THE WORLD OF THE CELL

Of Related Interest

From the Benjamin/Cummings Series in the Life Sciences

General Biology

N. A. Campbell
Biology, Fourth Edition (1996)
N. A. Campbell, L. G. Mitchell, and J. B. Reece
Biology: Concepts and Connections (1994)
J. Dickey
Laboratory Investigations for Biology (1995)
J. G. Morgan and M. E. B. Carter
Investigating Biology: A Laboratory Manual for Biology,
    Second Edition (1996)

Plant Biology

M. G. Barbour, J. H. Burk, and W. D. Pitts
Terrestrial Plant Ecology, Second Edition (1987)
J. D. Mauseth
Plant Anatomy (1988)
L. Taiz and E. Zeiger
Plant Physiology (1991)

Biochemistry

R. F. Boyer
Modern Experimental Biochemistry, Second Edition (1993)
C. K. Mathews and K. E. van Holde
Biochemistry, Second Edition (1996)
G. L. Sackheim
Chemistry for Biology Students, Fifth Edition (1995)

Molecular Biology

M. V. Bloom, G. A. Freyer, and D. A. Micklos
Laboratory DNA Science (1996)
L. E. Hood, I. L. Weissman, W. B. Wood, and J. H. Wilson
Immunology, Second Edition (1984)
J. D. Watson, N. H. Hopkins, J. W. Roberts, J. A. Steitz, and
    A. M. Weiner
Molecular Biology of the Gene, Fourth Edition (1987)

Microbiology

I. E. Alcamo
Fundamentals of Microbiology, Fourth Edition (1994)
R. M. Atlas and R. Bartha
Microbial Ecology: Fundamentals and Applications,
    Third Edition (1992)
J. G. Cappuccino and N. Sherman
Microbiology: A Laboratory Manual, Fourth Edition (1996)
T. R. Johnson and C. L. Case
Laboratory Experiments in Microbiology, Brief Version,
    Fourth Edition (1995)
G. J. Tortora, B. R. Funke, and C. L. Case
Microbiology: An Introduction, Fifth Edition (1995)

Human Anatomy and Physiology

E. N. Marieb
Essentials of Human Anatomy and Physiology,
    Fourth Edition (1994)
E. N. Marieb
Human Anatomy and Physiology, Third Edition (1995)
E. N. Marieb and J. Mallatt
Human Anatomy (1992)
A. P. Spence
Basic Human Anatomy, Third Edition (1991)

THE WORLD OF THE CELL

Third Edition

WAYNE M. BECKER
University of Wisconsin, Madison

JANE B. REECE
The Benjamin/Cummings Publishing Company

MARTIN F. POENIE
University of Texas, Austin

CONTRIBUTORS

John Raasch
University of Wisconsin, Madison
Chapter 13

Elizabeth Maynard Schaefer
Fremont, California
Chapters 19 and 25

Mary Jane Niles
University of San Francisco
Chapter 24

Timothy A. Ryan
Stanford University
Appendix

The Benjamin/Cummings Publishing Company

Menlo Park, California • Reading, Massachusetts
New York • Don Mills, Ontario • Wokingham, U.K.
Amsterdam • Bonn • Paris • Milan • Madrid • Sydney • Singapore • Tokyo
Seoul • Taipei • Mexico City • San Juan, Puerto Rico

Sponsoring Editor: **Catherine Pusateri**
Senior Developmental Editor: **Suzanne Olivier**
Associate Editor/Project Manager: **Kimberly Viano**
Editorial Assistants: **Clary Alward and Lisa Woo-Bloxberg**
Senior Production Editor: **Judith Hibbard**
Art Supervisor: **Kelly Murphy**
Photo Editors: **Kelli d'Angona-West and Lisa Lougee**
Cover Designer: **Yvo Riezebos Design**
Text Designer: **Cloyce Wall**
Copyeditor: **Betsy Dilernia**
Proofreader: **Janet E. deProsse**
Indexer: **Katherine Pitcoff**
Third Edition Artists: **Illustrious, Inc.**
Second Edition Artists: **Georg Klatt, Linda McVay,
Irene Imfeld**
First Edition Artists: **Greg Boren, Cyndie Clark-Huegel,
Michael Fornalski, Janet Hayes, Darwen and Vally
Hennings, Christina Jordan, Fran Milner, Kathy Monahan,
Judy Morley, Audre Newman, Carol Verbeeck**
Composition and Film Manager: **Lillian Hom**
Senior Manufacturing Coordinator: **Merry Free Osborn**
Manager of Visual Communications: **Don Kesner**
Composition and Film: **Interactive Composition
Corporation**
Cover Printer: **Phoenix Color Corp.**
Text Printer and Binder: **Von Hoffmann Press**

The cover art is a detail from the painting
Accent en Rose by Wassily Kandinsky. 1926.
Giraudon/Art Resource, NY/© 1996 Artists Rights Society,
NY/ADAGP, Paris, France.

Credits for photos, illustrations, and text appear following
the Appendix.

Library of Congress Cataloging-in-Publication Data

Becker, Wayne M.
    The World of the Cell / Wayne M. Becker, Jane B.
Reece, Martin F. Poenie; contributors, John Raasch, . . .
[et al]. — 3rd ed.
        p.   cm.
    Includes index.
    ISBN 0-8053-0880-6
    1. Cytology. 2. Molecular biology. I. Reece, Jane M.
II. Peonie, Martin F. III. Title.
QH581.2.B43   1995
574.87—dc20   95-25809
                                                        CIP

3 4 5 6 7 8 9 10 – VH – 99 98 97

The Benjamin/Cummings Publishing Company, Inc.
2725 Sand Hill Road
Menlo Park, California 94025

ABOUT THE AUTHORS

WAYNE M. BECKER teaches cell biology at the University of Wisconsin, Madison. His interest in textbook writing grew out of notes, outlines, and problem sets that he assembled for his students, culminating in *Energy and the Living Cell,* a paperback text on bioenergetics published in 1977, and *The World of the Cell,* the first edition of which appeared in 1986. He is a faculty member at the University  of Wisconsin, Madison, and earned all his degrees at that institution. All three degrees are in biochemistry, an orientation that is readily discernible in his textbooks. His research interests are in plant molecular biology, focusing specifically on the regulation of the expression of genes that encode enzymes of the photorespiratory pathway. His interests in teaching, learning, and research have taken him on sabbatical leaves at Harvard University, Edinburgh University, the University of Indonesia, the University of Puerto Rico, and Canterbury University in Christchurch, New Zealand.

JANE B. REECE has been affiliated with Benjamin/ Cummings since 1978, as an editor and, in recent years, as a coauthor of *Biology: Concepts and Connections,* with Neil Campbell and Lawrence Mitchell. At Benjamin/Cummings, she was sponsoring editor for Watson's *Molecular Biology of the Gene, 4th Edition,* and earlier editions of *The World of the Cell,* among other successful texts. She holds degrees  from Harvard University, Rutgers University, and the University of California, Berkeley, and has taught biology at community colleges in New Jersey and New York. Her research at Berkeley and as a postdoctoral fellow at Stanford University focused on genetic recombination in bacteria. She is the primary author for Part Four of this book.

MARTIN F. POENIE (Ph.D. Stanford University) is a research scientist and teacher at the University of Texas, Austin, where he is currently investigating the role of calcium and protein kinase C in T cell function. Dr. Poenie's commitment to teaching is reflected in the awards he has received, including Part-Time Teacher of the Year at Hartnell College, Salinas, California (1983), and the National Sci-  ence Foundation's Presidential Young Investigator Award (1987).

PREFACE

The World of the Cell is intended as a comprehensive introduction to cellular and molecular biology for students preparing for careers in biology, medicine, and related fields. Portions of this book began as lecture notes, problem sets, and exams in Biocore 303, a cell biology course at the University of Wisconsin, Madison, where one of us (WMB) teaches. These materials were expanded into the first edition of this textbook, published in 1986, and subsequently into the second edition, authored jointly by Wayne Becker and David Deamer of the University of California, Davis, and published in 1991. Heartened by the large number of users and by the responses of instructors and students alike, we have prepared this edition jointly, each of us bringing our own teaching and writing experience and our professional expertise to the venture in a way that we have found mutually complementary—and that we hope our readers will find beneficial.

Something Old and Something New

This book is not intended to be encyclopedic in its coverage. Our goal is to present the essential principles, processes, and methodology of cell biology as lucidly as possible. Recognizing the exceptionally rapid pace of discovery in cell and molecular biology during the past several years, we have sought to weave new knowledge and insights into the fabric of the text while remaining faithful to our central goal of focusing on the essentials of the discipline. Like the proverbial bride, this edition has "something old and something new," in the sense that we have tried to retain the features of the first two editions that readers have identified as "user-friendly," yet reorganizing and updating the material and adding new features that we hope will make the text even more useful.

Features we have retained from the first and second editions include an organization of subject matter that is readily adaptable to a great variety of course syllabi; careful and selective use of micrographs, accompanied by size bars to indicate magnification; problem sets intended to encourage not just recall but thoughtful application of information in each chapter; and boxed essays to provide further insights into selected topics. In addition, we have continued to make frequent use of overview figures, which outline complicated structures or processes in broad strokes before the details are examined more closely in the text and figures that follow. Overview figures provide context for students and help them appreciate and stay focused on the main points of a topic. Finally, we have as always paid careful attention to accuracy, consistency, vocabulary, and readability, hoping thereby to minimize confusion and maximize understanding for our readers.

New features that further enhance the usefulness of the text include the following:

• More color added to figures, facilitating an understanding of complex topics by the color coding of atoms, molecules, structures, pathways, and organelles, as appropriate.

• Molecular biology coverage updated and expanded, with material on recombinant DNA technology integrated into the appropriate chapters.

• Chapters on the cytoskeleton, motility, and nerve cell function substantially revised and updated.

• Coverage of signal transduction and cell-surface receptors significantly updated to reflect recent progress in this rapidly growing field of research.

• New chapter on cell junctions and extracellular structures added to enhance the coverage of these important topics.

Techniques and Methods

Throughout the text, we have tried to explain not only *what* we know about cells but also *how* we know what we know. Toward that end, we have included descriptions of experimental techniques and findings in every chapter, almost always in the context of the questions they address and in anticipation of the answers they provide. For example, equilibrium density centrifugation is introduced not in a chapter that describes a variety of methods for studying cells but in

Chapter 9, where it becomes important to our understanding of how lysosomes were originally distinguished from mitochondria and subsequently from peroxisomes as well. To help readers locate techniques out of context, an alphabetical **Guide to Techniques and Methods** appears on page x, with page references to particular techniques.

The only exception to the introduction of techniques in context is microscopy. The techniques of light and electron microscopy are relevant to so much of contemporary cell biology that they warrant special consideration as a self-contained unit. Accordingly, we include an **Appendix** devoted specifically to the principles and techniques of microscopy. Thoroughly updated for this edition, the Appendix is fully illustrated and cross-referenced at numerous points in the text.

In-Text Learning Aids

To enhance the effectiveness of this text as a learning tool, each chapter includes the following basic features:

• One or two **Boxed Essays** to help students better understand particularly important or intriguing aspects of cell biology. Some of the essays provide interesting historical perspectives on how science is done—the discovery of the double helix as described in Box 3A of Chapter 3, for example. Other essays are intended to help readers understand potentially difficult principles, such as the essay that uses the analogy of monkeys shelling peanuts to explain enzyme kinetics (Box 6A). Still others provide insights into contemporary techniques used by cell biologists, as exemplified by the description of studies on membrane asymmetry in Box 7A. Yet another role of the boxed essays is to illustrate clinical applications of research findings in cell and molecular biology, as illustrated by the discussion of membrane transport and cystic fibrosis in Box 8A.

• A list of **Key Terms** that includes the page number of the location at which each term first appears in boldface and is defined or described.

• A **Suggested Reading** list, with an emphasis on review articles and carefully selected research publications from the current or historical literature that motivated users are likely to find understandable. We have tried to avoid overwhelming readers with lengthy bibliographies of the original literature but have referenced articles that are especially relevant to the topics of the chapter.

• A **Problem Set**, reflecting our conviction that we learn science not just by reading or hearing about it, but by working with it. The problems are designed to emphasize understanding and application of the principles taught in the chapter, not just rote recall. Many of the problems are class-tested, having been selected from problem sets and exams we have used in our own courses. To maximize the usefulness of the problem sets, answers for all problems appear in the **Solutions Manual** described below. At the discretion of the instructor, this manual can be made available to students through the local bookstore or used by the instructor as a resource for homework and exam questions.

Supplementary Learning Aids

Supplementary learning aids that are available with this text include the following:

• A **Solutions Manual** containing detailed answers to all of the problems in the text. (ISBN 0-8053-0882-2)

• A set of 78 transparencies corresponding to selected figures from the text but with enlarged labels to enhance their usefulness in the classroom. (ISBN 0-8053-0881-4)

User Comments Warmly Welcome

The ultimate test of any textbook is how effectively it helps instructors teach and students learn. We welcome feedback from readers and will try to acknowledge all correspondence. Please send your comments, criticisms, and suggestions to the appropriate author, as follows:

Chapters 1-13, 24, and 25: Wayne M. Becker
Department of Botany
University of Wisconsin, Madison
Madison, Wisconsin 53706

Chapters 14-19: Jane B. Reece
Benjamin/Cummings Publishing, Inc.
Menlo Park, California 94025

Chapters 20-23: Martin F. Poenie
Department of Zoology
University of Texas, Austin
Austin, Texas 78712

ACKNOWLEDGMENTS

We want to acknowledge the contributions of the numerous people who have made this book possible. We are indebted especially to the many students whose words of encouragement catalyzed the writing of these chapters and whose thoughtful comments and criticisms have contributed much to whatever level of reader-friendliness the text may be judged to have.

Each of us owes a special debt of gratitude to our colleagues, from whose insights and understanding we have benefited greatly and borrowed freely. These include Ann Burgess, Millard Susman, Bill Sudgen, and Rick Eisenstein at the University of Wisconsin, Madison. We greatly appreciate the contributions of Tara Poenie, for her diligent library research, and Josh Trachtenberg (University of Texas, Austin), for his attentive examination of the chapters in Part Five. We also thank Mark Guyer at the National Center for Human Genome Research, Jim Weber at the Marshfield Clinic, and Bob Tjian at the University of California, Berkeley, for valuable information and advice.

We are especially grateful to colleagues who have brought their expertise to bear on the revision of selected chapters, including Mary Jane Niles (Chapter 24), John Raasch (Chapter 13), Beth Schaefer (Chapters 19 and 25), Timothy A. Ryan (Appendix), and Lisa Smit (Box 8A). We also acknowledge Peter Armstrong, John Carson, Ed Clark, David Deamer, Joel Goodman, David Gunn, Jeanette Natzle, David Speigel, Akif Uzman, and Karen Valentine for their contributions to the previous editions. In addition, we want to express our appreciation to the many colleagues who responded so generously to our requests for micrographs and other visual aids. One of us (WMB) is also especially grateful to Professor Juan Gonzalez Lagoa at the University of Puerto Rico–Mayagüez and Jack Heinemann at the University of Canterbury in Christchurch, New Zealand, for their gracious hospitality at the sabbatical institutions where significant portions of this book were revised.

The many reviewers listed below provided helpful criticisms and suggestions at various stages of manuscript development and revision. Their words of appraisal and counsel were gratefully received and greatly appreciated. Indeed, the extensive review process to which both this and the prior editions of the book have been exposed should be considered a significant feature of the book. Nonetheless, the final responsibility for what you read here remains ours, and you may confidently attribute to us any errors of omission or commission encountered in these pages.

We are also deeply indebted to the many people at The Benjamin/Cummings Publishing Company who made this venture a reality. Special recognition goes to Catherine Pusateri, Kimberly Viano, Lisa Woo-Bloxberg, Suzanne Olivier, Judith Hibbard, Betsy Dilernia, Kelly Murphy, Clary Alward, and Lisa Lougee, whose consistent encouragement, hard work, and careful attention to detail contributed much to the clarity of both the text and the art.

Finally, we are grateful beyond measure to our families and friends, without whose patience, understanding, and forbearance this book would never have been written.

Reviewers of the Third Edition:

L. Rao Ayyagari, Lindenwood College
Steve Benson, California State University, Hayward
Joseph J. Berger, Springfield College
Gerald Bergtrom, University of Wisconsin, Milwaukee
Frank L. Binder, Marshall University
R. B. Boley, University of Texas at Arlington
Edward M. Bonder, Rutgers, the State University of New Jersey
J. D. Brammer, North Dakota State University
Chris Brinegar, San Jose State University
P. Samuel Campbell, University of Alabama, Huntsville
George L. Card, The University of Montana
C. H. Chen, South Dakota State University
Jonathan Copeland, Georgia Southern University
Douglas Dennis, James Madison University
Elizabeth D. Dolci, Johnson State College
Michael P. Donovan, Southern Utah University
Diane D. Eardley, University of California, Santa Barbara
Robert M. Dores, University of Denver

Carl S. Frankel, Pennsylvania State University, Hazleton Campus
Carol V. Gay, Pennsylvania State University
Michael L. Gleason, Central Washington University
Karen F. Greif, Bryn Mawr College
Leah T. Haimo, University of California, Riverside
James P. Holland, Indiana University, Bloomington
Betty A. Houck, University of Portland
Kenneth C. Jones, California State University, Northridge
Martin A. Kapper, Central Connecticut State University
Lon S. Kaufman, University of Illinois, Chicago
Steven J. Keller, University of Cincinnati Main Campus
Joseph R. Koke, Southwest Texas State University
Roderick MacLeod, University of Illinois, Urbana-Champaign
Thomas D. McKnight, Texas A & M University
Robert L. Metzenberg, University of Wisconsin, Madison
Deborah B. Mowshowitz, Columbia University
Carl E. Nordahl, University of Nebraska, Omaha
Gilbert C. Pogany, Northern Arizona University
Ralph E. Reiner, College of the Redwoods
Gary Reiness, Lewis and Clark College
Adrian Rodriguez, San Jose State University
Donald H. Roush, University of North Alabama
David W. Scupham, Valparaiso University
Edna Seaman, University of Massachusetts, Boston
Diane C. Shakes, University of Houston
Sheldon S. Shen, Iowa State University
Randall D. Shortridge, State University of New York, Buffalo
Dwayne D. Simmons, University of California, Los Angeles
Richard D. Storey, Colorado College
Philip Stukus, Denison University
Elizabeth J. Taparowsky, Purdue University
Barbara J. Taylor, Oregon State University
Bruce R. Telzer, Pomona College

Reviewers of the First and Second Editions:

L. Rao Ayyagari, Lindenwood College
Margaret Beard, Columbia University
Paul Benko, Sonoma State University
Robert Blystone, Trinity University
Alan H. Brush, University of Connecticut, Storrs
Brower R. Burchill, University of Kansas
Ann B. Burgess, University of Wisconsin, Madison
Thomas J. Byers, Ohio State University
Edward A. Clark, University of Washington
Philippa Claude, University of Wisconsin, Madison
John M. Coffin, Tufts University School of Medicine
J. John Cohen, University of Colorado Medical School
Larry Cohen, Pomona College
David DeGroote, St. Cloud State
Aris J. Domnas, University of North Carolina, Chapel Hill
David R. Fromson, California State University, Fullerton
Stephen A. George, Amherst College
T. T. Gleeson, University of Colorado
Ursula W. Goodenough, Washington University
Thomas A. Gorell, Colorado State University
Marion Greaser, University of Wisconsin, Madison
Mark T. Groudine, Fred Hutchinson Cancer Research Center, University of Washington School of Medicine
Gary Gussin, University of Iowa
Laszlo Hanzely, Northern Illinois University
Bettina Harrison, University of Massachusetts, Boston
Lawrence Hightower, University of Connecticut, Storrs
Johns Hopkins III, Washington University
William R. Jeffery, University of Texas, Austin
Kwang W. Jeon, University of Tennessee
Patricia P. Jones, Stanford University
Robert Koch, California State University, Fullerton
Hal Krider, University of Connecticut, Storrs
William B. Kristan, Jr., University of California, San Diego
Frederic Kundig, Towson State University
Elias Lazarides, California Institute of Technology
John T. Lis, Cornell University
Robert Macey, University of California, Berkeley
Gary G. Matthews, State University of New York, Stony Brook
Richard Nuccitelli, University of California, Davis
Joanna Olmsted, University of Rochester
Alan Orr, University of Northern Iowa
Curtis L. Parker, Morehouse School of Medicine
Lee D. Peachey, University of Colorado
Howard Petty, Wayne State University
Susan Pierce, Northwestern University
Ralph Quatrano, Oregon State University
Gary Reiness, Pomona College
Edmund Samuel, Southern Connecticut State University
Robert D. Simoni, Stanford University
William R. Sistrom, University of Oregon
Barbara Y. Stewart, Swarthmore College
Antony O. Stretton, University of Wisconsin, Madison
Stephen Subtelny, Rice University
Millard Susman, University of Wisconsin, Madison
John J. Tyson, Virginia Polytechnic University
Akif Uzman, University of Texas, Austin
Fred D. Warner, Syracuse University
James Watrous, St. Joseph's University
Fred H. Wilt, University of California, Berkeley

GUIDE TO TECHNIQUES AND METHODS

The following techniques are important to cell biologists. Each technique is described in the text at the indicated location, in the context of its actual use by researchers.

BRIEF CONTENTS

DETAILED CONTENTS

two CELL STRUCTURE AND FUNCTION 79

4 CELLS AND ORGANELLES 80

5 BIOENERGETICS: THE FLOW OF ENERGY IN THE CELL 112

6 ENZYMES: THE CATALYSTS OF LIFE 138

7 MEMBRANES: THEIR STRUCTURE, FUNCTION, AND CHEMISTRY 167

13 PHOTOTROPIC ENERGY METABOLISM: PHOTOSYNTHESIS   374

four INFORMATION FLOW IN CELLS   407

14 THE STRUCTURAL BASIS OF CELLULAR INFORMATION: DNA, CHROMOSOMES, AND THE NUCLEUS   408

15 THE CELL CYCLE, DNA REPLICATION, AND MITOSIS   453

16 SEXUAL REPRODUCTION, MEIOSIS, AND GENETIC VARIABILITY   495

17 GENE EXPRESSION: I. THE GENETIC CODE AND TRANSCRIPTION   535

21 CELLULAR MOVEMENT: MOTILITY AND CONTRACTILITY   675

22 SIGNAL TRANSDUCTION MECHANISMS: I. ELECTRICAL SIGNALS IN NERVE CELLS   714

23 SIGNAL TRANSDUCTION MECHANISMS: II. MESSENGERS AND RECEPTORS   751

six SPECIAL TOPICS IN CELL BIOLOGY   783

24 CELLULAR ASPECTS OF THE IMMUNE RESPONSE   784

25 CELLULAR ASPECTS OF CANCER   817

APPENDIX: PRINCIPLES AND TECHNIQUES OF MICROSCOPY   841

PHOTO, ILLUSTRATION, AND TEXT CREDITS   869

INDEX   873

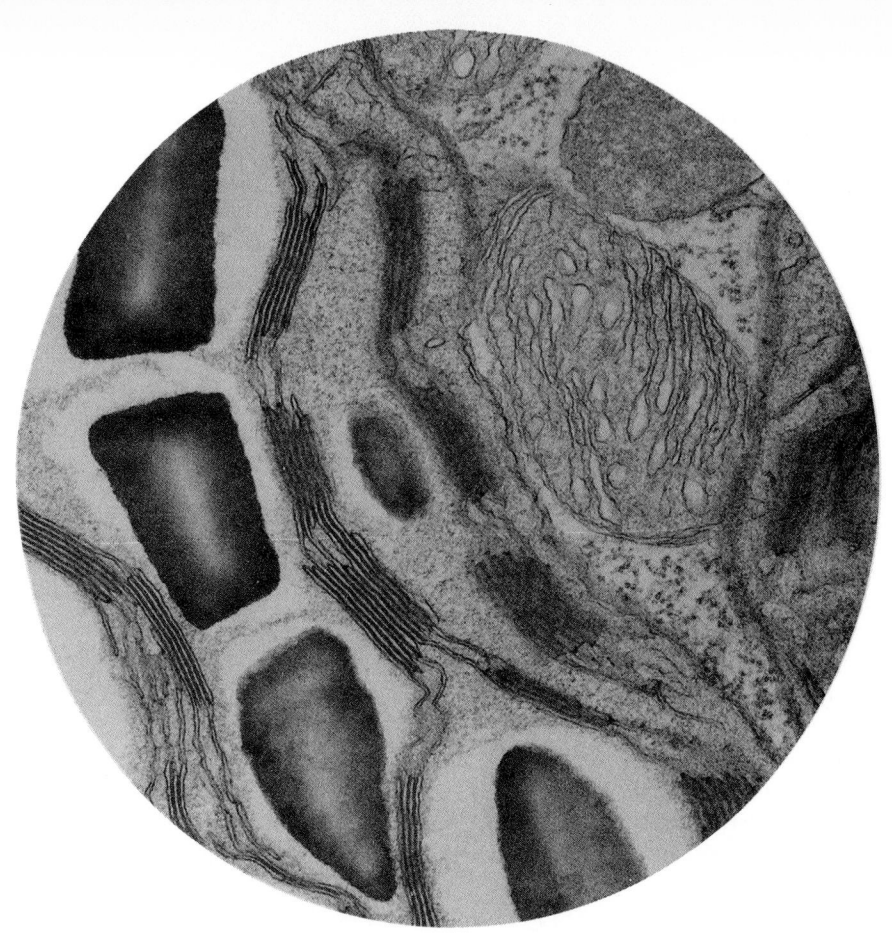

INTRODUCTION
TO THE CELL

1

THE WORLD OF THE CELL: A PREVIEW

The **cell** is the basic unit of biology. Every organism either consists of cells or is itself a single cell. Therefore, it is only as we understand the structure and function of cells that we can appreciate both the capabilities and the limitations of living organisms, whether animal, plant, or micro-organism.

We are in the midst of a revolution in biology that has brought with it tremendous advances in our understanding of how cells are constructed and how they carry out the intricate functions necessary for life. Particularly significant is the dynamic nature of the cell, as evidenced by its capacity to grow, reproduce, and become specialized and by its ability to respond to stimuli and to adapt to changes in its environment.

Cell biology itself is changing, as scientists from a variety of related disciplines focus their efforts on the common objective of understanding more adequately how cells work. The convergence of cytology, genetics, and biochemistry has made modern cell biology one of the most exciting and dynamic disciplines in contemporary biology.

In this chapter, we will look briefly at the beginnings of cell biology. Then we will consider the three main historical strands that have given rise to our current understanding of what cells are and how they function.

The Cell Theory: A Brief History

The story of cell biology started to unfold more than 300 years ago, as European scientists began to focus their crude microscopes on a variety of biological material ranging from tree bark to human sperm. One such scientist was Robert Hooke, Curator of Instruments for the Royal Society of London. Hooke examined a thin slice of cork cut with a penknife and saw a network of tiny boxlike compartments that reminded him of a honeycomb. He called these little compartments *cellulae,* a Latin term meaning "little rooms." It is from this word that we get our present-day term, *cell.*

Actually, what Hooke observed were not cells at all but the empty cell walls of dead plant tissue, which is what tree bark really is. However, Hooke would not have thought of his *cellulae* as dead because he did not understand that they could be alive! Although he noticed that cells in other plant tissues were filled with what he called "juices," he preferred to concentrate on the more prominent cell walls that he had first encountered.

Meanwhile, Antonie van Leeuwenhoek was making an amazing series of microscopic observations that would do much to lay the foundation for our appreciation of the cellular basis of life. Van Leeuwenhoek was a Dutch shopkeeper who devoted much of his spare time to the design of simple microscopes. He used his microscopes to examine almost anything he could get his hands on. He reported his observations to the Royal Society in a series of papers during the last quarter of the seventeenth century. His detailed reports attest to both the high quality of his lenses and his keen powers of observation.

Two factors restricted further understanding of the nature of cells. One was the limited resolution of the microscopes of the day, which even van Leeuwenhoek's superior instruments could push just so far. The second and probably more fundamental factor was the essentially descriptive nature of seventeenth-century biology. It was basically an age of observation, with little thought given to explaining the intriguing architectural details of biological materials that were beginning to yield to the probing lens of the microscope.

More than a century passed before the combination of improved microscopes and more experimentally minded microscopists resulted in a series of developments that culminated in an understanding of the importance of cells in biological organization. By the 1830s, improved lenses led to higher magnification and better resolution, such that structures only 1 micrometer (μm) apart could be resolved. (A *micrometer* is 10^{-6} m, or one-millionth of a meter; see Box 1A for a discussion of the units of measurement appropriate to cell biology.)

UNITS OF MEASUREMENT IN CELL BIOLOGY

The challenge of understanding cellular structure and organization is complicated by the problem of size. Cells and their organelles are so small that the units used to measure them are unfamiliar to many students and therefore often difficult to appreciate. The problem can be approached in two ways: by realizing that there are really only two units necessary to express the dimensions of most structures of interest to us, and by illustrating a variety of structures that can be appropriately measured with each of these units.

The **micrometer** (μm) is the most useful unit for expressing the size of cells and larger organelles. A micrometer (sometimes also called a **micron**) corresponds to one-millionth of a meter (10^{-6} m). In general, bacterial cells are a few micrometers in diameter, and the cells of plants and animals are 10- to 20-fold larger in any single dimension.

Larger organelles such as mitochondria and chloroplasts tend to have diameters or lengths of a few micrometers and are therefore comparable in size to whole bacterial cells. Smaller organelles are usually in the range of 0.2–1.0 μm. As a rule of thumb, if you can see it with a light microscope, you can probably express its dimensions conveniently in micrometers, since the resolution limit of the light microscope is about 0.20–0.35 μm. Figure 1A-1 illustrates a variety of structures that are usually measured in micrometers.

The **nanometer** (nm), on the other hand, is the unit of choice for molecules and subcellular structures that are too small or too thin to be seen with the light microscope. A nanometer is one-billionth of a meter (10^{-9} m). It therefore takes 1000 nanometers to equal 1 micrometer. As a benchmark on the nanometer scale, a ribosome has a diameter of about 25–30 nm. Other structures that can be measured

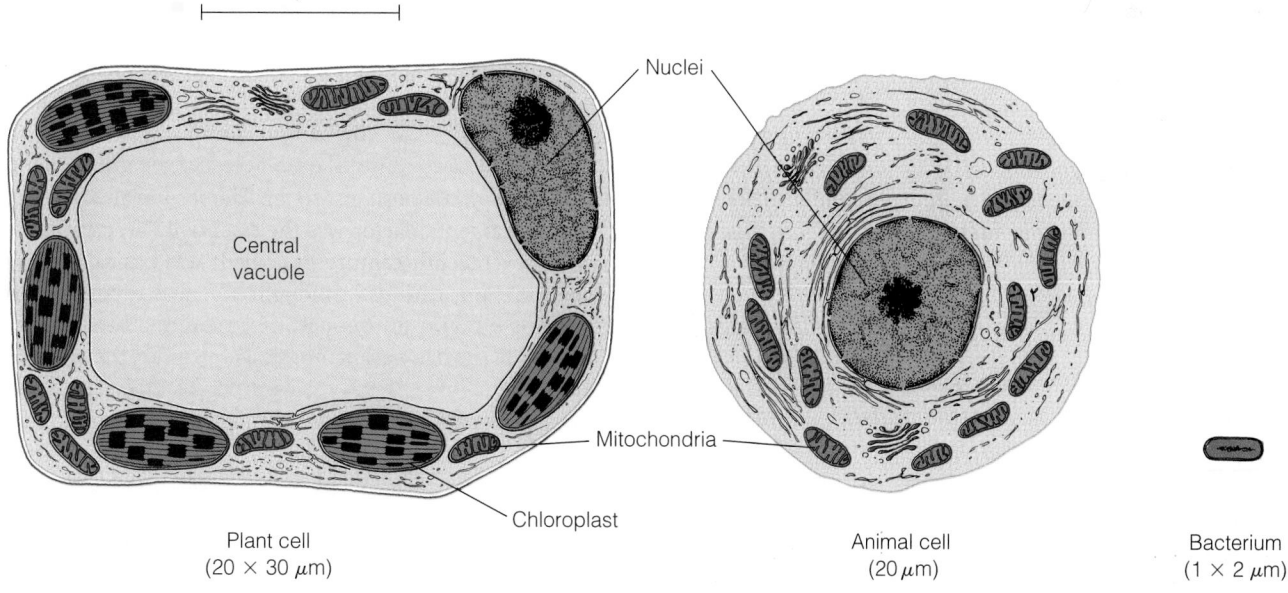

10 μm

Nuclei

Central vacuole

Mitochondria

Chloroplast

Plant cell
(20 × 30 μm)

Animal cell
(20 μm)

Bacterium
(1 × 2 μm)

Figure 1A-1   The World of the Micrometer. Structures with dimensions that can be measured conveniently in micrometers include almost all cells and some of the larger organelles.

conveniently in nanometers are microtubules, microfilaments, membranes, and DNA molecules. The dimensions of these structures are indicated in Figure 1A-2.

Another unit frequently used in cell biology is the **angstrom** (Å) which corresponds to 10^{-10} m or 0.1 nm. Molecular dimensions, in particular, are often expressed in angstroms. However, because the angstrom differs from the nanometer by only a factor of ten, it adds little flexibility to the expression of dimensions at the cellular level and will therefore not be used in this text. ■

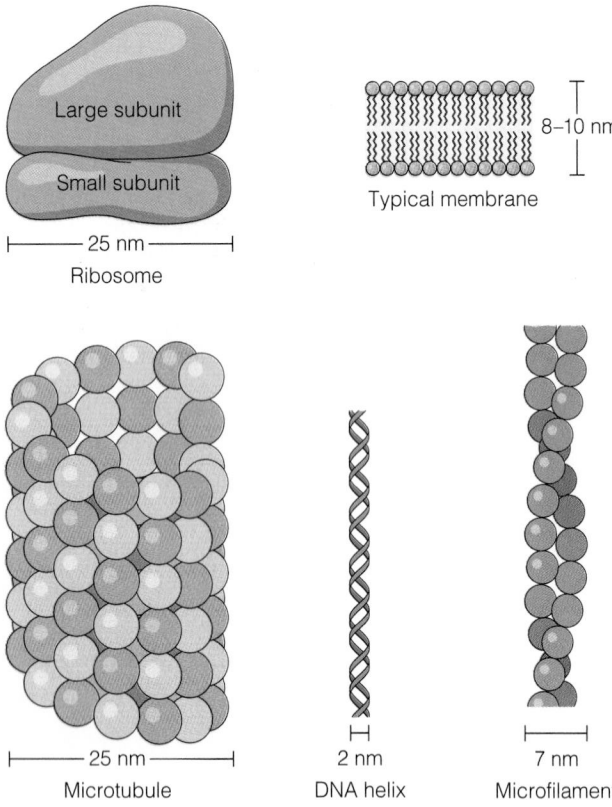

Figure 1A-2   The World of the Nanometer. Structures with dimensions that can be measured conveniently in nanometers include ribosomes, membranes, microtubules, the DNA double helix, and microfilaments. (Scale: 1 in = 20 nm.)

Aided by such improved lenses, the English botanist Robert Brown found that every plant cell he looked at contained a rounded structure, which he called a *nucleus*. In 1838, his German colleague Matthias Schleiden came to the important conclusion that all plant tissues were composed of cells and that an embryonic plant always arose from a single cell. Similar conclusions concerning animal tissue were reported only a year later by Theodor Schwann, thereby laying to rest earlier speculations that plants and animals might be structurally quite different. It is easy to understand how such speculations could have arisen. After all, plant cell walls provide conspicuous boundaries between cells that are readily visible even with a crude microscope, whereas individual animal cells, which lack cell walls, are much harder to distinguish in a tissue sample. It was only when Schwann examined animal cartilage cells that he became convinced of the fundamental similarity between plant and animal tissue, since cartilage cells, unlike most other animal cells, have boundaries that are well defined by thick deposits of collagen fibers.

It is to Schwann's credit that he was able to draw all these observations together into a single unified theory of cellular organization, illustrating the more interpretive atmosphere that was beginning to emerge. Moreover, Schwann's formulation has stood the test of time and continues to provide the basis for our own understanding of the importance of cells and cell biology.

As originally postulated by Schwann, the **cell theory** had two basic tenets:

1. All organisms consist of one or more cells.

2. The cell is the basic unit of structure for all organisms.

Less than 20 years later, a third tenet was added. This grew out of Brown's original description of nuclei, extended by Karl Nägeli to include observations on the nature of cell division. By 1855, Rudolf Virchow, a German physiologist, was able to conclude that cells arose in only one manner—by the division of other, preexisting cells. Virchow encapsulated this conclusion in the now-famous Latin phrase *omnis cellula e cellula*, which in translation becomes the third tenet of the modern cell theory:

3. All cells arise only from preexisting cells.

Thus, the cell is not only the basic unit of structure for all organisms but also the basic unit of reproduction. In other words, all of life has a cellular basis. No wonder, then, that an understanding of cells and their properties is so fundamental to a proper appreciation of all other aspects of biology.

The Emergence of Modern Cell Biology

Modern cell biology involves the weaving together of three distinctly different strands into a single cord. As the time line of Figure 1-1 illustrates, each of the strands had its own historical origins, and most of the intertwining has occurred

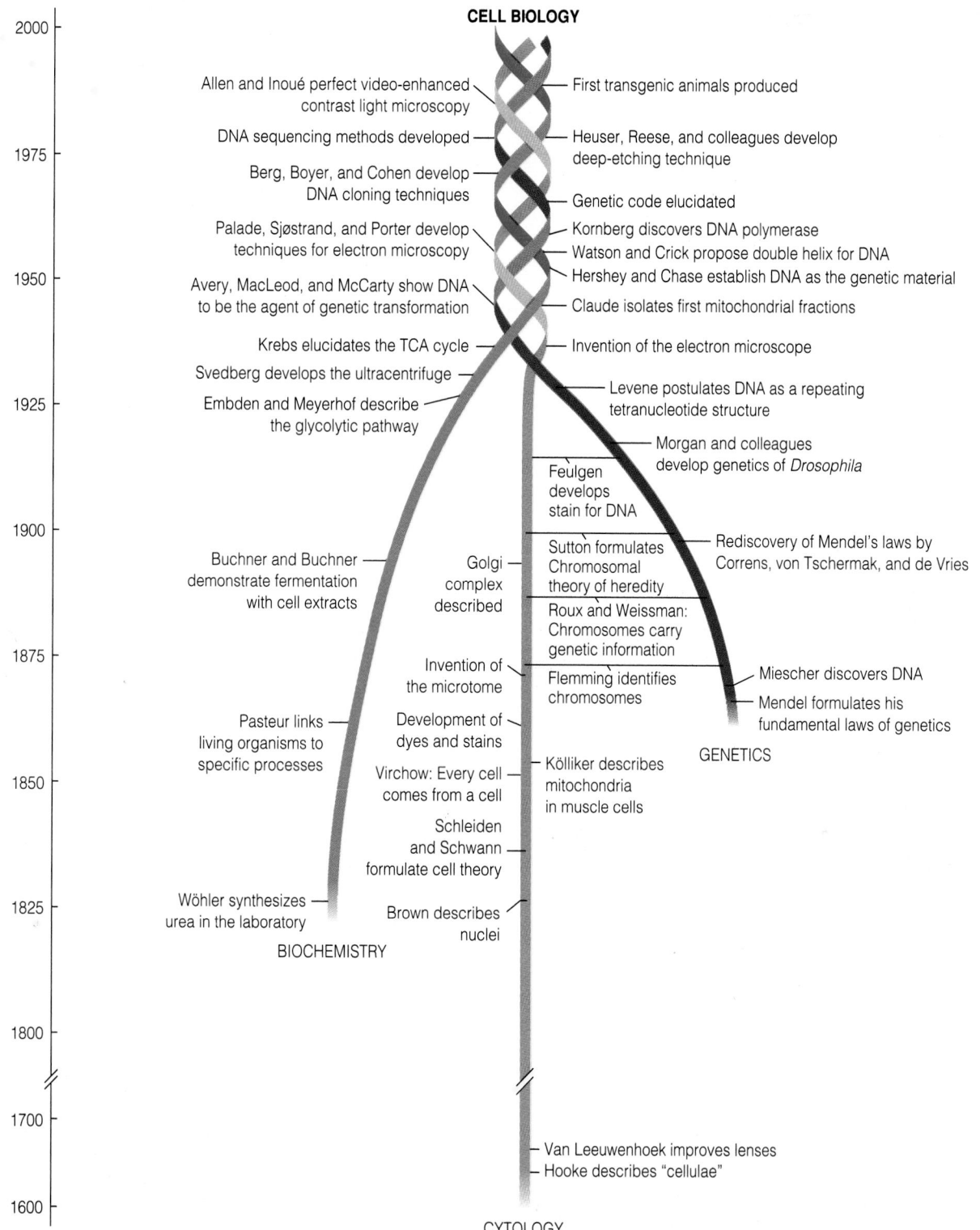

Figure 1-1 The Cell Biology Time Line. Although cytology, biochemistry, and genetics began as separate disciplines, they have increasingly merged since about the second quarter of the twentieth century.

only within the last 50 years. Each strand should be appreciated in its own right because each makes its own unique and significant contribution. The contemporary cell biologist must be adequately informed about all three strands, regardless of what his or her own immediate interests happen to be.

The first of these historical strands is **cytology,** which is concerned primarily with cellular structure. (The Greek prefix *cyto-* means "cell," as does the suffix *-cyte.*) As we have already seen, cytology had its origins more than three centuries ago and depended heavily on the light microscope for its initial impetus. More recently, the advent of electron microscopy and several related optical techniques has led to considerable additional cytological activity and understanding.

The second strand represents the contributions of **biochemistry** to our understanding of cellular function. Most of the developments in this field have occurred within the last 50 years, though again the roots go back much further. Especially important has been the development of techniques such as centrifugation and chromatography for the separation of cellular components and molecules.

The third strand is **genetics.** Here, the historical continuum stretches back more than a century to Gregor Mendel. Again, however, much of our present understanding has come within the last several decades. An especially important landmark on the genetic strand came with the realization that DNA (deoxyribonucleic acid) is the bearer of genetic information in most (though not all) life forms.

To understand present-day cell biology therefore means to appreciate its diverse roots and the important contributions that each of its component strands has made to our current understanding of what a cell is and what it can do. Each of the three historical strands of cell biology will be discussed briefly here, but a fuller appreciation of each is likely to come only as various aspects of cell structure, function, and genetics are explored in later chapters. In fact, much of the rest of the text can be thought of as a further development and integration of the several historical strands that are woven into modern cell biology.

The Cytological Strand

Strictly speaking, cytology is the study of cells. (Actually, the literal meaning of the Greek word *cytos* is "hollow vessel," which fits well with Hooke's initial impression of cells.) Historically, however, cytology has dealt primarily with cellular structure, mainly through the use of optical techniques.

The Light Microscope. The **light microscope** was the earliest tool of the cytologists and continues to play an important role in our elucidation of cellular structure. Light microscopy allowed cytologists to identify membrane-bounded structures such as *nuclei, mitochondria,* and *chloroplasts* within a variety of cell types. Such structures are called **organelles** ("little organs") and are prominent features of most plant and animal (but not bacterial) cells.

Other significant developments include the invention of the microtome in 1870 and the availability of various dyes and stains at about the same time. A **microtome** is an instrument for slicing thin sections of biological samples, usually after they have been dehydrated and embedded in paraffin or plastic. The technique enables rapid and efficient preparation of thin tissue slices of uniform thickness. The dyes that came to play so important a role in staining and identifying subcellular structures were primarily developed in the latter half of the nineteenth century by German industrial chemists working with coal tar derivatives.

Together with improved optics and more sophisticated lenses, these and related developments extended light microscopy as far as it could go—to the physical limits of resolution imposed by the wavelengths of visible light. The theoretical limit of resolution is $\lambda/2$, where λ is the wavelength of the light used to illuminate the sample. For visible light in the wavelength range of 400–700 nanometers (nm), the limit of resolution is about 200–350 nm. (A *nanometer* is 10^{-9} or one-billionth of a meter; 1 nm = 0.001 μm.) Figure 1-2 illustrates the useful range of the light microscope and compares its resolving power with that of the human eye and the electron microscope.

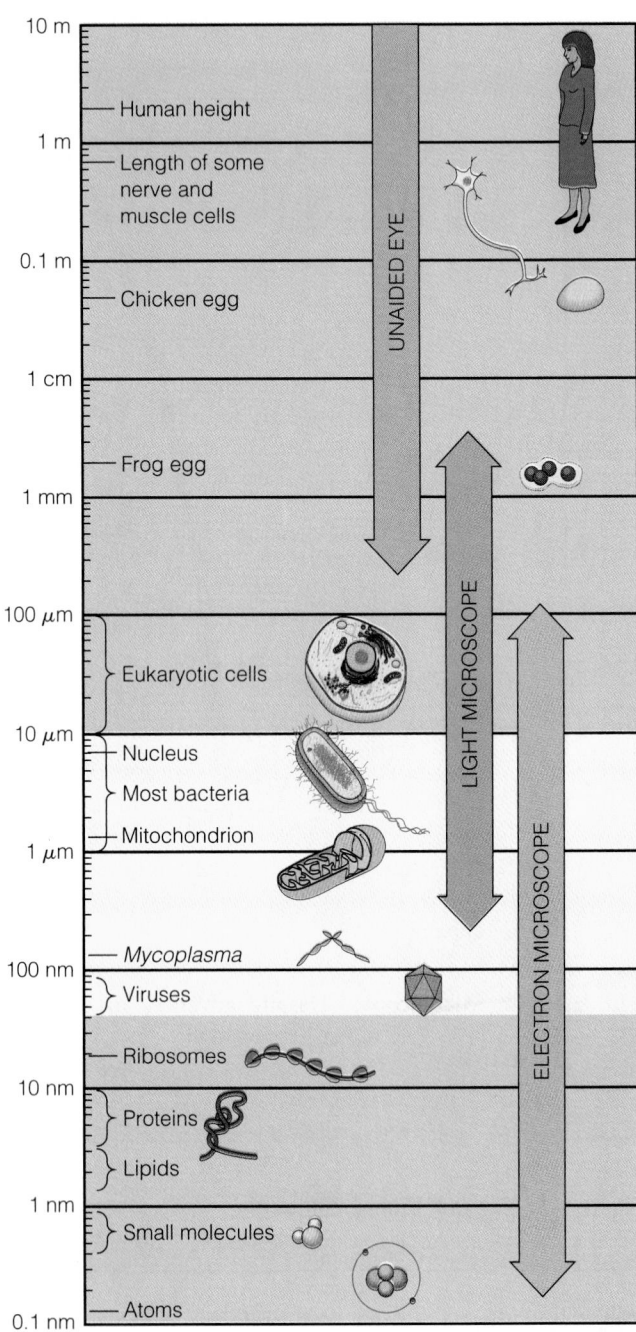

Figure 1-2   Resolving Power of the Human Eye, the Light Microscope, and the Electron Microscope. Notice that the vertical axis is on a logarithmic scale to accommodate the range of sizes shown.

For a more detailed discussion of microscopy, see the Appendix, which describes a variety of special optical techniques for visualizing living cells, including fluorescence microscopy, phase-contrast microscopy, and differential interference contrast microscopy. Because researchers can observe living cells directly using these techniques, they can avoid artifacts or distortions due to tissue fixation and sample preparation. *Fluorescence microscopy* enables researchers to detect specific proteins or other molecules that are made fluorescent by coupling them covalently to a fluorescent dye. By the simultaneous use of two or more such dyes, each coupled to a different kind of molecule, the distributions of different kinds of molecules can be followed in the same cell.

Phase-contrast and *differential interference contrast microscopy* make it possible to see living cells clearly and are widely used for this purpose. Both of these techniques enhance and amplify slight changes in the phase of transmitted light as it passes through a structure that has a different refractive index than the surrounding medium. Most modern light microscopes are equipped for phase-contrast and differential interference contrast in addition to the simple transmission of light, with conversion from one use to another accomplished by interchanging optical components.

Digital Video Microscopy. An exciting recent development in light microscopy is **digital video microscopy,** which makes use of video cameras and computer storage, and allows computerized image processing to enhance and analyze images. Attachment of a highly light-sensitive video camera to a light microscope makes it possible to observe cells for extended periods of time using very low levels of light. This *image intensification* is particularly useful for visualizing fluorescent molecules in living cells with a fluorescence microscope. For *computerized image processing,* the electronic signal from the video camera is processed by a computer in a variety of ways to compensate for optical flaws in the microscope and to enhance the contrast between objects and background. The enhanced contrast makes it possible to see small transparent objects that would otherwise have been impossible to distinguish from the background. For further details, see the Appendix.

The Electron Microscope. Despite advances in optical techniques and contrast enhancement, light microscopy is inevitably subject to the limit of resolution imposed by the wavelength of the light used to view the sample. Even the use of ultraviolet radiation with shorter wavelengths increases the resolution only by a factor of two.

A major breakthrough in resolving power came with the development of the **electron microscope.** The electron microscope was invented in Germany in 1932 and came into widespread biological use in the early 1950s, with G. E. Palade, F. S. Sjøstrand, and K. R. Porter among its most notable early users. In place of visible light and optical lenses, the electron microscope uses a beam of electrons that is deflected and focused by an electromagnetic field. Because the wavelength of electrons is so much shorter than that of photons of visible light, the limit of resolution of the electron microscope is much lower than that of the light microscope: about 0.1–0.2 nm for the electron microscope compared with about 200–350 nm for the light microscope.

However, for biological samples the practical limit of resolution is usually 2 nm or more. The difference is due to problems with specimen preparation and contrast. Nevertheless, the electron microscope has about 100 times more resolving power than the light microscope (see Figure 1-2). The useful magnification is also greater: up to 100,000-fold for the electron microscope, compared with about 1000- to 1500-fold for the light microscope. Figure 1-3 shows how much more structural detail can be seen when a cell is examined with an electron microscope than with a light microscope.

Because of the low penetration power of electrons, samples prepared for electron microscopy must be exceedingly thin. The instrument used for this purpose is called an **ultramicrotome.** It is equipped with a diamond knife and can cut sections as thin as 20 nm. Substantially thicker samples can also be examined by electron microscopy, but a much higher accelerating voltage is then required to increase the penetration power of the electrons adequately. Such a **high-voltage electron microscope** uses accelerating voltages up to several thousand kilovolts (kV), compared with the range of 50–100 kV common to most conventional instruments. Sections up to 1 μm thick can be studied with such a high-voltage instrument. This thickness allows organelles and other cellular structures to be examined in more depth.

Electron microscopy is discussed in detail in the Appendix. Specialized techniques described there include *freeze-fracturing, freeze-etching, negative staining,* and *scanning electron microscopy.* Scanning electron microscopy is an especially spectacular technique because of the sense of depth it gives to biological structures (Figure 1-4).

Electron microscopy has revolutionized our understanding of cellular architecture by making detailed ultrastructural investigations possible. Some organelles (such as nuclei or mitochondria) are large enough to be seen with a light microscope but can be studied in much greater detail with an electron microscope (see Figure 1-3). In addition, electron microscopy has revealed cellular structures that are too small to be seen at all with a light microscope. An example is the ribosome, the site of protein synthesis in all cells. Ribosomes escape detection entirely with the light microscope because they are only 25–30 nm in diameter, depending on the type of cell in which they are found. They are readily visualized with the electron microscope, however, because of its greater resolving power.

The Scanning Tunneling Microscope. Our ability to visualize biological structures took another potential leap forward recently when scientists in California obtained the first-ever three-dimensional images of an unaltered DNA molecule, with the structural features of the molecule shown in unprecedented detail. The images were obtained using a **scanning tunneling microscope,** a novel kind of microscope

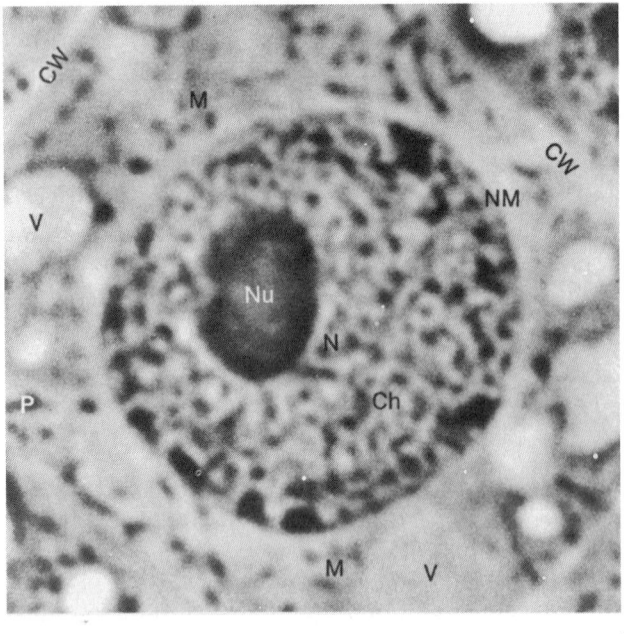

(a) Light microscopy · 50 μm

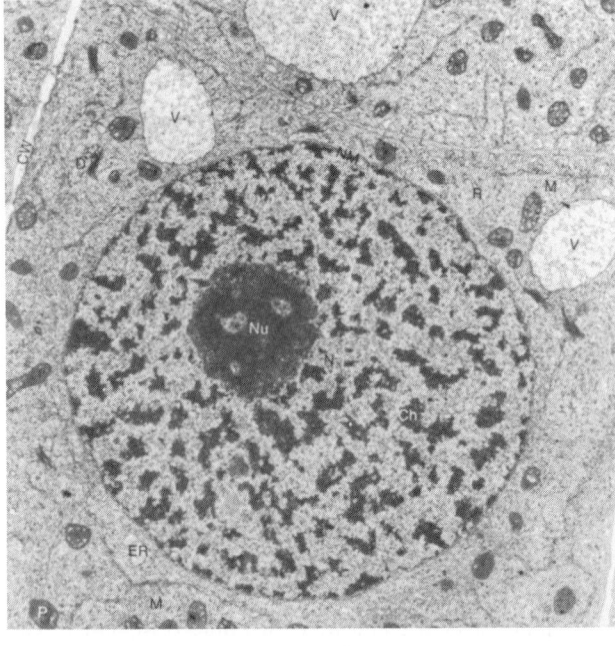

(b) Electron microscopy · 50 μm

Figure 1-3 Comparison of Resolving Power. Cells can be visualized by either **(a)** light microscopy or **(b)** electron microscopy, but much more detail is seen in the latter case because of the greater resolving power of the electron microscope. Both sections were cut from the same piece of onion root tissue. For light microscopy, a 1.5-μm section was cut and photographed with phase-contrast optics. For electron microscopy, a section of 0.03 μm was used. Labeled structures include the nucleus (N), nucleolus (Nu), chromatin (Ch), nuclear membranes (NM), plastids (P), mitochondria (M), endoplasmic reticulum (ER), vacuoles (V), ribosomes (R), and cell wall (CW). (750×. A magnification of 1000× is the upper limit for the light microscope but is a very low magnification for the electron microscope.)

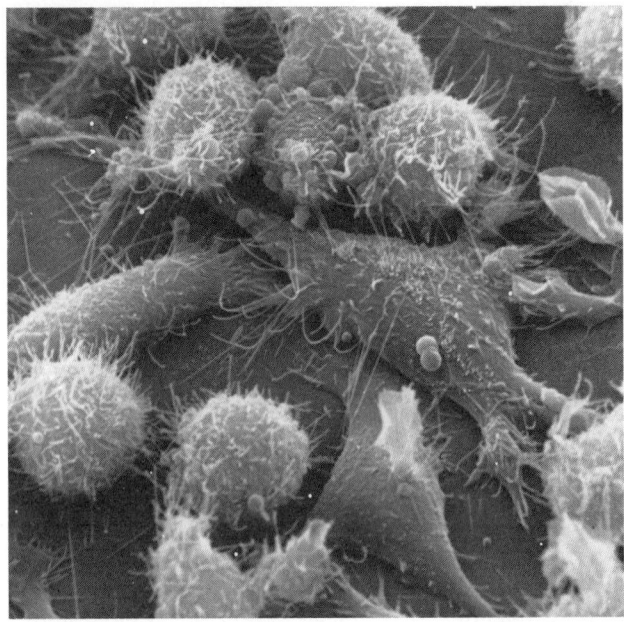

(a) Human neuroblastoma cells · 50 μm

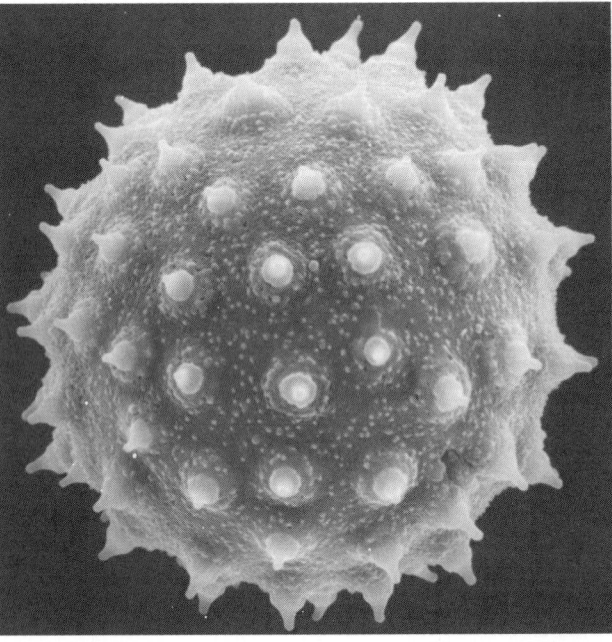

(b) Pollen grain · 10 μm

Figure 1-4 Scanning Electron Microscopy. A scanning electron microscope was used to visualize **(a)** cultured human neuroblastoma cells and **(b)** a pollen grain.

invented only six years earlier. This instrument has a magnifying power that may enable scientists to visualize directly the intricate atomic structure of complex biological molecules, particularly nucleic acids and proteins.

The scanning tunneling microscope features a sharp-tipped needle made of a platinum-rhodium alloy. The needle is attached to a tube-shaped piezoelectric ceramic, a material that expands and contracts in response to an electrical charge. When the tip of the needle is positioned very close to the surface of the sample, a current of electrons crosses the gap from needle to sample by means of a phenomenon of quantum mechanics known as *tunneling*. The *scanning* then follows: The tip scans across the sample, moving in and out to maintain the current at a constant level and thus tracing out the surface contours of the sample. The topographic data from the tip movements are then translated into three-dimensional images by a computer.

With its capacity to probe biological molecules, the scanning tunneling microscope may well revolutionize our understanding of the structure of these molecules. For example, the instrument might prove useful for studying the three-dimensional structures of proteins as well as the interactions between DNA and proteins.

The Biochemical Strand

At about the time when cytologists were starting to explore cellular structure with their microscopes, other scientists were making observations that began to explain and clarify cellular function. Much of what is now called biochemistry dates from a discovery reported by the German chemist Friedrich Wöhler in 1828. Wöhler was a contemporary (as well as fellow countryman) of Schleiden and Schwann. It is doubtful, however, that he would have thought he had anything in common with a botanist and a zoologist peering at plant and animal tissues through their crude microscopes. Nor is Wöhler likely to have imagined that he would one day be placed alongside these men in a history of cellular biology. Nevertheless, Wöhler revolutionized our thinking about biology and chemistry by demonstrating that urea, an organic compound of biological origin, could be synthesized in the laboratory from an inorganic starting material, ammonium cyanate:

$$NH_4^+ + {}^-OC\equiv N \longrightarrow H_2N-\overset{\overset{\displaystyle O}{\|}}{C}-NH_2$$

Ammonium ion   Cyanate ion   Urea

Previously, it had been widely held that living organisms were a world unto themselves, not governed by the laws of chemistry and physics that apply to the nonliving world. By showing that a compound made by living organisms—a "bio-chemical"—could be synthesized in a laboratory just like any other chemical, Wöhler helped to break down the conceptual distinction between the living and nonliving worlds and to dispel the notion that biochemical processes were somehow exempt from the laws of chemistry and physics.

Another major advance came about 40 years later, when Louis Pasteur linked the activity of living organisms to specific processes by showing that living yeast cells were needed to carry out the fermentation of sugar into alcohol. This observation was followed in 1897 by the finding of Eduard and Hans Buchner that fermentation could also take place with extracts from yeast cells—that is, the intact cells themselves were not required. Initially, such extracts were called "ferments," but gradually it became clear that the active agents in the extracts were specific biological catalysts that have since come to be called **enzymes.**

Significant progress in our understanding of cellular function came in the 1920s and 1930s as the biochemical pathways for fermentation and related cellular processes were elucidated. This was a period dominated by German biochemists such as Gustav Embden, Otto Meyerhof, Otto Warburg, and Hans Krebs. Several of these men have long since been immortalized by the pathways that have come to bear their names. For example, the *Embden-Meyerhof pathway* for glycolysis was a major research triumph of the early 1930s. It was followed shortly by the *Krebs cycle* (also known as the TCA cycle). Both of these pathways are important because of their role in the process by which cells extract energy from foodstuffs. At about the same time, the high-energy compound *adenosine triphosphate (ATP)* came to be recognized as the principal energy storage compound in most cells.

Biochemistry took a major step forward with the development of the **ultracentrifuge** as a means of separating subcellular structures and macromolecules on the basis of size, shape, and density. In many ways, the ultracentrifuge was as significant for biochemistry as the electron microscope was for cytology. In fact, both instruments were developed at about the same time, so the ability to see organelles and other subcellular structures with much greater resolution came almost simultaneously with the capacity to isolate and purify them. The ultracentrifuge was developed in Sweden by The Svedberg during the period 1925–1930 and was used initially for determining the sedimentation rates of proteins. Adaptation of centrifugation techniques for the isolation of subcellular fractions came in the early 1940s, largely through the pioneering work of Albert Claude.

With an enhanced ability both to see subcellular structures and to isolate them, cytologists and biochemists began to realize the extent to which their respective observations on cellular structure and function could complement each other, thereby laying the foundations for modern cell biology.

The Genetic Strand

The third strand in the cord of cell biology is genetics. Like the other two, this strand also has important roots in the nineteenth century. In this case, the strand begins with Gregor Mendel, whose contributions to science are well known. His studies with the pea plants he grew in the monastery garden must surely rank among the most famous experiments in all of biology. His findings were published in 1866, laying out the principles of segregation and independent assortment of the "hereditary factors" that we know today as **genes.** These were singularly important principles, destined to provide the foundation for what would eventually be known as Mendelian genetics. But Mendel was clearly a man ahead of his time. His work went almost unnoticed when it was first published and was not fully appreciated until its rediscovery nearly 35 years later.

As a prelude to that rediscovery, the role of the nucleus in the genetic continuity of cells came to be appreciated in the decade following Mendel's work. Shortly thereafter, **chromosomes** were identified by Walther Flemming as threadlike bodies seen in dividing cells. Flemming called the division process *mitosis,* from the Greek word for thread. Chromosome number soon came to be recognized as a distinctive characteristic of a species and was shown to remain constant from generation to generation. That the chromosomes themselves might be the actual bearers of genetic information was suggested by Wilhelm Roux as early as 1883 and was expressed more formally by August Weissman shortly thereafter.

With the role of the nucleus and chromosomes established and appreciated, the stage was set for the rediscovery of Mendel's initial observations. This came in 1900, when his studies were cited almost simultaneously by three plant geneticists working independently: Carl Correns in Germany, Ernst von Tschermak in Austria, and Hugo de Vries in Holland. Within three years, the **chromosome theory of heredity** was formulated by Walter Sutton, who was the first to link the chromosomal "threads" of Flemming with the "hereditary factors" of Mendel.

Sutton's theory proposed that the hereditary factors responsible for Mendelian inheritance are located on the chromosomes within the nucleus. This hypothesis received its strongest confirmation from the work of Thomas Hunt Morgan and his students at Columbia University during the first two decades of this century. They chose as their experimental species *Drosophila melanogaster,* the common fruit fly. By identifying a variety of morphological mutants of *Drosophila,* Morgan and his co-workers were able to link specific traits to specific chromosomes.

Meanwhile, the foundation for our understanding of the chemical basis of inheritance was also slowly being laid. An important milestone was the discovery of DNA by Johann Friedrich Miescher in 1869. Using such unlikely sources as salmon sperm and human pus from surgical bandages, Miescher isolated and described what he called "nuclein." But, like Mendel, Miescher was ahead of his time. It was about 75 years before the role of his nuclein as the genetic information of the cell came to be fully appreciated.

As early as 1914, DNA was implicated as an important component of chromosomes by the staining technique of Robert Feulgen, a method that is still in use today. But little consideration was given to the possibility that DNA could be the bearer of genetic information. In fact, that would have been considered quite unlikely in light of the apparently uninteresting structure of the mononucleotide constituents of DNA that were known by 1930. Until the middle of the century, it was widely held that genes were made up of proteins, since these were the only nuclear components that seemed to be able to account for the obvious diversity of genes.

A landmark experiment that clearly pointed to DNA as the genetic material was reported in 1944 by Oswald Avery, Colin MacLeod, and Maclyn McCarty. Their work focused on the phenomenon of genetic transformation in bacteria, to be discussed in Chapter 13. Their evidence was compelling, but the scientific community remained largely unconvinced of the conclusion. Just eight years later, however, a considerably more favorable reception was accorded the report of Alfred Hershey and Martha Chase that DNA, and not protein, enters a bacterial cell when it is infected by a bacterial virus.

Shortly thereafter, in 1953, James Watson and Francis Crick proposed their now-famous **double helix model** for DNA structure, with features that immediately suggested how replication and genetic mutations could occur. The *Watson-Crick model,* as it came to be known, catapulted DNA into prominence and launched an era of molecular genetics that has since revolutionized biology. In the process, the historical strand of genetics became intimately entwined with those of cytology and biochemistry, and the discipline of cell biology as we know it today came into being.

The Scientific Method

To become familiar with an area of science such as cell biology means, at least in part, to learn the facts about that subject. Even in this short introductory chapter, we have already encountered a number of facts about cell biology. When we say, for example, that "all organisms consist of one or more cells" or that "DNA is the bearer of genetic information," we recognize these statements as facts of cell biology. But we also recognize that the first of these statements was initially regarded as part of a theory and the second statement actually replaced an earlier misconception that genes were made of proteins.

Clearly, then, a scientific "fact" is a much more tenuous piece of information than our everyday sense of the word might imply. To a scientist, a "fact" is simply an attempt to state our best current understanding of a specific phenomenon and is only valid until it is revised or replaced by a better understanding. Box 1B explores the meaning of "facts" in biology and the **scientific method** by which new and better information becomes available.

1B

Further Insights

BIOLOGY, "FACTS," AND THE SCIENTIFIC METHOD

If asked what they expect to get out of a science textbook, most readers would probably reply that they intend to learn the facts relevant to the particular scientific area the book is about—cell biology, in the case of the text you are reading right now. If pressed to explain what a fact is, most people would probably reply that a fact is "something that we know to be true." That sense of the word agrees with the dictionary, since one of the definitions of *fact* is "a piece of information presented as having objective reality."

To a scientist, however, a fact is a much more tenuous piece of information than such a definition might imply. The "facts" of science are really just attempts to state our current understanding of the natural world around us, based on observations that we make and experiments that we do. As such, a given "fact" is only as sound as the observations or experiments on which it is based and can be modified or superseded at any time by a better understanding based on more careful observations or more discriminating experiments. As one scientist so aptly put it, truth to a researcher "is not a citadel of certainty to be defended against error; it is a shady spot where one eats lunch before tramping on" (White, 1968, p. 3).

Cell biology is rich with examples of "facts" that were once widely held but have since been superseded as cell biologists have "tramped on" to a better understanding of the phenomena those "facts" attempted to explain. As recently as the early nineteenth century, for example, it was widely held (i.e., regarded as fact) that living matter consisted of substances quite different from those in nonliving matter. According to this view, called *vitalism,* the chemical reactions that occurred within living matter did not follow the known laws of chemistry and physics but were instead directed by a "vital force." Then came Friedrich Wöhler's demonstration (in 1828) that the biological compound urea could be synthesized in the laboratory from an inorganic

compound, thereby undermining one of the "facts" of vitalism. The other "fact" was refuted by the work of Eduard and Hans Buchner, who showed (in 1897) that nonliving extracts from yeast cells could ferment sugar into ethanol. Thus, a view held as "fact" by generations of scientists was eventually discredited and replaced by the new "fact" that the components and reactions of living matter are not a world unto themselves, but follow all the laws of chemistry and physics.

For a more contemporary example, consider what we know about the energy needed to support life. Until recently, it was regarded as a fact that the sun is the ultimate source of all energy in the biosphere, such that every organism either uses solar energy directly (i.e., green plants, algae, and certain bacteria) or is a part of a food chain that is sustained by such photosynthetic organisms. Then came the discovery of *deep-sea thermal vents* and the thriving communities of organisms that live around them, none of which depends upon solar energy. Instead, these organisms depend on the bond energy of hydrogen sulfide (H_2S), which is extracted by bacteria that live around the thermal vents and used to synthesize organic compounds from carbon dioxide. These bacteria form the basis of food chains that include zooplankton (microscopic animals), worms, and other residents of the thermal vent environment.

The "facts" presented in biology textbooks such as this one are our best current attempts to describe and explain the workings of the biological world around us. They are subject to change whenever we become aware of new or better information.

How does new and better information become available? Scientists usually follow a systematic approach to new information called the *scientific method.* As Figure 1B-1 indicates, the scientific method begins as a researcher *makes observations,* either in the field or in a research laboratory.

Based on these observations and on knowledge gained in prior studies, the scientist *formulates a testable hypothesis,* a tentative explanation or model consistent with the observations and with prior knowledge that can be tested experimentally. Next, the investigator *designs a controlled experiment* to test the hypothesis by varying specific conditions while holding everything else as constant as possible. The scientist then *collects the data, interprets the results,* and *draws reasonable conclusions,* which obviously must be consistent not only with the results of this particular experiment but with prior knowledge as well.

Although the scientific method may seem quite formidable, perhaps even quite foreign, when described in this way or when diagrammed as in the figure here, it is second nature to a practicing scientist. It is, in fact, more a way of thinking than a set of procedures to be followed. Most likely, this is the way our ancestors explained and interpreted natural phenomena long before scientists were trained at universities—and long before students read essays about the scientific method!

For a good example of the scientific method in action, consider a recent report that points to a protein in cow's milk as a trigger of insulin-dependent diabetes. The research was conducted by Dr. Jutta Karjalainen and his col-

leagues at the Universities of Turku and Helsinki in Finland and at the Hospital for Sick Children in Toronto. As you read the next several paragraphs, refer to Figure 1B-1 and see if you can identify each of the elements of the scientific method in the work of Dr. Karjalainen and his colleagues.

In insulin-dependent diabetes, the body's immune system attacks and destroys the insulin-producing *beta cells* of the pancreas. The body is then unable to process sugars into energy because insulin is required for normal carbohydrate utilization, including the uptake of blood glucose into cells. Cow's milk was initially implicated as the possible culprit by experiments in which diabetes-prone rats did not get diabetes when fed a diet containing no cow's milk. Moreover, children who were breast-fed instead of receiving cow's milk showed a significantly lower risk of diabetes.

Armed with these facts and with preliminary data that suggested the milk protein *bovine serum albumin (BSA)* as the specific trigger molecule, the investigators formulated their hypothesis. They postulated that children with insulin-dependent diabetes produce antibodies against a specific segment of the BSA molecule and that these antibodies also recognize a specific protein, called *p69,* on the surface of the pancreatic beta cells, leading to the destruction of these cells. To test their hypothesis, the researchers used immunological techniques to look for antibodies against BSA in the blood of diabetic and nondiabetic children. They found that the diabetic children had a much higher level of anti-BSA antibodies in their blood than the nondiabetic children did. Furthermore, most of those antibodies were specific for the short segment of the BSA molecule that is almost identical to a portion of the p69 cell surface protein.

Based on these results, the investigators concluded that diabetic children have antibodies capable of recognizing a protein on the surface of their own pancreatic beta cells, and suggested that these antibodies may be the connection between cow's milk and insulin-dependent diabetes. This link has not yet been proved, but it appears likely that eliminating cow's milk from the diet of infants might dramatically reduce the incidence of insulin-dependent diabetes.

When illustrated by experiments such as this diabetes study, the scientific method sounds very neat and orderly. Not all scientific discoveries are made in this way, however. Many important advances in biology have come about more by accident than by plan. Alexander Fleming's discovery of penicillin in 1928 is a classic example. Fleming, a Scottish physician and bacteriologist, accidently left a culture dish of *Staphylococcus* bacteria uncovered, such that it was inadvertently exposed to contamination by other microorganisms. Fleming was about to discard the contam-

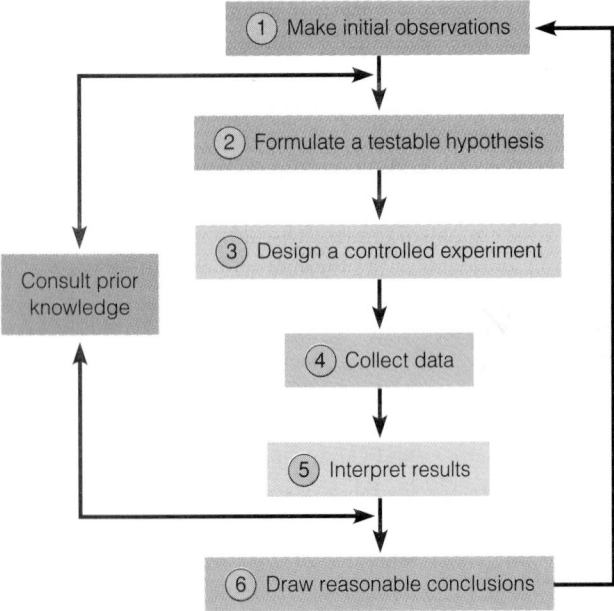

Figure 1B-1　The Scientific Method.

inated culture when he happened to notice some clear patches where the bacteria were not growing. Reasoning that the bacterial growth may have been inhibited by some contaminant in the air and recognizing how important an inhibitor of bacterial growth might be, Fleming kept the culture dish and began attempts to isolate and characterize the substance. The actual identification of penicillin and the demonstration that it was the product of a mold was left to others, but Fleming is credited with the initial discovery.

Boxes in subsequent chapters will acquaint you with further examples of apparently accidental discoveries. The discovery of a cellular structure called the *peroxisome* by Christian de Duve and his colleagues is certainly in this category, as you will learn in Chapter 9. Regardless of how accidental such discoveries may appear, however, it is almost always true that "chance favors the prepared mind." Behind the apparent "chance" of each such discovery is the "prepared mind" that has been trained to observe carefully and to think astutely.

As you proceed through this text, be on the outlook for applications of the scientific method such as that of Karjalainen and his colleagues, as well as examples of serendipitous discoveries such as those of Fleming and de Duve. You will find that regardless of the approach, the conclusions from each experiment add to our knowledge of how biological systems work and usually lead to more questions as well, continuing the cycle of scientific inquiry. And that's good news if you aspire to a career in research, because it's your best insurance that there will still be questions to answer when you are ready to begin. ∎

PERSPECTIVE

The biological world is a world of cells. All living organisms are made up of one or more cells, each of which came from a preexisting cell. Although the importance of cells in biological organization has been appreciated for about 150 years, the discipline of cell biology as we know it today is of much more recent origin. Modern cell biology has come about by the interweaving of three historically distinct strands—cytology, biochemistry, and genetics—which in their early development probably did not seem at all related. But the contemporary cell biologist must understand all three strands because they complement one another in the quest to learn what cells are and how they function.

KEY TERMS FOR SELF-TESTING

cell (p. 2)
cell biology (p. 2)

The Cell Theory: A Brief History

cell theory (p. 4)

The Emergence of Modern Cell Biology

cytology (p. 5)
biochemistry (p. 6)
genetics (p. 6)
light microscope (p. 6)
organelle (p. 6)

microtome (p. 6)
digital video microscopy (p. 7)
electron microscope (p. 7)
ultramicrotome (p. 7)
high-voltage electron microscope (p. 7)
scanning tunneling microscope (p. 7)
enzyme (p. 9)
ultracentrifuge (p. 9)
gene (p. 10)
chromosome (p. 10)

chromosome theory of heredity (p. 10)
double helix model (p. 10)

The Scientific Method

scientific method (p. 10)

Box 1A: Units of Measurement in Cell Biology

micrometer (micron) (p. 3)
nanometer (p. 3)
angstrom (p. 4)

PROBLEM SET

1-1. The Historical Strands of Cell Biology. For each of the following events, indicate whether it belongs mainly to the cytological (C), biochemical (B), or genetic (G) strand in the historical development of cell biology.

(a) Kölliker describes "sarcosomes" (now called mitochondria) in muscle cells (1857).

(b) Hoppe-Seyler isolates the protein hemoglobin in crystalline form (1864).

(c) Haeckel postulates that the nucleus is responsible for heredity (1868).

(d) Ostwald proves that enzymes are catalysts (1893).

(e) Muller discovers that X-rays induce mutations (1927).

(f) Davson and Danielli postulate a model for the structure of cell membranes (1935).

(g) Beadle and Tatum formulate the one gene–one enzyme hypothesis (1940).

(h) Claude isolates the first mitochondrial fractions from rat liver (1940).

(i) Lipmann postulates the central importance of ATP in cellular energy transactions (1940).

(j) Avery, MacLeod, and McCarty demonstrate that bacterial transformation is attributable to DNA, not protein (1944).

(k) Palade, Porter, and Sjøstrand each develop techniques for fixing and sectioning biological tissue for electron microscopy (1952–1953).

(l) Lehninger demonstrates that oxidative phosphorylation depends for its immediate energy source on the transport of electrons in the mitochondrion (1957).

1-2. More on Historical Strands. Several of the events shown in Figure 1-1 are associated with not one but two of the three strands that contribute to the historical development of cell biology. For each of the following pairs of strands, identify at least one event that is associated with both strands and indicate why you think it belongs to both.

(a) Biochemistry and cytology

(b) Genetics and biochemistry

(c) Genetics and cytology

1-3. Cell Sizes. To appreciate the differences in cell size illustrated in Figure 1A-1 on p. 3, consider the following specific examples. *Escherichia coli,* a typical bacterial cell, is cylindrical in shape, with a diameter of about 1 μm and a length of about 2 μm. As a typical animal cell, consider a human liver cell, which is roughly spherical in shape and has a diameter of about 20 μm. And for a typical plant cell, consider the columnar *palisade cells* located just beneath the upper surface of many plant leaves. These cells are cylindrical in shape, with a diameter of about 20 μm and a length of about 35 μm.

(a) Calculate the approximate volume of each of these three cell types in cubic micrometers. (Recall that $V = \pi r^2 h$ for a cylinder and that $V = 4\pi r^3/3$ for a sphere.)

(b) Approximately how many bacterial cells would fit in the internal volume of a human liver cell?

(c) Approximately how many liver cells would fit inside a palisade cell?

1-4. Sizing Things Up. To appreciate the sizes of the subcellular structures shown in Figure 1A-2 on p. 4, consider the following calculations.

(a) All cells and many subcellular structures are surrounded by a membrane. Assuming a typical membrane to be about 8 nm wide, how many such membranes would have to be aligned side by side before the structure could be seen with the light microscope? How many with the electron microscope?

(b) Ribosomes are the structures in cells on which the process of protein synthesis takes place. A human ribosome is a roughly spherical structure with a diameter of about 30 nm. How many ribosomes would fit in the internal volume of the human liver cell described in Problem 1-3 if the entire volume of the cell were filled with ribosomes?

(c) The genetic material of the *Escherichia coli* cell described in Problem 1-3 consists of a DNA molecule with a diameter of 2 nm and a total length of 1.36 mm. (The molecule is actually circular, with 1.36 mm as its circumference.) To be accommodated in a cell that is only a few micrometers long, this large DNA molecule is tightly coiled and folded into a *nucleoid* that occupies a small proportion of the internal volume of the cell. Calculate the smallest possible volume into which the DNA molecule could fit, and express that as a percentage of the internal volume of the bacterial cell that you calculated in Problem 1-3a.

1-5. A Question of Resolution. The light microscope was the earliest tool of the cytologist and still plays a vital role in the study of cells. A drawback of the light microscope is its *limit of resolution*—how close together two points can be and still be distinguished from each other—and hence, its *useful magnification*. (To answer the following questions, you may find it useful to consult the first several pages of the Appendix.)

(a) What is the major factor that determines the limit of resolution of a light microscope?

(b) Why does the use of ultraviolet light rather than visible (white) light increase the resolution of a light microscope by a factor of about two?

(c) What does it mean to say that the useful magnification of a light microscope is about 1000-fold?

(d) An image obtained with a light microscope can easily be magnified much more than 1000-fold by photographic enlargement. In what sense is any enlargement beyond 1000-fold called "empty magnification"?

1-6. The "Facts" of Life. Each of the following statements was once regarded as a biological fact but is now understood to be untrue. In each case, indicate why the statement was once thought to be true and why it is no longer considered a fact.

(a) Plant and animal tissues are constructed quite differently because animal tissues do not have conspicuous boundaries that divide them into cells.

(b) Living organisms are not governed by the laws of chemistry and physics as is nonliving matter, but are subject to a "vital force" that is responsible for the formation of organic compounds.

(c) Genes most likely consist of proteins because the only other likely candidate, DNA, is a relatively uninteresting molecule consisting of only four kinds of monomers (nucleotides) arranged in a relatively invariant repeating tetranucleotide sequence.

(d) The fermentation of sugar to alcohol can take place only if living yeast cells are present.

1-7. More "Facts" of Life. Each of the following statements was regarded as a biological fact until quite recently but is now either rejected or qualified to at least some extent. In each case, speculate on why the statement was once thought to be true, and then try to determine what evidence might have made it necessary to reject or at least to qualify the statement. (Note: This question requires a more intrepid sleuth than Problem 1-6 does, but chapter references are provided to aid in your sleuthing.)

(a) A biological membrane can be thought of as a protein-lipid "sandwich" consisting of an exclusively phospholipid interior coated on both sides with thin layers of protein. (Chapter 7)

(b) The enzymes required to catalyze the conversion of sugar into a compound called pyruvate invariably occur in the cyto-plasm of the cell, rather than being compartmentalized in membrane-enclosed structures. (Chapter 11)

(c) The mechanism by which the oxidation of organic molecules such as sugars leads to the generation of ATP involves a high-energy phosphorylated molecule as an intermediate. (Chapter 12)

(d) When carbon dioxide from the air is "fixed" (covalently linked) into organic form by photosynthetic organisms such as green plants, the first form in which the carbon atom of the CO_2 molecule appears is always the three-carbon compound 3-phosphoglycerate. (Chapter 13)

(e) DNA always exists as a duplex of two strands wound together into a right-handed helix. (Chapter 14)

(f) The genetic code that specifies how the information present in the DNA molecule is used to make proteins is universal in the sense that all organisms use the same code. (Chapter 17)

1-8. Cell Biology in 1875. Friedrich Miescher (1844–1895), Gregor Mendel (1822–1884), Louis Pasteur (1822–1895), and Rudolf Virchow (1821–1902) were European scientists (from Switzerland, Austria, France, and Germany, respectively) whose most important discoveries were made in the 20-year period from 1855 to 1875. Assume that the four men met at a scientific conference in 1875 to discuss their respective contributions to biology.

(a) What might the four scientists have found they had in common?

(b) What do you think Pasteur might have found most intriguing or relevant about Virchow's work? Explain your answer.

(c) What do you think Virchow might have found most intriguing or relevant about Mendel's work? Explain your answer.

(d) In whose work do you think Miescher might have been the most interested? Explain your answer.

1-9. Facts and Truth. An especially apt characterization of scientific fact was made by Lynn White, Jr. As quoted in Box 1B, White wrote that, to a scientist, truth "is not a citadel of certainty to be defended against error; it is a shady spot where one eats lunch before tramping on." Explain what White means by this statement. In what sense does it aptly describe scientific facts? How does his statement relate to the scientific method?

1-10. Pizza, Heartburn, and the Scientific Method. Although the scientific method may sound rather foreign when described in formal terms or when diagrammed as in Figure 1B-1 on p. 12, it is in reality not very different from the way most of us go about answering questions or solving problems. You probably use the scientific method frequently without even realizing it. Suppose, for example, that you have been experiencing heartburn quite often recently. By keeping track of your eating habits for a few weeks, you realize that the heartburn is most likely to occur on nights after you have eaten pizza for supper, especially if you have had your favorite pizza, with pepperoni, anchovies, and onions. You wonder if the heartburn is caused by eating pizza and, if so, which of the ingredients might be the culprit.

(a) Describe how you might go about determining whether the heartburn is due to the pizza and, if so, to which of the ingredients.

(b) Now compare your approach to the scientific method as shown in Figure 1B-1. How scientific was your method?

Suggested Reading

The Emergence of Modern Cell Biology

Bodenheimer, F. S. *The History of Biology.* London: Dawson, 1958.

Bradbury, S. *The Evolution of the Microscope.* New York: Pergamon Press, 1967.

Claude, A. The coming of age of the cell. *Science* 189 (1975): 433.

de Duve, C., and H. Beaufay. A short history of tissue fractionation. *J. Cell Biol.* 91 (1981): 293s.

Fruton, J. S. The emergence of biochemistry. *Science* 192 (1976): 327.

Gall, J. G., K. R. Porter, and P. Siekevitz, eds. Discovery in cell biology. *J. Cell Biol.* 91, part 3 (1981).

Hughes, A. *A History of Cytology.* New York: Abelard-Schumann, 1959.

Jacob, F. *The Logic of Life: A History of Heredity.* New York: Pantheon, 1973.

Judson, H. F. *The Eighth Day of Creation: Makers of the Revolution in Biology.* New York: Simon & Schuster, 1979.

Mirsky, A. E. The discovery of DNA. *Sci. Amer.* 218 (June 1968): 78.

Palade, G. E. Albert Claude and the beginning of biological electron microscopy. *J. Cell Biol.* 50 (1971): 5D.

Peters, J. A. *Classic Papers in Genetics.* Englewood Cliffs, NJ: Prentice-Hall, 1959. (Includes classic papers by G. Mendel and W. S. Sutton.)

Quastel, H. J. The development of biochemistry in the 20th century. *Canad. J. Cell Biol.* 62 (1984): 1103.

Stent, G. C. That was the molecular biology that was. *Science* 160 (1968): 390.

Stent, G. C., and R. Calendar. *Molecular Genetics: An Introductory Narrative,* 2d ed. New York: W. H. Freeman, 1978.

Watson, J. D. *The Double Helix.* New York: Atheneum, 1968.

Methods in Modern Cell Biology

Allen, R. D. New observations on cell architecture and dynamics by video-enhanced contrast optical microscopy. *Annu. Rev. Biophys. Biophys. Chem.* 14 (1985): 265.

Bracegirdle, B. Microscopy and comprehension: The development of understanding of the nature of the cell. *Trends Biochem. Sci.* 14 (1989): 464.

Bradbury, S. *An Introduction to the Optical Microscope.* Oxford, England: Oxford University Press, 1984.

Cooper, T. G. *The Tools of Biochemistry.* New York: Wiley, 1977.

de Duve, C. Exploring cells with a centrifuge. *Science* 189 (1975): 186.

de Duve, C., and H. Beaufay. A short history of tissue fractionation. *J. Cell Biol.* 9 (1981): 293s.

Dykstra, M. J. *Biological Electron Microscopy: Theory, Techniques, and Troubleshooting.* New York: Plenum, 1992.

Dykstra, M. J. *A Manual of Applied Techniques for Biological Electron Microscopy.* New York: Plenum, 1993.

Everhart, T. W., and T. L. Hayes. The scanning electron microscope. *Sci. Amer.* (January 1972): 54.

Flegler, S. L., J. W. Heckman, and K. L. Klomparens. *Scanning and Transmission Microscopy: An Introduction.* New York: W. H. Freeman, 1993.

Glauert, A. M. The high-voltage electron microscope in biology. *J. Cell Biol.* 63 (1974): 717.

Hunter, E., and M. Silver. *Practical Electron Microscopy.* New York: Cambridge University Press, 1993.

Inoue, S. *Video Microscopy.* New York: Plenum, 1986.

Olson, A. J., and D. S. Goodsell. Visualizing biological molecules. *Sci. Amer.* 267 (November 1992): 76.

Scheller, P. *Centrifugation in Biology and Medical Science.* New York: Wiley, 1981.

Slayter, E. M., and H. S. Slayter. *Light and Electron Microscopy.* New York: Cambridge University Press, 1992.

Sommerville, J., and U. Scheer, eds. *Electron Microscopy in Molecular Biology: A Practical Approach.* Washington, DC: IRL Press, 1987.

Spencer, M. *Fundamentals of Light Microscopy.* Cambridge, England: Cambridge University Press, 1982.

Tanaka, K. Scanning electron microscopy of intracellular structure. *Internat. Rev. Cytol.* 68 (1980): 97.

Watt, I. M. *The Principles and Practice of Electron Microscopy.* Cambridge, England: Cambridge University Press, 1985.

Wischnitzer, S. *Introduction to Electron Microscopy,* 3d ed. Elmsford, NY: Pergamon Press, 1981.

The Scientific Method

Braben, D. *To Be a Scientist: The Spirit of Adventure in Science and Technology.* New York: Oxford University Press, 1994.

Karjalainen, J., J. M. Martin, M. Knip, J. Ilonen, B. H. Robinson, E. Savilahti, H. K. Akerblom, and H.-M. Dorsch. A bovine albumin peptide as a possible trigger of insulin-dependent diabetes mellitus. *New England J. Med.* 327 (1992): 302.

White, L., Jr. *Machina ex Deo: Essays in the Dynamism of Western Culture.* Cambridge, MA: MIT Press, 1968.

2

THE CHEMISTRY OF THE CELL

Students just beginning in cell biology are sometimes surprised—occasionally even dismayed—to find that almost all courses and textbooks dealing with cell biology involve a substantial amount of chemistry. Yet biology in general and cell biology in particular depend heavily on both chemistry and physics. After all, cells and organisms follow all the laws of the physical universe, and biology is really just the study of chemistry in systems that happen to be alive. In fact, everything cells are and do has a molecular and chemical basis. Therefore, we can truly understand and appreciate cellular structure and function only when we can describe that structure in molecular terms and express that function in terms of chemical reactions and events.

Trying to appreciate cellular biology without a knowledge of chemistry would be like trying to appreciate a translation of Goethe without a knowledge of German. Most of the meaning would probably get through, but much of the beauty and depth of appreciation would be lost in the translation. For this reason, we will concentrate on the chemical background necessary for the cell biologist. Specifically, this chapter will focus on several principles that underlie much of cellular biology, preparing, in turn, for the next chapter, which focuses on the major classes of chemical constituents in cells.

The main points of this chapter can conveniently be structured around five principles:

1. *The importance of carbon.* The chemistry of cells is essentially the chemistry of carbon-containing compounds because the carbon atom has several unique properties that make it especially suitable as the backbone of biologically important molecules.

2. *The importance of water.* The chemistry of cells is also the chemistry of water-soluble compounds because the water molecule has several unique properties that make it especially suitable as the universal solvent of living systems.

3. *The importance of selectively permeable membranes.* Given that most biologically important molecules are

water-soluble, membranes that do not dissolve in water and are differentially permeable are very important both in defining cellular spaces and compartments and in controlling the movements of molecules and ions into and out of such spaces and compartments.

4. *The importance of synthesis by polymerization of small molecules.* Most biologically important molecules are either small, water-soluble organic molecules that can be transported across membranes or large macromolecules that cannot in general be so transported. Biological macromolecules are polymers formed by the linking together of similar or identical small molecules. The synthesis of macromolecules by polymerization of monomeric subunits is an important principle of cellular chemistry.

5. *The importance of self-assembly.* Proteins and other biological macromolecules made of repeating monomeric subunits are often capable of self-assembly into higher levels of structural organization. Self-assembly is possible because the information needed to specify the spatial configuration of the molecule is inherent in the linear array of monomers present in the polymer. Self-assembly is qualified, however, by the need in many cases for proteins called "molecular chaperones" that assist in the assembly process by inhibiting incorrect interactions that would lead to inactive structures.

Given these five principles, we can appreciate the main topics in cellular chemistry with which we need to be familiar before venturing further into our exploration of what it means to be a cell.

The Importance of Carbon

To study cellular molecules really means to study carbon-containing compounds. Almost without exception, molecules of importance to the cell biologist have a backbone, or skeleton, of carbon atoms linked together covalently. Actually, the study of carbon-containing compounds is the domain of **organic chemistry.** In its early days, organic chemistry was almost synonymous with biological chemistry because most of the carbon-containing compounds that were first investigated were obtained from biological sources (hence the word *organic*, acknowledging the organismal origins of the compounds). The terms have long since gone their separate ways, however, because organic chemists have now synthesized a bewildering variety of carbon-containing compounds that do not occur naturally (that is, in the biological world). Organic chemistry therefore includes all classes of carbon-containing compounds, whereas **biological chemistry** (**biochemistry** for short) deals specifically with the chemistry of living systems and is, as we have already seen, one of the several historical strands that form an integral part of modern cell biology.

The **carbon atom** (C) is the most important atom in bio-logical molecules. The diversity and stability of carbon-containing compounds are due to specific properties of the carbon atom and especially to the nature of the interactions of carbon atoms with one another as well as with a limited number of other elements found in molecules of biological importance.

The single most fundamental property of the carbon atom is its **valence** of four, which means that the outermost electron orbital of the atom lacks four of the eight electrons needed to fill it completely (Figure 2-1a). Since a complete outer orbital is required for the most stable chemical state of an atom, carbon atoms tend to associate with one another or with other electron-deficient atoms, allowing adjacent atoms to share a pair of electrons. For each such pair, one electron comes from each of the atoms. Atoms that share each other's electrons in this way are said to be joined together by a **covalent bond.** Carbon atoms are most likely to form covalent bonds with one another and with atoms of oxygen (O), hydrogen (H), nitrogen (N), and sulfur (S).

The electronic configurations of several of these atoms are shown in Figure 2-1a. Notice that, in each case, one or more electrons are required to complete the outer orbital.

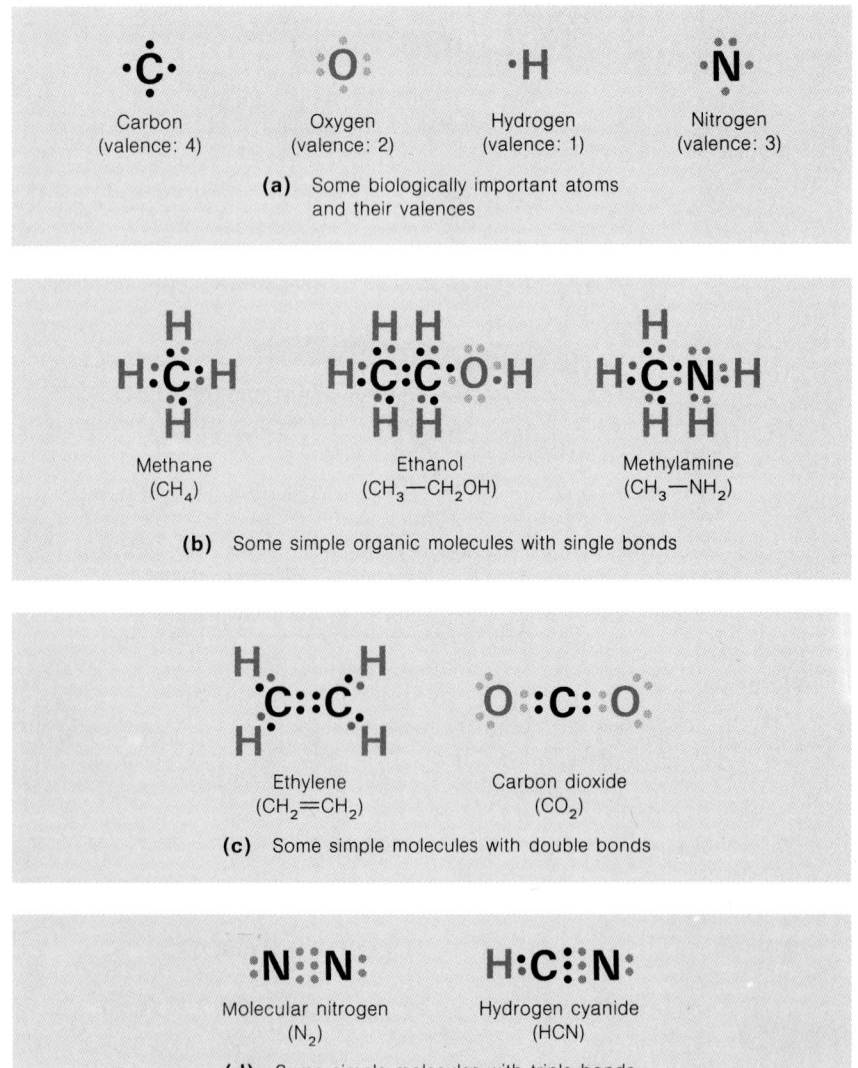

(a) Some biologically important atoms and their valences

(b) Some simple organic molecules with single bonds

(c) Some simple molecules with double bonds

(d) Some simple molecules with triple bonds

Figure 2-1 Electron Configurations of Some Biologically Important Atoms and Molecules. Electronic configurations are shown for **(a)** atoms of carbon, oxygen, hydrogen, and nitrogen and for simple organic molecules with **(b)** single bonds, **(c)** double bonds, and **(d)** triple bonds. Only electrons in the outermost electron orbital are shown. In each case, the two electrons positioned between adjacent atoms represent a shared electron pair, with one electron provided by each of the two atoms. Electrons are color-coded; those from carbon, oxygen, hydrogen, and nitrogen are black, light pink, dark pink, and gray, respectively. (All electrons are equivalent, of course; the color coding is simply to illustrate which electrons are contributed by each atom.)

The number of "missing" electrons corresponds in each case to the valence of the atom, which indicates, in turn, the number of covalent bonds the atom can form. Carbon, oxygen, hydrogen, and nitrogen are the lightest elements that form covalent bonds by sharing electron pairs. This lightness makes the resulting compounds especially stable because the strength of a covalent bond is inversely proportional to the atomic weights of the elements involved in the bond.

Because four electrons are required to fill the outer orbital of carbon, stable organic compounds have four covalent bonds for every carbon atom. Methane, ethanol, and methylamine are simple examples of such compounds, containing only **single bonds** between atoms (Figure 2-1b). Sometimes, two or even three pairs of electrons can be shared by two atoms, giving rise to **double bonds** or even **triple bonds**. Ethylene and carbon dioxide are examples of double-bonded compounds (Figure 2-1c). Triple bonds are found in molecular nitrogen (N_2) and hydrogen cyanide (Figure 2-1d).

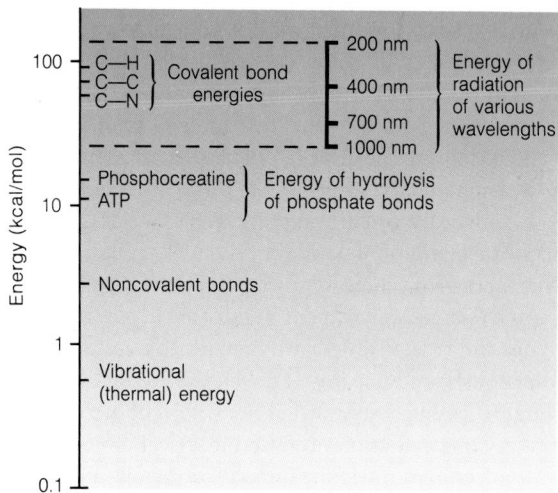

Figure 2-2 Energies of Biologically Important Transitions, Bonds, and Wavelengths of Electromagnetic Radiation. Note that energy is plotted on a logarithmic scale to accommodate the range of values shown.

Carbon-Containing Molecules Are Stable

As already implied, the stability of organic molecules is a property of the favorable electronic configuration of each carbon atom in the molecule. This stability is expressed in terms of **bond energy**—the amount of energy required to break 1 mole (6×10^{23}) of such bonds. Bond energies are usually expressed in *calories per mole (cal/mol)*, where a **calorie** is the amount of energy needed to raise the temperature of one gram of water one degree centigrade.*

It takes a great deal of energy to break a covalent bond. For example, the carbon-carbon (C—C) bond has a bond energy of 83 kilocalories per mole (kcal/mol). The bond energies for carbon-nitrogen (C—N), carbon-oxygen (C—O), and carbon-hydrogen (C—H) bonds are all in the same range: 70, 84, and 99 kcal/mol, respectively. Even more energy is required to break a carbon-carbon double bond (C=C; 146 kcal/mol) or a carbon-carbon triple bond (C≡C; 212 kcal/mol), so these compounds are even more stable.

We can appreciate the significance of these bond energies by comparing them with other relevant energy values, as shown in Figure 2-2. Most noncovalent bonds in biologically important molecules have energies of only a few kilocalories per mole, and the energy of thermal vibration is even lower—about 0.6 kcal/mol. Covalent bonds are much higher in energy and therefore very stable. In fact, specific enzyme-catalyzed reactions capable of quite high energy input are required to break such bonds during chemical reactions in cells.

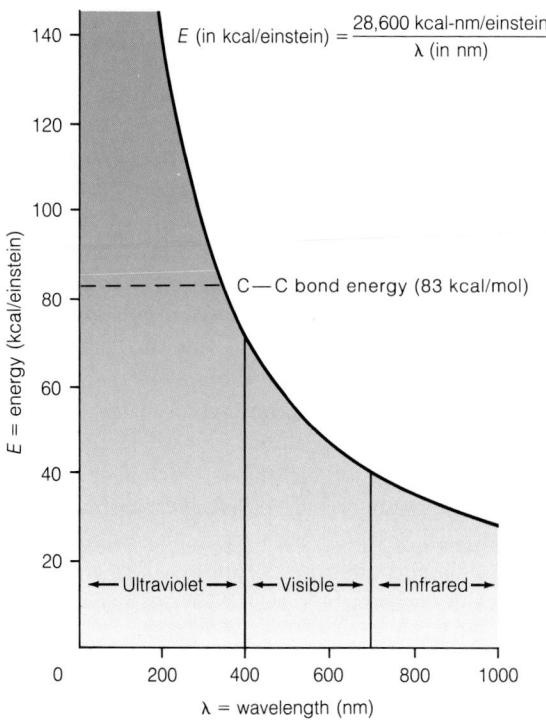

Figure 2-3 The Relationship Between Energy (E) and Wavelength (λ) for Electromagnetic Radiation. The dashed line marks the energy content of the C—C single bond (83 kcal/mol).

The "fitness" of the carbon-carbon bond for biological chemistry on Earth is especially clear when its energy is compared with that of solar radiation. As shown in Figure 2-3, there is an inverse relationship between the wavelength of electromagnetic radiation and its energy content. Specifically, the energy of electromagnetic radiation is related to

*Energy, heat, and work can be expressed either in *calories* and *kilocalories* or in *joules* and *kilojoules*. The joule (J) is the unit of choice among physicists and some biochemists; the calorie (cal) continues to be used in most cell biology texts, including this one. Conversion is easy: 1 cal = 4.184 J, or 1 J = 0.239 cal. Similarly, 1 kcal = 4.184 kJ, or 1 kJ = 0.239 kcal.

the wavelength by the equation $E = 28{,}600/\lambda$, where λ is the wavelength in nm, E is the energy in kilocalories per einstein, and 28,600 is a constant with the units kcal-nm/einstein. (An **einstein** is equal to 1 mole of photons.) Using this equation, you can readily calculate that the visible portion of sunlight (wavelengths of 400–700 nm) is lower in energy than the carbon-carbon bond. For example, green light with a wavelength of 500 nm has an energy content of about 57.2 kcal/einstein. The energy of green light is therefore well below the energies of covalent bonds (see Figure 2-2). If this were not the case, visible light would break covalent bonds spontaneously, and life as we know it would not exist.

Figure 2-3 suggests another important point: the hazard that ultraviolet radiation poses to biological molecules. At a wavelength of 300 nm, for example, ultraviolet light has an energy content of about 95.3 kcal/einstein, clearly enough to break carbon-carbon bonds spontaneously. It is this threat that has led to the current concern about pollutants that destroy the ozone layer in the upper atmosphere because the ozone layer filters out much of the ultraviolet radiation that would otherwise reach the Earth's surface.

Carbon-Containing Molecules Are Diverse

In addition to their stability, carbon-containing compounds are characterized by the great diversity of molecules that can be generated from relatively few different kinds of atoms. Again, this diversity is due to the tetravalent nature of the carbon atom and the resulting propensity of each carbon atom to form covalent bonds to four other atoms. Because one or more of these bonds can be to other carbon atoms, molecules consisting of long chains of carbon atoms can be built up. Ring compounds are also common. Further variety is possible by the introduction of branching and of double and single bonds into the carbon-carbon chains.

When only hydrogen atoms are used to complete the valence requirements of such linear or circular molecules, the resulting compounds are called **hydrocarbons** (Figure 2-4). Hydrocarbons are very important economically because gasoline and other petroleum products are mixtures of short-chain hydrocarbons such as *octane,* an eight-carbon compound (C_8H_{18}).

In biology, on the other hand, hydrocarbons play only a limited role because they are essentially insoluble in water, the universal solvent in biological systems. There is an important exception to this general rule, however: The interior of every biological membrane is a nonaqueous environment from which water and water-soluble compounds are excluded by the long hydrocarbon "tails" of phospholipid molecules that project into the interior of the membrane from either surface. This feature of membranes has important implications for their role as permeability barriers, as we will see shortly.

Most biological compounds contain, in addition to carbon and hydrogen, one or more atoms of oxygen and often nitrogen, phosphorus, or sulfur as well. These atoms are usually part of various **functional groups** that confer both

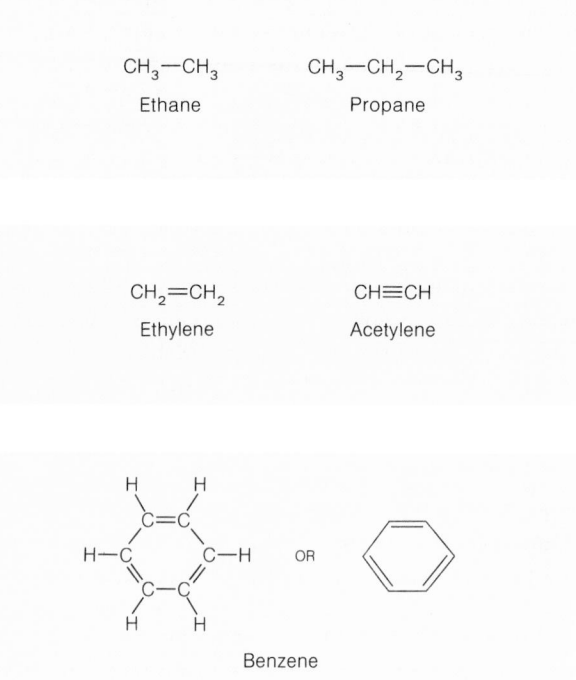

Figure 2-4   Some Simple Hydrocarbon Compounds.
Compounds in the top row have single bonds only, whereas those in the second row have double or triple bonds. The condensed structure shown for benzene is an example of the simplified structures that chemists frequently use for such compounds.

water solubility and chemical reactivity on the molecules of which they are a part. (Even the phospholipid molecules whose hydrocarbon tails contribute so importantly to the nonaqueous nature of the membrane interior contain atoms other than hydrogen and carbon.)

Some of the more common functional groups present in biological molecules are shown in Figure 2-5. Several of these groups are ionized or protonated at the near-neutral pH of most cells, including the negatively charged *carboxyl* and *phosphoryl* groups and the positively charged *amino* group. Other groups, such as the *hydroxyl, sulfhydryl, carbonyl,* and *aldehyde* groups, are uncharged at pH values near neutrality. However, they are much more polar than hydrocarbons and so cause a significant redistribution of electrons within the molecules to which they are attached, thereby conferring on these molecules greater water solubility and chemical reactivity.

Carbon-Containing Molecules Can Form Stereoisomers

Carbon-containing molecules are capable of still greater diversity because the carbon atom is a **tetrahedral** structure with *geometric symmetry* (Figure 2-6). When four different atoms or groups of atoms are bonded to the four corners of such a tetrahedral structure, two different spatial configurations are possible. Although both forms have the same struc-

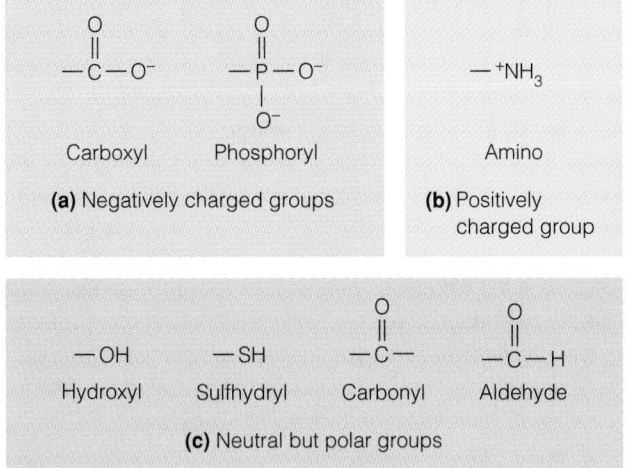

(a) Negatively charged groups

Carboxyl Phosphoryl

(b) Positively charged group

Amino

— OH — SH Carbonyl Aldehyde

Hydroxyl Sulfhydryl

(c) Neutral but polar groups

Figure 2-5 Some Common Functional Groups Found in Biological Molecules. Each functional group is shown in the form that predominates at the near-neutral pH of most cells. **(a)** The carboxyl and phosphoryl groups are ionized and therefore have a negative charge. **(b)** The amino group, on the other hand, is protonated and is therefore positively charged. **(c)** Hydroxyl, sulfhydryl, carbonyl, and aldehyde groups are uncharged at pH values near neutrality but are much more polar than hydrocarbons, thereby conferring greater polarity and hence greater water solubility on the organic molecules to which they are attached.

tural formula, they are not superimposable but are, in fact, mirror images of each other (see Figure 2-6). Such mirror-image forms of the same compound are called **stereoisomers.**

A carbon atom that has four different substituents is called an **asymmetric carbon atom.** Because two stereoisomers are possible for each asymmetric carbon atom, a compound with n asymmetric carbon atoms will have 2^n possible stereoisomers. As shown in Figure 2-7a, the three-carbon amino acid *alanine* has a single asymmetric carbon atom (in the center) and thus has two stereoisomers, called L-alanine and D-alanine. (Neither of the other two carbon atoms of alanine is an asymmetric carbon atom because one has three identical substituents and the other has two bonds to a single oxygen atom.) Both stereoisomers of alanine occur in nature, but only L-alanine is present as a component of proteins.

As an example of a compound with multiple asymmetric carbon atoms, consider the six-carbon sugar *glucose* shown in Figure 2-7b. Of the six carbon atoms of glucose, the four shown in pink are asymmetric. (Again, you should be able to figure out why the other two carbon atoms are not asymmetric.) With four asymmetric carbon atoms, the structure shown, D-glucose, is only one of 2^4, or 16, possible stereoisomers of the $C_6H_{12}O_6$ molecule. In this case, however, not all of the other possible stereoisomers exist in nature, mainly because some are energetically much less favorable than others.

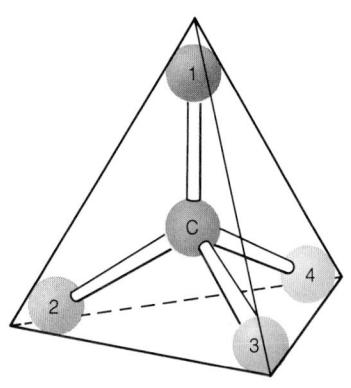
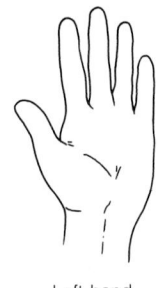
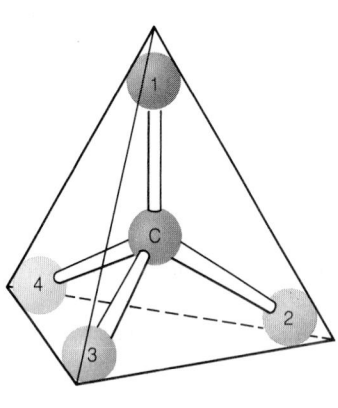
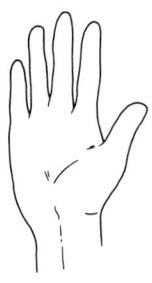

Figure 2-6 Stereoisomers. Stereoisomers of organic compounds occur when four different groups are attached to a tetrahedral carbon atom. Stereoisomers, like left and right hands, are mirror images of each other and cannot be superimposed on one another. (The dashed line down the center of the figure is the plane of the mirror.)

Left hand Right hand

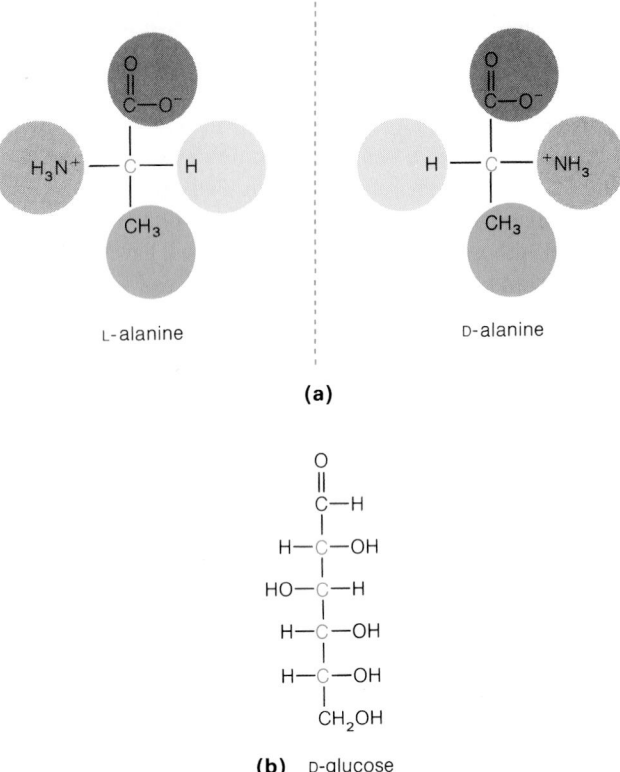

(a)

(b) D-glucose

Figure 2-7　Stereoisomers of Biological Molecules.
(a) The amino acid alanine has a single asymmetric carbon atom (in the center) and can therefore exist in two spatially different forms designated as L- and D-alanine. (The dashed line down the center of the figure is the plane of the mirror.) **(b)** The six-carbon sugar glucose has four asymmetric carbon atoms (pink), so D-glucose is just one of 16 (2^4) possible stereoisomers of the $C_6H_{12}O_6$ molecule.

The Importance of Water

Just as the carbon atom is uniquely important because of its role as the universal backbone of biologically important molecules, the water molecule commands special attention because of its indispensable role as the universal solvent in biological systems. Water is, in fact, the single most abundant component of cells and organisms. Typically, about 75–85% of a cell by weight is water, and many cells depend on an extracellular environment that is essentially aqueous as well. In some cases this is the body of water—whether an ocean, lake, or river—in which the cell or organism lives, while in other cases it may be body fluids in which the cell is suspended or with which the cell is bathed.

Water is indispensable for life as we know it. True, there are life forms that can go into a "holding action" and survive periods of severe water scarcity. Seeds of plants and spores of bacteria and fungi are clearly in this category; the moisture content of a dry seed is frequently as low as 10–20%. Some lower plants and animals, notably certain mosses, lichens, nematodes, and rotifers, can also undergo physiological adaptations that allow them to dry out and survive in a

highly desiccated form, sometimes for surprisingly long periods of time. Such adaptations are clearly an advantage in environments characterized by periods of drought. Yet all of these are at best holding actions, and resumption of normal activity always requires rehydration. Thus, an adequate water supply is needed to serve as the solvent system for the variety of activities that we associate with what it means to be alive.

To understand why water is so uniquely suitable for its role, we need to look at its chemical properties. The most critical attribute is clearly its *polarity*, because this property in turn accounts for its temperature-stabilizing capacity and its cohesiveness, both of which have important consequences for biological chemistry.

Water Molecules Are Polar

To understand the polar nature of water, we need to consider the shape of the molecule. As shown in Figure 2-8a, the water molecule is triangular rather than linear in shape, with the two hydrogen atoms bonded to the oxygen at an angle of 109.5° rather than 180°. It is not an overstatement to say that life as we know it depends critically on this angle, given the distinctive properties that the resulting asymmetry confers on the water molecule. Although the molecule as a whole is uncharged, the electrons tend to be unevenly distributed. The oxygen atom at the head of the molecule is **electronegative;** that is, it tends to draw electrons toward it, giving that end of the molecule a partial negative charge and leaving the other end of the molecule with a partial positive charge around the hydrogen atoms. This charge separation gives the water molecule its **polarity,** a property it shares with all

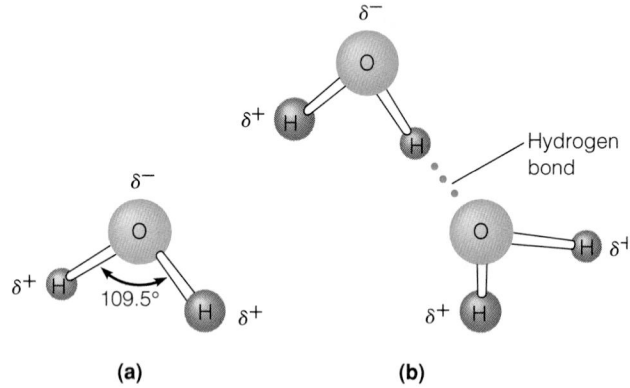

Figure 2-8　Polarity of Water Molecules.　(a) The water molecule is polar because it has an asymmetric charge distribution. The two hydrogen atoms are bonded to the oxygen at an angle of 109.5°. The oxygen atom bears a partial negative charge (δ^-; the Greek letter delta stands for "partial") and is thus the electronegative portion of the molecule. The two hydrogen atoms are electropositive; their end of the molecule has a partial positive charge (δ^+). **(b)** Since the electropositive hydrogen atom of one molecule is attracted to the electronegative oxygen atom of a neighboring molecule, adjacent water molecules tend to orient themselves so that weak hydrogen bonds form between them.

other molecules with an asymmetric internal distribution of charge. In the case of water, the polarity of the molecule has enormous consequences, accounting for the *cohesiveness*, the *temperature-stabilizing capacity*, and the *solvent properties* of water.

Water Molecules Are Cohesive

Because of their polarity, water molecules have an affinity for each other and tend to orient themselves spontaneously so that the electronegative oxygen atom of one molecule is associated with the electropositive hydrogen atoms of adjacent molecules. Each such association is called a *hydrogen bond* and is frequently represented by a dotted line, as in Figure 2-8b. Since each oxygen atom can bond to two hydrogens, and both of the hydrogen atoms can associate in this way with the oxygen atoms of adjacent molecules, liquid water is characterized by an extensive three-dimensional network of hydrogen-bonded molecules (Figure 2-9). The hydrogen bonds between adjacent molecules are constantly being broken and reformed, but on average, each molecule of water in the liquid state is hydrogen-bonded to about 3½ neighbor molecules at any given time. In ice, the hydrogen bonding is still more extensive, giving rise to a rigid, highly regular crystalline lattice with every oxygen hydrogen-bonded to hydrogens of two adjacent molecules and every water molecule therefore hydrogen-bonded to four neighboring molecules.

It is this tendency to form hydrogen bonds between adjacent molecules that makes water so highly *cohesive*. This cohesiveness accounts for the high *surface tension* of water, as well as for its high *boiling point*, high *specific heat*, and high *heat of vaporization*. The high surface tension of water causes the capillary action that enables water to move up the conducting tissues of plants and allows insects such as the water strider to move across the surface of a pond without sinking.

Water Has a High Temperature-Stabilizing Capacity

An important property of water that derives directly from the hydrogen bonding between adjacent molecules is the high specific heat that gives water its **temperature-stabilizing capacity. Specific heat** is the amount of heat a substance absorbs per gram to increase its temperature 1°C. The specific heat of water is 1.0 calorie per gram (cal/g). (This is, in fact, the way the *calorie* is defined.)

The heat capacity of water is much higher than that of most other liquids because of its extensive hydrogen bonding. Much of the energy that in other liquids would contribute directly to an increase in the motion of solvent molecules and therefore to an elevation in temperature is used instead to break hydrogen bonds between neighboring water molecules. In effect, by absorbing heat that would otherwise increase the temperature of the water more rapidly, hydrogen bonds buffer aqueous solutions against large changes in temperature. This capability is an important consideration

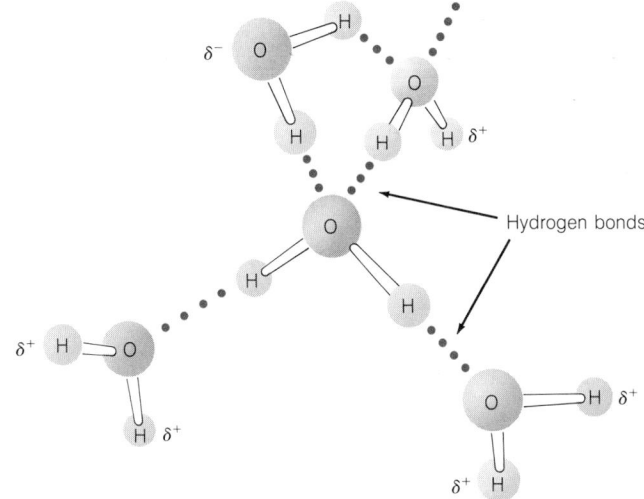

Figure 2-9 Hydrogen Bonding Between Water Molecules. The extensive association of water molecules with one another in either the liquid or the solid state is due to hydrogen bonds (dotted lines) between the electronegative oxygen atom of one water molecule and the electropositive hydrogen atoms of adjacent molecules. In ice, the resulting crystal lattice is regular and complete; every oxygen is hydrogen-bonded to hydrogens of two adjacent molecules. In water, some of the structure is disrupted, but much is retained.

for the cell biologist, because cells release large amounts of energy during metabolic reactions. This release of energy would pose a serious overheating problem for cells if it were not for the extensive hydrogen bonding and the resulting high specific heat of water molecules.

Water Is an Excellent Solvent

Probably the single most important property of water from a biological perspective is its excellence as a general solvent. A **solvent** is a fluid in which another substance, called the **solute,** can be dissolved. Water is an especially good solvent for biological purposes because of its remarkable capacity to dissolve a great variety of solutes and also because it is generally inert (i.e., does not react chemically with most solutes).

It is the polarity of water that makes it so useful as a solvent. Most of the molecules in cells are also polar and therefore interact electrostatically with water molecules, as do charged ions. Solutes that are polar and therefore dissolve readily in water are called **hydrophilic** ("water-loving"). Most small organic molecules found in cells are hydrophilic; examples are sugars, organic acids, and some of the amino acids. Nonpolar molecules are not very soluble in water and are accordingly termed **hydrophobic** ("water-fearing"). Among the more important hydrophobic compounds found in cells are the lipids and proteins of which membranes are made. Some molecules, as we shall see, have both hydrophobic and hydrophilic regions, so parts of the molecule have an affinity for an aqueous environment while other parts of the molecule do not.

To understand why polar substances dissolve so readily in water, first consider a salt such as sodium chloride (NaCl) (Figure 2-10). Because it is a salt, NaCl exists in crystalline form as a lattice of positively charged sodium ions (Na⁺) and negatively charged chloride ions (Cl⁻). For NaCl to dissolve in a liquid, solvent molecules must overcome the attraction of the oppositely charged Na⁺ cations and Cl⁻ anions for each other and involve them instead in interactions with the solvent molecules themselves. Because of their polarity, water molecules can form **spheres of hydration** around both Na⁺ and Cl⁻, thereby neutralizing their attrac-

tion for each other and lessening their likelihood of reassociation. As Figure 2-10a shows, the sphere of hydration around a cation such as Na⁺ involves water molecules clustered around the ion with their negative (oxygen) ends pointing toward it. For an anion such as Cl⁻, the orientation of the water molecules is reversed, with the positive (hydrogen) ends of the solvent molecules pointing in toward the ion (Figure 2-10b).

Some biological compounds are soluble in water because they exist as ions at the near-neutral pH of the cell and are therefore solubilized and hydrated like the ions of Figure 2-10. Compounds containing carboxyl, phosphoryl, or amino groups are in this category (see Figure 2-5). Most organic acids, for example, are almost completely ionized at a pH near 7 and therefore exist as anions that are kept in solution by spheres of hydration, just as the chloride ion of Figure 2-10b is. Amines, on the other hand, are usually protonated at cellular pH and thus exist as hydrated cations, like the sodium ion of Figure 2-10a.

More frequently, organic molecules have no net charge but are nonetheless hydrophilic because they have some regions that are positively charged and other regions that are negatively charged. Water molecules tend to cluster around such regions, and the resulting interactions between solute and water molecules keep the solute molecules from associating with one another. Compounds containing the hydroxyl, sulfhydryl, carbonyl, or aldehyde groups shown in Figure 2-5 are usually in this category.

Hydrophobic molecules, on the other hand, have no such polar regions and therefore show no tendency to interact electrostatically with water molecules. In fact, they actually disrupt the hydrogen-bonded structure of water and, for this reason, tend to be excluded by the water molecules. Hydrophobic molecules therefore tend to coalesce in an aqueous medium, associating with one another rather than with the water. This association is driven not so much by any specific affinity of the hydrophobic molecules for one another but by the strong tendency of water molecules to form hydrogen bonds and to exclude molecules that disrupt hydrogen bonding. As we will see later in the chapter, such associations of hydrophobic molecules (or parts of molecules) are a major driving force in the folding of molecules, the assembly of cellular structures, and the organization of membranes.

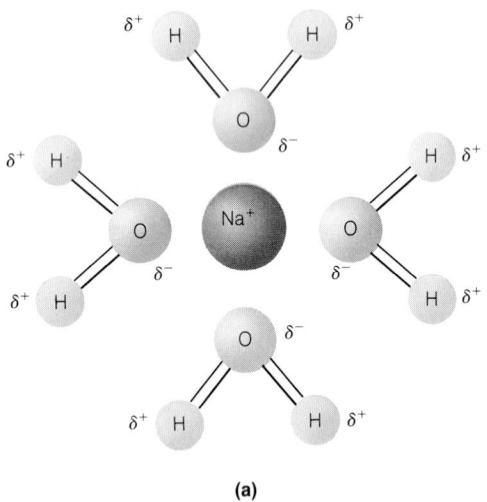

(a)

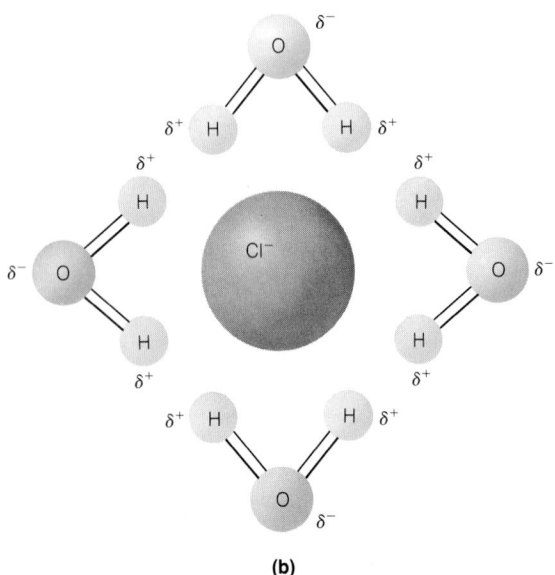

(b)

Figure 2-10   The Solubilization of Sodium Chloride.
Sodium chloride (NaCl) dissolves in water because of the formation of spheres of hydration around **(a)** the sodium ions and **(b)** the chloride ions. The oxygen atom and the sodium and chloride ions are drawn to scale.

The Importance of Selectively Permeable Membranes

Every cell and organelle needs some sort of physical barrier to keep its contents in and external materials out, as well as some means of controlling exchange between its internal environment and the extracellular environment. Ideally, such a barrier should be impermeable to most of the molecules and ions found in cells and their surroundings. Otherwise, substances could diffuse freely in and out, and the cell would not really have a defined content at all. On the other hand, the barrier cannot be completely impermeable, or else desired exchanges between the cell and its environment could not take place. Moreover, such a barrier must be insoluble in water so that it will not be dissolved by the aqueous medium of the cell. At the same time, it must be readily permeable to water because water is the basic solvent system of the cell and must be able to flow into and out of the cell as needed.

As you might expect, the membranes that surround cells and organelles satisfy these criteria admirably. A **membrane** is essentially a hydrophobic permeability barrier consisting of hydrophobic *phospholipids*, hydrophobic *proteins*, and (in the case of human and animal cells) *cholesterol*. Ac-

tually, the phospholipids and many of the proteins in the membrane are not simply hydrophobic; they have both hydrophilic and hydrophobic regions and are therefore referred to as **amphipathic molecules** (the Greek prefix *amphi-* means "of both kinds"). The amphipathic nature of membrane phospholipids is illustrated in Figure 2-11, which shows the structure of *phosphatidyl ethanolamine*, a prominent phospholipid in many kinds of membranes. (Phosphatidyl ethanolamine is an example of a *phosphoglyceride*, the major class of membrane phospholipids in most cells. For other examples of phosphoglycerides and other classes of membrane phospholipids, see Figures 3-26 and 3-27.) The distinguishing feature of amphipathic phospholipids is

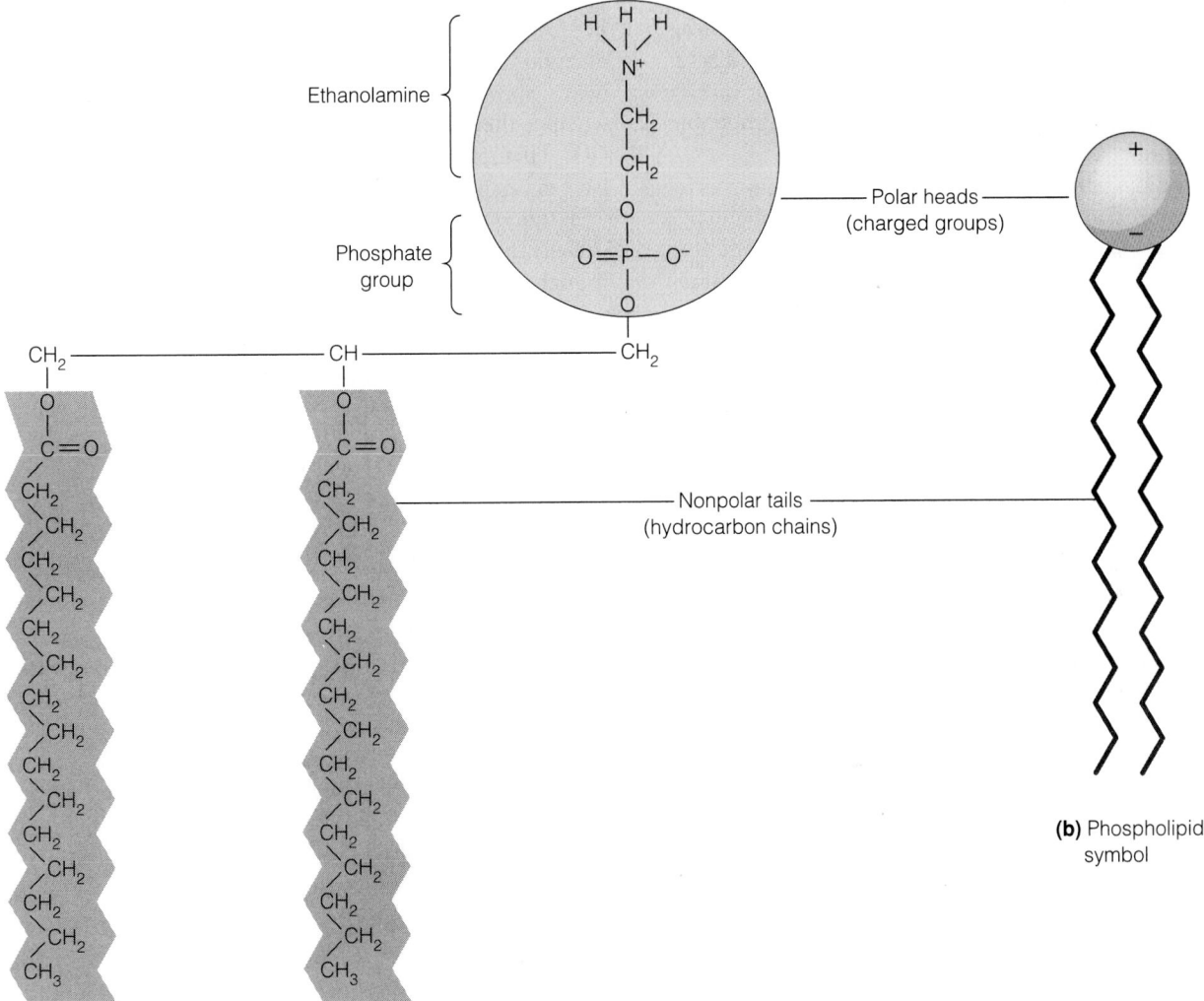

(a) Phospholipid structure

Figure 2-11 The Amphipathic Nature of Membrane Phospholipids. **(a)** A phospholipid molecule consists of two long nonpolar tails (gray) and a polar head (yellow). Shown here is phosphatidyl ethanolamine, an example of the phosphoglyceride class of membrane phospholipids. The polarity of the head of a phospholipid molecule results from a negatively charged phosphate group linked to a positively charged group—an amino group, in the case of phosphatidyl ethanolamine. Other

common phosphoglycerides with both a phosphate group and an amino group include phosphatidyl serine and phosphatidyl choline, the structures for which are shown in Figure 3-26. **(b)** A phospholipid molecule is often represented schematically by a circle for the polar head (notice the plus and minus charges) and two zigzag lines for the nonpolar hydrocarbon chains.

that each molecule consists of a polar "head" and two non-polar hydrocarbon "tails." The polarity of the hydrophilic head is due to the presence of a negatively charged phosphate group linked to a positively charged group—an amino group, in the case of phosphatidyl ethanolamine and most other phosphoglycerides.

The Membrane Bilayer

When exposed to an aqueous environment, amphipathic molecules undergo hydrophobic interactions. In a membrane, for example, phospholipids spontaneously arrange themselves so that their polar heads are facing outward toward the aqueous milieu but their hydrophobic tails are hidden from the water by interacting with the tails of other molecules oriented in the opposite direction. The resulting structure is the **phospholipid bilayer,** shown in Figure 2-12. The heads of both layers face outward and the hydrocarbon tails extend inward, forming the continuous hydrophobic interior of the membrane.

Every biological membrane has such a lipid bilayer as its basic structure. Each of the phospholipid layers is about 4–5 nm thick, so the bilayer has a width of about 8–10 nm. It is the phospholipid bilayer that gives membranes their characteristic "railroad track" appearance when seen with the electron microscope (Figure 2-13a). Apparently, the osmium used in preparing the tissue for electron microscopy reacts with the hydrophilic heads of the phospholipid molecules at both surfaces of the membrane but not with the hydrophobic tails in the interior of the membrane, giving the membrane its *trilaminar* (three-layered) appearance.

The structure of biological membranes is shown in Figure 2-13b. Embedded within or associated with the membrane lipid bilayer are various **membrane proteins.** These proteins are almost always amphipathic, and they orient themselves in the lipid bilayer accordingly. Hydrophobic regions of the protein associate with the interior of the membrane, whereas hydrophilic regions protrude into the aqueous environment at the surface of the membrane. Some proteins of the plasma membrane have carbohydrate side chains attached to their outer surfaces.

Depending on the particular membrane, the membrane proteins may play any of a variety of roles. Some are *trans-port proteins*, responsible for moving specific substances across an otherwise impermeable membrane. Others are *enzymes* that catalyze reactions associated with the specific membrane. Still others are the *receptors* on the outer surface of the cell membrane, the *electron transport intermediates* of the mitochondrial membrane, or the *chlorophyll-binding proteins* of the chloroplast.

Movement Across the Membrane

Because of its hydrophobic interior, a membrane is readily permeable to nonpolar molecules but is quite impermeable to most polar molecules and is highly impermeable to all ions. Because most cellular constituents are either polar or charged, they have little or no affinity for the membrane interior and are effectively prevented from entering or escaping from the cell or organelle. Very small molecules are an exception, however. Compounds with molecular weights below about 100 diffuse across membranes regardless of whether they are nonpolar (such as O_2) or polar (such as CO_2, ethanol, and urea). Thus, both carbon dioxide and oxygen can move readily into and out of cells and organelles, even though one is polar and the other is not. Water is an especially important example of a very small molecule that, although polar, diffuses very rapidly across membranes.

In contrast, even the smallest ions are very effectively excluded from the hydrophobic interior of the membrane. For example, a synthetic lipid bilayer is 10^8 to 10^9 times less permeable to such small cations as Na^+ or K^+ than it is to water. This striking difference is due to both the charge on an ion and the sphere of hydration that surrounds the ion.

Of course, it is essential that cells have ways of transferring ions such as Na^+ and K^+ as well as a wide variety of polar molecules across membranes that are not otherwise permeable to these substances. As already noted, membranes are equipped with **transport proteins** to serve this function (see Figure 2-13b). A transport protein is a specialized transmembrane protein that serves either as a *hydrophilic channel* through an otherwise hydrophobic membrane or as a *carrier* that binds a specific solute on one side of the membrane and then undergoes a conformational change to move the solute across the membrane.

Whether a channel or a carrier, each transport protein is specific for a particular molecule or ion (or, in some cases,

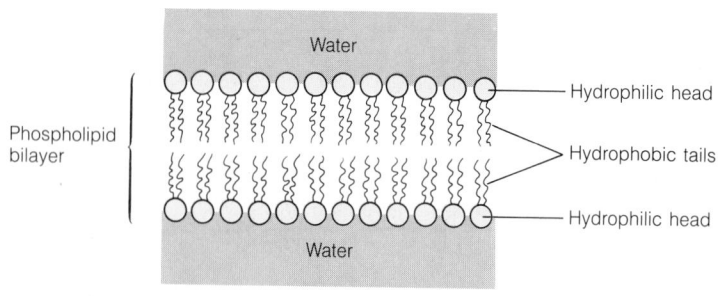

Figure 2-12 The Phospholipid Bilayer as the Basis of Membrane Structure.   Due to their amphipathic nature, phospholipids in an aqueous environment orient themselves in a double layer, with the hydrophobic tails (gray) buried on the inside and the hydrophilic heads (yellow) pointing toward the aqueous milieu on either side of the membrane.

for a class of closely related molecules or ions). Moreover, the activities of these proteins can be carefully regulated to meet cellular needs. As a result, biological membranes can best be described as **selectively permeable**: With the exception of very small molecules, the only molecules or ions that can move across a particular membrane are those for which the appropriate transport proteins are present in the membrane.

The Importance of Synthesis by Polymerization

For the most part, cellular structures such as ribosomes, chromosomes, membranes, flagella, and cell walls are made up of ordered arrays of linear polymers called **macromolecules.** (Some branching occurs, notably in the polysaccharides starch and glycogen, but the linearity of biological polymers is a good first approximation.) Examples of important macromolecules in cells include *proteins, nucleic acids* (both DNA and RNA), and *polysaccharides* such as starch, glycogen, and cellulose. Macromolecules are very important in both the function and the structure of cells. To understand the biochemical basis of cell biology therefore really means to understand macromolecules—how they are made, how they are assembled, and how they function.

The Importance of Macromolecules

The importance of macromolecules in cell biology is emphasized by the cellular hierarchy shown in Figure 2-14. The compounds of which most cellular structures are made are small, water-soluble *organic molecules* (level 1) that cells either obtain from other cells or synthesize from simple nonbiological molecules such as carbon dioxide, ammonia, or phosphate ions available from the environment. These organic molecules polymerize to form *biological macromolecules* (level 2) such as nucleic acids, proteins, or carbohydrates. Macromolecules are then assembled into a variety of *supramolecular structures* (level 3), which in turn are components of organelles and other subcellular structures (level 4) and hence of the cell itself (level 5).

One of the examples in Figure 2-14 is that of cell wall biogenesis (panels on right). In plants, a major component of the cell wall (level 3) is the polysaccharide cellulose (level 2). Cellulose is in turn a repeating polymer of the simple sugar glucose (level 1), formed by the plant cell from carbon dioxide and water in the process of photosynthesis.

From such examples, a general principle emerges: *The macromolecules that are responsible for most of the form and order characteristic of living systems are generated by the polymerization of small organic molecules.* This strategy of forming large molecules by joining smaller units in a repetitive

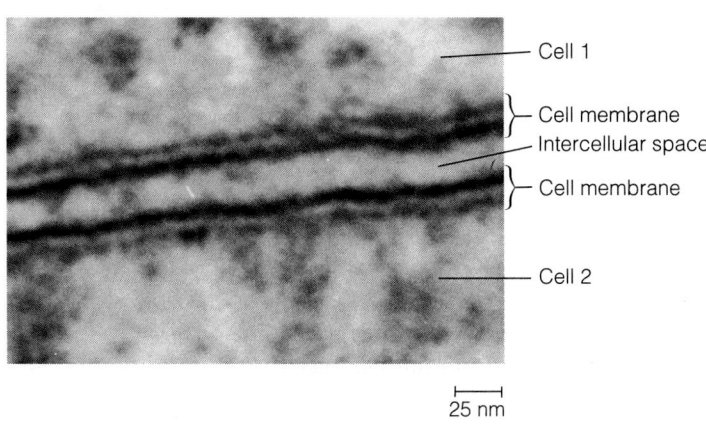

(a) Electron micrograph
of the membranes of
two adjacent cells

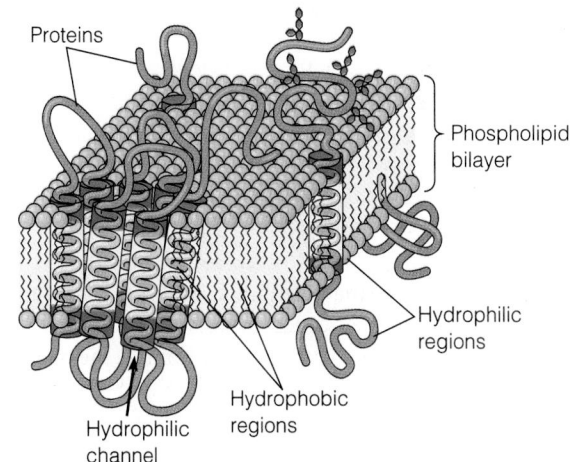

(b) Model of a membrane as
a phospholipid bilayer with
proteins embedded in it

Figure 2-13 Membranes and Membrane Structure.
(a) With the electron microscope, the cell membranes surrounding two adjacent cells each appear as a pair of dark bands. This characteristic trilaminar, or "railroad track," appearance is thought to be due to the association of osmium with the hydrophilic heads of the phospholipid molecules but not with their hydrophobic tails (TEM). **(b)** Biological membranes consist of amphipathic proteins embedded within a phospho-

lipid bilayer. Proteins are positioned in the membrane so that their hydrophobic regions (light purple) are located in the hydrophobic interior of the phospholipid bilayer and their hydrophilic regions (dark purple) are exposed to the aqueous milieu on either side of the membrane. The short chains attached to the upper surface of one of the membrane proteins represent carbohydrate side chains.

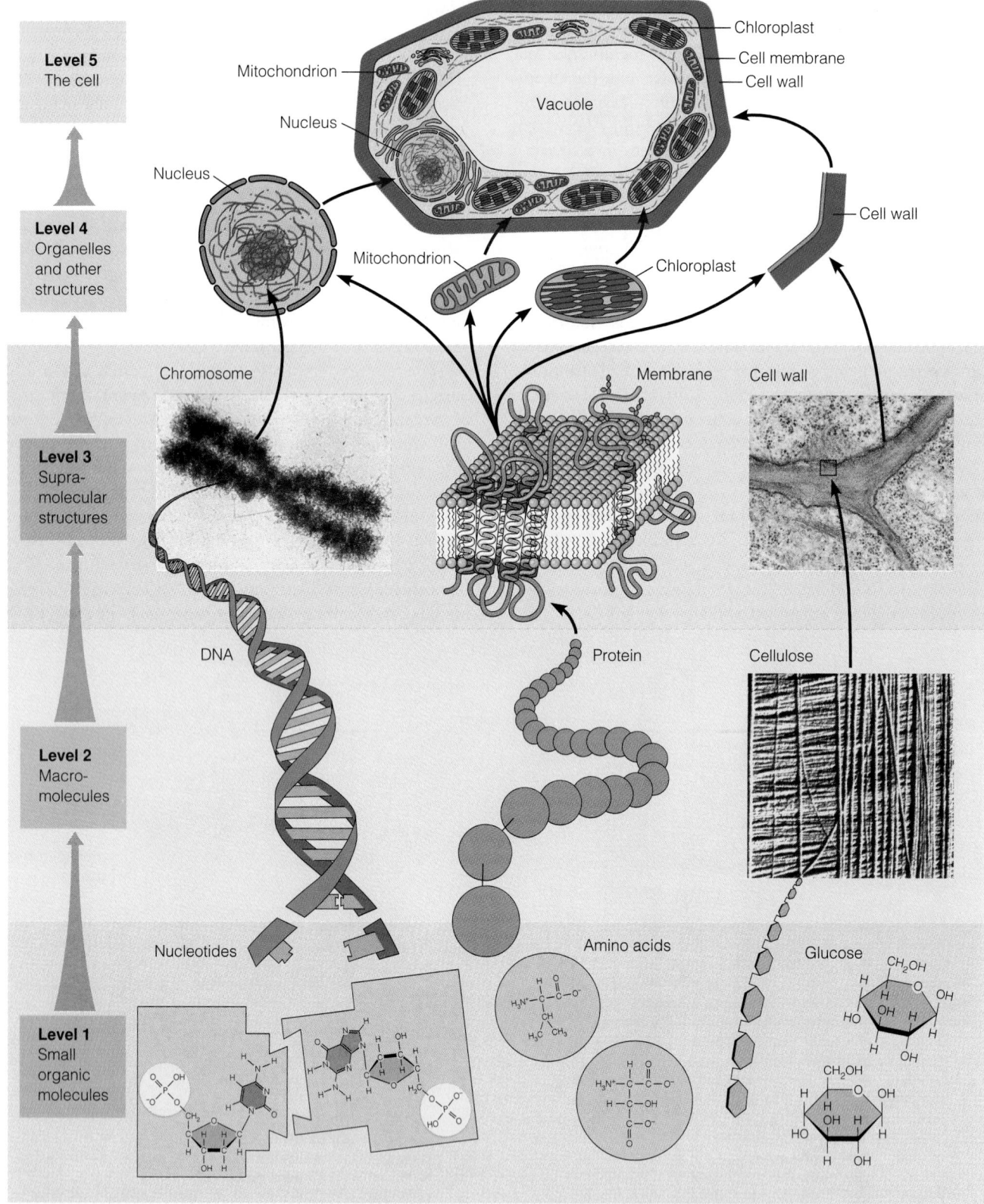

Level 5
The cell

Level 4
Organelles and other structures

Level 3
Supra-molecular structures

Level 2
Macro-molecules

Level 1
Small organic molecules

Chloroplast
Cell membrane
Cell wall

Mitochondrion

Vacuole

Nucleus

Nucleus

Mitochondrion

Chloroplast

Cell wall

Chromosome

Membrane

Cell wall

DNA

Protein

Cellulose

Nucleotides

Amino acids

Glucose

Figure 2-14   The Hierarchical Nature of Cellular Structures and Their Assembly.   Small organic molecules (level 1) are synthesized from simple inorganic substances and are polymerized to form macromolecules (level 2). The macromolecules then assemble into the supramolecular structures (level 3) that make up organelles and other subcellular structures (level 4) and, ultimately, the cell (level 5).

manner is illustrated in Figure 2-15. The importance of this strategy can hardly be overemphasized because it is a fundamental principle of cellular chemistry. The enzymes responsible for the catalysis of cellular reactions, the nucleic acids involved in the storage and expression of genetic information, the glycogen stored by your liver, and the cellulose that gives rigidity to a plant cell wall are all variations on the same design theme. Each is a macromolecule made by the linking together of small repeating units.

Examples of such repeating units, or **monomers,** are the *glucose* present in cellulose or glycogen, the *amino acids* needed to make proteins, and the *nucleotides* of which nucleic acids are made. In general, these are small, water-soluble organic molecules with molecular weights less than 350. They can be transported across most biological membranes, provided that appropriate transport proteins are present. By contrast, most macromolecules that are synthesized from these monomers cannot traverse membranes and therefore must be made in the cell or compartment in which they are needed. (We will encounter several exceptions to this rule when we get to messenger RNA and organellar proteins later in the text, but the generalization is nonetheless a good one.)

Kinds of Macromolecules

Table 2-1 lists the major kinds of macromolecules found in the cell, along with the number and kind of monomeric subunits required for polymer formation. The distinction between *informational macromolecules* and *storage* or *structural macromolecules* is important because the function of these biological polymers affects the number and order of different monomers we can expect to find in them.

Nucleic acids (both DNA and RNA) and proteins are called **informational macromolecules** because the order of the several kinds of nonidentical subunits they contain is nonrandom and highly significant to their function. Figure 2-16 schematically illustrates the structures of these two major classes of informational macromolecules. The order of the monomers (nucleotides in nucleic acids and amino acids in proteins) is genetically determined and carries important information that specifies the function or utilization of these macromolecules. For example, insulin (Figure 2-16b) is a *protein* because it is a string of amino acids linked together by peptide bonds, but it is the specific protein *insulin* because it is a particular string of amino acids in a

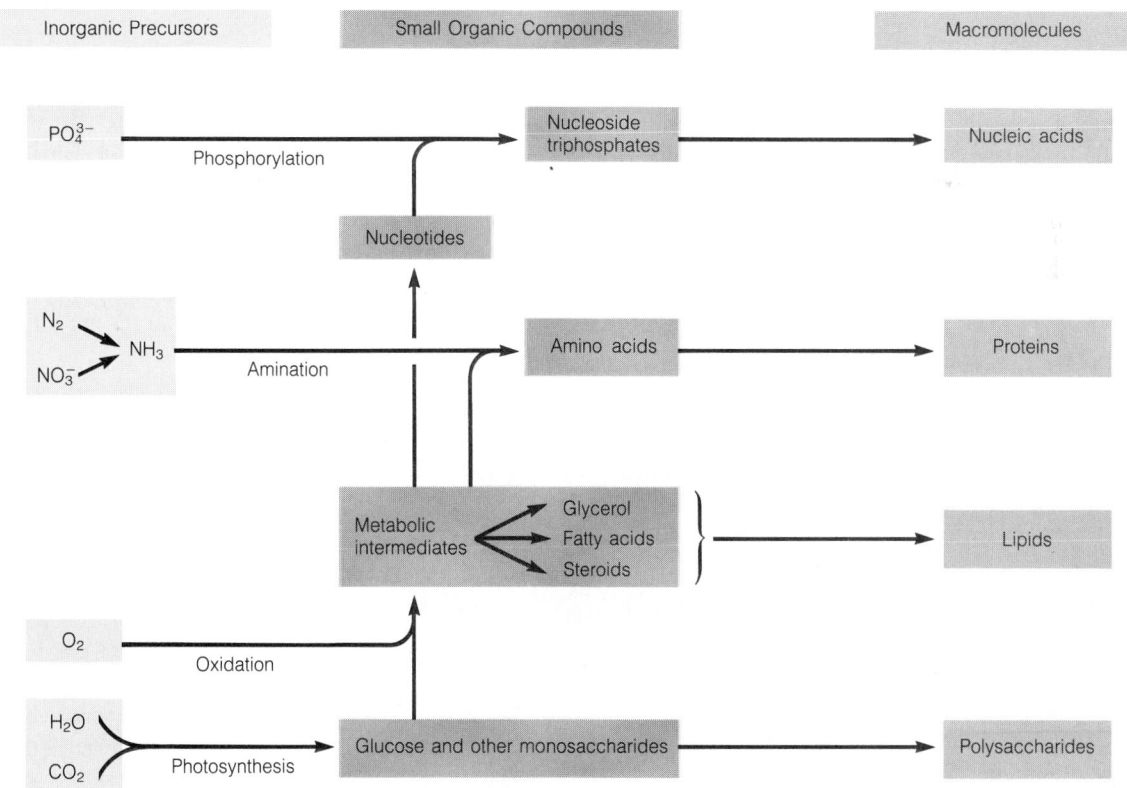

Figure 2-15   The Synthesis of Biological Macromolecules.
Simple inorganic precursors (left) react to form small organic molecules (center), which are then used in the synthesis of the macromolecules (right) of which most cellular structures consist.

Table 2-1 Biologically Important Macromolecules and Their Repeating Units

	Biological Polymer			
	Proteins	*Nucleic Acids*	*Polysaccharides*	
Kind of macromolecule	Informational (*structure*)	Informational	Storage	Structural
Examples	Enzymes, hormones, antibodies	DNA, RNA	Starch, glycogen	Cellulose
Repeating monomers	Amino acids	Nucleotides	Monosaccharides	Monosaccharides
Number of kinds of repeating units	20	4 in DNA; 4 in RNA	One or a few	One or a few

unique, specified order. In fact, the sequence of amino acids is so important that any variation in that sequence is likely to have a deleterious effect on the ability of the protein to perform its function.

For informational macromolecules, polymerization of smaller subunits is not just an economical way of building large molecules; it is an essential part of the role these macromolecules play in the cell. For example, the synthesis of nucleic acids by linkage of small repeating units (nucleotides) is important because it is in the specific order of the monomeric units that these molecules store and transmit the genetic information of the cell.

Polysaccharides, on the other hand, are not informational macromolecules. For them, the order of the monomeric units is enzymatically determined, carries no information, and is not essential to the function or utilization of the molecule in the sense that it is for proteins or nucleic acids. Instead, polysaccharides are either **storage macromolecules** or **structural macromolecules** and usually consist of a single kind of repeating monomer or a strictly repetitious sequence of a few different kinds of monomers. The

most familiar storage polysaccharides are the *starch* of plant cells and the *glycogen* found in animal cells. As shown in Figure 2-17a, both of these storage polysaccharides consist of a single repeating monomer, the simple sugar glucose. (The main difference between glycogen and starch lies in the occurrence and extent of branching, a feature that is not shown in Figure 2-17a but will be encountered in Chapter 3.)

The best-known example of a structural polysaccharide is the *cellulose* present in plant cell walls. Like starch and glycogen, cellulose also consists of glucose units, but the units are linked together by a somewhat different bond, as we will see in Chapter 3. The cell walls of some bacterial cells contain a somewhat more complicated kind of structural polysaccharide. In this case, the molecule consists of two different kinds of monomers, *N*-acetylglucosamine (GlcNAc) and *N*-acetylmuramic acid (MurNAc), as shown in Figure 2-17b. The two monomers occur in strictly alternating sequence, however, and carry no information. A further example of a structural polysaccharide is *chitin*, found in insect exoskeletons and crustacean shells. Chitin consists of GlcNAc units only.

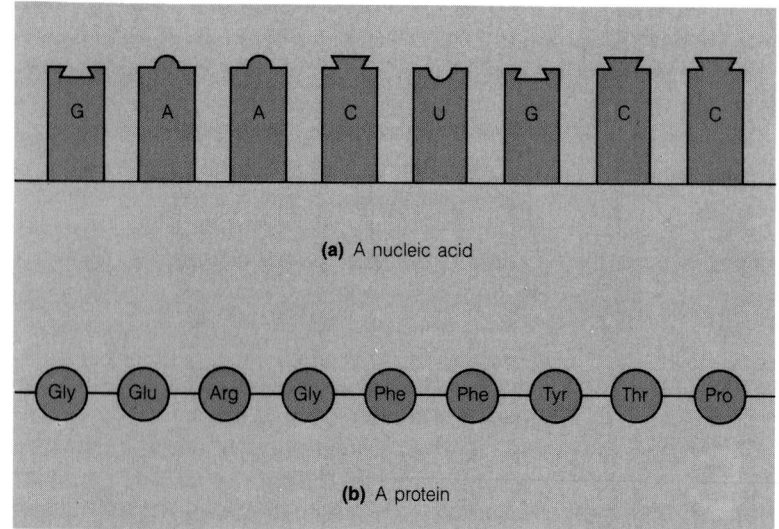

(a) A nucleic acid

(b) A protein

Figure 2-16 Informational Macromolecules. Informational macromolecules contain a variety of kinds of repeating units in a nonrandom, genetically determined order. The two main classes of informational macromolecules are nucleic acids (DNA and RNA) and proteins. **(a)** A portion of the nucleotide sequence of a specific RNA molecule, called 5S ribosomal RNA, from the bacterium *Escherichia coli*. The nucleotides present in RNA are the monophosphorylated forms of the ribonucleotides with the bases guanine (G), adenine (A), cytosine (C), and uracil (U); for the structure of a nucleotide, see Figure 3-12. **(b)** A portion of the amino acid sequence of a specific protein, human insulin. For the structures and three-letter codes of all 20 amino acids, see Figure 3-2 and Table 3-1.

The Synthesis of Macromolecules

In Chapter 3, we will look at each of the major kinds of biologically important polymers, called **biopolymers.** First, however, it will be useful to consider several important principles that underlie the polymerization processes by which these macromolecules arise. Although the chemistry of the monomeric units, and hence of the resulting polymers, differs markedly between such macromolecules as proteins, nucleic acids, and polysaccharides, the following basic principles apply in each case:

1. Biopolymers are always synthesized by the stepwise polymerization of similar or identical small molecules called monomers.

2. The addition of each monomeric unit occurs with the removal of a water molecule and is therefore termed a **condensation reaction.**

3. The monomeric units that are to be joined together must be present in an *activated form* before condensation can occur.

4. Activation usually involves coupling of the monomer to some sort of **carrier molecule** to form a high-energy **activated monomer.**

5. The energy to couple the monomer to the carrier molecule is provided by ATP or a related high-energy compound.

6. Because of the way in which they are synthesized, biopolymers have an inherent **directionality;** that is, the two ends of the polymer chain are chemically different from each other.

Because the elimination of water is essential in all biological polymerization reactions, a feature common to each of the monomeric units, regardless of their other chemical properties, is the presence of a reactive hydrogen (H) on at least one functional group and a reactive hydroxyl group (—OH) elsewhere on the molecule. This structural feature is depicted schematically in Figure 2-18a, which represents the monomeric units (M) simply as boxes but indicates the reactive hydrogen and hydroxyl group on each. For a given kind of polymer, the monomers may differ from one another in other aspects of their structure—indeed, they *must* differ from one another if the polymer is an informational macromolecule—but each monomer possesses the same kind of reactive hydrogen and hydroxyl group.

Figure 2-18a also shows the activation of monomers, an energy-requiring process. Regardless of the nature of the polymer, the addition of each monomer to a growing chain is always energetically unfavorable unless the incoming monomer is in an activated, energized form. Activation involves coupling of the monomer to some sort of carrier molecule (C), with the energy to drive this activation process provided by ATP or a closely related high-energy compound. A different kind of carrier molecule is used for each kind of polymer. For protein synthesis, amino acids are activated by linking them to carriers called *transfer RNA* (or *tRNA*) molecules, whereas polysaccharides are synthesized from sugar (often glucose) molecules that are activated by linking them to derivatives of nucleotides (*adenosine diphosphate* for starch, *uridine diphosphate* for glycogen).

Once activated, monomers are capable of reacting with each other in a condensation reaction that is followed or accompanied by the release of the carrier molecule from one of the two monomers (Figure 2-18b). Subsequent elongation of the polymer is a sequential, stepwise process, with a single activated monomeric unit added at a time, thereby lengthening the elongating polymer by one unit. Figure 2-18c illustrates the *n*th step in such a process, in which the next monomer unit is added to an elongating polymer that already contains *n* monomeric units.

The chemical nature of the energized monomers, the carrier, and the actual activation process differ for each biological polymer, but the general principle is always the same: *Polymer synthesis always involves activated monomers, and ATP or a similar source of energy is always required to activate the monomers by linking them to an appropriate carrier molecule.*

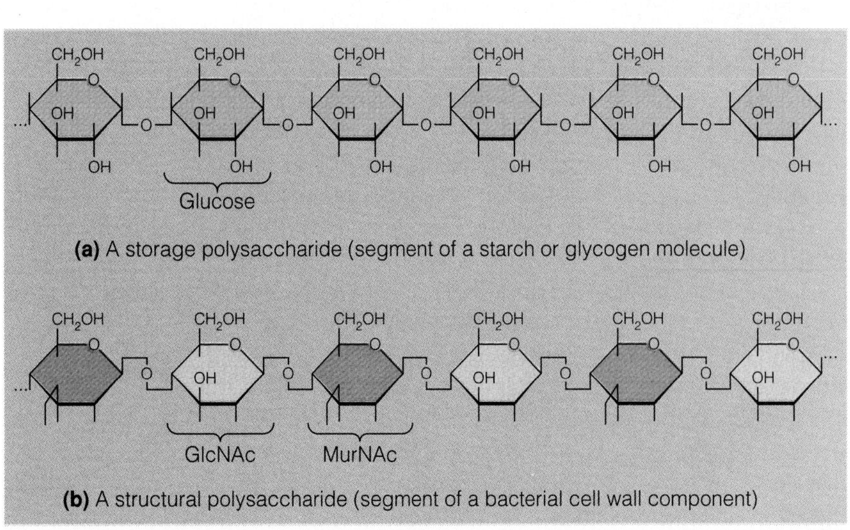

Figure 2-17 Storage and Structural Macromolecules. Storage and structural macromolecules contain one or a few kinds of repeating units in a strictly repetitious sequence. **(a)** A portion of the sequence of a linear segment of the storage polysaccharide starch or glycogen consisting of glucose units linked together by glycosidic bonds. **(b)** A portion of the sequence of a bacterial cell wall polysaccharide consisting of the sugar derivatives *N*-acetylglucosamine (GlcNAc) and *N*-acetylmuramic acid (MurNAc) in strict alternation. (For more detailed representations of the structures of glucose, starch, glycogen, GlcNAc, MurNAc, and a bacterial cell wall polysaccharide, see Figures 3-19, 3-21, and 3-23.)

(a) A storage polysaccharide (segment of a starch or glycogen molecule)

(b) A structural polysaccharide (segment of a bacterial cell wall component)

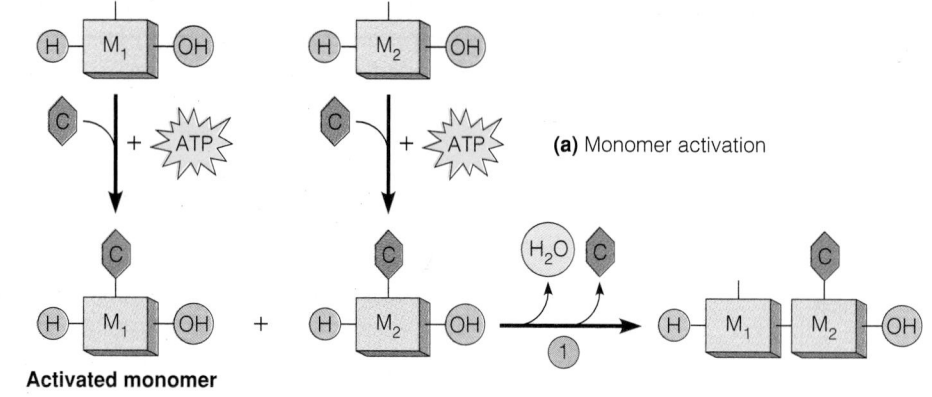

(a) Monomer activation

Activated monomer

(b) First step in biopolymer synthesis

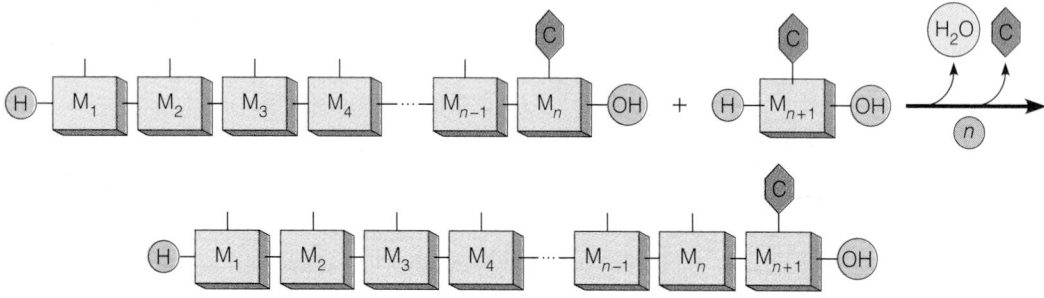

(c) Continuation of (*n*th step in) biopolymer synthesis

Figure 2-18   The Synthesis of Biopolymers.   Biopolymers are synthesized in a process that involves activation of monomeric units followed by their stepwise addition to the elongating polymer chain. Depending upon the polymer, the monomers may all be identical (the glucose units in starch or glycogen, for example) or may differ from one another (the amino acids in proteins or the nucleotides in nucleic acids), but the polymerization process is conceptually the same in each case. **(a)** Monomers (M_1, M_2, etc.) with reactive H and OH groups (shown in blue) are activated by coupling to the appropriate carrier molecule (C, shown in purple), using energy provided by ATP or a similar high-energy compound. **(b)** The first step in biopolymer synthesis involves the condensation of two activated monomers, accompanied or followed by the release of one of the carrier molecules. **(c)** The *n*th step in the polymerization process involves the addition of the next activated monomer (M_{n+1}) to a polymer that already consists of *n* monomeric units, with release of the carrier molecule bound to the *n*th unit.

The Importance of Self-Assembly

So far, we have seen that the macromolecules that characterize biological organization and function are polymers of small, hydrophilic organic molecules. The only requirements for polymerization are an adequate supply of the monomeric subunits, a source of energy, and, in the case of proteins and nucleic acids, sufficient information to specify the order in which the subunits (amino acids or nucleotides) are added. Still to be considered are the steps that lie beyond—the processes through which these biopolymers are organized into the supramolecular assemblies and organelles that are readily recognizable as cellular structures. Or, in terms of Figure 2-14, we need to ask how the macromolecules of level 2 are assembled into the higher-level structures of levels 3, 4, and 5.

Crucial to our understanding of these higher-level structures is the principle of **self-assembly,** which asserts that *the information required to specify the folding of macromolecules and their interaction to form more complicated structures with specific biological functions is inherent in the polymers themselves.* This principle says that once macromolecules are synthesized in the cell, their assembly into more complex structures occurs spontaneously, without further input of energy or information. As we will see shortly, this principle has been qualified somewhat by the recognition that proteins called *molecular chaperones* are needed in some, maybe even many, cases of protein folding to prevent incorrect molecular interactions that would lead to inactive structures. Even in such cases, however, the chaperone molecules do not provide additional steric information; they simply assist the assembly process by inhibiting interactions that would produce incorrect structures.

The Self-Assembly of Proteins

A useful prototype for understanding self-assembly processes is the coiling and folding necessary to form a functional three-dimensional protein from one or more linear chains of amino acids. Although the distinction is not always properly made, the immediate product of amino acid polymerization is not a protein but a **polypeptide.** To become a functional protein, one or more such linear polypeptide chains must coil and fold in a precise, predetermined manner to assume the unique three-dimensional structure necessary for biological activity.

The evidence for self-assembly of proteins comes largely from studies in which the "native," or natural, structure of a protein is disrupted by changing the environmental conditions. Such disruption, or unfolding, can be achieved by raising the temperature, by making the pH of the solution highly acid or highly alkaline, or by adding certain chemical agents such as urea or any of several alcohols. The unfolding of a polypeptide under such conditions is called **denaturation** because it results in the loss of the natural three-dimensional structure of the protein—and its function as well, such as the loss of catalytic activity, in the case of an enzyme.

When the denatured polypeptide is returned to conditions in which the native structure is stable, the polypeptide slowly undergoes **renaturation,** the return to its correct three-dimensional structure. In at least some cases, the renatured protein also regains its biological function—catalytic activity, in the case of an enzyme.

Figure 2-19a depicts the denaturation and subsequent renaturation of *ribonuclease,* the protein used by Calvin Anfinsen and his colleagues in their classic studies on protein self-assembly. ① When a solution of ribonuclease is heated, the protein will denature, resulting in an unfolded, randomly coiled polypeptide with freedom of rotation about the covalent bonds in both the polypeptide chain and the functional groups of the various amino acids. In this form, ribonuclease has no fixed shape and no catalytic activity. ② If the solution of denatured molecules is then allowed to cool slowly, the ribonuclease molecules will regain their original form and will again have catalytic activity. Thus, all of the information necessary to specify the three-dimensional structure of the ribonuclease molecule is inherent in its amino acid sequence. Similar results have been obtained from denaturation/renaturation experiments with other proteins and protein-containing structures, although renaturation is much more difficult, sometimes impossible, to demonstrate with larger, more complex proteins.

You may have noticed that the conditions for ribonuclease denaturation (Figure 2-19a, step 1) include not only heat but also the presence of a reducing agent. Similarly, renaturation (step 2) requires not just cooling but the presence of an oxidizing agent. The reducing agent is required because ribonuclease contains the amino acid cysteine at eight locations along the polypeptide chain, and each cysteine has a free sulfhydryl (—SH) group that is capable of forming a covalent *disulfide bond* (—S—S—) by reacting oxidatively with the sulfhydryl group of another cysteine elsewhere along the polypeptide chain. As shown in Figure 2-19, the native ribonuclease molecule has four such disulfide bonds, which confer additional stability upon the structure. To generate the fully denatured molecule, an agent is needed to reduce the

disulfide bonds to free sulfhydryl groups. To renature the molecule, an oxidizing agent is needed to reconstitute the disulfide bonds. Remarkably, each disulfide bond in the renatured molecule is formed from the same two sulfhydryl groups constituting that particular disulfide bond in the original undenatured molecule, even though the cysteines of which those sulfhydryl groups are a part may be quite distant from one another along the polypeptide chain.

Molecular Chaperones and Assisted Self-Assembly

Based on the ability of denatured proteins to return to their original configuration and to regain biological function as they do so, biologists have generally assumed that proteins and protein-containing structures self-assemble in cells as well. As shown in Figure 2-19b for ribonuclease, the self-assembly model envisions polypeptide chains coiling and folding spontaneously and progressively as polypeptides are synthesized in the cell. By the time the fully elongated polypeptide is released from the ribosome, it is thought to have attained a stable, predictable three-dimensional structure without any input of energy or information beyond the basic polymerization process. Furthermore, folding is assumed to be unique, in the sense that each polypeptide with the same amino acid sequence will fold in an identical, reproducible manner under the same conditions. Thus, the self-assembly model assumes that interactions occurring within and between polypeptides are all that is necessary for the biogenesis of proteins in their functional forms.

However, this model for self-assembly in vivo is based entirely on in vitro studies with isolated proteins, and even under in vitro conditions, not all proteins regain their native structure. Based on their work with one such protein (*ribulose bisphosphate carboxylase/oxygenase,* the multimeric enzyme in chloroplasts that captures carbon dioxide in the process of photosynthesis), John Ellis and his colleagues at Warwick University concluded that the self-assembly model may not be adequate for all proteins, at least not in its simplest formulation. In many cases, the interactions that drive protein folding may need to be assisted and controlled to reduce the probability of the formation of incorrect structures having no biological activity. For assembly processes with a low probability of incorrect interactions and nonfunctional structures (as is likely the case of the folding of a protein like ribonuclease, which consists of a single polypeptide chain), such control may not be needed. For more complex processes (such as the assembly of ribulose bisphosphate carboxylase/oxygenase from its 16 component subunits), control is essential to produce a sufficient number of correct structures for cellular needs.

This control of complex assembly processes is exerted by preexisting proteins called **molecular chaperones,** which facilitate the correct assembly of proteins and protein-containing structures but are not components of the assembled structures. Molecular chaperones that have been identified to date do not convey steric information either for polypeptide folding or for the assembly of multiple polypeptides into a single protein. Instead, they function by binding to specific structural features that are exposed only in the early stages of assembly, thereby inhibiting unproductive assembly pathways that would lead to incorrect structures. Commenting on the term *molecular chaperone*, Ellis and van der Vies observe that "the term chaperone is appropriate for this family of proteins because the role of the human chaperone is to prevent incorrect interactions between people, not to provide steric information for those interactions" (Ellis and van der Vies, 1991, p. 323).

The mode of action of molecular chaperones is best described as **assisted self-assembly.** We can therefore distinguish two types of self-assembly: *strict self-assembly*, for which no factors other than the structure of the polypep-

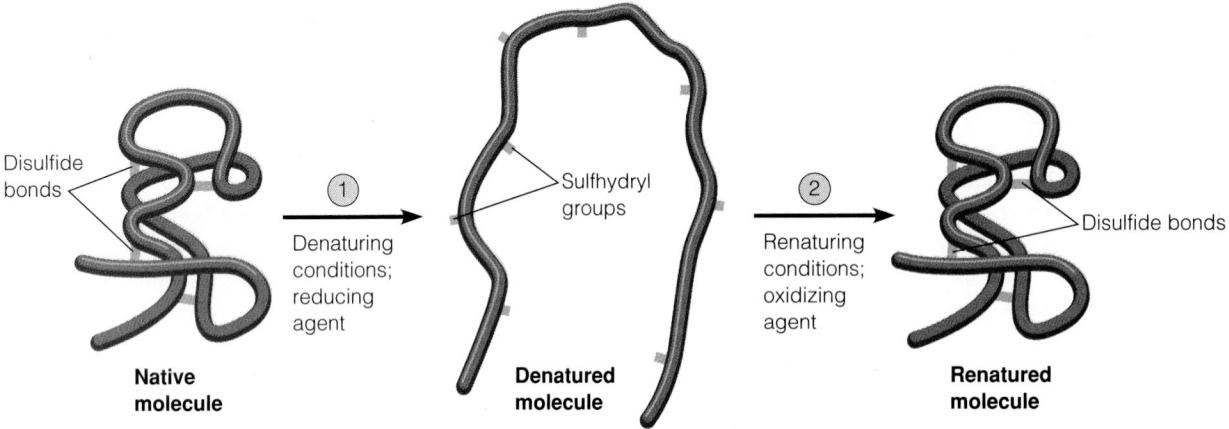

(a) Denaturation and renaturation of ribonuclease

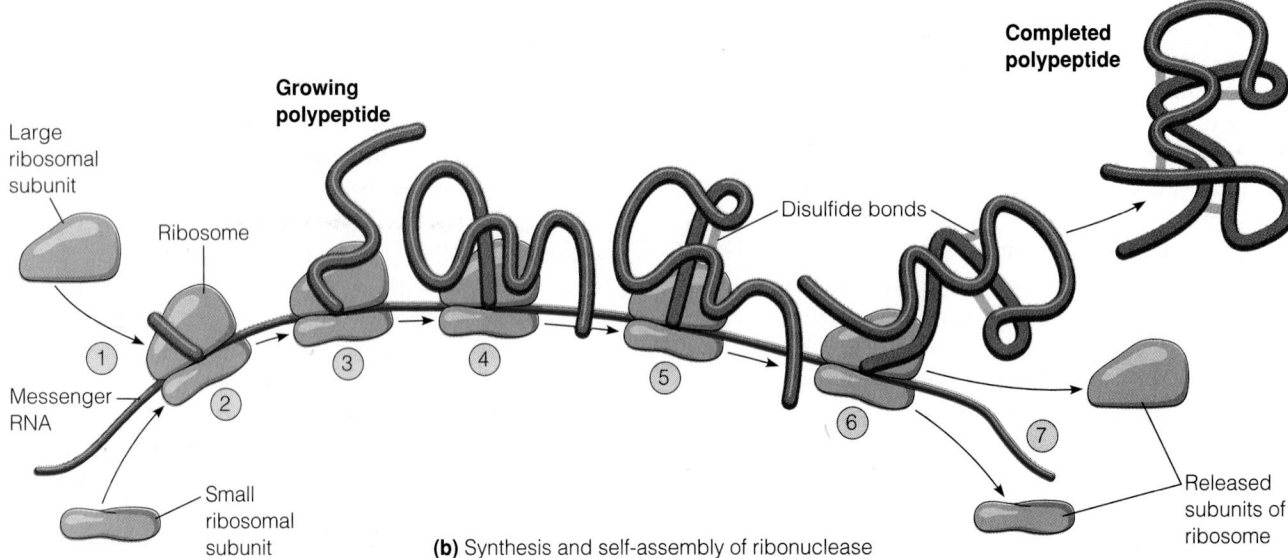

(b) Synthesis and self-assembly of ribonuclease

Figure 2-19 The Spontaneity of Polypeptide Folding.

(a) To demonstrate that the information for the three-dimensional configuration of a polypeptide resides in its amino acid sequence, ① the intact polypeptide can be exposed to denaturing conditions. The protein depicted here is the enzyme ribonuclease, which consists of a single polypeptide chain. Heat was used as the denaturing agent in Anfinsen's original experiment with ribonuclease, but a variety of chemical agents can be used also. This treatment results in a denatured molecule that has no fixed shape and has lost its biological function (enzyme activity, in the case of ribonuclease). ② Upon return to conditions that allow restoration of structure (gradual cooling or gradual removal of the denaturing agent), the polypeptide returns spontaneously to its original native configuration (renaturation) without the input of any additional information. (For polypeptides such as ribonuclease in which the native configuration of the molecule is stabilized by covalent disulfide bonds (—S—S—), complete denaturation requires the use of a reducing agent to reduce the disulfide bonds to their component sulfhydryl (—SH) groups. An oxidizing agent is then used to restore the disulfide bonds as the polypeptide is renatured.) **(b)** The same interactions that ensure the return of a denatured polypeptide to its native form are thought to act on the elongating polypeptide during its synthesis on a ribosome to bring about the progressive coiling and folding of the polypeptide into its unique three-dimensional configuration. In this drawing, synthesis of the polypeptide ribonuclease is initiated on the left, such that each successive ribosome from left to right has a larger portion of the total polypeptide chain completed.

tide itself are required for proper folding; and *assisted self-assembly,* in which the appropriate molecular chaperone is required to ensure that correct assembly will predominate over incorrect assembly.

The list of molecular chaperones has grown steadily since 1987, when Ellis and his colleagues first proposed the term. The first molecular chaperones studied were those of chloroplasts, mitochondria, and bacteria. Within a few years, however, chaperones were identified in other eukaryotic locations. Chaperone proteins are abundant under normal conditions and increase to still higher levels in response to stresses such as increased temperature—a condition called *heat shock*—or an increase in the cellular content of unfolded proteins. Many common chaperone proteins fall into two families, called *Hsp60* and *Hsp70.* ("Hsp" stands for *heat-shock protein,* and the numbers refer to the approximate molecular weights of the protein's polypeptide monomers, 60,000 and 70,000, respectively.) The proteins within each Hsp family are evolutionarily related, and they are found throughout the living world. Hsp70 proteins, for example, have been found in bacteria and in cells of a wide variety of eukaryotes, where they are present in several cellular locations.

In addition to their role in the folding of newly made polypeptides, chaperone proteins have other functions. Some of these relate to protein activity in the cell, others to protein transport into organelles (see Chapter 18). One important class of chaperones, the *nucleoplasmins,* are nuclear proteins that perform a variety of functions during such nuclear processes as DNA replication, transcription and the processing of RNA, and the transport of molecules into and out of the nucleus.

The Importance of Noncovalent Interactions in Protein Folding

Whether assisted by molecular chaperones or not, polypeptides clearly fold and self-assemble without the input of further energy or information, and the three-dimensional structure of a protein is remarkably stable once it has been attained. To understand the self-assembly of proteins (and, as it turns out, of other biological molecules and structures as well), we need to consider both the covalent and the noncovalent bonds that hold polypeptides and other macromolecules together.

Covalent bonds are easy to appreciate: Every protein or other macromolecule in the cell is held together by strong covalent bonds such as those discussed earlier in the chapter. A covalent bond forms whenever two atoms *share* electrons rather than gaining or losing electrons completely. Electron sharing is an especially prominent feature of the carbon atom, which has four electrons in its outer orbital and is therefore at the midpoint between the tendency to gain or lose electrons.

Covalent bonds not only link the monomers of a polypeptide together, they also stabilize the three-dimensional structure of many proteins. Specifically, sulfur-sulfur covalent bonds (—S—S—) play an important role in protein structure. Consider, for example, the ribonuclease molecule in Figure 2-19. As noted earlier, this molecule contains the amino acid cysteine at eight locations along the polypeptide chain. Each cysteine has a free sulfhydryl group that is capable of forming a covalent disulfide bond by reacting oxidatively with the sulfhydryl group of another cysteine located elsewhere along the polypeptide chain. The native ribonuclease molecule has four disulfide bonds, which confer structural stability upon the molecule. Many proteins are stabilized in this way by disulfide bonds, both within a single polypeptide and between the monomeric polypeptides of a multimeric protein.

As important as covalent bonds are in cellular chemistry, the complexity of molecular structure cannot be described in terms of covalent bonds alone. Most of the structures in the cell are held together by much weaker forces—the **noncovalent interactions** within and between proteins and other macromolecules.

In the case of proteins, opportunities for noncovalent interactions are especially diverse because proteins contain 20 different kinds of amino acids that vary greatly in the chemical properties of their respective side chains. As we will see in more detail in Chapter 3, some side chains are charged, others are uncharged but highly polar, and still others are essentially nonpolar, with little or no affinity for water. Given the chemical diversity of the amino acids, it is not surprising that a variety of noncovalent interactions contribute to the overall folding and assembly of a polypeptide. Four such interactions are especially important: electrostatic interactions, hydrogen bonds, van der Waals interactions, and the hydrophobic effect.

The role of **electrostatic interactions** (or *ionic bonds*) in polypeptide folding is easy to understand. Because the side chains of some amino acids are positively charged and the side chains of others are negatively charged, polypeptide folding is dictated in part by the tendency of charged groups to attract oppositely charged groups and to repel groups with the same charge. Several features of electrostatic interactions are particularly significant. The *strength* of such interactions allows them to exert an attractive force over distances greater than other noncovalent bonds. Moreover, the attractive force is *nondirectional,* so that electrostatic interactions are not limited to discrete angles, as in the case of covalent bonds. (Because electrostatic interactions depend upon both groups remaining charged, they will be disrupted if the pH value becomes so high or so low that either of the groups loses its charge. This loss of electrostatic interactions accounts in part for the denaturation that most proteins undergo at high or low pH.)

Hydrogen bonds also play a role in determining and stabilizing the three-dimensional structure of polypeptides (and other macromolecules as well). In the case of water molecules, a hydrogen bond forms between a covalently bonded hydrogen atom on one water molecule and a pair of nonbonding electrons of the oxygen atom on another molecule (see Figure 2-8b). In the case of a polypeptide, many of the amino acid side chains have functional groups that are either good hydrogen bond donors or good acceptors. Examples of good donors include the hydroxyl groups present on the side chains of several amino acids and the amino groups of others. The carbonyl and sulfhydryl groups present on the side chains of still other amino acids are good acceptors.

Electrostatic interactions and hydrogen bonds tend to draw parts of a protein molecule together, but when atoms or groups of atoms come so close together that their outer electron orbitals begin to overlap, they repulse each other mutually. This mutual repulsion rises so steeply as the distance between the two atoms or groups gets shorter that it acts as a very effective barrier to tighter packing closer than a specific distance. That distance defines the so-called **van der Waals radius** for a specific atom or group of atoms. Thus, each atom or group of atoms within the protein (or any other biological molecule) has its own inviolate space. Noncovalent attractive forces between different parts of the polypeptide tend to draw the molecule together into a compact shape, while the van der Waals radii for the various atoms and groups set the limits on how tight the molecular packing can be. Van der Waals radii for some common atoms and groups are given in Table 2-2. The value for a given atom or group serves as a measure of how close another atom or group can come. Van der Waals radii provide the basis for *space-filling models* of biological molecules, such as that for the protein *insulin* shown in Figure 2-20.

Yet another factor contributing to the folding of polypeptides is not really a bond or interaction but rather a tendency to *avoid* an interaction. As already noted, the side chains of the 20 different amino acids present in polypeptides vary greatly in their affinity for water. Some of the side chains are hydrophilic and are capable of forming hydrogen

bonds, not just with each other but also with the water molecules of the medium. Not surprisingly, such groups tend to seek out positions near the surface of the folded structure, where they can interact maximally with the surrounding water molecules. Other amino acids are hydrophobic. They are essentially nonpolar, cannot form hydrogen bonds, and tend therefore to gravitate toward the center of the polypeptide, where they interact with one another in an essentially nonaqueous milieu. This tendency of hydrophobic groups to avoid the aqueous surface of the polypeptide (or other molecule) is called the **hydrophobic effect.**

Thus, protein structure is, in part, the result of a balance between the tendency of hydrophilic groups to seek an aqueous environment near the surface of the molecule and that of the hydrophobic groups to minimize contact with water by associating with one another in the interior of the molecule. Clearly, if all or even most of the amino acids in a protein were hydrophobic, the protein would be virtually insoluble in water and would seek out a nonpolar environment. Recall that membrane proteins are localized in membranes for this very reason. Similarly, if all or most of the amino acids were hydrophilic, the polypeptide would most likely remain in a fairly distended, random shape, allowing maximum access of each amino acid to an aqueous environment. But precisely because most polypeptide chains contain both hydrophobic and hydrophilic amino acids (and in a specified order), parts of the molecule are drawn toward the surface while other parts are driven toward the interior.

Table 2-2 Van der Waals Radii of Some Atoms and Groups of Atoms

	Van der Waals Radius (nm)
Atoms	
Hydrogen (H)	0.12
Oxygen (O)	0.14
Nitrogen (N)	0.15
Carbon (C)	0.17
Sulfur (S)	0.18
Phosphorus (P)	0.19
Chemical groups	
Hydroxyl (—OH)	0.14
Amino (—NH₂)	0.15
methyl (—CH₃)	0.20

Figure 2-20   A Space-Filling Model of a Protein. All of the atoms and chemical groups in this model of the protein insulin are represented as spheres with the appropriate van der Waals radii. Notice how tightly the atoms pack against one another.

Overall, then, the stability of the folded structure of a polypeptide depends upon an interplay of four noncovalent factors: electrostatic interactions between charged amino acids, hydrogen bonding between side groups that are good donors and good acceptors, van der Waals radii that set limits on how tight the molecular packing can be, and the hydrophobic effect that drives nonpolar groups to the interior of the structure. The final configuration of the fully folded polypeptide is the result of these tendencies. Any one of these bonds or interactions is quite low in energy. A hydrogen bond, for example, is only about 5% as strong as a covalent bond. But because there are so many such bonds between the various side chains of the hundreds of amino acids that make up a typical protein, their aggregate strength is sufficient to confer great stability upon the configuration of the folded polypeptide.

These same factors are also important for structures that involve associations of polypeptides with each other. Such structures include the many proteins that consist of multiple polypeptide subunits as well as multiprotein complexes and the various filament- and tubule-containing structures responsible for contractility and motility. The assembly of these structures will almost certainly turn out to be dependent on the same kinds of noncovalent interactions as are involved in the folding and association of individual polypeptides. Ultimately, it appears that most, perhaps even all, of the interactions a polypeptide can undergo are intrinsic in its amino acid sequence and therefore specified by the gene that encodes that sequence.

The Self-Assembly of Other Cellular Structures

The same principle of self-assembly that accounts for the folding and interactions of polypeptides may also apply to more complex cellular structures. Many of the characteristic structures of the cell are complexes of two or more different kinds of polymers and therefore clearly involve interactions chemically distinct from those of polypeptide folding and association. However, the principle of self-assembly may nonetheless apply. For example, *ribosomes* contain both RNA and proteins; *membranes,* as we have already seen, are made up of both phospholipids and proteins; and even the *plant cell wall,* though composed mainly of cellulose fibrils, contains a small but apparently crucial protein component. Yet, despite the chemical differences between such polymers as proteins, nucleic acids, and polysaccharides, the interactions that drive these supramolecular assembly processes seem to be essentially the same as those that dictate the folding of individual protein molecules.

Tobacco Mosaic Virus: A Case Study in Self-Assembly

Some of the most definitive findings concerning the self-assembly of complex biological structures have come from studies with viruses. As we will learn in Chapter 4, a *virus* is a complex of proteins and nucleic acid, either DNA or RNA. A virus is not itself alive, but it can invade and infect a living cell and subvert the synthetic machinery of the cell for the production of more viruses. Inherent in the production of more viruses are the synthesis of the viral nucleic acid and viral proteins and their subsequent assembly into the mature **virion,** or viral particle. Studies of this assembly process have provided a wealth of information on structure and assembly that exceeds what we know about any other self-assembly system.

An especially good example is **tobacco mosaic virus (TMV),** a plant virus that has long been popular with molecular biologists. TMV is a rodlike particle about 18 nm in diameter and 300 nm in length. It consists of a single strand of RNA with about 6000 nucleotides and about 2130 copies of a single kind of polypeptide, the **coat protein,** each with 158 amino acids. The RNA molecule forms a helical core, with a cylinder of protein subunits clustered around it (Figure 2-21).

Heinz Fraenkel-Conrat and his colleagues carried out several important experiments that contributed significantly to our understanding of self-assembly. They separated TMV into its RNA and protein components and then allowed them to reassemble in vitro. Viral particles regenerated that were capable of infecting plant cells. This result was one of the first and most convincing demonstrations that the components of a complex biological structure can reassemble spontaneously into functional entities without external information. Especially interesting was the finding that the RNA from one strain of virus could be mixed with the protein component from another strain to form a hybrid virus that was also infective. As expected, the source of the RNA

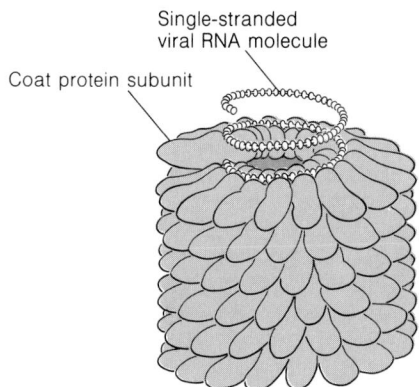

Figure 2-21   A Structural Model of Tobacco Mosaic Virus (TMV).   A single-stranded RNA molecule is coiled into a helix, surrounded by a coat consisting of 2130 identical protein subunits. Only a portion of the entire TMV virion is shown here, and several layers of protein have been omitted from the upper end of the structure to reveal the helical RNA molecule inside.

and not the protein determined the type of virus that was made by the infected cells.

The assembly process has since been studied in detail and is known to be surprisingly complex. The basic unit of assembly is a two-layered disk of coat protein, each layer consisting of 17 identical subunits arranged in a ring (Figure 2-22a). Each disk is initially a cylindrical structure but undergoes a conformational change that tightens it into a helical shape as it interacts with a short segment (about 102 nucleotides) of the RNA molecule (Figure 2-22b). This transition allows another disk to bind (Figure 2-22c), and each successive disk undergoes a conformational change from a cylinder to a helix and binds to another 102 bases of

the RNA. This disk-by-disk elongation process continues, the successively stacked disks creating a helical path for the RNA strand (Figure 2-22d). The process eventually gives rise to the mature virion, its RNA completely covered with coat protein.

The Limits of Self-Assembly

In many cases, the information required to specify the exact configuration of a cellular structure seems to lie entirely within the polymers that contribute to the structure. Such self-assembling systems achieve stable three-dimensional configurations without additional information input be-

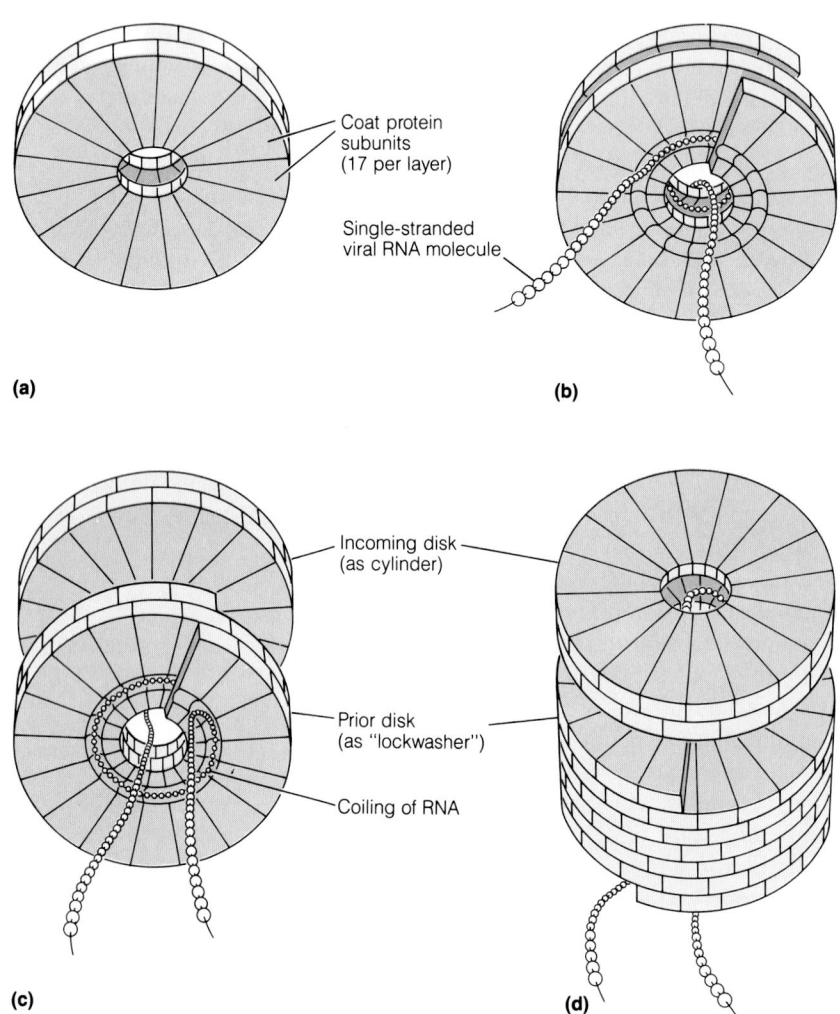

Figure 2-22 Self-Assembly of the Tobacco Mosaic Virus (TMV) Virion. **(a)** The unit of assembly is a double-layered disk with 17 protein subunits per layer. **(b)** Assembly begins as the viral RNA molecule associates with a disk, causing the RNA molecule to form a loop (involving 102 nucleotides) and the disk to change from a cylindrical to a helical conformation. **(c)** Another disk is then added, and another length of

RNA becomes associated with it. The prior disk, meanwhile, has changed from a cylindrical to a helical conformation that makes it look somewhat like a lockwasher. **(d)** Each incoming disk adds two more layers of protein subunits and causes another length of RNA to become associated with the structure in a spiral loop. The process continues until the entire RNA molecule is involved and the virion is complete.

cause the information content of the component polymers is adequate to specify the complete assembly process. Even in assisted self-assembly, the molecular chaperones provide no additional information.

However, there also appear to be some assembly systems that depend, in addition, on information supplied by a preexisting structure. In such cases, the ultimate structure arises not by assembling the components into a new structure but rather by ordering the components into the matrix of an existing structure. Examples of cellular structures that are routinely built up by adding new material to existing structures are membranes, cell walls, and chromosomes.

On the other hand, such structures are not yet sufficiently characterized to determine whether the presence of a preexisting structure is obligatory or whether, under the right conditions, the components might be capable of self-assembly. Evidence from studies with artificial membranes and with chromatin (isolated chromosomal components), for example, suggests that a preexisting structure, though routinely present in vivo, may not be an indispensable requirement for the assembly process. Additional insight will be necessary before we can say with certainty whether, and to what extent, external information is required or exploited in cellular assembly processes.

The Advantages of Hierarchical Assembly

Each of the assembly processes we have looked at exemplifies the basic cellular strategy of **hierarchical assembly** illustrated in Figure 2-14. Biological structures are almost always constructed in a hierarchical manner, with subassemblies acting as important intermediates en route from simple starting molecules to the end products of organelles, cells, and organisms. Consider how cellular structures are made.

First, large numbers of similar, or even identical, monomeric subunits are assembled by condensation into polymers. These polymers then aggregate spontaneously but specifically into characteristic multimeric units. The multimeric units can, in turn, give rise to still more complex structures and eventually to assemblies that are recognizable as distinctive subcellular structures. This hierarchical process has the double advantage of chemical simplicity and efficiency of assembly.

To appreciate the chemical simplicity, we need only recognize that almost all structures found in cells and organisms are synthesized from about 30 small precursor molecules, which George Wald has called the "alphabet of biochemistry." As Table 2-3 (page 41) indicates, this "alphabet" includes the 20 amino acids found in proteins, the 5 aromatic bases present in nucleic acids, 2 sugars, and 3 lipid molecules. We will encounter each of these in Chapter 3. Given these building blocks and the polymers that can be derived from them through just a few different kinds of condensation reactions, most of the structural complexity of life can be readily elaborated by hierarchical assembly into successively more complex structures.

The second advantage of hierarchical assembly lies in the "quality control" that can be exerted at each level of assembly, allowing defective components to be discarded at an early stage rather than being built into a more complex structure that would be more costly to reject and replace. Thus, if the wrong subunit has been inserted into a polymer at some critical point in the chain, that particular molecule may have to be discarded, but the cell will be spared the cost of synthesizing a more complicated supramolecular assembly or even a whole organelle before the defect is discovered. The story presented in Box 2A is intended to illustrate this basic principle.

Tempus Fugit and the Fine Art of Watchmaking

"Drat! Another defective watch!" With a look of disgust, Tempus Fugit the watchmaker tossed the faulty timepiece into the wastebasket and grumbled to himself, "That's two out of the last three watches I've had to throw away. What kind of a watchmaker am I, anyway?"

"A good question, Fugit," came a voice from the doorway. Tempus looked up to see Caveat Emptor entering the shop. "Maybe I can help you with it if you'll tell me a bit about how you make your watches."

"Nobody asked for your help, Emptor," growled Tempus testily, wishing fervently he hadn't been caught thinking out loud.

"Ah well, you'll get it just the same," continued Caveat, quite unperturbed. "Now tell me just how you make a watch and how long it takes you. I'd be especially interested in comparing your procedure with the way Pluribus Unum does it in his new shop down the street."

At the very mention of his competitor's name, Tempus groaned again. As much as possible, he avoided even thinking about E. Pluribus Unum and that disgustingly prosperous shop he had opened recently in the next block. "Unum!" he blustered. "What does he know about the fine art of watchmaking?"

"A good deal, apparently; probably more than you do," replied Caveat. "But tell me exactly how you go about it. How many steps does it take you per watch, and how often do you make a mistake?"

"It takes exactly 100 operations to make a watch. Every step has to be done exactly right, or the watch won't work. It's tricky business, but I've got my error rate down to 1%," said Tempus with at least a trace of pride in his voice. "I can make a watch from start to finish in exactly one hour, but I only make 36 watches each week because I always take Tuesday afternoons off to play darts."

"Well now, let's see," said Caveat, as he pulled out a pocket calculator. He mumbled to himself as he punched in a few numbers, then looked up and announced brightly, "That means, my good Mr. Fugit, that you only make about 13 watches per week that actually work. All the rest you have to throw away just like the one you pitched as I entered your shop."

"How did you know that, you busybody?" Tempus asked defensively, wondering how this know-it-all with his calculator had guessed his carefully kept secret.

"Elementary, my dear Watson," returned Caveat gleefully. "Simple probability is all it takes. You told me that each watch requires 100 operations and that there's a 99% chance that you'll get a given step right. That's 0.99 times itself 100 times, which comes out to 0.366. So only about 37% of your watches will be put together right, and you can't even tell anything about a given watch until it's finished and you test it. Want to know how that compares with Unum's shop?"

Tempus was about to protest that his name wasn't Watson and that he wasn't at all sure he wanted to know anything about that wretched Unum, but his tormentor hurried on with scarcely a pause for breath. "He can manage 100 operations per hour too, and his error rate is exactly the same as yours. He's also off every Tuesday afternoon, but he gets about 27 watches made every week. That's twice your output, Fugit! No wonder he's got so many watches in his window and so many customers at his door. I've heard that he's even thinking of expanding his shop. Want to know how he does it?"

Tempus was too depressed to even attempt a protest. And besides, although he wouldn't like to admit it, he really was dying to know how Unum managed it.

"Subunit assembly, Fugit, that's the answer! Subunit assembly! Instead of making each watch from scratch in 100 separate steps, Unum assembles the components into 10 pieces, each requiring 10 steps." Caveat began pushing calculator buttons again. "Let's see, he performs 10 operations with a 99% success rate for each, so I take 0.99 to the tenth power instead of the hundredth. That's 0.904, which means that about 90% of his subunits have no errors in them. So he spends about 33 hours each week making 330 subunits, throws the defective ones away, and still has about 300 to assemble into 30 watches during the last three hours on Friday afternoon. That takes 10 steps per watch, with the same error rate as before, so again he comes up with about 90% success. He has to throw away about 3 of his finished watches and ends up with about 27 watches to show for his efforts. Meanwhile, you've been working just as hard and just as accurately, yet you only make half as many watches. What do you have to say to that, my dear Fugit?"

Tempus buried his head in his hands as visions of E. Pluribus Unum and his subassembled watches danced through his mind. Finally, he managed a weary response. "Just one question, Emptor, and I'll probably hate myself for asking. But do you happen to know what Unum does on Tuesday afternoons?"

"I'm glad you asked," replied Caveat as he slipped his calculator back into his pocket and headed for the door. "But I don't think you'll be able to interest him in darts—he spends every Tuesday afternoon giving watchmaking lessons." ■

Table 2-3 The Thirty Most Common Small Molecules in Cells

Kind of Molecule	Number Present	Molecule Name		Role in Cell	Figure Number for Structures
Amino acid	20	Alanine	Leucine	Monomeric units of all proteins	3-2
		Arginine	Lysine		
		Asparagine	Methionine		
		Aspartate	Phenylalanine		
		Cysteine	Proline		
		Glutamate	Serine		
		Glutamine	Threonine		
		Glycine	Tryptophan		
		Histidine	Tyrosine		
		Isoleucine	Valine		
Aromatic base	5	Adenine	Thymine	Components of nucleic acids (DNA and RNA)	3-12
		Cytosine	Uracil		
		Guanine			
Sugar	2	Ribose		Component of nucleic acids	3-12
		Glucose		Energy metabolism; component of starch and glycogen	3-18
Lipid	3	Choline		Components of phospholipids	3-24
		Glycerol			3-25
		Palmitate			3-26

In summary, the basic molecular building blocks of the cell are small organic molecules that can be strung together by stepwise polymerization to form the macromolecules that are so important to cellular structure and function. About 30 different kinds of monomeric units account for most cellular structure. These monomeric units consist primarily of carbon, hydrogen, oxygen, nitrogen, and phosphorus, all readily available to the living world in the form of inorganic compounds such as carbon dioxide, water, ammonia, nitrate, and phosphate. Because of the presence of polar functional groups, most of these small organic molecules are quite water-soluble. The cell is able to retain these molecules within defined spaces because it has selectively permeable membranes with a hydrophobic interior that most polar molecules or ions cannot penetrate unless specific transport proteins are present in the membranes.

Polymerization into macromolecular form requires that the monomeric molecules be appropriately activated, usually at the expense of ATP. The resulting polymers can fold and coil spontaneously and then interact with one another in a unique, predictable manner to generate successively higher-order structures. The information needed for these assembly processes is inherent in the chemical nature of the monomeric units and the order in which chemically different monomers are strung together. In some (maybe even many) cases, proteins called molecular chaperones appear to mediate assembly, but they do so by inhibiting or preventing the formation of incorrect structure, not by providing additional steric information. This overall strategy of hierarchical assembly from subunits has the dual advantages of chemical simplicity and efficiency of assembly.

With these principles in mind, we are ready to proceed to Chapter 3. There, we will examine the major kinds of biological macromolecules and the chemical nature of the subunits from which they are synthesized.

KEY TERMS FOR SELF-TESTING

The Importance of Carbon

organic chemistry (p. 18)
biological chemistry (p. 18)
biochemistry (p. 18)
carbon atom (p. 18)
valence (p. 18)
covalent bond (p. 18)
single bond (p. 19)
double bond (p. 19)
triple bond (p. 19)
bond energy (p. 19)
calorie (p. 19)
einstein (p. 20)
hydrocarbon (p. 20)
functional group (p. 20)
tetrahedral (carbon atom) (p. 20)
stereoisomer (p. 21)
asymmetric carbon atom (p. 21)

The Importance of Water

electronegative (p. 22)
polarity (p. 22)
temperature-stabilizing capacity
  (p. 23)

specific heat (p. 23)
solvent (p. 23)
solute (p. 23)
hydrophilic (p. 23)
hydrophobic (p. 23)
sphere of hydration (p. 24)

The Importance of Selectively Permeable Membranes

membrane (p. 25)
amphipathic molecule (p. 25)
phospholipid bilayer (p. 26)
membrane protein (p. 26)
transport protein (p. 26)
selectively permeable (p. 27)

The Importance of Synthesis by Polymerization

macromolecule (p. 27)
monomer (p. 29)
informational macromolecule (p. 29)
storage macromolecule (p. 30)
structural macromolecule (p. 30)
biopolymer (p. 31)

condensation reaction (p. 31)
carrier molecule (p. 31)
activated monomer (p. 31)
directionality (p. 31)

The Importance of Self-Assembly

self-assembly (p. 32)
polypeptide (p. 33)
denaturation (p. 33)
renaturation (p. 33)
molecular chaperone (p. 33)
assisted self-assembly (p. 34)
noncovalent interaction (p. 35)
electrostatic interaction (p. 35)
hydrogen bond (p. 35)
van der Waals radius (p. 36)
hydrophobic effect (p. 36)
virion (p. 37)
tobacco mosaic virus (TMV) (p. 37)
coat protein (p. 37)
hierarchical assembly (p. 39)

PROBLEM SET

2-1. The Fitness of Carbon. Carbon has a number of properties that make it especially fit to play a key role in biological molecules. One way to appreciate these properties is to compare and contrast them with those of silicon, the element immediately beneath carbon in the periodic table. Contrast each of the following properties of silicon with the comparable property of carbon, and indicate, if appropriate, why carbon is a fitter element than silicon for biological purposes in that particular regard.

(a) Silicon is a larger atom, with an atomic weight of 28.

(b) Silicon combines with oxygen to form insoluble silicates or network polymers of silicon dioxide (quartz).

(c) Silicon does not readily form double or triple bonds.

(d) Silicon has four valence electrons and is quite abundant in the Earth's crust (28% vs. 0.19% for carbon).

(e) Polymers of silicon are not stable in water.

2-2. The Fitness of Water. For each of the following statements about water, decide whether the statement is true and describes a property that makes water a desirable component of cells (T); is true but describes a property that has no bearing on water as a cellular constituent (X); or is false (F).

(a) Water is a polar molecule and hence an excellent solvent for polar compounds.

(b) Water can be formed by the reduction of molecular oxygen (O_2).

(c) The density of water is less than the density of ice.

(d) The molecules of liquid water are extensively hydrogen-bonded to one another.

(e) Water does not absorb visible light.

(f) Water is odorless and tasteless.

2-3. Bond Energies. Of considerable importance in assessing the fitness of the bonds in organic molecules is the relationship between the amount of energy required to break these bonds and the amount of energy available in the solar radiation to which these bonds are exposed.

(a) Given that the carbon-carbon single bond is safe in green light (500 nm) but can be broken by the energy of ultraviolet light (300 nm, for example), where does the cutoff come? That is, what wavelength of light has just enough energy to break a carbon-carbon single bond?

(b) Would a carbon-nitrogen bond be more or less stable than a carbon-carbon single bond at this wavelength?

(c) What about a carbon-carbon double bond?

2-4. Stereoisomers. For each compound below, indicate how many stereoisomers exist, and draw the structure of each.

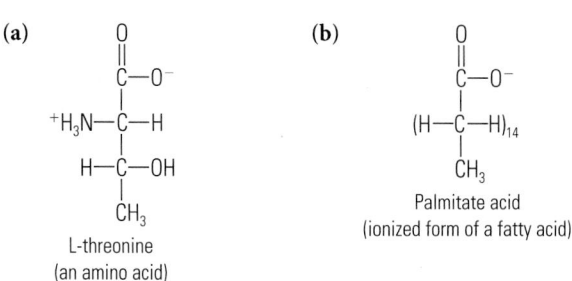

(a)

L-threonine
(an amino acid)

(b)

Palmitate acid
(ionized form of a fatty acid)

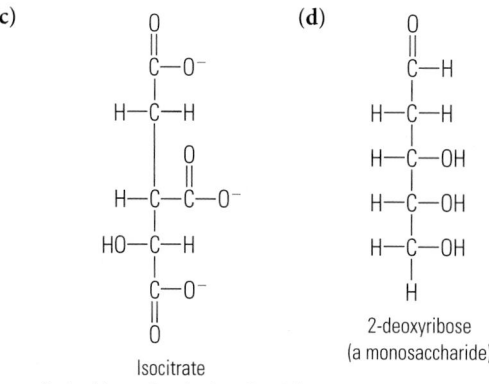

(c)

Isocitrate
(Ionized form of a tricarboxylic acid)

(d)

2-deoxyribose
(a monosaccharide)

2-5. Solubility Properties of Biological Molecules.

(a) Why is the organic compound hexadecane not a common chemical constituent of cells, whereas palmitate is?

$$CH_3—(CH_2)_{14}—CH_3$$
Hexadecane

$$CH_3—(CH_2)_{14}—C—O^-$$
Palmitate

(b) Which of the following compounds is more likely to occur in cells? Explain your answer.

(i)

(ii)

2-6. The Polarity of Water. Defend the assertion that all of life as we know it depends critically on the fact that the angle between the two hydrogen atoms in the water molecule is 109.5° and not 180°.

2-7. The Principle of Polymers. Polymers clearly play an important role in the molecular economy of the cell. What advantage does each of the following features of biopolymers have for the cell?

(a) A particular kind of polymer is formed using the same kind of condensation reaction to add each successive monomeric unit.

(b) The bonds between monomers are formed by the removal of water and are broken or cleaved by the addition of water.

2-8. TMV Assembly. Each of the following statements is an experimental observation concerning the reassembly of tobacco mosaic virus (TMV) virions from TMV RNA and coat protein subunits. In each case, state as carefully as possible a reasonable conclusion that can be drawn from the experimental finding.

(a) When RNA from a specific strain of TMV is mixed with coat protein from the same strain, infectious virions are formed.

(b) When RNA from strain A of TMV is mixed with coat protein from strain B, the reassembled virions are infectious, giving rise to strain A virus particles in the infected tobacco cells.

(c) Isolated coat protein monomers can polymerize into a virus-like helix in the absence of RNA.

(d) In infected plant cells, the TMV virions that form contain only TMV RNA and never any of the various kinds of cellular RNAs present in the host cell.

(e) Regardless of the ratio of RNA to coat protein in the starting mixture, the reassembled virions always contain RNA and coat protein in the ratio of three nucleotides of RNA per coat protein monomer.

2-9. The Alphabet and the Building Blocks. Understanding the great diversity of molecular structures in cells is simplified by the generalization that most of the larger molecules in a cell are made from the 30 small precursor molecules listed in Table 2-3 (the "alphabet") and consist primarily of the seven functional groups shown in Figure 2-5 (the "building blocks"). Although Table 2-3 provides only the names of the small precursor molecules, the structures of all 30 molecules can be found in Chapter 3.

(a) Six of the seven functional groups shown in Figure 2-5 are present in one or more of the 30 molecules from Table 2-3. Identify the six functional groups, and indicate for each at least one molecule in which that group is found.

(b) Which is the only functional group *not* present in any of the 30 molecules? Why do you suppose it is included as one of the seven most common functional groups in biological compounds when it is not found in any of the 30 most common precursor molecules?

(c) Identify at least five molecules from Table 2-3 that contain one or more functional groups *not* shown in Figure 2-5.

(d) If you were asked to make Figure 2-5 more inclusive by adding one more functional group to the seven already shown, which one would you choose? Explain your choice.

2-10. Jeopardy. Each of the following statements is an *answer;* indicate in each case what the *question* is.

(a) This component of solar radiation is sufficiently energetic to break carbon-carbon bonds.

(b) This property of water makes it possible for land animals to cool themselves by surface evaporation with minimum loss of body fluid.

(c) The next monomeric unit to be added to a biopolymer must be in this form in order for the condensation reaction to be energetically favorable.

(d) This principle of macromolecular biosynthesis minimizes the amount of genetic information needed and allows imperfect components to be discarded at several stages.

SUGGESTED READING

General References

Henderson, L. J. *The Fitness of the Environment.* Boston: Beacon Press, 1927; reprinted 1958.

Herriott, J., G. Jacobson, J. Marmur, and W. Parsom. *Papers in Biochemistry.* Reading, MA: Addison-Wesley, 1984.

Lambert, J. B. The shapes of organic molecules. *Sci. Amer.* 22 (January 1970): 58.

Stryer, L. *Biochemistry,* 3d ed. New York: W. H. Freeman, 1988.

Wald, G. The origins of life. *Proc. Natl. Acad. Sci. USA* 52 (1964): 595.

The Importance of Water

Eisenberg, D., and W. Kauzmann. *The Structure and Properties of Water.* Oxford, England: Oxford University Press, 1969.

Franks, F., ed. *Water—A Comprehensive Treatise,* vol. 4. New York: Plenum, 1975.

Scholander, P. F. Tensile water. *Amer. Sci.* 60 (1972): 584.

The Importance of Membranes

Bretscher, M. The molecules of the cell membrane. *Sci. Amer.* 253 (April 1985): 100.

Gennis, R. B. *Biomembranes.* New York: Springer-Verlag, 1989.

Jain, M., and R. Wagner. *Introduction to Biological Membranes,* 2d ed. New York: Wiley, 1988.

Kleinfeld, A. Current views of membrane structure. *Current Topics Membr. Transp.* 29 (1987): 1.

Robertson, R. N. *The Lively Membranes.* Cambridge, England: Cambridge University Press, 1987.

Yeagle, P. *The Membranes of Cells.* Orlando, FL: Academic Press, 1987.

The Importance of Self-Assembly

Anfinsen, C. B. Principles that govern the folding of protein chains. *Science* 181 (1973): 223.

Butler, P. J. G., and A. Klug. The assembly of a virus. *Sci. Amer.* 239 (November 1978): 62.

Craig, E. A. Chaperones: Helpers along the pathways to protein folding. *Science* 260 (1993): 1902.

Creighton, C. B. *The Proteins: Structure and Molecular Properties.* New York: W. H. Freeman, 1984.

Ellis, R. J. Molecular chaperones: The plant connection. *Science* 250 (1990): 954.

Ellis, R. J. The molecular chaperone concept. *Seminars in Cell Biol.* 1 (1990): 1.

Fraenkel-Conrat, H., and R. C. Williams. Reconstitution of active tobacco mosaic virus from its inactive protein and nucleic acid components. *Proc. Natl. Acad. Sci. USA* 41 (1955): 690.

Hendrick, J.P., and F.-U. Hartl. Molecular chaperone functions of heat-shock proteins. *Annu. Rev. Biochem.* 62 (1993): 349.

Klug, A. From macromolecules to biological assemblies. *Biosci. Rep.* 3 (1983): 395.

Lake, J., and C. F. Fox, eds. *Biological Recognition and Assembly.* New York: A. R. Liss, 1980.

Properties of Proteins

Dickerson, R. E., and I. Geis. *The Structure and Action of Proteins.* Menlo Park, CA: Benjamin/Cummings, 1969.

Doolittle, R. F. Proteins. *Sci. Amer.* 253 (October 1985): 88.

Fletterick, R. J., T. Schoer, and R. J. Matela. *Molecular Structure: Macromolecules in Three Dimensions.* Oxford, England: Blackwell Scientific, 1985.

Kendrew, J. C. The three-dimensional structure of a protein molecule. *Sci. Amer.* 205 (December 1961): 96.

Moore, S., and W. H. Stein. Chemical structures of pancreative ribonuclease and deoxyribonuclease. *Science* 180 (1973): 458.

Richardson, J. S. The anatomy and taxonomy of protein structure. *Adv. Protein Chem.* 34 (1981): 166.

Rupler, J. A., E. Gratton, and G. Careri. Water and globular proteins. *Trends Biochem. Sci.* 8 (1983): 18.

Schulz, G. E., and R. H. Schirmer. *Principles of Protein Structure.* New York: Springer-Verlag, 1979.

3

THE MACROMOLECULES OF THE CELL

In Chapter 2, we looked at some of the basic chemical principles of cellular organization. We saw that each of the major kinds of biopolymers—proteins, nucleic acids, and polysaccharides—consists of a relatively small number (from 1 to 20) of repeating monomeric units. These polymers are synthesized by condensation reactions in which activated monomers are linked together and water is removed. Once synthesized, the individual polymer molecules fold and coil spontaneously into stable, three-dimensional shapes. These folded molecules then associate with one another in a hierarchical manner to generate higher levels of structural complexity, usually without further input of energy or information.

We are now ready to examine the major kinds of biological macromolecules, focusing first on the chemical nature of the monomeric components and then on the synthesis and properties of the polymer itself. We begin our survey with proteins because they play such an important and widespread role in cellular structure and function. We then move on to nucleic acids and polysaccharides. The tour concludes with lipids, which do not quite fit the definition of a polymer but are important cellular components in their own right.

Proteins

Proteins are without a doubt the most important and ubiquitous macromolecules in the cell. All antibodies are proteins, as are most enzymes and many hormones. Connective tissue, muscle fibrils, cilia, flagella—all are made primarily or exclusively of proteins. Whether we are talking about carbon dioxide fixation in photosynthesis, oxygen transport in the blood, or the motility of a flagellated bacterium, we are dealing with processes that depend crucially on particular proteins with specific properties and functions.

Virtually everything a cell is or does depends on specific proteins, so it is important that we understand what proteins are and why they have the properties they do. Actually, we have picked up a few clues about proteins already. We know, for example, that proteins are linear polymers of amino acids joined in a genetically determined sequence. We also know that the properties of a particular protein depend on the proportions and sequence of the various amino acids it contains. In addition, we are aware that proteins are synthesized as polypeptide chains on "workbenches" called ribosomes under the genetic direction of a molecule of messenger RNA that is in turn derived by transcription of the genetic information stored in the DNA. The details of protein synthesis will not concern us until Chapter 18. At this point, it will suffice to look briefly at some general aspects of amino acid and protein chemistry.

The Monomers Are Amino Acids

There are 20 different kinds of **amino acids** present in proteins. Although most proteins contain all or most of the 20 amino acids, the proportions vary greatly between proteins, and no two proteins have the same amino acid sequence. Every amino acid has the basic structure illustrated in Figure 3-1, with a carboxyl group, an amino group, a hydrogen atom, and a so-called R group all attached to a single carbon atom. Except for glycine, for which the R group is just a hydrogen atom, all amino acids have at least one asymmetric carbon atom and therefore exist in two stereoisomeric forms, called D- and L-amino acids (see Figure 2-7a). Both kinds exist in nature, but only L-amino acids occur in proteins.

Because the carboxyl and amino groups in Figure 3-1 are common features of all amino acids, the specific properties of a particular amino acid obviously depend on the chemical nature of the R groups, which range from a single hydrogen atom to relatively complex aromatic groups. Shown in Figure 3-2 are the structures of the 20 L-amino acids found in proteins. The three-letter abbreviations given in parentheses for each amino acid are widely used by

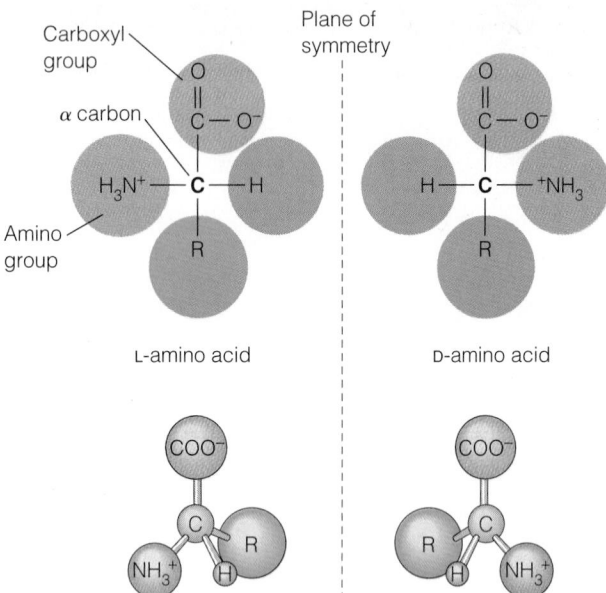

Figure 3-1 The Structure and Stereochemistry of an Amino Acid. Because the α carbon atom is asymmetric in all amino acids except glycine, most amino acids can exist in two isomeric forms, designated L and D and shown here as (top) conventional structural formulas and (bottom) ball-and-stick models. The L and D forms are stereoisomers, with the vertical dashed line as the plane of symmetry. Of the two forms, only L-amino acids are present in proteins.

amino acid to the growing chain by what is essentially a *condensation reaction*. This process is illustrated schematically in Figure 3-3 using ball-and-stick models. The —H and —OH groups that are removed during any condensation reaction come in this case from the amino group of one amino acid and the carboxyl group of the other, respectively. The general term for the resulting covalent linkage is an *amide bond,* but in the special case in which both reactants are amino acids, it is usually called a **peptide bond.** Notice that the chain of amino acids formed in this way has an intrinsic *directionality* because it terminates in an amino group at one end and a carboxyl group at the other end. The end with the amino group is called the **N-** (or **amino**) **terminus,** and the end with the carboxyl group is called the **C-** (or **carboxyl**) **terminus.**

Peptide bond formation is actually more complicated than is suggested by Figure 3-3 because both energy and information are needed for the addition of each amino acid. Energy is needed to activate the incoming amino acid and link it to a special kind of RNA molecule called a *transfer RNA*. Information is needed because the order of amino acids in the growing polypeptide chain is not random but is specified genetically. Ensuring that the correct amino acid is chosen for each successive addition to the growing chain depends on a recognition process between the transfer RNA

biochemists and molecular biologists. Table 3-1 lists these three-letter abbreviations and the corresponding one-letter abbreviations that are also commonly used.

Nine of these amino acids have nonpolar R groups and are therefore *hydrophobic* (Group A). As you look at their structures, you will notice the hydrocarbon nature of the R groups, with oxygen and nitrogen conspicuous by their absence. These are the amino acids that tend to seek out the interior of the molecule as a polypeptide folds into its three-dimensional shape. If a protein (or a region of the molecule) has a preponderance of hydrophobic amino acids, the whole protein will tend to avoid aqueous environments and will instead seek out hydrophobic locations, such as the interior of a membrane.

The remaining 11 amino acids are *hydrophilic*, with an R group that is either distinctly *polar* (Group B; notice the prominence of oxygen) or actually *charged* at the pH values characteristic of cells (Group C; notice the net negative or positive charge in each case). Hydrophilic amino acids tend to cluster on the surface of proteins, thereby maximizing their interaction with the polar water molecules in the surrounding environment.

The Polymers Are Polypeptides and Proteins

The process of stringing individual amino acids together into a linear polymer involves stepwise addition of each new

Table 3-1 Abbreviations of Amino Acids

Amino Acid	Three-Letter Abbreviation	One-Letter Abbreviation
Alanine	Ala	A
Arginine	Arg	R
Asparagine	Asn	N
Aspartate	Asp	D
Cysteine	Cys	C
Glutamate	Glu	E
Glutamine	Gln	Q
Glycine	Gly	G
Histidine	His	H
Isoleucine	Ile	I
Leucine	Leu	L
Lysine	Lys	K
Methionine	Met	M
Phenylalanine	Phe	F
Proline	Pro	P
Serine	Ser	S
Threonine	Thr	T
Tyrosine	Tyr	Y
Tryptophan	Trp	W
Valine	Val	V

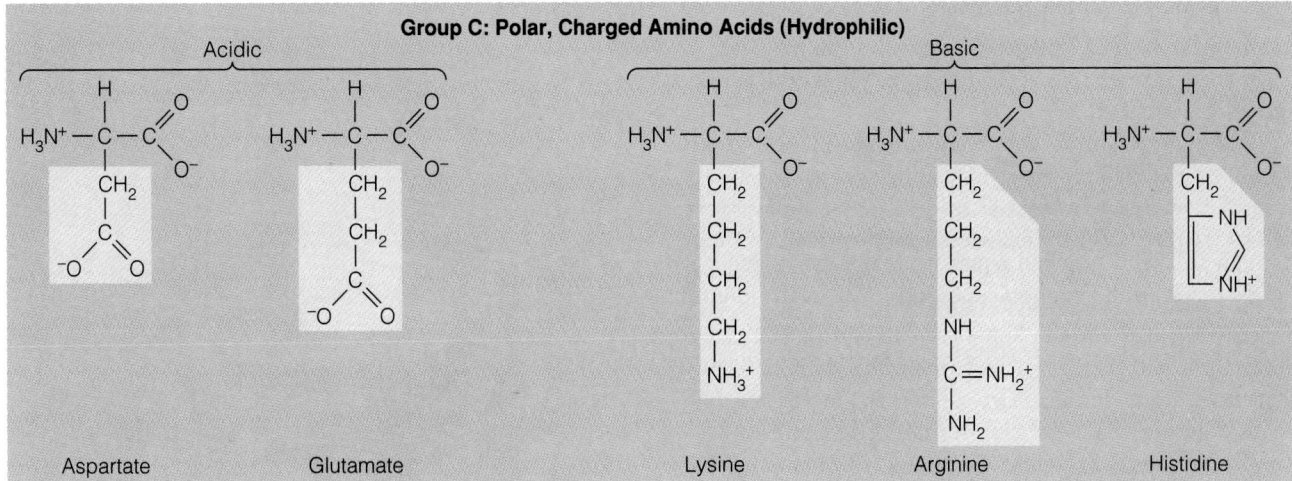

Figure 3-2 The Structures of the Twenty Amino Acids Found in Proteins. All amino acids have a carboxyl group and an amino group on the adjacent (α) carbon (purple), but each has its own distinctive R group (white). Those in Group A have nonpolar R groups and are therefore hydrophobic; notice the hydrocarbon nature of their R groups. The others are hydrophilic, either because the R group is polar in nature (Group B) or because the R group is protonated or ionized at cellular pH and therefore carries a formal electrostatic charge (Group C).

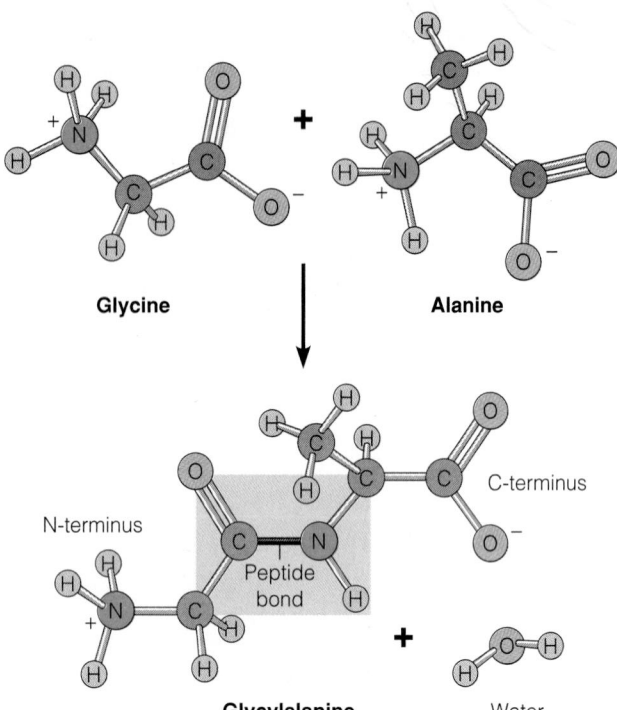

Figure 3-3 Peptide Bond Formation. Successive amino acids in a polypeptide are linked to one another by peptide bonds between the carboxyl group of one amino acid and the amino group of the next. Shown here is the formation of a peptide bond between the amino acids glycine and alanine.

that is linked to the amino acid and the *messenger RNA* that is bound to the ribosome. We will get to all of these details in due course. Here, we need only note that as each new peptide bond is formed, the growing chain of amino acids is lengthened by one and that the process requires both energy and information.

Although this process of elongating a chain of amino acids is often called *protein synthesis,* the term is not entirely accurate because the immediate product of amino acid polymerization is not a protein but a **polypeptide.** A protein is a polypeptide chain (or several such chains) that has attained a unique, stable, three-dimensional shape and is biologically active as a result. Some proteins consist of a single polypeptide and therefore achieve their final shape as a consequence of the folding and coiling that occur spontaneously as the chain is being formed. Such proteins are called **monomeric proteins.** (Be careful with the terminology here. On the one hand, a polypeptide is a *polymer,* with amino acids as its monomeric repeating units; on the other hand, such a polypeptide becomes the *monomer* of which proteins are made.) The enzyme ribonuclease, depicted in Figure 2-19, is an example of a monomeric protein. For such proteins, the term protein synthesis is appropriate for the polymerization because the functional protein forms spontaneously as the polypeptide is elongated.

Many other proteins, however, are **multimeric proteins,** consisting of two or more polypeptides. The *hemoglobin* that carries oxygen in your bloodstream is a multimeric protein because it contains four polypeptides, two called α chains and two called β chains (Figure 3-4). In such cases, protein synthesis involves not only elongation and folding of the individual polypeptide chains but also their subsequent interaction.

As discussed in Chapter 2, the folding of a polypeptide is driven by several noncovalent factors: *electrostatic interactions* between charged amino acids, *hydrogen bonding* between side groups that are good donors and others that are good acceptors, *van der Waals radii* that set limits on how tightly atoms or groups of atoms can be packed together, and the *hydrophobic effect* that drives nonpolar groups to the interior of the molecule. Not surprisingly, the association of two or more polypeptides to form a multimeric protein is controlled by these same factors. Thus, both the initial folding of individual polypeptides and their subsequent interaction to form a multimeric protein are driven by the same forces.

Protein Structure Depends on Amino Acid Sequence and Interactions

The structure of a protein is usually described in terms of four hierarchical levels of organization: the *primary, secondary, tertiary,* and *quaternary* structures (Table 3-2). Primary structure refers to the amino acid sequence, while the three higher levels of organization concern the interactions

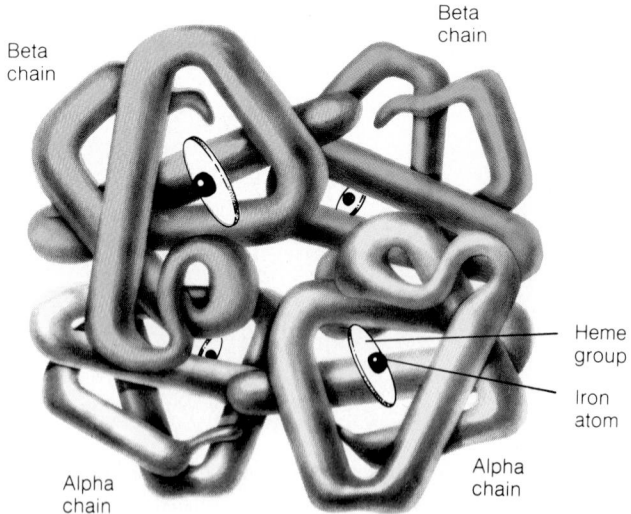

Figure 3-4 The Structure of Hemoglobin. Hemoglobin is a multimeric protein with four subunits (two α polypeptides and two β polypeptides). Each subunit contains a heme group with an iron atom. Each heme iron can bind a single oxygen molecule.

Table 3-2 Levels of Organization in Protein Structure

Level of Structure	Basis of Structure	Kinds of Bonds and Interactions Involved
Primary	Amino acid sequence	Covalent peptide bonds
Secondary	Folding into α helix, β sheet, or random coil	Hydrogen bonds
Tertiary	Three-dimensional folding of a single polypeptide chain	Hydrogen bonds, disulfide bonds, electrostatic interactions, hydrophobic effect
Quaternary	Association of two or more folded polypeptides to form a multimeric protein	Same as for tertiary structure

between the amino acid R groups that give the protein its characteristic **conformation,** or three-dimensional arrangement of atoms in space (Figure 3-5).

Secondary structure involves local interactions between contiguous amino acids along the chain; tertiary structure results from long-distance interactions between stretches of amino acids from different parts of the molecule; and quaternary structure concerns the interaction of two or more individual polypeptides to form a single multimeric protein. All are dictated by the primary structure, but each is important in its own right in the overall structure of the protein. Secondary and tertiary structures are obviously involved in determining the conformation of the individual polypep-

tide, while quaternary structure is relevant only for proteins consisting of more than one polypeptide.

Primary Structure. As already noted, the **primary structure** of a protein is just a formal designation for the amino acid sequence of the constituent polypeptide(s) (Figure 3-5a). When we describe the primary structure of a protein, we are simply specifying the order in which its amino acids appear from one end of the molecule to the other. By convention, amino acid sequences are always written from the N-terminus to the C-terminus of the polypeptide.

The first complete determination of the amino acid sequence of a protein was for the hormone *insulin,* reported in

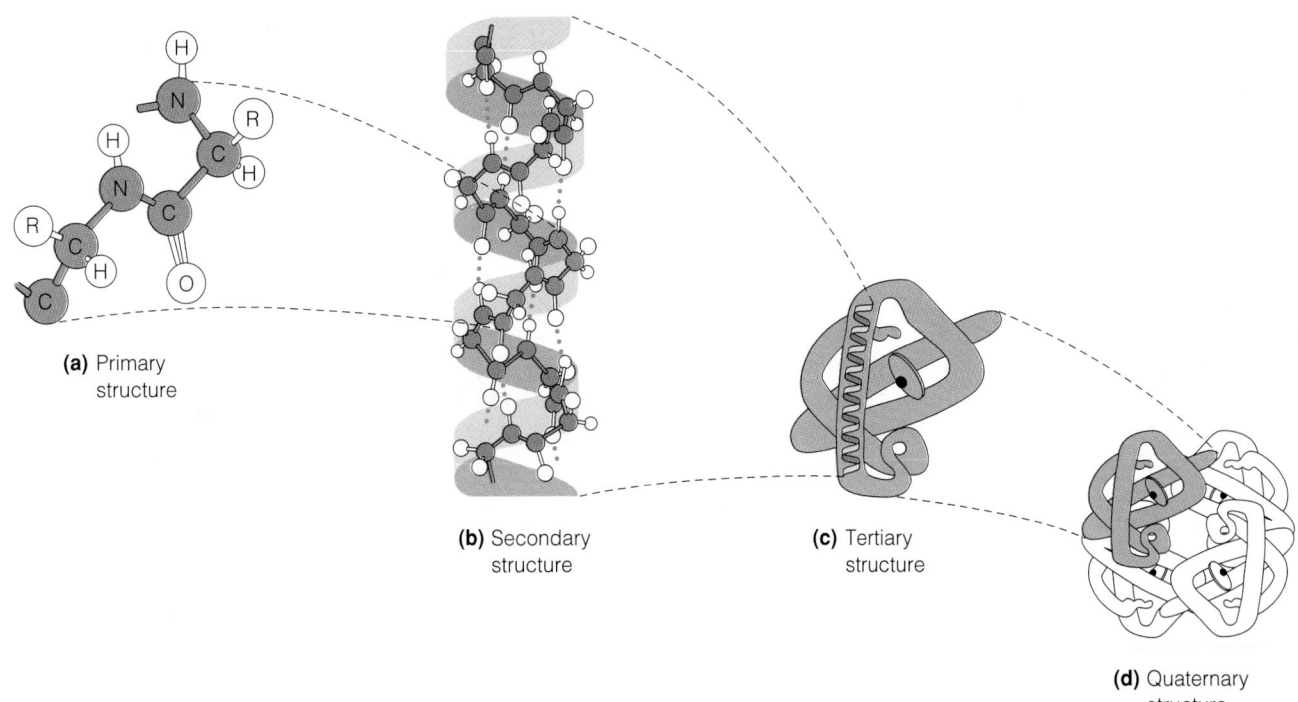

(a) Primary structure

(b) Secondary structure

(c) Tertiary structure

(d) Quaternary structure

Figure 3-5 The Four Levels of Organization of Protein Structure. **(a)** The primary structure of a protein consists of a sequence of amino acids linked together by peptide bonds. **(b)** The resulting polypeptide can then be coiled into an α helix, one form of secondary structure. **(c)** The α helix is a part of the tertiary structure of the folded, coiled polypeptide, which is itself one of the subunits that make up **(d)** the quaternary structure of the multimeric protein, in this case the oxygen-carrying protein hemoglobin.

1953 by Frederick Sanger, who eventually received a Nobel Prize for this important technical advance. The next protein to be sequenced was the enzyme *ribonuclease*. Since then, the sequencing of polypeptides has become an automated, routine procedure and has been successfully applied to hundreds of proteins. In fact, computerized data banks of polypeptide sequences are now available, making it easy to compare sequences and look for regions of similarity.

The primary structure of a protein is important both genetically and structurally. Genetically, it is significant because the amino acid sequence of the polypeptide derives directly from the order of nucleotides in the corresponding messenger RNA. The messenger RNA is in turn encoded by the DNA that represents the gene for this protein, so the primary structure of a protein is an inevitable result of the order of nucleotides in the DNA of the gene.

Of more immediate significance are the implications of the primary structure for higher levels of protein structure. This topic has already been covered to some extent because polypeptide folding and interaction were used as examples in our discussion of self-assembly in Chapter 2. In essence, all three higher levels of protein organization—secondary, tertiary, and quaternary structures—are direct consequences of the primary structure. This means that much, if not all, of the information necessary to specify how polypeptide chains will coil and fold and, if appropriate, how they will interact with one another is inherent in the amino acid sequence (and hence in the genetic information that dictates that sequence). Thus, if polypeptides are made that correspond in sequence to the α and β polypeptides of hemoglobin, they will assume the three-dimensional conformations appropriate to these respective subunits and then interact spontaneously in just the right way to form the $\alpha_2\beta_2$ tetramer that we recognize as hemoglobin.

Secondary Structure. The **secondary structure** of a protein involves local interactions between adjacent amino acids of a polypeptide chain (Figure 3-5b). These interactions result in two major structural patterns, referred to respectively as the α **helix** and β **sheet** (or β **pleated sheet**) conformations (Figure 3-6).

The α helix structure was proposed in 1951 by Linus Pauling and Robert Corey. As shown in Figure 3-6a, an α helix is spiral in shape, consisting of a backbone of peptide

Figure 3-6   The α Helix and β Sheet.
The α helix and β sheet are the two most important elements in the secondary structure of proteins. **(a)** The α helix resembles a coiled telephone cord, but with each turn of the coil stabilized by hydrogen bonds (blue dots) between the carbonyl and imino groups of one peptide bond and those of the peptide bonds just "below" and "above" it in the helix. The hydrogen bonds of an α helix are therefore within a single polypeptide chain and parallel to the polypeptide axis. **(b)** The β sheet resembles a pleated skirt, with successive atoms of each polypeptide chain located at folds of the pleats and with R groups of the amino acids jutting out on alternating sides of the sheet. The structure is stabilized by hydrogen bonds (blue dots) between the carbonyl and imino groups of peptide bonds in the adjacent polypeptides (or adjacent segments of the same polypeptide). For the structure shown here, hydrogen bonds are between adjacent polypeptide chains, but β sheets can also form within the same polypeptide when segments of the molecule are folded back alongside one another.

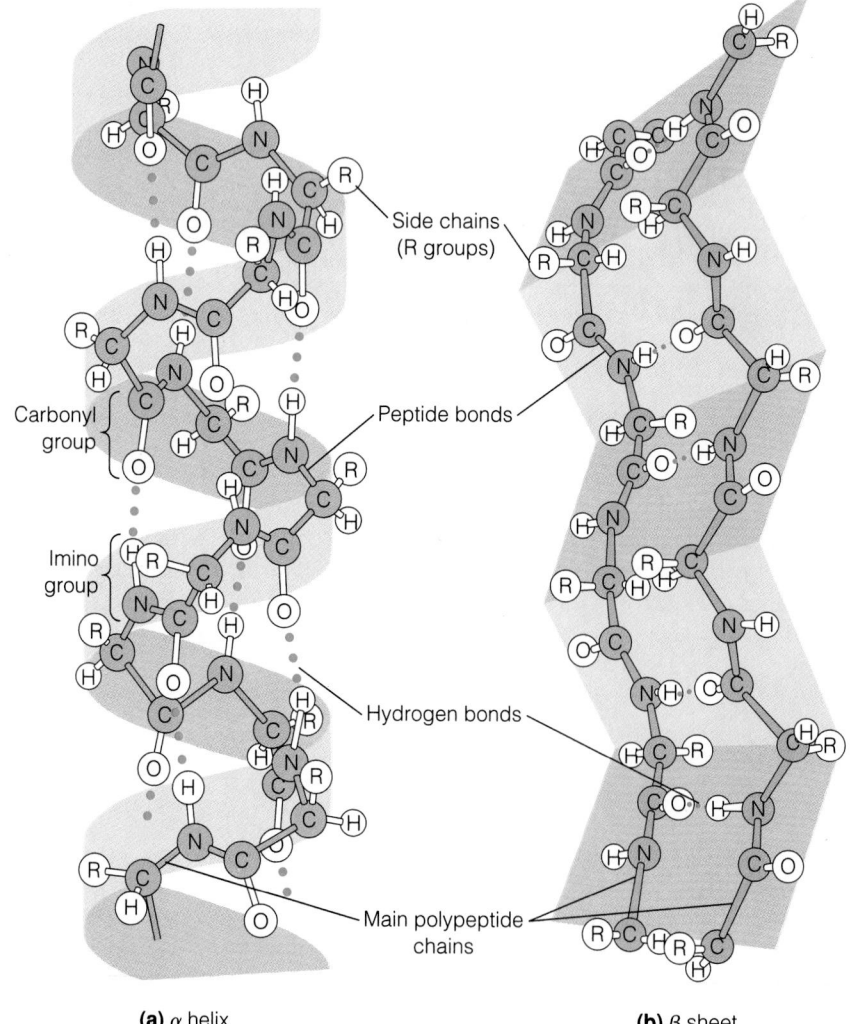

Side chains (R groups)

Carbonyl group

Imino group

Peptide bonds

Hydrogen bonds

Main polypeptide chains

(a) α helix

(b) β sheet

bonds with the specific R groups of the individual amino acids jutting out from it. A helical shape is common to repeating polymers, as we will see when we get to the nucleic acids and the polysaccharides. For the α helix, there are 3.6 amino acids per turn of the helix, bringing the peptide bonds of every fourth amino acid in close proximity. The distance between these juxtaposed peptide bonds is, in fact, just right for the formation of a hydrogen bond between the *imino group* (—NH—) of one peptide bond and the *carbonyl group* (—CO—) of the other, as shown in Figure 3-6a.

As a result, every peptide bond in the helix is hydrogen-bonded through its carbonyl group to the peptide bond immediately "below" it in the spiral and through its imino group to the peptide bond just "above" it. In each case, however, the hydrogen-bonded peptide bonds are separated in linear sequence by the three amino acids required to advance the helix far enough to allow the two bonds to be juxtaposed. These hydrogen bonds are all nearly parallel to the main axis of the helix and therefore tend to stabilize the spiral structure by holding successive turns of the helix together.

A major alternative to the α helix is the β sheet, also initially proposed by Pauling and Corey. As shown in Figure 3-6b, this structure is an extended sheetlike conformation with successive atoms in the polypeptide chain located at the "peaks" and "troughs" of the pleats. The R groups of successive amino acids jut out on alternating sides of the sheet.

Like the α helix, the β sheet is characterized by a maximum of hydrogen bonding. In both cases, all of the imino groups and carbonyl groups of the peptide bonds are involved. However, hydrogen bonding in an α helix is invariably intramolecular, between peptide bonds within the same polypeptide, whereas in the β sheet, hydrogen bonding is perpendicular to the plane of the sheet, linking peptide bonds in adjacent polypeptides or in juxtaposed segments of the same polypeptide.

Tertiary Structure. The **tertiary structure** of a protein can probably be best understood by contrasting it with secondary structure (Figure 3-5c). Secondary structure is a predictable, repeating conformational pattern that derives from the repetitive nature of the polypeptide because it involves hydrogen bonding between peptide bonds, the common structural elements along every polypeptide chain. If proteins contained only one or a few kinds of similar amino acids, virtually all aspects of protein conformation could probably be understood in terms of secondary structure, with only modest variations between proteins.

Tertiary structure comes about precisely because of the variety of amino acids present in proteins and the very different chemical properties of their R groups. In fact, tertiary structure depends almost entirely on interactions between the various R groups, regardless of where along the primary sequence they happen to come. Tertiary structure therefore reflects the nonrepetitive aspect of a polypeptide because it depends not on the carboxyl and amino groups common to all of the amino acids in the chain, but on the very feature that makes each amino acid distinctive—its R group.

Tertiary structure is neither repetitive nor readily predictable; it involves competing interactions between side groups with different properties. Hydrophobic R groups, for example, will associate with one another and spontaneously seek out a nonaqueous environment in the interior of the molecule, whereas polar amino acids will be drawn to the surface, by both their affinity for one another and their attraction to the polar water molecules in the surrounding milieu. As a result, the polypeptide chain will be folded, coiled, bent, and twisted into the **native conformation** that represents the most stable state for that particular sequence of amino acids.

Once the tertiary structure of a polypeptide has been achieved, it is stabilized and maintained by several types of bonds, both covalent and noncovalent. Noncovalent bonds that are important in this regard include *hydrogen bonds* between appropriate R groups, *electrostatic interactions* between charged R groups, and *hydrophobic interactions* between nonpolar R groups (see Table 3-2).

The most common covalent bond contributing to the stabilization of tertiary structure is the **disulfide bond** that is formed between two cysteines upon oxidation, as shown in Figure 3-7. The cysteines involved in a particular disulfide bond are usually quite distant from each other along the primary sequence but are juxtaposed by the folding process. Once formed, a disulfide bond confers considerable stability because of its covalent nature. A disulfide bond can be broken only by reducing it again and regenerating the two sulfhydryl groups of the cysteines.

The relative contributions of secondary and tertiary structures to the overall shape of a polypeptide vary from protein to protein and depend critically on the relative proportions and sequence of amino acids in the chain. Broadly speaking, proteins can be divided into two categories: *fibrous proteins* and *globular proteins*. **Fibrous proteins** have extensive secondary structure (either α helix or β sheet) throughout the molecule, giving them a highly ordered, repetitive structure. In general, secondary structure is much more important than tertiary interactions in determining the shape of such proteins, which often have an extended, filamentous structure. Especially prominent examples of fibrous proteins include silk *fibroin* and the *keratins* of hair and wool, as well as *collagen* (found in tendons and skin) and *elastin* (present in ligaments and blood vessels).

The amino acid sequence of each of these proteins favors a particular kind of secondary structure, which in turn confers a specific set of desirable mechanical properties on the protein. Fibroin, for example, consists mainly of long stretches of β sheet, with the polypeptide chains running parallel to the axis of the silk fiber. The most prevalent amino acids in fibroin are glycine, alanine, and serine. These amino acids have small R groups that pack together well (see Figure 3-2). The result is a silk fiber that is strong and relatively inextensible because the covalently bonded chains are stretched to nearly their maximum possible length.

Hair and wool fibers, on the other hand, consist of the protein α-keratin, which is almost entirely helical. The

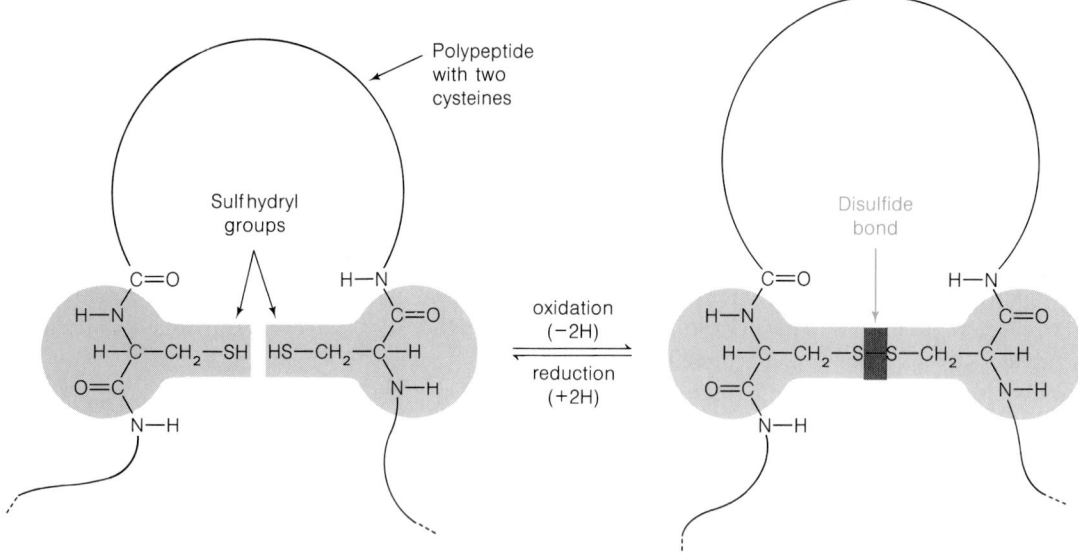

Figure 3-7 Disulfide Bond Formation. A disulfide bond is formed by the oxidation of sulfhydryl (—SH) groups of two cysteines (blue) located at different positions in a polypeptide chain (represented by the continuous line). A disulfide bond can be broken by reducing it to the two sulfhydryl groups from which it was originally formed. Disulfide bonds can also be formed between sulfhydryl groups in two separate polypeptide chains, thereby linking the two polypeptides covalently.

individual keratin molecules are very long and lie with their helix axes nearly parallel to the fiber axis. As a result, hair is quite extensible because stretching of the fiber is opposed not by the covalent bonds of the polypeptide chain, as in β sheets, but by the hydrogen bonds that stabilize the α-helical structure. The individual α helices in a hair are wound together to form a ropelike structure, as shown in Figure 3-8. First, three helices are wrapped around each other to form a *protofibril*. Eleven protofibrils are then combined to form a *microfibril*. Microfibrils, in turn, are packed together to form *macrofibrils*, bundles of which pass through and around the cells in a hair.

As important as fibrous proteins may be, they represent only a small fraction of the kinds of proteins present in most cells. Most of the proteins involved in cellular structure are **globular proteins,** so named because their polypeptide chains are folded into compact structures rather than extended filaments. The polypeptide chain of a globular protein is often folded locally into regions with α-helical or β-sheet structures, and these regions are themselves folded on one another to give the protein its compact, globular shape. This folding is possible because regions of α helix or β sheet are interspersed with irregularly structured regions (sometimes called *random coils*) that allow the polypeptide chain to loop and fold. Thus, every globular protein has its own unique tertiary structure, made up of secondary structural elements (helices and sheets) folded in a specific way especially suited to the protein's particular functional role.

Whether a specific segment of a polypeptide will form an α helix or a β sheet or neither depends on the amino acids present in that segment. For example, leucine, methio-

nine, and glutamate are strong "helix formers," whereas isoleucine, valine, and phenylalanine are strong "sheet formers." Both glycine and proline, the only cyclic amino acid, are "helix breakers" and are, in fact, largely responsible for the bends and turns in α helices. Such bends and turns usually occur at the surface of a polypeptide.

Figure 3-9 shows the tertiary structure of a typical globular protein, the enzyme ribonuclease. (Recall that ribonuclease was one of the first proteins to have its sequence determined. We also used it in Figure 2-19 to illustrate the denaturation and renaturation of a polypeptide and the spontaneity of its folding.) Two different conventional models are used to represent the structure of ribonuclease: the familiar ball-and-stick model, and a spiral-and-ribbon model, in which α-helical segments of the polypeptide are represented by spirals and β-sheet segments are depicted as ribbons with arrows pointing in the C-terminus direction. For clarity, most of the side chains of ribonuclease have been omitted in both models. The orange groups in Figure 3-9a are the four disulfide bonds that help to stabilize the three-dimensional structure of the ribonuclease model (see Figure 2-19a).

Globular proteins can be mainly α helical, mainly β sheet, or a mixture of both structures. These categories are illustrated in Figure 3-10 by the coat protein of tobacco mosaic virus (TMV), a portion of an immunoglobulin molecule, and a portion of the enzyme hexokinase, respectively. (In addition to illustrating aspects of tertiary structure, each of these proteins is important to us in its own right. We have already encountered the TMV coat protein in our discussion of self-assembly of viruses in Chapter 2. We will meet

Figure 3-8 **The Structure of Hair.** The main structural protein of hair is α-keratin, a fibrous protein with an α-helical shape. Three helices of α-keratin wrap into protofibrils, which then bond together to form microfibrils. Microfibrils have a characteristic "9 + 2" structure in which two central protofibrils are surrounded by nine others. Microfibrils, in turn, aggregate laterally to form macrofibrils. Bundles of macrofibrils pass through and around the cells in a hair.

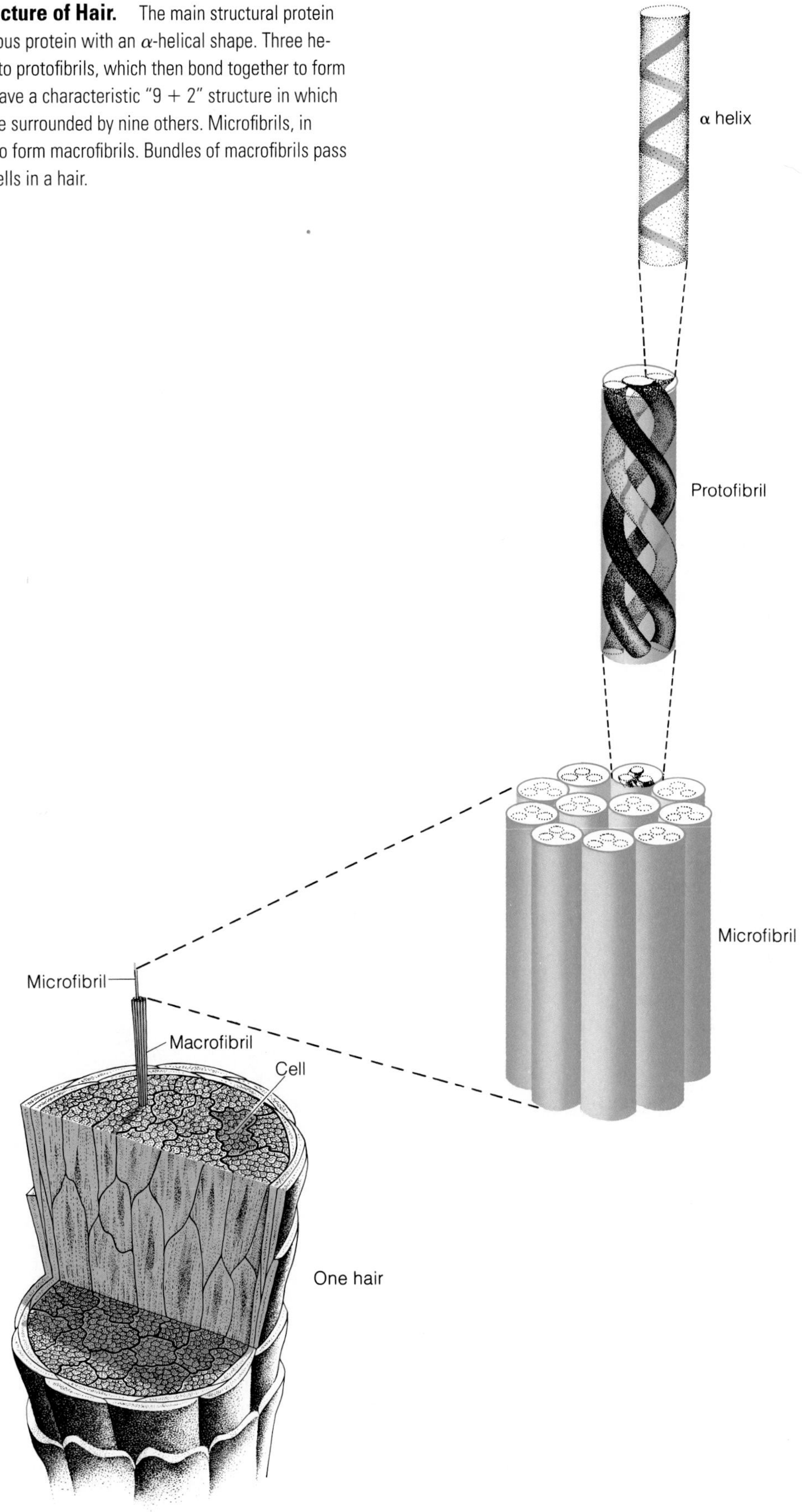

α helix

Protofibril

Microfibril

Microfibril

Macrofibril

Cell

One hair

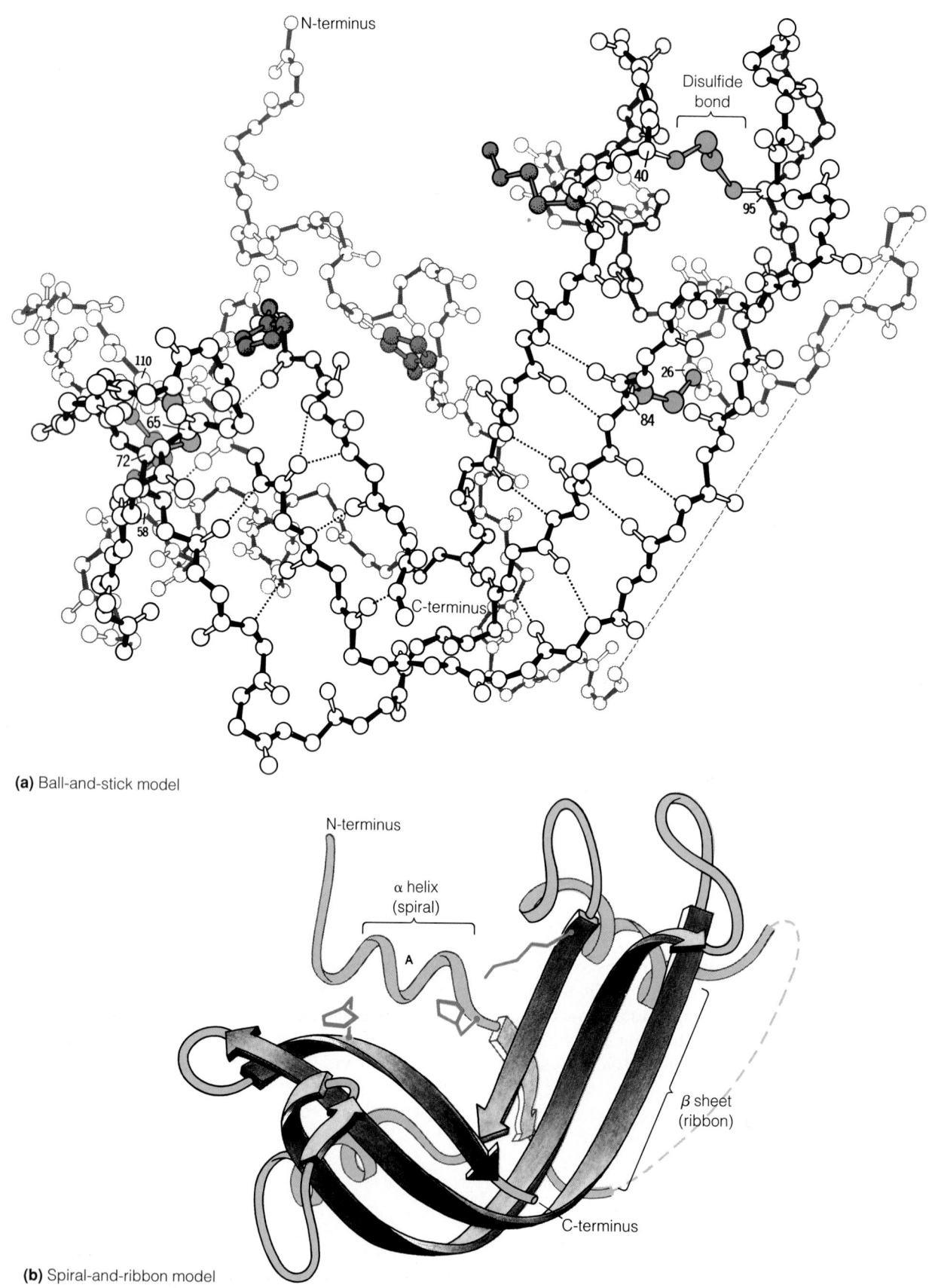

(a) Ball-and-stick model

(b) Spiral-and-ribbon model

Figure 3-9   The Three-Dimensional Structure of Ribonuclease.   Ribonuclease is a typical globular protein with significant α-helical and β-sheet segments. **(a)** A ball-and-stick model with disulfide bonds shown in orange and each of the relevant cysteines identified by its number in the amino acid sequence. The only side (R) groups shown are two histidines and a lysine that are important in the catalysis by this enzyme (purple). **(b)** An alternative model in which α-helical regions are represented as spirals and β-sheet regions as ribbons with arrows pointing in the direction of the C-terminus. Amino acid R groups and disulfide bonds have been omitted for clarity. Notice that the β-sheet structure is antiparallel (adjacent strands point in opposite directions) and highly twisted. Protein structure is determined by X-ray crystallography.

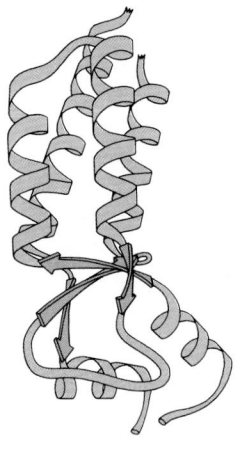 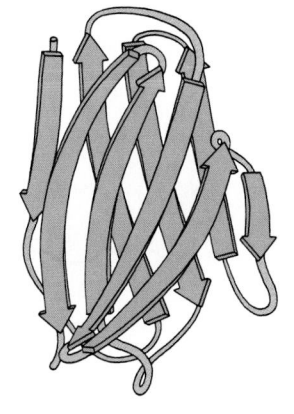 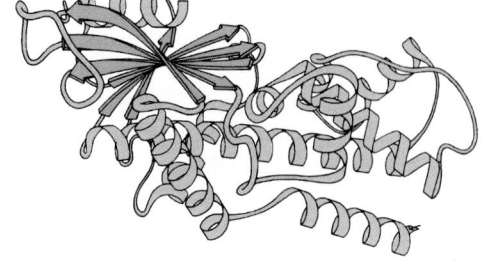

Tobacco mosaic coat protein

(a) Predominantly α helix

Immunoglobulin, V_2 domain

(b) Predominantly β sheet

Hexokinase, domain 2

(c) Mixed α helix and β sheet

Figure 3-10 Structures of Several Globular Proteins.
Shown here are proteins with different tertiary structures: **(a)** a predominantly α-helical structure (blue spirals), the coat protein of TMV; **(b)** a mainly β-sheet structure (orange ribbons with arrows), the V_2 domain of immunoglobulin; and **(c)** a structure that mixes α helix and β sheet, domain 2 of hexokinase. The immunoglobulin V_2 domain is an example of an antiparallel β-barrel structure, whereas the hexokinase domain 2 illustrates a twisted β sheet. (Green segments are random coils.)

hexokinase again when we consider the biological oxidation of glucose in Chapter 11, and we will encounter immunoglobulins in Chapter 24.)

Helical segments of globular proteins often consist of bundles of helices, as seen for the coat protein of tobacco mosaic virus (TMV) in Figure 3-10a. Segments with mainly β-sheet structure are usually characterized by a barrel-like configuration (Figure 3-10b) or by a twisted sheet (Figure 3-10c).

Most globular proteins consist of a number of segments called domains. A **domain** is a discrete, locally folded unit of tertiary structure, often containing regions of α helices and β sheets packed together compactly. A domain typically includes about 50–350 amino acids and usually has a specific function. Small globular proteins tend to be folded into a single domain, as shown in Figure 3-9b for ribonuclease, a relatively small protein. Large globular proteins usually have multiple domains. The portions of the immunoglobulin and hexokinase molecules shown in Figure 3-10b and c are, in fact, specific domains of these proteins (the V_2 domain of immunoglobulin and domain 2 of hexokinase, respectively).

Proteins having a common function (such as binding a specific metal ion or recognizing a specific molecule) usually have a common domain. Moreover, proteins with multiple functions usually have a separate domain for each function. Thus, domains can be thought of as the modular units from which globular proteins are constructed.

Before leaving the topic of tertiary structure, we should emphasize again the dependence of these higher levels of organization on the primary structure of the polypeptide. The significance of primary structure is exemplified especially well by the inherited condition **sickle-cell anemia.** People with this trait have hemoglobin molecules with normal α chains, but their β chains have a single amino acid that is abnormal. At a specific position in the chain, the glutamate normally present at that point is replaced by valine. The other 145 amino acids in the chain are correct, but this single substitution causes enough of a difference in the tertiary structure of the β chain that the oxygen-carrying capacity of the hemoglobin molecule is impaired significantly. Not all amino acid substitutions cause such dramatic changes in structure and function, but the example serves to underscore the crucial relationship between the amino acid sequence of a polypeptide and the final shape (and often the biological activity) of the molecule.

Quaternary Structure. The **quaternary structure** of a protein is the level of organization concerned with subunit interactions and assembly (Figure 3-5d). Quaternary structure therefore applies only to multimeric proteins. Many proteins are included in this category, however, particularly those with molecular weights above 50,000. Hemoglobin, for example, is a multimeric protein (see Figure 3-4). Some multimeric proteins contain identical polypeptide subunits; others, such as hemoglobin, contain several different kinds of polypeptides.

The bonds and forces that maintain quaternary structure are the same as those responsible for tertiary structure: hydrogen bonds, electrostatic interactions, hydrophobic interactions, and covalent disulfide bonds. Disulfide bonds may be either within a polypeptide chain or between chains. When they occur within a polypeptide, they stabilize tertiary structure; but when they occur between polypeptides, they help maintain quaternary structure. As for polypeptide folding, the process of assembly is spontaneous, with most, if not all, of the requisite information provided by the amino acid sequence of the individual polypeptide.

In some cases, a still higher level of assembly is possible, in the sense that two or more proteins (enzymes, usually) are organized into a **multiprotein complex,** with each protein involved sequentially in the same multistep process. An example of such a complex is an enzyme system called the *pyruvate dehydrogenase complex.* This complex catalyzes the oxidative removal of a carbon atom (as CO_2) from the three-carbon compound pyruvate (or pyruvic acid), a reaction that will be of interest to us when we get to Chapter 12. Three individual enzymes and five different coenzymes constitute a highly organized multienzyme complex that has a mass of 4.6 million daltons in bacterial cells and almost twice that size in the mitochondria of your liver cells. The pyruvate dehydrogenase complex is one of the best understood examples of how cells can achieve economy of function by ordering the enzymes that catalyze sequential reactions into a single multienzyme complex.

Nucleic Acids

Next we come to the **nucleic acids,** macromolecules of paramount importance to the cell because of their role in the storage, transmission, and expression of genetic information. Nucleic acids are linear polymers of nucleotides, strung together in a genetically determined order that is critical to their role as informational macromolecules. The two major types of nucleic acids are **DNA (deoxyribonucleic acid)** and **RNA (ribonucleic acid).** DNA and RNA differ in terms of both chemistry and their role in the cell. As the names suggest, RNA contains the five-carbon sugar **ribose** in each of its nucleotides, while DNA contains the closely related sugar **deoxyribose.** Functionally, DNA serves primarily as the repository of genetic information, whereas RNA molecules play several different roles in the expression of that information.

RNA is transcribed from DNA in several different forms, each of which plays a specific role in protein synthesis (Figure 3-11). **Messenger RNA (mRNA)** provides the information that dictates amino acid sequence during polypeptide synthesis. **Transfer RNA (tRNA)** directs the correct amino acid to the next site in a growing polypeptide chain. **Ribosomal RNA (rRNA),** the third major type of RNA, is an important constituent of the ribosomes that serve as the site of protein synthesis.

These roles of DNA and RNA as the bearers or mediators of genetic information will be considered in detail in Part Four. For now, we will focus on the chemistry of nucleic acids and the nucleotides of which they are composed.

The Monomers Are Nucleotides

Like proteins, nucleic acids are informational macromolecules and therefore contain nonidentical monomeric units

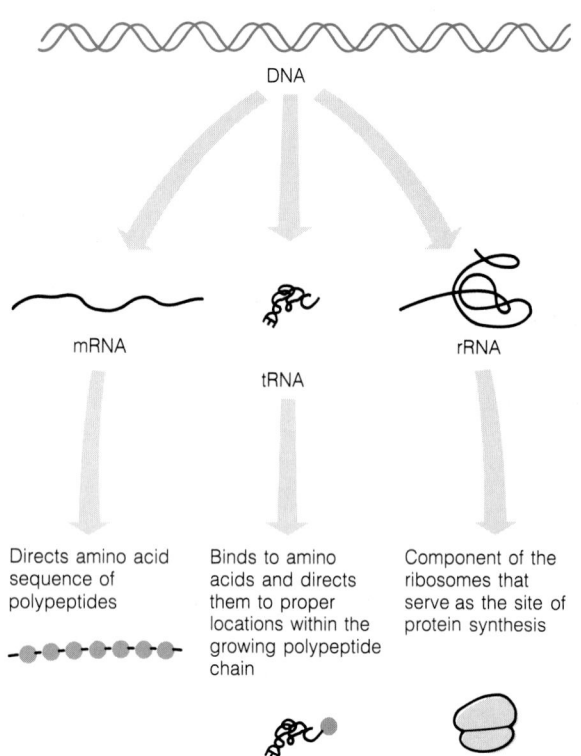

Figure 3-11   The Roles of DNA and RNA in Protein Synthesis. The genetic information in DNA is transcribed into molecules of messenger RNA (mRNA), transfer RNA (tRNA), and ribosomal RNA (rRNA). Each type of RNA plays a specific role in protein synthesis.

in a specified sequence. The monomeric units of nucleic acids are called **nucleotides.** Nucleotides exhibit less variety than amino acids; DNA and RNA each contain only four different kinds. (Actually, there is more variety than this suggests, especially in some RNA molecules; but virtually all variant structures found in nucleic acids represent one of the four basic nucleotides that has been chemically modified after insertion into the chain.)

As shown in Figure 3-12, each nucleotide consists of a five-carbon sugar, a phosphate group, and a nitrogen-containing aromatic base. The sugar is either D-ribose (for RNA) or D-deoxyribose (for DNA). The phosphate is joined by a **phosphoester bond** to the 5′ carbon of the sugar, and the base is attached at the 1′ carbon. The base may be either a **purine** or a **pyrimidine.** DNA contains the purines **adenine (A)** and **guanine (G)** and the pyrimidines **cytosine (C)** and **thymine (T).** RNA also has adenine, guanine, and cytosine but contains the pyrimidine **uracil (U)** in place of thymine.

If the phosphate is removed from a nucleotide, the remaining base-sugar unit is called a **nucleoside.** Each pyrimidine and purine may therefore occur as the free base, the nucleoside, or the nucleotide. The appropriate names for

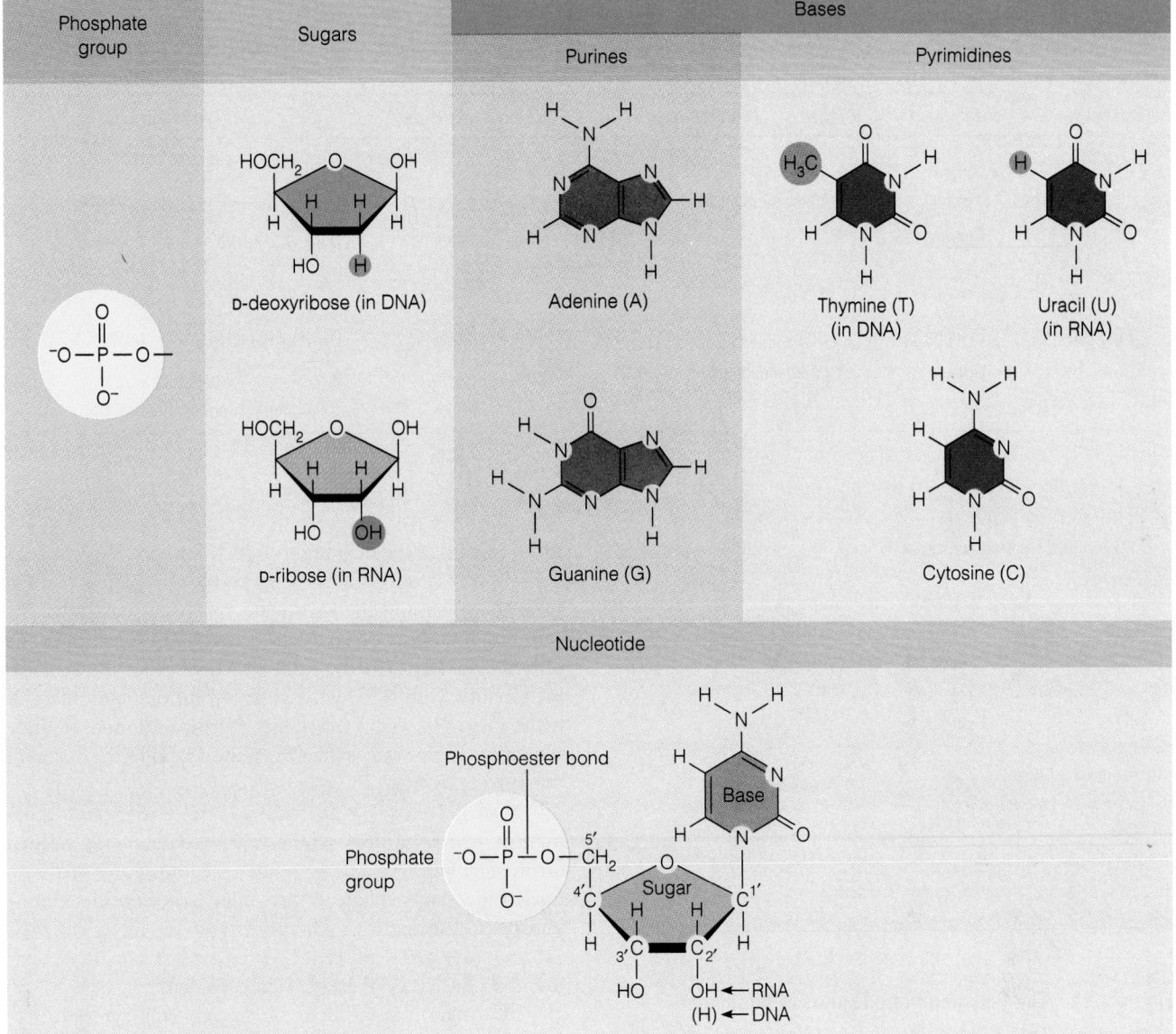

Figure 3-12 The Structure of a Nucleotide. In RNA, a nucleotide consists of the five-carbon sugar D-ribose with an aromatic nitrogen-containing base attached to the 1′ carbon and a phosphate group linked to the 5′ carbon by a phosphoester bond. (Carbon atoms in the sugar of a nucleotide are numbered from 1′ to 5′ to distinguish them from those in the base, which are numbered without the prime.) In DNA, the hydroxyl group on the 2′ carbon is replaced by a hydrogen atom, so the sugar is D-deoxyribose. The bases in DNA are the purines adenine (A) and guanine (G) and the pyrimidines thymine (T) and cytosine (C). In RNA, thymine is replaced by the pyrimidine uracil (U).

these compounds are given in Table 3-3. Notice that nucleotides and nucleosides containing deoxyribose are specified by a lowercase *d* preceding the letter that identifies the base.

As the nomenclature indicates, a nucleotide can be thought of as a **nucleoside monophosphate** because it is a nucleoside with a single phosphate group attached to it. This terminology can be readily extended to molecules with two or three phosphate groups attached to the 5′ carbon. For ex-

ample, the nucleoside adenosine (adenine plus ribose) can have one, two, or three phosphates attached and is designated accordingly as **adenosine monophosphate (AMP), adenosine diphosphate (ADP),** or **adenosine triphosphate (ATP).** The relationships among these compounds are shown in Figure 3-13.

You probably recognize ATP as the energy-rich compound used to drive a variety of reactions in the cell, including the activation of monomers for polymer formation. As

Table 3-3 The Bases, Nucleosides, and Nucleotides of RNA and DNA

	RNA		DNA	
Bases	Nucleoside	Nucleotide	Deoxynucleoside	Deoxynucleotide
Purines				
Adenine (A)	Adenosine	Adenosine monophosphate (AMP)	Deoxyadenosine	Deoxyadenosine monophosphate (dAMP)
Guanine (G)	Guanosine	Guanosine monophosphate (GMP)	Deoxyguanosine	Deoxyguanosine monophosphate (dGMP)
Pyrimidines				
Cytosine (C)	Cytidine	Cytidine monophosphate (CMP)	Deoxycytidine	Deoxycytidine monophosphate (dCMP)
Uracil (U)	Uridine	Uridine monophosphate (UMP)	—	—
Thymine (T)	—	—	Thymidine	Thymidine monophosphate (dTMP)

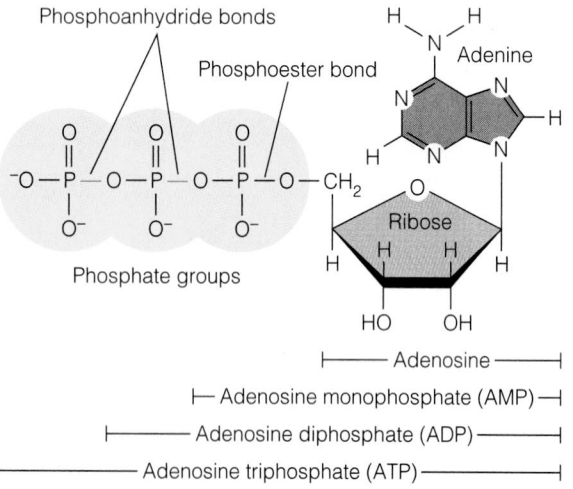

Figure 3-13   The Phosphorylated Forms of Adenosine. Adenosine occurs as the free nucleoside, the monophosphate (AMP), the diphosphate (ADP), and the triphosphate (ATP). The bond that links the first phosphate to the ribose of adenosine is a low-energy phosphoester bond, whereas the bonds that link the second and third phosphate groups to the molecule are higher-energy phosphoanhydride bonds.

this example suggests, nucleotides actually play two roles in cells: They are the monomeric units of nucleic acids, and they serve as intermediates in various energy-transferring reactions.

The Polymers Are DNA and RNA

Nucleic acids are linear polymers formed by linking each nucleotide to the next through a phosphate group, as shown in Figure 3-14. Specifically, the phosphate group already attached by a phosphoester bond to the 5′ carbon of one nucleotide becomes linked by a second phosphoester bond to

the 3′ carbon of the next nucleotide. In essence, this is a condensation reaction, with the —H and —OH groups coming from the sugar and the phosphate groups, respectively. The resulting linkage is called a 3′,5′ **phosphodiester bond.** The **polynucleotide** formed by this process has an intrinsic directionality, with a 5′ hydroxyl group on one end and a 3′ hydroxyl group on the other end. By convention, nucleotide sequences are always written from the 5′ end to the 3′ end of the polynucleotide.

Nucleic acid synthesis requires both energy and information, just as protein synthesis does. The energy needed for the formation of a phosphodiester bond is provided by a nucleoside triphosphate rather than a nucleoside monophosphate (nucleotide). The precursors for DNA synthesis are therefore dATP, dCTP, dGTP, and dTTP. For RNA synthesis, ATP, CTP, GTP, and UTP are needed.

Information is required for nucleic acid synthesis because successive incoming nucleotides must be added in a specific, genetically determined sequence. For this purpose, a preexisting molecule is used as a **template** to specify nucleotide order. For both DNA and RNA synthesis, the template is usually DNA. The essence of template-directed nucleic acid synthesis is that each incoming nucleotide is selected because its base can be recognized by (will interact with) the base of the nucleotide already present at that position in the template.

This recognition process depends on an important chemical feature of the purine and pyrimidine bases shown in Figure 3-15. These bases have carbonyl groups and nitrogen atoms capable of hydrogen bond formation under appropriate conditions. Furthermore, there exist complementary relationships between purines and pyrimidines that allow A to form two hydrogen bonds with T (or U) and G to form three hydrogen bonds with C, as shown in Figure 3-15. This pairing of A with T (or U) and G with C is a fundamental property of nucleic acids. Genetically, this **base pairing** provides a mechanism for nucleic acids to recognize one another, as we will see in Part Four. For now, however, let us concentrate on the structural implications.

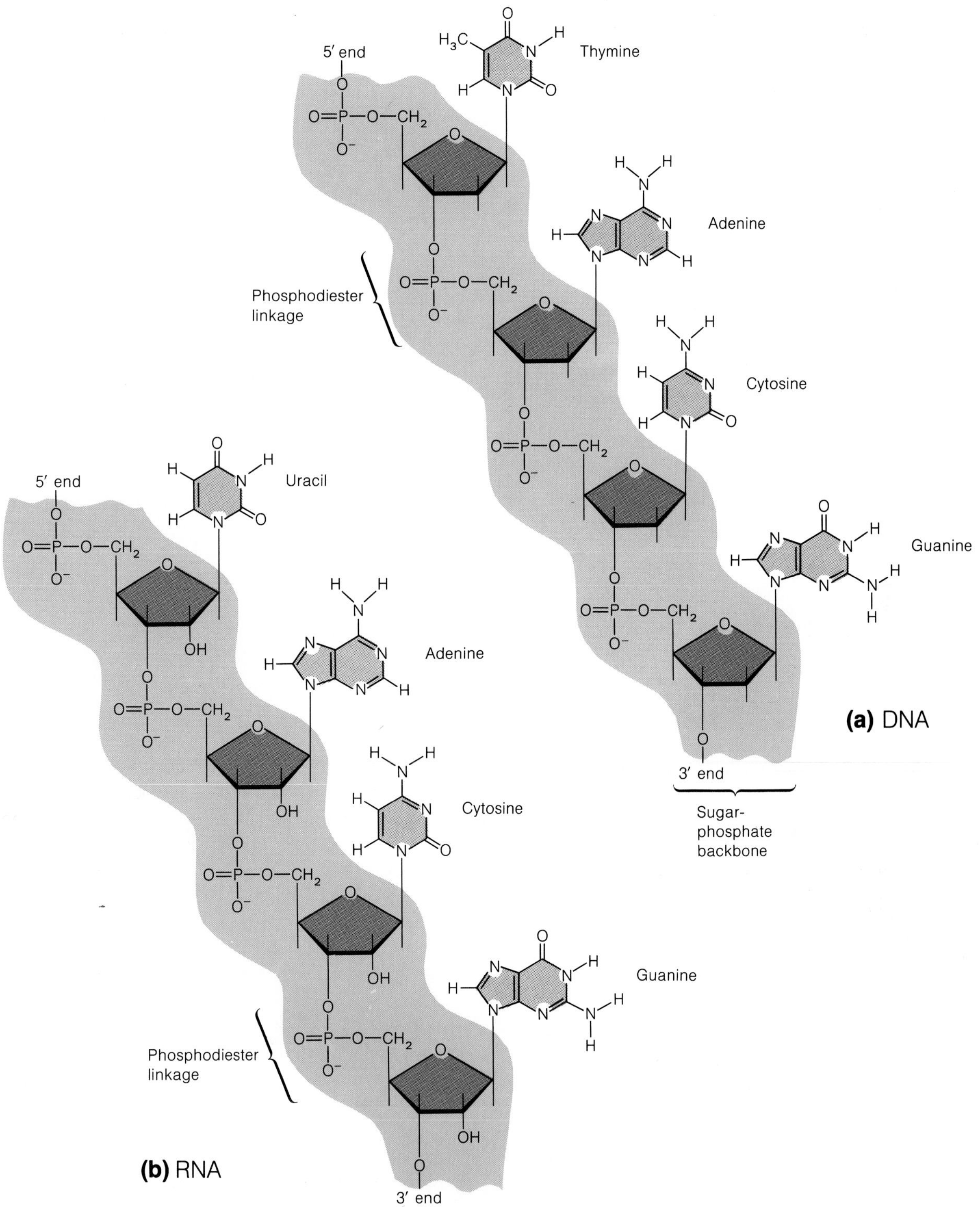

Figure 3-14 The Structure of Nucleic Acids. Nucleic acids consist of chains of nucleotides, each containing a sugar, a phosphate, and a base. The sugar is **(a)** deoxyribose in DNA and **(b)** ribose in RNA. Successive nucleotides in the chain are joined together by 3′,5′ phosphodiester linkages. The backbone of the chain (blue) is an alternating sugar-phosphate sequence, from which the bases (gray) jut out.

A DNA Molecule Is a Double-Stranded Helix

One of the most significant biological advances of the twentieth century came in 1953 in a two-page article in the scientific journal *Nature.* In the article, Francis Crick and James Watson postulated a double-stranded helical structure for DNA—the now-famous **double helix**—that not only accounted for the known physical and chemical properties of DNA but also suggested a mechanism for replication of the structure. Some highlights of this exciting chapter in the history of contemporary biology are related in Box 3A.

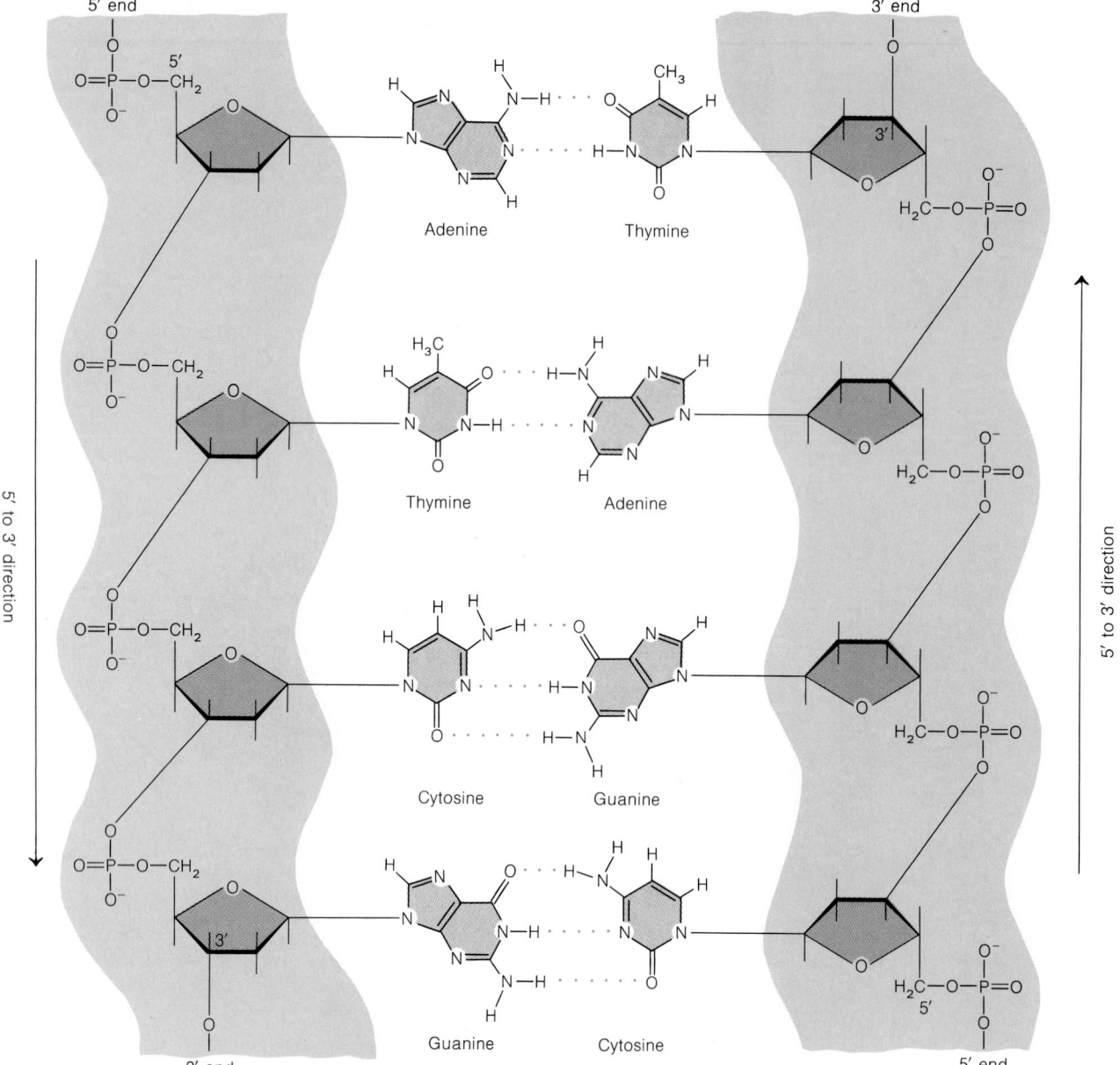

Figure 3-15   Hydrogen Bonding in Nucleic Acid Structure. The hydrogen bonds (blue dots) between adenine and thymine and between cytosine and guanine account for the AT and CG base pairing of DNA. If one or both strands were RNA instead, the pairing partner for adenine would be uracil (U).

Essentially, the double helix consists of two complementary chains of DNA twisted together around a common axis to form a right-handed helical structure that resembles a circular staircase (Figure 3-16). The two chains are oriented in opposite directions along the helix, one running in the 5′ → 3′ direction and the other in the 3′ → 5′ direction (see Figure 3-15). The backbone of each chain consists of alternating sugar and phosphate groups. The phosphate groups are charged and the sugar molecules contain polar hydroxyl groups, so it is not surprising that the sugar-

phosphate backbones of the two strands are on the outside of the DNA helix, where their interaction with the surrounding aqueous milieu can be maximized. The pyrimidine and purine bases, on the other hand, are aromatic compounds with less affinity for water. Accordingly, they are oriented inward, forming the base pairs that hold the two chains together.

To form a stable double helix, the two component strands must be not only *antiparallel* (running in opposite directions) but also *complementary*. By this we mean that

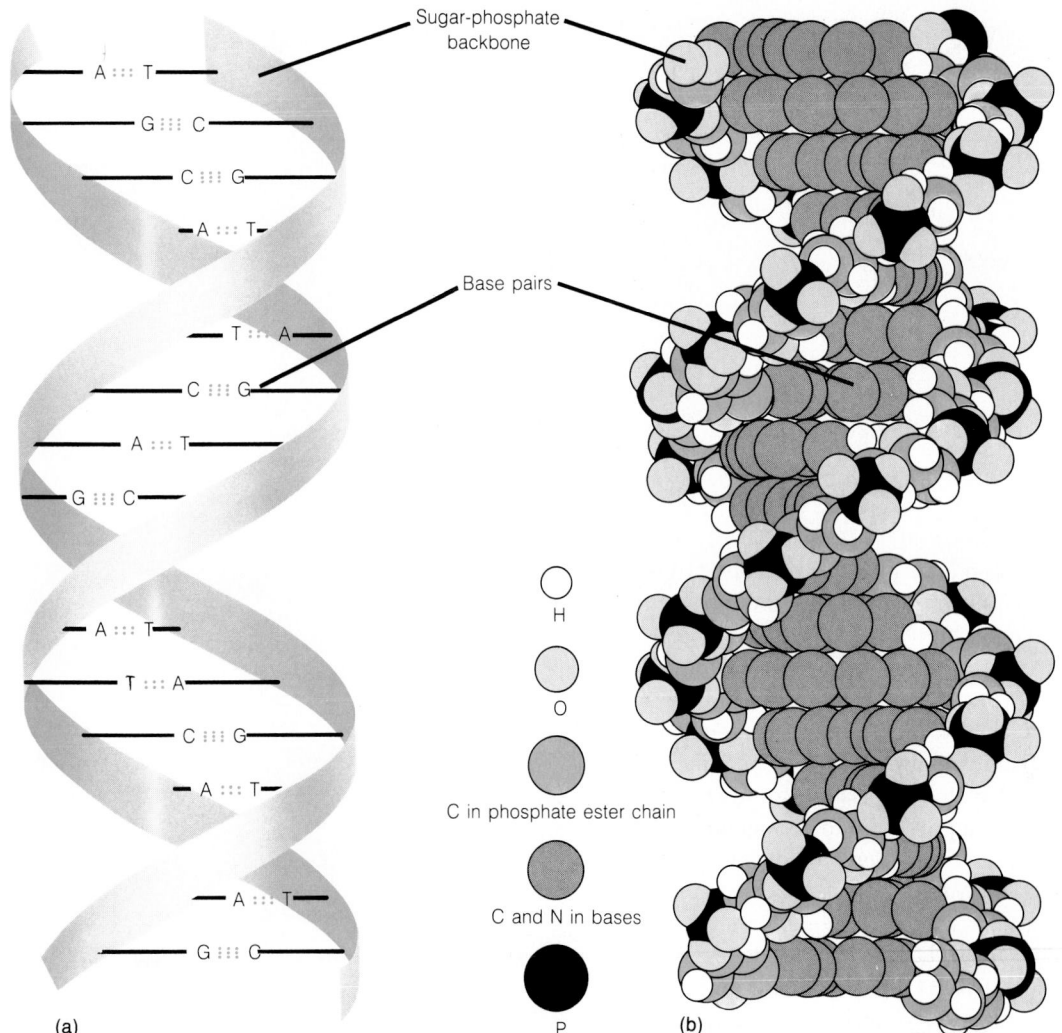

Sugar-phosphate backbone

Base pairs

H

O

C in phosphate ester chain

C and N in bases

P

(a)

(b)

Figure 3-16   The Structure of Double-Stranded DNA.
(a) A schematic representation of the double-helical structure of DNA. The continuously turning strips represent the sugar-phosphate backbones of the molecule, while the horizontal bars represent paired bases of the two strands. **(b)** A space-filling model of the DNA double helix.

each base in one strand can form specific hydrogen bonds with the base in the other strand directly across from it. From the pairing possibilities shown in Figure 3-15, this means that each A must be paired with a T, and each G with a C. In both cases, one member of the pair is a pyrimidine (T or C) and the other is a purine (A or G). The distance between the two sugar-phosphate backbones in the double helix is just sufficient to accommodate one of each kind of base. If we envision the sugar-phosphate backbones of the two strands as the sides of a circular staircase, then each step or rung of the stairway corresponds to a pair of bases held in place by hydrogen bonding (see Figure 3-16).

RNA structure also depends in part on base pairing, but this pairing is usually between complementary regions within the same strand and is much less extensive than the interstrand pairing of the DNA duplex. Of the various RNA species, secondary and tertiary structures are well understood only for the tRNA molecules, as we will see when we get to Chapter 17.

Polysaccharides

The next class of macromolecules we will consider is the **polysaccharides.** Unlike proteins and nucleic acids, polysaccharides play no informational role in the cell. In fact, they usually consist of a single kind of repeating unit, or sometimes a strictly alternating pattern of two kinds. As noted earlier, the major polysaccharides, at least in higher organisms, are the storage polysaccharides starch and glycogen and the structural polysaccharide cellulose. Each of these polymers contains the six-carbon sugar glucose as its single

Historical Perspectives

ON THE TRAIL OF THE DOUBLE HELIX

"I have never seen Francis Crick in a modest mood. Perhaps in other company he is that way, but I have never had reason so to judge him." With this typically irreverent observation as an introduction, James Watson goes on to describe in a very personal and highly entertaining way the events that eventually led to the discovery of the structure of DNA. The account, published in 1968 under the title *The Double Helix,* is still fascinating reading for the personal, unvarnished insights it provides into how an immense scientific discovery came about. Commenting on his reasons for writing the book, Watson observes in the preface that "there remains general ignorance about how science is 'done.' That is not to say that all science is done in the manner described here. This is far from the case, for styles of scientific research vary almost as much as human personalities. On the other hand, I do not believe that the way DNA came out constitutes an odd exception to a scientific world complicated by the contradictory pulls of ambition and the sense of fair play."

As portrayed in Watson's account, Crick and Watson are about as different from each other in nature and background as they could be. But there was one thing they shared, and that was an unconventional but highly productive way of "doing" science. They did little actual experimentation on DNA, choosing instead to draw heavily upon the research findings of others and to bring their own considerable ingenuities to bear building models and exercising astute insights and hunches (Figure 3A-1). And out of it all emerged, in a relatively short time, an understanding of the double-helical structure of DNA that has come to rank as one of the major scientific events of this century.

To appreciate their findings and their brilliance, we must first understand the setting in which Watson and Crick worked. The early 1950s was an exciting time in biology. It had been only a few years since Avery, MacLeod, and McCarty had published evidence on the genetic transformation of bacteria, but the work of Hershey and Chase that confirmed DNA as the genetic material had not yet appeared in print. Meanwhile, at Columbia University, Erwin

Figure 3A-1   James Watson (left) and Francis Crick (right) at work with their model of DNA.

Chargaff's careful chemical analyses had revealed that although the relative proportions of the four bases—A, T, C, and G—varied greatly from one species to the next, it was always the same for all members of a single species. Even more puzzling and portentous was Chargaff's second finding: For a given species, A and T always occurred in the same proportions, and so did G and C (that is, A = T and C = G).

The most important clues came from the work of Maurice Wilkins and Rosalind Franklin at King's College in London. Wilkins and Franklin were using the technique of X-ray diffraction to study DNA structure, and they took a rather dim view of Watson and Crick's strategy of model-building. X-ray diffraction is a useful tool for detecting regularly occurring structural elements in a crystalline substance because any structural feature that repeats at some fixed interval in the crystal contributes in a characteristic way to the diffraction pattern that is obtained. From Franklin's painstaking analysis of the diffraction patterns of DNA, it became clear that the molecule was long and thin, with some structural element being repeated every 0.34 nm and another being repeated every 3.4 nm. Even more intriguing, the molecule appeared to be some sort of helix.

This stirred the imaginations of Watson and Crick, because they had heard only recently of Pauling and Corey's α-helical structure for proteins. Working with models of the bases cut from stiff cardboard, Watson and Crick came to the momentous insight that DNA was also a helix but with an all-important difference: It was a *double* helix, with hydrogen-bonded pairing of purines and pyrimidines. The actual discovery is best recounted in Watson's own words:

When I got to our still empty office the following morning, I quickly cleared away the papers from my desk top so that I would have a large, flat surface on which to form pairs of bases held together by hydrogen bonds. Though I initially went back to my like-with-like prejudices, I saw all too well that they led nowhere. When Jerry [Donohue, an American crystallographer working in the same laboratory] came in I looked up, saw that it was not Francis, and began shifting the bases in and out of various other pairing possibilities. Suddenly I became aware that an adenine-thymine pair held together by two hydrogen bonds was identical in shape to a guanine-cytosine pair held together by at least two hydrogen bonds. All the hydrogen bonds seemed to form naturally; no fudging was required to make the two types of base pairs identical in shape. Quickly I called Jerry over to ask him whether this time he had any objections to my new base pairs.

When he said no, my morale skyrocketed, for I suspected that we now had the answer to the riddle of why the number of purine residues exactly equaled the number of pyrimidine residues. Two irregular sequences of bases could be regularly packed in the center of a helix if a purine always hydrogen-bonded to a pyrimidine. Furthermore, the hydrogen-bonding requirement meant that adenine would always pair with thymine, while guanine could pair only with cytosine. Chargaff's rules then suddenly stood out as a consequence of a double-helical structure for DNA. Even more exciting, this type of double helix suggested a replication scheme much more satisfactory than my briefly considered like-with-like pairing. Always pairing adenine with thymine and guanine with cytosine meant that the base sequences of the two intertwined chains were complementary to each other. Given the base sequence of one chain, that of its partner was automatically determined. Conceptually, it was thus very easy to visualize how a single chain could be the template for the synthesis of a chain with the complementary sequence.

Upon his arrival Francis did not get more than halfway through the door before I let loose that the answer to everything was in our hands. Though as a matter of principle he maintained skepticism for a few moments, the similarly shaped AT and GC pairs had their expected impact. His quickly pushing the bases together in a number of different ways did not reveal any other way to satisfy Chargaff's rules. A few minutes later he spotted the fact that the two glycosidic bonds (joining base and sugar) of each base pair were systematically related by a diad axis perpendicular to the helical axis. Thus, both pairs could be flip-flopped over and still have their glycosidic bonds facing in the same direction. This had the important consequence that a given chain could contain both purines and pyrimidines. At the same time, it strongly suggested that the backbones of the two chains must run in opposite directions.

The question then became whether the AT and GC base pairs would easily fit the backbone configuration devised during the previous two weeks. At first glance this looked like a good bet, since I had left free in the center a large vacant area for the bases. However, we both knew that we would not be home until a complete model was built in which all the stereochemical contacts were satisfactory. There was also the obvious fact that the implications of its existence were far too important to risk crying wolf. Thus I felt slightly queasy when at lunch Francis winged into the Eagle to tell everyone within hearing distance that we had found the secret of life.*

The rest is history. Shortly thereafter, the prestigious journal *Nature* carried an unpretentious two-page article entitled simply "Molecular Structure of Nucleic Acids: A Structure for Deoxyribose Nucleic Acid," by James Watson and Francis Crick. Though modest in length, that paper has had far-reaching implications, for the double-stranded model that Watson and Crick worked out in 1953 has proved to be correct in all its essential details, unleashing a revolution in the field of biology. ■

*Excerpted from *The Double Helix,* pp. 194–197. Copyright ©1968 James D. Watson. Reprinted with the permission of the author and Atheneum Publishers, Inc.

repeating unit, but they differ in both the nature of the bond between successive glucose units and the presence and extent of side branches on the chains.

The Monomers Are Monosaccharides

The repeating units of polysaccharides are simple sugars called **monosaccharides** (from the Greek *mono* meaning "single" and *saccharide* meaning "sugar"). A sugar can be defined as an aldehyde or ketone that has two or more hydroxyl groups. Thus, there are two categories of sugars: the **aldosugars,** with a terminal carbonyl group (Figure 3-17a); and the **ketosugars,** with an internal carbonyl group (Figure 3-17b). Within these categories, sugars are named generically according to the number of carbon atoms they contain. Most sugars have between three and seven carbon atoms and are therefore classified as a *triose* (three carbons), a *tetrose* (four carbons), a *pentose* (five carbons), a *hexose* (six carbons), or a *heptose* (seven carbons). We have already encountered two pentoses—the ribose of RNA and the deoxyribose of DNA.

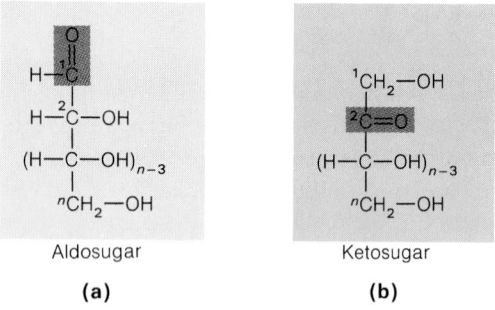

Aldosugar **(a)** Ketosugar **(b)**

Figure 3-17 Structures of Monosaccharides. **(a)** Aldosugars have a carbonyl group (gray) on carbon atom 1. **(b)** Ketosugars have a carbonyl group (gray) on carbon atom 2. The number of carbon atoms in a monosaccharide (*n*) varies from three to seven.

The single most common sugar in the biological world is the aldohexose D-glucose, represented by the formula $C_6H_{12}O_6$ and by the structure shown in Figure 3-18. The formula $C_nH_{2n}O_n$ is characteristic of sugars and gave rise to the general term **carbohydrate** because compounds of this sort were originally thought of as "hydrates of carbon"—$C_n(H_2O)_n$. The term persists, although in no sense can a carbohydrate be thought of as hydrated carbon atoms.

In keeping with the general rule for numbering carbon atoms in organic molecules, the carbons of glucose are numbered beginning with the most oxidized end of the molecule, the aldehyde group. Notice that glucose has four asymmetric carbon atoms (carbon atoms 2, 3, 4, and 5). There are therefore 2^4 different possible stereoisomers of the aldosugar $C_6H_{12}O_6$, but we will concern ourselves only with D-glucose, which is the most stable of the 16 isomers.

Figure 3-18a shows D-glucose as it appears in what chemists call a **Fischer projection,** with the —H and —OH groups intended to be projecting slightly out of the plane of the paper. This structure indicates that glucose is a linear molecule, and it is often a useful representation of glucose for pedagogic purposes. In reality, however, glucose exists in the cell in a dynamic equilibrium between the linear (or open-chain) configuration of Figure 3-18a and the ring form of Figure 3-18b. The ring form is the predominant structure because it is energetically more stable and therefore favored. The ring form results from the addition of the hydroxyl group on carbon atom 5 across the carbonyl group of carbon atom 1. Although the juxtaposition of carbon atoms 1 and 5 required for ring formation seems unlikely from the Fischer projection, it is actually favored by the tetrahedral nature of each carbon atom in the chain.

A more satisfactory representation of glucose is that shown in Figure 3-18c. The advantage of this **Haworth projection** is that it at least suggests the spatial relationship of different parts of the molecule and makes the spontaneous formation of a bond between carbon atoms 1 and 5 appear more likely. Any of the three representations of glucose

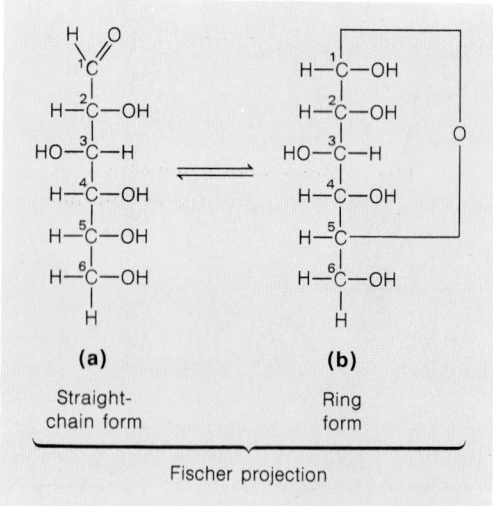

(a) Straight-chain form **(b)** Ring form

Fischer projection

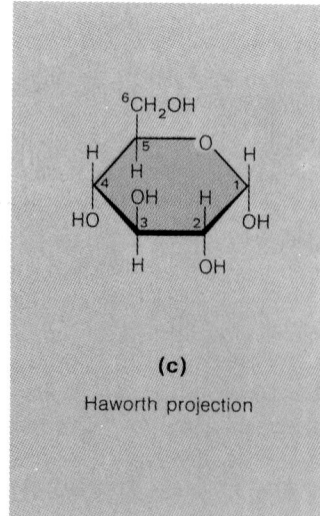

(c) Haworth projection

Figure 3-18 The Structure of Glucose. The glucose molecule can be represented by Fischer projections of **(a)** the straight-chain form or **(b)** the ring form of the molecule, as well as by **(c)** the Haworth projection of the ring form. In the Fischer projections, the —H and —OH groups are intended to be projecting slightly out of the plane of the paper. In the Haworth projection, carbon atoms 2 and 3 are intended to be jutting out of the plane of the paper, and carbon atoms 5 and 6 are behind the plane of the paper. The —H and —OH groups then project upward or downward, as indicated.

shown in Figure 3-18 is valid, but the Haworth projection is preferred because it indicates both the ring form and the spatial relationship of the carbon atoms.

Notice that formation of the ring structure results in the generation of one of two alternative forms of the molecule, depending on the spatial orientation of the hydroxyl group on carbon atom 1. These alternative forms of glucose are designated α and β. As shown in Figure 3-19, α-D-glucose has the hydroxyl group on carbon atom 1 pointing downward in the Haworth projection, and β-D-glucose has the hydroxyl group on carbon atom 1 pointing upward. Starch and glycogen both have α-D-glucose as their repeating unit, whereas cellulose consists of strings of β-D-glucose.

In addition to the free monosaccharide and the long-chain polysaccharides, glucose also occurs in **disaccharides,** which consist of two monosaccharide units linked covalently. Three common disaccharides are shown in Figure 3-20. *Maltose* consists of two glucose units linked together, whereas *lactose* (milk sugar) contains a glucose linked to a galactose and *sucrose* (common table sugar) has a glucose linked to a fructose. Both galactose and fructose will be discussed in more detail in Chapter 11, where the chemistry and metabolism of several sugars are considered.

Each of these disaccharides is formed by a condensation reaction in which two monosaccharides are linked together by the elimination of water. The resulting bond is called a **glycosidic bond** and is characteristic of linkages between sugars. In the case of maltose, both of the constituent glucose molecules are in the α form, and the glycosidic bond forms between carbon atom 1 of one glucose and carbon atom 4 of the other. This is called an α *glycosidic bond* because it involves a carbon atom 1 with its hydroxyl group in the α configuration. Lactose, on the other hand, is characterized by a β *glycosidic bond* because the hydroxyl group on carbon atom 1 of the galactose is in the β configuration. The distinction between α and β glycosidic bonds becomes especially critical when we get to the polysaccharides because both the three-dimensional configuration and the biological role of the polymer depend on the nature of the bond between the repeating monosaccharide units.

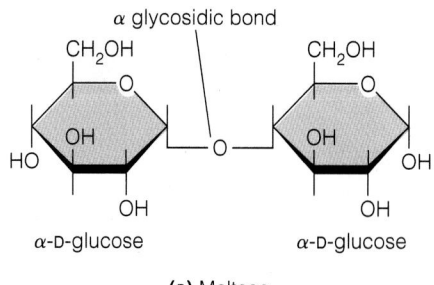

(a) Maltose

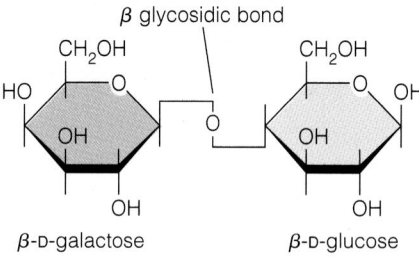

(b) Lactose

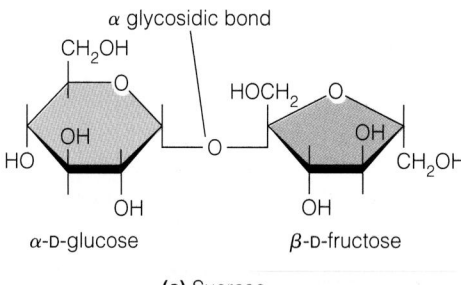

(c) Sucrose

Figure 3-20 Some Common Disaccharides. **(a)** Maltose (malt sugar) consists of two molecules of α-D-glucose linked by an α-glycosidic bond. **(b)** Lactose (milk sugar) consists of a molecule of β-D-galactose linked to a molecule of β-D-glucose by a β glycosidic bond. **(c)** Sucrose (table sugar) consists of a molecule of α-D-glucose linked to a molecule of β-D-fructose by an α glycosidic bond.

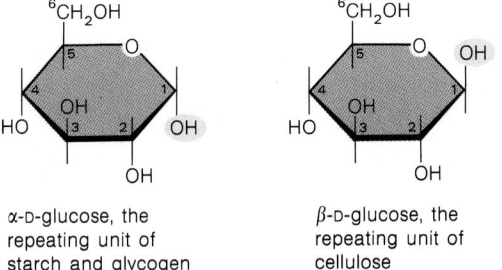

Figure 3-19 The Ring Forms of D-Glucose. The hydroxyl group on carbon atom 1 points downward in the α form and upward in the β form.

The Polymers Are Storage and Structural Polysaccharides

Polysaccharides perform either storage or structural functions in cells. The most familiar *storage polysaccharides* are the **starch** of plant cells (Figure 3-21a) and the **glycogen** of animal cells (Figure 3-21b). Both of these polymers consist of α-D-glucose units linked together by α glycosidic bonds. In addition to $\alpha(1 \rightarrow 4)$ bonds that link carbon atoms 1 and 4 of adjacent glucose units, these polysaccharides may contain occasional $\alpha(1 \rightarrow 6)$ linkages along the backbone, giving rise to side chains (Figure 3-21c). Storage polysaccharides can therefore be branched or unbranched polymers, depending on the presence or absence of $\alpha(1 \rightarrow 6)$ linkages.

Glycogen is highly branched, with $\alpha(1 \rightarrow 6)$ linkages occurring every 8 to 10 glucose units along the backbone

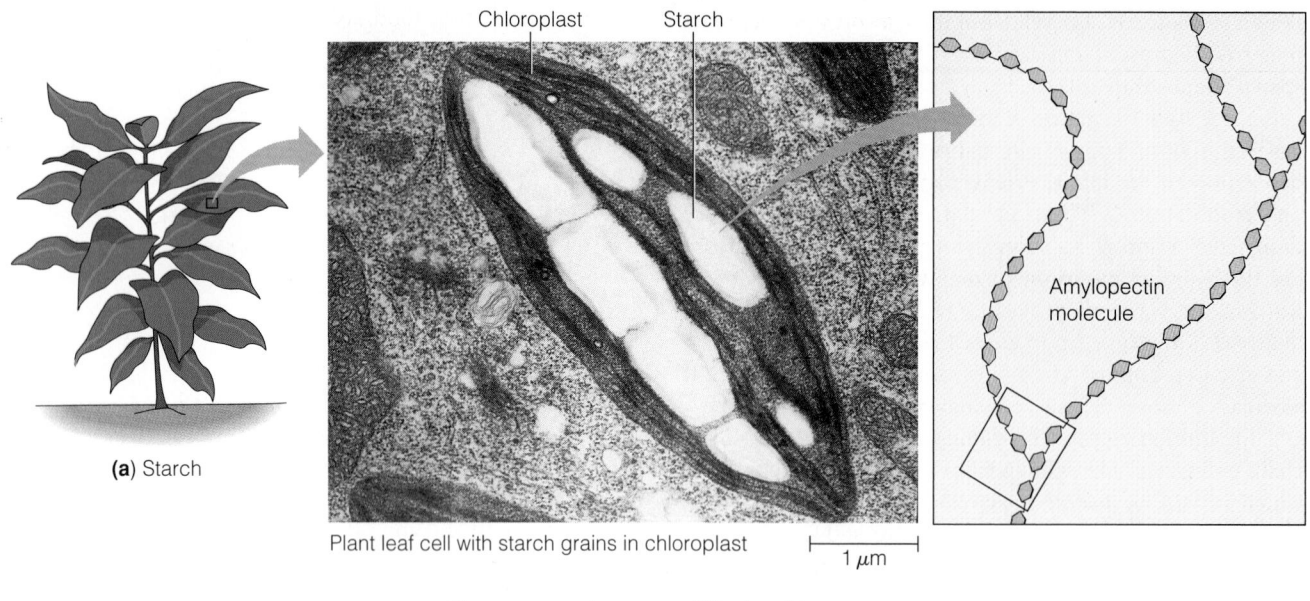

Plant leaf cell with starch grains in chloroplast

Chloroplast   Starch

Amylopectin molecule

1 μm

(a) Starch

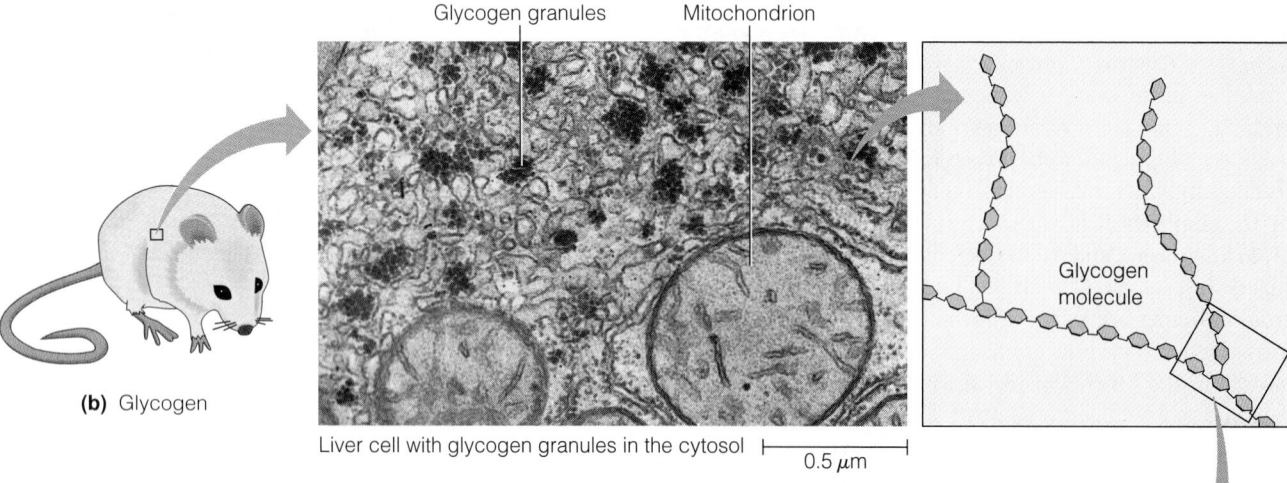

Liver cell with glycogen granules in the cytosol

Glycogen granules   Mitochondrion

Glycogen molecule

0.5 μm

(b) Glycogen

Figure 3-21   The Structure of Starch and Glycogen.

(a) The starch found in plant cells and **(b)** the glycogen found in animal cells are both storage polysaccharides composed of linear chains of α-D-glucose units, with or without occasional branch points (TEMs). Starch occurs in two forms: branched amylopectin, as shown in part a, and unbranched amylose (not shown). Glycogen occurs only as the branched form shown in part b. **(c)** The straight-chain portion of all three kinds of molecules consists of α-D-glucose units linked by $\alpha(1 \rightarrow 4)$ glycosidic bonds. In the case of amylopectin and glycogen, branch chains originate at $\alpha(1 \rightarrow 6)$ glycosidic bonds.

Side chain

$\alpha(1 \rightarrow 6)$ bond

$\alpha(1 \rightarrow 4)$ bond

(c) Glycogen or amylopectin structure

and giving rise to short side chains of about 8 to 12 glucose units (Figure 3-21b). Glycogen is stored mainly in the liver and in muscle tissue. In the liver it is used as a source of glucose to maintain blood sugar levels, whereas in muscle it serves as a fuel source to generate the ATP needed for muscle contraction.

Starch occurs both as unbranched **amylose** and as branched **amylopectin.** Like glycogen, amylopectin has $\alpha (1 \rightarrow 6)$ branches, but these occur less frequently along the backbone (once every 12 to 25 glucose units) and give rise to longer side chains (lengths of 20 to 25 glucose units are common) (see Figure 3-21a). Starch deposits are usually about 10–30% amylose and 70–90% amylopectin. Starch is stored in plant cells as *starch grains* within the plastids—either the *chloroplasts* that are the sites of carbon fixation and sugar synthesis in photosynthetic tissue or the *amyloplasts* that are specialized plastids for starch storage. The potato tuber, for example, is filled with starch-laden amyloplasts.

The best-known example of a *structural polysaccharide* is the **cellulose** found in plant cell walls (Figure 3-22). Cellulose is an important polymer quantitatively; more than half of the carbon in higher plants is present in cellulose! Like starch and glycogen, cellulose is also a polymer of glucose, but the repeating monomer is β-D-glucose and the linkage is therefore $\beta(1 \rightarrow 4)$. This linkage has structural consequences that we will get to shortly, but it also has nutritional implications. Mammals do not possess an enzyme that can hydrolyze a $\beta(1 \rightarrow 4)$ bond; therefore, mammals cannot

utilize cellulose as food. As a result, you can digest potatoes (starch) but not grass (cellulose). Animals such as cows and sheep might seem to be exceptions because they do eat grass and similar plant products. But they cannot cleave β glycosidic bonds either; they depend on the population of bacteria and protozoa in their rumen (part of their compound stomach) to do this for them. The microorganisms eat the cellulose, and the host animal then obtains the end-products of microbial digestion, now in a form the animal can use.

Although $\beta(1 \rightarrow 4)$-linked cellulose is quantitatively the most significant structural polysaccharide, others are also known. The celluloses of fungal cell walls, for example, contain either $\beta(1 \rightarrow 4)$ or $\beta(1 \rightarrow 3)$ linkages, depending on the species. The cell wall of many bacteria is somewhat more complex and contains two kinds of sugars, **N-acetylglucosamine (GlcNAc)** and **N-acetylmuramic acid (MurNAc).** As shown in Figure 3-23a, GlcNAc and MurNAc are derivatives of *β-glucosamine,* a glucose molecule with the hydroxyl group on carbon atom 2 replaced by an amino group. GlcNAc is formed by acetylation of the amino group, and MurNAc requires the further addition of a three-carbon lactyl group to carbon atom 3. The cell wall polysaccharide is then formed by the linking of GlcNAc and MurNAc in a strictly alternating sequence with $\beta(1 \rightarrow 4)$ bonds (Figure 3-23b). Figure 3-23c shows the structure of yet another structural polysaccharide, the **chitin** found in insect exoskeletons and crustacean shells. Chitin consists of Glc-NAc units only, joined by $\beta(1 \rightarrow 4)$ bonds.

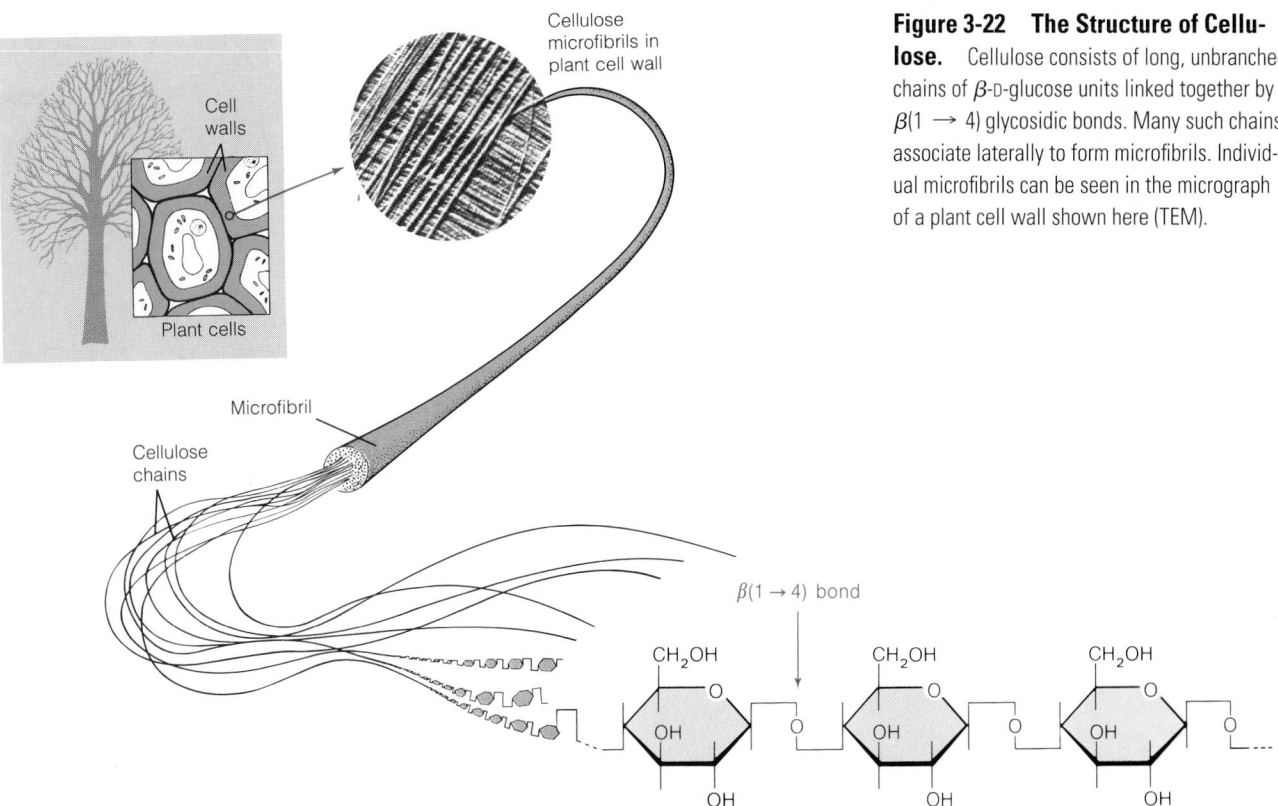

Cellulose microfibrils in plant cell wall

Cell walls

Plant cells

Microfibril

Cellulose chains

$\beta(1 \rightarrow 4)$ bond

CH$_2$OH

CH$_2$OH

CH$_2$OH

OH

OH

OH

OH

OH

OH

Figure 3-22 The Structure of Cellulose. Cellulose consists of long, unbranched chains of β-D-glucose units linked together by $\beta(1 \rightarrow 4)$ glycosidic bonds. Many such chains associate laterally to form microfibrils. Individual microfibrils can be seen in the micrograph of a plant cell wall shown here (TEM).

Polysaccharide Structure Depends on the Kinds of Glycosidic Bonds Involved

The distinction between the α and β glycosidic bonds of storage and structural polysaccharides has more than just nutritional significance. Because of the difference in linkages and therefore in the spatial relationship between successive glucose units, the two classes of polysaccharides differ markedly in secondary structure. The helical shape already established as a characteristic of both proteins and nucleic acids is also found in polysaccharides. Both starch and glycogen coil spontaneously into loose helices, but often the structure is not highly ordered because of the numerous side chains of amylopectin and glycogen.

Cellulose, by contrast, forms rigid, linear rods. These, in turn, aggregate laterally into **microfibrils** (see Figure 3-22).

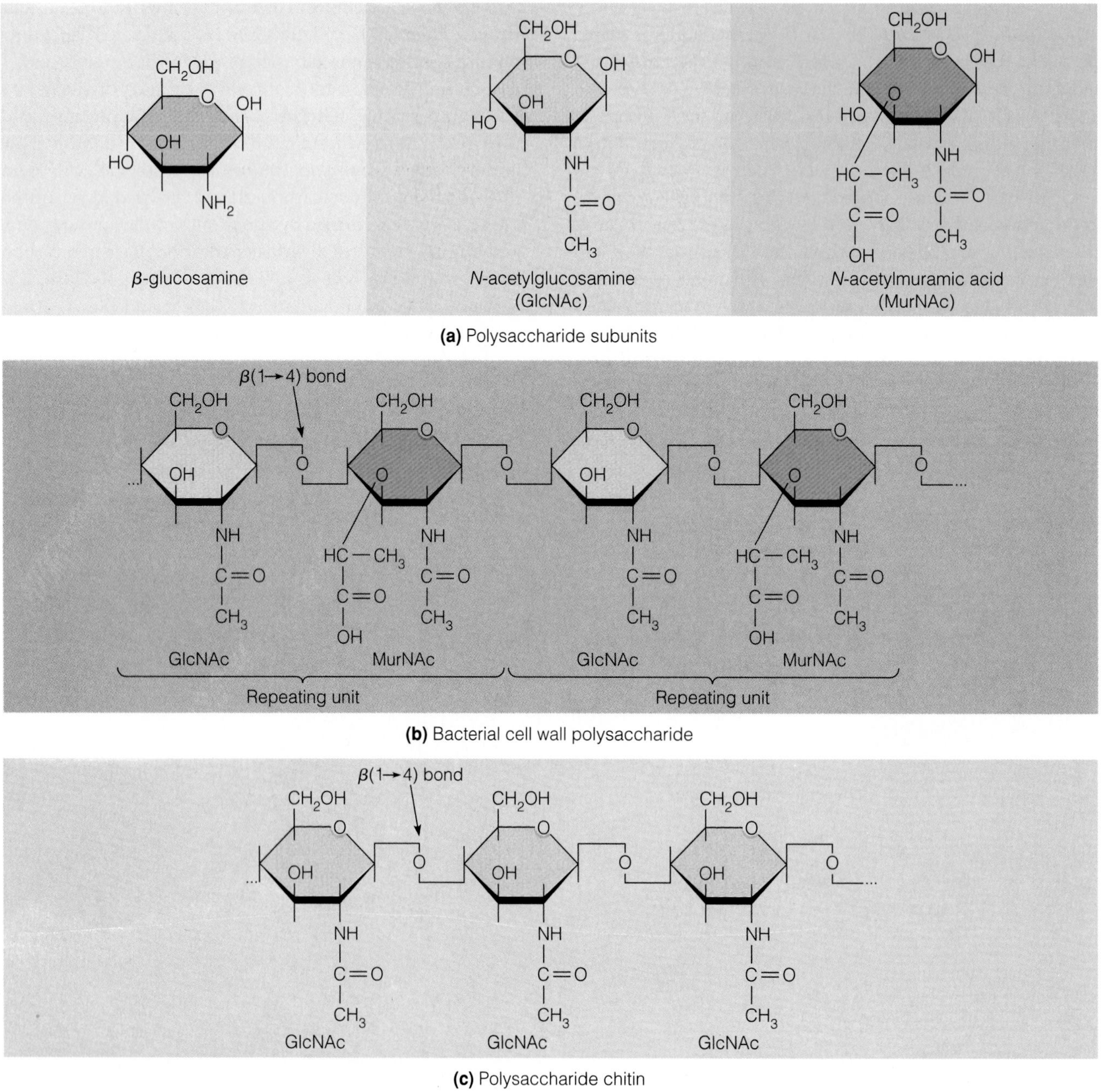

(a) Polysaccharide subunits

(b) Bacterial cell wall polysaccharide

(c) Polysaccharide chitin

Figure 3-23 Polysaccharides of Bacterial Cell Walls and Insect Exoskeletons. **(a)** The subunits glucosamine, N-acetylglucosamine (GlcNAc), and N-acetylmuramic acid (MurNAc). **(b)** A bacterial cell wall polysaccharide, consisting of alternating GlcNAc and MurNAc units. **(c)** The polysaccharide chitin found in insect exoskeletons and crustacean shells, with GlcNAc as its single repeating unit.

Microfibrils are about 25 nm in diameter and are composed of about 2000 cellulose chains. Plant and fungal cell walls consist of these rigid microfibrils of cellulose embedded in a **noncellulosic matrix** containing a rather variable mixture of several other polymers (*hemicellulose, pectin,* and *lignin*) and a protein called *extensin* that occurs only in the cell wall. Cell walls have been aptly compared to reinforced concrete, in which steel rods are embedded in the cement before it hardens to add strength. In cell walls, the cellulose microfibrils are the "rods" and the noncellulosic matrix the "cement."

Lipids

Strictly speaking, **lipids** do not qualify for inclusion in this chapter because they are not macromolecules and they are not formed by the kind of stepwise polymerization that gives rise to proteins, nucleic acids, and polysaccharides. Yet any discussion of cellular structure and chemical components would be incomplete without reference to this important group of molecules. Their inclusion here also seems reasonable in light of their frequent association with the macromolecules we have already discussed, especially proteins.

Lipids constitute a rather heterogeneous category of cellular components that resemble one another more in their solubility properties than in their chemical structures. The distinguishing feature of lipids is their hydrophobic nature. They have little, if any, affinity for water but are readily soluble in nonpolar solvents such as chloroform or ether. Accordingly, we can expect to find that they are rich in nonpolar hydrocarbon regions and have relatively few polar groups. Some lipids, however, are *amphipathic,* having both a polar and a nonpolar region. As we have already seen, this characteristic has profound implications for membrane structure. Because of their chemical heterogeneity, the lipids can be conveniently grouped into four separate categories: triglycerides, phospholipids, sphingolipids, and steroids.

Triglycerides Are Storage Lipids

The **triglycerides,** or true fats, consist of a glycerol molecule with three fatty acids linked to it by ester bonds. Their primary purpose in cells is to store energy, which makes them of special interest in our later discussion of energy metabolism in Part Three. As shown in Figure 3-24, **glycerol** is a

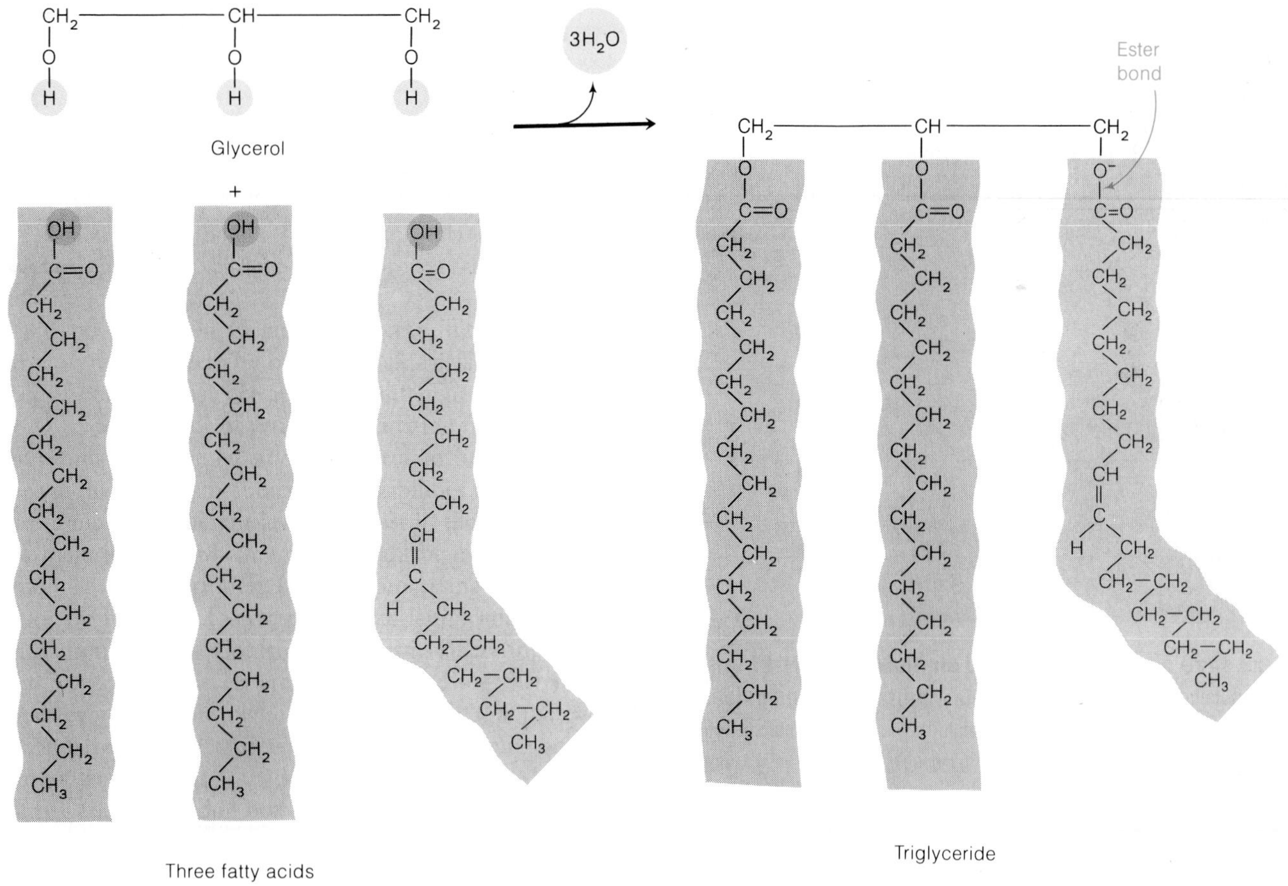

Figure 3-24   The Formation of a Triglyceride. Triglycerides are synthesized by the stepwise formation of ester bonds between three fatty acids and the three hydroxyl groups of glycerol. (The intermediates in the process are not shown; they are a monoglyceride and a diglyceride, which consist of glycerol esterified to one and two fatty acids, respectively.)

three-carbon alcohol with a hydroxyl group on each carbon. The **fatty acids** linked to these hydroxyl groups are generally long, unbranched hydrocarbon chains with a carboxyl group at one end. The fatty acid molecule is therefore amphipathic; the carboxyl group renders one end (often called the "head") polar, whereas the hydrocarbon "tail" is nonpolar. Fatty acids contain a variable, but usually even, number of carbon atoms. The usual range is from 12 to 24 carbon atoms per chain, with 16- and 18-carbon fatty acids especially common.

Table 3-4 summarizes the nomenclature of fatty acid chain length. Even numbers of carbon atoms are greatly favored because of the mode of fatty acid synthesis. Each molecule is generated by the stepwise addition of two-carbon units as acetyl coenzyme A, a carrier of acyl groups that we will consider in more detail in Chapter 12. Once added, each new two-carbon increment is then reduced to the hydrocarbon level.

Because they are highly reduced, fats yield a great deal of energy upon oxidation and are therefore compact and efficient forms of energy storage. We will see how to quantify the efficiency of storing energy as fat rather than as carbohydrate when we get to Part Three. For the present, simply note that a gram of fat contains more than twice as much usable energy as a gram of sugar or polysaccharide.

Table 3-4 also shows the variability in fatty acid structure due to the presence of double bonds between carbons. Fatty acids without double bonds are referred to as **saturated fatty acids** because every carbon atom in the chain has the maximum number of hydrogen atoms attached to it (Figure 3-25a). The general formula for a saturated fatty acid with n carbon atoms is $C_n H_{2n} O_2$. By contrast, **unsaturated fatty acids** contain one or a few double bonds. The presence of such sites of unsaturation affects the shape of the molecule and therefore the kinds of structures of which it can be a part. Saturated fatty acids have long straight tails that pack together well, whereas each double bond results in a bend or kink in the molecule that prevents tight packing (Figure 3-25b).

Fatty acids are linked to glycerol by **ester bonds** (see Figure 3-24; note again the principle of bond formation by the removal of water). *Monoglycerides* contain a single fatty acid, *diglycerides* have two, and *triglycerides* have each of the three hydroxyl groups of glycerol esterified to a fatty acid. The three fatty acids of a given triglyceride need not be identical; they can, and generally do, vary in both chain length and degree of saturation.

Triglycerides containing a preponderance of saturated fatty acids are usually solid or semisolid at room temperature and are called **fats.** Fats are prominent in the bodies of animals, as evidenced by the fat that you buy with most cuts of meat, by the large quantity of lard that is obtained as a by-product of the meat-packing industry, and by the widespread concern people have that they are "getting fat." In plants, most triglycerides are liquid at room temperature, as the term **vegetable oil** suggests. Because the fatty acids of oils are predominantly unsaturated, their hydrocarbon chains have kinks that prevent an orderly packing of the molecules. As a result, vegetable oils have lower melting temperatures than most animal fats. Soybean oil and corn oil are two familiar vegetable oils. Vegetable oils can be converted into solid products such as margarine and shortening by partial hydrogenation (saturation) of the double bonds, a process explored further in Problem 3-16 at the end of the chapter.

Table 3-4   Nomenclature of the Fatty Acids*

Number of Carbons	Number of Double Bonds	Common Name	Systematic Name	Formula
12	0	Laurate	*n*-dodecanoate	$CH_3(CH_2)_{10}COO^-$
14	0	Myristate	*n*-tetradecanoate	$CH_3(CH_2)_{12}COO^-$
16	0	Palmitate	*n*-hexadecanoate	$CH_3(CH_2)_{14}COO^-$
18	0	Stearate	*n*-octadecanoate	$CH_3(CH_2)_{16}COO^-$
20	0	Arachidate	*n*-eicosanoate	$CH_3(CH_2)_{18}COO^-$
16	1	Palmitoleate	*cis*-Δ^9-hexadecenoate	$CH_3(CH_2)_5CH{=}CH(CH_2)_7COO^-$
18	1	Oleate	*cis*-Δ^9-octadecenoate	$CH_3(CH_2)_7CH{=}CH(CH_2)_7COO^-$
18	2	Linoleate	*cis, cis*-Δ^9, Δ^{12}-octadecadienoate	$CH_3(CH_2)_4(CH{=}CHCH_2)_2(CH_2)_6COO^-$
18	3	Linolenate	All *cis*-$\Delta^9, \Delta^{12}, \Delta^{15}$-octadecatrienoate	$CH_3CH_2(CH{=}CHCH_2)_3(CH_2)_6COO^-$
20	4	Arachidonate	All *cis*-$\Delta^5, \Delta^8, \Delta^{11}, \Delta^{14}$-eicosatetraenoate	$CH_3(CH_2)_4(CH{=}CHCH_2)_4(CH_2)_2COO^-$

The common names, systematic names, and formulas are for the ionized (anionic) forms of the fatty acids because fatty acids exist primarily in the anionic form at the near-neutral pH of most cells. For the names and structures of the free fatty acids, simply replace the "-ate*" ending with "*-ic acid*" and substitute a hydrogen atom (H) for the negative charge in each case.

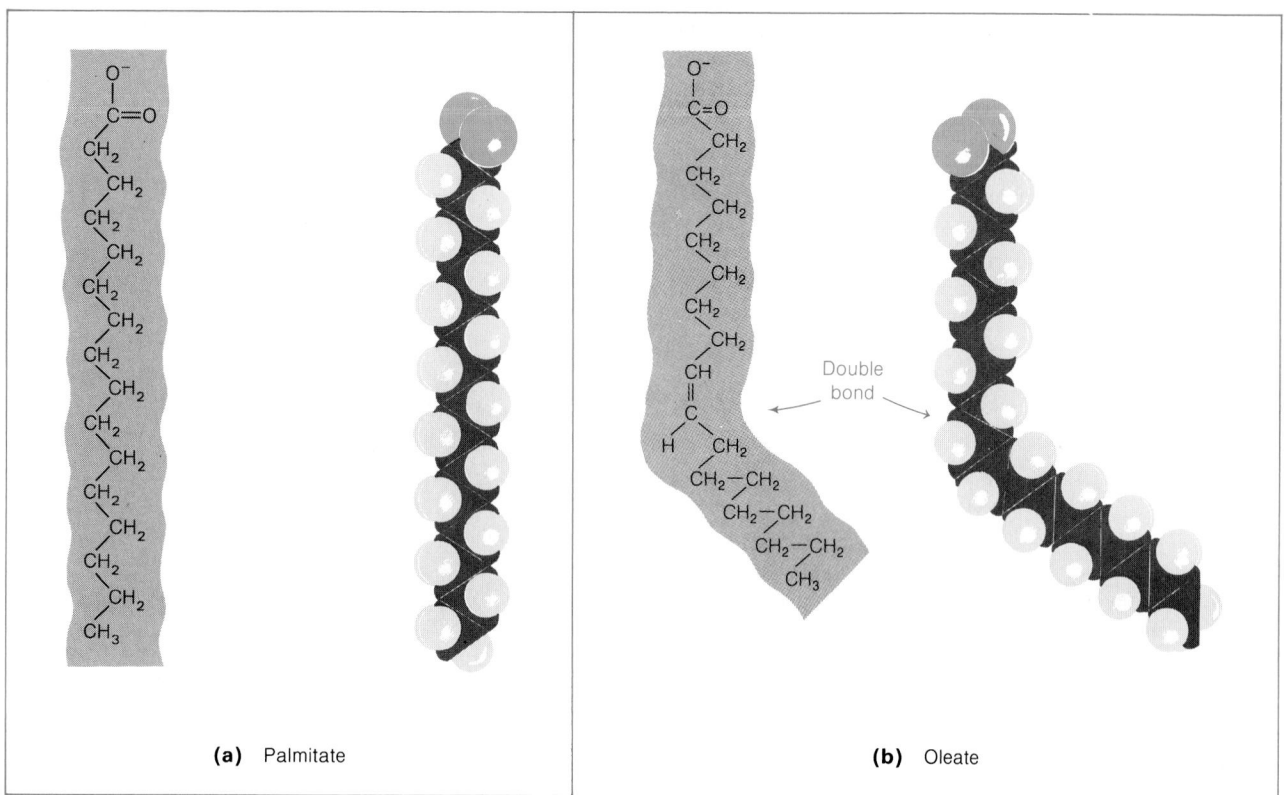

Figure 3-25 Structures of Saturated and Unsaturated Fatty Acids. **(a)** The saturated 16-carbon fatty acid palmitate. **(b)** The unsaturated 18-carbon fatty acid oleate. The space-filling models are intended to emphasize the overall shape of the molecules. Notice the kink that the double bond creates in the oleate molecule.

(a) Palmitate

(b) Oleate

Phospholipids Are Important in Membrane Structure

Phospholipids are lipids that contain one or more phosphate groups. They are similar to triglycerides in some chemical details but differ strikingly in their properties and their role in the cell. Phospholipids were mentioned in our discussion of membrane structure in Chapter 2 because they are critical to the bilayer structure found in all membranes (see Figure 2-12).

Membrane phospholipids resemble triglycerides chemically, as most of them contain a glycerol molecule esterified to two fatty acids. The difference comes at the third hydroxyl group, which in a phospholipid is occupied not by a third fatty acid but by a phosphate group. This basic structure is called a **phosphoglyceride.** The structures of some common phosphoglycerides are shown in Figure 3-26. Most membrane phospholipids are phosphoglycerides. (Exceptions include the sphingomyelins, which will be discussed shortly, and the glycolipids found in plants and some bacteria.)

The basic component of a phosphoglyceride is **phosphatidic acid,** which has just two fatty acids and a phosphate group (Figure 3-26a). Phosphatidic acid is a key intermediate in the synthesis of other phosphoglycerides but is itself not at all prominent in membranes. Instead, membrane phosphoglycerides invariably have, in addition, a small hydrophilic alcohol linked to the phosphate by an ester bond. The alcohol is usually *serine, ethanolamine, choline,* or *inositol* (Figure 3-26b–e). Except for inositol, these alcohols contain an amino group that is protonated and therefore charged at cellular pH. The presence of a negatively charged phosphate and a positively charged amine in juxtaposition makes these phosphoglycerides electrically neutral but gives them a highly polar head region.

The presence of a highly polar head and two long nonpolar chains gives the phosphoglycerides the characteristic amphipathic nature that is so critical to membrane structure. As we saw earlier, the fatty acids can vary considerably in both length and the presence and position of sites of unsaturation. In membranes, 16- and 18-carbon fatty acids are most common, and a typical phosphoglyceride molecule is likely to have one saturated and one unsaturated fatty acid. The length and the degree of unsaturation of fatty acid chains in membrane phospholipids profoundly affect membrane fluidity and can, in fact, be regulated by the cell to control this crucial membrane property.

Sphingolipids Are Also Found in Membranes

In addition to the phosphoglycerides, membranes of animal cells contain a class of lipids based not on glycerol but on the

Figure 3-26 Structures of Some Common Phosphoglycerides. (a) The starting point for phosphoglyceride synthesis is phosphatidic acid, to which several different polar alcohols can be added, including **(b)** serine, **(c)** ethanolamine, **(d)** choline, and **(e)** inositol. In each case, the phosphatidic acid backbone (gray) and the polar alcohol (pink) are linked by a phosphoester bond.

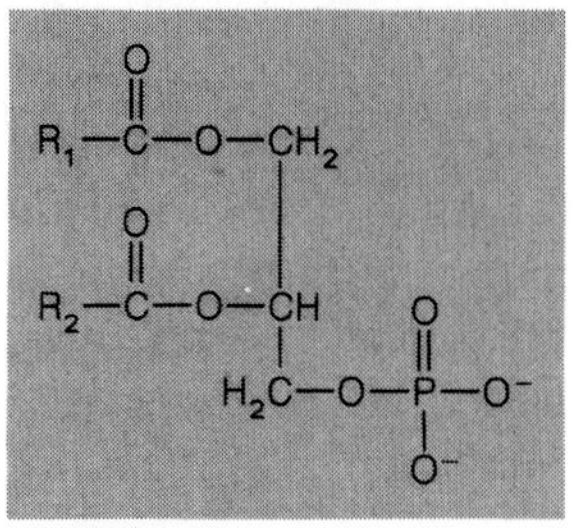

(a) Phosphatidic acid

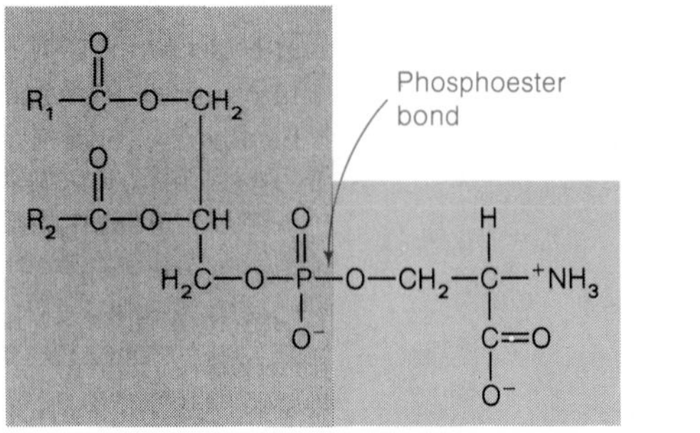

(b) Phosphatidyl serine

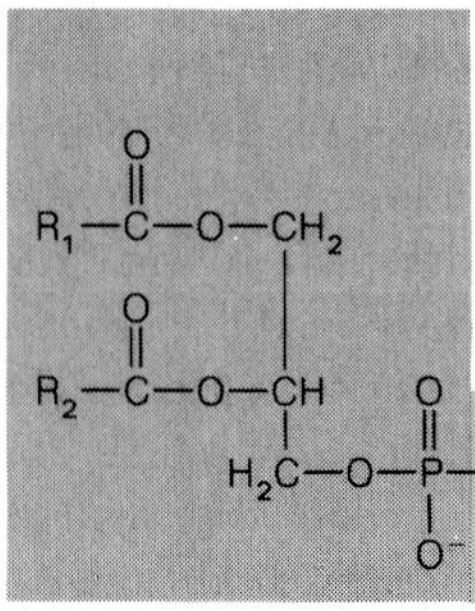

(c) Phosphatidyl ethanolamine

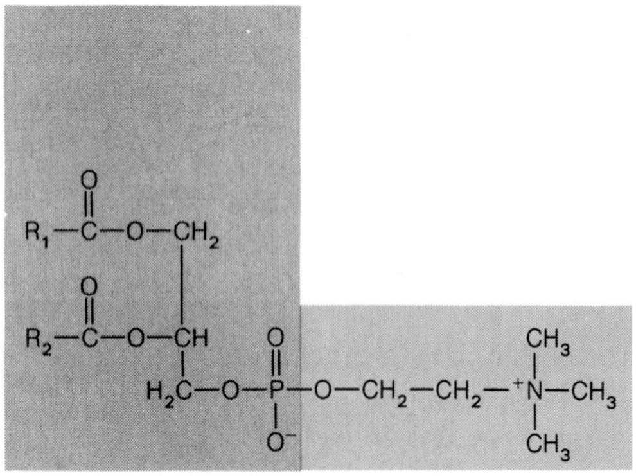

(d) Phosphatidyl choline

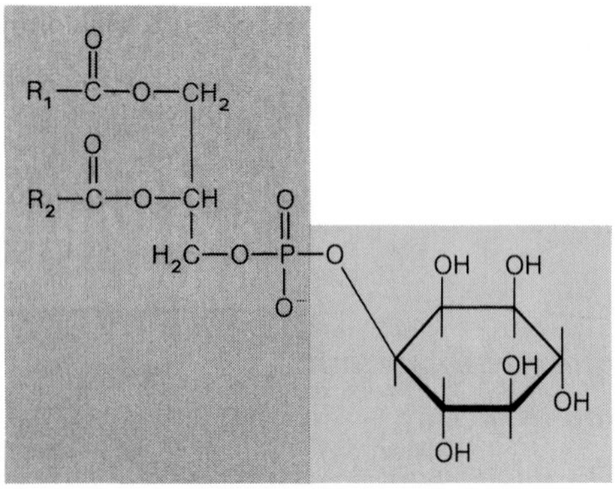

(e) Phosphatidyl inositol

amine alcohol **sphingosine.** As shown in Figure 3-27a, sphingosine has a long hydrocarbon chain with a single site of unsaturation near the polar end. Through its amine group, sphingosine can form an amide bond to a long-chain fatty acid. The resulting molecule is called a **ceramide** and consists of a polar region flanked by two long nonpolar tails (Figure 3-27b). Because of their common nonpolar nature, the two tails tend to bend around and associate with each other, giving the molecule a hairpin bend and a shape that approximates that of the phospholipids.

The hydroxyl group on carbon atom 1 of the sphingosine juts out from what is effectively the head of this hairpin molecule. A **sphingolipid** is formed when any of several polar groups becomes linked to this hydroxyl group (Figure 3-27c). Actually, a whole family of sphingolipids exists, differing only in the chemical nature of the polar group at-

tached to the hydroxyl group of the ceramide. The **sphingomyelins,** for example, contain phosphorylethanolamine or phosphorylcholine and therefore closely resemble phosphoglycerides in overall shape and the chemical nature of the polar head. The particular sphingolipid shown in Figure 3-27c is a sphingomyelin with phosphorylcholine linked to ceramide by a phosphoester bond.

Steroids Are Lipids with a Variety of Functions

The final category of lipids is the **steroids.** As you can see from the structures shown in Figures 3-28 and 3-29, steroids have little in common chemically with the other three categories of lipids. However, they share the common property

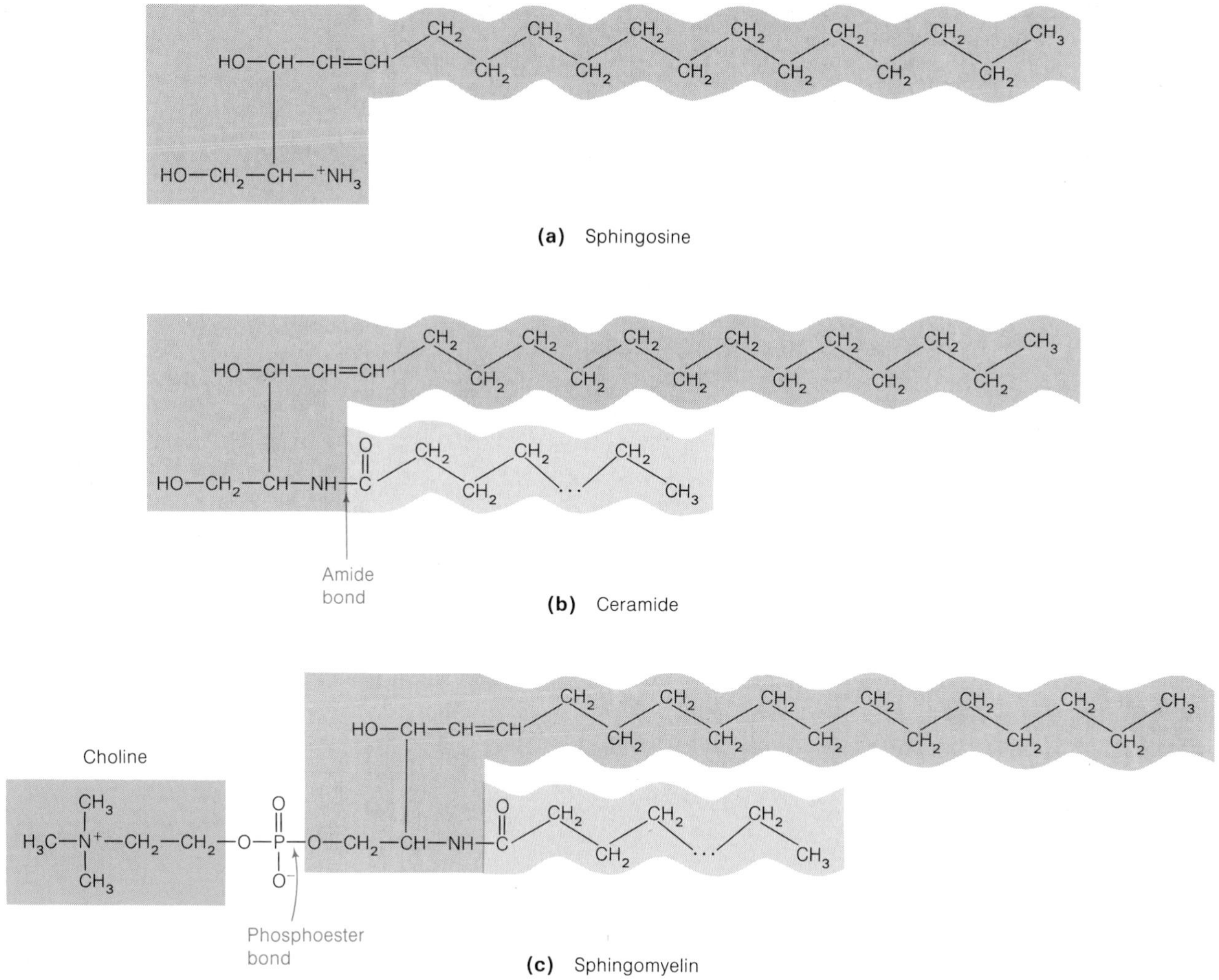

Figure 3-27   Structures of Some Common Sphingolipids.
(a) The starting point for sphingolipid synthesis is sphingosine.
(b) Attachment of a fatty acid to the amino group generates a ceramide.
(c) Linking choline to the terminal hydroxyl group of a ceramide by a

phosphoester bond converts the ceramide to a sphingomyelin, which is an example of a sphingolipid. Other sphingolipids can be formed by linking different polar groups to the hydroxyl group of the ceramide.

of being nonpolar and therefore hydrophobic. Most steroids contain four joined rings, but they differ in the number and positions of double bonds and functional groups. Steroids play a variety of roles in the cells of higher organisms but are not present in bacteria.

The most common steroid in animals is **cholesterol** (Figure 3-28). Cholesterol is very hydrophobic in nature, with only a hydroxyl group to confer a slightly hydrophilic nature on one end of the molecule. Because of its hydrophobicity, cholesterol is found primarily in membranes. It occurs in the plasma membrane of animal cells and in most of the membranes of organelles, except the inner membranes of mitochondria and chloroplasts.

Cholesterol is also the biosynthetic source of all **steroid hormones** (Figure 3-29), which include *progesterone,* the male and female *sex hormones,* the *glucocorticoids,* and the *mineralocorticoids.* The sex hormones include the *estrogens* produced by the ovaries of females and the *androgens* produced by the testes of males. *Estradiol* and *testosterone* are examples of estrogens and androgens, respectively. The glucocorticoids are a family of hormones that promote gluconeogenesis (synthesis of glucose) and suppress inflammation reactions. *Cortisol* is a glucocorticoid hormone. Mineralocorticoids such as *aldosterone* regulate ion balance by promoting the reabsorption of sodium, chloride, and bicarbonate ions in the kidney.

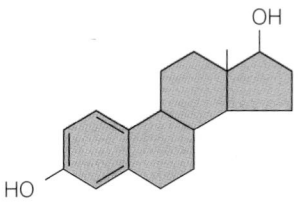

(a)

(b)

(c)

Figure 3-28   The Structure of Cholesterol.   Cholesterol consists of four joined rings, two methyl groups, and an eight-carbon side chain, with a total of 27 carbon atoms. **(a)** The structural formula for cholesterol. The letters A, B, and C denote six-membered rings, and D denotes a five-membered ring. **(b)** A three-dimensional skeletal model, with hydrogen atoms and sidegroups that project above the plane of the ring system represented by a solid wedge and those that project below the plane represented by a dashed wedge. **(c)** A space-filling model.

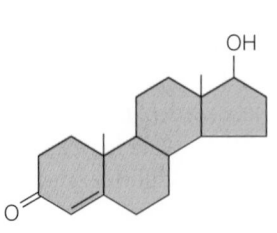

(a) Progesterone

(b) Estradiol

(c) Testosterone

(d) Cortisol

(e) Aldosterone

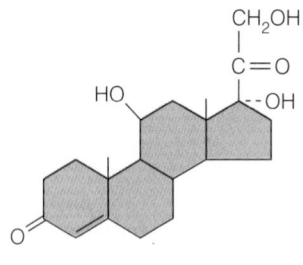

Figure 3-29   Structures of Several Common Steroid Hormones.   Among the many steroids that are synthesized from cholesterol are the hormones **(a)** progesterone; **(b)** estradiol, an estrogen; **(c)** testosterone, an androgen; **(d)** cortisol, a glucocorticoid; and **(e)** aldosterone, a mineralocorticoid.

Three kinds of polymers characterize most of cell structure: proteins, nucleic acids, and polysaccharides. Proteins and nucleic acids are informational macromolecules that depend directly or indirectly on genetic information to determine the order of subunits that is so critical to their roles in the cell. Polysaccharides, on the other hand, need no such information; they usually contain only a single kind of repeating unit and play storage or structural roles instead.

Proteins consist of linear chains of amino acids that differ markedly in the chemical properties of their R groups. The amino acid sequence, or primary structure, of a polypeptide is all-important because it contains most, if not all, of the information necessary to specify the local folding or orientation of the amino acid chain (secondary structure), the overall shape of the polypeptide (tertiary structure), and in the case of multimeric proteins, the further association with other polypeptides (quaternary structure). A major force behind protein folding and polypeptide interaction is the tendency of hydrophobic amino acids to avoid an aqueous environment. In addition, hydrogen bonds, electrostatic charges, and disulfide bonds are important in stabilizing protein structure.

Nucleic acids also attain a shape that is dictated by the chemical nature of their subunits. This is seen most strikingly in the DNA double helix, in which complementary base pairing (A with T, C with G) is stabilized by hydrogen bonding. The elucidation of the double-helical structure of DNA was one of the outstanding biological advances of the twentieth century.

Polysaccharides are chains of monosaccharides linked together by either α or β glycosidic bonds. The difference is critical because α linkages are readily digested by animals and are therefore suitable for storage polysaccharides such as starch or glycogen. By contrast, the β glycosidic bonds of structural polysaccharides such as cellulose and chitin are less readily digestible and give the molecule a rigid shape suitable to its function.

Lipids are not macromolecules but are included in this chapter because of their general importance as constituents of cells (especially membranes) and their frequent association with macromolecules, particularly proteins. Lipids differ substantially in chemical structure, but they all share the common property of solubility in organic solvents but not in water. The major classes of lipids include the triglycerides that make up fats and oils, the phospholipids and sphingolipids found in membranes, and the steroids, which perform a variety of functions in eukaryotic cells.

KEY TERMS FOR SELF-TESTING

Proteins

protein (p. 45)
amino acid (p. 45)
peptide bond (p. 46)
N- (amino) terminus (p. 46)
C- (carboxyl) terminus (p. 46)
polypeptide (p. 48)
monomeric protein (p. 48)
multimeric protein (p. 48)
conformation (p. 49)
primary structure (p. 49)
secondary structure (p. 50)
α helix (p. 50)
β sheet (β pleated sheet) (p. 50)
tertiary structure (p. 51)
native conformation (p. 51)
disulfide bond (p. 51)
fibrous protein (p. 51)
globular protein (p. 52)
domain (p. 53)
sickle-cell anemia (p. 53)
quaternary structure (p. 55)
multiprotein complex (p. 56)

Nucleic Acids

nucleic acid (p. 56)
DNA (deoxyribonucleic acid) (p. 56)
RNA (ribonucleic acid) (p. 56)
ribose (p. 56)
deoxyribose (p. 56)
messenger RNA (mRNA) (p. 56)
transfer RNA (tRNA) (p. 56)
ribosomal RNA (rRNA) (p. 56)
nucleotide (p. 56)
phosphoester bond (p. 56)
purine (p. 56)
pyrimidine (p. 56)
adenine (A) (p. 56)
guanine (G) (p. 56)
cytosine (C) (p. 56)
thymine (T) (p. 56)
uracil (U) (p. 56)
nucleoside (p. 56)
nucleoside monophosphate (p. 57)
adenosine monophosphate (AMP) (p. 57)
adenosine diphosphate (ADP) (p. 57)
adenosine triphosphate (ATP) (p. 57)
phosphodiester bond (p. 58)
polynucleotide (p. 58)
template (p. 58)
base pairing (p. 58)
double helix (p. 59)

Polysaccharides

polysaccharide (p. 61)
monosaccharide (p. 64)
aldosugar (p. 64)
ketosugar (p. 64)
carbohydrate (p. 64)
Fischer projection (p. 64)
Haworth projection (p. 64)
disaccharide (p. 65)
glycosidic bond (p. 65)
starch (p. 65)
glycogen (p. 65)
amylose (p. 67)
amylopectin (p. 67)
cellulose (p. 67)
N-acetylglucosamine (GlcNAc) (p. 67)

N-acetylmuramic acid (MurNAc)
  (p. 67)
chitin (p. 67)
microfibril (p. 68)
noncellulosic matrix (p. 69)

Lipids

lipid (p. 69)
triglyceride (p. 69)

glycerol (p. 69)
fatty acid (p. 70)
saturated fatty acid (p. 70)
unsaturated fatty acid (p. 70)
ester bond (p. 70)
fat (p. 70)
vegetable oil (p. 70)
phospholipid (p. 71)
phosphoglyceride (p. 71)

phosphatidic acid (p. 71)
sphingosine (p. 72)
ceramide (p. 72)
sphingolipid (p. 72)
sphingomyelin (p. 72)
steroid (p. 72)
cholesterol (p. 73)
steroid hormone (p. 73)

PROBLEM SET

3-1. Polymers and Their Properties. For each of the six biological polymers listed below, indicate which of the properties apply. Each polymer has multiple properties, and a given property may be used more than once.

Polymers

(a) Cellulose

(b) Messenger RNA

(c) Globular protein

(d) Amylopectin

(e) DNA

(f) Fibrous protein

Properties

1. Branched-chain polymer

2. Extracellular location

3. Aminoacyl tRNAs

4. Glycosidic bonds

5. Informational macromolecule

6. Peptide bond

7. *N*-acetylglucosamine

8. β linkage

9. Phosphodiester bond

10. Nucleoside triphosphates

11. Helical structure possible

12. Synthesis requires a template

3-2. Stability of Protein Structure. Several different kinds of bonds or interactions are involved in generating and maintaining the structure of proteins. List four or five such bonds or interactions, give an example of an amino acid that might be involved in each, and indicate which level(s) of protein structure might be generated or stabilized by that particular kind of bond or interaction.

3-3. Amino Acid Localization in Proteins. Amino acids tend to be localized either in the interior or on the exterior of a globular protein molecule, depending on their relative affinities for water.

(a) Classify each of the following amino acids as likely to be found in the interior, on the exterior, or at either location, and explain.

  valine   phenylalanine   glycine

  alanine   aspartate   lysine

(b) For each of the following pairs of amino acids, choose the one that is more likely to be found in the interior of a protein molecule, and explain why.

  alanine; glycine   glutamate; aspartate

  tyrosine; phenylalanine   methionine; cysteine

(c) Explain why cysteines with free sulfhydryl groups tend to be localized on the exterior of a protein molecule, whereas those involved in disulfide bonds are more likely to be buried in the interior of the molecule.

3-4. Myoglobin Versus Hemoglobin. Myoglobin and hemoglobin are both oxygen-binding proteins. Myoglobin is a monomeric protein found in muscle cells, whereas hemoglobin is tetrameric and is found in red blood cells. The tertiary structure of myoglobin is strikingly similar to that of both the α and β subunits of hemoglobin; yet when the primary structures are compared, myoglobin can be shown to have hydrophilic amino acids at several positions that in the hemoglobin chains are occupied by hydrophobic amino acids. Given the extent to which tertiary structure is thought to depend on primary structure, how can a relatively hydrophobic polypeptide such as the α or β subunit of hemoglobin have a tertiary structure very much like that of the relatively hydrophilic myoglobin?

3-5. Sickle-Cell Anemia. Sickle-cell anemia is a striking example of the drastic effect that a single amino acid substitution can have on the structure and function of a protein.

(a) Given the chemical nature of glutamate and valine, can you suggest why the substitution of the latter for the former at position 6 of the β chain would be especially deleterious?

(b) Suggest several other amino acids that would be much less likely to cause impairment of hemoglobin function if substituted for the glutamate at position 6 of the β chain.

(c) Can you see why in some cases two proteins could differ at a number of points in their amino acid sequence and still be very similar in structure and function? Explain.

3-6. Hair Versus Silk. The α-keratin of human hair is a good example of a fibrous protein with extensive α-helical structure. Silk fibroin is also a fibrous protein, but it consists primarily of β-sheet structure. Fibroin is essentially a polymer of alternating glycines and alanines, whereas α-keratin contains most of the common amino acids and has many disulfide bonds.

(a) If you were able to grab onto both ends of an α-keratin polypeptide and pull, you would find it to be both extensible (it can be stretched to about twice its length in moist heat) and elastic (when you let go, it will return to its normal length). In contrast, a fibroin polypeptide has essentially no extensibility, but it has great tensile strength. Explain these differences.

(b) Can you suggest why fibroin assumes a pleated sheet structure, whereas α-keratin exists as an α helix and even reverts spontaneously to a helical shape when it has been stretched artificially?

3-7. The "Permanent" Wave That Isn't. The "permanent" wave that your local beauty parlor offers depends critically on rearrangements in the extensive disulfide bonds of keratin that give your hair its characteristic shape. To change the shape of your hair (to give it a wave or curl), the beautician first treats your hair with a sulfhydryl reducing agent, then uses curlers or rollers to impose the desired artificial shape, and follows this by treatment with an oxidizing agent.

(a) What is the chemical basis of a "permanent"? Be sure to include the use of a reducing agent and an oxidizing agent in your explanation.

(b) Why do you suppose a "permanent" isn't permanent? (Explain why the wave or curl is gradually lost during the weeks following your visit to the beautician.)

(c) Can you suggest an explanation for naturally curly hair?

3-8. The Importance of Hydrogen Bonds. Hydrogen bonds play an important role in stabilizing the secondary structure of both proteins and nucleic acids. When a solution of either a protein or a nucleic acid is heated, one of the main effects of the thermal energy is to break hydrogen bonds; this is called *thermal denaturation*. For each statement below, decide for which, if any, of the following polymers the statement is true: the fibrous protein α-keratin (k); the silk protein fibroin (f); the DNA double helix (d).

(a) All of the hydrogen bonds involve one nitrogen atom.

(b) The hydrogen bonds are between units of the same polymer strand.

(c) The hydrogen bonds are perpendicular to the main axis of the polymer.

(d) The hydrogen bonds make the polymer more polar than it would otherwise be.

(e) The number of hydrogen bonds in a given length of the polymer does not depend on the particular amino acids or nucleotides present in that specific segment of the polymer.

(f) One of the effects of heating will be to separate polymer strands that are otherwise bonded to each other.

(g) The two ends of an individual heat-denatured polymer strand are likely to be much farther away from each other than they are in the fully hydrogen-bonded structure.

(h) Upon cooling under the appropriate conditions, the strands will return spontaneously to the original three-dimensional conformation.

(i) The renaturation process described in (h) will proceed much more slowly if the solution of denatured strands is diluted before being cooled.

3-9. The Size of DNA Molecules. The DNA double helix contains exactly ten nucleotide pairs per turn, and each turn adds 3.4 nm to the overall length of the duplex. The circular DNA molecule of *E. coli* has about 4 million nucleotide pairs and a molecular weight of about 2.8×10^9.

(a) What is the circumference of the *E. coli* DNA molecule? What problems might this pose for a cell that is only about 2 μm long?

(b) Human mitochondrial DNA is also a circular duplex, with a diameter of about 1.8 μm. About how many nucleotide pairs does it contain? If it takes about 1500 nucleotides of DNA to encode (specify the amino acid sequence of) a typical protein, how many proteins could mitochondrial DNA encode?

(c) What is the weight (in grams) of one nucleotide pair? How many nucleotide pairs are present in the 6 picograms (6×10^{-12} g) of double-stranded DNA present in the nucleus of most cells in your body?

(d) What is the total length of the 6 picograms of DNA present in a single nucleus?

3-10. Storage Polysaccharides. The only common examples of branched-chain polymers in cells are the storage polysaccharides glycogen and amylopectin. Both are degraded exolytically, which means by stepwise removal of terminal glucose units.

(a) Why might it be advantageous for a storage polysaccharide to have a branched-chain structure instead of a linear structure?

(b) Can you foresee any metabolic complications in the process of glycogen degradation? How do you think the cell handles this?

(c) Can you see why cells that must degrade amylose instead of amylopectin have enzymes capable of endolytic (internal) as well as exolytic cleavage of glycosidic bonds?

(d) Why do you suppose the structural polysaccharide cellulose does not contain branches?

3-11. Carbohydrate Structure. From the following descriptions of gentiobiose, raffinose, and a dextran, draw Haworth projections of each.

(a) *Gentiobiose* is a disaccharide found in gentians and other plants. It consists of two molecules of β-D-glucose linked to each other by a $\beta(1 \rightarrow 6)$ glycosidic bond.

(b) *Raffinose* is a trisaccharide found in sugar beets. It consists of one molecule each of α-D-galactose, α-D-glucose, and β-D-fructose, with the galactose linked to the glucose by an $\alpha(1 \rightarrow 6)$ glycosidic bond and the glucose linked to the fructose by an $\alpha(1 \rightarrow 2)$ bond.

(c) *Dextrans* are polysaccharides produced by some bacteria. They are polymers of α-D-glucose linked by $\alpha(1 \rightarrow 6)$ glycosidic bonds, with frequent $\alpha(1 \rightarrow 3)$ branching. Draw a portion of a dextran, including one branch point.

3-12. Reducing Sugars. A *reducing sugar* is one that will undergo the *Fehling reaction* shown in Figure 3-30 for the monosaccharide glucose. In this reaction, cupric ions (Cu^{2+}) are reduced in an alkaline solution to insoluble cuprous oxide (Cu_2O), with concomitant oxidation of the sugar to the corresponding acid—gluconic acid, in the case of glucose. (To understand the oxidation of glucose, draw the ring form of the sugar in its Fischer projection, convert it to the straight-chain form, and note the free aldehyde group on carbon atom 1 that can be oxidized to a carboxyl group.) The production of a red precipitate of Cu_2O indicates that the sugar being tested is a reducing sugar. The Fehling reaction, named for the scientist who devised it, was at one time used to test for excess sugar in the urine of people thought to have diabetes.

(a) Which of the disaccharides shown in Figure 3-20 are reducing sugars? Explain.

(b) Is either gentiobiose or raffinose a reducing sugar? (See Problem 3-11 for information on the structures of these sugars.)

Figure 3-30   The Fehling Reaction: A Test for Reducing Sugars.

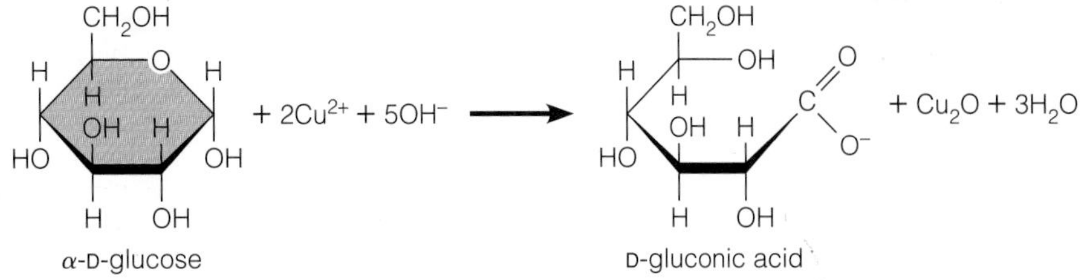

α-D-glucose $\quad + 2Cu^{2+} + 5OH^- \longrightarrow \quad$ D-gluconic acid $\quad + Cu_2O + 3H_2O$

3-13. The Polarity of Lipids. Arrange the following lipids in order of decreasing polarity, and explain your reasoning: cholesterol; fatty acid; phosphatidyl choline; triglyceride.

3-14. Phospholipids. Which would you expect to resemble a sphingomyelin molecule more closely: a molecule of phosphatidyl choline containing two palmitates as its fatty acid chains, or a phosphatidyl choline molecule with one palmitate and one oleate as its fatty acid chains? Explain.

3-15. Melting the Fat. Assume that you and your lab partner determined the melting temperature for each of the following fatty acids: arachidaic, linoleic, linolenic, oleic, palmitic, and stearic acids. Your partner recorded the melting points but neglected to note the specific fatty acid to which each value belongs. Can you assign each of the following melting temperatures (in °C) to the appropriate fatty acid? -11, $+5$, $+16$, $+63$, $+70$, $+76.5$.

3-16. Shortening. A popular brand of shortening has a label on the can that identifies the product as "partially hydrogenated soybean oil, palm oil, and cottonseed oil."

(a) What does the process of "partial hydrogenation" accomplish chemically?

(b) What did the contents of the can look like before it was partially hydrogenated?

(c) What is the physical effect of partial hydrogenation?

SUGGESTED READING

General References

Mathews, C. K., and K. E. van Holde. *Biochemistry,* 2d ed. Menlo Park, CA: Benjamin/Cummings, 1996.

Olson, A. J., and D. S. Goodsell. Visualizing biological molecules. *Sci. Amer.* 267 (November 1992): 76.

Stryer, L. *Biochemistry,* 3d ed. New York: W. H. Freeman, 1988.

Tanford, C. The hydrophobic effect and the organization of living matter. *Science* 200 (1978): 1012.

Watson, J. D., N. H. Hopkins, J. W. Roberts, J. A. Steitz, and A. M. Weiner. *Molecular Biology of the Gene,* 4th ed. Menlo Park, CA: Benjamin/Cummings, 1987.

Weinberg, R. A. The molecules of life. *Sci. Amer.* 253 (October 1985): 48.

Proteins

Anfinsen, C. B. Principles that govern the folding of protein chains. *Science* 181 (1973): 223.

Branden, C., and J. Tooze. *Introduction to Protein Structure.* New York: Garland, 1991.

Chothia, C. Principles that determine the structure of proteins. *Annu. Rev. Biochem.* 53 (1984): 537.

Chothia, C., and A. V. Finkelstein. The classification and origins of protein folding patterns. *Annu. Rev. Biochem.* 59 (1990): 1007.

Creighton, T. E. Disulfide bonds and protein stability. *Bioessays* 8 (1988): 57.

Creighton, T. E. *Proteins: Structure and Molecular Properties.* New York: W. H. Freeman, 1984.

Dickerson, R. E., and I. Geis. *Proteins: Structure, Function and Evolution.* Menlo Park, CA: Benjamin/Cummings, 1982.

Doolittle, R. F. Proteins. *Sci. Amer.* 253 (October 1985): 88.

Doolittle, R. F., and P. Bork. Evolutionarily mobile modules in proteins. *Sci. Amer.* 269 (October 1993): 50.

Karplus, M., and J. A. McCammon. The dynamics of proteins. *Sci. Amer.* 254 (April 1986): 42.

Pace, C. N. Conformational stability of proteins. *Trends Biochem. Sci.* 15 (1990): 14.

Richardson, J. S. The anatomy and taxonomy of protein structure. *Adv. Protein Chem.* 34 (1981): 167.

Sanger, F. Sequences, sequences, and sequences. *Annu. Rev. Biochem.* 57 (1988): 1.

Schulz, G. E., and R. H. Schirmer. *Principles of Protein Structure.* New York: Springer-Verlag, 1979.

Nucleic Acids

Cohen, J. S. DNA: Is the backbone boring? *Trends Biochem. Sci.* 5 (1980): 58.

Crick, F. H. C. The structure of the hereditary material. *Sci. Amer.* 194 (October 1954): 54.

Darnell, J. E., Jr. RNA. *Sci. Amer.* 253 (October 1985): 68.

Dickerson, R. E. The DNA helix and how it is read. *Sci. Amer.* 249 (December 1983): 94.

Felsenfeld, G. DNA. *Sci. Amer.* 253 (October 1985): 58.

Olby, R. *The Path to the Double Helix.* Seattle: University of Washington Press, 1974.

Portugal, F. H., and J. S. Cohen. *A Century of DNA: A History of the Discovery of the Structure and Function of the Genetic Substance.* Cambridge, MA: MIT Press, 1977.

Saenger, W. *Principles of Nucleic Acid Structure.* Berlin: Springer-Verlag, 1984.

Watson, J. D. *The Double Helix.* New York: Atheneum, 1968.

Watson, J. D., and F. H. C. Crick. Molecular structure of nucleic acids. A structure for deoxyribose nucleic acid. *Nature* 171 (1953): 737.

Carbohydrates and Lipids

Aspinwall, G. O., ed. *Polysaccharides,* vol. I. New York: Academic Press, 1982.

Davison, E. A. *Carbohydrate Chemistry.* New York: Holt, Rinehart, & Winston, 1967.

Ginsburg, V., and P. Robbins, eds. *Biology of Carbohydrates.* New York: Wiley, 1984.

Gurr, M. I., and A. T. James. *Lipid Biochemistry: An Introduction.* New York: Cornell University Press, 1971.

Hakomori, S. Glycosphingolipids. *Sci. Amer.* 254 (May 1986): 44.

Kobata, A. Structures and functions of the sugar chains of glycoproteins. *Europ. J. Biochem.* 209 (1992): 483.

Roehrig, K. L. *Carbohydrate Biochemistry and Metabolism.* Westport, CT: AUI Publishing, 1984.

Sharon, N. Carbohydrates. *Sci. Amer.* 243 (November 1980): 90.

Sharon, N., and H. Lis. Carbohydrates in cell recognition. *Sci. Amer.* 268 (January 1993): 82.

Storch, J., and A. M. Kleinfeld. The lipid structure of biological membranes. *Trends Biochem. Sci.* 10 (1982): 418.

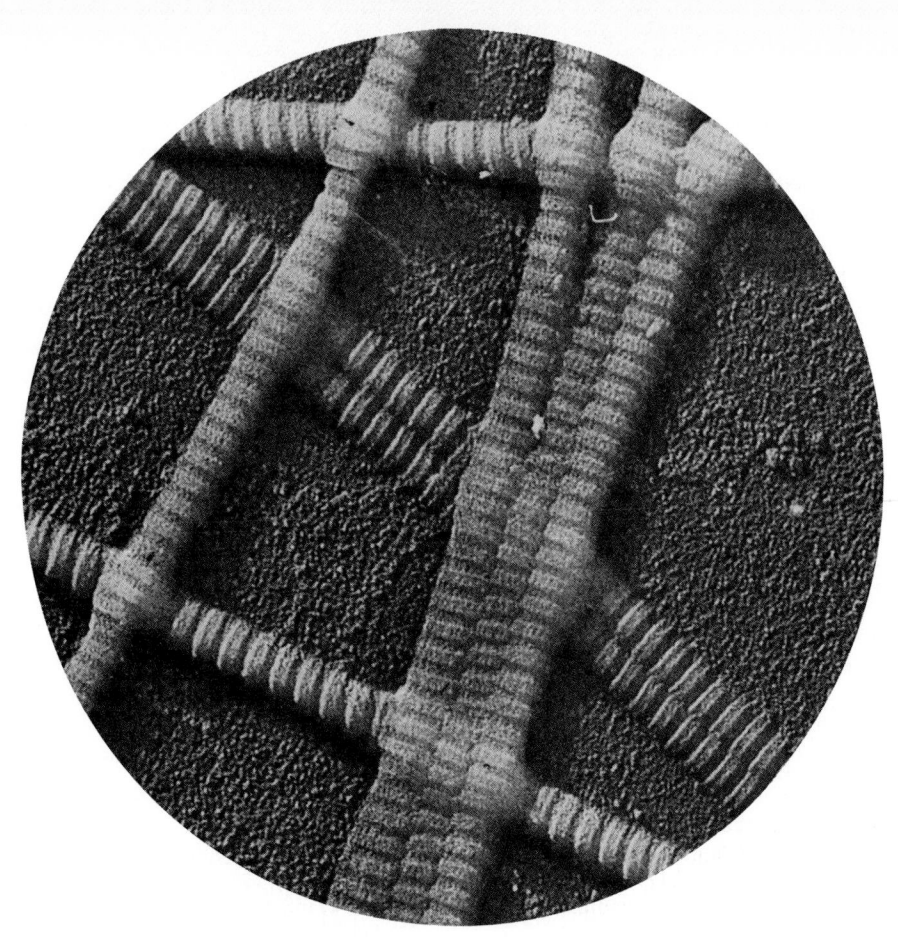

CELL STRUCTURE AND FUNCTION

4

CELLS AND ORGANELLES

In the previous two chapters, we encountered most of the major kinds of molecules found in cells, as well as some of the principles governing the assembly of these molecules into the supramolecular structures of which cells and their organelles are composed (see Figure 2-14). Now we are ready to focus our attention on cells and organelles directly.

Properties and Strategies of Cells

As we begin to consider what cells are and how they function, several general characteristics of cells quickly emerge. These include the sizes and shapes of cells, the classification of cells on the basis of their organizational complexity, and the specializations that cells undergo.

Cell Sizes and Shapes

Cells come in a variety of sizes, shapes, and forms. Some of the smallest bacterial cells, for example, are only about 0.2–0.3 μm in diameter—so small that about 30,000 such cells could fit side by side across the head of a thumbtack! At the other extreme are highly elongated nerve cells, which may extend one or more meters. Those in the neck and legs of a giraffe are dramatic examples. On the other hand, the oft-cited examples of bird eggs, especially ostrich eggs, are rather misleading because they are indeed single cells, but with most of their internal volume occupied by yolk—large deposits of stored food intended as nourishment for the developing embryo.

Most cells fall into a rather narrow and predictable range of sizes, however. Bacterial cells, for example, are usually about 1–5 μm in diameter, and most cells of higher plants and animals have dimensions in the range of 10–50 μm. (Recall that Box 1A describes the units used to express cellular dimensions and illustrates the relative sizes of cells and related structures.) The two most important factors that

limit cell size are the requirement for an adequate surface area/volume ratio and the need to maintain adequate concentrations of the substances and enzymes involved in various cellular processes.

The Surface Area/Volume Ratio. The main constraint on cell size is that set by the need to maintain an adequate **surface area/volume ratio**. Surface area is important because it is here that the exchange between the cell and its environment takes place. The internal volume of the cell determines the amount of nutrients that will have to be imported and the quantity of waste products that must be excreted, but the surface area effectively measures the amount of membrane available for such uptake and excretion.

The problem of maintaining adequate surface area arises because the volume of a cell increases with the cube of the cell's length or diameter, whereas its surface area only increases with the square. Consider, for example, the cube-shaped cells shown in Figure 4-1. The cell on the left is 20 μm on a side and has a volume of 8000 μm^3 ($V = s^3$, where $s = 20$ μm) and a surface area of 2400 μm^2 ($A = 6s^2$). The surface area/volume ratio is therefore 2400 μm^2/8000 μm^3, or 0.3 μm^{-1}. When this single large cell is divided into smaller cells, the total volume remains the same but the surface area increases. Thus, the surface area/volume ratio increases as the linear dimension of the cell decreases. The 1000 cells on the right, for example, still have a total volume of 8000 μm^3 (1000 $\times$ 2^3) but the total surface area is 24,000 μm^2 (1000 $\times$ 6 $\times$ 2^2), so the surface area/volume ratio is 24,000 μm^2/8000 μm^3, or 3.0 μm^{-1}.

This comparison illustrates a major constraint on cell size. As a cell increases in size, its surface area does not keep pace with its volume, and the necessary exchange of substances between the cell and its surroundings becomes more and more problematic. Cell size can therefore increase only over the range of values for which the membrane surface area is still adequate for the passage of materials into and out of the cell. Once the limiting surface area/volume ratio is reached, further increases in cell size would generate more

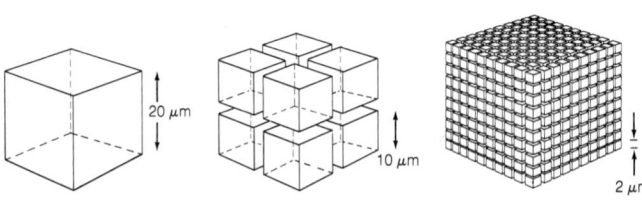

Surface area increases, volume stays the same

Length of one side	20 μm	10 μm	2 μm
Total surface area (height × width × number of sides × number of cubes)	2400 μm²	4800 μm²	24,000 μm²
Total volume (length × width × height × number of cubes)	8000 μm³	8000 μm³	8000 μm³
Surface area to volume ratio (surface area ÷ volume)	0.3	0.6	3.0

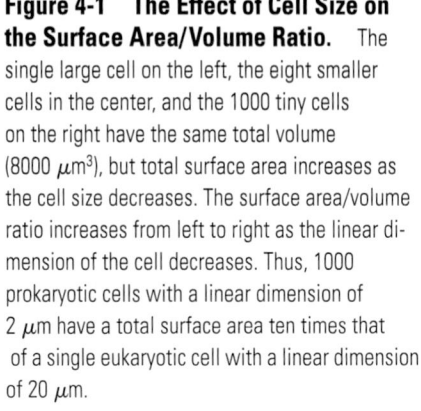

Figure 4-1 The Effect of Cell Size on the Surface Area/Volume Ratio. The single large cell on the left, the eight smaller cells in the center, and the 1000 tiny cells on the right have the same total volume (8000 μm³), but total surface area increases as the cell size decreases. The surface area/volume ratio increases from left to right as the linear dimension of the cell decreases. Thus, 1000 prokaryotic cells with a linear dimension of 2 μm have a total surface area ten times that of a single eukaryotic cell with a linear dimension of 20 μm.

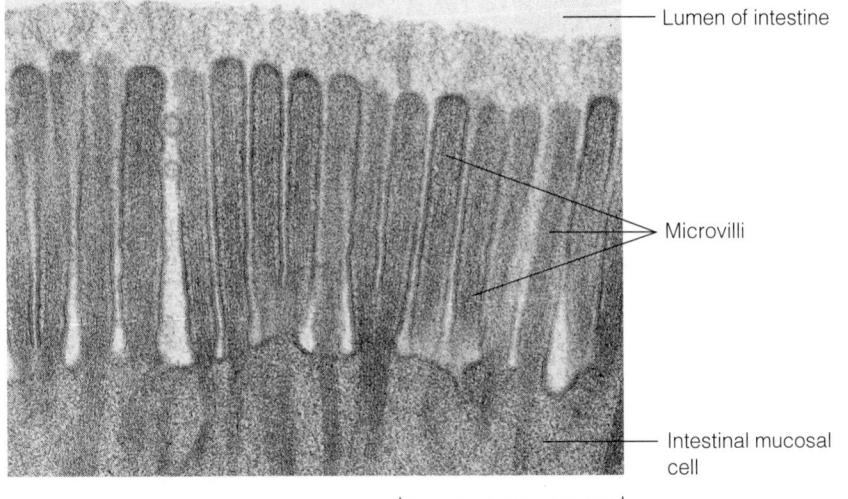

Lumen of intestine

Microvilli

Intestinal mucosal cell

0.5 μm

Figure 4-2 The Microvilli of Intestinal Mucosal Cells. Microvilli are fingerlike projections of the cell membrane that greatly increase the absorptive surface area of these cells (TEM).

cytoplasmic volume and therefore greater exchange requirements than could be met by the more modest increases in membrane surface area.

Some cells, particularly those that play a role in absorption, have additional characteristics that maximize their surface area. Effective surface area is most commonly increased by the inward folding or outward protrusion of the cell membrane. The cells that line your intestine, for example, contain many fingerlike projections called *microvilli* that greatly increase the effective membrane surface area and therefore the absorbing capacity of these cells (Figure 4-2).

Compartmentalization. Another limit on cell size is that imposed by the need to maintain adequate concentrations of the essential compounds and catalysts (enzymes) for the various processes cells must carry out. For a chemical reaction to occur in a cell, the appropriate reactants must collide with and bind to the surface of a particular enzyme. The frequency of such collisions will be greatly influenced by the concentrations of the reactants and of the enzyme itself. To maintain appropriate levels of reactants and enzymes as the size of a cell increases, the number of all such molecules must increase eightfold every time the three dimensions of the cell double. This increase taxes the synthetic capabilities of the cell.

The **compartmentalization** of activities within the cell is one solution to the problem of concentration. If all the enzymes and compounds involved in a specific process are localized within a specific region of the cell, a locally high concentration of those enzymes and compounds can be maintained in that region, rather than throughout the whole cell. This is what happens in plant and animal cells and is presumably the main reason they can be so much larger than bacterial cells and still function efficiently.

Plant and animal cells have a variety of **organelles,** internal compartments that are delineated by membranes and are highly specialized for specific functions. For example, the cells in a plant leaf have most of the enzymes, compounds, and pigments needed for photosynthesis compartmentalized together into structures called *chloroplasts.* Such cells can therefore maintain appropriately high concentrations of everything that is essential for photosynthesis within the chloroplasts without having to have similarly high levels of these substances elsewhere in the cell. In a similar way, other functions are localized within other compartments. This internal compartmentalization of specific functions makes it possible for the large cells of plants and animals to maintain locally high concentrations of the specific enzymes and compounds involved in particular cellular processes. Such processes can therefore proceed efficiently, even though as a whole such cells are orders of magnitude larger than bacterial cells.

Prokaryotes and Eukaryotes: An Organizational Dichotomy

With the advent of electron microscopy, biologists came to recognize two fundamentally different plans of cellular organization, the simpler one found in bacteria and the more complex one found in all other kinds of cells. Based on these structural differences, organisms can be divided into two broad groups, the **prokaryotes** (bacteria) and the **eukaryotes** (all other forms of life). The most fundamental distinction is that eukaryotic cells have a true, membrane-bounded nucleus (*eu-* is Greek for "true" or "genuine"; *karyon* means "nucleus"), whereas prokaryotic cells do not (*pro-* means "before," suggesting an evolutionarily earlier form of life).

Prokaryotes, in turn, are either *eubacteria* or *archebacteria.* The eubacteria ("true bacteria") include most present-day *bacteria* and the *cyanobacteria* (previously called *blue-green algae*). Although similar to eubacteria in cellular structure, archebacteria are as different from eubacteria as they are from eukaryotes in terms of molecular structure and biochemistry. They are regarded as modern descendants of an evolutionarily ancient form of prokaryote that differed fundamentally from the ancestors of present eubacteria. (*Arche-* is a Greek prefix meaning "ancient" or "original.") Present-day archebacteria include the *methanogens,* which produce methane gas (CH_4) from carbon dioxide; the *halophiles,* which can grow in the presence of high salt concentrations (up to 5.5 *M* NaCl); and the *thermacidophiles,* which thrive in acidic hot springs (pH as low as 2, temperatures as high as 80°C).

Eukaryotes represent a much broader spectrum of organisms. Whereas all prokaryotes are single-celled organisms, eukaryotes may be either unicellular or multicellular organisms. Included in this group are all *plants, animals, fungi,* and *protists* (single-celled eukaryotes such as *algae* and *protozoa*).

Prokaryotes and eukaryotes differ at the cellular level in important structural, biochemical, and genetic features. Some of these differences are summarized in Table 4-1 and discussed briefly here.

Table 4-1   A Comparison of Some Properties of Prokaryotic and Eukaryotic Cells

Property	Prokaryotic Cells	Eukaryotic Cells
Size*	Small (a few micrometers in length or diameter, in most cases)	Large (10–50 times the length or diameter of prokaryotes, in most cases)
Membrane-bounded nucleus	No	Yes
Organelles	No	Yes
Microtubules	No	Yes
Microfilaments	No	Yes
Intermediate filaments	No	Yes
Exocytosis and endocytosis	No	Yes
Mode of cell division	Cell fission	Mitosis and meiosis
Genetic information	DNA molecule complexed with relatively few proteins	DNA complexed with proteins (notably histones) to form chromosomes
Processing of RNA	Little	Much
Ribosomes**	Small (70S); 3 RNA molecules and 55 proteins	Large (80S); 4 RNA molecules and about 78 proteins

*The disparity in size between prokaryotic and eukaryotic cells indicated here is generally valid, but cells of both types vary greatly in size, with some overlap in size ranges.

**Ribosomes are characterized in terms of their *sedimentation coefficients* or *S values,* a measure of their sedimentation rate based on size and shape. Sedimentation coefficients are normally expressed in Svedberg units (S), where $1S = 1 \times 10^{-13}$ sec.

Eukaryotic Cells Have True Nuclei. As already noted, the most fundamental distinction between eukaryotes and prokaryotes is reflected in the nomenclature itself: Eukaryotic cells have a true nucleus and prokaryotic cells do not. Instead of being sequestered within a membrane-bounded nucleus, the genetic information of a prokaryotic cell is simply localized in a region of the cytoplasm called the *nucleoid*, without a membrane around it. Figure 4-3 compares these structural features of a typical prokaryotic cell and the nucleus of a representative eukaryotic cell. Structures within the nucleus include the *nucleolus*, the site of ribosome synthesis, and the DNA-bearing *chromosomes*, which in Figure 4-3b are dispersed throughout the semifluid *nucleoplasm*.

Eukaryotic Cells Segregate Functions with Internal Membranes. Internal membranes that compartmentalize specific functions are a general feature of eukaryotic cells. The presence of a nuclear membrane is just one example. As we already know, a variety of other internal membranes allow the eukaryotic cell to compartmentalize functions that in prokaryotic cells take place within the cytoplasm or on the plasma (cell) membrane. By contrast, prokaryotic cells contain few internal membranes and do not compartmentalize specific functions.

Examples of internal membrane systems in eukaryotic cells include the *endoplasmic reticulum*, the *Golgi complex*, and the membranes that surround and delimit organelles such as *mitochondria*, *chloroplasts*, *lysosomes*, and *peroxisomes*, as well as various kinds of *vacuoles* and *vesicles*. Each of these organelles has its own characteristic membrane (or pair of membranes, in the case of mitochondria and chloroplasts), similar to other membranes in basic structure but often with its own specific chemical composition and enzymes. Localized within each such organelle are the particular cellular functions for which the structure is specialized. Figure 4-4 illustrates several of these structural features in a cross-sectional view of a human Leydig cell. (*Leydig cells* are found in the testis and are responsible for synthesis of the steroid sex hormones.) We will meet each of these organelles later in this chapter and then return to each in its appropriate context in succeeding chapters.

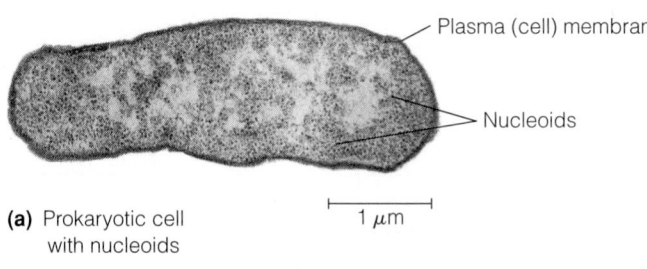

Plasma (cell) membrane

Nucleoids

1 μm

(a) Prokaryotic cell with nucleoids

Figure 4-3   The Storage of Genetic Information by Prokaryotic and Eukaryotic Cells.   **(a)** A bacterial cell with several nucleoids, each representing a single circular molecule of DNA (TEM). **(b)** The nucleus of a eukaryotic cell shown at the same magnification as the bacterial cell (TEM). The pair of membranes that encloses the nucleus is called the nuclear envelope. The semifluid nucleoplasm inside contains dispersed chromosomes, the DNA-bearing structures of the eukaryotic nucleus. The nucleolus is a structure within the eukaryotic nucleus that is involved in the synthesis of ribosomes.

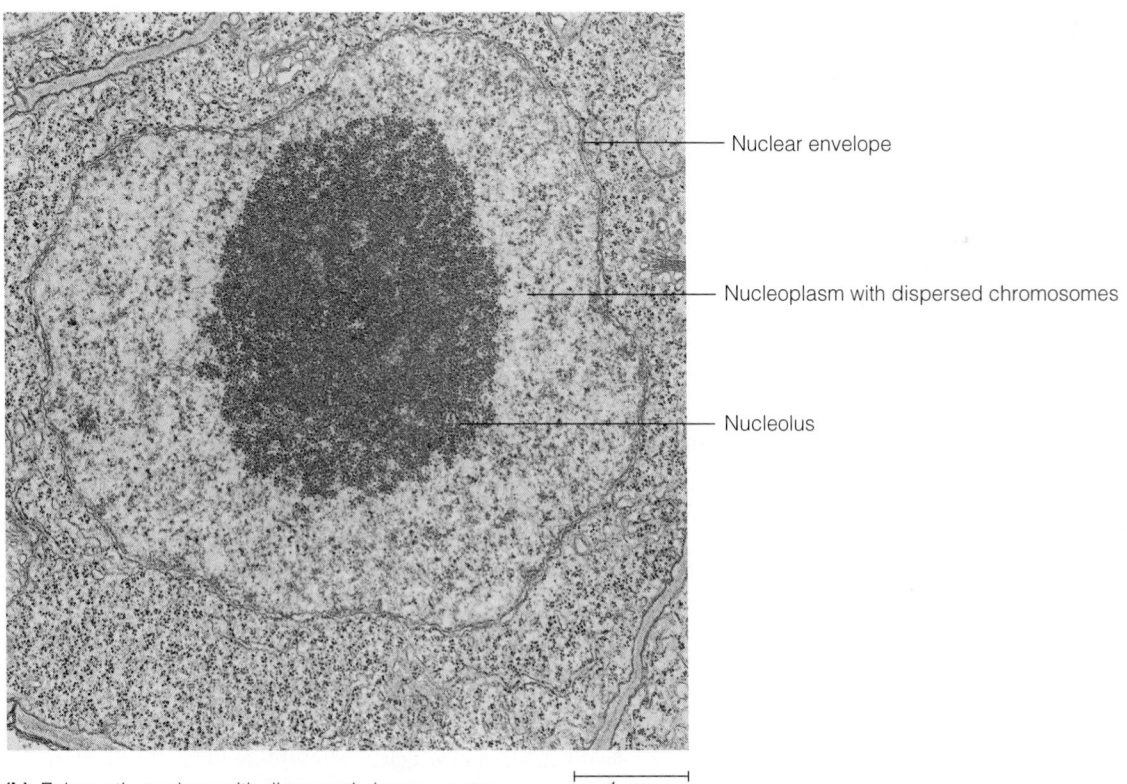

Nuclear envelope

Nucleoplasm with dispersed chromosomes

Nucleolus

(b) Eukaryotic nucleus with dispersed chromosomes

1 μm

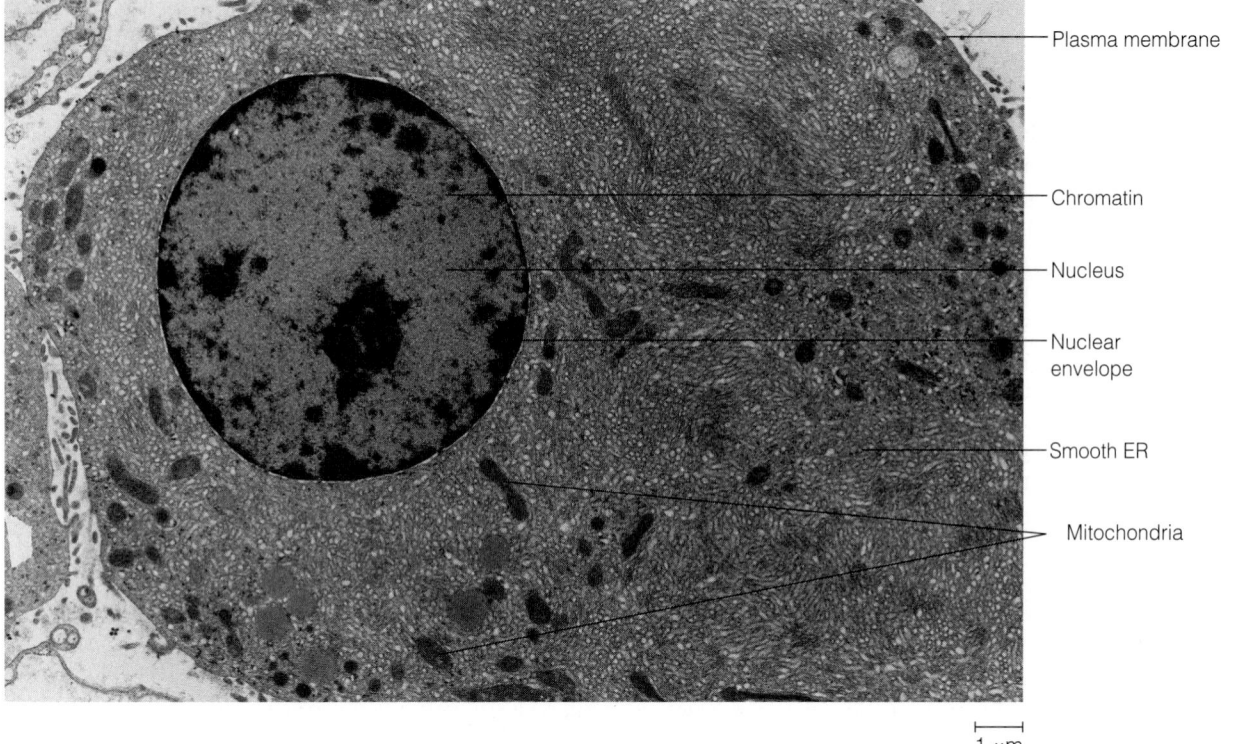

Plasma membrane

Chromatin

Nucleus

Nuclear envelope

Smooth ER

Mitochondria

1 μm

Figure 4-4   Structural Features of a Eukaryotic Cell.   Some of the major organelles present in eukaryotic cells are identified in this electron micrograph of a human Leydig cell seen in cross section (TEM). Other common features visible at this magnification but not identified in the figure include lysosomes, peroxisomes, and the Golgi complex. (Leydig cells are found in the testis and are characterized by extensive smooth ER, consistent with the production of steroid sex hormones.)

Eukaryotic Cells Have Tubules and Filaments. Also found in the cytoplasm of eukaryotic cells are several non-membranous structures that are involved in contraction, motility, and the establishment and support of cellular architecture. These include the *microtubules* found in the cilia and flagella of many cell types, the *microfilaments* of actin and myosin found in muscle fibrils and other structures involved in motility, and the *intermediate filaments,* which are especially prominent in parts of the cell that are subject to stress. Microtubules, microfilaments, and intermediate filaments are also involved in the *cytoskeletal framework* that imparts structure and elasticity to eukaryotic cells, as we will learn shortly and explore in more detail in Chapter 20.

Eukaryotic Cells Carry Out Exocytosis and Endocytosis. A further feature of eukaryotic cells is their ability to exchange materials between the membrane-bounded compartments within the cell and the exterior of the cell. This exchange is possible because of *exocytosis* and *endocytosis,* processes that are unique to eukaryotic cells. In endocytosis, portions of the plasma membrane invaginate and are pinched off to form membrane-bounded cytoplasmic vesicles containing substances that were previously on the outside of the cell. Exocytosis is essentially the reverse of this process; membrane-bounded vesicles inside the cell fuse with the plasma membrane and release their contents to the outside of the cell.

Eukaryotic Cells Organize Their DNA into Chromosomes. Another distinction between prokaryotes and eukaryotes becomes apparent when we consider the amount and organization of the genetic material. Prokaryotes characteristically contain amounts of DNA that might be described as "reasonable"; that is, we can account for much of the DNA in terms of known proteins for which the DNA serves as a genetic blueprint. Prokaryotic DNA is usually present in the cell as one or more circular molecules with which relatively few proteins are associated.

Though of "reasonable" size in a genetic sense, the circular DNA molecule of a prokaryotic cell is usually much longer than the cell itself. It therefore has to be folded and packed together tightly to fit into the nucleoid of the cell. For example, the common intestinal bacterium *Escherichia coli* is only about a micrometer or two long, yet it has a circular DNA molecule that is about 1300 μm in circumference. Clearly, a great deal of folding and packing is necessary to fit that much DNA into a small region of such a small cell. By way of analogy, it is roughly equivalent to packing about 60 feet (18 m) of very thin thread into a typical thimble.

But if DNA appears to pose a packaging problem for prokaryotic cells, consider the case of the poor eukaryotic cell! Although some of the lower eukaryotes (such as yeasts and fruit flies) contain only 10 to 50 times as much DNA as bacteria, most eukaryotic cells have at least a thousand times as much DNA as *E. coli.* It is tempting to label such amounts of DNA as "unreasonable" because we cannot at present assign any known function to much of it. But that is probably a more telling commentary on cell biologists than on cells.

Whatever the genetic function of such large amounts of DNA, the packaging problem is clearly acute. It is solved universally among eukaryotes by the organization of DNA into complex structures called **chromosomes** that contain at least as much protein as DNA. (The circular molecule of DNA in prokaryotic cells is sometimes also called a chromosome, but the convention in this text will be to use that term only for the structure in the eukaryotic nucleus.) It is as chromosomes that the DNA of eukaryotic cells is packaged, segregated during cell division, transmitted to daughter cells, and transcribed as needed into the molecules of RNA that are involved in protein synthesis. Figure 4-5 shows a chromosome from an animal cell as seen by high-voltage electron microscopy.

Eukaryotic Cells Segregate Genetic Information by Mitosis and Meiosis. A further contrast between prokaryotes and eukaryotes is the way they allocate genetic information to daughter cells upon division. Prokaryotic cells simply replicate their DNA and divide by a relatively simple process called *cell fission*, with one molecule of DNA going to each daughter cell. Eukaryotic cells also replicate their DNA, but they then use the more complex processes of *mitosis* and *meiosis* to distribute chromosomes equitably to daughter cells. The chromosome in Figure 4-5 was prepared from a cell that was undergoing mitosis. If the process had been allowed to proceed, the chromosome would have divided into two daughter chromosomes, each destined for one of the two daughter cells.

Eukaryotic Cells Express Their DNA Differently. The differences between prokaryotic and eukaryotic cells extend to the expression of genetic information. Eukaryotic cells tend to transcribe genetic information in the nucleus into much larger RNA molecules than they eventually use to direct protein synthesis in the cytoplasm. They depend on later processing and transport processes to deliver RNA molecules of the proper sizes to the cytoplasm.

By contrast, prokaryotes seem to transcribe very specific segments of genetic information into RNA messages, and little or no processing or selection appears to be either necessary or possible. In fact, the absence of a nuclear membrane makes it possible for new messenger RNA molecules to become involved in the process of protein synthesis even before they are themselves completely synthesized (Figure 4-6). Prokaryotes and eukaryotes also differ in the size and composition of the ribosomes used to synthesize proteins (see Table 4-1). We will return to this distinction in more detail later in the chapter.

Cell Specialization: The Unity and Diversity of Biology

In terms of structure and function, cells are characterized by both unity and diversity, as you can see in the typical animal and plant cells shown in Figures 4-7 and 4-8. By unity and diversity, we simply mean that all cells resemble one another in some ways, yet differ from one another in other ways. In

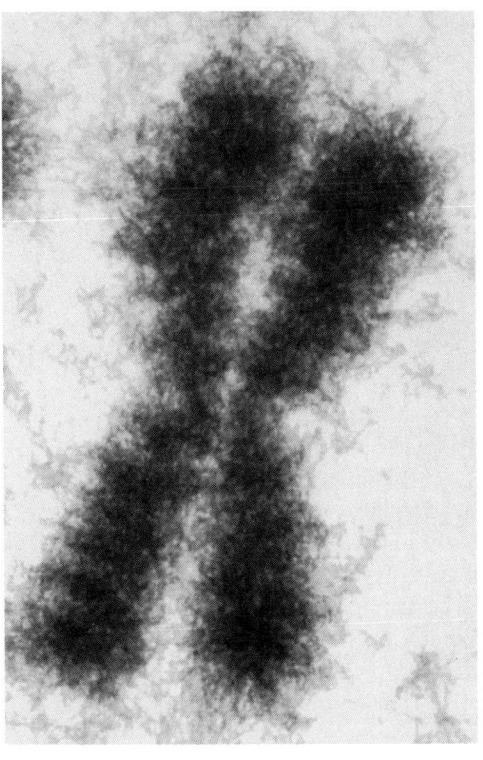

1 μm

Figure 4-5   A Eukaryotic Chromosome. This chromosome was obtained from a cultured Chinese hamster cell and visualized by high-voltage electron microscopy (HVEM). The cell was undergoing mitosis, and the chromosome is therefore highly coiled and condensed. If mitosis had been allowed to proceed, the chromosome would have divided into two daughter chromosomes, one destined for each of the two daughter cells that result from subsequent cell division.

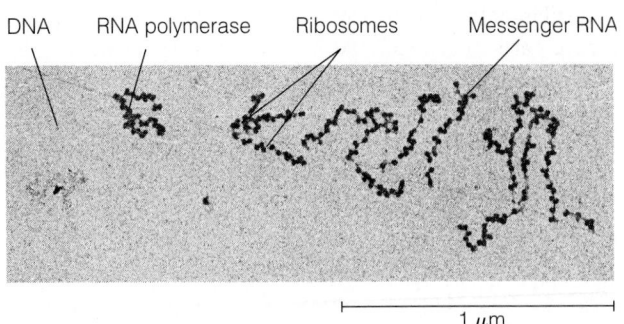

1 μm

Figure 4-6   The Expression of Genetic Information in a Prokaryotic Cell. This electron micrograph shows a small segment of a DNA molecule from a bacterial cell being transcribed into messenger RNA molecules by the enzyme RNA polymerase (TEM). As soon as the RNA becomes available, it associates with ribosomes, allowing protein synthesis to occur.

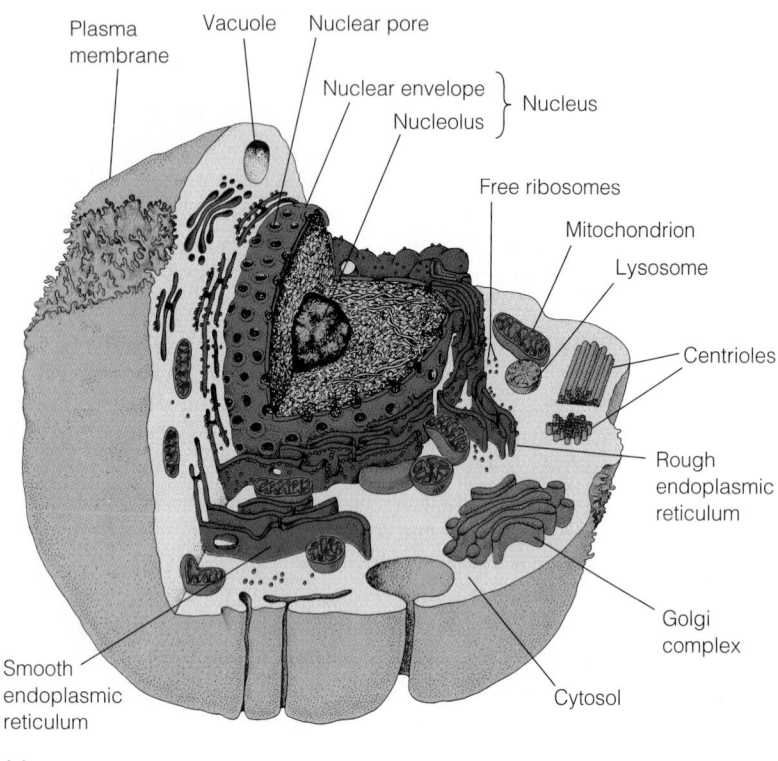

Plasma membrane

Vacuole

Nuclear pore

Nuclear envelope

Nucleolus

Nucleus

Free ribosomes

Mitochondrion

Lysosome

Centrioles

Rough endoplasmic reticulum

Golgi complex

Cytosol

Smooth endoplasmic reticulum

(a)

Figure 4-7   A Typical Animal Cell.
(a) A schematic drawing of an animal cell to provide perspective on the relative sizes and shapes of organelles and other subcellular structures. **(b)** A plasma cell (white blood cell), with several subcellular structures identified (TEM).

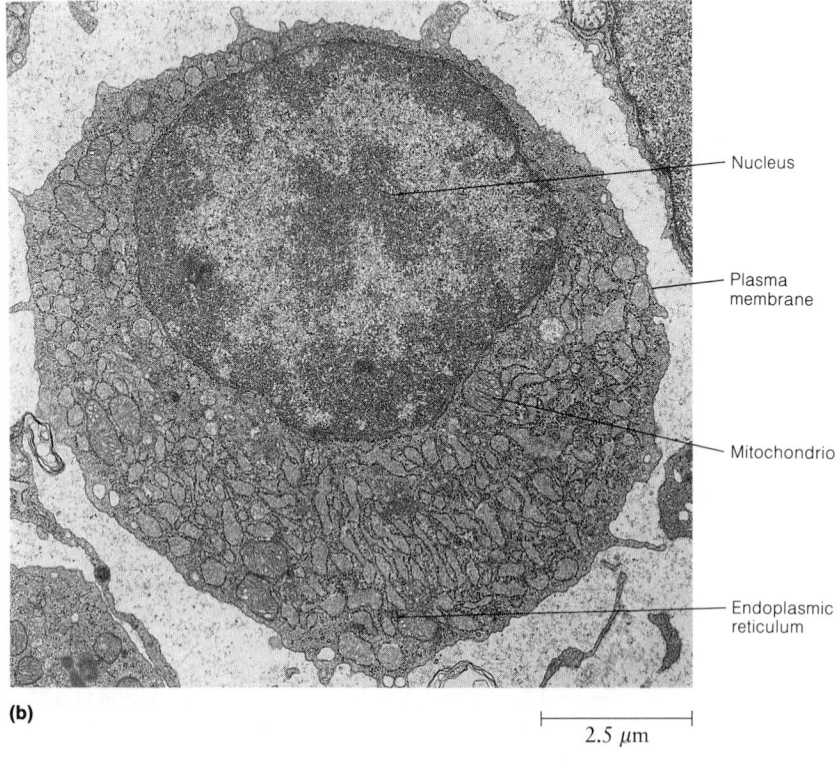

Nucleus

Plasma membrane

Mitochondrion

Endoplasmic reticulum

(b)

2.5 µm

Figure 4-8   A Typical Plant Cell.

(a) A schematic drawing of a plant cell. Compare this drawing with the animal cell in Figure 4-7a and notice that plant cells are characterized by the absence of lysosomes and the presence of chloroplasts, a cell wall, and a large central vacuole. **(b)** A cell from a *Coleus* leaf, with several subcellular structures identified (TEM).

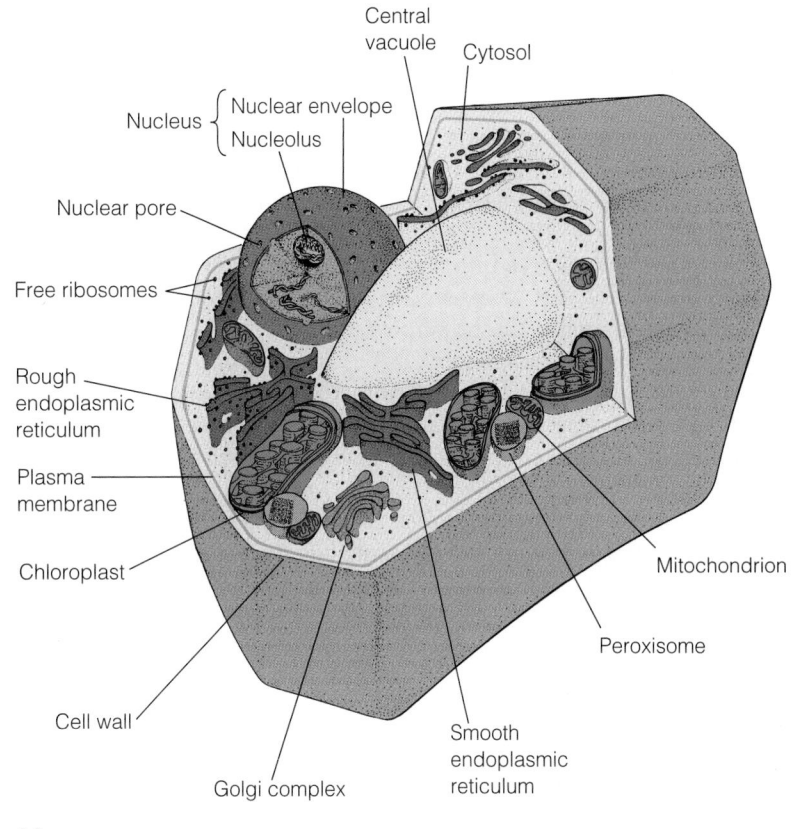

(a)

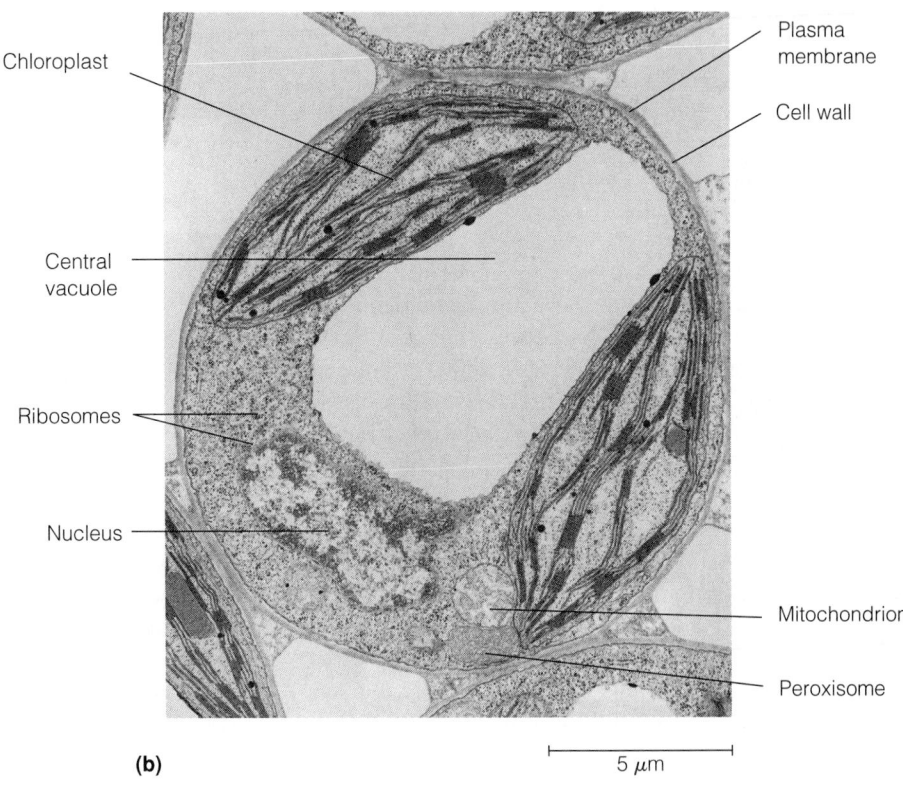

(b)

upcoming chapters, we will concentrate on aspects of structure and function common to most cell types, as these are the features of cells that are of the greatest general interest. We will find, for example, that virtually all cells oxidize sugar molecules for energy, transport ions across membranes, transcribe DNA into RNA, and undergo division to generate daughter cells. These, then, become topics of legitimate concern for us.

Much the same is true in terms of structural features. All cells are surrounded by a selectively permeable membrane, all have ribosomes for the purpose of protein synthesis, and all contain double-stranded DNA as their genetic information. Clearly, we can be confident that we are dealing with fundamental aspects of cellular organization and function when we consider processes and structures common to most, if not all, cells.

But sometimes our understanding of cellular biology is enhanced by considering not just the unity but also the diversity of cells—not just features common to most cells, but also features that are especially prominent in a particular cell type. For example, to understand how the process of protein secretion works, it would be an advantage to consider a cell that is highly specialized for that particular function. Cells from the human pancreas would be a good choice, for example, because they secrete large amounts of digestive enzymes, such as amylase and trypsin.

Similarly, to study functions known to occur in mitochondria, it would clearly be an advantage to select a cell type that is highly specialized in the energy-releasing processes that occur in the mitochondrion, since such a cell would probably have a lot of well-developed, highly active mitochondria. It was for this very reason, in fact, that Hans Krebs chose the flight muscle of the pigeon as the tissue with which to carry out the now-classic studies on the cyclic pathway of oxidative reactions that we know as the *tricarboxylic acid (TCA) cycle*.

Whenever we exploit the specialized functions of specific cell types to study a particular function, we are acknowledging the diversity of cell structure and function that arises primarily because of cellular specialization. Although we may not often realize it, we are also taking advantage of the multicellularity of many organisms, because it is usually only as part of a multicellular organism that a cell can afford to commit itself to a specialized function.

In general, unicellular organisms such as bacteria, protozoa, and some algae must be capable of carrying out any and all of the functions necessary for survival, growth, and reproduction and cannot afford to overemphasize any single function at the expense of others. Multicellular organisms, on the other hand, are characterized by a division of labor among tissues and organs that not only allows for, but actually depends on, specialization of structure and function. Whole groups of cells become highly specialized for a particular task, which then becomes their specific role in the overall economy of the organism.

The Eukaryotic Cell in Overview: Pictures at an Exhibition

From the preceding discussion, it should be clear that all cells carry out many of the same basic functions and have some of the same basic structural features. However, cells of eukaryotic organisms are far more complicated structurally than prokaryotic cells, primarily because of the organelles and other intracellular structures that eukaryotes use to compartmentalize various functions. The structural complexity of eukaryotic cells is illustrated by the typical animal and plant cells shown in Figures 4-7 and 4-8.

In reality, of course, there is no such thing as a truly "typical" cell; all eukaryotic cells have features that distinguish them from the particular cells shown in Figures 4-7 and 4-8. But most eukaryotic cells are sufficiently similar to warrant a general overview of their structural features.

In essence, a typical eukaryotic cell has at least four major structural features: a *plasma* (or *cell*) *membrane* to define its boundary and retain its contents, a *nucleus* to house the DNA that directs cellular activities, *membrane-bounded organelles* in which various cellular functions are localized, and the *cytoplasm* with its *cytoskeleton* of tubules and filaments. In addition, plant cells have a rigid *cell wall* external to the plasma membrane. Animal cells do not have a wall but are usually surrounded by an *extracellular matrix,* which consists primarily of *collagen* and polysaccharides called *proteoglycans.*

Our intention here is to look at each of these structural features in overview, as an introduction to cellular architecture. We are not yet ready to consider any of these features in detail; that will wait until later chapters, when we encounter the cellular processes in which the various organelles and other structures are involved. For the present, we will simply look at each structure as one might look at pictures at an exhibition. We will move through the gallery rather quickly, just to get a feel for the overall display, but with the intention of returning to each structure later for a careful and more detailed examination. (If you are interested in more information on a specific topic at this point, see the Reference Box on p. 107 for a list of the chapters in which particular cellular structures or research techniques are considered in detail.)

The Plasma Membrane

Our tour begins with the **plasma membrane** that surrounds every cell (Figure 4-9a). The plasma membrane defines the boundaries of the cell and ensures that its contents are retained. Like all biological membranes, the plasma membrane consists of phospholipids and proteins organized into two layers (Figure 4-9b). Each phospholipid molecule consists of two hydrophobic "tails" and a hydrophilic "head" and is therefore an *amphipathic molecule* (see Figures 2-11 and 2-12).

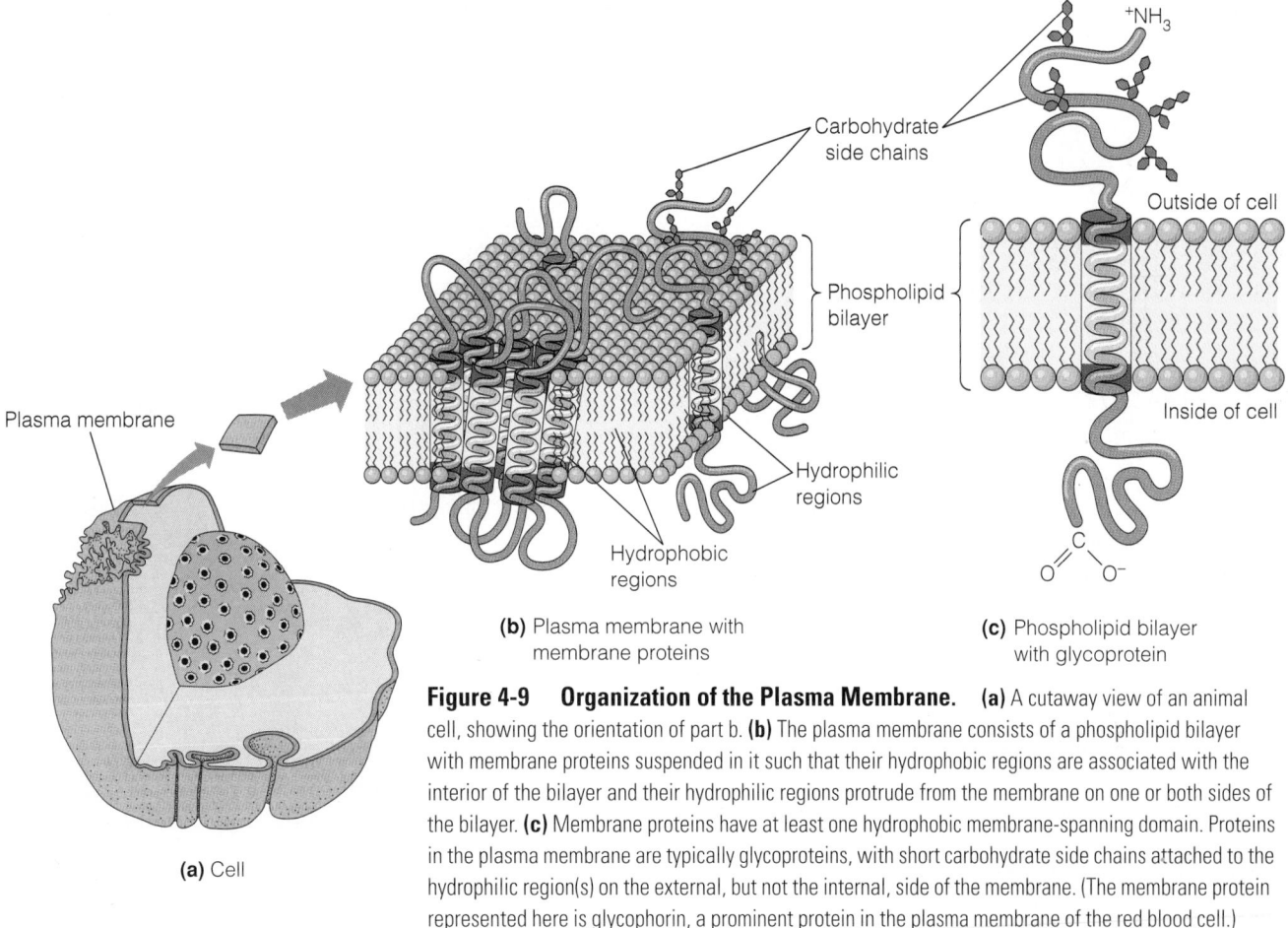

(b) Plasma membrane with membrane proteins

(c) Phospholipid bilayer with glycoprotein

Figure 4-9   Organization of the Plasma Membrane.   (a) A cutaway view of an animal cell, showing the orientation of part b. **(b)** The plasma membrane consists of a phospholipid bilayer with membrane proteins suspended in it such that their hydrophobic regions are associated with the interior of the bilayer and their hydrophilic regions protrude from the membrane on one or both sides of the bilayer. **(c)** Membrane proteins have at least one hydrophobic membrane-spanning domain. Proteins in the plasma membrane are typically glycoproteins, with short carbohydrate side chains attached to the hydrophilic region(s) on the external, but not the internal, side of the membrane. (The membrane protein represented here is glycophorin, a prominent protein in the plasma membrane of the red blood cell.)

(a) Cell

The phospholipid molecules orient themselves in the two layers of the membrane such that the "tails" of each molecule (the nonpolar hydrocarbon chains of the fatty acids) face inward, toward the other layer. The hydrophilic "head" of each molecule (containing a negatively charged phosphate group and positively charged amine) faces outward, either toward the inside or toward the outside of the cell, depending on the layer in which the lipid molecule is located (see Figure 4-9b). The resulting **phospholipid bilayer** is the basic structural unit of all membranes and serves as a permeability barrier to most water-soluble substances. The phospholipid bilayer appears as a pair of dark bands when viewed with the electron microscope (see Figure 2-13).

Membrane proteins are also amphipathic, with both hydrophobic and hydrophilic regions on their surface. They orient themselves in the membrane such that hydrophobic regions of the protein are located within the interior of the membrane, whereas hydrophilic regions protrude into the aqueous environment at the surface of the membrane. Most of the proteins with hydrophilic regions exposed on the external side of the plasma membrane have oligosaccharide (short carbohydrate) side chains attached to them and are therefore called *glycoproteins* (Figure 4-9c).

The proteins present in the plasma membrane play a variety of roles. Some are *enzymes,* which catalyze reactions known to be associated with the membrane. Others serve as *anchors* for structural elements of the cytoskeleton that we will encounter later in the chapter. Still others are *transport proteins,* responsible for moving specific substances (ions and hydrophilic solutes, usually) across the membrane. Membrane proteins are also important as *receptors* for specific chemical signals that impinge on the cell from its environment and need to be transmitted inward to the cell. Transport proteins and receptor proteins (and other membrane proteins as well) are *transmembrane proteins,* with hydrophilic regions protruding on both sides of the membrane, connected by one or more hydrophobic membrane-spanning domains.

The Nucleus

If we now enter the cell, one of the most prominent structures we encounter is the **nucleus** (Figure 4-10). The nucleus serves as the control center for the entire cell. Its interior is called the **nucleoplasm.** Here, separated from the rest of the cell by a membrane, are the DNA-bearing *chromosomes* of

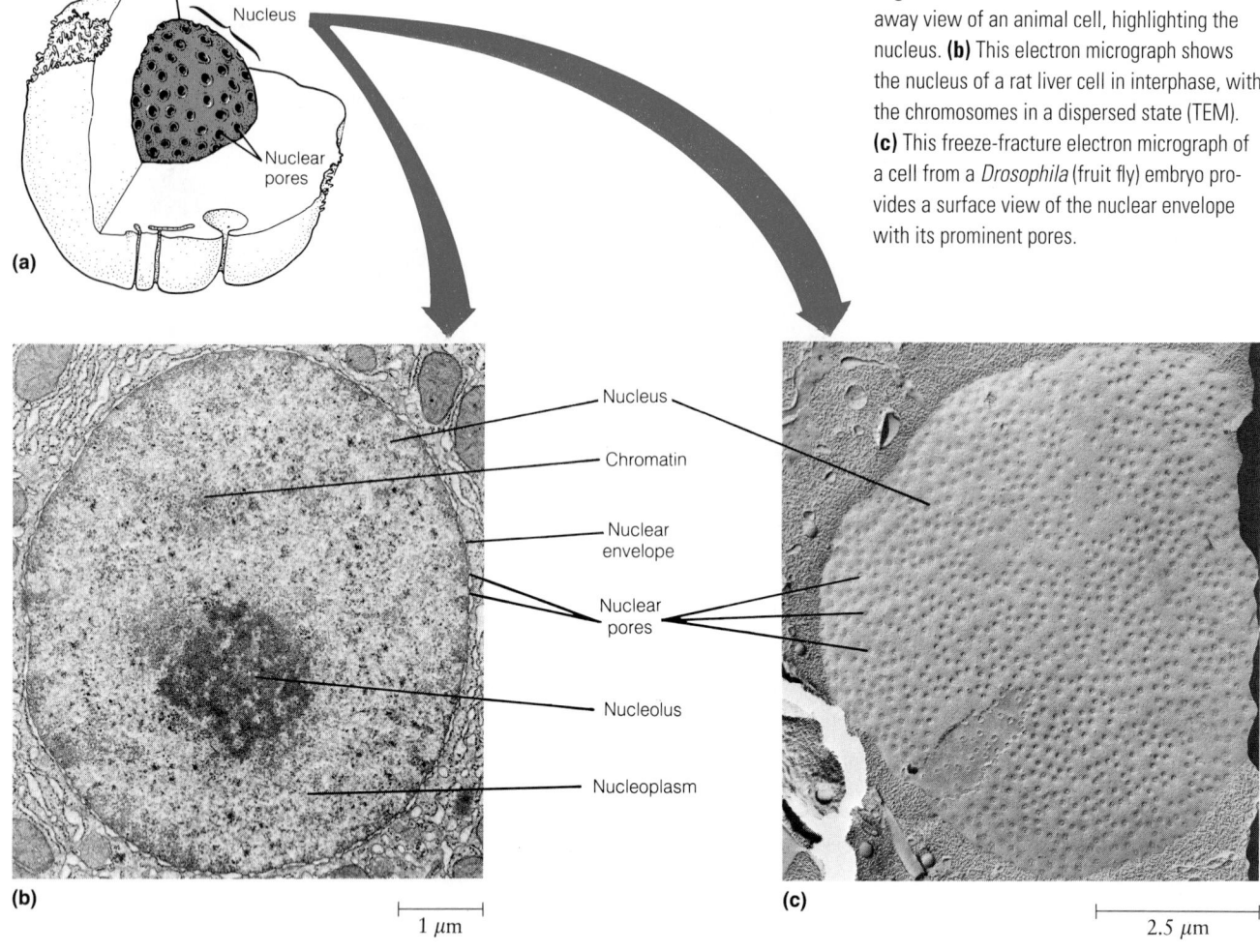

Figure 4-10   The Nucleus.   (a) A cutaway view of an animal cell, highlighting the nucleus. **(b)** This electron micrograph shows the nucleus of a rat liver cell in interphase, with the chromosomes in a dispersed state (TEM). **(c)** This freeze-fracture electron micrograph of a cell from a *Drosophila* (fruit fly) embryo provides a surface view of the nuclear envelope with its prominent pores.

(a)

Nucleus

Nuclear pores

Nucleus

Chromatin

Nuclear envelope

Nuclear pores

Nucleolus

Nucleoplasm

(b)

1 μm

(c)

2.5 μm

the cell. Actually, the membrane boundary around the nucleus consists of two membranes and is more properly called the **nuclear envelope.** Unique to the membranes of the nuclear envelope are numerous small openings called **pores** (Figure 4-10b and c). Each pore is a channel through which water-soluble molecules can move between the nucleus and cytoplasm. Ribosomes, messenger RNA molecules, chromosomal proteins, and enzymes needed for nuclear activities are also presumed to be transported across the nuclear envelope through its pores.

The number of chromosomes within the nucleus is characteristic of the species. It can be as low as two (in the sperm and egg cells of some grasshoppers, for example), or it can run into the hundreds. Chromosomes are most readily visualized during mitosis, when chromosomes in dividing cells are highly condensed and can easily be stained (see Figure 4-5). During the *interphase* between divisions, on the other hand, chromosomes are dispersed as **chromatin** and are not easy to visualize (see Figure 4-10b).

Also present in the nucleus are **nucleoli** (singular: **nucleolus**), structures responsible for the synthesis and assembly of the subunits that make up ribosomes. Nucleoli are usually associated with specific regions of particular chromosomes.

Intracellular Membranes and Organelles

The internal volume of the cell exclusive of the nucleus is called the *cytoplasm* and is occupied by membrane-bounded *organelles* and by the *cytosol* in which they are suspended. In this section, we will look at each of the major eukaryotic organelles. In a typical animal cell, these compartments make up almost half of the total internal volume of the cell.

The Mitochondrion.   Our tour of the eukaryotic organelles begins with a very prominent organelle, the **mitochondrion** (plural: **mitochondria**). The structure of the mitochondrion is shown in Figure 4-11. Mitochondria are large by cellular standards—up to a micrometer across and usually a few micrometers long. A mitochondrion is therefore comparable in size to a whole bacterial cell. The fact that most eukaryotic cells contain hundreds of mitochondria, each approximately the size of an entire bacterial cell, emphasizes again the great difference in size between prokaryotic and eukaryotic cells.

The mitochondrion is surrounded by two membranes, designated the **inner** and **outer mitochondrial membranes.** Both mitochondria and chloroplasts, which we will meet next, contain their own DNA and ribosomes and can there-

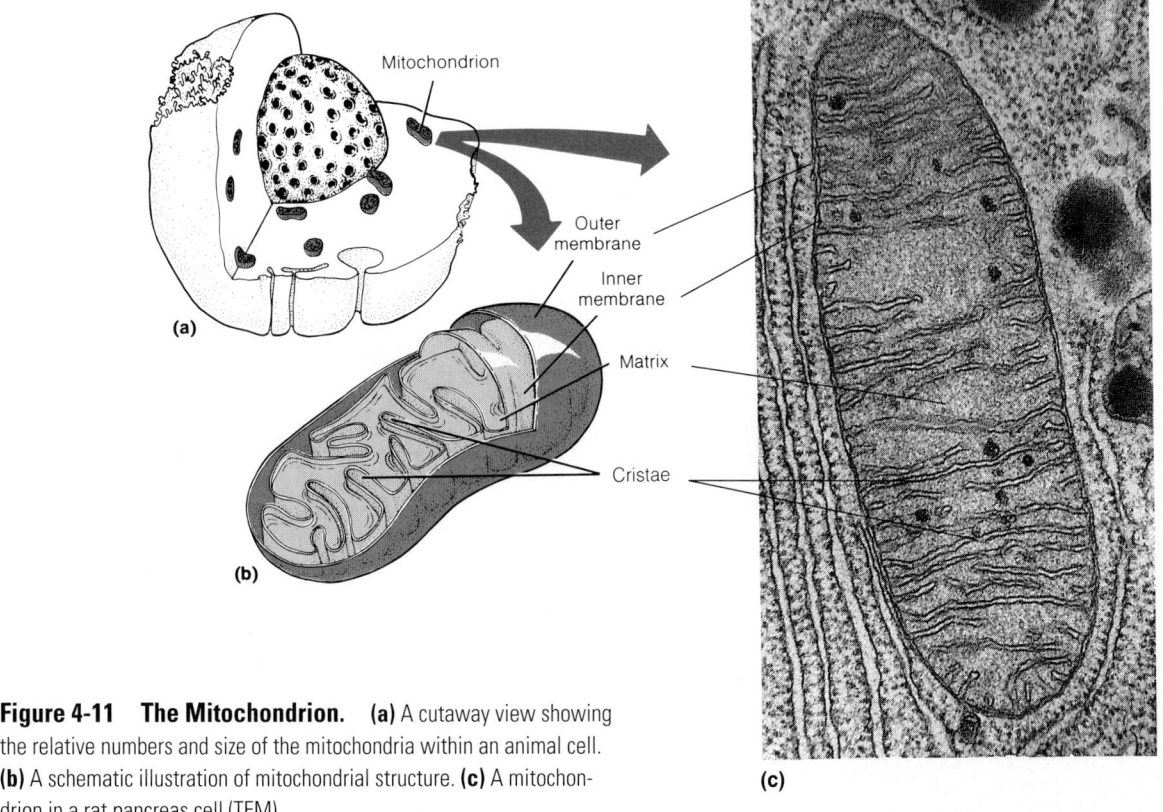

Figure 4-11 The Mitochondrion. **(a)** A cutaway view showing the relative numbers and size of the mitochondria within an animal cell. **(b)** A schematic illustration of mitochondrial structure. **(c)** A mitochondrion in a rat pancreas cell (TEM).

fore code for and synthesize some (though by no means all) of their own proteins.

Most of the chemical reactions involved in the oxidation of sugars and other cellular "fuel" molecules occur within the mitochondria. The purpose of these oxidative events is to extract energy from food and conserve as much of it as possible in the form of the high-energy compound *adenosine triphosphate (ATP)*. It is within the mitochondrion that the cell localizes most of the enzymes and intermediates involved in such important cellular processes as the tricarboxylic acid (TCA) cycle, fat oxidation, and ATP generation. Most of the intermediates involved in the transport of electrons from oxidizable food molecules to oxygen are located in or on the **cristae**, infoldings of the inner mitochondrial membrane. Other reaction sequences, particularly those of the TCA cycle and those involved in fat oxidation, occur in the semifluid **matrix** that fills the inside of the mitochondrion.

The number and location of mitochondria within a cell can often be related directly to their role in that cell. Tissues with an especially heavy demand for ATP as an energy source can be expected to have cells that are well endowed with mitochondria, and the organelles are usually located within the cell just where the energy need is greatest. This localization is illustrated by the sperm cell shown in Figure 4-12. As the drawing indicates, a sperm cell often has a single spiral mitochondrion wrapped around the central shaft, or *axoneme,* of the cell. The numerous mitochondrial profiles

seen along the length of the sperm are therefore multiple cross sections of the same mitochondrion. Notice how tightly the mitochondrion coils around the axoneme, just where the ATP is actually needed to propel the sperm cell. Muscle cells and cells that specialize in the transport of ions also have numerous mitochondria located strategically to meet the special energy needs of such cells (Figure 4-13).

The Chloroplast. Next, our gallery tour takes in the **chloroplast,** in many ways a close relative of the mitochondrion. A typical chloroplast is shown in Figure 4-14. Chloroplasts are large organelles, typically a few micrometers in diameter and 5–10 μm long. Chloroplasts are therefore substantially bigger than mitochondria and larger than any other structure in the cell except the nucleus. Like mitochondria, chloroplasts are surrounded by both an inner and an outer membrane. In addition, they have a third membrane system consisting of flattened sacs called **thylakoids** and the membranes, called **stroma lamellae** (or *stroma thylakoids*), that interconnect them. Thylakoids are stacked together to form the **grana** (singular: **granum**) that characterize most chloroplasts (Figure 4-14c and d).

Chloroplasts are the site of *photosynthesis,* the light-driven process whereby carbon dioxide and water are used to manufacture the sugars and other organic compounds from which all life is ultimately fabricated. Chloroplasts are found in leaves and other photosynthetic tissues of higher plants, as well as in all of the eukaryotic algae. Located

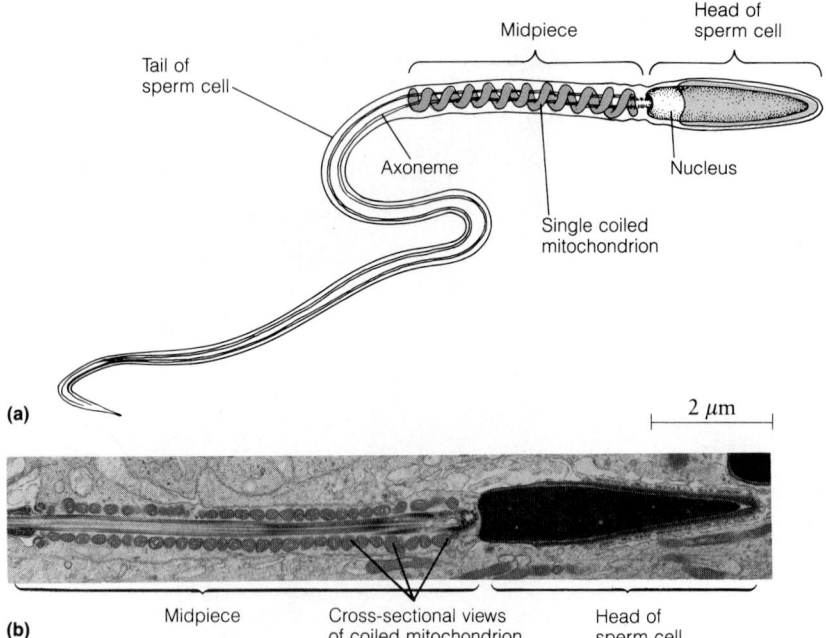

(a)

2 μm

(b)

Midpiece Cross-sectional views Head of
            of coiled mitochondrion sperm cell

Figure 4-12   Localization of the Mitochondrion Within a Sperm Cell. The single mitochondrion present in a sperm cell is coiled tightly around the axoneme of the tail, reflecting the localized need of the sperm tail (flagellum) for energy. **(a)** A schematic drawing of a sperm. **(b)** An electron micrograph of a sperm cell from a marmoset monkey (TEM).

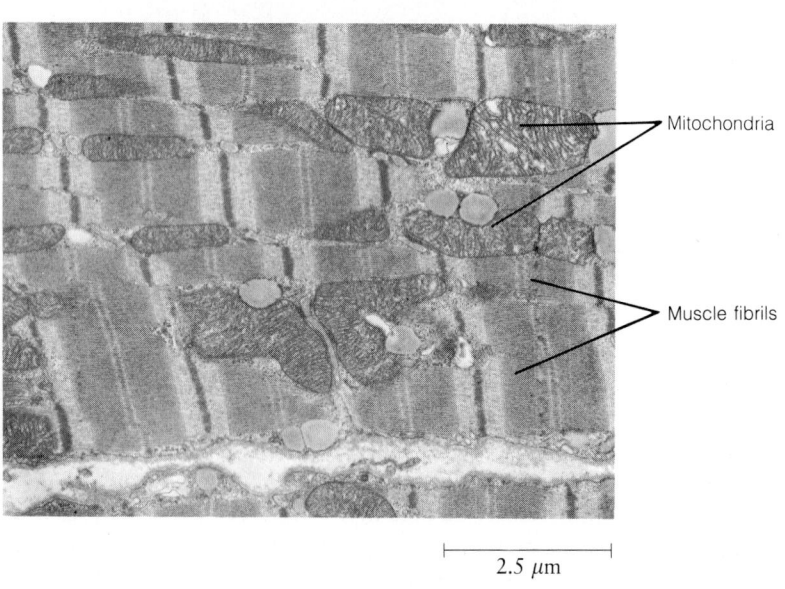

Mitochondria

Muscle fibrils

2.5 μm

Figure 4-13   Localization of Mitochondria Within a Muscle Cell. This electron micrograph of a muscle cell from a cat heart shows the intimate association of mitochondria with the muscle fibrils that are responsible for muscle contraction (TEM).

within the organelle are most of the enzymes, intermediates, and pigments (light-absorbing molecules) needed to "fix" carbon from carbon dioxide into organic form and convert it reductively into sugars. The reactions that depend directly on solar energy are localized in or on the thylakoid membrane system. Reactions involved in the initial trapping of carbon dioxide into organic form and its subsequent reduction and rearrangement into sugar molecules occur within the semifluid **stroma** that fills the interior of the chloroplast.

Although known primarily for their role in photosynthesis, chloroplasts are involved in a variety of other processes as well. An important example involves the reduction of nitrogen from the oxidation level of the nitrate (NO_3^-) that plants obtain from the soil to the oxidation level of ammonia (NH_3), the form of nitrogen required for protein synthesis. Furthermore, the chloroplast is only the most prominent example of a broader class of plant organelles, the **plastids.** Plastids perform a variety of functions in plant cells. *Chromoplasts,* for example, are pigment-containing plastids that are responsible for the characteristic coloration of flowers, fruits, and other plant parts. *Amyloplasts* are plastids that are specialized for starch storage.

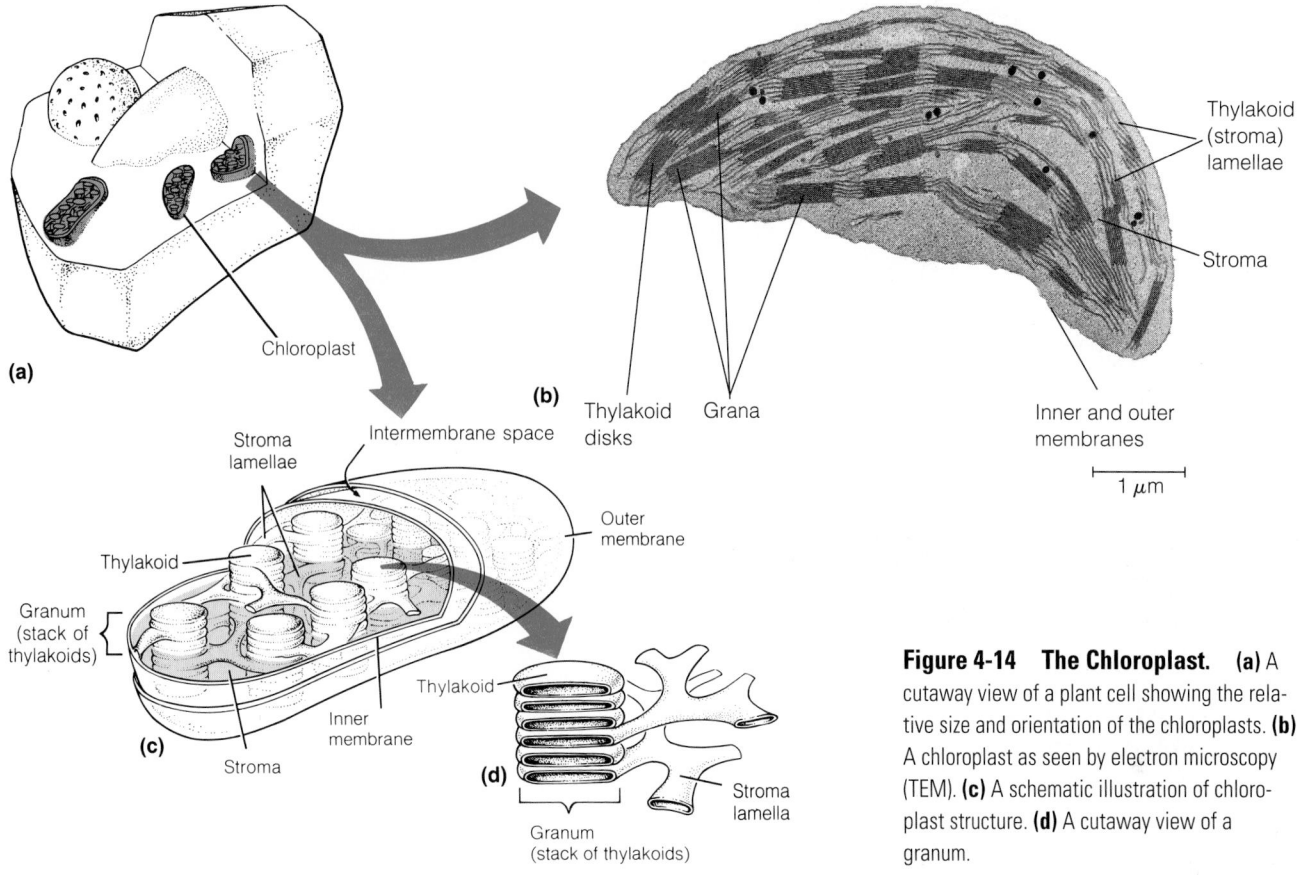

(a)

Chloroplast

(b) Thylakoid disks   Grana   Thylakoid (stroma) lamellae   Stroma

Inner and outer membranes

1 μm

Intermembrane space

Stroma lamellae

Thylakoid

Granum (stack of thylakoids)

Outer membrane

Thylakoid

Inner membrane

(c)   Stroma

(d)   Stroma lamella

Granum (stack of thylakoids)

Figure 4-14 The Chloroplast. (a) A cutaway view of a plant cell showing the relative size and orientation of the chloroplasts. **(b)** A chloroplast as seen by electron microscopy (TEM). **(c)** A schematic illustration of chloroplast structure. **(d)** A cutaway view of a granum.

The Endoplasmic Reticulum.   Extending throughout the cytoplasm of almost every eukaryotic cell is a network of membranes called the **endoplasmic reticulum, or ER** (Figure 4-15). The name sounds complicated, but *endoplasmic* just means "within the plasm" (of the cell), and *reticulum* is simply a fancy word for "network." The endoplasmic reticulum consists of tubular membranes and flattened sacs, or **cisternae,** that appear to be interconnected. The internal space enclosed by the ER membranes is called the **lumen.** The ER is continuous with the outer membrane of the nuclear envelope (Figure 4-15a and b). The space between the two nuclear membranes is therefore a part of the same compartment as the lumen of the ER.

The ER can be either *rough* or *smooth*. **Rough endoplasmic reticulum (rough ER)** appears "rough" in the electron microscope because it is studded with ribosomes on the side of the membrane that faces the cytoplasm (Figure 4-15d). These ribosomes are actively synthesizing proteins. Most of these proteins are transported into or across the membrane as they are synthesized, accumulating in completed form within either the membrane or the lumen of the ER. *Secretory proteins* (proteins destined to be exported from the cell) are synthesized in this way. They then make their way to the cell surface by a complex process that involves not only the rough ER but also the Golgi complex and secretory vesicles.

Not all proteins are synthesized on the rough ER, however. Much protein synthesis occurs on ribosomes that are not attached to the ER but are found free in the cytosol instead. In general, secretory proteins and membrane proteins are made by ribosomes on the rough ER, whereas proteins intended for use within the cytosol are made on free ribosomes.

The **smooth endoplasmic reticulum (smooth ER)** has no role in protein synthesis and hence no ribosomes. It therefore has a characteristic "smooth" appearance when viewed by electron microscopy (Figure 4-15c). Smooth ER is involved in the packaging of secretory proteins, as well as in the synthesis of lipids and steroids. The Leydig cell shown in Figure 4-4 produces steroid hormones in the testis and is characterized by an unusually extensive smooth ER. In addition, smooth ER is responsible for the inactivation and detoxification of drugs and other compounds that might otherwise be toxic or harmful to the cell.

The Golgi Complex.   Closely related to the smooth ER in both proximity and function is the **Golgi complex** (or *Golgi apparatus*), named after its Italian discoverer, Camillo Golgi. The Golgi complex consists of a stack of flattened vesicles, shown in Figure 4-16. The Golgi complex plays an important role in the processing and packaging of secretory proteins and in the synthesis of complex polysaccharides. Vesicles that arise by budding off the ER are accepted by the Golgi complex. Here, the contents of the vesicles (proteins, for the most part) and sometimes the vesicle membranes are

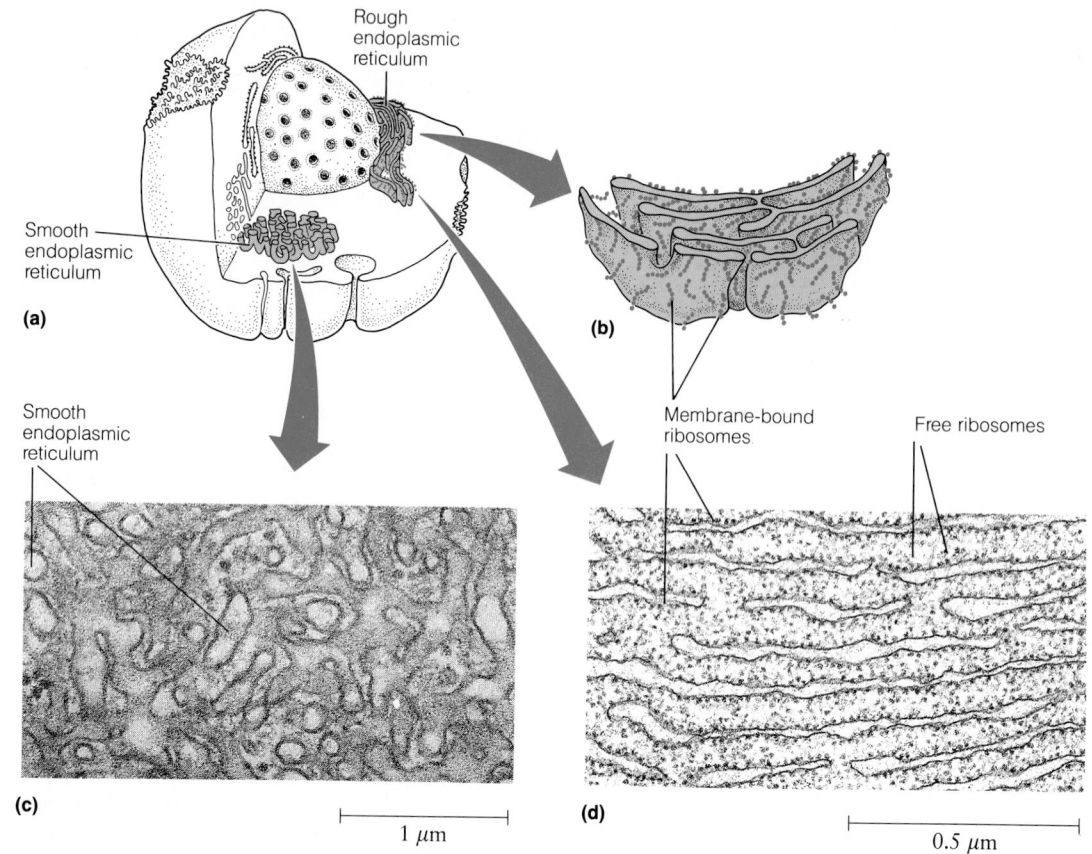

Figure 4-15   The Endoplasmic Reticulum.   (a) A cutaway view of a typical animal cell showing the location and relative size of the endoplasmic reticulum (ER). **(b)** A schematic illustration depicting the organization of the rough ER as layers of flattened membranes studded on their outer surface with ribosomes. **(c)** An electron micrograph of smooth ER in a cell from guinea pig testis (TEM). **(d)** An electron micrograph of rough ER in a rat pancreas cell (TEM); notice that ribosomes are either attached to the ER or free in the cytosol.

Figure 4-16   The Golgi Complex.   (a) A cutaway view showing the relative orientation and size of a Golgi complex within a cell. **(b)** An electron micrograph of a Golgi complex in a cell from a bean root tip (TEM). Notice the vesicles forming at the edges of the stack and the free vesicles that have presumably just arisen in this way. **(c)** A schematic drawing of a Golgi complex, showing vesicle formation by budding.

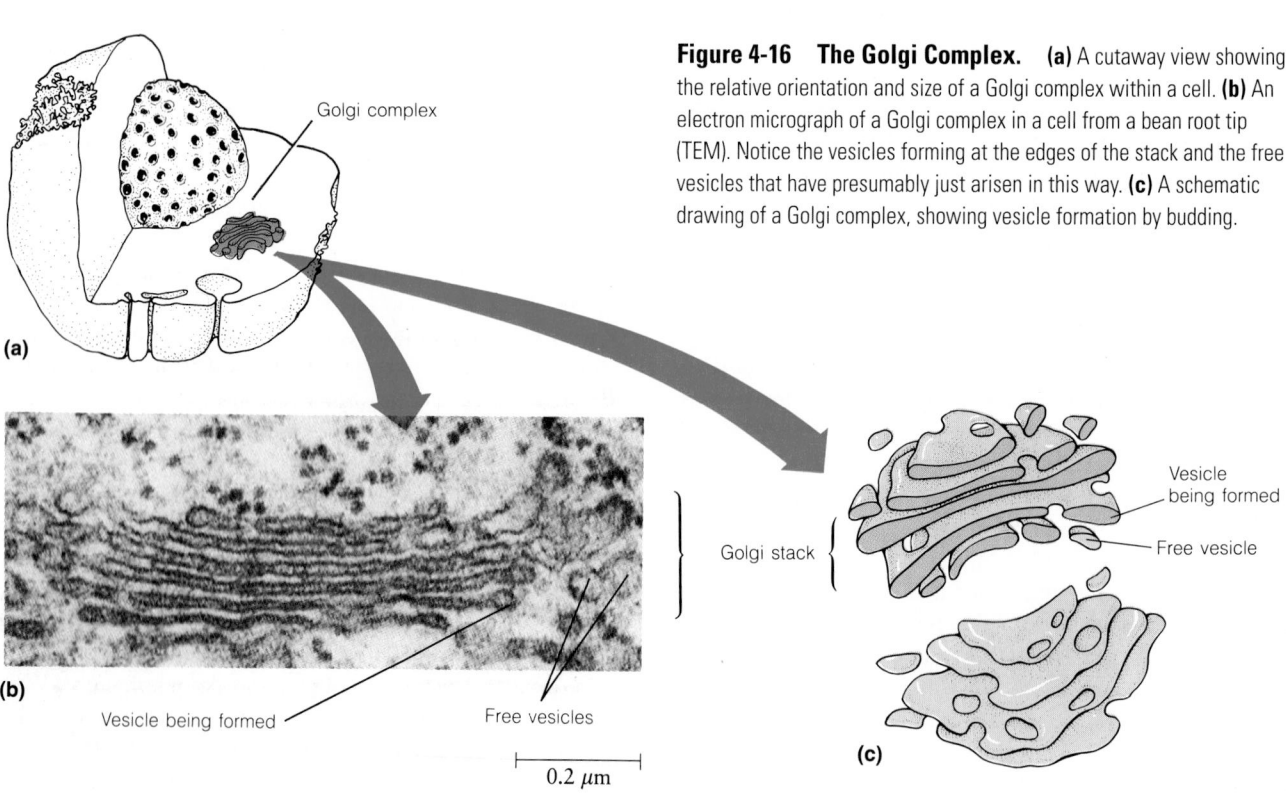

further modified and processed. The processed contents are then passed on to other components of the cell by means of vesicles that arise by budding off the Golgi complex (Figure 4-16c).

Most membrane proteins and secretory proteins are glycoproteins. The initial steps in *glycosylation* (sugar addition) take place within the lumen of the ER, but the process is usually completed within the Golgi complex. The Golgi complex should therefore be understood primarily as a processing station, with vesicles both fusing with it and arising from it. Almost everything that goes into it comes back out, but in a modified, packaged form, often ready for export from the cell.

Secretory Vesicles. Once processed by the Golgi complex, secretory proteins and other substances intended for export from the cell are packaged into **secretory vesicles.** The cells of your pancreas, for example, are likely to contain many such vesicles because the pancreas is responsible for the synthesis of several important digestive enzymes. These enzymes are synthesized on the rough ER, packaged by the Golgi complex, and then released from the cell via secretory vesicles, as shown in Figure 4-17. These vesicles move from the Golgi region to the plasma membrane that surrounds

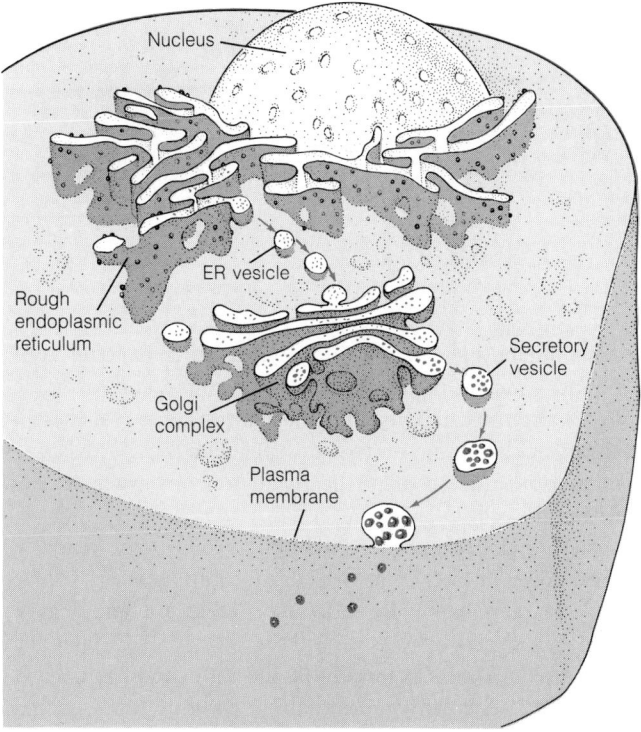

Figure 4-17 The Process of Secretion in Eukaryotic Cells.
Proteins to be packaged for export are synthesized on the rough ER, passed to the Golgi complex for processing, and eventually compartmentalized into secretory vesicles. These vesicles then make their way to the plasma membrane and fuse with it to release their contents to the exterior of the cell.

the cell. The vesicles then fuse with the plasma membrane and discharge their contents to the exterior of the cell by the process of exocytosis. The whole process of protein synthesis, processing, and export via the ER, the Golgi complex, and secretory vesicles will be considered in more detail in Chapter 9.

The Lysosome. The next picture at our cellular exhibition is of the **lysosome,** an organelle about 0.5–1.0 μm in diameter and surrounded by a single membrane (Figure 4-18). Lysosomes were discovered in the early 1950s by Christian de Duve and his colleagues. The story of that discovery is recounted in Box 4A, both to underscore the significance of chance observations when they are made by the right people and to illustrate the importance of new techniques to the progress of science. In this case, the new technique was *differential centrifugation,* which allows cellular contents to be fractionated according to size and density. Use of this technique led de Duve and his colleagues to the realization that an acid phosphatase initially thought to be located in the mitochondrion was in fact associated with a class of particles that had never been reported before. Along with acid phosphatase, these organelles contained several other hydrolytic enzymes. Because of its apparent role in cellular lysis, de Duve gave the new organelle the name *lysosome.*

Lysosomes are used by the cell as a means of storing *hydrolases,* enzymes capable of digesting specific biological molecules such as proteins, carbohydrates, or fats. It is important for cells to possess such enzymes, both to digest food molecules that the cell may acquire from its environment and to break down cellular constituents that are no longer needed. But it is also essential that such enzymes be carefully sequestered until actually needed, lest they digest cellular components that were not scheduled for destruction.

We are considering the lysosome at this point in our overview because of its relationship to the ER and the Golgi complex. Lysosomal enzymes are somewhat similar to secretory proteins in their synthesis and packaging. They are thought to be synthesized on the rough ER and transported to the Golgi, probably in an inactive form to prevent unwanted digestion of the structures through which they pass. Lysosomes then develop by budding off the ends of the Golgi cisternae. The resulting organelle is a *primary lysosome*—a lysosome containing hydrolytic enzymes but not yet engaged in digestive activity.

To initiate the digestion process, a lysosome must encounter and fuse with a membrane-bounded vacuole containing food particles. Such fused vacuoles are called *secondary lysosomes.* The hydrolytic enzymes of the secondary lysosome are then free to digest the contents of the vacuole, breaking them down to smaller and smaller components. Eventually, the digestion products are small enough to pass through the membrane out into the cytosol of the cell, where they can be utilized for the synthesis of macromolecules—recycling at the cellular level!

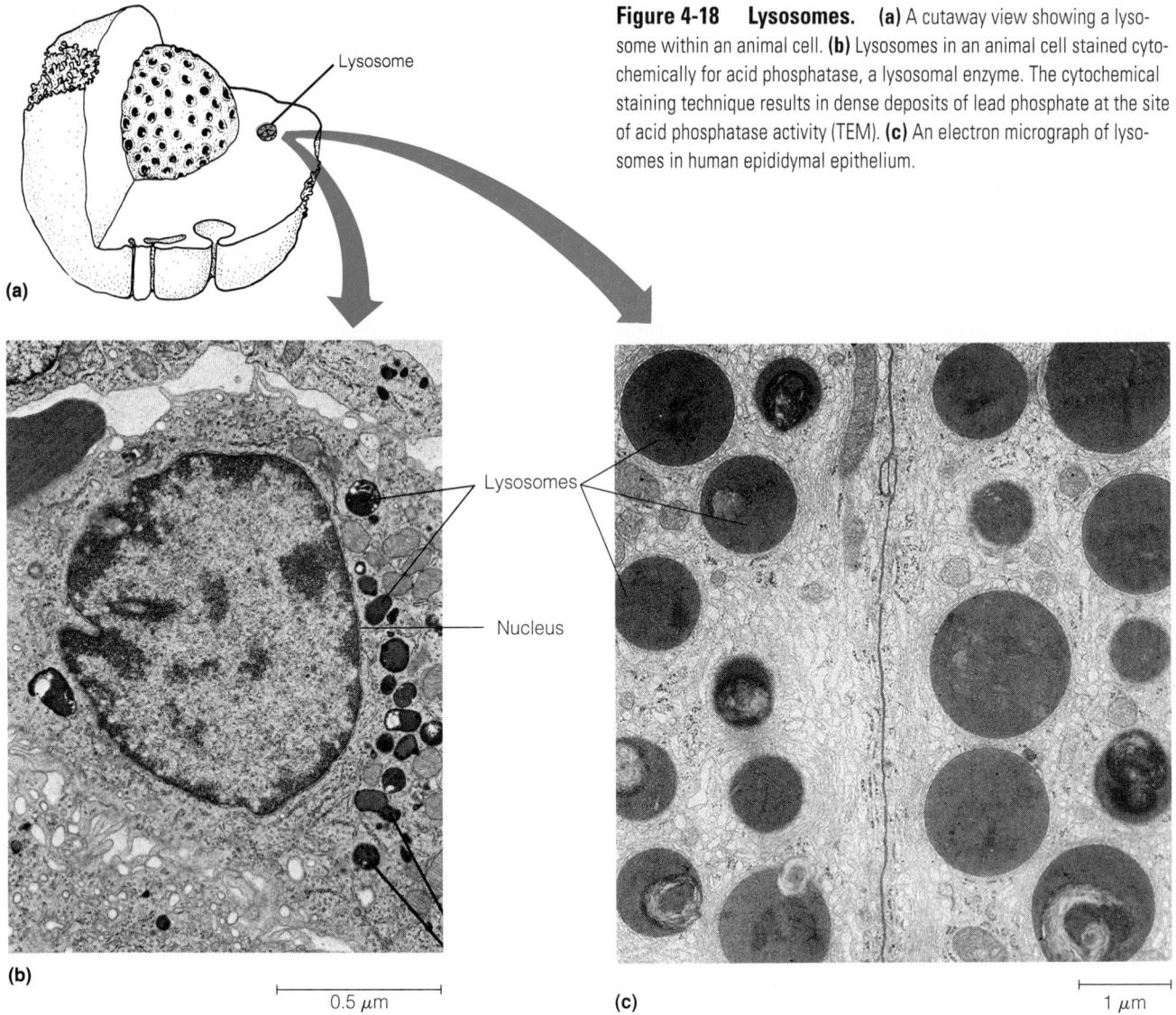

Lysosome

Figure 4-18   Lysosomes.   (a) A cutaway view showing a lysosome within an animal cell. **(b)** Lysosomes in an animal cell stained cytochemically for acid phosphatase, a lysosomal enzyme. The cytochemical staining technique results in dense deposits of lead phosphate at the site of acid phosphatase activity (TEM). **(c)** An electron micrograph of lysosomes in human epididymal epithelium.

(a)

Lysosomes

Nucleus

(b)

0.5 μm

(c)

1 μm

The Peroxisome.   The next organelle at our exhibition is the **peroxisome,** also called a *microbody.* Peroxisomes resemble lysosomes in size, mode of origin, and general lack of obvious internal structure. Like lysosomes, they are surrounded by a single, rather than a double, membrane. Peroxisomes are found in both plant and animal cells, as well as in fungi, protozoa, and algae. They perform several distinctive functions that differ with cell type but have the common property of both generating and degrading hydrogen peroxide (H_2O_2). Hydrogen peroxide is highly toxic to cells but can be decomposed into water and oxygen by the enzyme *catalase.* Eukaryotic cells protect themselves from the detrimental effects of hydrogen peroxide by packaging peroxide-generating reactions together with catalase in a single compartment, the peroxisome.

In animals, peroxisomes are found in most cell types but are especially prominent in liver and kidney cells (Figure 4-19). Beyond their role in the detoxification of hydrogen peroxide, animal peroxisomes may have several other func-

tions, including the detoxification of other harmful compounds (such as methanol, ethanol, formate, and formaldehyde) and the catabolism of unusual substances (such as D-amino acids). Some researchers speculate that peroxisomes may also be involved in the regulation of oxygen tension within the cell.

Animal peroxisomes also play a role in the oxidative breakdown of fatty acids, which are components of triglycerides and phosphoglycerides (see Table 3-4 and Figures 3-24, 3-25, and 3-26). As we will see in Chapter 12, fatty acid breakdown occurs primarily in the mitochondrion. However, fatty acids with more than 12 carbon atoms are oxidized relatively slowly by mitochondria. In the peroxisomes, on the other hand, fatty acids up to 22 carbon atoms long are oxidized rapidly. The long chains are degraded two carbon units at a time until they get to a length (10–12 carbon atoms) that can be handled efficiently by the mitochondria.

The vital role that peroxisomes play in breaking down long-chain fatty acids is underscored by the serious human

DISCOVERING ORGANELLES: THE IMPORTANCE OF CENTRIFUGES AND CHANCE OBSERVATIONS

Have you ever wondered how the various organelles within eukaryotic cells were discovered? There are almost as many answers to that question as there are kinds of organelles. In general, they were described by microscopists before their role in the cell was understood. As a result, the names of organelles usually reflect structural features rather than physiological roles. Thus, *chloroplast* simply means "green particle" and *endoplasmic reticulum* just means "network within the plasm (of the cell)."

Such is not the case for the *lysosome*, however. This organelle was the first to have its biochemical properties described before it had ever been reported by microscopists. Only after fractionation data had predicted the existence and properties of such an organelle were lysosomes actually observed in cells. A suggestion of its function is even inherent in the name given to the organelle, because the Greek root *lys-* means "to digest." (The literal meaning is "to loosen," but that's essentially what digestion does to chemical bonds!)

The lysosome is something of a newcomer on the cellular biology scene; it was not discovered until the early 1950s. The story of that discovery is fascinating because it illustrates how important chance observations can be, especially when made by the right people at the right time. The account also illustrates how significant new techniques can be, since the discovery depended on subcellular fractionation, a technique that at the time was still in its infancy.

The story begins in 1949 in the laboratory of Christian de Duve, who later received a Nobel Prize for this work. Like so many scientific advances, the discovery of lysosomes depended on a chance observation made by an astute investigator. Because of an interest in the effect of insulin on carbohydrate metabolism, de Duve was attempting at the time to pinpoint the cellular location of *glucose-6-phosphatase,* the enzyme responsible for the release of free glucose in liver cells. As a control enzyme (that is, one not involved in carbohydrate metabolism), de Duve happened to choose *acid phosphatase.*

De Duve first homogenized liver tissue and resolved it into several fractions by the new technique of *differential centrifugation,* which separates cellular components on the basis of differences in size and density. In this way, he was able to show that the glucose-6-phosphatase activity could be recovered with the microsomal fraction. (*Microsomes* are small vesicles that form from ER fragments when tissue is homogenized.) This in itself was an important observation because it helped establish the identity of microsomes, which at the time tended to be dismissed as fragments of mitochondria.

But the acid phosphatase results turned out to be still more interesting, even though they were at first quite puzzling. When de Duve and his colleagues assayed their liver homogenates for this enzyme, they found only a fraction of the expected activity. When assayed again for the same enzyme a few days later, however, the same homogenates had about ten times as much activity.

Speculating that he was dealing with some sort of activation phenomenon, de Duve subjected the homogenates to differential centrifugation to see with what subcellular fraction the phenomenon was associated. He and his colleagues were able to demonstrate that much of the acid phosphatase activity could be recovered in the mitochondrial fraction, and that this fraction showed an even greater increase in activity after standing a few days than did the original homogenates.

To their surprise, they then discovered that upon recentrifugation, this elevated activity no longer sedimented with the mitochondria but stayed in the supernatant. They went on to show that the activity could be increased and the enzyme solubilized by a variety of treatments, including harsh grinding, freezing and thawing, and exposure to detergents or hypotonic conditions. From these results, de Duve concluded that the enzyme must be present in some sort of membrane-bounded particle that could easily be ruptured to release the enzyme. Apparently, the enzyme could not be detected within the particle, probably because the membrane was not permeable to the substrates used in the enzyme assay.

Assuming that particle to be the mitochondrion, they continued to isolate and study this fraction of their liver homegenates. At this point, another chance observation occurred, this time because of a broken centrifuge. The unexpected breakdown forced one of de Duve's students to use an older, slower centrifuge, and the result was a mitochondrial fraction that had little or no acid phosphatase in it. Based on this unexpected finding, de Duve speculated that the mitochondrial fraction as they usually prepared it might in fact contain two kinds of organelles: the actual mitochondria, which could be sedimented with either centrifuge, and some sort of more slowly sedimenting particle that came down only in the faster centrifuge.

This led them to devise a fractionation scheme that allowed the original mitochondrial fraction to be subdivided into a rapidly sedimenting component and a slowly sedimenting component. As you might guess, the rapidly sedimenting component contained the mitochondria, as evidenced by the presence of enzymes known to be mitochondrial markers. The acid phosphatase, on the other hand, was in the slowly sedimenting component, along with several other hydrolytic enzymes, including ribonuclease, deoxyribonuclease, β-glucuronidase, and a protease. Each of these enzymes showed the same characteristic of increased activity upon membrane rupture, a property that de Duve termed *latency*.

By 1955, de Duve was convinced that these hydrolytic enzymes were packaged together in a previously undescribed organelle. In keeping with his speculation that this organelle was involved in intracellular lysis, he called it a *lysosome*.

Thus, the lysosome became the first organelle to be identified entirely on biochemical criteria. At the time, no such particles had been described by microscopy. But when de Duve's lysosome-containing fractions were examined with the electron microscope, they were found to contain membrane-bounded vesicles that were clearly not mitochondria and were in fact absent from the mitochondrial fraction. Knowing what the isolated particles looked like, microscopists were then able to search for them in fixed tissue. As a result, lysosomes were soon identified and reported in a variety of animal tissues. Within six years, then, the organelle that began as a puzzling observation in an insulin experiment became established as a bona fide feature of most animal cells. ■

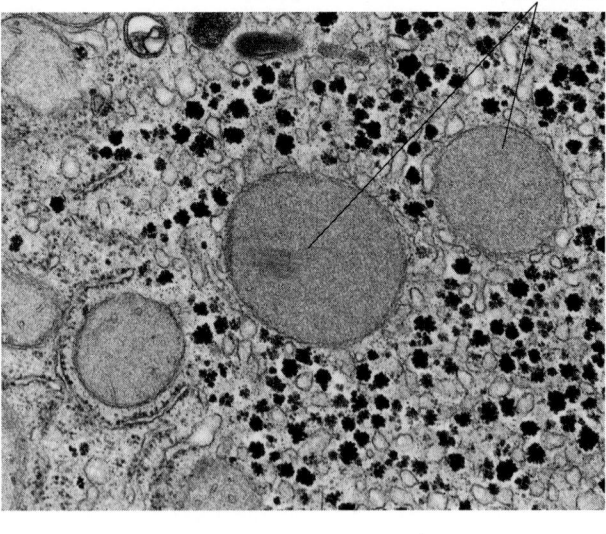

Peroxisomes

1 μm

Figure 4-19 Animal Peroxisomes. Several peroxisomes can be seen in this cross-section of a liver cell (TEM). Peroxisomes are found in most animal cells but are especially prominent features of liver and kidney cells.

diseases that result when one or more peroxisomal enzymes involved in fatty acid degradation are defective or absent. One such disease is adrenoleukodystrophy (ALD), a sex-linked (male-only) disorder that leads to profound neurological debilitation and eventually to death. ALD is the disease that afflicted Lorenzo Odone, the boy in the motion picture *Lorenzo's Oil*.

The best-understood metabolic roles of peroxisomes occur in plant cells. During the germination of fat-storing seeds, specialized peroxisomes called **glyoxysomes** play a key role in the conversion of stored fat into carbohydrate. In photosynthetic tissue, **leaf peroxisomes** are prominent because of their role in *photorespiration*, a light-dependent pathway that detracts from the efficiency of photosynthesis in many plants by "unfixing" some of the carbon that is fixed by the chloroplasts (to be discussed further in Chapter 13). The photorespiratory pathway is an example of a cellular process that involves several organelles. Some of the enzymes that catalyze the reactions in this sequence occur in the peroxisome, whereas others are located in the chloroplast or the mitochondrion. This mutual involvement in a common cellular process is suggested by the intimate association of peroxisomes with mitochondria and chloroplasts in many leaf cells, as shown in Figure 4-20.

Vacuoles. Cells also contain a variety of other membrane-bounded organelles called **vacuoles.** In animal cells, vacuoles are frequently used for temporary storage or transport. Some protozoa, for example, take up food particles or other materials from their environment by a process called *phagocytosis* ("cell eating"). Phagocytosis is a form of endocytosis that involves an inpocketing of the plasma mem-

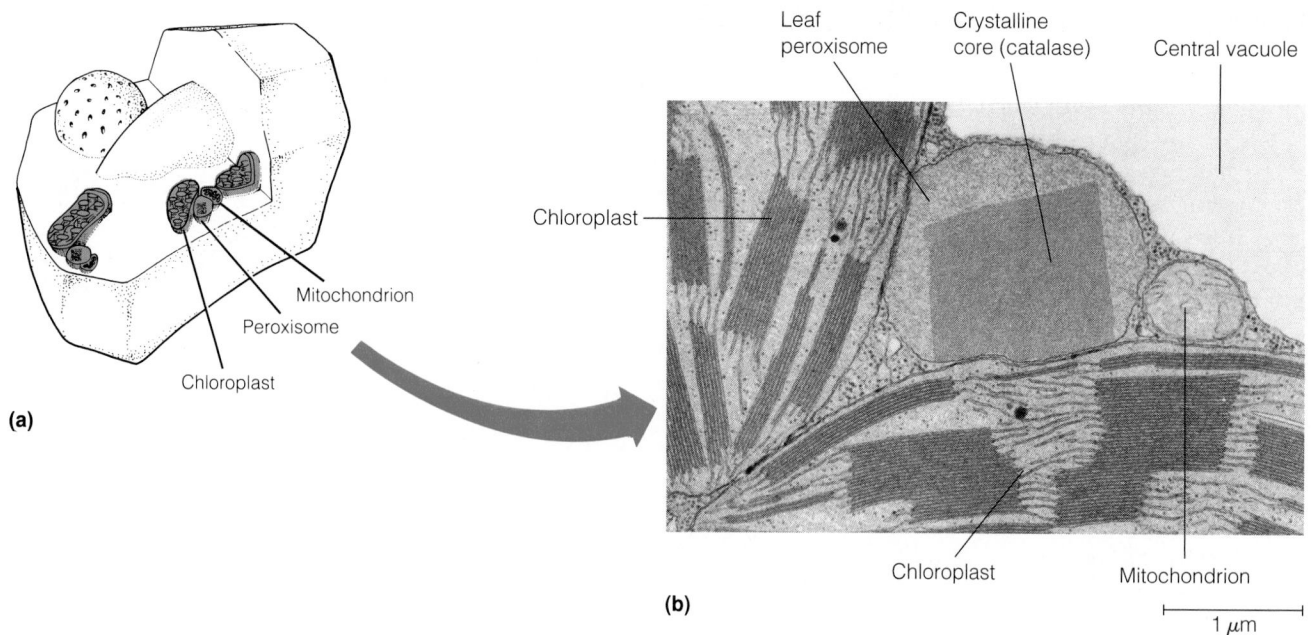

Figure 4-20   A Leaf Peroxisome and Its Relationship to Other Organelles in a Plant Cell. **(a)** A cutaway view showing a peroxisome, a mitochondrion, and a chloroplast within a plant cell. **(b)** A peroxisome in close proximity to chloroplasts and to a mitochondrion within a tobacco leaf cell (TEM). This is probably a functional relationship because all three organelles participate in the process of photorespiration. The crystalline core frequently observed in leaf peroxisomes is the enzyme catalase, which catalyzes the decomposition of hydrogen peroxide into water and oxygen.

brane around the desired substance, followed by a pinching-off process that internalizes the membrane-bounded particle as a vacuole.

The term *vacuole* is also used with plant cells, but usually in reference to the large **central vacuole** that characterizes most mature plant cells (Figure 4-21). Although the central vacuole may play a limited role in storage and appears also to be capable of a lysosomelike function in intracellular digestion, its real importance lies in the maintenance of the *turgor pressure* of the plant cell.

As you already know, a plant cell is surrounded by a rigid cell wall. The central vacuole is like an inflatable sphere in the center of the cell that is "pumped up" with liquid, pressing the rest of the cellular constituents out against the cell wall and thereby maintaining the turgor pressure characteristic of nonwilted plant tissue. The limp, flaccid appearance associated with wilting comes about when the central vacuole does not provide adequate pressure, allowing the cell (and hence the tissue) to go limp. We can easily demonstrate this by placing a piece of crisp celery in salt water. The high concentration of salt on the outside of the cells will cause water to move out of the cells; the turgor pressure will then decrease, and the tissue will quickly become flaccid and lose its crispness.

Ribosomes.   The last portrait in our gallery of organelles is the **ribosome,** which serves as the site of protein synthesis. Strictly speaking, a ribosome should not be considered an organelle because it is not bounded by a membrane. But it is convenient to include ribosomes at this time because ribosomes, like organelles, are the focal point for a specific cellular activity—in this case, protein synthesis. Unlike true organelles, ribosomes are found in both eukaryotic and prokaryotic cells. Even here, however, the dichotomy between the two basic cell types manifests itself in that prokaryotic and eukaryotic ribosomes differ characteristically in size and in the number and kinds of protein and RNA molecules they contain.

Compared to membrane-bounded organelles, ribosomes are tiny structures. The ribosomes of eukaryotic cells have a diameter of about 30 nm, and those of prokaryotic cells are slightly smaller. An electron microscope is therefore required to visualize ribosomes (see Figures 4-6 and 4-15). To appreciate how small ribosomes are, consider that more than 350,000 ribosomes could fit inside a typical bacterial cell, with room to spare!

Another way to express the size of such a small particle is to refer to its **sedimentation coefficient.** The sedimentation coefficient of a particle or macromolecule is a measure of how rapidly the particle sediments in an ultracentrifuge and is expressed in *Svedberg units (S)*. Sedimentation coefficients are widely used to indicate relative size, especially for large macromolecules such as proteins and nucleic acids and small particles such as ribosomes. Ribosomes from eukaryotic cells have sedimentation coefficients of about 80S; those from prokaryotic cells are about 70S (see Table 4-1).

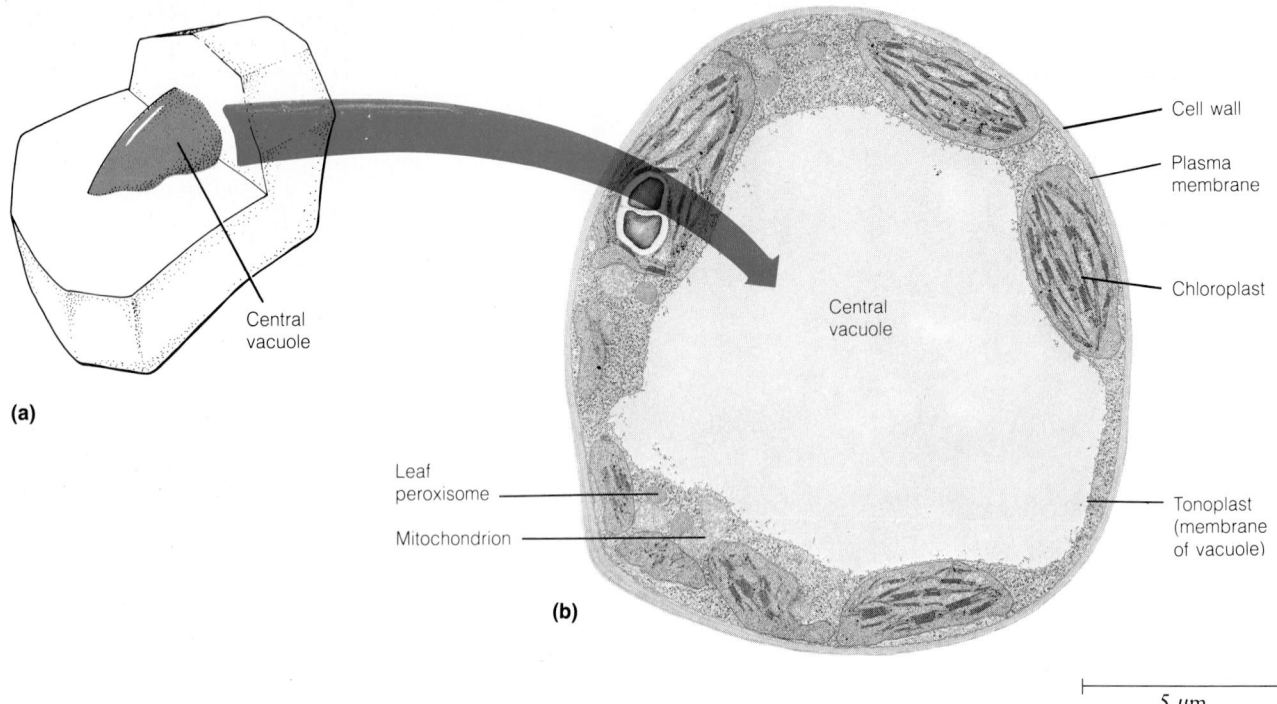

(a)

(b)

5 μm

Figure 4-21   The Central Vacuole in a Plant Cell.
(a) A cutaway view showing the central vacuole in a plant cell. **(b)** An electron micrograph of a bean leaf cell with a large central vacuole (TEM). The central vacuole occupies much of the internal volume of the cell, with the cytoplasm sandwiched into a thin sphere between the vacuole and the plasma membrane. The membrane of the vacuole is called the tonoplast.

A ribosome consists of two subunits differing in size, shape, and composition. In eukaryotic cells, the **large** and **small ribosomal subunits** have sedimentation coefficients of about 60S and 40S, respectively (Figure 4-22). For prokaryotic ribosomes, the corresponding values are about 50S and 30S. (Note that the sedimentation coefficients of the subunits do not add up to that of the intact ribosome; this is because sedimentation coefficients depend critically on both size and shape and are therefore not linearly related to molecular weight.) In both eukaryotic and prokaryotic cells, ribosomal subunits are synthesized and assembled separately in the cell but come together for the purpose of making proteins.

Ribosomes are far more numerous than most other cellular structures. Prokaryotic cells usually contain thousands of ribosomes, and eukaryotic cells may have hundreds of thousands or even millions of them. Ribosomes are also found in both chloroplasts and mitochondria, where they function in organelle-specific protein synthesis. Although the ribosomes of these eukaryotic organelles differ in size and composition from the ribosomes found in the cytoplasm of the same cell, they are strikingly similar to those found in bacteria and cyanobacteria.

This similarity is particularly striking when the nucleotide sequences of ribosomal RNA (rRNA) from mitochondria and chloroplasts are compared with those of prokaryotic rRNAs. These similarities support the **endosym-**

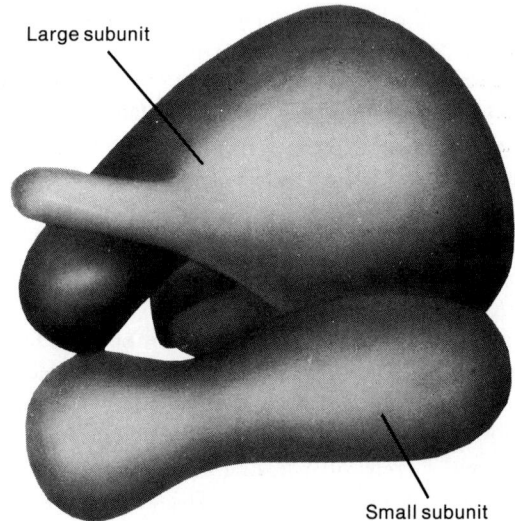

Figure 4-22   Structure of a Eukaryotic Ribosome.   Each ribosome is made up of a large subunit and a small subunit that join together when they attach to a messenger RNA and begin to make a protein. The large subunit of a eukaryotic ribosome consists of about 45 different protein molecules and 3 molecules of RNA. The small subunit has about 33 proteins and 1 molecule of RNA. The fully assembled ribosome of a eukaryotic cell has a diameter of about 30 nm. The ribosomes and ribo-somal subunits of prokaryotic cells are slightly smaller than those of eukaryotic cells and consist of their own distinctive protein and RNA molecules.

biont theory for the evolutionary origins of mitochondria and chloroplasts. This theory, developed most fully by Lynn Margulis, proposes that both of these organelles originated from prokaryotes that gained entry to, and established a symbiotic relationship within, the cytoplasm of ancestral cells (sometimes called *protoeukaryotic* cells) that eventually gave rise to present-day eukaryotic cells. (In biology, *symbiosis* involves the intimate living together of two organisms or cells for the mutual benefit of both.) Specifically, the endosymbiont theory postulates that both mitochondria and chloroplasts arose from specific prokaryotic ancestors—bacteria and cyanobacteria, respectively. Although still speculative, the endosymbiont theory is the most widely accepted explanation of the origin of eukaryotic organelles.

The Cytoplasm, Cytosol, and Cytoskeleton

The **cytoplasm** of a eukaryotic cell consists of that portion of the interior of the cell not occupied by the nucleus. Thus, the cytoplasm includes organelles such as the mitochondria; it also includes the **cytosol,** the semifluid substance in which the organelles are suspended. In a typical animal cell, the cytoplasm occupies more than half of the total internal volume of the cell. Many cellular activities take place in the cytoplasm, including the synthesis of proteins, the synthesis of fats, and the initial steps in the release of energy from sugars.

In the early days of cell biology, the cytosol was regarded as a rather amorphous, gel-like substance, and its proteins were thought to be soluble and freely diffusible. However, several new techniques have done much to change this view greatly. We now know that the cytosol of eukaryotic cells, far from being a structureless fluid, is permeated by an intricate three-dimensional array of interconnected filaments and tubules called the **cytoskeleton,** or the **cytoskeletal network** (Figure 4-23). Specifically, the cytoskeleton consists of a network of microtubules, microfilaments, and intermediate filaments, all of which are unique to eukaryotes.

As the name suggests, the cytoskeleton is an internal framework that gives a eukaryotic cell its distinctive shape and high level of internal organization. This elaborate array of filaments and tubules forms a highly structured yet very dynamic matrix that not only helps establish and maintain shape but also plays important roles in cell movement and cell division. As we will see in later chapters, the filaments and tubules that make up the cytoskeleton function in the contraction of muscle cells, the beating of cilia and flagella, the movement of chromosomes during cell division, and in some cases the locomotion of the cell itself.

In addition, the cytoskeleton serves as a framework for positioning and actively moving organelles within the cytosol. The same may be true of ribosomes and enzymes. Some researchers estimate that up to 80% of the proteins of the cytosol are not freely diffusible but are instead associated with the cytoskeleton. Even water, which accounts for about 70% of the cell volume, may be influenced by the cytoskeleton. It has been estimated that as much as 20–40% of the

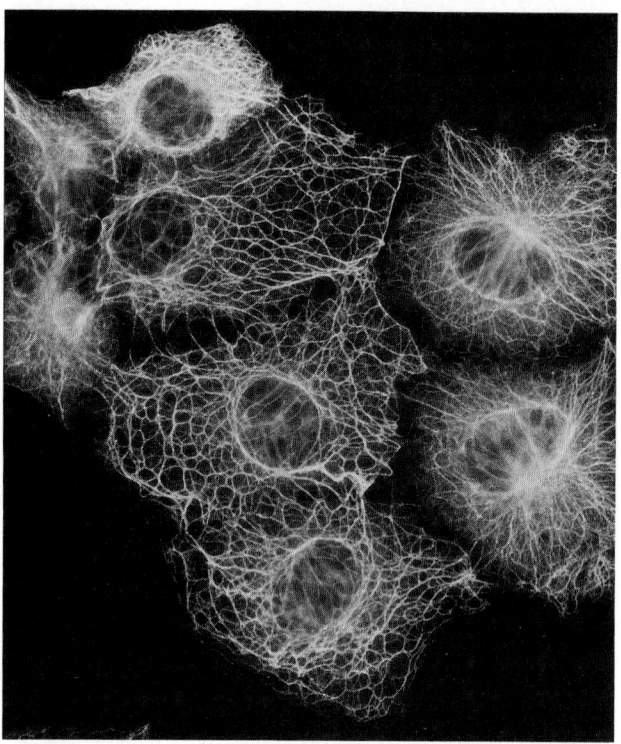

———| 25 μm |———

Figure 4-23 The Cytoskeleton. The cytoskeleton consists of a network of tubules and filaments that gives the cell shape, anchors organelles and directs their movement, and enables the whole cell to move or to change its shape. This micrograph shows the cytoskeleton of human amnion epithelial cells, as revealed by the technique of immunofluorescence microscopy using antibodies against keratin proteins found in the intermediate filaments of the cytoskeleton. Notice that these filaments surround, but do not extend into, the space occupied by the nucleus.

water in the cytosol may be bound to the filaments and tubules of the cytoskeleton.

The three major structural elements of the cytoskeleton are *microtubules, microfilaments,* and *intermediate filaments.* These structures occur only in eukaryotic cells. They can be visualized by phase-contrast and immunofluorescence microscopy and by electron microscopy. Some of the structures in which they are found (such as cilia, flagella, or muscle fibrils) can even be seen by ordinary light microscopy. Microfilaments and microtubules are best known for their roles in contraction and motility. In fact, these roles were appreciated well before it became clear that the same structural elements are also integral parts of the pervasive network of filaments and tubules that gives cells their characteristic shape and structure.

Chapter 20 provides a detailed description of the cytoskeleton, followed in Chapter 21 by a discussion of microtubule- and microfilament-mediated contraction and motility. We will also encounter microtubules and microfilaments in Chapter 15 because of their roles in chromosome separation and cell division, respectively. Here, we will focus on the structural features of the three major components of the cytoskeleton: microtubules, microfilaments, and intermediate filaments.

Microtubules. Of the structural elements found in the cytoskeleton, **microtubules (MTs)** are the largest. A well-known MT-based cellular structure is the *axoneme* of cilia and flagella, the appendages responsible for motility of eukaryotic cells. We have already encountered an example of such a structure; the axoneme of the sperm tail in Figure 4-12 consists of microtubules. MTs also form the *spindle fibers* that separate chromosomes prior to cell division, as we will see in Chapter 15.

In addition to their involvement in motility and chromosome movement, MTs also play an important role in the organization of the cytoplasm. They contribute to the polarity and overall shape of the cell, the spatial disposition of its organelles, and the distribution of microfilaments and intermediate filaments. Examples of the diverse phenomena that are governed by MTs include the asymmetric shapes of animal cells, the plane of cell division in plant cells, the ordering of filaments during muscle development, and the positioning of mitochondria around the axoneme of motile appendages.

As shown in Figure 4-24a, MTs are straight, hollow cylinders with an outer diameter of about 25 nm and an inner diameter of about 15 nm. The wall of the microtubule consists of longitudinal arrays of *protofilaments*, usually 13 arranged side by side around the hollow center, or *lumen*. Each protofilament is a linear polymer of *tubulin* molecules. Tubulin is a dimeric protein, consisting of two similar but distinct polypeptide subunits, *α-tubulin* and *β-tubulin*. All of the tubulin dimers in each of the protofilaments are oriented in the same direction, such that all of the subunits face

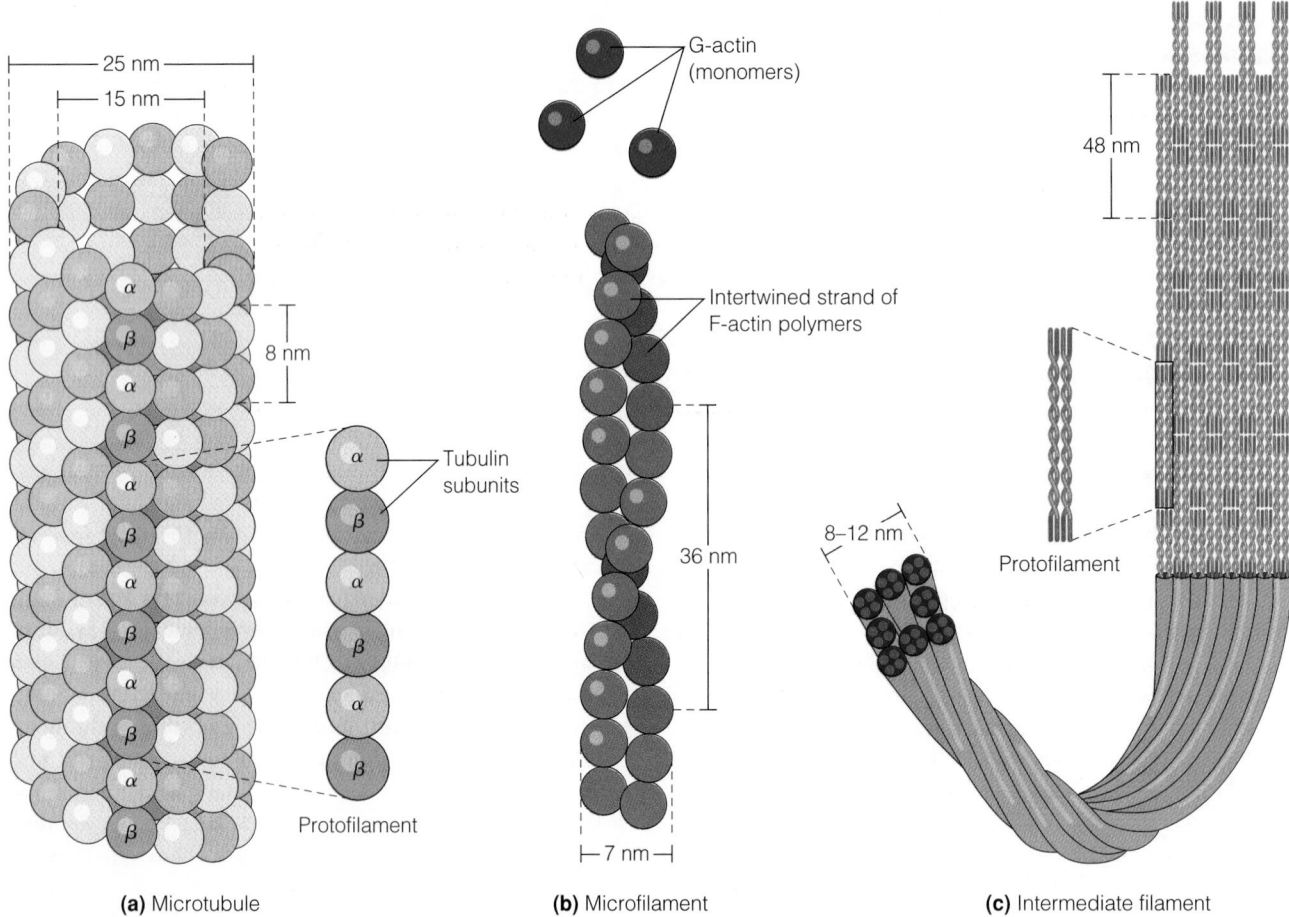

(a) Microtubule   **(b)** Microfilament   **(c)** Intermediate filament

Figure 4-24 Structures of Microtubules, Microfilaments, and Intermediate Filaments. **(a)** A schematic diagram of a microtubule, showing 13 protofilaments arrayed longitudinally forming a hollow cylinder with an outer diameter of about 25 nm and an inner diameter of about 15 nm. Each protofilament is a polymer of tubulin dimers, each about 8 nm long. All dimers are oriented in the same direction, thereby accounting for the polarity of the protofilament and hence of the whole microtubule. **(b)** A schematic diagram of a microfilament, showing a strand of F-actin twisted into a right-handed double helix with a diameter of about 7 nm. A half-turn of the helix occurs every 36 nm. The F-actin polymer consists of monomers of G-actin, all oriented in the same direction to give the microfilament its inherent polarity. **(c)** A schematic diagram of an intermediate filament. The structural unit is the tetrameric protofilament, consisting of two pairs of coiled polypeptides, with a length of about 48 nm. Protofilaments assemble by end-to-end and side-to-side alignment, forming an intermediate filament having a diameter of 8–12 nm that is thought to be eight protofilaments thick at any point.

the same end of the microtubule. This uniform orientation gives the MT an inherent *polarity*. The polarity of microtubules has important implications for their assembly, as we will see in Chapter 20.

Microfilaments. Microfilaments (MFs) are much thinner than microtubules. They have a diameter of about 7 nm, which makes them the smallest of the major cytoskeletal components. MFs are best known for their role in the contractile fibrils of muscle cells; we will meet them again in Chapter 21. However, microfilaments are involved in a variety of other cellular phenomena as well. They can form connections with the plasma membrane and thereby influence *locomotion, amoeboid movement,* and *cytoplasmic streaming,* a cyclic or back-and-forth flow of cytoplasm seen in a variety of algal, plant, and animal cells. MFs also produce the *cleavage furrows* that divide the cytoplasm of animal cells after chromosomes have been separated by the spindle fibers. In addition, MFs contribute importantly to the development and maintenance of cell shape.

MFs are polymers of the protein *actin* (Figure 4-24b). Actin is synthesized as a monomer called *G-actin* (G for globular). G-actin monomers polymerize reversibly into long, double-helical strands of *F-actin* (F for filamentous), each strand about 4 nm wide. A microfilament therefore consists of a strand of F-actin wrapped around itself to form a right-handed double helix with a diameter of about 7 nm (Figure 4-24b). Like microtubules, microfilaments are polar structures; all of the subunits are oriented in the same direction. This polarity influences the direction of MF elongation; assembly usually proceeds more readily at one end of the growing microfilament, whereas disassembly is favored at the other end.

Intermediate Filaments. Intermediate filaments (IFs) comprise the third structural element of the cytoskeleton. As their name suggests, IFs have a diameter of about 8–12 nm, larger than the diameter of microfilaments but smaller than that of microtubules. Intermediate filaments are the most stable and the least soluble constituents of the cytoskeleton. Because of this stability, some researchers regard IFs as a scaffold that supports the entire cytoskeletal framework. Intermediate filaments are also thought to have a tension-bearing role in some cells because they often occur in areas that are subject to mechanical stress.

In contrast to microtubules and microfilaments, intermediate filaments differ in their composition from tissue to tissue. Based on biochemical and immunological criteria, IFs from animal cells can be grouped into six classes. A specific cell type usually contains only one or sometimes two classes of IF proteins. Because of this tissue specificity, animal cells from different tissues can be distinguished on the basis of the IF proteins present. This *intermediate filament typing* serves as a diagnostic tool in medicine.

Despite their heterogeneity of size and chemical properties, all IF proteins share common structural features. They all have a central rodlike segment that is remarkably similar from one IF protein to the other. Flanking the central region

of the protein are N-terminal and C-terminal segments that differ greatly in size and sequence, presumably accounting for the functional diversity of these proteins.

Figure 4-24c shows a possible model for IF structure. The basic structural unit is a dimer of two intertwined IF polypeptides. Two such dimers align laterally to form a tetrameric *protofilament*. Protofilaments then interact with each other to form an intermediate filament that is thought to be eight protofilaments thick at any point, with protofilaments probably joined end to end in an overlapping manner.

Outside the Cell: The Extracellular Matrix and the Cell Wall

So far, we have considered the plasma membrane that surrounds every cell, the nucleus and cytoplasm within, and the variety of organelles, membrane systems, and filaments found in the cytoplasm of most eukaryotic cells. While it may seem that our tour of the cell is complete, most cells are also characterized by extracellular structures formed from materials that the cells transport outward across the plasma membrane. For animal cells, these structures are called the **extracellular matrix (ECM)** and consist primarily of *collagen fibrils* and polysaccharides called *proteoglycans*. For plant cells, the extracellular structure is the rigid **cell wall,** which consists mainly of *cellulose microfibrils* embedded in a matrix of other polysaccharides and small amounts of protein.

This difference between plant and animal cells is in keeping with the lifestyles of these two broad groups of eukaryotes. Plants are generally *nonmotile,* a lifestyle that is compatible with the rigidity cell walls confer on an organism. Animals, on the other hand, are usually *motile,* an essential feature for an organism that needs not only to find food but to escape becoming food for other organisms. Accordingly, their cells are not encased in rigid walls but are instead surrounded by an elastic network of collagen fibrils.

The extracellular matrix of animal cells forms a variety of structures, depending on the cell type. The primary function of the ECM is support, but the kinds of extracellular materials and the patterns in which they are deposited may regulate such diverse processes as cell motility and migration, cell division, cell recognition and adhesion, and cell differentiation during embryonic development. The main constituents of animal extracellular structures are the collagen fibers and the network of proteoglycans that surround them. The collagens are a group of generally insoluble glycoproteins that contain large amounts of glycine and the hydroxylated forms of lysine and proline. In vertebrates, collagen is such a prominent part of tendons, cartilage, and bone that it is the single most abundant protein of the animal body.

Figure 4-25 illustrates the prominence of the wall as a structural feature of a typical plant cell. Notice that neighboring cells, though separated by the wall, are actually connected by numerous cytoplasmic bridges, called **plasmodesmata** (singular: **plasmodesma**), which pass through the

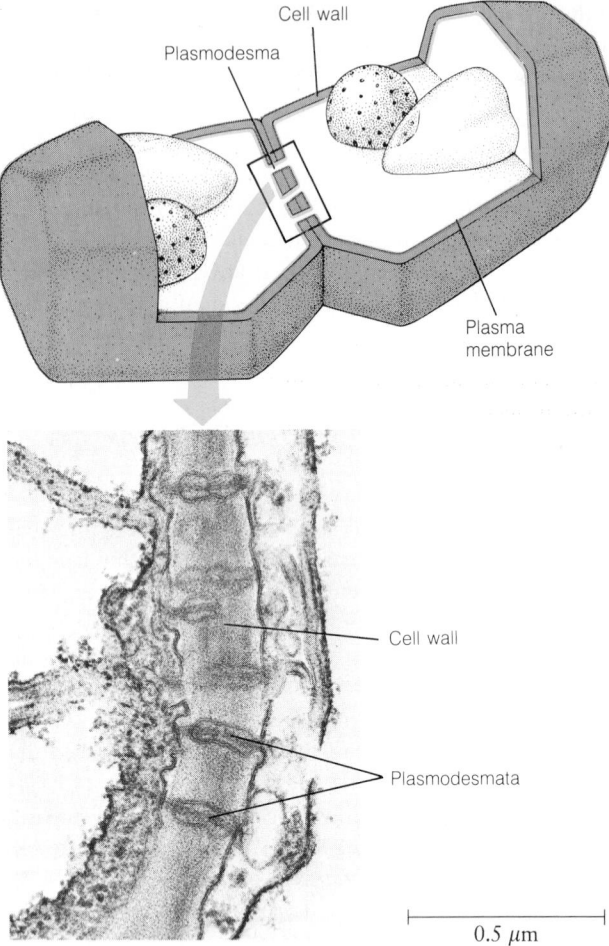

Figure 4-25   The Plant Cell Wall. The wall surrounding a plant cell consists of rigid microfibrils of cellulose embedded in a noncellulosic matrix of proteins and sugar polymers. Notice how the neighboring cells are connected by plasmodesmata (TEM).

wall. The plasma membranes of adjacent cells are continuous through each plasmodesma, such that the channel is membrane-lined. The diameter of a typical plasmodesma is large enough to allow water and small solutes to pass freely from cell to cell. Most of the cells of the plant are interconnected in this way.

Animal cells can also communicate with each other. But instead of plasmodesmata they have intercellular connections called **gap junctions** that are specialized for the transfer of material between the cytoplasms of adjacent cells. Two other types of intercellular junctions are also characteristic of animal cells. **Tight junctions** hold cells together so tightly that the transport of substances through the spaces between the cells is effectively blocked. **Desmosomes** also link adjacent cells, but for the purpose of connecting them tightly into sturdy yet flexible sheets. Each of these types of junctions is discussed in detail in Chapter 10.

Most prokaryotes are also surrounded by an extracellular structure called a cell wall. However, bacterial cell walls

consist not of cellulose but mainly of *peptidoglycans*. Peptidoglycans are built up from a backbone of repeating units of *N*-acetyglucosamine (GlcNAc) and *N*-acetylmuramic acid (MurNAc), amino sugars we encountered in Chapter 3 (see Figure 3-23). In addition, bacterial cell walls contain a variety of other constituents, including some that are unique to each of the major structural groups of bacteria identified by their response to the *Gram stain*. The Gram stain is named for Hans Christian Gram, the Danish bacteriologist who devised it. Bacterial cells are stained with a solution of crystal violet and iodine, then destained with alcohol or acetone. Gram-positive cells are resistant to destaining and remain deep blue in color, whereas gram-negative cells are rapidly decolorized.

Cell wall constituents unique to gram-negative bacteria include *lipopolysaccharides, lipoproteins,* and a class of proteins called *porins*. Compounds found only in gram-positive bacteria include the *teichoic acids,* which consist of three- or five-carbon alcohols (glycerol and ribitol, respectively) linked together by phosphate groups.

Viruses, Viroids, and Prions

Before concluding this preview of cellular biology, we need to look at several kinds of agents that invade cells, subverting normal cellular functions and often killing their unwilling hosts. These include the *viruses* and two other agents about which we know much less, the *viroids* and *prions*.

Viruses

Viruses are subcellular parasites that are incapable of a free-living existence but that invade and infect cells and redirect their synthetic machinery toward the production of more viruses. Viruses cannot carry on all of the functions required for independent existence and must therefore depend for most of their needs on the cells they invade.

Viruses are responsible for many diseases in humans, animals, and plants and are therefore important in their own right. They are also very significant research tools for cell and molecular biologists because they are much less complicated than cells. The *tobacco mosaic virus (TMV)* that we encountered in Chapter 2 (see Figures 2-21 and 2-22) is a good example of a virus that is important both economically and scientifically—economically because of the threat it poses to tobacco and other crop plants that it can infect, and scientifically because it is so amenable to laboratory study. Fraenkel-Conrat's studies on TMV self-assembly described in Chapter 2 are good examples of the usefulness of viruses to cell biologists.

Viruses are sometimes named for the diseases they cause. Poliovirus, influenza virus, herpes simplex virus, and TMV are several common examples. Other viruses have more cryptic laboratory names such as T4, Qβ, λ, or Epstein-Barr virus. Viruses that infect bacterial cells are called **bacteriophages,** or often just **phages** for short. Bacteriophages and other viruses will figure prominently in our discussion of molecular genetics in Part Four because of the striking genetic similarities between viruses and cells. In terms of the storage, expression, and transmission of genetic information, viruses seem to follow most of the same rules and use many of the same mechanisms that cells do, yet they are simpler than cells and much easier to manipulate. Viruses are therefore very useful in the study of genetics at the molecular level, a point to which we will return in Chapter 14. (For an advance glimpse into the importance of phages in genetic research, see Box 14A.)

Viruses range in size from about 25 nm to 300 nm. The smallest viruses are therefore about the size of a ribosome, whereas the largest ones are about one-quarter the diameter of a typical bacterial cell. Each virus has its own characteristic shape, as shown in Figure 4-26.

Despite their morphological diversity, viruses are chemically quite simple. Most viruses consist of little more than a *coat* (or *capsid*) of protein surrounding a *core* that contains one or more molecules of either RNA or DNA, depending on the type of virus. The simplest viruses, such as TMV, have a single nucleic acid molecule surrounded by a capsid consisting of proteins of a single type (see Figure 2-21). More complex viruses have cores that contain several nucleic acid molecules and capsids consisting of several (or even many) different kinds of proteins. Some viruses are surrounded by a membrane that is derived from the plasma membrane of the host cell in which the viral particles are made and assembled. Such viruses are called *enveloped viruses;* the virus that causes AIDS is an example.

The question is sometimes asked whether or not viruses are living. The answer depends crucially on what we mean by "living," and it is probably worth pondering only to the extent that it helps us more fully understand what viruses are—and what they are not. The most fundamental properties of living things are *motility, irritability* (perception of, and response to, environmental stimuli), and the *ability to reproduce.* Viruses clearly do not satisfy the first two criteria. Outside their host cells, viruses are inert and inactive. They can, in fact, be isolated and crystallized almost like a chemical compound. It is only in an appropriate host cell that a virus becomes functional, undergoing a cycle of synthesis and assembly that gives rise to more viruses.

Even the ability of viruses to reproduce has to be qualified carefully. A basic tenet of the cell theory is that cells arise only from preexisting cells, but this is not true of viruses. No virus can give rise to another virus by any sort of self-duplication process. Rather, the virus must subvert the metabolic and genetic machinery of the host cell, reprogramming it for synthesis of the proteins necessary to package the DNA or RNA molecules that arise by copying the genetic information of the parent virus.

It is only in a genetic sense that one can think of viruses as living at all. Another fundamental property of living things is the capability of specifying and directing the ge-

netic composition of progeny—an ability that viruses clearly possess. It is probably most helpful to think of viruses as "quasi-living," satisfying part but not all of the basic definition of life.

Viroids

As simple as viruses are, even simpler agents are known that can infect eukaryotic cells (though apparently not prokaryotic cells, as far as is presently known). The **viroids** found in some plant cells represent one class of such agents. Viroids are small, circular RNA molecules, rather like an RNA virus without its capsid. They are, in fact, the smallest known infectious agents. The RNA circles are only about 300–400 nucleotides long and are replicated in the host cell despite the fact that they do not code for any protein.

It is not yet known how viroids are transmitted from one host to another, since they apparently do not occur in a free form. Most likely, they pass from one plant cell to another only when the surfaces of adjacent cells are damaged so that there is no membrane barrier for the RNA molecules to cross.

Viroids are responsible for diseases of several crop plants, including potatoes and tobacco. A viroidal disease that has severe economic consequences is *cadang-cadang disease* of the coconut palm. It is not yet clear how viroids cause disease. They may enter the nucleus and interfere with the transcription of DNA into RNA. Alternatively, they may interfere with the subsequent processing required of most eukaryotic RNA transcripts.

Prions

Prions represent another class of infectious agents about which little is yet known. The term was coined to describe *proteinaceous infective particles* that are thought to be responsible for neurological diseases such as scrapie in sheep and goats and kuru in humans. *Scrapie* is so named because infected animals rub incessantly against trees or other objects, scraping off most of their wool in the process. *Kuru* is a degenerative disease of the central nervous system originally reported among native peoples in New Guinea. Patients with this or other prion-based diseases suffer initially from mild physical weakness and dementia, but the effects slowly become more severe, and the diseases are eventually fatal.

The proteinaceous particles to which the name refers are the only agents that have thus far been isolated from infected individuals. The best guess at present is that the prion protein is a variant product of a normal cellular gene, capable of subverting the transcription or processing of the normal mRNA product of that gene to produce more of the infectious (prion) form instead. Both the normal and variant forms of the prion protein are found on the surfaces of neurons, suggesting that the protein may somehow affect the receptors that detect nerve signals.

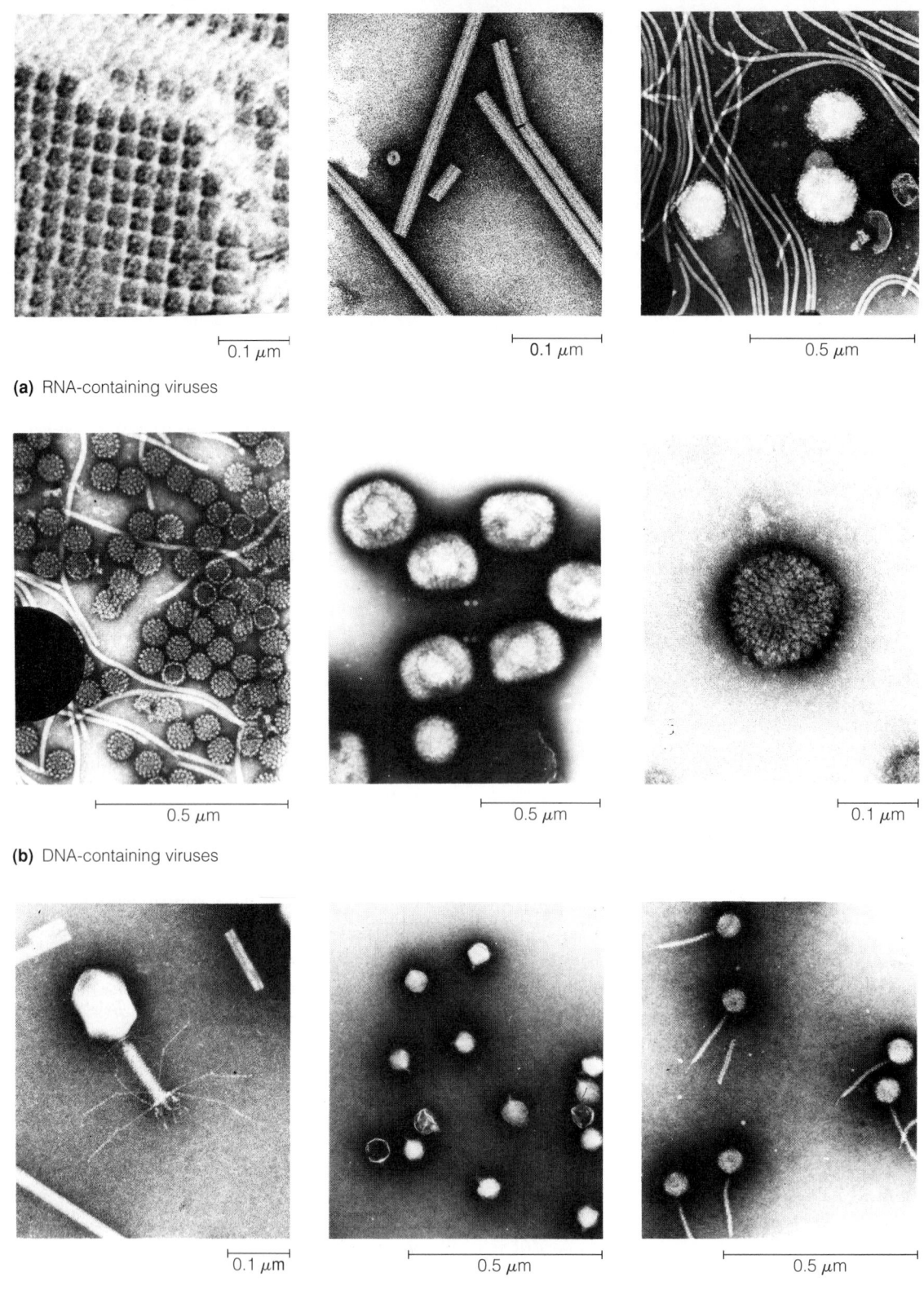

(a) RNA-containing viruses

(b) DNA-containing viruses

(c) DNA-containing bacteriophages

Figure 4-26 Sizes and Shapes of Viruses. These electron micrographs illustrate the morphological diversity of viruses. **(a)** RNA-containing viruses (left to right): polio, tobacco mosaic, and Rous sarcoma viruses. **(b)** DNA-containing viruses (left to right): papilloma, vaccinia, and herpes simplex viruses. **(c)** DNA-containing bacteriophages (left to right): T4, T7, and λ. (All TEMs.)

PERSPECTIVE

A fundamental distinction in biology is the one between the prokaryotes (eubacteria and archebacteria) and the eukaryotes (animals, plants, fungi, and protists). Prokaryotic cells are relatively small and structurally less complex than eukaryotic cells, lacking most of the internal membrane systems and organelles of the latter. A plasma membrane and ribosomes are the only two structural features common to both cell types. All other organelles are found only in eukaryotic cells, where they play indispensable roles in the compartmentalization of function.

Eukaryotic cells have at least four major structural features: a plasma membrane that defines the boundaries of the cell and retains its contents, a nucleus that houses the cell's DNA, membrane-bounded organelles, and the cytosol with its cytoskeleton of tubules and filaments. In addition, plant cells almost always have a rigid cell wall, and animal cells are usually surrounded by an extracellular matrix of collagen and proteoglycans.

The nucleus is surrounded by a double membrane called the nuclear envelope. The chromosomes within the nucleus contain most of the DNA of the cell.

Mitochondria play an important role in the oxidation of food molecules to release energy, which is used to make ATP by ADP phosphorylation. Chloroplasts trap solar energy and use it to "fix" carbon from carbon dioxide into organic form and convert it to sugar. Mitochondria and chloroplasts are surrounded by a double membrane and have an extensive system of internal membranes in which most of the components involved in ATP generation are embedded.

The endoplasmic reticulum is an extensive network of membranes that are either rough (studded with ribosomes) or smooth. Rough ER is responsible for the synthesis of secretory and membrane proteins, whereas smooth ER is involved in lipid synthesis and drug detoxification. Proteins synthesized on the rough ER are further processed and packaged in the Golgi complex and are then transported to the surface of the cell by secretory vesicles.

Lysosomes contain hydrolytic enzymes and are involved in cellular digestion. Peroxisomes are often about the same size as lysosomes but function in the generation and degradation of hydrogen peroxide. Animal peroxisomes play an important role in the catabolism of long-chain fatty acids. In plants, specialized peroxisomes are involved in the conversion of stored fat into carbohydrate during seed germination and in the process of photorespiration.

Although not membrane-bounded, ribosomes are usually considered organelles. Ribosomes serve as sites of protein synthesis in both prokaryotic and eukaryotic cells and also in mitochondria and chloroplasts. The striking similarities between mitochondrial and chloroplast ribosomes and those of bacteria and cyanobacteria, respectively, lend strong support to the endosymbiont theory that these organelles are of prokaryotic origin.

The cytoskeleton is an extensive network of microtubules, microfilaments, and intermediate filaments that gives eukaryotic cells their distinctive shapes. The cytoskeleton is also important in cellular motility and contractility, topics of later chapters.

Also to be encountered in later chapters are the viruses, which satisfy some, but not all, of the basic criteria of living things. Viruses are important both as infectious agents that cause diseases in humans, animals, and plants and as laboratory tools, particularly for geneticists. Viroids and prions are infectious agents that are even smaller (and less well understood) than viruses. Viroids are small RNA molecules, whereas prions are thought to be abnormal proteins of normal cellular genes.

Reference Box

For more detailed information about cellular structures, see the following chapters:

Plasma membrane	Chapter 7	Cytoskeleton	Chapter 20
Nucleus	Chapter 14	Cell wall and extracellular matrix	Chapter 10
Mitochondrion	Chapter 12	*For more detailed information about techniques used to study cellular structures, see the following chapters and the Appendix:*	
Chloroplast	Chapter 13		
Endoplasmic reticulum	Chapter 9	Differential centrifugation	Chapter 9
Golgi complex	Chapter 9	Equilibrium density centrifugation	Chapters 9 and 15
Lysosome	Chapter 9	Light microscopy	Appendix
Peroxisome	Chapter 9	Electron microscopy	Appendix
Ribosome	Chapter 18	Autoradiography	Chapter 13 and Appendix

KEY TERMS FOR SELF-TESTING

Properties and Strategies of Cells

surface area/volume ratio (p. 80)
compartmentalization (p. 81)
organelle (p. 82)
prokaryote (p. 82)
eukaryote (p. 82)
chromosome (p. 85)

The Plasma Membrane

plasma membrane (p. 88)
phospholipid bilayer (p. 89)

The Nucleus

nucleus (p. 89)
nucleoplasm (p. 89)
nuclear envelope (p. 90)
pore (p. 90)
chromatin (p. 90)
nucleolus (p. 90)

Intracellular Membranes and Organelles

mitochondrion (p. 90)
inner mitochondrial membrane (p. 90)
outer mitochondrial membrane (p. 90)
crista (p. 91)

matrix (p. 91)
chloroplast (p. 91)
thylakoid (p. 91)
stroma lamella (p. 91)
granum (p. 91)
stroma (p. 92)
plastid (p. 92)
endoplasmic reticulum (ER) (p. 93)
cisterna (p. 93)
lumen (p. 93)
rough endoplasmic reticulum (rough ER) (p. 93)
smooth endoplasmic reticulum (smooth ER) (p. 93)
Golgi complex (p. 93)
secretory vesicle (p. 95)
lysosome (p. 95)
peroxisome (p. 96)
glyoxysome (p. 98)
leaf peroxisome (p. 98)
vacuole (p. 98)
central vacuole (p. 99)
ribosome (p. 99)
sedimentation coefficient (p. 99)
large ribosomal subunit (p. 100)
small ribosomal subunit (p. 100)
endosymbiont theory (p. 100)

The Cytoplasm, Cytosol, and Cytoskeleton

cytoplasm (p. 101)
cytosol (p. 101)
cytoskeleton (cytoskeletal network) (p. 101)
microtubule (MT) (p. 102)
microfilament (MF) (p. 103)
intermediate filament (IF) (p. 103)

Outside the Cell: The Extracellular Matrix and the Cell Wall

extracellular matrix (ECM) (p. 103)
cell wall (p. 103)
plasmodesma (p. 103)
gap junction (p. 104)
tight junction (p. 104)
desmosome (p. 104)

Viruses, Viroids, and Prions

virus (p. 104)
bacteriophage (phage) (p. 105)
viroid (p. 105)
prion (p. 105)

PROBLEM SET

4-1. Prokaryotes and Eukaryotes. Indicate whether each of the following statements is true (T) or false (F). If false, reword the statement to make it true.

(a) Eukaryotic cells are in most cases larger than prokaryotic cells.

(b) Some cells are large enough to be seen with the unaided eye.

(c) Prokaryotic cells possess none of the following features: mitochondria, membrane-bounded nucleus, plasma membrane, microtubules.

(d) The surface area/volume ratio is generally greater for a prokaryotic cell than for a eukaryotic cell.

(e) The ribosomes found in the mitochondria of your muscle cells are more like those of the bacteria in your intestine than they are like the ribosomes in the cytoplasm of your muscle cells.

(f) Because prokaryotic cells have neither mitochondria nor chloroplasts, they cannot carry out either ATP synthesis or photosynthesis.

4-2. That's About the Size of It. To get some feeling for the differences in size of various cellular structures, it is useful to compare the structures on a macroscopic scale. Listed below is a variety of structures, with the approximate dimensions of each. To compare their dimensions on a macroscopic scale, assume that each struc-

ture has been magnified a millionfold, using a scale such that 1 nm is represented by 1 mm. On this scale, a prokaryotic ribosome has a diameter of 25 mm (about 1 in) and is therefore the size of a large marble. Convert each of the other dimensions to this macroscopic scale, and suggest a physical object that has approximately the same dimensions.

(a) Eukaryotic ribosome: 30 nm in diameter

(b) Microtubule: 25 nm $\times$ 1 μm

(c) Microfilament: 7 $\times$ 200 nm

(d) Peroxisome: 0.5 μm in diameter

(e) Mitochondrion: 1 $\times$ 2 μm

(f) Chloroplast: 2 $\times$ 8 μm

(g) Nucleus: 6 μm in diameter

(h) Liver cell: 20 μm in diameter

(i) Chicken egg: 4 $\times$ 6 cm

(j) Adult human: 1.8 m tall

4-3. Cellular Specialization. Each of the cell types listed here is a good example of a cell that is specialized for a specific function. Match each cell type in list A with the appropriate function from list B, and explain why you matched each as you did.

List A	List B
(a) Pancreatic cell	Cell division
(b) Cell from flight muscle	Absorption
(c) Palisade cell from leaf	Motility
(d) Cell of intestinal lining	Photosynthesis
(e) Nerve cell	Secretion
(f) Bacterial cell	Transmission of electrical impulses

4-4. Cellular Structure. Indicate whether each of the following cellular structures or components is found in animal cells (A), bacterial cells (B), and/or plant cells (P).

(a) Chloroplasts   (g) Golgi complex

(b) Cell wall   (h) Central vacuole

(c) Microtubules   (i) Thylakoids

(d) DNA   (j) Ribosomes

(e) Nuclear envelope   (k) Lipid bilayers

(f) Nucleoli   (l) Actin

4-5. Matching. For each of the structures in list A, choose the single term from list B that matches best, and explain your choice.

List A

(a) Plant cell wall   (f) Rough ER

(b) Cyanobacterium   (g) Smooth ER

(c) Lysosome   (h) Plasma membrane

(d) Nuclear envelope   (i) Virus

(e) Nucleolus   (j) Viroid

List B

Bacteriophage	Pores
Hydrolases	Ribosome synthesis
RNA	Glycoprotein
Cellulose	Eukaryote
Photorespiration	Svedberg
Lipid synthesis	Glyoxysomes
Muscle cells	Granum
Peptidoglycan	Interphase
Prokaryote	Secretory proteins

4-6. Sentence Completion. Complete each of the following statements about cellular structure in ten words or less.

(a) If you were shown an electron micrograph of a section of a cell and were asked to identify the cell as plant or animal, one thing you might do is . . .

(b) A slice of raw apple placed in a concentrated sugar solution will . . .

(c) A cellular structure that is visible with an electron microscope but not with a light microscope is . . .

(d) Ribosomes are not true organelles because . . .

(e) One reason why it might be difficult to separate lysosomes from peroxisomes by centrifugation techniques is that . . .

4-7. Telling Them Apart. Suggest a way to distinguish between the two elements in each of the following pairs.

(a) Bacterial cell vs. a protist of the same size

(b) Rough ER vs. smooth ER

(c) Animal peroxisome vs. leaf peroxisome

(d) Primary lysosome vs. secondary lysosome

(e) Virus vs. viroid

(f) Microfilament vs. intermediate filament

(g) Polio virus vs. herpes simplex virus

(h) Eukaryotic ribosome vs. prokaryotic ribosome

4-8. Structural Relationships. For each pair of structural elements, indicate with an A if the first element is a constituent part of the second, with a B if the second element is a constituent part of the first, and with an N if they are separate structures with no particular relationship to each other.

(a) Mitochondrion; crista

(b) Golgi complex; nucleus

(c) Cytoplasm; cytoskeleton

(d) Cell wall; extracellular matrix

(e) Nucleolus; nucleus

(f) Smooth ER; ribosome

(g) Lipid bilayer; plasma membrane

(h) Peroxisome; thylakoid

(i) Chloroplast; granum

4-9. Organellar Cooperation. For each pair of organelles or other structures, indicate a single cellular function in which both organelles play a role, and indicate (if possible) the role of each organelle in the function.

(a) Chloroplast; leaf peroxisome

(b) Rough ER; Golgi complex

(c) Nucleus; ribosome

(d) Mitochondrion; cytosol

(e) Secretory vesicle; plasma membrane

4-10. The Palisade Cell: A Look Inside. Just under the upper surface of many plant leaves is a layer of columnar *palisade cells,* the site of much photosynthesis. A typical palisade cell is cylindrical in shape, with a diameter of 20 μm and a length of 35 μm. In round numbers, such a cell might contain 200 mitochondria, 40 chloroplasts, 100 peroxisomes, 2 million ribosomes, and 1 nucleus. The dimensions of each of these organelles are given in Problem 4-2. Except for the region occupied by the nucleus, the cytoplasm of the palisade cell is restricted to a 2.5-μm layer just beneath the plasma membrane because of the central vacuole, which, like the cell itself, is roughly cylindrical in shape.

(a) Calculate the proportion of the total internal volume of the cell that is occupied by each of these populations of organelles. Consider the cell, the vacuole, the chloroplasts, and the mitochondria to be cylindrical in shape ($V = \pi r^2 h$) and the nucleus, peroxisomes, and ribosomes to be approximately spherical ($V = 4\pi r^3/3$).

(b) What proportion of the total internal volume of the cell is not accounted for by the named organelles? What other major structural features must be accommodated in this remaining cytoplasmic volume?

4-11. Protein Synthesis and Secretion. Although we will not encounter protein synthesis and secretion in detail until later chapters, you already have enough information about these processes to place in order the seven events that are now listed randomly. Order events 1–7 so that they represent the correct sequence of events corresponding to steps a–g, tracing a typical secretory protein from the initial transcription (readout) of the relevant genetic information in the nucleus to the eventual secretion of the protein from the cell by exocytosis.

Transcription → (a) → (b) → (c) →
(d) → (e) → (f) → (g) → Secretion

(1) The protein is partially glycosylated within the lumen of the ER.

(2) The secretory vesicle arrives at and fuses with the plasma membrane.

(3) The RNA transcript is transported from the nucleus to the cytoplasm.

(4) The final sugar groups are added to the protein in the Golgi complex.

(5) As the protein is synthesized, it passes across the ER membrane into the lumen of a cisterna.

(6) The enzyme is packaged into a secretory vesicle and released from the Golgi complex.

(7) The RNA message associates with a ribosome and begins synthesis of the desired protein on the surface of the rough ER.

4-12. Disorders at the Organelle Level. Each of the following medical problems involves a disorder in the function of an organelle or other cell structure. In each case, identify the organelle or structure involved, and indicate whether it is likely to be underactive or overactive.

(a) A girl inadvertently consumes cyanide and dies almost immediately because ATP production ceases.

(b) A boy is diagnosed with adrenoleukodystrophy (ALD), characterized by an inability of his body to break down very long-chain fatty acids.

(c) A smoker develops lung cancer and is told that the cause of the problem is a population of cells in her lungs that are undergoing mitosis at a much greater rate than is normal for lung cells.

(d) A young man learns that he is infertile because his sperm are nonmotile.

(e) A young child dies of Tay-Sachs disease because her cells lack the hydrolase that normally breaks down a membrane component called ganglioside G_{M2}, which therefore accumulated in the membranes of her brain.

SUGGESTED READING

Properties and Strategies of Cells

Bonner, J. T. *Cells and Societies.* Princeton, NJ: Princeton University Press, 1966.

de Duve, C., and H. Beaufay. A short history of tissue fractionation. *J. Cell Biol.* 91 (1981): 293s.

Dixon, B. *Power Unseen: How Microbes Rule the World.* New York: W. H. Freeman, 1994.

Fawcett, D. W. *The Cell: Its Organelles and Inclusions,* 2d ed. Philadelphia: Saunders, 1981.

Stanier, R., E. Adelberg, and J. Ingraham. *The Microbial World,* 4th ed. Englewood Cliffs, NJ: Prentice-Hall, 1976.

Thomas, L. *Lives of a Cell: Notes of a Biology Watcher.* New York: Viking Press, 1974.

Valentine, J. W. The evolution of multicellular plants and animals. *Sci. Amer.* 239 (September 1978): 140.

Watson, J. D., N. H. Hopkins, J. W. Roberts, J. A. Steitz, and A. M. Weiner. *Molecular Biology of the Gene,* 4th ed. Menlo Park, CA: Benjamin/Cummings, 1987.

Prokaryotic and Eukaryotic Cells

Angert, E. R., K. D. Clements, and N. R. Pace. The largest bacterium. *Nature* (1993): 239.

Carlile, M. Prokaryotes and eukaryotes: Strategies and successes. *Trends Biochem. Sci.* 7 (1982): 128.

Cavalier-Smith, T. The origin of eukaryotic and archaebacterial cells. *Ann. N.Y. Acad. Sci.* 503 (1987): 17.

Doolittle, W. F. Archaebacteria coming of age. *Trends Genet.* 1 (1985): 268.

Dyer, B. D., and R. Obar, eds. *The Origin of Eukaryotic Cells.* New York: Van Nostrand Reinhold, 1985.

Margulis, L. *Symbiosis in Cell Evolution.* New York: W. H. Freeman, 1981.

Margulis, L., and K. V. Schwartz. *Five Kingdoms.* New York: W. H. Freeman, 1982.

Sogin, M. L. Giants among the prokaryotes. *Nature* 362 (1993): 207.

Vidal, G. The oldest eukaryotic cells. *Sci. Amer.* 250 (February 1984): 48.

The Plasma Membrane

Bretscher, M. S. The molecules of the cell membrane. *Sci. Amer.* 253 (October 1985): 100.

Robertson, R. N. *The Lively Membranes.* New York: Cambridge University Press, 1983.

Yeagle, P. *The Membranes of Cells.* Orlando, FL: Academic Press, 1987.

The Nucleus

Hoffman, M. The cell's nucleus shapes up. *Science* 259 (1993): 1257.

Miller, O. L. The nucleolus, chromosomes, and visualization of genetic activity. *J. Cell Biol.* 91 (1981): 15.

Newport, J. W., and D. J. Forbes. The nucleus: Structure, function, and dynamics. *Annu. Rev. Biochem.* 56 (1987): 535.

Intracellular Membranes and Organelles

Bainton, D. The discovery of lysosomes. *J. Cell Biol.* 91 (1981): 66s.

Barinaga, M. Secrets of secretion revealed. *Science* 260 (1993): 487.

Bielka, H., ed. *The Eukaryotic Ribosome.* New York: Springer-Verlag, 1982.

Bogorad, L. Chloroplasts. *J. Cell Biol.* 91 (1981): 256s.

Bretscher, M. S., and M. C. Raff. Mammalian plasma membranes. *Nature* 258 (1975): 43.

de Duve, C. The lysosome. *Sci. Amer.* 208 (May 1973): 64.

de Duve, C. Microbodies in the living cell. *Sci. Amer.* 248 (May 1983): 74.

Dustin, P. Microtubules. *Sci. Amer.* 243 (August 1980): 66.

Ernster, L., and G. Schatz. Mitochondria: A historical review. *J. Cell Biol.* 91 (1981): 227s.

Farquhar, M., and G. Palade. The Golgi apparatus (complex)—(1945–1981): From artifact to center stage. *J. Cell Biol.* 91 (1981): 77s.

Fawcett, D. W. *The Cell: Its Organelles and Inclusions,* 2d ed. Philadelphia: Saunders, 1981.

Hoober, J. K. *Chloroplasts.* New York: Plenum, 1984.

Lake, J. A. The ribosome. *Sci. Amer.* 245 (August 1981): 84.

Novikoff, A. The endoplasmic reticulum: A cytochemist's view (a review). *Proc. Nat. Acad. Sci.* 73 (1976): 2781.

Palade, G. Intracellular aspects of the process of protein synthesis. *Science* 189 (1975): 347.

Roland, J. C., A. Szølløsi, and D. Szølløsi. *Atlas of Cell Biology.* Boston: Little, Brown, 1977.

Tolbert, N. E., and E. Essner. Microbodies: Peroxisomes and glyoxysomes. *J. Cell Biol.* 91 (1981): 271s.

Tzagoloff, A. *Mitochondria.* New York: Plenum, 1982.

Valle, D., and J. Gartner. Penetrating the peroxisome. *Nature* 361 (1993): 682.

Wolfe, S. L. *Cell Ultrastructure.* Belmont, CA: Wadsworth, 1985.

Wool, I. G. The structure and function of eukaryotic ribosomes. *Annu. Rev. Biochem.* 48 (1979): 719.

The Cytoplasm and the Cytoskeleton

Fulton, A. B. How crowded is the cytoplasm? *Cell* 30 (1980): 345.

Luby-Phelps, K., D. L. Taylor, and F. Lanni. Probing the structure of the cytoplasm. *J. Cell Biol.* 102 (1986): 2015.

Schliwa, M. *The Cytoskeleton.* New York: Springer-Verlag, 1986.

Outside the Cell: The Extracellular Matrix and the Cell Wall

Albersheim, P. The wall of growing plant cells. *Sci. Amer.* 232 (April 1975): 80.

Hay, E. D., ed. *Cell Biology of Extracellular Matrix.* New York: Plenum, 1982.

Hay, E. D. Extracellular matrix. *J. Cell Biol.* 91 (1981): 205s.

Luft, J. H. The structure and properties of the cell surface coat. *Int. Rev. Cytol.* 45 (1976): 291.

Viruses, Viroids, and Prions

Joklik, W. K., ed. *Virology,* 2d ed. Norwalk, CT: Appleton & Lange, 1985.

Prusiner, S. B. The prion diseases. *Sci. Amer.* 272 (January 1995): 48.

Reisner, D., and H. H. Gross. Viroids. *Annu. Rev. Biochem.* 54 (1985): 531.

Weissman, C. Prions: Sheep disease in human clothing. *Nature* 352 (1989): 298.

Weissman, C. A "unified theory" of prion propagation. *Nature* 352 (1991): 679.

5

BIOENERGETICS:
THE FLOW OF ENERGY IN THE CELL

Have you ever considered how downright improbable you are? Your body contains approximately 100 trillion cells, all integrated and functioning together as a harmonious whole. Each of those cells contains trillions of molecules, every one in turn made up of dozens, hundreds, or thousands of atoms, all carefully organized in a manner that is strikingly nonrandom. As you contemplate the grandeur of that organization, consider how utterly improbable it all is—how very unlikely that so many atoms of carbon, hydrogen, oxygen, nitrogen, and other elements would ever come together and be ordered into the molecules, structures, cells, tissues, and organs that make up your body.

What you are contemplating is the thermodynamic improbability that the order of the human body (or any other biological entity) could come into being spontaneously or, for that matter, could be maintained in such a highly ordered state once it had come into being. On the contrary, things in nature usually proceed from an ordered state to a less ordered one, not the other way around. For example, the carbon, hydrogen, and oxygen atoms of which this page is made tend to exist as simple inorganic molecules such as carbon dioxide and water instead of the complex cellulose microfibrils it takes to make paper. If you doubt that tendency, just touch a match to a piece of paper and you will quickly be reminded of how strong that tendency really is (Figure 5-1a)! Similarly, you know that a drop of dye solution placed in a beaker of water will spontaneously diffuse until the dye molecules are randomized in the beaker (Figure 5-1b).

In each of our examples, it is obvious that the reverse process is highly improbable. No matter how long you wait, it is extremely unlikely that molecules of carbon dioxide and water will spontaneously reassemble and regenerate paper, and it is equally improbable that the dye molecules in the beaker will spontaneously reconvene at some localized point once they have been dispersed throughout the solution. So, too, with your body and with every cell in it: Your body is a highly ordered structure composed of large numbers of atoms that could not reasonably be expected to assemble spontaneously.

What, then, does it take to bring about the highly improbable order of biological systems? What is it that allows such complex organization to happen? Or, to put it another way, why are you possible if you are so improbable?

The Importance of Energy

The answer to our question has two components: information and energy. You are already familiar with these two important concepts; we encountered a dual requirement for information and energy in our discussion of informational macromolecules in Chapter 3. Recall that the assembly of amino acids into polypeptides or nucleotides into nucleic acids requires both a source of energy (to activate the incoming monomeric unit) and a source of information (to specify which of several alternative subunits should be added at a specific point in the chain). The same is true for biological systems in general and for cells in particular. However improbable a structure may be because of its order, it can always be generated if sufficient energy and information are available.

Energy and information are, in other words, two indispensable prerequisites for the existence of life. Order can be brought about, maintained, and even extended in biological systems provided that adequate information and energy are available. The information is required to specify what form that order should take, and the energy is needed to drive the reactions and processes that lead to the order.

We can summarize the discussion thus far by saying that you are improbable because you are so highly ordered, but you are nonetheless possible because of the information available to you (in the DNA of your cells) and the copious quantities of energy at your disposal (in the bond energies of the food you eat).

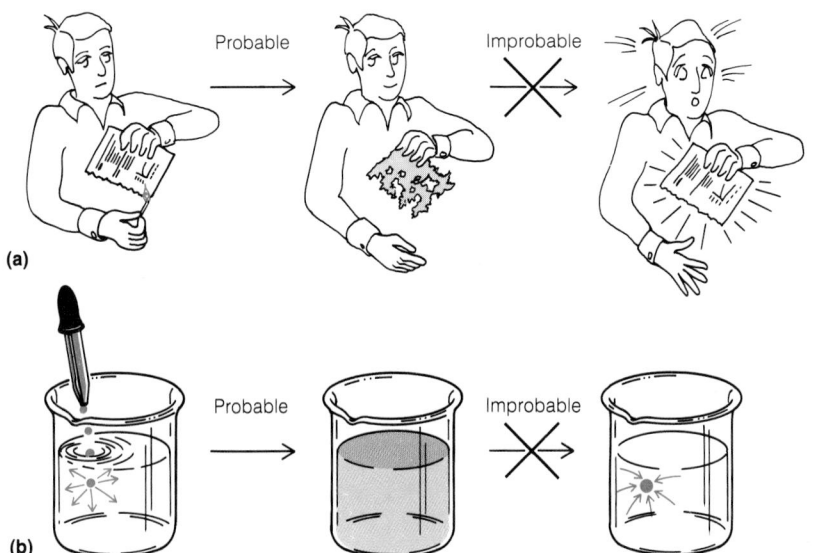

Figure 5-1   Probable and Improbable Processes.   Prior experience with familiar processes such as **(a)** the burning of paper or **(b)** the diffusion of dye molecules allows us to predict that such events are probable and take place spontaneously once initiated, while their reversal is highly improbable and never occurs spontaneously. Probable, spontaneous events are associated with the evolution of heat and/or with a greater randomness of the components of the system.

Considering their importance to cells, it is not surprising that the topics of *information* and *energy* are given considerable attention by cell biologists. Each is treated as a major theme in this text. Part Three deals with energy flow in detail, and Part Four is devoted to the topic of information flow.

The Need for Energy

All living systems require an ongoing supply of energy. Before discussing why cells need energy, it might be useful to consider what we mean by energy. Usually, energy is defined as the capacity to do work. But that turns out to be a somewhat circular definition because work is frequently defined in terms of energy changes. A more useful definition would be that **energy** *is the ability to cause specific changes.* Since life is characterized first and foremost by change, this definition underscores the total dependence of all forms of life on the continuous availability of energy.

Now that we have defined energy in this way, we recognize that asking about cellular needs for energy really means inquiring into the kinds of changes that cells must effect, that is, the cellular activities that give rise to change. Six categories of change come to mind (Figure 5-2).

Synthetic Work (Changes in Chemical Bonds).   An important activity of virtually every cell at all times is the work of **biosynthesis,** which results in the formation of new bonds and the generation of new molecules. This activity is especially obvious in a population of growing cells, where it can be shown that additional molecules are being synthesized if the cells are increasing in size or number or both. But **synthetic work** is required to maintain structures just as surely as it is needed to generate them originally. Most existing structural components of the cell are in a state of constant *turnover.* The molecules that make up the structure are continuously being degraded and replaced.

Why might cells expend energy to synthesize large molecules, only to break them up again in the turnover process? The answer has to do with molecular damage. As discussed earlier, larger molecules tend to be inherently unstable, and both physical and chemical processes are continuously damaging proteins and nucleic acids. For example, exposure to the ultraviolet component of sunlight can damage the DNA of skin cells. If the damage is too great, the cells cannot recover, and sunburn results. The skin renews itself by "turning over" the dead cells, sloughing them off and replacing them with new tissues. Something similar happens at the molecular level, except that the damaged molecules are broken down by specialized enzymes, then replaced by the ongoing synthetic processes.

In terms of the hierarchy of cellular structure shown in Figure 2-14, almost all of the energy that cells require for biosynthetic work is used to make energy-rich organic molecules from simpler starting materials and to activate such organic molecules for incorporation into macromolecules. Higher levels of structural complexity usually occur by spontaneous self-assembly, without further energy input. Of the two energy-requiring levels, synthesis of the small organic molecules actually requires much more energy than their subsequent polymerization into macromolecules. As we will see in Chapter 13, it takes 18 molecules of ATP (and another 12 molecules of a reduced coenzyme called nicotinamide adenine dinucleotide phosphate) to make one molecule of glucose out of carbon dioxide and water in a photosynthetic cell. Thereafter, only two additional high-energy bonds are required to link the glucose onto the growing chain of a starch molecule. Therefore, the work of biosynthesis is essentially the work of fabricating sugars, amino acids, nucleotides, and lipid molecules from simpler starting compounds.

Figure 5-2 Several Kinds of Biological Work. The six major categories of biological work are shown here. **(a)** Synthetic work is illustrated by the process of photosynthesis, **(b)** mechanical work by the contraction of a weight lifter's muscles, and **(c)** concentration work by the uptake of molecules into a cell against a concentration gradient. **(d)** Electrical work is represented by the membrane potential of mitochondria (shown being generated by active proton transport), **(e)** heat production is illustrated in skunk cabbage, a plant that melts its way through spring snow, and **(f)** bioluminescence is depicted by the courtship of fireflies.

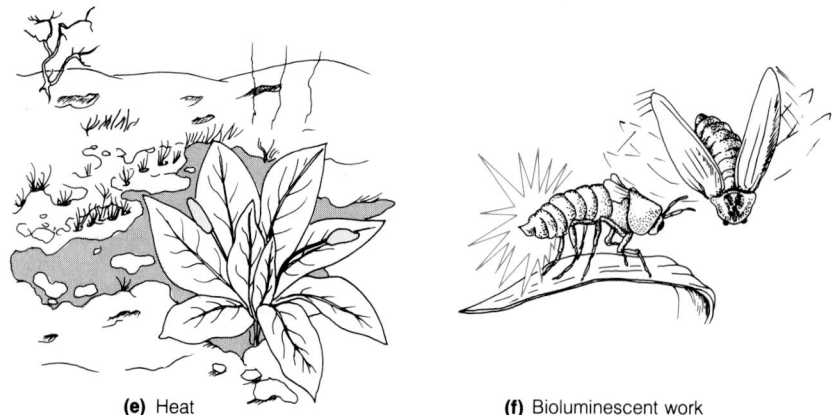

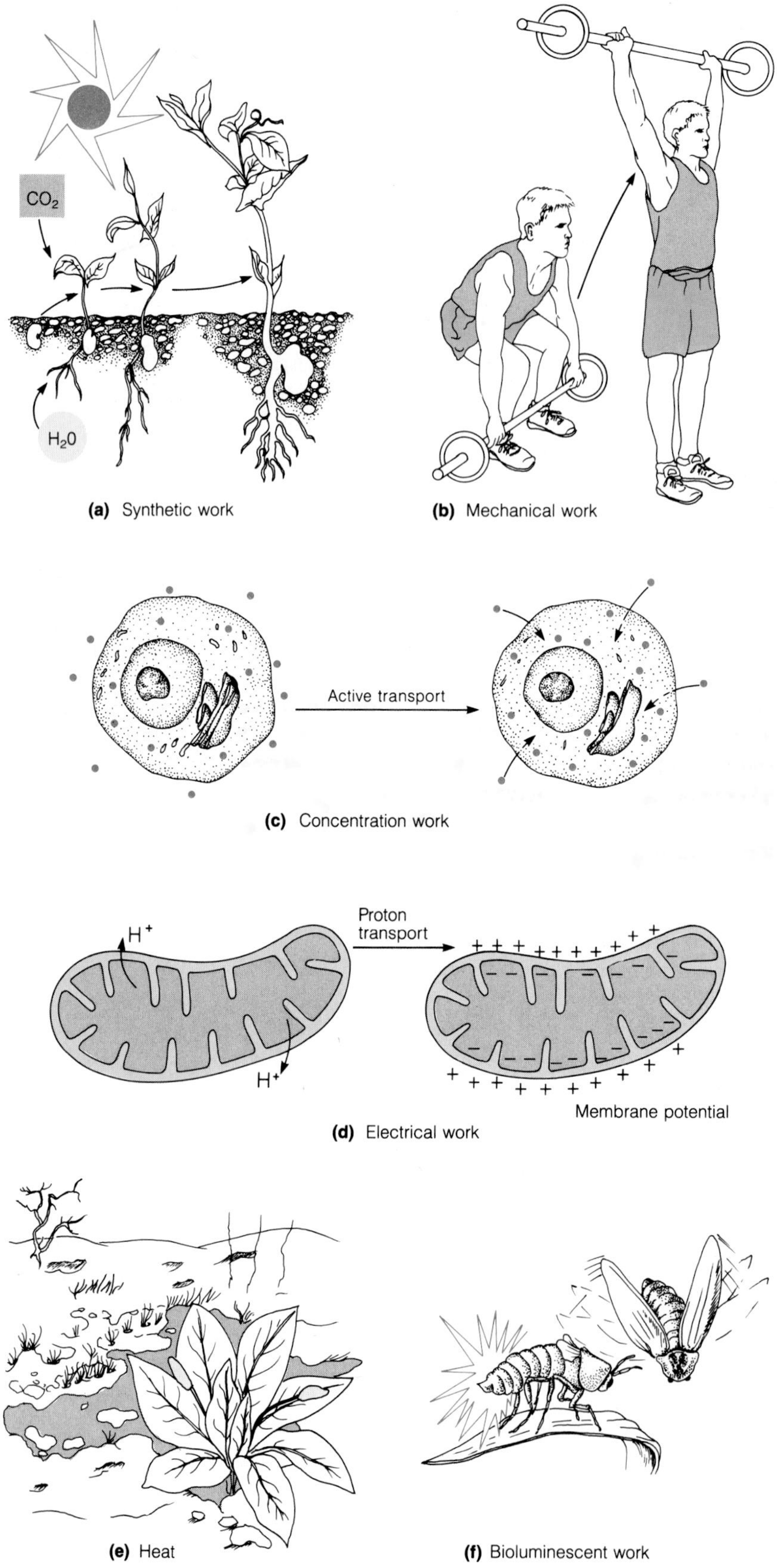

CO_2

H_2O

(a) Synthetic work

(b) Mechanical work

Active transport

(c) Concentration work

H^+

Proton transport

H^+

Membrane potential

(d) Electrical work

(e) Heat

(f) Bioluminescent work

Mechanical Work (Changes in Location or Orientation). **Mechanical work** involves a physical change in the position or orientation of a cell or some part of it. An especially good example is the movement of a cell with respect to its environment. This movement requires the presence of some sort of motile appendage such as a flagellum or a cilium. Many prokaryotic cells propel themselves through the environment, as in the case of the flagellated bacterium in Figure 5-3. Sometimes, however, it is the environment that moves past the cell, as when the ciliated cells that line your trachea beat upward to sweep inhaled particles back to the mouth or nose, thus protecting the lungs. Muscle contraction is another good example of mechanical work, involving not just a single cell but a large number of muscle cells (Figure 5-4). Other examples of mechanical work that occur within the cell include the movement of chromosomes along the spindle fibers during mitosis, the streaming of cytoplasm, and the movement of a ribosome along a strand of messenger RNA.

Concentration Work (Changes in Concentration Across Membranes). Less conspicuous than either of the previous two categories but every bit as important to the cell is the work of moving molecules or ions against a concentration gradient. In fact, in a resting state, about two-thirds of your energy consumption is used for the ongoing work of concentrating ions across membranes. The purpose of **concentration work** is either to accumulate substances within a cell or subcellular compartment, or to remove by-products of cellular activity that cannot be further utilized by the cell and indeed might be toxic if allowed to build up in the cell or compartment. Examples of concentration work include

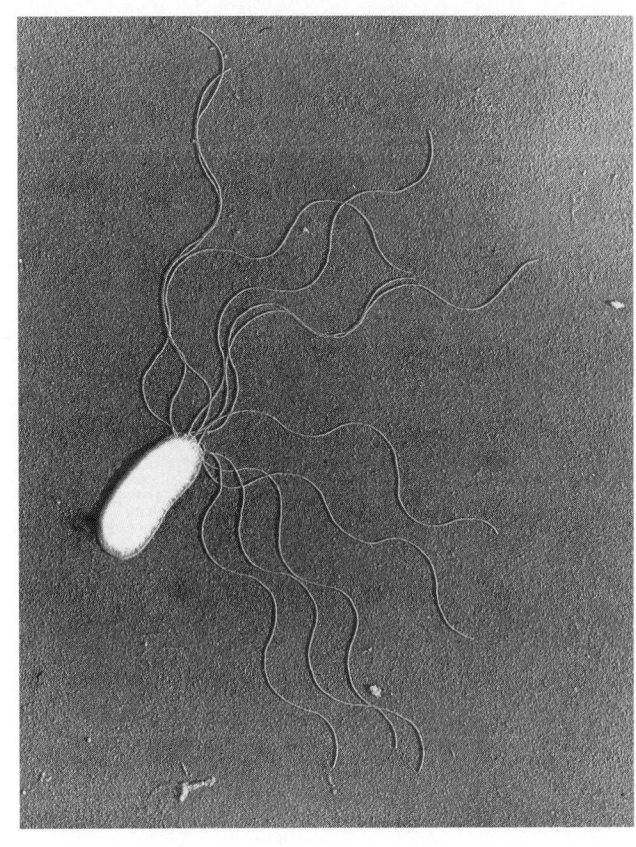

1 μm

Figure 5-3 A Flagellated Bacterium. The whipping motion of bacterial flagella is an example of mechanical work, providing motility for certain bacterial species (TEM).

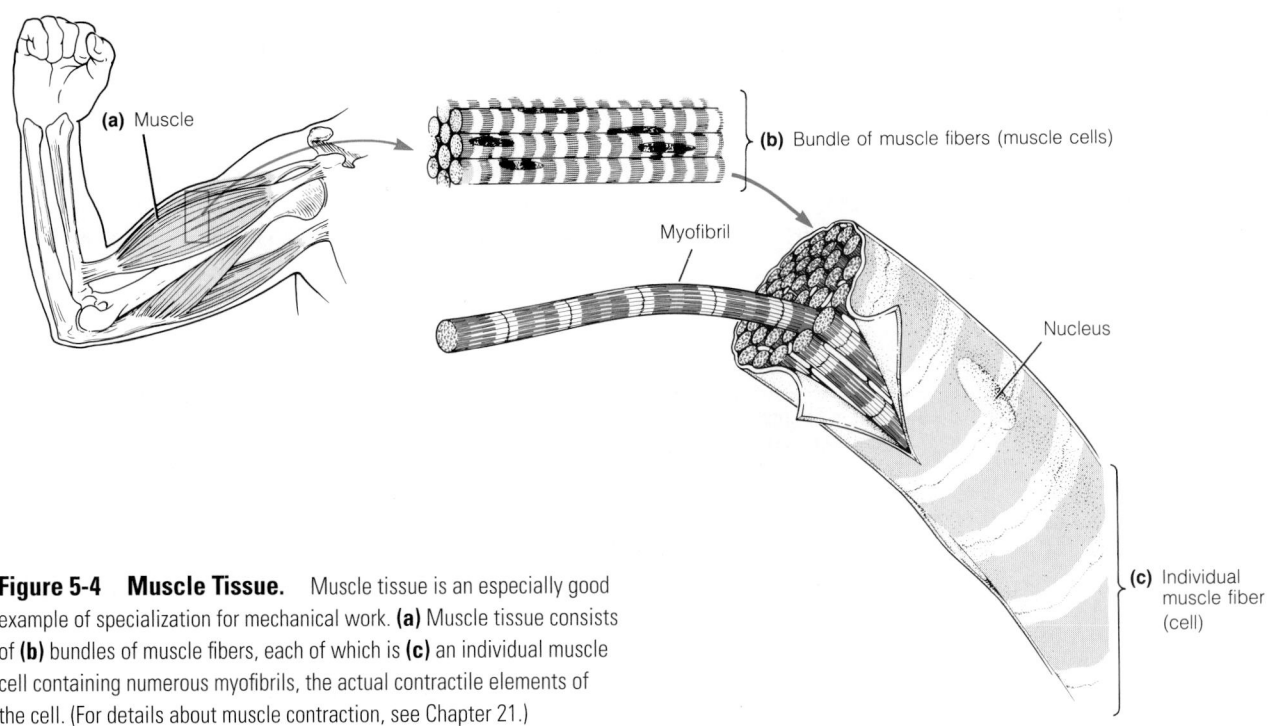

(a) Muscle

(b) Bundle of muscle fibers (muscle cells)

Myofibril

Nucleus

(c) Individual muscle fiber (cell)

Figure 5-4 Muscle Tissue. Muscle tissue is an especially good example of specialization for mechanical work. **(a)** Muscle tissue consists of **(b)** bundles of muscle fibers, each of which is **(c)** an individual muscle cell containing numerous myofibrils, the actual contractile elements of the cell. (For details about muscle contraction, see Chapter 21.)

the pumping of sodium and potassium ions across plasma membranes and the light-driven accumulation of protons within the chloroplasts of a plant cell.

Electrical Work (Movement of Ions Across Membranes). **Electrical work** is often considered a specialized case of concentration work because it also involves movement across membranes. In this case, however, the species that is translocated is a charged ion, and the result is not just a change in concentration but also the establishment of an electrical potential across the membrane. Every membrane has some characteristic potential that is generated in this way. An electrochemical gradient of protons across the mitochondrial or chloroplast membrane is essential to the production of ATP in both respiration (Chapter 12) and photosynthesis (Chapter 13). Electrical work is also important in the mechanism whereby impulses are conducted in nerve and muscle cells. An especially dramatic example of electrical work is found in *Electrophorus electricus,* the electric eel. The electric organ of *Electrophorus* consists of layers of cells called *electroplaxes,* each of which can generate a membrane potential of about 150 millivolts (mV). Because the electric organ contains thousands of such cells arranged in series, the eel can develop potentials of several hundred volts.

Heat and Work.   It is easy to forget about **heat** because living organisms do not use heat as a form of energy in the same way that a steam engine does. But heat is, in fact, a major use of energy in all *homeotherms* (animals that regulate their body temperature independent of the environment). In fact, as you read these lines, about two-thirds of your metabolic energy is being used just to stay warm. How can two-thirds of our energy be used for heat when we noted earlier that about two-thirds of our energy is used for ion transport? The answer is straightforward: The energy of ATP hydrolysis is used to pump ions, and when ATP is hydrolyzed to ADP and phosphate, heat is released as a by-product. This is the heat that warm-blooded animals use to maintain body temperature near 37°C, where their metabolism is most efficient. The relationship between work and heat energy is demonstrated when you get hot from exercise

or shiver from the cold. You should now be able to understand the relationship between these seemingly unrelated processes and ATP hydrolysis, which powers muscle contraction.

Bioluminescent Work (the Production of Light).   To be complete, we must also include the production of light, or **bioluminescence,** as yet another way in which energy is used by cells. The light produced by bioluminescent organisms is generated by the reaction of ATP with specific luminescent compounds and is usually pale blue. This is a much more specialized kind of energy use than the other five categories, and for present purposes we can leave it to the fireflies, luminous toadstools, dinoflagellates, deep-sea fish, and other creatures that live in its strange, cold light.

Using Energy: Phototrophs and Chemotrophs

The main forms of energy in our environment are listed in Table 5-1. It is immediately clear from the table that sunlight represents a major source of energy at the Earth's surface. All organisms (and therefore all cells) can be classified into one of two groups, which differ in the way they utilize energy sources.

The first group consists of organisms capable of capturing light energy by means of photosynthetic pigment systems, then storing the energy in the form of chemical bonds of organic molecules like glucose. Such organisms are called **phototrophs** (literally, "light-feeders") and include plants, algae, and certain groups of bacteria that are capable of photosynthesis.

These chemical bonds provide an energy source that can be used by a second group of organisms called **chemotrophs** (literally, "chemical-feeders") because they require the intake of chemical compounds such as carbohydrates, fats, and proteins. All animals, protists, and fungi and most bacteria are chemotrophs. The chemical energy is released by two methods. In the first, the chemical bonds are simply broken, releasing some of the stored energy. The general term for this process is *fermentation.* The more specific term *glycolysis* is used to describe the breakdown of the glu-

Table 5-1   Sources of Energy on the Earth's Surface

Source of Energy	Amount of Energy (cal/cm^2 per year)	Notes
Solar radiation	2.6×10^5	Includes infrared and ultraviolet light.
Electrical discharge	4.0	Lightning is the main source.
Radioactivity	0.8	Occurs to a depth of about 1 km in the Earth's crust.
Heat (volcanic)	0.13	
Chemical energy (photosynthesis)	100	
Chemical energy (stored)	—	A total of about 10^{26} calories is stored in coal, oil, and other geological deposits.

cose molecule during the fermentation process. The second method involves oxidation, in which electrons are transferred from the chemical bonds to molecular oxygen or other electron acceptor. This process is often called *respiration*. During respiration, essentially all the energy stored in chemical bonds is released.

A point that is often not appreciated about phototrophs is that although they can utilize solar energy when it is available, they are also capable of functioning as chemotrophs and, in fact, do so whenever they are not illuminated. Most higher plants are really a mixture of phototrophic and chemotrophic cells. A plant root cell, for example, though part of an obviously phototrophic organism, is in most cases incapable of carrying out photosynthesis and is every bit as chemotrophic as an animal cell.

The Flow of Energy in the Biosphere

So far we have seen that both chemotrophs and phototrophs depend on their environment for the energy they need, but differ in the forms of energy they can use. Chemotrophs require organic molecules, while phototrophs trap solar radiation and transduce it into chemical bond energies.

The flow of energy through the biosphere is depicted in Figure 5-5. Solar energy is trapped by phototrophs and used to convert carbon dioxide and water into more complex (and more reduced) cellular materials in the process of photosynthesis. As we will see in Chapter 13, the immediate products of photosynthetic carbon fixation are sugars, but in a sense we can consider the entire phototrophic organism to be the "product" of photosynthesis because every carbon atom in every molecule of that organism is derived from carbon dioxide that is fixed into organic form by the photosynthetic process.

Chemotrophs, on the other hand, are unable to use solar energy directly and depend on the chemical energy of oxidizable molecules. The energy needs of the chemotrophs can be met either *anaerobically* (in the absence of oxygen) by a variety of fermentation processes or *aerobically* (in the presence of oxygen) by the complete oxidation of chemical compounds in the process of respiration. Chemotrophs therefore depend completely on energy that has been packaged into the bonds of fermentable or oxidizable food molecules by phototrophs. A world composed only of chemotrophs would last only so long as food supplies held out, for even though we live on a planet that is flooded each day with solar energy, it is in a form that we cannot use to meet our energy needs.

Both phototrophs and chemotrophs use energy to carry out work—that is, to effect the various kinds of changes we

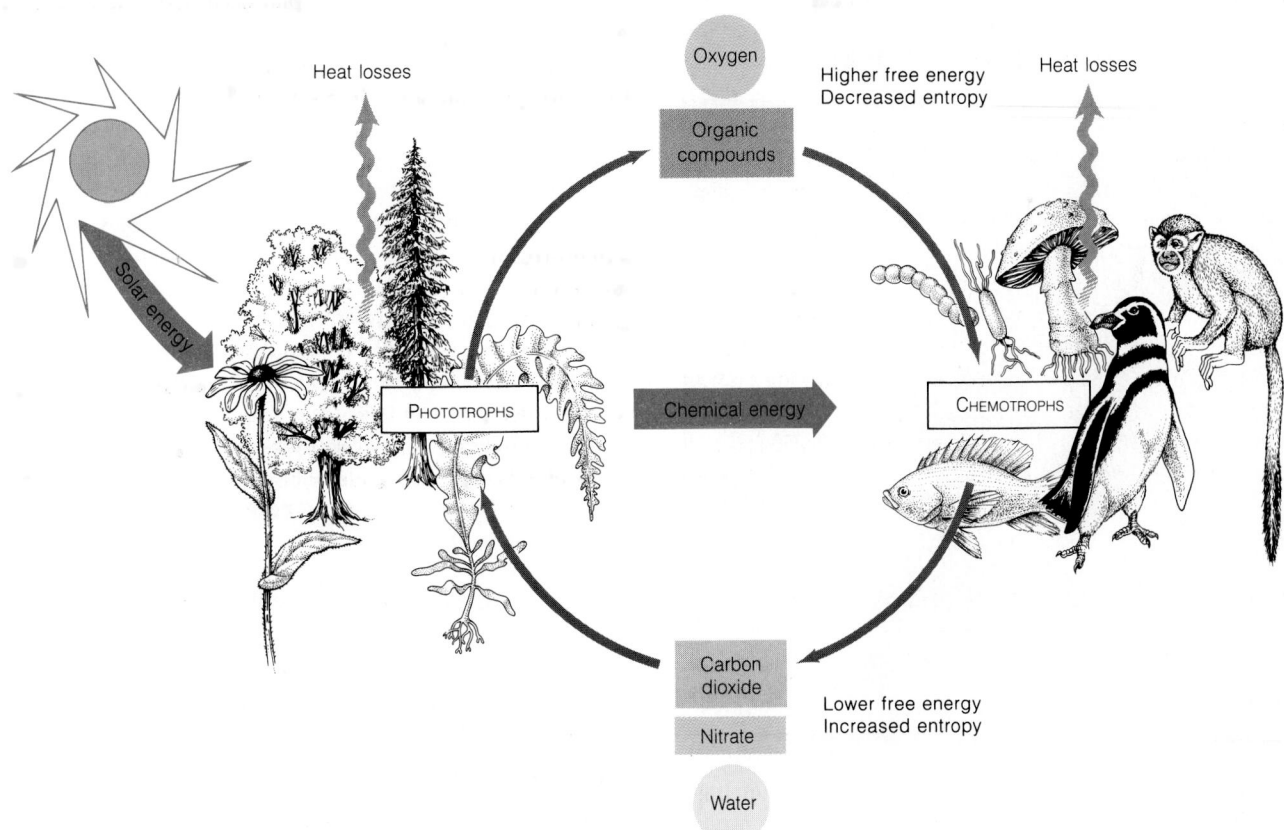

Figure 5-5   The Flow of Energy Through the Biosphere.
The energy in the biosphere originates in the sun and contributes eventually to the ever-increasing entropy of the universe. Accompanying the uni-directional flow of energy from phototrophs to chemotrophs is a cyclic flow of matter between the two groups of organisms.

have already catalogued. In the process, two kinds of losses occur (see Figure 5-5). One of the principles of energy conversion is that no chemical or physical process occurs with 100% efficiency; some energy is lost as heat. In fact, most processes that involve the conversion of energy from one form to another actually dissipate more energy as heat than they succeed in converting to the desired form. An electric light bulb, for example, generates more heat than light.

As we will see in Part Three, biological processes are remarkably efficient in energy conversion. *Heat losses* are nonetheless inevitable in every biological energy transaction. Sometimes the heat that is liberated during cellular processes is put to good use. As discussed earlier, warm-blooded animals use heat to maintain the body temperature at some constant level, usually well above ambient. Some plants use metabolically generated heat to melt overlying snow (Figure 5-2e) or to attract pollinators (Figure 5-6). But in general, the heat is simply dissipated into the environment and lost.

Even more fundamental is the *increase in entropy* that accompanies cellular activities. We will get to that in more detail shortly; here, we can simply note that every process or reaction that occurs anywhere in the universe always does so in such a way that the total entropy, or disorder, in the universe is increased. This change in entropy occurs at the expense of energy that might otherwise have been available to do useful work and is therefore an inevitable "sink" into which energy is lost. Just as the ultimate source of all energy in the biosphere is the sun, the ultimate fate of all energy in the biosphere is to become randomized in the universe as increased entropy or randomness.

Viewed on a cosmic scale, there is a continuous, massive, and unidirectional flow of energy from its source in the nuclear fusion reactions of the sun to its eventual sink, the entropy of the universe. We here in the biosphere are the transient custodians of an almost infinitesimally small portion of that energy, but it is precisely that small but critical fraction of energy and its flow through living systems that is of concern to us. The flow begins with green plants, which use light energy to drive electrons energetically "uphill" into new chemical bonds. This energy is then released by both plants and animals in "downhill" fermentative or oxidative reactions. This flux of energy through living matter—from the sun, to phototrophs, to chemotrophs, to heat—drives the molecular machinery of all life processes.

The Flow of Matter in the Biosphere

Accompanying the flow of energy in the biosphere is a corresponding flow of matter or mass because the energy in cells and organisms is almost inevitably stored and transferred as chemical bond energies of organic molecules. Although energy enters the biosphere unaccompanied by matter (that is, as photons of light) and leaves the biosphere similarly unaccompanied (as heat losses and increases in entropy) while it is passing through the biosphere, energy exists primarily in the form of chemical bond energy. As a result, the flow of energy in the biosphere is coupled to a correspondingly immense flow of matter.

Whereas energy flows unidirectionally from the sun through phototrophs to chemotrophs, matter flows in cyclic fashion between the two groups of organisms (see Figure 5-5). During respiration, aerobic chemotrophs take in organic nutrients from their surroundings, usually by ingesting phototrophs or other chemotrophs that have in turn eaten phototrophs. These nutrients are oxidized to carbon dioxide and water, low-energy molecules that are returned to the environment. Those molecules then become the raw materials that phototrophic organisms use to make new organic molecules photosynthetically, returning oxygen to the environment in the process.

In addition, there is an accompanying cycle of nitrogen. Phototrophs obtain nitrogen from the environment in inorganic form (often as nitrate from the soil, in some cases as N_2 from the atmosphere), convert it into ammonia, and use it in the synthesis of amino acids, proteins, nucleotides, and nucleic acids. Eventually, these molecules, like other components of phototrophic cells, are consumed by chemotrophs. The nitrogen is then converted back into ammonia and

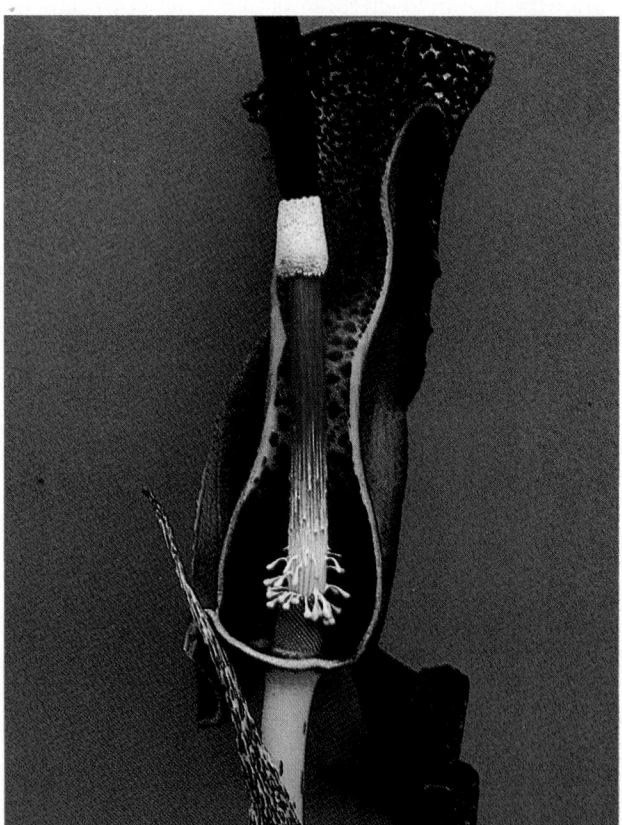

Figure 5-6   Voodoo Lily, a Plant That Depends on Metabolically Generated Heat to Attract Pollinators.   The voodoo lily *(Sauromatum guttatum)* warms certain parts of its flowers. The plant is pollinated by flies, which apparently mistake the flowers for dead meat. The flower emits odors that help attract the flies, and heating helps disperse the smelly gases. (This is a cutaway view of a flower.)

eventually to nitrate, under the influence of soil microorganisms.

Carbon dioxide, oxygen, nitrogen, and water thus cycle continuously between the phototrophic and chemotrophic worlds, always entering the chemotrophic sphere as energy-rich compounds and leaving again in an energy-poor form. The two great groups of organisms can therefore be thought of as living in symbiotic relationship with each other, with a cyclic flow of matter and a unidirectional flow of energy as components of that symbiosis.

On to Cellular Energetics

When we deal with the overall macroscopic flux of energy and matter through living organisms, we find cellular biology interfacing with ecology. Ecologists are very much concerned with cycles of energy and nutrients, with the roles of various species in these cycles, and with environmental factors that affect the flow. At the cellular level, our ultimate concern is how the flux of energy and matter we have been considering on a macroscopic scale can be expressed and explained on a molecular scale in terms of energy transactions and the chemical processes that occur within cells. We therefore leave the macroscopic cycles to the ecologist and turn our attention to the reactions that occur within individual cells of bacteria, plants, and animals to account for those cycles. First, however, we must acquaint ourselves with the physical principles underlying energy transactions, and for that we turn to the topic of bioenergetics.

Bioenergetics

The principles that govern energy flow are incorporated in an area of science that the physical chemist calls **thermodynamics.** Although the prefix *thermo-* suggests that the term is limited to heat (and that is indeed its historical origin), thermodynamics also takes into account other forms of energy and processes that convert energy from one form to another. Specifically, thermodynamics concerns the laws governing the energy transactions that inevitably accompany most physical processes and all chemical reactions. **Bioenergetics,** in turn, can be thought of as applied thermodynamics—that is, it concerns the application of thermodynamic principles to reactions and processes in the biological world.

Energy, Systems, Heat, and Work

As we have already seen, it is useful to define energy not simply as the ability to do work but specifically as the ability to cause change. Without energy, all processes would be at a standstill, including those that we associate with living cells.

Energy exists in a variety of forms, many of them of interest to biologists. Think, for example, of the energy represented by a ray of sunlight, a teaspoon of sugar, a moving flagellum, an excited electron, or the concentration of ions or small molecules within a cell or an organelle. These phenomena are diverse, but they are all governed by certain basic principles of energetics.

Energy is distributed throughout the universe, and for some purposes it is necessary to consider the total energy of the universe, at least in a theoretical way. Usually, however, we are interested not in the whole universe but only in a small portion of it. We might, for example, be concerned with a reaction or process occurring in a beaker of chemicals, in a cell, or in a block of metal. By convention, the restricted portion of the universe that one wishes to consider at the moment is called the **system,** and all the rest of the universe is referred to as the **surroundings.** Sometimes the system has a natural boundary, such as a glass beaker or a cell membrane. In other cases, the boundary between the system and its surroundings is a hypothetical one used only for convenience of discussion, such as the imaginary boundary around one mole of glucose molecules in a solution.

Systems can be either open or closed, depending on whether or not they can exchange energy with their surroundings (Figure 5-7). A **closed system** is sealed from its environment and can neither take in nor release energy in any form. An **open system,** on the other hand, can have energy added to it or removed from it. As we will see later, the improbable levels of organization that biological systems routinely display are possible only because cells and organisms are open systems, capable of both the uptake and the release of energy. Specifically, biological systems require a constant, large-scale influx of energy from their surroundings both to attain and to maintain the levels of complexity characteristic of them. That is why plants need sunlight and you need food.

Whenever we talk about a system, we have to be careful to specify the state of the system. A system is said to be in a specific **state** if each of its variable properties (such as temperature, pressure, and volume) is held constant at a specified value. In such a situation, the total energy content of the system, while not directly measurable, has some unique value. If such a system then changes from one state to another as a result of some interaction between the system and its surroundings, the change in its total energy is determined uniquely by the initial and final states of the system and is not affected at all by the mechanism by which the change occurs or the intermediate states through which the system may pass. This is a very useful property because it allows energy changes to be determined from a knowledge of the initial and final states only.

The problem of keeping track of system variables and their effect on energy changes can be simplified if one or more of the variables are held constant. Fortunately, this is the case with most biological reactions, because they usually occur in dilute solutions within cells that are at approximately the same temperature and pressure during the entire course of the reaction. These environmental conditions, as well as the cell volume, are generally slow to change compared to the speed of biological reactions. This means that three of the most important system variables that physical chemists usually concern themselves with—temperature,

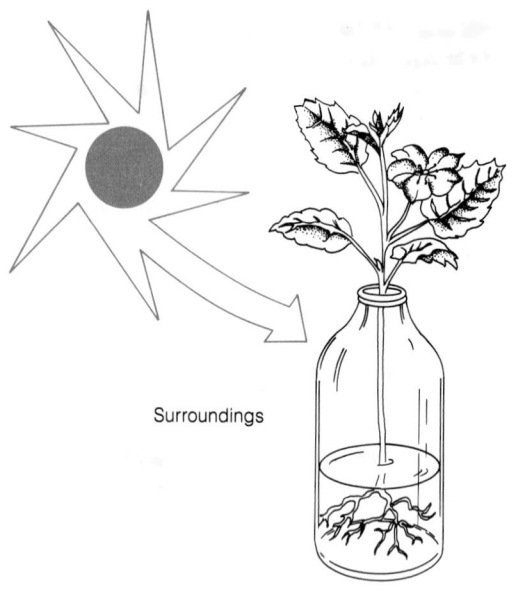

(a) Open system

(b) Closed system

Figure 5-7 Open and Closed Systems. A system is that portion of the universe under consideration. The rest of the universe is called the surroundings of the system. **(a)** An open system can exchange energy with its surroundings, whereas **(b)** a closed system cannot. The open system can use incoming energy to increase its orderliness, thereby decreasing its entropy. The closed system tends toward equilibrium and increases its entropy.

---

pressure, and volume—are essentially constant for biological reactions.

The exchange of energy between a system and its surroundings occurs in two ways: as heat and as work. Heat is energy transfer from one place to another as a result of a temperature difference between the two places. Spontaneous transfer always occurs from the hotter place to the colder place. Heat is an exceedingly useful form of energy for many machines and other devices designed to accomplish mechanical work. However, it has only limited biological utility because most biological systems operate under conditions of either fixed or only minimally variable temperature. Such **isothermal** systems lack the temperature gradients required to convert heat into other forms of energy. As a result, heat is not a useful source of energy for cells.

In biological systems, **work** is the application of energy from one place or form to another place or form to drive any process other than heat flow. For example, work is performed when the muscles in your arm expend chemical energy to lift this book, when a corn leaf uses light energy to synthesize sugar, or when an electrical eel draws on the ion concentration gradients of its electroplax tissue to deliver a shock. It is the amount of useful energy available to do cellular work that we will be primarily interested in when we begin calculating energy changes associated with specific reactions that cells carry out.

To quantify energy changes during chemical reactions or physical processes, we need units in which energy can be expressed. The most common units are the *calorie* and the *joule.* In biological chemistry, energy changes are usually expressed in terms of the **calorie (cal),** which we have already defined as the amount of energy required to warm 1 gram of water 1 degree centigrade (specifically, from 14.5° to 15.5°C) at a pressure of 1 atmosphere. (Again, note that the unit of energy measurement, like the very term *thermodynamics,* is based on heat but is applied generally to all forms of energy.) An alternative energy unit, the **joule (J),** is preferred by physicists and is used in some biochemistry texts. Conversion is easy: 1 cal = 4.184 J, or 1 J = 0.239 cal.

Energy changes are often measured on a per-mole basis, and the most common form in which we will encounter energy units in biological chemistry will be as calories (or sometimes kilocalories) per mole (cal/mol or kcal/mol). (Be careful to distinguish between the *calorie* as defined here and the *Calorie* that is often used to express the energy content of foods. The nutritional Calorie is represented with a capital C and is really a *kilocalorie* as defined here.)

Conservation of Energy: The First Law of Thermodynamics

Much of what we understand about the principles governing energy flow can be summarized by the three laws of thermodynamics. Of these, only the first and second laws are of particular relevance to the cell biologist. The **first law** is called the *law of conservation of energy.* Simply put, it states that *energy can be converted from one form to another but can never be created or destroyed.* (If you are familiar with the conversion of mass to energy that occurs in nuclear reactions, you will recognize that a more accurate statement would take both mass and energy into account. For purposes of biological chemistry, however, the law is adequate as stated.)

Applied to the universe as a whole or to a closed system, the first law means that the total amount of energy present in all forms must be the same before and after any process or reaction occurs. Applied to an open system, such as a cell, the first law says that during the course of any reaction or process, the total amount of energy that leaves the system must be exactly equal to the energy that enters the system minus any energy that remains behind and is therefore stored within the system:

$$\text{energy out} = \text{energy in} - \text{energy stored} \qquad \textbf{(5-1)}$$

Or, by simple rearrangement,

$$\text{energy stored} = \text{energy in} - \text{energy out} \qquad \textbf{(5-2)}$$

The total energy stored within a system is called the **internal energy** of the system, represented by the symbol E. We are not usually concerned with the actual value of E for a system because that value cannot be measured directly. However, it is possible to measure the *change in internal energy*, ΔE, that occurs during a given process. ΔE is the difference in internal energy of the system before and after the process:

$$\Delta E = E_2 - E_1 \qquad \textbf{(5-3)}$$

Equation 5-3 is valid for all physical processes or chemical reactions under any conditions. For biological reactions and processes, however, we are usually more interested in the change in **enthalpy,** or *heat content*. Enthalpy is represented by the symbol H and is related to the internal energy E as follows:

$$H = E + PV \qquad \textbf{(5-4)}$$

For reactions at constant pressure, ΔH, the *change in enthalpy,* is therefore related to ΔE as follows:

$$\Delta H = \Delta E + P\Delta V \qquad \textbf{(5-5)}$$

For our purposes, equation 5-5 can be simplified because most biological reactions occur without any significant change in volume. This means that ΔV is zero (or at least negligible), and equation 5-5 becomes simply

$$\Delta H = \Delta E \qquad \textbf{(5-6)}$$

Thus, biologists routinely determine changes in heat content for reactions of interest, confident that the values are valid estimates of ΔE.

The enthalpy change that accompanies a specific reaction is simply the difference in the heat content between the reactants and the products of the reaction:

$$\Delta H = H_{\text{products}} - H_{\text{reactants}} \qquad \textbf{(5-7)}$$

If the heat content of the products is less than that of the reactants, ΔH will be *negative* and the reaction is said to

be *exothermic* (Figure 5-8a). If the heat content of the products is greater than that of the reactants, ΔH will be *positive* and the reaction is *endothermic* (Figure 5-8b). Thus, the ΔH value for a reaction is simply a measure of the heat that is either produced by or taken up by that reaction (for negative and positive values, respectively) as it occurs under conditions of constant temperature and pressure.

For a biological example of such an energy transaction, consider the oxidation of glucose to carbon dioxide and water:

$$C_6H_{12}O_6 + 6O_2 \longrightarrow 6CO_2 + 6H_2O + \text{energy} \qquad \textbf{(5-8)}$$

You may recognize this as the summary equation for the process of *respiration,* whereby aerobic chemotrophs obtain energy from the sugar glucose. (Most of the cells in your body are carrying out this process right now.) The reaction is therefore highly relevant to cellular energy flow, in addition to illustrating several points about ΔH.

By combusting glucose under controlled conditions in the laboratory, we can show that 673,000 cal (673 kcal) of energy are liberated for every mole of glucose that is oxidized. This means that the products (carbon dioxide and water) have a lesser heat content than the reactants (glucose and oxygen). Reaction 5-8 therefore has a ΔH of -673 kcal/mol, where the negative sign indicates a *decrease* in enthalpy (liberation of heat) as glucose is oxidized to carbon dioxide and water (Figure 5-9a).

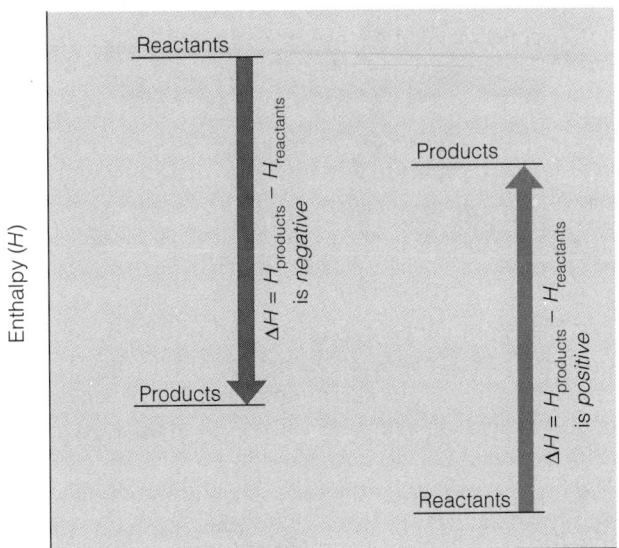

(a) *Exothermic* reaction (*decrease* in enthalpy)

(b) *Endothermic* reaction (*increase* in enthalpy)

Figure 5-8   Changes in Enthalpy for Chemical Reactions. The change in enthalpy, or heat content, that accompanies a chemical reaction is the difference in the heat content between the reactions and the products. **(a)** If the products are lower in enthalpy than the reactants, the reaction involves a *decrease* in enthalpy ($\Delta H < 0$) and is exothermic. **(b)** If the products are higher in enthalpy than the reactants, the reaction involves an *increase* in enthalpy ($\Delta H > 0$) and is endothermic.

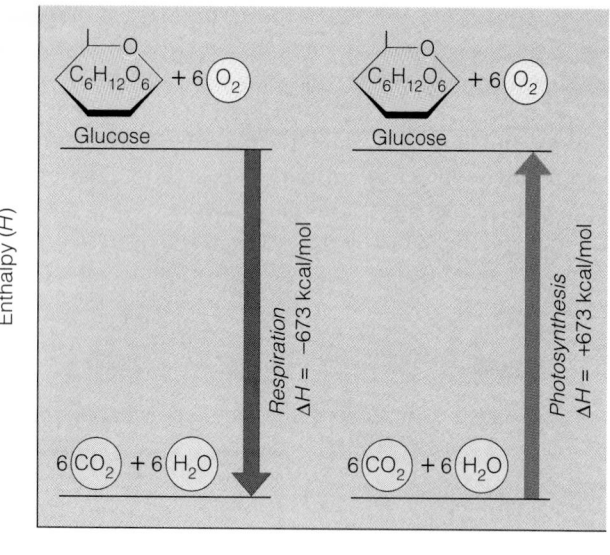

(a) Respiration
(exothermic)

(b) Photosynthesis
(endothermic)

Figure 5-9   Changes in Enthalpy for Respiration and Photosynthesis.   (a) In respiration, glucose is oxidized to carbon dioxide and water with the release of 673 kcal of energy per mole of glucose oxidized. The enthalpy change, ΔH, is therefore -673 kcal/mol, and the process is exothermic. **(b)** In photosynthesis, glucose is synthesized from carbon dioxide and water with the release of oxygen. This process consumes 673 kcal of energy per mole of glucose synthesized. ΔH is therefore $+673$ kcal/mol, and the process is endothermic.

Now consider the reverse reaction, which turns out to be a summary equation for photosynthesis:

$$6CO_2 + 6H_2O + \text{energy} \longrightarrow C_6H_{12}O_6 + 6O_2 \qquad \textbf{(5-9)}$$

If this reaction is carried out under the same conditions used earlier to measure ΔH for glucose oxidation, it can be shown that the ΔH value for glucose synthesis is $+673$ kcal/mol, where the positive sign indicates an *increase* in internal energy as carbon dioxide and water are converted to glucose (Figure 5-9b).

The important point is that ΔH values for reactions 5-8 and 5-9 are equal in magnitude but opposite in sign. The increase in energy as glucose is synthesized is matched exactly by the decrease in energy as glucose is oxidized. Energy is neither created nor destroyed, just as the first law predicts. The first law of thermodynamics, in other words, allows us to calculate energy changes with confidence because it assures us that energy can always be accounted for.

How to Know Which Way It Will Go: The Second Law of Thermodynamics

So far, all that thermodynamics has been able to tell us is that energy is conserved whenever a process or reaction occurs—that all of the energy going into a system must either be stored within the system or released again to the surroundings. We have seen the usefulness of ΔH as a measure

of how much the total enthalpy of a system would change if a given process were to occur, but we have no way as yet of predicting whether and to what extent the process will in fact occur under the prevailing conditions.

We have, at least in some cases, an intuitive feeling that some reactions or processes are possible, whereas others are not. Returning to the examples of Figure 5-1, we are somehow very sure that if we were to set a match to paper, it would burn. The oxidation of cellulose to carbon dioxide and water is, in other words, a possible reaction. Or, to use more precise terminology, it is a *thermodynamically spontaneous* reaction. Furthermore, there is clearly a directionality to the process, for we are equally convinced that the reverse process will not occur—that if we were to stand around clutching the charred remains, the paper would not spontaneously reassemble in our hands. We have, in other words, a feeling for both the possibility and the directionality of cellulose oxidation.

You can probably think of other processes for which you can make such thermodynamic predictions with equal confidence. We know, for example, that drops of dye diffuse in water, that ice cubes melt at room temperature, and that sugar dissolves in water, and we can therefore label these as thermodynamically spontaneous events. But if we ask why we recognize them as such, the answer has to do with repeated prior experience. We have seen paper burn, ice cubes melt, and sugar dissolve often enough to know intuitively that these are processes that really occur, and with such predictability that we can label them as spontaneous, provided only that we know the conditions.

However, when we move from the world of familiar physical processes to the realm of chemical reactions in cells, we quickly find that we cannot depend on prior experience to guide us in our predictions. Consider, for example, the conversion of glucose-6-phosphate into fructose-6-phosphate:

$$\text{glucose-6-phosphate} \rightleftharpoons \text{fructose-6-phosphate} \qquad \textbf{(5-10)}$$

Glucose is a six-carbon aldosugar and fructose is its keto equivalent (see Figure 3-17). Both can form a phospho-ester bond between a phosphoric acid (phosphate) molecule and the hydroxyl group on carbon 6 of the sugar, giving rise to the phosphorylated compounds. Reaction 5-10 therefore involves the interconversion of a phosphorylated aldosugar to the corresponding phosphorylated ketosugar, as shown in Figure 5-10.

This particular interconversion is a significant reaction in all cells. It is, in fact, the second step in an important and universal reaction sequence called the glycolytic pathway. In addition to illustrating an important thermodynamic principle, therefore, reaction 5-10 introduces a bit of cellular chemistry that will come in handy later on. For now, however, just focus on the reaction from a thermodynamic point of view, and ask yourself what predictions you can make about the likelihood of glucose-6-phosphate being converted into fructose-6-phosphate. You will probably be at a loss to make any predictions at all. We know what will happen with

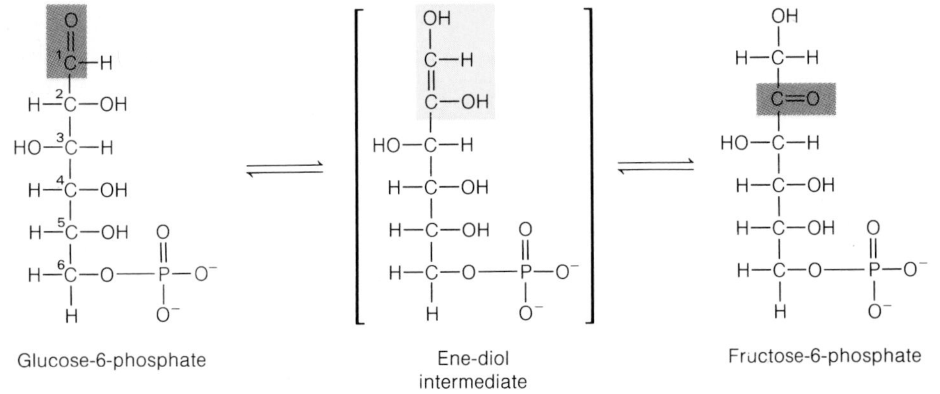

Glucose-6-phosphate   Ene-diol intermediate   Fructose-6-phosphate

Figure 5-10   **The Interconversion of Glucose-6-Phosphate and Fructose-6-Phosphate.**   This reaction involves the interconversion of the phosphorylated forms of an aldosugar (glucose) and a ketosugar (fructose). It is catalyzed by an enzyme called phospho- glucoisomerase and is readily reversible. The reaction proceeds via an intermediate called an ene-diol because it has a carbon-carbon double bond ("ene") with two alcohol groups ("diol") attached.

burning paper and melting ice, but we lack the familiarity and prior experience with phosphorylated sugars even to make an intelligent guess. Clearly, what we need is a reliable means of determining whether a given physical or chemical change can occur under specific conditions without having to rely on prior experience, familiarity, or intuition.

Thermodynamics provides us with exactly such a measure of spontaneity in the **second law,** or the *law of thermodynamic spontaneity.* Moreover, the second law allows us to predict not only in what direction a reaction will proceed under specified conditions but also how far from equilibrium it lies, how much energy the reaction will release or consume as it proceeds, and how the energetics of the reaction will be affected by specific changes in the conditions.

The two specific thermodynamic parameters that measure thermodynamic spontaneity are *entropy* and *free energy.* These concepts are sometimes difficult to understand because they are somewhat abstract. We will therefore limit our discussion here to their use in determining what kinds of changes can occur in biological systems. For further help, see Box 5A for an essay that uses jumping beans to introduce the concepts of internal energy, entropy, and free energy.

Entropy.   Although we cannot experience **entropy** directly, we can get some feel for it by considering it to be a measure of *randomness* or *disorder.* Entropy is represented by the symbol S. For any system, the *change in entropy, ΔS,* represents a change in the degree of randomness or disorder of the components of the system. As an example, consider again the drop of dye in Figure 5-1. As the dye molecules diffuse through the water, they become ever more random in their location in the beaker, and the entropy of the system therefore increases. Similarly, the combustion of paper involves an increase in entropy because the carbon, oxygen, and hydrogen atoms of cellulose are much more randomly distributed in space once they are converted to carbon dioxide and water. Entropy also increases as ice melts or as a volatile solvent such as gasoline is allowed to evaporate.

Entropy Change as a Measure of Thermodynamic Spontaneity.   How can any of this help predict what changes will occur in a cell? There is a very important link between spontaneous events and entropy changes because, whenever a process occurs in nature, the randomness or disorder of the universe (that is, the entropy of the universe) invariably increases. This is one of two alternative ways to state the second law of thermodynamics (Box 5B, page 128). According to this formulation, *the value of $\Delta S_{universe}$ is positive for every real process or reaction.* Processes that would lead to a decrease in the entropy of the universe simply do not occur. Remembering that entropy is a measure of randomness, we can rephrase the second law to state that *the universe becomes more random with every reaction that occurs; reactions that would make it less random are never observed.*

We have to keep in mind, however, that this formulation of the second law pertains to the universe as a whole and may not apply to the specific system under consideration. Every real process, without exception, must be accompanied by an increase in the entropy of the universe, but for a given system the entropy may increase, decrease, or stay the same as the result of a specific process. For example, the oxidation of glucose to carbon dioxide and water (reaction 5-8) is spontaneous under standard conditions (25°C, a pressure of 1 atm, and pH 7.0) and is accompanied by an *increase* in the system entropy ($\Delta S_{system} = +43.6$ cal/mol-degree). On the other hand, the freezing of water at -0.1°C is also a spontaneous event, yet it involves a *decrease* in the system entropy ($\Delta S_{system} = -0.5$ cal/mol-degree). This makes sense when you consider the greater ordering of water molecules in ice crystals. Thus, while the change in entropy of the universe is a valid measure of the spontaneity of a process, the change in entropy of the system is not.

To understand how the entropy of the universe can increase during a process while the entropy of the system decreases, we need only realize that the decrease in entropy of the system can be accompanied by an equal or even greater increase in the entropy of the surroundings. On the basis of

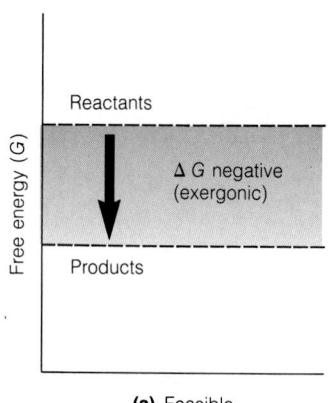

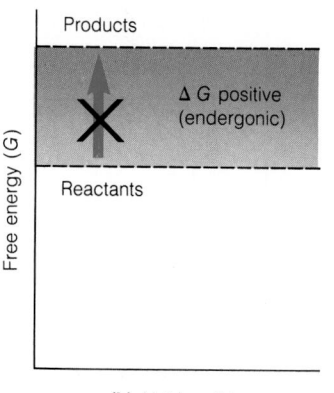

(a) Feasible
(thermodynamically spontaneous)

(b) Not feasible
(thermodynamically nonspontaneous)

Figure 5-11 ΔG as a Measure of Thermodynamic Spontaneity.
(a) Every reaction or process for which ΔG is negative is thermodynamically spontaneous and therefore feasible. **(b)** Any reaction or process for which ΔG is positive would yield products with a greater free energy content than the reactants and is therefore thermodynamically impossible. However, if the reaction is coupled (linked mechanistically) to an exergonic reaction with a sufficiently negative ΔG to ensure that the overall ΔG for the coupled reaction is negative, the first reaction can be "driven" by the second reaction. In cells, the exergonic reaction of ATP hydrolysis is often used in this manner.

the second law, such local increases in order (decreases in entropy) must be offset by an even greater decrease in the order (increase in entropy) of the surroundings. This is, in fact, the situation that normally prevails within living cells, which by their growth and reproduction produce striking local increases in order.

This means, however, that the second law, stated in terms of entropy of the universe, is of limited value in predicting the spontaneity of biological processes, for it would require keeping track of changes that occur not only within the system but also in its surroundings. Far more convenient would be a parameter that would enable prediction of the spontaneity of reactions from a consideration of the system alone.

Free Energy. As you might guess, a measure of spontaneity for the system alone does in fact exist. It is called **free energy** and is represented by the symbol G after Willard Gibbs, who first developed the concept. Because of its predictive value and its ease of calculation, the free energy function is one of the most useful thermodynamic concepts in biology. One could even make the case that our entire discussion of thermodynamics so far has really been a way of getting us to free energy, because it is here that the usefulness of thermodynamics for cell biologists becomes apparent.

Like most other thermodynamic functions, free energy is defined only in terms of mathematical relationships. But for biological systems at constant pressure, volume, and temperature, the **free energy change, ΔG,** is related to the changes in enthalpy and entropy as follows:

$$\Delta H = \Delta G + T\Delta S \qquad \textbf{(5-11)}$$

or

$$\Delta G = \Delta H - T\Delta S \qquad \textbf{(5-12)}$$

where ΔH is the change in enthalpy, ΔG the change in free energy, ΔS the change in entropy, and T the temperature of the system in kelvin units (K = °C + 273).

Free Energy Change as a Measure of Thermodynamic Spontaneity. Free energy is an exceptionally useful concept as a readily measurable indicator of spontaneity. As we shall see shortly, ΔG for a reaction can be readily calculated from the equilibrium constant for the reaction and from easily measurable system variables, such as the concentrations of reactants and products. Once determined, ΔG provides exactly what we have been looking for: a measure of the spontaneity of a reaction based solely on the properties of the system in which the reaction is occurring.

Specifically, every spontaneous reaction is characterized by a *decrease* in the free energy of the system ($\Delta G_{system} < 0$) just as surely as it is characterized by an *increase* in the entropy of the universe ($\Delta S_{universe} > 0$). This is true because with the temperature and pressure held constant, ΔG for the system is related to ΔS for the universe in a simple but inverse way. This gives us a second, equally valid way of expressing the second law: *All reactions that occur spontaneously result in a decrease in the free energy content of the system* (Figure 5-11a).

Such reactions are called **exergonic,** which means *energy-yielding.* Note carefully that the reference is specifically to the change in free energy and not to the change in total internal energy, which may be negative, positive, or zero for a spontaneous reaction and is therefore *not* a valid measure of thermodynamic spontaneity. Conversely, any process that would result in an increase in the free energy of the system is called **endergonic** (*energy-requiring*) and cannot proceed under the conditions used to calculate ΔG (Figure 5-11b). If, however, conditions are altered (as, for example, by increasing the concentrations of the reactants or decreasing the concentrations of the products), it is sometimes possible to convert an otherwise endergonic reaction into an exergonic reaction, thereby rendering it spontaneous. As you will see later, this fact is fundamental to our understanding of biosynthetic reactions.

The Meaning of Spontaneity. Before looking at how we can actually calculate and use ΔG as a measure of thermody-

5A

F u r t h e r   I n s i g h t s

JUMPING BEANS AND FREE ENERGY

If you are finding the concepts of free energy, entropy, and equilibrium constants difficult to grasp, perhaps a simple analogy might help.[*] For this we will need an imaginary supply of jumping beans, which are really seeds of certain Mexican shrubs, with larvae of the moth *Laspeyresia salitans* inside. Whenever the larvae inside the seed wiggle about, the seeds wiggle, too. They don't really jump in the usual sense of the word, although certain wasp larvae can in fact cause the galls they inhabit (called fleaseeds) to leap several centimeters from a standing start! The jumping action probably serves to get the larvae out of direct sunlight, which could heat them to lethal temperatures.

The Jumping Reaction

For purposes of illustration, imagine that we have some high-powered jumping beans in two chambers separated by a low partition, as shown below. Notice that the chambers have the same floor area and are at the same level, although we will want to vary both of these properties shortly. As soon as we place a handful of jumping beans in chamber 1, they begin jumping about randomly. Although most of the beans jump only to a modest height most of the time, occasionally one of them, in a burst of ambition, gives a more energetic leap, surmounting the barrier and falling into chamber 2. We can write this as the *jumping reaction:*

Beans in chamber 1 ⇌ Beans in chamber 2

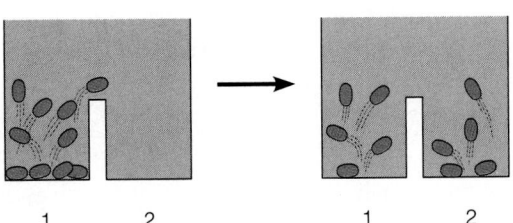

[*]We are indebted to Princeton University Press for permission to use this analogy, first developed by Harold F. Blum in the book *Time's Arrow and Evolution* (3d ed., 1968), pp. 17–26.

We will imagine this to be a completely random event, happening at irregular, infrequent intervals. Occasionally, one of the beans that has reached chamber 2 will happen to jump back into chamber 1, which is the *back reaction.* At first, of course, there will be more beans jumping from chamber 1 to chamber 2 because there are more beans in chamber 1, but things will eventually even out so that, on average, there will be the same number of beans in both compartments. The system will then be at *equilibrium.* Beans will still continue to jump between the two chambers, but the numbers jumping in both directions will be equal.

The Equilibrium Constant

Once our system is at equilibrium, we can count up the number of beans in each chamber and express the results as the ratio of the number of beans in chamber 2 to the number in chamber 1. This is simply the *equilibrium constant* K_{eq} for the jumping reaction:

$$K_{eq} = \frac{\text{number of beans in chamber 2 at equilibrium}}{\text{number of beans in chamber 1 at equilibrium}}$$

For the specific case shown above, the numbers of beans in the two chambers are equal at equilibrium, so the equilibrium constant for the jumping reaction under these conditions is 1.0.

Enthalpy Change (ΔH)

Now suppose that the level of chamber 1 is somewhat higher than that of chamber 2, as shown in the next diagram. Jumping beans placed in chamber 1 will again tend to distribute themselves between chambers 1 and 2, but this time a higher jump is required to get from 2 to 1 than from 1 to 2, so the latter will occur more frequently. As a result, there will be more beans in chamber 2 than in chamber 1 at equilibrium, and the equilibrium constant will therefore be greater than 1.

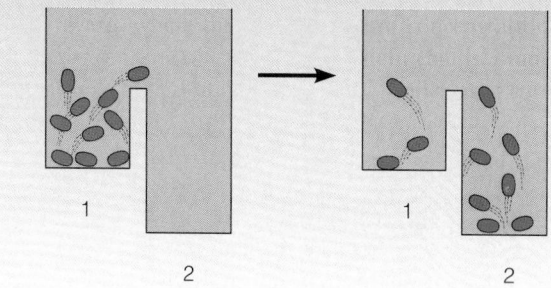

The relative heights of the two chambers can be thought of as measures of the *enthalpy,* or *heat content* (H), of the chambers, such that chamber 1 has a higher H value than chamber 2, and the difference between them is represented by ΔH. Since it is a "downhill" jump from chamber 1 to chamber 2, it makes sense that ΔH has a negative value for the jumping reaction from chamber 1 to chamber 2. Similarly, it seems reasonable that ΔH for the reverse reaction should have a positive value because that jump is "uphill."

Entropy Change (ΔS)

So far, it might seem as if the only thing that can affect the equilibrium distribution of beans between the two chambers is the difference in enthalpy, ΔH. But that is only because we have kept the floor area of the two chambers constant. Imagine instead the situation shown below, where the two chambers are again at the same height, but chamber 2 now has a greater floor area than chamber 1. The probability of a bean finding itself in chamber 2 is therefore correspondingly greater, so there will be more beans in chamber 2 than in chamber 1 at equilibrium, and the equilibrium constant will be greater than 1 in this case also. This means that the equilibrium position of the jumping reaction has been shifted to the right, even though there is no change in internal energy at all.

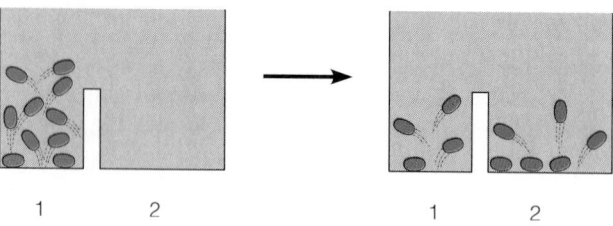

The floor area of the chambers can be thought of as a measure of the *entropy,* or *randomness,* of the system, S,

and the *difference* between the two chambers can be represented by ΔS. Since chamber 2 has a greater floor area than chamber 1, the entropy change is positive for the jumping reaction as it proceeds from left to right under these conditions. Note that for ΔH, negative values are associated with favorable reactions, while for ΔS, favorable reactions are indicated by positive values.

Free Energy Change (ΔG)

So far, we have encountered two different factors that affect the distribution of beans: the difference in levels of the two chambers (ΔH) and the difference in floor area (ΔS). Moreover, it should be clear that neither of these factors by itself is an adequate indicator of how the beans will be distributed at equilibrium because a favorable (negative) ΔH could be more than offset by an unfavorable (negative) ΔS, and a favorable (positive) ΔS could be more than offset by an unfavorable (positive) ΔH. You should, in fact, be able to design chamber conditions that illustrate both of these situations, as well as situations in which ΔH and ΔS tend to reinforce rather than counteract each other.

Clearly, what we need is a way of summing these two effects algebraically to see what the net tendency will be. The new measure we come up with is called the *free energy change,* ΔG, and is defined so that *negative* values correspond to favorable (i.e., thermodynamically spontaneous) reactions and *positive* values represent unfavorable reactions. Thus, ΔG should have the *same* sign as ΔH (since a negative ΔH is also favorable) but the *opposite* sign from ΔS (since for ΔS a positive value is favorable). In terms of real-life thermodynamics, the expression for ΔG in terms of ΔH and ΔS is

$$\Delta G = \Delta H - T\Delta S$$

(Notice that the temperature dependence of ΔS is the only feature of this relationship that cannot be readily explained by our model, unless we assume that the effect of changes in room size is somehow greater at higher temperatures.)

ΔG *and the Capacity to Do Work*

You should be able to appreciate the difficulty of suggesting a physical equivalent for ΔG because it represents an algebraic sum of entropy and energy changes, which may either reinforce or partially offset each other. But as long as ΔG is negative, beans will continue to jump from chamber 1 to chamber 2, whether driven primarily by changes in entropy, internal energy, or both. This means that if some sort of bean-powered "bean wheel" is placed between the two chambers as shown below, the movement of beans from

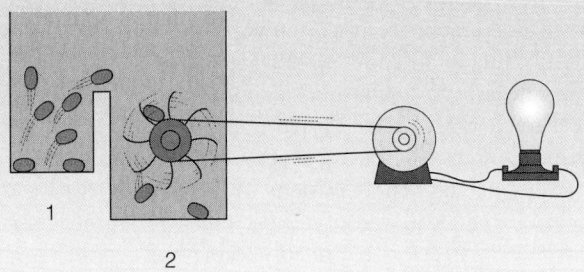

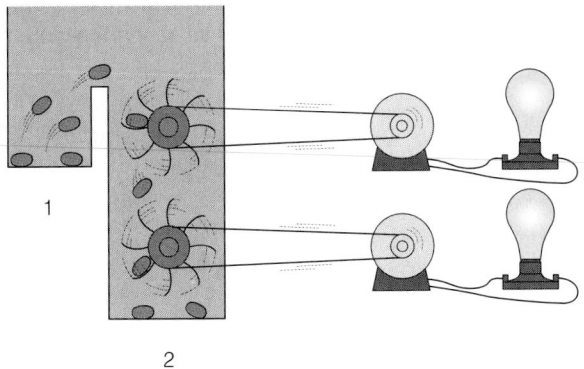

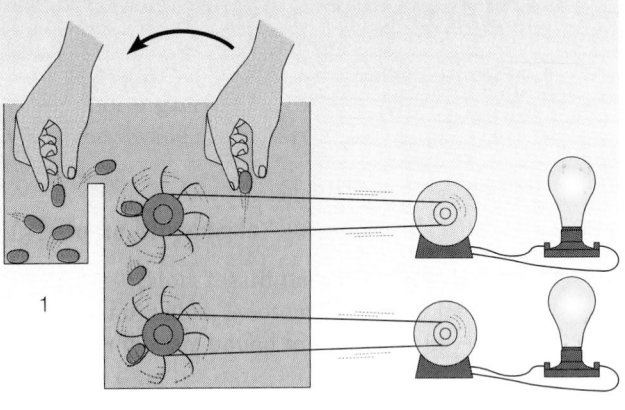

one chamber to the other can be harnessed to do work until equilibrium is reached, at which point no further work is possible. Furthermore, the greater the difference in free energy between the two chambers (that is, the more highly negative ΔG is), the more work the system can do.

Thus, ΔG is first and foremost a measure of the capacity of a system to do work under specified conditions. You might, in fact, want to think of ΔG as free energy in the sense of energy that is free or available to do useful work. Moreover, if we contrive to keep ΔG negative by continuously adding beans to chamber 1 and removing them from chamber 2, we have a dynamic *steady state*, a condition that

effectively harnesses the inexorable drive to achieve equilibrium. Work can then be performed continuously by beans that are forever jumping toward equilibrium but that never actually reach it.

Looking Ahead

To anticipate the transition from the thermodynamics of this chapter to the kinetics of the next, begin thinking about the *rate* at which beans actually proceed from chamber 1 to chamber 2. Clearly, ΔG measures how much energy will be released if beans do jump, but it says nothing at all about the rate. That would appear to depend critically on how high the barrier between the two chambers is. Label this the *activation energy barrier,* and then contemplate means by which you might get the beans to move over the barrier more rapidly. You will probably be helped by the knowledge that the larvae inside the seeds wiggle more vigorously if they are warmed. ■

Historical Perspectives

ENERGY AND ENTROPY: THE GREEK CONNECTION

Sometimes it requires genius to see the obvious. Rudolph Clausius was just 28 when he published his first scientific paper in Berlin, in 1850. It concerned heat as an energy source, and his conclusion was that "Heat of itself cannot pass from a colder to a hotter body." This seems so simple to us now, with the understanding that heat is a form of random kinetic energy that is transferred between atoms and molecules through collision and radiation. Yet in 1850, heat was still a mysterious entity that scientists and engineers were struggling to understand.

In later work, Clausius went on to a second, equally powerful insight, which is not so obvious even today. The word "energy" is from the Greek, with the sense of "containing work." In other words, something has "energy" if it can cause something else to move or change in some way. (One definition of work is moving a mass over a distance; moving a piano is work!) Clausius noticed that when energy was used to do work, he could never get an exact balance between the energy used and the work done. For instance, when heat was used to make steam and the steam energy was transformed into mechanical work by a steam engine, measurements showed that the work done was always less than the energy put into the system. This strange effect needed a descriptive term. Clausius reasoned that if *energy* meant "work content," he would coin the word *entropy* from the Greek words for "transformation content": The greater the number of transformations energy went through, the greater the entropy.

As these ideas developed further, it became clear that this concept represented a universal law, now known as the second law of thermodynamics, in which entropy is related to the relative disorder of a system. As Clausius summed it up: "The energy of the universe is constant; the entropy of the universe tends towards a maximum." One of Murphy's laws also sums it up nicely: "If something can go wrong, it will." ∎

namic spontaneity, we need to make sure we understand what is—and what is not—meant by the term *spontaneous*. As used in thermodynamics and bioenergetics, **spontaneity** is the ability of a reaction to have a negative ΔG and therefore the capability of proceeding in the direction indicated. However, spontaneity tells us only that the reaction *can* go; it says nothing at all about whether it *will* go. A reaction can have a negative ΔG value and yet not actually proceed to any measurable extent at all. The combustion of paper is a good example. The cellulose of paper obviously burns spontaneously once ignited, consistent with a highly negative ΔG value of -686 kcal/mol of glucose units. Yet in the absence of a match, paper is reasonably stable and would require thousands of years to oxidize. Thus, ΔG can really tell us only whether a reaction or process is thermodynamically possible—whether it has the *potential* for occurring. Whether an exergonic reaction will in fact proceed depends not only on its favorable (negative) ΔG but also on the availability of a mechanism or pathway to get from the initial state to the final state. Usually, an initial input of activation energy is required as well, such as the heat energy from the match that was used to ignite the piece of paper.

Thermodynamic spontaneity is therefore a necessary but insufficient criterion for determining whether a reaction will actually occur. In Chapter 6, we will have an opportunity to explore the subject of reaction rates in the context of enzyme-catalyzed reactions. For the moment, we need only note that when we designate a reaction as thermodynamically spontaneous, we simply mean that it is an energetically feasible event that will liberate free energy if and when it actually takes place.

Understanding ΔG

Our final task in this chapter will be to understand how ΔG is calculated and how it can then be used to assess the thermodynamic feasibility of reactions under specified conditions. For that we come back to the reaction that converts glucose-6-phosphate into fructose-6-phosphate (reaction 5-10) and ask what we can learn about the spontaneity of the conversion in the direction written (from left to right). Prior experience and familiarity provide no clues here, nor is it obvious how the entropy of the universe would be affected if the reaction were to proceed. Clearly, we need to be able to calculate ΔG and to determine whether it is positive or negative under the particular conditions we specify for the reaction.

The Equilibrium Constant as a Measure of Directionality

One means of assessing whether a reaction can proceed in a given direction under specified conditions involves the **equilibrium constant K_{eq}**, which is the ratio of product concentrations to reactant concentrations at equilibrium. For the general reaction in which A is converted reversibly into B, the equilibrium constant is simply the ratio of the equilibrium concentrations of A and B:

$$A \rightleftharpoons B \qquad \text{(5-13)}$$

$$K_{eq} = \frac{[B]_{eq}}{[A]_{eq}} \qquad \text{(5-14)}$$

where $[A]_{eq}$ and $[B]_{eq}$ are the concentrations of A and B, in moles per liter, when reaction 5-13 is at equilibrium at 25°C. Given the equilibrium constant for a reaction, you can easily tell whether a specific mixture of products and reactants is at equilibrium, and, if not, how far the reaction is away from equilibrium and in which direction it must proceed to reach equilibrium.

For example, the equilibrium constant for reaction 5-10 at 25°C is known to be 0.5. This means that at equilibrium there will be one-half as much fructose-6-phosphate as glucose-6-phosphate, regardless of the actual magnitudes of the concentrations:

$$K_{eq} = \frac{[\text{fructose-6-phosphate}]_{eq}}{[\text{glucose-6-phosphate}]_{eq}} = 0.5 \qquad \text{(5-15)}$$

If the two compounds are present in any other concentration ratio, the reaction will not be at equilibrium and will tend toward (be thermodynamically spontaneous in the direction of) equilibrium. Thus, a ratio of concentrations less than K_{eq} means that there is too little fructose-6-phosphate present, and the reaction will tend to proceed to the right to generate more fructose-6-phosphate at the expense of glucose-6-phosphate. Conversely, a ratio greater than K_{eq} indicates that the relative concentration of fructose-6-phosphate is too high, and the reaction will tend to proceed to the left.

Figure 5-12 illustrates this concept for the interconversion of A and B (reaction 5-13), showing the relationship between the free energy of the reaction and how far the concentrations of A and B are from equilibrium. (Note that K_{eq} is assumed to be 1.0 in this illustration; for other values of K_{eq}, the curve would be the same but the numbers along the X axis would be different.) The point of Figure 5-12 is clear: The free energy is lowest at equilibrium and increases as the system is displaced from equilibrium in either direction. Moreover, if we know how the ratio of prevailing concentrations compares with the equilibrium concentration ratio, we can predict in which direction a reaction will tend to proceed *and* how much free energy will be released as it does so. Thus, the tendency toward equilibrium provides the driving force for every chemical reaction, and a comparison of prevailing and equilibrium concentration ratios provides one measure of that tendency.

Calculation of ΔG

It should come as no surprise that ΔG is really just a means of calculating how far from equilibrium a reaction lies under

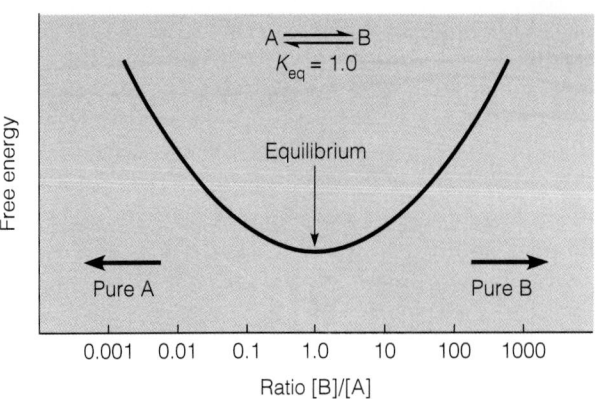

Figure 5-12 Free Energy and Chemical Equilibrium. The amount of free energy available from a chemical reaction depends on how far the components are from equilibrium. This is illustrated here for a reaction that interconverts A and B and has an equilibrium constant, K_{eq}, of 1.0. The free energy of the system increases as the [B]/[A] ratio changes on either side of the equilibrium point. (For a reversible reaction with an equilibrium constant other than 1.0, the shape of the curve would be the same, but the curve would be centered at the K_{eq} value rather than at 1.0.)

specified conditions and how much energy will be released as the reaction proceeds toward equilibrium. Nor should it be surprising that both the equilibrium constant and the prevailing concentrations of reactants and products are needed to calculate ΔG. For reaction 5-13, the equation relating these variables is as follows:

$$\Delta G = RT \ln \frac{[B]_{pr}}{[A]_{pr}} - RT \ln \frac{[B]_{eq}}{[A]_{eq}}$$

$$= RT \ln \frac{[B]_{pr}}{[A]_{pr}} - RT \ln K_{eq} \qquad \text{(5-16)}$$

where ΔG is the free energy change, in cal/mol, under the specified conditions; R is the gas constant (1.987 cal/mol-K); T is the temperature in Kelvins (use 25°C = 298 K unless otherwise specified); ln is the natural log (i.e., to the base e); $[A]_{pr}$ and $[B]_{pr}$ are the prevailing concentrations of A and B in moles per liter; $[A]_{eq}$ and $[B]_{eq}$ are the equilibrium concentrations of A and B in moles per liter; and K_{eq} is the equilibrium constant at the standard temperature of 298 K (25°C).

More generally, for a reaction in which a molecules of reactant A combine with b molecules of reactant B to form c and d molecules, respectively, of products C and D,

$$a\text{A} + b\text{B} \rightleftharpoons c\text{C} + d\text{D} \qquad \text{(5-17)}$$

ΔG is calculated as

$$\Delta G = RT \ln \frac{[C]_{pr}^{c}[D]_{pr}^{d}}{[A]_{pr}^{a}[B]_{pr}^{b}} - RT \ln K_{eq} \qquad \text{(5-18)}$$

where all the constants and variables are as previously defined.

Returning to reaction 5-10, assume that the prevailing concentrations of glucose-6-phosphate and fructose-6-phosphate in a cell are 10 μM (10×10^{-6} M) and 1 μM (1×10^{-6} M), respectively, at 25°C. Since the ratio of prevailing product concentrations to reactant concentrations is 0.1 and the equilibrium constant is 0.5, there is clearly too little fructose-6-phosphate present relative to glucose-6-phosphate for the reaction to be at equilibrium. The reaction should therefore tend toward the right (in the direction of fructose-6-phosphate generation), and we should expect ΔG to have a negative value for the reaction in that direction.

The actual value for ΔG is calculated as follows:

$$\begin{aligned}
\Delta G &= (1.987 \text{ cal/mol-K})\,(298 \text{ K}) \ln \frac{[1 \times 10^{-6}\,\text{M}]}{[10 \times 10^{-6}\,\text{M}]} \\
&\quad - (1.987 \text{ cal/mol-K})\,(298 \text{ K}) \ln (0.5) \\
&= (592 \text{ cal/mol}) \ln (0.1) - (592 \text{ cal/mol}) \ln (0.5) \\
&= (592 \text{ cal/mol})\,(-2.303) - (592 \text{ cal/mol})\,(-0.693) \\
&= -1364 \text{ cal/mol} + 410 \text{ cal/mol} \\
&= -954 \text{ cal/mol} \qquad\qquad\qquad\qquad \textbf{(5-19)}
\end{aligned}$$

Notice that our expectation of a negative ΔG is confirmed, and we now know exactly how much free energy is liberated upon the spontaneous conversion of 1 mole of glucose-6-phosphate into 1 mole of fructose-6-phosphate under the specified conditions. (As we will see later, the free energy liberated in an exergonic reaction can be either "harnessed" to do work or lost as heat.)

It is important to understand exactly what this calculated value for ΔG means and under what conditions it is valid. Because it is a thermodynamic parameter, ΔG can tell us whether a reaction is thermodynamically possible as written, but it says nothing about the rate or mechanism. It simply says that if the reaction does occur, it will be to the right and will liberate 954 calories of free energy for every mole of glucose-6-phosphate that is converted to fructose-6-phosphate, *provided* that the concentrations of both the reactant and the product are maintained at the initial values (10 μM and 1 μM, respectively) throughout the course of the reaction.

More generally, ΔG is a measure of thermodynamic spontaneity for a reaction in the direction in which it is written (from left to right), at the specified concentrations of reactants and products. In a beaker or test tube, this requirement for constant reactant and product concentrations means that reactants must be added continuously and products must be removed continuously. In the cell, each reaction is part of some metabolic pathway, and its reactants and products are maintained at fairly constant, nonequilibrium concentrations by the reactions that precede and follow it in the sequence.

The Standard State and the Standard Free Energy Change

Because it is a thermodynamic parameter, ΔG is independent of the actual mechanism or pathway for a reaction, but it depends crucially on the conditions under which the reaction occurs. A reaction characterized by a large decrease in free energy under one set of conditions may have a much smaller (but still negative) ΔG or may even go in the opposite direction under a different set of conditions. The melting of ice, for example, depends on temperature; it proceeds spontaneously above 0°C but goes in the opposite direction (freezing) below that temperature. It is therefore important to identify the conditions under which a given measurement of ΔG is made.

By convention, biochemists have agreed on certain arbitrary conditions to define the **standard state** of a system for convenience in reporting, comparing, and tabulating free energy changes in chemical reactions. For systems consisting of dilute aqueous solutions, these are usually a standard temperature of 25°C (298 K), a pressure of 1 atmosphere, and all products and reactants present in their most stable forms at a standard concentration of 1 mol/L (or a pressure of 1 atmosphere for gases). The only common exception to this standard concentration rule is water. The concentration of water in a dilute aqueous solution is approximately 55.5 M and does not change significantly during the course of reactions, even when water is itself a reactant or product. By convention, biochemists do not include the concentration of water in calculations of free energy changes, even though the reaction may indicate a net consumption or production of water.

In addition to standard conditions of temperature, pressure, and concentration, biochemists also frequently specify a standard pH of 7.0 because most biological reactions occur at or near neutrality. The concentration of hydrogen ions (and of hydroxyl ions as well) is therefore 10^{-7} M, so the standard concentration of 1.0 M does not apply to H^+ or OH^- ions when a pH of 7.0 is specified. Values of K_{eq}, ΔG, or other thermodynamic parameters determined or calculated at pH 7.0 are always written with a prime (as K'_{eq}, $\Delta G'$, and so on) to indicate this fact.

Energy changes for reactions are usually reported in standardized form as the change that would occur if the reactions were run under the standard conditions. More precisely, the standard change in any thermodynamic parameter refers to the conversion of a mole of a specified reactant to products or the formation of a mole of a specified product from the reactants under conditions where the temperature, pressure, pH, and concentrations of all relevant species are maintained at their standard values. (As for the more general case, the maintenance of concentrations at 1.0 M implies that reactants are added as they are used up and that products are removed as they are formed.)

The free energy change calculated under these conditions is called the **standard free energy change,** designated $\Delta G^{\circ\prime}$, where the superscript ($\circ$) refers to standard conditions of temperature, pressure, and concentration and the prime ($\prime$) emphasizes that the standard hydrogen ion concentration for biochemists is 10^{-7} M, not 1.0 M.

It turns out that $\Delta G^{\circ\prime}$ bears a simple relationship to the equilibrium constant K'_{eq} (Figure 5-13). This relationship

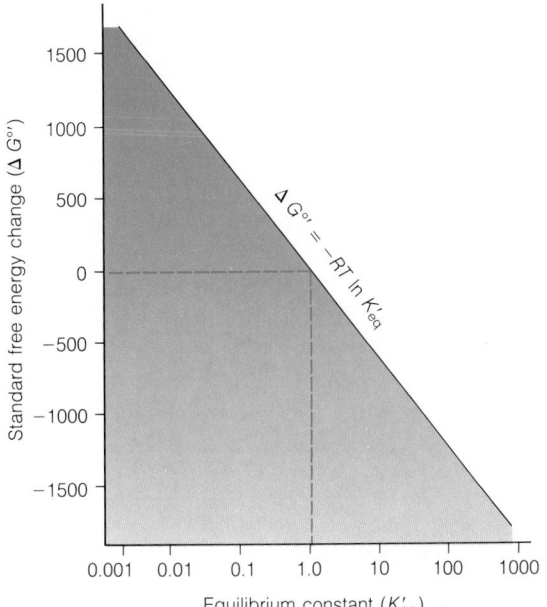

Figure 5-13 The Relationship Between $\Delta G^{\circ\prime}$ and K'_{eq}.
The standard free energy change and the equilibrium constant are related by the equation $\Delta G^{\circ\prime} = -RT \ln K'_{eq}$. Note that the equilibrium constant is plotted on an exponential scale. Note also that when the equilibrium constant is 1.0, the standard free energy change is zero.

can readily be seen by rewriting equation 5-18 with primes and then assuming standard concentrations for all reactants and products. All concentration terms are now 1.0 and the logarithm of 1.0 is zero, so the first term in the general expression for $\Delta G^{\circ\prime}$ is eliminated, and what remains is an equation for $\Delta G^{\circ\prime}$, the free energy change under standard conditions:

$$\Delta G^{\circ\prime} = -RT \ln K'_{eq} \qquad (5\text{-}20)$$

In other words, $\Delta G^{\circ\prime}$ can be calculated directly from the equilibrium constant, provided that the latter has also been determined under the same standard conditions of temperature, pressure, and pH. This, in turn, allows equation 5-18 to be expressed in somewhat simpler form as

$$\Delta G' = RT \ln \frac{[C]^c_{pr}[D]^d_{pr}}{[A]^a_{pr}[B]^b_{pr}} + \Delta G^{\circ\prime} \qquad (5\text{-}21)$$

At the standard temperature of 25°C (298 K), the term RT becomes $(1.987)(298) = 592$ cal/mol, so equations 5-20 and 5-21 can be rewritten as follows, in what are the most useful formulas for our purposes:

$$\Delta G^{\circ\prime} = -592 \ln K'_{eq} \qquad (5\text{-}22)$$

$$\Delta G' = \Delta G^{\circ\prime} + 592 \ln \frac{[C]^c_{pr}[D]^d_{pr}}{[A]^a_{pr}[B]^b_{pr}} \qquad (5\text{-}23)$$

Summing Up: The Meaning of $\Delta G'$ and $\Delta G^{\circ\prime}$

Equations 5-22 and 5-23 summarize the most important contribution of thermodynamics to biochemistry and cell biology—a means of assessing the feasibility of a chemical reaction based on the prevailing concentrations of products and reactants and a knowledge of the equilibrium constant. Equation 5-22 expresses the relationship between the standard free energy change $\Delta G^{\circ\prime}$ and the equilibrium constant K'_{eq} and enables us to calculate the free energy change that would be associated with any reaction of interest if all reactants and products were maintained at a standard concentration of 1.0 M.

If K'_{eq} is greater than 1.0, $\ln K'_{eq}$ will be positive and $\Delta G^{\circ\prime}$ will be negative, and the reaction can proceed to the right under standard conditions. This makes sense because if K'_{eq} is greater than 1.0, products will predominate over reactants at equilibrium. A predominance of products can only be achieved from the standard state by the conversion of reactants to products, so the reaction will tend to proceed spontaneously to the right. Conversely, if K'_{eq} is less than 1.0, $\Delta G^{\circ\prime}$ will be positive and the reaction cannot proceed to the right. Instead, it will tend toward the left because $\Delta G^{\circ\prime}$ for the reverse reaction will have the same numerical value but will be opposite in sign. This is in keeping with the small value for K'_{eq}, which specifies that reactants are favored over products (that is, the equilibrium lies to the left).

The $\Delta G^{\circ\prime}$ values are convenient both because of the ease with which they can be determined from the equilibrium constant and because they provide a uniform convention for reporting free energy changes. But bear in mind that a $\Delta G^{\circ\prime}$ value is an arbitrary standard in that it refers to an arbitrary state specifying conditions of concentration that cannot be achieved with most biologically important compounds. It is therefore useful for standardized reporting, but it is not a valid measure of the thermodynamic spontaneity of reactions as they occur under real conditions.

For that purpose we need $\Delta G'$, which provides a direct measure of how far from equilibrium a reaction is at the concentrations of reactants and products that actually prevail in the cell or other system of interest (equation 5-23). Therefore, $\Delta G'$ is the most useful measure of thermodynamic spontaneity. If it is negative, the reaction in question is thermodynamically spontaneous and can proceed as written under the conditions for which the calculations were made. Its magnitude serves as a measure of how much free energy will be liberated as the reaction occurs. This, in turn, determines the maximum amount of work that can be performed on the surroundings, provided a mechanism is available to conserve and use the energy as it is liberated.

A positive $\Delta G'$, on the other hand, indicates that the reaction cannot occur in the direction written under the conditions for which the calculations were made. Such reactions can sometimes be rendered spontaneous, however, by changes in the concentrations of products or reactants. For the special case where $\Delta G' = 0$, the reaction is clearly at equilibrium, and no net energy change accompanies the

Table 5-2 The Meaning of $\Delta G^{\circ\prime}$ and $\Delta G'$

The Meaning of $\Delta G^{\circ\prime}$

$\Delta G^{\circ\prime}$ *Negative* ($K'_{eq} > 1.0$)	$\Delta G^{\circ\prime}$ *Positive* ($K'_{eq} < 1.0$)	$\Delta G^{\circ\prime} = 0$ ($K'_{eq} = 1.0$)
Products predominate over reactants at equilibrium at standard temperature, pressure, and pH.	Reactants predominate over products at equilibrium at standard temperature, pressure, and pH.	Products and reactants are present equally at equilibrium at standard temperature, pressure, and pH.
Reaction goes spontaneously to the right under standard conditions.	Reaction goes spontaneously to the left under standard conditions.	Reaction is at equilibrium under standard conditions.

The Meaning of $\Delta G'$

$\Delta G'$ *Negative*	$\Delta G'$ *Positive*	$\Delta G' = 0$
Reaction is thermodynamically feasible as written under conditions for which $\Delta G'$ was calculated.	Reaction is not feasible as written under the conditions for which $\Delta G'$ was calculated.	Reaction is at equilibrium under the conditions for which $\Delta G'$ was calculated.
Work can be done by the reaction under conditions for which $\Delta G'$ was calculated.	Energy must be supplied to drive the reaction under the conditions for which $\Delta G'$ was calculated.	No work can be done nor is energy required by the reaction under the conditions for which $\Delta G'$ was calculated.

conversion of reactant molecules into product molecules or vice versa. These features of K'_{eq}, $\Delta G^{\circ\prime}$, and $\Delta G'$ are summarized in Table 5-2.

Free Energy Change: Sample Calculations

To illustrate the calculation and utility of $\Delta G'$ and $\Delta G^{\circ\prime}$, we return once more to the interconversion of glucose-6-phosphate and fructose-6-phosphate (reaction 5-10). We already know that the equilibrium constant for this reaction under standard conditions of temperature, pH, and pressure is 0.5 (equation 5-15). This means that if the enzyme that catalyzes this reaction in cells is added to a solution of glucose-6-phosphate at 25°C, 1 atmosphere, and pH 7.0 and the solution is incubated until no further reaction occurs, both fructose-6-phosphate and glucose-6-phosphate will be present in an equilibrium ratio of 0.5. (Note that this ratio is independent of the actual starting concentration of glucose-6-phosphate and could have been achieved equally well by starting with any concentration of fructose-6-phosphate or any mixture of both, in any starting concentrations.)

The standard free energy change $\Delta G^{\circ\prime}$ can be calculated from K'_{eq} as follows:

$$
\begin{aligned}
\Delta G^{\circ\prime} &= -RT \ln K'_{eq} = -592 \ln K'_{eq} \\
&= -592 \ln (0.5) = -592 \, (-0.693) \\
&= +410 \text{ cal/mol}
\end{aligned}
\tag{5-24}
$$

The positive value for $\Delta G^{\circ\prime}$ is therefore another way of expressing the fact that the reactant (glucose-6-phosphate) is the predominant species at equilibrium. A positive $\Delta G^{\circ\prime}$ value also means that under standard conditions of concentration, the reaction is nonspontaneous (thermodynamically impossible) in the direction written. In other words, if we begin with both glucose-6-phosphate and fructose-6-phosphate present at concentrations of 1.0 M, no net conversion of glucose-6-phosphate to fructose-6-phosphate can occur.

As a matter of fact, given an appropriate catalyst, the reaction will proceed to the *left* under standard conditions because the $\Delta G^{\circ\prime}$ for the reaction in that direction is -410 cal/mol. Fructose-6-phosphate will therefore be converted into glucose-6-phosphate until the equilibrium ratio of 0.5 was reached. Alternatively, if both species were added or removed continuously as necessary to maintain the concentrations of both at 1.0 M, the reaction would proceed continuously and spontaneously to the left (assuming the presence of a catalyst), with the liberation of 410 cal of free energy per mole of fructose-6-phosphate converted to glucose-6-phosphate. In the absence of any provision for conserving this energy, it would be dissipated as heat.

In a real cell, it is extremely unlikely that either of these phosphorylated sugars would ever be present at a concentration even approaching 1.0 M. In fact, experimental values for the actual concentrations of these substances in human red blood cells are as follows:

[glucose-6-phosphate]: 83 μM (83 $\times$ 10^{-6} M)
[fructose-6-phosphate]: 14 μM (14 $\times$ 10^{-6} M)

Using these values, we can calculate the actual $\Delta G'$ for the interconversion of these sugars in red blood cells as follows:

$$\Delta G' = \Delta G^{\circ \prime} + 592 \ln \frac{\text{[fructose-6-phosphate]}_{pr}}{\text{[glucose-6-phosphate]}_{pr}}$$

$$= +410 + 592 \ln \frac{(14 \times 10^{-6})}{(83 \times 10^{-6})}$$

$$= +410 + 592 \ln (0.169)$$

$$= +410 + 592 (-1.78) = +410 - 1054$$

$$= -644 \text{ cal/mol} \qquad \textbf{(5-25)}$$

The negative value for $\Delta G'$ means that the conversion of glucose-6-phosphate into fructose-6-phosphate is thermodynamically possible under the conditions of concentration actually prevailing in red blood cells and that the reaction will yield 644 cal of free energy per mole of reactant converted to product. Thus, the conversion of reactant to product is thermodynamically impossible under standard conditions, but the red blood cell maintains these two phosphorylated sugars at concentrations adequate to offset the positive $\Delta G^{\circ \prime}$, thereby rendering the reaction possible. This adaptation, of course, is essential if the red blood cell is to be successful in carrying out the glucose-degrading process of glycolysis of which this reaction is a part.

Life and the Steady State

As this chapter has emphasized, the driving force in all reactions is their tendency to move toward equilibrium. Indeed, $\Delta G^{\circ \prime}$ and $\Delta G'$ are really nothing more than convenient means of quantifying how far and in what direction from equilibrium a reaction lies under the specific conditions dictated by standard or prevailing concentrations of products and reactants. But to understand how cells really function, we must appreciate the importance of reactions that move toward equilibrium without ever achieving it. At equilibrium, the forward and backward rates become the same for a reaction, there is no net flow of matter in either direction,

and, most importantly, no further energy can be extracted from the reaction because $\Delta G'$ is zero for a reaction at equilibrium.

For all practical purposes, then, a reaction at equilibrium is a reaction that has stopped. But a living cell is characterized by reactions that are continuous, not stopped. A cell at equilibrium would be a dead cell. We might, in fact, define life as a continual struggle to maintain a myriad of cellular reactions in positions far from equilibrium because at equilibrium no net reactions are possible, no energy can be released, no work can be done, and the thermodynamically improbable order of the living state cannot be maintained.

Thus, life is possible only because living cells maintain themselves in a **steady state** far from thermodynamic equilibrium. The levels of glucose-6-phosphate and fructose-6-phosphate found in red blood cells illustrate this point. As we have seen, these compounds are maintained in the cell at steady-state concentrations far from the equilibrium condition predicted by the K'_{eq} value of 0.5. In fact, the levels are so far from equilibrium concentrations that the conversion of glucose-6-phosphate to fructose-6-phosphate can occur continuously in the cell, even though the equilibrium state has a positive $\Delta G^{\circ \prime}$ and actually favors glucose-6-phosphate. The same is true of most reactions and pathways in the cell. They can proceed and can be harnessed to perform various kinds of cellular work because reactants, products, and intermediates are maintained at steady-state concentrations far from the thermodynamic equilibrium.

This state, in turn, is possible only because a cell is an open system and receives large amounts of energy from its environment. If the cell were a closed system, all its reactions would gradually run to equilibrium and the cell would come inexorably to a state of minimum free energy, after which no further changes could occur, no work could be accomplished, and life would cease. The steady state so vital to life is possible only because the cell is able to take up energy continuously from its environment, whether in the form of light or preformed organic food molecules. This continuous uptake of energy and the accompanying flow of matter make possible the maintenance of a steady state in which all the reactants and products of cellular chemistry are kept far enough from equilibrium to ensure that the thermodynamic drive toward equilibrium can be harnessed by the cell to perform useful work, thereby maintaining and extending its activities and structural complexity.

We will focus on how this is accomplished in later chapters. In the next chapter, we will look at principles of enzyme catalysis that determine the rates of cellular reactions (that convert the "can go" of thermodynamics into the "will go" of kinetics). Then we will be ready to move on to Part Three and the functional metabolic pathways that result from series of such reactions acting in concert.

The thermodynamically improbable level of order that exists in cells is possible only because of the availability of energy from the environment. Cells require energy to carry out various kinds of work, including synthesis, movement, concentration, charge separation, the generation of heat, and bioluminescence. The energy needed for these processes comes either from the sun or from the bonds of oxidizable organic molecules such as carbohydrates, fats, and proteins. Since chemotrophs feed directly or indirectly on phototrophs, there is a unidirectional flow of energy through the biosphere, with the sun as the ultimate source and entropy and heat losses as the eventual fate of all the energy that moves through living systems.

The flow of energy through cells is governed by the laws of thermodynamics. The first law specifies that energy can change form but must always be conserved. The second law provides a measure of thermodynamic spontaneity, although this means only that a reaction can go and says nothing about whether it will actually go, or at what rate.

Spontaneous processes are always accompanied by an *increase* in the entropy of the universe and by a *decrease* in the free energy of the system. The latter is a far more practical indicator of spontaneity because it can be calculated readily from the equilibrium constant and the prevailing concentrations of reactants and products.

Cells obtain the energy they need to carry out their activities by maintaining the many reactants and products of the various reaction sequences at steady-state concentrations far from equilibrium, thereby allowing the reactions to move exergonically toward equilibrium without ever actually reaching it. A negative $\Delta G'$ is a necessary prerequisite for a reaction to proceed, but it does not guarantee that the reaction will actually occur at a reasonable rate. To assess that, we must know more about the reaction than just its thermodynamic status. We need to know whether an appropriate catalyst is on hand and at what rate the reaction can occur in the presence of the catalyst. In other words, we need the enzymes that will be discussed in Chapter 6.

KEY TERMS FOR SELF-TESTING

The Importance of Energy

energy (p. 113)
biosynthesis (p. 113)
synthetic work (p. 113)
mechanical work (p. 115)
concentration work (p. 115)
electrical work (p. 116)
heat (p. 116)
bioluminescence (p. 116)
phototroph (p. 116)
chemotroph (p. 116)

Bioenergetics

thermodynamics (p. 119)
bioenergetics (p. 119)

system (p. 119)
surroundings (p. 119)
closed system (p. 119)
open system (p. 119)
state (p. 119)
isothermal (p. 120)
work (p. 120)
calorie (cal) (p. 120)
joule (J) (p. 120)
first law of thermodynamics (p. 120)
internal energy (E) (p. 121)
enthalpy (H) (p. 121)
second law of thermodynamics (p. 123)
entropy (S) (p. 123)

free energy (G) (p. 124)
free energy change (ΔG) (p. 124)
exergonic (p. 124)
endergonic (p. 124)
spontaneity (p. 128)

Understanding ΔG

equilibrium constant (K_{eq}) (p. 128)
standard state (p. 130)
standard free energy change ($\Delta G^{o\prime}$) (p. 130)

Life and the Steady State

steady state (p. 133)

PROBLEM SET

5-1. Solar Energy. Although we are concerned at present with a global energy crisis, we actually live on a planet that is flooded continuously with an extravagant amount of energy in the form of solar radiation. Every day, year in and year out, solar energy arrives at the upper surface of the Earth's atmosphere at the rate of 1.94 cal/min per square centimeter of cross-sectional area (the *solar energy constant*).

(a) Assuming the cross-sectional area of the Earth to be about 1.28×10^{18} cm^2, what is the total annual amount of incoming energy?

(b) A substantial portion of that energy, particularly in the wavelength ranges below 300 nm and above 800 nm, never reaches the Earth's surface. Can you suggest what happens to it?

(c) Of the radiation that reaches the Earth's surface, only a small proportion is actually trapped photosynthetically by phototrophs. (You can calculate the actual value in Problem 5-2). Why do you think the efficiency of utilization is so low?

5-2. Photosynthetic Energy Transduction. The amount of energy trapped and the volume of carbon converted to organic form by photosynthetic energy transducers are mind-boggling: about 5×10^{16} g of carbon per year over the entire Earth's surface.

(a) Assuming that the average organic molecule in a cell has about the same proportion of carbon as glucose does, how many grams of organic matter are produced annually by carbon-fixing phototrophs?

(b) Assuming that all the organic matter in part a is glucose (or any molecule with an energy content equivalent to that of glucose), how much energy is represented by that quantity of organic matter? Assume that glucose has an energy content (free energy of combustion) of 3.8 kcal/g.

(c) Refer to the answer for Problem 5-1a. What is the average efficiency with which the radiant energy incident on the upper atmosphere is trapped photosynthetically on the Earth's surface?

(d) What proportion of the net annual phototrophic production of organic matter calculated in part a do you think is consumed by chemotrophs each year?

5-3. Energy Conversion. Most cellular activities involve the conversion of energy from one form to another. For each of the following cases, give a biological example and explain the significance of the conversion.

(a) Chemical energy into mechanical energy

(b) Chemical energy into radiant energy

(c) Radiant energy into chemical energy

(d) Chemical energy into electrical energy

(e) Chemical energy into the potential energy of a concentration gradient

5-4. Enthalpy, Entropy, and Free Energy. The oxidation of glucose to carbon dioxide and water is represented by the following reaction whether the oxidation occurs by combustion in the laboratory or by biological oxidation in living cells:

$$C_6H_{12}O_6 + 6O_2 \rightleftharpoons 6CO_2 + 6H_2O \qquad \textbf{(5-26)}$$

When combustion is carried out under controlled conditions in the laboratory, the reaction is highly exothermic, with an enthalpy change (ΔH) of -673 kcal/mol. As you will discover in Chapter 11, ΔG for this reaction at 25°C is -686 kcal/mol, so the reaction is also highly exergonic.

(a) Explain in your own words what the ΔH and ΔG values mean. What do the negative signs mean in each case?

(b) What does it mean to say that the difference between the ΔG and ΔH values is due to entropy?

(c) Without doing any calculations, would you expect ΔS (entropy change) for this reaction to be positive or negative? Explain your answer.

(d) Now calculate ΔS for this reaction at 25°C. Does the calculated value agree in sign with your prediction in part c?

(e) What are the values of ΔG, ΔH, and ΔS for the reverse of the above reaction as carried out by a photosynthetic algal cell that is using CO_2 and H_2O to make $C_6H_{12}O_6$?

5-5. The Equilibrium Constant. The following reaction is one of the steps in the glycolytic pathway, which we will encounter again in Chapter 11. You should recognize it already, however, because we used it as an example earlier (reaction 5-10):

$$\text{glucose-6-phosphate} \rightleftharpoons \text{fructose-6-phosphate} \qquad \textbf{(5-27)}$$

The equilibrium constant for this reaction at 25°C is 0.5.

(a) Assume that you incubate a solution containing 0.15 M glucose-6-phosphate (G6P) overnight at 25°C with the enzyme phosphoglucomutase that catalyzes the above reaction. How many millimoles of fructose-6-phosphate (F6P) will you recover from 10 mL of the incubation mixture the next morning, assuming you have an appropriate chromatographic procedure to separate F6P from G6P?

(b) What answer would you get for part a if you had started with a solution containing 0.15 M F6P instead?

(c) What answer would you expect for part a if you had started with a solution containing 0.15 M G6P but forgot to add phosphoglucomutase to the incubation mixture?

(d) Would you be able to answer the question in part a if you had used 15°C as your incubation temperature instead of 25°C? Why or why not?

5-6. Calculating $\Delta G^{\circ\prime}$ and $\Delta G^{\prime}$. Like the reaction in Problem 5-5, the conversion of 3-phosphoglycerate to 2-phosphoglycerate is an important cellular reaction because it is one of the steps in the glycolytic pathway (see Chapter 11):

$$\text{3-phosphoglycerate} \rightleftharpoons \text{2-phosphoglycerate} \qquad \textbf{(5-28)}$$

If the enzyme that catalyzes this reaction is added to a solution of 3-phosphoglycerate at 25°C and pH 7.0, the equilibrium ratio between the two species will be 0.165:

$$K^{\prime}_{eq} = \frac{[\text{2-phosphoglycerate}]}{[\text{3-phosphoglycerate}]} = 0.165 \qquad \textbf{(5-29)}$$

Experimental values for the actual steady-state concentrations of these compounds in human red blood cells are 61 μM for 3-phosphoglycerate and 4.3 μM for 2-phosphoglycerate.

(a) Calculate $\Delta G^{\circ\prime}$. Explain in your own words what this value means.

(b) Calculate $\Delta G^{\prime}$. Explain in your own words what this value means.

(c) If conditions in the cell change such that the concentration of 3-phosphoglycerate remains fixed at 61 μM but the concentration of 2-phosphoglycerate begins to rise, how high can the 2-phosphoglycerate concentration get before reaction 5-28 will cease because it is no longer thermodynamically feasible?

5-7. Hydrolysis of Glucose-6-Phosphate. The hydrolysis of glucose-6-phosphate to glucose and phosphate is an important step in glycogen catabolism (see Figure 9-11). The free glucose formed by this reaction is released into the blood for transport to cells in need of energy.

$$\text{glucose-6-phosphate} + H_2O \longrightarrow \text{glucose} + \text{phosphate} \qquad \textbf{(5-30)}$$

$\Delta G^{\circ\prime}$ for this reaction is -3.3 kcal/mol at 25°C and pH 7.0.

(a) Calculate $\Delta G^{\prime}$ at 25°C when the concentration of glucose-6-phosphate is 20 μM and the glucose and phosphate concentrations are 5 mM each. Assume the appropriate enzyme is present to catalyze this reaction.

(b) In which direction would you expect the reaction to proceed under the conditions defined in part a?

5-8. Backward or Forward? The interconversion of dihydroxyacetone phosphate (DHAP) and glyceraldehyde-3-phosphate (G3P) is a part of both the glycolytic pathway (see Chapter 11) and the Calvin cycle for photosynthetic carbon fixation (see Chapter 13):

$$\text{(5-31)}$$

The value of $\Delta G^{\circ\prime}$ for this reaction is $+1.8$ kcal/mol at 25°C. In the glycolytic pathway, this reaction goes to the *right*, converting DHAP to G3P. In the Calvin cycle, this reaction proceeds to the *left*, converting G3P to DHAP.

(a) In which direction does the equilibrium lie? What is the equilibrium constant at 25°C?

(b) In which direction does this reaction tend to proceed under standard conditions? What is $\Delta G^\prime$ for the reaction *in that direction*?

(c) In the glycolytic pathway, this reaction is driven to the right because G3P is consumed by the next reaction in the sequence, thereby maintaining a low G3P concentration. What will $\Delta G^\prime$ be (at 25°C) if the concentration of G3P is maintained at 1% of the DHAP concentration (i.e., if [G3P]/[DHAP] = 0.01)?

(d) In the Calvin cycle, this reaction proceeds to the left. How low must the [G3P]/[DHAP] ratio be to ensure that the reaction is exergonic by at least -1.0 kcal/mol (at 25°C)?

5-9. Succinate Oxidation. The oxidation of succinate to fumarate occurs as one of the reactions in the tricarboxylic acid (TCA) cycle, an important component of chemotrophic energy metabolism (see Chapter 12). The two hydrogen atoms that are removed from succinate are accepted by a coenzyme molecule called flavin adenine dinucleotide (FAD), which is thereby reduced to $FADH_2$:

$$\text{(5-32)}$$

The value of $\Delta G^{\circ\prime}$ for this reaction is 0 cal/mol.

(a) If you start with a solution containing 0.01 M each of succinate and FAD and add an appropriate amount of the enzyme that catalyzes this reaction, will any fumarate be formed? If so, calculate the resulting equilibrium concentrations of all four species. If not, explain why not.

(b) Answer part a assuming that 0.01 M $FADH_2$ is also present initially.

(c) Assuming that the steady-state conditions in a cell are such that the $FADH_2$/FAD ratio is 5 and the fumarate concentration is 2.5 μM, what steady-state concentration of succinate

would be necessary to maintain $\Delta G^\prime$ for succinate oxidation at -1.5 kcal/mol?

5-10. Proof of Additivity. A useful property of thermodynamic parameters such as $\Delta G^\prime$ or $\Delta G^{\circ\prime}$ is that they are additive for sequential reactions. Assume that $K^\prime_{AB}$, $K^\prime_{BC}$, and $K^\prime_{CD}$ are the respective equilibrium constants for reactions 1, 2, and 3 of the following sequence:

$$A \underset{\text{Reaction 1}}{\rightleftharpoons} B \underset{\text{Reaction 2}}{\rightleftharpoons} C \underset{\text{Reaction 3}}{\rightleftharpoons} D \qquad \text{(5-33)}$$

(a) Prove that the equilibrium constant $K^\prime_{AD}$ for the overall conversion of A to D is the *product* of the three component equilibrium constants:

$$K^\prime_{AD} = K^\prime_{AB} \cdot K^\prime_{BC} \cdot K^\prime_{CD} \qquad \text{(5-34)}$$

(b) Prove that the $\Delta G^{\circ\prime}$ for the overall conversion of A to D is the *sum* of the three component $\Delta G^{\circ\prime}$ values:

$$\Delta G^{\circ\prime}_{AD} = \Delta G^{\circ\prime}_{AB} + \Delta G^{\circ\prime}_{BC} + \Delta G^{\circ\prime}_{CD} \qquad \text{(5-35)}$$

(c) Prove that the $\Delta G^\prime$ values are similarly additive.

5-11. Utilizing Additivity. The additivity of thermodynamic parameters discussed in Problem 5-10 applies not just to sequential reactions in a pathway, but to *any* reactions or processes. Moreover, it also applies to subtraction of reactions. Use this information to answer the following questions.

(a) The phosphorylation of glucose using inorganic phosphate (abbreviated P_i) is endergonic ($\Delta G^{\circ\prime} = +3.3$ kcal/mol), whereas the dephosphorylation (hydrolysis) of ATP is exergonic ($\Delta G^{\circ\prime} = -7.3$ kcal/mol).

$$\text{glucose} + P_i \rightleftharpoons \text{glucose-6-phosphate} + H_2O \quad \text{(5-36)}$$

$$\text{ATP} + H_2O \rightleftharpoons \text{ADP} + P_i \qquad \text{(5-37)}$$

Use this information to write a reaction for the phosphorylation of glucose by the transfer of a phosphate group from ATP, and calculate the $\Delta G^{\circ\prime}$ value for the reaction.

(b) Phosphocreatine is used by your muscle cells to store energy. The dephosphorylation of phosphocreatine, like that of ATP (reaction 5-37), is a highly exergonic reaction with $\Delta G^{\circ\prime} = -10.3$ kcal/mol:

$$\text{phosphocreatine} + H_2O \rightleftharpoons \text{creatine} + P_i \quad \text{(5-38)}$$

Use this information to write a reaction for the transfer of phosphate from phosphocreatine to ADP to generate creatine and ATP, and calculate the $\Delta G^{\circ\prime}$ value for the reaction.

5-12. Protein Folding. In Chapter 2, you learned that a polypeptide in solution usually folds into its proper three-dimensional shape spontaneously. The driving force for this folding is the tendency to achieve the thermodynamically most favored conformation. A folded polypeptide can be induced to unfold (i.e., will undergo denaturation) if the solution is heated or made acidic or alkaline. The denatured polypeptide is a random structure, with many possible conformations.

(a) What is the sign of ΔG for the folding process? What about for the unfolding (denaturation) process?

(b) What is the sign of ΔS for the folding process? What about for the unfolding (denaturation) process?

(c) Will the contribution of ΔS to the free energy change be positive or negative?

5-13. Thermodynamic Logic. Because most biochemical reactions occur with little or no change in pressure, volume, or temperature, biochemists express the free energy change, ΔG, in terms of either ΔH (change in enthalpy) or ΔE (change in internal energy):

$$\Delta G = \Delta E - T\Delta S \qquad \text{(5-39)}$$

$$\Delta G = \Delta H - T\Delta S \qquad \text{(5-40)}$$

(a) Which of these equations, if either, is likely to be more exact for a system in which pressure, volume, and temperature are held rigorously constant? Explain your answer.

(b) Which of these equations, if either, is likely to be more exact for a biological system in which pressure, volume, and temperature are approximately, but not rigorously, constant? Explain your answer.

5-14. A Special Challenge. Consider two consecutive reactions in the TCA cycle, a metabolic pathway that you will encounter in Chapter 12. In reaction 5-41, a four-carbon compound called malate is oxidized to oxaloacetate with concomitant reduction of a coenzyme molecule called nicotinamide adenine dinucleotide (NAD). In reaction 5-42, oxaloacetate combines with acetate, a two-carbon compound carried into the reaction by a carrier molecule called coenzyme A (CoA), to form citrate, a six-carbon compound, releasing the CoA in the process.

$$\text{malate} + \text{NAD}_{oxid} \rightleftharpoons \text{oxaloacetate} + \text{NAD}_{red} \qquad \text{(5-41)}$$

$$\text{oxaloacetate} + \text{acetyl CoA} + \text{H}_2\text{O} \rightleftharpoons \text{citrate} + \text{CoA} \qquad \text{(5-42)}$$

The $\Delta G^{o\prime}$ values for these reactions at 25°C are +7.1 kcal/mol and −8.0 kcal/mol, respectively. Assume that these reactions are occurring in a bacterial cell growing at 25°C and that the concentrations of the reactants and products are as follows:

[malate]: 0.4 mM	[citrate]: 3.0 mM
[CoA]: 1.0 mM	[acetyl CoA]: 3.0 mM
[NAD$_{oxid}$]/[NAD$_{red}$]: 100	

(a) What is the maximum concentration of oxaloacetate permissible in the bacterial cell under these conditions, given that reaction 5-41 must proceed spontaneously to the right for the TCA cycle to be functional?

(b) For the concentration of oxaloacetate calculated in part a, determine $\Delta G'$ for reaction 5-42. Does this reaction also proceed spontaneously to the right?

(c) Why must the concentration of oxaloacetate within the bacterial cell be kept very low?

SUGGESTED READING

General References

Barrow, G. M. *Physical Chemistry for the Life Sciences,* 2d ed. New York: McGraw-Hill, 1981.

Eisenberg, D., and D. Crothers. *Physical Chemistry with Applications to the Life Sciences.* Menlo Park, CA: Benjamin/Cummings, 1979.

Mathews, C. K., and K. E. van Holde, *Biochemistry,* 2d ed. Menlo Park, CA: Benjamin/Cummings, 1996.

van Holde, K. E. *Physical Biochemistry,* 2d ed. Englewood Cliffs, NJ: Prentice-Hall, 1985.

Wood, W. B., J. H. Wilson, R. M. Benbow, and L. E. Hood. *Biochemistry: A Problems Approach,* 2d ed. Menlo Park, CA: Benjamin/Cummings, 1981.

Zubay, G. *Biochemistry,* 2d ed. New York: Macmillan, 1988.

Bioenergetics

Baker, J. J. W., and G. E. Allen. *Matter, Energy, and Life: An Introduction to Chemical Concepts,* 4th ed. Reading, MA: Addison-Wesley, 1981.

Christensen, H. N., and R. A. Cellarius. *Introduction to Bioenergetics.* Philadelphia: Saunders, 1972.

Harold, F. M. *The Vital Force: A Study of Bioenergetics.* New York: W. H. Freeman, 1986.

Racker, E. From Pasteur to Mitchell: A hundred years of bioenergetics. *Fed. Proc.* 39 (1980): 210.

Energy Flow

Gates, D. M. The flow of energy in the biosphere. *Sci. Amer.* 224 (September 1971): 88.

Miller, G. T., Jr. *Energetics, Kinetics, and Life—An Ecological Approach.* Belmont, CA: Wadsworth, 1971.

Free Energy and Entropy

Blum, H. F. *Time's Arrow and Evolution,* 3d ed. Princeton, NJ: Princeton University Press, 1968.

Hess, B., and M. Markus. Order and chaos in biochemistry. *Trends Biochem. Sci.* 12 (1987): 45.

Hill, T. L. *Free Energy Transduction in Biology.* New York: Academic Press, 1977.

Morowitz, H. J. Entropy anyone? *Hosp. Pract.* 16 (1981): 114.

Morowitz, H. J. *Entropy for Biologists: An Introduction to Thermodynamics.* New York: Academic Press, 1970.

6

ENZYMES: THE CATALYSTS OF LIFE

I n Chapter 5, we encountered $\Delta G'$, the change in free energy, and saw its importance as an indicator of thermodynamic spontaneity. Specifically, the *sign* of $\Delta G'$ tells us whether a reaction is possible in the indicated direction, and the *magnitude* of $\Delta G'$ indicates how much energy can be obtained (or must be invested) as the reaction proceeds in that direction under the conditions for which $\Delta G'$ was calculated. At the same time, we were careful to note that because it is a thermodynamic parameter, $\Delta G'$ can provide us with no clue as to whether a feasible reaction will actually take place, or at what rate. In other words, $\Delta G'$ tells us only whether a reaction *can* go but says nothing at all about whether it actually *will* go. For that distinction we need to know not just the direction and energetics of the reaction, but something about the mechanism and rate as well.

This brings us to the topic of **enzyme catalysis,** because virtually all cellular reactions or processes are mediated by protein (or, in certain cases, RNA) catalysts called **enzymes.** The only reactions that occur at any appreciable rate in a cell are those for which the appropriate enzymes are present and active. Thus, enzymes almost always spell the difference between "can go" and "will go" for cellular reactions. It is only as we explore the nature of enzymes and their catalytic properties that we begin to understand how reactions that are energetically feasible actually take place in cells and how the rates of such reactions are controlled.

In this chapter, we will first examine why thermodynamically spontaneous reactions do not usually occur at appreciable rates without a catalyst, and then we will look at the role of enzymes as specific biological catalysts. We will also see how the rate of an enzyme-catalyzed reaction is affected by the amount of substrate available to it, as well as some of the ways in which reaction rates are regulated to meet the needs of the cell.

Activation Energy and the Metastable State

If you stop to think about it, you are already familiar with many reactions that are thermodynamically feasible yet do not occur to any appreciable extent. An obvious example from Chapter 5 is the oxidation of glucose (see reaction 5-8.) This reaction (or series of reactions, really) is highly exergonic ($\Delta G^{\circ\prime} = -686$ kcal/mol) and yet does not take place on its own. In fact, glucose crystals or a glucose solution can be exposed to the oxygen in the air indefinitely, and little or no oxidation occurs. The cellulose in the paper on which these words are printed is another example—and so, for that matter, are you, consisting as you do of an improbable and complex collection of thermodynamically unstable molecules.

Not nearly as familiar but equally important to cellular chemistry are the myriad thermodynamically feasible reactions in cells that could go, but seem not to on their own. As an example, consider the high-energy molecule adenosine triphosphate (ATP), which has a highly favorable $\Delta G^{\circ\prime}$ (-7.3 kcal/mol) for the hydrolysis of its terminal phosphate group to form the corresponding diphosphate (ADP) and inorganic phosphate (P_i):

$$\text{ATP} + \text{H}_2\text{O} \rightleftharpoons \text{ADP} + \text{P}_i \qquad \textbf{(6-1)}$$

The reaction is very exergonic under standard conditions and is even more so under the conditions that prevail in cells. Yet despite the highly favorable free energy change, this reaction occurs only slowly on its own, so that ATP remains stable for several days when dissolved in pure water. This property turns out to be shared by many biologically important molecules and reactions, and it is important to understand why.

Activation Energy

Molecules that should react with one another often do not because they lack sufficient energy. For every reaction, there is a specific **activation energy** (E_A), which is the minimum amount of energy that two molecules must have in order to react. Figure 6-1a shows the activation energy required for the reaction between ATP and H_2O.

In addition to $\Delta G°'$, which measures the difference in free energy between reactants and products (-7.3 kcal/mol, in this case), there is also a characteristic activation energy that a given pair of ATP and H_2O molecules must have before a collision between them will be successful, leading to a reaction. The actual rate of a reaction is always proportional to the fraction of molecules that have an energy content equal to or greater than E_A. When in solution at room temperature, molecules of ATP and water move about readily, each possessing a certain amount of energy at any instant. As Figure 6-1b shows, the energy distribution among molecules will be bell-shaped; some molecules will have very little energy, some will have a lot, and most will be somewhere near the average. The important point is that only those with enough energy to exceed the *activation energy barrier* are capable of reacting at a given instant.

The Metastable State

For most biologically important reactions at normal cellular temperatures, the activation energy is sufficiently high that the proportion of molecules possessing that much energy at any instant is extremely small. Accordingly, the rates of uncatalyzed reactions in cells are very low, and most molecules appear to be stable even though they are potential reactants in thermodynamically favored reactions. They are, in other words, thermodynamically unstable but kinetically stable, lacking adequate kinetic energy to exceed the activation energy barrier.

Such seemingly stable molecules are said to be in a **metastable state**. For cells and cell biologists, high activation energies and the resulting metastable state of cellular constituents are crucial, because life by its very nature is a system maintained in a steady state a long way from equilibrium. If it were not for the metastable state, all reactions would proceed quickly to equilibrium, and life as we know it would be impossible. Life, then, depends critically on the high activation energies that prevent most cellular reactions from occurring at appreciable rates in the absence of a suitable catalyst.

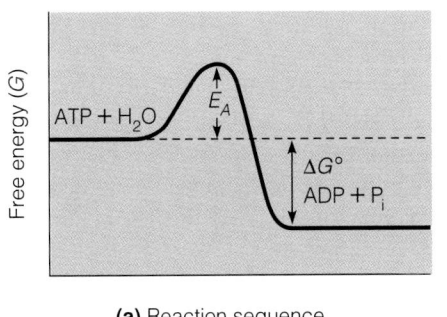

(a) Reaction sequence

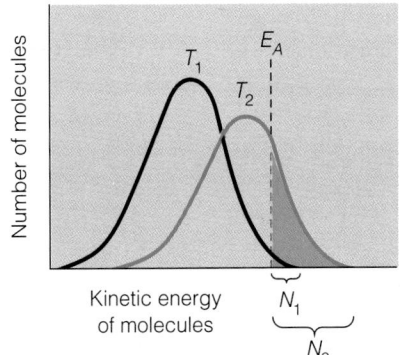

(b) Thermal activation

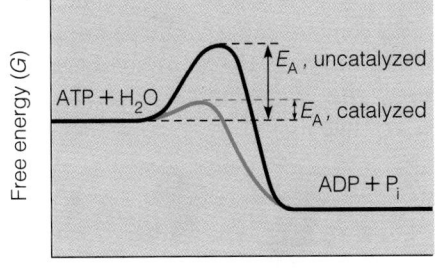

(c) Reaction sequence with catalyst

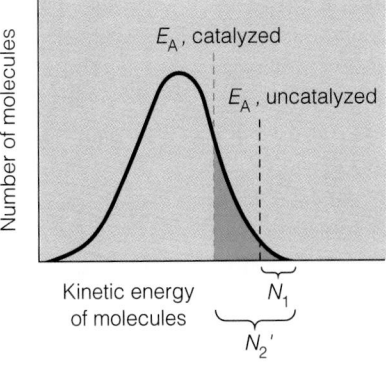

(d) Catalytic activation

Figure 6-1 The Effect of Catalysis on Activation Energy and Number of Molecules Capable of Reaction. **(a)** The activation energy E_A is the minimum amount of kinetic energy reactant molecules (here ATP and H_2O) must possess to permit collisions leading to product formation. After reactants overcome the activation energy barrier and enter into a reaction, the products have less free energy by the amount $\Delta G°$. **(b)** The number of molecules N_1 that have sufficient energy to exceed the activation energy barrier (E_A) and collide successfully can be increased to N_2 by raising the temperature from T_1 to T_2. **(c)** Alternatively, the activation energy can be lowered by a catalyst, thereby **(d)** increasing the number of molecules from N_1 to N_2'.

Overcoming the Activation Energy Barrier

Important as it is to the maintenance of the metastable state, the activation energy requirement is nonetheless a barrier that must be overcome if desirable reactions are to proceed at reasonable rates. Since the energy content of a given molecule must exceed E_A before that molecule is capable of undergoing reaction, the only way a reaction involving metastable reactants can be made to proceed at an appreciable rate is to increase the proportion of molecules with sufficient energy. This can be achieved either by increasing the average energy content of all molecules or by lowering the activation energy requirement.

One way to increase the energy content of the system is by the input of heat. As Figure 6-1b illustrates, simply increasing the temperature of the system from T_1 to T_2 will increase the kinetic energy of the average molecule, thereby ensuring a greater number of reactive molecules (N_2 instead of N_1). Thus, the hydrolysis of ATP could be facilitated by heating the solution, giving each ATP and water molecule more energy.

Sometimes, only a brief initial increase in temperature is enough to trigger a thermodynamically feasible reaction, because the reaction, once under way, liberates enough energy to activate additional molecules, thereby becoming self-sustaining. Lighting a match illustrates this point. Once initiated, the combustion of the match head generates so much heat that it sustains itself. An electric spark sometimes provides the same sort of initial energy. For example, the reaction of gasoline (a hydrocarbon) and oxygen in air is potentially a violently exergonic reaction, yet the activation energy is sufficiently high that the hydrocarbon and oxygen molecules can exist side by side in the metastable state. An electric spark, however, provides the energy to overcome the barrier, and the reaction becomes self-sustaining as the two gases combine explosively. When appropriately confined in a cylinder with a movable piston, this reaction provides the energy source for the gasoline engine.

The problem with using a match or spark or a continuously elevated temperature is that such approaches are incompatible with life, because biological systems require a relatively constant temperature. Cells would find it extremely difficult to function if activation energy requirements for specific reactions required touching a match to the molecules in question! They are basically *isothermal* (constant-temperature) systems and require isothermal methods to solve the activation problem. Moreover, we can anticipate a later topic if we note that this approach would also suffer from a lack of specificity because the input of heat would result in many activation energy barriers being indiscriminately overcome. Yet the essence of successful regulation of cellular activities lies in the ability to facilitate specific reactions under specific conditions while leaving other metastable molecules undisturbed.

The alternative to thermal activation is to lower the activation energy requirement, thereby making it statistically more likely that a greater proportion of molecules will have sufficient energy to collide successfully and undergo reaction. This is possible because, unlike the free energy change for a given reaction, the activation energy depends not only on the initial and final states but also on the mechanism. If a reaction depends on random collision between molecules, a relatively high energy is almost always required to ensure reactivity. But if the reactants can be carefully ordered on some sort of surface in a way that brings potentially reactive portions of adjacent molecules into close juxtaposition, their interaction will be greatly favored and the activation energy effectively reduced.

Providing such a reactive surface is the task of a **catalyst**, an agent that enhances the rate of a reaction by lowering the energy of activation (Figure 6-1c), thereby ensuring that a higher proportion of the molecules are energetic enough to undergo reaction without the input of heat (Figure 6-1d). A primary feature of a catalyst is that it is not permanently changed or consumed as the reaction proceeds. It simply provides a suitable surface and environment to facilitate the reaction.

For a specific example of catalysis, consider the decomposition of hydrogen peroxide (H_2O_2) into water and oxygen:

$$2H_2O_2 \rightleftharpoons 2H_2O + O_2 \qquad \textbf{(6-2)}$$

This is a thermodynamically favored reaction, yet hydrogen peroxide exists in a metastable state because of the high activation energy of the reaction. However, if we add a small number of ferric ions (Fe^{3+}) to a hydrogen peroxide solution, the decomposition reaction proceeds about 30,000 times faster than without the ferric ions. Clearly, Fe^{3+} is a catalyst for this reaction, lowering the activation energy (as shown in Figure 6-1c) and thereby ensuring that a significantly greater proportion (30,000-fold more) of the hydrogen peroxide molecules possess adequate energy to decompose at the existing temperature without the input of added energy.

The biological solution to hydrogen peroxide breakdown, however, is not the addition of ferric ions but the enzyme *catalase*, an iron-containing protein. In the presence of catalase, the reaction proceeds about 100,000,000 times faster than the uncatalyzed reaction. Catalase contains iron atoms bound in chemical structures called *porphyrins*, thus taking advantage of inorganic catalysis within the context of a protein molecule. This combination is a much more effective catalyst for hydrogen peroxide decomposition than ferric acids by themselves. This example underscores the importance of enzymes as catalysts and brings us to the main theme of this chapter.

Enzymes as Biological Catalysts

Regardless of their chemical nature, all catalysts share the following three basic properties.

1. A catalyst increases the rate of a reaction by lowering the activation energy requirement, thereby allowing thermodynamically feasible reactions to occur at a reasonable rate without thermal activation.

2. A catalyst acts by forming transient complexes with substrate molecules, ordering them in a manner that facilitates their interaction.

3. A catalyst changes only the *rate* at which equilibrium is achieved; it has no effect on the *position* of the equilibrium. This means that a catalyst can enhance the rate of exergonic reactions but cannot somehow drive an endergonic reaction. Catalysts, in other words, are not thermodynamic genies.

These properties are common to all catalysts, organic and inorganic alike. In terms of our example, they apply equally to ferric ions and to catalase molecules. However, biological systems rarely use inorganic catalysts. Instead, essentially all catalysis in cells is carried out by organic molecules (proteins, in most cases) called *enzymes.* Because enzymes are organic molecules, they are much more specific than inorganic catalysts, and their activities can be regulated much more carefully.

Enzymes as Proteins

The capacity of cellular extracts to catalyze chemical reactions has been known since the fermentation studies of Eduard and Hans Buchner in 1897. In fact, one of the first terms for what we now call enzymes was *ferments.* However, it was not until 1926 that a specific enzyme, *urease,* was crystallized (from jack beans by James B. Sumner) and shown to be a protein. (Urease catalyzes the decomposition of urea, NH_2—CO—NH_2, to ammonia and carbon dioxide. Its crystallization from a common plant not only established the protein nature of enzymes but also drove the final nail in the coffin of *vitalism,* the belief that biochemical reactions could only occur in living cells through the agency of a "vital force.")

Even then, it took a while for biochemists and enzymologists to appreciate that the "ferments" they were studying were, in fact, protein catalysts, and that to understand them as catalysts really meant understanding their structure and function as proteins. (Since the early 1980s, biologists have recognized that, in addition to proteins, certain RNA molecules, called *ribozymes,* also have catalytic activity. Ribozymes will be discussed in a subsequent section. Here, we will consider enzymes as proteins—which, in fact, most are.)

The Active Site. One of the most important concepts to emerge from our understanding of enzymes as proteins is the active site. Every enzyme, regardless of the reaction it catalyzes or the details of its structure, contains somewhere within its tertiary configuration a characteristic cluster of amino acids forming the **active site** where the actual catalytic event, for which that enzyme is responsible, occurs. Usually, the active site is an actual groove or pocket with chemical and structural properties that accommodate the intended substrate with high specificity. Figure 6-2 shows

computer graphic models of the enzymes *lysozyme* and *carboxypeptidase A.* The three-dimensional structure of the active site can be appreciated by the precise fit of the substrate molecule into pockets produced by the characteristic folding of polypeptide chains. (Lysozyme is an enzyme that breaks the glycosidic bond between *N*-acetylglucosamine [GlcNAc] and *N*-acetylmuramic acid [MurNAc] groups in the peptidoglycan of bacterial cell walls [see Figure 3-23], thereby causing the bacterial cells to lyse [break open] and die. Carboxypeptidase A is an enzyme that breaks down polypeptides by removing one amino acid at a time from the C-terminus of the polypeptide.)

The amino acids that make up the active site of an enzyme are not usually contiguous to one another along the primary sequence of the protein. Instead, they are brought together in just the right conformation by the specific three-dimensional folding of the polypeptide chain. For the carboxypeptidase molecule, for example, the active site consists of a tightly bound zinc ion bonded to the side chains of the histidine at position 69, the glutamate at position 72, and another histidine at position 196 of the polypeptide chain. This configuration is shown in Figure 6-3a. When the substrate for the enzyme binds to the active site, the structure of the enzyme changes, bringing the arginine at position 145, the glutamate at position 270, and the tyrosine at position 248 into close proximity with the substrate. Each of these amino acids plays a critical role in the actual catalytic process (Figure 6-3b).

For carboxypeptidase, then, 6 amino acids out of a total chain length of 307 are actually involved at the active site. This is typical of most enzymes, with the active site usually involving about 5% of the surface area of the enzyme. Also typical is the involvement of amino acids from distant positions along the primary structure of the polypeptide. This involvement underscores the importance of the overall tertiary structure of the protein; only as an enzyme molecule attains its stable three-dimensional conformation are the specific amino acids brought together to constitute the active site.

Of the 20 different amino acids that make up proteins, only a few are actually involved in the active sites of the many proteins that have been studied. In most cases, these are the amino acids cysteine, histidine, serine, aspartate, glutamate, and lysine. All of these can participate in binding or bonding of the substrate to the active site during the catalytic process, and several (histidine, aspartate, and glutamate) also serve as donors or acceptors of protons.

Some enzymes consist not only of one or more polypeptide chains but of specific nonprotein components as well. These components are called **prosthetic groups** and are usually either small organic molecules or metal ions, such as the iron in catalase. Frequently, they function as electron acceptors, because none of the amino acid side chains is a good electron acceptor. Where present, prosthetic groups are located at the active site and are indispensable for the catalytic activity of the enzyme. Carboxypeptidase A is an example of an enzyme with a prosthetic group, containing a zinc atom at its active site (see Figure 6-3a).

Ball-and-stick model

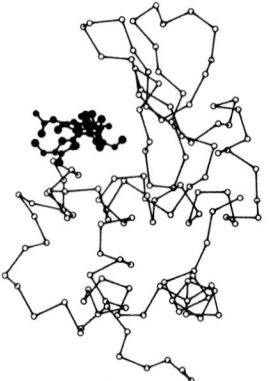

Space-filling model

Substrate
bound to
active site

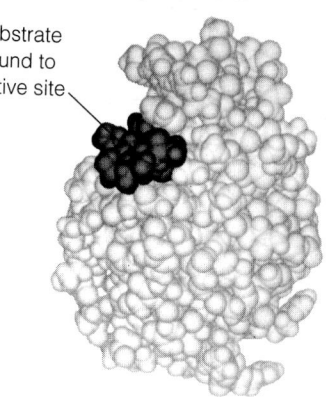

(a) Lysozyme

Figure 6-2   Molecular Structures of Lysozyme and Carboxypeptidase A.
(a) The enzyme lysozyme is represented as a ball-and-stick model (left), showing only the backbone of the protein as deduced by X-ray diffraction studies of protein crystals, and as a space-filling model (right) generated by computer. The molecule shown here is lysozyme from chicken eggs. A substrate molecule is shown in the active site, which appears as a cleft in the side of the enzyme molecule.
(b) The enzyme carboxypeptidase A is represented as a ball-and-stick model (left) and a space-filling model (right), with a substrate molecule bound to the active site. For a detailed view of the active site, see Figure 6-3.

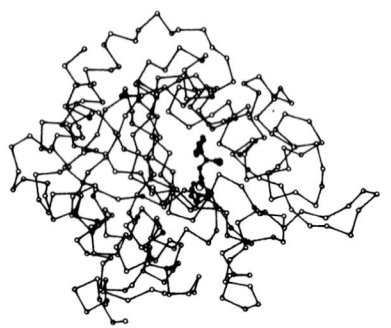

Substrate
bound to
active site

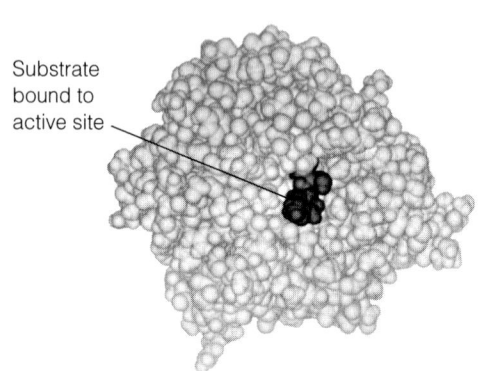

(b) Carboxypeptidase A

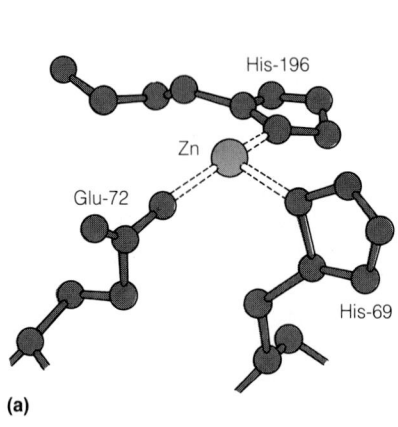

(a)

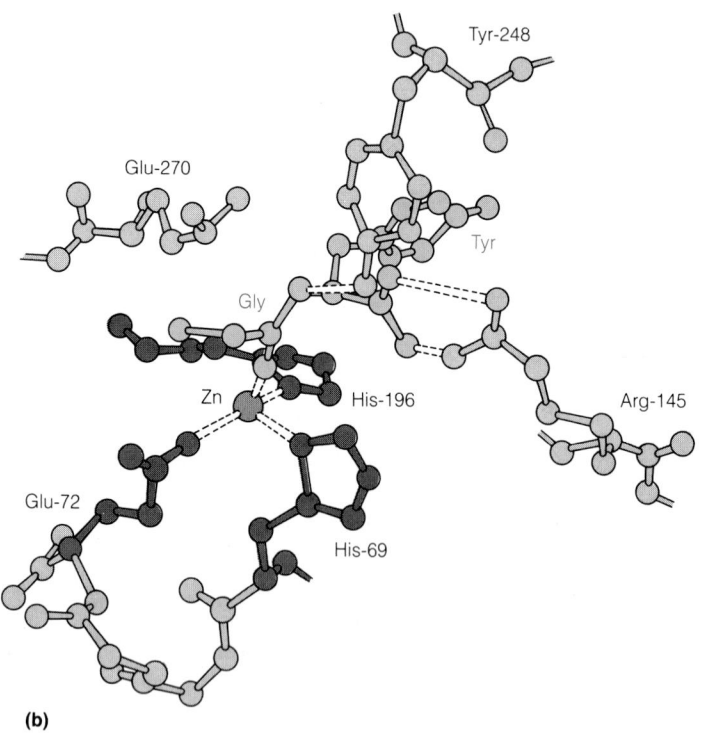

(b)

Figure 6-3   The Active Site of Carboxypeptidase A.
(a) The unoccupied active site of carboxypeptidase A consists of a zinc ion coordinated to two histidines and a glutamate. **(b)** Upon substrate binding (using an artificial peptide, shown in pink), the active site is induced to close in on the substrate, bringing an arginine, a glutamate, and a tyrosine into close proximity with the substrate.

Enzyme Specificity. A consequence of the structure of the active site is that enzymes display a high degree of **substrate specificity,** evidenced by an ability to discriminate between very similar molecules. Specificity is probably one of the most characteristic properties of living systems, and enzymes are especially dramatic examples of biological specificity.

We can illustrate their specificity by comparing enzymes with inorganic catalysts. Most inorganic catalysts are quite nonspecific in that they will act on a variety of compounds that share some general chemical feature. Consider, for example the *hydrogenation* of (addition of hydrogen to) an unsaturated C=C bond:

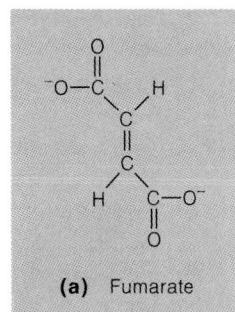

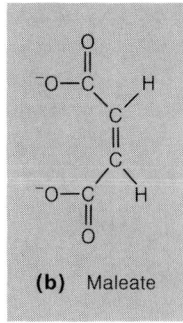

Figure 6-4   The Stereoisomers (a) Fumarate and (b) Maleate.

$$R-\overset{\overset{\displaystyle H}{|}}{C}=\overset{\overset{\displaystyle H}{|}}{C}-R' + H_2 \xrightarrow[\text{Pt or Ni}]{} R-\overset{\overset{\displaystyle H}{|}}{\underset{\underset{\displaystyle H}{|}}{C}}-\overset{\overset{\displaystyle H}{|}}{\underset{\underset{\displaystyle H}{|}}{C}}-R' + H_2 \qquad \textbf{(6-3)}$$

This reaction can be carried out in the laboratory using a platinum (Pt) or nickel (Ni) catalyst, as indicated. These inorganic catalysts are very nonspecific, however; they can be used to hydrogenate a wide variety of unsaturated compounds. In fact, nickel or platinum is used commercially to hydrogenate polyunsaturated vegetable oils in the manufacture of solid cooking fats or shortenings. Regardless of the exact structure of the unsaturated compound, it can be effectively hydrogenated in the presence of nickel or platinum.

By way of contrast, consider the biological example of hydrogenation involved in the conversion of fumarate to succinate, a reaction we will encounter again in Chapter 12:

$$\text{Fumarate} + 2H^+ + 2e^- \rightleftharpoons \text{Succinate} \qquad \textbf{(6-4)}$$

This particular reaction is catalyzed in cells by the enzyme *succinate dehydrogenase* (so named because it normally functions in the opposite direction during energy metabolism). This dehydrogenase, like most enzymes, is highly specific. It will not add or subtract hydrogens from any compounds except those shown in reaction 6-4. In fact, this particular enzyme is so specific that it will not even recognize maleate, a geometric stereoisomer of fumarate (Figure 6-4).

Not all enzymes are quite this specific; some accept a number of closely related substrates, and others accept any of a whole group of substrates as long as they possess some common structural feature. Such **group specificity** is seen most often with enzymes involved in the synthesis or degradation of polymers. The purpose of carboxypeptidase A is to degrade polypeptide chains from the carboxyl end; thus, it makes sense for the enzyme to accept any of a wide variety of polypeptides as substrate, since it would be needlessly extravagant of the cell to require a separate enzyme for every different peptide bond that has to be hydrolyzed in polypeptide degradation.

In general, however, enzymes are highly specific with respect to substrate, such that a cell must possess almost as many different kinds of enzymes as it has reactions to catalyze. For a typical cell, this means that several thousand different kinds of enzymes are necessary to carry out its full metabolic program. At first, that may seem wasteful in terms of proteins to be synthesized, genetic information to be stored and read out, and enzyme molecules to have on hand in the cell. But you should also be able to see the tremendous regulatory possibilities this suggests, a point we will return to later.

Enzyme Diversity and Nomenclature. Given the specificity of enzymes and the large number of reactions that occur within a cell, it is not surprising that thousands of different enzymes have been identified. This enormous diversity of enzymes and enzymatic functions led initially to a variety of schemes for naming enzymes as they were discovered and characterized. Some were given names indicative of the substrate; *ribonuclease, protease,* and *amylase* are examples. Others, such as *succinic dehydrogenase* and *phosphoglucoisomerase,* were named to describe their function. Still other enzymes have names that provide little indication of either substrate or function. *Trypsin, catalase,* and *lysozyme* are enzymes in this category.

The proliferation of common names for enzymes and the resulting confusion eventually prompted the International Union of Biochemistry to appoint an Enzyme Commission (EC) charged with devising a rational system for naming enzymes. The EC system is represented in Table 6-1. Enzymes are divided into six major classes based on their general functions, with subgroups used to define their functions more precisely. The six major classes are *oxidoreductases*, *transferases*, *hydrolases*, *lyases*, *isomerases*, and *ligases*. Table 6-1 provides an example of each class, using enzymes that catalyze reactions to be discussed later in the text.

The EC system assigns every known enzyme a four-part number. For example, EC 3.4.17.1 is the number for carboxypeptidase A. The first three numbers define the major class, subclass, and sub-subclass, and the final number is the serial number assigned to the enzyme when it was added to the list. Thus, carboxypeptidase A is the first entry in the 17th sub-subclass of the fourth subclass of class 3 (hydrolases). The list of enzymes that have been classified in this way grows continuously as new enzymes are discovered and characterized.

Sensitivity to Temperature.
In addition to their specificity and diversity, enzymes are also characterized by their sensitivity to temperature. This temperature dependence is not usually a practical concern for enzymes in the cells of mammals or birds because these organisms are *homeotherms*, capable of regulating body temperature independent of the environment. However, in organisms such as lower animals, plants, protists, and bacteria, the dependence of enzyme activity on temperature is very significant because these organisms are often subjected to considerable ranges in temperature.

Within limits, the rate of an enzyme-catalyzed reaction increases with temperature because the greater kinetic energy of both enzyme and substrate molecules ensures more frequent collisions, thereby increasing the likelihood of correct substrate binding. At some point, however, further increases in temperature are counterproductive because the enzyme molecule begins to denature. Hydrogen bonds break, hydrophobic interactions change, and the structural integrity of the active site is disrupted, causing a loss of activity.

The temperature range over which an enzyme denatures varies greatly from enzyme to enzyme and especially from organism to organism. Figure 6-5a contrasts the temperature dependence of a typical enzyme from the human body with that of a typical enzyme from a thermophilic bacterium. Not surprisingly, the reaction rate of a human enzyme is maximum at about 37°C, which is normal body temperature. The sharp decrease in activity thereafter reflects the progressive denaturation of the enzyme molecules. Most enzymes of homeotherms are inactivated by temperatures above about 50–55°C. For such enzymes, heat inactivation is clearly a laboratory phenomenon only. However, some enzymes are remarkably sensitive to heat and are denatured and inactivated at lower temperatures—in some cases, even by body temperatures encountered in people with high fevers.

On the other end of the spectrum, some enzymes retain activity at unusually high temperatures. The right-hand

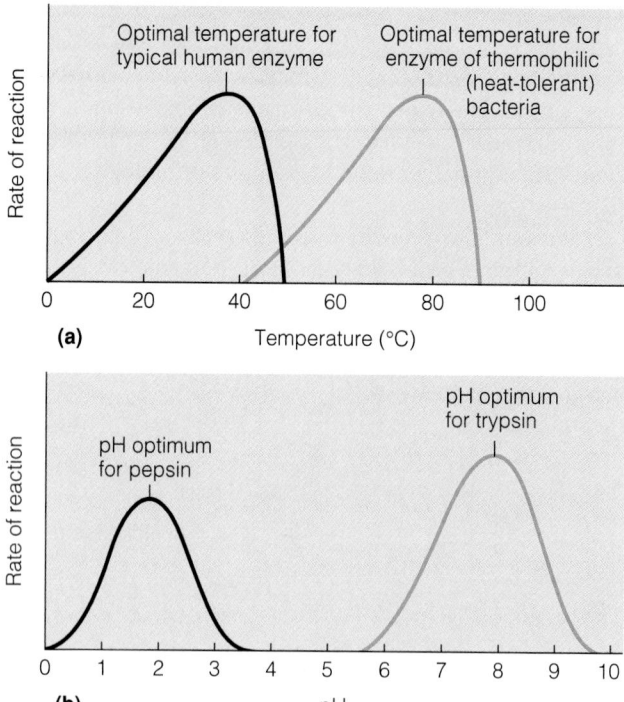

Figure 6-5 The Effect of Temperature and pH on the Reaction Rate of Enzyme-Catalyzed Reactions. **(a)** The dependence of reaction rate on temperature for a typical human enzyme (black) and a typical enzyme from a thermophilic bacterium (blue). The reaction rate is maximized at the optimal temperature, which is about 37°C (body temperature) for the human enzyme and about 75°C (the temperature of a typical hot spring) for the bacterial enzyme. The initial increase in activity with temperature is due to the greater kinetic energy of enzyme and substrate molecules at higher temperatures. Eventually, however, an inactivating effect is seen as further increases in temperature result in thermal denaturation of the enzyme protein. **(b)** The dependence of reaction rate on pH for the gastric enzyme pepsin (black) and the intestinal enzyme trypsin (blue). The reaction rate is maximized at the optimal pH, which is about 2.0 for pepsin and about 8.0 for trypsin. The pH optimum for an enzyme corresponds to the proton concentration at which ionizable groups on both the enzyme and the substrate molecules are in the most favorable form for maximum reactivity. Changes in pH away from the optimum usually reflect titration of charged groups on the enzyme, the substrate, or both. The pH optimum of an enzyme usually reflects the pH of the intracellular compartment or other environment in which the enzyme is active.

curve in Figure 6-5a depicts the temperature dependence of an enzyme from one of the *thermacidophilic archebacteria* mentioned in Chapter 4. These organisms thrive in acidic hot springs at temperatures as high as 80°C! Clearly, each of the thousands of different kinds of enzymes such organisms need for normal cellular activity must be capable of functioning at temperatures that would denature most of the enzymes in the cells of other organisms, your own included.

Sensitivity to pH.
Enzymes are also sensitive to pH, with many enzymes active only within a pH range of about 3–4 pH units. This pH dependence is usually due to the presence

Table 6-1 The Major Classes of Enzymes with an Example of Each*

Class	Reaction Type	Example — Enzyme Name	Example — Reaction Catalyzed
1. Oxidoreductases	Oxidation-reduction reactions	Alcohol dehydrogenase (EC 1.1.1.1) (oxidation with NAD^+)	$CH_3-CH_2-OH \xrightarrow[\text{NAD}^+ \quad \text{NADH + H}^+]{}$ Ethanol $\rightarrow$ $CH_3-\overset{\overset{\displaystyle O}{\|}}{C}-H$ Acetaldehyde
2. Transferases	Transfer of functional groups from one molecule to another	Glycerokinase (EC 2.4.3.2) (phosphorylation)	$HO-CH_2-\overset{\overset{\displaystyle OH}{\|}}{CH}-CH_2-OH \xrightarrow[\text{ATP} \quad \text{ADP}]{} HO-CH_2-\overset{\overset{\displaystyle OH}{\|}}{CH}-CH_2-O-PO_3^{2-}$ Glycerol phosphate
3. Hydrolases	Hydrolytic cleavage of one molecule into two molecules	Carboxypeptidase A (EC 3.4.17.1) (peptide bond cleavage)	$-NH-\overset{R_{n-1}}{\underset{}{CH}}-\overset{\overset{\displaystyle O}{\|}}{C}-NH-\overset{R_n}{\underset{}{CH}}-\overset{\overset{\displaystyle O}{\|}}{C}-O^-$ C-terminus of polypeptide $\xrightarrow{H_2O}$ $-NH-\overset{R_{n-1}}{\underset{}{CH}}-\overset{\overset{\displaystyle O}{\|}}{C}-O^- + H_3N^+-\overset{R_n}{\underset{}{CH}}-\overset{\overset{\displaystyle O}{\|}}{C}-O^-$ Shortened polypeptide / C-terminal amino acid
4. Lyases	Removal of a group from, or addition of a group to, a molecule with rearrangement of electrons	Pyruvate decarboxylase (EC 4.1.1.1) (decarboxylation)	$CH_3-\overset{\overset{\displaystyle O}{\|}}{C}-\overset{\overset{\displaystyle O}{\|}}{C}-O^- + H^+$ Pyruvate $\longrightarrow$ $CH_3-\overset{\overset{\displaystyle O}{\|}}{C}-H + CO_2$ Acetaldehyde
5. Isomerases	Movement of a functional group within a molecule	Maleate isomerase (EC 5.2.1.1) (cis-trans isomerization)	Maleate $\rightleftharpoons$ Fumarate
6. Ligases	Joining of two molecules to form a single molecule	Pyruvate carboxylase (EC 6.4.1.1) (carboxylation)	$CH_3-\overset{\overset{\displaystyle O}{\|}}{C}-\overset{\overset{\displaystyle O}{\|}}{C}-O^- + CO_2$ Pyruvate $\xrightarrow[\text{ATP} \quad \text{ADP} + P_i]{}$ $^-O-\overset{\overset{\displaystyle O}{\|}}{C}-CH_2-\overset{\overset{\displaystyle O}{\|}}{C}-C-O^-$ Oxaloacetate

*This system for classifying, naming, and assigning numbers to enzymes was devised by the Enzyme Commission of the International Union of Biochemistry.

of one or more charged amino acids at the active site, with activity usually dependent on having such groups present in a specific form, either charged or uncharged. For example, the active site of carboxypeptidase A involves the carboxyl groups from each of two glutamates, located at positions 72 and 270 of the protein (see Figure 6-3). These carboxyl groups must be present in the charged (ionized) form, so the enzyme becomes inactive if the pH is decreased to the point at which the glutamate carboxyl groups on most of the enzyme's molecules are protonated and therefore uncharged.

As you might expect, the pH dependence of an enzyme usually reflects the environment in which that enzyme is normally active. Figure 6-5b shows the pH dependence of two protein-degrading enzymes found in the human digestive tract. Pepsin is present in the stomach, where the pH is usually about 2, whereas trypsin is secreted into the small intestine, which has a pH of about 8. Both enzymes are active over a range of almost 4 pH units but differ greatly in their pH optima, consistent with the conditions in their respective locations within the body.

Enzymes as Catalysts

Because of the precise "fit" between the active site of an enzyme and its substrates, enzymes are highly effective as catalysts. This effectiveness can be seen in the much greater extent to which reaction rates are catalyzed by enzymes compared with inorganic catalysts. Typically, enzyme-catalyzed reactions proceed 10^8 to 10^{10} times faster than uncatalyzed reactions, compared with a stimulation of 10^3 to 10^4 times for inorganic catalysts. As you might guess, most of the interest in enzymes focuses on the active site, where binding, activation, and chemical transformation of the substrate occur.

Substrate Binding. Initial contact between the active site of an enzyme and a potential substrate molecule depends on random collision. Once in the groove, or pocket, of the active site, however, the substrate molecules are bound temporarily to the enzyme surface in just the right orientation to one another and to specific catalytic groups on the enzyme to facilitate reaction. Substrate binding usually involves hydrogen bonds or ionic bonds (or both) to charged amino acids. These are generally weak bonds, but several bonds may hold a single molecule in place. The strength of the bonds between an enzyme and a substrate molecule is often in the range of 3–12 kcal/mol, less than one-tenth the strength of a single covalent bond. Substrate binding is therefore readily reversible.

For a long time, enzymologists regarded the active site as a rigid structure. They likened the fit of a substrate into the active site to that of a key into a lock, an analogy first suggested in 1890 by the German biochemist Emil Fischer. This *lock-and-key model* explained enzyme specificity but did little to enhance our understanding of the catalysis itself. This understanding came with a more recent view of enzyme-substrate interaction called the **induced-fit model.**

This model, first proposed by Daniel Koshland in 1958, assumes that the initial binding of the substrate molecule(s) at the active site induces a *conformational change* in the shape of the enzyme molecule and hence in the configuration of the active site, as shown in Figure 6-6. This change positions the proper reactive groups of the enzyme optimally for the catalytic reaction in which they are involved, thereby enhancing the likelihood of that reaction. Specifically, the conformational change brings into the active site amino acid side chains that are critical to the catalytic process but are not in the immediate area of the active site in the uninduced conformation. In many cases, these are acidic or basic groups that promote catalysis. In the case of carboxypeptidase A, for example, substrate binding brings three critical amino acids—an arginine, a glutamate, and a tyrosine—into the active site (see Figure 6-3).

Evidence that such conformational changes actually occur upon binding of substrate has come from X-ray diffraction studies of crystallized proteins. Specifically, X-ray crystallography is used to determine the shape of an enzyme molecule with and without substrate bound to the active site. Figure 6-7 illustrates the conformational changes that take place upon substrate binding for *lysozyme*, the enzyme shown in Figure 6-2a, and for *hexokinase*, an enzyme involved in the glycolytic pathway for glucose catabolism that we will encounter in Chapter 11. For both enzymes, a noticeable change takes place in the conformation of the protein molecule in response to substrate binding. In the case of hexokinase, for example, the binding of substrate (a molecule of D-glucose) causes two domains of the enzyme to fold toward each other, closing the binding site cleft about the substrate (Figure 6-7b).

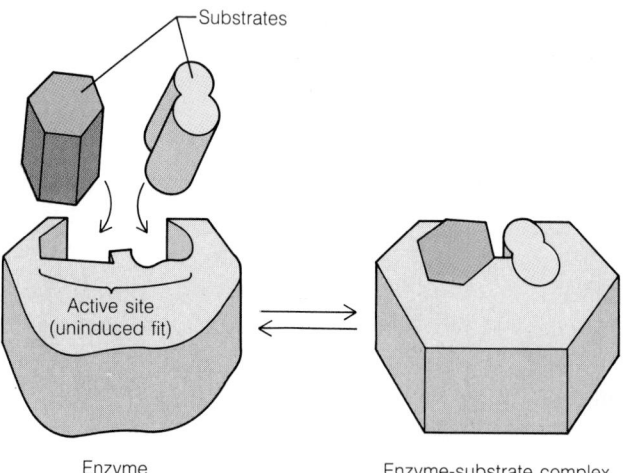

Figure 6-6   The Induced-Fit Model for Enzyme-Substrate Interaction.   According to the induced-fit model, binding of a substrate molecule to the active site of an enzyme induces a conformational change in the enzyme molecule that positions the proper reactive groups at the active site optimally for the catalytic reaction. In addition, the tighter enzyme-substrate fit distorts one or more of the bonds of the substrate, thereby making it more susceptible to catalytic attack.

Figure 6-7 The Conformational Change in Enzyme Structure Induced by Substrate Binding. Shown here are space-filling models for the enzymes **(a)** lysozyme and **(b)** hexokinase and their substrates (an artificial peptidoglycan and a molecule of D-glucose, respectively). In both cases, substrate binding induces a conformational change in the enzyme that is detectable by X-ray diffraction analysis. In the case of hexokinase, the two major domains of the enzyme are shaded differently to distinguish them.

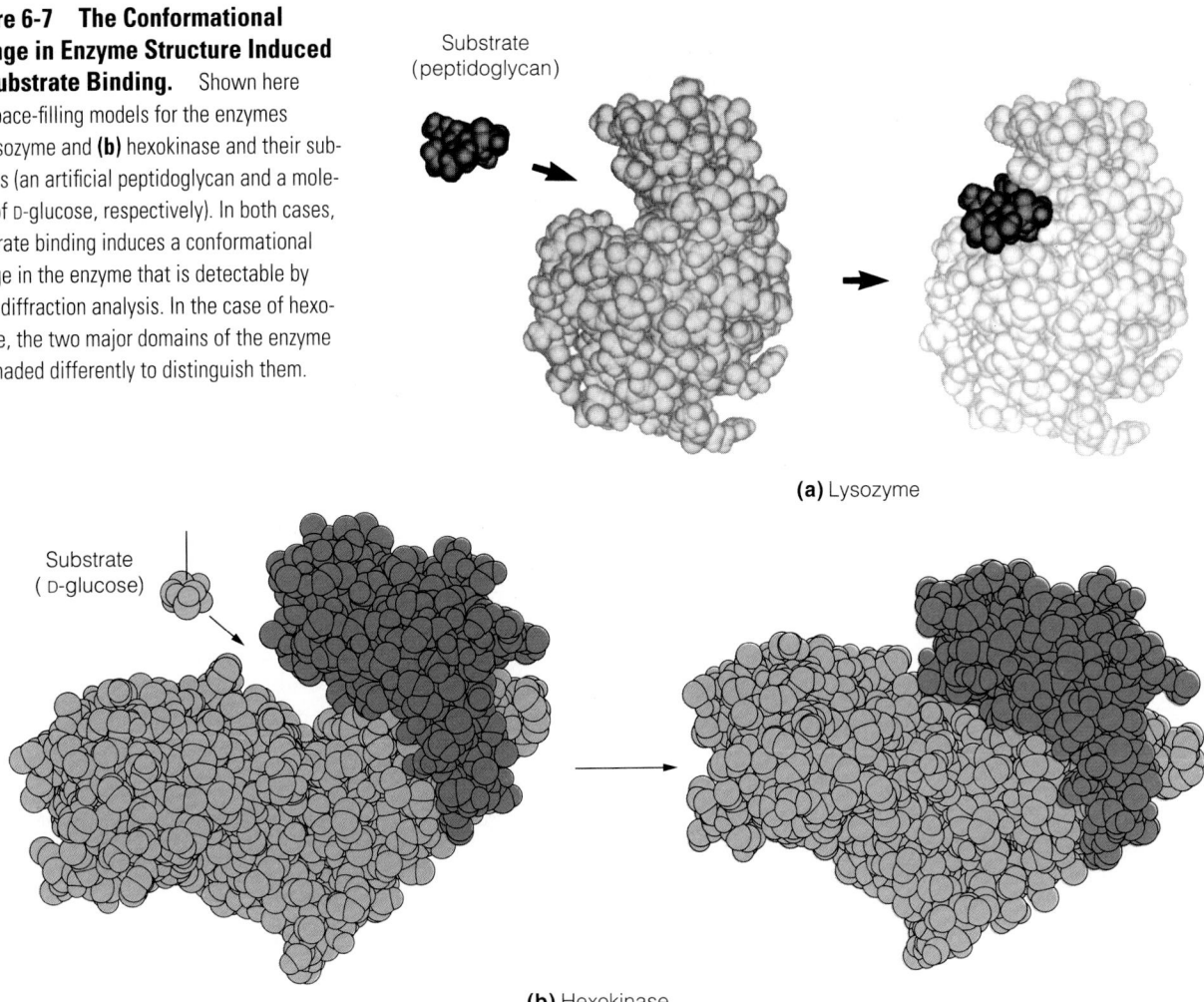

Substrate (peptidoglycan)

(a) Lysozyme

Substrate (D-glucose)

(b) Hexokinase

The conformational change upon substrate binding can result in remarkably large displacements of amino acid groups within the protein molecule. In the case of carboxypeptidase A, for example, substrate binding causes the tyrosine at position 248 (see Figure 6-3) to move 1.2 nm, a distance equal to about a quarter of the diameter of the enzyme molecule!

Substrate Activation. The role of the active site is not just to recognize and bind the appropriate substrate but also to *activate* it by subjecting it to the right chemical environment for catalysis. For a given enzyme-catalyzed reaction, **substrate activation** may involve one or more means of activation. Three of the most common mechanisms are as follows:

1. The change in enzyme conformation induced by initial substrate binding to the active site not only causes better complementarity and a tighter enzyme-substrate fit but also distorts one or more of its bonds, thereby weakening the bond and making it more susceptible to catalytic attack.

2. The enzyme may also accept or donate protons, thereby increasing the chemical reactivity of the substrate. This is similar to the chemist's approach of changing the pH of the reaction mixture to enhance the rate of organic or inorganic reactions in the laboratory. For enzymes, the "pH change" is effected by acidic or basic groups at or near the active site. This accounts for the importance of charged amino acids in active-site chemistry, which in turn explains why enzyme activity is so often dependent on pH (see Figure 6-5b).

3. As a further means of substrate activation, enzymes may also accept or donate electrons, thereby forming temporary covalent bonds between the enzyme and its substrate. This mechanism requires that the substrate have a region that is either electropositive (electron-deficient) or electronegative (electron-rich) and that the active site of the enzyme have one or more groups of opposite polarity. In one case, the electronegative side group of an appropriate amino acid at the active site donates electrons to the electropositive region of the substrate, in what is called a **nucleophilic substitution** reaction (Figure 6-8a). The hydroxyl group of serine, the sulfhydryl group of cysteine, and the indole group of histidine are all active nucleophilic attacking groups. Alternatively, the electronegative group may be on the substrate and the electropositive group on

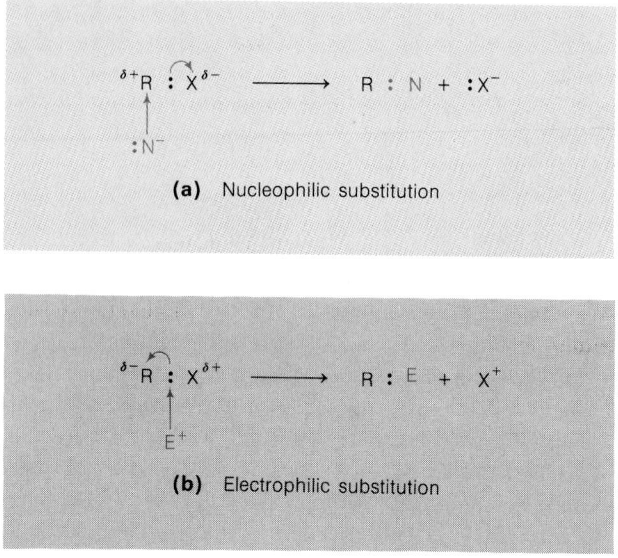

Figure 6-8   Substitution Reactions in the Mechanism of Action of Some Enzymes.   **(a)** Nucleophilic substitution, with N^- as the nucleophilic attacking group on the enzyme and X^- as the leaving group on the substrate. **(b)** Electrophilic substitution, with E^+ as the electrophilic attacking group on the enzyme and X^+ as the leaving group on the substrate.

the enzyme, and the reaction is then an **electrophilic substitution** (Figure 6-8b). Enzymes capable of electrophilic attack invariably require a prosthetic group at the active site because none of the amino acid side groups is sufficiently electrophilic. Metal ions are strongly electropositive and therefore serve well in this role.

The Catalytic Event.   So far, we've seen that the initial random collision of a substrate molecule with the active site of an enzyme induces a change in the enzyme conformation that tightens the fit between the substrate molecule and the active site. In the process, the substrate is positioned correctly at the active site, and one or more of its bonds are distorted and weakened to facilitate catalytic attack.

The catalytic event then occurs, usually involving the formation of temporary covalent bonds between the substrate and one or more specific amino acid side chains at the active site. When this process of binding, activation, and catalysis has been completed, the product or products of the reaction are released from the active site. The active site then returns to its original state, ready for another round of catalysis. This entire sequence of events takes place in a sufficiently short time to allow hundreds or even thousands of such reactions to occur per second at a single active site!

Enzyme Kinetics

So far, our discussion of enzymes has been basically descriptive. We have dealt with the activation energy requirement that prevents thermodynamically feasible reactions from occurring, and with catalysts as a means of reducing the activation energy and thereby facilitating such reactions. We have also encountered enzymes as biological catalysts and have examined their structure and function in some detail. Certainly, the only reactions likely to occur in cells at reasonable rates are those for which specific enzymes are on hand, and the metabolic capabilities of cells are therefore effectively specified by the enzymes that are present.

Still lacking, however, is a means of assessing the actual rates at which enzyme-catalyzed reactions will proceed, as well as an appreciation for the factors that influence reaction rates. The mere presence of the appropriate enzyme in a cell may not necessarily ensure that a given reaction will occur at an adequate rate unless we can also be assured that cellular conditions are favorable for enzyme activity. We have already seen how factors such as temperature and pH can affect enzyme activity. Now we are ready to appreciate how critically enzyme activity also depends on the concentrations of substrates, products, and inhibitors that prevail in the cell. In addition, we will see how at least some of these effects can be defined quantitatively.

It is as we turn our attention to these quantitative aspects of enzyme catalysis that we encounter the field of **enzyme kinetics.** The word *kinetics* is of Greek origin (*kinetikos,* meaning "moving"). Applied to chemical reactions, kinetics concerns reaction rates and the manner in which those rates are influenced by a variety of factors, but especially by concentrations of substrates, products, and inhibitors. Most of our attention will focus on the effects of substrate concentration on the kinetics of enzyme-catalyzed reactions. Our considerations will be restricted to initial reaction rates, measured over a period of time during which the substrate concentration has not yet decreased enough to affect the rate and the accumulation of product is still too small to cause any measurable back reaction. Although this is oversimplified compared to the real-life situation, it nonetheless allows us to understand some important principles of enzyme kinetics.

Enzyme kinetics can seem quite complex at first. To help you understand the basic concepts, Box 6A uses an analogy in which enzymes acting on substrate molecules are likened to a roomful of monkeys shelling peanuts. You may find it useful to turn to the analogy at this point and then come back to this section.

Michaelis-Menten Kinetics

It has long been known that the rate of an enzyme-catalyzed reaction increases with increasing substrate concentration, but in a manner such that each additional increment of substrate results in a smaller increase in reaction rate. More precisely, the relationship between the initial reaction rate (or velocity v) and the substrate concentration [S] can be shown experimentally to be that of a hyperbola, as illustrated in Figure 6-9. An important property of this hyperbolic relationship is that, as [S] tends toward infinity, v tends toward

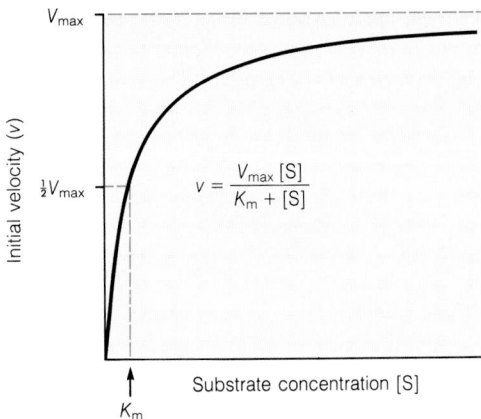

Figure 6-9 The Relationship Between Reaction Velocity and Substrate Concentration. For an enzyme-catalyzed reaction that follows Michaelis-Menten kinetics, the initial velocity tends toward an upper-limiting velocity V_{max} as the substrate concentration [S] tends toward infinity. The Michaelis constant K_m corresponds to that substrate concentration at which the reaction is proceeding at one-half of the maximum velocity.

an upper limiting value that depends on the number of enzyme molecules and can be increased only by adding more enzyme.

The inability of higher and higher substrate concentrations to increase the reaction velocity beyond a finite upper value is called **saturation.** This property is a fundamental characteristic of enzyme-catalyzed reactions. Catalyzed reactions always become saturated at high substrate concentrations, whereas uncatalyzed reactions do not.

Much of our understanding of the hyperbolic relationship between [S] and v is owed to the pioneering work of two German enzymologists, Leonor Michaelis and Maud Menten. In 1913, they postulated a general theory of enzyme action that has turned out to be basic to the quantitative analysis of almost all aspects of enzyme kinetics. To understand their approach, consider one of the simplest possible enzyme-catalyzed reactions, in which a substrate S is converted into a product P:

$$S \xrightarrow[\text{Enzyme (E)}]{} P \qquad \textbf{(6-5)}$$

According to the Michaelis-Menten hypothesis, the enzyme E that catalyzes this reaction first reacts with the substrate S, forming the transient enzyme-substrate complex ES, which then undergoes the actual catalytic reaction to form free enzyme and product P, as shown in the following sequence:

$$E_f + S \underset{k_2}{\overset{k_1}{\rightleftharpoons}} ES \underset{k_4}{\overset{k_3}{\rightleftharpoons}} E_f + P \qquad \textbf{(6-6)}$$

where E_f is the free form of the enzyme, S is the substrate, ES is the enzyme-substrate complex, P is the product, and k_1, k_2, k_3, k_4 are the rate constants for the indicated reactions.

Starting with this model and several simplifying assumptions, Michaelis and Menten arrived at the following relationship between the velocity of an enzyme-catalyzed reaction and the substrate concentration:

$$v = \frac{V_{max}\,[S]}{K_m + [S]} \qquad \textbf{(6-7)}$$

Here [S] is the initial substrate concentration, v is the initial reaction velocity at that substrate concentration, and V_{max} and K_m are kinetic parameters. This is the **Michaelis-Menten equation,** the central relationship of enzyme kinetics. Problem 6-12 at the end of the chapter gives you an opportunity to derive the Michaelis-Menten equation yourself. To do so, you will need to assume, as they did, that the reverse reaction between E_f and P is negligible, that the rate of product formation (k_3) is slow compared to the rates of ES formation (k_1) and redissociation (k_2), and that the total concentration of enzyme, E_t, is the sum of the free and complexed forms (that is, $[E_t] = [E_f] + [ES]$).

The Meaning of V_{max} and K_m

To appreciate the implications of this relationship between v and [S] and to examine the meaning of the parameters V_{max} and K_m, we can consider three special cases of substrate concentration: very low substrate concentration, very high substrate concentration, and the special case of $[S] = K_m$.

Case 1: Very Low Substrate Concentration ($[S] \ll K_m$). At very low substrate concentration, [S] becomes negligibly small compared to the constant K_m in the denominator of the Michaelis-Menten equation, so we can write

$$v = \frac{V_{max}\,[S]}{K_m + [S]} \cong \frac{V_{max}\,[S]}{K_m} \qquad \textbf{(6-8)}$$

Thus, at very low substrate concentration, the initial reaction velocity is roughly proportional to the substrate concentration. This is therefore the *first-order region* of the Michaelis-Menten plot. As long as the substrate concentration is much lower than the K_m value, the velocity of an enzyme-catalyzed reaction increases almost linearly with substrate concentration.

Case 2: Very High Substrate Concentration ($[S] \gg K_m$). At very high substrate concentration, K_m becomes negligibly small compared to [S] in the denominator of the Michaelis-Menten equation, so we can write

$$v = \frac{V_{max}\,[S]}{K_m + [S]} \cong \frac{V_{max}\,[S]}{[S]} = V_{max} \qquad \textbf{(6-9)}$$

This relationship means that, at very high substrate concentrations, the velocity of an enzyme-catalyzed reaction is independent of variation in [S] and is therefore approximately constant. This is therefore the *zero-order region* of the

6A

Further Insights

MONKEYS AND PEANUTS

I f you found the Mexican jumping beans helpful in understanding free energy in Chapter 5, you might appreciate an approach to enzyme kinetics based on the analogy of a roomful of monkeys ("enzymes") shelling peanuts ("substrates"), with the peanuts present in varying abundance. Try to understand each step first in terms of monkeys shelling peanuts, then in terms of an actual enzyme-catalyzed reaction.

The Peanut Gallery

For our model, we need a troop of ten monkeys, all equally adept at finding and shelling peanuts. We shall assume that the monkeys are too full to eat any of the peanuts they shell, but nonetheless have an irresistible compulsion to go on shelling. To make the model a bit more rigorous, we should insist that the peanuts are a new hybrid variety that can be readily stuck back together again and that the monkeys are just as likely to put the peanuts back in the shells as they are to take them out. But these considerations need not concern us here; we are interested only in the initial conditions in which all the peanuts start out in their shells.

Next, we need the Peanut Gallery, a room of fixed floor space with peanuts scattered equally about on the floor. The amount of peanuts will be varied as we proceed, but in all cases there will be vastly more peanuts than monkeys in the room. Moreover, because we know the number of peanuts and the total floor space, we can always calculate the "concentration" (more accurately, the density) of peanuts in the room. In each case, the monkeys start out in an adjacent room. To start an assay, we simply open the door and allow the eager monkeys to enter the Peanut Gallery.

The Shelling Begins

Now we are ready for our first assay. We start with an initial peanut concentration of one peanut per square meter, and we assume that at this concentration of peanuts, the average

monkey spends 9 seconds looking for a peanut to shell and 1 second shelling it. This means that each monkey requires 10 seconds per peanut and can thus shell peanuts at the rate of 0.1 peanut per second. And since there are ten monkeys in the Gallery, the rate (let's call it the velocity v) of peanut-shelling for all the monkeys is 1 peanut per second at this particular concentration of peanuts (which we will call [S] to remind ourselves that the peanuts are really the substrate of the shelling action). All of this can be tabulated as follows:

[S] = Concentration of peanuts (peanuts/m^2)	1
Time required per peanut:	
To find (sec/peanut)	9
To shell (sec/peanut)	1
Total (sec/peanut)	10
Rate of shelling:	
Per monkey (peanut/sec)	0.10
Total (v) (peanut/sec)	1.0

The Peanuts Become More Abundant

For our second assay, we herd all the monkeys back into the waiting room, sweep up the debris, and arrange peanuts about the Peanut Gallery at a concentration of 3 peanuts per square meter. Since peanuts are now three times more abundant than previously, the average monkey should find a peanut three times more quickly than before, such that the time spent finding the average peanut is now only 3 seconds. But each peanut, once found, still takes 1 second to shell, so the total time per peanut is now 4 seconds and the velocity of shelling is 0.25 peanut per second for each monkey, or 2.5 peanuts per second for the roomful of monkeys. This generates another column of entries for our data table:

[S] = Concentration of peanuts (peanuts/m²)	1	3
Time required per peanut:		
To find (sec/peanut)	9	3
To shell (sec/peanut)	1	1
Total (sec/peanut)	10	4
Rate of shelling:		
Per monkey (peanut/sec)	0.10	0.25
Total (v) (peanut/sec)	1.0	2.5

What Happens to v as [S] Continues to Increase?

To find out what eventually happens to the velocity of peanut-shelling as the peanut concentration in the room gets higher and higher, all you need do is extend the data table by assuming ever-increasing values for [S] and calculating the corresponding v. For example, you should be able to convince yourself that a further tripling of the peanut concentration (from 3 to 9 peanuts/m²) will bring the time required per peanut down to 2 seconds (1 second to find and another second to shell), which will result in a shelling rate of 0.5 peanut per second for each monkey, or 5.0 peanuts per second overall.

Already you should begin to see a trend. The first tripling of peanut concentration increased the rate 2.5-fold, but the next tripling only resulted in a further doubling of the rate. There seems, in other words, to be a diminishing return on additional peanuts. You can see this clearly if you choose a few more peanut concentrations and then plot v on the y-axis (suggested scale: 0–10 peanuts/sec) versus [S] on the x-axis (suggested scale: 0–100 peanuts/m²). What you should find is that the data generate a hyperbolic curve that looks strikingly like Figure 6-9. And if you look at your data carefully, you should see that the reason your curve continues to "bend over" as [S] gets higher (i.e., the reason you get less and less additional velocity for each further increment of peanuts): The shelling time is fixed and therefore becomes a more and more prominent component of the total processing time per peanut as the finding time gets smaller and smaller. You should also appreciate that it is this fixed shelling time that ultimately sets the upper limit on the overall rate of peanut processing, because even when [S] is infinite (i.e., in a world flooded with peanuts), there will still be a finite time of 1 second required to process each peanut.

Finally, you should realize that there is something special about the peanut concentration at which the finding time is exactly equal to the shelling time (it turns out to be 9 peanuts/m²); this is the point along the curve at which the rate of peanut processing is exactly one-half of the maximum rate. In fact, it is such an important benchmark along the concentration scale that you might even be tempted to give it a special name, particularly if your name were Michaelis and you were monkeying around with enzymes instead of peanuts! ■

Michaelis-Menten plot. As long as the substrate concentration is much higher than the K_m value, the velocity is essentially unaffected by changes in substrate concentration, remaining approximately constant at values close to V_{max}.

This, then, provides us with a definition of V_{max}, one of the two kinetic parameters in the Michaelis-Menten equation: V_{max} is the **maximum velocity,** or the upper limiting value, to which the initial reaction velocity v tends as the substrate concentration [S] approaches infinity. In other words, V_{max} is the velocity at saturating substrate concentrations. Under these conditions, every enzyme molecule is occupied in the actual process of catalysis, since the substrate concentration is so high that another substrate molecule arrives at the active site almost as soon as a product molecule is released.

V_{max} is therefore an upper limit determined by (1) the time required for the actual catalytic reaction plus subsequent release of product from the surface of each enzyme molecule, and (2) how many such enzyme molecules are present. Because the actual reaction rate is fixed, the only way that V_{max} can be increased is to increase enzyme concentration. In fact, V_{max} is linearly proportional to the amount of enzyme present, as shown in Figure 6-10.

Case 3: [S] = K_m. So far, we have seen the reason for first-order reaction kinetics at low substrate concentrations and for zero-order kinetics at high concentrations. We have also formulated a definition for V_{max} but have yet to discover the meaning of the second kinetic parameter, K_m. Note, however, that whatever its meaning, K_m appears to have something to do with determining how low the substrate concentration must be to ensure first-order kinetics or, alternatively, how high a concentration is required to ensure zero-order kinetics. Thus, K_m seems to be some sort of benchmark on the concentration scale that determines how

high is high and how low is low. To explore its meaning more precisely, consider the special case where [S] is exactly equal to K_m. Under these conditions, the Michaelis-Menten equation can be written as follows:

$$v = \frac{V_{max}[S]}{K_m + [S]} = \frac{V_{max}[S]}{2[S]} = \frac{V_{max}}{2} \qquad \textbf{(6-10)}$$

This equation provides us with the definition we have been looking for: K_m is that specific substrate concentration at which the reaction proceeds at one-half of its maximum (upper-limiting) velocity. This specific concentration is a fixed value for a given enzyme catalyzing a specific reaction under specified conditions and is called the **Michaelis constant** (hence the designation K_m) after the enzymologist who first elucidated its meaning.* Figure 6-9 illustrates the meaning of both V_{max} and K_m.

The Double-Reciprocal Plot

A classic Michaelis-Menten plot of v versus [S] is a faithful representation of the dependence of velocity on substrate concentration, but it is not an especially useful tool for the quantitative determination of the key kinetic parameters K_m and V_{max}. Its hyperbolic shape makes it difficult to extrapolate accurately to infinite substrate concentration, as would be required to determine the critical parameter V_{max}, and if V_{max} is not known accurately, K_m cannot be determined. This problem is readily apparent from Figure 6-9, since it would be difficult to estimate V_{max} accurately if it were not already sketched in, and without V_{max}, K_m cannot easily be estimated either.

To circumvent this problem and provide a more useful graphic approach, H. Lineweaver and D. Burk converted the hyperbolic relationship of the Michaelis-Menten equation into a linear function by inverting both sides of equation 6-7 and simplifying the resulting expression into the form of an equation for a straight line:

$$\frac{1}{v} = \frac{K_m + [S]}{V_{max}[S]} = \frac{K_m}{V_{max}[S]} + \frac{[S]}{V_{max}[S]}$$

$$= \frac{K_m}{V_{max}} \left[\frac{1}{[S]} \right] + \frac{1}{V_{max}} \qquad \textbf{(6-11)}$$

Equation 6-11 is the **Lineweaver-Burk equation.** If it is plotted as $1/v$ versus $1/[S]$, as in Figure 6-11, the resulting **double-reciprocal plot** is linear, with a y-intercept of $1/V_{max}$, an x-intercept of $-1/K_m$, and a slope of K_m/V_{max}. (You should be able to convince yourself of these intercept values by setting first $1/[S]$ and then $1/v$ equal to zero in equation 6-11 and solving for the other value.) Therefore, once the double-reciprocal plot has been constructed, V_{max} can be de-

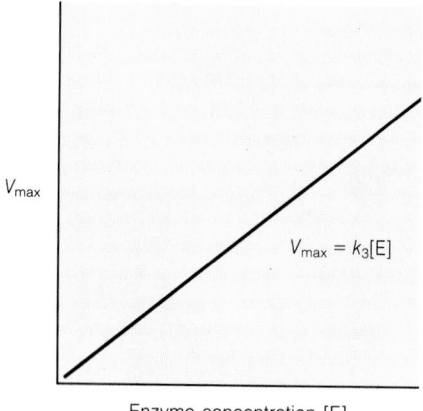

V_{max}

$V_{max} = k_3[E]$

Enzyme concentration [E]

Figure 6-10   The Linear Relationship Between V_{max} and Enzyme Concentration.   The linear increase in reaction velocity with enzyme concentration provides the basis for determining enzyme concentrations experimentally.

*K_m is often considered a measure of the affinity of an enzyme for its substrate, or, in effect, the dissociation constant for ES. In fact, however, it is a ratio of rate constants ($K_m = [k_2 + k_3]/k_1$), as you will discover if you choose to work Problem 6-12 at the end of the chapter.

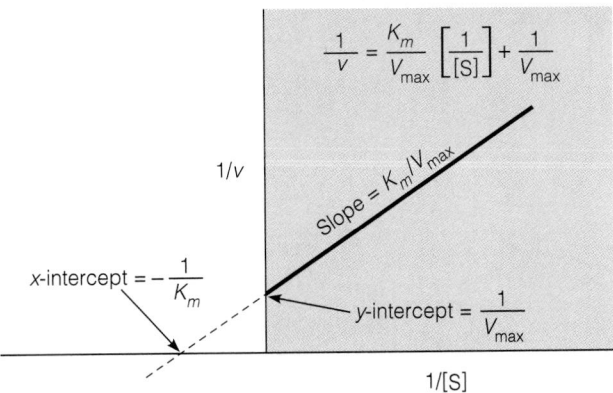

Figure 6-11   The Lineweaver-Burk Double-Reciprocal Plot. The reciprocal of the initial velocity, $1/v$, is plotted as a function of the reciprocal of the substrate concentration, $1/[S]$. K_m can be calculated from the x-intercept and V_{max} from the y-intercept.

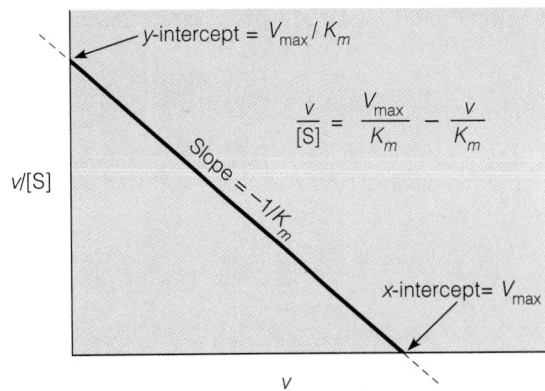

Figure 6-12   The Eadie-Hofstee Plot. The ratio $v/[S]$ is plotted as a function of v. K_m can be determined from the slope and V_{max} from the x-intercept.

termined directly from the reciprocal of the y-intercept and K_m from the negative reciprocal of the x-intercept. Furthermore, the slope can be used to check both values.

Thus, the Lineweaver-Burk plot is useful because it confirms by its linearity that the reaction in question is following Michaelis-Menten kinetics, and it allows determination of the parameters V_{max} and K_m without the complication of a hyperbolic shape. It also serves as a useful diagnostic in the analysis of enzyme inhibition because the several different kinds of reversible inhibitors affect the shape of the plot in characteristic ways.

The Lineweaver-Burk equation has some limitations, however. The main problem with it is that a long extrapolation is often necessary to determine K_m, with resulting uncertainty in the result. Moreover, the data points that are often the most crucial in determining the slope of the curve are those the farthest from the y-axis, and those points represent the samples with the lowest substrate concentrations, the lowest levels of enzyme activity, and hence the most uncertain values.

To circumvent these disadvantages, several alternatives to the Lineweaver-Burk equation have come into use to linearize kinetic data. One such alternative is the **Eadie-Hofstee equation,** which is represented graphically as a plot of v versus $v/[S]$. As Figure 6-12 illustrates, V_{max} is determined from the x-intercept and K_m from the slope of this plot. (To explore the Eadie-Hofstee plot and another alternative to the Lineweaver-Burk plot further, see Problem 6-13 at the end of this chapter.)

Determining K_m and V_{max}: An Example

To illustrate the value of the double-reciprocal plot in determining V_{max} and K_m, consider a specific example involving the enzyme hexokinase, as illustrated in Figures 6-13 and 6-14. Hexokinase is an important enzyme in cellular energy metabolism because it catalyzes the first reaction in the

glycolytic pathway. Using ATP as a source of both the phosphate group and the energy needed for the reaction, hexokinase catalyzes the phosphorylation of glucose on carbon atom 6:

$$\text{glucose} + \text{ATP} \underset{\text{hexokinase}}{\overset{}{\rightleftharpoons}} \text{glucose-6-phosphate} + \text{ADP} \quad \textbf{(6-12)}$$

To analyze this reaction kinetically, one must determine the initial velocity at each of several substrate concentrations. When an enzyme has two substrates, the usual approach is to vary the concentration of one substrate at a time, holding that of the other one constant at a sufficiently high (near-saturating) level to ensure that it does not become rate-limiting. Care must also be taken to ensure that the velocity determination is made before product accumulates to the point that the back reaction becomes significant.

In the experimental approach shown in Figure 6-13, glucose is the variable substrate, as ATP is present at a saturating concentration in each tube. Of the nine reaction mixtures set up for this experiment, one is designated the reagent blank (B) because it contains no glucose. The other eight tubes contain graded levels of glucose ranging from 0.05 to 0.40 mM. With all tubes prepared and maintained at some favorable temperature (25°C is often used), the reaction in each is initiated by the addition of a fixed amount of hexokinase.

The rate of product formation can then be determined either by continuous spectrophotometric monitoring of the reaction mixture (provided that one of the reactants or products absorbs light of a specific wavelength) or by allowing the reaction mixture to incubate for some short, fixed period of time, followed by chemical assay for either substrate depletion or product accumulation. In the case of the hexokinase reaction, the latter procedure is used because there is no direct photometric means of detecting any of the products or reactants.

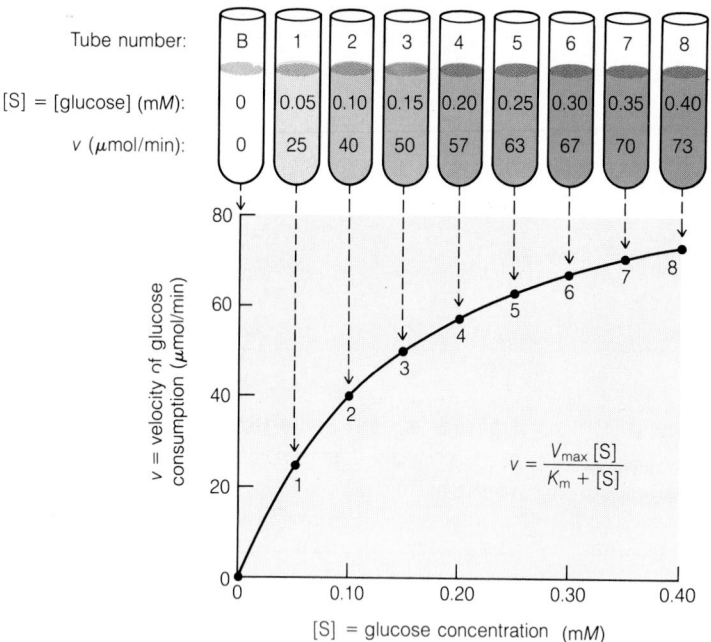

Figure 6-13 **Experimental Procedure for Studying the Kinetics of the Hexokinase Reaction.** Test tubes containing graded concentrations of glucose and saturating levels of ATP are incubated with a standard amount of hexokinase, and the initial rate of product appearance, v, is plotted as a function of substrate concentration [S]. The curve is hyperbolic, approaching V_{max} as the substrate concentration gets higher and higher. For the double-reciprocal plot derived from these data, see Figure 6-14.

As Figure 6-13 indicates, the initial velocity in tubes 1–8 ranged from 25 to 73 μmol of glucose consumed per minute, with no detectable reaction in the blank. (If any glucose consumption were noted in the blank, the values of tubes 1–8 would have to be corrected for that amount of noncatalytic reaction.) When these reaction velocities are plotted as a function of glucose concentration, the eight data points generate a hyperbolic curve like that in Figure 6-13. Although the data of Figure 6-13 are idealized for illustrative purposes, most kinetic data generated by this approach do, in fact, fit a hyperbolic curve unless the enzyme has some special properties that cause departure from Michaelis-Menten kinetics.

The hyperbolic curve of Figure 6-13 illustrates the need for some means of linearizing the analysis because neither

V_{max} nor K_m can be determined from the values as plotted, even though the data are known to be idealized. This need is met by the linear double-reciprocal plot of Figure 6-14. To obtain the data plotted here, reciprocals were calculated for each value of [S] and v from Figure 6-13. Thus, the [S] values of 0.05–0.40 mM generate reciprocals of 20–2.5 mM^{-1}, and the v values of 25–73 μmol/min give rise to reciprocals of 0.04–0.014 min/μmol. Because these are reciprocals, the data point for tube 1 is farthest from the origin, and each successive tube is represented by a point closer to the origin.

When these data points are connected by a straight line, the y-intercept is found to be 0.01 min/μmol and the x-intercept is -6.7 mM^{-1}. From these intercepts, we can calculate that $V_{max} = 1/0.01 = 100$ μmol/min and $K_m = -(1/-6.7) = 0.15$ mM. If we now go back to the Michaelis-

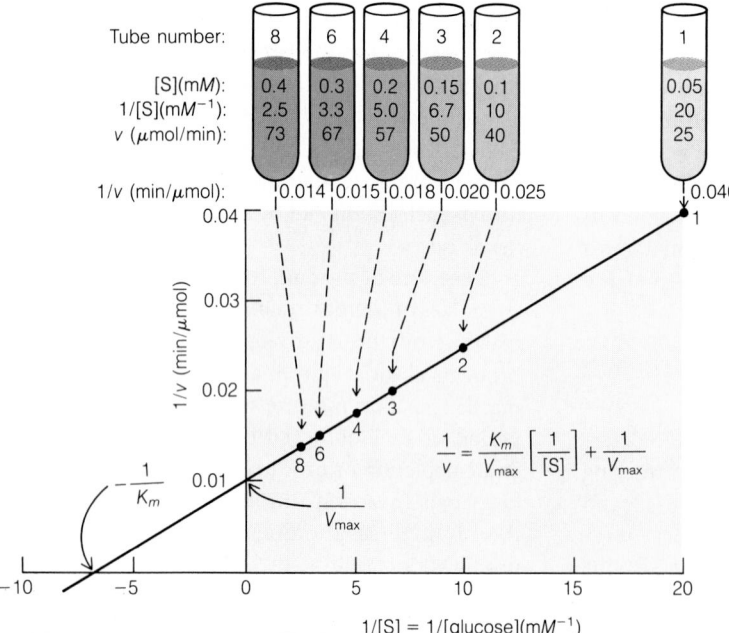

Figure 6-14 **Double-Reciprocal Plot for the Hexokinase Data of Figure 6-13.** For each test tube, $1/v$ and $1/[S]$ were calculated from the data of Figure 6-13, and $1/v$ was then plotted as a function of $1/[S]$. The y-intercept of 0.01 corresponds to $1/V_{max}$, so V_{max} is 100 μmol/min. The x-intercept of -6.7 corresponds to $-1/K_m$, so K_m is 0.15 mM.

Menten plot of Figure 6-13, we can see that both of these values are eminently reasonable because we can readily imagine that the plot is rising hyperbolically to a maximum of 100 µmol/min. Moreover, the graph reaches one-half of this value at a substrate concentration of 0.15 mM, which turns out to be the data point for tube 3. This, of course, is the K_m of hexokinase for glucose, often written $K_{m, glucose}$.

The enzyme also has a K_m value for the other substrate, $K_{m, ATP}$, but that would have to be determined by varying the ATP concentration while holding the glucose concentration constant. Interestingly, hexokinase phosphorylates not only glucose but also other hexoses, and has a distinctive K_m value for each. The K_m for fructose, for example, is 1.5 mM, which means that it takes ten times more fructose than glucose to sustain the reaction at one-half of its maximum velocity.

Though somewhat simplified and idealized, this is the approach that enzymologists take in studying the kinetics of enzyme-catalyzed reactions. Their analyses are often more complicated than this, and they almost always use a computer to calculate and plot double-reciprocal data and to determine K_m and V_{max} values, but the basic approach is the same.

Enzyme Inhibition

Thus far, we have assumed that the only substances in cells that affect the activities of enzymes are their substrates. However, enzymes are also influenced by *products, alternative substrates, substrate analogues, drugs, toxins,* and an especially important class of regulators called *allosteric effectors.* Most of these substances have an inhibitory effect on enzyme activity, reducing (or sometimes even abolishing) the reaction rate with the desired substrate.

This **inhibition** of enzyme activity is important for several reasons. First and foremost, enzyme inhibition plays a vital role as a control mechanism in cells. As discussed in the next section, many enzymes are subject to regulation by specific small molecules other than their substrates. Enzyme inhibition is also important in understanding the mode of action of drugs and poisons, which frequently exert their effects by inhibiting specific enzymes. Inhibitors are also useful to enzymologists as tools in their studies of reaction mechanisms. Especially important in this latter case are *substrate analogues,* compounds that resemble the real substrate closely enough to bind to the active site but are then chemically unable to undergo reaction.

Inhibitors may be either *reversible* or *irreversible.* An **irreversible inhibitor** binds to the enzyme *covalently,* causing an irrevocable loss of catalytic activity. Not surprisingly, irreversible inhibitors are usually toxic to cells. Ions of heavy metals are often irreversible inhibitors, as are alkylating agents and nerve gas poisons. This is, in fact, the mode of action of many *insecticides* and *nerve gases* and the reason they are so toxic. These substances bind irreversibly to *acetylcholinesterase,* an enzyme that is vital to the transmission of nerve impulses (see Chapter 22). Inhibition of acetylcholinesterase activity leads to rapid paralysis of vital func-

tions and therefore to death. One such nerve gas is *diisopropyl fluorophosphate,* which binds covalently to the hydroxyl group of a critical serine at the active site of the enzyme, thereby rendering the enzyme molecule permanently inactive.

Many natural toxins are also irreversible inhibitors of enzymes. For example, the alkaloid *physostigmine,* a natural constituent of calabar beans, is toxic to animals because it is a potent inhibitor of acetylcholinesterase. The antibiotic *penicillin* is an irreversible inhibitor of serine-containing enzymes involved in bacterial cell wall synthesis. Penicillin is therefore effective in treating bacterial infections because it prevents the bacterial cells from forming cell walls.

In contrast, a **reversible inhibitor** binds to an enzyme in a dissociable manner, such that the bound and free forms of the inhibitor exist in equilibrium with each other. We can represent such binding as follows, with E as the enzyme, I as the inhibitor, and EI as the enzyme-inhibitor complex:

$$E + I \rightleftharpoons EI \qquad \text{(6-13)}$$

Clearly, the fraction of the enzyme that is available to the cell in active form depends on the concentration of the inhibitor and the strength of the enzyme-inhibitor complex.

The two most common forms of reversible inhibitors are called *competitive inhibitors* and *noncompetitive inhibitors.* A competitive inhibitor binds to the active site of the enzyme but cannot be processed by the enzyme. Such an inhibitor therefore competes with substrate molecules for the same site on the enzyme (Figure 6-15a) and reduces enzyme activity to the extent that the active sites of the enzyme molecules have inhibitor molecules rather than substrate molecules bound to them at any point in time.

A noncompetitive inhibitor, on the other hand, binds to the enzyme surface at a location other than the active site but in a manner that nonetheless inhibits catalytic activity (Figure 6-15b). Binding of a noncompetitive inhibitor to an enzyme reduces enzyme activity without affecting substrate binding because the inhibitor and the substrate do not compete for the same binding sites.

Enzyme Regulation

To understand the role of enzymes in cellular function, it is important to recognize that it is rarely in the cell's best interest to allow an enzyme to function at an indiscriminately high rate. Instead, the rates of enzyme-catalyzed reactions and the biochemical sequences of which they are a part must be continuously adjusted to keep them finely tuned to the needs of the cell. An important aspect of that adjustment lies in the cell's ability to control enzyme activities with specificity and precision.

We have already encountered a variety of regulatory mechanisms, including changes in substrate (and product) concentrations, alterations in pH, and the presence and concentration of inhibitors. Regulation that depends directly on the interactions of substrates and products with the enzyme

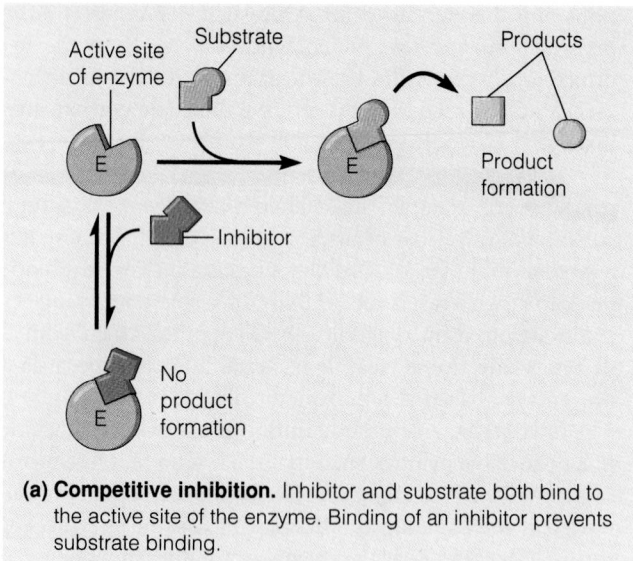

(a) Competitive inhibition. Inhibitor and substrate both bind to the active site of the enzyme. Binding of an inhibitor prevents substrate binding.

(b) Noncompetitive inhibition. Inhibitor and substrate bind to different sites. Binding of an inhibitor distorts the enzyme and decreases the likelihood of substrate binding.

Figure 6-15   Mode of Action of Competitive and Noncompetitive Inhibitors. Both competitive and noncompetitive inhibitors (red) bind reversibly to the enzyme (E), thereby inhibiting its activity.

is called **substrate-level regulation.** As the Michaelis-Menten equation makes clear, increases in substrate concentration result in higher reaction rates (see Figure 6-9). Conversely, increases in product concentration reduce the rate at which substrate is converted to product. (This inhibitory effect of product concentration is why *v* needs to be identified as the *initial* reaction velocity in the Michaelis-Menten equation, as given by equation 6-7. If a significant amount of product is already present, or accumulates during the course of the reaction, the equation becomes more complex than the simple form we have considered.)

As an example of substrate-level regulation, consider the phosphorylation of glucose to generate glucose-6-phosphate (reaction 6-12). This reaction is the first step in the glycolytic pathway, which will be discussed in Chapter 11. The enzyme *hexokinase* that catalyzes this reaction is inhibited by its product, glucose-6-phosphate. If utilization of glucose-6-phosphate is blocked for any reason, it will accumulate, inhibiting the hexokinase reaction and slowing down the further entry of glucose into the pathway.

Substrate-level regulation is an important control mechanism in cells, but it is not sufficient for the regulation of most reactions or reaction sequences. For most pathways, enzymes are regulated by other mechanisms as well. Two of the most important mechanisms are *allosteric regulation* and *covalent modification*. These mechanisms allow cells to turn enzymes on or off or to fine-tune their reaction rates by modulating enzyme activities appropriately.

Almost invariably, an enzyme that is regulated by such a mechanism catalyzes a reaction that represents the first step of a multistep sequence. By increasing or reducing the rate at which the first step functions, the whole sequence is effec-

tively controlled. Pathways that are regulated in this way include those required to break down large molecules (such as sugars, fats, or amino acids) to extract energy from them, as well as pathways that lead to the synthesis of substances needed by the cell.

We will discuss allosteric regulation and covalent modification in an introductory manner here, with the intention of returning to these mechanisms as we encounter specific examples in later chapters.

Allosteric Regulation

The single most important control mechanism whereby the rates of enzyme-catalyzed reactions are adjusted to meet cellular needs is **allosteric regulation.** To understand this mode of regulation, consider the pathway shown below, by which a cell converts some precursor A into some final product P via a series of intermediates B, C, and D, in a sequence of reactions catalyzed respectively by enzymes E_1, E_2, E_3, and E_4:

$$A \xrightarrow{E_1} B \xrightarrow{E_2} C \xrightarrow{E_3} D \xrightarrow{E_4} P$$

Product P could, for example, be an amino acid needed by the cell for protein synthesis, and A could be some common cellular component that serves as the starting point for the specific reaction sequence leading to P.

If allowed to proceed at a constant, unrestrained rate, this pathway has the capacity to convert large amounts of A to P, with possible adverse effects resulting from a depletion of A or an excessive accumulation of P (or both). Clearly, the best interests of the cell are served when the pathway is functioning not at its maximum rate or even some constant rate, but at a rate that is carefully tuned to the cellular need for P. Somehow, the enzymes of this pathway must be responsive to the cellular level of the product P in somewhat the same way that a furnace needs to be responsive to the temperature of the rooms it is intended to heat.

In the latter case, a thermostat provides the necessary regulatory link between the furnace and its "product," heat. In the case of enzymes, the desired regulation is possible because the product P is a specific inhibitor of E_1, the enzyme that catalyzes the first reaction in the sequence. This phenomenon is called **feedback** (or **end-product**) **inhibition** and is represented by the dashed arrow that connects the product P to enzyme E_1 in the following pathway:

$$A \xrightarrow[E_1]{} B \xrightarrow[E_2]{} C \xrightarrow[E_3]{} D \xrightarrow[E_4]{} P \qquad \textbf{(6-15)}$$

Feedback inhibition of E_1 by P

Feedback inhibition is one of the most common mechanisms used by cells to ensure that reaction sequences are adjusted to cellular needs.

Figure 6-16 provides a specific example of such a pathway—the five-step sequence whereby the amino acid *isoleucine* is synthesized from *threonine*, another amino acid. In this case, the first enzyme in the pathway, *threonine deaminase*, is regulated by the concentration of isoleucine within the cell. If isoleucine is being used by the cell (to synthesize proteins, most likely), the isoleucine concentration will be low. Under these conditions, threonine deaminase is active and the pathway functions to produce more isoleucine, thereby meeting the ongoing need for this amino acid. If the need for isoleucine decreases, isoleucine will begin to accumulate in the cell, and the increase in its concentration will lead to a decrease in the activity of threonine deaminase and hence to a reduced rate of isoleucine synthesis.

How can the first enzyme in a pathway (i.e., enzyme E_1 in reaction sequence 6-15) be sensitive to the concentration of a substance P that is neither its substrate nor its immediate product? Or, in terms of Figure 6-16, how can the activity of threonine deaminase be responsive to the concentration of isoleucine, when isoleucine differs sufficiently from threonine in structure that it is unlikely to be recognized by the active site of threonine deaminase?

The answer to this question was first proposed in 1963 by Jacques Monod and his fellow scientists Jean-Pierre Changeux and François Jacob. Although based initially on inconclusive data, their model was quickly substantiated and went on to become the foundation for our understanding of *allosteric regulation*. The term *allosteric* derives from the Greek for "another shape (or state)," thereby indicating that all enzymes capable of allosteric regulation can exist in two different states. In one of the two forms, the enzyme has a high affinity for its substrate(s), whereas in the other form, it has little or no affinity for its substrate. Enzymes with this property are called **allosteric enzymes.** The two different forms of an allosteric enzyme are readily interconvertible and are, in fact, in equilibrium with each other. Obviously,

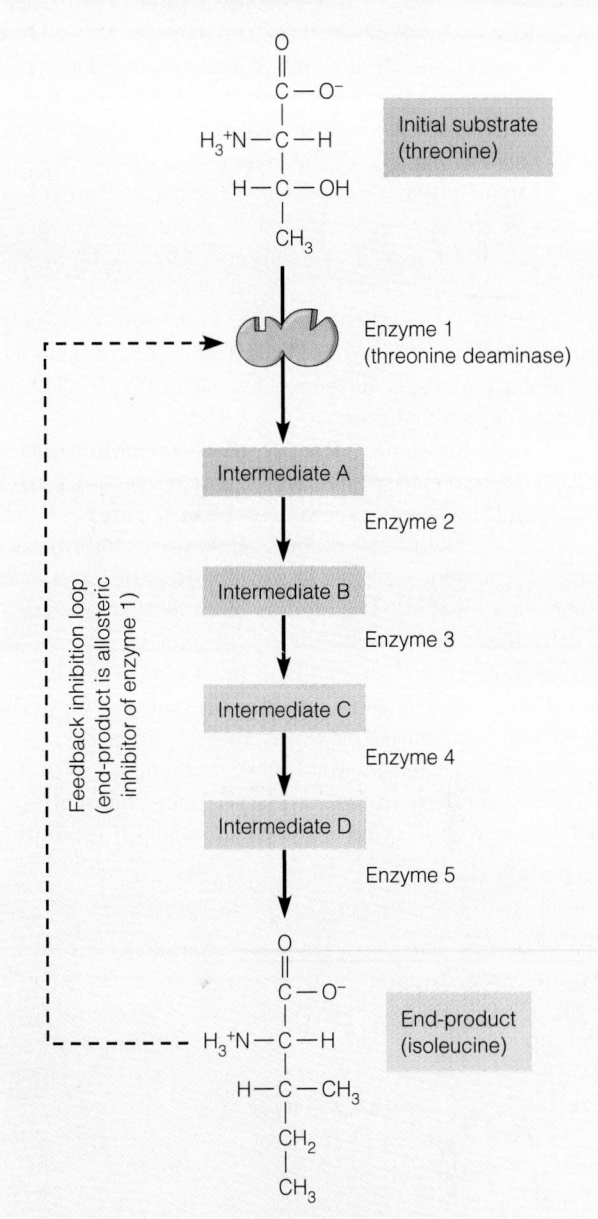

Figure 6-16   Allosteric Regulation of Enzyme Activity.
A specific example of feedback inhibition is seen in the pathway by which the amino acid isoleucine is synthesized from threonine, another amino acid. The first enzyme in the sequence, threonine deaminase, is allosterically inhibited by isoleucine.

the reaction rate is high when the enzyme is in its high-affinity form and low or even zero when the enzyme is in its low-affinity form.

Whether the active or inactive form of an allosteric enzyme is favored depends on the cellular concentration of the appropriate regulatory substance, called an **allosteric effector.** In the case of isoleucine synthesis, the allosteric enzyme is threonine deaminase and the allosteric effector is isoleucine. More generally, an allosteric effector is a small organic molecule that regulates the activity of an enzyme for which it is neither the substrate nor the immediate product.

An allosteric effector influences enzyme activity by binding to one of the two interconvertible forms of the enzyme, thereby stabilizing it in that state. The effector binds to the enzyme because of the presence on the enzyme surface of an **allosteric** (or **regulatory**) **site** that is distinct from the active site at which the catalytic event occurs. Thus, a distinguishing feature of all allosteric enzymes (and other allosteric proteins, as well) is the presence on the surface of an *active site* to which the substrate binds and an *allosteric site* to which the effector binds. In fact, some allosteric enzymes have multiple allosteric sites, each capable of recognizing a different effector.

An effector may be either an **allosteric inhibitor** or an **allosteric activator,** depending on the effect it has when bound to the allosteric site on the enzyme (Figure 6-17). The binding of an allosteric inhibitor shifts the equilibrium between the two forms of the enzyme to favor the low-affinity state (Figure 6-17a). The binding of an allosteric activator, on the other hand, shifts the equilibrium in favor of the high-affinity state (Figure 6-17b). In either case, binding of the effector to the allosteric site stabilizes the enzyme in one of its two interconvertible forms, thereby either decreasing or increasing the likelihood of substrate binding.

Most allosteric enzymes are large, multisubunit proteins with an active site or an allosteric site on each subunit. In fact, the active sites and allosteric sites are usually on different subunits of the protein, referred to as **catalytic subunits** and **regulatory subunits,** respectively (see Figure 6-17). This means, in turn, that the binding of effector molecules to the allosteric sites affects not just the shape of the regulatory subunits but that of the catalytic subunits as well.

Furthermore, many allosteric enzymes (and other multisubunit enzymes) display a property known as **cooperativity.** This means that as the multiple catalytic sites on the enzyme become successively occupied by substrate molecules, the affinity of the remaining sites increases. Since the multiple catalytic sites are on separate subunits, the effect of substrate binding at one site must be transmitted to other, as yet unfilled sites by conformational changes in the molecule.

Covalent Modification

In addition to allosteric regulation, many enzymes are also subject to control by **covalent modification.** In this form of regulation, the activity of an enzyme is affected by the addition or removal of specific chemical groups. Common modifications include the addition of phosphate groups, methyl groups, acetyl groups, or derivatives of nucleotides. Some of these modifications can be reversed, whereas others cannot. In each case, the effect of the modification is to activate or to inactivate the enzyme, or at least to adjust its activity upward or downward.

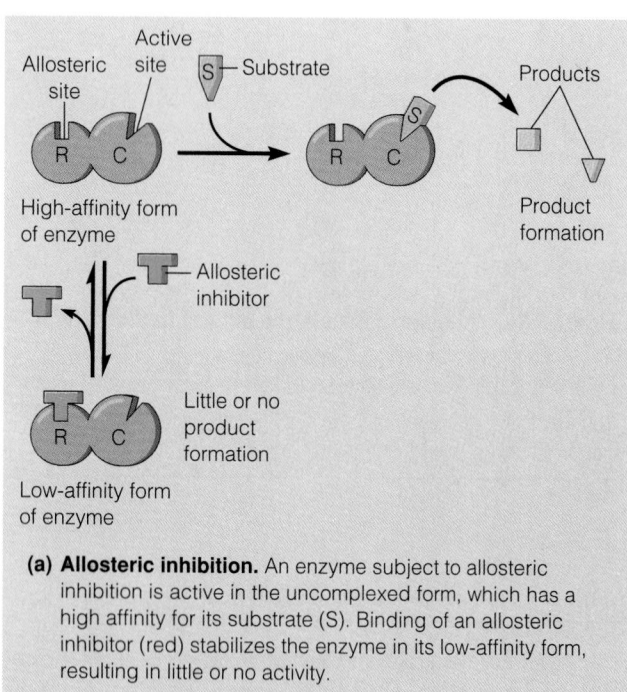

(a) Allosteric inhibition. An enzyme subject to allosteric inhibition is active in the uncomplexed form, which has a high affinity for its substrate (S). Binding of an allosteric inhibitor (red) stabilizes the enzyme in its low-affinity form, resulting in little or no activity.

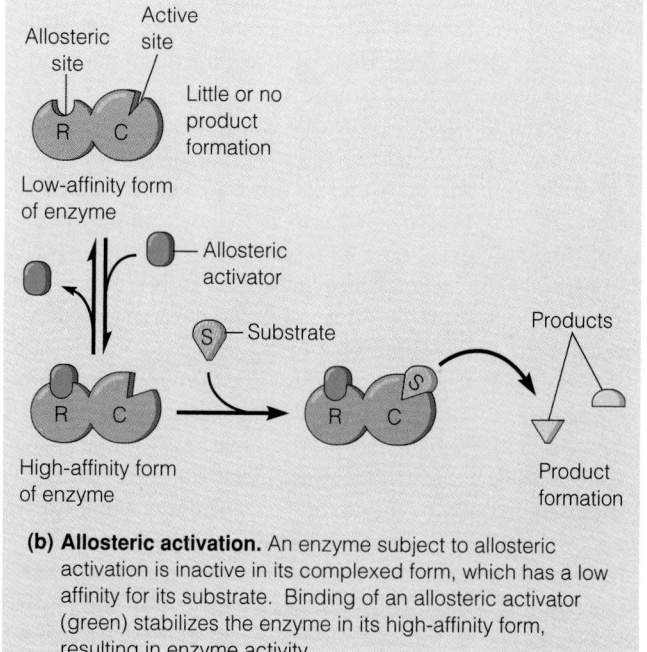

(b) Allosteric activation. An enzyme subject to allosteric activation is inactive in its complexed form, which has a low affinity for its substrate. Binding of an allosteric activator (green) stabilizes the enzyme in its high-affinity form, resulting in enzyme activity.

Figure 6-17 Mechanisms of Allosteric Inhibition and Activation. An allosteric enzyme consists of one or more catalytic subunits (C) and one or more regulatory subunits (R), each with an active site or an allosteric site, respectively. The enzyme exists in two forms, one with a high affinity for its substrate (and therefore a high likelihood of product formation) and the other with a low affinity (and a correspondingly low likelihood of product formation).

Phosphorylation/Dephosphorylation. One of the most frequently encountered and best understood covalent modifications involves the reversible addition of phosphate groups. The addition of phosphate groups is called **phosphorylation** and occurs most commonly by transfer of the phosphate group from ATP to the hydroxyl group of a serine, threonine, or tyrosine in the protein. Enzymes that catalyze the phosphorylation of other enzymes (or of other proteins) are called *protein kinases*. The reversal of this process, **dephosphorylation,** involves the removal of a phosphate group from a phosphorylated protein, catalyzed by enzymes called *phosphoprotein phosphatases*.

This mode of regulation is illustrated by *glycogen phosphorylase*, an enzyme found in skeletal muscle cells (Figure 6-18). This enzyme breaks down glycogen by successive removal of glucose units as glucose-1-phosphate (Figure 6-18a). Regulation of this dimeric enzyme is achieved in part by the presence in muscle cells of two interconvertible

forms of the enzyme, an active form called *phosphorylase a* and an inactive form called *phosphorylase b* (Figure 6-18b). When glycogen breakdown is required in the muscle cell, the inactive *b* form of the enzyme is converted into the active *a* form by the addition of a phosphate group to a particular serine on each of the two subunits of the phosphorylase molecule. This reaction is catalyzed by *phosphorylase kinase* and results in a conformational change of phosphorylase to the active form. When glycogen breakdown is no longer needed, the phosphate groups are removed from phosphorylase *b* by a specific phosphatase. *Glycogen synthase*, the enzyme that adds glucose units onto the glycogen chain, responds in the opposite manner: It is inactive in the phosphorylated form and is activated by dephosphorylation.

In addition to regulation by the phosphorylation/dephosphorylation mechanism shown in Figure 6-18, glycogen phosphorylase is also an allosteric enzyme, inhibited by

Figure 6-18   The Regulation of Glycogen Phosphorylase by Phosphorylation. **(a)** Glycogen phosphorylase is a dimeric enzyme in muscle cells that releases glucose units from glycogen molecules as glucose-1-phosphate, which can then be used by the muscle cells as an energy source. **(b)** Glycogen phosphorylase is regulated in part by a phosphorylation/dephosphorylation mechanism. The inactive form of the enzyme, phosphorylase *b*, can be converted to the active form, phosphorylase *a*, by the transfer of phosphate groups from ATP to a particular serine on each of the two subunits of the enzyme molecule. This phosphorylation reaction is catalyzed by the enzyme phosphorylase kinase. Removal of the phosphate groups by a specific phosphatase enzyme returns the phosphorylase molecule to the inactive *b* form.

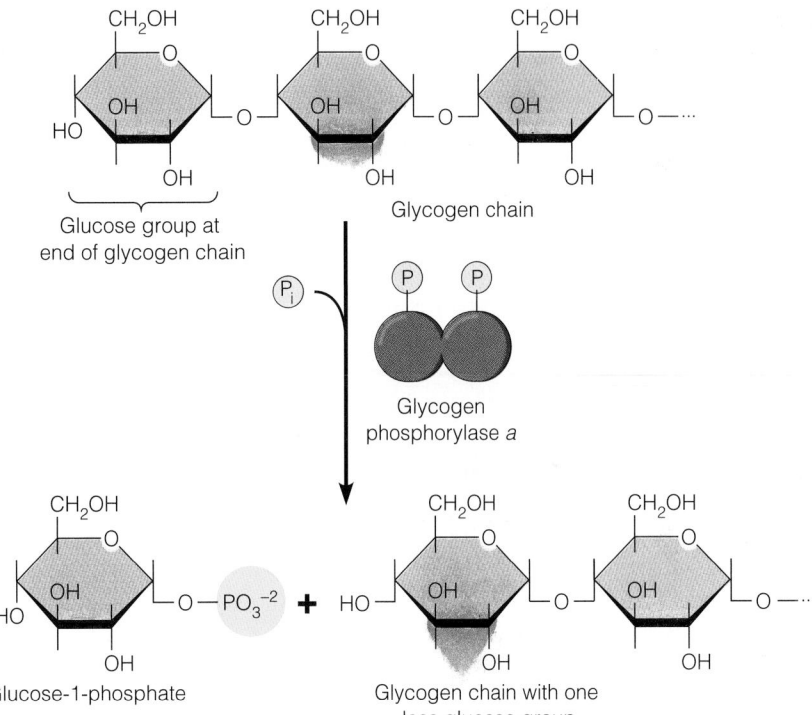

(a) The reaction catalyzed by glycogen phosphorylase

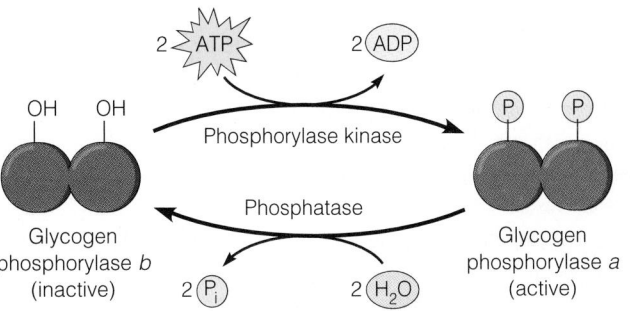

(b) Regulation of glycogen phosphorylase

glucose and ATP and activated by AMP. If a hormonal signal arrives that triggers the phosphorylation and hence the activation of the phosphorylase in a muscle cell that still has an adequate supply of glucose, the glucose will inhibit the enzyme allosterically until it is actually needed. On the other hand, muscle cells with a low level of glucose will benefit immediately from the conversion of phosphorylase to the active form.

The existence of two levels of regulation for glycogen phosphorylase illustrates an important aspect of enzyme regulation. Many enzymes are controlled by two or more regulatory mechanisms, thereby enabling the cell to make appropriate responses to a variety of situations.

Proteolytic Cleavage. A different kind of covalent activation of enzymes involves the one-time, irreversible removal of a portion of the polypeptide chain by an appropriate proteolytic (protein-degrading) enzyme. This kind of modification, called **proteolytic cleavage,** is exemplified especially well by the *proteolytic enzymes* of the pancreas, which include trypsin, chymotrypsin, and carboxypeptidase. Synthesized in the pancreas, these enzymes are secreted into the duodenum of the stomach in response to a hormonal signal. These proteases, along with the pepsin of the stomach and other proteolytic enzymes secreted by the cells of the intestine, can digest almost all ingested proteins into free amino acids, which can then be absorbed by the intestinal epithelial cells.

Pancreatic proteases are not synthesized in their final, active form; that would likely cause problems for the cells of the pancreas, which must protect themselves against their own proteolytic enzymes. Instead, each of these enzymes is synthesized as a slightly longer, catalytically inactive molecule called a *zymogen.* Zymogens must themselves be cleaved proteolytically to yield active enzymes. Several such events are shown in Figure 6-19. For example, trypsin is synthesized initially as a zymogen called *trypsinogen.* When trypsinogen reaches the duodenum, it is activated by the removal of a hexapeptide (a string of six amino acids) from its N-terminus by the action of *enterokinase,* a membrane-bound protease produced by the duodenal cells. The active trypsin then activates the other zymogens by specific proteolytic cleavages.

RNA Molecules as Enzymes: Ribozymes

Prior to the early 1980s, it was thought that all enzymes were proteins. Indeed, that statement was regarded as one of the fundamental truths of cellular biology and was found in every textbook. This chapter is no exception; all the discussion to this point has been about protein catalysts. Cell biologists became convinced that all enzymes were proteins because every enzyme isolated in the 55 years following Sumner's purification of urease in 1926 turned out to be a protein. But biology is full of surprises, and we now know that the statement needs to be revised to include RNA molecules as well.

The first evidence came in 1981, when Thomas Cech and his colleagues discovered an apparent exception to the rule. They were studying the splicing of an internal segment from a specific ribosomal RNA precursor (pre-rRNA) in *Tetrahymena thermophila,* a single-celled eukaryote. (As you will learn in Chapter 18, many eukaryotic RNA molecules require the removal of one or more internal segments called *introns* or *intervening sequences* [IVS] before they become functional in the cell. The removal process involves the excision of the IVS and a splicing together of the two pieces of the original molecule at the excision site.) In the course of their work, the researchers made the remarkable observation that the process apparently proceeded without the presence of proteins! Describing their attempt to study intron excision in vitro, Cech later wrote:

It turned out that some of the small molecules—notably magnesium ion and any of several forms of the nucleotide guanosine—were required for the reaction to proceed. To our great surprise, however, the nuclear extract containing the enzymes was not. We were forced to conclude either that the enzymatic activity came from a protein bound so tightly to the RNA that we were unable to strip it off or that the

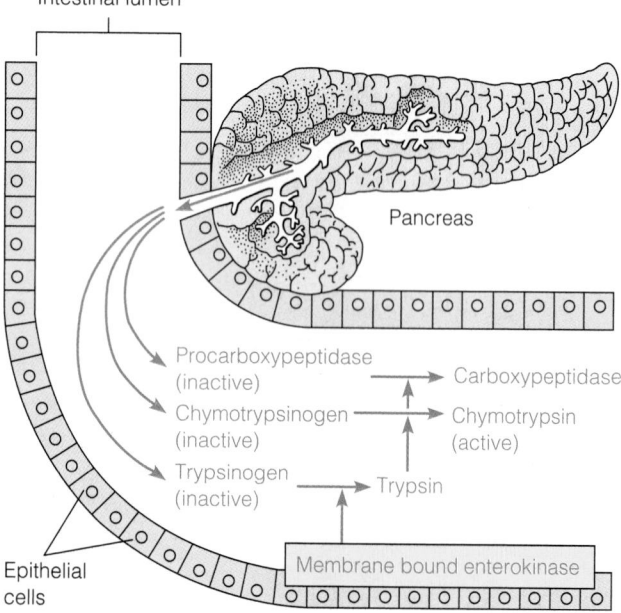

Figure 6-19 Activation of Pancreatic Zymogens by Proteolytic Cleavage. Pancreatic proteases are synthesized and secreted into the small intestine as inactive precursors called zymogens. Procarboxypeptidase, trypsinogen, and chymotrypsinogen are zymogens. Activation of trypsinogen to trypsin requires removal of a six-peptide segment by the enzyme enterokinase, a duodenal enzyme. Trypsin then activates other zymogens by proteolytic cleavage. (Procarboxypeptidase is activated by a single cleavage event, whereas the activation of chymotrypsinogen is a somewhat more complicated two-step process, the details of which are not shown here.)

RNA was catalyzing its own splicing. Given the deeply rooted nature of the idea that all biological catalysts are proteins, the hypothesis of catalysis by RNA was not easy to accept. (Cech, 1987, p. 1533.)

But further experimentation bore out their initial conclusion: The removal of a 413-nucleotide IVS from the *Tetrahymena* pre-rRNA is catalyzed by the pre-rRNA molecule itself.

Figure 6-20 shows the excision process, which involves ① folding of the unspliced pre-rRNA molecule to form a loop, ② attack by a hydroxyl group of a free guanosine nucleotide that acts as a cofactor, ③ cleavage and splicing of the rRNA molecule with release of the IVS, and ④ further autocatalytic cleavage of the IVS molecule to remove 19 more nucleotides. The fully processed IVS is itself a ribozyme, capable of shortening or elongating small oligonucleotides.

One might argue that the rRNA molecule in Figure 6-20 fails to satisfy the definition of a catalyst, which requires that the catalyst itself is not altered in the reaction process. However, another RNA-based catalyst discovered at about the same time in the laboratory of Sidney Altman overcomes this objection. The enzyme, called *ribonuclease P*, cleaves transfer RNA precursors (pre-tRNAs) to yield functional RNA molecules. (In this case, a terminal segment of the RNA molecule is removed rather than an internal IVS as for pre-rRNA processing.)

It had been known for some time that ribonuclease P consisted of a protein component and an RNA component, but it was generally assumed that the active site was on the protein component. By isolating the components and studying them separately, however, Altman and his colleagues showed unequivocally that the isolated protein component was completely inactive, whereas the RNA component was capable of catalyzing the specific cleavage of tRNA precursors on its own and was not itself altered in the process. Furthermore, the RNA-catalyzed reaction followed Michaelis-Menten kinetics, further evidence that the RNA component was acting like a true enzyme. (The protein component enhances activity but is not required for either substrate binding or cleavage.)

The significance of these findings was recognized by the Nobel Prize that Cech and Altman shared in 1989 for their discovery of RNA-based catalysts, or **ribozymes.** Since these initial discoveries, additional examples of ribozymes have been reported. Especially significant are recent reports that ribosomes may have active sites involving RNA. Ribosomes can be thought of as very large enzymes because they catalyze the formation of the peptide bonds that add successive amino acids to a growing polypeptide chain (see Figure 2-19b). Ribosomes consist of both protein and RNA molecules, and it had long been assumed that the active sites involved in the polymerization process were on the protein molecules. However, recent evidence from the laboratory of Harry Noller indicates that the proteins can be extracted from ribosomes without destroying the ability of the ribosomes to catalyze peptide bond formation.

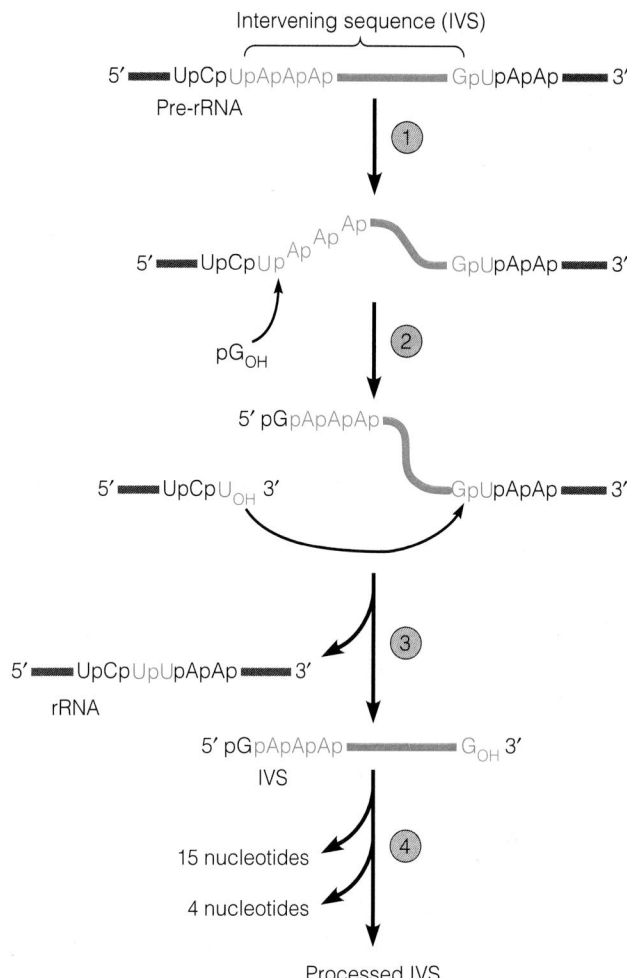

Figure 6-20 Self-Excision and Splicing of the Intervening Sequence from *Tetrahymena* Pre-rRNA. A ribosomal RNA precursor (pre-rRNA) molecule from *Tetrahymena* contains an intervening sequence (IVS) that is capable of catalyzing its own excision from the pre-rRNA molecule. The excision and splicing process occurs in four steps. ① The pre-rRNA molecule forms a loop at the IVS region. ② The IVS loop is attacked by a hydroxyl group of a free guanosine nucleotide, pG, that functions as a cofactor. ③ The pre-rRNA molecule is cleaved at the 5′ end of the IVS segment, with the addition of G to the IVS. The uridine at the 3′ end of the other rRNA fragment then attacks the 3′ end of the IVS, splicing the two rRNA pieces together and releasing the IVS. ④ The IVS then undergoes further autocatalytic cleavage to remove 19 more nucleotides by cleaving first a 15-nucleotide segment and then a 4-nucleotide piece.

Although most biologists were initially astonished by the discovery of ribozymes, in retrospect there is no reason why RNA molecules should not be able to function as enzymes. Like proteins, RNA molecules can take on complex tertiary structure, and that is the prerequisite for catalytic function in both cases.

The discovery of ribozymes has markedly changed the way we think about the origin of life on Earth. For many years, scientists have speculated that the first catalytic macromolecules must have been amino acid polymers resembling proteins. But this concept immediately ran into difficulty because there was no obvious way for a primitive protein to carry information or to replicate itself, two primary attributes of life. However, if the first catalysts were RNA rather than protein molecules, it becomes conceptually easier to imagine a system of RNA molecules acting both as catalysts and as replicating systems capable of transferring information between generations.

PERSPECTIVE

We come full circle and return to the theme raised in the introduction to this chapter—namely, that thermodynamics allows us to assess the feasibility of a reaction but says nothing about the likelihood that the reaction will actually occur at a reasonable rate in the cell. To ensure that the activation energy requirement is met, a catalyst is required, which in biological systems is always an enzyme. All protein enzymes are chains of amino acids in a genetically programmed sequence that are sensitive to temperature and pH. They are also exquisitely specific, either for a single specific substrate or for a class of compounds. The actual catalytic process takes place at the active site, a critical cluster of amino acids responsible for substrate binding and activation and for the actual chemical reaction. Binding of the appropriate substrate at the active site induces a more stringent fit between enzyme and substrate, thereby facilitating substrate activation.

An enzyme-catalyzed reaction proceeds via an enzyme-substrate intermediate and follows Michaelis-Menten kinetics, characterized by a hyperbolic relationship between the initial reaction velocity and the substrate concentration. The upper limit on velocity is called V_{max}, and the substrate concentration needed to reach one-half of the maximum velocity is termed the Michaelis constant, K_m. The hyperbolic relationship between v and [S] can be linearized by a double-reciprocal equation and plot, from which V_{max} and K_m can be determined graphically or by computer analysis.

Enzyme activity is influenced not only by substrate availability but also by products, alternative substrates, substrate analogues, drugs, and toxins, most of which have an inhibitory effect. Inhibition may be either reversible or irreversible, with the latter category involving covalent bonding of the inhibitor to the enzyme surface. A reversible inhibitor, on the other hand, binds to an enzyme in a dissociable manner, either at the active site (competitive inhibition) or elsewhere on the enzyme surface (noncompetitive inhibition).

Enzymes must be regulated to adjust their activity levels to cellular needs. Substrate-level regulation involves the effects of substrate and product concentrations on the reaction rate. Additional control mechanisms include allosteric regulation and covalent modification. Most allosterically regulated enzymes catalyze the first step in a reaction sequence and are multisubunit proteins with multiple catalytic subunits and multiple regulatory subunits. Each of the catalytic subunits has an active site that recognizes substrates and products, whereas each regulatory subunit has one or more allosteric sites that recognize specific effector molecules. A given effector may either inhibit or activate the enzyme, depending upon which form of the enzyme is favored by effector binding. The most common covalent modifications include phosphorylation and dephosphorylation, as exemplified by the enzyme glycogen phosphorylase, and proteolytic cleavage, as occurs in the activation of the zymogen forms of proteolytic enzymes secreted by the pancreas.

Although it was long thought that all enzymes were proteins, we now recognize the catalytic properties of certain RNA molecules called ribozymes. These include some rRNA molecules that are able to catalyze the removal of their own intervening sequences, RNA components of enzymes that also contain protein components, and perhaps even the RNA components of assembled ribosomes. The discovery of ribozymes has changed the way we think about the origin of life on Earth because RNA molecules, unlike proteins, are capable of replicating themselves.

KEY TERMS FOR SELF-TESTING

enzyme catalysis (p. 138)
enzyme (p. 138)

Activation Energy and the Metastable State

activation energy (E_A) (p. 139)

metastable state (p. 139)
catalyst (p. 140)

Enzymes as Biological Catalysts

active site (p. 141)
prosthetic group (p. 141)

substrate specificity (p. 143)
group specificity (p. 143)
induced-fit model (p. 146)
substrate activation (p. 147)
nucleophilic substitution (p. 147)
electrophilic substitution (p. 148)

PROBLEM SET

6-1. The Need for Enzymes. You are now in a position to appreciate the difference between the thermodynamic feasibility of a reaction and the likelihood that it will actually proceed.

(a) Many reactions that are thermodynamically possible do not occur at an appreciable rate because of an activation energy requirement. In molecular terms, what does this mean?

(b) One way to meet this requirement is by an input of heat, which in some cases need only be an initial, transient input. Give an example, and explain what this accomplishes in molecular terms.

(c) An alternative solution is to lower the activation energy. What does it mean in molecular terms to say that a catalyst lowers the activation energy of a reaction?

(d) Organic chemists often use inorganic catalysts such as nickel, platinum, or cations in their reactions, whereas cells use proteins called enzymes. What advantages can you see to the use of enzymes? Can you think of any disadvantages?

6-2. Activation Energy. As shown in reaction 6-2, hydrogen peroxide, H_2O_2, decomposes to H_2O and O_2. The activation energy, E_A, for the uncatalyzed reaction at 20°C is 18 kcal/mol. The reaction can be catalyzed by colloidal platinum ($E_A = 13$ kcal/mol) or by the enzyme *catalase* ($E_A = 7$ kcal/mol).

(a) Draw an activation energy diagram for this reaction under catalyzed and uncatalyzed conditions, and explain what it means for the activation energy to be lowered from 18 to 13 kcal/mol by platinum but from 18 to 7 kcal/mol by catalase.

(b) Suggest two properties of catalase that make it a more suitable intracellular catalyst than platinum.

(c) Suggest yet another way in which the rate of hydrogen peroxide decomposition can be accelerated. Is this a suitable means of increasing reaction rates within cells? Why or why not?

(d) Recall from Chapter 4 that the catalase present in eukaryotic cells is localized within the peroxisomes, along with any of several H_2O_2-generating enzymes. Given the toxicity of hydrogen peroxide to the cell, explain in terms of enzyme kinetics why it is advantageous to have H_2O_2-generating enzymes and catalase compartmentalized together within a membrane-bounded organelle.

6-3. Properties of Enzymes. For each of the following, indicate with a C if it is a property of enzymes as catalysts, with a P if it is a property of enzymes as proteins, and with an N if it is not a valid property of an enzyme.

(a) Contains an active site that usually represents less than 5% of the surface area of the molecule.

(b) Increases the rate of a chemical reaction by reducing the standard free energy change for the reaction.

(c) Can speed up the rate of exergonic but not endergonic reactions.

(d) Complexes transiently with reactant molecules.

(e) Is sensitive to variations in pH of the milieu.

(f) Displays a high degree of substrate specificity.

(g) Distorts substrate molecules, thereby weakening bonds and increasing the likelihood of breaking bonds.

(h) Is usually very heat-stable.

6-4. Temperature and pH Effects. Figure 6-5 illustrates enzyme activities as a function of temperature and pH. In general, the activity of a specific enzyme is highest at the temperature and pH that are characteristic of the environment in which the enzyme functions in vivo.

(a) Suggest the biological significance of these characteristic temperature and pH enzyme activity curves.

(b) Explain the shape of each curve in terms of the major chemical or physical factors that affect enzyme activity.

(c) Figure 6-21 is a graph of enzyme activity versus pH for three enzymes: pepsin, papain, and cholinesterase. For each example, suggest the adaptive advantage, to the enzyme or organism, of having the enzyme activity curve shown, and explain the activity in terms of enzyme structure.

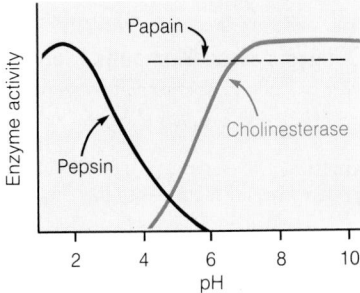

Figure 6-21   The pH Dependence of Several Enzymes.
The differing pH requirements of enzymes are shown here for the enzymes pepsin, papain, and cholinesterase. Pepsin is a protease found in the stomach, papain is a protease from papaya fruit, and cholinesterase, or acetylcholinesterase, is important in the transmission of nerve signals.

6-5. Enzyme Specificity. All enzymes are highly specific in both the reaction they catalyze and their choice of substrate. Substrate specificity can be for a particular molecule or class of molecules.

(a) A proteolytic enzyme such as trypsin can usually degrade a variety of polypeptide chains, whereas a dehydrogenase is usually absolutely specific for a particular substrate. Explain.

(b) Subtilisin is a bacterial protease that can cleave any peptide bond, regardless of the specific amino acids involved. Trypsin, on the other hand, splits peptide bonds only on the carboxyl side of lysine and arginine groups. What differences might you expect in the active sites of these two enzymes?

(c) Compounds that are sufficiently similar in structure to a bona fide substrate to allow them to bind to the active site of an enzyme but that cannot then undergo the reaction catalyzed by that enzyme are usually highly effective competitive inhibitors of enzyme activity. Explain.

6-6. Michaelis-Menten Kinetics. Figure 6-22 represents a Michaelis-Menten plot for a typical enzyme, with initial reaction plotted as a function of substrate concentration. Three regions of the curve are identified by the letters A, B, and C. For each of the statements that follow, indicate with a single letter which one of the three regions of the curve fits the statement best. A given letter can be used more than once.

(a) The active site of an enzyme molecule is occupied by substrate most of the time.

(b) The active site of an enzyme molecule is free most of the time.

(c) The range of substrate concentration in which most enzymes usually function in normal cells.

(d) Includes the point $(K_m, V_{max}/2)$.

(e) Reaction velocity is limited mainly by the number of enzyme molecules present.

(f) Reaction velocity is limited mainly by the number of substrate molecules present.

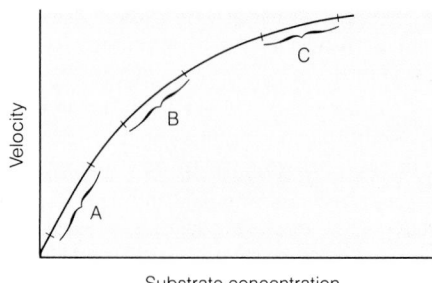

Figure 6-22   Analysis of the Michaelis-Menten Plot.

6-7. Enzyme Kinetics. The enzyme β-galactosidase catalyzes the hydrolysis of the disaccharide lactose into its component monosaccharides:

$$\text{lactose} + \text{H}_2\text{O} \xrightarrow{\;\;\beta\text{-galactosidase}\;\;} \text{glucose} + \text{galactose} \quad \textbf{(6-16)}$$

To determine V_{max} and K_m of β-galactosidase for lactose, the same amount of enzyme (1 μg per tube) was incubated with a series of lactose concentrations under conditions where product concentrations remained negligible. At each lactose concentration, the initial reaction velocity was determined by assaying for the amount of

lactose remaining at the end of the assay. The following data were obtained:

Lactose concentration (mM)	Rate of lactose consumption (μmol/min)
1	10.0
2	16.7
4	25.0
8	33.3
16	40.0
32	44.4

(a) Why is it necessary to specify that "product concentrations remained negligible" during the course of the reaction?

(b) Plot v (rate of lactose consumption) versus [S] (lactose concentration). Why is it that when the lactose concentration is doubled, the increase in velocity is always less than twofold?

(c) On the same graph as part b, plot the results you would expect if each tube contained only 0.5 μg of enzyme. Explain.

(d) Calculate $1/v$ and $1/[S]$ for each entry on the data table, and plot $1/v$ versus $1/[S]$.

(e) Determine K_m and V_{max} from your double-reciprocal plot.

6-8. More Enzyme Kinetics. The reaction sequence in equation 6-17 (page 165) is a part of the tricarboxylic acid (TCA) cycle, which you will encounter in Chapter 12. The enzyme fumarase converts the four-carbon compound fumarate into malate by the addition of water across the double bond. The enzyme malate dehydrogenase (MDH) then catalyzes the oxidation of malate to oxaloacetate, with a coenzyme called nicotinamide adenine dinucleotide (NAD) as the electron acceptor. The K_m value of fumarase for fumarate is 5×10^{-6} M (5 μM).

(a) At what fraction of its maximum velocity will fumarase proceed in a test tube containing a negligible concentration of malate with fumarate present at a concentration of (i) 0.5 μM, (ii) 5 μM, or (iii) 50 μM?

(b) In each of two test tubes (A and B), MDH is incubated with malate in the presence of excess NAD^+ (i.e., the coenzyme concentration is nonlimiting). The starting concentrations of malate are 0.1 mM and 0.2 mM for test tubes A and B, respectively, and the concentrations of NADH and oxaloacetate remain negligible during the experiment. The reaction in test tube B is found to have an initial reaction velocity (v_B) that is 1.5 times the initial velocity in tube A (v_A). What is the K_m of MDH for malate?

(c) Why is it necessary to specify in part b that v_A and v_B are *initial* reaction velocities?

(d) How would you proceed if you wished to determine the K_m value of MDH for its other substrate, NAD^+?

6-9. Still More Enzyme Kinetics. Shown in Figure 6-23 are Lineweaver-Burk plots for the enzyme hexokinase (see reaction 6-12) as isolated from a normal (nonmutant) strain of the bacterium *Escherichia coli* and two variants of the enzyme as isolated from mutant *E. coli* strains that are defective in glucose utilization.

(a) Indicate whether the K_m and V_{max} values for mutant 1 and mutant 2 are higher than, lower than, or the same as the corresponding values for the normal enzyme.

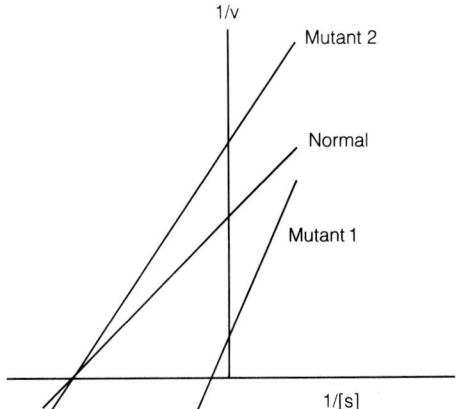

Fumarate — Fumarase (H₂O) → Malate — MDH (NAD⁺ → NADH, H⁺) → Oxaloacetate **(6-17)**

(The chemical structures shown:)

$$\text{Fumarate} \xrightarrow[\text{Fumarase}]{H_2O} \text{Malate} \xrightarrow[\text{MDH}]{NAD^+ \quad NADH, H^+} \text{Oxaloacetate} \qquad (6\text{-}17)$$

(b) If the normal form of the enzyme can phosphorylate glucose at the rate of 40 μmol/min when the glucose concentration is 0.1 mM and the ATP concentration is nonlimiting, which of the two mutants is likely to require a higher concentration of glucose to achieve the same reaction velocity?

(c) With the added information that $V_{\max}/K_m$ is a measure of the efficiency of the reaction (i.e., $V_{\max}/K_m$ is high for an efficient enzyme), determine whether mutant 1 is more or less efficient than the normal enzyme. What about mutant 2?

Figure 6-23   Lineweaver-Burk Plots for Normal and Mutant Enzymes.

6-10. Competitive Inhibition. A distinguishing characteristic of a competitive inhibitor is that it binds reversibly to the active site of the enzyme, thereby competing with the substrate for the active site and reducing the rate at which the substrate is processed. To understand competitive inhibition, consider that the monkeys in Box 6A have access not only to real peanuts but also to plastic peanuts that look just like real peanuts. A monkey will therefore pick up plastic peanuts and real peanuts indiscriminately, but only the latter can be shelled. The initial reaction velocity v is expressed in terms of real peanuts shelled per minute. Decide whether each of the following statements is true (T) or false (F) of a roomful of monkeys confronted with a fixed "concentration" of plastic peanuts and variable "concentrations" of real peanuts. If false, reword the statement to make it true.

(a) At a fixed concentration of plastic peanuts, the reaction velocity achieved for a specific concentration of real peanuts will be less than would be observed in the absence of the plastic peanuts.

(b) If the concentration of plastic peanuts is increased, the concentration of real peanuts necessary to reach a specific reaction velocity will be higher also.

(c) The K_m value of the monkeys for real peanuts will be higher when determined in the presence of a fixed concentration of plastic peanuts than in their absence.

(d) The $V_{\max}$ value for a specific number of monkeys will be the same in the presence or absence of plastic peanuts, because any finite concentration of plastic peanuts would have no effect on reaction velocity if the concentration of real peanuts were infinite.

(e) The Lineweaver-Burk double-reciprocal plot for data obtained in the presence of a fixed concentration of plastic peanuts will have the same y-intercept but a steeper slope than it would if the data had been obtained in the absence of the plastic peanuts.

6-11. Biological Relevance. Explain the biological relevance of each of the following observations concerning enzyme regulation.

(a) When you need a burst of energy, the hormones epinephrine and glucagon are secreted into your bloodstream and circulated to your muscle cells, where they initiate a cascade of reactions that leads to the phosphorylation of the inactive (b) form of glycogen phosphorylase, thereby converting the enzyme into the active (a) form.

(b) Even in the a form, glycogen phosphorylase is allosterically inhibited by a high concentration of glucose or ATP within a specific muscle cell.

(c) Your pancreas synthesizes and secretes the proteolytic enzyme carboxypeptidase in the form of an inactive precursor called procarboxypeptidase, which is activated as a result of proteolytic cleavage by the enzyme trypsin in the duodenum of your small intestine.

6-12. Derivation of the Michaelis-Menten Equation. For the enzyme-catalyzed reaction in which a substrate S is converted into a product P (see reaction 6-5), velocity can be defined as the disappearance of substrate or the appearance of product per unit time:

$$v = \frac{-d[S]}{dt} = \frac{+d[P]}{dt} \qquad (6\text{-}18)$$

Beginning with this definition and restricting your consideration to the initial stage of the reaction when [P] is essentially zero, derive the Michaelis-Menten equation (see equation 6-7). The following points may help you in your derivation:

(a) Begin by expressing the rate equations for $d[S]/dt$, $d[P]/dt$, and $d[ES]/dt$ in terms of concentrations and rate constants.

(b) Assume a steady state at which the enzyme-substrate complex of reaction 6-6 is being broken down at the same rate at which it is being formed, such that the net rate of change, $d[\text{ES}]/dt$, is zero.

(c) Note that the total amount of enzyme present, E_t, is the sum of the free form E_f plus the amount of complexed enzyme ES: $\text{E}_t = \text{E}_f + \text{ES}$.

(d) When you get that far, note that $V_{\max}$ and K_m can be defined as follows:

$$V_{\max} = k_3[\text{E}_t] \qquad K_m = \frac{k_2 + k_3}{k_1}$$

6-13. Linearizing Michaelis and Menten. In addition to the Lineweaver-Burk plot, two other straight-line forms of the Michaelis-Menten equation are sometimes used. The Eadie-Hofstee plot is a graph of $v/[\text{S}]$ versus v, and the Hanes-Woolf plot graphs $[\text{S}]/v$ versus $[\text{S}]$.

(a) In both cases, show that the equation being graphed can be derived from the Michaelis-Menten equation by simple arithmetic manipulation.

(b) In both cases, indicate how K_m and $V_{\max}$ can be determined from the resulting graph.

(c) Make a Hanes-Woolf plot, with the intercepts and slope labeled as for the Lineweaver-Burk and Eadie-Hofstee plots (see Figures 6-11 and 6-12). Can you suggest why the Hanes-Woolf plot is the most statistically satisfactory of the three?

S UGGESTED R EADING

General References

Colowick, S. P., and N. O. Kaplans, eds. *Methods in Enzymology.* New York: Academic Press, 1970– (ongoing series).

Creighton, T. E. *Proteins.* New York: W. H. Freeman, 1983.

Fresht, A. *Enzyme Structure and Mechanism,* 2d ed. New York: W. H. Freeman, 1985.

Mathews, C. K., and K. E. van Holde. *Biochemistry,* 2d ed. Menlo Park, CA: Benjamin/Cummings, 1996.

Stryer, L. *Biochemistry,* 4th ed. New York: W. H. Freeman, 1995.

Webb, E. *Enzyme Nomenclature.* Orlando, FL: Academic Press, 1984.

Zubay, G. *Biochemistry,* 3d ed. Dubuque, IA: W. C. Brown, 1993.

Structure and Function of Enzymes

Bernard, S. A. *The Structure and Function of Enzymes.* Menlo Park, CA: Benjamin/Cummings, 1968.

Cori, C. F. James B. Sumner and the chemical nature of enzymes. *Trends Biochem. Sci.* 6 (1981): 194.

Dickerson, R. E. and I. Geis. *The Structure and Action of Proteins.* Menlo Park, CA: Benjamin/Cummings, 1969.

Ferdinand, W. *The Enzyme Molecule.* New York: Wiley, 1976.

Knowles, J. R. Tinkering with enzymes: What are we learning? *Science* 236 (1987): 1252.

Kraut, J. How do enzymes work? *Science* 242 (1988): 533.

Srere, P. A. Why are enzymes so big? *Trends Biochem. Sci.* 9 (1984): 387.

Mechanisms of Enzyme Catalysis

Bennet, W. S., and T. A. Steitz. Glucose-induced conformational change in yeast hexokinase. *Proc. Nat. Acad. Sci. USA* 75 (1978): 4848.

Neurath, H. Proteolytic enzymes, past and present. *Fed. Proc.* 44 (1985): 2907.

Stroud, R. M. A family of protein-cutting proteins. *Sci. Amer.* 231 (July 1974): 74.

Walsh, C. *Enzymatic Reaction Mechanisms.* New York: W. H. Freeman, 1979.

Enzyme Regulation

Changeux, J. P. The control of biochemical reactions. *Sci. Amer.* 212 (April 1965): 36.

Cohen, P. *Control of Enzyme Activity.* New York: Chapman and Hall, 1976.

Hammes, G. C., and C. W. Wu. Regulation of enzyme activity. *Science* 172 (1971): 105.

Koshland, D. E., Jr. Control of enzyme activity and metabolic pathways. *Trends Biochem. Sci.* 9 (1984): 155.

Pardee, A. B. Control of metabolic reactions by feedback inhibition. *Harvey Lect.* 64 (1971): 59.

Schirmer, T., and P. R. Evans. Structural basis of the allosteric behavior of phosphofructokinase. *Nature* 343 (1990): 140.

RNA as an Enzyme

Altman, S. Enzymatic cleavage of RNA by RNA. *Biosci. Reporter* 10 (1990): 317.

Cech, T. R. RNA as an enzyme. *Sci. Amer.* 255 (November 1986): 64.

Cech, T. R. The chemistry of self-splicing RNA and RNA enzymes. *Science* 236 (1987): 1532.

Cedergren, R. RNA—the catalyst. *Biochem. Cell Biol.* 68 (1990): 903.

Noller, H.F., and V. Hoffarth, and L. Zimniak. Unusual resistance of peptidyl transferase to protein extraction procedures. *Science* 256 (1992): 1416.

7

MEMBRANES: THEIR STRUCTURE, FUNCTION, AND CHEMISTRY

A n essential feature of every cell is the presence of **membranes** that define the boundaries of the cell and delineate its various internal compartments. Even the casual observer of electron micrographs is likely to be struck by the prominence of the membranes around and within cells, especially those of eukaryotic organisms (Figure 7-1). We encountered membranes and membrane-bounded organelles in an introductory manner in Chapter 4; now we are ready to look at membrane structure and function in greater detail.

In this chapter, we will examine the molecular structure of membranes and explore the multiple roles that membranes play in the life of the cell. In Chapter 8, we will consider the transport of solutes across membranes. We will discuss several specific membrane systems of eukaryotic cells in Chapter 9, and in Chapter 10 we will encounter several kinds of junctions between cells as well as the extracellular matrix and walls that are exterior to the plasma membrane.

The Functions of Membranes

We begin our discussion of biological membranes by asking about the functions they perform. As Figure 7-2 illustrates, the membranes of a cell perform five main related yet distinct functions: They define the boundaries of the cell and delineate its compartments, serve as loci of specific functions, provide for and regulate the movement of substances into and out of the cell and its compartments, contain the receptors required for the detection of external signals, and provide mechanisms for cell-to-cell communication. We will look briefly at each of these functions and then consider in detail the structural and chemical features that make membranes so well suited for these roles.

① Delineation and Compartmentalization

One of the most obvious roles of membranes is to delineate the boundaries of the cell and its compartments. The interior of the cell must be physically separated from the surrounding environment not only to keep desirable substances in the cell, but also to keep undesirable substances out. Membranes serve this purpose well because the hydrophobic interior of the membrane is an effective barrier for hydrophilic molecules and ions. For the cell as a whole, the **plasma** (or **cell**) **membrane** is the permeability barrier. (The term *plasma* was originally used to describe the contents of a cell; we now use the term *cytoplasm* instead, but the outer limiting membrane is still called the plasma membrane.)

In addition to the plasma membrane, various **intracellular membranes** serve to compartmentalize function within eukaryotic cells. For example, the *mitochondrion* is the specific intracellular compartment that contains the enzymes, substrates, and other molecules required for the process of *aerobic respiration*, while the *chloroplast* contains the enzymes, pigments, and intermediates needed for *photosynthesis*. Without such compartments, eukaryotic life as we know it would not exist.

② Localization and Organization of Function

In addition to delineating cells and their components, membranes and membranous compartments have specific functions associated with them because the molecules and structures responsible for those functions are either embedded in or localized on membranes. One of the best ways to characterize a specific membrane, in fact, is to describe the specific enzymes, transport proteins, pigments, and other molecules associated with it.

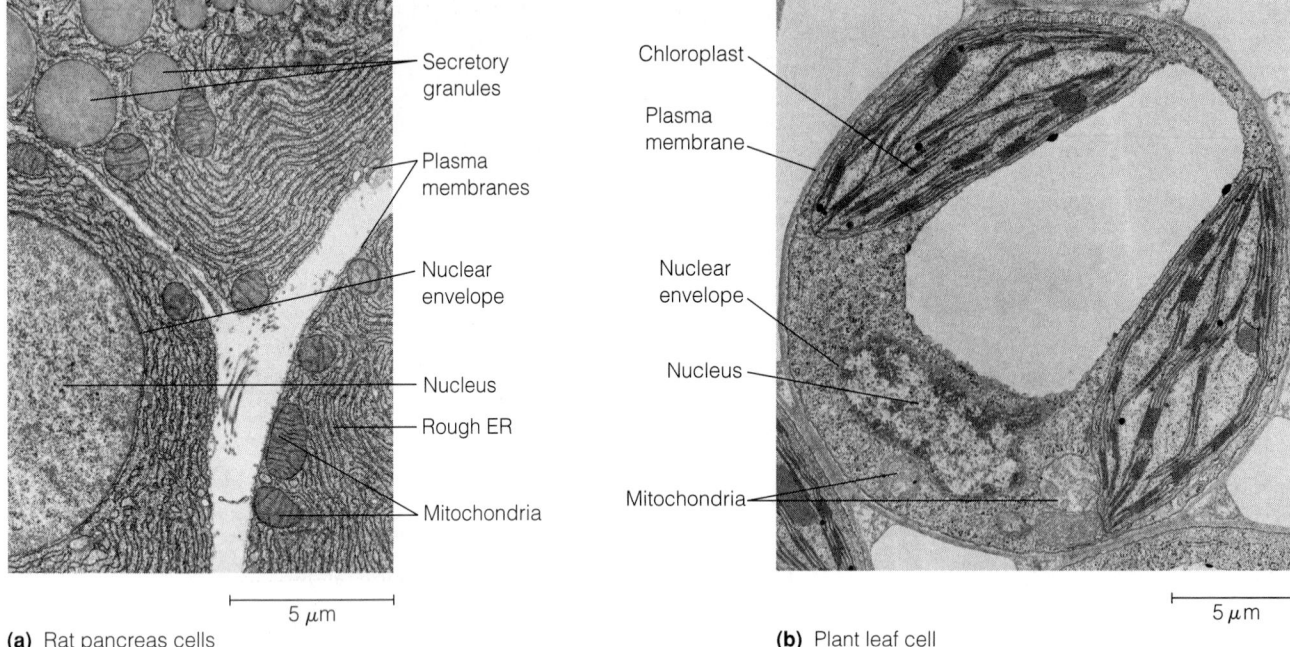

(a) Rat pancreas cells

(b) Plant leaf cell

Figure 7-1   The Prominence of Membranes Around and Within Eukaryotic Cells. Among the structures of eukaryotic cells that involve membranes are the plasma membrane, nucleus, chloroplasts, mitochondria, endoplasmic reticulum (ER), and secretory granules. These structures are shown here in **(a)** portions of three cells from a rat pancreas and **(b)** a plant leaf cell (TEMs).

The localization of specific functions to membranes is exemplified by the roles of the inner membrane of the mitochondrion and the thylakoid membranes of the chloroplast in the processes of respiration and photosynthesis, respectively—topics we will explore in detail in Chapters 12 and

13. Membrane associations are also evident for other cellular components, including the many enzymes known to be localized in or on the membranes of organelles such as the endoplasmic reticulum, Golgi complex, lysosomes, and peroxisomes.

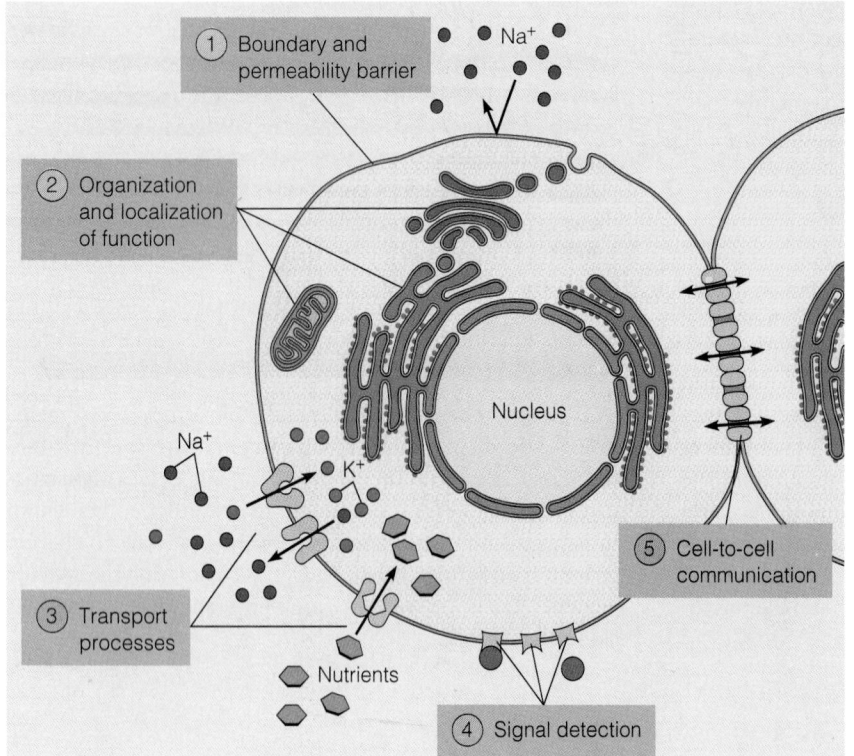

Figure 7-2   Functions of Membranes. Membranes ① define the boundaries of the cell and its organelles, ② serve as loci for specific functions, ③ provide for and regulate transport processes, ④ contain the receptors needed to detect external signals, and ⑤ provide mechanisms for cell-to-cell contact and communication.

Such enzymes are often useful as *markers* during the isolation of organelles from suspensions of disrupted cells. For example, the lysosome is the only organelle that contains the enzyme acid phosphatase. When lysosomes are isolated from other cell components by successive purification steps, acid phosphatase activity can be used as a **marker enzyme,** enabling the investigator to determine both the distribution of lysosomes among various fractions and the purity of the final preparation.

③ *Regulation of Transport*

Another function of membranes is to provide for and regulate the movement of substances into and out of cells and their organelles. Nutrients, ions, gases, water, and other substances are taken up by the cell, and various products and wastes must be removed.

As we will see in Chapter 8, the modes of entry for various substances differ. Many substances move in the direction dictated by their concentration gradients (or, in the case of ions, by the combined effect of the concentration and charge gradients across the membrane). This process, called *passive transport,* does not require energy because movement is "down" the gradient. Passive transport occurs by two different modes. Molecules such as water, oxygen, and carbon dioxide can cross membranes by *simple diffusion* because of their small size. Other molecules, such as ethanol and cholesterol, are able to diffuse across membranes because of their nonpolar nature. Larger, more polar molecules such as sugars and amino acids require special *transport proteins* to facilitate their passage across the membrane.

Alternatively, substances can be transported against, or "up," a concentration gradient and/or a charge gradient, with ATP as the usual energy source. This process, called *active transport,* is carried out by membrane-bound enzyme systems that are commonly referred to as *pumps.* Nutrient molecules such as sugars and amino acids are transported by this mechanism, as are ions such as sodium, potassium, calcium, chloride, and hydrogen (protons).

Even molecules as large as proteins can gain entry to cells by being engulfed and incorporated into vesicles, a process called *endocytosis.* And within cells, proteins that are synthesized on the endoplasmic reticulum or in the cytosol can be secreted from the cell or imported into specific membrane-bounded organelles such as chloroplasts and mitochondria.

④ *Detection and Transmission of Signals*

Cells receive information from their environment, usually in the form of electrical or chemical signals that impinge upon the outer surface of the cell. The various hormones present in your circulatory system at this moment are examples of such signals, as are growth-promoting substances and neurotransmitters. These signals can cause changes in either the nature or the rate of cellular activities. Frequently, in fact, the information that impinges on a cell initiates new patterns or cycles of activities, such as cell division or differentiation.

The plasma membrane plays a key role in **signal transduction,** a term that includes both the detection of specific signals on the outer surface of the cell and the specific responses that a particular signal triggers within the cell. In some cases, the signal molecules are taken up by the cell and act internally. The hormone estrogen is an example. Because estrogen is a steroid (see Figure 3-29b), it is nonpolar in nature and can cross membranes readily. Estrogen therefore gains entry to the cell and interacts with regulatory proteins within the cell.

In most cases, however, the impinging signal molecules do not enter the cell but instead bind to specific **receptors** on the outer surface of the plasma membrane. Binding of such substances, called *ligands*, is followed by specific chemical events on the inner surface of the membrane, thereby generating internal signals called *second messengers*. Membrane receptors therefore allow cells to recognize, transmit, and respond to a variety of specific chemical signals, a topic we will consider in Chapter 23.

⑤ *Cell-to-Cell Communication*

Membranes also provide a means of communication between adjacent cells. Although textbooks often depict cells as separate entities, many cells in multicellular organisms are in contact with one another via direct cytoplasmic connections. This **cell-to-cell communication** is provided by *gap junctions* in animal cells and by *plasmodesmata* in plant cells.

All of these functions—compartmentalization, localization of function, transport, signal detection, and intercellular communication—depend on the chemical composition and structural features of membranes. It is to these topics that we now turn.

Membrane Structure: A Historical Perspective

In Chapter 2, a membrane was described as a semifluid "sea" of phospholipids with proteins "floating" in it. This contemporary concept can best be appreciated if we look first at how our present understanding of membrane structure developed. As we do so, you may also gain some insight into how such developments come about, as well as a greater respect for the diversity of approaches and techniques that are often important for advancing our understanding of biological phenomena. Figure 7-3 presents a chronology of membrane studies that began over a century ago with the understanding that lipid layers are a part of membrane structure and continue today with investigations of membrane proteins, some of which we will encounter in this chapter. Refer to Figure 7-3 as you read the sections that follow.

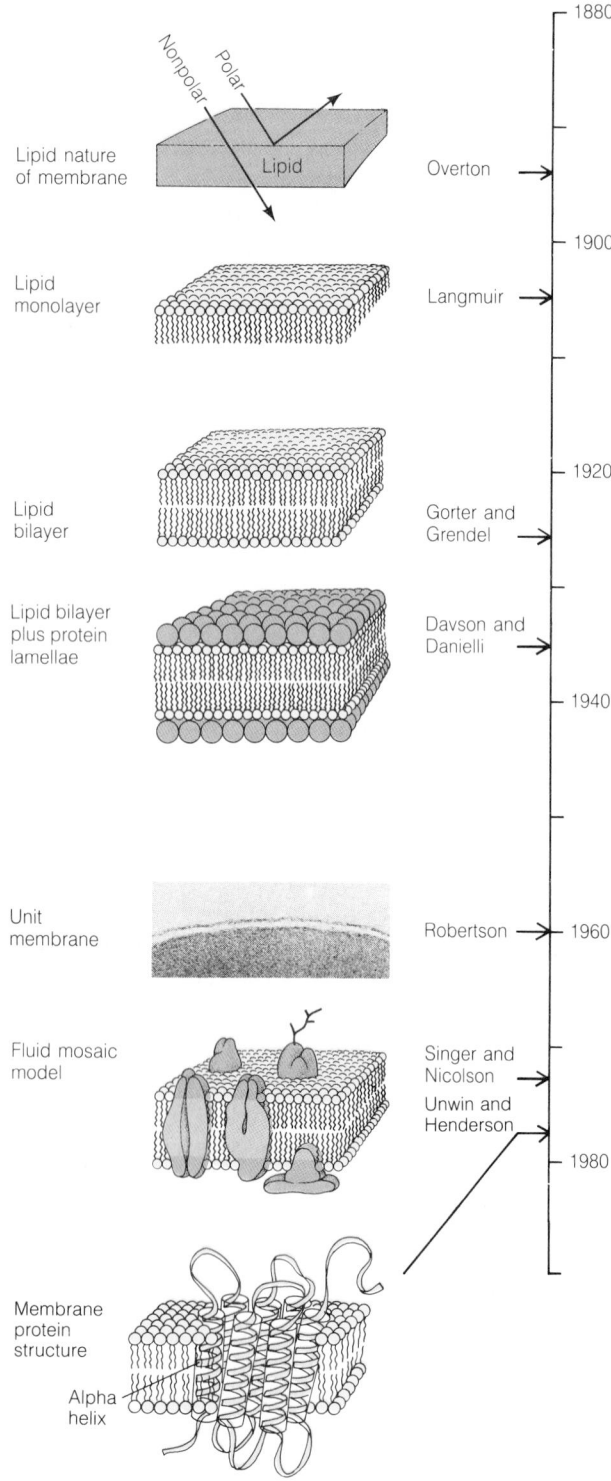

Lipid nature of membrane — Overton — 1880

Lipid monolayer — Langmuir — 1900

Lipid bilayer — Gorter and Grendel — 1920

Lipid bilayer plus protein lamellae — Davson and Danielli

Unit membrane — Robertson — 1960

Fluid mosaic model — Singer and Nicolson

Unwin and Henderson — 1980

Membrane protein structure

Alpha helix

Figure 7-3   Timeline for the Development of Our Understanding of Membrane Chemistry and Structure.

Overton and Langmuir: The Importance of Lipids

A good starting point for our historical overview is the pioneering work of Charles Overton in the 1890s. Overton was aware that cells seemed to be enveloped by some sort of selectively permeable layer that allowed the passage of some substances but not others. He reasoned that the ability of a substance to cross the membrane might be related in some way to its chemical affinity for the membrane. This, in turn, suggested that he might be able to learn something about cell membranes by determining the nature of substances that could move across membranes easily.

Working with cells of plant root hairs, Overton found that lipid-soluble substances penetrated readily into cells, whereas water-soluble substances did not. In fact, he found a good correlation between the *lipophilic* ("lipid-loving") nature of a substance and the ease with which it could enter cells. From his studies, Overton concluded that lipids were present on the cell surface as some sort of "coat." He even suggested that cell membranes were probably mixtures of cholesterol and lecithin, an insight that later proved to be remarkably foresighted.

A second important advance came about a decade later through the work of Irving Langmuir, who studied the behavior of lipids by spreading them out as a thin layer on a water surface of a so-called *Langmuir trough* (Figure 7-4a). He dissolved the lipids in benzene and layered samples of the benzene-lipid solution onto the surface of the water. As the benzene evaporated, the molecules were left as a **lipid monolayer** on the water surface, with their hydrophilic heads in the water and their hydrophobic tails in the air (Figure 7-4b and c). Langmuir's lipid monolayer became the basis for further thought about membrane structure in the early years of the twentieth century.

Gorter and Grendel: The Lipid Bilayer

The next major step came in 1925 when two Dutch physiologists, E. Gorter and F. Grendel, read Langmuir's papers and thought that his approach might help answer a question regarding the surface of the red blood cells, or *erythrocytes,* with which they worked. Overton's earlier research had shown that lipid was present on the cell surface as some sort of coat. But how many lipid layers were in the coat? Gorter and Grendel could see in their microscope that the human erythrocyte resembled a flattened sphere about 7 μm in diameter. They could therefore determine the surface area of a known volume of erythrocytes by counting the cells and multiplying by 100 μm^2, their best estimate of the surface area of an erythrocyte based on its apparent diameter. They then extracted the lipids from a known number of erythrocytes, spread the lipids out as a monolayer using Langmuir's method, and compared the area of the lipid monolayer with the total surface area of the erythrocytes.

When they did so, Gorter and Grendel found that the area of the monolayer of lipid on the water was about twice the estimated total surface area of the erythrocyte. They therefore concluded that the erythrocyte plasma membrane consisted of *two* monolayers of lipid, giving rise to the important concept of the **lipid bilayer.** (As it later turned out, Grendel and Gorter made two major errors, underestimating by about one-third both the surface area of the red blood cell and the amount of lipid present in its plasma membrane. Fortunately, these errors canceled each other out, so

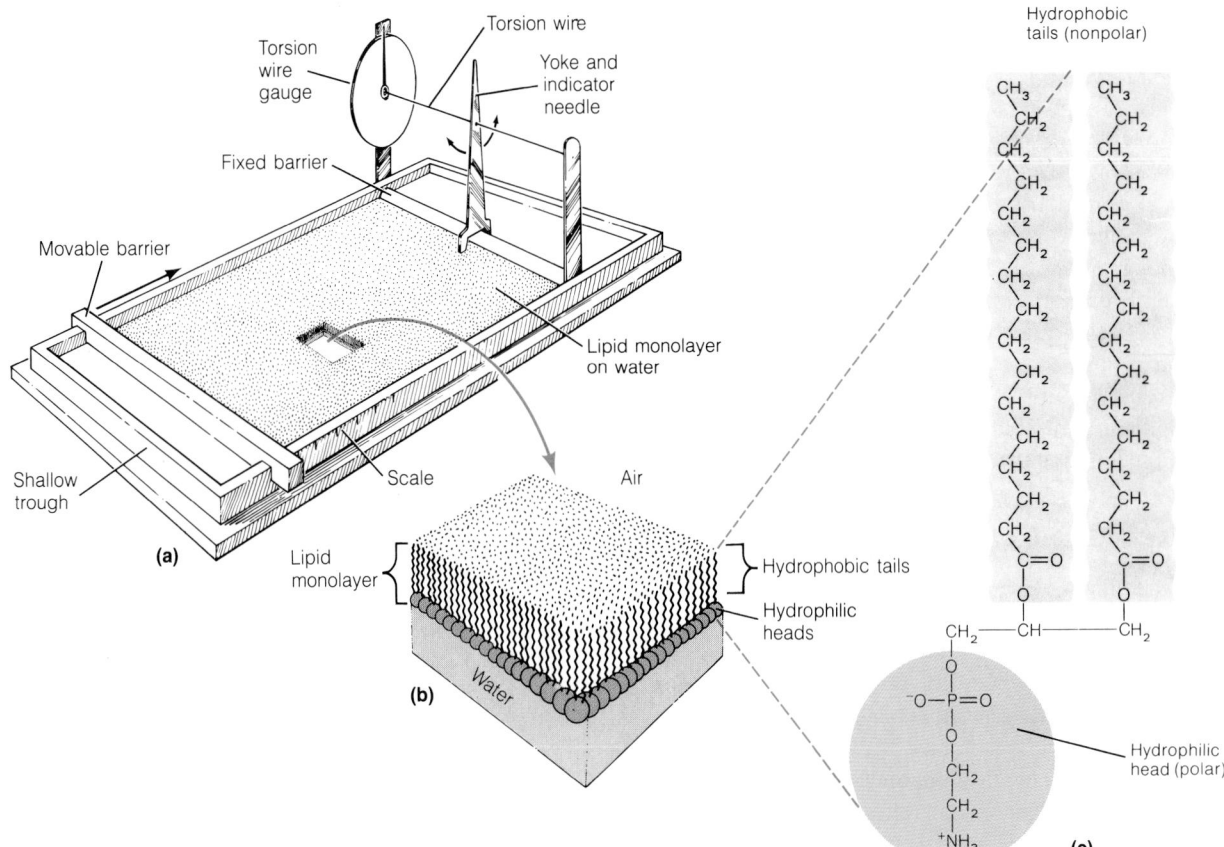

Figure 7-4 The Langmuir Trough. **(a)** The trough consists of a tray equipped with a movable barrier that can be used to push a phospholipid monolayer up against a fixed barrier, the latter equipped with a torsion wire gauge to allow the monolayer to be compressed to a uniform extent each time. **(b)** The orientation of the phospholipids in the mono-layer is such that the hydrophilic heads protrude downward into the water and the hydrophobic tails stick upward into the air. **(c)** The structure of a phospholipid molecule, indicating the hydrophobic tails (gray) and the hydrophilic, polar head (pink).

their conclusion was correct even though their data were not. For a chance to repeat their calculations and discover the source of their errors, see Problem 7-3 at the end of the chapter.)

Assuming a bilayer structure, Grendel and Gorter reasoned that it would be thermodynamically favorable for the nonpolar hydrocarbon chains of each layer to face inward, away from the aqueous milieu on either side of the membrane. The polar hydrophilic groups would then face outward, toward the aqueous environment. Their experiment and their conclusions were momentous because this work represented the first attempt to understand membranes at the molecular level. Moreover, the lipid bilayer that they envisioned became the basic underlying assumption for each successive refinement in our understanding of membrane structure.

Davson and Danielli: The Importance of Proteins

It soon became clear that the lipid bilayer, though an essential feature of membrane structure, could not explain all the properties of membranes, particularly those related to *solute permeability* and *electrical resistance.* The permeability of

membranes to solute molecules often depends on quite subtle chemical differences. For example, glucose gets into most cells readily but galactose, a sugar molecule of similar size and shape, does not. It was not easy for early membrane biologists to imagine how a lipid bilayer could be capable of such **differential permeability.** At about the same time, however, biologists were becoming aware that enzymes have the kind of specificity that seemed to be needed to distinguish between molecules with very similar properties. Cell biologists therefore began to wonder whether membrane components with enzymelike binding sites might be present in the membrane to facilitate glucose transport. Similarly, biological membranes are much more permeable to ions than artificial lipid bilayers are. For example, the leakage rate of potassium ions across artificial bilayers is measured in days, whereas the same amount of leakage across erythrocyte membranes occurs in an hour or so.

To explain these additional features, Hugh Davson and James Danielli invoked the presence of proteins in membranes, proposing in 1935 that biological membranes consist of lipid bilayers coated on both sides with layers, or *lamellae,* of proteins. Thus, the original **Davson-Danielli model** was in essence a protein-lipid "sandwich." The model was modified somewhat in the years that followed, however. Particularly notable was the suggestion, made in 1954, that hydrophilic proteins might penetrate into the membrane in places to provide a *polar pore* through what was otherwise a very hydrophobic bilayer. The protein could then account for the permeability and resistivity characteristics that were not easily explained in terms of the lipid bilayer alone. Specifically, the lipid interior accounted for the hydrophobic properties of membranes, and the protein components explained their hydrophilic properties.

The real significance of the model, however, was its recognition of the importance of proteins in membrane structure. This feature more than any other made the Davson-Danielli "sandwich" the basis for much subsequent research on membrane structure.

Robertson: The Unit Membrane

All of the membrane structures discussed so far were developed as models of the plasma membrane. With the advent of electron microscopy in the 1950s, however, it became clear that most subcellular organelles of eukaryotic cells are bounded by membranes as well. Furthermore, when membranes were stained with osmium, a heavy metal, and examined closely at high magnification, they were found to have extensive regions of trilaminar "railroad track" structure that appeared as two dark lines separated by an unstained space, with an overall thickness of about 7.5 nm (Figure 7-5). These considerations led J. David Robertson to suggest the first general structure for all membranes, the **unit membrane.** In this model, membranous components of the cell were regarded as a continuous lipid bilayer with functional proteins adhering to the surface.

When first proposed, the unit membrane structure appeared to agree remarkably well with the Davson-Danielli model: The two dark lines were equated with the outer protein layers, which appeared electron-dense because of their affinity for heavy metal stains, while the space in between was thought to correspond to the hydrophobic part of the lipid molecules, which presumably did not take up stain. This latter assumption was supported by experiments showing essentially the same staining pattern for membranes with and without prior extraction of the lipid. Later, however, it was found that the same "railroad track" pattern was seen when artificial lipid bilayers were stained with osmium, even though such bilayers contained no protein. (We now know that osmium reacts at several sites in membranes—with unsaturated lipids in the bilayer as well as with proteins on the membrane surface.) This finding was to take on special significance, because continuous outer layers of proteins are not part of the contemporary model of membrane structure, as we are about to see.

Singer and Nicolson: The Fluid Mosaic Model

An attractive feature of the unit membrane model was its apparent universality: Membranes from widely different types of cells and organelles showed the same characteristic

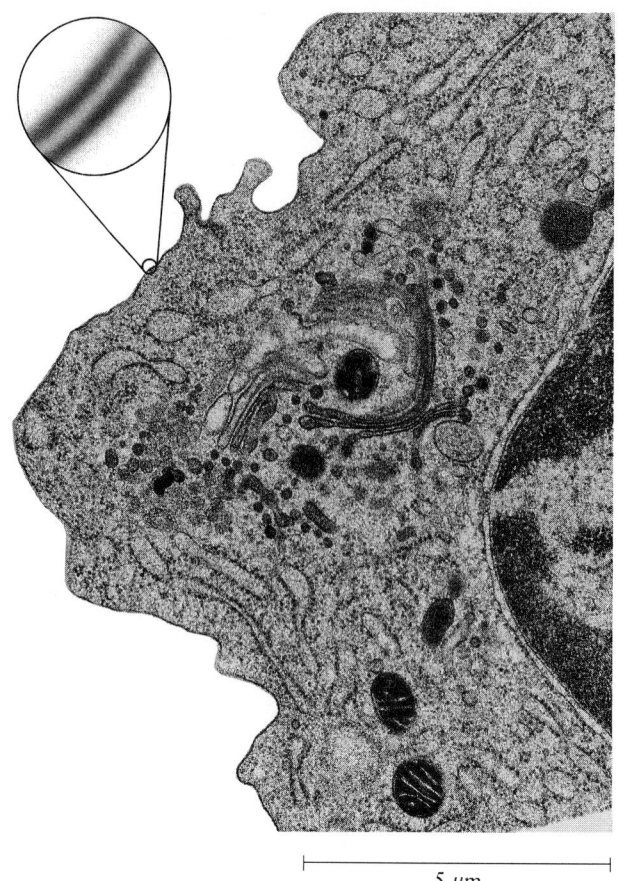

Figure 7-5 Ultrastructure of Cell Membranes. An electron micrograph of fixed and stained cells reveals ultrastructural features such as membranes, the cytoskeleton, and nucleoprotein complexes. At higher magnification (inset), most of the membranes have the basic trilaminar "railroad track" appearance that gave rise to the unit membrane concept of Robertson (TEM).

5 μm

features when examined by electron microscopy. However, the model did not readily account for the distinctiveness of different kinds of membranes. Depending on their source, membranes vary considerably in chemical composition and especially in the ratio of protein to lipid (Table 7-1). The protein/lipid ratio can be as high as 3 or more in some bacterial cells and as low as 0.23 for the myelin sheath that serves as a membranous electrical insulation around nerve axons (see Chapter 22). Even the two membranes of the mitochondrion differ significantly: The protein/lipid ratio is about 1.2 for the outer membrane and about 3.5 for the inner membrane, which contains all the enzymes and proteins related to electron transport and ATP synthesis.

As more membranes were studied, it became increasingly difficult to rationalize such variations in protein content with the unit membrane model, because the width and appearance of the "rails" simply did not vary correspondingly. Furthermore, as membrane proteins were isolated and studied, it became apparent that most of them were *globular proteins* with sizes and shapes that were inconsistent with the concept of thin layers or sheets of proteins on the two

Table 7-1 Protein, Lipid, and Carbohydrate Content of Biological Membranes

Membrane	Approximate Percent by Weight			Protein/Lipid Ratio
	Protein	Lipid	Carbohydrate	
Plasma membrane				
Human erythrocyte	49	43	8	1.14
Mammalian liver cell	54	36	10	1.50
Amoeba	54	42	4	1.29
Myelin sheath of nerve axon	18	79	3	0.23
Nuclear envelope	66	32	2	2.06
Endoplasmic reticulum	63	27	10	2.33
Golgi complex	64	26	10	2.46
Chloroplast lamella	70	30	0	2.33
Mitochondrial outer membrane	55	45	0	1.22
Mitochondrial inner membrane	78	22	0	3.54
Gram-positive bacterium	75	25	0	3.00

surfaces of the membrane. These findings, along with others on membrane fluidity, caused biologists to rethink membrane structure and led eventually to the **fluid mosaic model,** which now has widespread support and acceptance.

The fluid mosaic model was first proposed in 1972 by S. Jonathan Singer and Garth Nicolson. The model has two key features, both implied by its name. The basic lipid bilayer structure of earlier models is retained, but membrane proteins are not thought of as continuous sheets on the membrane surface. Rather, they are recognized as discrete globular entities that associate with the membrane to an extent dictated by their affinity for the hydrophobic interior of the lipid bilayer (Figure 7-6). The membrane is therefore envisioned as a *mosaic* of proteins discontinuously embedded in a lipid bilayer.

Also important is the *fluid* nature that the model attributes to the membrane. Rather than being rigidly locked in place, the lipid components of a membrane are in constant motion, capable of *lateral diffusion* (movement parallel to the membrane surface). Many membrane proteins are also able to move laterally within the membrane, although some proteins are anchored to cytoskeletal structures beneath the membrane and are therefore limited in their mobility.

The fluid mosaic model recognizes two broad categories of membrane proteins that differ in their affinity for the hydrophobic interior of the membrane and therefore in the extent to which they interact with the lipid bilayer (see Figure 7-6). **Integral membrane proteins** have one or more hydrophobic regions with an affinity for the hydrophobic interior of the lipid bilayer. These proteins are therefore intimately associated with the membrane and cannot be readily removed. However, such proteins also have one or more hydrophilic regions that extend outward from the membrane into the aqueous phase on one or both sides of the membrane. Because of the affinity of their hydrophobic re-

gions for the lipid bilayer, integral membrane proteins are difficult to isolate and study by standard protein purification techniques designed for relatively water-soluble proteins. Treatment with a detergent is usually required to solubilize integral membrane proteins.

In contrast, **peripheral membrane proteins** lack discrete hydrophobic sequences and therefore do not penetrate into the lipid bilayer. Instead, they associate with the membrane surfaces through weak electrostatic forces, binding either to the polar heads of membrane lipids or to the hydrophilic portions of integral proteins that extend out of the membrane. Peripheral proteins are much more readily removed from the membrane than integral proteins and can usually be solubilized by aqueous salt solutions, without resorting to detergents.

Peripheral membrane proteins are the elements of the fluid mosaic model that most resemble the sheets of surface proteins originally postulated by Davson and Danielli. However, peripheral proteins are individual molecules and are globular in nature, rather than forming a continuum of protein strands as envisioned in earlier membrane models.

Unwin and Henderson: The Structure of Membrane Proteins

The final illustration in the timeline of Figure 7-3 depicts an important property of membrane proteins that cell biologists began to understand in the 1970s: Each of the hydrophobic sequences of an integral membrane protein is a **transmembrane segment** that spans the lipid bilayer, anchoring the protein to the membrane and holding it in proper alignment within the lipid bilayer. The example in Figure 7-3 is *bacteriorhodopsin*, the first membrane protein for which this structural feature was shown. Bacteriorhodopsin is a membrane protein that serves as a light-

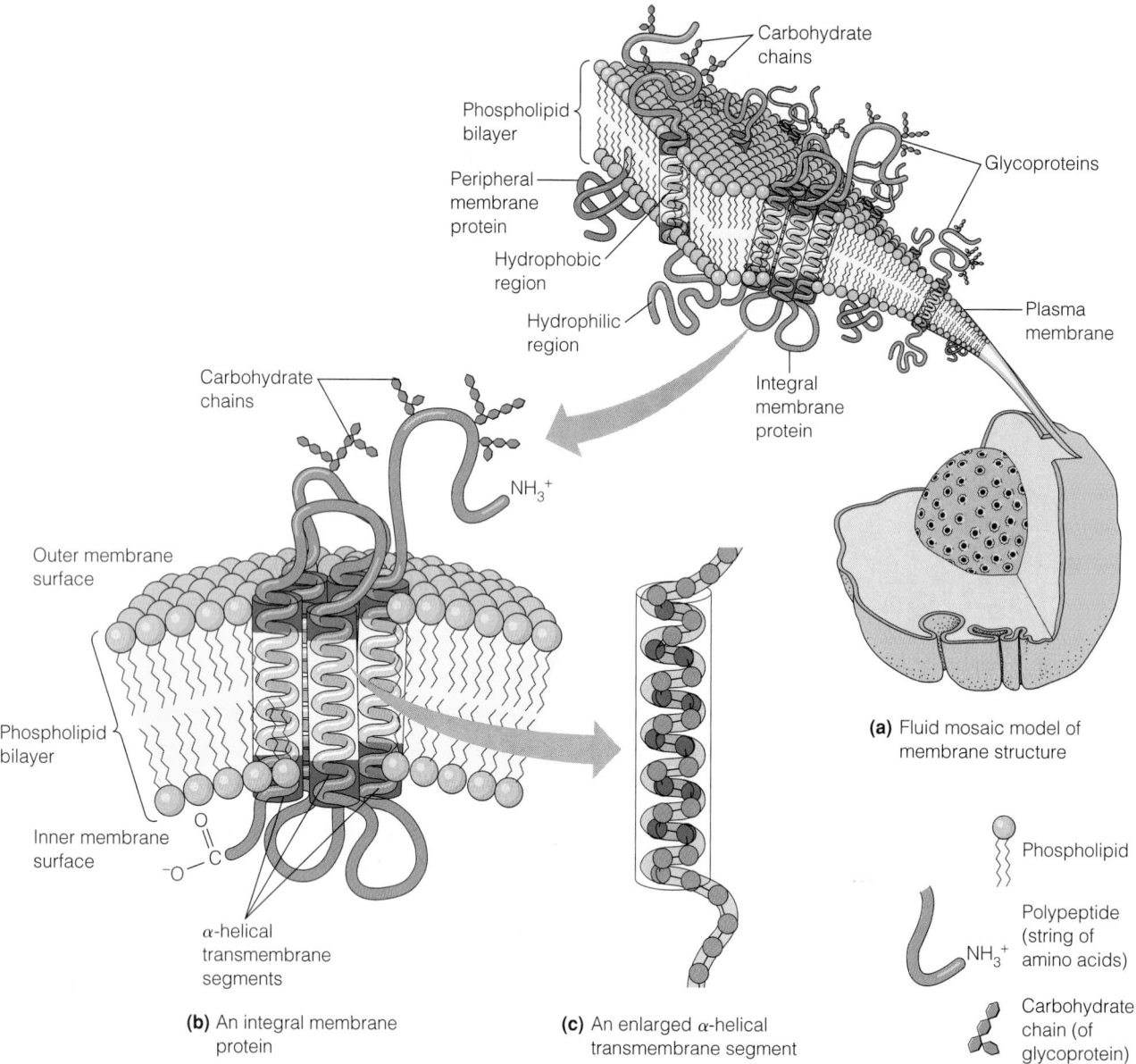

Carbohydrate chains

Phospholipid bilayer

Peripheral membrane protein

Hydrophobic region

Hydrophilic region

Glycoproteins

Plasma membrane

Integral membrane protein

Carbohydrate chains

NH_3^+

Outer membrane surface

Phospholipid bilayer

Inner membrane surface

α-helical transmembrane segments

(b) An integral membrane protein

(c) An enlarged α-helical transmembrane segment

(a) Fluid mosaic model of membrane structure

Phospholipid

Polypeptide (string of amino acids)

NH_3^+

Carbohydrate chain (of glycoprotein)

7-6   The Fluid Mosaic Model of Membrane Structure.
(a) Sketch of a membrane according to the fluid mosaic model, showing hydrophobic and hydrophilic regions. Singer and Nicolson's model envisions the membrane as a fluid bilayer of lipids with a mosaic of associated proteins that are either integral or peripheral parts of the membrane. Integral membrane proteins have one or more hydrophobic transmembrane segments, usually with an α-helical conformation, that anchor them to the hydrophobic interior of the membrane. Hydrophilic segments

extend outward on one or both sides of the membrane. Peripheral membrane proteins are associated with the membrane surface by weak electrostatic forces that bind them to the polar head groups of membrane phospholipids or to hydrophilic regions of integral membrane proteins.
(b) An integral membrane protein with multiple α-helical transmembrane segments. Many integral membrane proteins have carbohydrate side chains attached to the hydrophilic segments that protrude from the outer membrane surface. **(c)** An α-helical transmembrane segment.

activated proton pump in bacteria of the genus *Halobacterium.* (The halobacteria are *halophilic,* or salt-loving, organisms that not only thrive in a high-salt environment but actually require it. A halobacterium can use solar energy to transport protons across its plasma membrane because of the bacteriorhodopsin molecules in its plasma membrane. To detect light, bacteriorhodopsin has as part of its structure a molecule of *retinal,* the same pigment molecule used to capture light energy in the human eye.)

Bacteriorhodopsin molecules in the membrane are arranged in a crystalline lattice, a property that enabled Nigel Unwin and Richard Henderson to determine the three-dimensional structure of bacteriorhodopsin and its orientation in the membrane. Their remarkable finding, reported in 1975, was that bacteriorhodopsin consists of a single peptide chain folded back and forth across the lipid bilayer a total of seven times (Figure 7-7). Each of the seven transmembrane segments of the protein is a closely packed

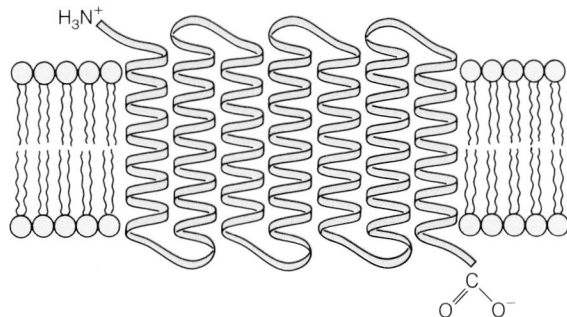

Figure 7-7 The Structure of Bacteriorhodopsin. Bacteriorhodopsin was the first integral membrane protein for which the detailed three-dimensional structure was determined. It occurs in the plasma membrane of bacteria in the genus *Halobacterium* and has seven transmembrane segments organized into a proton channel. The protein has 248 amino acids, about 70% of which are part of the transmembrane segments. The C-terminus and N-terminus of the protein have short hydrophilic segments that protrude on the inner and outer surfaces of the bacterial plasma membrane, respectively. Short hydrophilic segments also link each of the transmembrane segments.

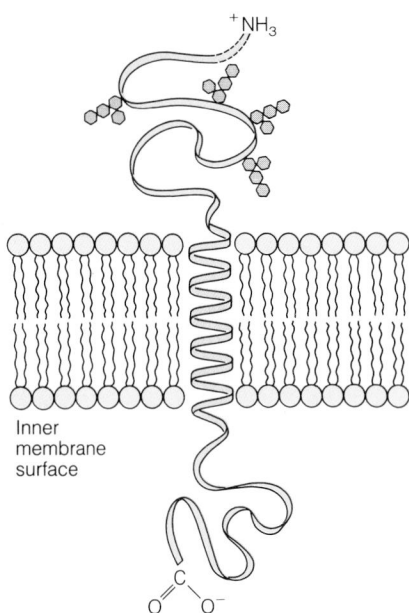

Figure 7-8 The Structure of Glycophorin. Glycophorin is an integral membrane protein in the red blood cell (erythrocyte) plasma membrane. It has a single transmembrane segment that consists entirely of hydrophobic amino acids. The N-terminus protrudes on the outer surface and the C-terminus on the cytoplasmic surface. Glycophorin is a glycoprotein, with 16 carbohydrate groups on its surface.

α helix composed mainly of hydrophobic amino acids. Successive transmembrane segments are linked to each other by short loops of hydrophilic amino acids that extend into or through the polar surfaces of the membrane.

We now recognize this pattern as a characteristic of all integral membrane proteins. Such proteins invariably have hydrophobic amino acids clustered in one or more segments of about 20–25 amino acids each, separated by stretches of hydrophilic amino acids. Some membrane proteins have a single transmembrane segment, with a hydrophilic C-terminus extending out of the membrane on one side and a hydrophilic N-terminus protruding on the other side. An example of this structure is *glycophorin*, a protein found in the plasma membrane of the red blood cell (Figure 7-8). Other proteins have multiple transmembrane segments, ranging from two or three to 20 or more such segments. Regardless of the number, these segments are arranged in the membrane in a manner that facilitates the specific function of the protein. In the case of bacteriorhodopsin, for example, the seven transmembrane segments are positioned in the membrane to facilitate the light-activated pumping, or active transport, of protons across the membrane, a topic to which we will return in Chapter 8.

The Fluid Mosaic Model: Evidence and Implications

Almost from the moment Singer and Nicolson proposed it, the fluid mosaic model revolutionized the way scientists thought about membrane structure. The model launched a new era in membrane research that has not only confirmed the basic model but has further augmented and extended it. As a result, the fluid mosaic model is now the generally accepted view of membrane structure. It is therefore important for us to examine the essential features of the model in detail. These features include the nature of the lipid bilayer, the fluidity of membrane lipids, the relationship of membrane proteins to the bilayer, the mobility of proteins within the bilayer, and the asymmetric distribution of membrane components.

We will explore each of these features in turn, focusing on both the supporting evidence and the implications of each feature for membrane function. First, however, we will look at a particular membrane that is especially popular with membrane biologists—the plasma membrane of the **erythrocyte,** the red blood cell responsible for oxygen transport between the heart, lungs, and body tissues. Figure 7-9 shows several human erythrocytes as seen by scanning electron microscopy.

The Erythrocyte Membrane

The plasma membrane of the human erythrocyte has been one of the most widely studied membranes since the time of Gorter and Grendel, mainly because of the availability of erythrocytes and the ease with which pure plasma membrane preparations can be obtained from them. Availability is easy to understand: Red blood cells can be prepared readily, either from outdated blood provided by blood banks or

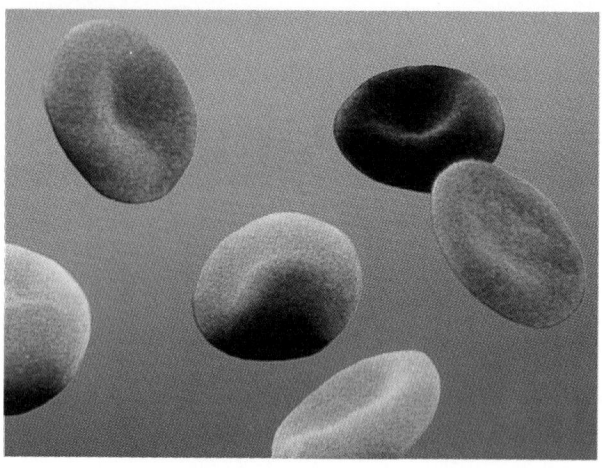

5 μm

Figure 7-9 Electron Micrograph of Human Erythrocytes.
An erythrocyte is a small disk-shaped cell that is concave on both surfaces. The diameter is typically about 7 μm. A mammalian erythrocyte contains no nucleus or other organelles. It is essentially a bag of hemoglobin, which is all an erythrocyte needs to transport oxygen from the lungs to other body tissues and carbon dioxide back to the lungs (SEM).

from freshly drawn blood supplied by human volunteers. Equally important, an erythrocyte is a highly specialized cell that has few, if any, organelles, so very pure plasma membrane preparations can be obtained without contamination by organellar membranes, as often occurs with plasma membrane preparations from other cell types. (Erythrocytes begin life with a full complement of internal structures but lose organelles progressively as they differentiate in the bone marrow prior to being released into the circulatory system. In mammals, though not in birds, even the nucleus is lost during the differentiation process.)

A recent model of the erythrocyte plasma membrane is shown in Figure 7-10. The major integral membrane proteins are glycophorin and an *anion channel protein* (originally called "band 3" because of its electrophoretic mobility; see Figure 7-18). Glycophorin is involved in cell recognition, whereas the anion channel protein facilitates the exchange of bicarbonate and chloride ions across the membrane, a function essential to the role of the erythrocyte in CO_2 transport. As shown in Figure 7-8, glycophorin has a single hydrophobic transmembrane segment with hydrophilic domains on both sides of the membrane. Both polypeptides of the dimeric anion channel protein have at least six transmembrane segments.

The main peripheral proteins of the erythrocyte plasma membrane are *spectrin, ankyrin, actin,* and a protein called *band 4.1* because of its electrophoretic mobility. Each of these proteins is localized to the inner surface of the plasma membrane (see Figure 7-10). Molecules of ankyrin link the anion channel protein to tetramers of α and β spectrin, $(\alpha\beta)_2$, thereby anchoring the membrane to the cytoskeletal network beneath the membrane surface. (*Ankyrin* is derived from the Greek word for "anchor.") The free ends of adjacent spectrin tetramers are held together by short chains of actin with the assistance of the band 4.1 protein.

The erythrocyte plasma membrane is a useful and well-characterized example of membrane organization and has served membrane biologists well as a model. However, it differs from most other plasma membranes in several significant ways. It is, for instance, a remarkably simple and homogenous membrane, containing relatively few kinds of proteins distributed more or less uniformly in the plane of the membrane. Most other membranes contain a greater diversity of proteins, which are often unevenly distributed in the membrane.

The erythrocyte membrane is also unusual in its many sites of attachment to the cytoskeleton, which forms a shell that underlies the entire plasma membrane (see Figure

Figure 7-10   Structure of the Erythrocyte Plasma Membrane.   The two major integral membrane proteins are glycophorin and the dimeric anion channel protein. Glycophorin has a single transmembrane segment, whereas each of the polypeptides of the anion channel protein spans the membranes at least six times. Ankyrin is a peripheral membrane protein that anchors the membrane to the cytoskeleton by linking the anion channel protein to the underlying strands of tetrameric spectrin, $(\alpha\beta)_2$. The free ends of adjacent spectrin tetramers are held together by short chains of actin and an additional membrane protein known (from its electrophoretic mobility) as band 4.1.

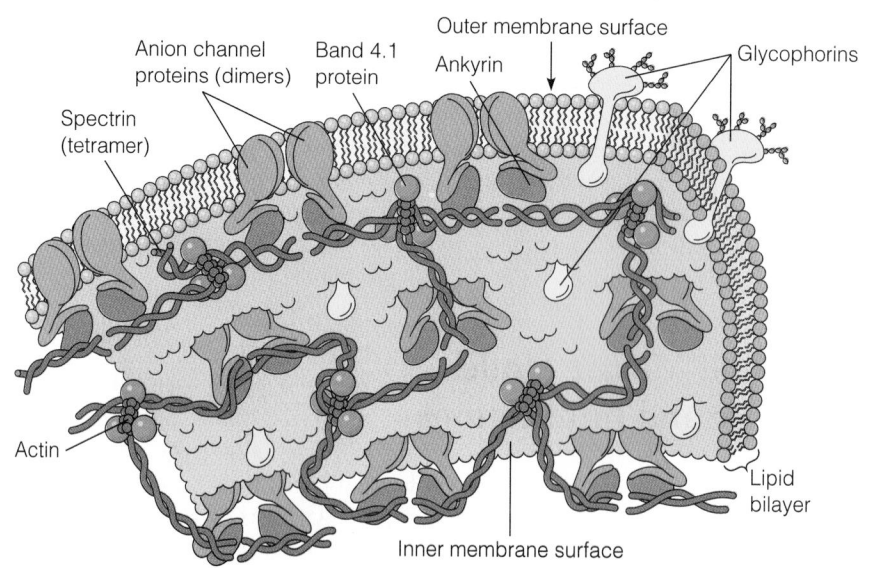

Spectrin (tetramer) · Anion channel proteins (dimers) · Band 4.1 protein · Outer membrane surface · Ankyrin · Glycophorins · Actin · Lipid bilayer · Inner membrane surface

7-10). By contrast, the plasma membrane of a more typical mammalian cell is not anchored as extensively to the cytoskeleton. The numerous connections between the erythrocyte plasma membrane and the underlying cytoskeleton give the membrane great strength and resilience, which are important properties for a cell that often has to squeeze through capillaries that are much thinner than its diameter of 7 μm. These properties also make the erythrocyte quite durable. A typical red blood cell exists for about four months in the bloodstream, traveling about 300 miles as it makes several hundred thousand trips between the heart, lungs, and body tissues, delivering billions of oxygen molecules every minute across a membrane only two lipid molecules thick!

Membrane Lipids: Chemistry and Properties

With the erythrocyte plasma membrane as our example, we can now turn to some of the specific features and implications of the fluid mosaic model. The most obvious property of Singer and Nicolson's model is its retention of the lipid bilayer initially proposed by Gorter and Grendel, though with a greater diversity and fluidity of lipid components than early investigators are likely to have recognized.

Table 7-2 lists the major membrane lipids and their main functions. As we already know from Chapters 2 and 3, the most prominent components of membranes are **phospholipids** and **sphingolipids** (see Figures 2-12, 3-26, and 3-27). The membranes of most animal cells also contain significant amounts of **cholesterol** (see Figure 3-28). Cholesterol is not found in membranes of prokaryotic or plant cells, however, and is also absent from the inner membranes of both mitochondria and chloroplasts.

Analysis of Lipids: Thin-Layer Chromatography. An important technique for the analysis of lipid membranes is thin-layer chromatography (TLC), depicted schematically in Figure 7-11. This technique is capable of resolving different kinds of lipids based on their relative affinities for a hydrophilic *stationary phase* and a hydrophobic *mobile phase*. The stationary phase is usually a thin layer of silicic acid on a glass or metal plate, while the mobile phase is a

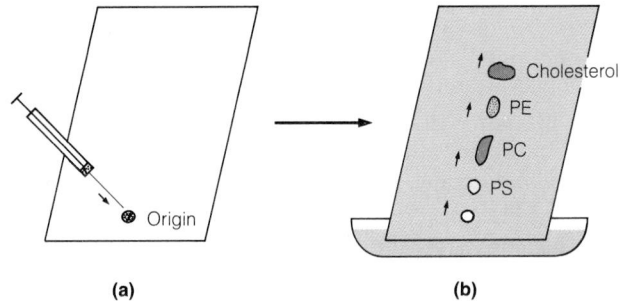

Figure 7-11 Use of Thin-Layer Chromatography in the Analysis of Membrane Lipids. Thin-layer chromatography (TLC) is a useful technique for the analysis of membrane lipids. **(a)** Lipids are extracted from a membrane preparation with a mixture of organic solvents, and a sample is spotted onto a small area of a glass or metal plate coated with a thin layer of silicic acid. When the sample has dried at the origin, the plate is dipped into a solvent system, usually a mixture of chloroform, methanol, and water. As the solvent moves up the plate by capillary action, the lipids are separated according to their polarity: Nonpolar lipids such as cholesterol do not adhere strongly to the silicic acid and move further up the plate, while more polar lipids remain closer to the origin. **(b)** When the solvent front nears the top, the plate is removed from the solvent system and the lipids are eluted and identified. The pattern shown is that for lipids from the erythrocyte plasma membrane. The main components are cholesterol, phosphatidyl ethanolamine (PE), phosphatidyl choline (PC), and phosphatidyl serine (PS).

Table 7-2 Classes and Functions of Membrane Lipids

Lipid Class	Source	Function
Major phospholipids		
Phosphatidyl choline	Present in most membranes	Form bilayers that provide barriers
Phosphatidyl ethanolamine	Present in most membranes	to diffusion of polar solutes
Phosphatidyl serine	Present in most membranes	
Minor phospholipids		
Cardiolipin	Mitochondrial inner membrane	Activates cytochromes
Phosphatidyl inositol	Present in most membranes	Source of inositol trisphosphate
Sphingolipids	Most mammalian cell membranes, particularly those of nervous tissue	Barrier function; activates certain enzymes
Glycolipids	Major lipid in thylakoid membranes of chloroplasts	Barrier function
Cholesterol	Most animal cell membranes	Reduces bilayer permeability; modulates membrane fluidity

mixture of appropriate solvents. To analyze membrane lipids, the lipids are first extracted from a membrane preparation using a mixture of organic solvents. A sample of the extract is applied to one end of a silicic acid–coated plate by spotting the extract onto a small area called the *origin* and allowing the solvent to evaporate (Figure 7-11a).

The edge of the plate is then dipped into a solvent system that typically consists of chloroform, methanol, and water. As the solvent moves past the origin and up the plate by capillary action, the lipids are separated according to their polarity. Nonpolar lipids such as cholesterol have little affinity for the silicic acid (the stationary phase) and therefore move up the plate with the solvent system (the mobile phase). More polar lipids such as phospholipids interact more strongly with the silicic acid, which slows their movement. In this way, the various lipids are separated progressively as the leading edge of the mobile phase continues to move up the plate. When the leading edge, or *solvent front,* approaches the top, the plate is removed from the solvent system and dried, and the separated lipids are then eluted (recovered from the plate by dissolving them in a solvent such as chloroform) and identified.

Figure 7-11b shows the TLC pattern seen for the lipids of the erythrocyte plasma membrane. The main components of this membrane are phospholipids (55%) and cholesterol (25%). The most prominent phospholipids in the erythrocyte membrane are *phosphatidyl ethanolamine* (*PE*), *phosphatidyl choline* (*PC*), and *phosphatidyl serine* (*PS*), the structures of which are shown in Figure 3-26. (The erythrocyte plasma membrane also contains phosphatidyl inositol and sphingolipids, but these components are present in lesser amounts and are less likely to be detected by thin-layer chromatography. These lipids are nonetheless important membrane components and are, in fact, more prominent in other membranes. Sphingolipids, for example, are especially prominent in nervous tissue.)

Fatty Acids and Membrane Properties. Fatty acids are a part of all membrane lipids except cholesterol. They are essential to membrane structure because their long hydrocarbon tails form an effective hydrophobic barrier to the diffusion of polar solutes. Most fatty acids in membranes are 14–24 carbon atoms in length, with 16- and 18-carbon fatty acids especially common. This size range appears to be optimal for bilayer formation because chains with fewer than 12 carbons are unable to form a stable bilayer. Thus, the thickness of membranes (about 8–10 μm) is dictated primarily by the chain length of the fatty acids required for bilayer stability.

In addition to differences in length, the fatty acids found in membrane lipids also vary considerably in the presence and number of double bonds. Table 7-3 shows the structures of several fatty acids that are especially common in membrane lipids. *Palmitate* and *stearate* are saturated fatty acids with 16 and 18 carbon atoms, respectively. *Oleate* and *linoleate* are 18-carbon unsaturated fatty acids with one and two double bonds, respectively. Other common unsaturated fatty acids in membranes are *linolenate* with 18 car-

bons and three double bonds and *arachidonate* with 20 carbons and four double bonds.

All unsaturated fatty acids in membranes are in the *cis* configuration, resulting in a sharp bend, or kink, in the hydrocarbon chain at every double bond. Because of the nonlinearity of their side chains, fatty acids with double bonds do not pack tightly in the membrane, a feature with considerable implications for membrane fluidity, as we are about to see.

Membrane Fluidity: The Motility of Membrane Lipids

An important aspect of membrane function is the **fluidity of the lipid bilayer** and the resulting freedom of motion of membrane components that the term implies. Lipid molecules are much smaller than proteins and are therefore especially mobile. A phospholipid molecule, for example, has a molecular weight of about 800 and can travel the entire length of a bacterial cell in about a second! Many, though not all, membrane proteins are also free to move laterally within the membrane. However, proteins typically move much more slowly than lipids, in part because they are much larger molecules with molecular weights 20 to 100 times those of phospholipids, but also because they tend to establish connections with other membrane proteins and with underlying cytoskeletal structures.

The mobility of membrane lipids can be demonstrated experimentally by a technique called **fluorescence recovery after photobleaching** (Figure 7-12, page 180). The investigator tags, or labels, lipid (or protein) molecules in the membrane of a living cell by covalently linking molecules of a fluorescent dye to them. A high-intensity laser beam is then used to photobleach the fluorescence of the dye in a tiny spot (a few square micrometers) on the cell surface. When the cell surface is examined with a fluorescence microscope, a dark, nonfluorescent spot is seen on the membrane initially. Within seconds, however, the edges of the spot become fluorescent as bleached lipid molecules diffuse out of the laser-treated area and fluorescent lipid molecules from adjoining regions of the membrane diffuse in. Eventually, the spot is indistinguishable from the rest of the cell surface. This technique not only demonstrates that membrane lipids are in a fluid rather than a static state but also provides a direct means of measuring the lateral movement of specific molecules.

Membrane Fluidity and Temperature. As you might guess, membrane fluidity changes with temperature, decreasing as the temperature drops and increasing as it rises. In fact, we know from studies with artificial lipid bilayers that every lipid bilayer has a characteristic **transition temperature** at which it gels ("freezes") when cooled and becomes fluid again ("melts") when warmed. This change in the state of the membrane is called a **phase transition** and somewhat resembles the change that butter or margarine undergoes upon heating or cooling. Not surprisingly,

Table 7-3 Structures of Some Common Fatty Acids

Name of Fatty Acid	Number of Carbon Atoms	Number of Double Bonds	Structural Formula	Space-Filling Model
Saturated				
Palmitate	16	0		
Stearate	18	0		
Unsaturated				
Oleate	18	1		
Linoleate	18	2		

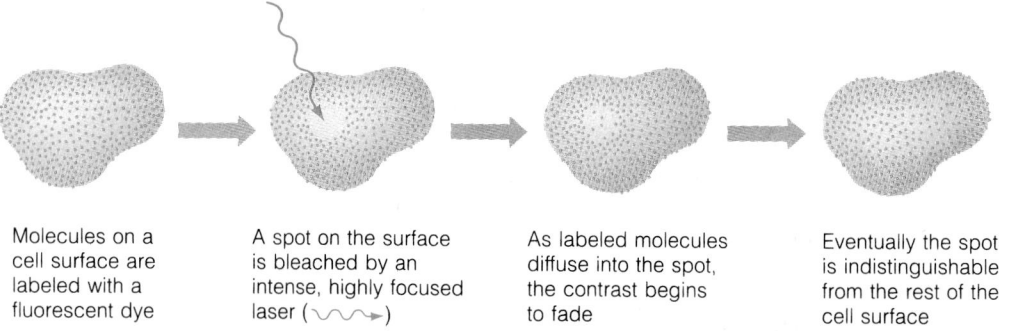

Molecules on a cell surface are labeled with a fluorescent dye

A spot on the surface is bleached by an intense, highly focused laser (〰️➤)

As labeled molecules diffuse into the spot, the contrast begins to fade

Eventually the spot is indistinguishable from the rest of the cell surface

Figure 7-12 Demonstration of Lipid Mobility Within Membranes by Fluorescence Recovery After Photobleaching.

natural membranes function normally only if they are maintained above the transition temperature, in the fluid rather than the gel state.

The fluidity of a membrane depends primarily on the kinds of lipids present. The three most important aspects of lipid composition that affect membrane fluidity are the length of the fatty acids present, the degree of unsaturation of their side chains (that is, the number of double bonds present), and, for animal membranes, the amount of cholesterol in the membrane.

The temperature at which a saturated fatty acid "melts" depends on its chain length, with transition temperatures ranging from 32°C to 76°C for saturated fatty acids with 10 and 20 carbon atoms, respectively (Figure 7-13a). Sites of unsaturation affect melting temperature even more markedly. For fatty acids with 18 carbon atoms, the transi-

tion temperatures are 70, 16, 5, and −11°C for zero, one, two, and three double bonds, respectively (Figure 7-13b). As a result, membranes containing many saturated chains have transition temperatures between 40° and 50°C, whereas membranes with mainly unsaturated chains "melt" at about 0°C or lower.

The effect of unsaturation on membrane fluidity is so dramatic because the kinks that double bonds introduce into fatty acids prevent the hydrocarbon chains from fitting together snugly. Membrane lipids with saturated fatty acids pack together tightly, thereby optimizing their van der Waals interactions (Figure 7-14a), whereas lipids with unsaturated fatty acids do not (Figure 7-14b). The lipids of most plasma membranes contain fatty acids that vary in both chain length and degree of unsaturation. In fact, the variability is often intramolecular because membrane lipids commonly

Figure 7-13 The Effect of Chain Length and Number of Double Bonds on the Transition Temperature of Fatty Acids. The transition temperature of fatty acids **(a)** increases with chain length for saturated fatty acids and **(b)** decreases with the number of double bonds for fatty acids with a fixed chain length (18 carbon atoms, in the example shown here).

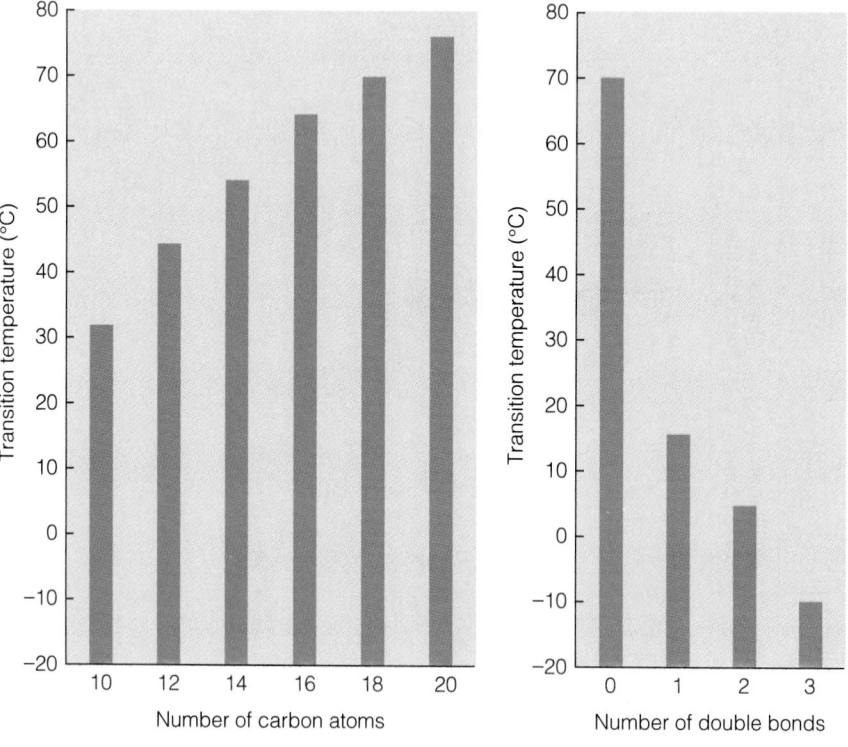

(a) Effect of chain length on the transition temperature of saturated fatty acids

(b) Effect of unsaturation on the transition temperature of fatty acids with 18 carbon atoms

contain one saturated and one unsaturated fatty acid. This property keeps most membrane lipids in the fluid state at physiological temperatures.

The Effects of Cholesterol on Membrane Fluidity. For animal cells, membrane fluidity is also affected by the presence of cholesterol. The plasma membranes of most animal cells contain large amounts of cholesterol—up to 50% of the total membrane lipid on a molar basis. Cholesterol molecules are usually found in both layers of the plasma membrane, but a given molecule is localized to one of the two layers (Figure 7-15a). The molecule orients itself in the layer with its single hydroxyl group—the most polar part of the molecule—close to the polar head group of a neighboring phospholipid molecule, where it can form a hydrogen bond

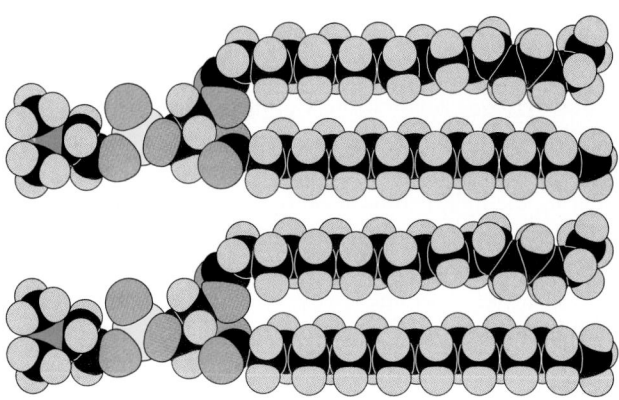

(a) Lipids with saturated fatty acids pack together well in the membrane

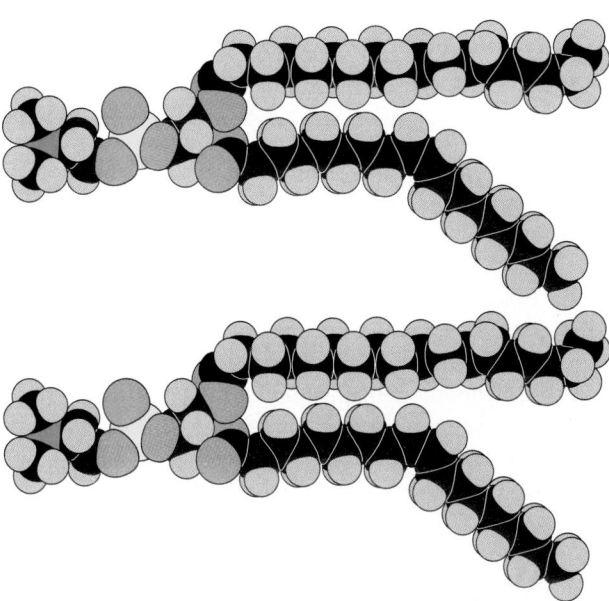

(b) Lipids with a mixture of saturated and unsaturated fatty acids do not pack together well in the membrane

Figure 7-14 The Effect of Unsaturated Fatty Acids on the Packing of Membrane Lipids. **(a)** Membrane lipids with no unsaturated fatty acids fit together well because the fatty acid chains are parallel to each other. **(b)** Membrane lipids with one or more unsaturated fatty acids do not fit together well because the *cis* double bonds cause bends in the chains that interfere with packing. Each of the structures shown is a phosphatidyl choline molecule with either two 18-carbon saturated fatty acids or two 18-carbon fatty acids, one saturated and the other with one double bond.

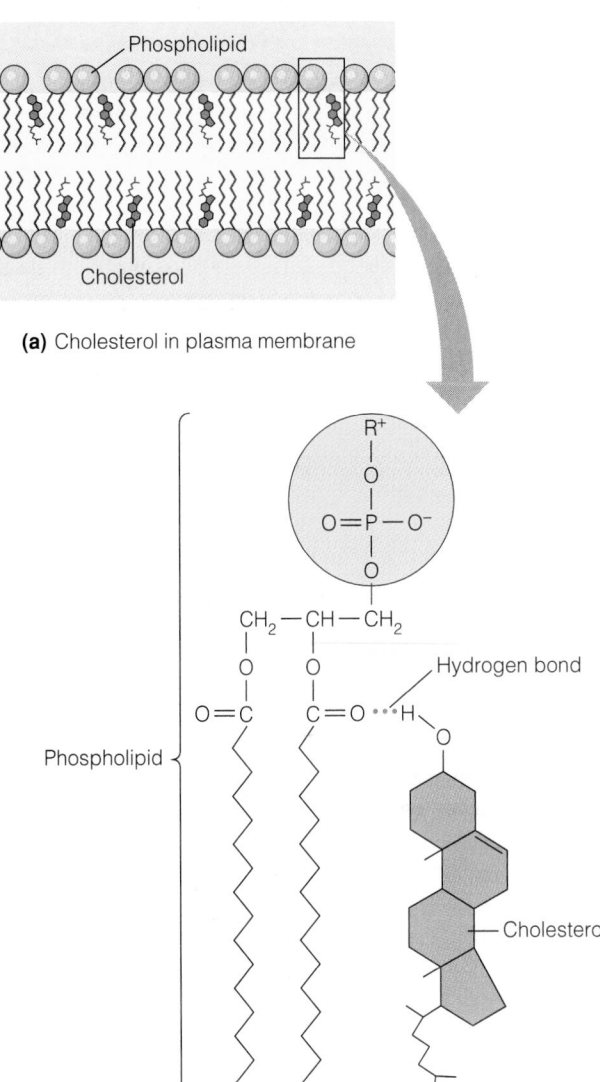

(a) Cholesterol in plasma membrane

(b) Bonding of cholesterol to phospholipid

Figure 7-15 Orientation of Cholesterol Molecules in a Lipid Bilayer. **(a)** Cholesterol molecules are present in both lipid layers in the plasma membranes of most animal cells, but a specific molecule is localized to one of the two layers. **(b)** Each molecule orients itself in the lipid layer such that its single hydroxyl group is close to the polar head group of a neighboring phospholipid molecule, where it can form a hydrogen bond with the oxygen of the ester bond between the glycerol backbone and a fatty acid. The steroid rings and hydrocarbon side group of the cholesterol molecule interact with adjacent hydrocarbon chains of the membrane phospholipids.

with the oxygen of the ester bond between the glycerol backbone and a fatty acid of the phospholipid molecule (Figure 7-15b). The rigid hydrophobic steroid rings and the hydrocarbon side chain of the cholesterol molecule interact with the portions of adjacent hydrocarbon chains that are closest to the phospholipid head groups.

This intercalation of cholesterol molecules into the lipid monolayers results in reduced membrane fluidity at higher temperatures. However, cholesterol also effectively prevents the hydrocarbon chains of phospholipids from aggregating as temperature is decreased, thereby reducing the tendency of membranes to "freeze" upon cooling. Thus, cholesterol has the dual effect of *decreasing* membrane fluidity at high temperatures and *increasing* it at low temperatures.

In addition to its effects on membrane fluidity, cholesterol also decreases the permeability of a lipid bilayer to ions and small polar molecules. It probably does so by filling in spaces between hydrocarbon chains of membrane phospholipids, thereby plugging small channels through which ions and small molecules might otherwise pass. In general, a lipid bilayer containing cholesterol is less permeable to ions and small molecules than a bilayer lacking cholesterol.

Cholesterol is also thought to enhance both the flexibility and mechanical stability of the phospholipid bilayer. The importance of cholesterol in maintaining membrane stability can be demonstrated with lines of mutant animal cells that are unable to synthesize cholesterol due to a genetic defect. Such cells *lyse* (break open) if cultured in the absence of cholesterol. If cholesterol is added to the culture medium, however, the cells take up the cholesterol and incorporate it into the plasma membrane, thereby stabilizing the cells and enabling them to survive.

The Regulation of Membrane Fluidity. Most organisms, whether prokaryotic or eukaryotic, are able to regulate membrane fluidity, primarily by changing the lipid composition of the membranes. This ability is especially important for *poikilotherms*—organisms such as bacteria, fungi, protists, plants, and cold-blooded animals that cannot regulate their own temperature. For these organisms, changes in the external temperature affect the fluidity of membranes significantly. Since lipid fluidity decreases with temperature, membranes of these organisms would "freeze" upon cooling if the organism had no way to compensate for the decrease in temperature. (You may have experienced this effect even though you are a *homeotherm*, or warm-blooded organism; on chilly days, your fingers and toes can get so cold that the membranes of sensory nerve endings cease to function, resulting in temporary numbness.) At high temperatures, on the other hand, the lipid bilayers of poikilotherms become so fluid that they no longer serve as an effective permeability barrier. For example, most cold-blooded animals are paralyzed by temperatures much above 45°C, probably because certain nerve cell membranes become so leaky to ions that overall nervous function is disabled.

The relationship between temperature, fatty acid composition, and membrane fluidity can be seen clearly with

bacteria of the genus *Acholeplasma.* These bacteria cannot manufacture their own fatty acids but must instead use whatever is available in the environment. Their membranes therefore take on the physical characteristics of the fatty acids available to the bacteria at the time. If the bacteria have access to unsaturated fatty acids, their membranes remain fluid and the bacteria grow readily. If only saturated fatty acids are available, however, the bacteria grow only until their membrane lipids contain so many saturated hydrocarbon chains that the membranes undergo a phase transition to the gel state. If the temperature is then raised to induce a phase transition back to the fluid state, the bacteria are again able to grow.

Homeoviscous Adaptation: Compensating for the Effects of Temperature Changes. Fortunately, most poikilotherms are able to compensate for temperature changes by altering the lipid composition of their membranes, thereby regulating membrane fluidity. This capability is called **homeoviscous adaptation** because the main effect of such regulation is to keep the viscosity of the membrane approximately the same despite the change in temperature. Consider, for example, what happens when bacterial cells are transferred from a warmer to a cooler environment. In some species of the genus *Micrococcus*, a drop in temperature results in an increase in the proportion of 16-carbon rather than 18-carbon fatty acids in the plasma membrane. In this case, the desired increase in membrane viscosity is effected by an increase in the activity of an enzyme that removes two terminal carbons from 18-carbon hydrocarbon tails.

In other bacterial species, adaptation to temperature changes involves an alteration in the extent of unsaturation of membrane fatty acids rather than in their length. In the common intestinal bacterium *Escherichia coli*, for example, a decrease in temperature triggers the synthesis of a *desaturase* enzyme that introduces double bonds into the hydrocarbon chains of fatty acids. As these unsaturated fatty acids are incorporated into the plasma membrane, they decrease the transition temperature of the membrane, thereby ensuring that the membrane remains fluid at the lower temperature. If the temperature is then increased, the gene encoding the desaturase enzyme becomes inactive again, and existing molecules of the enzyme are inhibited allosterically. As a result, fatty acids without double bonds are synthesized and incorporated into the plasma membrane, thereby keeping the membrane from becoming too fluid and losing its effectiveness as a permeability barrier.

Homeoviscous adaptation also occurs in yeasts and plants. In these organisms, changes in membrane fluidity with temperature appear to depend on the increased solubility of oxygen in the cytoplasm at lower temperatures. Oxygen is a substrate for the desaturase enzyme system involved in the generation of unsaturated fatty acids. With more oxygen available at lower temperatures, unsaturated fatty acids are synthesized at a greater rate and membrane fluidity increases, thereby offsetting the temperature effect. This capability has great agricultural significance because plants that are able to adapt in this way are cold-hardy (resistant to chilling) and can therefore be grown in colder environments.

Poikilothermic animals also adapt to lower temperatures by increasing the proportion of unsaturated fatty acids in their membranes. These organisms also increase the proportion of cholesterol in the membrane, thereby decreasing the interaction between hydrocarbon chains and reducing the tendency of the membrane to "freeze."

Although homeoviscous adaptation is of greatest general relevance to poikilothermic organisms, it is also important to mammals that hibernate because the body temperature often drops substantially as the animal enters hibernation—a decrease of more than 30°C for some rodents. An animal entering hibernation adapts to this change by synthesizing a greater proportion of unsaturated fatty acids as its body temperature falls.

Membrane Proteins: The "Mosaic" Part of the Model

Having looked in some detail at the "fluid" aspect of the fluid mosaic model, we come now to the "mosaic" part. This means that we come to membrane proteins, because the rev-olutionary feature of the Singer-Nicolson model was to envision the membrane as a *mosaic* of proteins suspended and floating in a lipid bilayer. Earlier in the chapter, we noted that prior models could not readily account for the different amounts of protein present in various membranes, nor could they easily accommodate the evidence that membrane proteins were globular molecules rather than thin sheets. Now we are ready to consider in more detail the specific evidence concerning the nature of membrane proteins.

The Membrane as a Mosaic of Proteins: Evidence from Freeze-Fracture Microscopy. Especially compelling evidence for the fluid mosaic model came from studies in which artificial bilayers and natural membranes were prepared for electron microscopy by the **freeze-fracture technique** described in the Appendix. If a bilayer or membrane (or a cell containing membranes) is frozen quickly and the frozen sample is then subjected to a sharp blow from a microtome knife, the resulting fracture frequently follows the plane between the two layers of the membrane because the nonpolar interior of the bilayer is the path of least resistance through the frozen specimen (see Figure A-30). As a result, the bilayer is split into its inner and outer layers, revealing the inner surface of each layer (Figure 7-16).

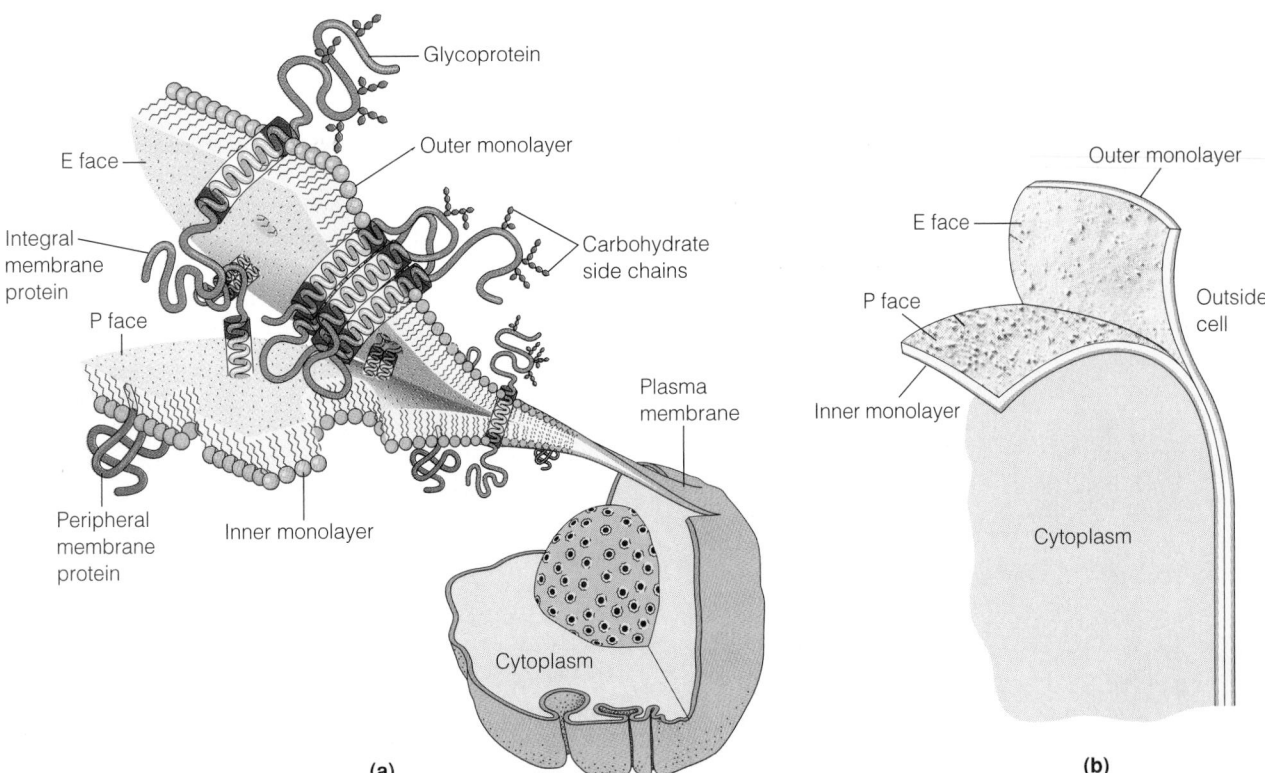

Figure 7-16   Freeze-Fracture Analysis of a Membrane.
(a) Sketch of a freeze-fractured membrane in which the fracture plane has passed through the hydrophobic interior of the membrane, revealing the inner surfaces of the two monolayers. Hydrophobic regions of proteins are shown in light purple and hydrophilic regions in dark purple. Integral membrane proteins that remain with the outer monolayer are seen on the E (exterior) face, whereas those that remain with the inner monolayer are seen on the P (protoplasmic) face. **(b)** Sketch of a freeze-fractured membrane, with electron micrographs of the E and P faces from the plasma membrane of a mouse kidney tubule cell superimposed on the art (TEMs). The P face of the membrane is studded with intramembranous particles, whereas the E face is relatively smooth.

Electron micrographs of membranes prepared by the freeze-fracture technique are entirely consistent with the bilayer structure of membranes. More significant, however, is the evidence from such studies that proteins are actually suspended within membranes. Whenever a fracture plane splits the membrane into its two layers, particles having the size and shape of globular proteins can be seen adhering to one or the other of the inner membrane surfaces, called the *E* (for *exterior*) and *P* (for *protoplasmic*) *faces*. Moreover, the abundance of such particles correlates well with the known protein content of the particular membrane under investigation. Thus, membranes from the myelin sheath of nerve cell axons have a low protein/lipid ratio (0.23; see Table 7-1) and show almost no particles when subjected to freeze-fracture, whereas the inner mitochondrial membrane has a high protein/lipid ratio (3.5) and a high density of intramembranous particles, especially on the inner lipid layer.

Still more convincing evidence that the particles seen in this way are really proteins came from work by David Deamer and Daniel Branton, who used the freeze-fracture technique to examine artificial bilayers with and without added protein. Bilayers formed from pure phospholipids showed no evidence of particles on their interior surfaces (Figure 7-17a). When proteins were added to the artificial bilayers, however, particles similar to those seen in freeze-fractured natural membranes were visible (Figure 7-17b).

Analysis of Membrane Proteins: SDS-Polyacrylamide Gel Electrophoresis. As noted earlier, peripheral membrane proteins are bound to the membrane by weak electrostatic interactions with either the polar head groups of membrane lipids or the hydrophilic portions of integral membrane proteins. Peripheral proteins can therefore be ex-

tracted from the membrane by salt solutions. Integral membrane proteins, on the other hand, can be solubilized only by the use of organic solvents or detergents that dissolve the membrane.

A useful technique for studying membrane proteins is **SDS-polyacrylamide gel electrophoresis,** illustrated in Figure 7-18. Membrane fragments are first solubilized with the anionic detergent *sodium dodecyl sulfate* (*SDS*), which disrupts most protein-protein and protein-lipid associations. The proteins denature, unfolding into stiff polypeptide rods that cannot refold because their surfaces are coated with negatively charged detergent molecules. The solubilized, SDS-coated polypeptides are then applied to the top of a gel of polyacrylamide and a potential difference is applied across the gel, such that the bottom of the gel is the positively charged anode. Because the polypeptides are coated with negatively charged SDS molecules, they migrate down the gel toward the anode. The polyacrylamide gel can be thought of as a fine meshwork that impedes the movement of large molecules more than that of small molecules. As a result, polypeptides move down the gel at a rate that is inversely related to their size.

When the smallest polypeptides approach the bottom of the gel, the process is terminated and the gel is stained with a dye (usually *Coomassie brilliant blue*) that binds to polypeptides and makes them visible. The particular polypeptide profile shown as ⑦ in Figure 7-18 is for the membrane proteins of human erythrocytes. Bands are numbered from the largest to the smallest polypeptides. Proteins mentioned in this chapter are also identified by name. (Glycophorin is also a major protein of the erythrocyte but does not show up here because it is a glycoprotein and does not stain well with the dye that was used.)

Figure 7-17 Comparison of Lipid Bilayers With and Without Added Proteins. This figure compares the appearance of **(a)** artificial lipid bilayers with **(b)** artificial bilayers to which proteins were added. Both samples were prepared by freeze-fracture electron microscopy, in which biological specimens are first frozen in liquid nitrogen and then fractured and shadowed with platinum. The white lines in the artificial membranes of part a represent individual lipid bilayers in a multilayered specimen, and the gray regions show where single bilayers have split to reveal smooth surfaces. In contrast, the artificial membrane of part b shows large numbers of globular particles in the fracture surface. They represent the proteins that were added to the membrane preparation. Similar particles are observed when natural membranes are freeze-fractured (TEMs).

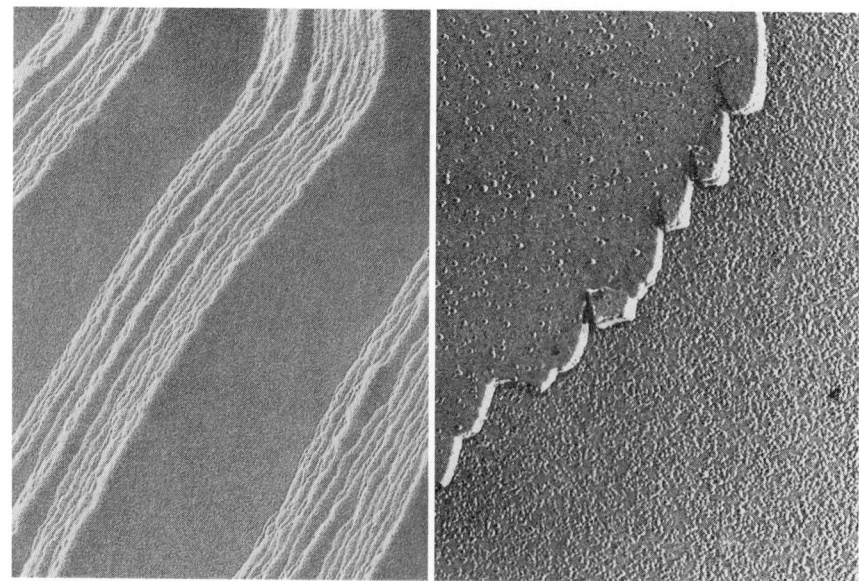

0.1 μm

(a) Artificial bilayers without proteins

(b) Artificial bilayers with proteins

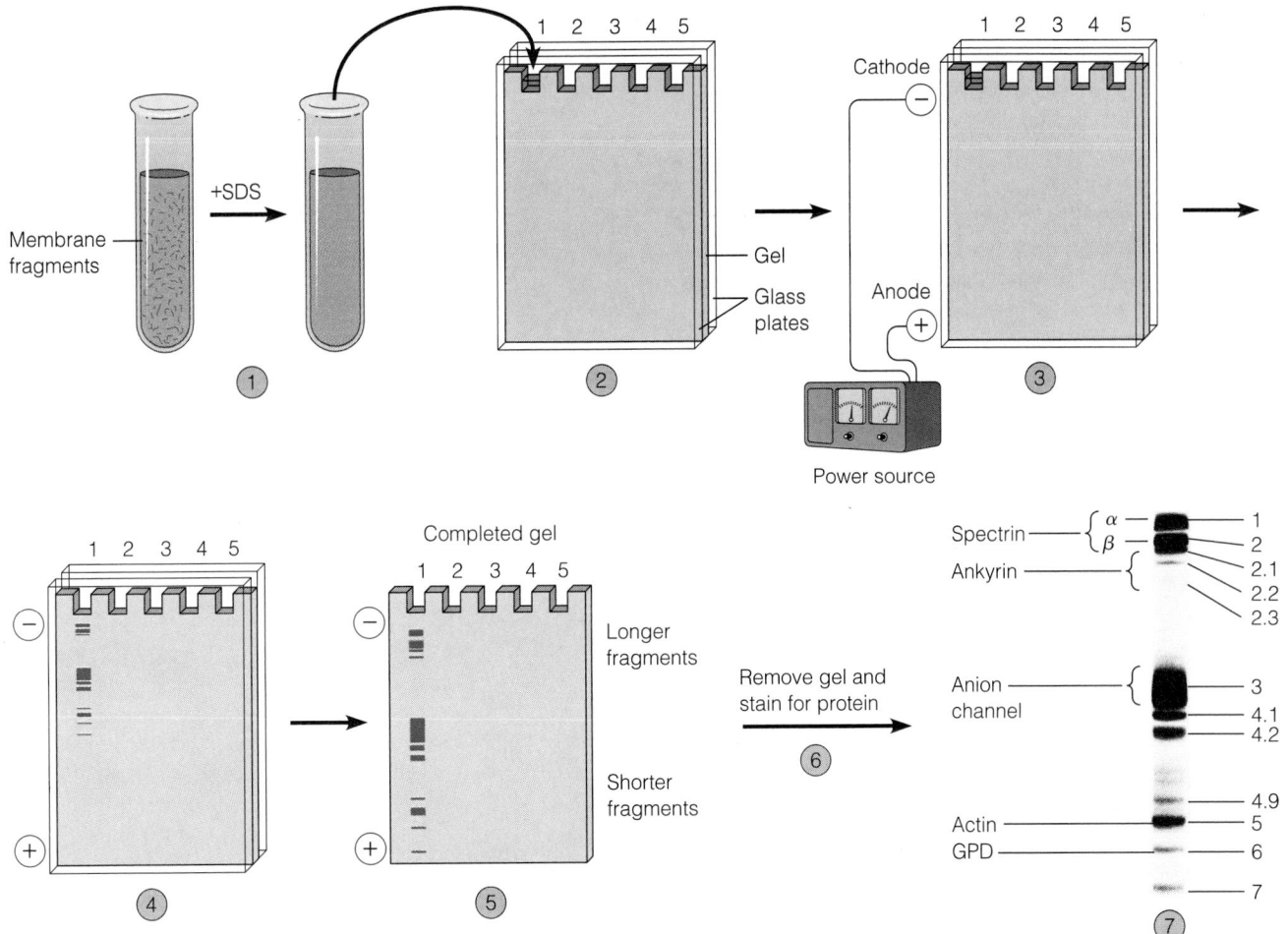

Figure 7-18 **SDS-Polyacrylamide Gel Electrophoresis of Membrane Proteins.** Membrane fragments are suspended in buffer and ① treated with sodium dodecyl sulfate (SDS). ② A small sample of the solubilized polypeptides is then applied to one of the wells at the top of a gel of polyacrylamide that has been polymerized and cross-linked between two glass plates. ③ A potential difference is applied across the gel, ④ causing the protein molecules to migrate toward the far end of the gel. ⑤ The polypeptides move down the gel at rates that are inversely related to their respective sizes. ⑥ When the smallest polypeptides are near the bottom, the gel is removed from between the glass plates and stained with a dye that binds to polypeptides and makes them visible. (The bands shown in color in steps 4 and 5 are not actually visible on the gel until the gel has been removed and stained.) ⑦ The particular polypeptide profile shown here is for the membrane proteins of the human erythrocyte. (GPD = glyceraldehyde-3-phosphate dehydrogenase.)

Glycoproteins. In addition to lipids and proteins, most membranes contain small but significant amounts of *carbohydrates*. The plasma membrane of the human erythrocyte, for example, contains about 52% protein, 40% lipid, and 8% carbohydrate by weight. A small proportion of membrane carbohydrate is present as *glycolipids,* membrane lipids with short oligosaccharide chains attached to them, but most of it is in the form of **glycoproteins**—membrane proteins with carbohydrate chains linked covalently to amino acid side chains. As Figure 7-19 shows, the linkages are of two types: *N-linked carbohydrates* are attached to the amino group on the side chain of asparagine, whereas *O-linked carbohydrates* are usually bonded to the hydroxyl groups of either serine or threonine. In some cases, O-linked carbohydrates are bonded to the hydroxyl group of either hydroxylysine or hydroxyproline which are derivatives of the amino acids lysine and proline, respectively.

The carbohydrate chains found in glycoproteins form straight or branched chains ranging in length from 2 to about 60 sugar units. The most common sugars involved are *galactose, mannose, N-acetylglucosamine,* and *sialic acid* (Figure 7-20a). Figure 7-20b shows the carbohydrate group of glycophorin, one of the integral membrane glycoproteins in the erythrocyte plasma membrane (see Figure 7-8). Glycophorin has 16 carbohydrate groups on the portion of the molecule (the N-terminus) that extends outward from the membrane; of these, 1 is N-linked and 15 are O-linked. Notice that both branches of the carbohydrate group terminate in a negatively charged sialic acid. Because of these anionic groups on their surfaces, erythrocytes repel each other, thereby reducing blood viscosity.

Glycoproteins are most prominent in plasma membranes, where they are always positioned such that the carbohydrate groups are found only on the external surface of

Figure 7-19   Linkage of Carbohydrate Groups to Membrane Proteins. Carbohydrate groups may be linked to membrane proteins in two different ways. **(a)** N-linked carbohydrates are bonded to the amino group of asparagine side chains. **(b)** O-linked carbohydrates are bonded to the hydroxyl group of either serine or threonine side chains. **(c)** In some cases, O-linked carbohydrates bind to the hydroxyl group of the modified amino acids hydroxylysine and hydroxyproline.

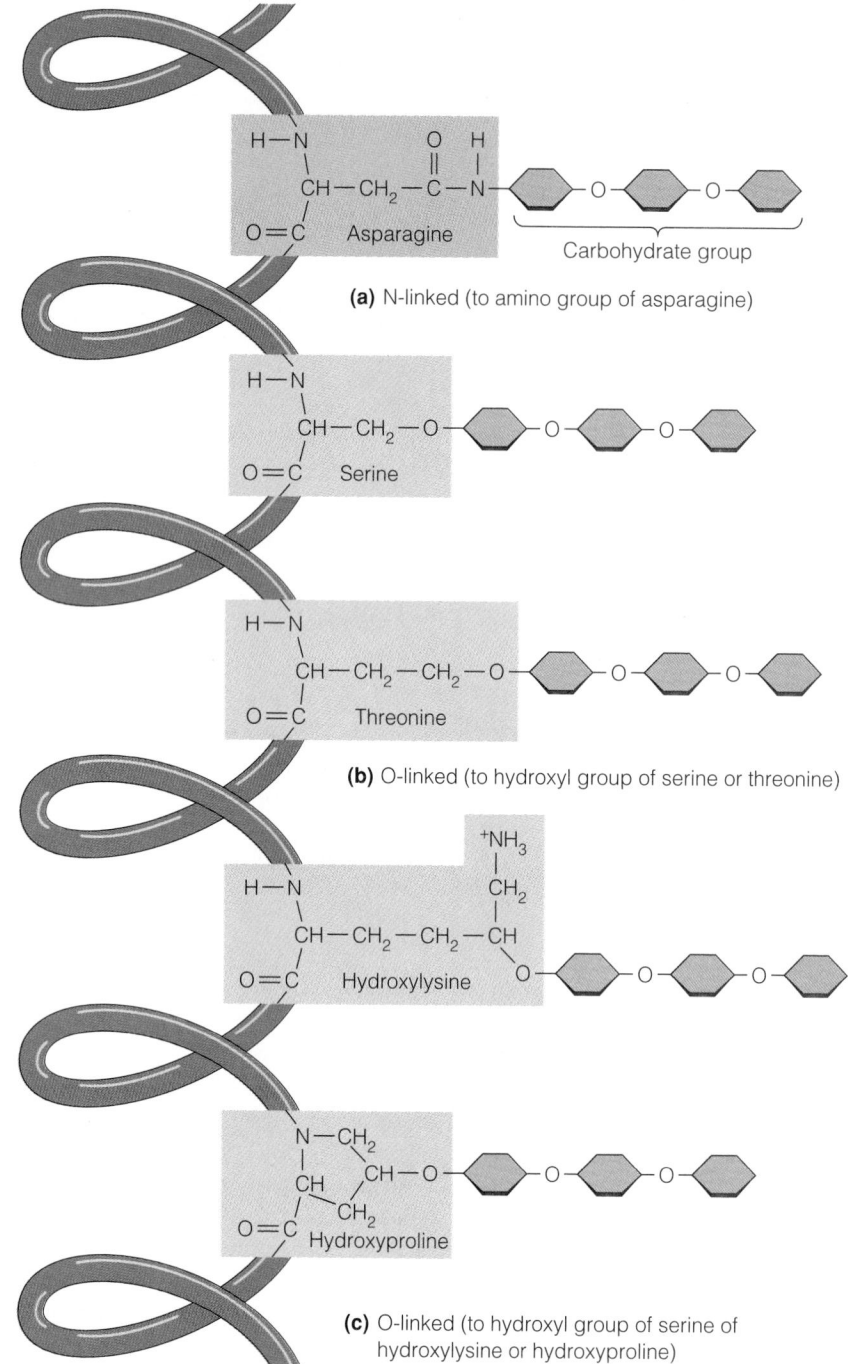

(a) N-linked (to amino group of asparagine)

(b) O-linked (to hydroxyl group of serine or threonine)

(c) O-linked (to hydroxyl group of serine of hydroxylysine or hydroxyproline)

the cell membrane. This arrangement has been shown experimentally by using *lectins*, plant proteins that bind specific sugar groups very avidly. For example, *wheat germ agglutinin*, a lectin found in wheat embryos, binds very specifically to oligosaccharides that terminate in *N*-acetylglucosamine, whereas *concanavalin A*, a lectin from beans, recognizes mannose groups in internal positions. Investigators visualize these lectins by linking them to *ferritin*, an iron-containing protein that shows up as a very electron-dense spot when viewed with an electron microscope. When such ferritin-linked lectins are used as probes to localize the

oligosaccharide chains of membrane glycoproteins, binding is always specifically to the outer surface of the membrane. More techniques for studying membrane glycoproteins are described in Box 7A (pages 191–192).

In many animal cells, the carbohydrate groups of the glycoproteins on the plasma membrane form a surface coat called the **glycocalyx.** Figure 7-21 shows the prominent glycocalyx of an intestinal epithelial cell. The carbohydrate groups on the cell surface are important as components of the recognition sites of membrane receptors such as those involved in binding extracellular signal molecules, in

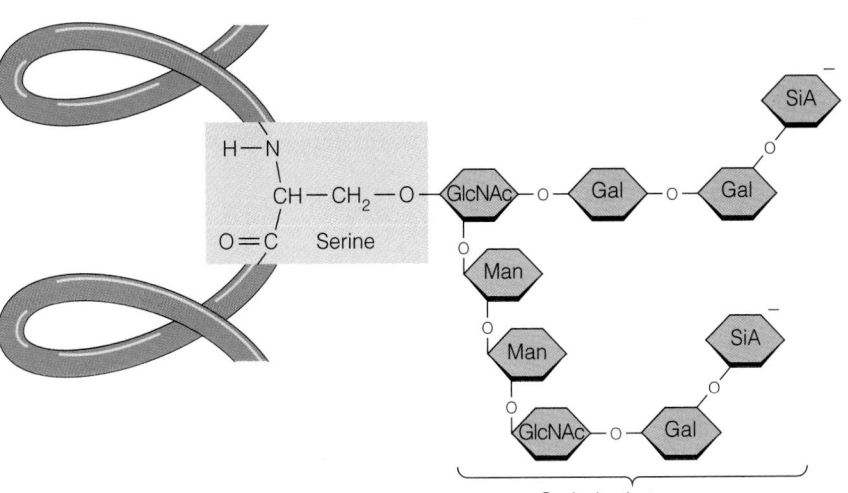

(a) Common sugars found in glycoproteins

(b) The carbohydrate group of glycophorin

Carbohydrate group

Figure 7-20 The Carbohydrates Found in Glycoproteins. **(a)** The four most common carbohydrates in glycoproteins are galactose (Gal), mannose (Man), N-acetylglucosamine (GlcNAc), and sialic acid (SiA). **(b)** For example, glycophorin has 16 carbohydrate groups linked to the hydrophilic portion of the protein on the outer surface of the erythrocyte plasma membrane. Each carbohydrate group consists of three units of Gal and two units each of GlcNAc, Man, and SiA. The sialic acid groups are negatively charged.

antibody-antigen reactions, and in intercellular adhesion to form tissues. The carbohydrate groups probably also protect the proteins from digestion by extracellular proteases and may contribute to membrane stability as well.

Functions of Membrane Proteins. What functions do membrane proteins perform? Not surprisingly, the answers are the same as those to the question about membrane function at the beginning of the chapter. Membrane proteins have been isolated and shown to have a variety of functions. In some cases, it has even been possible to identify the functions of specific proteins that can be visualized as membrane particles by freeze-fracture analysis.

Some of the proteins in membranes are enzymes, which accounts for the localization of specific functions to specific membranes. For example, one of the peripheral proteins of the erythrocyte plasma membrane is glyceraldehyde-3-phosphate dehydrogenase, an enzyme involved in the process whereby the erythrocyte uses blood glucose for energy. Each of the organelles in eukaryotic cells is characterized by its own distinctive set of membrane-bound enzymes, as we will see in coming chapters. Other membrane proteins are

intermediates in energy transduction processes, such as the cytochromes and iron-sulfur proteins found in the inner membrane of the mitochondrion that are involved in electron transport.

Still other membrane proteins are involved in solute transport across membranes. These include *transport proteins*, which facilitate the movement of nutrients such as sugars and amino acids across membranes, and *channel proteins*, which provide hydrophilic passageways through otherwise hydrophobic membranes. Also in this category are *transport ATPases*, which use the energy of ATP to pump ions across membranes, as we will see in more detail in Chapter 8.

Some membrane proteins are *receptors* involved in the recognition and mediation of specific chemical signals that impinge on the surface of the cell. Hormones, neurotransmitters, and growth-promoting substances are examples of chemical signals that interact with specific receptors in or on the membrane of target cells. Usually, the binding of a hormone or other signal molecule to the appropriate receptor on the membrane surface triggers some sort of intracellular response that then elicits the desired effect.

Figure 7-21 The Glycocalyx of an Intestinal Epithelial Cell. This electron micrograph of a cat intestinal epithelial cell shows the microvilli that are involved in absorption and the glycocalyx on the cell surface. The glycocalyx on this cell is about 150 nm thick and consists primarily of oligosaccharide chains about 1.2–1.5 nm in diameter (TEM).

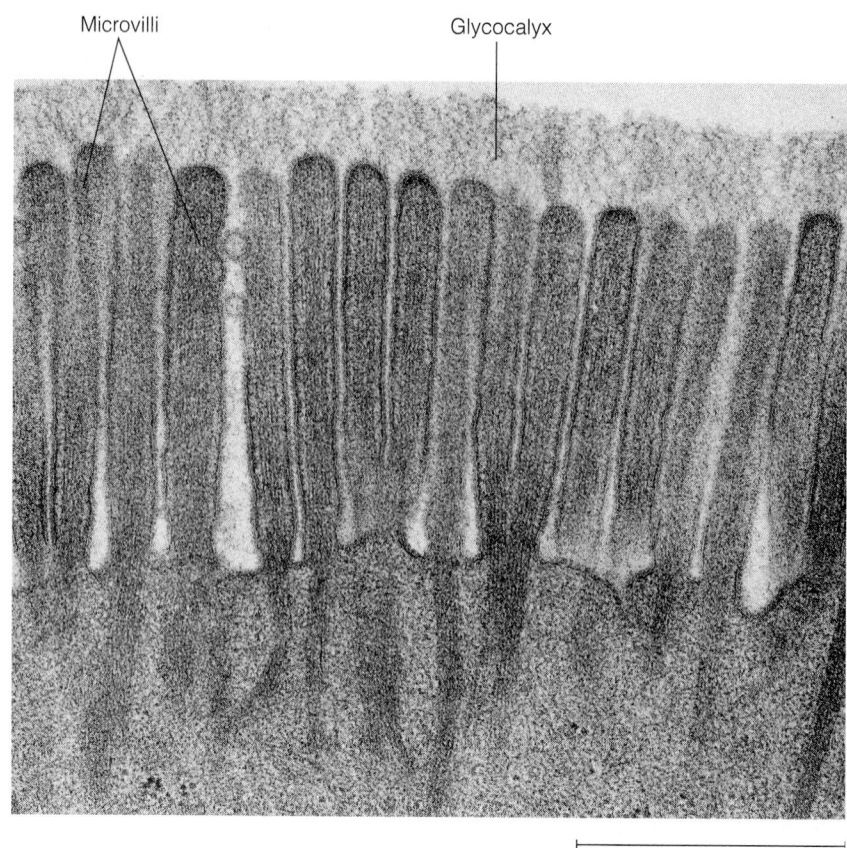

Microvilli

Glycocalyx

0.5 μm

Proteins involved with cell-to-cell communication include those that form the structures called *connexons* at *gap junctions* between animal cells and those that make up the *desmotubules* of *plasmodesmata* between plant cells. These structures will be discussed in Chapter 10.

A final group of membrane-associated proteins are those with structural roles in stabilizing and shaping the cell membrane. Examples include spectrin, ankyrin, and actin in the erythrocyte (see Figure 7-10) and a protein called clathrin, which is present in cells such as the hepatocytes of liver tissue. Clathrin has the ability to coat membrane surfaces, as will be discussed in detail in Chapter 9.

The Mobility of Membrane Proteins

The concept of membrane fluidity applies not just to membrane lipids but also to the proteins embedded within those lipids. Particularly convincing evidence for the mobility of membrane proteins came from *cell fusion* experiments and from studies of a phenomenon called *patching and capping* in white blood cells.

Cell Fusion. The mobility of membrane proteins was demonstrated by **cell fusion** experiments such as that shown in Figure 7-22 from the work of David Frye and Michael Edidin. These researchers took advantage of two powerful

Figure 7-22 Demonstration of the Mobility of Membrane Proteins by Cell Fusion. ① Cells from two different species (human and mouse) are tagged with fluorescent antibodies that react with membrane proteins and ② induced to fuse by exposure to Sendai virus. The proteins from the two parent cells are localized to two separate halves of the hybrid cell initially but ③ begin to intermix within minutes and ④ are randomly distributed in the plasma membrane within 40 minutes.

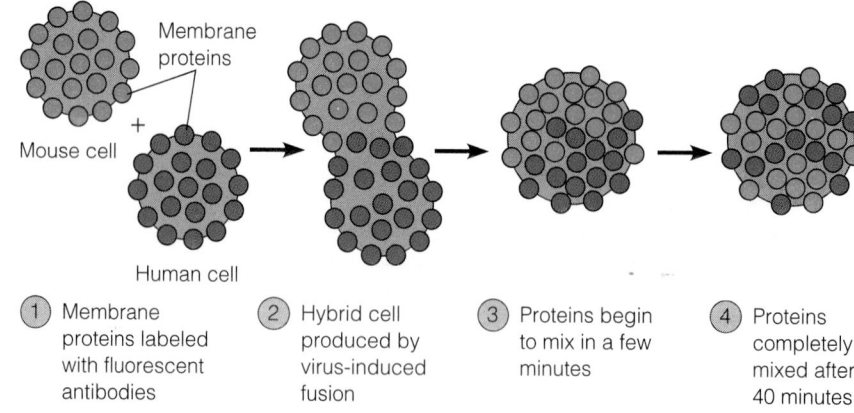

Membrane proteins

Mouse cell

+

Human cell

① Membrane proteins labeled with fluorescent antibodies

② Hybrid cell produced by virus-induced fusion

③ Proteins begin to mix in a few minutes

④ Proteins completely mixed after 40 minutes

techniques, one that enabled them to fuse cells from two different species and another that made it possible for them to tag the surfaces of cells with fluorescent dye molecules. Their approach was to tag mouse and human cells with **fluorescent antibodies** against plasma membrane proteins specific to the two species. They then fused the two types of cells and observed the hybrid human/mouse cells with a fluorescence microscope to see what happened to the plasma membrane proteins from the two parent cells.

The antibodies needed for this kind of experiment are prepared by injecting small quantities of purified membrane proteins from human and mouse cells into separate experimental animals. (Rabbits or goats are commonly used for this purpose.) The animals respond immunologically to the foreign protein by producing antibodies (blood proteins called *immunoglobulins;* see Chapter 24) that react specifically with the membrane proteins used in the injections. After the antibodies have been isolated from the blood of the animal, fluorescent dyes are covalently linked to them so that the protein/antibody/dye complexes can be visualized with a fluorescence microscope.

Frye and Edidin tagged human and mouse cells with antibodies that had two different kinds of fluorescent dyes linked to them so that the two types of cells could be readily distinguished by the characteristic fluorescence of the dye molecules bound to their membrane proteins. They then exposed the tagged cells to Sendai virus, an agent known to cause fusion of eukaryotic cells, even those from different species. When the cells were first fused, the membrane proteins from the mouse cell were localized on one half of the plasma membrane of the hybrid cell, whereas those from the human cell were restricted to the other half. In a few minutes, however, the proteins from the two parent cells began to intermix, and within 40 minutes they were randomly distributed in the plasma membrane of the hybrid cell. Based on these results, Frye and Edidin concluded that the inter-mixing of the fluorescent proteins resulted from lateral diffusion of the human and mouse proteins through the fluid bilayer.

Patching and Capping. As further proof that membrane proteins move laterally within the membrane, consider the phenomenon of **patching and capping,** as seen in lymphocytes (Figure 7-23). A *lymphocyte* is a white blood cell that responds to a foreign protein by producing antibody molecules, which are present on the plasma membrane of the lymphocyte. One kind of foreign protein that will elicit antibody production in an animal is an antibody from another species. If mouse antibodies are injected into a rabbit, for example, the rabbit lymphocytes will produce anti-mouse antibodies. These antibodies can then be isolated from the blood of the rabbit and linked covalently to a fluorescent dye so they can be visualized by fluorescence microscopy.

When rabbit anti-mouse antibodies are mixed with mouse lymphocytes, the tagged antibodies will be spread evenly over the surface of the plasma membrane initially. Soon, however, clusters and then patches of antibody molecules form. (Clustering occurs because antibodies are *multivalent;* that is, each molecule of rabbit antibody can react with multiple mouse antibodies on the plasma membrane, and each membrane antibody can react with more than one rabbit antibody molecule.) This clustering leads to the formation of "patches" of rabbit anti-mouse antibody molecules linked to membrane-bound mouse antibody molecules. The patches then aggregate as a dense "cap" on one side of the cell, where they are taken into the cell by the process of endocytosis. Clearly, the process of patching and capping could not occur if proteins were not free to move laterally within the membrane.

Capping in lymphocytes occurs much more rapidly than does intermixing of membrane proteins in hybrid cells. The process shown in Figure 7-22 requires about 40

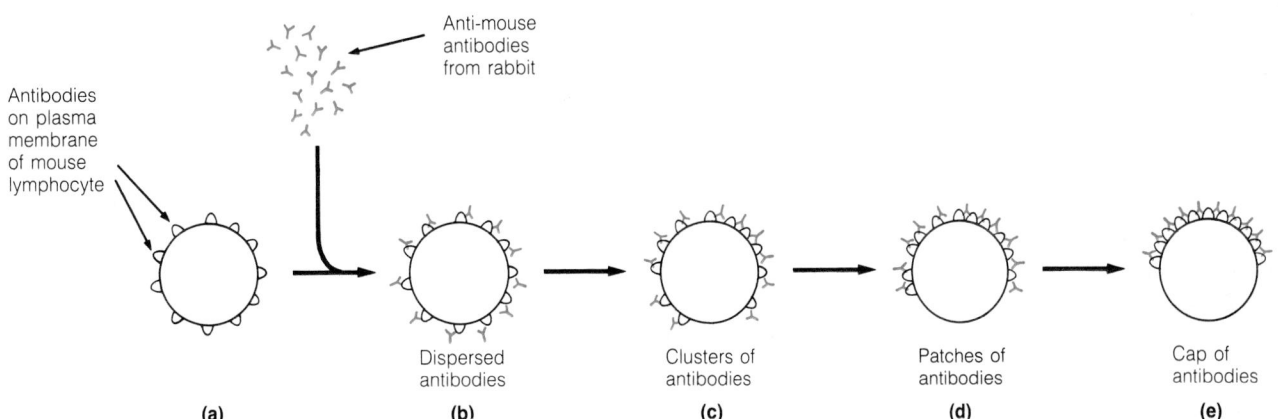

Figure 7-23   Demonstration of the Mobility of Membrane Proteins by Patching and Capping.   **(a)** Antibodies are normally dispersed randomly on the external surface of the plasma membrane of a mouse lymphocyte. **(b)** When rabbit anti-mouse antibodies are initially mixed with the mouse lymphocytes, the rabbit antibodies bind to the mouse antibodies and appear dispersed on the membrane surface of the lymphocyte. **(c)** Because the rabbit antibodies are capable of binding to more than one mouse antibody, antibody complexes begin to cluster on the surface membrane. **(d)** These clusters aggregate into large patches and **(e)** accumulate as a cap on one side of the membrane, where they are taken into the cell by endocytosis.

minutes, whereas the tagged proteins shown in Figure 7-23 are swept into caps within minutes. The capping process is so rapid because the proteins are actively moved through the membrane rather than diffusing randomly. The process requires ATP and is mediated by microfilaments attached to the cytoplasmic side of membrane proteins.

Membrane Asymmetry

Most membrane lipids are highly asymmetric in their distribution between the inner and outer monolayers of the bilayer. **Membrane asymmetry** is established during membrane biogenesis by the insertion of different lipids, or different proportions of the various lipids, into the two monolayers. Once established, asymmetry is usually maintained for hours or days because the movement of the hydrophilic heads of lipids through the hydrophobic interior of the membrane is thermodynamically unfavorable. Such "flip-flopping," or **transverse diffusion,** can be shown to occur in membrane lipids but is relatively rare. For instance, a typical phospholipid molecule undergoes flip-flop less than once a week in a pure phospholipid bilayer. This contrasts strikingly with the **rotation** of phospholipid molecules about their long axis and with the **lateral diffusion** of phospholipids in the plane of the membrane, which is so rapid that a lipid molecule can traverse the length of a bacterial cell in about a second. Figure 7-24 illustrates these three types of lipid movements. (The rate of phospholipid flip-flop is greater in some natural membranes than in artificial bilayers, apparently because some membranes have proteins that catalyze the transverse diffusion of lipid molecules.)

The asymmetry of membrane lipids includes differences both in the kinds of lipid present and in the degree of unsaturation of the fatty acids in the phospholipid molecules. For instance, most of the phosphatidyl choline in the erythrocyte plasma membrane is present in the outer of the two monolayers, whereas most of the phosphatidyl ethanolamine and phosphatidyl serine is in the inner monolayer. Cholesterol, on the other hand, is present in the two monolayers in approximately equal amounts. Especially significant is the localization of all glycolipids in the outer layer of the plasma membrane of an animal cell, such that the carbohydrate groups of these molecules protrude from the outer membrane surface, forming part of the glycocalyx.

Membrane proteins are also asymmetrically localized. A peripheral protein is by definition associated with one or the other of the membrane surfaces and, once in place, cannot readily move across the membrane from one surface to the other. Integral membrane proteins are also asymmetrically

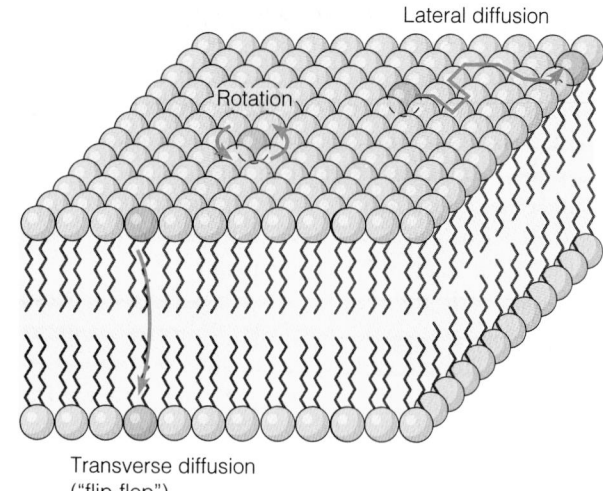

Figure 7-24 Movements of Phospholipid Molecules Within Membranes. A phospholipid molecule is capable of three kinds of movement in a membrane: rotation about its long axis, lateral diffusion by exchanging places with neighboring molecules in the same monolayer, and transverse diffusion, or "flip-flop," from one monolayer to the other. In a pure phospholipid bilayer at 37°C, a typical lipid molecule exchanges places with neighboring molecules about ten million times per second and can move laterally at a rate of about several micrometers per second. By contrast, the frequency with which an individual lipid molecule flip-flops from one layer to the other ranges from less than once a week in a pure phospholipid bilayer to once every few hours in some natural membranes. The difference is apparently due to the presence in some membranes of proteins that catalyze the transverse diffusion of phospholipid molecules.

oriented in the membrane, often held in place by their association with peripheral proteins. The highly ordered protein complexes that make up the respiratory assemblies of the inner mitochondrial membrane (see Chapter 12) and the photosystems of the chloroplast thylakoids (see Chapter 13) are dramatic examples of structures that contribute to membrane asymmetry. The asymmetric distribution of membrane proteins arises at the time of protein insertion into the membrane and is then preserved by the thermodynamic restrictions on transverse diffusion.

As noted earlier, glycoproteins are oriented asymmetrically in the plasma membrane of animal cells such that their carbohydrate groups, like those of glycolipids, are located exclusively on the outer surface. Box 7A provides some insights into the ingenious ways in which the asymmetry of membrane proteins has been demonstrated and studied.

Contemporary Techniques

RED BLOOD CELLS, MEMBRANES, AND INGENUITY

Do you ever stop to wonder how we know the things we know about cells? Consider, for example, the simple assertion that membranes are usually asymmetric, having some proteins associated with the external lipid layer, others with the internal layer, and still others extending into or even through both layers. How do we know these things? And how good is the evidence? In answering these questions, we will encounter some ingenious techniques and gain some fascinating insights into the way cell biology is done.

To explore membrane asymmetry, we will need some red blood cells, the enzymes lactoperoxidase (LP) and galactose oxidase (GO), some radioactive iodine (^{125}I), and the radioactively labeled reducing agent ^{3}H-BH$_4$ (tritiated borohydride). In addition, we will use the technique of SDS-polyacrylamide gel electrophoresis described in Figure 7-18. Finally, we will need some means of finding out which polypeptide bands on the polyacrylamide gels are radioactively labeled.

LP is useful because it is an enzyme capable of linking radioactive iodine atoms to tyrosine groups on a protein, provided that the reaction is carried out in the presence of hydrogen peroxide, H$_2$O$_2$. This means that a membrane protein can be labeled with ^{125}I, provided only that at least part of the molecule is exposed to the LP enzyme and that one or more tyrosine groups are present in that part of the molecule.

GO and tritiated borohydride are useful for labeling carbohydrate side chains on membrane proteins because GO can oxidize galactose (and galactosamine) groups, which then become labeled with ^{3}H when they are subsequently reduced with tritiated borohydride. Because LP and GO are too large to penetrate into or cross a membrane, they label only those carbohydrate chains or portions of protein molecules that are accessible on the outside of the membrane.

For example, when intact red blood cells are incubated with LP in the presence of ^{125}I and peroxide and the membrane proteins are then extracted and analyzed by SDS-polyacrylamide gel electrophoresis, only a few of the polypeptide bands on the gel can be shown to be radioactive.

This is because the other bands represent polypeptides that do not protrude from the external surface of the cell and therefore are not accessible to the LP enzyme.

Similarly, any carbohydrate chains on the external surface can be labeled by treating intact red blood cells with GO (to oxidize galactose groups), followed by exposure to labeled borohydride (to reduce the galactose groups again, introducing labeled hydrogen atoms in the process). As you might guess from what you already know about the location of glycoproteins in membranes, all of the carbohydrate side chains of the membrane proteins are labeled by this procedure.

For comparison, if you first rupture the red blood cell membrane by subjecting the cell to hypotonic shock, the LP or GO enzymes will have access to both sides of the membrane (because the cell is no longer intact and the enzyme molecules are therefore not excluded from the interior of the cell). Now what you find on electrophoresis is that virtually all of the proteins from the LP-treated membranes are labeled with ^{125}I. This indicates that even the most hydrophobic of membrane proteins have some portion that protrudes from the membrane enough to be accessible to the LP enzyme. On the other hand, GO-borohydride treatment of disrupted membranes labels no more polypeptides than were labeled with intact cells, further strengthening the conclusion that all the carbohydrate side chains are on the external membrane surface.

The cleverest experiments, though, are those designed to label only the proteins (or carbohydrates, if there were any) that are exposed on the inner, but not the outer, surface of the red blood cell membrane. To understand how this might be done, you must know that once the cell is disrupted, the membrane can be fragmented into smaller pieces, which tend to seal again spontaneously, forming empty vesicles. The resealing process turns out to be random with respect to membrane orientation, so some of the vesicles are right-side-out, but others have the membrane inside-out. The two kinds of vesicles are sufficiently different in properties to allow them to be separated, so both kinds can then be treated separately with either LP or GO.

As you might expect, the right-side-out vesicles show the same labeling response as intact cells. More revealing are the inside-out vesicles, because these now allow specific labeling of inside proteins only. It is interesting that almost all the polypeptides on the gels that did not become labeled when intact cells were exposed to LP and radioactive iodine are labeled when inside-out vesicles are treated. Apparently, almost every membrane protein protrudes from the membrane on one surface or the other. In fact, a few of the red blood cell membrane proteins are labeled with ^{125}I whether the inside-out or the right-side-out vesicles are used. What conclusion can we draw from that observation?

From these and similar experiments, a clear picture emerges of the distribution of proteins within the red blood cell membrane. Most of the proteins are associated with the inner lipid layer and contain no carbohydrate chains. All glycoproteins are located in the outer lipid layer, with the carbohydrate side chain invariably protruding into the external environment. Several proteins extend all the way through the membrane and can be labeled with LP on both inner and outer surfaces, as well as with GO and borohydride on the outer surface.

These results were confirmed by another approach that uses the freeze-fracture technique. Imagine that you could somehow shrink to molecular size and watch what happens when a cell is frozen in liquid nitrogen and then broken open to be examined by the freeze-fracture method of electron microscopy. You are near a plasma membrane, and all the molecules have been cooled to a temperature of about −200°C. Instead of the chaotic motion of the fluid state, with molecules rushing around at high velocities, colliding, vibrating, and rotating billions of times per second, every molecule is now fixed in place, shivering a little, and only occasionally changing places with a neighbor. The lipids and proteins of the membrane are held together by hydrogen bonds and electrostatic forces between ions and look just like the static image of the fluid mosaic model. In the interior of the lipid bilayer, only weak van der Waals forces hold lipid tails together.

Suddenly you hear in the distance a loud thundering sound, and a huge metallic blade can be seen approaching, crashing through the ice and frozen cytoplasm. As it passes overhead, an immense crack descends into the cell and breaks open the membrane, easily parting the lipid bilayer along the weakly bonded tails. A brief flash of heat is felt as some of the fracture energy is turned into vibrational molecular energy, and the lipid chains melt and oscillate for a few microseconds like miniature tuning forks. What happens to the membrane proteins? Are they ripped apart by the fracture process, leaving broken peptide bonds in their amino acid chains, or do they simply pop out of the lipid like little corks?

Although we cannot shrink to molecular size to answer this question, it can be approached through biochemical analysis of membrane composition following freezing and fracturing. Knute Fischer, at the University of California, San Francisco, found that red blood cell membranes adhered tightly to a glass surface if it was coated with polylysine, a strongly cationic polymer of the amino acid lysine. The membranes could then be frozen together with a thin layer of water and the glass split away from the ice. This left behind on the glass a layer of half-membranes that resulted from splitting of the frozen preparation, just as occurs in the freeze-fracture method. When the membranes were analyzed by microanalytical techniques for lipids, cholesterol was found to be approximately evenly divided between the inner and outer leaflets of the lipid bilayer. Phosphatidyl choline was mostly in the outer leaflet, while phosphatidyl ethanolamine and phosphatidyl serine composed the inner leaflet. The protein composition was then analyzed by gel electrophoresis. Every protein could be accounted for, even those known to be transmembrane in orientation, and no broken fragments were seen. The surprising conclusion is that proteins in frozen membranes easily break away from the lipid and ice during fracturing. Therefore, the particles observed in freeze-fracture images represent complete protein molecules, probably with some associated lipids, and are not fragments of proteins pulled apart like pieces of taffy.

Once again, it is a judicious combination of clever techniques and ingenious thinking that makes such experiments—and hence such findings—possible. ■

Cells need membranes to define and compartmentalize space, regulate the flow of materials and information, and mediate recognition and interaction between cells. Our current understanding of membrane structure represents the culmination of almost a century of studies, beginning with the recognition that lipids are an important membrane component and moving progressively from a lipid monolayer to a phospholipid bilayer. Once proteins were recognized as important components, Davson and Danielli proposed their "sandwich" model, with the lipid bilayer surrounded on both sides by layers of proteins. Their model was acknowledged as the basis of the unit membrane but was later discredited because it could not account for key features of membrane proteins.

In its place, the fluid mosaic model has emerged as the generally accepted description of membrane structure. According to this model, proteins with varying affinities for the hydrophobic membrane interior float in a sea of phospholipids (and cholesterol, in animal cells). Each of these proteins contains one or more domains of predominantly hydrophobic amino acids that anchor the protein to the membrane. In many cases, these are transmembrane segments, such that portions of the protein are exposed on both sides. Membrane phospholipids and proteins are free to move within the plane of the membrane unless they are specifically anchored to structures beneath the cell. Lipids and proteins are not free to move between the two monolayers of a membrane, however. Once inserted into a monolayer, they remain in that layer. As a result, most membranes are characterized by an asymmetry between the two monolayers.

Membrane proteins function as enzymes, transport molecules, electron carriers, and receptor sites for chemical signals such as neurotransmitters and hormones. Many proteins in the plasma membrane are glycoproteins, with carbohydrate side chains that invariably protrude from the membrane on the external side. These carbohydrate side chains play important roles as recognition markers on the cell surface.

With our understanding of the structure and function of membranes acquired in this chapter, we are now ready to explore the mechanisms by which molecules and ions are transported across membranes. The transport of small molecules and ions across membranes will be discussed in Chapter 8, followed in Chapter 9 by a consideration of the mechanisms whereby membrane-bounded vesicles are involved in the uptake and release of macromolecules and even larger substances.

KEY TERMS FOR SELF-TESTING

membrane (p. 167)

The Functions of Membranes

plasma (cell) membrane (p. 167)
intracellular membrane (p. 167)
marker enzyme (p. 169)
signal transduction (p. 169)
receptor (p. 169)
cell-to-cell communication (p. 169)

Membrane Structure: A Historical Perspective

lipid monolayer (p. 170)
lipid bilayer (p. 170)
differential permeability (p.171)
Davson-Danielli model (p. 171)

unit membrane (p. 172)
fluid mosaic model (p. 173)
integral membrane protein (p. 173)
peripheral membrane protein (p. 173)
transmembrane segment (p. 173)

The Fluid Mosaic Model: Evidence and Implications

erythrocyte (p. 175)
phospholipid (p. 177)
sphingolipid (p. 177)
cholesterol (p. 177)
thin-layer chromatography (TLC) (p. 177)
fluidity (p. 178)
fluorescence recovery after photo-bleaching (p. 178)

transition temperature (p. 178)
phase transition (p. 178)
homeoviscous adaptation (p. 182)
freeze-fracture technique (p. 183)
SDS-polyacrylamide gel electrophoresis (p. 184)
glycoprotein (p. 185)
glycocalyx (p. 186)
cell fusion (p. 188)
fluorescent antibody (p. 189)
patching and capping (p. 189)
membrane asymmetry (p. 190)
transverse diffusion (p. 190)
rotation (of phospholipid molecules) (p. 190)
lateral diffusion (p. 190)

PROBLEM SET

7-1. Functions of Membranes. For each of the following statements, indicate which one of the five general membrane functions (compartmentalization, localization of function, regulation of transport, detection of signals, or cell-to-cell communication) the statement seems to illustrate.

(a) The plasma membrane of a muscle cell is excitable and capable of conducting an action potential.

(b) The membrane of a plant root cell has an ion pump that exchanges phosphate inward for bicarbonate outward.

(c) When cells are disrupted and fractionated into subcellular components, the enzyme cytochrome *c* reductase is recovered with the endoplasmic reticulum fraction.

(d) Cells of multicellular organisms carry tissue-specific glycoproteins on their outer surface that are responsible for cell-cell adhesion.

(e) Membranes are composed primarily of phospholipids and hydrophobic proteins.

(f) Photosystems I and II are embedded in the thylakoid membrane of the chloroplast.

(g) All of the acid phosphatase in a mammalian cell is found within the lysosomes.

(h) The mitochondrial membrane is impermeable to ATP but contains an ATP-ADP carrier that couples outward ATP movement to inward ADP movement.

(i) Insulin does not enter a target cell, but instead binds to a specific membrane receptor on the external surface of the membrane, thereby activating the enzyme adenylate cyclase on the inner membrane surface.

(j) Adjacent plant cells frequently exchange cytoplasmic components through channels called plasmodesmata.

7-2. Elucidation of Membrane Structure. Each of the following observations played an important role in enhancing our understanding of membrane structure. Explain the significance of each, and indicate in what decade in the timeline of Figure 7-3 the observation was most likely made.

(a) When a membrane is observed in the electron microscope, both of the thin, electron-dense lines are about 2 nm thick, but they are often distinctly different from each other in appearance.

(b) Ethylurea penetrates much more readily into a membrane than does urea, and diethylurea penetrates still more readily.

(c) The addition of phospholipase to living cells causes rapid digestion of the lipid bilayers of the membranes, which suggests that the enzyme has ready access to the membrane phospholipids.

(d) When artificial lipid bilayers are subjected to freeze-fracture, no particles are seen on either face.

(e) The electrical resistivity of artificial lipid bilayers is several orders of magnitude greater than that of real membranes.

(f) When artificial lipid bilayers are fixed, stained with osmium, and viewed in the electron microscope, they have the same "railroad track" appearance as natural membranes.

(g) Some membrane proteins can be readily extracted with 1 *M* NaCl, while others require the use of an organic solvent or a detergent.

(h) When halobacteria are grown in the absence of oxygen, they produce a purple pigment that is embedded in their plasma membranes and has the ability to pump protons outward when illuminated. If the purple bacteria membranes are isolated and viewed by freeze-fracture electron microscopy, they are found to contain patches of crystallized particles.

7-3. Gorter and Grendel Revisited. Gorter and Grendel's classic conclusion that the plasma membrane of the human erythrocyte consists of a lipid bilayer was based on the following observations: (i) the lipids that they extracted with acetone from 4.74×10^9 erythrocytes formed a monolayer 0.89 m^2 in area when spread out on a water surface; and (ii) the surface area of one erythrocyte was about 100 μm^2, according to their measurements.

(a) Show from these data how they came to the conclusion that the erythrocyte membrane is a bilayer.

(b) We now know that the surface area of a human erythrocyte is about 145 μm^2. Explain how Gorter and Grendel could have come to the right conclusion when one of their measurements was only about two-thirds of the correct value.

7-4. Lipid, Protein, and Bilayers in Erythrocytes. The membrane of a human erythrocyte has a surface area of about 145 μm^2 and is about 8 nm thick. It contains about 0.52 picogram of lipid and 0.60 picogram of protein. (A picogram is 1×10^{-12} g.) Assume equimolar amounts of cholesterol and phospholipids, with molecular weights of 386 and about 800, respectively. In a monolayer, each phospholipid molecule occupies a surface area of 0.55 nm^2 per molecule, and each cholesterol molecule occupies 0.38 nm^2 per molecule.

(a) Assuming an average molecular weight of 50,000 for membrane proteins, how many protein molecules are there in a single erythrocyte membrane?

(b) What is the ratio of lipid molecules to protein molecules in the erythrocyte membrane?

(c) What proportion of the total surface area of the membrane is occupied by the lipids in the bilayer?

7-5. Lateral and Transverse Diffusion. Lateral diffusion occurs readily and rapidly for the lipids and many of the proteins in a membrane, but neither lipids nor proteins can undergo transverse diffusion very readily.

(a) Explain this difference between lateral and transverse diffusion.

(b) What is the consequence for the cell of free lateral diffusion of most membrane components?

(c) What would be the consequence for the cell if transverse diffusion were also to occur readily?

7-6. Temperature and Membrane Composition. Which of the following responses are likely to be seen when a bacterial culture growing at 37°C is transferred to a culture room maintained at 25°C?

(a) Initial decrease in membrane fluidity.

(b) Gradual replacement of shorter fatty acids by longer fatty acids in the membrane phospholipids.

(c) Gradual replacement of stearate by oleate in the membrane phospholipids.

(d) Enhanced rate of synthesis of unsaturated fatty acids.

(e) Incorporation of more cholesterol into the membrane.

7-7. Glycophorin: The "Sugar Carrier." The word *glycophorin* is derived from Greek roots meaning "to carry sugar." To understand why one of the major integral membrane proteins of the erythrocyte membrane is so named, consider the following:

(a) The protein consists of a single polypeptide with 131 amino acids. Assuming that amino acids have an average molecular weight of 110, what is the molecular weight of the glycophorin polypeptide?

(b) Glycophorin contains 16 carbohydrate side chains, each with three galactose units ($C_6H_{12}O_6$), two mannose units ($C_6H_{12}O_6$), two glucosamine units ($C_6H_{13}O_5N$), and two sialic acid units ($C_{11}H_{19}O_9N$). (Note: One molecule of water is lost as each of these units is linked to the carbohydrate side chain and as the chain is linked to the protein.) Determine the molecular weight of each unit in a carbohydrate side chains, then calculate the total molecular weight of the 16 carbohydrate side chains.

(c) What is the total molecular weight of the glycophorin polypeptide plus its 16 carbohydrate side chains? What percentage of the total molecular weight do the 16 carbohydrate side chains represent?

(d) Can you see why this glycoprotein was given the name glycophorin?

7-8. Membrane Fluidity and Temperature. The effects of temperature and lipid composition on membrane fluidity are often studied by using artificial membranes containing only one or a few kinds of lipids and no proteins. Assume that you and your lab partner have made the following artificial membranes:

Membrane 1: Made entirely from phosphatidyl choline with saturated 16-carbon fatty acids.

Membrane 2: As for membrane 1, except that each of the 16-carbon fatty acids has a single *cis* double bond.

Membrane 3: As for membrane 1, except that each of the saturated fatty acids has only 14 carbon atoms.

After determining the transition temperatures for three samples representing each of the membranes, you discover that your lab partner failed to record the membranes to which the samples correspond. The three values you determined are $-36°$, $23°$, and $41°C$. Can you assign each of these transition temperatures to the correct artificial membrane? Explain your reasoning.

7-9. Membrane Potpourri. Explain each of the following observations as succinctly as possible.

(a) Proteins A and B both have a molecular weight of 33,000 and consist of a single polypeptide chain with 240 hydrophilic amino acids and 60 hydrophobic amino acids, yet protein A is an integral membrane protein and protein B is a soluble protein.

(b) When the photobleaching experiment described in Figure 7-12 is carried out using plasma membranes from myoblasts (precursors to muscle cells) with fluorescence-tagged surface glycoproteins, it takes much longer for the dark, nonfluorescent spot to become fluorescent again than when phospholipids are labeled with a fluorescent dye. Moreover, only 55% of the protein fluorescence eventually returns to the laser-bleached spot.

(c) In the legs of a reindeer, the membranes of cells near the hoof have a higher relative amount of unsaturated fatty acids than the membranes of cells in the rest of the leg.

7-10. Anesthesia and Membrane Fluidity. Many of the substances used as anesthetics in medicine are nonpolar, which makes them soluble in the hydrophobic interior of the membrane and increases the fluidity of lipid bilayers. Suggest how such substances might function in disrupting the transmission of nerve impulses and thereby eliminating sensations of pain.

7-11. Inside or Outside? From Box 7A, we know that exposed regions of membrane proteins can be labeled with ^{125}I by the lactoperoxidase (LP) reaction, whereas carbohydrate side chains of membrane glycoproteins can be labeled with ^3H by oxidation of galactose groups with galactose oxidase (GO), followed by reduction with tritiated borohydride (^3H-BH_4). Noting that both LP and GO are too large to penetrate into the interior of an intact cell, explain each of the following observations made with intact erythrocytes.

(a) When intact cells are incubated with LP in the presence of ^{125}I and peroxide and the membrane proteins are then extracted and analyzed on SDS-polyacrylamide gels, several of the bands on the gel are found to be radioactive.

(b) When intact cells are incubated with GO and then reduced with ^3H-BH_4, several of the bands on the gel are found to be radioactive.

(c) All of the proteins of the plasma membrane that are known to contain carbohydrates are labeled by the GO/^3H-BH_4 method.

(d) None of the proteins of the erythrocyte plasma membrane that are known to be devoid of carbohydrate is labeled by the LP/^{125}I method.

(e) If the erythrocytes are ruptured before the labeling procedure, the LP procedure labels virtually all of the major membrane proteins.

7-12. Inside-Out Membranes. It is technically possible to prepare sealed vesicles from erythrocyte membranes in which the original orientation of the membrane is inverted. Such vesicles have what was originally the cytoplasmic side of the membrane facing outward.

(a) What results would you expect if such inside-out vesicles were subjected to the GO/borohydride procedure described in Problem 7-11?

(b) What results would you expect if such inside-out vesicles were subjected to the LP/^{125}I procedure of Problem 7-11?

(c) What conclusion would you draw if some of the proteins that become labeled by the LP/^{125}I method of part b were among those that had been labeled when intact cells were treated in the same way in part a of Problem 7-11?

7-13. An Inside Job. Knowing that it is possible to prepare inside-out vesicles from erythrocyte plasma membranes as described in Problem 7-12, can you think of a way to label a transmembrane protein with ^3H on one side of the membrane and with ^{125}I on the other side?

SUGGESTED READING

Membrane Structure and Function

Branton, D., and R. Park. *Papers on Biological Membrane Structure.* Boston: Little, Brown, 1968.

Bretscher, M. The molecules of the cell membrane. *Sci. Amer.* 253 (April 1985): 100.

Gennis, R. B. *Biomembranes: Molecular Structure and Function.* New York: Springer-Verlag, 1989.

Houslay, M. D., and K. K. Stanley. *Dynamics of Biological Membranes.* New York: Wiley, 1982.

Jain, M., and R. Wagner. *Introduction to Biological Membranes,* 2d ed. New York: Wiley, 1988.

Kleinfeld, A. Current views of membrane structure. *Curr. Topics Membr. Transp.* 29 (1987): 1.

Lipowsky, R. The conformation of membranes. *Nature* 249 (1991): 475.

Petty, H. R. *Molecular Biology of Membranes.* New York: Plenum, 1993.

Vance, D. E., and J. E. Vance. *Biochemistry of Lipids and Membranes.* Menlo Park, CA: Benjamin/Cummings, 1985.

The Fluid Mosaic Model

Helenius, A., and K. Simons. Solubilization of membranes by detergents. *Biochim. Biophys. Acta* 415 (1975): 29.

Singer, S. M. The molecular organization of membranes. *Annu. Rev. Biochem.* 43 (1974): 805.

Singer, S. J., and G. L. Nicolson. The fluid mosaic model of the structure of cell membranes. *Science* 175 (1972): 720.

Storch, J., and A. M. Kleinfeld. The lipid structure of biological membranes. *Trends Biochem. Sci.* 10 (1985): 418.

Yeagle, P. L. Cholesterol and the cell membrane. *Biochim. Biophys. Acta* 822 (1985): 267.

The Erythrocyte and Its Plasma Membrane

Golan, D. E., and W. Veatch. Lateral mobility of band 3 in the human erythrocyte membrane studied by fluorescence photobleaching recovery: Evidence for control by cytoskeletal interactions. *Proc. Natl. Acad. Sci. USA* 77 (1980): 2537.

Kopito, R. R., and H. F. Lodish. Primary structure and transmembrane orientation of the murine anion exchange protein. *Nature* 316 (1985): 234.

Viitala, J., and J. Jarnefelt. The red cell surface revisited. *Trends Biochem. Sci.* 10 (1985): 392.

Membrane Proteins

Byers, T. J., and D. Branton. Visualization of the protein associations in the erythrocyte membrane skeleton. *Proc. Natl. Acad. Sci. USA* 82 (1985): 6153.

Jennings, M. L. Topology of membrane proteins. *Annu. Rev. Biochem.* 58 (1989): 999.

Khorana, H. G. Bacteriorhodopsin: A membrane protein that uses light to translate photons. *J. Biol. Chem.* 163 (1988): 7439.

Lodish, H. F., and J. E. Rothman. The assembly of cell membranes. *Sci. Amer.* 240 (January 1979): 48.

Paulson, J. C. Glycoproteins: What are the sugar chains for? *Trends Biochem. Sci.* 14 (1989): 272.

Singer, S. J. The structure and insertion of integral proteins in membranes. *Annu. Rev. Cell Biol.* 6 (1990): 247.

Rothman, J. E., and J. Lenard. Membrane asymmetry. *Science* 195 (1977): 743.

Unwin, N., and R. Henderson. The structure of proteins in biological membranes. *Sci. Amer.* 250 (February 1984): 78.

Von Heijne, G. Transcending the impenetrable: How proteins come to terms with membranes. *Biochim. Biophys. Acta* 947 (1988): 307.

Membrane Fluidity

DePetris, S., and M. C. Raff. Normal distribution, patching, and capping of lymphocyte surface immunoglobulin studied by electron microscopy. *Nature New Biol.* 241 (1979): 5163.

Frye, L. D., and M. Edidin. The rapid intermixing of cell surface antigens after formation of mouse-human heterokaryons. *J. Cell Sci.* 7 (1970): 319.

Jacobson, K., E. Elson, D. Koppel, and W. Webb. Fluorescence photobleaching in cell biology. *Nature* 295 (1982): 283.

Jacobson, K., A. Ishihara, and R. Inman. Lateral diffusion of proteins in membranes. *Annu. Rev. Physiol.* 49 (1987): 163.

Lenaz, G. Lipid fluidity and membrane protein dynamics. *Biosci. Sci. Rep.* 7 (1987): 823.

Quinn, P. J., F. Joo, and L. Vigh. The role of unsaturated lipids in membrane structure and stability. *Prog. Biophys. Mol. Biol.* 53 (1989): 71.

Thompson, G. Membrane acclimation by unicellular organisms in response to temperature change. *J. Bioenerg. Biomembr.* 21 (1989): 43.

TRANSPORT ACROSS MEMBRANES: OVERCOMING THE PERMEABILITY BARRIER

In Chapter 7, we focused on membrane structure and chemistry. Membranes were described as barriers to the free diffusion of molecules or ions, thereby keeping desirable substances within the cell and undesirable substances out. Within eukaryotic cells, membranes also define specific subcellular compartments by confining the appropriate molecules, ions, enzymes, and other factors needed for specific functions.

However, it is inadequate to think of membranes simply as permeability barriers. Crucial to the proper functioning of a cell or organelle is the ability to overcome the permeability barrier for specific molecules and ions so that they can be moved into and out of the cell or organelle selectively. Membranes, in other words, not only serve as barriers to the indiscriminate movement of substances into and out of cells and organelles, but must also provide for the controlled passage of specific molecules and ions. In this chapter, we will explore the ways in which substances are moved selectively across membranes and the significance of such transport processes to the life of the cell.

Cells and Transport Processes

An essential feature of every cell and subcellular compartment is the ability to accumulate a variety of substances at concentrations that are often strikingly different from those in the surrounding milieu. Some of these substances are proteins and other macromolecules, which are moved into and out of cells and organelles by mechanisms we will consider in later chapters. Specifically, Chapter 9 includes a discussion of *endocytosis* and *exocytosis*, bulk uptake and transfer processes whereby substances are moved into and out of cells as part of a volume of fluid enclosed within membrane-bounded vesicles. Mechanisms for the *secretion* of proteins from cells and for the *import* of proteins into organelles are discussed in Chapter 18.

As important as these topics are, most of the substances that move across membranes are not macromolecules or fluids but *ions* and *small organic molecules*. Some of the more common ions that are transported across membranes are sodium (Na^+), potassium (K^+), calcium (Ca^{2+}), chloride (Cl^-), and hydrogen ions (H^+). Most of the small organic molecules are *metabolites*—substrates, intermediates, and products in the various metabolic pathways. For most such substances, the concentration inside the cell is markedly higher than outside. In fact, life can be defined as a state not only of improbable structural order but also of improbable concentrations. Very few cellular reactions or processes could occur at reasonable rates if they had to depend on the concentrations at which essential substrates are present in the cell's surroundings.

The importance of solute transport is underscored by human diseases that result from the malfunctioning of specific transport systems. For example, cystic fibrosis is a serious genetic disease characterized by the accumulation in the lungs of unusually thick mucus that often leads to pneumonia and other lung disorders. Until recently, we knew that people with cystic fibrosis have abnormal perspiration but the cause was not known. Now we understand that the underlying problem is an inability to secrete chloride ions and that the actual genetic defect is in a specific plasma membrane protein required for chloride secretion. Recent developments in our understanding of cystic fibrosis at the molecular level are reported in Box 8A (page 200), along with prospects for the use of gene therapy as a treatment.

A central aspect of cell function, then, is **transport:** the ability to move ions and organic molecules across membranes selectively. Figure 8-1 illustrates a variety of such transport processes. Except for a few small or nonpolar molecules, the transport of ions and metabolites across membranes is possible only because of the presence of membrane proteins that can recognize substances with great specificity and mediate their movement across the membrane. It is to

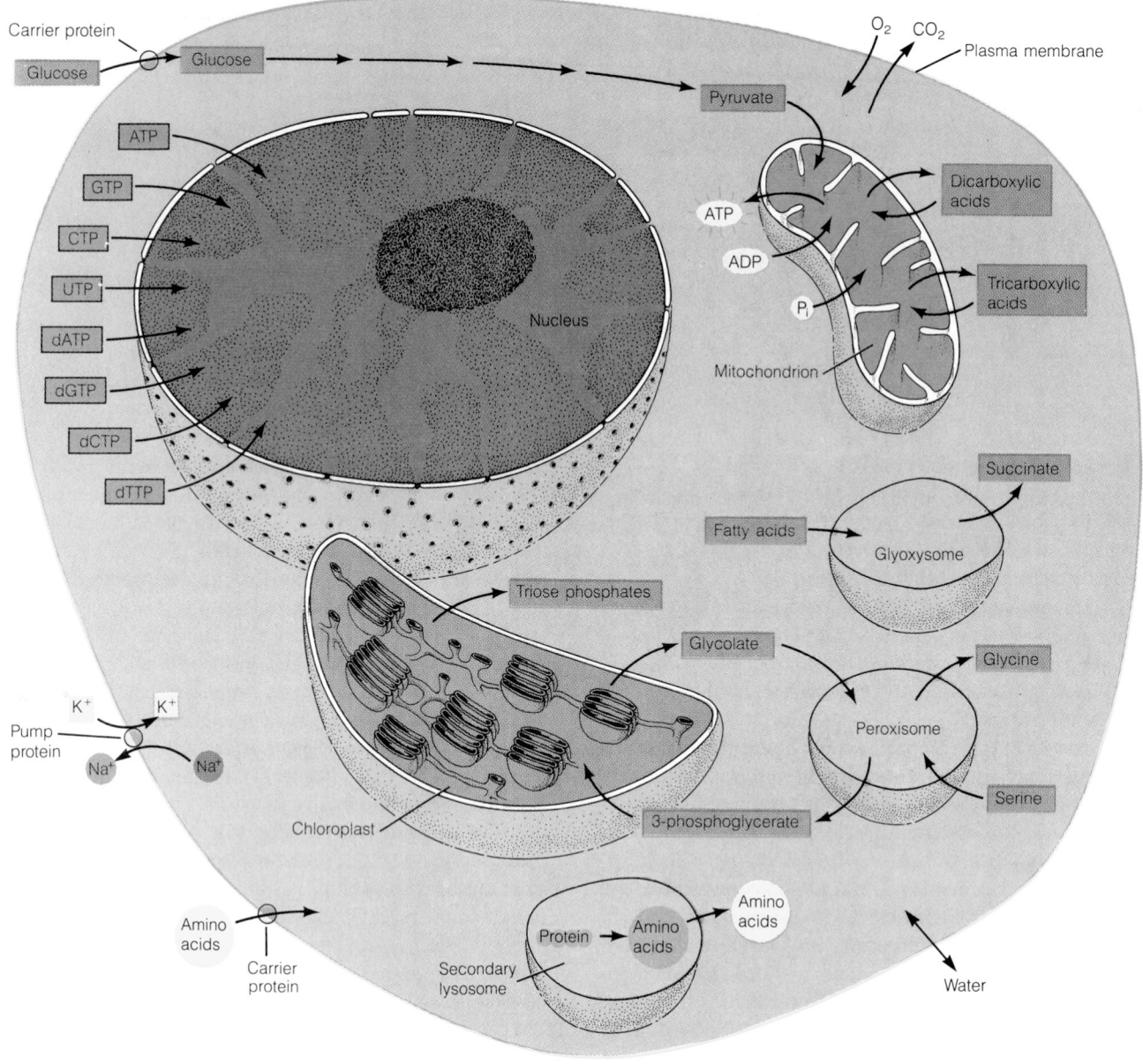

Figure 8-1 Transport Processes Within a Composite Eukaryotic Cell. The substances shown in this composite plant/animal cell are some of the many kinds of molecules and ions that are transported across the membranes of eukaryotic cells. Yellow dots indicate specific transport sites.

these transport processes and the proteins involved that we now turn our attention.

Categories of Transport

When we consider the transport of small organic molecules and inorganic ions across membranes, three categories of transport come readily to mind. The three use similar mechanisms but accomplish different end results.

Cellular transport involves the exchange of materials between the cell and its environment. Included in this category are the uptake of nutrients and other raw materials across the plasma membrane and the removal of wastes and secretory products. The uptake of glucose from the bloodstream into the cells of your body is an example of cellular transport, as is the inward pumping of potassium ions accompanied by the outward pumping of sodium ions that most animal cells carry out continuously. **Intracellular transport,** on the other hand, involves the movement of substances across membranes of organelles within the cell and is therefore a eukaryotic phenomenon. It includes the traffic of molecules and ions into and out of such organelles as the nucleus, mitochondrion, chloroplast, lysosome, peroxisome, Golgi complex, and endoplasmic reticulum.

Transcellular transport, as the name implies, does not simply move a substance into the cell from the outside, but moves it in one side and out the other, thereby accomplishing net transport across the cell. In multicellular organisms, transcellular transport occurs through cell layers that act as semipermeable barriers. Examples include the epithelial cells that line your gastrointestinal tract and the cells of a plant root that are responsible for the absorption of water and mineral salts.

Mechanisms of Membrane Transport

Membranes represent significant barriers to the free movement of *solutes* (i.e., ions or molecules in solution) into and out of cellular compartments. Sometimes the barrier is necessary for cell function, such as the maintenance of ion gradients required for nerve cell function. In other cases, the barrier must be overcome so that a solute can move into or out of the cell or intracellular compartment.

Transport of solutes across a membrane can be either passive or active, depending on the energy requirement of the process. **Passive transport** does not require energy. It occurs because of the tendency for dissolved molecules to move, or *diffuse*, from higher to lower concentrations, as shown in Figure 8-2a. Most passive transport depends on specialized membrane proteins to facilitate the passage of ions or polar molecules across the hydrophobic interior of the membrane. However, some molecules can diffuse across the membrane unaided because they are sufficiently small or nonpolar, or both. Examples of such molecules are propanol, ethanol, urea, CO_2, O_2, and water. Whether unaided or facilitated by a membrane protein, passive transport is always *exergonic* ($\Delta G < 0$).

In contrast, **active transport** always requires the input of energy because such processes are *endergonic* ($\Delta G > 0$) and would not be thermodynamically feasible in the absence

of an energy source. Active transport is catalyzed by specialized membrane proteins that are often referred to as "pumps." Because of the energy input, active transport can result in the generation of a concentration gradient across the membrane (Figure 8-2b).

Active transport of ions is an especially important process. Recall from Chapter 5 that more than two-thirds of the energy your body expends in the resting state is used to maintain gradients of ions such as H^+, K^+, Na^+, and Ca^{2+} across the membranes of your cells. Because ions have an electrical charge, the transport process and resulting concentration gradient can produce an electrical voltage, or *membrane potential*, across the membrane. The combined concentration gradient and associated membrane potential are referred to as an *electrochemical gradient*. The stored energy of such gradients is used to drive a variety of processes in cells. In mitochondria and chloroplasts, for example, the stored energy of an electrochemical gradient of protons is used to drive the synthesis of ATP during the processes of cellular respiration (see Chapter 12) and photosynthesis (see Chapter 13), respectively. In nerve cells, gradients of K^+ and Na^+ are responsible for the action potential of nerve cell membranes and for the transmission of nerve impulses, topics we will explore in Chapter 22.

With these concepts in mind, we are now ready for a more detailed discussion of passive and active transport processes. As in Chapter 7, we will use the erythrocyte, or red blood cell, as an example at several points. Vital to the erythrocyte's role in providing oxygen to body tissues is the permeability of its plasma membrane to O_2 and CO_2—and to bicarbonate ion (HCO_3^-), as it turns out—as well as to glucose, which serves as its energy source. These transport processes are depicted in Figure 8-3 (page 202) and described in the next section.

Figure 8-2 Passive and Active Transport. Passive and active transport processes differ in their effects on concentration gradients. **(a)** Passive transport moves *toward* equilibrium, driven by the diffusion of a solute *down* its concentration gradient. **(b)** Active transport moves *away* from equilibrium, using energy to pump solute *up* its concentration gradient.

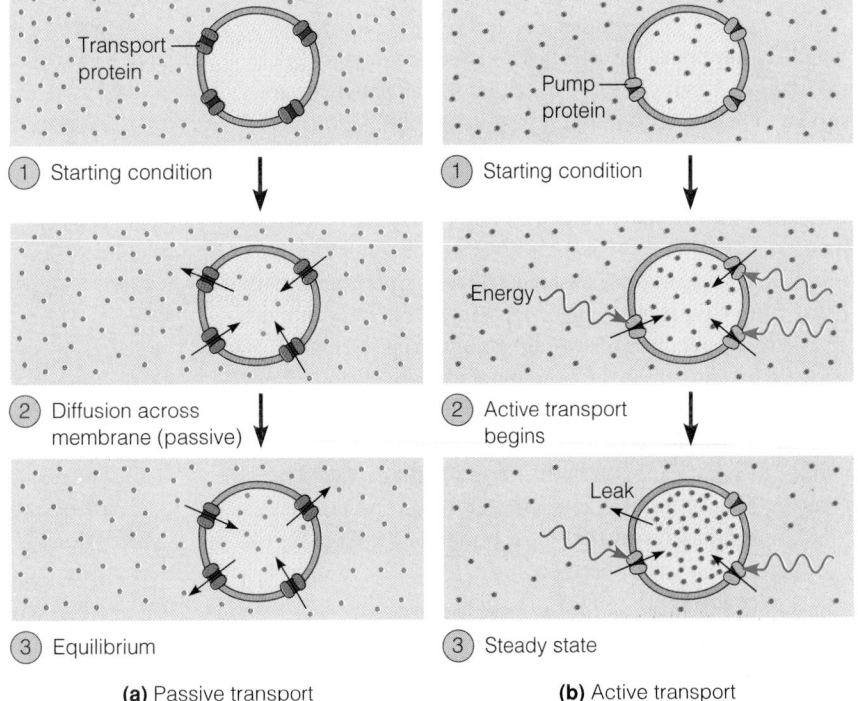

(1) Starting condition

(2) Diffusion across membrane (passive)

(3) Equilibrium

(a) Passive transport

(1) Starting condition

(2) Active transport begins

(3) Steady state

(b) Active transport

8A

Clinical Applications

MEMBRANE TRANSPORT, CYSTIC FIBROSIS, AND THE PROSPECTS FOR GENE THERAPY

Transport proteins located in the plasma membrane play critical roles in facilitating and controlling the movement of molecules and ions into and out of cells. To remain healthy, our bodies depend on the proper functioning of many such membrane proteins. If any of these proteins is defective, the movement of a particular ion or molecule across cell membranes is likely to be impaired, and disease may result.

An example that has attracted the attention of researchers and doctors alike is **cystic fibrosis (CF),** a disease that has recently been shown to result from genetic defects in a transport protein in the plasma membrane. Cystic fibrosis is a common genetic disease that typically affects children and young adults. The parts of the body that are most noticeably affected are the lungs, pancreas, and sweat glands. Complications in the lungs are the most severe medical problems because they are difficult to treat and can become life-threatening. The airways of a CF patient are often obstructed with abnormally thick mucus and are vulnerable to chronic bacterial infections.

Using the tools of molecular and cellular biology, researchers have recently achieved a better understanding of this disease. During the 1980s, cells from CF patients were shown to be defective in the secretion of chloride ions (Cl^-). The cells that line the lungs of people without CF secrete Cl^- ions in response to a substance called *cyclic AMP*, whereas cells from CF patients do not. (Cyclic AMP is a form of AMP that is involved in a variety of regulatory roles in cells; for its structure, see Figure 23-8.) Experiments with tissue from CF patients suggested that this difference might be due to a defect in a membrane protein that normally serves as a channel for the movement of chloride ions across the membrane.

Many symptoms of CF can be explained by faulty Cl^- secretion (Figure 8A-1a). In the lungs of an unaffected person (top panel), chloride ions are secreted from the cells that line the airways and enter the lumen of the passage normally. (A lumen is the space inside a passage or duct.) The movement of Cl^- out of the cell and into the lumen provides the driving force for the concurrent movement of sodium ions into the lumen. Osmotic pressure causes water

to follow the sodium and chloride ions, resulting in the secretion of a dilute salt solution. The water that moves into the lumen in this way provides vital hydration to the mucus lining of the air passages. In the cells of a person with cystic fibrosis, Cl^- ions cannot exit into the lumen, so sodium ions and water do not either (bottom panel). As a result, the mucus is insufficiently hydrated, a condition that favors bacterial growth. Despite this understanding of the link between chloride ion transport and cystic fibrosis, however, scientists were unable to isolate the Cl^- channel protein associated with CF based on their knowledge of its function.

An exciting breakthrough in CF research came in 1989 when investigators in the laboratories of Francis Collins at the University of Michigan and of Lap-Chee Tsui and John Riordan at the University of Toronto isolated the gene that is defective in CF patients. The gene encodes a protein called the **cystic fibrosis transmembrane conductance regulator (CFTR).** The sequence of nucleotides in the gene was determined using methods that are described in Chapter 14 (see Figure 14-14). Knowing the nucleotide sequence of the gene, scientists were able to predict the amino acid sequence and the structure of the CFTR protein. As shown in Figure 8A-1b, the protein is thought to have two sets of *transmembrane domains* that anchor the protein in the plasma membrane and two *nucleotide binding folds* that serve as binding sites for ATP, which provides the energy to drive transport of chloride ions across the membrane. In addition, the protein has a large cytoplasmic domain called the *regulatory domain*, which has several serine hydroxyl groups that can be reversibly phosphorylated. The CFTR protein has since been shown to function as a chloride channel in cells, and channel function is known to be affected when the phosphorylation sites in the regulatory domain are changed as a result of a mutation in the CFTR gene.

By sequencing the CFTR genes from CF patients, investigators quickly determined that the most common mutation causes the deletion of a single amino acid in the first nucleotide binding fold. The question of how this mutation causes CF remained unanswered until researchers examined the location of CFTR in cells with and without the

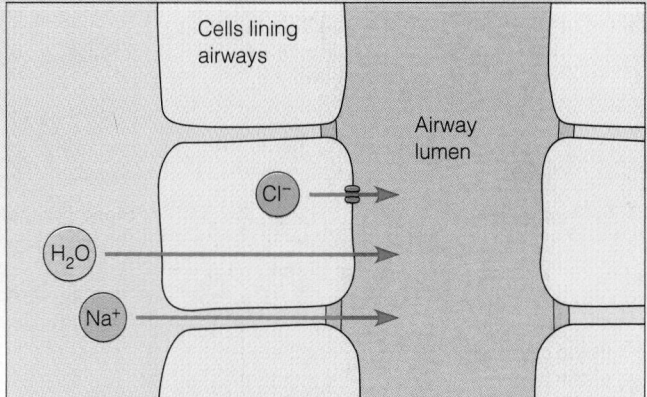

Normal cells lining airways

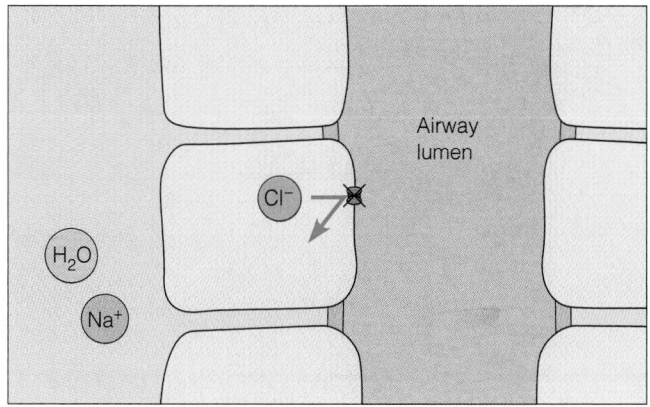

Cells of a person with cystic fibrosis

(a) Normal and faulty chloride ion secretion

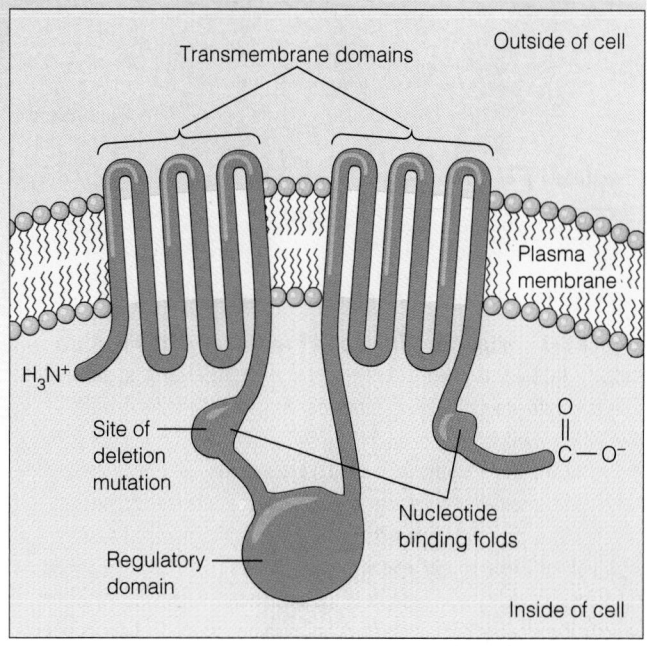

(b) CFTR protein

Figure 8A-1   Cystic Fibrosis and Chloride Ion Secretion.
(a) Cystic fibrosis is caused by a defect in the secretion of chloride ions in cells lining the lungs, leading to insufficient hydration and the promotion of bacterial growth. **(b)** The cystic fibrosis transmembrane conductance regulator (CFTR) is an integral membrane protein that functions as a chloride ion channel. The most common mutation found in cystic fibrosis patients causes the deletion of a single amino acid in one of the two nucleotide binding folds of the CFTR protein.

mutation. Normal CFTR was found in the cell membrane, as predicted. In contrast, mutant CFTR was not detected in the plasma membrane.

The most likely explanation at present is that normal CFTR is synthesized on the rough endoplasmic reticulum (ER), moves through the Golgi complex, and is eventually inserted in the plasma membrane by a route that is explained in the next chapter. Mutant CFTR, on the other hand, is apparently trapped in the ER, perhaps because it is folded improperly. It is therefore recognized as a defective protein and degraded. Consequently, CFTR is not present in the plasma membrane of CF cells, chloride ion secretion cannot take place, and disease results.

Armed with information about the gene and the protein associated with CF, researchers are now actively trying to develop new treatments or perhaps even a cure for the disease. A promising approach is **gene therapy,** in which a normal copy of a gene is introduced into affected cells of the body—lung cells, in the case of CF. Investigators would like to direct normal copies of the CFTR gene into the cells that line the airways of CF patients. These cells should then be able to synthesize a correct CFTR protein, which, unlike mutant CFTR, would be located in the plasma membrane, thereby allowing proper Cl⁻ secretion and correcting the disease. No such cure is currently available, but experiments and clinical trials are under way. ■

This box was written by Lisa S. Smit, University of Michigan.

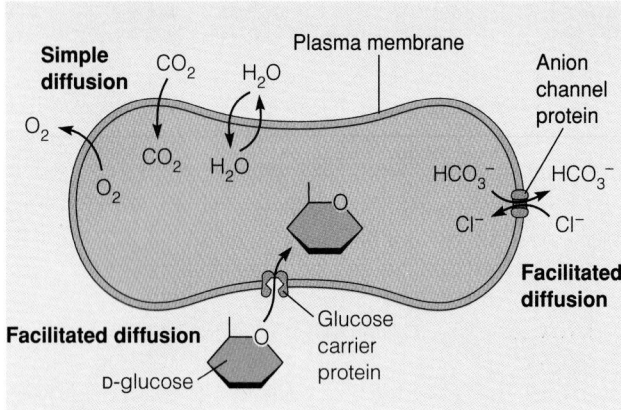

(a) Basic transport processes in an erythrocyte

Figure 8-3   Important Transport Processes of the Erythrocyte.   (a) Depicted here are several transport processes that are vital to erythrocyte function. *Simple diffusion:* Oxygen and carbon dioxide cross the plasma membrane by simple diffusion in directions dictated by their concentrations on the inside and outside of the cell. *Facilitated diffusion of glucose:* The movement of glucose across the membrane is facilitated by a transport protein that recognizes only D-glucose and the D-isomers of a few closely related sugars. *Facilitated diffusion of anions:* An anion channel protein present in the plasma membrane facilitates the reciprocal exchange of chloride (Cl^-) and bicarbonate (HCO_3^-) ions across the membrane. **(b)** The directions in which oxygen, carbon dioxide, and bicarbonate ions move across the plasma membrane of the erythrocyte depend on the location of the erythrocyte in the body. Top: In the capillaries of body tissues, where the concentration of CO_2 is high and that of O_2 is low relative to the concentrations of these gases within the erythrocytes, O_2 is released by hemoglobin and diffuses outward to meet tissue needs. CO_2 diffuses inward and is converted to bicarbonate by carbonic anhydrase, a cytosolic enzyme. Bicarbonate ions are transported outward by the anion channel protein, balanced by an inward movement of chloride ions to maintain charge balance. Carbon dioxide therefore returns to the lungs as bicarbonate ions in both the blood plasma and the cytoplasm of the erythrocytes. Bottom: In the capillaries of the lungs, where the concentration of O_2 is high and the concentration of CO_2 is low, O_2 diffuses inward and binds to hemoglobin. Bicarbonate moves inward from the blood plasma, accompanied by an outward movement of chloride ions. Incoming bicarbonate is converted into CO_2, which diffuses out of the erythrocytes and into the cells that line the capillaries of the lungs.

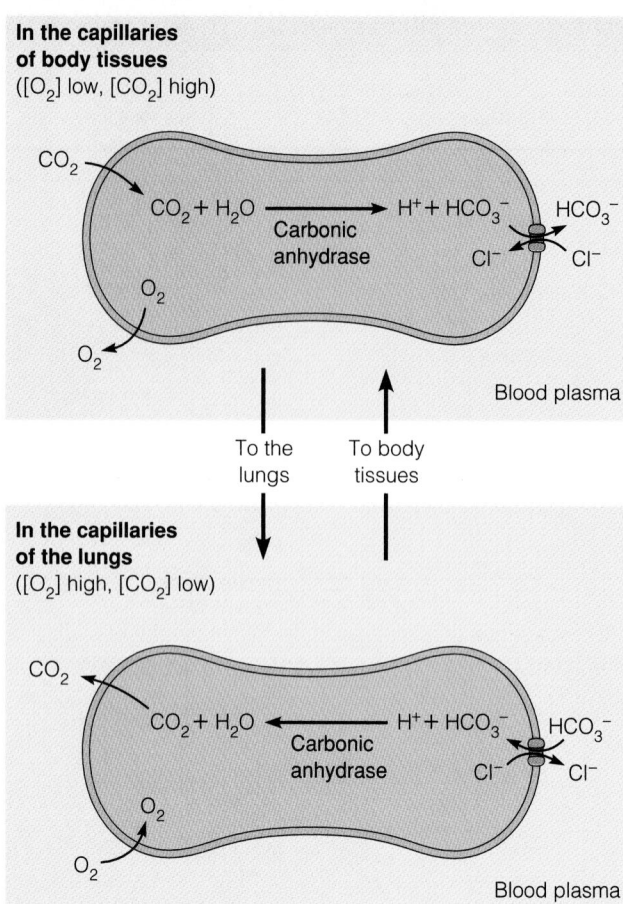

(b) Erythrocyte transport processes in the body

Passive Transport: Simple and Facilitated Diffusion

Passive transport is very common in cells; many molecules and ions move exergonically across membranes into and out of cells and organelles. Passive transport can be thought of as a diffusional process, in which a flux, or flow, of molecules or ions occurs continuously across a membrane in both directions, with net movement in the direction dictated by the concentration (or, for ions, the electrochemical) gradient.

Some substances are small enough or nonpolar enough to cross membranes unaided, a process called **simple diffusion.** In simple diffusion, a small, relatively nonpolar molecule that is in aqueous solution on one side of the membrane simply permeates, or moves into, the phospholipid bilayer, diffuses passively across the bilayer, and emerges on the other side, again in aqueous solution.

Oxygen is an example of a molecule that moves into and out of cells by simple diffusion (Figure 8-3a). This property of oxygen enables erythrocytes in the circulatory system to take up oxygen in the lungs and release it again in body tissues without the involvement of membrane proteins (Figure 8-3b). In the capillaries of body tissues, where the concentration of oxygen is low, oxygen is released from hemoglobin and diffuses passively from the cytoplasm of the erythrocytes into the blood plasma and from there into the cells that line the capillaries (Figure 8-3b, top panel). In the lungs, the opposite occurs: Oxygen diffuses from the inhaled air in the lungs, where its concentration is higher, into the cytoplasm of the erythrocytes, where its concentration is lower (Figure 8-3b, bottom panel). Carbon dioxide is also able to cross membranes by simple diffusion; not surprisingly, its movement across the erythrocyte membrane is reciprocal to that of oxygen.

Most substances in cells are too large or too polar to cross the membrane by simple diffusion, however. They can move into and out of cells and organelles only with the assistance of specialized membrane proteins. This process is called **facilitated diffusion** because the solute still diffuses from higher to lower concentration but its movement is facilitated by a specific *transport protein* in the membrane. As an example of facilitated diffusion, consider the movement of glucose across the plasma membrane of an erythrocyte (or almost any other cell in your body, for that matter). The concentration of glucose is higher in the blood than in the erythrocyte, so the transport of glucose across the plasma membrane of the cell is passive. However, glucose is too large and too polar to diffuse across the membrane unaided; a transport protein is required to facilitate its inward or outward movement (see Figure 8-3a).

We will discuss facilitated diffusion in some detail because of its importance to cells. First, however, we will consider simple diffusion, focusing on membrane permeability and diffusion rates.

Simple Diffusion and Membrane Permeability

Simple diffusion across a membrane is only possible for solutes that are readily permeable in the membrane. To determine what governs membrane permeability and how permeable a membrane is to different solutes, scientists frequently use membrane models. An important advance in understanding the barrier properties of membranes came in 1965, when Alec Bangham and his colleagues reported that membrane lipids could be extracted and dispersed as **liposomes,** small vesicles about 0.1 μm in diameter consisting of closed spherical lipid bilayers. Bangham found that it was possible to trap solutes such as potassium ions in the liposomes as they form, and then measure the rate at which the solute escapes by diffusion across the liposome bilayers.

The results were remarkable: Ions such as potassium and sodium were trapped in the vesicles for days, whereas small molecules such as water exchanged so rapidly that the rates could hardly be measured. The inescapable conclusion was that lipid bilayers represent the primary permeability barrier of all membranes. Small molecules such as water could pass through the barrier by simple diffusion, but sodium and potassium ions could hardly pass at all.

Over the years, this type of experiment has been repeated for a variety of lipid bilayer systems, both synthetic and natural, and with thousands of different solutes. From these studies, several general rules have emerged that let us predict with considerable confidence how permeable a membrane will be to various solutes. Table 8-1 summarizes these rules and gives examples of solutes to which bilayers are relatively permeable and relatively impermeable. The rules suggest that three primary properties of solutes must be taken into consideration: relative size, relative polarity, and whether the solute is an ion.

Relative Size. Generally speaking, lipid bilayers are more permeable to smaller molecules than to larger molecules. The smallest molecules relevant to cell function are water, oxygen, and carbon dioxide. Membranes are sufficiently permeable to these molecules that no specialized transport processes are required to get them into and out of cells. Even these small molecules do not move across membranes freely, however. Water molecules, for example, diffuse across a bilayer 10,000 times slower than they move when allowed to diffuse freely in the absence of a membrane! The best way to think about the passive transport of small molecules is that their diffusion is strongly hindered by the presence of a lipid bilayer but that they occasionally permeate the bilayer and diffuse randomly to the other side, where they leave the membrane again.

The size rule holds for molecules up to about the size of glucose ($C_6H_{12}O_6$; 180 daltons). Glycerol ($C_3H_8O_3$; 92 daltons) and ethanol (CH_3CH_2OH; 46 daltons) are able to diffuse across membranes at reasonable rates but glucose is not. Most cells therefore include specialized proteins in their plasma membranes to facilitate the entry of glucose and other large nutrient molecules, as we will see in the next section.

Polarity. In general, lipid bilayers are relatively permeable to nonpolar molecules and less permeable to polar molecules. This difference is because nonpolar molecules dissolve more readily in the hydrophobic phase of the lipid bilayer, which increases their probability of crossing the membrane barrier. As discussed in Chapter 7, the relationship that Overton discovered between the lipid solubility of various solutes and their membrane permeability was one of the

Table 8-1 Factors Governing Diffusion Across Lipid Bilayers

	Examples		
Rule	*More Permeable*	*Less Permeable*	**Permeability Ratio***
1. Size rule: bilayer more permeable to smaller molecules	H_2O (Water)	$H_2N — CO — NH_2$ (Urea)	10^2:1
2. Polarity rule: bilayer more permeable to nonpolar molecules	$CH_3 — CH_2 — CH_2 — OH$ (Propanol)	$HO — CH_2 — CHOH — CH_2 — OH$ (Glycerol)	10^3:1
3. Ionic rule: bilayer highly impermeable to ions	O_2 (Oxygen)	OH^- (Hydroxide ion)	10^3:1

*Ratio of diffusion rate for the more permeable solute to the less permeable solute.

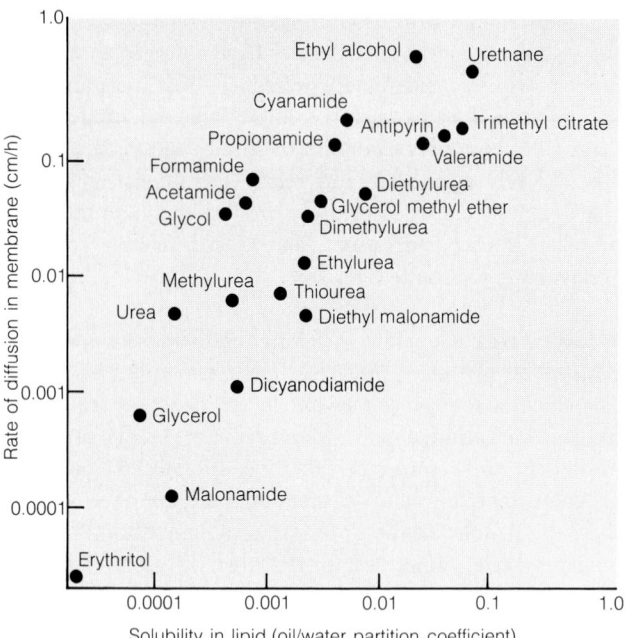

Figure 8-4   The Relationship Between Hydrophobicity and Rate of Diffusion Across a Membrane.   The rates at which various solutes cross the plasma membrane of the alga *Chara* are plotted against the hydrophobicity of the solutes. A logarithmic scale is used for both axes to accommodate the range of values. The *x*-axis is K, the oil-water partition coefficient, defined as the ratio of the equilibrium concentrations of solute X in olive oil versus water (K = $[X]_{eq,oil}/[X]_{eq,water}$). The *y*-axis is the diffusion rate of the solutes in the plasma membrane of *Chara*, or, equivalently, the rate at which the solute enters the algal cell. The diffusion rate is expressed in cm/hour (1cm/h = 2.78 μm/sec).

first indications that a nonpolar hydrophobic phase might be present in membranes.

Figure 8-4 illustrates the relationship between the hydrophobicity of a solute as measured by its relative solubility in a nonpolar solvent (olive oil) and its rate of diffusion across a membrane. In general, the more hydrophobic, or nonpolar, a substance is, the more readily and rapidly it can move across a membrane. Note that a polar compound like malonamide has a relatively low membrane permeability. However, if it is made less polar by adding ethyl groups to form diethyl malonamide, both its solubility in lipid and its membrane permeability increase. A similar effect can be seen as ethyl groups are added to urea to generate ethylurea and diethylurea.

Ion Permeability.   As a general rule, lipid bilayers are not very permeable to ions. This is because a great deal of energy (about 40 kcal/mol) is required to move ions from an aqueous environment into a nonpolar environment. The impermeability of membranes to ions is very important to cell activity, because nerve cells, mitochondria, and chloroplasts must maintain large ion gradients in order to function. On the other hand, membranes must also allow ions to cross the barrier under certain circumstances, as in the conduction of nerve cell action potentials. As we shall see, this task is carried out by special proteins embedded in the lipid bilayer.

Simple Diffusion: Direction and Rate

So far, we have provided only a qualitative understanding of simple diffusion. We can be more quantitative by considering the thermodynamic and kinetic properties of the process (Table 8-2). Thermodynamically, simple diffusion is always an exergonic process, requiring no input of metabolic energy. Individual molecules diffuse randomly in both directions, but net flux is always down the concentration gradient for the molecule in question.

Kinetically, a key feature of simple diffusion is that the net rate of transport for a specific substance is directly proportional to the concentration difference for that substance across the membrane over a broad range of concentrations. For the diffusion of substance X from the outside to the inside of a cell or organelle, the expression for the rate, or velocity *v*, of diffusion through the membrane is:

$$v_{inward} = P([X]_{outside} - [X]_{inside})\qquad(8\text{-}1)$$

where v_{inward} is the rate of inward diffusion (in moles per second per cm^2 of membrane surface) while $[X]_{inside}$ and $[X]_{outside}$ are the concentrations of substance X on the inside and outside of the cell, respectively. P is the *permeability coefficient*, an experimentally determined parameter that depends on the thickness and viscosity of the membrane, the size, shape, and polarity of X, and the solubility of X in the membrane.* Table 8-3 lists permeability coefficients for several ions and molecules in the plasma membrane of the human erythrocyte and in an artificial membrane.

As equation 8-1 indicates, simple diffusion is characterized by a linear relationship between the rate of diffusion of the solute across the membrane and the concentration gradient, with no evidence of saturation at high concentrations (Figure 8-5, black line; page 206). Simple diffusion differs in this respect from facilitated diffusion, which is subject to saturation and generally follows Michaelis-Menten kinetics, as we will see shortly. Simple diffusion can therefore be distinguished from facilitated diffusion on a kinetic basis, as indicated in Table 8-2.

We can summarize simple diffusion by noting that it applies only to molecules such as ethanol, O_2, and CO_2, which are small enough and/or nonpolar enough to cross membranes at a reasonable rate without the aid of a transport protein. Simple diffusion always proceeds exergonically in the direction dictated by the concentration gradient, with a linear, nonsaturating relationship between the diffusion rate and the concentration gradient.

*Equation 8-1 is a modified version of *Fick's first law of diffusion*, which is usually written as J = $-D\Delta C/\Delta x$, where J is the flux density (flux per unit area), D is the diffusion constant (usually expressed as cm^2/sec), ΔC is the difference in concentration between two regions separated by a distance Δx (the difference in concentration across a membrane of thickness Δx, in the case of membrane transport). The permeability coefficient, P, is related to D for the diffusion of substance X in the lipid bilayer by the expression P = KD/L, where L is the thickness of the membrane and K is the partition coefficient for X between oil and water, as defined in the legend for Figure 8-4.

Table 8-2 Properties of Passive and Active Transport

Properties		Passive Transport		Active Transport
		Diffusion	*Facilitated*	
Solutes transported	Examples			
Small nonpolar	Oxygen	Yes	No	No
Large nonpolar	Fatty acids	Yes	No	No
Small polar	Water	Yes	No	No
Large polar	Glucose	No	Yes	Yes
Ions	N^+, K^+, Ca^{2+}	No	Yes	Yes
Thermodynamic properties				
Direction relative to electrochemical gradient		Down	Down	Up
Effect on entropy		Increased	Increased	Decreased
Metabolic energy required		No	No	Yes
Intrinsic directionality		No	No	Yes
Kinetic properties				
Carrier-mediated		No	Yes	Yes (pump)
Michaelis-Menten kinetics		No	Yes	Yes
Competitive inhibition		No	Yes	Yes

Facilitated Diffusion

Simple diffusion explains how oxygen and other small non-polar molecules cross the lipid bilayer of membranes. However, most substances required by cells are polar or charged and move across membranes at appreciable rates only if cells and organelles have specific means of facilitating that movement. Facilitated diffusion therefore differs from simple diffusion in that specific **transport proteins** are required to effect the passage of particular molecules or ions across an otherwise impermeable membrane. Facilitated diffusion uses the same driving force as simple diffusion: the movement of solutes from a region of higher concentration to a region of lower concentration. The role of transport proteins is simply to facilitate the diffusion of polar or charged solutes across an otherwise impermeable barrier.

Transport proteins are invariably integral membrane proteins embedded in the membranes of cells and organelles. These proteins have at least two different *domains*, one to anchor the protein within the appropriate membrane and the other to recognize, bind, and translocate the solute.

Many transport proteins have additional domains as well, serving such diverse functions as interaction with cytoskeletal proteins and the recognition of regulatory molecules.

Transport proteins are sometimes also called **permeases.** This term is especially apt because the suffix *-ase* suggests a similarity between transport proteins and enzymes that turns out to be valid. Like an enzyme-catalyzed reaction, the process of facilitated diffusion always involves an initial binding of the solute to a specific site on a protein surface, a process mediated on the protein surface, a subsequent release of product, and a reduction in the activation energy of the reaction due to the involvement of the protein catalyst. Another property that transport proteins share with enzymes is specificity. Like enzymes, transport proteins are highly specific, often for a single compound or small groups of closely related compounds—and sometimes even for a specific stereoisomer.

As might be expected from the analogy with enzymes, permeases become saturated as the concentration of the transportable solute is raised, because the number of transport proteins is limited and each has some finite maximum

Table 8-3 Permeability Coefficients for Some Ions and Molecules

Membrane	Permeability Coefficient (cm/second)				
	K^+	*Na^+*	*Cl^-*	*Glucose*	*Water*
Human erythrocyte (plasma membrane)	2.4×10^{-10}	1.0×10^{-10}	1.4×10^{-4}*	2×10^{-5}*	5×10^{-3}
Phosphatidyl serine (artificial membrane)	$<9 \times 10^{-13}$	$<2 \times 10^{-13}$	1.5×10^{-11}	4×10^{-10}	5×10^{-3}

*Chloride ions and glucose are transported into the erythrocyte by facilitated diffusion due to the presence of the appropriate transport proteins in the plasma membrane; all other molecules and ions move by simple diffusion only.

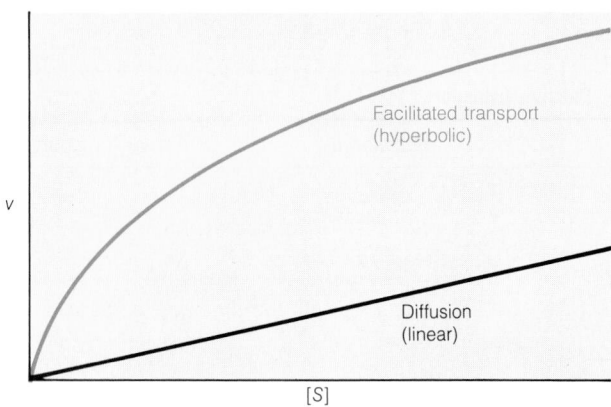

Figure 8-5 Comparison of the Kinetics of Simple Diffusion and Facilitated Diffusion. For simple diffusion of solutes across a membrane, the relationship between v, the rate of diffusion, and the solute concentration [S] is linear (black line). For facilitated diffusion, the relationship is hyperbolic, in accord with Michaelis-Menten kinetics (blue line). For simplicity, the initial solute concentration is assumed to be [S] on one side of the membrane and zero on the other side.

velocity at which it can function. As a result, permease-facilitated transport, like enzyme catalysis, follows Michaelis-Menten kinetics, with an upper limiting velocity and a Michaelis constant corresponding to the concentration of transportable solute needed to achieve one-half of the maximum rate of transport. A plot of transport rate versus solute concentration is therefore hyperbolic for facilitated diffusion instead of linear as for simple diffusion (see Figure 8-5, blue line). As we have already noted, this difference is an important means of distinguishing between simple diffusion and facilitated diffusion (see Table 8-2).

An example of facilitated diffusion is the transport of glucose from the blood into erythrocytes (see Figure 8-3a). Like many other cells in the body, erythrocytes can obtain the glucose they need only because of the presence of glucose transport proteins in the plasma membrane of each cell.

The Mechanism of Transport Protein Function

Transport proteins function by binding to the desired solute molecules in such a way as to shield the polar groups of the solute from the nonpolar interior of the membranes. As you might expect, transport proteins are an important topic of contemporary research. In no case, however, is the actual mechanism of transport known. At one time, it was thought that such proteins functioned as *carriers*, binding the solute on one side of the membrane and then either diffusing across the membrane or flipping within it to bring the solute to the other side of the membrane, where release could occur. Most membrane biochemists agree that these mechanisms are not tenable in light of what we now know about membrane chemistry. Your own understanding of membrane asymmetry from Chapter 7 should enable you to appreciate how unlikely such transverse movements would be.

Instead, transport proteins are thought to be transmembrane proteins that form solute-specific hydrophilic channels through the membrane, thereby allowing specific polar or charged solutes to move from one side to another. The hydrophilic channel is formed as folding of the polypeptide(s) of the transport protein brings hydrophilic amino acids together such that they line a pore through the protein. The channel accounts not only for the ability of the protein to transport the solute across the membrane but also for the specificity of the process.

The Alternating Conformation Model

The molecular mechanism for facilitated diffusion is not yet known. One possibility is the **alternating conformation model** proposed by S. Jonathan Singer. According to this model, a transport protein alternates between two *conformational states*, such that the solute binding site of the protein is open or accessible first to one side of the membrane and then to the other. Binding of a solute molecule or ion to the protein on one side of the membrane triggers a conformational change in the protein that transfers the solute across the membrane, thereby opening the binding site to the other side of the membrane.

Figure 8-6 illustrates this model using the inward transport of glucose as an example. The cycle begins when a molecule of glucose collides with the binding site of a transport protein that is open to the outside of the membrane. Binding of glucose causes the protein to shift to its alternative conformation, with the binding site now facing the other side of the membrane. The conformational change facilitates the release of glucose on the other side of the membrane. Upon release of the solute, the protein returns to its original conformation, with the binding site again facing outward.

The example shown in Figure 8-6 is for inward transport, but the process is readily reversible because the transport protein is presumed to function equally well in both directions. A transport protein is really just a gate in an otherwise impenetrable wall, and, like most gates, it facilitates traffic in either direction. Individual solute molecules may be transported either inward or outward, with the net direction of solute movement through the membrane determined by the relative concentrations of the solute on the two sides of the membrane. If the concentration is higher outside, net flow will be inward; if the higher concentration exists inside, net flow will be outward.

Specificity of Transport Proteins

An important property of transport proteins that is only implied by Figure 8-6 is their *specificity*. Like enzymes, transport proteins are specific for only one or a few structurally related solutes. For example, the transport protein that facilitates the movement of glucose into erythrocytes recognizes no other substances except for a few closely related monosaccharides, including galactose and mannose. Moreover,

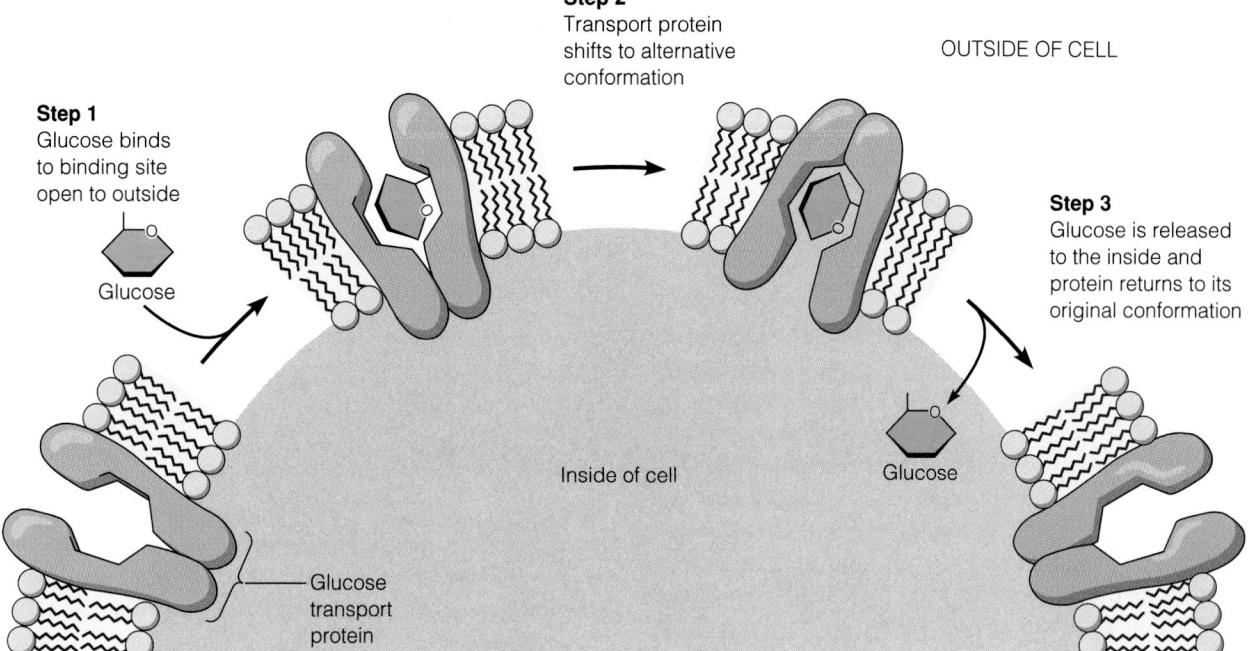

Step 1
Glucose binds to binding site open to outside

Glucose

Step 2
Transport protein shifts to alternative conformation

OUTSIDE OF CELL

Step 3
Glucose is released to the inside and protein returns to its original conformation

Inside of cell

Glucose

Glucose transport protein

Figure 8-6 The Alternating Conformation Model for Facilitated Diffusion.   According to this model, facilitated diffusion involves a transmembrane protein that provides a hydrophilic channel for a specific polar solute and is capable of alternating between two conformations, one open to the outside of the cell and the other to the inside. The transport process for the glucose transport protein of the erythrocyte plasma membrane is shown here in three steps, arranged around the periphery of a cell. ① With the transport protein in the conformation that exposes the binding site to the outside of the cell, a molecule of D-glucose collides with and binds to the binding site. ② Binding of glucose causes the transport protein to shift to its alternative conformation, with the binding site now open to the inside of the cell. ③ As glucose is released from the binding site to the inside, the transport protein reverts to its original conformation, ready for a further transport cycle.

the protein is *stereospecific:* It accepts the D- but not the L- isomers of these sugars. This specificity is presumably a result of the precise stereochemical fit between the solute and its binding site on the transport protein.

A further similarity with enzymes is that transport proteins are often subject to *competitive inhibition* by molecules or ions that are structurally related to the intended "substrate." In the case of glucose transport, for example, the activity of the transport protein toward glucose is competitively inhibited by the alternative monosaccharides that it also accepts.

Ionophores as Models of Transport Proteins

Additional insights into the nature of membrane transport have come from studies with **ionophores**, substances that greatly increase the permeability of membranes for specific ions. (Most ionophores are antibiotics produced by bacteria to kill other microorganisms. They do so by making the membranes of the target organism so permeable, or leaky, to ions that it becomes impossible for the organism to maintain the gradients of H^+, Na^+, K^+, and other ions that are essential to life.)

Some ionophores are *channel formers,* providing a hydrophilic channel through an otherwise hydrophobic membrane. Such ionophores are of great interest as models of the transport proteins involved in facilitated diffusion. Other ionophores are *ion carriers,* which are hydrophobic on the outside but have a hydrophilic cavity inside, thereby enabling an ion to be ferried across the interior of the membrane in a hydrophilic environment. The relevance of ionophores to our understanding of membrane transport proteins is explored further in Box 8B (page 209).

Examples of Facilitated Transport

As Figure 8-1 indicates, numerous solutes must be transported across the various membranes of cells. In many cases, the transport is by facilitated diffusion. We will consider several examples here, including the transport of specific solutes into and out of the erythrocyte (see Figure 8-3).

Facilitated Transport of Glucose.   The movement of glucose into an erythrocyte is an example of facilitated transport across the plasma membrane (see Figure 8-3a). The level of glucose in the blood is usually in the range of 65–90 mg/100 mL, or about 3.6–5.0 mM. The erythrocyte (or almost any cell in contact with the blood, for that matter) can meet its need for glucose by equipping its plasma membrane with a glucose transport protein and keeping its intracellular glucose concentration well below 3.6 mM.

That, in fact, is precisely what most cells do. As soon as glucose crosses the cell membrane, it is phosphorylated to glucose-6-phosphate by the enzyme *hexokinase*, with ATP as the phosphate donor and energy source:

$$\text{Glucose} \xrightarrow[\text{ATP}\quad\text{ADP}]{} \text{Glucose-6-phosphate} \xrightarrow[\text{Glycolytic pathway}]{} \longrightarrow \longrightarrow \longrightarrow \text{Pyruvate}$$

The hexokinase reaction is the first step in the *glycolytic pathway*, to be discussed in Chapter 11. This pathway occurs in the cytosol of almost all cells and is represented here by the sequence of arrows that connects glucose to pyruvate. Because the hexokinase reaction is highly exergonic ($\Delta G^{\circ\prime} = -4.0$ kcal/mol), the concentration of glucose within the cell is kept low, thereby ensuring continued uptake by facilitated diffusion. For many mammalian cells, the glucose concentration inside the cell is in the range of 0.5–1.0 mM, about 15–20% of the glucose level in the blood outside the cell.

The phosphorylation of glucose also has the effect of "locking" glucose in the cell because the plasma membrane of the erythrocyte does not have a transport protein for glucose-6-phosphate. This is, in fact, a general strategy for retaining molecules within the cell because most eukaryotic cells do not have transport proteins for phosphorylated compounds in their membranes. For example, all of the intermediates in the glycolytic pathway between glucose and pyruvate are phosphorylated compounds for which no transport proteins are present in the plasma membrane of an erythrocyte or any other cell.

Facilitated Diffusion Across the Membranes of Organelles. Facilitated diffusion is also the mechanism whereby many metabolites move into and out of organelles. Examples shown in Figure 8-1 are the entry of pyruvate into the mitochondrion and the transport of triose phosphates out of the chloroplast.

Frequently, transport proteins in the membranes of organelles move two different but related substances in opposite directions, with the inward transport of one substance coupled with the outward movement of the other. Several of the mitochondrial transport proteins in Figure 8-1 function in this way. The *ATP-ADP transport protein*, for example, couples outward ATP movement with the inward movement of ADP. Since ATP is generated in the mitochondrion and used in the cytosol, the ATP concentration is usually higher inside the organelle, while the ADP concentration is higher outside. The result is *coupled transport*, with ATP moving out and ADP moving in. (The ATP-ADP transport protein is an exception to the generalization that membranes do not contain transport proteins for phosphorylated compounds.) The *dicarboxylate* and *tricarboxylate transport proteins* function in the same manner, carrying out the passive reciprocal exchange of their respective solutes into and out of the mitochondrion in response to the prevailing internal and external concentrations of each.

The Erythrocyte Anion Channel Protein. Another example of coupled passive transport is the *anion channel protein* found in the plasma membrane of the red blood cell (see Figure 8-3a). This integral membrane protein facilitates the exchange of chloride (Cl^-) and bicarbonate (HCO_3^-) ions across the plasma membrane by providing a channel, or *ion pore*, through which these ions can pass. Because of the presence of this protein in the membrane, the permeability coefficient for chloride is about 10 million times greater for the erythrocyte plasma membrane than for an artificial membrane made of phosphatidyl serine (see Table 8-3). Moreover, the channel is very selective, exchanging bicarbonate for chloride in a 1:1 ratio and accepting no other anion.

The rapid flux of bicarbonate ions that the anion channel protein makes possible is essential to the role of the erythrocyte in the transport of waste CO_2 from body tissues to the lungs (see Figure 8-3b). In gaseous form, CO_2 is not very soluble in aqueous solutions such as cytoplasm or blood plasma. However, as rapidly as CO_2 diffuses into the erythrocyte, it is converted into bicarbonate anions by *carbonic anhydrase*, a cytosolic enzyme. The bicarbonate anions are then transported outward, with chloride ions moving inward to maintain charge balance. In the lungs, the process is reversed: Bicarbonate anions enter the erythrocytes in exchange for chloride ions and are converted into CO_2 by carbonic anhydrase. The CO_2 then diffuses out of the erythrocyte and into the cells that line the capillaries of the lung. This scheme makes use of both the blood plasma and the total cytoplasmic volume of the erythrocytes to carry bicarbonate to the lungs. The quantity of bicarbonate carried in this way is sufficient to keep pace with the generation of CO_2 in the body.

Active Transport: Energy and Gradients

Before continuing to a discussion of active transport, we will summarize the properties of passive transport and compare them to active transport (see Table 8-2). Passive transport refers to movement of a solute through a membrane in the direction dictated by the existing gradient of concentration or charge—that is, *down* the concentration or electrochemical gradient. The value of ΔG is therefore always negative and the process always exergonic.

Active transport, on the other hand, involves the movement of substances into or out of cells or organelles *against* a concentration or electrochemical gradient. The value of ΔG is therefore always positive and the process always endergonic. As a result, the pumps involved in active transport must provide not only for translocation of the desired solute molecules across the membrane but also for coupling of that translocation to an energy-yielding reaction. Often, though not always, energy is provided by the hydrolysis of a high-energy phosphate bond. For example, the inward pumping of potassium and concomitant outward pumping of sodium characteristic of most animal cells is driven by ATP. In other

Further Insights

IONOPHORES AND THE STUDY OF MEMBRANES

icroorganisms synthesize various antibiotics that are as much a boon to cell biologists and membrane biochemists as they are a bane to cells and membranes. These include *ionophores,* small molecules that greatly increase the permeability, or leakiness, of membranes to cations such as K^+, Na^+, and H^+. Though obviously detrimental to membrane function in the cell, such antibiotics have proved exceedingly useful in studying transport phenomena.

Ionophores fall into two categories, depending on how they facilitate ion movement across membranes. Some are **channel formers.** They organize themselves within the membrane to form a hydrophilic channel for the ion in much the way that transport proteins function. Ions enter the channel on one side of the membrane and pass through it to the other side, as illustrated in Figure 8B-1. *Gramicidin* is an example of a channel-forming antibiotic (Figure 8B-2).

Other ionophores function by binding ions on one side of the membrane and then diffusing across the membrane,

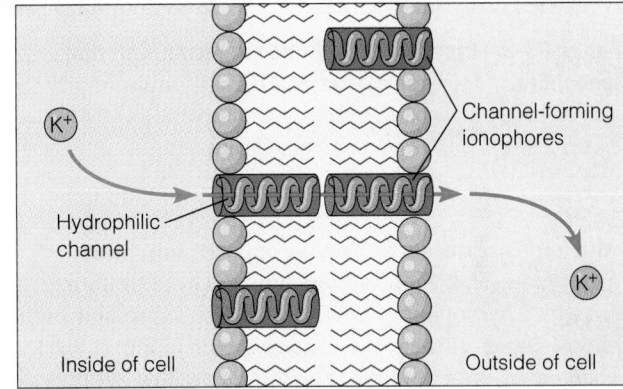

Figure 8B-1   Mechanism of Action of a Channel-Forming Ionophore.   A channel former assumes a helical conformation in the membrane, with a hydrophilic channel, or pore, in the center. The channel serves as a conduit for ions, thereby making the membrane leaky and discharging ion gradients across the membrane.

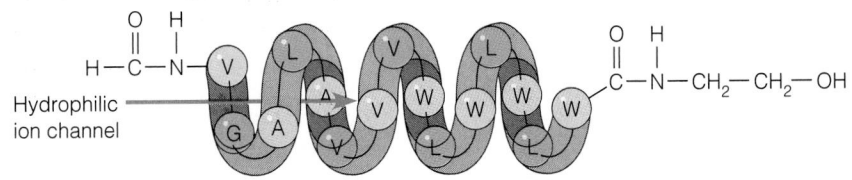

(a) Primary structure of gramicidin

(b) Helical conformation of gramicidin within a membrane

Figure 8B-2   Structure of Gramicidin.   **(a)** Gramicidin is an antibiotic with eight L-amino acids (light purple) and seven D-amino acids (dark purple) in strictly alternating order. **(b)** In the membrane, gramicidin assumes a helical shape, with the hydrophobic side chains facing the surrounding phospholipids and a hydrophilic ion channel on the inside of the helix. The amino acids and their abbreviations are alanine (Ala, A), glycine (Gly, G), leucine (Leu, L), tryptophan (Trp, W), and valine (Val, V).

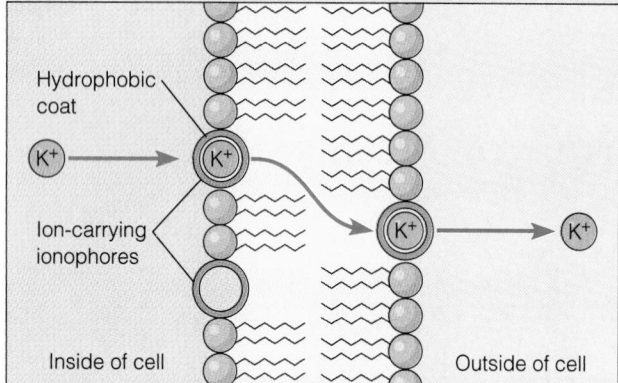

Figure 8B-3 Mechanism of Action of an Ion-Carrying Ionophore. An ion-carrying ionophore is a mobile molecule within the membrane that surrounds an ion with a hydrophobic coat and carries the ion with it as it diffuses across the membrane.

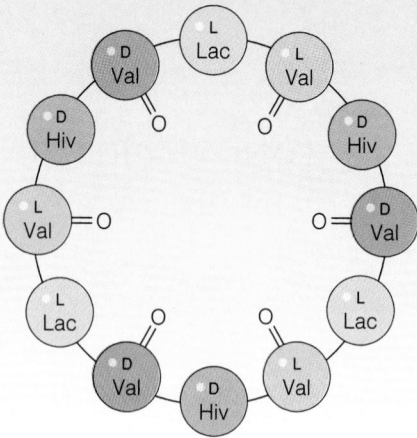

(a) Primary structure of valinomycin

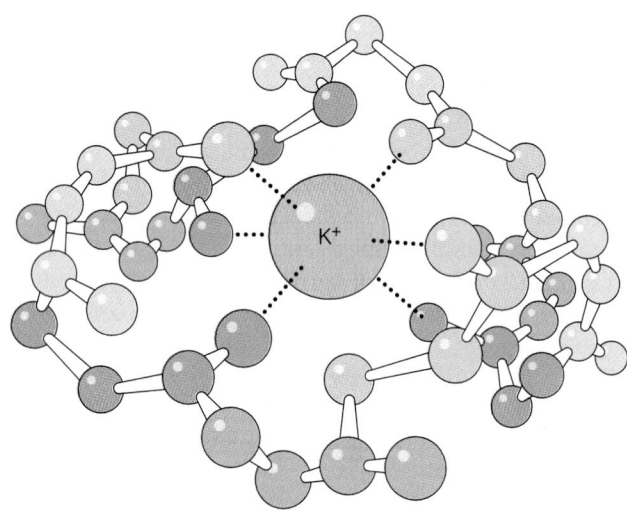

(b) Spherical formation of valinomycin within a membrane

Figure 8B-4 Structure of Valinomycin. **(a)** Valinomycin is a cyclic antibiotic in which D- and L-valines (D-Val and L-Val) alternate with L-lactate (L-Lac) and D-hydroxyisovalerate (D-Hiv). The carboxyl oxygens of the six valines project into the center of the ring. **(b)** In the membrane, the molecule assumes a conformation with an inner cavity that is lined with oxygens and nitrogens and is the right size to accommodate a K+ ion.

with ion release occurring on the opposite side. Such membrane-soluble **ion carriers** function by surrounding an ion with a hydrophobic "coat," making the ion soluble in the phospholipids of the membrane, as shown in Figure 8B-3. *Valinomycin* is an example of an ion carrier that has proved highly useful in the study of membrane transport (Figure 8B-4).

It is easy to understand how ion carriers can be distinguished from channel formers, once you realize that a carrier moves across the membrane with the ion but a channel former does not. Recall that as the temperature of a membrane is lowered, the membrane phospholipids undergo a thermal transition from a highly fluid state to a much more rigidly ordered state. This has little effect on the function of a channel-forming antibiotic, because the ionophore does not have to move within the membrane. But the change in membrane fluidity greatly affects an ion carrier because its activity depends on diffusion through the membrane. Figure 8B-5 illustrates the difference in temperature dependence of the two alternative modes of ionophore function.

Gramicidin, a representative channel-forming ionophore, is a polypeptide with 15 amino acids. This remarkable substance was discovered in the late 1930s by René Dubos, who found that it blocked the growth of pathogenic gram-positive bacteria, hence its name. In fact, a good case can be made that gramicidin, not penicillin, was the first antibiotic to be discovered and used to treat bacterial infections. Unlike the polypeptides of proteins, gramicidin consists of L- and D-amino acids in strict alternation (see Figure 8B-2a). When inserted into a membrane, the gramicidin polypeptide assumes a helical shape, with hydrophobic groups on the outside in contact with membrane phospholipids. Polar groups are on the inside of the helix, forming a

hydrophilic "lining" for the 0.4-nm channel down the center of the helix.

A single gramicidin helix spans only one-half of the membrane and essentially "floats" in one of the two monolayers that form the phospholipid bilayers. When a helix in one monolayer lines up with another such helix in the other monolayer, the two polypeptides associate end to end, forming a continuous channel that spans the membrane, as

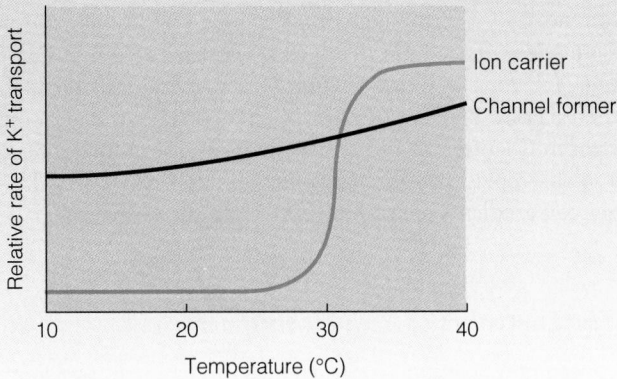

Figure 8B-5 Temperature Dependence of Ion-Carrying and Channel-Forming Ionophores. The rate of transmembrane transport of K$^+$ by ionophores is shown as a function of temperature for ion carriers (blue) and channel formers (black).

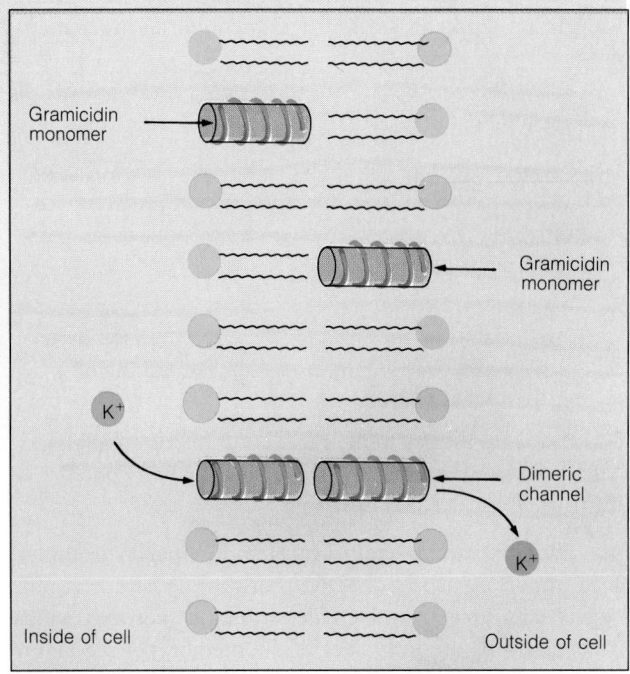

Figure 8B-6 Mode of Action of Gramicidin. Gramicidin monomers diffuse laterally in both monolayers of the membrane. When monomers line up as an end-to-end dimer, they form a transient channel through which potassium ions can flow.

shown in Figure 8B-6. This channel exists only as long as the two monomers remain aligned (about a second, usually), but during that time, potassium ions can pass through the channel at a rate exceeding 10 million ions per second! Membrane biochemists are especially fascinated with channel-forming antibiotics such as gramicidin because of the attractive model they provide for transport proteins, all of which are thought to function by providing hydrophilic channels through what is otherwise a hydrophobic membrane.

Other ionophores function by providing a hydrophilic environment for an ion as it traverses a membrane, but in this case the antibiotic literally surrounds the ion and moves across the membrane with it. Such ionophores are doughnut-shaped, with hydrophobic hydrocarbon groups on the outside and oxygen atoms (six to eight, usually) projecting into the central cavity. These oxygen atoms bind the desired ion coordinately, stabilizing it within the cavity as

the ionophore diffuses across the membrane. In the case of valinomycin (see Figure 8B-4), the oxygen atoms come from the carbonyl groups of the six valines in the molecule, whereas the methyl and isopropyl side chains make up the hydrocarbon periphery. Because of its specificity for potassium ions, valinomycin is a powerful nervous system poison. Can you understand why? ■

cases, the immediate driving force for active transport is an ion gradient, usually of protons or sodium ions. Pyruvate transport into the mitochondrion, for example, is coupled to the inward movement of protons; as a result, it depends on both the existence and the magnitude of an electrochemical proton gradient across the mitochondrial inner membrane.

Active transport performs three major functions in cells and organelles:

1. It makes possible the uptake of energy-rich molecules and other essential nutrients from the environment or surrounding fluid, even when their concentrations in the environment are very low.

2. It allows various substances, such as secretory products, waste materials, and sodium ions to be removed from the cell or organelle, even when the concentration outside is greater than that inside.

3. It enables the cell to maintain constant, optimal internal concentrations of inorganic ions, particularly potassium, calcium, and hydrogen ions.

The ability to maintain a constant internal environment is an important aspect of active transport. Passive transport results in a greater and greater similarity between solute concentrations on both sides of the membrane (see Figure 8-2a). Active transport, on the other hand, represents a means of creating deliberate concentration differences across membranes (see Figure 8-2b) and leads to a nonequilibrium steady state without which life as we know it would be impossible.

The Directionality of Active Transport

Another important distinction between active and passive transport concerns the direction of transport with respect to the inside and outside of the membrane. Passive transport is inherently *nondirectional* with respect to the membrane; solute can move either inward or outward, depending entirely on the prevailing concentration or electrochemical gradient. Active transport, on the other hand, has an intrinsic **directionality.** An active transport system that transports a solute across a membrane in one direction will not be able to transport that solute actively in the other direction. Active transport is therefore said to be a *unidirectional*, or *vectorial*, process. However, in the absence of an energy supply, some carriers that normally function in active transport can also be used to facilitate passive transport of the same solutes. In such cases, passive transport can occur in either direction, as dictated by the gradient. Clearly, carriers that show this property are not inherently vectorial; directionality is imposed on them by the energy-yielding system to which they are coupled when involved in active transport.

Under certain conditions, active transport can actually be driven in reverse. In muscle cells, for example, calcium ions are transported across the *sarcoplasmic reticulum* (*SR*) membrane by a specific **transport ATPase.** When these membranes are isolated as membranous vesicles, they maintain their ability to transport calcium, using one molecule of ATP to drive two calcium ions inward. After a high internal calcium concentration is produced, the vesicles can be quickly placed in a reaction medium containing ADP and P_i but no Ca^{2+}. Surprisingly, under these conditions ATP is *synthesized* in the ratio in which it was used to pump Ca^{2+}: one ATP for every two calcium ions that leak out through the ATPase. The importance of this concept will become clear in Chapters 12 and 13, when we discuss mitochondrial and chloroplast membranes, which use the energy of hydrogen ion gradients to synthesize ATP in a similar manner.

The Energetics of Active Transport

We can now go on to a more detailed discussion of transport energetics. Every transport event is an energy transaction; energy is either released as transport occurs or is required to drive transport. To understand the energetics of transport, we must recognize that two different factors may be involved. For uncharged solutes, the only variable is the concentration gradient across the membrane, and transport is either "downhill" (passive) or "uphill" (active). For charged solutes, however, there may be both a concentration gradient and a potential gradient across the membrane, and the two may either reinforce each other or oppose each other, depending on the charge on the ion and the direction of transport. We will first look at the transport of uncharged substances and then consider the additional complication that arises when charged species are moved across membranes.

Transport of Uncharged Solutes. For solutes with no net charge, we are concerned only with the concentration gradient across the membrane. We can therefore treat the transport process as a simple chemical reaction and calculate ΔG as we would for any other reaction.

The general "reaction" for the transport of molecules of solute X from the outside of a membrane-bounded compartment to the inside can be represented as

$$X_{outside} \longrightarrow X_{inside} \qquad (8\text{-}2)$$

From Chapter 5, we know that the free energy change for this reaction can be written as

$$\Delta G = \Delta G^\circ + RT \ln [X]_{inside}/[X]_{outside} \qquad (8\text{-}3)$$

where ΔG is the free energy change, ΔG° is the standard free energy change, R is the gas constant (1.987 cal/mol-K), T is the absolute temperature, and $[X]_{inside}$ and $[X]_{outside}$ are the prevailing concentrations of X on the inside and outside of the membrane, respectively. However, the equilibrium constant K_{eq} for the transport of an uncharged solute is always 1, because at equilibrium, the solute concentrations on the two sides of the membrane will be the same:

$$K_{eq} = [X]_{inside}/[X]_{outside} = 1.0 \qquad (8\text{-}4)$$

This means that $\Delta G°$ is always zero:

$$\Delta G° = -RT \ln K_{eq} = -RT \ln(1) = 0 \quad \text{(8-5)}$$

So the expression for ΔG of inward transport of an uncharged solute simplifies to

$$\Delta G_{inward} = +RT \ln [X]_{inside}/[X]_{outside} \quad \text{(8-6)}$$

Notice that if $[X]_{inside}$ is less than $[X]_{outside}$, then ΔG will be negative, indicating that the inward transport of substance X is exergonic and may occur spontaneously, as would be expected for passive transport down a concentration gradient. But if $[X]_{inside}$ is greater than $[X]_{outside}$, inward transport of X will be against the concentration gradient and the amount of energy required to drive the transport is indicated by the positive value of ΔG.

As an example, suppose that the concentration of lactose within a bacterial cell is to be maintained at 10 mM, while the external lactose concentration is only 0.20 mM. The energy requirement for the inward transport of lactose at 25°C can be calculated from equation 8-6 as

$$
\begin{aligned}
\Delta G_{inward} &= +RT \ln[\text{lactose}]_{inside}/[\text{lactose}]_{outside} \\
&= +(1.987)(273 + 25) \ln(0.010)/(0.0002) \\
&= +592 \ln(50) = +2316 \text{ cal/mol} \\
&= +2.32 \text{ kcal/mol} \quad \text{(8-7)}
\end{aligned}
$$

Clearly, this is an energy requirement that can be met easily by coupling the transport to ATP hydrolysis ($\Delta G°' = -7.3$ kcal/mol), which is one of the several sources of energy that bacterial cells can use for lactose uptake.

As written, equation 8-6 applies to inward transport. For outward transport, the positions of X_{inside} and $X_{outside}$ are simply interchanged within the logarithm. As a result, the numeric value of ΔG remains the same, but the sign is changed. As for any other process, a transport reaction that is exergonic in one direction will be endergonic to the same degree in the opposite direction. To help you keep the signs straight, the equations for calculating ΔG of inward and outward transport of uncharged solutes are summarized in Table 8-4.

Transport of Charged Solutes. For charged solutes, we need to take both the concentration gradient and the potential (electrical charge) gradient into account. The latter is often important because most biological membranes have a significant **membrane potential (V_m)** that makes them negative on one side and positive on the other. The membrane potential is indicated in Figure 8-7 by the negative charges associated with the inner surface of the plasma membrane. V_m is expressed either in volts (V) or millivolts (mV). For animal cells, the plasma membrane potential is usually in the range -60 to -90 mV. In bacterial and plant cells, the plasma membrane potential is significantly more negative, often about -150 mV in bacteria and between -200 and

Table 8-4 Calculation of ΔG for the Transport of Charged and Uncharged Solutes

Reaction:

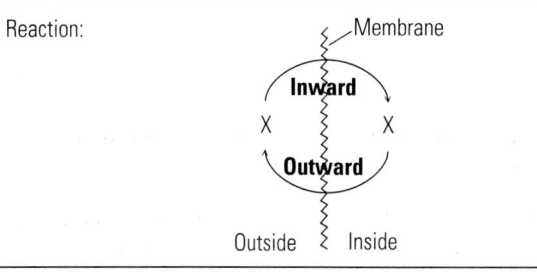

ΔG for Transport of Uncharged Solutes:

$$\Delta G_{inward} = +RT \ln \frac{[X]_{inside}}{[X]_{outside}}$$

$$\Delta G_{outward} = +RT \ln \frac{[X]_{outside}}{[X]_{inside}}$$

$R = 1.987$ cal/mol-K
$T = K = °C + 273$

ΔG for Transport of Charged Solutes:

$$\Delta G_{inward} = +RT \ln \frac{[X]_{inside}}{[X]_{outside}} + zFV_m$$

$$\Delta G_{outward} = +RT \ln \frac{[X]_{outside}}{[X]_{inside}} - zFV_m$$

z = charge on ion
$F = 23,062$ cal/mol-V
V_m = membrane potential (in volts)

-300 mV in plants. Notice the convention: The V_m value indicates how negative (or positive, in the case of a plus sign) the *inside* of the membrane is compared to the *outside*.

The membrane potential obviously has no effect on uncharged solutes (Figure 8-7a and b), but it affects the energetics of ion transport significantly. Because it is almost always negative, the membrane potential *favors* the inward movement of cations (Figure 8-7c) and *opposes* their outward movement (Figure 8-7d). Conversely, anions move *against* the membrane potential when they are transported inward (Figure 8-7e) but *with* it when they are transported outward (Figure 8-7f). The net effect of both the concentration gradient and the potential gradient for an ion is called the electrochemical gradient for that ion.

Both components of the electrochemical gradient must be considered when we are determining the energetics of ion transport. To calculate ΔG for the transport of ions therefore requires an equation with two terms, one to express the effect of the concentration gradient across the membrane and the other to take the membrane potential into account.

If we let X^z represent a solute with a charge z, then we can calculate ΔG for the inward transport of X^z as

$$\Delta G_{inward} = +RT \ln[X]_{inside}/[X]_{outside} + zFV_m \quad \text{(8-8)}$$

where R, T, and $[X]$ are defined as before and z is the charge on X (such as 1, +2, −1, −2), F is the Faraday constant

Figure 8-7 The Effect of Negative Membrane Potential on Inward and Outward Movement of Molecules and Ions. These diagrams illustrate the effect of membrane potential on the transport of **(a, b)** uncharged solutes, **(c, d)** cations, and **(e, f)** anions across a membrane with a negative membrane potential. In each pair, inward transport is shown on the left (a, c, e) and outward transport on the right (b, d, f).

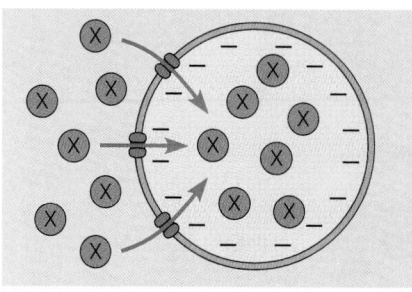

(a) Membrane potential has no effect on inward transport of uncharged solutes (zFV_m is zero)

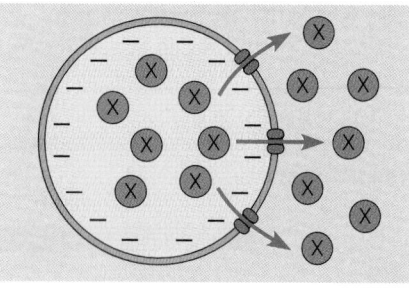

(b) Membrane potential has no effect on outward transport of uncharged solutes (zFV_m is zero)

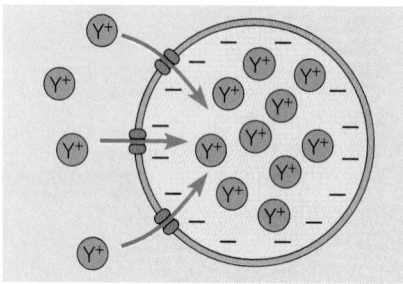

(c) Membrane potential favors inward transport of cations because of charge attraction (zFV_m is negative)

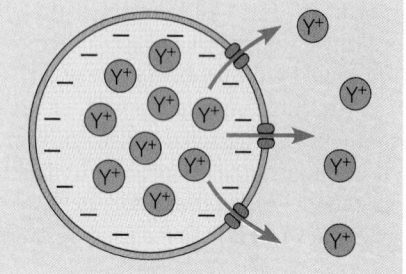

(d) Membrane potential opposes outward transport of cations because of charge attraction (zFV_m is negative)

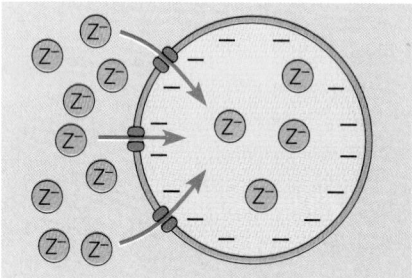

(e) Membrane potential opposes inward transport of anions because of charge repulsion (zFV_m is positive)

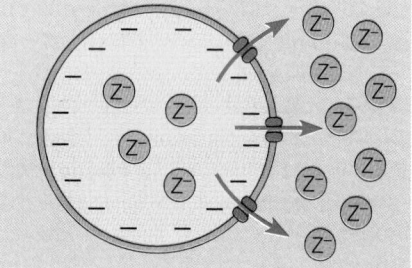

(f) Membrane potential favors outward transport of anions because of charge repulsion (zFV_m is positive)

(23,062 cal/mol-V), and V_m is the membrane potential, in volts. For the outward transport of X, ΔG has the same value as for inward transport but is opposite in sign, so we can write

$$\Delta G_{outward} = -\Delta G_{inward}$$
$$= -RT \ln[X]_{inside}/[X]_{outside} - zFV_m$$

Or, by interchanging terms within the logarithm,

$$\Delta G_{outward} = +RT \ln[X]_{outside}/[X]_{inside} - zFV_m \qquad \text{(8-9)}$$

To help you in such calculations, these equations for inward and outward transport of charged solutes are included along with those for the transport of uncharged solutes in Table 8-4, which summarizes the thermodynamic properties of each of these processes.

Mechanisms of Active Transport

Most membranes of living cells carry out one or more active transport processes, summarized in Table 8-5. Here we will consider several of the basic underlying mechanisms. This is an area of biology that has elicited a great deal of interest and excitement but has only recently begun to yield definitive answers. Much of the initial difficulties lay in the hydrophobic nature of most transport proteins and the problems involved in extracting these proteins from membranes in an active form. A related difficulty has been the lack of adequate assay systems once such proteins were isolated. But at least for several systems, these problems have been solved, either through the use of closed membrane vesicles that retain transport activity or by the incorporation

Table 8-5 Examples of Active Transport

Transported Solute	Membrane	Transport Protein	Comments
Hydrogen ion	*Halobacterium*	Bacteriorhodopsin	Light-dependent pump
	Chloroplast	Reaction centers	Light-dependent pump
	Mitochondrion	Respiratory complexes	Uses energy of electron transport
	Plasma membrane	E_1-E_2 ATPase	Regulates pH of cell
	Lysosome	E_1-E_2 ATPase	Maintains acid interior of organelle
Sodium and potassium ions	Plasma membrane	E_1-E_2 ATPase	Maintains gradient across plasma membrane of animal cells
Calcium ions	Sarcoplasmic reticulum	E_1-E_2 ATPase	Regulates muscle contraction
	Endoplasmic reticulum	E_1-E_2 ATPase	Regulates cytosolic Ca^{2+} concentration
	Plasma membrane	E_1-E_2 ATPase	Regulates cytosolic Ca^{2+} concentration
Nutrients, metabolites	Plasma membrane	Cotransport proteins	Uses energy of Na^+ or H^+ gradients

of membrane proteins into artificial phospholipid vesicles (the *liposomes* described earlier) so that transport activity is reconstituted. Because of these and other technical advances, substantial progress has been made toward understanding the molecular basis of active transport.

Properties of Active Transport Mechanisms

An important principle that has emerged from various studies is that not all active transport occurs by the same mechanism. Instead, several mechanisms are involved, differing primarily in terms of whether or not another solute is transported concomitantly, the source of energy used to drive the process, and whether or not the solute is chemically modified (phosphorylated) during transport. As a further source of complexity, many substrates are known to be transported into the same cell by several different mechanisms. This multiplicity of uptake mechanisms is observed in all organisms but is most striking in bacteria. *Escherichia coli*, for example, has at least five different systems for transporting galactose into the cell.

We will consider two general properties of active transport systems first and then look at several specific examples. Throughout this discussion, the term *pump* will be used to refer to the various mechanisms responsible for the active transport of molecules and ions across membranes. This terminology is consistent with the scientific literature and textbook usage, but no functional analogy with mechanical pumps is intended. On the contrary, mechanical pumps invariably effect a mass flow from one location to another, whereas membrane pumps selectively transport specific components from one mass to another.

Simple Transport Versus Cotransport. Some solutes are moved actively across membranes by **simple active transport**—a unidirectional pumping of the species of interest,

unaccompanied by the flow of any other substance in either direction. An example of simple active transport is the *calcium pump* found in muscle cells. As we will see in Chapter 21, calcium ions play an important role in the regulation of muscle contraction. Within the muscle cell, calcium is sequestered in the SR, as described earlier. When the cell is stimulated by a nerve impulse, calcium is released into the cytoplasm, triggering muscle contraction. As the cell returns to its resting state, the calcium is quickly returned to the SR by an ATP-driven calcium pump. This pump requires only calcium ions and ATP (plus catalytic amounts of Mg^{2+} to activate the ATPase). It therefore represents a simple active transport system, accomplishing the unidirectional transport of a single solute.

More frequently, however, coupled mechanisms are involved, such that two different solutes are moved simultaneously across a membrane in a tightly coupled manner. This process is called **cotransport,** which implies an obligatory comcomitant passage of two different solutes across the membrane. The two cotransported solutes may move either in the same direction, **symport,** or in opposite directions, **antiport,** as shown in Figure 8-8.

In some cases, both solutes are moved against their concentration gradient (or, for ions, their electrochemical gradient), and the energy source must therefore be adequate to pump both solutes "uphill." This is the case with the sodium-potassium pump, which we will consider shortly. In other cases, the transmembrane gradients of the two solutes are such that exergonic transport of one solute down its concentration (or electrochemical) gradient can be used to transport the other species actively against its concentration gradient. Clearly, the gradient of the passively transported solute must exceed that of the actively transported solute; otherwise the coupled process will not be exergonic. In animal cells, the active uptake of sugars, amino acids, and other metabolites is coupled to the passive inward movement of sodium ions in a process called *sodium cotransport*. Because of the large outside-to-inside sodium gradient that most

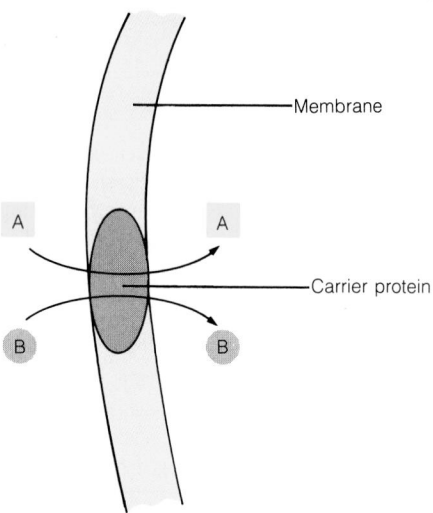

(a) Symport: Both solutes move in the same direction

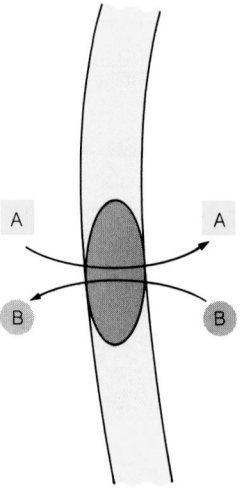

(b) Antiport: Solutes move in opposite directions

Figure 8-8 Cotransport. Cotransport involves the obligatory coupled transport of two different solutes, shown here as A and B, across a membrane. The two solutes may move either **(a)** in the same direction (symport) or **(b)** in opposite directions (antiport).

cells maintain across the plasma membrane, inward sodium transport is highly exergonic and can be used to drive the active symport of organic solutes.

Energy Sources. The energy required for active transport is provided either by hydrolysis of a high-energy phosphate bond or by cotransport of a second species. Not surprisingly, ATP is the compound of choice for most reaction mechanisms that couple to phosphate bond hydrolysis. However, some transport systems use other compounds instead. For example, one of the mechanisms used for sugar transport by bacterial cells depends on phosphoenolpyruvate as its energy source.

Cotransport almost always depends on an electrochemical gradient of either sodium ions or protons to drive the active symport of the desired solute, which is usually a small organic molecule. For animal cells, sodium cotransport is a common option. Bacteria, fungi, and plants, on the other hand, usually depend instead on an electrochemical proton gradient as the driving force for coupled transport. In some cases, the proton gradient is harnessed directly by means of a proton symport mechanism. In other instances, coupling is indirect, involving symport or antiport with some other ion, usually sodium, the electrochemical gradient for which is in turn maintained by a proton-driven pump.

Several Active Transport Mechanisms

Having considered some of the general features of active transport systems, we are now ready to look at four specific mechanisms, two in animal cells and two in bacteria. In each case, we will ask what kinds of solutes are transported, what the driving force is, and how the energy source is coupled to the transport mechanism. We will look first at the *sodium-potassium pump* present in all animal cells, because this is currently the best-understood transport system. Then we will consider two alternative modes for the uptake of organic molecules: *cotransport* (using sodium cotransport in animal cells as our example) and *phosphorylating transport* in bacterial cells. Finally, we will explore the concept of *proton transport* in the bacteriorhodopsin system, in anticipation of the important role that proton gradients play in respiration and photosynthesis, discussed in Chapters 12 and 13, respectively.

The Sodium-Potassium Pump of Animal Cells. A characteristic feature of most cells is a high intracellular level of potassium ions and a low intracellular level of sodium ions. This property is essential for the well-being of the cell, because potassium is required for a number of vital cellular processes, including ribosome function and activation of a variety of enzymes, and sodium often inhibits these functions. In addition, the resulting gradients of potassium and sodium ions are essential for the transmission of nerve impulses (see Chapter 22) and as the driving force for cotransport.

Potassium concentration is usually maintained at about 100–150 mM inside most animal cells, whereas external levels of potassium are generally much lower than that and may fluctuate widely. Conversely, the intracellular concentration of sodium is about 10–15 mM, which is considerably less than that in the surrounding medium (Table 8-6). Both the inward pumping of potassium ions and the outward pumping of sodium ions are therefore energy-requiring processes.

The pump responsible for this process was discovered in the mid-1950s and represented the first documented case of active transport. The pump uses ATP as its energy source and is therefore an example of a *transport ATPase*. Specifically, the **sodium-potassium pump,** as this ATPase is called, couples the exergonic hydrolysis of ATP to the inward trans-

Table 8-6 Concentrations of Ions on the Inside and Outside of a Typical Animal Cell

Ion	Concentration Inside (mM)	Concentration Outside (mM)	Concentration Ratio	
			$[Ion]_{inside}/[Ion]_{outside}$	$[Ion]_{outside}/[Ion]_{inside}$
K^+	140.0	4.0	35	0.03
Na^+	12.0	145.0	0.083	12.08
Mg^{2+}	0.8	1.5	0.53	1.88
Ca^{2+}	<0.001	1.8	<0.0005	1800.00
Cl^-	4.0	116.0	0.035	29.0
HCO_3^-	12.0	29.0	0.41	2.42

port of potassium ions and the outward transport of sodium ions. The sodium-potassium pump requires both potassium and sodium ions for activation, differing in this respect from other potassium-activated proteins in the cell, which are almost all inhibited by sodium.

The sodium-potassium pump is present in the plasma membrane of virtually all animal cells but has been studied in the greatest detail in red blood cells. Like other active transport systems, the sodium-potassium pump is a vectorial system: It has inherent directionality in that potassium is pumped only inward and sodium is pumped only outward. In fact, sodium and potassium ions activate the ATPase only on the side of the membrane from which they are transported—sodium from the inside, potassium from the outside.

Because the gradients against which sodium and potassium are pumped by animal cells seldom exceed 50:1 (see Table 8-6), the energy requirement per ion moved is relatively low, usually less than 2 kcal/mol ($\Delta G = +RT \ln(50) = 2.32$ kcal/mol at 25°C). This suggests that the hydrolysis of a single ATP molecule can drive the transport of several ions at a time, which is what actually happens in cells. The stoichiometry apparently varies somewhat with cell type, but for red blood cells, three sodium ions are moved out and two potassium ions are moved in per molecule of ATP hydrolyzed.

Figure 8-9 is a schematic illustration of the sodium-potassium pump consistent with available evidence. The pump is a tetrameric transmembrane protein, with two α and two β subunits. The α subunits are catalytically active. They have binding sites for ATP and sodium ions on the cytoplasmic (inner) side and for potassium ions on the external side of the membrane. Because the β subunits are glycosylated, they are located on the extracellular side. The function of the β subunits is not yet clear.

The sodium-potassium pump is an allosteric protein, with two alternative conformational states, referred to as E_1 and E_2. E_1 is thought to be "open" to the inside of the cell and to have a high affinity for sodium ions, whereas E_2 is regarded as "open" to the outside, with a high affinity for potassium ions. Phosphorylation of the enzyme, a sodium-

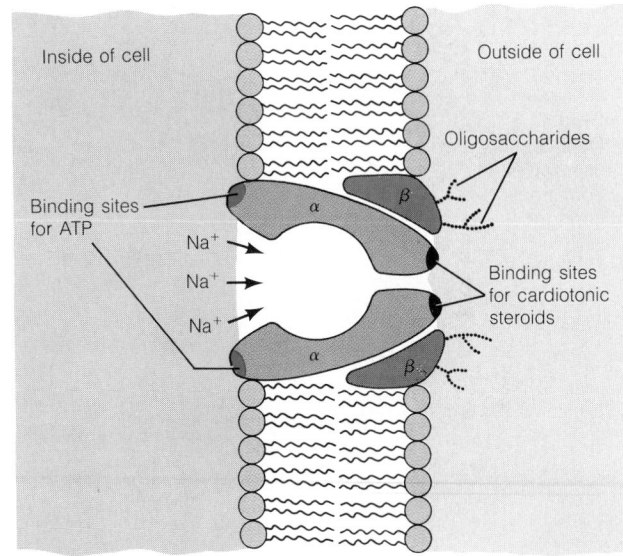

Figure 8-9   The Sodium-Potassium Pump.   The sodium-potassium pump found in most animal cells consists of two α and two β subunits. The α subunits are transmembrane proteins, with binding sites for ATP on the cytoplasmic side and for cardiotonic steroids on the extracellular side. The β subunits are located on the outer side of the membrane and are glycosylated. The pump is shown in the E_1 conformation, with the ion binding sites facing the inside of the cell. Binding of sodium ions causes a conformational change to the E_2 form, which opens to the outside.

triggered event, stabilizes it in the E_2 form. Dephosphorylation, on the other hand, is triggered by potassium and stabilizes the enzyme in the E_1 form. Other ions transported by E_1-E_2 *ATPases* include Ca^{2+} in sarcoplasmic reticulum of muscle tissue and H^+ in membranes such as lysosomes, secretory granules, and the plasma membrane.

As illustrated in Figure 8-10, the actual transport mechanism probably involves an initial binding of three sodium ions to E_1 on the inner side of the membrane (step 1). The binding of sodium ions triggers phosphorylation of the

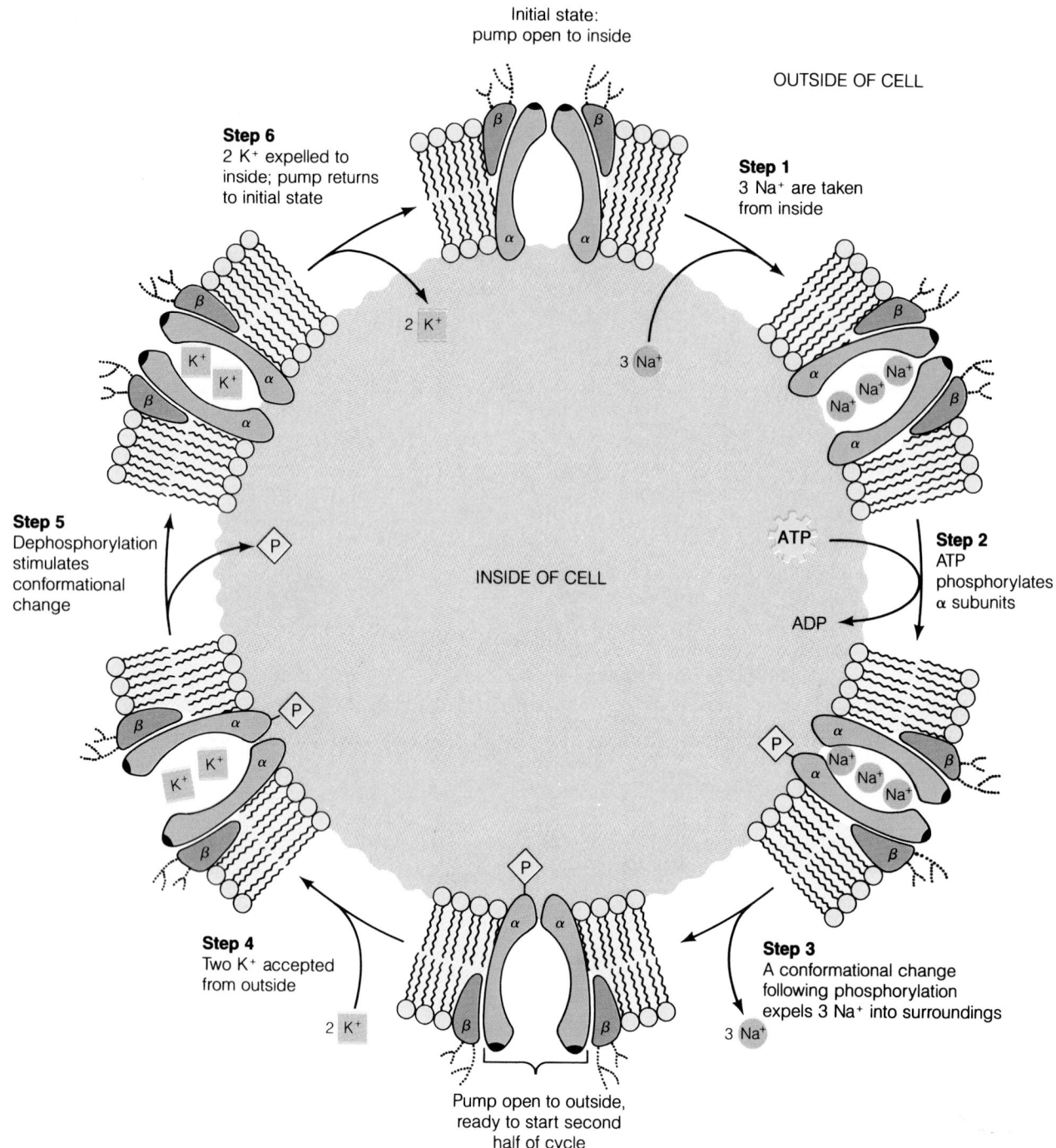

Figure 8-10 A Model Mechanism for the Sodium-Potassium Pump. The transport process is shown here in six steps arranged around the periphery of a cell. The outward transport of sodium ions is coupled to the inward transport of potassium ions, both against their respective electrochemical gradients. The driving force is provided by ATP hydrolysis, which is required for phosphorylation of the α subunit of the pump. Selective transport of sodium ions and potassium ions in opposite directions is possible because sodium ions activate the ATPase only on the inner surface of the membrane, whereas potassium ion activation of the dephosphorylation reaction occurs only on the outer surface. E_1 and E_2 are the conformation states of the protein with the channel open to the inside (top of figure) and to the outside (bottom of figure) of the cell, respectively. As depicted here, the binding of sodium ions (step 1) precedes phosphorylation of the α subunit (step 2), but these steps may occur in the reverse order. Pumps that work this way are called E_1-E_2 ATPases.

enzyme by ATP (step 2), leading to a conformational change to E_2. As a result, the bound sodium ions are translocated through the membrane to the external surface, where they are released to the outside (step 3). Then potassium ions from the outside bind to the α subunits (step 4), triggering dephosphorylation and a return to the original conformation (step 5). In the process, the potassium ions are translocated to the inner surface, where they dissociate, leaving the carrier ready to accept more sodium ions (step 6).

The sodium-potassium pump is sensitive to inhibition by a class of poisonous plant steroids that includes compounds such as ouabain and digitalis. These inhibitors are quite toxic; *ouabain*, for example, is used by indigenous South Americans to poison their arrowheads because of its effectiveness in killing prey. In carefully controlled doses, however, these compounds are of great medical importance because of their profound effects on the function of the heart. For this reason, they are called *cardiotonic steroids*. *Digitalis* is often the drug of choice in the treatment of congestive heart failure because it causes an increase in intracellular calcium concentration, thereby stimulating muscle contraction in heart cells.

The binding sites for the cardiotonic steroids are located on the external side of the α subunits, as shown in Figure 8-9. These steroids are thought to inhibit the sodium-potassium pump by blocking the dephosphorylation reaction (step 5 of Figure 8-10). The demonstration that ouabain blocks the transport of both sodium and potassium as well as ATP hydrolysis was crucial confirmation that all three of these processes are, in fact, catalyzed by a single protein.

The sodium-potassium pump is not only the best-understood transport system but also one of the most important. In addition to maintaining the appropriate intracellular concentrations of both potassium and sodium, it is responsible for maintenance of the membrane potential that exists across the plasma membrane. As we will see in Chapter 22, the membrane potential plays an important part in the transmission of nerve signals. The sodium-potassium pump assumes still more significance when we take into account the vital role that sodium plays in the inward transport of organic substrates, a topic we now come to as we consider sodium cotransport.

Sodium Cotransport of Organic Molecules in Animal Cells. A unifying feature that has emerged from studies of the active uptake of sugars, amino acids, and other organic molecules by animal cells is that the inward transport of such molecules is often coupled obligatorily to the concomitant inward cotransport of sodium ions. It is, in fact, only as we consider **sodium cotransport** that we appreciate the full significance of the continuous outward movement of sodium accomplished by the sodium-potassium pump. It is the steep electrochemical gradient of sodium ions maintained by this pump that serves as the driving force for the inward transport of a variety of sugars and amino acids. The uptake of such compounds by animal cells is therefore not

coupled directly to ATP hydrolysis but is powered instead by the sodium gradient. Ultimately, of course, such uptake is still dependent on ATP, since the sodium-potassium pump is itself an ATP-driven system.

The pump proteins involved in sodium-driven transport have not yet been isolated in purified form, but several of their properties can be predicted from what is already known about the process in which they are involved. Clearly, the pump must have two kinds of sites, one specific for sodium ions and the other specific for the solute to be cotransported. Furthermore, the binding of sodium to its site almost certainly affects the affinity of the other site for the desired solute, because the solute should bind when sodium binds and dissociate when sodium dissociates if its transport is to follow the sodium gradient faithfully.

To visualize the mechanism whereby sodium-driven uptake of a compound like glucose might occur, we need an allosteric membrane protein with sites for sodium and the appropriate solute. Figure 8-11 shows such a pump protein, with glucose as the transported solute. When sodium binds to its site on the outside of the membrane, the conformation of the protein changes, giving the glucose binding site a high affinity for glucose (step 1). Binding of glucose (step 2) presumably causes a further conformational change in the protein that results in the translocation of both sodium and glucose to the inner surface of the membrane (step 3). There, the sodium dissociates in response to the low intracellular sodium concentration (step 4). This causes glucose to be ejected from its site regardless of the cellular glucose level (step 5), probably because the affinity of that site for glucose is greatly reduced on dissociation of the sodium ion. The protein then reverts to its original conformation, returning the empty binding sites to the outer surface of the membrane (step 6).

Sodium cotransport is a common mechanism for the uptake of organic substrates by animal cells. Other organisms also use cotransport as a means of uptake for sugars, amino acids, and other solutes, but they usually rely on a proton gradient rather than a sodium gradient to drive the uptake. Bacterial cells, for example, make extensive use of proton cotransport to drive the uptake of organic solutes, with the proton gradient maintained, in turn, by electron transport. Fungi and plants also utilize proton symport for the uptake of organic solutes, but these organisms depend on an ATP-driven proton pump for the generation and maintenance of the electrochemical proton gradient.

Transport by Modification: Phosphorylating Transport in Bacterial Cells. The several types of transport we have examined so far differ in mechanism and in the immediate energy source, but they share the property that a given molecule or ion is transported across a membrane against its concentration (or electrochemical) gradient without being chemically changed in any way. Such is not universally the case, however. An alternative is **transport by modification,** in which the uptake of a solute is accomplished in a way that couples transport of the solute to its chemical modification.

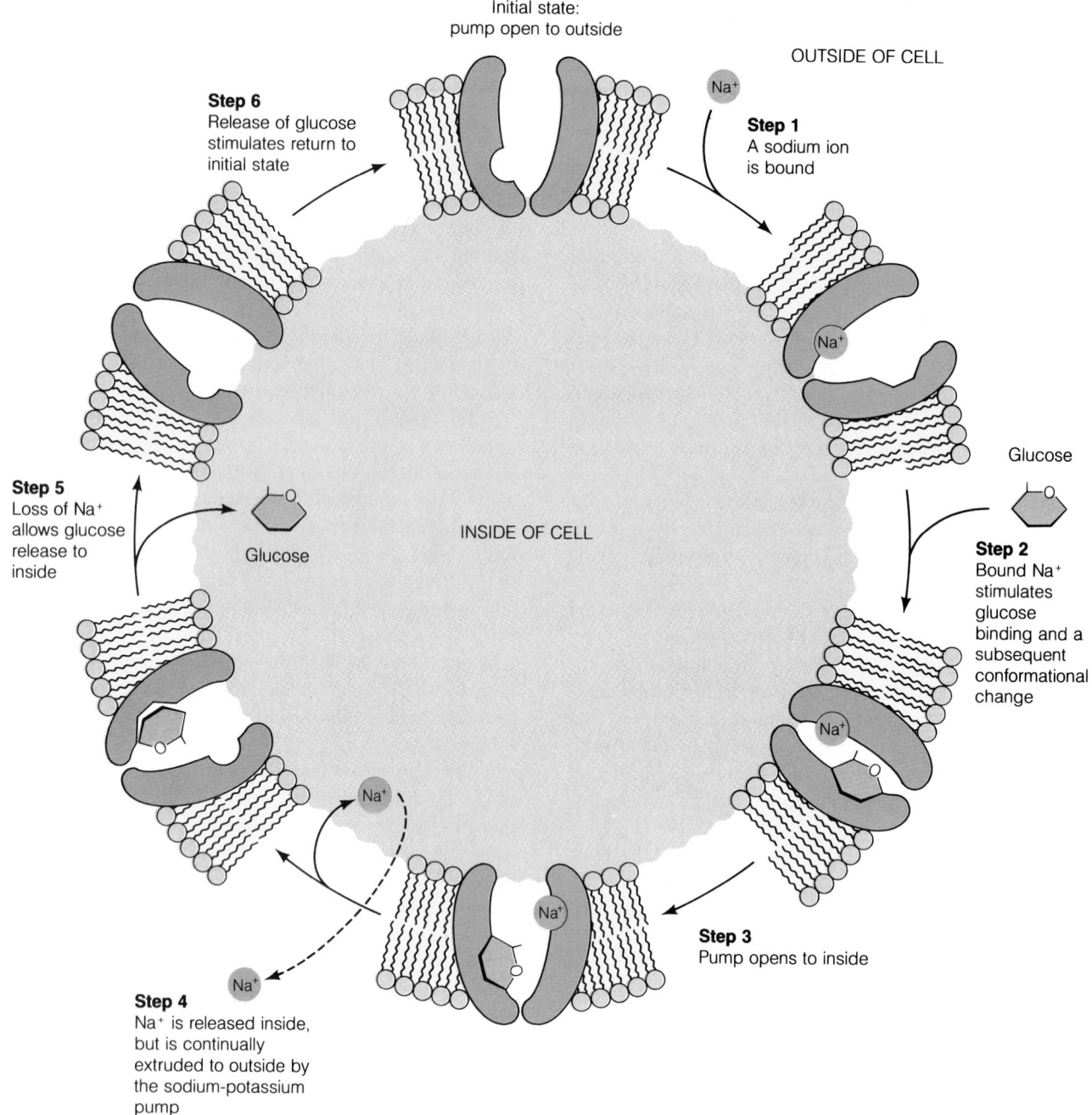

Figure 8-11   A Model Mechanism for Sodium Cotransport. The transport process is shown here in six steps arranged around the periphery of a cell. The inward transport of an organic solute such as glucose *against* its concentration gradient is driven by the concomitant inward transport of sodium *down* its electrochemical gradient. The sodium gradient is in turn maintained by the continuous outward extrusion of sodium ions (dashed arrow) by the sodium-potassium pump of Figure 8-10.

The best-understood example of such a mechanism is the *phosphotransferase system* responsible for sugar uptake in *Escherichia coli* and other bacteria. This system uses the high-energy phosphate group of a compound called *phosphoenolpyruvate (PEP)* to phosphorylate the sugar molecule (Figure 8-12). The process is termed **phosphorylating transport** because the sugar is phosphorylated as an inherent part of the uptake mechanism. In the case of glucose, for example, the sugar is taken up at the outer membrane surface as free glucose but is released at the inner membrane surface as glucose-6-phosphate:

$$\text{glucose} + \text{PEP} \longrightarrow \text{glucose-6-phosphate} + \text{pyruvate} \quad \textbf{(8-10)}$$

Because the plasma membrane contains no transport proteins for glucose-6-phosphate or other sugar phosphates, these intermediates remain inside the cell.

As Figure 8-12 shows, the phosphotransferase system responsible for this process is complex. It involves four com-

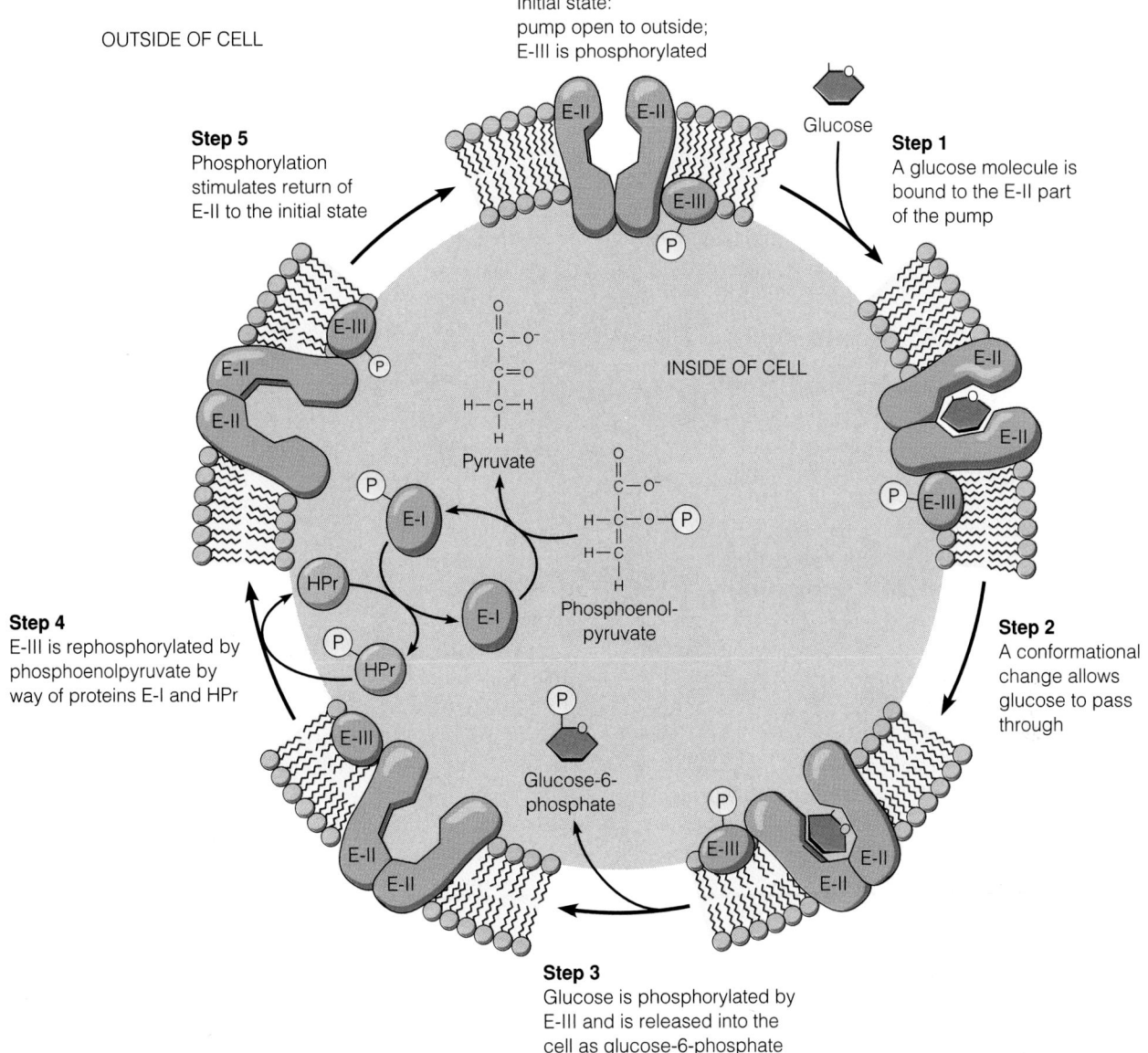

OUTSIDE OF CELL

Initial state:
pump open to outside;
E-III is phosphorylated

Glucose

Step 5
Phosphorylation
stimulates return of
E-II to the initial state

Step 1
A glucose molecule is
bound to the E-II part
of the pump

INSIDE OF CELL

E-I

HPr

Pyruvate

Phosphoenol-
pyruvate

Step 4
E-III is rephosphorylated by
phosphoenolpyruvate by
way of proteins E-I and HPr

Step 2
A conformational
change allows
glucose to pass
through

Glucose-6-
phosphate

Step 3
Glucose is phosphorylated by
E-III and is released into the
cell as glucose-6-phosphate

Figure 8-12   A Model Mechanism for Phosphorylating Transport of Sugars in Bacterial Cells.   The transport process is shown here in five steps arranged around the periphery of a cell. The inward transport of a solute such as glucose is driven by phosphorylation of the solute on the inner surface of the membrane, with phosphoenolpyruvate (PEP) as both the phosphate donor and the energy source.

Enzymes II and III (E-II and E-III) are solute-specific membrane proteins. Enzyme I and HPr are nonspecific soluble proteins that transfer the high-energy phosphate group of PEP to specific E-III proteins, which then transfer the phosphate group to the specific sugar that binds to the associated E-II protein.

ponents, designated *enzyme I* (*E-I*), *enzyme II* (*E-II*), *enzyme III* (*E-III*), and *HPr*, a small heat-stable protein. E-II is the actual transport protein. It is an integral membrane protein with a domain that forms a transmembrane channel and catalyzes phosphorylation of the solute. There are probably as many kinds of E-II proteins in the membrane as there are different sugars to be transported by this mechanism because the specificity of sugar recognition seems to be a property of E-II.

At the beginning of the transport process, the appropriate E-II protein is open to the outside of the cell. The mole-

cule to be transported inward (glucose, in the example shown) binds to the E-II protein at the outer membrane surface (step 1), resulting in a conformational change in E-II. As the sugar is translocated inward, presumably through the hydrophilic transmembrane channel of E-II (step 2), it receives a phosphate group from E-III, a peripheral membrane protein that interacts with E-II (step 3). The phosphate group is added to the carbon atom 6 of glucose, forming glucose-6-phosphate. Phosphorylation of the sugar reduces its affinity for the binding site of E-II, resulting in the release of glucose-6-phosphate into the cytoplasm.

Although the phosphate group transferred to glucose came from E-III, it is actually supplied by phosphoenolpyruvate and reaches E-III via E-I and HPr (step 4). As Figure 8-12 suggests, E-I and HPr are both soluble proteins, having no association with the membrane. Finally, E-II reverts to its original conformation (step 5), ready for the entry of another molecule of solute.

Like E-II, E-III is specific for a particular sugar. Thus, there is a separate set of E-II and E-III proteins for each sugar to be transported. E-I and HPr, however, are nonspecific proteins that serve as the common phosphorylating system for every E-III.

The Bacteriorhodopsin Proton Pump.

The final active transport system we will consider is the simplest because nothing more is involved than a small integral membrane protein called the **bacteriorhodopsin proton pump.** As discussed in Chapter 7, bacteriorhodopsin is a light-activated proton pump found in the plasma membrane of halophilic (salt-loving) bacteria belonging to the genus *Halobacterium*. Halobacteria produce bacteriorhodopsin under anaerobic conditions as a means of trapping light energy, which they use to pump protons outward across the plasma membrane. The *electrochemical proton gradient* that is thereby established can be used to generate ATP.

This transport process is one of the most significant recent discoveries in membrane biology. In part, this is because bacteriorhodopsin was the first membrane protein to have its three-dimensional structure and orientation determined (by Unwin and Henderson; see Figure 7-3). In addition, however, the bacteriorhodopsin transport system provided convincing evidence in support of the current model for proton-driven ATP generation.

The light-absorbing pigment, or *chromophore*, of bacteriorhodopsin is *retinal*, a carotenoid derivative related to vitamin A. (Retinal also serves as the visual pigment in the retina of your eyes.) Because of retinal, bacteriorhodopsin is bright purple in color and is the reason that halobacteria are also called *purple photosynthetic bacteria* (Figure 8-13a).

Bacteriorhodopsin appears in the plasma membrane of the *Halobacterium* cell as colored patches called *purple membrane* (Figure 8-13b).

Bacteriorhodopsin is an integral membrane protein with seven α-helical membrane-spanning segments that are oriented in the membrane to form an overall cylindrical shape (Figure 8-13c). The retinal molecule is present in the all-*trans* form and is covalently linked to the side chain of the lysine at position 216, forming a positively charged *Schiff base* (Figure 8-13d). When the retinal absorbs a photon of light, the photoactivated bacteriorhodopsin molecule is capable of transferring one or two photons from the inside to the outside of the cell. The details of the proton transfer process are not yet known, even though bacteriorhodopsin has been successfully isolated and incorporated into artificial membranes in functional form.

One possible mechanism for proton pumping is shown in Figure 8-14. According to this model, the absorption of an incoming photon of light causes one of the double bonds of all-*trans* retinal to isomerize to the *cis* conformation (step 1). This change decreases the ability of the nitrogen atom to bind protons. A proton is therefore released to the outside of the cell (step 2), permitting the double bond to relax to the *trans* form (step 3). The nitrogen again binds a proton, but always from the inside of the cell (step 4). Because protons are always drawn from the inside and released to the outside of the cell, the process leads to a vectorial pumping of protons from inside to outside. This pumping results in an electrochemical proton gradient across the plasma membrane that is then used to drive ATP synthesis.

Energy-dependent proton pumping is one of the most basic concepts in cellular energetics. Proton pumping occurs in all bacteria, mitochondria, and chloroplasts and represents the driving energy of life on Earth because it is an absolute requirement for the efficient synthesis of ATP. The mechanisms underlying the generation of proton gradients and the use of the energy of such gradients will be discussed in Part Three.

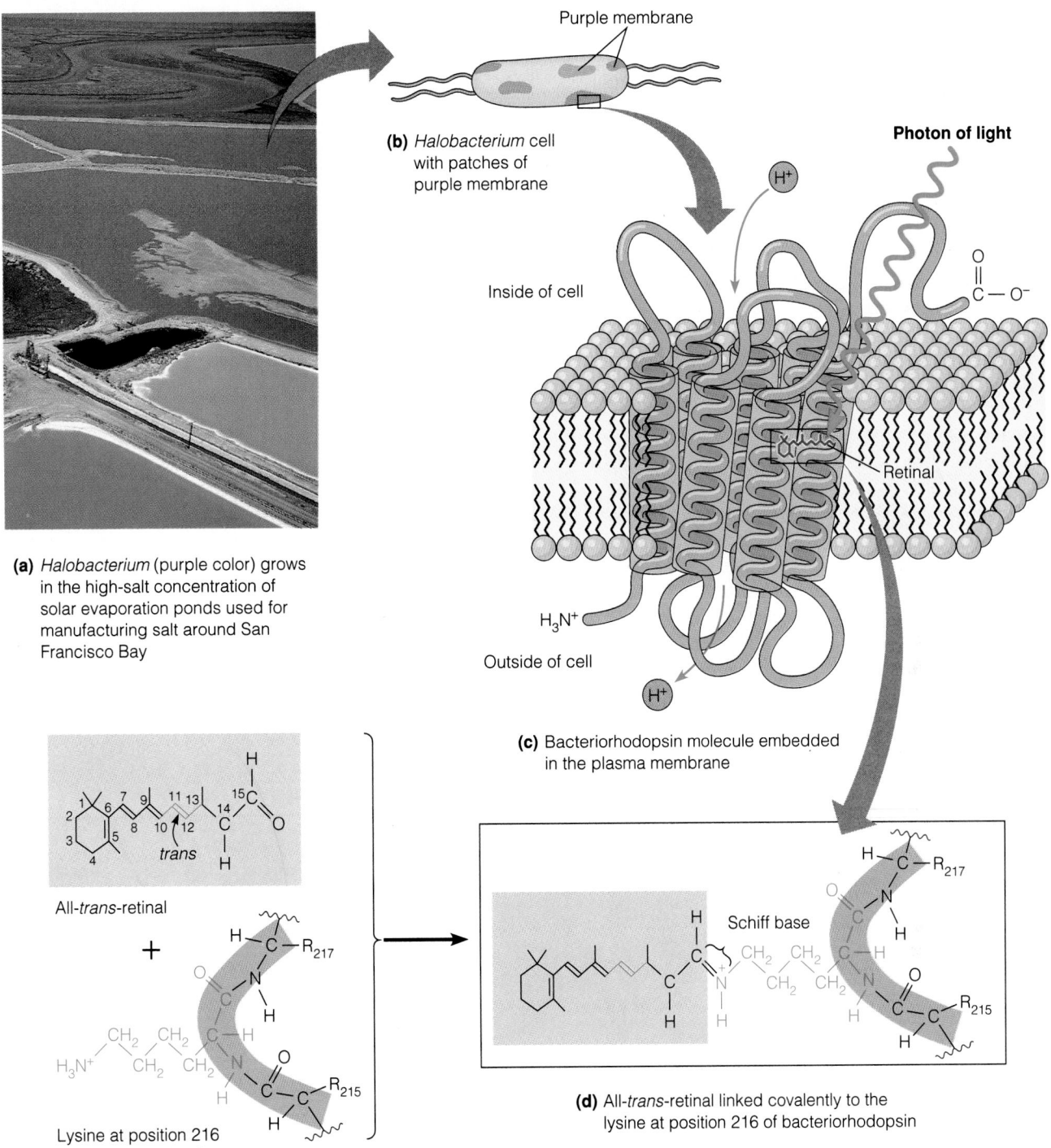

(b) *Halobacterium* cell with patches of purple membrane

(c) Bacteriorhodopsin molecule embedded in the plasma membrane

(a) *Halobacterium* (purple color) grows in the high-salt concentration of solar evaporation ponds used for manufacturing salt around San Francisco Bay

All-*trans*-retinal

Lysine at position 216 of bacteriorhodopsin

(d) All-*trans*-retinal linked covalently to the lysine at position 216 of bacteriorhodopsin

Figure 8-13   The Bacteriorhodopsin Proton Pump of Halobacteria. **(a)** Bacteria belonging to the genus *Halobacterium* are characterized by a purple color that is due to the protein bacteriorhodopsin. **(b)** Bacteriorhodopsin is a light-activated proton pump that is present in the plasma membrane of *Halobacterium* cells as bright purple patches known as purple membrane. **(c)** The seven α-helical transmem-brane segments of bacteriorhodopsin are separated by short nonhelical segments and are oriented in the membrane to form an overall cylindrical shape. **(d)** The chromophore, all-*trans*-retinal, is linked as a Schiff base to the lysine at position 216 in the seventh transmembrane segment of the protein.

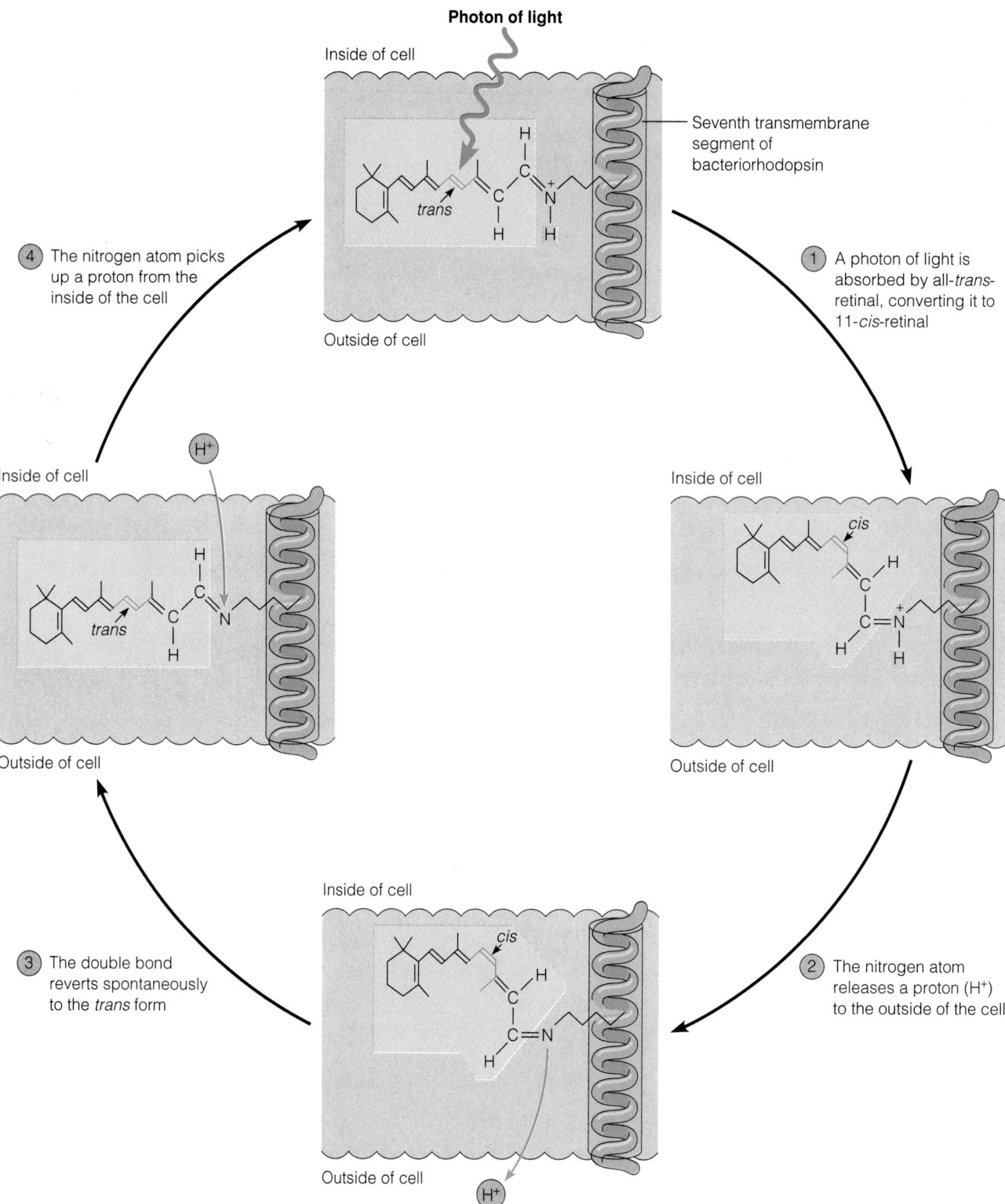

Photon of light

Inside of cell

Seventh transmembrane
segment of
bacteriorhodopsin

trans

Outside of cell

④ The nitrogen atom picks
up a proton from the
inside of the cell

① A photon of light is
absorbed by all-*trans*-
retinal, converting it to
11-*cis*-retinal

Inside of cell

H⁺

trans

Outside of cell

Inside of cell

cis

Outside of cell

③ The double bond
reverts spontaneously
to the *trans* form

② The nitrogen atom
releases a proton (H⁺)
to the outside of the cell

Inside of cell

cis

Outside of cell

H⁺

**Figure 8-14   A Model Mechanism for the Bacterio-
rhodopsin Proton Pump.**   A possible mechanism for light-
activated proton pumping by bacteriorhodopsin is shown here in four
steps. Protons are always released to the outside of the cell (step 2) but
drawn from the inside of the cell (step 4), thereby accounting for the outward pumping of protons that generates an electrochemical proton gradient across the plasma membrane of illuminated cells.

The selective transport of molecules and ions across membrane barriers ensures that the right substances are moved into and out of the right compartments at the right times, in the right amounts, and at the right rates. Except for molecules that are quite small and/or nonpolar, transport requires integral membrane proteins to facilitate or drive the movement of solutes across the membrane. Transport is either passive or active, depending on whether or not the process requires metabolic energy.

Passive transport does not require energy and can be thought of as a diffusional process in which flux occurs continuously across membranes, with net movement in the direction dictated by the concentration gradient. Simple diffusion is adequate for molecules such as ethanol, oxygen, carbon dioxide and water, to which the lipid bilayer barrier is relatively permeable. For most molecules and all ions, however, movement across the membrane is facilitated by transport proteins that provide solute-specific mechanisms for transport through an otherwise impermeable barrier. Because such transport proteins are finite in number and in velocity of transport, they become saturated at high solute concentrations and follow Michaelis-Menten kinetics.

Active transport requires energy and is carried out by enzymelike pump proteins embedded in the membrane. The energy is provided by light, ion gradients, or by the hydrolysis of high-energy phosphate bonds (of ATP, usually). Because of the energy input, concentration or electrochemical gradients of the transported solutes can be produced. Most cells maintain electrochemical gradients of sodium, potassium, calcium, and hydrogen ions.

The energy of such gradients can be used in turn to drive secondary transport processes, called cotransport. Cotransport uses the exergonic movement of sodium or hydrogen ions down their electrochemical gradient to drive the concomitant transport of the desired solute up its concentration gradient. Animal cells typically use sodium ions for this purpose and depend on the sodium-potassium pump to maintain the high outside-to-inside sodium gradient that this process requires. Bacterial, plant, and fungal cells use an electrochemical proton gradient instead, with outward proton pumping driven either by electron transport or by an ATP-powered proton pump. Most transport mechanisms move the desired solute across the membrane unaltered, but in some bacteria, the inward transport of solutes is coupled to their chemical modification, with phosphorylation of sugars being the best-studied example. Bacteriorhodopsin, the light-driven proton pump of halobacteria, is of special interest because it clearly illustrates the use of an ion gradient to drive the synthesis of ATP, a topic to which we will return in Chapter 12.

Membrane transport is an area of intense current interest. Recent technical advances in the study of membrane proteins and the reconstitution of transport activity in vitro promises further rapid progress. We may soon have a better understanding of the mechanisms whereby solutes are translocated from one side of a membrane to another, as well as the specific means by which such translocation is coupled to the appropriate energy source.

KEY TERMS FOR SELF-TESTING

Cells and Transport Processes

transport (p. 197)
cellular transport (p. 198)
intracellular transport (p. 198)
transcellular transport (p. 198)
passive transport (p. 199)
active transport (p. 199)

Passive Transport: Simple and Facilitated Diffusion

simple diffusion (p. 202)
facilitated diffusion (p. 203)
liposome (p. 203)
transport protein (p. 205)
permease (p. 205)
alternating conformation model (p. 206)

ionophore (p. 207)

Active Transport: Energy and Gradients

directionality (p. 212)
transport ATPase (p. 212)
membrane potential (V_m) (p. 213)

Mechanisms of Active Transport

simple active transport (p. 215)
cotransport (p. 215)
symport (p. 215)
antiport (p. 215)
sodium-potassium pump (p. 216)
sodium cotransport (p. 219)
transport by modification (p. 219)

phosphorylating transport (p. 220)
bacteriorhodopsin proton pump (p. 222)

Box 8A: Membrane Transport, Cystic Fibrosis, and the Prospects for Gene Therapy

cystic fibrosis (CF) (p. 200)
cystic fibrosis transmembrane conductance regulator (CFTR) (p. 200)
gene therapy (p. 201)

Box 8B: Ionophores and the Study of Membranes

channel former (p. 209)
ion carrier (p. 210)

PROBLEM SET

8-1. True or False? Indicate whether each of the following statements about membrane transport is true (T) or false (F). If false, reword the statement to make it true.

(a) Facilitated diffusion of an uncharged solute occurs only from a compartment of lower solute concentration to a compartment of higher solute concentration.

(b) Active transport is always driven by the hydrolysis of high-energy phosphate bonds.

(c) The K_{eq} value for the diffusion of polar molecules out of the cell is less than 1 because membranes are essentially impermeable to such molecules.

(d) All active transport processes are coupled to reactions that are exergonic under prevailing conditions.

(e) The permeability coefficient for a particular solute is likely to be orders of magnitude lower if a transport protein for that solute is present in the membrane.

(f) Plasma membranes have few, if any, transport proteins that are specific for phosphorylated compounds.

(g) Carbon dioxide and bicarbonate anions usually move in the same direction across the plasma membrane of an erythrocyte.

(h) Treatment of an animal cell with an inhibitor that is specific for the sodium-potassium pump is not likely to affect the uptake of glucose by sodium cotransport.

8-2. Telling Them Apart. From the list of properties below, indicate which one(s) can be used to distinguish between each of the following pairs of transport mechanisms.

(a) Simple diffusion; facilitated diffusion

(b) Facilitated diffusion; active transport

(c) Simple diffusion; active transport

(d) Simple active transport; cotransport

(e) Symport; antiport

Properties

1. Directions in which two transported solutes move.

2. Direction in which solute moves relative to its concentration or electrochemical gradient.

3. Kinetics of solute transport.

4. Requirement for metabolic energy.

5. Requirement for simultaneous transport of two solutes.

6. Intrinsic directionality.

7. Competitive inhibition.

8-3. Mechanisms of Transport. For each of the following statements, answer with a D if the statement is true of simple diffusion, with an F if it is true of facilitated diffusion, and with an A if it is true of active transport. Any, all, or none (N) of the choices may be appropriate for a given statement.

(a) Requires the presence of an integral membrane protein.

(b) Depends primarily on solubility properties of the substance.

(c) Doubling the concentration gradient of the molecule to be transported will double the rate of transport over a broad range of concentrations.

(d) Applies only to transcellular processes.

(e) Work is done during the transport process.

(f) Applies only to small, nonpolar solutes.

(g) Applies only to ions.

(h) Transport can occur in either direction across the membrane, depending on the prevailing concentration gradient.

(i) $\Delta G° = 0$.

(j) Michaelis constant can be calculated.

8-4. Evidence of Active Transport. The red blood cells in your body maintain internal concentrations of sodium and potassium ions that are significantly different from those in blood plasma. However, these gradients of ion concentration can be abolished if isolated red blood cells are treated in any of the following ways. In each case, explain how the effectiveness of the treatment shows that the outward movement of sodium ions and the inward movement of potassium ions are energy-requiring processes.

(a) If isolated red blood cells are treated with poisons that inhibit specific reactions involved in energy metabolism, the concentrations of potassium ions and sodium ions across the plasma membrane will slowly come to equilibrium.

(b) If isolated red blood cells are placed on ice, sodium and potassium ions will equilibrate across the plasma membrane, but normal ion gradients will be restored if the chilled cells are returned to 37°C.

(c) If isolated red blood cells are held for an extended period of time at 37°C in a medium without an energy source, they will eventually begin to leak potassium ions outward and sodium ions inward. However, the outward movement of sodium ions and inward movement of potassium ions can be restored by the addition of glucose to the medium.

(d) If valinomycin is added to isolated red blood cells, the K^+ gradient across the membrane will reach equilibrium much more rapidly than the Na^+ gradient.

8-5. Potassium Transport. Most of the cells of your body pump potassium ions inward to maintain an internal potassium concentration that is 35 times the external concentration.

(a) What is ΔG (at 37°C) for the transport of potassium ions into a cell that maintains no membrane potential across its plasma membrane?

(b) For a nerve cell with a membrane potential of -60 mV, what is ΔG for the inward transport of potassium ions at 37°C?

(c) What is the maximum number of potassium ions that can be pumped inward by the hydrolysis of one ATP molecule if the ATP/ADP ratio in the cell is 5:1 and the inorganic phosphate concentration is 10 mM? (Assume $\Delta G' = -7.3$ kcal/mol for ATP hydrolysis.)

8-6. Ion Gradients and ATP Synthesis. The ion gradients maintained across the plasma membranes of most cells play a very significant role in cellular energetics. Ion gradients are often either generated by the hydrolysis of ATP or used to make ATP by the phosphorylation of ADP.

(a) Cite an example in which ATP is used to generate and maintain an ion gradient. What is another way that an ion gradient can be generated and maintained?

(b) Cite an example in which an ion gradient is used to make ATP. What is another use to which ion gradients are put?

(c) Assume the concentration gradient for Na^+ ions given in Table 8-6 and a membrane potential of -90 mV. Can a cell use ATP hydrolysis to drive the outward transport of Na^+ ions on a 2:1 basis (two sodium ions translocated per ATP hydrolyzed) if the ATP/ADP ratio is 5, the inorganic phosphate concentration is 50 mM, and the temperature is 37°C? What about on a 3:1 basis? Explain your answers.

(d) Assume that a bacterial cell maintains a proton gradient across its plasma membrane such that the pH inside the cell is 8.0 when the outside pH is 7.0. Can the cell use the proton gradient to drive ATP synthesis on a 1:1 basis (one ATP synthesized per proton translocated) if the membrane potential is -180 mV, the temperature is 25°C, and the ATP, ADP, and inorganic phosphate concentrations are as in part c? What about on a 1:2 basis? Explain your answers.

8-7. Sodium Transport. A marine protozoan is known to pump sodium ions outward by a simple ATP-driven sodium pump that operates independently of potassium. The intracellular concentrations of ATP, ADP, and P_i are 20 mM, 2 mM, and 1 mM, respectively, and the membrane potential is -75 mV.

(a) Assuming that the pump transports three sodium ions outward per molecule of ATP hydrolyzed, what is the lowest internal sodium ion concentration that can be maintained at 25°C when the external sodium ion concentration is 150 mM?

(b) If you were dealing with a neutral molecule rather than an ion, would your answer for part a be higher or lower, assuming all conditions remained the same? Explain.

8-8. Sodium Cotransport as a Mechanism of Amino Acid Uptake. The epithelial cells that line your small intestine are thought to take up amino acids from the gut by a process of cotransport, in which the electrochemical gradient of sodium ions across the plasma membrane drives the uptake of amino acids against their concentration gradients. The following values were obtained at 37°C:

pH of epithelial cells $= 7.0$

$V_m =$ membrane potential $= -65$ mV (inside negative)

$[Na^+]_{inside} = 7.5$ mM $[Na^+]_{outside} = 105$ mM

$[glycine]_{inside} = 15.0$ mM $[glycine]_{outside} = 0.10$ mM

$[aspartate]_{inside} = 22.5$ mM $[aspartate]_{outside} = 0.15$ mM

The structures of glycine and aspartate at pH 7.0 are as follows:

Glycine   Aspartate

(a) Calculate ΔG for the movement of 1 mole-equivalent of sodium ions into the epithelial cells.

(b) Calculate ΔG for the movement of 1 mole of glycine into the epithelial cells. What is the minimum number of sodium ions that must be cotransported with one glycine molecule to drive the uptake?

(c) Calculate ΔG for the movement of 1 mole of aspartate into the epithelial cells. What is the minimum number of sodium ions that must be cotransported with one aspartate molecule to drive the uptake?

(d) Explain why the number of sodium ions required for amino acid cotransport is different for glycine (part b) and aspartate (part c).

(e) Do you think the diffusion of aspartate back out through the epithelial cell membrane would be a problem? Why or why not? What about that of glycine?

8-9. The Case of the Acid Stomach. The gastric juice in your stomach has a pH of 2.0. This acidity is due to the secretion of hydrogen ions into the stomach by the epithelial cells of the gastric mucosa. Epithelial cells have an internal pH of 7.0 and a membrane potential of -70 mV (inside negative) and function at body temperature (37°C).

(a) What is the concentration gradient of hydrogen ions across the epithelial membrane?

(b) Calculate the free energy change associated with the secretion of 1 mole of protons into gastric juice at 37°C.

(c) Do you think that hydrogen transport can be driven by ATP hydrolysis at the ratio of one molecule of ATP per hydrogen ion transported?

(d) If hydrogen ions were free to move back into the cell, calculate the membrane potential that would be required to prevent them from doing so.

8-10. Calcium Pump of Sarcoplasmic Reticulum. Muscle cells use calcium ions to regulate the contractile process. Calcium is both released and taken up by the sarcoplasmic reticulum (SR). The release of calcium from the SR activates muscle contraction, and ATP-driven calcium uptake causes the muscle cell to relax afterward. When muscle tissue is disrupted by homogenization, the SR forms small vesicles called *microsomes* that maintain their ability to take up calcium. In the experiment shown in Figure 8-15, a reaction medium was prepared to contain 5 mM Mg^{2+}-ATP and 0.1 M KCl at pH 7.5. An aliquot of SR microsomes containing 1.0 mg protein was added to 1 mL of the reaction mixture, followed by 0.4 μmol of calcium. Two minutes later, a calcium ionophore was added. ATPase activity was monitored during the additions, with the results shown in the figure.

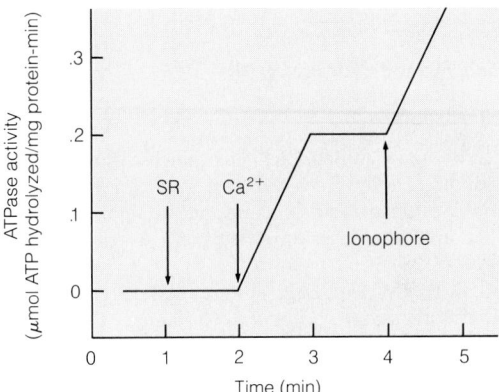

Figure 8-15   Calcium Uptake by the Sarcoplasmic Reticulum.

(a) What is the ATPase activity, calculated as micromoles of ATP hydrolyzed per milligram of protein per minute?

(b) The ATPase is calcium-activated, as shown by the increase in ATP hydrolysis when the calcium was added and the decrease in hydrolysis when all the added calcium was taken up into the vesicles 1 minute after it was added. How many calcium ions are taken up for each ATP hydrolyzed?

(c) The final addition is an ionophore that carries calcium ions across membranes. Why does ATP hydrolysis begin again?

8-11. Ionophores. Gramicidin is a channel-forming ionophore that allows the passage of K^+ ions when monomers in the two phospholipid monolayers of the membrane line up transiently (see Figure 8B-6, p. 211). When the conductance of a membrane containing a very low concentration of gramicidin is measured, the conductance changes up and down in quantum steps. Explain.

8-12. Inverted Vesicles. An important advance in transport research was the development of methods for making closed membrane vesicles that retain the activity of certain transport systems. One such system uses resealed vesicles from red blood cell membranes, in which the orientation of membrane proteins may be either the same as in the intact erythrocyte (right-side-out) or inverted (inside-out). Such vesicles have demonstrable ATP-driven sodium-potassium pump activity. By resealing the vesicles in one medium and then placing them in another, it is possible to have ATP, sodium, and potassium present inside the vesicle, present outside the vesicle, or not present at all.

(a) Suggest one or two advantages that such vesicles might have compared to intact red blood cells for studying the sodium-potassium pump. Can you think of any possible disadvantages?

(b) For the inverted vesicles, indicate whether each of the following should be present inside the vesicle (I), outside the vesicle (O), or not present at all (N) in order to demonstrate ATP hydrolysis: Na^+, K^+, ATP.

(c) If you plot the rate of ATP hydrolysis as a function of time after initiating transport in such inverted vesicles, what sort of a curve would you expect to obtain?

8-13. Ouabain Inhibition. Ouabain is a very specific inhibitor of the active transport of sodium ions out of the cell and is therefore a valuable tool in studies of membrane transport mechanisms. Which of the following processes in your own body would you expect to be sensitive to inhibition by ouabain? Explain in each case.

(a) Passive transport of glucose into a muscle cell.

(b) Active transport of dietary phenylalanine across the intestinal mucosa.

(c) Uptake of potassium ions by red blood cells.

(d) Active uptake of lactose by the bacteria in your intestine.

SUGGESTED READING

General References

Bonting, S. L., and J. J. dePont, eds. *Membrane Transport*. Amsterdam: Elsevier/North Holland, 1981.

Gennis, R. B. *Biomembranes: Molecular Structure and Function*. New York: Springer-Verlag, 1989.

Graves, J. S., ed. *Regulation and Development of Membrane Transport Processes*. New York: Wiley, 1985.

Martinosi, A. N., ed. *Membranes and Transport*, vol. 1. New York: Plenum, 1982.

Stein, W. D. *Channels, Carriers, and Pumps: An Introduction to Membrane Transport*. New York: Academic Press, 1990.

Stein, W. D. *Transport and Diffusion Across Cell Membranes*. Orlando, FL: Academic Press, 1986.

Passive Transport

Gould, G. W., and G. I. Bell. Facilitative glucose transporters: An expanding family. *Trends Biochim Sci.* 15 (1990): 18.

Jennings, M. L. Oligomeric structure and the anion transport function of human erythrocyte band 3 protein. *J. Membrane Biol.* 80 (1984): 105.

Lienhard, G. E., J. W. Slot, D. E. James, and M. M. Mueckler. How cells absorb glucose. *Sci. Amer.* 266 (January 1992): 86.

Pessin, J. E., and G. I. Bell. Mammalian facilitative glucose transporter family. *Annu. Rev. Physiol.* 54 (1992): 911.

Widdas, W. F. Old and new concepts of the membrane transport for glucose in cells. *Biochim. Biophys. Acta* 947 (1988): 386.

Active Transport

Baldwin, S. A., and P. J. F. Henderson. Homologies between sugar transporters from eukaryotes and prokaryotes. *Annu. Rev. Physiol.* 51 (1989): 459.

Carafoli, E., and M. Chiesi. Calcium pumps in the plasma and intracellular membranes. *Curr. Topics Cell. Regul.* 32 (1992): 209.

Henderson, P. J. F. Proton-linked sugar transport systems in bacteria. *J. Bioenerg. Biomembr.* 22 (1990): 571.

Hobbs, A. S., and R. W. Alberts. The structure of proteins involved in active membrane transport. *Annu. Rev. Biophys. Bioeng.* 9 (1980): 259.

Rossier, B. C., K. Geering, and J. P. Krahenbuhl. Regulation of the sodium pump: How and why. *Trends Biochem. Sci.* 10 (1987): 483.

Stein, W. D., ed. *Ion Pumps: Structure, Function, and Regulation*. New York: A. R. Liss, 1988.

Stekhoven, F. S., and S. L. Bonting. Transport adenosine triphosphatases: Properties and function. *Physiol. Rev.* 61 (1981): 1.

Tanford, C. Mechanism of free energy coupling in active transport. *Annu. Rev. Biochem.* 52 (1983): 379.

Specific Transport Mechanisms

Cantley, L. C. Structure and mechanism of the (Na,K)-ATPase. *Current Topics Bioenerg.* 11 (1981): 201.

Dills, S. S., A. Apperson, M. R. Schmidt, and M. H. Saier, Jr. Carbohydrate transport in bacteria. *Microbiol. Rev.* 44 (1980): 385.

Glynn, I. M. The Na^+-K^+ transporting adenosine triphosphatase. In *The Enzymes of Biological Membranes*, 2d ed. (A. Martinosi, ed.). New York: Plenum, 1985.

Horisberger, J.-D., V. Lemas, J. P. Krahenbuhl, and B. C. Rossier. Structure-function relationships of the Na,K-ATPase. *Annu. Rev. Physiol.* 53 (1991): 565.

Kimmich, G. A. Membrane potentials and the mechanism of intestinal Na^+-dependent sugar transport. *J. Membrane Biol.* 114 (1990): 1.

Jay, D., and C. Cautley. Structural aspects of the red cell anion exchange protein. *Annu. Rev. Biochem.* 55 (1986): 311.

Jones, M. N., and J. K. Nickson. Monosaccharide transport proteins of the human erythrocyte membrane. *Biochim. Biophys. Acta* 650 (1981): 1.

Postma, P. W., and J. W. Lengeler. Phosphoenolpyruvate: Carbohydrate phosphotransferase system of bacteria. *Microbiol. Rev.* 49 (1985): 232.

Scott, D. M. Sodium cotransport systems: Cellular, molecular and regulatory aspects. *Bioessays* 7 (1987): 71.

Trachtenberg, M. C., D. J. Packey, and T. Sweeney. In vivo functioning of the Na,K-activated ATPase. *Curr. Topics Cell. Regul.* 19 (1981): 217.

Box 8A: Membrane Transport, Cystic Fibrosis, and the Prospects for Gene Therapy

Collins, F. S. Cystic fibrosis: Molecular biology and therapeutic implications. *Science* 256 (1992): 774.

Collins, F. S., J. R. Riordan, and L.-C. Tsui. The cystic fibrosis gene: Isolation and significance. *Hospital Practice* 25 (1990): 47.

Lindahl, R., and R. Parry. Molecular medicine: A primer for clinicians. Part IV: Cystic fibrosis and the power and limitations of molecular medicine. *South Dakota J. Med.* 46 (1993): 393.

Sferra, T. J., and F. S. Collins. The molecular biology of cystic fibrosis. *Annu. Rev. Med.* 44 (1993): 133.

9

INTRACELLULAR COMPARTMENTS:
THE ENDOPLASMIC RETICULUM, GOLGI COMPLEX,
LYSOSOMES, AND PEROXISOMES

The study of eukaryotic cells is essentially the exploration of an elaborate series of membrane-bounded compartments in which most cellular activities take place. Whether we consider the storage and readout of genetic information, the secretion of proteins, the digestion of food particles, or the breakdown of long-chain fatty acids, the locale for the process within a eukaryotic cell is in each case a membrane-bounded organelle. To appreciate the eukaryotic cell fully, therefore, is to understand the prominent role that intracellular membranes play and the compartmentalization of function that these membranes make possible. (Because prokaryotes have few internal membranes, most of the discussion in this chapter does not apply to them.)

We met each of the major organelles in Chapter 4; now we are ready to begin considering them in more detail. Some organelles will be encountered in coming chapters, when we discuss cellular functions localized to the mitochondrion (respiration; Chapter 12), the chloroplast (photosynthesis; Chapter 13), the nucleus (information storage and readout; Chapter 14), and the ribosome (protein synthesis; Chapter 18).

In this chapter, we will consider the endoplasmic reticulum, the Golgi complex, the lysosome, and the peroxisome. Before doing so, however, we will take note of several techniques that have contributed significantly to our present understanding of these and other organelles, and then look at three centrifugation techniques in detail.

Microscopy and Centrifugation: Indispensable Techniques of Contemporary Cell Biology

We owe much of our current understanding of eukaryotic membranes and organelles to insights provided by microscopists and biochemists, and especially to studies made pos-

sible by the development of the electron microscope and the ultracentrifuge. These instruments became available to biologists at about the same time and quickly became indispensable tools for the study of cells, organelles, membranes, and macromolecules.

The **electron microscope** came into use by cell biologists in the early 1950s, making possible detailed ultrastructural (high-resolution) studies that rapidly revolutionized our understanding of cellular architecture. Electron microscopy enabled cell biologists not only to study known structures such as the mitochondrion and nucleus in unprecedented detail but also to describe and explore cellular structures that had never been seen before, including ribosomes and components of the cytoskeleton.

At about the same time that electron microscopy was enabling cell biologists to see cellular structures with extraordinary resolution, the development of the **ultracentrifuge** made it possible for biochemists to isolate and separate subcellular structures and macromolecules. Developed initially to determine the sedimentation rates of proteins, the ultracentrifuge proved equally useful as a means of isolating and purifying the organelles and other cellular structures that electron microscopy had so recently revealed.

The approaches of the microscopist and the biochemist are complementary because the microscopist usually describes cells in terms of their *structure*, whereas the biochemist focuses on *function*. For a detailed discussion of microscopy in general and electron microscopy in particular, see the Appendix. Here, we will consider centrifugation as an indispensable tool for the isolation and purification of specific organelles and macromolecules, a procedure commonly referred to as subcellular fractionation.

Centrifugation and Subcellular Fractionation

Centrifugation is a useful means of isolating and purifying cellular components because most organelles and macromolecules differ from one another significantly in *size* and *density* (and sometimes also in *shape*, especially in the case of macromolecules). These properties, as well as the density and viscosity of the surrounding medium, determine the rate at which a specific particle will move when it is subjected to a centrifugal force by being spun at high speed in a centrifuge. Because of their inherent difference in size, shape, and/or density, the various organelles and other cellular structures can usually be separated, or *fractionated*, from one another by centrifugation. This process, called **subcellular fractionation,** enables cell biologists to isolate and purify specific cellular components, which can then be studied in vitro.

For their pioneering work in this area, Albert Claude, George Palade, and Christian de Duve shared a Nobel Prize in 1974. Claude was instrumental in developing *differential centrifugation* as a way of isolating specific organelles. Palade was quick to use this new technique in studies of the endoplasmic reticulum and the Golgi complex. This enabled him to establish the roles of these organelles in the synthesis, processing, and secretion of proteins, as we will learn later in this chapter. De Duve, in turn, was the discoverer of both lysosomes and peroxisomes. In each case, subcellular fractionation and electron microscopy played vital roles in the detection and characterization of the new class of organelles. De Duve's discovery of lysosomes depended on *differential centrifugation* (see Box 4A). His discovery of peroxisomes, detailed later in this chapter, depended specifically on *equilibrium density centrifugation*, a powerful technique for resolving organelles and macromolecules from one another based on density differences.

We will discuss each of these techniques, beginning with a consideration of centrifuges and sample preparation in general.

Centrifuges. In essence, a centrifuge consists of a rotor that is driven by an electric motor. The rotor holds tubes that contain solutions or suspensions of the particles to be separated. Rotors are of two basic types: *Fixed-angle rotors* have wells in which the tubes are maintained at a specific angle (Figure 9-1a), whereas *swinging-bucket rotors* have hinges that allow the buckets to swing out as the rotor spins (Figure 9-1b). Many centrifuges have a refrigerated chamber to maintain the sample at a specified temperature, usually just above freezing. Centrifugation at very high speeds (above about 20,000 rpm) requires an *ultracentrifuge* that has a vacuum system to reduce heating due to air friction and armor plating around the chamber to contain the rotor in the event of an accident. Some ultracentrifuges are capable of speeds up to 80,000 revolutions per minute (rpm), reaching forces up to 500,000 times the force of gravity (*g*).

Sample Preparation. Separation of cellular components by centrifugation requires that the tissue first be disrupted, or *homogenized*, usually in a cold isotonic solution (0.25 *M* sucrose is often used for this purpose). Disruption can be achieved by forcing cells through a narrow orifice, by subjecting the tissue to ultrasonic vibration or osmotic shock,

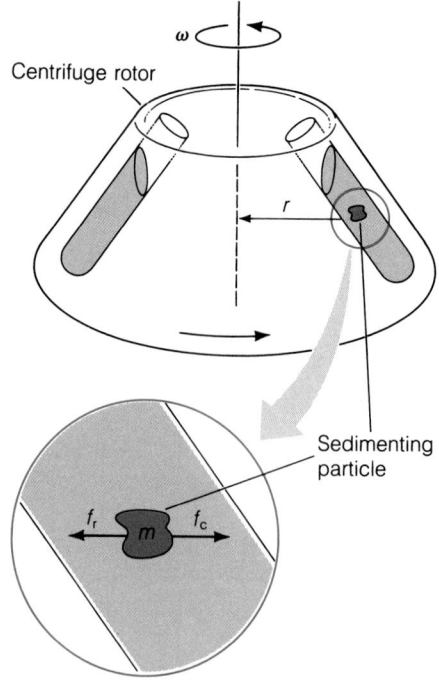

(a) Fixed-angle rotor

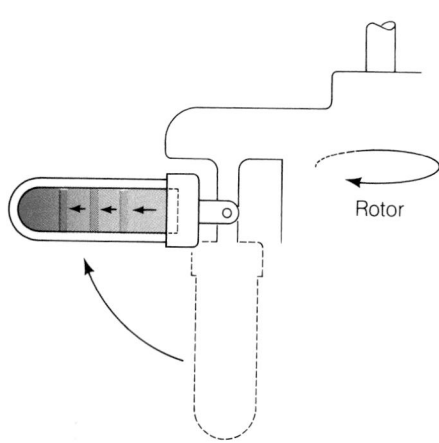

(b) Swinging-bucket rotor

Figure 9-1 Centrifuge Rotors. A centrifuge consists of a rotor driven by an electric motor, usually within a refrigerated chamber. The rotor either **(a)** holds the tubes at a fixed angle or **(b)** has hinged buckets to allow the tubes to swing out and be parallel to the centrifugal force field during the centrifugation process. For a particle (a macromolecule or organelle, usually) sedimenting in a centrifugal force field, the centrifugal force F_c acting on the particle is given by the equation $F_c = m\omega^2 r$, where m is the mass of the particle (in g/cm^3), ω is the angular velocity of the rotor (in radians/sec), and r is the distance from the axis of rotation to the particle (in cm).

or by grinding in a mortar and pestle, sometimes in the presence of glass beads or other abrasive material. Depending on the specific organelles to be isolated, a detergent may also be used to solubilize membranes. The resulting **homogenate** is a suspension of organelles, molecules, and other cellular components. If the tissue has been disrupted gently enough, many of the organelles and other structures in the homogenate are likely to remain intact and retain their original biochemical functions.

Centrifugation Techniques

Because they differ greatly in size and density, most organelles can be at least partially resolved from one another. The three most common techniques for this purpose are differential centrifugation, density gradient centrifugation, and equilibrium (or buoyant) density centrifugation. The latter two techniques are also applicable for macromolecules.

Differential Centrifugation. **Differential centrifugation** is an effective means of subcellular fractionation because organelles differ from one another so much in size and weight that they move, or *sediment*, at very different rates in response to centrifugal force. As shown schematically in Figure 9-2, particles that are large and/or dense (purple spheres) sediment rapidly, those that are intermediate in size and/or density (blue spheres) sediment less rapidly, while smaller and/or less dense particles (black dots) sediment still more slowly.

Thus, differential centrifugation takes advantage of differences in *sedimentation rate* to separate organelles. In fact, one way to express the size of an organelle or molecule is in terms of its *sedimentation coefficient*, which measures how rapidly the particle sediments when subjected to centrifugation. Sedimentation coefficients are normally expressed in *Svedberg units* (S) and are widely used to indicate relative sizes of organelles and macromolecules. (A sedimentation coefficient actually has the units of seconds; the Svedberg unit (S) is defined such that $1\ S = 1 \times 10^{-13}$ sec.)

Differential centrifugation is illustrated in Figure 9-3. The tissue of interest is first homogenized, usually in a cold isotonic solution. Subcellular fractions are then isolated by subjecting first the homogenate and then subsequent supernatant fractions to successively higher centrifugal forces and/or to longer centrifugation times. The *supernatant* is the clarified homogenate that remains after particles of a given size and density are removed as a *pellet* by the centrifugation process. In each case, the supernatant from one step is decanted, or poured off, into a new centrifuge tube and then returned to the centrifuge and subjected to greater centrifugal force to obtain the next pellet. In successive steps, the pellets are enriched in unbroken cells and debris; mitochondria, lysosomes, and peroxisomes; ER and other membrane fragments; and free ribosomes and large macromolecules. The material in each pellet can be resuspended and used for biochemical studies, if desired. The final supernatant contains only soluble cellular components and is called the *cytosol*. (An ultracentrifuge is needed to achieve the forces required for steps 4 and 5, whereas steps 2 and 3 are usually carried out in a centrifuge that is refrigerated but without a vacuum system or armored plating.)

Each of the fractions obtained in this way is enriched for the respective organelles but is also likely to be contaminated with other organelles and cellular components. Most of the contaminants in a specific pellet can usually be removed by resuspending the pellet and repeating the centrifugation procedure.

Density Gradient Centrifugation. **Density gradient** (or **rate-zonal**) **centrifugation** is a variation of the differential centrifugation technique. Instead of being uniformly distributed throughout the solution or suspension at the start, the particles to be separated are present initially as a thin layer on top of a solution containing a solute that increases

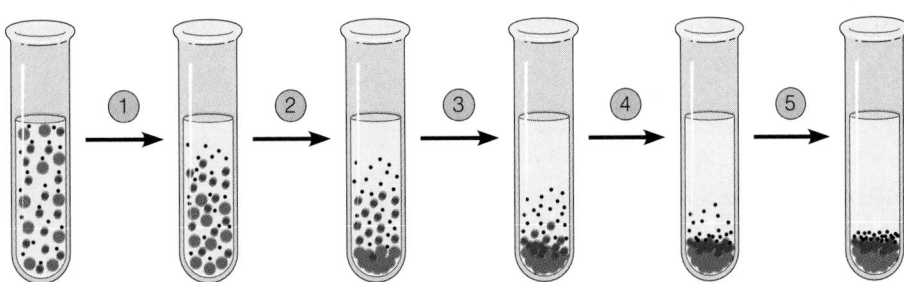

- ● Particles with large sedimentation coefficients
- ● Particles with intermediate sedimentation coefficients
- · Particles with small sedimentation coefficients

Figure 9-2 Differential Centrifugation. Differential centrifugation is a means of separating particles based on differences in size, shape, and/or density that result in different rates of sedimentation in response to centrifugal force. These effects are illustrated here by three kinds of particles that differ significantly in size. A solution containing the particles is subjected to a fixed centrifugal force for five successive time intervals (circled numbers). Particles that are large and/or dense (purple spheres) sediment rapidly in response to centrifugal force. Particles that are intermediate in size and/or density (blue spheres) sediment less rapidly, while smaller and/or less dense particles (black dots) sediment still more slowly.

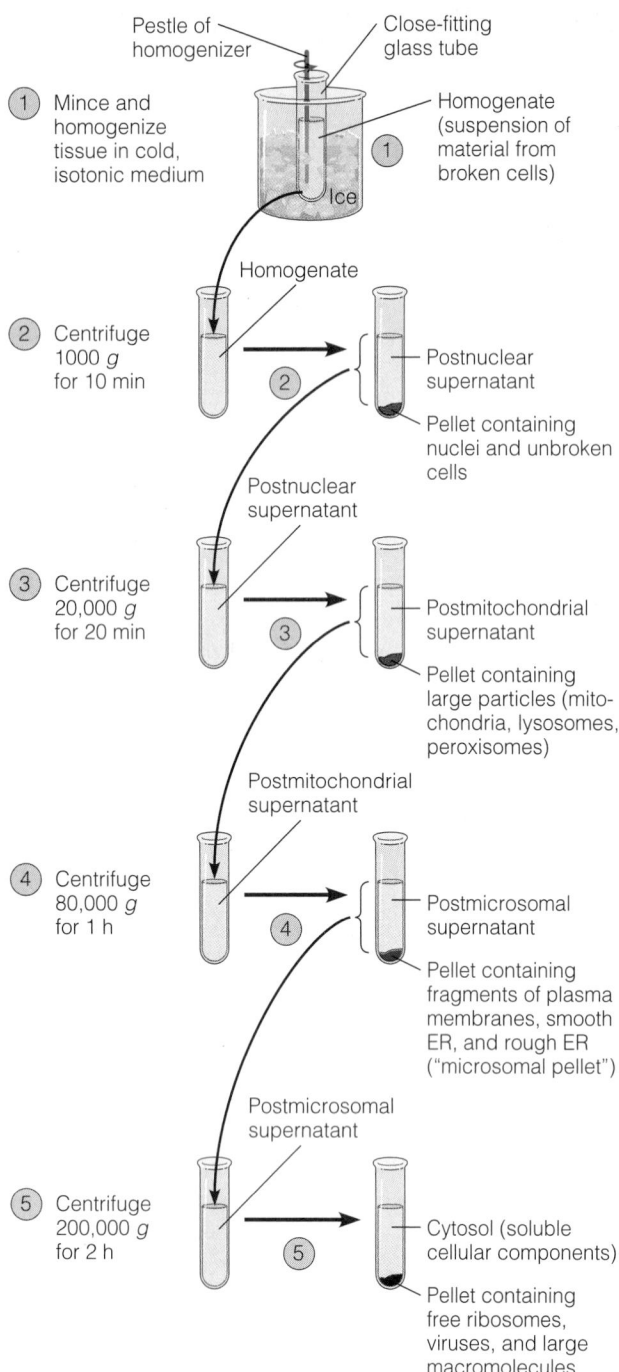

Figure 9-3 Differential Centrifugation as a Means of Isolating Organelles. The use of differential centrifugation in subcellular fractionation is shown here. ① Tissue is first minced and homogenized in a cold isotonic solution. ② Nuclei and unbroken cells are recovered by centrifugation at low speed, followed by ③ recovery of large organelles such as mitochondria, lysosomes, and peroxisomes at a higher speed. ④ The postmitochondrial supernatant is centrifuged at a significantly higher centrifugal force to recover microsomes (fragments of rough ER) and other membrane fragments. ⑤ To sediment free ribosomes, viral particles, and large macromolecules, the postmicrosomal supernatant is centrifuged at a still higher centrifugal force. In each case, the acceleration due to centrifugation, $\omega^2 r$, is expressed as a multiple of g, the acceleration due to gravity (980 cm/sec²). The final supernatant is called the cytosol.

Labels within figure:

① Mince and homogenize tissue in cold, isotonic medium

Pestle of homogenizer

Close-fitting glass tube

Homogenate (suspension of material from broken cells)

Ice

Homogenate

② Centrifuge 1000 g for 10 min

Postnuclear supernatant

Pellet containing nuclei and unbroken cells

Postnuclear supernatant

③ Centrifuge 20,000 g for 20 min

Postmitochondrial supernatant

Pellet containing large particles (mitochondria, lysosomes, peroxisomes)

Postmitochondrial supernatant

④ Centrifuge 80,000 g for 1 h

Postmicrosomal supernatant

Pellet containing fragments of plasma membranes, smooth ER, and rough ER ("microsomal pellet")

Postmicrosomal supernatant

⑤ Centrifuge 200,000 g for 2 h

Cytosol (soluble cellular components)

Pellet containing free ribosomes, viruses, and large macromolecules

in concentration, and hence in density, from the top to the bottom of the tube. When subjected to a centrifugal force, particles that differ in size and/or density move downward as discrete bands, or zones, that migrate at different rates.

This process is illustrated schematically in Figure 9-4. The largest and/or densest particles (purple spheres) move into the gradient as a rapidly sedimenting band, while particles that are intermediate in size and/or density (blue spheres) sediment less rapidly, and the smallest particles (black dots) move still more slowly. Because of the gradient of solute in the tube, the particles at the leading edge of each band continually encounter a slightly more dense solution and are therefore slowed slightly in their sedimentation. As a result, each band is kept very compact, maximizing the resolution of different-sized particles.

Centrifugation is stopped after the bands of interest have moved sufficiently far into the gradient to be resolved from each other but before any of the bands reaches the bottom of the tube. Stopping at this point is essential, because the density of the solution is *less* than the density of any of the particles, even at the very bottom of the tube. If centrifugation were continued too long, the bands would reach the bottom of the tube one after another, piling up on top of each other and negating the very purpose of the process.

Density gradient centrifugation is widely used for separating both organelles and macromolecules. Figure 9-5 (page 234) illustrates the separation of mitochondria and lysosomes by this technique. The tissue of interest is homogenized in a cold isotonic solution, and a pellet containing primarily mitochondria and lysosomes is prepared as in Figure 9-3 (steps 1–3). (For simplicity, we will assume that the tissue contains few peroxisomes.) The pellet is resuspended and the suspension is then layered onto a sucrose solution that increases in concentration and density from the top to the bottom of a plastic or celluloid tube. (A glass tube cannot be used because the tube needs to be punctured to collect the bands; see step 4).

Upon centrifugation, the mitochondria move into the gradient as a band that sediments more rapidly than the band of lysosomes because mitochondria are both larger and denser than lysosomes. After a suitable length of time, the centrifuge is stopped, the bottom of the tube is punctured, and fractions are collected. Because the mitochondria move faster than the lysosomes and are therefore closer to the bottom of the tube when it is punctured, a series of mitochondria-enriched fractions will be collected first, followed by a series of lysosome-enriched fractions. By assaying each of the fractions for marker enzymes that are unique to either mitochondria or lysosomes, the fractions containing these organelles can be readily identified and the extent of cross-contamination determined.

Equilibrium Density Centrifugation. The third major centrifugation technique is called **equilibrium** (or **buoyant**) **density centrifugation**. This procedure also uses a gradient of solute concentration and density, but in this case the solute is sufficiently concentrated such that the density gra-

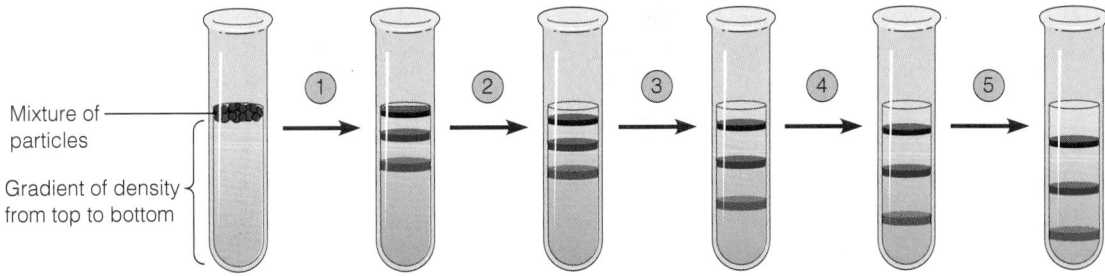

Mixture of particles

Gradient of density from top to bottom

- Particles with large sedimentation coefficients
- Particles with intermediate sedimentation coefficients
· Particles with small sedimentation coefficients

Figure 9-4 Density Gradient Centrifugation. Density gradient centrifugation is a means of separating particles (organelles or macromolecules) based on differences in size, shape, and/or density that result in different rates of sedimentation in response to centrifugal force. These effects are illustrated here by three kinds of particles that differ significantly in size. A concentrated mixture of the three kinds of particles is layered onto a solution of a solute that increases in concentration (and hence in density) from the top to the bottom of a celluloid tube (but is lower in density than any of the particles even at the bottom of the tube). The tube is then subjected to a fixed centrifugal force for five successive time intervals (circled numbers). Particles that are large and/or dense (purple spheres) move into the gradient as a rapidly sedimenting band. Particles that are intermediate in size and/or density (blue spheres) sediment less rapidly, forming a more slowly moving band, while smaller particles (black dots) form a band that moves still more slowly. The centrifugation process is stopped after the bands of interest have moved sufficiently far into the gradient to be resolved from each other but before any of the bands reach the bottom of the tube. To collect the bands as separate fractions, a hole is punched in the bottom of the tube, as shown in Figure 9-5.

dient spans the range of densities of the organelles or macromolecules to be separated on it. For organelles, a gradient of *sucrose* is often used, and the density range is 1.10–1.30 g/cm³, corresponding to a sucrose concentration range of about 0.75–2.3 *M*. For macromolecules, on the other hand, *cesium chloride* (*CsCl*) is usually the solute of choice. (For a classic experiment in which CsCl density gradients were used to resolve double-stranded DNA molecules containing the ^{14}N versus ^{15}N isotopes of nitrogen (a density difference of about 1%), as well as to detect hybrid [$^{14}N/^{15}N$] DNA molecules at an intermediate density, see Figure 15-3.)

Figure 9-6 illustrates the use of equilibrium density centrifugation for the separation of organelles on a sucrose gradient. The tissue of interest is first homogenized in a cold isotonic solution (0.25 *M* sucrose), and a pellet containing primarily mitochondria, lysosomes, and peroxisomes is prepared as in Figure 9-3, steps 1–3. The pellet is resuspended in 0.25 *M* sucrose, and the suspension is layered onto a sucrose gradient that increases in density from 1.10 to 1.30 g/cm³, a range that includes the densities of all three organelles.

Upon centrifugation, the organelles move into the gradient until each reaches its *equilibrium,* or *buoyant, density*—the point in the gradient at which the density of the sucrose is exactly equal to the density of the organelle. At its buoyant density, usually represented by ρ (the Greek letter rho), an organelle has no net force acting on it, so it moves no further. Given enough time, all organelles will reach their characteristic buoyant density positions in the gradient and will remain there. When the organelles are at their equilibrium positions, the centrifuge is stopped, the bottom of the tube is punctured, and fractions are collected.

Because of their differing densities, the three classes of organelles are recovered in different fractions.

The Organelles

Having considered centrifugation techniques in some detail, we are now ready to discuss the properties and roles of four specific organelles that have been isolated and studied using these and other techniques. Each of these organelles is highlighted in Figure 9-7 (page 235). We will focus first on the endoplasmic reticulum and the Golgi complex. Then, we will move on to lysosomes and peroxisomes, organelles with specific properties and functions of their own. In each case, we will ask how the organelle was discovered and how it is studied, what we know about its structure, and what sorts of functions take place in or on it.

The Endoplasmic Reticulum

The **endoplasmic reticulum (ER)** is one of the most pervasive features of eukaryotic cells. Unlike more prominent organelles such as the mitochondrion and chloroplast, however, the ER cannot usually be seen with the light microscope. In the late nineteenth century it was noted that some eukaryotic cells, particularly those involved in secretion, contained regions that stained intensely with basic dyes. These regions were called *ergastoplasm*, but their significance remained in doubt until the advent of electron microscopy in the 1950s. With the tremendous increase in resolving power that this technique allowed, it became possible for the first time to visualize the elaborate network of

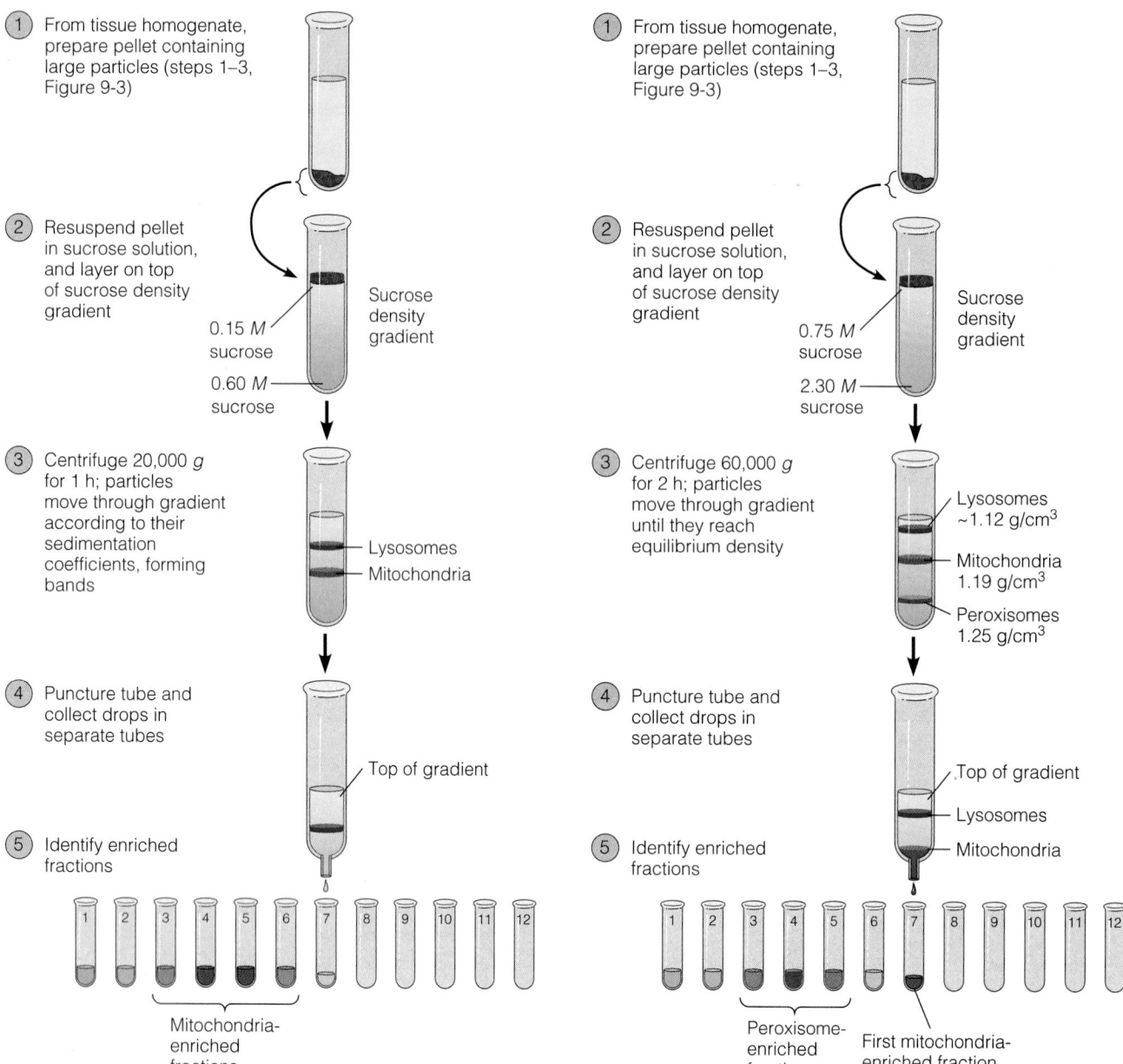

Figure 9-5 Density Gradient Centrifugation as a Means of Isolating Organelles. The use of density gradient centrifugation in subcellular fractionation is illustrated here for the separation of mitochondria and lysosomes, based on the larger size and greater density of mitochondria. ① The tissue of interest is first minced and homogenized as in Figure 9-3, step 1, in a cold isotonic solution (0.25 *M* sucrose). A pellet containing primarily mitochondria and lysosomes is then prepared by the centrifugation procedure also shown in Figure 9-3, steps 2 and 3 (for simplicity, assume that the tissue contains few peroxisomes). ② The pellet is resuspended in a cold sucrose solution, and the suspension is then layered onto a sucrose solution that increases in density from the top to the bottom of a celluloid tube. ③ Upon centrifugation, the mitochondria sediment as a band that moves more rapidly than the band of lysosomes. After a suitable length of time, the centrifuge is stopped, ④ the bottom of the tube is punctured, and fractions are collected as the sucrose solution drips out of the bottom of the tube. ⑤ Contents of individual fractions are then identified, usually by assaying for specific marker enzymes.

Figure 9-6 Equilibrium Density Centrifugation as a Means of Isolating Organelles. The use of equilibrium (or buoyant) density centrifugation in subcellular fractionation is illustrated here for the separation of mitochondria, lysosomes, and peroxisomes based on the differing densities of these organelles. ① The tissue of interest is minced and homogenized in a cold isotonic solution (0.25 *M* sucrose), and a pellet containing primarily mitochondria, lysosomes, and peroxisomes is prepared. ② The pellet is resuspended in a cold sucrose solution, and the suspension is layered onto a sucrose solution that increases in density from 1.10 g/cm³ (0.75 *M*) to 1.30 g/cm³ (2.30 *M*). ③ Upon centrifugation, the organelles move into the gradient until each is at its equilibrium density. The centrifuge is then stopped, ④ the bottom of the tube is punctured, and fractions are collected as the sucrose solution drips out of the bottom of the tube. ⑤ Contents of individual fractions are then identified, usually by assaying for specific marker enzymes.

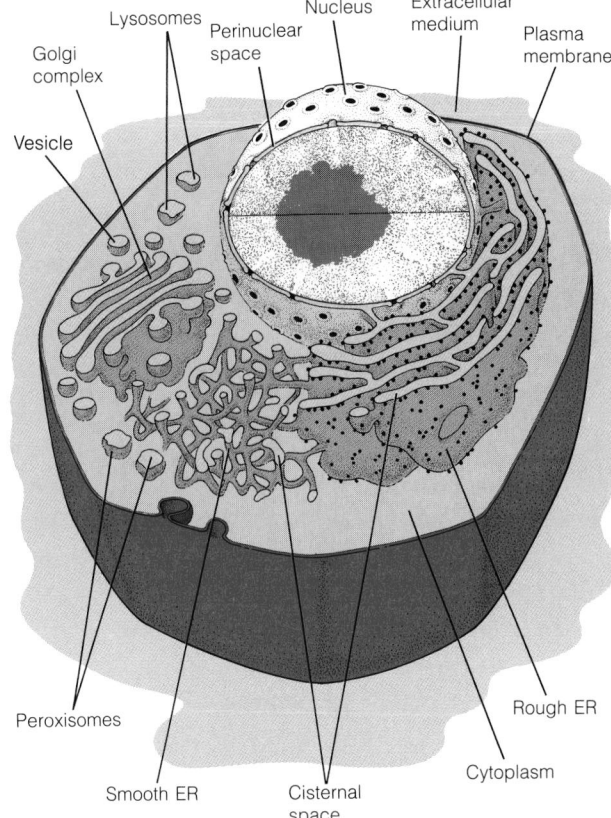

Figure 9-7 The Main Cellular Structures Discussed in This Chapter. This chapter focuses on the endoplasmic reticulum, the Golgi complex, lysosomes, and peroxisomes, in that order. Also relevant to the discussion are the plasma membrane and the nucleus with its envelope of two membranes.

intracellular membranes that came to be called the endoplasmic reticulum.

Although the name sounds formidable, it is, in fact, quite descriptive. *Endoplasmic* simply means "within the (cyto)plasm," and *reticulum* is a Latin word meaning "network." (Actually, the term was originally coined because of the extent to which the ER resembled a *reticule*, a netted handbag with a drawstring that was popular in the late nineteenth century.)

The ER, then, is an interconnecting membranous network of vesicles, tubules, and flattened sacs. These sacs are called **cisternae,** and the internal volume that they define is called the **cisternal space.** The cisternal space of the ER is continuous with the *perinuclear space* between the two membranes of the *nuclear envelope* and can communicate with the internal spaces of the Golgi complex and lysosomes by means of *vesicles* that shuttle between these structures. Moreover, these internal spaces also communicate with the exterior of the cell as vesicles fuse with, and form from, the plasma membrane. This system of intracellular spaces is separated from the surrounding **cytosol** by the membranes of these structures. The ER, Golgi complex, lysosomes, and nuclear envelope, along with the plasma membrane and the

vesicles that interconnect these structures, comprise the **endomembrane system** of the eukaryotic cell. These relationships are illustrated in Figure 9-8.

The ER Membrane

Our knowledge of the endoplasmic reticulum comes largely from studies of hepatocytes (liver cells) because these cells are relatively accessible and can be obtained in substantial quantities. The properties of hepatocyte ER can usually be generalized to other cells, but these cells also have a few specialized functions that we will point out as we go along.

ER membranes are only about 5-6 nm thick, which makes them noticeably thinner than the plasma membrane. ER membranes also differ significantly from plasma membranes in their higher protein/lipid ratios and in their lipid composition. Data for the plasma and ER membranes of hepatocytes are shown in Table 9-1. Notice that the protein/lipid ratio is about 1.5:1 for the plasma membrane but 2.3:1 for the ER. ER membranes also have a much higher relative content of phosphoglycerides and correspondingly lower levels of cholesterol, sphingomyelin, and especially glycolipids.

Much of the protein content can be accounted for in terms of enzymes known to be associated with the ER. Some of the best-known enzymes of the ER include cytochrome b_5 reductase, cytochrome c reductase, and glucose-6-phosphatase. Because these enzymes are unique to the endoplasmic reticulum, they are often used as markers to identify and keep track of the ER in subcellular fractionation experiments. Other prominent proteins include cytochrome b_5 and cytochrome P-450, both of which are components of an *electron transport chain* present in the ER of hepatocytes. The function of this transport chain is to transfer electrons stepwise from a reduced coenzyme called nicotinamide dinucleotide phosphate (NADPH) to oxygen, thereby activating the oxygen for hydroxylation reactions involved in detoxification processes. (We will encounter electron transport chains in more detail as part of our discussion of energy metabolism in Chapters 12 and 13.)

Two Types of Endoplasmic Reticulum

The ER actually consists of two different kinds of membranous networks that can be readily distinguished from each other by the presence or absence of membrane-bound ribosomes. **Rough endoplasmic reticulum (rough ER)** is characterized by the presence of ribosomes on the outer surface of the membrane. **Smooth endoplasmic reticulum (smooth ER)** is "smooth" because of the absence of ribosomes. (It is, by the way, the RNA in the ribosomes of rough ER that reacts strongly with the basic dyes originally used to identify the ergastoplasm seen with the light microscope. Ergastoplasm, in other words, turned out to be rough ER.)

Rough and smooth ER can also be distinguished morphologically. Rough ER membranes usually consist of large, flattened sheets, whereas smooth ER membranes are gener-

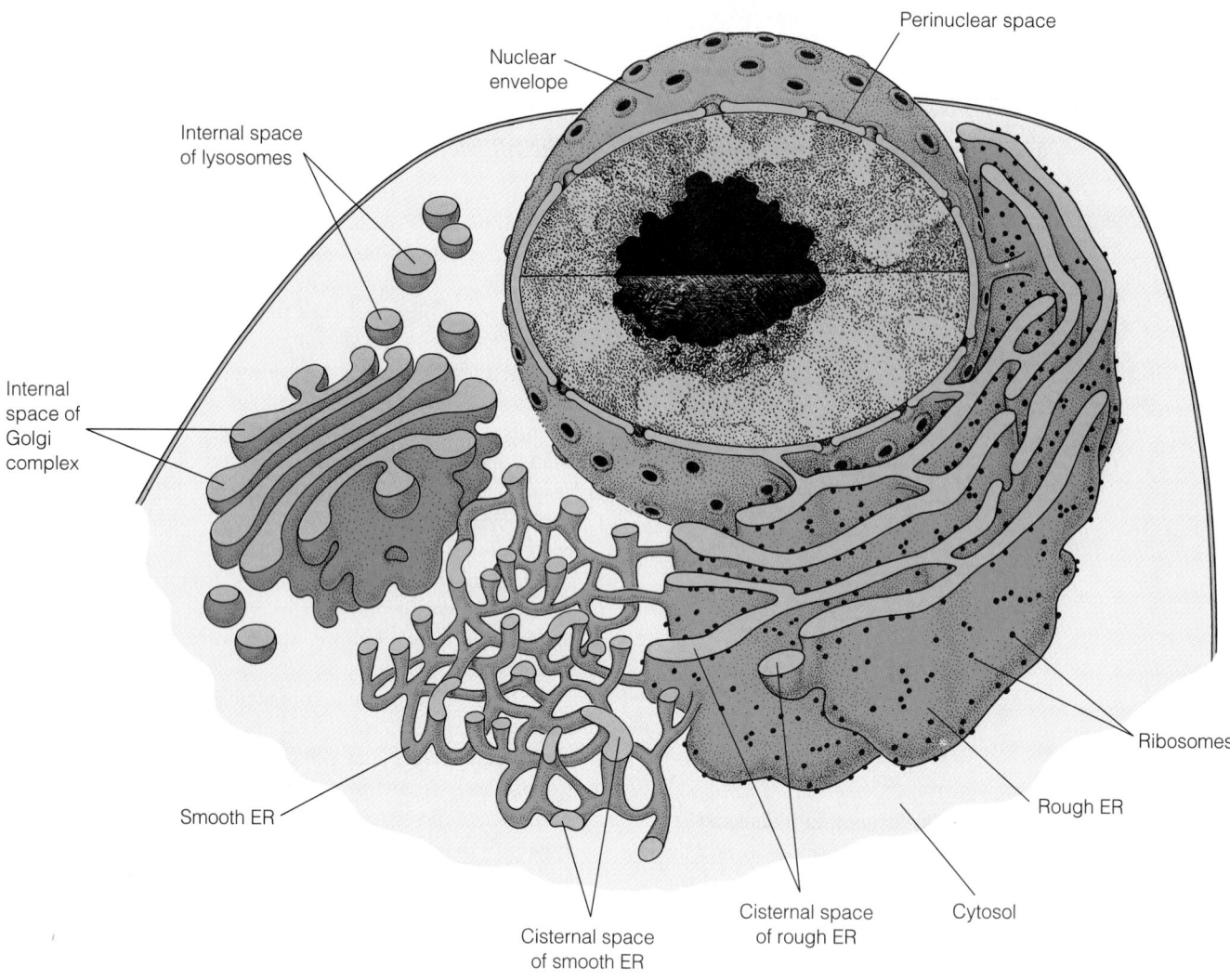

Figure 9-8   The Cisternal Space of the ER.   The cisternal space of the ER consists of the internal volume of the cisternae of the rough ER and smooth ER and is continuous with the perinuclear space between the two membranes of the nuclear envelope. The cisternal space is linked to the internal spaces of the Golgi complex and lysosomes by vesicles that shuttle between these compartments. This network of intracellular cisternae and spaces is separated from the surrounding cytosol by the membranes of these structures, which comprise the endomembrane system of the eukaryotic cell.

Table 9-1   Composition of the Plasma and ER Membranes of Rat Hepatocytes (Liver Cells)

Membrane Components	Plasma Membrane	ER Membrane
Membrane components (as % of membrane by weight)		
Carbohydrate	10	10
Protein	54	62
Total lipid	36	27
Protein/lipid ratio	1.5:1	2.3:1
Membrane lipids (as % of total lipids by weight)		
Phosphoglycerides*	35	62
Cholesterol	17	6
Sphingomyelin	19	5
Glycolipids	7	Trace
Other lipids	22	27

*Includes phosphatidyl choline, phosphatidyl ethanolamine, and phosphatidyl serine.

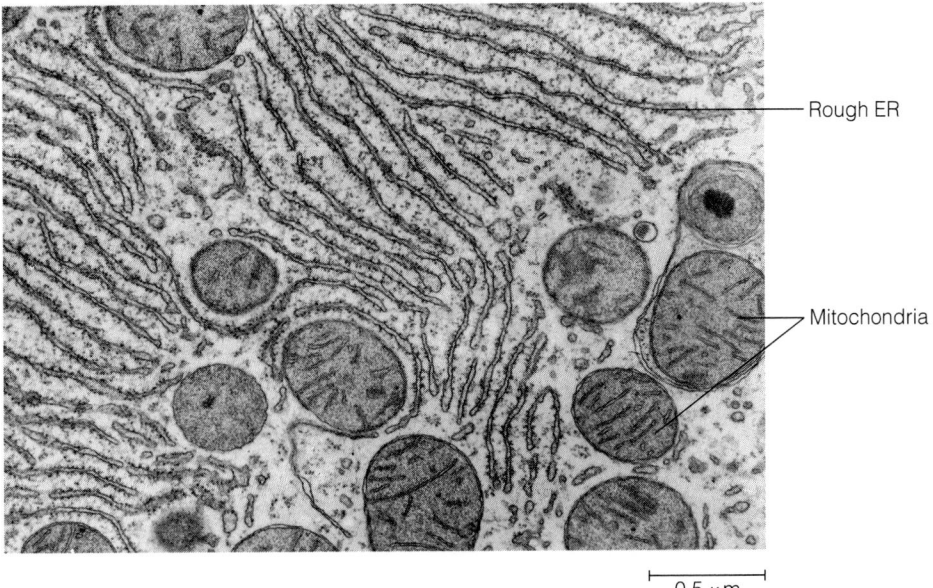

Rough ER

Mitochondria

0.5 μm

Figure 9-9   Rough ER.   This electron micrograph shows rough ER in a human liver cell. The rough ER is studded with ribosomes. Also visible in this micrograph are several mitochondria (TEM).

ally more tubular in shape. In most cells, components of rough and smooth ER appear to be continuous.

Both types of ER are present in most eukaryotic cells, but considerable variation is seen in the relative amounts, depending on the activities of the cell. Cells characterized by the synthesis of secretory proteins tend to have a very prominent rough ER network, as shown in Figure 9-9 for a liver cell. This is due to the involvement of rough ER in the process of protein synthesis (which, as you have probably guessed, has something to do with the presence of all those ribosomes on the rough ER surface). Smooth ER, on the other hand, is involved in the synthesis of steroid hormones and is therefore a prominent feature of steroid hormone-producing cells (Figure 9-10).

When tissue is homogenized for subcellular fractionation, the ER membranes break into smaller fragments, which then seal spontaneously into vesicles called **microsomes.** When microsomes are prepared by differential centrifugation, fractions can be isolated with and without attached ribosomes, depending on whether the membrane originated from the rough or smooth ER. Such preparations have proved tremendously useful in exploring both types of ER, but keep in mind that microsomes do not exist as such in the cell; they are simply an artifact of the fractionation process.

Smooth Endoplasmic Reticulum

Smooth ER is involved in several different cellular processes, including hydroxylation reactions, drug detoxification, carbohydrate metabolism, and synthesis of neutral fats, phospholipids, and steroids. Especially important in this latter

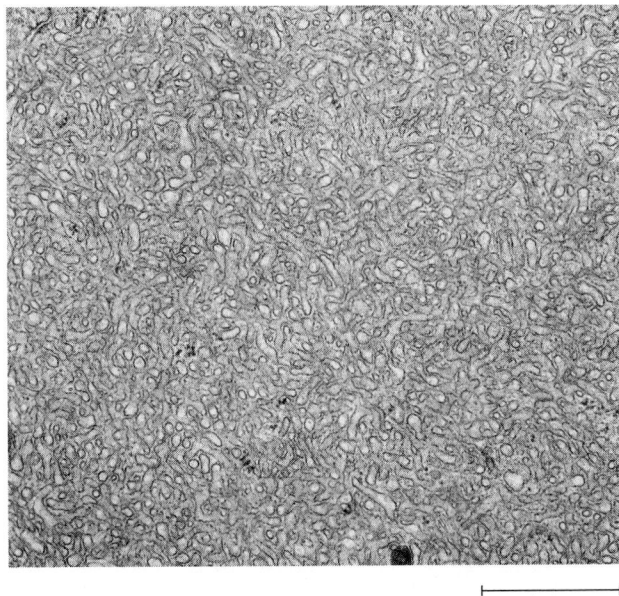

1 μm

Figure 9-10   Smooth ER.   This electron micrograph shows smooth ER in a Leydig cell from a rat testis. The prominence of smooth ER reflects the role of Leydig cells in the biosynthesis of steroids. Steroid synthesis is important in the testes because the sex hormones are steroids (TEM).

category are the male and female sex hormones and the steroid hormones of the adrenal cortex.

Hydroxylation Reactions.   The smooth ER of hepatocytes plays an important role in a variety of **hydroxylation reactions,** in which molecular oxygen (O_2) is used to generate hydroxyl groups, with electrons supplied by NADPH, the reduced coenzyme mentioned earlier. One such hydroxylation reaction converts the amino acid phenylalanine into tyrosine, the hydroxylated equivalent:

Phenylalanine $+ O_2 + $ NADPH $+ H^+ \longrightarrow$ Tyrosine $+ H_2O + $ NADP$^+$

(9-1)

All such hydroxylation reactions require both NADPH and oxygen. One atom of the oxygen molecule is used to hydroxylate the substrate, and the other is reduced to water. Enzymes that carry out such hydroxylation reactions are called *mono-oxygenases* or *mixed-function oxidases*. Although reaction 9-1 suggests a direct transfer of electrons from NADPH to oxygen, the transfer actually involves an electron transport chain found in smooth ER, with cytochrome P-450 as the terminal component. It is the reduced form of P-450 that actually activates oxygen for hydroxylation.

Drug Detoxification and Carcinogenesis. The cytochrome P-450 found in smooth ER is important in **drug detoxification,** especially in the liver. The mode of detoxification usually involves hydroxylation, which increases the solubility of the drug in water. This alteration is critical because hydrophobic compounds tend to be membrane-soluble and are therefore retained in the body, whereas water-soluble compounds are more easily flushed away by the blood and subsequently excreted from the body. Barbiturate drugs, for example, can be detoxified by hydroxylation enzymes associated with smooth ER. This can be demonstrated by injecting the sedative phenobarbital into a rat. One of the most striking effects is a rapid increase in the level of barbiturate-detoxifying enzymes in the liver, accompanied by a dramatic proliferation of smooth ER.

This process enhances removal of the drug from the body, but at the same time means that increasingly higher doses of the drug must be administered to achieve the same sedative effect in habitual users of phenobarbital. Furthermore, the mono-oxygenase that is induced by phenobarbital is of such broad specificity that it can hydroxylate and therefore solubilize a variety of other drugs, including such useful agents as antibiotics, narcotics, steroids, and anticoagulants. As a result, the chronic use of barbiturates decreases the effectiveness of a host of other clinically useful drugs.

Another hydroxylase present in smooth ER appears to have a *carcinogenic* (cancer-causing) effect. The enzyme, called *aryl hydrocarbon hydroxylase*, is thought to be involved in converting potentially carcinogenic compounds into their chemically active forms. This enzyme complex contains cytochrome P-448, a molecule closely related to

cytochrome P-450. (The numbers in cytochrome names refer to the wavelength, in nm, at which the cytochrome absorbs light maximally.) Significantly, cigarette smoke is a potent inducer of aryl hydrocarbon hydroxylase.

Glycogen Catabolism. Yet another function of the smooth ER of hepatocytes is in the *catabolism*, or breakdown, of stored glycogen, as evidenced by the presence of the enzyme *glucose-6-phosphatase* in the smooth ER membrane. This enzyme catalyzes the removal of the phosphate group from glucose-6-phosphate to form free glucose:

$$\text{glucose-6-phosphate} + H_2O \longrightarrow \text{glucose} + P_i \quad \textbf{(9-2)}$$

To understand the importance of this phosphatase, we need to appreciate both the way in which liver cells store glycogen and the reason and mechanism for its subsequent breakdown.

A major role of the liver is to keep the level of glucose in the blood relatively constant. The liver stores glucose as glycogen and releases it as needed by the body, especially between meals and in response to muscular activity. Liver glycogen is stored as granules associated with smooth ER (Figure 9-11a). The mobilization of liver glycogen is hormonally mediated. This process involves the stimulation of a membrane-bound enzyme called *adenylate cyclase*, resulting in a transient elevation in the intracellular level of *cyclic AMP* (*cAMP*). (Cyclic AMP is a form of AMP that is involved in the regulation of a variety of cellular processes, as we will see in more detail in Chapter 23.) The cAMP elevation triggers a complex cascade of events, leading eventually to the activation of *glycogen phosphorylase*, an enzyme that cleaves glycogen into glucose-1-phosphate units (Figure 9-11b). Glucose-1-phosphate can readily be converted to glucose-6-phosphate in the cytosol by an enzyme called *phosphoglucomutase*.

To leave the liver cell and enter the bloodstream, however, the glucose-6-phosphate must be converted to free glucose, because membranes are generally impermeable to phosphorylated sugars. The glucose-6-phosphatase in the smooth ER membrane removes the phosphate group from glucose-6-phosphate, thereby allowing the glucose to move out of the liver cell into the blood for transport to cells in the body that are in need of energy (see Figure 9-11b). Significantly, glucose-6-phosphatase activity is present in liver, kidney, and intestinal cells but not in muscle or brain cells. Muscle and brain cells retain glucose-6-phosphate and use it to meet their high energy needs.

Rough Endoplasmic Reticulum

As already noted, rough ER is characterized by the presence of ribosomes. Rough ER ribosomes are always located on the side of the membrane that faces the cytosol, and they are responsible for the protein synthesis associated with the rough ER. Proteins made on the rough ER are usually destined either for export from the cell as secretory products or for incorporation into one of the several membrane systems

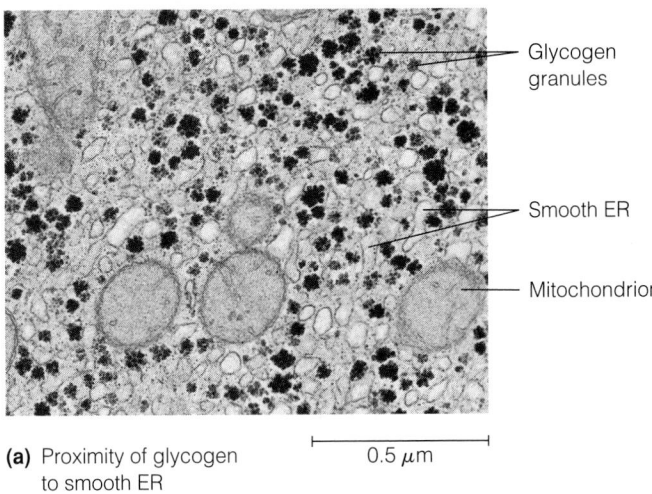

(a) Proximity of glycogen to smooth ER

0.5 μm

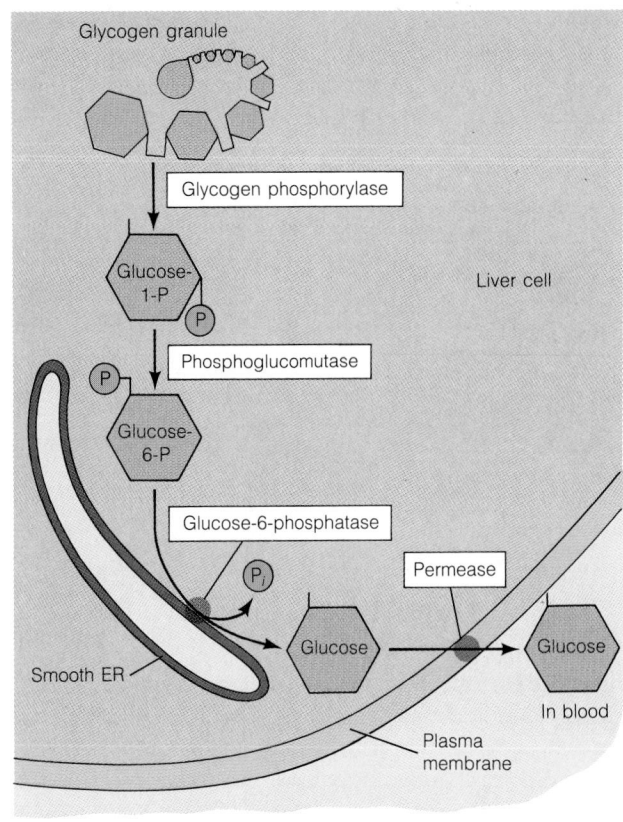

Glycogen granule

Glycogen phosphorylase

Glucose-1-P

Phosphoglucomutase

Glucose-6-P

Glucose-6-phosphatase

Permease

Smooth ER

Glucose

Glucose

In blood

Plasma membrane

Liver cell

Figure 9-11   The Role of Smooth ER in the Breakdown of Liver Glycogen. **(a)** This electron micrograph of a monkey liver cell shows numerous granules of glycogen in close association with smooth ER (TEM). **(b)** The breakdown of liver glycogen involves the stepwise removal of glucose units as glucose-1-phosphate, and then the conversion of glucose-1-phosphate to glucose-6-phosphate by enzymes in the cytosol. Removal of the phosphate group requires the action of glucose-6-phosphatase, an enzyme associated with the smooth ER membrane. Free glucose is then transported out of the liver cell by a permease in the plasma membrane.

(b) Process of glycogen breakdown in liver

derived directly or indirectly from the ER. In addition to protein synthesis, rough ER is also the site of the initial steps in the addition of sugar groups to glycoproteins.

Investigators are not in full agreement about how ribosomes are attached to the rough ER membrane, nor is it entirely clear that there is a single mechanism. However, for integral membrane proteins and for secretory proteins that are inserted into the cisternal space of rough ER as they are synthesized, it has been convincingly established that the ribosomes are bound to the rough ER membrane in part by the growing polypeptide chain. It is likely that one or more specialized ER proteins are also involved, since ribosomes remain loosely attached to the ER even when protein synthesis is not taking place. The mechanism of protein synthesis and the role of the ER in protein processing will be discussed in detail in Chapter 18. Here, we turn our attention to the Golgi complex, a component of the endomembrane system that is intimately linked to the ER in its functions.

The Golgi Complex

The **Golgi complex** (sometimes also called the *Golgi apparatus*) derives its name from Camillo Golgi, the Italian biologist who first described this structure in 1898. He reported that nerve cells soaked in osmium tetroxide showed a heavy metal deposit in a threadlike network surrounding the nucleus. The same staining reaction was demonstrated with other heavy metals and a variety of cell types. The results were inconsistent, however, and no cellular structure could be identified that might explain the staining. As a result, the nature (the very existence, in fact) of the Golgi complex remained highly controversial until the 1950s, when its existence was confirmed by electron microscopy. Since then, we have come to understand much about the Golgi complex, its intimate relationship with the ER, and its role in the chemical modification, packaging, and sorting of proteins that are shuttled to it from the ER.

Golgi Structure

The Golgi complex consists of a series of flattened, membrane-bounded cisternae (also called *saccules*), disk-shaped sacs that are stacked together as shown in Figure 9-12a. Such a series of cisternae is called a **Golgi stack** in animal cells and a **dictyosome** in plant cells. These structures can be readily visualized by electron microscopy (Figure 9-12b). Usually, there are 4–8 cisternae per stack, though the Golgi stacks of some lower organisms can have several dozen cisternae. The size and number of Golgi stacks vary with cell type and also with the metabolic activity of the cell. Some cells have one large stack, whereas others (particularly those that are especially active in secretion) have hundreds or even thousands of Golgi stacks. The intracisternal spaces of the Golgi complex are part of the network of internal spaces of the endomembrane system shown in Figure 9-8.

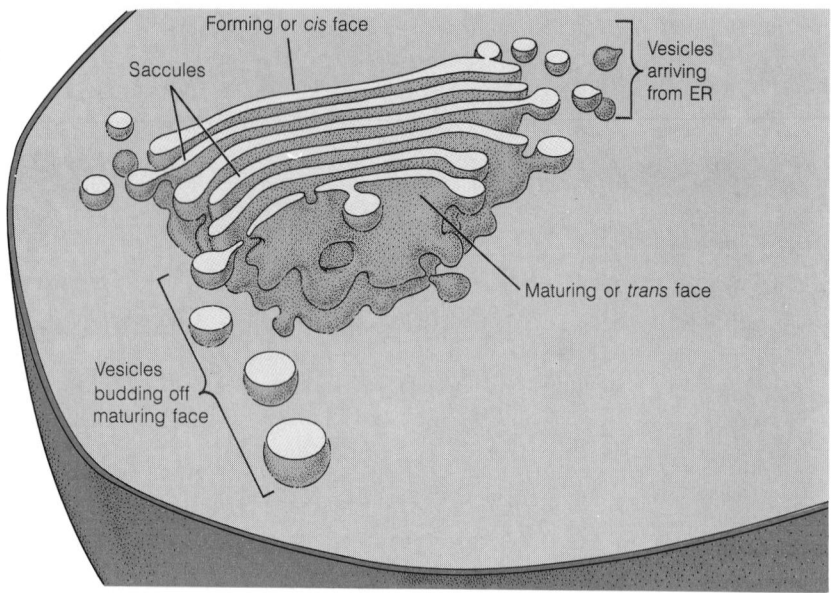

Forming or *cis* face

Saccules

Vesicles arriving from ER

Maturing or *trans* face

Vesicles budding off maturing face

(a) A Golgi stack in an animal cell

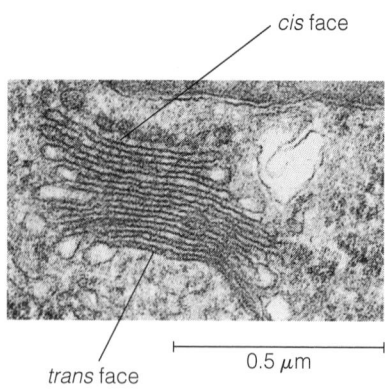

cis face

trans face

0.5 μm

(b) A dictyosome in an algal cell

Figure 9-12 Golgi Structure. Each Golgi stack, or dictyosome, consists of a small number of flattened cisternae (saccules) stacked together. **(a)** On the *cis* (forming) face, vesicles arriving from the ER fuse to form new cisternae. On the *trans* (maturing) face, vesicles arise by budding off the tips of the cisternae. Secretory vesicles and primary lysosomes are formed in this way. **(b)** This electron micrograph shows a dictyosome from an algal cell (TEM).

The Prominence of Vesicles. The static view of Golgi stacks that we get from micrographs such as Figure 9-12b is very misleading; the Golgi complex is actually a very dynamic, active structure. As evidence of its activity, a Golgi stack is typically surrounded by numerous **vesicles** of various sizes (see Figure 9-12a). These vesicles shuttle proteins from the ER to the Golgi, between the cisternae of the Golgi stack, and from the Golgi to various destinations in the cell, including secretory vesicles, lysosomes, and the plasma membrane. Moreover, the edges, or margins, of the Golgi cisternae are characteristically enlarged and filled with granular material. This material consists mainly of proteins, many of them glycosylated, that are moving to, from, or within the cisternae by means of vesicles that bud off and fuse with the edges of the cisternae.

The vesicles involved in protein transfer between the ER, Golgi, and related organelles are called *coated vesicles* because of the presence on their surface of a coat, or layer, of a bristlelike, polyhedral protein lattice. The main protein in the lattice is either *clathrin* or *coatomer*, depending on the function of the vesicle. Coated vesicles are a common feature of cellular processes that involve the transfer or exchange of substances between specific membrane-bounded compartments of eukaryotic cells. We will discuss coated vesicles later in the chapter, when we encounter clathrin-coated membranes again in the process of endocytosis.

The Faces of the Golgi Stack. Each Golgi stack has two distinct sides, or *faces*. The *cis* (or **forming**) face is oriented toward the rough ER with which most Golgi stacks are intimately associated. At the *cis* face, **transition vesicles** containing newly synthesized proteins arrive continuously from the ER. Meanwhile, on the ***trans* (or maturing) face** of the Golgi stack, vesicles bud off the tips of the cisternae continuously, carrying processed proteins to their various destinations. The sacs between the two faces comprise the *medial cisternae* of the Golgi stack, in which much of the processing of proteins occurs. Traffic between successive cisternae of a Golgi stack is mediated by **shuttle vesicles** that bud off one cisterna and fuse with another, usually the next cisterna in the *cis*-to-*trans* sequence. In addition, "back traffic" also occurs, primarily to facilitate the return of ER-specific proteins that are inadvertently passed from the ER to the *cis* face of a nearby Golgi stack.

The two faces of a Golgi stack are biochemically and functionally distinct. Cytochemical and immunological staining techniques have shown that specific enzymes and receptor proteins are concentrated within the cisterna on the *cis* face of the stack, whereas other proteins are localized primarily in the cisterna on the *trans* face. This biochemical polarity is illustrated in Figure 9-13, which shows several Golgi stacks in an epithelial cell of the rat epididymis. The cell section has been stained to detect the receptor protein for the carbohydrate mannose-6-phosphate, using an antibody against the receptor protein as a probe. Clearly, the receptor protein for this carbohydrate is concentrated in the cisterna on the *cis* face of each Golgi stack. (As we will see later in the chapter, mannose-6-phosphate is a distinctive component of glycoproteins that are destined to become lysosomal enzymes. Mannose-6-phosphate is added to these proteins in the ER, and the presence of the mannose-6-phosphate receptor in the *cis* cisterna of the Golgi stack enables the Golgi

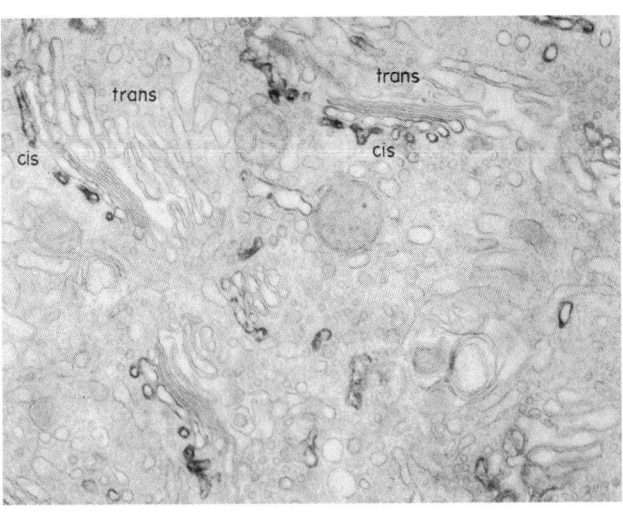

Figure 9-13 Immunochemical Staining of Golgi Stacks.
This electron micrograph shows several Golgi stacks in a cell from the absorptive epithelium of the rat epididymis. The *cis* and *trans* faces of two stacks are labeled. The cell was subjected to an immunochemical technique to stain receptor proteins that are specific for the sugar mannose-6-phosphate. The receptors for this sugar are concentrated in the cisterna on the *cis* face of the Golgi stack. Thus, glycoproteins with mannose-6-phosphate groups that arrive at the Golgi complex will be recognized and processed on the *cis* face of the stack (TEM).

complex to recognize such proteins and to target them eventually to the lysosomes.)

The two faces of a Golgi stack also differ in the kind of coated vesicles associated with them. Vesicles that bud off the *trans* cisternae are coated with clathrin, whereas vesicles that bud off the ER or the *cis* or *medial* cisternae of the Golgi stack are coated with coatomer. Thus, coatomer-coated vesicles shuttle proteins between the ER and Golgi cisternae, while clathrin-coated vesicles mediate the transport of proteins destined for secretory vesicles, lysosomes, or the plasma membrane.

Roles of the ER and Golgi Complex in Protein Glycosylation

Much of the processing of proteins that occurs within the ER and Golgi complex involves the addition and restructuring of carbohydrate side chains of glycoproteins. Conceptually, we can think of the glycosylation process in two phases, the initial glycosylation event and the subsequent modification of the sugar side chain.

Core Glycosylation: Addition of the Core Oligosaccharide.
The initial stage of glycosylation occurs in the ER, where a single kind of *N-linked oligosaccharide* is added to the terminal amino group of specific asparagines in the protein to be glycosylated (or in a nascent polypeptide; in some cases,

glycosylation occurs as a protein is being synthesized). This process is called **core glycosylation.**

Invariably, the sugar that is directly linked to asparagine is *N-acetylglucosamine* (GlcNAc), a modified sugar that we encountered in Chapter 7 (see Figure 7-20). Despite the variety of oligosaccharides found in glycoproteins, all of the sugar groups added to proteins in the ER have a common high-mannose core structure consisting of two GlcNAc units, three glucose units, and nine mannose units. This **core oligosaccharide** is built up by the sequential addition of monosaccharide units to an activated lipid carrier called *dolichol phosphate*. This process is illustrated in Figure 9-14. The dolichol phosphate molecule is shown in part a and the stepwise addition of monosaccharides to form the core oligosaccharide is depicted in part b. The completed core oligosaccharide is then transferred as a block to an asparagine side chain of a polypeptide, as shown in part c.

Terminal Glycosylation: Processing of the Core Oligosaccharide. Initial processing of the core oligosaccharide occurs while the glycoprotein is still in the ER. Specifically, the three glucose units and one of the mannose units are usually removed. Further processing of glycoproteins then takes place in the Golgi complex as the proteins move from the *cis* face through the medial cisternae to the *trans* face of the Golgi stack. This sequence of events is called **terminal glycosylation** to distinguish it from the initial core glycosylation that occurs in the ER.

As already noted, terminal glycosylation always involves the removal of some of the sugars of the core oligosaccharide that was added in the ER. In some cases, no further processing is necessary. The resulting *high-mannose oligosaccharides* retain the two GlcNAc units and most of the mannose groups present in the core oligosaccharide. In other cases, *complex oligosaccharides* are generated by the addition of other sugars, including galactose, sialic acid, and sometimes fucose. (For the structures of these sugars, see Figures 7-20a and 10-13.) As it exits the Golgi complex, a specific glycoprotein may contain only high-mannose oligosaccharides, only complex oligosaccharides, or both.

Given the role of the Golgi complex in glycosylation, it is not surprising that two of the most important kinds of enzymes present in Golgi stacks are *glycosyl transferases* and *glucan synthetases*. The transferases are instrumental in attaching sugar groups to proteins, and the synthetases catalyze the formation of polysaccharides from sugar nucleotides. These and other enzymes are distributed within the Golgi stack in a manner consistent with their role in the glycosylation process, such that each cisterna in the stack contains its own distinctive set of processing enzymes.

The Role of the Golgi Complex in Protein Sorting Within the Endomembrane System

In addition to its role in terminal glycosylation, the Golgi complex also mediates the flow of proteins from the ER to various destinations within the endomembrane system.

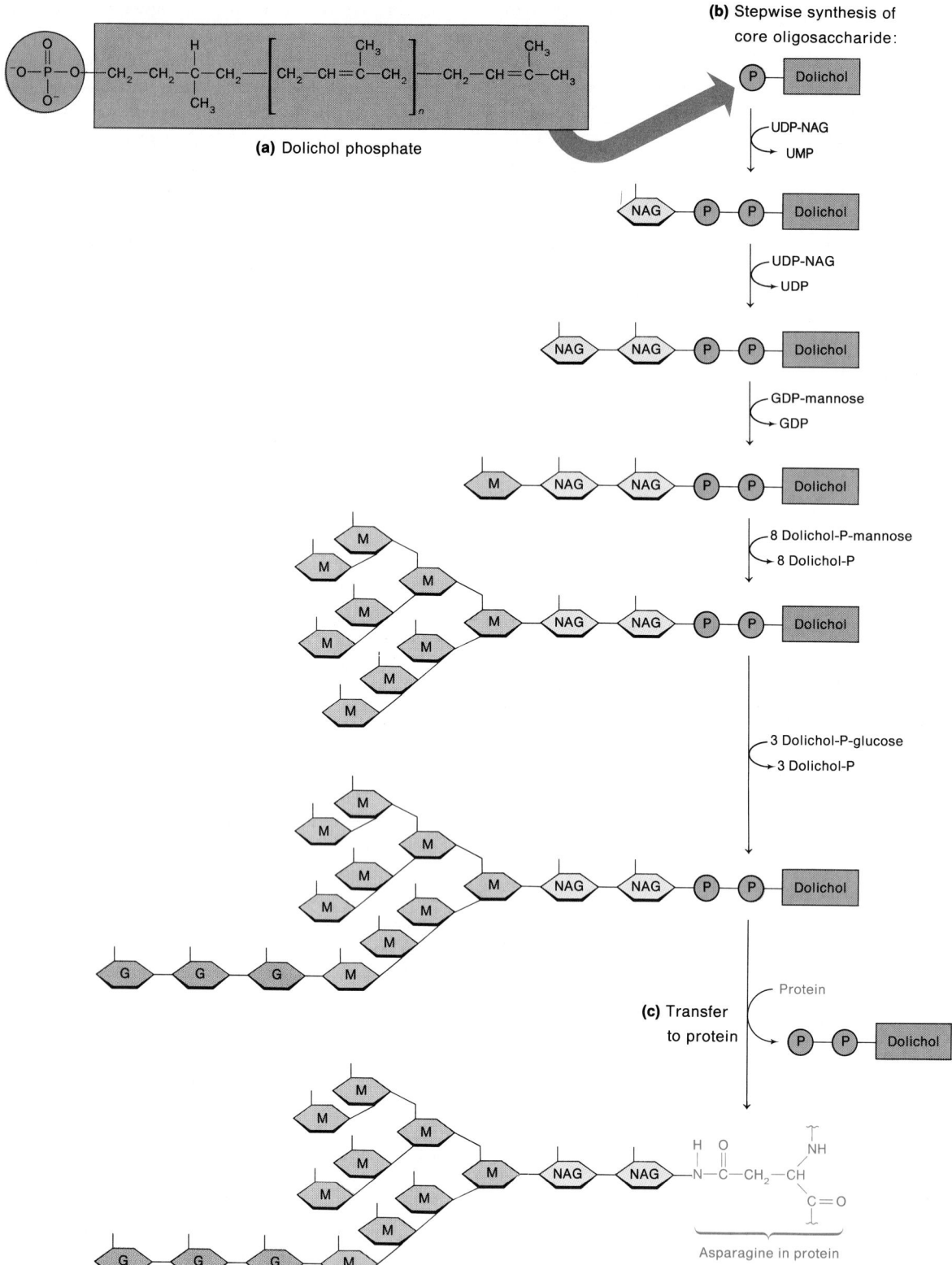

Figure 9-14 Glycosylation of Proteins in the ER.

(a) Dolichol phosphate, the carrier of oligosaccharide units in glycosylation reactions. (b) Stepwise synthesis of the high-mannose core oligosaccharide by the successive addition of N-acetylglucosamine (GlcNAc), mannose (M), and glucose (G) units to the growing chain, with UDP-GlcNAc, GDP-mannose, and dolichol phosphate-glucose as the sugar donors. (c) When the core oligosaccharide is complete, it is transferred from dolichol phosphate to the asparagine side chain of the protein that is to be glycosylated.

Proteins that are synthesized on the rough ER and sorted in the Golgi complex include *secretory proteins* destined for release to the exterior of the cell, *hydrolytic enzymes* intended for packaging into lysosomes, *Golgi proteins* that are retained within specific cisternae of the Golgi stack, and *integral membrane proteins* that either remain a part of the Golgi membrane or, more commonly, become incorporated into the plasma membrane of the cell, the membranes of lysosomes, or the membranes of the nuclear envelope (with which the ER is contiguous; see Figure 9-7). The vesicle traffic involved in these pathways is illustrated in Figure 9-15. In some cells, the Golgi complex is also involved in the processing of proteins that enter the cell from the outside by endocytosis, a process described later in the chapter.

The "default pathway" of Figure 9-15 is the route whereby proteins pass from the ER via the Golgi complex to vesicles that move directly to the cell surface and fuse with the plasma membrane. As we will learn shortly, this process is called *constitutive secretion* to distinguish it from *regulated secretion*, which is not a default pathway. A protein that is synthesized in the ER will be constitutively secreted from the cell unless it contains or acquires a specific "tag" that routes it in another direction. Depending on the destination, this tag may be either a short but specific amino acid sequence or a specific oligosaccharide side chain.

Thus, proteins that are constitutively secreted to the outside of the cell reach their destination by default; all other ER-synthesized proteins must be specifically diverted

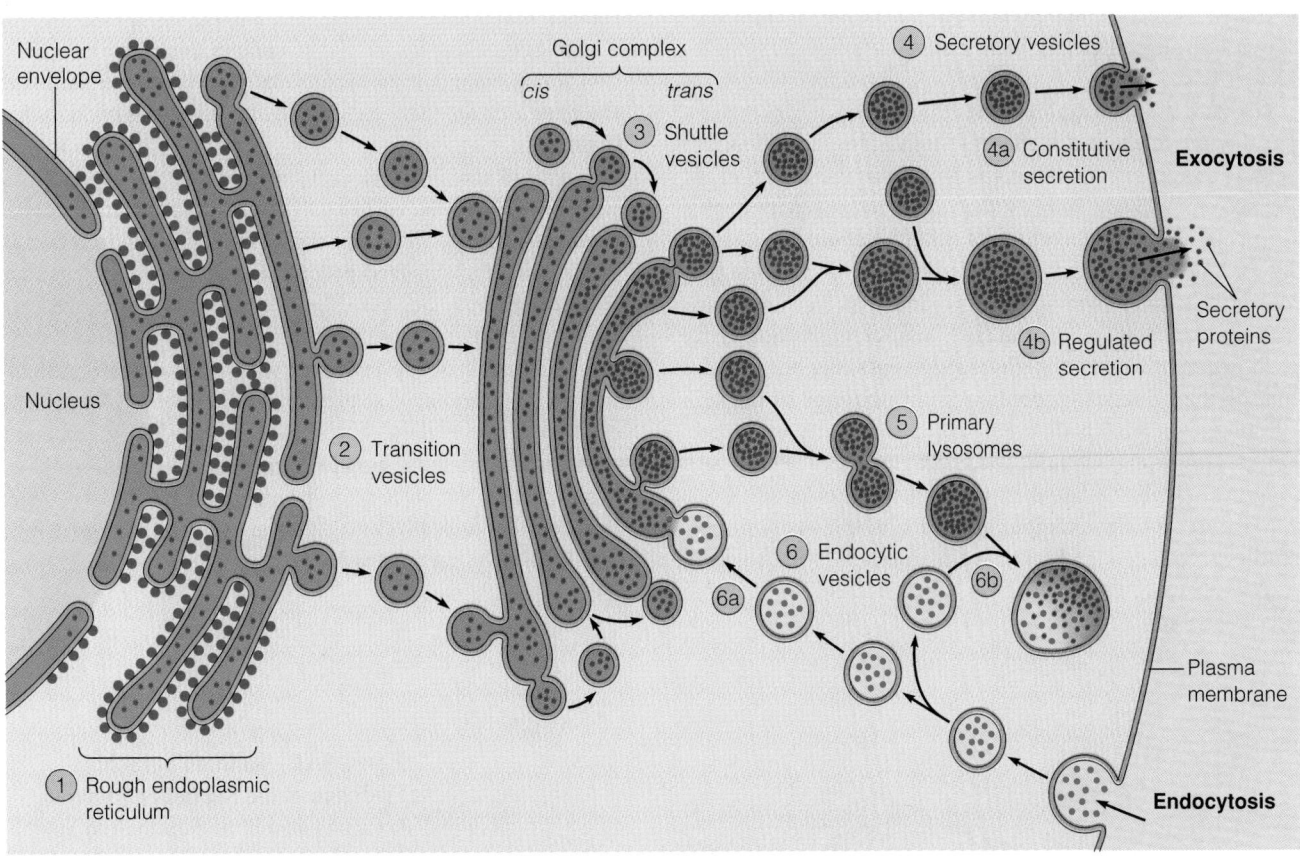

Figure 9-15   Protein and Vesicle Traffic to and from the Golgi Complex.   Shown here are the several routes whereby proteins move via vesicles from the ER through the Golgi complex to any of several destinations, including lysosomes and secretory vesicles. ① Proteins are synthesized by ribosomes bound to the outer surfaces of rough ER and the nuclear envelope with which the ER is continuous. Initial glycosylation events occur within the ER cisternae. ② Transition vesicles carry newly synthesized and glycosylated proteins to the *cis* face of a nearby Golgi stack. ③ Proteins then move through the cisternae of the Golgi stack from the *cis* to the *trans* face by means of shuttle vesicles that bud from one cisterna and fuse with the next cisterna in the stack.

Within the Golgi cisternae, glycoproteins are processed further, and proteins are progressively concentrated for packaging into vesicles. At the *trans* face, vesicles bud off to form either ④ secretory vesicles or ⑤ primary lysosomes, depending on their protein content. Secretory vesicles move to the plasma membrane, where they discharge their proteins by exocytosis either ④ₐ constitutively or ④♭ in a regulated manner, in response to the appropriate signal. ⑥ Proteins and other material are taken into the cell by endocytosis, forming endocytotic vesicles that fuse either ⑥ₐ with cisternae of the Golgi complex or ⑥♭ with primary lysosomes, depending on the nature of the endocytosed material.

from this pathway for retention in the ER or in the Golgi complex, for incorporation into the plasma membrane or the nuclear envelope, for packaging into lysosomes or specialized secretory vesicles, or for return to the ER, in the case of proteins that are meant to remain in the ER but escape inadvertently.

Secretory Pathways

Integral to the cellular traffic patterns shown in Figure 9-15 are the **secretory pathways** by which proteins move from the ER through the Golgi complex to **secretory vesicles** that discharge their contents to the exterior of the cell. The concerted roles of the ER and the Golgi complex in secretion were first demonstrated by George Palade and his colleagues, who injected rabbits with a radioactive amino acid and then used autoradiography to trace the fate of the radioactivity through the cells in the parotid salivary gland, a secretory tissue.

Figure 9-16 illustrates the results of such an experiment. For the first few minutes after injection of the labeled amino acid, the radioactive label is found only in the rough ER of the secretory tissue. Shortly thereafter, radioactivity begins to appear in the Golgi complex, peaking at about 30–40 min. By 30 min, radioactivity can be detected in vesicles that Palade called *condensing vacuoles* (and that we now know bud off the *trans* face of the Golgi complex). Radioactivity then begins to accumulate in *secretory granules*, the vesicles that discharge secretory proteins to the exterior of the cell. Although not shown in the figure, radioactivity eventually appears in the extracellular space, proving that the secretory granules actually release their contents to the outside.

Based on Palade's initial experiments and many similar studies since then, the secretory pathway shown in Figure 9-15 is now understood in considerable detail. Specifically, we can trace the flow of proteins in this sequence:

rough ER → transition vesicles → *cis* cisternae of the Golgi → medial cisternae → *trans* cisternae → secretory vesicles → fusion of vesicles with plasma membrane and discharge of secretory proteins to the exterior of the cell.

Although the distinction was not recognized by early investigators, we now distinguish two different modes of exocytotic secretion by eukaryotic cells. **Constitutive secretion** involves the continuous discharge of vesicles at the membrane surface, whereas **regulated secretion** entails release at intervals, usually in response to a signal from outside the cell. Examples of constitutive secretion include the continuous release of mucus by certain cells that line your intestine as well as the secretion of the glycoproteins of the extracellular matrix, a process we will encounter in Chapter 10. The release of insulin from the acinar cells of your pancreas, on the other hand, is an example of regulated secretion, as is the release of digestive enzymes from other cells of the same organ.

All eukaryotic cells carry out constitutive secretion, which we identified earlier as the default pathway for proteins that traverse the Golgi complex. Regulated secretion is a much more specialized process, found mainly in cells that secrete proteins such as digestive enzymes, hormones, or neurotransmitters. The information needed to direct a protein to a vesicle of the regulated pathway is probably inherent in the amino acid sequence of the protein. Most likely, each secretory protein that is to be released in a regulated manner has a specific signal (a short stretch of amino acids, presumably) that targets the protein to a regulated vesicle. Proteins that lack this and other signals are packaged into constitutive vesicles instead.

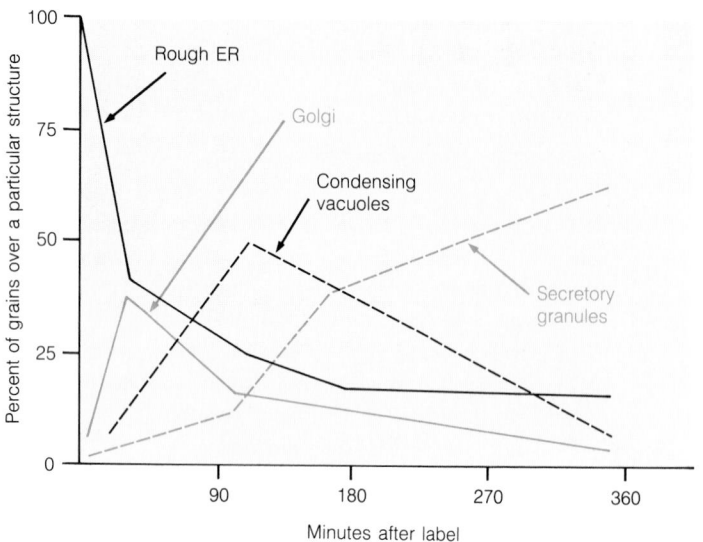

Figure 9-16   Autoradiographic Evidence of the Secretory Pathway. To trace the path of newly synthesized protein in the cell, George Palade and his colleagues injected rabbits with a radioactively labeled amino acid and used autoradiography to determine the distribution of radioactivity within the parotid gland cells at various times during the first six hours after injection. Initially, most of the radioactivity (measured as silver grains on the autoradiogram) was in the rough ER. Then it moved through the Golgi complex to vesicles called condensing vacuoles and finally to mature secretory granules for secretion by exocytosis.

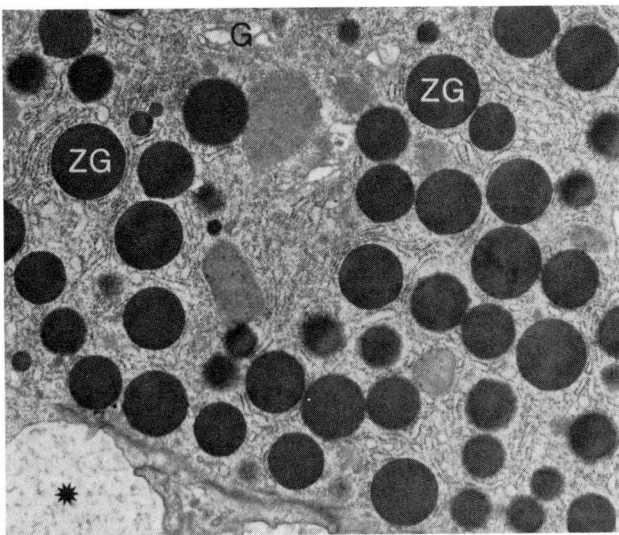

Figure 9-17 Zymogen Granules. This electron micrograph of an acinar (secretory) cell from the exocrine pancreas of a rat illustrates the prominence of zymogen granules (ZG), also called secretory granules. Zymogen granules are usually concentrated in the region of the cell between the Golgi complex (G) from which they arise and the portion of the plasma membrane bordering the acinar lumen (*) into which the contents of the granules are eventually discharged by exocytosis (TEM).

Whereas vesicles containing constitutively secreted proteins move continuously and directly to the plasma membrane for immediate release, vesicles involved in regulated secretion accumulate in the cell before their eventual discharge. Such vesicles form by budding from *trans* Golgi cisternae as immature secretory vesicles, which then undergo a maturation process that involves concentration of the proteins (a process called *condensation*) and frequently also the proteolytic processing of the proteins. The mature vesicles then move to the site of secretion and wait near the plasma membrane for the signal that eventually triggers their release in a process called *exocytosis*.

The vesicles involved in regulated secretion are usually quite large and contain much more highly concentrated proteins than do vesicles of the constitutive pathway. Such large, dense vesicles are usually called **secretory granules** (or, sometimes, **zymogen granules**). Figure 9-17 is an electron micrograph of a secretory cell from the pancreas of a rat. The prominent zymogen granules (ZG) are concentrated in the region of the cell between the Golgi stacks from which they arise and the portion of the plasma membrane bordering the lumen into which the contents of the granules are eventually discharged.

Outward and Inward Transport: Exocytosis and Endocytosis

The process whereby secretory granules discharge their contents to the exterior of the cell is called *exocytosis*. The complementary process by which cells internalize materials that were previously outside the cell is called *endocytosis*. Both of these processes serve as means of transporting macromolecules and other substances across the plasma membrane of the eukaryotic cell. We will consider exocytosis first because it is the final step in a secretory pathway that began with the ER and Golgi complex.

Exocytosis

In **exocytosis,** proteins sequestered within a vesicle are released to the outside of the cell as the membrane of the vesicle fuses with the plasma membrane. Many different kinds of proteins are exported from both animal and plant cells by exocytosis. Animal cells secrete digestive enzymes, protein and peptide hormones, mucus, and milk proteins in this manner. Plant cells secrete mainly proteins that are associated with the cell wall, including both enzymes and structural proteins.

The process of exocytosis is illustrated schematically in Figure 9-18. As noted earlier, cellular products destined for secretion are first concentrated and packaged into secretory vesicles (or granules). These vesicles then move to the cell surface, where the membrane of the vesicle fuses with the plasma membrane. This ruptures the plasma membrane, thereby allowing discharge of the vesicle contents to the outside of the cell. In the process, the membrane of the vesicle becomes integrated into the plasma membrane, with the *inner* surface of the vesicle becoming the *outer* (extracellular) surface of the plasma membrane.

The mechanism underlying the movement of exocytotic vesicles to the cell surface is not yet clear. Current evidence points to the involvement of microtubules in both exocytotic and endocytotic vesicle movement. In some cells, for example, vesicles appear to move from the Golgi to the plasma membrane along "tracks" of microtubules that are oriented parallel to the direction of vesicle movement. Moreover, vesicle movement stops when cells are treated with *colchicine*, a plant alkaloid that binds to tubulin monomers and prevents their assembly into microtubules.

The Likely Role of Calcium as a Trigger of Exocytosis. The discharge of regulated vesicles is triggered by a specific signal from outside the cell. In most cases, the signal is a substance such as a hormone or a neurotransmitter that is known to bind to specific receptors on the cell surface, thereby activating the synthesis or release of a *second messenger* within the cell. (We will discuss second messengers and their role in intracellular signaling pathways in Chapter 23.) Calcium ions are probably involved in triggering the discharge process because most substances that activate regulated secretion share the property of elevating the intracellular calcium concentration. Moreover, the secretory vesicles in many types of cells that carry out regulated secretion can be triggered to release their contents by experimental manipulation of the intracellular calcium content. In the case of pancreatic cells, for example, mature secretory granules can

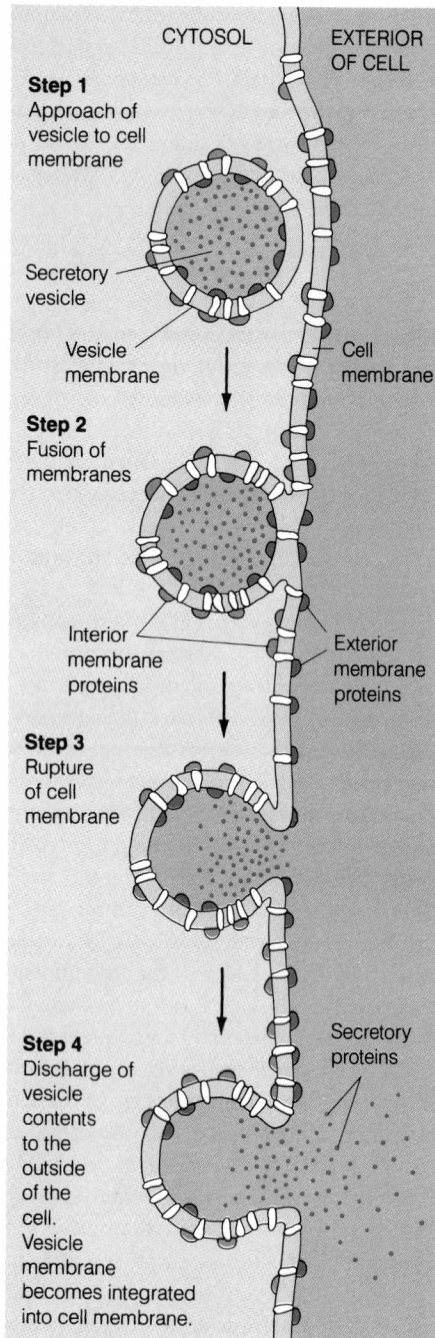

Step 1
Approach of
vesicle to cell
membrane

CYTOSOL

EXTERIOR
OF CELL

Secretory
vesicle

Vesicle
membrane

Cell
membrane

Step 2
Fusion of
membranes

Interior
membrane
proteins

Exterior
membrane
proteins

Step 3
Rupture
of cell
membrane

Step 4
Discharge of
vesicle
contents
to the
outside
of the
cell.
Vesicle
membrane
becomes
integrated
into cell
membrane.

Secretory
proteins

Figure 9-18   Exocytosis.   Exocytotic discharge of the contents of a secretory vesicle or granule to the exterior of the cell involves ① approach of the vesicle to the plasma membrane, ② fusion of the two membranes, ③ rupture of the plasma membrane, and ④ release of the vesicle contents to the outside of the cell. Notice that the vesicle membrane becomes integrated into the plasma membrane, with internal and external membrane components correctly positioned.

be induced to discharge their contents by microinjection of calcium into the cells.

The specific role of calcium in triggering the release of regulated vesicles is not yet understood. As we will see in Chapter 23, calcium is known to activate *protein kinases,* en-

zymes that phosphorylate specific kinds of proteins. It may be that an elevation in the intracellular calcium concentration leads to the activation of protein kinases whose target proteins are components of the vesicle membrane or perhaps the plasma membrane.

Polarized Secretion.   In many cases, exocytosis of specific proteins is limited to a specific surface of the cell. For example, the secretory cells that line your intestine release digestive enzymes only on the side of the cell that faces the interior of the intestine. Exocytosis also occurs on the opposite side of the cell, but a completely different set of proteins is secreted on that side. This phenomenon, called **polarized secretion,** is also seen in nerve cells, which secrete neurotransmitter molecules only at junctions with other nerve cells.

Proteins destined for polarized secretion are apparently sorted into vesicles with specific recognition properties—capable, perhaps, of binding specifically to localized recognition sites on the plasma membrane. Studies with the protozoan *Tetrahymena* suggest that the plasma membrane contains specific recognition sites for membrane fusion. These sites involve rosettes of protein particles that can be seen by freeze-fracture microscopy.

Endocytosis

In addition to exocytosis, many eukaryotic cells also carry out **endocytosis.** In this process, the cell internalizes materials by the progressive infolding of a small segment of the plasma membrane, which then pinches off to form an **endocytotic vesicle** containing the ingested substance or particle. In this way, substances that were previously outside the cell are entrapped within vesicles and brought into the cell (Figure 9-19). In a formal sense, endocytosis is the reverse of exocytosis, but the mechanisms are somewhat different, as we will see shortly.

In terms of membrane flow, endocytosis and exocytosis clearly have opposite effects. Endocytosis internalizes small portions of the plasma membrane as vesicles, whereas exocytosis adds to the plasma membrane by the fusion of vesicles with it. The magnitude of the resulting membrane exchange can be very impressive, especially in cells that carry out both processes actively. The secretory cells in your pancreas, for example, require only about 90 min to recycle an amount of membrane equal to the whole surface area of the cell. Cultured macrophages (large white blood cells) are even faster, replacing an amount of membrane equivalent to the plasma membrane in about 30 min. And cells of the slime mold *Dictyostelium* accomplish the same feat in about 20 min!

The term *endocytosis* encompasses several different processes that vary in the nature and quantity of material taken up and the mechanism employed. Usually, a distinction is made between bulk-phase endocytosis, phagocytosis, and receptor-mediated endocytosis. In all three cases, how-

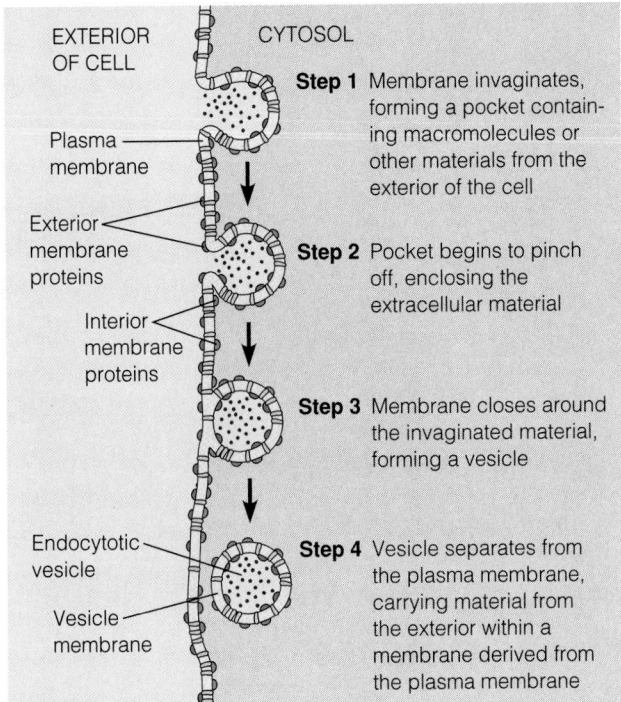

EXTERIOR OF CELL   CYTOSOL

Plasma membrane

Exterior membrane proteins

Interior membrane proteins

Endocytotic vesicle

Vesicle membrane

Step 1 Membrane invaginates, forming a pocket containing macromolecules or other materials from the exterior of the cell

Step 2 Pocket begins to pinch off, enclosing the extracellular material

Step 3 Membrane closes around the invaginated material, forming a vesicle

Step 4 Vesicle separates from the plasma membrane, carrying material from the exterior within a membrane derived from the plasma membrane

Figure 9-19 Endocytosis. Endocytotic uptake of materials from the exterior of the cell involves ① invagination of the plasma membrane and ② gradual pinching off of the invagination, followed by ③ closing of the membrane around the invaginated material to form ④ an endocytotic vesicle. (For clarity, the clathrin coats at the site of invagination and around the endocytotic vesicle have been omitted from this diagram. See Figure 9-20 for the involvement of clathrin.)

ever, the substances that are internalized remain separated from the cytosolic space by the membrane of the vesicle.

Bulk-Phase Endocytosis. Bulk-phase endocytosis involves the uptake of fluid from the medium in a nonspecific manner. Whatever molecules happen to be in the surrounding medium are taken into the cell indiscriminately. Bulk-phase endocytosis occurs in many, perhaps all, types of eukaryotic cells and has been studied especially in macrophages, leukocytes (another kind of white blood cell), kidney cells, intestinal epithelial cells, and plant root cells. In contrast to the other forms of endocytosis, bulk-phase uptake seems to proceed at a relatively constant rate in most eukaryotic cells and may in some cell types be the main mechanism whereby the cell compensates for the membrane segments that are continuously added to the plasma membrane by exocytosis.

Phagocytosis. Phagocytosis is characterized by the uptake of large substances, including aggregates of macromolecules, parts of other cells, and even whole microorganisms or other cells. As you might guess, phagocytosis is used as a means of obtaining food by many single-celled eukaryotes, especially amoebas and ciliated protozoa. (The term *phagocytosis* comes, in fact, from Greek roots that mean "cellular

eating.") The materials these organisms take up from their environment are internalized as large endocytotic vesicles called *phagocytotic vacuoles*. These vacuoles, also called *phagosomes*, then fuse with lysosomes, forming large vesicles in which the ingested material is digested.

Phagocytosis is not found widely among higher eukaryotes, however. In your body, for example, only macrophages and other white blood cells called neutrophils are able to take in materials by phagocytosis. Moreover, their purpose is defensive rather than nutritional: These cells ingest and digest microorganisms that gain entry to the body. In addition, macrophages play a scavenging role, ingesting damaged cells and cellular debris from the circulatory system.

Receptor-Mediated Endocytosis.   Current research interest focuses especially on **receptor-mediated endocytosis (RME),** a process for the efficient uptake of macromolecules for which the cell has specific receptors on its surface. RME is the main mechanism for the specific uptake of most macromolecules by eukaryotic cells. Depending on the cell type, mammalian cells take in hormones, growth factors, enzymes, serum proteins, antibodies, and even some viruses and bacterial toxins by this mechanism. Box 9A (page 250) describes the uptake by receptor-mediated endocytosis of low-density lipoproteins (LDL), the means by which cholesterol gains entry to the cells in your body. An interest in familial hypercholesterolemia, a hereditary predisposition to high blood cholesterol levels and hence to atherosclerosis and heart disease, led Michael S. Brown and Joseph L. Goldstein to the discovery of RME, for which they shared a Nobel Prize in 1986.

For every different kind of substance that a cell can internalize by RME, **receptors** specific for that substance must be present on the cell surface. Each such receptor is a transmembrane protein that is oriented in the plasma membrane with the binding site for a specific substance protruding from the external side of the membrane. For example, each of your liver cells has in its plasma membrane many thousands of receptors for an iron-transporting protein called *transferrin,* and each such receptor has a binding site for transferrin exposed on the exterior surface. Substances such as transferrin that bind to a specific site on a receptor molecule are called **ligands.** Receptors are free to move laterally in the membrane in both the presence and the absence of bound ligand.

RME is illustrated in Figure 9-20a. The process begins with the binding of ligand molecules to their respective receptors on the outer surface of the plasma membrane. As the receptor-ligand complexes diffuse laterally in the membrane, they encounter specialized membrane regions called *coated pits,* which serve as localized sites of collection and internalization for receptor-ligand complexes. The actual internalization process involves the invagination of the pit, which appears to be facilitated by the presence of the protein coat on the cytoplasmic surface. Invagination continues until the pit pinches off from the plasma membrane as an endocytotic vesicle within the cytoplasm of the cell.

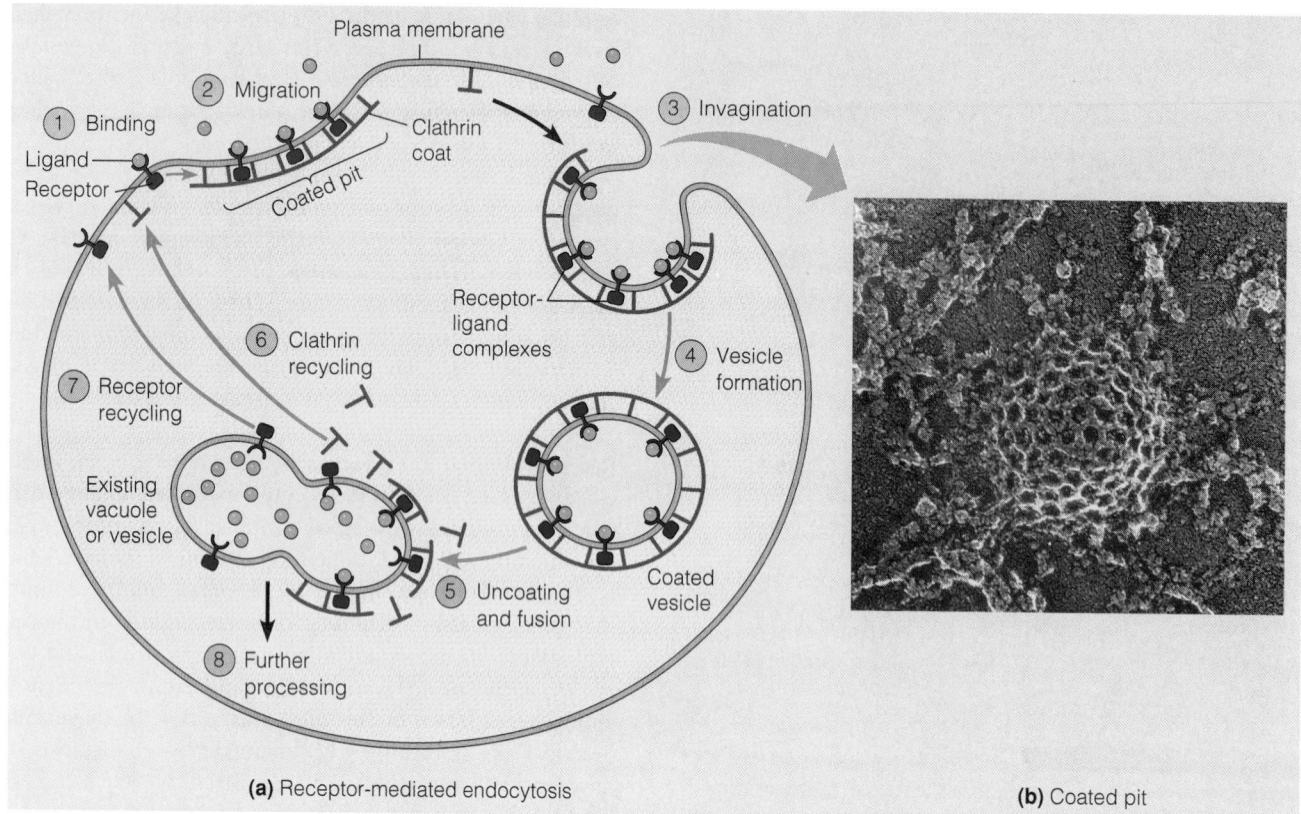

Figure 9-20 Receptor-Mediated Endocytosis. (a) In this schematic diagram of receptor-mediated endocytosis (RME), ① the molecules to be internalized bind to specific receptors on the surface of the plasma membrane. ② Receptor-ligand complexes migrate to coated pits, ③ where invagination is facilitated by the clathrin coat on the cytoplasmic surface of the membrane. ④ The resulting endosome is a coated vesicle, which ⑤ quickly loses all or most of its clathrin coat, thereby enabling it to fuse with the membrane of an internal vesicle, vacuoles, or other struc-

ture. ⑥ In most cases, clathrin is recycled to the plasma membrane, ⑦ as are many of the receptors. ⑧ Vesicles containing internalized molecules are then left to their appropriate intracellular fates, as dictated by the nature of the molecules they contain. **(b)** This electron micrograph of a freeze-etched cell shows clathrin-coated pits on the inner surface of the plasma membrane of a cultured human fibroblast cell, as visualized by freeze-fracture microscopy. The cagelike structure of the pits is due to the clathrin coat on the inner surface of each pit (TEM).

The protein coat that gives coated pits their name occurs on the inner (cytoplasmic) surface of the membrane. The coat consists primarily of clathrin, the same protein that forms the bristlelike lattice around coated vesicles that bud from the *trans* cisternae of the Golgi complex. The role of clathrin in coated pits and vesicles is discussed in the next section. Figure 9-20b shows a coated pit on the inner surface of the plasma membrane of a cultured human fibroblast cell. In a typical mammalian cell, the coated pits occupy about 2% of the total surface area of the plasma membrane.

Thus, macromolecules taken up by RME bind to specific receptors on the cell surface, collect as receptor-ligand complexes in coated pits, and enter the cell as clathrin-coated vesicles. The speed and scope of the process are impressive: A coated pit usually invaginates within a minute or so of being formed, and up to 2500 such coated pits invaginate per minute in a cultured fibroblast cell! Figure 9-21 shows the progressive formation of a coated vesicle from a coated pit as particles of yolk protein are taken up by the maturing oocyte (egg cell) of a chicken.

Once a coated pit has invaginated and become internalized as a coated vesicle, the clathrin coat is released almost immediately, leaving an uncoated endocytotic vesicle that is able to fuse with other vesicles or membrane surfaces. At least some of the clathrin coat protein is thought to be recycled back to the plasma membrane, as are many of the receptors and some of the lipid of the vesicle membrane. Recycling of receptor components appears to depend on acidification of the endocytotic vesicle, mediated by an ATP-dependent proton pump in the vesicle membrane. Typically, the pH decreases from about 7 to about 5 or 5.5. The acid pH is apparently required to decrease the affinity of the receptor-ligand complexes, thereby freeing receptors to be recycled to the plasma membrane. Endocytotic vesicles may remain roughly spherical or may form elongated or even branched structures.

Further processing of the endocytotic vesicle depends on the intended fate of the internalized molecules. Some vesicles fuse to form larger vesicles, whereas others give rise to smaller vesicles that pinch off from the initial vesicle. In

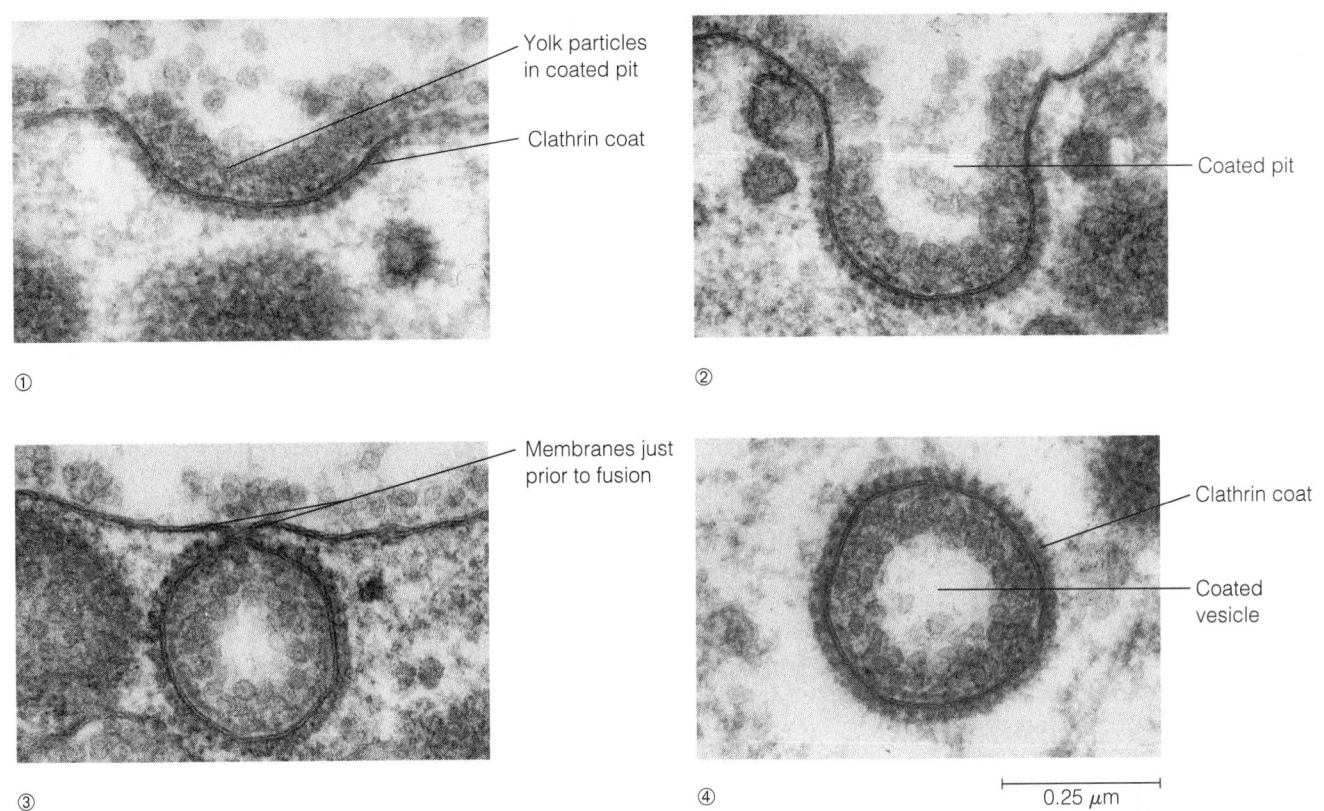

Figure 9-21   Receptor-Mediated Endocytosis of Yolk Protein by a Chicken Oocyte.   This series of electron micrographs illustrates the formation of a coated vesicle from a coated pit during receptor-mediated endocytosis. ① Yolk particles accumulate in a coated pit, which appears initially as a shallow invagination of the plasma membrane with a clathrin coat on its inner surface. ② A deeper coated pit containing several free particles in addition to those adhering to the membrane. ③ The final stage in the formation of a coated vesicle, just prior to fusion of the membranes at the neck of the invagination. ④ A coated vesicle that has just formed at the plasma membrane and that still has its clathrin coat intact (TEMs).

the process, the substances they contain are delivered either to the Golgi complex or to lysosomes, depending on the nature of the ligand. In each case, delivery of the substance occurs as endocytotic vesicles (or smaller vesicles derived from them) fuse with the appropriate membranous structure within the cell. Those substances that are routed to the Golgi complex become incorporated into cisternae of the Golgi, where they either remain or are further sorted to the ER, the plasma membrane, or the perinuclear space. Endocytotic vesicles that contain substances intended for degradation deliver their contents to lysosomes for digestion, as will be described later in the chapter.

The Role of Coated Pits and Vesicles in Cellular Transport Processes

From the preceding discussion, it is clear that **coated pits** are an integral part of receptor-mediated endocytosis. And as we noted earlier, **coated vesicles** mediate the movement of proteins from the ER through the various cisternae of the Golgi complex to secretory vesicles, lysosomes, and the plasma membrane. A common feature of both coated pits and coated vesicles is the presence on the membrane surface of a bristlelike, polyhedral protein lattice. For coated vesicles, the main protein is either *clathrin* or *coatomer*, whereas clathrin is always the main protein of coated pits. Thus, coated membranes are a common feature of many cellular processes involving the transfer or exchange of substances between specific membranous compartments of eukaryotic cells.

Coated Vesicles

Coated vesicles were first reported in 1964 by Thomas Roth and Keith Porter, who described their involvement in the selective uptake of yolk protein by developing mosquito oocytes. Since then, coated vesicles have been shown to play a vital role in diverse cellular processes. Examples are the delivery of newly synthesized molecules from the ER or Golgi complex to lysosomes, the plasma membrane, and other destinations within the cell; the uptake of substances destined for lysosomal digestion; the recycling of membranes at neuromuscular junctions; and the transport of immunoglobulins from maternal blood to fetal blood (a

9A

Clinical Applications

CHOLESTEROL, THE LDL RECEPTOR, AND RECEPTOR-MEDIATED ENDOCYTOSIS

Receptor-mediated endocytosis (RME) is a highly efficient pathway for the uptake of specific macromolecules by eukaryotic cells. We now know of several dozen different kinds of macromolecules that can be taken up by this means, each recognized by its own specific receptor in the plasma membrane of the appropriate cell types. If we consider the discovery of RME, however, we focus on a specific receptor—and, as it turns out, on a health issue that many of us are concerned about: the level of cholesterol in our blood.

As you may already know, one of the primary factors that predisposes a person to heart attacks is an abnormally high level of cholesterol in the blood serum, a condition called *hypercholesterolemia*. Because of its insolubility in the aqueous serum, cholesterol tends to be deposited on the inside walls of blood vessels. These deposits build up over time, forming the *atherosclerotic plaques* that cause *atherosclerosis*, commonly known as hardening of the arteries. Ultimately, the plaques may block the flow of blood through the vessels, causing strokes and heart attacks.

Although high blood cholesterol levels and atherosclerosis are often linked to dietary intake of cholesterol and fatty acids, some people are genetically predisposed to high blood cholesterol levels and hence to atherosclerosis and heart disease. This hereditary predisposition, called *familial hypercholesterolemia* (*FH*), is especially debilitating to homozygous individuals—that is, to those who inherited a defective FH gene from both parents. These people have grossly elevated levels of serum cholesterol (about 650–1000 mg/100 mL of blood serum, compared to the normal range of about 130–260 mg/100 mL), and develop atherosclerosis early in life, often leading to death from heart disease before the age of 20. People who are heterozygous—those with one defective and one normal copy of the gene—are affected less severely, but they nonetheless have elevated serum cholesterol levels (about 250–500 mg/100 mL) and are at high risk for heart attacks in their thirties and forties.

The link between FH and RME came about because of a study of FH by M. S. Brown and J. L. Goldstein that was begun in 1972 and led not only to the discovery of RME but also to Nobel Prizes for both scientists in 1986. Brown and Goldstein began by culturing fibroblast cells from FH patients in the laboratory and showing that such cells synthesized cholesterol at abnormally high rates compared to normal cells. Their next key observation was that normal cells also synthesized cholesterol at abnormally high rates when they were deprived of the **low-density lipoproteins (LDL)** that were usually present in the culture medium. LDL is the form in which cholesterol is transported in the blood and taken up into cells.

LDL is one of several classes of *blood lipoproteins*, which are classified according to their density. Basically, a lipoprotein is a droplet of oil that consists of a monolayer of phospholipid and cholesterol molecules and one or more protein molecules, with the lipids oriented such that their polar head groups face the aqueous medium on the outside and their nonpolar tails extend into the interior of the droplet (Figure 9A-1). LDL is the class of lipoproteins with the highest cholesterol content: Free cholesterol and cholesterol esters make up more than half of the LDL particle by weight! The esterified form of cholesterol has a long-chain fatty acid linked to it, which makes it highly hydrophobic. Esterified cholesterol molecules (about 1500 per particle) therefore cluster in the interior of the droplet, whereas the free, or unesterified, cholesterol molecules (about 500 per particle) are found mainly in the lipid monolayer (Figure 9A-1).

In addition to phospholipids and cholesterol, each LDL particle has a single molecule of a large protein called *apoprotein B-100* embedded in its lipid monolayer. This protein is crucial to our understanding of the difference in the response of FH and normal cells to the level of LDL in the medium. The ability of normal fibroblasts to maintain an appropriately low rate of cholesterol synthesis in the presence of LDL suggested to Brown and Goldstein that LDL was involved in the transport of cholesterol into the cell, where the cholesterol then regulated its own synthesis by *allosteric* (or *feedback*) *inhibition* (see Figure 6-17). FH fibroblasts, on the other hand, synthesized cholesterol at a high rate regardless of whether LDL was present in the medium, suggesting that these cells might be defective in LDL-dependent cholesterol uptake.

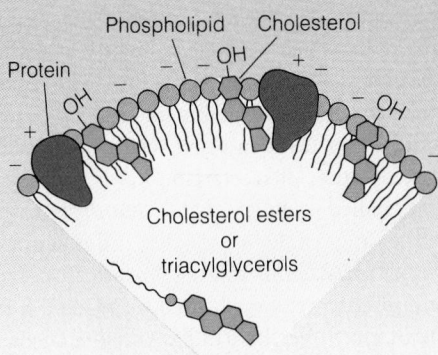

Figure 9A-1   Lipoprotein Structure.   A typical lipoprotein consists of a monolayer of phospholipid and cholesterol molecules surrounding a very hydrophobic interior, a portion of which is shown here. One or more protein molecules are embedded in the monolayer. Cholesterol molecules esterified to long-chain fatty acids are highly hydrophobic and tend to cluster in the interior of the lipoprotein particle. Lipoproteins differ from one another in density, depending on the relative amounts of lipid and protein present. (Lipids are much less dense than proteins, so the density of the particle is inversely related to the abundance of lipids.) The particular lipoprotein shown here is low-density lipoprotein (LDL), with a density of 1.02–1.06 g/mL. A typical LDL particle contains about 800 phospholipid and 500 free (unesterified) cholesterol molecules in the lipid monolayer and about 1500 esterified cholesterol molecules in the interior. The single protein molecule, called apoprotein B-100, has a molecular weight of about 500,000 and is embedded in the lipid monolayer. The protein provides structural organization to the particle and mediates the binding of the LDL to the LDL receptors on the surfaces of cells.

Based on their observations, Brown and Goldstein postulated that the uptake of cholesterol into cells requires the action of a specific receptor on the cell surface and that this receptor is absent or defective in FH patients. In a brilliant series of experiments, these investigators and their colleagues demonstrated the existence of an LDL-specific membrane protein, called the **LDL receptor,** and showed that it recognizes the apoprotein B-100 molecule that is present in every LDL particle. They also showed that the cells from FH patients either lacked this protein entirely or had receptor molecules that were defective in any of several ways.

To visualize the LDL particles, these scientists conjugated, or linked, them to molecules of ferritin, a protein that binds iron atoms. Because iron atoms are electron-dense, they appear as dark dots in the electron microscope (Figure 9A-2). Using this technique, Brown and Goldstein showed that the ferritin-conjugated LDL particles bound to the surface of the cell and clustered at specific locations (Figure 9A-2a). We now recognize these sites as *coated pits*, localized regions of the plasma membrane characterized by the presence of *clathrin* on the cytoplasmic side of the membrane

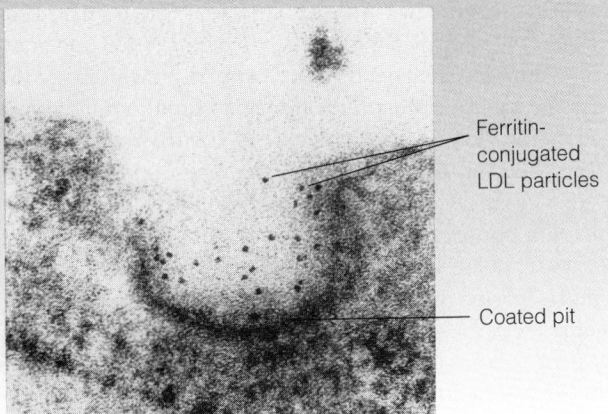

(a) Particles bind to coat pit

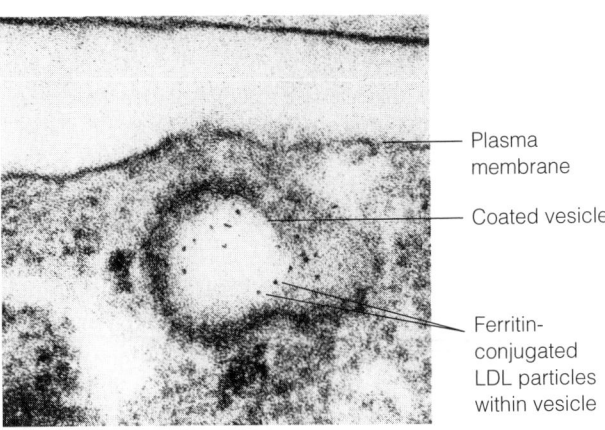

(b) Pit forms coated vesicle

Figure 9A-2   Visualization of LDL Binding.   LDL particles can be visualized for electron microscopy by conjugating them with ferritin, an iron-binding protein that allows the LDL-ferritin complex to be seen in the electron microscope because of the density of the iron atoms bound to the ferritin. **(a)** LDL-ferritin conjugates, visible as dark dots, bind to a coated pit on the surface of a cultured human fibroblast cell. **(b)** LDL-ferritin conjugates are internalized within a coated vesicle that is formed as the coated pit region invaginates beneath the plasma membrane and pinches off (TEMs).

and by the accumulation of membrane-bound receptor-ligand complexes on the exterior of the membrane.

Dark dots were also seen on the inside of vesicles that formed by invagination and the pinching off of coated pits (Figure 9A-2b). The receptors, in other words, not only bound the LDL on the cell surface but were also apparently involved in the internalization of LDL within vesicles. In short, these workers had discovered a new mechanism by which cells can take up macromolecules from their environment. And since it was an endocytotic process involving specific receptors, Brown and Goldstein gave it the name by which we know it today—*receptor-mediated endocytosis.* ∎

process called *transcytosis*). Coated vesicles have even been implicated in the uptake of viruses into cells and in the transport of viral proteins from the ER to the plasma membrane. In fact, coated vesicles may well be involved in all exchanges between membranes, accounting not only for the mechanism of the interaction but perhaps also for the selectivity.

Clathrin-coated vesicles are primarily involved in the selective transport of specific proteins from the *trans* cisternae of the Golgi complex as well as the internalization of receptor-ligand complexes from the plasma membrane by endocytosis. Coatomer-coated vesicles, on the other hand, facilitate nonselective transport from the ER to the Golgi complex and between Golgi cisternae. We will focus here on clathrin-coated vesicles because we know far more about them. (Recent evidence suggests that there may be a third class of coat vesicles, with the protein *caveolin* as its coat. However, we know too little about their function to discuss them further.)

Clathrin-Coated Vesicles

As isolated from a variety of tissues, clathrin-coated vesicles have diameters of about 50–250 nm, with each vesicle surrounded by a coat that consists mainly of pentagonal and hexagonal units of **clathrin,** a polypeptide with a molecular weight of about 180,000. As Figure 9-22 illustrates, the clathrin coat seems to form a latticelike "basket" around the vesicle. (The term *clathrin* comes from *clathratus*, the Latin word for "lattice.") The clathrin lattice is flexible, accommodating itself readily to changes in vesicle size during budding, probably by varying the number of hexagonal units.

The vesicles dissociate readily into membranous components and soluble clathrin complexes. These, in turn, can reassemble spontaneously under appropriate conditions. Clathrin complexes can even be induced to reassemble in the absence of vesicle membranes, resulting in empty shells called clathrin "cages." Such reassembly occurs remarkably fast—within seconds, under the right conditions. The ease with which it can be assembled and disassembled is probably an important feature of the clathrin coat because fusion of the underlying membrane with the membrane of another structure seems to require partial or complete uncoating of the vesicle.

Structure of the Clathrin Lattice: The Triskelion. In 1981, Ernst Ungewickell and Daniel Branton succeeded in visualizing the structural units of the clathrin lattice and showed that they exist as trimers called **triskelions** (Figure 9-23a). Each triskelion is a three-legged structure, consisting of three clathrin polypeptides radiating from a central vertex (Figure 9-23b). Each leg is bent slightly near the middle and has a globular domain at its tip. The triskelion is thought to be the unit of assembly for clathrin coats.

In addition to clathrin itself, clathrin-coated vesicles contain several other characteristic kinds of polypeptides. One group, called *clathrin light chains*, has molecular weights in the range of 30,000–36,000 and is about as abundant in clathrin triskelions as are the clathrin polypeptides. This stoichiometric relationship suggests that a triskelion probably contains three clathrin light chains, one associated with each of the clathrin legs of the triskelion. Antibodies against clathrin light chains bind to the legs of the triskelion near its center, suggesting that the light chains are associated with the inner half of each triskelion leg, as indicated in Figure 9-23b.

The *adaptins* are a third class of proteins found in clathrin coats. These multisubunit proteins are required to attach the clathrin coat to the membrane and to recognize the specific transmembrane receptor proteins needed to ensure that the appropriate macromolecules will bind to the membrane surface and be internalized as the membrane invaginates.

By examining negatively stained preparations of partly reassembled clathrin structures, B. M. F. Pearse and her colleagues were able to suggest a model for the assembly of clathrin triskelions into the characteristic hexagons and pentagons of coated pits and vesicles (Figure 9-23c). According to their model, one clathrin triskelion is located at each polyhedral vertex, with each polypeptide leg extending along one of the legs of two neighboring trimers. (Only modest differences in the angles of the legs of the triskelion are necessary to accommodate pentagonal, hexagonal, or other lattice patterns.) This arrangement of triskelions into overlapping networks allows a maximum amount of longitudinal contact between clathrin polypeptides. Such contacts may confer the mechanical strength and flexibility needed when a vesicle buds off a membrane.

Figure 9-22 Clathrin-Coated Vesicles. Clathrin-coated vesicles are involved in a variety of transport and shuttle processes in eukaryotic cells. Each such vesicle is surrounded by a latticelike "basket" or "cage" of clathrin molecules. **(a)** Electron micrographs of clathrin cages that have been negatively stained with uranyl acetate. Arrows point to edges that exhibit a clear double- or triple-lined nature due to the overlap of the legs of adjacent clathrin molecules, as shown in Figure 9-23 (TEMs). **(b)** An interpretive drawing of a clathrin cage.

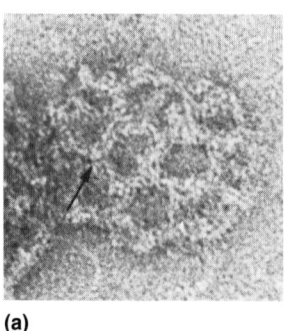
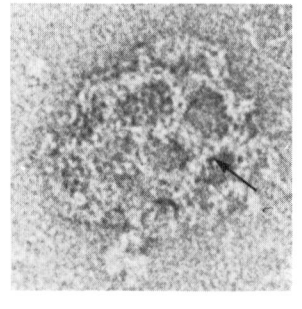
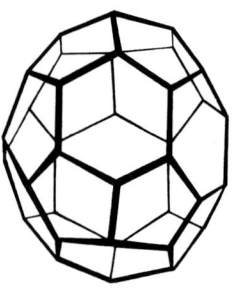

(a)

50 nm

(b)

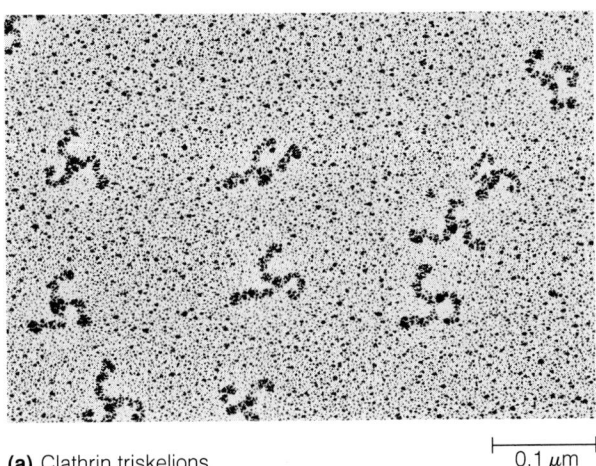

(a) Clathrin triskelions

0.1 μm

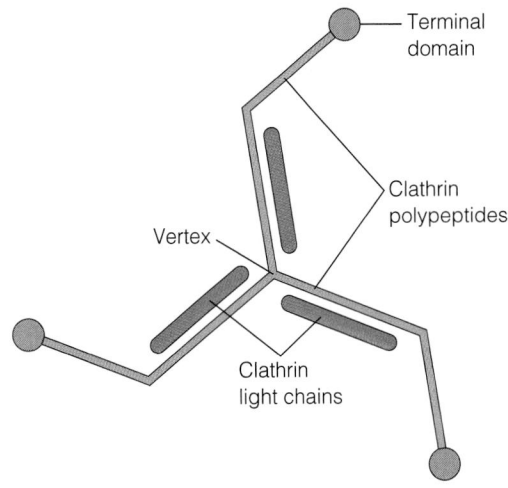

(b) Structure of clathrin triskelion

Pentagon

Hexagon

(c) Model for assembly of clathrin triskelions

Figure 9-23   Clathrin Triskelions.
(a) This micrograph shows individual triskelions of clathrin isolated from calf brain (TEM). **(b)** Each triskelion consists of three clathrin polypeptides radiating from a central vertex, with a terminal beadlike domain at the tip of each triskelion leg and a clathrin light chain bound to the inner half of each leg. (Not shown are the adaptins that are also found in clathrin coats.) **(c)** Under appropriate conditions, clathrin triskelions assemble into the pentagonal and hexagonal structures characteristic of coated pits and vesicles. According to the model illustrated here, one clathrin triskelion is located at each polyhedral vertex, with each polypeptide leg extending along one of the legs of two neighboring triskelions.

Clathrin Assembly and Bud Formation. The assembly of clathrin triskelions on a membrane surface appears to provide the force that drives the formation of a bud at that site. In the case of receptor-mediated endocytosis, for example, the plasma membrane is thought to invaginate inward in response to the assembly of the clathrin coat on the inner side of the membrane. However, we do not yet know what initiates the process at a particular site on the membrane, nor do we understand how the invaginated bud eventually pinches off to form an endocytotic vesicle. Moreover, clathrin assembly is not always required for bud formation because phagocytosis also involves membrane invagination, yet it occurs without any involvement of clathrin.

Some mechanism is also clearly required to *uncoat* a clathrin-coated membrane. Moreover, the uncoating must be done in a regulated manner because, in most cases, the clathrin coat on a membrane segment remains intact as long as the membrane is part of a coated pit or a bud, but the coat dissociates rapidly once the vesicle is formed. Unlike assembly of the clathrin coat, which appears to occur spontaneously without an energy requirement, dissociation of the coat is an energy-requiring process. Dissociation is driven by the energy of ATP hydrolysis—about 3 ATP molecules per triskelion—mediated perhaps by an *uncoating ATPase*.

The specificity necessary to ensure that the dissociation is limited to vesicles may be provided by differences in calcium concentration within the cell. Calcium is known to bind to clathrin light chains, which appears to destabilize the clathrin coat. Because calcium is pumped out of the cell by calcium pumps in the plasma membrane, the calcium concentration within the cell is the lowest in the immediate vicinity of the membrane. Thus, coated pits are likely to be in a microenvironment with a sufficiently low calcium concentration to ensure the stability of clathrin coats. When a coated vesicle buds off the membrane, however, it is likely to move inward, where the higher calcium concentration may lead to the binding of calcium to the light chains. Binding of calcium may then destabilize the coat, rendering it susceptible to the action of the uncoating ATPase.

Biosynthesis of Membrane Proteins and Lipids

Inherent in the secretory processes we discussed earlier is a unidirectional flow of membrane from the ER through the Golgi complex to the plasma membrane. Every time a secretory granule fuses with the plasma membrane and discharges its contents by exocytosis, another bit of membrane that originated with the rough ER becomes a part of the plasma membrane. This membrane flow suggests that the rough ER membrane is the precursor to both the Golgi membranes and the plasma membrane of eukaryotic cells. Studies of lipid biosynthesis and their fate within the cell indicate that rough ER may be the source of other cellular membranes as well, including the membranes of lysosomes and both inner and outer membranes of the mitochondrion.

Because rough ER is the source of most (perhaps even all) membranes in the eukaryotic cell, it clearly must be the site of synthesis not just of secretory proteins but of membrane components as well. This appears to be true of membrane proteins and phospholipids alike.

Biosynthesis and the Processing of Membrane Proteins

Earlier in the chapter we emphasized secretory proteins, but only because these are easiest to approach experimentally. In fact, most of the proteins a cell makes are not destined for secretion; they are targeted for intracellular sites such as membranes. Nonetheless, evidence suggests that the processing of membrane proteins is similar to that already described for secretory proteins. The ribosomal complex becomes anchored to the rough ER, just as it does for a secretory protein. At this point, however, an important difference is noted. Instead of passing through the ER membrane into the lumen, as secretory proteins do during synthesis, membrane proteins remain anchored in the membrane as they are synthesized.

Glycosylation of membrane proteins is also similar to the processing of secretory proteins. Glycosylation is begun in the rough ER and completed in the Golgi complex. Experiments with radioactively labeled sugars show a progressive movement of label from the Golgi to the cell surface. Glycosylation of proteins in the ER and Golgi occurs on the luminal side (i.e., the inner surface) of the membrane only. Because this side is topologically equivalent to the exterior surface of the plasma membrane, it is easy to see why all plasma membrane glycoproteins are found on the extracellular side of the membrane.

Biosynthesis of Membrane Lipids

The ER, both rough and smooth, is the site of synthesis of most membrane lipids, including phospholipids and cholesterol. In fact, most of the enzymes needed for the synthesis of the various membrane phospholipids are found nowhere else in the cell. Labeling experiments with radioactive choline confirm the role of rough ER in phospholipid biosynthesis.

Lipids synthesized in the ER become incorporated into all the membranes of the cell, not just the plasma membrane. For most membranes, such as those of the Golgi complex and lysosomes, this transfer of lipids (and proteins as well) probably occurs by means of small **transport vesicles** that pinch off from rough ER and fuse with target membranes. Transport vesicles are probably coated vesicles similar to those involved in the transport of secretory proteins from rough ER to the Golgi complex.

The transfer of phospholipids from the ER to the mitochondrion poses a unique problem because mitochondria do not grow by fusion with vesicles synthesized elsewhere in the cell. Instead, special **phospholipid transfer proteins** mediate the transfer of phospholipid molecules into the mitochondrial membrane. Each such protein recognizes a specific kind of phospholipid, removes it from one membrane, and adds it to the other. These transfer proteins may be involved in the movement of phospholipids from the ER to other membrane systems of the cell as well, possibly including the plasma membrane.

Lysosomes and Cellular Digestion

Lysosomes are derived from the ER and the Golgi complex but are distinctively different in their function in the cell. Basically, a **lysosome** is an organelle that contains digestive enzymes capable of degrading all major classes of biological macromolecules. These enzymes are needed to both degrade materials brought into the cell from the outside and digest cellular molecules and structures that are damaged or no longer needed. We will first look at the organelles themselves and then consider the digestive processes in which they are involved, as well as some of the disease conditions that result from lysosomal malfunction.

The Discovery of Lysosomes

As previously noted, lysosomes were discovered in the early 1950s by Christian de Duve and his colleagues. Differential centrifugation led the researchers to realize that an acid phosphatase initially thought to be located in the mitochondrion was in fact associated with a class of particles that had never been reported before. Along with acid phosphatase, these organelles contained several other hydrolytic enzymes, including β-glucuronidase, ribonuclease, deoxyribonuclease, and a protease. Because of its apparent role in cellular lysis, de Duve called the new organelle a *lysosome.*

Only after its existence had been predicted, its properties described, and its enzyme content specified was the lysosome actually observed in the electron microscope and recognized as a regular constituent of most animal cells. Final confirmation came from cytochemical staining reactions capable of localizing acid phosphatase and other lysosomal enzymes to specific structures that can be seen in the electron microscope, as illustrated in Figure 9-24. Lysosomes vary considerably in size and shape but are in general about 0.5

μm in diameter. They are surrounded by a single membrane and have an electron-dense, granular matrix when visualized in the electron microscope.

The list of lysosomal enzymes has expanded considerably since de Duve's original work, but all known lysosomal enzymes have the common property of being *acid hydrolases*—hydrolytic enzymes with a pH optimum around 5. The list includes at least five phosphatases, four proteases, two nucleases, six lipases, twelve glycosidases, and an arylsulfatase. Taken together, these lysosomal enzymes are capable of digesting all the major classes of biological molecules. No wonder, then, that they are sequestered together in a single kind of organelle, away from the rest of the cell, with which they would quickly wreak havoc.

Lysosome Biogenesis

Most investigators regard the Golgi complex as the site of lysosome formation. Lysosomal enzymes are synthesized on ribosomes that are bound to the rough ER and transferred

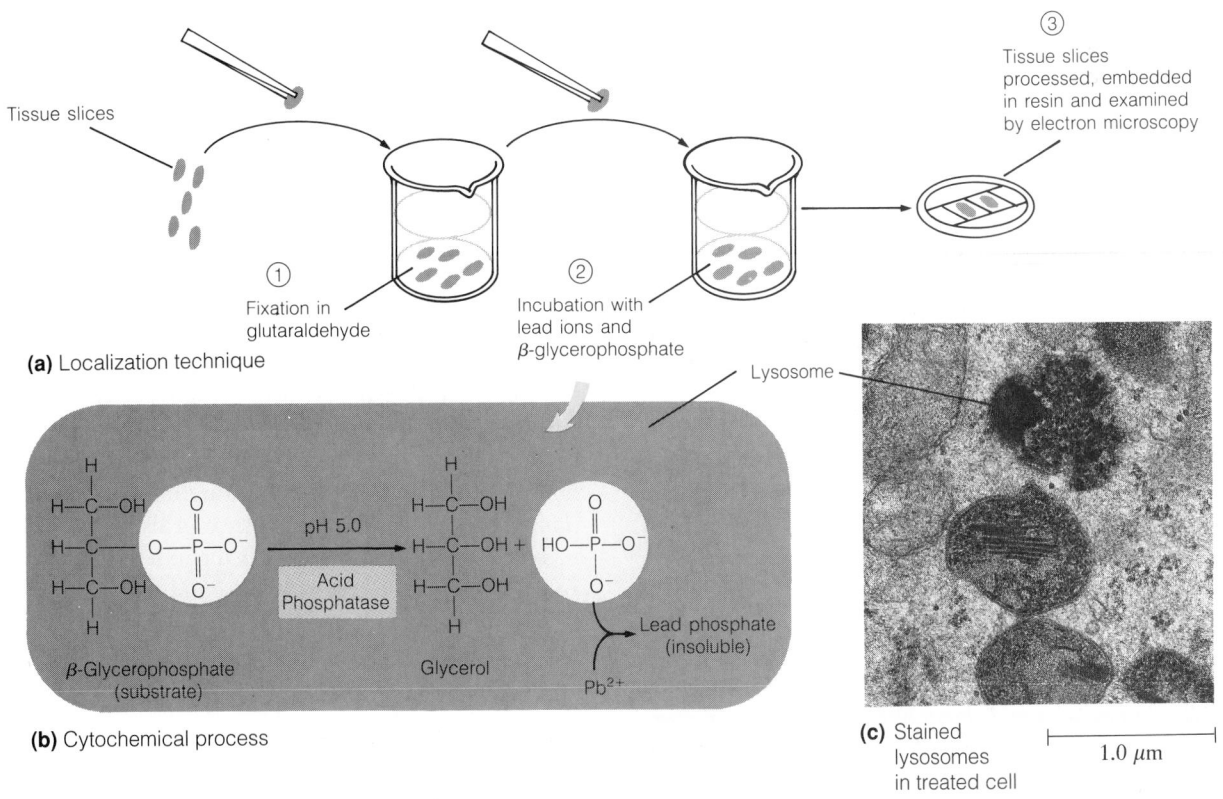

Figure 9-24   Cytochemical Localization of Acid Phosphatase, a Lysosomal Enzyme.   (a) ① Thin sections of tissue are fixed in glutaraldehyde and ② incubated at pH 5.0 in a medium containing β-glycerophosphate (a substrate for the enzyme acid phosphatase) and a soluble lead salt (lead nitrate, commonly). ③ The tissue slices are then embedded in resin and examined in the electron microscope.
(b) During the incubation, the acid phosphatase within the lysosomes cleaves the phosphate from the β-glycerophosphate substrate, leaving free glycerol and phosphate anions. The phosphate anions react with lead ions to form lead phosphate, which is insoluble and therefore precipitates at the site of enzyme activity. Since the lead phosphate is electrondense, microscopy reveals the site of lead phosphate formation and hence the location of the acid phosphatase within the cell. **(c)** Cells treated as in part a contain lysosomes that appear darkly stained in the electron microscope, indicating the presence of lead phosphate. Thus, acid phosphatase must be located in lysosomes (TEM).

to its lumen before being transported to the Golgi. Lysosomes then develop from clathrin-coated vesicles that bud off the *trans*-most cisternae of the Golgi complex.

Targeting Hydrolases to the Lysosomes

Enzymes destined for the lysosomes are glycoproteins, containing oligosaccharides that are added in the ER and processed in the Golgi complex. However, lysosomal enzymes are distinguishable from other glycoproteins by the presence of an unusual oligosaccharide containing mannose-6-phosphate. This unique feature of lysosomal enzymes appears to serve as a recognition marker, or "address," that targets all such proteins to the lysosomes. The Golgi membrane, in turn, contains receptors that recognize this special oligosaccharide and ensure that all enzymes with its address are packaged into lysosomes. Recall from Figure 9-13 that mannose-6-phosphate receptors are concentrated in the cisternae on the *cis* face of the Golgi stack. This localization is consistent with the role of these receptors in the detection of glycoproteins with the address, as they arrive at the Golgi complex from their site of synthesis on the rough ER.

Strong support for this concept of hydrolase targeting by an oligosaccharide address came from studies of a human genetic disorder called *I-cell disease*. Fibroblast cells from patients with this disease synthesize all the expected hydrolases in culture, but they release the enzymes into the medium instead of incorporating them into their lysosomes. The distinguishing feature of these hydrolases is that they lack the oligosaccharide that contains mannose-6-phosphate. Apparently, I-cell disease involves a genetic defect at some step in the process of adding the correct oligosaccharide address to all the lysosomal enzymes. As a result, the enzymes are secreted by default, rather than being targeted to lysosomes.

Cellular Digestion

The organelle formed as we have just described is called a **primary lysosome**—a lysosome with its full complement of lytic enzymes, but not yet engaged in digestive activity. A primary lysosome, in other words, is a collection of as-yet-unused digestive enzymes packaged together in a way that protects the cell from their activities until they are required for specific hydrolytic functions. These functions include the digestion of foreign materials, often brought into the cell for that express purpose, the breakdown of cellular components that are damaged in some way, and the removal of compounds that are no longer needed.

The digestive processes in which lysosomal enzymes are involved can be distinguished by both the site of their activity and the origin of the materials that are digested, as shown in Figure 9-25. Usually, the site of activity is intracellular, though in some cases lysosomes may release their enzymes to the outside of the cell by exocytosis. The materials to be digested are often of extracellular origin, but important processes are also known to involve lysosomal digestion of

cellular components or even whole cells or tissues. As a result, lysosomes play a role in such diverse cellular activities as nutrition, defense, recycling of cellular components, differentiation, and even cell death. The specific processes in which lysosomes are involved are called phagocytosis, autophagy, autolysis, and extracellular digestion. These are illustrated in Figure 9-25 as pathways A, B, C, and D, respectively. We will now examine each of these processes in detail.

Phagocytosis: Lysosomes in Nutrition and Defense.   One of the most important functions of lysosomal enzymes is the degradation of foreign materials brought into the cell by the process of endocytosis. Among the endocytotic processes contributing to the uptake of foreign substances destined for lysosomal digestion are *bulk-phase* and *receptor-mediated endocytosis* (see Figures 9-19 and 9-20). In most cells, the majority of endocytotic vesicles formed by these processes ultimately fuse with lysosomes to initiate digestion of the endocytosed macromolecules.

In addition, some cells can take up much larger particles, including whole cells, by *phagocytosis*, a process seen most prominently among single-celled heterotrophs. These organisms routinely engulf and internalize food particles and smaller organisms for intracellular digestion. Phagocytosis is also used by some primitive animals, notably flatworms, coelenterates, and sponges, as a means of obtaining nutrients. In higher organisms, phagocytosis is usually restricted to specialized cells called **phagocytes.** Your body, for example, contains two classes of white blood cells that function as phagocytes: *macrophages* and *neutrophils* (also called *polymorphonuclear leukocytes*). These types of cells engulf and digest invading microorganisms or foreign materials in the bloodstream. Phagocytosis therefore serves a nutritional purpose among lower organisms but becomes a means of defense in higher animals.

Phagocytosis has been studied most extensively in the amoeba. The cell surface of most amoebas is covered with either fine hairs or a coat of glycosaminoglycans. The function is the same in either case—a means of adsorbing and trapping food particles or smaller organisms. Contact with such an object appears to trigger the onset of phagocytosis. Folds of membrane gradually surround and engulf the particle, and a *food cup* forms as the membrane invaginates and folds around the particle. Eventually, this pinches off to form a free **phagocytic vacuole** within the cell, as pathway A of Figure 9-25 shows.

Once inside the cell, phagocytic vacuoles move toward the Golgi region, where they encounter and fuse with the primary lysosomes that form from the *trans* cisternae of the Golgi complexes in that part of the cell. (Obviously, such fusions must be highly selective because the fusion of lysosomes with any membrane-bounded structures other than phagocytic vacuoles could have disastrous consequences for the cell.) Such fused vacuoles are called **secondary lysosomes**. Because they contain both digestive enzymes and digestible foodstuffs, secondary lysosomes are the site of digestive activity. Soluble products of digestion, such as sugars,

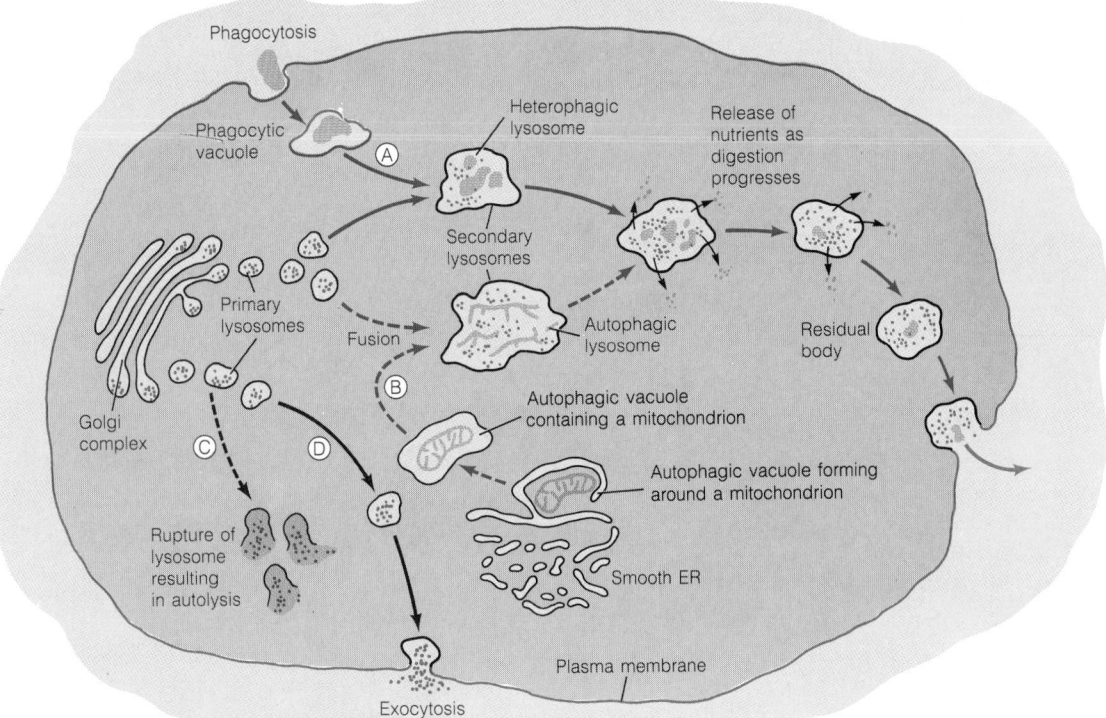

Figure 9-25   The Formation of Primary and Secondary Lysosomes and Their Role in Cellular Digestive Processes. Shown in this composite cell are the major processes in which lysosomes are involved. The pathways depicted are Ⓐ phagocytosis, Ⓑ autophagy, Ⓒ autolysis, and Ⓓ extracellular digestion.

amino acids, and nucleotides, cross the membrane and are used as a source of nutrients by the cell. Secondary lysosomes are capable of fusing with further phagocytic vacuoles, thereby acquiring further material to digest. As a result, they vary considerably in size, appearance, content of digestible material, and stage of digestion.

Eventually, however, only indigestible material remains in the vacuole, which becomes a **residual body** as digestion ceases. In protozoa, residual bodies routinely fuse with the plasma membrane and expel their contents to the outside by exocytosis, as illustrated in Figure 9-25. In vertebrates, there appears to be no such mechanism, so residual bodies accumulate in the cytoplasm. This accumulation is thought to contribute to cellular aging, particularly in long-lived cells such as those of the nervous system.

Autophagy: The Original Recycling System. A second major role of lysosomes is the breakdown of cellular components and structures that have become damaged or are no longer needed. Most cellular organelles are in a state of dynamic flux, with new organelles being synthesized and old organelles being destroyed continuously. The digestion of old or unwanted organelles or other cell structures is called **autophagy,** which literally means "self-eating." Autophagy is illustrated in Figure 9-25 as pathway B.

Autophagy begins when an organelle or other structure becomes wrapped in membranes that are thought to arise from the ER. The resulting vesicle is called an **autophagic**

vacuole. It is often possible to see identifiable remains of cellular structures in these vacuoles, as shown in Figure 9-26. The fate of an autophagic vacuole closely parallels that of a phagocytic vacuole. In both cases, fusion with a primary lysosome gives rise to a secondary lysosome, in which the digestive action of the lytic enzymes is unleashed. To distinguish between secondary lysosomes, we refer to those containing substances of extracellular origin as **heterophagic lysosomes,** while those with materials of intracellular origin are called **autophagic lysosomes.**

Autophagy occurs at varying rates in most cells most of the time, but it is especially prominent in certain developmental situations. During the maturation of a red blood cell, for example, virtually all of the intracellular contents must be destroyed, including all the mitochondria. This is accomplished by autophagic digestion. A marked increase in autophagy is also noted in cells stressed by starvation. Presumably, the process represents a rather desperate attempt on the part of the cell to continue to provide for its energy needs, even if it has to consume its own structures to do so.

Autolysis: Cellular Self-Destruction. Normally, the hydrolytic enzymes of the lysosome are kept safely packaged within the membranes of the organelle to prevent unwanted degradation of the cell itself. Materials destined for degradation, whether originating from inside the cell or from outside, are first carefully enclosed in vacuoles and then brought in contact with the digestive enzymes by fusion of

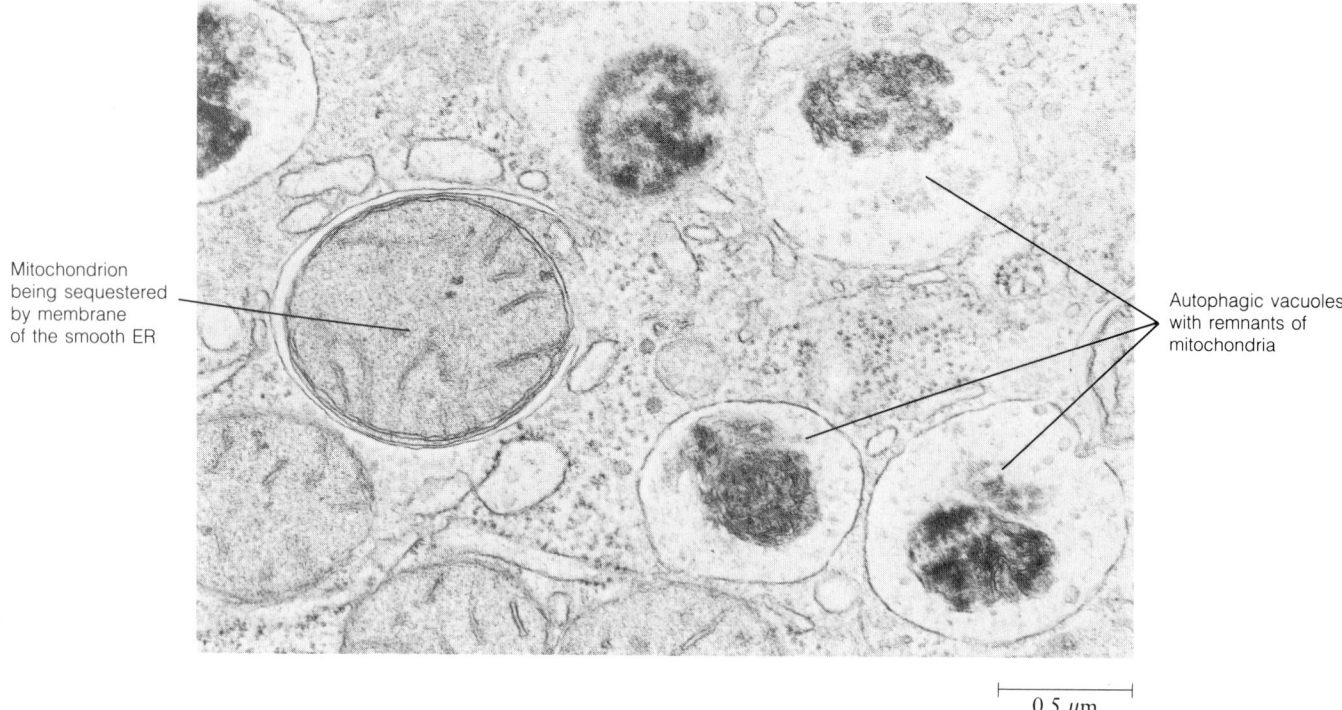

Mitochondrion being sequestered by membrane of the smooth ER

Autophagic vacuoles with remnants of mitochondria

0.5 μm

Figure 9-26 Autophagic Digestion. Earlier and later stages of autophagic digestion are shown here in a rat liver cell. On the left is the formation of an autophagic vacuole, as a mitochondrion is sequestered by membrane derived from the smooth ER. On the right are several autophagic vacuoles with remnants of mitochondria in them (TEM).

the vacuoles with primary lysosomes. At no point in the process are the lytic enzymes allowed to escape from behind the membranes that confine them. If they do, as sometimes occurs in injured red blood cells, the released enzymes rapidly digest and destroy the cell. Under most conditions, this would be disastrous.

Such is not always the case, however. During the development of many multicellular organisms, there are specific processes in which such self-destruction, or **autolysis,** plays an important role. Wherever the removal of specific cells or tissues is necessary to the development of a particular structure or organ system, autolysis is the means by which the unwanted cells are destroyed. The creation of individual digits (fingers and toes) from the initially webbed hands and feet of a human embryo by the selective removal of interdigital cells is an example of such *programmed cell death* during development. The progressive loss of its tail as a tadpole undergoes metamorphosis is another case in which lysosomally mediated autolysis of cells is part of a normal developmental process. The mechanism appears to involve the deliberate release of digestive enzymes from the lysosomes, giving the organelle its reputation as a "suicide bag" in such cases. Autolysis appears in Figure 9-25 as pathway C.

Extracellular Digestion. Most of the digestive processes involving lysosomal enzymes occur intracellularly, either within vacuoles (phagocytosis and autophagy) or within the

cell as a whole (autolysis). In rare cases, however, lysosomes may discharge their enzymes to the outside of the cell by exocytosis, resulting in **extracellular digestion** (see Figure 9-25, pathway D). One such example occurs during fertilization of animal eggs. The head of the sperm releases lysosomal enzymes capable of degrading chemical barriers that would otherwise keep the sperm from penetrating the egg surface. Moreover, certain inflammatory diseases, such as rheumatoid arthritis, appear to result from the release of lysosomal enzymes into the joints. The steroid hormones cortisone and hydrocortisone are thought to be effective antiinflammatory agents because of their role in stabilizing lysosomal membranes and thereby inhibiting enzyme release.

Lysosomal Storage Diseases

The important role of lysosomes in cellular turnover is clearly seen in diseases that result from the deficiency of specific lysosomal enzymes. There are, in fact, about 40 such **lysosomal storage diseases,** each characterized by the undesirable accumulation and storage of excessive amounts of specific substances, usually polysaccharides or lipids. The cells in which this accumulation occurs are greatly impaired in function, if not destroyed. Muscle weakness, skeletal deformities, and mental retardation commonly result, often with a fatal outcome.

In each case, a specific lysosomal enzyme is absent or deficient. The first of these storage diseases to be explained was *type II glycogenosis*, in which young children accumulate excessive amounts of glycogen in the liver, heart, and muscles and eventually die. The problem turned out to be a deficiency of the lysosomal enzyme α-glucosidase, which catalyzes glycogen hydrolysis in normal cells.

Two of the best-known lysosomal storage diseases are *Hurler syndrome* and *Hunter syndrome*. Both involve defects in the degradation of acid glycosaminoglycans. When sweat gland cells from a patient with Hurler syndrome were observed in the electron microscope, large numbers of atypical vacuoles were seen that stained for both acid phosphatase and undegraded acid glycosaminoglycans. These are thought to be aberrant secondary lysosomes with contents that cannot be degraded.

Mental retardation is so common a feature of the lysosomal storage diseases because of the impaired metabolism of glycolipids, which are important in brain tissue and in the sheaths of nerve cell axons. One particularly well-known example is *Tay-Sachs disease*, which occurs mainly among people of eastern European Jewish origin. The condition is inherited as a recessive trait. Afflicted children show rapid mental deterioration after about six months of age and usually die within the first three years. The disease results from the accumulation in the brain of a particular kind of glycolipid called a *ganglioside*. The missing lysosomal enzyme in this case is β-*N*-acetylhexosaminidase, which is responsible for cleaving the terminal *N*-acetylgalactosamine from the carbohydrate portion of the ganglioside. Lysosomes from children suffering from Tay-Sachs disease are filled with membrane fragments containing undegraded gangliosides.

Peroxisomes

Like lysosomes, **peroxisomes** are organelles that are bounded by a single membrane and are usually somewhat smaller than mitochondria—though with considerable variation in size, depending on the tissue in which they are found. Unlike lysosomes, however, peroxisomes are not derived from the endoplasmic reticulum and are therefore not a part of the endomembrane system that includes the other organelles in this chapter. Peroxisomes are found ubiquitously in eukaryotic cells but are especially prominent in mammalian kidney and liver cells, in algae and photosynthetic cells of plants, and in germinating seedlings of plant species that store fat in their seeds.

Regardless of where they occur, peroxisomes are characterized by the presence of *catalase*, an enzyme that plays a vital role in the breakdown of hydrogen peroxide. Hydrogen peroxide (H_2O_2) is a potentially toxic compound that is formed in a variety of oxidative reactions catalyzed by enzymes called *oxidases*. These enzymes are localized to peroxisomes along with catalase. Thus, the generation and degradation of H_2O_2 occur within the same organelle, protecting

other parts of the cell from exposure to this toxic compound. Before discussing the functions of peroxisomes further, however, we will look first at how peroxisomes were discovered and how they can be distinguished from other organelles when viewed with an electron microscope.

The Discovery of Peroxisomes

As noted earlier, de Duve and his colleagues discovered both lysosomes and peroxisomes. The discovery of lysosomes depended critically on the difference in the rates at which mitochondria and lysosomes sediment when subjected to centrifugation (see Chapter 4). Because lysosomes are smaller and less dense than mitochondria, they sediment more slowly when subjected to centrifugation. This difference explains the separation of acid phosphatase (and other hydrolytic enzymes) from mitochondrial marker enzymes that de Duve and his colleagues achieved using density gradient centrifugation of rat liver homogenates (see Figure 9-5).

The discovery of peroxisomes also depended on the apparently anomalous behavior of specific enzymes, although a different centrifugation technique was used in this case. During the course of their early studies on lysosomes, de Duve and his colleagues became aware of at least one enzyme, *urate oxidase*, that seemed to be associated with their lysosomal fractions, yet was not an acid hydrolase. Moreover, this enzyme differed somewhat from known lysosomal enzymes in its behavior during differential centrifugation, which also kept de Duve from identifying it as a lysosomal enzyme.

The clue that eventually allowed de Duve and his associates to assign urate oxidase to a new class of previously undescribed organelles was provided by the technique of equilibrium density centrifugation (Figure 9-27; see also Figure 9-6). By using a gradient of sucrose concentration, the researchers found that urate oxidase from rat liver was recovered in a region of the gradient having a somewhat different density (1.25 g/cm^3) from that of other organelles, notably mitochondria (1.19 g/cm^3) and lysosomes (about 1.20–1.24 g/cm^3). However, the density differences were too small and the range of lysosome densities too broad to allow the new class of organelles to be separated adequately from lysosomes under normal conditions (Figure 9-27a). By using an experimental trick, however, de Duve was able to achieve good separation. The trick was based on the chance observation that animals injected with the detergent Triton WR-1339 accumulate the detergent preferentially in their lysosomes, giving these organelles a much lower buoyant density than normal (about 1.10–1.14 g/cm^3). By using liver homogenates from detergent-treated rats, de Duve was able to separate urate oxidase cleanly from both mitochondrial and lysosomal marker enzymes (Figure 9-27b).

Once this separation was achieved, additional enzymes were shown to be present in the urate oxidase-containing fractions, including catalase and D-amino acid oxidase. This

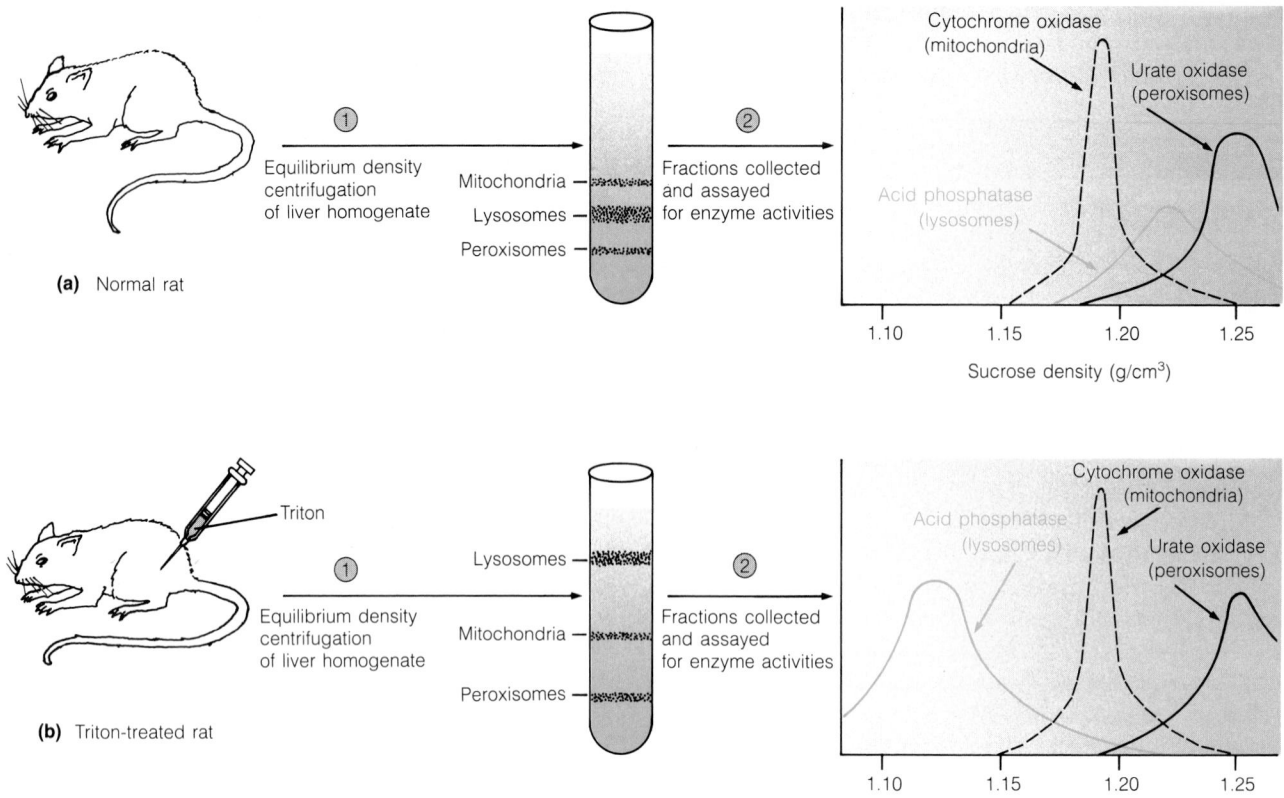

Figure 9-27   Separation of Lysosomes from Peroxisomes by Equilibrium Density Centrifugation.   (a) For organelles obtained from the liver of a normal rat and centrifuged to equilibrium on a sucrose density gradient, lysosomes (marker enzyme: *acid phosphatase;* density 1.22g/cm³) have a range of densities intermediate between those of mitochondria (marker enzyme: *cytochrome oxidase;* density 1.19 g/cm³) and peroxisomes (marker enzyme: *urate oxidase;* density 1.25 g/cm³), making resolution of the three organelles difficult. (b) For organelles obtained from the liver of a rat that had previously been treated with the detergent Triton WR-1339, the lysosomes band at a much lower density (about 1.10–1.14 g/cm³), allowing them to be resolved from mitochondria and peroxisomes, the densities of which remain unchanged.

latter enzyme, like urate oxidase, is a H_2O_2-forming enzyme, and catalase, of course, degrades H_2O_2. Because of its apparent involvement in hydrogen peroxide, the new organelle became known as a *peroxisome.* Other enzymes have since been recognized as peroxisomal in their location, and it is now clear that the enzyme complement of the organelle varies significantly from species to species and sometimes from organ to organ, or from one developmental stage to another within the same organ. However, the presence of catalase and one or more hydrogen peroxide-generating oxidases remains a distinguishing characteristic of peroxisomes.

Peroxisomes as Catalase-Containing Microbodies

Once peroxisomes had been identified and isolated biochemically, electron microscopy was employed to verify the existence of organelles with the expected properties, first in isolated peroxisomal fractions from density gradients and then in intact cells. Peroxisomes turned out to be the functional equivalents of organelles that had been seen earlier in electron micrographs of both animal and plant cells. Because their function was unknown at the time, these organelles were simply called **microbodies.** In both plant and animal cells, microbodies are usually about 0.2–2.0 μ in diameter, are surrounded by a single membrane, and generally have a finely granular *matrix* (interior of the organelle). Figure 9-28 shows the appearance of microbodies (peroxisomes) in a rat liver cell.

As can be seen in Figure 9-28, animal peroxisomes often contain a distinctive crystalline core, which usually consists of urate oxidase in crystalline form. Crystalline cores are often present in the peroxisomes of plant leaves also but usually consist of catalase instead (Figure 9-29). When such cores are present, it is easy to identify microbodies as peroxisomes, since urate oxidase and catalase are two of the enzymes by which peroxisomes are defined. In the absence of a crystalline core, however, it is not always easy to spot peroxisomes ultrastructurally.

A useful technique in such cases is a cytochemical test for catalase called the *diaminobenzidine (DAB) reaction.* This assay depends on the ability of catalase to oxidize DAB to a

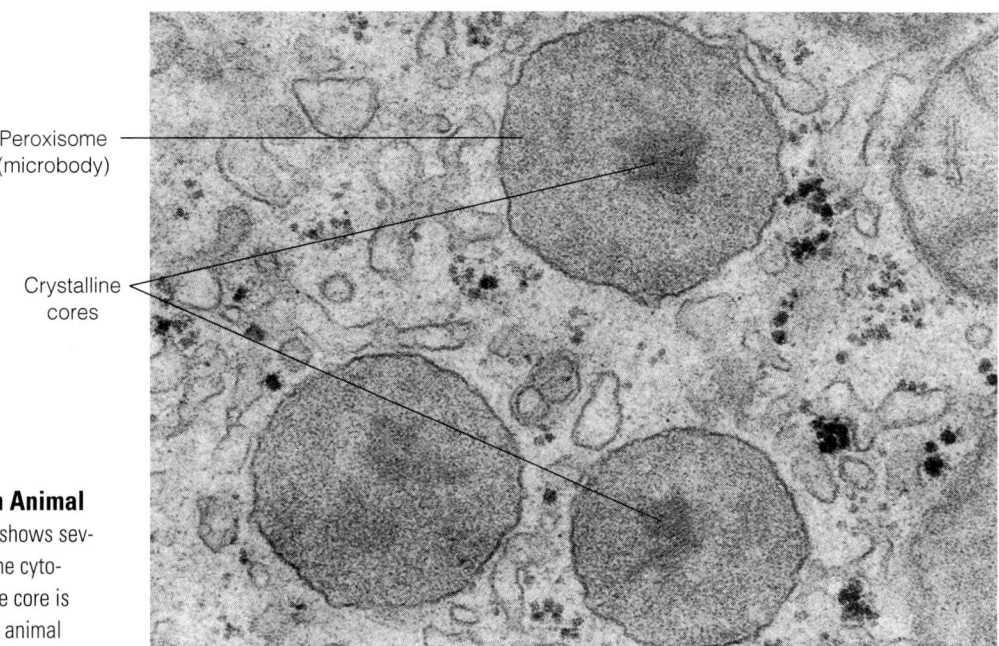

Figure 9-28 Peroxisomes in Animal Cells. This electron micrograph shows several peroxisomes (microbodies) in the cytoplasm of a rat liver cell. A crystalline core is readily visible in each microbody. In animal microbodies, the cores are almost always crystalline urate oxidase (TEM).

Peroxisome (microbody)

Crystalline cores

0.5 μm

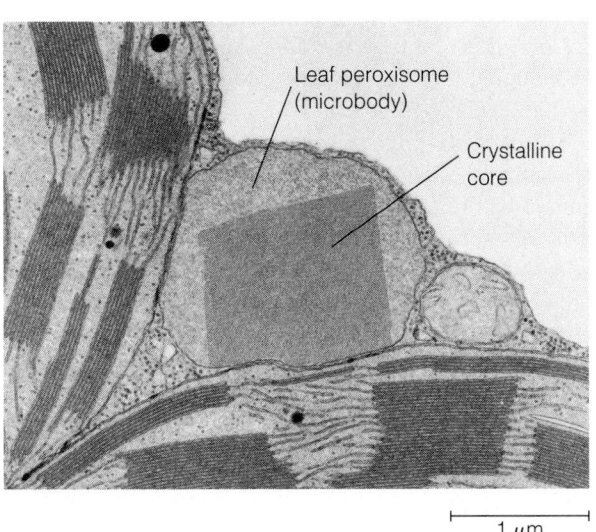

Leaf peroxisome (microbody)

Crystalline core

1 μm

Figure 9-29 Peroxisomes in Plant Cells. The tobacco leaf cell shown in this electron micrograph contains a prominent peroxisome (microbody) with a crystalline core. In plant microbodies, the cores are almost always crystalline catalase (TEM).

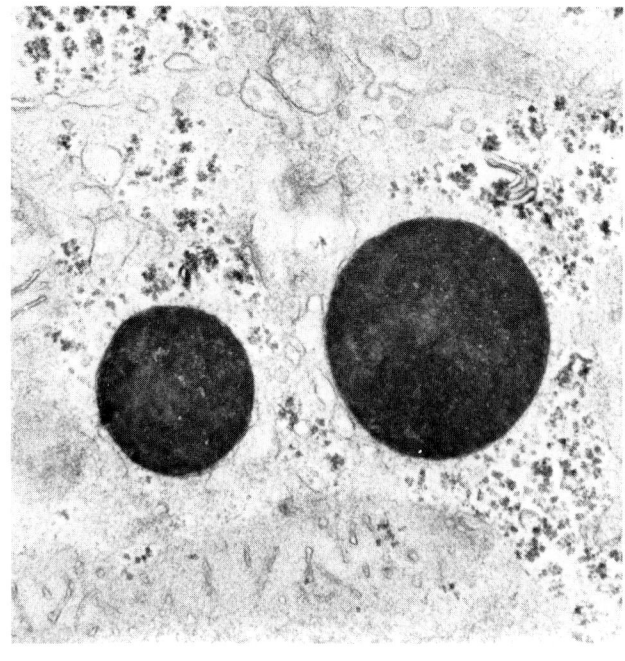

0.5 μm

Figure 9-30 Cytochemical Localization of Catalase in Animal Peroxisomes by the DAB Reaction. Diaminobenzidine (DAB) is oxidized by catalase to a polymeric form that causes electron-dense osmium atoms to be deposited in peroxisomes of tissue treated with osmium tetroxide (OsO_4). The principle of this assay is similar to that of the cytochemical test for acid phosphatase described in Figure 9-24. Shown here are rat liver peroxisomes stained with the DAB reaction (TEM); compare with the unstained peroxisomes from the same tissue shown in Figure 9-28.

polymeric form that causes deposition of electron-dense osmium atoms when the tissue is treated with osmium tetroxide (OsO_4). The resulting electron-dense deposits can be readily seen in cells from stained tissue (Figure 9-30). In animal peroxisomes, the entire internal volume often stains intensely with DAB, indicating that catalase exists as a soluble enzyme uniformly distributed throughout the matrix of the organelle; compare the DAB-treated organelles

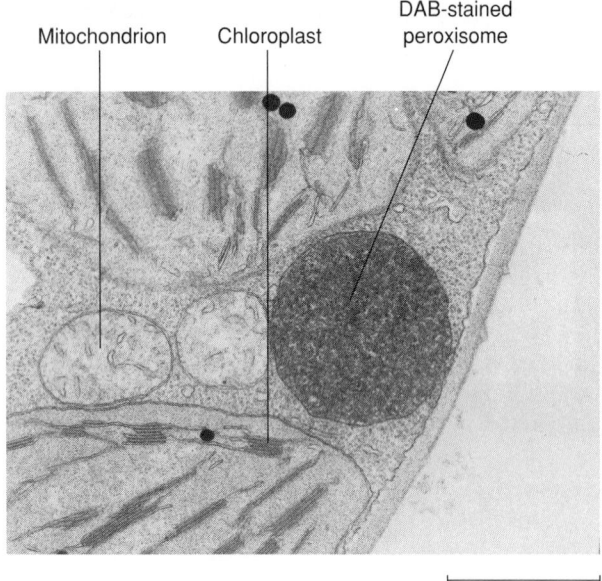

Mitochondrion   Chloroplast   DAB-stained peroxisome

1 μm

Figure 9-31   Cytochemical Localization of Catalase in Plant Peroxisomes.   Shown here is a tobacco leaf cell similar to that shown in Figure 9-29 but stained with DAB as described in the legend to Figure 9-30. Note that DAB stains the crystalline cores intensely, identifying them as crystalline catalase (TEM).

in Figure 9-30 with unstained peroxisomes from the same tissue in Figure 9-28). In plant leaf cells, DAB treatment stains the crystalline cores of the peroxisomes preferentially (Figure 9-31), thereby identifying these cores as crystalline catalase. Because catalase is the single enzyme present in all peroxisomes and does not routinely occur in any other organelle, the DAB reaction is a highly reliable and highly specific means of identifying organelles unequivocally as peroxisomes.

The Occurrence and Functions of Animal Peroxisomes

Peroxisomes occur widely in animals, plants, algae, and at least some fungi. In animals, peroxisomes are most prominent in liver and kidney tissue; in fact, they were initially thought to be unique to these organs. It is now recognized, however, that peroxisomes occur widely in most animal tissues, though not as prominently as in the liver and kidney. The roles of peroxisomes in animal cells are not yet as well understood as they are in plants. At least five possible functions have been suggested, each of which may be significant in at least some types of cells.

Hydrogen Peroxide Metabolism.   The most obvious role for peroxisomes in animal cells is the detoxification of hydrogen peroxide that is facilitated by having H_2O_2-generating oxidases packaged in the same organelle with catalase. The oxidases in peroxisomes vary considerably in

the specific reactions they catalyze, but they all share the property of transferring electrons from their respective substrates to molecules of oxygen (O_2) to form H_2O_2. Using RH_2 to represent an oxidizable substrate, the general reaction catalyzed by oxidases can be written as:

$$RH_2 + O_2 \longrightarrow R + H_2O_2 \qquad \text{(9-3)}$$

The hydrogen peroxide formed in this manner is broken down by catalase in one of two ways. Usually, catalase functions in what is called its *catalitic mode*, in which one molecule of H_2O_2 is oxidized to oxygen and a second is reduced to water:

$$2H_2O_2 \longrightarrow O_2 + 2H_2O \qquad \text{(9-4)}$$

Dividing reaction 9-4 by 2 and adding it to reaction 9-3 yields a summary reaction for the overall process:

$$RH_2 + \tfrac{1}{2}O_2 \longrightarrow R + H_2O \qquad \text{(9-5)}$$

Alternatively, catalase can function in its *peroxidatic mode*, in which hydrogen peroxide is reduced to water using electrons derived from an organic donor:

$$R'H_2 + H_2O_2 \longrightarrow R' + 2H_2O \qquad \text{(9-6)}$$

(The prime on the R group indicates that this substrate is likely to be different than the substrate in reaction 9-3.) The corresponding summary reaction in this case is:

$$RH_2 + R'H_2 + O_2 \longrightarrow R + R' + 2H_2O \qquad \text{(9-7)}$$

The result is the same in either case: Hydrogen peroxide is degraded without ever leaving the peroxisome. Given the toxicity of hydrogen peroxide (which is the main active ingredient in a variety of disinfectants), it makes good sense for the enzymes responsible for peroxide generation to be compartmentalized together with the catalase that catalyzes its degradation. Indeed, catalase is the most abundant protein in most peroxisomes, representing up to 15% of the total protein content of the organelle. In this way, virtually every molecule of hydrogen peroxide generated by the oxidases will encounter a molecule of catalase almost immediately and will be promptly degraded.

Detoxification of Harmful Compounds.   In its peroxidatic mode (reaction 9-6), catalase can use as its electron donor a variety of substances, including methanol, ethanol, formic acid, formaldehyde, nitrites, and phenols. Because all of these compounds are harmful to cells, it has been suggested that their oxidative detoxification by catalase may be a vital peroxisomal function. The prominent peroxisomes of liver and kidney cells are thought to be important in such **detoxification reactions.**

Oxidation of Fatty Acids.   Peroxisomes in many animal tissues have the enzymes necessary to oxidize fatty acids. This process, called *β oxidation*, also occurs in the mitochondrion and will be encountered again in Chapter 12.

About 25–50% of fatty acid oxidation in animal tissues occurs in peroxisomes, with the remainder localized in mitochondria. Peroxisomal β oxidation appears to be especially important in the catabolism (breakdown) of fatty acids with especially long carbon chains.

Metabolism of Nitrogen-Containing Compounds. In the peroxisomes of most animals except primates, the enzyme *urate oxidase* (also called *uricase*) is needed to oxidize urate, a purine formed during the catabolism of nucleic acids and some proteins. Urate oxidation is an important step in the elimination of nitrogenous wastes in these animals. Oxygen is required and hydrogen peroxide is formed:

$$\text{urate} + O_2 \longrightarrow \text{allantoin} + H_2O_2 \qquad \textbf{(9-8)}$$

The H_2O_2 is immediately degraded by catalase, and the allantoin is further metabolized and excreted by the organism, either as allantoic acid or, in the case of fish and amphibians, as urea.

Other peroxisomal enzymes are also involved in nitrogen metabolism. These include the *aminotransferases*, enzymes that catalyze the transfer of amino groups (—NH$_2$) from amino acids to α-keto acids:

$$
\underset{\text{Amino acid}}{\overset{^+NH_3}{\underset{|}{R—\overset{|}{C}—COO^-}}} + \underset{\alpha\text{-keto acid}}{\overset{O}{\underset{\|}{R'—\overset{\|}{C}—COO^-}}} \rightleftharpoons
$$

$$
\underset{\alpha\text{-keto acid}}{\overset{O}{\underset{\|}{R—\overset{\|}{C}—COO^-}}} + \underset{\text{Amino acid}}{\overset{^+NH_3}{\underset{|}{R'—\overset{|}{C}—COO^-}}} \qquad \textbf{(9-9)}
$$

These enzymes play important roles in the synthesis and degradation of amino acids by enabling amino groups to be moved from one molecule to another.

The Breakdown of Unusual Substances. Some of the substrates for peroxisomal oxidases are rare compounds for which the cell may have no other degradative pathways. D-amino acids may be in this category because they are not recognized by enzymes involved in the degradation of the L-amino acids found in proteins. In some fungi, the peroxisomes contain enzymes that are also involved in the breakdown of unusual substances. In this case, the enzymes are capable of metabolizing *alkanes*, short-chain hydrocarbon compounds found in oil and other petroleum products. Such fungi may turn out to be useful for cleaning up oil spills that would otherwise contaminate the environment.

The Occurrence and Functions of Plant Peroxisomes

In plants and algae, peroxisomes are involved in several specific aspects of energy metabolism and will therefore be discussed in more detail in Part Three. Here, we will simply introduce the several kinds of plant peroxisomes and describe their functions briefly.

Leaf Peroxisomes. Cells of leaves and other photosynthetic tissue are characterized by the presence of large, prominent **leaf peroxisomes,** which are often seen in close contact with chloroplasts and mitochondria (see Figure 9-29). This spatial proximity probably reflects the mutual involvement of these three organelles in *photorespiration,* a light-dependent process that occurs only in photosynthetic tissue. Photorespiration involves the uptake of O_2 and release of CO_2 and thereby decreases the net amount of carbon converted into organic form by photosynthetic tissue. Several of the key enzymes of the photorespiratory pathway are localized to leaf peroxisomes, including several specific aminotransferases. Because of the link between photorespiration and photosynthesis, we will discuss leaf peroxisomes in Chapter 13.

Glyoxysomes. Another functionally distinct type of plant peroxisome occurs transiently in seedlings of plant species that store carbon and energy reserves in the seed as fat (triglycerides, mainly; see Figure 3-24). In such species, stored triglycerides are mobilized and converted to sucrose during early postgerminative development by a sequence of events that includes β oxidation of fatty acids as well as a pathway known as the *glyoxylate cycle*. All of the enzymes needed for these processes are localized to specialized peroxisomes called **glyoxysomes.** Glyoxysomes are found only in the tissues in which the fat is stored (endosperm or cotyledons, depending on the species) and are present only for the relatively short period of time (a week or two, in most cases) required for the seedling to deplete its supply of stored fat. Glyoxysomes have been reported to appear again in senescing (aging) tissues of some plant species, presumably to degrade lipids derived from the membrane of the senescent cells. However, it is not yet clear how important their involvement is in senescence. Because of their role in the oxidative breakdown of fat, we will encounter glyoxysomes in Chapter 12.

Other Kinds of Plant Peroxisomes. In addition to their presence in tissues that carry out either photorespiration or fat oxidation, peroxisomes are found in other plant tissues as well. For example, another kind of specialized peroxisome is present in *nodules*, the structures on plant roots in which plant cells and certain bacteria cooperate in the fixation of atmospheric nitrogen (i.e., in the conversion of N_2 into organic form). The peroxisomes in these cells are involved in the processing of fixed nitrogen.

Peroxisome Biogenesis

As for other organelles in the cell, peroxisomes increase in number as cells grow and divide, a process called *biogenesis.*

In the case of lysosomes, biogenesis occurs by budding off the endoplasmic reticulum. On the other hand, new mitochondria and chloroplasts are formed by the division of preexisting organelles. Although peroxisomes were once thought to form by budding off the ER, most investigators now agree that their biogenesis occurs by the division of preexisting peroxisomes.

This mode of biogenesis raises the question of how the enzymes and other proteins present in the matrix or membrane of the organelle get to these locations as new peroxisomes are formed. The answer is that proteins destined for the peroxisome are synthesized not on the membrane-bound ribosomes of rough ER but on ribosomes in the cytosol and then enter preexisting peroxisomes as full-length polypeptides. The passage of polypeptides across the peroxisome membrane is almost certainly mediated by specific membrane proteins, but they have not yet been identified.

This mode of protein import into organelles is called *posttranslational protein import* (or *uptake*) because it occurs after *translation* (synthesis of proteins based on the sequence information in a messenger RNA molecule) is complete. Mitochondria, chloroplasts, and nuclei also import proteins posttranslationally, as we will learn in Chapter 18.

Figure 9-32 illustrates the uptake of membrane components and matrix enzymes from the cytosol into a peroxisome, and the formation of new peroxisomes by division of the preexisting organelle. The protein depicted in the figure is catalase, a tetrameric protein with a heme group bound to each subunit. The subunits are synthesized individually on cytosolic ribosomes, imported into the peroxisome and assembled, along with heme, into the active tetrameric enzyme.

For posttranslational uptake to work, each protein destined for a specific organelle must have some sort of signal that directs, or targets, the protein to the correct organelle. Presumably, such a signal functions by recognizing specific receptors or other features on the surface of the appropriate membrane. The signal in each case is a sequence of amino acids, which differs in sequence, length, and location for proteins targeted to different organelles. For chloroplast and mitochondrial proteins, for example, the targeting signal is located at the N-terminus of the molecule and ranges in length from 30 to almost 100 amino acids in chloroplasts and up to 70 amino acids in mitochondria.

In contrast, the signal that targets at least some peroxisomal proteins to their destination is found at or near the C-terminus of the molecule and consists of just three amino acids. The most common sequence is serine-lysine-leucine, or SKL, although a limited number of other amino acids are possible at each of the three locations. Thus, we expect each of the four subunits of catalase shown in Figure 9-32 to have the SKL sequence (or one of the acceptable variants) at or near its carboxyl terminus.

The work of S. Subramani and others has shown that a protein which would normally be targeted to the peroxi-

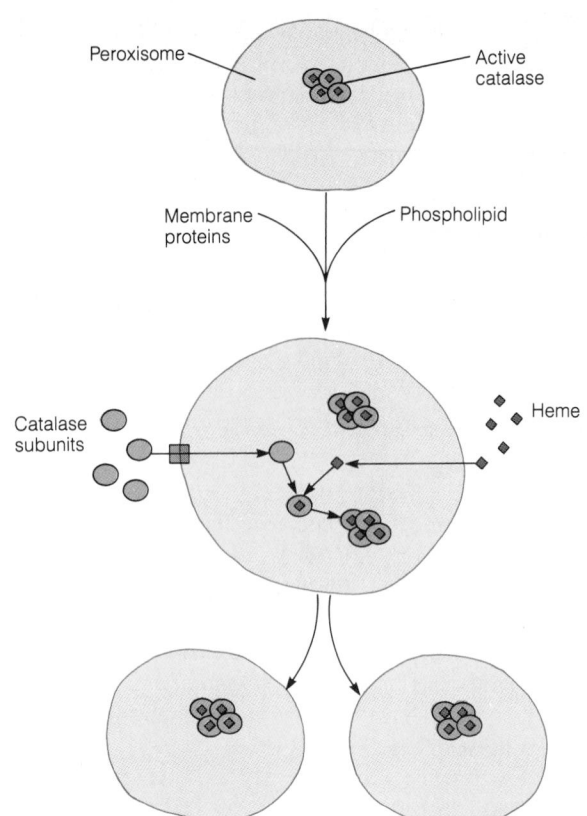

Figure 9-32 Biogenesis of Peroxisomes. Peroxisomes arise from preexisting peroxisomes rather than from the ER-Golgi system. Top: Membrane proteins, lipids, and matrix enzymes are added to preexisting peroxisomes from sources in the cytosol. Shown here are subunits of catalase, a tetrameric protein, being transported into a peroxisome in a process mediated by an as-yet-unidentified receptor protein or mechanism in the membrane of the peroxisome. Most peroxisomal polypeptides have at or near their carboxyl terminus a distinctive tripeptide sequence, usually SKL, that is both necessary and sufficient for targeting to the peroxisome. Bottom: New peroxisomes are formed by the division of preexisting organelles, though neither the signal to divide nor the mechanism whereby division occurs is understood as yet.

some will remain in the cytosol if its SKL sequence is removed experimentally. Conversely, the addition of the SKL sequence to a protein normally found in the cytosol will direct that protein to the peroxisome instead. Thus, the SKL sequence (or one of several acceptable variants) is both *necessary* and *sufficient* to direct proteins to peroxisomes. Moreover, a peroxisomal targeting sequence in one species can be recognized in other species as well, even when the species are as evolutionarily diverse from one another as plants, yeast, insects, and animals. As a particularly intriguing example, the gene for *luciferase*, a peroxisomal enzyme in fireflies, was transferred genetically into cells of yeast and plants, resulting in the synthesis of luciferase and its insertion into the peroxisomes of these cells (Figure 9-33)!

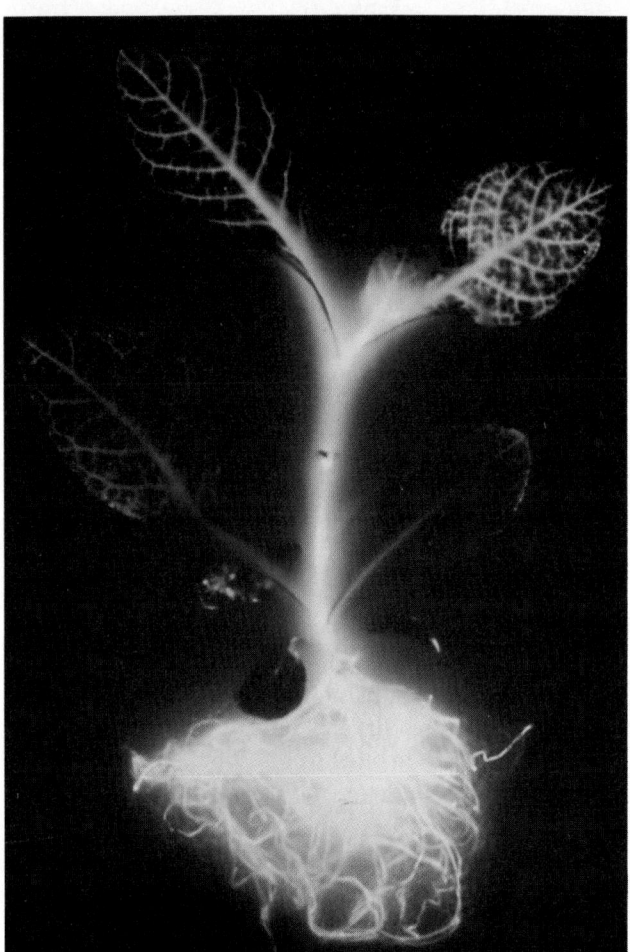

Figure 9-33 A Tobacco Plant Expressing the Gene for Firefly Luciferase. The gene for luciferase, the enzyme that enables fireflies to emit flashes of light, was used to transform tobacco cells, from which whole tobacco plants were then grown. As the photograph shows, the firefly enzyme is expressed in cells in most parts of the plant, causing the tobacco plant to glow in the dark. When cells of the genetically transformed tobacco plant were examined carefully, the luciferase was found to be present in the peroxisomes, the same organelle in which it is found in fireflies.

PERSPECTIVE

To appreciate the eukaryotic cell fully means to understand the prominent role that intracellular membranes play. Especially prevalent within most eukaryotic cells is an elaborate system of membranes and organelles derived either directly or indirectly from the endoplasmic reticulum (ER). The ER itself is an interconnecting network of vesicles, tubules, and flattened sacs that separates the cisternal space (interior of the ER and related compartments) from the surrounding cytosol. The rough ER is studded with ribosomes and is the site of the synthesis of proteins destined for secretion from the cell, for inclusion within lysosomes, or for insertion into specific membranes—including those of the ER, the Golgi complex, the nuclear envelope, and the plasma membrane. Both rough ER and smooth ER synthesize various lipids, while smooth ER is the site of other processes, including steroid synthesis, hydroxylation reactions, drug detoxification, and carbohydrate metabolism.

The Golgi complex plays key roles in protein glycosylation and sorting. The Golgi stacks are in intimate association with the ER by means of transition vesicles that bud off the ER and fuse with the cisternae on the *cis*, or forming, face of the Golgi stack. Proteins move through the Golgi stacks toward the *trans*, or maturing, face, transported from one cisterna to the next by shuttle vesicles. Vesicles that bud off the *trans* face carry the processed proteins to their final destinations within the cell. Among the proteins processed by the Golgi complex are those intended for secretion. Secretory proteins are released from the cell by exocytosis, which may be either constitutive (continuous) or regulated (in response to a specific signal). In a sense, endocytosis is the opposite process, involving the uptake of extracellular substances by an infolding of the plasma membrane. Mechanisms of endocytosis include bulk-phase endocytosis, phagocytosis, and receptor-mediated endocytosis. The latter mechanism is highly specific because it involves the recognition and binding of specific ligands to receptors on the cell surface. Once the receptor-ligand complex has formed, endocytosis occurs at clathrin-coated pits and results in the formation of clathrin-coated vesicles. Once inside the cell, the vesicle rapidly loses its clathrin coat and recycles membrane and receptors back to the cell surface. The basic unit of structure of clathrin-coated membranes is the triskelion.

Lysosomes and peroxisomes share many properties, including size and the presence of a single bounding membrane. Moreover, both organelles were discovered by de Duve, based on the seemingly anomalous behavior of specific enzymes upon centrifugation of tissue homogenates. The two organelles differ, however, in their origin within the cell and in the enzymes they contain. Lysosomes form by budding off the Golgi complex, whereas peroxisomes increase in number by the division of preexisting peroxisomes. As the name suggests, lysosomes contain hydrolytic enzymes and are involved in a variety of cellular digestive processes. In some cases, primary lysosomes containing acid hydrolases fuse with vacuoles containing digestible material from either outside the cell (phagocytosis) or inside the cell (autophagy). In other cases, primary lysosomes fuse with the plasma membrane and release their contents to the outside of the cell by exocytosis. Yet another mode of lysosomal function involves autolysis, or self-digestion, of the cell by the intracellular rupture of primary lysosomes.

Peroxisomes are named for the hydrogen peroxide that is generated and degraded within the organelle by oxidases and catalase, respectively. In animal cells, peroxisomes perform several functions, including the detoxification of harmful substances, oxidation of fatty acids, and metabolism of nitrogen-containing compounds. In plants, peroxisomes have distinctive roles in the conversion of stored fat into carbohydrate (glyoxysomes) and in photorespiration (leaf peroxisomes).

In this chapter, we have covered several major membranous structures of eukaryotic cells, including the ER, the Golgi complex, lysosomes, peroxisomes, exocytotic and endocytotic vesicles, and the plasma membrane itself. In each case, we have recognized the unique collection of proteins that characterizes each of these structures and the specific functions each serves. In coming chapters, we will discuss other important intracellular structures, including the mitochondrion (Chapter 12), the chloroplast (Chapter 13), and the nucleus (Chapter 14). First, however, we will consider the junctions that link cells together into tissues and the extracellular structures that are found outside the plasma membrane. We are, in other words, about to look "beyond the cell."

KEY TERMS FOR SELF-TESTING

Microscopy and Centrifugation

electron microscope (p. 229)
ultracentrifuge (p. 229)
centrifugation (p. 229)
subcellular fractionation (p. 230)
homogenate (p. 230)
differential centrifugation (p. 231)
density gradient (rate-zonal) centrifugation (p. 231)
equilibrium (buoyant) density centrifugation (p. 232)

The Endoplasmic Reticulum

endoplasmic reticulum (ER) (p. 233)
cisterna (p. 235)
cisternal space (p. 235)
cytosol (p. 235)
endomembrane system (p. 235)
rough endoplasmic reticulum (rough ER) (p. 235)
smooth endoplasmic reticulum (smooth ER) (p. 235)
microsome (p. 237)
hydroxylation reaction (p. 237)
drug detoxification (p. 238)

The Golgi Complex

Golgi complex (p. 239)

Golgi stack (p. 239)
dictyosome (p. 239)
vesicle (p. 240)
cis (forming) face (p. 240)
transition vesicle (p. 240)
trans (maturing) face (p. 240)
shuttle vesicle (p. 240)

Roles of the ER and Golgi Complex in Protein Glycosylation

core glycosylation (p. 241)
core oligosaccharide (p. 241)
terminal glycosylation (p. 241)

Secretory Pathways

secretory pathway (p. 244)
secretory vesicle (p. 244)
constitutive secretion (p. 244)
regulated secretion (p. 244)
secretory (zymogen) granule (p. 245)

Outward and Inward Transport: Exocytosis and Endocytosis

exocytosis (p. 245)
polarized secretion (p. 246)
endocytosis (p. 246)
endocytotic vesicles (p. 246)
bulk-phase endocytosis (p. 247)

phagocytosis (p. 247)
receptor-mediated endocytosis (RME) (p. 247)
receptor (p. 247)
ligand (p. 247)

The Role of Coated Pits and Vesicles in Cellular Transport Processes

coated pit (p. 249)
coated vesicle (p. 249)
clathrin (p. 252)
triskelion (p. 252)

Biosynthesis of Membrane Proteins and Lipids

transport vesicle (p. 254)
phospholipid transfer protein (p. 254)

Lysosomes and Cellular Digestion

lysosome (p. 254)
primary lysosome (p. 256)
phagocyte (p. 256)
phagocytic vacuole (p. 256)
secondary lysosome (p. 256)
residual body (p. 257)
autophagy (p. 257)

PROBLEM SET

9-1. Subcellular Fractionation. Shown in Figure 9-34 is a two-dimensional diagram depicting the densities (in g/cm³) and sedimentation coefficients (S values; note the logarithmic scale) of various organelles, molecules, and viruses. In each case, the oval-shaped area delimits the usual ranges of density and S values for a particular particle. Thus, mitochondria have densities in the range of 1.15–1.20 g/cm³ and S values in the range of 0.2–1.0 × 10⁵, whereas DNA molecules have densities of about 1.70–1.72 g/cm³ and S values in the range of 10–80. Use Figure 9-34 and the centrifugation techniques described in this chapter to answer the following questions.

(a) What molecules or organelles have the lowest densities? The highest densities? The lowest S values? The highest S values?

(b) How does the density of a ribosome compare with the densities of RNA and protein molecules? Given what you know about the structure of a ribosome, can you explain this observation?

(c) Given that most viruses consist of DNA or RNA complexed with protein molecules, why is the sedimentation coefficient of a typical virus greater than that of DNA, RNA, or protein molecules individually?

(d) Can differential centrifugation be used to separate nuclei from mitochondria effectively? Why or why not? Can this technique be used to separate peroxisomes from mitochondria? Why or why not?

(e) Can equilibrium density centrifugation be used to separate microsomes from mitochondria effectively? Why or why not? Can this technique be used to separate RNA from DNA? Why or why not?

(f) Devise a separation scheme whereby peroxisomes, lysosomes, and mitochondria are first separated from the other contents of a homogenate and then resolved into three separate fractions.

9-2. Vesicle Match-Up. For each of the vesicles listed below, indicate which of the listed properties apply. Each vesicle has multiple properties, and a given property may be used more than once (or not at all).

Vesicles

(a) Coated vesicle   (d) Transition vesicle

(b) Secretory vesicle   (e) Transport vesicle

(c) Shuttle vesicle   (f) Zymogen granule

Properties

1. Transports proteins from the ER to the Golgi.

2. Exterior surface may be covered with a layer of the protein clathrin or coatomer.

3. Transports proteins from the plasma membrane to the ER or the Golgi complex.

4. Contains primarily acid hydrolases.

5. Transports protein from one Golgi cisterna to the next.

6. Transports membrane lipids from the ER to any of several target membranes.

7. Also called secretory granules in some cases.

8. Buds or pinches off the rough ER.

9. Capable of fusing with the plasma membrane.

9-3. Compartmentalization of Function. Each of the following processes is associated with a specific eukaryotic organelle. In each case, identify the organelle, and indicate the more general cellular function of which the process is a part. In some cases, more than one organelle may be appropriate.

(a) Hydroxylation of phenobarbital

Figure 9-34 Densities and Sedimentation Coefficients of Organelles, Macromolecules, and Viruses.

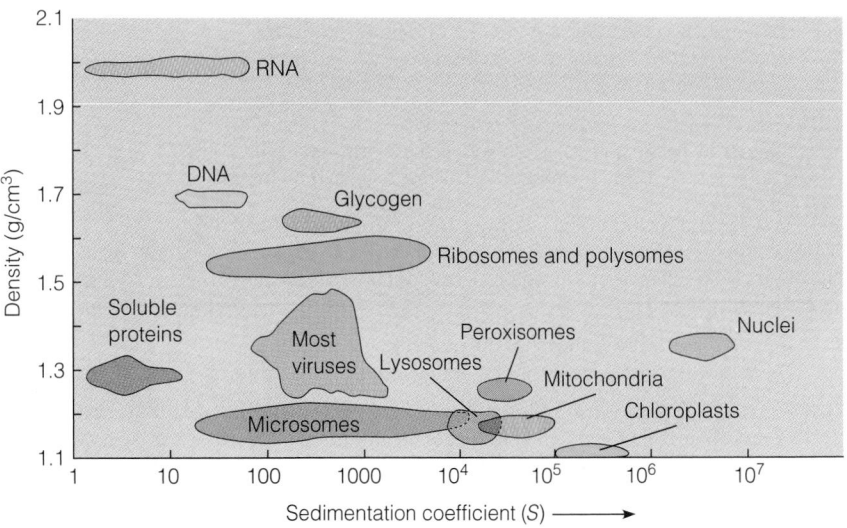

(b) Synthesis of insulin

(c) Synthesis of testosterone, a male sex hormone

(d) Glycosylation of proteins

(e) Detoxification of toxic compounds

(f) Digestion of old or unwanted organelles

9-4. Endoplasmic Reticulum. Label each of the statements below with an S if it is true of smooth ER only, with an R if it is true of rough ER only, with an RS if it is true of both, and with an N if it is true of neither.

(a) Consists of about 70% protein and 30% lipid by weight.

(b) Is studded with ribosomes on its outer surface.

(c) Is involved in the breakdown of glycogen.

(d) Is involved in the detoxification of drugs.

(e) Is the site of the synthesis of secretory proteins.

(f) Tends to be tubular in shape.

(g) Can be seen only with the electron microscope.

(h) Usually consists of flattened sacs.

9-5. Synthesis of Integral Membrane Proteins. In addition to their role in cellular secretion, rough ER and the Golgi complex are also responsible for the synthesis of integral membrane proteins. Specifically, glycoproteins that are found so commonly in the outer phospholipid layer of many plasma membranes are synthesized by this route.

(a) In a series of diagrams, depict the synthesis and glycosylation of glycoproteins of the plasma membrane.

(b) Explain why the sugar groups of membrane glycoproteins are always located on the outer surface of the plasma membrane.

(c) What assumptions about membrane asymmetry did you make in order to draw the diagrams of part a and answer the question of part b?

9-6. Interpreting the Data. Each of the following statements summarizes the results of a specific experiment relating to exocytosis or endocytosis. In each case, explain the relevance of the experiment and its result to our understanding of these processes.

(a) Addition of the drug colchicine to cultured fibroblast cells causes vesicle movement to cease.

(b) Certain cells of the pituitary gland secrete laminin continuously but secrete adrenocorticotropic hormone only in response to specific signals.

(c) Certain cells in the adrenal gland of a rat can be induced to secrete granules containing adrenalin when their internal calcium concentration is increased experimentally.

(d) When the gene for insulin, a protein secreted in a regulated manner by pancreatic cells, is introduced into pituitary cells, insulin is secreted by the pituitary cells in a regulated manner.

9-7. Lysosomal Enzymes. The following are experimental observations about lysosomes and their enzymes. In each case, explain the observation in terms of known properties of the organelle and its enzymes.

(a) In most classical procedures for differential centrifugation, lysosomes are found mainly in the "mitochondrial" fraction.

(b) Substantially higher acid hydrolase activities are detected in homogenates of animal tissues if the homogenization is performed in distilled water instead if isotonic (0.25 M) sucrose solution.

(c) Lysosomal nucleases, phospholipase, and proteases are synthesized by ribosomes bound to the membranes of the rough ER but do not digest either the ribosomes or the ER membrane.

(d) Lysosomes can be stained cytochemically by incubating a tissue section with a phosphorylated substrate in the presence of lead ions.

(e) Most of the lysosomal enzymes are latent—that is, they are inactive against their substrates as long as the lysosomal membrane remains intact.

9-8. Cellular Digestion. For each of the following statements, indicate with the appropriate letter code(s) the specific digestion process(es) of which the statement is true: autolysis (AL); autophagy (APh); phagocytosis (PhC); or extracellular digestion (ED). A given statement may be true of one, some, all, or none of these processes.

(a) Occurs within vacuoles called secondary lysosomes.

(b) Involves acid hydrolases.

(c) Important in certain developmental processes.

(d) Essential for sperm penetration during fertilization.

(e) Serves as a source of nutrients within the cell.

(f) May involve exocytosis.

(g) May involve endocytosis.

(h) Material digested is of intracellular origin.

(i) Involves fusion of primary lysosomes with the plasma membrane.

(j) Involves fusion of primary lysosomes with a vacuole.

(k) Involves no fusion of primary lysosomes.

(l) Also found in prokaryotic cells.

9-9. Silicosis and Asbestosis. *Silicosis* is a debilitating miner's disease that results from the uptake of silica particles (such as sand or glass) by macrophages in the lungs. *Asbestosis* is a similarly serious disease caused by the inhalation of asbestos fibers. In both cases, the particles or fibers are found in secondary lysosomes in the cells, and fibroblasts (collagen-secreting cells) in the lungs are stimulated to deposit nodules of collagen fibers, leading to reduced lung elasticity, impaired breathing, and eventually to death.

(a) How do you think the fibers or particles get into the secondary lysosomes?

b) What effect do you think fiber or particle accumulation has on the secondary lysosomes?

(c) How might you explain the death of silica-containing or asbestos-containing cells?

(d) What do you think happens to the silica particles or asbestos fibers when such a cell dies? Can you see how cell death could continue almost indefinitely, even after further exposure to silica dust or asbestos fibers has ceased?

(e) Cultured fibroblast cells will secrete collagen and produce connective tissue fibers after the addition of material from a culture of lung macrophages that have been exposed to silica particles. What does this tell you about the deposition of collagen nodules in the lungs of silicosis patients?

9-10. Lysosomal Storage Diseases. Despite a bewildering variety of symptoms, lysosomal storage diseases have a number of properties in common. For each of the following statements, respond with an A if you would expect the property to be common to all (or most) lysosomal storage diseases, with an S if you would expect it to be true of a specific lysosomal storage disease, or with an N if you do not consider it to be true of lysosomal storage diseases at all.

(a) Results in accumulation of excessive quantities of glycogen.

(b) Is a rare heritable disorder.

(c) Symptoms include mental retardation and enlargement of the liver and spleen.

(d) Results in proliferation of catalase-containing organelles.

(e) Results from the genetic deficiency of an acid hydrolase.

(f) Leads to death in early infancy, usually before the age of six months.

(g) Results from inability to regulate synthesis of acid glycosaminoglycans.

(h) Results in accumulation of secondary lysosomes in the cell.

9-11. Peroxisomal Properties. For each of the following statements, indicate whether the statement is true of all (A), some (S), or none (N) of the kinds of peroxisomes described in this chapter, and explain your answer.

(a) Surrounded by a single membrane.

(b) Contain acid hydrolases.

(c) Capable of catabolizing fatty acids.

(d) Contain crystals of either urate oxidase or catalase.

(e) Stain positively in the DAB reaction.

(f) Capable of catabolizing pyruvate.

(g) Import proteins posttranslationally.

(h) Contain the genes for luciferase.

9-12. Peroxisomal Metabolism. Answer each of the following questions based on the information in this chapter.

(a) Complete the following reaction for the peroxisomal oxidation of D-alanine as catalyzed by D-*amino acid oxidase*, a peroxisomal enzyme:

$$\overset{\overset{+\text{NH}_3}{|}}{\text{CH}_3-\text{CH}-\text{COO}^-} + \underline{\quad} + \underline{\quad} \longrightarrow$$

D-alanine

$$\overset{\overset{O}{||}}{\text{CH}_3-\text{C}-\text{COO}^-} + \text{NH}_3 + \underline{\quad}$$

Pyruvate acid   Ammonia

(b) Complete the following reaction for the detoxification of formate acid by catalase:

$$\text{HCOO}^- + \underline{\quad} \longrightarrow \text{CO}_2 + \underline{\quad}$$

Formate acid

Does this reaction represent the catalatic or the peroxidatic function of catalase?

(c) Assume that the oxidation of D-alanine and the detoxification of formate are occurring concomitantly in a peroxisome. Write a summary equation for the two processes.

(d) Comparing the reactants and the products of the reaction you wrote for part c and knowing that pyruvate acid can be used by most cells as a source of energy, what can you say about the importance of peroxisomes to the well-being of the cell in which these reactions are occurring?

9-13. Enzymes and Organelles. Most organelles are identified and described in terms of their enzyme content. Both the lysosome and the peroxisome were discovered because specific enzymes displayed properties that could not be easily reconciled with their presence in already-known structures.

(a) What criteria would you invoke to identify a new organelle unequivocally, based on enzymes contained within that organelle?

(b) Would your criteria have been adequate to identify lysosomes in a mitochondrial preparation? Explain your answer.

(c) Would your criteria have been adequate to identify peroxisomes in a lysosomal preparation? Explain your answer.

S U G G E S T E D   R E A D I N G

General References

Alberts, B., D. Bray, J. Lewis, M. Raff, K. Roberts, and J. D. Watson. *Molecular Biology of the Cell*, 3d ed. New York: Garland, 1994.

Gruenberg, J., and M. J. Clague. Regulation of intracellular membrane transport. *Curr. Opin. Cell Biol.* 4 (1992): 593.

Lodish, H., D. Baltimore, A. Berk, S. L. Zipursky, P. Matsudaira, and J. Darnell. *Molecular Cell Biology*, 3d ed. New York: Scientific American Books, 1995.

Pryer, N. K., L. J., Wuesthube, and R. Schekman. Vesicle-mediated protein sorting. *Annu. Rev. Biochem.* 61 (1992): 471.

Watson, J. D., N. Hopkins, J. Roberts, J. Steitz, and A. Weiner. *Molecular Biology of the Gene*, 4th ed. Menlo Park, CA: Benjamin/Cummings, 1987.

Centrifugation and Subcellular Fractionation

Claude, A. The coming of age of the cell. *Science* 189 (1975): 433.

Cooper, T. G. *The Tools of Biochemistry*. New York: Wiley, 1977.

de Duve, C. Exploring cells with a centrifuge. *Science* 189 (1975): 186.

de Duve, C., and H. Beaufay. A short history of tissue fractionation. *J. Cell. Biol.* 91 (1981): 293s.

Scheeler, P. *Centrifugation in Biology and Medical Science*. New York: Wiley, 1981.

The Endoplasmic Reticulum and the Golgi Complex

Bretcher, M. S., and S. Munro. Cholesterol and the Golgi apparatus. *Science* 261 (1993): 1280.

Farquhar, M. G. Progress in unraveling pathways of Golgi traffic. *Annu. Rev. Cell Biol.* 1 (1985): 447.

Hicke, L., and R. Schekman. Molecular machinery required for protein transport from the ER to Golgi complex. *Bioessays* 12 (1990): 253.

Kelley, R. B. Microtubules, membrane traffic, and cell organization. *Cell* 61 (1990): 5.

Mellman, I., and K. Simons. The Golgi complex: *In vitro veritas?* *Cell* 68 (1992): 829.

Pelham, H. R. B. Control of protein export from the ER. *Annu. Rev. Cell Biol.* 5 (1989): 1.

Rose, J. K., and R. W. Doms. Regulation of protein export from the ER. *Annu. Rev. Cell Biol.* 4 (1988): 257.

Rothman, J. E. The compartmental organization of the Golgi apparatus. *Sci. Amer.* 253 (October 1985): 74.

Exocytosis and Endocytosis

Almers, W. Exocytosis. *Annu. Rev. Physiol.* 52 (1990): 607.

Bruzzone, R. The molecular basis of enzyme secretion. *Gastro-enterology* 99 (1990): 1157.

Burgess, L., and R. B. Kelly. Constitutive and regulated secretion of proteins. *Annu. Rev. Cell Biol.* 3 (1987): 243.

Burgoyne, R. D., and A. Morgan. Regulated exocytosis. *Biochem. J.* 293 (1993): 305.

Dautry-Varsat, A., and H. F. Lodish. How receptors bring proteins and particles into cells. *Sci. Amer.* 250 (May 1984): 52.

Goldstein, J. L., M. S. Brown, R. G. W. Anderson, D. W. Russell, and W. J. Schneider. Receptor-mediated endocytosis: Concepts emerging from the LDL receptor system. *Annu. Rev. Cell Biol.* 1 (1985): 1.

Gordon, J. I. Targeting of proteins into the eukaryotic secretory pathway. *Bioessays* 12 (1990): 479.

Hong, W., and B. L. Tang. Protein trafficking along the exocytotic pathway. *Bioessays* 15 (1993): 231.

Pastan, I., and M. C. Willingham. *Endocytosis.* New York: Plenum, 1985.

Plattner, H. Regulation of membrane fusion during exocytosis. *Internat. Rev. Cytol.* 119 (1989): 197.

Tartakoff, M. *The Secretory and Endocytotic Paths: Mechanisms and Specificity of Vesicular Traffic in the Cell Cytoplasm.* New York: Wiley, 1987.

van Deurs, B., O. W. Petersen, S. Olsnes, and K. Sandvig. The ways of endocytosis. *Internat. Rev. Cytol.* 117 (1989): 131.

Watts, C., and M. Marsh. Endocytosis: What goes in and how? *J. Cell Sci.* 103 (1992): 1.

Clathrin and Coated Vesicles

Brodsky, F. M. Living with clathrin: Its role in intracellular membrane traffic. *Science* 242 (1988): 1396.

Brodsky, F. M., B. L. Hill, S. A. Acton, I. Nathke, D. H. Wong, W. Ponnambalam, and P. Parham. Clathrin light chains: Arrays of protein motifs that regulate coated-vesicle dynamics. *Trends Biochem. Sci.* 16 (1991): 208.

Crowther, R. A., and B. M. F. Pearse. Assembly and packing of clathrin into coats. *J. Cell. Biol.* 91 (1981): 790.

Keen, J. H. 1990. Clathrin assembly proteins and the organization of the coated membrane. *Adv. Cell Biol.* 3 (1990): 153.

Morris, S., S. Ahle, and E. Ungewickell. Clathrin-coated vesicles. *Curr. Opin. Cell Biol.* 1 (1989): 684.

Pearse, B. M., and R. A. Crowther. Structure and assembly of coated vesicles. *Annu. Rev. Biophys. Chem.* 16 (1987): 49.

Pearse, B. M., and M. S. Robinson. Clathrin, adaptors, and sorting. *Annu. Rev. Cell Biol.* 6 (1990): 151.

Robinson, D. G., and H. Depta. Coated vesicles. *Annu. Rev. Plant Physiol. Plant Mol. Biol.* 39 (1988): 53.

Robinson, M. S. Coated vesicles and protein sorting. *J. Cell Sci.* 87 (1987): 203.

Trowbridge, I. S. Endocytosis and signals for internalization. *Curr. Opin. Cell Biol.* 3 (1991): 634.

Ungewickell, E., and D. Branton. Triskelions: The building blocks of clathrin coats. *Trends Biochem. Sci.* 7 (1982): 358.

Lysosomes

Bainton, D. The discovery of lysosomes. *J. Cell. Biol.* 91 (1981): 66s.

Dahms, N. M., P. Lobel, and S. Kornfeld. Mannose-6-P receptors and lysosomal enzyme targeting. *J. Biol. Chem.* 264 (1989): 12115.

de Duve, C. The lysosome. *Sci. Amer.* 208 (May 1963): 64.

Fukuda, M. Lysosomal membrane glycoproteins: Structure, biosynthesis and intracellular trafficking. *J. Biol. Chem.* 266 (1991): 21327.

Holtzman, E. *Lysosomes.* New York: Academic Press, 1989.

Kornfeld, S., and I. Mellman. The biogenesis of lysosomes. *Annu. Rev. Cell Biol.* 5 (1989): 483.

Neufeld, E. F. Lysosomal storage diseases. *Annu. Rev. Biochem.* 60 (1991): 257.

Pfeffer, S. R. Mannose-6-P receptors and their role in targeting proteins to lysosomes. *J. Membrane Biol.* 103 (1988): 7.

von Figura, K. Molecular recognition and targeting of lysosomal proteins. *Curr. Opin. Cell Biol.* 3 (1991): 642.

Watts, R. W. E., and D. A. Gibbs. *Lysosomal Storage Diseases.* Philadelphia: Taylor and Francis, 1986.

Peroxisomes

Borst, P. Peroxisome biogenesis revisited. *Biochim. Biophys. Acta* 1008 (1989): 1.

de Duve, C. Microbodies in the living cell. *Sci. Amer.* 248 (May 1983): 74.

de Duve, C. The peroxisome: A new cytoplasmic organelle. *Proc. Roy. Soc. London Ser. B* 173 (1969): 71.

Fahimi, H. D., and H. Sies, eds. *Peroxisomes in Biology and Medicine.* Heidelberg: Springer-Verlag, 1987.

Lazarow, P. B. Genetic approaches to studying peroxisome biogenesis. *Trends Biochem. Sci.* 3 (1993): 89.

Lazarow, P. B. Peroxisome biogenesis. *Curr. Opin. Cell Biol.* 1 (1989): 630.

Lazarow, P. B., and Y. Fujiki. Biogenesis of peroxisomes. *Annu. Rev. Cell Biol.* 1 (1985): 489.

Subramani, S. Targeting of proteins into the peroxisomal matrix. *J. Membrane Biol.* 125 (1992): 99.

Tolbert, N. E., and E. Essner. Microbodies: Peroxisomes and glyoxysomes. *J. Cell Biol.* 91 (1981): 271s.

10

BEYOND THE CELL: JUNCTIONS, ADHESION, AND EXTRACELLULAR STRUCTURES

I n the preceding chapters, we have been discussing cells as if they exist in isolation and as if they "end" at the plasma membrane. These assumptions are not true of most cells, however. Many kinds of cells—including most of those in your own body—spend all of their lives linked to neighboring cells, and almost all cells have some sort of structure exterior to the plasma membrane. We will explore both of these aspects in this chapter. First, we will discuss the various kinds of *junctions* that bind cells together into multicellular organisms and the specialized functions they perform. Then we will consider the structures that are typically found exterior to the plasma membrane, including the *extracellular matrix* of animal cells and the *cell walls* that enclose plant and bacterial cells.

Cell Junctions

Unicellular organisms, by definition, have no permanent associations between cells; each cell is an entity unto itself. Multicellular organisms, on the other hand, have specific means of joining cells in permanent associations to form tissues and organs. Usually, these means involve specialized modifications of the plasma membrane at the point where two cells come together. Such specialized structures are called **cell junctions.**

Several general types of junctions are recognized, differing mainly in the role of the junction and in how tightly the cells are appressed at the point of contact. In plants, cells are often connected by special structures called plasmodesmata, as we will see later. In animals, the three most common kinds of junctions, summarized in Table 10-1, are adhesive junctions, tight junctions, and gap junctions.

Adhesive junctions hold cells together in fixed positions within tissues and are especially prominent in epithelial tissue. *Tight junctions* function mainly as seals between two cell-lined compartments that prevent the flow of molecules and ions through the extracellular space. Tight junctions

keep contents such as gastric juices, urine, or other body fluids in their proper compartments. *Gap junctions* form open channels between cells, thereby allowing ions and small molecules to move directly from one cell to another. This free flow of ions and molecules means that adjacent cells are in direct chemical and electrical communication with each other. Smooth muscle and heart muscle are examples of tissues characterized by gap junctions.

Adhesive Junctions

Two kinds of **adhesive junctions** occur widely in animal tissues. The two types differ primarily in the kinds of cytoskeletal filaments with which they are associated. *Desmosomes* are attachment points for intermediate filaments, whereas *adherens junctions* are connection sites for actin microfilaments.

Desmosomes. **Desmosomes** are buttonlike points of tight adhesion between adjacent cells in a body tissue. This adhesion gives the tissue structural integrity and enables the cells to function together as a single unit. Desmosomes are found in many tissues but are especially abundant in skin, heart muscle, and the tissue that forms the neck of the uterus. A common feature of these tissues is that they must withstand considerable mechanical stress. Desmosomes form early in embryonic development and play an important role in maintaining cell position during development.

The structure of a typical desmosome is shown in Figure 10-1 (page 273). The plasma membranes of the two adjacent cells are parallel to each other and are separated by about 25–35 nm. The extracellular space between the two membranes is called the **desmosome core.** On either side of the desmosome, a thick layer of dense material called a **plaque** is found just beneath the plasma membrane. The desmosome core is filled with filaments and granules of glycoproteins called *desmocollins,* which are most likely the

Table 10-1 Junctions Between Animal Cells

Type of Junction	Function	Features	Intermembrane Space	Associated Structures
Adhesive junction				
Desmosome	Cell adhesion	Localized points of attachment	25–35 nm	Intermediate filaments (called tonofilaments)
Adherens junction	Cell adhesion	Localized points and continuous zones of attachment	25–35 nm	Actin microfilaments
Tight junction	Sealing spaces between membranes	Membranes joined along ridges	None	Tight junction elements
Gap junction	Exchange of ions and molecules between cells	Connexons (trans-membrane proteins with 1.5 nm pores)	2–3 nm	Connexons in one membrane align with those in another to form channels between cells

agents that bind the two plasma membranes together. Another family of glycoproteins, the *desmogleins*, appears to function as transmembrane links between the plaque on the inside of the plasma membrane and the core on the outside. These proteins interact through their extracellular domains to hold the adjacent membranes together. The binding activity of the desmogleins requires Ca^{2+} ions, which may account for the tendency of desmosomes (and hence, cells) to separate if excised tissue is bathed in a solution with a very low calcium concentration.

The plaques on either side of the desmosome contain at least two major proteins, *desmoplakins I and II*. Filamentous structures called **tonofilaments** extend inward from the plaque, anchoring the desmosome in the cytoplasm. These are *intermediate filaments* with a diameter of about 8–12 nm (see Figure 4-22c). The main component of tonofilaments is a protein called *keratin, desmin,* or *vimentin*, depending on the cell type. Because each desmosome serves as an anchoring site for filaments in the cells on both sides of the junction, the intermediate filaments of adjacent cells are linked indirectly through these junctions. In this way, the intermediate filaments within individual cells form what is, in effect, a continuous network throughout the tissue.

Adherens Junctions. Adherens junctions are found in heart muscle and in the thin layers of tissue that line body cavities and cover body organs. Some adherens junctions are small, buttonlike points of attachment, whereas others form a continuous **adhesion belt** around each of the cells in a sheet of tissue. Adherens junctions are quite similar to desmosomes in both structure and function but have more loosely structured plaques and **actin microfilaments** rather than intermediate filaments. The presence of actin suggests that these junctions may play a role in actin-based motility, a

topic we will consider in Chapter 21. In heart muscle cells, for example, actin microfilaments of adherens junctions are continuous with the internal microfilaments responsible for the cell contractions of the heartbeat. Actin microfilaments are attached to the plasma membrane by means of a protein called *vinculin.* As for desmosomes, the integrity of adherens junctions is calcium-dependent: If the extracellular calcium concentration is decreased significantly, the junctions dissociate.

Tight Junctions

Tight junctions function as seals between two cell-lined compartments, preventing the flow of fluids through the extracellular space. As the name implies, a tight junction involves very close contact between the plasma membranes of adjacent cells, with no space at all in between (Figure 10-2a, page 274). Tight junctions form a continuous belt around the cells that line a body cavity such that the spaces between cells are tightly sealed. As a result, the space on one side of the junction plane is effectively separated from the space on the other side, thereby ensuring that fluids cannot pass from one side to the other.

Tight junctions are especially prominent in intestinal epithelial cells, which must form an effective barrier so that liquid from the intestine cannot pass across the epithelial layer. Tight junctions are also abundant in the ducts and cavities of glands, such as the liver and pancreas, that connect with the digestive tract, as well as the urinary bladder, where they perform the obvious function of ensuring that the urine stored in the bladder does not seep out between cells.

The role of tight junctions in preventing unwanted exchange between body compartments is illustrated by the

Figure 10-1   The Desmosome.   (a) An electron micrograph of a desmosome joining two cells in the skin of a newt (TEM). **(b)** A schematic diagram of a desmosome. The distance between cells in the desmosome region is 25–35 nm, about that for a nonjunction region. The desmosome core between the two membranes is filled with glycoproteins called desmocollins. The plaque on the cytoplasmic side of the membrane contains proteins called desmoplakins and is anchored in the cytoplasm by tonofilaments that consist of keratin, desmin, or vimentin, depending on the cell type. Desmogleins are transmembrane proteins that serve as links between the plaque on the inside of the membrane and the core on the outside.

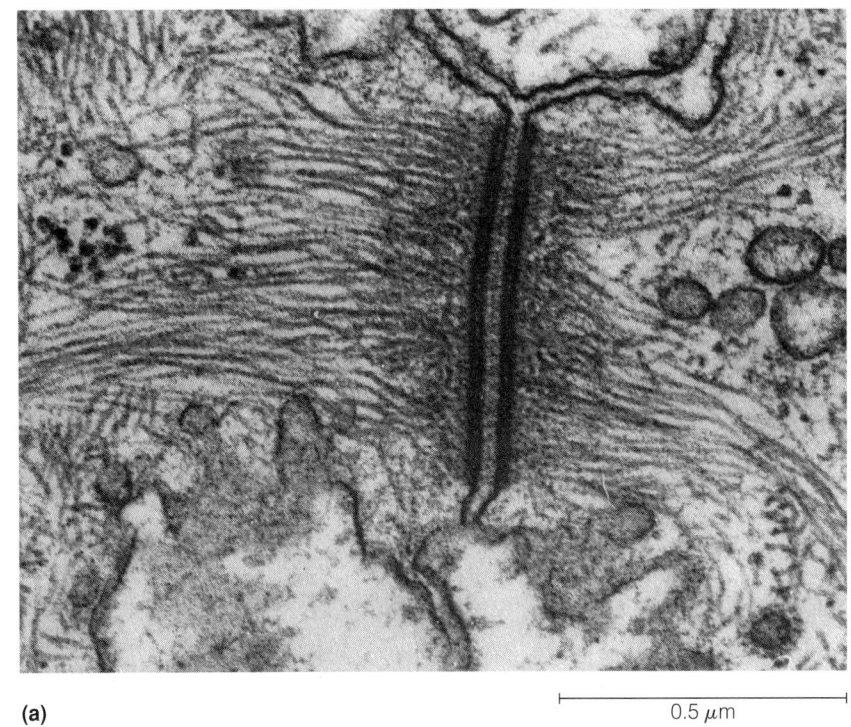

(a)

0.5 μm

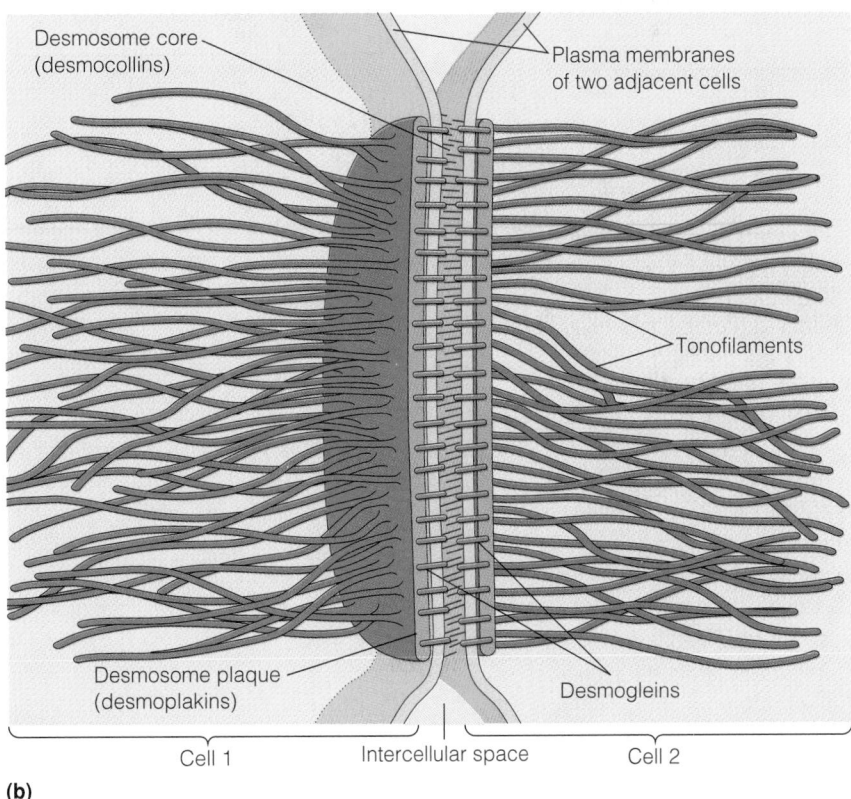

Desmosome core (desmocollins)

Plasma membranes of two adjacent cells

Tonofilaments

Desmosome plaque (desmoplakins)

Desmogleins

Cell 1

Intercellular space

Cell 2

(b)

electron micrograph of pancreatic tissue shown in Figure 10-2b. Exocrine (secretory) cells surround the lumen, or cavity, into which secretory proteins stored in the zymogen granules are secreted. Adjacent exocrine cells are linked by tight junctions, thereby ensuring that the lumen is isolated from the intercellular space and hence from the space on the other side of the cells.

Although tight junctions fuse the membranes of adjacent cells together very effectively, the membranes are not actually in close contact over broad areas. Rather, they are

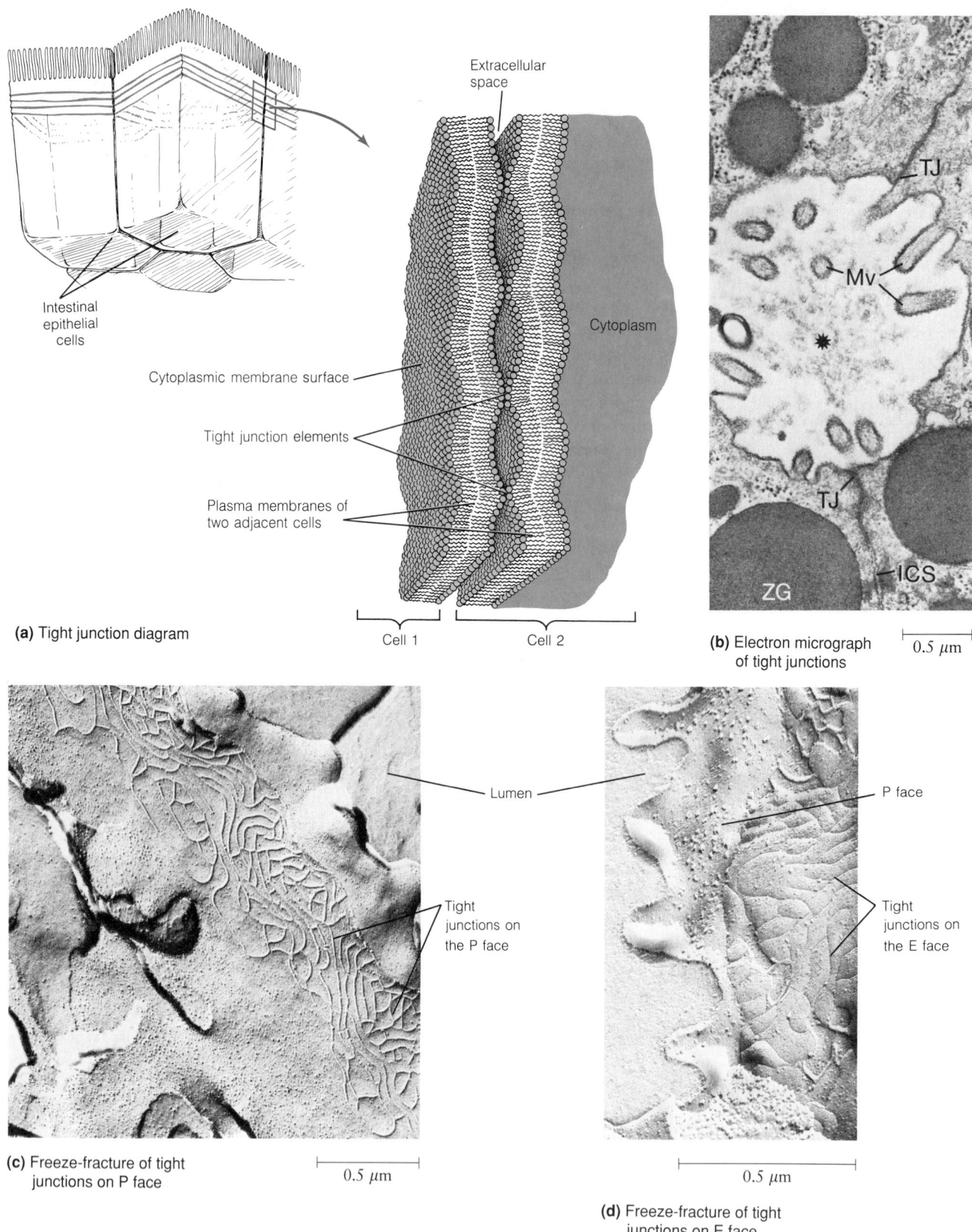

(a) Tight junction diagram

Extracellular space

Cytoplasm

Cytoplasmic membrane surface

Tight junction elements

Plasma membranes of two adjacent cells

Intestinal epithelial cells

Cell 1

Cell 2

(b) Electron micrograph of tight junctions

0.5 μm

(c) Freeze-fracture of tight junctions on P face

Lumen

Tight junctions on the P face

0.5 μm

(d) Freeze-fracture of tight junctions on E face

P face

Tight junctions on the E face

0.5 μm

Figure 10-2   The Tight Junction.   (a) A schematic representation of a tight junction between two cells of the intestinal epithelium. Tight junction elements are fused ridges at which the two plasma membranes are joined together tightly. **(b)** An electron micrograph of a lumen in a rat pancreas (TEM). TJ = tight junction; ICS = intercellular space;

Mv = short microvilli; ZG = zymogen granule; * = lumen. Tight junctions between cells in a frog bladder, as revealed by the freeze-fracture technique, appear as **(c)** raised strands on the protoplasmic (P) face and **(d)** shallow grooves on the exterior (E) face of the membrane (TEMs). The lumen is the cavity of the bladder.

fused along sharply defined ridges (see Figure 10-2a). The result is rather like placing two pieces of corrugated metal together so that their ridges are aligned and then fusing the two pieces lengthwise along each ridge of contact. The fused ridges eliminate the extracellular space and effectively seal the junction. Each such ridge is called a **tight junction element.** Not surprisingly, there appears to be a good correlation between the number of such elements across the junction and the tightness of the seal that the junction makes.

Tight junctions can be seen especially well in tissue prepared for electron microscopy by the freeze-fracture technique (Figure 10-2c and d). Depending on the fracture plane, tight junction elements appear as raised strands (part c) or as shallow grooves (part d). Notice that the ridges in part c form an interlocking network that extends across the junction.

In addition to preventing the movement of fluids, ions, and molecules across cell layers, tight junction elements also block the lateral movement of lipids and proteins within the membrane bilayers. Lipid movement is blocked in the outer monolayer only, but the movement of integral membrane proteins is blocked entirely. As a result, different kinds of integral membrane proteins can be maintained in the portions of a plasma membrane on opposite sides of a tight junction belt.

Consider, for example, the need for localization of integral membrane proteins in the epithelial cells that line your small intestine. These cells carry out the **transcellular transport** of nutrients such as glucose from the intestinal lumen on one side of the cell into the bloodstream on the other side (Figure 10-3). In these epithelial cells, the integral membrane proteins, or "pumps," responsible for the active uptake of glucose from the intestinal tract are restricted to the *apical surface* (the cell surface that faces the lumen of the intestine), whereas proteins needed for the passive movement of glucose from the cell into the extracellular fluid are present only on the *basolateral surface* (the cell surface that faces the circulatory system). In this way, glucose is transported directionally across the epithelial cells from the intestinal tract into the bloodstream. This transcellular transport would not be possible if both types of transport proteins occurred on both surfaces of the cell. Thus, tight junctions not only seal the spaces between adjacent epithelial cells so that transported molecules such as glucose cannot diffuse back into the lumen; they also prevent transport proteins on the apical surface from moving laterally in the membrane to the basolateral surface, and vice versa.

The molecular structure of tight junctions is not yet known. However, the ability of tight junctions to block membrane fluidity makes it likely that the ridges visible in Figure 10-2c consist of tightly packed integral membrane proteins that prevent other membrane proteins from moving laterally across the ridge. If so, tight junctions may be points at which adjacent plasma membranes are held together by tightly packed transmembrane junctional proteins that make contact across the intercellular space to create the desired seal.

Gap Junctions

The **gap junction** is the most common type of junction between animal cells. It is a region at which the plasma membranes of two cells are aligned and brought into intimate contact, with a gap of only 2–3 nm in between. The basic function of a gap junction is to provide a point of communication between two adjacent cells through which ions and small molecules can pass from one cell to the other. This free flow of ions and molecules means that adjacent cells are in direct chemical and electrical communication with each other. Smooth muscle and heart muscle are examples of tissues characterized by gap junctions.

Gap junction structure is illustrated in Figure 10-4 (page 277). At a gap junction, the two plasma membranes from adjacent cells are joined by tightly packed, hollow cylinders called **connexons.** A single gap junction may consist of just a few or thousands of clustered connexons. Each connexon is a circular assembly of six subunits of the protein *connexin.* The assembly spans the membrane and protrudes into the space, or *gap,* between the two cells (Figure 10-4a). Each connexon has a diameter of about 7 nm and a hollow center that forms a very thin hydrophilic channel through the membrane. The channel is about 1.5 nm in diameter at its narrowest point, just large enough to allow the passage of ions and small molecules.

When connexons in the plasma membranes of two adjacent cells are aligned, the cylinders in the two membranes meet end to end, forming direct channels of communication between the two cells that can be seen in the electron microscope (Figure 10-4b and c). The passage of solutes from cell to cell can be demonstrated experimentally by injecting a dye into one cell in a row of cells that are connected by gap junctions and then observing dye movement with a light microscope. The dye molecules move from cell to cell much more rapidly than would be expected if the molecules had to move into and out of the cells across their plasma membranes. By using molecules and ions of different sizes, W. R. Loewenstein and his colleagues have shown that gap junctions allow the passage of solutes with molecular weights up to about 1200. Included in this range are such common substances as monosaccharides, amino acids, and nucleotides.

Gap junctions are also the sites of lesser electrical resistance than is normally found across the plasma membrane, which means that ions can move freely through the connexons of gap junctions. Gap junctions are important in heart muscle and smooth muscle because they provide for the intercellular flow of potassium ions that is responsible for muscle activation. Thus, both electrical and chemical communication between adjacent cells is specifically facilitated by the direct cytoplasmic connections that exist through the channels of gap junction proteins.

Gap junction permeability can be changed rapidly by experimental manipulation of the pH or the Ca^{2+} concentration of the coupled cells. This suggests that cells may be able to open and close their gap junctions and thus control intercellular communication. Regulation is probably made

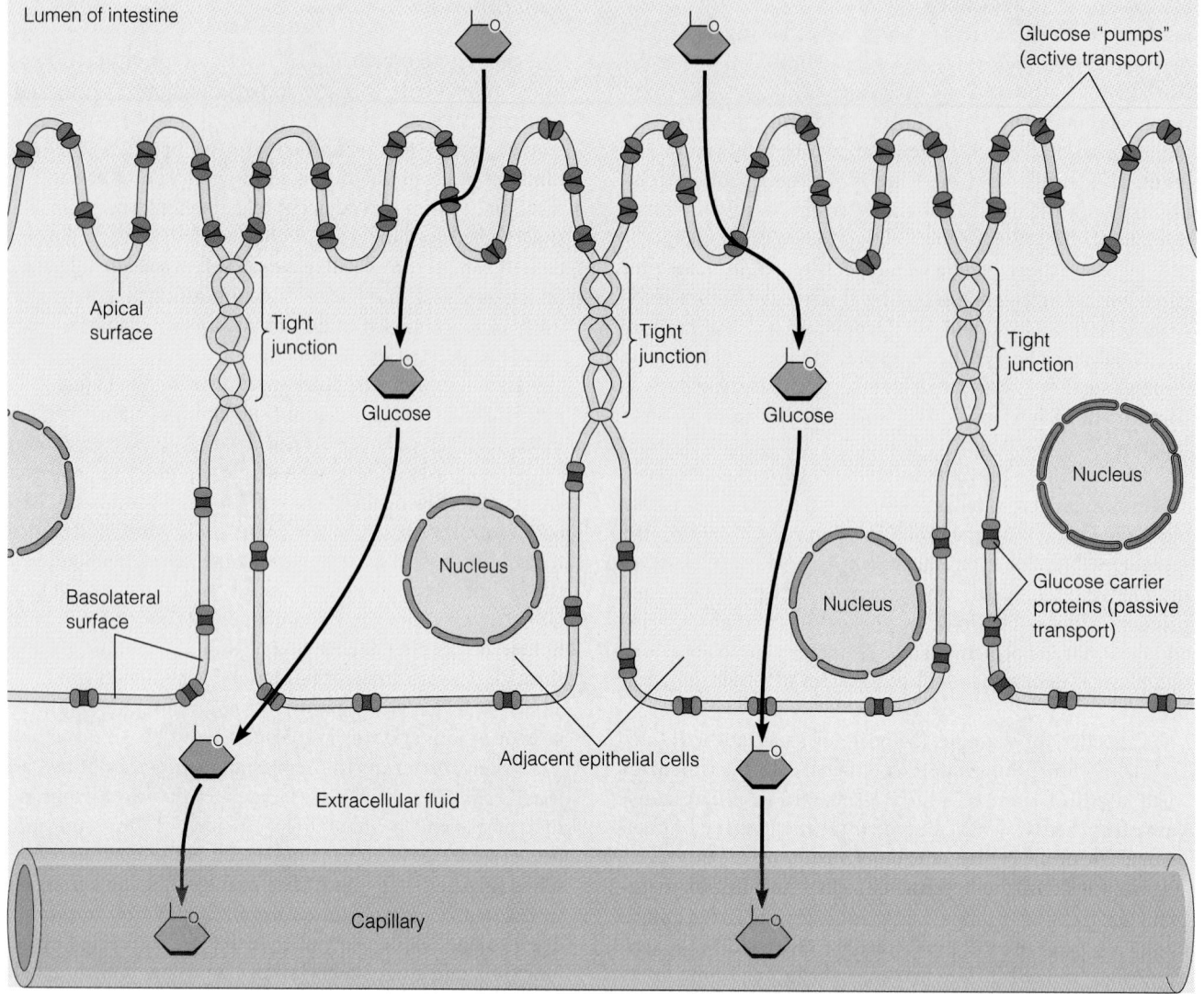

Figure 10-3 The Transport of Glucose Across the Intestinal Epithelium. The epithelial cells that line the small intestine need tight junctions to prevent the leakage of fluid from the intestinal lumen into the circulatory system on the other side of the epithelium, and to restrict membrane proteins to either the apical surface or the basolateral surface of the cells. Illustrated here is the transcellular (or transepithelial) transport of glucose from the lumen of the intestine (low glucose content) into the cytoplasm of epithelial cells (high glucose content) and then via the extracellular fluid into a capillary of the bloodstream. Glucose is actively transported into the epithelial cell by glucose "pump" proteins that are localized to the apical surface of the cell, then moves out of the cell by facilitated diffusion across the basolateral surface. This transcellular transport of glucose is possible because the tight junctions between adjacent cells restrict glucose pump proteins to the apical surface and glucose carrier proteins to the basolateral surface.

possible by conformational changes in the proteins that make up the connexons. Gap junctions tend to be open at low Ca^{2+} or low H^+ concentration and to close as the Ca^{2+} or proton concentration is raised.

Beyond the Membrane: Extracellular Structures

Few, if any, cells "end" at the plasma membrane. Most have some sort of structure that is external to the membrane but is nonetheless an integral part of the cell, both structurally and functionally. These **extracellular structures** consist of substances that are secreted by the cell. The chemical nature of these substances differs among organisms, but the extracellular structures of most eukaryotes have a common theme: They contain long, rigid fibers embedded in an amorphous, hydrated matrix of branched molecules that are usually glycoproteins or polysaccharides. In addition to holding the fibers in place, this gel-like network of glycoproteins or polysaccharides also impedes the flow of water molecules, thereby resisting compression. As Table 10-2 indicates, animal cells have an *extracellular matrix* that consists primarily of fibers of *collagen* embedded in a network of

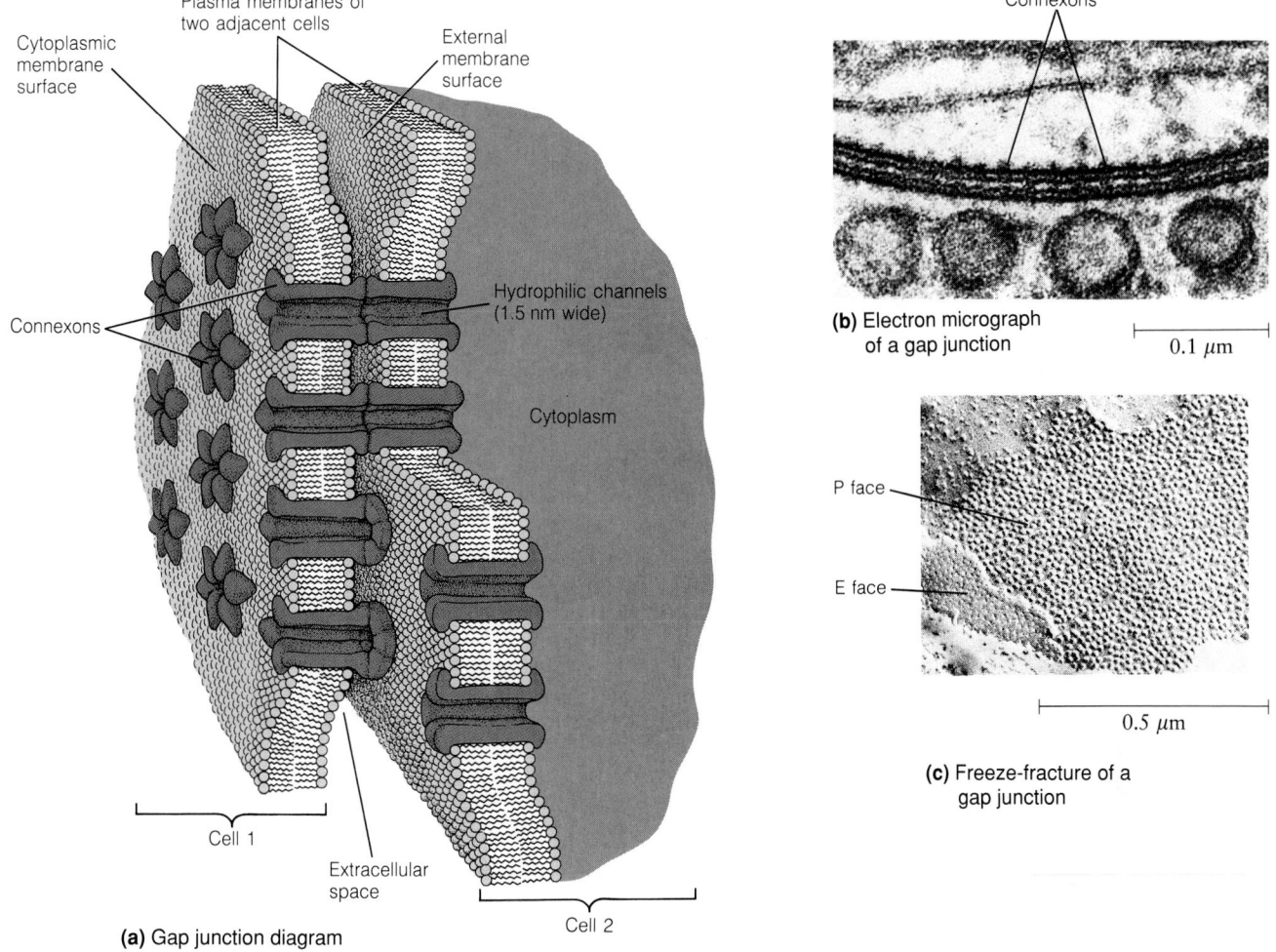

(a) Gap junction diagram

(b) Electron micrograph of a gap junction

0.1 μm

(c) Freeze-fracture of a gap junction

0.5 μm

Figure 10-4   The Gap Junction.   (a) A schematic representation of a gap junction. A gap junction consists of a large number of hydrophilic channels formed by the alignment of connexons in the plasma membranes of two adjoining cells. **(b)** An electron micrograph of a gap junction between two adjacent nerve cells (TEM). The connexons that extend through the membranes are visible here as beadlike projections spaced about 17 nm apart on either side of the membrane-membrane junction. **(c)** A gap junction as revealed by the freeze-fracture technique (TEM). The junction appears as an aggregation of intramembranous particles on the protoplasmic (P) face and as a series of pits on the exterior (E) face of the membrane.

glycoproteins called *proteoglycans*. In plants, fungi, algae, and prokaryotes, the extracellular structure is called a *cell wall*. In higher plants, the cell wall consists primarily of fibers of *cellulose* embedded in a complex network of polysaccharides called *hemicelluloses* and *pectins*.

The cell walls of plants and other organisms confer rigidity on the cells they encase and protect the cells from physical damage and attack by viruses and infectious organisms. In animals, the extracellular matrix forms a wide variety of structures and performs a diversity of functions,

Table 10-2   Extracellular Structures of Eukaryotic Cells

Kind of Organism	Extracellular Structure	Structural Fiber	Components of Hydrated Matrix	Adhesive Molecules
Animals	Extracellular matrix (ECM)	Collagen and elastin	Proteoglycans	Fibronectins and laminin
Plants	Cell wall	Cellulose	Hemicelluloses, pectins, and extensins	—

depending on the tissue or organ. The kinds and patterns of extracellular materials that are laid down during development regulate cellular processes as diverse as division, motility, differentiation, and adhesion.

We will look at the extracellular structures of both animal and plant cells, noting especially their molecular composition and the contributions they make to the structural and functional properties of cells. Bacterial cell walls are discussed in Box 10A.

Animal Cells: The Extracellular Matrix

The **extracellular matrix (ECM)** of animal cells consists mainly of long, semicrystalline fibers of collagen embedded in an elastic network of proteoglycans, a very diverse family of glycoproteins. The considerable variety in the properties of the ECM from different tissues results not only from differences in the types of collagen and the kinds of proteoglycans present but also from variations in the ratio of collagen to proteoglycans.

Collagen. The **collagen** that makes up the fibers of the ECM is the most abundant protein in your body. It is secreted by cells called *fibroblasts* that occur throughout most

tissues. Collagen is an especially prominent component of *connective tissue* such as bone, cartilage, tendons, and ligaments. In the absence of collagen, cells in these and other tissues would not have sufficient adhesive strength to maintain a given form. Collagen is actually a large family of closely related glycoproteins, all of which form fibers with high tensile strength. Collagens have a high content of the amino acids glycine, hydroxylysine, and hydroxyproline. (For the structures of hydroxylysine and hydroxyproline, see Figure 7-19c). The carbohydrate component of collagen makes up about 10% of the collagen molecule by weight and consists primarily of short chains of glucose and galactose units in alternation.

In electron micrographs of most animal tissues, collagen fibers can be seen in bundles throughout the extracellular matrix. The structure of one such fiber is illustrated in Figure 10-5. Each fiber is composed of many *macrofibrils*, each consisting of many *microfibrils*. A microfibril, in turn, is made up of many molecules of **tropocollagen,** which consist of three polypeptide chains twisted together into a rigid, right-handed triple helix about 270 nm in length and 1.5 nm in diameter. A typical collagen fiber contains about 10,000 molecules of tropocollagen per micron of length, which represents about 270 tropocollagen molecules in cross section.

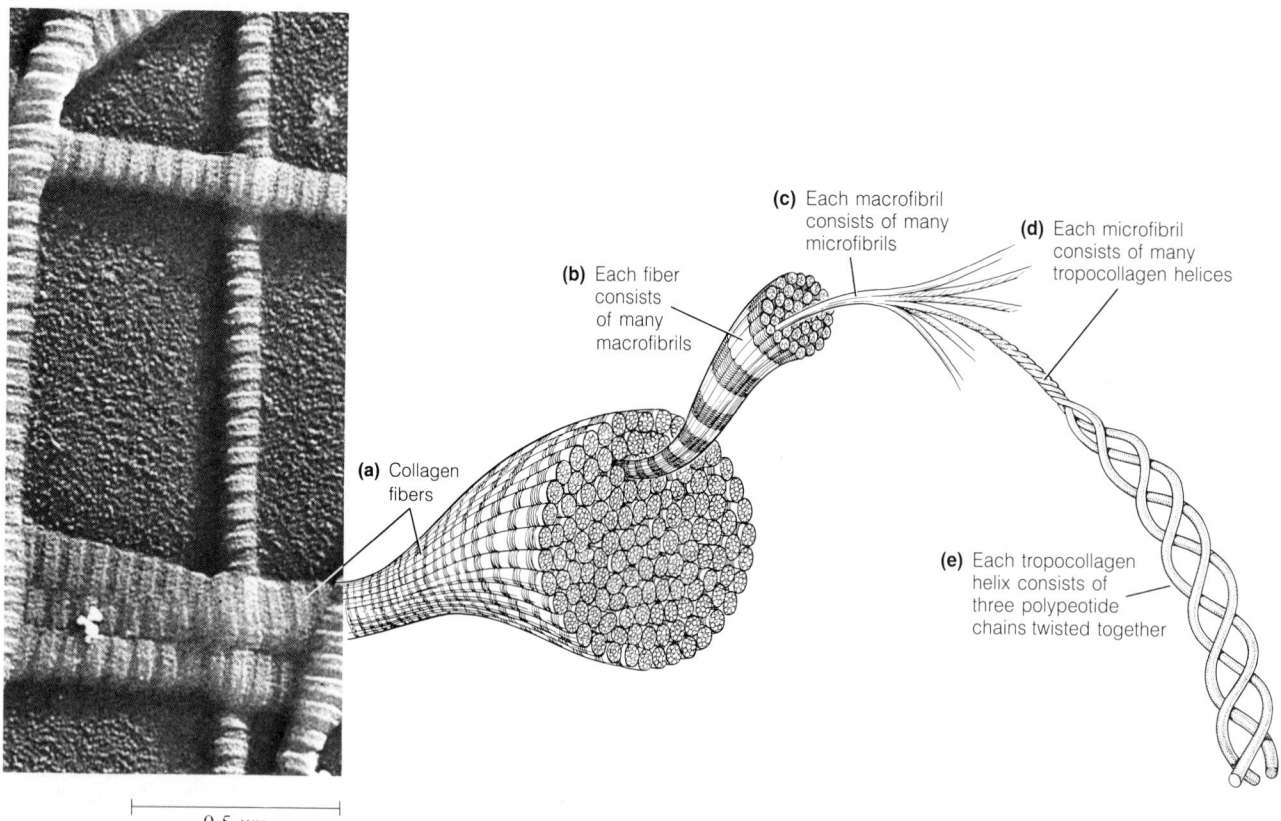

(c) Each macrofibril consists of many microfibrils

(d) Each microfibril consists of many tropocollagen helices

(b) Each fiber consists of many macrofibrils

(a) Collagen fibers

(e) Each tropocollagen helix consists of three polypeotide chains twisted together

0.5 μm

Figure 10-5   The Structure of Collagen.   (a) Collagen fibers as seen in the scanning electron microscope. **(b, c)** A collagen fiber contains macrofibrils, each of which consists of microfibrils. **(d)** Every microfibril is a tiny bundle of tropocollagen molecules, **(e)** each of which contains three polypeptides called α chains twisted together into a right-handed helix. The repeating bands visible on the fibers in the SEM in part a reflect the regular but offset manner in which tropocollagen helices associate laterally, forming microfibrils.

Clinical Applications

BACTERIAL CELL WALLS, THE GRAM STAIN, PENICILLIN, AND TEARS

Like plant cells, a bacterial cell is also surrounded by a cell wall, consisting of complex polysaccharides, that provides support and protection to the cell. The **bacterial cell wall** protects the cell from bursting due to osmotic pressure and from invasion by viruses. The bacterial cell wall also serves as a permeability barrier and plays a role in cell-cell recognition and adhesion.

That is the extent of the similarities with plant cell walls, however; bacterial cell walls differ strikingly from the cell walls of plants. In fact, the chemical nature of the cell wall differs among bacteria as well. This difference is the basis for classifying bacteria into two major groups based on their response to the **Gram stain,** a colored complex of iodine and the dye crystal violet. (The stain is named for Hans Christian Gram, a Danish bacteriologist who devised it in the nineteenth century.) The chemical basis for the differential staining is not known, but the technique is very straightforward. Bacterial cells are fixed by heating, stained with the iodine-crystal violet mixture, and then treated with a decolorizing agent (alcohol or acetone, usually). *Gram-positive bacteria* retain the stain and appear deep blue-purple, whereas *Gram-negative bacteria* do not and therefore appear reddish (Figure 10A-1). *Staphylococcus aureus* is a representative gram-positive species, while *Escherichia coli* is an example of a gram-negative bacterium. The critical difference between the two groups of bacteria is the chemical nature of the cell wall, but the stain is retained within the cell, not just by the cell wall. Somehow, the cell wall around a gram-positive bacterium prevents destaining of the cell itself.

The key structural differences between the two types of cell walls are shown in Figure 10A-2 for *S. aureus* (gram-positive) and *E. coli* (gram-negative). The cell wall of a gram-positive bacterium is a multilayered, cross-linked network of **peptidoglycans** (polysaccharides and peptides) located external to the plasma membrane of the cell (Figure 10A–2a). These peptidoglycans consist of long chains of *N*-acetylglucosamine (*GlcNAc*) and *N*-acetylmuramic acid (*MurNAc*) in strictly alternating order, cross-linked by short peptides. The peptide cross-links consist of a tetrapeptide

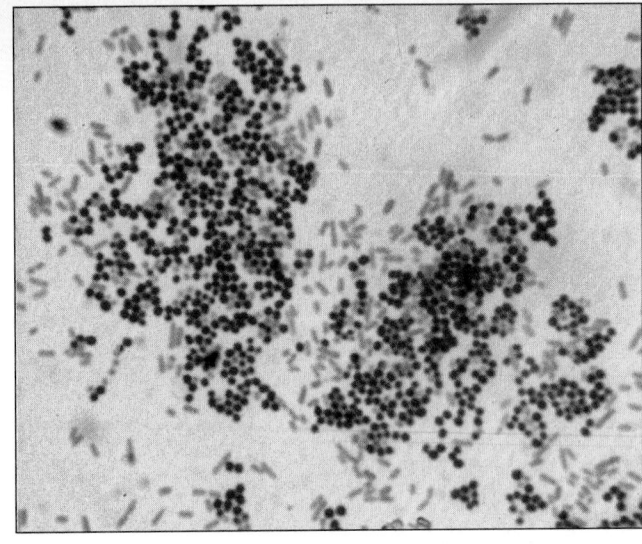

1 μm

Figure 10A-1 Gram-Stained Bacterial Cells. Shown here are gram-positive *Staphylococcus aureus* (blue-purple) and gram-negative *Escherichia coli* (reddish) cells (LM).

group on each MurNAc that is joined to adjacent tetrapeptides, usually by a chain of five glycines. In addition to the peptidoglycans, the cell wall of gram-positive bacteria also contains *teichoic acids*, which are polymers of glycerol or a 5-carbon alcohol called ribitol. The teichoic acid polymers, usually 5–30 units long, protrude from the membrane through the wall and are cross-linked to the peptidoglycan backbone. The result is a covalently cross-linked network that can be thought of as a single enormous molecule surrounding the cell (see Figure 10A-2a). Typically, the cell wall of a gram-positive bacterium consists of about 20–40 peptidoglycan layers and is about 50 nm thick.

The cell wall of a gram-negative bacterium also contains peptidoglycans but is much thinner, consisting typically of just one or a few peptidoglycan layers and covered by an *outer membrane* (Figure 10A-2b). In this case, the

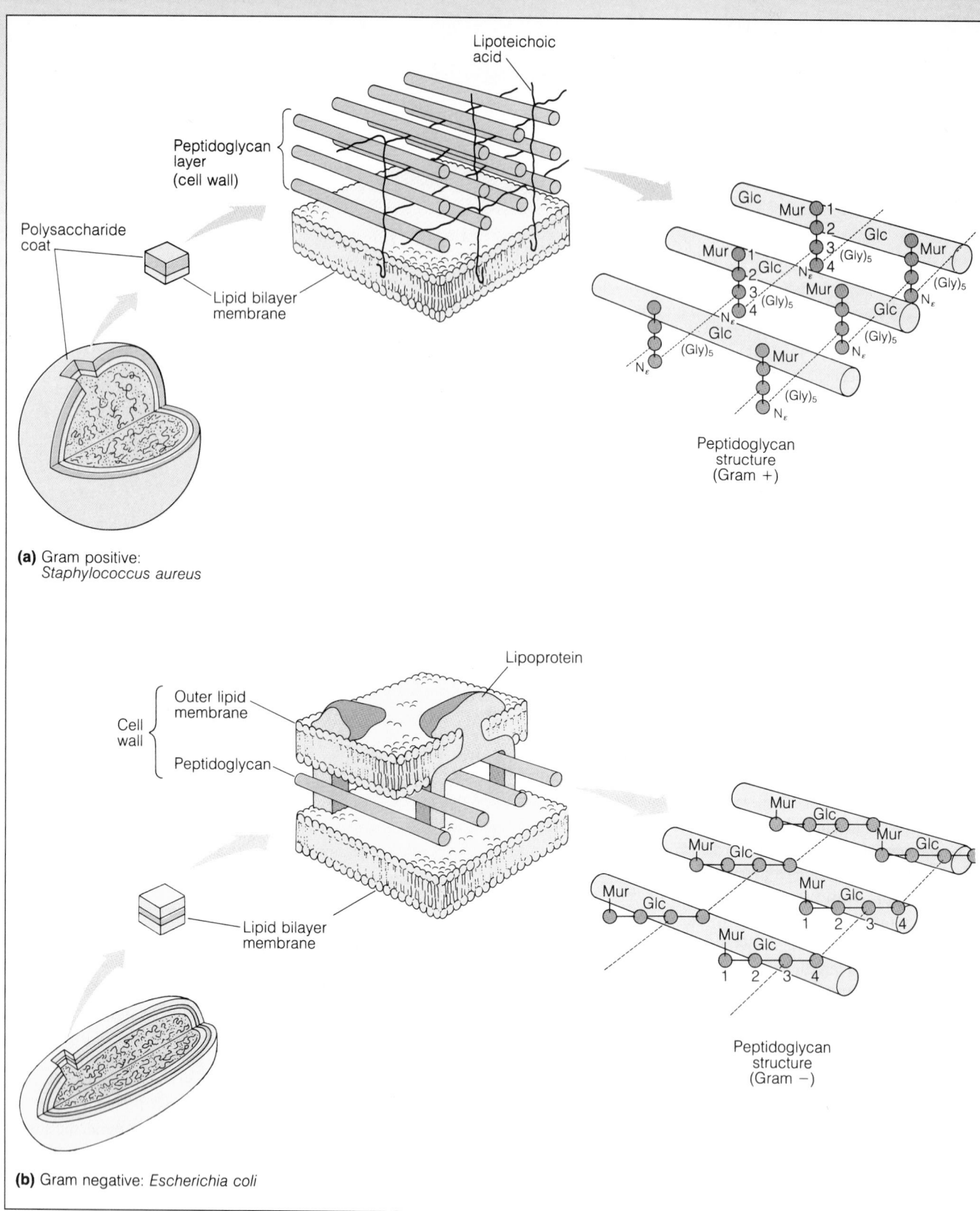

Figure 10A-2   Bacterial Cell Walls.   (a) The cell wall of gram-positive *Staphylococcus aureus*. The wall consists of 20–40 layers of a cross-linked network of peptidoglycans. The polysaccharide is a polymer of *N*-acetylglucosamine (GlcNAc or Glc) and *N*-acetylmuramic acid (MurNAc or Mur) in strictly alternating sequence. Attached to the MurNAc units are tetrapeptides linked together by pentaglycine peptide chains. Lipoteichoic acid molecules protrude from the plasma membrane through the wall and are linked to the peptidoglycans. **(b)** The cell wall of gram-negative *Escherichia coli*. The wall consists of one or a few layers of peptidoglycans bounded on the inner side by the plasma membrane of the cell and on the outside by an outer membrane. The periplasmic space between the two membranes contains periplasmic proteins that shuttle amino acids and sugars from the porin channels in the outer membrane to transport proteins in the plasma membrane.

tetrapeptide chains attached to MurNAc units are joined to one another directly by a peptide bond rather than by a pentaglycine link. Because the cell wall of a gram-negative bacterium is bounded by the plasma membrane on one side and the outer membrane on the other, its location is described as the *periplasmic space*. The thin cell wall is porous enough to allow molecules as large as proteins to move within the periplasmic space. Especially important in this regard are the *periplasmic proteins* that act as shuttles for amino acids and sugars. These solutes cross the outer membrane through protein channels called *porins* and are then shuttled across the wall by the periplasmic proteins, which deliver them to transport proteins in the plasma membrane.

The synthesis and assembly of bacterial cell walls is obviously a complex process. The cell wall appears to grow by the addition of new peptidoglycan material on the inside of the old. Most of the enzymes required for this process are localized to the plasma membrane. Because the cell wall is vital to the viability of the bacterium and is synthesized by enzymes that are unique to prokaryotes, cell wall synthesis is a common target of *antibiotics*, drugs that are used to prevent and control the growth of pathogenic bacteria. A widely used antibiotic is *penicillin*, a compound produced by the mold *Penicillium notatum*. Penicillin is effective as an antibacterial drug because it inhibits the enzymes that join the peptide side chains together after peptidoglycan assembly is complete. Without the side chains, the cell wall cannot withstand the normal osmotic pressure of the bacterial cell, resulting in the bursting, or *lysis*, of the treated cells.

Naturally occurring antibiotics of another class act in a quite different way. These are the *lysozymes*, a family of enzymes present in such diverse sources as egg white, bacteriophage, and your own tears. Rather than interfering with cell wall synthesis, lysozymes digest the peptidoglycan layer directly by catalyzing the hydrolysis of the glycoside links between the GlcNAc and MurNAc units in the polysaccharides. As a result, the cell wall is dissolved, leading to lysis and death of the cell. So the next time you shed a tear, take a moment to appreciate the effective defense system it represents! ■

The individual polypeptides are called **α chains.** Every third amino acid in each α chain is glycine, which makes the triple helix possible because the spacing of the glycines places them in the axis of the helix, and glycine is the only amino acid that is small enough to fit in the interior of the triple helix. There are about two dozen different kinds of α chains in vertebrates, each with its own unique amino acid sequence. These associate in various combinations to form at least 14 different types of collagen molecules, each of which is found in specific tissues (Table 10-3). Types I, II, and III are found in connective tissue and are the most abundant forms of collagen α chains. Type I alone makes up about 90% of the collagen in the human body, comprising most of the collagen of skin, bones, and tendons. The other two prominent forms of collagen are found in cartilage (type II) and in skin and tendons (type III). Type IV collagen is found mainly in *basal laminae*, the thin, sheetlike mats of extracellular material that underlie layers of epithelial cells. Most of the other forms (types V–XIV) are minor constituents of specific connective tissues.

When fibers containing type I, II, or III tropocollagen molecules are examined with an electron microscope, visible bands, or *striations*, are seen (see Figure 10-5a). These striations reflect the regular but offset manner in which tropocollagen helices associate laterally to form microfibrils. The molecules are aligned in parallel but staggered rows. Molecules in adjacent rows overlap by about one-quarter of their lengths, such that the repeat distance of about 67 nm corresponds to one-quarter of the length of the individual molecules. Only types I, II, and III collagen are known to have striations. Type IV collagen is known to form very fine, unstriated fibrils. The structures of other collagen types is not yet known.

Although collagen provides support to cells in tissues, its fibrous, rodlike structure is not particularly suited to the elastic qualities required by tissues such as skin, lungs, and the intestine, which undergo continuous shape changes throughout life. Elasticity is provided by a second extracellular matrix protein, appropriately called **elastin.** Elastin forms a relatively loose, flexible framework of cross-linked chains.

The important roles of collagen and elastin are clearly demonstrated during aging. Over time, collagen becomes increasingly cross-linked and inflexible, and elastin is lost from tissues like skin. As a result, older people often find that their bones and joints are less flexible, and their skin becomes wrinkled and returns to its original shape only slowly after being deformed (by gentle pinching, for instance).

Proteoglycans. The hydrated, gel-like network in which the collagen fibers of the ECM are enmeshed consists primarily of **proteoglycans** (see Table 10-2). The proteoglycans (sometimes also called *mucoproteins*) are a family of glycoproteins that are remarkably high in carbohydrate—up to 95% by weight! Proteoglycans vary greatly in size, depending on the molecular weight of the polypeptide chain (ranging from about 10,000 to over 500,000) and the number and length of the carbohydrate chains (1–200 per molecule, with

Table 10-3   Types of Collagen and Their Occurrence and Structure

Type	Occurrence	Structure*
I	Skin, bones, tendons, ligaments; cornea of eye	Striated fibrils
II	Cartilage, intervertebral discs; vitreous humor of eye	Striated fibrils
III	Skin, tendons; blood vessel walls; uterine walls	Striated fibrils
IV	Basal laminae	Fine, unstriated fibrils
V	Cornea of eye; interstitial tissues	Fine fibrils (unstriated?)
VI	All connective tissues (minor component only)	Fine fibrils (unstriated?)
VII–XIV	Occurrence varies with type; most are minor components of cartilage or tendons	Fine fibrils (unstriated?)

*Fibrils of types I, II, and III have striated patterns that can be readily seen with an electron microscope, whereas type IV collagen is not striated. Types V–XI have molecular structures that are consistent with striation, but no such patterns have yet been seen. Types XII, XIII, and XIV have molecular structures that make striations unlikely.

an average length of about 800 monosaccharide units). Most proteoglycans have molecular weights in the range of 0.25–3 million.

The main carbohydrate components of proteoglycans are the **glycosaminoglycans** (also called *mucopolysaccha-rides*). Most of the glycosaminoglycans, or GAGs, are O-linked to the hydroxyl groups of serine side chains, though some are O-linked to threonine or N-linked to asparagine. GAGs are characterized by repeating disaccharide units, as illustrated in Figure 10-6 for three of the most common

Figure 10-6   The Disaccharide Repeating Units of Several Common Glycosaminoglycans (GAGs).
Top: chondroitin sulfate; center: keratan sulfate; and bottom: hyaluronic acid—three common extracellular glycosaminoglycans (GAGs) found in the extracellular matrix of animal cells. In chondroitin sulfate and hyaluronic acid, the disaccharide repeating unit consists of glucuronic acid (GlcUA), present in the ionized form, and an amino sugar—either *N*-acetylgalactosamine (GalNAc) or *N*-acetylglucosamine (GlcNAc). For keratan sulfate, the repeating unit consists of galactose (Gal) and GlcNAc.

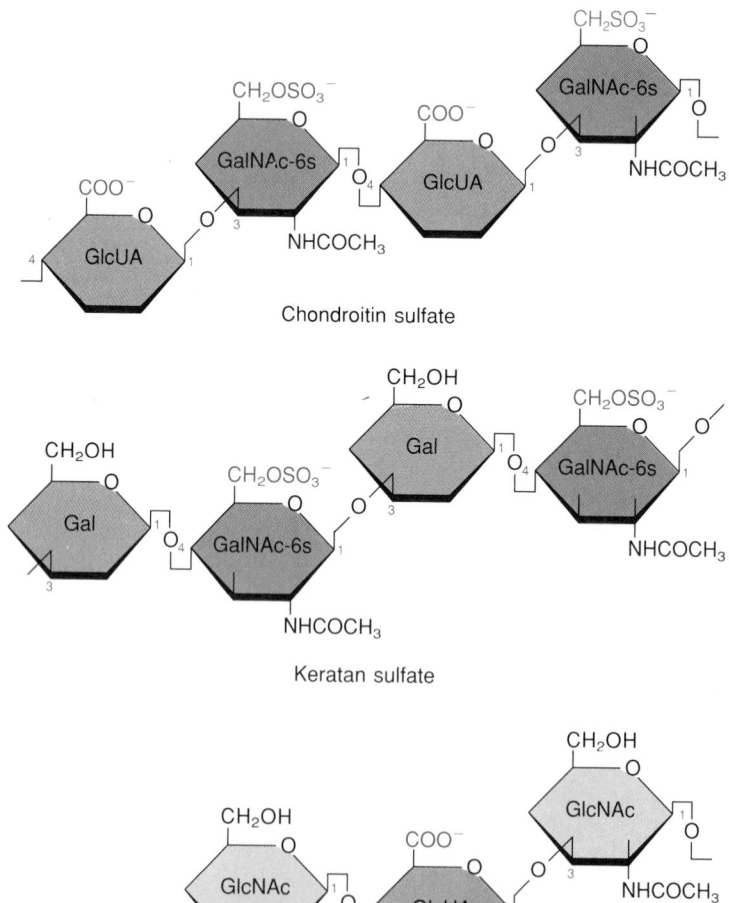

Chondroitin sulfate

Keratan sulfate

Hyaluronic acid

types. In each case, one of the two disaccharides in the repeating unit is always an amino sugar, with *N-acetylglucosamine* (*GlcNAc*) and *N-acetylgalactosamine* (*GalNAc*) as the two most common examples. The other repeating unit is usually a sugar or a sugar acid, commonly *galactose* (*Gal*) or *glucuronic acid* (*GlcUA*). In most cases, the amino sugar has one or more sulfate groups attached; of the three common GAGs shown in Figure 10-6, only one—*hyaluronic acid*—is unsulfated. Because of the many sulfate and carboxyl groups, GAGs are very acidic.

Most proteoglycans have the general structure shown in Figure 10-7, with numerous GAG chains attached along the length of the polypeptide core to form molecules with molecular weights in the hundreds of thousands. In cartilage, even larger complexes form by the noncovalent association of proteoglycans with long molecules of hyaluronic acid (Figure 10-8). A single such complex can have a molecular weight that exceeds 10^8. These complexes are then linked directly to collagen fibers to make up the fiber/network structure of the ECM.

An important role of proteoglycans is to trap water molecules, thereby slowing their flow. In effect, proteoglycan networks are like extracellular sponges that can hold remarkable quantities of water—up to 50 times their weight, in fact! Because of their high water content, these networks are quite resistant to compression and regain their shape quickly if distorted. Much of the elasticity of cartilage is due to this capability of proteoglycans to absorb and hold water molecules.

What Holds the Extracellular Matrix in Place?

The ECM is held in place by linkages of the proteoglycans to the plasma membrane. In some cases, the proteoglycans are

Figure 10-7 General Proteoglycan Structure. A proteoglycan molecule consists of a core polypeptide with numerous short GAG chains attached along its length and radiating outward. **(a)** The gray circles represent the amino acids of the core polypeptide, while the colored hexagons represent the sugars or sugar acids of the GAG chains. **(b)** This enlargement shows the first two sugar units of a keratan sulfate chain attached to the hydroxyl group of a serine in the polypeptide. Keratan sulfate consists of *N*-acetylglucosamine and galactose in strict alternation.

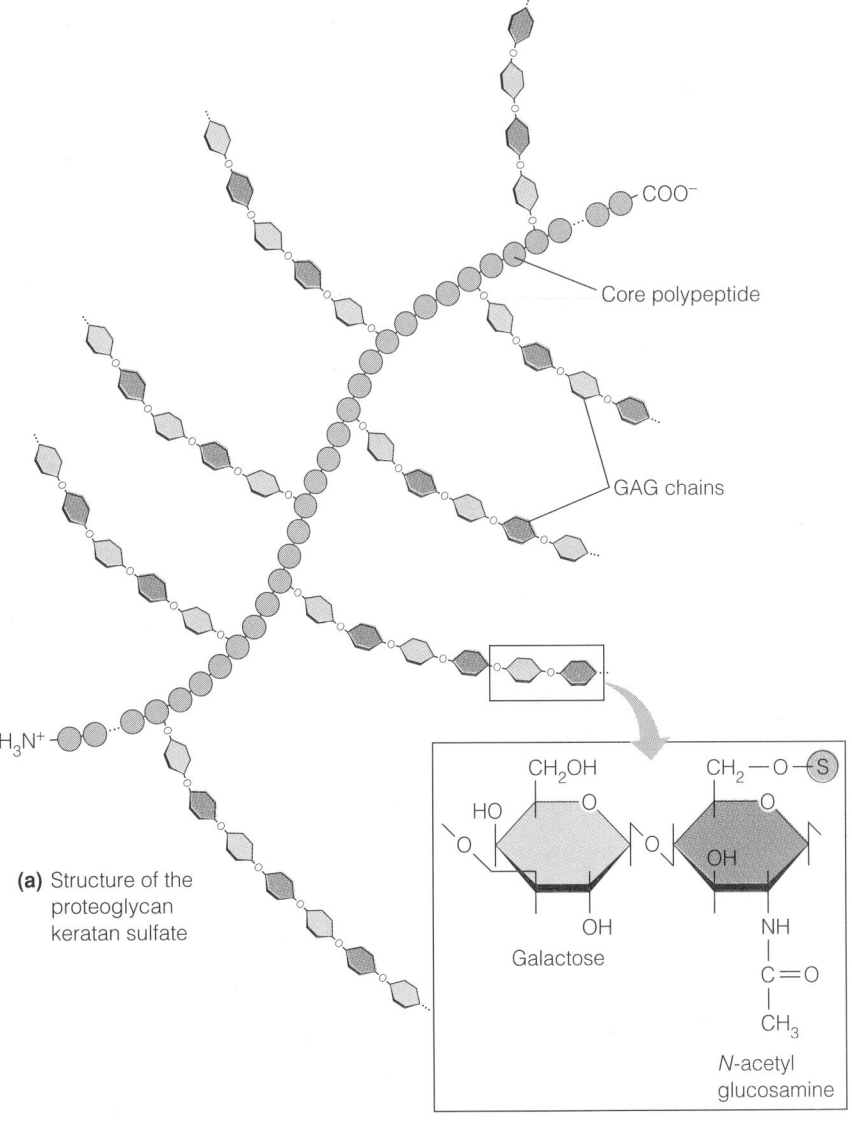

(a) Structure of the proteoglycan keratan sulfate

(b) Structure of GAG chain subunits

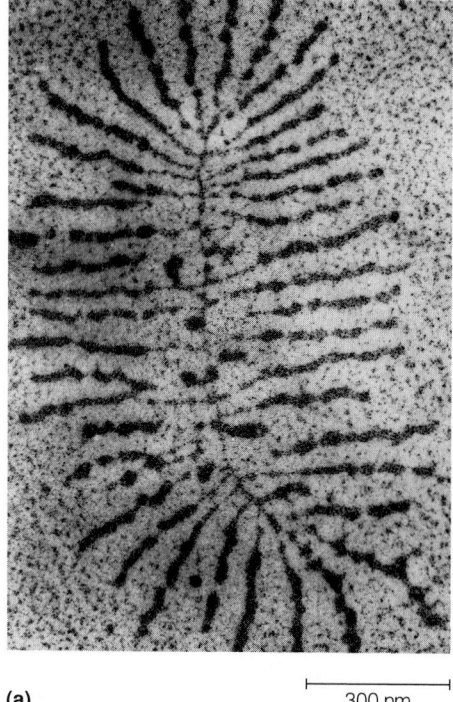

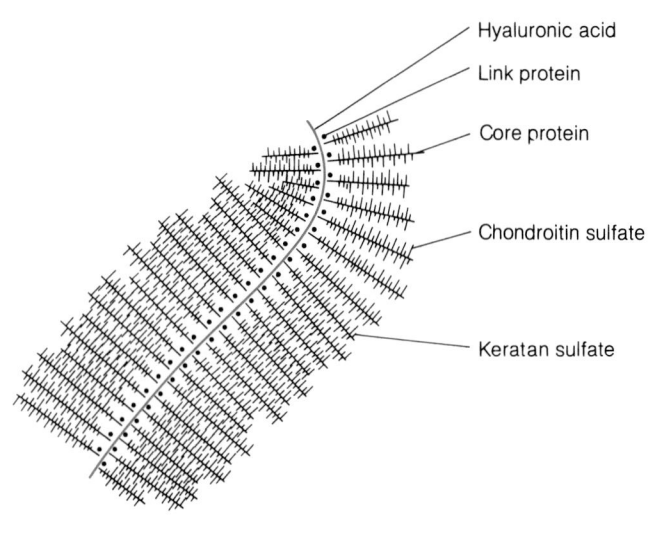

Hyaluronic acid

Link protein

Core protein

Chondroitin sulfate

Keratan sulfate

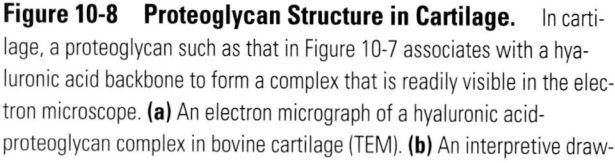

(a)   300 nm   **(b)**

Figure 10-8   Proteoglycan Structure in Cartilage.   In carti-lage, a proteoglycan such as that in Figure 10-7 associates with a hya-luronic acid backbone to form a complex that is readily visible in the elec-tron microscope. **(a)** An electron micrograph of a hyaluronic acid-proteoglycan complex in bovine cartilage (TEM). **(b)** An interpretive draw-ing of the structure shown in part a. Core polypeptides are attached non-covalently to a long hyaluronic acid molecule, with short keratan sulfate and chondroitin sulfate chains linked covalently along the length of the polypeptides.

themselves integral components of the plasma membrane, with their core polypeptides embedded within the membrane. In other cases, proteoglycans are linked covalently to membrane phospholipids. Alternatively, either the proteo-glycans or the collagen may bind to specific *receptor proteins* on the outer surface of the plasma membrane.

These direct links between the ECM and the plasma membrane are reinforced by a family of **adhesive glyco-proteins** that bind proteoglycans and collagen molecules to each other and to receptors on the membrane surface. The two most common kinds of adhesive glycoproteins are the *fibronectins* and *laminin*, while many of the membrane re-ceptors belong to a family of transmembrane proteins called *integrins*.

Fibronectins.   **Fibronectins** are a family of closely related adhesive glycoproteins that are widely distributed through-out the animal kingdom. A fibronectin molecule consists of two large polypeptides linked near their carboxyl ends by a pair of disulfide bonds (Figure 10-9). The two polypeptides are similar but nonidentical. Each has about 2500 amino acids and is folded into a series of globular domains con-nected by short, flexible segments of the polypeptide chain.

To study the function of the globular domains, re-searchers cleaved the polypeptide chains of fibronectin at

specific sites, isolated individual domains, and analyzed them for binding activity. From such studies, it is clear that some of the domains recognize and bind specific compo-nents of the ECM, including several types of collagen (I, II, and IV) and specific proteoglycans. Other domains recog-nize and bind to cell surface receptors. The receptor binding activity of these domains has been localized to a specific tetrapeptide sequence, RGDS (arginine-glycine-aspartate-serine). This *RGDS sequence* is a common motif among ex-tracellular adhesive proteins and is recognized by integrins such as the *fibronectin receptor*, to be discussed shortly.

Because of their multiple binding sites, fibronectin mol-ecules can link the various components of the ECM to each other as well as to receptors on the cell surface. Furthermore, fibronectin molecules can also bind to each other, further enhancing their capacity to form stable networks of ECM components and to anchor such networks to the cell surface.

In addition to (or perhaps as a consequence of) their role in cross-linking and anchoring the ECM, fibronectins are important in the maintenance of cell shape. Evidence for this role comes primarily from studies of cells maintained as laboratory cultures. Such cultured cells usually require a coating of fibronectins to assume and maintain the appear-ance they would have in an intact organism. A characteristic of many kinds of cancer cells is their inability to synthesize

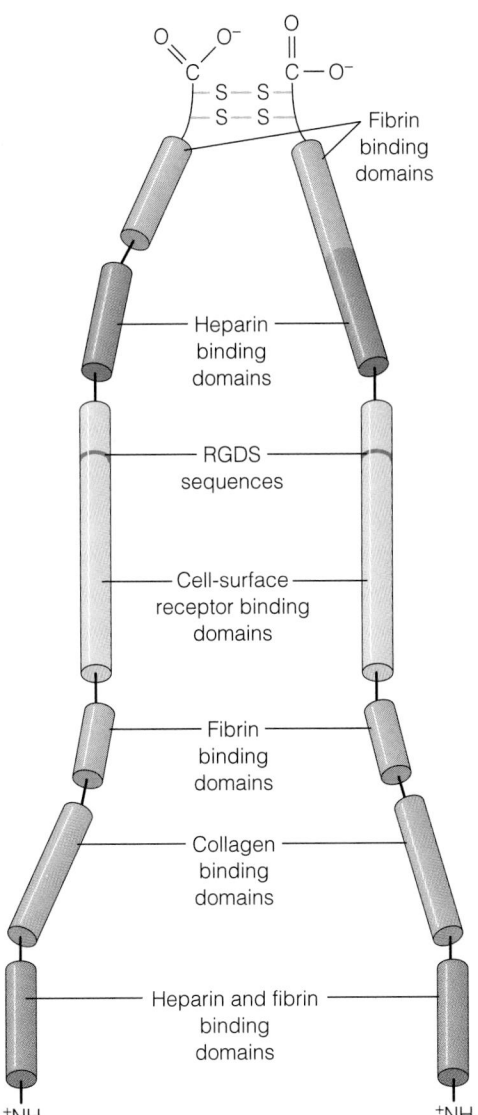

Figure 10-9 Fibronectin Structure. A fibronectin molecule consists of two nearly identical polypeptide chains held together by two disulfide bonds near their carboxyl ends. Each polypeptide chain is folded into a series of six globular domains linked by short, flexible segments. The globular domains have binding sites for ECM components or for specific receptors on the cell surface. The receptor-binding domain contains the tetrapeptide sequence RGDS (arginine-glycine-aspartate-serine) that is recognized by fibronectin receptors. In addition to the binding activities shown, fibronectin also has binding sites for heparan sulfate, hyaluronic acid, and gangliosides (glycosphingolipids that contain sialic acid groups).

fibronectins. In fact, their loss of ability to make fibronectins often correlates well with their loss of normal cell shape and with the detachment of the ECM. Moreover, if such cells are supplied with fibronectin, they often return to their normal shape and recover their ability to bind ECM.

A soluble form of fibronectin is present in the blood and other body fluids. This form, called *plasma fibronectin*,

is involved in blood clotting and perhaps also in wound healing and phagocytosis. Fibronectin promotes blood clotting because it has several binding domains that recognize *fibrin*, the blood-clotting protein (see Figure 10-9).

Laminin.   Unlike the fibronectins, **laminin** is a single glycoprotein rather than a family of related proteins. It is found mainly in basal laminae and is therefore much more restricted in occurrence than the fibronectins. **Basal laminae** are thin sheets, or mats, of specialized extracellular material that underlie layers of epithelial cells as well as the endothelial cells of blood vessels, supporting the overlying cells and separating them from the connective tissue that lies beneath or around them. Basal laminae also surround muscle cells, fat cells, and the *Schwann cells* that form myelin sheaths around nerve cells. In the glomerulus of the kidney, the basal lamina functions as a selective filter, regulating the movement of macromolecules from the blood into the urine. Basal laminae perform a number of other important functions. They determine cell polarity, organize proteins in the membranes of adjacent cells, and are also involved in the induction of cellular differentiation and the regeneration of injured tissue.

Despite these differences in function and specific molecular composition from tissue to tissue, all forms of basal laminae contain type IV collagen, proteoglycans (mainly heparan sulfate), and laminin. Fibronectins may also be present, but laminin is the most abundant adhesive glycoprotein in basal laminae. Laminin is thought to be localized mainly on the surface of the lamina that faces the overlying epithelial cells, where it helps bind the cells to the lamina. Fibronectins, on the other hand, are located on the other side of the lamina, where they help anchor cells of the connective tissue.

Laminin is a very large protein, with a molecular weight of at least 850,000. It consists of three long polypeptides, identified as α, β_1, and β_2. Disulfide bonds hold the polypeptide chains together in the shape of a cross, with part of the long arm wound into a three-stranded coiled coil (Figure 10-10). Like fibronectins, laminin has several domains, which represent binding sites for type IV collagen and for heparin and heparan sulfate, as well as for laminin receptor proteins on the surface of the plasma membranes of the overlying cells. Laminin molecules can also bind to each other, thereby increasing their capacity to form extensive ECM networks in the basal laminae.

Integrins.   Fibronectins and laminin can bind components of the ECM to the plasma membranes of cells because the membranes have specific *receptor glycoproteins* on their surfaces that recognize and bind to specific regions of the fibronectin or laminin molecule. These receptors belong to a family of transmembrane proteins called **integrins,** which differ from most other receptor proteins (such as those we will meet in Chapter 23) in their higher concentrations on the cell surface and their lower affinity for their respective *ligands* (the specific molecules they recognize and bind).

Figure 10-10   Laminin Structure.
A laminin molecule consists of three large polypeptides, α, β_1, and β_2, linked by disulfide bonds into a crosslike structure. A portion of the long arm consists of a three-stranded coiled coil. The molecule has several functional domains. Those on the ends of the α chain bind to organ-specific cell-surface receptors, whereas those at the ends of the two arms of the cross are specific for type IV collagen. The cross-arms also contain laminin-laminin binding sites, thereby enabling laminin molecules to bind to each other and form large aggregates.

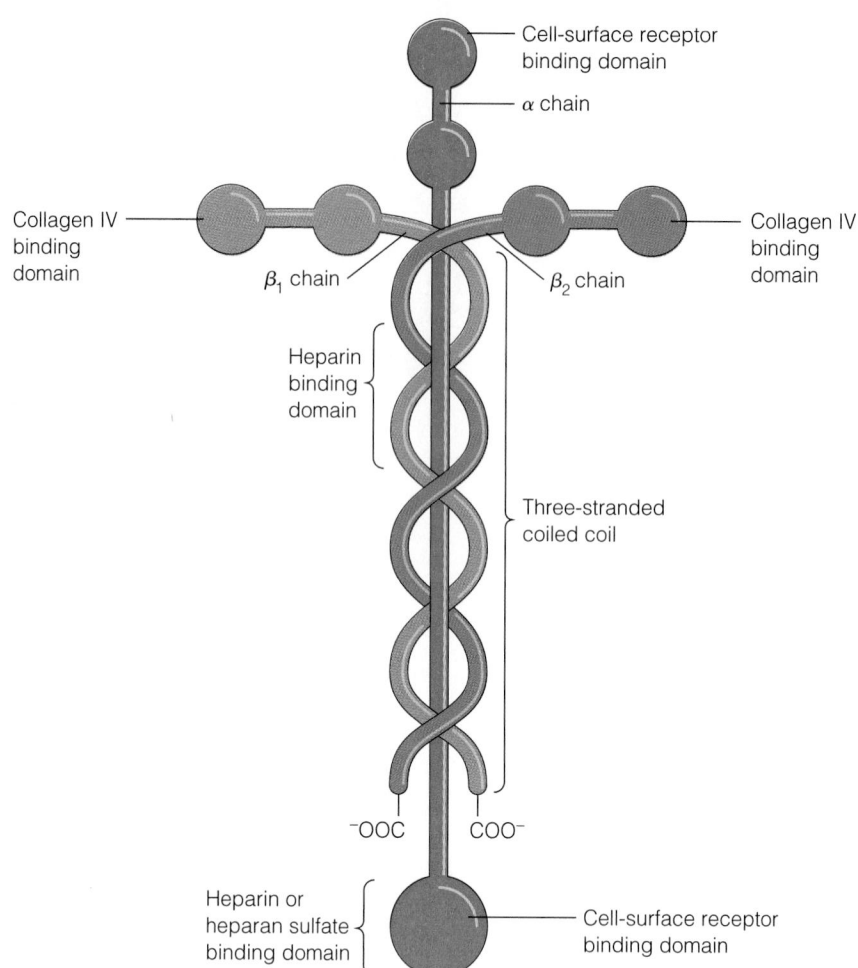

Integrins consist of two large transmembrane polypeptides, the α and β subunits, that are associated with each other noncovalently (Figure 10-11a). Integrins differ from one another in their binding specificities and in the sizes of their subunits (molecular weight ranges: 110,000–140,000 for the α subunit; 85,000–91,000 for the β subunit). The α and β subunits interact on the outer surface to form the binding site for the adhesive glycoprotein, with most of the binding specificity apparently dependent on the α subunit. On the cytoplasmic side of the membrane, many integrins have binding sites for specific molecules of the cytoskeleton, thereby enabling the cytoskeleton and the ECM to communicate across the plasma membrane.

The **fibronectin receptor** shown in Figure 10-11a is the best-characterized ECM receptor. It was initially identified as a membrane glycoprotein that bound to a fibronectin *affinity column* (a chromatographic column with fibronectin molecules immobilized on it) and could be eluted from the column with a solution of a small peptide containing the RGDS binding sequence. The receptor has a binding site for fibronectin on the outer surface of the plasma membrane. On the inner surface, the receptor has a binding site for *talin*, a cytoskeletal protein involved in the attachment of actin microfilaments to the plasma membrane. In this way, the ECM is effectively linked to the cytoskeleton.

The affinity of the talin binding site for its ligand depends on the phosphorylation state of a crucial tyrosine in the talin binding site of the receptor (Figure 10-11b). When the tyrosine is not phosphorylated, the binding site has a high affinity for talin. However, when the tyrosine is phosphorylated (by a protein kinase that is specific for tyrosine), the affinity of the binding site for talin is decreased, leading to dissociation of talin and therefore to a break in the link between the fibronectin in the ECM on the outside of the cell and the actin microfilaments that are a part of the cytoskeleton on the inside.

Many of the integrins recognize the RGDS sequence in the specific ECM glycoproteins that they bind. However, other parts of the glycoprotein molecule must also be recognized by the integrin binding site because integrins display a greater specificity of glycoprotein binding than can be accounted for by the RGDS sequence alone. Laminin can apparently be recognized by several different kinds of receptors, some of which resemble the fibronectin receptor in structure and properties, while others differ significantly from the integrins. An example of the latter is a laminin receptor that recognizes the amino acid sequence YIGSR (tyrosine-isoleucine-glycine-serine-arginine).

Most of the attachments between the ECM and the plasma membrane of an animal cell are more or less perma-

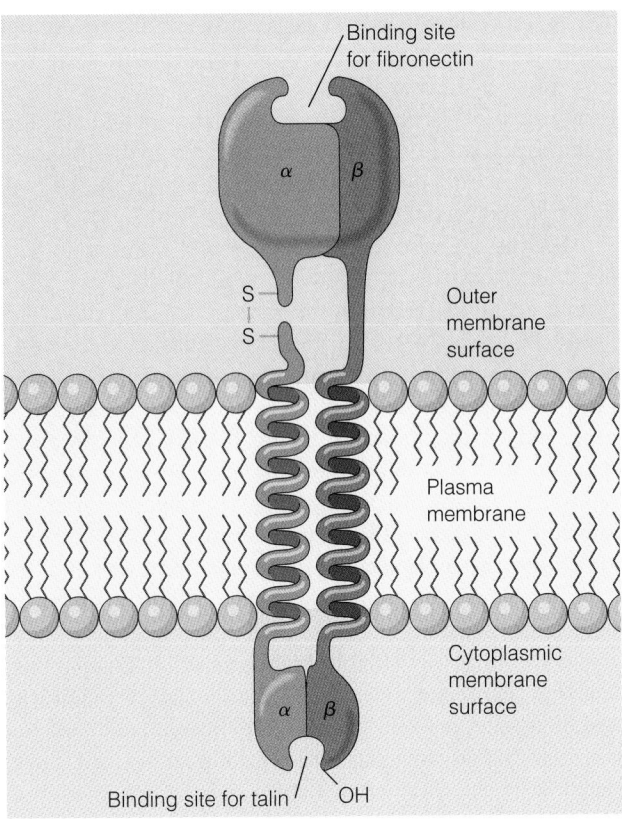

(a) The fibronectin receptor

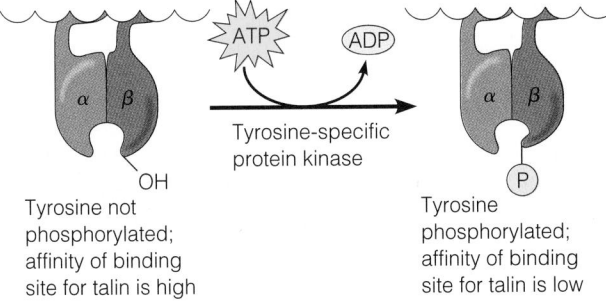

Tyrosine not phosphorylated; affinity of binding site for talin is high

Tyrosine phosphorylated; affinity of binding site for talin is low

(b) Phosphorylation of tyrosine at the talin binding site reduces affinity for talin

Figure 10-11 An Integrin: The Fibronectin Receptor.
(a) An integrin consists of α and β subunits, transmembrane polypeptides that associate with each other noncovalently. The α and β subunits interact on the outer membrane surface to form the binding site for the ligand. On the cytoplasmic side of the membrane, integrins have binding sites for cytoskeletal molecules. The integrin shown here is the fibronectin receptor, which has a binding site for fibronectin on the outer surface and a binding site for talin on the cytoplasmic side of the membrane. In this and several other integrins, the α subunit is split into two segments held together by a disulfide bond. Both the α and β subunits are glycosylated on the exterior side, although the sugar side chains are not shown here. **(b)** The talin binding site on the cytoplasmic surface of the fibronectin receptor has a high affinity for talin when the hydroxyl group of a specific tyrosine is not phosphorylated. Phosphorylation of the hydroxyl group (by a tyrosine-specific protein kinase) causes a distortion of the binding site that greatly reduces its affinity for talin.

nent, especially if the cell is part of a mature tissue. During embryonic development, however, ECM/membrane links can change as part of the developmental process, particularly in the case of cells that migrate within the developing embryo. Such cells appear to move along tracks of extracellular material, with which they continually make and break connections as they migrate.

Plant Cells: The Cell Wall

The rigid **cell walls** that surround almost all plant cells play important roles in supporting and defining the shape of plant tissues, and in protecting the cells from both mechanical injury and invasion by microorganisms, especially fungi and bacteria. Plant cell walls are remarkably sturdy, enabling a plant cell to withstand the considerable *turgor pressure* that is exerted because of the uptake of water by the cell. Turgor pressure is vital to plants because it accounts for much of the rigidity of plant tissues and provides the driving force behind cell expansion.

Cell walls are of two types. The wall that is laid down during cell division is called the **primary cell wall.** Primary cell walls are characterized by a flexibility and extensibility that enables them to expand in response to cell enlargement and elongation. The primary cell wall consists mainly of cellulose fibers embedded in a gel-like polysaccharide matrix that is about 60% water by weight.

As a cell reaches its final size and shape, a much thicker and more rigid **secondary cell wall** may form by the deposition of additional cell wall material on the inner surface of the primary wall. The secondary cell wall is much more complex than the primary wall, in both structure and chemical composition. The secondary wall usually contains more cellulose than the primary wall and may have a high content of *lignin*, a major component of wood. Deposition of a secondary cell wall renders a cell very inextensible and therefore specifies the final size and shape of the cell definitively.

Like the extracellular matrix of animal cells, plant cell walls consist predominantly of long fibers embedded in a network of branched molecules. Instead of collagen and proteoglycans, however, plant cell walls contain *cellulose microfibrils* enmeshed in a network of polysaccharides and protein (see Table 10-2). In a typical cell wall, cellulose microfibrils represent about 40% of the cell wall on a dry-weight basis, with the polysaccharides and protein accounting for the other 60%.

Cellulose Microfibrils. Recall from Chapter 3 that **cellulose** is an unbranched polymer of β-D-glucose units linked together by $\beta(1\rightarrow4)$ bonds (see Figure 3-22). Because of its presence in cell walls, cellulose is the most abundant organic macromolecule on Earth. Cellulose molecules are long, ribbonlike structures that are stabilized by intramolecular hydrogen bonds. Those in primary walls have 500–6000 glucose units, while those in secondary walls are generally larger, having some 10,000–14,000 glucose units per molecule. Many such molecules associate laterally and form the **microfibrils** found in cell walls. The cellulose molecules in a microfibril all have the same polarity and are tightly cross-linked by hydrogen bonds between adjacent molecules. Typically, microfibrils from the cell walls of higher plants are about 60–70 molecules thick and have a diameter of about 25 nm, which makes them large enough to be seen readily in the electron microscope (Figure 10-12).

Molecules of the Cell Wall Matrix. The matrix of the cell wall in which the cellulose microfibrils are embedded is a complex network of branched polysaccharides and proteins. The two main types of polysaccharides are the *hemicelluloses* and the *pectins*. Most of the proteins are glycoproteins called *extensins*. On a dry-weight basis, hemicelluloses typically make up about 20% of the cell wall, pectins account for another 25–30%, and extensins represent about 10%. The highly branched structures of hemicelluloses and pectins give these molecules the property of trapping water. As a result, the cell wall has a very gel-like nature that makes it resistant to compression.

Despite the name, **hemicelluloses** are chemically and structurally distinct from cellulose. The hemicelluloses are a heterogenous group of polysaccharides, but they all consist of a long, linear chain of a single kind of sugar with short side chains that usually contain several different kinds of sugars, including the hexoses glucose, galactose, and fucose and the pentoses xylose and arabinose (Figure 10-13a). The sugar units in the backbone form hydrogen bonds along the surface of the cellulose microfibrils.

Depending on the species in which they are found, hemicelluloses have as their backbone a $\beta(1{\rightarrow}4)$-linked polymer of either glucose or xylose. In monocotyledonous plants, the main hemicellulose is a polymer of xylose called *xylan*. To this backbone are attached side chains that contain

xylose and arabinose as well as galactose. In dicotyledonous plants, on the other hand, the main hemicellulose is *xyloglucan*, which is a linear polymer of glucose with side chains that consist primarily of xylose but also contain galactose, arabinose, and fucose. Figure 10-13b depicts the proposed structure for a segment of the xyloglucan from the cell wall of a typical dicotyledonous plant (Figure 10-13c).

Pectins are also branched polysaccharides, but with backbones that consist mainly of negatively charged galacturonic acid units, mixed in some cases with rhamnose (Figure 10-14a, page 290). The side chains attached to the backbone contain some of the same monosaccharides found in hemi-celluloses, including glucose, galactose, xylose, arabinose, and fucose. Shown in Figure 10-14b is the proposed structure for a segment of the *rhamnogalacturonan* from the cell wall of a dicot. The strings of galacturonic acid units are interrupted at intervals by rhamnose units, which introduce kinks into the molecule and serve as attachment sites for *neutral pectins* that cross-link rhamnogalacturonan molecules to hemicelluloses.

Because of their highly branched structure and their negative charge, pectins such as rhamnogalacturonan trap and bind water molecules. As a result, pectins have a gel-like consistency that ranges from very fluid to very rigid, depending on the chemical structure and physical properties of the specific pectin molecules that are present. (It is because of their gel-forming capacity that pectins are added to fruit juice in the process of making jelly.) Negatively charged pectins also bind cations avidly. Calcium is an especially important cation because it cross-links pectin molecules to other components of the cell wall.

In addition to hemicelluloses and pectins, cell walls also contain glycoproteins called **extensins.** The polypeptide backbone of the extensins is rich in the amino acids serine, lysine, and hydroxyproline. In fact, hydroxyproline accounts for up to 30% of the amino acids in a typical extensin molecule. Because of the high proportion of lysine, extensin molecules have a net positive charge and therefore a high affinity for the negatively charged pectin molecules.

Extensins contain numerous short oligosaccharides, most of which are attached to the hydroxyl groups of serine and hydroxyproline. As shown in Figure 10-15 (page 290), oligosaccharide side chains consist primarily of 1–4 units of arabinose or galactose. Although short, the oligosaccharide side chains are so numerous that they account for about two-thirds of the extensin molecule by weight.

Despite the name, extensins are not actually very extensible or flexible; they are rigid, rodlike molecules that are tightly woven into the complex polysaccharide network of the cell wall. In fact, they are so integral a part of the cell wall matrix that attempts to extract them chemically usually result in the loss of cell wall structure. Extensins are least abundant in the cell walls of actively growing tissues and most abundant in the cell walls of tissues that provide mechanical support to the plant. Extensins are initially de-

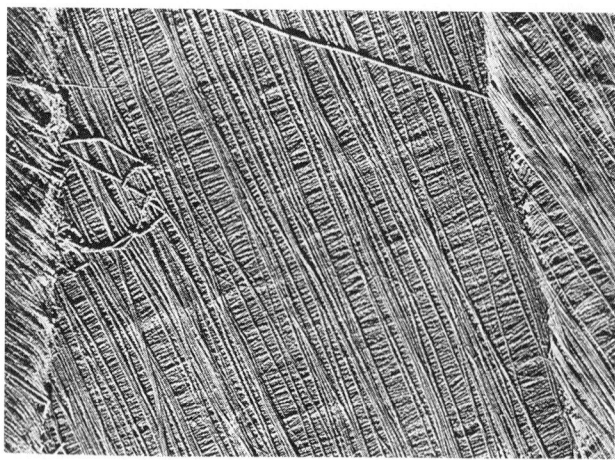

0.25 µm

Figure 10-12 The Structure of Cellulose Microfibrils. This electron micrograph shows individual cellulose microfibrils in the cell wall of a green alga. Each microfibril has a diameter of about 25 nm and consists of many cellulose molecules aligned laterally (TEM).

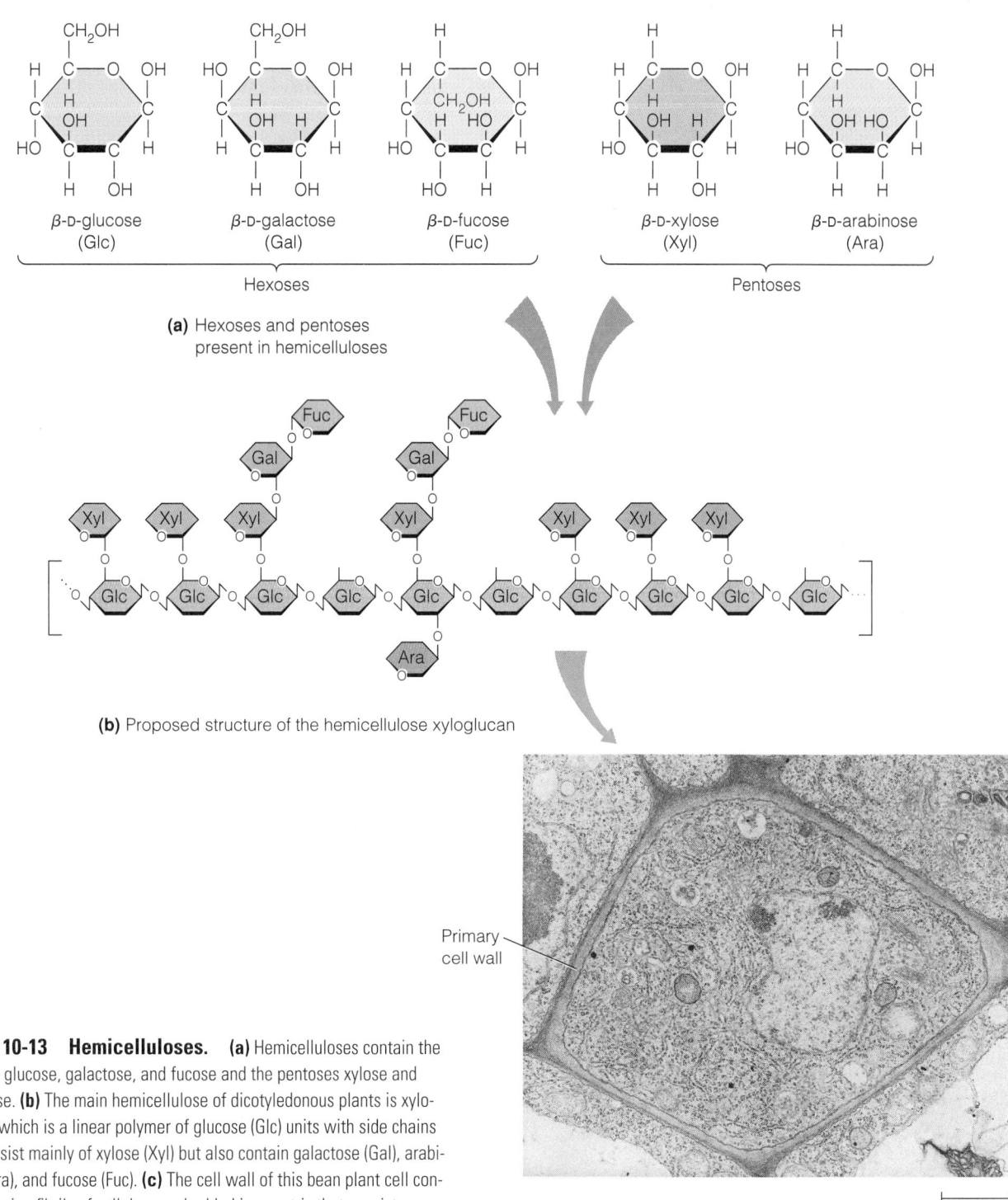

(a) Hexoses and pentoses present in hemicelluloses

(b) Proposed structure of the hemicellulose xyloglucan

Primary cell wall

(c) Xyloglucans are found in the cell walls of dicots, such as this bean cell

1 μm

Figure 10-13 Hemicelluloses. **(a)** Hemicelluloses contain the hexoses glucose, galactose, and fucose and the pentoses xylose and arabinose. **(b)** The main hemicellulose of dicotyledonous plants is xyloglucan, which is a linear polymer of glucose (Glc) units with side chains that consist mainly of xylose (Xyl) but also contain galactose (Gal), arabinose (Ara), and fucose (Fuc). **(c)** The cell wall of this bean plant cell consists of microfibrils of cellulose embedded in a matrix that consists mainly of xyloglucans (TEM).

posited in the cell wall in a soluble form. Once deposited, however, extensin molecules become covalently cross-linked to one another (primarily through tyrosine side groups) and to lignin.

Lignin is the final cell wall component we will consider here. As noted earlier, lignins are found primarily in the sec-ondary cell walls of mature tissue. They are an especially prominent component of wood (the Latin word for which is *lignum*). The lignins are very insoluble polymers of aromatic alcohols that form an extensive cross-linked network with other cell wall constituents, especially the extensins. Three of the major alcohols found in lignin are *p-coumaryl alcohol,*

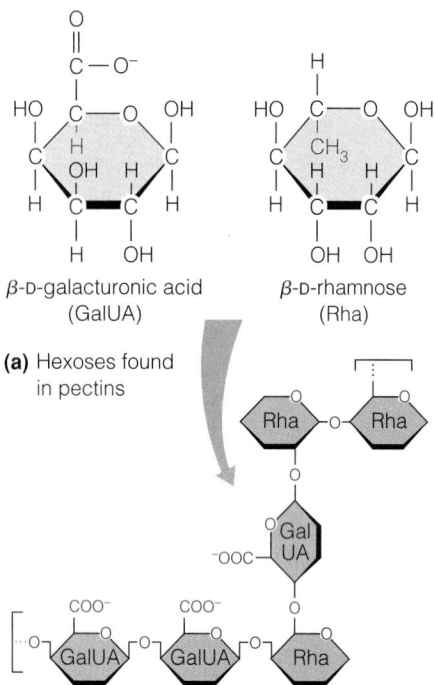

(a) Hexoses found in pectins

(b) Proposed structure of the pectin rhamnogalacturonans

Figure 10-14 Pectins. **(a)** Pectins consist primarily of galacturonic acid and rhamnose. In each case, the form of the sugar present in hemicellulose is the β-D isomer. **(b)** Most pectins have a backbone of galacturonic acid and rhamnose units and are therefore called rhamnogalacturonans. Each rhamnose introduces a kink in the molecule and serves as a potential cross-linking site. Not shown in the diagram are the numerous side chains consisting of glucose, galactose, xylose, arabinose, and fucose that are attached to the backbone.

coniferyl alcohol, and *sinapyl alcohol* (Figure 10-16a). Coniferyl alcohol is present in the lignins of gymnosperms, and the other two alcohols are characteristic of angiosperm lignins. These alcohols are deposited in the cell wall and become linked covalently by the action of the enzyme peroxidase (Figure 10-16b), forming large networks of insoluble polymers and giving the tissue the structural strength we associate with wood.

The Cell Wall as a Permeability Barrier.   The presence of a wall around a plant cell poses a significant permeability barrier, though only for large molecules. For water, gases, ions, and small water-soluble molecules, the cell wall has little or no resistance; ions and small molecules diffuse through the wall readily. In fact, the plasma membrane is usually more of a barrier than the cell wall to the passage of these substances. For macromolecules, however, the cell wall is a significant barrier. In a typical cell wall, the diameter of the pores is about 5 nm, big enough to allow the passage of globular molecules with molecular weights up to about 20,000. The passage of molecules larger than this size range is greatly or even completely impeded. The traffic into and out of cells is therefore restricted to ions and small water-soluble molecules. Not surprisingly, the hormones that serve as intercellular signals in plants are without exception small molecules, with molecular weights well below 1000.

Plasmodesmata: Bridging the Barrier

Recognizing that every plant cell is surrounded by a plasma membrane and a cell wall with pores no more than 5 nm in diameter, you may think it hopeless to expect any sort of in-

Figure 10-15 The Side Chains of Extensin.   An extensin molecule consists of a polypeptide backbone that is rich in the amino acids serine (Ser), lysine (Lys), and hydroxyproline (HPro). Linked to the serine and hydroxyproline side chains are numerous short oligosaccharides consisting of 1–4 units of arabinose (Ara) or galactose (Gal).

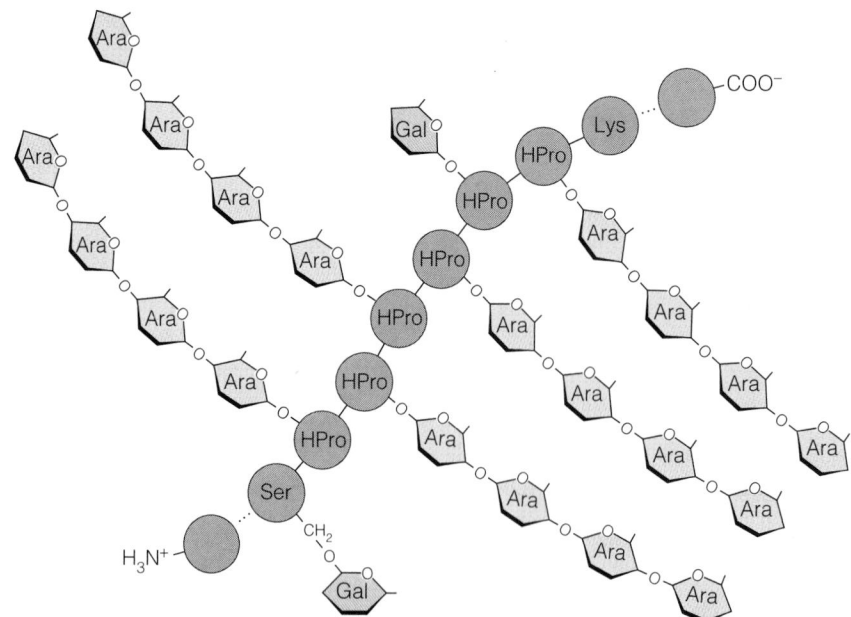

Figure 10-16 The Alcohols Present in Lignins. (a) Lignins are highly insoluble polymers of aromatic alcohols derived from the amino acid phenylalanine. The three main alcohols are p-coumaryl alcohol, coniferyl alcohol, and sinapyl alcohol. Within the cell wall, these aromatic alcohols polymerize to form extensively cross-linked networks. (b) An example of the numerous reactions that lead to the formation of lignin polymers and cross-links is the oxidative reaction between the alcohol group of one alcohol molecule and the double bond of another molecule. This reaction requires hydrogen peroxide and is catalyzed by the enzyme peroxidase. The reaction mechanism is more complicated than shown here; peroxidase actually removes hydrogens from alcohol groups to form very reactive radicals, which then react to form the covalent bonds that link alcohol units into polymers.

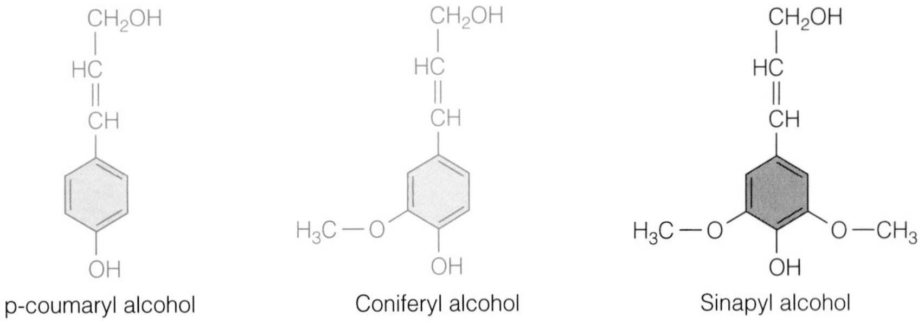

(a) Structures of the main alcohols present in lignins

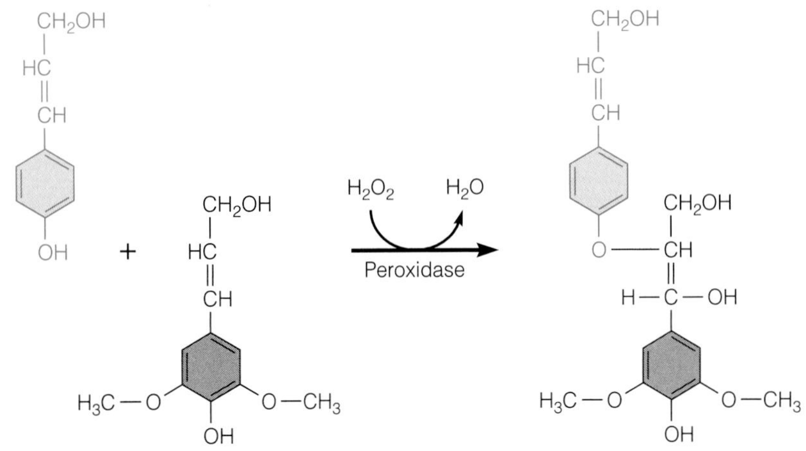

(b) Example of the linkage of these alcohols by peroxidase

tercellular communication such as that afforded by the gap junctions of animal cells. But plasmodesmata accomplish this very purpose. As shown in Figure 10-17, **plasmodesmata** are cytoplasmic channels through relatively large pores in the cell wall, allowing fusion of the plasma membranes from two adjacent cells. Each plasmodesma is therefore lined with plasma membrane common to the two connected cells. A plasmodesma is cylindrical in shape, with the cylinder narrower in diameter at both ends. The channel diameter varies from about 20 to about 200 nm. A single tubular structure, the **desmotubule,** usually lies in the central channel of the plasmodesma. Endoplasmic reticulum (ER) cisternae often approach the plasmodesmata on either side of the cell wall and may actually make connections between cells, possibly by means of the desmotubule.

The ring of cytoplasm between the desmotubule and the membrane that lines the plasmodesma is called the **annulus.** The annulus is thought to provide cytoplasmic continuity between adjacent cells, thereby allowing molecules to pass from one cell to the next without the stringent size limits that are otherwise set by the cell wall. The plasmodesmata

therefore provide for continuity of the plasma membrane, the ER, and the cytoplasm between adjacent cells.

Even after cell division and deposition of new cell walls between the two daughter cells, cytoplasmic continuities are maintained between the daughter cells by plasmodesmata that pass through the newly formed walls. In fact, most plasmodesmata are formed at the time of cell division, when the new cell wall is being formed. Minor changes may occur later, but the number and location of plasmodesmata are largely fixed at the time of division.

In many respects, plasmodesmata appear to be similar to gap junctions in function. They reduce electrical resistance between adjacent cells by about fiftyfold, compared to cells that are completely separated by plasma membranes. When studied experimentally, the movement of ions between adjacent cells (measured as current flow) is proportional to the number of plasmodesmata that connect the cells. As noted earlier for gap junctions, calcium appears to regulate movement through the plasmodesmata, because traffic between cells can be decreased significantly by the injection of Ca^{2+} into cells.

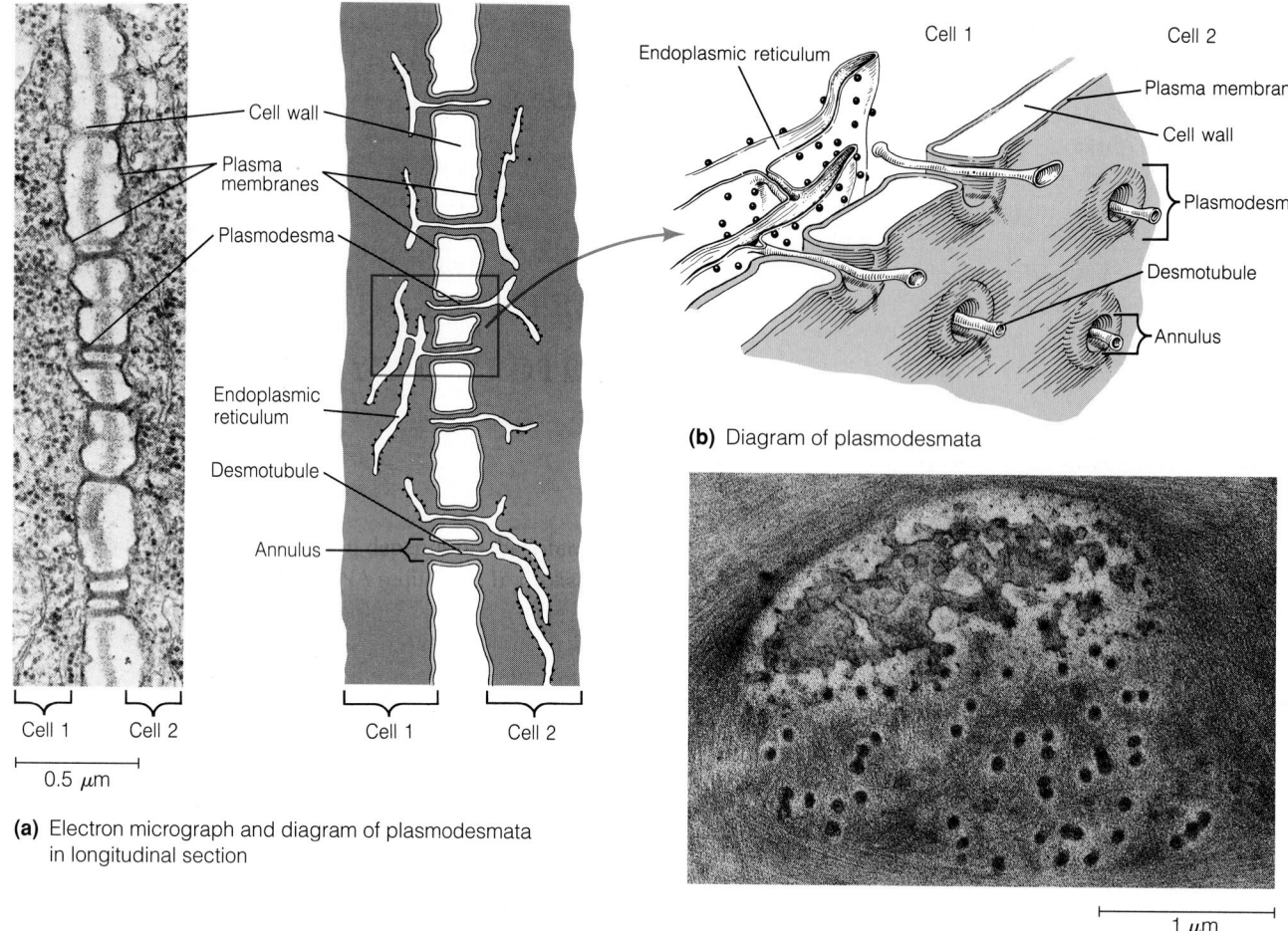

(a) Electron micrograph and diagram of plasmodesmata in longitudinal section

(b) Diagram of plasmodesmata

(c) Electron micrograph of plasmodesmata in cross section

Figure 10-17 Plasmodesmata. A plasmodesma is a channel or pore through the cell wall between two adjacent plant cells, allowing cytoplasmic exchange between the cells. The plasma membrane of one cell is continuous with that of the other cell at each plasmodesma. Most plasmodesmata have at the center a narrow cylindrical desmotubule derived from the ER that appears to be continuous with the ER of both cells. Between the desmotubule and the plasma membrane that lines the plasmodesma is a narrow ring of cytoplasm called the annulus. **(a)** This electron micrograph and diagram show the cell wall between two adjacent root cells of timothy grass, with numerous plasmodesmata (TEM). **(b)** A diagrammatic view of a cell wall with numerous plasmodesmata. **(c)** This electron micrograph shows many plasmodesmata in cross section (TEM).

PERSPECTIVE

Most of the cells in multicellular organisms are in close and ongoing association with neighboring cells. The junctions that link animal cells together are of three general types. Adhesive junctions hold cells together in fixed positions within tissues. Tissues such as skin, heart muscle, and the tissue at the neck of the uterus that must withstand considerable mechanical stress have especially prominent adhesive junctions called desmosomes. Tight junctions function mainly as seals between two cell-lined compartments, preventing the leakage of fluid (such as gastric juice, urine, or blood) and the molecules and ions in the fluid. The membranes of cells linked by tight junctions are fused along sharply defined ridges called tight junction elements. Gap junctions form open channels between cells, allowing direct chemical and electrical communication between cells. By using dye molecules and ions of different sizes, researchers have shown that gap junctions allow the passage of solutes with molecular weights up to about 1200. This direct communication enables the cells of your heart to beat in unison.

Both plant and animal cells have extracellular structures that consist of long, rigid fibers embedded in an amorphous, hydrated matrix of branched molecules. In the case of animal cells, the extracellular matrix (ECM) consists of collagen (and elastin) fibers embedded in a network of proteoglycans, especially glycosaminoglycans. The ECM is held in place by proteoglycans that are themselves integral membrane proteins and/or are linked covalently to membrane phospholipids. In addition, proteoglycans or the collagen may bind to specific receptor proteins on the outer surface of the plasma membrane. These direct links are reinforced by adhesive glycoproteins, especially fibronectins and laminin, that bind to receptor glycoproteins called integrins.

For plants, the primary cell wall consists mainly of cellulose fibers embedded in a complex network of hemicelluloses and pectins. The secondary cell wall that forms as a cell reaches its final size and shape is reinforced with lignins, a major component of wood. Plasmodesmata are membrane-lined cytoplasmic channels between adjacent plant cells that allow chemical and electrical communication rather like that facilitated by the gap junctions between animal cells. The flow of molecules and ions between cells appears to be regulated in both plasmodesmata and gap junctions.

KEY TERMS FOR SELF-TESTING

Cell Junctions

cell junction (p. 271)
adhesive junction (p. 271)
desmosome (p. 271)
desmosome core (p. 271)
plaque (p. 271)
tonofilament (p. 272)
adherens junction (p. 272)
adhesion belt (p. 272)
actin microfilament (p. 272)
tight junction (p. 272)
tight junction element (p. 275)
transcellular transport (p. 275)
gap junction (p. 275)
connexon (p. 275)

Beyond the Membrane: Extracellular Structure

extracellular structure (p. 276)

Animal Cells: the Extracellular Matrix

extracellular matrix (ECM) (p. 278)
collagen (p. 278)
tropocollagen (p. 278)
α chain (p. 281)
elastin (p. 281)
proteoglycan (p. 281)
glycosaminoglycan (p. 282)
adhesive glycoprotein (p. 284)
fibronectin (p. 284)
laminin (p. 285)
basal lamina (p. 285)
integrin (p. 285)
fibronectin receptor (p. 286)

Plant Cells: The Cell Wall

cell wall (p. 287)
primary cell wall (p. 287)

secondary cell wall (p. 287)
cellulose (p. 287)
microfibril (p. 287)
hemicellulose (p. 288)
pectin (p. 288)
extensin (p. 288)
lignin (p. 289)
plasmodesma (p. 291)
desmotubule (p. 291)
annulus (p. 291)

Box 10A: Bacterial Cell Walls, the Gram Stain, Penicillin, and Tears

bacterial cell wall (p. 279)
Gram stain (p. 279)
peptidoglycan (p. 279)

PROBLEM SET

10-1. Cellular Junctions and Plasmodesmata. Indicate whether each of the following statements is true of adhesive junctions (A), tight junctions (T), gap junctions (G), and/or plasmodesmata (P). A given statement may be true of any, all, or none of these structures.

(a) Associated with filaments that confer either contractile or tensile properties.

(b) Sites of true membrane fusion are restricted to abutting ridges of adjacent membranes.

(c) Involve peptidoglycan chains with *N*-acetylglucosamine repeating units.

(d) Contain hexagonal particles with a central opening or core.

(e) Membranes of two adjacent cells sealed tightly together.

(f) Allow the exchange of metabolites between the cytoplasms of two adjacent cells.

(g) Found in animal cells but not in plant cells.

10-2. The Function of the Junction: Completing the Thought. Complete each of the following statements about cell junctions in 25 words or less.

(a) Tight junctions are prominent features of organs such as the stomach and kidneys because . . .

(b) Adhesive junctions are prominent features of epithelial tissue because . . .

(c) Gap junctions are prominent features of muscle tissue because . . .

(d) Of the three major types of animal cell junctions, the plasmodesmata of plant tissue are most like the . . .

10-3. Junction Proteins. Indicate whether each of the following proteins or structures is a component of adhesive junctions (A), tight junctions (T), gap junctions (G), or plasmodesmata (P), and briefly describe the role the protein plays in the junction.

(a) connexin **(c)** desmocollins **(e)** desmoplakins

(b) vinculin **(d)** desmotubule **(f)** annulus

10-4. Beyond the Membrane: ECM and Cell Walls. Compare and contrast the extracellular material (ECM) of animal cells with the walls around plant cells.

(a) What basic organizational principle underlies both ECM and cell walls?

(b) What are the chemical constituents in each case?

(c) What functional roles do the ECM and cell wall share in common?

(d) What functional roles are unique to the ECM? To the cell wall?

10-5. Holding the ECM in Place. According to our current understanding, the ECM around animal cells is held in place by several different kinds of protein-mediated linkages.

(a) What are the two main kinds of linkages? How could you distinguish between them experimentally, based on the proteins involved?

(b) What is the role of the adhesive glycoproteins?

(c) Briefly explain how the various domains of the fibronectin molecule (see Figure 10-9) or the laminin molecule (see Figure 10-10) were identified experimentally.

10-6. Plant Cell Walls. Distinguish between the terms in each of the following pairs with respect to the structure of the plant cell wall, and briefly indicate the significance of each.

(a) Primary wall; secondary wall

(b) Cellulose; hemicellulose

(c) Xylan; xyloglucan

(d) Extensin; lignin

(e) Desmotubule; annulus

SUGGESTED READING

Cell Junctions

Bennett, M., and D. Spray, eds. *Gap Junctions*. Cold Spring Harbor, NY: Cold Spring Harbor Laboratory, 1985.

Da Silva, P. P., and B. Kachar. On tight-junction structure. *Cell* 28 (1982): 441.

Evans, W. H. Communication between cells. *Nature* 283 (1980): 521.

Hertzberg, E., T. S. Lawrence, and N. B. Gilula. Gap junctional communication. *Annu. Rev. Physiol.* 43 (1981): 479.

Robards, A. W., and W. J. Lucas. Plasmodesmata. *Annu. Rev. Plant Physiol. Plant Mol. Biol.* 41 (1990): 585.

Unwin, P. N. T., and G. Zampighi. Structure of the junction between communicating cells. *Nature* 283 (1980): 545.

The Extracellular Matrix of Animal Cells

Burgeson, R. E. New collagens, new concepts. *Annu. Rev. Cell Biol.* 4 (1988): 551.

Caplan, A. I. Cartilage. *Sci. Amer.* 251 (October 1984): 84.

Fransson, L.-A. Structure and function of cell-associated proteoglycans. *Trends Biochem. Sci.* 12 (1987): 406.

Hay, E. D., ed. *Cell Biology of Extracellular Matrix*. New York: Plenum, 1981.

Hay, E. D. Extracellular matrix. *J. Cell Biol.* 91 (1981): 205s.

Kjellen, L., and U. Lindahl. Proteoglycans: Structures and interactions. *Annu. Rev. Biochem.* 60 (1991): 443.

McDonald, J. A. Extracellular matrix assembly. *Annu. Rev. Cell Biol.* 4 (1988): 183.

Scott, J. E. Supramolecular organization of extracellular matrix of glycosaminoglycans, in vitro and in the tissues. *FASEB J.* 6 (1992): 2639.

van der Rest, M., and R. Garrone. Collagen family of proteins. *FASEB J.* 5 (1991): 2814.

Vuorio, E., and de Crombrugghe. The family of collagen genes. *Annu. Rev. Biochem.* 59 (1990): 837.

Weber, K., and M. Osborn. The molecules of the cell matrix. *Sci. Amer.* 253 (October 1985): 110.

The Plant Cell Wall

Casab, G. I., and J. E. Varner. Cell wall proteins. *Annu. Rev. Plant Physiol.* 329 (1988): 321.

Gunning, B. E. S., and R. L. Overall. *Biosci.* 33 (1983): 260.

Hayashi, T. Zyloglucans in the primary cell wall. *Annu. Rev. Plant Physiol. Plant Mol. Biol.* 40 (1989): 139.

Knox, P. Emerging patterns of organization at the plant cell surface. *J. Cell Sci.* 96 (1990): 557.

McNeil, M., A. G. Darvill, S. C. Fry, and P. Albersheim. Structure and function of the primary cell walls of plants. *Annu. Rev. Biochem.* 53 (1984): 625.

Robards, A. W., and W. J. Lucas. Plasmodesmata. *Annu. Rev. Plant Physiol. Plant Mol. Biol.* 41 (1990): 369.

Varner, J. E., and L.-S. Lin. Plant cell wall architecture. *Cell* 56 (1989): 231.

The Bacterial Cell Wall

Beveridge, T. J., and L. L. Graham. Surface layers of bacteria. *Microbiol. Rev.* 55 (1991): 684.

Ferris, F. G., and T. J. Beveridge. Functions of bacterial cell surface structures. *Biosci.* 35 (1985): 172.

Scherrer, R. Gram's staining reaction, gram types and cell walls of bacteria. *Trends Biochem. Sci.* 9 (1984): 242.

Schockman, G. D., and J. F. Barnett. Structure, function, and assembly of cell walls of gram-positive bacteria. *Annu. Rev. Microbiol.* 37 (1983): 27.

ENERGY FLOW IN CELLS

CHEMOTROPHIC ENERGY METABOLISM: GLYCOLYSIS AND FERMENTATION

I n this and the following two chapters, we will be discussing energy flow in cells, focusing especially on the sources of energy available to the biological world and the processes of energy conversion and utilization in cells. As we do so, we will draw on foundations that have been laid in previous chapters. In Chapter 5, for example, we considered the thermodynamic principles that govern all reactions and processes in cells. We saw that $\Delta G'$, the free energy change, determines in which direction a reaction will proceed and how much free energy can be derived from (or must be put into) the system during the reaction.

In Chapter 6, we learned that almost all reactions in cells are catalyzed by enzymes and that the rate of an enzyme-catalyzed reaction depends on a variety of factors, including temperature, pH, and the concentrations of substrates and products. In Chapters 7 and 8, we encountered the membranes that define cell boundaries and compartments, and we learned how energy is used to transport various solute molecules across those boundaries. And in Chapters 9 and 10, we considered a variety of cellular processes and reactions that also require energy.

Still unanswered, however, is the question of where the energy comes from and how cells gain access to it. We will answer this question by looking at the ways in which cells obtain the energy they need to drive cellular reactions and processes. In this chapter and the next, we will consider how *chemotrophs* obtain energy from the food they engulf or ingest, focusing especially on the oxidative breakdown of sugar molecules. Then, in Chapter 13, we will discuss the process by which *phototrophs,* such as the algae and green plants, tap the solar radiation that is the ultimate energy source for almost all living organisms.

Metabolic Pathways

When we encountered enzymes in Chapter 6, we considered only individual chemical reactions catalyzed by individual enzymes functioning in isolation. But that is not the way

cells operate. To accomplish any major task, a cell requires a series of reactions occurring in an organized sequence. This, in turn, requires many enzymes because most enzymes catalyze only a single reaction, and many such reactions are usually needed to accomplish a major biochemical operation.

When we consider all of the chemical reactions that occur within a cell, we are talking about **metabolism.** The overall metabolism of a cell consists, in turn, of a large number of specific **metabolic pathways,** each of which accomplishes a particular task. From a biochemist's perspective, life at the cellular level can be defined as a network of integrated and carefully regulated metabolic pathways, each contributing to the sum of activities a cell must carry out.

Metabolic pathways are of two general types. Pathways concerned with the synthesis of cellular components are called **anabolic pathways** (from the Greek prefix *ana-,* meaning "up"), whereas those involved in the breakdown of cellular constituents are called **catabolic pathways** (from the Greek prefix *kata-,* meaning "down"). Anabolic pathways usually involve a substantial increase in molecular order (and therefore a local decrease in entropy) and are *endergonic* (energy-requiring). Catabolic pathways, by contrast, are *exergonic* (energy-liberating) because they involve a decrease in atomic order (increase in entropy). Catabolic pathways play two roles in cells: They release energy, and they give rise to the small organic molecules, or *metabolites,* that are the building blocks for biosynthesis. The energy is needed in part to drive anabolic pathways in which specific metabolites are used to synthesize the macromolecules and other cellular components needed for cell structure and function.

As we will see shortly, catabolism can be carried out either in the presence of oxygen (*aerobic* conditions) or in the absence of oxygen (*anaerobic* conditions). The energy yield is much greater in the presence of oxygen, which probably explains the preponderance of aerobic organisms in the world. However, anaerobic catabolism is also important, both for organisms in environments that are always devoid of oxygen and for organisms and cells that are temporarily deprived of oxygen.

ATP: The Universal Energy Coupler

The anabolic reactions of cells are responsible for the growth and repair processes characteristic of all living systems, while catabolic reactions release the energy needed to drive the anabolic reactions and to carry out other kinds of cellular work. The efficient linking, or *coupling*, of energy-yielding processes to energy-requiring processes is therefore crucial to cell function. This coupling is made possible by specific kinds of molecules that conserve the energy derived from exergonic reactions, which is then available whenever and wherever energy is needed. In the biological world, the molecule that is used most commonly as an energy intermediate is the phosphorylated compound **adenosine triphosphate (ATP)**. ATP is, in other words, the energy currency in the biological realm. Because ATP is involved in almost every cellular energy transaction, it is essential that we understand its structure and function.

ATP Structure and Function

Recall from Chapter 3 that ATP is a complex molecule containing the aromatic base *adenine*, the five-carbon sugar *ribose*, and a chain of three phosphate groups linked to each other by **phosphoanhydride bonds** and to the ribose by a **phosphoester bond** (see Figure 3-13). The compound formed by the linking of adenine and ribose is called *adenosine*. Adenosine may occur in the cell in the unphosphorylated form or with one, two, or three phosphates attached to carbon atom 5 of the ribose, forming *adenosine monophosphate* (*AMP*), *diphosphate* (*ADP*), and *triphosphate* (*ATP*), respectively.

The ATP molecule is well suited for its role as an intermediate in cellular energy metabolism because of the unstable nature of the phosphoanhydride bond that links the third (outermost) phosphate to the second. (The bond between the first and second phosphates has the same properties and is important in some reactions, but at present we are mainly interested in the terminal phosphoanhydride bond.) To say that this bond is *unstable* means that its hydrolysis is an exergonic reaction. (*Hydrolysis* is the reaction whereby a bond in a molecule is broken by reaction with water. Two products are formed; one receives an —H from the water molecule and the other gets an —OH group. In the case of ATP, the hydrolysis reaction can be catalyzed by any of several *ATPases* in the cell.) Thus, the reaction of ATP with water to form ADP and inorganic phosphate (P_i) is exergonic, with a standard free energy change ($\Delta G^{\circ\prime}$) of about -7.3 kcal/mol (Figure 11-1, reaction 1). The reverse reaction, whereby ATP is synthesized from ADP and P_i with the loss of a water molecule, is correspondingly endergonic, with a $\Delta G^{\circ\prime}$ of about $+7.3$ kcal/mol (Figure 11-1, reaction 2).

Biochemists sometimes refer to bonds such as the phosphoanhydride bonds of ATP as "high-energy" or "energy-rich" bonds, a convention introduced in 1941 by Fritz Lipmann, a leading bioenergetics researcher of the time. Unfortunately, this terminology often conveys the erroneous impression that the bond somehow "contains" energy that can be "released." In fact, of course, all chemical bonds *require* energy to be broken and *release* energy when they form. What we really mean by "high-energy bond" is that free energy is released when the bond is hydrolyzed. And that, in turn, means that more energy is released as the new bonds are formed, between the —H and —OH groups of the water molecule and the two products of the reaction, than is required to break the so-called "high-energy bond." The energy is therefore a characteristic of the reaction in which the molecule is involved and *not* of a particular bond

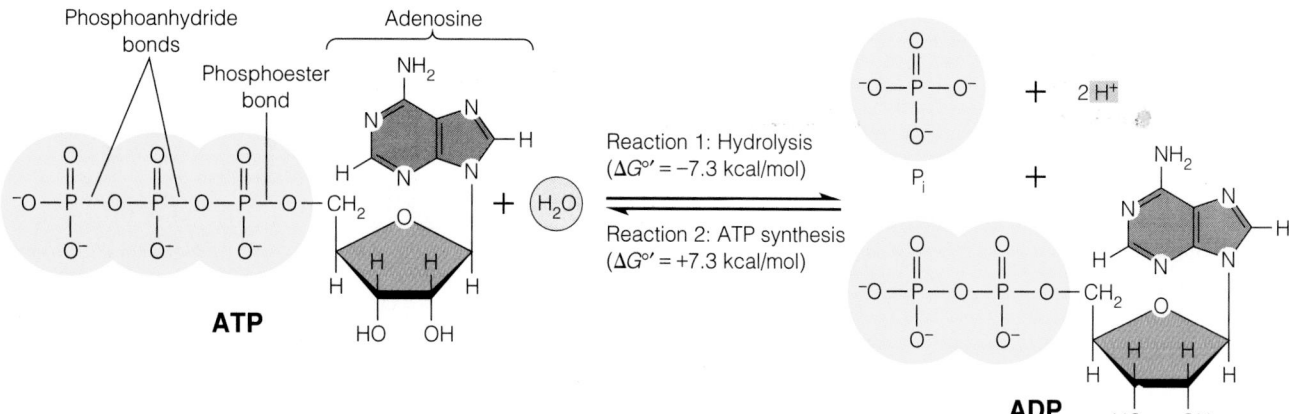

Figure 11-1   Hydrolysis and Synthesis of ATP. ATP is the most common intermediate in cellular energy transactions. Reaction 1: The hydrolysis of ATP to ADP and inorganic phosphate (P_i) is a highly exergonic reaction, with a standard free energy change of about -7.3 kcal/mol. (The two protons shown as products of the reaction result from the ionization of the ADP and P_i after hydrolysis.) Reaction 2: The synthesis of ATP by phosphorylation of ADP is a highly endergonic reaction, with a standard free energy change of about $+7.3$ kcal/mol.

within that molecule. Or, to borrow words from J. C. Morris, "It is about as sensible to try to identify a 'high-energy bond' in a 'high-energy compound' as it is to attempt to attribute the tone of a violin to any one of its structural members." Thus, to call ATP or any other molecule a "high-energy compound" should always be understood as a shorthand way of saying that the molecule possesses one or more bonds—the phosphoanhydride bonds, in the case of ATP— the hydrolysis of which is strongly exergonic.

What is it about the ATP molecule that makes the hydrolysis of its phosphoanhydride bonds so exergonic? The answer to this question has two parts: Hydrolysis of ATP to ADP is exergonic because of charge repulsion between the adjacent negatively charged phosphate groups and because of resonance stabilization of both products of hydrolysis (ADP and inorganic phosphate).

Charge repulsion is easy to understand. By way of analogy, imagine holding two magnets together with like poles touching. The like poles repel each other, and you need to make an effort (put energy into the system) to force them together. If you let go, the magnets spring apart, releasing the energy. Now consider the three phosphate groups of ATP. As Figure 11-1 shows, each group bears at least one negative charge at the near-neutral pH of the cell, and these negative charges tend to repel one another, thereby straining the covalent bond linking the phosphate groups together. Because of this strain, less energy is needed to break the bond than would otherwise be required.

A second, even more important contribution to ATP bond energy is **resonance stabilization.** To understand this phenomenon, we need to realize that groups like the carboxylate or phosphate groups, though formally written with one double bond and one or more single bonds to oxygen, in reality have an unshared electron pair that is delocalized

over all of the bonds to oxygen (Figure 11-2). The true structure of the carboxylate and phosphate groups is actually an average of the contributing structures called a **resonance hybrid,** in which the extra electrons are delocalized over all possible bonds. As Figure 11-2 shows, the resonance hybrid involves delocalization of electrons over two carbon-oxygen bonds in the case of the carboxylate group and over four phosphorus-oxygen bonds in the case of the inorganic phosphate ion (P_i). (The Greek letter δ is used in resonance hybrids to indicate a partial charge on an atom.)

When the electrons of such a compound are maximally delocalized in this way, the molecule is in its most stable (lowest-energy) configuration and is said to be *resonance-stabilized.* Consider now what happens if one (or more) of the oxygen atoms of such an anion becomes involved in a covalent bond to an organic compound: The oxygen atom is no longer available for electron delocalization (Figure 11-3). Now, the extra electrons are delocalized over fewer than the maximum number of oxygen atoms, and the molecule is consequently locked into a higher-energy (less delocalized) configuration. This means that any chemical bond to a carboxylate or phosphate group will result in a compound that is higher in energy than the free carboxylate or phosphate group itself. Specifically, it means that esters, phosphoesters, anhydrides, and phosphoanhydrides are higher-energy compounds than the alcohol, carboxylate, and phosphate compounds they yield upon hydrolysis and will therefore liberate energy if they are hydrolyzed to these products.

For esters, only a moderate amount of energy is liberated upon hydrolysis, because only one of the two products—the acid—can undergo resonance stabilization. For anhydrides, however, both products can be resonance-stabilized, so the hydrolysis reaction is highly exergonic. In addition, both products of anhydride hydrolysis are charged

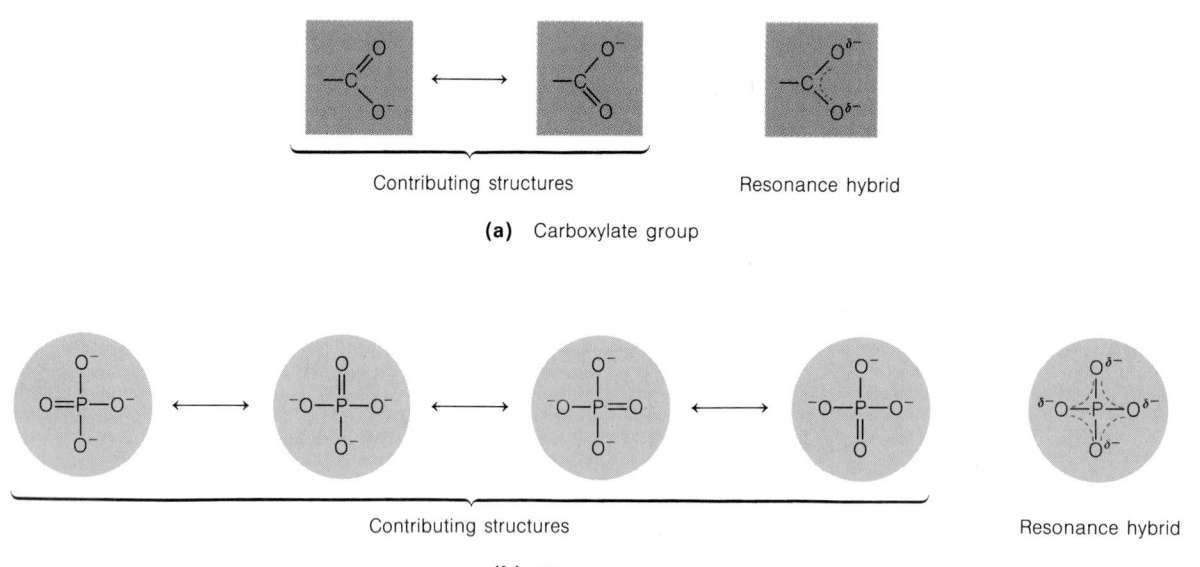

Contributing structures   Resonance hybrid

(a) Carboxylate group

Contributing structures   Resonance hybrid

(b) Phosphate ion

Figure 11-2 Resonance Stabilization. Resonance stabilization is an important feature of both **(a)** the carboxylate group and **(b)** the phosphate ion. In each case, the extra electron pair that is shown as formally localized to one of the oxygen bonds of the contributing structures is in fact delocalized over all available C—O or P—O bonds, as shown in the resonance hybrid structures on the right. (Delocalized electrons are shown by pink dashed lines, and the Greek letter δ indicates a partial negative charge on each of the oxygen atoms.)

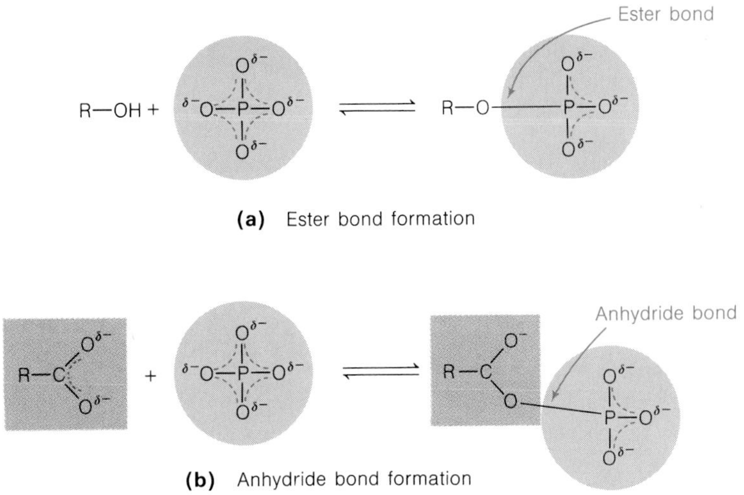

Figure 11-3 Decreased Resonance Stabilization of the Phosphate Group Upon Bond Formation.

Involvement of a phosphate ion in either **(a)** an ester bond or **(b)** an anhydride bond results in decreased opportunity for electron delocalization. As a result, the ester or anhydride product is a higher-energy compound than the phosphate group and the alcohol or carboxylic acid involved in bond formation. An ester bond has a lower free energy of hydrolysis because hydrolysis (reversal of the reaction in part a) results in increased electron delocalization for only one of the two products. An anhydride bond has a higher energy of hydrolysis because hydrolysis (reversal of the reaction in part b) leads to increased electron delocalization for both of the products and relieves charge repulsion as well. (The R refers to the rest of the molecule; only the hydroxyl or carboxyl group is shown.)

(a) Ester bond formation

(b) Anhydride bond formation

and therefore repel each other, which is not the case with ester hydrolysis. In general, esters and phosphoesters have free energies of hydrolysis around -3 to -4 kcal/mol, whereas the hydrolysis of anhydrides and phosphoanhydrides releases roughly twice that much free energy. ATP illustrates this difference well: Hydrolysis of either of the phosphoanhydride bonds that link its phosphate groups together has a standard free energy change of about -7.3 kcal/mol, whereas hydrolysis of the phosphoester bond that links the first (innermost) phosphate group to the ribose has a $\Delta G^{\circ\prime}$ of about -3.6 kcal/mol:

$$\text{ATP} + \text{H}_2\text{O} \longrightarrow \text{ADP} + \text{P}_i \qquad \Delta G^{\circ\prime} = -7.3 \text{ kcal/mol} \tag{11-1}$$

$$\text{ADP} + \text{H}_2\text{O} \longrightarrow \text{AMP} + \text{P}_i \qquad \Delta G^{\circ\prime} = -7.3 \text{ kcal/mol} \tag{11-2}$$

$$\text{AMP} + \text{H}_2\text{O} \longrightarrow \text{adenosine} + \text{P}_i \quad \Delta G^{\circ\prime} = -3.6 \text{ kcal/mol} \tag{11-3}$$

Thus, ATP and ADP are both "higher-energy compounds" than AMP, to use the shorthand of biochemists.

If anything, the $\Delta G^{\circ\prime}$ value of -7.3 kcal/mol underestimates the free energy change associated with the hydrolysis of ATP to ADP under biological conditions. As we learned in Chapter 5, the actual free energy change, $\Delta G^{\prime}$, depends on the prevailing concentrations of reactants and products (see equation 5-21). For the hydrolysis of ATP (reaction 11-1), $\Delta G^{\prime}$ is calculated as

$$\Delta G^{\prime} = \Delta G^{\circ\prime} + RT \ln [\text{ADP}] [\text{P}_i]/[\text{ATP}] \tag{11-4}$$

In most cells, the ATP/ADP ratio is significantly greater than 1:1, often in the range of about 5:1. As a result, $\Delta G^{\prime}$ is more highly negative than -7.3 kcal/mol—usually in the range of -10 to -14 kcal/mol. (Can you use equation 11-4 to convince yourself that $\Delta G^{\prime}$ is in this range when the ATP/ADP ratio is about 5:1? What must you assume about $[\text{P}_i]$?)

ATP as an Intermediate in Cellular Energy Transactions

Although scientists often refer to ATP as a high-energy compound, we really should think of it as an *intermediate-energy* compound because that is where it actually ranks in the overall spectrum of phosphorylated compounds in the cell. Figure 11-4 makes this point clear by ranking some of the more common phosphorylated intermediates in cellular energy metabolism according to their $\Delta G^{\circ\prime}$ values. The values along the y-axis are negative, so the compounds closest to the top of the figure have the most negative $\Delta G^{\circ\prime}$ values for hydrolysis of the phosphate group. This means that, under standard conditions, any compound is capable of exergonically phosphorylating any compound below it (i. e., one that has a less negative $\Delta G^{\circ\prime}$ value) but none of the compounds above it. Thus, ATP can phosphorylate glucose, but not pyruvate. Similarly, ATP can be formed from ADP by the transfer of a phosphate group from phosphoenolpyruvate (PEP), but not from glucose-6-phosphate. The following reactions, in other words, have negative $\Delta G^{\circ\prime}$ values that can be predicted from Figure 11-4 and calculated as shown:

$$\text{glucose} + \text{ATP} \longrightarrow \text{glucose-6-phosphate} + \text{ADP}$$
$$\begin{aligned}\Delta G^{\circ\prime}_{\text{net}} &= \Delta G^{\circ\prime}_{\text{donor}} - \Delta G^{\circ\prime}_{\text{acceptor}} \\ &= -7.3 - (-3.3) = -4.0 \text{ kcal/mol}\end{aligned} \tag{11-5}$$

$$\text{PEP} + \text{ADP} \longrightarrow \text{pyruvate} + \text{ATP}$$
$$\Delta G^{\circ\prime}_{\text{net}} = -14.8 - (-7.3) = -7.5 \text{ kcal/mol} \tag{11-6}$$

As we will discover shortly, reactions 11-5 and 11-6 are the first and final reactions, respectively, in an important catabolic pathway called *glycolysis*, which is the starting point for chemotrophic energy metabolism in virtually all forms of life.

Figure 11-4   Standard Free Energies of Hydrolysis for Some Common Phosphorylated Compounds Involved in Cellular Energy Metabolism. Compounds such as phosphoenolpyruvate, 1,3-bisphosphoglycerate, and phosphocreatine are considered high-energy compounds because hydrolysis of the phosphate group is highly exergonic, with $\Delta G°'$ values more negative than -10 kcal/mol. Compounds such as glucose-1-phosphate, glucose-6-phosphate, and glycerol phosphate are considered low-energy compounds because hydrolysis of the phospho-ester bond is less exergonic, with $\Delta G°'$ values in the range -2 to -5 kcal/mol. ATP is an intermediate-energy compound, with a $\Delta G°'$ value of about -7.3 kcal/mol. Based on $\Delta G°'$ values, the phosphorylated form of any compound can transfer its phosphate group exergonically to the unphosphorylated form of any compound that lies *below* it on the figure (i.e., that has a less negative $\Delta G°'$ value).

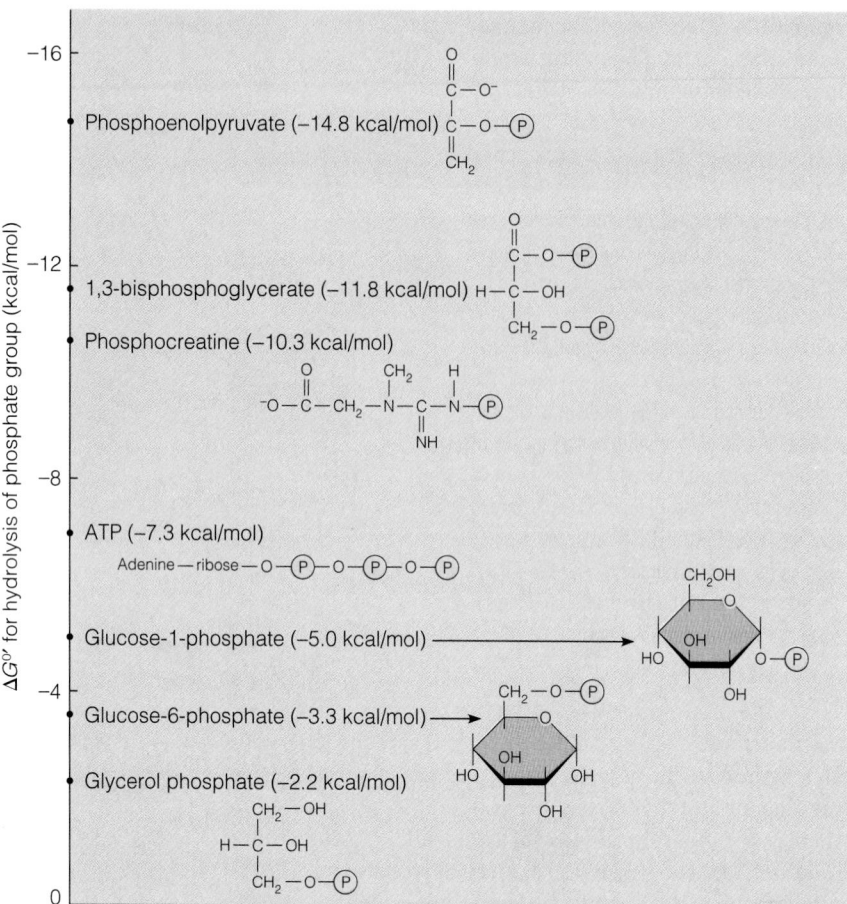

Reactions such as 11-5 and 11-6 involving the movement of a chemical group from one molecule to another are called **group transfer reactions.** Group transfer reactions represent one of the most common processes in cellular metabolism, and the phosphate group is one of the most frequently transferred groups, especially in energy metabolism.

The most important point to understand from Figure 11-4 is that the ATP/ADP pair occupies a crucial *intermediate* position in terms of bond energies. This means that ATP can serve as a phosphate *donor* in biological reactions and that its dephosphorylated form, ADP, can serve as a phosphate *acceptor* because there are compounds both above and below the ATP/ADP pair in energy. Thus, the important role of ATP in energy metabolism really depends on the hydrolysis of its terminal phosphate group being not a high-energy but an intermediate-energy reaction.

In summary, the ATP/ADP pair represents a reversible means of conserving, transferring, and releasing energy within the cell (Figure 11-5). ATP is the "charged," or higher-energy, form, whereas ADP is the "discharged," or lower-energy, form. As catabolic processes occur in the cell, whether anaerobically (Figure 11-5a) or aerobically (Figure 11-5b), the energy-liberating reactions in the sequence are coupled to the ATP/ADP system such that the available free energy drives the formation of ATP from ADP. The energy

of ATP hydrolysis then provides the driving force for the variety of processes (such as biosynthesis, active transport, charge separation, and muscle contraction) that are essential to life and require the input of energy. Thus, we are dealing with an energy-conserving system that is charged during the catabolism of nutrients (or during the photosynthetic trapping of solar energy, as we will see in Chapter 13) and discharged during the performance of cellular work.

Chemotrophic Energy Metabolism

Now we have the essential concepts in hand to take up the main theme of this chapter and the next. We are ready to discuss **chemotrophic energy metabolism**—the reactions and pathways by which cells catabolize nutrients such as carbohydrates, fats, and proteins, conserving as ATP some of the free energy that is released in the process. Or, to put it more personally, we will be looking at the specific metabolic processes by which the cells of your own body make use of the food you eat to meet your energy needs. We begin our discussion by considering *oxidation*, because much of chemotrophic energy metabolism involves oxidative reactions.

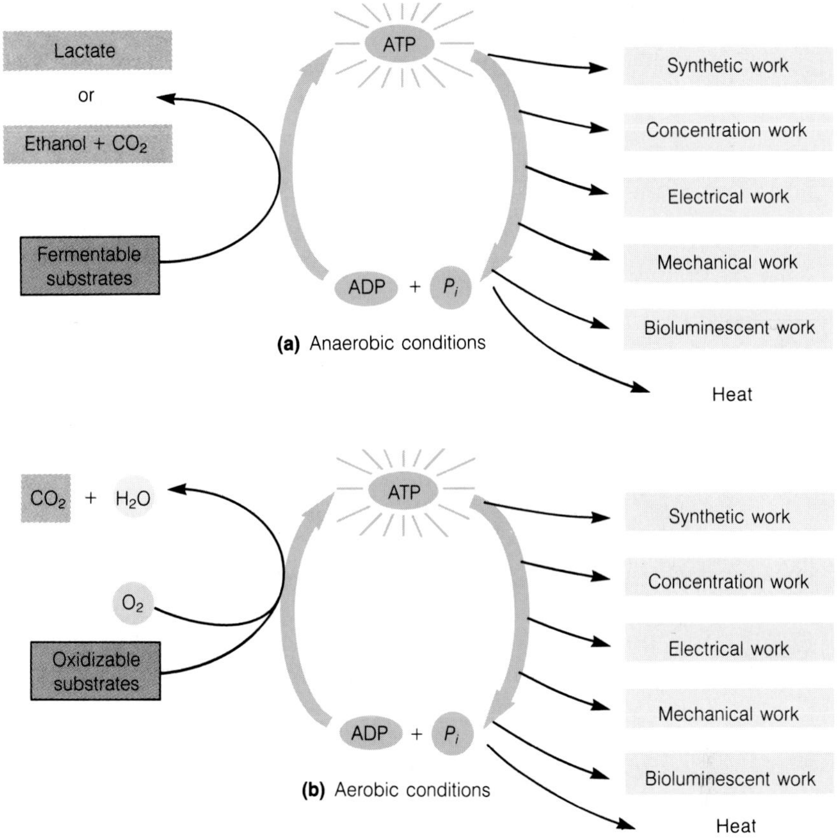

Figure 11-5 The ATP/ADP System as a Means of Conserving and Releasing Energy Within the Cell. ATP is generated during the oxidative catabolism of nutrients (left side) and is used to do cellular work (right side). **(a)** Under anaerobic or hypoxic (oxygen-deficient) conditions, ATP is generated by *fermentation,* with lactate as the most common end-product in some organisms and ethanol plus carbon dioxide as the most common end-products in other organisms. **(b)** Under aerobic conditions, ATP is generated by *respiration,* as oxidizable nutrients are catabolized completely to carbon dioxide and water.

Biological Oxidation

To say that nutrients such as carbohydrates, fats, or proteins are sources of energy for cells is another way of saying that these are *oxidizable organic compounds* and that their oxidation is highly exergonic. Recall from chemistry that **oxidation** is the removal of electrons. Thus, ferrous ion (Fe^{2+}) is oxidizable because it readily gives up an electron:

$$Fe^{2+} \longrightarrow Fe^{3+} + e^- \qquad \textbf{(11-7)}$$

In organic and biological chemistry, the definition is exactly the same. The only difference is that the oxidation of organic molecules frequently involves the removal of both electrons and hydrogen ions (protons), so that the process is also one of **dehydrogenation.** Consider, for example, the oxidation of ethanol to the corresponding aldehyde:

$$CH_2-CH_2-OH \xrightarrow{\text{oxidation}} \underset{\text{Acetaldehyde}}{CH_2-\overset{\overset{\text{H}}{|}}{C}=O} + 2e^- + 2H^+ \quad \textbf{(11-8)}$$

ethanol

Electrons are removed, so this is clearly an oxidation. But protons are liberated as well, and an electron plus a proton is the equivalent of a hydrogen atom. Therefore, what happens, in effect, is the removal of the equivalent of two hydrogen atoms:

$$\underset{\text{ethanol}}{CH_2-CH_2-OH} \xrightarrow[\text{(dehydrogenation)}]{\text{oxidation}} CH_2-\overset{\overset{\text{H}}{|}}{C}=O + 2[H] \quad \textbf{(11-9)}$$

Thus, for cellular reactions involving organic molecules, oxidation is almost always manifested as a dehydrogenation reaction. Many of the enzymes that catalyze oxidative reactions in cells are in fact called *dehydrogenases.*

None of the above reactions can take place in isolation, of course; the electrons must be transferred to another molecule, which is *reduced* in the process. **Reduction is defined as the addition of electrons,** but in biological reactions the electrons are frequently accompanied by protons and the overall reaction is therefore a **hydrogenation:**

$$CH_2-\overset{\overset{\text{H}}{|}}{C}=O + 2[H] \xrightarrow[\text{(hydrogenation)}]{\text{reduction}} CH_2-CH_2-OH \quad \textbf{(11-10)}$$

Reactions 11-9 and 11-10 both illustrate the further general feature that biological oxidation-reduction reactions almost always involve two-electron (and therefore two-proton) transfers.

As written, reactions 11-9 and 11-10 are only *half-reactions,* representing an oxidation and a reduction event, respectively. In real reactions, however, oxidation and reduction always take place simultaneously. Any time an oxidation occurs, a reduction occurs as well because the electrons (and protons) removed from one molecule must be added to another molecule. Despite the way reactions 11-9 and 11-10 are written, hydrogen atoms are never actually released into solution as such but are transferred to another molecule instead.

The Role of Coenzymes in Biological Oxidations

In chemotrophic energy metabolism, the ultimate acceptor of electrons is usually, though not always, oxygen. (We will encounter some alternative acceptors in Chapter 12, including H_2 and H_2S.) Rarely, however, are electrons ever passed directly from an oxidizable substrate to oxygen. Instead, the immediate electron acceptor in most biological oxidations (and the immediate electron donor in most biological reductions) is any of several coenzymes. In general, **coenzymes** are small molecules that function along with enzymes (hence the name), usually by serving as a carrier of electrons or functional groups. As we will see shortly, coenzymes actually function as substrates in the reactions in which they participate. However, they are continuously recycled within the cell, so the relatively low intracellular concentration of a given coenzyme is adequate to meet the needs of the cell for that particular coenzyme.

The most common coenzyme involved in energy metabolism is **nicotinamide adenine dinucleotide (NAD⁺).** Its structure is shown in Figure 11-6. Despite its formidable name and structure, its function is very straightforward. NAD⁺ serves as an electron acceptor by adding two electrons and one proton to a carbon atom on its aromatic ring, thereby generating the reduced form, NADH, with the concomitant release of a proton into the medium:

$$NAD^+ + 2[H] \longrightarrow NADH + H^+ \qquad \textbf{(11-11)}$$

As a nutritional note, the nicotinamide of NAD⁺ is a derivative of *niacin*, which we recognize as a *B vitamin*—one of a family of water-soluble compounds essential to the diet of humans and other vertebrates that cannot synthesize these compounds for themselves. In Chapter 12, we will meet two other coenzymes that also have B vitamin derivatives as part of their structures. It is precisely because these compounds are components of indispensable coenzymes that they are essential in the diet of any organism unable to manufacture its own supply of them.

Glucose as a Substrate

We are interested in oxidation because it is the means by which chemotrophs meet most of their energy needs. Many different kinds of substances serve as substrates for biological oxidation. Some microorganisms can use reduced inorganic compounds (such as reduced forms of iron, sulfur, or nitrogen) as their energy sources. These organisms play important roles in the inorganic economy of the biosphere and utilize rather specialized oxidative reactions. But most chemotrophs depend on organic food molecules, with carbohydrates, fats, and proteins as the three major categories.

To simplify our discussion initially and to provide a unifying metabolic theme, we will concentrate on the biological oxidation of the six-carbon sugar **glucose** ($C_6H_{12}O_6$). Glucose is a good choice for several reasons. In many vertebrates, including humans, glucose is the main sugar in the blood, and hence the main energy source for the organism. Blood glucose comes primarily from dietary carbohydrates such as sucrose or starch and from the breakdown of stored glycogen. Glucose is therefore an especially important molecule for you personally, a point that is underscored in Box 11A.

Glucose is important to plants because it is the monosaccharide released upon starch breakdown and also makes

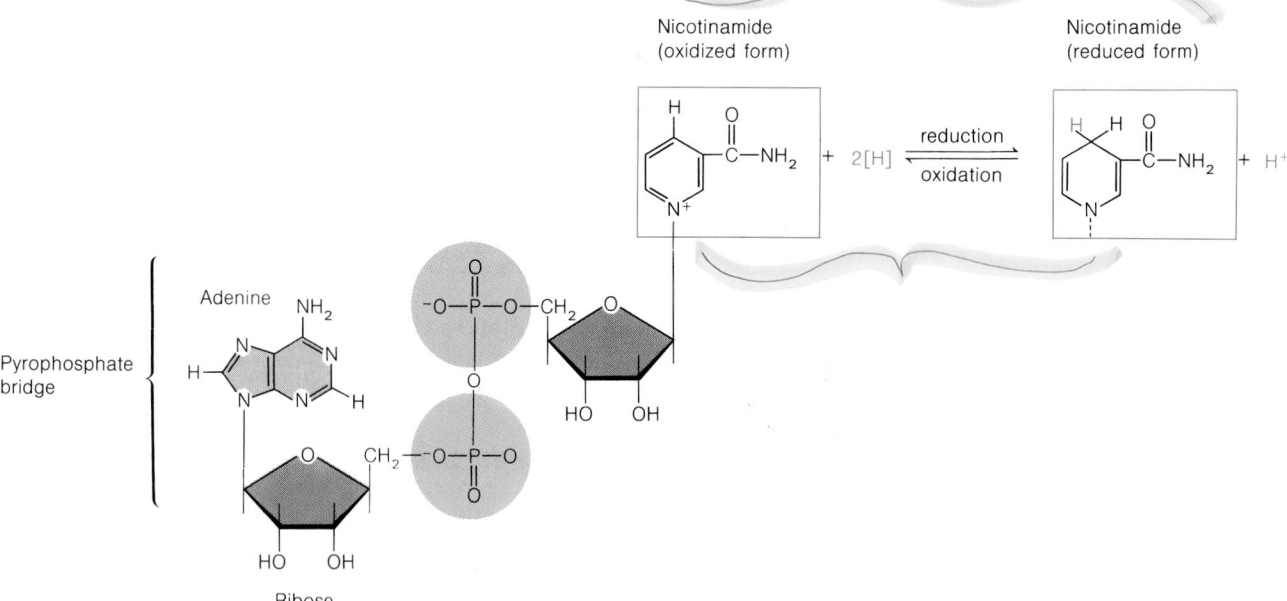

Figure 11-6 The Structure of NAD⁺ and Its Oxidation and Reduction. The portion of the coenzyme outlined in pink is nicotinamide, a B vitamin. The hydrogen atoms derived from an oxidizable substrate are also shown in pink. When NAD⁺ is used as an electron acceptor, two electrons and one proton from the oxidizable substrate are transferred to one of the carbon atoms of nicotinamide, and the other proton is released into solution.

F u r t h e r   I n s i g h t s
"WHAT HAPPENS TO THE SUGAR?"

While it's important to acquire an academic understanding of processes like glycolysis and gluconeogenesis, it is also important to appreciate what that knowledge means for you as a person—or as an organism, to use a biological term. Will you, for example, be able to use the information in this chapter and the next to understand more adequately how your body meets its energy needs and what it does with the nutrients you eat?

To put it more specifically, can you relate what you are learning in these chapters to what the cells in your body are doing with the food you had for breakfast this morning? As you add sugar to your coffee or cereal, can you answer the question, "What happens to the sugar?" This brief essay will address these questions and might help you appreciate the relevance of these chapters if you can see more immediately how the information they contain applies to your daily life.

To keep the topic manageable, let's focus on your bowl of cereal and ask what your body does with the sugar you sprinkle on it, as well as with the sugar in the milk and the carbohydrate in the cereal itself (Figure 11A-1). We are, in other words, considering the disaccharides *sucrose* (from the sugar bowl) and *lactose* (in the milk) and the polysaccharide *starch* in the cereal. First we'll follow the sugars and starch through your digestive tract, then consider the glucose in your blood and the several ways the cells in different parts of your body use it.

Let's start with a spoonful of cereal you've just eaten. The sucrose and lactose remain intact until they reach your small intestine, but the digestion of starch begins in your mouth because saliva contains *salivary amylase,* an enzyme that splits starch into smaller polysaccharides. (Theoretically, amylase can hydrolyze starch to the disaccharide maltose, but food is not usually in the mouth long enough for digestion to proceed that far.) Further digestion occurs in your small intestine, where *pancreatic amylase* completes the breakdown of starch to maltose. (Note, by the way, that the intestinal digestion of starch involves simple hydrolysis of the glycosidic bonds between glucose units. This contrasts with the phosphorolytic cleavage that occurs during

Figure 11A-1   What Happens to the Sugar?   Is this a question you ever ponder as you sprinkle sugar on your breakfast cereal?

glycogen breakdown in liver or muscle cells and during starch breakdown in plants.)

The maltose generated from starch is hydrolyzed to glucose in your intestine by the enzyme *maltase.* Maltase is one of a family of *intestinal disaccharidases,* each specific for a different disaccharide. The lactose from the milk and the sucrose that you sprinkled on the cereal are hydrolyzed by other members of this family—*lactase* and *sucrase,* respectively. Lactose yields one molecule each of glucose and galactose, whereas sucrose is hydrolyzed to one molecule each of glucose and fructose (see Figure 11-11). (In some people, intestinal lactase disappears gradually after age 4 or so, when milk drinking usually decreases. If such people ingest milk or other dairy products, they are likely to experience cramps and diarrhea, a condition called *lactose intolerance.*)

Glucose, galactose, and fructose molecules are absorbed by intestinal epithelial cells. These cells are well suited to the task because of the numerous *microvilli* that project into the lumen of the intestine, thereby greatly increasing the absorptive surface of the cell. Moreover, only two layers of epithelial cells separate nutrients in the lumen of your intestine from the blood in your capillaries. Some sugars, such as fructose, move across the plasma membrane of an epithelial cell by facilitated diffusion, because the concentrations of these sugars are lower in capillary blood than in the intestinal lumen. Glucose, however, is moved by active transport because of its high concentration in blood. (Other, less abundant nutrients such as vitamins and amino acids are also pumped actively from the intestine into the bloodstream.)

Sugars such as fructose and galactose are transported by your bloodstream to the various tissues of your body. These sugars are eventually absorbed by body cells and converted to intermediates in the glycolytic (and therefore also the gluconeogenic) pathway, as shown in Figure 11-11. The pathway for galactose utilization is more complex than that of most other simple sugars, with five reactions required to convert a molecule of galactose into glucose-6-phosphate. (A genetic defect in this pathway may result in an inability to metabolize galactose, resulting in high levels of galactose in the blood and high levels of galactose-1-phosphate in tissues. This disorder, called *galactosemia,* has serious consequences, including mental retardation. Not surprisingly, it occurs most commonly in infants because the major dietary source of galactose is milk. The symptoms can be prevented or alleviated by removing milk and dairy products from the diet, provided that the condition is detected early.)

The main sugar in the blood, of course, is glucose. Its concentration in your blood a few hours after a meal is probably about 80 mg% (80 mg per 100 mL of blood, or about 4.4 mM). The level may rise to 120 mg% (6.6 mM) shortly after you've eaten, and it decreases somewhat if you wait longer than usual between meals. In general, however, blood glucose level is maintained within rather narrow limits. In fact, maintenance of blood glucose level is one of the most important regulatory functions in your body, particularly for proper functioning of your nervous system, including your brain. Your blood glucose level is under the control of several hormones, including *insulin, glucagon, epinephrine,* and *norepinephrine.*

Once in your bloodstream, glucose has four main fates: It can be catabolized to CO_2 and H_2O by aerobic respiration, converted to lactate, used to synthesize the storage polysaccharide glycogen, or converted via glycolysis and pyruvate oxidation to acetyl CoA, from which body fat is made. Aerobic respiration is the most common fate of blood glucose because most of the tissues in your body function aerobically most of the time. Your brain is particularly noteworthy as an aerobic organ. It needs large amounts of energy to maintain the membrane potentials essential for the transmission of nerve impulses and normally depends solely on glucose to meet this need. In fact, your brain needs about 120g of glucose per day, which is about 15% of your total energy consumption. When you are at rest, your brain accounts for about 60% of your glucose usage. The brain is also a heavy user of oxygen—about 20% of your total oxygen consumption! Moreover, the brain has no significant stores of glycogen, so the supply of both oxygen and glucose must be continuous. Even a short interruption of either has dire consequences. Your heart has similar requirements because it is also a completely aerobic organ and has little or no energy reserves. The supply of oxygen and fuel molecules must therefore be constant, though the heart, unlike the brain, can use a variety of fuel molecules, including glucose, lactate, and fatty acids.

In addition to aerobic respiration in any of a wide variety of tissues, glucose in your blood can also be catabolized anaerobically, especially in red blood cells and in skeletal muscle cells. Red blood cells have no mitochondria and cannot carry out aerobic respiration. They depend exclusively on glycolysis to meet their energy needs. Unlike heart muscle, skeletal muscle can function in either the presence or the absence of oxygen. When you exert yourself strenuously, the ATP demand of your muscle cells temporarily exceeds the capacity of your circulatory system to supply oxygen, so the rate of glycolysis exceeds that of aerobic respiration, and excess pyruvate is converted to lactate. Lactate generated in this way is released into the blood and taken up not only by your heart for use as fuel but also by gluconeogenic tissues, especially your liver. When lactate molecules enter a liver cell, they are reoxidized to pyruvate, which is then used to make glucose by gluconeogenesis (see Figure 11-13). The glucose is returned to the bloodstream, where it can be taken up by muscle (or any other) cells again.

Skeletal muscle is the main source of blood lactate, and the liver is the primary site of gluconeogenesis, so a cycle is set up, as shown in Figure 11A-2. Lactate produced by glycolysis in hypoxic (oxygen-deficient) muscle cells is transported via the blood to the liver. There, gluconeogenesis converts the lactose to glucose, which is released into the blood. This process is called the *Cori cycle* because it was originally described by Carl and Gerti Cori. The next time you are resting after strenuous exercise, think of what is happening: The lactate your muscle cells have just released into the bloodstream is being taken up by liver cells and converted back to glucose. The reason you are probably breathing heavily is to provide the oxygen your body needs

Figure 11A-2 The Cori Cycle: The Link Between Glycolysis in Muscle Cells and Gluconeogenesis in the Liver. Skeletal muscle derives much of its energy from glycolysis, especially during periods of strenuous exercise. The lactate produced in this way is transported by the bloodstream to the liver, where it is reoxidized to pyruvate. Pyruvate is used as substrate for the gluconeogenic pathway, generating glucose that is then returned to the blood. This cyclic process is called the Cori cycle.

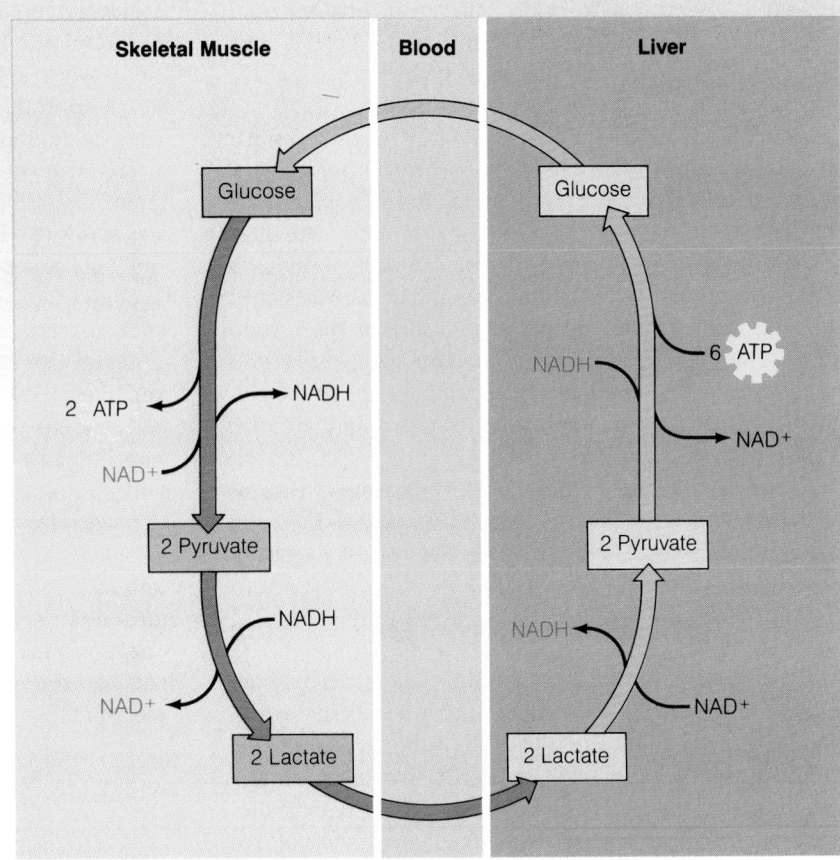

to return your muscle cells to aerobic condition, and to generate all the ATP and GTP needed for gluconeogenesis in your liver and for rebuilding body glycogen stores.

Glycogen storage is, in fact, the third significant fate of blood glucose. Glycogen is stored primarily in the cells of your liver and skeletal muscle. Muscle glycogen is used to supply glucose during times of strenuous exertion, when the glycolytic rate of the muscle cells temporarily exceeds the capability of the circulatory system to deliver blood glucose. Liver glycogen, on the other hand, is used as a source of glucose when the liver is stimulated hormonally to release glucose into the bloodstream to maintain the blood glucose level.

The fourth possible fate of blood glucose is its use for the synthesis of body fat. The route to fat is via pyruvate to acetyl CoA, just as in the initial phase of aerobic respiration. Whenever you eat more food than your body needs for energy and the biosynthesis of other molecules, excess glucose is oxidized to acetyl CoA and shunted into the synthesis of triglycerides. These are then stored as body fat, especially in *adipose tissue*, specialized for this purpose. Thus, your body

has three main sources of energy at all times: the glucose in your blood, the glycogen in your liver and skeletal muscle cells, and the triglycerides stored in adipose tissue.

To conclude, let's come back to our original question: "What happens to the sugar?" Hopefully, you're in a better position to answer that question now. All the glucose (and other sugars) in your body come originally from the food you eat—either as monosaccharides directly, or from the breakdown of disaccharides and polysaccharides in your intestinal tract. The ultimate fate of that glucose is oxidation to CO_2 and water, which you then exhale and excrete. But in the meantime, glucose molecules can circulate in your bloodstream or be stored as glycogen in liver or muscle cells. In its circulating form, glucose can be oxidized immediately by aerobic tissues such as the brain, it can be converted to lactate and become a part of the Cori cycle, or it can be used to synthesize glycogen or fat for storage.

It may look like just a modest spoonful of sugar as you sprinkle it on your cereal or add it to your coffee, but treat it with respect—it plays an important role in the energy metabolism of all of the cells in your body! ■

up one-half of the disaccharide sucrose, the major transport sugar in most plants. Moreover, the catabolism of most other energy-rich substances in plants, animals, and microorganisms alike begins with their conversion into one or another of the intermediates in the pathway for glucose catabolism. Rather than looking at the fate of a single compound, then, we are considering a metabolic pathway that is at the very heart of chemotrophic energy metabolism.

Chemically, glucose is an *aldohexose*, a six-carbon sugar that has a terminal carbonyl group when the molecule is shown in its straight-chain form. Recall from Chapter 3 that the glucose molecule can be represented by any of the three structures shown in Figure 3-18. The *Haworth projection* is the most satisfactory of the three because it indicates both the ring form and the spatial relationship of the carbon atoms. However, we will use the straight-chain *Fischer projection* because the chemistry of the initial steps of the glycolytic pathway is somewhat easier to understand with this model.

Glucose is a good potential source of energy because its oxidation is an exergonic process, with a $\Delta G^{o'}$ of -686 kcal/mol for the complete combustion of glucose to carbon dioxide and water:

$$C_6H_{12}O_6 + 6O_2 \longrightarrow 6CO_2 + 6H_2O \qquad \textbf{(11-12)}$$

As a thermodynamic parameter, $\Delta G^{o'}$ is unaffected by route and will therefore have the same value whether the oxidation is by direct combustion with all of the energy released as heat or by biological oxidation, with maximum conservation of energy as ATP. Thus, oxidation of the sugar molecules in a marshmallow will release the same amount of free energy whether you burn the marshmallow over a campfire or eat it and catabolize the sugar molecules in your body. Biologically, however, the distinction is critical: Uncontrolled combustion would be incompatible with life, whereas controlled, stepwise oxidation mediated by a series of enzyme-catalyzed reactions represents an isothermal process that can be coupled to ATP generation and subjected to careful regulation.

In this and the following chapter, therefore, we will consider the biological processes whereby the energy of the oxidizable bonds of glucose is released in ways that ensure the conservation of as much of that energy as possible in the form of ATP, and under conditions compatible with life. We will also discuss the means by which cells continuously adjust the rate of glucose oxidation and synthesis to meet actual cellular needs for ATP.

Respiration Versus Fermentation: The Difference Oxygen Makes

To speak of glucose (or any other compound) as an oxidizable substrate assumes the availability of some sort of electron acceptor, without which there can be no oxidation. The electron acceptor for most organisms is molecular oxygen (O_2). Notice, for example, that reaction 11-12 assumes the availability of oxygen. Access to the full 686 kcal/mol of free energy in glucose is possible only with oxygen as the electron acceptor and hence requires aerobic conditions. The complete oxidation of glucose to carbon dioxide and water in the presence of oxygen is called **aerobic respiration.** Aerobic respiration, or *respiration* for short, is a complex, multistep process, to be discussed in detail in Chapter 12.

Even in the absence or scarcity of oxygen, most organisms can still extract limited amounts of energy from substrates such as glucose but with no net oxidation of the substrate. They do so by means of *glycolysis*, a pathway that does not require oxygen. Instead, electrons are removed from an intermediate at one step in the pathway but are returned to another intermediate later in the same pathway. Each such anaerobic process is called a **fermentation** and is identified in terms of the principal end-product. In some animal cells and many bacteria, the end-product is *lactate,* so the process of anaerobic glucose catabolism is called *lactate fermentation.* In most plant cells and in microorganisms such as yeast, the process is termed *alcoholic fermentation* because the end-product is ethanol, an *alcohol.* (As we will learn in Chapter 12, some organisms can carry out *anaerobic respiration,* using electron acceptors other than oxygen. Unlike fermentation, anaerobic respiration involves a net transfer of electrons to an acceptor that is not an intermediate in the pathway and therefore results in the net oxidation of the substrate.)

Aerobic and Anaerobic Organisms

Organisms can be classified in terms of their need for and use of oxygen as an electron acceptor in energy metabolism. Many organisms have an absolute requirement for oxygen and are called **strict,** or **obligate, aerobes.** You look at such an organism every morning in the mirror. **Strict,** or **obligate, anaerobes,** on the other hand, cannot use oxygen as an electron acceptor under any conditions. Not surprisingly, such organisms occupy environments from which oxygen is generally excluded, as in puncture wounds or the sludge at the bottoms of ponds. Most strict anaerobes are bacteria, including those responsible for gangrene, food poisoning, and methane production.

Facultative organisms are those that can function under either aerobic or anaerobic conditions. Given the availability of oxygen, most facultative organisms carry out the full respiratory process, but they can switch to a fermentative mode if oxygen is limiting or absent. Many bacteria and fungi are facultative organisms. Some cells or tissues of otherwise aerobic organisms can function anaerobically if required to do so. Your muscle cells are an example; they normally function aerobically but switch to lactate fermentation whenever the oxygen supply becomes limiting (as during periods of strenuous exercise, for example).

The remainder of this chapter is devoted mainly to the anaerobic generation of ATP by the fermentation of glucose, with lactate and alcohol as the main products of interest.

Chapter 12 will follow with a discussion of aerobic energy metabolism. It seems appropriate to consider fermentation processes first because the process of glycolysis is common to both fermentation and aerobic respiration. Thus, by beginning with fermentation, we will not only be considering the ways in which energy can be extracted from glucose without net oxidation; we will also be laying the foundation for the aerobic processes of the next chapter.

Glycolysis and Fermentation: ATP Generation with No Net Oxidation

Whether an obligate or facultative anaerobe, any organism or cell that meets its energy needs by fermentation carries out energy-yielding oxidative reactions without using oxygen as an electron acceptor. The six-carbon glucose molecule is split into two three-carbon molecules, followed by what is in effect an intramolecular oxidation-reduction reaction. The overall process is sufficiently exergonic to generate two ATP molecules per molecule of glucose fermented. This is the maximum possible energy yield without access to oxygen or to an alternative electron acceptor.

Glycolysis

The process of **glycolysis,** also called the **glycolytic pathway,** is a ten-step reaction sequence that occurs in all organisms. This pathway is common to both aerobic and anaerobic glucose metabolism because *pyruvate,* the three-carbon molecule to which it leads, can be either *reduced* to lactate or ethanol under anaerobic conditions or *oxidized* further if oxygen is available. Historically, the glycolytic pathway was the first major metabolic sequence to be elucidated. Most of the decisive work was done in the 1930s by the German biochemists Gustav Embden, Otto Meyerhof, and Otto Warburg. In fact, an alternative name for the glycolytic sequence is the *Embden-Meyerhof pathway.*

An Overview. The glycolytic pathway is shown in the context of fermentation in Figure 11-7 and will also appear as a component of aerobic respiration in the next chapter. Each of the ten reactions in the pathway is numbered (Gly-1 through Gly-10), and the enzymes that catalyze the reactions are identified in the center of the figure. The essence of the process is suggested by its very name because the term *glycolysis* derives from two Greek roots: *glykos,* meaning "sweet," and *lysis,* meaning "loosening" or "splitting." Literally, glycolysis is the loosening, or splitting, of something sweet—the starting sugar. The splitting occurs at reaction Gly-4 in Figure 11-7 because it is at that point that the six-carbon sugar is cleaved into two three-carbon molecules. One of these molecules, *glyceraldehyde-3-phosphate,* turns out to be the only oxidizable molecule in the whole pathway.

The oxidation of that molecule in reaction Gly-6 involves NAD^+ as the electron acceptor. Under anaerobic conditions, the NADH is used as the source of electrons for the reductive fermentation step at the end of the pathway, thereby regenerating the oxidized form of the coenzyme. At two points in the glycolytic sequence (Gly-7 and Gly-10), specific reactions are sufficiently exergonic to drive the generation of ATP from ADP. These reactions represent the energy payoff of the process because they are the only ATP-yielding steps in the whole pathway as it functions under anaerobic conditions.

The important features of the glycolytic pathway are therefore the sugar-splitting reaction for which the sequence is named, the oxidative event that generates NADH, and the two specific steps at which the reaction sequence is coupled to ATP generation. These features will be emphasized as we consider the overall pathway in three phases: the preparatory and cleavage steps (Gly-1 through Gly-5); the oxidative sequence, which includes the first ATP-generating event (Gly-6 and Gly-7); and the second ATP-generating sequence (Gly-8 through Gly-10).

Phase 1: Preparation and Cleavage. Reaction Gly-4 is the step at which the cleavage of a six-carbon compound to two three-carbon molecules occurs. It is useful to focus on the substrate for that reaction, *fructose-1,6-bisphosphate,* and ask how it can be formed from glucose, because that will explain the first three preparatory steps in the sequence. The crucial feature of fructose-1,6-bisphosphate is clearly the presence of two phosphate groups, one on each terminal carbon. Looking at glucose, it is easy to see how phosphorylation can take place on carbon atom 6, because the hydroxyl group there can be readily linked to a phosphate group to form a *phosphoester.* That is, in fact, what happens in reaction Gly-1, converting glucose into *glucose-6-phosphate.*

ATP provides not only the phosphate group but also the driving force that renders the phosphorylation reaction strongly exergonic ($\Delta G^{\circ\prime} = -4.0$ kcal/mol), making it essentially irreversible in the direction of glucose phosphorylation. Notice, by the way, that the bond being formed is a *phosphoester bond,* while the bond by which the terminal phosphate is linked to ATP is a *phosphoanhydride bond.* This difference is what makes the transfer of the phosphate group from ATP to glucose exergonic (see Figure 11-4 and reaction 11-5). The enzyme that catalyzes this first reaction is called *hexokinase;* as the name suggests, it is not specific for glucose but catalyzes the phosphorylation of other hexoses (six-carbon sugars) as well. (Liver cells contain an additional enzyme, *glucokinase,* that phosphorylates only glucose.)

The carbonyl group on carbon atom 1 of the glucose molecule is not as readily phosphorylated as the hydroxyl group on carbon atom 6. But in the next reaction, the aldo-sugar is converted to the corresponding ketosugar, *fructose-6-phosphate* (reaction Gly-2), with a hydroxyl group on carbon atom 1. That hydroxyl group can then be phosphorylated, yielding the doubly phosphorylated sugar,

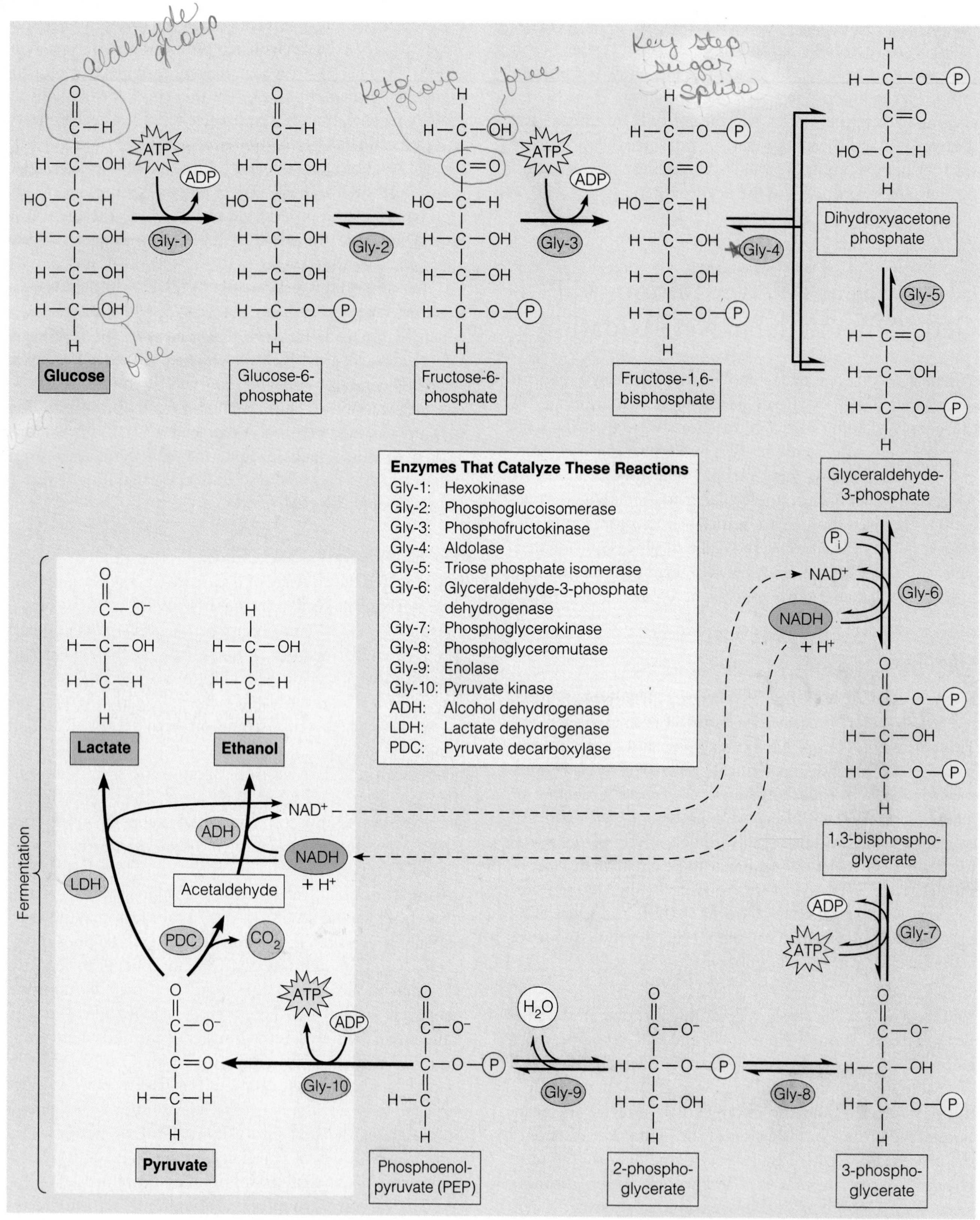

Figure 11-7 The Glycolytic Pathway from Glucose to Pyruvate, with Two Fermentation Alternatives.   Glycolysis is a sequence of ten reactions in which glucose (or any of several related sugars) is catabolized to pyruvate, with a single oxidative reaction (Gly-6) and two ATP-generating steps (Gly-7 and Gly-10). In the absence of oxygen or another electron acceptor, the NADH generated by reaction Gly-6 is reoxidized by transferring its electrons to pyruvate. The most common products of glucose fermentation are lactate or ethanol plus carbon dioxide. The enzymes that catalyze these reactions are identified at the center of the figure. In most organisms, these enzymes occur in the cytosol of the cell. In some protozoans, however, they are compartmentalized in membrane-bounded organelles called *glycosomes*.

fructose-1,6-bisphosphate (reaction Gly-3). Again, the energy difference between the anhydride bond of ATP and the phosphoester bond on the glucose molecule renders the reaction highly exergonic and essentially irreversible in the glycolytic direction ($\Delta G^{\circ\prime} = -3.4$ kcal/mol). This reaction is catalyzed by *phosphofructokinase*, an especially important enzyme in the regulation of glycolysis, as we will see later.

Next comes the actual cleavage reaction from which glycolysis derives its name. Fructose-1,6-bisphosphate is split reversibly by *aldolase* to yield two trioses (three-carbon sugars) called *dihydroxyacetone phosphate* and *glyceraldehyde-3-phosphate* (reaction Gly-4). The two trioses formed in Gly-4 have the same relationship to each other as do glucose-6-phosphate and fructose-6-phosphate: One is an aldose and the other is a ketose. It is therefore not surprising that dihydroxyacetone phosphate and glyceraldehyde-3-phosphate are readily interconvertible (reaction Gly-5). Since only the latter of these compounds is directly oxidizable in the next phase of glycolysis, interconversion of the two trioses enables dihydroxyacetone phosphate to be catabolized simply by conversion to glyceraldehyde-3-phosphate. We can summarize this first phase of the glycolytic pathway as follows:

$$\text{glucose} + 2\text{ATP} \longrightarrow 2 \text{ glyceraldehyde-3-phosphate} + 2\text{ADP}$$
(11-13)

Phase 2: Oxidation and ATP Generation. So far, five of the ten steps of glycolysis have been accounted for, and the original glucose molecule has been doubly phosphorylated and cleaved into two interconvertible trioses. The energy yield is negative thus far, however; two molecules of ATP have been consumed per molecule of glucose up to this point. Now, however, the ATP debt is about to be repaid with interest as we encounter the two energy-yielding phases of glycolysis. In the first sequence (reactions Gly-6 and Gly-7), ATP production is linked directly to an oxidative event, while in the second case (Gly-8 through Gly-10), a highly unstable form of the pyruvate molecule serves as the driving force behind ATP generation.

The oxidation of glyceraldehyde-3-phosphate to the corresponding acid, *3-phosphoglycerate,* is highly exergonic—sufficiently so, in fact, to drive both the reduction of the coenzyme NAD⁺ (Gly-6) and the phosphorylation of ADP with inorganic phosphate, P_i (Gly-7). Historically, this was the first example of a reaction sequence in which the coupling of ATP generation to an oxidative event was understood. Because of its usefulness as a prototype for understanding such coupling, this two-reaction sequence is illustrated in detail in Figure 11-8. The figure focuses especially on the mechanism of action of *glyceraldehyde-3-phosphate dehydrogenase*, the enzyme that catalyzes the actual oxidative reaction (Gly-6).

The important features of this highly exergonic reaction are the involvement of NAD⁺ as the electron acceptor and the coupling of the oxidation to the formation of a high-

energy, doubly phosphorylated intermediate, *1,3-bisphosphoglycerate.* The phosphoanhydride bond on carbon atom 1 of this intermediate has such a highly negative $\Delta G^{\circ\prime}$ of hydrolysis (-11.8 kcal/mol; see Figure 11-4) that the transfer of the phosphate to ADP, catalyzed by the enzyme *phosphoglycerate kinase*, is a very exergonic reaction. ATP generation by the direct transfer to ADP of a high-energy phosphate group from a phosphorylated substrate such as 1,3-bisphosphoglycerate is called **substrate-level phosphorylation.** This mode of ATP synthesis is distinct from *oxidative phosphorylation*, in which phosphorylation of ADP is driven by the exergonic transfer of electrons from reduced coenzymes to oxygen, as we will see in Chapter 12.

To summarize the substrate-level phosphorylation of reactions Gly-6 and Gly-7, we can write an overall reaction with a stoichiometry that accounts for one of the two glyceraldehyde-3-phosphate molecules generated from each glucose molecule in the first phase of glycolysis:

$$\text{glyceraldehyde-3-P} + \text{NAD}^+ + \text{ADP} + P_i \longrightarrow$$
$$\text{3-phosphoglycerate} + \text{NADH} + \text{H}^+ + \text{ATP} \quad \text{(11-14)}$$

Keep in mind that each reaction in the glycolytic pathway beyond glyceraldehyde-3-phosphate occurs *twice* per starting molecule of glucose, thereby accounting for both molecules of triose phosphate generated in the cleavage reaction (Gly-4). This means that, on a per-glucose basis, two molecules of NADH need to be reoxidized. It also means that the initial investment of two ATP molecules in phase 1 is recovered here in phase 2, so the net ATP yield is now zero.

Thus far, then, seven of the ten reactions of glycolysis have been used to convert one molecule of glucose into two molecules of 3-phosphoglycerate, but there is as yet no net ATP generation. That comes in the final phase of the pathway.

Phase 3: Pyruvate Formation and ATP Generation. Generating another molecule of ATP at the expense of 3-phosphoglycerate depends on the phosphate group on carbon atom 3. Yet at this stage, the phosphate group is linked to the carbon atom by a phosphoester bond with an unpromisingly low free energy ($\Delta G^{\circ\prime} = -3.3$ kcal/mol). In this last phase of the glycolytic pathway, this phosphoester bond is converted to a *phosphoenol bond*, which has a highly negative free energy change for hydrolysis ($\Delta G^{\circ\prime} = -14.8$ kcal/mol; see Figure 11-4). This increase in the amount of free energy released upon hydrolysis involves a rearrangement of internal energy within the molecule. To accomplish this, the phosphate group of 3-phosphoglycerate is moved to the adjacent carbon atom, forming *2-phosphoglycerate* (reaction Gly-8). Water is then removed from 2-phosphoglycerate by the enzyme *enolase* (reaction Gly-9), thereby generating *phosphoenolpyruvate (PEP).*

If you look carefully at the structure of PEP (see Figure 11-7, Gly-9), you will notice that, unlike the phosphoester bonds of either 3- or 2-phosphoglycerate, the phosphoenol bond of PEP has what we might define as a distinguishing

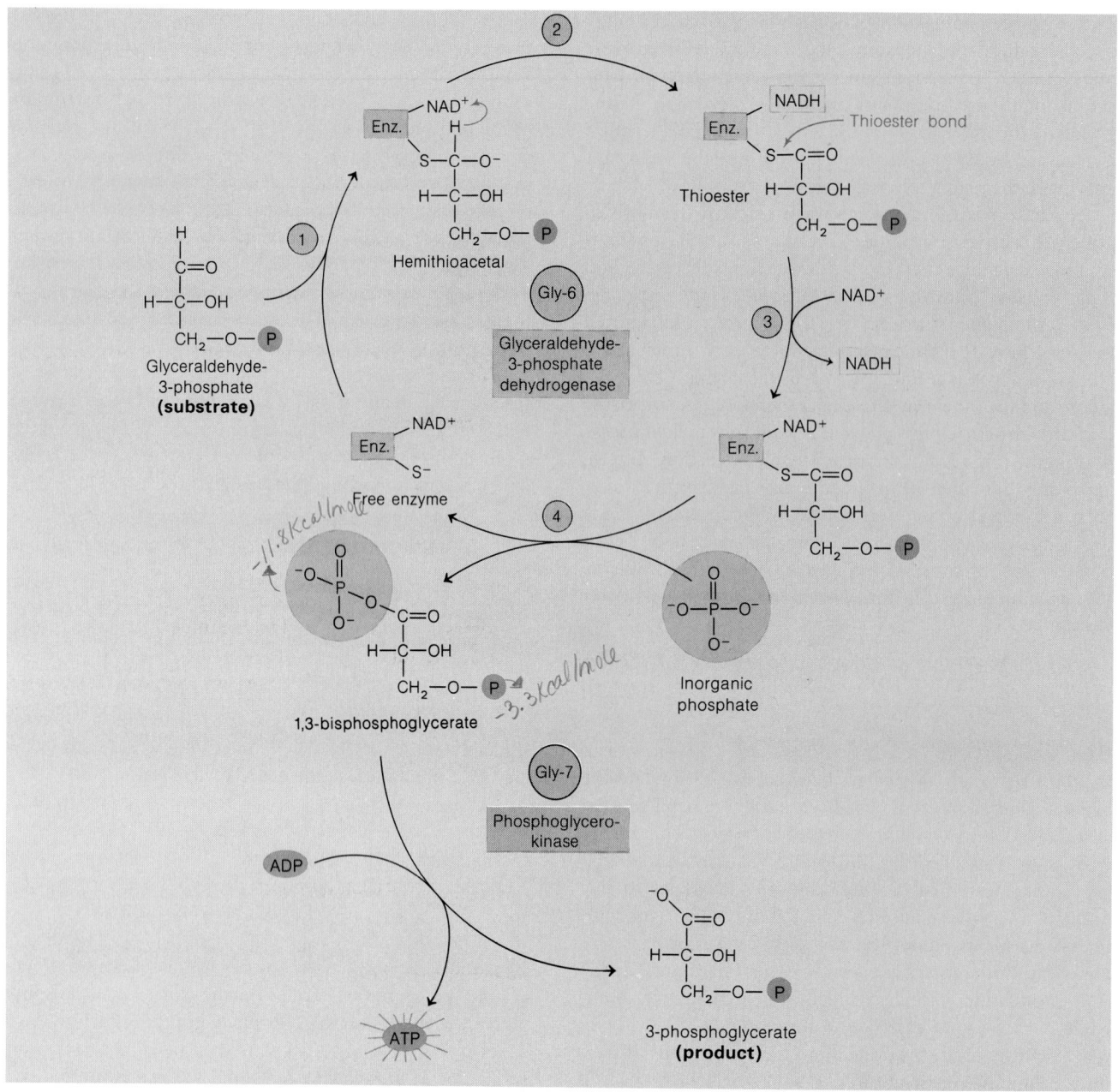

Figure 11-8   Substrate-Level Phosphorylation: A Detailed Mechanism for Reactions Gly-6 and Gly-7 of the Glycolytic Pathway.   ATP is generated by substrate-level phosphorylation in reactions Gly-6 and Gly-7, a two-reaction sequence in which glyceraldehyde-3-phosphate (G3P) is oxidized to 3-phosphoglycerate (PGA) with 1,3-bisphosphoglycerate (BPG) as an intermediate. ① The reaction sequence begins with the covalent binding of G3P to a sulfhydryl group at the active site of the enzyme glyceraldehyde-3-phosphate dehydrogenase, forming a *hemithioacetal*. ② The sulfur bond of the hemithioacetal is oxidized to a thioester by an enzyme-bound molecule of NAD^+. ③ The reduced NADH is displaced from the enzyme surface by a molecule of NAD^+. ④ Inorganic phosphate attacks the thioester bond, forming BPG

and displacing it from the enzyme surface. BPG, a high-energy phosphoanhydride, then binds to the active site of phosphoglycerokinase, where the high energy of hydrolysis of the phosphoanhydride bond is used to drive the synthesis of ATP, releasing PGA as the product. The essential feature of the sequence is that a thermodynamically unfavorable reaction, the formation of an anhydride between a carboxylic acid and inorganic phosphate, is driven by a thermodynamically favorable reaction, the oxidation of an aldehyde. The two reactions are coupled by an enzyme-bound thioester intermediate, which preserves much of the free energy that would otherwise have been released as heat in the oxidation reaction.

characteristic of a high-energy phosphate bond: a phosphate group on a carbon atom with a double bond. In fact, hydrolysis of the phosphoenol bond of PEP is one of the most exergonic hydrolytic reactions known in biological systems.

To understand why PEP is such a high-energy compound compared to most other phosphorylated intermediates in the cell, consider Figure 11-9. Like many keto compounds, pyruvate can exist in either the enol or the keto form, but the equilibrium of the interconversion greatly favors the latter (Figure 11-9a). This means that the keto form of pyruvate is much more stable, whereas the enol form is a highly unstable, thermodynamically unlikely configuration. Thus, when pyruvate is "trapped" in the enol form by chemical constraints that do not allow its transition to the keto form, the molecule will be highly unstable.

Such is the case with the PEP generated in reaction Gly-9. When water is removed from 2-phosphoglycerate, the product is pyruvate "locked" in the enol form by the presence of a phosphate group on carbon atom 2 that prevents transition (or *tautomerization*) to the more stable keto form (Figure 11-9b). PEP is therefore higher in energy than most other phosphorylated compounds; in addition to the usual decrease in free energy when the extra electron pair becomes maximally delocalized over the P—O bonds of the phosphate group, reversion of the pyruvate from the enol form to the keto form is also exergonic.

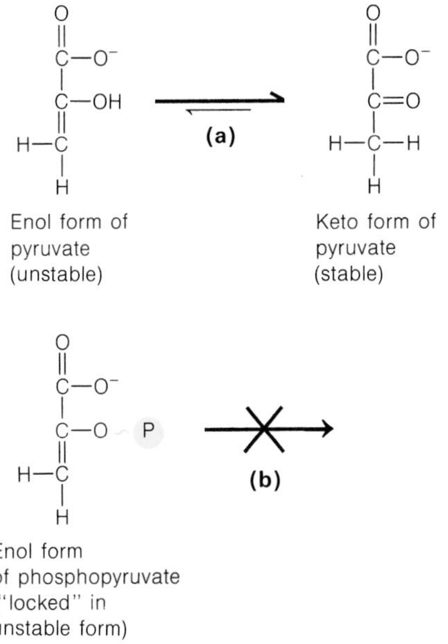

Enol form of pyruvate (unstable)

(a)

Keto form of pyruvate (stable)

Enol form of phosphopyruvate ("locked" in unstable form)

(b)

Figure 11-9 The Thermodynamic Instability of Phosphoenolpyruvate. **(a)** As a keto compound, pyruvate can exist in either the keto or the enol form, but the equilibrium greatly favors the keto form. Conversion of the enol form to the keto form is therefore thermodynamically very favorable. **(b)** Phosphoenolpyruvate, which is the form generated in reaction Gly-9, is in the thermodynamically unfavorable enol configuration and cannot undergo conversion to the more stable keto form because of the phosphate group covalently linked to carbon atom 2. Release of this phosphate group is highly exergonic because the pyruvate molecule is then free to assume the more stable keto form.

To say that the phosphate bond of PEP has a highly negative free energy of hydrolysis is to make the last step in the sequence, reaction Gly-10, entirely reasonable because this reaction involves the transfer of that phosphate to ADP, generating another molecule (or, on a per-glucose basis, two more molecules) of ATP. That transfer, catalyzed by the enzyme *pyruvate kinase,* is highly exergonic ($\Delta G^{\circ\prime} = -7.5$ kcal/mol; see Figure 11-4 reaction 11-6) and is therefore essentially irreversible in the direction of pyruvate and ATP formation.

To summarize this third phase of glycolysis, we can write an overall reaction for pyruvate formation, again using stoichiometry that accounts for one of the two trioses derived from glucose:

$$\text{3-phosphoglycerate} + \text{ADP} \longrightarrow \text{pyruvate} + H_2O + \text{ATP}$$

$$\textbf{(11-15)}$$

In Summary. The ATP initially invested in reactions Gly-1 and Gly-3 was recouped in the first phosphorylation event (Gly-7), so the two molecules of ATP formed per molecule of glucose by the second phosphorylation event (Gly-10) represent the net ATP yield of the glycolytic pathway. This becomes clear when we add up the three reactions that summarize the three phases of the pathway (reactions 11-13, 11-14, and 11-15) with the latter two reactions multiplied by two to account for both triose molecules generated in reaction 11-13. The result is an overall expression for the pathway from glucose to pyruvate:

$$\text{glucose} + 2NAD^+ + 2ADP + 2P_i \xrightarrow{\text{Gly-1 through Gly-10}} 2 \text{ pyruvate} + 2NADH + 2H^+ + 2ATP + 2H_2O \quad \textbf{(11-16)}$$

This sequence is exergonic in the direction of pyruvate formation. Under typical intracellular conditions in your own body, for example, $\Delta G'$ for the overall pathway from glucose to pyruvate is about -20 kcal/mol.

The glycolytic pathway is one of the most universal metabolic pathways known. Virtually all cells possess the ability to convert glucose to pyruvate, with some of the energy of the oxidative event in the pathway trapped in the form of two molecules of ATP per molecule of glucose. What happens next, however, usually depends on the availability of oxygen because catabolism beyond pyruvate is different under aerobic conditions than under the anaerobic conditions we are presuming at present.

Pyruvate as a Branching Point

Pyruvate occupies a key position as a branching point in chemotrophic energy metabolism (Figure 11-10). Its fate depends on the kind of organism involved and whether oxygen is available. In the presence of oxygen, pyruvate undergoes further oxidation to a molecule called *acetyl coenzyme A* (*acetyl CoA*), and the glycolytic pathway becomes just the first of several major segments of *aerobic respiration* (Figure 11-10a). As we will see in Chapter 12, this results in the complete oxidation of pyruvate to carbon dioxide, with the generation of much more ATP than is otherwise possible.

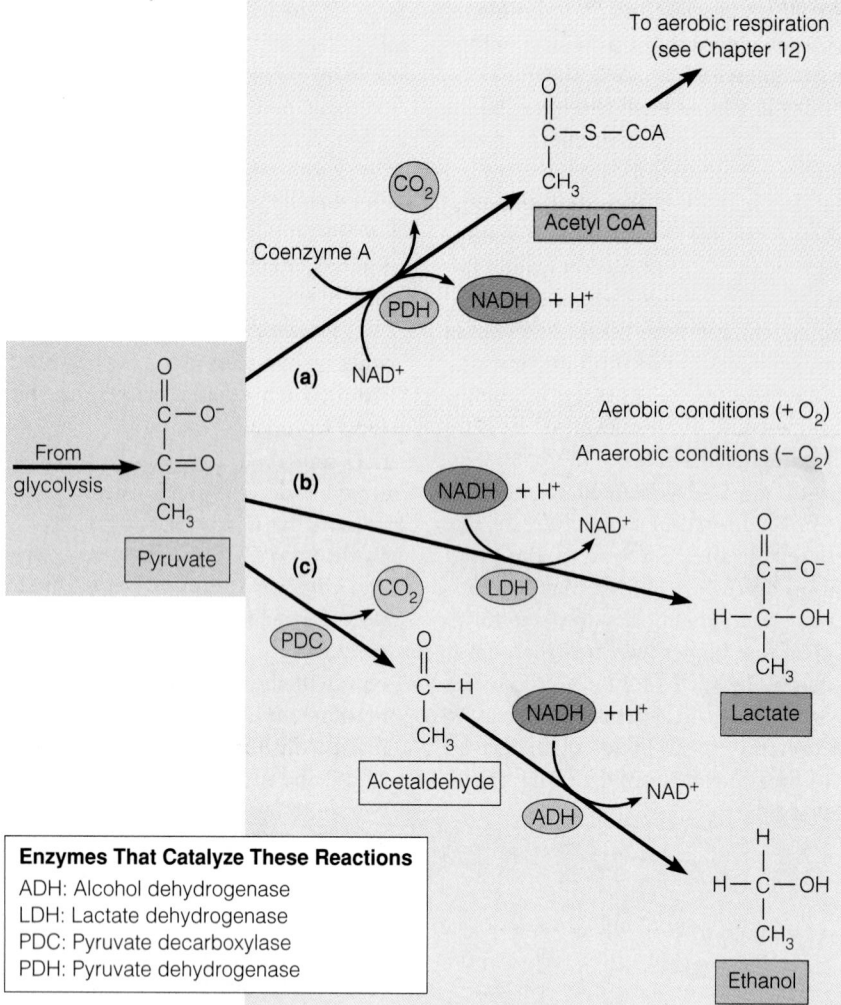

Figure 11-10   The Fate of Pyruvate Under Aerobic and Anaerobic Conditions.   The fate of pyruvate depends on the organism involved and whether oxygen is available. **(a)** Under aerobic conditions, most organisms convert pyruvate to an activated form of acetate in a reaction that involves both oxidation (with NAD^+ as the electron acceptor) and decarboxylation (liberation of a carbon atom as CO_2). The activated acetate is bound to the carrier coenzyme A as acetyl coenzyme A (acetyl CoA). Acetyl CoA then becomes the substrate for aerobic respiration (see Chapter 12). Under anaerobic or hypoxic conditions, pyruvate serves as the electron acceptor for the oxidation of NADH to NAD^+, thereby regenerating the oxidized form of the coenzyme required in reaction Gly-6 of glycolysis. The most common products of pyruvate reduction are **(b)** lactate (in most animal cells and many bacteria) or **(c)** ethanol and CO_2 (in many plant cells and in yeasts and other microorganisms). The enzymes that catalyze these reactions are identified at the bottom of the figure. In eukaryotic cells, all of the reactions of aerobic respiration beyond pyruvate occur within the mitochondrion.

An important feature of glycolysis, however, is that it can also take place in the absence of oxygen. Under these conditions, no further oxidation of pyruvate is possible, no acetyl coenzyme A is formed, and no additional ATP can be generated. Instead, the energy needs of the cell are met by the modest ATP yield of the glycolytic pathway. Rather than being oxidized, pyruvate is reduced because it serves as an acceptor molecule for the electrons (and protons) that must continually be removed from NADH to provide for the continued regeneration of NAD^+. As Figure 11-10 illustrates, the most common products of pyruvate reduction are lactate (part b) or ethanol and carbon dioxide (part c).

Fermentation Pathways: Lactate and Alcoholic Fermentation

As usually defined, the glycolytic pathway ends with pyruvate. Fermentative processes cannot end there, however, because of the need to regenerate NAD^+, the oxidized form of the coenzyme. As reaction 11-16 indicates, the conversion of glucose to pyruvate involves the stoichiometric reduction of NAD^+: One molecule of NAD^+ is reduced to NADH per molecule of pyruvate generated. Coenzymes are present in cells at only modest concentrations, however, and cannot be consumed stoichiometrically. Thus, the reduction of NAD^+

to NADH that occurs during glycolysis must be accompanied by the concurrent reoxidation of NADH to NAD^+, or glycolysis would quickly come to a halt.

In the presence of oxygen, NADH is reoxidized by the transfer of its electrons (and proton) to oxygen, as we will see in Chapter 12. Under anaerobic conditions, however, the electrons are transferred to an acceptor that is itself generated by the anaerobic pathway. Regardless of the particular end-products, fermentative pathways always terminate in some sort of reductive reaction sequence that regenerates NAD^+ from NADH by the transfer of electrons to an acceptor molecule that is a part of the pathway. For the fermentation of glucose (or other sugars), pyruvate is the molecule that accepts electrons from NAD^+. Inspection of its structure reveals why: Pyruvate has a carbonyl group that can be readily reduced to a hydroxyl group (see Figure 11-10). It is therefore not surprising that the two most common pathways for fermentation use pyruvate as the electron acceptor, converting it either to lactate or to CO_2 and ethanol.

Lactate Fermentation. The fermentation process that terminates in lactate is called **lactate fermentation.** As Figure 11-10b indicates, lactate is generated by the direct transfer of electrons from NADH to the carbonyl group of pyruvate, reducing it to the hydroxyl group of lactate. On a per-glucose basis, this can be represented as follows:

$$2 \text{ pyruvate} + 2\text{NADH} + 2\text{H}^+ \longrightarrow 2 \text{ lactate} + 2\text{NAD}^+$$
$$\text{(11-17)}$$

This reaction is readily reversible; in fact, the enzyme that catalyzes it is called *lactate dehydrogenase* because of its ability to catalyze the oxidation, or dehydrogenation, of lactate to pyruvate.

Because the glycolytic pathway generates NADH and pyruvate on an equimolar basis, both of the NADH molecules generated per glucose are reoxidized at the expense of pyruvate, as shown in reaction 11-17. Thus, the overall reaction for the metabolism of glucose to lactate under anaerobic conditions is as follows:

$$\text{glucose} + 2\text{ADP} + 2\text{P}_i \longrightarrow 2 \text{ lactate} + 2\text{ATP} + 2\text{H}_2\text{O}$$
$$\text{(11-18)}$$

Lactate fermentation is the major energy-yielding pathway in many anaerobic bacteria and in animal cells operating under anaerobic or hypoxic (oxygen-deficient) conditions. Lactate fermentation is very important commercially because the production of cheese, yogurt, and other dairy products depends on microbial fermentation of lactose, the main sugar found in milk.

A more personal example of lactate fermentation involves your own muscles during periods of strenuous exertion. Whenever muscle cells use oxygen faster than it can be supplied by the circulatory system, the cells become temporarily hypoxic. Under these conditions, some, or even most, of the pyruvate is reduced to lactate instead of being oxidized to acetyl CoA. (The resulting local buildup of lac-

tate is what causes the muscle pain that sometimes accompanies vigorous exercise.) The lactate produced in this way is transported by the circulatory system from the muscle to the liver. There it is converted to glucose again by the process of *gluconeogenesis*. As we will see later in this chapter, gluconeogenesis is essentially the reverse of lactate fermentation but with several critical differences that enable the pathway to proceed exergonically in the direction of glucose formation. The relationship between glycolysis and gluconeogenesis in your own body was explored in Box 11A, page 303.

Alcoholic Fermentation. **Alcoholic fermentation** also involves both NADH and pyruvate, but with different end-products. Plant cells carry out alcoholic fermentation when under anaerobic conditions (in water-logged roots, for example), as do yeasts and other microorganisms. In this case, the pyruvate is first decarboxylated to a two-carbon compound, acetaldehyde, which then serves as the electron acceptor. Acetaldehyde reduction gives rise to ethanol, the alcohol for which the process is named. This reductive sequence actually involves two separate events (pyruvate decarboxylation and subsequent acetaldehyde reduction) catalyzed by two separate enzymes, *pyruvate decarboxylase* and *alcohol dehydrogenase*, respectively (see Figure 11-10c). The overall reaction can be summarized as follows:

$$2 \text{ pyruvate} + 2\text{NADH} + 2\text{H}^+ \longrightarrow 2 \text{ ethanol} + 2\text{CO}_2 + 2\text{NAD}^+$$
$$\text{(11-19)}$$

By adding this reductive step to the overall equation for glycolysis (reaction 11-16), we arrive at the following summary equation for alcoholic fermentation:

$$\text{glucose} + 2\text{ADP} + 2\text{P}_i \longrightarrow 2 \text{ ethanol} + 2\text{CO}_2 + 2\text{ATP} + 2\text{H}_2\text{O}$$
$$\text{(11-20)}$$

Alcoholic fermentation has considerable economic significance because fermentation as carried out by yeast cells is a key process in the baking, brewing, and winemaking industries. For the baker, carbon dioxide is the important end-product. The yeast cells that are added to bread dough function anaerobically, generating both CO_2 and ethanol. Carbon dioxide becomes entrapped within the mass of dough, causing it to rise, and the alcohol is driven off harmlessly during the subsequent baking process, contributing to the pleasant aroma of baking bread.

For the brewer, both CO_2 and ethanol are essential; ethanol makes the product an alcoholic beverage, and CO_2 accounts for the carbonation. (In modern brewing technology, carbon dioxide is added artificially, however.) In the making of wine, on the other hand, interest focuses specifically on the ethanol; the CO_2 simply helps produce an anaerobic environment favorable to fermentation by the yeast. (Amateur winemakers should note the importance of maintaining anaerobic conditions, because contaminating bacteria can convert sugar to acetic acid if oxygen is available, and the result may be a keg of vinegar instead of wine!)

Other Fermentation Pathways. Although lactate and ethanol are the fermentation products of greatest economic significance, they by no means exhaust the microbial repertoire with respect to fermentation. In *propionate fermentation,* for example, bacteria convert pyruvate reductively to propionate (CH_3—CH_2—COO^-), an important reaction in the production of Swiss cheese. Many bacteria that cause food spoilage do so by *butylene glycol fermentation.* Other fermentation processes yield butyrate (the cause of rancid butter), acetone, or isopropyl alcohol. All these reactions, however, are just metabolic variations on the common theme of reoxidizing NADH by the transfer of electrons to an organic acceptor.

The Energetics of Fermentation

An essential feature of every fermentative process is that no external electron acceptor is involved and no net oxidation occurs. In both lactate and alcoholic fermentation, for example, the NADH generated by the single oxidative step of glycolysis (reaction Gly-6) is reoxidized stoichiometrically in the final reaction of the sequence (reactions 11-17 and 11-19, respectively). Because no net oxidation occurs, fermentation processes are characterized by a modest ATP yield—two molecules of ATP per molecule of glucose, in the case of either lactate or alcoholic fermentation.

Most of the free energy of the glucose molecule is still present in the lactate or ethanol molecules. In the case of lactate fermentation, for example, the two lactate molecules produced from every glucose molecule account for most of the 686 kcal of free energy present per mole of glucose because the complete aerobic oxidation of lactate has a $\Delta G^{o\prime}$ of −319.5 kcal/mol. In other words, about 93% (2 × 319.5/686 × 100%) of the original free energy content of glucose is still present in the lactate molecules. Lactate fermentation is therefore able to tap only about 7% (47 kcal/mol) of the free energy potentially available from glucose.

Although the energy yield from lactate fermentation is low, the available free energy is conserved very efficiently as ATP. The ATP yield, of course, is two molecules of ATP per molecule of glucose. Based on standard free energy changes (admittedly somewhat arbitrary), these two molecules of ATP represent 2 × 7.3 = 14.6 kcal/mol, corresponding to an efficiency of energy conservation of about 30% (14.6/47 × 100%). If anything, this underestimates the efficiency of fermentation, because actual $\Delta G^\prime$ values for ATP hydrolysis under cellular conditions are usually substantially more negative than −7.3 kcal/mol, often in the range of −10 to −14 kcal/mol. Based on these values, the two molecules of ATP represent at least 20 kcal/mol, which means that the efficiency of energy conservation most likely exceeds 40%.

Alternative Substrates for Glycolysis

Thus far, we have assumed glucose to be the starting point for glycolysis and therefore, by implication, for all of cellular energy metabolism. Glucose is certainly a major substrate for both fermentation and respiration in a variety of organisms and tissues. It is not the only such substrate, however; there are many organisms and tissues within organisms for which glucose is not very important at all. So it is useful to ask what some of the major alternatives to glucose are and how they are handled by cells.

One principle quickly emerges: Regardless of the chemical nature of the alternative substrates, they are converted in the fewest possible steps into an intermediate in the mainstream pathway for glucose catabolism. Most carbohydrates, for example, are converted to intermediates in the glycolytic pathway. To emphasize this point, we will briefly consider two classes of alternative substrates: other sugars and storage carbohydrates.

Catabolism of Sugars

Many sugars other than glucose are available to cells, depending on the food sources of the organism in question. Most of them are either monosaccharides (usually hexoses or pentoses), or disaccharides that can be readily hydrolyzed into the component monosaccharides. Ordinary table sugar (sucrose), for example, is a disaccharide consisting of the hexoses glucose and fructose, and milk sugar (lactose) contains glucose and galactose (see Figure 3-20). In addition to glucose, fructose, and galactose, mannose is another relatively common dietary hexose.

Figure 11-11 illustrates the reaction sequences that bring various carbohydrates into the glycolytic pathway. In general, disaccharides are hydrolyzed into their component monosaccharides, and each monosaccharide is converted to a glycolytic intermediate in one or a few steps. Glucose and fructose enter most directly because their conversion requires only phosphorylation on carbon atom 6, catalyzed by the enzyme hexokinase. Hexokinase can also phosphorylate mannose, and the resulting mannose-6-phosphate undergoes conversion (by the enzyme *phosphomannoisomerase)* to fructose-6-phosphate, a glycolytic intermediate. The entry of galactose requires a somewhat more complex reaction sequence, the details of which are explored in Problem 11-9 at the end of the chapter.

Phosphorylated pentoses can also be channeled into the glycolytic pathway, but only after being converted to hexose phosphates. That conversion is accomplished by a metabolic sequence called the *phosphogluconate pathway,* which shuffles carbon atoms between 3-, 4-, 5-, 6-, and 7-carbon sugars. Thus, the typical cell has metabolic capabilities to convert most naturally occurring sugars (and a variety of other compounds as well) to one or another of the glycolytic intermediates for further catabolism under anaerobic or aerobic conditions.

Catabolism of Storage Polysaccharides

Although glucose is the immediate substrate for both fermentation and respiration in many cells and tissues, it is not present in cells to any large extent as the free monosaccharide. Instead, it occurs in the form of *storage polysaccharides,*

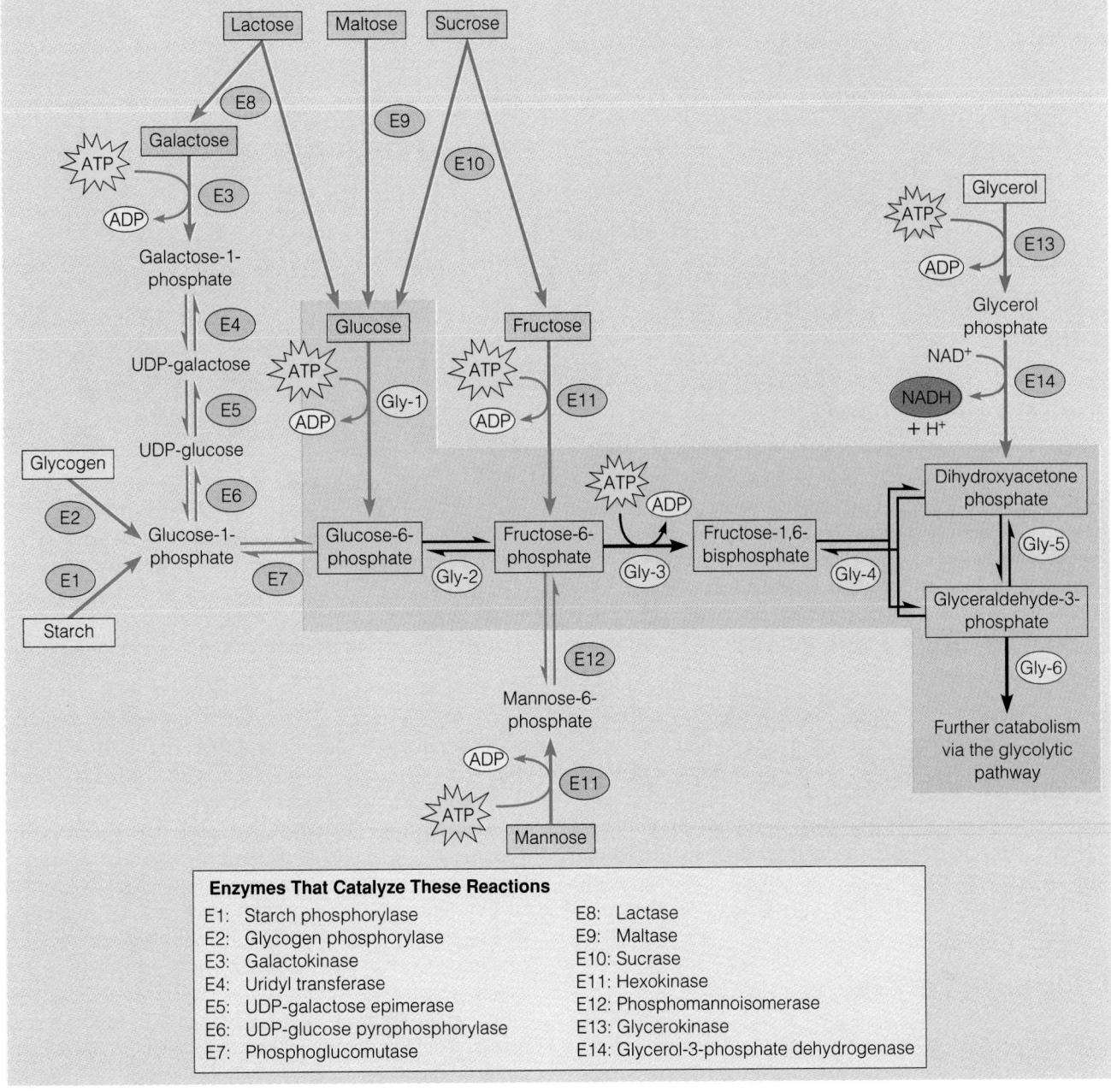

Enzymes That Catalyze These Reactions

E1: Starch phosphorylase	E8: Lactase
E2: Glycogen phosphorylase	E9: Maltase
E3: Galactokinase	E10: Sucrase
E4: Uridyl transferase	E11: Hexokinase
E5: UDP-galactose epimerase	E12: Phosphomannoisomerase
E6: UDP-glucose pyrophosphorylase	E13: Glycerokinase
E7: Phosphoglucomutase	E14: Glycerol-3-phosphate dehydrogenase

Figure 11-11   Carbohydrate Catabolism by the Glycolytic Pathway. Carbohydrate substrates that can be metabolized by the common means of conversion to an intermediate in the glycolytic pathway include the hexoses galactose, glucose, fructose, and mannose; the disaccharides lactose, maltose, and sucrose; the polysaccharides glycogen and starch; and the three-carbon compound glycerol. The conversion reactions are shown by blue arrows. In each case, the enzyme that catalyzes the reaction is indicated by a circled E and is identified at the bottom of the figure. The first six reactions of the glycolytic pathway are highlighted in blue-green; for the names of the enzymes that catalyze these reactions, see Figure 11-7. In some cases, other enzymes or reaction sequences may be involved, depending on the organism and tissue.

most commonly *starch* in plants and *glycogen* in animals. As indicated in Figure 11-12, these storage polysaccharides are mobilized by a process called *phosphorolytic cleavage*. Inorganic phosphate is used to break the α (1 → 4) bond between successive glucose units, liberating the glucose monomers as glucose-1-phosphate. Both glycogen and starch are cleaved in this manner, primarily by the enzymes

glycogen phosphorylase and *starch phosphorylase*, respectively. The glucose-1-phosphate liberated by phosphorylase activity can be converted by the enzyme phosphoglucomutase to glucose-6-phosphate, which is then catabolized by the glycolytic pathway.

Notice that glucose stored in polymerized form enters the glycolytic pathway as glucose-6-phosphate, without the

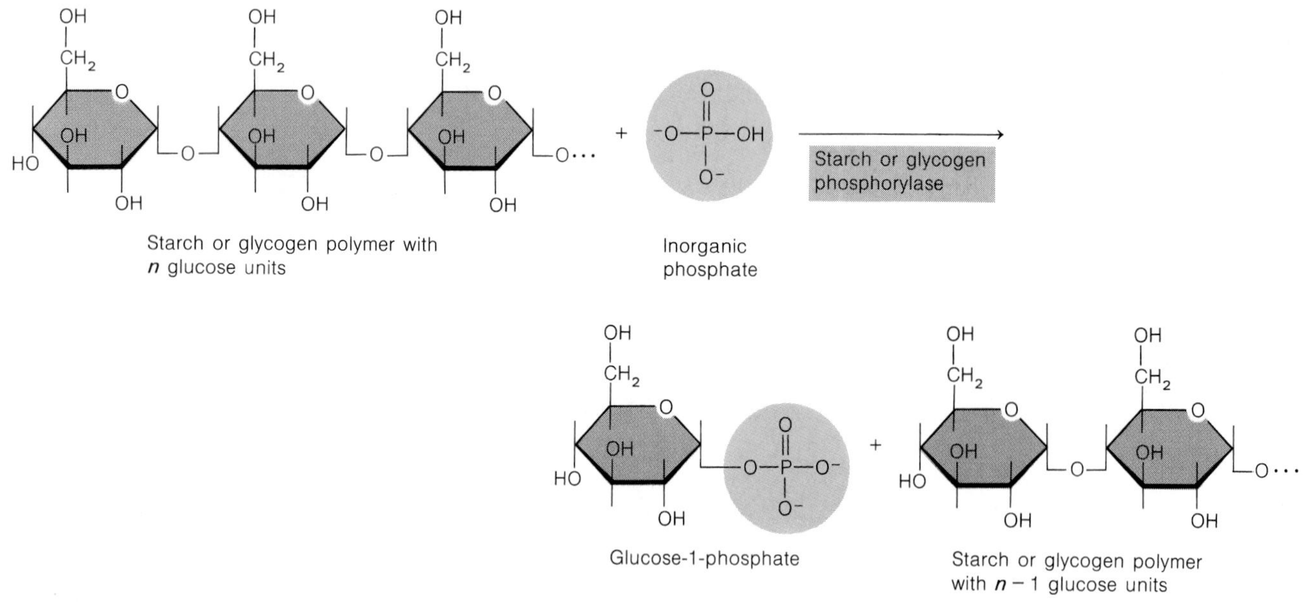

Figure 11-12 Phosphorolytic Cleavage of Storage Polysaccharides. Terminal glucose units of storage polysaccharides such as starch or glycogen are liberated as glucose-1-phosphate by phosphorolytic cleavage. The glucose-1-phosphate is then converted to glucose-6-phosphate and catabolized by the glycolytic pathway, as shown in Figure 11-7.

input of the ATP that would be required for the initial phosphorylation of the free sugar. Consequently, the overall energy yield for glucose is greater by one molecule of ATP when it is catabolized from the polysaccharide level than when it is catabolized with the free sugar as the starting substrate. (This is not a case of getting something for nothing, however, because energy is required in the polymerization process whereby glucose units are added to the growing starch or glycogen chain during polysaccharide synthesis.)

Gluconeogenesis

Cells are able not only to catabolize glucose and other carbohydrates as sources of energy but also to synthesize sugars and polysaccharides to meet cellular or organismal needs. The process of glucose synthesis is called **gluconeogenesis,** literally the *genesis,* or formation, of "new glucose." More specifically, gluconeogenesis is defined as the process by which animal cells synthesize glucose (and other carbohydrates) from three-carbon and four-carbon precursors that are usually noncarbohydrate in nature. The most common starting material is *pyruvate,* or the *lactate* into which pyruvate is converted under anaerobic conditions.

Figure 11-13 compares gluconeogenesis and glycolysis. The two pathways share much in common; in fact, seven of the reactions in the gluconeogenic pathway occur by simple reversal of the corresponding reactions in glycolysis (reactions Gly-2 and Gly-4 through Gly-9). In each case, the same enzyme is used in both directions. Not all of the steps of the gluconeogenic pathway are simply the reversal of glycolytic reactions, however. As Figure 11-13 illustrates, three of the

reactions of the glycolytic pathway—the first, third, and tenth—are accomplished by other means in the direction of gluconeogenesis. In fact, these differences illustrate well an important principle of cellular metabolism: *Biosynthetic pathways are never simply the reversal of the corresponding catabolic pathways.* This principle is based on the energy requirements in each direction. For a metabolic pathway to be thermodynamically favorable in a specific direction, it must be sufficiently exergonic in that direction. That certainly is true of glycolysis; recall that the overall sequence from glucose to pyruvate as summarized by reaction 11-7 has a $\Delta G'$ value of about -20 kcal/mol under typical intracellular conditions in the human body. Clearly, then, $\Delta G'$ for the reverse process would be about $+20$ kcal/mol, making glucose synthesis by the direct reversal of glycolysis highly endergonic and therefore thermodynamically impossible.

Gluconeogenesis is possible because the three most exergonic reactions in the glycolytic pathway (Gly-1, Gly-3, and Gly-10) do not simply run in reverse in the gluconeogenic direction. Instead, the gluconeogenic pathway has **bypass reactions** at each of those three sites—alternative reactions that effectively circumvent the glycolytic reactions that would be the most difficult to drive in the reverse direction. In fact, the three reactions of the glycolytic pathway that are bypassed in gluconeogenesis in Figure 11-13 are the only three that are shown as unidirectional in Figure 11-7. In each of these three instances, the bypass reactions of gluconeogenesis circumvent the irreversibility of the glycolytic step. In the case of both Gly-1 and Gly-3, the requirement for ATP synthesis in the reverse direction is bypassed by a simple hydrolytic reaction, catalyzed by *glucose-6-phosphatase* and *fructose-1,6-bisphosphatase,* respectively (see

Figure 11-13 Pathways for Glycolysis and Gluconeogenesis Compared.

The pathways for glycolysis (left) and gluconeogenesis (right) have nine intermediates and seven enzyme-catalyzed reactions in common. The three essentially irreversible reactions of the glycolytic pathway (in blue-green) are circumvented in gluconeogenesis by four bypass reactions (in yellow). As a catabolic pathway, glycolysis is inherently exergonic, capable of yielding 2 ATP per glucose. Gluconeogenesis, on the other hand, is an anabolic pathway, requiring the coupled hydrolysis of six phosphoanhydride bonds (four from ATP, two from ADP) to drive it in the direction of glucose formation. (The enzymes of the glycolytic pathway are identified in Figure 11-7.)

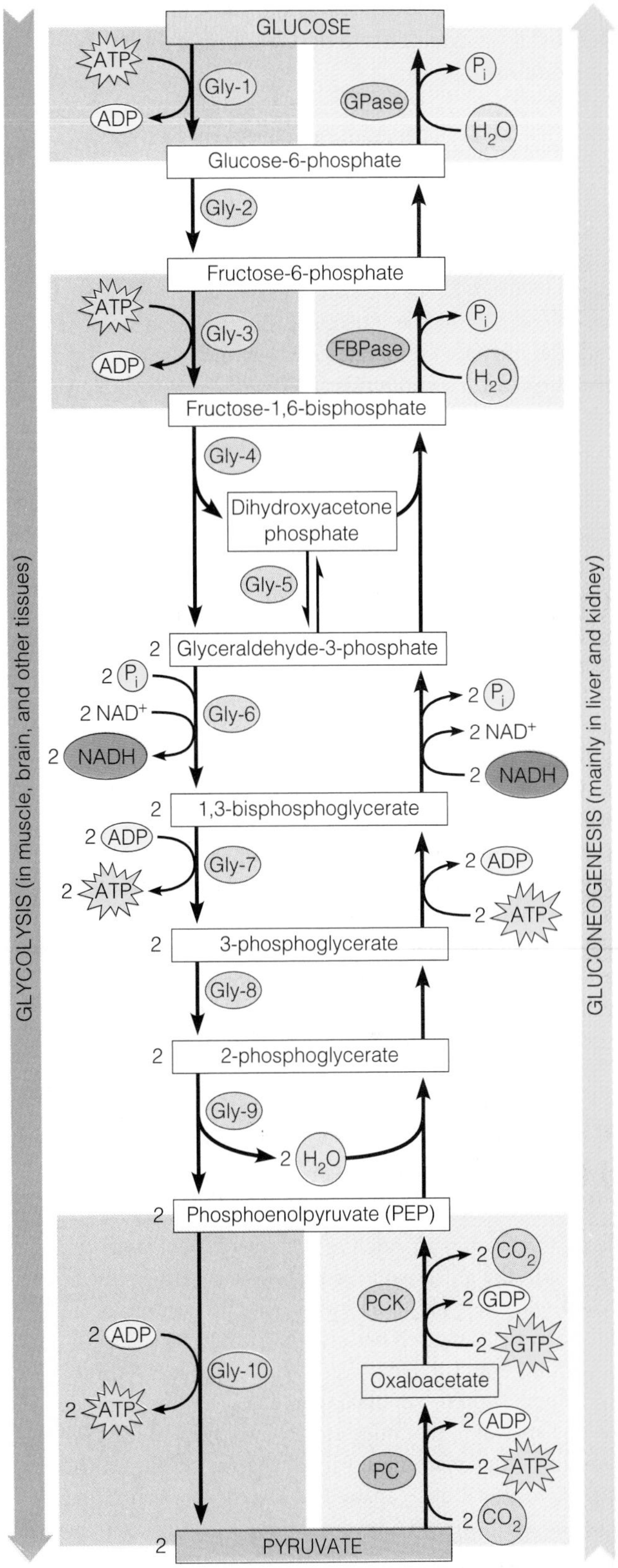

Enzymes That Catalyze the Bypass Reactions of Gluconeogenesis

PC: Pyruvate carboxylase
PCK: Phosphoenolpyruvate carboxykinase
FBPase: Fructose-1,6-bisphosphatase
GPase: Glucose-6-phosphatase

Figure 11-13). Notice how effectively this simple metabolic ploy overcomes the thermodynamic hurdle. In the case of the interconversion of glucose and glucose-6-phosphate, for example, the reaction is exergonic in the glycolytic direction because of the input of an ATP molecule. In other words, the reaction is driven in that direction by the difference in energies of hydrolysis of the phosphoanhydride bond of ATP and the phosphoester bond of glucose-6-phosphate ($\Delta G^{\circ\prime}$ values of -7.3 and -3.3 kcal/mol, respectively; see Figure 11-4). And in the gluconeogenic direction, exergonicity is ensured by the simple hydrolysis of the phosphoester bond, which provides a driving force of 3.3 kcal/mol under standard conditions.

The third site of irreversibility in the glycolytic pathway, reaction Gly-10, is bypassed in gluconeogenesis by a two-reaction sequence (see Figure 11-13). Both of these reactions are driven by the hydrolysis of a phosphoanhydride bond, from ATP in one case and from the related compound GTP in the other. (GTP is the abbreviation for guanosine triphosphate; see Figure 3-12 for the structure of guanine.) The first of these two steps in the gluconeogenic direction involves the addition of CO_2 to pyruvate (a *carboxylation* reaction) to form a four-carbon compound called *oxaloacetate,* which we will meet again in the next chapter. In the second step, the carboxyl group is removed (a *decarboxylation* reaction), and a phosphate group is added to form phosphoenolpyruvate (PEP). In this case, both the phosphate group and the energy are provided by GTP, which is energetically the equivalent of ATP. (GTP and ATP have identical $\Delta G^{o'}$ values for hydrolysis of their respective terminal phosphate groups.) The enzymes that catalyze these reactions are called *pyruvate carboxylase* and *PEP carboxykinase,* respectively.

What these bypass reactions accomplish is clear when the glycolytic and gluconeogenic pathways are compared directly (see Figure 11-13). Glycolysis uses 2 ATPs but generates 4 ATPs, for a net yield of 2 ATP molecules formed per molecule of glucose catabolized. Gluconeogenesis, on the other hand, requires 4 ATPs and 2 GTPs per glucose, or the equivalent of 6 ATP molecules consumed per molecule of glucose synthesized. The difference between the two pathways of 4 ATP molecules per glucose represents enough energy to ensure that gluconeogenesis proceeds at least as exergonically in the direction of glucose synthesis as glycolysis does in the direction of glucose breakdown.

The Regulation of Glycolysis and Gluconeogenesis

If cells have enzymes to catalyze the reactions of both the glycolytic and the gluconeogenic pathways, it seems reasonable to ask what keeps both pathways from proceeding at the same time in the same cell because that would be a very futile metabolic exercise. How, we may ask, do cells "know" whether they should be making glucose or breaking it down? The answer to that question has two parts, at least for humans and other mammals. Part of the answer lies in the differences in metabolic capabilities of cells from different parts of the body. Box 11A, page 303, took up this topic by tracing the fate of sugar molecules in the human body, including glycolysis in muscle (and other) cells and gluconeogenesis in the liver and kidney. Here, we will focus on the other part of the answer by considering how glycolysis and gluconeogenesis are regulated within the same cell.

Like all metabolic pathways, glycolysis and gluconeogenesis are regulated to ensure that they function at rates that are responsive to cellular and organismal needs for their products—ATP and glucose, respectively. Not surprisingly, glycolysis and gluconeogenesis are regulated in a reciprocal, or inverse, manner: Intracellular conditions known to stimulate one pathway usually have an inhibitory effect on the other. In addition, glycolysis is closely coordinated with other major pathways of energy generation and utilization in the cell, especially the pathways involved in aerobic respiration that we will be considering in Chapter 12.

Most of the regulation of glycolysis and gluconeogenesis involves one or both of two major control mechanisms, *allosteric regulation* and *hormonal regulation.* Allosteric regulation operates at the level of enzyme activity and is a strictly intracellular mechanism. Hormonal regulation, on the other hand, is an intrinsically organismal mechanism because the initiating signal is a hormone produced in another, often distant, part of the body. We will consider allosteric regulation here, focusing specifically on liver cells, which possess the enzymes for both glycolysis and gluconeogenesis. Hormonal regulation will also be mentioned, but only briefly because it will be discussed in detail in the context of *signal transduction* in Chapter 23.

Allosteric Regulation of Glycolysis and Gluconeogenesis

Recall from Chapter 6 that **allosteric regulation** of enzyme activity involves the interconversion of an enzyme between two forms, one of which is catalytically active (or more active), whereas the other is inactive (or less active). Whether a specific enzyme molecule is in its active or inactive form depends on whether a specific *allosteric effector* is bound to the *allosteric site,* and whether that effector is an *allosteric activator* or an *allosteric inhibitor* (see Figure 6-19). Every allosterically regulated enzyme has one or more *catalytic subunits* with an active site on each and one or more *regulatory subunits* containing the allosteric site(s) to which the effector(s) bind.

Figure 11-14 shows the key regulatory enzymes of the glycolytic and gluconeogenic pathways and the allosteric effectors that regulate each enzyme. Each allosteric effector is identified as either an activator $(+)$ or an inhibitor $(-)$ of the enzyme(s) to which it binds, based primarily on studies with liver cells. Several points quickly become apparent from the figure. Notice, for instance, that each of the regulatory enzymes is unique to its pathway, thereby making it possible for each pathway to be regulated independently of the other. Notice also the reciprocal nature of the regulation of the two pathways: AMP and acetyl CoA, the two effectors to which both pathways are sensitive, have opposite effects in the two directions. AMP, for example, *activates* glycolysis but *inhibits* gluconeogenesis.

Moreover, the effects of the regulatory agents "make sense"—that is, they are invariably in the direction one would predict, based on an understanding of the role each pathway plays in the cell. Consider, for example, the effects

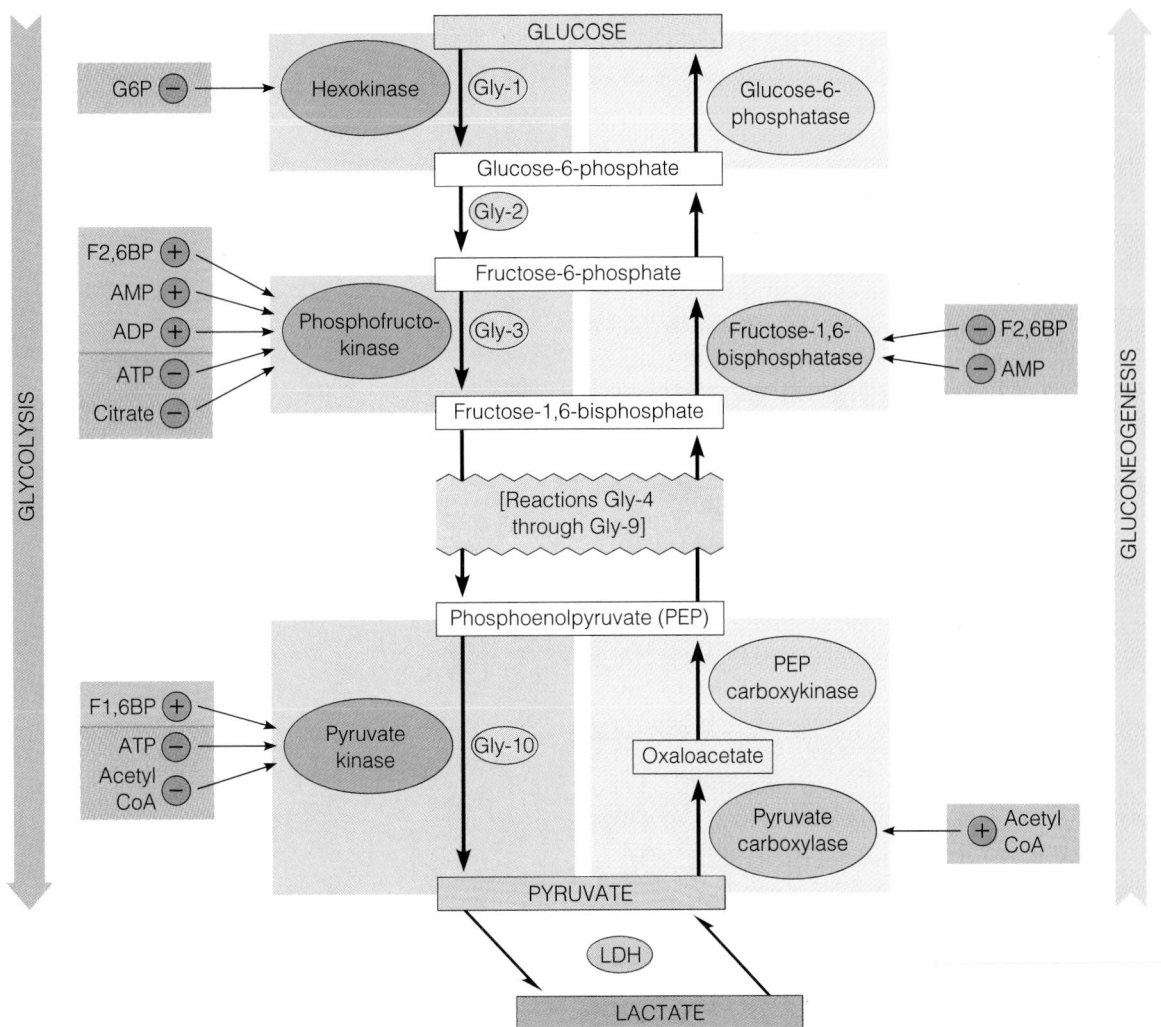

Figure 11-14 The Regulation of Glycolysis and Gluconeogenesis. Glycolysis and gluconeogenesis are regulated in a reciprocal manner. In both cases, regulation involves allosteric activation (+) or inhibition (−) of enzymes that catalyze reactions unique to the pathway; reactions common to both pathways (Gly-2 and Gly-4 through Gly-9) are therefore not shown. For glycolysis, the key regulatory enzymes are those that catalyze the three essentially irreversible reactions unique to this pathway (red). For gluconeogenesis, two of the four bypass enzymes (green) that are unique to this pathway are the main sites of allosteric regulation. Allosteric regulators include acetyl CoA, AMP, ATP, citrate, fructose-1,6-bisphosphate (F1,6B), fructose-2,6-bisphosphate (F2,6BP), and glucose-6-phosphate (G6P). Acetyl CoA and citrate are important intermediates in aerobic respiration. F2,6BP is synthesized by PFK-2, as shown in Figure 11-15.

of ATP, ADP, and AMP, which can be thought of as the "charged," "discharged," and "very discharged" forms of the adenosine phosphates, with 2, 1, and 0 phosphoanhydride bonds, respectively. When the concentration of ATP is low and the ADP and/or AMP concentrations are high, the cell is clearly low on energy, so it is reasonable for ADP and AMP to activate glycolysis. Conversely, as the ATP concentration increases and the ADP and/or AMP concentrations decrease, the stimulatory effects of AMP and ADP on glycolysis lessen, and the inhibitory effect of ATP on both phosphofructokinase and pyruvate kinase comes into play, reducing the rate of glycolysis appropriately.

In the case of phosphofructokinase (PFK), it may be surprising to find that ATP is an allosteric inhibitor of an enzyme for which it is also a required substrate. This seems to create a contradiction of effects because increases in substrate concentration should increase the rate of an enzyme-catalyzed reaction, yet increasing the concentration of an allosteric inhibitor should render the enzyme less active. This apparent contradiction is readily explained, however, because the active site of PFK has a high affinity—a low K_m—for ATP, whereas the affinity of the effector site for ATP is lower—that is, the K_m is higher. Thus, at low ATP concentrations, binding occurs at the catalytic site but not to any appreciable extent at the allosteric site, so most of the PFK molecules remain in the active form and glycolysis proceeds. As the ATP concentration increases, however, binding is promoted at the effector site, converting more and more of the PFK molecules to the inactive form and thereby serving effectively as a throttle for the whole glycolytic sequence.

Both pathways shown in Figure 11-14 are also subject to allosteric regulation by compounds involved in aerobic respiration. As we will learn in the next chapter, acetyl CoA and citrate are key players in an aerobic pathway called the *tricarboxylic acid cycle*. (They are, in fact, a substrate and product, respectively, of the first reaction of that sequence.) High concentrations of acetyl CoA and citrate indicate that the cell is well supplied with substrate for the next phase of respiratory metabolism beyond pyruvate. Thus, it is not surprising to find that both of these compounds have an inhibitory effect on a glycolytic enzyme, thereby decreasing the rate at which pyruvate is formed. Similarly, the stimulatory effect of acetyl CoA on a gluconeogenic enzyme is consistent with the availability of pyruvate for conversion to glucose.

Fructose-2,6-Bisphosphate as a Regulator of Glycolysis and Gluconeogenesis

Although each of the preceding effects plays a significant role in the regulation of glycolysis and gluconeogenesis, research findings since 1980 have established **fructose-2,6-bisphosphate (F2,6BP)** as the most important regulator of both pathways, especially in liver cells. As Figure 11-14 shows, F2,6BP activates the glycolytic enzyme (PFK) that phosphorylates fructose-6-phosphate and inhibits the gluconeogenic enzyme (FBPase) that catalyzes the reverse reaction.

The formation, dephosphorylation, and main effects of F2,6BP in liver cells are depicted in Figure 11-15. F2,6BP is synthesized by ATP-dependent phosphorylation of fructose-6-phosphate, the same reaction that gives rise to fructose-1,6-bisphosphate at reaction Gly-3 in the glycolytic pathway. However, synthesis of F2,6BP is catalyzed by a separate form of phosphofructokinase, called **phosphofructokinase-2 (PFK-2)** to distinguish it from PFK-1, the glycolytic enzyme. The activity of PFK-2 depends on the phosphorylation status of one of its subunits. When that subunit is in the phosphorylated form, the activity of the enzyme is decreased (because of an increase in its K_m for fructose-6-phosphate). Conversely, removal of the phosphate group increases its activity.

The phosphorylation of PFK-2 by ATP is catalyzed by a *protein kinase* (see Figure 11-15). The activity of this enzyme depends, in turn, on *cyclic AMP (cAMP)*, a form of AMP known to be a key intermediate in many cellular signal transduction pathways (Figure 11-16). Another enzyme activity, initially called *fructose-2,6-bisphosphatase*, was found to cleave the phosphate group from F2,6BP, converting the compound back to fructose-6-phosphate (see Figure 11-15). This activity was also shown to be regulated by cAMP-stimulated phosphorylation, but in this case phosphorylation *increases* the activity. We now understand that both of these activities are, in fact, properties of the same enzyme, PFK-2. Because it has two separate catalytic activities, PFK-2

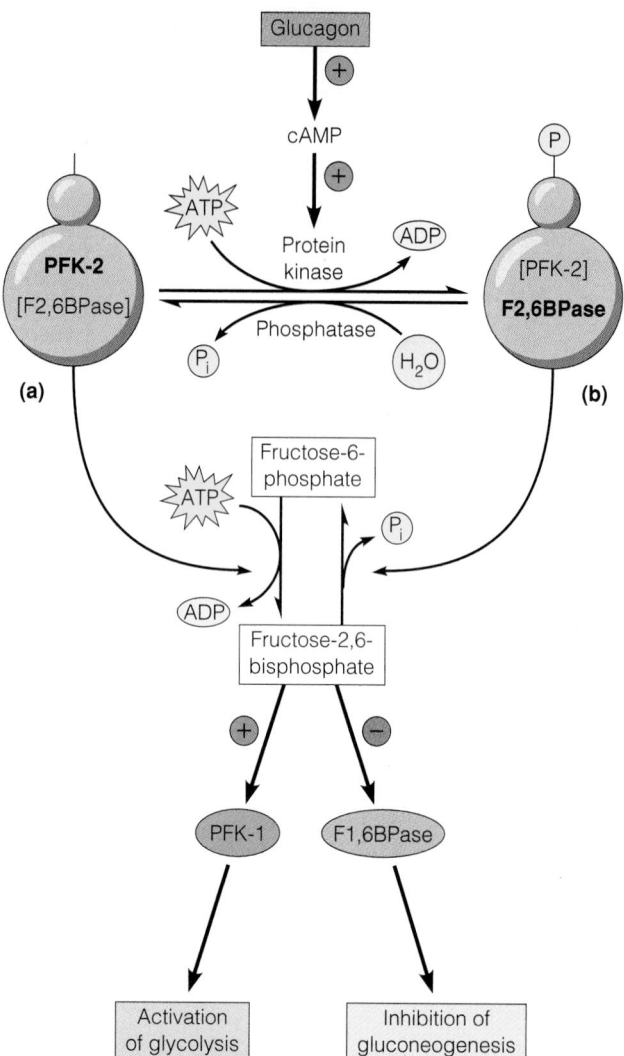

Figure 11-15 The Regulatory Role of PFK-2. The enzyme phosphofructokinase-2 (PFK-2) is a bifunctional enzyme, with two different catalytic activities. **(a)** In its unphosphorylated form, the enzyme catalyzes the phosphorylation of fructose-6-phosphate to form fructose-2,6-bisphosphate. **(b)** In its phosphorylated form, the enzyme catalyzes the hydrolysis of the phosphate group from carbon atom 2 of fructose-2,6-bisphosphate. Fructose-2,6-bisphosphate is an allosteric activator of the glycolytic enzyme PFK-1 and an allosteric inhibitor of the gluconeogenic enzyme fructose-1,6-bisphosphatase. PFK-2 is phosphorylated by a protein kinase under the control of cyclic AMP (cAMP) and is dephosphorylated by a phosphatase. In turn, cAMP is regulated by the hormones glucagon and epinephrine. The effect of glucagon is to increase the cAMP level, thereby stimulating the phosphorylation of PFK-2 and converting it to the form with fructose-2,6-bisphosphatase activity. The resulting decrease in the concentration of fructose-2,6-bisphosphate eliminates the activation of glycolysis and alleviates the inhibition of gluconeogenesis.

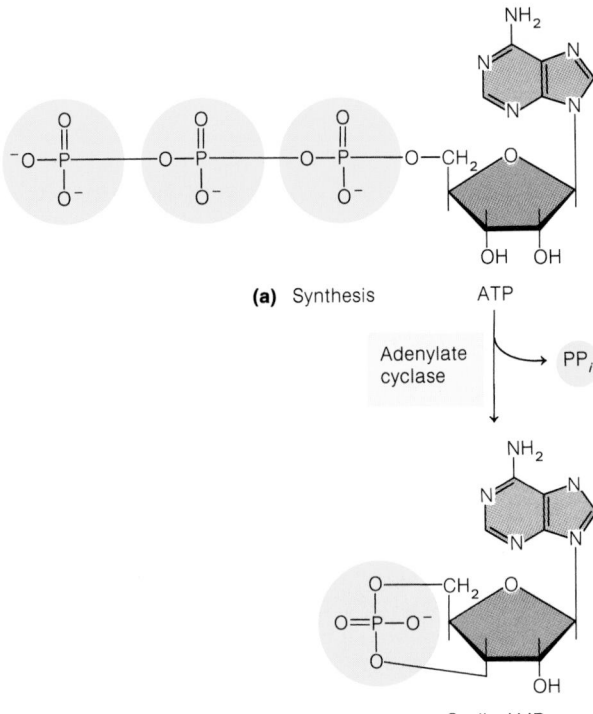

(a) Synthesis   ATP

Adenylate cyclase → PP$_i$

Cyclic AMP

Figure 11-16   The Structure of Cyclic AMP. Cyclic AMP (adenosine-3′, 5′-cyclic monophosphate, abbreviated cAMP) is a form of AMP in which the phosphate group is linked covalently to both carbon atom 3 and carbon atom 5, forming a ring. Cyclic AMP is formed from ATP by the action of the enzyme adenylate cyclase. Cyclic AMP plays a key role in intracellular mediation of hormonal and other signals that impinge upon cells from the outside.

is called a *bifunctional enzyme.* When the subunit with the phosphorylation site is in the nonphosphorylated form, the bifunctional enzyme is active as a kinase, forming F2,6BP from fructose-6-phosphate (see Figure 11-15a). When the subunit is phosphorylated, the enzyme is active as a phosphatase, cleaving F2,6BP to fructose-6-phosphate (see Figure 11-15b).

Thus, cAMP affects the F2,6BP concentration in two ways: It inactivates the kinase activity and stimulates the phosphatase activity. Both of these effects tend to *decrease* the concentration of F2,6BP in the cell. This change leads, in turn, to lessened stimulation of PFK-1 and lessened inhibition of fructose-1,6-bisphosphatase, thereby decreasing the glycolytic flux and increasing the gluconeogenic flux.

The effects of cAMP shown in Figure 11-15 provide a link to the topic of hormonal regulation because the cAMP level in liver cells is controlled primarily by the hormones *glucagon* and *epinephrine (adrenalin)*. These hormones cause an increase in cAMP concentration, thereby stimulating gluconeogenesis and consequently raising the blood glucose level. Moreover, the increase in cAMP also stimulates a regulatory cascade that *increases* the rate of glycogen breakdown and therefore the blood glucose level. Not surprisingly, the effect of cAMP on glycogen synthesis is just the opposite: Whether triggered by glucagon or epinephrine, an increase in liver cAMP concentration leads to a *decrease* in the rate of glycogen formation. For further details on hormonal regulation and the role of cAMP in mediating hormonal effects, see the discussion on hormonal signal transduction in Chapter 23.

Metabolic pathways in cells are either anabolic (synthetic) or catabolic (degradative). The latter reactions provide the energy necessary to drive the former. ATP is a useful intermediate for this purpose because its terminal anhydride bond has a free energy of hydrolysis that allows ATP to serve as a donor, and ADP to serve as an acceptor, of phosphate groups. Most chemotrophs derive the energy needed for ATP generation from the catabolism of organic nutrients such as carbohydrates, fats, or proteins. They do so either by fermentative processes in the absence of oxygen or by respiratory metabolism, which is usually, though not always, an aerobic process.

Using glucose as a prototype substrate, catabolism under both anaerobic and aerobic conditions begins with the glycolytic pathway, a ten-step sequence of reactions in which glucose is converted to pyruvate with the net production of two molecules of ATP per molecule of glucose. In the absence of oxygen, the reduced coenzyme NADH generated during glycolysis must be reoxidized at the expense of pyruvate, leading to fermentation end-products such as lactate or ethanol and carbon dioxide. This severely limits the extent to which the free energy content of the glucose molecule can be released, but the 7% or so that is available is conserved as ATP quite efficiently. Although usually written with glucose as the starting substrate, the glycolytic sequence is also the mainstream pathway for the catabolism of a variety of related sugars, as well as for the utilization of the glucose-1-phosphate derived by phosphorolytic cleavage of storage polysaccharides such as starch or glycogen.

Gluconeogenesis is, in a sense, the "opposite" of glycolysis because it is the pathway whereby animal cells synthesize glucose (and other carbohydrates) from three-carbon and four-carbon noncarbohydrate starting materials such as pyruvate. However, the gluconeogenic pathway is not just glycolysis in reverse. The two pathways share seven enzyme-catalyzed reactions in common, but in gluconeogenesis the three most exergonic reactions of glycolysis are bypassed by reactions that render the pathway exergonic in the gluconeogenic direction by the input of energy from ATP and GTP.

Like other metabolic pathways, glycolysis and gluconeogenesis are carefully regulated to ensure that the rate of product formation (ATP and glucose, respectively) is carefully tuned to actual cellular needs. Both allosteric and hormonal regulation are involved. The enzymes that are subject to allosteric regulation catalyze reactions unique to the respective pathways. These enzymes are regulated by one or more effectors, which include ATP, ADP, and AMP, as well as acetyl CoA and citrate, key intermediates in aerobic respiration. In animal cells, the most important allosteric regulator of both glycolysis and gluconeogenesis is fructose-2,6-bisphosphate, the concentration of which depends on the relative kinase and phosphatase activities of the bifunctional enzyme PFK-2. The function of PFK-2 is regulated, in turn, by the hormones glucagon and epinephrine, mediated by the intracellular concentration of cyclic AMP.

As complex as glycolysis may seem upon first encounter, it represents the simplest mechanism by which glucose can be degraded in dilute solution at temperatures compatible with life and with a large portion of the free energy yield conserved as ATP. Coupled to an appropriate reductive sequence to regenerate the coenzyme NAD^+, glycolysis serves the cell well under anaerobic conditions, meeting energy needs despite the absence of oxygen. All we have seen so far, however, pales in comparison with the potential for energy release and conservation in the presence of oxygen, for aerobic respiration is the capstone of bioenergetics and the mainspring of cellular energy metabolism for most chemotrophic forms of life. For that, we take the NADH and the pyruvate provided by glycolysis and proceed to the process of aerobic respiration, the subject of Chapter 12.

KEY TERMS FOR SELF-TESTING

Metabolic Pathways

metabolism (p. 296)
metabolic pathway (p. 296)
anabolic pathway (p. 296)
catabolic pathway (p. 296)

ATP: The Universal Energy Coupler

adenosine triphosphate (ATP) (p. 297)
phosphoanhydride bond (p. 297)
phosphoester bond (p. 297)
charge repulsion (p. 298)

resonance stabilization (p. 298)
resonance hybrid (p. 298)
group transfer reaction (p. 300)

Chemotrophic Energy Metabolism

chemotrophic energy metabolism (p. 300)
oxidation (p. 301)
dehydrogenation (p. 301)
reduction (p. 301)
hydrogenation (p. 301)

coenzyme (p. 302)
nicotinamide adenine dinucleotide (NAD^+) (p. 302)
glucose (p. 302)
aerobic respiration (p. 306)
fermentation (p. 306)
strict (obligate) aerobe (p. 306)
strict (obligate) anaerobe (p. 306)
facultative organism (p. 306)

PROBLEM SET

11-1. "High-Energy Bonds." When first introduced by Fritz Lipmann in 1941, the term "high-energy bond" was considered a useful concept for describing the energetics of biochemical molecules and reactions. However, use of the term can lead to confusion when relating ideas about cellular energy metabolism to those of physical chemistry. To check out your own understanding, indicate whether each of the following statements is true (T) or false (F). If false, reword the statement to make it true.

(a) Energy is stored in special high-energy bonds in molecules such as ATP and is released when these bonds are broken.

(b) Energy is always released whenever a covalent bond is formed and is always required to break a covalent bond.

(c) To a physical chemist, "high-energy bond" means a very stable bond that requires a lot of energy to break, whereas to a biochemist, the term is likely to mean a bond that releases a lot of energy upon hydrolysis.

(d) The terminal phosphate of the ATP is a "high-energy" phosphate that takes its high energy with it when it is hydrolyzed.

(e) Phosphoester bonds are "low-energy bonds" because they require less energy to break than the "high-energy bonds" of phosphoanhydrides.

(f) The term "high-energy molecule," if used at all, should be thought of as a characteristic of the reaction in which the molecule is involved and not a property of a particular bond within that molecule.

11-2. The History of Glycolysis. Following are several observations that led to the elucidation of the glycolytic pathway. In each case, suggest a metabolic basis for the observed effect, and explain the significance of the observation for the elucidation of the pathway.

(a) Alcoholic fermentation in yeast extracts requires a heat-labile fraction originally called *zymase* and a heat-stable fraction (*cozymase)* that is necessary for the activity of zymase.

(b) Alcoholic fermentation does not take place in the absence of inorganic phosphate.

(c) In the presence of iodoacetate, a known inhibitor of glycolysis, fermenting yeast extracts accumulate a doubly phosphorylated hexose.

(d) In the presence of fluoride ion, another known glycolytic inhibitor, fermenting yeast extracts accumulate two phosphorylated three-carbon acids.

11-3. What's Your Reaction? Figure 11-4 indicates the free energy of hydrolysis for a variety of phosphorylated compounds commonly found in cells. Use the information in the figure and your familiarity with $\Delta G^{\circ\prime}$ and $\Delta G^\prime$ from Chapter 5 to answer the following questions:

(a) For each of the following compounds, indicate whether the transfer of its phosphate group to ADP to form ATP would be exergonic or endergonic, assuming a temperature of 25°C and conditions such that $\Delta G^\prime = \Delta G^{\circ\prime}$: phosphoenolpyruvate; glucose-1-phosphate; phosphocreatine; and glycerol phosphate.

(b) If equimolar concentrations of ADP, ATP, 1,3-bisphosphoglycerate, and 3-phosphoglycerate are mixed together at 25°C and an appropriate amount of the enzyme phosphoglycerokinase is added, what reaction will occur, and how exergonic will it be in the direction written?

(c) If equimolar concentrations of ADP, phosphoenolpyruvate, and glucose are mixed together at 25°C with appropriate amounts of the enzymes pyruvate kinase and hexokinase, what reaction(s) do you predict will occur? Explain your answer.

(d) If the mixture in part c is allowed to come to equilibrium, what compounds will be present in the equilibrium mixture? Explain your answer.

(e) The conversion of glycerol to glycerol phosphate requires the transfer of a phosphate group from a compound with a suitable $\Delta G^\prime$ for its hydrolysis. In most cells, the donor is ATP. Would 1,3-bisphosphoglycerate (BPG) be a possible donor also? If you answer no, explain why not. If you answer yes, suggest a reason why this reaction does not actually occur in cells. (Base your reason on bioenergetics.)

(f) Under conditions such that $\Delta G^\prime = \Delta G^{\circ\prime}$, would the transfer of a phosphate group from 1,3-bisphosphoglycerate to ADP be able to drive the synthesis of glycerol phosphate from free glycerol and inorganic phosphate, assuming the existence of an appropriate coupling mechanism? Making the same assumptions, would phosphate transfer from BPG to ADP be able to drive the synthesis of glucose-1-phosphate from free glucose and inorganic phosphate? Explain your answer in both cases.

11-4. Glycolysis in 25 Words or Less. Complete each of the following statements about the glycolytic pathway in 25 words or less:

(a) Although the brain is an obligately aerobic organ, it still depends on glycolysis because . . .

(b) Although one of its reactions is an oxidation, glycolysis can proceed in the absence of oxygen because . . .

(c) What happens to the pyruvate generated by the glycolytic pathway depends on . . .

(d) If you bake bread or brew beer, you are dependent on glycolysis for . . .

(e) Two organs in your body that can use lactate are . . .

(f) The synthesis of glucose from lactate in a liver cell requires more molecules of nucleoside triphosphates (ATP and GTP) than are formed during the catabolism of glucose to lactate in a muscle cell because . . .

11-5. Energetics of Carbohydrate Utilization. The anaerobic fermentation of free glucose has an ATP yield of 2 ATP molecules per molecule of glucose. For glucose units in a glycogen molecule, the yield is 3 ATP molecules per molecule of glucose. The corresponding value for the disaccharide sucrose is 2 ATP molecules per molecule of monosaccharide if the sucrose is eaten by an animal, but 2.5 ATP molecules per molecule of monosaccharide if the sucrose is metabolized by a bacterium.

(a) Explain why the glucose units present in glycogen have a higher ATP yield than free glucose molecules.

(b) What is the likely mechanism for sucrose breakdown in the gut of an animal to explain the energy yield of 2 ATP molecules per molecule of monosaccharide?

(c) Based on what you know about the process of glycogen breakdown, suggest a mechanism for bacterial sucrose metabolism that is consistent with an energy yield of 2.5 ATP molecules per molecule of monosaccharide.

(d) What energy yield (in molecules of ATP per molecule of monosaccharide) would you predict for the bacterial catabolism of raffinose, a trisaccharide?

11-6. Glucose Phosphorylation. The direct phosphorylation of glucose by inorganic phosphate is a thermodynamically unfavorable reaction:

glucose + P$_i$ $\longrightarrow$ glucose-6-P + H$_2$O

$$\Delta G^{\circ\prime} = +3.3 \text{ kcal/mol} \quad \textbf{(11-21)}$$

In the cell, this is accomplished by coupling the reaction to the hydrolysis of ATP, a highly exergonic reaction:

ATP + H$_2$O $\longrightarrow$ ADP + P$_i$

$$\Delta G^{\circ\prime} = -7.3 \text{ kcal/mol} \quad \textbf{(11-22)}$$

Typical concentrations of these intermediates in yeast cells are as follows:

[glucose-6-phosphate]: 0.08 mM [ATP]: 1.8 mM
[P$_i$]: 1.0 mM [ADP]: 0.15 mM

(a) What minimum concentration of glucose would have to be maintained in a yeast cell for direct phosphorylation (reaction 11-21) to be thermodynamically spontaneous? Is this physiologically reasonable? (Assume a temperature of 25°C for all calculations.)

(b) What is the overall equation for the coupled (ATP-driven) phosphorylation reaction? What is its $\Delta G^{\circ\prime}$ value?

(c) What minimum concentration of glucose would have to be maintained in a yeast cell for the coupled reaction to be thermodynamically spontaneous? Is this physiologically reasonable?

(d) By about how many orders of magnitude is the minimum required glucose concentration reduced when the phosphorylation of glucose is coupled the hydrolysis of ATP?

(e) Assuming a yeast cell to have a glucose concentration of 5.0 mM, what is $\Delta G^\prime$ for the coupled phosphorylation reaction?

11-7. Ethanol Intoxication and Methanol Toxicity. The enzyme alcohol dehydrogenase was mentioned in this chapter because of its role in the final step of alcoholic fermentation. However, the enzyme also occurs commonly in aerobic organisms, including humans. The ability of the human body to catabolize the ethanol in alcoholic beverages depends on the presence of alcohol dehydrogenase in the liver. One of the effects of ethanol intoxication is a dramatic decrease in the NAD$^+$ concentration in liver cells, which decreases the aerobic utilization of glucose. Methanol

(sometimes called "wood alcohol"), on the other hand, is not just an intoxicant; it is a deadly poison because of the toxic effect of the formaldehyde to which it is converted in the liver.

(a) Why does ethanol consumption lead to a reduction in NAD$^+$ concentration and to a decrease in aerobic respiration?

(b) Most of the unpleasant effects of hangovers result from an accumulation of acetaldehyde and its metabolites. Where does the acetaldehyde come from?

(c) The medical treatment for methanol poisoning usually involves administration of large doses of ethanol. Why is this treatment effective?

11-8. Propionate Fermentation. Although lactate and ethanol are the best-known products of fermentation, other pathways are also known, some with important commercial application. Swiss cheese production, for example, depends on the bacterium *Propionibacterium freudenreichii*, which converts pyruvate to propionate (CH$_3$—CH$_2$—COO$^-$). Fermentation of glucose to propionate always generates at least one other product as well.

(a) Why is it not possible to devise a scheme for the fermentation of glucose with propionate as the sole end-product?

(b) Suggest an overall scheme for propionate production that generates only one additional product, and indicate what that product might be.

(c) If you know that Swiss cheese production actually requires both propionate and carbon dioxide and that both are produced by *Propionibacterium* fermentation, what else can you now say about the fermentation process that this bacterium carries out?

11-9. Galactose Metabolism. The glycolytic pathway is usually written with glucose as the starting substrate because it is the single most important sugar for most organisms. But cells can utilize a variety of sugars. The diet of young mammals, for example, consists almost entirely of milk, which contains as its principal carbohydrate the disaccharide lactose. When lactose is hydrolyzed in the intestine, it yields one molecule each of the hexoses glucose and galactose, so the cells of these animals (and of human babies) must metabolize just as much galactose as glucose. Galactose is metabolized by phosphorylation and conversion to glucose. The reaction sequence is a bit complicated, though, because the conversion to glucose (an *epimerization* reaction on carbon atom 4) occurs while the sugar is attached to the carrier *uridine diphosphate (UDP)*, a close relative of ADP. The reactions are as follows:

galactose + ATP $\longrightarrow$ galactose-1-phosphate + ADP   **(11-23)**

galactose-1-phosphate + UDP-glucose $\longrightarrow$
glucose-1-phosphate + UDP-galactose   **(11-24)**

UDP-galactose $\longrightarrow$ UDP-glucose   **(11-25)**

glucose-1-phosphate $\longrightarrow$ glucose-6-phosphate   **(11-26)**

(a) Write a reaction for the overall conversion of galactose to glucose-6-phosphate.

(b) How do you think the $\Delta G^{\circ\prime}$ value for the overall reaction of part a compares with that for the hexokinase reaction (reaction Gly-1)?

(c) If you know that the epimerase reaction (reaction 11-25) has an absolute requirement for the coenzyme NAD$^+$ and involves 4-ketoglucose as an enzyme-bound intermediate, can you suggest a reaction sequence to explain the conversion of galactose to glucose? (Use Haworth projections in this case.)

(d) One form of the congenital disease *galactosemia* is caused by a genetic absence of the enzyme that catalyzes reaction 11-24. The symptoms of galactosemia, which include mental disorders and cataracts of the eye, are thought to result from high levels of galactose in the blood and from an intracellular accumulation of galactose-1-phosphate. Why does this seem a reasonable hypothesis?

11-10. Glycolysis and Gluconeogenesis. As Figure 11-13 indicates, gluconeogenesis is accomplished by what is essentially the reverse of the glycolytic pathway but with bypass reactions in place of the first, third, and tenth reactions in glycolysis.

(a) Explain why it is not possible to accomplish gluconeogenesis by a simple reversal of every reaction in glycolysis.

(b) Write an overall reaction for gluconeogenesis that is comparable to reaction 11-16 for glycolysis.

(c) Explain why gluconeogenesis requires the input of six molecules of nucleoside triphosphates (4 ATPs and 2 GTPs) per molecule of glucose synthesized when glycolysis only yields two molecules of ATP per molecule of glucose.

(d) Assuming concentrations of ATP, ADP, and P_i such that $\Delta G'$ for hydrolysis of ATP is about –10 kcal/mol, what is the approximate $\Delta G'$ value for the overall reaction for gluconeogenesis that you wrote in part b?

(e) With all of the enzymes for glycolysis and gluconeogenesis present in a liver cell, how does the cell "know" whether it should be synthesizing or catabolizing glucose at any point in time?

11-11. Energy Charge. A useful way to express the energy status of a cell is in terms of the relative concentrations of AMP, ADP, and ATP. Specifically, the concentrations of these three nucleotides are used to determine the *energy charge* of the cell, which is a measure of the relative abundance of the high-energy forms, ATP and ADP:

$$\text{energy charge} = \frac{2[\text{ATP}] + [\text{ADP}]}{2[\text{ATP}] + 2[\text{ADP}] + 2[\text{AMP}]} \quad \textbf{(11-27)}$$

Most cells maintain an energy charge of about 0.9.

(a) Why does AMP appear in the denominator but not in the numerator?

(b) Why does ATP have a weighting factor twice that of ADP in the numerator?

(c) What is the lowest theoretical value that the energy charge can have? What would be true of the concentrations of AMP, ADP, and ATP at that value? Is it likely that a cell would ever have an energy charge that low?

(d) What is the highest theoretical value that the energy charge can have? What would be true of the concentrations of AMP, ADP, and ATP at that value? Is it likely that a cell would ever have an energy charge that high?

(e) If the AMP concentration is about one-tenth that of ADP, what is the ATP/ADP ratio in a cell that has an energy charge of 0.9?

(f) How does the ATP/ADP ratio that you calculated in part e affect the $\Delta G'$ value for the hydrolysis of ATP, assuming a $\Delta G^{\circ\prime}$ value of –7.3 kcal/mol at 25°C?

(g) How does the allosteric regulation of glycolysis and gluconeogenesis render these pathways sensitive to the energy charge of the cell?

11-12. Arsenate Poisoning. Arsenate ($HAsO_4^{2-}$) is a potent poison in almost all living systems. Among other effects, arsenate is known to uncouple the phosphorylation event from the oxidation of glyceraldehyde-3-phosphate. This uncoupling occurs because the enzyme involved, glyceraldehyde-3-phosphate dehydrogenase, can utilize arsenate instead of inorganic phosphate, forming glycerate-1-arseno-3-phosphate. This product is a highly unstable compound that immediately undergoes nonenzymatic hydrolysis into glycerate-3-phosphate and free arsenate.

(a) In what sense might arsenate be called an *uncoupler* of substrate-level phosphorylation?

(b) Why is arsenate such a toxic substance for an organism that depends critically on glycolysis to meet its energy needs?

(c) Can you think of other reactions that are likely to be uncoupled by arsenate in the same way as the glyceraldehyde-3-phosphate dehydrogenase reaction?

11-13. Enzyme Tests in Clinical Medicine. Lactate dehydrogenase (LDH) catalyzes the reduction of pyruvate to lactate in anaerobic metabolism. LDH activity in the blood is elevated following a heart attack, and the degree of elevation can be used to diagnose the severity of tissue damage. Other soluble enzymes of cells, such as creatine phosphokinase and the transaminases, show similar patterns. Why might there be a relationship between a heart attack and metabolic enzymes circulating in the blood?

SUGGESTED READING

General References

Abeles, R. H., P. A. Frey, and W. P. Jencks. *Biochemistry.* Boston: Jones and Bartlett, 1992.

Beitner, R., ed. *Regulation of Carbohydrate Metabolism,* vols. I and II. Boca Raton, FL: CRC Press, 1985.

Lehninger, A. L., D. L. Nelson, and M. M. Cox. *Principles of Biochemistry,* 2d ed. New York: Worth, 1993.

Mathews, C. K., and K. E. van Holde. *Biochemistry,* 2d ed. Menlo Park, CA: Benjamin/Cummings, 1996.

Stryer, L. *Biochemistry,* 4th ed. New York: W. H. Freeman, 1995.

ATP Generation

Bridger, W. A., and J. F. Henderson. *Cell ATP.* New York: Wiley, 1983.

Gayford, C. ATP: A coherent view for school advanced level studies in biology. *J. Biol. Educ.* 20 (1986): 27.

Hinkle, P. C., and R. E. McCarty. How cells make ATP. *Sci. Amer.* 238 (March 1978): 104.

Lipmann, F. Metabolic generation and utilization of phosphate bond energy. *Adv. Enzymol.* 18 (1941): 99.

Lipmann, F. *Wanderings of a Biochemist.* New York: Wiley, 1971.

Morris, J. C. *A Biologist's Physical Chemistry,* 2d ed. London: Edward Arnold, 1974.

Schlenk, F. The ancestry, birth and adolescence of ATP. *Trends Biochem. Sci.* 12 (1987): 367.

Westheimer, F. Why nature chose phosphates. *Science* 235 (1987): 1173.

Glycolysis and Fermentation

Erecinska, M., and D. F. Wilson. Regulation of cellular energy metabolism. *J. Membr. Biol.* 70 (1982): 1.

Fothergill-Gilmore, L. A. The evolution of the glycolytic pathway. *Trends Biochem. Sci.* 11 (1986): 47.

Masters, C. J., S. Reid, and M. Don. Glycolysis—new concepts in an old pathway. *Molec. Cell. Biochem.* 76 (1987): 3.

Ovadi, J. Old pathway—new concept. Control of glycolysis by metabolite-modulated dynamic enzyme associations. *Trends Biochem. Sci.* 13 (1988): 486.

Gluconeogenesis

Hers, G. A., and L. Hue. Gluconeogenesis and related aspects of glycolysis. *Annu. Rev. Biochem.* 52 (1983): 617.

Regulation of Glycolysis and Gluconeogenesis

Pilkis, S. J., and D. K. Granner. Molecular physiology of the regulation of hepatic gluconeogenesis and glycolysis. *Annu. Rev. Physiol.* 54 (1992): 885.

Pilkis, S. J., M. R. El-Maghrabi, and T. H. Claus. Hormonal regulation of hepatic gluconeogenesis and glycolysis. *Annu. Rev. Biochem.* 57 (1988): 755.

12

CHEMOTROPHIC ENERGY METABOLISM: AEROBIC RESPIRATION

In the previous chapter, we learned that some cells meet their energy needs by anaerobic fermentation, either because they are strict anaerobes or because they are facultative cells functioning temporarily in the absence of oxygen. However, we also noted that fermentation processes yield only modest amounts of energy. In the absence of an external electron acceptor—one that is not itself a part of the glycolytic pathway—the electrons that are removed from one three-carbon organic compound (glyceraldehyde-3-phosphate) are eventually transferred to another three-carbon compound (pyruvate), and the difference in energy is such that only two molecules of ATP can be generated per molecule of glucose. In short, fermentation is a feasible way of meeting energy needs, but the ATP yield is low and the process is energetically wasteful because the cell has access to only a limited portion of the total free energy that is potentially available from the oxidizable molecules it uses as substrates. In addition, fermentation processes always result in the accumulation of waste products such as ethanol or lactate that are toxic to the cells as they accumulate—unless, of course, they are removed and metabolized elsewhere in the environment (or in the organism, as in the Cori cycle, shown in Figure 11A-2). Furthermore, fermentation products cannot be used as starting materials by phototrophs, thereby failing to complete the chemotrophic portion of the cyclic flow of matter shown in Figure 5-5.

Respiration: Maximizing ATP Yields

All of this changes dramatically when we come to **respiratory metabolism,** or **respiration** for short. With an external electron acceptor available, complete substrate oxidation becomes possible, and ATP yields are much higher. As a formal definition, *respiration is the flow of electrons, through or within a membrane, from reduced coenzymes to an electron acceptor, usually accompanied by the generation of ATP.* We will get to the membrane and ATP generation parts of the definition later in the chapter. For now, let's focus on the reduced coenzymes and electron acceptors. We have already encountered NADH as the reduced coenzyme generated by the glycolytic catabolism of sugars or related compounds. As we will see shortly, two other coenzymes, *FAD* (for *flavin adenine dinucleotide*) and *coenzyme Q* (or *ubiquinone*), also collect the electrons that are removed from oxidizable organic substrates and pass them to the ultimate electron acceptor via a series of electron carriers.

For many organisms, including the authors and readers of this textbook, the ultimate electron acceptor is *oxygen*, the reduced form of the acceptor is *water*, and the overall process is called **aerobic respiration.** However, a variety of other acceptors are also used, especially by bacteria. Examples of alternative acceptors and their reduced forms are *elemental sulfur* (S/H_2S), *protons* (H^+/H_2), and *ferric ions* (Fe^{3+}/Fe^{2+}). Respiratory processes that involve electron acceptors such as these require no oxygen and are therefore examples of **anaerobic respiration.** Anaerobic respiratory processes play important roles in the cycling of elements such as sulfur, hydrogen, and iron in the environment and contribute significantly to the overall energy economy of the biosphere. Here, however, we will focus on aerobic respiration because it is the mainspring of energy metabolism in the aerobic world of which we and all other higher organisms are a part.

At this point, attention focuses on the *mitochondrion* (Figure 12-1a) because *most aerobic ATP production in eukaryotic cells takes place within this organelle* (Figure 12-1b). For example, the complete aerobic oxidation of glucose by a muscle cell releases enough free energy to drive the synthesis of 36 molecules of ATP per molecule of glucose, and most of these ATPs (all but the two produced by glycolysis) are generated within the mitochondrion. For fats, the proportion of ATP produced in the mitochondrion is even higher.

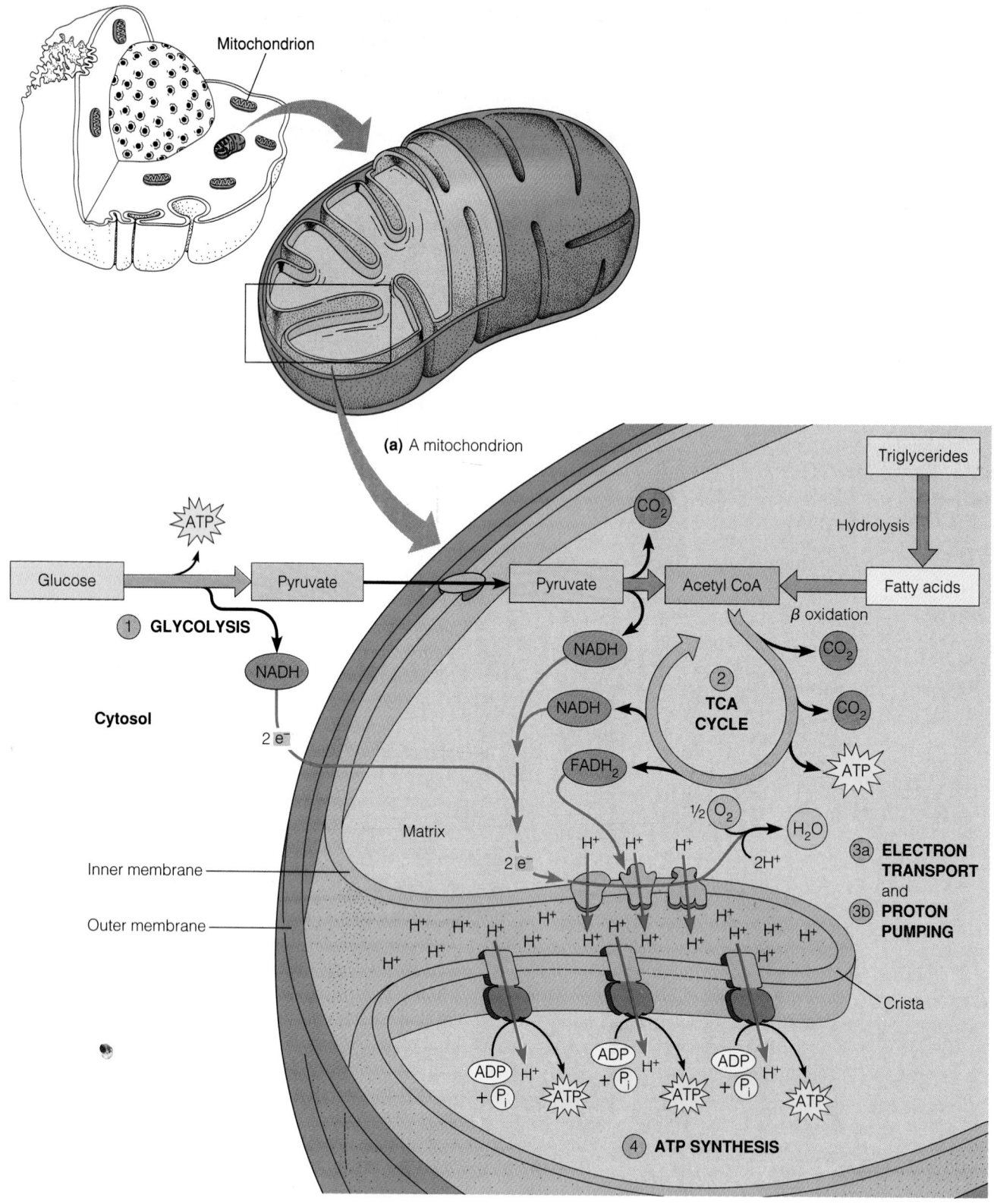

(a) A mitochondrion

(b) Localization of aerobic respiration within the mitochondrion

Figure 12-1   The Role of the Mitochondrion in Aerobic Respiration.   **(a)** The mitochondrion plays a central role in aerobic respiration; most respiratory ATP production in eukaryotic cells occurs in this organelle. **(b)** Oxidation of glucose and other sugars begins in the cytosol with ① glycolysis (stage 1), producing pyruvate. Pyruvate is transported across the inner mitochondrial membrane and is oxidized within the matrix to acetyl CoA, the primary substrate of ② the TCA cycle (stage

2). An alternative source of acetyl CoA is β oxidation of fatty acids. ③ Electron transport (stage 3a) is coupled to proton pumping (stage 3b), with the energy of electron transport conserved as an electrochemical proton gradient across the inner membrane of the mitochondrion (or across the plasma membrane, in the case of prokaryotes). The energy of the proton gradient is used in part to drive ④ the synthesis of ATP from ADP and inorganic phosphate (stage 4).

Aerobic Respiration: An Overview

With oxygen available as the ultimate electron acceptor, we can begin with the pyruvate generated by glycolysis and ask how pyruvate molecules are catabolized and how the free energy released in the process is conserved by the generation of ATP. As we know from Chapter 11, the end-products of aerobic respiration are carbon dioxide and water (see reaction 11-12). These are the very molecules with which photosynthesis begins, as required for the cyclic flow of matter between the chemotrophic and phototrophic worlds. Moreover, the energy yield for aerobic respiration is remarkably higher than for fermentation. Instead of just 2 ATP molecules per molecule of glucose, aerobic respiration has the potential of generating up to 38 ATPs per glucose in prokaryotes and either 36 or 38 ATPs per glucose in eukaryotes, depending on the cell type. (These are maximum possible ATP yields, assuming that the free energy released during respiration is used only for ATP synthesis. As we will see later in this chapter, ATP yields are typically somewhat lower because the cell uses at least some of the energy for other purposes.)

The oxygen that makes all of this possible serves as the ultimate, or terminal, electron acceptor, thereby providing a means for the continuous reoxidation of NADH and other reduced coenzymes. These coenzyme molecules accept electrons during the stepwise oxidation of the organic intermediates derived from pyruvate (as well as from other oxidizable substrates) and transfer these electrons to oxygen via a sequence of membrane-bound electron carriers. Aerobic respiration therefore involves oxidative pathways in which electrons are removed from organic substrates and transferred to coenzyme carriers, as well as concomitant processes whereby the reduced (electron-bearing) coenzymes are reoxidized by the transfer of electrons to oxygen, accompanied indirectly by the generation of ATP.

The Stages of Aerobic Respiration

Aerobic respiration can be considered in four stages, two concerned with coenzyme-mediated oxidative processes and two involving coenzyme reoxidation and the generation of ATP—driven, as it turns out, by the energy stored in a transmembrane gradient of protons. These stages are shown in Figure 12-1b. (In cells, of course, all these processes occur continuously and simultaneously. Their division into four arbitrary "stages" may be a useful framework for our discussion, but keep in mind that none of these stages functions in isolation; each is an integral part of the overall respiratory process.)

Stage 1 is the *glycolytic pathway* that we encountered in Chapter 11. The function of glycolysis is the same under aerobic and anaerobic conditions: the conversion of glucose to pyruvate. But the fate of the pyruvate is different in the presence of oxygen. Instead of being used as an electron acceptor as in fermentation, pyruvate is further oxidized and enters *stage 2*, the *tricarboxylic acid (TCA) cycle*. This cyclic pathway oxidizes incoming carbon atoms to CO_2 and conserves the

energy as reduced coenzyme molecules, which are high-energy compounds in their own right.

Stage 3 involves (a) *electron transport*, or transfer, from reduced coenzymes to oxygen, coupled to (b) the *active transport, or "pumping," of protons* across a membrane. The transfer of electrons from coenzymes to oxygen is exergonic and occurs stepwise via a sequence of membrane-bound electron carriers called the *electron transport system*. The exergonic transfer of electrons between carrier molecules within the membrane drives the pumping of protons across the membrane in which the carriers are embedded. This active transport of protons generates and maintains an *electrochemical gradient of protons* across the membrane. In *stage 4*, the energy of the proton gradient is used to drive *ATP synthesis*. Recall from Chapter 11 that this mode of oxygen-dependent ATP synthesis is called *oxidative phosphorylation* to distinguish it from *substrate-level phosphorylation*, which takes place at one step in the glycolytic pathway—and, as it turns out, in one of the reactions of the TCA cycle as well.

Our goal in this chapter is to understand the processes identified as stages 2, 3, and 4 in Figure 12-1b. Specifically, we want to understand (1) what happens to pyruvate (and other oxidizable substrates such as fats and amino acids) under aerobic conditions; (2) how coenzymes and other electron carriers mediate the exergonic transfer of electrons from oxidizable substrates to oxygen; (3) how the energy of electron transfer events is used to maintain the electrochemical proton gradient; and (4) how the energy of that gradient drives ATP synthesis. We will begin our consideration of aerobic energy metabolism with a close look at the mitochondrion because of the prominent role this organelle plays in eukaryotic energy metabolism. After a discussion of mitochondrial structure and function, we will also note the prokaryotic alternative to mitochondrial compartmentalization.

The Mitochondrion: Where the Action Takes Place

Launching our discussion of aerobic respiration with a consideration of the **mitochondrion** is appropriate because most of aerobic energy metabolism in eukaryotic cells takes place within this organelle. Thus, most of the ATP generation of most animal cells occurs in the mitochondria, as does much of the nonphotosynthetic energy metabolism of plant and algal cells. No wonder, then, that the mitochondrion is called the "energy powerhouse" of the eukaryotic cell.

Mitochondria have been known and studied for more than a century. As early as 1850, the German biologist Rudolph Kölliker described the presence of what he called "ordered arrays of particles" in muscle cells. When these particles were isolated, they were found to swell in water, suggesting the presence of a limiting membrane with osmotic activity. Various names were given to such particles in early work, but the term *mitochondrion* (meaning "thread-like granule"), first introduced in 1898, gradually replaced other names and is now the universally recognized term.

Evidence suggesting a role for the organelle in oxidative events began to accumulate early in the present century. In 1913, for example, Otto Warburg showed that cellular oxygen consumption was associated with particles obtained by filtering tissue homogenates. However, most of our understanding of the role of mitochondria in energy metabolism dates from the development of differential centrifugation, pioneered by Albert Claude (see Figure 9-3). Intact mitochondria were first isolated by this technique in 1948 and were subsequently shown by Eugene Kennedy, Albert Lehninger, and others to be capable of carrying out all the reactions of the TCA cycle, electron transport, and oxidative phosphorylation.

Occurrence and Size of Mitochondria

Mitochondria are found in virtually all aerobic cells of eukaryotes and are prominent features of many cell types when examined by electron microscopy (Figure 12-2). Mitochondria are present in both chemotrophic and phototrophic cells and are therefore common to both plants and animals. Their occurrence in phototrophic cells reminds us that photosynthetic organisms are capable of respiration, a capability on which they depend to meet energy and carbon needs during periods of darkness—and at all times in nonphotosynthetic tissue such as the roots of plants.

The number of mitochondria per cell is highly variable, ranging from one or a few per cell in many protists, fungi, and algae (and in some mammalian cells as well) up to several hundred or even a few thousand per cell in some tissues of higher plants and animals. Mammalian liver cells, for example, contain about 500–1000 mitochondria each, though only a small fraction of these are seen in a typical thin section prepared for electron microscopy, such as that shown in Figure 12-2.

After the nucleus, the mitochondrion is the largest organelle in most animal cells. (In plant cells, the chloroplasts and some of the vacuoles are also larger than the mitochondria.) Typically, a mitochondrion has a diameter in the range 0.5–1.0 μm and a length of several micrometers (more rarely, up to 10 μm). A mitochondrion may therefore be similar in size to an entire bacterial cell.

Localization of Mitochondria

The crucial role of the mitochondrion in meeting cellular ATP needs is often reflected in the localization of mitochondria within the cell. Frequently, mitochondria are clustered within the cell in regions of most intense metabolic activity,

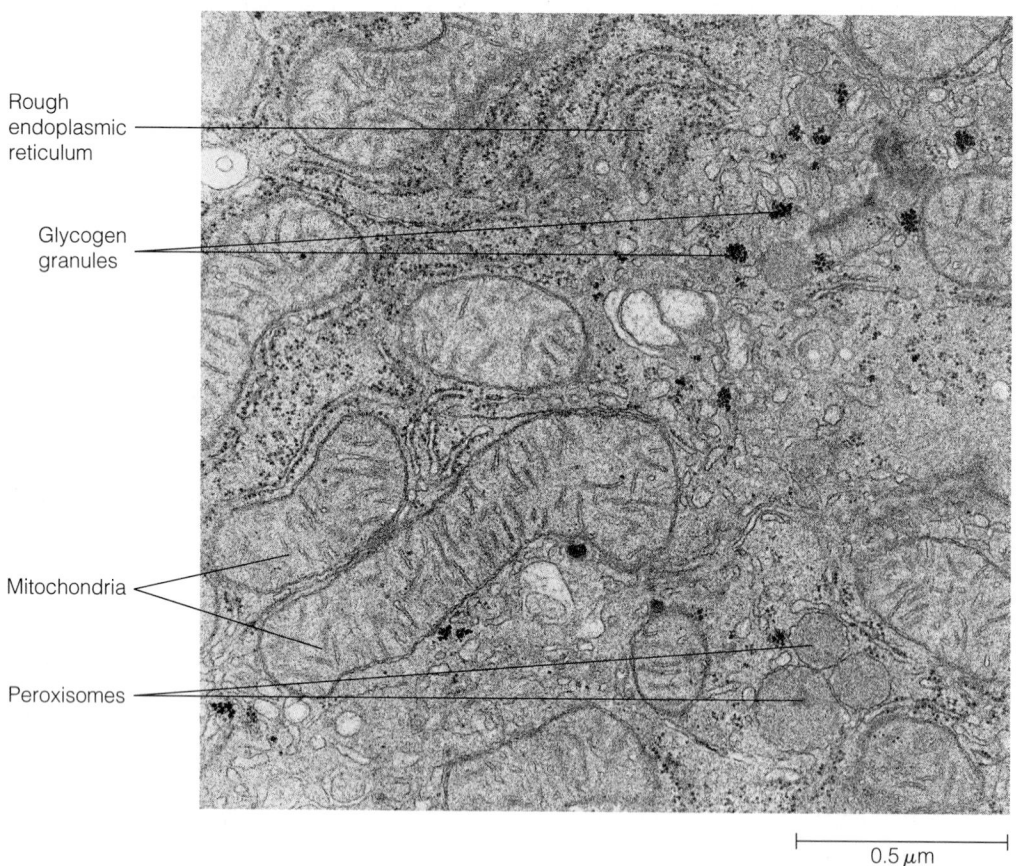

Rough endoplasmic reticulum

Glycogen granules

Mitochondria

Peroxisomes

0.5 μm

Figure 12-2 The Prominence of Mitochondria. Mitochondria are prominent features in this electron micrograph of a rat liver cell. Other cellular components shown include the rough endoplasmic reticulum, glycogen granules, and peroxisomes (TEM).

where the ATP need is greatest. An especially good example is in muscle cells (see Figure 4-13). Just as Kölliker originally observed, the mitochondria in such cells are organized in rows along the fibrils responsible for contraction. Presumably, this association minimizes the distance that ATP molecules must diffuse to get from their site of generation in the mitochondrion to the site of ATP utilization in the contracting fibrils. A similar localization of mitochondria occurs in sperm tails (see Figure 4-12), in flagella and cilia, and at the base of kidney tubule cells, where exchange with the blood is most rapid.

Structural Features of Mitochondria

The structure of a typical mitochondrion is shown in Figure 12-3. A distinctive feature is the presence of two membranes, called the inner and outer membranes. The **outer membrane** is smooth, has no folds, and is readily permeable to solutes with molecular weights below about 5000. This size limit excludes proteins and other macromolecules but allows the passage of small molecules and ions. Consequently, the **intermembrane space** between the inner and outer membranes is essentially continuous with the cytosol with respect to the ions and small molecules that are pertinent to mitochondrial function.

In contrast to the outer membrane, the **inner membrane** of the mitochondrion presents a permeability barrier to most solutes. Not surprisingly, the inner membrane contains the transport proteins needed to facilitate the movement of solutes such as pyruvate, fatty acids, ATP, ADP, and inorganic phosphate across the membrane. The inner membrane is also the locale of the protein complexes involved in electron transport and ATP synthesis.

The inner membrane has many distinctive infoldings called **cristae** (singular: **crista**) that greatly increase its surface area. Because of its large surface area, the inner membrane is able to accommodate large numbers of the protein complexes needed for electron transport and ATP synthesis, thereby enhancing the capacity of the mitochondrion for ATP generation. Figure 12-4 illustrates structural details of the inner and outer mitochondrial membranes as revealed by the freeze-fracture technique described in the Appendix. Notice especially the high density of protein particles associated with both fracture faces of the inner membrane. These represent the transmembrane portions of proteins involved in solute transport, electron transport, and ATP synthesis.

The prominence of cristae within the mitochondrion frequently reflects the relative metabolic activity of the cell or tissue in which the organelle is located. Heart, kidney, and muscle cells have high respiratory activities, and their mitochondria have correspondingly large numbers of prominent cristae. The flight muscles of birds are especially high in respiratory activity and have mitochondria that are exceptionally well endowed with cristae. Plant cells, by contrast, are generally characterized by lower rates of respiratory activity and by fewer cristae within their mitochondria.

The interior of the mitochondrion is filled with a semifluid **matrix.** Within the matrix are many of the enzymes involved in mitochondrial function as well as the DNA molecules and ribosomes that give the organelle the ability to make some of its own proteins. In most mammals, the mitochondrial genome consists of a circular DNA molecule that

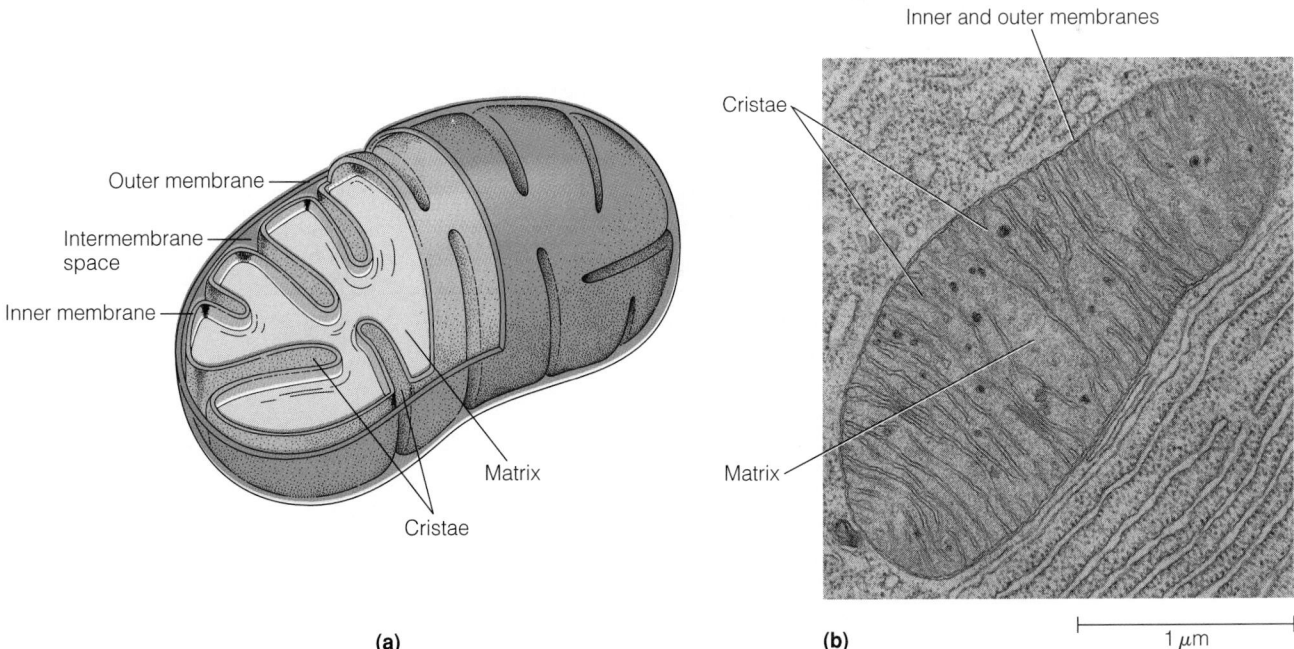

(a)

(b)

Figure 12-3 Mitochondrial Structure. **(a)** Mitochondrial structure is illustrated schematically in this cutaway view. **(b)** A mitochondrion of a bat pancreas cell as seen by electron microscopy (TEM). Note the prominent cristae.

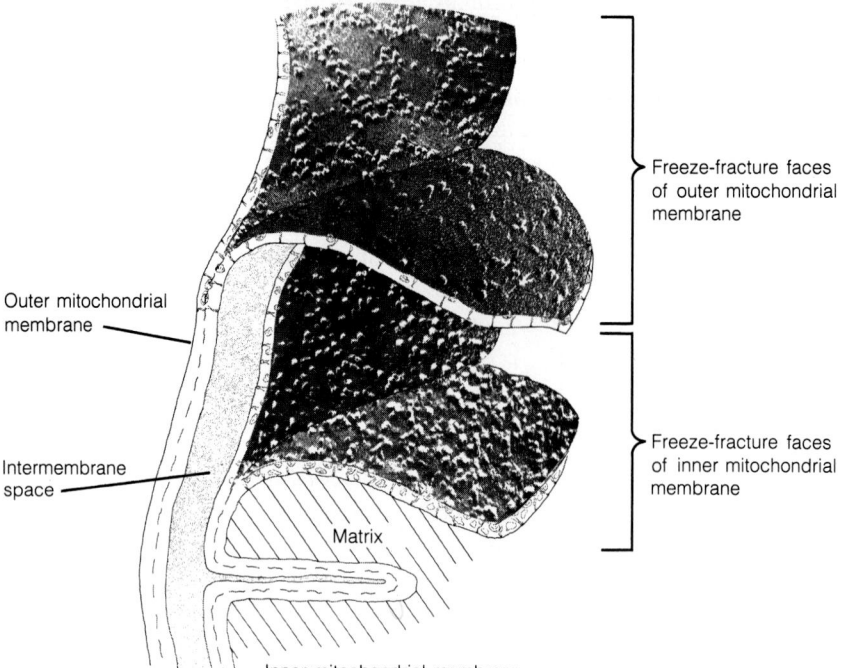

Figure 12-4 Structure of the Inner and Outer Mitochondrial Membranes. When the inner and outer mitochondrial membranes are subjected to freeze-fracturing, each of the membranes splits along its hydrophobic interior, separating each membrane into two fracture faces. Segments of electron micrographs of the two fracture faces for both inner and outer membranes are superimposed on a schematic diagram of the freeze-fractured membranes to illustrate the density of protein particles in each membrane.

Labels on figure: Freeze-fracture faces of outer mitochondrial membrane; Outer mitochondrial membrane; Freeze-fracture faces of inner mitochondrial membrane; Intermembrane space; Matrix; Inner mitochondrial membrane

has a molecular weight of about 11 million and encodes about a dozen polypeptides.

Localization of Respiratory Functions in the Mitochondrion

Specific functions and pathways have been localized within the mitochondrion by disruption of the organelle and fractionation of the various components. Table 12-1 lists some of the main functions localized to each of the compartments of the mitochondrion.

Enzymes and other proteins that are readily solubilized upon disruption are assumed to be either in the matrix or only loosely associated with the membrane. By this criterion, most of the mitochondrial enzymes involved in pyruvate oxidation, in the TCA cycle, and in the catabolism of fatty acids and amino acids are matrix enzymes. On the other hand, most of the intermediates in the electron trans-

Table 12-1 Localization of Metabolic Functions Within the Mitochondrion

Membrane or Compartment	Functions
Outer membrane	Phospholipid synthesis Fatty acid desaturation Fatty acid elongation
Inner membrane	Electron transport Oxidative phosphorylation Transport of metabolites
Intermembrane space	Phosphorylation of nucleotides
Matrix	Pyruvate oxidation TCA cycle β oxidation of fats DNA replication RNA synthesis (transcription) Protein synthesis (translation)

port system are integral components of the inner membrane, where they are organized into large complexes. Protruding from the inner membrane are knoblike spheres called F₁ complexes or **particles** (Figure 12-5). Each F₁ particle is a complex of proteins and is attached by a short protein neck to an **F$_o$ complex**, an assembly of hydrophobic proteins embedded within the inner membrane. Individual F₁ complexes can be seen in Figure 12-5a, an electron micrograph taken at high magnification using a technique called *negative staining*. (As described in the Appendix, negative staining results in a light image against a dark background

and is the preferred method for examining very small objects.)

The combination of an F₁ complex linked to an F$_o$ complex is called an *F$_o$F₁ complex*. Functionally, the F$_o$F₁ complex is called an *ATP synthase* because it is the structure that is responsible for ATP generation in the mitochondrion—and, as it turns out, in prokaryotic cells and in chloroplasts as well. In each of these settings, ATP generation by F$_o$F₁ complexes is driven by an electrochemical gradient of protons across the membrane to which the F$_o$F₁ complexes are anchored.

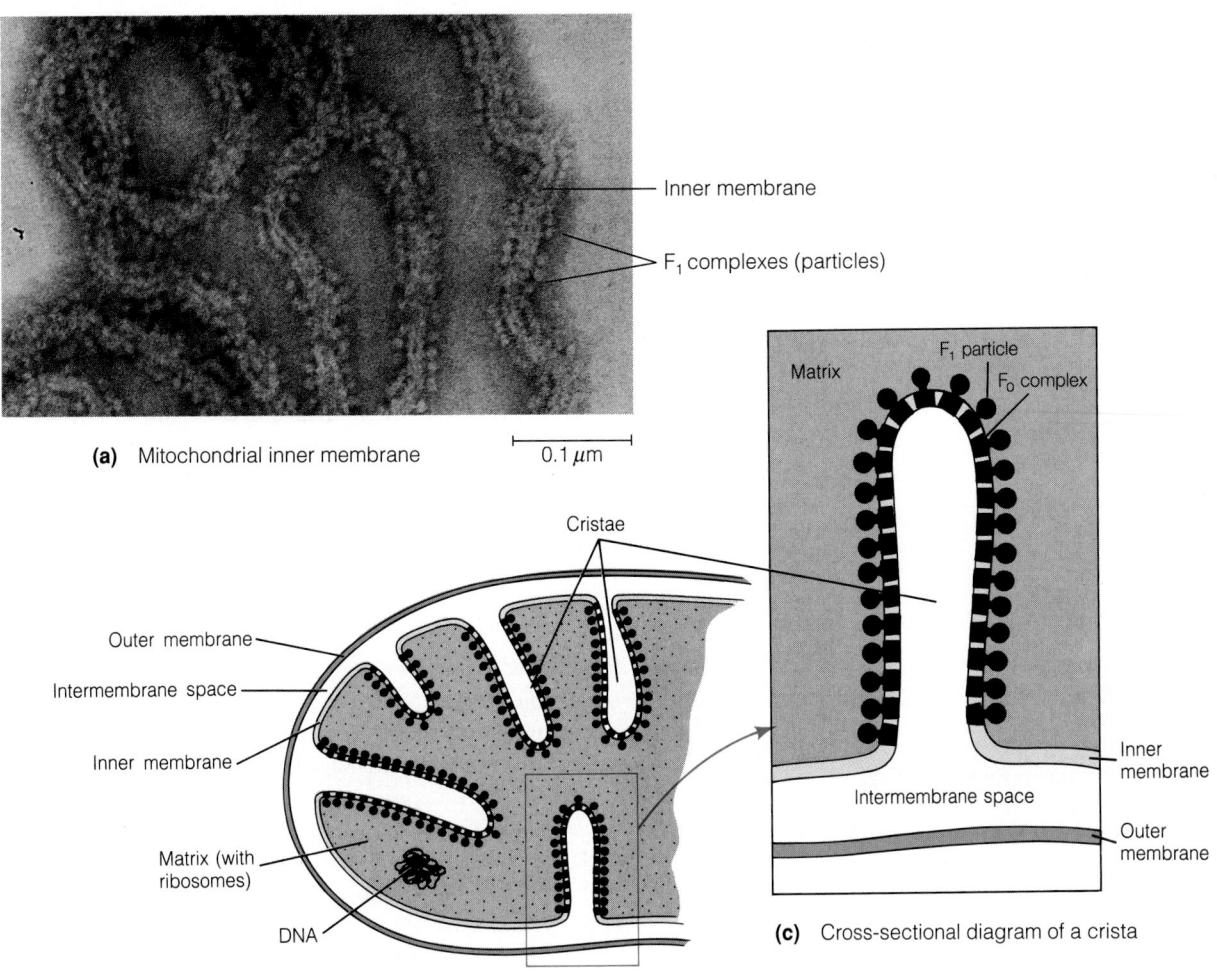

(a) Mitochondrial inner membrane

0.1 μm

(b) Cross-sectional diagram of a mitochondrion

(c) Cross-sectional diagram of a crista

Figure 12-5   The F₁ and F$_o$ Complexes of the Inner Mitochondrial Membrane.   **(a)** This electron micrograph was prepared by negative staining to show the spherical F₁ complexes that line the matrix side of the inner membrane of a bovine heart mitochondrion (TEM). **(b)** Cross section of a mitochondrion, showing major structural features.

(c) An enlargement of a single crista, showing the F₁ complexes (particles) that project from the inner membrane on the matrix side and the F$_o$ complexes embedded in the inner membrane. Each F₁ complex is attached to an F$_o$ complex by a short protein stalk. Together, an F$_o$F₁ pair constitutes a functional ATP synthase complex.

Localization of Respiratory Functions in the Prokaryotic Cell

Prokaryotes do not have mitochondria, yet many prokaryotic cells are capable of aerobic respiration. Where in the prokaryotic cell are the various components of respiratory metabolism localized? Essentially, the *cytoplasm* and *plasma membrane* of a prokaryotic cell perform the same functions as the mitochondrial matrix and inner membrane, respectively. In the prokaryotic cell, therefore, most of the enzymes of the TCA cycle and of fatty acid and amino acid catabolism are found in the cytoplasm, whereas the electron transport proteins are located in the plasma membrane.

The F_oF_1 complex is also localized to the plasma membrane in prokaryotes, with the F_o component embedded in the membrane and the F_1 component protruding from the membrane into the cytoplasm. Figure 12-6 is an electron micrograph of bacterial F_oF_1 complexes reconstituted in phospholipid vesicles. F_1 complexes can be clearly seen protruding from the outer surface of the vesicle. Because of the ease with which they can be isolated, bacterial membranes are often used as the starting material in studies of F_oF_1 complexes. (Notice, by the way, that the F_1 complexes of the vesicles shown in Figure 12-6 protrude *outward* from the membrane of the vesicle rather than *inward*, as they do when located on either the plasma membrane of the bacterial cell or the inner membrane of the mitochondrion. This reversed orientation is caused by the inside-out manner in which small F_oF_1-bearing membrane pieces seal into vesicles when F_oF_1 complexes are reconstituted in this way.)

Because of the absence of mitochondria in the prokaryotic cell, glycolysis and the TCA cycle are localized to the *same* compartment, the cytoplasm, in prokaryotes rather than to two *separate* compartments, the cytosol and the mitochondrial matrix, as in eukaryotes. As we will see later in this chapter, this difference explains why the maximum ATP yield per glucose is somewhat variable in eukaryotic cells, depending on the tissue and conditions.

The Tricarboxylic Acid Cycle: Oxidation in the Round

Having considered localization of respiratory functions in mitochondria and in prokaryotic cells, we are now ready to discuss the several stages of aerobic respiration. To do so, we will return to the eukaryotic context and follow a molecule of pyruvate across the inner membrane of the mitochondrion to see what fate awaits it inside.

In the presence of oxygen, pyruvate is oxidized fully to carbon dioxide in a cyclic process that is a central feature of energy metabolism in almost all aerobic chemotrophs. An important intermediate in this cyclic series of reactions is *citrate*, which has three carboxylic acid groups and is therefore a *tricarboxylic acid*. For this reason, this pathway is usually called the **tricarboxylic acid (TCA) cycle.** Other names for it are the *citrate* (or *citric acid*) *cycle* or the *Krebs cycle*, in honor of Hans Krebs, in whose laboratory this metabolic sequence was elucidated in the 1930s.

The TCA cycle is shown in Figure 12-7. In broad outline, the cycle begins with *acetyl coenzyme A* (*acetyl CoA*), its only real substrate. (As we will see shortly, acetyl CoA consists of a two-carbon acetate group linked to a carrier called *coenzyme A*. Coenzyme A was discovered by Fritz Lipmann, who shared a Nobel Prize with Krebs in 1953 for their work on aerobic respiration.) Acetyl CoA arises either by oxidative decarboxylation of pyruvate or by the stepwise oxidative breakdown of fatty acids. Regardless of its origin, acetyl CoA transfers its acetate group to a four-carbon acceptor called *oxaloacetate*, thereby generating the *citrate* for which the cycle was originally named. In a cyclic series of reactions, citrate is subjected to two successive decarboxylations and several oxidations, leaving a four-carbon compound from which the starting oxaloacetate is regenerated.

Each "round" of TCA cycle activity involves the entry of two carbons (as the acetate from acetyl CoA), the release of two carbons as carbon dioxide, and the regeneration of oxaloacetate. Oxidation occurs at five steps: four in the cycle itself and one in the reaction that converts pyruvate to acetyl CoA. In each case, electrons are accepted by coenzyme molecules. The substrate for the TCA cycle is therefore acetyl CoA, and the products are carbon dioxide, reduced coenzymes, and a molecule of ATP, as it turns out.

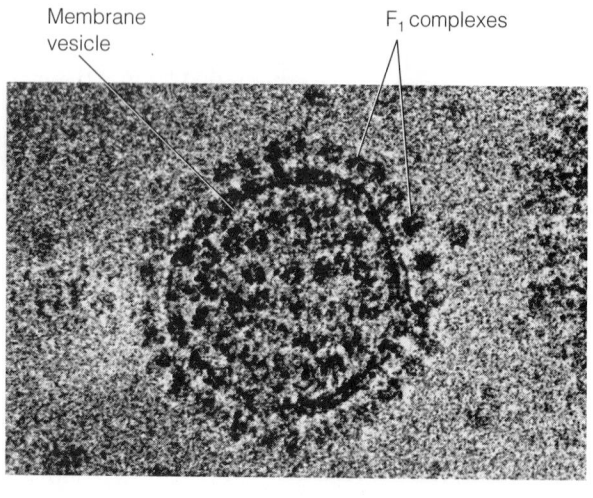

Membrane vesicle

F_1 complexes

50 nm

Figure 12-6 Bacterial F_1 Complexes. This electron micrograph shows bacterial F_oF_1 complexes reconstituted in phospholipid vesicles, with F_1 complexes lining the outer surface of the vesicle. Bacterial F_oF_1 complexes are frequently used in studies of ATP synthase because of the ease with which these membranes can be isolated. (The membrane of the vesicle is inside-out because of the way in which vesicles form from small pieces of F_oF_1-bearing membrane. In an intact bacterial cell, the F_1 complexes are located on the inner surface of the plasma membrane, projecting into the interior of the cell.)

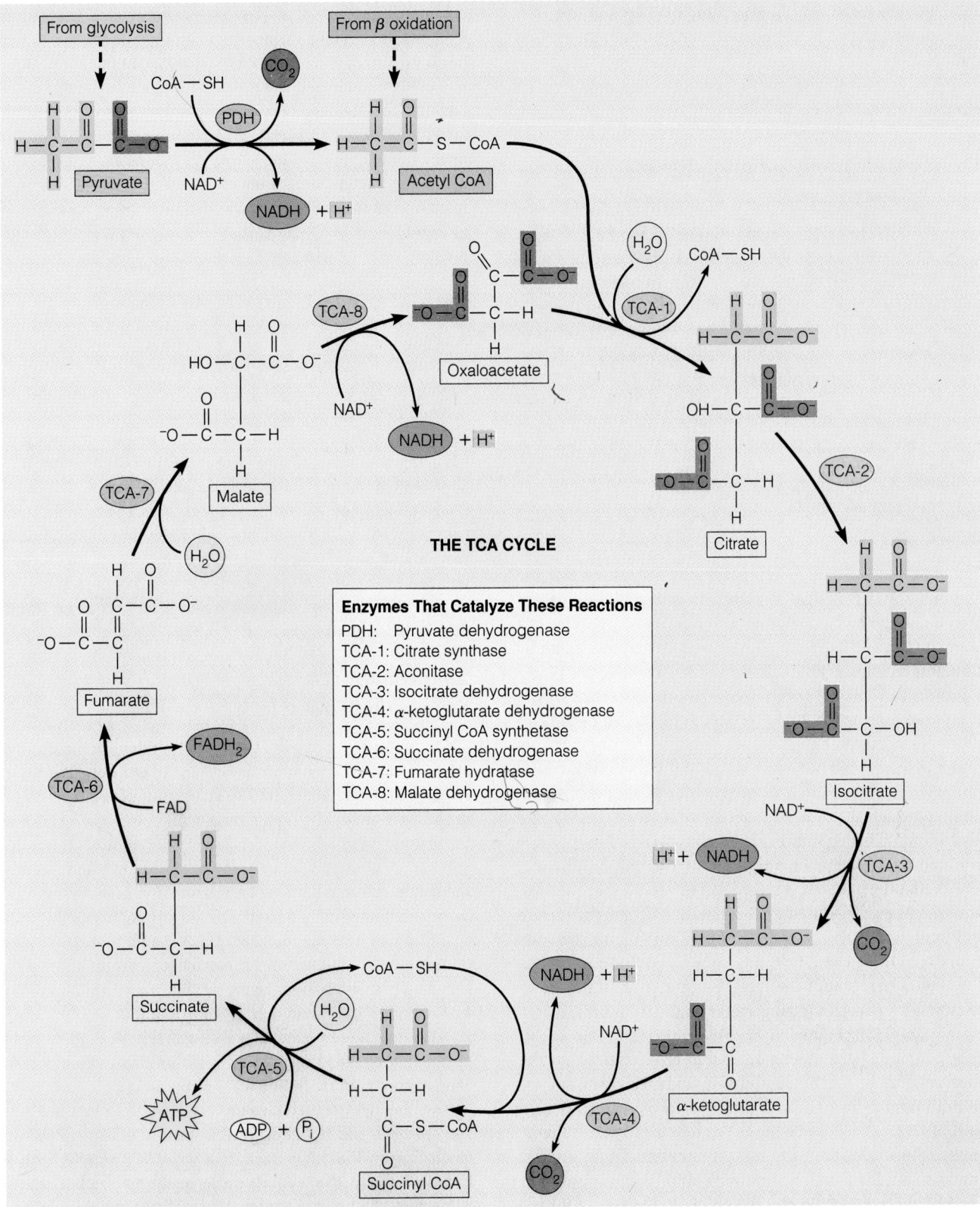

Figure 12-7 The Tricarboxylic Acid (TCA) Cycle. The two carbon atoms of pyruvate that enter the cycle via acetyl CoA are shown in pink in citrate and subsequent molecules until they are randomized by the symmetry of the fumarate molecule. The carbon atom of pyruvate that is lost as carbon dioxide is shown in gray, as are the two carboxyl groups of oxaloacetate that give rise to carbon dioxide in steps TCA-3 and TCA-4. Five of the reactions are oxidations, with NAD⁺ as the electron acceptor in four reactions (PDH, TCA-3, TCA-4, and TCA-8) and FAD as the electron acceptor in one case (TCA-6). The generation of ATP shown in reaction TCA-5 is characteristic of bacterial cells and plant mitochondria; in animal mitochondria, GTP is formed instead, but the reactions are energetically equivalent because the terminal phosphoanhydride bonds of GTP and ATP are identical in free energy of hydrolysis and are readily interconvertible.

With this brief overview in mind, let's look at the TCA cycle in more detail, focusing on what happens to the carbon molecules that enter as acetyl CoA and how the energy released by each of the oxidations is conserved as high-energy electrons of reduced coenzymes.

Oxidative Decarboxylation of Pyruvate to Acetyl Coenzyme A

Acetyl CoA is the starting point for the TCA cycle, but from Chapter 11 we know that the glycolytic pathway ends with pyruvate, not acetyl CoA. To get from pyruvate to acetyl CoA requires the activity of the *pyruvate dehydrogenase (PDH) complex*. This complex consists of multiple copies of three different enzymes, each of which plays a role in the *oxidative decarboxylation* of pyruvate:

(12-1)

This reaction is called a *decarboxylation* because one of the carbons of pyruvate (carbon atom 1) is liberated as carbon dioxide. As a result, carbon atoms 2 and 3 of pyruvate become, respectively, carbon atoms 1 and 2 of acetate. In addition, this reaction is an *oxidation* because two electrons (and a proton as well) are transferred from the substrate to the coenzyme acceptor, NAD^+. The electrons carried by NADH represent potential energy that is tapped to drive ATP synthesis when the NADH is reoxidized by the electron transport system. The oxidation of pyruvate occurs on carbon atom 2, which is oxidized from an α-keto to a carboxylic acid group. This oxidation is possible because of the concomitant elimination of carbon atom 1 as carbon dioxide. The oxidation is highly exergonic ($\Delta G^{\circ\prime} = -7.5$ kcal/mol), with the free energy used to energize, or activate, the acetate molecule for further metabolism by linking it to the sulfhydryl group of one of the enzymes in the PDH complex, followed by transfer of the acetyl group from the enzyme to **coenzyme A (CoA)**, forming **acetyl CoA.**

As shown in Figure 12-8, CoA is a complicated molecule containing the B vitamin *pantothenic acid.* (Like the nicotinamide of NAD^+, pantothenic acid is classified as a vitamin because humans and other vertebrates need it as a part of an essential coenzyme but cannot synthesize it themselves.) To understand the role of coenzyme A in cellular metabolism, focus on the free sulfhydryl, or *thiol*, group at the end of the molecule. It is this thiol group that can form a *thioester bond* with organic acids such as acetate. The thiol group is sufficiently important to the function of the molecule that coenzyme A is sometimes abbreviated not just as CoA but as CoA-SH. Compared to an ester bond, a thioester bond is a higher-energy bond because significantly more free energy is released upon its hydrolysis.* Just as NAD^+ is adapted for transfer of electrons, so coenzyme A is well suited as a carrier of acetate and other acyl groups. (The "A" in the name comes, in fact, from its role as an *acyl* carrier.) Thus, the acetyl group that is transferred to coenzyme A from one of the enzymes of the PDH complex (represented in Figure 12-8 as E-SH) is in a higher-energy, or activated, form.

Entry of Acetate into the TCA Cycle

As Figure 12-7 illustrates, the TCA cycle itself begins with the entry of acetate in the form of acetyl CoA. With each round of TCA cycle activity, two carbon atoms enter in organic form (as acetate) and two carbon atoms leave in inorganic form (as carbon dioxide). In the first reaction (TCA-1), the acetate group of acetyl CoA is added onto oxaloacetate to form *citrate*, with the condensation driven by the free energy of hydrolysis of the thioester bond. The reaction is catalyzed by the enzyme *citrate synthase*. Acetate has two carbon atoms and oxaloacetate has four, so citrate is a six-carbon tricarboxylic acid. (The colored carbon atoms in Figure 12-7 will help you keep track of the incoming atoms during subsequent steps in the cycle.)

The Oxidative Decarboxylation Steps of the TCA Cycle

Since the function of the TCA cycle is to oxidize the carbon atoms of the incoming acetate group completely, it should not be too surprising to find that four out of the eight steps in the cycle are oxidations. This is evident in Figure 12-7 because four steps (TCA-3, TCA-4, TCA-6, and TCA-8) involve coenzymes that enter in the oxidized form and leave in the reduced form. Each of these reactions is catalyzed by a *dehydrogenase* that is specific for the particular substrate. The first two of these reactions, TCA-3 and TCA-4, are also decarboxylation steps; one molecule of carbon dioxide is eliminated in each step, reducing the number of carbons first from six to five and then from five to four.

Prior to these events, however, reaction TCA-2 converts citrate to the related compound *isocitrate*. If you look closely at the structures of these two compounds in Figure 12-7, you will see why this conversion is essential. Citrate is a *tertiary alcohol*, and tertiary alcohols are not easily oxidizable. Isocitrate, on the other hand, is a *secondary alcohol*, with a hydroxyl group that can be quite easily oxidized. The enzyme that catalyzes the interconversion of citrate and isocitrate is called *aconitase*. Aconitase actually carries out successive dehydration and rehydration reactions, in which the

---

*Having learned in Chapter 11 that esters are relatively low-energy compounds compared to anhydrides, you may wonder why a thioester is a high-energy compound that has an even more negative standard free energy change ($\Delta G^{\circ\prime} = -7.5$ kcal/mol) than does ATP. Resonance stabilization provides the answer here, as it did in for the instability of anhydrides in Chapter 11. Although not mentioned there, the C—O bond of an ester has a partial double-bond character because of

the partial overlap of π electrons, thereby allowing the ester to resonate between two forms. By contrast, the larger atomic size of S compared to O reduces the π-electron overlap between C and S, thereby eliminating any significant contribution of the C=S structure to resonance stabilization. A thioester is therefore thermodynamically less stable than an ester, which results in a more highly negative $\Delta G^{\circ\prime}$ for hydrolysis.

Figure 12-8   Structure of Coenzyme A and Acetyl CoA Formation.   The portion of the coenzyme indicated in pink is pantothenic acid, a B vitamin. Formation of a thioester bond between CoA and an acetyl group generates acetyl coenzyme A. The acetyl group is formed by the oxidative decarboxylation of pyruvate (reaction PDH in Figure 12-7) or by β oxidation of fatty acids and is transferred to CoA from one of the enzymes of the pyruvate dehydrogenase complex, to which it is bound as a thioester after the oxidation of pyruvate.

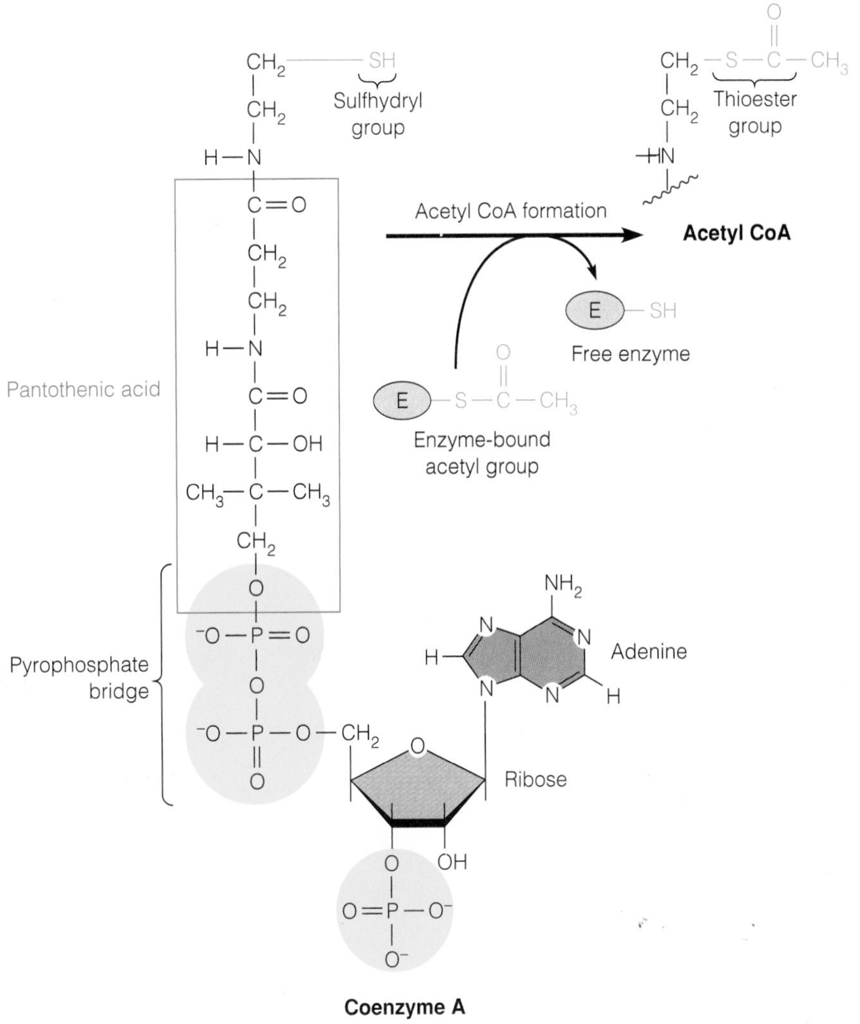

Coenzyme A

elements of water are removed from citrate to generate an unsaturated intermediate called aconitate (not shown), to which the —H and —OH of water are added back again, but in opposite positions.

The hydroxyl group of isocitrate is now the target of the first oxidation, or dehydrogenation, of the cycle. Isocitrate is oxidized by the enzyme *isocitrate dehydrogenase* to a six-carbon α-keto compound called oxalosuccinate (not shown), with NAD^+ as the usual electron acceptor. Oxalosuccinate is unstable and immediately undergoes decarboxylation to the five-carbon compound α-ketoglutarate. This reaction, TCA-3, is therefore one of the two decarboxylation steps of the cycle.

The second decarboxylation event occurs in the next step, reaction TCA-4. This reaction is also an oxidation, with NAD^+ as the electron acceptor. To understand this reaction, compare the structure of α-ketoglutarate with that of pyruvate. Both compounds are α-keto acids, so it should not be surprising that the mechanism of oxidation is the same for both, complete with decarboxylation (from pyruvate to acetate in one case, and from α-ketoglutarate to succinate in the other) and linkage of the oxidized product to coenzyme A as a thioester. Thus, α-ketoglutarate is oxidized to *succinyl*

CoA, a four-carbon compound with a thioester bond. The enzyme that catalyzes this reaction is called *α-ketoglutarate dehydrogenase*.

The ATP-Generating Step of the TCA Cycle

Since we are already halfway around the cycle, let's pause a moment to take stock. Note that the carbon balance of the cycle is already satisfied: Two carbon atoms entered as acetyl CoA, and two carbon atoms have now been lost as carbon dioxide. (But notice carefully from Figure 12-7 that the two carbon atoms that leave in a given cycle are not the *same* two that entered in reaction TCA-1 of that cycle. Instead, the two CO_2 molecules arise from what were initially the two carboxyl carbons of oxaloacetate.) We have also encountered two of the four oxidation reactions of the TCA cycle and therefore have two molecules of NADH that will require re-oxidation via the electron transport system. In addition, we recognize succinyl CoA as a compound that, like acetyl CoA, has a thioester bond, the hydrolysis of which is highly exergonic.

Unlike acetyl CoA, however, succinyl CoA is not condensed onto another molecule, so the energy of the thioester bond is not needed for that purpose. Instead, the energy is used to generate a molecule of ATP, with the release of free *succinate*. In fact, it is because succinyl CoA is formed rather than free succinate in reaction TCA-4 that the energy of α-ketoglutarate oxidation is conserved in a form suitable for ATP production. The direct generation of ATP shown in Figure 12-7 for reaction TCA-5 is characteristic of bacterial cells and plant mitochondria; in animal mitochondria, the high-energy intermediate formed is not ATP but guanosine triphosphate (GTP), a closely related compound (see Figure 3-12). The two alternatives are energetically equivalent, however, because the terminal phosphoanhydride bonds of GTP and ATP are identical in free energy of hydrolysis. Moreover, the terminal phosphate group of GTP can be readily transferred to ADP by a mitochondrial enzyme that exchanges phosphate groups among nucleotides. Thus, the net result of succinyl CoA hydrolysis is the generation of one molecule of ATP in either case.

The Final Oxidative Reactions of the TCA Cycle

Of the remaining three steps in the TCA cycle, two are oxidations. In reaction TCA-6, the succinate formed in the previous step is oxidized to *fumarate*. This reaction is unique in that both electrons come from carbon atoms, generating a C=C double bond. (All the other oxidations we have

encountered thus far involve the removal of electrons from adjacent carbon and oxygen atoms, generating a C=O double bond.) The initial electron acceptor for this dehydrogenation is not NAD^+ but another coenzyme called **flavin adenine dinucleotide (FAD)**. Like NAD^+ and coenzyme A, FAD contains a B vitamin as part of its structure—*riboflavin*, in this case (Figure 12-9). FAD accepts two protons and two electrons, so the reduced form is written as $FADH_2$. Note that FAD accepts electrons on nitrogen atoms (Figure 12-9, arrows), whereas NAD^+ accepts electrons on carbon atoms (see Figure 11-6). This feature has implications for the ATP yield when the two coenzymes are oxidized. As we will see later, the maximum ATP yield upon coenzyme oxidation is about three for NADH but only about two for $FADH_2$.

Reaction TCA-6 has a further unique feature: The enzyme that catalyzes it, *succinate dehydrogenase*, does not occur as a soluble protein in the matrix like the other enzymes of the TCA cycle. Instead, it is an integral membrane protein, embedded in the inner membrane of the mitochondrion (or in the plasma membrane, in the case of prokaryotes). Furthermore, the FAD coenzyme, unlike NAD^+, is tightly bound to the enzyme, passing its electrons to *coenzyme Q*, a membrane-soluble quinone that we will meet later in our discussion of the electron transport chain. (Some scientists point out that FAD is really a prosthetic

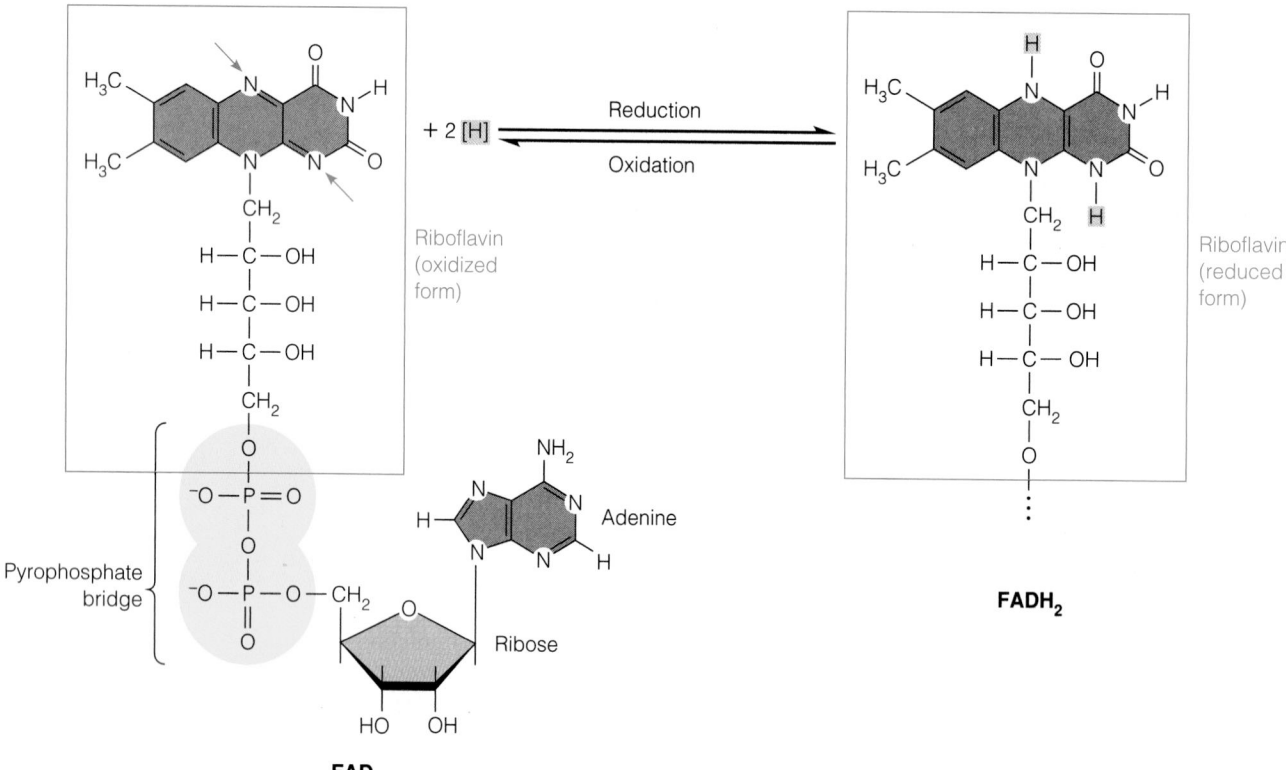

Figure 12-9   Structure of FAD and Its Oxidation and Reduction.   The portion of the coenzyme indicated in pink is riboflavin, a

B vitamin. The arrows point to the two nitrogen atoms of riboflavin that acquire one proton and one electron each when FAD is reduced to $FADH_2$.

group and that coenzyme Q ought to be regarded as the real electron acceptor for the succinate dehydrogenase reaction. The oxidized and reduced forms of the coenzyme for reaction TCA-6 would then be represented as CoQ and $CoQH_2$, respectively. However, most textbooks continue to represent FAD as the electron acceptor for this reaction, a convention observed in Figure 12-7 also.)

In the next step of the cycle, the double bond of fumarate is hydrated, producing *malate* (reaction TCA-7, catalyzed by the enzyme *fumarate hydratase*). Because fumarate is a symmetric molecule, the hydroxyl group of water has an equal chance of adding to either of the internal carbon atoms. As a result, the colored carbon atoms of Figure 12-7, which were used to keep track of the most recent acetate group to enter the cycle, are randomized at this step between the "upper" and "lower" two carbon atoms of malate and are therefore not color-coded from this point on.* In reaction TCA-8, the hydroxyl group of malate becomes the target of the final oxidation in the cycle. Again, NAD^+ serves as the electron acceptor, and the product is the corresponding keto compound, *oxaloacetate*.

The TCA Cycle in Summary

With the regeneration of oxaloacetate, one cycle is complete. We can summarize what has been accomplished by noting the following properties of the TCA cycle:

1. Acetate enters the cycle as acetyl CoA by condensation onto a four-carbon acceptor molecule to form citrate, a six-carbon compound.

2. Decarboxylation occurs at two steps in the cycle so that the input of two carbons as acetate is balanced by the loss of two carbons as carbon dioxide.

3. Oxidation occurs at four steps, with NAD^+ as the electron acceptor in three cases and enzyme-bound FAD as the electron acceptor in one case.

4. ATP is generated at one point, with GTP as an intermediate in animal cells.

5. The cycle is completed upon regeneration of the original acceptor.

By summing the eight component reactions of the TCA cycle as shown in Figure 12-7, we arrive at the following summary reaction:

$$acetyl\ CoA + 3H_2O + 3NAD^+ + FAD + ADP + P_i \longrightarrow$$
$$2CO_2 + 3NADH + 3H^+ + FADH_2 + CoA\text{-}SH + ATP + H_2O$$
$$(12\text{-}2)$$

(2x)

*The discerning chemist will note that the citrate of reaction TCA-2 is also a symmetric molecule, and its hydroxyl group should, by the same argument, be moved either "up" or "down" the molecule, thereby randomizing carbon atoms at an early stage in the cycle. However, aconitase, the enzyme that catalyzes reaction TCA-2, is capable of distinguishing between the two ends of the molecule and moves the hydroxyl group in one direction only. By contrast, fumarate hydratase, the enzyme responsible for reaction TCA-7, cannot distinguish between the two inner carbon atoms.

Because the cycle must, in effect, occur twice to metabolize both of the acetyl CoA molecules derived from a single molecule of glucose, the summary reaction on a per-glucose basis can be obtained by doubling all the coefficients in reaction 12-2. If we then add to this reaction the summary reactions for glycolysis through pyruvate (reaction 11-16) and for the oxidative decarboxylation of pyruvate to acetyl CoA (reaction 12-1, also multiplied by 2), we arrive at the following summary reaction for the entire sequence from glucose through the TCA cycle:

$$glucose + 6H_2O + 10NAD^+ + 2FAD + 4ADP + 4P_i \longrightarrow$$
$$6CO_2 + 10NADH + 10H^+ + 2FADH_2 + 4ATP + 4H_2O$$
$$(12\text{-}3)$$

As you consider this summary reaction, two points may strike you: how modest the ATP yield is thus far, and how many coenzyme molecules are reduced during the oxidation of glucose. With only four ATP molecules generated per glucose molecule, we have as yet little evidence for the substantially greater ATP yield that is supposed to be characteristic of respiratory metabolism. However, we also recognize the need to regenerate the oxidized forms of the coenzymes, since only catalytic quantities of these compounds are present in the mitochondrion. As you may well guess, there is a connection between the two observations. The reoxidation of the 12 reduced coenzyme molecules shown on the right side of reaction 12-3 is a highly exergonic process that provides the energy needed to drive the synthesis of the majority of the ATP molecules produced during the complete oxidation of glucose.

For the release of that energy, we must look to the remaining stages of respiratory metabolism—electron transport and oxidative phosphorylation. Before doing so, however, we will consider several additional features of the TCA cycle: its centrality in energy metabolism, its regulation, and its role in other metabolic pathways.

The Centrality of the TCA Cycle

It is important to understand the central role of the TCA cycle in all of aerobic energy metabolism. Thus far, we have regarded glucose as the main substrate for respiration. True, a variety of alternative carbohydrate substrates were considered in Chapter 11 (see Figure 11-11), but most summary reactions written for chemotrophic energy metabolism assume glucose as the starting compound, and reaction 12-3 is no exception. In a sense, the assumption is very reasonable, given the importance of glucose as an energy source for the chemotrophic world. However, it is also important to note the role of other substrates in cellular energy metabolism and the centrality of the TCA cycle in the catabolism of a variety of alternative fuel molecules, especially fats and proteins. Far from being a minor pathway for the catabolism of a single sugar, the TCA cycle represents the mainstream of aerobic energy metabolism.

Fat as a Source of Energy. When we first encountered fats in Chapter 3, we noted their role in energy storage and observed that they are highly reduced compounds that liberate more energy per gram upon oxidation than do carbohydrates. For this reason, fats are an important long-term energy storage form for many organisms. Storage polysaccharides such as starch and glycogen are important as localized, mobilizable energy reserves, but for long-term energy stores, fats are often used. Fat reserves are especially important in hibernating animals and migrating birds and also represent a common (though by no means the only) form in which energy and carbon are stored by plants in their seeds. Fats are well suited for this storage function because they allow a maximum number of calories to be stored with minimum of volume and weight, a feature of obvious significance for both animal motility and seed dispersal.

Most fat is stored as deposits of **triglycerides,** which are neutral triesters of *glycerol* and long-chain *fatty acids* (see Figure 3-24). Triglycerides are obtained either by mobilization of stored fat or, in the case of animals, from the diet. In either case, catabolism of triglycerides begins with their hydrolysis to glycerol and free fatty acids. The glycerol is channeled into the glycolytic pathway by oxidative conversion to dihydroxyacetone phosphate, as shown in Figure 11-11. The fatty acids are oxidatively degraded by a sequential, stepwise process that involves the successive removal of two-carbon units as acetyl CoA. Thus, the fatty acids derived from fats, like the pyruvate derived from carbohydrates, are oxidatively converted into acetyl CoA, which is then further catabolized by the TCA cycle. Moreover, the enzymes of fatty acid oxidation are localized to the mitochondrion in many (though not all) eukaryotic cells, so the acetyl CoA derived from fats is usually generated and catabolized within the same cellular compartment (see Figure 12-1b).

The sequential process of fatty acid catabolism to acetyl CoA is called **β oxidation** because the initial oxidative event in each successive cycle occurs on the carbon atom in the β position of the fatty acid (i.e., the second carbon from the carboxylic acid group). The process involves successive cycles of oxidative attack on the fatty acid, as shown in Figure 12-10. Each cycle begins with oxidation of the β carbon, results in the release of two carbon atoms as acetyl CoA, and leaves the fatty acid shortened by two carbon atoms but ready for another round of oxidation.

In brief, β oxidation of a fatty acid starts with an activation step in which the energy of ATP is used to form the *fatty acyl CoA* derivative (reaction FA-1 in Figure 12-10). This activated form of the fatty acid is then oxidized by a series of four reactions, three of which (FA-2, FA-3, and FA-4) parallel exactly the sequence whereby succinate is converted into oxaloacetate in the TCA cycle (see reactions TCA-6, TCA-7, and TCA-8 in Figure 12-7). Just as that series of reactions begins with the FAD-mediated dehydrogenation of succinate, here fatty acyl CoA is dehydrogenated to the corresponding *α,β unsaturated acyl CoA*, with FAD as the electron acceptor (reaction FA-2). Not surprisingly, the enzyme

that catalyzes this reaction, fatty acyl dehydrogenase, is an integral membrane protein with a molecule of FAD bound tightly to it. In the next step (reaction FA-3), water is added across the double bond, generating *β-hydroxyl fatty acyl CoA* just as malate is formed from fumarate.

The hydroxyl group is then oxidized in an NAD$^+$-dependent reaction to form the corresponding *β-keto acid* (reaction FA-4). This compound is not very stable; in the final step of the sequence, another molecule of coenzyme A is taken up and the *β*-keto acid is split into acetyl CoA and an acyl CoA compound that is two carbon atoms shorter than the original molecule (reaction FA-5).

The newly formed acyl CoA compound is identical to the activated starting compound that served as the substrate for reaction FA-2, except that it is two carbons shorter. This new acyl CoA molecule then undergoes the same series of reactions represented by reactions FA-2 through FA-5, with another two-carbon unit removed as acetyl CoA. Thus, *β* oxidation is a cyclic, repetitive process in which oxidation continues down the fatty acid backbone, removing two carbon atoms at a time. The repetitive nature of *β* oxidation is illustrated in Figure 12-10 for a fatty acid with eight carbon atoms.

Protein as a Source of Energy. Proteins are not regarded primarily as energy sources because they have more fundamental roles in the cell—as enzymes, permeases, hormones, and receptors, for example. But proteins, too, can be catabolized to generate ATP if necessary. In animals, protein catabolism is prominent under conditions of fasting or starvation, or when the dietary intake of proteins exceeds the need for amino acids. In plants, catabolism of proteins is especially important during the germination of protein-storing seeds and in the senescence, or aging, of leaves. In addition, all cells undergo metabolic turnover of most proteins and protein-containing structures, and the amino acids to which the proteins are degraded can either be recycled into proteins or degraded oxidatively to yield energy.

As shown in Figure 12-11 (page 342), protein catabolism begins with hydrolysis of the peptide bonds that link amino acids together in the polypeptide chain. The process is called **proteolysis,** and the enzymes responsible for it are called *proteases*. The products of proteolytic digestion are small peptides and free amino acids. Further digestion of peptides is catalyzed by *peptidases*, which either hydrolyze internal peptide bonds (*endopeptidases*) or remove successive amino acids from the end of the peptide (*exopeptidases*). Exopeptidases are in turn either *aminopeptidases* or *carboxypeptidases,* depending on the end of the peptide from which digestion proceeds.

Free amino acids, whether ingested as such or obtained by the digestion of proteins, can be catabolized for energy. The pathways by which they are degraded illustrate the general principle that alternative substrates are converted to intermediates of mainstream catabolism in as few steps as possible. In spite of their number and chemical diversity, all

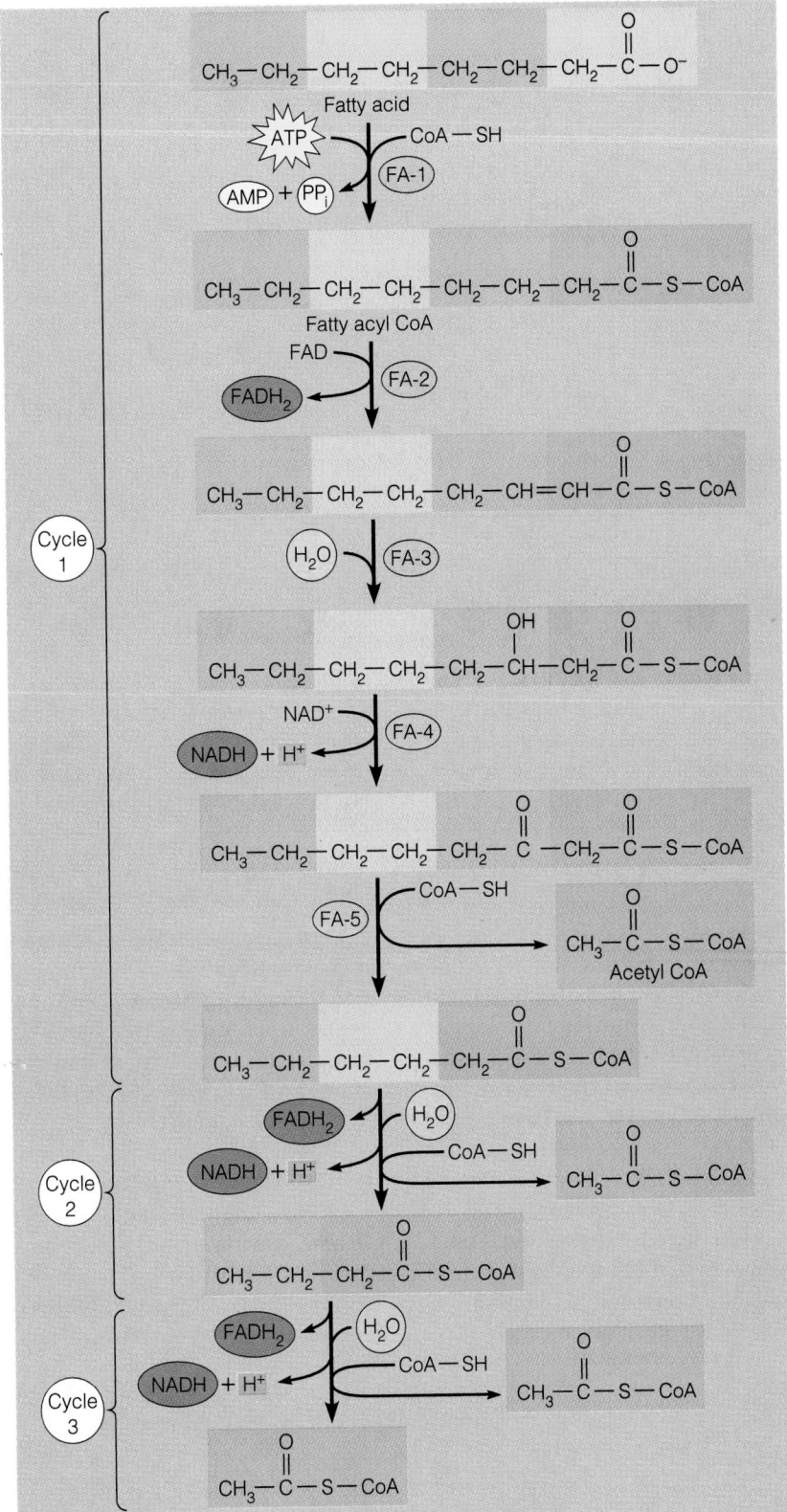

Figure 12-10 The Process of β Oxidation. Most fatty acids have an even number of carbon atoms ranging from $n = 10$ to $n = 24$ and must undergo $(n/2) - 1$ cycles of β oxidation to be catabolized to $n/2$ molecules of acetyl CoA. Reaction numbers and chemical details are shown for the first cycle of β oxidation only. The specific fatty acid shown here is octanoic acid (octanoate), a saturated fatty acid that has eight carbon atoms and therefore requires three cycles of β oxidation, yielding three molecules of NADH, three molecules of FADH$_2$, and four molecules of acetyl CoA. The acetyl CoA molecules are then subject to further catabolism by the TCA cycle.

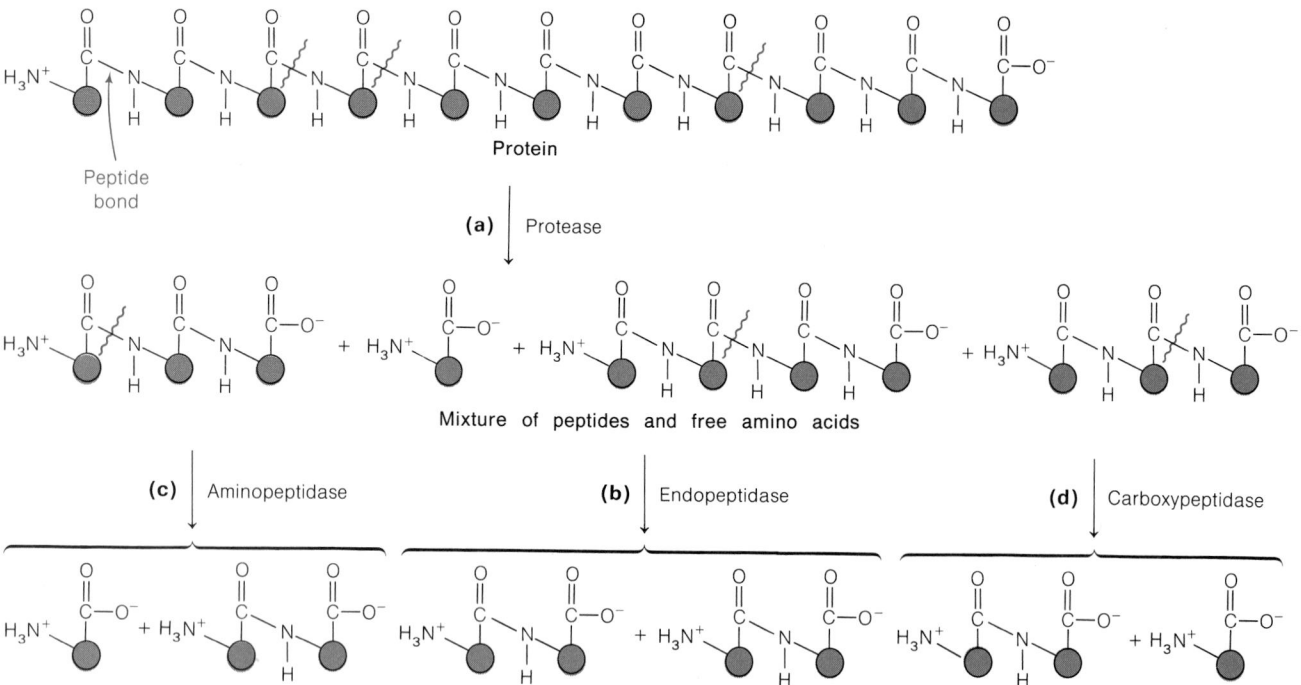

Figure 12-11   Hydrolysis of Proteins and Peptides.   Amino acids are represented by blue circles; only peptide bonds are shown in detail. **(a)** The protein shown at the top is digested to a mixture of peptides and free amino acids by a protease that hydrolyzes peptide bonds at the sites indicated by the wavy lines. Peptides are further digested either by **(b)** internal cleavage (endopeptidases) or by the successive removal of amino acids from either **(c)** the amino end (aminopeptidases) or **(d)** the carboxyl end (carboxypeptidases) of the chain. Eventually, proteins and polypeptides are degraded in this way to free amino acids, which can then be further catabolized by the cell.

these pathways eventually lead to a few key intermediates in the TCA cycle, notably acetyl CoA, α-ketoglutarate, oxaloacetate, fumarate, and succinyl CoA. Most pathways for amino acid catabolism begin with removal of the amino group, either by transamination or by direct oxidative deamination (Figure 12-12). **Transamination** involves the transfer of an amino group from an amino acid to an α-keto acid acceptor (Figure 12-12a). The *deamination* of the amino acid is therefore accompanied by the *amination* of the α-keto acid. Transamination is a common means of shifting amino groups between carbon skeletons and is used by the cell not only in degradative processes but also in synthetic pathways. Transamination does not accomplish the net removal of nitrogen from the pool of organic molecules,

as must happen if net catabolism of amino acids is to occur. It does, however, allow the transfer of amino groups to several common carbon skeletons, especially glutamate. The amino group can then be liberated as free ammonia by a process called **oxidative deamination** (Figure 12-12b). All amino acids can undergo transamination, but only a few can be oxidatively deaminated.

Of the 20 amino acids found in proteins, three give rise to TCA cycle intermediates or precursors directly upon transamination or deamination. These are the amino acids alanine, aspartate, and glutamate, which are converted to pyruvate, oxaloacetate, and α-ketoglutarate, respectively (Figure 12-13). All the other amino acids require more complicated pathways, often with many intermediates. You will

Figure 12-12   Transamination and Oxidative Deamination.   **(a)** Transamination of an amino acid involves transfer of the amino group to an α-keto acid acceptor. **(b)** In oxidative deamination, the amino group is liberated as NH_4^+ or NH_3, depending on the pH. All amino acids can undergo transamination, but only a few can be oxidatively deaminated.

$$R_1 - \underset{\underset{H}{|}}{\overset{+NH_3}{\overset{|}{C}}} - \overset{O}{\overset{||}{C}} - O^- \ + \ R_2 - \overset{O}{\overset{||}{C}} - \overset{O}{\overset{||}{C}} - O^- \ \rightleftharpoons \ R_1 - \overset{O}{\overset{||}{C}} - \overset{O}{\overset{||}{C}} - O^- \ + \ R_2 - \underset{\underset{H}{|}}{\overset{+NH_3}{\overset{|}{C}}} - \overset{O}{\overset{||}{C}} - O^-$$

(a)   Transamination

$$R - \underset{\underset{H}{|}}{\overset{+NH_3}{\overset{|}{C}}} - \overset{O}{\overset{||}{C}} - O^- \ + \ NAD^+ \ + \ H_2O \ \longrightarrow \ R - \overset{O}{\overset{||}{C}} - \overset{O}{\overset{||}{C}} - O^- \ + \ NH_4^+ \ + \ NADH \ + \ H^+$$

(b)   Oxidative deamination

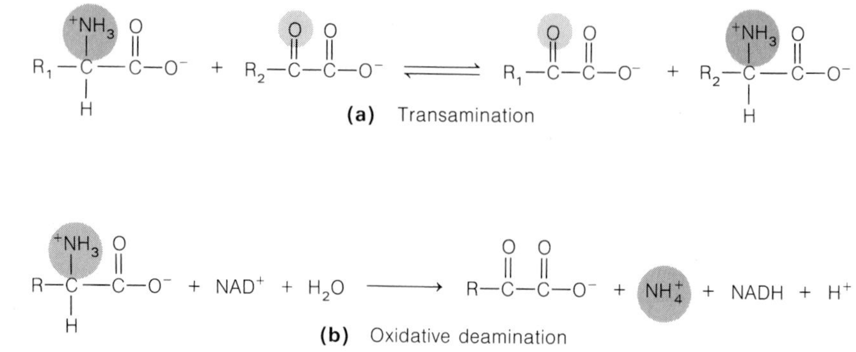

Figure 12-13 Interconversion of Several Amino Acids and Their Cognate Keto Acids in the TCA Cycle.
The amino acids **(a)** alanine, **(b)** aspartate, and **(c)** glutamate can be reversibly transaminated into the corresponding α-keto acids: pyruvate, oxaloacetate, and α-ketoglutarate, respectively. Each of these keto acids is an intermediate in the TCA cycle, a portion of which is shown by dashed lines to provide the metabolic context for these transamination reactions. In each case, the carbon atom that is reversibly transaminated is shown in blue, as are the amino and keto groups. (For the mechanism of transamination, see Figure 12-12a; note specifically that the conversion of an amino acid to the cognate keto acid requires the concomitant conversion of another α-keto acid to its cognate amino acid.) These transamination reactions perform the catabolic function of converting amino acids to TCA-cycle intermediates and the anabolic function of converting TCA-cycle intermediates to amino acids for use in protein synthesis. The degradative and synthetic pathways for other amino acids are less direct but also lead to and from intermediates in respiratory metabolism.

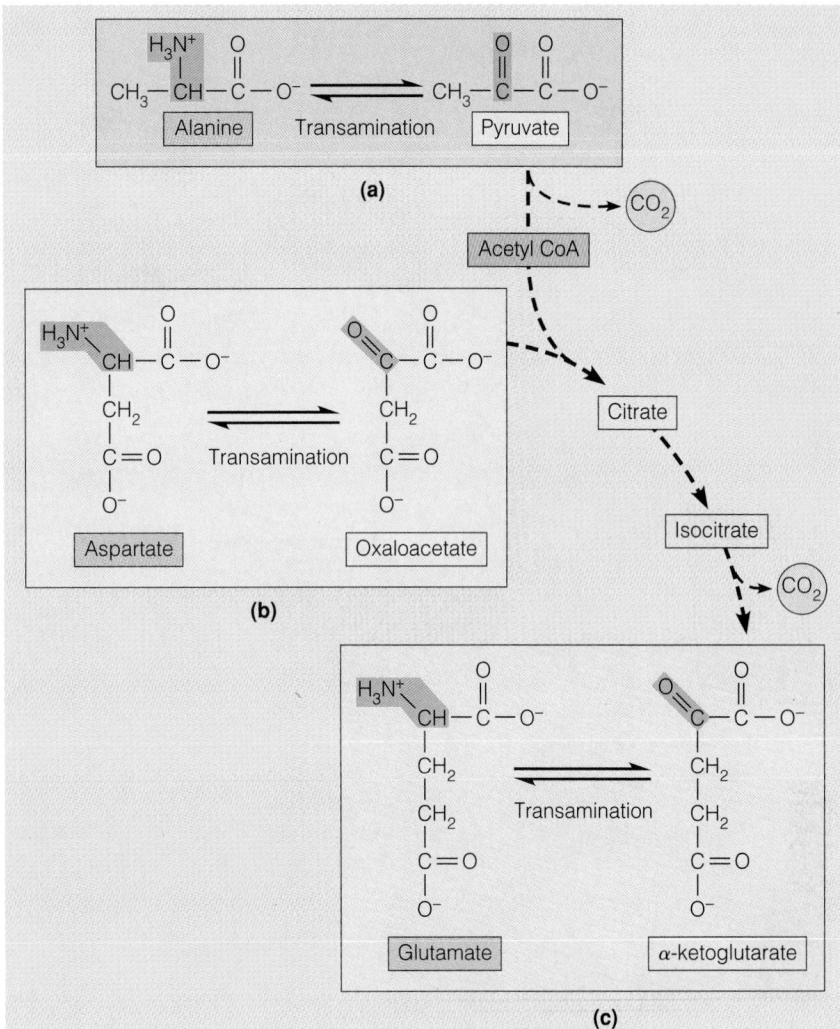

probably encounter these pathways at some future point, most likely in a biochemistry course. When you do, be sure to note how many of them have end-products that are TCA cycle intermediates, because that will further emphasize the centrality of the TCA cycle for all of cellular energy metabolism, regardless of the starting substrate.

The Regulation of TCA Cycle Activity

Like all metabolic pathways, the TCA cycle must be carefully regulated to ensure that its level of activity reflects cellular needs for its products. Regulation can be either *kinetic* or *allosteric*, depending on whether the regulatory agent is a substrate or product of the reaction or an allosteric effector of the enzyme. The pyruvate dehydrogenase complex is also regulated by an additional mechanism to control the availability of acetyl CoA, the cycle's substrate. The main sites of regulation are indicated in Figure 12-14, with kinetic regulators and allosteric regulators shown in different colors.

Several of the dehydrogenases are subject to kinetic regulation by the intramitochondrial $[NAD^+]/[NADH]$ ratio because these enzymes require NAD^+ as a substrate and re-

lease NADH as a product. Whenever the NAD^+ concentration decreases significantly or the NADH concentration increases significantly, the activities of these enzymes are inhibited by either a shortage of NAD^+ or an excess of NADH, or both. Thus, the cycle is highly sensitive to the relative levels of the reduced (high-energy) and oxidized (low-energy) forms of its key coenzyme. Conditions that decrease the $[NAD^+]/[NADH]$ ratio can limit the activities of these dehydrogenases, leading to a reduction in TCA cycle activity. Citrate synthase, the enzyme responsible for citrate formation from oxaloacetate and incoming acetyl CoA (TCA-1), also appears to be regulated kinetically because the acetyl CoA concentration fluctuates significantly with the metabolic status of the cell, and the concentration of oxaloacetate is frequently in the regulatory range also.

The TCA cycle is also subject to allosteric regulation. Both isocitrate dehydrogenase (TCA-3) and α-ketoglutarate dehydrogenase (TCA-4) are allosterically inhibited by NADH. In addition, α-ketoglutarate dehydrogenase is allosterically inhibited by succinyl CoA. Isocitrate dehydrogenase is allosterically activated by ADP, rendering the cycle sensitive to the ATP/ADP status of the cell.

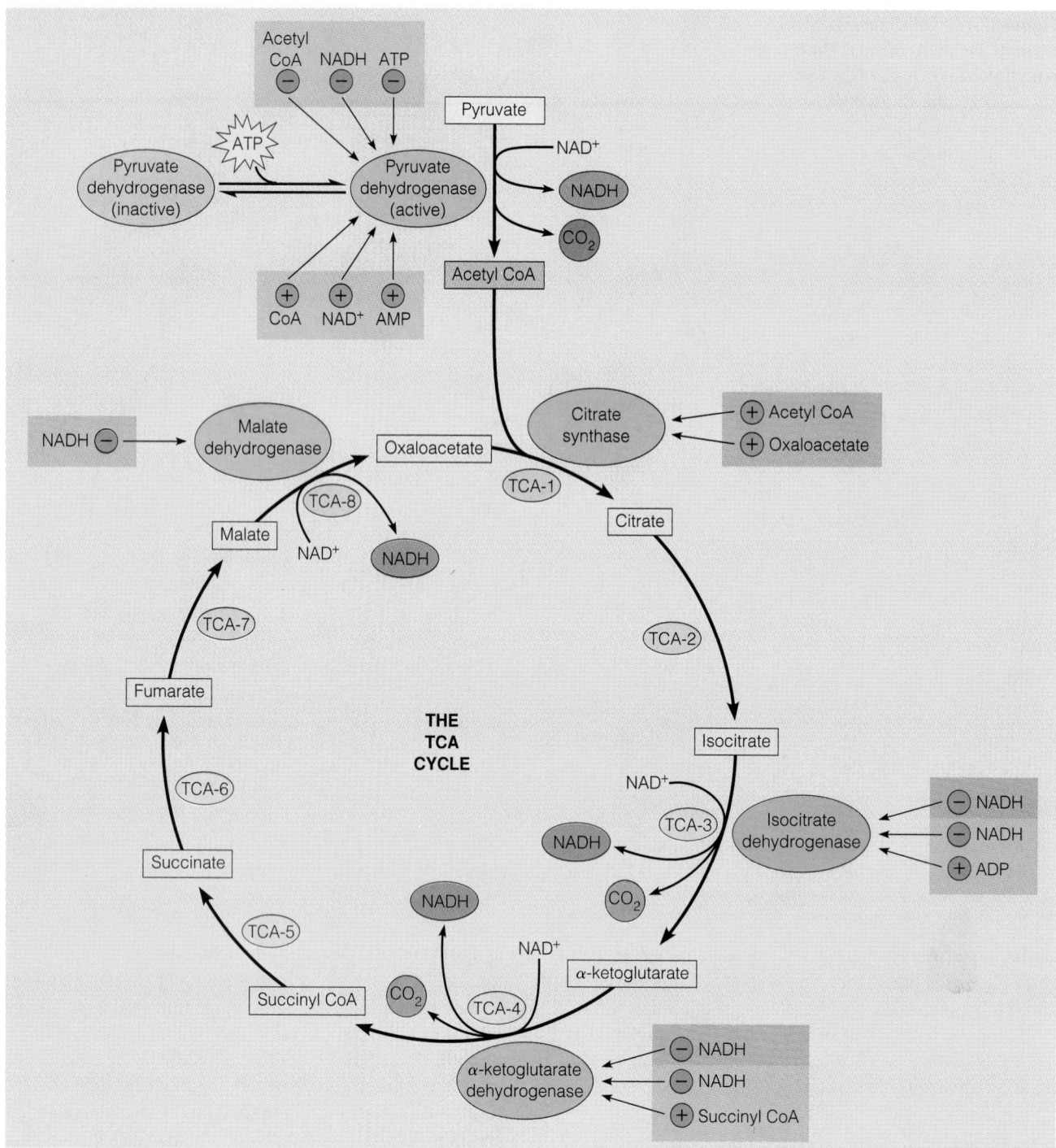

Figure 12-14 Regulation of the TCA Cycle. The TCA cycle is shown here in outline form, with regulatory enzymes highlighted. Major regulatory effects are indicated as either activation (+) or inhibition (−). Kinetic regulators are highlighted in blue, including stimulatory effects of acetyl CoA and oxaloacetate and inhibitory effects of NADH. Allosteric effectors are highlighted in purple, including CoA, NAD⁺, AMP, and ADP as activators and acetyl CoA, NADH, and ATP as inhibitors. In addition to its allosteric effect on the active form of pyruvate dehydrogenase, ATP promotes the phosphorylation of this enzyme, converting it reversibly to an inactive form.

The overall availability of acetyl CoA is determined primarily by the activity of the pyruvate dehydrogenase complex (see reaction 12-1). Components of this multienzyme complex are allosterically inhibited by acetyl CoA, NADH, and ATP and/or allosterically activated by CoA, NAD⁺, and AMP. In addition, the enzyme complex is deactivated by an ATP-dependent phosphorylation when the ATP level in the mitochondrial matrix rises and is reactivated by dephosphorylation when the ATP level falls again. As a result of these multiple control mechanisms, the generation of acetyl CoA is sensitive to the [acetyl CoA]/[CoA] and [NADH]/[NAD$^+$] ratios within the mitochondrion and to the mitochondrial ATP status as well.

In addition to these regulatory effects on reactions of the TCA cycle, feedback control from the cycle to the gly-colytic pathway is provided by the inhibitory effects of cit-rate and acetyl CoA on phosphofructokinase and pyruvate kinase, respectively (see Figure 11-14).

The Amphibolic Role of the TCA Cycle

Catabolism is clearly the major function of the TCA cycle, given its role in the oxidation of acetyl CoA to carbon diox-ide, with concomitant conservation of free energy as re-duced coenzymes. Strictly speaking, in fact, acetyl CoA is the only substrate the TCA cycle can accept; it is not possible for the cycle to accomplish the net intake and catabolism of any other substance. Yet in most cells, there is a considerable flow of four-, five-, and six-carbon intermediates into and out of the cycle. This flux takes place in addition to the cata-bolic function of the cycle. These side reactions replenish the supply of intermediates in the cycle as needed, as well as providing for the synthesis of compounds derived from any of several intermediates in the cycle. Because the TCA cycle can function both in a catabolic mode and as a source of precursors for anabolic pathways, it is called an **amphibolic pathway** (from the Greek *amphi*, meaning "both").

The TCA cycle is involved in a variety of anabolic processes. For example, the three transamination reactions shown in Figure 12-13 convert α-keto intermediates into the amino acids pyruvate, aspartate, and glutamate. These amino acids are constituents of proteins, so the TCA cycle is indirectly involved in protein synthesis by providing several of the amino acids required for the process. Other amphi-bolic precursors in the cycle include succinyl CoA and cit-rate. Succinyl CoA is the starting point for the biosynthesis of heme, while citrate can be transported out of the mito-chondrion and used as a source of acetyl CoA for the syn-thesis of fatty acids in the cytosol.

In each of these cases, intermediates of the cycle are drawn off for biosynthetic purposes. This, in turn, dictates the need for mechanisms that can replenish the intermedi-ates in the cycle as needed. The most important replenish-ment reactions are those in which oxaloacetate is formed by carboxylation of either pyruvate or phosphoenolpyruvate (PEP). Several such possibilities are shown in Figure 12-15. Transamination reactions can also serve as a source of TCA intermediates by functioning in the direction of α-ketoglutarate and oxaloacetate synthesis at the expense of the analogous amino acids (see Figure 12-13).

By a variety of such side reactions, the TCA cycle ac-quires a metabolic versatility beyond its primary catabolic role. Even anaerobic organisms use portions of the TCA pathway for synthetic purposes, despite the lack of TCA cycle activity as such in the absence of oxygen. The common feature of these amphibolic roles is that no cyclic function is involved. One intermediate is removed for synthetic pur-poses, another may be replenished by a side reaction, and only the portion of the cycle that links the two intermediates actually operates.

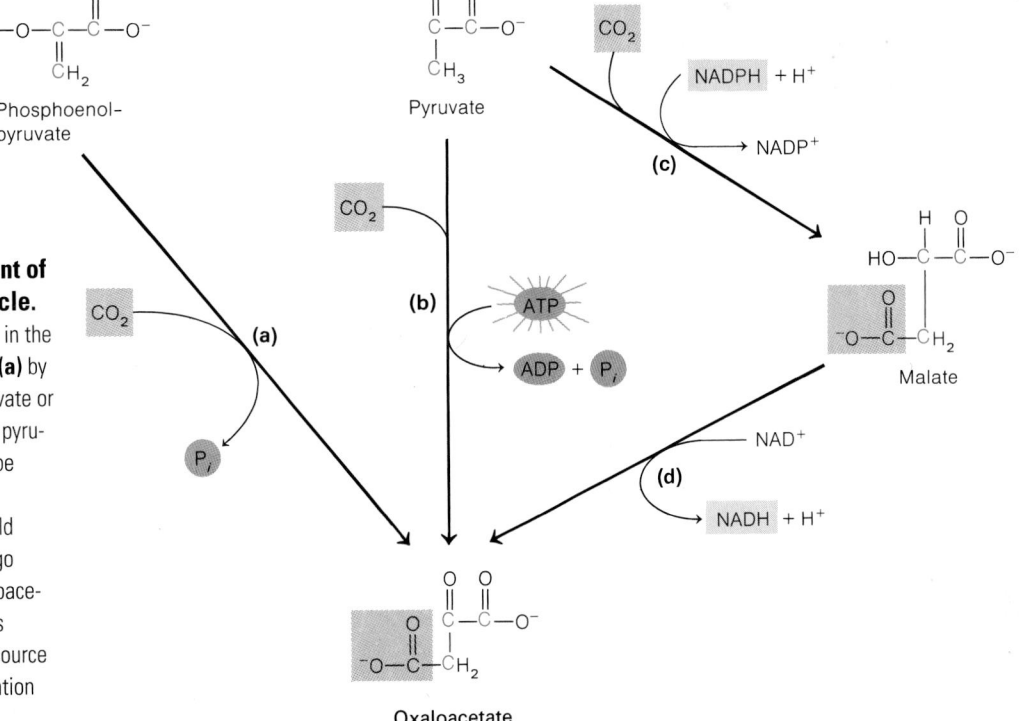

Figure 12-15 Replenishment of Oxaloacetate in the TCA Cycle. The concentration of oxaloacetate in the mitochondrion can be augmented **(a)** by carboxylation of phosphoenolpyruvate or **(b)** by ATP-driven carboxylation of pyru-vate. Alternatively, pyruvate may be **(c)** reductively carboxylated in an NADPH-dependent reaction to yield malate, which can then **(d)** undergo NAD-dependent oxidation to oxaloace-tate. If the amino acid aspartate is present, it can also be used as a source of oxaloacetate by the transamination reaction shown in Figure 12-13b.

The Glyoxylate Cycle: Converting Acetyl CoA to Sugar

A pathway that is related to the TCA cycle but that performs a very specialized function is the *glyoxylate cycle* described in Box 12A. The glyoxylate cycle shares several reactions with the TCA cycle but with the critical difference that it is an anabolic rather than a catabolic pathway. Specifically, it lacks the two decarboxylating reactions of the TCA cycle, using instead a reaction sequence capable of synthesizing the four-carbon compound succinate from two acetate molecules that enter as acetyl CoA (see Figure 12A-2). The succinate is then converted to pyruvate, from which sugars can be synthesized by gluconeogenesis.

The glyoxylate cycle enables organisms to synthesize sugars from two-carbon compounds such as acetate, a metabolic capability that most organisms do not possess. (The TCA cycle includes four-carbon intermediates that can be used for gluconeogenesis, but the *net* conversion of acetate carbons to glucose is not possible because every cycle of activity takes in two carbons as acetate and releases two carbons as CO_2.) The glyoxylate cycle also makes possible the conversion of stored fat to carbohydrate, with acetyl CoA as the critical intermediate. This capability is vital to seed germination in those plant species that store significant amounts of carbon reserves in their seeds as fats. In the seedlings of such species, both β oxidation and the glyoxylate cycle take place in organelles called *glyoxysomes* (see Figure 12A-3).

Electron Transport: Electron Flow from Coenzymes to Oxygen

Having considered the first two stages of aerobic respiration, glycolysis and the TCA cycle, let's pause briefly to ask what has been achieved thus far in terms of energy metabolism. As reaction 12-3 indicates, chemotrophic energy metabolism through the TCA cycle accounts for the synthesis of four ATP molecules per glucose, only twice the yield for glycolysis alone. This modest enhancement in ATP yield hardly seems to justify the metabolic jungle we have been through. Where, one might ask, is all the rest of the free energy? And when will we get to the substantially greater ATP yield that is supposed to be characteristic of respiration?

The answer is that the free energy is right there in reaction 12-3, represented by the reduced coenzyme molecules NADH and $FADH_2$. As we will see shortly, large amounts of free energy are released when reduced coenzymes are reoxidized by molecular oxygen. Such coenzymes are therefore high-energy compounds in their own right. Put more rigorously, reduced coenzymes carry high-energy electrons with potential energy that can be tapped to drive ATP synthesis.

Reduced coenzymes are major players in all of chemotrophic energy metabolism. As we have seen, the catabolism of glucose involves six different oxidations: one during glycolysis, one in the conversion of pyruvate to acetyl CoA, and the other four as part of the TCA cycle (see Figure 12-7). In the process, all six carbon atoms of the glucose molecule are oxidized to carbon dioxide, and twelve pairs of electrons are transferred to NAD^+ (in five of the six reactions) or to FAD (in one reaction). Similarly, the oxidation of fatty acids involves the reduction of one molecule each of NAD^+ and FAD per cycle of oxidation (Figure 12-10) and of additional reduced coenzymes as the resulting acetyl CoA molecules are oxidized further in the TCA cycle. Amino acids are also catabolized by oxidative pathways that use these same coenzymes as electron acceptors. Clearly, much of the energy released by the oxidative catabolism of carbohydrates, fats, or proteins is stored transiently in the high-energy reduced coenzymes generated in the process. Moreover, these oxidative events can be sustained only if oxidized coenzyme molecules continue to be available as electron acceptors. That, in turn, depends on the continuous reoxidation of reduced coenzymes by the transfer of electrons to the terminal acceptor—O_2, in the case of aerobic respiration.

The process of coenzyme reoxidation by the transfer of electrons to oxygen is called **electron transport.** Electron transport is the third stage of respiratory metabolism (see Figure 12-1b). The accompanying process of ATP synthesis (stage 4) will be discussed later in this chapter. Keep in mind, however, that electron transport and ATP synthesis are not isolated processes but are integral parts of respiratory metabolism, intimately linked to each other by the electrochemical proton gradient that is both the "product" of electron transport and the source of the energy that drives ATP synthesis.

The role of the proton gradient as the energy link between electron transport and ATP synthesis raises two key questions: How is the energy of electron transport used to establish and maintain a gradient of protons across the inner membrane of the mitochondrion (or the plasma membrane, in the case of prokaryotic cells)? And how can the free energy of the gradient be used to drive the phosphorylation of ADP? The answers to these questions will occupy our attention for the remainder of this chapter. First we will consider the oxidation of reduced coenzymes, focusing on the free energy released in the process and on the sequence of electron carriers that link coenzyme oxidation to the reduction of oxygen, the terminal electron acceptor. Then we will discuss maintenance and use of the electrochemical proton gradient and its role in ATP synthesis.

Coenzyme Oxidation

Since electron transport involves oxidation of the coenzymes NADH and $FADH_2$ with molecular oxygen as the electron acceptor, we can write summary reactions as follows:

$$NADH + H^+ + \tfrac{1}{2}O_2 \longrightarrow NAD^+ + H_2O \qquad \textbf{(12-4)}$$

$$FADH_2 + \tfrac{1}{2}O_2 \longrightarrow FAD + H_2O \qquad \textbf{(12-5)}$$

Electron transport therefore accounts not only for the reoxidation of coenzymes and the consumption of oxygen but

Further Insights

THE GLYOXYLATE CYCLE, GLYOXYSOMES, AND SEED GERMINATION

P lant species that store substantial carbon and energy reserves in their seeds as *fats* face a special metabolic challenge when their seeds germinate: the conversion of the stored fat to *sucrose*, the circulating form of carbon and energy in the growing seedling. Many plant species are in this category, including such well-known oil-bearing species as soybeans, peanuts, sunflowers, castor beans, and maize. The fat consists mainly of triglycerides and is stored as *lipid bodies*, either in the cotyledons of the plant embryo or in the surrounding endosperm. The electron micrograph in Figure 12A-1 shows the prominence of lipid bodies in the cotyledon of a cucumber seedling. The advantage of storing fat rather than carbohydrate is clear when you consider that one gram of triglyceride contains about 2.25 times as much energy as one gram of carbohydrate. This difference enables fat-storing species to pack the greatest amount of carbon and calories into the least amount of space. But it also

means that such species must be able to convert the stored fat into sugar when the seeds germinate.

The conversion of fat to sugar is not possible for most organisms. Many organisms readily convert sugars and other carbohydrate to fat—some of us all too readily, in fact! But most eukaryotic organisms cannot carry out the reverse process. Yet for the seedlings of fat-storing plant species, the conversion of storage triglycerides to sucrose is essential because it is mainly as sucrose that carbon and energy reserves reach the growing shoot and root tips of the developing seedling.

The metabolic pathways that make this conversion possible are *β oxidation*, with which you should already be familiar, and the **glyoxylate cycle,** with which you probably are not. The function of *β* oxidation is to degrade the stored fat to acetyl CoA (see Figure 12-10). The acetyl CoA then enters the glyoxylate cycle (Figure 12A-2), a five-step cyclic

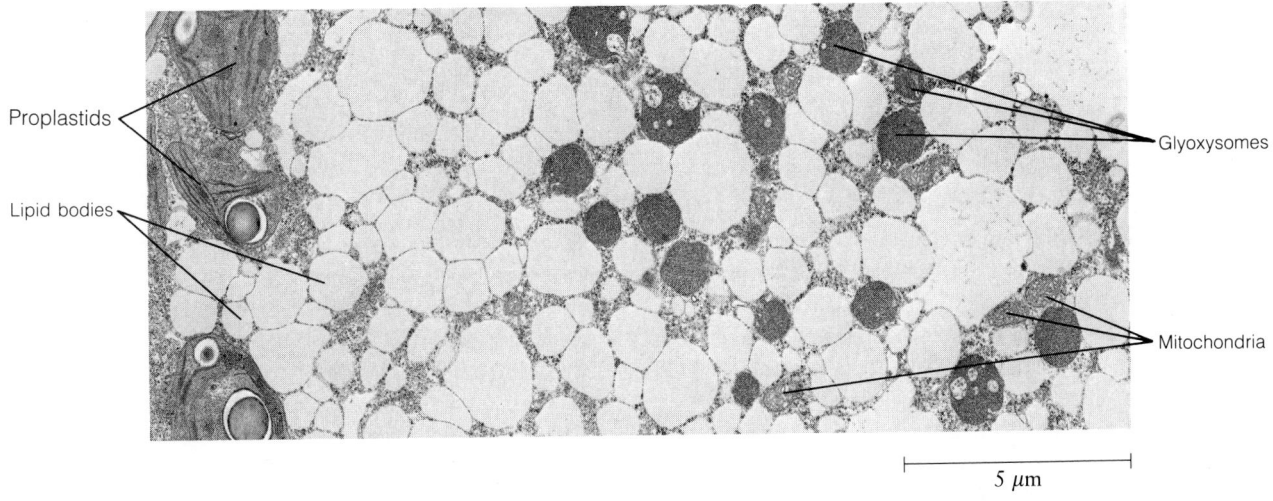

Proplastids

Lipid bodies

Glyoxysomes

Mitochondria

5 μm

Figure 12A-1   The Association of Glyoxysomes and Lipid Bodies in Fat-Storing Seedlings.   Shown here is a cell from the cotyledon of a cucumber seedling during early postgerminative development. The glyoxysomes and mitochondria involved in fat mobilization and gluconeogenesis are intimately associated with the lipid bodies in which the fat is stored. Evidence that the cotyledon is not yet photosynthetically active and is therefore still heterotrophic in its nutritional mode can be seen in the presence of proplastids instead of mature chloroplasts (TEM).

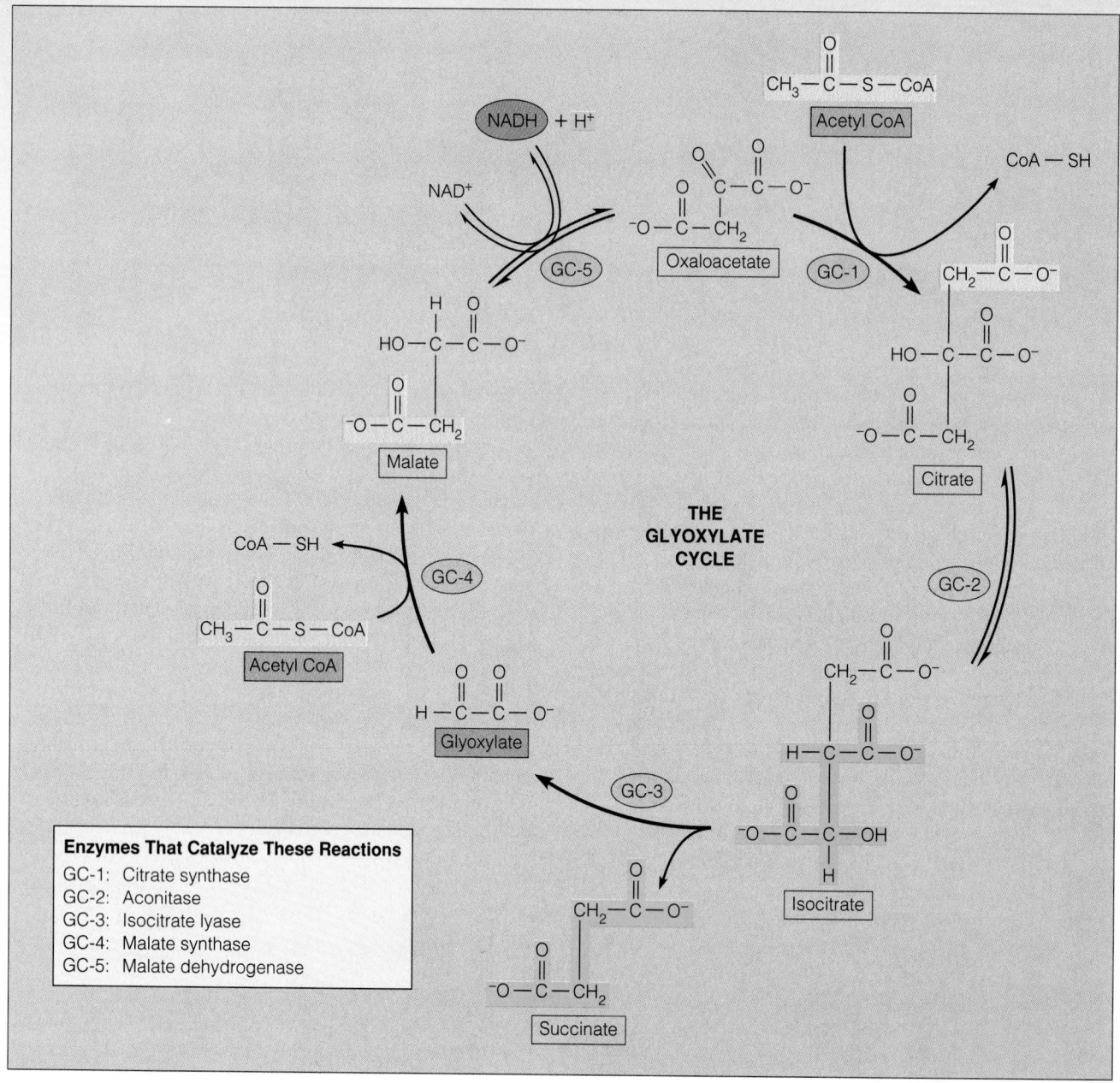

Figure 12A-2 The Glyoxylate Cycle. The glyoxylate cycle is a five-reaction sequence that converts acetyl coenzyme A into succinate. Acetyl CoA enters the cycle at two steps (reactions GC-1 and GC-4), and succinate, a four-carbon molecule, is released upon cleavage of isocitrate in reaction GC-3. The atoms shown in yellow in the citrate and malate molecules are those derived from acetyl CoA in reactions GC-1 and GC-4, respectively, and the blue atoms in isocitrate are those that give rise to succinate in reaction GC-3. Reactions GC-1, GC-2, and GC-5 also occur in the TCA cycle, whereas reactions GC-3 and GC-4 are unique to the gly-oxylate cycle. The cycle is named for glyoxylate, the molecule that is generated and consumed by reactions GC-3 and GC-4, respectively. In the seedlings of fat-storing plant species such as soybean, peanut, maize, and castor bean, acetyl CoA is derived from β oxidation of fatty acids. In bacteria and eukaryotic microorganisms capable of growing on two-carbon substrates such as ethanol or acetate, acetyl CoA is generated from acetyl phosphate, which is formed by ATP-dependent phosphorylation of free acetate.

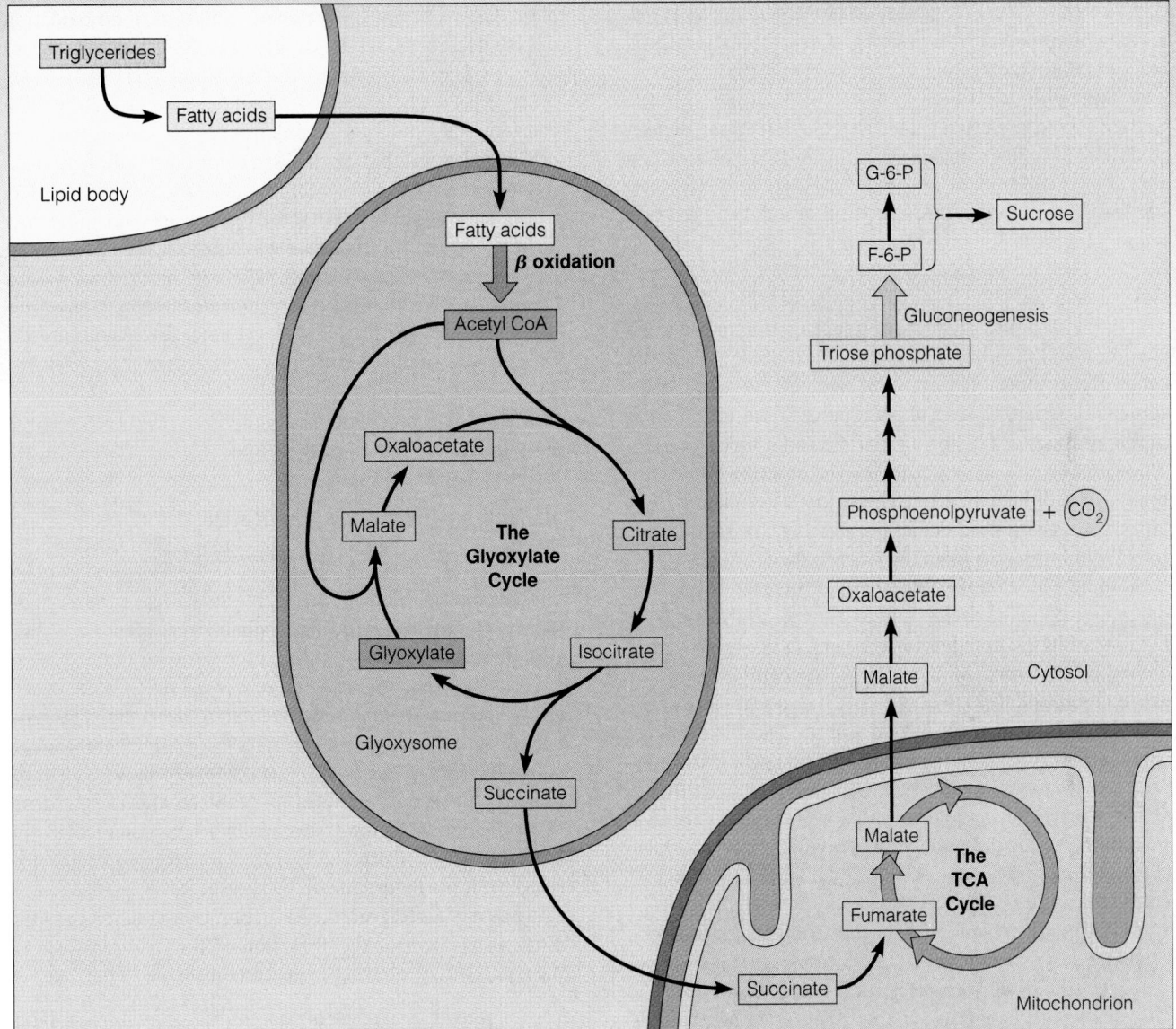

Figure 12A-3 The Glyoxylate Cycle and Gluconeogenesis in Fat-Storing Seedlings. Seedlings of fat-storing plant species are capable of converting stored fat into sugar. Fatty acids derived from hydrolysis of storage triglycerides are oxidized to acetyl CoA by the process of β oxidation. Acetyl CoA is then converted into succinate by the glyoxylate cycle. All the enzymes of β oxidation and the glyoxylate cycle are located in the glyoxysome. Conversion of succinate to oxaloacetate occurs within the mitochondrion, whereas the further metabolism of oxaloacetate via phosphoenolpyruvate (PEP) to hexoses and hence to sucrose takes place in the cytosol.

pathway that is named for one of its intermediates, the two-carbon keto acid *glyoxylate*. The glyoxylate cycle is related to the TCA cycle, with which it shares three reactions in common. There is a critical difference, however: The glyoxylate cycle bypasses the two decarboxylation reactions of the TCA cycle at which carbon dioxide would otherwise be evolved. Instead, the glyoxylate cycle has a strategic two-reaction sequence that enables it to take in not one but two molecules of acetyl CoA per cycle, generating succinate, a four-carbon compound.

As Figure 12A-2 illustrates, carbon enters the glyoxylate cycle as acetyl CoA at reactions GC-1 and GC-4. In the first reaction, the acetate group of acetyl CoA is transferred to oxaloacetate (four carbons, or 4C) to form citrate (6C), as in the TCA cycle. In step GC-4, acetyl CoA reacts with glyoxylate (2C) to form malate (4C). Carbon leaves the cycle as succinate in reaction GC-3. Thus, the glyoxylate cycle is *anabolic* (carbon enters as two-carbon molecules and leaves as a four-carbon molecule), whereas the TCA cycle is *catabolic* (carbon enters as a two-carbon molecule but leaves as two CO_2 molecules).

In the seedlings of fat-storing species, the enzymes of β oxidation and the glyoxylate cycle are localized in organelles called **glyoxysomes.** Recall from Chapter 9 that a glyoxysome is a specialized kind of plant peroxisome, found only in the seedlings of fat-storing species (and sometimes in senescing leaves). Glyoxysomes are visible in the electron micrograph of Figure 12A-1. The intimate association of glyoxysomes with lipid bodies presumably facilitates the transfer of fatty acids from the latter to the former.

Figure 12A-3 (page 349) puts all of the relevant metabolism together in a cellular context. Storage triglycerides are hydrolyzed in the lipid bodies, releasing fatty acids. These are transported into the glyoxysome, where they are degraded by β oxidation to acetyl CoA. The acetyl CoA is converted to succinate by the enzymes of the glyoxylate cycle. The succinate moves to the mitochondrion, where it is converted via fumarate to malate by a reaction sequence that is a part of the TCA cycle. (Notice that mitochondria are also present in the cucumber cotyledon in Figure 12A-1.) Malate is then transported to the cytosol and oxidized to oxaloacetate, which is decarboxylated to form phosphoenolpyruvate (PEP). PEP serves as the starting point for gluconeogenesis, also a cytosolic pathway. Gluconeogenesis leads to the formation of the phosphorylated monosaccharides glucose-6-phosphate (G-6-P) and fructose-6-phosphate (F-6-P), from which sucrose can be synthesized.

The route from stored triglycerides to sucrose is obviously quite complex, involving enzymes located in lipid bodies, glyoxysomes, mitochondria, and the cytosol. But it is the metabolic lifeline on which the seedlings of all fat-storing plant species depend. And it consolidates for us much of the metabolism we've been learning in Chapters 11 and 12, including gluconeogenesis, the TCA cycle, β oxidation—and now the glyoxylate cycle as well. ■

also for the formation of water, which we recognize as the reduced form of oxygen and one of the two end-products of aerobic energy metabolism. (The other end-product, of course, is the carbon dioxide generated in the TCA cycle.)

The most important aspect of reactions 12-4 and 12-5 is the large amount of free energy released upon oxidation of NADH and $FADH_2$ by oxygen. The $\Delta G^{o\prime}$ values for reactions 12-4 and 12-5 are −52.4 kcal/mol and −45.9 kcal/mol, respectively, making it clear that the oxidation of a coenzyme is a highly exergonic process. It should therefore not surprise you that electrons are not passed directly from reduced coenzymes to oxygen because such direct transfers would almost certainly liberate excessive amounts of energy as heat. Rather, the transfer is accomplished as a multistep process that involves a series of reversibly oxidizable electron carriers functioning together in what is called an **electron transport system.** In this way, the total free energy difference between reduced coenzymes and oxygen is parceled out among a series of electron transfers and is released in increments, maximizing the opportunity for energy conservation and, indirectly, for ATP generation.

Reduction Potentials: The Key to Understanding Electron Transport

Understanding the sequence of intermediates in the electron transport system requires an acquaintance with the **reduction potential, E',** a convention used to quantify the *electron transfer potential*, or reducing power, of redox pairs at pH 7.0. A **reduction-oxidation (redox) pair** consists of two molecules or ions that are interconvertible by the loss or gain of electrons. For example, NAD^+ and NADH constitute a redox pair, as do the ferric (Fe^{3+}) and ferrous (Fe^{2+}) forms of iron. By convention, redox pairs are represented in the form Fe^{3+}/Fe^{2+}, with the oxidized form given first, separated from the reduced form by a slash.

For any redox pair, the reduction potential refers to the *half-cell reaction* for the reduction of the oxidized form of the pair. Thus, the half-cell reaction for the Fe^{3+}/Fe^{2+} pair is:

$$Fe^{3+} + e^- \longrightarrow Fe^{2+} \tag{12-6}$$

(Half-cell reactions such as the above do not occur in isolation, of course, because every reduction must be accompanied by an oxidation. However, the concept of half-cell reactions is very useful because it provides us with a means of thinking about the tendency of a reduction to occur and enables us to quantify that tendency as an E' value.)

A reduction potential is a measure of the affinity that the oxidized form of a redox pair has for electrons or, in other words, how strongly the half-reaction tends to proceed in the direction of reduction. A redox pair has a *positive E'* if its oxidized form has a high affinity for electrons (i.e., is a good electron acceptor) and a *negative E'* if the oxidized form has a low affinity for electrons (i.e., is a poor electron acceptor). For example, the E' value for reaction 12-6 is highly positive, meaning that Fe^{3+} is a good electron acceptor and the reaction tends to proceed in the direction written. Alternatively, negative E' values can be thought of as a measure of how good an electron *donor* the reduced form of a redox pair is: The more highly negative the value, the better the donor. The E' for the $NAD^+/NADH$ pair (i.e., for the reduction of NAD^+ to NADH) is highly negative, indicating that NADH is a good electron donor.

The Standard Reduction Potential. The **standard reduction potential, E_0',** is the reduction potential for a redox pair under standard conditions (25°C, 1 M concentration, 1 atmosphere pressure, and pH 7.0), expressed in volts (V). The standard reduction potentials for the half-cell reactions of a number of biologically important redox pairs are given in Table 12-2. For a brief description of how these values are determined experimentally, see Figure 12-16.

By convention, the H^+/H_2 redox pair is used as a standard and is assigned the value 0.00 V (Table 12-2, boldface line). For a redox pair to have a positive standard reduction potential means that, under standard conditions, the oxidized form of the pair is a better electron acceptor, and therefore a better oxidizing agent, than H^+ (or, alternatively,

that the reduced form of the pair is not as good an electron donor, and hence a poorer reducing agent, than H_2). Conversely, a negative reduction potential means that the oxidized form of the pair has less affinity for electrons than H^+ (or, alternatively, that the reduced form is a better reducing agent than H_2).

The E_0' values in Table 12-2 are valid only under standard conditions. To calculate E', the reduction potential under nonstandard conditions, an additional term is needed that takes into account the actual concentrations of the oxidized and reduced forms of the redox pair. The equation is:

$$E' = E_0' + (RT/nF) \ln ([\text{oxidized form}]/[\text{reduced form}]) \quad \text{(12-7)}$$

where E_0' is the standard reduction potential, R is the gas constant (1.987 cal/mol-K), T is the temperature in Kelvin units, n is the number of electrons transferred in the half-cell reaction (1 or 2; see Table 12-2), and F is the Faraday constant (23,062 cal/mol-V).

The redox pairs of Table 12-2 are arranged in order, with the most negative E_0' values (i.e., the best electron donors and hence the strongest reducing agents) at the top.

Table 12-2 Standard Reduction Potentials for Redox Pairs of Biological Relevance

Half-Cell Reaction	E_0'	n
	Volts	
acetate + CO_2 + $2H^+$ + $2e^-$ → pyruvate + H_2O	−0.70	2
succinate + CO_2 + $2H^+$ + $2e^-$ → α-ketoglutarate + H_2O	−0.67	2
acetate + $2H^+$ + $2e^-$ → acetaldehyde + H_2O	−0.60	2
3-phosphoglycerate + $2H^+$ + $2e^-$ → glyceraldehyde-3-P + H_2O	−0.55	2
α-ketoglutarate + CO_2 + $2H^+$ + $2e^-$ → isocitrate	−0.38	2
NAD^+ + H^+ + $2e^-$ → NADH	−0.32	2
FMN + $2H^+$ + $2e^-$ → $FMNH_2$	−0.30	2
1,3-bisphosphoglycerate + $2H^+$ + $2e^-$ → glyceraldehyde-3-P	−0.29	2
S + $2H^+$ + $2e^-$ → H_2S	−0.23	2
acetaldehyde + $2H^+$ + $2e^-$ → ethanol	−0.20	2
pyruvate + $2H^+$ + $2e^-$ → lactate	−0.19	2
FAD + $2H^+$ + $2e^-$ → $FADH_2$	−0.18*	2
oxaloacetate + $2H^+$ + $2e^-$ → malate	−0.17	2
fumarate + $2H^+$ + $2e^-$ → succinate	−0.03	2
$2H^+$ + $2e^-$ → H_2	**0.00****	**2**
CoQ + $2H^+$ + $2e^-$ → $CoQH_2$	+0.04	2
cytochrome b (Fe^{3+}) + e^- → cytochrome b (Fe^{2+})	+0.07	1
cytochrome c (Fe^{3+}) + e^- → cytochrome c (Fe^{2+})	+0.25	1
cytochrome a (Fe^{3+}) + e^- → cytochrome a (Fe^{2+})	+0.29	1
cytochrome a_3 (Fe^{3+}) + e^- → cytochrome a_3 (Fe^{2+})	+0.55	1
Fe^{3+} + e^- → Fe^{2+}	+0.77	1
$\frac{1}{2} O_2$ + $2H^+$ + $2e^-$ → H_2O	+0.816	2

E_0' values are determined at pH 7.0 and 25°C, relative to the standard hydrogen half-cell. For a two-electron reaction, a difference in E_0' of 0.10 corresponds to $\Delta G^{\circ\prime}$ value of −4.6 kcal/mol.

In the far right column, n represents the number of electrons transferred in the half-cell reaction.

*The value for the $FAD/FADH_2$ pair presumes the free coenzyme; when bound to a flavoprotein, the coenzyme has an E_0' value in the range of from 0.0 to +0.3 V, depending on the specific protein.

** This is the standard hydrogen half-cell; it requires that $[H^+]$ = 1.0 M and therefore specifies pH 0.0. At pH 7.0, the value for the $2H^+/H_2$ pair is −0.42 V.

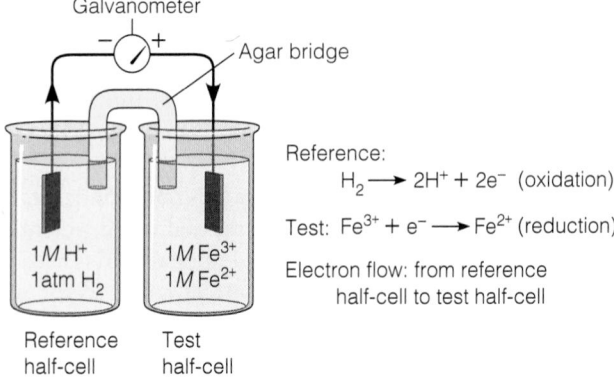

Reference:

$$H_2 \longrightarrow 2H^+ + 2e^- \quad \text{(oxidation)}$$

Test: $Fe^{3+} + e^- \longrightarrow Fe^{2+}$ (reduction)

Electron flow: from reference
half-cell to test half-cell

(a) Fe^{3+}/Fe^{2+} reduction potential

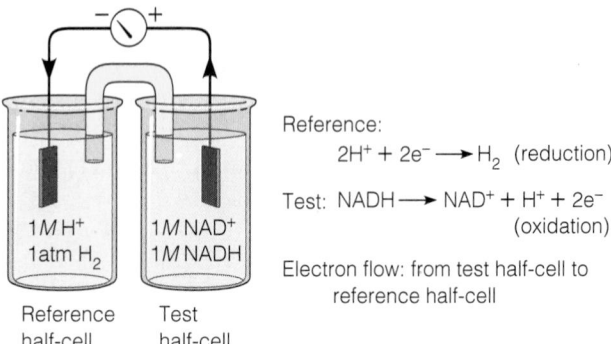

Reference:

$$2H^+ + 2e^- \longrightarrow H_2 \quad \text{(reduction)}$$

Test: $NADH \longrightarrow NAD^+ + H^+ + 2e^-$
(oxidation)

Electron flow: from test half-cell to
reference half-cell

(b) $NAD^+/NADH$ reduction potential

Figure 12-16   Determination of Standard Reduction Potentials. Standard reduction potentials are determined using an electrochemical cell that consists of two half-cells, each containing the oxidized and reduced forms of a redox pair under standard conditions. The reference half-cell (on the left) contains protons at a concentration of 1 M and hydrogen gas (H_2) at a pressure of 1 atmosphere. The test half-cell (on the right) contains the oxidized and reduced forms (both at 1 M) of the redox pair under study. An agar bridge connects the two half-cells to maintain electrical neutrality. The galvanometer that links the two half-cells measures the electromotive force (emf), which is the difference between the reduction potentials in the reference and test half-cells. Electrons may flow either away from or toward the reference half-cell, depending on whether H_2 has a greater or lesser tendency to lose electrons than the reduced form of the redox pair in the test half-cell. **(a)** When the Fe^{3+}/Fe^{2+} redox pair is in the test half-cell, as shown here, electrons flow *from* the reference half-cell *to* the test half-cell because H_2 loses electrons more readily than does Fe^{2+}. As a result, H_2 will be oxidized, Fe^{3+} ions will be reduced, and the galvanometer will record a *positive* emf: $+0.77$ V. The Fe^{3+}/Fe^{2+} redox pair therefore has a *positive* standard reduction potential. **(b)** When the test half-cell contains the $NAD^+/NADH$ redox pair, electron flow will be *from* the test half-cell *to* the reference half-cell because NADH is a better electron donor than H_2. Under these conditions, NADH will be oxidized, H^+ will be reduced, and the galvanometer will record a *negative* emf: -0.32 V. The $NAD^+/NADH$ redox pair therefore has a *negative* standard reduction potential.

Any redox pair shown in Table 12-2 can undergo a redox reaction with any other pair. The direction of such a reaction under standard conditions can be predicted by inspection, because the reduced form of any pair will spontaneously reduce the oxidized form of any pair below it on the table. Thus, NADH will spontaneously reduce oxaloacetate or pyruvate, but not α-ketoglutarate or acetate.

The tendency of the reduced form of one pair to reduce the oxidized form of another pair can be quantified by determining $\Delta E_0'$, the *difference* in the E_0' values between the two pairs:

$$\Delta E_0' = E_{0, \text{acceptor}}' - E_{0, \text{donor}}' \qquad \textbf{(12-8)}$$

As you may have already guessed, $\Delta E_0'$ is a measure of thermodynamic spontaneity for the redox reaction between any two redox pairs under standard conditions. The spontaneity of a redox reaction can therefore be expressed as either $\Delta G^{\circ\prime}$ or $\Delta E_0'$. The sign convention for $\Delta E_0'$ is the opposite of that for $\Delta G^{\circ\prime}$, however: A thermodynamically feasible reaction is one with a *negative* $\Delta G^{\circ\prime}$ but a *positive* $\Delta E_0'$. For example, $\Delta E_0'$ for the transfer of electrons from NADH to O_2 (reaction 12-4) is calculated as follows, with NADH as the donor and O_2 as the acceptor:

$$\Delta E_0' = E_{0, \text{acceptor}}' - E_{0, \text{donor}}' = +0.816 - (-0.32) = +1.136 \text{ V}$$
$$\textbf{(12-9)}$$

The $\Delta E_0'$ value for the reaction is positive, so the transfer of electrons from NADH to O_2 is thermodynamically spontaneous under standard conditions. (It is, in fact, the difference in reduction potentials between the $NAD^+/NADH$ and O_2/H_2O redox pairs that drives the maintenance of the electrochemical proton gradient and, hence, most of the ATP synthesis of aerobic respiration.)

The Relationship Between $\Delta E_0'$ and $\Delta G^{\circ\prime}$ Because both $\Delta G^{\circ\prime}$ and $\Delta E_0'$ are measures of thermodynamic spontaneity, we should expect some sort of mathematical relationship between them. The relationship is linear but with a change of sign because of the difference in conventions for $\Delta G^{\circ\prime}$ and $\Delta E_0'$. For any oxidation-reduction reaction, $\Delta G^{\circ\prime}$ is related to E_0' by the equation

$$\Delta G^{\circ\prime} = -nF\Delta E_0' \qquad \textbf{(12-10)}$$

where n is the number of electrons transferred and F is the Faraday constant as previously defined. For example, the reaction of NADH with oxygen (reaction 12-4) involves the transfer of two electrons, so $\Delta G^{\circ\prime}$ for the reaction can be calculated as:

$$\Delta G^{\circ\prime} = -nF\Delta E_0' = -2(23,062)(+1.136) = -52,400 \text{ cal/mol}$$
$$= -52.4 \text{ kcal/mol} \qquad \textbf{(12-11)}$$

The Electron Transport System

We are now ready to consider the electron transport system in detail. To do so, we will ask three questions: (1) What are the major electron carriers in the system? (2) What is the sequence of these carriers in the electron transport system? (3) How are these carriers organized in the membrane to ensure that the flow of electrons from reduced coenzymes to oxygen is coupled to the pumping of protons across the membrane, thereby maintaining the electrochemical proton gradient on which ATP synthesis depends?

What Are the Electron Carriers in the System?

The carriers that make up the electron transport system include flavoproteins, iron-sulfur proteins, cytochromes, copper-containing proteins, and a quinone called coenzyme Q. Except for coenzyme Q, all these carriers are proteins with specific prosthetic groups capable of being reversibly oxidized and reduced. Most of the events of electron transport occur within membranes, so it is not surprising that, except for cytochrome *c*, all these carriers are hydrophobic molecules. We will look briefly at their chemistry and then see how they are ordered into a sequence.

Flavoproteins. Several membrane-bound **flavoproteins** participate in electron transport, using either *flavin adenine dinucleotide (FAD)* or *flavin mononucleotide (FMN)* as the prosthetic group. FMN is essentially the flavin–containing half of the FAD molecule shown in Figure 12-9. An example of a flavoprotein is *NADH dehydrogenase*, an FMN-containing protein that is part of the protein complex responsible for the transfer of electrons from NADH in the matrix to coenzyme Q in the membrane. Another example, already familiar to us from the TCA cycle, is the enzyme *succinate dehydrogenase*, which has FAD as its prosthetic group.

Unlike the other enzymes of the TCA cycle, succinate dehydrogenase is an integral membrane protein. It is, in fact, the complex that transfers electrons from succinate via its FAD prosthetic group to coenzyme Q and is more properly referred to as *succinate-coenzyme Q reductase*. As we will see shortly, both NADH dehydrogenase and succinate-coenzyme Q reductase are major structural components of the electron transport system as it functions in the membrane.

Iron-Sulfur Proteins. **Iron-sulfur proteins,** also called *nonheme iron proteins*, are a family of proteins containing **iron-sulfur (Fe/S) centers** that consist of iron and sulfur atoms complexed with cysteine groups of the protein. The iron and sulfur atoms are usually present in the iron-sulfur centers in equimolar amounts, with 2Fe/2S and 4Fe/4S as the most common forms. The iron atoms of the iron-sulfur centers are the actual electron acceptors and donors of the iron-sulfur proteins. Each iron atom alternates between the Fe^{3+} (ferric) and Fe^{2+} (ferrous) states as the centers are reduced and oxidized by one-electron transfers. Unlike NAD^+, FAD, FMN, and CoQ, iron-sulfur centers do not pick up and release protons as they cycle between the oxidized and reduced states. Iron-sulfur proteins are the most numerous intermediates in the electron transport system. At least a dozen different Fe/S centers are known to be involved in the mitochondrial transport system.

Cytochromes. Like the iron-sulfur proteins, **cytochromes** also contain iron, but as part of a porphyrin prosthetic group called *heme*. The basic structure of heme is shown in Figure 12-17. There are at least five different cytochromes in

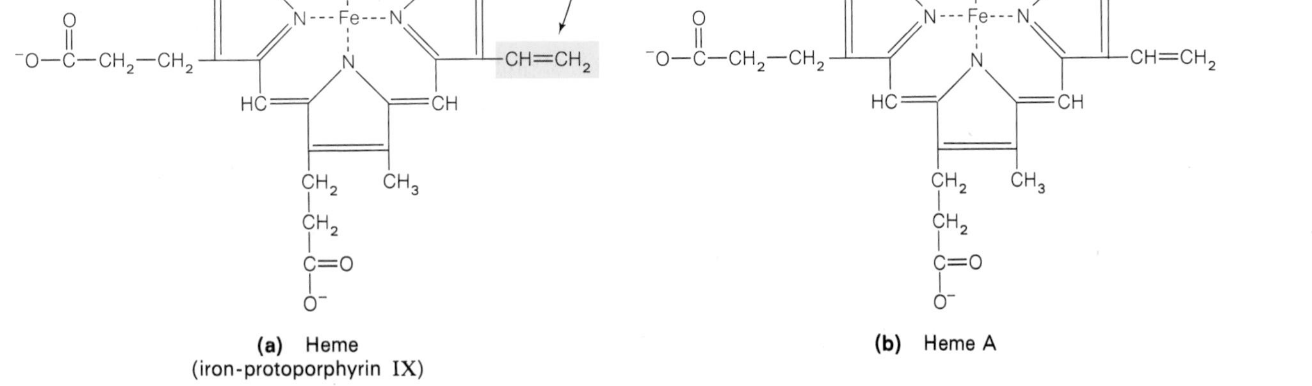

(a) Heme
(iron-protoporphyrin IX)

(b) Heme A

Figure 12-17 The Structure of Heme. **(a)** Heme (iron-protoporphyrin IX), as found in cytochromes *b*, *c*, and *c₁* **(b)** Heme A, as found in cytochromes *a₁* and *a₃*. The heme of cytochromes *c* and *c₁* is covalently attached to the protein by thioether bonds between the sulfhydryl groups of two cysteines in the protein and the vinyl ($—CH{=}CH_2$) groups of the heme (highlighted in yellow).

the electron transport system, designated as cytochromes *b*, *c*, *c₁*, *a*, and *a₃*. Cytochromes *b*, *c*, and *c₁* all contain *iron-protoporphyrin IX*, the form of heme found in hemoglobin (Figure 12-17a), whereas cytochromes *a* and *a₃* contain a modified prosthetic group called *heme A* (Figure 12-17b). Like the iron of the iron-sulfur complex, the iron of the heme prosthetic group is reversibly oxidizable and serves as the electron acceptor for the cytochromes. Both cytochromes and iron-sulfur proteins are therefore one-electron carriers. Cytochromes *b*, *c₁*, *a*, and *a₃* are integral membrane proteins, the latter two occurring together at the end of the transport sequence as a complex called *cytochrome c oxidase*. Cytochrome *c*, on the other hand, is relatively hydrophilic and is only loosely associated with the inner surface of the membrane. Moreover, cytochrome *c* is not a part of a large complex and can therefore diffuse much more rapidly in the plane of the membrane, a property that is important to its role in transferring electrons between protein complexes.

Copper-Containing Proteins. **Copper-containing proteins** are also involved in mitochondrial electron transport, where they occur in conjunction with cytochromes *a* and *a₃*. The *copper (Cu) centers* of these proteins consist of copper ions that alternate between the Cu^{2+} (cupric) and Cu^+ (cuprous) states as they accept and donate single electrons. A copper center appears to be a single copper atom bound to the heme group of the cytochrome, where it forms a bimetallic **iron-copper (Fe/Cu) center** with an iron atom. The iron-copper center plays a critical role in binding an O_2 molecule as it picks up the four electrons that are needed to reduce the oxygen molecule to two molecules of water. (A nutritional note: If you have ever wondered why you require the elements iron and copper in your diet, their roles in electron transport are part of the reason.)

Coenzyme Q. **Coenzyme Q (CoQ)** is the only nonprotein component of the electron transport system. Coenzyme Q is a *quinone*; its structure is shown in Figure 12-18. Because of its ubiquitous occurrence in nature, coenzyme Q is also known as *ubiquinone*. Figure 12-18 also illustrates the reversible reduction, in two one-electron steps, from the quinone form (CoQ) via the *semiquinone* form (CoQH) to the *dihydroquinone* form (CoQH₂). Unlike most of the proteins of the electron transport system, coenzyme Q is not part of a complex. Instead, the nonpolar interior of the inner mitochondrial membrane (or of the plasma membrane, in the case of prokaryotes) contains a large pool of CoQ molecules that transfer electrons between other carriers. CoQ molecules are the most abundant electron carriers in the membrane. As we will see shortly, coenzyme Q occupies a central position in the electron transport system, serving as a collection point for electrons from the reduced prosthetic groups of FMN- and FAD-linked dehydrogenases in the membrane.

Note that coenzyme Q accepts not only electrons but also protons when it is reduced and that it releases both elec-

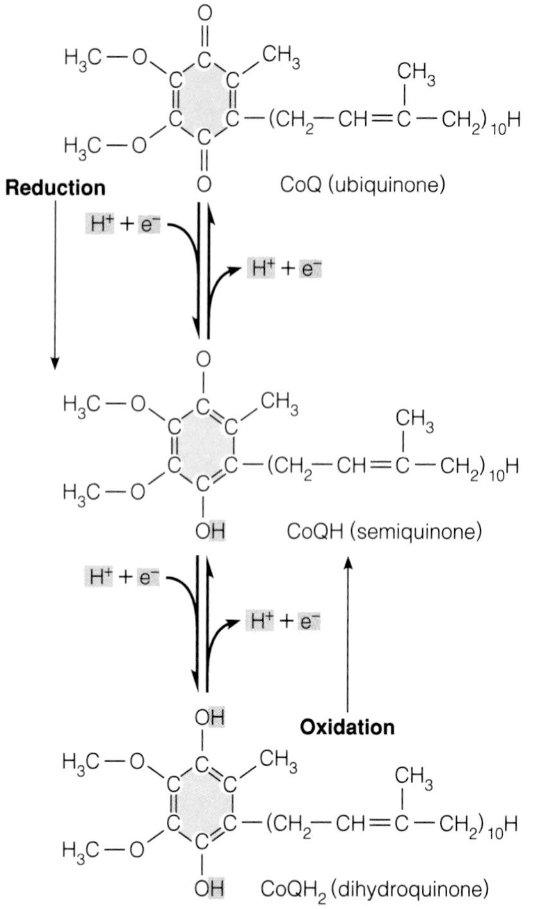

Figure 12-18   Oxidized and Reduced Forms of Coenzyme Q. Coenzyme Q (also called ubiquinone) accepts both electrons and protons as it is reversibly reduced in two one-electron steps to form first CoQH (the semiquinone form) and then CoQH₂ (the dihydroquinone form).

trons and protons when it is oxidized. Now imagine a closed membrane vesicle with coenzyme Q embedded in the membrane. Assume that whenever CoQ is reduced to CoQH₂, it always accepts protons from *inside* the vesicle. The reduced form, CoQH₂, then diffuses across the membrane to the outer surface, where it is oxidized to CoQ, with the protons ejected to the *outside* of the vesicle. Under these conditions, protons would be picked up by coenzyme Q on the inside and delivered to the outside, thereby providing a proton pump coupled to electron transport. As we will see, this mode of proton pumping is thought to be one of the mechanisms whereby mitochondria, chloroplasts, and prokaryotes establish and maintain the electrochemical proton gradients that are used to store the energy of electron transport.

What Is the Sequence of the Carriers?

The respiratory electron transport system consists of several FMN- and FAD-linked dehydrogenases, iron-sulfur proteins with a total of twelve or more Fe/S centers, five cytochromes, copper proteins with Fe/Cu centers, and a pool of coenzyme Q molecules that exist in the oxidized (CoQ), partially reduced (CoQH), or reduced (CoQH₂) states. The next question concerns the order in which these carriers function. Not surprisingly, the sequence is determined by the relative reduction potentials of the components.

Figure 12-19 shows most of the major components of the electron transport system from free NADH ($E_0' = -0.32$ V) and the FADH$_2$ of succinate dehydrogenase ($E_0' = -0.18$ V) to oxygen ($E_0' = +0.816$ V), ordered according to their standard reduction potentials (E_0' values as obtained from Table 12-2). The standard reduction potential scale is shown on the left. The scales on the right are for $\Delta E_0'$ and $\Delta G^{\circ\prime}$ calculated according to equations 12-8 and 12-10, respectively, for a two-electron reaction with O$_2$ as the electron acceptor.

We will consider the organization of these electron carriers in the next section. For the moment, simply notice that the position of each carrier is determined by its standard re-

duction potential. The electron transport system, in other words, consists of a series of chemically diverse electron carriers, with their order of participation in electron transfer determined by their relative reduction potentials. Notice also that electron transfer from NADH or FADH$_2$ on the left to O$_2$ on the right is spontaneous and exergonic, with some of the transfers between successive carriers characterized by quite large differences in E_0' values and hence by large changes in free energy.

How Are the Carriers Organized in the Membrane?

Although there are many electron carriers involved in the electron transport system, they are not present in the membrane as separate entities. Instead, all the carriers except

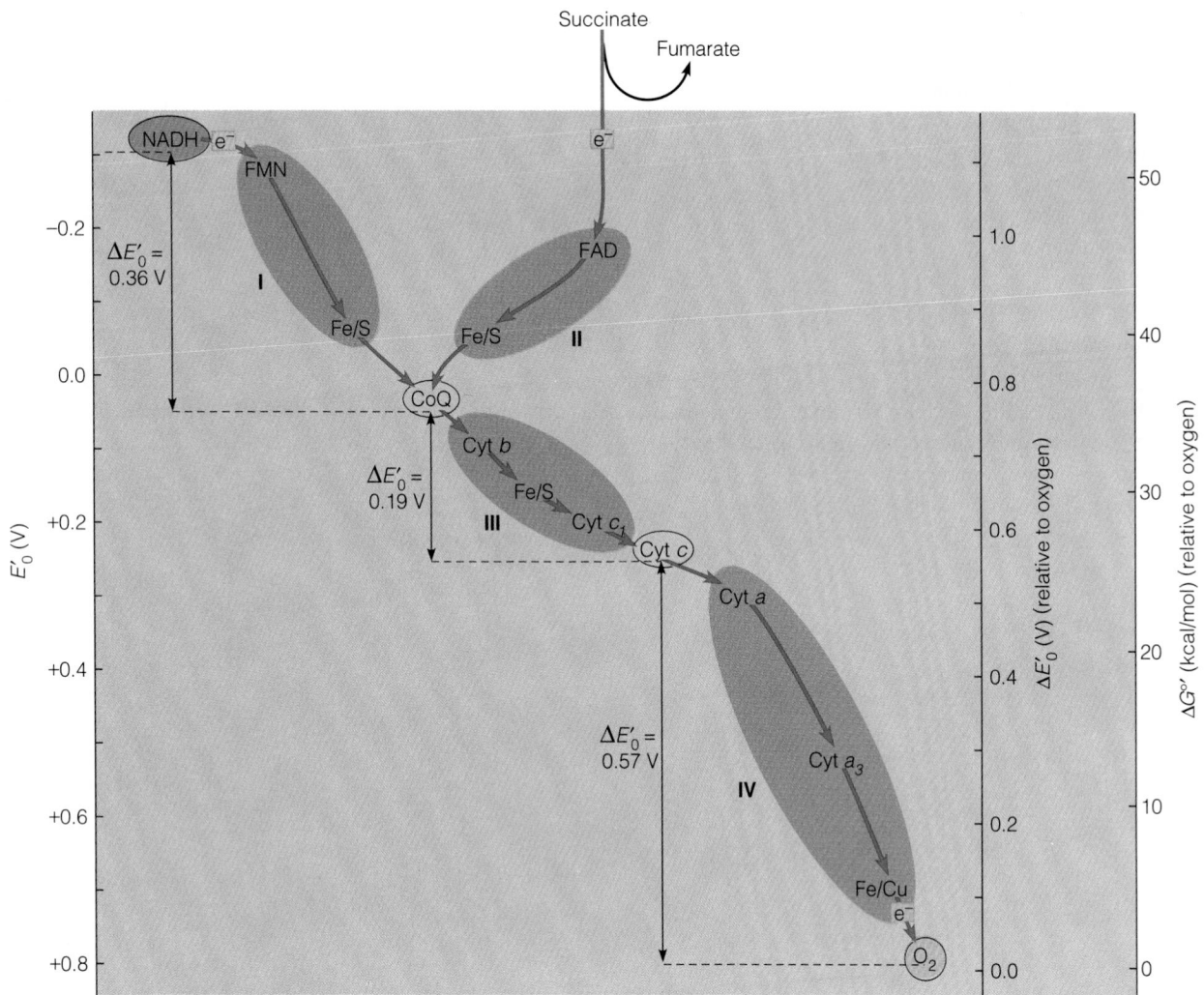

Figure 12-19 The Energetics of Electron Transport. The major intermediates in the transport of electrons from NADH (-0.32 V) and FADH$_2$ (-0.18 V) to oxygen ($+0.816$ V) are positioned vertically according to their energy levels, as measured by their standard reduction potentials (E_0', left axis). On the right axes are the $\Delta E_0'$ and $\Delta G^{\circ\prime}$ values relative to oxygen (i.e., the changes in the standard reduction potential and the standard free energy change for the transfer of electrons to O$_2$).

The pink lines trace the exergonic flow of electrons through the system. The Roman numerals identify the four respiratory complexes, with the major electron carriers of each complex enclosed within brown ovals. The vertical lines indicate the change in E_0' across complexes I, III, and IV. In each case, the change in reduction potential provides the energy that drives proton pumping by these three complexes (but not by complex II). For details of the respiratory complexes, see Figure 12-20.

NAD$^+$, coenzyme Q, and cytochrome c are organized into four different kinds of **respiratory complexes.** These complexes are identified in Figure 12-19 by the brown ovals and Roman numerals, and are shown in more detail in Figure 12-20.

Especially significant in Figure 12-19 are the large decreases in reduction potential across complexes I, III, and IV indicated by the vertical arrows, because these changes in reduction potential provide the energy needed to maintain the electrochemical proton gradient on which ATP synthesis depends. Each of these three complexes (but not complex II) acts as an energy transducer, using the free energy change of electron transfer, as measured by $\Delta E'$ across the complex, to pump protons outward across the membrane.

Table 12-3 summarizes some of the properties of these complexes. *Complex I* transfers electrons from NADH to coenzyme Q and is called **NADH dehydrogenase.** *Complex II* transfers to CoQ the electrons derived from succinate oxi-

dation in reaction TCA-6. This complex is properly called **succinate-coenzyme Q reductase,** though it is often also referred to by its common name, *succinate dehydrogenase.* Similar but separate complexes are required to transfer electrons to coenzyme Q from other FAD-linked dehydrogenases, such as that involved in the oxidation of fatty acids (reaction FA-2 of Figure 12-10). *Complex III* is called **coenzyme Q-cytochrome c reductase** because it accepts electrons from coenzyme Q and passes them to cytochrome c. *Complex IV* transfers electrons from cytochrome c to oxygen and is called **cytochrome c oxidase.** Each of these complexes consists of multiple polypeptide subunits and has as its prosthetic groups flavin nucleotides (complexes I and II), cytochromes (complexes III and IV), iron-sulfur centers (complexes I–III) and/or copper centers (complex IV).

Of all the electron-transferring intermediates involved in aerobic respiration, only cytochrome a_3, at the end of the transport system, is a *terminal oxidase,* capable of direct

Figure 12-20 Major Components of the Respiratory Complexes. The four respiratory complexes are indicated by brown ovals, with the major electron carriers in each complex shown in blue-green ovals. Coenzyme Q and cytochrome c are small, mobile intermediates that transfer electrons between the several complexes as they collide with them in the membrane. The dashed arrow indicates the site at which electron transport is inhibited by rotenone, a potent insecticide that is the subject of Problem 12-10.

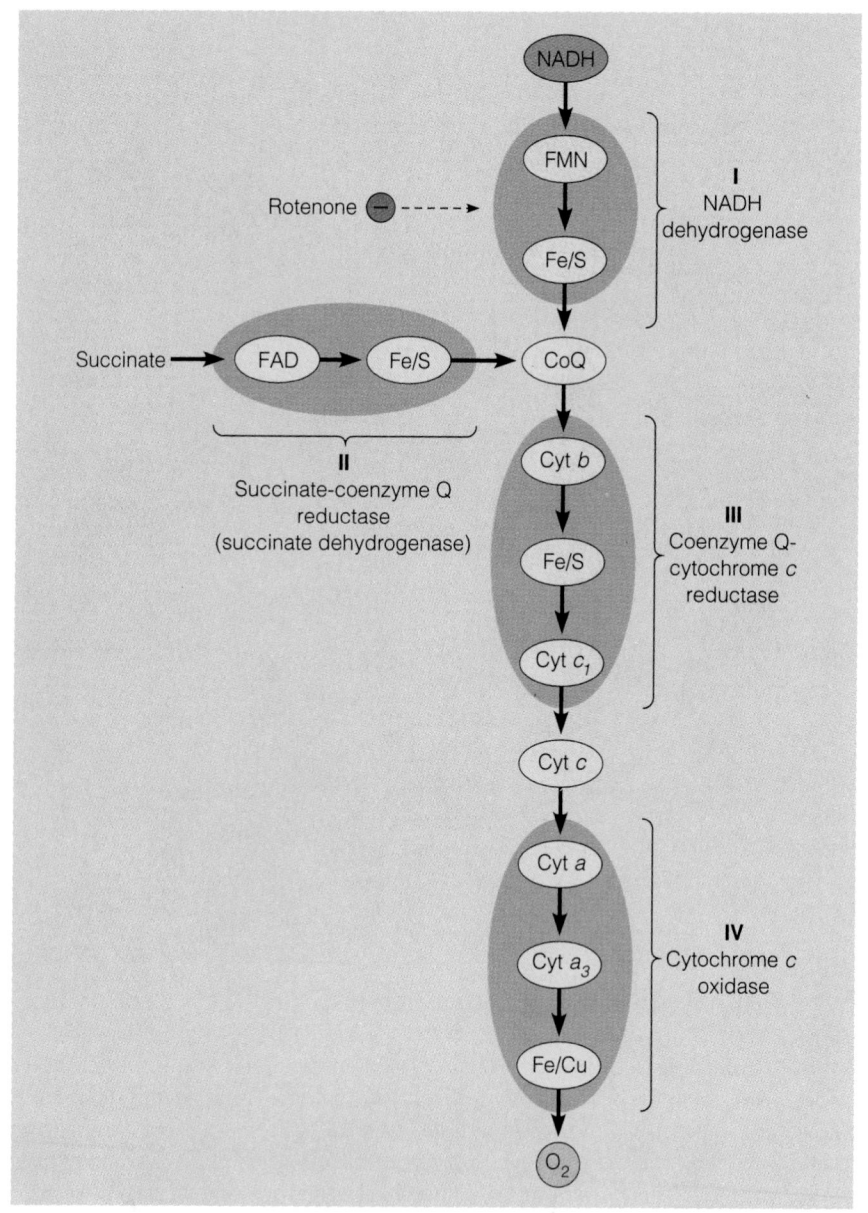

Table 12-3 Properties of the Mitochondrial Respiratory Complexes

| Respiratory Complex | | Number of Polypeptides | Prosthetic Groups | Electron Flow | | Proton Transport? |
Number	Name			Accepted from	Passed to	
I	NADH dehydrogenase (NADH-coenzyme Q reductase)	22–26	1 FMN 6–9 Fe/S centers	NADH	Coenzyme Q	Yes
II	Succinate-coenzyme Q reductase (succinate dehydrogenase)	4–5	1 FAD 3 Fe/S centers	Succinate (via enzyme-bound FAD)	Coenzyme Q	No
III	Coenzyme Q-cytochrome c reductase (cytochrome b-c_1 complex)	8–10	2 cytochrome b 1 cytochrome c_1 1 Fe/S center	Coenzyme Q	Cytochrome c	Yes
IV	Cytochrome c oxidase	9	1 cytochrome a 1 cytochrome a_3 2 Cu centers (as Fe/Cu centers with cytochrome a_3)	Cytochrome c	Oxygen (O_2)	Yes

transfer of electrons to oxygen. Therefore, almost every electron extracted from any oxidizable organic molecule anywhere in an aerobic cell passes eventually through cytochrome a_3 because this carrier is the critical link between aerobic respiration and the oxygen that makes it all possible.

Figure 12-21 places the three complexes needed for NADH oxidation (complexes I, III, and IV) in their proper eukaryotic context, the inner mitochondrial membrane. (Complex II is not included because it is not involved in NADH oxidation.) Also shown in Figure 12-21 is the outward pumping of protons *across* the membrane that accompanies electron transport *within* the membrane. Each of these complexes represents one site of proton pumping, consistent with the drops in reduction potentials across these complexes noted in Figure 12-19. The respiratory complexes are very numerous in the inner mitochondrial membrane, as are the F_oF_1 complexes that are responsible for ATP synthesis. One estimate puts the numbers of complexes I, II, III, and IV in the inner membrane of a single liver mitochondrion at about 2600, 5500, 5500, and 15,600, respectively, in addition to about 15,000 F_oF_1 complexes.

Unlike some integral membranes of the plasma membrane, the protein complexes of the mitochondrial inner membrane are mobile, free to diffuse within the plane of the membrane. This diffusion can be visualized by freeze-fracture analysis. In one study, Charles Hackenbrock placed inner mitochondrial membranes in an electrical field and found that the field caused the freeze-fracture particles (representing the protein complexes) to accumulate at one end of the membrane by a kind of electrophoresis. When the electrical field was turned off, the particles returned to a random distribution within seconds, clearly demonstrating

Figure 12-21 The Flow of Electrons Through Respiratory Complexes I, III, and IV, with Concomitant Directional Proton Pumping. Electrons flow spontaneously from NADH to O_2 via respiratory complexes I, III, and IV, with the transfers between I and III and between III and IV mediated by coenzyme Q (here labeled Q) and cytochrome c, (here labeled C), respectively. Electron flow through each carrier is coupled to the outward pumping of protons against the electrochemical gradient. The number of protons pumped outward per electron pair transferred through the complex is shown as *n* to indicate present uncertainty concerning the stoichiometry of proton pumping. When needed for energy calculations in this chapter, the value of *n* is assumed to be 3.

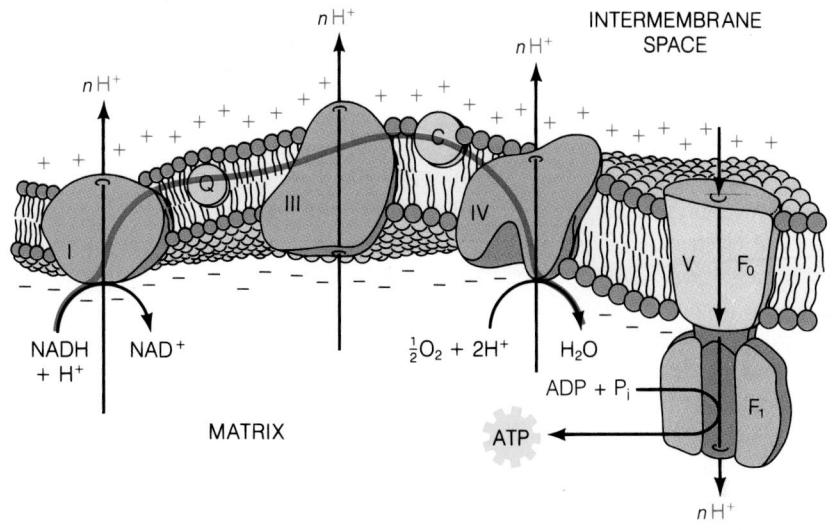

their freedom to diffuse in a fluid lipid bilayer. The results of this and similar studies make it clear that the respiratory complexes are *not* lined up in the membrane in the orderly fashion often seen in textbook diagrams, but exist in the membrane as random arrays of mobile complexes.

As Figure 12-20 indicates, NADH, coenzyme Q, and cytochrome *c* are key links in the electron transfer process. NADH links the system to the dehydrogenase (oxidation) reactions of the TCA cycle and to most other oxidation reactions in the matrix, while coenzyme Q and cytochrome *c* carry electrons between the respiratory complexes. Coenzyme Q accepts electrons from both complexes I and II, and is therefore the "funnel" that collects electrons from virtually every oxidation reaction in the cell. Coenzyme Q and cytochrome *c* are both relatively small molecules that can diffuse rapidly in the plane of the membrane. They are also quite numerous—about 10 molecules of cytochrome *c* and 50 molecules of ubiquinone for every complex I, according to one estimate. Because of their abundance and mobility, these carriers collide with the major complexes frequently enough to account for the observed rates of electron transfer in actively respiring mitochondria.

The electron transport system is sometimes referred to as a "chain," which is an apt description if we understand it to mean that the net flow of electrons is from NADH dehydrogenase to cytochrome oxidase because each of the carriers in the sequence has a higher affinity for electrons (i.e., a more negative reduction potential) than the previous carrier. The transport process is also a "chain" in the sense that each carrier in the sequence can only interact with the component from which it receives electrons and the component to which it donates electrons. The word "chain" is *not* appropriate, however, if it suggests any sort of fixed location or ordered sequence of carriers within the membrane. Instead, each of the complexes exists and diffuses randomly within the membrane, as do the coenzyme Q and cytochrome *c* molecules that shuttle electrons between them.

Oxidative Phosphorylation and the Electrochemical Proton Gradient

So far, we have learned that coenzymes are reduced during the oxidative events of the first two stages of aerobic respiration (stages 1 and 2 of Figure 12-1b). We have also seen that reduced coenzymes are reoxidized by the exergonic transport of electrons to oxygen via a membrane-bound system of reversibly oxidizable intermediates (stage 3a). Now we are ready to discuss the final stages of respiration: the processes by which the free energy released during electron transport is used to generate a proton gradient and the energy of the gradient is then used to drive ATP synthesis. In terms of Figure 12-1b, we are about to consider stage 3b (proton pumping) and stage 4 (ATP synthesis) of aerobic respiration.

Because this means of ATP production depends on phosphorylation events that are coupled to oxygen-dependent electron transport, the process is called **oxidative phosphorylation** to distinguish it from *substrate-level phosphorylation,* which occurs as an integral part of a specific reaction in a metabolic pathway. (The generation of ATP at reactions Gly-7 and Gly-10 of the glycolytic pathway and at reaction TCA-5 of the TCA cycle are examples of substrate-level phosphorylation.)

Mechanistically, oxidative phosphorylation is far more complex than substrate-level phosphorylation. Oxidative phosphorylation has, in fact, been a confusing and highly controversial topic for much of its 50-year history, prompting Efraim Racker, a respected researcher in this field, to comment at one point, "Anyone who is not confused about oxidative phosphorylation just doesn't understand the situation." Much of the confusion and controversy focused on the mechanism responsible for the actual coupling of electron transport to ATP generation. Now, however, there is good general agreement that the crucial link between electron transport and ATP production is an **electrochemical proton gradient** that is established by the directional pumping of protons across the membrane in which electron transport is occurring. We will come to the details of this model shortly, but first we need to consider the evidence that ATP synthesis is really coupled to electron transport.

Coupling of ATP Synthesis to Electron Transport

Normally, electron transport is tightly coupled to phosphorylation. By **coupling,** we mean not only that ATP synthesis depends critically on electron flow but also that electron flow is possible only when ATP can be synthesized. Coupling is an important regulatory mechanism because it ensures that the electron transport system—and therefore all of aerobic respiration—is dependent on, and responsive to, ADP and ATP concentrations in the mitochondrion and hence to the energy status of the cell.

If two processes can be coupled, one might imagine that they could also be uncoupled. In fact, chemical compounds called *uncouplers* exist and have proved to be useful tools in the study of oxidative phosphorylation mechanisms. A classic uncoupler is *2,4-dinitrophenol (DNP).* If DNP is added to respiring mitochondria that are actively synthesizing ATP, the rates of coenzyme oxidation and oxygen consumption increase suddenly, and ATP synthesis drops precipitously. The DNP has *uncoupled* electron transport from ATP synthesis. In a sense, this is like putting your car's automatic transmission in neutral: The engine speeds up somewhat, yet forward motion ceases. Any mechanism that attempts to explain the coupling between electron transport and ATP synthesis must also account for the action of uncoupling agents.

Respiratory Control. Because electron transport is coupled to ATP synthesis, the availability of ADP regulates the rate of oxidative phosphorylation and therefore of electron transport. This is called **respiratory control,** and its physiological significance is easy to appreciate: Electron transport and ATP generation will be favored when ADP concentration is high (i.e., when ATP concentration is low) and inhibited when ADP concentration is low (when ATP concentration is high). Oxidative phosphorylation is therefore regulated by cellular ATP needs, such that electron flow from organic fuel molecules to oxygen is adjusted to the energy needs of the cell. This regulatory mechanism becomes apparent during exercise, when the accumulation of ADP in muscle tissue causes an increase in electron transport rates, followed by a dramatic rise in the need for oxygen.

Sites of ATP Synthesis. Much research in the field of oxidative phosphorylation focused on the question of where along the electron transport system the actual events occur that account for the coupling of ATP generation to electron flow. In the early 1940s, Severo Ochoa showed that a fixed relationship usually exists between the number of moles of ATP generated and the number of moles of oxygen atoms consumed in respiration. This relationship is expressed as the **P/O ratio:** The number of molecules of ATP generated as a pair of electrons passes along the sequence of carriers from reduced coenzyme (or other electron donor) to oxygen. For many years, the P/O ratio was thought to be about 3 for the mitochondrial oxidation of NADH and about 2 for the oxidation of enzyme-bound $FADH_2$ or $FMNH_2$. Recent work on the mechanism of oxidative phosphorylation has called these numbers into question, however. Specifically, the realization that electron transport is coupled to ATP synthesis by an electrochemical proton gradient has reopened the question of the true P/O ratios for oxidative phosphorylation.

We will assume the historical P/O ratios of 3 for NADH and 2 for $FADH_2$ but with the recognition that these values are not as firmly established as was once thought. They are in any case reasonable numbers, considering the energetics of the system. As noted earlier, the $\Delta G^{o\prime}$ value for the oxidation of NADH by molecular oxidation is -52.4 kcal/mol. Assuming conditions such that $\Delta G'$ for this reaction is similar to the value under standard conditions, and recognizing that $+10$ kcal/mol is a reasonable value for the phosphorylation of ADP to ATP, a P/O ratio of 3 would mean that NADH oxidation drives ATP synthesis with an efficiency of almost 60% ($3 \times 10/52.4 = 0.57$).

Early evidence supporting a P/0 ratio of 3 came from a variety of ingenious experiments in which artificial electron donors and acceptors and specific inhibitors of electron transport were used to ensure that electrons flowed only through a selected portion of the sequence. From such studies, three sites along the electron transport sequence were linked to ATP synthesis, one between NADH and coenzyme Q, the second between coenzyme Q and cytochrome *c*, and the third between cytochrome *c* and oxygen. There is, in other words, one ATP-linked event associated with each of the complexes (I, III, and IV) that transfer electrons from NADH to oxygen. Confirmation of these findings was provided by Efraim Racker and his colleagues, who succeeded in incorporating each of the three complexes into synthetic

phospholipid vesicles along with the mitochondrial ATP-synthesizing system. Upon addition of the appropriate oxidizable substrate, such vesicles are capable of generating one molecule of ATP per pair of electrons that passes through the complex. We now understand that each of these three complexes (but not complex II) serves as a proton pump, thereby contributing to the proton gradient that drives ATP synthesis (see Figure 12-21).

The Chemiosmotic Model

How can ATP generation, a dehydration reaction, be tightly coupled to electron transport, which involves the sequential oxidation and reduction of various protein complexes in a lipid bilayer? The answer to this question is provided by the **chemiosmotic coupling model,** first proposed in 1961 by Peter Mitchell, a British biochemist. According to this model, the exergonic transfer of electrons between and within the respiratory complexes is accompanied by the unidirectional pumping of protons across the membrane in which the transport system is localized, either the inner mitochondrial membrane or the prokaryotic plasma membrane. The electrochemical proton gradient that is generated and maintained in this way provides the driving force for ATP synthesis by the F_oF_1 ATP synthase. Figure 12-21 illustrates both the outward pumping of protons that accompanies electron transport through complexes I, III, and IV and the proton-driven synthesis of ATP by the F_oF_1 complex.

The essential feature of the chemiosmotic model is that *the link between electron and ATP formation is an electrochemical potential across a membrane.* It is, in fact, this feature that gives the model its name: The "chemi" part of the term refers to the chemical processes of oxidation and electron transfer and the "osmotic" part comes from the Greek word *osmos,* which means "to push"—to push protons across the membrane, in this case.

The chemiosmotic model has turned out to be exceptionally useful not only because of the very plausible explanation it provides for coupled ATP generation but also because of its pervasive influence on the way we think about energy conservation in biological systems. To understand the chemiosmotic model, we will look briefly at each of its major features: the existence of a transmembrane electrochemical proton gradient, the unidirectional pumping of protons that accounts for the gradient, and the capability of the gradient to drive ATP synthesis.

The Transmembrane Electrochemical Proton Gradient. According to the chemiosmotic hypothesis, the energy liberated during electron transport is conserved as an electrochemical proton gradient across the membrane in which the electron transport intermediates are embedded. The term *electrochemical* is intended to imply the two components of the gradient: an "electro" component and a "chemical" component, both of which are induced (or at least increased) by the pumping of protons across the membrane. The "chemical" component is the *concentration gradient* of protons across the membrane, whereas the "electro" component is the *membrane potential* (V_m). The concentration gradient is, in effect, a pH gradient, because pH is simply a measure of proton concentration. The membrane potential arises because of an unequal charge distribution across the membrane, with the inside of the inner membrane *negative* with respect to the outside.

This electrochemical gradient exerts a force called the **proton motive force, or pmf**. The pmf, measured in volts, is analogous to the *electromotive force*, or *emf*, that is used to describe the electrical potential produced by electrons. In fact, both pmf and emf are measured in volts. If you think about the terms, a *motive force* is simply the force needed to *move* something—protons in one case, electrons in the other. The two fundamental concepts of chemiosmosis are (1) that a pmf is produced by a proton pump coupled to electron transport and (2) that the energy available in the pmf is then used to drive the enzymatic synthesis of ATP.

To understand how the pmf drives ATP synthesis, we need to recognize that the F_oF_1 complex thus far identified as an ATP synthase also has ATPase (ATP-hydrolyzing) activity. In fact, the F_oF_1 complex was originally characterized as an ATPase because its F_1 component catalyzes the hydrolysis of ATP when dissociated from the F_o component. The proton motive force drives ATP synthesis by reversing the ATPase activity of the F_oF_1 complex so that instead of ATP being hydrolyzed to ADP and P_i, the elements of water are removed from ADP and P_i to produce ATP.

The idea that ATP can be synthesized by reversing an ATPase enzyme is best understood by remembering that, in thermodynamics, any reaction has an equilibrium point and is therefore reversible in principle. The pmf-dependent reversibility of the F_oF_1 ATPase activity is illustrated in Figure 12-22. In part a, an imaginary lipid vesicle is shown with one F_oF_1 complex embedded in its membrane. ATP is presented to the vesicle, and the ATPase activity of the complex begins

Figure 12-22 The Chemiosmotic Mechanism of ATP Synthesis. According to the chemiosmotic theory, ATP synthesis involves the reversal of ATP hydrolysis, so that ATP is produced from ADP and P_i. The process is catalyzed by an F_oF_1-ATPase complex anchored to the membrane. Such an ATPase is shown here within a simple, membrane-bounded vesicle. **(a)** As ATP is hydrolyzed by the ATPase complex, the energy is used to pump protons outward and is stored as the potential energy of the electrochemical proton gradient. **(b)** After a short time, the magnitude of the proton gradient becomes so large that no further net pumping of protons is possible and no additional net ATP hydrolysis occurs. The system comes to a steady state, indicated by the bidirectional arrows. **(c)** If a respiratory complex is introduced into the membrane, a means is provided for increasing the proton gradient independent of the F_oF_1-ATPase activity. **(d)** As electron transport through the respiratory complex continues to drive the outward pumping of protons, the resulting electrochemical proton gradient becomes large enough to drive the ATPase reaction in reverse, synthesizing ATP from ADP and P_i. The proton gradient is measured as pmf (proton motive force). The pmf is the sum of the concentration gradient of protons (pH gradient) and the electrical potential (membrane potential) that is produced when positively charged protons are pumped across the membrane.

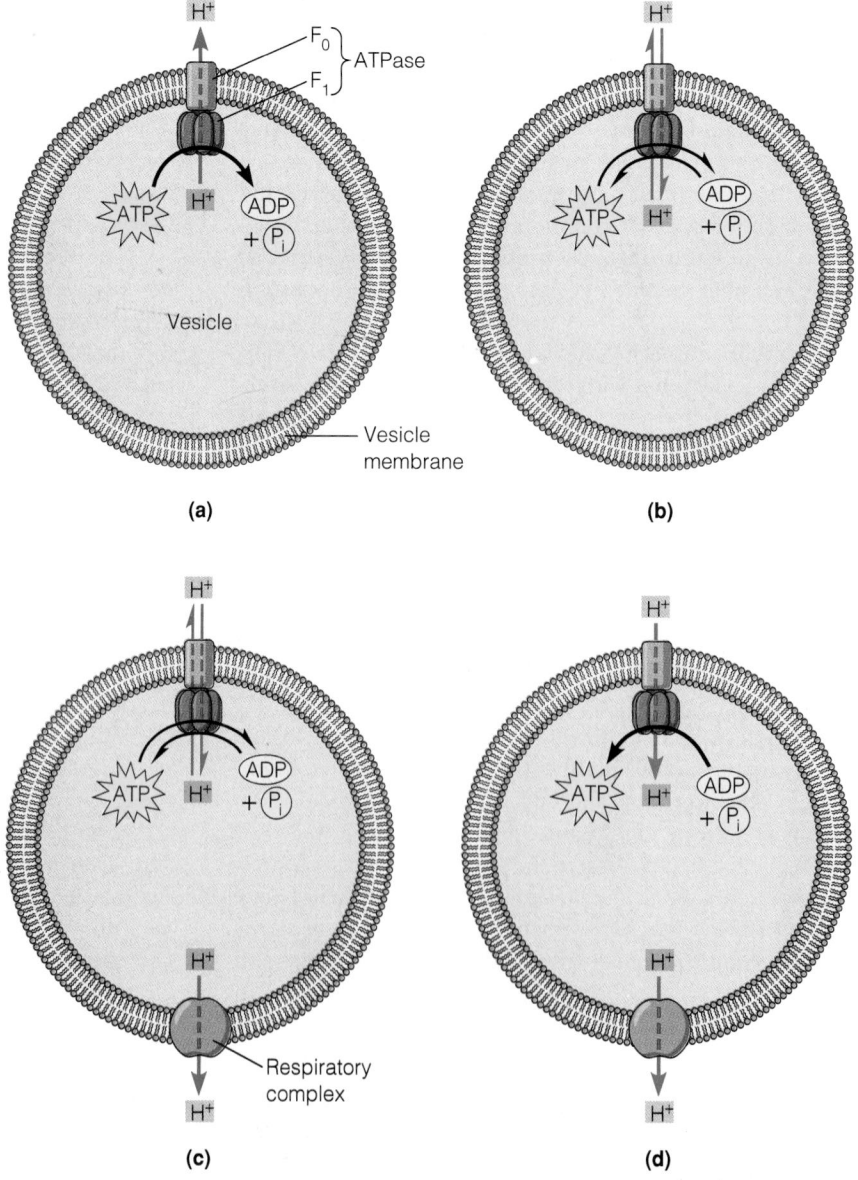

to hydrolyze it to ADP and inorganic phosphate. An integral part of Mitchell's original model was the concept that such ATPases have the ability to pump protons, using ATP as an energy source. In the illustration here, a pmf is produced across the membrane as protons are pumped across it, with the magnitude of the pmf determined by the pH gradient and the membrane potential. Thus, the energy of ATP hydrolysis in such a system is conserved by the pH gradient and the membrane potential that are generated in the process.

As ATP hydrolysis and proton pumping continue, a point will eventually be reached at which such a large proton gradient has been generated that no more proton pumping is possible and no further net hydrolysis of ATP can occur (Figure 12-22b). This is analogous to a water turbine pump that has driven water so far up a standpipe that no more water can be pumped against the back pressure. To continue the analogy, imagine that water is poured into the top of the standpipe at this point, further increasing the back pressure. Under these conditions, the pump will begin to run backward. If there were some way to put a thermodynamic "back pressure" on the ATPase, it would also run backward, synthesizing ATP from ADP and inorganic phosphate.

How can such a back pressure be exerted? The answer is to have an alternative mechanism for pumping protons out of the lipid vesicle—that is, some means of proton pumping that does not depend on ATP. As you may have already guessed, the alternative system is a proton pump coupled to electron transport. By adding a respiratory complex to our imaginary vesicle (Figure 12-22c), we introduce a means of maintaining the proton gradient that is independent of ATPase activity. As protons are pumped outward by the respiratory complex, the proton gradient becomes so great that the ATPase begins to function in reverse, synthesizing ATP by the phosphorylation of ADP (Figure 12-22d).

The model in Figure 12-22, although shown for an imaginary vesicle with a single F_oF_1 complex and a single respiratory complex, is the mechanism whereby electron transport drives ATP synthesis in aerobic respiration. The model is broadly applicable in biology, underlying the relationship between ATP and ion gradients in a wide variety of systems. The chemiosmotic model explains ATP synthesis in both mitochondrial and prokaryotic respiration and is also the mechanism for photosynthetic ATP synthesis by both eukaryotic chloroplasts and prokaryotes (see Chapter 13). Moreover, it illustrates the important point that *all such membrane-bound F_oF_1 complexes can function in either direction, depending on the membrane potential, the relative concentrations of ATP and ADP, and the concentration of protons (or other ions) across the membrane in which the complexes are embedded.* Thus, a pool of ATP molecules and an electrochemical ion gradient are two alternative, mutually interconvertible energy sources.

The Unidirectional Pumping of Protons. A basic concept of the chemiosmotic model is that the establishment and maintenance of an electrochemical proton gradient occurs through **unidirectional pumping** of protons across the membrane. The first definitive evidence in support of this prediction came in 1963, when André Jagendorf and Geoffrey Hind discovered that isolated spinach chloroplasts pump protons inward upon illumination. Similar observations were soon made in mitochondrial and bacterial membranes, except that the protons were pumped in the opposite direction, from inside to out, and the energy source was respiration rather than light.

The mechanism whereby the transfer of electrons from one carrier to another is coupled to the directional transport of protons is not yet understood. There may even be more than one such mechanism. An attractive possibility is that discussed earlier for coenzyme Q. As we already know, coenzyme Q picks up a proton from the medium whenever it accepts an electron and releases the proton again when it passes the electron on. Coenzyme Q can move freely within the membrane, so it could easily accept electrons from complex I or II on the *inner* surface of the membrane and pass them to complex III at the *outer* surface, thereby moving one proton across the membrane for every electron transferred. Because coenzyme Q is known to associate with complex III, the mechanism already discussed may underlie proton pumping by this complex.

In the case of the other proton-pumping complexes (I and IV), current evidence suggests that the pumping mechanism may involve allosteric changes in protein conformation. Suppose, for example, that a specific protein in one of these complexes is an allosteric protein with two different conformational states. In one conformational state, the protein might spontaneously bind a proton on the *inside* of the membrane. Transfer of one or more electrons through the complex might then provide the energy needed for the protein to assume an alternative conformational state, capable of releasing the proton on the *outside* of the membrane. The protein could then return spontaneously to its initial conformation, ready to bind another proton on the inside. Thus, the pumping of protons would be achieved by conformational changes in a protein within a respiratory complex, rather than by the physical movement of a carrier molecule, as suggested for coenzyme Q.

The Energetics of Proton Pumping

The energy represented by an electrochemical proton gradient can be quantified by determining the free energy required to pump protons against the gradient or, equivalently, by measuring the free energy liberated when protons are allowed to move "down" the gradient. Knowing that the concentration (or pH) gradient and the membrane potential (V_m) are both components of the gradient, we should not be surprised to find out that the equation used to calculate the $\Delta G'$ for proton pumping has two terms, involving the concentration gradient and the membrane potential, respectively. For a system such as the vesicle shown in Figure

12-22, the "reaction" of outward proton pumping can be written as:

$$H^+_{inside} \longrightarrow H^+_{outside} \tag{12-12}$$

From our consideration of the energetics of ion transport in Chapter 8 and specifically from equation 8-9, we know that $\Delta G'$ for reaction 12-12 can be calculated as:

$$\Delta G'_{outward} = RT \ln ([H^+]_{outside}/[H^+]_{inside}) - nFV_m \tag{12-13}$$

where V_m is the membrane potential across the inner membrane of the mitochondrion (or the plasma membrane of a prokaryotic cell), and all the other terms are as previously defined. Because pH is defined as $-\log[H^+]$, we can convert the logarithm in the first term of equation 12-13 from the base e to the base 10 and express the gradient in pH units instead:

$$\begin{aligned} \Delta G'_{outward} &= 2.303\, RT \log ([H^+]_{outside}/[H^+]_{inside}) - nFV_m \\ &= 2.303\, RT (\log [H^+]_{outside} - \log [H^+]_{inside}) - nFV_m \\ &= 2.303\, RT (pH_{inside} - pH_{outside}) - nFV_m \end{aligned} \tag{12-14}$$

For a sample calculation, consider a typical mitochondrion actively involved in aerobic respiration at 37°C. Such a mitochondrion might have a membrane potential of -0.14 V (more negative on the inside than on the outside because of the outward pumping of protons) and a pH in the matrix that is 1.4 units higher than that of the surrounding cytosol. (Typical values might be pH 7.4 in the cytosol and pH 8.8 in the matrix.) The free-energy change associated with the outward pumping of one mole of protons can be calculated as follows:

$$\begin{aligned} \Delta G'_{outward} &= 2.303\, RT (pH_{inside} - pH_{outside}) - nFV_m \\ &= (2.303)(1.987)(273+37)(8.8-7.4) - (1)(23,062)(-0.14) \\ &= 1985 + 3228 = 5213 \text{ cal/mol} = +5.2 \text{ kcal/mol} \end{aligned} \tag{12-15}$$

Thus, the outward pumping of protons under these conditions requires about 5.2 kcal of free energy per mole of protons. The inward movement of protons will be correspondingly exergonic: -5.2 kcal/mol.

ATP Synthesis: Putting It All Together

We are now ready to take up the fourth, and final, stage of aerobic respiration: ATP synthesis. We have seen how the energy of an oxidizable substrate such as glucose is transferred to reduced coenzymes during the oxidation reactions of glycolysis and the TCA cycle (see Figure 12-1b, stages 1 and 2) and then used to generate an electrochemical proton gradient across the inner membrane of the mitochondrion (see Figure 12-1b, stage 3) or across the plasma membrane of a prokaryotic cell. Now we can ask how the energy of that gradient is harnessed to drive ATP synthesis. For that, we return to the F_1 particles that can be seen along the inner surfaces of cristae (see Figure 12-5a) and prokaryotic membranes (see Figure 12-6) and ask about the evidence that these spheres are capable of ATP synthesis.

F_oF_1 as the ATP Synthase Complex

Key evidence concerning the role of the F_1 particles came from studies carried out by Efraim Racker and his colleagues to test the prediction of the chemiosmotic hypothesis that a reversible, proton-translocating ATPase is present in membranes that are capable of coupled ATP synthesis. Beginning with intact mitochondria (Figure 12-23a), Racker and his colleagues found that they could disrupt the mitochondria in such a way that fragments of the inner membrane formed small vesicles, which they called *submitochondrial particles* (Figure 12-23b). Like the intact mitochondrial membrane from which they were derived, these submitochondrial particles were capable of carrying out electron transport and ATP synthesis. (In effect, these particles are real-world examples of the imaginary vesicles shown in Figure 12-22.) By subjecting these submitochondrial particles to mechanical agitation or enzyme treatment, Racker was able to dislodge the F_1 structures from the membranous vesicles (Figure 12-23c). A mixture of the F_1 particles and membranous vesicles was capable of electron transport but not ATP synthesis. However, the mixture had ATPase activity.

When the F_1 particles and the membranous vesicles were separated from each other by centrifugation, the membranous fraction could still carry out electron transport but could no longer synthesize ATP; the two functions had become "uncoupled" (Figure 12-23d). The isolated F_1 particles, on the other hand, were not capable of either electron transport or ATP synthesis but had ATPase activity (Figure 12-23e). The ATP-generating capability of the membranous fraction was restored by adding the F_1 particles back to the membranes, suggesting that the spherical projections are an important part of the ATP-generating complex of the membrane. These particles were therefore referred to as *coupling factors*, and are now known to be the structures responsible for the ATP-synthesizing activity of the inner mitochondrial membrane.

As Figures 12-21 and 12-22 have already indicated, F_1 is only part of the **ATP synthase** complex. F_1 is attached by a short stalk to F_o, which is embedded in the inner mitochondrial membrane at the base of the stalk. F_o serves as the **proton translocator,** the channel through which protons flow when the electrochemical gradient across the membrane is being used to drive ATP synthesis by F_1. Thus, the functional unit is the $\mathbf{F_oF_1}$ **complex:** The F_o component provides a channel for the exergonic flow of protons from the outside to the inside of the membrane, thereby tapping into the pmf, or driving force, of the electrochemical proton gradient, and the F_1 component carries out the actual synthesis of ATP, driven by the energy of the proton gradient.

Polypeptide Composition of the F_oF_1 Components. Table 12-4 presents the polypeptide composition of the bacterial F_oF_1 complex, and Figure 12-24 (page 364) illustrates their

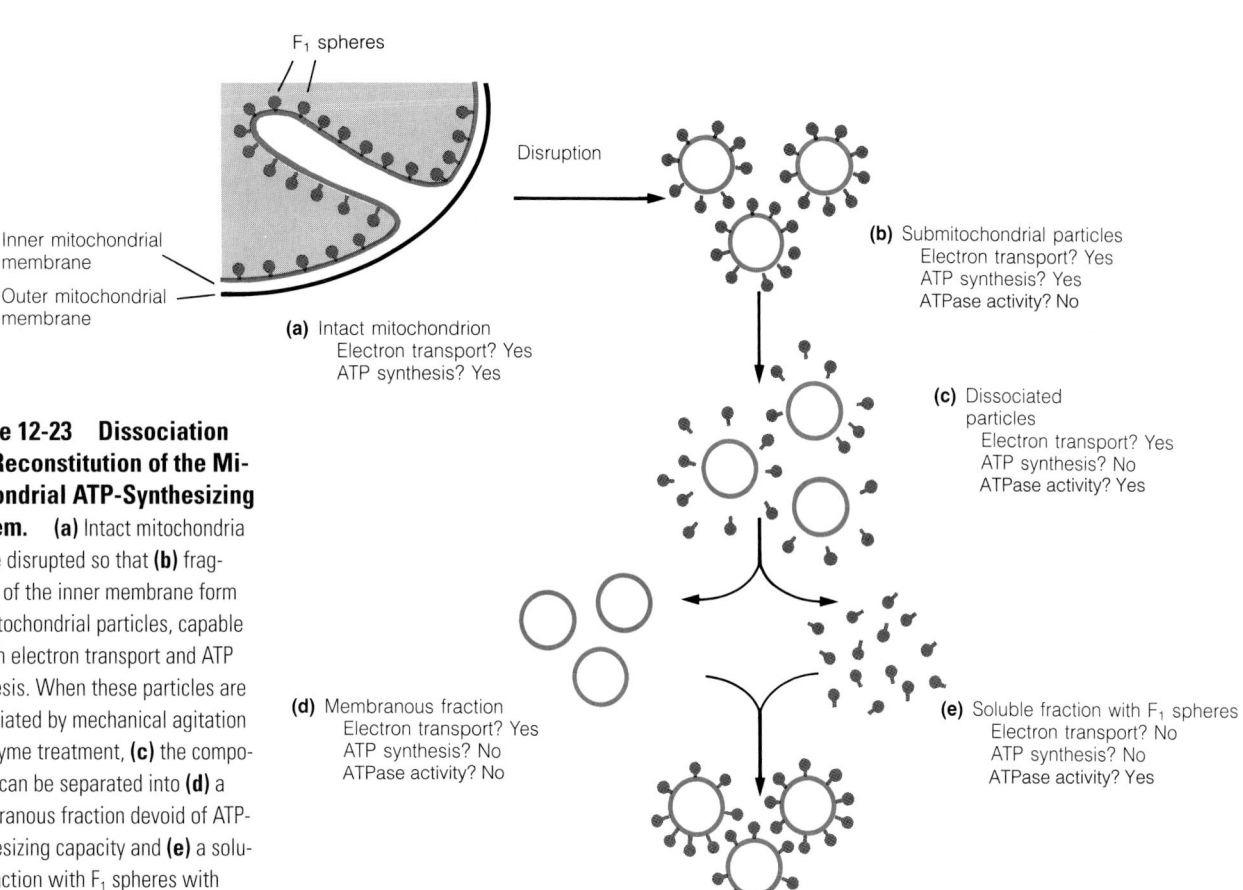

Figure 12-23 Dissociation and Reconstitution of the Mitochondrial ATP-Synthesizing System. (a) Intact mitochondria can be disrupted so that (b) fragments of the inner membrane form submitochondrial particles, capable of both electron transport and ATP synthesis. When these particles are dissociated by mechanical agitation or enzyme treatment, (c) the components can be separated into (d) a membranous fraction devoid of ATP-synthesizing capacity and (e) a soluble fraction with F_1 spheres with ATPase activity. (f) Mixing the two fractions reconstitutes the structure and restores ATP synthase activity.

F_1 spheres

Disruption

Inner mitochondrial membrane

Outer mitochondrial membrane

(a) Intact mitochondrion
Electron transport? Yes
ATP synthesis? Yes

(b) Submitochondrial particles
Electron transport? Yes
ATP synthesis? Yes
ATPase activity? No

(c) Dissociated particles
Electron transport? Yes
ATP synthesis? No
ATPase activity? Yes

(d) Membranous fraction
Electron transport? Yes
ATP synthesis? No
ATPase activity? No

(e) Soluble fraction with F_1 spheres
Electron transport? No
ATP synthesis? No
ATPase activity? Yes

(f) Reconstituted particles
Electron transport? Yes
ATP synthesis? Yes
ATPase activity? No

Table 12-4 Polypeptide Composition of the *E. coli* F_oF_1- ATP Synthase (ATPase)*

Structure	Polypeptide	Molecular Weight**	Number Present	Function
F_1	α	52,000	3	ATP/ADP binding site; promotes activity of β subunit
	β	55,000	3	Catalytic site for ATP hydrolysis and synthesis
Stalk	γ	31,000	1	Stabilizes assembly of the F_oF_1 complex
	δ	19,000	1	Main component of stalk; required for F_oF_1 assembly
	ϵ	15,000	1	Binds to δ subunit; required for F_oF_1 assembly
F_o	a	30,000	1	Stabilizes proton channel
	b	17,000	2	Binds to δ subunit; stabilizes proton channel
	c	8,000	10[†]	Forms proton channel

*The mitochondrial F_oF_1 complex is similar to the bacterial complex but with one additional polypeptide in F_1 and seven additional polypeptides in F_o.

**The molecular weights of the three components of the *E. coli* F_oF_1 are about 321,000 for F_1 ($\alpha_3\beta_3$), 64,000 for the stalk ($\gamma\delta\epsilon$), and 144,000 for F_o (ab_2c_{10}). The total molecular weight for the assembled complex ($\alpha_3\beta_3\gamma\delta\epsilon ab_2c_{10}$) is therefore about 530,000.

[†]Estimates of the number of c subunits in the functional F_oF_1 complex from *E. coli* range from 6 to 10, with 10 regarded as the most likely number.

Source: Data from Futai and Kanazawa, 1983.

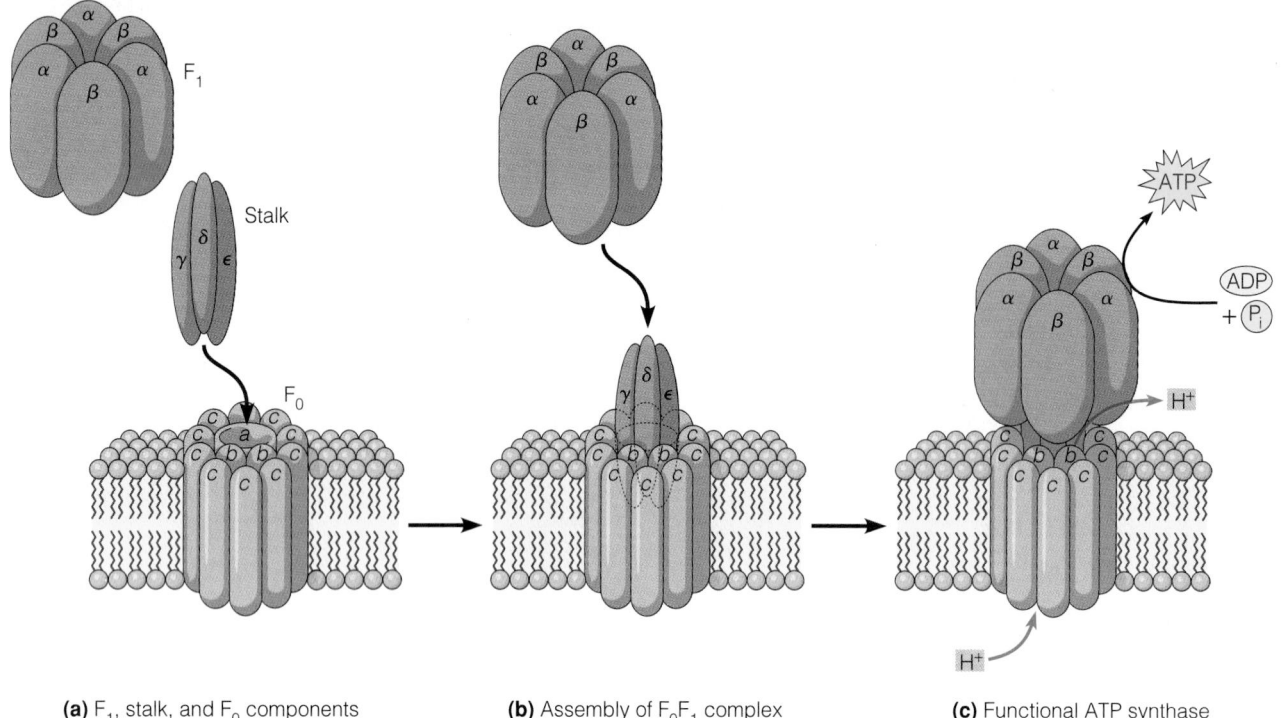

(a) F_1, stalk, and F_0 components

(b) Assembly of F_0F_1 complex

(c) Functional ATP synthase

Figure 12-24 F_1 and F_0 Components of the Bacterial ATP-Synthesizing Enzyme. F_1 is a spherical projection on the inner mitochondrial membrane that is responsible for ATP generation. F_0 is embedded in the lipid bilayer of the inner membrane and provides a channel for proton translocation from the cytosolic to the matrix side of the membrane. **(a)** The F_1 particle is a complex of three α and three β subunits. The F_0 complex consists of one *a* subunit, two *b* subunits, and six to ten *c* subunits, the latter arranged in a circle to form a cylindrical channel through the membrane. **(b)** F_1 and F_0 are attached through a stalk that consists of γ, δ, and ϵ subunits. **(c)** The assembled F_0F_1-ATP synthase is capable of using the energy of proton flux through the F_0 complex to drive ATP synthesis by the β subunits of the F_1 particle.

assembly into the functional complex. The F_1 headpiece consists of three α and three β polypeptides. The catalytic site for ATP synthesis and hydrolysis is located on the β subunit; the α subunit serves as an ATP/ADP binding site, thereby promoting the catalytic activity of the β subunit. The stalk consists of three polypeptides: γ, δ, and ϵ. The δ and ϵ subunits are required for assembly of the F_0F_1 complex, and the γ subunit is thought to stabilize the assembled structure. In addition, the γ subunit may regulate the movement of protons through the channel in the F_0 structure. The F_0 complex consists of polypeptides *a*, *b*, and *c*, with one *a* subunit, two *b* subunits, and six to ten *c* subunits present in the functional complex. The *c* polypeptides are thought to be organized in a circle, forming the proton channel through the membrane. The *a* and *b* subunits apparently stabilize the proton channel. In addition, subunit *b* binds to the δ polypeptide of the stalk, thereby anchoring the F_1/stalk structure to the F_0 base.

The Mechanism of ATP Synthesis by F_0F_1.   Despite years of intense research, we do not yet understand the mechanism whereby the exergonic flow of protons through F_0 drives the otherwise endergonic phosphorylation of ADP to ATP by F_1. A clue to the mechanism may lie in the extensive conformational changes that are observed in the F_0F_1 complex when ATP synthesis is occurring. A likely explanation is that the inward movement of protons through the channel F_0 causes a conformational change in one or more of the F_0 subunits, which is then transmitted through the stalk to the F_1 complex, activating the phosphorylation of ADP.

The Dependence of ATP Synthesis on the Electrochemical Proton Gradient

If the electrochemical proton gradient is really the critical link between electron transport and the phosphorylation of ADP, it should be possible to show not only that the gradient is generated during electron transport but also that the gradient is both necessary and sufficient to account for ATP generation. A crucial test of this came in 1966, when Jagendorf and his colleagues demonstrated that a pH gradient imposed on isolated chloroplasts was adequate to cause ATP synthesis in the absence of light. In these experiments, chloroplasts were first incubated in an acidic buffer for several hours so that internal membranous compartments contained substantial amounts of buffer at pH 4. When the equilibrated chloroplasts were quickly transferred to a solution at pH 8 with ADP and inorganic phosphate present, a pH gradient of 4 units was produced, and a burst of ATP synthesis was observed as the pH gradient was discharged. In subsequent studies, investigators showed that the energy of experimentally imposed membrane potentials could also be used to synthesize ATP in mitochondria. These results strongly supported the chemiosmotic concept that an electrochemical proton gradient across a membrane can drive ATP generation.

The dynamics of the proton motive force are illustrated in Figure 12-25. The electron transport system pumps protons outward across the inner membrane of the mitochondrion, and the resulting electrochemical gradient then drives ATP generation by means of the F_oF_1 complexes associated with the same membrane. Presuming only that the number of protons that must pass through the F_o channel to cause one phosphorylation event is the same as the number of protons extruded by the passage of a pair of electrons through one of the respiratory assemblies (3, by our earlier

assumption), then the P/O ratios of 3 for NADH and 2 for $FADH_2$ are understandable.

In addition, it now becomes clear how uncouplers might work. Compounds like DNP are actually weak organic acids that are soluble in membranes. If DNP is present when a proton gradient begins to build up from pump activity, DNP molecules can continually associate with protons on one side of the membrane, carry them across, and release the protons on the other side. If enough DNP is present, proton pumps associated with electron transport may not be able to keep up with this constant "leak" and therefore cannot produce a proton gradient of sufficient magnitude to drive ATP synthesis. The result is that the membranes are *uncoupled*: Electron transport continues, but ATP cannot be synthesized because the DNP carries the protons back across the membrane, thereby discharging the gradient as fast as it is formed.

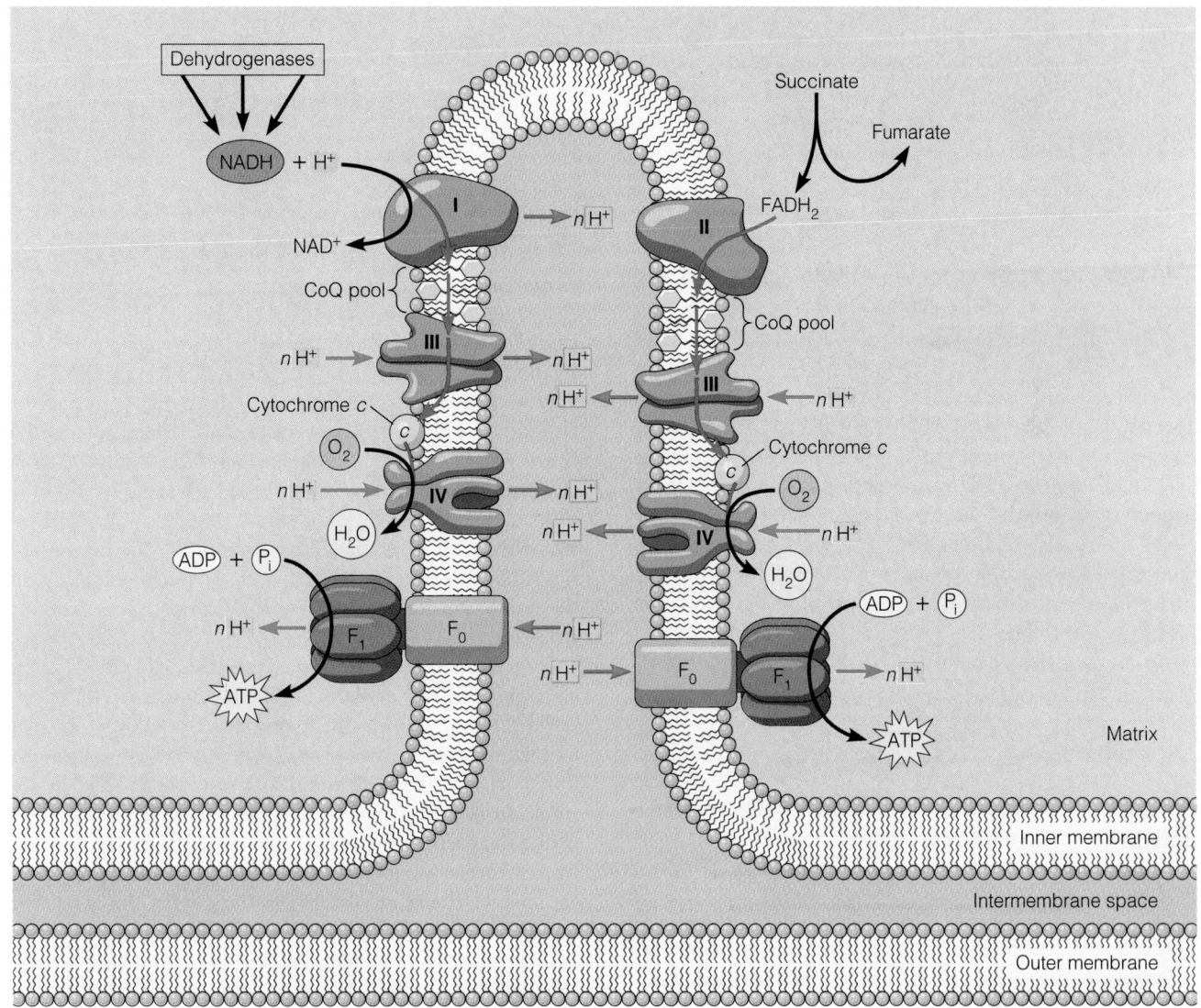

Figure 12-25 Dynamics of the Electrochemical Proton Gradient. The respiratory complexes I, II, III, and IV are integral components of the inner mitochondrial membrane. Complexes I, III, and IV (but not complex II) couple the exergonic flow of electrons (pink line) through the complex with the outward pumping of protons (blue) across the membrane. The resulting electrochemical proton gradient then drives ATP synthesis by F_1 as protons are translocated back across the membrane by the F_o complex, which is also embedded in the inner membrane.

A Summary of Aerobic Respiration

To summarize aerobic respiration, we need to review the role of each of the components. As the glycolytic pathway and the TCA cycle (or other catabolic pathways, such as β oxidation) proceed, coenzymes are continuously reduced. These reduced coenzymes represent a storage form of much of the free energy of substrate oxidation—energy that can be tapped and released as the coenzymes are themselves reoxidized by the electron transport system. As electrons are transported from NADH or $FADH_2$ to oxygen, they pass through several respiratory complexes. Three of the four complexes are sites at which the energy of electron transport is conserved by being coupled to the directional pumping of protons across the membrane. The resulting electrochemical gradient exerts a pmf that serves as the driving force for ATP synthesis. Under most conditions, a steady-state pmf will be maintained across the membrane, with the transfer of electrons from coenzymes to oxygen carefully and continuously adjusted so that the outward pumping of protons balances the inward flux of protons necessary to synthesize ATP at the desired rate.

The ATP Yield of Aerobic Respiration

Now we can return to the question of the total ATP yield per molecule of glucose under aerobic conditions. Recall from reaction 12-3 that the complete oxidation of glucose to carbon dioxide results in the generation of 4 molecules of ATP by substrate-level phosphorylation, with most of the remaining free energy of glucose oxidation stored in the 12 coenzyme molecules—10 of NADH and 2 of $FADH_2$. In prokaryotes and some eukaryotic cells, electrons from NADH pass through all 3 ATP-generating complexes of the electron transport system, yielding 3 ATP molecules per molecule of coenzyme. Electrons from $FADH_2$, on the other hand, traverse only two of the three complexes, yielding only 2 ATP molecules per molecule of coenzyme. The maximum theoretical ATP yield obtainable upon reoxidation of the 12 coenzyme molecules formed per glucose can therefore be represented as follows:

$$10NADH + 10H^+ + 5O_2 + 30ADP + 30P_i \longrightarrow$$
$$10NAD^+ + 10H_2O + 30ATP + 30H_2O \qquad \textbf{(12-16)}$$

$$2FADH_2 + O_2 + 4ADP + 4P_i \longrightarrow$$
$$2FAD + 2H_2O + 4ATP + 4H_2O \qquad \textbf{(12-17)}$$

Summing the above reactions gives us an overall reaction for electron transport and ATP synthesis:

$$10NADH + 10H^+ + 2 FADH_2 + 6O_2 + 34ADP + 34P_i \longrightarrow$$
$$10NAD^+ + 2FAD + 12H_2O + 34ATP + 34H_2O \qquad \textbf{(12-18)}$$

Addition of reaction 12-18 to the summary reaction for glycolysis and the TCA cycle (reaction 12-3) leads to the following overall expression for the maximum theoretical ATP yield upon the complete aerobic respiration of glucose or other hexoses:

$$C_6H_{12}O_6 + 6O_2 + 38ADP + 38P_i \longrightarrow$$
$$6CO_2 + 38ATP + 44H_2O \qquad \textbf{(12-19)}$$

This summary reaction is valid for most prokaryotic cells and for some eukaryotic cells. Depending on the cell type, however, the maximum ATP yield for a eukaryotic cell may be 36 instead of 38.

Before leaving reaction 12-19 and the aerobic energy metabolism that it summarizes, it may be useful to address several questions that are often asked concerning the number of water molecules and ATP molecules in the reaction. We will consider three specific questions, with the hope that each will further enhance your understanding of aerobic respiration.

1. *Why does reaction 12-19 have so many water molecules on the right?* Summary reactions such as reaction 12-19 are commonly written with only $6H_2O$ on the right-hand side. However, doing so ignores the 38 water molecules that result when ADP and P_i condense to form ATP. Keep in mind that, for every glucose molecule that is catabolized by aerobic respiration, electron transport generates 12 water molecules, the TCA cycle consumes 6 water molecules, and ATP synthesis produces another 38 water molecules. The net result is the production of 44 molecules of H_2O per molecule of glucose, as shown in reaction 12-19. Factoring out the water molecules that arise from ATP formation, we can rewrite reaction 12-19 as:

$$C_6H_{12}O_6 + 6O_2 \xrightarrow[\text{38ADP + 38P}_i \quad \text{38ATP + 38H}_2\text{O}]{} 6CO_2 + 6H_2O \qquad \textbf{(12-19a)}$$

2. *Why does the maximum theoretical ATP yield in eukaryotic cells vary between 36 and 38 ATPs per glucose?* Recall that when glucose is catabolized aerobically in a eukaryotic cell, glycolysis gives rise to 2 molecules of NADH per glucose in the cytosol, while the catabolism of pyruvate generates another 8 molecules of NADH in the matrix of the mitochondrion. This spatial distinction is important because the inner membrane of the mitochondrion does not have a carrier protein for NADH or NAD^+, so NADH cannot enter the mitochondrion to deliver its electrons to the transport system directly. Instead, the electrons and H^+ ions are passed inward by one of several *electron shuttle systems* that differ in the number of ATP molecules formed per NADH molecule oxidized.

An **electron shuttle system** consists of one or more electron carriers that can be reversibly reduced, with transport proteins present in the membrane for both the oxidized and the reduced forms of the carrier. In the case of electron transport into the mitochondrion, cytosolic NADH passes its electrons and proton to a carrier molecule, thereby oxidizing the coenzyme so that it can again be used as an electron acceptor in glycolysis or other cytosolic processes. Meanwhile, the reduced form of the carrier is transported into the mitochondrion, where it is oxidized by a mitochondrial enzyme. The difference in ATP yield among eukaryotic cells comes about because the oxidation of the carrier molecule within the mitochondrion may involve the transfer of electrons to either NAD^+ or FAD, depending on the cell type.

In liver, kidney, and heart cells, electrons from cytosolic NADH are transferred into the mitochondrion by means of a complex electron shuttle that uses NAD$^+$ as the electron acceptor on the matrix side. This shuttle is freely reversible and is therefore only useful in cells that maintain a higher [NADH]/[NAD$^+$] ratio in the cytosol than in the mitochondrion, such that inward electron transport is exergonic. In such cells, the oxidation of NADH to NAD$^+$ in the cytosol is accompanied by the generation of NADH from NAD$^+$ in the matrix. Electrons derived from cytosolic NADH therefore pass through all three ATP-generating complexes of the transport system and are therefore capable of generating up to three molecules of ATP per molecule of coenzyme.

In skeletal muscle, brain, and other tissues, the [NADH]/[NAD$^+$] ratio is sometimes lower in the cytosol than in the mitochondrion. Cells in these tissues therefore deliver electrons to the transport system by means of a shuttle mechanism that uses FAD rather than NAD$^+$ as the mitochondrial electron acceptor. This mechanism, called the **glycerol-3-phosphate/dihydroxyacetone phosphate shuttle,** or often just the **glycerol phosphate shuttle,** is shown in Figure 12-26. The glycerol phosphate shuttle involves cytosolic and mitochondrial forms of the enzyme *glycerol phosphate dehydrogenase (GPDH)*, which reversibly oxidizes glycerol-3-phosphate to dihydroxyacetone phosphate. The cytosolic form of the enzyme transfers electrons from NADH to dihydroxyacetone phosphate, reducing it to glycerol-3-phosphate. The glycerol-3-phosphate then

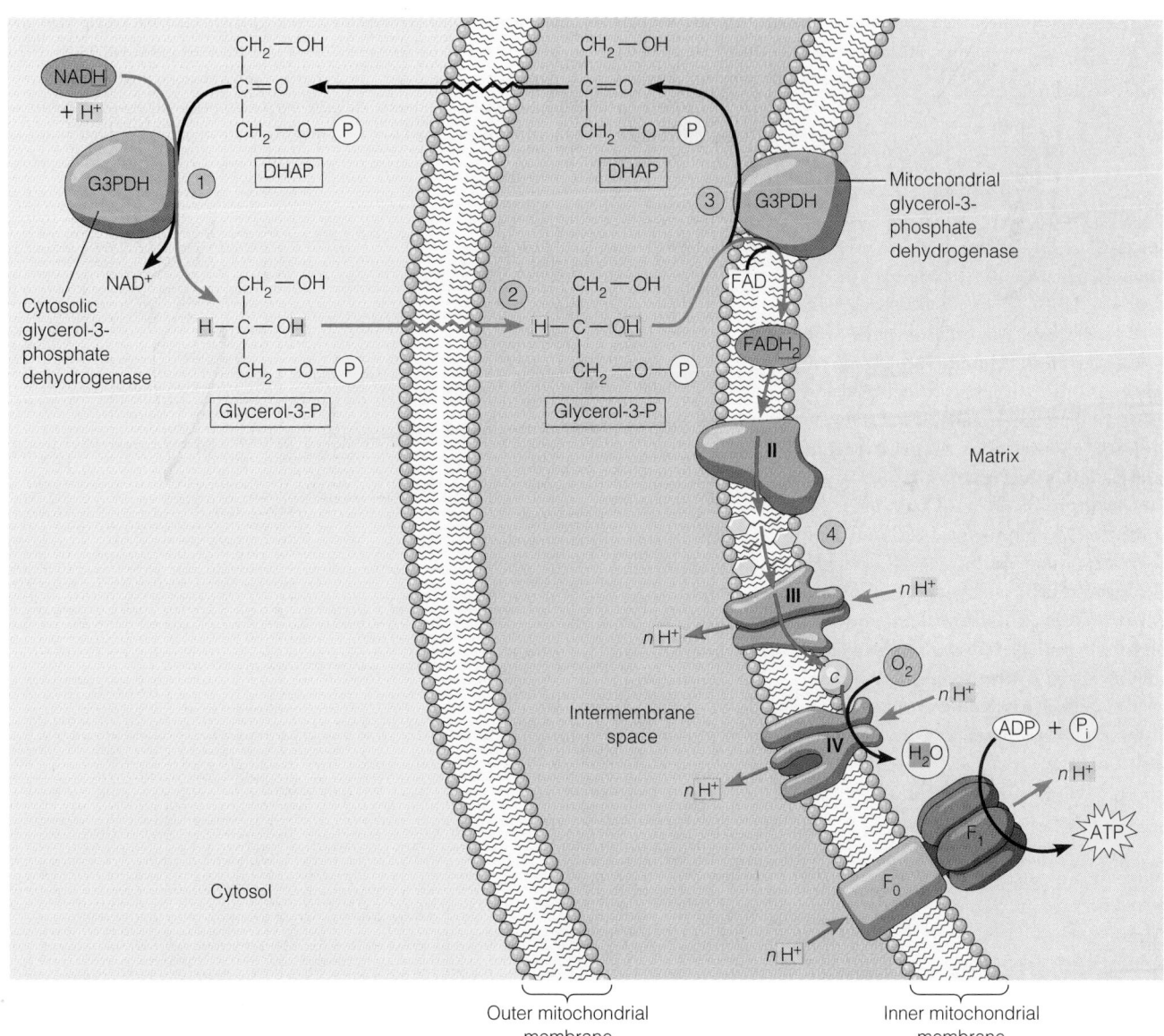

Figure 12-26 The Transport of Electrons from Cytosolic NADH into the Mitochondrion: The Glycerol-3-Phosphate/Dihydroxyacetone Phosphate Shuttle. ① The cytosolic enzyme glycerol-3-phosphate dehydrogenase (G3PDH) uses electrons (and protons) from NADH to reduce dihydroxyacetone phosphate (DHAP) to glycerol-3-phosphate (glycerol-3-P), which ② can move freely across the outer mitochondrial membrane. ③ Glycerol-3-P is then reoxidized to DHAP by an FAD-linked G3PDH in the inner membrane. Because NADH is a more energy-rich compound than FADH$_2$, the inward transport of electrons is exergonic, driven by the difference in the reduction potentials for the two coenzymes. The "cost" of this inward transport is a reduced ATP yield, because ④ electrons from the FADH$_2$ generated by the mitochondrial G3PDH pass through only two proton-pumping respiratory complexes (III and IV) en route to oxygen.

passes into the mitochondrion, where it is reoxidized to dihydroxyacetone phosphate by an FAD-dependent GPDH located on the outer face of the inner membrane. Because this membrane-bound enzyme uses FAD instead of NAD^+ as its electron acceptor, the electrons are transferred directly to coenzyme Q. As a result, these electrons bypass the first energy-conserving site of the transport system, generating only two molecules of ATP instead of three, and reducing the maximum theoretical yield by one ATP per cytosolic NADH and therefore by two ATP molecules per molecule of glucose.

This shuttle may seem inherently wasteful because the oxidation of NADH to NAD^+ in the cytosol leads to the generation of $FADH_2$ rather than NADH in the mitochondrion, with a consequent reduction in the maximum theoretical ATP yield. However, the difference in E_0' between NADH and $FADH_2$ provides the driving force necessary for inward electron transport when the $[NADH]/[NAD^+]$ is lower in the cytosol than in the mitochondrion. In fact, the difference

in E_0' values is sufficiently large that the shuttle is essentially irreversible and functions effectively even when the $[NADH]/[NAD^+]$ ratio in the cytosol is very low. This shuttle is therefore especially prominent in cells with low levels of cytosolic NADH.

3. *Why is the ATP yield of coenzyme oxidation referred to as the "maximum theoretical ATP yield"?* The wording is a reminder that yields of 36 or 38 ATP molecules per molecule of glucose are possible *only* if we assume that the energy of the electrochemical proton gradient is used solely to drive ATP synthesis. Such an assumption may be useful for the calculation of maximum possible yields, but it is not otherwise realistic because the energy of the proton gradient provides the driving force not only for ATP synthesis but also for other energy-requiring reactions and processes. For example, some of the energy of the proton gradient is used to drive the transport of various metabolites and ions across the membrane. Several of these transport processes are illustrated in Figure 12-27. Depending on the relative concentra-

Figure 12-27 Major Transport Systems of the Inner Mitochondrial Membrane. The major transport proteins localized in the inner mitochondrial membrane are shown here in schematic form. **(a)** The *pyruvate carrier* cotransports pyruvate and protons inward, driven by the proton gradient. The **(b)** *dicarboxylate* and **(c)** *tricarboxylate carriers* exchange organic acids across the membrane, with the direction of transport dependent on the relative concentrations of tricarboxylic and dicarboxylic acids on the inside and outside of the inner membrane, respectively. **(d)** The *ATP-ADP carrier* exchanges ATP outward for ADP inward, and **(e)** the *phosphate carrier* couples the inward movement of phosphate with the outward movement of hydroxyl ions, which are neutralized by protons in the intermembrane space. (For the mechanism of inward electron transport, see Figure 12-26).

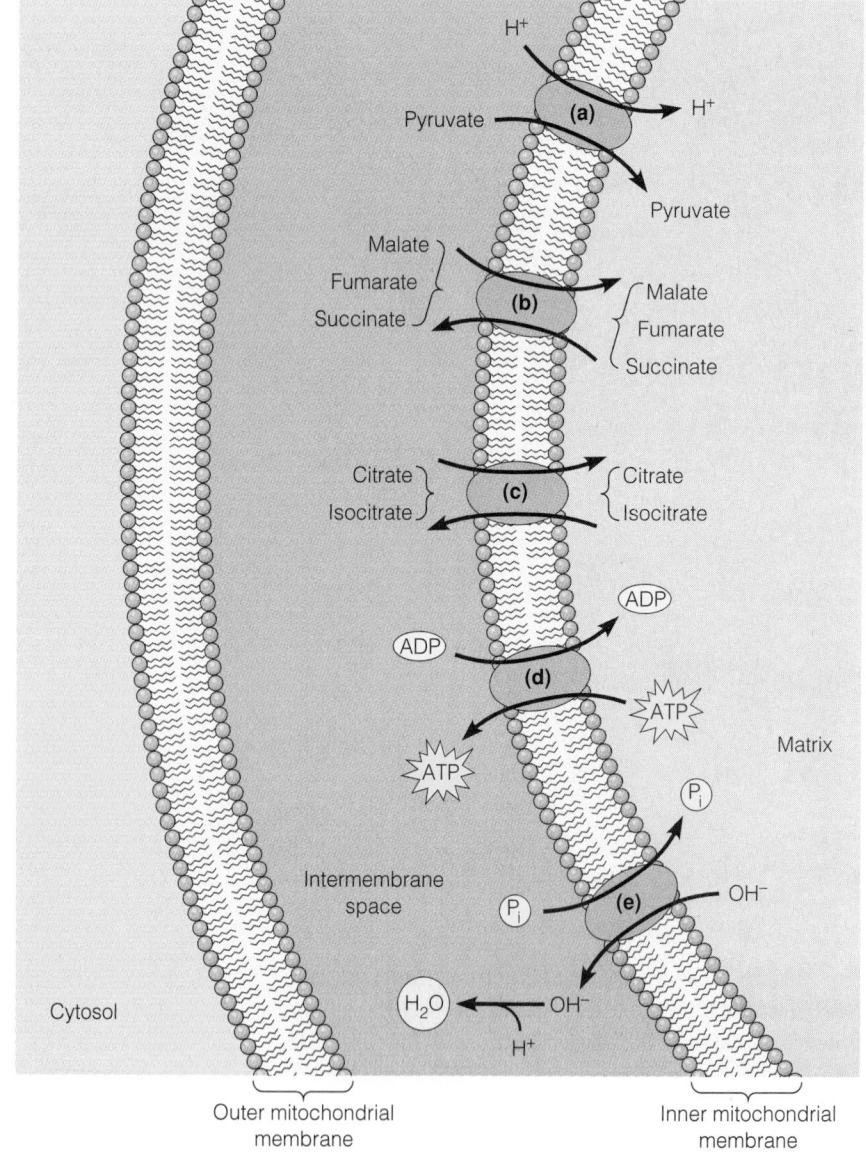

tions of pyruvate, fatty acids, amino acids, and TCA-cycle intermediates in the cytosol and the mitochondrial matrix, variable amounts of energy may be needed to ensure that the mitochondrion has adequate supplies of oxidizable substrates and TCA-cycle intermediates. Moreover, the inward transport of phosphate ions needed for ATP synthesis is accompanied by the concomitant outward movement of hydroxyl ions, which are neutralized by protons in the intermembrane space, thereby drawing on the proton gradient also. Problem 12-9 (at the end of the chapter) provides an opportunity to consider how the use of the proton gradient to drive transport processes affects the actual ATP yield of aerobic respiration.

The Efficiency of Aerobic Respiration

To determine the overall efficiency of ATP production by aerobic respiration, we must ask what proportion of the energy of glucose oxidation is preserved in the 36 or 38 molecules of ATP generated per molecule of glucose. The $\Delta G^{\circ\prime}$ value for ATP hydrolysis is about -7.3 kcal/mol, but $\Delta G'$ is typically in the range of -10 to -14 kcal/mol. Assuming a conservative estimate of 10 kcal/mol, the 36–38 moles of ATP generated by aerobic respiration of 1 mole of glucose in an aerobic cell correspond to about 360–380 kcal of energy conserved per mole of glucose oxidized. This is an efficiency of about 52–55%, well above that obtainable with the most efficient machines.

PERSPECTIVE

Compared to fermentative processes, aerobic respiration gives the cell access to much more of the free energy that is liberated by the oxidation of organic substrates such as sugars, fats, and proteins. The complete catabolism of carbohydrates begins with the glycolytic pathway, but the pyruvate that is formed is passed into the mitochondrion, where it is oxidatively decarboxylated to acetyl CoA. The acetyl CoA is then oxidized fully by enzymes of the TCA cycle. Fatty acids are alternative substrates for energy metabolism in many cells. Their catabolism begins with β oxidation to acetyl CoA, which then enters the TCA cycle. Proteins can also be used as energy sources, particularly under conditions of fasting or starvation, or when the diet contains protein in excess of amino acid requirements. In such cases, proteins are degraded to amino acids, each of which is then catabolized to one or more end-products that enter either the glycolytic pathway or the TCA cycle.

Reduced coenzymes (NADH and FADH$_2$) are reoxidized by an electron transport system that consists of respiratory complexes, large multiprotein assemblies embedded in the inner mitochondrial membrane (or, in the case of prokaryotes, in the plasma membrane). The respiratory complexes are free to move laterally within the membrane. Key intermediates in the electron transport system are coenzyme Q and cytochrome c, which transfer electrons between the several complexes. Oxygen is the ultimate electron acceptor, and water is the final product.

Of the four main kinds of respiratory complexes, three (complexes I, III, and IV) couple the transfer of electrons to the outward pumping of protons. This establishes an electrochemical proton gradient that is the driving force for ATP generation. The ATP-synthesizing system consists of a proton translocator, F$_o$, embedded in the membrane, and an ATP synthase, F$_1$, a knoblike structure that projects from the inner membrane on the matrix side. Because the F$_o$F$_1$ complex can function as either an ATPase or an ATP synthase, the electrochemical proton gradient and ATP are, in effect, interconvertible forms of stored energy.

Mitochondria are the site of respiratory metabolism in eukaryotic cells and are prominent organelles in most eukaryotic cells in terms of both size and numbers. They are usually several micrometers long and range in abundance from one or a few up to hundreds or even a few thousand per cell. A mitochondrion is surrounded by two membranes. The inner membrane has many infoldings called cristae, which greatly increase the surface area of the membrane and hence its ability to accommodate the numerous respiratory complexes, F$_o$F$_1$ complexes, and transport proteins needed for respiratory function. Specific carriers are required to effect the inward transport of pyruvate, fatty acids, and other organic molecules across the inner membrane of the organelle. ATP transport outward is coupled to the inward movement of ADP and the concurrent inward movement of phosphate ions is coupled to the outward movement of hydroxyl ions, which is driven by the proton gradient. The electrons of coenzyme molecules that undergo reduction in the cytosol must be passed inward to the electron transport system by specific shuttle mechanisms because the inner membrane is not permeable to the coenzymes themselves.

This, then, is aerobic energy metabolism. No transistors, no mechanical parts, no noise, no pollution—and all done in units of organization that require an electron microscope to visualize. Yet the process goes on continuously in living cells with a degree of integration, efficiency, fidelity, and control that we can scarcely understand well enough to appreciate fully, let alone aspire to reproduce in our test tubes.

KEY TERMS FOR SELF-TESTING

Respiration: Maximizing ATP Yields

respiratory metabolism (respiration) (p. 327)
aerobic respiration (p. 327)
anaerobic respiration (p. 327)

The Mitochondrion: Where the Action Takes Place

mitochondrion (p. 329)
outer membrane (p. 331)
intermembrane space (p. 331)
inner membrane (p. 331)
crista (p. 331)
matrix (p. 331)
F_1 complex (particle) (p. 333)
F_o complex (p. 333)

The Tricarboxylic Acid Cycle: Oxidation in the Round

tricarboxylic acid (TCA) cycle (p. 334)
coenzyme A (CoA) (p. 336)
acetyl CoA (p. 336)
flavin adenine dinucleotide (FAD) (p. 338)
triglyceride (p. 340)
β oxidation (p. 340)
proteolysis (p. 340)

transamination (p. 342)
oxidative deamination (p. 342)
amphibolic pathway (p. 345)

Electron Transport: Electron Flow from Coenzymes to Oxygen

electron transport (p. 346)
electron transport system (p. 350)
reduction potential (E') (p. 350)
reduction-oxidation (redox) pair (p. 350)
standard reduction potential (E_0') (p. 351)
flavoprotein (p. 353)
iron-sulfur protein (p. 354)
iron-sulfur (Fe/S) center (p. 353)
cytochrome (p. 353)
copper-containing protein (p. 354)
iron-copper (Fe/Cu) center (p. 354)
coenzyme Q (CoQ) (p. 354)
respiratory complex (p. 356)
NADH dehydrogenase (p. 356)
succinate-coenzyme A reductase (p. 356)
coenzyme Q–cytochrome c reductase (p. 356)
cytochrome c oxidase (p. 356)

Oxidative Phosphorylation and the Electrochemical Proton Gradient

oxidative phosphorylation (p. 358)
electrochemical proton gradient (p. 358)
coupling (p. 358)
respiratory control (p. 358)
P/O ratio (p. 359)
chemiosmotic coupling model (p. 359)
proton motive force (pmf) (p. 360)
unidirectional pumping (p. 361)

ATP Synthesis: Putting It All Together

ATP synthase (p. 362)
proton translocator (p. 362)
F_oF_1 complex (p. 362)
electron shuttle system (p. 366)
glycerol-3-phosphate/dihydroxyacetone phosphate (glycerol phosphate) shuttle (p. 367)

Box 12A: The Glyoxylate Cycle, Glyoxysomes, and Seed Germination

glyoxylate cycle (p. 347)
glyoxysome (p. 350)

PROBLEM SET

12-1. Localization of Molecules and Functions Within the Mitochondrion. Indicate whether you would expect to find each of the following molecules or functions in the matrix (M), the inner membrane (IM), the outer membrane (OM), the intermembrane space (IS), or not in the mitochondrion at all (N).

(a) Coenzyme A

(b) Coenzyme Q

(c) Malate dehydrogenase

(d) Succinate dehydrogenase

(e) Fatty acyl CoA dehydrogenase

(f) Fatty acid elongation

(g) Dicarboxylate carrier

(h) Conversion of lactate into pyruvate

(i) ATP synthase

(j) Accumulation of a high proton concentration

12-2. Localization of Molecules and Functions Within the Prokaryotic Cell. Repeat Problem 12-1, but indicate where in a prokaryotic cell you would expect to find each of the molecules or functions. Choices: cytoplasm (C), plasma membrane (PM), exterior of cell (EX), or not present at all (N).

12-3. True or False. Indicate whether each of the following statements is true (T) or false (F). If false, reword the statement to make it true.

(a) The orderly flow of carbon through the TCA cycle is possible because each of the enzymes of the cycle is embedded in the inner mitochondrial membrane in such a manner that their order in the membrane is the same as their sequence in the cycle.

(b) Thermodynamically, acetyl CoA should be capable of driving the phosphorylation of ADP (or GDP), just as succinyl CoA does.

(c) Respiration is an aerobic process in all organisms because oxygen is the only known electron acceptor for reoxidation of coenzymes.

(d) Because of the cyclic nature of the pathway, the TCA cycle can continue to operate even though the $\Delta G'$ value for one or more of its reactions is positive.

(e) Unlike NAD^+, the coenzyme FAD tends to be tightly associated with the dehydrogenase enzymes that use it as an electron acceptor.

(f) Nine cycles of β oxidation are required to degrade the 18-carbon fatty acid oleate completely to acetyl CoA.

12-4. Completing the Pathway. In each of the following cases, complete the pathway by indicating the structures and order of intermediates.

(a) The conversion of citrate to α-ketoglutarate by reactions TCA-2 and TCA-3 involves as intermediates not only isocitrate but also molecules identified (but not shown in Figure 12-7) as *aconitate* and *oxalosuccinate*. Illustrate the pathway from citrate to α-ketoglutarate by showing the structures and order of all three intermediates.

(b) Synthesis of the amino acid glutamate can be effected from pyruvate and alanine by a metabolic sequence that illustrates the amphibolic role of the TCA cycle. Devise such a pathway, assuming the availability of whatever additional enzymes may be needed.

Glutamate

Alanine

(c) If pyruvate-2-^{14}C (pyruvate with the middle carbon atom radioactively labeled) is provided to actively respiring mitochondria, most of the radioactivity will be incorporated into citrate. Trace the route whereby radioactively labeled carbon atoms are incorporated into citrate, and indicate where in the citrate molecule the label will first appear.

(d) If pyruvate-1-^{14}C (pyruvate with the carboxyl carbon atom radioactively labeled) is provided to actively respiring mitochondria, most of the radioactivity will be released as $^{14}CO_2$ but a small amount will be incorporated into citrate, where it will first appear as the carbon atom of one of the terminal carboxyl groups. Trace the route by which some radioactively labeled carbon atoms are incorporated into citrate, and show how the terminal carboxyl group becomes labeled.

12-5. Regulation of Catabolism. Explain the advantage to the cell of each of the following regulatory mechanisms.

(a) Isocitrate dehydrogenase (reaction TCA-3) is allosterically activated by ADP.

(b) The dehydrogenases that oxidize isocitrate, α-ketoglutarate, and malate (reactions TCA-3, TCA-4, and TCA-8) are all allosterically inhibited by NADH.

(c) Pyruvate dehydrogenase (reaction 12-1) is inactivated by ATP.

(d) Phosphofructokinase (Figure 11-7, reaction Gly-3) is allosterically inhibited by citrate.

(e) Pyruvate carboxylase (Figure 12-5b) is allosterically activated by acetyl CoA.

12-6. Lethal Synthesis. The leaves of *Dichapetalum cymosum*, a South African plant, are very poisonous. Animals that eat the leaves have convulsions and usually die shortly thereafter. One of the most pronounced effects of poisoning is a marked elevation in citrate concentration and a blockage of the TCA cycle in many organs of the affected animal. The toxic agent in the leaves of the plant is fluoroacetate, but the actual poison in the tissues of the animal is fluorocitrate. If fluoroacetate is incubated with purified enzymes of the TCA cycle, it has no inhibitory effect.

Fluoroacetate

Fluorocitrate

(a) Why might you expect fluorocitrate to have an inhibitory effect on one or more of the TCA cycle enzymes if incubated with the purified enzymes in vitro, even though fluoroacetate has no such effect?

(b) Which enzyme in the TCA cycle do you suspect is affected by fluorocitrate? Give two reasons for your answer.

(c) How do you suppose fluoroacetate gets converted to fluorocitrate?

(d) Why is this phenomenon referred to as *lethal synthesis*?

12-7. The Energetics of Malate Oxidation. Malate dehydrogenase is the enzyme that catalyzes the oxidation of malate to oxaloacetate in reaction TCA-8 of the TCA cycle, with NAD^+ as the electron acceptor.

(a) Without doing any calculations, predict from Table 12-2 in which direction reaction TCA-8 should be thermodynamically spontaneous under standard conditions.

(b) Calculate $\Delta E_0'$ and $\Delta G^{\circ\prime}$ for reaction TCA-8 in the direction required for TCA cycle activity. What do the values of these parameters tell you about the reaction? What is it that makes this reaction possible under cellular conditions when $\Delta G^{\circ\prime}$ is so unfavorable?

(c) Again, without doing any calculations, predict whether the reactions TCA-3 and TCA-6 will be thermodynamically feasible in the direction required for TCA cycle activity under standard conditions. Why is it not possible to make a similar prediction for reaction TCA-4 without any calculations?

12-8. Oxidation of Cytosolic NADH. In some eukaryotic cells, the NADH generated by glycolysis in the cytosol is reoxidized by the glycerol-3-phosphate/dihydroxyacetone phosphate shuttle shown in Figure 12-26.

(a) Write balanced reactions for the reduction of dihydroxyacetone phosphate (DHAP) to glycerol-3-phosphate (glycerol-3-P) by cytosolic NADH and for the oxidation of glycerol-3-P to DHAP by the FAD-linked dehydrogenase in the inner membrane of the mitochondrion.

(b) Add the two reactions in part a to obtain a summary reaction for the transfer of electrons from cytosolic NADH to mitochondrial FAD. Calculate $\Delta E_0'$ and $\Delta G^{\circ\prime}$ for this reaction. Is the inward movement of electrons thermodynamically feasible under standard conditions?

(c) Write a balanced reaction for the reoxidation of $FADH_2$ by coenzyme Q within the inner membrane, assuming that CoQ is reduced to $CoQH_2$. Calculate $\Delta E_0'$ and $\Delta G^{\circ\prime}$ for this reaction. Is this transfer thermodynamically feasible under standard conditions?

(d) Write a balanced reaction for the transfer of electrons from cytosolic NADH to mitochondrial CoQ, and calculate $\Delta E_0'$ and $\Delta G^{\circ\prime}$ for this reaction. Is this transfer thermodynamically feasible under standard conditions?

(e) Assume that the [NADH]/[NAD$^+$] ratio in the cytosol is 5.0 and that the [CoQH$_2$]/[CoQ] ratio in the inner membrane is 2.0. What is $\Delta G'$ for the reaction in part d at 25°C and pH 7.0?

(f) Is $\Delta G'$ for the inward transfer of electrons from NADH to CoQ affected by the ratio of the reduced to the oxidized forms of the enzyme-bound FAD in the inner membrane? Why or why not?

12-9. Multiple Uses of the Electrochemical Proton Gradient.

Most of our discussion in this chapter focused on the use of the electrochemical proton gradient to drive the synthesis of ATP. In particular, the conclusion that the electrochemical proton gradient across the inner membrane of the mitochondrion can generate 34 molecules of ATP per molecule of glucose (reaction 12-18) assumes that the energy of the proton gradient is not tapped for any other purpose. Actually, however, the energy of the proton gradient is also used to drive other processes, notably the transport of ions and metabolites across the inner membrane. Assume (1) that two protons are required to drive the symport of each molecule of pyruvate into the mitochondrion; (2) that the coupled outward transport of ATP and inward transport of ADP is exergonic; and (3) that the inward movement of phosphate anions (P$_i$) is coupled on an equimolar basis to the outward transport of hydroxyl ions, each of which is neutralized by reaction with a proton on the outer surface of the inner membrane. Assume further that each of the respiratory complexes pumps three protons outward across the inner membrane per pair of electrons transported, and that ATP synthesis by the F_oF_1 complex is driven by the inward movement of three protons per ATP.

(a) Under these conditions, what is the total number of protons pumped outward by the electron transport that accompanies the aerobic catabolism of one molecule of glucose? How many ATP molecules could be synthesized per glucose if the energy of the proton gradient were used for no other purpose?

(b) How many pyruvate molecules must be transported into the mitochondrion per molecule of glucose catabolized? How many protons are required for this purpose?

(c) How many ATP molecules can be synthesized per glucose by the remaining protons, recalling that the inward transport of phosphate is coupled to the outward movement of hydroxyl ions, which are then neutralized by protons?

(d) What is the total ATP yield per molecule of glucose under these conditions? Write a balanced reaction to show the aerobic catabolism of glucose accompanied by the generation of the appropriate number of ATP molecules.

12-10. Rotenone Poisoning.

Rotenone is a potent insecticide and fish poison. As shown in Figure 12-20, its mode of action is to block electron transport from the FMN of NADH dehydrogenase to coenzyme Q.

(a) Why do fish and insects die after digesting rotenone?

(b) Would you expect the use of rotenone as an insecticide to be a potential hazard to other forms of animal life? Explain.

(c) Would you expect the use of rotenone as a fish poison to be a potential hazard to aquatic plants that might be exposed to the compound? Explain.

12-11. Mitochondrial Transport.

For aerobic respiration, a variety of substances must be in a state of flux across the inner mitochondrial membrane. Assuming a cell in which glucose is the sole energy source, indicate for each of the following substances whether you would expect a net flux across the membrane and, if so, how many molecules will move in which direction per molecule of glucose catabolized?

(a) Pyruvate	**(g)** NADH
(b) Oxygen	**(h)** FADH$_2$
(c) ATP	**(i)** Oxaloacetate
(d) ADP	**(j)** Water
(e) Acetyl CoA	**(k)** Electrons
(f) Glycerol-3-phosphate	**(l)** Protons

12-12. Glyoxysomal Function.

The glyoxysomes of fat-storing seedlings contain all the enzymes necessary to degrade fatty acids completely to acetyl CoA and to synthesize succinate from the resulting acetyl CoA. Fatty acid oxidation (Figure 12-10) begins with ATP-dependent formation of fatty acyl CoA thioester (reaction FA-1). Each cycle of β oxidation generates one molecule each of FADH$_2$ (FA-2) and NADH (FA-4), culminating in the release of two carbons as acetyl CoA (FA-5). In the glyoxysomes, FADH$_2$, but not NADH, is reoxidized by the direct, oxidase-mediated transfer of electrons to oxygen, generating hydrogen peroxide and no ATP. Synthesis of succinate from the acetyl CoA occurs by means of the glyoxylate cycle (Figure 12A-2). The further steps required to convert succinate to hexoses (and then to sucrose) occur elsewhere in the cell (Figure 12A-3).

(a) What are the main products of β oxidation of a fatty acid in the glyoxysome? What happens to each?

(b) How many molecules of succinate are produced from a single molecule of palmitate ($C_{16}H_{32}O_2$)? How many molecules of FADH$_2$ and of NADH are generated in the process?

(c) How many hexose molecules can be formed from the succinate of part b? Where in the cell does this process occur? What other products are formed in the conversion of succinate to hexose?

(d) How many molecules of sucrose can be produced from a single molecule of palmitate? How many of the original 16 carbon atoms of palmitate eventually appear in sucrose? What happens to the others?

(e) Assuming a quantitative conversion of fat to carbohydrate, how many grams of sucrose can a fat-storing seedling produce from 1 gram of stored palmitate?

(f) In addition to fatty acids and succinate, what other substances would you predict have to be transported across the glyoxysomal membrane? In which direction would you expect these substances to move?

SUGGESTED READING

General References

Abeles, R. H., P. A. Frey, and W. P. Jencks. *Biochemistry.* Boston: Jones and Bartlett, 1992.

Cramer, W. A., and D. B. Knaff. *Energy Transduction in Biological Membranes.* New York: Springer-Verlag, 1991.

Lemasters, J., C. Hackenbrock, R. Thurman, and H. Wetserhoff. *Integration of Mitochondrial Function.* New York: Plenum, 1989.

Mathews, C. K., and K. E. van Holde. *Biochemistry,* 2d ed., Menlo Park, CA.: Benjamin/Cummings, 1996.

Nicholls, D. G. *Bioenergetics: An Introduction to the Chemiosmotic Theory.* New York: Academic Press, 1982.

Racker, E. *A New Look at Mechanisms in Bioenergetics.* New York: Academic Press, 1976.

Racker, E. From Pasteur to Mitchell: A hundred years of bioenergetics. *Federation Proc.* 39 (1980): 210.

Stryer, L. *Biochemistry,* 4th ed. New York: W. H. Freeman, 1995.

Mitochondrial Structure

Attardi, G., and G. Schatz. Biogenesis of mitochondria. *Annu. Rev. Cell Biol.* 4 (1988): 289.

Ernster, L., and G. Schatz. Mitochondria: An historical overview. *J. Cell Biol.* 91 (1981): 227s.

Hahn-Bereiter, J. Behavior of mitochondria in the living cell. *Internat. Rev. Cytol.* 122 (1990): 1.

Quagliariello, E., E. C. Slater, F. Palmieri, C. Saccone, and A. M. Kroon, eds. *Achievements and Perspectives of Mitochondrial Research.* New York: Elsevier, 1985.

Rosamond, J. Structure and function of mitochondria. *Internat. Rev. Cytol. Suppl.* 17 (1987): 121.

Srere, P. A. The structure of the mitochondrial inner membrane matrix compartment. *Trends Biochem. Sci.* 7 (1982): 375.

Tzagoloff, A. *Mitochondria.* New York: Plenum, 1982.

von Heinje, G. Why mitochondria need a genome. *FEBS Lett.* 198 (1986): 1.

The Tricarboxylic Acid Cycle

Cammack, R. FADH$_2$ as a "product" of the citric acid cycle. *Trends Biochem. Sci.* 12 (1987): 377.

Kornberg, H. L. Tricarboxylic acid cycles. *Bioessays* 7 (1987): 236.

Krebs, H. A. The history of the tricarboxylic acid cycle. *Perspect. Biol. Med.* 14 (1970): 154.

Krebs, H. A., and A. Martin. *Reminiscences and Reflections.* New York: Oxford University Press, 1981.

Electron Transport and Oxidative Phosphorylation

Anraku, Y. Bacterial electron transport chains. *Annu. Rev. Biochem.* 57 (1988): 101.

Babcock, G. T., and M. Wikstrom. Oxygen activation and the conservation of energy in cell respiration. *Nature* 356 (1992): 301.

Beinert, H. 1990. Recent developments in the field of Fe-S proteins. *FASEB J.* 4 (1990): 2483.

Brand, N., and M. Murphy. Control of electron flux through the respiratory chain in mitochondria and cells. *Biol. Rev.* 62 (1987): 141.

Capaldi, R. A. Structure and function of cytochrome *c* oxidase. *Annu. Rev. Biochem.* 59 (1990): 569.

Chan, S.I., and M. Li. Cytochrome *c* oxidase: Understanding nature's design of a proton pump. *Biochemistry* 29 (1990): 1.

Cooper, C. E., P. Nicholls, and J. A. Freedman. Cytochrome *c* oxidase: Structure, function, and membrane topology of the polypeptide subunits. *Biochim. Cell Biol.* 69 (1991): 586.

Ferguson, S. J. The ups and downs of P/O ratios. *Trends Biochem. Sci.* 11 (1986): 351.

Futai, M., and H. Kanazawa. Structure and function of proton-translocating adenosine triphosphatase ($F_o F_1$): Biochemical and molecular biological approaches. *Microbiol. Rev.* 47 (1983): 285.

Futai, M., T. Noumi, and M. Maeda. ATP synthase (H^+-ATPase): Results by combined biochemical and molecular biology approaches. *Annu. Rev. Biochem.* 58 (1989): 111.

Hackenbrock, C. R. Lateral diffusion and electron transfer in the mitochondrial inner membrane. *Trends Biochem. Sci.* 6 (1981): 151.

Lenas, G. Role of mobility of redox components in the inner mitochondrial membrane. *J. Membr. Biol.* 104 (1989): 193.

Mitchell, P. Coupling of phosphorylation to electron and hydrogen transfer by a chemi-osmotic type of mechanism. *Nature* 191 (1961): 144.

Mitchell, P. Keilin's respiratory chain concept and its chemiosmotic consequences. *Science* 206 (1979): 1148.

Ohnishi, T. Structure of the succinate-ubiquinone oxidoreductase (complex IV). *Current Topics Bioenerget.* 15 (1987): 37.

Ragan, C. I. Structure of the NADH-ubiquinone reductase (complex I). *Current Topics Bioenerget.* 15 (1987): 1.

Senior, A. E. The proton-translocating ATPase of *E. coli. Annu. Rev. Biophys. Biophys. Chem.* 19 (1990): 7.

Slater, E. C. The mechanism of the conservation of energy of biological oxidations. *Europ. J. Biochem.* 166 (1988): 489.

Weiss, H. Structure of the mitochondrial ubiquinol-cytochrome *c* reductase (complex II). *Current Topics Bioenerget.* 15 (1987): 67.

Weiss, H., T. Friedrich, G. Hofhaus, and D. Preis. The respiratory-chain NADH dehydrogenase (complex I) of mitochondria. *Europ. J. Biochem.* 197 (1991): 563.

Wilson, M. T., and D. Bickar. Cytochrome oxidase as a proton pump. *J. Bioenerget. Biomembr.* 23 (1991): 755.

Ysern, X., L. M. Amzel, and P. L. Pedersen. ATP synthesis: Structure of the F_1-moiety and its relationship to function and mechanism. *J. Bioenerget. Biomembr.* 20 (1988): 423.

Regulation of Respiratory Metabolism

Erecinska, M., and D. F. Wilson. Regulation of cellular energy metabolism. *J. Membr. Biol.* 70 (1982): 1.

Martin, B. R. *Metabolic Regulation: A Molecular Approach.* Oxford, England: Blackwell Scientific, 1987.

13

PHOTOTROPHIC ENERGY METABOLISM: PHOTOSYNTHESIS

In the two preceding chapters, we studied fermentation and respiration, two chemotrophic solutions to the universal problem of meeting the energy and carbon needs of a living cell. Although a few prokaryotic chemotrophs can acquire energy, and even reduced carbon, from inorganic molecules, most chemotrophs depend on organic substrates. This creates a secondary problem: Left alone, most chemotrophs would perish shortly after they oxidized the Earth's limited supply of reduced carbon to carbon dioxide and water. Moreover, the first organisms affected would be those dependent on aerobic respiration, because molecular oxygen (O_2), like reduced carbon, is an exhaustible resource.

Nature's ultimate solution to the energy and carbon drain from the biosphere is **photosynthesis.** Nearly all life on Earth is sustained by a cascade of energy that arrives at the planet as sunlight. **Phototrophs** convert solar energy to chemical energy in the form of ATP. For one group of phototrophs, the halobacteria described in Chapter 7, this is the extent of their photosynthetic ability. The halobacteria are an example of **photoheterotrophs,** organisms that acquire energy from sunlight but depend on organic sources of reduced carbon. Most other phototrophs—plants, algae, and most photosynthetic bacteria—are photoautotrophs. **Photoautotrophs** use solar energy to drive the synthesis of energy-rich organic molecules from simple inorganic starting materials—carbon dioxide and water.* Moreover, members of one group of photoautotrophs, appropriately called *oxygenic phototrophs,* release molecular oxygen as a by-product of photosynthesis. Thus, phototrophs replenish reduced carbon in the biosphere and oxygen in the atmosphere, completing the cyclic flow of energy and carbon introduced in Chapter 5.

In this chapter, we will study two general aspects of photosynthesis: how photoautotrophs capture solar energy

and convert it to chemical energy, and how that energy is used to transform energy-poor carbon dioxide and water into energy-rich organic molecules, such as carbohydrates.

An Overview of Photosynthesis

In photoautotrophs, photosynthesis may be divided into two major processes: energy transduction and carbon assimilation (Figure 13-1). During the **energy transduction reactions,** light energy is converted to chemical energy in the form of ATP and the coenzyme NADPH. To emphasize the importance of light, the energy transduction reactions are also called the *light-dependent reactions* of photosynthesis. ATP and NADPH generated by the energy transduction reactions subsequently provide energy and reducing power for the **carbon assimilation reactions.** During this process, fully oxidized carbon atoms from carbon dioxide are *fixed*—covalently attached—to organic acceptor molecules and then reduced and rearranged to form the carbohydrates and other organic compounds required for building a living cell. Unfortunately, the carbon assimilation reactions are traditionally referred to as the *light-independent reactions* or even the *dark reactions* of photosynthesis; these terms are not that accurate because carbon assimilation is far from light-independent.

A family of green pigment molecules called *chlorophyll* has a key role in every photoautotroph's energy transduction pathway. While a variety of pigments found in photoautotrophs absorb light energy, only chlorophyll converts solar energy to chemical energy as it donates photoexcited electrons to organic acceptor molecules. From chlorophyll, photoexcited electrons flow energetically downhill through an electron transport system. Like electron transport in mitochondria, the flow of electrons is coupled to unidirectional proton pumping, which stores energy in an electrochemical proton gradient that drives an ATP synthase. The light-dependent synthesis of ATP by this process is called **photophosphorylation.**

*Some phototrophs are *facultative* phototrophs, which can function as either photoautotrophs or photoheterotrophs, depending on the availability of nutrients. Plants are *obligate* photoautotrophs; they *must* acquire carbon from carbon dioxide.

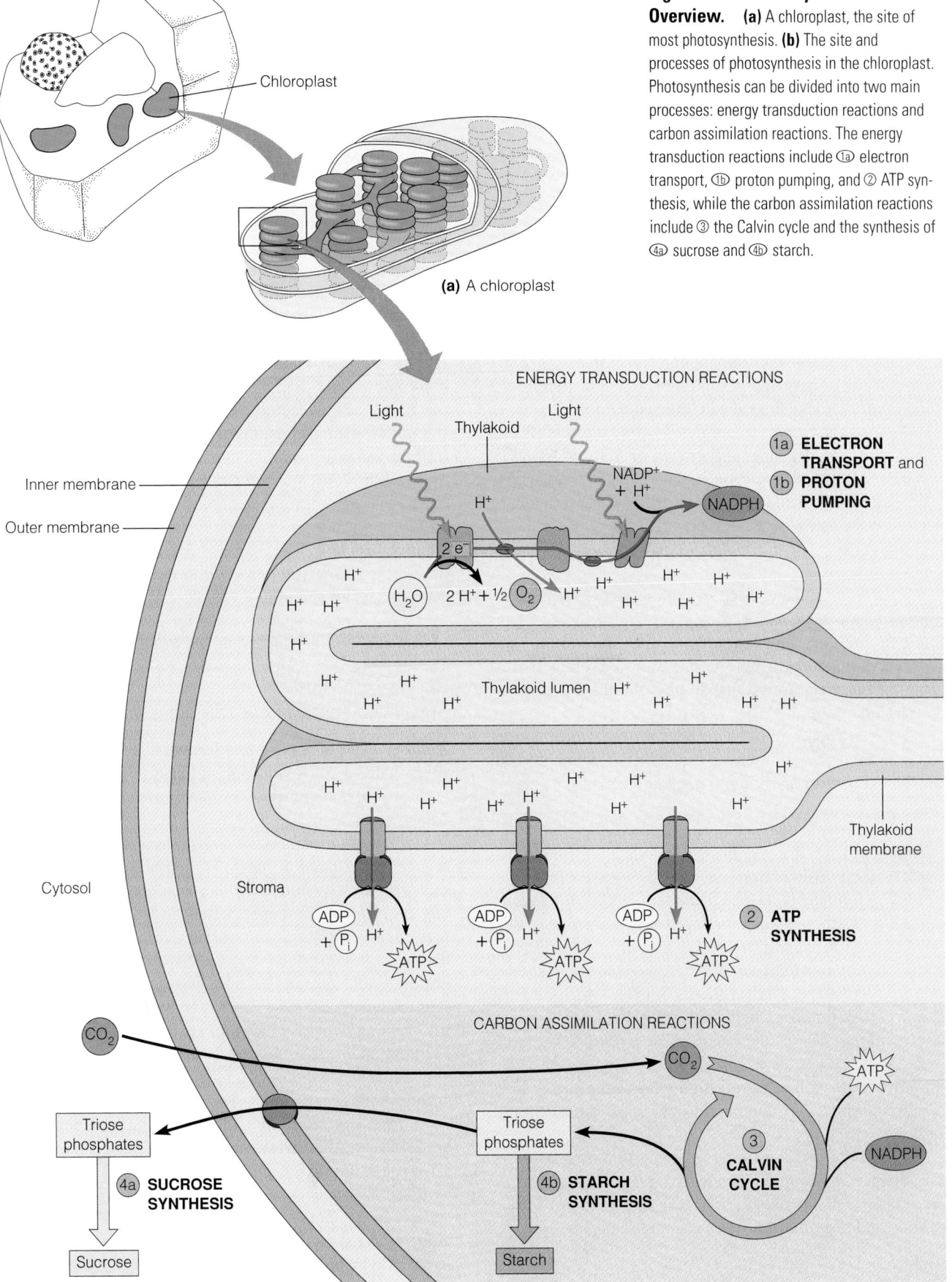

Figure 13-1 Photosynthesis: An Overview. **(a)** A chloroplast, the site of most photosynthesis. **(b)** The site and processes of photosynthesis in the chloroplast. Photosynthesis can be divided into two main processes: energy transduction reactions and carbon assimilation reactions. The energy transduction reactions include ①a electron transport, ①b proton pumping, and ② ATP synthesis, while the carbon assimilation reactions include ③ the Calvin cycle and the synthesis of ④a sucrose and ④b starch.

(a) A chloroplast

ENERGY TRANSDUCTION REACTIONS

Light

Thylakoid

Light

$NADP^+ + H^+$

NADPH

1a ELECTRON TRANSPORT and
1b PROTON PUMPING

Inner membrane

Outer membrane

$2 e^-$

H_2O $2 H^+ + \frac{1}{2} O_2$

H^+

Thylakoid lumen

Thylakoid membrane

Cytosol

Stroma

ADP $+ P_i$ H^+ ATP

ADP $+ P_i$ H^+ ATP

ADP $+ P_i$ H^+ ATP

2 ATP SYNTHESIS

CARBON ASSIMILATION REACTIONS

CO_2

CO_2

ATP

Triose phosphates

Triose phosphates

③ CALVIN CYCLE

NADPH

4a SUCROSE SYNTHESIS

4b STARCH SYNTHESIS

Sucrose

Starch

(b) Localization of photosynthesis within the chloroplast

To incorporate fully oxidized carbon atoms from carbon dioxide into organic molecules, photoautotrophs also need NADPH, a reduced form of the coenzyme nicotinamide adenine dinucleotide phosphate ($NADP^+$). In **oxygenic phototrophs**—plants, algae, and cyanobacteria—light energy absorbed by chlorophyll and other pigment molecules drives the movement of electrons from water, with a very positive reduction potential, to *ferredoxin,* with a very negative reduction potential. From ferredoxin, electrons may then flow exergonically to $NADP^+$, thereby generating NADPH. In **anoxygenic phototrophs**—green and purple bacteria—compounds with less positive reduction potentials, such as sulfide (HS^-), thiosulfate ($S_2O_3^{2-}$), or succinate, serve as electron donors. In this case, the light-dependent generation of reduced coenzyme may be less direct, with ATP generated by photophosphorylation driving the movement of electrons from donors to $NADP^+$. In both oxygenic and anoxygenic phototrophs, the light-dependent generation of NADPH is called **photoreduction.**

Most of the energy accumulated within photosynthetic cells by the light-dependent generation of ATP and NADPH is rapidly consumed by carbon assimilation pathways, where it drives carbon dioxide fixation and reduction. A general reaction for the complete process may be written as:

$$\text{light} + CO_2 + 2H_2A \longrightarrow [CH_2O] + 2A + H_2O \quad \textbf{(13-1)}$$

where H_2A is a suitable electron donor, $[CH_2O]$ represents an organic molecule with carbon at the oxidation level of an aldehyde, and A is the oxidized form of the electron donor. By expressing photosynthesis in this way, we avoid perpetuating the incorrect notion that all phototrophs use water as an electron donor.

When we focus on oxygenic phototrophs, which do use water as an electron donor, we can rewrite reaction 13-1 in the following more specific and familiar form:

$$\text{light} + 3CO_2 + 6H_2O \longrightarrow C_3H_6O_3 + 3O_2 + 3H_2O \quad \textbf{(13-2)}$$

Later we will see that the primary products of photosynthetic carbon assimilation are actually two triose phosphates, *glyceraldehyde-3-phosphate* and *dihydroxyacetone phosphate.* Carbon assimilation generally diverges at this point, as the triose phosphates enter a variety of biosynthetic pathways. The most important pathways for our consideration are sucrose and starch synthesis. Sucrose, which conveys energy and reduced carbon from photosynthetic to nonphotosynthetic cells, is the major transport carbohydrate in most plant species. And starch, which accumulates when photosynthetic carbon assimilation exceeds the energy and carbon demands of a photoautotroph, is the major storage carbohydrate.

Our model organisms for studying photosynthesis in this chapter are the most familiar oxygenic phototrophs—green plants.* Before we study photosynthetic energy transduction and carbon assimilation in more detail, however, we

---

*At least one plant, an unusual orchid, lacks chlorophyll, the green pigment essential for photosynthesis. This particular plant lives underground and derives all of its energy and reduced carbon from organic compounds in the soil.

will look at the structure and function of the chloroplast. In every eukaryotic phototroph, most of the events of photosynthesis are localized in this organelle.

The Chloroplast: A Photosynthetic Organelle

In plants and algae, the primary events of photosynthetic energy transduction and carbon assimilation are confined to specialized organelles called **chloroplasts.** The prominence of chloroplasts within a photosynthetic cell is clearly illustrated in the electron micrograph of a *Coleus* leaf cell shown in Figure 13-2a. But an electron microscope is not necessary for observing chloroplasts. Because they are generally large (about 1–5 μm wide and 5–10 μm long) and opaque, chloroplasts were described and studied early in the history of cell biology. In fact, the earliest descriptions go back to the work of Antonie van Leeuwenhoek and Nehemiah Grew in the seventeenth century, at the very dawn of microscopic observation. A mature leaf cell may contain 20–100 chloroplasts, the precise number depending on the plant species and habitat, while an algal cell typically contains only one or a few chloroplasts. The shapes of these organelles vary from the simple flattened spheres common in plants to the more elaborate forms found in green algae. A cell of the filamentous green alga *Spirogyra,* for example, contains one or more ribbon-shaped chloroplasts, shown in Figure 13-2b.

Structural Features of Chloroplasts

A closer view of a typical chloroplast is shown in Figure 13-3a (page 378). A chloroplast, like a mitochondrion, has both an **outer membrane** and an **inner membrane,** often separated by a narrow **intermembrane space** (Figure 13-3b). The outer membrane is freely permeable to most small organic molecules. But the inner membrane forms a significant permeability barrier, with transport proteins controlling the flow of most metabolites between the intermembrane space and the stroma. (Three important metabolites that are able to diffuse freely across the inner membrane, however, are carbon dioxide, water, and oxygen.) The **stroma,** enclosed by the inner membrane, is a gel-like matrix teeming with enzymes for carbon, nitrogen, and sulfur assimilation. Because the stroma also contains DNA, mRNA, and ribosomes, chloroplasts, like mitochondria, are **semiautonomous organelles.**

Unlike mitochondria, chloroplasts have a third membrane system, the **thylakoids,** seen in Figure 13-3c and d. Thylakoids are flat, saclike structures suspended in the stroma and usually arranged in stacks called **grana** (singular: **granum**). Grana, which resemble stacks of coins, are interconnected by a network of longer thylakoids called **stroma thylakoids.** Essentially all of the photosynthetic pigments, the enzymes required for the photoreactions, the carriers involved in electron transport, and the proteins that

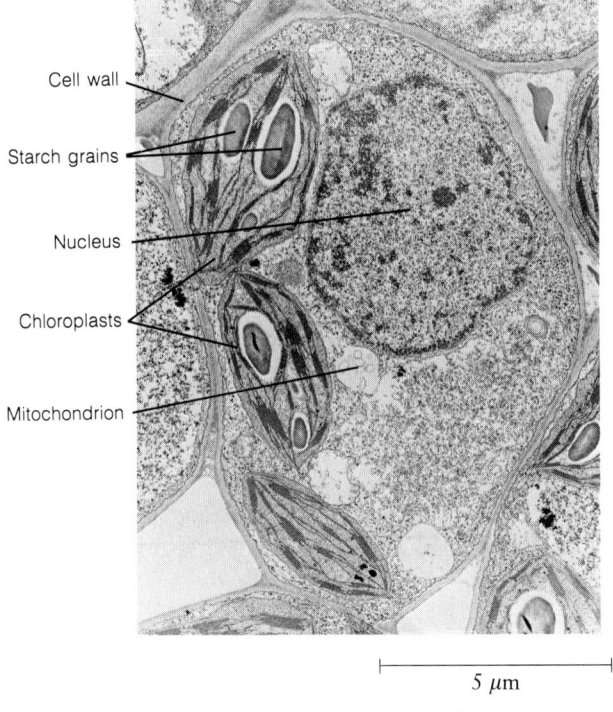

(a) Chloroplasts in a plant leaf cell

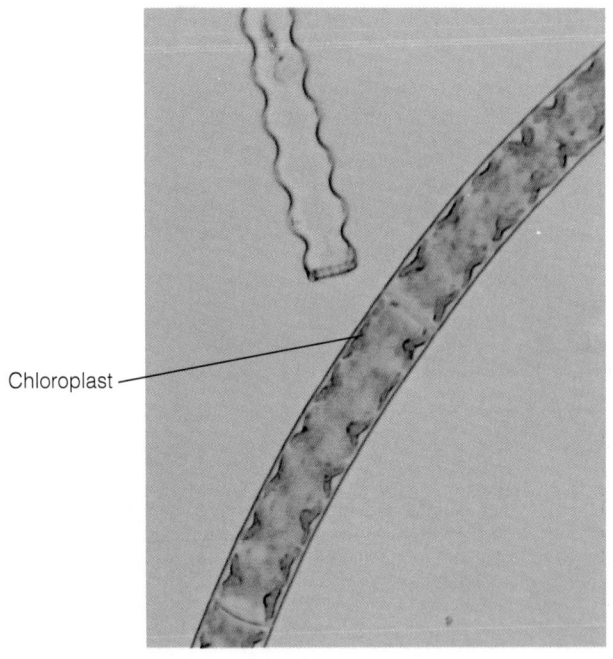

(b) Chloroplasts in an algal cell

Figure 13-2 Chloroplasts. (a) The prominence of chloroplasts in leaf cells of a plant is illustrated by this electron micrograph of a parenchyma cell from a *Coleus* leaf. The cell contains many chloroplasts, three of which are seen in this particular cross section. The presence of large starch grains in the chloroplasts indicates that the cell was photosynthetically active just prior to fixation for electron microscopy (TEM). **(b)** This light micrograph reveals the unusual ribbon-shaped chloroplasts of the filamentous green alga *Spirogyra*.

couple electron transport to proton pumping and ATP synthesis are localized in or on the thylakoid membranes. Regions where one thylakoid membrane within a granum touches another are called *appressed regions.* Two of the photosynthetic complexes described later, the photosystem I and ATP synthase complexes, are restricted to *nonappressed regions.*

The thylakoid membranes might arise from invaginations of the inner membrane during chloroplast development. However, they are not physically contiguous with the inner membrane. Moreover, their lipid and protein content is clearly different from that of either the outer or the inner membrane. Electron micrographs of serial sections suggest that the grana and stroma thylakoids enclose a single compartment, the **thylakoid lumen.** Separation of the lumen from the stroma by the thylakoid membrane plays an important role in the generation of an electrochemical proton gradient and the synthesis of ATP, because protons pumped into the lumen during light-driven electron transport drive ATP synthesis as they return to the stroma.

Photosynthetic prokaryotes do not have chloroplasts. However, in some prokaryotes, such as the cyanobacteria, the plasma membrane is folded inward and forms *photosynthetic membranes,* structures analogous to the thylakoids (Figure 13-4). Indeed, to some extent cyanobacteria appear to be free-living chloroplasts.

Photosynthetic Energy Transduction

To understand how light energy is converted to chemical energy within a chloroplast or bacterium, we require some knowledge of the nature of light (electromagnetic radiation) and its interaction with molecules. We often regard light as a wave; the visible portion of the electromagnetic spectrum, for example, consists of wavelengths ranging from about 380 to 750 nm. However, light actually behaves as a stream of discrete particles called **photons,** each photon carrying an indivisible packet, or **quantum,** of energy. The precise quantum of energy carried by a photon is inversely related to the photon's wavelength (see Figure 2-3). A photon of ultraviolet or blue light, for example, has a shorter wavelength and carries a larger quantum of energy than a photon of red or infrared light.

When a **pigment** (light-absorbing molecule) such as chlorophyll absorbs a photon, the photon's energy is transferred to an electron, which is energized from its *ground state* in a low-energy orbital to an *excited state* in a higher-energy orbital. This event is called **photoexcitation** and is the first step in photosynthesis. The photon's quantum of energy must be identical to the difference in energy between the ground-state orbital and the excited-state orbital of the electron or absorption will not occur. Because electron orbitals are characterized by discrete energy levels and every

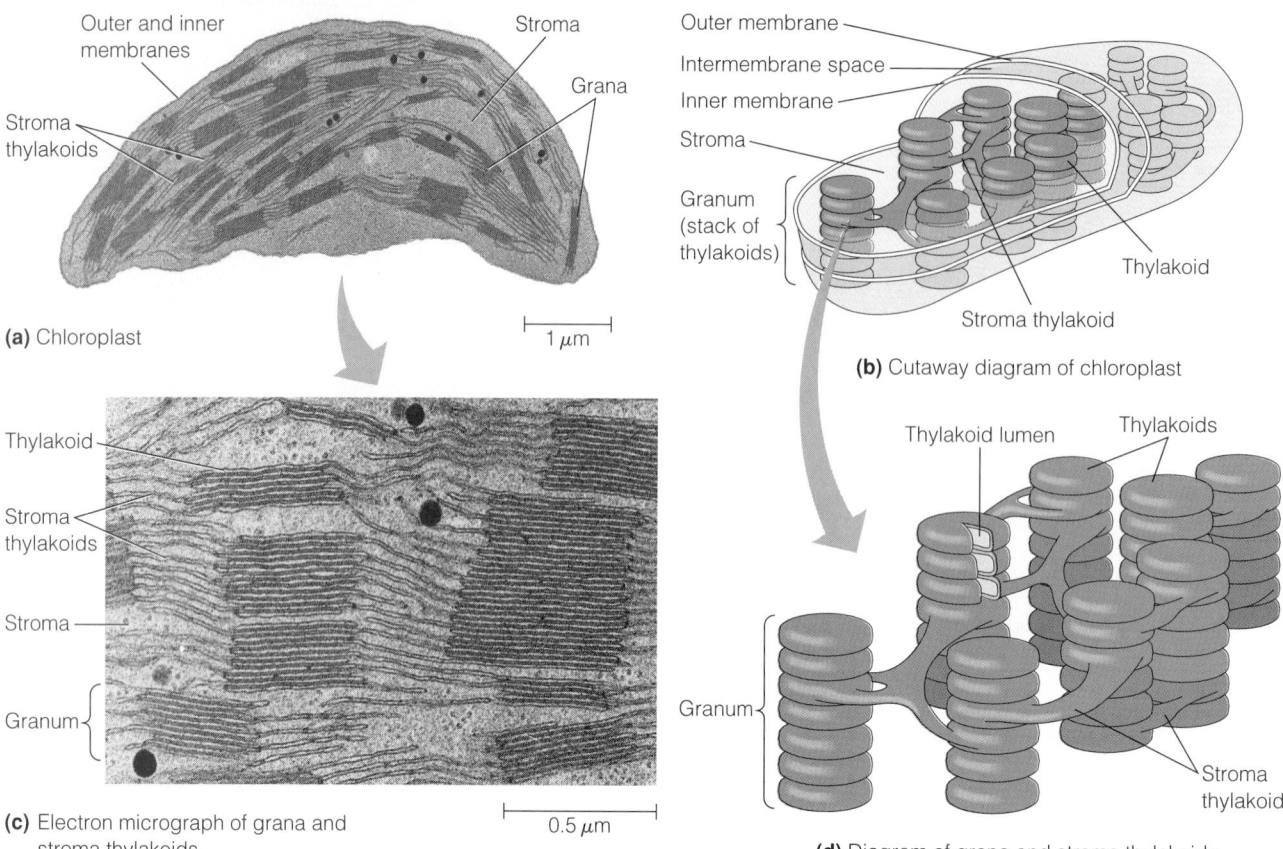

(a) Chloroplast

Outer and inner membranes

Stroma thylakoids

Stroma

Grana

1 μm

Thylakoid

Stroma thylakoids

Stroma

Granum

(c) Electron micrograph of grana and stroma thylakoids

0.5 μm

Outer membrane

Intermembrane space

Inner membrane

Stroma

Granum (stack of thylakoids)

Thylakoid

Stroma thylakoid

(b) Cutaway diagram of chloroplast

Thylakoid lumen

Thylakoids

Granum

Stroma thylakoids

(d) Diagram of grana and stroma thylakoids

Figure 13-3   Structural Features of a Chloroplast.
(a) An electron micrograph of a chloroplast from a leaf of timothy grass (*Phleum pratense*) (TEM). **(b)** A diagram showing the three-dimensional structure of a typical chloroplast. **(c)** A more highly magnified electron micrograph of the chloroplast in part a, showing the arrangement of grana and stroma thylakoids (TEM). **(d)** A diagram depicting the continuity of the thylakoid membranes, the arrangement of thylakoids into stacks called grana, and the stroma thylakoids that interconnect the grana. The thylakoid membranes enclose a space called the thylakoid lumen.

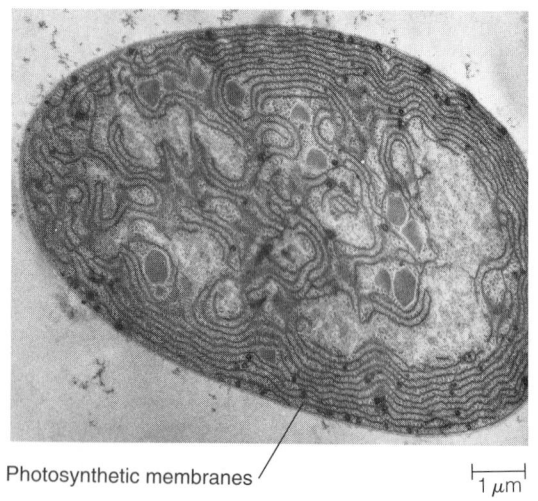

Photosynthetic membranes

1 μm

Figure 13-4   Photosynthetic Membranes of a Cyanobacterium.   This electron micrograph of a thin section of *Anabaena azollae* reveals the extensive folded membranes of cyanobacterial cells that resemble the thylakoids of chloroplasts (TEM).

pigment has a different electron configuration, each one displays a characteristic **absorption spectrum.** The wavelengths absorbed by a molecule correspond to the orbital transitions that electrons in the molecule can undergo. The absorption spectra of several common pigments found in photosynthetic organisms are shown in Figure 13-5, along with the spectrum of solar radiation reaching the Earth's surface.

A photoexcited electron in a pigment molecule is very unstable and has several potential fates. Most frequently, excitation energy is lost as heat, often concomitant with the release of a photon carrying a smaller quantum of energy (fluorescent emission), as the electron returns to its ground state in a low-energy orbital. Alternatively, most of the excitation energy is conserved when it excites an electron in an adjacent pigment molecule by **resonance transfer,** or the electron itself is passed to another molecule by **photochemical reduction.** Resonance transfer and photochemical reduction are important events in photosynthesis, the latter being essential for converting light energy to chemical energy.

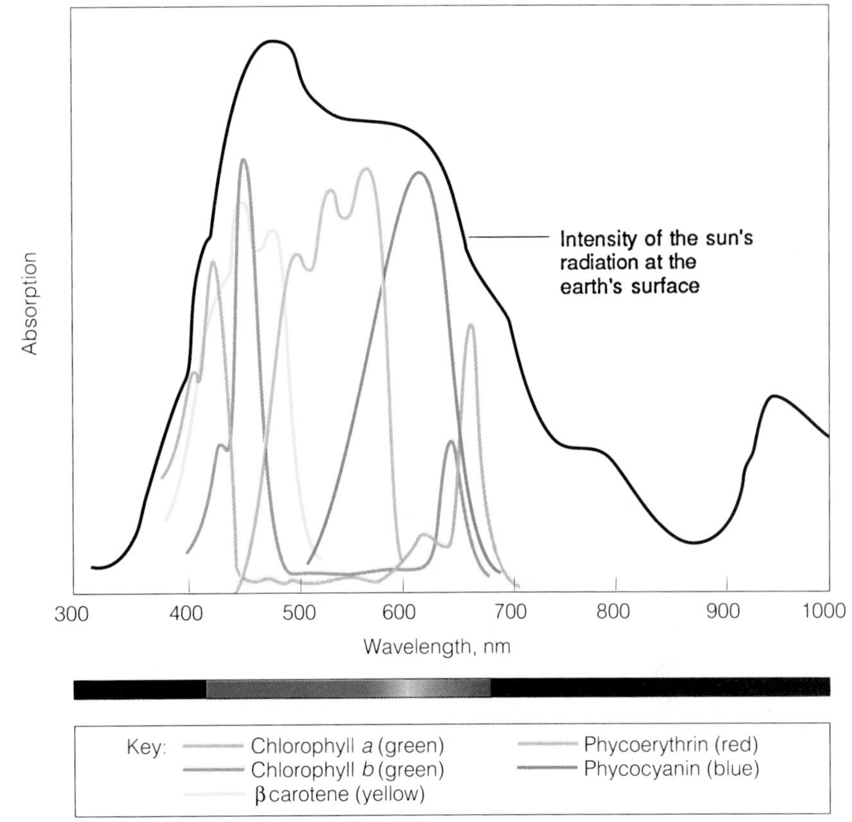

Figure 13-5 Absorption Spectra of Common Plant Pigments. The absorption spectra of various pigments are compared with the spectral distribution of solar energy reaching the Earth's surface. The absorption spectrum of lutein, a carotenoid, resembles that of β-carotene. The overlapping absorption spectra of various chlorophylls and accessory pigments effectively covers most of the spectrum of sunlight that reaches the Earth's surface.

Intensity of the sun's radiation at the earth's surface

Absorption

Wavelength, nm

Key:
Chlorophyll *a* (green)
Chlorophyll *b* (green)
β carotene (yellow)
Phycoerythrin (red)
Phycocyanin (blue)

Figure 13-6 The Structures of Chlorophyll *a* and *b*. Chlorophyll *b* has a formyl (CHO) group at the indicated position, whereas chlorophyll *a* has a methyl (CH₃) group. The bacteriochlorophyll found in anoxygenic phototrophs has a saturated carbon-carbon bond at the location indicated by the arrow.

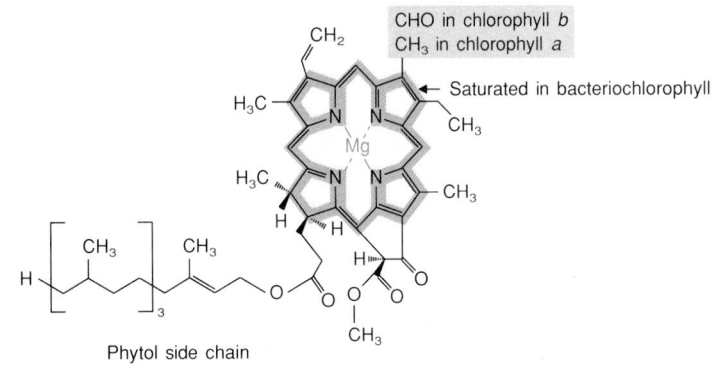

CHO in chlorophyll *b*
CH₃ in chlorophyll *a*
Saturated in bacteriochlorophyll

Phytol side chain

Chlorophyll: The Energy Transduction Pigment

Found in all photosynthetic organisms except the halobacteria, **chlorophyll** is life's primary link to sunlight. Moreover, only specific forms of chlorophyll can pass a photoexcited electron to another molecule in an oxidation-reduction reaction that traps light energy and converts it to chemical energy. The structures of both chlorophyll *a* and chlorophyll *b* are shown in Figure 13-6. All plants and green algae contain both chlorophyll *a* and *b*, which enables them to absorb more photons than either pigment alone would. Other photosynthetic eukaryotes supplement chlorophyll *a* with either chlorophyll *c* (in brown algae, diatoms, and dinoflagellates) or *d* (in red algae).

The magnesium ion (Mg^{2+}) in the chlorophyll molecule affects the electron distribution in the porphyrin ring of the molecule, ensuring that a variety of higher-energy orbitals are available and that more than a few specific wavelengths of light can be absorbed. Chlorophyll *a*, for example, has a broad absorption spectrum, with maxima at about 420 and 660 nm (see Figure 13-5). The phytol side chain is strongly hydrophobic and interacts with lipids of thylakoid or cyanobacterial membranes.

The precise absorption spectrum of a chlorophyll molecule is modified by structural changes and by the presence of specific *chlorophyll binding proteins*. Chlorophyll *b*, for example, is distinguished by a *formyl* (CHO) group in place of one of the *methyl* (CH₃) groups on the porphyrin ring (see Figure 13-6). This minor alteration shifts the absorption maxima toward the center of the visible spectrum, as shown in Figure 13-5.

Bacteriochlorophyll is a subfamily of chlorophyll molecules restricted to anoxygenic phototrophs and characterized by a site of saturation not found in other chlorophylls (Figure 13-6, arrow). This and other structural changes shift the absorption maxima of bacteriochlorophylls toward the near-ultraviolet and far-red regions of the spectrum. While this enables such organisms to avoid competition with plants, green algae, and cyanobacteria, it also prevents them from exploiting water as an electron donor for photosynthesis. Unlike other chlorophylls, bacteriochlorophylls are not strong enough oxidants to accept electrons from water.

Accessory Pigments Expand Access to Solar Energy

Most photosynthetic organisms also contain **accessory pigments**, which absorb photons that cannot be captured by chlorophyll. This feature enables organisms to collect energy from a much larger portion of the sunlight reaching the Earth's surface, as shown in Figure 13-5. Two types of accessory pigments are **carotenoids** and **phycobilins.** Two carotenoids that are abundant in the thylakoid membranes of most plants and green algae are *β-carotene* and *lutein.* When sufficiently abundant and not masked by chlorophyll, these pigments give a yellow or orange tint to leaves. With absorption maxima between 420 and 480 nm, carotenoids absorb photons from a broader range of blue wavelengths of light than the various forms of chlorophyll do. Phycobilins are found only in red algae and cyanobacteria. Two common examples are *phycoerythrin* and *phycocyanin.* Phycoerythrin absorbs photons from the blue, green, and yellow regions of the spectrum, enabling red algae to utilize the dim light that penetrates an ocean's surface water. Phycocyanin, on the other hand, absorbs photons from the orange region of the spectrum and is characteristic of cyanobacteria living close to the surface of a lake or on land. The adaptation of phototrophs to specific conditions of illumination often reflects differences in the amounts and properties of accessory pigments they contain.

Photosystems and Light-Harvesting Complexes

Chlorophyll molecules, accessory pigments, and associated proteins are organized into functional units called **photosystems**, which are localized to thylakoid or photosynthetic bacterial membranes. The chlorophyll molecules are anchored to the membranes by long hydrophobic side chains, while *chlorophyll-binding proteins* stabilize the arrangement of the chlorophyll within a photosystem and modify the absorption spectra of specific chlorophyll molecules. Other proteins bind components of electron transport systems or catalyze oxidation-reduction reactions.

Most of a photosystem's pigments serve solely as light-gathering **antenna pigments**, absorbing photons and passing the energy to a neighboring chlorophyll molecule or accessory pigment by resonance transfer. The photochemical events that drive electron flow and proton pumping do not begin until energy reaches the **reaction center** of a photosystem, where two special chlorophyll molecules are localized. Only these special chlorophyll molecules can catalyze the conversion of solar energy to chemical energy.

The absorption of a photon and the transfer of energy to the reaction center of a photosystem are illustrated in Figure 13-7. Each chlorophyll molecule and accessory pigment within a photosystem has a characteristic absorption maximum, depending on its structure and immediate chemical environment. Following the laws of thermodynamics, electrons always flow exergonically from antenna pigments that absorb light of the shortest wavelengths (i.e., those with electrons that can be excited by the largest quanta of energy) to those that absorb light of the longest wavelengths (i.e., those with electrons that can be excited by the smallest quanta of energy). Thus, pigments are arranged to funnel energy toward the reaction center of a photosystem. The chlorophyll molecules within the reaction center absorb light of the longest wavelength and therefore act as a *sink.* Within a single nanosecond after absorption of a photon by any pigment within a photosystem, an electron at the reac-

Figure 13-7   The Transfer of Energy to the Reaction Center of a Photosystem.
Light energy absorbed by antenna pigments is passed from one molecule to another by resonance transfer until it reaches a special chlorophyll molecule at the reaction center. Energy is captured when a photoexcited electron is transferred to a specific organic acceptor molecule.

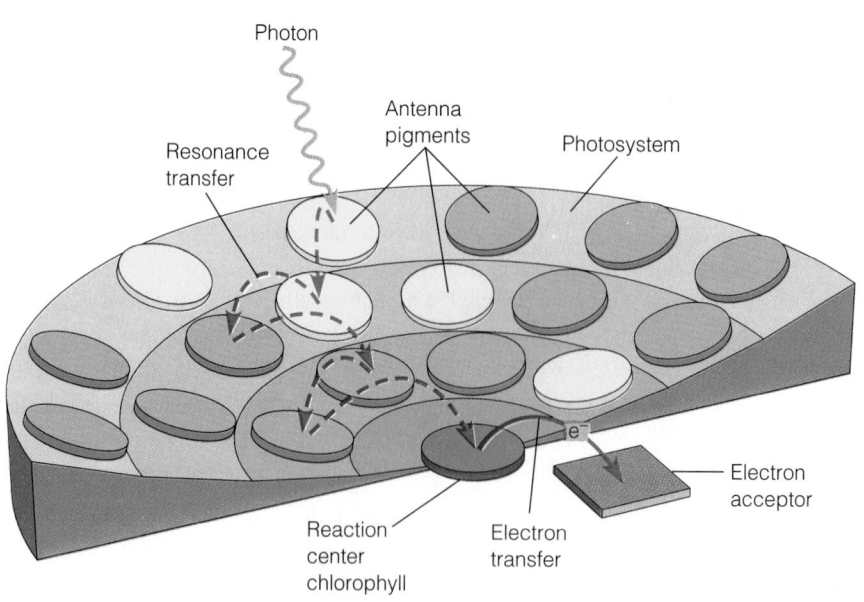

tion center is excited and passed to an acceptor molecule positioned at the "top" of an electron transport system. The oxidized chlorophyll, with a positive charge referred to as an *electron hole*, is then a strong oxidant, ready to accept an electron from a suitable donor.

A photosystem is generally associated with one or more **light-harvesting complexes** (**LHCs**), which, like photosystems, collect light energy. However, LHCs do not contain reaction centers; instead, they pass the energy they collect to a nearby photosystem by resonance transfer. Plants and green algae have LHCs composed of about 100–250 chlorophyll *a* and *b* molecules, along with carotenoids and pigment-binding proteins. Red algae and cyanobacteria have a different type of light-harvesting complex, called a **phycobilisome**, which contains phycobilins rather than chlorophyll and carotenoids. Together, a photosystem and its associated LHCs are referred to as a **photosystem complex**.

Oxygenic Phototrophs Have Two Types of Photosystems

In the 1940s, Robert Emerson and his colleagues at the University of Illinois uncovered evidence for the existence of two separate photoreactions during photoreduction in oxygenic phototrophs. Initially, they observed a dramatic drop, at a wavelength of about 690 nm, in the **action spectrum** for photosynthesis by the green alga *Chlorella* (Figure 13-8). (An action spectrum for photosynthesis depicts the relationship between photosynthetic efficiency and wavelength of light.) Emerson's group considered this *red drop* odd because *Chlorella* actually contains chlorophyll molecules that absorb a significant amount of light at wavelengths above 690 nm. When they supplemented the longer wavelengths of light with a shorter wavelength (about 650 nm), the red drop was less severe, as shown in Figure 13-8. Indeed, photo-

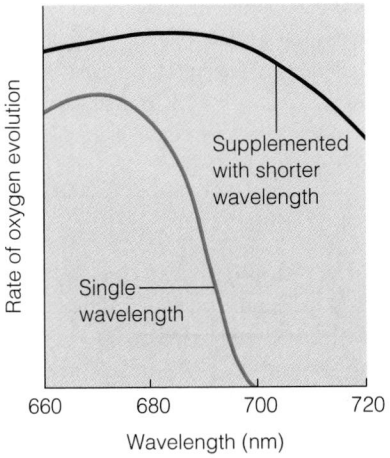

Figure 13-8 The Emerson Enhancement Effect. When cells of the green alga *Chlorella* are illuminated by light of a single wavelength, there is a dramatic drop in the action spectrum (measured by oxygen evolution) above a wavelength of about 690 nm. This red drop is less severe when long wavelengths of light are supplemented with a shorter wavelength.

synthesis driven by a combination of short and long wavelengths of red light exceeded the sum of activities obtained with either wavelength alone. This synergistic phenomenon became known as the **Emerson enhancement effect**.

We now know that the Emerson enhancement effect in oxygenic phototrophs is the result of two distinct photosystems working in concert. **Photosystem I** (**PS I**), with an absorption maximum of 700 nm, absorbs both short and long wavelengths of red light, while **photosystem II** (**PS II**), with an absorption maximum of 680 nm, absorbs only short wavelengths of red light. As we are about to see, each electron that passes from water to NADP⁺ must be photoexcited twice, once by PS I and once by PS II. When illumination is restricted to wavelengths above 690 nm, photosystem II is not active and photosynthesis is severely impaired. As an indication of their specific absorption maxima, the special chlorophyll molecules within the reaction centers of PS I and PS II are designated **P700** and **P680**, respectively.

Photoreduction (NADPH Synthesis) in Oxygenic Phototrophs

Two distinct photosystems in oxygenic phototrophs are apparently essential for the efficient conservation of energy as photoexcited electrons, donated by water, are transferred to NADP⁺ to form NADPH. As shown in Figure 13-9, the complete photoreduction pathway includes several components. Absorption of light energy by each photosystem boosts electrons to the "top" of an electron transport system. As the electrons flow exergonically from photosystem II, a portion of their energy is conserved in an electrochemical proton gradient across the thylakoid (or cyanobacterial) membrane. From photosystem I, electrons are passed to *ferredoxin (Fd)* and then to **nicotinamide adenine dinucleotide phosphate** (**NADP⁺**), conserving reducing power in the form of NADPH. As indicated in Figure 13-10 (page 383), NADP⁺ differs from NAD⁺ by having an additional phosphate group attached to its adenosine. NADP⁺ is the coenzyme of choice for a large number of anabolic pathways, while NAD⁺ tends to be involved in catabolic pathways.

The Photosystem II Complex and the Oxidation of Water

Our detailed tour of the photoreduction pathway begins at photosystem II, which must reduce a *plastoquinone* to a *plastoquinol*, using electrons from water. The PS II reaction center, part of which is shown in Figure 13-11 (page 383), includes two large proteins, D1 and D2, that bind not only chlorophyll but also components of an electron transport system. (Not shown in Figure 13-11 are about 30 additional chlorophyll *a* and a few β-carotene molecules that generally surround the reaction center and enhance photon collection.)

Reduction does not depend on the direct absorption of a photon at the reaction center, or even by a nearby pigment.

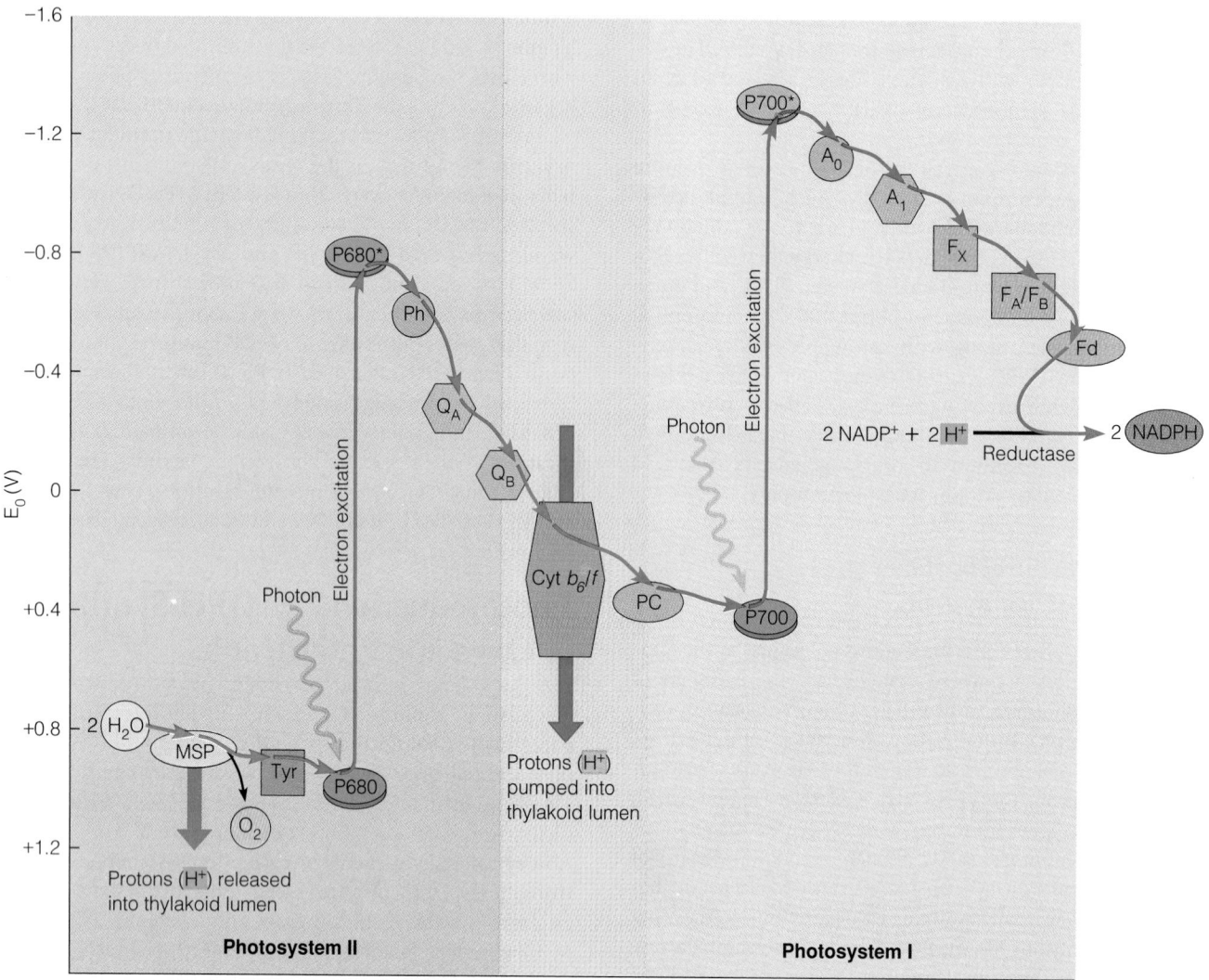

Figure 13-9 Noncyclic Electron Flow in Oxygenic Phototrophs.   Components of the electron transport system that are part of PS II include manganese stabilizing protein (MSP), a tyrosine residue (Tyr) on protein D1 (see Figure 13-11), a pair of special chlorophyll *a* molecules (P680), pheophytin (Ph), and two plastoquinones (Q_A and Q_B). Components of PS I include a pair of special chlorophyll *a* molecules (P700), a modified chlorophyll *a* molecule (A_0), phylloquinone (A_1), and three iron-sulfur centers (F_X, F_A, and F_B). The photosystems are linked by a cytochrome b_6/f complex, which couples electron transport (the path highlighted in pink) to the pumping of protons into the thylakoid lumen, and by plastocyanin (PC). P680 and P700 are unusual because the absorption of photons alters their redox potentials (vertical axis), enabling them to accept electrons from a donor with a highly positive redox potential and donate electrons to an acceptor with a highly negative redox potential, thereby capturing energy. A reductase catalyzes the transfer of electrons from ferredoxin (Fd) to $NADP^+$. For a model showing the orientation of the various components described here within a thylakoid membrane, see Figure 13-11.

In plants and green algae, photosystem II may be associated with a specific light-harvesting complex, **LHC-II,** containing about 250 chlorophyll and numerous carotenoid molecules. Energy captured by antenna pigments of PS II or of an associated LHC-II is funneled to the reaction center by resonance transfer (see Figure 13-7). When the energy reaches the reaction center, it lowers the reduction potential of a P680 molecule to about -0.80 V (see Figure 13-9). A photoexcited electron is then passed to *pheophytin (Ph)* in an oxidation-reduction reaction. Pheophytin is a modified chlorophyll *a* molecule with two protons in place of the magnesium ion. The charge separation between oxidized

P680 and reduced pheophytin traps the electron, preventing it from returning to its ground state in the P680 molecule.

Next, the electron is passed to Q_A, a specific **plastoquinone** tightly bound to protein D2 (see Figure 13-11). Plastoquinone is similar to *coenzyme* Q, a component of the mitochondrial electron transport system described in Chapter 12. Another plastoquinone, Q_B, is tightly bound to protein D1, but only until it receives two electrons, one at time, from Q_A. Reduced Q_B picks up two protons from the stroma and joins a pool of **plastoquinol**, Q_BH_2, in the lipid phase of the photosynthetic membrane. Q_BH_2, which is a mobile electron carrier, can then convey two electrons and two pro-

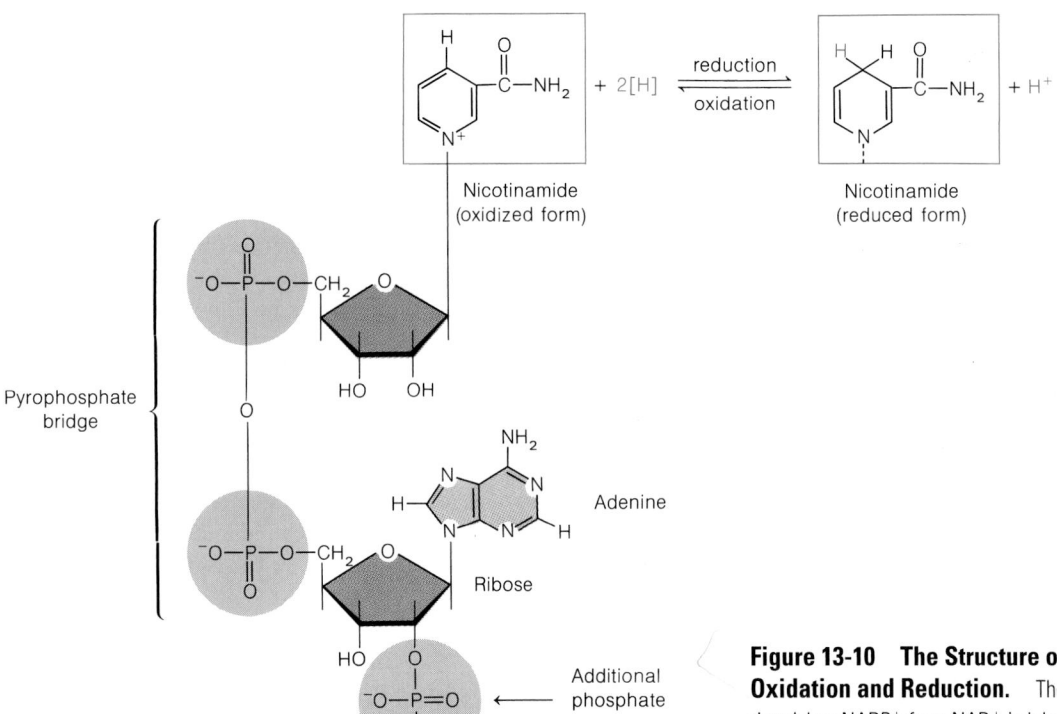

Figure 13-10 The Structure of NADP⁺ and Its Oxidation and Reduction. The extra phosphate that distinguishes NADP⁺ from NAD⁺ is labeled. NADPH is the coenzyme of choice in most reductive reactions in anabolic pathways.

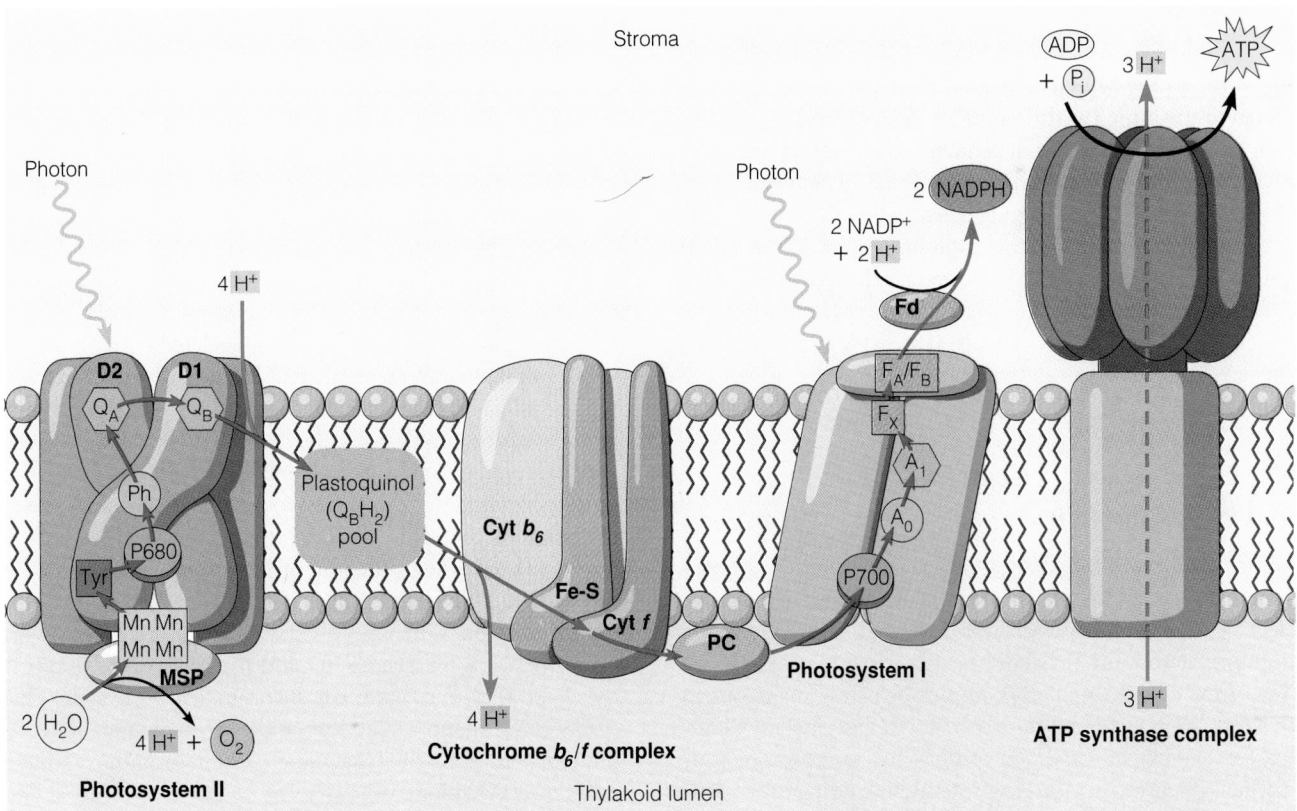

Figure 13-11 A Model of the Orientation of the Major Energy Transduction Complexes Within the Thylakoid Membrane. This diagram shows the possible arrangement of PS II, the cytochrome b_6/F complex, PSI, and the ATP synthase complex within the thylakoid membrane. The path highlighted in pink depicts the probable flow of electrons from H_2O to NADP⁺. PS II includes at least two major proteins, D1 and D2, that bind components of the electron transport system. The cluster of four manganese atoms at the base of PS II are part of the water-splitting complex. The cytochrome b_6/f complex includes two cytochromes (b_6 and f) and an iron-sulfur protein (Fe-S). (See Figure 13-9 for a general key to symbols.) The stoichiometry shown is for the passage of four electrons from H_2O to NADPH.

tons to the *cytochrome b_6/f complex*. Because a chlorophyll molecule catalyzes the transfer of only one electron per photon absorbed, the formation of one mobile plastoquinol molecule depends on two sequential photoreactions at the same reaction center:

$$2 \text{ photons} + Q_B + 2H^+_{\text{stroma}} \longrightarrow Q_BH_2 \qquad \textbf{(13-3)}$$

To absorb photons repeatedly, oxidized P680 must be reduced each time an electron is lost to plastoquinone. Photosystem II includes a **water-splitting complex**, an assembly of proteins and manganese ions that catalyzes the oxidation of water to molecular oxygen. Two water molecules donate four electrons, one at a time, by way of a tyrosine on protein D1, to oxidized P680 (see Figures 13-9 and 13-11). In the process, four protons and one oxygen molecule are released in the thylakoid lumen. The protons contribute to an electrochemical proton gradient across the membrane, and oxygen diffuses out of the chloroplast (or cyanobacterium). Because the complete oxidation of two water molecules to molecular oxygen depends on four photoreactions, the net reaction catalyzed by PS II can be summarized as:

$$4 \text{ photons} + 2H_2O + 2Q_B + 4H^+_{\text{stroma}} \longrightarrow$$
$$4H^+_{\text{lumen}} + O_2 + 2Q_BH_2 \qquad \textbf{(13-4)}$$

The protons removed from the stroma are actually still in transit to the lumen as part of Q_BH_2, while the protons added to the lumen at this point are derived from the oxidation of water. The light-dependent oxidation of water to protons and molecular oxygen, called *water photolysis*, probably appeared in cyanobacteria between 2 and 3 billion years ago, thereby permitting the exploitation of water as an abundant electron donor. The oxygen released by the process dramatically changed the Earth's early atmosphere, which did not originally contain free oxygen, and allowed the development of aerobic respiration.

The Cytochrome b_6/f Complex and Proton Pumping

Before electrons carried by Q_BH_2 are delivered to photosystem I for another boost in energy, they flow through an electron transport system coupled to unidirectional proton pumping across the thylakoid membrane. This happens by way of the **cytochrome b_6/f complex**, which is analogous to the mitochondrial *respiratory complex III* described in Chapter 12. The cytochrome b_6/f complex is composed of four distinct integral membrane proteins, including two cytochromes and an iron-sulfur protein.

Q_BH_2 diffuses within the lipid phase of the thylakoid membrane from PS II to a cytochrome b_6/f complex. There, it donates two electrons, one at a time, via *cytochrome b_6* or the iron-sulfur protein, to *cytochrome f*. Oxidation of Q_BH_2 releases two protons into the thylakoid lumen and enables

Q_B to return to the pool of plastoquinone available for accepting electrons from photosystem II.*

Cytochrome *f* donates electrons to a copper-containing protein called **plastocyanin (PC)**. Plastocyanin, like plastoquinol, is a mobile electron carrier. Unlike plastoquinol, however, it is a peripheral membrane protein found in the thylakoid lumen and carries only one electron at a time. Starting with the two Q_BH_2 molecules generated at photosystem II (see reaction 13-4), the net reaction catalyzed by the cytochrome b_6/f complex can be summarized as:

$$2Q_BH_2 + 4 \text{ plastocyanin (Cu}^{2+}) \longrightarrow$$
$$2Q_B + 4H^+_{\text{lumen}} + 4 \text{ plastocyanin (Cu}^+) \qquad \textbf{(13-5)}$$

Thus, four photoreactions at photosystem II add a total of eight protons to the thylakoid lumen: four from the oxidation of water to oxygen and four carried to the lumen by two plastoquinol molecules. Each reduced plastocyanin now diffuses to a PS I complex, where it can donate an electron to an oxidized P700 molecule.

The Photosystem I Complex and the Reduction of Ferredoxin

The task of photosystem I is to transfer photoexcited electrons to *ferredoxin*, using reduced plastocyanin as an electron donor. As indicated in Figure 13-11, the reaction center of PS I includes a third chlorophyll *a* molecule, referred to as A_0, instead of a pheophytin molecule. Other components of the reaction center include a *phylloquinone* (A_1) and three *iron-sulfur centers* that form an electron transport system linking A_0 to ferredoxin. As in photosystem II, the reaction center of photosystem I is surrounded by additional pigment molecules—between 50 and 100 chlorophyll and several β-carotene molecules—to enhance photon collection.

The light-harvesting complex associated with photosystem I in plants and green algae, **LHC-I**, contains fewer antenna molecules than LHC-II does—about 100 chlorophyll and a few carotenoid molecules. As in the PS II complex, energy captured by antenna pigments in the photosystem or in an associated LHC is funneled to a chlorophyll molecule at the reaction center. In photosystem I, however, the energy lowers the reduction potential of the reaction center chlorophyll, P700, to about -1.30 V (see Figure 13-9). A photoexcited electron is then rapidly passed to A_0, and charge separation between oxidized P700 and the reduced A_0 prevents the electron from returning to its original ground state. This oxidation-reduction reaction and the analogous one in photosystem II are the true *photoreactions* of photosynthetic energy transduction.

*According to several models, the oxidation of Q_BH_2 is actually accompanied by the transfer of four protons into the lumen by a mechanism called *Q-cycling*. The evidence is equivocal, however, so we will leave this mechanism out of our discussion.

From A_0, electrons flow exergonically through the electron transport system to ferredoxin, the final electron acceptor for photosystem I. **Ferredoxin** is a mobile iron-sulfur protein found in the chloroplast stroma and, as we will see later, an important reductant in several metabolic pathways, including those for nitrogen and sulfur assimilation. To transfer electrons repeatedly to ferredoxin, oxidized P700 must be reduced each time an electron is lost. The donor, of course, is reduced plastocyanin. Starting with the four reduced plastocyanin molecules generated at the cytochrome b_6/f complex (reaction 13-5), the net reaction catalyzed by PS I can be summarized as:

$$4 \text{ photons } + 4 \text{ plastocyanin (Cu}^+) + 4 \text{ ferredoxin (Fe}^{3+}) \longrightarrow$$
$$4 \text{ plastocyanin (Cu}^{2+}) + 4 \text{ ferredoxin (Fe}^{2+}) \qquad \textbf{(13-6)}$$

Ferredoxin-NADP$^+$ Reductase

The final step in the photoreduction pathway is the transfer of electrons from ferredoxin to NADP$^+$, thereby providing the NADPH essential for photosynthetic carbon assimilation. This transfer is catalyzed by the enzyme **ferredoxin-NADP$^+$ reductase,** a peripheral membrane protein found on the stromal side of the thylakoid membrane. Starting with the four reduced ferredoxin generated at photosystem I (see reaction 13-6), we obtain the following:

$$4 \text{ ferredoxin (Fe}^{2+}) + 2\text{NADP}^+ + 2\text{H}^+_{\text{stroma}} \longrightarrow$$
$$4 \text{ ferredoxin (Fe}^{3+}) + 2\text{NADPH} \qquad \textbf{(13-7)}$$

Notice that the reduction of one NADP$^+$ molecule consumes two electrons, each from a single reduced ferredoxin. While not actually moving protons from the stroma to the lumen, this reaction nonetheless raises the pH of the stroma and contributes to the electrochemical proton gradient across the thylakoid membrane.

Functioning together within the chloroplast, the various components of the electron transport system provide a continuous, unidirectional flow of electrons from water to NADP$^+$, as indicated in Figures 13-9 and 13-11. This is referred to as **noncyclic electron flow,** primarily to distinguish it from the *cyclic electron flow* we will encounter shortly. The net result of the complete light-dependent oxidation of two water molecules to molecular oxygen can be obtained by summing reactions 13-4 through 13-7, assuming ferredoxin does not donate electrons to other pathways:

$$4 \text{ photons at PS II} + 4 \text{ photons at PS I} + 2\text{H}_2\text{O} + 6\text{H}^+_{\text{stroma}}$$
$$+ 2\text{NADP}^+ \longrightarrow 8\text{H}^+_{\text{lumen}} + \text{O}_2 + 2\text{NADPH} \qquad \textbf{(13-8)}$$

For every eight photons absorbed, two NADPH molecules are generated. Furthermore, the electron transport system for photoreduction is coupled to a mechanism for unidirectional proton pumping across the thylakoid membrane from the stroma to the lumen. Thus, solar energy is captured and stored in two forms: the reductant NADPH and an electrochemical proton gradient.

Photophosphorylation (ATP Synthesis) in Oxygenic Phototrophs

Because the thylakoid membrane is virtually impermeable to protons, a substantial electrochemical proton gradient can develop across the membrane in illuminated chloroplasts. It is not unusual, for example, for the thylakoid lumen to have a pH of 4.5, while the stroma pH is in the range 8.0–8.5. As described in Chapter 12 for mitochondria, we can calculate the total proton motive force across the thylakoid membrane, or the equivalent free-energy change ($\Delta G'$), for the movement of protons from the lumen to the stroma by summing a concentration (pH) term and a membrane potential (V_m) term (see equation 12-14). In mitochondria, the pH difference across the inner membrane is only about 1.0–1.5 units, so the membrane potential term is usually more significant than the concentration term. In chloroplasts, however, magnesium ions readily diffuse from the lumen to the stroma, maintaining charge neutrality across the thylakoid membrane. Thus, nearly all of the proton motive force is due to the light-induced pH gradient, which can be as steep as 3.5 or 4 units. This difference corresponds to a $\Delta G'$ value of about -5.1 kcal/mol of protons moving from the lumen to the stroma.

The ATP Synthase Complex

In chloroplasts, as in mitochondria, the movement of protons from an acidic to a basic compartment is coupled to the synthesis of ATP by an **ATP synthase complex**. The ATP synthase complexes found in chloroplasts, also designated **CF$_o$CF$_1$ complexes,** are remarkably similar to the F$_o$F$_1$ complexes of mitochondria described in Chapter 12. The **CF$_1$** component is a hydrophilic assembly of polypeptides that protrudes from the stromal side of the thylakoid membrane and contains the catalytic site for ATP synthesis. Like the mitochondrial F$_1$ and its stalk, CF$_1$ is composed of five distinct polypeptides with a stoichiometry of $\alpha_3\beta_3\gamma\delta\epsilon$.

The other half of the chloroplast ATP synthase complex, the **CF$_o$** component, is an assembly of polypeptides embedded in the thylakoid membrane, where it forms a hydrophilic channel between the lumen and the stroma. CF$_o$ is the **proton translocator** through which protons flow under the "pressure" of the proton motive force. How this movement is coupled to ATP synthesis by the CF$_1$ component is not clearly understood, but it appears that about one ATP molecule is generated for every three protons that pass through the channel. The reaction catalyzed by the ATP synthase complex can therefore be summarized as:

$$3\text{H}^+_{\text{lumen}} + \text{ADP} + \text{P}_i \longrightarrow 3\text{H}^+_{\text{stroma}} + \text{ATP} + \text{H}_2\text{O} \qquad \textbf{(13-9)}$$

This is quite reasonable in terms of thermodynamics. The $\Delta G'$ value for the synthesis of ATP within the chloroplast stroma is usually 10–14 kcal/mol, while the energy available is about 15 kcal per 3 moles of protons passing through the ATP synthase complex (5.1 kcal/mol × 3 mol = 15.3 kcal). Thus, the flow of four electrons through the noncyclic pathway shown in Figures 13-9 and 13-11 not only generates two NADPH molecules but leads indirectly to the formation of almost three ATP molecules from ADP and P_i (4 electrons × 2 protons/electron × 1 ATP/3 protons = 2.7 ATP).

Recall from Chapter 6 that enzymes change only the *rate* at which equilibrium between the reactants and products of a chemical reaction is achieved; they have no effect on the *position* of the equilibrium. Within an illuminated chloroplast, the net free energy change favors ATP synthesis; net $\Delta G'$ is between −1 and −5 kcal/mol. But what happens in the dark, when the light-induced proton gradient is no longer maintained? If the CF_oCF_1 complex remained active, it would become an *ATPase*, consuming ATP as it actively transported protons into the lumen. Not surprisingly, chloroplasts have developed systems for regulating ATP synthase.

Independent of the flow of protons through the CF_o component, catalytic activity of the chloroplast ATP synthase appears to depend on the actual establishment of a proton gradient across the thylakoid membrane. In the dark, deterioration of the light-induced proton gradient substantially reduces enzyme activity but does not completely inactivate it. A second mechanism, to be described later, links the activation of ATP synthase to the presence of reduced ferredoxin in the stroma. In the absence of photoreduction, there is not enough reduced ferredoxin to maintain the chloroplast ATP synthase in an active state.

Cyclic Photophosphorylation

Having seen how NADPH and ATP are generated by light-driven electron flow, we are near the end of our discussion of the photosynthetic energy transduction reactions. But before we move on to carbon assimilation, we must consider how phototrophs might balance NADPH and ATP synthesis to meet the precise energy needs of a living cell. Notice that noncyclic electron flow leads to the generation of slightly less than three ATP for every two NADPH molecules. Because both compounds are consumed by a variety of metabolic pathways, which may be more or less active at any given time, it is very unlikely that a photosynthetic cell will always require ATP and NADPH in the precise ratio generated by noncyclic electron flow.

When NADPH consumption is low, an optional pathway may divert the reducing power generated at photosystem I into ATP synthesis rather than NADP⁺ reduction. As shown in Figure 13-12, the reduced ferredoxin generated by photosystem I can return electrons to a cytochrome b_6/f complex instead of donating them to NADP⁺. By a mechanism that probably involves plastoquinol, the exergonic flow of electrons from reduced ferredoxin to plastocyanin is coupled to proton pumping, thereby contributing to the proton motive force across the thylakoid membrane. At least three protons are moved from the stroma to the lumen for every

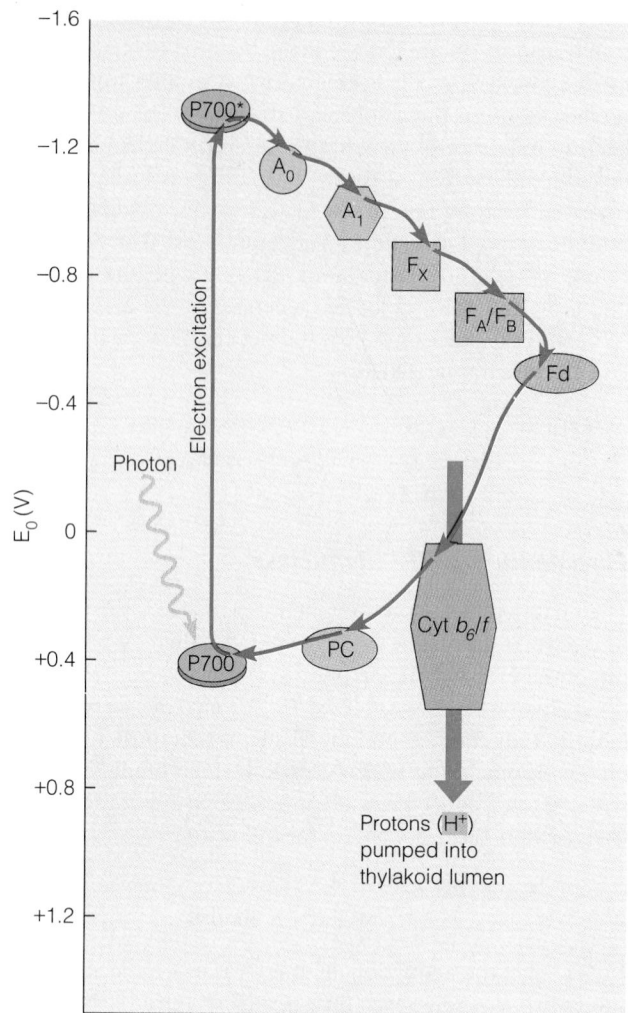

Figure 13-12 Cyclic Electron Flow. Cyclic electron flow through photosystem I enables oxygenic phototrophs to regulate the ratio of ATP to NADPH within photosynthetic cells. When the concentration of NADP⁺ is low (i.e., when the concentration of NADPH is high), ferredoxin (Fd) donates electrons to the cytochrome b_6/f complex. Electrons then return to P700 via plastocyanin (PC). Because this cyclic electron flow is coupled to unidirectional proton pumping across the thylakoid membrane, excess reducing power is channeled into ATP synthesis. (See Figure 13-9 for a general key to symbols.)

two electrons, but the precise number is not known. From plastocyanin, electrons return to an oxidized P700 molecule in photosystem I, completing a closed circuit and allowing P700 to absorb another photon. This is referred to as **cyclic electron flow,** and the ATP synthesis that it supports is called *cyclic photophosphorylation*. No water is oxidized and no oxygen is released, because the flow of electrons from photosystem II is bypassed.

The Complete Energy Transduction System

The model shown in Figure 13-11 represents the entire photosynthetic energy transduction system within a thylakoid membrane. The essential features of the complete system for the transfer of electrons from water to $NADP^+$ concomitant with ATP synthesis can be summarized in terms of the following component parts:

1. *Photosystem II complex:* An assembly of chlorophyll molecules, accessory pigments, and proteins, with a sufficiently positive reduction potential to allow a reaction center chlorophyll to accept electrons exergonically from water ($E_0 = +0.816$ V). Photoexcited from about $+0.9$ to -0.8 V by a single photon, photosystem II has a sufficiently negative reduction potential to donate electrons to plastoquinone ($E_0 \approx 0.1$ V), a mobile electron carrier.

2. *Photosystem I complex:* A second assembly of chlorophyll molecules, accessory pigments, and proteins, with a sufficiently positive reduction potential to allow a reaction center chlorophyll to accept electrons exergonically from plastocyanin ($E_0 = +0.360$ V). Photoexcited from about $+0.48$ to -1.3 V by a single photon, photosystem I has a sufficiently negative reduction potential to donate electrons to ferredoxin ($E_0 = -0.420$ V), a stromal protein.

3. *Cytochrome b_6/f complex:* An electron transport system linking plastoquinol, the electron shuttle for photosystem II, with plastocyanin, the electron shuttle for photosystem I. The flow of electrons through the cytochrome b_6/f complex is coupled to unidirectional proton pumping across the thylakoid membrane, thereby establishing and maintaining an electrochemical proton gradient. An option for cyclic electron flow allows a cytochrome b_6/f complex to accept electrons from ferredoxin as well as plastoquinol, allowing for the generation of additional ATP, as needed.

4. *Ferredoxin-$NADP^+$ reductase:* An enzyme on the stromal side of the thylakoid membrane that catalyzes the transfer of electrons from two reduced ferredoxin proteins to a single $NADP^+$ molecule ($E_0 = -0.32$ V). The NADPH

generated is an essential reducing agent in anabolic pathways, particularly carbon assimilation.

5. *ATP synthase complex (also called the CF_oCF_1 complex):* A proton channel and ATP synthase that couples the exergonic flow of protons, from the thylakoid lumen to the stroma, to the synthesis of ATP from ADP and P_i. Like NADPH, the ATP accumulates in the stroma, where it provides energy for carbon assimilation.

Within a chloroplast, both noncyclic and cyclic pathways of electron flow are operative, thereby providing much flexibility in the relative amounts of ATP and NADPH that are generated. ATP can be produced on a close to equimolar basis with respect to NADPH if the noncyclic pathway is operating alone, or it can be generated in any desired excess simply by shunting more and more of the electrons from ferredoxin to the cyclic rather than the noncyclic side of photosystem I.

Localization of Complexes in the Thylakoid Membrane

The complexes we just described are not necessarily contiguous within a membrane. In fact, components appear to be segregated to regions of the thylakoids, as illustrated in Figure 13-13. Photosystem II complexes are localized primarily in appressed regions of stacked thylakoids (grana), while photosystem I and ATP synthase complexes are localized exclusively in nonappressed regions of stacked thylakoids and in stroma thylakoids. This allows NADPH and ATP to be generated on the stromal side of the membrane. Cytochrome b_6/f complexes are present in roughly equal concentration in both stacked and stroma thylakoids. Plastoquinol and plastocyanin, of course, are mobile electron carriers that can diffuse over long distances, linking the non-

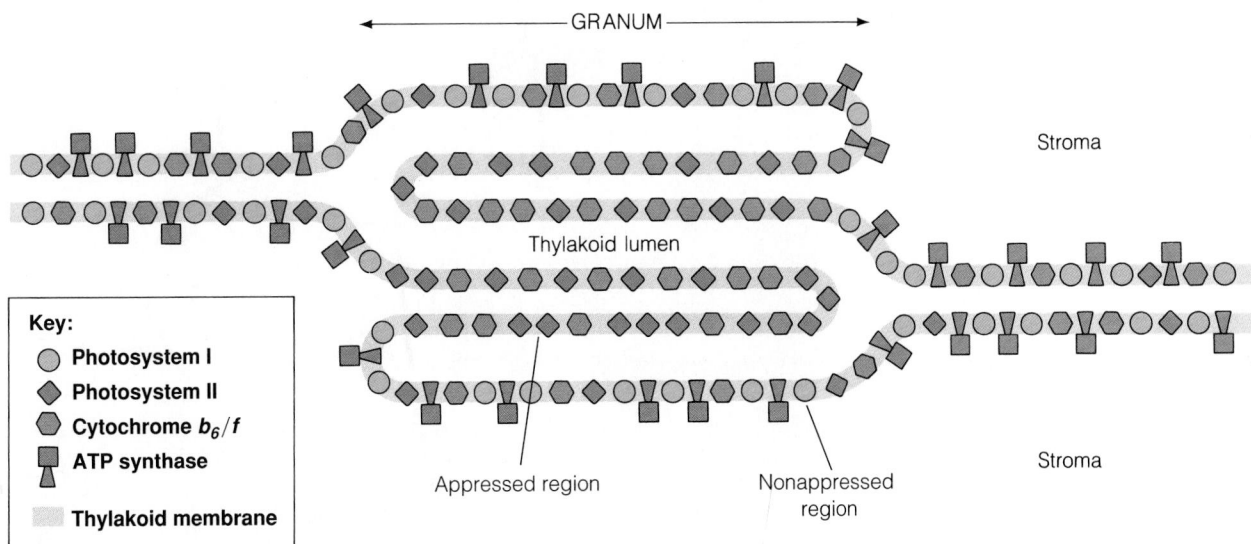

Figure 13-13   Localization of the Four Major Complexes of Photosynthetic Energy Transduction. The complexes are localized to different regions of the stacked thylakoids (grana) and stroma thy-

lakoids. The noncontiguous components are linked by plastoquinol, which is soluble in the lipid phase of the thylakoid membrane, and plastocyanin, which is a peripheral membrane protein in the lumen.

contiguous components into a complete system. As already noted, plastoquinol is confined to the lipid phase of the thylakoid membrane, while plastocyanin is a peripheral membrane protein on the lumen side.

A Photosynthetic Reaction Center from a Purple Bacterium

Much of our knowledge of photosynthetic reaction centers and the chemistry of capturing light energy has come from studies of reaction center complexes isolated from photosynthetic bacteria. The stage was set in the late 1960s when Roderick Clayton successfully purified the reaction center complex of a purple bacterium, *Rhodopseudomonas sphaeroides*. More recently, Hartmut Michel, Johann Deisenhofer, and Robert Huber crystallized a reaction center complex from a different purple bacterium, *Rhodopseudomonas*

viridis, and determined its molecular structure by X-ray crystallography. (For a discussion of X-ray crystallography, see the Appendix.) Michel and his colleagues not only provided the first detailed look at how pigment molecules are arranged to capture light energy, they were the first group to crystallize any membrane protein complex at all. For their exciting contributions, Michel, Deisenhofer, and Huber shared a 1988 Nobel Prize.

As shown in Figure 13-14, the reaction center of *R. viridis* includes four protein subunits. The first subunit is a cytochrome *c* molecule bound to the periplasmic surface of the bacterial membrane. (The periplasm is the space between the plasma membrane and cell wall of a bacterium.) The second and third subunits, called L and M, span the membrane and stabilize a total of four bacteriochlorophyll *b* molecules, two bacteriopheophytin molecules, and two quinones. Subunits L and M are homologous to proteins D1 and D2, respectively, of photosystem II in oxygenic phototrophs. A fourth subunit, called H, is bound to the cyto-

Figure 13-14 A Model of the Photosynthetic Reaction Center from a Purple Bacterium. The structure for a reaction center of *Rhodopseudomonas viridis* was deduced from X-ray diffraction measurements. The cytochrome (blue-green) with four heme groups (yellow) lies in the periplasm outside the bacterial plasma membrane. Subunits L and M span the membrane and bind components for light harvesting and electron transfer. Subunit H lies mostly on the cytosolic side of the membrane.

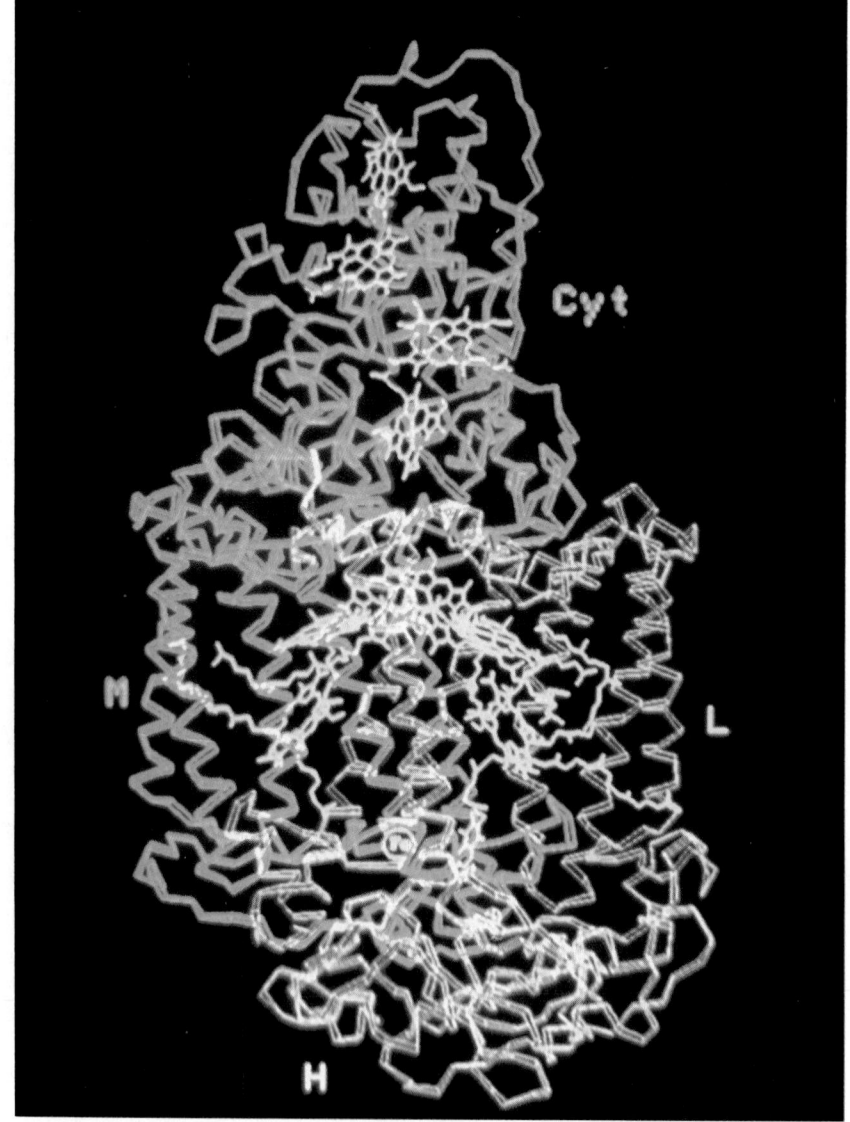

plasmic surface of the membrane and is homologous to a photosystem II protein called CP43. Electron flow through this bacterial photosystem and a second membrane protein complex, the *cytochrome b/c_1 complex*, resembles the flow of electrons through photosystem II and the cytochrome b_6/f complex of oxygenic phototrophs, with one major difference: The bacterial photosystem does not include a water-splitting complex.

Two of the bacteriochlorophyll *b* molecules, each designated *P960* to indicate their absorption maxima, have a direct role in catalyzing the light-dependent transfer of an electron to an electron transport system. Absorption of a photon, either directly or by accessory pigments, lowers the reduction potential of P960 from about $+0.5$ to -0.7 V. The photoexcited electron is immediately transferred to bacteriopheophytin, thereby stabilizing charge separation. From bacteriopheophytin, the electron flows exergonically to a tightly bound quinone and then to a second quinone. After accepting two electrons, the reduced quinone diffuses to a cytochrome b/c_1 complex, which then passes electrons to cytochrome *c*. This event is coupled to unidirectional proton pumping across the bacterial membrane from the cytoplasm to the periplasm. Cytochrome *c* then returns the electron to oxidized P960. As in chloroplasts and mitochondria, the electrochemical proton gradient across the membrane drives an ATP synthase.

Notice that the flow of electrons described above is cyclic, with no net gain of reducing power. How, then, does this particular phototroph generate reductant? In *R. viridis*, the cytochrome b/c_1 complex and cytochrome *c* accept electrons from donors such as hydrogen sulfide, thiosulfate, or succinate. And ATP generated by cyclic photophosphorylation is used to push electrons energetically uphill from the cytochrome b/c_1 complex or cytochrome *c* to NAD^+. The cyclic flow of electrons described above is therefore very different than the cyclic flow through photosystem I in oxygenic phototrophs. The bacterial photosystem is most accurately described as a somewhat simplified version of photosystem II in plants, green algae, and cyanobacteria, the main product of the reaction center being a reduced quinone.

Photosynthetic Carbon Assimilation: The Calvin Cycle

With information about chloroplast structure and photosynthetic energy transduction in mind, we are now prepared to look closely at photosynthetic carbon assimilation. More specifically, we will look at the primary events of carbon assimilation: the initial fixation and reduction of carbon dioxide to form simple three-carbon carbohydrates. As mentioned above, the primary products of carbon assimilation then enter a variety of metabolic pathways, the most important being sucrose and starch synthesis.

The fundamental pathway for the movement of inorganic carbon into the biosphere is the **Calvin cycle,** which is found in all oxygenic and most anoxygenic phototrophs.

The pathway is named after Melvin Calvin, who received a Nobel Prize in 1961 for the work he and his colleagues Andrew Benson and James Bassham did to elucidate the process. The exciting sequence of events that led to their discoveries is described in Box 13A.

In plants and algae, the Calvin cycle is confined to the chloroplast stroma, where the ATP and NADPH generated by photosynthetic energy transduction reactions accumulate. In plants, the carbon dioxide that is to be assimilated generally enters the leaves through special pores called **stomata** (singular: **stoma**), allowing gas exchange between the atmosphere and interior of a leaf. Once inside the leaf, carbon dioxide diffuses into **mesophyll cells** and, in over 95% of plant species, travels unhindered into the chloroplast stroma, where the Calvin cycle begins. In a relatively small number of species, however, the Calvin cycle is preceded by one of two types of preliminary carboxylation/decarboxylation pathways that can enhance photosynthetic efficiency. After we look at the Calvin cycle in detail, we will consider how it may be augmented by additional pathways in some plant species.

For convenience, the Calvin cycle can be considered as a sequence of three stages (Figure 13-15):

1. The carboxylation of the organic acceptor molecule *ribulose-1,5-bisphosphate*, followed immediately by hydrolysis to generate two molecules of *3-phosphoglycerate*.

2. The reduction of 3-phosphoglycerate from the oxidation level of a carboxylic acid to that of an aldehyde, forming the triose phosphate *glyceraldehyde-3-phosphate*.

3. The regeneration of the original acceptor molecule, thereby allowing continuous carbon assimilation.

Because the first detectable product of the Calvin cycle is a three-carbon molecule, 3-phosphoglycerate, an alternative name for this metabolic sequence is the **C_3 pathway**. As we look at each stage, keep in mind that each one is actually an integral part of the complete pathway presented in Figure 13-15 and that all three function continuously in illuminated chloroplasts.

Carboxylation of Ribulose-1,5-Bisphosphate

Our tour of the Calvin cycle begins with the covalent attachment of carbon dioxide to the carbonyl carbon of ribulose-1,5-bisphosphate (reaction CC-1 in Figure 13-15). Carboxylation of this five-carbon carbohydrate should generate a six-carbon product, but such a molecule has never been isolated. As Calvin and his colleagues discovered, the first detectable product of carbon dioxide fixation by this pathway is a three-carbon molecule, 3-phosphoglycerate. Presumably, the six-carbon compound exists only as an enzyme-bound intermediate, and the carboxylation step is followed immediately by hydrolysis of this intermediate to yield two molecules of 3-phosphoglycerate. The newly fixed carbon dioxide, which is highlighted in Figure 13-15, appears as a carboxyl group on one of the two 3-phosphoglycerate molecules.

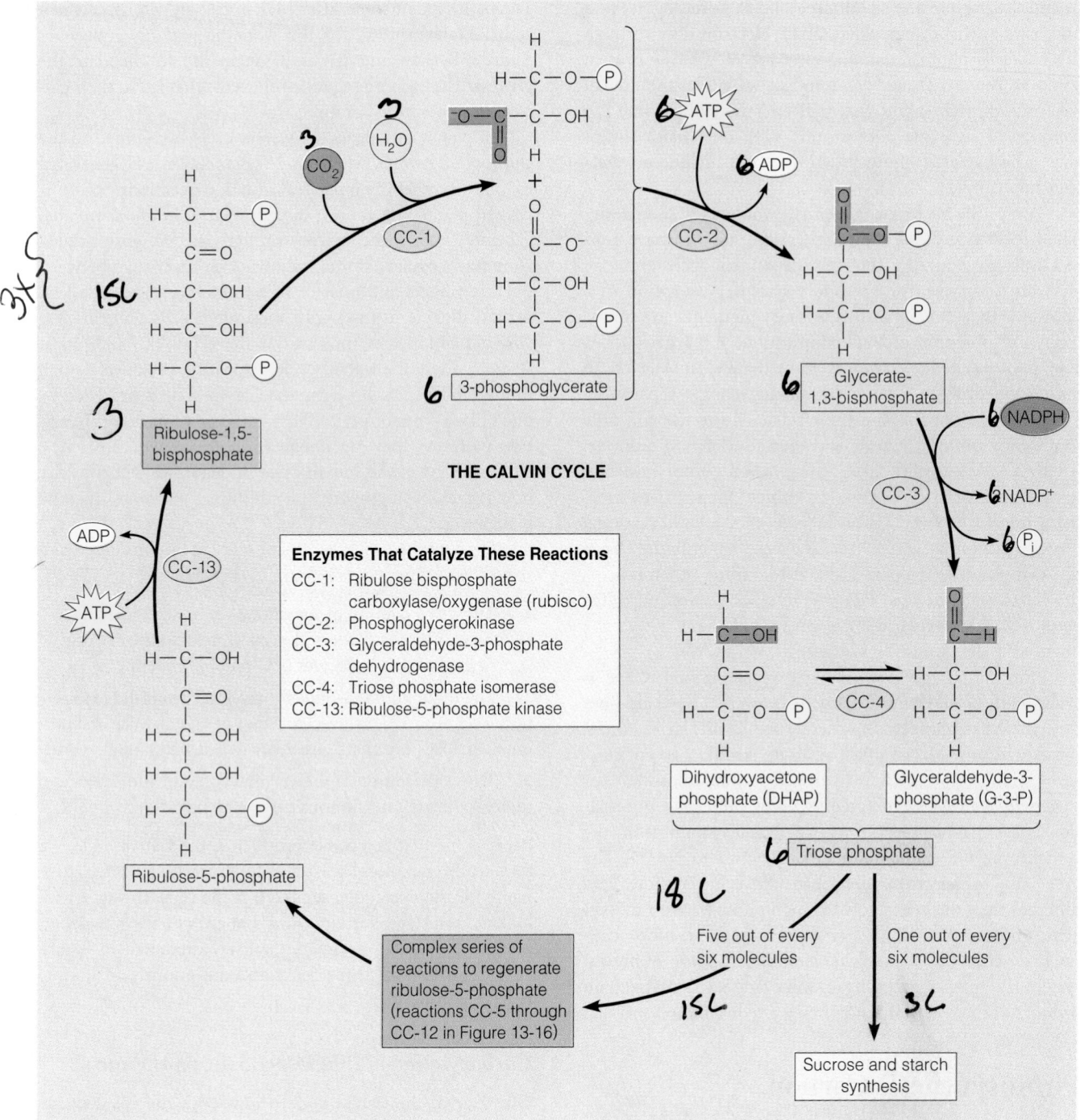

Figure 13-15 The Calvin Cycle for Photosynthetic Carbon Assimilation. Reactions CC-1, CC-2, and CC-3 provide for the initial fixation and reduction of carbon dioxide, generating the interconvertible trioses glyceraldehyde-3-phosphate (G-3-P) and dihydroxyacetone phosphate (DHAP). On the average, one out of every six triose molecules is used in the synthesis of sucrose, starch, or other organic molecules. Five out of every six triose molecules are used to generate ribulose-5-phosphate by the complex series of reactions summarized in Figure 13-16 (reactions CC-5 through CC-12). The ribulose-5-phosphate is then phosphorylated at the expense of ATP in reaction CC-13 to form the acceptor molecule with which the sequence began. The enzymes that catalyze these reactions are identified in the center of the figure.

The enzyme that catalyzes the capture of carbon dioxide and formation of 3-phosphoglycerate is called **ribulose bisphosphate carboxylase/oxygenase** (**rubisco**). This relatively large enzyme (about 560,000 Da) is unique to phototrophs and is found in all photosynthetic organisms except for a few photosynthetic bacteria. Considering its essential role in carbon dioxide fixation for virtually the entire biosphere, it is hardly surprising that rubisco has the distinction of being the most abundant protein on the planet. About 10–25% of soluble leaf protein is rubisco, and one estimate

Historical Perspectives

CARBON-14, PAPER CHROMATOGRAPHY, AND THE CALVIN CYCLE

One of the most reliable marks of scientific genius is the ability to bring new techniques to bear on old problems in a way that pushes the frontiers of science ahead in quantum leaps. An especially good example of this can be seen in the pioneering work of Melvin Calvin and his colleagues at the University of California, Berkeley—work that eventually led to the elucidation of the cycle that now bears Calvin's name. It had long been known that atmospheric carbon dioxide could somehow be incorporated into organic form by green plants, but the mechanism behind that "somehow" was hidden in a maze of enzyme-catalyzed biochemical reactions. This was the problem that Calvin, along with Andrew Benson and James Bassham, set out to solve. One measure of their eventual success was a Nobel Prize, later awarded to Calvin for this work. Aspiring scientists would do well to take note of their approach, for much of their success depended on the ingenuity with which they were able to tackle the old problem of carbon dioxide fixation with several powerful new techniques.

Their story also illustrates the impact that political and social developments have on science, because the success of their approach depended on the use of ^{14}C, a radioactive isotope of carbon that had just become available in the 1940s as a result of research undertaken in connection with World War II. The basic approach in Calvin's laboratory was to expose illuminated cells of the green alga *Chlorella* to $^{14}CO_2$ for a short period of time, and then extract and analyze the various molecules in the algal cells to see in which ones the labeled carbon atoms appeared. For that analysis, they successfully adapted the newly developed technique of two-dimensional *paper chromatography*, which essentially enables researchers to resolve compounds of different solubilities in one or both of two solvent systems from each other, based on their differential mobility as the solvent is allowed to move up a sheet of filter paper. To locate the resolved spots in which the ^{14}C was localized, they used yet another new technique called *autoradiography*. This technique depends on the ability of decaying radioactive atoms

such as ^{14}C to darken an X-ray film if the film is pressed tightly against the paper chromatogram for a sufficient period of time.

Calvin and his colleagues grew the *Chlorella* cells in illuminated chambers into which radioactive carbon dioxide was injected. After the desired incubation period (measured usually in seconds or minutes), they opened a valve at the bottom of the chamber and collected the algae in a beaker of hot alcohol. The alcohol served both to kill the algae and to extract soluble organic compounds. Next, the researchers spotted extracts of the cells onto a sheet of filter paper and separated the various compounds by chromatography, first in one dimension with one solvent system, then in a second dimension (at 90° to the first) with a second solvent system. They exposed the dried chromatogram to X-ray film to locate the radioactive compounds, which were then identified by chemical analysis.

Many of the molecules in the cell were found to be radioactively labeled when the algae were incubated with $^{14}CO_2$ for more than a few minutes. This is hardly surprising because every organic compound in the cell is synthesized from photosynthetically fixed carbon. What they really wanted to know, of course, is which molecules were labeled *first* upon exposure to the $^{14}CO_2$. For this, Calvin and his colleagues used very short incubation times. Even with an incubation time of only 30 seconds, for example, the chromatogram turned out to have a large number of radioactive spots, as shown in Figure 13A-1a. With shorter and shorter incubation times, however, they detected fewer and fewer radioactive compounds (Figure 13A-1b). By 5 seconds, only about a half-dozen spots were found (Figure 13A-1c), and with incubation times of 2 seconds or less, a single labeled compound predominated. This turned out to be 3-phosphoglycerate, a compound that had earlier been identified as one of the intermediates in glycolysis. By painstakingly identifying each of the additional compounds that became labeled with slightly longer incubations and by determining where within each such molecule the labeled carbon was

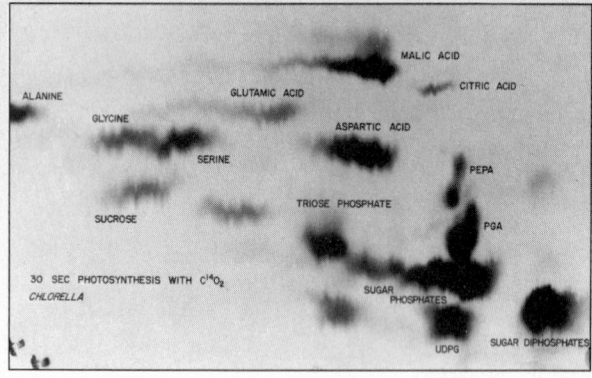

(a) 30 sec incubation period

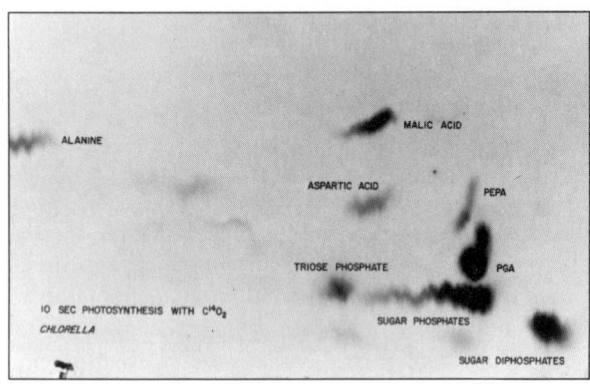

(b) 10 sec incubation period

located, Calvin and his colleagues were eventually able to piece together the complex pathway of photosynthetic carbon metabolism.

Something of the perspiration that inevitably accompanies inspiration in any definition of genius can be sensed in Calvin's own words when he noted, in his Nobel laureate lecture, that the data on which their conclusions were based came mainly from "the number, position, and intensity—that is, radioactivity—of the blackened areas. The paper ordinarily does not print out the names of these compounds, unfortunately, and our principal chore for the succeeding ten years was to properly label those blackened areas on the film" (Calvin, 1962, p. 879).

Perspiration, then, to be sure—ten years of it, by Calvin's report. But inspiration as well, for without the foresight to bring together radioactive carbon, paper chromatography, and X-ray autoradiography, there would have been no blackened areas to label, no pathway to name, and no prize to claim. ∎

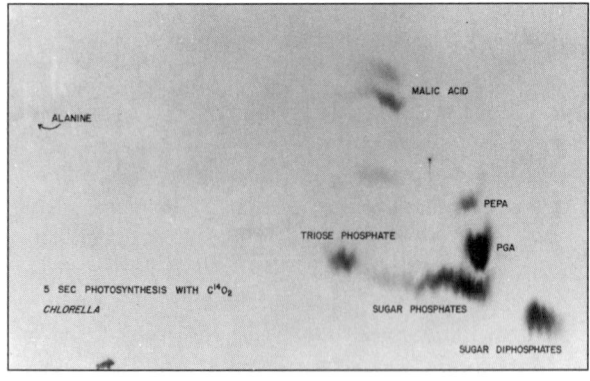

(c) 5 sec incubation period

Figure 13A-1   Autoradiographic Identification of Radioactive Compounds in *Chlorella* Cells After Exposure to
$^{14}CO_2$.   Algal cells were incubated with radioacitve CO_2 (as the bicarbonate ion, $H^{14}CO_3^-$) for **(a)** 30 sec, **(b)** 10 sec, and **(c)** 5 sec and were then killed by being plunged into hot alcohol. Extracts were then chromatographed, and radioactive compounds were localized by exposure of X-ray film to the paper chromatogram. Radioactive compounds were identified by chemical analysis and cochromatography of known compounds.

puts the total amount of rubisco on the Earth at 40 million tons, or almost 15 lbs for each living person! After we look at the next two stages of the Calvin cycle, we will return to rubisco and consider regulation of the enzyme's *carboxylase* activity, as well as problems created by its *oxygenase* activity.

Reduction of 3-Phosphoglycerate

The 3-phosphoglycerate molecules formed upon carbon dioxide fixation are reduced to glyceraldehyde-3-phosphate by a sequence of reactions that is essentially the reverse of the oxidative sequence of glycolysis (reactions Gly-6 and Gly-7 in Figure 11-7), except that the coenzyme involved is NADPH, not NADH. The steps for NADPH-mediated reduction are shown in Figure 13-15 as reactions CC-2 and CC-3. In the first reaction, *phosphoglycerokinase* catalyzes the transfer of a phosphate group from ATP to 3-phosphoglycerate. This reaction generates an activated intermediate, glycerate-1,3-bisphosphate. In the second reaction, *glyceraldehyde-3-phosphate dehydrogenase* catalyzes the transfer of two electrons from NADPH to glycerate-1,3-bisphosphate, reducing the molecule to glyceraldehyde-3-phosphate.

Some accounting is in order at this point. For every carbon dioxide molecule that is fixed by rubisco (reaction CC-1), two 3-phosphoglycerate molecules are generated. The reduction of both of these molecules to glyceraldehyde-3-phosphate (reactions CC-2 and CC-3) requires the hydrolysis of two ATP molecules and the oxidation of two NADPH molecules. This is for a net gain of only one carbon atom, however. The net synthesis of one triose phosphate molecule requires the fixation and reduction of *three* carbon dioxide molecules to maintain carbon balance and will therefore consume *six* ATP and six NADPH:

$$\text{3 ribulose bisphosphate} + 3CO_2 + 3H_2O + 6ATP + 6NADPH \longrightarrow$$
$$\text{6 glyceraldehyde-3-phosphate} + 6ADP + 6P_i + 6NADP^+$$

(13-10)

The glyceraldehyde-3-phosphate produced by reactions CC-1 through CC-3 enters an interconvertible pool of glyceraldehyde-3-phosphate and dihydroxyacetone phosphate. Both types of triose phosphate are required for continuation of the Calvin cycle, as well as for the synthesis of sucrose and starch and for other metabolic pathways. The conversion of glyceraldehyde-3-phosphate to dihydroxyacetone phosphate (reaction CC-4 in Figure 13-15) is catalyzed by *triose phosphate isomerase*, which maintains a balanced pool of the two compounds.

Regeneration of Ribulose-1,5-Bisphosphate

One out of every six triose phosphate molecules—the net gain of the Calvin cycle—is used for the synthesis of sucrose, starch, or other molecules. The five remaining triose phosphates are required within the Calvin cycle for the regeneration of the starting compound. For each triose phosphate that leaves the cycle, a total of three ribulose-1,5-bisphosphate molecules must be regenerated (see reaction 13-10).

The sequence of reactions for converting five three-carbon molecules into three five-carbon molecules is presented in Figure 13-16 as reactions CC-5 through CC-10. This process is accomplished by three types of enzymes. The condensation of dihydroxyacetone phosphate with an aldosugar is catalyzed by *aldolases* (CC-5 and CC-8). One of the phosphate groups on the product of each condensation reaction is then removed by a hydrolytic reaction catalyzed by *phosphatases* (CC-6 and CC-9). And two-carbon units are transferred from ketosugars to aldosugars by *transketolases* (CC-7 and CC-10). The result is three pentose phosphates, but not ribulose-1,5-bisphosphate.

The pentose phosphates, two molecules of *xylulose-5-phosphate* and one molecule of *ribose-5-phosphate*, are readily converted into *ribulose-5-phosphate* by *isomerases* (reactions CC-11 and CC-12). Then, each ribulose-5-phosphate molecule is finally phosphorylated by *ribulose-5-phosphate kinase* to form the ribulose-1,5-bisphosphate that accepts carbon dioxide in reaction CC-1. Thus, regeneration of ribulose-1,5-bisphosphate from dihydroxyacetone phosphate (DHAP) and glyceraldehyde-3-phosphate (G-3-P) consumes three ATP molecules, and the regeneration phase of the Calvin cycle may be summarized as:

$$\text{2 DHAP} + \text{3 G-3-P} + 3ATP + 2H_2O \longrightarrow$$
$$\text{3 ribulose bisphosphate} + 3ADP + 2P_i$$

(13-11)

The Complete Calvin Cycle

To summarize primary carbon assimilation by the Calvin cycle, we can simply add reactions 13-10 and 13-11, which encompass all the chemistry of the cycle from carbon dioxide fixation and reduction through triose phosphate synthesis and regeneration of the organic acceptor molecule. The resulting summary reaction is:

$$3CO_2 + 9ATP + 6NADPH + 5H_2O \longrightarrow$$
$$\text{glyceraldehyde-3-phosphate} + 9ADP + 6NADP^+ + 8P_i$$

(13-12)

Note the requirements for ATP and NADPH: The Calvin cycle consumes nine ATP molecules and six NADPH molecules for every three-carbon carbohydrate synthesized, or three ATP and two NADPH for each carbon dioxide molecule fixed. As noted earlier, the product of the Calvin cycle is actually an interconvertible pool of glyceraldehyde-3-phosphate and dihydroxyacetone phosphate that may be diverted to various metabolic pathways.

Because triose phosphates generated by the Calvin cycle are consumed by metabolic pathways in the cytosol as well as the chloroplast stroma, there must be a mechanism for transporting them across the chloroplast inner membrane. The most abundant protein in the chloroplast inner membrane is a *phosphate translocator* that catalyzes the exchange of dihydroxyacetone phosphate, glyceraldehyde-3-phosphate, or 3-phosphoglycerate in the stroma for P_i in the cytosol. This *antiport system* (see Figure 8-10b) ensures that triose phosphates will not be exported unless P_i returns to the stroma. Moreover, the phosphate translocator is suffi-

Figure 13-16 Regeneration Sequence of the Calvin Cycle. Three molecules of glyceraldehyde-3-phosphate (G-3-P) and two molecules of dihydroxyacetone phosphate (DHAP) enter from the left. Four basic reactions—catalyzed by aldolases, phosphatases, transketolases, or isomerases—transform the five three-carbon compounds into three five-carbon compounds (reactions CC-5 through CC-10) and thence into the common intermediate ribulose-5-phosphate (reactions CC-11 and CC-12). The final step of regeneration, phosphorylation of ribulose-5-phosphate to form ribulose-1,5-bisphosphate, is shown in Figure 13-15 (reaction CC-13).

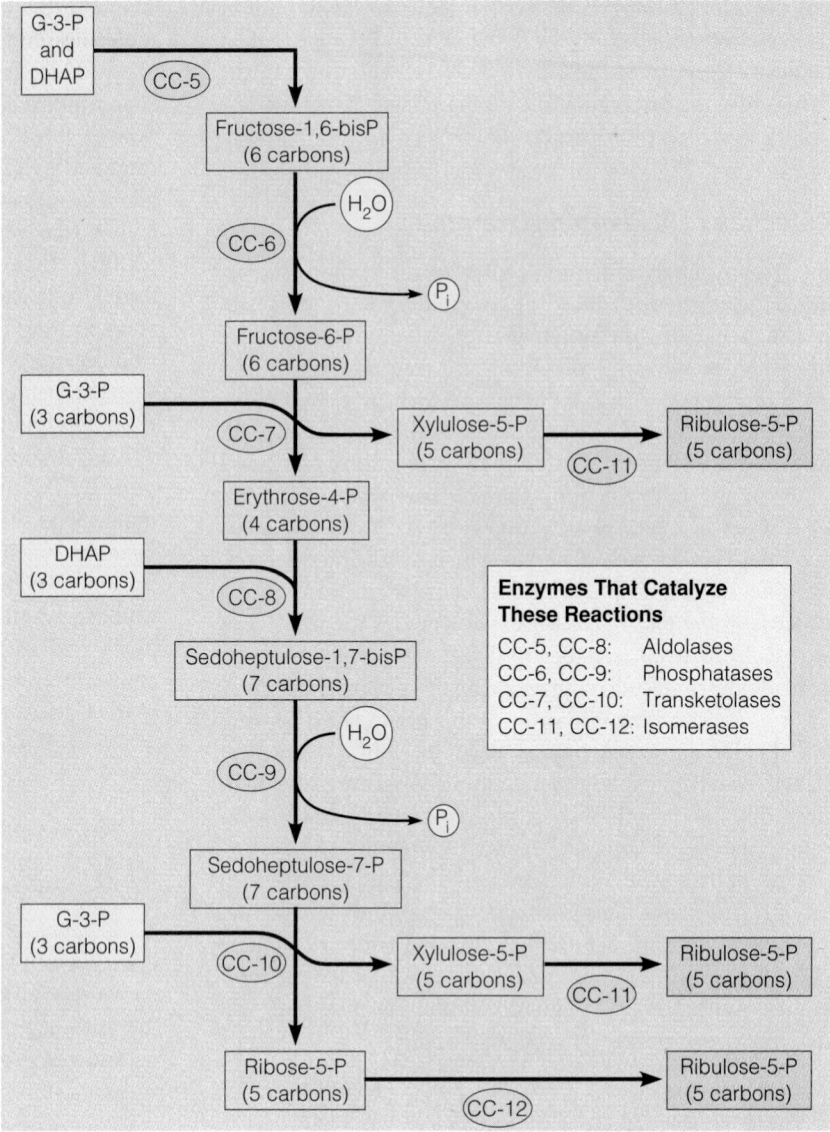

ciently specific to prevent other intermediates of the Calvin cycle from leaving the stroma. Within the cytosol, triose phosphates may be used for sucrose synthesis, as we will describe below, or they may enter the glycolytic pathway to provide ATP and NADH in the cytosol. The triose phosphates remaining in the stroma are generally used for starch synthesis, which is also described below.

Regulation of the Calvin Cycle

In the dark, all phototrophs essentially become chemotrophs. To meet the steady demand for energy and carbon, they must tap the surplus of carbohydrates that accumulate when light is available. Such activity would be futile, however, if the Calvin cycle were not rendered inoperable while the glycolytic and other pathways consume storage carbohydrates. Remember that all biological pathways run at less than 100% thermodynamic efficiency because some energy is inevitably lost as heat and entropy. Channeling energy from carbohydrates into ATP or NADH synthesis and then

back into carbohydrate synthesis would eventually drain all energy and carbon reserves. It is not surprising, then, that phototrophs have several regulatory systems to ensure that the Calvin cycle does not operate unless light is available.

The first level of control, appropriate for enzymes unique to a specific pathway, is the regulation of gene expression. Enzymes that are important only for light-dependent carbon dioxide fixation and reduction are generally not present in plant tissues not exposed to light. Plastids in root cells, for example, contain less than 1% of the amount of rubisco activity found in the chloroplasts of leaf cells.

Enzymes of the Calvin cycle are also regulated by metabolites. Because the photosynthetic energy transduction reactions and the Calvin cycle are localized in the same compartment—the cytoplasm of a photosynthetic bacterium or the chloroplast stroma of a plant or algal cell—there is ample opportunity for metabolites of one process to affect the other. Consider the changes that occur in the chloroplast stroma as light drives the movement of electrons from water to ferredoxin. As protons are pumped from the stroma to the lumen, the pH of the stroma rises from about 7.2 (a typical value in the dark) to about 8.5. Simultaneously, magnesium ions diffuse from the lumen to the stroma (to maintain charge neutrality across the thylakoid membrane), and the concentration of magnesium ions in the stroma rises about tenfold. Eventually, high levels of reduced ferredoxin, NADPH, and ATP also accumulate. Each of these factors serves as a signal that can affect enzymes of the Calvin cycle as well as enzymes of other metabolic pathways.

Three enzymes unique to the Calvin cycle are logical points for metabolic control. Rubisco is an obvious candidate because it catalyzes the first reaction of the Calvin cycle. The others are *sedoheptulose bisphosphatase* and *ribulose-5-phosphate kinase*, which have roles in regenerating the organic acceptor molecule ribulose-1,5-bisphosphate (see Figure 13-16). Rubisco, sedoheptulose bisphosphatase, and ribulose-5-phosphate kinase are all stimulated by a high pH and a high concentration of magnesium ions. In the dark, when pH and magnesium ion levels in the stroma decline, these enzymes are less active. Each one is not only unique to the Calvin cycle but catalyzes an essentially irreversible reaction. Recall from Chapter 11 that the glycolytic pathway is also regulated at sites of unique and irreversible reactions. This is a common theme in metabolism.

Another system for regulating enzyme activity, shown in Figure 13-17, depends on ferredoxin. During illumination, electrons donated by water are used to reduce ferredoxin. An enzyme called *ferredoxin-thioredoxin reductase* then catalyzes the transfer of electrons from ferredoxin to *thioredoxin*, another mobile electron carrier. Thioredoxin affects an enzyme by reducing *disulfide (S-S)* bonds to *sulfhydryl (SH)* groups, which can cause the protein to undergo a conformational change. Sedoheptulose bisphosphatase and ribulose-5-phosphate kinase contain disulfide bonds that can be reduced by thioredoxin. For these enzymes, the resulting conformational change stimulates activity. In the dark, when no reduced ferredoxin is available, the sulfhydryl groups spontaneously reoxidize to disulfide bonds, thereby inactivating the enzymes. The CF_oCF_1 ATP synthase described earlier is also affected in this way by the thioredoxin system.

Because the Calvin cycle produces three-carbon molecules that can enter the glycolytic pathway directly, the breakdown of complex carbohydrates is not necessary in the light. Not surprisingly, the same mechanisms that activate enzymes of the Calvin cycle inactivate enzymes of degradative pathways. One example is *phosphofructokinase*, the most important control point in glycolysis. While the Calvin cycle is operating in the light, inhibition of phosphofructokinase ensures that the early steps of the glycolytic pathway are bypassed, thereby preventing the development of another potential futile cycle.

Photosynthetic Energy Transduction and the Calvin Cycle

Now we are ready to write an overall reaction for photosynthesis that takes both energy transduction and carbon assimilation into account. Considering the needs of the Calvin cycle and assuming no diversion of ATP or NADPH to other pathways, the requirements are clear. According to reaction 13-12, the synthesis of one molecule of glyceraldehyde-3-phosphate (or of dihydroxyacetone phosphate) requires 9 ATP and 6 NADPH molecules. This demand can be met by the flow of 12 electrons through the noncyclic pathway, which provides all 6 of the NADPH molecules (3 × reaction 13-8) and 8 of the 9 ATP molecules (8 × reaction 13-9), plus the flow of 2 electrons through the cyclic pathway, which will provide at least one more ATP molecule (1 × reaction 13-9). Thus, the net reaction for energy transduction is:

$$26 \text{ photons} + 9ADP + 9P_i + 6NADP^+ \longrightarrow$$
$$3O_2 + 9ATP + 6NADPH + 3H_2O \qquad \textbf{(13-13)}$$

The absorption of 26 photons is accounted for by 12 photoexcitation events at photosystem I and another 12 at photosystem II during noncyclic electron flow and 2 more photoexcitation events at photosystem I during cyclic electron flow.

By adding the summary reaction for the Calvin cycle (reaction 13-12) to the net reaction for energy transduction (reaction 13-13), we obtain the following overall expression:

$$26 \text{ photons} + 3CO_2 + 5H_2O + P_i \longrightarrow$$
$$\text{glyceraldehyde-3-phosphate} + 3O_2 + 3H_2O \qquad \textbf{(13-14)}$$

Because the phosphate group is generally removed by hydrolysis when glyceraldehyde-3-phosphate is incorporated

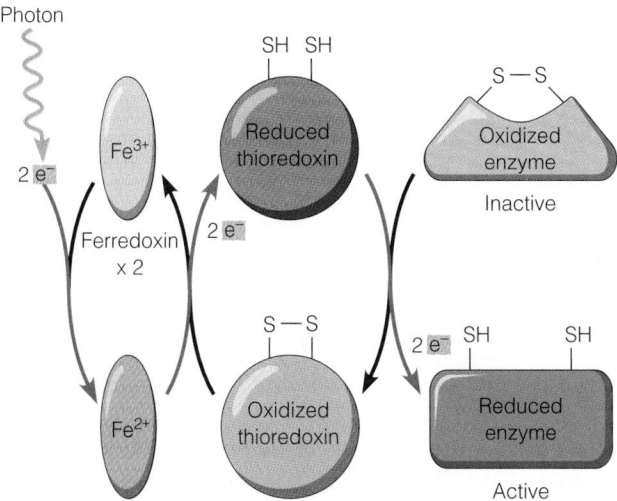

Figure 13-17 Thioredoxin-Mediated Activation of a Calvin Cycle Enzyme. Two reduced ferredoxin molecules generated by the photoreduction pathway donate electrons to thioredoxin, an intermediate electron carrier. Thioredoxin subsequently activates a target enzyme by reducing a disulfide bond to two sulfhydryl groups. In other cases, thioredoxin may inactivate enzymes.

into more complex organic compounds, we may rewrite reaction 13-14 as:

$$26 \text{ photons} + 3CO_2 + 6H_2O \longrightarrow$$
$$\text{glyceraldehyde} + 3O_2 + 3H_2O \qquad \textbf{(13-15)}$$

This reaction is almost identical to the net photosynthetic reaction (reaction 13-2) we introduced near the beginning of the chapter. Now, however, we are in a much better position to understand some of the photochemical and metabolic complexity behind what might otherwise appear to be a simple reaction. Moreover, we have replaced the vague terms "light" and "$C_3H_6O_3$," with a specific number of photons and a primary product of carbon dioxide fixation.

This information enables us to calculate the *maximum* efficiency of photosynthetic energy transduction and carbon assimilation. For red light, assuming a wavelength of 670 nm, 26 photons represent about 26×43 kcal/einstein (an einstein is a mole of photons), or a total of 1118 kcal of energy. Because glyceraldehyde differs in free energy from carbon dioxide and water by 343 kcal/mol, the maximum efficiency of photosynthetic energy transduction is about 31% ($343/1118 = 0.31$). Although an efficiency of this order might be observed in the laboratory with isolated chloroplasts or algal cells, photosynthesis under field conditions is far less efficient. After we look at carbohydrate synthesis, we will consider a few of the factors that may limit the overall productivity of a photosynthetic organism.

Carbohydrate Synthesis

As we have already learned, the Calvin cycle generates an interconvertible pool of glyceraldehyde-3-phosphate and dihydroxyacetone phosphate, both of which can be transported out of the chloroplast by a phosphate translocator located in the chloroplast inner membrane. These triose phosphates are, in turn, the starting points for starch synthesis within the chloroplast and for sucrose synthesis in the cytosol (Figure 13-18). We will consider both pathways, each of which begins with the formation of hexose phosphates.

Hexose Phosphate Formation

The key hexose phosphate required for both starch and sucrose synthesis is glucose-1-phosphate, which is formed from triose phosphates as shown in Figure 13-18, reactions S-1 through S-4. These reactions occur both in the cytosol (c) and in the chloroplast stroma (s) because hexoses and hexose phosphates cannot cross the chloroplast inner membrane and must therefore be synthesized separately in each compartment. The first several steps in the sequence resemble reactions we encountered in our discussion of gluconeogenesis in Chapter 11 (see Figure 11-12). The first step in triose phosphate utilization is a condensation reaction that generates fructose-1,6-bisphosphate (reaction S-1). As in gluconeogenesis, the pathway is rendered exergonic by the

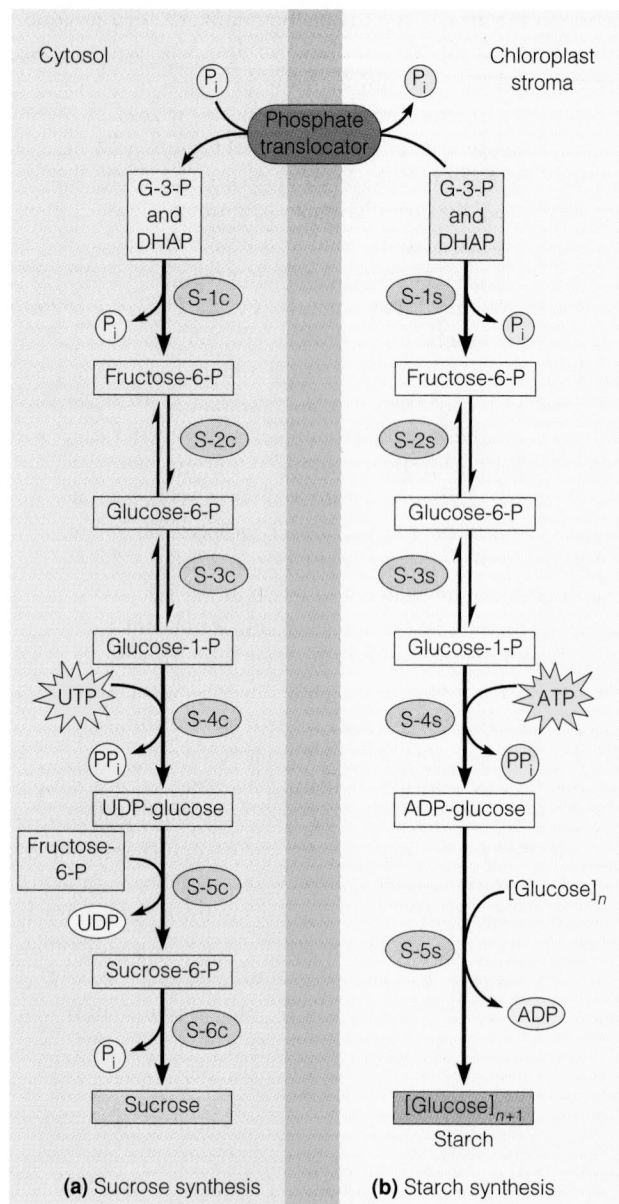

Enzymes That Catalyze These Reactions

S-1: Aldolase and
         fructose-1,6-bisphosphatase
S-2: Phosphoglucoisomerase
S-3: Phosphoglucomutase
S-4c: UDP-glucose pyrophosphorylase
S-5c: Sucrose phosphate synthase
S-6c: Sucrose phosphatase
S-4s: ADP-glucose pyrophosphorylase
S-5s: Starch synthase

Figure 13-18 The Synthesis of Sucrose and Starch from Products of the Calvin Cycle.   The triose phosphates glyceraldehyde-3-phosphate (G-3-P) and dihydroxyacetone phosphate (DHAP) from the chloroplast stroma are exchanged, via a phosphate translocator, for inorganic phosphate from the cytosol. **(a)** Sucrose synthesis is confined to the cytosol, while **(b)** starch synthesis occurs only in the chloroplast stroma. The enzymes and isozymes that catalyze these reactions are restricted to either the cytosol (c) or the stroma (s).

hydrolytic removal of a phosphate group from fructose-1, 6-bisphosphate to form fructose-6-phosphate (reaction S-2). The enzyme that catalyzes this reaction, fructose-1, 6-bisphosphatase, exists in two forms, one in the cytosol and the other in the chloroplast stroma. The multiple forms of such an enzyme are called **isoenzymes**—physically distinct proteins that catalyze the same reaction. The fructose-6-phosphate can then be converted, via glucose-6-phosphate, to glucose-1-phosphate (reactions S-3 and S-4). Again, separate isoenzymes catalyze each of these reactions in the cytosol and in the chloroplast stroma.

Textbooks often depict glucose as the end-product of photosynthetic carbon assimilation. But this portrayal is more a definition of convenience for writing summary reactions than a metabolic fact, as very little free glucose actually accumulates in photosynthetic cells. Most glucose is converted to either transport carbohydrates such as sucrose or storage carbohydrates such as starch. In reality, the designation of an end-product for photosynthetic carbon assimilation becomes rather arbitrary once you get past the pool of triose phosphates produced by the Calvin cycle. Indeed, we could regard the whole phototrophic organism—plant, algal cell, or bacterium—as the "end-product" of photosynthesis.

Sucrose Formation

Recall from Chapter 3 that sucrose is a disaccharide consisting of one molecule each of glucose and fructose linked by a glycosidic bond (see Figure 3-20). Sucrose is of interest because it is the major transport carbohydrate in most plant species. Moreover, in some species, such as sugar beets and sugar cane, sucrose also serves as a storage carbohydrate. As shown in Figure 13-18a, sucrose synthesis is localized in the cytosol of a photosynthetic cell. Triose phosphates exported from the chloroplast stroma and not consumed by other metabolic pathways are converted to fructose-6-phosphate and glucose-1-phosphate. Glucose from glucose-1-phosphate is then activated by the reaction of glucose-1-phosphate with *uridine triphosphate* (*UTP*), generating *UDP-glucose* (reaction S-4c). Finally, the glucose is transferred to fructose-6-phosphate to form the phosphorylated disaccharide *sucrose-6-phosphate*, and hydrolytic removal of the phosphate group then generates free sucrose (reactions S-5c and S-6c). (In some plant species, glucose may be transferred directly from UDP-glucose to free fructose.) The sucrose is exported from leaves by way of vascular bundles and conveys energy and reduced carbon to nonphotosynthetic tissues of the organism.

Like the Calvin cycle, sucrose synthesis is carefully controlled to prevent conflict with degradation pathways. *Cytosolic* fructose-1,6-bisphosphatase, for example, is inhibited by *fructose-2,6-bisphosphate*, a metabolite introduced in Chapter 11 as an important regulator of glycolysis and gluconeogenesis in liver cells. In plant cells, fructose-2,6-bisphosphate accumulates when high levels of fructose-6-phosphate and P_i (signals that indicate low sucrose de-

mand) or low levels of 3-phosphoglycerate and dihydroxy-acetone phosphate (signals that indicate high triose phosphate demand) are present. Another control point for sucrose synthesis is *sucrose phosphate synthase*, the enzyme that catalyzes the transfer of glucose from UDP-glucose to fructose-6-phosphate. This enzyme is stimulated by glucose-6-phosphate and inhibited by sucrose-6-phosphate, UDP, and P_i.

Starch Formation

Starch synthesis in plant cells is confined to plastids, which we encountered in Chapter 4. In photosynthetic plant cells, starch synthesis is generally restricted to chloroplasts, which are essentially photosynthetic plastids. When sufficient energy and carbon are available to meet the metabolic needs of a plant, the excess triose phosphates within the chloroplast stroma are converted to glucose-1-phosphate, which is then used for starch synthesis. As shown in Figure 13-18b, glucose is activated by the reaction of glucose-1-phosphate with ATP, generating *ADP-glucose* (reaction S-4s). The activated glucose is then transferred directly to a growing starch chain by *starch synthase*, leading to elongation of the polysaccharide (reaction S-5s). The two forms of starch commonly found in plants are amylose and amylopectin, as described in Chapter 3. As shown in Figure 13-2, starch may accumulate in large storage granules within the chloroplast stroma. When photosynthesis is limited by darkness or other factors, starch is degraded to triose phosphates, which can then enter glycolysis or be converted to sucrose in the cytosol and exported from the cell.

Like the Calvin cycle and sucrose synthesis, starch synthesis is carefully controlled. Chloroplast fructose-1,6-bisphosphatase, which channels fructose-1,6-bisphosphate into glucose and starch biosynthesis in the stroma, is activated by the same thioredoxin system that affects enzymes of the Calvin cycle. This regulation ensures that starch synthesis occurs only when there is sufficient illumination for photoreduction. The key enzyme for regulation, however, is *ADP-glucose pyrophosphorylase*, which catalyzes the activation of glucose and commits it to starch synthesis. ADP-glucose pyrophosphorylase is stimulated by glyceraldehyde-3-phosphate and inhibited by P_i. Thus, when triose phosphates are diverted to the cytosol and ATP is hydrolyzed to ADP and P_i (signals that indicate high energy demand), starch synthesis is blocked.

Other Photosynthetic Assimilation Pathways

Photosynthesis encompasses more than carbon dioxide fixation and carbohydrate synthesis. In plants and algae, the ATP and NADPH generated by photosynthetic energy transduction reactions is directly consumed by a variety of other anabolic pathways found in chloroplasts. Additional examples of carbon metabolism within chloroplasts include the synthesis of fatty acids, chlorophyll, and carotenoids. Moving beyond carbon metabolism, several key steps of nitrogen and sulfur assimilation are also localized in chloroplasts. The reduction of nitrite (NO_2^-) to ammonia (NH_3), for example, is catalyzed by a reductase enzyme in the chloroplast stroma, with reduced ferredoxin serving as an electron donor. The ammonia is then channeled into amino acid and nucleotide synthesis, portions of which also occur in chloroplasts. Furthermore, much of the reduction of sulfate (SO_4^{2-}) to sulfide is catalyzed by enzymes in the chloroplast stroma. In this case, both ATP and reduced ferredoxin provide energy and reducing power. The sulfide, like ammonia, may then be used for amino acid synthesis.

The Decrease in Photosynthetic Efficiency due to Rubisco's Oxygenase Activity

The primary reaction catalyzed by rubisco is its *carboxylase* activity—the addition of carbon dioxide and water to ribulose-1,5-bisphosphate, forming two molecules of 3-phosphoglycerate. However, rubisco can also catalyze a reaction in which ribulose-1,5-bisphosphate is split with the uptake of oxygen, rather than carbon dioxide:

Ribulose-1,5-bisphosphate

3-phosphoglycerate

+ Phosphoglycolate   **(13-16)**

The result of this *oxygenase* activity is one three-carbon product, 3-phosphoglycerate, and one two-carbon product, **phosphoglycolate.** Because phosphoglycolate cannot be used during the next step of the Calvin cycle, it appears to be a wasteful diversion of material from carbon assimilation. Furthermore, alternative functions for rubisco's oxygenase activity have not been clearly demonstrated. Indeed, not only does the production of phosphoglycolate appear to waste energy and carbon, the accumulation of phosphoglycolate may kill a plant by inhibiting the triose phosphate isomerase that maintains a balance of glyceraldehyde-3-phos-

phate and dihydroxyacetone phosphate in the chloroplast stroma.

Considering that rubisco has a much lower affinity for oxygen than for carbon dioxide and that the oxygenation reaction proceeds much more slowly than the carboxylation reaction, one might expect that the oxygenation reaction is not a serious problem. However, the low carbon dioxide and high oxygen concentrations in the Earth's atmosphere—about 0.035% and 21%, respectively—lead to significant competition between carbon dioxide and oxygen for binding to rubisco's catalytic site. Within a typical leaf or algal cell exposed to the atmosphere, the low CO_2/O_2 ratio leads to one oxygenation of ribulose-1,5-bisphosphate for every two or three carboxylation events, seriously reducing photosynthetic efficiency and generating large amounts of phosphoglycolate.

Plants living in hot, arid environments with intense illumination are particularly affected by rubisco's oxygenase activity, since such conditions may further lower the CO_2/O_2 ratio in the chloroplast stroma. Although the solubilities of both carbon dioxide and oxygen decrease as temperature increases, the solubility of carbon dioxide declines more rapidly, thereby lowering the CO_2/O_2 ratio in solution. Another problem occurs when plants respond to drought by closing their stomata during the day to reduce water loss. When the stomata are closed, carbon dioxide cannot enter the leaf. Without a steady supply of carbon dioxide for assimilation, the concentration of carbon dioxide in leaf cells may decline. Moreover, water photolysis continues to generate oxygen, which accumulates because it cannot diffuse out of the leaf. Intense sunlight will exacerbate this problem by increasing the rate of water photolysis, which depends on the absorption of light that drives noncyclic electron flow and photoreduction.

Efforts to reduce rubisco's oxygenase activity through alteration of the amino acid sequence of the enzyme have failed. According to one theory, the oxygenase activity is an evolutionary relic from a time when oxygen did not make up a large part of the Earth's atmosphere, and it cannot be eliminated without seriously compromising the carboxylase function. Not even natural selection appears to be up to the task of altering this enzyme. Instead, phototrophs that depend on rubisco have developed three alternative strategies for coping with the enzyme's apparently wasteful oxygenase activity. We will consider each briefly.

The Glycolate Pathway: Photorespiration

In all photosynthetic plant cells, phosphoglycolate generated by rubisco's oxygenase activity is channeled into the **glycolate pathway.** This *salvage pathway* disposes of phosphoglycolate and returns up to 75% of the reduced carbon present in phosphoglycolate to the Calvin cycle as 3-phosphoglycerate. Because the glycolate pathway is characterized by the light-dependent uptake of oxygen and evolution of carbon dioxide, it is also referred to as **photorespiration.**

Several steps of the glycolate pathway are localized in a specific type of peroxisome called a **leaf peroxisome.** Recall from Chapter 9 that peroxisomes are membrane-bounded microbodies containing oxidase enzymes that generate hydrogen peroxide. The potentially destructive hydrogen peroxide is then eliminated by another peroxisomal enzyme, catalase, which degrades it to water and oxygen. Because of their essential role in the glycolate pathway, leaf peroxisomes are found not only in leaf cells but in all photosynthetic plant tissues.

A typical leaf peroxisome from a mesophyll cell is shown in the electron micrograph of Figure 9-29. The core within the matrix of the organelle is crystalline catalase, a common feature of peroxisomes. You will also notice the close proximity of the leaf peroxisome to a chloroplast and a mitochondrion. This association is frequently found in photosynthetic plant cells and may reflect the functional relationship between the three organelles, because one or more steps of the glycolate pathway occur in each of these organelles (Figure 13-19).

In the chloroplast, the phosphoglycolate generated by rubisco is rapidly dephosphorylated by a phosphatase attached to the stromal side of the chloroplast inner membrane (reaction GP-1). The product is glycolate, which diffuses to a nearby leaf peroxisome where an *oxidase* converts

it to glyoxylate (reaction GP-2). Presumably, the need to pass glycolate from the chloroplast to the peroxisome dictates the close juxtaposition of these organelles in photosynthetic cells. As you might expect of a peroxisomal process, the oxidation of glycolate is accompanied by the uptake of oxygen and the generation of hydrogen peroxide, which is immediately degraded by catalase (reaction GP-C). During the next reaction in the peroxisome, an *aminotransferase* catalyzes the transfer of an amino group from glutamate to glyoxylate, forming glycine (reaction GP-3).

Glycine diffuses from the leaf peroxisome to a mitochondrion, where two enzyme activities working in series—a *decarboxylase* and a *hydroxymethyl transferase*—convert two glycine molecules to a single serine, concomitant with the generation of NADH and the release of carbon dioxide and ammonia (reaction GP-4). Rubisco's oxygenase activity therefore leads not only to a loss of carbon, but to a loss of nitrogen. To prevent the depletion of nitrogen reserves, the ammonia must be reassimilated at the expense of ATP and reductant.

Back in the peroxisome, another aminotransferase removes the amino group from serine, generating hydroxypyruvate (reaction GP-5). A *reductase*, using NADH as an

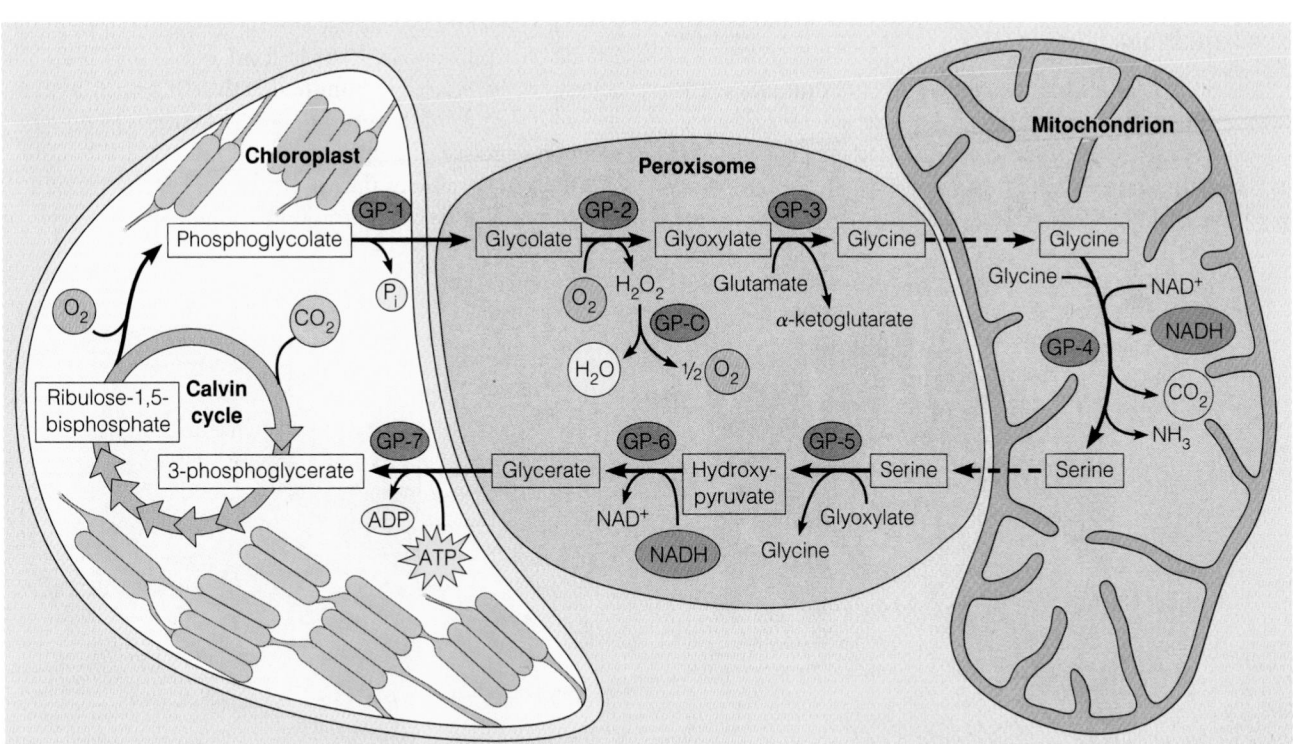

Figure 13-19 The Glycolate Pathway. Glycolate arises as a result of the oxygenase activity of ribulose-1,5-bisphosphate carboxylase (rubisco). The immediate product is phosphoglycolate, which is converted to free glycolate by a phosphatase localized in the chloroplast membrane (reaction GP-1). Free glycolate diffuses out of the chloroplast stroma and is metabolized by a five-step pathway (GP-2 through GP-6) that occurs partially in the peroxisome and partially in the mitochondrion. Glycerate then diffuses into the chloroplast and is phosphorylated to form 3-phosphoglycerate (GP-7), which enters the Calvin cycle. The oxygen uptake and carbon dioxide evolution characteristic of photorespiration occur in the peroxisome (reaction GP-2) and mitochondrion (reaction GP-4), respectively.

Enzymes That Catalyze These Reactions

GP-1: Phosphoglycolate phosphatase
GP-2: Glycolate oxidase
GP-3: Glutamate: glyoxylate aminotransferase
GP-4: Glycine decarboxylase and serine hydroxymethyl transferase
GP-5: Serine: glyoxylate aminotransferase
GP-6: Hydroxypyruvate reductase
GP-7: Glycerate kinase
GP-C: Catalase

electron donor, then reduces hydroxypyruvate to glycerate (reaction GP-6). Finally, glycerate is phosphorylated by a *kinase* in the cytosol or chloroplast stroma to generate 3-phosphoglycerate (reaction GP-7), the key intermediate of the Calvin cycle.

What is the benefit of this long salvage pathway, winding through several organelles? Three out of every four carbon atoms that exit the Calvin cycle as part of phosphoglycolate are recovered as 3-phosphoglycerate. Without this pathway, phosphoglycolate would not only accumulate to toxic levels, but triose phosphates essential for the regeneration of ribulose-1,5-bisphosphate and the continuation of the Calvin cycle would be depleted. In terms of energy and reduced carbon, phosphoglycolate metabolism is quite expensive. For every three carbon atoms salvaged, an ammonia molecule must be reassimilated at the expense of one ATP and two reduced ferredoxin molecules, and the glycerate generated by reaction GP-5 must be phosphorylated at the cost of one ATP molecule (see Figure 13-19). With rubisco's apparently unavoidable oxygenase activity, however, photorespiration is a net gain for the plant. Just consider the value of the three carbon atoms salvaged: 9 ATP and 6 NADPH were consumed when they were originally fixed and reduced.

The Hatch-Slack Cycle

As mentioned earlier, plants in hot, arid environments with intense illumination are particularly affected by rubisco's oxygenase activity. In some cases, the potential for energy and carbon drain through photorespiration is so overwhelming that plants must depend on adaptive strategies for solving the problem. One general approach is confining rubisco to cells that contain a high concentration of carbon dioxide, thereby minimizing the inherent oxygenase activity of rubisco.

In many tropical grasses, including economically important plants such as maize, sorghum, and sugar cane, an enhanced CO_2 concentration is achieved by a short carboxylation/decarboxylation sequence referred to as the **Hatch-Slack cycle,** after M. D. Hatch and C. R. Slack, two plant physiologists who played key roles in the elucidation of the pathway. Because the immediate product of carbon dioxide fixation by the Hatch-Slack cycle is the four-carbon organic acid oxaloacetate, the pathway is also called the C_4 **pathway,** and plants containing the Hatch-Slack cycle are referred to as C_4 **plants.** This term distinguishes such plants from C_3 **plants,** in which the immediate product of carbon dioxide fixation is the three-carbon compound 3-phosphoglycerate.

To appreciate the advantage of the Hatch-Slack cycle, we must first consider the arrangement of the Hatch-Slack and Calvin cycles within the leaf of a C_4 plant. As shown in Figure 13-20, C_4 plants, unlike C_3 plants, have within their leaves two distinct types of photosynthetic cells: *mesophyll cells* and *bundle sheath cells*. The first steps of carbon dioxide fixation within a C_4 plant are accomplished by the Hatch-Slack cycle in mesophyll cells, which are exposed to the carbon dioxide and oxygen that enter a leaf through its stomata. The carbon dioxide that is fixed in mesophyll cells is subsequently released in **bundle sheath cells,** which are relatively isolated from the atmosphere. The entire Calvin cycle, including rubisco, is confined to chloroplasts in the bundle sheath cells. Because of the activity of the Hatch-Slack cycle, the carbon dioxide concentration in bundle sheath cells may

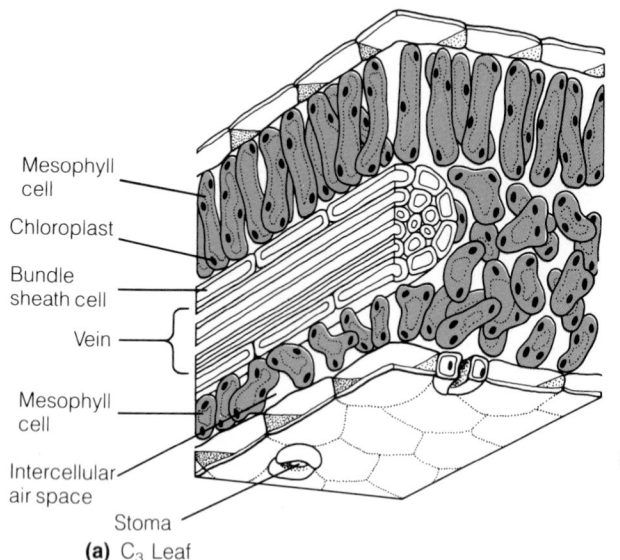

(a) C_3 Leaf

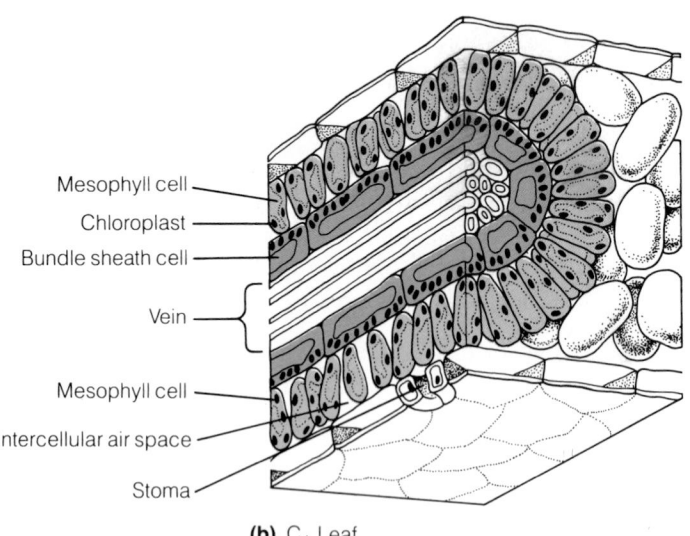

(b) C_4 Leaf

Figure 13-20   Structural Differences Between Leaves of C_3 and C_4 Plants.   **(a)** In C_3 plants, the Calvin cycle occurs in mesophyll cells. **(b)** In C_4 plants, the Calvin cycle is confined to bundle sheath cells, which are relatively isolated from atmospheric carbon dioxide and oxygen. C_4 plants utilize the Hatch-Slack cycle for collecting carbon dioxide

in mesophyll cells and concentrating it in bundle sheath cells. The bundle sheath cells surround the vascular bundles (veins) of the leaf, which carry carbohydrates to other parts of the plant. This concentric arrangement is called Kranz (German for "halo" or "wreath") anatomy and is essential to the photosynthetic efficiency of C_4 plants.

be as much as ten times the level in the atmosphere, strongly favoring rubisco's carboxylase activity and minimizing its oxygenase activity.

As detailed in Figure 13-21, the Hatch-Slack cycle begins with the carboxylation of *phosphoenolpyruvate* to form oxaloacetate (reaction HS-1). Both phosphoenolpyruvate and oxaloacetate should be familiar to you, since the same carboxylation reaction was encountered earlier as one of the means of replenishment of oxaloacetate for the TCA cycle (see Figure 12-15). Carboxylation is catalyzed by a specific cytosolic form of *phosphoenolpyruvate carboxylase*, which is particularly abundant in mesophyll cells of C_4 plants. This carboxylase not only lacks rubisco's oxygenase activity but is an excellent "scavenger" for carbon dioxide. In other words, it has a high affinity for its substrate—*bicarbonate* (HCO_3^-)—and operates very efficiently even when the concentration of bicarbonate is quite low. (Bicarbonate forms when carbon dioxide dissolves in water; its concentration therefore reflects the availability of carbon dioxide gas.)

In one variation of the Hatch-Slack pathway, the oxaloacetate generated by phosphoenolpyruvate carboxylase is rapidly converted to malate by an *NADPH-dependent dehydrogenase* (reaction HS-2 in Figure 13-21). Malate is a stable four-carbon acid that carries carbon from mesophyll cells to chloroplasts of bundle sheath cells, where decarboxylation occurs (reaction HS-3). The liberated carbon dioxide is then refixed and reduced by the Calvin cycle. Because the decarboxylation of malate is accompanied by the generation of NADPH, the Hatch-Slack cycle also conveys reducing power from mesophyll to bundle sheath cells. This might limit the demand for noncyclic electron flow from water to $NADP^+$ in the bundle sheath cells, thereby minimizing the formation of oxygen and further favoring rubisco's carboxylase activity.

The pyruvate that remains after malate decarboxylation diffuses to a mesophyll cell, where it is phosphorylated at the expense of ATP to regenerate phosphoenolpyruvate (reaction HS-4), the original carbon dioxide acceptor of the Hatch-Slack cycle. Thus, the overall process is cyclic, and the net result is a "feeder system" that captures carbon dioxide in mesophyll cells and passes it to the Calvin cycle in bundle sheath cells. The Hatch-Slack cycle is not a substitute for the Calvin cycle, but simply a preliminary carboxylation/decarboxylation sequence that concentrates CO_2 in the bundle sheath cells.

Because ATP is hydrolyzed to AMP in reaction to HS-4, the actual cost of moving carbon from mesophyll to bundle sheath cells is equivalent to two ATP molecules per carbon dioxide molecule. Carbon assimilation within a C_4 plant therefore consumes a total of five ATP molecules per carbon atom, rather than the three required in C_3 plants. In an environment that would otherwise enhance rubisco's oxygenase activity, however, the energy required to prevent phosphoglycolate formation may be far less than the energy that would otherwise be lost through photorespiration.

When temperatures exceed about 30°C, the photosynthetic efficiency of a C_4 plant exposed to intense sunlight may be twice that of a C_3 plant. While the higher efficiency is largely due to the reduced photorespiration in the C_4 plant and enhanced photorespiration in the C_3 plant, other factors are also important. In a C_3 plant, photosynthesis is often limited by the low atmospheric concentration of carbon dioxide, not by the availability of sunlight. In a C_4 plant, however, the Hatch-Slack cycle actively concentrates carbon dioxide in bundle sheath cells, where the Calvin cycle is localized, enabling the plant to take advantage of higher levels of illumination.

Enrichment of carbon dioxide in the vicinity of rubisco by the Hatch-Slack pathway confers at least one additional advantage on C_4 plants. Because phosphoenolpyruvate is an efficient scavenger of carbon dioxide, gas exchange through the stomata of C_4 plants can be substantially reduced to conserve water without adversely affecting photosynthetic efficiency. As a result, C_4 plants may assimilate over twice as much carbon as a C_3 plants for each unit of water transpired. This adaptation makes C_4 plants suitable for regions of periodic drought, such as tropical savannas.

Although fewer than 1% of plant species investigated depend on the Hatch-Slack cycle, the pathway is of particular interest because many economically important species are in this group. Moreover, C_4 plants such as maize and sugar cane are characterized by net photosynthetic rates that are often two or three times those of C_3 plants such as cereal grains. Little wonder, then, that crop physiologists and plant breeders have devoted so much attention to C_4 species and to the question of whether it is possible to improve on the relatively inefficient carbon dioxide fixation pathway of the C_3 plants. Some scenarios of genetic engineering even envision the genetic "conversion" of C_3 plants to C_4 plants.

Crassulacean Acid Metabolism: CAM Plants

Some plant species that live in deserts, salt marshes, and other environments where access to water is severely limited contain a preliminary carbon dioxide fixation pathway closely related to the Hatch-Slack cycle. The sequence of reactions is similar, but these plants segregate the carboxylation and decarboxylation reactions by *time* rather than by *space*. Because the pathway was first recognized in the family Crassulaceae, it is called **crassulacean acid metabolism (CAM),** and plants that take advantage of CAM photosynthesis are called **CAM plants.** CAM photosynthesis has been found in about 4% of plant species investigated, including many succulents, cacti, and orchids.

CAM plants, unlike most C_3 and C_4 plants, generally open their stomata only at night, when the atmosphere is relatively cool and moist. As carbon dioxide diffuses into mesophyll cells, it is assimilated by the first two steps of a pathway similar to the Hatch-Slack cycle, and accumulates as malate. Instead of being exported from mesophyll cells, however, the malate is stored in large vacuoles, which become very acidic. The process of moving malate into vacuoles consumes ATP, but it is necessary as a means of protecting cytosolic enzymes from a large drop in pH at night.

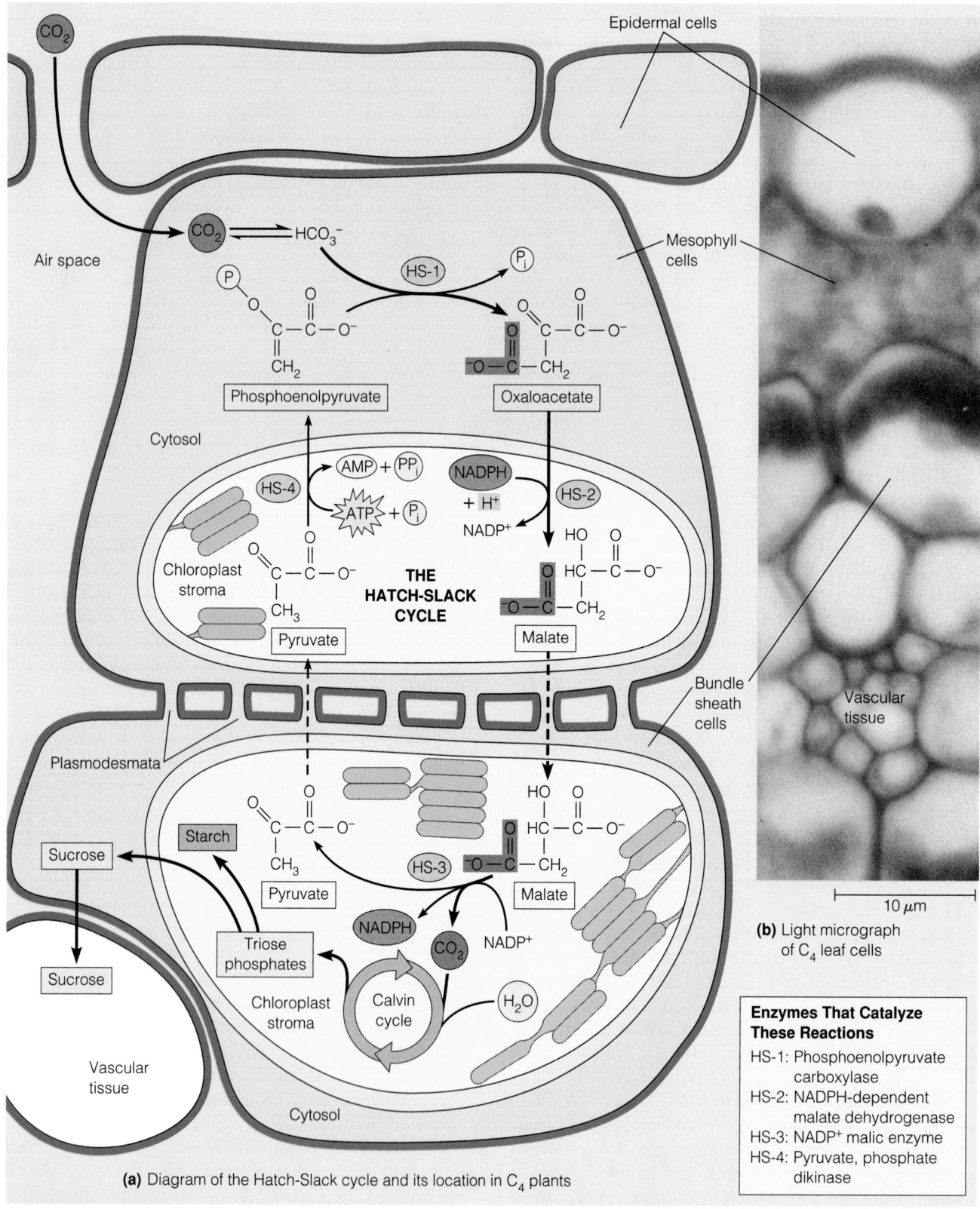

(a) Diagram of the Hatch-Slack cycle and its location in C$_4$ plants

Enzymes That Catalyze These Reactions

HS-1: Phosphoenolpyruvate carboxylase
HS-2: NADPH-dependent malate dehydrogenase
HS-3: NADP$^+$ malic enzyme
HS-4: Pyruvate, phosphate dikinase

(b) Light micrograph of C$_4$ leaf cells

10 μm

Figure 13-21 Localization of the Hatch-Slack Cycle Within Different Cells of a C$_4$ Leaf. **(a)** Carbon dioxide fixation in C$_4$ plants initially occurs by the Hatch-Slack cycle within the mesophyll cells. (The path of incoming carbon is indicated by heavy black arrows.) Malate is then passed inward to the bundle sheath cells, where it is decarboxylated. The carbon dioxide is refixed by the Calvin cycle, eventually yielding sucrose, which passes into the adjacent vascular tissue for transport to other parts of the plant. The enzyme that catalyzes the phosphorylation of pyruvate, pyruvate phosphate dikinase, is unique to the Hatch-Slack cycle. The enzymes that catalyze the Hatch-Slack reactions are listed in the box at the lower right. **(b)** The particular C$_4$ leaf shown is that of maize, *Zea mays* (LM).

During the day, CAM plants close their stomata to conserve water. The malate then diffuses from the vacuoles to the cytosol, where the Hatch-Slack cycle continues. Carbon dioxide released by the decarboxylation of malate diffuses into the chloroplast stroma, where it is refixed and reduced by the Calvin cycle. The high carbon dioxide and low oxygen concentrations established when light is available for generating ATP and NADPH strongly favor rubisco's carboxylase activity and minimize the loss of carbon through photorespiration. Notice that the carboxylation of phosphoenolpyruvate and decarboxylation of malate occur in the same compartment. Because of this, the activity of phosphoenol-pyruvate carboxylase in CAM plants must be strictly inhibited during the day to prevent a futile cycle from developing.

With their remarkable ability to conserve water, CAM plants may assimilate over 25 times as much carbon as C_3 plants for each unit of water transpired! Moreover, some CAM plants display a process called *CAM idling*, whereby the plant keeps its stomata closed both night *and* day. Carbon dioxide is simply recycled between photosynthesis and respiration, with virtually no loss of water. Such plants will not, of course, display a net gain of carbohydrate. This ability, however, may enable them to survive droughts lasting up to several months.

PERSPECTIVE

Photosynthesis is the single most vital metabolic process for virtually all forms of life on Earth, because all of us, whatever our immediate sources of energy, ultimately depend on the energy radiating from the sun. Photosynthesis involves both energy transduction reactions and carbon assimilation reactions. During the energy transduction reactions, photons of light are absorbed by chlorophyll or accessory pigment molecules within the thylakoid or photosynthetic bacterial membranes, and the energy is rapidly passed to one of two special chlorophyll molecules at the reaction center of a photosystem. There, the energy is used to excite and eject an electron, which, in the case of photosystem I of oxygenic phototrophs, is passed via ferredoxin to $NADP^+$, generating the NADPH required for carbon dioxide fixation and reduction.

The source of electrons in oxygenic phototrophs is water. Electron transfer from water to $NADP^+$ depends on two photosystems acting in series, with photosystem II responsible for the oxidation of water and photosystem I responsible for the reduction of $NADP^+$. In plants, electron flow between the two photosystems (or in cyclic fashion around photosystem I) passes through a cytochrome b_6/f complex, where it drives the pumping of protons into the thylakoid lumen. The resulting proton motive force across the thylakoid membrane is due largely to the pH differential and is used to drive ATP synthesis by the CF_1 particles that protrude outward from the thylakoid membranes into the stroma of the chloroplast.

In the stroma, ATP and NADPH are used for the fixation and reduction of carbon dioxide into organic form by enzymes of the Calvin cycle. In C_3 plants, carbon dioxide is fixed directly onto ribulose-1,5-bisphosphate by rubisco, generating 3-phosphoglycerate. In C_4 and CAM plants, carbon dioxide is fixed by a preliminary carboxylation/decarboxylation pathway that concentrates it within a photosynthetic cell—either in a different cell or at a different time of day—for subsequent assimilation by the Calvin cycle. The eventual product of carbon dioxide fixation in each case is glyceraldehyde-3-phosphate, which can be converted to a second triose phosphate called dihydroxyacetone phosphate. Some of these triose phosphate molecules are used for the synthesis of more complex carbohydrates, such as sucrose or starch, or as sources of energy for other metabolic pathways. The remainder must be used for regenerating the acceptor molecule with which the Calvin cycle began. The net synthesis of one triose phosphate molecule requires the fixation of three carbon dioxide molecules and uses nine ATP and six NADPH molecules. A combination of noncyclic and cyclic electron flow ensures that the ratio of ATP to NADPH within a photosynthetic cell meets the metabolic demands imposed not only by carbon assimilation but also by other pathways, including those involved in nitrogen and sulfur assimilation.

This transduction of solar energy into chemical energy is crucial to the continued existence of the biological world. All the energy stored in organic molecules on which chemotrophs depend represents the energy of sunlight, originally trapped within the molecules of organic compounds during photosynthesis. We have not yet discovered anything particularly unique about the carbon metabolism of photoautotrophs, because carbon dioxide fixation (carboxylation) occurs in animals and other nonphotosynthetic organisms. What is remarkable about photosynthetic organisms is their ability to carry out sustained *net* fixation and reduction of carbon dioxide using solar energy to drive what would otherwise be a highly endergonic process. Only phototrophs can utilize sunlight to extract electrons from such poor (electropositive) donors as water and use them to reduce carbon atoms in carbon dioxide to the level of an organic compound. And they can do so because of the photochemical events that are initiated whenever light of the appropriate wavelength is absorbed by chlorophyll, a remarkable molecule that has transformed the biosphere of an entire planet—Earth.

KEY TERMS FOR SELF-TESTING

photosynthesis (p. 374)
phototroph (p. 374)
photoheterotroph (p. 374)
photoautotroph (p. 374)

An Overview of Photosynthesis

energy transduction reaction (p. 374)
carbon assimilation reaction (p. 374)
photophosphorylation (p. 374)
oxygenic phototroph (p. 376)
anoxygenic phototroph (p. 376)
photoreduction (p. 376)

The Chloroplast: A Photosynthetic Organelle

chloroplast (p. 376)
outer membrane (p. 376)
inner membrane (p. 376)
intermembrane space (p. 376)
stroma (p. 376)
semiautonomous organelle (p. 376)
thylakoid (p. 376)
granum (p. 376)
stroma thylakoid (p. 376)
thylakoid lumen (p. 377)

Photosynthetic Energy Transduction

photon (p. 377)
quantum (p. 377)
pigment (p. 377)
photoexcitation (p. 377)
absorption spectrum (p. 378)
resonance transfer (p. 378)
photochemical reduction (p. 378)
chlorophyll (p. 379)
bacteriochlorophyll (p. 380)
accessory pigment (p. 380)
carotenoid (p. 380)
phycobilin (p. 380)

Photosystems and Light-Harvesting Complexes

photosystem (p. 380)
antenna pigment (p. 380)
reaction center (p. 380)
light-harvesting complex (LHC) (p. 381)
phycobilisome (p. 381)
photosystem complex (p. 381)

Oxygenic Phototrophs Have Two Types of Photosystems

action spectrum (p. 381)
Emerson enhancement effect (p. 381)
photosystem I (PS I) (p. 381)
photosystem II (PS II) (p. 381)
P700 (p. 381)
P680 (p. 381)

Photoreduction (NADPH Synthesis) in Oxygenic Phototrophs

nicotinamide adenine dinucleotide phosphate ($NADP^+$) (p. 381)
LHC-II (p. 382)
plastoquinone (p. 382)
plastoquinol (p. 382)
water-splitting complex (p. 384)
cytochrome b_6/f complex (p. 384)
plastocyanin (PC) (p. 384)
LHC-I (p. 384)
ferredoxin (p. 385)
ferredoxin-$NADP^+$ reductase (p. 385)
noncyclic electron flow (p. 385)

Photophosphorylation (ATP Synthesis) in Oxygenic Phototrophs

ATP synthase complex (CF_oCF_1 complex) (p. 385)
CF_1 (p. 385)
CF_o (p. 385)
proton translocator (p. 385)
cyclic electron flow (p. 386)

Photosynthetic Carbon Assimilation: The Calvin Cycle

Calvin cycle (p. 389)
stoma (p. 389)
mesophyll cell (p. 389)
C_3 pathway (p. 389)
ribulose bisphosphate carboxylase/ oxygenase (rubisco) (p. 392)

Carbohydrate Synthesis

isoenzyme (p. 397)

The Reduction of Photosynthetic Efficiency by Rubisco's Oxygenase Activity

phosphoglycolate (p. 398)
glycolate pathway (p. 398)
photorespiration (p. 398)
leaf peroxisome (p. 0398)
Hatch-Slack cycle (p. 400)
C_4 pathway (p. 400)
C_4 plant (p. 400)
C_3 plant (p. 400)
bundle sheath cell (p. 400)
crassulacean acid metabolism (CAM) (p. 401)
CAM plant (p. 401)

PROBLEM SET

13-1. True, False, or ? Identify whether each of the following statements is true (T), false (F), or indecisive(?), if you cannot tell whether the statement is true or false on the basis of the information provided.

(a) Although sometimes called the light-independent reactions of photosynthesis, photosynthetic carbon assimilation actually depends indirectly on light and will not continue to function very long in the dark.

(b) Sucrose is synthesized in the chloroplast stroma and exported from photosynthetic cells to provide energy and reduced carbon for nonphotosynthetic plant cells.

(c) If the time interval between $^{14}CO_2$ exposure and the extraction of intermediates is short enough, the first organic compound to be radioactively labeled in a plant leaf will be 3-phosphoglycerate.

(d) The energy requirement expressed as ATP consumed per molecule of carbon dioxide fixed is higher for a C_3 plant than for a C_4 plant.

(e) The ultimate electron donor in the photosynthetic generation of NADPH is water.

(f) The enzyme rubisco is atypical in that it exhibits two different enzymatic activities, depending on conditions.

13-2. The Advantage of the Hatch-Slack Cycle. A C_4 plant may be more efficient than a C_3 plant at fixing carbon dioxide, an advantage that becomes more evident as the carbon dioxide concentration decreases.

(a) Explain in your own words why a C_4 plant may be inherently more efficient at carbon fixation than a C_3 plant.

(b) Why would this advantage be more apparent at an atmospheric carbon dioxide concentration of 0.004% than at the normal level of about 0.04%?

(c) If a C_4 plant and a C_3 plant are grown under constant illumination in a sealed container with an initial carbon dioxide concentration of 0.04%, the C_4 plant will eventually kill the C_3 plant. Explain why.

13-3. The Role of Sugar. A plant can be viewed as a system that uses solar energy to make ATP and NADPH, which then drive the synthesis of carbohydrates in the leaves. At least one carbohydrate, sucrose, is translocated to nonphotosynthetic parts of the plant (roots, stems, flowers, fruits) for use as an energy source. Thus, ATP is used to make sucrose, and the sucrose is then used to make ATP. It would seem simpler for the plant just to make ATP and translocate the ATP itself directly to other parts of the plant, thereby completely eliminating the need for a Calvin cycle, a glycolytic pathway, and a TCA cycle (and making life a lot easier for cell biology students in the process). Why do you suppose this is not the way a plant manages its energy economy? You should be able to think of at least two major reasons.

13-4. Effects on Photosynthesis. Assume that you have an illuminated suspension of *Chlorella* cells carrying out photosynthesis in the presence of 0.1% carbon dioxide and 20% oxygen. What will be the short-term effects of the following changes in conditions on the levels of 3-phosphoglycerate (PGA) and ribulose-1,5-bisphosphate (RuBP)? Explain your answer in each case.

(a) Carbon dioxide concentration is suddenly reduced 1000-fold.

(b) Light is restricted to green wavelengths (510–550 nm).

(c) An inhibitor of photosystem II is added.

(d) Oxygen concentration is reduced from 20% to 1%.

13-5. Overall Reactions. Eukaryotic phototrophs use water as an electron donor, but most prokaryotic phototrophs depend instead on a variety of inorganic and organic electron donors. Reaction 13-1 is intended as a general reaction to cover all cases. Some bacteria use H_2S as their source of electrons, generating elemental sulfur instead of molecular oxygen.

(a) Write a balanced overall reaction for photosynthesis based on H_2S as the electron donor and a triose phosphate as the end-product.

(b) Why are photosynthetic bacteria able to use H_2S but not H_2O as an electron donor?

(c) Do you think photosynthetic bacteria are able to generate reduced coenzyme (either NADPH or NADH) by oxidizing ferrous ions to ferric ions? (The E_0 value for the Fe^{3+}/Fe^{2+} redox pair is $+0.77$ V.)

13-6. Energy Flow in Photosynthesis. A portion of the solar energy that arrives at the surface of a plant leaf is eventually converted to chemical energy and appears in the chemical bonds of carbohydrates that are generally regarded as the end-products of photosynthesis. In between the photons and the carbohydrate molecules, however, the energy exists in a variety of forms. Trace the flow of energy from photon through ATP to starch molecule, assuming the wavelength of light to be in the absorption range of one of the accessory pigments rather than that of chlorophyll.

13-7. The Mint and the Mouse. A prominent name in the early history of research in photosynthesis is that of Joseph Priestley, a British clergyman. In 1771, Priestley wrote these words:

One might have imagined that since common air is necessary to vegetable as well as to animal life, both plants and animals had affected it in the same manner; and I own that I had that expectation when I first put a sprig of mint into a glass jar standing inverted in a vessel of water; but when it had continued growing there for some months, I found that the air would neither extinguish a candle, nor was it at all inconvenient to a mouse which I put into it.

Explain the basis of Priestley's observations, and indicate their relevance to the early understanding of the nature of photosynthesis.

13-8. The Hill Reaction. A highly significant advance in our understanding of photosynthesis came in 1937 when Robert Hill showed that isolated chloroplasts, though not capable of fixing carbon dioxide, were able to produce molecular oxygen. This only occurred if the chloroplasts were illuminated and provided with an artificial electron acceptor. (Ferricyanide was the acceptor of choice.) Which of the following statements are valid conclusions from Hill's experiment? Some, all, or none may be correct.

(a) The oxygen generated during photosynthesis apparently does not come from carbon dioxide, as was believed earlier.

(b) These findings are in accord with a proposal made six years earlier by C. B. van Niel that the oxygen evolved during photosynthesis comes from water.

(c) Reaction 13-2 would be less indicative of what happens during photosynthesis in eukaryotic phototrophs if three molecules of water were subtracted from each side of the reaction.

13-9. Photophosphorylation. ATP generation in higher plants can occur as the result of either cyclic or noncyclic electron flow. To sustain the ratio of three ATP to two NADPH molecules required by the Calvin cycle, the ratio of cyclic to noncyclic electron flow would have to be about 1:6. But other processes also consume ATP and NADPH. For each of the following conditions, indicate whether you would expect the ratio of cyclic to noncyclic flow to be higher than 1:6, unchanged, or lower than 1:6.

(a) The reduction of nitrite to ammonia (NH_3), which occurs in the chloroplast stroma and consumes NADPH but not ATP.

(b) Extensive active transport across the chloroplast inner membrane, which consumes ATP but not NADPH.

(c) In some plants, the Hatch-Slack cycle conveys carbon dioxide from mesophyll cells to bundle sheath cells.

(d) Treatment of a plant with DCMU, a herbicide that blocks electron transfer from Q_A to Q_B in photosystem II.

(e) Treatment of a plant with ferricyanide, which accepts electrons from the "bottom" of the electron transport system, becoming reduced to ferrocyanide. (The E_0 for the ferricyanide/ferrocyanide redox pair is $+0.4$ V.)

13-10. Photosynthetic Efficiency. Earlier, we estimated the maximum photosynthetic efficiency for the conversion of red light (wavelength of 670 nm), carbon dioxide, and water to glyceraldehyde. Under laboratory conditions, a photosynthetic organism might convert 31% of the light energy striking it to chemical bond energy of organic molecules. In reality, however, photosynthetic efficiency is far lower—closer to 5% or less. Considering a plant growing in a natural environment, suggest three or four reasons for this discrepancy.

13-11. Chloroplast Structure. Where in the chloroplast is each of the following substances or processes located? Be as specific as possible.

(a) Reduction of 3-phosphoglycerate

(b) Cyclic electron flow

(c) Carotenoid molecules

(d) Photophosphorylation

(e) Proton pumping

(f) P700

(g) Transketolase

(h) Plastoquinol

13-12. Metabolite Transport Across Membranes. For each of the following metabolites, indicate whether you would expect it to be in steady-state flux across one or more membranes in a photosynthetically active chloroplast, and, if so, indicate which membrane(s) the metabolite must cross.

(a) Carbon dioxide

(b) Protons

(c) Electrons

(d) Starch

(e) Glyceraldehyde-3-phosphate

(f) NADPH

(g) ATP

13-13. Crassulacean Acid Metabolism. A CAM plant uses a pathway very similar to the Hatch-Slack cycle for preliminary carbon dioxide fixation. Trace the flow of carbon from the atmosphere to glyceraldehyde-3-phosphate within a CAM plant. How does this minimize water loss by such plants?

SUGGESTED READING

General References

Hall, D. O., and K. K. Rao. *Photosynthesis*, 5th ed. New York: Cambridge University Press, 1994.

Hatch, M. D., and N. K. Boardman, eds. *Photosynthesis. The Biochemistry of Plants*, Vol. 10. New York: Academic Press, 1987.

Lawlor, D. W. *Photosynthesis: Molecular, Physiological and Environmental Processes*, 2d ed. Essex, England: Longman Scientific & Technical, 1993.

Mathews, C. K., and K. E. van Holde. *Biochemistry*, 2d ed. Menlo Park, CA: Benjamin/Cummings, 1996.

Preiss, J., ed. *Carbohydrates. The Biochemistry of Plants*, Vol. 14. New York: Academic Press, 1988.

Salisbury, F. B., and C. W. Ross. *Plant Physiology*, 4th ed. Belmont, CA: Wadsworth, 1992.

Staehelin, L. A., and C. J. Arntzen, eds. *Photosynthesis III: Photosynthetic Membranes and Light Harvesting Systems. Encyclopedia of Plant Physiology*, New Series, Vol. 19. Berlin: Springer-Verlag, 1986.

Tobin, A. K., ed. *Plant Organelles*. New York: Cambridge University Press, 1992.

Youvan, D. C., and B. L. Marrs. Molecular mechanisms of photosynthesis. *Sci. Amer.* 249 (June 1987): 42.

The Chloroplast

Bogorad, L. Chloroplasts. *J. Cell Biol.* 91 (1981): 256s.

Cramer, W. A., W. R. Widger, R. G. Herrmann, and A. Trebst. Topography and function of thylakoid membrane proteins. *Trends Biochem. Sci.* 10 (1985): 125.

Haliwell, B. *Chloroplast Metabolism: The Structure and Function of Chloroplasts in Green Leaf Cells*. Oxford, England: Clarendon Press, 1981.

Hoober, J. K. *Chloroplasts*. New York: Plenum, 1984.

Miller, K. R., and M. K. Lyon. Do we really know why chloroplast membranes stack? *Trends Biochem. Sci.* 10 (1985): 219.

Photosynthetic Energy Transduction

Andréaesson, L.-H., and T. Vänngard. Electron transport in photosystems I and II. *Annu. Rev. Plant Physiol. Plant Molec. Biol.* 39 (1988): 379.

Blankenship, R. E., and R. C. Prince. Excited state redox potentials and the Z scheme of photosynthesis. *Trends Biochem. Sci.* 10 (1985): 382.

Deisenhofer, J., and H. Michel. The photosynthetic reaction center from the purple bacterium *Rhodopseudomonas viridis*. *Science* 245 (1989): 1463.

Glazer, A. N., and A. Melis. Photochemical reaction centers: Structure, organization, and function. *Annu. Rev. Plant Physiol.* 38 (1987): 11.

Goodwin, T. W., ed. *Plant Pigments*. London: Academic Press, 1988.

Govindjee and W. J. Coleman. How plants make oxygen. *Sci. Amer.* 262 (February 1990): 50.

Malkin, R. Cytochrome bc_1 and b_6f complexes of photosynthetic membranes. *Photosynth. Res.* 33 (1992): 121.

The Calvin Cycle and Carbohydrate Synthesis

Buchanan, B. B. Carbon dioxide assimilation in oxygenic and anoxygenic photosynthesis. *Photosynth. Res.* 33 (1992): 147.

Buchanan, B. B. The ferredoxin/thioredoxin system: A key element in the regulatory function of light in photosynthesis. *BioSci.* 34 (1984): 378.

Ellis, R. J. The most abundant protein in the world. *Trends Biochem. Sci.* 4 (1979): 241.

Ellis, R. J., and J. C. Gray, eds. Ribulose bisphosphate carboxylase-oxygenase. *Philos. Trans. R. Soc. London Ser. B* 313 (1986): 303.

Flügge, U. I., and H. W. Heldt. Metabolite translocators of the chloroplast envelope. *Annu. Rev. Plant Physiol. Plant Molec. Biol.* 42 (1991): 129.

Sharkey, T. D. Evaluating the role of rubisco regulation in photosynthesis of C3 plants. *Philos. Trans. R. Soc. London Ser. B* 323 (1989): 435.

Photorespiration, C$_4$ Plants, and CAM Plants

Bazzaz, F. A., and E. D. Fajer. Plant life in a CO_2-rich world. *Sci. Amer.* 266 (January 1992): 68.

Bowes, G. Facing the inevitable: Plants and increasing CO_2. *Annu. Rev. Plant Physiol. Plant Molec. Biol.* 44 (1993): 309.

Hatch, M. D. C4 photosynthesis: An unlikely process full of surprises. *Plant Cell Physiol.* 33 (1992): 333.

Heber, U., and G. H. Krause. What is the physiological role of photorespiration? *Trends Biochem. Sci.* 4 (1979): 32.

Huang, A. H. C., R. N. Trelease, and T. S. Moore, Jr. *Plant Peroxisomes*. New York: Academic Press, 1983.

Ogren, W. L. Photorespiration: Pathways, regulation, and modification. *Annu. Rev. Plant Physiol.* 35 (1984): 415.

Ting, I. P. Crassulacean acid metabolism. *Annu. Rev. Plant Physiol.* 36 (1985): 595.

Box 13A: Carbon-14, Paper Chromatography, and the Calvin Cycle

Calvin, M. The path of carbon in photosynthesis. *Science* 135 (1962): 879.

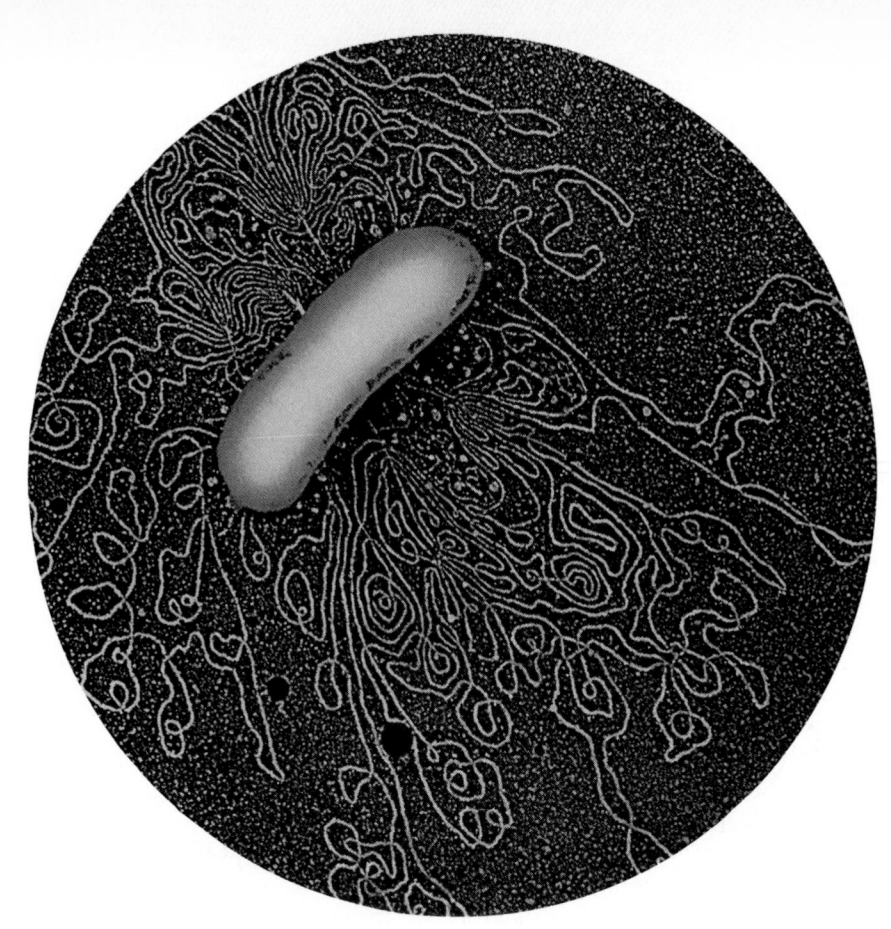

INFORMATION FLOW
IN CELLS

14

The Structural Basis of Cellular Information: DNA, Chromosomes, and the Nucleus

Implicit in our earlier discussions of cellular structure and function has been a sense of predictability, order, and control. We have come to expect that organelles and other cellular structures will have a predictable appearance and function, that metabolic pathways and other cellular activities will proceed in an orderly fashion in specific intracellular locations, and that all of this will be carried out in a carefully controlled, highly efficient, and heritable manner.

Such expectations express our confidence that cells have "instructions" that specify their structure, dictate their functions, and regulate their activities, and that these instructions can be passed on faithfully to daughter cells. Over a hundred years ago, the Augustinian monk Gregor Mendel worked out rules accounting for inheritance patterns he observed with pea plants, although he had little inkling of the cellular or molecular basis for these rules. Now we know that an organism's inherited instructions reside in the genetic information of each of its cells as DNA, and we can tell a coherent genetic story starting with this molecule.

But first let's step back and use Figure 14-1 to preview how DNA functions as instructional information in cells and, at the same time, how this unit of chapters on information flow in cells is organized. The information carried by cellular DNA flows both between generations of cells and within each individual cell. As Figure 14-1a indicates, the information carried in a eukaryotic cell's DNA is passed on to daughter cells by the processes of DNA replication and mitosis. The DNA is first duplicated (replication), and then the two copies are distributed to the daughter cells when the cell divides. This and the next two chapters focus on the structures and events associated with this aspect of information flow. This chapter covers the structure of DNA and its organization, along with proteins, in structures called chromosomes; it also discusses the nucleus, the organelle that houses the chromosomes of eukaryotic cells. Chapter 15 discusses DNA replication and cell division, and Chapter 16 considers the cellular and molecular bases of information flow between

generations of sexually reproducing organisms (including Mendel's work and its chromosomal basis).

Figure 14-1b outlines the instructional role of DNA information *within* a cell. The two main processes here are called *transcription* and *translation*. In transcription, RNA is synthesized. In translation, sequences of nucleotides in RNA determine the sequences of amino acids in proteins. (The diagrams here show eukaryotic cells, which carry out transcription within a membrane-bounded nucleus, transport the messenger RNA to the cytoplasm, and carry out translation there.) It is the particular proteins synthesized by a cell that most directly determine the cell's structure and the functions it performs. Transcription and translation constitute the expression of genetic information. These processes are the subjects of Chapters 17–19.

We begin this chapter with Johann Friedrich Miescher, the Swiss biochemist who discovered the chemical we call DNA.

The Chemical Nature of the Genetic Material

At the same time that Mendel was deducing the basic rules of inheritance, Johann Friedrich Miescher was investigating the chemical substance that would turn out to be the genetic material, DNA. Miescher published his first findings in 1869, just a few years after Mendel's initial report of his findings on the heritability of specific traits in peas and just a few years before the cytologist Walther Flemming would first see chromosomes as he studied dividing cells under the microscope.

Pus, Fish Sperm, and the Discovery of DNA

Miescher was interested in studying the chemistry of the nucleus, which most scientists guessed was the site of the cell's

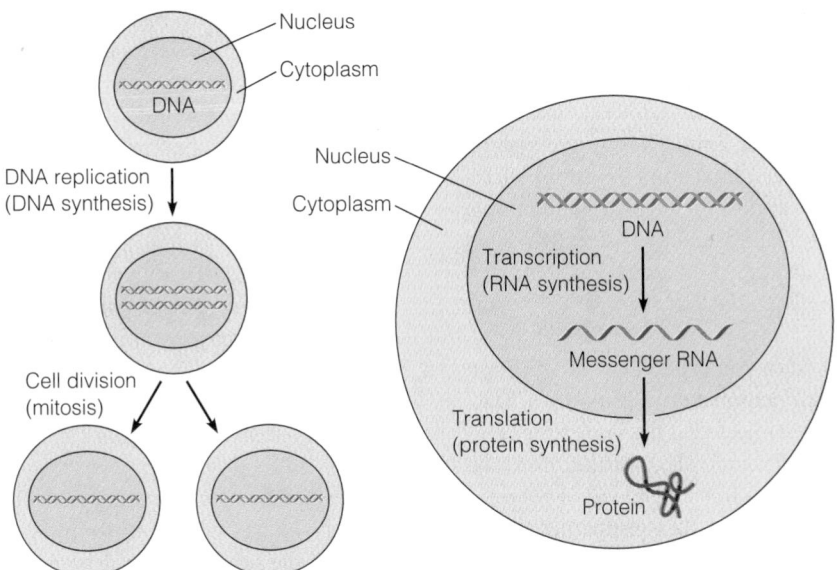

(a) The flow of genetic information between generations of cells

(b) The flow of genetic information within a cell: the expression of genetic information

Figure 14-1   The Flow of Information in Cells.   The diagrams here feature eukaryotic cells, but DNA replication, cell division, transcription, and translation are processes that occur in prokaryotic cells as well. **(a)** Genetic information encoded in DNA molecules is passed on to successive generations of cells by DNA replication and cell division (in eukaryotic cells, by means of mitosis). The DNA is first duplicated and then divided equally between the two daughter cells. In this way, each daughter cell is assured of having the same genetic information as the cell from which it arose. **(b)** Within each cell, genetic information encoded in the DNA is expressed through the processes of transcription (RNA synthesis) and translation (protein synthesis). Transcription involves the use of selected segments of DNA as templates for the synthesis of messenger RNA and other RNA molecules. Translation is the process whereby amino acids are joined together in a sequence dictated by the sequence of nucleotides in messenger RNA.

genetic material. In his initial experiments, Miescher isolated nuclei from the white blood cells present in pus recovered from surgical bandages. By treating the nuclei with alkali, he was able to prepare an extract containing a chemically unusual substance that he called "nuclein," which we now know to have been largely DNA. Miescher then went on to study "nuclein" from a more pleasant source, salmon sperm. (Fish sperm may seem a somewhat unusual source material, until we realize that the nucleus accounts for more than 90% of the mass of a typical sperm cell. Sperm cells are therefore rich in DNA, and fish sperm is a widely used source of DNA.) From sperm nuclei, Miescher isolated a purer form of "nuclein" and found it to consist of unusually large acidic molecules, rich in phosphorus. He also isolated a basic substance he called *protamine*. We now recognize protamine as a class of proteins unique to sperm nuclei; they are small proteins containing large amounts of the basic amino acids lysine and arginine. Ironically, the implications of this finding—that nuclei contain proteins as well as "nuclein"—contributed to an 80-year delay in the recognition of DNA's role in heredity.

From Miescher's day until the middle of the twentieth century, most scientists believed that nuclear proteins, rather than "nuclein," were the genetic material. The basic chemical structures of both protein and nucleic acid were worked out by the early 1900s. Protein was perceived to be the more complex kind of substance and was therefore considered the more logical candidate for genes. Nucleic acid, by contrast, was widely believed to be a simple polymer consisting of a single tetranucleotide sequence (e.g., ATCG) repeated over and over, hence lacking the variability expected of a genetic molecule. Nucleic acid was thought to be merely a scaffold for genes made of protein. This view prevailed until two lines of experiments resolved the matter definitively in favor of DNA as the genetic material, as we describe in the next two sections.

DNA as the Transforming Principle of "Pneumococcus"

In 1928, the British physician Frederick Griffith was studying a pathogenic strain of a bacterium, then called "pneumococcus," that causes a fatal pneumonia in animals. When the bacteria (now called *Streptococcus pneumoniae*) are grown on a solid agar medium in the laboratory, these pathogenic bacteria produce colonies that are smooth and shiny, because of the mucous, polysacccharide coat that each cell secretes. Such strains are designated S (smooth) strains, to distinguish them from R (rough) strains, which lack the ability to synthesize the surrounding mucous capsule and which therefore produce colonies with a rough surface.

Significantly, the R strains of pneumococcus are non-pathogenic. When injected into a mouse, for example, these bacteria are readily killed by the animal's immune system, and the mouse usually recovers. Injection of S-strain bacteria, on the other hand, leads almost inevitably to pneumonia and the death of the mouse. Pathogenicity is directly related to the presence of the S strain's polysaccharide coat; it is this coat that protects the bacterial cell from attack by the mouse's immune system.

Griffith showed that mice could be killed by infection with cells of an R strain of pneumococcus, but only if heat-killed cells of an S strain were injected at the same time (Figure 14-2d; parts a–c are the controls). Furthermore, he could recover live S-type bacteria from the blood of the dead mouse (part e)! Griffith concluded that nonpathogenic R bacteria could be converted into pathogenic S bacteria by something present in the heat-killed S bacteria that were coinjected. He called the phenomenon **transformation** and referred to the active (though still unknown) substance in the S cells as the "transforming principle."

Griffith's discoveries set the stage for 14 years of work by Oswald Avery and his colleagues at Rockefeller Institute in New York. These researchers pursued the investigation of bacterial transformation to its logical conclusion by asking which component of the heat-killed S bacteria was actually responsible for the transforming activity. They fractionated

cell-free extracts of S bacteria and found that only the nucleic acid fraction was capable of causing transformation. Moreover, the activity was specifically eliminated by treatment with deoxyribonuclease, an enzyme that degrades DNA. This and other evidence convinced them that the transforming substance of pneumococcus was DNA. Avery, Colin MacLeod, and Maclyn McCarty published this conclusion in 1944.

This was the first rigorously documented assertion that DNA could carry genetic information. Some of the excitement of that discovery and a glimpse into Avery's appreciation of its implications can be found in a letter that Avery wrote to his brother Roy in May, 1943. Here is an excerpt from that letter:

For the past two years, first with MacLeod and now with Dr. McCarty, I have been trying to find out what is the chemical nature of the substance in the bacterial extract which induces this specific change. The crude extract of Type III [the pathogenic, S strain of bacteria] is full of capsular polysaccharide, . . . carbohydrate, nucleoproteins, free nucleic acids of both the yeast [RNA] and thymus [DNA] type, lipids, and other cell constituents. Try to find in the complex mixtures the active principle! Try to isolate and chemically identify the particular substance that will by itself, when brought into contact with the R cell . . . cause it to elaborate Type III capsular polysaccharide and to acquire all the aristocratic distinctions of the same specific type of cells

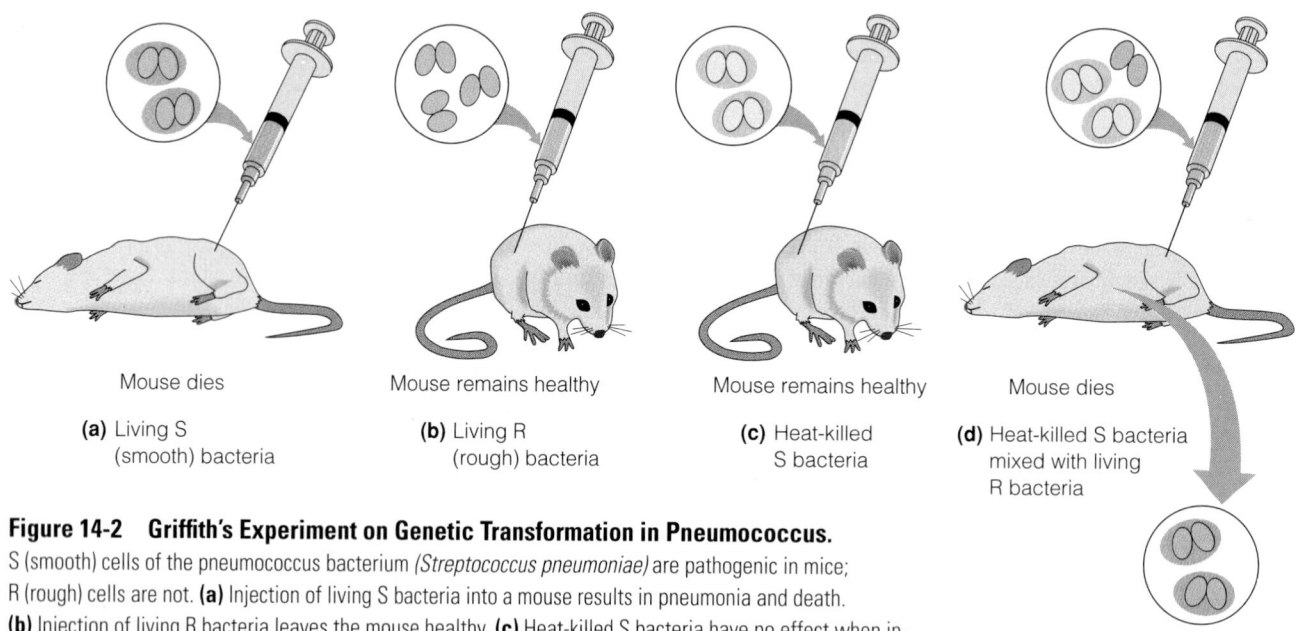

Mouse dies

Mouse remains healthy

Mouse remains healthy

Mouse dies

(a) Living S (smooth) bacteria

(b) Living R (rough) bacteria

(c) Heat-killed S bacteria

(d) Heat-killed S bacteria mixed with living R bacteria

(e) Living S bacteria in blood from dead mouse

Figure 14-2   Griffith's Experiment on Genetic Transformation in Pneumococcus.
S (smooth) cells of the pneumococcus bacterium *(Streptococcus pneumoniae)* are pathogenic in mice; R (rough) cells are not. **(a)** Injection of living S bacteria into a mouse results in pneumonia and death. **(b)** Injection of living R bacteria leaves the mouse healthy. **(c)** Heat-killed S bacteria have no effect when injected alone. **(d)** When a mixture of living R bacteria and heat-killed S bacteria is injected, the result is pneumonia and death. **(e)** The finding that living pathogenic S-type bacteria could be recovered from the blood of the mouse in part d suggested to Griffith that some chemical factor from the heat-killed S cells was able to cause a heritable change (transformation) of nonpathogenic R bacteria into pathogenic S bacteria. The chemical factor was later identified as DNA.

as that from which the extract was prepared! Some job, full of headaches and heartbreaks. But at last perhaps we have it.

. . . If we prove to be right—and of course that is a big if—then it means that both the chemical nature of the inducing stimulus is known and the chemical structure of the substance produced is also known, the former being thymus nucleic acid [DNA], the latter Type III polysaccharide, and both are thereafter reduplicated in the daughter cells and after innumerable transfers without further addition of the inducing agent and the same active and specific transforming substance can be recovered far in excess of the amount originally used to induce the reaction. Sounds like a virus—may be a gene. But with mechanisms I am not now concerned. One step at a time and the first step is what is the chemical nature of the transforming principle? Some one else can work out the rest. Of course the problem bristles with implications. It touches the biochemistry of . . . [DNA molecules,] which are known to constitute the major part of chromosomes but have been thought to be alike regardless of origin and species. It touches genetics, enzyme chemistry, cell metabolism and carbohydrate synthesis. But today it takes a lot of well documented evidence to convince anyone that the sodium salt of deoxyribose nucleic acid, protein free, could possibly be endowed with such biologically active and specific properties and that is the evidence we are now trying to get. lt is lots of fun to blow bubbles but it is wiser to prick them yourself before someone else tries to.*

Though the experiments of Avery and his colleagues were rigorous, the assignment of a genetic role to DNA did not meet with immediate acceptance. Skepticism was due in part to the persistent, widespread conviction that DNA lacked the necessary complexity for such a role. In addition, many scientists questioned whether genetic information in bacteria had anything to do with heredity in other organisms. Most remaining doubts were alleviated eight years later, however, when DNA was also shown to be the genetic material of a virus, the bacteriophage T2.

DNA as the Genetic Material of Viruses

Bacteriophages—or **phages,** for short—are viruses that infect bacteria. They have been objects of genetic research since the 1930s. As a result, a great deal is known about phages, and much of our understanding of molecular genetics has come from experiments with these viruses. Box 14A (pages 413–416) describes the anatomy and replication cycle of some phages and highlights some of the advantages of bacteriophages for genetic studies.

*DNA as the Transforming Principle of Pneumococcus. Excerpt from a letter written by Oswald Avery to his brother Roy, capturing some of the excitement and implications of Avery's findings on the transforming principle in "Pneumococcus." Reproduced by R. D. Hotchkiss in *Phage and the Origins of Molecular Biology*, J. Cairns, G. S. Stent, and J. D. Watson, eds., Cold Spring Harbor, New York: Cold Spring Harbor Laboratory, 1966. Reprinted by permission.

Bacteriophage T2 is one of the most thoroughly studied of the phages that infect the well-studied bacterium *Escherichia coli*. That was, in fact, a distinction T2 already enjoyed in 1952, when Alfred Hershey and Martha Chase, at the Cold Spring Harbor Laboratory in New York, devised an experiment to determine the chemical nature of this phage's genetic material. Hershey and Chase took advantage of the fact that the proteins of T2, like most proteins, contain the element sulfur (in the amino acids methionine and cysteine) but not phosphorus, and that DNA contains phosphorus (in its sugar-phosphate backbone) but not sulfur. Hershey and Chase were therefore able to prepare two batches of T2 phage particles (as intact phages are called) with different kinds of radioactive labeling. In one batch, the phage proteins were labeled with the radioactive isotope ^{35}S; in the other batch, the phage DNA was labeled with the isotope ^{32}P.

By using radioactive isotopes in this way, Hershey and Chase were able to trace the fates of both protein and DNA during the infection process (Figure 14-3a). The experiment began with adsorption (attachment) of the radioactive phage particles to the bacterial cells and the transfer of genetic information from the phage particles into the host cell. Hershey and Chase found that the empty protein coats (or phage "ghosts") could be effectively removed from the surface of the bacterial cells by whirling the suspension in an ordinary kitchen blender and recovering the bacterial cells by centrifugation. They then measured the radioactivity in the pellet of bacteria at the bottom of the tube and in the supernatant liquid.

The critical observation made by Hershey and Chase was that most (80%) of the ^{35}S could be dislodged from the bacterial cells but most (65%) of the ^{32}P remained with the cells (Figure 14-3b). They also showed that, if the infected bacteria are resuspended in fresh liquid and incubated longer, the ^{32}P gets transferred to some of the offspring phage particles, but the ^{35}S does not. These findings led them to the conclusion that DNA, not protein, is the actual genetic material of phage T2.

Thus, the Hershey-Chase experiment further strengthened the case for DNA as the genetic material. This experiment received a much warmer welcome than Avery's work on bacterial transformation. The main reason seems to have been simply the passage of time and the accumulation of additional, circumstantial evidence since Avery's 1944 publication. Perhaps most important was evidence that DNA was indeed variable enough to serve as the genetic material. This evidence came from studies of the base composition of DNA, as we describe next.

DNA Base Composition and Chargaff's Rules

Despite the generally lukewarm reaction Avery, MacLeod, and McCarty received in 1944, their work was an important influence on several other scientists. Among these was Erwin Chargaff of Columbia University, who focused on the base composition of DNA. Between 1944 and 1952, Chargaff and

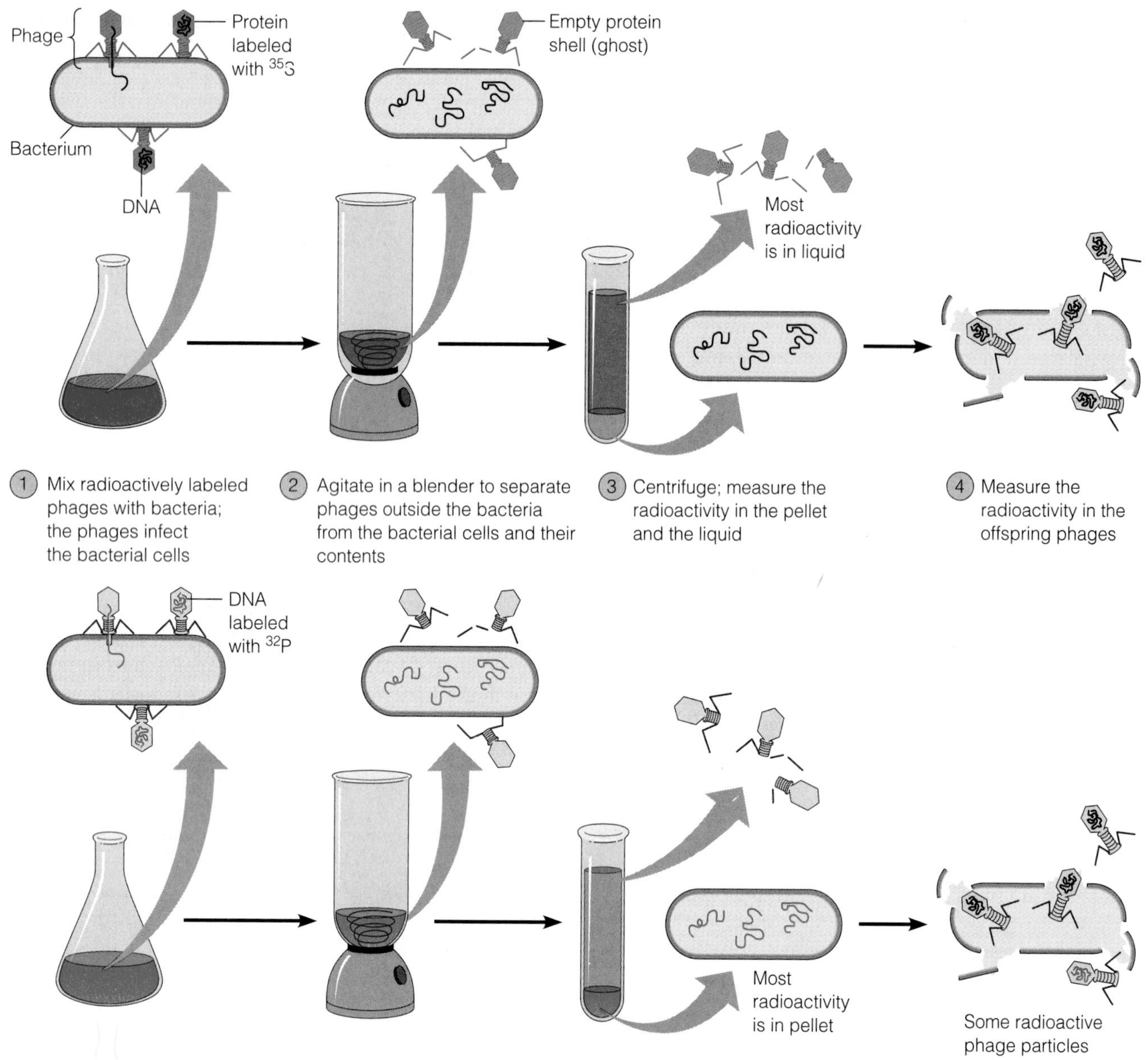

Phage

Protein labeled with ^{35}S

Empty protein shell (ghost)

Bacterium

DNA

Most radioactivity is in liquid

① Mix radioactively labeled phages with bacteria; the phages infect the bacterial cells

② Agitate in a blender to separate phages outside the bacteria from the bacterial cells and their contents

③ Centrifuge; measure the radioactivity in the pellet and the liquid

④ Measure the radioactivity in the offspring phages

DNA labeled with ^{32}P

Most radioactivity is in pellet

Some radioactive phage particles

(a) The Hershey-Chase experiment

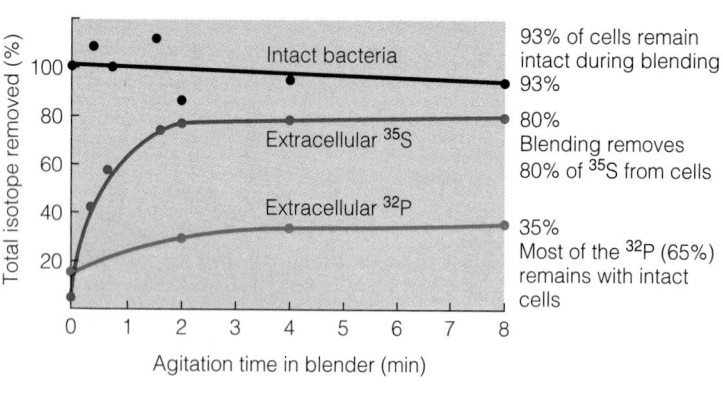

93% of cells remain intact during blending 93%

80% Blending removes 80% of ^{35}S from cells

35% Most of the ^{32}P (65%) remains with intact cells

(b) Experimental data from part a, step 3

Figure 14-3   The Hershey-Chase Experiment: DNA as the Genetic Material of Phage T2.   (a) ① T2 radioactively labeled with either ^{35}S (to label protein) or ^{32}P (to label DNA) is used to infect duplicate batches of bacteria. The phages adsorb to the cell surface and inject their DNA. ② Agitation of the infected cells in a blender dislodges most of the ^{35}S from the cells in phage ghosts, whereas most of the ^{32}P remains with the cells. ③ Immediate centrifugation causes the cells to form a pellet in the tube; any free phage particles, including ghosts, remain in the supernatant liquid. The radioactivity of the pellet and supernatant are measured separately. ④ When the cells in each pellet are incubated further, the phage DNA within them dictates the synthesis and eventual release of new phage particles. Some of these phages contain ^{32}P in their DNA (because the old, labeled phage DNA is packaged into some of the new particles), but none contain ^{35}S in their coat proteins. **(b)** The graph shows the extent to which ^{35}S and ^{32}P are removed from the intact cells at step 3, as a function of time in the blender. More than 90% of the cells remain intact, even after 8 minutes of blending, and even a few minutes of blending is enough to remove most (80%) of the ^{35}S, while leaving most (65%) of the ^{32}P with the cells.

14A

PHAGES: THE BEST-UNDERSTOOD GENETIC SYSTEMS

From its inception in the mid-nineteenth century, genetics has drawn upon a wide variety of organisms for its experimental materials. Initially, attention focused on plants and animals, such as Mendel's peas and the fruit flies popularized by later investigators. Around 1940, however, bacteria and viruses came into their own, providing geneticists with experimental systems that literally revolutionized the science by bringing it to the molecular level.

Bacteriophages have been especially important. Bacteriophages, or phages, for short, are viruses that infect bacterial cells. It is easy to obtain huge numbers of phage particles in a short time, which greatly facilitates screening for mutants—phages with heritable variations—thereby making it possible for geneticists to identify particular genes. In addition, phage genetic material—its "genome," which may be DNA or RNA—is usually much smaller than that of even the simplest bacteria, making it particularly amenable to physical, chemical, and genetic analysis. As a result, more is known about the storage, transmission, and expression of genetic information in bacteriophages than in any other kind of genetic system.

Some of the most thoroughly studied phages are the T2, T4, and T6 (the so-called T-even) bacteriophages, which infect the bacterium *E. coli*. The three T-even phages have similar structures, which are quite elaborate. T4 is shown in Figure 14A-1. The *head* of the phage is a protein capsule that is shaped like a hollow icosahedron (a 20-sided object), and it is filled with DNA. The head is attached to a protein *tail*, which consists of a hollow *tail core* surrounded by a contractile *tail sheath* and terminating in a hexagonal *baseplate*, to which six *tail fibers* are attached.

Figure 14A-2a depicts the main events in the replication cycle of the T4 phage. The drawings are not to scale; the bacterium is proportionately larger, as the electron micrograph indicates (Figure 14A-2b). The process begins with the adsorption of a phage particle to the wall of a bacterial cell. When the phage collides with the cell, it "squats" so that its baseplate attaches to a specific receptor protein in the wall. Next, the tail sheath contracts, driving the hollow tail core through the cell wall. The core forms a needle through which the DNA is injected into the bacterium.

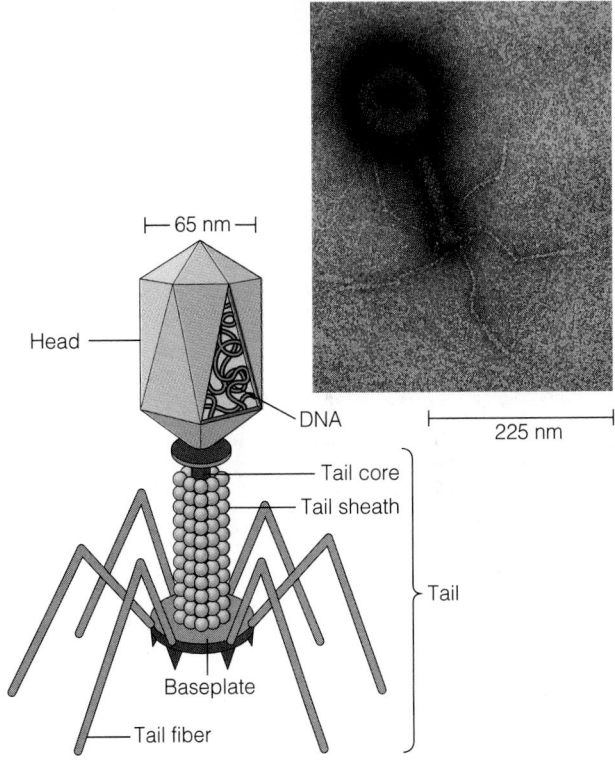

Figure 14A-1   The Structure of Bacteriophage T4.   The drawing identifies the main structural components of this phage, not all of which are visible in the micrograph (TEM).

Once the DNA of the bacteriophage has gained entry to the bacterial cell, the genetic information of the phage is transcribed and translated. This gives rise to a few key proteins that subvert the metabolic machinery of the host cell for the phage's benefit, which is usually its own rapid multiplication. Since the phage consists simply of a DNA molecule surrounded by a protein coat (its *capsid*), most of the metabolic activity in the infected cell is channeled toward the replication of phage DNA and the synthesis of capsid proteins.

(Continued on page 415)

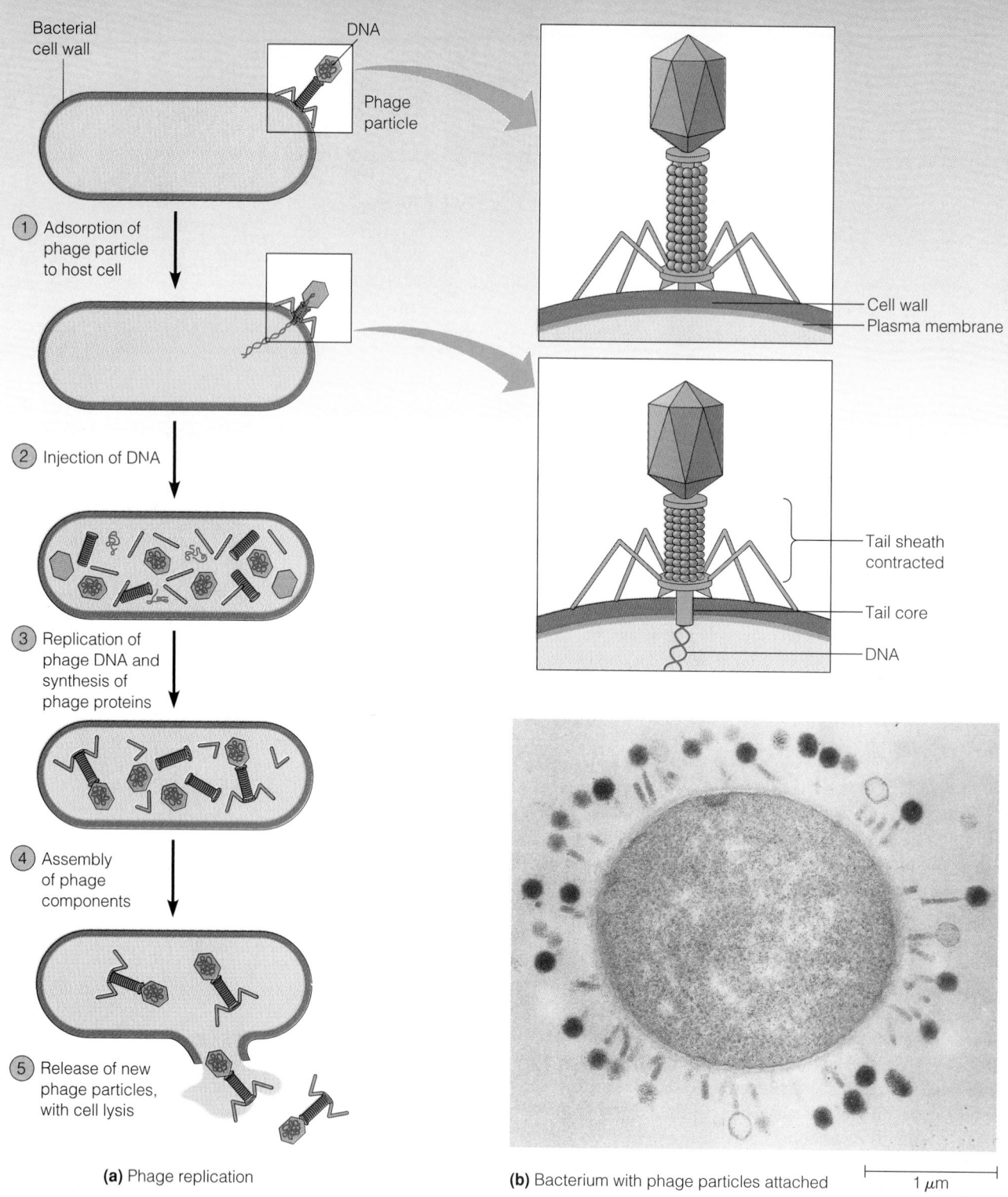

(a) Phage replication

(b) Bacterium with phage particles attached

1 μm

Figure 14A-2   Replication of a T-Even Phage.   (a) The replication cycle of a T-even phage begins when a phage particle ① becomes adsorbed to the surface of a bacterial cell and ② injects its DNA into the cell. ③ The phage DNA replicates in the host cell and codes for the production of phage proteins. ④ These components assemble into new phage particles. ⑤ Eventually, the host cell lyses, releasing offspring phage particles that can infect additional bacteria. This replication process is typical of the lytic growth of many phages. **(b)** Electron micrograph of a bacterium with phage particles attached to its surface (TEM).

The phage DNA and capsid proteins then self-assemble into hundreds of new phage particles. Within about half an hour, the infected cell lyses (breaks open), releasing the new phage particles into the medium. Each of these can now infect another bacterial cell, making it possible to obtain enormous populations of phage—as many as 10^{11} phage particles per milliliter in infected bacterial cultures.

To determine the number of phage particles in a sample, a measured volume is diluted appropriately and mixed with a few drops of a bacterial culture growing in liquid medium, to allow adsorption of the phages to the bacteria. The mixture is then spread onto solid (agar-containing) nutrient medium in a Petri dish. Upon incubation, the bacteria multiply to produce a dense "lawn" of cells on the surface of the nutrient medium. Wherever a virus particle has infected a bacterial cell, however, a clear spot is visible in the lawn, because the bacterial cells there have been killed by the multiplying phage population. Such clear spots are called **plaques.** The number of plaques on the bacterial lawn represents the number of phage particles in the original phage-bacterium mixture, provided only that the initial number of phages was small enough to ensure that each gives rise to a separate plaque. Figure 14A-3 shows plaques formed by T4 bacteriophage on a lawn of *E. coli* cells.

The course of events shown in Figure 14A-2a is called *lytic growth* and is characteristic of a **virulent phage.** Lytic growth results in the lysis of the host cell and the production of many progeny phage particles. In contrast, a **temperate phage** can either produce lytic growth, just as a virulent phage does, or integrate its DNA into the bacterial chromosome without causing any immediate harm to the host cell. An especially well-studied example of a temperate phage is bacteriophage λ (lambda), which, like the T-even phages, infects *E. coli* cells.

In the integrated state, the **lysogenic state,** the DNA of the temperate phage is called a **prophage.** The prophage is replicated along with the bacterial DNA, often through many generations of host cells (Figure 14A-4). During this

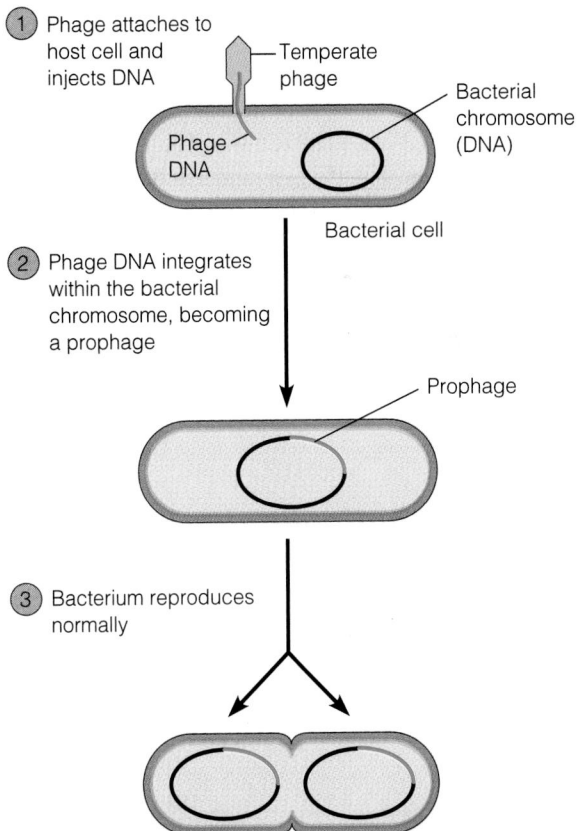

Figure 14A-3   Phage Plaques on a Lawn of Bacteria.
Clear phage plaques have formed on a lawn of *E. coli* infected with phage T4. The ratio of phage particles to cells in the inital mixture was adjusted so that the offspring of each parental phage particle would produce a single plaque. Thus, the number of plaques in the Petri dish is a measure of the phage concentration in the original phage sample.

Figure 14A-4   Propagation of a Prophage Within a Bacterial Chromosome.   The DNA injected by a temperate phage can become integrated into the DNA of the bacterial chromosome. The integrated phage DNA, called a prophage, is replicated along with the bacterial DNA each time the bacterium reproduces.

time, the bacteriophage genes, though potentially lethal to the host, are inactive, or *repressed*. Under certain conditions, however, the prophage DNA is excised from the bacterial chromosome and again enters a lytic cycle, producing progeny phage particles and lysing the host cell.

One reason bacteriophages have been so attractive to geneticists is the small size of their genomes. This makes it relatively easy to identify all or most of their genes and to understand the genetic organization and regulation of the genome in great detail. Bacteriophage λ, for example, has as its genome a single DNA molecule with a molecular weight of 32 million. This DNA codes for fewer than 60 genes, compared to the several thousand genes in a bacterium such as *E. coli*. Other phages are still smaller. A phage called φX174, for example, has only 1.8 million daltons of DNA, encoding only 9 genes (φ is the Greek letter *phi*). This phage is also noteworthy because its DNA is single-stranded rather than double-stranded as in all organisms and many other viruses. It is a single-stranded circle of 5375 nucleotides and was the first complete genome to be sequenced, a feat accomplished in 1977 by Frederick Sanger and his colleagues.

Largely because of their simple genomes, their rapidity of multiplication, and the enormous numbers of progeny that can be produced in a small volume of culture medium, bacteriophages are among the best-understood of all "organisms." They have proved exceedingly useful as model systems in our continuing quest to understand the much more complex genomes of true organisms. ■

his colleagues used quantitative chromatographic methods to separate and analyze the individual purines (adenine and guanine) and pyrimidines (cytosine and thymine) present in the DNA isolated from a variety of organisms. From their analyses came two important discoveries. First, they showed clearly that DNA from the same species always had the same percentage of each of the four bases, and that the percentage did not vary with individual, tissue, age, nutritional state, or environment. However, the base composition of DNA varied from species to species. In general, DNA preparations from closely related species were found to have similar base compositions, whereas those from very different species were likely to have quite different base compositions.

At least as significant was their observation that for all the DNA samples they examined, the number of adenines was equal to the number of thymines (A = T), and the number of guanines was equal to the number of cytosines (G = C). This result meant that the number of purines always equaled the number of pyrimidines (A + G = C + T). The significance of these equivalencies, known as **Chargaff's rules,** was an enigma and remained so until the double-helical model of DNA was established.

The Structure of DNA

The experiments of Avery and his colleagues with the bacterial "transforming principle" and of Hershey and Chase with bacteriophage infection provided strong biological support for DNA as the genetic material. The clincher, however, came in 1953, when James Watson and Francis Crick published their model of DNA as a double helix. We described the structure of the double helix in Chapter 3 and its discovery in Box 3A, but we return to it now for review and some further details.

Watson, Crick, and the Double Helix

In 1952, Watson and Crick were among a small number of scientists who were convinced not only that DNA was the genetic material but also that knowledge of its three-dimensional structure would provide valuable clues as to how it functioned. Working at Cambridge University in England, they approached the puzzle by building wire models of possible structures. DNA had been known for years to be a long polymer with a backbone of repeating deoxyribose and phosphate units, and a nitrogenous base attached to each sugar. Watson and Crick were also aided in their model building by knowing that the specific forms in which the bases A, G, C, and T exist at physiological pH permit specific hydrogen bonds to form between pairs of them. The crucial experimental evidence, however, came from an X-ray diffraction pattern of DNA determined by Rosalind Franklin, working in the laboratory of Maurice Wilkins at King's College in London. Franklin's picture told Watson and Crick that DNA was a helical model made up of two strands—a double helix. Watson and Crick then put this information together with what they already knew to come up with the model shown in Figure 3A-1, p. 62. In their model, the sugar-phosphate backbones are on the outside of the helix, and the bases face inward toward the center of the helix, forming the "steps" of the "circular staircase" that the structure resembles.

Figure 14-4a illustrates several structural features of the Watson-Crick double helix. The helix is right-handed, meaning that it curves "upward" to the right (notice that this is true even if you turn the diagram upside down). It contains ten nucleotide pairs per turn and advances 0.34 nm per nucleotide pair. Consequently, each complete turn of the helix adds 3.4 nm to the length of the molecule. The diameter of the helix is 2 nm. This distance turns out to be too small for two purines and too great for two pyrimidines, but it accommodates a purine and a pyrimidine well, consistent with Chargaff's rules. Pyrimidine-purine pairing, in other words, was necessitated by steric considerations. The two strands are held together by hydrogen bonding between the bases on opposite strands. In addition, because the bases are flat and relatively nonpolar, hydrophobic interactions tend to hold the stack of base pairs together, providing additional stability to the double helix.

Also shown in Figure 14-4a are the **major and minor grooves** that result from the way the two strands are twisted around each other. These are significant in the interactions of a variety of molecules with the DNA helix.

A further important structural feature of DNA is the *antiparallel* orientation of the two strands, as reviewed in Figure 14-4b (see also Chapter 3). As you move along one strand in a given direction, successive nucleotides are linked together by phosphodiester bonds that have a $5' \rightarrow 3'$ orientation, where the $5'$ and $3'$ refer to carbon atoms of the sugar unit. The complementary nucleotides of the other strand, however, are joined by bonds with a $5' \rightarrow 3'$ orientation that runs in the *opposite* direction. In other words, the two strands are upside down with respect to each other. The differing polarities (directions) of the two strands is a feature that has important implications for both replication and transcription of double-stranded DNA, as we shall see in Chapters 15 and 17.

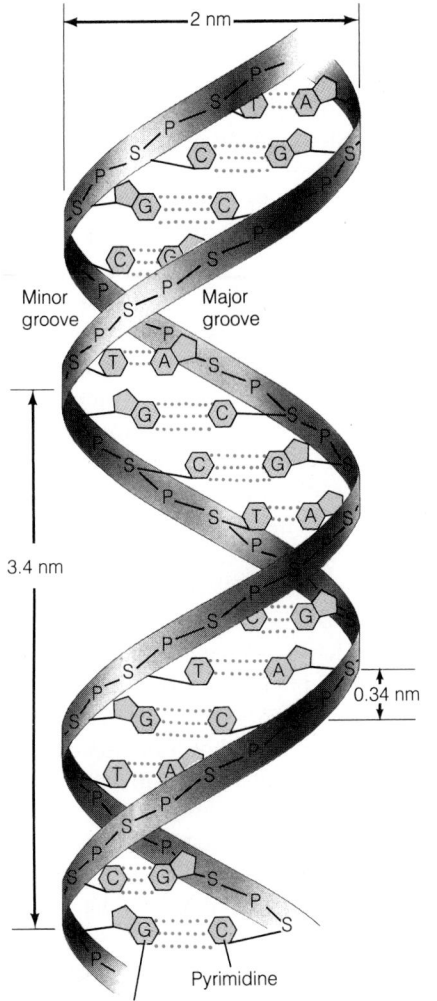

(a) Double helix

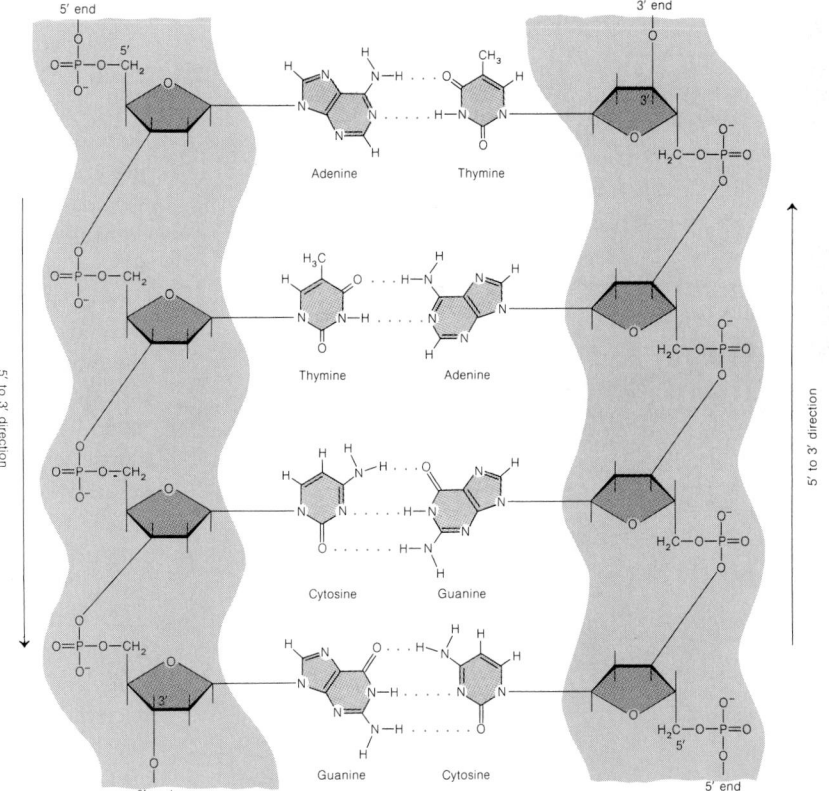

(b) Antiparallel orientation of strands

Figure 14-4   The DNA Double Helix.   (a) This schematic illustration shows the sugar-phosphate chains of the DNA backbone, the complementary base pairs, the major and minor grooves, and several important dimensions. A = adenine, G = guanine, C = cytosine, T = thymine, P = phosphate, and S = sugar (deoxyribose). **(b)** One of the strands of a DNA duplex (untwisted helix) is oriented $5' \rightarrow 3'$ in one direction, whereas its complement has a $5' \rightarrow 3'$ orientation in the opposite direction. This diagram also shows the hydrogen bonds that connect the bases in AT and GC pairs.

Alternative Forms of DNA

The right-handed Watson-Crick helix is an idealized version of what is called **B-DNA** (Figure 14-5a). Naturally occurring B-DNA double-helices, however, are flexible molecules that often deviate from this ideal, with exact shapes and dimensions dependent on the local nucleotide sequence. Furthermore, although B-DNA is undoubtedly the main form of DNA, other forms might also exist in cells, perhaps in short segments interspersed in molecules that are mostly B-DNA. The alternative forms that are leading candidates for this are A-DNA and Z-DNA. **A-DNA** has a right-handed helical configuration that is shorter and thicker than equivalent B-DNA. It can be induced in vitro by dehydration of B-DNA but is not known to exist in vivo. As Figure 14-5b shows, **Z-DNA** is a *left-handed* double helix. Its name derives from the zigzag pattern of its sugar-phosphate backbone, and it is longer and thinner than B-DNA. The Z form is more likely to arise in DNA where there are alternating

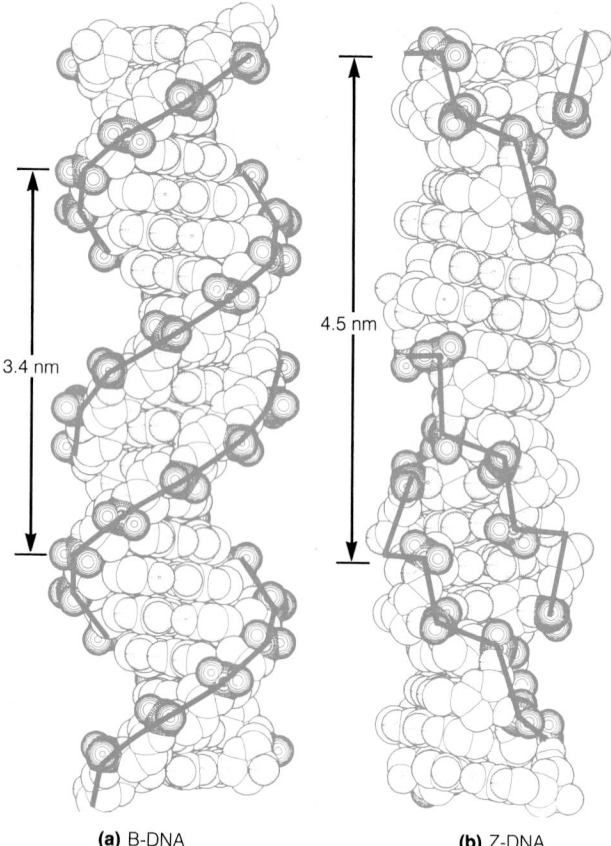

(a) B-DNA   **(b)** Z-DNA

Figure 14-5   Alternative Forms of DNA.   (a) B-DNA has a right-handed helix and a highly regular sugar-phosphate backbone, as originally predicted by Watson and Crick. (A-DNA, not shown, is shorter and thicker than B-DNA, with larger major grooves and smaller minor grooves.) **(b)** Z-DNA has a left-handed configuration and an irregular backbone, as traced by the heavy lines that zigzag along the strands from phosphate to phosphate.

purines and pyrimidines, or cytosines with extra methyl groups (which do occur in chromosomal DNA; see Chapter 19). But despite great excitement about Z-DNA immediately following its discovery in 1979, the amount of Z-DNA in cells and its biological significance, if any, are still unknown.

Supercoiled DNA

In addition to the helical configuration typical of all DNA molecules, a DNA molecule can be twisted upon itself to form a new, higher-order helix: **supercoiled DNA.** Although now known to be an important property of DNA molecules in all organisms, supercoiling was first identified in the DNA of certain small viruses that have circular DNA molecules—molecules that exist as closed loops. Electron microscopic observations have established that DNA molecules from many sources are circular. In addition to many viruses, bacteria generally have circular DNA, as do the mitochondria and chloroplasts of eukaryotic cells. The bacterium *E. coli*, for example, has a circular DNA molecule with a contour length of about 1.6 mm. For some viruses, such as bacteriophage λ, the DNA molecule exists alternately in linear and circular forms, during different parts of the life cycle.

Supercoiling is easiest to study in circular DNA molecules. A DNA molecule can go back and forth between the supercoiled state and the nonsupercoiled, or *relaxed,* state. To understand the basic idea, you might perform the following exercise. Start with a length of rope consisting of two strands twisted together into a right-handed coil; this is the equivalent of a relaxed, linear DNA molecule. Just joining the ends of the rope together changes nothing; the rope is now circular but still in a relaxed state. But if before sealing the ends you first give the rope an extra right-handed twist (i.e., another twist in the direction in which the strands are already entwined about each other), the rope becomes overwound and is thrown into what is called a *positive supercoil.* Conversely, if the rope is given a left-handed twist before sealing (i.e., twisted in the direction opposite to that in which it is wound), it becomes underwound and is thrown into a *negative supercoil.*

In the same manner, a relaxed DNA molecule can be converted to a positive supercoil by a right-handed twist and to a negative supercoil by a left-handed twist (Figure 14-6a). Circular DNA molecules found in nature, including those of bacteria, viruses, and eukaryotic organelles, are invariably negatively supercoiled. (Figure 14-6b shows a relaxed circular DNA molecule from a phage and a similar molecule with negative supercoils.) Moreover, as mentioned earlier, supercoiling is not limited to circular DNA molecules. It also occurs in linear DNA, provided only that the ends of the molecule are not free to rotate. (Try this with the rope.) At any given time, significant portions of the linear DNA in the chromatin of eukaryotic cells may be supercoiled, and when the chromatin condenses to form compact chromosomes during mitosis, the DNA is virtually all highly supercoiled, as we shall see later. Supercoiling makes DNA more com-

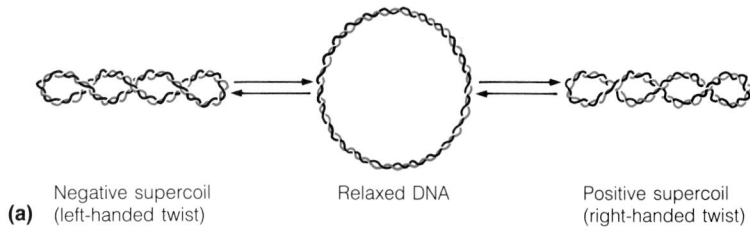

(a)   Negative supercoil   Relaxed DNA   Positive supercoil
(left-handed twist)   (right-handed twist)

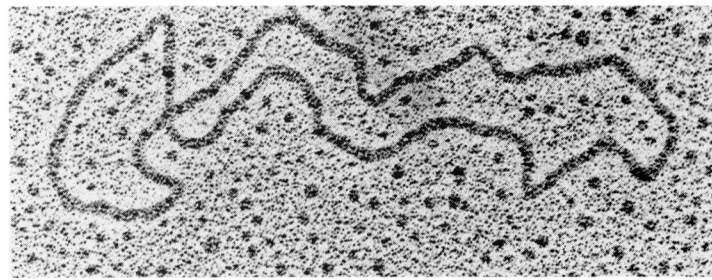

Figure 14-6 Interconversion of Relaxed and Supercoiled DNA. **(a)** A schematic diagram of a relaxed circular DNA molecule and its conversion to supercoiled forms with a left-handed twist (negative supercoiling) or a right-handed twist (positive supercoiling). **(b)** Electron micrographs of circular molecules of DNA from a bacteriophage called PM2, with a relaxed molecule on the top and a supercoiled molecule on the bottom (TEMs).

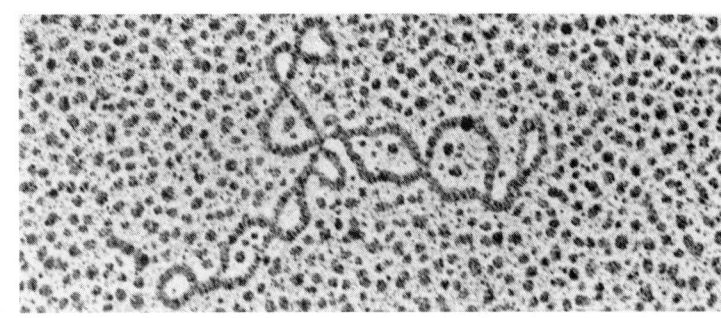

(b)

pact. In the laboratory, this compactness causes the DNA to sediment more rapidly during centrifugation and to move faster during gel electrophoresis.

By influencing both the spatial organization and the energy state of DNA, supercoiling affects the ability of a DNA molecule to interact with other molecules. Positive supercoiling favors tighter winding of the double helix and therefore reduces opportunities for interaction. Conversely, negative supercoiling tends to unwind the double helix, thereby increasing access to its strands for proteins involved in DNA replication or transcription.

DNA molecules that differ only in their state of supercoiling are called **topological isomers** of one another. Correspondingly, the enzymes that carry out the interconversion of relaxed and supercoiled forms of DNA are called **topoisomerases.** Most topoisomerases are classified as *type I* or *type II.* Both types catalyze the stepwise relaxation of supercoiled DNA, but type I enzymes relax one coil at a time, whereas type II enzymes relax two coils at a time. Topoisomerases function by attaching to a supercoiled duplex and producing a transient break, called a *nick,* in either one strand (type I) or both strands (type II). The two strands can then rotate around each other, after which the nicks are resealed. This relaxation reaction requires no addition of en-

ergy, suggesting that the phosphodiester bond energy that is liberated when a strand is nicked must somehow be conserved and used to form the new bond.

DNA gyrase is a type II topoisomerase that can induce as well as relax supercoiling. DNA gyrase is one of several enzymes required for DNA replication, as we will see in Chapter 15. It can relax the positive supercoiling that results from partial unwinding of a double helix, or it can actively introduce negative supercoils that promote strand separation, allowing the access of other proteins involved in the DNA replication process. DNA gyrase requires ATP to generate supercoiling but not to relax an already supercoiled molecule.

Topoisomerases are involved in many other DNA transactions. The particular balance of different topoisomerases present at a particular place and time presumably determines the state of the DNA there.

DNA Denaturation and Renaturation

Because the two strands of the DNA duplex are held together only by relatively weak, noncovalent bonds, they can be readily separated under appropriate conditions. Strand separation is an integral part of both the replication of DNA and the synthesis of RNA, as we will see in coming chapters. Strand separation can also be induced experimentally, resulting in **DNA denaturation;** the reverse process, which reestablishes a double helix, is called **DNA renaturation.** The ability to denature and renature nucleic acids in vitro has many scientific applications; for example, as we will discuss later, it is the basis for the use of nucleic acid "probes" to identify and isolate particular genes.

One way to denature DNA in the laboratory is to raise the temperature. If this is done slowly, the DNA retains its double-stranded, or native, state until some critical temperature is reached, at which point the duplex quickly denatures, or "melts," into its component strands. The melting process can easily be monitored because of the difference in absorbance between double-stranded and single-stranded DNA. All DNA absorbs ultraviolet light, with an absorption maximum around 260 nm (Figure 14-7). (RNA has a similar absorption profile; the absorbance at 260 nm is routinely used to determine concentrations of both DNA and RNA.) When the temperature of a DNA solution is raised, the absorbance remains relatively constant until the duplex structure melts into its component strands. The melting occurs over a very small temperature range, and as the strands separate, the absorbance of the solution increases rapidly because of the higher intrinsic absorption of single-stranded DNA. Figure 14-8 shows the absorbance of a DNA solution undergoing this process of thermal denaturation.

The temperature at which one-half of the absorbance change has been achieved is called the **DNA melting temperature (T_m).** The melting temperature is a sensitive measure of the base composition of DNA. GC pairs, which are held together by three hydrogen bonds, are more resistant to separation than AT pairs, which have only two (see Figure 14-4b). The melting temperature generally increases linearly

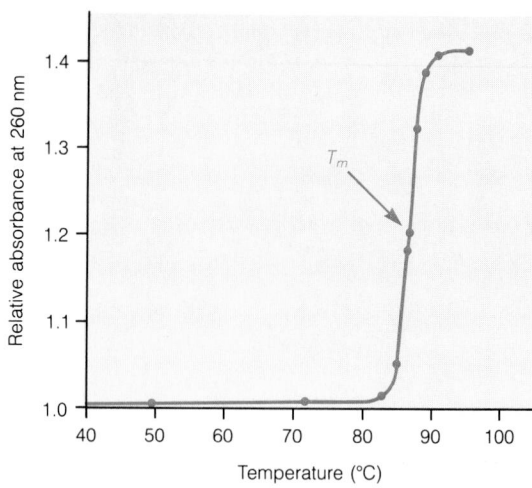

Figure 14-8 A Thermal Denaturation Profile for DNA. As the temperature of a solution of double-stranded (native) DNA is raised, a point is reached at which the heat energy causes the DNA to denature rapidly. The conversion to single strands is accompanied by a characteristic increase in the absorbance of light at 260 nm. The temperature at which the midpoint of this increase occurs is called the melting temperature, T_m. For the sample shown, the T_m value is about 86°C.

with the proportion of GC base pairs in the DNA (Figure 14-9).

Denatured DNA will renature if conditions are favorable for the reestablishment of hydrogen bonding. The renaturation process is illustrated in Figure 14-10. DNA strands collide randomly until a small region of one strand encounters a complementary region of another strand and hydrogen bonds form between base pairs. This initial *nucleation event* is then followed by a "zipping-up" of the two strands, in both directions. Because the nucleation event depends on random collision, it is the rate-limiting step in the

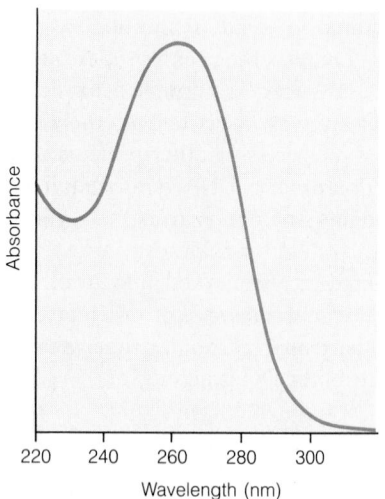

Figure 14-7 Ultraviolet Absorption Spectrum of DNA.
Nucleic acids absorb ultraviolet light with an absorption maximum at a wavelength of about 260 nm because of the absorption by the pyrimidine and purine bases.

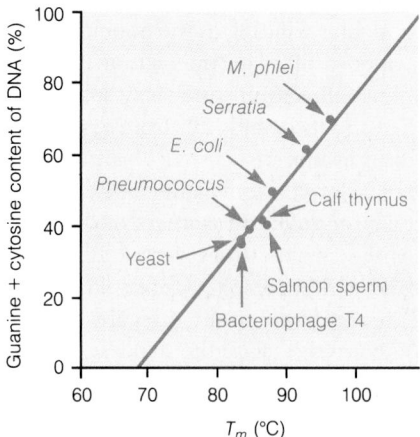

Figure 14-9 The Dependence of Melting Temperature on DNA Base Composition. The melting temperature of DNA increases linearly with its G+C content, as illustrated by the relationship between T_m and G+C content for DNA samples from a variety of organisms.

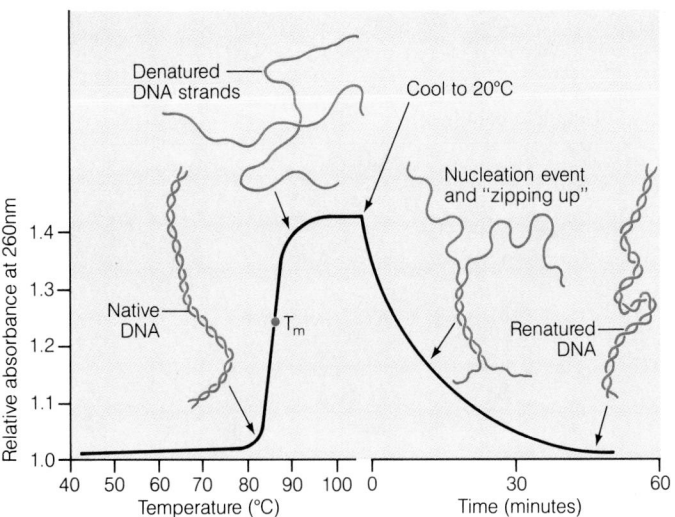

Figure 14-10 DNA Denaturation and Renaturation. If a solution of native (double-stranded) DNA is heated slowly under carefully controlled conditions, the DNA "melts" over a narrow temperature range, with an increase in absorbance at 260 nm. When the solution is allowed to cool, the separated DNA strands reassociate with kinetics that depend on the initial concentration. Complementary strands collide randomly in the nucleation event, followed by a rapid "zipping up" of adjacent nucleotide pairs. The reassociation requires varying amounts of time, depending on both the DNA concentration in the solution and the length of the DNA strands.

reassociation process. The frequency of collision is in turn dependent on the number of strands in solution, so the overall process is concentration-dependent. Because the time required for complete renaturation also depends on the lengths of the DNA strands, researchers were able to use renaturation experiments to make some of the first estimates of relative lengths of the DNA of viruses, bacteria, and eukaryotic cells.

The Organization of DNA in Genomes

So far, we have considered several chemical and physical properties of DNA. But as cell biologists, we are primarily interested in its importance to the cell. We therefore want to know how much DNA cells have, how and where they store it, and how they access the genetic information it contains. We begin by inquiring about the amount of DNA present, because that determines the maximum amount of information that the cell can possibly contain.

The **genome** of an organism or virus consists of the DNA (or RNA, in the case of some viruses) that makes up one complete copy of all the genetic information of that organism or virus. For many viruses and prokaryotes, the genome consists of one linear or circular DNA molecule, or a small number of them. Eukaryotic cells have a nuclear genome, a mitochondrial genome, and, in the case of plants and algae, a chloroplast genome as well. The nuclear genome is usually defined to consist of the DNA molecules present in a haploid set of chromosomes. (As we will explore in more detail in Chapter 16, a *haploid* set of chromosomes consists of one representative of each type of chromosome, whereas a *diploid* set consists of two copies of each type of chromosome, one copy from the mother and one from the father. Sperm and egg cells each have a haploid set of chromosomes, whereas most other eukaryotic cells have a diploid chromosome set.)

Genome Size

Genome size is usually expressed in numbers of nucleotide pairs, commonly referred to as base pairs (bp). The circular DNA molecule that constitutes the genome of an *E. coli* cell, for example, contains about 4.7 million base pairs. Table 14-1 presents the approximate genome sizes of a variety of viruses and organisms, both as base pairs per haploid genome and as fractions or multiples of the size of the *E. coli* genome. Even with the limited number of species included in the table, there is a spread of almost eight orders of magnitude in genome size, from 5×10^3 base pairs in the case of virus SV40 to 10^{11} base pairs for the nuclear genome of *Trillium*, a common spring wildflower.

Table 14-1 also gives the actual physical length of the DNA. For the genomes shown, this ranges from less than 2 μm in the case of SV40 to about 34 m (more than 100 feet!) for *Trillium*. For viruses and bacteria, the genome is usually a single molecule, but for eukaryotes the nuclear genome is always dispersed among a number of chromosomes and is therefore never present as a single molecule. (However, mitochondrial and chloroplast genomes are single, usually circular DNA molecules, resembling those of bacteria.)

Broadly speaking, genome size increases with the complexity of the organism (Figure 14-11). Viruses contain enough nucleic acid to code for only a few or a few dozen proteins, bacteria can specify a few thousand proteins, and eukaryotic cells have enough DNA—theoretically—to encode millions of proteins. Upon closer examination of such data, however, many puzzling features emerge. Notably, there are great variations in genome size among eukaryotic species that do not seem to correlate with any known differences in organismal complexity. Some amphibians and plants, for example, have inordinately large genomes—tens or even hundreds of times larger than those of other amphibians or plants, or of mammalian species. *Trillium*, for example, is a member of the lily family, with no obvious need for exceptional amounts of genetic information. Yet its

Table 14-1 DNA Content of Selected Genomes

Species	Genome Size (Base Pairs)	Relative Size (*E. coli* = 1)	Length of DNA (mm)
Viruses			
SV40	5×10^3	0.001	0.0017
T7	4×10^4	0.01	0.014
T2	2×10^5	0.04	0.068
Prokaryotes			
Mycoplasma	3×10^5	0.06	0.10
Bacillus	3×10^6	0.6	1.02
E. coli	4.7×10^6	1	1.6
Animals			
Fruit fly	2×10^8	40	68
Chicken	2×10^9	400	680
Human	3×10^9	600	1,000
Fungi			
Yeast	2×10^7	4	6.8
Plants			
Peas	5×10^9	1,000	1,600
Trillium	1×10^{11}	20,000	34,000

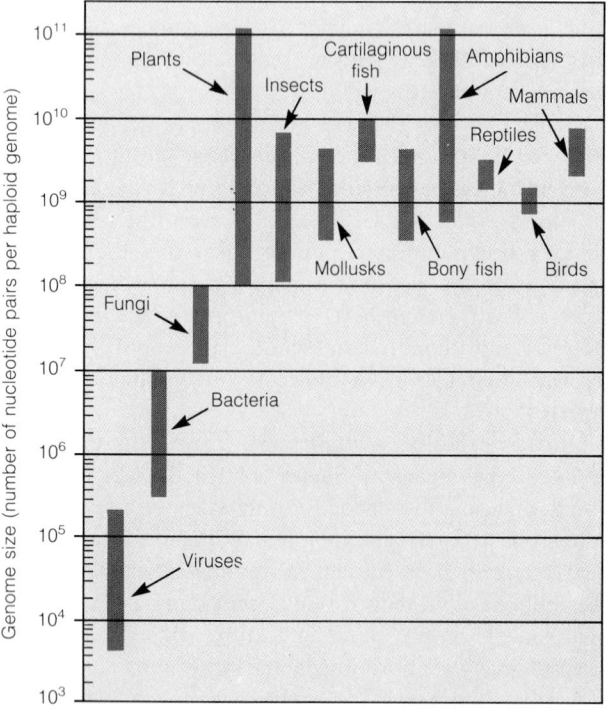

Figure 14-11 The Relationship Between Genome Size and Type of Organism. For each group of organisms shown, the bar represents the approximate range in genome size measured as the number of nucleotide pairs per haploid genome.

genome size is more than 10 times that of peas and 20 times that of humans. We have no idea what it does with all that DNA. In the final analysis, genome size is probably much less important than the number and identity of functional genes and the noncoding DNA sequences that help control their expression. As we discuss shortly, most eukaryotic genomes carry large amounts of DNA of no known function.

Restriction Enzymes as Tools for Studying the DNA of Genomes

If the hereditary similarities and differences observed among organisms derive from their DNA, then surely we can expect the study of DNA itself to yield important biological insights. Clues to a myriad of mysteries—from the control of gene expression within a cell to the evolution of new species—are undoubtedly to be found in the nucleotide sequences of genomic DNA. Most DNA molecules, however, are far too large to be studied intact. In fact, until the early 1970s, DNA was the most difficult biological molecule to analyze biochemically. Eukaryotic DNA seemed especially intimidating, given the size of the genomes of most eukaryotes, and no method was known for cutting DNA at specific sites to yield reproducible fragments. The prospect of ever being able to identify, isolate, sequence, or manipulate specific eukaryotic genes seemed unlikely. Yet in less than a decade, DNA became one of the easiest biological molecules to work with.

This breakthrough was made possible largely by the discovery of **restriction enzymes.** Restriction enzymes are enzymes, isolated from bacteria, that make cuts in foreign

DNA molecules at specific sites. (Technically, these enzymes are "endonucleases," because they act within [*endo-*] a nucleic acid.) The cutting action of such an enzyme generates a specific set of DNA pieces called **restriction fragments**. The site at which a given restriction enzyme cleaves double-stranded DNA consists of a segment usually four to six nucleotide pairs long, called the **recognition sequence**. For example, below is the sequence recognized by a widely used *E. coli* enzyme called *Eco*RI:

$$\downarrow$$
$$5'—G—A—A—T—T—C—3'$$
$$3'—C—T—T—A—A—G—5'$$
$$\uparrow$$

The arrows indicate where *Eco*RI cuts the DNA. Note that this restriction enzyme, like many, makes a *staggered* cut in the double-stranded DNA molecule. (Box 14B, page 428, describes the biological role of restriction enzymes in bacteria and the characteristics of recognition sequences.)

The frequency with which a restriction enzyme recognition sequence occurs along a typical DNA molecule is such that a restriction enzyme cleaves most DNA into fragments ranging in length from a few hundred to a few thousand base pairs. Fragments of these lengths are far more amenable to further manipulation than the enormously long DNA molecules from which they are generated.

The Separation of Restriction Fragments by Gel Electrophoresis. Digestion of a DNA sample by a specific restriction enzyme results in a collection of restriction fragments of different sizes. To determine the number and lengths of the different restriction fragments produced from each original DNA molecule and to make preparations of individual fragments for further study, one must be able to separate the fragments from each other. The technique of choice for this purpose is **gel electrophoresis**, essentially the same method used for the separation of proteins and polypeptides (see Figure 7-18). In fact, the procedure for DNA is even simpler than for proteins, because DNA molecules have an inherent negative charge (due to their phosphate groups) and therefore have no need for treatment with a negatively charged detergent to make them move toward the anode.

For small pieces of DNA (less than about 500 nucleotides), gels of *polyacrylamide* may be used, just as for electrophoresis of proteins and polypeptides. In fact, fragments that differ in length by as little as a single nucleotide can be separated from each other on such gels. For larger DNA fragments, however, the pores in polyacrylamide gels are too small to allow passage. Such fragments are therefore separated on more porous gels of *agarose*, a polysaccharide. Improved agarose gels are now often used for smaller fragments, too.

Figure 14-12 illustrates the separation of restriction fragments of different sizes using gel electrophoresis. DNA

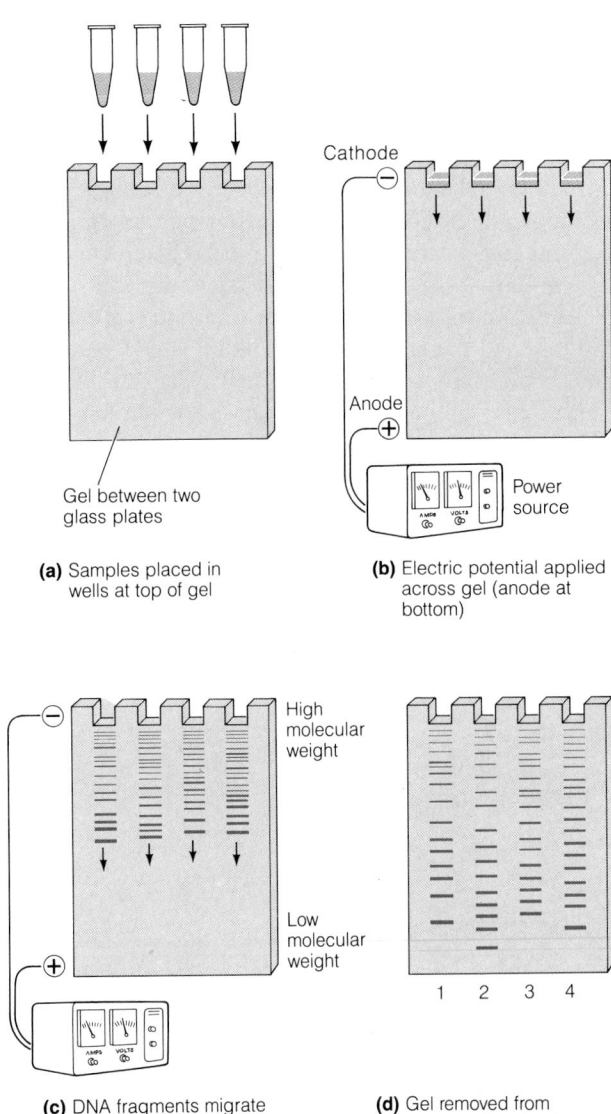

(a) Samples placed in wells at top of gel

(b) Electric potential applied across gel (anode at bottom)

(c) DNA fragments migrate toward the anode at a rate inversely related to their size

(d) Gel removed from plates and stained with ethidium bromide to visualize bands

Figure 14-12   Gel Electrophoresis of DNA.   (a) The four test tubes contain mixtures of DNA fragments produced by incubating DNA aliquots with four different restriction enzymes. To fractionate a DNA preparation containing fragments of various sizes, a small sample of the preparation is applied to the top of a gel of either polyacrylamide (for small DNA fragments) or agarose (for large DNA fragments). The gel is formed between two glass plates, and DNA solutions are placed in the wells at the top. **(b)** An electrical potential of several hundred volts is then applied across the gel, such that the anode (the positive electrode) is at the bottom of the gel and the cathode (the negative electrode) is at the top. **(c)** The DNA fragments in the applied sample migrate toward the anode at a rate that is inversely related to their size; the smaller fragments migrate more rapidly down the gel than do the larger ones.
(d) After a predetermined amount of time, the gel is removed and stained with a dye such as ethidium bromide, which binds to the DNA fragments and causes them to fluoresce under ultraviolet light. Alternatively, autoradiography can be used to locate the DNA bands on the gel, provided that the DNA is radioactively labeled.

samples are first digested with the desired restriction enzyme; in the figure, four different restriction enzymes are used. A small portion of each digestion mixture is then placed in a separate well at the top of the gel (or at one end of the gel, if it is positioned horizontally). An electrical potential of several hundred volts is applied across the gel, with the anode, the positive electrode, at the bottom (or other end) of the gel. Because of the negative charge of their phosphate groups, DNA fragments migrate toward the anode. Smaller fragments are able to move through the gel with relative ease and therefore migrate rapidly, while larger fragments move more slowly. The current is left on until the fragments are well spaced out on the gel. The result is a separation of DNA fragments based on their size.

DNA fragments on the gel can be visualized either by staining or, in the case of radioactively labeled DNA, by autoradiography. A common staining technique involves soaking the gel in the dye *ethidium bromide*, which binds to DNA fragments and fluoresces pink when exposed to ultraviolet light. If the DNA fragments contain a radioisotope, their locations on the gel can be determined with photographic film. When the film is developed, the resulting autoradiogram will be marked wherever the radioactivity in a DNA fragment has exposed the film. Once the locations of the various fragments are identified, specific fragments can be eluted (washed) from gels for further study. Large amounts of specific fragments can be readily prepared by cloning or by the polymerase chain reaction (PCR), which we discuss in the next chapter.

Restriction Mapping Techniques. How does a researcher determine how a set of restriction fragments were ordered in the original DNA molecule? One method makes use of the fact that the restriction enzyme does not cut all the recogni-

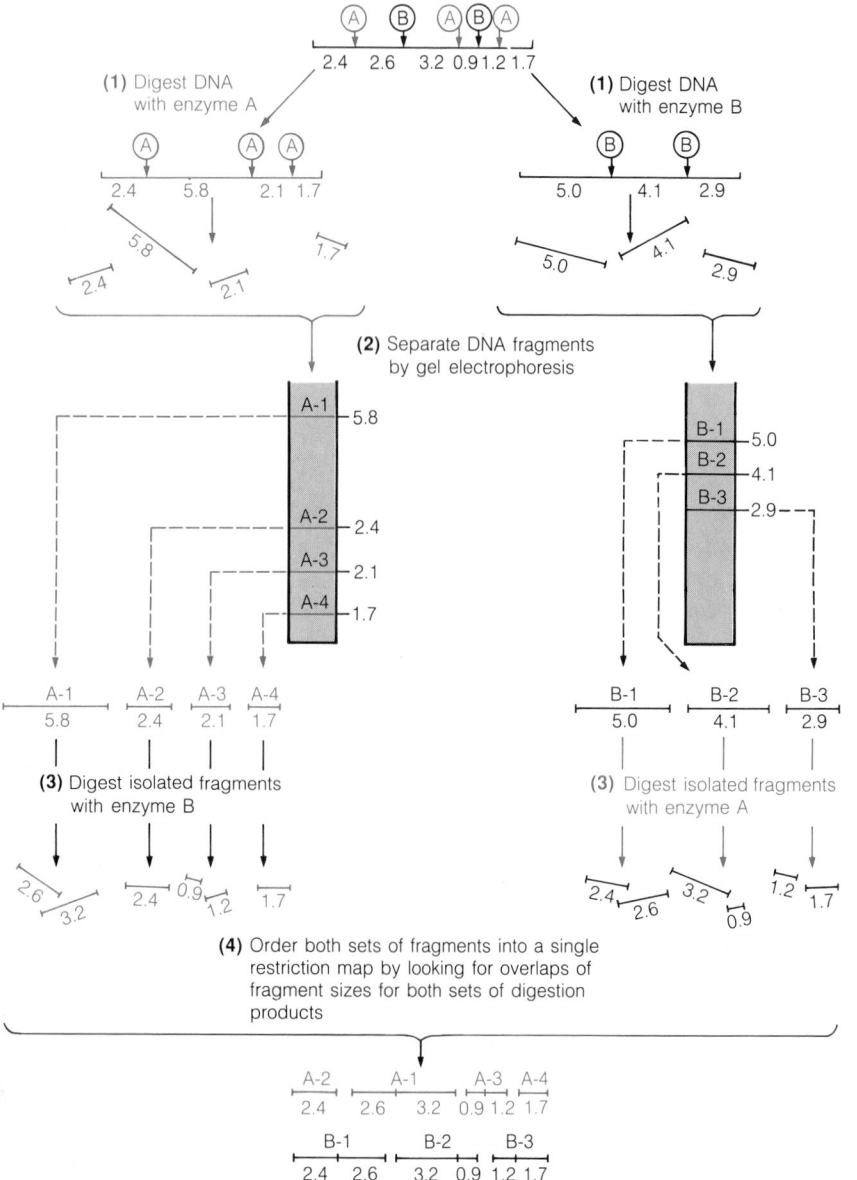

Figure 14-13 Restriction Mapping. The DNA molecule shown at the top contains 12,000 base pairs. Based on an assumption that the four kinds of nucleotides are present in equal numbers, such a molecule should contain an average of $12,000/4^6 = 2.93$ recognition sites for a restriction enzyme with a six-nucleotide recognition sequence. As indicated, the molecule contains three recognition sites for enzyme A and two sites for enzyme B. (The numbers between recognition sites represent thousands of base pairs.) ① Upon digestion with these enzymes, the molecule is cleaved at three points by enzyme A (left, blue) or at two points by enzyme B (right, black). The fragments generated in this way are of characteristic length and are specific for the restriction enzyme. ② The DNA fragments are separated according to size by gel electrophoresis on polyacrylamide or agarose. The fragments are usually numbered in order from largest to smallest. ③ Each of the fragments can be recovered from the gel and digested with the other restriction enzyme. In some cases (such as fragments A-1, A-3, and all three B fragments), further cleavage occurs, because the fragment contains a recognition site for the second enzyme. In other instances (fragments A-2 and A-4), this is not the case. The number and sizes of such fragments should be the same for both digestions, but the fragments occur in different permutations because the order of digestion is different (A followed by B in one case, B followed by A in the other). ④ The two sets of fragments can then be ordered into a single restriction map by a trial-and-error process of looking for overlaps in fragment lengths for the two sets of digestion products. If ambiguities are found, they can frequently be resolved by use of a third restriction enzyme.

tion sequences simultaneously. By stopping the reaction at various times prior to completion, and then determining the sizes of the fragments existing at each time, the fragments can often be put in order.

Another method, which provides more elaborate restriction maps, involves sequential (and complete) digestion of the starting DNA with two or more restriction enzymes. The sizes of the restriction fragments produced by the various treatments are then compared. This method of restriction mapping is illustrated in Figure 14-13, using two restriction enzymes designated A and B. Because of its specificity, a given restriction enzyme will always cleave a specific DNA molecule into the same number of fragments, and each such fragment will have a characteristic length. The length of each fragment can be determined by gel electrophoresis. In general, two restriction enzymes with different recognition sequences will generate two quite different sets of fragments, although the sum of the fragment lengths will of course be the same in both cases.

All the fragments generated by one enzyme are then isolated—for example, by cutting the gel into slices and soaking each slice in a buffer solution. The fragments are then individually subjected to digestion by the second enzyme, producing two series of smaller fragments. Each such fragment is generated twice in the procedure, once when the fragments generated by enzyme A are further digested with enzyme B, and again when the fragments generated by enzyme B are further digested with enzyme A. It is therefore often possible to determine the order of both sets of fragments by looking for overlapping pieces common to two fragments, one in each set. The resulting **restriction map** depicts the location of each cleavage site with respect to the others in the original DNA molecule. Ambiguous cases can usually be resolved by the use of a third enzyme, which generates yet another distinctive pattern of fragments.

Restriction maps have been constructed for the genomes of phages, eukaryotic viruses, bacteria, and organelles, and for extensive segments of eukaryotic chromosomes as well.

DNA Sequencing

Now that we know how to prepare reproducible fragments of manageable length from very long DNA molecules, it becomes possible to determine nucleotide sequence. At about the same time that the methods for preparing restriction fragments were developed, two methods were devised for the rapid determination of the nucleotide sequence of DNA fragments. One method was devised by the Americans Allan Maxam and Walter Gilbert, the other by the British biochemist Frederick Sanger and his colleagues. The Maxam-Gilbert method, called the *chemical method,* is based on the use of (nonprotein) chemicals that cleave DNA preferentially at specific bases, whereas the Sanger method, called the *enzymatic method* or *dideoxy method,* is based on the use of a *DNA polymerase* enzyme and special, chain-terminating

*di*deoxyribonucleotides. Because it is Sanger's method that is the basis of almost all DNA sequencing today, we describe that method here.

For the enzymatic sequencing technique, a single-stranded DNA fragment is used as a template for the in vitro synthesis of complementary strands of all possible lengths. The new DNA strands are labeled with either a radioactive isotope or a fluorescent dye. When these labeled strands are subsequently fractionated on a polyacrylamide gel, the nucleotide sequence of the DNA can be deduced from the positions of the bands on the gel. The key to the procedure is the use of 2′,3′-dideoxyribonucleotides, which are identical to the nucleotides the cell normally uses for DNA synthesis except that the 3′-carbon of the sugar (as well as the 2′-carbon) lacks an —OH group.

Figure 14-14 diagrams the enzymatic sequencing procedure, using as the DNA under study a strand only twelve nucleotides in length (rather than the hundreds of nucleotides that would be more typical). The starting material, a solution containing many copies of the single-stranded DNA molecule to be sequenced, is divided among four test tubes. To each tube are also added DNA polymerase, the four deoxyribonucleotides that are the usual precursors of DNA synthesis (dATP, dCTP, dGTP, and dTTP), and a short, single-stranded piece of DNA (oligonucleotide) that is complementary to the 3′ end of the DNA strand to be sequenced. This oligonucleotide serves as a necessary *primer* for the initiation of DNA synthesis, and in the procedure shown here, it is radioactively labeled (at its 5′ end) to facilitate later detection of the new DNA. In addition, each of the four tubes receives a small amount of *one* of the four dideoxyribonucleotides ddATP, ddCTP, ddTTP, and ddGTP.

When the four reaction mixtures are incubated, the DNA polymerase catalyzes the attachment of nucleotides, one by one, to the 3′ end of the primer, producing a DNA strand complementary to the template. Wherever the template has an A, the DNA polymerase inserts a T in the new strand, and so forth, according to the base-pairing rules. Most of the nucleotides inserted are the normal deoxyribonucleotides, because these are present at high concentration in the reaction mixture. However, every so often, at random, a dideoxyribonucleotide is inserted instead of its normal equivalent. The dideoxyribonucleotide terminates the chain, because it lacks the 3′—OH needed for attachment of the next nucleotide. Eventually, a mixture of radioactive strands of various lengths is generated in each tube. If you examine step 2 of the figure, you will see the expected results. DNA strands of all possible lengths have been produced. Tube 1 contains all the possible strands ending in A, tube 2 all those ending in C, and so on.

Next, samples from each of the four tubes are subjected to electrophoresis in adjacent lanes of a polyacrylamide gel, and autoradiography is then used to make visible the bands containing radioactively labeled DNA. The sequence of the complete new strand can now be deduced by considering the bands present in each of the four lanes. First, look at lane

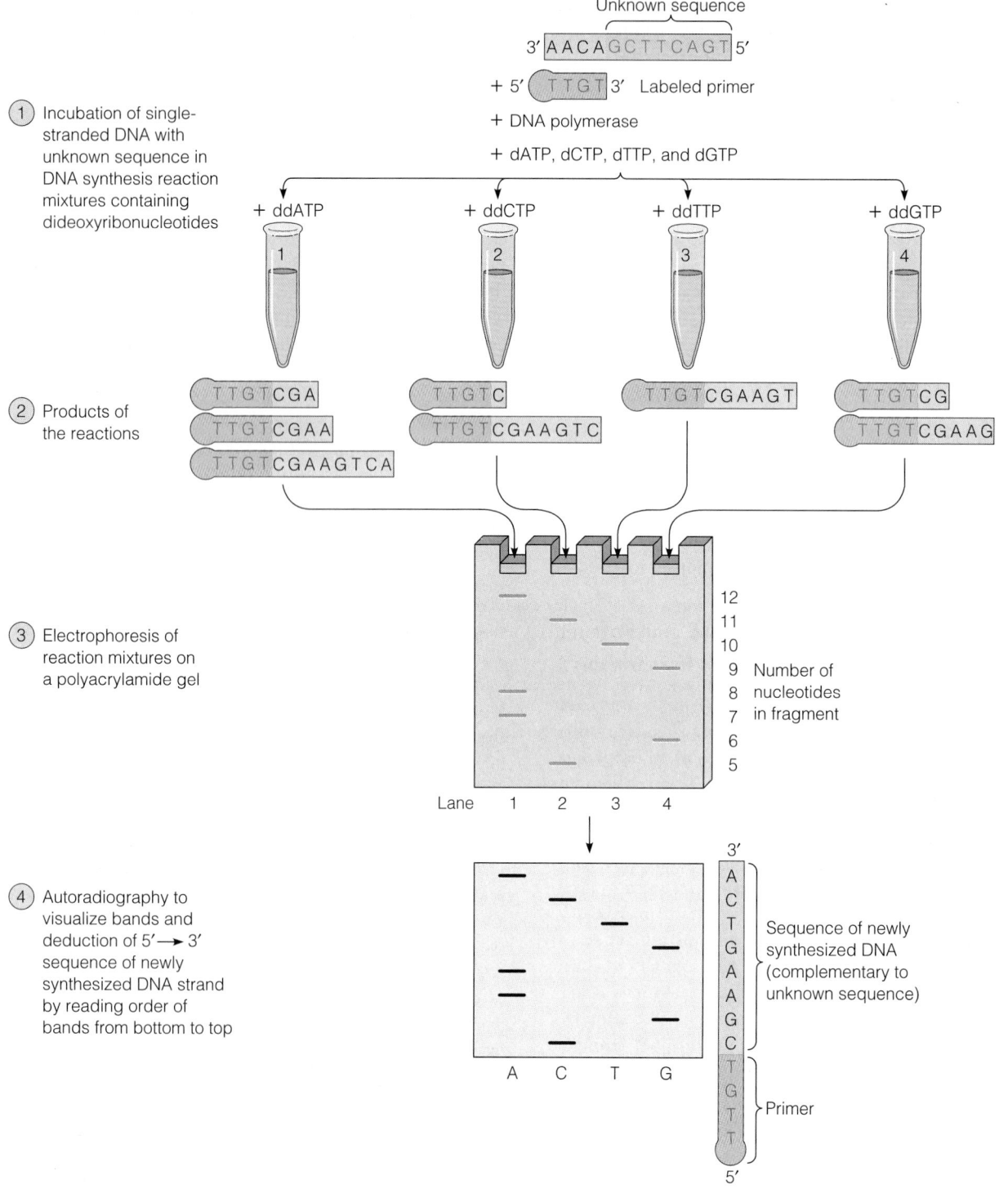

Figure 14-14 DNA Sequencing. The enzymatic (Sanger) sequencing technique is illustrated here for a short (12-nucleotide) strand of DNA. ① Equal portions of the single-stranded fragment are incubated in similar DNA synthesis reaction mixtures, all of which contain a DNA primer (complementary to the 3′ end of the fragment to be sequenced), the enzyme DNA polymerase, and the four usual precursor deoxyribonucleotides for DNA synthesis. In addition, each reaction tube contains a small amount of one of the four nucleotides in its dideoxy form (dd-). In the version of the procedure shown here, the DNA primer is radioactively labeled, at its 5′ end. ② The expected products from each reaction tube.

③ Samples from each of the four tubes are then subjected to electrophoresis in adjacent lanes of a polyacrylamide gel. ④ The gel exposes photographic film, making visible the radioactively labeled oligonucleotides in each of the four lanes. The sequence of the newly synthesized strand can then be deduced directly from the autoradiogram by reading the banding pattern upward, recording in order the letter of the lane (A, C, T, or G) in which each successive band appears. The sequence of the starting strand is simply the complement of this strand. (In practice, the initial DNA fragment would likely be hundreds of nucleotides long.)

1, which contains the pieces terminated by ddATP (Figure 14-14, step 3). Three bands appear in lane 1, at positions corresponding to oligonucleotides 7, 8, and 12 nucleotides long. This indicates that there must be adenines at the 7th, 8th, and 12th positions of the new strand, counting from its 5′ end. By similar logic, we can establish the positions of the cytosines, thymines, and guanines. By analyzing the banding patterns in the four lanes, we deduce the sequence of the complete, newly synthesized strand.

Notice that by starting at the bottom of the autoradiogram (Figure 14-14, step 4) and recording the letter (A, C, T, or G) of the lane in which each successive band appears, we can read the nucleotide sequence of the new DNA strand directly. It is obviously a very attractive feature of the technique that it is so easy to deduce the sequence once the autoradiogram is in hand. Once the sequence of the new strand is determined, of course, the sequence of the original strand is also known because the sequences of the two strands are complementary.

Most of the sequencing procedure can now be carried out quickly and automatically by DNA sequencing machines. As an alternative to labeling the new DNA by using radioactive primers, most modern machines use dideoxyribonucleotides tagged with fluorescent dyes, a different color for each of the four types. This approach allows the four reactions to be performed in a single tube; the machine distinguishes the four types of new strands by the wavelength (color) of their fluorescence.

From Genes to Genomes

The significance of the techniques we have just been exploring can scarcely be overestimated. DNA sequencing is now so routine and so automated that thousands of nucleotides of DNA can be sequenced in a single day. Complete sequences are already available for many genes from microbes, plants, and animals (including humans); and computerized data banks have been established to store and analyze sequence information. When the gene encoding a particular, unsequenced polypeptide has been isolated (by techniques to be described in Chapter 16), it is usually easier and faster to sequence the DNA and deduce the amino acid sequence of the polypeptide from the nucleotide sequence of the DNA than to sequence the polypeptide directly.

Increasingly, sequencing techniques are being applied not just to individual genes, but to whole genomes. The first genome to be completely sequenced was, not surprisingly, that of a virus, ϕX174, a phage with a single-stranded DNA genome only 5375 bases long. Other small viral genomes and several chloroplast and mitochondrial genomes followed next (the smallest mitochondrial genomes are about 16,000 base pairs; chloroplast genomes are typically about 120,000 bp). Much larger genomes are now being tackled as well, including those of several organisms important in biological research. These include the bacterium *E. coli* (genome size 4.7×10^6 bp), the plant *Arabidopsis thaliana*

(7×10^7 bp, one of the smallest plant genomes known), the roundworm *Caenorhabditis elegans* (1×10^8 bp), and the fruit fly *Drosophila melanogaster* (1.7×10^8 bp). At the time of this writing, several of the 16 chromosomes in a haploid cell of bakers' yeast, *Saccharomyces cerevisiae*, have been completely sequenced, and the entire genome (1.4×10^7 bp) is expected to be completed by 1996.

To us as human beings, of course, the ultimate challenge of DNA sequencing is the human genome. How awesome a challenge is that one? To answer this question, we need to consider the amount of DNA involved. But how do we comprehend the 3 billion base pairs in the haploid nuclear genome? We can describe it as about a thousandfold more DNA than is present in an *E. coli* cell, or more than a hundred thousand times more DNA than most viruses have, but even these comparisons fail to capture the magnitude of the task. One way to comprehend the challenge is to note that, at the time of this writing, state-of-the-art laboratories each sequence only 1–2 million bases a year; at this rate it would take a single lab 1500–3000 years to sequence the entire human genome!

The sequence of the human genome is a major goal of the **Human Genome Project,** a cooperative, international effort that is currently planned to involve about 15 years of intense work. The project also encompasses a rough "mapping" of the human genome by locating specific genes and other sites ("markers") along the DNA on each human chromosome and analysis of the genomes of several other species, as mentioned above. Comparative analysis of the DNA of various species will not only be interesting in itself but will also help in interpreting the human data.

The wisdom of allocating large research funds to a mechanical sequencing of human DNA is controversial. Much of the genomic DNA of eukaryotes does not code for protein and has no other known function. (This DNA is often called "junk DNA"; one type, repeated DNA, is discussed in the next section.) Many biologists believe that research efforts should be directed primarily toward particular genes, arguing that a focused approach is most likely to yield scientifically interesting and useful results. For example, we know that alterations at specific sites in the human genome are involved in many diseases, and focusing on the DNA alterations (mutations) that cause these diseases should be the quickest route to significant progress in prevention and treatment. In fact, in the past few years, researchers studying a number of important genetic diseases have succeeded in identifying and sequencing the genes responsible. They have identified, among others, the genes involved in cystic fibrosis, Duchenne muscular dystrophy, Huntington's disease, and several kinds of leukemia.

The Human Genome Project also raises ethical controversies. The potential exists for the abuse of genetic information about individuals—for example, genetic discrimination by insurance companies, employers, or even government agencies. Furthermore, detailed knowledge of the human genome will increase the potential for using recombinant

(Continued on page 430)

14B

A CLOSER LOOK AT RESTRICTION ENZYMES
AND RESTRICTION FRAGMENTS

Restriction enzymes are endonucleases that are present in most bacterial cells. These enzymes protect the bacterial cell from foreign DNA molecules, particularly those of bacteriophages, that may invade the cell. In fact, restriction enzymes get their name from their ability to *restrict* the ability of foreign DNA to take over the transcription and translation machinery of the bacterial cell.

To protect its own DNA from being degraded, the bacterial cell has enzymes that add methyl groups ($—CH_3$) to specific nucleotides that its own restriction enzymes would otherwise recognize. Once methylated, the nucleotides cannot be recognized by the restriction enzymes, and the bacterial DNA is therefore not digested by the restriction enzymes present in the same cell. In other words, restriction enzymes are part of a **restriction/methylation system** in the bacterial cell: Foreign DNA is degraded by the restriction enzymes, and the bacterial genome is protected by prior methylation.

Restriction enzymes are named after the bacteria from which they are obtained. In each case, the name is derived by combining the first letter of the genus with the first two letters of the species. The particular strain of the bacterium is also indicated in the name, and if two or more enzymes have been isolated from the same species, the enzymes are numbered (using Roman numerals) in order of discovery. Thus, the first restriction enzyme isolated from *E. coli* strain R is designated *Eco*RI, whereas the third enzyme isolated from *Hemophilus aegyptius* is called *Hae*III.

Restriction enzymes are specific for double-stranded DNA and always cleave both of the strands. Every restriction enzyme recognizes a specific DNA sequence, which usually consists of four or six (but may be eight or more) nucleotide pairs. The recognition sequence is therefore a characteristic of the particular restriction enzyme and dictates where the enzyme will cleave the DNA molecule. For example, the enzyme *Hae*III recognizes the tetranucleotide sequence GGCC and cleaves the DNA double helix as shown in Figure 14B-la. Table 14B-1 indicates the recognition sequences for a number of common restriction enzymes.

The recognition sequences for most restriction enzymes are *palindromes*, which simply means that each such sequence reads the same in either direction. (The English word "radar" is a palindrome, for example.) The palindromic nature of a recognition sequence is due to the twofold rotational symmetry of the sequence: When rotated 180° in the plane of the paper, the sequence reads the same as it did before rotation. In other words, the recognition sequence has the same order of nucleotides on both strands but is read in opposite directions on the strands because of their antiparallel orientation.

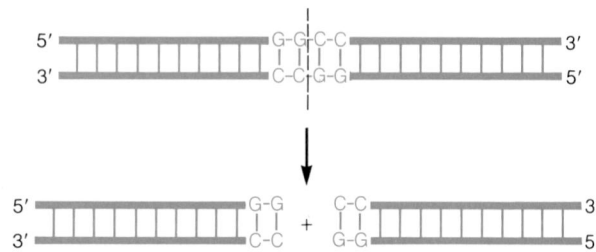

(a) Blunt-ended cleavage by the enzyme *Hae*III

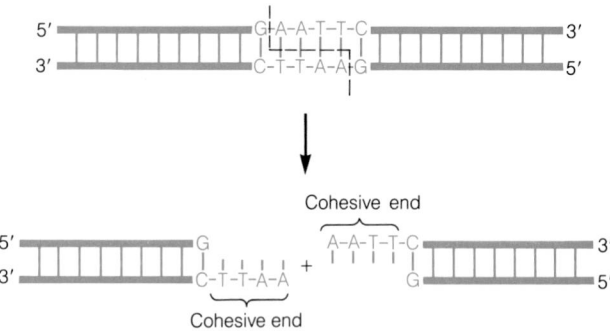

(b) Staggered cleavage by the enzyme *Eco*RI

Figure 14B-1   Cleavage of DNA by Restriction Enzymes.
(a) Some enzymes, such as *Hae*III, cut both strands of DNA at the same point, generating blunt-ended fragments. **(b)** Many other enzymes, such as *Eco*RI, make staggered cuts, leading to fragments with complementary cohesive ends (sticky ends). A genetic "engineer" can use such cohesive ends for joining DNA fragments from different sources, as we will explain in Chapter 16.

The frequency with which a specific recognition sequence is likely to occur within a DNA molecule is statistically predictable. For a DNA molecule containing equal amounts of the four bases (A, T, C, and G), we can predict that, on the average, a recognition site with four nucleotide pairs will occur once every 256 (i.e., 4^4) nucleotide pairs, whereas the likely frequency of a six-nucleotide sequence is once every 4096 (i.e., 4^6) nucleotide pairs. Restriction enzymes therefore cleave most DNA into pieces that vary in length from a few hundred to a few thousand nucleotide pairs—gene-sized pieces, essentially. Such pieces are called *restriction fragments*. Because of the specificity of the recognition sequence, a particular restriction enzyme always cleaves a particular DNA molecule in a predictable manner, generating a reproducible set of restriction fragments.

Some restriction enzymes cut both strands at the same point, generating restriction fragments with *blunt ends*.

*Hae*III has such a cleavage pattern, as we have already seen (Figure 14B-la). Many other restriction enzymes cleave the two strands in an offset (staggered) manner, generating short, single-stranded tails on both fragments. *Eco*RI is an example of such an enzyme; it recognizes the hexanucleotide sequence GAATTC and cuts the DNA molecule in an offset manner, leaving a —TTAA tail on both fragments (Figure 14B-lb). The restriction fragments generated by enzymes with this staggered cleavage pattern always have **cohesive ends,** also called **sticky ends.** These terms derive from the fact that the single-stranded tail on the end of each such fragment can base-pair with the tail at either end of any other fragment generated by the same enzyme, causing the fragments to stick to one another by hydrogen bonding. Enzymes that generate such fragments are particularly useful because they are an important means of creating recombinant DNA molecules, as we will see in Chapter 16. ■

Table 14B-1 Some Common Restriction Enzymes and Their Recognition Sequences

Enzyme	Source Organism	Recognition Sequence*
*Ava*I	*Anabena variabilis*	5′ C–Py–C–G–Pu–G 3′ G–Pu–G–C–Py–C
*Bam*HI	*Bacillus amyloliquefaciens*	5′ G–G–A–T–C^m–C 3′ C–C^m–T–A–A–G
*Eco*RI	*Escherichia coli*	5′ G–A–A–T–T–C 3′ C–T–T–A–A–G
*Hae*III	*Hemophilus aegyptius*	5′ G–G–C–C 3′ C–C–G–G
*Hind*III	*Hemophilus influenzae*	5′ A^m–A–G–C–T–T 3′ T–T–C–G–A–A^m
*Pst*I	*Providencia stuartii* 164	5′ C–T–G–C–A–G 3′ G–A–C–G–T–C
*Pvu*I	*Proteus vulgaris*	5′ C–G–A–T–C–G 3′ G–C–T–A–G–C
*Pvu*II	*Proteus vulgaris*	5′ C–A–G–C–T–G 3′ G–T–C–G–A–C
*Sal*I	*Streptomyces albus* G	5′ G–T–C–G–A–C 3′ C–A–G–C–T–G

*The arrows within the recognition sequence indicate the points at which the restriction enzyme cuts the two strands of the DNA molecule. Py = Pyrimidine (C or T), Pu = Purine (G or A), A^m = N^6-methyladenosine, C^m = 5-methylcytosine.

DNA techniques (see Chapter 16) to *alter* genomes—gene therapy—not only to treat selected body tissues that may be malfunctioning, but also to change germ lines (sperm and eggs). What use to make of these abilities and how they should be regulated are clearly questions that concern not only the scientific community but human society as a whole.

Repeated Sequences in Eukaryotic Genomes

We noted earlier that many restriction enzymes cleave DNA at specific sequences of four to six base pairs. A specific sequence of that length would be expected to occur only once in several thousand base pairs if the four kinds of DNA nucleotides were randomly distributed, and most DNA fragments produced by such cleavage would be a few thousand nucleotides in length. Much DNA does, in fact, end up in fragments of that length after digestion with a restriction enzyme, but a surprising amount of eukaryotic DNA has been found to give rise to much shorter fragments. This could only mean that eukaryotic genomes have **repeated sequences,** short sequences of nucleotides that are repeated many times in the genome.

The existence of repeated sequences in DNA, particularly eukaryotic DNA, was actually first discovered in DNA-renaturation studies. It is now clear that all known eukaryotic genomes contain repeated DNA sequences. The proportion of the genome represented by such sequences varies greatly among species, as does the number of times a sequence is repeated. Now we can explain, at least in part, the mystery of the seemingly excess amount of DNA in species such as *Trillium:* This organism contains a relatively high proportion of repeated DNA sequences. Using the sequencing techniques we described earlier, researchers have been able to determine the order of nucleotides in various repeating sequences and to classify them into two main categories (Table 14-2).

One major category of repeated DNA is now usually called **tandemly repeated DNA.** This type of DNA makes up 10–15% of most mammalian genomes, and the repeated unit may be as few as 1 bp or as many as 2000 bp. However, for much of this DNA, the repeated unit is fewer than 10 bp; consequently, this subcategory of DNA is often called **simple-sequence repeated DNA.** The key feature shared by all tandemly repeated DNA is that the multiple copies are arranged next to each other in a row—that is, tandemly. For example, the following would be a typical segment of tandemly repeated DNA with a simple repeated unit (showing one strand only):

$$\ldots \text{GTTACGTTACGTTACGTTACGTTAC} \ldots$$

The number of repetitions of the GTTAC unit at a given site in the genome could be as high as several hundred thousand.

Tandemly repeated DNA of the simple-sequence variety was originally called **satellite DNA** because, in many cases, distinctive base composition causes it to appear in a "satellite" band separate from the rest of the genomic DNA in centrifugation procedures that fractionate by density (equilibrium density centrifugation in a cesium chloride gradient; see Figure 15-3). Recall Chargaff's discovery that, although all double-stranded DNA has equal amounts of purines and pyrimidines, the proportions of particular purines and pyrimidines vary with the source of the DNA. Since adenine and guanine differ slightly in molecular weight, as do cytosine and thymine, the densities of DNAs with different base compositions will differ. In the procedures that reveal satellite bands of DNA, the genomic DNA is cleaved to short lengths before centrifugation, so that segments of DNA of different densities are free to migrate to different positions in the centrifuge tube.

What could be the function of simple-sequence segments of DNA? We know that it is not usually transcribed,

---

Table 14-2 Categories of Repeated Sequences in Eukaryotic DNA

I. Tandemly repeated DNA, including simple-sequence repeated DNA (satellite DNA)

10–15% of most mammalian genomes is this type of DNA

Length of each repeated unit:	1–2000 bp; typically 5–10 bp for simple-sequence repeated DNA
Number of repetitions per genome:	10^2–10^5
Arrangement of repeated units:	Tandem
Total length of satellite DNA at each site:	
Regular satellite DNA:	10^5–10^7 bp
Minisatellite DNA:	10^2–10^5 bp
Microsatellite DNA:	10^1–10^2 bp

II. Interspersed repeated DNA

25–40% of most mammalian genomes is this type of DNA

Length of each repeated unit:	10^2–10^4 bp
Arrangement of repeated units:	Scattered within the genome
Number of repetitions per genome:	10^1–10^6; "copies" not identical

---

because the nucleotide sequences that are known to signal the beginnings of genes are usually absent. One hypothesis is that highly repeated sequences are responsible for imparting special physical properties to some portions of the DNA. In most eukaryotes, the regions of chromosomes called **centromeres,** which play an important role in the distribution of chromosomes in cell division (see Chapter 15), are particularly rich in satellite DNA, and it is possible that these repeated sequences provide a specialized structure for the centromere chromatin. **Telomeres,** the ends of chromosomes, also have simple-sequence repeats. Their function seems to be to protect the chromosomes from degradation at their vulnerable ends and from harm due to shrinkage with each round of replication (see Figure 15-12). Human telomeres each have 250–1500 copies of the sequence TTAGGG. This sequence has been highly conserved over hundreds of millions of years of evolution; all vertebrates have the identical sequence, and even unicellular eukaryotes have similar sequences. Apparently, these sequences are absolutely critical to the survival of an organism.

Minisatellite DNA is a term often used for simple-sequence DNA that occurs in relatively short stretches, in the range of 10^2–10^5 bp at each site. *Microsatellite DNA* (in which the repeated unit is only 1–4 bp) typically has only about 20–200 bp per site, although there may be numerous sites in the genome with the same sequence. (In contrast, a stretch of regular satellite DNA is 10^5–10^7 bp in length.) Certain microsatellite and minisatellite DNA segments are extremely useful in the laboratory for **DNA typing,** also called **DNA fingerprinting.** This procedure, described more fully in Box 14C (page 446), uses gel electrophoresis of DNA restriction fragments to compare the nucleotide sequences at particular sites, or "markers," in the genomes of two or more individuals. It is a means of individual identification that is as accurate as conventional fingerprinting.

Within the past few years, medical researchers have made the surprising discovery that several genetic diseases affecting the nervous system are traceable to excessive numbers of repeated trinucleotide units within an otherwise normal gene. An example of such a disease is *Huntington's disease,* a devastating neurological disease that strikes in middle age and is invariably fatal. The normal version of the Huntington's gene contains the trinucleotide CAG tandemly repeated 11–34 times. The genes of afflicted individuals, however, may have as many as 100 copies of the repeated unit. Other neurological diseases resulting from the same phenomenon include *fragile X syndrome,* which is a major cause of mental retardation, and *myotonic dystrophy,* which affects the muscles. For some of these genes, the repeated sequence is in a region of the gene that is not translated; for others, it is translated into a polypeptide segment consisting of a long string of the same amino acid. In both kinds of cases, the normal function of the repeated sequences is not known, nor is it known how the extra copies lead to neurological dysfunction.

Besides tandemly repeated DNA, the other main category of repeated sequence is called **interspersed repeated DNA.** Rather than being clustered in tandem arrangements, the repeated units of this type of DNA are scattered about the genome. A single unit tends to be hundreds or even thousands of base pairs long, and its dispersed "copies," which may number in the hundreds of thousands, are similar but usually not identical to each other. Interspersed repeated DNA makes up 25–40% of most mammalian genomes.

In humans and other primates, a large portion of the interspersed repeated DNA consists of a family of similar sequences called the *Alu* **family,** so named because the first ones identified all contained a recognition sequence for the restriction enzyme *Alu*I. A single Alu unit is about 300 bp long, and the hundreds of thousands of Alu units in the human genome add up to at least 5% of the DNA. The Alu sequence turns out to be similar to the nucleotide sequence of a special type of RNA involved in protein synthesis on the rough ER (as a component of the signal-recognition particle, discussed in Chapter 18). Furthermore, many Alu units can actually be transcribed into RNA in the cell. However, this Alu RNA is quickly broken down, and there is no indication that it has any function. The function, if any, of Alu sequences remains unknown.

Although the cellular roles of interspersed repeated sequences are a mystery, we do know something about how the large numbers and variable locations of these sequences come to be. Most interspersed repeated sequences, at least in mammals, seem to be a kind of molecular parasite resembling *transposable elements.* These are segments of DNA that can move around the genome, leaving copies of themselves wherever they alight. We discuss one type of transposable element in Chapter 17 (see Box 17A).

We conclude this section on the organization of DNA in genomes by deferring discussion of some of the most interesting discoveries to come from molecular genetics and DNA sequencing—the organization of the DNA that constitutes *genes.* We will take up this topic in Chapter 17.

DNA Packaging

The amount of DNA that a cell must accommodate is awesome, even for organisms with genomes of modest size. Figure 14-15 illustrates the logistical problem posed by the amount of DNA that must fit in a bacterium, *E. coli.* As you can see, the total length of DNA greatly exceeds the dimensions of the space into which it must fit. The typical *E. coli* cell is about 1 μm in diameter and 2 μm in length, yet it must accommodate a (circular) DNA molecule with a length of about 1600 μm—enough DNA to encircle the cell more than 400 times! Somehow all this DNA must be efficiently packaged to fit into the cell yet still remain readily accessible

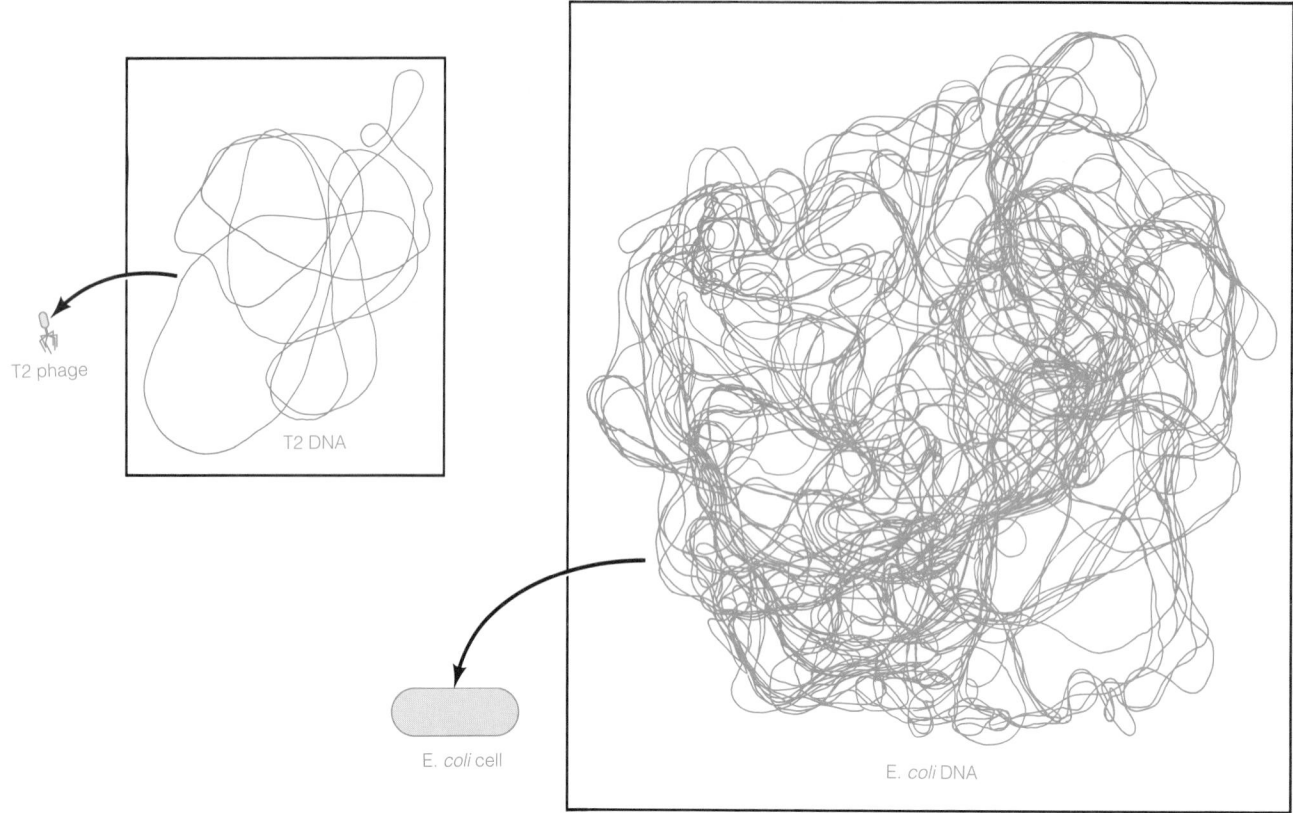

Figure 14-15 DNA Packaging in Viruses and Bacteria.
The bacteriophage T2, the bacterium *E. coli,* and the DNA molecule that must fit in each are shown to scale. Here the actual dimensions are mag-nified 10,000-fold; 1 cm in the figure represents 1 μm. On this scale, the T2 DNA molecule is 68 cm long, and the *E. coli* genome is 1600 cm.

to the cellular machinery for replication, transcription of specific genes, and other processes involving DNA.

Eukaryotic cells face an even greater challenge. A typical human liver cell, for example, has a diameter of about 35 μm and contains enough DNA to wrap around the cell more than 15,000 times. If it were included to scale in Figure 14-15, the cell would have a diameter of 35 cm, and the DNA would have a total length of about 20,400 m—slightly more than 12 miles. Clearly, DNA packaging is an important topic for all forms of life. We will look first at how prokaryotes organize their DNA and then consider the eukaryotic answer to the same problem.

Packaging of DNA in Prokaryotes

In marked contrast to the chromosomes of the eukaryotic cell, the genome of prokaryotes such as *E. coli* was once thought to be a "naked" DNA molecule lacking any elaborate organization and with only trivial amounts of protein associated with it. Now, however, we know that the organization of the bacterial genome within the cell is more like the eukaryotic case than previously realized. Bacterial ge-neticists usually refer to the structure that contains the main bacterial genome as the **bacterial chromosome.**

Packaging of the Bacterial Chromosome. The DNA mole-cule of the bacterial chromosome is highly folded and coiled in a complex but orderly manner. The molecule is negatively supercoiled and seems be folded into a num-ber of loops (see Figure 14-15). These loops average 50,000–100,000 bp, about the same size as similar looped domains in eukaryotic chromatin. The loops of bacterial DNA seem to be held in place by RNA and protein mole-cules. Evidence for a structural role for RNA in the bacterial chromosome is that treatment with ribonuclease, an enzyme that degrades RNA, releases some of the loops, although it does not relax the supercoiling. Nicking the DNA with a topoisomerase, on the other hand, relaxes the supercoiling but does not disrupt the loops.

The supercoiled DNA that forms the loops is actually organized into beadlike packets containing small, basic pro-tein molecules, analogous to the histones of eukaryotic chromatin (discussed in the next section). There is evidence that the DNA wraps around a core of protein.

Thus, from what we know so far, the bacterial chromosome consists of supercoiled DNA complexed with small, basic proteins and folded into looped domains. The compact structure that results is visible with the electron microscope as a tangled mass of threadlike material within the cell. The microscopist's term for this mass of material—the bacterial chromosome—is the **nucleoid.**

Bacterial Plasmids. In addition to its "chromosome," a bacterial cell may contain one or more plasmids. **Plasmids** are relatively small, circular molecules of DNA that carry genes both for their own replication and, often, for one or more cellular functions (usually nonessential ones). Most plasmids are supercoiled, giving them a condensed, compact form. Although plasmids replicate autonomously, the replication is usually in sufficient synchrony with the replication of the bacterial chromosome to ensure a roughly comparable number of plasmids from one generation to the next. In *E. coli* cells, three classes of plasmids are recognized. *F (fertility) factors* are involved in the process of conjugation, a sexual process we will discuss later. *R (resistance) factors* carry genes that confer drug resistance on the bacterial cell. *Col (colicinogenic) factors* allow the bacterium to secrete *colicins,* compounds that kill other bacteria lacking the col factor. In addition, some strains of *E. coli* contain *cryptic plasmids,* which have no known function.

Packaging of DNA in Eukaryotic Chromosomes

When we turn from prokaryotic cells to eukaryotic cells, DNA packaging becomes more complicated. First, substantially greater amounts of DNA are involved. Each eukaryotic chromosome is thought to have a single, very long, linear DNA molecule. In human cells, for example, just *one* of these DNA molecules may be 10 cm or more long. Second, greater structural complexity is introduced by the association of eukaryotic DNA with greater masses and numbers of proteins. The DNA with its associated proteins is called **chromatin.** It is important to realize that the number of "levels" of packaging, and hence the degree of compaction, of eukaryotic chromatin varies with the stage in the life cycle of the cell. The chromatin is most tightly packed during cell division, when discrete chromosomes are visible with the microscope; at other times, the degree of compaction is generally much less, and the chromatin appears as one diffuse mass. However, even when the chromatin is in its most diffuse state, it varies in its degree of compaction along the chromosome. We will start our discussion of eukaryotic DNA packaging with its basic structural unit, the nucleosome, which is found at all stages of the cell cycle. In the following sections, we use the term *chromosome* in its most general sense, not limiting it to the highly compacted structures that are microscopically visible during cell division.

The Nucleosome. The basic structural unit of eukaryotic chromosomes is a DNA-protein complex called the **nucleosome.** Nucleosomes consist of DNA associated with molecules of basic proteins called **histones.** The mass of histones in chromatin is approximately equal to the mass of DNA, except in the chromosomes of sperm cells, where the histones are replaced by protamines (discovered by Miescher, as you may recall). There are five major classes of histones: H1, H2A, H2B, H3, and H4. Histone H1, which is the largest at about 220 amino acids, is especially rich in lysine. H2A and H2B are slightly lysine-rich; and H3 and H4 are rich in arginine instead. These basic amino acids give the histones a positive charge. Their association with DNA, which is negatively charged, is therefore stabilized by ionic bonds. Chromatin also contains a small but highly significant proportion of acidic proteins, called **nonhistone chromosomal proteins.**

When the nuclei of nondividing (i.e., interphase) cells are isolated and washed free of all soluble substances, only the chromatin remains. This chromatin is almost completely unfolded and uncoiled. With the electron microscope, it is seen to consist of structural units, the nucleosomes, which repeat relatively regularly along the chromatin fiber, looking rather like beads on a string (Figure 14-16). Each nucleosome bead, or *core particle,* is about 11 nm in diameter and consists of eight histone molecules associated with 146 nu-

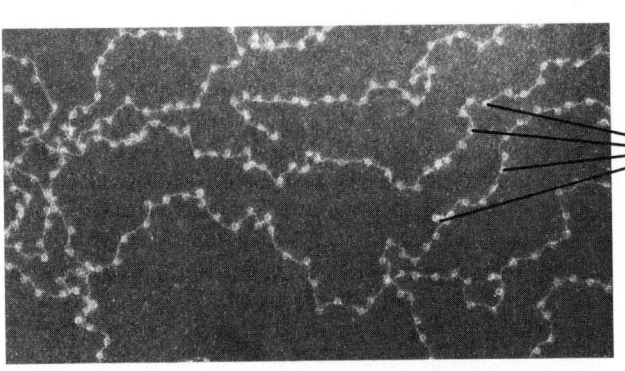

Nucleosome core particles

0.25 μm

Figure 14-16 Nucleosomes. The core particles of nucleosomes appear as beadlike structures spaced at regular intervals along eukaryotic chromatin fibers, as shown in this electron micrograph of chromatin from the red blood cells of a chicken. A nucleosome is defined to include both a core particle and the stretch of DNA that connects to the next core particle (TEM).

cleotide pairs of DNA. The histones include two molecules each of histones H2A, H2B, H3, and H4. As Figure 14-17 shows, these histones form a core around which the DNA is wrapped in a negative supercoil of almost two full turns.

As Figure 14-17 and Figure 14-18a show, nucleosome beads are separated from each other by a length of **nucleosome linker DNA** (also called *spacer DNA*), at least in isolated chromatin. The amount of linker DNA between successive nucleosome beads varies but averages about 50–60 nucleotides, and nucleosomes are usually defined to include it; therefore, a typical nucleosome can be said to include about 200 bp of DNA. Treatment of chromatin fibers with deoxyribonuclease results in preferential digestion of the linker DNA and the release of free nucleosome beads.

The positioning of nucleosome beads along the DNA depends on two factors. First, because severe bending of the DNA must occur for it to wrap tightly around the histone core, nucleosome positioning depends on the local flexibility of the DNA molecule, which in turn is influenced by the local sequence of bases. Second, nucleosome formation requires an absence of other proteins bound to the DNA. Because DNA binding proteins generally bind to specific base sequences, this factor, too, ultimately depends on DNA base sequence. However, the histone core does not interact directly with DNA bases. Thus, the positions of nucleosomes are usually not at precise points along the DNA.

Higher Levels of Chromatin Structure. Histone H1 is not a part of the nucleosomal structure but plays an essential role in further packaging of the beads-on-a-string structure to form a fiber roughly 30 nm in thickness, called the **30-nm chromatin fiber.** It has been established that one H1 molecule binds to each nucleosome in forming the 30-nm chro-

matin fiber, but the fiber's detailed structure is not known. One model is shown in Figure 14-18b. This is a coil, sometimes called a *solenoid,* with 6 nucleosomes per turn; both the linker DNA and the H1 molecules are in its interior. Recent studies suggest that the structure of the 30-nm fiber in the living cell may be much less uniform along the fiber's length than this model proposes and that it may be more like a zigzag ribbon than a coil. It is not yet known how much of the chromatin in a typical interphase cell is in the form of a 30-nm fiber.

The next level of chromatin packaging seems to be the formation of **looped domains** averaging 50,000–100,000 bp. The loop structures are held together by nonhistone proteins, which form a chromosomal axis. Such loops can be clearly seen in electron micrographs of chromosomes isolated from dividing cells and treated to remove all the histones and most of the nonhistone proteins (Figure 14-18c). Loops can also be seen in micrographs of intact chromosomes from certain specialized cells that are not in the process of dividing. (See the discussions of polytene chromosomes in Chapter 19.) In these cases, the chromatin loops turn out to contain "active" regions of DNA, that is, DNA that is being transcribed. It makes sense that active DNA would be less tightly packed than inactive DNA, allowing easier access by the proteins of the cellular machinery.

Even in a metabolically active interphase cell, however, significant amounts of the chromatin may be further compacted (Figure 14-18d). The degree of folding in such a cell probably varies over a continuum. Segments of chromatin so highly compacted that they show up as dark spots in micrographs are called **heterochromatin,** to distinguish them from the more extended, diffuse **euchromatin.** The DNA in heterochromatin is, not surprisingly, unlikely to be transcriptionally active.

As a cell prepares to undergo division, *all* of its chromatin becomes highly condensed, forming microscopically distinguishable chromosomes. Because the chromosomes have recently duplicated, each consists of two copies, called *chromatids.* Each chromatid is about 700 nm thick (Figure 14-18e).

The extent of DNA compaction can be quantified by the **DNA packing ratio,** the length of the linear DNA molecule divided by the length of the fiber or chromosome into which it is packaged. The initial coiling of the DNA around the histone cores of the nucleosomes reduces the length by a factor of about seven, and formation of the 30-nm fiber results in a further sixfold condensation. The packing ratio of the 30-nm fiber is therefore about 42. Further folding and coiling brings the overall packing ratio of typical euchromatin to about 750.

For heterochromatin and the chromosomes of dividing cells, the packing ratio is still higher. For example, a typical human chromosome from a dividing cell is about 4 or 5 μm long, yet contains about 75 mm of DNA. The overall packing ratio is therefore somewhere in the range of 15,000–20,000.

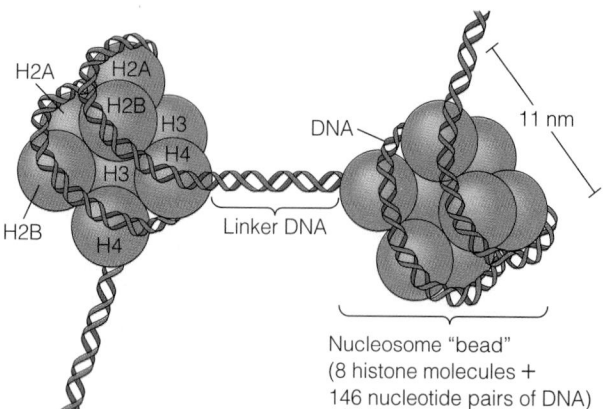

Figure 14-17 A Closer Look at Nucleosome Structure.
Each nucleosome consists of 8 histone molecules (two each of histones H2A, H2B, H3, and H4) associated with 146 nucleotide pairs of DNA and a stretch of linker DNA about 50 nucleotide pairs in length. The diameter of the nucleosome "bead," or core particle, is about 11 nm.

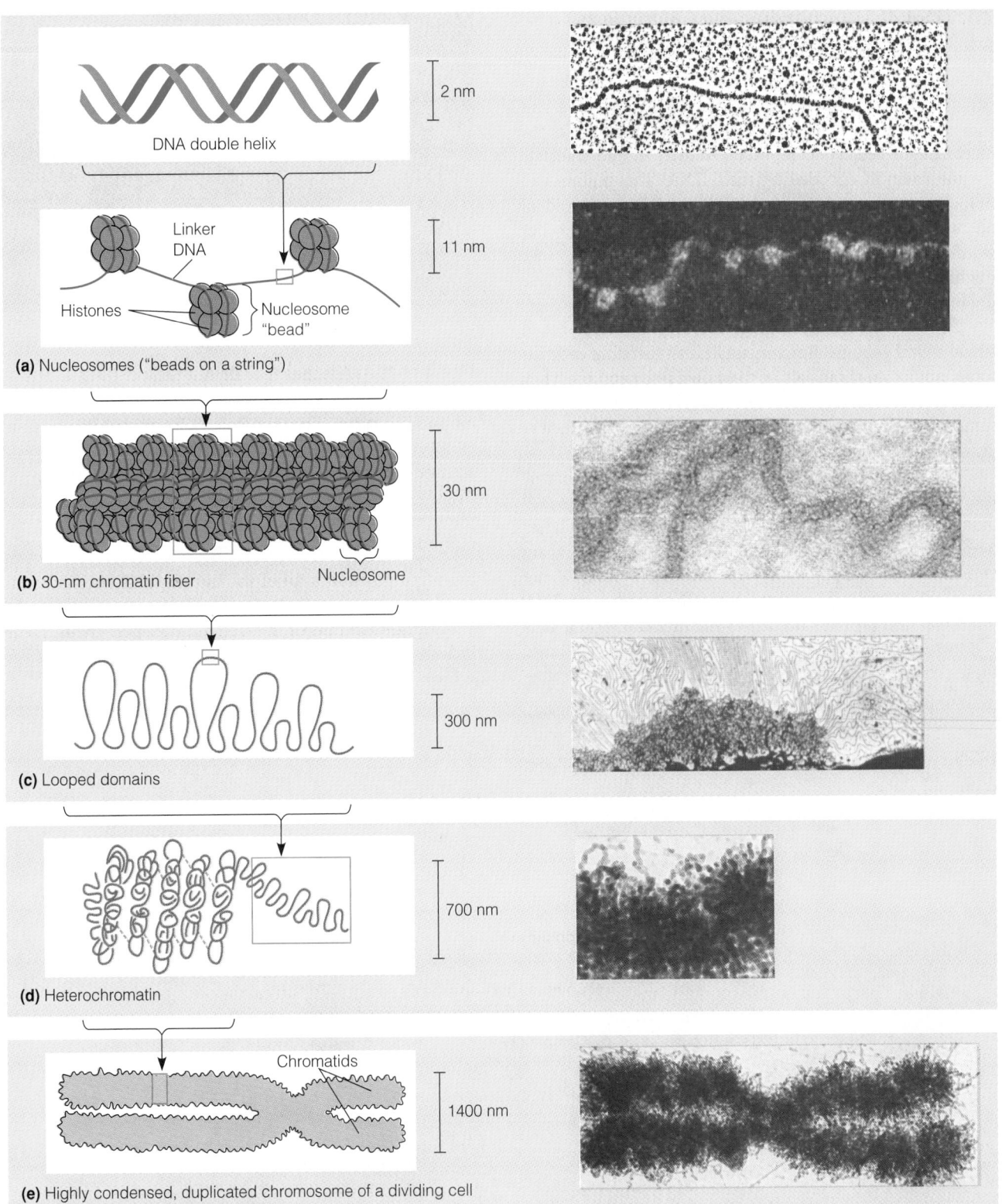

(a) Nucleosomes ("beads on a string")

Linker DNA

Histones

Nucleosome "bead"

2 nm

DNA double helix

11 nm

(b) 30-nm chromatin fiber

Nucleosome

30 nm

(c) Looped domains

300 nm

(d) Heterochromatin

700 nm

(e) Highly condensed, duplicated chromosome of a dividing cell

Chromatids

1400 nm

Figure 14-18 Levels of Chromatin Packing. These diagrams and TEMs show a current model for progressive stages of DNA coiling and folding, culminating in the highly compacted chromosome of a dividing cell. **(a)** "Beads on a string," an extended configuration of nucleosomes formed by the association of DNA with four types of histones. **(b)** The 30-nm chromatin fiber, shown here as a tightly wound coil. The fifth histone, H1, may be located in the interior of the fiber. **(c)** Looped do- mains of 30-nm fibers, visible in the TEM here because a mitotic chromo- some has been experimentally unraveled. **(d)** Heterochromatin, highly folded chromatin that is visible as discrete spots even in interphase cells. **(e)** A duplicated chromosome (two attached chromatids) from a dividing cell, with all the DNA of the chromosome in the form of very highly com- pacted heterochromatin.

The DNA of Mitochondria and Chloroplasts

Not all of the DNA of a eukaryotic cell is present in the nucleus. Nuclear DNA certainly accounts for most of the genetic information of the cell, but both mitochondria and chloroplasts contain some DNA of their own, as well as the machinery necessary to replicate, transcribe, and translate the information encoded by their DNA. The genomes of mitochondria and chloroplasts consist of DNA devoid of histones and are usually circular (Figure 14-19). In other words, they are similar to the genomes of prokaryotes, as we might expect from the likely endosymbiotic origin of these organelles (see Chapter 4).

In both organelles, the genome is small, comparable in size to a viral genome. Both organelles can therefore code for some, but by no means all, of their own polypeptides. They are therefore semiautonomous organelles, able to specify some of their polypeptides but dependent on the nuclear genome to encode most of them.

The genome of the human mitochondrion, for example, consists of a circular DNA molecule containing 16,569 base pairs and having a contour length of about 5 μm. It has been completely sequenced, and all of its 37 genes are known. The RNA and polypeptides encoded in this DNA are just a small fraction (about 5%) of the number of RNA molecules and proteins needed by the mitochondrion. This is nonetheless a vital genetic contribution, however, because these products include the RNA molecules present in mito-

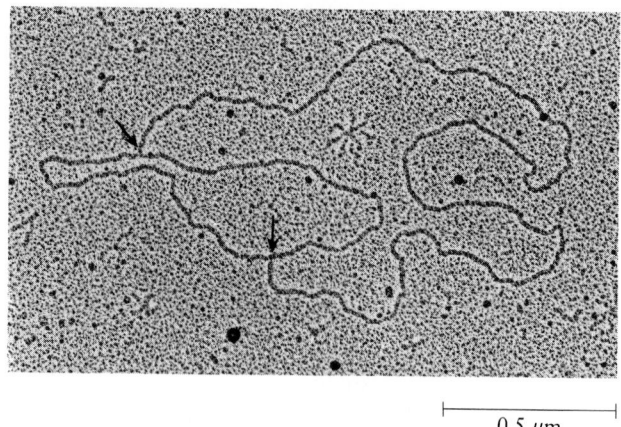

Figure 14-19 Mitochondrial DNA. Mitochondrial DNA from most organisms is circular, as seen in this electron micrograph of a mitochondrial DNA molecule from a rat liver cell. The molecule was caught in the act of replication; the arrows indicate the points at which replication was proceeding when the molecule was fixed for electron microscopy (TEM).

chondrial ribosomes, all of the transfer RNA molecules required for mitochondrial protein synthesis, and 13 of the polypeptide subunits of the electron transport chain. These include subunits of NADH dehydrogenase, cytochrome *b*, cytochrome *c* oxidase, and ATP synthase (Figure 14-20).

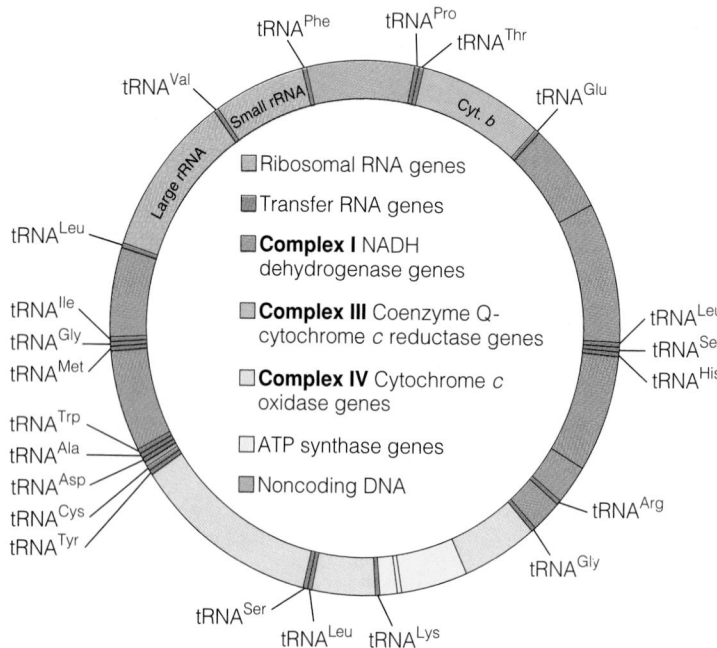

Figure 14-20 The Genome of the Human Mitochondrion.
The double-stranded DNA molecule of the human mitochondrion is circular and contains 16,569 nucleotide pairs. This genome codes for large and small ribosomal RNA molecules, transfer RNA (tRNA) molecules (each identified by a superscript with the three-letter abbreviation for the amino acid it carries), and subunits of a number of the proteins that make up the mitochondrial electron transport chain complexes. The tRNA genes are very short because the RNA molecules they encode each contain only about 75 nucleotides. Notice that there are two tRNA genes for leucine and two for serine; they code for slightly different versions of tRNAs for these amino acids. The mitochondrial genome is extremely compact, with little noncoding DNA between genes.

The size of the mitochondrial genome varies considerably with species. Mammalian mitochondria typically have about 16,500 bp of DNA, but yeast mitochondria have about five times that much, and plant mitochondria have even larger amounts. It is not at all clear, however, that larger mitochondrial genomes necessarily code for correspondingly more polypeptides. A comparison of yeast and human mitochondrial DNA molecules, for example, suggests that most of the additional DNA present in the yeast mitochondrion consists of long noncoding sequences.

Chloroplasts typically contain circular DNA molecules with about 120,000 bp, encoding about 120 genes. In addition to ribosomal and transfer RNAs and polypeptides involved in protein synthesis, the chloroplast genome also codes for a number of polypeptides specifically involved in photosynthesis. These include several proteins that are part of photosystems I and II and one of the two subunits present in ribulose-1,5-bisphosphate carboxylase, the carbon-fixing enzyme of the Calvin cycle (see Figure 13-16).

Interestingly, most polypeptides known to be encoded by the mitochondrial or chloroplast genome are part of multimeric proteins that also contain subunits encoded by the nuclear genome. In other words, most organelle proteins that contain subunits encoded within the organelle are hybrids, consisting of some polypeptides encoded and synthesized within the organelle and others encoded by the nuclear genome and synthesized by cytoplasmic ribosomes. This raises intriguing questions as to how the polypeptides synthesized in the cytoplasm enter the organelle and how their uptake and availability are synchronized with the synthetic activities of the organelle. We will return to these questions in Chapter 18.

The Nucleus

So far, we have considered DNA as the genetic material of the cell, the genome as the DNA available to the cell of a particular species, and the chromosome as the physical means of packaging almost all the DNA within the cell. Now we come to the **nucleus,** the site within the eukaryotic cell where the chromosomes are localized and replicated and where the DNA they contain is selectively transcribed. The nucleus is therefore both the repository of almost all the genetic information for the cell and the control center for the expression of that information.

The nucleus is one of the most prominent and characteristic features of eukaryotic cells (Figure 14-21). Recall that the term *eukaryon* means "true nucleus." The very essence of a eukaryote is its membrane-bounded nucleus, which compartmentalizes the activities of the cell's genome—both replication and transcription—from the rest of cellular

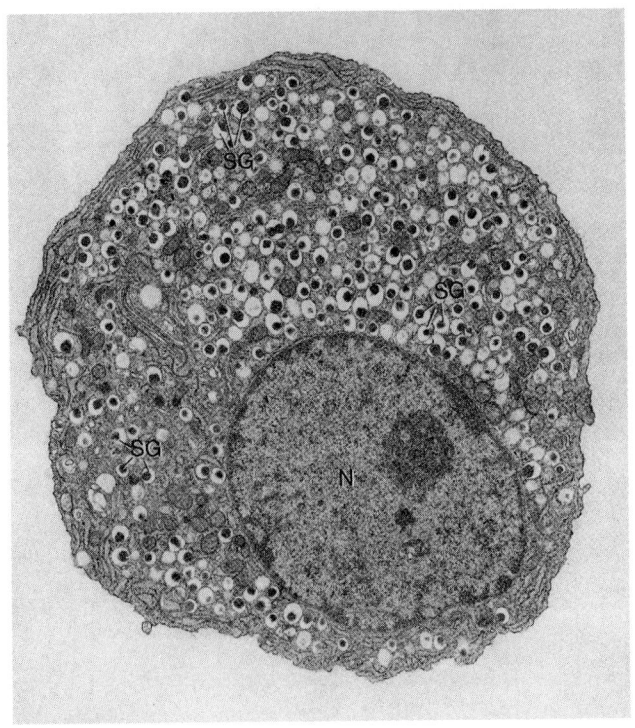

(a) Animal cell nucleus   5 μm

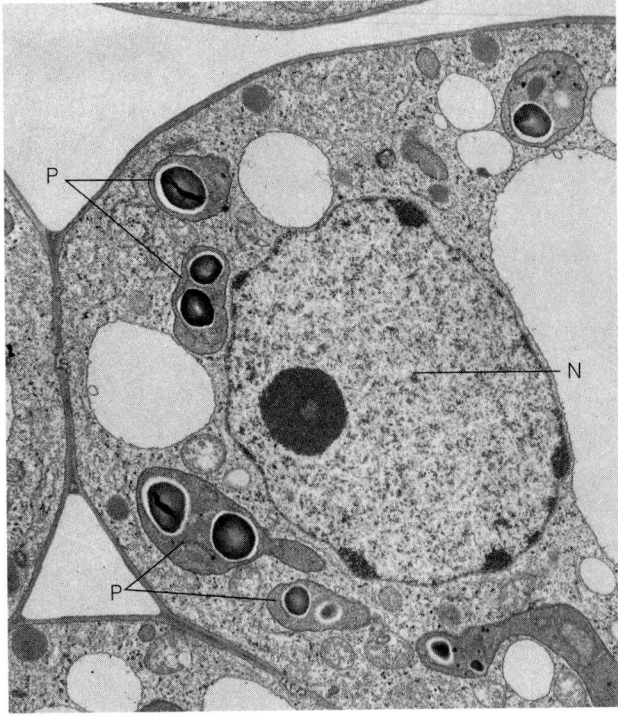

(b) Plant nucleus   5 μm

Figure 14-21 The Nucleus. The nucleus is a prominent structural feature in most eukaryotic cells. **(a)** The nucleus (N) of an animal cell. This is an insulin-producing cell from a rat pancreas; hence, the prominence of secretory granules (SG) in the cytoplasm. **(b)** The nucleus (N) of a plant cell. This is a cell from a soybean root nodule. The prominence of plastids (P) reflects their role in the storage of starch granules (TEMs).

metabolism. In the following discussion, we focus first on the envelope of membranes that is the boundary of the nucleus. Then we turn our attention to several other aspects of the nucleus, including the pores that perforate the envelope, the structural fibers inside the nucleus, the localization of the chromatin, and the nucleolus. Figure 14-22 provides an overview of the nuclear structures.

The Nuclear Envelope

The existence of a membrane around the nucleus was suggested in the late nineteenth century, primarily on the basis of observed osmotic properties of the nucleus, and was confirmed by phase-contrast microscopy during the first half of the present century. Further investigation gradually led to the realization that the nucleus is surrounded by a *double* membrane, with a space between the two phospholipid bilayers. Additional insight came with the advent of electron microscopy, which verified the presence of a double membrane and added much to our understanding of its structure.

As shown in Figure 14-22, the **nuclear envelope** consists of two unit membranes, the inner and outer nuclear membranes, with a **perinuclear space** between them. Each membrane is about 7–8 nm thick and has the same trilamellar appearance as most other cellular membranes. The inner nuclear membrane is lined with a network of fibers called the *nuclear lamina* (discussed later in this chapter). The outer nuclear membrane is continuous with the endoplasmic reticulum and is thought to derive from the ER. Like the membrane of the rough ER, the outer membrane is often studded on its outer surface with ribosomes engaged in protein synthesis. In electron micrographs of some cells, intermediate filaments of the cell's cytoskeleton are seen to extend outward from the outer membrane into the cytoplasm, possibly anchored on the other end to other organelles or the plasma membrane. These filaments probably position the nucleus firmly within the cell.

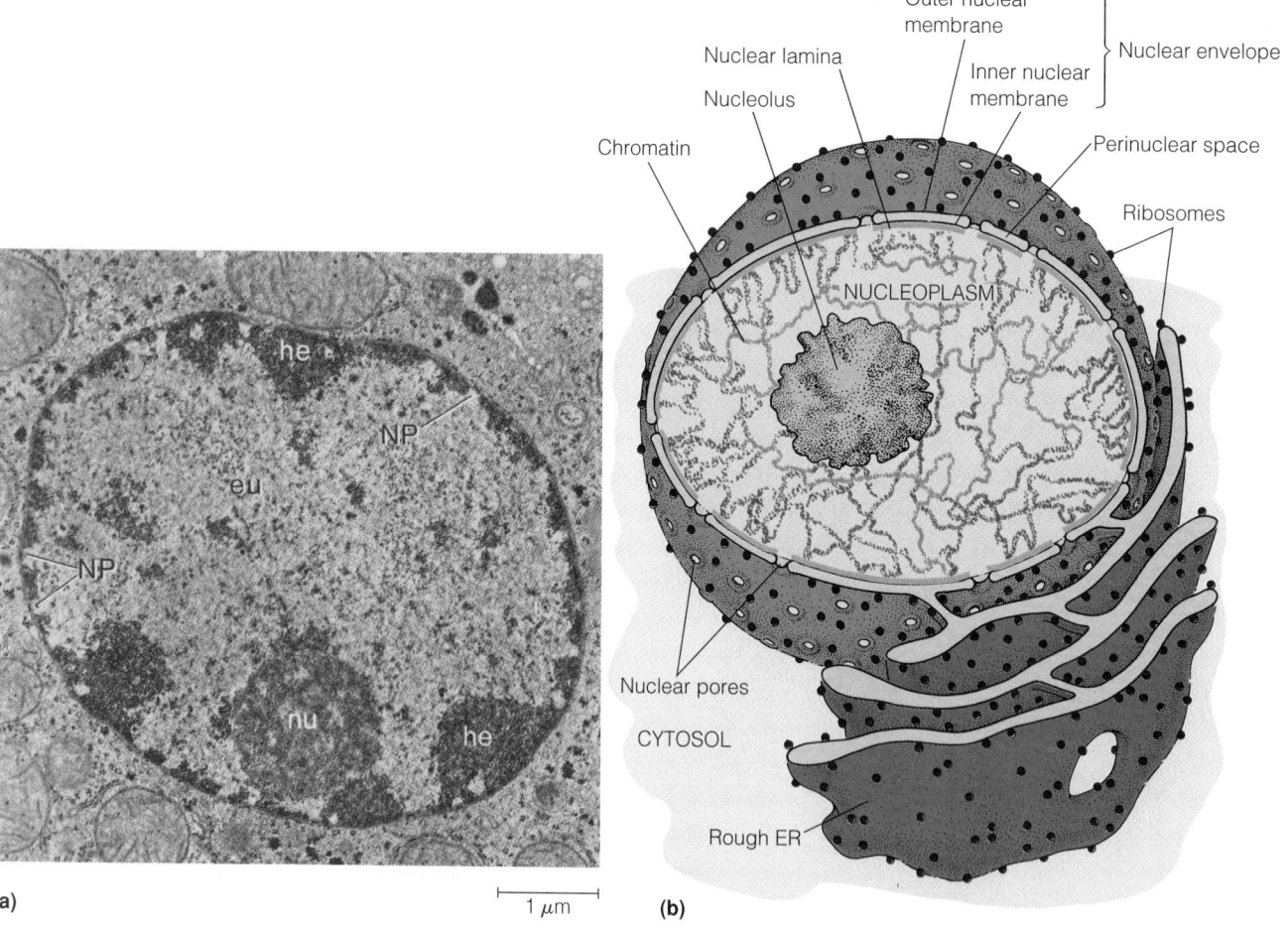

(a)

1 μm

(b)

Figure 14-22 The Structure of the Nucleus. **(a)** An electron micrograph of the nucleus from a mouse liver cell, with prominent structural features labeled (TEM). The nuclear envelope is a double membrane perforated by nuclear pores (NP). Internal structures include the nucleolus (nu), euchromatin (eu), and heterochromatin (he). **(b)** A drawing of a typical nucleus. Structural features included here but not visible in the micrograph include the nuclear lamina, ribosomes on the outer nuclear membrane, and the continuity between the outer nuclear membrane and the rough ER.

The perinuclear space between the two membranes is a gap of about 20–40 nm. This fluid-filled compartment seems to be continuous with the cisternae of the ER. Little is known of the biochemical composition of the perinuclear fluid or the significance of this compartment.

The most distinctive feature of the nuclear envelope is the presence of numerous **nuclear pores,** small cylindrical channels that extend through both membranes, providing direct contact between the cytosol and the **nucleoplasm,** the interior of the nucleus. Nuclear pores are readily visible when the nuclear envelope is examined by freeze-fracture microscopy (Figure 14-23).

The density of pores (number per unit surface area of the nuclear envelope) varies greatly with cell type and activity. A typical mammalian nucleus has about 3000–4000 pores, a density of about 10–20 pores per square micrometer. Researchers studying nuclear pores often use oocytes (egg cells) of *Xenopus laevis,* the South African clawed toad, because these have very large nuclei with an unusually high pore density; a single nucleus may contain more than 10 million pores! (Presumably all these pores are needed to handle the heavy transenvelope traffic occurring in this metabolically hyperactive cell.)

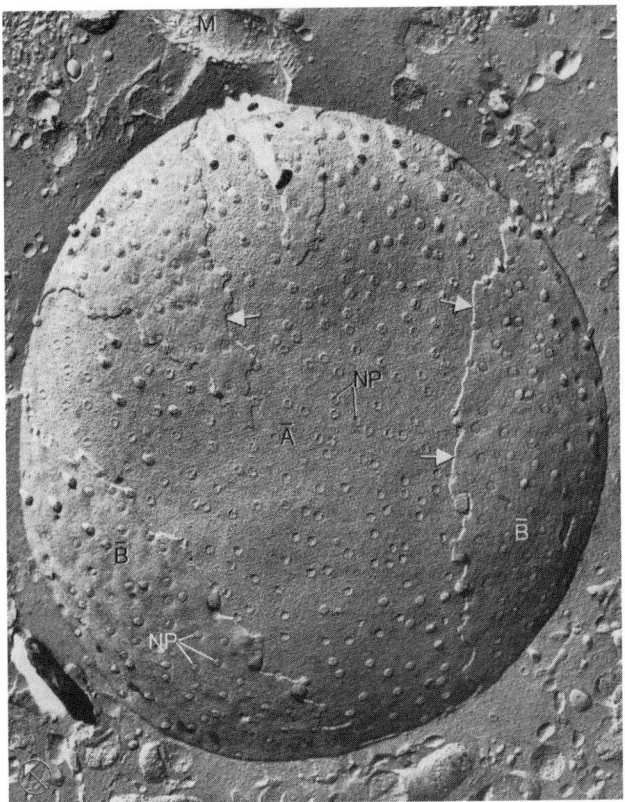

1 μm

Figure 14-23 Nuclear Pores. Numerous nuclear pores (NP) are visible in this freeze-fracture micrograph of the nuclear envelope of an epithelial cell from a rat kidney. The fracture plane reveals faces of both the inner membrane (A) and the outer membrane (B). The ridges to which the arrows point represent the perinuclear space delimited by the two membranes (TEM).

At each pore, the inner and outer nuclear membranes are fused. The two membranes of the nuclear envelope are therefore physically continuous, although they differ distinctively in their biochemical properties and functional roles. Because the outer nuclear membrane is continuous with both the ER and the inner membrane, both membranes of the nuclear envelope can exchange material directly with the ER membrane, thereby allowing the nuclear envelope to expand and contract as needed.

Nuclear pores represent much more than points of continuity between the two membranes. They have a structural complexity that is still being unraveled, and they control the transport of key molecules between the cytoplasm and the nucleus. Because of the importance of this transport role, we will look at the protein complex that forms the nuclear pores in some detail.

The Nuclear Pore Complex

A nuclear pore is formed by the fusion of the two membranes of the nuclear envelope and is lined with an intricate protein structure called the **nuclear pore complex (NPC).** The diameter of the entire pore complex is about 120 nm. It has an overall mass of some 120 million daltons (Da) and may consist of 100 or more different kinds of polypeptide subunits.

The morphology of nuclear pores in living cells is not yet entirely understood. Their appearance in electron micrographs seems to vary somewhat with the way the sample is prepared and with the type of cell. In electron micrographs, the most striking feature of the pore complex is the octagonal arrangement of its subunits. Micrographs such as the one in Figure 14-24a show rings of eight subunits arranged in an octagonal pattern. In other views, the eight subunits are seen to protrude on both the cytoplasmic and the nucleoplasmic sides of the envelope. Notice that central granules can be seen in some of the nuclear pore complexes in Figure 14-24a. Although these granules were once thought to be particles in transit through the pores, recent evidence supports the idea that they are an integral part of the pore complex. Apparently, they are easily lost during the preparation of samples for microscopy.

In recent years, increasingly sophisticated biochemical and genetic techniques have supplemented electron microscopy as tools for delving into the mysteries of the nuclear pore. Figure 14-24b and c show a current model for pore structure. These illustrations distinguish by color several major components of the pore complex, each consisting of a number of polypeptides. The pore complex as a whole is shaped somewhat like a wheel lying on its side within the nuclear envelope. Two parallel rings, outlining the rim of the wheel, each consist of the eight subunits seen in electron micrographs. Eight spokes (shown in green) extend from the rings to the wheel's hub (dark pink), which is the "central granule" seen in many EMs. This granule is now usually called the **transporter** because it functions to move macromolecules across the nuclear envelope. Proteins extending

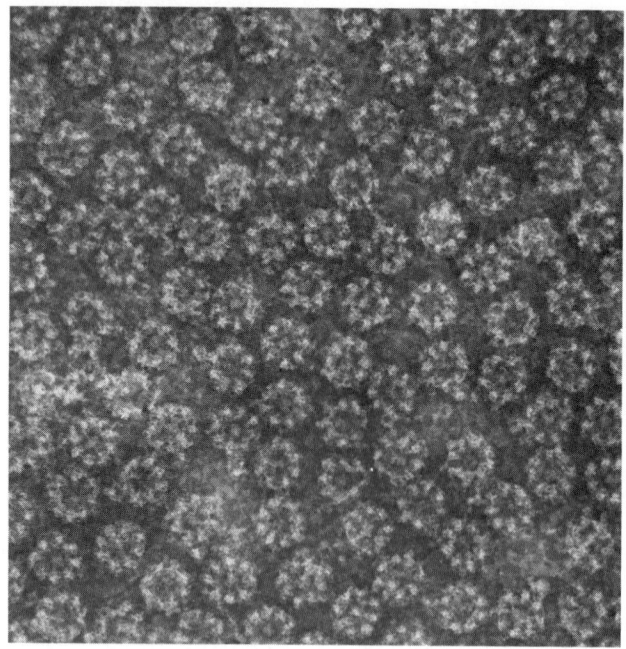

(a) Nuclear envelope showing nuclear pores

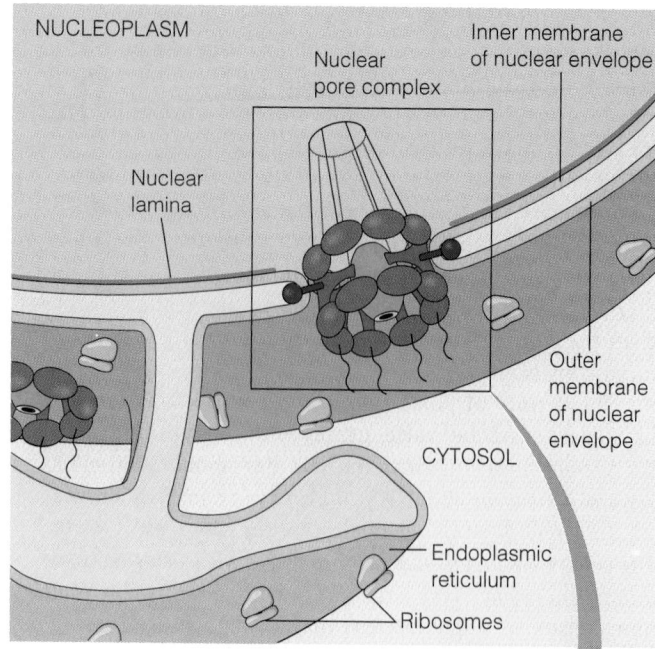

(b) Location of nuclear pores in nuclear membrane

Figure 14-24 The Structure of the Nuclear Pore.
(a) Negative staining of an oocyte nuclear envelope reveals the octagonal pattern of the nuclear pore complexes. This nuclear envelope is from an oocyte of the newt *Taricha granulosa* (TEM).
(b) A nuclear pore is formed by the fusion of the inner and outer nuclear membranes and is lined by an intricate protein structure called the nuclear pore complex (NPC). **(c)** A current model for the detailed structure of the nuclear pore complex. The structure is roughly wheel-shaped and has octagonal symmetry. Two parallel rings, each consisting of eight subunits (dark purple), outline the rim of the wheel. Eight spokes (green) are shaped so that they connect the two rings (two of the spokes have been omitted from the drawing). The spokes extend to the central transporter (dark pink) at the hub of the wheel; the transporter may be a diaphragmlike structure that opens and closes to allow the passage of particles of different sizes. Proteins extending from the rim-and-spoke assembly into the perinuclear space (pale purple) presumably help anchor the complex within the nuclear envelope. Fibers extend above and below the complex, with the ones on the nucleoplasmic side forming a basket of unknown function. The nuclear pore complex may consist of as many as 100 distinct polypeptides.

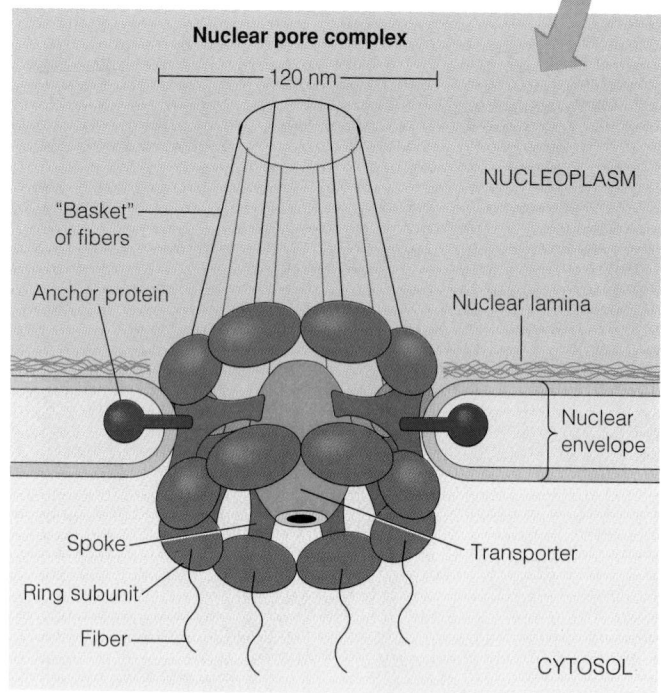

(c) Nuclear pore structure

from the rim into the perinuclear space (pale purple) are thought to help anchor the pore complex to the envelope. In addition, fibers extend from the rings into the cytosol and nucleoplasm, the ones on the nucleoplasm side forming a basket (sometimes called a "cage" or "fishtrap"). We do not yet know the function of this basket. However, we do know something about the general mechanisms of nuclear pore function, as discussed in the next section.

Transport Across the Nuclear Envelope

The nuclear envelope is both a solution to one "problem" and the source of another. As a means of restricting chromosomes and their activities to one part of the cell, it is an example of the general eukaryotic strategy of compartmentalization. Presumably, it is advantageous for a nucleus to possess a barrier that keeps chromosomes in and organelles

such as ribosomes, mitochondria, lysosomes, and microtubules out. For example, it is useful for newly synthesized RNA to be secluded, because, as we will discuss in Chapter 17, the RNA transcribed from eukaryotic DNA usually needs to be processed further before it can function properly in the cytoplasm. The nuclear envelope protects the immature RNA from being acted upon by cytoplasmic organelles or enzymes.

At the same time as it protects the chromosomes and immature RNA molecules from exposure to the cytoplasm, however, the nuclear envelope creates for the eukaryotic cell several formidable transport problems that are unknown in prokaryotes. All the enzymes and other proteins required for chromosome replication and transcription of DNA in the nucleus must be imported from the cytoplasm, and all the RNA molecules and partially assembled ribosomes needed for protein synthesis in the cytoplasm must be obtained from the nucleus (Figure 14-25). Nature's solution to these

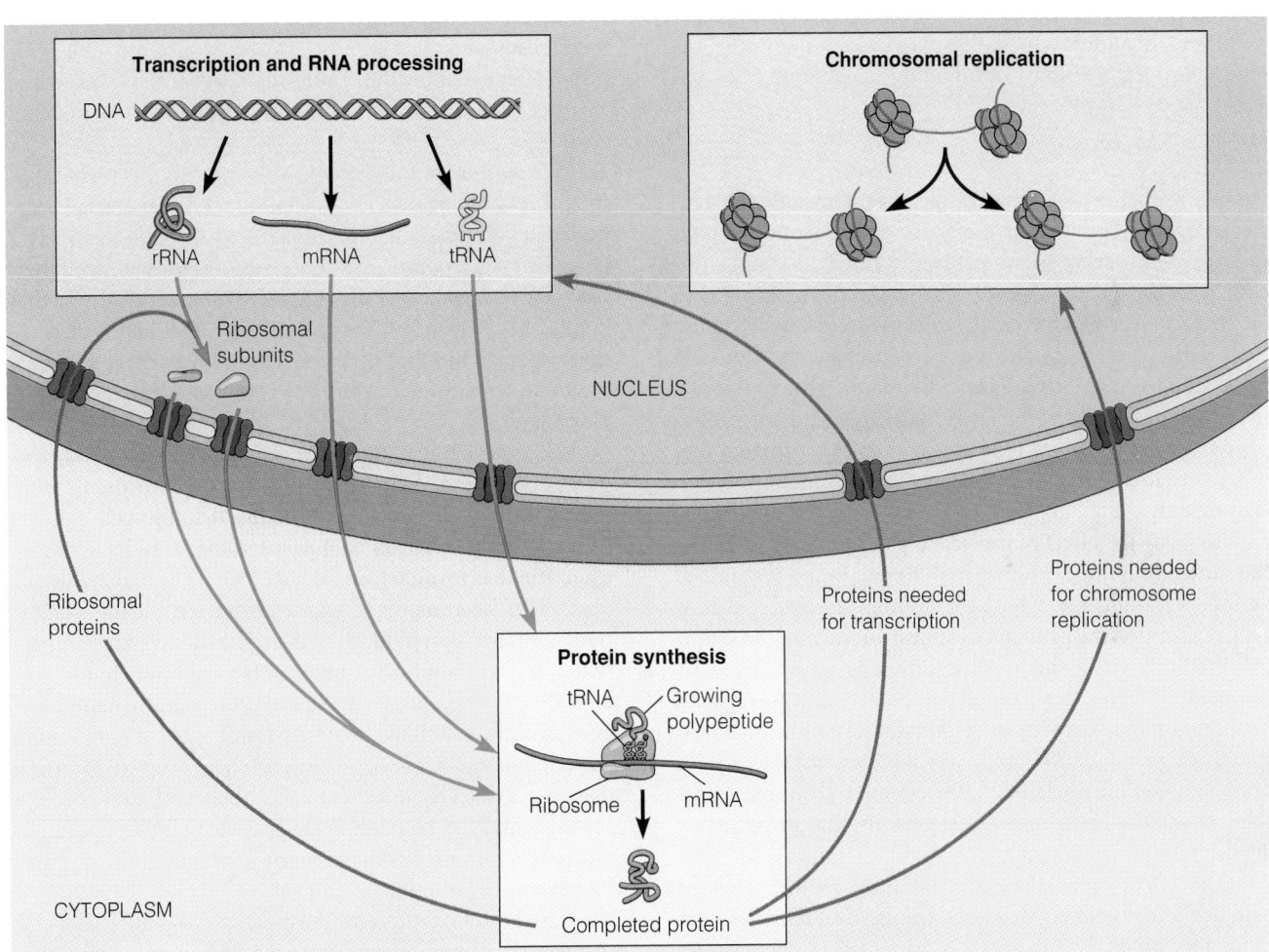

Figure 14-25 Macromolecular Transport Into and Out of the Nucleus. Because eukaryotic cells store their genetic information in the nucleus but synthesize proteins in the cytoplasm, all the proteins needed in the nucleus must be transported inward from the cytoplasm (purple arrows), and all the RNA molecules and ribosomal subunits needed for protein synthesis in the cytoplasm must be transported outward from the nucleus (red arrows). The three kinds of RNA molecules required for protein synthesis are ribosomal RNA (rRNA), messenger RNA (mRNA), and transfer RNA (tRNA).

problems of macromolecular transport has been the evolution of a eukaryotic nuclear envelope with pores, which mediate virtually all transport across the envelope.

To get some idea of how much traffic must travel through the nuclear pores, consider the flow of ribosomal subunits from the nucleus to the cytoplasm. Ribosomes are partially assembled in the nucleus as two classes of subunits, each of which is a complex of RNA and proteins. These subunits move to the cytoplasm and, when needed for protein synthesis, are combined into functional ribosomes consisting of one of each type of subunit. An actively growing mammalian cell can easily be synthesizing 20,000 ribosomal subunits per minute. We already know that such a cell has about 3000–4000 nuclear pores, so ribosomal subunits must be transported to the cytosol at a rate of about 5–6 subunits per minute per pore. Traffic in the opposite direction is, if anything, even heavier. When chromosomes are being replicated, histones are needed at the rate of about 300,000 molecules per minute. The rate of inward movement must therefore be about 100 histone molecules per minute per pore! And, in addition to all this macromolecular traffic, the pores allow the transport of numerous small molecules and ions.

Passive Transport of Small Molecules Through the Nuclear Pores. The nuclear pores provide aqueous channels that apparently serve as a direct connection between the nuclear interior and the cytosol. The channels are thought to be freely permeable to small molecules and ions, because such substances cross the nuclear envelope very quickly when they are injected into cells. Thus, the nucleoside triphosphates required for DNA and RNA synthesis probably diffuse freely through the pores, as do other small molecules needed for metabolic pathways that function within the nucleus.

To study the effect of molecular size on transport across the nuclear envelope, investigators have injected proteins of various sizes into the cytoplasm of cells and observed how long it takes for the proteins to appear in the nucleus. Many proteins can cross the nuclear envelope, presumably by means of its pores, but passage becomes more difficult with increasing molecular weight. A globular protein with a molecular weight of 20,000 takes only a few minutes to equilibrate between the nucleus and the cytoplasm, but most proteins of 60,000 Da or more seem barely able to penetrate at all.

The effect of particle size has been further assessed using colloidal gold particles of various sizes. Particles with diameters up to 5–6 nm can be detected in the nucleus within minutes; those with diameters of about 9 nm take several hours; and larger particles do not enter at all. These and other transport measurements indicate that the aqueous channels in the nuclear pore complexes are about 9 nm in diameter. Until recently, researchers generally assumed that there was one such channel in each pore complex. Now, however, it is thought that there may be eight 9-nm channels

on the periphery of the pore complex, between the spokes, and perhaps an additional 9-nm channel at the center of the transporter.

Active Transport of Macromolecules and Ribosomal Subunits Through the Nuclear Pores. Many of the proteins involved with DNA packaging, replication, and transcription are small enough to pass through a 9-nm-wide channel. Histones, for example, have molecular weights of 21,000 or less and should therefore passively traverse the nuclear pores with little problem. Some nuclear proteins are very large, however. The polymerases involved in both DNA and RNA synthesis, for example, have subunits with molecular weights in excess of 100,000, which would presumably not fit through a 9-nm orifice. Messenger RNA poses a challenge, too, because mRNA molecules are believed to leave the nucleus complexed with proteins in the form of **ribonucleoprotein particles**—complexes of RNA and protein—that are quite large. The ribonucleoprotein particles that are the ribosomal subunits must also be exported to the cytoplasm after assembly in the nucleus. Clearly, the transport of all these particles through the nuclear pores is a significant challenge.

We do not yet fully know the answer to this puzzle, but in recent years, evidence has accumulated that certain large molecules and particles are transported through the nuclear pores by an active process. This process is probably different from active transport across single membranes. But, like that kind of active transport, it does require ATP, and it does require specific binding of the transported particle to membrane proteins, which in this case are part of the pore complex.

The mechanism of active transport through nuclear pores is best understood for the transport of proteins from the cytosol to the nucleus. Proteins that the cell actually transports from cytosol to nucleus all seem to have one or more **nuclear localization signals (NLS),** which enable the protein to be recognized and transported by the nuclear pore complex. A typical nuclear localization signal is a particular sequence of 8–30 amino acids somewhere in the protein; the sequence varies with the protein but usually contains proline as well as the positively charged (basic) amino acids lysine and arginine. Gold particles of up to 26 nm in diameter are transported into the nucleus if they are first coated with NLS-containing polypeptides. Thus, the maximum diameter for *active* transport across the nuclear envelope seems to be 26 nm.

The process of protein transport into the nucleus seems to have at least three steps (Figure 14-26). In step 1, the NLS-containing protein is conveyed to a nuclear pore. Much evidence suggests that this stage involves the binding of a receptor protein *in the cytosol* to the NLS, followed by movement of the protein complex to a pore, perhaps with the pore's cytoplasmic fibers somehow serving as tracks. The protein complex then binds to the nuclear pore complex (step 2). In the ATP-dependent step, the NLS-containing protein (at

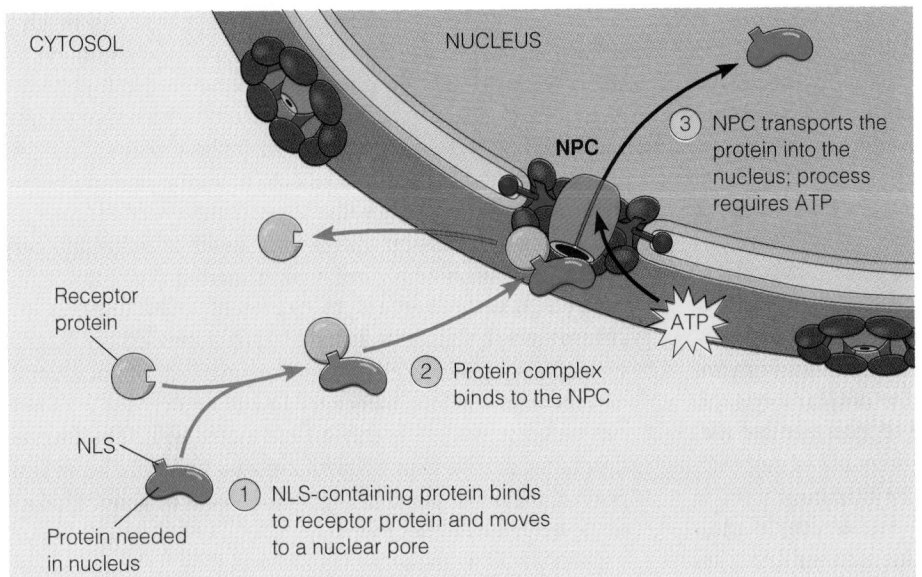

Figure 14-26 A Proposed Mechanism for the Transport of Proteins from the Cytoplasm into the Nucleus. Proteins destined for use in the nucleus contain short stretches of amino acids called nuclear localization sequences (NLS) that target them for transport through the nuclear pores. ① The NLS-containing protein is thought first to bind to a receptor protein in the cytosol and to move to a pore. ② The nuclear protein/receptor protein complex then binds to the nuclear pore complex (NPC). ③ Finally, with the hydrolysis of ATP, the NPC's transporter (dark pink) opens and the nuclear protein passes through it into the nucleus.

least) is transported through the envelope by the transporter at the center of the pore complex (step 3). Some investigators hypothesize that the transporter operates like the diaphragm of a camera, opening and closing to allow the passage of particles of different sizes.

For the transport of material out of the nucleus, an equally selective mechanism seems to operate. In further experiments with colloidal gold, investigators have coated gold particles with RNA molecules (e.g., transfer RNA), injected the coated particles into the nucleus, and observed the transport of the particles through pores into the cytoplasm. The *direction* of transport is specific to the type of molecule: In most cases, RNA is only exported, whereas proteins designed for function in the nucleus are only imported. However, a given pore can carry out transport in either direction.

Structural Fibers Within the Nucleus

During the past two decades, investigators have amassed evidence for a structural network of fibers, analogous to the cytoskeleton of the cytoplasm, that extends throughout the nucleus. Presumably this network, usually called the **nuclear matrix,** helps maintain the shape of the nucleus and provide an organizing scaffold for the interphase chromatin. The existence of the nuclear matrix has not been accepted by all cell biologists. The fibers are visible only in certain micrographs (Figure 14-27), leading skeptics to raise the question of whether they are artifacts introduced in sample preparation. In recent years, however, additional evidence from studies of eukaryotic DNA replication and transcription have bolstered the case for the nuclear matrix and have sug-

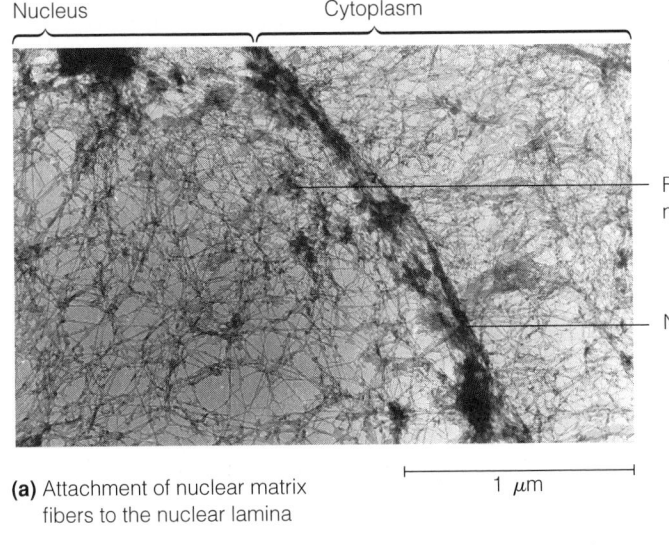

(a) Attachment of nuclear matrix fibers to the nuclear lamina

1 μm

Figure 14-27 The Nuclear Matrix and the Nuclear Lamina. **(a)** This electron micrograph of part of a mammalian cell nucleus shows a branched network of nuclear matrix filaments traversing the nucleus.

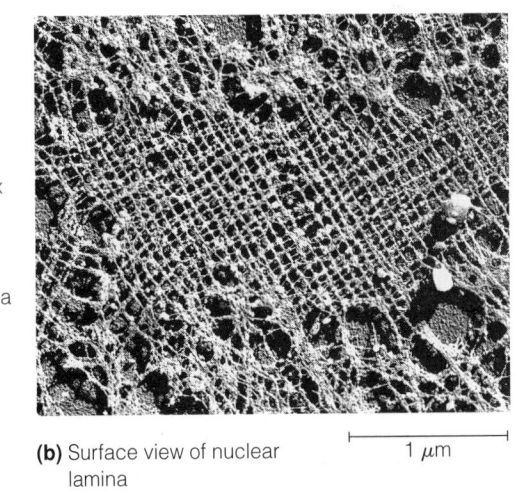

Fibers of the nuclear matrix

Nuclear lamina

(b) Surface view of nuclear lamina

1 μm

These filaments seem attached to the nuclear lamina, the dense layer of filaments that lines the nucleoplasm side of the nuclear envelope. **(b)** A surface view of the nuclear lamina of a frog oocyte (TEMs).

gested that it plays a role in organizing the DNA for orderly replication and perhaps even for the localization and control of transcription. Also, there is evidence that fibers of the nuclear matrix may provide tracks along which mRNA moves from the interior of the nucleus to the pores on its periphery. In another intriguing line of research, key differences have been reported between the nuclear matrix proteins of normal cells and those of cancerous cells.

In any case, the existence of a more limited network of protein fibers within the nucleus has been well established. This is the **nuclear lamina,** a thin, dense meshwork of fibers, in a layer 10–40 nm thick, that lines the inner nuclear membrane and presumably helps support the nuclear envelope. The nuclear lamina consists of a type of intermediate filament made of proteins called *lamins* (discussed in more detail in Chapter 20). At least some of the filaments seem to anchor to proteins of the inner nuclear membrane. In addition, the nuclear lamina may provide attachment sites for chromatin, a topic addressed in the next section. The chromatin-binding sites on the lamina seem to be located away from the immediate vicinity of nuclear pores, perhaps to ensure unobstructed access to the pores for transport purposes.

The Localization of Chromatin in the Interphase Nucleus

As discussed earlier, the chromatin is in a highly extended state in the interphase between cell divisions. One might guess that the chromatin threads corresponding to the cell's individual chromosomes were randomly distributed and highly intertwined within the nucleus. Surprisingly, this seems not to be the case. Instead, the chromatin of each chromosome apparently has its own discrete domain. This idea was first proposed in 1885, but evidence that it is true in a variety of cells awaited the techniques of modern molecular biology. Recently, using nucleic acid probes specific for individual chromosomes, several research groups have demonstrated that each chromosome occupies a compact domain within the nucleus. The chromosome positions do not seem to be fixed, however. They vary from cell to cell of the same organism and seem to change during a cell's life cycle, perhaps reflecting changes in gene activity on the different chromosomes.

The nuclear lamina may help organize the interphase cell's chromatin by binding certain segments of it to specific sites on the nuclear envelope. The segments of chromatin that bind in this way are highly compacted—that is, they are heterochromatin. In electron micrographs, this material appears as a dark irregular layer around the nuclear periphery, as you saw in Figure 14-22a. Most of it seems to be the type called **constitutive heterochromatin,** which exists in a highly condensed form at virtually all times in all cells of the organism. The DNA of constitutive heterochromatin consists of simple-sequence repeated DNA, short nucleotide sequences that repeat tandemly and are not transcribed. The

two major chromsosomal regions that appear as constitutive hetero-chromatin are the centromere and the telomere. In many cases, it is chromosomal telomeres, the highly repeated DNA found at the ends of chromosomes, that are attached to the nuclear envelope in interphase cells.

In contrast to constitutive heterochromatin, **facultative heterochromatin** varies with the particular activities the cell is carrying out. Thus, it differs from tissue to tissue and can even vary from time to time within a given cell. Facultative heterochromatin appears to represent chromosomes or chromosomal regions that have become specifically inactivated in a specific cell type. The amount of facultative heterochromatin is usually very low in embryonic cells but can be substantial in highly differentiated cells. The formation of facultative heterochromatin may therefore be an important means of inactivating entire blocks of genetic information during development.

The Nucleolus

The remaining structural feature of the eukaryotic nucleus is the **nucleolus,** the ribosome factory of the cell. Nucleoli are large, prominent structures present in every eukaryotic nucleus. The existence of the nucleolus has been known since 1774, but it was not until the 1960s that cell biologists determined its function and began to correlate its function with its microscopic appearance. The nucleolus does not have a membrane. In electron micrographs, it appears to consist of a *fibrillar component* and a *granular component* (Figure 14-28). The "fibrils" are basically DNA and RNA. The DNA, in the form of loops of unraveled chromatin, carries genes for ribosomal RNA (rRNA), the RNA component of ribosomes. The fibrillar component is especially dense in areas of the nucleolus where transcription is going on and RNA is present. The RNA is rRNA that is being synthesized and processed. The granules are rRNA molecules being packaged with proteins (imported from the cytoplasm) into ribosomal subunits. As mentioned earlier, the ribosomal subunits will be exported through the nuclear pores to the cytoplasm.

A stretch of DNA carrying rRNA genes is called a **nucleolus organizer region (NOR).** These genes are present in multiple copies in all genomes and are thus an important example of repeated DNA that carries genetic information. The number of copies of rRNA genes varies greatly, depending on the species, but animal cells often contain hundreds of copies and plant cells usually contain thousands. The multiple copies are grouped into one or more clusters, in which the gene copies are tandemly arranged, and clusters of rRNA genes may be found on more than one chromosome. A single nucleolus does not always correspond to a single NOR, however. The human genome, for instance, contains five NORs per haploid chromosome set, or ten per diploid nucleus, each located near the tip of a different chromosome. But instead of ten separate nucleoli, the typical human nucleus contains a single large nucleolus represent-

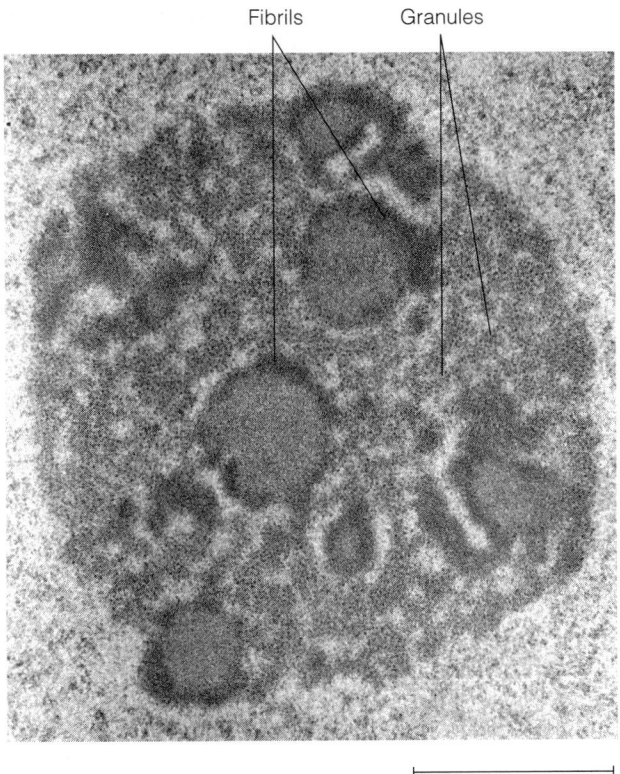

Fibrils   Granules

1 μm

Figure 14-28 The Nucleolus. The nucleolus is a prominent in-tranuclear structure. It is a mass of fibrils and granules. The fibrils are DNA and rRNA; the granules are nascent ribosomal subunits. Shown here is a nucleolus of a spermatogonium, a cell that gives rise to sperm cells (TEM).

ing a gathering of loops of chromatin from ten separate chromosomes. Not surprisingly, nucleoli stain intensely for ribonucleoprotein and become heavily labeled when the cell is exposed to radioactive precursors of RNA (Figure 14-29).

The size of the nucleolus is correlated with its level of activity. In cells having a high rate of protein synthesis and hence a need for many ribosomes, the nucleolus can account for 20–25% of the total volume of the nucleus. In less active cells, it is much smaller. The main difference is in the amount of granular component present. Cells that are producing many ribosomes transcribe, process, and package large quantities of rRNA and have higher steady-state levels of partially complete ribosomal subunits on hand in the nucleolus, thus accounting for the prominent granular component.

The nucleolus disappears during mitosis, at least in the cells of higher plants and animals. As the cell approaches division, the condensation of chromatin into compact chromosomes is accompanied by the shrinkage and then disappearance of the nucleoli. With our current knowledge of the nucleolus's composition and function, this makes perfect sense: The extended chromatin loops of the nucleolus cease being transcribed as they are coiled and folded, and any remaining rRNA and ribosomal protein molecules presumably disperse or are degraded. Then, when mitosis is complete, the chromatin uncoils, the NORs loop out again, and rRNA synthesis resumes. This is the only point in the cell cycle at which the multiple NORs of the human genome are apparent. As rRNA synthesis begins again, ten tiny nucleoli become visible, one near the tip of each of ten chromosomes. As these nucleoli enlarge, they quickly fuse into the single large nucleolus characteristic of the interphase human nucleus.

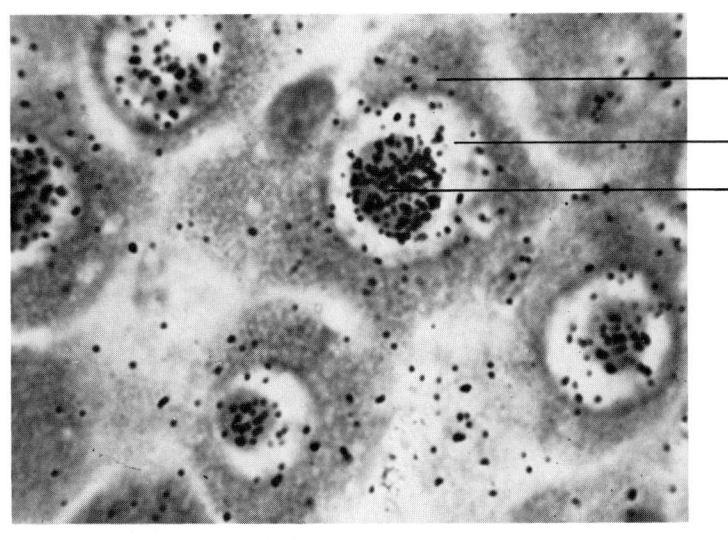

Cytoplasm

Nucleus

Nucleolus

10 μm

Figure 14-29 The Nucleolus as a Site of RNA Synthesis. To demonstrate the role of the nucleolus in RNA synthesis, a rat was injected with ³H-cytidine, a radioactively labeled RNA precursor. Five hours later, liver tissue was removed and subjected to autoradiography. The black spots over the nucleoli in this autoradiograph indicate that the ³H is concentrated in the nucleoli (TEM).

Contemporary Techniques

DNA FINGERPRINTING

The specificity of the restriction fragment patterns produced when DNA is digested with restriction enzymes has been exploited for a number of purposes, ranging from basic research into the organization of eukaryotic genomes to the diagnosis of genetic diseases and the solving of violent crimes. For medical and legal applications, the focus is on the differences in restriction fragment patterns of DNA from different people. Scattered throughout the human genome (as well as those of other organisms) are many small variations (called *polymorphisms*) in nucleotide sequences that can be detected because they affect the lengths of fragments produced by a restriction enzyme. These

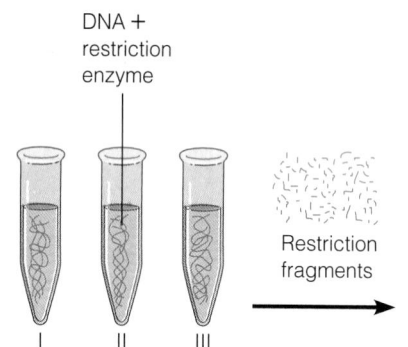

① Restriction fragment preparation. DNA is extracted from white blood cells taken from individuals I, II, and III. A restriction enzyme is added to the three samples of DNA to produce restriction fragments.

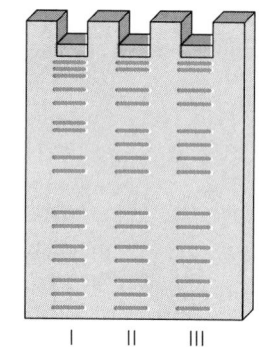

② Gel electrophoresis. The mixtures of restriction fragments from each sample are separated by electrophoresis. Each sample forms a characteristic pattern of bands. (There would be many more bands than are shown here.)

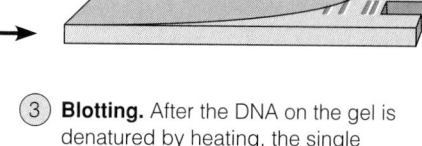

③ Blotting. After the DNA on the gel is denatured by heating, the single strands are transferred onto special paper by blotting.

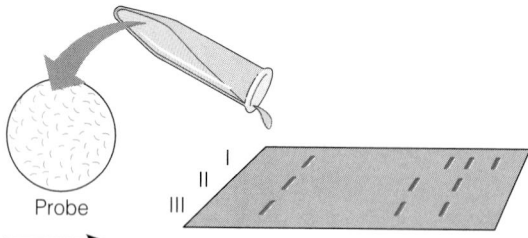

④ Radioactive probe. A solution of radioactive probe is added to the paper blot. The probe is a single-stranded DNA molecule that is complementary to the DNA of interest. The probe attaches only to bands containing complementary DNA, by base pairing.

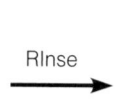

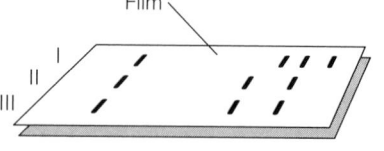

⑤ Autoradiography. After the excess probe is rinsed off, a sheet of photographic film is laid over the paper blot. The radioactivity in the bound probe exposes the film to form an image corresponding to specific DNA bands—the bands containing DNA that base-pairs with the probe.

Figure 14C-1   DNA Fingerprinting by RFLP Analysis.

differences are called **restriction fragment length polymorphisms (RFLPs).** Because of these variations, different individuals will have different sets of RFLP markers. Restriction fragment band patterns revealed by gel electrophoresis can provide a many-banded "DNA fingerprint" that is unique to an individual. In practical use, however, *DNA fingerprinting* (also called *DNA typing*) is carried out in such a way that only a small, selected subset of restriction fragment bands is examined.

Figure 14C-1 outlines the main steps of DNA fingerprinting by RFLP analysis, as it might be used for diagnosis of a hereditary disease whose genetic basis is known. For this example, imagine that individuals I, II, and III are members of a family in which the disease-causing gene is common. It is known that individual I carries the "bad" gene but individual II does not, and we want to find out the status of the gene for their child, individual III.

The key step in DNA fingerprinting procedure is called **Southern blotting** (after E. M. Southern, who developed it in 1975), but the fingerprinting procedure actually involves several different techniques, as Figure 14C-1 indicates. Step

1 is the restriction-enzyme digestion of DNA from the three individuals. (The DNA is usually obtained from the white blood cells in a blood sample.) Step 2 is the electrophoresis of the resulting restriction fragments on a gel. Because the DNA from each person represented an entire genome, there would be many hundreds of thousands of bands if all were made visible. Here is where Southern's stroke of brilliance comes in, with a technique that enables us to single out the bands of interest. In the Southern blotting process, step 3, a special kind of "blotter" paper (nitrocellulose or nylon) is pressed against the completed gel, allowing the DNA fragments to transfer to the paper, where the pattern of bands is replicated. In step 4, a **radioactive probe** is used to label the bands of interest. Such a probe is simply a preparation of radioactive single-stranded DNA (or RNA) that has a base sequence complementary to the DNA of interest—in this case, the DNA of the potentially disease-causing gene. The probe attaches to complementary DNA by base pairing (the same process that occurs when denatured DNA renatures). In step 5, the bands that the probe binds are made visible by autoradiography. The results indicate that the child's version of the gene matches that of the *healthy* parent, individual II.

The autoradiograph in Figure 14C-2 illustrates the use of DNA fingerprinting in a murder case. Here RFLP analysis makes it clear that blood found on the defendant's clothes came from the victim, strongly implicating the defendant in the murder. Another application of DNA fingerprinting in the legal area is the determination of paternity or maternity.

Recently, for most DNA fingerprinting purposes, scientists have turned away from classical RFLP analysis. They now use certain kinds of tandemly repeated DNA for DNA fingerprinting, usually minisatellite or microsatellite DNA. Repeated DNA sequences are chosen for which the number of tandemly repeated copies at a particular site in the genome varies from person to person; these are called **variable number tandem repeats (VNTRs).** The most useful VNTRs are relatively short in overall length but still have many copy number variants in the human population. With this type of polymorphism, the DNA of three suspects in a murder-rape case, for example, might all have different numbers of copies of the dinucleotide sequence TG at a particular place in the genome; one might have 19 copies, the second 21 copies, and the third 32 copies. The differences in restriction fragment patterns would then come from the differences in the number of repeated (TG) units between two restriction cutting sites, rather than from the absence or presence of a cutting site, as for traditional RFLPs. Furthermore, the newer methods also involve the use of the polymerase chain reaction (PCR), a powerful technique that will be described in Chapter 15. ■

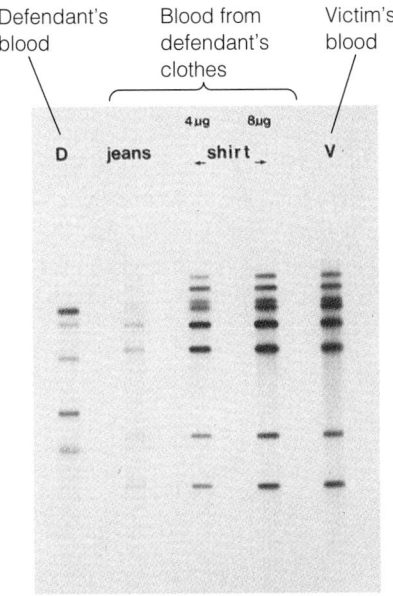

Figure 14C-2   DNA Fingerprints from a Murder Case.   DNA was isolated from bloodstains on the defendant's clothes and compared, by RFLP analysis, with DNA from the defendant and DNA from the victim. The band pattern for the bloodstain DNA matches that for the victim, showing that the blood on the defendant's clothes came from the victim. (Courtesy of Cellmark Diagnostics, Germantown, MD.)

Our knowledge of DNA dates back to the early studies of Miescher, but it was not until the mid-twentieth century that experiments with pneumococcal bacteria and bacteriophage T2 made a strong case that DNA was the primary genetic material. This work was followed closely by Watson and Crick's elucidation of the double-helical structure of DNA, one of the major landmarks in twentieth-century biology. Their discovery has matured into the discipline of molecular biology, which has had a profound effect on our understanding of life at all levels. Molecular biologists working with bacteria have developed a number of powerful tools for the study of genetics. Chief among these are restriction enzymes. These enzymes, isolated from bacteria, can be used to cut very long DNA molecules into reproducible fragments that are short enough to be manipulated easily in the laboratory. In this chapter, we have described how the use of these enzymes makes it feasible to determine the nucleotide sequence of DNA. Later, in Chapter 16, we will see how restriction enzymes are also the key to making recombinant DNA.

The DNA that makes up one complete set of an organism's genetic information is called its genome; for some viruses, the genome is RNA. For most viruses and prokaryotes, the genome consists of one DNA molecule or a small number of them. Eukaryotes have a nuclear genome divided among multiple chromosomes, with one, very long DNA molecule per chromosome. Eukaryotes also have a mitochondrial genome and, in the case of plants, a chloroplast genome as well. One of the most striking features of the nuclear genomes of eukaryotes, especially multicellular organisms, is the large fraction of DNA that does not contain coded information for RNA or protein synthesis. Much of this DNA consists of repeated nucleotide sequences. Very little is known about the functions of repeated-sequence DNA, but it is thought that at least some of it plays a structural role for the chromosome.

The physical length of the DNA molecules present in cells (and even viruses) necessitates considerable packaging.

In both prokaryotes and the nuclei of eukaryotic cells, the DNA is complexed with proteins to form chromatin, but the ingredients and organization of the eukaryotic chromatin are much more elaborate. The basic structural unit of the eukaryotic chromosome is the nucleosome, which consists of a short length of DNA wrapped around a core of eight histone molecules. Higher levels of chromatin packaging are still not well understood. Stretches of nucleosomes ("beads on a string") seem to coil or fold to form a 30-nm chromatin fiber, which can then loop and fold further. The more highly compacted the DNA, the less likely it is to be transcriptionally active in the cell. In nondividing (interphase) cells that are actively transcribing DNA, much of the chromatin is in a relatively extended, highly diffuse form called euchromatin. However, other portions of chromatin are in a highly condensed form called heterochromatin. During cell division, all the chromatin becomes highly compacted, forming discrete chromosomes visible with the light microscope.

The eukaryotic chromosomes are localized within the nucleus. The double-membraned nuclear envelope is perforated with nuclear pores that allow two-way passage of materials between the nucleoplasm and the cytosol. Within each pore is an elaborate protein structure called the nuclear pore complex. Molecules and ions up to about 9 nm in diameter diffuse passively through aqueous channels in the pore complex; larger particles are actively transported through it. The pores thus control the inward movement of proteins used in the nucleus and the outward movement of RNA and ribosomal subunits. We have known for several decades that the nuclear structure called the nucleolus carries out ribosomal RNA synthesis and the assembly of ribosomal subunits. Now we are learning that the other activities of the nucleus—DNA replication and the production of messenger RNA—may also occur within discrete regions of this master organelle.

KEY TERMS FOR SELF-TESTING

The Chemical Nature of the Genetic Material

transformation (p. 410)
bacteriophage (phage) (p. 411)
Chargaff's rules (p. 416)

The Structure of DNA

major and minor grooves of DNA (p. 417)
B-DNA (p. 418)
A-DNA (p. 418)

Z-DNA (p. 418)
supercoiled DNA (p. 418)
topological isomer (p. 419)
topoisomerase (p. 419)
DNA gyrase (p. 419)

PROBLEM SET

14-1. The Genetic Material. Label each of the following statements concerning the chemical nature of the genetic material with a B if the statement is most appropriately dated to the period *before* 1944, with an I if it belongs to the *interim* period 1944–1952, with an A if it is most appropriate to the period *after* 1952, and with an N if it was *never* a widely held concept.

(a) DNA may not be simply a repeating tetranucleotide sequence after all.

(b) The genetic material in higher organisms is most likely protein.

(c) DNA is the genetic information in both bacteria and their phages.

(d) DNA is chemically too simple to be considered the genetic information of any cell.

(e) Nuclein is an important component of virtually every bacterial cell, even though we do not yet know what it does.

(f) DNA may well be the genetic material of bacterial cells, but it is still an open question what that means for higher organisms.

(g) Smooth (S) strains of pneumococcus are capable of converting nonpathogenic rough (R) strains into pathogenic S strains, but we do not yet know which component of the S cells effects this transformation.

(h) Once adsorbed onto a host cell, a bacteriophage injects its proteins into the bacterium.

14-2. Prior Knowledge. Virtually every experiment performed by biologists builds upon knowledge that has resulted from prior experiments.

(a) Of what significance to Avery and his colleagues was the finding (made in 1932 by J. L. Alloway) that the same kind of transformation of R cells into S cells that Griffith observed to occur in mice could also be demonstrated in vitro with cultured pneumococcus cells?

(b) Of what significance to Hershey and Chase was the following suggestion (made in 1951 by R. M. Herriott)? "A virus may act like a little hypodermic needle full of transforming principles; the virus as such never enters the cell; only the tail contacts the host and perhaps enzymatically cuts a small hole through the outer membrane and then the nucleic acid of the virus head flows into the cell."

(c) Of what significance to Watson and Crick were the data of their colleagues at Cambridge that the specific forms in which A, G, C, and T exist at physiologic pH permit the formation of specific hydrogen bonds?

(d) How did the findings of Hershey and Chase help explain an earlier report (by T. F. Anderson and R. M. Herriott) that bacteriophage T2 loses its ability to reproduce when it is burst open osmotically by suspending the viral particles in distilled water prior to their addition to a bacterial culture?

14-3. DNA Structure. Carefully inspect the double-stranded DNA molecule shown here, and notice that it has two-fold rotational symmetry. Label each of the following statements as T if true and F if false.

$$3'-A-G-C-G-C-T-A-T-A-G-C-G-C-T-5'$$
$$5'-T-C-G-C-G-A-T-A-T-C-G-C-G-A-3'$$

(a) There is no way to distinguish the right end of the double helix from the left end.

(b) If a solution of these molecules were heated to denature them, every single-stranded molecule in the solution would be capable of annealing with every other molecule.

(c) If the molecule were cut at its midpoint into two halves, it would be possible to distinguish the left half from the right half.

(d) If the two single strands were separated from each other, it would not be possible to distinguish one strand from the other.

(e) In a single strand from this molecule, it would be impossible to determine which is the 3' end and which is the 5' end.

14-4. Restriction Mapping of DNA. The genome of a newly discovered bacteriophage is a DNA molecule 10,500 nucleotide pairs in length. One sample of this DNA has been incubated with restriction enzyme X and another sample with restriction enzyme Y. The lengths (in thousands of base pairs) of the restriction fragments produced by the two enzymes have been determined by gel electrophoresis to be:

Enzyme X: Fragment X-1 = 4.5; X-2 = 3.6; X-3 = 2.4
Enzyme Y: Fragment Y-1 = 5.2; Y-2 = 3.8; Y-3 = 1.5

Next, the fragments from the enzyme X reaction are isolated and treated with enzyme Y, and the fragments from the enzyme Y reaction are treated with enzyme X. The results are:

X fragments treated with Y: X-1 $\longrightarrow$ 4.5 (unchanged)
                              X-2 $\longrightarrow$ 2.1 + 1.5
                              X-3 $\longrightarrow$ 1.7 + 0.7
Y fragments treated with X: Y-1 $\longrightarrow$ 4.5 + 0.7
                              Y-2 $\longrightarrow$ 2.1 + 1.7
                              Y-3 $\longrightarrow$ 1.5 (unchanged)

Draw a restriction map of the phage DNA, indicating the positions of all the enzyme X and enzyme Y restriction sites and the lengths of DNA between them.

14-5. DNA Sequencing. You have isolated the DNA fragment shown in Problem 14-3 but do not know its complete sequence. From knowledge of the specificity of the restriction enzyme used to prepare it, you know the first four bases at the left end and have prepared a radioactively labeled single-stranded DNA primer of sequence 5' T–C–G–C 3'. Explain how you would determine the rest of the sequence using Sanger's enzymatic method. Draw the gel pattern that would be observed, and indicate the base sequence of the DNA in each radioactive spot.

14-6. Genome Size. The bacterium *E. coli* has as its genome a circular DNA molecule containing about 4.7 million nucleotide pairs. The bacterial cell itself can be thought of as a cylinder with a diameter of 1 μm and a length of 2 μm.

(a) What is the length of the DNA molecule?

(b) If the molecule existed as a fully extended circle, what would its radius be? Do you think it is likely that it exists as a fully extended circle in the bacterial cell? Explain.

(c) Assuming a molecular weight of 330 for the average nucleotide, what is the molecular weight of the molecule? What is its actual weight in grams?

(d) Knowing that the density of DNA is about 1.7 g/cm³, calculate the volume that the circular *E. coli* DNA molecule would occupy if tightly packed together. What proportion of the total internal volume of the *E. coli* cell does this represent?

(e) At what rate (in nucleotides added per second) must the replication of *E. coli* DNA proceed in a culture with a generation time of 20 minutes?

(f) At what rate (in revolutions per second) must the double helix of DNA unwind to sustain the rate of replication calculated in part e?

14-7. DNA Melting. Shown in Figure 14-30 are the melting curves for two DNA samples that were thermally denatured under the same conditions.

(a) What conclusion can you draw concerning the base compositions of the two samples? Explain.

(b) How might you explain the steeper slope of the melting curve for sample A?

(c) Formamide and urea are agents known to form hydrogen bonds with pyrimidines and purines. What effect, if any, would the inclusion of a small amount of formamide or urea in the incubation mixture have on the melting curves?

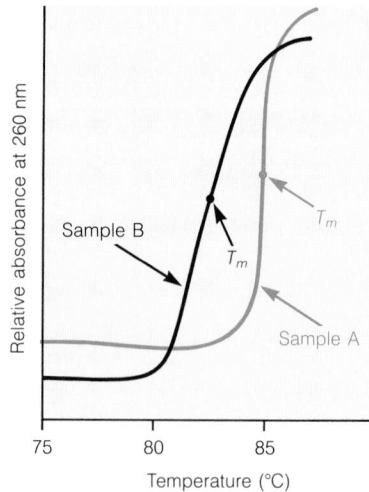

Figure 14-30 Thermal Denaturation Profiles for Two DNA Samples.

14-8. Nuclear Structure and Function. Indicate the implications for nuclear structure or function of each of the following experimental observations.

(a) Sucrose crosses the nuclear envelope so rapidly that its rate of movement cannot be accurately measured.

(b) Colloidal gold particles with a diameter of 5.5 nm equilibrate rapidly between the nucleus and cytoplasm when injected into an amoeba, but gold particles with a diameter of 15 nm do not.

(c) Nuclear pore complexes sometimes stain heavily for ribonucleoprotein.

(d) If gold particles up to 26 nm in diameter are coated with a polypeptide containing a nuclear localization signal (NLS) and are then injected into the cytoplasm of a living cell, they are transported into the nucleus. If they are injected into the nucleus, however, they remain there.

(e) Many of the proteins of the nuclear envelope appear from electrophoretic analysis to be the same as those found in the endoplasmic reticulum.

(f) Ribosomal proteins are synthesized in the cytoplasm but are packaged with rRNA into ribosomal subunits in the nucleus.

(g) If nucleoli are irradiated with a microbeam of ultraviolet light, synthesis of ribosomal RNA is inhibited.

(h) Treatment of nuclei with the nonionic detergent Triton X-100 dissolves away the nuclear envelope but leaves an otherwise intact nucleus.

14-9. Nuclear Pores. At one stage in its development, the oocyte of *Xenopus laevis* has a diameter of 0.3 mm and a nuclear surface area of 200,000 μm^2. The pore density is 50 pores per square micrometer, and each pore has a diameter of about 100 nm.

(a) What is the total number of nuclear pores on the nuclear envelope of this oocyte?

(b) Approximately how far apart are the pores, assuming an even distribution over the envelope?

(c) What proportion of the total surface area of the nucleus do the pores occupy? Is that a high or low value, compared to most other cell types?

(d) Why do you suppose an oocyte is so well endowed with nuclear pores?

14-10. Nucleoli. Indicate whether each of the following statements is true (T) or false (F). If false, reword the statement to make it true.

(a) Nucleoli are membrane-bounded structures present in the eukaryotic nucleus.

(b) The fibrils seen in electron micrographs of nucleoli are primarily RNA and DNA.

(c) The DNA of nucleoli carries the cell's tRNA genes, which are present in clusters of multiple copies.

(d) A single nucleolus always corresponds to a single nucleolus organizer region (NOR).

(e) Nucleoli become heavily labeled when radioactive ribonucleotides are provided to the cell.

(f) In animals and plants, the disappearance of nucleoli during mitosis correlates with cessation of ribosome synthesis.

SUGGESTED READING

General References

Alberts, B., D. Bray, J. Lewis, M. Raff, K. Roberts, and J. D. Watson. *Molecular Biology of the Cell,* 3d ed. New York: Garland, 1994.

Cairns, J., G. S. Stent, and J. D. Watson, eds. *Phage and the Origins of Molecular Biology.* Cold Spring Harbor, NY: Cold Spring Harbor Laboratory, 1966.

Judson, H. F. *The Eighth Day of Creation: Makers of the Revolution in Biology.* New York: Simon & Schuster, 1979.

Lewin, B. *Genes V.* New York: Oxford University Press, 1994.

Lodish, H., D. Baltimore, A. Berk, S. L. Zipursky, P. Matsudaira, and J. Darnell. *Molecular Cell Biology,* 3d ed. New York: Scientific American Books, 1995.

Singer, M., and P. Berg. *Genes and Genomes.* Mill Valley, CA: University Science Books, 1990.

Watson, J. D., M. Gilman, J. Witkowski, and M. Zoller. *Recombinant DNA: A Short Course,* 2d ed. New York: Scientific American Books, 1992.

Watson, J. D., N. Hopkins, J. Roberts, J. Steitz, and A. Weiner. *Molecular Biology of the Gene,* 4th ed. Menlo Park, CA: Benjamin/Cummings, 1987.

The Chemical Nature of the Genetic Material

Avery, O. T., C. M. MacLeod, and M. McCarty. Studies on the chemical nature of the substance inducing transformation of pneumococcal types. Induction of transformation by a desoxyribonucleic acid fraction isolated from *Pneumococcus* Type III. *J. Exp. Med.* 79 (1944): 137.

Chargaff, E. Preface to a grammar of biology: A hundred years of nucleic acid research. *Science* 172 (1971): 637.

Hershey, A. D., and M. Chase. Independent functions of viral protein and nucleic acid in growth of bacteriophage. *J. Gen. Physiol.* 36 (1952): 39.

The Structure of DNA

Bauer, W. R., F. H. C. Crick, and J. H. White. Supercoiled DNA. *Sci. Amer.* 243 (July 1980): 118.

Dickerson, R. E., H. R. Drew, B. N. Conner, R. M. Wing, A. V. Fratini, and M. L. Kopka. The anatomy of A-, B-, and Z-DNA. *Science* 216 (1982): 475.

Maxam, A. M., and W. Gilbert. A new method of sequencing DNA. *Proc. Natl. Acad. Sci. USA* 74 (1977): 560.

Sanger, F. Determination of nucleotide sequences in DNA. *Science* 214 (1981): 1205.

Wang, J. C. DNA topoisomerases. *Sci. Amer.* 247 (July 1982): 94.

Watson, J. D. *The Double Helix.* New York: Atheneum, 1968.

Watson, J. D., and F. H. C. Crick. Genetical implications of the structure for deoxyribonucleic acid. *Nature* 171 (1953): 964.

Watson, J. D., and F. H. C . Crick. Molecular structure of nucleic acids: A structure for deoxyribose nucleic acid. *Nature* 171 (1953): 737.

The Organization of DNA in Genomes

Britten, R. J., and D. E. Kohne. Repeated sequences of DNA. *Science* 161 (1968): 529.

Collins, F., and D. Galas. A new five-year plan for the U.S. human genome project. *Science* 262 (1993): 43.

DNA Technology in Forensic Science. Washington, DC: National Academy Press, 1992.

Gall, J. G. Chromosome structure and the C-value paradox. *J. Cell Biol*. 91 (1981): 3s.

Genome issue [an editorial and 6 articles on the status of genome research and the Human Genome Project, as of September 1994]. *Science* 265 (1994): 1991, 2031–2070.

Jelinek, W. R., and C. W. Schmid. Repetitive sequences in eukaryotic DNA and their expression. *Annu. Rev. Biochem.* 51 (1982): 813.

Moyzis, R. K. The human telomere. *Sci. Amer.* 265 (August 1991): 48.

Nowak, R. Mining treasures from "junk DNA" (Research News). *Science* 263 (1994): 608.

Roberts, R. Restriction and modification enzymes and their recognition sequences. *Nucleic Acids Res.* 10 (1982): r117.

Smith, H. O. Nucleotide sequence specificity of restriction endonucleases. *Science* 205 (1979): 455.

DNA Packaging

Adolph, K. W., ed. *Chromosomes and Chromatin,* vols. 1–3. Boca Raton, FL: CRC Press, 1988.

Borst, P., L. A. Grivell, and G. S. P. Groot. Organelle DNA. *Trends Biochem. Sci.* 9 (1984): 128.

Grivell, L. A. Mitochondrial DNA. *Sci. Amer.* 248 (March 1983):78.

Horowitz, R. A., D. A. Agard, J. W. Sedat, and C. L. Woodcock. The three-dimensional architecture of chromatin in situ: Electron tomography reveals fibers composed of a continuously variable zig-zag nucleosomal ribbon. *J. Cell Biol.* 125 (1994): 1.

Kornberg, R. D., and A. Klug. The nucleosome. *Sci. Amer.* 244 (February 1981): 52.

Swedlow, J. R., D. A. Agard, and J. W. Sedat. Chromosome structure inside the nucleus. *Curr. Opin. Cell. Biol.* 5 (1993): 412.

The Nucleus

Forbes, D. J. Structure and function of the nuclear pore complex. *Annu. Rev. Cell Biol.* 8 (1992): 495.

Hoffman, M. The cell's nucleus shapes up (Research News). *Science* 259 (1993): 1257.

Jordan, E.-C., and C. A. Cullis, eds. *The Nucleolus.* Cambridge, England: Cambridge University Press, 1982.

Maul, G. G., ed. *The Nuclear Envelope and the Nuclear Matrix.* New York: A. R. Liss, 1982.

Miller, O. L., Jr. The nucleolus, chromosomes, and visualization of genetic activity. *J. Cell Biol.* 91 (1981): 15s.

Newmeyer, D. D. The nuclear pore complex and nucleocytoplasmic transport. *Curr. Opin. Cell. Biol.* 5 (1993): 395.

Newport, J. W., and D. J. Forbes. The nucleus: Structure, function and dynamics. *Annu. Rev. Biochem.* 56 (1987): 535.

Spector, D. L. Macromolecular domains within the cell nucleus. *Annu. Rev. Cell Biol.* 9 (1993): 265.

15

THE CELL CYCLE, DNA REPLICATION, AND MITOSIS

A fundamental property of living organisms is their capacity to grow. At the cellular level, this is accomplished by the synthesis of additional molecules of proteins, nucleic acids, carbohydrates, lipids, and other cellular constituents. As a cell grows, the plasma membrane must expand to allow the internal volume of the cell to increase. However, a cell cannot continue to enlarge indefinitely because there is a concomitant decrease in its surface area/volume ratio and hence in its capacity for effective exchange with the environment (recall Figure 4-1). For this reason, cell growth must be accompanied by **cell division,** whereby one cell gives rise to two **daughter cells.**

For single-celled organisms, cell division achieves an increase in the total number of individuals in a population. In multicellular organisms, cell division either increases the number of cells in the organism, leading to growth of the organism, or replaces cells that have died. In an adult human, for example, about two million stem cells in bone marrow divide every second to maintain a constant number of red blood cells in the body.

An important feature of the division process is its genetic fidelity. Both daughter cells are faithful genetic duplicates of the parent cell, containing the same (or almost the same) DNA sequences. Therefore, in the generation of two daughter cells from a single parent cell, all the genetic information present in the nucleus of the parent cell must be duplicated and carefully parceled out to the daughter cells. For eukaryotic cells, there is a particular sequence of events that extends from the "birth" of a daughter cell until the division of that cell into two. This sequence is called the **cell cycle.**

An Overview of the Cell Cycle

To early cell biologists studying eukaryotic cells with the microscope, the most dramatic events in the life of a cell were those associated with the end of the cycle, the actual process of cell division. The division phase, called the **M (mitotic) phase,** consists of two overlapping processes, in which the nucleus divides first and the cytoplasm second. Nuclear division is called **mitosis,** and the subsequent division of the cytoplasm to produce two daughter cells is called **cytokinesis** (Figure 15-1a).

The stars of the mitotic drama are, of course, the chromosomes. The beginning of mitosis is marked by the condensation (coiling and folding) of the cell's chromatin, until the chromosomes are thick enough to be individually discernable under the microscope. Already duplicated, each chromosome consists of two identical **sister chromatids,** still joined at their centromeres. As the chromosomes become visible, the nuclear envelope breaks apart into fragments. Then, in a stately ballet guided by the microtubule fibers of the *mitotic spindle,* the sister chromatids separate and—each now a full-fledged chromosome—move to opposite ends of the cell. By this time, cytokinesis has usually begun. Nuclear membranes envelop the two groups of daughter chromosomes as cell division is completed.

While visually striking, the events of the mitotic phase account for only a relatively small proportion of the total cell cycle; for a mammalian cell growing in culture, the mitotic phase takes less than an hour, only about 5% of the total cycle time. The typical cell spends most of its time in the growth phase between divisions, called **interphase.** It is during interphase that the cell's metabolism is most active. Most cellular contents are synthesized continuously during interphase, so that cell mass increases gradually as the cell approaches division. However, DNA synthesis and the assembly of DNA and histones into chromatin are restricted to a limited portion of interphase, called the **S phase** (S for synthesis). The "gaps" between S phase and M phase are called the **G1 phase** and the **G2 phase,** as indicated in Figure 15-1b.

How long is the cell cycle? The cells in a multicellular organism divide at varying rates, but most studies of the cell cycle have been done with cells in culture, where the length of the cycle is similar for different cell types. We can easily determine the length of the cell cycle—the **generation**

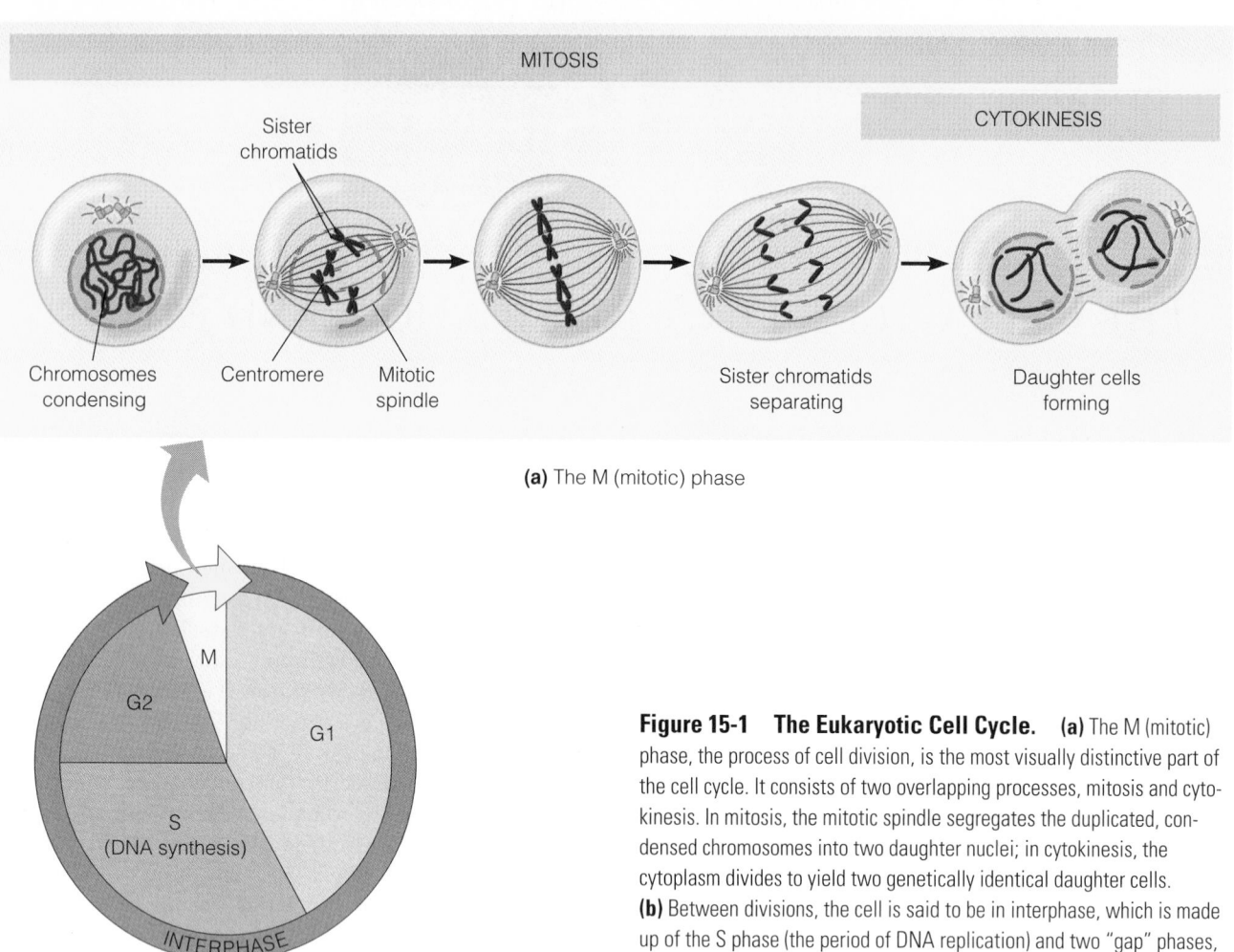

Sister chromatids

Chromosomes condensing   Centromere   Mitotic spindle   Sister chromatids separating   Daughter cells forming

(a) The M (mitotic) phase

M

G2

G1

S
(DNA synthesis)

INTERPHASE

(b) The cell cycle

Figure 15-1   The Eukaryotic Cell Cycle.   (a) The M (mitotic) phase, the process of cell division, is the most visually distinctive part of the cell cycle. It consists of two overlapping processes, mitosis and cytokinesis. In mitosis, the mitotic spindle segregates the duplicated, condensed chromosomes into two daughter nuclei; in cytokinesis, the cytoplasm divides to yield two genetically identical daughter cells. **(b)** Between divisions, the cell is said to be in interphase, which is made up of the S phase (the period of DNA replication) and two "gap" phases, called G1 and G2. The cell continues to grow throughout interphase, a time of high metabolic activity.

time—for cultured cells either by counting the cells at intervals under a microscope or by monitoring the cell mass with a spectrophotometer. For mammalian cells in culture, the total cycle usually takes about 18–24 hours.

Once the total length of the cycle is known, the length of specific phases can also be determined. The S and M phases are easiest to measure. To determine the length of the S phase, we can expose cells to a radioactively labeled DNA precursor (usually ³H-thymidine) for a short period of time and then examine them by autoradiography. The fraction of cells with exposed silver grains over their nuclei represents the fraction of cells that were somewhere in S phase when the radioactive compound was available. When this fraction is multiplied by the total length of the cell cycle, the result is an estimate of the average length of the S phase. For mammalian cells in culture, this fraction is often around 0.33, and the S phase is therefore about 6–8 hours in length.

Similarly, the length of the M phase can be estimated by multiplying the generation time by the fraction of the cells that are actually in mitosis at any point in time. This fraction is called the **mitotic index.** The mitotic index for cultured cells is often about 0.03–0.05.

G1 and G2 must be determined by less direct methods, because there is no easy way at present to label or identify cells in these phases specifically. The length of G1 is quite variable, depending on the cell type. Within a multicellular organism, some cells may spend only minutes or hours in G1, whereas others spend weeks, months, or years. During G1, a major "decision" is made as to whether and when the cell is to divide again. Cells that are arrested in G1 for long periods are often said to be in a **G0 state.** Some cells in G0 are destined never to divide again; most of the nerve cells in your body are in this state. A similar kind of arrest in G2 also occurs in some cells. In general, however, G2 is shorter than G1 and is much more uniform in duration among the different cell types of an organism.

The various phases of the cell cycle have been measured in many types of cultured cells. For typical mammalian cells, G1 lasts about 8–10 hours, S is completed in about 6–8 hours, G2 takes about 4–6 hours, and mitosis requires an hour or less, usually about 30–45 minutes.

Now that we have looked at the entire cell cycle in overview, we will consider its workings in more detail. Because DNA replication is, in a sense, the "purpose" of the cell cycle, we will start by examining the S phase. Then we will discuss the details of how the replicated genetic information is distributed into daughter cells in the M phase, mitosis and cytokinesis. Finally, we will have more to say about phases G1 and G2 when we outline our current understanding of the regulation of the cell cycle.

DNA Replication

One of the most significant features of the double-helical model of DNA is that it immediately suggests a mechanism for DNA replication. In fact, a month after Watson and Crick published their now-classic paper postulating a double helix for DNA, they followed it with an equally important paper suggesting how such a base-paired structure might duplicate itself. Here, in their own words, is the basis of that suggestion:

Now our model for deoxyribonucleic acid is, in effect, a pair of templates, each of which is complementary to the other. We imagine that prior to duplication the hydrogen bonds are broken, and the two chains unwind and separate. Each chain then acts as a template for the formation onto itself of a new companion chain, so that eventually we shall have two pairs of chains, where we only had one before. Moreover, the sequence of the pairs of bases will have been duplicated exactly (Watson and Crick, 1953, p. 966).

The model they proposed for DNA replication is shown in Figure 15-2. The essence of their suggestion is that one of

Figure 15-2 The Watson-Crick Model of DNA Replication. In 1953, Watson and Crick proposed that the DNA double helix replicates semiconservatively, using a model like this one to illustrate the principle. The double-stranded helix unwinds, and each parent strand serves as a template for the synthesis of a complementary daughter strand, assembled according to the base-pairing rules. A, T, C, and G stand for the adenine, thymine, cytosine, and guanine nucleotides. A pairs with T, and G with C. The hydrogen bonds that join the complementary bases are shown as short dotted lines.

the two strands of every daughter DNA molecule is derived from the parent molecule, whereas the other strand is newly synthesized. This is called **semiconservative replication,** because half of the parent molecule is retained by each daughter molecule.

Proof of the Semiconservative Model

Within five years of its publication, the Watson-Crick model for DNA replication was tested and proved correct by Matthew Meselson and Franklin Stahl. The ingenuity of their contribution lay in the method they devised, in collaboration with Jerome Vinograd, for distinguishing semiconservative replication from other possibilities. Their approach was to grow prokaryotic cells (*E. coli*) in a medium containing ^{15}N, a heavy (but nonradioactive) isotope of nitrogen. Because all the DNA synthesized by *E. coli* cells under these conditions will have the ^{15}N isotope instead of ^{14}N, it will have a higher density than ordinary DNA. After many generations of growth on the ^{15}N-containing medium, the cells were transferred to a medium containing only ^{14}N, and the density of the DNA was examined after successive cycles of replication.

The density of DNA molecules was assessed by **equilibrium density centrifugation,** a technique we encountered earlier in its application to the separation of organelles (see Figure 9-6). Briefly, the technique allows organelles or macromolecules to be separated on the basis of differences in buoyant density by centrifugation in an appropriate solution that forms a gradient increasing in density from the top of the tube to the bottom. In response to the centrifugal force, the particles migrate "down" the tube (actually, they move *outward*, away from the axis of rotation) until they reach a density equal to their own. They then remain at this equilibrium density and can be recovered as a band at that position in the tube after centrifugation.

A useful solute for equilibrium density centrifugation of macromolecules such as DNA is cesium chloride (CsCl), because it is highly soluble, very dense, and chemically inert toward organic molecules. The DNA to be analyzed is mixed with an initially homogeneous solution of cesium chloride of an appropriate density. The solution is then centrifuged at high speed for a relatively long time (with a modern centrifuge, 80,000 rpm for 8 hours, for example).

As a gradient of cesium chloride establishes itself during the centrifugation, the DNA molecules float "up" or sink "down" within the gradient to reach their equilibrium density positions. The density of a DNA molecule and thus the exact position where it comes to rest depends on its base composition and on whether it is single-stranded or double-stranded DNA, but most DNA molecules band in the range of 1.65–1.75 g/cm^3.

Meselson and Stahl were able to separate the DNA of ^{15}N-grown cells from that of ^{14}N-grown cells because the difference in density between "heavy" (^{15}N-containing) and "light" (^{14}N-containing) DNA is sufficiently great to allow the two kinds of DNA to be resolved as separate bands in ce-

sium chloride solution (Figure 15-3). This density difference made possible the critical experiment shown in Figure 15-4, where equilibrium density centrifugation was used to produce a banding pattern of DNA from cells that were first grown in ^{15}N (Figure 15-4a) and then transferred to ^{14}N for one or more additional cycles of replication.

What result would be predicted for a semiconservative mechanism of DNA replication? After one replication cycle in ^{14}N, each DNA molecule present would consist of one ^{15}N strand (the old strand) and one ^{14}N strand (the new strand); the overall density would be intermediate between heavy DNA and light DNA. (To address the predictions of other models for replication, see Problem 15-3 at the end of the chapter.)

The experimental results of Meselson and Stahl clearly supported the semiconservative model. After one replication cycle in the ^{14}N medium, a single band of DNA was observed in the cesium chloride gradient, with a density exactly halfway between that of ^{15}N DNA and that of ^{14}N DNA (Figure 15-4b). Because they saw no band at the density of heavy DNA, Meselson and Stahl concluded that the original, double-stranded parental DNA was not preserved as such in the replication process. Similarly, the absence of a band at the density of light DNA indicated that no daughter DNA molecules consisted exclusively of newly synthesized nucleotides. Instead, it appeared that a part of every daughter DNA molecule was newly synthesized, while another part was derived from the parent molecule. In fact, the density exactly midway between those of ^{14}N DNA and ^{15}N DNA meant that the hybrid DNA molecules were one-half parental and one-half newly synthesized, just as predicted by the Watson-Crick model of semiconservative replication.

The data from cells allowed to grow on ^{14}N for additional generations provided further confirmation. After two cycles of DNA replication, for example, Meselson and Stahl saw two equal bands, one at the hybrid density of the previous cycle and one at the density of purely ^{14}N DNA (Figure 15-4c). As the figure illustrates, this is also consistent with a semiconservative mode of replication.

From these findings, Meselson and Stahl concluded "that the nitrogen of a DNA molecule is divided equally between two physically continuous subunits; that, following duplication, each daughter molecule receives one of these; and that the subunits are conserved through many duplications" (1958, p. 682). Further experimentation was then required to prove that the "physically continuous subunits" into which DNA is partitioned are indeed separate DNA strands. In an extension of the cesium chloride technique, $^{14}N/^{15}N$ hybrid DNA was separated into single strands, and these were shown to be either all heavy or all light, demonstrating that the semiconservative partitioning of nitrogen reported by Meselson and Stahl extends over long stretches of DNA. Meanwhile, other researchers had used radioactive labeling and autoradiography to look at the process of DNA replication in eukaryotic chromosomes. In the end, Watson and Crick were proven right—DNA is replicated semiconservatively, and in all organisms.

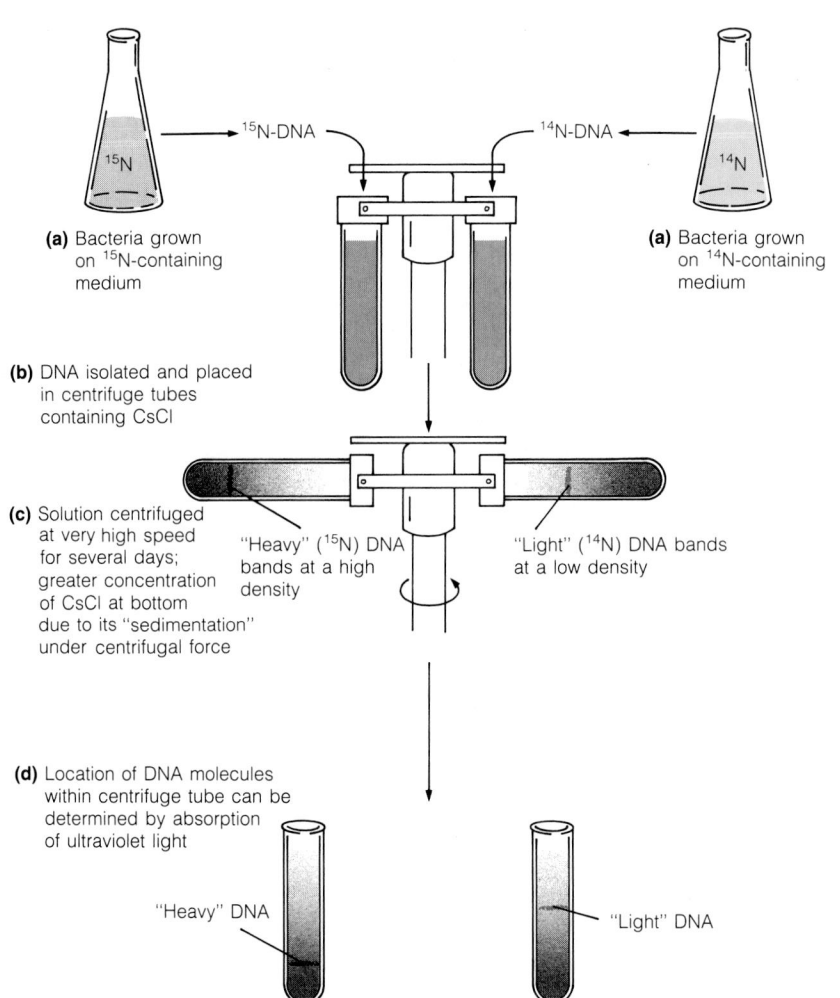

Figure 15-3   Equilibrium Density Centrifugation in DNA Analysis. Equilibrium density centrifugation can be used to distinguish between heavy (^{15}N-containing) and light (^{14}N-containing) DNA. **(a)** If bacterial cells are grown on either ^{15}N or ^{14}N medium for many generations, we can distinguish the DNA from the two cultures by **(b)** placing it in tubes containing cesium chloride at the appropriate concentration and **(c)** centrifuging the tubes at a very high speed until the DNA reaches its equilibrium, or buoyant, density. Heavy DNA bands at a higher density than does light DNA because of the presence of the ^{15}N atoms in its structure. **(d)** After centrifugation, the DNA bands can be visualized by their absorption of ultraviolet light. The density difference (about 1%) is sufficient not only to resolve the two bands, but also to detect hybrid DNA molecules of an intermediate density, as in Figure 15-4c.

(a) Bacteria grown on ^{15}N-containing medium

^{15}N-DNA

^{14}N-DNA

(a) Bacteria grown on ^{14}N-containing medium

(b) DNA isolated and placed in centrifuge tubes containing CsCl

(c) Solution centrifuged at very high speed for several days; greater concentration of CsCl at bottom due to its "sedimentation" under centrifugal force

"Heavy" (^{15}N) DNA bands at a high density

"Light" (^{14}N) DNA bands at a low density

(d) Location of DNA molecules within centrifuge tube can be determined by absorption of ultraviolet light

"Heavy" DNA

"Light" DNA

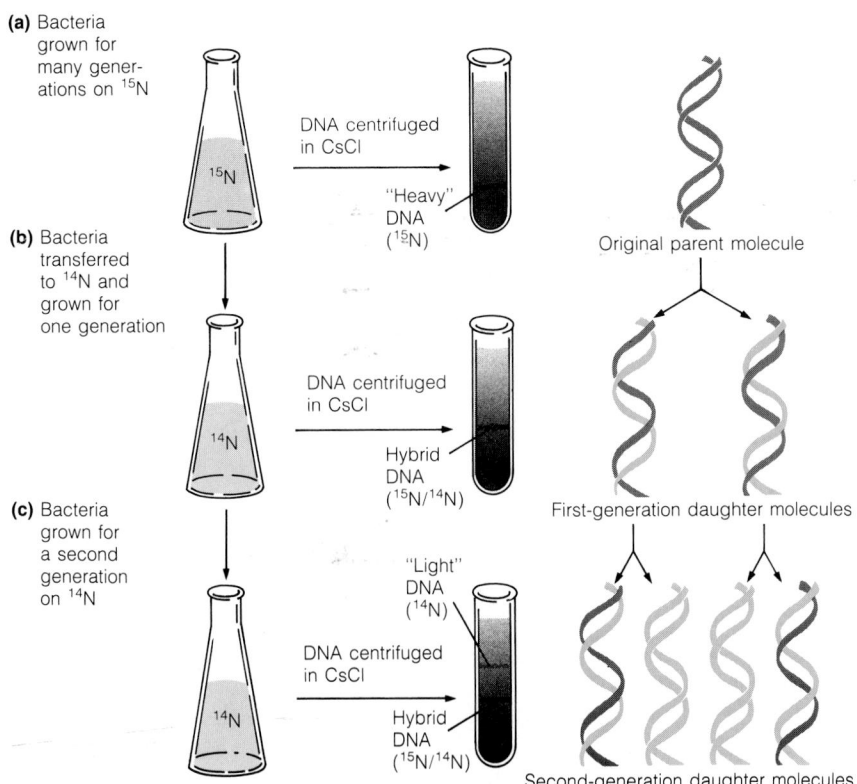

(a) Bacteria grown for many generations on ^{15}N

(b) Bacteria transferred to ^{14}N and grown for one generation

(c) Bacteria grown for a second generation on ^{14}N

DNA centrifuged in CsCl

"Heavy" DNA (^{15}N)

DNA centrifuged in CsCl

Hybrid DNA ($^{15}N/^{14}N$)

DNA centrifuged in CsCl

"Light" DNA (^{14}N)

Hybrid DNA ($^{15}N/^{14}N$)

Original parent molecule

First-generation daughter molecules

Second-generation daughter molecules

Figure 15-4   Semiconservative Replication of Density-Labeled DNA. Meselson and Stahl **(a)** grew bacteria for many generations on an ^{15}N-containing medium and then transferred the cells to an ^{14}N-containing medium for **(b)** one or **(c)** two further cycles of replication. In each case, DNA was extracted from the cells and centrifuged to equilibrium on cesium chloride, as described in Figure 15-3. Bacterial cultures appear on the left, cesium chloride gradients in the center, and schematic illustrations of the DNA molecules on the right. Dark-gray strands contain ^{15}N, whereas blue strands are synthesized with ^{14}N.

The Mechanism of DNA Replication

Conceptually, the overall process of DNA replication is as simple as the Watson-Crick model implied. The same cannot be said of the actual mechanism, however. DNA replication has been shown to be a complex process requiring a number of enzymes and other proteins and even the participation of RNA! We will first look at some general features of the replicative process and then focus on the details of the mechanisms involved. Throughout, we will make frequent reference to *E. coli,* as DNA replication is currently best understood in this prokaryotic organism. However, studies of mammalian viruses such as SV40 and of the yeast *Saccharomyces cerevisiae* are now revealing important details about DNA replication in eukaryotes. DNA replication seems to be a drama whose plot and molecular actors are basically similar in prokaryotic and eukaryotic cells. This is perhaps not surprising for such a fundamental process—one that must have arisen very early in the evolution of life.

The first direct visualization of DNA replication was provided by John Cairns, working with *E. coli.* He grew cells for varying lengths of time in a medium containing the DNA precursor ^{3}H-thymidine and then used autoradiography to look for DNA molecules caught in the act of replication. One such molecule is seen in the autoradiograph of Figure 15-5a. The two forklike structures indicated by the arrows represent the sites at which the DNA duplex was being replicated. These **replication forks** are now known to result from a replication process that begins at a specific

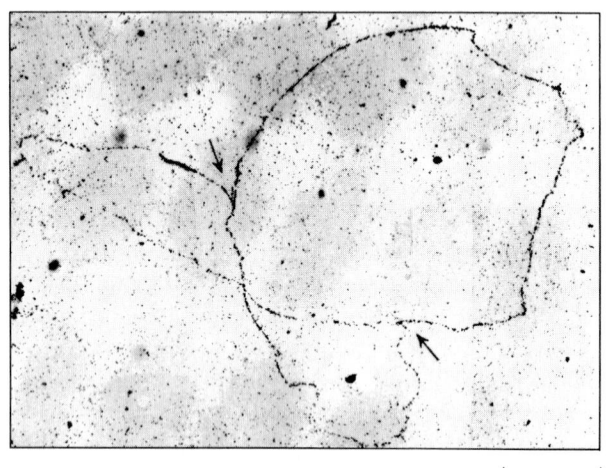

(a) *E. coli* DNA replication

0.25 μm

Figure 15-5 Replication of Circular DNA. **(a)** This autoradiograph shows an *E. coli* DNA molecule caught in the act of replication. The bacterium from which the molecule came had been grown in a medium containing ^{3}H-thymidine, thereby ensuring that the DNA molecule could be visualized by autoradiography. The arrows point to the two replication forks. **(b)** Replication of a circular DNA molecule begins at a single origin and proceeds bidirectionally around the circle, with the two replication forks moving in opposite directions. The new strands are shown in blue. The replication process generates intermediates that resemble the Greek letter theta (θ), from which this type of replication derives its name.

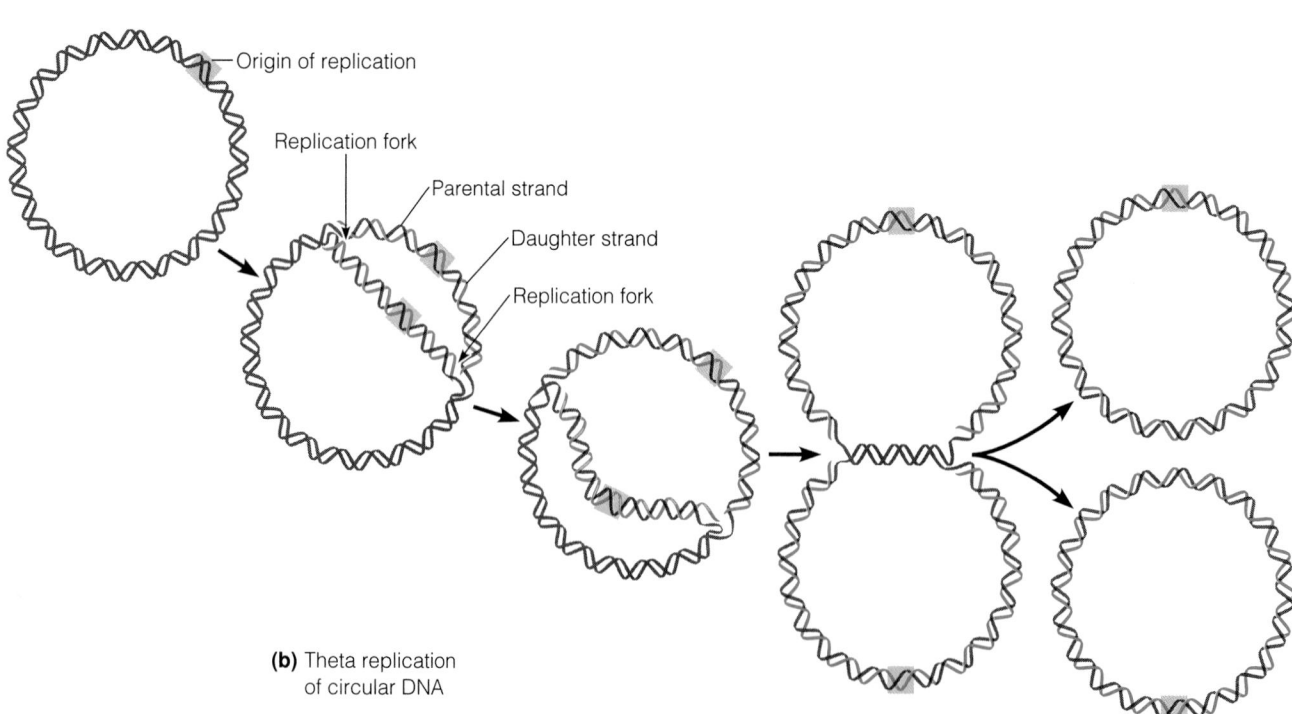

Origin of replication

Replication fork

Parental strand

Daughter strand

Replication fork

(b) Theta replication
of circular DNA

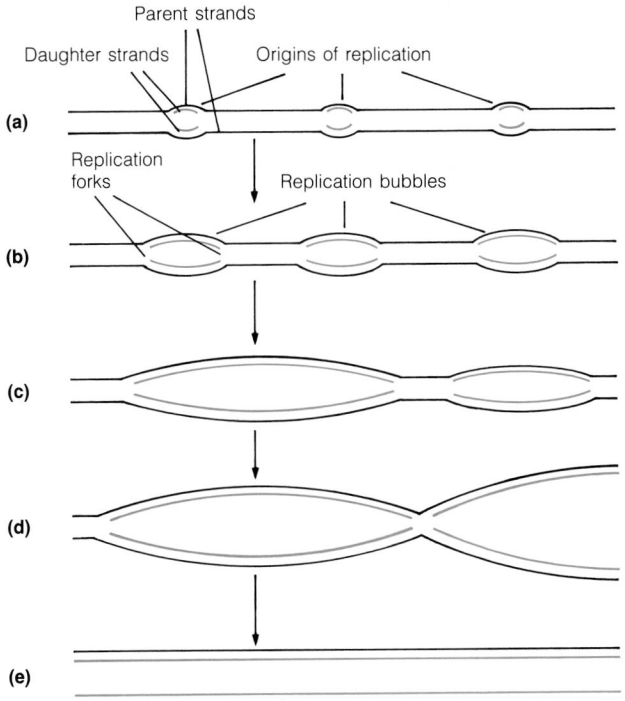

Parent strands
Daughter strands
Origins of replication

(a)

Replication forks
Replication bubbles

(b)

(c)

(d)

(e)

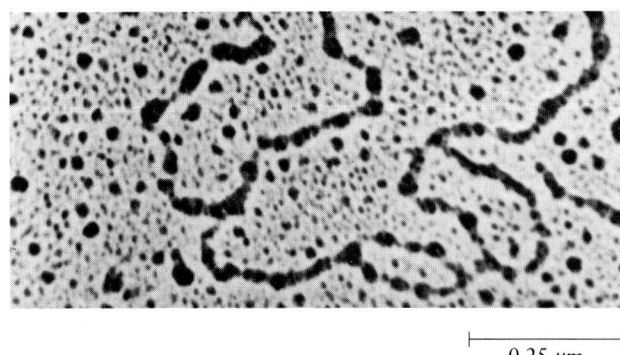

0.25 μm

Figure 15-6   Multiple Origins of Replication in Eukaryotic DNA. Replication of the linear eukaryotic DNA molecules is initiated at numerous origins along the DNA; the timing of initiation is specific for each cluster of origins. **(a)** Replication bubbles form at origins. **(b)** The bubbles grow as the replication forks move along the DNA in both directions from each origin. The micrograph shows three replication bubbles in DNA from cultured Chinese hamster cells (TEM). **(c)** Eventually, individual bubbles meet and fuse. **(d)** A Y-shaped structure forms as a replication fork reaches the end of a DNA molecule. **(e)** When all bubbles have fused, replication is complete, and the two daughter molecules separate.

point and moves along the DNA, unwinding the helix and copying both strands as it goes.

DNA replication is initiated at a defined sequence of nucleotides called an **origin of replication.** An **initiator protein** binds to this sequence. (In *E. coli*, 20–40 molecules of a single polypeptide bind at the origin.) Using energy from ATP, the initiator protein slightly unwinds the double helix. This allows the rest of the DNA replication machinery access to single-stranded DNA. We now know that in most cases, DNA replication is bidirectional: Two replication forks are formed at an origin, and replication proceeds simultaneously in both directions. The forks are at the ends of a "replication bubble" that grows in size as replication continues bidirectionally. For circular DNA, this process is called **theta replication** because it produces intermediates that look like the Greek letter theta (θ), as you can see in Figure 15-5b. Most circular DNA molecules are replicated in this way, starting from a single origin. Theta replication is used not only for bacterial genomes such as *E. coli*'s, but also for the circular DNAs of plasmids, mitochondria, chloroplasts, and some viruses.

The linear DNA molecules of certain other viruses and of all eukaryotic chromosomes are also replicated bidirectionally. However, while linear viral DNAs have a single origin of replication, the very large DNA molecules of eukaryotic chromosomes may have hundreds or even thousands of origins.

Figure 15-6 depicts the replication process for a small length of a eukaryotic DNA molecule. Replication bubbles form at multiple origins and grow as replication proceeds in both directions. The electron micrograph shows a segment of eukaryotic DNA with three replication bubbles. Eventu-

ally, adjacent bubbles fuse to generate larger bubbles, and Y-shaped intermediates form as a bubble reaches the end of a molecule. Finally, the parent strands separate completely, leaving two double-stranded daughter molecules, each with one parental strand and one new strand.

The individual unit of replication is called a **replicon.** A replicon is thus a self-replicating segment of a chromosome that includes an origin from which replication proceeds bidirectionally. For viruses and prokaryotes, the entire genome is usually a single replicon. A eukaryotic chromosome, on the other hand, consists of a large number of replicons, each about 50,000–300,000 nucleotide pairs in length. Not all replicons begin replicating at the same time during the S phase of the eukaryotic cell cycle. Clusters of 20–80 replicons seem to be activated together. Certain clusters always replicate early and others always late in the S phase. Neither the regulation of replicon activation nor the significance of the timing sequence is yet understood, but it is clear that all the replicons of a eukaryotic genome must be replicated before mitosis can start.

DNA Polymerases

Enzymes capable of adding successive nucleotides to a growing DNA strand are called **DNA polymerases.** Virtually all known DNA polymerases require a template, all use deoxyribonucleoside triphosphates as their substrates, and all catalyze the covalent bonding of nucleotides to the 3′-hydroxyl end of the growing polynucleotide chain. Each successive nucleotide is linked to the growing chain by a phosphoester bond between the phosphate group on its 5′

carbon and the hydroxyl group on the 3′ carbon of the nucleotide added in the previous step (Figure 15-7). Chain elongation therefore always occurs at the 3′ end of a DNA strand, and the strand grows in the 5′ → 3′ direction.

Multiple DNA polymerases have been found in both prokaryotic and eukaryotic cells. (Table 15-1 lists the main DNA polymerases used in DNA replication, along with other key proteins involved in the process.) In *E. coli*, three DNA polymerases have been described, of which DNA polymerases I and III have been studied the most thoroughly. DNA polymerase I was discovered first (by Arthur Kornberg and his colleagues, in 1956) and is the most abundant DNA

Figure 15-7 The Directionality of DNA Synthesis. Addition of the next nucleotide to a growing DNA strand is catalyzed by DNA polymerase and always occurs at the 3′ end of the strand. A phosphoester bond is formed between the 3′ hydroxyl group of the terminal nucleotide and the 5′ phosphate of the incoming deoxyribonucleoside triphosphate (here dTTP), extending the growing chain by one nucleotide, liberating pyrophosphate (PP$_i$), and leaving the 3′ end of the strand with a free hydroxyl group to accept the next nucleotide.

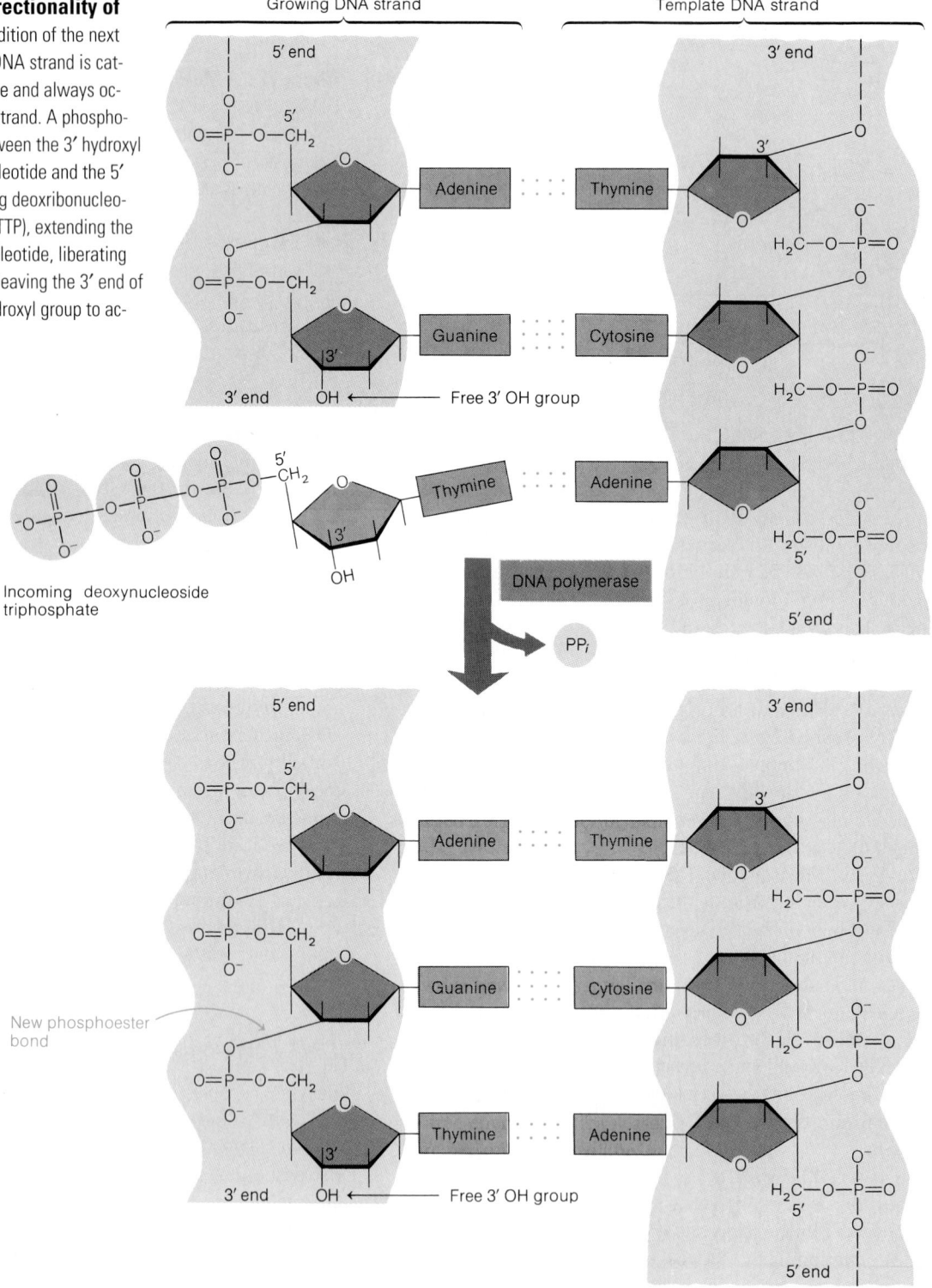

Table 15-1 Important DNA Replication Proteins

Protein	Cell Type*	Activities and/or Functions
DNA polymerase I	Prok.	DNA synthesis; 3′ → 5′ exonuclease (for proofreading); 5′ → 3′ exonuclease; removes and replaces RNA primers used in DNA replication (also functions in excision repair of damaged DNA)
DNA polymerase III	Prok.	DNA synthesis; 3′ → 5′ exonuclease (for proofreading); used in synthesis of both DNA strands
DNA polymerase α	Euk.	DNA synthesis; 3′ → 5′ exonuclease (for proofreading); used in lagging-strand synthesis
DNA polymerase δ	Euk.	DNA synthesis; associated 3′ → 5′ exonuclease (for proofreading); used in leading-strand synthesis
Primase	Both	RNA synthesis; makes RNA oligonucleotides that are used as primers for DNA synthesis
Helicase	Both	Unwinds double-stranded DNA
Single-strand binding protein (SSB)	Both	Binds to single-stranded DNA; stabilizes strands of unwound DNA in an extended configuration that facilitates access by other proteins
DNA topoisomerase (type I and type II)	Both	Makes single-strand cuts (type I) or double-strand cuts (type II) in DNA; induces and/or relaxes DNA supercoiling; can serve as swivel to prevent overwinding ahead of the DNA replication fork; can separate linked DNA circles at the end of DNA replication
DNA gyrase	Prok.	Type II DNA topoisomerase that serves as a swivel to relax supercoiling ahead of the DNA replication fork in *E. coli*
DNA ligase	Both	Makes covalent bonds to join together adjacent DNA strands, including the Okazaki fragments in lagging-strand DNA synthesis and the new and old DNA segments in excision repair of DNA
Initiator protein	Both	Binds to origin of replication and initiates unwinding of DNA double helix
Telomerase	Euk.	Using an integral RNA molecule as template, synthesizes DNA for extension of telomeres (sequences at ends of chromosomal DNA)

* Cell type(s) where found. Prok. = prokaryotic; Euk. = eukaryotic.

polymerase in *E. coli*. However, the main DNA replication enzyme in *E. coli* cells is **DNA polymerase III.** This enzyme was first discovered in mutant cells that had lost most (all but 2%) of their polymerase I activity—but had no noticeable impairment in their ability to grow.

DNA polymerase III is an especially complicated enzyme. The complete enzyme, called the *holoenzyme,* consists of ten different polypeptide subunits. Although we do not yet know the exact roles of all the polymerase III subunits, we do know something about the key roles of some of them. As is true for most DNA polymerases, the catalytic activities of polymerase III are known to include not only DNA synthesis but also a proofreading function, which checks the newly synthesized DNA and removes nucleotides that do not properly base-pair with the template. This proofreading activity, associated with one particular subunit (epsilon, ϵ), is called a 3′ → 5′ exonuclease. An **exonuclease** is an enzyme that degrades nucleic acid (usually DNA) from one end, rather than making internal cuts, as *endo*nucleases do. A 3′ → 5′ exonuclease is one that clips off nucleotides start-

ing at the 3′ end of the molecule and proceeding toward the 5′ end. Coupled with the 5′ → 3′ polymerase activity of polymerase III, the 3′ → 5′ exonuclease activity allows the enzyme to excise an incorrect nucleotide and insert the correct nucleotide in its place in the growing chain. This proofreading activity greatly enhances the accuracy of DNA replication.

E. coli's **DNA polymerase I,** on the other hand, functions primarily as a repair enzyme, although one of its "repair" functions is thought to play an important role in normal DNA replication. DNA polymerase I is a single polypeptide with three enzymatic activities. Like polymerase III, it has a 5′ → 3′ polymerase activity and a 3′ → 5′ exonuclease activity used for proofreading. In addition, a discrete domain of the protein has a 5′ → 3′ exonuclease activity that plays an important role in a repair process for removing and replacing damaged bases in DNA. The 5′ → 3′ exonuclease activity is also involved in removing small pieces of RNA that are used as primers in DNA replication, as we will see shortly. Finally, **DNA polymerase II** of *E. coli* seems to play a role in DNA repair, though it is not essential for cell viability if polymerases I and III are present.

In eukaryotic cells, five different DNA polymerases have been found to date; they are named with Greek letters. **DNA polymerase α** (alpha) and **DNA polymerase δ** (delta) seem to be the key enzymes for DNA synthesis in the nucleus, together playing a role equivalent to that of DNA polymerase III in *E. coli*. **DNA polymerase γ** (gamma) is found only in mitochondria and is the main polymerase for the replication of mitochondrial DNA. Polymerase ϵ (epsilon) is essential for cell division in yeast, and it is also found in human cells, but its function is still obscure, as is the function of polymerase β (beta).

DNA polymerases from diverse sources often contain stretches of amino acids that are strikingly similar in sequence, a structural indication of how much DNA replication processes have been conserved over billions of years of evolutionary history. For example, enzymes corresponding in function to eukaryotic polymerase α, isolated from sources as diverse as bacteriophages, mammalian viruses, bacterial plasmids, yeast, and human cells, contain similar amino acid sequences. Mysteriously, the corresponding *E. coli* enzyme, polymerase III, lacks these sequences.

In addition to their biological importance, DNA polymerases have important uses in biotechnology. Box 15A (pages 468–469) describes a technique called the **polymerase chain reaction (PCR),** in which a special type of DNA polymerase is the prime tool. Used for the rapid amplification of tiny samples of DNA, PCR is a powerful adjunct to the DNA fingerprinting method discussed in Chapter 14.

Discontinuous Synthesis

When it became clear that DNA replication always involves the addition of nucleotides to the 3′ end of each growing nucleotide chain, a conceptual problem became evident. Given the antiparallel orientation of the two strands in a DNA molecule, continuous nucleotide addition along both strands of a replication fork would require synthesis in the 5′ → 3′ direction on one strand and in the 3′ → 5′ direction on the other strand. But all known DNA polymerases function only in the 5′ → 3′ direction.

The dilemma was resolved when it was realized that synthesis at each replication fork is *continuous* in the direction of fork movement for one strand but *discontinuous* in the opposite direction for the other strand (Figure 15-8). The two daughter strands can therefore be distinguished based on their mode of growth. The **leading strand** is the one to which nucleotides can be added continuously, because it is growing in the 5′ → 3′ direction as the replication fork advances. The **lagging strand,** on the other hand, grows by synthesis of short pieces in the reverse direction, followed by joining of the pieces. In prokaryotes, the same DNA polymerase, polymerase III, is the main enzyme used for DNA synthesis on both strands. In eukaryotes, polymerase α seems to be the main DNA polymerase for synthesis of the lagging strand, while polymerase δ works on the leading strand.

The short pieces that serve as replication intermediates for the lagging strand were first described by Reiji Okazaki and his colleagues and are therefore called **Okazaki fragments.** Each fragment grows until it meets the adjacent fragment. Okazaki fragments are about 1000–2000 nucleotides long in viral and bacterial systems, but only about one-tenth this length in eukaryotic cells. Okazaki fragments are eventually joined together by an enzyme called **DNA ligase,** which catalyzes the formation of phosphoester bonds between the nucleotides at the 3′ and 5′ ends of adjacent fragments.

The Role of RNA in Priming DNA Replication

A surprising feature of Okazaki fragments came to light soon after their discovery. When first synthesized, each fragment has a short piece of RNA (3–10 nucleotides) at its 5′ end. In fact, the little piece of RNA is synthesized first and

Figure 15-8   Directions of DNA Synthesis at a Replication Fork. Because DNA polymerases function only in the 5′ → 3′ direction, synthesis at each replication fork is continuous in the direction of fork movement for the leading strand but discontinuous in the opposite direction for the lagging strand. Discontinuous synthesis involves short intermediates called Okazaki fragments, which are 1000–2000 nucleotides long in bacteria and about 100–200 nucleotides long in eukaryotic cells. The fragments are later joined together by the enzyme DNA ligase. Parent DNA is shown in dark blue, newly synthesized DNA in lighter blue. Here and in subsequent figures, arrowheads indicate the direction in which the nucleic acid chain is being elongated.

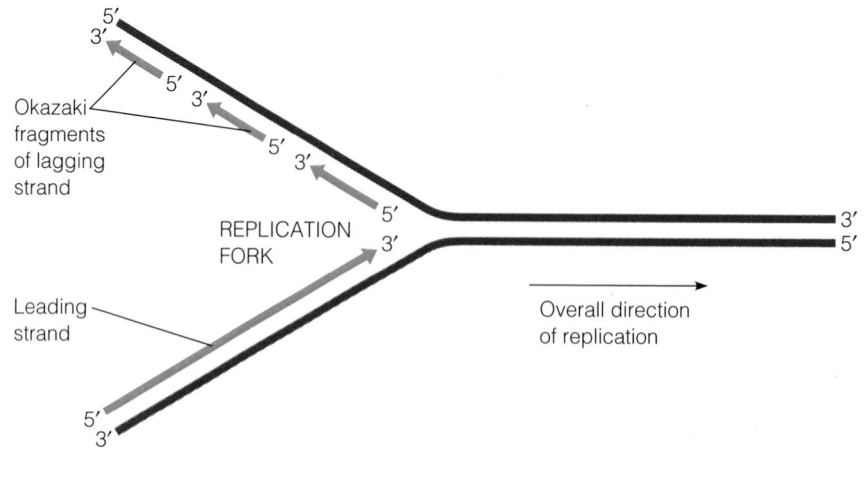

serves as a **primer** for DNA synthesis—an oligonucleotide onto which deoxyribonucleotides are added. These RNA primers are synthesized by a DNA-dependent RNA polymerase called **primase,** as Figure 15-9 illustrates. Primase is a specific kind of RNA polymerase that is involved only in the process of DNA replication. Like all other RNA polymerases, and unlike all DNA polymerases, primases can *initiate* a new polynucleotide strand complementary to a template strand; they do not themselves require a primer.

Primases preferentially recognize and transcribe into RNA a specific short sequence of DNA nucleotides. In *E. coli,* the primase is relatively inactive unless accompanied by six other proteins, forming a complex called a **primosome.** One important function contributed by the other primo-

some proteins is the unwinding of the parental DNA, as we shall discuss shortly. Other functions seem to include roles in assembling the primosome and recognizing the target sequence. The situation in eukaryotic cells is slightly different, so the term *primosome* is not used. The eukaryotic primase is not as closely associated with unwinding proteins, but it is very tightly bound to DNA polymerase α, the main lagging-strand DNA polymerase.

Once the RNA primer is available, DNA synthesis can proceed, with DNA polymerase III (or DNA polymerase α in eukaryotes) adding successive deoxyribonucleotides to the 3′ end of the primer (Figure 15-9, step 2). Polymerization continues until the enzyme reaches the adjacent Okazaki fragment. No longer needed, the RNA section is now removed by a 5′ → 3′ exonuclease, and DNA nucleotides are polymerized to fill its place. In *E. coli,* both these actions are thought to be carried out by DNA polymerase I (Figure 15-9, step 3). Adjacent fragments are then linked together covalently by DNA ligase.

In the absence of a preexisting nucleotide chain to which deoxyribonucleotides can be added, DNA synthesis simply does not occur. Thus, the synthesis of RNA primers is an integral part of almost all DNA replication. This is true for both the lagging strand and the leading strand, although for the leading strand a primer is only needed at an origin of replication. We will focus on the initiation of DNA replication at an origin after considering another important aspect of DNA replication, the unwinding of DNA.

Unwinding the DNA

For DNA replication to proceed, the DNA must unwind to expose the single strands to the enzymes responsible for copying them. This aspect of DNA replication is known to involve at least three kinds of proteins with distinct functions (Figure 15-10).

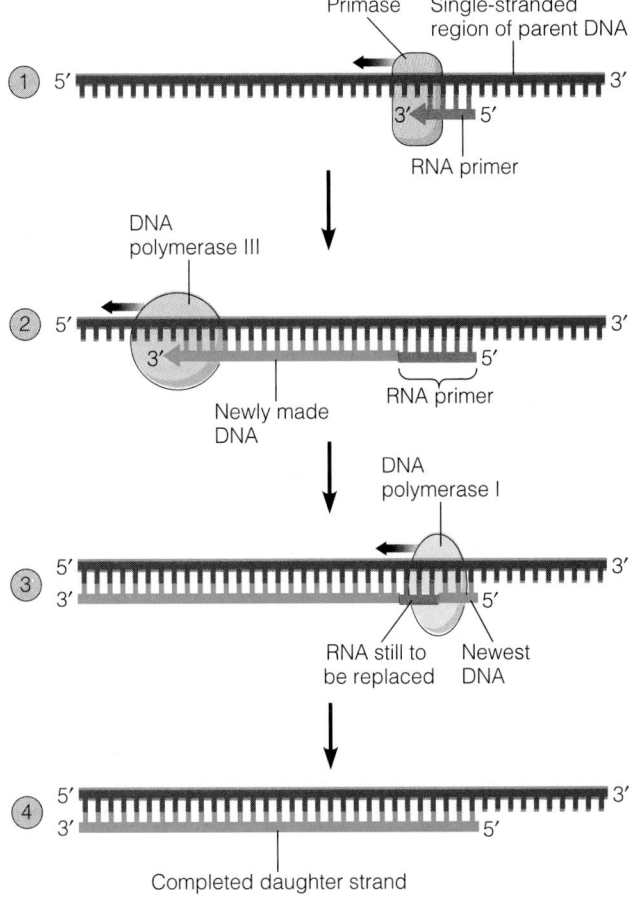

Figure 15-9 The Role of RNA Primer in DNA Replication. DNA synthesis is initiated with a short RNA primer in both prokaryotes and eukaryotes. This figure shows the process as it occurs in *E. coli.* ① The primer (pink) is synthesized by primase, an RNA polymerase that uses a single strand of DNA as its template. In *E. coli,* the primase is part of a protein complex called a primosome (not shown here). ② Once the short stretch of RNA is available, DNA polymerase III uses it as a primer to initiate DNA synthesis, which proceeds in the 5′ → 3′ direction. ③ The RNA primer is eventually removed by the 5′ → 3′ exonuclease activity of DNA polymerase I, which replaces ribonucleotides with deoxyribonucleotides as it proceeds. ④ The daughter strand is complete.

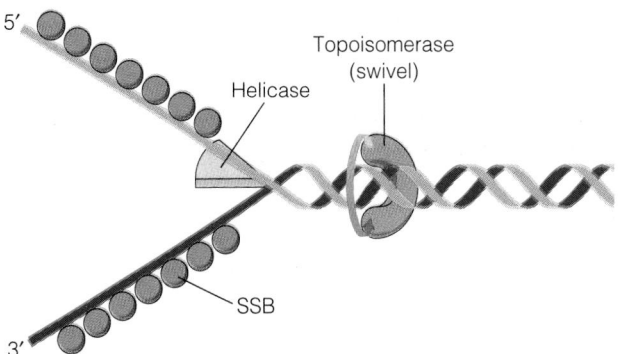

Figure 15-10 Proteins Involved in Unwinding the DNA at the Replication Fork.   Three types of proteins are required for this aspect of DNA replication. The actual unwinding proteins are the helicases; the principal one in *E. coli,* which is part of the primosome, operates 5′ → 3′ along the template for the lagging strand, as shown here. Single-strand binding proteins (SSB) stabilize the unwound DNA in an extended position. A topoisomerase forms a swivel ahead of the replication fork; in *E. coli,* this topoisomerase is DNA gyrase.

The proteins directly responsible for unwinding the DNA double helix are called **helicases.** These are enzymes that use the energy of ATP to unwind the DNA in advance of the replication fork, breaking the hydrogen bonds as they go. In *E. coli*, at least two different helicases seem to be involved in DNA replication; one attaches to the lagging-strand template and moves in a $5' \rightarrow 3'$ direction, and the other attaches to the leading-strand template and moves $3' \rightarrow 5'$. Both are part of the primosome, but the $5' \rightarrow 3'$ helicase is the more important for unwinding the DNA at the replication fork. In eukaryotic cells, less is known about the helicase(s) involved in DNA replication.

The unwinding associated with DNA replication would create an intolerable amount of supercoiling and possibly tangling in the rest of the DNA were it not for the actions of **DNA topoisomerases,** which we introduced in Chapter 14. These enzymes can form swivels in the DNA by making and then quickly resealing single- or double-stranded breaks in the double helix. Of the ten or so topoisomerases found in *E. coli*, the one that seems to be most important is DNA gyrase, a type II topoisomerase (it cuts both DNA strands). Using the energy of ATP, gyrase can make negative supercoils and thereby relax positive ones. It is chiefly gyrase that serves as the swivel to prevent overwinding (positive supercoiling) of the DNA ahead of the replication fork. In addition, gyrase has special roles in both the initiation and the completion of DNA replication in *E. coli*—in opening up the double helix at the origin of replication and in separating the linked circles of daughter DNA at the end. The situation in eukaryotic cells is not yet well understood, although topoisomerases of both types have been isolated.

Once strand separation has begun, molecules of **single-strand binding protein (SSB)** quickly attach to the exposed single strands at the replication fork, in such a way that they do not cover the nitrogenous bases. The SSB molecules hold the separated strands in a semiextended position that makes them more accessible to the DNA replication machinery. Once a particular segment of DNA is replicated, the SSB molecules fall off and are recycled, attaching to the next single-stranded segment.

Putting It All Together: DNA Replication in Summary

Figure 15-11 reviews the highlights of what we currently understand about the mechanics of DNA replication in *E. coli* cells (although without showing the actual three-dimensional arrangement of the DNA strands). Starting at the origin of replication, we gradually build up the machinery at the replication fork, until, at step 7, seven different proteins are involved. It is important to keep in mind that these and a number of additional proteins are actually all closely associated in one big complex. This complex, called a **replisome,** is about the size of a ribosome. The replisome machine is powered by the hydrolysis of nucleoside triphos-

phates. These include both the nucleoside triphosphates used by the polymerases and primase as building blocks for polynucleotide synthesis and the ATP used in the functioning of several other DNA-replication proteins: initiator protein, helicase, gyrase, and ligase.

Much remains to be learned about DNA replication, especially in eukaryotic cells. The great length and elaborate folding of eukaryotic DNA molecules pose special challenges for its replication. How are the many replication origins coordinated, and how is their activation linked to other key events in the cell cycle? How exactly are the histones and other chromosomal proteins removed from the replicating DNA and then readded in forming the two new chromatin strands? Researchers have learned that, as the replication fork passes through a nucleosome, the old histone octamers are distributed randomly to the two daughter DNA molecules, with new histone octamers assembling among them. But when and how are other chromosomal proteins added, what controls the higher levels of chromatin packing, and how is all this accomplished without tangling the chromatin? Research is underway on these and other difficult questions, with varying rates of progress. Meanwhile, an answer has emerged to one of the most baffling questions related to DNA replication in eukaryotes, the end-replication problem.

The Problem of DNA Ends: DNA Synthesis at Telomeres

If we continue the process summarized in Figure 15-11 for the circular genome of *E. coli*, or for any other circular DNA molecule, we will eventually complete the circle. The leading strand can simply continue to grow $5' \rightarrow 3'$ until its $3'$ end touches the $5'$ end of the lagging strand coming around in the other direction. For the lagging strand, the very last bit of DNA to be synthesized, the replacement for the RNA primer of the last Okazaki fragment, can be added to the free $3'$-OH end of the leading strand coming around in the other direction.

For linear DNA, the fact that *DNA polymerases can only add nucleotides to the $3'$-OH end of a preexisting polynucleotide* creates a serious problem, which is illustrated in Figure 15-12 (page 466). When a growing lagging strand (lighter blue) reaches the end of the molecule, there is no way to polymerize the last few DNA nucleotides. After the last RNA primer, located at or near the end of the template strand, is removed by a $5' \rightarrow 3'$ exonuclease, there is no way to fill in the gap with DNA, because there is no $3'$-OH end ahead of it to which deoxyribonucleotides can be added. As a result, each round of replication produces shorter and shorter daughter DNA molecules. Clearly, if this trend continued indefinitely, we would not be here today!

Viruses with linear DNA genomes solve this problem in a variety of ways. In some cases, such as bacteriophage λ, the linear DNA injected into the cell forms a closed circle before

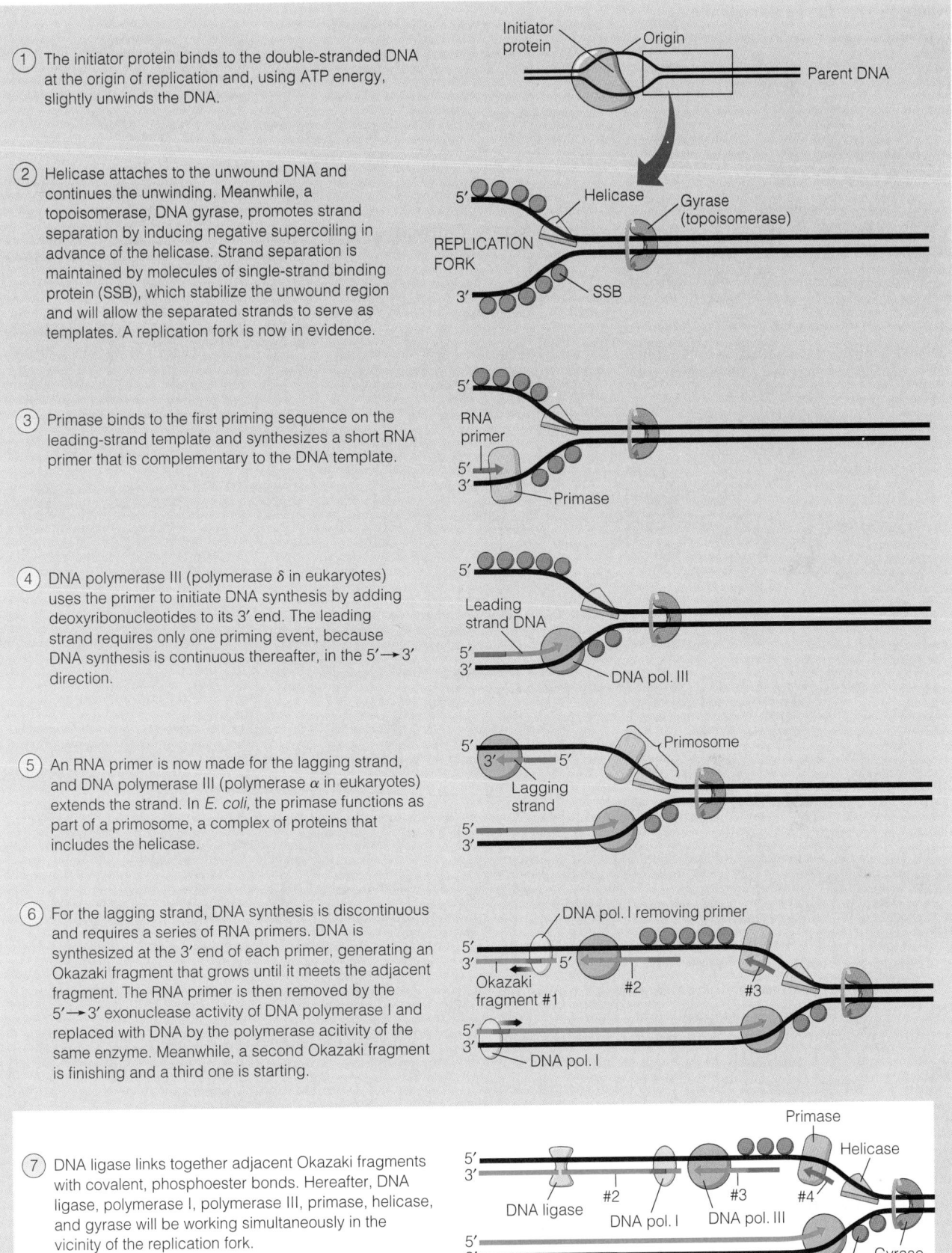

① The initiator protein binds to the double-stranded DNA at the origin of replication and, using ATP energy, slightly unwinds the DNA.

② Helicase attaches to the unwound DNA and continues the unwinding. Meanwhile, a topoisomerase, DNA gyrase, promotes strand separation by inducing negative supercoiling in advance of the helicase. Strand separation is maintained by molecules of single-strand binding protein (SSB), which stabilize the unwound region and will allow the separated strands to serve as templates. A replication fork is now in evidence.

③ Primase binds to the first priming sequence on the leading-strand template and synthesizes a short RNA primer that is complementary to the DNA template.

④ DNA polymerase III (polymerase δ in eukaryotes) uses the primer to initiate DNA synthesis by adding deoxyribonucleotides to its 3′ end. The leading strand requires only one priming event, because DNA synthesis is continuous thereafter, in the 5′→3′ direction.

⑤ An RNA primer is now made for the lagging strand, and DNA polymerase III (polymerase α in eukaryotes) extends the strand. In E. coli, the primase functions as part of a primosome, a complex of proteins that includes the helicase.

⑥ For the lagging strand, DNA synthesis is discontinuous and requires a series of RNA primers. DNA is synthesized at the 3′ end of each primer, generating an Okazaki fragment that grows until it meets the adjacent fragment. The RNA primer is then removed by the 5′→3′ exonuclease activity of DNA polymerase I and replaced with DNA by the polymerase acitivity of the same enzyme. Meanwhile, a second Okazaki fragment is finishing and a third one is starting.

⑦ DNA ligase links together adjacent Okazaki fragments with covalent, phosphoester bonds. Hereafter, DNA ligase, polymerase I, polymerase III, primase, helicase, and gyrase will be working simultaneously in the vicinity of the replication fork.

Figure 15-11   A Summary of DNA Replication in Bacteria.
Starting with the initiation event at the replication origin, this figure depicts DNA replication in E. coli in seven steps. Two replication forks move in opposite directions from the origin, but only one fork is illustrated for steps 2–7. The various proteins shown here as separate entities are actually closely associated (along with others) in a single large complex called a replisome. The primase and helicase are particularly closely bound, within a primosome. Parental DNA is shown in black, newly synthesized DNA in blue, and RNA in pink. This series of diagrams does not attempt to show the topological arrangement of the DNA strands.

Figure 15-12 The End-Replication Problem. For a linear DNA molecule, such as that of a eukaryotic chromosome, the usual DNA replication machinery is unable to replicate the ends. As a result, with each round of replication, the DNA molecules will get shorter, with potentially disastrous consequences for the cell. In this diagram, the initial parent DNA strands are black, daughter DNA strands are blue, and RNA primers are pink. For simplicity, we show only one origin of replication and, in the last two steps, only the shortest of the progeny molecules. (The lagging-strand daughter DNA is shown in a lighter blue in the first three steps to make a point unrelated to the end-replication problem—that each daughter strand is leading at one end and lagging at the other. This is apparent in this figure because it shows the entire replicating molecule, with both replication forks.)

① DNA replication is initiated at the origin; the replication bubble grows as the two replication forks move in opposite directions

② Finally only one primer (pink) remains on each daughter DNA molecule

③ The last primers are removed by a 5′ → 3′ exonuclease, but no DNA polymerase can fill the resulting gaps because there is no 3′–OH available to add nucleotides to

④ Each round of replication generates shorter and shorter DNA molecules

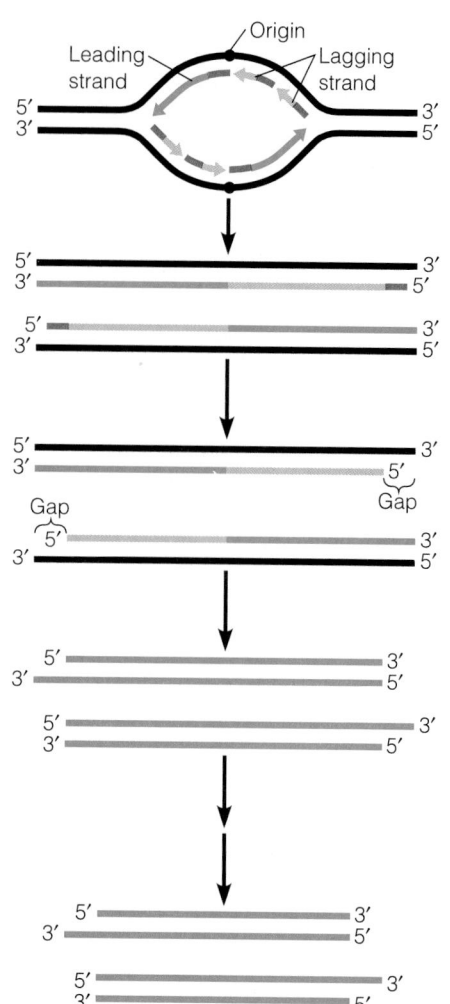

it replicates. In other cases, the viruses use more exotic reproduction strategies, although the DNA polymerases always progress 5′ to 3′, adding nucleotides to a 3′ end.

The strategy employed for eukaryotic chromosomal DNA involves the special end-sequences called telomeres. Recall from Chapter 14 that **telomeres,** the protective ends of the linear DNA molecules of eukaryotic chromosomes, consist of simple-sequence repeated DNA. The repeated unit in human DNA, which is typical, is TTAGGG. The number of repetitions in a telomere varies between 100 and 1000 or so. This variation and the fact that the 3′ end of the telomere extends as a single strand 12–16 nucleotides beyond the 5′ end are clues to the telomere role. The ends of the DNA usually *do* shorten with replication of the molecule, but the noncoding sequences of the telomeres can protect the coding DNA from degradation through many rounds of replication. And, on occasion, the telomeres are lengthened by the action of a special DNA polymerase called

telomerase. This protein actually includes an RNA molecule containing a sequence that serves as a template for telomere extension. Figure 15-13 depicts the action of the telomerase of the ciliated protozoan *Tetrahymena,* the first telomerase to be isolated.

Evidence for telomerases has been found in all eukaryotes studied to date. In multicellular organisms, telomerases may be a crucial factor in the longevity of tissues and perhaps even the organism as a whole. In most (if not all) somatic tissues—tissues other than those that give rise to the "germ cells," sperm and eggs—the chromosomes' telomeres get shorter and shorter with each cell division. For example, in humans, the telomeres of skin cells and blood cells diminish by about 15–40 bp each year. By contrast, sperm telomeres increase in length with age. This and mounting evidence from the measurement of telomerase levels suggest that the genes encoding telomerase are normally turned on only in germ line cells, and not in somatic cells. Presumably,

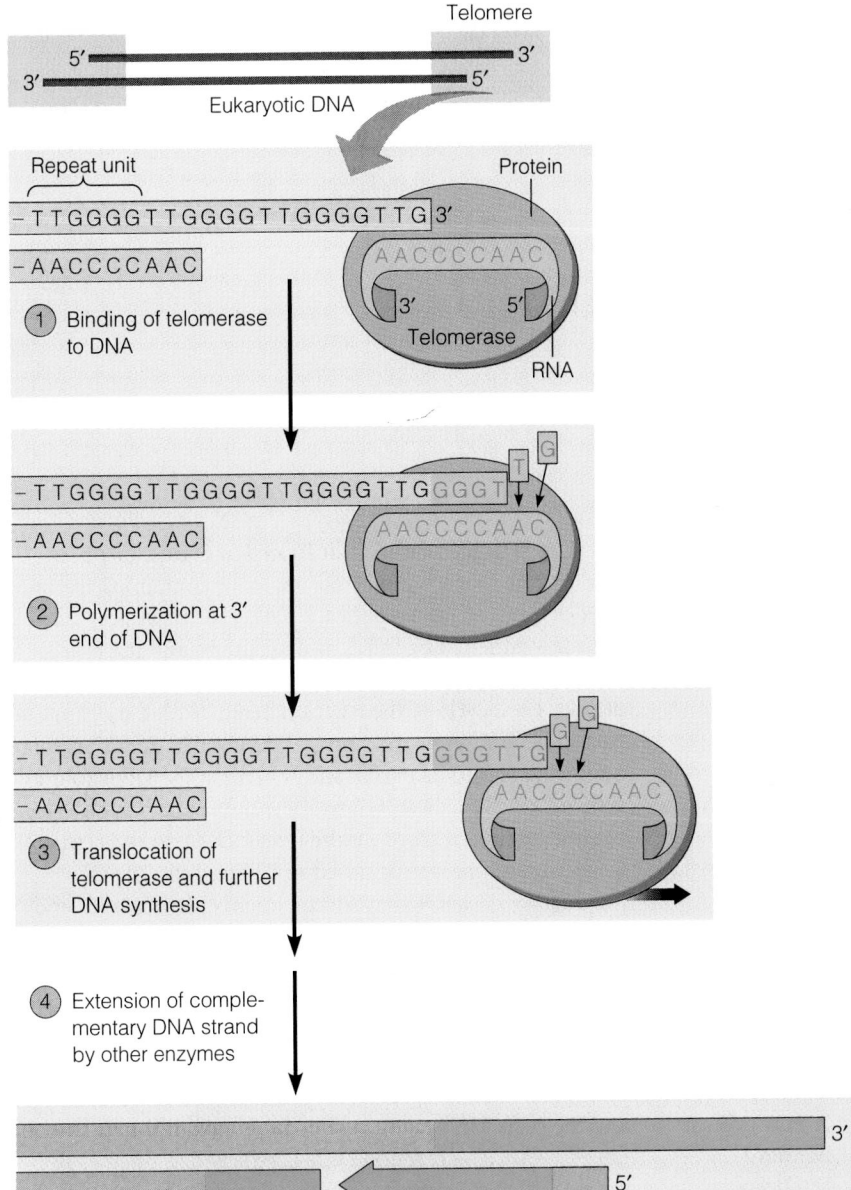

Telomere

5′ ————————— 3′
3′ ————————— 5′
Eukaryotic DNA

Repeat unit   Protein

- T T G G G G T T G G G G T T G G G G T T G 3′

- A A C C C C A A C

A A C C C C A A C
3′   5′
Telomere
RNA

① Binding of telomerase to DNA

- T T G G G G T T G G G G T T G G G G T T G G G G T

- A A C C C C A A C

A A C C C C A A C

② Polymerization at 3′ end of DNA

- T T G G G G T T G G G G T T G G G G T T G G G G T T G

- A A C C C C A A C

A A C C C C A A C

③ Translocation of telomerase and further DNA synthesis

④ Extension of complementary DNA strand by other enzymes

3′

5′

Figure 15-13 The Extension of Telomeres by Telomerase. Telomeres are stretches of repeated DNA at the very ends of eukaryotic chromosomes. This figure focuses on one end of a DNA molecule from *Tetrahymena*, for which the telomeric repeat unit is TTGGGG (those of other species are very similar). The 3′ end of the DNA extends beyond the 5′ end and is the substrate for the action of the telomerase enzyme. Telomerase is composed of both protein and RNA; the RNA portion of *Tetrahymena* telomerase is 159 nucleotides long and contains a 9-base sequence complementary to 1½ telomeric repeat units. As outlined here, the telomerase ① binds to the 3′ end of the telomere, positioning itself so that there is base pairing between part of the 9-base RNA sequence and the last few DNA bases. ② Next it catalyzes the elongation of the DNA, with the remainder of the 9-base RNA sequence serving as the template. Like all DNA polymerases, telomerase adds nucleotides to the 3′ end of a preexisting polynucleotide. ③ The telomerase then translocates (moves) along the DNA strand in the 3′ direction and repeats step 2 and step 3 several times. ④ Meanwhile, the standard DNA replication machinery synthesizes a lagging strand complementary to the strand elongated by telomerase. The result is a lengthened telomere.

if a line of somatic cells went through enough rounds of cell division, the telomeres on one or more chromosomes could entirely disappear, leading to the erosion of coding DNA and premature cell death. It is possible that the telomeres at birth, about 10,000 bp in humans, may be long enough to endure more than a lifetime of replicative neglect. But it is also possible that telomeres are a limiting factor in determining an organism's life span.

Interestingly, active telomerase has recently been found in human somatic cells that are cancerous. Cells from a tumor that is large enough to be visible typically have unusually short telomeres, as one would expect for cells that

have undergone an unusually large number of cell divisions. Progressive shortening would presumably lead eventually to self-destruction of the cancer, unless telomerase became available to stabilize telomere length. This is exactly what seems to have happened in the cancer cells studied. A survey of a large number of human cell samples, including over 100 from tumors, found telomerase activity in almost all the cancer cell samples (and in "immortal" strains of cultured cells) but in none of the samples from normal somatic tissues. If telomerase is indeed an important factor in the progression of most or all cancers, it may provide a useful target for both cancer diagnosis and anticancer drug therapy.

Contemporary Techniques

THE PCR REVOLUTION

The ability to work with minuscule amounts of DNA is proving valuable in a wide range of endeavors, from paleontology to criminology. In Chapter 14, we described how RFLP analysis can be used to identify and characterize particular sequences contained within a very small amount of DNA, as little as 1 μg, the amount in a small drop of blood (see Box 14C). But sometimes even that amount of DNA may not be available. In such cases, another method, polymerase chain reaction (PCR), can come to the rescue. With PCR, one can rapidly replicate, or "amplify," selected DNA segments in a test tube. In only a few hours, PCR can make millions or even billions of copies of a specific DNA sequence, to produce enough material for RFLP analysis, DNA sequencing, or other uses. Like RFLP analysis, PCR is often in the news in connection with the solving of violent crimes.

The complicated, multiprotein system that the cell uses for DNA replication in vivo is not required for the PCR method; neither origins of replication, nor DNA unwinding proteins, nor the apparatus for lagging-strand synthesis are involved. The keys to the simplicity of PCR as practiced today are an unusual DNA polymerase and the realization that synthetic primers can be used to set up a chain reaction to produce an exponentially growing population of specific DNA molecules. For this insight, biochemist Kary Mullis received a Nobel Prize.

The PCR technique requires the preliminary preparation of single-stranded DNA oligonucleotides to serve as primers for DNA synthesis. The primers are generally 15–20 nucleotides long and correspond to sequences at the two ends of the DNA segment to be amplified. (If sequences that naturally flank the sequence of interest are not known, artificial ones can be attached in vitro.) The DNA polymerase used for PCR today was first isolated from the bacterium *Thermus aquaticus,* a denizen of thermal hot springs where the waters are normally 70–80°C. The optimal temperature for the enzyme, called *Taq* polymerase, is 72°C, and it is stable at even higher temperatures—a property that makes possible the automation of PCR.

The PCR technique is illustrated in Figure 15A-1. Included in the reaction mixture are DNA containing the sequence targeted for amplification, the *Taq* DNA polymerase, the synthetic oligonucleotide primers, and the four deoxyribonucleoside triphosphates (dATP, dTTP, dCTP, and dGTP). Each cycle of the reaction begins with a short period of heating to near boiling (95°C) to denature the DNA double helix into its two strands (step 1). The DNA solution is then cooled to about 50° to allow the primers to bind to complementary regions on the longer DNA strands (step 2). The temperature is then raised to 72°, and the *Taq* DNA polymerase goes to work, adding nucleotides to the 3′ end of the primer (step 3). The specificity of the primers ensures the selective copying of the stretches of template DNA downstream from the primers.

Notice that this basic cycle (steps 1–3) results in a doubling of the targeted DNA segment. The reaction mixture can now be heated again to melt the new double helices, and the cycle can be repeated to double the DNA again (steps 1′–3′). The reaction cycle is repeated as often as necessary, with each cycle doubling the amount of DNA from the previous cycle. After the third cycle, more and more of the product DNA molecules will be of a uniform length that consists only of the targeted sequence (like two of the molecules in the last line of the figure). Because heating to 95°C does not destroy the *Taq* polymerase, there is no need to add fresh enzyme for each round of the cycle.

In most cases, 20–30 reaction cycles are sufficient to produce the desired quantity of DNA. Since the theoretical amplification accomplished by n cycles is 2^n, 20 cycles can yield amplification of a millionfold or more ($2^{20} = 1,048,576$), and 30 cycles over a billionfold ($2^{30} = 1,073,741,824$). Each cycle takes only about 5 minutes, and the process can be automated to complete 20 or more cycles in a few hours. This amount of time is considerably shorter than the several days required for amplifying DNA by cloning it in bacteria, a method to be discussed in Chapter 16. Furthermore, PCR can be used with as little as one molecule of DNA, and it does not require that the starting sample of DNA be puri-

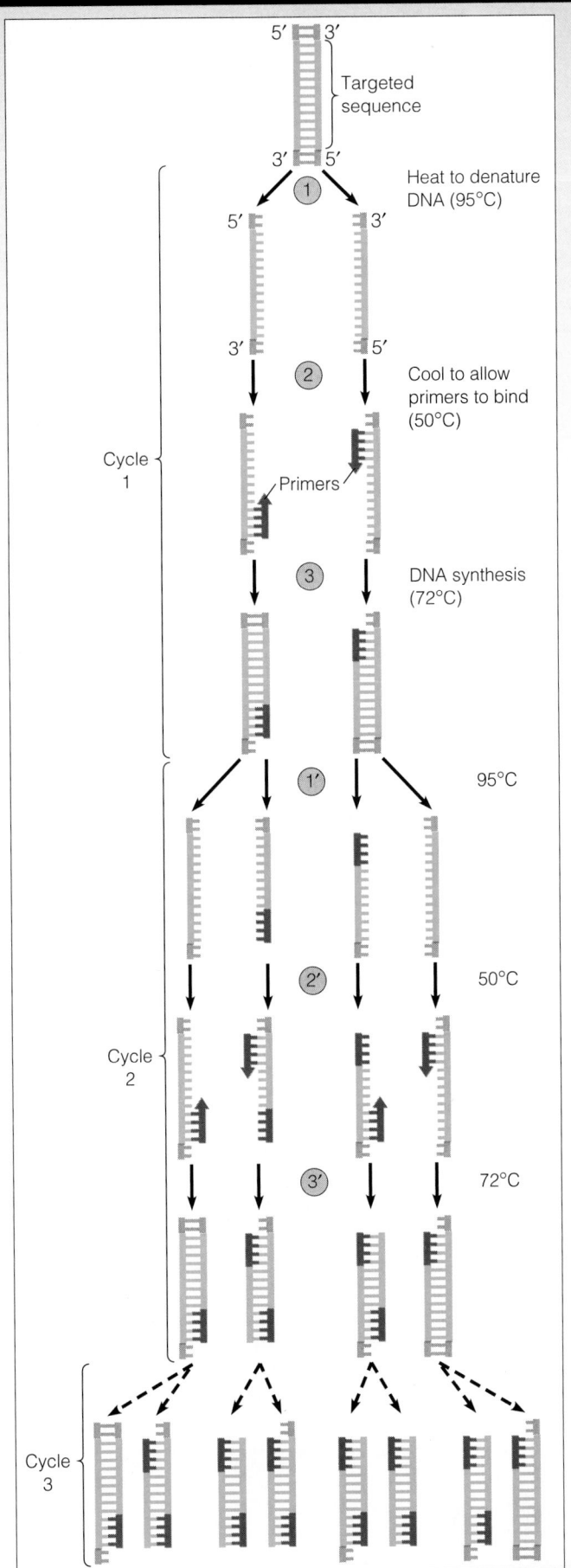

fied. In fact, the method's sensitivity and power can cause problems. A few contaminating DNA molecules (such as from skin cells shed by a lab technician) could be amplified along with the DNA of interest, yielding misleading results. In a legal case, such an error could obviously lead to grave injustice, and for this reason the courts are proceeding cautiously in allowing the introduction of PCR evidence.

Nevertheless, with the proper precautions and controls, PCR is proving extremely valuable. As an aid in evolution research, it has been used to amplify DNA fragments recovered from ancient Egyptian mummies, a 40,000-year-old woolly mammoth frozen in a glacier, and a 30-million-year-old plant fossil. In medical diagnosis, PCR has been used to amplify DNA from single embryonic cells for rapid prenatal diagnosis, and it has made possible the detection of viral genes in cells infected with HIV and certain other viruses. Perhaps most importantly for the long run, PCR has revolutionized basic research in molecular genetics, by allowing easy amplification of particular genes or sequences from among the 100,000 or so genes in mammalian genomes. ■

Figure 15A-1   DNA Amplification Using the Polymerase Chain Reaction.   See the description in the box text. PCR works best when the DNA segment to be amplified—the region flanked by the two primers—is 50–2000 nucleotides long.

DNA Repair

The faithfulness with which DNA sequences are maintained from one generation of cells to the next requires not only that DNA be replicated very accurately, but also that provision be made for repairing the many "spontaneous" changes to DNA that result from interactions with its environment. DNA interacts with many other macromolecules in the cell, with small molecules and ions, and often with various kinds of radiation. Some spontaneous changes to DNA are actually desirable because such changes, called **mutations,** provide the genetic variability that is the raw material of evolu-

tion. The mutation rate is very low, however; by some estimates, an average gene accumulates one mutation every 200,000 years. But the rate at which spontaneous DNA damage occurs is really far greater than the mutation rate suggests, because most damage is repaired shortly after it occurs and therefore does not affect the genetic record of the organism.

Kinds of DNA Damage

DNA can be damaged by a variety of chemical and physical agents, but most spontaneous changes result from the

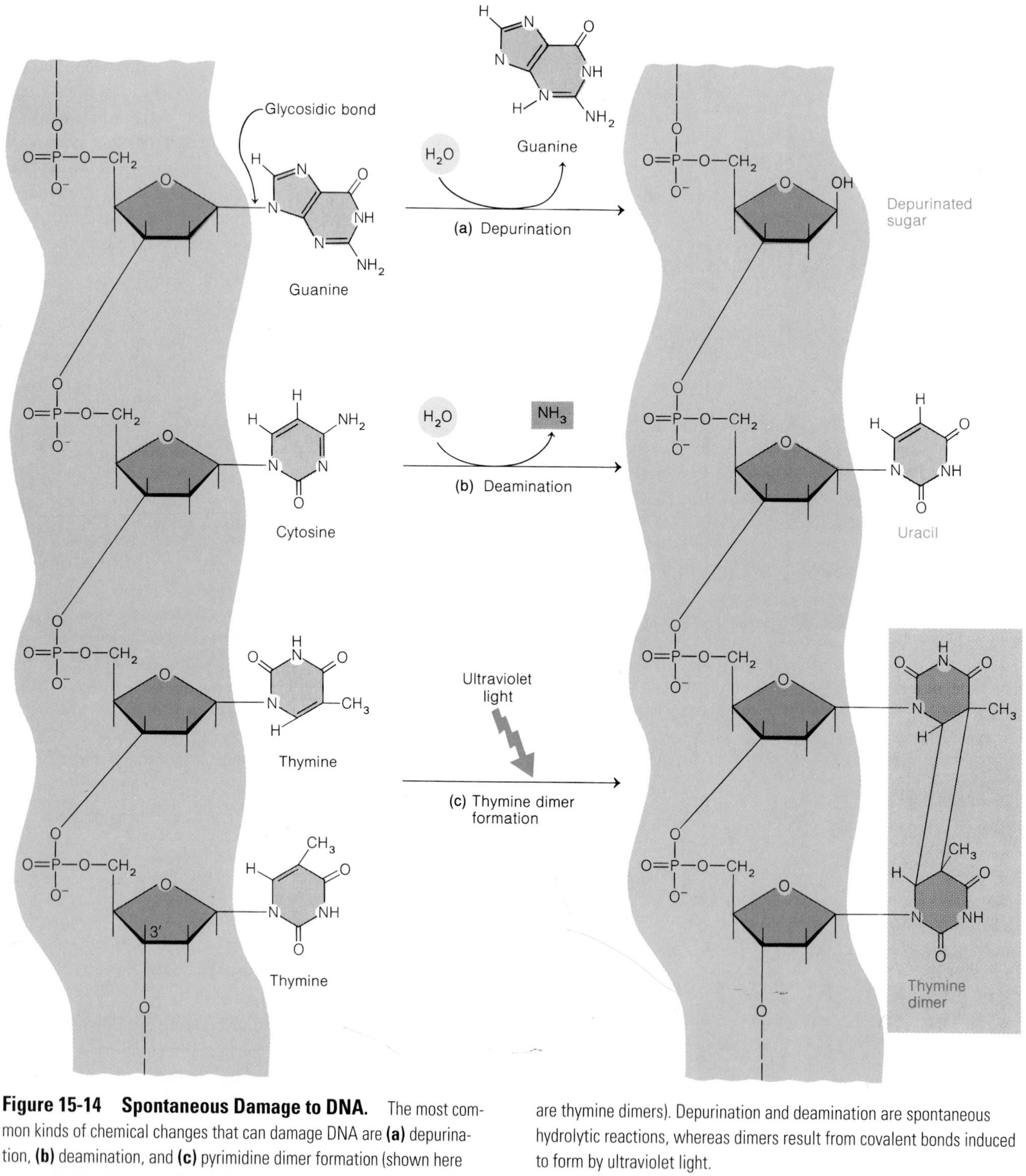

Figure 15-14 Spontaneous Damage to DNA. The most common kinds of chemical changes that can damage DNA are **(a)** depurination, **(b)** deamination, and **(c)** pyrimidine dimer formation (shown here are thymine dimers). Depurination and deamination are spontaneous hydrolytic reactions, whereas dimers result from covalent bonds induced to form by ultraviolet light.

depurination of purine nucleotides, the deamination of cytosines and other bases, and the formation of pyrimidine dimers (Figure 15-14). The first two categories are generally due to random thermal interactions between DNA and the water molecules around it; these changes are spontaneous hydrolytic reactions. Pyrimidine dimer formation, on the other hand, is induced by ultraviolet light.

Depurination involves the loss of a purine base (either adenine or guanine) by spontaneous hydrolysis of the glycosidic bond that links it to deoxyribose (Figure 15-14a). This glycosidic bond is intrinsically unstable and is in fact so susceptible to random thermal cleavage that the DNA in a human cell may lose thousands of purine bases every day.

A second common source of DNA damage is the spontaneous **deamination** of cytosine, adenine, or guanine—that is, the loss of an amino group (—NH_2). Of the three bases, cytosine is most susceptible to deamination, giving rise to uracil (Figure 15-14b). Like depurination, deamination is a hydrolytic reaction, usually caused by random collision of a water molecule with the bond that links the amino group of the base to the pyrimidine or purine ring. The rate of damage to the DNA in a human cell by this means is about 100 deaminations per day.

If a DNA strand with missing purines or deaminated bases is not repaired, an erroneous base sequence may be propagated when the strand serves as a template in the next round of DNA replication. For example, where a cytosine has been converted to a uracil by deamination, the uracil will direct the insertion of an adenine on the opposite strand, rather than guanine, the correct base. The ultimate result of this now heritable change in base sequence may be a significant change in the amino acid sequence and function of a protein.

Pyrimidine dimer formation is yet another form of spontaneous damage to DNA. It is promoted by ultraviolet light from the sun and involves the formation of covalent bonds between two adjacent pyrimidine bases, usually two thymines (Figure 15-14c). Both replication and transcription are blocked by such dimers, presumably because the enzymes carrying out these functions cannot cope with the resulting bulge in the DNA.

Repair Mechanisms

Perhaps not surprisingly for a function so important to the health and survival of the organism, many different enzymes and mechanisms have evolved for repairing damaged DNA. In *E. coli* alone, almost 100 genes are known to play roles in DNA repair; many of these are also involved in DNA replication and/or recombination. In some cases, DNA damage can be directly reversed. Usually, however, the repair process involves cutting out and replacing the damaged nucleotides. This type of repair, called **excision repair,** occurs in three main steps, as shown in Figure 15-15. The first step is the excision of the defective nucleotide(s) from one strand of the duplex. Special enzymes called **repair endonucleases** recog-

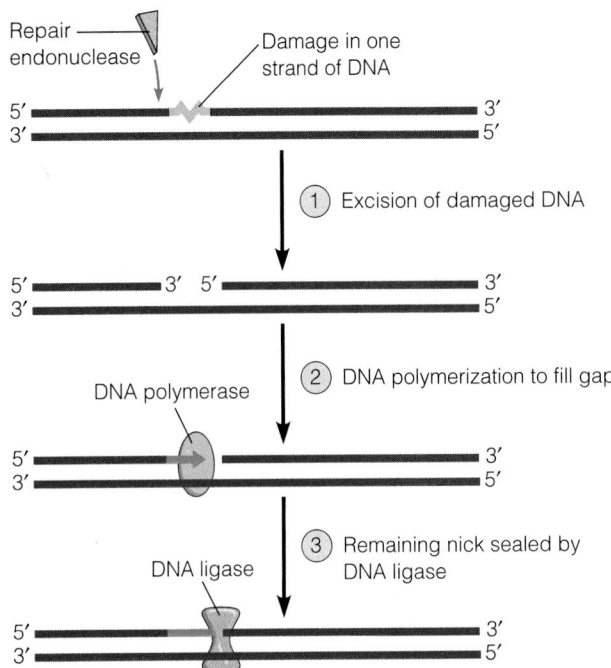

Figure 15-15   General Scheme for Excision Repair of DNA Damage.   The three steps shown here are common to all types of excision repair of DNA with damage in one strand. ① The damaged part of the DNA strand, and possibly some DNA on either side of it, is cut out (excised) from the double helix. An endonuclease, which nicks the DNA near the damage, is crucial for this step; a helicase and/or exonuclease may help remove the damaged segment. ② A DNA polymerase fills in the gap by adding nucleotides to the strand's 3′ end; DNA polymerase I plays this role in *E. coli.* ③ The remaining nick is sealed by DNA ligase.

nize the defect and nick the phosphodiester backbone of the DNA near it. Additional enzymes may be required to fully remove the defective nucleotide(s). For example, a helicase might unwind a stretch of damaged DNA between two endonuclease nicks, to free it from the DNA double helix; alternatively, an exonuclease might attach to an end created by a single nick and chew away the damaged strand one nucleotide at a time. The second step in excision repair is the replacement of the missing nucleotides with the correct ones by a DNA polymerase—in *E. coli,* usually DNA polymerase I. The nucleotide sequence of the complementary strand ensures correct base insertion, just as it does in DNA replication. Then, in the third step, DNA ligase seals the remaining nick in the repaired strand by forming the missing phosphoester bond.

Excision repair pathways can be classified into three main types, each carried out by a somewhat different set of proteins. The three types of repair are called mismatch repair, base excision repair, and nucleotide excision repair. **Mismatch repair** targets errors made during DNA replication, when incorrect nucleotides are sometimes successfully inserted into a growing strand of DNA. Specific proteins of the mismatch repair system recognize such sites in DNA and can discern which of the two strands is the "wrong" one. In this type of repair, a single nick by a repair endonuclease is followed by the successive removal of nucleotides from the defective strand by an exonuclease that starts at the nick. The importance of this repair system was highlighted recently with the discovery that one of the most common hereditary cancers, *hereditary nonpolyposis colon cancer* (*HNPCC*), can result from mutations in the genes encoding proteins of the mismatch repair system.

Base excision repair targets single damaged bases in the DNA. The detection of deaminated bases, for example, is accomplished by specific **DNA glycosylases,** enzymes that recognize a specific deaminated base and remove it from the DNA molecule by cleaving the bond between the base and the sugar to which it is attached. The sugar with the missing base is then recognized by a repair endonuclease that detects depurination. This repair endonuclease breaks the phosphodiester backbone on one side of the sugar lacking a base; another enzyme completes the removal of the sugar-phosphate.

For the removal of pyrimidine dimers and other large, bulky lesions in DNA, the cell uses the third repair pathway, called **nucleotide excision repair (NER).** This repair system recognizes major distortions of the double helix. The NER endonuclease (often called an excision nuclease, or excinuclease) makes two nicks, one on either side of the distortion. Then a helicase binds to the stretch of DNA between the nicks (12 nucleotides long in *E. coli,* 29 in humans) and unwinds it, freeing it from the rest of the DNA. Finally, as in the other two types of repair, the gap left by the excision step is filled in by DNA polymerase and sealed by DNA ligase. The NER system is the most versatile of the three repair systems discussed here; it recognizes and acts on a huge number of kinds of damage that cannot otherwise be eliminated from DNA. As for mismatch repair, the importance of NER is underscored by the plight of people with mutations in one of the repair genes. Individuals with the disease *xeroderma pigmentosum* carry a mutation in one of the genes of the NER system (it can be any of seven genes). Consequently, they lack one of the dozen or so proteins needed for excision repair of skin cell damage caused by the UV radiation in sunlight. As a result, sunlight readily kills skin cells in these people and invariably causes skin cancers.

At the same time that we distinguish among the three main types of DNA repair, it is important to realize that some of the proteins involved in repair are used in more than one repair system. In addition, more and more "repair" proteins are turning out to play roles in other important cellular processes, including DNA replication, the control of the cell cycle, and the transcription of genes. In the words of the journal *Science,* which selected DNA repair enzymes as "molecule of the year" for 1994, "this new research sketches a more integrated picture of the cell, in which all processes relating to DNA are coordinated and the same molecular toolkit is used for different tasks."

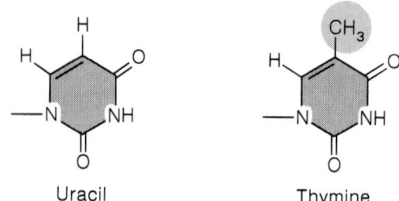

Figure 15-16   Uracil and Thymine Compared.

Why DNA Contains Thymine Instead of Uracil

Until recently, it was not clear why DNA contains thymine instead of the uracil found in RNA. Both bases pair with adenine, although thymine contains a methyl group not present on uracil (Figure 15-16). In terms of energy, it would make more sense for DNA to contain uracil because the methylation step in the synthesis of thymine nucleotides is energetically expensive. But now that we understand how deamination damage is repaired, we also understand why thymine, rather than uracil, is present in DNA.

The deamination of cytosine generates uracil (see Figure 15-14b), which is then detected and removed by uracil-DNA glycosylase, a DNA repair enzyme that recognizes only uracil. If uracil were present as a normal component of DNA, it is not clear how such uracils would be distinguished from those generated by the accidental deamination of cytosines. The methyl group on thymine is apparently the "label" that tells the glycosylase to leave it alone, thereby ensuring that deaminated cytosines can be replaced without causing other changes in the DNA molecule.

Nuclear and Cell Division

Having seen how the DNA in the nucleus is replicated and repaired, we turn now to the processes whereby the two copies of each chromosome that have been generated during the prior S phase are separated from each other and partitioned into daughter cells. These processes are mitosis and cytokinesis.

The Stages of Mitosis

Mitosis has been known and studied for a century, but only in the past 25 years has significant progress been made toward understanding the mitotic process at the molecular level. We will begin by surveying the morphological changes that occur in a cell as it undergoes mitosis; later we will examine the underlying molecular mechanisms.

Morphologically, mitosis can be described as a series of five phases, based primarily on the appearance and behavior of the chromosomes. As with any dynamic process, we must remember that the division into phases is somewhat arbitrary and that the phases are primarily a convenience for studying and describing the process.

The five phases of mitosis are *prophase, prometaphase, metaphase, anaphase,* and *telophase.* (An alternative term for prometaphase is simply "late prophase.") The micrographs and schematic diagrams of Figure 15-17 illustrate the phases as they appear in a typical animal cell. The micrographs of Figure 15-18 depict the sequence of mitotic stages as they appear during division of a plant cell. As you follow the events of each phase, keep in mind that the overall purpose of mitosis is to ensure that each of the two daughter nuclei receives precisely one copy of each duplicated chromosome. Note that we defer discussion of the molecular mechanisms of mitosis until later in the chapter.

Prophase. Toward the end of G2, the chromosomes start to condense from the extended, highly diffuse form of interphase chromatin to the dense, coiled structures characteristic of mitosis. Although the transition from interphase to mitotic prophase is not sharply defined, a cell is considered to be in **prophase** when the chromosomes have condensed to the point of being visible as threads in the light microscope. Each chromosome has duplicated during the preceding S phase and now consists of two sister chromatids (Figure 15-17a). Sister chromatids are tightly attached to each other at a constricted region, the **centromere,** which corresponds to a particular stretch of the chromosome's DNA. As the chromosomes condense, the nucleolus (or nucleoli) gradually disappears.

Meanwhile, outside the nucleus, another important organelle has sprung into action. This is the **centrosome,** an amorphous cloud of material that, in most animal cells, surrounds a pair of **centrioles.** Because they are lacking in some mitotic cells, including all plant cells and fungal cells, centrioles cannot be essential for mitosis, and their function in the centrosome remains a mystery.* However, the centriole structure—a cylinder made of microtubules—is related to the centrosome's role in the cell. The centrosome is a cellular organizing center for microtubules. During interphase, the microtubules of the cytoskeleton originate there. During prophase, the cytoskeletal microtubules disassemble, and their tubulin subunits (see Figure 4-24a) start to reassemble to form the **mitotic spindle,** the apparatus that will distribute chromosomes to the daughter cells. The centrosome, including its two centrioles, has duplicated during interphase, and in prophase the two centrosomes are seen to be moving apart from each other. Radiating from them are microtubules that are growing to form the mitotic spindle. The starburst of microtubules in the immediate vicinity of a centrosome is called an *aster.* The terms centrosome and aster were first used for animal cells but are now used more broadly.

Prometaphase. The start of **prometaphase** is marked by the fragmentation of the nuclear envelope into membranous vesicles, allowing the mitotic spindle to enter the nuclear area. Eventually the two centrosomes are at opposite poles of the cell (Figure 15-17b). On each chromosomal centromere, proteins assemble to form a protein-DNA complex called a **kinetochore;** thus there are two kinetochores on each chromosome, one on each chromatid. The two kinetochores face in opposite directions, as the diagram shows. Some of the spindle microtubules "capture" (attach to) kinetochores; others interact with microtubules coming from the other centrosome. Forces exerted within the assembly of microtubules throw the chromosomes into agitated motion and gradually move them toward the center of the cell.

Metaphase. The cell is said to be in **metaphase** when the chromosomes, which are now maximally condensed, are all aligned at the **metaphase plate,** the plane equidistant between the two poles of the mitotic spindle (Figure 15-17c). The cell seems to pause at metaphase, which can last 20 minutes out of a total mitotic phase of an hour. While all the spindle microtubules are essentially identical in composition, they are classified into three types, based on the kinds of structures with which their tips interact. The ones attached to kinetochores are called **kinetochore microtubules;** those that interact with microtubules from the opposite pole of the cell are called **polar microtubules;** and the shorter ones that form the asters at each pole are called **astral microtubules.** At least some of the astral microtubules seem to interact with proteins lining the plasma membrane. (A cluster of closely associated spindle microtubules that looks like a single unit in the light microscope is often referred to as a spindle "fiber.")

The chromosomes appear to be relatively stationary at metaphase, but this appearance is misleading. Actually, sister chromatids are already being actively tugged toward their respective poles. They appear stationary because the forces acting on them are equal in magnitude and opposite in direction. The chromatids are the prizes in a tug of war between two equally strong opponents. (We will discuss what is known about the sources of these forces shortly.)

Anaphase. Usually the shortest phase of mitosis, **anaphase** typically lasts only a few minutes. The two sister chromatids of each metaphase chromosome abruptly separate and move toward opposite poles of the cell (Figure 15-17d). All chromatid pairs start to separate at the same time, with the daughter chromosomes being pulled toward the spindle poles at a rate of about 1 μm/min. Anaphase is characterized by two different but simultaneous kinds of movements, called anaphase A and anaphase B. In **anaphase A,** the chromo- somes move toward the spindle poles as the kinetochore microtubules shorten. In **anaphase B,** the poles themselves move away from each other, as the polar microtubules lengthen.

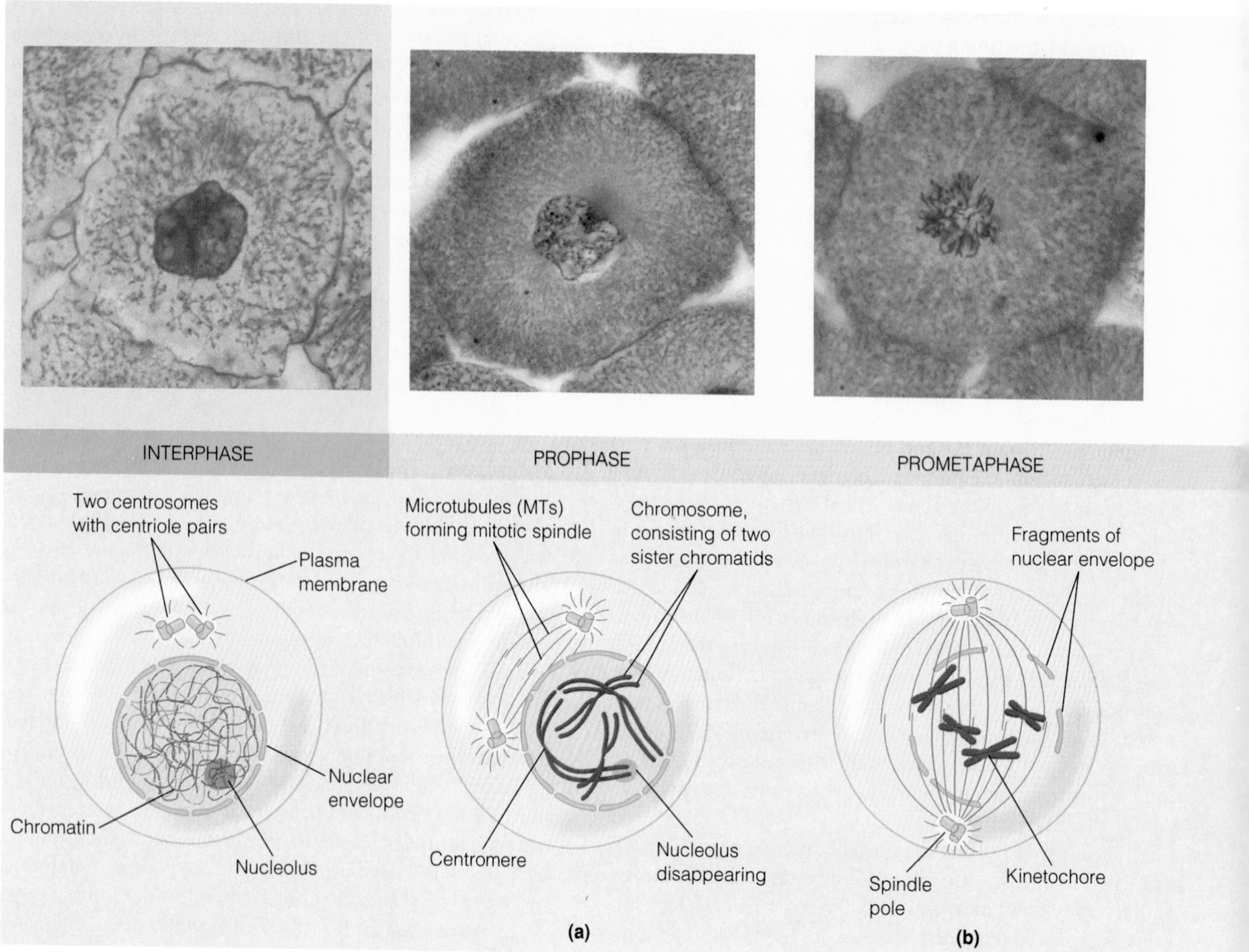

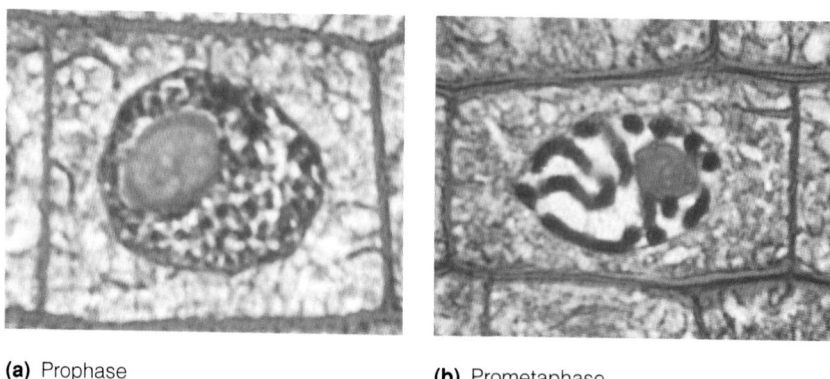

INTERPHASE

Two centrosomes with centriole pairs

Plasma membrane

Chromatin

Nuclear envelope

Nucleolus

PROPHASE

Microtubules (MTs) forming mitotic spindle

Chromosome, consisting of two sister chromatids

Centromere

Nucleolus disappearing

(a)

PROMETAPHASE

Fragments of nuclear envelope

Spindle pole

Kinetochore

(b)

Figure 15-17 The Phases of Mitosis in an Animal Cell.
The micrographs show mitosis in cells from a fish embryo (LMs). The mitotic spindle, including asters, is visible in the metaphase and anaphase
micrographs. At this low magnification (about 600X), however, we see

Figure 15-18 The Phases of Mitosis in a Plant Cell. These micrographs show mitosis in cells of an onion root (LMs).

(a) Prophase

(b) Prometaphase

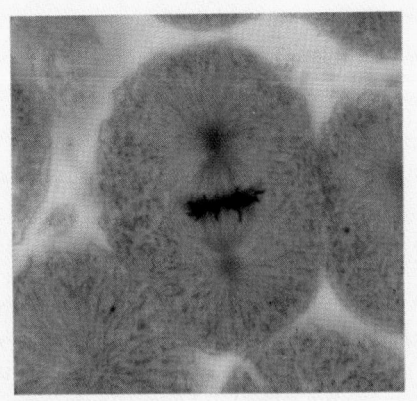
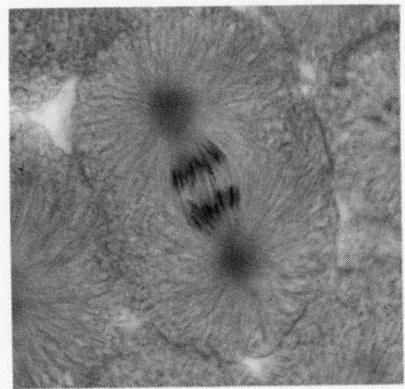
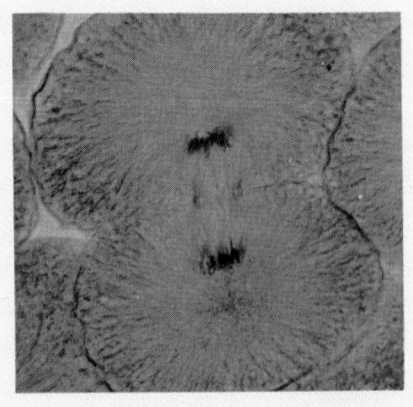

25 μm

| METAPHASE | ANAPHASE | TELOPHASE AND CYTOKINESIS |

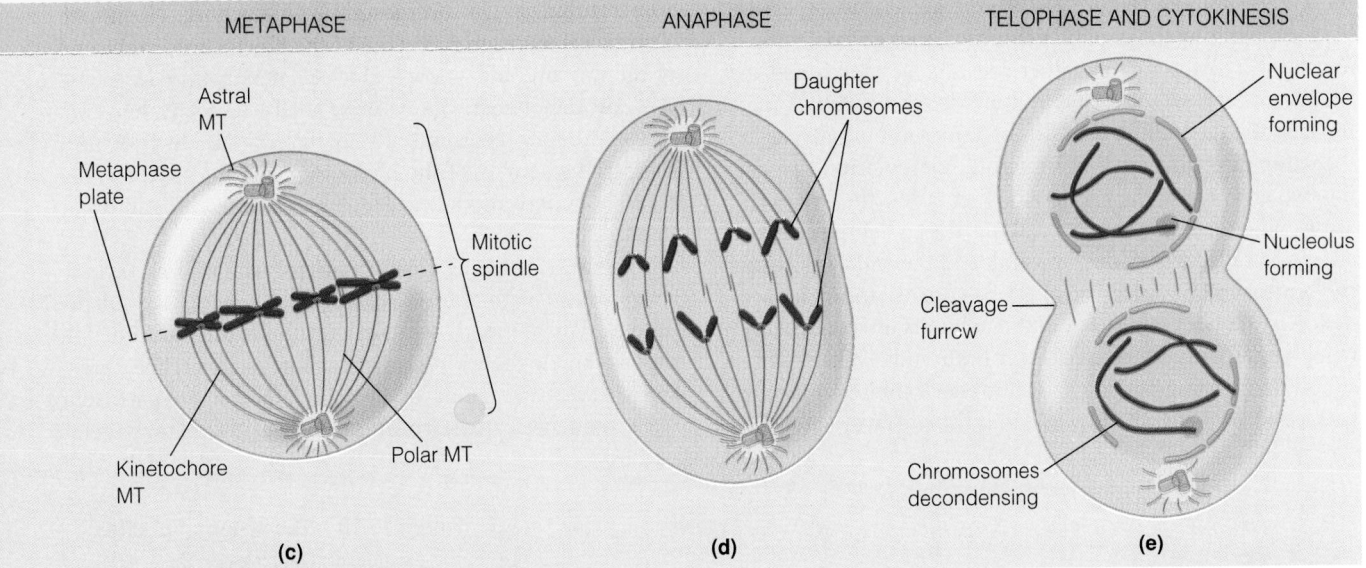

Astral MT

Metaphase plate

Mitotic spindle

Kinetochore MT

Polar MT

(c)

Daughter chromosomes

(d)

Nuclear envelope forming

Nucleolus forming

Cleavage furrow

Chromosomes decondensing

(e)

spindle "fibers," rather than individual microtubules; each fiber consists of a number of microtubules. The drawings are schematic and include details not visible in the micrographs; for simplicity, only four chromosomes are drawn. MT = microtubule.

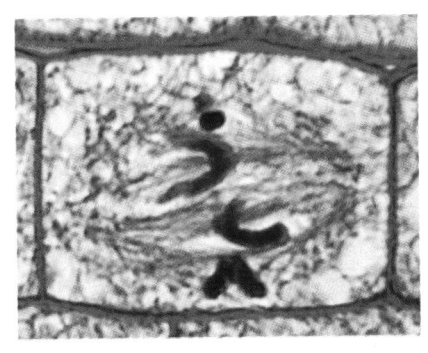

(c) Metaphase

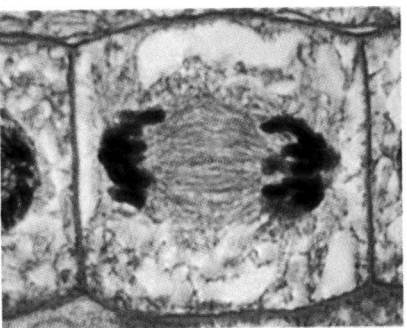

(d) Anaphase

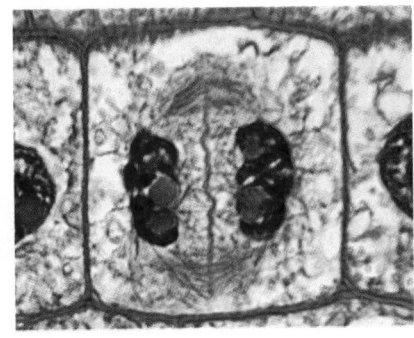

(e) Telophase

25 μm

Telophase. By the beginning of **telophase,** the daughter chromosomes have arrived at the poles of the spindle (Figure 15-17e). During telophase, the chromosomes uncoil and revert to the extended form and homogeneous appearance of interphase chromatin. Nucleoli develop at the nucleolar organizing sites on the DNA. The spindle disassembles, and nuclear envelopes form around the two groups of daughter chromosomes, completing the mitotic process. At the same time, the cell is usually undergoing cytokinesis, to produce two daughter cells.

The Mechanisms of Mitosis

Any discussion of the mechanisms of mitosis must focus on the mitotic spindle (Figure 15-19), the apparatus responsible for separating the daughter chromosomes and partitioning them into the two daughter cells. As we mentioned in Chapter 4 (and will describe in more detail in Chapter 20), the fact that the tubulin subunits of a microtubule all face in the same direction gives the microtubule polarity; that is, the two ends are slightly different (Figure 15-20). The initiating end—for a spindle microtubule, the end at the centrosome—is called the minus (−) end, and the end away from the centrosome is called the plus (+) end. Microtubules are *dynamic* structures, in that tubulin subunits are continually being added and subtracted from both ends. When more subunits are being added than subtracted, the microtubule gets longer. In general, the plus end is the site favored for the addition of tubulin subunits and the minus end for subunit removal, but, as we shall see, the removal of subunits from the plus end can be important. Growth in microtubule length comes mostly from the more rapid addition of subunits to the plus end.

Spindle Assembly and Chromosome Attachment. In late prophase, microtubule activity speeds up dramatically, and the frequency of initiation of new microtubules at the centrosome increases. This results in the prominent asters that we see around the centrosomes during mitosis. More importantly, once the nuclear envelope has disintegrated at the start of prometaphase, the increased microtubule activity increases the probablility of contact between the plus end of a microtubule and a chromosome's kinetochore. When such a contact is made, the presence of the kinetochore at the end of the microtubule—now a *kinetochore microtubule*—slows down its depolymerization. As we shall see shortly, however, it would be more accurate to say that the kinetochore now *controls* the activity of the plus end of the microtubule, because polymerization and depolymerization can still occur there.

Figure 15-21 is an electron micrograph of a metaphase chromosome with two sets of attached kinetochore microtubules. The plus ends of the microtubules are embedded in the two kinetochores. Each kinetochore is a platelike, three-layered structure made of proteins attached to particular DNA sequences at the chromosome's centromere. Kineto-

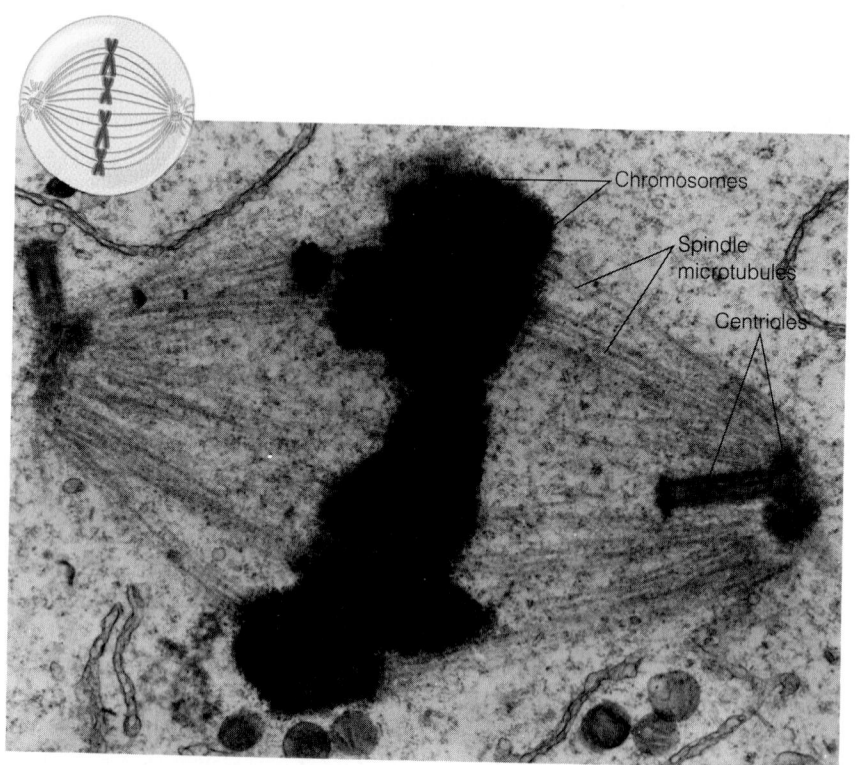

Chromosomes

Spindle microtubules

Centrioles

Figure 15-19 The Mitotic Spindle. This electron micrograph shows the mitotic spindle of a metaphase cell from a rooster. The chromosomes appear as a single mass, aligned at the metaphase phate. The centrioles at the two poles of the spindle and the spindle between the poles are clearly visible. As in Figure 15-17, we see spindle fibers rather than individual microtubules (TEM).

1 μm

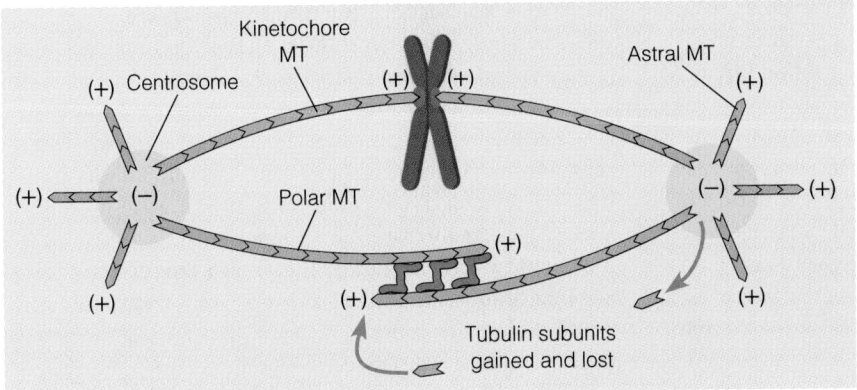

Figure 15-20 Micotubule Polarity in the Mitotic Spindle. This diagram shows only a few representatives of the many microtubules making up a spindle. The orientation of the tubulin subunits constituting a microtubule (MT) make the two ends of the MT different. The minus end is at the initiating centrosome; the plus end points away from the centrosome. MTs lengthen by adding tubulin subunits and shrink by losing subunits. In general, lengthening is due to addition at the plus ends and shortening to loss at the minus ends, but subunits can also be removed from the plus ends. The red structures between the plus ends of the polar MTs shown here represent proteins that crosslink them (see Figure 15-22a).

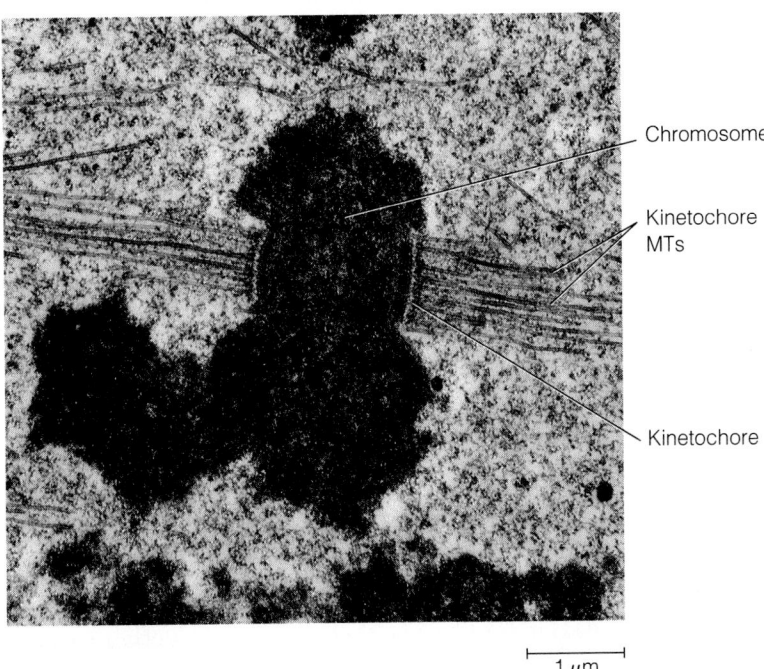

1 μm

Figure 15-21 Kinetochores and Their Microtubules. The striped structures on either side of this metaphase chromosome are its kinetochores, each associated with one of the two sister chromatids. Numerous kinetochore MTs are attached to each kinetochore. The two sets of microtubules come from opposite poles of the cell (TEM).

chores of different species vary in size. In yeast cells, for example, they are small and bind only one spindle microtubule each. In mammalian cells, they are much larger and each binds 30–40 microtubules, as shown in Figure 15-21.

Because the two kinetochores on a chromosome are back-to-back, they usually become attached to microtubules coming from opposite centrosomes. (The orientation of each chromosome is random; either kinetochore can end up facing either pole.) Opposing forces acting on the two sets of kinetochore microtubules move the chromsome to the metaphase plate.

Meanwhile, other growing spindle microtubules—the polar microtubules—are making direct contact with polar microtubules coming from the opposite centrosome. When the plus-end regions of two microtubules of opposite polar-

ity start to overlap, crosslinking proteins attach to them. Like the kinetochores on the kinetochore microtubules, this crosslinking stabilizes the polar microtubules. Thus we can picture a barrage of microtubules rapidly shooting out from each centrosome during late prophase. The ones that successfully hit a kinetochore or a microtubule of opposite polarity are stabilized; the others retreat by disassembling.

At metaphase, the chromosomes are held at the metaphase plate, but careful microscopic study of living cells reveals that they continue to make small jerking movements. Tubulin subunits continue to be added to the plus ends and subtracted from the minus ends of the spindle microtubules, although the addition and subtraction now balance. If chromatid pairs are experimentally separated, the microtubule ends at the kinetochores start to disassemble rapidly, and the chromosomes quickly move toward the poles, as in anaphase. This is evidence that the "stationary" situation at metaphase is due to a counterbalancing of opposing forces.

The natural separation of the sister chromatids at the onset of anaphase is triggered by a chemical signal, although exactly how this occurs is not yet known. One possibility is that a proteolytic enzyme involved in cell cycle regulation (a cyclin-degradation enzyme; see Figure 15-31) may cleave a protein that holds the sister chromatids together. Recent experiments suggest that kinetochores not yet attached to microtubules may provide some sort of chemical signal that prevents anaphase from starting.

In the process called anaphase A, the kinetochore microtubules rapidly disassemble at their plus ends, in addition to slow disassembly at their centrosome ends. The kinetochore microtubules thus get shorter and shorter, and the daughter chromosomes get dragged, centromere first, toward the poles. At the same time, in anaphase B, the polar microtubules are lengthening, and the cell is elongating as its poles are being pushed apart. To understand what is happening at the molecular level, we need to look at another aspect of mitosis, its *molecular motors*.

Molecular Motors. It would be appealing to conclude that formation of the mitotic spindle and the mitotic movements of chromosomes result solely from the elongation and shortening of spindle microtubules. But this is not the case. In recent years, cell biologists have discovered a number of **motor proteins** that seem to play active roles in mitosis. Like motor proteins involved in cellular motility and contractility (see Chapter 21), these proteins use energy from ATP to change shape in such a way that they exert force and can cause attached structures to move. Mitosis seems to involve at least three kinds of motor proteins, which operate at the plus ends of the three types of spindle microtubules (Figure 15-22a).

Motor proteins are now believed to play an important role at the kinetochore. Here, during anaphase, molecules of a kinetochore protein attached to the microtubule may move the kinetochore along the microtubule in the minus direction, exposing the plus end, which is progressively disassembled. In this model, the motor creates the driving force, and the shortening of the microtubules is a secondary event (Figure 15-22b).

The proteins that stabilize antiparallel pairs of polar microtubules by crosslinking them are also thought to act as molecular motors that push the overlapping microtubules in opposite directions, toward their poles of origin. This tends to push the poles of the spindle apart and may help keep both polar microtubules and kinetochore microtubules under appropriate tension in prometaphase and metaphase. The activity of these motor molecules may stabilize the spindle as a whole in anaphase. In anaphase B, this activity may be the primary force that elongates the cell; the lengthening of the polar microtubules may be secondary. (This situation is analogous to the model given above for the anaphase activity at kinetochore microtubules, although there it is the *shortening* of microtubules that is secondary.) Figure 15-22c shows electron microscopic evidence for the sliding of overlapping polar microtubules in anaphase B.

Another motor-produced force may be involved in the separation of the centrosomes during both spindle formation and anaphase B. In addition to the outward push by interacting polar microtubules, there may be an outward *pull* mediated by protein motors that attach the plus ends of astral microtubules to the *cell cortex*, a layer of actin microfilaments lining the inner surface of the plasma membrane. (The cell cortex is discussed in Chapter 20.)

The relative contributions of these pushing and pulling forces seem to differ with the species of organism. The pushing (sliding) of microtubules against adjacent ones of opposite polarity is particularly important in diatoms and yeasts. Pulling at the asters, on the other hand, is the main force operating in the cells of certain other fungi. In vertebrate cells, both mechanisms are probably operative, although astral pulling may play a greater role, especially during spindle formation.

Cytokinesis

Nuclear division and cytoplasmic division are not always linked events. Some fungi and algae, for example, undergo many rounds of nuclear division unaccompanied by cytoplasmic division, resulting in large multinucleated cells. In most cases, however, cytokinesis accompanies or closely follows mitosis, such that each of the daughter nuclei acquires its own cytoplasm and becomes a separate cell. Cytokinesis usually starts during late anaphase or telophase, as the nuclear envelope and nucleoli are re-forming and the chromosomes are decondensing.

Cytokinesis in Animal Cells. Cytokinesis occurs by quite different mechanisms in plant and animal cells. In animals, cytoplasmic division is called **cleavage** and has been studied most extensively in fertilized eggs of species such as frogs and sea urchins. Cleavage begins as a slight indentation or puckering of the cell surface, which deepens into a **cleavage furrow** that encircles the cell, as shown in Figure 15-23 (page 480) for a fertilized frog egg. The plane of the cleavage furrow is always perpendicular to the long axis of the mitotic spindle, thereby ensuring that the two sets of chromosomes will be segregated into the two daughter cells. The furrow continues to deepen until opposite surfaces make contact and the cell is split in two.

Cleavage depends on a beltlike bundle of actin microfilaments called the **contractile ring,** which forms in the cell cortex just inside the plasma membrane during early anaphase. Examination of the contractile ring with an electron microscope reveals large numbers of microfilaments oriented with their long axes parallel to the furrow. As cleavage progresses, the ring of microfilaments tightens around the cytoplasm, like a belt around the waist.

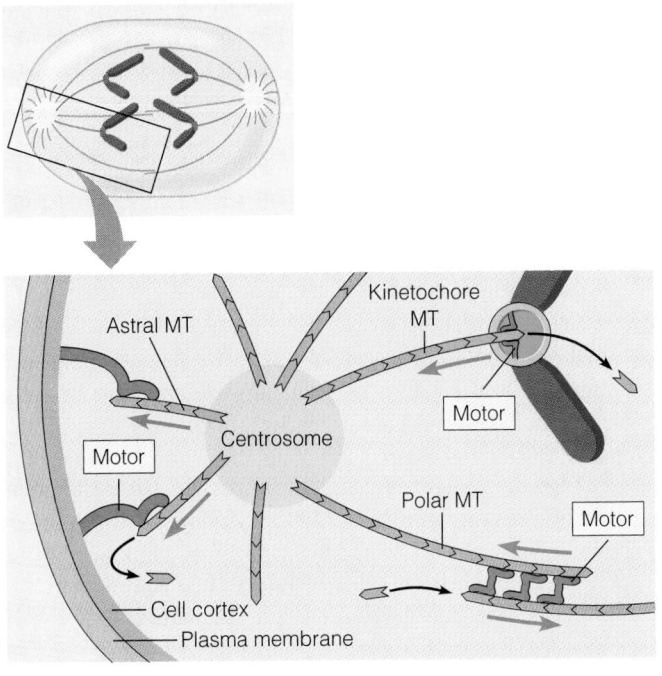

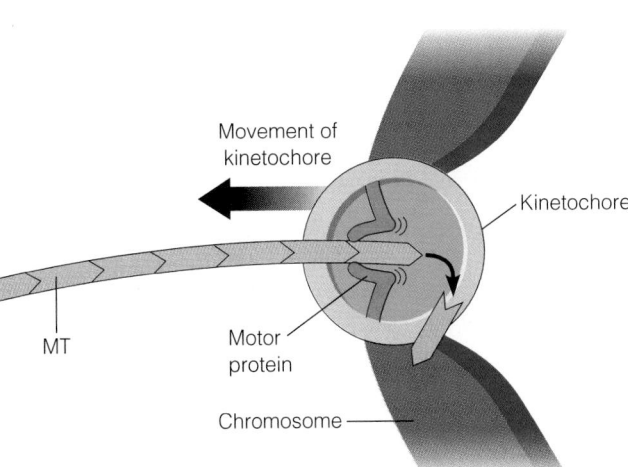

(a) Three motors

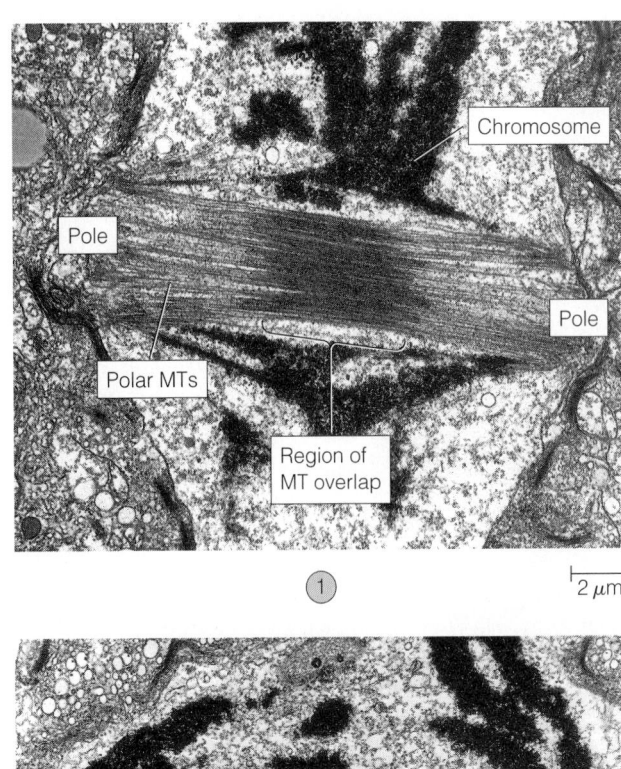

①   2 μm

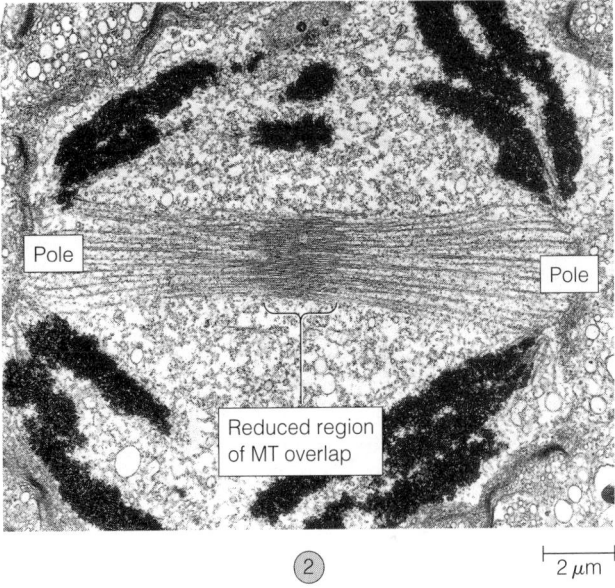

②   2 μm

(b) Closeup at kinetochore

(c) Movement of polar microtubules during anaphase

Figure 15-22 Mitotic Motors. **(a)** A model for mitotic movement based on three main types of molecular motors. These protein motors (shown in red) are thought to be associated with the plus ends of each of the three types of microtubules: kinetochore MTs, polar MTs, and astral MTs. The kinetochore motor is part of the kinetochore itself; the polar MT motor is made of proteins crosslinking the MTs; and the astral MT motor links the plus ends of astral MTs to the cell cortex. Pink arrows indicate the directions of movement generated by the motors. For both kinetochore MTs and astral MTs, the motor's operation is associated with loss of tubulin subunits at the plus ends; for polar MTs there is, in anaphase, a net gain of subunits, which leads to the lengthening of the spindle as the poles of the cell move farther apart. **(b)** A closer view of

the kinetochore motor, showing a single MT. Energized by ATP, the motor proteins move in such a way that they "walk" toward the minus end of the MT, dragging the rest of the kinetochore and its attached chromosome along with them. Simultaneously, the MT, while remaining attached to the kinetochore, loses tubulin subunits from its tip. As a result, the kinetochore MT shortens and the chromosome is pulled toward one pole of the cell. **(c)** These two electron micrographs show evidence for the motorized sliding of polar MTs during anaphase. ① During metaphase and early anaphase, the polar MTs from opposite ends of the cell overlap significantly. ② Later in anaphase, the amount of overlap has been reduced, due to the sliding of the MTs against each other (TEMs).

The force needed to tighten the contractile ring and divide the cytoplasm is thought to be generated by the interaction of the actin microfilaments with molecules of the protein *myosin*. Myosin is the protein with which actin interacts

in muscle tissue to produce the ratcheted sliding of protein filaments that accounts for muscle contraction (see Chapter 21). Unlike actin microfilaments, myosin filaments cannot be seen directly in the contractile ring, but their presence

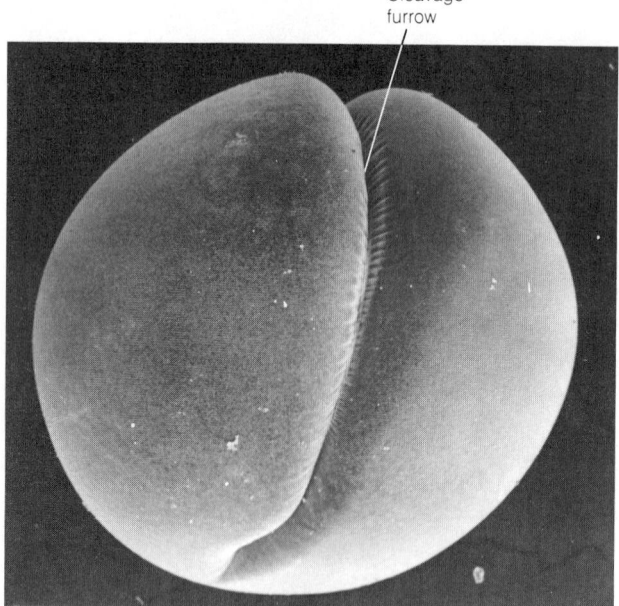

Cleavage furrow

Figure 15-23   Cytokinesis in an Animal Cell. An electron micrograph of a frog zygote (fertilized egg) caught in the act of dividing. The cleavage furrow is clearly visible as an inward constriction of the plasma membrane. Within the cell, mitosis is nearly complete, so the cleavage furrow will separate the two sets of chromosomes as it continues to constrict the membrane (SEM).

⊢──⊣
100 μm

and then reassemble in the daughter cells, thereby helping to ensure a roughly even division of cell components.

Cytokinesis in Plant Cells. In cells of higher plants, cytokinesis occurs by a different mechanism, presumably because of the presence of a cell wall. Plant cells divide not by pinching the cytoplasm in half with a contractile ring, but by assembling a plasma membrane and a cell wall between the two daughter nuclei (Figure 15-24). A cell wall in the process of formation is called a **cell plate.**

The cell plate is formed as small membranous vesicles filled with cell wall precursors align themselves along the midplane of the cell, usually during late anaphase or early telophase. These vesicles are derived from dictyosomes (the plant cell's Golgi stacks). They appear to be guided to the midplane by microtubules that are derived from the polar microtubules and that are oriented perpendicular to the developing cell plate. The parallel array of microtubules forms the **phragmoplast,** an open, cylindrical structure restricted initially to the central region of the cell. Fusion of the vesicles forms a large, flattened sac called the *early cell plate.* The contents of the vesicles assemble to form the noncellulose components of the primary cell wall, which expands outward as clusters of microtubules and vesicles form at the lateral edges of the advancing cell plate.

Eventually, contact is made with the original cell wall, and the two daughter cells are separated from each other. The new cell wall is then completed by deposition of cellulose microfibrils. The plasmodesmata that provide channels

can be deduced from studies using labeled antibodies to myosin. The filaments in the contractile ring apparently constrict progressively in a manner that involves the sliding of actin and myosin filaments; ATP provides the necessary energy.

The contractile ring of dividing animal cells provides one of the most dramatic examples of the transitory nature of actin-myosin structures in most nonmuscle cells and the rapidity with which these structures can be assembled and disassembled. Apparently, polymerization of actin occurs just before the initial indentation becomes visible, and the entire structure is dismantled again shortly after cytokinesis is complete. The actin monomers needed to assemble the microfilaments of the contractile ring are obtained by disassembly of the actin filaments of the cytoskeleton, just as the tubulin needed for spindle microtubules is derived from cytoskeletal microtubules.

The intermediate filaments are the only major components of the cytoskeleton that are not disassembled during cell division. Instead, the network of intermediate filaments elongates during mitosis and is parceled out into the two daughter cells as cleavage proceeds. All the organelles and other components of the cell are similarly segregated into the two daughter cells during division, so that daughter cells usually contain about the same numbers of components. Large organelles such as the endoplasmic reticulum and the Golgi complex fragment into small vesicles early in mitosis

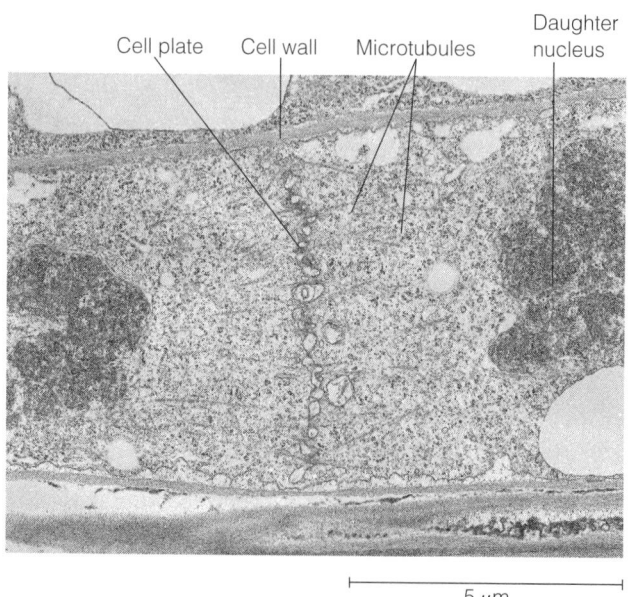

Cell plate   Cell wall   Microtubules   Daughter nucleus

⊢─────────────⊣
5 μm

Figure 15-24   Cytokinesis and Cell Plate Formation in a Plant Cell. This electron micrograph shows a plant cell at late telophase. The daughter nuclei with their sets of chromosomes are partially visible as the dark material on the far right and far left of the micrograph, and the developing cell plate is seen as a line of vesicles in the midregion of the cell. Microtubules are oriented perpendicular to the cell plate. The cell is from *Acer saccharinum,* the sugar maple (TEM).

of continuity between the cytoplasms of adjacent plant cells are also present in the cell plate and the new wall as it forms.

Variations in the Cell Cycle

As we mentioned early in this chapter, eukaryotic cells do not always proceed continuously through predictable cycles of growth and division, with G1, S, G2, and M following one another in uninterrupted progression and with every nuclear division accompanied by cytokinesis. Such is often the case, of course, particularly in growing organisms or cultured cells that have not run out of nutrients or space. But many variations are also possible, especially in terms of the relative length of time spent in various phases of the cycle and in the immediacy with which mitosis and cytokinesis are coupled.

Variations in Cell Cycle Length

Some of the most common variations in the cell cycle in vivo involve differences in generation time between different cell types. Within the same organism, some cells divide at approximately the same rate as cells in culture, but others differ greatly, depending on their role in the organism (Figure 15-25). Some cells divide rapidly and continuously

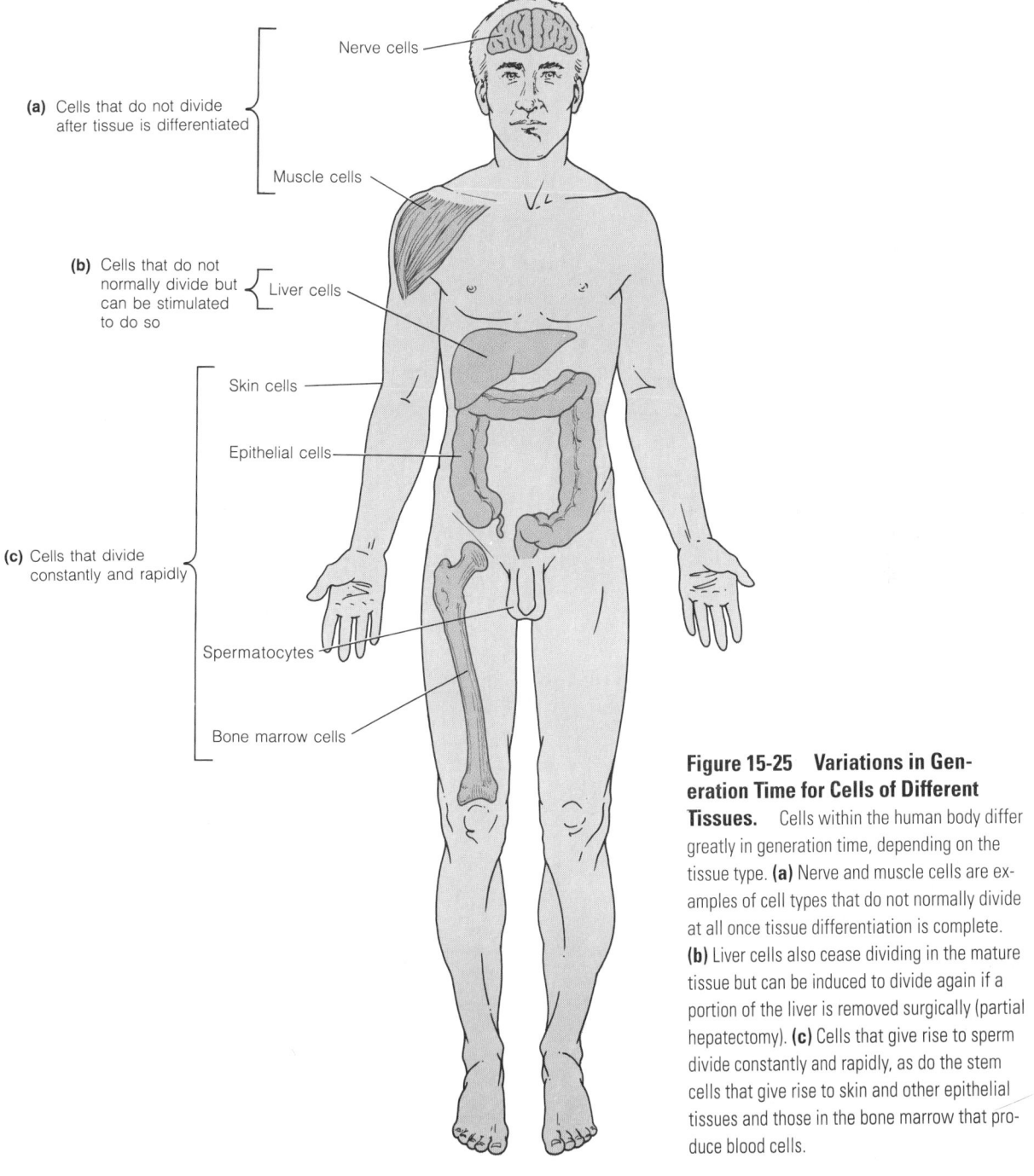

Figure 15-25 Variations in Generation Time for Cells of Different Tissues. Cells within the human body differ greatly in generation time, depending on the tissue type. **(a)** Nerve and muscle cells are examples of cell types that do not normally divide at all once tissue differentiation is complete. **(b)** Liver cells also cease dividing in the mature tissue but can be induced to divide again if a portion of the liver is removed surgically (partial hepatectomy). **(c)** Cells that give rise to sperm divide constantly and rapidly, as do the stem cells that give rise to skin and other epithelial tissues and those in the bone marrow that produce blood cells.

(a) Cells that do not divide after tissue is differentiated

(b) Cells that do not normally divide but can be stimulated to do so

(c) Cells that divide constantly and rapidly

Nerve cells
Muscle cells
Liver cells
Skin cells
Epithelial cells
Spermatocytes
Bone marrow cells

throughout the life of the organism as a means of replacing cells that are lost or destroyed during the normal functioning of the organism. Included in this category are the cells that lead to sperm formation and the precursor cells, called **stem cells,** that give rise to blood cells, skin cells, and the epithelial cells that line the inner surfaces of body organs such as the lungs and intestines. Human stem cells may have generation times as short as 8 hours.

Cells of slow-growing tissues, on the other hand, may have generation times of several days or more, and some cells, such as those of nerve or muscle tissue, do not divide at all. Still other cell types do not divide under normal conditions but can be induced to begin dividing again by an appropriate stimulus. Liver cells are in this category; they do not normally proliferate in the mature liver but can be induced to do so if a portion of the liver is removed surgically. Lymphocytes (white blood cells) are another example; when exposed to a foreign protein, they begin dividing as part of the immune response, as we will see in Chapter 24.

Most of these variations in generation time involve differences in G1, although S and G2 can also vary somewhat. Cells that divide very slowly can spend days, months, or even years in the offshoot of G1 called G0, whereas cells that divide very rapidly have almost no G1 phase at all. In fact,

some cells even begin DNA synthesis before mitosis is complete, eliminating G1 entirely.

The embryos of insects, amphibians, and certain other nonmammalian animals are dramatic examples of very short cell cycles, with no G1 phase and a very short S phase. During early embryonic development in amphibians such as the frog *Xenopus laevis,* for instance, cell division can take less than 30 minutes, even though the normal length of the cell cycle in adult tissues is about 20 hours. Under these conditions, the S phase is completed in less than 3 minutes, at least 100 times faster than in adult tissues. The incredible rate of DNA synthesis needed to sustain such a rapid cell cycle is possible because virtually all replicons are active at the same time, in contrast to the sequential activation seen in adult tissues. In addition, the average replicon length decreases, because new replicons are induced at this time.

Furthermore, these embryonic cells have little or no need to synthesize cellular components other than DNA because the fertilized egg is a very large cell with enough cytoplasm to sustain many rounds of cell division. Each round of division subdivides the initial cytoplasm into smaller cells, until the cell size characteristic of adult tissues is reached (Figure 15-26). During the early cleavage divisions of *Xenopus* embryos, for example, not only is G1 lacking but G2 is

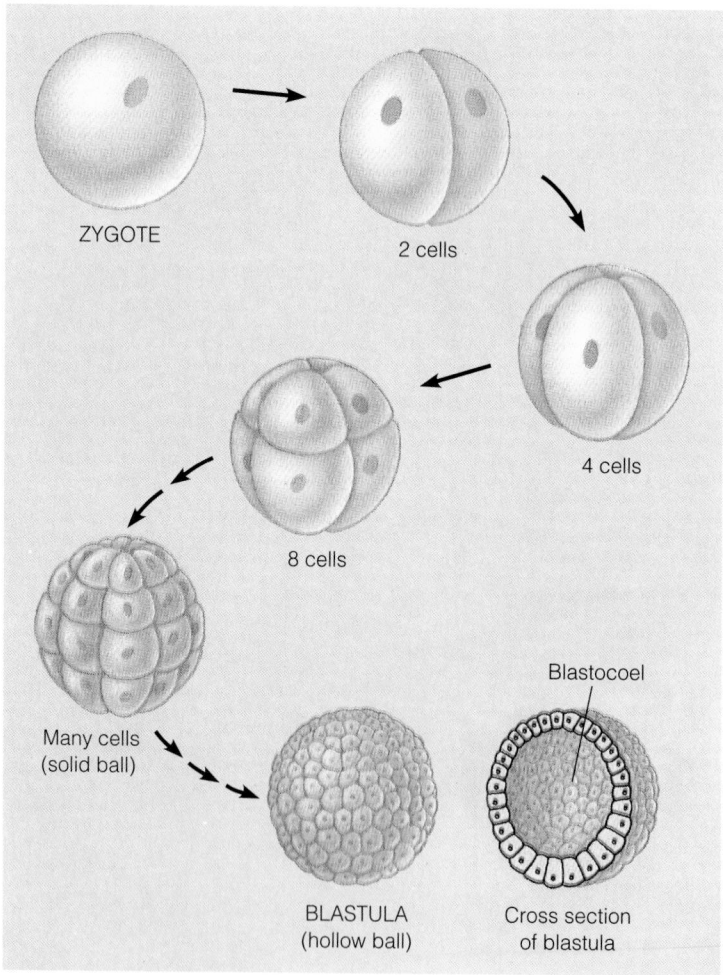

Figure 15-26 Cleavage of a Fertilized Egg into Progressively Smaller Cells. Amphibian eggs are very large, with enough cytoplasm to sustain many rounds of cell division after fertilization. Each round of division during early development parcels the cytoplasm into smaller cells. Eventually, a hollow ball of small cells called the blastula is produced, with a cavity called the blastocoel.

unusually short, so that cells go almost directly from DNA synthesis to mitosis and back to DNA synthesis. In fact, the S phase in such cells begins even before mitosis is complete. From such examples, we know that cell growth during the G1 and G2 phases is not an absolute prerequisite for cell division, even though growth and division are usually coupled processes.

Variations in Timing of Mitosis and Cytokinesis

For some multinucleate cells, such as the fungal and algal cells already mentioned and the skeletal muscle cells of vertebrates, the multinucleate condition is permanent. In other situations, however, the multinucleate state is only a temporary phase in the organism's development. This is the case, for example, in the development of a plant seed tissue called endosperm in the cereal grains. Here nuclear division occurs for a time unaccompanied by cytokinesis, generating many nuclei in a common cytoplasm. Successive rounds of cytokinesis then occur without mitosis, walling off the many nuclei into separate endosperm cells. A similar process occurs in developing insect eggs. The fertilized egg undergoes mitosis but not cytokinesis and soon consists of hundreds of nuclei in the same cytoplasm; later, cytokinesis catches up.

Regulation of the Cell Cycle

The variability in generation time for cells of the same organism tells us that the cell cycle must somehow be regulated. The molecular basis of this regulation is a subject of intense interest, not only for understanding the life cycles of normal cells but also for understanding how cancer cells manage to escape normal control mechanisms. Now one of the hottest areas of biological research, cell cycle regulation is beginning to reveal its underlying molecular mechanisms. We will begin our discussion with a look at the general concept of cell-cycle checkpoints and some of the early experimental evidence for the nature of their control.

Cell Cycle Checkpoints

Evidence acquired decades ago pointed to a particular point in G1 as critical for regulation of the mammalian cell cycle. We have already seen that G1 is the phase that varies most among cell types. Moreover, mammalian cells that have stopped dividing are almost always arrested during the G1 phase. For example, we can stop or slow down the process of cell division in cultured cells by allowing the cells to run out of either nutrients or space or by adding inhibitors of vital processes such as protein synthesis. In all such cases, the cells are arrested in G1.

These findings suggest that when a cell leaves G1 and enters the S phase, it is committed to completing the cycle. Therefore, the release of cells from G1 appears to be a critical control mechanism. More specifically, early researchers identified a point of no return in late G1, which they called

the *restriction point* (Figure 15-27). Cells that have passed this point are committed to division, whereas those that have not passed this point can remain in G1 indefinitely, in the resting state called the G0 state. As the previously mentioned experiments demonstrated, the ability to pass the restriction point can be heavily influenced by factors in the cell's environment.

Later, research with other types of cells revealed two other points of no return, and all three are now generally termed **cell cycle checkpoints.** At the end of G2 is a major checkpoint that controls the cell's entry into mitosis (M phase). At the G2 checkpoint, certain kinds of cells can enter a resting state analogous to G0. Within M phase, at metaphase, is a third checkpoint, which somehow determines whether all the chromosomes are properly attached to the spindle before allowing anaphase to begin. The relative importance of the G1 and G2 checkpoints varies with the organism and cell type. For example, the G1 checkpoint is the more important checkpoint in the budding yeast *Saccharomyces cerevisiae* (where it is called "Start"), as it is in most cells of multicellular organisms. However, the G2 checkpoint is the more important one in, for example, the mitotic divisions of a fertilized frog egg and in the yeast *Schizosaccharomyces pombe* (called a *fission yeast* because it reproduces by dividing evenly in two, rather than by budding). A cell's behavior at a checkpoint is influenced both by preceding events in the cell cycle (such as DNA replication) and by factors in the cell's environment (such as nutrients or

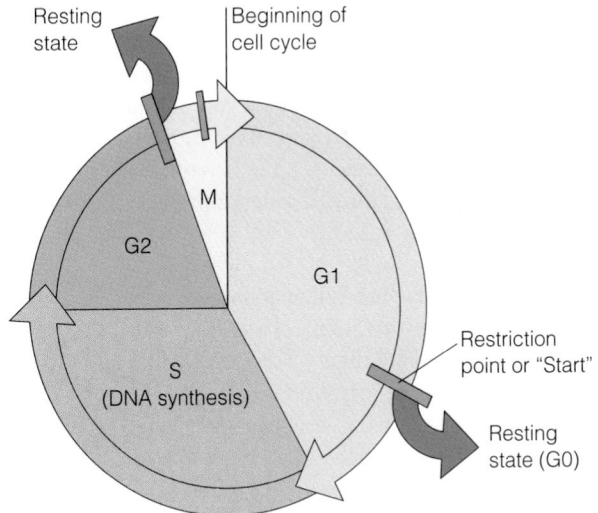

Figure 15-27 Cell-Cycle Checkpoints. The red rectangles mark three checkpoints in the eukaryotic cell cycle, points where the cell "decides" whether or not to proceed through the rest of the cycle. The decision is based on chemical signals reflecting both the cell's internal state and its external environment. The two main checkpoints that have been studied are near the end of G1 (also called the restriction point in mammalian cells and "Start" in yeast cells) and at the end of G2. If conditions are not satisfactory at these checkpoints, the cell exits the cycle and goes into a resting state (called G0 for the G1 checkpoint). There is also a checkpoint during the M phase, at metaphase.

hormones). Whatever the influence, its effects are mediated by cellular proteins, activating or inhibiting one another in chains of interactions that can be quite elaborate. As we will see later, however, there is an underlying unity in the molecular strategies of cell cycle regulation.

Early Evidence for Chemical Regulation of the Cell Cycle

As in many areas of scientific inquiry, early attempts to determine how the cell cycle is regulated were hampered by the difficulty of distinguishing between causal relationships and simple correlations. Just because an event usually happens at a specific point in the cell cycle does not necessarily mean it is involved in regulation of the cycle. For example, it was long thought that the critical regulatory factor was simply the ratio of cytoplasmic mass to nuclear mass and that when the cytoplasm reached a certain size, the cell would divide. It is certainly true that cell division is usually correlated with an increase in cytoplasmic mass, but there is no evidence to suggest a necessary causal relationship. Indeed, the cleavage of a fertilized egg into many smaller cells without accompanying cell growth seems to contradict such a suggestion. Moreover, there appears to be no validity to the suggestion that the transition from G1 to S is controlled by the availability of the DNA replication enzymes.

Between 1970 and 1975, it became clear that specific chemical signals present in the cytoplasm were responsible for moving the cell cycle past the G1 and G2 checkpoints—that is, for triggering DNA replication (S phase) and mitosis (M phase). Some of the first strong evidence for this came from experiments in which two cultured mammalian cells in different phases of the cell cycle were fused to form a single cell with two nuclei, a **heterokaryon.** As Figure 15-28a indicates, if one of the original cells is in S phase and the other is in G1, the G1 nucleus in the heterokaryon immediately enters S phase, as though a signal present in the cyto-plasm of the first cell triggers the S phase events. Similarly, if a cell undergoing mitosis is fused with another cell in any stage of its cell cycle, even G1, the second nucleus is immediately driven into the preparatory steps for mitosis, including condensation of dispersed interphase chromatin into visible chromosomes, spindle formation, and fragmentation of the nuclear envelope. If the second cell was in G1, the condensed chromosomes will be unduplicated (Figure 15-28b).

More direct evidence for a mitosis-inducing chemical signal came from experiments with frog eggs. In the frog, the oocyte, an egg cell precursor, is arrested in G2 until hormones stimulate meiosis. (Meiosis is the variation of mitosis that halves the number of chromosomes in egg or sperm production.) The oocyte proceeds through most of the phases of meiosis but is arrested in M phase—in metaphase of the second of two meiotic divisions. It is now a "mature" egg cell, capable of being fertilized. Because frog oocytes and eggs are very large, about 1 mm in diameter, it is easy to transfer cytoplasm between them with a fine pipette. In a crucial experiment, it was shown that if cytoplasm taken from a mature egg cell is injected into the cytoplasm of an oocyte, the oocyte immediately begins meiosis (Figure 15-29). The hypothetical cytoplasmic chemical that induces this oocyte "maturation" was dubbed **maturation-promoting factor (MPF).** It was quickly established that MPF also induces mitosis of fertilized frog eggs (cleavage).

MPF-like activities have since been found in the cytoplasms of a broad range of eukaryotes, including yeasts, marine invertebrates, and mammals. Furthermore, the mitosis-inducing factors have proven to be very similar in all these organisms. For example, in yeast cells with a defective or missing MPF gene, the human version of the gene can substitute perfectly well, despite the fact that the last ancestor common to yeasts and humans probably lived about 3 billion years ago! Through these and other kinds of experiments, investigators learned a lot about the MPF activity, even before the MPF protein was purified in 1988.

Figure 15-28 Cell Fusion Evidence for the Role of Cytoplasmic Chemical Signals in Cell Cycle Regulation. Important information can be obtained from experiments in which cells at two different points in the cell cycle are induced to fuse, forming a single cell with two nuclei, a heterokaryon. Cell fusion can be brought about by any of several methods, including the addition of certain viruses or polyethylene glycol, or the application of a brief electrical pulse, which causes plasma membranes to destabilize momentarily (electroporation). **(a)** If cells in S phase and G1 phase are fused, DNA synthesis begins in the original G1 nucleus, suggesting that a substance that activates S phase is present in the S phase cell. **(b)** If a cell in M phase is fused with one in any other phase, the latter cell immediately enters mitosis. If the cell was in G1, the condensed chromosomes that appear have *single* chromatids.

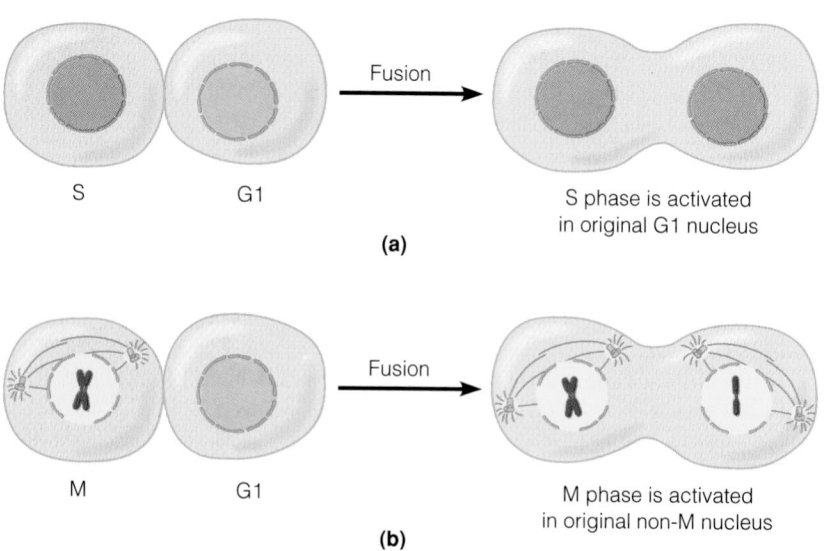

S   G1   Fusion   S phase is activated in original G1 nucleus

(a)

M   G1   Fusion   M phase is activated in original non-M nucleus

(b)

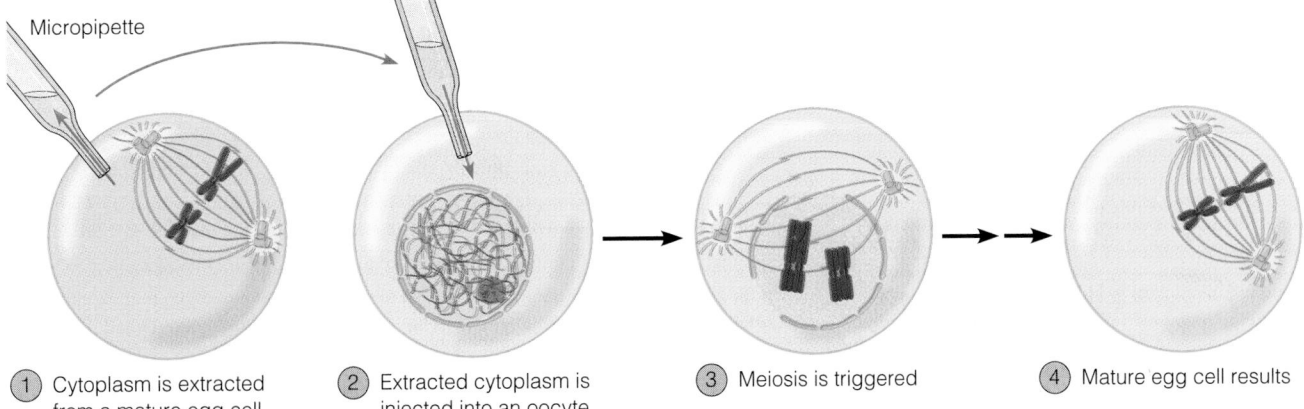

(1) Cytoplasm is extracted from a mature egg cell

(2) Extracted cytoplasm is injected into an oocyte

(3) Meiosis is triggered

(4) Mature egg cell results

Figure 15-29   Evidence for the Existence of MPF. In nature, hormones act on frog oocytes to trigger entry into meiosis and their development into mature frog eggs, which are arrested (until fertilization) in metaphase of the second meiotic division. The experiment shown here, performed by Y. Masui and C. L. Markert in 1971, established the existence of a cytoplasmic substance that induces this process; they called it maturation-promoting factor (MPF). (1) In their experiment, they used a micropipette to remove cytoplasm from a mature egg cell, arrested in metaphase of the second meiotic division, and (2) inject it into an oocyte arrested in G2 of interphase. (3) The oocyte then entered meiosis and (4) became a mature egg cell. This experimental procedure could now be used as an assay for the detection and eventual isolation of MPF. The hormones that trigger oocyte maturation in the frog were presumed to act by stimulating the synthesis or activation of MPF.

The Molecular Basis of Cell Cycle Regulation

The study of cell cycle regulation entered a molecular era in 1988. This new era was brought about by the merging of results from two main lines of research, the physiological/biochemical study of developing frog eggs and the genetic study of yeasts. Their status as single-celled microbes makes yeasts particularly useful model organisms for studying many aspects of eukaryotic cell biology. Intensive research on the genetics of yeast cell cycles had begun in the late 1960s, just a few years before MPF was discovered.

Working with *S. cerevisiae*, geneticist Leland Hartwell undertook a search for mutants that were "stuck" at some point in the cell cycle. Most such mutants would be difficult or impossible to work with, because their blocked cell cycle would prevent them from reproducing. But Hartwell was able to use a powerful strategy of microbial genetics, focusing his search on **conditional mutants.** These are mutants whose defect is apparent only under certain conditions—in this case, at temperatures above the normal range for the organism. A yeast cell with such a **temperature-sensitive mutation** in a gene required for cell cycle operation reproduces normally at 20–23°C but poorly or not at all at 35–37°C. The mutant can thus be grown at the lower ("permissive") temperature for genetic and biochemical study. How can the mutant behave normally under permissive conditions? Presumably the protein encoded by the mutated gene is close enough to the normal gene product to function at the lower temperature, while the increased thermal energy at higher temperatures disrupts its active conformation (the molecular shape needed for function) more readily than that of the normal protein.

In this way, Hartwell and his colleagues identified many genes involved in the cell cycle of *S. cerevisiae* and established the points in the cell cycle at which their products functioned. Predictably, some of these genes turned out to encode DNA replication proteins, but others seemed to function in cell cycle regulation. A breakthrough discovery was made by Paul Nurse and his colleagues, who carried out similar research with the fission yeast *Schizosaccharomyces pombe.* They identified a gene called *cdc2* whose activity was essential for the initiation of mitosis—that is, for passing the G2 checkpoint. The acronym *cdc* stands for cell division cycle. (Some guidelines on gene names are given in Box 15B.) The *cdc2* gene turned out to be essentially identical to a *S. cerevisiae* gene that Hartwell's group had called *CDC28* and to have counterparts in all eukaryotic cells. (It was Nurse who showed that the human version of the gene could "rescue" mutant yeast cells.) In tribute to the importance of Nurse's discovery, the protein encoded by such a gene is often called a **Cdc2 protein,** regardless of the organism where it is found.

This yeast research came together with the frog egg research when it was established that the Cdc2 protein was one of two proteins making up MPF. Researchers were now primed to unravel the mysteries of the G2 checkpoint.

The Cdc2 Protein, a Protein Kinase. The Cdc2 protein is a **protein kinase,** an enzyme that catalyzes the transfer of a phosphate group from ATP to certain other proteins. Phosphorylation by ATP is a major theme in cell biochemistry and is the usual mechanism by which ATP functions to activate molecules, both large and small. In earlier chapters, you have seen that phosphorylation of glucose activates it for glycolysis (see Figure 11-8) and that phosphorylation and dephosphorylation of the protein of the sodium-potassium pump causes the shape change that allows Na^+ and K^+ to

A FEW NOTES ABOUT GENE NAMES

Somewhat confusingly, the systems of nomenclature and typography used for genes are not entirely standardized, but differ with the organism. The names of genes are almost always italicized, but there is variation in the use of uppercase versus lowercase letters and the use of numbers versus letters. Here we describe how genes are written for the organisms mentioned in this chapter.

Yeast Genes

For fission yeasts, such as *Schizosaccharomyces pombe,* the gene name is written in lowercase, and numbers are used to distinguish between genes involved in the same area of cellular function. For example, *cdc1* and *cdc2* are two different genes involved in the cell division cycle. As illustrated by the "cdc" in this example, acronyms or abbreviations of terms relating to the gene's function are usually the basis for the gene name. For budding yeasts, such as *Saccharomyces cerevisiae,* gene names are similar to those for fission yeasts, but they are written with uppercase letters. For example, *CDC28* is the name of the *S. cerevisiae* gene that corresponds to *S. pombe*'s *cdc2*. When referring to the protein product encoded by a yeast gene, the usual designation is the same as the gene name, but with only the first letter capitalized and without italics; thus the protein encoded by *cdc2* is called Cdc2.

Bacterial Genes

For bacterial genes, lowercase letters are used for the part of the gene name indicating the area of gene function and uppercase letters to identify the particular gene. For example, *dnaA* and *dnaB* are two genes involved in *E. coli* DNA replication. The protein encoded by the *dnaA* gene is referred to as the DnaA protein, as with yeast.

Human Genes

For "higher" organisms such as animals and plants, a variety of conventions are employed. Human genes generally follow the convention for budding yeasts, although the length of the gene name is more variable. For example, *BRCA2* is one of two recently identified genes associated with susceptibility to breast cancer. ■

cross the plasma membrane (see Figure 8-12). The sodium-potassium pump protein has its own innate ability to hydrolyze ATP. However, in many other cases the phosphorylation of a protein is catalyzed by a separate protein—that is, a protein kinase. The phosphorylation of proteins by kinases, and their dephosphorylation by enzymes called *phosphatases,* is turning out to be a common cellular mechanism for regulating protein activity. And it is a mechanism that is used many times over in regulating the cell cycle.

The Roles of Cyclins. When one looks for the Cdc2 protein at different points in the cell cycle, it is found to be present continuously at about the same concentration. However, it is not continuously active as a trigger for mitosis (or meiosis). Its MPF activity rises rapidly during G2 phase, peaks during the first half of M phase, and then suddenly drops (Figure 15-30). MPF's activity is regulated by the second component of MPF, a protein called a **cyclin.** As the name suggests, cyclins are a class of proteins whose level in the cell oscillates; they are found in all eukaryotic cells. Not coincidentally, the cyclin level in a cell correlates with MPF activity level. MPF activity starts its climb when the cyclin level reaches a critical threshold.

The oscillation of cyclin level is unusual. Most other major cell proteins, like Cdc2, exist at a relatively constant concentration during the cell cycle because synthesis and any degradation occur at a constant rate as the cell grows. Unlike these proteins, cyclin is synthesized at a rate that allows cyclin accumulation to outpace the cell's growth rate—until M phase, when cyclin degradation markedly increases and destroys most of the cyclin present.

The Cdc2-cyclin complex plays a multifaceted role in stimulating mitosis—it has all the activities earlier attributed to active MPF. In fact, in the scientific literature, "MPF" or "active MPF" continues to be used to mean the Cdc2-

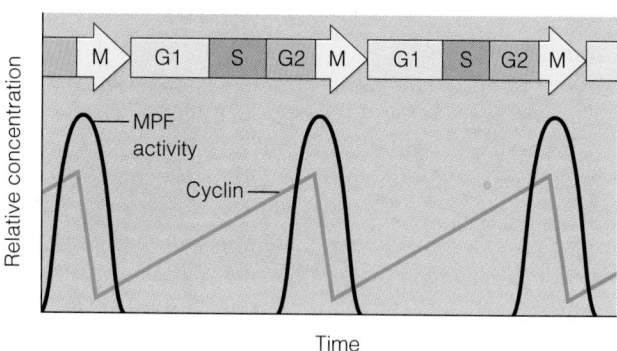

Figure 15-30 Fluctuating Levels of Cyclin and MPF Activity During the Cell Cycle. Cellular levels of cyclin rise during interphase (G1, S, and G2), then fall abruptly during M phase. The peaks of MPF activity (assayed by testing for its ability to stimulate mitosis) and cyclin concentration correspond, although the rise in MPF activity is not significant until a threshold concentration of cyclin is reached. Active MPF has been found to consist of a combination of cyclin protein and the Cdc2 protein. Cdc2 itself is present at a constant concentration (not shown on the graph), because the amount of Cdc2 protein increases at a rate corresponding to the overall growth of the cell.

cyclin complex that stimulates mitosis. As summarized by items 1–3 at the top of Figure 15-31, MPF (the Cdc2-cyclin complex) induces chromosome condensation, the assembly of the mitotic spindle, and the breakdown of the nuclear envelope. Only in the last of these three cases is much known about the molecular mechanism involved: The Cdc2 phosphorylates (and stimulates other kinases to phosphorylate) the *lamin* proteins of the *nuclear lamina,* to which the inner nuclear membrane is attached (see Figure 14-24); phosphorylation causes the lamins to dissociate from each other, and pieces of the nuclear envelope follow suit. By the time mitosis is well under way, yet another activity of the Cdc2-cyclin complex becomes important: It activates proteolytic enzymes that cause its own demise by degrading cyclin, including both cyclin bound to Cdc2 and free cyclin. The Cdc2 protein is recycled.

What about the G1 checkpoint, the restriction point identified decades ago in mammalian cells? In yeast, in which the G1 checkpoint is called Start, the go-ahead signal is also given by a Cdc2-cyclin complex, although the cyclin is a different one. In cells of vertebrates, including both frogs and humans, there is a whole family of different cyclins and also a family of proteins more or less similar to the Cdc2 protein. (Recently, a second Cdc2-like protein has also been found to be involved in the yeast cell cycle.) The generic term for a member of the Cdc2 protein family is **cyclin-dependent protein kinase (Cdk).** The various types of Cdk and cyclins act in different combinations at different stages of the animal cell cycle. The details are still being determined, but current evidence supports the involvement of the animal cell's Cdc2 along with cyclins B and A at the G2 checkpoint, and Cdk proteins called Cdk2, Cdk4, and Cdk5, along with cyclins E and D (several kinds), at the G1 checkpoint. Cdk2-cyclin A seems to be important during S phase. The different cyclins are made during different phases of the cell cycle.

You may be wondering about the third checkpoint mentioned earlier, the M-phase checkpoint where the decision is made whether or not to separate the metaphase chromatids and initiate anaphase. Here neither a new cyclin nor a new Cdk seems to be involved. Instead, the onset of anaphase appears to be triggered by proteolytic enzymes activated by the Cdc2-cyclin complex. However, cyclin breakdown (and the concomitant inactivation of the Cdc2-cyclin complex itself) is, surprisingly, not the anaphase-triggering event. In an experiment using an in vitro system based on frog egg extracts, a nondegradable form of cyclin B was added, creating a nondegradable, continuously active Cdc2-cyclin complex. Although this complex prevented mitosis from proceeding to completion, sister chromatid separation did occur. This result suggests that the proteolytic enzymes that normally attack cyclin must also attack other key pro-

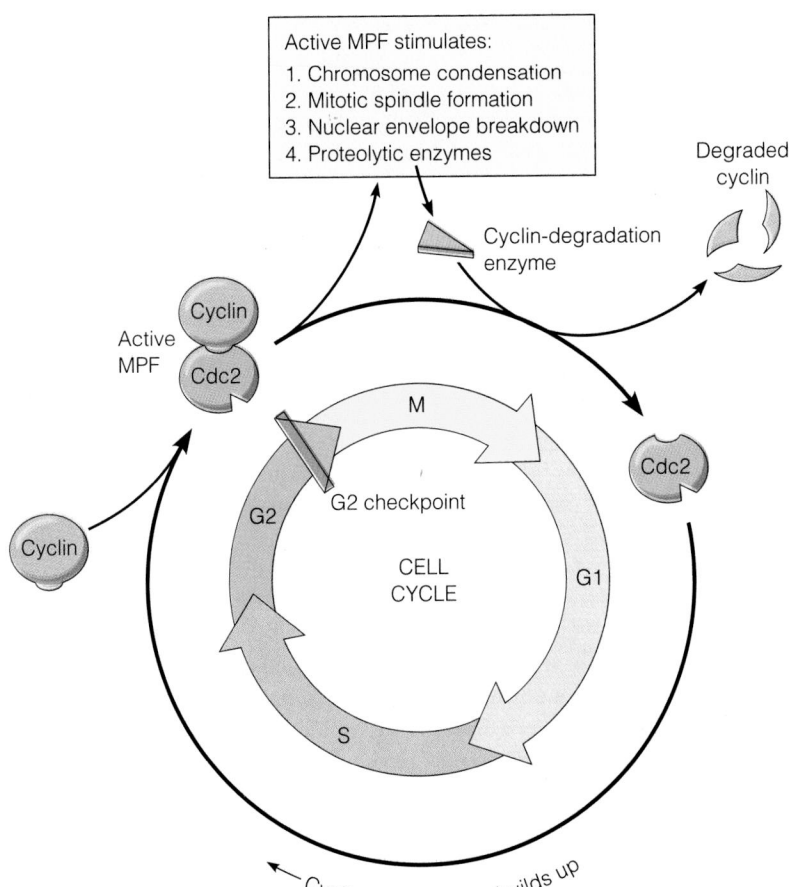

Figure 15-31   The Cdc2 Activity Cycle.
This diagram illustrates the activation and inactivation of the Cdc2 protein during the cell cycle. In G1, S, and G2, Cdc2 is made at a steady rate as the cell grows, while the cyclin concentration gradually increases. Cdc2 and cyclin form a complex, "active MPF," that drives the cell cycle past the G2 checkpoint and stimulates the mitotic events listed. By activating a proteolytic enzyme that degrades cyclin, the Cdc2-cyclin complex brings about its own demise, allowing the completion of mitosis and entry into G1 of the next cell cycle. Cdc2, sometimes called "inactive MPF," is recycled.

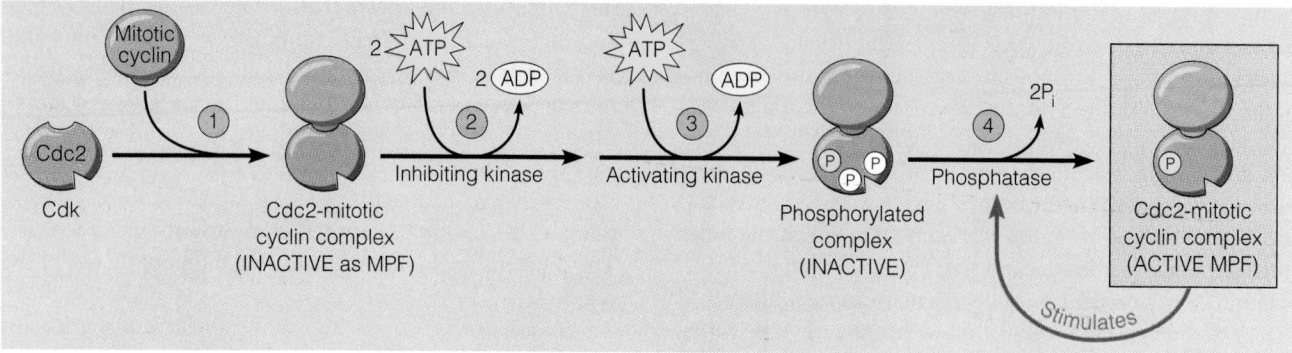

Figure 15-32 Phosphorylation and Dephosphorylation in the Activation of a Cdk-Cyclin Complex. The series of reactions shown here for the formation of active MPF (Cdc2-mitotic cyclin) was worked out using a combination of data from yeasts and frog eggs. ① The Cdc2 and mitotic cyclin proteins form an inactive complex. ② An inhibiting kinase adds two phosphate groups (white) to the com- plex, which block its active site. ③ An activating kinase phosphorylates a third site on the complex (yellow phosphate). ④ A phosphatase removes the inhibiting phosphate groups, converting the complex to a singly phos- phorylated form, which is active as MPF. This active MPF in turn stimu- lates the phosphatase to produce additional active MPF. The result is a burst of MPF activity.

teins, perhaps including proteins required for holding sister chromatids together.

Regulation of Cdk-Cyclin Complexes by Other Kinases. Unfortunately for students of this subject, there are addi- tional levels of complexity in the regulation of the cell cycle—making up, along with Cdk proteins and cyclins, the chains of activating and inactivating proteins we mentioned earlier. Fortunately, the reactions catalyzed by these proteins have a common theme: phosphorylation and dephosphory- lation. To put it another way, most of these other proteins are protein kinases and phosphatases.

Figure 15-32 indicates the main proteins and the four main reactions involved in the formation of active MPF dur- ing G2, starting with the Cdc2 protein. This scheme is prob- ably similar to what happens at the G1 checkpoint, also. The initial complex formed by the joining of the Cdc2 protein and the MPF (mitotic) cyclin is inactive in the cell; to trigger mitosis, the complex requires the addition of a phosphate group on a particular amino acid of Cdc2 (Thr-161). In the figure, this phosphate is highlighted with yellow. It is added by a specific kinase, which the figure calls "activating kinase." But before that enzyme acts, another *inhibiting* kinase phos- phorylates the protein in two other places (Thr-14 and Tyr-15), such that its active site is blocked. So the last step in the activation sequence is actually the removal of the inhibiting phosphates by a phosphatase enzyme. The extra phosphory- lation and dephosphorylation steps provide other points in the pathway where the process is subject to control by other factors. In addition, a positive feedback loop is involved: The active form of Cdc2-cyclin activates more and more phosphatase.

At either checkpoint, the Cdk protein and/or the cyclin protein may need to be further modified by additional se- quences of reactions before the final Cdk-cyclin complex is fully active. The details of these processes may vary with the organism and, in a multicellular organism, with the cell type. Furthermore, various environmental influences, such as nutrients and hormones, may help determine which cy- clins accumulate and at what rate. Most cells have many lay- ers of cell cycle control.

Putting It All Together: The Cell Cycle Regulation Ma- chine. Figure 15-33 is a generalized and simplified sum- mary of the operation of the molecular machine that regu- lates the eukaryotic cell cycle, as currently understood. Although much of what we know to date comes from re- search on the G2 checkpoint in frogs and yeasts, most cell cycle decisions are probably controlled in a similar way, with the key molecules being protein kinases and cyclins.

The cell cycle machine can be described in terms of two fundamental, interacting mechanisms. One mechanism is an autonomous clock, which on its own goes through a fixed cycle over and over again. The molecular basis of this clock is the synthesis and degradation of cyclins, which occur in a rhythmic fashion. The other mechanism adjusts the clock as needed, by providing feedback from the cell's internal and external environments. This mechanism makes use of cyclin- dependent kinases and additional proteins that, directly or indirectly, interact with cyclins. Many of the additional proteins are themselves protein kinases or phosphatases. It is this part of the cell cycle machine that transmits informa- tion about the state of the cell's metabolism—including DNA replication—and about conditions outside the cell. Energy required to activate the machine is supplied by ATP.

Even though the Cdk-cyclin core of the cell cycle regu- latory machine has been identified, we are still in the dark about many aspects of cell cycle regulation. How exactly do the Cdk-cyclin complexes influence cell cycle events? That is, what are the actual substrates in vivo for Cdk-cyclin kinase activity? Researchers have a number of candidates, but only lamins have thus far been proven to qualify. Only with more information about substrates will we be able to determine to what extent the different Cdk-cyclins control fundamentally

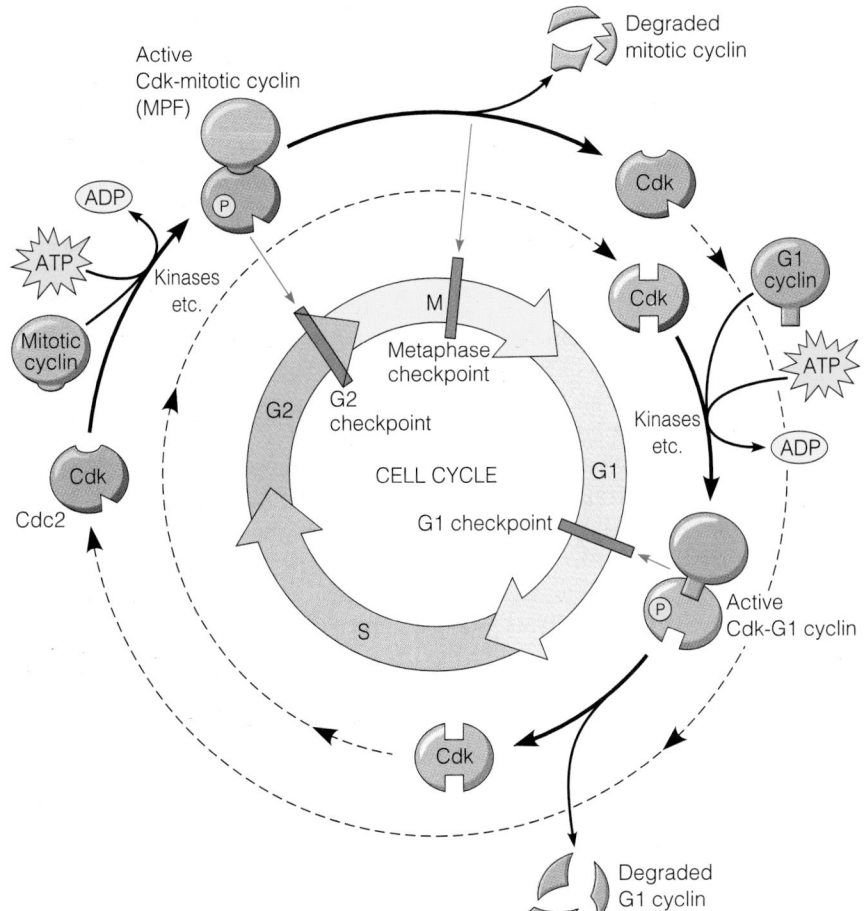

Active
Cdk-mitotic cyclin
(MPF)

ADP

ATP

Kinases
etc.

Mitotic
cyclin

Cdc2

Cdk

Degraded
mitotic cyclin

Cdk

Cdk

G1
cyclin

ATP

Kinases
etc.

ADP

M

Metaphase
checkpoint

G2
checkpoint

G2

CELL CYCLE

G1

G1 checkpoint

S

Active
Cdk-G1 cyclin

Cdk

Degraded
G1 cyclin

Figure 15-33   A General Model for Cell Cycle Regulation.   According to this model, which is based mostly on studies of the G2 checkpoint, the triggering proteins at both the G1 and the G2 checkpoints are specific protein complexes made of cyclin and Cdk (cyclin-dependent kinase), whose phosphorylation of other (still unknown) proteins somehow induces progression of the cycle. Activation of the Cdk-cyclin complexes themselves requires their phosphorylation by other kinases, as well as appropriate dephosphorylation by phosphatases. Different cyclins and, in most eukaryotes, different Cdk proteins are used at different points in the cell cycle. The particular Cdk and cyclin that combine during G2 are Cdc2 and mitotic cyclin. Except for this case, almost nothing is yet known about what controls the synthesis and degradation of these molecules.

distinct processes or, perhaps, cell-type specific versions of the same processes. Not only do we need to know more about the cell cycle in the systems that are already under intensive study, such as yeasts and mammalian cells; we also need to learn more about the cells of other organisms—plants, for instance. Such studies are likely to reveal that at least some of the key cell cycle proteins are also involved in regulating other aspects of cell metabolism. Already it is known that yeast has a Cdk (called PHO85) that, in combination with various cyclins, participates in both the cell cycle and phosphate metabolism.

At the other end of the chain of cell cycle control is the issue of how growth-promoting or growth-inhibiting signals coming from outside the cell connect with the cell cycle machinery. Hence the study of signal transduction pathways, which we treat in Chapter 23, is intimately interconnected with the study of the cell cycle. In addition, the topic of Chapter 25—cancer, which can be described as a genetic disease of the cell cycle—is closely related to signal transduction and the cell cycle. When the normal cell cycle is disturbed by abnormal molecules and events, normal cell growth may turn cancerous. More and more, therefore, there is convergence of research on the cell cycle per se and research on the cellular basis of cancer. Thus, we will return to the cell cycle at a number of points in later parts of this book.

PERSPECTIVE

The eukaryotic cell cycle, which stretches from a cell's "birth" by cell division to the completion of its own division, is divided into five main phases. For historical reasons, the phases are defined in relation to the two most dramatic aspects of the cycle, cell division and DNA replication. The cell division phase is called the M (mitotic) phase, and the period of DNA synthesis is called the S phase. The "gaps" between M and S, and between S and M, are called G1 and G2. Interphase, consisting of G1, S, and G2, is a time of continuous cell growth and active metabolism, and it typically takes up 95% of the cycle time. Cultured mammalian cells usually have cell cycle (generation) times of 18–24 hours, but in multicellular organisms, cells differ greatly in generation time, ranging from stem cells that divide rapidly and continuously to differentiated cells that do not normally divide at all.

The central biochemical process in cell reproduction is the replication of its genetic material. DNA replicates by a semiconservative mechanism in which the two strands unwind and each serves as a template for the synthesis of a complementary strand. A complex of many different proteins carries out the process. The characteristics of cellular DNA synthesis are largely determined by the capabilities of DNA polymerases. Because these enzymes can polymerize only in the $5' \rightarrow 3'$ direction, DNA synthesis is continuous in the overall direction of replication along one new DNA strand (the leading strand) but discontinuous in the opposite direction along the other (lagging) strand. And since DNA polymerases can add nucleotides only to the $3'$ end of a preexisting nucleotide chain, the start of each new chain is primed by the synthesis of a short stretch of RNA. The Okazaki fragments forming the lagging strand are joined by DNA ligase. Helicases, topoisomerases, and single-strand binding proteins are involved in unwinding the double helix. In *E. coli* cells, DNA polymerase III is the chief replication polymerase, and DNA polymerase I is used to remove and replace the RNA primers. DNA polymerase I is also involved in excision repair of DNA damage caused by deamination, depurination, or pyrimidine dimer formation. Other DNA polymerases are used in eukaryotes, but the overall scheme of DNA replication is very similar to that of prokaryotes.

The problem of replicating the ends of the linear DNA molecules of eukaryotic chromosomes is solved by the presence of telomerases, RNA-containing enzymes. The telomerase RNA serves as a primer for the extension of the telomeric repeated sequences at the ends of DNA.

In the M phase, both nucleus and cytoplasm divide. Mitosis is the process by which the duplicated chromosomes are parceled out into two daughter nuclei. It is subdivided into five phases: prophase, metaphase, prometaphase, anaphase, and telophase. During prophase, replicated chromosomes condense as sister chromatids, joined at the centromere. Meanwhile, the cell's two centrosomes move apart, as tubulin polymerizes to form the microtubules (MTs) of the mitotic spindle. In prometaphase, the nuclear envelope fragments, and the chromosomes' kinetochores are captured by spindle MTs (kinetochore MTs). The chromosomes are moved toward the equator of the cell. At metaphase, the chromosomes are arrayed at the metaphase plate. Anaphase begins with the separation of the sister chromatids and continues with their movement, as daughter chromosomes, toward the poles of the cell. In the process, the kinetochore MTs shorten, the polar MTs lengthen, and the cell starts to elongate. During telophase, the separated chromosomes decondense, and a nuclear envelope is re-formed around each daughter nucleus.

Although the forces responsible for chromosome movement in mitosis are not yet completely understood, motor proteins play key roles. Such proteins, which change shape when phosphorylated, are believed to be involved in mitosis in at least three places. Motor proteins at the kinetochore apparently move the chromosome along the kinetochore MTs toward the pole, with local disassembly of the MTs at their kinetochore ("plus") ends. Similar proteins push polar MTs against each other, and others pull astral MTs toward the plasma membrane at the cell poles. Both movements probably contribute to cell elongation during anaphase and telophase.

The division of the cytoplasm, cytokinesis, usually begins before mitosis is complete. In animal cells, microfilaments form a cleavage furrow, which progressively constricts the cell at the midline and eventually separates the cytoplasm into two daughter cells. In plant cells, a cell wall forms through the middle of the parent cell.

Regulation of the eukaryotic cell cycle is a topic of much current research. Complicated cascades of interacting proteins are undoubtedly involved, but at the core of the molecular machinery for cell cycle regulation seem to be cyclin-dependent protein kinases (Cdk proteins) and their partners, the cyclins. An attractive model suggests that passage of the cell cycle past two major checkpoints, in G1 and at the end of G2, is controlled by specific Cdk-cyclin complexes. Phosphorylation and dephosphorylation seem to be the main chemical reactions by which cell cycle regulators are activated and deactivated, although most of the substrates of phosphorylation by Cdk itself are still unknown. The rhythmic buildup and degradation of cyclins serves as a clocklike mechanism underlying the cell cycle, but the cycle is subject to important modification at the checkpoints, mediated by chemical signals reflecting conditions inside and outside the cell. Further research on these signal pathways is expected to lead to a greater understanding of the cancerous cell, as well as the normal cell.

KEY TERMS FOR SELF-TESTING

replication fork (p. 458)
origin of replication (p. 459)
initiator protein (p. 459)
theta replication (p. 459)
replicon (p. 459)
DNA polymerase (p. 459)
DNA polymerases I, II, and III
  (p. 461)
exonuclease (p. 461)
DNA polymerases α, γ, and δ (p. 462)
polymerase chain reaction (PCR)
  (p. 462)
leading strand (p. 462)
lagging strand (p. 462)
Okazaki fragment (p. 462)
DNA ligase (p. 462)
primer (p. 463)
primase (p. 463)
primosome (p. 463)
helicase (p. 464)
DNA topoisomerase (p. 464)
single-strand binding protein (SSB)
  (p. 464)
replisome (p. 464)
telomere (p. 466)
telomerase (p. 466)

DNA Repair

mutation (p. 470)
depurination (p. 471)
deamination (p. 471)
pyrimidine dimer formation (p. 471)
excision repair (p. 471)
repair endonuclease (p. 471)
mismatch repair (p. 471)
base excision repair (p. 472)
DNA glycosylase (p. 472)
nucleotide excision repair (NER)
  (p. 472)

Nuclear and Cell Division

prophase (p. 473)
centromere (p. 473)
centrosome (p. 473)
centriole (p. 473)
mitotic spindle (p. 473)
prometaphase (p. 473)
kinetochore (p. 473)
metaphase (p. 473)
metaphase plate (p. 473)
kinetochore microtubule (p. 473)
polar microtubule (p. 473)
astral microtubule (p. 473)

anaphase (p. 473)
anaphase A and B (p. 473)
telophase (p. 476)
motor protein (p. 478)
cleavage (p. 478)
cleavage furrow (p. 478)
contractile ring (p. 478)
cell plate (p. 480)
phragmoplast (p. 480)

Variations in the Cell Cycle

stem cell (p. 482)

Regulation of the Cell Cycle

cell cycle checkpoint (p. 483)
heterokaryon (p. 484)
maturation-promoting factor (MPF)
  (p. 484)
conditional mutant (p. 485)
temperature-sensitive mutation
  (p. 485)
Cdc2 protein (p. 485)
protein kinase (p. 485)
cyclin (p. 486)
cyclin-dependent protein kinase (Cdk)
  (p. 487)

PROBLEM SET

15-1. The Cell Cycle. Indicate whether each of the following statements is true of the G1 phase of the cell cycle, the S phase, the G2 phase, or the M phase. A given statement may be true of any, all, or none of the phases.

(a) The amount of nuclear DNA in the cell doubles.

(b) The nuclear envelope breaks into fragments.

(c) Sister chromatids separate from each other.

(d) Cells that will never divide again are likely to be arrested in this phase.

(e) The primary cell wall of a plant cell forms.

(f) Chromosomes are present as diffuse, extended chromatin.

(g) This phase is part of interphase.

(h) Mitotic cyclin is at its lowest level.

(i) A Cdk protein is present in the cell.

(j) A cell cycle checkpoint has been identified in this phase.

15-2. The Mitotic Index and the Cell Cycle. The mitotic index is a measure of the amount of mitotic activity in a cell. It is calculated as the percentage of cells in mitosis at any one time. Assume that upon examination of a sample of 1000 cells, you find 30 cells in prophase, 20 in prometaphase, 20 in metaphase, 10 in anaphase, 20 in telophase, and 900 in interphase. Of those in interphase, 400 are found (by microspectrophotometric analysis after staining the cells with a DNA-specific stain) to have an X amount of DNA, 200 to have a 2X amount, and 300 cells to be somewhere in between.

Autoradiographic analysis indicates that the G2 phase lasted 4 hours.

(a) What is the mitotic index for this population of cells?

(b) Specify the proportion of the cell cycle spent in each of the following phases: prophase, prometaphase, metaphase, anaphase, telophase, G1, S, G2.

(c) What is the total length of the cell cycle?

(d) What is the actual amount of time (in hours) spent in each of the phases of part b ?

(e) To measure the G2 phase, radioactive thymidine (a DNA precursor) is added to the culture at some time *t*, and samples of the culture are analyzed autoradiographically for labeled nuclei at regular intervals thereafter. What specific observation would have to be made to assess the length of the G2 phase?

(f) What proportion of the interphase cells would you expect to exhibit labeled nuclei in autoradiographs prepared shortly after exposure to the labeled thymidine? (Assume a labeling period just long enough to allow the thymidine to get into the cells and begin to be incorporated into DNA.)

15-3. Meselson and Stahl Revisited. Although the Watson-Crick structure for DNA seemed to suggest a semiconservative model for DNA, at least two other models are conceivable. In a *conservative model,* the parental DNA double helix remains intact and a second, all-new copy is made. In a *dispersive model,* each strand of both daughter molecules contains a mixture of old and newly synthesized segments.

(a) Starting with one parental double helix, sketch the progeny molecules for two rounds of replication according to each of these alternative models. Use one color for the original parent strands and another color for all the DNA synthesized thereafter (as is done in Figure 15-4 for the semiconservative model).

(b) Indicate the distribution of DNA bands that Meselson and Stahl would have found on their cesium chloride gradients after one and two rounds of replication, for each of the alternative models.

15-4. DNA Replication. Sketch a replication fork of bacterial DNA in which one strand is being replicated discontinuously and the other is being replicated continuously. List six different enzyme activities associated with the replication process, identify the function of each, and indicate on your sketch where each would be located on the replication fork. Identify, in addition, the following features on your sketch: DNA template, RNA primer, Okazaki fragments, and single-strand binding protein.

15-5. More DNA Replication. Following are the results of five experiments carried out to determine the mechanism of DNA replication in the hypothetical organism *Fungus mungus*. For each experiment, indicate whether the results support (S), refute (R), or have no bearing (NB) on the hypothesis that this fungus replicates its DNA by the same mechanism as that known for *E. coli*. Explain your reasoning in each case.

(a) Neither of the two DNA polymerases of *F. mungus* appears to have an exonuclease activity.

(b) Replicating DNA from *F. mungus* shows discontinuous synthesis on both strands of the replication fork.

(c) Some of the DNA sequences from *F. mungus* are present in multiple copies per genome, whereas other sequences are unique.

(d) Short fragments of *F. mungus* DNA isolated during replication contain both ribose and deoxyribose.

(e) *F. mungus* cells are grown in the presence of the heavy isotopes ^{15}N and ^{13}C for several generations and then grown for one generation in normal (^{14}N, ^{12}C) medium; then DNA is isolated from these cells and denatured. The single strands yield a single band in a cesium chloride density gradient.

15-6. Still More DNA Replication. DNA replication seems an extremely complicated process. Perhaps it evolved from some simpler process in which there were not distinct mechanisms for "leading strand" and "lagging strand" synthesis. In this primitive process, there may only have been DNA replication of the leading-strand type, which is less complicated than the lagging-strand mechanism. For example, DNA could have been replicated by a mechanism that involved the synthesis of one daughter strand at a time:

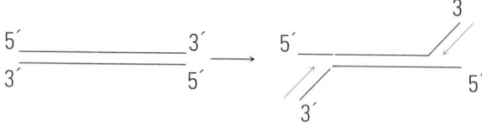

There is something wrong with this picture, however. All DNA synthesis requires a primer. Devise a simple model for DNA synthesis that would allow complete daughter molecules of DNA to be produced by a leading-strand replication mechanism. By simple model, we mean one that does not require more than one unifunctional enzyme in addition to the primitive DNA polymerase. You may assume that the primordial slime in which the DNA is repli-

cating contains an adequate supply of deoxyribonucleoside triphosphates (dNTPs). (There are numerous possible solutions to this problem.)

15-7. The Minimal Chromosome. To enable it to be transmitted intact from one cell generation to the next, the linear DNA molecule of a eukaryotic chromsome must have appropriate nucleotide sequences making up three special kinds of regions: origins of replication (at least one), a centromere, and two telomeres. What would happen if such a chromosomal DNA molecule somehow lost

(a) all of its origins of replication?

(b) all of the DNA constituting its centromere?

(c) one of its telomeres?

15-8. DNA Damage and Repair. Indicate whether each of the following statements is true of depurination (DP), deamination (DA), or pyrimidine dimer formation (DF). A given statement may be true of any, all, or none of these processes.

(a) This process is caused by spontaneous hydrolysis of a glycosidic bond.

(b) This process is induced by ultraviolet light.

(c) This can happen to guanine but not cytosine.

(d) This can happen to thymine but not adenine.

(e) This can happen to thymine but not cytosine.

(f) Repair involves a DNA glycosylase.

(g) Repair involves an endonuclease.

(h) Repair involves DNA ligase.

(i) Repair depends on the existence of separate copies of the genetic information in the two strands of the double helix.

(j) Repair depends on cleavage of both strands of the double helix.

15-9. Nonstandard Purines and Pyrimidines. Shown in Figure 15-34a are three nonstandard nitrogenous bases that are formed by the deamination of naturally occurring bases in DNA.

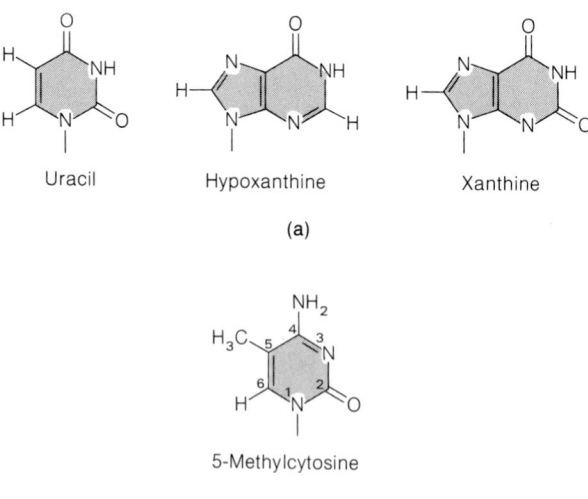

Figure 15-34 Structures of Several Nonstandard Purines and Pyrimidines.

(a) Indicate which base in DNA must be deaminated to form each of these bases.

(b) Why are there only three bases shown, when DNA contains four bases?

(c) Why is it important that none of the bases shown in Figure 15-34a occurs naturally in DNA?

(d) Figure 15-34b shows 5-methylcytosine, a pyrimidine that arises naturally in DNA when cellular enzymes methylate cytosine. Why is the presence of this base in the DNA sequence likely to increase the probability of a mutation at that site?

15-10. Prophase of Mitosis. Based on what you know about nucleoli from Chapter 14, why do the nucleoli disappear during prophase? That is, what is happening at the molecular level to cause their disappearance?

15-11. Chromosome Movement in Mitosis. It is possible to mark the microtubules of a spindle by photobleaching with a laser microbeam (Figure 15-35). When this is done, chromosomes move *toward* the marked area during anaphase. Are the following statements consistent or inconsistent with this experimental result?

(a) Microtubules move chromosomes by disassembling at the spindle poles.

(b) Chromosomes move by disassembling microtubules at their kinetochore ends.

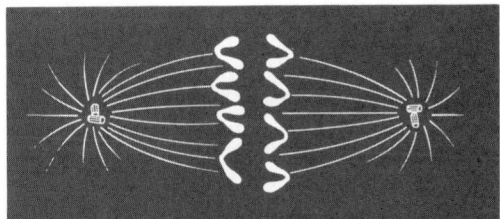

Microtubules are stained with a fluorescent antibody during anaphase.

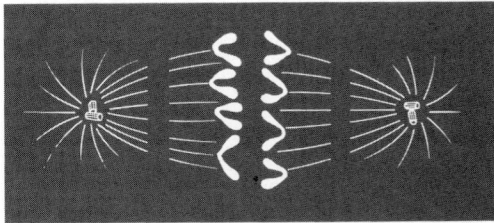

A laser microbeam is used to bleach the fluorescent marker (dark bands).

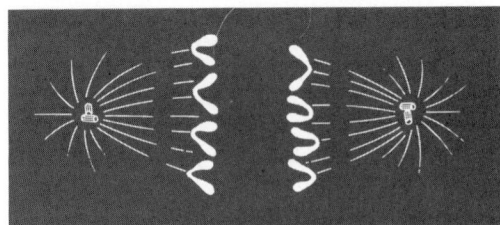

The chromosomes are observed to move toward the dark bands.

Figure 15-35   The Use of Laser Photobleaching to Study Chromosome Movement During Mitosis.

(c) Chromosomes are moved by a springlike elastic property of microtubules.

(d) Chromosomes are moved along microtubules by a motor protein.

15-12. More Cell Cycle. For each of the following pairs of phases from the cell cycle, indicate how you could tell which of the two phases a specific cell is in.

(a) G1 and G2

(c) G2 and M

(b) G1 and S

(d) G1 and M

15-13. Cell Cycle Variations. A normal diploid cell from the insect *Imagineria diploides* contains ten chromosomes and 0.02 ng of DNA in its nucleus. Elevated amounts of DNA are found in certain *Imagineria* cells, as follows:

Cell Type	DNA/Cell	Nuclei/Cell	Chromosomes/ Nucleus
Embryo	20 ng	1000	10
Liver	20 ng	1	10,000
Salivary gland	20 ng	1	10

(a) Explain how the elevated DNA content of each of these cell types might have arisen.

(b) Which of these cell types would you expect to have the shortest generation time? Explain.

(c) Would you expect the salivary gland cells to be dividing actively? Explain.

(d) How might you induce the liver cells to begin dividing again?

15-14. Cell Cycle Regulation. One approach to the study of cell cycle regulation has been to fuse cultured cells that are at different stages of the cell cycle and observe the effect of the fusion on the nuclei of the fused cells. When cells in G1 were fused with cells in S, the nuclei from the G1 cells were observed to begin DNA replication earlier than they would have if they had not been fused. In fusions of cells in G2 and S, however, nuclei continued their previous activities, apparently uninfluenced by the fusion. Fusions between mitotic cells and interphase cells always led to chromatin condensation in the nonmitotic nuclei. Based on these results, identify each of the following statements about cell cycle regulation as probably true (T), probably false (F), or not possible to conclude from the data (NP).

(a) The activation of DNA synthesis may result from the positive action of one or more cytoplasmic factors.

(b) The transition from S to G2 may result from the presence of a cytoplasmic factor that inhibits DNA synthesis.

(c) The transition from G2 to mitosis may result from the presence in the G2 cytoplasm of one or more factors that induce chromatin condensation.

(d) G1 is not an obligatory phase of all cell cycles.

(e) The transition from mitosis to G1 appears to result from the disappearance or inactivation of a cytoplasmic factor present during M phase.

SUGGESTED READING

General References

Alberts, B., D. Bray, J. Lewis, M. Raff, K. Roberts, and J. D. Watson. *Molecular Biology of the Cell*, 3d ed. New York: Garland, 1994.

Kornberg, A., and T. A. Baker. *DNA Replication*, 2d ed. New York: W. H. Freeman, 1992.

Lodish, H., D. Baltimore, A. Berk, S. L. Zipursky, P. Matsudaira, and J. Darnell. *Molecular Cell Biology*, 3d ed. New York: Scientific American Books, 1995.

Murray, A., and T. Hunt. *The Cell Cycle: An Introduction*. New York: W. H. Freeman, 1993.

DNA Replication

Blackburn, E. H. Telomerases. *Annu. Rev. Biochem.* 61 (1992): 111.

Cech, T. R. Chromosome end games (Perspectives). *Science* 266 (1994): 387.

Counter, C. M., H. W. Hirte, S. Bacchetti, and C. B. Harley. Telomerase activity in human ovarian carcinoma. *Proc. Natl. Acad. Sci. USA* 91 (1994): 2900.

DePamphilis, M. L. Origins of DNA replication that function in eukaryotic cells. *Curr. Opin. Cell Biol.* 5 (1993): 434.

Joyce, C. M., and T. A. Steitz. Function and structure relationships in DNA polymerases. *Annu. Rev. Biochem.* 63 (1994): 777.

Kim, N. W., et al. Specific association of human telomerase activity with immortal cells and cancer. *Science* 266 (1994): 2011.

Marx, J. Chromosome ends catch fire (Research News). *Science* 265 (1994): 1656.

McHenry, C. S. DNA polymerase III holoenzyme. *J. Biol. Chem.* 266 (1991): 1912.

Meselson, M., and F. W. Stahl. The replication of DNA in *E. coli*. *Proc. Natl. Acad. Sci. USA* 44 (1958): 671.

Moyzis, R. K. The human telomere. *Sci. Amer.* 265 (August 1991): 48.

Ogawa, T., and R. Okazaki. Discontinuous DNA replication. *Annu. Rev. Biochem.* 49 (1980): 421.

Wang, J. C. DNA topoisomerases. *Sci. Amer.* 247 (January 1982): 94.

Wang, T. S.-F. Eukaryotic DNA polymerases. *Annu. Rev. Biochem.* 60 (1991): 513.

Watson, J. D., and F. H. C. Crick. Genetical implications of the structure of deoxyribonucleic acid. *Nature* 171 (1953): 964.

DNA Repair

Barnes, D. E., T. Lindahl, and B. Sedgwick. DNA Repair. *Curr. Opin. Cell Biol.* 5 (1993): 424.

Culotta, E., and D. E. Koshland, Jr. DNA repair works its way to the top (Molecule of the Year). *Science* 266 (1994): 1926. In the same issue, see Perspectives articles by A. Sancar (p. 1954), P. C. Hanawalt (p. 1957), and P. Modrich (p. 1959). In the previous issue, see Research News article by J. Marx (p. 728).

Sancar, A. Z., and G. B. Sancar. DNA repair enzymes. *Annu. Rev. Biochem.* 57 (1988): 29.

Nuclear and Cell Division

Ault, J. G., and C. L. Rieder. Centrosome and kinetochore movement during mitosis. *Curr. Opin. Cell Biol.* 6 (1994): 41.

Fuller, M. T., and P. G. Wilson. Force and counterforce in the mitotic spindle. *Cell* 71 (1992): 547.

Glover, D. M., C. Gonzalez, and J. W. Raff. The centrosome. *Sci. Amer.* 268 (June 1993): 62.

Hyam, A. A., K. Middleton, M. Centola, T. J. Mitchison, and J. Carbon. Microtubule-motor activity of a yeast centromere-binding protein complex. *Nature* 359 (1992): 533.

Kellogg, D. R., M. Moritz, and B. M. Alberts. The centrosome and cellular organization. *Annu. Rev. Biochem.* 63 (1994): 639.

Rose, M. D., Biggins, S., and L. L. Satterwhite. Unravelling the tangled web at the microtubule-organizing center. *Curr. Opin. Cell Biol.* 5 (1993): 105.

Schulman, I., and K. S. Bloom. Centromeres: An integrated protein/DNA complex required for chromosome movement. *Annu. Rev. Cell Biol.* 7 (1991): 311.

Wadsworth, P. Mitosis: Spindle assembly and chromosome motion. *Curr. Opin. Cell Biol.* 5 (1993): 123.

Regulation of the Cell Cycle

Forsburg, S. L., and P. Nurse. Cell cycle regulation in the yeasts *Saccharomyces cerevisiae* and *Schizosaccharomyces pombe*. *Annu. Rev. Cell Biol.* 7 (1991): 227.

Hartwell, L. H., and M. B. Kastan. Cell cycle control and cancer. *Science* 266 (1994): 1821.

Holloway, S. L., M. Glotzer, R. W. King, and A. W. Murray. Anaphase is initiated by proteolysis rather than by the inactivation of maturation-promoting factor. *Cell* 73 (1993): 1393.

Hunt, T. Braking the cycle. *Cell* 75 (1993): 839.

Marx, J. How cells cycle toward cancer (Research News). *Science* 263 (1994): 319.

Marx, J. Researchers find new role for cell cycle proteins (Research News). *Science* 263 (1994): 1093.

Murray, A. W., and M. W. Kirschner. What controls the cell cycle? *Sci. Amer.* 264 (March 1991): 56.

Nigg, E. A. Targets of cyclin-dependent protein kinases. *Curr. Opin. Cell Biol.* 5 (1993): 187.

Norbury, C., and P. Nurse. Animal cell cycles and their control. *Annu. Rev. Biochem* 61 (1992): 441.

Pines, J. Arresting developments in cell-cycle control. *TIBS* 19 (1994): 143.

Reed, S. I. The role of p34 kinases in the G1 to S-phase transition. *Annu. Rev. Cell Biol.* 8 (1992): 529.

Sherr, C. J. Mammalian G_1 cyclins. *Cell* 73 (1993): 1959.

Solomon, M. J. Activation of the various cyclin/cdc2 protein kinases. *Curr. Opin. Cell Biol.* 5 (1993): 180.

Staiger, C., and J. Doonan. Cell division in plants. *Curr. Opin. Cell Biol.* 5 (1993): 226.

16

SEXUAL REPRODUCTION, MEIOSIS, AND GENETIC VARIABILITY

The process of mitotic cell division discussed in the preceding chapter is responsible for the proliferation of eukaryotic cells, leading either to more organisms or to more cells per organism. Because mitosis involves the segregation of identical chromatids, the daughter cells of every mitotic division are genetically identical, or very nearly so.

Mitotic division is an excellent way to perpetuate specific genetic traits faithfully, and it is the basis of all **asexual reproduction** in eukaryotes. In asexual reproduction, new individuals are generated by a single parent organism, either unicellular or multicellular. Asexual reproduction is widespread in nature. The details of the process, although always involving mitosis, differ with the organism (Figure 16-1, page 496). Examples are *mitotic division* of unicellular organisms, *budding* of offspring as outgrowths from a multicellular parent's body, and *regeneration* of whole organisms from pieces of a parent organism. In plants, organisms can even be regenerated from single cells isolated from the adult plant.

Evolutionarily speaking, asexual reproduction can be an efficient and successful mode of perpetuating a species. As long as a population's environment remains constant, asexual reproduction is adequate for maintaining the survival of the population. If a population is already well adapted to its environment, the genetic predictability of asexual reproduction fits well with the predictability of a static environment. If the environment changes, however, such a population may be ill suited to cope with the new conditions. In a changing environment, a sexually reproducing population will usually have an advantage, as we now discuss.

Sexual Reproduction

The fundamental characteristic of asexual reproduction is that all progeny are genetically similar to the single parent organism from which they arise mitotically. **Sexual reproduction**, on the other hand, involves the mixing of genetic information from two parent organisms and results in off-spring that are genetically dissimilar, both from each other and from the parents. Moreover, the offspring are unpredictably dissimilar; that is, we cannot anticipate exactly what combination of genetic information a particular offspring will receive from its two parents. Since most plants and animals—and even many microorganisms—reproduce sexually, this type of reproduction must provide some distinct advantages.

Advantages of Sexual Reproduction

The major advantage of sexual reproduction over asexual reproduction is that the sexual process can combine in a single individual desirable genetic changes that have arisen in separate individuals. In this way, sexual reproduction can generate great variety among the individuals making up a population.

Genetic modification ultimately depends on the occurrence of *mutations,* unpredictable alterations in the DNA that involve changes in the nucleotide sequence. These changes can result from either base changes or from the rearrangement of DNA segments. Mutations are rare events, and beneficial mutations are even rarer. When a beneficial mutation occurs, however, it is clearly to the advantage of the species to preserve the mutation in the population. It can be even more advantageous to combine several desirable mutations in a single individual—and therein lies the fundamental advantage of sexual reproduction. Although mutations occur in both sexual and asexual species, only sexual reproduction can readily bring together beneficial mutations that originally occurred in two separate individuals.

Sexual reproduction brings about a reshuffling of genetic information in each new generation, thereby providing a large range of individual variations within a given population, variations that include all sorts of combinations of mutations. It follows that at least some of these variations will confer advantages on some members of the population. Of course, other members of the population will be less suited for survival. For the species as a whole, however, it is clearly

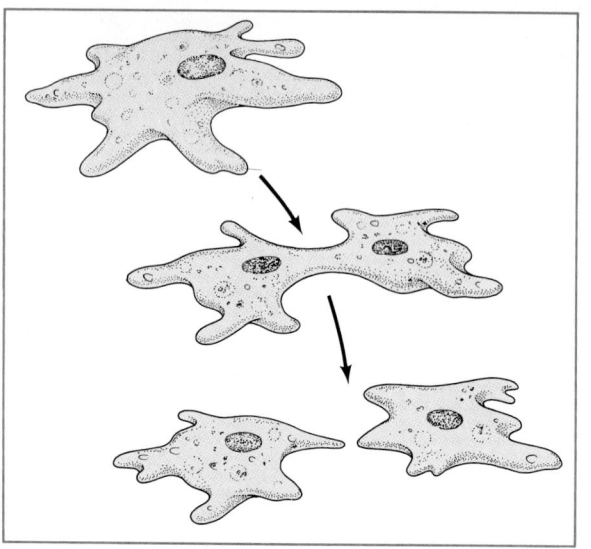

(a) Mitotic division of an amoeba

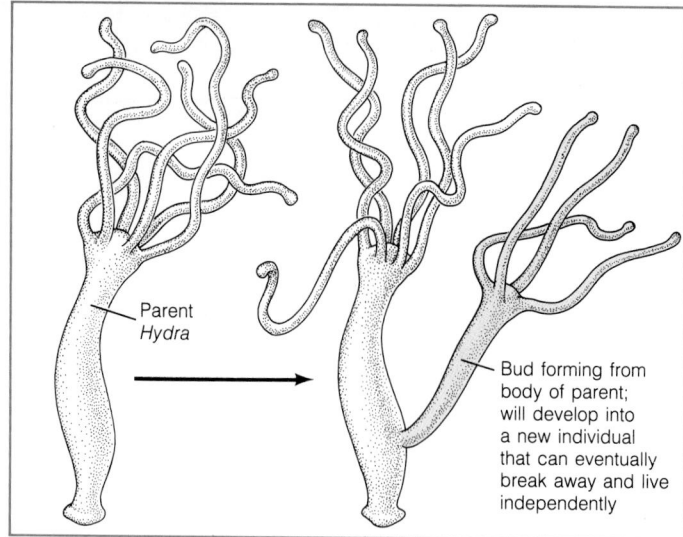

(b) Budding in the freshwater polyp *Hydra*

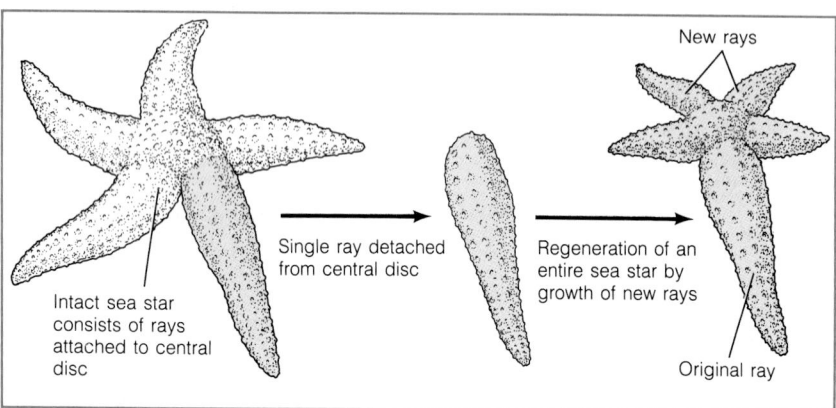

(c) Regeneration of a sea star from a detached ray of the parent organism

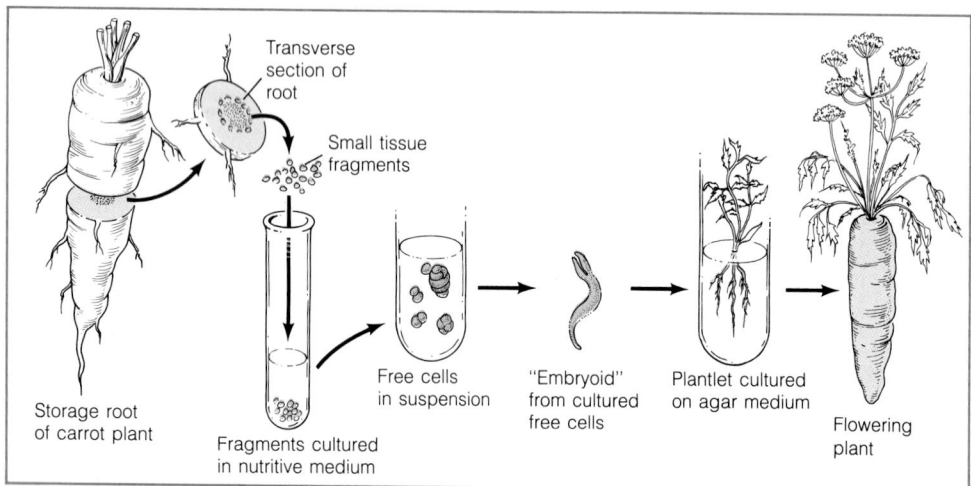

(d) Regeneration of an entire carrot plant from cultured cells obtained from a section of root tissue

Figure 16-1   Asexual Reproduction of Eukaryotes.   In asexual reproduction, new individuals are generated from a single parent by a mitotic process. Examples of asexual reproduction are **(a)** simple mitotic division of an amoeba cell; **(b)** budding of offspring as outgrowths from the body of a multicellular parent organism in *Hydra,* a simple freshwater animal (of the phylum Cnidaria); **(c)** regeneration of an entire sea star (genus *Linckia*) from a single ray (arm) that was detached from a parent organism; and **(d)** regeneration of an entire carrot plant from a single cell obtained from an adult plant.

preferable that there be a large amount of genetic variation within populations, so that some members will be able to survive when the environment changes.

Sexual Reproduction and the Diploid Organism

By definition, sexual reproduction involves the combination in an individual offspring of genetic information from two parents. At some point in its life cycle, every sexually reproducing species has cells with two sets of chromosomes, one set inherited from each parent. The corresponding chromosomes, one from each set, are called **homologous chromosomes**. Two homologous chromosomes carry the same lineup of genes, although, for a given gene, the two versions may be slightly different in nucleotide sequence. Not surprisingly, then, homologous chromosomes generally look exactly alike when viewed with a microscope. For many species, there is a major exception, however—the **sex chromosomes,** chromosomes that determine whether an individual is male or female. Some organisms have sex chromosomes of very different appearance. Male mammals, for example, have one *X* chromosome and one *Y* chromosome, which is much smaller. (Female mammals have two *X* chromosomes.) However, parts of the *X* and *Y* chromosomes are actually homologous, and in sexual reproduction, the chromosomes behave as homologues.

A cell or organism with two sets of chromosomes is said to be **diploid** (from the Greek word *diplous,* meaning "double") and contains two copies of its genome. A cell or organism with a single set of chromosomes, and therefore a single copy of its genome, is **haploid** (from the Greek word *haplous,* meaning "single"). By convention, the haploid chromosome number for a species is designated *n* (or 1*n*) and the diploid number 2*n*.

Let's now focus on a **gene locus** (plural: **loci**), a particular place on a chromosome (corresponding to a particular part of its DNA sequence) where a certain gene is located. For the sake of simplicity, let's assume that this gene is present in only one copy along the genomic DNA (not multiple copies, like the histone genes mentioned in Chapter 14). Let's also assume that the gene controls a single, clear-cut characteristic—or *character,* as geneticists usually say—in the organism. A diploid organism will have two copies of this gene, which may be either identical or slightly different. Different versions of a gene are called **alleles** of the gene, and the combination of alleles present in an organism determines what the organism will be like with respect to the character controlled by the gene. In garden peas, for example, the alleles at a single gene locus determine seed color, which may be green or yellow (Figure 16-2). An organism with two identical alleles for a given gene is said to be **homozygous** for the gene or, in informal usage, for the trait that allele determines. (The word *trait* is used for a particular variant of a character, such as a trait of green seeds, where seed color is the character.) Thus, a pea plant that inherited the same alleles for yellow seed color from both of its parents is said to be homozygous for (yellow) seed color. An organism with two different alleles for a gene is said to be **heterozygous** for that gene or for the character it determines. A pea plant with one allele that specifies yellow seed color and a second allele that specifies green seed color is therefore heterozygous for seed color.

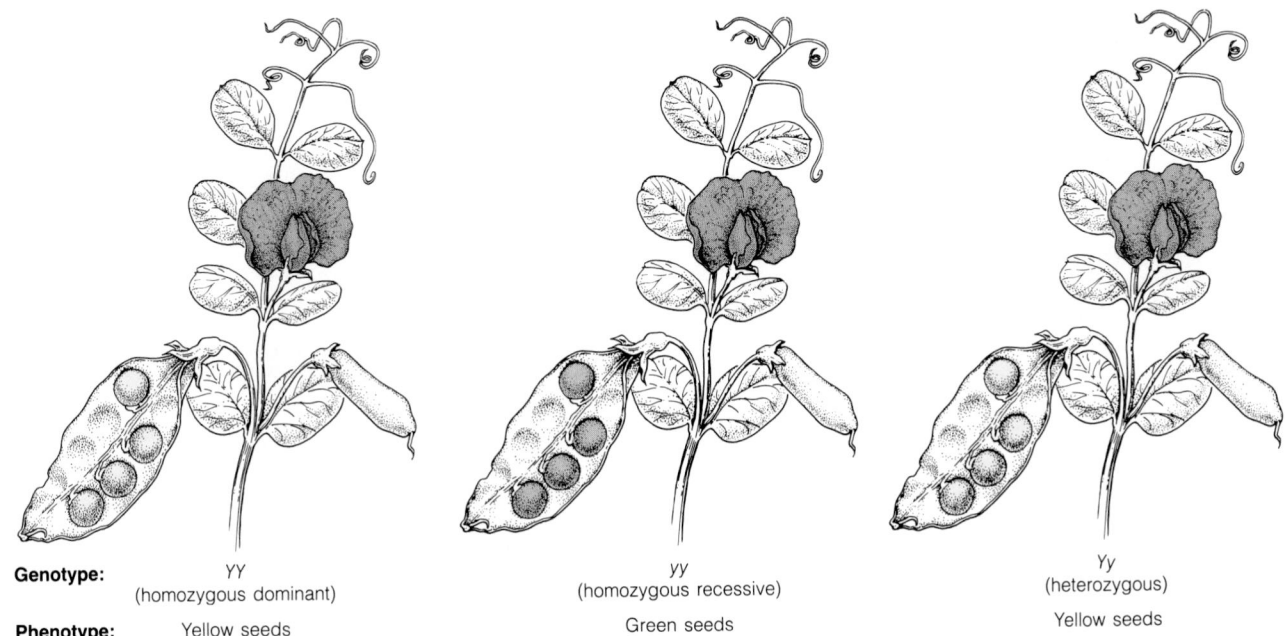

Genotype:	*YY*	*yy*	*Yy*
	(homozygous dominant)	(homozygous recessive)	(heterozygous)
Phenotype:	Yellow seeds	Green seeds	Yellow seeds

Figure 16-2 Genotype and Phenotype. In garden peas, seed color (phenotype) can be either yellow or green. The seed-color alleles are *Y* (yellow, dominant) or *y* (green, recessive; shown here as gray). Be-cause the pea plant is a diploid organism, its genetic makeup (genotype) for seed color may be homozygous dominant (*YY*), homozygous recessive (*yy*), or heterozygous (*Yy*).

In a heterozygous individual, one of the two alleles is often **dominant** and the other **recessive**. These terms are relative, simply conveying the idea that it is the dominant allele that determines the trait that appears in a heterozygous individual. "Dominant" and "recessive" are also used to refer to the trait in question. Consider, for example, the genetic character of seed color in peas. In this case, yellow seed color is dominant over green seed color; therefore, pea plants that are heterozygous for the seed-color gene will have yellow seeds. By the convention we use here, a dominant allele is designated by an uppercase letter that stands for the trait, and a corresponding recessive allele is represented by the same letter in lowercase. In both cases, italics are used. Thus, the alleles for seed color in peas are represented by *Y* for yellow and *y* for green, because yellow is the dominant trait. As Figure 16-2 illustrates, a pea plant can have a genotype for the seed color gene that is homozygous for the dominant allele (*YY*), homozygous for the recessive allele (*yy*), or heterozygous (*Yy*).

It is important to distinguish between the **genotype**, or genetic makeup of an organism, and its **phenotype**, or the physical expression of its genotype (see Figure 16-2). The phenotype of an organism can often be directly determined by inspection. Genotype, on the other hand, can be *directly* determined only from studying the organism's DNA. Usually, genotype is deduced from indirect evidence, including the organism's phenotype and information about its parents and/or offspring. All organisms having the same phenotype do not necessarily have identical genotypes. In the example of Figure 16-2, pea plants with yellow seeds (phenotype) can be either *YY* or *Yy* (genotype). Figure 16-3 summarizes the genetic terminology introduced so far.

Diploidy, the diploid state, is a necessary feature within the life cycle of sexually reproducing species. It is a distinct advantage because of the greater genetic flexibility it allows. In a sense, a diploid cell contains a spare set of genes that is available for mutation and genetic innovation. Changes in a spare copy of a gene, in other words, will not usually threaten the survival of the organism even if the mutation is deleterious to the original function of that particular gene.

The Role of Gametes

The hallmark of sexual reproduction is that it brings together in a single individual genetic information contributed by two parents. Because sexual reproduction results in offspring that are diploid, the contribution from each parent must be haploid. Haploid cells that are specialized for sexual reproduction are called **gametes,** and the process that produces them is **gametogenesis.** Biologists distinguish between male and female individuals on the basis of the gametes they produce. Gametes produced by the female parent, called **eggs** or **ova** (singular: **ovum**), are specialized for the storage of nutrients; they are nonmotile and relatively large. Male gametes, called **sperm** (or *spermatozoa*), are small and flagellated; they are specialized for motility. The union of a sperm and an egg is called **fertilization.** The fertilized egg that results, also called a **zygote,** is diploid, having received one chromosome set from the sperm and a homologous set from the egg. In the life cycles of multicellular organisms, fertilization is followed by **development**—a series of mitotic divisions and progressive specialization of various groups of cells to form a multicellular embryo, and eventually an adult.

We must note that the distinction between male and female parents is not universal. Among fungi and protists, the gametes produced by the two parents may be identical in appearance, although different at the molecular level. Such gametes are said to differ in **mating type.** The union of two gametes requires that they be of different mating types, but the number of possible mating types in a species may be greater than two—in some cases, more than ten!

Figure 16-3 Some Genetic Terminology. This diagram shows a homologous pair of chromosomes from a diploid cell; the chromosomes are the same size and shape and carry genes for the same characters (characteristics), in the same order. The site of a gene on a chromosome is called a gene locus. The particular versions of a gene—alleles—found at comparable gene loci on homologous chromosomes may be identical (giving the organism a genotype that is homozygous for that gene) or different (making the genotype heterozygous). If one allele of a gene is dominant and the other recessive, the heterozygous organism exhibits a dominant phenotype with respect to the character in question. A recessive phenotypic trait is observed only if the genotype is homozygous for the recessive allele. Red and blue are used here and in most later figures to distinguish the different parental origins of the two chromosomes.

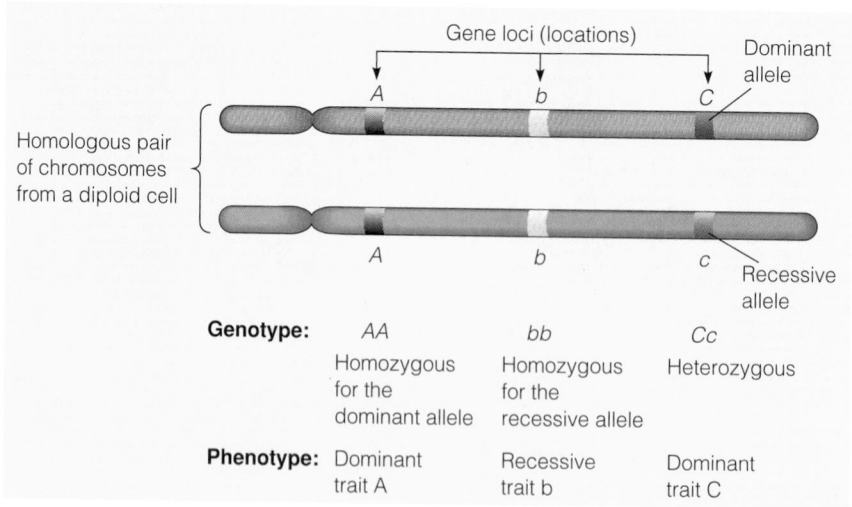

Meiosis

Gametes are haploid cells and therefore cannot be generated from diploid precursors by mitosis because mitotic division results in daughter cells that are genetically identical to the parent cell. If gametes were formed mitotically from diploid cells, both sperm and egg would have a diploid chromosome number, and the zygote that resulted from their fusion would be *tetraploid*, having *four* homologous sets of chromosomes. In fact, the chromosome number would then double every generation, which is clearly not the case. Thus, another kind of division must occur to maintain the constancy of chromosome number from generation to generation. That process, first described in the 1880s, is meiosis.

Meiosis can be described as two successive nuclear divisions following a single duplication of the chromosomes. The products are four daughter nuclei (each in a separate daughter cell, usually) with only one set of chromosomes per nucleus. Figure 16-4 outlines the principle of the meiotic process starting with a diploid cell with only four chromosomes ($2n = 4$). A single period of DNA replication is followed by two cell divisions, **meiosis I** and **meiosis II,** which result in the formation of four haploid daughter cells.

Meiosis and the Life Cycle

Both meiosis and fertilization are indispensable components of the life cycle of every sexually reproducing organism, because the doubling of chromosome number that takes place at fertilization is balanced by the halving that occurs during meiosis. As a result, the life cycle of sexual organisms is divided into two phases: a diploid ($2n$) phase and a haploid ($1n$) phase. The diploid phase begins at fertilization and extends until meiosis, whereas the haploid phase is initiated at meiosis and ends in fertilization.

Organisms vary greatly in the relative prominence of the haploid and diploid phases of their life cycles (Figure 16-5). Bacteria reproduce asexually (though by a process that is simpler than mitosis) and therefore do not have a diploid phase in their life cycle (Figure 16-5a). As we will learn later, some bacteria have a means of transmitting DNA from one cell to another that is sometimes termed "sexual." However, such transmission of genetic information is not an inherent part of the bacterial life cycle, and it usually involves only a portion of the genome; therefore, it does not qualify as true sexual reproduction.

Most fungi are also primarily haploid organisms, but they can reproduce sexually. Their life cycle thus includes a diploid phase that begins with gamete fusion (the fungal equivalent of fertilization) and ends with meiosis (Figure 16-5b). However, meiosis usually occurs almost immediately after gamete fusion, so the diploid phase is very short, and, accordingly, only a very small fraction of fungal nuclei are diploid at any one time. Fungal gametes develop, without meiosis, from cells that are already haploid.

Mosses and ferns are probably the best examples of organisms in which both the haploid and diploid phases are prominent features of the life cycle. Every species of these

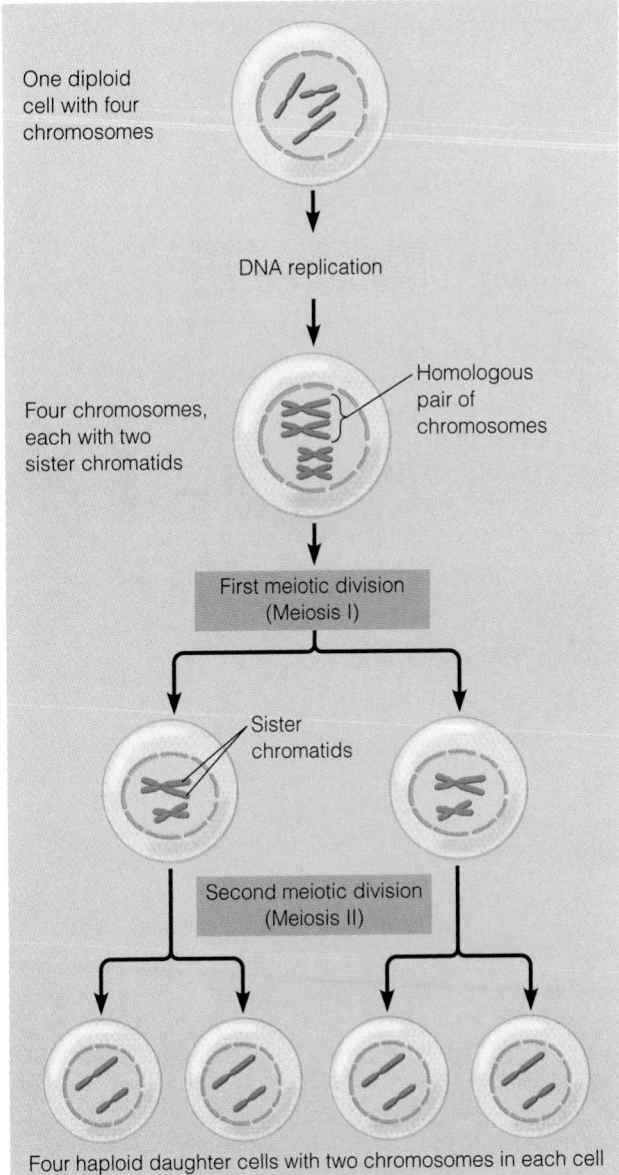

Figure 16-4 The Principle of Meiosis.   Meiosis involves a single round of DNA replication (chromosome duplication) in a diploid cell followed by two successive cell division events. In this example, the diploid cell has only four chromosomes, which can be grouped into two homologous pairs. After DNA replication, each chromosome consists of two sister chromatids. In the first meiotic division (meiosis I), homologous chromosomes separate, but sister chromatids remain attached. In the second meiotic division (meiosis II), sister chromatids separate, resulting in four haploid daughter cells with two chromosomes each. Notice that each haploid cell has one chromosome from each homologous pair that was present in the diploid cell.

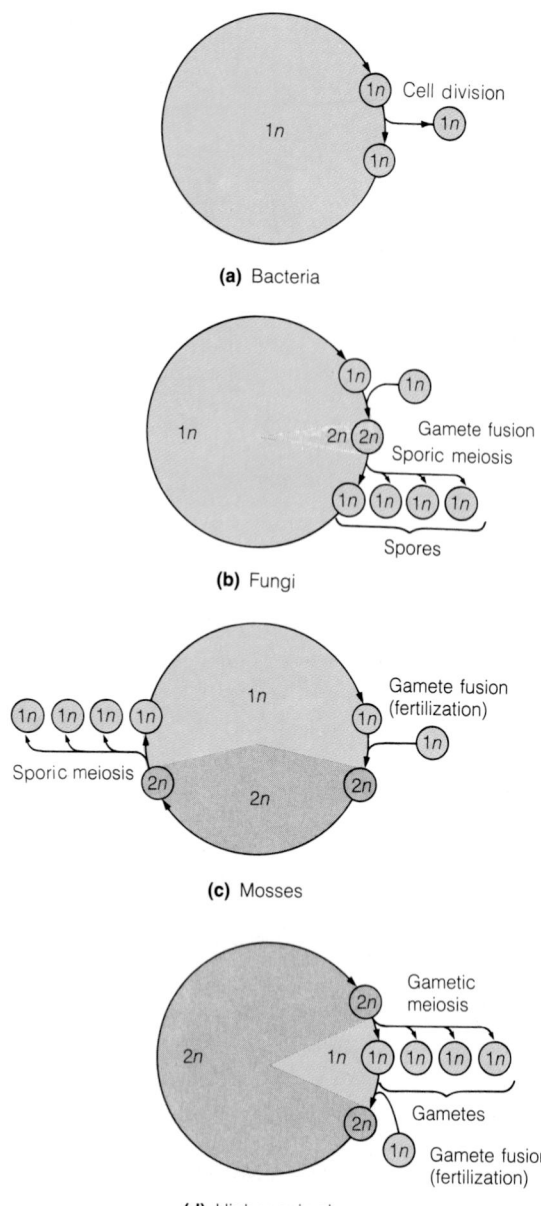

(a) Bacteria

(b) Fungi

(c) Mosses

(d) Higher animals

Figure 16-5 Types of Life Cycles. The relative prominence of the haploid (1*n*) and diploid (2*n*) phases of the life cycle differ greatly, depending on the organism. **(a)** Bacteria exist exclusively in the haploid state. **(b)** Most fungi exemplify a life form that is predominantly haploid but has a brief diploid phase. Because the products of meiosis in fungi are haploid spores, this type of meiosis is called sporic meiosis. The spores give rise to haploid cells, some of which later become gametes (without meiosis). **(c)** Mosses (and ferns, as well) alternate between haploid and diploid forms, both of which are significant components in the life cycles of these organisms. Sporic meiosis produces haploid spores, which in this case grow into haploid plants. Eventually, some of the haploid plant's cells differentiate into gametes. (In seed plants, such as conifers and flowering plants, the haploid forms of the organism are vestigial, each consisting of only a small number of cells.) **(d)** Higher animals are the best examples of organisms that are predominantly diploid, with only the gametes representing the haploid phase of the life cycle. Animals are said to have a gametic meiosis, since the immediate products of meiosis are haploid gametes.

plants has two alternative, morphologically distinct, multicellular forms, one haploid and the other diploid (Figure 16-5c). For mosses, the haploid form of the organism is larger and more prominent, and the diploid form is a modest, rather inconspicuous structure. For ferns, it is the other way around. In both cases, gametes develop from preexisting haploid cells.

Organisms that alternate between haploid and diploid multicellular forms in this way are said to display an **alternation of generations** in their life cycles. In addition to mosses and ferns, eukaryotic algae and all seed plants also have an alternation of diploid and haploid generations in their life cycles. In all such organisms, the products of meiosis are always **haploid spores,** which, after germination, give rise by mitotic cell division to the haploid form of the plant or alga. The haploid form in turn produces the gametes by specialization of cells that are already haploid. The gametes, upon fertilization, give rise to the diploid form. Because the diploid form produces spores, it is called a **sporophyte** ("spore producing plant"). The haploid form produces gametes and is therefore called a **gametophyte.** All plants have an alternation of generations. Among the seed plants, however, the sporophyte generation is predominant. In the angiosperms (flowering plants), for example, the gametophyte generation is an almost vestigial structure located in the flower (*male gametophytes* in the flower's anthers, the *female gametophyte* in the carpel).

The best examples of a life cycle dominated almost completely by the diploid form are found among the animals (Figure 16-5d). In such organisms, including humans, meiosis gives rise not to spores but to gametes directly, so the haploid phase of the life cycle is represented only by the gametes. Meiosis in such species is called *gametic meiosis* to distinguish it from the *sporic meiosis* that occurs in spore-producing organisms with an alternation of generations. Meiosis is thus gametic in animals and sporic in plants.

The Process of Meiosis

Wherever it occurs in the life cycle, the meiotic cell cycle is always characterized by an initial chromosome duplication followed by two successive division events. As shown in Figure 16-6 (page 502), in prophase of meiosis I the duplicated chromosomes are linked together in homologous pairs. This pairing of homologous chromosomes, called **synapsis,** is unique to meiosis; at all other times, including mitosis, the chromosomes of a homologous pair behave completely independently within the cell. The paired homologous chromosomes, each pair consisting of two pairs of sister chromatids, are called **bivalents** (a synonymous term, which emphasizes the total of four chromatids, is *tetrad*). After aligning at the metaphase plate in the middle of the cell, the bivalents split apart in such a way that each pair of sister chromatids goes to a different pole. Consequently, each daughter nucleus from the first meiotic division contains one of the two chromosomes of each bivalent. Meiosis II follows almost immediately, resulting in the separation of sister chromatids in a manner that is analogous to the mitotic process. The events unique to meiosis occur in meiosis I: the synapsis of homologous chromosomes and their subsequent segregation to different daughter cells. (As we'll see shortly, genetic recombination also takes place during meiosis I.)

Each meiotic division consists of the same basic stages as mitosis, although cell biologists do not usually distinguish prometaphase as a separate phase. Thus the meiotic phases are *prophase, metaphase, anaphase,* and *telophase.* Prophase I is much longer and more complicated than mitotic prophase, and prophase II, by contrast, tends to be very short. There is often no interphase between meiotic divisions, and the chromosomes may never decondense. If there is an interphase II, it is usually very short and—most important—does not include DNA replication. The interphase prior to meiosis I, on the other hand, is similar to mitotic interphase in that the DNA replicates and the chromosomes double.

After studying the meiosis diagrams in Figure 16-6, look at the micrographs in Figure 16-7 (page 504), which illustrate the phases of meiosis as they occur in the male grasshopper. Then continue with the text, where we examine the stages of meiosis in more detail.

Meiosis I

The first meiotic division results in the segregation of homologous chromosomes to different daughter cells. This is the feature of meiosis that is of greatest genetic significance because it is at this stage in the life cycle of the organism that the two alleles for each gene part company. It is this separation of alleles that makes possible the eventual "re-combination" of different pairs of alleles at fertilization, a type of genetic recombination. Also of great significance are events in meiosis I that result in another type of genetic recombination, involving the physical exchange of parts of DNA molecules. Changes in DNA molecules resulting from the exchange of segments of DNA from two different sources is **genetic recombination** as molecular biologists use the term (and as we use it in this book). As we discuss shortly, while the chromosomes are synapsed during prophase I of meiosis, DNA is exchanged between homologous chromosomes.

Prophase I. Prophase I is a particularly long and complex phase. Based on what they observed with the light microscope, early cytologists divided prophase I into five stages: leptotene, zygotene, pachytene, diplotene, and diakinesis (Figure 16-8, page 505). The **leptotene** stage is characterized by the start of chromosome condensation; the chromosomes first appear as very long, threadlike structures, just as in mitosis. At **zygotene**, however, homologous chromosomes have begun to pair laterally (synapse) to form bivalents. Keep in mind that each such bivalent contains four chromatids. Bivalent formation is of great genetic significance because it is between the nonsister (homologous)

chromatids of bivalents that crossing over occurs. **Crossing over** is the physical exchange of parts of homologous chromosomes that accounts for genetic recombination.

By the **pachytene** stage, synapsis is complete, and the bivalents have become still shorter and thicker. At the **diplotene** stage, the chromosomes are still more highly condensed, and sister chromatids remain tightly apposed all along their lengths. However, the homologous chromosomes of each bivalent begin to separate from each other, particularly near the centromere. As a result of this partial separation, **chiasmata** (singular: **chiasma**) become visible. As implied by its name—derived from the Greek letter *chi* (X)—a chiasma looks like an X, formed by two (nonsister) chromatids that are in contact. Each chiasma results from a crossover event that occurred earlier between the two chromatids, one from each of the two homologous chromosomes. A chiasma is therefore the cytological manifestation of crossing over.

In some organisms—female mammals, for instance—the chromosomes decondense during diplotene, transcription resumes, and the cells may "take a break" from meiosis for a prolonged period, sometimes years. However, with the onset of **diakinesis,** the final stage of prophase I, the chromosomes recondense to their maximally compacted state. Now the centromeres of the homologous chromosomes separate further, and the chiasmata eventually are the only remaining attachments between the homologues. Diakinesis is also the stage at which the the nucleoli completely disappear, the spindle forms, and the nuclear envelope fragments. Thus, it corresponds to prometaphase in mitosis (see Chapter 15).

With the advent of modern tools and techniques, especially the electron microscope, cytologists have been able to refine our picture of what happens during prophase I. They have discovered that what holds the homologous chromosomes in tight apposition in synapsis is an elaborate protein structure called the **synaptonemal complex.** This structure looks something like a zipper. As indicated at the bottom of Figure 16-8, certain elements of the complex (light purple) start to attach to individual chromosomes during leptotene, but the elements that actually join homologous chromosomes together (dark purple) develop at zygotene. The synaptonemal complex is fully formed during pachytene. It disassembles in diplotene, allowing the homologous chromosomes to separate except where they are joined by chiasmata. The synaptonemal complex appears to be essential for crossing over to occur, and some electron micrographs reveal additional protein complexes, called *recombination nodules,* that are thought to mediate the crossing-over process. Figure 16-9 (page 505) shows the synaptonemal complex as it appears at two magnifications under the electron microscope, as well as an enlarged diagram of the complex.

Metaphase I. At metaphase I, the bivalents, attached to spindle microtubules by their kinteochores, are lined up on the metaphase plate. The kinetochores of sister chromatids

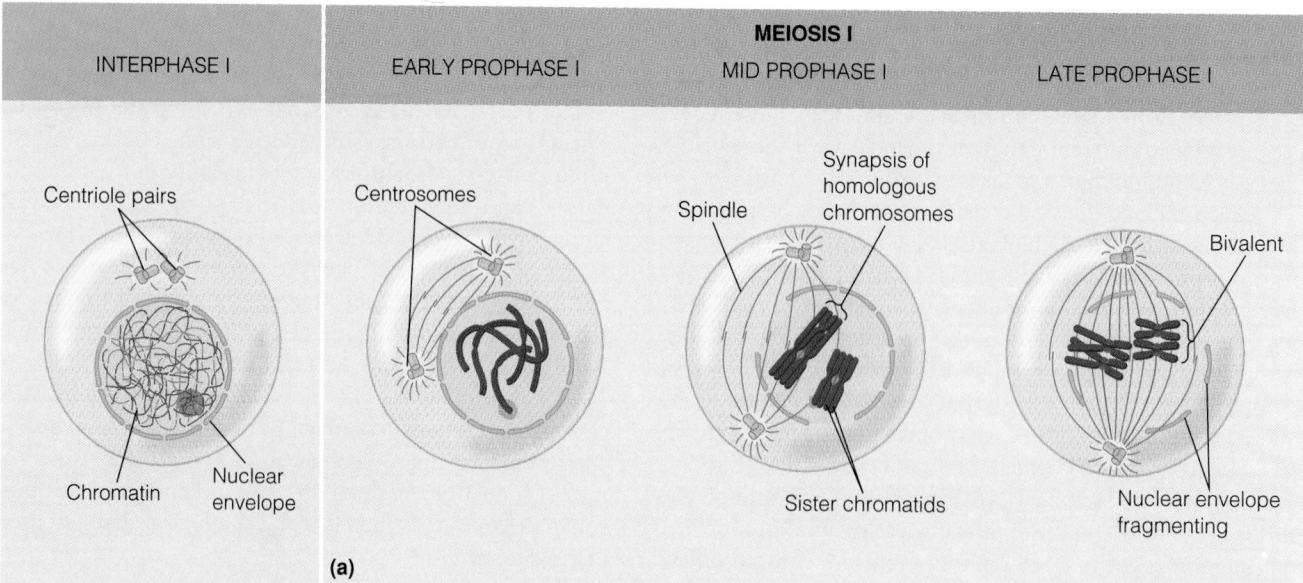

| INTERPHASE I | EARLY PROPHASE I | MEIOSIS I | |
| | | MID PROPHASE I | LATE PROPHASE I |

Centriole pairs

Centrosomes

Spindle

Synapsis of homologous chromosomes

Bivalent

Chromatin

Nuclear envelope

Sister chromatids

Nuclear envelope fragmenting

(a)

Figure 16-6 Meiosis in an Animal Cell.   Meiosis consists of two successive divisions, called meiosis I and II, with no intervening DNA synthesis or chromosome duplication. **(a)** During prophase I, the chromosomes (duplicated during the previous S phase) condense and the two centrosomes migrate to opposite poles of the cell. Each chromosome (four, in this example) consists of two sister chromatids. Homologous chromosomes pair to form bivalents. **(b)** Bivalents become aligned at the metaphase plate (metaphase I). **(c)** Homologous chromosomes separate during anaphase I, but sister chromatids remain attached at the centromere.

(d) Telophase and cytokinesis follow. Although not illustrated here, there may then be a short interphase (interphase II). In meiosis II, **(e)** chromosomes recondense (prophase II), **(f)** a second metaphase plate forms (metaphase II), and **(g)** sister chromatids at last separate (anaphase II). **(h)** After telophase II and cytokinesis, the result is four haploid daughter cells, each containing one chromosome of each homologous pair. Prophase I is a complicated process shown in more detail in Figure 16-8. Meiosis in plants is similar, except for the absence of centrioles and the mechanism of cytokinesis, which involves formation of a cell plate.

are side by side (in many species appearing as a single mass) and face the same pole of the cell; the side-by-side kinetochores of the homologous chromatid pair face the opposite pole. The number of bivalents is equal to the haploid chromosome number for the species. The positioning of *pairs* of homologous chromosomes at the metaphase I plate is a crucial distinction between meiosis I and mitosis, where such pairing does not occur (Figure 16-10, page 506). An important point about metaphase I is that the bivalents are randomly oriented on the metaphase plate, in the sense that, for each bivalent, either the maternal or paternal homologue may face a given pole of the cell.

Note that, at this stage, only chiasmata hold together the homologous chromosomes. If, for some reason, prophase I had occurred without crossing over, and hence without chiasma formation, the chromosomes might not arrange themselves properly on the metaphase I plate, and homologous chromosomes might not separate properly at anaphase I. This is exactly what happens in meiosis of mutant yeast cells that have deficiencies in genetic recombination.

Anaphase I.   At the start of anaphase I, the kinetochores of the two chromosomes (sister chromatid pairs) making up each bivalent begin to move toward opposite poles of the cell, pulling the homologous chromosomes of each bivalent

apart. The sister chromatids of each chromosome are apparently released from whatever glue was holding them in close apposition, allowing the recombinant chromatids to go to a single pole. (Otherwise, chromatids involved in chiasmata might be torn apart!) The chiasmata migrate toward the ends of the chromosomes and finally "resolve"; that is, the chromatids that were linked by chiasmata separate completely. Again, notice the distinction between meiosis and mitosis (see Figure 16-10). In mitotic anaphase, sister chromatids separate from each other. In anaphase I of meiosis, homologous chromosomes separate, but sister chromatids remain together.

Telophase I and Cytokinesis.   The movement of homologous chromosomes to opposite poles is completed during telophase I. Which member of a homologous pair ends up at which pole is determined entirely by how the chromosomes happened to be oriented on the metaphase plate at metaphase I. In any case, at telophase I, two nuclear envelopes form, each containing only one (duplicated) chromosome of each homologous pair. The cytoplasm then splits to produce two daughter cells; and, although their chromosomes are duplicated, these cells are haploid. As mentioned earlier, the chromosomes may not decondense before meiosis II begins.

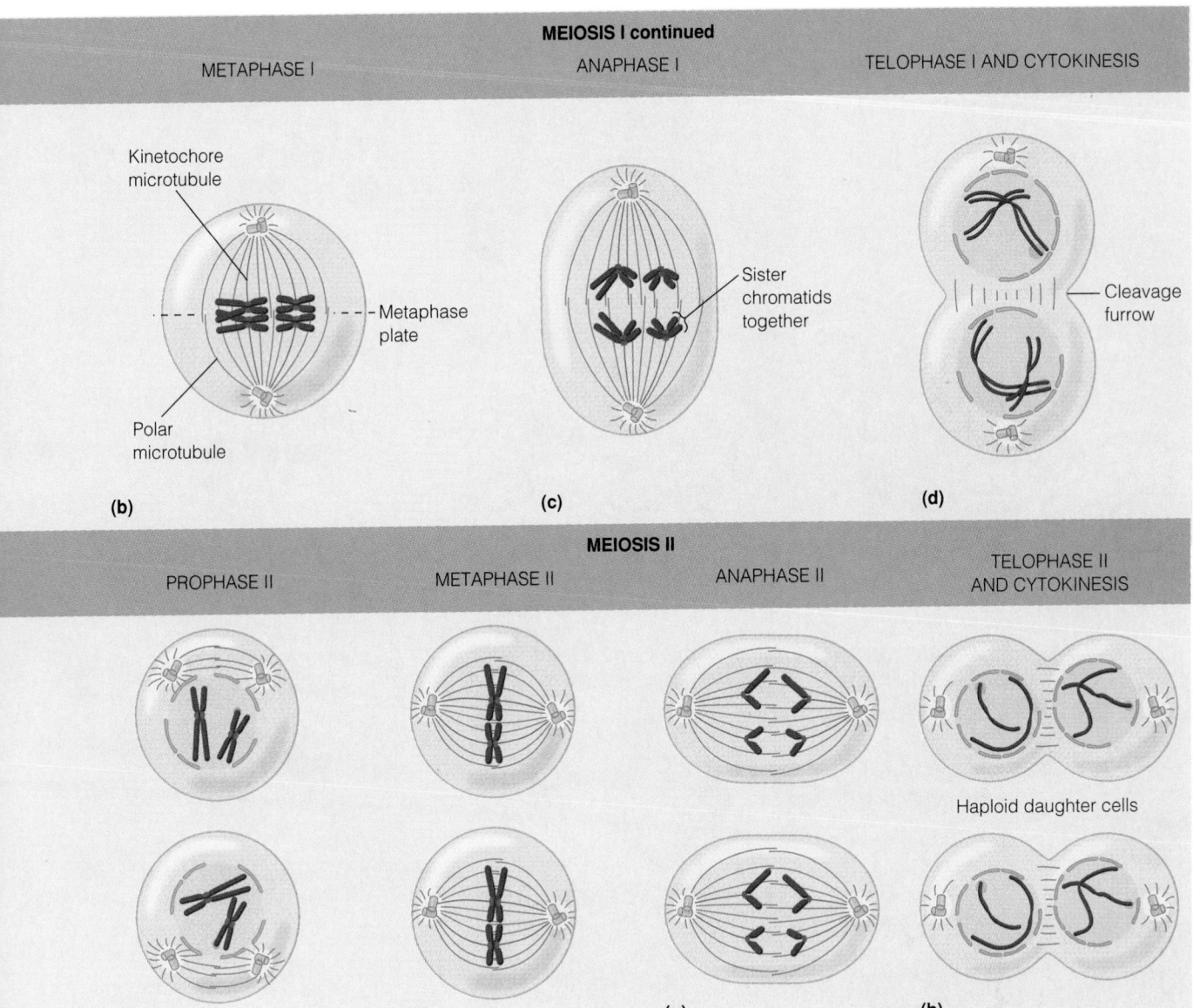

Kinetochore microtubule

Metaphase plate

Polar microtubule

(b)

Sister chromatids together

(c)

Cleavage furrow

(d)

MEIOSIS II

PROPHASE II   METAPHASE II   ANAPHASE II   TELOPHASE II AND CYTOKINESIS

Haploid daughter cells

(e)   (f)   (g)   (h)

Meiosis II

The second meiotic division usually follows almost immediately; the interphase between the two divisions is either very short or nonexistent, depending on the species. Chromosomes are already duplicated as meiosis II begins, but only one set of chromosomes is present because of the prior separation of homologues. Prophase II is very brief. If detectable at all, it is much like mitotic prophase. Metaphase II also resembles the equivalent stage in mitosis, except that only half as many chromosomes are present on the metaphase plate. The kinetochores of sister chromatids are now distinct structures that face opposite poles of the cell. Anaphase II follows, effecting the separation of sister chromatids. The new daughter chromosomes move to opposite poles. This movement is completed in telophase II, and a nuclear envelope re-forms around each of the haploid nu-

clei. (At this point, you may want to study Figure 16-10 in its entirety to review the similarities and differences between meiosis and mitosis.)

The Products of Meiosis

The process of meiosis is fundamentally similar in all organisms, but other aspects of gametogenesis vary. Figure 16-11 (page 507) is a schematic depiction of gametogenesis in animals. In male animals, meiosis leads to four haploid cells of the same size (Figure 16-11a). These cells subsequently differentiate into sperm cells by losing most of their cytoplasm and acquiring long flagella and other specialized structures. In female animals, however, only one of the four haploid cells survives and gives rise to a functional egg cell (Figure 16-11b). This is because the two cell divisions are very

(Continued on page 507)

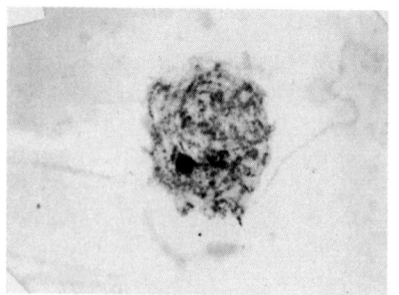

Prophase I: leptotene

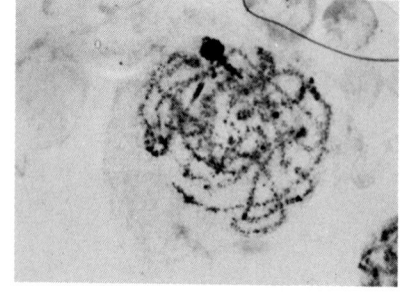

Prophase I: zygotene

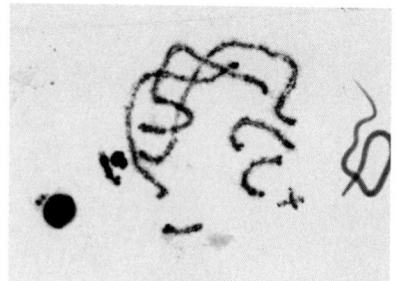

Prophase I: pachytene

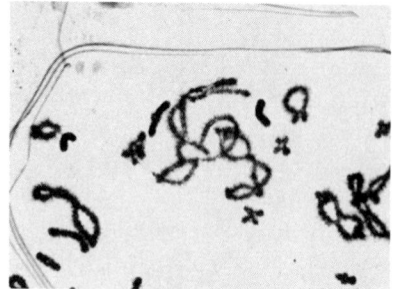

Prophase I: diplotene

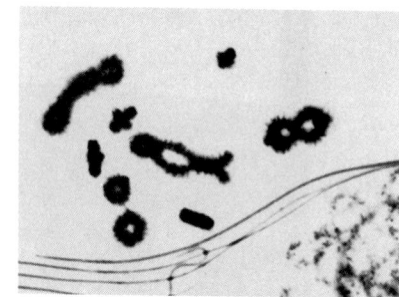

Prophase I: diakinesis

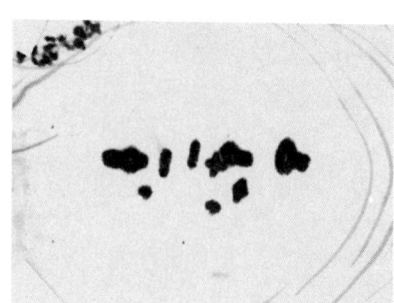

Metaphase I

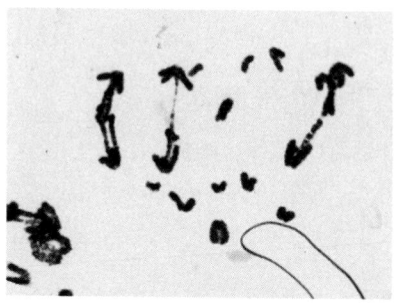

Anaphase I

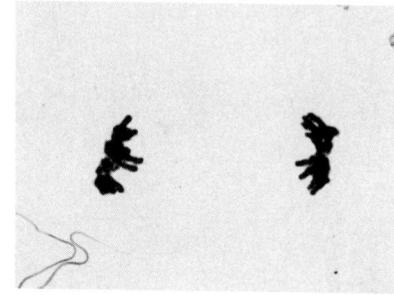

Telophase I

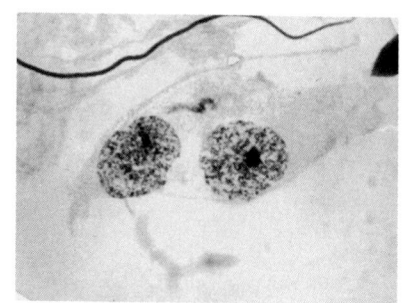

Interphase II

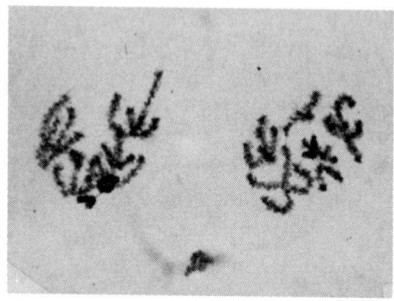

Prophase II

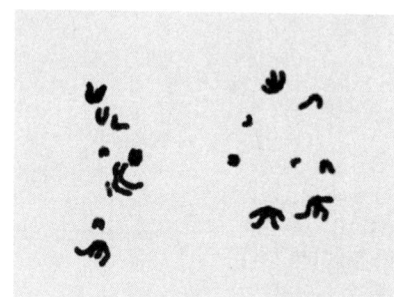

Metaphase II

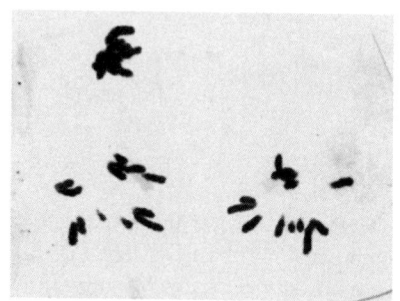

Anaphase II

Figure 16-7   Meiosis in the Male Grasshopper. These light micrographs illustrate meiosis as it occurs in the testis of a male grasshopper, *Chorthippus parallelus*. Each of the phases of meiosis I and meiosis II is shown, including the five stages of prophase I. ⊢——————⊣
25μm

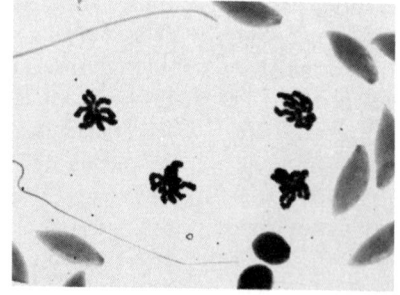

Telophase II

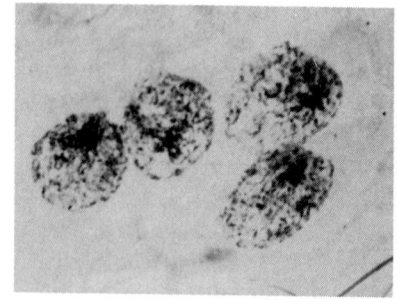

Daughter cells (interphase)

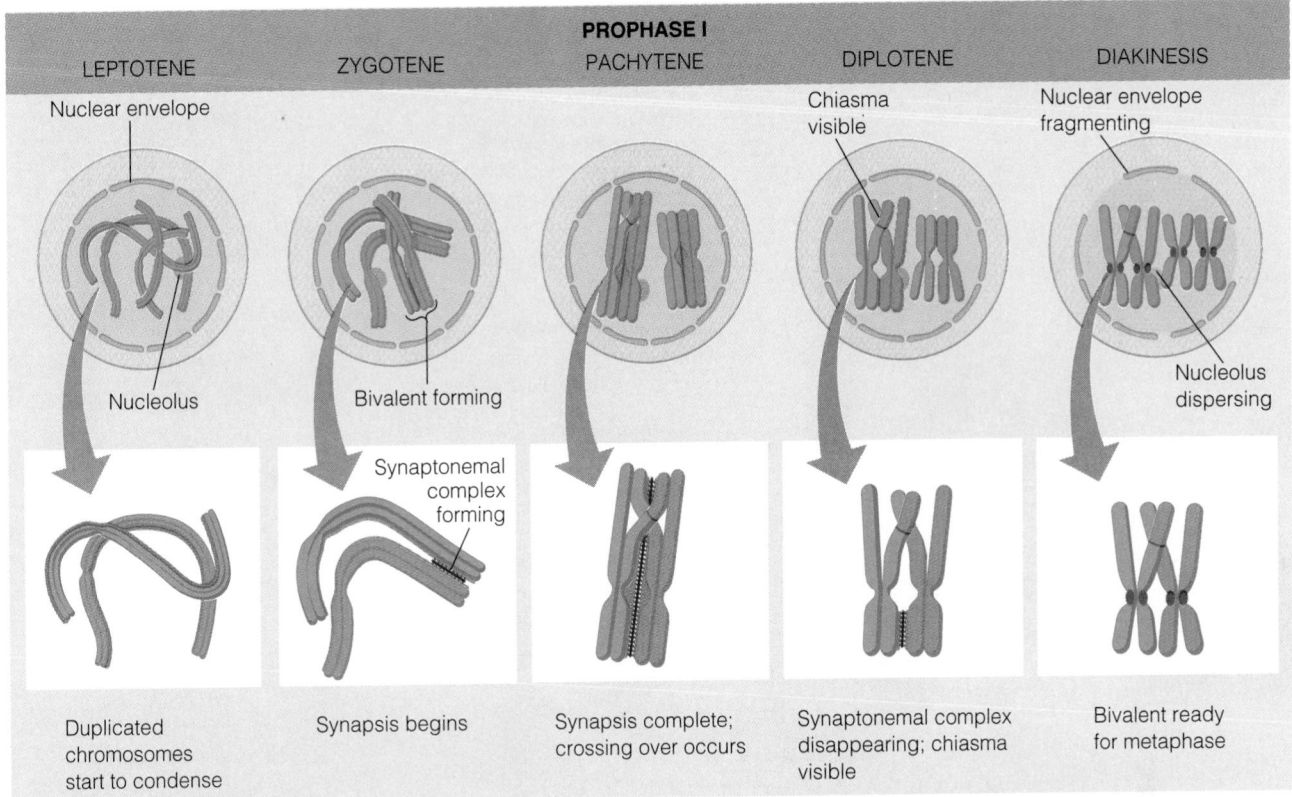

PROPHASE I				
LEPTOTENE	ZYGOTENE	PACHYTENE	DIPLOTENE	DIAKINESIS

Nuclear envelope

Nucleolus

Bivalent forming

Synaptonemal complex forming

Chiasma visible

Nuclear envelope fragmenting

Nucleolus dispersing

Duplicated chromosomes start to condense

Synapsis begins

Synapsis complete; crossing over occurs

Synaptonemal complex disappearing; chiasma visible

Bivalent ready for metaphase

Figure 16-8   Meiotic Prophase I. Prophase I is a complicated process that is subdivided into five stages. The top diagrams depict the cell nucleus at each stage, for a diploid cell with a total of four chromosomes (two homologous pairs). The bottom diagrams focus in on a single homologous pair in greater detail, revealing the formation and subsequent disappearance of the synaptonemal complex, a protein structure that holds homologous chromosomes in close lateral apposition during pachytene. Red and blue distinguish the paternal and maternal chromosomes of each homologous pair; the synaptonemal complex is shown in shades of purple.

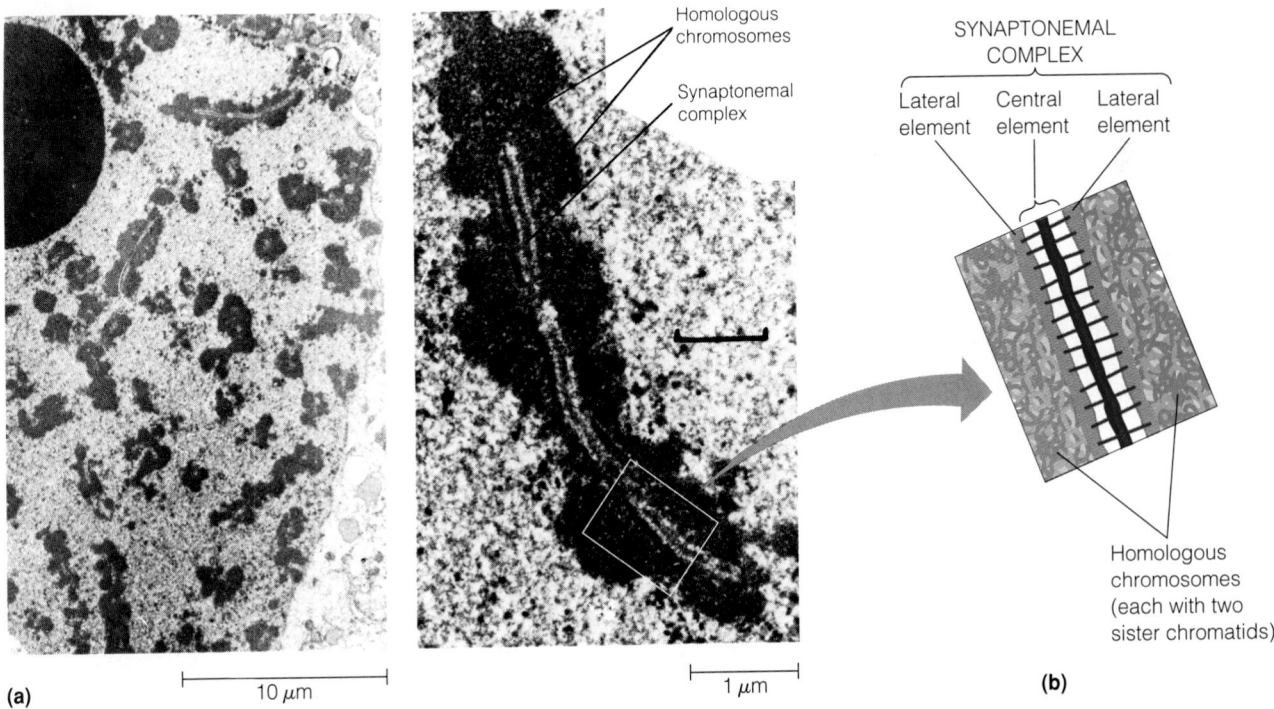

Homologous chromosomes

Synaptonemal complex

SYNAPTONEMAL COMPLEX

Lateral element   Central element   Lateral element

Homologous chromosomes (each with two sister chromatids)

10 µm

1 µm

(a)

(b)

Figure 16-9   The Synaptonemal Complex.   (a) These electron micrographs show, at two magnifications, synaptonemal complexes in the nuclei of cells from a lily. The cells are at the pachytene stage of prophase I (TEMs). **(b)** The diagram identifies the complex's *lateral elements* (light purple), which seem to form on the chromosomes during leptotene, and its *central* (or *axial*) *element* (dark purple), which starts to appear during zygotene and "zips" the homologous chromosomes together. At pachytene, the homologues are held tightly together all along their lengths.

MEIOSIS I

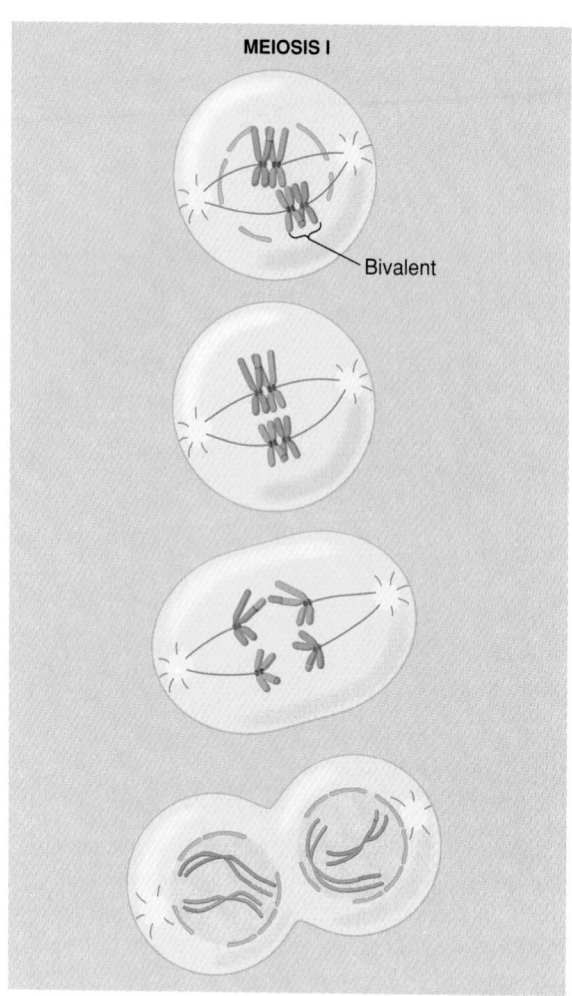

Bivalent

Prophase

Each condensing chromosome has two chromatids. In meiosis I, homologous chromosomes synapse, forming a bivalent. Crossing over occurs between nonsister chromatids, producing chiasmata. In mitosis, each chromosome acts independently.

Metaphase

In meiosis I, the bivalents align at the metaphase plate. In mitosis, individual chromosomes align at the metaphase plate.

Anaphase

In meiosis I, chromosomes (not chromatids) separate. In mitosis, chromatids separate.

Telophase and Cytokinesis

MITOSIS

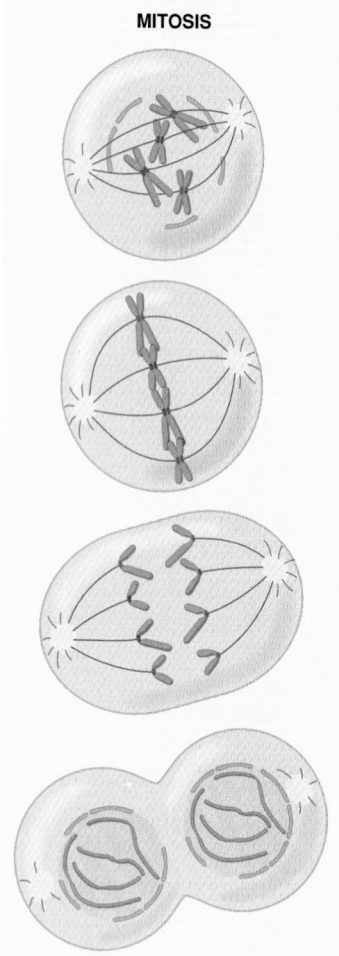

Result of mitosis: two cells, each with the same number of chromosomes as the original cell

MEIOSIS II

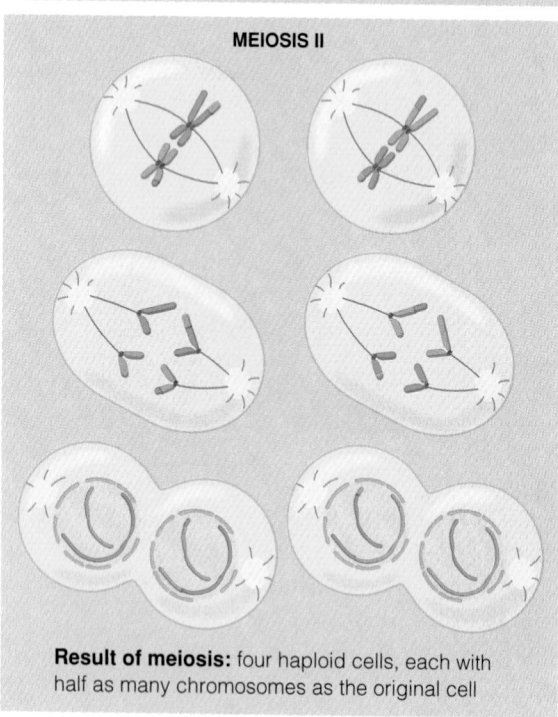

In meiosis II, sister chromatids separate.

Result of meiosis: four haploid cells, each with half as many chromosomes as the original cell

Figure 16-10   Meiosis and Mitosis Compared.   Meiosis and mitosis are both preceded by DNA replication, resulting in two sister chromatids per chromosome at prophase. Meiosis includes two nuclear (and cell) divisions, halving the chromosome number to the haploid level. Moreover, during the elaborate prophase of the first meiotic division, homologous chromosomes synapse, and crossing over occurs between nonsister chromatids. Mitosis involves only a single division, producing two diploid nuclei (and, usually, cells), each with the same number of chromosomes as the original cell. In mitosis, the homologous chromosomes behave independently; at no point do they come together in pairs.

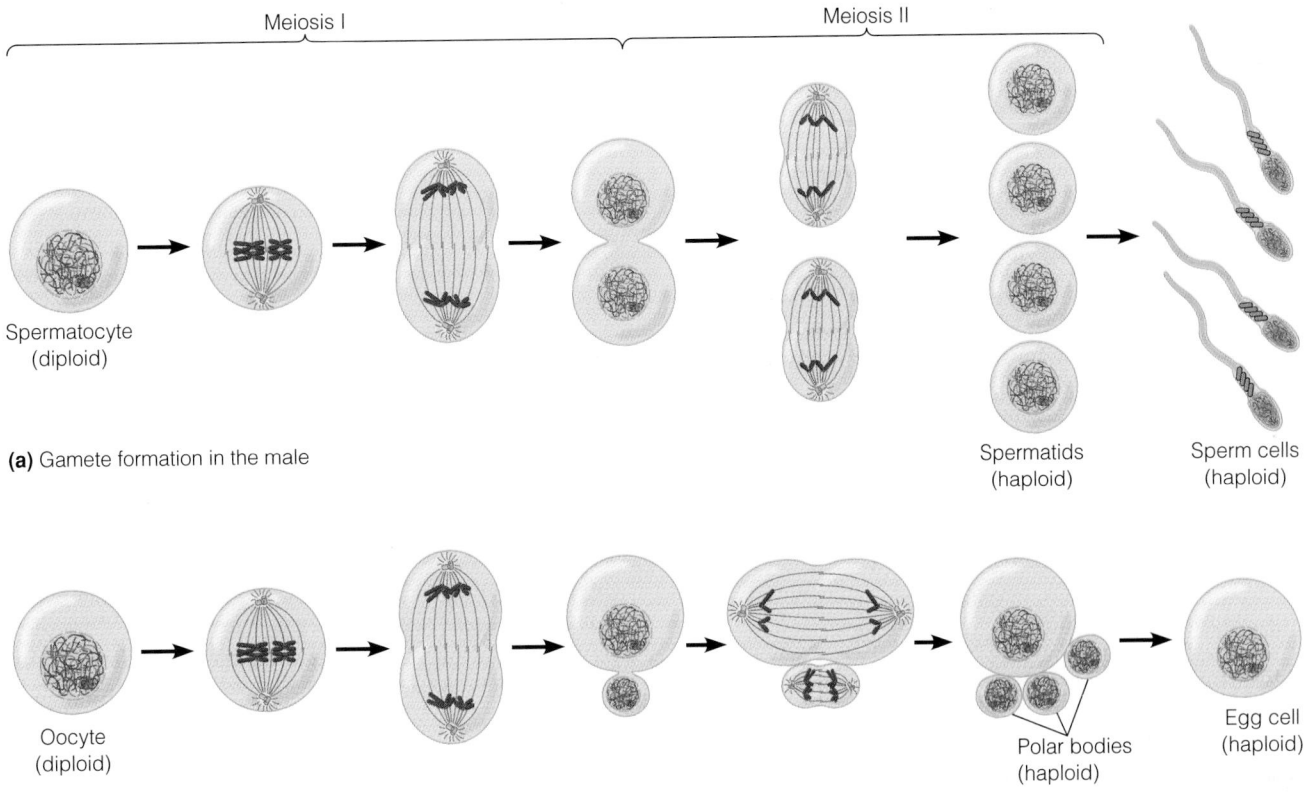

Meiosis II

Spermatocyte
(diploid)

Spermatids
(haploid)

Sperm cells
(haploid)

(a) Gamete formation in the male

Oocyte
(diploid)

Polar bodies
(haploid)

Egg cell
(haploid)

(b) Gamete formation in the female

Figure 16-11   Gamete Formation.   (a) In the male, all four haploid products of meiosis are retained and mature into sperm. **(b)** In the female, both meiotic divisions are asymmetric, forming one large egg cell and three (in some cases, only two) small cells called polar bodies that do not give rise to functional gametes. Although not indicated here, the mature egg cell has usually grown much larger than the oocyte from which it arose.

unequal, with only one of the four daughter cells receiving the bulk of the cytoplasm of the original diploid cell. The other three, smaller cells, called **polar bodies,** degenerate. In the case of the egg cell, by the time meiosis is completed, the gamete is fully "mature," and it may even have been fertilized (as in frogs). The differentiation of the egg cell involves the synthesis of specialized coat layers and a great amount of cytoplasm and nucleoplasm. A human egg cell, for example, has a diameter of about 100 μm, giving it a volume more than a hundred times larger than that of the diploid oocyte from which it arose. And consider the gigantic size of a bird egg!

Highly specialized cellular structures play important roles in the fertilization process. In animals, enzymes released from a membrane-bounded sac called the *acrosome,* at the front end of the sperm, dissolve the egg cell's outer coat, and species-specific proteins on the sperm's surface bind to receptors on the egg cell. After fusion of the plasma membranes of sperm and egg, the sperm nucleus enters the egg and fuses with the egg nucleus, generating a diploid cell, the zygote. Fusion of the plasma membranes triggers important changes in the animal egg. The fertilized egg quickly develops a barrier to the entry of additional sperm and simultaneously undergoes a tremendous burst of metabolic

activity in preparation for embryonic development. Although not described here, fertilization in plants involves similarly complex cellular and biochemical events.

The Significance of Meiosis

As we have pointed out, meiosis preserves the chromosome number in sexually reproducing organisms. If it were not for meiosis, gametes would have as many chromosomes as all the other cells in the body, and the chromosome number would double with every new generation.

Equally important, however, is the role meiosis plays in generating genetic diversity in a sexually reproducing population. Meiosis is significant as the point in the flow of genetic information at which a variety of combinations of chromosomes (and the alleles they carry) are assembled in gametes for potential passage to offspring. Although each gamete produced has exactly one of each homologous pair of chromosomes present in the organism's diploid cells, a gamete's particular combination of chromosomes is only one of many possibilities. The assortment of paternal and maternal chromosomes in a given gamete is random. It results from the random orientation of bivalents at metaphase I, in which the paternal and maternal homologues of each pair can face either pole, independent of the orientations of the other bivalents.

Furthermore, the crossing over that occurs between the synapsed homologous chromosomes during prophase I generates even more diversity among the gametes. Crossing over enables the gametes to have many more combinations of alleles than the independent assortment of homologous chromosomes allows. We will return to crossing over (recombination) later in the chapter. Now we turn to the historic experiments that first revealed the genetic consequences of chromosome segregation and independent assortment during meiosis. These experiments were carried out by Gregor Mendel, before chromosomes were even known to exist.

Genetic Variability: Segregation and Assortment of Alleles

Mendel's Experiments

Most students of biology have heard of Gregor Mendel and the classic genetics experiments he conducted in a monastery garden. Mendel's findings, first published in 1865, laid the foundation for what we now know as *Mendelian genetics*. Working with the common garden pea, Mendel chose seven readily identifiable characters of pea plants and selected in each case two strains (varieties) of plants that displayed alternative forms of the character. (As mentioned earlier, geneticists use the term *character* to refer to a characteristic of the organism under study.) For example, seed color was one character Mendel chose, because he had one strain of peas that had yellow seeds and another that had green seeds (see Figure 16-2).

Mendel's experimental approach was straightforward. He first established that each of the fourteen plant strains chosen was **true-breeding** upon self fertilization. To continue with our seed-color example, Mendel showed that plants grown from his yellow seeds produced only yellow seeds, and plants grown from his green seeds produced only green seeds. These results are depicted in Figure 16-12a. We now know that these strains of plants were homozygous for a gene that controls seed color.

Mendel then *cross*-fertilized his plants to produce **hybrids** for each pair of true-breeding strains, hoping to deduce from his results the principles that govern the inheritance of such traits. The results must have seemed mystifying at first: In every case, the resulting progeny—called the **F₁ generation**, where F₁ stands for first filial—regularly showed one or the other of the parental traits, but never both. In other words, one parental trait was always dominant and the other was always recessive. In the case of seed color, for example, all the F₁ plants had yellow seeds (Figure 16-12b), indicating that yellow seed color is dominant. Notice in the figure that the original parental plants are called the **P₁ generation**.

The next summer, Mendel allowed all the F₁ hybrids to self-fertilize. For each of the seven characters under study, he

Figure 16-12 Genetic Analysis of Seed Color in Pea Plants. Genetic crosses were performed starting with two true-breeding strains of pea plants, one having yellow seeds and one having green seeds. The yellow-seed trait is dominant (allele *Y*); the green-seed trait is recessive (allele *y*). The resulting phenotypes of the progeny are shown on the left and their genotypes on the right. (The genotypes were deduced later.) **(a)** The parent stocks are homozygous for either the dominant (*YY*) or recessive (*yy*) trait and breed true upon self-fertilization. **(b)** When crossed, the parent stocks yield F₁ plants (hybrids) that are all heterozygous (*Yy*) and therefore show the dominant trait. **(c)** Upon self-fertilization, the F₁ plants produce yellow and green seeds in the F₂ generation in a ratio of 3:1. See Figures 16-13 and 16-14 for further analyses.

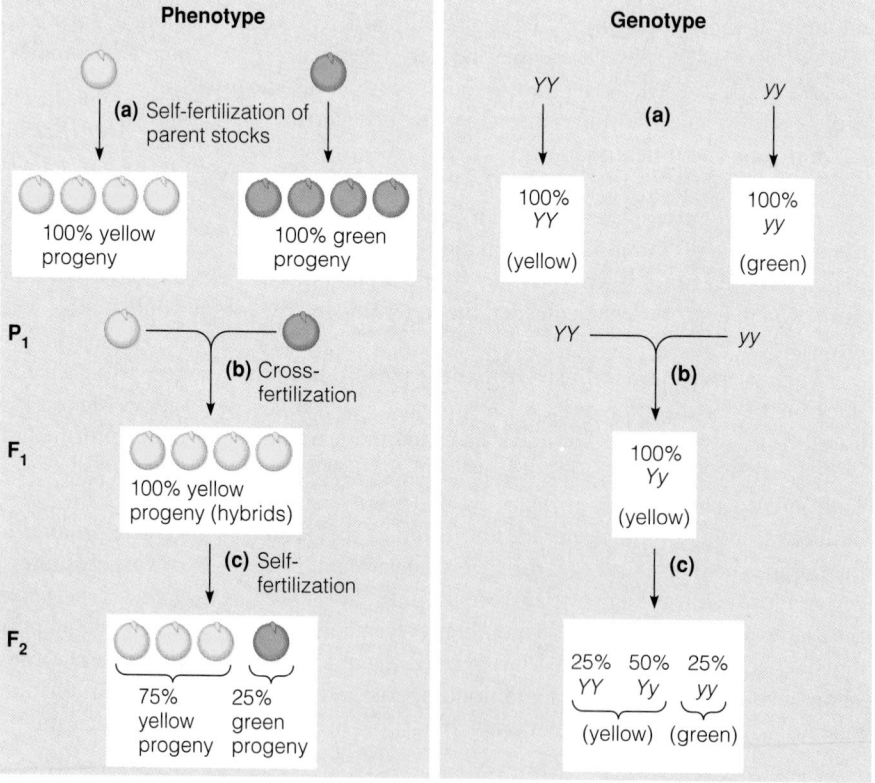

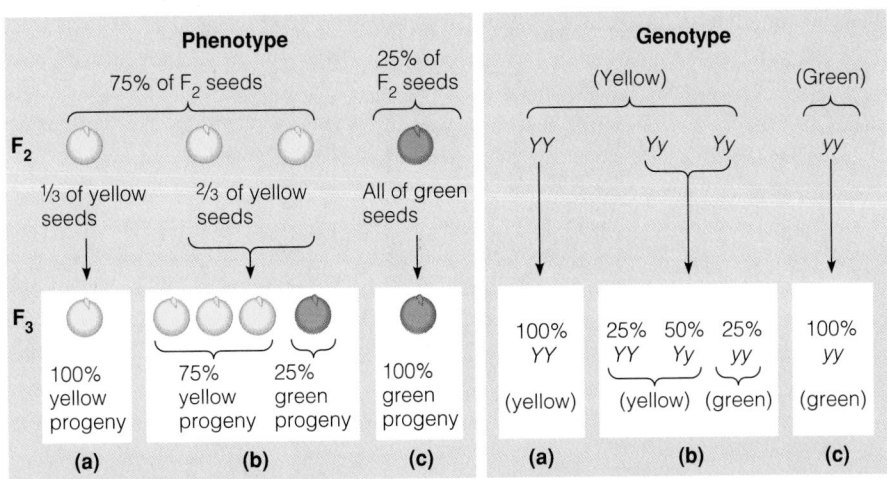

Figure 16-13   Analysis of F₂ Pea Plants by Self-Fertilization. The F₂ plants of Figure 16-12 were analyzed through self-fertilization. The phenotypic results are shown on the left and the genotypes (deduced later) on the right. **(a)** One-third of the yellow F₂ progeny of Figure 16-12 (25% of the total F₂ progeny) breed true for yellow seed color upon self-fertilization, because they are genotypically *YY.*

(b) Two-thirds of the yellow F₂ progeny (50% of the total F₂ progeny) yield yellow and green seeds upon self-fertilization, in a ratio of 3:1 (just as the F₁ plants of Figure 16-12 did upon self-fertilization). **(c)** All the green F₂ seeds (25% of the total F₂ progeny) breed true for green seed color upon self-fertilization, because they are genotypically *yy.*

made the same surprising observation: The recessive trait that had seemingly been lost in the F₁ generation appeared among the progeny in the next generation, the **F₂ generation.** Moreover, for each of the seven characters, the ratio of dominant to recessive phenotypes in the progeny was always about 3:1. In the case of seed color, for instance, plants grown from the yellow F₁ seeds produced about 75% yellow seeds and 25% green (Figure 16-12c).

This outcome, of course, was quite different from what had been obtained when the true-breeding yellow seeds of the parent strain had been self-fertilized. Clearly, there was an important difference between the yellow seeds of the P₁ stock and the yellow seeds of the F₁ generation. They looked alike, but the former bred true, whereas the latter did not.

Next, Mendel investigated the F₂ plants through self-fertilization (Figure 16-13). The F₂ plants that showed the recessive trait (green, in the case of seed color) gave the simplest results: They always bred true (Figure 16-13c), suggesting that they were genetically identical to the green-seeded P₁ strain with which Mendel had begun. F₂ plants with the dominant trait yielded a more complex pattern, however. One-third bred true for the dominant trait (Figure 16-13a) and therefore seemed to be identical to the P₁ plants with the dominant (yellow-seed) trait. However, the other two-thirds of the yellow-seeded F₂ plants produced progeny with both dominant and recessive phenotypes, in a 3:1 ratio (Figure 16-13b)—the same ratio that had arisen from the F₁ self-fertilization.

The consistency with which Mendel obtained these results led him to conclude that genetic information specifying the recessive trait must somehow be present in at least some of the hybrid seeds and plants, even though the trait is not always displayed. This conclusion was consistent with

results from another set of experiments, in which F₁ hybrids were crossed with the original parent strains, a process called **backcrossing** (Figure 16-14). The F₁ hybrids backcrossed to the dominant parent strain always produced progeny with the dominant trait (Figure 16-14a). When backcrossed to the recessive parent, though, they yielded a mixture of plants with dominant and recessive traits, in a ratio of 1:1 (Figure 16-14b). Moreover, the dominant progeny from the latter cross behaved just like the F₁ hybrids: Upon self-fertilization, they gave rise to a 3:1 mixture of phenotypes (Figure 16-14c), and upon backcrossing to the recessive parent, they yielded a 1:1 ratio of dominant to recessive progeny (Figure 16-14d). (An alternative way of diagramming crosses, called the *Punnett square,* is shown in Problem 16-7 at the end of the chapter.)

Mendel's Laws

After a decade of careful work, Mendel came to several important conclusions, which have since come to be known as **Mendel's laws of inheritance.** The first principle to emerge from Mendel's work was that phenotypic traits are determined by discrete "factors" that are present in most organisms as pairs of "determinants." Today, we call these "factors" *genes* and "determinants" *alleles* (alternative forms of genes). Mendel's conclusion seems almost self-evident to us, but it was an important assertion in his day. At that time, most scientists favored a *blending theory of inheritance,* one that viewed traits such as yellow and green seed color rather like cans of paint that are poured together to yield intermediate results. Other investigators had described the nonblending nature of inheritance before Mendel, but without the accompanying data and mathematical analysis that were

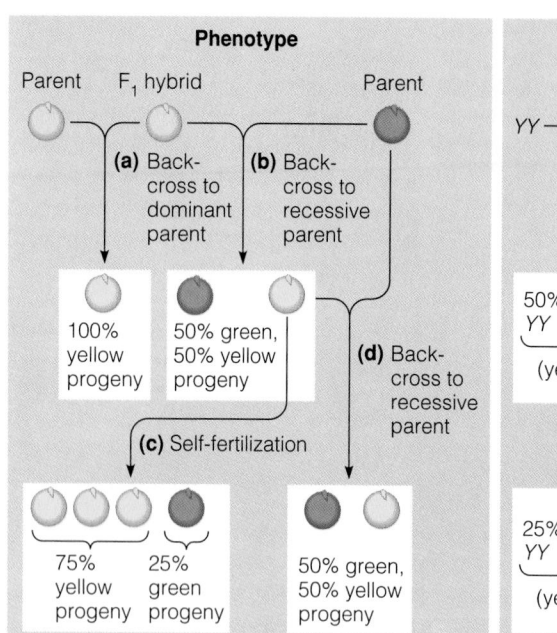

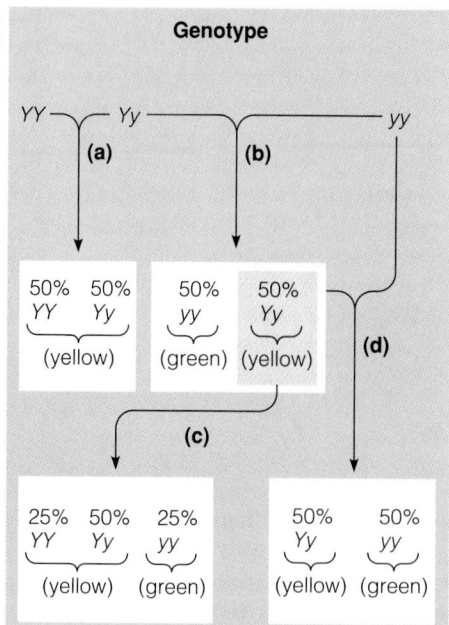

Figure 16-14 Analysis of F₁ Hybrids by Backcrossing. The F₁ hybrids of Figure 16-12 were analyzed by backcrossing to the parent (P₁) strains. The phenotypic results are shown on the left and the genotypes (deduced later) on the right. **(a)** Upon backcrossing of the F₁ hybrid (*Yy*) to the dominant parent (*YY*), all the progeny have yellow seeds, because the genotype is either *YY* or *Yy*. **(b)** Backcrossing to the recessive parent (*yy*) yields yellow (*Yy*) or green (*yy*) seeds in equal proportions. **(c)** These yellow-seeded progeny will give rise, upon self-fertilization, to a 3:1 mixture of yellow and green seeds (just as with the F₁ hybrids in Figure 16-12). **(d)** Backcrossing of the yellow-seeded progeny to the homozygous recessive parent again yields a 1:1 mixture of yellow and green seeds, as in the backcross of part b.

Mendel's special contribution. Mendel's breakthrough is especially impressive when we recall that he formulated his theory before anyone had seen chromosomes.

The Law of Segregation. Of singular importance to the development of genetics was Mendel's conclusion regarding how genes are parceled out during gamete formation. According to the **law of segregation,** *the two alleles of a gene are distinct entities that segregate, or separate from each other, in the formation of gametes.* In other words, the two alleles retain their identities even when present together in a hybrid organism, and they emerge in later generations unchanged.

The Law of Independent Assortment. In addition to the crosses already described, each of which focused on a single pair of alleles, Mendel also studied *multifactor* crosses, between plants that differed in *several* characters. In addition to differing in seed color, for example, the plants he crossed might differ in seed shape and flower position. As in his single-factor crosses, he used parent plants that were true-breeding (homozygous) for the characters he was testing and generated F₁ hybrids heterozygous for each character. He then self-fertilized these hybrids and determined the frequency with which the dominant and recessive forms of the several characters appeared among the progeny.

Mendel found that all possible combinations of traits appeared in the F₂ progeny, and he concluded that all possible combinations of the different alleles must have been present among the F₁ gametes. Furthermore, from the proportions in which the various phenotypes were found in the F₂ generation, Mendel was also able to deduce that all possible combinations of alleles occured in the gametes with equal frequencies. In other words, *the two alleles of a gene segregate independently of the alleles of other genes.* This is the **law of independent assortment,** another cornerstone of genetics. (Later it would be shown that this law only applies to genes on different chromosomes or very far apart on the same chromosome.)

Discovery of the Cytological Basis of Inheritance

Mendel's findings lay dormant in the scientific literature until 1900, when his paper was rediscovered almost simultaneously by three other European biologists. In the meantime, much had been learned about the cytological basis of inheritance. By 1875, for example, microscopists had identified chromosomes with the help of stains produced by the developing aniline dye industry. At about the same time, fertilization was shown to involve the fusion of two parent nuclei, one contained within the egg and the other contributed by the sperm.

The first suggestion that chromosomes might be the bearers of genetic information was made in 1883. Within ten years, chromosomes had been studied in dividing cells and had been seen to split longitudinally into two apparently identical daughter chromosomes. This led to the realization that the number of chromosomes per cell remained constant during the development of an organism. With the invention of better optical systems, more detailed analysis of chromosomes became possible. Mitotic cell division was shown to involve the movement of identical daughter chromosomes to opposite poles, thereby ensuring that daughter cells would have exactly the same complement of chromosomes as their parent cell.

Against this backdrop came the rediscovery of Mendel's paper, followed almost immediately by three crucial studies that established chromosomes as the carriers of Mendel's factors. The investigators were Edward Montgomery, Theodor Boveri, and Walter Sutton. Montgomery's contribution was to recognize the existence of homologous chromosomes. From careful observations of insect chromosomes, he concluded that in most cells the chromosomes could be grouped into pairs, with one member of the pair of maternal origin and the other of paternal origin. Furthermore, the two chromosomes of each type synapsed in the "reduction division" (now called meiosis) of gamete formation, a process that had been reported a decade or so earlier. To this, Boveri added the observation, based on his studies of fertilization in sea urchins, that chromosomes have functional individuality.

Sutton, meanwhile, was studying meiosis in the grasshopper. In 1902, he made the important observation that the orientation of each pair of homologous chromosomes (bivalent) on the metaphase plate at metaphase I is purely a matter of chance. Any homologous pair, in other words, may lie with the maternal or paternal chromosome toward either pole, regardless of the positions of other pairs. Many different combinations of maternal and paternal chromosomes are therefore possible in the gametes that an individual produces.

The Chromosome Theory of Inheritance

In the period 1902–1903, Sutton put all these observations together into a coherent theory describing the role of chromosomes in inheritance. Sutton's theory can be summarized as five main points:

1. Nuclei of all cells except those of the germ line (sperm and eggs) contain two sets of homologous chromosomes, one set of maternal origin, the other of paternal origin.

2. Chromosomes retain their individuality and are genetically continuous throughout the life cycle of an organism.

3. Maternal and paternal homologues synapse during meiosis, move to opposite poles of the division spindle, and are thus separated in the process.

4. The two sets of homologous chromosomes in a diploid are equivalent in a functional sense, each carrying one complement of genetic determinants (genes).

5. The maternal and paternal members of different homologous pairs segregate independently at meiosis.

Thus, chromosomes came to be appreciated as the physical basis for understanding how Mendel's factors could be carried, transmitted, and segregated. The presence of two sets of homologous chromosomes in the cell parallels Mendel's suggestion of two determinants for each phenotypic trait. Also, the segregation of homologous chromosomes during the meiotic process of gamete formation gives the basis for Mendel's law of segregation; the random, independent way homologous pairs orient at metaphase I and subsequently segregate accounts for his law of independent assortment.

Figures 16-15 and 16-16 illustrate the chromosomal basis of Mendel's laws, using examples from Mendel's pea

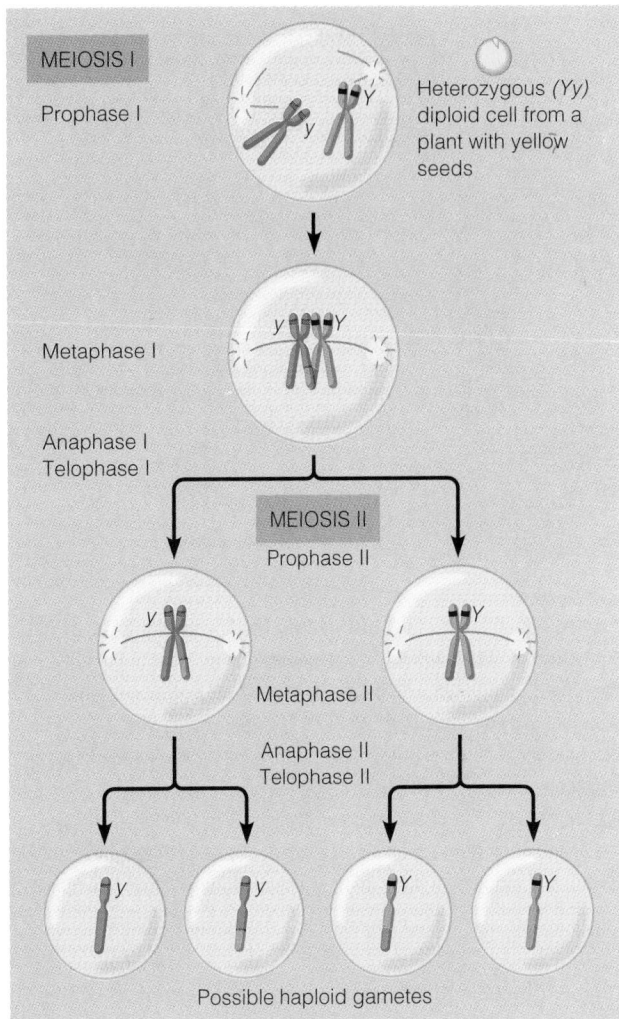

Figure 16-15   The Meiotic Basis for Mendel's Law of Segregation. Segregation of seed color alleles during meiosis is illustrated for the case of a pea plant heterozygous for this character. Peas have seven pairs of chromosomes, but only the homologous pair bearing the seed color alleles is shown here. During the first meiotic division (meiosis I), homologous chromosomes (each consisting of two sister chromatids) pair during prophase I, align as a pair on the metaphase plate at metaphase I, and segregate into separate cells at anaphase I and telophase I. In meiosis II, sister chromatids segregate to different daughter cells. The result is four haploid daughter cells, each of which has one allele for seed color.

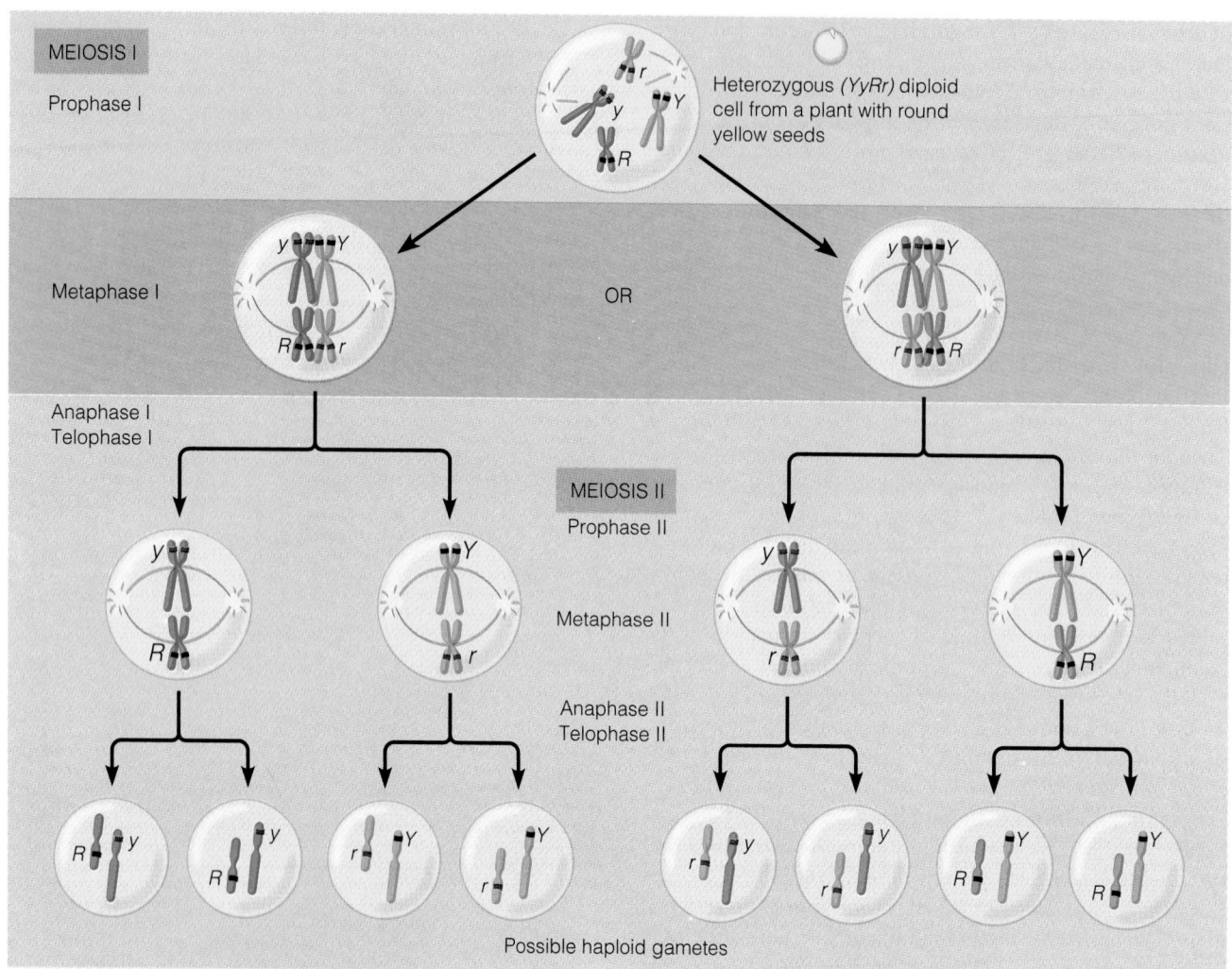

MEIOSIS I

Prophase I

Heterozygous (YyRr) diploid cell from a plant with round yellow seeds

Metaphase I

OR

Anaphase I
Telophase I

MEIOSIS II

Prophase II

Metaphase II

Anaphase II
Telophase II

Possible haploid gametes

Figure 16-16   The Meiotic Basis for Mendel's Law of Independent Assortment.   Independent assortment of the alleles of two genes on different chromosomes is illustrated by meiosis in a pea plant heterozygous for seed color (*Yy*) and seed shape (*Rr*). Pea plants have seven pairs of chromosomes, but only the two pairs bearing the seed color alleles and the seed shape alleles are shown. During meiosis, the segregation of the seed color alleles occurs independently of the segregation of the seed shape alleles. The basis of this independent assortment is found at metaphase I, when the homologous pairs (biva-lents) align on the metaphase plate. Each bivalent can face in either direction. Consequently, there are two alternative situations: Either the paternal homologues (blue) can both face the same pole of the cell, with the maternal homologues (red) facing the other pole, or they can face different poles. The arrangement at metaphase I determines which homologues subsequently go to which daughter cells. Since the alternative arrangements occur with equal probability, all the possible combinations of the alleles therefore occur with equal probability in the gametes.

plants. The basis for Mendel's law of segregation is shown in Figure 16-15 for the alleles governing seed color in a heterozygous pea plant with genotype *Yy*. (Peas have seven pairs of chromosomes, but only the pair bearing the alleles for seed color is shown.) During meiosis, the two homologous chromosomes, each with two sister chromatids, synapse during prophase I, align together on the metaphase plate at metaphase I, and then segregate into separate daughter cells. The second meiotic division separates sister chromatids, so that each haploid cell ends up with only one allele for seed color, either *Y* or *y*.

The basis for the law of independent assortment is shown in Figure 16-16 for the chromosomes carrying the genes for seed color (alleles *Y* and *y*) and seed shape (alleles *R* and *r*, where *R* stands for round seeds, a dominant trait, and *r* for wrinkled seeds, a recessive trait). The specific reason for independent assortment is that there are two possible, equally likely arrangements of the chromosome pairs at metaphase I. Of the *YyRr* cells that undergo meiosis, half will produce gametes like the four at the bottom left of the figure, and half will produce gametes like the four at the bottom right. Therefore, the *YyRr* plant will produce equal numbers of gametes of the eight types.

As we will discuss shortly, the law of independent assortment is *guaranteed* to hold only for genes on *different* chromosomes. It is remarkable that Mendel happened to choose seven independently assorting characters in an organism that has only seven pairs of chromosomes. Mendel was also fortunate that each of the seven characters was controlled by a single gene (pair of alleles). Perhaps he concentrated on pea strains that, in preliminary experiments, gave him the most consistent and comprehensible results.

The Molecular Basis of Chromosome Homology

What does it mean to say that the members of a homologous pair of chromosomes are "closely similar" but may carry "different versions" (alleles) of their genes? Simply put, it means that the sequences of nucleotides on their DNA molecules are almost but not entirely identical. Thus, in general, the homologous chromosomes carry the same genes in exactly the same order. Occasional differences in sequence along the DNA molecules are the molecular basis of the differences between alleles of the same gene. Such differences arise through mutation, and the different alleles for a gene that we find in a population—whether of pea plants or people—originally arose as changes (mutations) in an ancestral gene. Differences in phenotype between two individuals (in the same environment) result from differences in their alleles (genotype). The alleles are usually "expressed" by transcription into RNA and translation into proteins. It is ultimately the functioning of proteins that creates an organism's phenotype. As we will see in Chapter 18, a change as small as a single base pair in a gene can lead to a changed protein that is different enough to cause an observable change in the organism's phenotype—in fact, it can be lethal. (Furthermore, as we saw in Chapter 14, small sequence differences in noncoding DNA, while usually not affecting phenotype, can be useful as markers for RFLP analysis; see Box 14C).

The underlying similarity in nucleotide sequence between DNA molecules—*DNA homology*—is probably the basis of the appropriate pairing of homologous chromosomes in meiosis. In any case, DNA homology is essential for normal crossing over between homologous chromosomes. One popular model for synapsis holds that the correct alignment of homologous chromosomes is brought about before completion of the synaptonemal complex by some sort of base-pairing interaction between matching regions of DNA in the two chromosomes. Then, after completion of the synaptonemal complex, DNA recombination is completed at these (and perhaps other) sites. According to this model, crossing over and synaptonemal complex formation are mutually dependent—an idea that is supported by mutational studies with yeast.

Genetic Variability: Recombination and Crossing Over

The segregation and independent assortment of members of homologous pairs of chromosomes result in the independent assortment of the alleles of genes carried on the different chromosomes, as we saw in Figure 16-16. Figure 16-17a

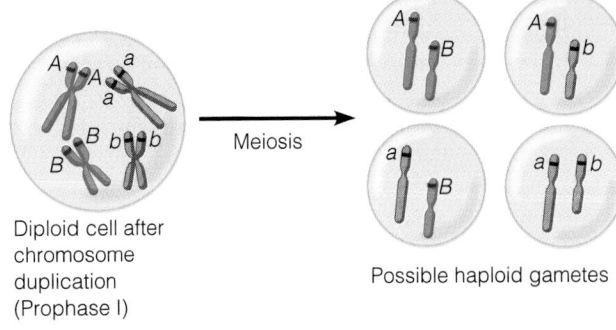

(a) Unlinked genes assort independently

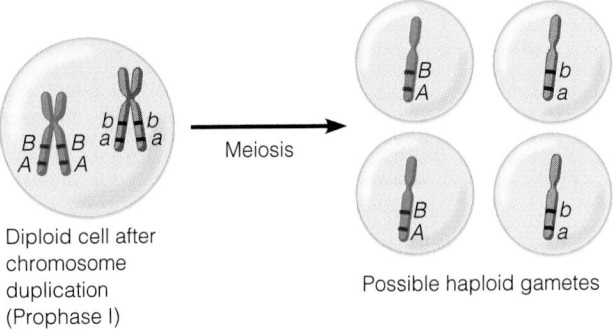

(b) Linked genes end up together in the absence of crossing over

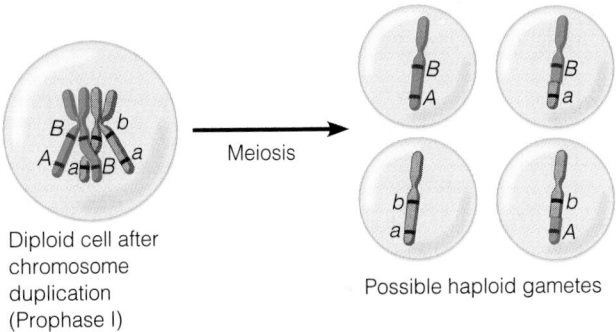

(c) Linked genes do not always end up together if crossing over occurs

Figure 16-17   Segregation and Assortment of Linked and Unlinked Genes.   In this figure, A and a are alleles of one gene, and B and b are alleles of another gene. **(a)** Alleles of genes on different chromosomes segregate and assort independently during meiosis; allele A is as likely to occur in a gamete with allele B as it is with allele b. **(b)** Alleles of genes on the same chromosome remain linked during meiosis in the absence of crossing over; in this case, allele A will occur routinely in the same gamete with allele B, but not with allele b. **(c)** Alleles of genes on the same chromosome can become interchanged when crossing over takes place, so that allele A occurs not only with allele B but also with allele b, and so forth.

summarizes the outcome of meiosis for a generalized version of the same situation. Here we have a diploid organism that is heterozygous for two genes on nonhomologous chromosomes, with the allele pairs called *Aa* and *Bb*. Meiosis in such an organism will produce gametes in which allele *A* is just as likely to occur with allele *B* as it is with allele *b*, and *B* is just as likely to occur with *A* as it is with *a*. But what happens if genes *A* and *B* are on the same chromosome? In that case, alleles *A* and *B* will routinely occur together in the same gamete, as will alleles *a* and *b* (Figure 16-17b).

However, even for genes on the same chromosome, some scrambling of alleles occurs because of the phenomenon of crossing over, which results in recombination. This process, which occurs while homologues are synapsed during prophase I of meiosis, involves an interchange of genetic material between homologous chromosomes. Recombination therefore contributes further to the genetic variability seen in sexually reproducing species by providing a mechanism for rearranging genetic material *within* a pair of homologous chromosomes to create chromosomes with new combinations of alleles (Figure 16-17c).

Recombination was originally discovered in studies with the fruit fly, *Drosophila melanogaster*, conducted by Thomas H. Morgan and his colleagues beginning around 1910. We will therefore turn to Morgan's work to understand recombination, where we encounter the concept of gene linkage.

Linked Genes and Linkage Groups

The fruit flies used by Morgan and his colleagues had certain advantages over Mendel's pea plants as objects of genetic study, not the least of which was the fly's relatively brief generation time, about two weeks (versus months for pea plants). Unlike Mendel's peas, however, the fruit flies did not come with a ready made variety of phenotypes and genotypes. Whereas Mendel was able to purchase seed stocks of different true-breeding varieties, the only type of fruit fly available to Morgan initially was what has come to be known as the **wild type,** or "normal" organism. Morgan and his colleagues therefore had to generate variants—that is, mutants—for their genetic experiments.

Morgan and his coworkers bred large numbers of flies and selected mutant individuals, flies having detectable phenotypic modifications that were heritable. (Later, X-irradiation was used to enhance the mutation rate, but in their early work Morgan's group depended entirely on spontaneous mutations.) Within five years, they were able to identify about 85 different mutants, each carrying a mutation in a different gene. Each of these mutants could be propagated as a laboratory stock and used for matings as needed.

One of the first things Morgan and his colleagues realized as they began analyzing their mutants was that, unlike the "factors" Mendel studied in peas, the genes identified by the fruit fly mutations did not all assort independently. In-

stead, some genes behaved as if they were linked together, and the new combinations of alleles predicted by Mendel were infrequent or even nonexistent. In fact, it was soon recognized that the fruit fly genes could be classified into four **linkage groups.** Each group consisted of a collection of **linked genes,** genes that were usually inherited together. Morgan quickly realized that the number of linkage groups correlated exactly with the number of different chromosomes in the organism, since the haploid chromosome number for *Drosophila* is four. The conclusion was obvious: Each chromosome is the physical basis for a specific linkage group.

Recombination, Crossing Over, and Chiasma Formation

Although the genes they discovered could all be ordered into linkage groups, Morgan and his colleagues found that the linkage within such groups was incomplete. Most of the time, genes that were known to be linked (and therefore on the same chromosome) assorted together, as would be expected. Sometimes, however, two or more such traits would appear in the offspring in nonparental combinations. This phenomenon of less-than-complete linkage was called *recombination* because the different alleles for the genes appeared in new and unexpected associations in the offspring.

To explain such recombinant offspring, Morgan proposed that homologous chromosomes could exchange segments, presumably by some sort of breakage-and-fusion event, as illustrated in Figure 16-18a. By this process, which Morgan termed *crossing over*, a particular allele or group of alleles initially present on one member of a homologous pair of chromosomes could be transferred to the other chromosome in a reciprocal manner.

In the example of Figure 16-18a, two homologous chromosomes, one with alleles *A* and *B* and the other with alleles *a* and *b*, lie side by side at synapsis. Portions of an *AB* chromatid and a nonsister *ab* chromatid then exchange positions, producing two recombinant chromatids, one with the alleles *A* and *b* and the other with the alleles *a* and *B*. Each of the four chromatids ends up in a different gamete at the end of the second meiotic division, so the products of meiosis will include two *parental* gametes and two *recombinant* gametes, assuming a single crossover event (Figure 16-18b).

We now know that crossing over occurs during the pachytene stage of meiotic prophase I, at a time when sister chromatids are packed tightly together and it is hard to see what is happening. Later in prophase, as the chromatids begin to separate at diplotene, each of the four chromatids in a bivalent can be identified as belonging to one or the other of the two homologues. Wherever crossing over has occurred between nonsister chromatids, the two homologues remain attached to each other, forming an X-shaped chiasma. As mentioned earlier, a *chiasma* is the cytological manifestation of a crossover event that occurred earlier in prophase.

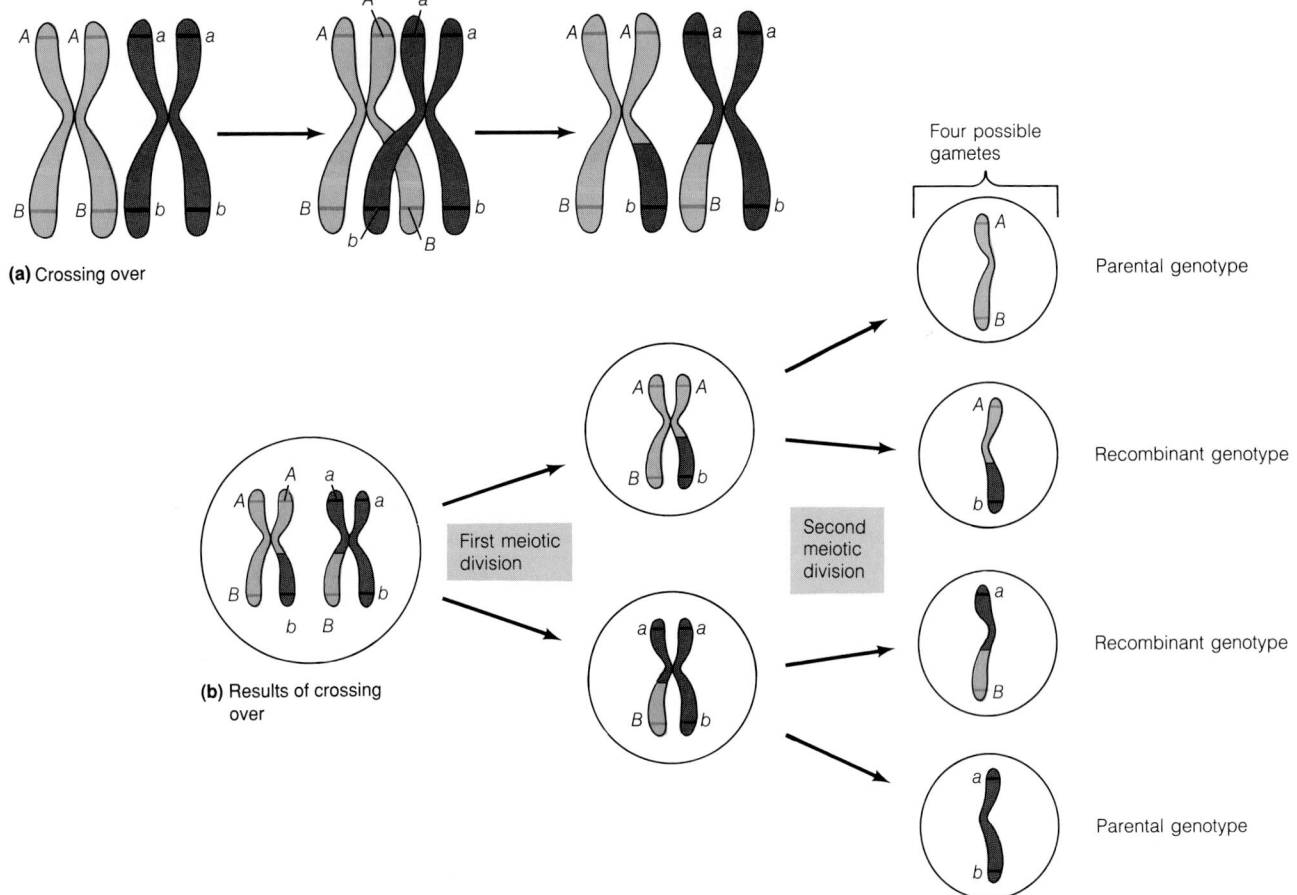

(a) Crossing over

(b) Results of crossing over

First meiotic division

Second meiotic division

Four possible gametes

Parental genotype

Recombinant genotype

Recombinant genotype

Parental genotype

Figure 16-18 Crossing Over and Recombination.   As in Figure 16-17, *A* and *a* are alleles of one gene, and *B* and *b* are alleles of another gene. **(a)** Two homologous chromosomes, one bearing alleles *A* and *B* and the other alleles *a* and *b*, are paired at prophase I. Two nonsister chromatids undergo crossing over, causing portions of each to exchange places. The result is two recombinant chromatids, with alleles *A* and *b* on one chromatid and *a* and *B* on the other. **(b)** Each chromatid ends up in a separate gamete, two of which have the parental genotypes (*AB* and *ab*) and two the recombinant genotypes (*Ab* and *aB*).

At the first meiotic metaphase, homologous chromosomes are almost always held together by at least one chiasma (if not, they may not segregate properly). Many bivalents contain multiple chiasmata. Human bivalents, for example, usually contain two or three chiasmata per bivalent, because multiple crossover events routinely occur between paired homologues. To be of genetic significance, crossing over must involve nonsister chromatids. In some species, exchanges between sister chromatids can also occur, but such exchanges are of no genetic consequence because sister chromatids are genetically identical.

Recombination Frequency and Gene Mapping

Eventually, it became clear to Morgan and others that the frequency of occurrence of recombinant progeny differed for different pairs of genes within the various linkage groups, but that for a specific pair of genes it was remarkably constant. This led to the suggestion that the frequency of observed recombination between two genes might be a measure of how far the genes are from each other along the chromosome. If the probability of crossing over is the same at every point along a chromatid (an assumption that appears to be justified much of the time), then genes located very close to each other would be less likely to become separated by a crossover event between them than would genes that are far apart.

It was, in fact, quickly realized that the frequency of recombination, expressed as the percentage of progeny that are recombinant, was a useful means of roughly quantifying distance between genetic markers. This insight led Alfred Sturtevant to suggest in 1911 that recombination data could be used to determine the sequence of genes along the chromosomes of *Drosophila*. Thus, a chromosome was now being viewed as a linear string of genes that could be ordered and spaced along its length on the basis of recombination data.

The sequencing and spacing of genes on a chromosome on the basis of recombinant frequencies is called **genetic mapping.** In the construction of such maps, recombinant frequency becomes the map distance. If, for example, alleles *A* and *B* of Figure 16-18b appear among the progeny in their parental combination 85% of the time and in the recombinant combination 15% of the time, we conclude that the two genes are linked (are on the same chromosome) and are 15 **map units** apart. This approach has been used to map the chromosomes of many species of plants and animals, and for bacteria and viruses, as well. However, because bacteria and viruses do not reproduce sexually, the methods used to generate recombinants are different.

Recombination in Bacteria and Viruses

From what we have discussed so far, we might expect genetic recombination to be restricted to organisms that reproduce sexually, since recombination depends on crossing over between homologous chromosomes. As we have seen, sexual reproduction brings with it an opportunity for recombination once every generation, because the necessary juxtaposition of homologous chromosomes is an intrinsic part of the meiotic process. Viruses and prokaryotes, on the other hand, might seem to be poor candidates for recombination. They reproduce asexually and have haploid genomes, with no obvious mechanisms for regularly bringing genomes from two different parents together.

Yet viruses and bacteria are also capable of genetic recombination. In fact, recombination data allowed the extensive mapping of viral and bacterial genomes well before the advent of modern DNA technology. To understand how recombination can occur despite a haploid genome, we need to examine the several mechanisms by which two haploid genomes, or portions of genomes, can be brought together within the same cell.

Bacteriophages: Recombination During Co-infection

Bacteriophages have been especially valuable to our understanding of genes and gene mapping. Much of our earliest understanding of genes at the molecular level, as well as the vocabulary in which we express that information, comes from experiments with phages, particularly with the T-even phages and phage λ (see Box 14A).

Recombinant bacteriophages can arise if a single bacterial cell is simultaneously infected by genetically different versions of the same phage. Two infecting T4 phages, for example, might have slightly different DNA sequences, such that they have different alleles for some of their genes. Figure 16-19 illustrates a laboratory experiment in which co-infection by two genotypically different T4 phages is used to create recombinant phage progeny. The two parent phage

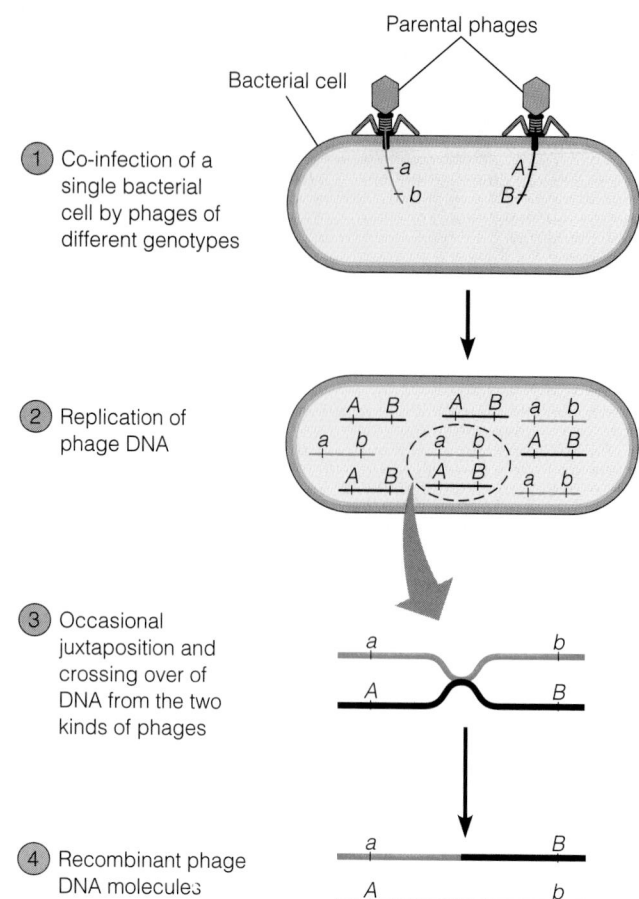

Figure 16-19 Genetic Recombination of Bacteriophages.
① Two parental phage stocks with different genotypes are used to co-infect a bacterial cell, thereby ensuring the simultaneous presence of both kinds of phage genomes in a single bacterial cell. One of the parental phages carries the mutant alleles *a* and *b*, while the other has alleles *A* and *B*. ② As the phage DNA molecules replicate, ③ they occasionally become juxtaposed in a way that allows crossing over and recombination to occur, ④ giving rise to recombinant phage molecules. The frequency of occurrence of recombinant genotypes provides a measure of the distance between markers. (As in other cases, biologists determine the genotypes of the progeny phage particles from the phenotypes they display. An example of a phage phenotype is the size of the plaques it forms on lawns of bacteria.)

stocks are mixed together, and bacteria are then infected with enough phage particles that virtually every bacterium is infected with some of both genotypes. The host cell therefore contains complete copies of the genomes of both parent phages. As the phages replicate in the bacterial cell, their DNA molecules occasionally become juxtaposed in ways that allow recombination to occur. Recombinant progeny are generated with a frequency that depends on the distance between the genes under study, just as in diploid organisms. In general, the farther apart the genes, the greater the likelihood of recombination between them.

In phage recombination, the crossing over involves relatively short, naked DNA molecules, rather than chromatids.

The simplicity of the situation has facilitated research on the molecular mechanisms of recombination and the proteins that catalyze the process. It is clear that general recombination of phage DNA involves the precise alignment of homologous DNA molecules at the region of crossing over—a requirement that presumably holds for the recombination of eukaryotic and prokaryotic DNA as well.

Bacteria: Recombination by Transformation or Transduction

We have already encountered one means of getting DNA from one bacterial cell into another. Recall from Chapter 14 the experiments with rough and smooth pneumococcus bacteria that led Frederick Griffith to postulate the existence of some sort of *transforming principle,* subsequently identified as DNA by Oswald Avery and his colleagues. The ability of a bacterial cell to take up DNA and to incorporate segments of that DNA into its own genome is called **transformation** (Figure 16-20a).

Although initially described for a single genetic trait in a specific bacterium, transformation is now recognized as a general mechanism whereby some (though by no means all) kinds of bacteria can acquire genetic information from other cells, genetic information they can then pass on to progeny. Transformation can occur naturally whenever bacterial cells have access to DNA from other cells, but it occurs

more readily with some bacteria than with others. For example, cells of *Streptococcus pneumoniae* ("pneumococcus") are easily transformed, but appropriate conditions for transformation with *E. coli* were not discovered until recently.

In **transduction,** DNA is brought into a bacterial cell by a DNA-containing bacteriophage. Most of the time, phages contain only their own DNA, but some phages occasionally incorporate some host cell DNA into their progeny particles. When a phage particle carrying host cell DNA infects another bacterium, the phage injects the fragment of DNA as usual, acting in effect like a syringe carrying DNA from one bacterial cell to the next (Figure 16-20b). Phages capable of carrying host cell DNA from one cell to another are called *transducing phages.*

Bacteriophage P1, which infects *E. coli,* is a transducing phage that has been especially useful for gene mapping in that bacterium. The amount of DNA that will fit into the protein coat of a bacteriophage is small compared to the size of the bacterial genome. Two genes—or, more generally, genetic markers (particular nucleotide sequences)—must therefore be close together for both of them to be carried into a bacterial cell by a single phage particle. This is the basis of *cotransductional mapping,* in which the proximity of one marker to another is determined by quantifying the frequency with which the markers accompany each other in a transducing phage particle. The closer two markers are, the more likely they are to be cotransduced into a bacterial cell. When such

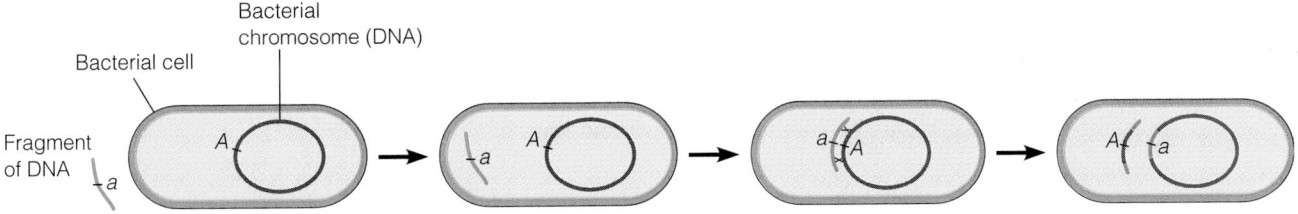

(a) Transformation of a bacterial cell by exogenous DNA

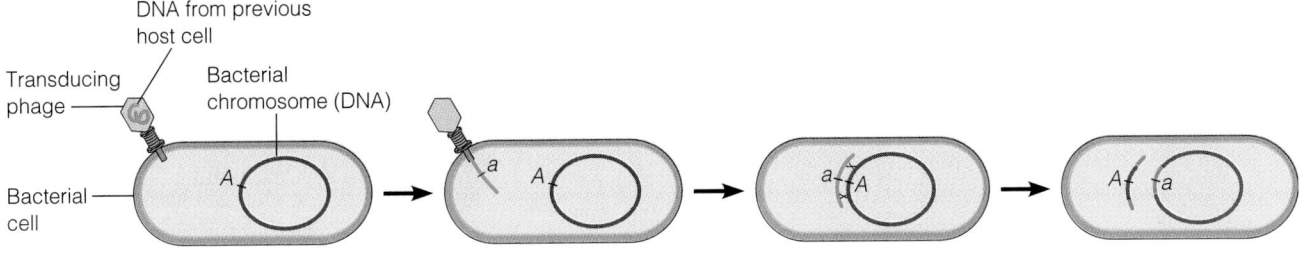

(b) Transduction of a bacterial cell by a transducing phage

Figure 16-20 Transformation and Transduction in Bacterial Cells. **(a)** Transformation involves uptake by the bacterial cell of exogenous DNA, which occasionally becomes integrated into the bacterial genome by two crossover events (indicated by X). The exogenous DNA will be detectable in progeny cells only if integrated into the bacterial chromosome, because the fragment of DNA initially taken up does not normally have the capacity to replicate itself autonomously in the cell.

(The main exception is an intact plasmid.) **(b)** Transduction involves the introduction of exogenous DNA into a bacterial cell by a phage. Once injected into the host cell, the DNA can become integrated into the bacterial genome in the same manner as in transformation. In both cases, linear fragments of DNA that end up outside the bacterial chromosome are eventually degraded by nucleases. In this figure, the letters *A* and *a* represent alleles of the same gene.

studies are carried out with the transducing phage P1, it is found that the markers cannot be cotransduced if they are separated on the bacterial DNA by more than about 10^5 nucleotide pairs. This finding agrees with the observation that the P1 phage has a genome of about this size.

Sex and the Single Cell: Conjugation in Bacteria

In addition to transformation and transduction, some bacteria can also transfer DNA from one cell to another by a process called **conjugation.** As the name suggests, conjugation resembles a mating in that one bacterium is clearly identifiable as the donor (often called a "male") and another as the recipient ("female"). Furthermore, the entire bacterial genome can be transferred, although this is very rare. The existence of conjugation was postulated in 1946 by Joshua Lederberg and Edward L. Tatum, who were the first to show that recombination could occur in bacteria. They also established that physical contact between two cells was necessary for conjugation to take place. We now understand that conjugation involves the directional transfer of DNA from the donor bacterium to the recipient bacterium. Recombinant cells always arise from the recipient, not from the donor.

The F Factor. The ability of an *E. coli* cell to act as a donor in conjugation is due to a piece of DNA called the **F factor** (F for fertility), which can exist either as an independent, replicating plasmid in the cell or as a segment of DNA within the bacterial chromosome. Donor bacteria that carry the F plasmid are designated F^+, and recipient cells, which usually lack the F factor completely, are designated F^-. Donor cells grow long, hairlike projections called **sex pili** (singular: **pilus**) on their surface (Figure 16-21a). The end of a sex pilus specifically adheres to the surface of a recipient

cell. This connection facilitates the formation of a transient cytoplasmic **mating bridge,** through which DNA can be transferred from the donor cell to the recipient cell (Figure 16-21b).

When an F factor is present in a donor cell as a plasmid, a copy of it is transferred very quickly and efficiently during conjugation, converting the recipient cell from F^- to F^+, as shown in Figure 16-22a. Transfer always begins at a point on the plasmid called its **origin of transfer,** represented in the figure by an arrowhead. As the arrowhead implies, the transfer process has an inherent directionality. Although conjugation involves transfer of an F factor to the F^- cell, the original donor cell does not lose its F factor in the process; the transfer of the F factor is coupled to its replication, so that an F plasmid is left behind in the donor cell. Mixing an F^+ population of bacteria with an F^- population can therefore result in a population of cells that is entirely F^+. "Maleness" is in a sense infective, and the F factor is the infective agent.

Hfr Cells and Bacterial Chromosome Transfer. So far, we have seen that donor and recipient bacterial cells are generally defined in terms of the presence or absence of the F factor, and we have seen that the F factor is transmissible by conjugation. But how do recombinant bacterial progeny arise by this means?

Although usually present as a plasmid, the F factor occasionally becomes integrated into the bacterial chromosome, as shown in Figure 16-22b. (Integration occurs by crossing over between short DNA sequences on the chromosome and similar sequences on the F factor.) The integration process converts an F^+ cell into an **Hfr cell,** which is capable of producing a *h*igh *f*requency of *r*ecombination in further matings because it can now transfer *genomic* DNA during conjugation.

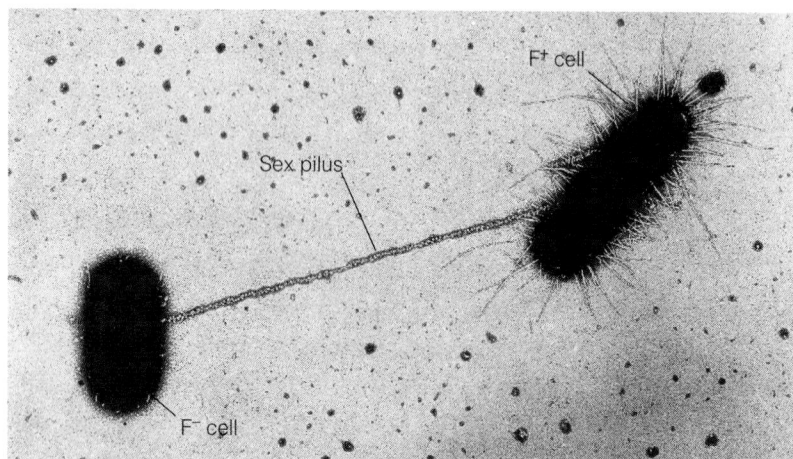

(a) Sex pilus

$\overline{1\ \mu m}$ **(b)** Mating bridge $\overline{0.5\ \mu m}$

Figure 16-21   The Cellular Apparatus for Bacterial Conjugation.   **(a)** The donor bacterial cell on the right, an F^+ cell, has numerous slender appendages, called pili, on its surface. Some of these pili are sex pili, including the very long pilus leading to the other cell, an

F^- cell. Made of protein encoded by a gene on the F factor, sex pili enable a donor cell to attach to a recipient cell. **(b)** Subsequently, a cytoplasmic mating bridge forms, through which DNA is passed from the donor cell to the recipient cell (TEMs).

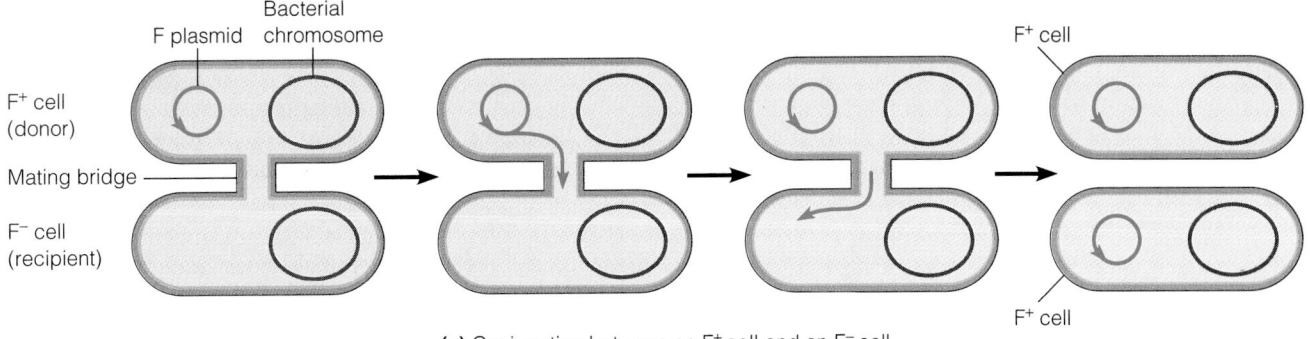

(a) Conjugation between an F⁺ cell and an F⁻ cell

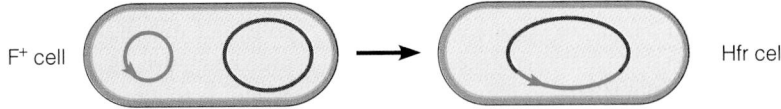

(b) Conversion of an F⁺ cell into an Hfr cell by integration of the F factor into the bacterial chromosome

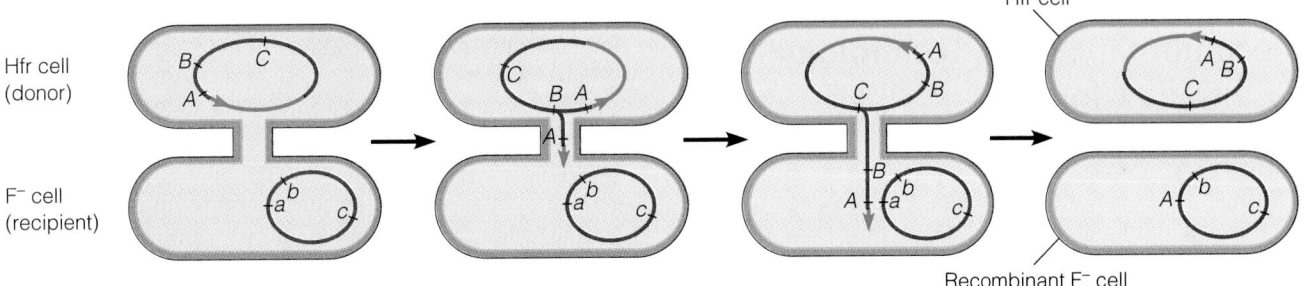

(c) Conjugation between an Hfr cell and an F⁻ cell

Figure 16-22 DNA Transfer by Bacterial Conjugation.
(a) Conjugation between an F⁺ donor bacterium and an F⁻ recipient involves the transfer of a copy of the F factor plasmid from the donor to the recipient, thereby converting the F⁻ cell to an F⁺ cell. Transfer of the F plasmid occurs through a mating bridge and always begins at the F factor's origin of transfer, indicated by the arrowhead. **(b)** Conversion of an F⁺ cell into an Hfr cell by integration of the F factor into the bacterial chromosome. **(c)** Conjugation between an Hfr donor cell and an F⁻ cell involves the transfer of a copy of the Hfr genome through a mating bridge into the F⁻ cell, beginning with the origin of transfer on the integrated F factor. Transfer is usually incomplete, because cells rarely remain in mating contact long enough for the entire bacterial chromosome to be transferred. Once inside the F⁻ cell, parts of the Hfr DNA can undergo recombination with the DNA of the F⁻ cell, just as in transformation or transduction (see Figure 16-20). Here, uppercase letters represent alleles carried by the Hfr; lowercase letters represent the corresponding alleles carried by the F⁻ cell. In the last step, allele *A* from the Hfr becomes recombined into the F⁻ cell's DNA in place of its *a* allele.

When an Hfr donor bacterium is mated to an F⁻ cell, DNA is transferred into the recipient cell. Instead of transferring just the F factor, however, the Hfr bacterium transfers at least part (and occasionally all) of its chromosomal DNA (retaining a copy, as in F⁺ DNA transfer). Transfer begins at the origin of transfer within the integrated F factor and proceeds in a direction dictated by the orientation of the F factor within the chromosome (Figure 16-22c). Notice that the DNA is transferred in a linear form, with a small part of the F factor at the leading end and the remainder at the trailing end.

Because the F factor is split in this way during the transfer process, only the recipient cells that receive a complete bacterial chromosome from the Hfr donor actually become Hfr cells themselves. Transfer of the whole chromosome is extremely rare, because it takes about 90 minutes. Usually, mating contact is spontaneously disrupted before transfer is complete, leaving the recipient cell with only a portion of the Hfr chromosome. As a result, genes located close to the origin of transfer on the Hfr chromosome ("downstream" of the origin of transfer) are much more likely to be transmitted to the recipient cell than are genes farther away.

Once part of the Hfr chromosome has been introduced into a recipient cell by conjugation, it can recombine with regions of the recipient's DNA that are homologous (similar) in sequence. The recombinant bacterial chromosomes that arise in this way contain some genetic information from the donor cell and some from the recipient. Only the Hfr DNA that successfully integrates by recombination into the recipient chromosome survives in the recipient cell and its progeny. Hfr DNA that is not integrated, as well as F⁻ DNA that leaves the chromosome in the recombination event, is eventually degraded by nuclease enzymes.

The correlation between the position of a gene on the bacterial chromosome and its likelihood of transfer can be used to map genes with respect to the origin of transfer and therefore with respect to one another. For example, if gene *A* of an Hfr is transferred in conjugation 95% of the time, gene *B* 70% of the time, and gene *C* 55% of the time, then the sequence of these genes is *A-B-C*, with gene *A* closest to the origin of transfer. Moreover, since the daughter cells of the recipient bacterium are recombinants, they can be used for genetic analysis. Typically, a cross is made between an Hfr strain and an F⁻ strain that differ in two or more genetic properties. After conjugation has taken place, the cells are plated on a nutrient medium on which recombinants can grow but "parent" strains cannot, thereby allowing the recombinants to be detected and recombinant frequencies calculated.

Recombinant DNA Technology and Gene Cloning

We have seen that genetic recombination is a naturally occurring process in which the underlying molecular process is the combining of pieces of DNA from two different sources into a single molecule, which thereafter carries a new combination of genes. In nature, the two sources of DNA are usually organisms of the same species. In animals and plants, for instance, it is an individual's two parents who were the original sources of the DNA that recombines during meiosis. Naturally occurring recombinant DNA usually differs from the parental DNA molecules only in its assortment of alleles; the fundamental identities and sequences of genes remain the same.

In the laboratory, there are no such limitations. Since the development of **recombinant DNA technology** in the 1970s, scientists have had at their disposal a collection of techniques for making and replicating recombinant DNA in the laboratory. Arising from basic research on the molecular biology of bacteria, these techniques enable researchers to manipulate and study genes from both prokaryotes and eukaryotes with greater ease and precision than was earlier thought possible. This is truly a new era in biology. Any segment of DNA can now be excised from any genome and spliced together with any other piece of DNA.

An essential aspect of recombinant DNA technology is the ability to replicate—or "clone"—specific pieces of DNA in order to prepare large enough amounts for research and other uses. This cloning is done by splicing the DNA of interest to the DNA of a genetic element, called a **cloning vector,** that can replicate autonomously when introduced into a cell grown in culture—usually, a bacterium such as *E. coli.* The cloning vector can be a plasmid or the DNA of a virus, usually a bacteriophage; in either case, the vector's DNA "passenger" is copied every time it replicates. In this way, it is possible to generate large quantities of specific genes or other DNA segments, and of their protein products as well, if the "passenger" genes are transcribed and translated in proliferating cells that carry the vector.

To appreciate the importance of recombinant DNA technology to basic research, we need to grasp the magnitude of the problem that biologists faced as they tried to study the genomes of eukaryotic organisms. As we have mentioned, most of our initial understanding of information flow in cells came from studies with bacteria and viruses. The genomes of bacteria, phages, and other viruses were mapped and analyzed in great detail. This work involved the use of ingenious genetic methods, but most of these were applicable only to bacteria and viruses. Until a couple of decades ago, therefore, investigators despaired of ever being able to understand and manipulate eukaryotic genomes to the same extent. The sizes of eukaryotic genomes were especially intimidating. The typical eukaryote has at least 10,000 times as much DNA as the best-studied phages—truly an awesome haystack in which to find a gene-sized needle. But the advent of recombinant DNA technology made possible an explosion of discoveries in eukaryotic genetics.

There are now many excellent books devoted to the latest details of recombinant DNA technology and how it can be applied to both prokaryotic and eukaryotic DNA. Here we will simply outline the most basic techniques, concentrating primarily on the use of plasmid cloning vectors.

Recombinant DNA Molecules

Much of what we call recombinant DNA technology was made possible by the discovery of *restriction enzymes,* which were discussed in Chapter 14 (see Box 14B). In the laboratory, these enzymes play a crucial role in the isolation of specific DNA segments and the joining of these segments to other pieces of DNA. Restriction enzymes that make staggered cuts in DNA are especially useful to molecular geneticists, because they generate single-stranded **cohesive ends** also called **sticky ends** that provide a simple means for joining DNA fragments obtained from different sources. In essence, any two DNA fragments generated by the same restriction enzyme can be joined together. As the only property needed for the fusion of fragments is the complementarity of their cohesive ends, novel combinations of DNA are readily made.

Figure 16-23 illustrates the general approach for splicing together DNA fragments from two different sources. DNA molecules from the two sources are first treated with a restriction enzyme known to generate fragments with cohesive ends. The fragments from the two sources are then mixed under conditions that favor base pairing between the complementary cohesive ends of the fragments. Once joined in this way, the DNA fragments are covalently sealed together by DNA ligase, an enzyme that is normally involved in DNA replication and repair (see Chapter 15). The product is a **recombinant DNA molecule,** consisting of DNA originating from two different sources.

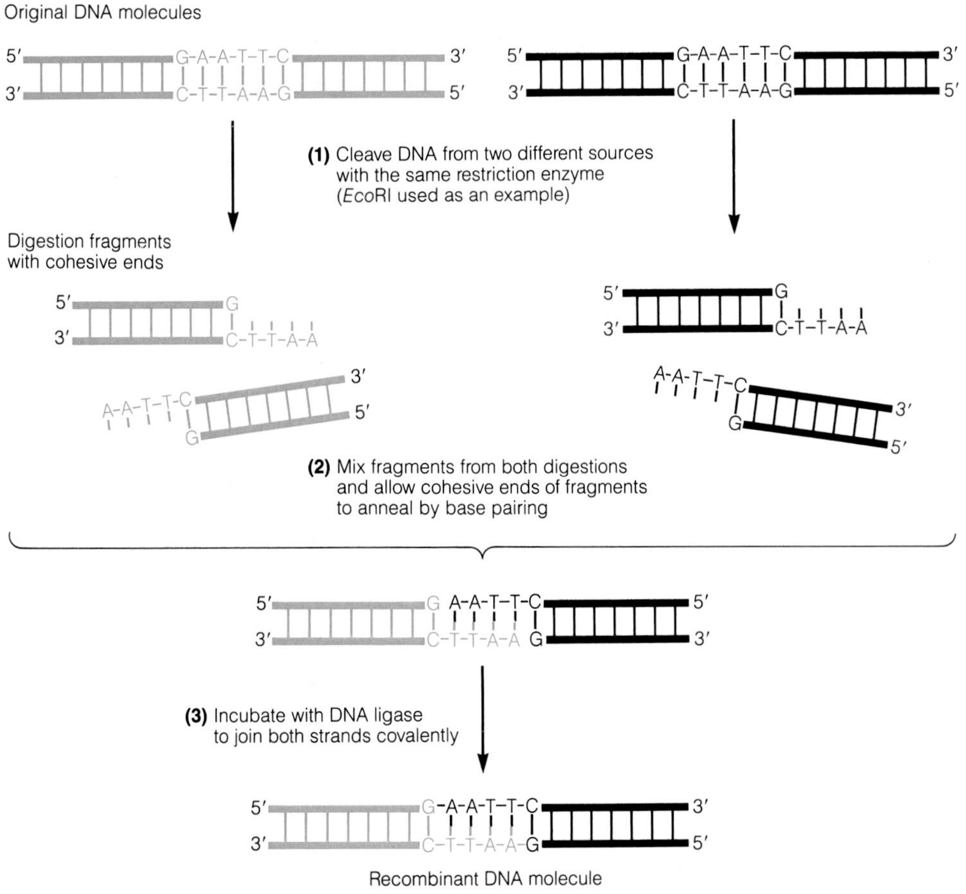

Original DNA molecules

5′ —[|||||]—G-A-A-T-T-C—[|||||]— 3′   5′ —[|||||]—G-A-A-T-T-C—[|||||]— 3′
3′ —[|||||]—C-T-T-A-A-G—[|||||]— 5′   3′ —[|||||]—C-T-T-A-A-G—[|||||]— 5′

(1) Cleave DNA from two different sources
with the same restriction enzyme
(*Eco*RI used as an example)

Digestion fragments
with cohesive ends

5′ —[|||||]—G   3′
3′ —[|||||]—C-T-T-A-A

                    A-A-T-T-C—[|||||]— 3′
                    G—[|||||]— 5′

5′ —[|||||]—G   3′
3′ —[|||||]—C-T-T-A-A

                    A-A-T-T-C—[|||||]— 3′
                    G—[|||||]— 5′

(2) Mix fragments from both digestions
and allow cohesive ends of fragments
to anneal by base pairing

5′ —[|||||]—G A-A-T-T-C—[|||||]— 5′
3′ —[|||||]—C-T-T-A-A G—[|||||]— 3′

(3) Incubate with DNA ligase
to join both strands covalently

5′ —[|||||]—G-A-A-T-T-C—[|||||]— 3′
3′ —[|||||]—C-T-T-A-A-G—[|||||]— 5′

Recombinant DNA molecule

Figure 16-23 The Generation of Recombinant DNA Molecules. ① A restriction enzyme (such as *Eco*RI) that generates cohesive ends is used to cleave DNA molecules from two different sources, making two sets of fragments with complementary cohesive ends. ② Mixing these fragments allows cohesive ends of fragments from both sets to join by base pairing of the complementary ends. ③ Upon incubation with DNA ligase, the strands are covalently sealed. Like the example shown here, some of the DNA molecules formed in this way will be recombinant molecules containing segments from both of the original sources.

The combined use of restriction enzymes and DNA ligase allows any two (or more) pieces of DNA to be spliced together, regardless of their origins. A piece of human DNA, for example, can be fused to bacterial or phage DNA just as easily as it can be linked to another piece of human DNA. It is possible, in other words, to form recombinant DNA molecules that have never existed in nature before, without any regard for the natural barriers that otherwise limit genetic recombination to genomes of the same or closely related species. Therein lies the power of recombinant DNA technology.

DNA Cloning

As mentioned earlier, a crucial feature of recombinant DNA technology is the ease with which a desired segment of DNA, usually a segment containing a specific gene, can be introduced into a bacterial cell as part of a self-replicating cloning vector. If the vector is a plasmid, it will continue to make copies of itself as the bacterium proliferates. By identifying bacterial colonies that contain the desired DNA and then growing up masses of such colonies, one can obtain large quantities of the desired DNA. The process of generating many copies of specific DNA fragments is called **DNA cloning,** or *gene cloning,* a less general term. (In biology, a *clone* is a population of organisms derived from a single an-

cestor and hence homogeneous, and a *cell clone* is a population of cells all derived from a single cell. By analogy, a *DNA clone* is a population of DNA molecules derived from the replication of a single molecule and hence identical to one another.)

The details vary, but DNA cloning almost always involves five main stages: (1) insertion of the DNA into a cloning vector; (2) introduction of the vector into cultured cells, usually bacteria; (3) amplification of the recombinant-DNA vector in the bacteria; (4) selection of all cell (or phage) clones that contain recombinant DNA (a stage that may be carried out simultaneously with stage 3); and (5) identification of the specific clones containing the DNA of interest. Figure 16-24 gives an overview of these stages, as performed using bacteria and a plasmid cloning vector. We will refer to this figure as we consider each stage in turn.

Insertion of DNA into a Cloning Vector.   The first stage in cloning a desired piece of DNA is its insertion into an appropriate cloning vector, either a plasmid or a bacteriophage. Most vectors used for DNA cloning are themselves recombinant DNA molecules, designed specifically for this purpose. For example, the DNA genome of the strains of bacteriophage λ commonly used as vectors have had nonessential genes removed, to make room in the phage head for spliced-in DNA. Plasmids used as cloning vectors

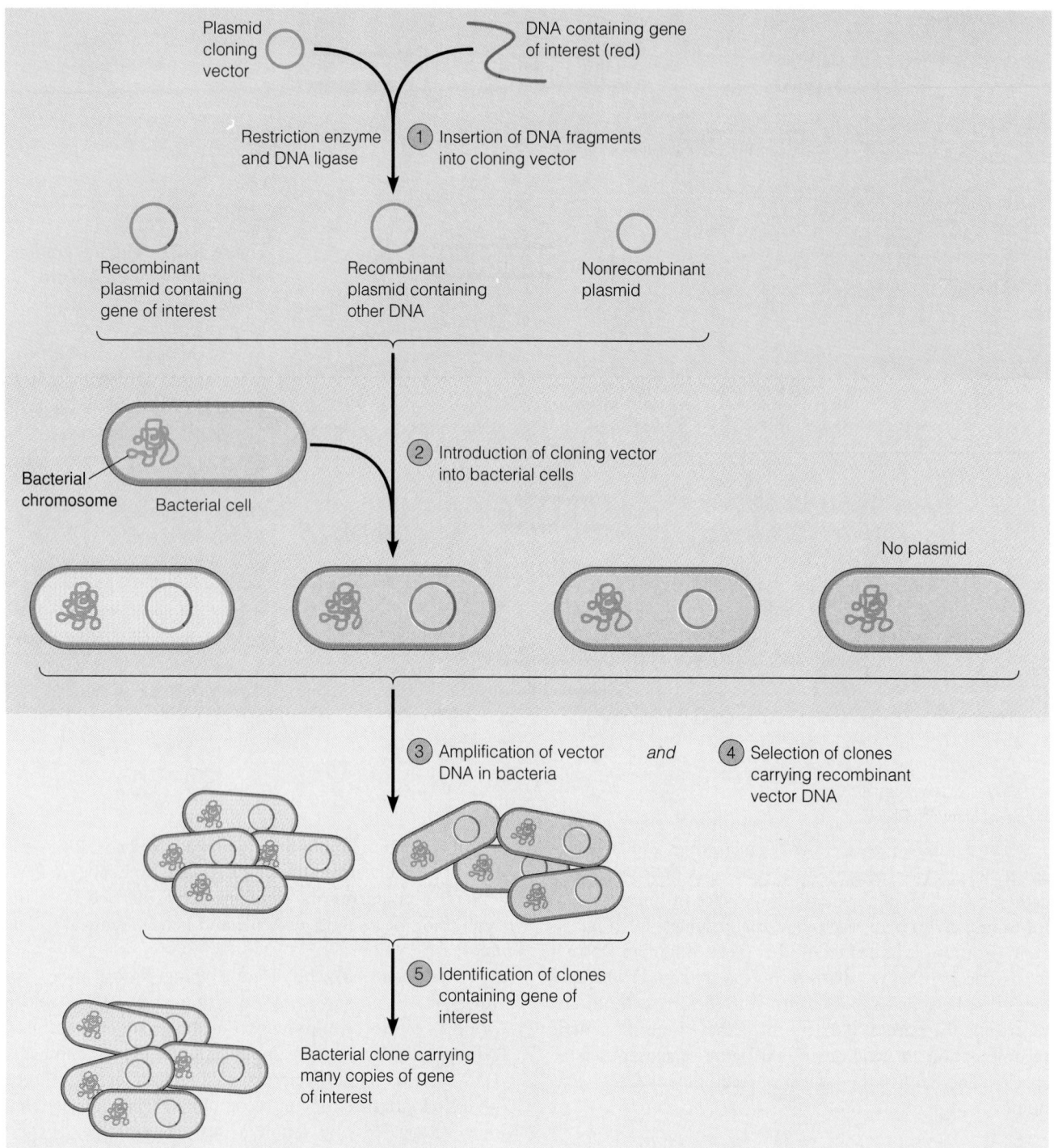

Figure 16-24 An Overview of DNA Cloning in Bacteria, Using a Plasmid Vector. The plasmids most widely used as cloning vectors are typically about 1/1000 the size of the bacterial chromosome.

usually carry genes that confer antibiotic resistance on their host cells, and they often have a variety of restriction sites (that is, restriction enzyme recognition sequences). The antibiotic-resistance genes can facilitate the selection stage (stage 4) in two different ways, as we will see, and the presence of multiple kinds of restriction sites allows the plasmid to incorporate DNA fragments prepared with a variety of alternative restriction enzymes.

A classic vector for DNA cloning in *E. coli* is a plasmid called *pBR322*, shown schematically in Figure 16-25a. This plasmid carries genes that confer resistance to the antibiotics ampicillin (amp^R) and tetracycline (tet^R) and contains restriction sites for ten different restriction enzymes (arrows). Bacteria containing this plasmid can be easily identified by their ability to grow on culture medium containing one or both antibiotics. Notice that some of the restriction sites lie *within* an antibiotic-resistance gene. Integration of foreign DNA at such an intragene site will create a recombinant plasmid with one of its antibiotic resistance genes disrupted and hence inactivated. For example, suppose a DNA frag-

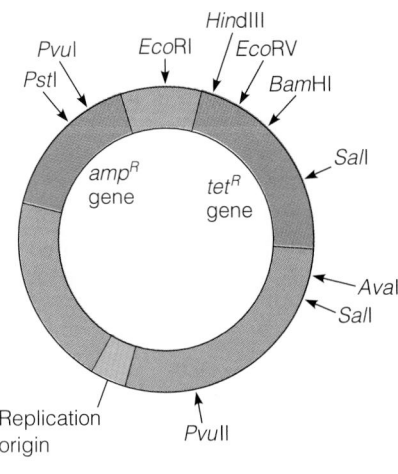

(a) Plasmid pBR322

Figure 16-25 Preparation of a Recombinant Plasmid Vector Using Plasmid pBR322. **(a)** pBR322 is a plasmid of *E. coli* that is widely used as a cloning vector; it has 4362 base pairs. The plasmid has been genetically engineered to include genes that make its "host" cell resistant to the antibiotics ampicillin (*amp^R*) and tetracycline (*tet^R*). Arrows indicate the sites of restriction enzyme recognition sequences. Inserting DNA at one of the restriction sites that lies within an antibiotic-resistance gene will inactivate the gene and render the plasmid incapable of making a host cell resistant to that antibiotic. **(b)** Insertion of foreign DNA into the plasmid. ① The plasmid is cleaved with a restriction enzyme known to recognize a single site, in this case one that is within the *tet^R* gene. ② The same enzyme is used to cleave a "foreign" DNA molecule containing a gene of interest, thereby generating fragments that have the same cohesive ends as the linearized plasmid DNA. ③ The fragments of foreign DNA are then incubated with the linearized plasmid DNA under conditions that favor base pairing between cohesive ends. Among the expected products will be plasmid molecules recircularized by base pairing with a single fragment of foreign DNA, and some of these will contain the gene of interest. ④ Incubation with DNA ligase covalently seals such a recombinant molecule, generating a recombinant plasmid vector carrying the gene of interest. Cells carrying such plasmids will be resistant to ampicillin but sensitive to tetracycline.

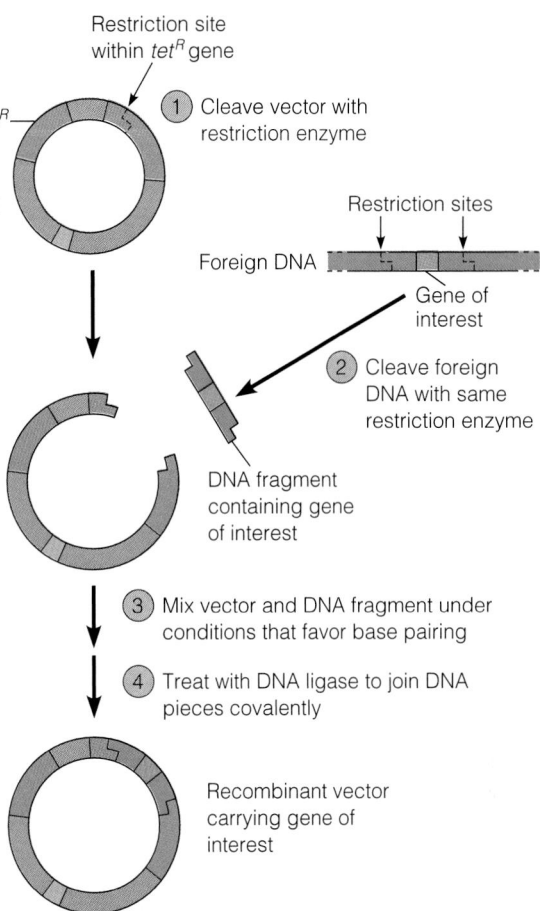

(b) Preparation of recombinant plasmid vector

ing the desired fragment of foreign DNA. In practice, however, a variety of DNA products will be present, including nonrecombinant plasmids and recombinant plasmids containing other fragments generated by the action of the restriction enzyme.

Introduction of the Recombinant Vector into Bacterial Cells. Once foreign DNA has been inserted into a cloning vector, the recombinant DNA molecules must be put into bacterial cells where they can replicate. If the vector is a plasmid, the recombinant DNA molecules are introduced into bacterial cells by the process of transformation, which in recombinant DNA procedures goes by the special term **transfection**. If the cloning vector is a phage such as λ, the recombinant phage DNA is usually packaged into phage particles in vitro and injected into bacterial cells by the natural infection process.

Amplification of Recombinant Vector DNA in Bacteria. Once they have taken up the recombinant vectors, the bacteria are plated out on a nutrient medium. Now the recombinant DNA will be replicated, or *amplified*. In the case of a plasmid vector, the bacteria proliferate and form colonies, each derived from a single cell. As the bacteria multiply, the recombinant plasmids also replicate, producing an enormous number of circular DNA molecules containing foreign DNA fragments.

ment generated by the enzyme *Hin*dIII were inserted at the single *Hin*dIII site on the plasmid. Because the *Hin*dIII site lies within the *tet^R* gene, that gene would be inactivated. Thus, a cell containing this recombinant plasmid would be sensitive to tetracycline, although resistant to ampicillin.

Figure 16-25b illustrates how a specific gene of interest contained in "foreign" DNA is inserted into a plasmid cloning vector, using plasmid pBR322 as the vector and a restriction enzyme that cleaves the plasmid at a single site Here the enzyme used is one, such as *Hin*dIII, that cuts the *tet^R* gene. Incubation with the restriction enzyme opens the plasmid at that site. The same restriction enzyme is used to cleave the DNA molecule containing the gene to be cloned. The sticky-ended fragments of foreign DNA are incubated with the linearized vector molecules under conditions that favor base pairing, followed by treatment with DNA ligase to link the molecules covalently. To keep the diagram simple, Figure 16-25b shows only the recombinant plasmid contain-

In the case of a phage vector such as phage λ, the bacteria are plated out soon after infection. Within each infected cell, the recombinant phage DNA replicates and is packaged into a large number of new phage particles, which are released upon lysis of the cell. Through further cycles of phage replication and cell lysis on the culture plate, the original infected bacterium produces a clear plaque upon a bacterial lawn (see Figure 14A-4, p. 415). The millions of phage particles in each plaque contain identical molecules of recombinant phage DNA.

Selection of Colonies Carrying Recombinant Vector DNA. An essential part of DNA cloning is to select for bacterial colonies (or phage plaques) that contain recombinant DNA. For plasmid vectors such as pBR322, the method of selection depends on the plasmid's antibiotic-resistance genes. For example, bacteria carrying the recombinant plasmids generated in Figure 16-25b will all be resistant to the antibiotic ampicillin, since all plasmids have an intact ampicillin-resistance gene. The amp^R gene is a **selectable marker,** which allows only the cells carrying plasmids to proliferate into colonies on culture medium containing ampicillin (the medium "selects for" the growth of the ampicillin-resistant cells). Thus, at least this part of the selection stage usually proceeds simultaneously with stage 3.

Not all the ampicillin-resistant cells will carry *recombinant* plasmids, however. Fortunately, cells containing recombinant plasmids can be identified because they are *not* resistant to the antibiotic tetracycline. In cells containing recombinant plasmids, the tet^R gene has been inactivated by the insertion of a DNA fragment in the middle of it.

Identification of Colonies Containing the DNA of Interest. After amplification by the multiplication of bacteria or phage, the cloned DNA fragments can be recovered by isolating vector DNA and digesting it with the same restriction enzyme used initially. In most cases, however, many different DNA fragments will have been cloned, only one or a few of which are relevant to the desired research. In Figure 16-24, for example, the bacterial colonies present on the Petri dishes at the end of stage 4 are likely to contain at least as many different kinds of fragments as there are restriction enzyme recognition sites in the DNA used in stage 1. The final stage in any recombinant DNA procedure is therefore the screening of the colonies (or phage plaques) to identify those that contain the specific DNA fragment of interest. This is frequently the most difficult stage of DNA cloning.

The technique used to screen colonies of bacteria depends on what the researcher knows about the gene being cloned. If something is known about the nucleotide sequence of the gene of interest, the researcher can employ a **nucleic acid probe,** a radioactively labeled, single-stranded molecule of DNA or RNA that is used to identify a desired DNA sequence by base pairing with it. (In Box 14C, we saw such a probe used to identify restriction fragment bands in Southern blotting.) The researcher prepares a labeled oligonucleotide corresponding to part of the DNA sequence of interest and uses it to tag the colonies that contain complementary DNA. Figure 16-26 outlines how this *colony hybridization technique* can be used to screen for colonies carrying the DNA of interest.

Another approach focuses on the protein encoded by a gene of interest. If this protein is known and has been purified, labeled antibodies against it can be prepared and used as probes to check bacterial colonies for the presence of the protein. Alternatively, the *function* of the protein can be assayed in some way; for example, an enzyme could be tested

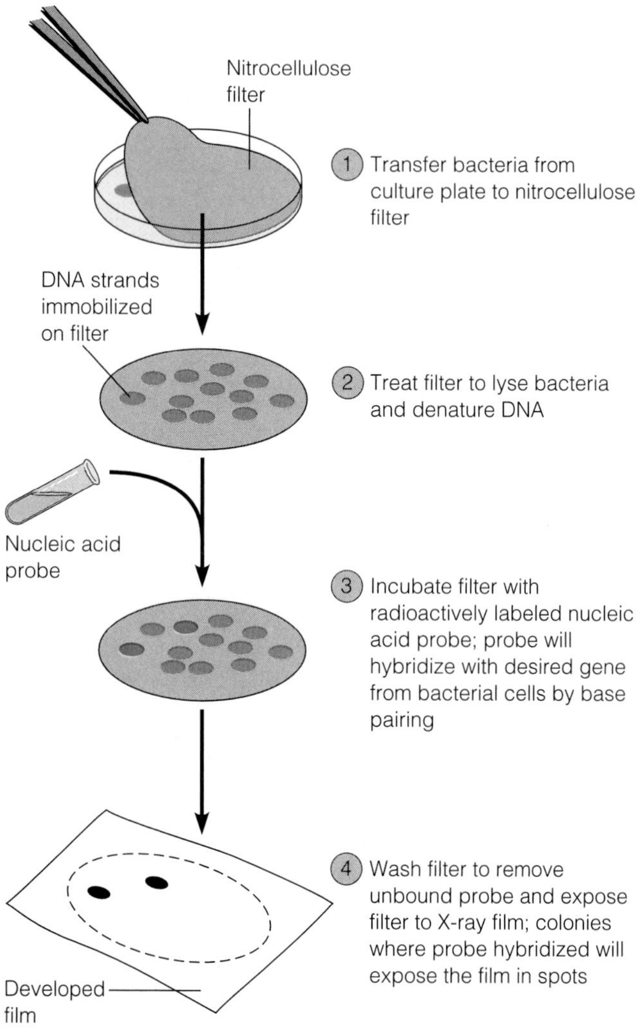

Nitrocellulose filter
1 Transfer bacteria from culture plate to nitrocellulose filter
DNA strands immobilized on filter
2 Treat filter to lyse bacteria and denature DNA
Nucleic acid probe
3 Incubate filter with radioactively labeled nucleic acid probe; probe will hybridize with desired gene from bacterial cells by base pairing
4 Wash filter to remove unbound probe and expose filter to X-ray film; colonies where probe hybridized will expose the film in spots
Developed film

Figure 16-26 The Colony Hybridization Technique. This technique is used to screen bacterial colonies for the presence of DNA that is complementary in sequence to a nucleic acid probe. ① Bacterial colonies are transferred from the surface of an agar culture plate onto a nitrocellulose filter. ② The filter is treated with detergent to lyse the bacteria and alakali (NaOH) to denature their DNA. ③ The filters are then incubated with molecules of the nucleic acid probe—radioactively labeled, single-stranded DNA or RNA—which attach by base pairing to any complementary DNA present on the filter. ④ The filter is rinsed and subjected to autoradiography, which will make visible only those colonies containing DNA that is complementary in sequence to the probe. The base pairing between strands of nucleic acid from two sources, such as the probe nucleic acid and the target DNA here, is called *nucleic acid hybridization.*

for its catalytic activity, or a receptor protein for its ability to bind a ligand. Protein-screening methods obviously depend on the ability of the bacterial cells to produce a foreign protein encoded by a cloned gene and will fail to detect cloned genes that are not expressed in the host cell. However, special *expression vectors* can be used to increase the likelihood that bacteria will transcribe eukaryotic genes properly and in large amounts. Expression vectors contain the special DNA sequences that signal the bacterial cell to perform these processes.

DNA Libraries

The cloning of foreign DNA in bacterial cells is now a routine procedure. In principle, two different approaches are used with respect to the DNA starting material. The procedure described above is the "shotgun" approach. In this approach, the whole genome of an organism (or some substantial portion of the genome) is cleaved into a large number of restriction fragments, and these are then introduced randomly into a large number of bacterial cells (or phage particles), such that the entire genome is represented in the collection. Such a collection of cloned DNA is called a **genomic library** because it contains cloned fragments representing most, if not all, of the genome. Of course, the DNA cuts made by a restriction enzyme do not respect gene boundaries, and some genes will be divided among two or more restriction fragments.

Genomic libraries of eukaryotic DNA are valuable resources from which specific genes can be isolated, provided only that a sufficiently sensitive identification technique is available. Once a rare bacterial colony that contains the desired DNA fragment has been identified, it can be grown on a nutrient medium to generate as many copies of the fragment as may be needed. Amplifications of a billionfold or more are easily achieved using the procedures already described. As mentioned, however, identifying the right colonies for further amplification is often the most difficult part of the cloning procedure.

A smaller, more selective DNA library can be prepared from **complementary DNA (cDNA).** This is DNA that is synthesized in vitro using messenger RNA (mRNA) as a template. Although ordinary cellular transcription is in the DNA → RNA direction, animal viruses called *retroviruses* have an enzyme called **reverse transcriptase** that catalyzes the synthesis of DNA on an RNA template. The researcher isolates mRNA from a batch of cells that are expressing the gene of interest, synthesizes single strands of cDNA by reverse transcription, converts the DNA into double-stranded form using DNA polymerase, adds single-stranded tails, and inserts it into a suitable cloning vector (Figure 16-27).

The collection of clones derived from an mRNA preparation as just described constitutes a **cDNA library.** Such a library can be screened by the same techniques used with a genomic library. Unlike a genomic library, however, a cDNA library will contain only those DNA sequences that are transcribed into RNA—presumably, the active genes in the cells or tissue from which the mRNA was prepared.

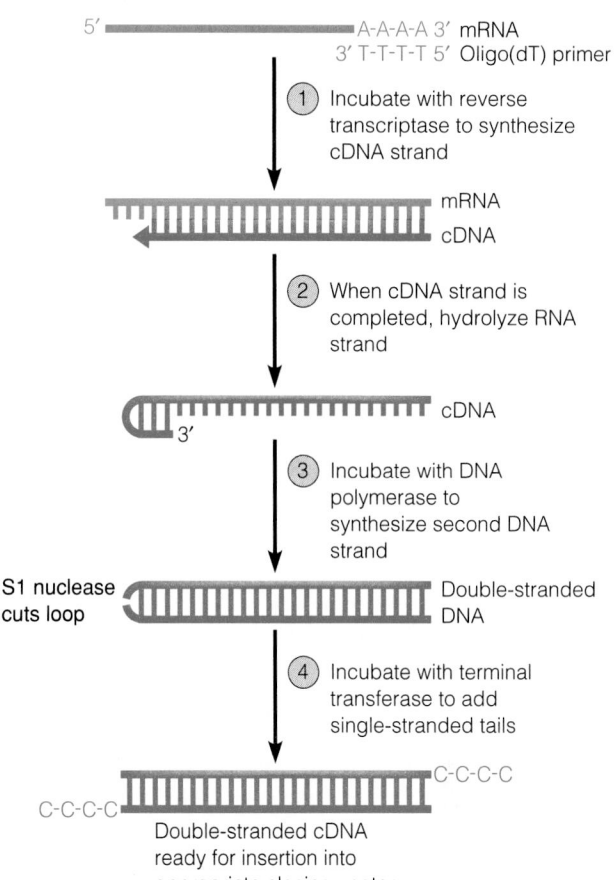

Figure 16-27 Preparation of Complementary DNA (cDNA) for Cloning. ① Messenger RNA is incubated with reverse transcriptase, which uses the mRNA as a template for synthesis of a complementary DNA (cDNA) strand. Oligo(dT), a short chain of thymine deoxyribonucleotides, can be used as a primer, because eukaryotic mRNA always has a stretch of adenine nucleotides at its 3′ end. ② The resulting mRNA-cDNA hybrid is treated with alkali or an enzyme to hydrolyze the RNA, leaving the single-stranded cDNA. ③ DNA polymerase can now synthesize the complementary DNA strand. The looped-around 3′ end of the first DNA strand can often be used as a primer. An enzyme called S1 nuclease is then used to cleave the loop. ④ For efficient insertion in a cloning vector, the double-stranded DNA must have single-stranded tails that are complementary to those of the vector. These can be added by incubation with terminal transferase, an enzyme that adds nucleotides one at a time to the ends of the molecule. If short stretches of cytosine nucleotides, for example, are added to the cDNA and short stretches of guanine nucleotides are added in the same way to a linearized cloning vector, recombinant molecules can be generated, as shown in Figure 16-25b. (As an alternative to step 4, short synthetic "linker" molecules containing a variety of restriction sites can be ligated to the ends of both the cDNA and a blunt-ended cloning vector. The linkers are then cleaved with a restriction enzyme that generates cohesive ends.)

In addition to being limited to transcribed genes, a cDNA library has another important advantage as a starting point for the cloning of eukaryotic genes. Using mRNA to make cDNA guarantees that the cloned genes will contain only gene-coding sequences, without the noncoding interruptions that are common in eukaryotic genes (see Chapter 17). Eukaryotic cells remove the noncoding segments, called *introns,* from RNA transcripts before translation starts, but prokaryotic cells do not have the cellular machinery to do this. Intron DNA can be so extensive that the overall length of a eukaryotic gene is too unwieldy for recombinant DNA manipulation. Using cDNA eliminates this problem. Furthermore, bacteria cannot possibly make the correct protein product of an intron-containing eukaryotic gene unless the introns have been removed—as they are in cDNA.

Yeast Artificial Chromosomes (YACs)

DNA cloning using the vectors we have mentioned so far is a very powerful methodology, but it has an important limitation: the relatively small amount of foreign DNA that can be successfully cloned in a single vector molecule. Certain phage λ vectors specially constructed to accommodate large pieces of DNA can carry up to 30,000 bp, but many eukaryotic genes are much longer than this. Moreover, many eukaryotic genes may not be able to function normally without the presence of adjoining DNA consisting, perhaps, of regulatory sequences or other genes. For the study of such genes and gene complexes, it is clearly desirable to be able to clone very long DNA segments. And for genome-mapping projects, the availability of clones of long DNA segments is extremely valuable; the more DNA per clone, the fewer the number of clones needed to cover the entire genome.

Fortunately, long segments of DNA can now be readily cloned using a type of vector developed in the 1980s, the **yeast artificial chromosome (YAC).** These new vectors are "minimalist" eukaryotic chromosomes; they are DNA molecules containing all the elements essential for normal chromosome replication and segregation to daughter cells, and very little else. As you might be able to guess from your knowledge of chromosome replication and segregation (see Chapter 15), a eukaryotic chromosome needs three crucial kinds of DNA sequences: (1) an origin of DNA replication, (2) two telomeres (to allow periodic extension of the shrinking ends by telomerase); and (3) a centromere (to ensure proper attachment, via a kinetochore, to spindle microtubules in cell division). If yeast-compatible versions of these elements are combined with foreign DNA and the resulting YAC is put into a yeast cell, the YAC will be replicated and segregated to daughter cells with each round of cell division, just like a natural chromosome. And, under appropriate conditions, its foreign genes may be expressed.

Figure 16-28 outlines the construction of a typical YAC. The YAC cloning vector is a small circular DNA molecule (actually a plasmid) that carries the three essential DNA ele-

Figure 16-28 Construction of a Yeast Artificial Chromosome (YAC). The YAC cloning vector is a circular DNA molecule with nucleotide sequences specifying an origin of DNA replication (*ORI*), a centromere (*CEN*), two telomeres (*TEL*), and two selectable markers. It has two recognition sequences for the restriction enzyme *Bam*HI and one for *Eco*RI. Digestion of the YAC vector with both restriction enzymes produces two linear DNA fragments that together contain all the essential sequences, as well as the fragment that connected the *Bam*HI sites, which is of no further use. The fragment mixture is incubated with fragments from light digestion of foreign DNA with *Eco*RI (light digestion generates large DNA fragments because not all the restriction sites are cut), and the resulting recombinant strands are sealed with DNA ligase. Among the products will be artificial chromosomes—YACs—carrying foreign DNA, as shown at the bottom of the figure. After yeast cells are transformed with the products of the procedure, the colonies of cells that have received complete YACs can be identified by the properties conferred by the two selectable markers.

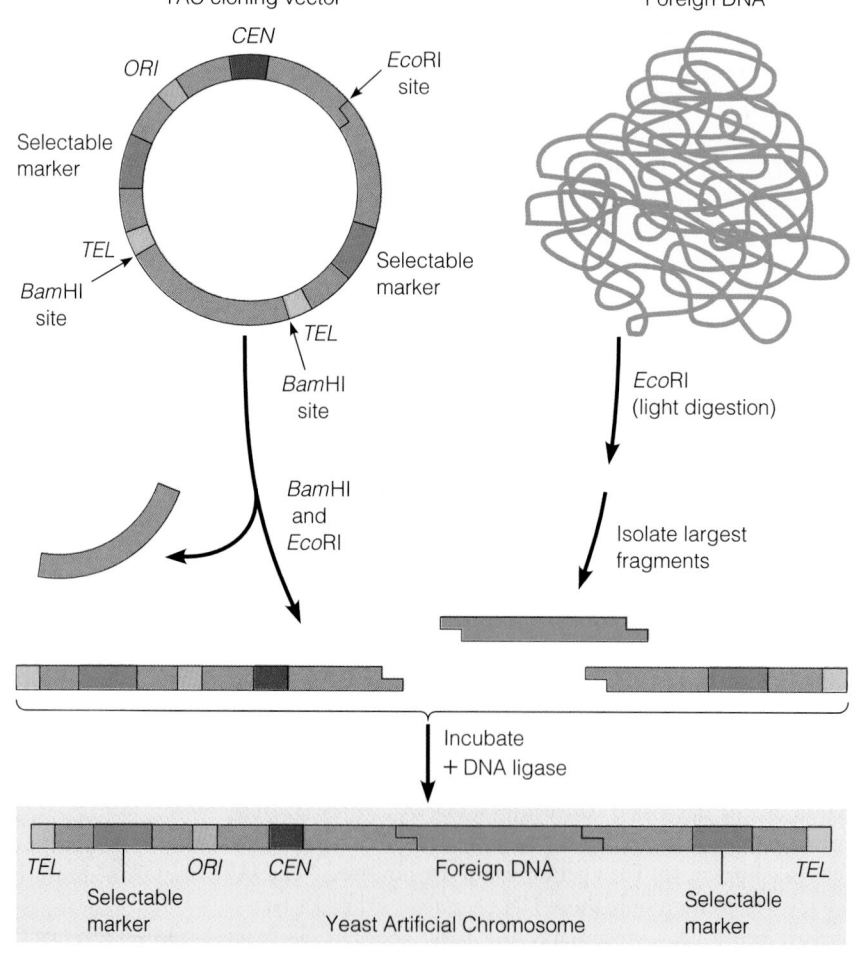

ments. (In yeast, an origin of DNA replication is called *ORI*, centromere DNA is called *CEN*, and telomere DNA is called *TEL*.) In addition, the YAC vector shown here carries two genes that can function as selectable markers and three restriction sites, one for *Eco*RI and two for another restriction enzyme, called *Bam*HI. The vector DNA is digested with both enzymes to generate linear fragments, and these are incubated with fragments of foreign DNA that has been digested with *Eco*RI. After covalent sealing by DNA ligase, the products include a variety of YACs, carrying different fragments of foreign DNA. The YACs are introduced by transformation into yeast cells whose cell walls have been removed. The presence of two different selectable markers in the original vector makes it easy for the researcher to select for yeast cells containing YACs with both chromosomal "arms." The diagram in the figure is not to scale: The YAC vector alone is only about 10,000 bp, but the inserted foreign DNA can be 300,000 to 1.5 million bp. In fact, YACs must carry at least 50,000 bp to be reliably replicated and segregated.

Genetic Engineering

Although the size of DNA fragment that can be cloned in bacterial cells is limited, it is still true that virtually any DNA sequence can be cloned in bacteria and that bacteria can express at least some foreign genes. Furthermore, bacteria are extremely easy to grow in huge numbers. Very early in the development of recombinant DNA technology, attention turned to the possibility of using bacteria to produce useful gene products on a commercial scale. Especially attractive was the possibility of using bacteria to make large amounts of mammalian proteins such as hormones, growth factors, and interferons (antiviral chemicals)—all substances that animals produce in only very tiny amounts. One of the first mammalian genes to be expressed in bacteria was that for rat insulin, an achievement reported in 1978. Since then, the gene for the human version of this hormone has also been cloned and expressed in bacteria, as has the gene for human growth hormone. Both of these hormones have been approved for use in treating human patients in the United States and are being produced commercially (Figure 16-29).

These hormones were among the first success stories of **genetic engineering,** the application of recombinant DNA technology to practical problems, primarily in medicine and agriculture. The availability of human insulin produced by genetically engineered bacteria was good news to the two million diabetics in the United States, especially since the only insulin otherwise available, which is obtained from animal sources, causes adverse reactions in some people. Similarly, the availability of human growth hormone is a great benefit to children born with hypopituitarism, a syndrome in which the pituitary gland fails to provide an adequate natural supply of the hormone. Other valuable products now being made by genetic engineering are proteins used in treating burns, heart attacks, anemia, hemophilia, and some

Figure 16-29   Commercial Production of Human Growth Hormone.   The bottles in this photograph contain human growth hormone, produced using recombinant DNA techniques. The bottles represent the final stage in a manufacturing process that makes human growth hormone commercially available for treating children with growth hormone deficiencies.

cancers. Medical uses of recombinant DNA technology also include the production of vaccines and the development of diagnostic tests for genetic disorders and for some infectious diseases (such as AIDS). Whenever possible, bacteria are used for the manufacturing process, but for some gene products yeast cells or cultured mammalian cells must be used.

For certain genetic diseases traceable to a single defective gene, such as cystic fibrosis (see Box 8A) and several disorders affecting the blood and immune system, medical scientists are already carrying out *human gene therapy* on an experimental basis. The goal of such therapy is to transplant a functioning allele for the gene into cells that will then manufacture the normal gene product in amounts large enough to alleviate the condition. Ideally, the normal allele would be inserted into cells that multiply throughout a person's life. For diseases affecting the blood and immune system, bone marrow cells, which include the *stem cells* that give rise to all blood cells, are prime candidates. Most of the gene therapy trials now in progress are yielding promising preliminary results, suggesting that eventually genetic engineering may provide means for correcting, rather than just treating, many genetic diseases. The ultimate cure, of course, would be to use genetic engineering to replace malfunctional alleles in the germ line, the tissues that give rise to sperm and eggs. This approach would eliminate the disease in future generations. However, it raises serious ethical and social questions.

Increasingly, nonhuman eukaryotes are serving as targets of genetic engineering. This field was pioneered with yeasts, which can take up DNA by transformation and which have plasmids as well. Plants and animals are also amenable to genetic engineering. The "supermouse" described in Box 16A was an early and visually dramatic example in the animal kingdom of a **transgenic organism,** an organism that carries genes from another species in all its cells, including its germ line.

Exciting results of genetic engineering in higher organisms are also being reported by plant molecular biologists, who enjoy the special advantage of being able to regenerate adult plants of many species from single cells grown in tissue culture (see Figure 16-1d). The vector of choice in most cases is a plasmid carried by the bacterium *Agrobacterium tumefaciens.* In nature, this bacterium is a pathogenic microorganism that causes tumors called *crown galls* in certain plant species. The tumors are induced by the plasmid, which is called the **Ti plasmid** (Ti stands for tumor-inducing). The significant feature of this plasmid for genetic engineering is that a segment of its DNA, called the *T DNA region,* integrates into the genome of the host plant cell. This property has been exploited by plant molecular biologists, who have developed strategies for genetically engineering plants using the Ti plasmid.

The general approach for creating transgenic plants using the Ti plasmid is shown in Figure 16-30. The Ti plasmid is isolated from *Agrobacterium* cells, and the desired DNA (a specific gene, most commonly) is inserted into the T DNA region of the plasmid using standard recombinant DNA techniques. (This step disrupts the tumor-inducing genes.) The plasmid is put back into *Agrobacterium,* and the genetically engineered bacterium is then used to infect plant cells growing in culture. When the recombinant plasmid enters the plant cell, its T DNA becomes stably integrated into the plant genome and is passed on to both daughter cells at every cell division. From such cells, plants are eventually regenerated that contain the recombinant T DNA—and therefore the desired foreign gene—in all of their cells. The foreign gene will now be inherited by progeny plants just like any other gene. Plant genetic engineering holds great promise as a means of increasing agricultural production because such diverse traits as herbicide resistance, stress tolerance, quality of seed proteins and lipids, and nitrogen fixation are potentially amenable to genetic modification.

In summary, recombinant DNA technology seems destined not only to enhance our understanding of the eukaryotic genome, but to influence profoundly the practices of medicine and agriculture as well.

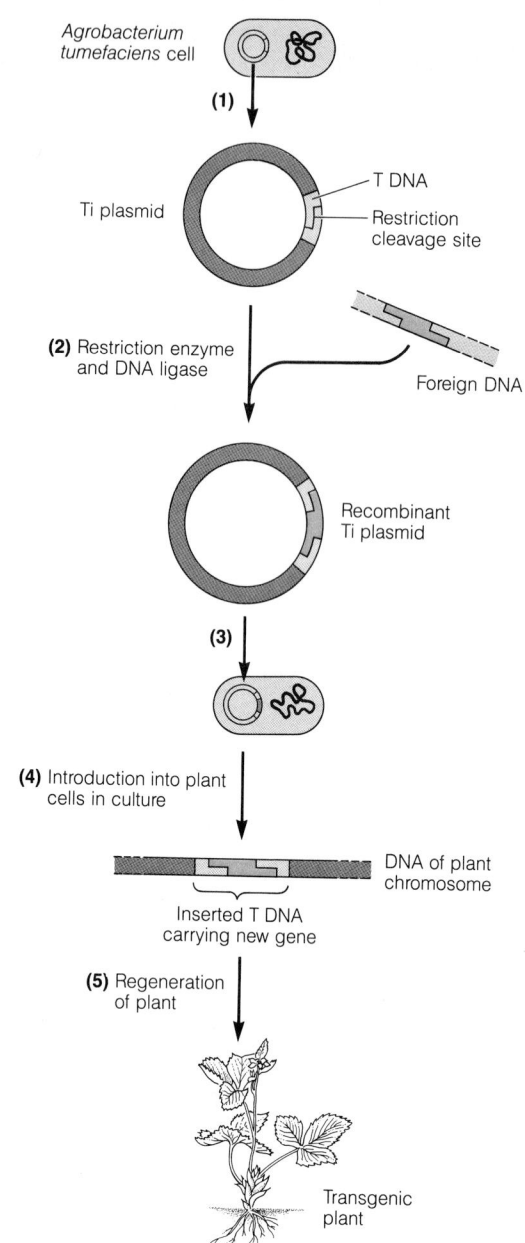

Figure 16-30 **The Molecular Basis for Generating a Transgenic Plant.** Most transgenic plants are generated using the Ti plasmid from *Agrobacterium tumefaciens.* ① The Ti plasmid is isolated from *Agrobacterium* cells, ② subjected to standard recombinant DNA procedures to insert the desired DNA into the T DNA region of the plasmid, and then ③ put back into *Agrobacterium.* ④ Cultured plant cells are infected with bacteria containing the recombinant plasmid, and ⑤ these plant cells are then used to regenerate whole plants. The resulting transgenic plants contain the recombinant T DNA region stably integrated into the genome of every cell.

Historical Perspectives

SUPERMOUSE, AN EARLY TRANSGENIC TRIUMPH

A DNA fragment containing the gene of rat growth hormone was microinjected into the pronuclei of fertilized mouse eggs. Of 21 mice that developed from these eggs, seven carried the gene and six of these grew significantly larger than their littermates. (Palmiter et al., 1982, p. 611)

With these words, a team of investigators led by Richard Palmiter and Ralph Brinster reported how a genetic trait can be introduced experimentally into mice without going through the usual procedure of "breeding," a process of sexual reproduction followed by selection of desired traits. The researchers injected rat growth hormone genes into fertilized mouse eggs, and from one of these eggs developed a "supermouse" weighing almost twice as much as its littermates (Figure 16A-1). The accomplishment was heralded as a significant breakthrough because it proved the feasibility of applying genetic engineering to animals, with all the scientific and practical consequences such engineering is likely to have.

What Palmiter, Brinster, and their colleagues did to create the supermouse is an intriguing story that begins with the isolation of the structural gene for growth hormone (GH) from a library of rat DNA, using techniques similar to those described in this chapter. The cloned GH gene from which the regulatory region had been deleted was then fused to the regulatory portion of a mouse gene, the gene that codes for *metallothionein* (*MT*). MT is a small metal-binding protein that is normally present in most mouse tissues and appears to be involved in regulating the level of zinc in the animal. The advantage of fusing the MT gene to the GH gene is that the expression of the MT gene can be specifically induced (turned on) by zinc.

To make multiple copies of the MT-GH hybrid gene, it was cloned in *E. coli* cells using a plasmid as a cloning vector. After the recombinant plasmid DNA was recovered from the bacterial cells, the MT-GH region was excised by digestion of the DNA with two restriction enzymes, each of

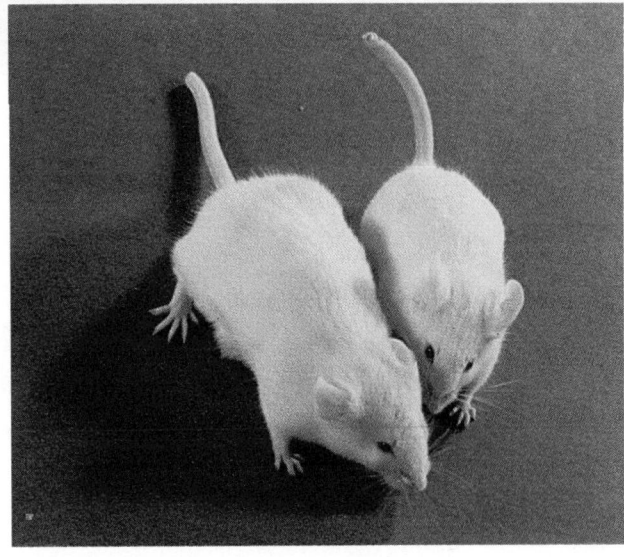

Figure 16A-1   Genetic Engineering in Mice.   "Supermouse" (on the left) is significantly larger than its littermate because it was engineered to carry and express at high levels the gene for rat growth hormone.

which has a unique cleavage site at one end of the desired DNA fragment. This fragment consisted of about 5000 nucleotide pairs and remained linear because it had two different end sequences and therefore could not circularize.

About 600 copies of this fragment were microinjected into fertilized mouse eggs, in a volume of about 2 picoliters (0.000002 μL!). The DNA was injected into the male pronucleus, the haploid sperm nucleus that has not yet fused with the haploid egg nucleus. (The success rate for integration and retention of the MT-GH gene had been found to be higher when the male pronucleus was used than when the DNA was injected into either the female pronucleus or the cytoplasm.) Of the 170 fertilized eggs that were injected and implanted back into the reproductive tracts of foster mothers, 21 animals developed. Seven of them turned out to be transgenic mice with MT-GH genes present in their cells. In at least one case, a transgenic mouse transmitted the MT-GH gene faithfully to about half of its offspring, suggesting that the gene was integrated stably into one of its chromosomes.

Because the growth hormone gene was linked to an MT gene regulatory region, the hybrid gene, if present in a mouse, could be turned on simply by giving the mice a high level of zinc in their drinking water. In this way, mice carrying the gene could be made to produce GH continually at high levels. Three kinds of evidence revealed the expression of the rat GH genes. When mouse liver tissue was assayed for the presence of messenger RNA for GH, the results indicated about 800–3000 mRNA molecules per liver cell. Moreover, elevated levels of growth hormone were found in the blood: Four of the transgenic mice had blood GH levels that were 100–800 times higher than those of their nonengineered littermates! Of course, the most obvious evidence that the genes were expressed was that the transgenic

mice grew significantly faster and got significantly bigger. They weighed about twice as much as normal mice, and during the period of maximum sensitivity to growth hormone (3 weeks to 3 months), they grew three to four times faster than normal littermates.

This dramatic experiment proved that it was possible to introduce isolated, cloned genes into higher organisms and get those genes integrated into the genome, where they could be expressed and passed on to offspring. As the authors pointed out when the report was published, "this approach has implications for studying the biological effects of growth hormone, as a way to accelerate animal growth, as a model for gigantism (a human growth abnormality caused by growth hormone), as a means of correcting genetic diseases, and as a method of farming valuable products." Whether (and how fast) all these possibilities become realities remains, of course, to be seen. But in the years since supermouse's creation, rapid progress has been made in most of these areas. Supermouse, it seems, was just the beginning. ■

PERSPECTIVE

Asexual reproduction is based on mitosis and results in offspring that are genetically identical to the single parent. By contrast, sexual reproduction involves two parents and leads to a mixture of parental traits in the offspring. Sexual reproduction allows populations to respond adaptively to environmental changes, enables desirable mutations to be combined in a single individual, and promotes genetic flexibility by maintaining a diploid genome.

The life cycle of every sexually reproducing, eukaryotic species includes both a haploid phase and a diploid phase. Haploid gametes are generated directly or indirectly by meiosis and fuse at fertilization to restore the diploid chromosome number. Meiosis consists of two successive cell divisions without an intervening duplication of chromosomes. In the first division (meiosis I), homologous chromosomes separate and segregate into the two daughter cells. In the second division (meiosis II), sister chromatids separate, and four haploid daughter cells are produced. Meiosis differs from mitosis not only in the ploidy level of the daughter cells but also in the synapsis of homologous chromosomes during prophase of the first meiotic division. It is during synapsis that crossing over occurs between nonsister chromatids, leading to genetic recombination. The chias-

mata observed during meiosis I are morphological manifestations of crossover events.

Mendel's laws describe the genetic consequences of chromosome behavior during meiosis, even though chromosomes had not yet been discovered at the time of Mendel's experiments. According to his laws, maternal and paternal alleles segregate into different gametes during meiosis, and the alleles of genes on separate chromosomes assort independently of one another. The enormous genetic variability among an organism's gametes arises in part from the independent assortment of chromosomes during anaphase I and in part from the recombination that occurs during prophase I.

The frequency with which recombination is observed between genetic markers on the same chromosome is a measure of the distance between the two markers and can be used to map genes in both eukaryotes and prokaryotes (and in viruses, as well). Although prokaryotes do not reproduce sexually, when part of the bacterial chromosome is transferred between cells by transformation, transduction, or conjugation, recombination between the transferred DNA and the recipient's DNA ensues, and the recombinant cells can be identified and counted.

The development of recombinant DNA technology has made it possible to combine in vitro DNA from any two (or more) sources into a single molecule of recombinant DNA. By combining a gene of interest with a cloning vector, either a plasmid or phage DNA, the gene can be cloned (amplified) in bacterial cells. In this way, large amounts of specific genes or their protein products can be prepared for research or applied purposes. Recombinant DNA technology has made possible the detailed analysis and manipulation of eukaryotic genomes, including the human genome. At the same time, applications of the technology are revolutionizing medicine and agriculture.

KEY TERMS FOR SELF-TESTING

asexual reproduction (p. 495)

Sexual Reproduction

sexual reproduction (p. 495)
homologous chromosomes (p. 497)
sex chromosome (p. 497)
diploid (p. 497)
haploid (p. 497)
gene locus (p. 497)
allele (p. 497)
homozygous (p. 497)
heterozygous (p. 497)
dominant allele or trait (p. 498)
recessive allele or trait (p. 498)
genotype (p. 498)
phenotype (p. 498)
gamete (p. 498)
gametogenesis (p. 498)
egg (p. 498)
sperm (p. 498)
fertilization (p. 498)
zygote (p. 498)
development (of an organism) (p. 498)
mating type (p. 498)

Meiosis

meiosis (p. 499)
meiosis I (p. 499)
meiosis II (p. 499)
alternation of generations (p. 499)
haploid spore (p. 499)
sporophyte (p. 499)
gametophyte (p. 499)
synapsis (p. 499)
bivalent (p. 499)
genetic recombination (p. 501)
leptotene (p. 501)
zygotene (p. 501)
crossing over (p. 501)
pachytene (p. 501)
diplotene (p. 501)
chiasma (p. 501)
diakinesis (p. 501)
synaptonemal complex (p. 501)
polar body (p. 507)

Genetic Variability: Segregation and Assortment of Alleles

true-breeding (plant strain) (p. 508)
hybrid (p. 508)
F_1 and P_1 generations (p. 508)
F_2 generation (p. 509)
backcrossing (p. 509)
Mendel's laws of inheritance (p. 509)
law of segregation (p. 510)
law of independent assortment (p. 510)

Genetic Variability: Recombination and Crossing Over

wild type (p. 514)
linkage group (p. 514)
linked genes (p. 514)
genetic mapping (p. 516)
map unit (p. 516)

Recombination in Bacteria and Viruses

transformation (p. 517)
transduction (p. 517)
conjugation (p. 518)
F factor (p. 518)
sex pilus (p. 518)
mating bridge (p. 518)
origin of transfer (p. 518)
Hfr cell (p. 518)

Recombinant DNA Technology and Gene Cloning

recombinant DNA technology (p. 520)
cloning vector (p. 520)
cohesive (sticky) end (p. 520)
recombinant DNA molecule (p. 520)
DNA cloning (p. 521)
transfection (p. 523)
selectable marker (p. 524)
nucleic acid probe (p. 524)
genomic library (p. 525)
complementary DNA (cDNA) (p. 525)
reverse transcriptase (p. 525)
cDNA library (p. 525)
yeast artificial chromosome (YAC) (p. 526)

Genetic Engineering

genetic engineering (p. 527)
transgenic organism (p. 528)
Ti plasmid (p. 528)

16-1. The Truth About Sex. For each of the following statements, indicate with an S if it is true of sexual reproduction, with an A if it is true of asexual reproduction, with a B if it is true of both, and with an N if it is not true of either.

(a) Traits from two different parents can be combined in a single offspring.

(b) Each generation of offspring is genetically identical to the previous generation.

(c) Mutations get propagated to the next generation.

(d) Some offspring in every generation will be less suited for survival than the parents, but others may be better suited.

(e) Mitosis is involved in the life cycle.

16-2. Ordering the Phases of Meiosis. Shown in Figure 16-31 are drawings of several phases of meiosis in an organism, labeled A through F.

(a) What is the diploid chromosome number in this species?

(b) Place the six phases in chronological order, and name each one.

(c) Between which two phases do homologous centromeres separate?

(d) Between which two phases does recombination occur?

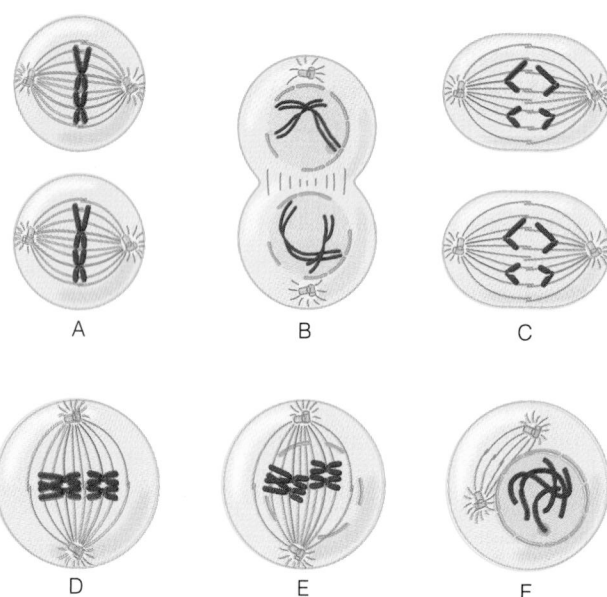

Figure 16-31   Six Phases of Meiosis to Be Ordered and Identified.

16-3. Telling Them Apart. Briefly describe how you might distinguish between each of the following pairs of phases in the same organism:

(a) Metaphase of mitosis and metaphase I of meiosis.

(b) Metaphase of mitosis and metaphase II of meiosis.

(c) Metaphase I and metaphase II of meiosis.

(d) Telophase of mitosis and telophase II of meiosis.

(e) Pachytene and diplotene stages of meiotic prophase I.

16-4. Your Centromere Is Showing. Suppose you have a diploid organism in which all the chromosomes contributed by the sperm have cytological markers on their centromeres that allow you to distinguish them visually from the chromosomes contributed by the egg.

(a) Would you expect all the somatic cells of the organism to have equal numbers of maternal and paternal centromeres? Explain.

(b) Would you expect equal numbers of maternal and paternal centromeres in each gamete produced by that individual? Explain.

16-5. How Much DNA? Let X be the amount of DNA present in the gamete of an organism that has a diploid chromosome number of 4. Assuming all chromosomes to be of approximately the same size, how much DNA (X, 2X, 1/2X, and so on) would you expect in each of the following?

(a) A zygote immediately after fertilization

(b) A single sister chromatid

(c) A daughter cell following mitosis

(d) A single chromosome following mitosis

(e) A nucleus in mitotic prophase

(f) The cell during metaphase II of meiosis

(g) One bivalent

16-6. Meiotic Mistakes and Down Syndrome. Down syndrome is a genetic disorder associated with heart and respiratory defects and varying degrees of mental retardation. It usually results from the presence of an entire extra chromosome (chromosome 21) in each of the diploid cells of an afflicted person. The extra chromosome, in turn, usually derives from a mistake in meiosis during the formation of the egg or sperm by one of the parents. What kind of mistake in meiosis would lead to a gamete with an extra copy of one of the chromosomes?

16-7. Punnett Squares as Genetic Tools. A *Punnett square* is a diagram representing all possible outcomes of a genetic cross. The genotypes of all possible gametes from the male and female parents are arranged along two adjacent sides of a square, and each box in the matrix is then used to represent the genotype resulting from the union of the two gametes at the heads of the intersecting rows. By the law of independent assortment, all possible combinations are equally likely, so the frequency of a given genotype among the boxes represents the frequency of that genotype among the progeny of the genetic cross represented by the Punnett square.

Figure 16-32 shows the Punnett squares for two crosses of pea plants. The genetic characters involved are seed color (where *Y* is the allele for yellow seeds and *y* for green seeds) and seed shape (where *R* is the allele for round seeds and *r* for wrinkled seeds). Punnett square a represents a one-factor cross between parent plants that are both heterozygous for seed color ($Yy \times Yy$). Punnett square b is a two-factor cross between plants that are heterozygous for both seed color (*Yy*) and seed shape (*Rr*).

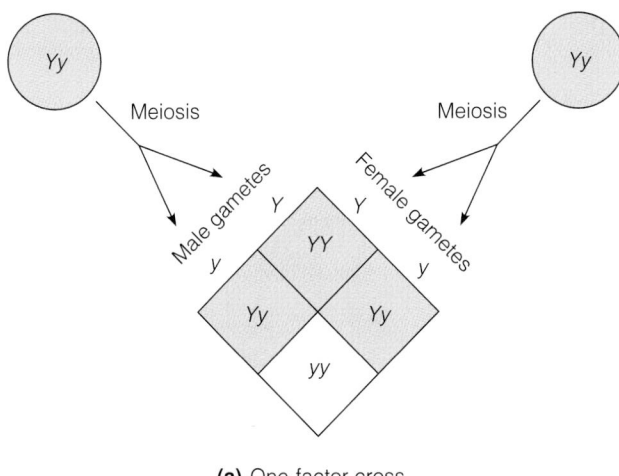

(a) One-factor cross

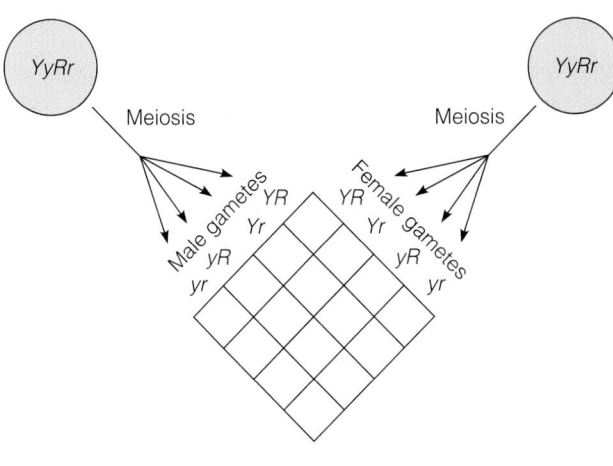

(b) Two-factor cross

Figure 16-32   Punnett Squares.

(a) Using the Punnett square of Figure 16-32a, explain the 3:1 phenotypic ratio that Mendel observed for the offspring of such a cross.

(b) Explain why the Punnett square of Figure 16-32b is a 4 × 4 matrix with 16 genotypes. In general, what is the mathematical relationship between the number of heterozygous allelic pairs in the parents and the number of different kinds of gametes?

(c) How does the Punnett square of Figure 16-32b reflect Mendel's law of independent assortment?

(d) Complete the Punnett square of Figure 16-32b by writing in each of the possible progeny genotypes. How many different genotypes will be found in the progeny? In what ratios?

(e) For the case of Figure 16-32b, how many different phenotypes will be found in the progeny? In what ratios?

16-8. More Punnett Squares. Three of the characters that Mendel chose to study in peas were flower position, stem length, and seed shape. Each of these characters is controlled by a gene that is located on a different chromosome, and each has a dominant and a recessive expression, as follows:

Character	Dominant Trait	Recessive Trait
Flower position	Axial (*A*)	Terminal (*a*)
Stem length	Long (*L*)	Short (*l*)
Seed shape	Round (*R*)	Wrinkled (*r*)

Using a Punnett square as described in Problem 16-7, answer the following:

(a) If you allow a plant that is heterozygous for all three characters to self fertilize, what proportion of the offspring would you expect to be homozygous for all three characters?

(b) If you cross a plant that is heterozygous for all three characters with one that is heterozygous for the first two but homozygous for round seeds, what proportion of the offspring would you expect to be homozygous for all three characters?

(c) What would have to be true of the genetic makeup of the two parent plants if you found that a particular genetic cross yielded no offspring that were homozygous for all three characters?

(d) Of what significance is it that these three characters are controlled by genes located on three different chromosomes?

16-9. Sex Linkage. In addition to discovering gene linkage (the fact that genes located on the same chromosome tend to be inherited together), T. H. Morgan's group discovered sex linkage. A gene is said to be *sex-linked* if it is located on a sex chromosome—the *X* chromosome in mammals, for instance. Sex-linked genes display patterns of inheritance that are different from those observed for genes located on nonsex chromosomes. A classic example of a sex-linked gene in humans is a gene with an aberrant allele that can cause hemophilia, a serious genetic disease characterized by failure of the body's blood-clotting mechanism. This gene is located on the *X* chromosome (and not on the *Y* chromosome, which has very few genes), and the hemophilia allele is recessive to the normal allele. Recalling that females have two *X* chromosomes and males have an *X* and a *Y*, answer the following questions.

(a) Using *h* for the hemophilia allele and *H* for the corresponding normal allele, give the possible genotype(s) for (i) a female hemophiliac, (ii) a male hemophiliac, (iii) a female with normal blood clotting, and (iv) a male with normal blood clotting.

(b) Suppose a man with normal blood clotting marries a woman with normal blood clotting *whose father was a hemophiliac.* Predict the ratios of genotypes and phenotypes among their children. (You might want to use a Punnett square.)

(c) Now look at the sexes of the offspring of the marriage in part b. What fraction of the girls are predicted to be hemophiliacs? What fraction of the boys?

16-10. Recombinant DNA Technology. A researcher wants to study an important protein involved in human gametogenesis. Because the protein is very difficult to prepare in sufficient quantities from human cells, she decides to clone its gene so that, if all goes well, she can use bacteria to make large batches of the protein. The amino acid sequence of the protein's single polypeptide chain has already been established. Briefly explain how she might clone the desired gene.

SUGGESTED READING

General References

Alberts, B., D. Bray, J. Lewis, M. Raff, K. Roberts, and J. D. Watson. *Molecular Biology of the Cell,* 3d ed. New York: Garland, 1994.

Griffiths, A. J. F., J. H. Miller, D. T. Suzuki, R. C. Lewontin, and W.M. Gelbart. *An Introduction to Genetic Analysis,* 5th ed. New York: W. H. Freeman, 1993.

Watson, J. D., N. Hopkins, J. Roberts, J. Steitz, and A. Weiner. *Molecular Biology of the Gene,* 4th ed. Menlo Park, CA: Benjamin/Cummings, 1987.

Advantages of Sexual Reproduction

Crow, J. F. The importance of recombination. In *The Evolution of Sex: An Examination of Current Ideas* (R. E. Michod and B. R. Levin, eds.). Sunderland, MA: Sinauer, 1988.

Meiosis

Chandley, A. C. Meiosis in man. *Trends Genet.* 4 (1988): 79.

Hawley, R. S., and T. Arbel. Yeast genetics and the fall of the classical view of meiosis. *Cell* 72 (1993): 301.

John, B. *Meiosis.* New York: Cambridge University Press, 1990.

Murray, A., and T. Hunt. *The Cell Cycle: An Introduction.* New York: W. H. Freeman, 1993.

Roeder, G. S. Chromosome synapsis and genetic recombination: Their roles in meiotic chromosome segregation. *Trends Genet.* 6 (1990): 385.

Mendel's Experiments

Fincham, J. R. S. Mendel—now down to the molecular level (News and Views). *Nature* 343 (1990): 208.

Mendel, G., H. de Vries, C. Correns, and E. Tschermak. The birth of genetics. *Genetics 35* (1950, Suppl.): 1. (Original papers in English translation.)

Sturtevant, A. H. *A History of Genetics.* New York: Harper & Row, 1965.

Recombinant DNA Technology

Abelson, J. and E. Butz, eds. Recombinant DNA. *Science* 209 (1980): 1317 (special issue).

Burke, D. T., G. F. Carle, and M. V. Olson. Cloning of large segments of exogenous DNA into yeast by means of artificial chromosome vectors. *Science* 236 (1987): 806.

Drlica, K. *Understanding DNA and Gene Cloning.* New York: Wiley, 1984.

Szostak, J. W. *In vitro* genetics. *Trends Biochem. Sci.* 17 (1992): 89.

Watson, J. D., M. Gilman, J. Witkowski, and M. Zoller. *Recombinant DNA: A Short Course,* 2d ed. New York: Scientific American Books, 1992.

Watson, J., and J. Tooze. *The DNA Story: A Documentary History of Gene Cloning.* New York: W. H. Freeman, 1981.

Genetic Engineering

Barinaga, M. Gene therapy: Step taken toward improved vectors for gene transfer (Research News). *Science* 266 (1994): 1326.

Boyd, A. L., and D. Samid. Review: Molecular biology of transgenic animals. *J. Anim. Sci.* 71 (S3) (1993): 1.

Capecchi, R. M. The new mouse genetics: Altering the genome by gene targeting. *Trends Genet.* 5 (1989): 70.

Chilton, M. D. A vector for introducing new genes into plants. *Sci. Amer.* (June 1983): 50.

Mulligan, R. C. The basic science of gene therapy. *Science* 260 (1993):926.

Palmiter, R. D., and R. L. Brinster. Germ line transformation in mice. *Annu. Rev. Genet.* 20 (1986): 465.

Palmiter, R. D., R. L. Brinster, R. E. Hammer, M. E. Trumbauer, M. G. Rosenfeld, N. C. Birnberg, and R. M. Evans. Dramatic growth of mice that develop from eggs microinjected with metallothionein-growth hormone fusion genes. *Nature* 300 (1982): 611.

Pennisi, E. Mouse of a different YAC. *Science News* 143 (1993): 360.

Stone, R. Agricultural biotechnology: Large plots are next test for transgenic crop safety (News and Comment). *Science* 266 (1994): 1472.

Westphal, H. Transgenic mammals and biotechnology. *FASEB J.* 3 (1989): 117.

17

GENE EXPRESSION:
I. THE GENETIC CODE AND TRANSCRIPTION

S o far, we have described DNA as the genetic material of cells and organisms.
We have come to understand its structure,
chemistry, and replication, as well as the way
it is packaged into chromosomes and
parceled out to daughter cells at mitotic and
meiotic cell divisions. Now we are ready to explore how DNA is expressed—that is, how this
coded information is used by the cell to specify
RNA and protein molecules. Our discussion of this important subject is divided among three chapters. This chapter
deals with the nature of the genetic code and the basic
processes of RNA synthesis or transcription. Chapter 18 describes translation, how protein is synthesized. Finally,
Chapter 19 elaborates on transcription and translation by
focusing on the *regulation* of gene expression. To put all
these topics in context, we start here with a general overview
of the roles of DNA, RNA, and protein in gene expression.

The Central Dogma of
Molecular Biology

As we mentioned at the start of this unit (Chapter 14), the
flow of genetic information in cells is generally from DNA to
RNA to protein. The DNA (more precisely, a segment of one
DNA strand) first serves as a template for the synthesis of an
RNA molecule, and, in most cases, the RNA molecule then
directs the synthesis of a particular protein. (In other cases,
the RNA is the final product of gene expression and functions as such within the cell.) The principle of unidirectional
genetic information flow through these three important
macromolecules is known as the **central dogma of molecular biology,** a term coined by Francis Crick. The central
dogma is usually summarized as follows:

$$\text{DNA} \xrightarrow{\text{replication}} \text{DNA} \xrightarrow{\text{transcription}} \text{RNA} \xrightarrow{\text{translation}} \text{protein}$$

Thus, the central dogma involves the processes
of *DNA replication, transcription* of the information carried by DNA into the form of
RNA, and *translation* of the information encoded in the RNA into protein.

RNA that is translated into protein is
called **messenger RNA (mRNA)** because, in
essence, it carries a genetic message from the
DNA to the ribosomes, where protein synthesis actually occurs. In addition to mRNA, two other classes of
RNA are required for protein synthesis. **Ribosomal RNA
(rRNA)** molecules are integral components of the ribosomes, and molecules of **transfer RNA (tRNA)** serve as the
keys for translating the coded nucleotide sequence of messenger RNA into the amino acid sequence of protein. Note
that rRNA and tRNA are RNAs that are not themselves
translated into protein; for their genes, expression ends at
the RNA level. The involvement of all three major classes of
RNA in the overall flow of information from DNA to protein is outlined in Figure 17-1.

The synthesis of RNA using DNA as a template is called
transcription to emphasize that this phase of gene expression is simply a transfer of information from DNA to RNA.
Both are nucleic acids, so the basic "language" remains the
same. By contrast, the process of protein synthesis is called
translation because it involves a language change, from the
nucleotide sequence of the mRNA to the amino acid sequence of the protein.

During the years since Crick's first statement of the central dogma, it has been elaborated in various ways. Many
viruses with RNA genomes synthesize RNA on an RNA template. Other viruses with RNA genomes, such as HIV, carry
out **reverse transcription,** whereby their RNA is used as a
template for DNA synthesis—a "backward" flow of genetic
information. (Box 17A, page 538, discusses these viruses and
also the involvement of reverse transcription in the rearrangement of certain DNA sequences in cellular DNA.)
However, the central dogma remains the main operating
principle for genetic information flow in the life of the cell.

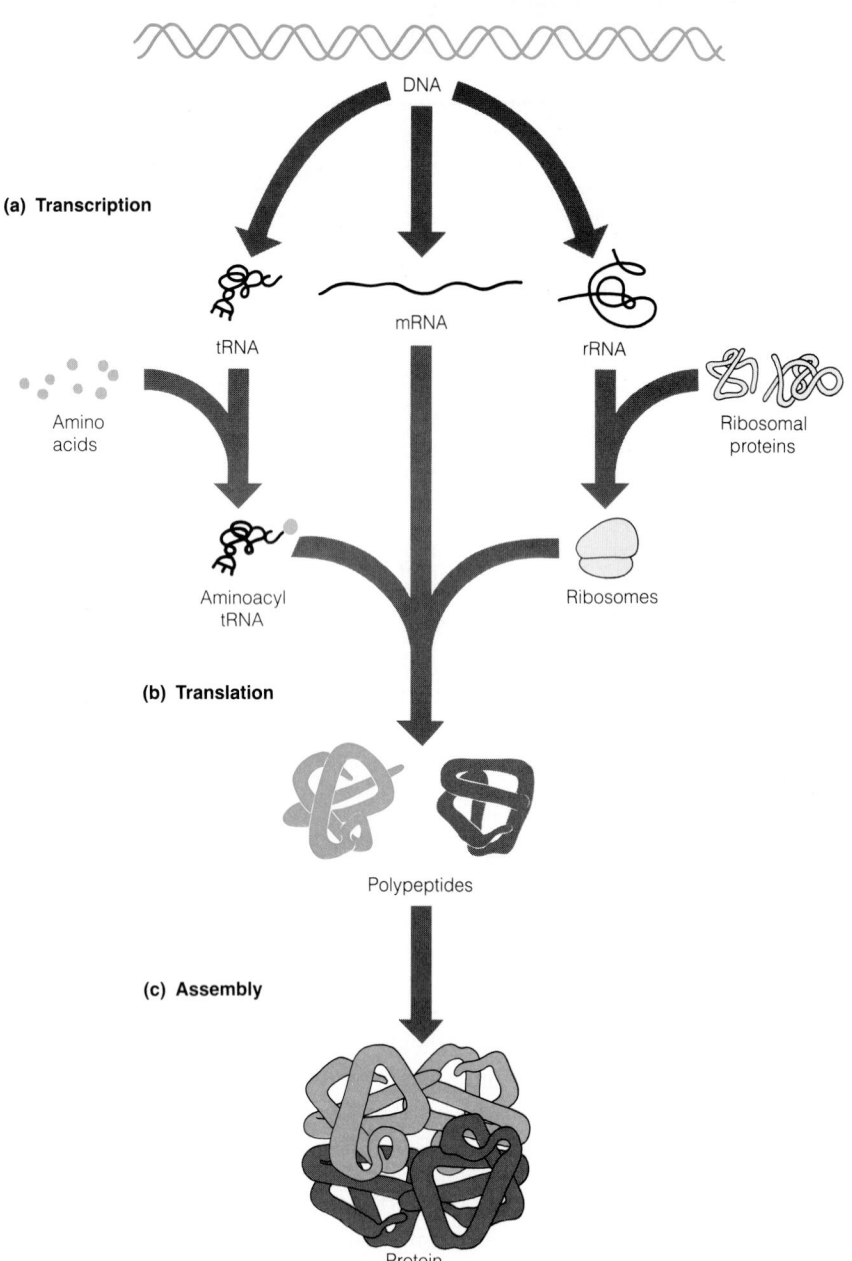

Figure 17-1   RNAs as Intermediates in the Flow of Genetic Information.
All three major classes of RNA—tRNA, mRNA, and rRNA—are **(a)** synthesized by transcription of the appropriate DNA sequences (genes) and **(b)** involved in the subsequent process of translation (polypeptide synthesis). The appropriate amino acids are brought to the mRNA and ribosome by tRNA. A tRNA molecule carrying an amino acid is called an aminoacyl tRNA. Polypeptides then fold and **(c)** assemble into functional proteins. The specific polypeptides shown here are the globin chains of the protein hemoglobin. For simplicity, this figure omits many details that will be described in this chapter and the next.

The Genetic Code

We now go on to look at the **genetic code,** the set of rules that governs the relationship between the order of nucleotides in DNA and the order of amino acids in proteins. In that relationship lies the essence of gene expression, and the cracking of the genetic code is one of the real landmarks of twentieth-century biology.

One of the definitions of a code is a system of symbols used to represent assigned and often secret meanings. Consider, for example, the International Morse Code (Figure 17-2), a system of dots, dashes, and spaces used primarily to send messages by telegraph and shortwave radio. To the uninitiated, the coded message in Figure 17-2a is just a series of dots and dashes, with no obvious meaning. But given ac-

cess to a set of equivalency rules for the Morse code (Figure 17-2b), anyone can convert the sequence of dots and dashes into a string of letters, and the message becomes recognizable as an English sentence (Figure 17-2c). Thus, a sequence of dots and dashes can carry useful information, but it must be translated into letters of the alphabet before the message makes sense to the reader.

So, too, with the genetic code. To the uninitiated, the message shown in Figure 17-3a is just a series of nucleotides in an mRNA molecule, with no obvious meaning. But given access to the genetic code (Figure 17-3b; see also Figure 17-7), anyone can convert the sequence of purines and pyrimidines into a string of amino acids, and the message becomes recognizable as a polypeptide (Figure 17-3c). Thus, a nucleotide sequence can contain useful information for

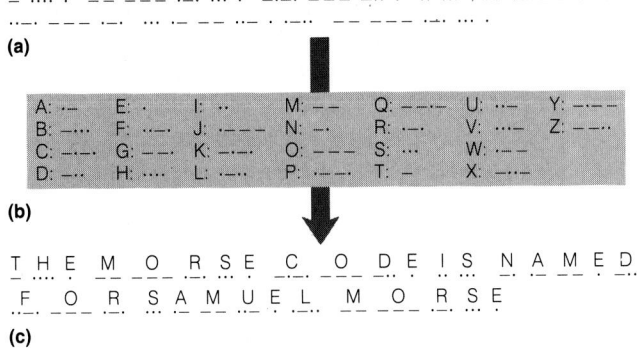

(a)

A: ·—	E: ·	I: ··	M: ——	Q: ——·—	U: ··—	Y: —·——
B: —···	F: ··—·	J: ·———	N: —·	R: ·—·	V: ···—	Z: ——··
C: —·—·	G: ——·	K: —·—	O: ———	S: ···	W: ·——	
D: —··	H: ····	L: ·—··	P: ·——·	T: —	X: —··—	

(b)

T H E   M O R S E   C O D E   I S   N A M E D
F O R   S A M U E L   M O R S E

(c)

Figure 17-2   The Morse Code. The International Morse Code is a system of dots, dashes, and spaces used to send messages by telegraph or shortwave radio. **(a)** A message written as a sequence of dots and dashes has no obvious meaning until **(b)** a set of equivalency rules is used to convert the sequence into **(c)** a recognizable English sentence—in this case, the first several words of an encyclopedia entry on the Morse code.

protein synthesis, but it must be translated into an amino acid sequence in order to make sense to the cell.

What is needed in both cases, of course, is a knowledge of the appropriate code—the set of rules that determines how many and which dots and dashes (or purines and pyrimidines) correspond to which letters (or amino acids). The encoded message can then be translated into sentences that make sense to the reader (or into proteins that the cell can use). In the case of the Morse code, one has only to take the trouble to learn the code, and all messages are readily

understandable. Most codes, however, are not public; they are specifically intended to convey secret messages that cannot be understood by outsiders unless they manage to crack the code.

In a sense, the genetic code was a secret code, and all of us were outsiders until the code was solved in the early 1960s. In fact, the genetic code was a secret in a double sense: Before scientists could figure out the exact relationship between the nucleotide sequence of a DNA molecule and the amino acid sequence of a protein, they first had to become aware that such a relationship existed at all. That awareness arose from the discovery that mutations in DNA could lead to changes in proteins.

Mutations Cause Changes in Proteins

The link between mutations and proteins was first detected experimentally by George Beadle and Edward Tatum, in their research with the common bread mold, *Neurospora crassa*. They induced mutations in *Neurospora* by X-ray treatment and then isolated nutritional mutants that could grow only if specific supplementary substances were added to the growth medium. They were able to infer that in each such mutant a particular enzyme-catalyzed step leading to the synthesis of a specific compound was blocked. There was, in other words, a one-to-one correspondence between a genetic mutation and the lack of a specific enzyme required in a biochemical pathway. From these findings, Beadle and Tatum formulated the *one gene–one enzyme hypothesis.* Their hypothesis was thoroughly confirmed in later studies and is now more precisely expressed as **one gene–one polypeptide.**

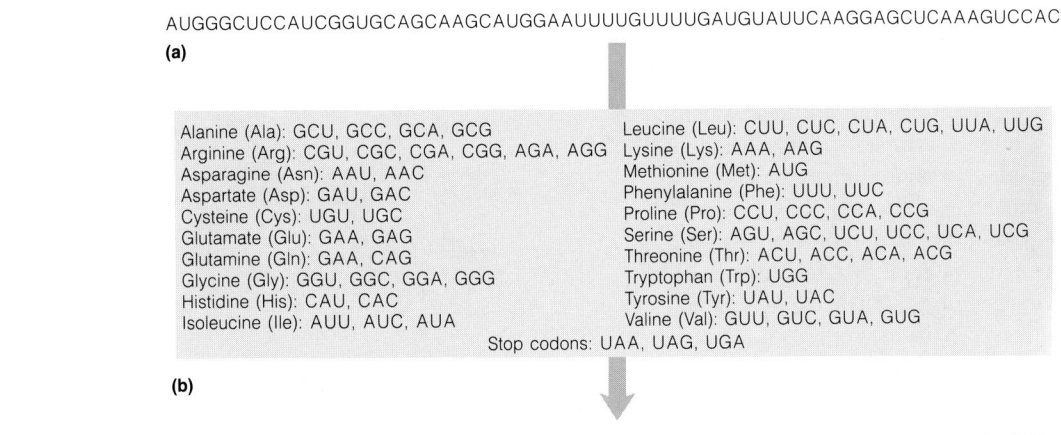

AUGGGCUCCAUCGGUGCAGCAAGCAUGGAAUUUUGUUUUGAUGUAUUCAAGGAGCUCAAAGUCCAC

(a)

Alanine (Ala): GCU, GCC, GCA, GCG	Leucine (Leu): CUU, CUC, CUA, CUG, UUA, UUG
Arginine (Arg): CGU, CGC, CGA, CGG, AGA, AGG	Lysine (Lys): AAA, AAG
Asparagine (Asn): AAU, AAC	Methionine (Met): AUG
Aspartate (Asp): GAU, GAC	Phenylalanine (Phe): UUU, UUC
Cysteine (Cys): UGU, UGC	Proline (Pro): CCU, CCC, CCA, CCG
Glutamate (Glu): GAA, GAG	Serine (Ser): AGU, AGC, UCU, UCC, UCA, UCG
Glutamine (Gln): GAA, CAG	Threonine (Thr): ACU, ACC, ACA, ACG
Glycine (Gly): GGU, GGC, GGA, GGG	Tryptophan (Trp): UGG
Histidine (His): CAU, CAC	Tyrosine (Tyr): UAU, UAC
Isoleucine (Ile): AUU, AUC, AUA	Valine (Val): GUU, GUC, GUA, GUG

Stop codons: UAA, UAG, UGA

(b)

5 ′ - AUG - GGC - UCC - AUC - GGU - GCA - GCA - AGC - AUG - GAA - UUU - UGU - UUU - GAU - GUA - UUC - AAG - GAG - CUC - AAA - GUC - CAC - 3 ′

Met - Gly - Ser - Ile - Gly - Ala - Ala - Ser - Met - Glu - Phe - Cys - Phe - Asp - Val - Phe - Lys - Glu - Leu - Lys - Val - His -

(c)

Figure 17-3   A First Look at the Genetic Code. The genetic code is a system of purines and pyrimidines used to send messages from the genome to the ribosomes. **(a)** A message written as a sequence of nucleotides in an mRNA molecule has no obvious meaning, until **(b)** a set of equivalency rules for the genetic code is used to convert the sequence into **(c)** the amino acid sequence of a recognizable polypeptide—in this case, the first 22 of the 385 amino acids in ovalbumin, the major protein of egg white. (The genetic code is shown in more conventional form in Figure 17-7.)

Further Insights

REVERSE TRANSCRIPTION, RETROVIRUSES, AND RETROTRANSPOSONS

Most transcription in the cell goes on as described in the central dogma, with DNA serving as a template for RNA synthesis. However, RNA can also serve as a template for DNA polymerization, a process called *reverse transcription,* which is catalyzed by the enzyme *reverse transcriptase.* This enzyme was first discovered by Howard Temin and David Baltimore in certain viruses with RNA genomes. Viruses that carry out reverse transcription are called *retroviruses.* Examples of retroviruses include some important pathogens, such as the human immunodeficiency virus (HIV, the AIDS virus) and a number of viruses that cause cancers in animals.

Retroviruses

Figure 17A-1 depicts the reproductive cycle of a typical retrovirus. In the virus particle, two copies of the RNA genome are enclosed within a protein capsid, which is enclosed, in turn, within a membranous envelope. Each RNA molecule has a molecule of reverse transcriptase attached to it. ① The virus first binds to the surface of the host cell, and its envelope fuses with the plasma membrane, releasing the capsid and its contents into the cytoplasm. Once inside the cell, the reverse transcriptase catalyzes the synthesis of ② a strand of DNA complementary to the viral RNA (the reverse transcription step) and then ③ a second DNA strand complementary to the first. The result is a double-stranded DNA version of the viral genome. ④ This DNA enters the cell nucleus and integrates into the host cell's DNA, much as the DNA genome of a lysogenic phage integrates into the DNA of the bacterial chromosome (see Box 14A). Like a prophage, the integrated viral genome, called a *provirus,* is replicated every time the cell replicates its own DNA. ⑤ Transcription of the proviral DNA (by cellular enzymes) produces many RNA transcripts, which function in two ways. ⑥ First, they serve as mRNA, dictating the synthesis of the three viral proteins: capsid protein, envelope protein, and reverse transcriptase. ⑦ Second, they are packaged with the viral proteins into new virus particles. ⑧ The new viruses "bud" from the cell membrane without necessarily killing the cell.

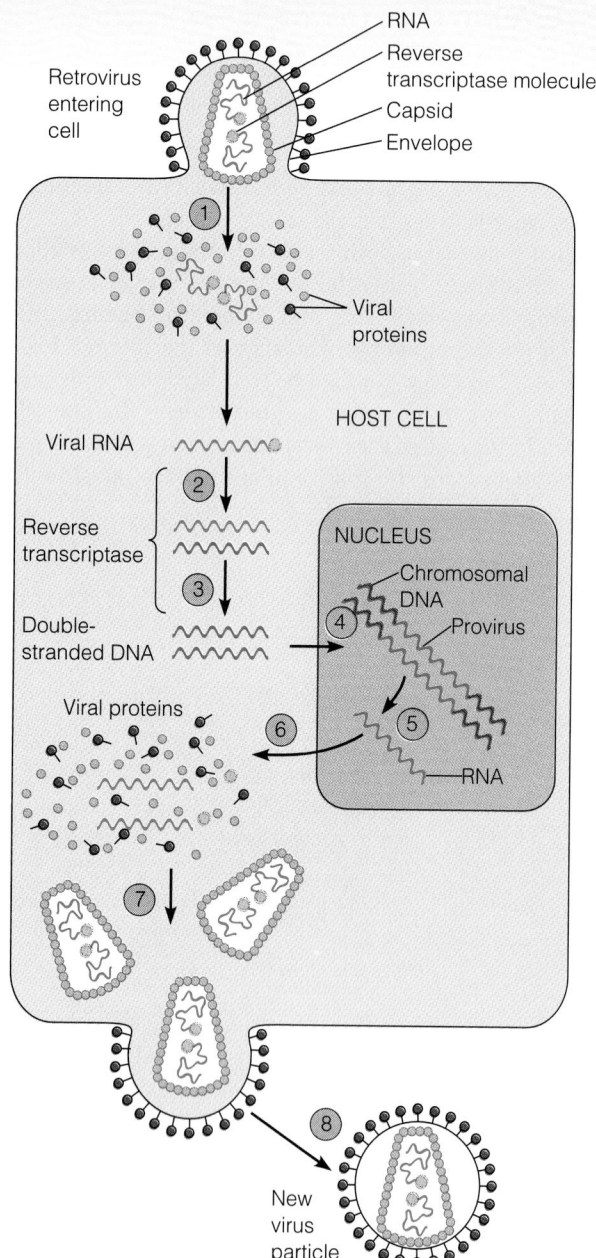

Figure 17A-1 The Reproductive Cycle of a Retrovirus.

The ability of a retroviral genome to integrate into host cell DNA helps explain how some retroviruses can cause cancer. These viruses, called RNA tumor viruses, are of two types. Viruses of the first class carry a cancer-causing gene—an *oncogene*—in their genomes, along with their genes for viral proteins. The oncogene is actually a mutated version of a normal cellular gene, usually a version of one of the proteins used to regulate the cell cycle. For example, the oncogene carried by the Rous sarcoma virus (a chicken virus that was the first RNA tumor virus to be discovered) is a modified version of a cellular gene for a protein kinase. The protein product of the viral gene is hyperactive and is not controllable by the cell in the normal way. As a result, cells expressing this gene proliferate wildly, producing cancerous tumors called *sarcomas*. RNA tumor viruses of the second type do not themselves carry oncogenes, but the integration of their genomes into cellular DNA can mutate the DNA in such a way that a normal cellular gene acquires oncogene activity. (We discuss oncogenes in greater detail in Chapter 25.)

Retrotransposons

Reverse transcription also takes place in normal eukaryotic cells in the absence of viral infection. Much of it involves genetic elements within cellular DNA called *retrotransposons*. These are nucleotide sequences that can move, or *transpose*, themselves from one site to another within the genomic DNA by a mechanism involving reverse transcrip-

tion. The mechanism is essentially identical to part of the retrovirus life cycle, with the reverse transcriptase often encoded in the retrotransposon DNA. As outlined in Figure 17A-2, the transposition process ① starts with transcription of the retrotransposon DNA by the usual cellular machinery. ② Translation of part of the RNA yields reverse transcriptase, which ③ catalyzes the synthesis of double-stranded DNA from the retrotransposon RNA. ④ The free retrotransposon DNA can then insert itself into the genome at some other location. In this way, the retrotransposon can spread to many places in the genome. (Steps 3 and 4 in Figure 17A-2 correspond to steps 2–4 in Figure 17A-1.)

Retrotransposons are probably present in all eukaryotic organisms. Moreover, although they transpose themselves only rarely, they can attain very high numbers of copies within a genome. Over 10% of the human genome, for example, is known to consist of retrotransposons. We encountered one type of these in Chapter 14—the *Alu family* of sequences. Alu sequences are only 300 bp long and do not encode a reverse transcriptase, but, using a reverse transcriptase encoded elsewhere in the genome, they have sent copies of themselves throughout the genomes of humans and other primates. The haploid genome of humans has about half a million Alu sequences, making up 5% of human DNA! Apparently molecular "parasites," such sequences are not known to have any function other than their own propagation. The question of what keeps retrotransposons from completely overrunning the genome still awaits an answer. ■

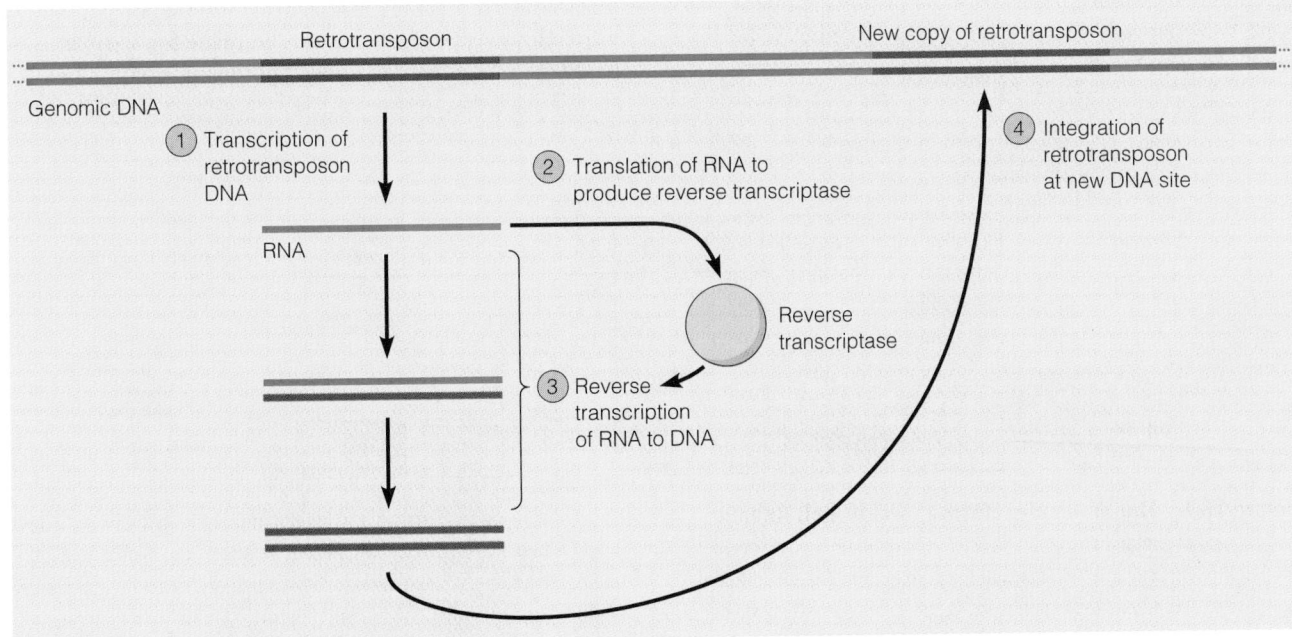

Figure 17A-2　Movement of a Retrotransposon.

A better understanding of the relationship between gene and protein developed as it became clear that mutations often result in changes in the amino acid sequence of specific polypeptides. The first evidence for this connection came from studies of *sickle-cell anemia,* a severe human blood disease characterized by the unusual "sickle" shape that the red blood cells of afflicted people often assume. This condition was traced to a difference in the oxygen-carrying protein *hemoglobin,* which consists of four polypeptide chains, two α chains and two β chains ($\alpha_2\beta_2$).

In 1957, Vernon Ingram showed that normal hemoglobin and sickle-cell hemoglobin have identical α chains but have β chains that differ in a single amino acid. Normal hemoglobin has glutamic acid (glutamate) at the sixth amino acid position of the β chain, whereas sickle-cell hemoglobin (hemoglobin S) has valine at that position (Figure 17-4). That single change (which results from the change of a single base pair in the DNA) is enough to affect the way in which hemoglobin molecules pack into red blood cells. Normal hemoglobin is jellylike in its consistency, but hemoglobin S tends to form a kind of crystal when it delivers oxygen to and picks up carbon dioxide from tissues. The crystalline array deforms the cell into a sickled shape, which then blocks blood flow in capillaries, leading to the symptoms of the disease.

From this and other examples, it became apparent that genes must somehow specify the amino acid sequences of proteins. This relationship was even more clearly established by further genetic studies, such as those of Charles Yanofsky on the enzyme tryptophan synthetase of *Escherichia coli.* This enzyme consists of two polypeptides, A and B, encoded by the genes *trpA* and *trpB,* respectively. Yanofsky identified many bacterial mutants lacking tryptophan synthetase activity and mapped their mutations within the genes. He found that most of the mutations could be ordered linearly within the *trpA* gene. He also discovered that a number of the mutants produced altered forms of polypeptide A that were enzymatically inactive but could nonetheless be isolated and sequenced.

From the sequence data for many such mutants, Yanofsky was able to show that each of the mutant polypeptides differed from the wild type by a single amino acid and that different mutants had different amino acid substitutions. When he compared the location of each altered amino acid in the polypeptide with the genetic map position of the mutation that caused that alteration, he found a close correlation. This correlation led Yanofsky to conclude that there is a **colinearity** between a gene and its polypeptide, further substantiating the relationship between the nucleotide sequence of DNA and the amino acid sequence of proteins. In retrospect, it was fortunate that Yanofsky was studying bacterial genes. As we will see later, most eukaryotic genes contain noncoding sequences interspersed among the coding regions of the gene, and therefore do not usually show quantitative colinearity between a gene and its polypeptide product.

The Nature of the Genetic Code

Given a sequence relationship between DNA and proteins, the next question concerned the number of nucleotides necessary to specify each amino acid. In the Morse code, more than one symbol is clearly required per letter, because only three different kinds of symbols—dots, dashes, and spaces—are used to represent 26 letters, 10 numbers, and various punctuation marks. In fact, the number of symbols per letter varies, ranging from a single dot for the letter E to various combinations of four symbols for some of the less common letters, such as X, Y, and Z. Fortunately, the genetic code is more orderly, and the number of nucleotides per amino acid is always three. Thus, three nucleotide pairs in the double-stranded DNA genome are required to specify each amino acid in a polypeptide.

Most of the initial evidence suggesting a triplet code was genetic, though biochemical confirmation quickly followed, once it became possible to synthesize polypeptides in vitro. However, a good case can be made for a triplet code based on logic alone. We know that the information in DNA

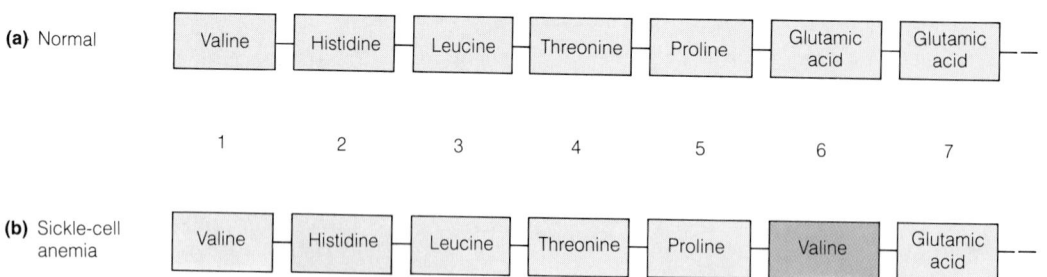

(a) Normal

Valine — Histidine — Leucine — Threonine — Proline — Glutamic acid — Glutamic acid

1   2   3   4   5   6   7

(b) Sickle-cell anemia

Valine — Histidine — Leucine — Threonine — Proline — Valine — Glutamic acid

Figure 17-4 The Molecular Basis of Sickle-Cell Anemia.

The β chain of human hemoglobin consists of 146 amino acids. Shown here are the first seven amino acids of the β chain from **(a)** an individual with normal (wild-type) hemoglobin and **(b)** an individual with sickle-cell anemia. The only difference between the two polypeptides is the substitution of a valine for glutamic acid (glutamate) at the sixth position. This single amino acid difference is sufficient to impair hemoglobin function and cause sickle-cell anemia in individuals homozygous for the sickle-cell allele.

must reside in the sequence of the four nucleotides that constitute the DNA: A, T, G, and C. These are the only "letters" of the DNA alphabet. Therefore, there must be more than one letter for each "word" in the message, because the DNA language has to contain at least 20 words, one for each amino acid. A doublet code would not be adequate, as four nucleotides taken two at a time can generate only $4^2 = 16$ different combinations.

With three letters per word, however, the number of different words that can be produced with an alphabet of just four letters is $4^3 = 64$. This number appears to be more than adequate, and it raises the obvious question of whether every one of the 64 possible combinations is actually used. The answer is yes, although a few of them have a special "punctuation" function instead of coding for an amino acid, as we will see later. First, however, we will look at the genetic evidence for the triplet nature of the code. And for that, we need to become acquainted with frameshift mutations.

Analysis of Frameshift Mutations by Crick and Brenner

In 1961, Francis Crick, Sydney Brenner, and their colleagues deduced the triplet nature of the genetic code by studying the effects of the mutagenic chemical *proflavin* on the bacteriophage T4. Their work is well worth considering, not just because of the critical genetic evidence it provided concerning the nature of the code but also because of the ingenuity of the deductive reasoning that was necessary to understand the significance of their findings.

Proflavin is one of several *acridine dyes* used as **mutagens** (mutation-inducing agents) in genetic research. The acridines are interesting mutagens because they act by causing the addition or deletion of single nucleotide pairs in the DNA. Sometimes, mutants generated by acridine treatment

of a wild-type virus or organism appear to revert to the wild type when treated with more of the same type of mutagen. Closer examination, however, shows that this apparent reversion is not a true reversal of the original event, but the acquisition of a second mutation that maps very close to the first. In fact, the two mutations are always in the same gene, and if they are separated by genetic recombination, each still gives rise to the mutant phenotype.

These mutations display an interesting kind of arithmetic. If the first is called a plus ($+$) mutation, then the second can be called a minus ($-$) mutation. By itself, each changes the phenotype, but when they occur close together as a double mutation, they cancel each other out.

This finding can be explained using the analogy in Figure 17-5. Line 1 is a wild-type "gene," written in a language that uses three-letter words. When we "translate" the line by starting at the beginning and reading three letters at a time, the message of the gene is readily comprehensible. A plus mutation is the addition of a single letter within the message (line 2). That change may seem minor, but since the message is always read three letters at a time, the insertion of an extra letter early in the sequence means that all the remaining letters are read out of phase. There is, in other words, a shift in the *reading frame*, and the result is a garbled message from the point of the insertion onward. A minus mutation can be explained in a similar way because the deletion of a single letter also causes the reading frame to shift, resulting in another garbled message (line 3). Such **frameshift mutations** are characteristic effects of the acridine dyes and other mutagens that cause the insertion or deletion of individual nucleotide pairs.

Individually, plus and minus mutations always change the reading frame and garble the message. When a plus and a minus mutation occur in the same gene, however, they can cancel out each other's effect, particularly if they are located

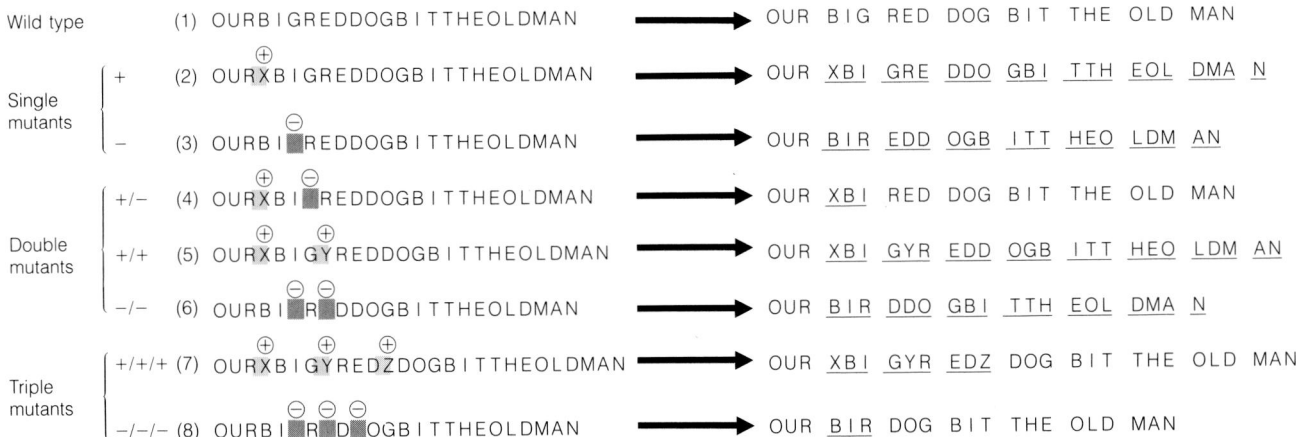

Figure 17-5 Frameshift Mutations.   The effect of frameshift mutations can be illustrated with an English sentence. The wild-type sentence (line 1) consists of three-letter words. When read in the correct frame, it is fully comprehensible. The insertion (line 2) or deletion (line 3) of a single letter shifts the reading frame and garbles the message from that point onward. (Garbled words due to shifts in the reading frame are underscored.) Double mutants containing a deletion that "cancels" a prior insertion have a restored reading frame from the point of the second mutation onward (line 4). However, double insertions (line 5) or double deletions (line 6) produce garbled messages. Triple insertions (line 7) or deletions (line 8) garble part of the message but restore the reading frame with the net addition or deletion of a single word.

in close proximity. The insertion caused by the plus mutation compensates for the deletion caused by the minus mutation, and the message is intelligible from that point on (line 4). Notice, however, that double mutations with either two additions (+/+; line 5) or two deletions (−/−; line 6) do not compensate in this way. They stay out of phase for the remainder of the message.

Crick and Brenner made similar observations with the T4 mutants they generated with proflavin. They were able to isolate revertants from their minus mutants and show that these always contained a plus mutation at a site different from, but close to, the original mutation. New plus and minus mutations could, in fact, be localized to a variety of positions within the gene, and double +/− mutants could be constructed in many combinations. Many of these had wild-type (or more properly, *pseudo-wild-type)* phenotypes. But when Crick and Brenner generated +/+ or −/− double mutants by recombination, no wild-type phenotypes were ever seen.

Crick and Brenner also constructed triple mutants of the same types (+/+/+ or −/−/−) and found, probably to their surprise, that many of these did have wild-type phenotypes. This finding, of course, can be readily understood by consulting lines 7 and 8 of Figure 17-5: The reading frame at the beginning and end of the message remains the same when three letters are either added or removed. The portion of the message between the first and third mutations will be garbled, but provided these are sufficiently close to each other, enough of the sentence may remain to convey an intelligible message.

Keep in mind, however, that Crick and Brenner did not have Figure 17-5 to assist them. Their ability to deduce the correct explanation from their analysis of acridine-induced mutations is an especially inspiring example of the careful, often ingenious reasoning that almost always accompanies significant advances in science.

The Genetic Code Is a Triplet Code

Because the wild-type phenotype was often maintained in the presence of three nucleotide-pair additions (or deletions) but not in the presence of one or two, Crick and Brenner concluded that the nucleotides making up a DNA strand must be read in groups of three (or perhaps multiples thereof). They reasoned that the reading of a message must begin at a specific starting place on the DNA (to ensure the proper reading frame) and must then proceed three nucleotides at a time, with each such triplet translated into the appropriate amino acid, until the end of the message is reached. The genetic code, in other words, is a *triplet code.*

Adding or deleting one nucleotide pair shifts the reading frame of the message from that point onward, and a second, similar change shifts the reading frame yet again. From the first mutation onward, the message is garbled. After a third change of the same type, however, the original reading frame is back again, and the only segment of the message that is translated incorrectly is the segment between the first

and third mutations. Such errors can apparently be tolerated in many proteins, provided that the affected region is short and the change in amino acid sequence in that region does not destroy protein function. This is why the individual mutations in a triple mutant with wild-type phenotype map so closely together. Subsequent sequencing of "wild-type" polypeptides from such triple mutants confirmed the slightly altered sequences of amino acids that one would predict.

The Genetic Code Is Degenerate

From the fact that so many of their triple mutants were viable, Crick and Brenner drew an additional conclusion: Most of the 64 possible nucleotide triplets must specify amino acids, even though proteins have only 20 different kinds of amino acids. If only 20 of the 64 possible combinations of nucleotides "made sense" to the cell, the chances of a meaningless triplet appearing in the out-of-phase stretches would be high. Such triplets would surely interfere with protein synthesis, and frameshift mutants would "revert" only rarely.

Because they detected such revertants frequently and found that any plus mutation had a fairly high probability of suppressing any minus mutation if the two were close together, Crick and Brenner reasoned that most of the 64 possible triplets must code for amino acids. This, in turn, told them that the genetic code is **degenerate**—that is, a given amino acid may be specified by more than one nucleotide triplet. (The term *degenerate* has a foreign sound to many biologists; it was actually borrowed from quantum mechanics and probably reflects the influence of physics on some of the early investigators in the emerging field of molecular biology.)

The Words of the Genetic Code Are Nonoverlapping

A further conclusion from Crick and Brenner's work is that the genetic code is *nonoverlapping,* as we have, in fact, been assuming. An *overlapping* code would be one in which the reading frame was advanced only one or two nucleotides at a time along a DNA strand, so that each nucleotide was used two or three times. But, in fact, when the reading frame is advanced three nucleotides at a time, each nucleotide is used only once, and the code is thus nonoverlapping.

Figure 17-6 compares a nonoverlapping code with an overlapping code in which the reading frame advances only one nucleotide at a time. With such an overlapping code, the insertion or deletion of a single nucleotide pair in the gene would lead to the insertion or deletion of one amino acid at one point in the polypeptide and would change several adjacent amino acids but would not affect the reading frame of the remainder of the gene. (You might want to explore for yourself what the consequences would be of an overlapping code in which the reading frame advanced two nucleotides at a time.) If the code were overlapping, Crick and Brenner

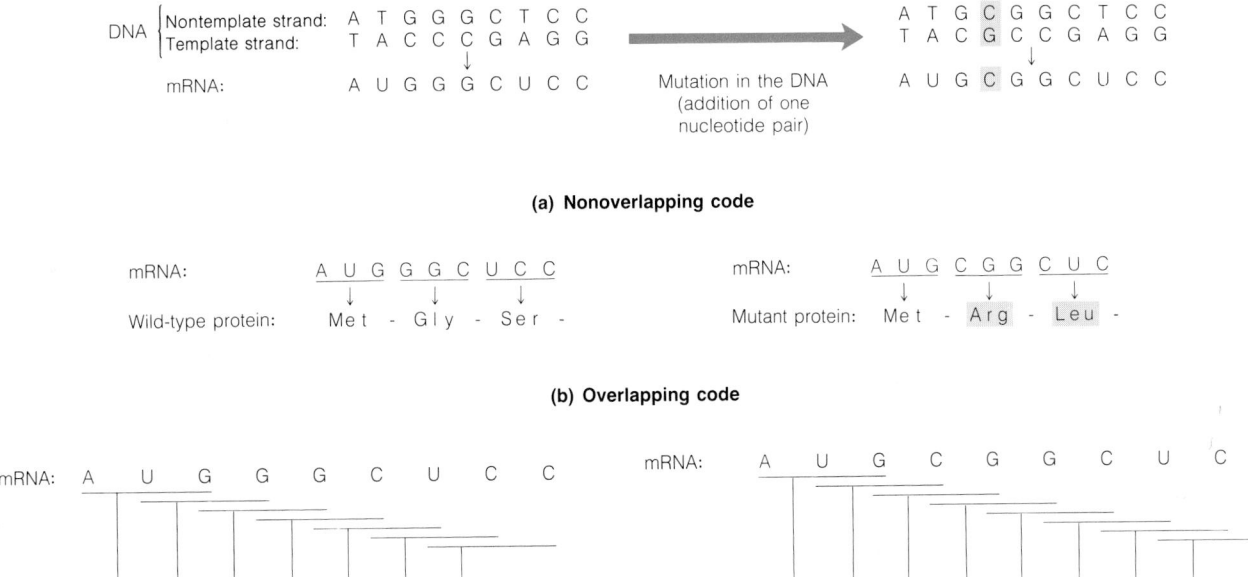

Figure 17-6 Effect of the Insertion of a Single Nucleotide Pair on Proteins Encoded by Overlapping and Nonoverlapping Genetic Codes. One strand of the DNA duplex at the top, called the *template strand,* is transcribed into the 9-nucleotide segment of mRNA shown, according to the same base-pairing rules used in DNA replication. (The complementary DNA strand, with a sequence essentially identical to that of the mRNA, is called the *nontemplate strand* or, because it carries essentially the same base sequence as the mRNA, the *coding strand.*) **(a)** With a nonoverlapping code, the reading frame advances three nucleotides at a time, and this mRNA segment is therefore read as three successive triplets, coding for the amino acids methionine, glycine, and serine. (See Figure 17-3 or 17-7 for amino acid coding rules.) If the DNA duplex is mutated by insertion of a single nucleotide pair (the CG pair in the top blue box), the mRNA will have an additional nucleotide (also in blue). This insertion alters the reading frame beyond that point, so that the remainder of the mRNA is read incorrectly and all amino acids

are wrong. In the example shown, the insertion occurs near the beginning of the message, and the only similarity between the wild-type protein and the mutant protein is the first amino acid (methionine). **(b)** In one type of overlapping code, the reading frame advances only one nucleotide at a time. The wild-type protein will therefore contain three times as many amino acids as would a protein generated from the same mRNA using a nonoverlapping code. Insertion of a single nucleotide pair in the DNA again results in an mRNA molecule with one extra nucleotide. However, in this case the effect of the insertion on the protein is modest; two amino acids in the wild-type protein are replaced by three different amino acids in the mutant protein, but the remainder of the protein is normal. The frameshift mutations that Crick and Brenner found in their studies with the mutagen proflavin would not have been observed if the genetic code were overlapping. Accordingly, their data indicated the code to be nonoverlapping.

would not have observed their frameshift mutations. Thus, their results clearly indicated the nonoverlapping nature of the code: Each nucleotide is a part of one, and only one, triplet.

Interestingly, although the genetic code is always translated in a nonoverlapping way, there are cases where a particular segment of DNA is translated in more than one reading frame. For example, certain viruses with very small genomes have overlapping genes, as was first discovered in 1977 for the phage ϕX174. In this phage's DNA, one gene is completely embedded within another gene, and, to complicate matters further, a third gene overlaps them both! The three genes are translated in different reading frames. Presumably, it is highly advantageous for the phage to keep its genome as small as possible. Other instances of overlapping genes are found in bacteria. As we will see in Chapter 19, some prokaryotic mRNAs encode more than one

polypeptide; in such cases, genes sometimes overlap by a few nucleotides at their boundaries.

From Triplets to Codons

From the publication of Crick and Brenner's historic findings in 1961, it took only five years for the meaning of each of the 64 triplets to be elucidated. We are about to look at how that was done. First, however, it is important to note that the genetic code as we usually describe it refers not to the order of nucleotide pairs in double-stranded DNA, but to the order of nucleotides in the single-stranded mRNA molecules that are actually used in protein synthesis.

As indicated in Figure 17-6a, transcription of DNA into RNA proceeds by a complementary base-pairing mechanism similar to DNA replication, except that only one of the two DNA strands is involved. The strand of the DNA duplex that serves as the template for RNA synthesis is called the **template strand.** The nontemplate DNA strand, although not itself directly involved in transcription, is by convention usually called the **coding strand,** because it is identical in sequence to the single-stranded RNA transcript that carries the coded message, except that the RNA transcript has uracil where the coding strand has thymine.

After transcription, the genetic information represented as triplets of deoxyribonucleotides in DNA is present in mRNA as triplets of ribonucleotides. These triplets are called **codons** because they are the actual coding units read by the translational machinery during protein synthesis. A DNA nucleotide sequence coding for a polypeptide is always expressed in terms of triplet codons in mRNA molecules. The four bases present in RNA are the purines adenine (A) and guanine (G) and the pyrimidines cytosine (C) and uracil (U), so the 64 codons of the genetic code consist of all possible combinations of these four "letters" taken three at a time. And since mRNA molecules are synthesized in the $5' \rightarrow 3'$ direction (like DNA) and are translated starting at the 5' end, codons are always written in the $5' \rightarrow 3'$ order.

Codon Assignments

The year 1961 was especially important in the history of the genetic code, for it brought not only the publication of Brenner and Crick's findings but also the development of the first of several methods for assigning specific codons to particular amino acids. Pioneered by Marshall Nirenberg and J. Heinrich Matthei, the original technique depended on the ability of an enzyme called *polynucleotide phosphorylase* to make synthetic mRNA of predictable nucleotide composition. Unlike the enzymes involved in cellular transcription, polynucleotide phosphorylase does not use a template but simply assembles available nucleotides randomly into a linear chain.

Homopolymers as mRNA. If polynucleotide phosphorylase were given access to all four ribonucleotides (ATP, GTP, CTP, and UTP), all possible codons would be made randomly, and no useful information could be obtained. But by using only one or two nucleotides, a restricted number of codons can be generated. The simplest RNA molecule results when a single kind of nucleotide is used, because the only possible product is a *homopolymer* consisting of a single repeating nucleotide. Not surprisingly, the first synthetic mRNA molecules that Nirenberg and Matthei made were homopolymers, and the first codon assignments were therefore easy.

For example, when polynucleotide phosphorylase was incubated with UTP as the sole precursor, the product was a homopolymer of uracil, called poly(U). Nirenberg and Matthei then added this synthetic mRNA to an in vitro protein-synthesizing system and showed that the product was a polypeptide containing phenylalanine as its only amino acid. Poly(U), in other words, codes for polyphenylalanine. From this observation, Nirenberg and Matthei de-

duced that the sequence UUU on mRNA is read as a signal for phenylalanine insertion during protein synthesis. They therefore made the first codon assignment: UUU = phenylalanine. Synthesis of other homopolymers quickly revealed that AAA codes for lysine and CCC for proline. [Poly(G) turned out not to be a good messenger because of unexpected structural complications and therefore could not be tested.]

The Copolymer Challenge. After the homopolymers were tested, the method was extended to more complex polymers, but it quickly proved to be less definitive because of the random choice of nucleotides by polynucleotide phosphorylase. Consider the **copolymers,** polynucleotides produced from mixtures of two nucleotides. For example, the copolymer synthesized by incubating polynucleotide phosphorylase with the precursors CTP and ATP contains Cs and As, but in no predictable order. Such a copolymer contains eight different codons: CCC, CCA, CAC, ACC, AAC, ACA, CAA, and AAA. When the copolymer was used to direct protein synthesis in vitro, the polypeptides that were produced contained six of the 20 possible amino acids. The codons for two of these (CCC and AAA) were known from the homopolymer studies, but the other four could not be unambiguously assigned.

Further progress depended on an alternative means of codon assignment devised by Nirenberg's group. Instead of using long polymers, they synthesized all possible codons (i.e., 64 very short mRNA molecules, each only three nucleotides long) and used these in binding studies to see which amino acid binds to the ribosome in response to each codon. With this approach, they were able to determine the majority of the codon assignments.

Meanwhile, a refined method of polymer synthesis had been devised in the laboratory of H. Gobind Khorana. Khorana's approach was similar to that of Nirenberg and Matthei, but with the important difference that the polymers he synthesized had defined sequences. Thus, he could produce a synthetic mRNA molecule with the strictly alternating sequence UAUA. . . . Such an RNA copolymer has only two codons, UAU and AUA, and they alternate in strict sequence. Knowing that the polypeptide product contained only tyrosine and isoleucine, Khorana was able to narrow the choices for UAU and AUA to these two amino acids. When the results obtained with various such synthetic polymers were combined with the findings of Nirenberg's binding studies, most of the codons could be assigned unambiguously.

Codon Assignments Complete. By 1966, just five years after the first codon was identified, all 64 codons had been assigned—that is, the entire genetic code had been worked out. Codon assignments are shown in Figure 17-7, in the conventional format. The elucidation of the code confirmed several properties that had been deduced earlier from indirect evidence. All 64 codons are in fact used in translation of mRNA; 61 of the codons specify the addition of specific

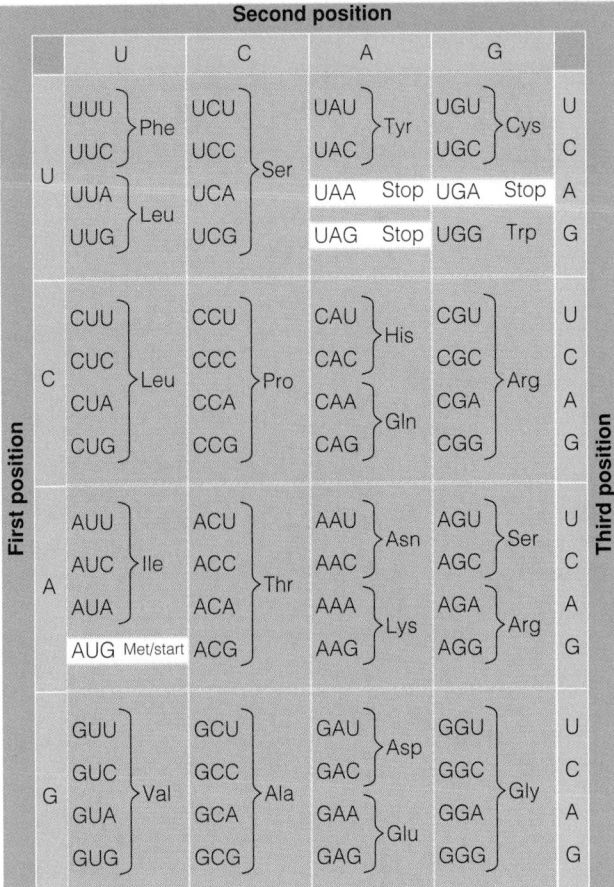

Second position

	U	C	A	G	
U	UUU ⎤ Phe UUC ⎦ UUA ⎤ Leu UUG ⎦	UCU ⎤ UCC ⎥ Ser UCA ⎥ UCG ⎦	UAU ⎤ Tyr UAC ⎦ UAA Stop UAG Stop	UGU ⎤ Cys UGC ⎦ UGA Stop UGG Trp	U C A G
C	CUU ⎤ CUC ⎥ Leu CUA ⎥ CUG ⎦	CCU ⎤ CCC ⎥ Pro CCA ⎥ CCG ⎦	CAU ⎤ His CAC ⎦ CAA ⎤ Gln CAG ⎦	CGU ⎤ CGC ⎥ Arg CGA ⎥ CGG ⎦	U C A G
A	AUU ⎤ AUC ⎥ Ile AUA ⎦ AUG Met/start	ACU ⎤ ACC ⎥ Thr ACA ⎥ ACG ⎦	AAU ⎤ Asn AAC ⎦ AAA ⎤ Lys AAG ⎦	AGU ⎤ Ser AGC ⎦ AGA ⎤ Arg AGG ⎦	U C A G
G	GUU ⎤ GUC ⎥ Val GUA ⎥ GUG ⎦	GCU ⎤ GCC ⎥ Ala GCA ⎥ GCG ⎦	GAU ⎤ Asp GAC ⎦ GAA ⎤ Glu GAG ⎦	GGU ⎤ GGC ⎥ Gly GGA ⎥ GGG ⎦	U C A G

(Left edge label: **First position**. Right edge label: **Third position**.)

Figure 17-7   The Genetic Code. The code "words" are three-letter codons present in the nucleotide sequence of mRNA, as read in the 5′ → 3′ direction. Letters represent the nucleotide bases uracil (U), cytosine (C), adenine (A), and guanine (G). Each codon specifies either an amino acid or a stop signal. To decode a codon, read down the left edge for the first letter, then across the grid for the second letter, and then down the right edge for the third letter. For example, the codon AUG represents methionine. (As we shall see in Chapter 18, AUG is also a start signal.)

amino acids to the polypeptide that is being synthesized. The remaining three codons (UAA, UAG, and UGA) are **stop signals** that instruct the cell to terminate the polypeptide. The degenerate nature of the code is clear, in that many of the amino acids are specified by more than one codon. There are, for example, two codons for histidine (His), four for threonine (Thr), and six for leucine (Leu). Also clear from Figure 17-7 is the *unambiguous* nature of the code: Every codon has one and only one meaning.

The (Near) Universality of the Genetic Code

A final property of the genetic code that is worth noting is its near universality. No matter what the organism, the same codons almost always stand for the same amino acids (or stop signals), suggesting that the genetic code was estab-

lished very early in the history of life and has remained virtually constant over billions of years of evolutionary history. Exceptions to the standard genetic code are known for only three kinds of genomes: mitochondrial genomes, certain bacterial genomes, and the nuclear genomes of some protozoa. In common with chloroplasts, mitochondria contain their own DNA and carry out both transcription and translation. However, the mitochondria of some organisms, including mammals and yeasts, use a genetic code that differs from the standard one for several codons. For example, the codon UGA, which is usually a stop codon, is translated as tryptophan in mammalian and yeast mitochondria. On the other hand, AGA is a stop codon in mammalian mitochondria, although in most other systems (including yeast mitochondria) it stands for arginine. Bacteria called mycoplasmas also use a few codons in a nonstandard way. Finally, some protozoa, such as the ciliate *Tetrahymena*, use a slightly deviant genetic code in translating their nuclear genomes. It is not clear when all these variations of the code originated, or what advantages (if any) they provide. But except for these relatively rare cases, all organisms studied so far, both prokaryotes and eukaryotes, use the same genetic code.

Transcription in the Prokaryotic Cell

Having looked in detail at the way in which the amino acid sequence of a protein is specified genetically, we now come to the first of the two major processes in the flow of genetic information from DNA to protein: the transcription of the nucleotide sequence of DNA into that of RNA. RNA is chemically similar to DNA, except that it contains ribose instead of deoxyribose as its sugar and has the base uracil in place of thymine. RNA is almost always a single-stranded molecule, though usually with some intramolecular double-strandedness. As in other areas of molecular genetics, the fundamental principles of RNA synthesis were first elucidated in bacteria, where the molecules and mechanisms are relatively simple. For that reason, we will start with transcription in prokaryotes.

Bacterial RNA Polymerase, the Transcription Enzyme

The enzyme responsible for catalyzing the synthesis of RNA from a DNA template is called **RNA polymerase.** Bacterial cells have a single kind of RNA polymerase, which synthesizes all three major classes of RNA—mRNA, tRNA, and rRNA. The enzymes from different bacteria are quite similar, and the RNA polymerase from *E. coli* has been especially well characterized. It is a large protein (about 480,000 Da), consisting of two α subunits, two β subunits that differ enough to be identified as β and β′, and a dissociable subunit called the **sigma (σ) factor.** The sigma subunit ensures that RNA synthesis is initiated at the right place on the DNA strand by promoting the attachment of the polymerase to

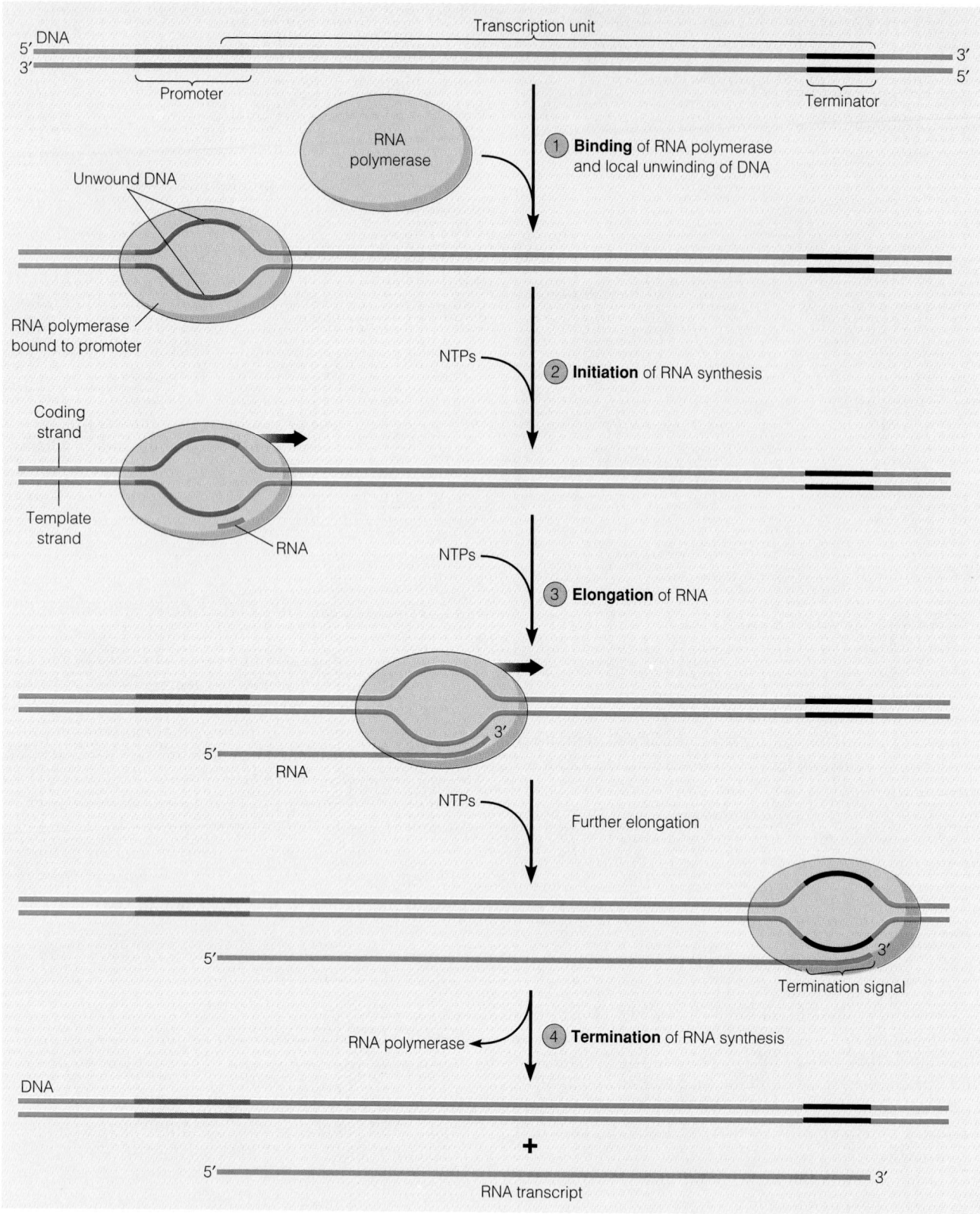

Figure 17-8   An Overview of Transcription.   Transcription of DNA occurs in four main stages: ① binding of RNA polymerase to DNA at a promoter, ② initiation of transcription on the template DNA strand (RNA synthesis), ③ subsequent elongation of the RNA chain, and ④ eventual termination of transcription, accompanied by the release of RNA polymerase and the completed RNA product from the DNA tem-

plate. RNA polymerase moves along the template strand of the DNA in the 3′ → 5′ direction, and the RNA molecule grows in the 5′ → 3′ direction. The general scheme shown in this figure holds for both prokaryotic and eukaryotic transcription. NTPs (ribonucleoside triphosphate molecules) = ATP, GTP, CTP, and UTP.

specific DNA sequences, called *promoters*, found at the beginnings of genes. It is thus essential to the specificity of transcription. The **core enzyme** ($\alpha_2\beta\beta'$) lacks the sigma subunit but is nonetheless competent to carry out RNA synthesis. However, the **holoenzyme** ($\alpha_2\beta\beta'\sigma$, the core enzyme plus sigma) is required to ensure correct initiation. Bacteria contain a variety of sigma factors, which can be utilized to recognize particular categories of genes, as we will discuss in Chapter 19.

The Stages of Transcription

Transcription is the synthesis of an RNA molecule that is complementary in base sequence to a strand of DNA. Figure 17-8 (on facing page) is an overview of the process that holds for both prokaryotes and eukaryotes. It shows a single **transcription unit,** a segment of DNA that is transcribed as a single, continuous RNA molecule. RNA polymerase binds to the DNA duplex at the promoter and brings about a local unwinding of the double helix, a separation of the strands. Using one of the strands as a template, RNA polymerase initiates the synthesis of an RNA chain. The polymerase then moves along the DNA, unwinding the double helix and elongating the RNA chain as it goes. The enzyme catalyzes the polymerization of nucleotides in an order determined by their base pairing to the DNA template. Eventually, the polymerase transcribes a sequence called a *terminator*. At that point, the now completed RNA molecule is released, and the enzyme dissociates from the template.

Transcription is a complicated process but can be thought of in four stages: binding, initiation, elongation, and termination. We will now look at each in detail, as it occurs in *E. coli.*

Binding. In bacteria, the *binding* stage starts with the attachment of the RNA-polymerase holoenzyme to DNA at a **promoter,** a specific sequence that determines where RNA synthesis starts and which DNA strand is used as a template.

Each transcription unit has a promoter located near the beginning of the sequence that is actually to be transcribed. The end of a transcribed sequence where the bacterial promoter is located is said to be upstream of the transcribed sequence. In other words, **upstream** is in the direction of the 5′ end of the DNA coding strand (or the 5′ end of the RNA transcript). In *E. coli*, as mentioned above, the sigma subunit of RNA polymerase is responsible for promoter recognition and therefore for specificity of binding.

The DNA segments of 40 or so bp that constitute bacterial promoters differ greatly among transcription units. How, then, does a single kind of RNA polymerase recognize them all? Enzyme recognition and binding, it turns out, depend only on several very short sequences spaced appropriately within the DNA; the identities of the nucleotides making up the rest of the promoter are irrelevant for this purpose.

Figure 17-9 shows a typical prokaryotic promoter and highlights its essential features. By convention, the base sequences are given as they occur on the coding strand. The point where transcription will begin, called the *startpoint,* is almost always a purine and often an adenine. Approximately 10 bases upstream of the startpoint is the 6-base sequence TATAAT, called the **−10 sequence** or the *Pribnow box,* after its discoverer. By convention, the nucleotides are numbered from the startpoint (= +1), with positive numbers to the right (downstream) and negative ones to the left (upstream). The "−1" nucleotide is immediately upstream of the startpoint (there is no "0"). At or near the −35 position is the 6-base sequence TTGACA, called the **−35 sequence.**

The −10 sequence and the −35 sequence (and their positions relative to the startpoint) have been conserved in evolution, but they are not identical in all prokaryotic promoters, or even in all the promoters in a single genome. For example, the −10 sequence in the promoter for one of the *E. coli* tRNA genes is TATGAT, whereas the −10 sequence for a group of genes needed for lactose breakdown in the same

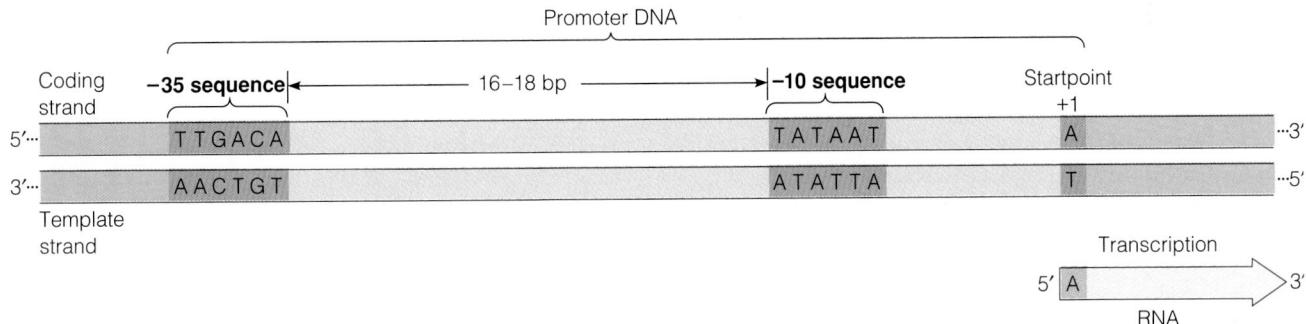

Figure 17-9   A Typical Prokaryotic Promoter. The DNA promoter region in prokaryotes is a stretch of about 40 bp adjacent to and including the transcription startpoint. By convention, the critical DNA sequences are given as they appear on the coding strand (the nontemplate strand, which corresponds in sequence to the RNA transcript). The essential features of the promoter are the startpoint (designated +1 and usually an A), the 6-nucleotide −10 sequence, and the 6-nucleotide −35

sequence. As their names imply, the two key sequences are located approximately at 10 nucleotides and 35 nucleotides, respectively, on the 5′ side of the startpoint. The sequences shown here are consensus sequences, meaning that they have the most commonly found base at each position. The numbers of nucleotides separating the consensus sequences from each other and from the startpoint are important for promoter function, but the identity of these nucleotides is not.

17B

Contemporary Techniques

THE FOOTPRINTING METHOD FOR
IDENTIFYING PROTEIN-BINDING SITES ON DNA

The initiation of transcription depends on the interactions of proteins with specific DNA sequences. In prokaryotic cells, the RNA polymerase (holoenzyme) directly recognizes and binds to the promoter sequence at the start of a gene. In eukaryotic cells, certain transcription factors must bind to the promoter before an RNA polymerase can bind. In all cells, as we shall discuss in Chapter 19, the *regulation* of transcription operates primarily at the level of transcription initiation, and it depends on the interaction of still other proteins with other sites on the DNA. Thus, the researcher seeking to understand transcription needs to know about both transcriptional proteins and the DNA sequences to which some of them bind.

A technique called *DNA footprinting* is widely used to identify sites on DNA where DNA-binding proteins attach. Footprinting can be used to identify any DNA site that binds a protein specifically, as long as the protein binds tightly enough. The underlying principle is that the binding of a protein to a particular DNA sequence should protect that sequence from degradation by enzymes or chemicals. A widely used version of footprinting, outlined in Figure 17B-1, uses an endonuclease called DNAse I as the degradative agent. Unlike restriction endonucleases, this enzyme is not sequence-specific but attacks the bonds between nucleotides more or less at random.

In the figure here, the starting material is a piece of DNA known to contain a binding site for a particular DNA-binding protein (in a real case, the DNA might be several hundred base pairs long). Radioactive phosphate (indicated by the stars) has been added to the 5′ ends of the DNA by

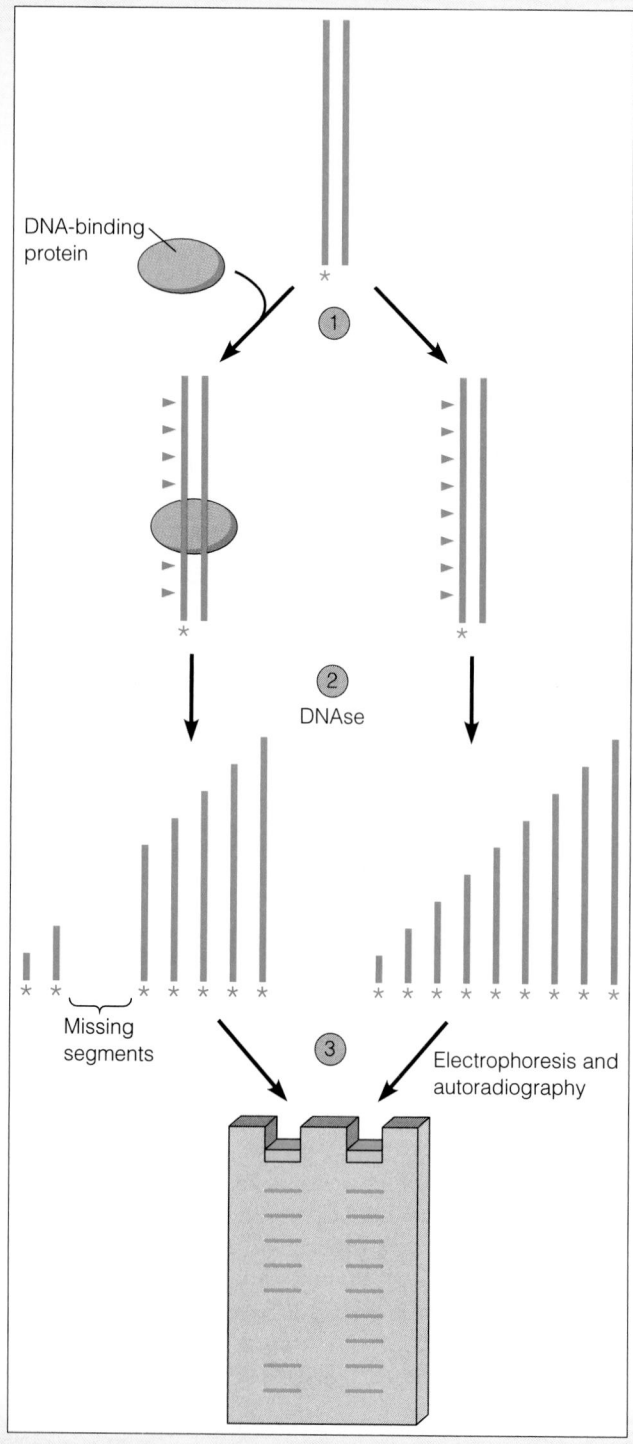

Figure 17B-1   DNAse Footprinting as a Tool to Identify DNA Sites That Bind Specific Proteins.

reacting the DNA with ^{32}P-labeled ATP and the enzyme polynucleotide kinase. ① An aliquot of the end-labeled DNA is mixed with the DNA-binding protein under study; another aliquot, without the added protein, serves as the control. ② Both aliquots are incubated for a short time with a low concentration of DNAse I—conditions where most of the DNA molecules will be cleaved only once. The arrowheads indicate eight possible cleavage sites in the DNA starting material. ③ The two incubation mixtures are submitted to electrophoresis in adjacent lanes of a polyacrylamide gel and visualized by autoradiography. The fragments resulting from the DNAse reactions are revealed as two ladders of bands. The control lane has nine bands spaced uniformly, because every possible cleavage site has been cut. However, the other lane is missing some of the bands, because the protein bound to the DNA during DNAse treatment protected some of the cleavage sites. The blank region in the left-hand lane is the "footprint" that identifies the location and length of the DNA sequence in contact with the DNA-binding protein. Often, a third aliquot of the starting DNA is sequenced and electrophoresed alongside the footprinting lanes. In this way, the exact position and sequence of the DNA protein-binding site can be readily determined.

Footprinting experiments like the one described here have revealed, for example, that when the bacterial RNA polymerase binds to a promoter, it protects a segment of DNA extending from about 40 nucleotides upstream of the transcriptional startpoint to almost 20 nucleotides past that point. ∎

organism is TATGTT, and the sequence given in Figure 17-9 is TATAAT. The particular hexanucleotides shown in the figure are **consensus sequences,** which consist of the most common and hence presumably the preferred nucleotides at each position within the sequence. Deviations from the consensus sequences are likely to affect promoter function. Mutations toward the consensus sequence can enhance promoter activity, whereas mutations away from the consensus sequence are likely to reduce the strength of the promoter and may even eliminate promoter activity entirely.

Molecular biologists use various techniques to identify the particular DNA sequences that constitute promoters. One method takes advantage of the fact that promoters, by definition, are specifically bound by particular proteins—in the case we are discussing here, the holoenzyme of bacterial RNA polymerase. This method, called **DNA footprinting,** can be used to identify any DNA sequence that is specifically bound by a protein (Box 17B). Footprinting depends on the fact that the attachment of the protein to the DNA protects the nucleotide sequence at that site from enzymatic (or chemical) attack.

The binding stage of prokaryotic transcription concludes with the unwinding of the DNA in the promoter region. Approximately a turn and a half of the double helix, about 15–18 bp, are opened up around the startpoint.

Initiation. Once an RNA polymerase molecule is bound to a promoter and the DNA double helix has been opened, *initiation* of RNA synthesis is possible. One of the two exposed segments of single-stranded DNA serves as the template for the synthesis of RNA from incoming ribonucleoside triphosphate molecules (NTPs). The DNA strand that carries the key promoter sequences given above determines which way the RNA polymerase faces, and the enzyme's orientation determines, in turn, which DNA strand it transcribes. As soon as the first two NTPs are hydrogen-bonded with the complementary nucleotides at the DNA startpoint, RNA polymerase catalyzes the formation of a phosphodiester bond between the 3′ OH hydroxyl group of the first NTP and the 5′-phosphate of the second, with the removal of pyrophosphate, PP$_i$. The polymerase remains at the promoter as additional nucleotides are added one by one, until the chain is about nine nucleotides long. At that point, the sigma factor detaches from the polymerase, and initiation is complete.

Elongation. Chain *elongation* (Figure 17-10) now occurs as the RNA polymerase moves along the DNA molecule, untwisting the helix bit by bit and adding one complementary NTP at a time to the growing RNA. The enzyme moves along the template DNA strand from the 3′ to the 5′ end. Because nucleotide pairing is antiparallel, the RNA strand is elongated in the 5′ → 3′ direction as each successive nucleotide is added to the 3′ end of the growing chain. (This is the same direction in which DNA strands are synthesized in

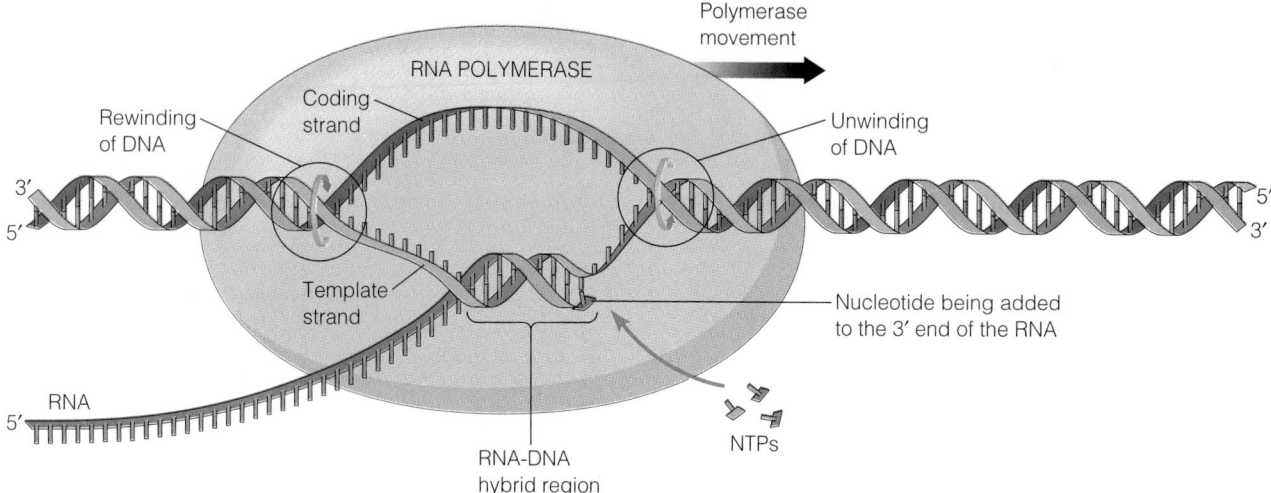

Figure 17-10   A Closeup of the Prokaryotic Elongation Complex.   During elongation, RNA polymerase binds to about 30 bp of DNA (recall that each complete turn of the DNA double helix is about 10 bp). At any given moment, about 18 bp of DNA are unwound, and the most recently synthesized RNA is still hydrogen-bonded to the DNA, forming a short RNA-DNA hybrid. This hybrid is probably about 12 bp long, but it may be shorter. The total length of growing RNA bound to the enzyme and/or DNA is about 25 nucleotides. This diagram is drawn roughly to scale.

DNA replication.) As the RNA chain grows, the most recently added nucleotides remain base-paired with the DNA template strand, forming a short RNA-DNA hybrid. Most evidence suggests that this hybrid is about 12 bp long, although it may be shorter. (Figure 17-10 is drawn roughly to scale.) In any case, the polymerase continues to move on, rewinding the DNA behind and unwinding the DNA ahead as it goes.

Termination.   *Termination* of transcription occurs after the RNA polymerase has transcribed a special sequence in the DNA called the **terminator.** The RNA transcript of the terminator is the actual **termination signal.** In prokaryotes, the termination signal is commonly an RNA sequence that gives rise to a **hairpin** structure followed by a string of Us (Figure 17-11). The sequence within the hairpin must be rich in GC base pairs, but the exact sequence is less significant than the self-complementarity that allows the RNA to fold spontaneously into a hairpin. The string of Us is apparently what triggers dissociation between the RNA polymerase and the DNA.

Not all transcription units in *E. coli* have termination signals like the one in Figure 17-11, however. Termination for these other transcription units requires a protein called **rho (ρ) factor.** Rho seems to recognize an RNA sequence 50–90 bases long preceding the termination point, but the sequence does not have a string of Us and is not necessarily able to form a hairpin. Whether termination is rho-dependent or not, it results in the release of the completed RNA molecule and of the core RNA polymerase. The core polymerase is then available to bind sigma factor again and reinitiate synthesis at another promoter.

Transcription in the Eukaryotic Cell

Transcription in eukaryotic cells occurs in the same four stages described in Figure 17-8, but the eukaryotic process is more complex than the process in prokaryotes. The main differences are as follows:

•   *Three different RNA polymerases* transcribe the nuclear DNA of eukaryotes. Each synthesizes one or more particular kinds of RNA.

•   *Eukaryotic promoters* are more varied than prokaryotic promoters. Not only are there different types of promoters for the three RNA polymerases, but there is great variation within each type, especially among the ones for protein-coding genes. Furthermore, some eukaryotic promoters are actually located *downstream* from the transcription start-point.

•   Many *transcription factors* are involved in the binding of eukaryotic RNA polymerase to DNA. Unlike the prokaryotic sigma factor, these proteins are not considered part of the polymerase. Most of them must bind to DNA *before* the polymerase can bind to the promoter and initiate transcription. Thus, transcription factors, rather than the RNA polymerase itself, determine the specificity of transcription in eukaryotes. In this chapter, we limit our discussion to the class of factors that are essential for the transcription of all genes transcribed by an RNA polymerase. We defer discussion of the regulatory class of transcription factors, which operate in combinations that vary from gene to gene, until Chapter 19.

•   *Protein-protein interactions* are very important in the first stage of eukaryotic transcription. Although some tran-

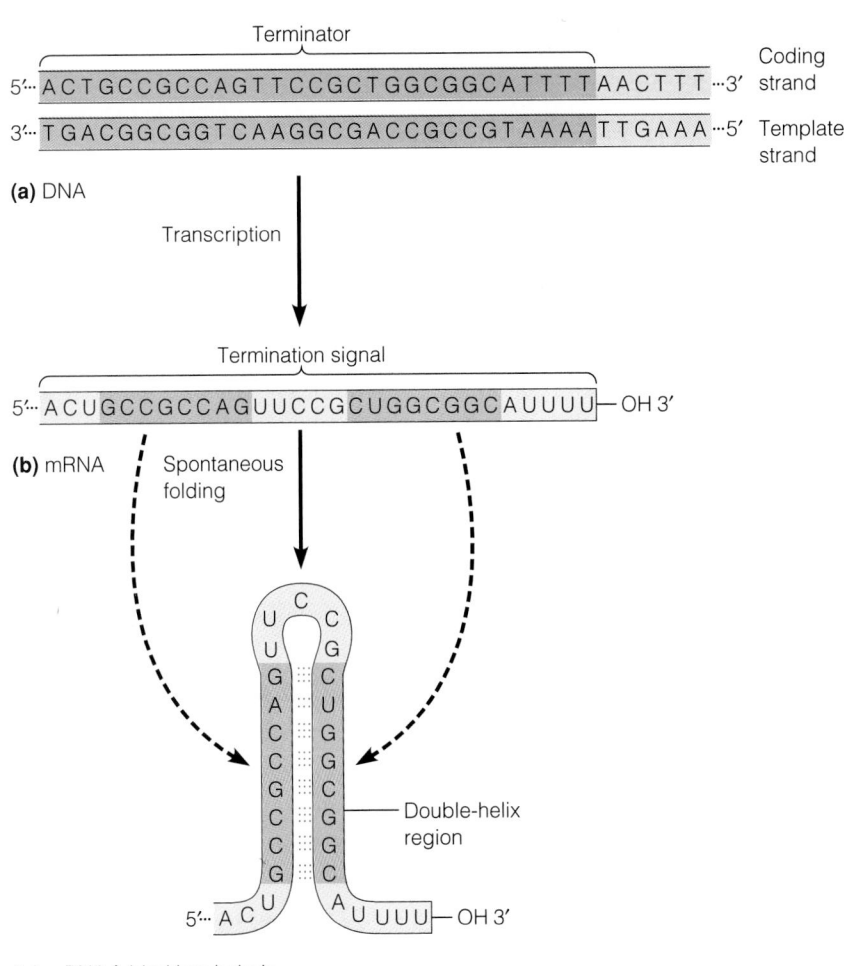

(a) DNA

Terminator

```
5'··· A C T G C C G C C A G T T C C G C T G G C G G C A T T T T A A C T T T ···3'   Coding strand
3'··· T G A C G G C G G T C A A G G C G A C C G C C G T A A A A T T G A A A ···5'   Template strand
```

Transcription

Termination signal

```
5'··· A C U G C C G C C A G U U C C G C U G G C G G C A U U U U ── OH 3'
```

(b) mRNA

Spontaneous folding

```
          U   C
       U        C
       G        G
       A        C
       C        U
       C        G
       G        G
       C        C
       C        G
       G        G
   5'··· A C U        A U U U U ── OH 3'
```

Double-helix region

(c) mRNA folded into hairpin

Figure 17-11   A Prokaryotic Termination Signal. **(a)** A DNA segment carrying the termination signal for one of the transcription units of *E. coli*. The DNA sequence for the termination signal is called the terminator. **(b)** The 3' end of the mRNA, showing the sequence of the termination signal itself. It consists of two complementary regions (darker pink) and a short string of Us at the very end. **(c)** The RNA transcript folds spontaneously into a hairpin structure, stabilized by hydrogen bonds between the complementary nucleotides in the two highlighted regions of part b. Termination seems to be determined by the hairpin structure itself rather than by any particular nucleotide sequence, because a variety of self-complementary sequences can serve as termination signals. However, the dissociation of the RNA from the DNA and the enzyme seems to require the string of Us.

scription factors bind directly to DNA, many others attach to other proteins—either other factors or the polymerase itself.

• *Termination* of transcription is often less important than RNA cleavage in determining the 3' end of the RNA product.

As you probably know already, another major difference between eukaryotic and prokaryotic RNA production is the extensive *processing of RNA* that occurs during and after the actual transcription process in eukaryotes. We will discuss RNA processing later in the chapter. Now, we will examine in more detail the aspects of transcription unique to eukaryotes, starting with their RNA polymerases.

Eukaryotic RNA Polymerases

Table 17-1 summarizes some of the properties of the three RNA polymerases that function in the nucleus of the eukaryotic cell, as well as two others found in mitochondria and chloroplasts. The nuclear enzymes are designated RNA polymerases I, II, and III. As the table indicates, these en-

zymes differ in their location within the nucleus and in the kinds of RNA they make. In addition, the nuclear polymerases can be distinguished from each other in the laboratory based on their sensitivity to *α-amanitin,* a deadly toxin produced by the mushroom *Amanita phalloides* (known commonly as the "death cap" or "destroying angel").

RNA polymerase I is found in the nucleolus. It is responsible for the synthesis of an RNA molecule that is the precursor to the three largest of the four RNAs of eukaryotic ribosomes (28S rRNA, 18S rRNA, and 5.8S rRNA). This enzyme is insensitive to *α-amanitin*. Its association with the nucleolus is understandable, as the nucleolus is the site of ribosomal RNA synthesis and ribosomal assembly, as we discussed in Chapter 14. Because cells need large amounts of rRNA, RNA polymerase I actually accounts for most cellular RNA synthesis.

RNA polymerase II is found in the nucleoplasm and synthesizes the precursors to mRNA, the RNA that codes for protein. In addition, it synthesizes most of the *snRNAs*, small RNAs involved in posttranscriptional RNA processing. Thus, polymerase II is responsible for the synthesis of the greatest variety of RNA molecules. The enzyme is highly

Table 17-1 Properties of Eukaryotic RNA Polymerases

RNA Polymerase	Location	Main Products	α-Amanitin Sensitivity
I	Nucleolus	Precursor for 28S rRNA, 18S rRNA, and 5.85 rRNA	Resistant
II	Nucleoplasm	Pre-mRNA (hnRNA) and most snRNA	Very sensitive
III	Nucleoplasm	Pre-tRNA, 55 rRNA, and other small RNAs	Moderately sensitive*
Mitochondrial	Mitochondrion	Mitochondrial RNA	Resistant
Chloroplast	Chloroplast	Chloroplast RNA	Resistant

*In mammals.

sensitive to α-amanitin, which explains the toxicity of this compound to humans and other animals. (Figure 17-12 is a computer graphic of RNA polymerase II from yeast.)

RNA polymerase III is also a nucleoplasmic enzyme, but it is responsible for the synthesis of a variety of small RNAs, including tRNA precursors and the smallest species of ribosomal RNA, 5S rRNA. The RNA polymerase III of mammals is sensitive to α-amanitin, but only at higher levels of the toxin than those required to inhibit RNA polymerase II. (The equivalent enzymes of some other organisms, such as insects and yeasts, are insensitive to α-amanitin.)

Structurally, RNA polymerases I, II, and III are somewhat similar to each other and also to prokaryotic core RNA polymerase. The three eukaryotic enzymes are all quite large, with multiple polypeptide subunits and molecular weights around 500,000. RNA polymerase II, for example, has at least 10 polypeptides, of at least 8 different types. The three biggest subunits are evolutionarily related to the prokaryotic RNA polymerase subunits α, β, and β'. The three smallest subunits lack that relationship but are also found in RNA polymerases II and III. The RNA polymerases of mitochondria and chloroplasts resemble the prokaryotic enzyme closely, as you might expect from the probable ori-

gins of these organelles as endosymbiotic bacteria (see Chapter 4). Like bacterial RNA polymerase, they are resistant to α-amanitin.

Eukaryotic Promoters and Their Binding by Transcription Factors

The most characteristic features of eukaryotic transcription have been revealed by studying the promoter regions of eukaryotic genes and the various proteins that attach to them during the binding stage of transcription.

Eukaryotic Promoters. As mentioned earlier, the promoters where eukaryotic RNA polymerases bind are extremely varied, even more so than prokaryotic promoters. However, the promoters used by the RNA polymerases of the eukaryotic nucleus can be classified into three main types, one for each type of polymerase. Figure 17-13 shows typical features of the three types of promoters.

The promoter used by RNA polymerase I—that is, the promoter of the transcription unit that gives rise to the biggest species of rRNA—has two parts (Figure 17-13a). The part called the **core promoter** actually extends into the nu-

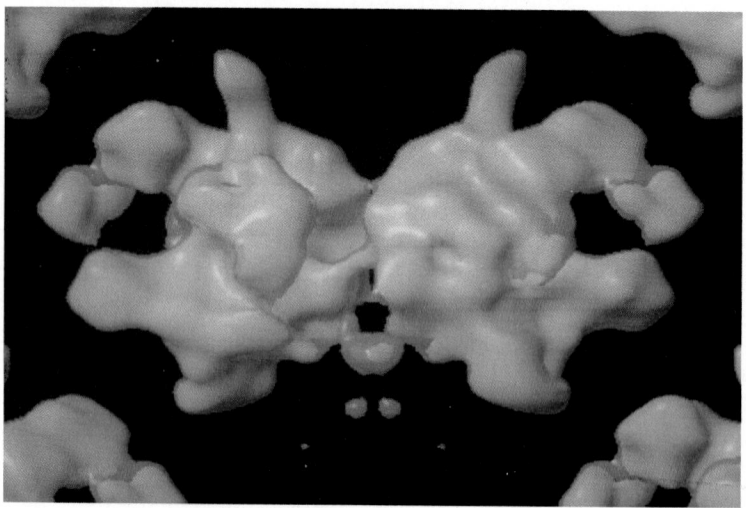

Figure 17-12 RNA Polymerase II. This is a computer model showing the overall shape of RNA polymerase II of yeast. Two molecules of the polymerase appear here. Each molecule of RNA polymerase II consists of a number of different polypeptides.

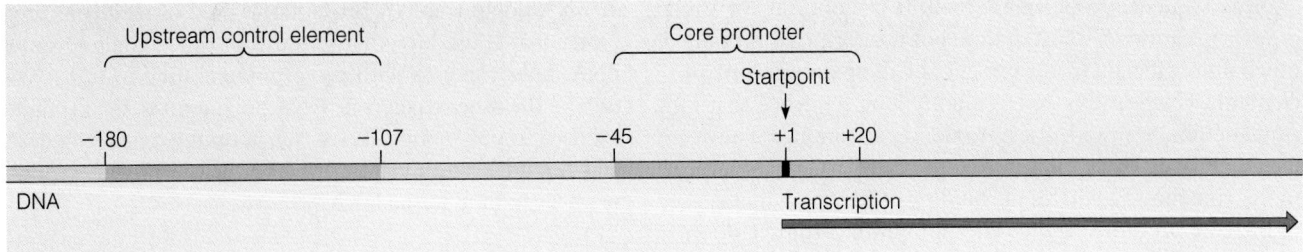

(a) Promoter for RNA polymerase I

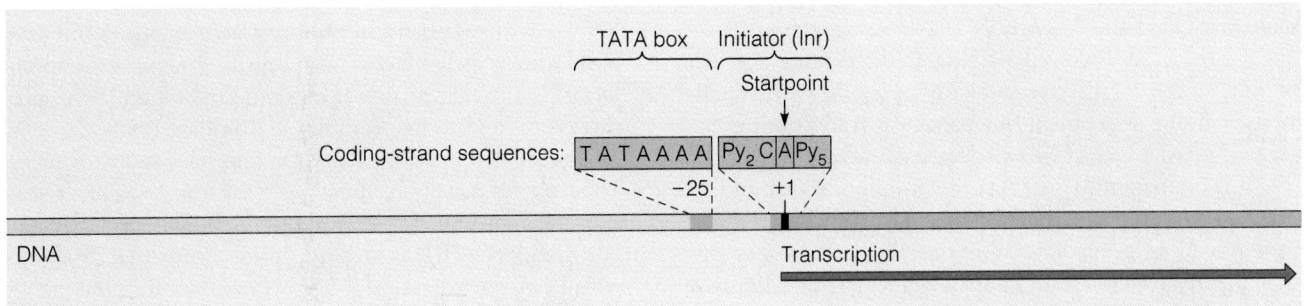

(b) Core promoter for RNA polymerase II

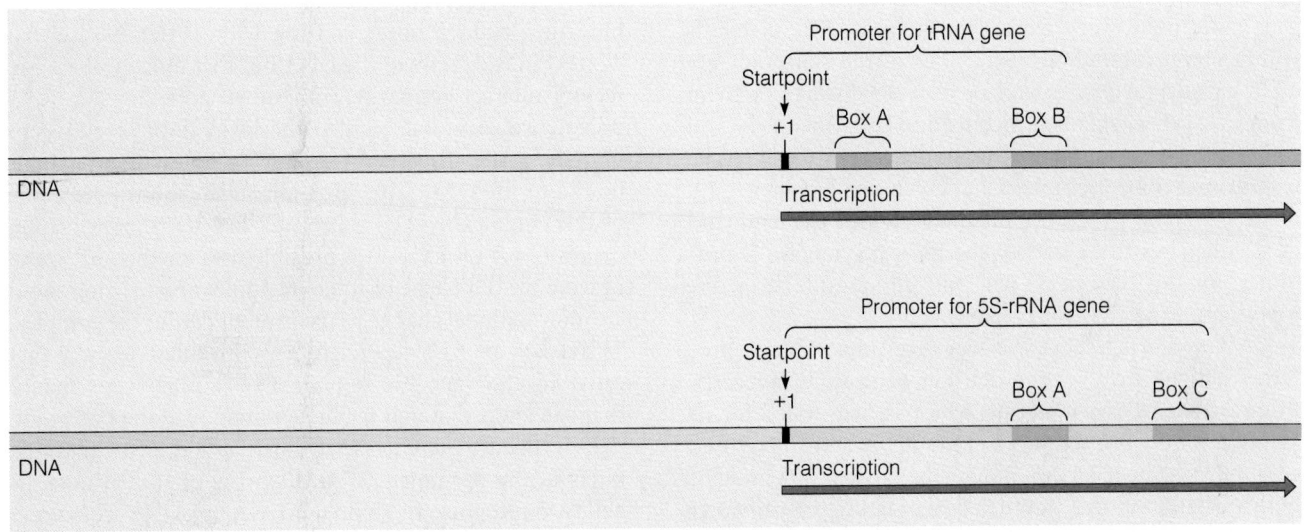

(c) Two types of promoters for RNA polymerase III

Figure 17-13 Typical Eukaryotic Promoters Used by RNA Polymerases I, II, and III.   **(a)** The promoter for RNA polymerase I has two parts, a core promoter surrounding the startpoint and an upstream control element. After the binding of appropriate transcription factors to both parts, the RNA polymerase binds to the core promoter. **(b)** The typical promoter for RNA polymerase II has a short initiator sequence, consisting mostly of pyrimidines (Py), and usually a TATA box about 25 bases upstream from the startpoint. This type of promoter (with or without the TATA box) is often called a polymerase II core promoter,

because for most genes, a variety of upstream control elements (not shown here) also play important roles in the initiation of transcription. **(c)** The promoters for RNA polymerase III vary in structure, but the ones for tRNA genes and 5S rRNA genes are located entirely downstream of the startpoint, within the transcribed sequence. Boxes A, B, and C are DNA consensus sequences, each about 10 bp long. In tRNA genes, about 30–60 bp of DNA separate boxes A and B; in 5S rRNA genes, about 10–30 bp separate boxes A and C.

cleotide sequence to be transcribed. It is sufficient for the proper initiation of transcription, but transcription is made much more efficient by the presence of an **upstream control element,** which in this case is a fairly long sequence that is similar (though not identical) to the core promoter. The attachment of transcription factors to both parts of the promoter somehow facilitates the binding of RNA polymerase I to the core promoter and enables it to initiate transcription at the startpoint.

A typical promoter used by RNA polymerase II is shown in Figure 17-13b. It has a short **initiator (Inr)** sequence surrounding the startpoint (which is often an A, as in prokaryotes) and usually, at around −25, a short sequence called the **TATA box,** which has a consensus sequence of TATA followed by two or three more As. The binding stage of transcription by RNA polymerase II usually starts with the attachment of a particular transcription factor to the TATA box, as we will describe below.

It is important to note that the promoter shown in Figure 17-13b would function at only a *basal* (low) level. Most protein-coding genes have additional short sequences further upstream—upstream control elements—that improve the promoter's efficiency. Some of these upstream elements occur with many different genes; examples of common ones include the *CAAT box* (consensus sequence GCCCAATCT in animals and yeasts) and the *GC box* (consensus sequence GGGCGG). The locations of these elements relative to a gene's startpoint vary from gene to gene. The elements within about a hundred nucleotides of the startpoint are often called *proximal control elements* to distinguish them from *enhancer* elements, which tend to be farther away and can even be located downstream of the gene. We will return to proximal control elements and enhancers in Chapter 19. Some researchers consider proximal control elements and even enhancers to be part of the RNA polymerase II promoter; for this reason, we will generally use the term *core promoter* for the TATA–Inr region.

A functional test for the sequences important in promoter activity involves the deletion of specific sequences from a cloned DNA molecule, which is then tested for its ability to direct transcription, either in vitro or in cultured cells. When the gene for β-globin (the β chain of hemoglobin) was tested in this way, deletion of either the promoter TATA box or a CAAT box upstream element reduced the level of transcription at least tenfold.

For transcribing genes for tRNA and 5S rRNA, RNA polymerase III uses promoters that are entirely *downstream* of the transcription unit's startpoint. The promoters for tRNA and 5S rRNA genes are different, but in both cases the consensus sequences fall into two blocks of about 10 bp each (Figure 17-13c). The tRNA-gene promoter has consensus sequences called box A and box B. The promoters for 5S rRNA genes have box A (positioned farther from the startpoint than in tRNA-gene promoters) and another critical sequence, called box C. (Not shown in the figure is a third type

of RNA polymerase III promoter, a kind of *upstream* promoter that is used for the synthesis of other kinds of small RNA molecules.) As with the promoters used in transcription by the other eukaryotic RNA polymerases, the promoters used in polymerase III transcription must be recognized and bound by transcription factors before the RNA polymerase can bind to the DNA.

The Role of Basal Transcription Factors in the Function of RNA Polymerase II. We now take a closer look at the important role of transcription factors in eukaryotes, using the RNA polymerase II transcription system as our example. A **basal transcription factor** (also called a *general transcription factor*) is a protein that is always required for an RNA polymerase to bind to its promoter and initiate RNA synthesis. As already mentioned, eukaryotic cells have many of them. Their names usually include "TF" (for transcription factor) and a roman numeral identifying the polymerase they aid. For example, TFIID is a transcription factor that functions with RNA polymerase II; "D" was used because the names TFIIA, TFIIB, and TFIIC had already been taken.

Figure 17-14 illustrates the series of steps that is believed to constitute the binding stage of transcription for RNA polymerase II. We show five basal transcription factors, although others are probably also involved. The basal factors bind in a defined order, starting with TFIID. Notice that TFIID binds directly to the DNA but that most of the other factors interact primarily with other proteins. The RNA polymerase does not bind to the DNA until several steps into the series. Eventually, a large complex of proteins is bound to the DNA. The initiation stage requires *activation* of this complex by ATP hydrolysis. Exactly what activation entails is not yet clear. Two possible uses for the ATP are to unwind the double helix at the startpoint and to bring about a conformational change in the protein part of the complex. TFIIH has an ATP-dependent protein kinase activity that can phosphorylate the polymerase, causing it to change shape in a way that may facilitate initiation at the startpoint.

TFIID, the polymerase II transcription factor that first binds to the promoter DNA, is worthy of special note. Its ability to recognize and bind the TATA box DNA sequence is conferred by one of its subunits, called **TATA-binding protein (TBP).** Surprisingly, TBP's role is not limited to TATA-containing polymerase II promoters. TBP is also essential for transcription initiation at promoters lacking TATA boxes, including promoters used by RNA polymerases I and III. Depending on the type of promoter, TBP associates with different proteins, and for TATA-less promoters, much of TBP's specificity is probably derived from protein-protein interactions. TBP's similarity from species to species shows that it has been highly conserved in evolution. Recently, the detailed three-dimensional structure of this important protein has been worked out (Figure 17-15, page 556).

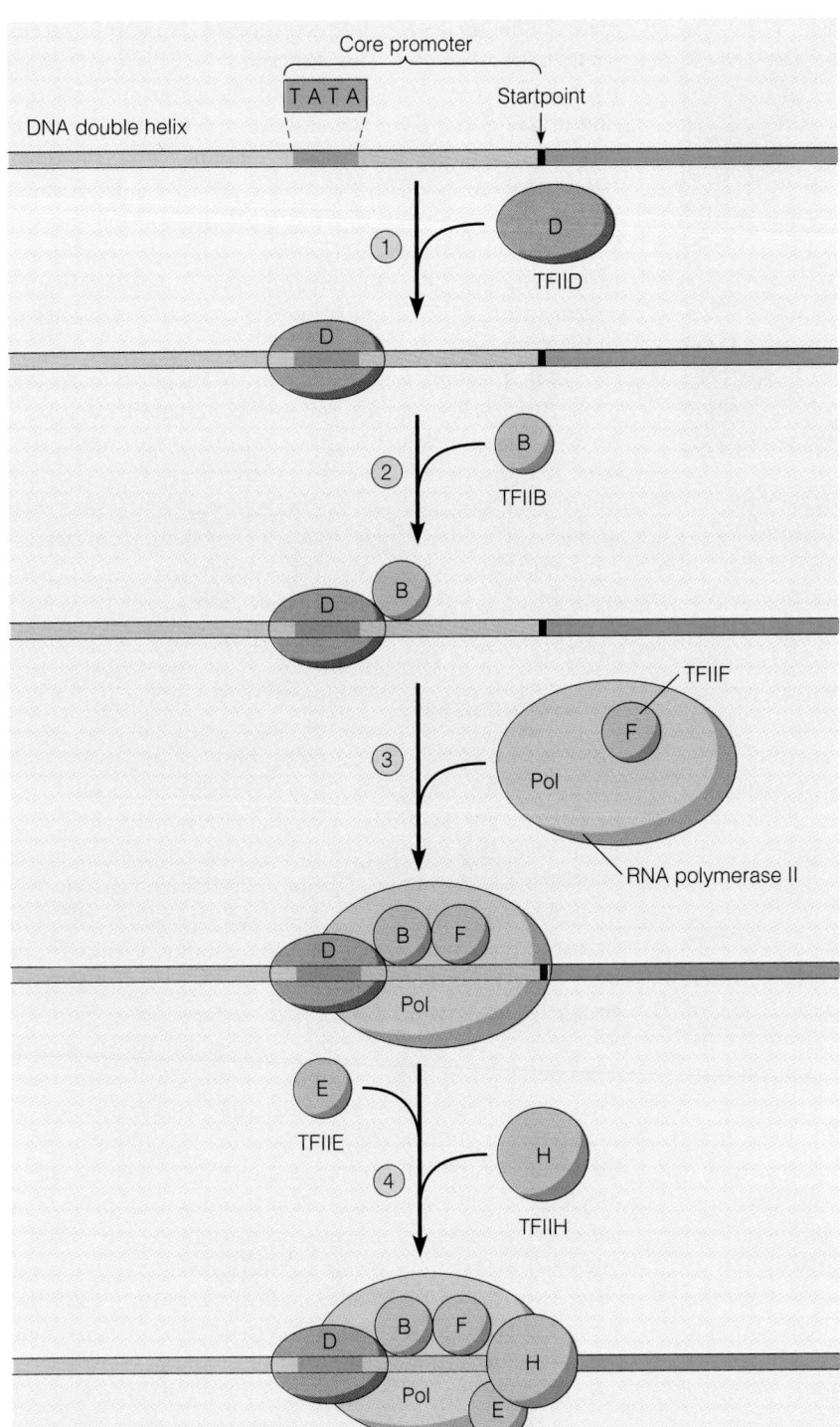

Figure 17-14 Steps in the Binding Stage of Eukaryotic Transcription by RNA Polymerase II. The figure outlines the sequential binding of five basal transcription factors (called TFII_, where _ is a letter identifying the particular factor) and the polymerase. ① TFIID binds to the TATA box on the DNA, followed by ② the binding of TFIIB. ③ The resulting complex is now bound by the polymerase, to which TFIIF has already attached. ④ The initiation complex is completed by the addition of TFIIE and TFIIH. After an activation step requiring ATP hydrolysis, the polymerase can initiate transcription at the startpoint. In addition to the basal transcription factors shown in this figure, two others, TFIIA and TFIIJ, are probably involved—TFIIA joining the complex after TFIID, and TFIIJ after TFIIH.

The numerous basal transcription factors are all potential targets for the regulation of transcription, and the recognition of other upstream elements in the DNA by other factors adds additional levels of complexity. Thus, the separation between our description of the fundamentals of transcription in this chapter and our discussion of the regulation of transcription in Chapter 19 is an artificial one.

Termination of Eukaryotic Transcription

Much less is known about the termination of eukaryotic transcription than about its initiation. RNA polymerase I termination involves recognition of an 18-nucleotide termination signal on the RNA by a protein factor. Termination signals for RNA polymerase III are also known; the signals

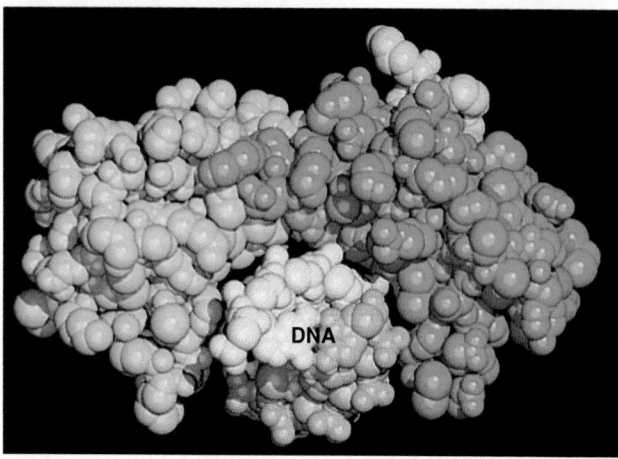

Figure 17-15 TATA-Binding Protein (TBP) Attached to DNA. In this computer graphic model, TBP is shown bound to DNA (white and gray, viewed looking down its axis). TBP differs from most DNA-binding proteins in that it interacts with the minor groove of DNA, rather than the major groove, and imparts a sharp bend to the DNA. The TBP molecule shown here is from the plant *Arabidopsis thaliana,* but TBP has been highly conserved in evolution. Dark and light blue differentiate the two, symmetrical domains of the polypeptide; light green is used for its nonconserved N-terminal segment. When TBP is bound to DNA, other transcription-factor proteins can interact with the convex surface of the TBP "saddle." TBP is required for transcription initiation on all types of eukaryotic promoters.

always include a short run of Us (as in prokaryotic termination signals), and no ancillary protein factors are needed for their recognition. Hairpin structures are apparently not involved in termination by either polymerase I or polymerase III.

For RNA polymerase II, almost nothing is known about termination signals. Most transcripts destined to become mRNA are cleaved at a specific site before transcription is actually terminated. The cleavage site is 10–35 nucleotides downstream from an AAUAAA sequence in the RNA. Termination may not occur until hundreds or even thousands of nucleotides are downstream of that site, but the RNA made after the cleavage step is quickly degraded. The cleavage site is also the site for the addition of a *poly(A) tail,* a string of adenine nucleotides that is found at the 3′ end of almost all eukaryotic mRNAs. The addition of the poly(A) tail is one aspect of RNA processing, our next topic.

RNA Processing

The immediate product of transcription, called a **primary transcript,** frequently must undergo chemical changes before it can function in the cell. We use the term **RNA processing** to mean all the chemical modifications necessary to generate the final RNA product from the primary transcript that serves as its precursor. Processing usually involves the hydrolytic removal of portions of the primary transcript, and it may also include the addition or chemical modification of specific nucleotides. For example, methylation of bases or ribose groups is a common way that individual nucleotides are modified. In addition to chemical changes to the RNA, other posttranscriptional events are often necessary before the RNA can function, such as association with specific proteins or (in eukaryotes) passage from the nucleus to the cytoplasm.

In this section, we look at some of the most important properties of rRNA, tRNA, and mRNA and at the processing involved in the production of each from the primary products of transcription. The term "RNA processing" is most often associated with eukaryotic systems, but prokaryotes also process some of their RNA. We include both eukaryotic and prokaryotic RNA in the discussion below.

Ribosomal RNA

Ribosomal RNA is by far the most abundant and most stable form of RNA in most cells. This can easily be demonstrated by extracting total cellular RNA and fractionating it according to size, either by sedimentation through sucrose gradients or by electrophoresis on polyacrylamide gels. Most of the RNA detectable on the gradient or gel is either rRNA or tRNA. In most cells, rRNA represents about 70–80% of the total RNA, tRNA represents about 10–20%, and mRNA accounts for less than 10%.

Eukaryotic cells have four species of rRNA, usually identified by their sedimentation rates in centrifugation (see Chapter 9). Table 17-2 lists the sedimentation coefficients (S values) for all the different types of rRNA. In eukaryotes, the smaller of the two ribosomal subunits has a single rRNA molecule with a sedimentation coefficient of about 18S. The larger of the two subunits contains three molecules of RNA, one of about 28S (as low as 25S in some species) and the other two of about 5.8S and 5S. In prokaryotes, only three species of rRNA are present: a 16S molecule associated with the small subunit and molecules of 23S and 5S associated with the large subunit.

Of the four species of eukaryotic rRNA, the three larger ones (28S, 18S, and 5.8S) are encoded by a single long transcription unit (Figure 17-16b), which is transcribed by RNA polymerase I in the nucleolus. Within this transcription unit, the sequences for the three rRNAs are separated by segments of DNA called *transcribed spacers.* The linkage of the three genes in a single transcription unit guarantees that the cell will make these three rRNAs in equal quantities. Furthermore, most eukaryotic genomes have multiple copies of this rRNA transcription unit, facilitating the production of the large amounts of these RNAs needed by the cell. The

Table 17-2 Properties of Ribosomal RNA (rRNA)

Source	Ribosomal Subunit	rRNA Sedimentation Coefficient	Nucleotides
Prokaryotic cells	Large (50S)	23S	2900
		5S	120
	Small (30S)	16S	1540
Eukaryotic cells	Large (60S)	25–28S	≤4700
		5.8S	160
		5S	120
	Small (40S)	18S	1900

copies are arranged in one or more tandem arrays (Figure 17-16a). The human (haploid) genome, for example, has 150–200 copies of the transcription unit, divided among five chromosomes. *Nontranscribed spacers* separate the transcription units within each cluster.

The gene for 5S rRNA constitutes a separate transcription unit, which is transcribed by RNA polymerase III. It, too, occurs in multiple copies arranged in long, tandem arrays. However, the 5S-RNA genes are not usually located near the genes for the bigger rRNAs and so are not usually part of the nucleolus.

As indicated in Figure 17-16c, the primary transcript for the three large eukaryotic rRNAs is processed by cleavage at the rRNA-spacer junctions to release the mature rRNAs (Figure 17-16d). (The spacer sequences are degraded.) The enzyme(s) that cut the RNA are not known, but the processing seems to occur in the nucleolus at the same time as the assembly of the RNA with proteins (and 5S rRNA) to form the ribosomal subunits.

The primary transcript for the large rRNAs in most mammalian cells contains about 13,000 nucleotides and has a sedimentation coefficient of about 45S. The three rRNA molecules that are generated from it use only about 52% of the original RNA. The remaining 48% (about 6200 nucleotides) consists of the transcribed spacer sequences that are excised during the processing steps. The rRNA precur-

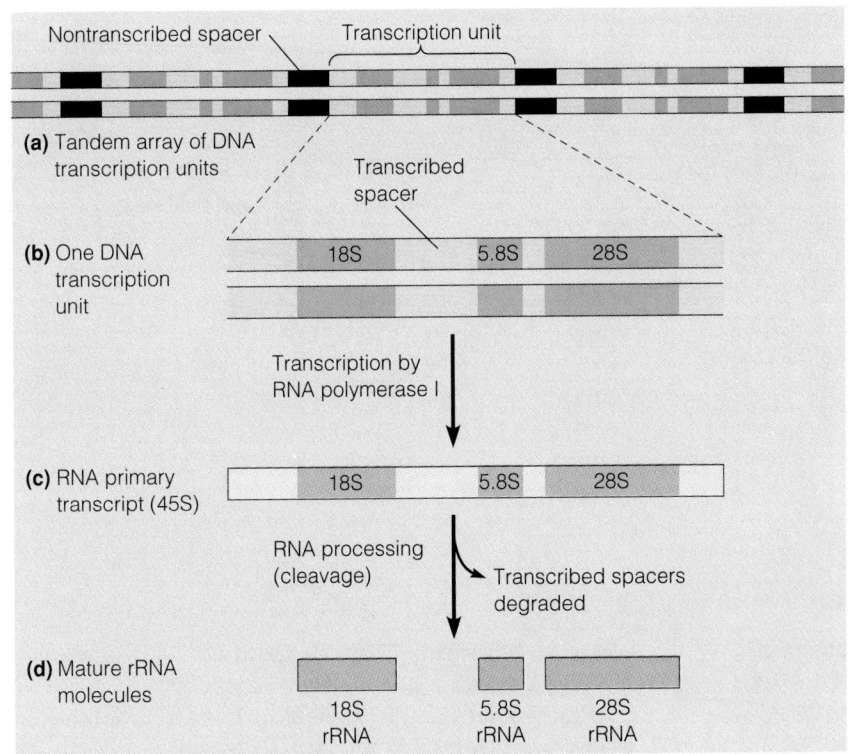

(a) Tandem array of DNA transcription units

(b) One DNA transcription unit

Transcription by RNA polymerase I

(c) RNA primary transcript (45S)

RNA processing (cleavage)

Transcribed spacers degraded

(d) Mature rRNA molecules

18S rRNA   5.8S rRNA   28S rRNA

Figure 17-16 Eukaryotic rRNA Genes and the Processing of the Primary Transcript. **(a)** The eukaryotic transcription unit that includes the genes for the three largest rRNAs occurs in multiple copies, arranged in tandem arrays. Nontranscribed spacers (black) separate the units. **(b)** Each transcription unit includes the genes for the three rRNAs (darker blue) and four transcribed spacers (lighter blue). **(c)** The transcription unit is transcribed by RNA polymerase I into a single long precursor with a sedimentation coefficient of about 45S. **(d)** RNA processing yields mature 18S, 5.8S, and 28S rRNA molecules. RNA cleavage actually occurs in a series of steps. The order of steps varies with the species and cell type, but the final products are always the same three types of rRNA molecules.

sors of some other eukaryotes contain smaller amounts of spacer sequences, but in all cases, the precursor RNA is larger than the aggregate size of the three rRNA molecules made from it. Thus, at least some processing is always necessary.

In bacterial cells, also, rRNA formation involves the processing of precursor forms that are larger than the end-products. *E. coli,* for example, has seven rRNA transcription units (*operons,* as we shall see in Chapter 19), which are scattered about its genome. Each transcription unit includes sequences for all three prokaryotic rRNAs—23S rRNA, 16S rRNA, and 5S rRNA—and, in addition, several tRNA genes. The processing of the primary transcripts involves two sets of enzymes, one for the rRNAs and one for the tRNAs.

Transfer RNA

Cells make a variety of tRNA molecules, with specificities for recognizing different codons on mRNA and carrying different amino acids. However, all tRNA molecules share a common general structure, as illustrated in Figure 17-17. A mature tRNA molecule has only 70–90 nucleotides, some of which are chemically modified. Self-complementary base sequences within the single-stranded molecule cause it to form a secondary structure with several hairpins, shown in part b of the figure. Molecular biologists call the tRNA secondary structure a cloverleaf structure. Most tRNAs have four base-paired regions, as shown here; in some, there is a fifth one at the *variable loop.* Each of these regions is a short stretch of RNA double helix. In the three-dimensional tertiary structure, the molecule is folded so that the overall shape is like an L (see Figure 18-3).

Like ribosomal RNA, transfer RNA is synthesized in a precursor form in both eukaryotic and prokaryotic cells. Processing of tRNA precursors involves several distinctly different events, as Figure 17-17 indicates for yeast tyrosine tRNA. At the 5′ end, a short *leader sequence* (of 16 nucleotides, in this case) is removed. At the 3′ end, the two terminal nucleotides of the precursor are removed and replaced with the trinucleotide sequence CCA, which is a dis-

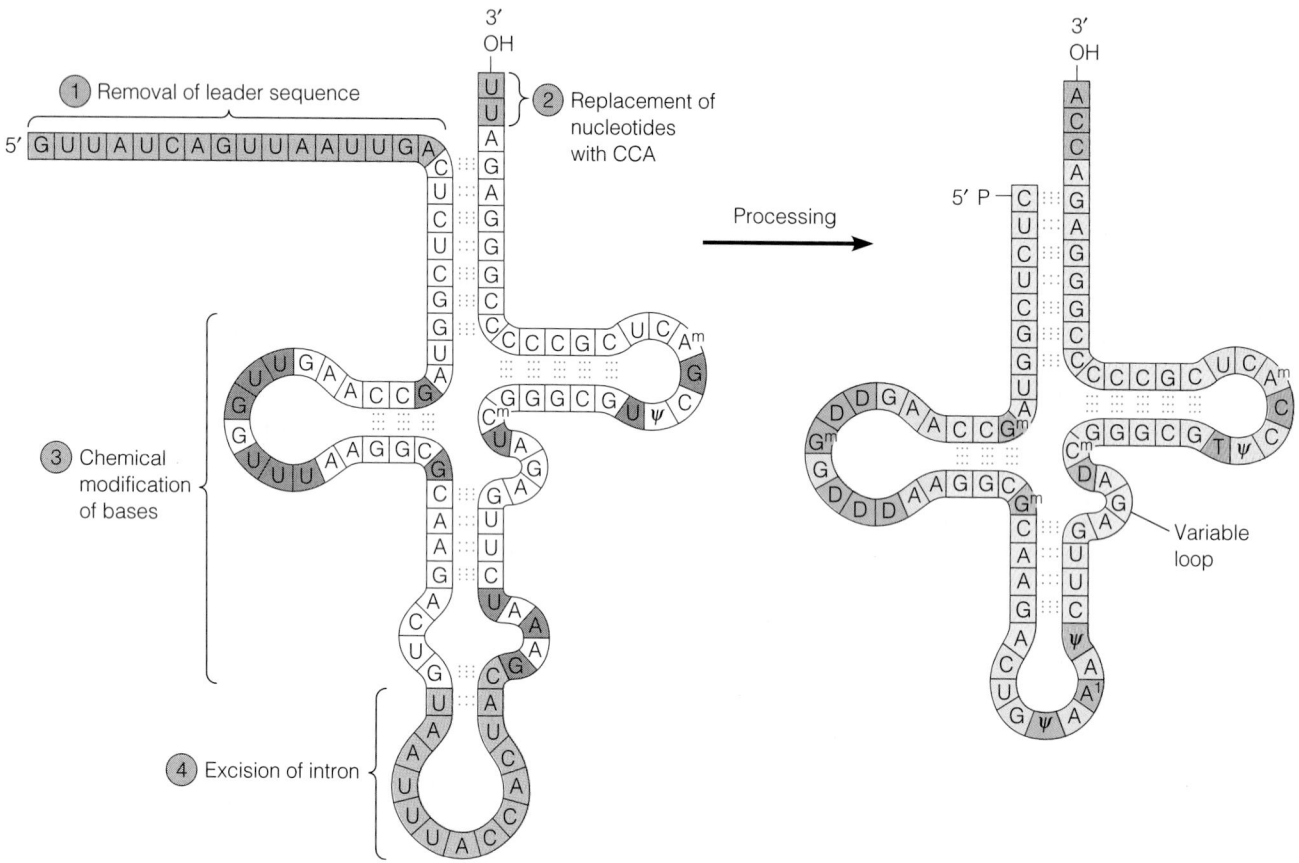

(a) Primary transcript (precursor) for yeast tyrosine tRNA

(b) Mature tRNA, secondary structure

Figure 17-17 Processing and Secondary Structure of Transfer RNA. **(a)** Every tRNA gene is transcribed as a precursor that must be processed into a mature tRNA molecule. This is the primary transcript for yeast tyrosine tRNA. Processing for this tRNA involves ① removal of the leader sequence at the 5′ end, ② replacement of two nucleotides at the 3′ end by the sequence CCA (with which all mature

tRNA molecules terminate), ③ chemical modification of certain bases (grey), and ④ excision of an intron. **(b)** The mature tRNA in a flattened, cloverleaf representation, which clearly shows the base pairing between self-complementary stretches in the molecule. Modified bases (darker colors) are abbreviated as A^m for methyladenine, G^m for methylguanine, D for dihydrouracil, T for ribothymine, and ψ for pseudouridine.

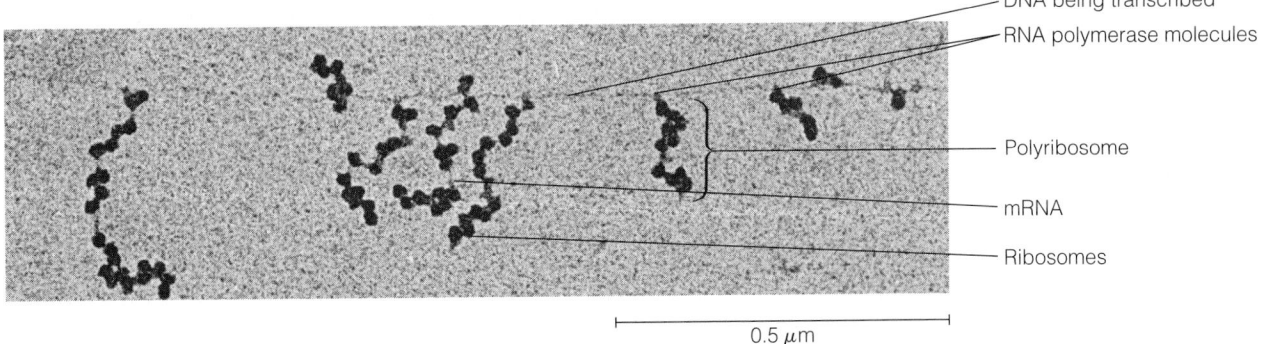

Figure 17-18 **Coupling of Transcription and Translation in Bacterial Cells.** This electron micrograph shows *E. coli* DNA being transcribed by RNA polymerase molecules that are moving from right to left. Attached to each polymerase molecule is a strand of mRNA still in the process of being transcribed. The large dark particles attached to each growing mRNA strand are ribosomes that are actively translating the partially complete mRNA. (The polypeptides being synthesized are not visible.) A cluster of ribosomes attached to a single mRNA strand is called a polyribosome (TEM).

tinguishing characteristic of all functional tRNA molecules. (Some tRNAs already have CCA in their primary transcripts and therefore do not require modification at the 3′ end.) In addition, a number of bases are modified chemically, particularly by methylation. Transfer RNA molecules are noted for the presence of a high proportion of modified bases.

The processing of yeast tyrosine tRNA is also characterized by the removal of an internal 14-nucleotide sequence, although the transcripts for most tRNAs do not require that kind of excision. An internal segment of an RNA transcript that must be removed to form the functional RNA product is called an RNA **intron,** or *intervening sequence.* We will consider introns in more detail in the discussion of mRNA processing because they are an almost universal feature of mRNA precursors in eukaryotic cells. For the present, simply note that some eukaryotic tRNA precursors contain introns and that these must be eliminated during processing by a precise mechanism that cuts and splices the precursor molecules at exactly the same location every time.

An intron-containing tRNA precursor is cut and spliced in two steps, catalyzed by two separate enzymes, an RNA endonuclease and an RNA ligase. These enzymes are very similar from species to species, even among species that are phylogenetically distant from one another. In an experiment that demonstrated this point vividly, cloned yeast genes for the tyrosine tRNA shown in Figure 17-17 were microinjected into eggs of *Xenopus laevis,* the South African clawed toad. Despite the long evolutionary divergence between fungi and amphibians, the yeast genes were transcribed in the toad eggs and were processed properly, including the removal of the 14-nucleotide intron.

Messenger RNA

Prokaryotic mRNA is, in almost all cases, an exception to the generalization that RNA requires processing before it can be used by the cell. Most bacterial mRNA is synthesized in a form that is ready for translation immediately, even before the entire RNA molecule is completed. Moreover, transcription in a bacterial cell is not separated by a membrane barrier from the ribosomes responsible for translation, so bacterial mRNA molecules in the process of elongation by RNA polymerase often have ribosomes already associated with them. The electron micrograph in Figure 17-18 shows this coupling of transcription and translation in a bacterial cell.

In eukaryotic cells, on the other hand, transcription occurs in the nucleus, but translation takes place in the cytoplasm. The two processes are therefore separated in both time and space. In addition, substantial chemical processing in the nucleus is necessary to convert primary transcripts into mature mRNA molecules that are ready for transport to the cytoplasm and translation.

Although each eukaryotic mRNA transcription unit encodes only one polypeptide, the primary transcripts are often very long, typically ranging from about 2000 to 20,000 nucleotides. This size heterogeneity is reflected in the term **heterogeneous nuclear RNA (hnRNA),** which has been applied to the nonribosomal, nontransfer RNA found in eukaryotic nuclei. Most hnRNA is thought to be mRNA molecules or their precursors, *pre-mRNA.* Some of this pre-mRNA has already been modified by the addition of 5′ caps and 3′ tails, as described next.

5′ Caps and 3′ Poly(A) Tails. Most eukaryotic mRNA molecules bear distinctive modifications at both ends. All mRNA molecules contain a modified nucleotide at their 5′ end, called a **5′ cap.** At their 3′ end, most mRNA molecules have a long stretch of adenine ribonucleotides called a **poly(A) tail.**

The 5′ cap is just a guanosine that is methylated at position 7 of the purine ring structure and is "on backward"; that is, the bond joining it to the RNA is a 5′ → 5′ pyrophosphate linkage, instead of the usual 3′ → 5′ bond. This distinctive feature of eukaryotic mRNA is added to the primary transcript shortly after initiation of RNA synthesis (Figure 17-19).

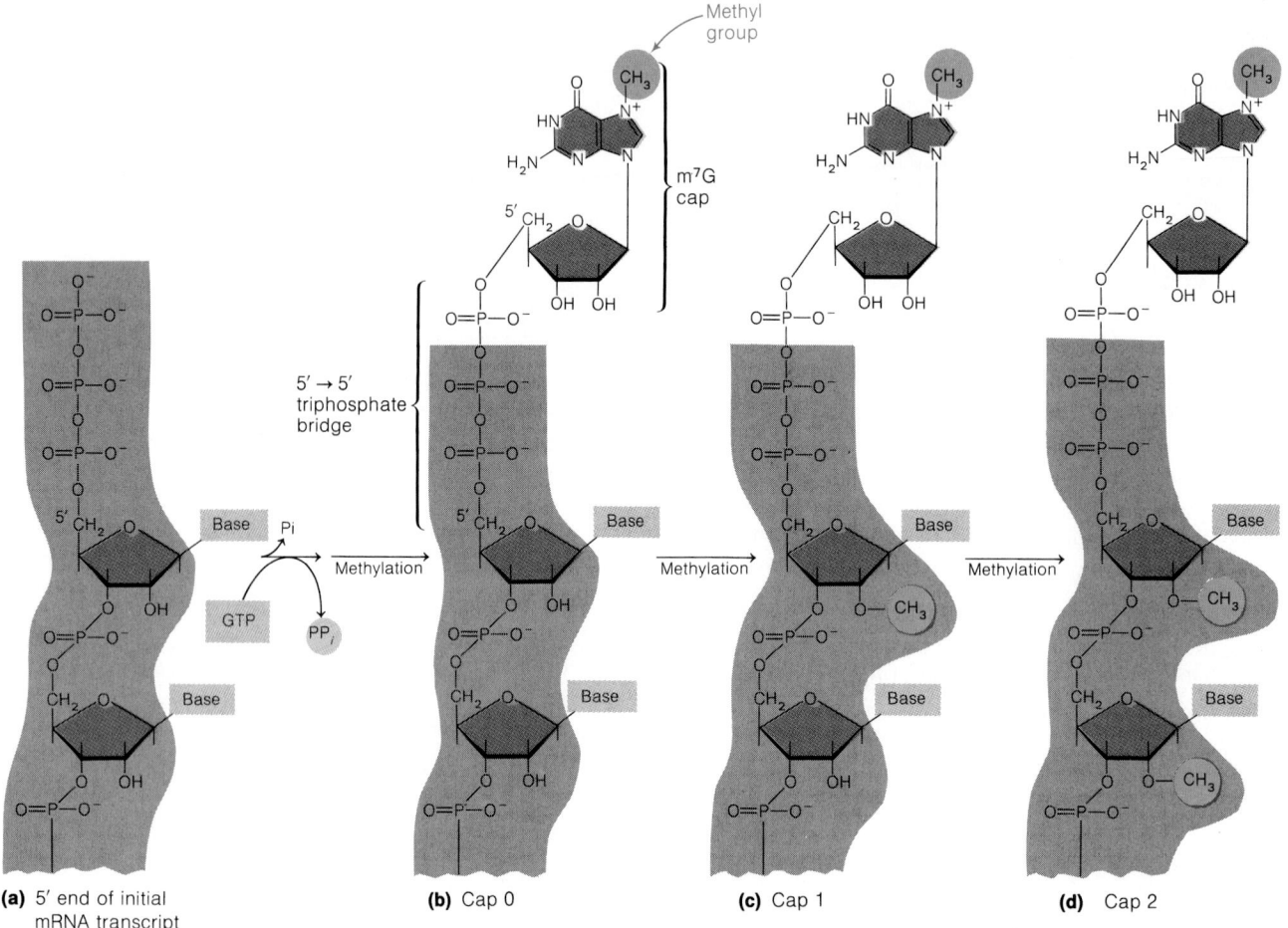

(a) 5′ end of initial mRNA transcript

(b) Cap 0

(c) Cap 1

(d) Cap 2

Figure 17-19 Capping of Eukaryotic Pre-mRNA. The processing of eukaryotic mRNA molecules involves both the addition of an m⁷G cap at the 5′ end of the initial transcript and one or more subsequent methylation events. **(a)** The initial transcript has a triphosphate group at its 5′ end but is unmodified. **(b)** Cap 0 is generated by the addition of a methylated guanosine (m⁷G) cap with a 5′ → 5′ linkage. **(c)** Cap 1 is formed by the addition of a methyl group on the ribose of the adjacent nucleotide. **(d)** Cap 2 requires methylation of the ribose of the next nucleotide as well.

If the only methyl group is on the N-7 of the guanine, the structure is called *cap 0* (Figure 17-19b). However, adjacent nucleotides can be methylated on their ribose groups, generating *cap 1* (next nucleotide only) or *cap 2* (next two nucleotides methylated on the ribose), as shown in Figure 17-19c and d. These cap structures protect the mRNA from degradation by nucleases that attack RNA at the 5′ end, thereby contributing to the stability of the mRNA. In addition, the cap is important in positioning the mRNA on the ribosome for the initiation of translation.

Poly(A) tails are found on most eukaryotic mRNA molecules. In animal cells, only the mRNAs for the major histones lack them. The tails are usually 100–200 nucleotides in length. It is clear that they are added after transcription, because the DNA strands coding for such mRNA molecules do not contain corresponding stretches of thymine nucleotides (T). The enzyme responsible for the sequential addition of the adenine nucleotides is called *poly(A) polymerase* (or *polyadenylate polymerase*). To be polyadenylated, a primary transcript must have the sequence AAUAAA. This signal triggers cleavage of the primary transcript at a site about 10–35 nucleotides downstream from there and the addition of adenine nucleotides to the 3′ end. Figure 17-20 shows an RNA transcript with both a cap and a tail.

The poly(A) tail seems to have several functions. Like the 5′ cap, it protects the mRNA from nuclease attack, and it seems also to help the ribosome recognize the mRNA as a molecule to be translated. In addition, it is apparently recognized by specific proteins that are involved in exporting the mRNA from the nucleus. In the laboratory, poly(A) tails can be used to isolate mRNA from the more prevalent rRNA and tRNA. The bulk RNA extracted from cells is passed through a column packed with particles coated with poly(dT), single strands of DNA consisting solely of thymine nucleotides. Molecules with poly(A) tails bind to the poly(dT), while others pass through. The polyadenylated mRNA is then eluted from the column.

The Discovery of Introns. In eukaryotic cells, the precursors for most mRNAs (and for some tRNAs and rRNAs)

5' Cap
m⁷G—P—P—P—[A A U A A A   ]—(A₁₀₀₋₂₀₀)—OH
                                                                3' Poly(A) tail
                              10—35 nucleotides

Figure 17-20   Pre-mRNA with Cap and Tail.   The 5′ m⁷G cap is added to the primary RNA transcript early in the elongation stage. Later, about 10–35 nucleotides downstream from an AAUAAA sequence in the transcript, the RNA is cleaved and a string of 100–200 adenine nu-cleotides is added to the end. This is the 3′ poly(A) tail. Almost all eukaryotic mRNAs have caps and tails, which help protect the message from degradation and serve as tags for identification by the ribosome.

---

contain *introns*, sequences that are a part of the primary transcript but not of the functional RNA. The discovery of introns in 1977 was a great surprise to biologists. It had been known for many years that in bacteria, the order of amino acids in a polypeptide correlates exactly with a sequence of contiguous nucleotide pairs in the DNA. This relationship was demonstrated in the colinearity experiments of Charles Yanofsky and others in the early 1960s and was confirmed by direct comparisons of nucleotide and amino acid sequences as rapid methods became available for sequencing DNA and proteins. Biologists naturally assumed that the same would turn out to be true for eukaryotes, once it became possible to isolate and sequence specific segments of eukaryotic DNA.

It was therefore a shock when several research groups reported almost simultaneously in 1977 that at least some eukaryotic genes did not follow this pattern, but were in fact interrupted by "extra" sequences of nucleotides—introns—that were not represented in either the functional mRNA or the protein product. Electron micrographs of hybrids between mRNA molecules and segments of DNA known to contain the genes coding for these mRNA molecules showed striking visual evidence for introns. To understand this evidence, let's first examine what happens with a protein-coding gene *lacking* introns. Figure 17-21a depicts the expected result when a prokaryotic mRNA is allowed to hybridize with double-stranded DNA corresponding to the gene from which the mRNA was transcribed. The mRNA hybridizes to the template strand of the DNA, leaving the other, displaced strand as one single-stranded DNA loop that is readily identifiable as such in the electron microscope.

If eukaryotic genes were contiguous sequences of coding nucleotides, they would be expected to show the same profile. However, when such experiments were actually done with eukaryotic mRNA and genes for proteins such as human β globin and chick ovalbumin, the surprising result was that *multiple* loops were seen (Figure 17-21b). This pattern is now known to be due to the presence within the gene of multiple introns that are not present in the mRNA and therefore do not participate in hybrid formation. The prominent loops are double-stranded intron DNA. Because the segments of the coding sequence, called *exons*, are individually very short, the single-stranded DNA loops displaced by mRNA are barely visible.

Once they had been reported for a few genes, introns began popping up everywhere, especially as restriction en-zyme techniques and DNA sequencing began to be applied to a wide variety of eukaryotic genes. To date, introns have been found in most protein-coding genes of multicellular eukaryotes; the histone genes are notable exceptions. Unicellular eukaryotes have many genes completely lacking introns, but in most of the living world, "split" genes seem to be the rule.

The genes of vertebrates display a remarkable variety with respect to the number and size of introns they contain, as well as the total proportion that the introns represent of the whole gene. The human β-globin gene, for example, has only two introns, one of 120 bp and the other of 550 bp. Together, these account for about 40% of the total length of the gene. More commonly in mammals, most of the gene is

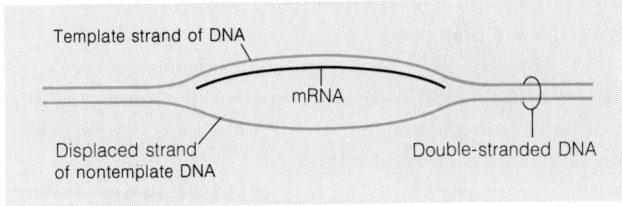

Template strand of DNA

mRNA

Displaced strand of nontemplate DNA

Double-stranded DNA

(a)

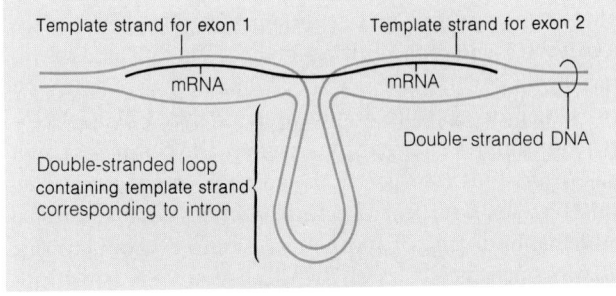

Template strand for exon 1

Template strand for exon 2

mRNA

mRNA

Double-stranded DNA

Double-stranded loop containing template strand corresponding to intron

(b)

Figure 17-21   Demonstration of Introns in Protein-Coding Genes.   Molecules of mature mRNA were allowed to hydrogen-bond with the DNA (genes) from which they had been transcribed. The resulting hybrid molecules were then examined under an electron microscope. The diagrams here show the results. **(a)** Hybridization of a prokaryotic mRNA molecule with DNA from its gene, which lacks introns. The displaced strand of DNA forms a wide loop. **(b)** Hybridization of a eukaryotic mRNA molecule with its gene, which has one intron. The prominent loop is double-stranded DNA corresponding to the intron. The loops formed by single DNA strands displaced by the mRNA are not readily discernible in micrographs because the two coding segments (exons) are relatively quite short.

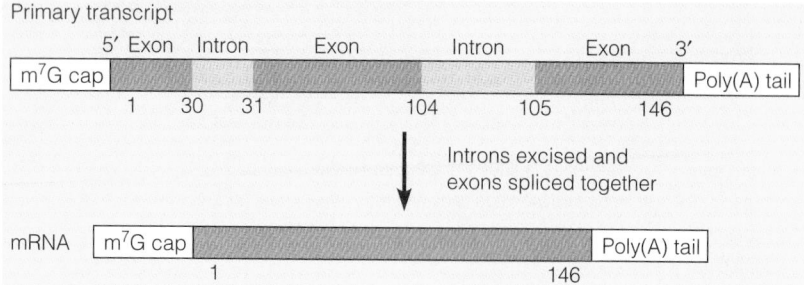

Figure 17-22 An Overview of RNA Splicing.
(a) The capped and tailed primary transcript of the human gene for β globin. Exons are shown in dark pink and introns in light pink. The numbers refer to the codons. **(b)** The mature mRNA that results from RNA splicing. The introns have been excised and the exons joined together to form a molecule with a continuous coding sequence. The cell's ribosomes will translate this message into a polypeptide of 146 amino acids.

made up of introns. An extreme example is the human dystrophin gene, an altered form (allele) of which causes Duchenne muscular dystrophy. This gene is over 2 million bp long and has more than 50 introns, representing over 99% of the DNA! To date, this gigantic gene holds the record for size.

RNA Splicing. To produce a functional molecule of mRNA from a primary transcript containing introns, the eukaryotic cell must somehow remove the introns and splice together the remaining RNA segments, which are called **exons** (because they are *expressed*). The entire process of cutting out the introns and rejoining the exons is termed **RNA splicing.** As an example, Figure 17-22 shows the starting material (primary transcript) and end-product (mRNA) from the β-globin gene.

Clearly, the RNA splicing mechanism must be very precise, because a shift of a single nucleotide would alter the reading frame and render the mRNA useless. The mechanism depends on specific base sequences at the junctions between introns and exons. As Figure 17-23 indicates, the sequence at the 5′ end of the intron, called the *5′ splice site*, is different from the one at the intron's 3′ end, the *3′ splice site*. These sequences can vary at certain base positions, but the GU at the 5′ end of the intron and the AG at its 3′ end are invariant.

The process of RNA splicing is catalyzed by an RNA-protein complex called a **spliceosome.** The complex is built up from four smaller RNA-protein complexes called **snRNPs** (small nuclear ribonucleoproteins) and additional proteins. Each snRNP (pronounced "snurp") contains one or two small molecules of RNA, called **snRNA** (small nuclear RNA). There are five snRNAs in the spliceosome: U1, U2, U4, U5, and U6. U4 and U6 snRNAs are together in one

snRNP, attached to each other by a stretch of base pairing. Recognition of the pre-mRNA splice sites by snRNPs is based on base pairing between snRNAs and pre-mRNA.

The spliceosome assembles on an mRNA precursor, usually while the primary transcript is still being made (Figure 17-24). The snRNPs—called U1, U2, U5, and U4/U6, after their snRNAs—join the growing spliceosome in a defined order. The completed spliceosome is almost as big as a ribosome. The actual splicing occurs in two steps (steps 2 and 3 in the figure). First, cleavage occurs at the 5′ splice site when the 2′ OH of a specific adenosine near the 3′ end of the intron attacks the G at the 5′ end and covalently attaches to it by a 2′–5′ phosphodiester bond. Thus, the adenosine forms a branch point within the intron RNA. In the next step, the 3′ splice site is cleaved, and the two exons are joined together. In this step, the newly formed 3′ end of exon 1 attacks at the 3′ splice site. The result is a mature mRNA with a continuous coding sequence and an excised RNA intron in the form of a lariat. Although not shown in the figure, both the primary transcript and the mature mRNA are associated along their lengths with non-snRNP proteins, forming **RNP particles** that are something like DNA nucleosomes. The RNA remains associated with these proteins until it enters the cytoplasm, where it joins with other proteins and the ribosomal subunits.

The detailed chemical mechanism of RNA splicing by spliceosomes is still being investigated. There is little doubt, however, that snRNAs play a role in the catalytic process, as well as in spliceosome assembly and splice-site recognition. The idea of a catalytic role for snRNAs arose from the discovery of self-splicing RNA introns, our next topic.

Self-Splicing RNA Introns. Not all RNA splicing involves snRNPs and spliceosomes. Certain genes have *self-splicing*

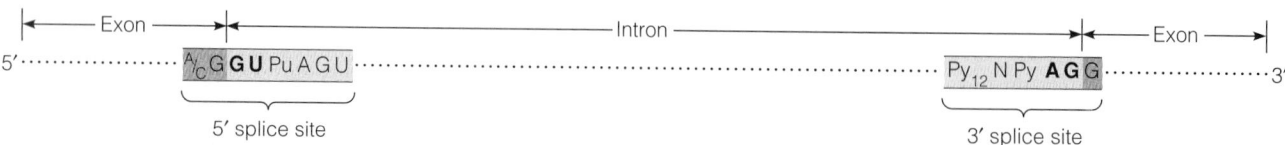

Figure 17-23 Consensus Sequences at Intron-Exon Splice Sites. Pre-mRNA molecules of higher eukaryotes have the nucleotide sequences shown here at their intron-exon junctions. As indicated, alternative bases are found at a number of positions within the sequences.

However, the GU at the 5′ end of the intron and the AG at its 3′ end are invariant. A/C = A or C; Pu = purine; Py = pyrimidine; N = any of the four RNA bases.

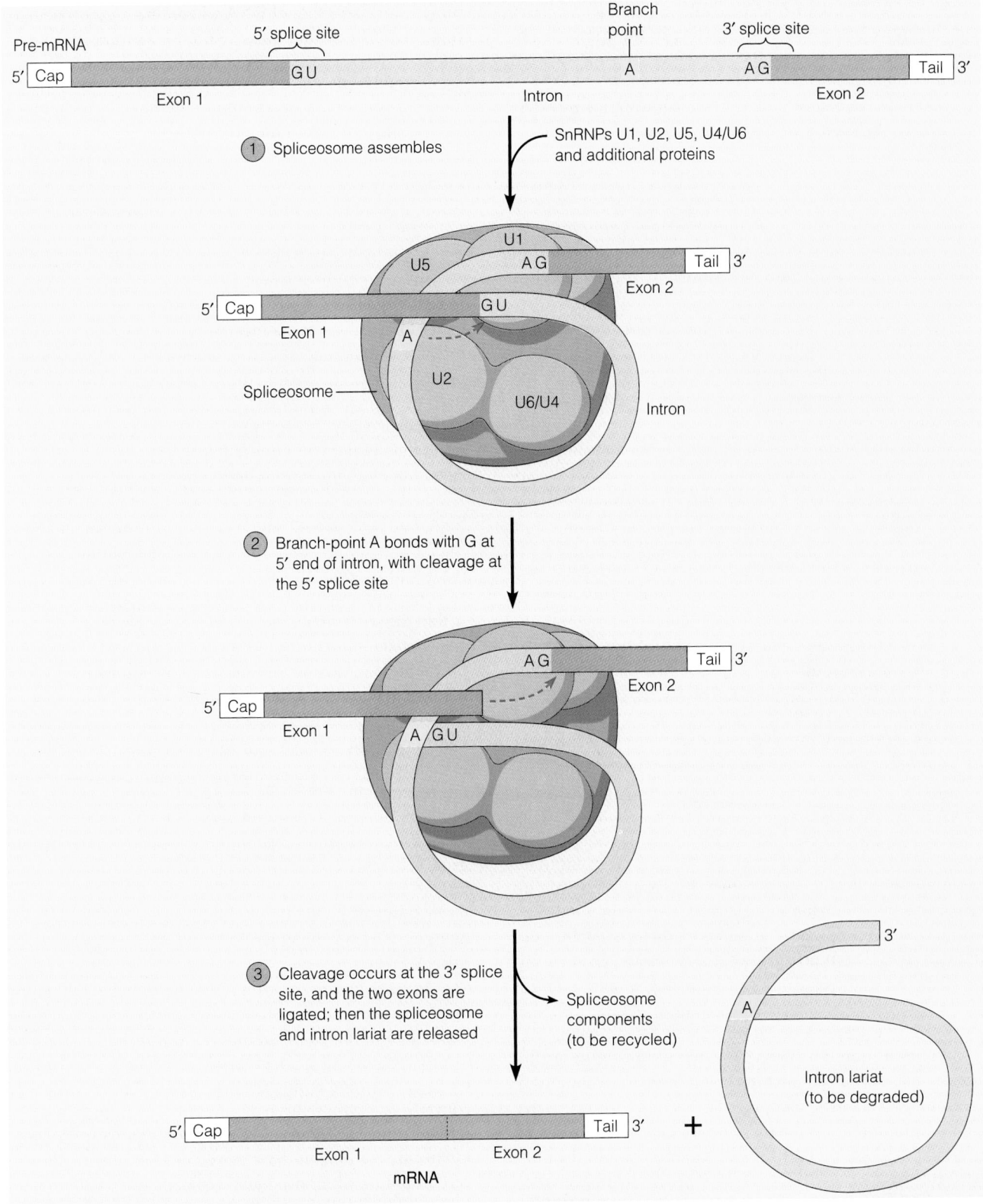

Figure 17-24   Assembly and Function of the Spliceosome.
The spliceosome is an RNA-protein complex that splices intron-containing pre-mRNA transcripts in the eukaryotic nucleus. The substrate here is a molecule of pre-mRNA with one intron and two exons. ① The spliceosome assembles on the pre-mRNA by the successive addition of four smaller RNA-protein complexes, called snRNPs, and other proteins. The U1 and U2 snRNPs bind to the DNA first. Part of the U1 RNA molecule (not shown) base-pairs with the 5′ splice site, and another part probably base-pairs with the 3′ splice site, thus bringing the two intron-exon junctions close together at the start. A specific adenosine (yellow), about 25 bases from the 3′ end of the intron, is now ready to attack (dashed arrow) the G at the 5′ end of the intron. ② The adenosine forms a 2′–5′ phosphodiester bond with the G, with cleavage of the intron at its 5′ splice site. The adenosine thus becomes a branch-point in the intron. The 3′ end of exon 1 is now free to attack (dashed arrow) the other intron-exon junction. ③ The 3′ end of exon 1 covalently attaches (ligates) to the 5′ end of exon 2, as the second RNA cleavage occurs. The spliceosome is released from the mRNA, along with the intron, which has the form of a lariat. The lariat is degraded, and the spliceosome components are recycled.

RNA introns. The RNA transcript of such a gene can carry out the entire process of RNA splicing in the complete absence of protein (e.g., in a test tube); the intron RNA itself catalyzes the process.

There are two classes of self-splicing RNA introns. The larger class, called *Group I introns,* are found in the RNA transcripts of certain mitochondrial genes of fungi, of some chloroplast tRNA genes, and of rRNA genes in *Tetrahymena, Physarum* (a slime mold), some fungi, and chloroplasts. In addition, Group I introns are found in three pre-mRNAs of bacteriophage T4—one of only a few examples of introns in prokaryotic systems. Group I RNA introns are excised as linear pieces of RNA.

Group II introns, on the other hand, excise themselves as lariats, just as in the spliceosome mechanism. Moreover, an adenosine within the Group II RNA intron forms the branch point of the lariat. In present-day organisms, Group II introns are found only in fungal mitochondria. But biologists think that today's prevailing splicing mechanism probably evolved from such a system, with the intron RNA's catalytic role being taken over by snRNA molecules of the spliceosome.

Why Introns?   The burning question, of course, is *why.* Why do nearly all genes in multicellular eukaryotes have so much DNA that seems to serve no coding function? Why, in generation after generation of cells, is so much energy invested in making segments of DNA molecules—and of RNA transcripts—that are apparently destined only for the splicing scrap heap?

One possibility involves *alternative splicing* of pre-mRNA. Many cases are known where introns and RNA splicing allow the cell to generate several different mRNAs, and proteins, from the same sort of primary transcript. This flexibility is possible because the junction sequences for different introns are the same, thereby allowing the exons to be assembled in different combinations by juxtaposition of different junctions. Having a single gene encode several different polypeptides is clearly an efficient usage of DNA, perhaps partially counterbalancing the apparent inefficiency involved in the presence of long noncoding sequences in the genome. Moreover, alternative splicing provides a way of regulating gene expression, as we will discuss in Chapter 19.

An even more interesting role for introns is an evolutionary one. It is likely that introns hasten the evolution of new and potentially useful proteins. For many proteins, the exons correspond to separate polypeptide *domains,* units of polypeptide folding that have discrete functions in the protein. For example, the three exons of the β-globin gene correspond to different structural and functional regions of the polypeptide. This kind of situation suggests that protein-coding genes with multiple exons may have been assembled during evolution from what were originally separate entities. The presence of introns could help new proteins evolve in two ways, both of which depend upon the fact that introns

provide long stretches of DNA where "incorrect" genetic recombination can take place without harming coding sequences. First, crossing over within introns of different genes could lead to the creation of genes with new combinations of exons—exon shuffling. Second, recombination within other combinations of introns could easily produce duplicates of particular exons within a single gene. These exons might continue as exact duplicates, or one might mutate to a sequence that produces a new activity in the polypeptide.

Many biologists believe that introns are actually relics of ancient unicellular organisms that were the ancestors of all today's organisms, both prokaryotes and eukaryotes. Evidence for this hypothesis includes the presence of introns in modern archaebacteria and in some of the tRNA genes of cyanobacteria. Presumably, during billions of years of evolution, evolutionary pressure for a streamlined genome has been particularly strong in unicellular organisms, with the result that modern bacteria and, to a large extent, unicellular eukaryotes lack the introns originally present.

RNA Editing.   About ten years ago, molecular biologists were surprised by the discovery of yet another type of mRNA processing. In this kind of processing, called **RNA editing,** nucleotides are inserted, removed, or changed within the coding sequence of the RNA. The result is usually an altered translational reading frame and often new polypeptide initiation or termination codons. The first-discovered and best-studied examples of RNA editing occur in the mitochondrial mRNAs of trypanosomes, which are parasitic protozoa. In these mRNAs, the editing involves the insertion and deletion of multiple uracil nucleotides at various points in the mRNA. It was discovered that the information for this editing is located in small RNA molecules called *guide RNAs,* which are apparently encoded by mitochondrial genes separate from the mRNA genes. In one proposed editing mechanism, hydrogen bonding links short complementary regions of the guide RNA and mRNA, and nearby sequences of Us in the guide RNA are then spliced into the mRNA.

Another important type of mRNA editing is found in many of the mitochondrial and chloroplast mRNAs of higher plants. In these cases, nucleotides are neither inserted nor deleted, but Cs are converted to Us (and vice versa) by deamination (and amination) reactions. Similar nucleotide conversions have also been discovered in a few mRNAs transcribed from *nuclear* genes of mammals. And, very recently, nucleotide conversions have been found in *transfer RNA* of certain protozoal mitochondria. In none of these situations is the mechanism known or is it known where the information specifying the conversions resides.

In all its manifestations, RNA editing seems relatively rare. However, its existence provides a reason to be cautious in inferring either polypeptide or RNA sequences from genomic DNA sequences!

Useful Properties of mRNA Metabolism

Before ending this chapter, we should note two general aspects of mRNA metabolism that contribute to mRNA's usefulness in the cell. These are the rapid turnover of mRNA and its ability to amplify genetic information.

Most mRNA has a high *turnover rate*—that is, the rate at which molecules are degraded and then replaced with newly synthesized versions. In this respect, mRNA contrasts with the other major forms of RNA in the cell, rRNA and tRNA, which are notable for their stability. Because of its more rapid turnover, mRNA accounts for most of the transcriptional activity in many cells, even though it represents only a small fraction of the total RNA content of the cell. Turnover is usually measured in terms of the *half-life* of the molecule in question: the length of time required for degradation of 50% of the molecules present at a given time. The mRNA molecules of bacterial cells generally have half-lives only minutes long. The mRNA molecules of eukaryotic cells, on the other hand, usually have half-lives of hours or sometimes even days.

The synthesis of a particular mRNA can be initiated again and again from the same stretch of DNA. This repeated synthesis affords an important opportunity for *amplification* of the genetic message. If DNA were used directly in protein synthesis, the number of protein molecules that could be translated from it within a given time period would be strictly limited by the rate of polypeptide synthesis. But in a system using mRNA as an intermediate, multiple copies of the information can be made, and each of these can in turn be used to direct the synthesis of many copies of the protein product.

As an especially dramatic example of this amplification effect, consider the synthesis of *fibroin*, the major protein of silk. The haploid genome of the silkworm has only one copy of the fibroin gene, but about 10^4 copies of fibroin mRNA are transcribed from the two copies of the gene in each diploid cell of the silk gland. Each of these mRNA molecules, in turn, directs the synthesis of about 10^5 fibroin molecules, resulting in the production of about 10^9 molecules of fibroin per cell—all within the 4-day period it takes the worm to make its cocoon! Without mRNA as an intermediate, the genome of the silkworm would need 10^4 copies of the fibroin gene (or about 40,000 days!) to make a cocoon.

Significantly, most genes that encode proteins are present in only one or a few copies per haploid genome. Genes that code for rRNA and tRNA, however, are always present in multiple copies in the genome. It is advantageous for the cell to have many copies of genes whose final products are RNA (rather than protein), because in this case there is no opportunity for amplifying each gene's effect by repeated translation.

PERSPECTIVE

The expression of genetic information is one of the fundamental activities of all cells. Instructions stored in the DNA of the chromosomes must first be transcribed and processed into molecules of mRNA, rRNA, and tRNA. These RNAs then play specific roles in the synthesis of proteins, with the nucleotide sequence of the mRNA providing the information that actually dictates the order of amino acids in the polypeptide product. The mRNA sequence is read as triplet codons. The genetic code specifies which amino acid corresponds to each of the triplet codons. The code is unambiguous, nonoverlapping, degenerate, and nearly universal.

Transcription is the synthesis of RNA by a mechanism that depends on complementary base pairing between incoming nucleotides and the template strand of the DNA. The process of transcription can be divided into four stages: (1) binding of RNA polymerase, the transcription enzyme, to the DNA at the promoter region; (2) initiation of RNA synthesis; (3) elongation of the RNA chain; and (4) termi-nation. Prokaryotic cells have only a single kind of RNA polymerase, but eukaryotic cells have several kinds, each used for making one or more different kinds of RNA: rRNA, mRNA, tRNA, mitochondrial DNA, or chloroplast DNA. The DNA promoter sequences used by the three nuclear RNA polymerases are distinctive, and, in contrast with the situation in bacteria, recognition of eukaryotic promoters is primarily the responsibility of transcription factors that are separate from the polymerase. Protein-protein interactions are thus much more important for eukaryotic transcription than for the prokaryotic process.

Once synthesized, most RNA transcripts must be processed to generate functional RNA molecules. Prokaryotic mRNA, which is translated as it is being made, is the major exception. RNA processing can involve cleavage of multigene transcription units, removal of noncoding sequences, addition of special structural features at the 3′ and/or the 5′ end, and chemical modification of specific nucleotides. The splicing of eukaryotic RNA, in which introns are excised and exons rejoined, is a particularly elaborate type of processing. Splicing of most eukaryotic mRNA precursors is carried out by spliceosomes, which consist of RNA and protein, but in some other cases, splicing is catalyzed solely by intron RNA. It is likely that introns and RNA splicing—in particular, self-splicing—are relics from primordial cells that were ancestral to all cells on Earth today.

KEY TERMS FOR SELF-TESTING

The Central Dogma of Molecular Biology

central dogma of molecular biology (p. 535)
messenger RNA (mRNA) (p. 535)
ribosomal RNA (rRNA (p. 535)
transfer RNA (tRNA) (p. 535)
transcription (p. 535)
translation (p. 535)
reverse transcription (p. 535)

The Genetic Code

genetic code (p. 536)
one gene–one polypeptide (p. 537)
colinearity (p. 540)
mutagen (p. 541)
frameshift mutation (p. 541)
degenerate code (p. 542)
template strand (p. 544)
coding strand (p. 544)
codon (p. 544)
copolymer (p. 544)
stop signal (p. 545)

Transcription in the Prokaryotic Cell

RNA polymerase (p. 545)
sigma (σ) factor (p. 545)
core enzyme (of bacterial RNA polymerase) (p. 547)
holoenzyme (p. 547)
transcription unit (p. 547)
promoter (p. 547)
upstream (p. 547)
−10 sequence (p. 547)
−35 sequence (p. 547)
consensus sequence (p. 549)
DNA footprinting (p. 549)
terminator (p. 550)
termination signal (p. 550)
hairpin (p. 550)
rho (ρ) factor (p. 550)

Transcription in the Eukaryotic Cell

RNA polymerases I, II, and III (pp. 551–552)
core promoter (p. 552)
upstream control element (p. 554)

initiator (Inr) (p. 554)
TATA box (p. 554)
basal transcription factor (p. 554)
TATA-binding protein (TBP) (p. 554)

RNA Processing

primary transcript (p. 556)
RNA processing (p. 556)
intron (p. 559)
heterogeneous nuclear RNA (hnRNA) (p. 559)
5′ cap (p. 559)
poly(A) tail (p. 559)
exon (p. 562)
RNA splicing (p. 562)
spliceosome (p. 562)
snRNP (p. 562)
snRNA (p. 562)
RNP particle (p. 562)
RNA editing (p. 564)

PROBLEM SET

17-1. Codes and Coding. One way to think about the genetic code is to compare it with the Morse code (see Figure 17-2). Explain each of the following terms as it relates to genetic coding, and indicate with an M if it is true of the Morse code and with a G if it is true of the genetic code.

(a) Degenerate

(b) Unambiguous

(c) Triplet

(d) Universal

(e) Nonoverlapping

17-2. The Genetic Code in a T-Even Phage. A portion of a polypeptide produced by bacteriophage T4 was found to have the following sequence of amino acids:

. . . Lys-Ser-Pro-Ser-Leu-Asn-Ala . . .

Deletion of a single nucleotide in one location on the T4 DNA template strand with subsequent insertion of a different nucleotide nearby changed the sequence to:

. . . Lys-Val-His-His-Leu-Met-Ala . . .

(a) What was the nucleotide sequence of the portion of mRNA that encoded this portion of the original polypeptide?

(b) What was the nucleotide sequence of the mRNA encoding this portion of the mutant polypeptide?

(c) Can you determine which nucleotide was deleted and which was inserted? Explain your answer.

17-3. Frameshift Mutations. The mutants listed below each have a different mutant form of the gene encoding protein X. Each mutant gene contains one or more nucleotide insertions (+) or deletions (−) of the type caused by acridine dyes. Assume that all the mutations are located very near the beginning of the gene for protein X. In each case, indicate with an "OK" if you would expect the mutant protein to be nearly normal and with a "Not OK" if you would expect it to be obviously abnormal.

(a) −   (d) +/−/+   (g) +/+/+/−/+

(b) −/+   (e) +/−/+/−   (h) −/−/+/+/−/−

(c) −/−   (f) +/+/+   (i) −/−/−/−/−/−

17-4. Life with an Overlapping Code. Assume that the genetic code of all forms of life on planet QB9 consists of overlapping triplets of nucleotides, such that the translation apparatus shifts only one nucleotide at a time. Thus, the nucleotide sequence ABCDEF would be read on Earth as two codons (ABC, DEF) but on QB9 as four codons (ABC, BCD, CDE, DEF) and the start of two more. For each of the following kinds of mutations, briefly describe the effect it would have on the amino acid sequence of the coded protein (i) on Earth and (ii) on QB9. (Assume that the mutation occurs near the middle of the gene.)

(a) Single-nucleotide substitution

(b) Single-nucleotide deletion

(c) Deletion of three consecutive nucleotides

17-5. The Case of the Economical Phage. As mentioned in the chapter, bacteriophage ϕX174 is marvelously economical with its genetic information in that it contains some *overlapping* genes. For example, gene D is entirely contained within gene E, as shown in the diagram below. However, the nucleotides constituting gene D are read in a different reading frame from those in gene E. With reference to the diagram, label each of the following statements as either true (T) or false (F).

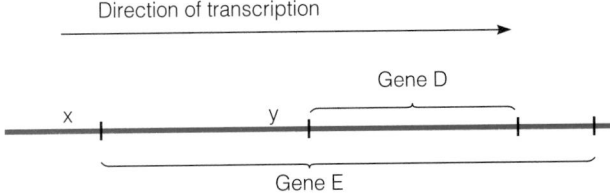

(a) A point mutation that causes an amino acid substitution in the D protein is likely also to cause an amino acid substitution in the E protein.

(b) Some mutations in gene D that result in the creation of new stop signals in that gene will also create new stop signals in gene E. (Mutations that create new stop signals, called *nonsense mutations,* are discussed further in Chapter 18.)

(c) In cells infected with a nonmutant form of the virus, some protein molecules will be made that correspond to the E protein at one end and to the D protein at the other end.

(d) The deletion of a single nucleotide in the region between x and y is more likely to affect the E protein than the D protein.

17-6. RNA Polymerases and Promoters. For each of the following statements about RNA polymerases, indicate with a B if the statement is true of the bacterial enzyme and with a I, II, or III if it is true of the respective eukaryotic RNA polymerase. A given statement may be true of any, all, or none of these enzymes.

(a) The enzyme is insensitive to α-amanitin.

(b) The enzyme catalyzes an exergonic reaction.

(c) All the primary transcripts must be processed before being used in translation.

(d) The enzyme may sometimes be found attached to an RNA molecule that also has ribosomes bound to it.

(e) The enzyme synthesizes rRNA.

(f) Transcription factors must bind to the promoter before the polymerase can bind.

(g) The enzyme adds a poly(A) sequence to mRNA.

(h) The enzyme moves along the DNA template strand in the $3' \rightarrow 5'$ direction.

(i) The enzyme synthesizes a product that is likely to acquire a 5′ cap.

(j) All promoters used by the enzyme lie mostly upstream of the transcriptional startpoint and are only partially transcribed.

(k) The specificity of transcription by the enzyme is determined by a subunit of the holoenzyme.

17-7. RNA Processing. The three major classes of RNA found in the cytoplasm of a typical eukaryotic cell are rRNA, tRNA, and mRNA. For each, indicate the following:

(a) Two or more kinds of processing to which that RNA has almost certainly been subjected.

(b) A processing activity unique to that RNA species.

(c) A processing activity that you would also expect to find for the same species of RNA from a bacterial cell.

17-8. Spliceosomes. The RNA processing carried out by spliceosomes in the eukaryotic nucleus involves a number of different kinds of protein and RNA molecules. For each of the following six components of the splicing process, indicate whether it is protein (P), RNA (R), or both (PR). Then briefly explain how each of the six fits into the process.

(a) snRNA

(b) spliceosome

(c) snRNP

(d) splice sites

(e) lariat

17-9. Antibiotic Inhibitors of Transcription. Rifamycin and actinomycin D are two antibiotics derived from the bacterium *Streptomyces*. Rifamycin binds to the β subunit of *E. coli* RNA polymerase and interferes with the formation of the first phosphodiester bond in the RNA chain. Actinomycin D binds to double-stranded DNA by intercalation (slipping in between neighboring base pairs).

(a) Which of the four stages in transcription would you expect rifamycin to affect primarily?

(b) Which of the four stages in transcription would you expect actinomycin D to affect primarily?

(c) Which of the two inhibitors is more likely to affect RNA synthesis in cultured human liver cells?

(d) Which of the two inhibitors would be more useful for an experiment in which it is necessary to block the initiation of new RNA chains without interfering with the elongation of chains that are already being synthesized?

17-10. Copolymer Analysis. In their initial attempts to determine codon assignments, Nirenberg and Matthei first used homopolymers and then used copolymers synthesized by the enzyme polynucleotide phosphorylase. This enzyme adds nucleotides randomly to the growing chain, but in proportion to their presence in the incubation mixture. By varying the ratio of precursor molecules in the synthesis of copolymers, Nirenberg and Matthei were able to deduce base compositions (but usually not actual sequences) of the codons that code for various amino acids. Suppose you carry out two polynucleotide phosphorylase incubations, with UTP and CTP present in both, but in different ratios. In incubation A the precursors are present in equimolar concentrations, but in incubation B there is three times as much UTP as CTP. The copolymers generated in both incubation mixtures are then used in an in vitro protein-synthesizing system, and the resulting polypeptides are analyzed for amino acid composition.

(a) What are the eight possible codons represented by the nucleotide sequences of the resulting copolymers in both incubation mixtures? What amino acids do these codons code for?

(b) For every 64 codons in the copolymer formed in incubation A, how many of each of the eight possible codons would you expect on the average? How many for incubation B?

(c) What can you say about the expected frequency of occurrence of the possible amino acids in the polypeptides obtained upon translation of the copolymers from incubation A? What about the polypeptides that result from translation of the incubation B copolymers?

(d) Explain what sort of information can be obtained by this technique.

(e) Would it be possible by this technique to determine that codons with 2 Us and 1 C code for phenylalanine, leucine, and serine?

(f) Would it be possible by this technique to decide which of the three codons with 2 Us and 1 C (UUC, UCU, CUU) correspond to each of the three amino acids mentioned in part e?

(g) Suggest a way to assign the three codons of part f to the appropriate amino acids of part e.

SUGGESTED READING

General References

Alberts, B., D. Bray, J. Lewis, M. Raff, K. Roberts, and J. D. Watson. *Molecular Biology of the Cell,* 3d ed. New York: Garland, 1994.

Gesteland, R. F., and J. F. Atkins, eds. *The RNA World.* Cold Spring Harbor, NY: Cold Spring Harbor Laboratory, 1993.

Griffiths, A., J. Miller, D. Suzuki, R. Lewontin, and W. Gelbart. *An Introduction to Genetic Analysis,* 5th ed. New York: W. H. Freeman, 1993.

Lewin, B. *Genes V.* New York: Oxford University Press, 1994.

McKnight, S. L. and K. R. Yamamoto, eds. *Transcriptional Regulation.* Cold Spring Harbor, NY: Cold Spring Harbor Laboratory, 1992.

Singer, M., and P. Berg. *Genes and Genomes.* Mill Valley, CA: University Science Books, 1990.

Stent, G. S. *Molecular Genetics: An Introductory Narrative.* San Francisco: W. H. Freeman, 1971.

Watson, J. D., N. H. Hopkins, J. W. Roberts, J. A. Steitz, and A. M. Weiner. *Molecular Biology of the Gene,* 4th ed. Menlo Park, CA: Benjamin/Cummings, 1987.

The Central Dogma

Rennie, J. DNA's new twists. *Sci. Amer.* 266 (March 1992): 122.

Varmus, H. Reverse transcription. *Sci. Amer.* 30 (September 1987): 56.

Weiner, A. M., P. L. Deininger, and A. Estratiadis. Nonviral retroposons: Genes, pseudogenes, and transposable elements generated by the reverse flow of genetic information. *Annu. Rev. Biochem.* 55 (1986): 631.

The Genetic Code

Crick, F. H. C. The genetic code. *Sci. Amer.* 207 (October 1962): 66.

Crick, F. H. C. The genetic code III. *Sci. Amer.* 215 (October 1966): 55.

Khorana, H. G. Nucleic acid synthesis in the study of the genetic code. In *Nobel Lectures: Physiology or Medicine (1963–1970),* 341. New York: American Elsevier, 1973.

Nirenberg, M. W. The genetic code II. *Sci. Amer.* 208 (March 1963): 80.

Transcription in the Prokaryotic Cell

Das, A. Control of transcription termination by RNA-binding proteins. *Annu. Rev. Biochem.* 62 (1993): 893.

Kainz, M., and J. Roberts. Structure of transcription elongation complexes in vivo. *Science* 255 (1992): 838.

McClure, W. Mechanism and control of transcription initiation in prokaryotes. *Annu. Rev Biochem.* 54 (1985): 171.

Nudler, E., A. Goldfarb, and M. Kashlev. Discontinuous mechanism of transcription elongation. *Science* 265 (1994): 793.

Ross, W. et al. A third recognition element in bacterial promoters: DNA binding by the a subunit of RNA polymerase. *Science* 262 (1993): 1407.

Transcription in the Eukaryotic Cell

Buratowski, S. The basics of basal transcription by RNA polymerase II. *Cell* 77 (1994): 1.

Comai, L., et al. Reconstitution of transcription factor SL1: Exclusive binding of TBP by SL1 or TFIID subunits. *Science* 266 (1994): 1966.

Conaway, R. C., and J. W. Conaway. General initiation factors for RNA polymerase II. *Annu. Rev. Biochem.* 62 (1993): 161.

Goodrich, J. A., and R. Tjian. TBF-TAF complexes: Selectivity factors for eukaryotic transcription. *Curr. Opin. Cell Biol.* 6 (1994): 403.

Nikolov, D. B., et al. Crystal structure of TFIID TATA-box binding protein. *Nature* 360 (1992): 40.

Tjian, R. Molecular machines that control genes. *Sci. Amer.* 272 (February 1995): 54.

Young, R. A. RNA polymerase II. *Annu. Rev. Biochem.* 60 (1991): 689.

Zawel, L., and D. Reinberg. Initiation of transcription by RNA polymerase II: A multi-step process. *Prog. Nucl. Acid Res. Mol. Biol.* 44 (1993): 67.

RNA Processing

Cattaneo, R. Different types of messenger RNA editing. *Annu. Rev. Genet.* 25 (1991): 71.

Dreyfuss, G., et al. hnRNP proteins and the biogenesis of mRNA. *Annu. Rev. Biochem.* 62 (1993): 289.

Green, M. R. Biochemical mechanisms of constitutive and regulated pre-mRNA splicing. *Annu. Rev. Cell Biol.* 7 (1991): 559.

Nilsen, T. W. RNA-RNA interactions in the spliceosome: Unraveling the ties that bind. *Cell* 78 (1994): 1.

Seiwert, S. D., and K. Stuart. RNA editing: Transfer of genetic information from gRNA to precursor mRNA in vitro. *Science* 266 (1994): 114.

Spector, D. L. Macromolecular domains within the cell nucleus. *Annu. Rev. Cell Biol.* 9 (1993): 265.

Steitz, J. A. Snurps. *Sci. Amer.* 258 (June 1988): 58.

Wise, J. A. Guide to the heart of the spliceosome (Perspectives). *Science* 262 (1993): 1978. (See also three related research articles in the same issue.)

Yang, J.-H., R. Cedergren, and B. Nadal-Ginard. Catalytic activity of an RNA domain derived from the U6-U4 RNA complex. *Science* 263 (1994): 77.

18

Gene Expression:
II. Protein Synthesis and Sorting

In the previous chapter, we took gene expression from DNA to RNA, including both transcription and the processing of the RNA transcript. For genes encoding ribosomal RNA and transfer RNA (and certain other small RNA molecules), RNA is the ultimate expression of the gene. But for the thousands of other genes in an organism's genome, the ultimate gene product is protein. This chapter describes how the messenger RNA produced from protein-coding genes is translated into polypeptides, how polypeptides become functional proteins, and how proteins reach the destinations where they carry out their functions.

Translation, the first and most important phase of protein synthesis, is inherently complicated, involving a change in "language" from the nucleotide order in an mRNA molecule to the amino acid sequence of a polypeptide. In essence, the order of nucleotides, read as triplet codons, specifies the order in which incoming amino acids are added to a growing polypeptide chain. Ribosomes serve as the intracellular sites of translation, and RNA molecules are the agents that ensure insertion of the correct amino acids at each position in the polypeptide. We will start by surveying the cell's cast of characters for performing translation, before describing in detail the steps of the process.

Translation: The Cast of Characters

The cellular apparatus that translates mRNA into polypeptide includes the *tRNA* molecules that align amino acids in the correct order along the mRNA template, the *aminoacyl-tRNA synthetases* that attach amino acids to their appropriate tRNA molecules, and a number of *protein factors* that participate at several stages in the translation process. The *ribosomes* play a central role. They orient the mRNA and amino acid–carrying tRNAs in the proper relation to each

other so that the genetic code is read accurately, and they catalyze the formation of the peptide bonds that link the amino acids into a polypeptide.

Ribosomes

As we described in Chapter 4, ribosomes are particles made of RNA and protein that are found in the cytoplasm of both prokaryotic and eukaryotic cells, as well as in the matrix of mitochondria and the stroma of chloroplasts. In the cytoplasm, they exist both free in the cytosol and bound to the membrane of the endoplasmic reticulum and nuclear envelope. Ribosomes are often considered organelles, but they differ from true organelles in not being surrounded by a membrane. Prokaryotic and eukaryotic ribosomes are similar in structure, but they are not identical. Prokaryotic ribosomes are less massive (2.5×10^6 instead of 4.2×10^6 Da), contain fewer proteins, have smaller RNA molecules (and one fewer), and are sensitive to different inhibitors of protein synthesis.

Electron microscopic techniques have revealed the shape of the prokaryotic ribosome shown in Figure 18-1. Like all ribosomes, it consists of two dissociable subunits of different sizes. A complete prokaryotic ribosome, with a sedimentation coefficient of about 70S, consists of a 30S subunit and a 50S subunit; its eukaryotic equivalent is an 80S particle made up of a 40S subunit and a 60S subunit. Table 18-1 lists some of the properties of prokaryotic and eukaryotic ribosomes and their subunits. The RNA and protein components of each ribosomal subunit self-assemble spontaneously, both in vivo and in vitro. However, the two subunits only come together when mRNA is present, as we will see.

Functionally, ribosomes have sometimes been called the "workbenches" of protein synthesis, but their very active role in polypeptide synthesis makes "factory" a more apt label. Furthermore, the earlier view that the role of rRNA was simply to provide a structural scaffold for functional ribosomal proteins is now known to be false. Today we realize that the reverse is probably true—that rRNA performs most or all of the ribosome's key functions.

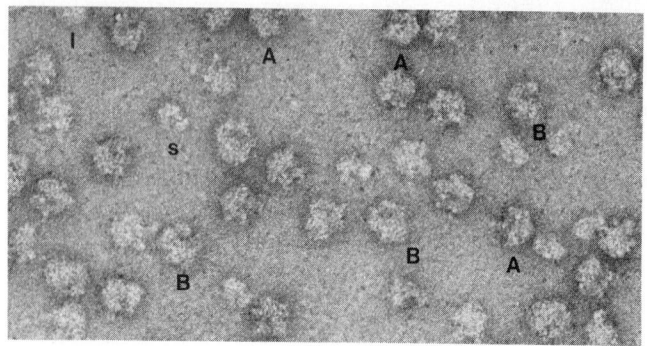

(a) Prokaryotic ribosomes and free subunits

50 nm

Figure 18-1 The Prokaryotic Ribosome. **(a)** This electron micrograph includes both complete ribosomes and individual subunits (TEM). "A" and "B" indicate two different views of the complete ribosomes; "l" and "s" indicate free subunits, large and small respectively. **(b)** These models are based on such micrographs. The prokaryotic ribosome is about 25 nm in diameter. (The eukaryotic ribosome is roughly similar in shape and about 30 nm in diameter.)

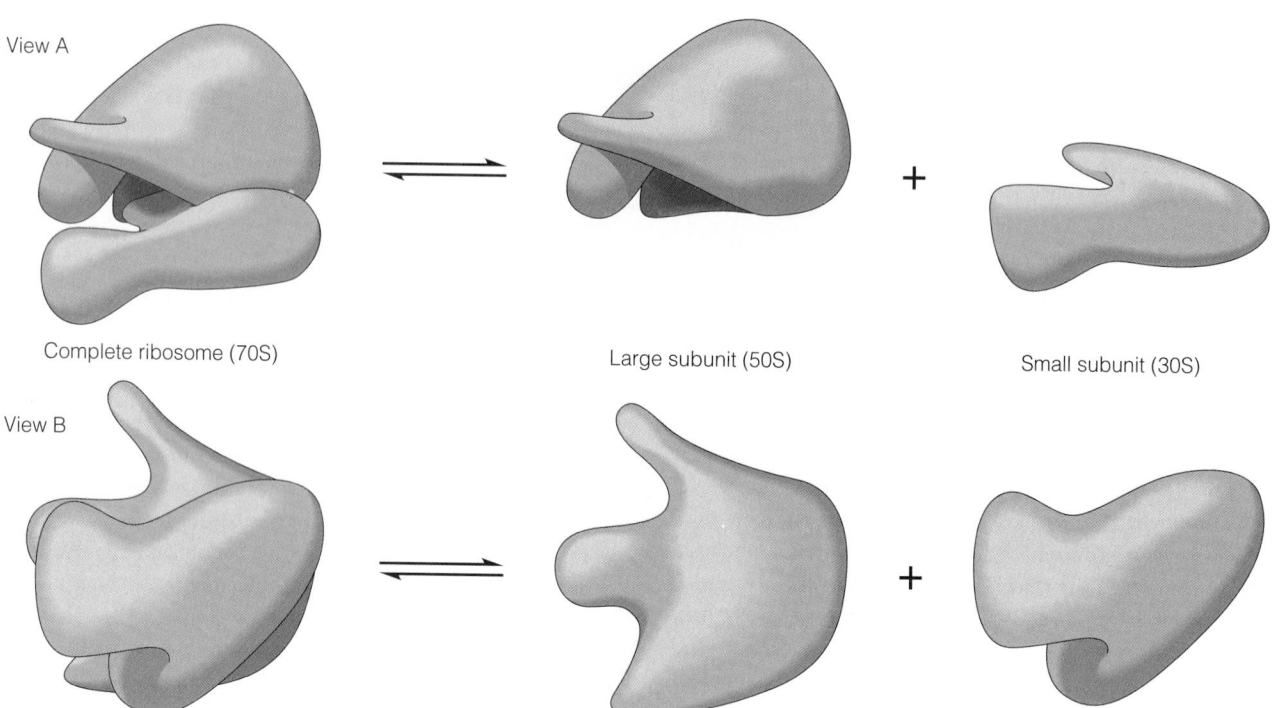

View A

Complete ribosome (70S)   Large subunit (50S)   Small subunit (30S)

View B

(b) Two views of the prokaryotic ribosome and its subunits

Table 18-1   Properties of Prokaryotic and Eukaryotic Ribosomes

| Source | Size of Ribosomes | | Subunit | Subunit Size | | Subunit Proteins | Subunit RNA | |
	S Value*	Mol. Wt.		S Value	Mol. Wt.		S Value	Nucleotides
Prokaryotic cells	70S	2.5×10^6	Large	50S	1.6×10^6	34	23S	2900
							5S	120
			Small	30S	0.9×10^6	21	16S	1540
Eukaryotic cells	80S	4.2×10^6	Large	60S	2.8×10^6	About 45	(25–)28S	≤4700
							5.8S	160
							5S	120
			Small	40S	1.4×10^6	About 33	18S	1900

*If you are surprised that the S values of the subunits do not add up to that of the ribosome, recall that an S value is a measure of the velocity at which a particle sediments upon centrifugation and is only indirectly related to the mass of the particle.

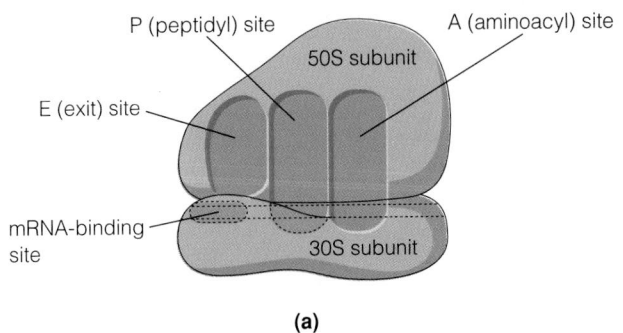

(a)

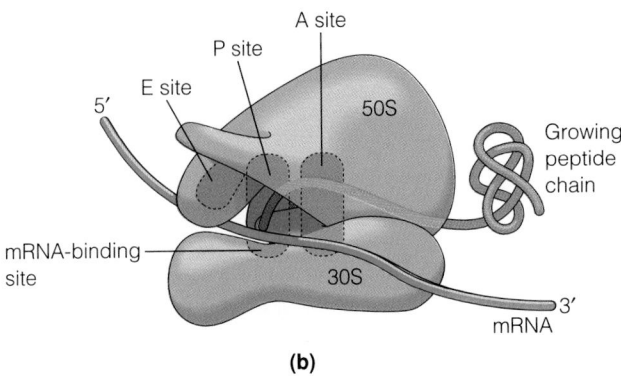

(b)

Figure 18-2   Important Binding Sites on the Prokaryotic Ribosome. The A (aminoacyl) and P (peptidyl) sites are cavities on the ribosome where charged (amino acid–carrying) tRNA molecules bind during polypeptide synthesis. The more recently discovered E (exit) site is the site from which discharged tRNAs leave the ribosome. The mRNA-binding site binds a particular nucleotide sequence near the 5′ end of the mRNA, positioning the mRNA in the proper position for the translation of its first codon. **(a)** The diagrammatic representation of a ribosome that is used in this chapter. **(b)** A more realistic representation. The binding sites are all located at or near the interface between the large and small subunits.

Four specific sites on the ribosome are particularly important for protein synthesis (Figure 18-2). These are an **mRNA-binding site** and three sites where tRNA can bind: an **A (aminoacyl) site,** where the incoming amino acid–carrying tRNA usually binds; a **P (peptidyl) site,** where the tRNA carrying the growing polypeptide chain resides; and the **E (exit) site,** from which tRNAs leave the ribosome after they have discharged their amino acids. How these sites function in the process of translation will become clear in a few pages.

Transfer RNA

Because amino acids themselves are unable to recognize different nucleotide codons along the mRNA template, they must be brought to the correct codons by other molecules, which Francis Crick originally termed *adaptors*. Transfer RNA molecules are these adaptors, the intermediaries between the nucleotide sequence of the messenger and the

amino acid sequence of the polypeptide. Appropriate to this role, tRNA molecules have two kinds of specificity. Each tRNA molecule recognizes a specific codon on the mRNA, and each carries an amino acid that is specifically chosen to match that codon according to the dictates of the genetic code.

As Figure 18-3 indicates, the amino acid is always linked by an ester bond to the 3′ OH group of the nucleotide (always an A) at one end of the molecule. (Selection of the appropriate amino acid for a given tRNA is the responsibility of the aminoacyl-tRNA synthetase enzyme that catalyzes formation of the ester bond; we discuss these enzymes shortly.) The amino acid that can attach to a given tRNA is indicated by a superscript. The tRNA molecules specific for alanine, for example, are identified as tRNA^{Ala}. When a tRNA has its amino acid attached, it is called an **aminoacyl-tRNA** (e.g., alanyl-tRNA^{Ala}). The tRNA is said to be in its *charged* form, and the amino acid is said to be *activated*.

The other aspect of tRNA specificity involves recognition of the appropriate codon. This is the role of the **anticodon,** a trinucleotide sequence located on one of the loops of the tRNA, as Figure 18-3 shows. The anticodon is always complementary to the codon, which it recognizes by base pairing. Take careful note of the convention used in representing codons and anticodons: Codons on mRNA molecules are usually written in the 5′ → 3′ direction, whereas anticodons on tRNA molecules are usually represented in the 3′ → 5′ orientation. Thus, one of the codons for alanine is 5′-GCC-3′ and the corresponding anticodon is 3′-CGG-5′.

Since the genetic code has 61 different codons that specify amino acids, you might expect to find 61 different tRNA molecules involved in protein synthesis, each responsible for recognizing a different codon. However, the number of different tRNA molecules is actually significantly less than 61, a situation made possible by the fact that some tRNA molecules are able to recognize several different codons. This, in turn, is possible largely because the complementary binding of codon to anticodon is not as strict as base pairing is in other contexts, such as the DNA duplex.

Specifically, there is some flexibility in the binding of the base in the third position (the 5′ end) of the anticodon to that in the complementary position (the 3′ end) of the codon. Crick referred to this flexibility as **"wobble"** when he first postulated it, and the term has stuck. The wobble rules are shown in Table 18-2. Notice that wobble is allowed with G or U in the third position of the anticodon, but not with C or A.

Notice also that tRNA molecules contain **inosine (I),** an unusual base not found in DNA or in other RNA molecules. (Actually, "inosine" refers to the nucleoside, the base-ribose unit.) The structure of inosine is shown in Figure 18-4a (page 573). Inosine is often present in the third position of the anticodon. In *E. coli,* in fact, A is never found in the third position of anticodons but is almost always modified to I. Inosine is the "wobbliest" of all third-position bases, since it can pair with U, C, or A (Figure 18-4b). For example, a tRNA with the anticodon UAI could recognize the codons AUU,

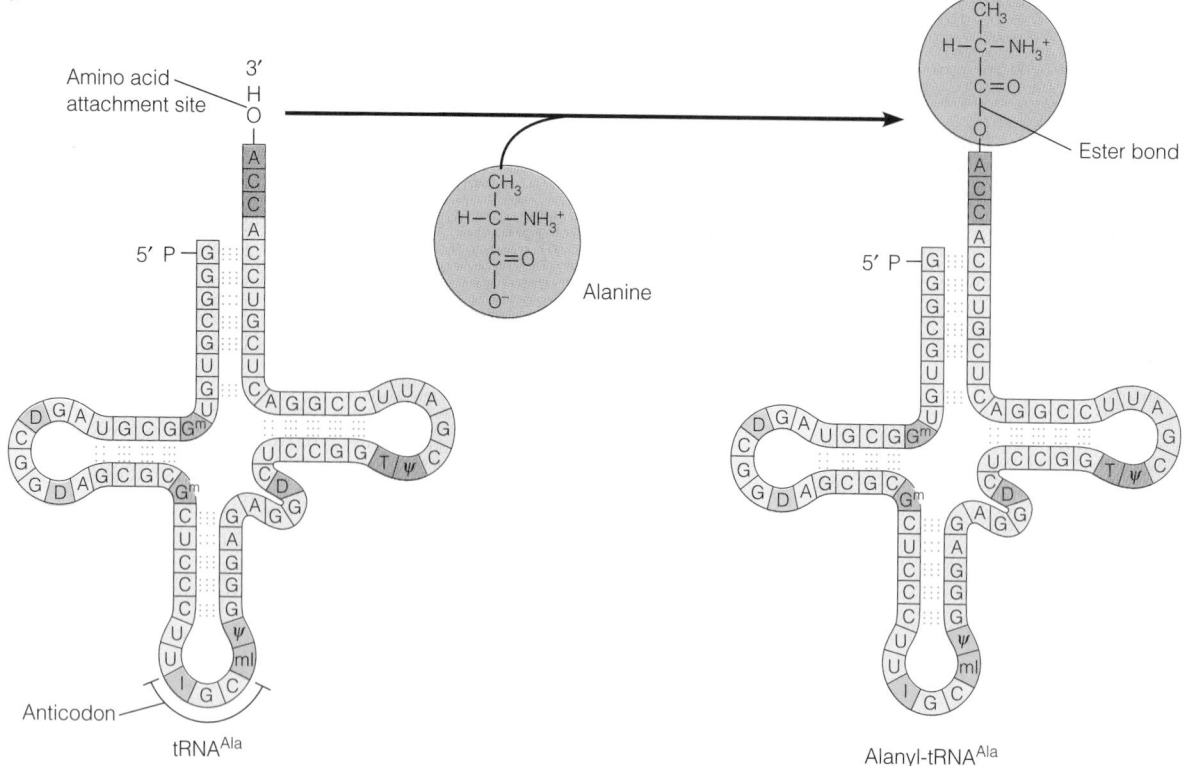

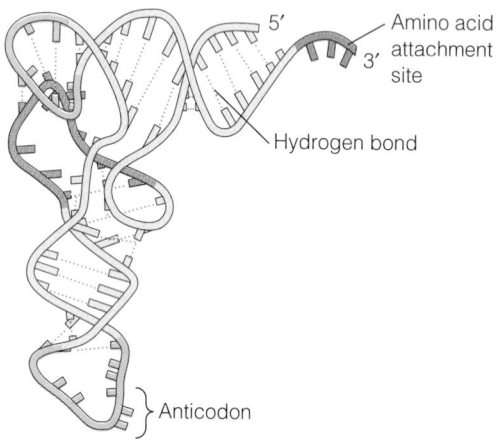

(a) Secondary structure of tRNA, before and after amino acid attachment

(b) Tertiary structure of tRNA

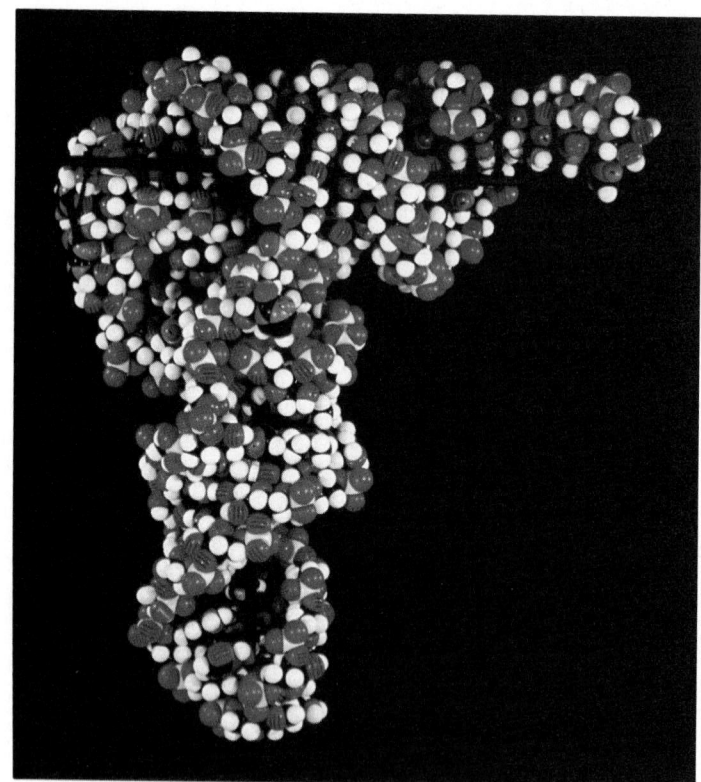

Figure 18-3 Sequence, Structure, and Aminoacylation of a tRNA. **(a)** Yeast alanine tRNA, like all tRNA molecules, contains three major loops, four base-paired regions, an anticodon triplet, and a 3′ terminal sequence of CCA, to which the appropriate amino acid can be attached by an ester bond. Modified bases are dark colored, and their names (as nucleosides) are abbreviated I for inosine, mI for methylinosine, D for dihydrouridine, T for ribothymidine, ψ for pseudouridine, and Gᵐ for methylguanosine. (For the significance of an inosine in the anticodon, see Figure 18-4 and the wobble rules of Table 18-2.) **(b)** In the L-shaped tertiary structure of tRNA, the

amino acid attachment site is at one end and the anticodon at the other. Color coding of the tRNA loops correlates the secondary (two-dimensional) structures of part a and the tertiary (three-dimensional) structure on the left in part b. The tRNA model on the right in part b shows the individual atoms. The structure shown here is characteristic of all tRNAs.

AUC, and AUA, which code for the amino acid isoleucine (Ile; see Figure 17-7).

As a result of wobble, fewer tRNA molecules are required for some amino acids than the number of codons that specify those amino acids. In the case of isoleucine, a cell can translate all three codons with a single tRNA molecule provided only that the tRNA has UAI as its anticodon. Similarly, the six codons for the amino acid leucine (UUA, UUG, CUU, CUC, CUA, and CUG) require only three tRNA molecules because of wobble.

Aminoacyl-tRNA Synthetases

The enzymes responsible for linking amino acids to their cognate tRNA molecules are called **aminoacyl-tRNA synthetases.** There are 20 different aminoacyl-tRNA synthetases in the cell, one for each of the 20 amino acids used in protein synthesis. The same enzyme can recognize all the tRNAs for the same amino acid, when there is more than one.

The exquisite specificity of the synthetase enzymes ensures the joining of the correct amino acid to each tRNA with a high degree of accuracy. The linkage is an ester bond, and the reaction involves the hydrolysis of ATP to AMP and pyrophosphate:

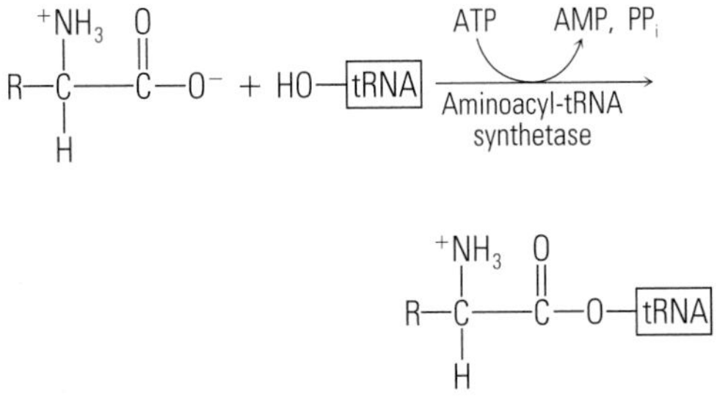

Table 18-2   The Wobble Rules for Codon-Anticodon Pairing

Normal Base-Pairing Rules	
Base Present in Anticodon of tRNA	Base Recognized in Codon of mRNA
C	G
A	U
U	A
G	C

Wobble Base-Pairing Rules	
Base Present in 5′ Position of tRNA Anticodon	Bases Recognized in 3′ Position of mRNA Codon
C	G
A	U
U	A, G
G	C, U
I*	U, C, A

*I stands for inosine; for its structure, see Figure 18-4a.

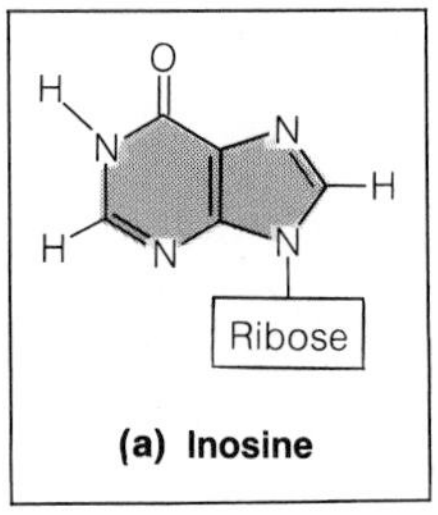

(a) Inosine

Figure 18-4   The Structure of Inosine and Its Base-Pairing Possibilities.
(a) Inosine (I) is an unusual nucleoside found in tRNA molecules. **(b)** Its structure allows it to base-pair with U, C, or A, such that tRNA molecules with an I at the 5′ position of the anti-codon can recognize codons that end with any of these three nucleotides. For example, the tRNA shown in Figure 18-3 can recognize the codons GCU, GCC, or GCA, all of which code for alanine.

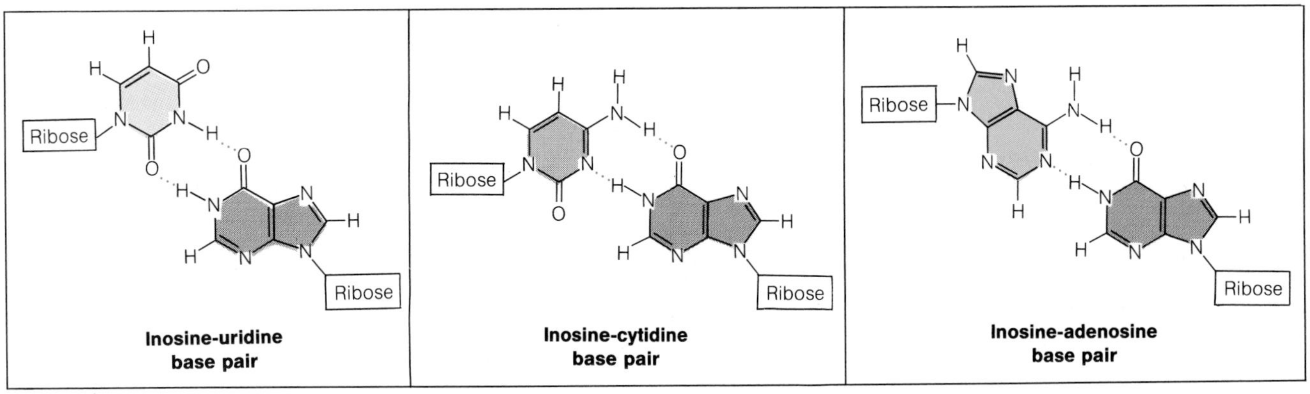

| Inosine-uridine base pair | Inosine-cytidine base pair | Inosine-adenosine base pair |

(b)

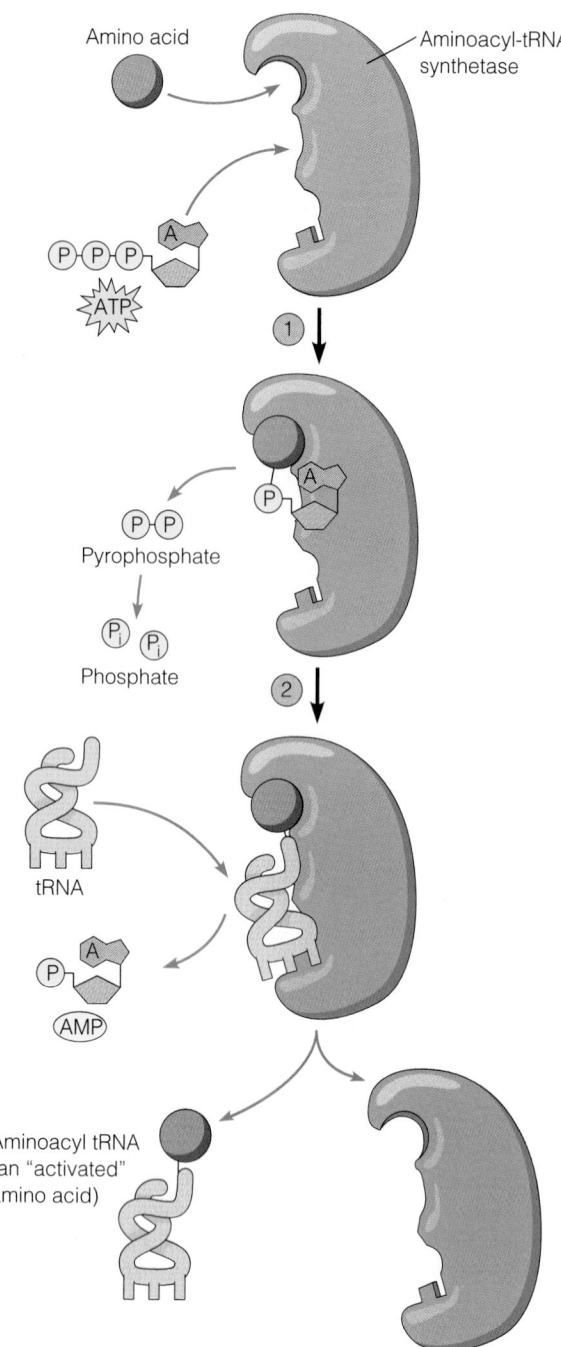

(a) Process of amino acid activation by aminoacyl-tRNA synthetase

Figure 18-5   Aminoacyl-tRNA Synthetase.   (a) In two chemical steps, this enzyme catalyzes the formation of an ester bond between the carboxyl group of an amino acid and the 3′ OH of the appropriate tRNA. ① The amino acid and a molecule of ATP enter the active site of the enzyme. Simultaneously, ATP loses pyrophosphate, and the resulting AMP bonds covalently to the amino acid. The pyrophosphate is hydrolyzed to $2P_i$. ② The tRNA covalently bonds to the amino acid, displacing the AMP, and the aminoacyl tRNA is released from the enzyme. **(b)** A molecule of tRNA binding to an enzyme molecule. This computer graphic model of a tRNA molecule (red and yellow) bound to a molecule of aminoacyl-tRNA synthetase (blue) shows their true relative sizes. The green molecule is ATP. The amino acid is not shown; it would be less than half the size of the ATP.

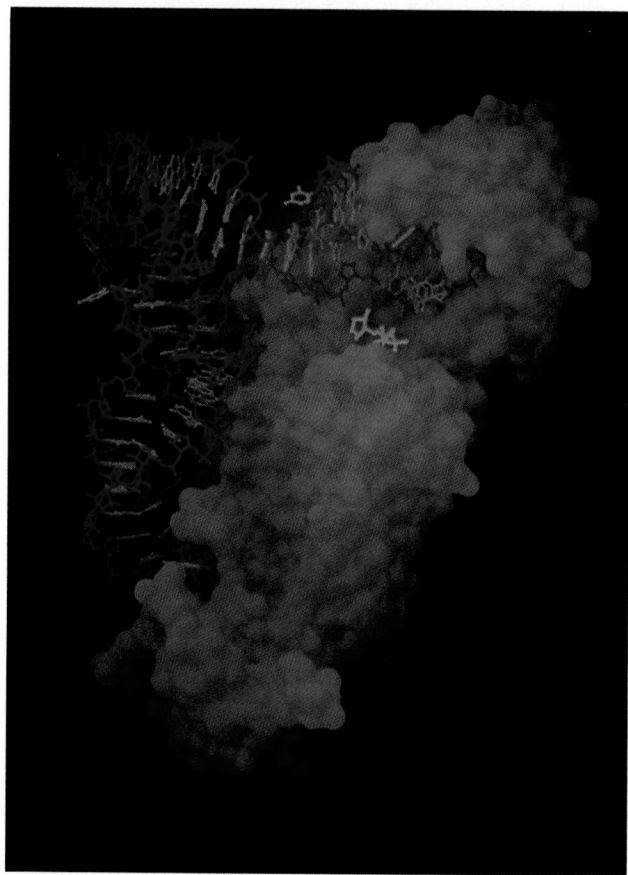

(b) A model of tRNA bound to aminoacyl-tRNA synthetase

Figure 18-5a outlines the steps by which this reaction occurs. The driving force for the reaction is actually provided by the hydrolysis of pyrophosphate to $2P_i$.

In the product, aminoacyl tRNA, the ester bond linking the amino acid to the tRNA is said to be a "high-energy" bond. This simply means that hydrolysis of this bond releases sufficient energy to drive formation of the peptide bond that will eventually join the amino acid to a growing polypeptide chain. The process of aminoacylation of a tRNA molecule is therefore also called **amino acid activation**, be-

cause it not only links an amino acid to its proper tRNA but also activates it for subsequent peptide bond formation.

Protein Factors

In addition to the aminoacyl-tRNA synthetase proteins and the protein constituents of the ribosomes, translation also requires the participation of a number of protein factors. Some are involved in initiation of the translation process, others in elongation of the polypeptide, and still others in

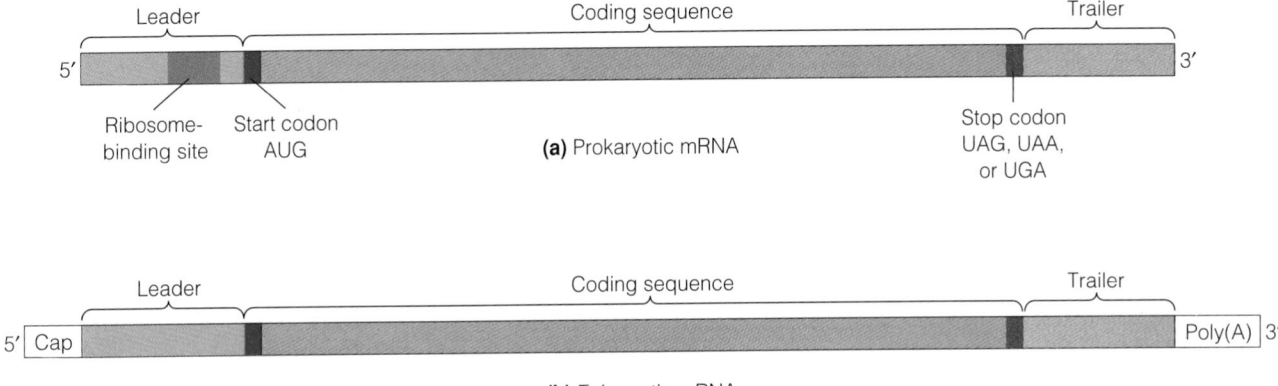

Leader — Coding sequence — Trailer

5'

Ribosome-
binding site

Start codon
AUG

(a) Prokaryotic mRNA

Stop codon
UAG, UAA,
or UGA

3'

Leader — Coding sequence — Trailer

5' Cap

Poly(A) 3'

(b) Eukaryotic mRNA

Figure 18-6 Messenger RNA. **(a)** A prokaryotic mRNA molecule encoding a single polypeptide has the features shown here. (A polycistronic prokaryotic mRNA would generally have a set of these features for each gene.) **(b)** A eukaryotic mRNA molecule has, in addition, a 5' cap and a 3' poly(A) tail. It lacks a ribosome-binding site (a nucleotide sequence also called a Shine-Dalgarno sequence, after its discoverers).

termination. We will describe the functions of these factors during our discussion of the steps of translation.

mRNA

The heart of messenger RNA is, of course, its message—the sequence of nucleotides that encode a polypeptide. However, an mRNA molecule also has sequences at either end that are not translated, as shown in Figure 18-6. The nontranslated sequence at the 5' end of mRNA is called the **leader,** because it precedes the **start codon,** the first codon to be translated. The start codon is usually AUG; it can be GUG in eukaryotes. The nontranslated sequence at the 3' end is the **trailer,** and it follows the **stop codon,** which can be UAG, UAA, or UGA. The leader and trailer sequences range from about 30 to several hundred nucleotides in length. While these sequences are not themselves translated, their presence is essential for translation of the message. In addition, eukaryotic mRNAs have 5' caps and 3' poly(A) tails, as we saw in Chapter 17.

In eukaryotes, each mRNA molecule generally encodes a single polypeptide. In prokaryotes, however, some mRNAs encode several polypeptides, usually with related functions in the cell. The clusters of genes that give rise to such mRNA molecules are thus single transcription units, called *operons;* they are regulated coordinately, as we will discuss in Chapter 19. On polygenic mRNAs, the genes are often separated by spacer regions.

The Stages of Translation

Like transcription, translation is easier to understand when it is broken down into stages. The major stages in translation are initiation, elongation of the polypeptide chain, and chain termination. Figure 18-7 provides an overview of the three stages. In *initiation,* the mRNA, the two ribosomal subunits, and a tRNA carrying the first amino acid of the polypeptide to be synthesized come together to form a single, large complex. The anticodon of the tRNA hydrogenbonds to the start codon of the message. During *elongation,* other tRNAs bring succeeding amino acids to the complex,

one by one. These are attached sequentially to the first amino acid, and the polypeptide chain elongates. As each amino acid is added, the mRNA moves with respect to the ribosome, so that the next codon is positioned for translation. This is equivalent to saying that the ribosome moves along the mRNA. Eventually, the ribosome reaches a stop codon, which is recognized by a *release factor.* This interaction triggers the *termination* of translation, and the components of the translational apparatus come apart, releasing the completed polypeptide.

The three stages of translation are similar in prokaryotes and eukaryotes. The descriptions that follow focus mainly on the details of the prokaryotic system, which is relatively well understood. From what is known to date, the aspects of translation that are unique to prokaryotes or eukaryotes are mostly confined to the initiation stage, as we describe in the next section.

Initiation

Prokaryotic Initiation. The initiation of translation in prokaryotes is illustrated in Figure 18-8 (page 577). It has three steps. First, three **initiation factors (IF)**—called IF1, IF2, and IF3—bind to the small ribosomal subunit, with GTP attaching to IF2. Next, a molecule of charged tRNA and the mRNA bind to the ribosomal subunit. The tRNA is usually a special one, called **initiator tRNA.** In bacteria, it is tRNA$^{f\text{-Met}}$, carrying **N-formylmethionine (f-Met),** a modified methionine with a formyl group on its nitrogen atom (Figure 18-9, page 577). N-formylmethionine is the amino acid with which almost every polypeptide is initiated in prokaryotes. (Eukaryotes also have a special initiator tRNA, and it carries methionine, but the amino acid is not formylated in these organisms.) IF2 (with its bound GTP) helps position the N-formylmethionyl tRNA$^{f\text{-Met}}$ in the P site of the 30S ribosomal subunit. This is the only aminoacyl tRNA that can bind to the P site.

The anticodon of tRNA$^{f\text{-Met}}$ base-pairs with the initiation codon, AUG, on the mRNA. This AUG is brought into the correct position when the mRNA binds to the 30S

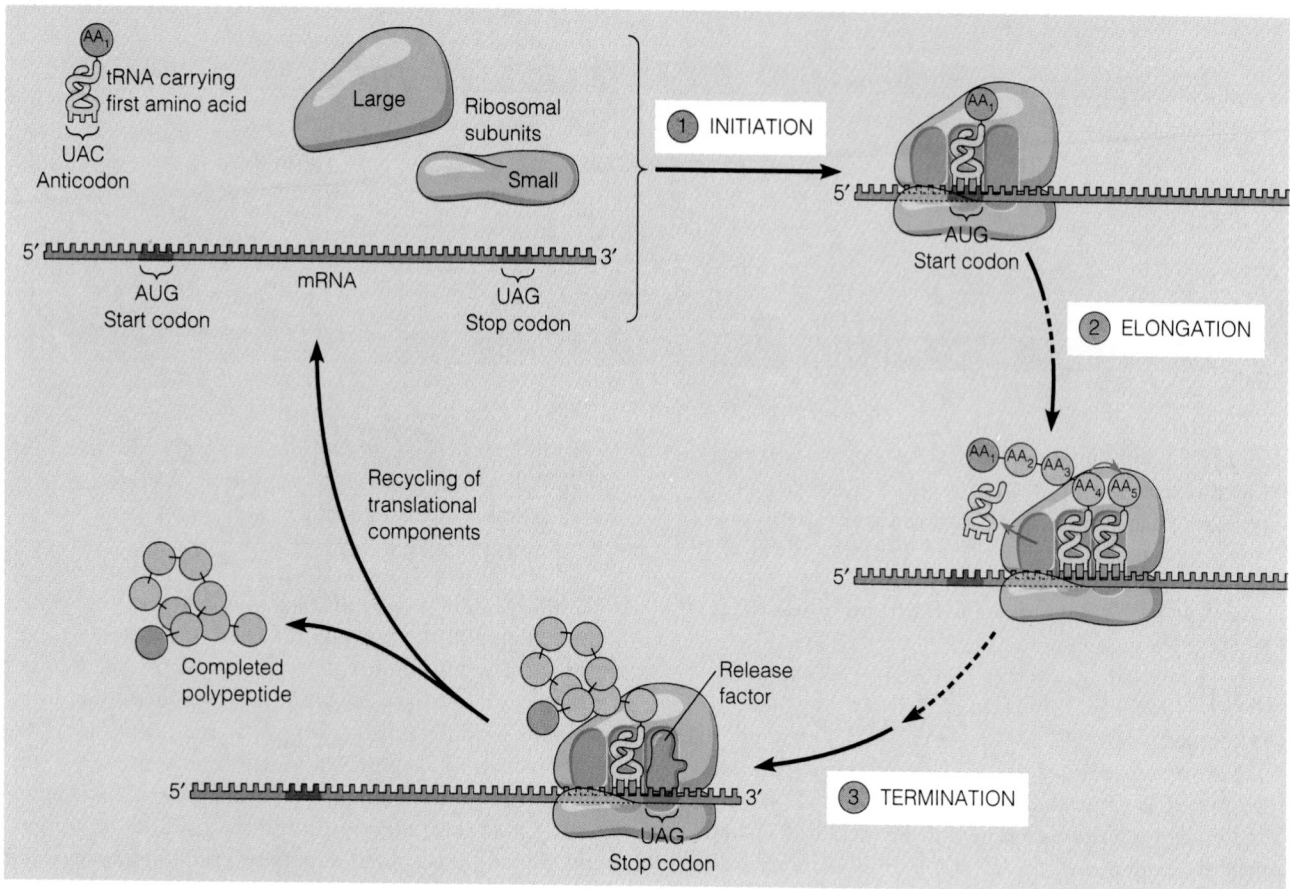

Figure 18-7 An Overview of Translation. Translation occurs in three stages. ① In initiation, the components of the translational apparatus come together with a molecule of mRNA. A tRNA carrying the first amino acid binds to the start codon. ② In elongation, amino acids are conveyed to the mRNA by tRNAs and are added, one by one, to a growing polypeptide chain. ③ In termination, a stop codon on the mRNA is recognized by a protein release factor, and the translational apparatus comes apart, releasing a completed polypeptide.

ribosomal subunit by means of the mRNA's **ribosome-binding site** (also known as the **Shine-Dalgarno sequence,** after its discoverers). This is a nucleotide sequence on the mRNA consisting of a string of 3–9 purine nucleotides (often AGGA) located slightly upstream of the initiation codon. The purine sequence base-pairs with a pyrimidine-rich sequence at the 3′ end of 16S rRNA, the ribosome's *mRNA-binding site* referred to earlier. It is the ability of the initiation region of the mRNA to base-pair simultaneously with both 16S rRNA and tRNA$^{f\text{-Met}}$ that distinguishes the start AUG from AUG codons elsewhere in the mRNA.

The **30S initiation complex** formed in this way is joined, in the final step of initiation, by a free 50S ribosomal subunit, generating the **70S initiation complex** needed for the growth of a polypeptide. As indicated in step 3 of Figure 18-8, binding of the 50S subunit is driven by hydrolysis of the GTP that was bound to IF2. By now, all three initiation factors have been released.

We can summarize the complicated process of prokaryotic initiation by representing the three steps of Figure 18-8 as reactions:

30S subunit + IF1 + IF3 + IF2 + GTP →
$\qquad$ 30S–IF1–IF3–IF2-GTP complex

30S–IF1–IF3–IF2-GTP complex + f-Met-tRNA$^{f\text{-Met}}$ + mRNA →
$\qquad$ 30S initiation complex + IF1 + IF3

30S initiation complex + 50S subunit →
$\qquad$ 70S initiation complex + IF2 + GDP + P$_i$

30S subunit + f-Met-tRNA$^{f\text{-Met}}$ + mRNA + 50S subunit + GTP →
$\qquad$ **70S initiation complex** + GDP + P$_i$

Eukaryotic Initiation. The initiation process in eukaryotes involves different initiation factors (called *eIF* in eukaryotes) and a somewhat different pathway for assembling the initiation complex. The initiation factor *eIF2*, with GTP already attached, binds to the initiator aminoacyl tRNA, methionyl tRNA^{Met}, *before* the tRNA binds to the small ribosomal subunit. The resulting small-subunit complex then binds to the 5′ end of the mRNA, recognizing the 5′ cap; there is no mRNA sequence corresponding to the ribosome-binding site on prokaryotic mRNA. The small subunit, with the initiator tRNA in tow, slides along the mRNA and usually begins translation at the first AUG triplet it comes to.

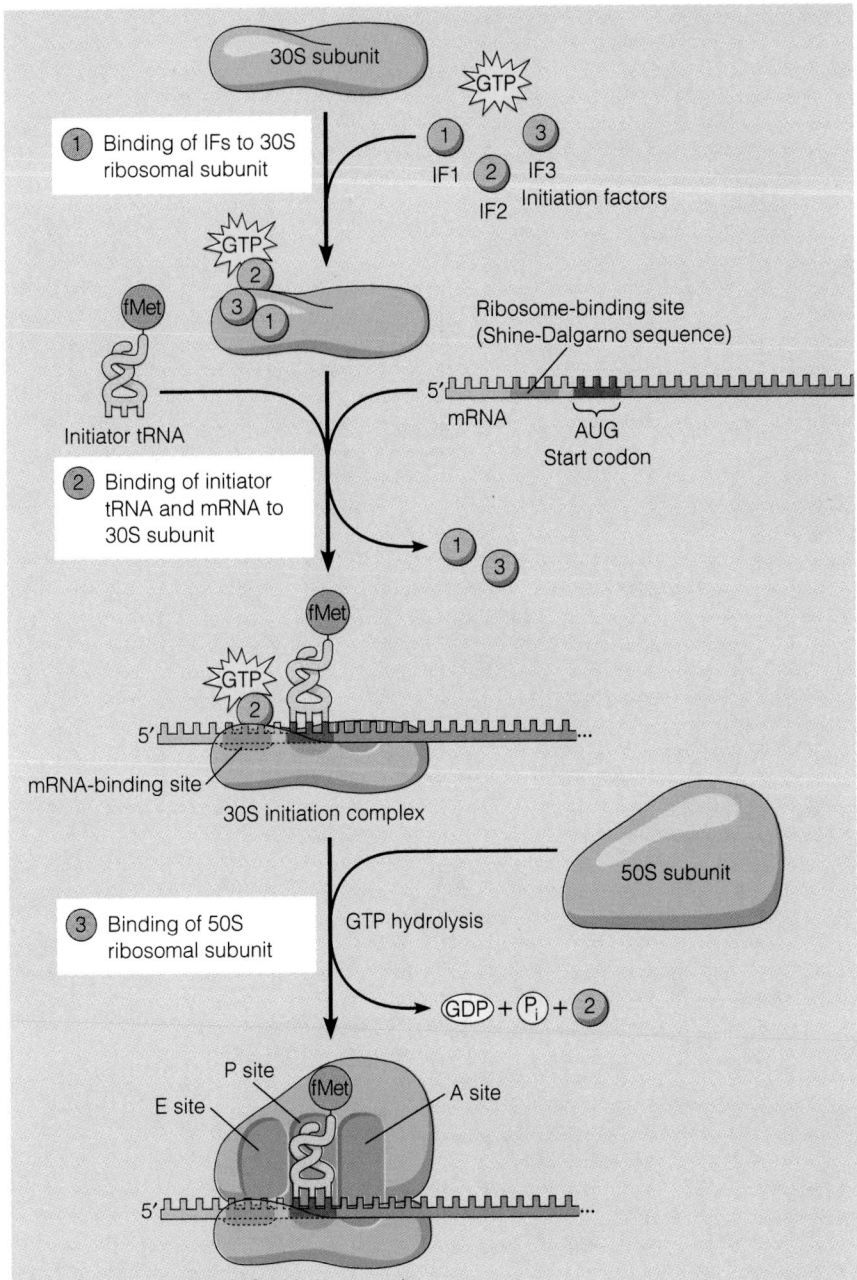

Figure 18-8   Initiation of Translation in Prokaryotes. The formation of the 70S translation initiation complex occurs in three steps. ① Three initiation factors (IF) and GTP bind to the small ribosomal subunit. ② The initiator aminoacyl tRNA and mRNA are attached. The mRNA-binding site is on the 16S rRNA of the small ribosomal subunit. ③ The large ribosomal subunit joins the complex. The resulting 70S initiation complex has f-Met-tRNA$^{\text{f-Met}}$ residing in the ribosome's P site.

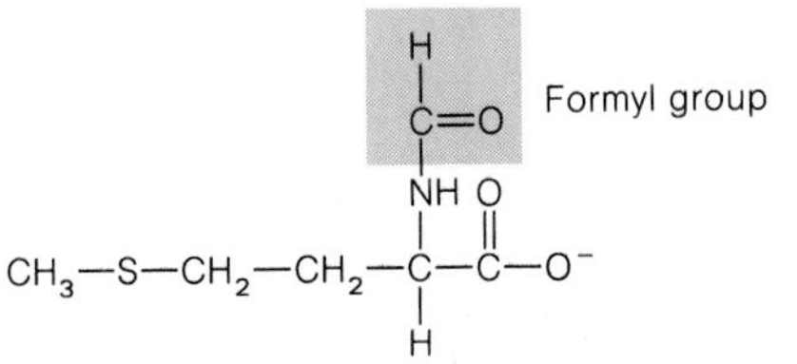

Formyl group

Figure 18-9   The Structure of N-Formylmethionine.
N-formylmethionine (f-Met) is the modified amino acid with which every polypeptide is initiated in prokaryotes.

The nucleotides on either side of the eukaryotic start codon seem to be involved in its recognition; ACCAUGG is a common start sequence, where the underlined triplet is the actual start codon. When the tRNA base-pairs with the start codon, the large ribosomal subunit joins the complex.

Chain Elongation

Once initiated, a polypeptide chain is elongated by successive additions of amino acids. Each such addition involves the formation of a new peptide bond. Elongation proceeds in prokaryotes via the series of steps shown in Figure 18-10. First, an incoming aminoacyl tRNA base-pairs with the

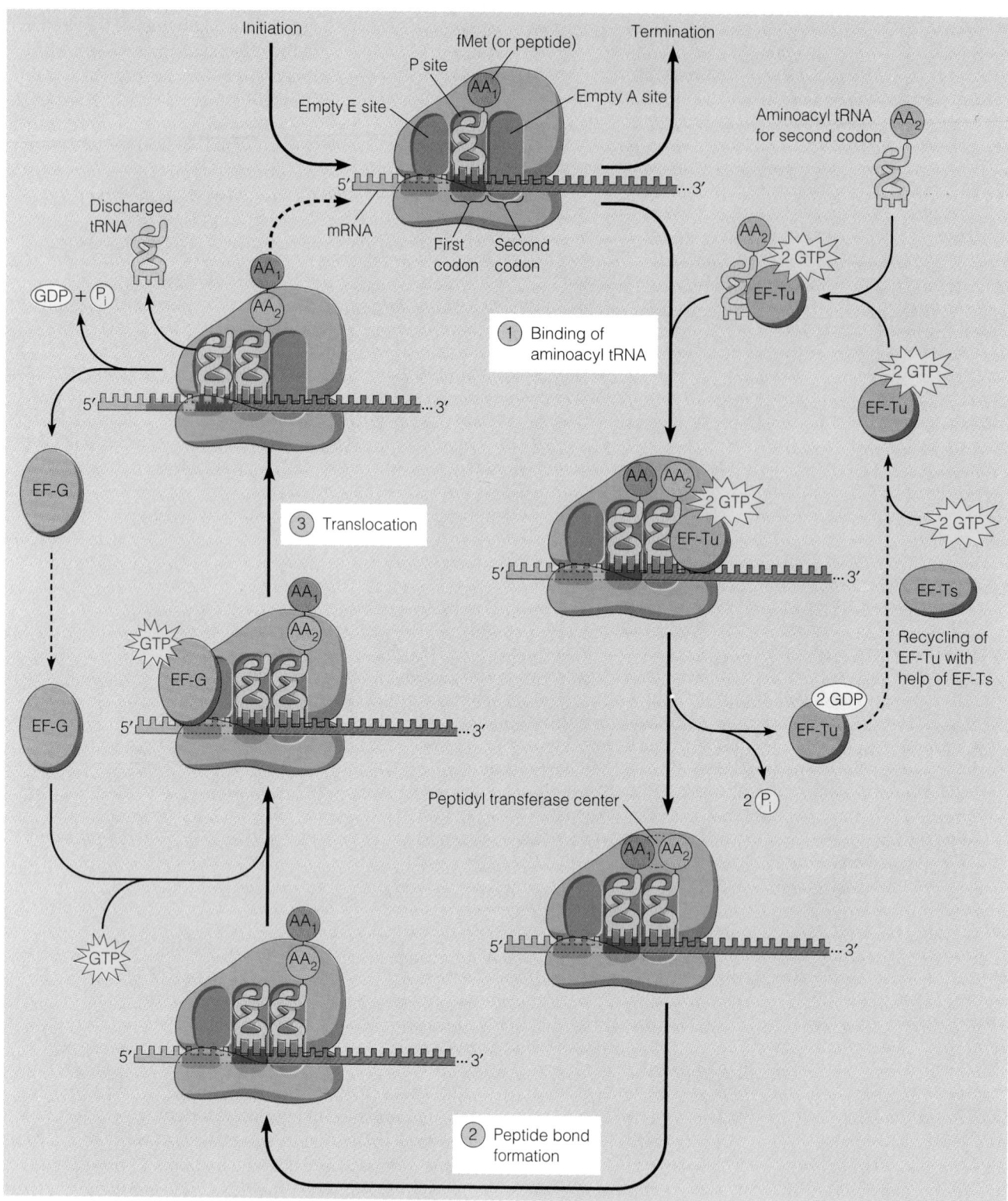

**Figure 18-10 Polypeptide Chain Elongation in Prokary-
otes.** Chain elongation during protein synthesis requires the presence
of a peptidyl tRNA or, in the first elongation cycle (as shown here), an
f-Met-tRNA^{f-Met} at the peptidyl (P) site. ① Elongation begins with the
binding of the second aminoacyl tRNA at the aminoacyl (A) site. The
tRNA is escorted to the ribosomal A site by elongation factor EF-Tu,
which also carries two bound GTP. As the tRNA binds, the GTP is hydrolyzed, and EF-Tu is released. EF-Ts helps recycle the EF-Tu. ② A peptide bond is formed between the carboxyl group of the f-Met (or, in a later
cycle, of the terminal amino acid) at the P site and the amino group of the
newly arrived amino acid at the A site. This reaction is catalyzed by the
peptidyltransferase activity of the 23S-rRNA molecule in the large ribosomal subunit. ③ After EF-G–GTP binds to the ribosome, the tRNA carrying the elongated peptide translocates from the A site to the P site.
Meanwhile, GTP is simultaneously released and hydrolyzed, and the discharged tRNA moves from the P site to the E (exit) site and leaves the
ribosome. As the peptidyl tRNA translocates, it takes the mRNA along
with it. Consequently, the next mRNA codon is moved into the A site,
which is open for the next aminoacyl tRNA. This cycle of events is repeated for each amino acid that is added.

mRNA codon present in the ribosome's A site and binds there. In the first round of elongation, the codon in the A site is the one immediately downstream (i.e., to the 3′ side) of the start codon. The binding of the new aminoacyl tRNA requires the elongation factors EF-Tu and EF-Ts and is driven by the hydrolysis of 2GTP to 2GDP and 2P$_i$. Note that, from now on, every incoming aminoacyl tRNA binds first to the A (aminoacyl) site—hence the site's name.

The role of **EF-Tu,** with two bound molecules of GTP, is to convey the aminoacyl tRNA to the A site of the ribosome. (Although not shown as such in the figure, the EF-Tu may be a dimer of two proteins, each of which binds one GTP.) As the aminoacyl tRNA is transferred to the ribosome, the GTP is hydrolyzed, and EF-Tu–2GDP is released. The role of **EF-Ts** is to regenerate EF-Tu–2GTP for the next round of the cycle. Thus the aminoacyl-tRNA binding sequence can be summarized as follows (abbreviating aminoacyl tRNA as AA-tRNA):

EF-Tu + 2GTP + AA-tRNA $\longrightarrow$ EF-Tu-2GTP–AA-tRNA

EF-Tu-2GTP–AA-tRNA + 70S complex $\longrightarrow$
           70S complex–AA-tRNA + EF-Tu–2GDP + 2P$_i$

EF-Tu–2GDP $\xrightarrow{\text{EF-Ts}}$ EF-Tu + 2GDP

70S complex + AA-tRNA + 2GTP $\longrightarrow$
         70S complex–AA-tRNA + 2GDP + 2P$_i$

With f-Met-tRNA$^{\text{f-Met}}$ now at the P site and the second aminoacyl tRNA at the A site, the stage is set for the second major step of elongation: peptide bond formation between the activated carboxyl group of f-Met and the amino group of the second amino acid. The result of this chemical reaction is that *two* amino acids are now attached to the second tRNA; that is, this tRNA now carries a dipeptide. (Either the initial methionine or the amino acid linked to it will become the N-terminus of the completed polypeptide.)

Peptide bond formation is catalyzed by **peptidyl transferase,** an enzymatic activity inherent in the 23S rRNA of the large ribosomal subunit. (The 23S rRNA is a *ribozyme,* an enzyme made entirely of RNA.) The energy to drive peptide bond formation is provided by hydrolysis of the "high-energy" ester bond by which f-Met was attached to its tRNA.

To prepare for the next major step in elongation, an elongation factor called EF-G, with a bound GTP, attaches to the ribosome. In the final step, **translocation,** the dipeptidyl tRNA moves from the A site to the P (peptidyl) site. At the same time, the uncharged tRNA$^{\text{f-Met}}$ moves to the E (exit) site and is then released from the ribosome. The energy for translocation is supplied by the hydrolysis of the GTP, as EF-G leaves the ribosome. As the peptidyl tRNA translocates, it remains hydrogen-bonded to the mRNA and pulls the mRNA along with it. As a result, the mRNA advances by three nucleotides. (That the translocation itself moves the mRNA, rather than some separate ribosomal mechanism, was demonstrated by the use of mutant tRNA molecules with *four*-nucleotide anticodons. These tRNAs hydrogen-bond to four nucleotides on the mRNA, and when translocation occurs, the mRNA advances by four nucleotides rather than the usual three.)

The movement of the mRNA positions its third codon in the A site, and the ribosome is now set to receive the next aminoacyl tRNA and repeat the cycle shown in the figure. Each successive amino acid is added in this way, as the mRNA is read in the 5′ → 3′ direction. The amino end of the growing polypeptide is thought to pass out of the ribosome through a tunnel in the 50S subunit, and the chain usually starts to fold while it is being elongated. Polypeptide synthesis is very rapid; in a growing *E. coli* cell, a polypeptide of 400 amino acids can be made in 10 seconds.

Chain Termination

The elongation process depicted in Figure 18-10 continues in cyclic fashion, reading one codon after another and adding successive amino acids to the polypeptide chain, until one of the three possible stop codons (UAG, UAA, or UGA) on the mRNA arrives at the ribosome's A site (Figure 18-11). There are no tRNA molecules that recognize these

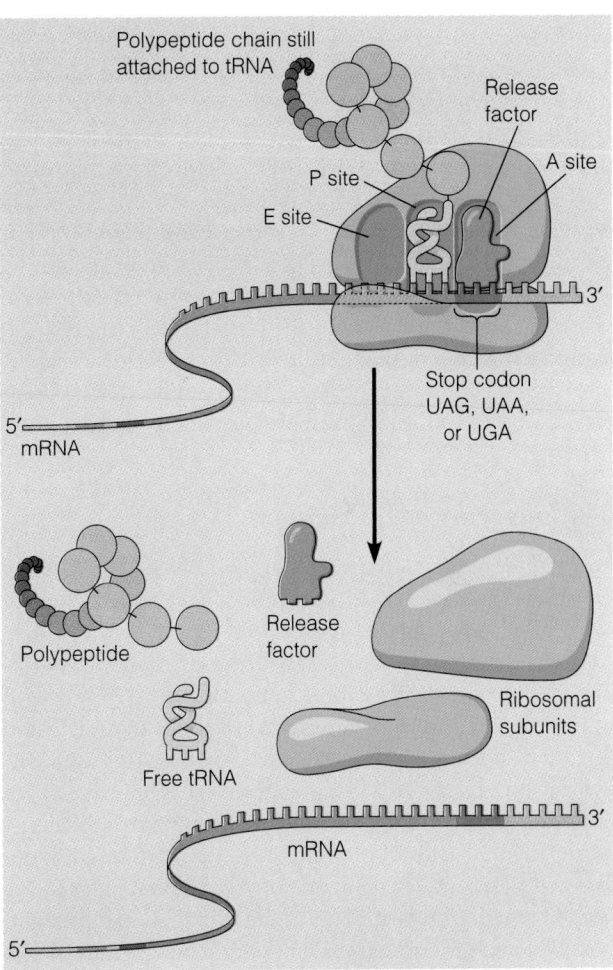

Figure 18-11 Termination of Translation. When a stop codon—UAG, UAA, or UGA—arrives at the A site, it is recognized and bound by a protein release factor. This protein causes the polypeptide to be transferred to a molecule of water, causing its release from the tRNA and the dissociation of the other components of the elongation complex.

codons. Instead, stop codons are recognized by proteins called **release factors,** which terminate translation by triggering the release of the completed polypeptide from the peptidyl tRNA. The cleavage is hydrolytic; in essence, the polypeptide transfers to a water molecule instead of to an activated amino acid. This produces a free carboxyl group at the end of the polypeptide, its C-terminus. As the polypeptide is released, the other components of the complex come apart. They are now available for reuse in a new initiation complex.

Polypeptide Folding

A polypeptide starts to fold into its secondary and tertiary structure while it is still being synthesized. As discussed in Chapters 2 and 3, the primary sequence of a protein is sufficient to specify its three-dimensional structure, and, indeed, many polypeptides can spontaneously fold into their proper tertiary conformations in vitro. However, inside the cell, protein folding is aided by proteins called **molecular chaperones** (see Chapter 2). Several evolutionarily unrelated families of chaperone proteins are found throughout the living world, from bacteria to the various compartments of eukaryotic cells. As mentioned in Chapter 2, two of the most widely occurring chaperone families are called *Hsp70* and *Hsp60.* (The "Hsp" comes from the original isolation of these proteins as "heat-shock proteins" selectively expressed in bacterial cells subjected to metabolic stress.) These two types of chaperones operate by somewhat different mechanisms, but they both have ATPase activity and function by ATP-dependent cycles of binding and releasing their protein substrates. It is the release of the protein that requires ATP hydrolysis. In recent years, it has become clear that chaperone proteins are involved in other important cellular processes; we shall see an example later in this chapter.

The Energetics of Translation

Polypeptide elongation involves the hydrolysis of at least five phosphoanhydride bonds per amino acid added. Two of these bonds are supplied by the ATP that is hydrolyzed to AMP in the aminoacyl-tRNA synthetase reaction. The remainder are supplied by three molecules of GTP: two used in binding the incoming aminoacyl tRNA at the A site and the other in translocation. Assuming that each phosphoanhydride bond has a $\Delta G^{o\prime}$ (standard free energy) value of 7.3 kcal/mol, the five bonds represent a standard free energy input of 36.5 kcal/mol of amino acid inserted. In addition, for each polypeptide chain, one or two additional GTPs are needed, one in the formation of the initiation complex and possibly one for termination. Thus, for a polypeptide of 100 amino acids, the $\Delta G^{o\prime}$ value for synthesis may be as high as 3665 kcal/mol. Clearly, protein synthesis is a very expensive process energetically; in fact, it accounts for a substantial fraction of the total energy budget of most cells. If we also consider the energy required to synthesize the messenger RNA and the components of the translational apparatus, and perhaps the use of ATP by chaperone proteins as well, the cost of protein synthesis becomes even greater.

It is important to note that, in translation, GTP does not function as an ATP-like energy donor. There is no evidence that its hydrolysis is used directly in the formation of any covalent bond. Instead, the function of GTP is apparently to induce conformational changes in the translation factors by binding to them and releasing from them. These shape changes, in turn, allow the translation factors to bind (noncovalently) to and release from the ribosome. Hydrolysis of the GTP on a ribosome-bound translation factor serves to release the factor from the ribosome. Moreover, hydrolysis of the GTP attached to EF-Tu apparently contributes to the accuracy of translation by playing a role in a proofreading mechanism that ejects incorrect aminoacyl tRNAs that enter the A site.

A Summary of Translation

At this point, you may want to refer back to Figure 18-7 for a summary of the translation process. Translation is a mechanism for converting the information in strings of RNA codons to a chain of amino acids linked by peptide bonds. As the ribosome reads the mRNA codon by codon in the $5' \rightarrow 3'$ direction, successive amino acids are brought into place by complementary base pairing between the codons on the mRNA and the anticodons on specific aminoacyl-tRNA molecules. When a stop codon is encountered, the completed polypeptide is released, and the mRNA and ribosomal subunits become available for further use.

In reviewing the process of translation, you will notice that RNA molecules play especially important roles. The mRNA plays a central role, of course, as the carrier of the genetic message. The tRNA molecules serve as the "adaptors" that link the amino acids to the appropriate codons. Last but not least, the rRNA molecules have multiple functions. Not only do they serve as structural components of the ribosomes, but one (the 16S rRNA of the small subunit) provides the binding site for incoming mRNA, and another (23S rRNA of the large subunit) catalyzes the formation of the peptide bond. Additional roles for rRNA are likely to be discovered in the near future. As Box 18A relates, the active roles of rRNA may be a vestige of the ancient origins of translation.

Most messages are read by many ribosomes at once, one following closely behind the other. As mentioned in the last chapter, such clusters of ribosomes and mRNA molecules are called **polyribosomes** (see Figure 17-18). By allowing the synthesis of many polypeptides from the same message simultaneously, cells maximize the efficiency of mRNA utilization.

Further Insights

Ribosomal RNA and the Origins of Translation

In a 1991 review article, Harry F. Noller, a preeminent ribosome researcher, offered the following thoughts on the origins of protein synthesis:

It has been argued that the design of the translation apparatus is imprinted with and reflects its evolution, and to understand one is to understand the other. The sheer complexity of even the simplest bacterial translational systems, which require well over a hundred different species of macromolecules, makes it clear that protein synthesis could not have evolved in a single step. The crucial role of translation, linking genotype with phenotype, only compounds the difficulty of imagining how it could have evolved. A striking characteristic of the translational apparatus is that it is rich in RNA. Francis Crick suggested that this is because the RNA components were part of the primitive machinery for protein synthesis, and wondered whether the original ribosomes were made entirely of RNA. The "chicken-or-the-egg" question of whether proteins or nucleic acids (or both) arose prior to the evolution of translation was addressed by Leslie Orgel, who pointed out a key problem: the unlikelihood of "reverse translation" of a polypeptide se-quence into a polynucleotide sequence. Thus, if any proteins existed prior to the evolution of translation, they are no longer with us.

A way out of the RNA-protein paradox has been provided by Thomas Cech and his colleagues, the discoverers of catalytic RNA. If RNA molecules can have enzymatic activity, one can imagine the existence of an RNA-based genetic system, in which genotype and phenotype are embodied in the same molecule. It can then be postulated that within such an RNA-based system, a primitive translational apparatus evolved. But this somewhat plausible sequence of events implies evolutionary selection for a relatively complex system, even in its simplest conceptual form. That is, the RNA-based system could not have "known" in advance that it was preparing for a more powerful protein-based system during the time that the precursors to tRNA, mRNA, and rRNA were evolving. More likely, the translational apparatus emerged from the modification of some pre-existing functional RNA molecules; the questions is, what was their function? ■

Source: Excerpted from Noller, H. Ribosomal RNA and translation. *Annu. Rev. Biochem.* 60 (1991): 195–196.

Nonsense Mutations and Suppressor tRNA

Mutations sometimes convert a codon that codes for an amino acid into a stop codon. Figure 18-12 illustrates such a case, in which the change of a single nucleotide pair in the DNA results in the replacement of a lysine codon in the mRNA (Figure 18-12a) with the stop signal UAG (Figure 18-12b). Such mutations normally lead to the production of short, nonfunctional peptides that end at the stop codon, as Figure 18-12b shows. A mutation that converts an amino acid–coding codon into a stop codon is called a **nonsense mutation.** This kind of mutation was first studied in the bacteriophage T4.

Nonsense mutations in vital genes are usually lethal, but phages that have such mutations can nonetheless grow in certain strains of bacteria. The bacteria seem to "suppress" the normal chain-terminating effect of the mutation. This is possible because such **suppressor strains** of bacteria have a mutant tRNA that recognizes what would otherwise be a stop codon and inserts an amino acid at that point (Figure 18-12c). In the example shown, the codon UAG is read as a codon for tyrosine by a mutant tRNA that has an altered anticodon. The inserted amino acid is almost always different from the amino acid that would be present at that position in the wild-type protein, but the crucial feature of suppression is that chain termination is averted and a full-length polypeptide can be made.

A tRNA molecule that somehow negates the effect of a mutation is called a **suppressor tRNA.** As you might imagine, suppressor tRNAs exist that negate the effects of various types of mutations in addition to nonsense mutations (see Problem 18-7 at the end of the chapter). For the cell to survive, suppressor tRNAs must be relatively inefficient; otherwise, the protein-synthesizing apparatus would produce too many abnormal proteins. An overly efficient nonsense suppressor, for example, would fail to terminate many proteins correctly.

Knowledge of the genetic code and the mechanism of translation also helps us understand the effects of other kinds of mutations. Box 18B (page 584) summarizes the major types of mutations and the terminology used to describe them.

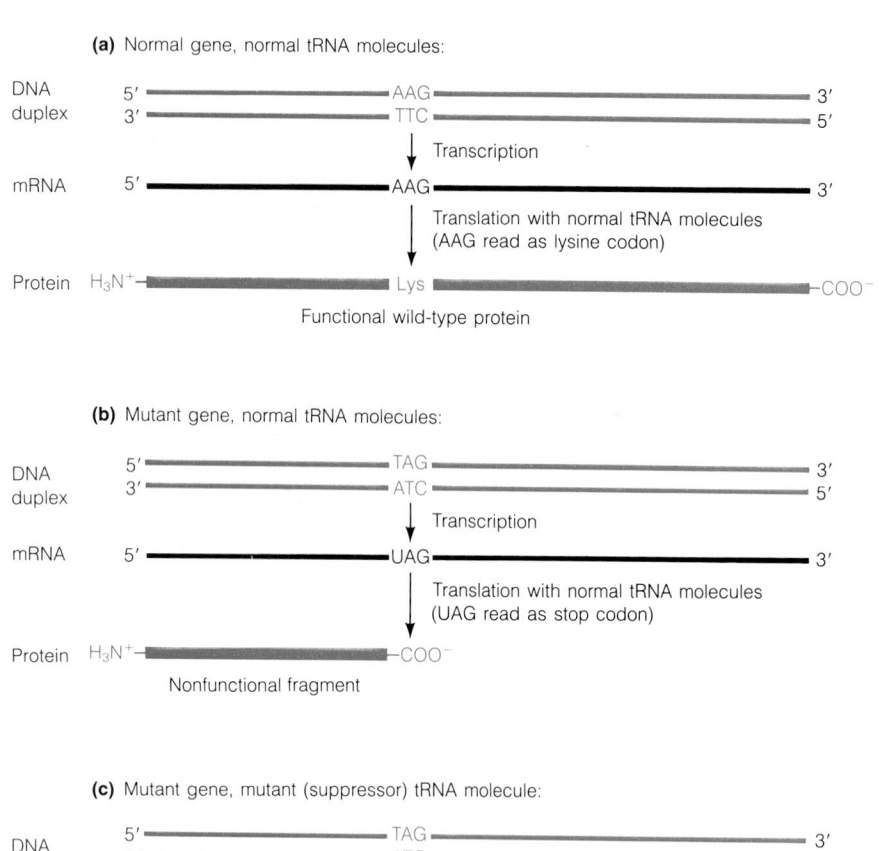

(a) Normal gene, normal tRNA molecules:

(b) Mutant gene, normal tRNA molecules:

(c) Mutant gene, mutant (suppressor) tRNA molecule:

Figure 18-12   Nonsense Mutations and Suppressor tRNAs.   (a) A wild-type (normal) gene is transcribed into an mRNA molecule that contains the codon AAG at one point. Upon translation, this codon specifies the amino acid lysine (Lys) at one point in the functional, wild-type protein. **(b)** If a mutation occurs in the DNA that changes the AAG codon in the mRNA to UAG, the UAG codon will be read as a stop signal, and the translation product will be a short, nonfunctional polypeptide. Mutations of this sort are called nonsense mutations. **(c)** In the presence of a mutant tRNA molecule that reads UAG as an amino acid codon instead of a stop signal, an amino acid will be inserted, and polypeptide synthesis will continue. In the example shown, UAG is read as a codon for tyrosine because the mutant tyrosine tRNA has as its anticodon 3'-AUC-5' instead of the usual 3'-AUG-5' (which recognizes the tyrosine codon 5'-UAC-3'). The resulting protein will be mutant because a lysine has been replaced by a tyrosine at one point along the chain. The protein may still be functional, however, if its biological activity is not adversely affected by the amino acid substitution.

Posttranslational Processing

Like RNA, polypeptides may be modified after synthesis. The N-formyl group at the amino end of the prokaryotic polypeptide chain is always removed. Moreover, the methionine to which it was attached is often removed also, as is the methionine that starts eukaryotic polypeptides. As a result, relatively few mature polypeptides have methionine at their N-terminus, even though they started out that way. Sometimes, whole blocks of amino acids are removed from the polypeptide. Some enzymes, for example, are synthesized as inactive precursors and must be activated by the removal of a specific sequence at one end or the other. The transport of proteins across membranes also may involve the removal of a terminal *signal sequence,* as we will see shortly. Some polypeptides have *internal* stretches of amino acids removed to produce the active protein. For instance, insulin is synthesized as a single polypeptide and then processed to remove an internal segment; the two end segments remain linked by disulfide bonds between cysteine residues in the active hormone (Figure 18-13).

Other common processing events include chemical modifications of individual amino acid groups—by methylation, phosphorylation, or acetylation reactions, for example. In addition, a polypeptide may undergo glycosylation (the addition of carbohydrate side chains; see Chapter 9), binding of prosthetic groups, and association with other polypeptides into functional multimeric proteins or supramolecular complexes.

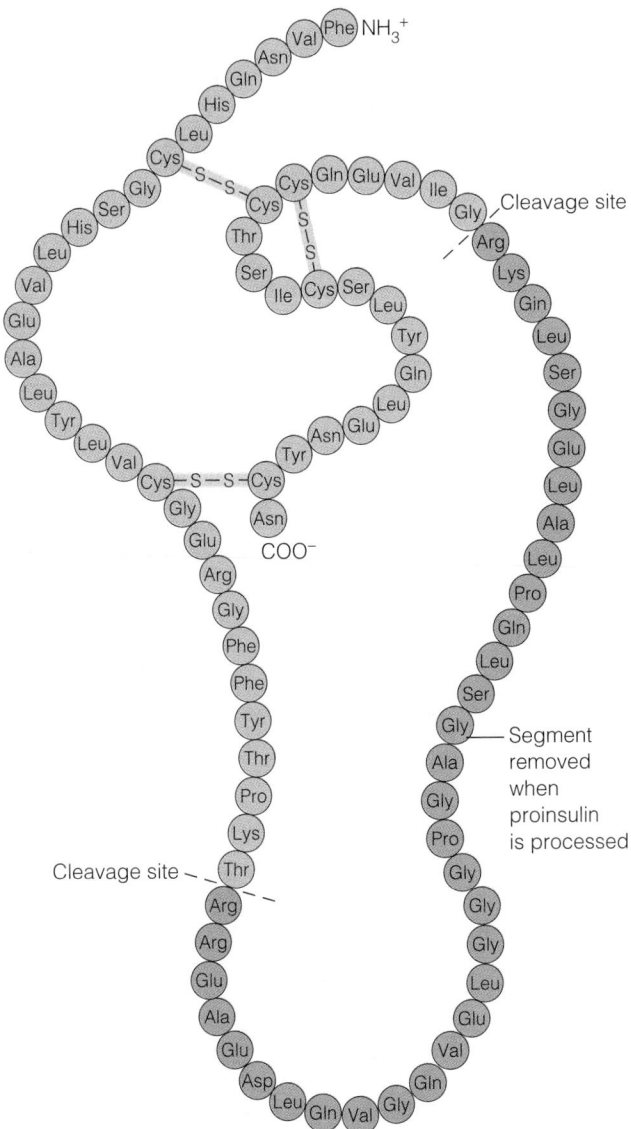

Figure 18-13   Processing of the Polypeptide Precursor of Insulin. The protein hormone insulin is synthesized in a precursor form called proinsulin, which is illustrated here. (Each circle represents an amino acid.) Proinsulin is converted to the active hormone by the enzymatic removal of a long internal section of polypeptide (blue). The two remaining chains (purple) continue to be covalently connected by disulfide bonds (gold) connecting cysteine residues in insulin.

Protein Targeting and Sorting

So far, we have considered the flow of genetic information from genes to proteins, including the intricacies of transcription, RNA processing, translation, and polypeptide processing. Not yet explored, however, are the mechanisms needed to ensure that each protein synthesized within the cell reaches its correct intracellular destination. Think for a moment about a typical eukaryotic cell with its diversity of organelles, each containing its own unique set of proteins. Such a cell is likely to have billions of protein molecules, representing at least 10,000 different kinds of polypeptides, each of which must find its way to the appropriate location.

A limited number of these polypeptides are encoded by the genome of the mitochondrion (and, for plant cells, by the chloroplast genome as well), but most are encoded by nuclear genes and are synthesized by a process that begins in the cytosol. Each of these polypeptides must then be directed to the right cellular compartment and must therefore have some sort of molecular "address" that ensures its delivery to the proper location. As our final topic for this chapter, we will now discuss this process: **protein targeting and sorting.**

We can begin by grouping the various compartments of eukaryotic cells into three categories: (1) the cytosol; (2) the mitochondria, chloroplasts, peroxisomes (and related organelles), and the interior of the nucleus; (3) the endomembrane system, which includes the endoplasmic reticulum and all other organelles with which it is in contact. Included in the endomembrane system are the Golgi complex, lysosomes, and secretory vesicles (see Chapter 9), and also the nuclear envelope and the plasma membrane. Recall that the interiors of all the compartments in the endomembrane system can communicate with one another and with the outside of the cell either directly or by means of membranous vesicles.

18B

A MUTATION PRIMER

In its broadest sense, the term *mutation* refers to any change in the sequence of nucleotides in a genome. Now that we have examined the processes of transcription and translation, we can understand the effects of a number of different kinds of mutations. Limiting our discussion to protein-coding genes, let's consider here some of the main types of mutations and their impact on the polypeptide encoded by the mutant gene. Figure 18B-1 shows examples of the types of mutations described here.

In this and the previous chapter, we have encountered several types of mutation in which the change to the DNA involves only one or a few base pairs. At the beginning of

Chapter 17, for instance, we mentioned the genetic allele that causes sickle-cell anemia when present homozygously. This allele originated as the result of a mutation called a *base-pair substitution*. In this case, an AT pair was substituted for a TA pair in the wild-type DNA. In the mRNA transcribed from the mutant allele, a GUA codon replaces a GAA, and, in the polypeptide (β-globin), a valine replaces a glutamic acid. This single amino acid change, resulting from a single base-pair change, is enough to change significantly the conformation of the β-globin and, in turn, the hemoglobin tetramer. Such a base-pair substitution mutation is also called a *missense mutation*, because the mutated

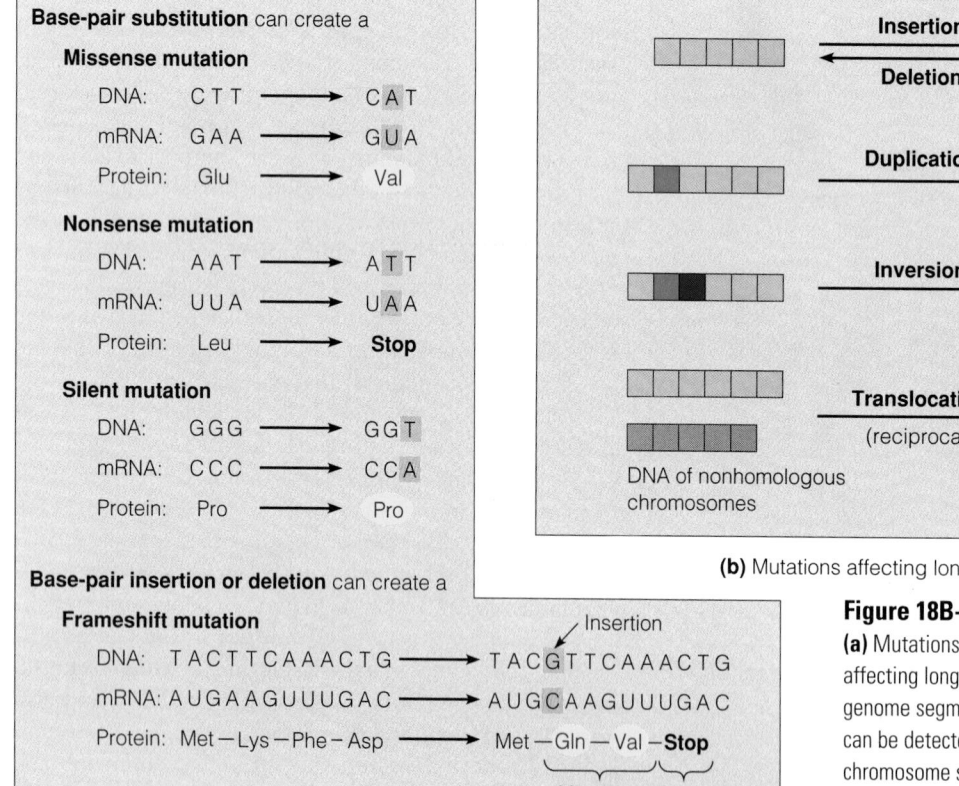

Base-pair substitution can create a

Missense mutation

DNA:	C T T	⟶	C A T
mRNA:	G A A	⟶	G U A
Protein:	Glu	⟶	Val

Nonsense mutation

DNA:	A A T	⟶	A T T
mRNA:	U U A	⟶	U A A
Protein:	Leu	⟶	**Stop**

Silent mutation

DNA:	G G G	⟶	G G T
mRNA:	C C C	⟶	C C A
Protein:	Pro	⟶	Pro

Base-pair insertion or deletion can create a

Frameshift mutation

DNA: TACTTCAAACTG ⟶ TACGTTCAAACTG ↙ Insertion
mRNA: AUGAAGUUUGAC ⟶ AUGCAAGUUUGAC
Protein: Met—Lys—Phe—Asp ⟶ Met—Gln—Val—**Stop**
　　　　　　　　　　　　　　　　　　　　　　Missense Nonsense

(a) Mutations affecting one base pair

(b) Mutations affecting long DNA segments

Insertion / Deletion / Duplication / Inversion / Translocation (reciprocal) / DNA of nonhomologous chromosomes

Figure 18B-1   Types of Mutations.
(a) Mutations affecting one base pair. **(b)** Mutations affecting long DNA segments. These may involve genome segments so large that their rearrangements can be detected microscopically as rearrangements of chromosome segments. Although sometimes called *chromosomal mutations*, such alterations are now usually referred to as *chromosomal aberrations*, a clearer term.

codon continues to code for an amino acid—but the "wrong" one.

A single amino acid change (or even a change in several amino acids) does not always affect the protein's function in a major way. As long as the protein's three-dimensional conformation remains relatively unchanged, biological activity may be unaffected. Substitution of one amino acid for another of the same type—for example, valine for isoleucine—is especially unlikely to affect protein function.

Alternatively, a base-pair substitution can create a *nonsense mutation;* that is, it can change an amino acid codon to one of the stop codons. As a result, the translation machinery will terminate the polypeptide prematurely. Unless the nonsense mutation is very close to the end of the message or a suppressor tRNA is present, the polypeptide is not likely to be functional. Both nonsense and missense codons also result from the *base-pair insertions* and *deletions* that cause *frameshift mutations.*

Silent mutations are base-pair substitutions that change the nucleotide sequence without changing the meaning of the message. Here the "mutant" polypeptide is exactly the same as the wild type. The nature of the genetic code actually minimizes the effects of single base-pair mutations. For example, changing the third base of a codon often produces a new codon that is a synonym of the original one.

Not all mutations involve the substitution, insertion, or deletion of one or a few base pairs. Many involve long stretches of DNA, sometimes even microscopically detectable segments of eukaryotic chromosomes. In addition to insertions and deletions of long DNA segments, there are several other categories of such mutations. In a *duplication,* a section of DNA is tandemly repeated. In an *inversion,* a segment of chromosome is cut out and reinserted in its original position but in reversed direction. A *translocation* involves the movement of a DNA segment from its normal location in the genome to another place, on the same chromosome or a different one. Because these large-scale mutations may or may not affect the expression of many genes, they have a wide range of phenotypic effects, from no effect at all to lethality.

When we think about the potential effects of mutations, it is useful to remember that genes have important noncoding components and that these, too, can be mutated in ways that seriously affect the gene product. A mutation in a promoter, for example, can result in more or less frequent transcription of the gene. Even a mutation in an intron can affect the gene product in a major way if it touches a critical part of a splice-site sequence.

Finally, mutations in genes that encode regulatory proteins—that is, proteins that control the expression of other genes—can have far-reaching effects on many other proteins. We will discuss this topic in Chapter 19. ■

As Figure 18-14 indicates, polypeptides encoded by nuclear genes follow one of two divergent pathways, each with several alternative destinations. The templates for all such polypeptides are mRNA molecules made in the nucleus, and polypeptide synthesis begins on free ribosomes in the cytosol (Figure 18-14a). Shortly thereafter, the two pathways diverge. Ribosomes involved in the synthesis of polypeptides destined for the endomembrane system become attached to the membrane of the ER early in the translational process. Such polypeptides are transferred across (or, in the case of integral membrane proteins, inserted into) the membrane of the ER as synthesis proceeds (Figure 18-14b). The transfer of polypeptides into the ER is called **cotranslational import** because uptake (or membrane insertion) is coupled directly to the translational process. The conveyance of such proteins from the ER to their final destinations is carried out by various vesicles and the Golgi complex, as discussed in Chapter 9 (see Figure 9-16).

Polypeptides destined for the cytosol, nuclear interior, mitochondria, chloroplasts, or peroxisomes follow an alternative pathway (Figure 18-14c). Ribosomes involved in the synthesis of these polypeptides remain free in the cytosol, unattached to any membrane. The polypeptides are released from the ribosomes after synthesis and either remain in the cytosol or are taken up by the appropriate organelle. The uptake by organelles of completed polypeptides is called **posttranslational import.** In the case of the nucleus, polypeptides enter through the nuclear pores, as discussed in Chapter 14 (see Figure 14-26). Polypeptide entrance into mitochondria, chloroplasts, and peroxisomes occurs by a different kind of mechanism, as we shall see.

In the next section, we look at cotranslational import, including in our discussion both soluble proteins and membrane proteins. Then we turn to posttranslational import of proteins into mitochondria and chloroplasts.

Cotranslational Import of Polypeptides into the ER

Proteins destined for the endomembrane system or for secretion from the cell are synthesized on ribosomes that become attached to the membrane of the rough ER shortly after translation starts. We focus first on soluble proteins. These include the secretory proteins and also the soluble proteins that end up inside the cisternae of organelles such as the Golgi complex, lysosomes, and the ER itself.

The Signal Mechanism. A model for explaining the cotranslational transfer of secretory polypeptides into the lumen (cisternal space) of the rough ER was put forward in 1971 by Gunter Blobel and David Sabatini. Their model was called the **signal hypothesis** because it proposed that some sort of signal was required to single out these proteins from the many proteins destined to remain in the cytosol. Blobel and Sabatini suggested that the first segment of the polypeptide to be synthesized, the amino end (N-terminus), is a **signal peptide** (or **signal sequence**) that directs the

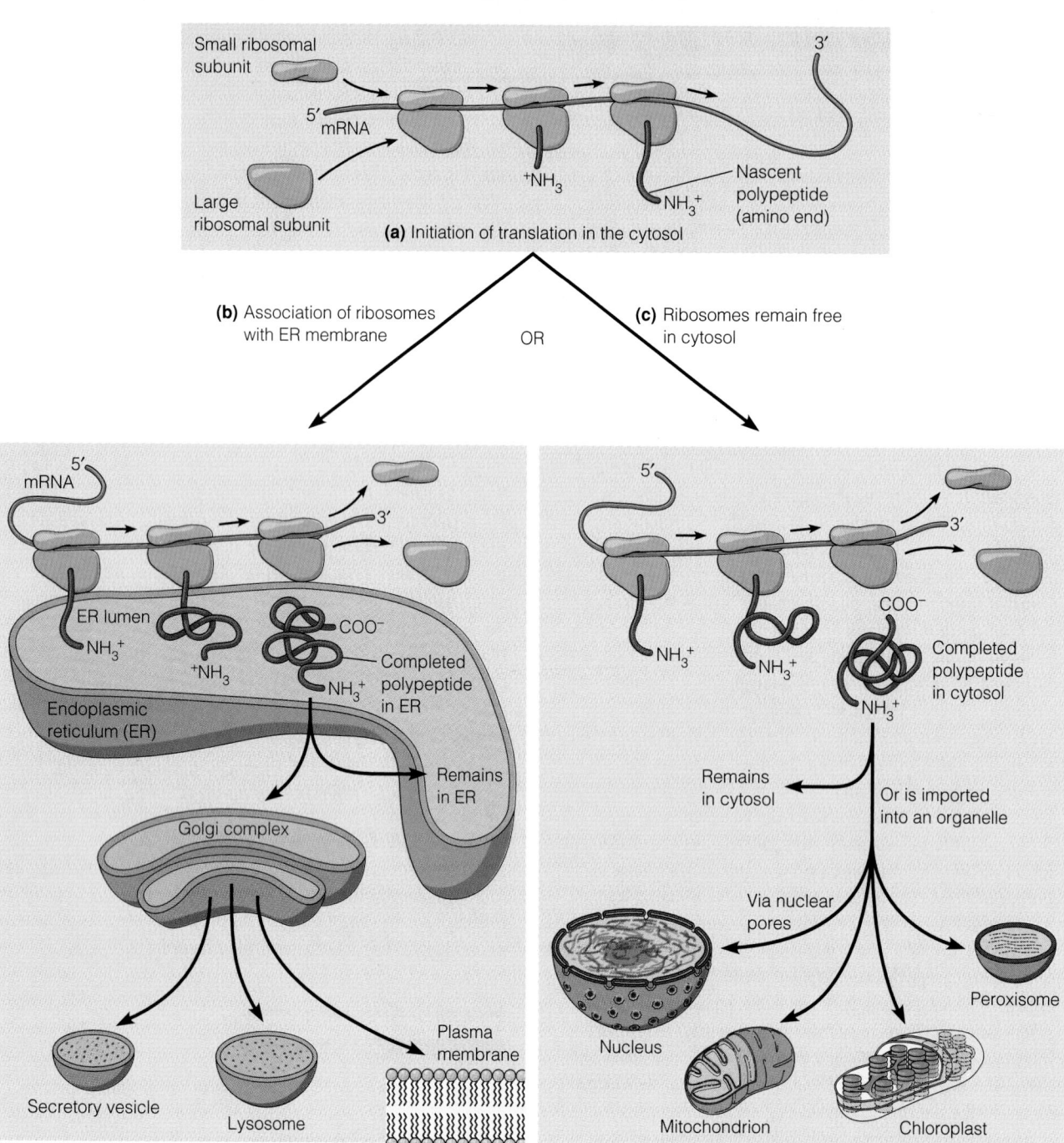

Figure 18-14   Intracellular Sorting of Proteins.   (a) Synthesis of all polypeptides encoded by nuclear genes begins in the cytosol. The large and small ribosomal subunits associate with each other and with the 5′ end of an mRNA molecule, forming a functional ribosome that starts making the polypeptide. When the nascent polypeptide is about 30 amino acids long, it enters one of two alternative pathways. **(b)** If destined for any of the compartments of the endomembrane system, the nascent polypeptide becomes associated with the ER membrane and is transferred across the membrane into the lumen (cisternal space) of the ER as synthesis continues. This process is cotranslational import. The completed polypeptide then either remains in the ER or is transported via various vesicles and the Golgi complex to another final destination. (Integral membrane proteins are inserted into the ER membrane as they are made, rather than into the lumen.) **(c)** If the polypeptide is destined for the cytosol or for import into the nucleus, mitochondria, chloroplasts, or peroxisomes, its synthesis continues in the cytosol. When the polypeptide is complete, it is released from the ribosome and either remains in the cytosol or is transported into the appropriate organelle by posttranslational import. Polypeptide uptake by the nucleus occurs via the nuclear pores, using a mechanism different from that involved in posttranslational uptake by other organelles.

ribosome-mRNA-polypeptide complex to the surface of the rough ER. The complex anchors at a protein "dock" on the ER surface, they proposed. Then, as the polypeptide chain lengthens, it progressively crosses the ER membrane and enters the ER lumen.

Over the past 25 years, the details of the signal hypothesis have been modified and elaborated, but the main idea has stood the test of time. Sequencing of many signal peptides has established that they have several characteristic features. They are typically 15–30 amino acids long, with a few positively charged amino acids at their amino end and a string of 6–12 hydrophobic amino acids in the middle of the sequence; at the carboxyl end (C-terminus) of the peptide is a sequence that constitutes the cleavage site. The charged end may promote interaction with the hydrophilic exterior of the ER membrane, and the hydrophobic region may facilitate interaction of the signal peptide with the membrane's lipid interior. In any case, it is now established that only peptides with such signal sequences are capable of being inserted into or across the membrane of the rough ER as synthesis proceeds. In fact, when recombinant DNA methods are used to give signal sequences to polypeptides that do not usually have them, the recombinant polypeptides are directed to the ER.

Once the existence of signal sequences wes established, it quickly became clear from further studies that nascent polypeptides must become attached to the ER membrane before very much of the polypeptide has emerged from the ribosome. If translation were allowed to continue without attachment to the ER, the folding of the growing polypeptide chain might bury the signal peptide. To understand what prevents this from happening, we need to look at the signal mechanism in further detail.

Contrary to the original signal hypothesis, the signal peptide does not itself initiate direct contact with the ER. Instead, the contact is mediated by a **signal-recognition particle (SRP),** which first recognizes and binds to the signal sequence of the nascent polypeptide and then binds to an **SRP receptor** on the ER membrane. At first, the SRP was thought to be purely protein (the P in its name originally stood for protein). Later, however, the SRP was shown to consist of six different polypeptides complexed with a 300-nucleotide (7S) molecule of RNA (Figure 18-15). The protein components have three main active sites: one that recognizes and binds to the signal peptide, one that binds to the SRP receptor, and one that interacts with the ribosome to block further translation. Thus, it is the SRP that blocks translation until docking occurs. (Beyond its structural role in the particle, the SRP RNA is also speculated to play a role in recognizing the translation complex.)

Figure 18-16 illustrates a current version of the signal mechanism for cotranslational import, showing the roles of the signal peptide, the SRP, and the SRP receptor, along with several other components. As the signal sequence of a nascent polypeptide emerges from the ribosome, SRP (shown in green) binds to it and blocks further translation (step 1). The block persists until the ribosomal complex

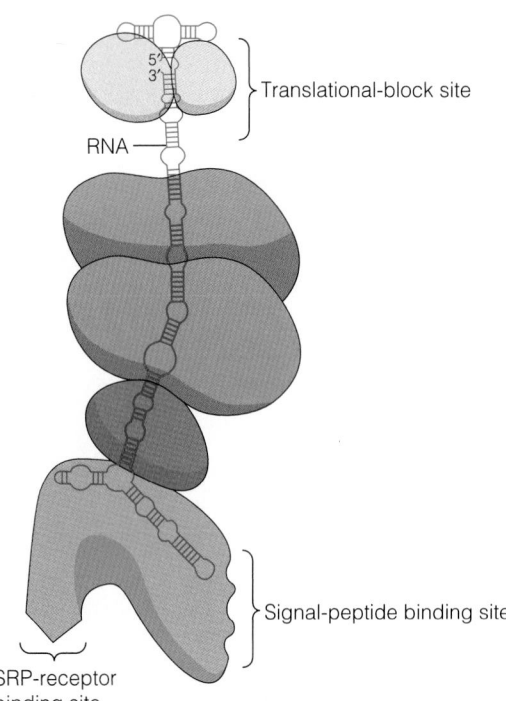

Figure 18-15   The Signal-Recognition Particle (SRP). The SRP is an elongated particle, about 25 nm by 5 nm, that consists of a 300-nucleotide (7S) molecule of RNA and six different polypeptides. The relative sizes, shapes, and arrangement of the polypeptides shown here, and the details of the self-complementary structure of the RNA, are only approximate. SRP has three main activities: binding to signal peptides, blocking translation, and binding to the SRP receptor on the ER membrane. These activities are associated with different protein domains of the particle, as indicated.

contacts the membrane. The SRP soon "docks" the ribosomal complex onto the ER by binding to the SRP receptor (step 2). In the model shown in the figure, the SRP receptor is located close to several other membrane proteins that are important for cotranslational import. One is a ribosome receptor that helps hold the ribosome in place, one forms a pore through which the growing polypeptide can enter the ER lumen, and one is an enzyme called signal peptidase.

Next, GTP binds to the SRP–SRP-receptor complex, unblocking translation and causing transfer of the signal peptide to the pore protein (step 3). GTP is then hydrolyzed and the products released, along with the SRP (step 4). With the signal peptide remaining bound at or near the pore protein, the polypeptide elongates and translocates into the ER lumen, in the form of a loop (step 5). No energy source beyond the GTP used in polypeptide elongation is needed to drive the translocation. When polypeptide synthesis is completed and the entire polypeptide has been translocated, the **signal peptidase** clips off the signal peptide, thus creating a free amino end for the polypeptide (step 6). Simultaneously, the ribosomal subunits dissociate from the mRNA, leaving the newly synthesized polypeptide

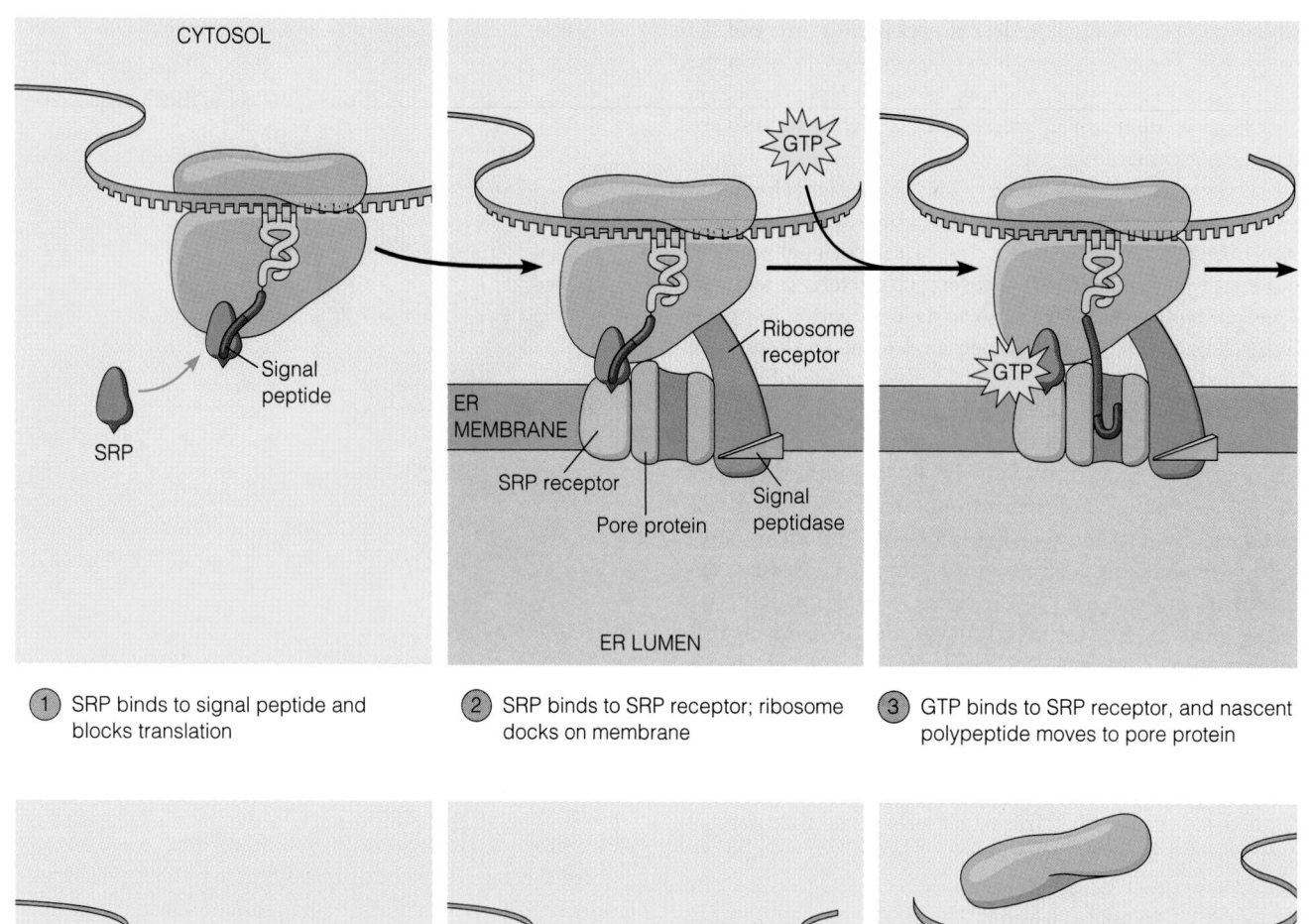

① SRP binds to signal peptide and blocks translation

② SRP binds to SRP receptor; ribosome docks on membrane

③ GTP binds to SRP receptor, and nascent polypeptide moves to pore protein

④ GTP is hydrolyzed and SRP is released

⑤ Polypeptide elongates and translocates into the ER lumen

⑥ Polypeptide is completed and cleaved by signal peptidase, releasing it from the membrane

Figure 18-16   A Model for the Signal Mechanism of Co-translational Import.   This figure shows a current model for the signal mechanism. It is now well established that the growing polypeptide translocates through an aqueous pore created by one or more membrane proteins. The complex of membrane proteins that carry out translocation is sometimes called the *translocon*. The most recent evidence suggests that the ribosome fits tightly across the cytoplasmic side of the pore and that the ER-lumen side is somehow closed off until the polypeptide is about 70 amino acids long (after step 4). In any case, after step 6, the ribosome is released and the pore closes completely (not shown).

sequestered within the ER. The polypeptide folds into its final tertiary form and, in some cases, associates with other polypeptides to form a quaternary structure.

An important aspect of this final stage for most proteins made on the ER—both soluble proteins and membrane proteins—is their glycosylation. As described in Chapter 9, in the ER, short side chains of sugars are added to certain amino acid residues, chiefly asparagine. For some proteins, the oligosaccharides are further processed in the Golgi complex.

The default pathway for soluble proteins made on the ER leads to secretion from the cell (see Figure 9-16). However, certain specific sugar side chains added to polypeptides or intrinsic short amino acid sequences (signal peptides) within polypeptides target proteins to final destinations in the ER-related organelles. For example, the **KDEL signal**, which targets a polypeptide for the ER itself, is the following sequence of four amino acids at the polypeptide's C-terminus: -Lys-Asp-Glu-Leu-COO$^-$. (The letters K, D, E, and L are the one-letter abbreviations for those four amino acids.) For soluble proteins with other destinations, the main sorting site is the Golgi complex.

Insertion of Transmembrane Proteins into the ER Membrane. Proteins that end up as integral membrane proteins in the endomembrane system must first become inserted in the membrane of the ER. These proteins have one or more hydrophobic regions located inside the membrane. Mechanisms similar to the one by which soluble proteins reach the ER lumen ensure that membrane proteins insert properly into the ER membrane. We focus here on the simplest case, proteins with only a single transmembrane region; but the principles involved extend to proteins with more complicated configurations.

Researchers postulate two main mechanisms by which transmembrane proteins become inserted into the ER membrane. Both involve an *internal*, hydrophobic signal sequence. This peptide differs from the signal peptide at the amino end of a polypeptide targeted for the ER lumen in that it lacks a cleavage site.

Figure 18-17 illustrates the two postulated mechanisms. Some polypeptides have both an N-terminal signal peptide and an internal **stop-transfer peptide** (Figure 18-17a). When the terminal signal peptide gives the "start-transfer" signal by attaching to the pore complex, the nascent polypeptide starts across the ER membrane as in the soluble polypeptide case. However, when the stop-transfer peptide reaches the membrane, it halts there, and translocation stops. Translation continues to completion, but the rest of the polypeptide remains on the cytosolic side of the ER membrane. When the terminal signal peptide is cleaved off, the result is a transmembrane protein with its amino end in the ER lumen and its carboxyl end in the cytosol. The hydrophobic stop-transfer peptide stays anchored in the membrane even after departure of the translocational apparatus.

Other membrane proteins lack a terminal signal peptide but have an internal signal peptide that serves as both a "start-transfer" signal and a membrane anchor (Figure 18-17b). In this case, the start-transfer signal peptide remains permanently in the membrane. Its orientation at insertion determines which terminus of the polypeptide ends up in the ER lumen and which in the cytosol.

Transmembrane proteins with multiple membrane regions probably insert in the ER membrane by means of a number of start-transfer and stop-transfer signal peptides.

Posttranslational Import of Proteins

In contrast to the cotranslational import of proteins into the ER, proteins destined for the nuclear interior, mitochondrion, chloroplast, or peroxisome are imported into these organelles after translation has been completed. Targeting of proteins to peroxisomes is discussed in Chapter 9. As discussed in Chapter 14, the posttranslational import of polypeptides by the nucleus depends on peptide sequences called *nuclear localization signals*, and entrance into the nucleus is an ATP-requiring process mediated by the nuclear pore complexes (see Figure 14-26). Here we focus on protein import into mitochondria and chloroplasts, which occurs by a mechanism involving signal sequences similar to those used in cotranslational import.

The Import of Polypeptides into Mitochondria and Chloroplasts. Although mitochondria and chloroplasts both contain their own DNA and protein-synthesizing machinery, they encode and synthesize relatively few of the polypeptides that they require. Most of the proteins in these organelles, like all of the proteins in the nucleus and the peroxisome, are encoded by nuclear genes and synthesized on cytosolic ribosomes (see Figure 18-14c). (The polypeptides encoded by the organellar genome are located mainly in the mitochondrial inner membrane or in the thylakoid membrane of the chloroplast. Almost without exception, these polypeptides are subunits of multimeric proteins, with one or more of the other subunits encoded by the nuclear genome and imported from the cytosol.)

Polypeptides intended for the mitochondrion or chloroplast are completed and released from the ribosome before import but are then taken up by the organelle within minutes. The targeting signal for such polypeptides is a signal peptide that is often called a **transit peptide.** Like the signal peptide of ER-targeted polypeptides, the transit peptide is located at the amino end of the polypeptide. Once inside the mitochondrion or chloroplast, the transit peptide is removed by a **transit peptidase** present inside the organelle. Removal of the transit peptide usually occurs before the transport process is complete.

The transit peptides of mitochondrial or chloroplast polypeptides usually consist of both hydrophobic and hydrophilic amino acids. The presence of positively charged amino acids is critical, although the secondary structure of the peptide may be more important than the specific amino

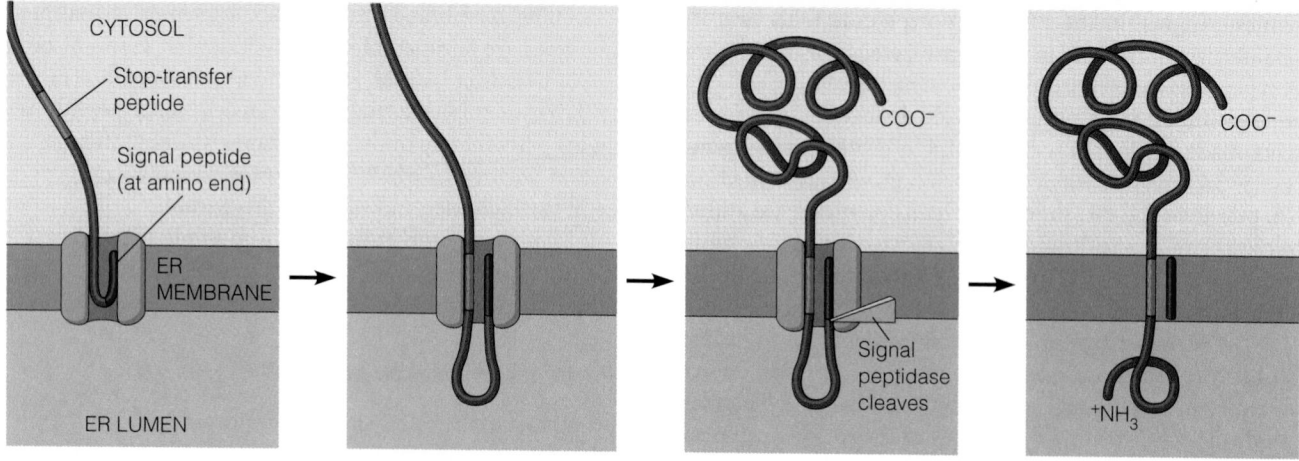

(a) Polypeptide with an internal stop-transfer peptide and a terminal signal peptide

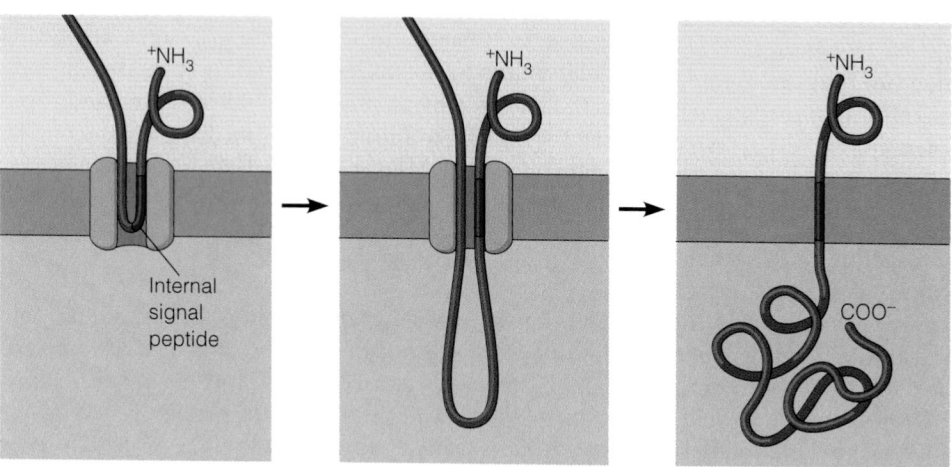

(b) Polypeptide with only an internal signal peptide

Figure 18-17 Cotranslational Insertion of Transmembrane Proteins into the ER Membrane.   This figure shows two mechanisms postulated for the insertion of integral membrane proteins having a single transmembrane segment. For clarity, the SRP, ribosome, and most other parts of the translocational apparatus have been omitted. **(a)** Insertion of a polypeptide with both a terminal signal peptide and an internal stop-transfer peptide. The terminal peptide is eventually cut off, leaving a transmembrane protein with its N-terminus in the ER lumen and its C-terminus in the cytosol. **(b)** Insertion of a polypeptide with only a single, internal signal peptide, which both starts polypeptide transfer and anchors itself permanently in the membrane. The amino-carboxyl orientation of the completed protein depends on the orientation of the signal peptide when it first inserts into the translocation apparatus.

acids. For example, some mitochondrial transit peptides have positively charged amino acids interspersed with hydrophobic amino acids in such a way that, when the peptide is coiled into an alpha helix, most of the positively charged amino acids are on one side of the helix and the hydrophobic amino acids on the other. Such an amphipathic helix may somehow be important for moving the polypeptide across the membrane.

Initial recognition of their respective transit peptides by the mitochondrion or chloroplast involves transit peptide receptors in the outer membrane. Once bound to the outer membrane, a polypeptide apparently passes through both the inner and outer membrane in rapid succession, presumably at a **contact site,** where the two membranes are close together.

Evidence for this model comes not only from electron microscopy, which reveals many sites of apparent contact between the two membranes, but also from biochemical experiments, in which a cell-free mitochondrial import system is incubated on ice to trap polypeptides in the act of translocation (Figure 18-18). The low temperature causes the translocation process to halt soon after it starts. At this point, the polypeptides have already had their transit peptides removed by the transit peptidase enzyme in the mitochondrial matrix. However, they can still be attacked by exogenously added proteolytic enzymes. These results indicate that polypeptides transiently span both membranes during import. The amino end is clearly within the matrix of the mitochondrion while the rest of the molecule is still on the outside of the organelle.

Polypeptides are almost certainly in an unfolded state when they are imported into the mitochondrion or chloroplast. It is unlikely that a polypeptide could maintain its three-dimensional conformation while crossing a membrane, especially when one considers the wide range of polypeptide sizes. It is equally unlikely that any sort of pore in the membrane could be large enough to allow a fully folded polypeptide through without disrupting such essential membrane functions as the maintenance of an electrochemical gradient. Recent research has revealed that the polypeptides targeted for the mitochondrion or chloroplast are held in a partially unfolded state by bound *chaperone proteins,* proteins like the ones that help many newly synthesized polypeptides fold correctly. Moreover, other chaperone molecules may actually help drive the translocation.

Figure 18-19 shows a current model for the chaperone-mediated posttranslational import of polypeptides by mitochondria. During polypeptide synthesis, cytosolic chaperones of the Hsp70 class bind to the polypeptide, keeping it in a loosely folded state that will allow translocation. The transit peptide at the amino end of the polypeptide binds to a receptor protein on the outer membrane of the mitochondrion. The chaperone molecules are then released, with ATP hydrolysis, as the polypeptide is translocated into the mitochondrial matrix. The transit peptide is clipped off in the matrix. As the rest of the polypeptide enters the matrix, *mitochondrial* Hsp70 molecules bind to it temporarily; their release, too, requires ATP hydrolysis. The ATP hydrolysis associated with mitochondrial chaperone release is likely to be what actually drives the translocation. Finally, mitochondrial Hsp60 chaperone molecules help some imported polypeptides achieve their fully folded conformations.

Both chloroplasts and mitochondria are known to require energy for the import of polypeptides. In the case of mitochondria, import requires not only the hydrolysis of ATP, but also an electrochemical gradient across the inner membrane. The electrochemical gradient seems to be necessary only for the binding and penetration of the transit peptide. Once this step has occurred, experimental abolition of the membrane potential does not interfere with the rest of the transfer process. Chloroplasts, on the other hand, maintain an electrochemical gradient across the thylakoid membrane, but not across the inner membrane. Presumably, their energy needs for import into the stroma are met by ATP alone.

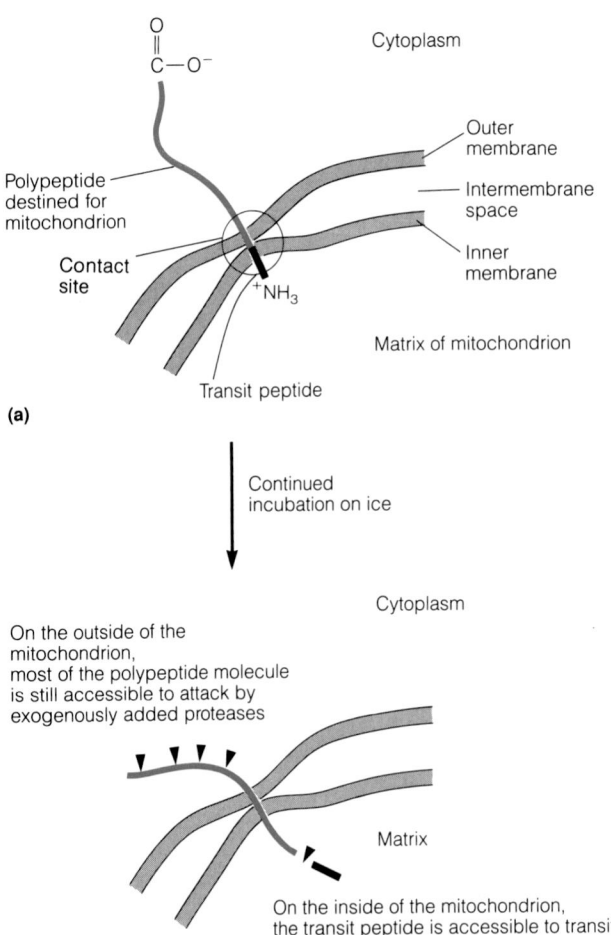

(a)

Continued incubation on ice

On the outside of the mitochondrion, most of the polypeptide molecule is still accessible to attack by exogenously added proteases

On the inside of the mitochondrion, the transit peptide is accessible to transit peptidase

(b)

Figure 18-18 Experimental Demonstration That Polypeptides Span Both Mitochondrial Membranes During Import. (a) To demonstrate that polypeptides being imported into the mitochondrion span both membranes at the same time, a cell-free import system is incubated on ice, instead of at the usual temperature of 37°C. At low temperature, polypeptides can start to penetrate the mitochondrion, but their translocation then stalls. (b) Under these conditions, the transit peptide is cleaved by the transit peptidase present in the matrix, indicating that the amino end of the polypeptide is within the mitochondrion. At the same time, most of the polypeptide molecule is readily accessible to attack by exogenously added proteolytic enzymes on the outside of the mitochondrion. Therefore, we conclude that the polypeptide must span both membranes transiently during import, presumably at a contact site between the two membranes.

Targeting Polypeptides to Organellar Compartments. Because of the structural complexity of mitochondria and chloroplasts, proteins to be imported from the cytoplasm must be targeted not only to the right organelle, but also to the appropriate compartment within the organelle. Mitochondria have four compartments: the outer membrane, the intermembrane space, the inner membrane, and the matrix

Figure 18-19   A Model for the Post-translational Import of Polypeptides into the Mitochondrion.   Like cotranslational import into the ER, posttranslational import into a mitochondrion involves a signal peptide (called a transit peptide in this case), a membrane receptor, pore-forming membrane proteins, and a peptidase. However, in the mitochondrion, the membrane receptor recognizes the signal sequence directly, without the intervention of a cytosolic SRP. Furthermore, chaperone proteins play several crucial roles in the mitochondrial process: They keep the polypeptide partially unfolded after synthesis in the cytosol so that binding of the transit peptide and translocation can occur (steps 1–3); they drive the translocation itself by binding to and releasing from the polypeptide *within* the matrix, an ATP-requiring process (step 4); and, in many cases, they help the polypeptide fold into its final conformation (step 5). The chaperones included here are cytosolic and mitochondrial versions of Hsp70 (light and dark blue, respectively) and a mitochondrial Hsp60 (not illustrated).

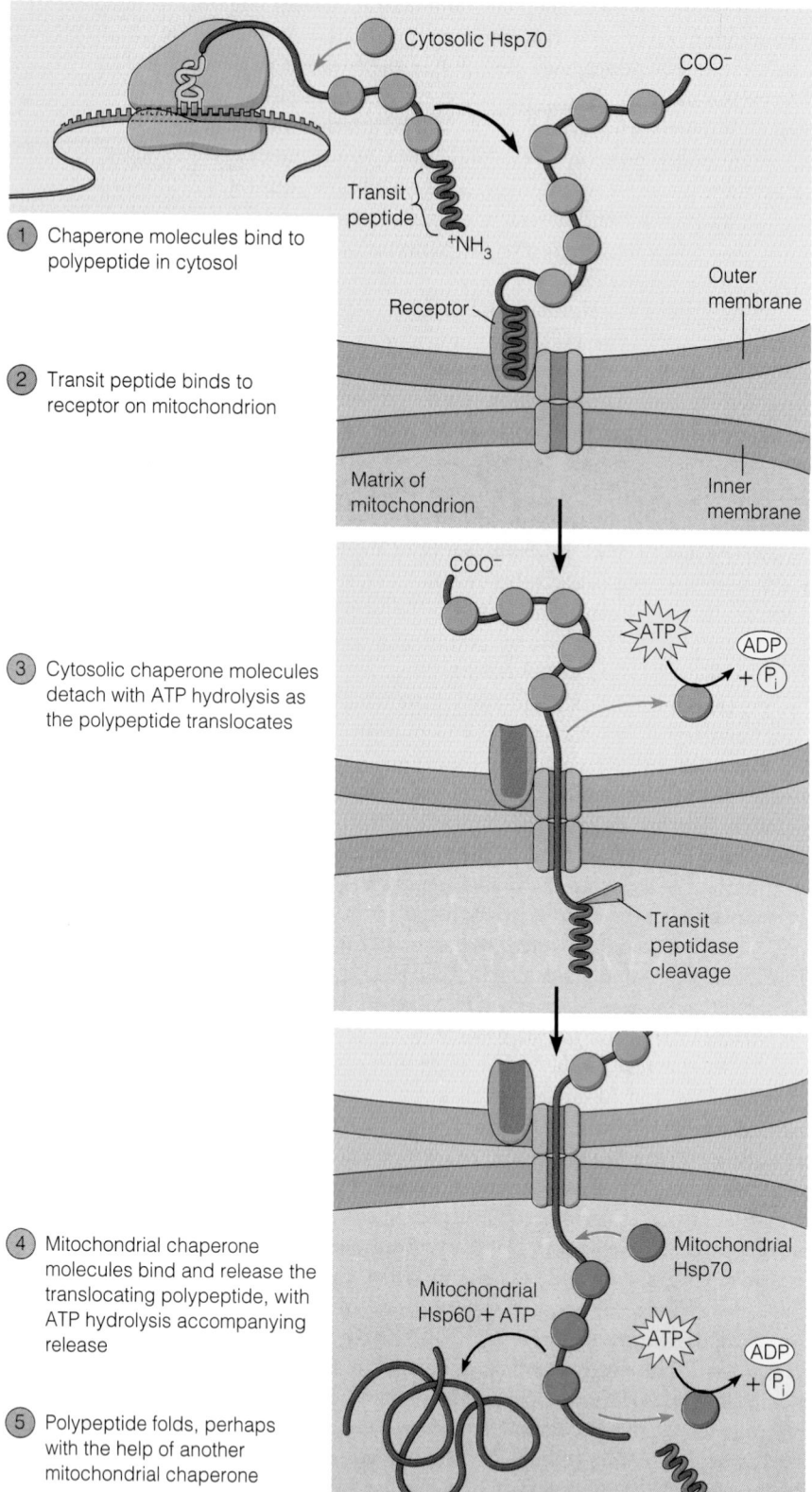

① Chaperone molecules bind to polypeptide in cytosol

② Transit peptide binds to receptor on mitochondrion

③ Cytosolic chaperone molecules detach with ATP hydrolysis as the polypeptide translocates

④ Mitochondrial chaperone molecules bind and release the translocating polypeptide, with ATP hydrolysis accompanying release

⑤ Polypeptide folds, perhaps with the help of another mitochondrial chaperone

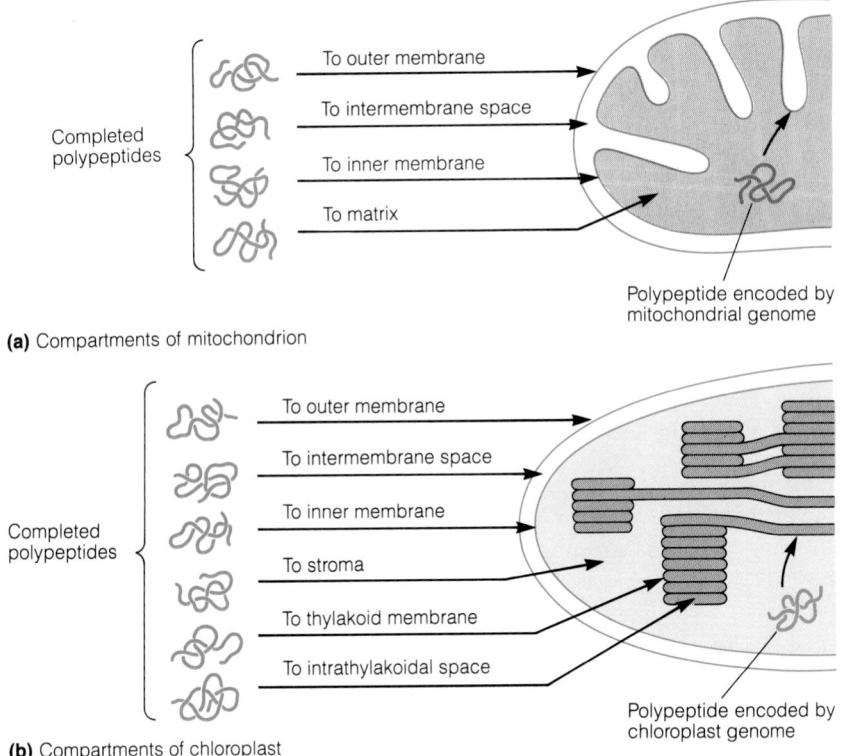

(a) Compartments of mitochondrion

To outer membrane
To intermembrane space
To inner membrane
To matrix

Completed polypeptides

Polypeptide encoded by mitochondrial genome

To outer membrane
To intermembrane space
To inner membrane
To stroma
To thylakoid membrane
To intrathylakoidal space

Completed polypeptides

Polypeptide encoded by chloroplast genome

(b) Compartments of chloroplast

Figure 18-20   Compartments of the Mitochondrion and Chloroplast.
Polypeptides encoded by nuclear genes and synthesized on cytosolic ribosomes are targeted to each of the compartments of the mitochondrion and chloroplast. **(a)** Mitochondria have four compartments: the outer membrane, the intermembrane space, the inner membrane, and the matrix. Polypeptides encoded by the mitochondrial genome and synthesized within the organelle are targeted mainly to the inner membrane. **(b)** Chloroplasts have six compartments: the outer membrane, the intermembrane space, the inner membrane, the stroma, the thylakoid membrane, and the thylakoid lumen (intrathylakoidal space). Polypeptides encoded by the chloroplast genome and synthesized within the organelle are targeted mainly to the thylakoid membrane.

(Figure 18-20a). Chloroplasts have four similar compartments (with the stroma substituted for matrix), and two additional compartments as well: the thylakoid membrane and the thylakoid lumen (Figure 18-20b). Notice that a polypeptide may have to cross one, two, or even three membranes to reach its final destination.

Given the structural complexity of both organelles, it is probably not surprising that at least some mitochondrial and chloroplast polypeptides require more than one signal to arrive at their proper destinations. For example, targeting of a polypeptide to the intermembrane space or to the inner membrane of the mitochondrion requires both a transit peptide to direct the polypeptide to the mitochondrion and a second, very hydrophobic signal sequence to target it to its final destination.

As Figure 18-21 indicates, when the transit peptide of a polypeptide imported into a mitochondrion is clipped off, a second signal sequence is revealed. This sequence is highly hydrophobic and resembles the signal peptide required for cotranslational transport into the ER. This sequence targets the polypeptide to the inner mitochondrial membrane, in the same way that mitochondrially encoded polypeptides are known to be inserted into this membrane. With its hydrophobic signal peptide anchored in the inner membrane, the rest of the polypeptide is then transported across the inner membrane into the intermembrane space (or into the inner membrane itself, if that is its final destination). Once in the intermembrane space, the polypeptide may be separated from its signal peptide by another peptidase, or the signal peptide may remain part of the final protein. In either case, the signal peptide is left in the inner membrane.

In some—perhaps many—cases, proteins destined for the mitochondrial membrane may be translocated by a slightly different pathway, in which most of the polypeptide chain never enters the matrix. Instead, the second, hydrophobic signal sequence remains in the inner membrane, and the rest of the polypeptide is deposited directly into the intermembrane space (and/or membrane). Such a mechanism is similar to those described earlier for insertion of transmembrane proteins into the ER membrane.

Multiple signals and transport steps are also required to direct some chloroplast polypeptides to their final destination. Polypeptides intended for insertion into (or transport across) the thylakoid membrane, for example, must first be targeted to the chloroplast and transported into the stroma, presumably crossing the inner and outer membranes at a contact site. In the stroma, the transit peptide used for this first step is cleaved from the polypeptide, unmasking a hydrophobic thylakoid signal peptide. This hydrophobic sequence is then used to insert the polypeptide into the thylakoid membrane, in the same way that chloroplast-encoded polypeptides are known to be inserted into this membrane. A similar mechanism is thought to be involved in targeting polypeptides to the thylakoid lumen, except that in this case, the polypeptide moves across the thylakoid membrane instead of simply inserting into it.

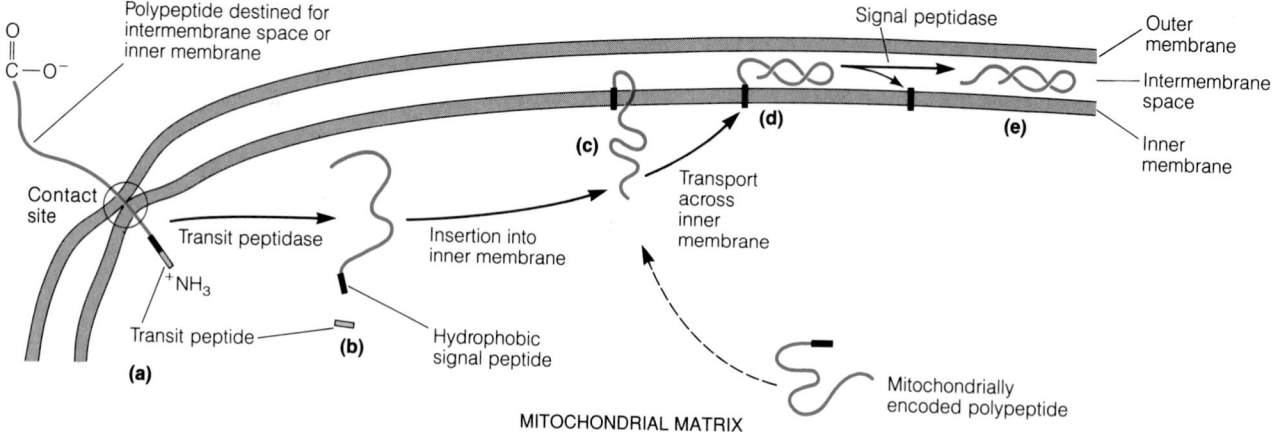

Figure 18-21   Targeting of Polypeptides to the Intermembrane Space or the Inner Membrane of the Mitochondrion.
In the mechanism shown here, polypeptides synthesized on cytosolic ribosomes but destined for either the intermembrane space or the inner membrane of the mitochondrion require two separate targeting sequences, both located at the amino end. **(a)** The polypeptide is directed to a contact (translocation) site on the mitochondrion by a positively charged or amphipathic transit peptide. **(b)** Cleavage of the transit peptide by a peptidase in the mitochondrial matrix uncovers a second signal sequence, which is highly hydrophobic. **(c)** This second signal peptide causes the polypeptide to be inserted into the inner membrane in the same way that mitochondrially encoded polypeptides are targeted to this membrane. **(d)** The remainder of the polypeptide is then moved across the membrane into the intermembrane space (or into the inner membrane, if that is its final destination). **(e)** Cleavage by a second peptidase can release the polypeptide into the intermembrane space, leaving the signal peptide behind in the inner membrane.

Translation is the process by which polypeptides are synthesized on ribosomes in the cell. The cellular machinery of translation is dominated by RNA molecules of various kinds. Messenger RNA determines the order of amino acids in the polypeptide, tRNA brings the amino acids to the ribosome, and rRNA helps position the mRNA on the ribosome and catalyzes peptide bond formation. In addition, a number of protein factors trigger specific events during the initiation, elongation, and termination stages of the process. GTP binding and hydrolysis drive the necessary conformational changes in the protein factors. The specificity required to link the right amino acids to the right tRNA molecules is a property of the aminoacyl-tRNA synthetases that catalyze these reactions. After the mRNA, ribosomal subunits, and initiator aminoacyl tRNA come together to form the initiation complex, other aminoacyl tRNAs recognize successive codons on the mRNA and add their amino acids to the growing polypeptide chain. Chain termination occurs when one of the stop codons is encountered, and the completed polypeptide is then released from the ribosome.

Knowledge of the genetic code and the details of the translation process enable us to understand how nonsense mutations cause their deleterious effects, and also how they can be suppressed by compensating mutations in tRNA. The phenotypic effect of a mutation that changes an amino acid codon to a stop codon can be overcome if a tRNA mutated in its anticodon reads the stop codon as an amino acid.

Proteins reach their final destinations in the cell by two main pathways, both of which involve polypeptide targeting and sorting. The general strategy is that newly made polypeptides have special sequences of amino acids that serve as "targeting" signals; proteins selectively recognize and bind to the signals, thus "sorting" the polypeptides. In one pathway, proteins destined for the endomembrane system or secretion from the cell are cotranslationally imported into the

ER. (The endomembrane system includes the ER, the Golgi complex, lysosomes, various kinds of vesicles, the nuclear envelope, and the plasma membrane.) The signal peptide that targets these polypeptides to the ER is on the amino end of the nascent polypeptide. An SRP in the cytosol binds to the signal peptide and then to an SRP receptor on the ER membrane, docking the ribosome-mRNA-peptide complex on the membrane. Polypeptide synthesis then proceeds, translocating the growing polypeptide across the ER membrane through a protein pore. The signal peptide is clipped off by a signal peptidase, leaving the remaining polypeptide to fold into its final three-dimensional shape. Polypeptides that insert into the ER membrane have internal signal peptides, instead of or in addition to a terminal signal peptide. Most proteins made on the ER are glycosylated; certain of these oligosaccharide side chains also serve as targeting signals for other parts of the ER system.

In the other sorting pathway, proteins destined for the nuclear interior, mitochondria, chloroplasts, or peroxisomes are synthesized on cytosolic ribosomes (as are proteins that remain in the cytosol) and are then imported posttranslationally into the targeted organelle. Polypeptides targeted to the nucleus contain a peptide sequence called a nuclear lo-calization signal and enter that organelle through the nuclear pores, in an ATP-dependent process (see Chapter 14). Polypeptides are targeted to the mitochondrion or chloroplast by an amphipathic transit peptide at the amino end. Polypeptides appear to be transported into the organelle at contact sites where the inner and outer membranes of the organelle are close together. In this pathway, receptor proteins in the outer membrane recognize the transit peptide directly. The energy needed to transport the unfolded polypeptide into the mitochondrion is provided by ATP hydrolysis associated with the release of chaperone proteins from the polypeptide and by the electrochemical gradient across the inner membrane. In the case of the chloroplast, unfolding and transport are driven by ATP hydrolysis alone. Mitochondria and chloroplasts have multiple compartments (four and six, respectively) to which polypeptides may be targeted. Import into some compartments requires multiple signal peptides at the amino end of the polypeptide; these peptides are cleaved successively as the polypeptide moves first into the matrix (or, in the case of the chloroplast, the stroma) and then to its final location within the organelle.

KEY TERMS FOR SELF-TESTING

Translation: The Cast of Characters

mRNA-binding site (p. 571)
A (aminoacyl) site (p. 571)
P (peptidyl) site (p. 571)
E (exit) site (p. 571)
aminoacyl tRNA (p. 571)
anticodon (p. 571)
wobble (p. 571)
inosine (I) (p. 571)
aminoacyl-tRNA synthetase (p. 573)
amino acid activation (p. 574)
leader (on mRNA) (p. 575)
start codon (p. 575)
trailer (on mRNA) (p. 575)
stop codon (p. 575)

The Stages of Translation

initiation factor (IF) (p. 575)

initiator tRNA (p. 575)
N-formylmethionine (f-Met) (p. 575)
ribosome-binding site (Shine-
    Dalgarno sequence) (p. 576)
30S initiation complex (p. 576)
70S initiation complex (p. 576)
EF-Tu (p. 579)
EF-Ts (p. 579)
peptidyl transferase (p. 579)
translocation (p. 579)
release factor (p. 580)
molecular chaperone (p. 580)
polyribosome (p. 580)

**Nonsense Mutations and
Suppressor tRNA**

nonsense mutation (p. 582)
suppressor strain (p. 582)

suppressor tRNA (p. 582)

Protein Targeting and Sorting

protein targeting and sorting (p. 583)
cotranslational import (p. 585)
posttranslational import (p. 585)
signal hypothesis (p. 585)
signal peptide (signal sequence)
    (p. 585)
signal-recognition particle (SRP)
    (p. 587)
SRP receptor (p. 587)
signal peptidase (p. 587)
KDEL signal (p. 589)
stop-transfer peptide (p. 589)
transit peptide (p. 589)
transit peptidase (p. 589)
contact site (p. 590)

18-1. The Genetic Code and Two Human Hormones. Below is the actual sequence of a small stretch of human DNA:

3′ AATTATACACGATGAAGCTTGTGACAGGGTTTCCAATCATTAA 5′
5′ TTAATATGTGCTACTTCGAACACTGTCCCAAAGGTTAGTAATT 3′

(a) What are the two possible RNA molecules that could be transcribed from this DNA?

(b) Only one of these two RNA molecules can actually be translated. It is, in fact, the mRNA for the hormone vasopressin. What is the apparent amino acid sequence for vasopressin? (The genetic code is given in Figure 17-7.)

(c) In its active form, vasopressin is a nonapeptide (i.e., it has nine amino acids) with cysteine at the N-terminus. How can you explain this in light of your answer to part b?

(d) A related hormone, oxytocin, has the following amino acid sequence:

Cys-Tyr-Ile-Glu-Asp-Cys-Pro-Leu-Gly

Where and how would you change the DNA that codes for vasopressin so that it would code for oxytocin instead? Does your answer suggest a possible evolutionary relationship between the genes for vasopressin and oxytocin?

18-2. Tracking a Series of Mutations. The following diagram shows the amino acids that result from mutations in the codon for a particular amino acid in a bacterial polypeptide:

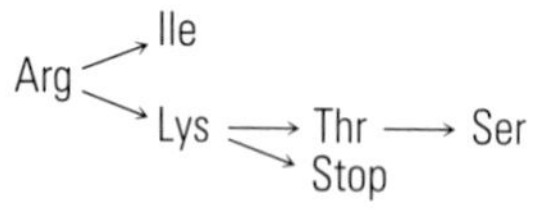

(a) Assume that each arrow denotes a single base-pair substitution in the bacterial DNA. Referring to the genetic code in Figure 17-7, determine the most likely codons for each of the amino acids and the stop signal in the above diagram.

(b) Starting with a population of the mutant carrying the nonsense mutation, another mutant is isolated in which the premature stop signal is suppressed. Assuming wobble does not occur and assuming a single base change in the tRNA anticodon, what are all the possible amino acids that might be found in this mutant at the amino acid position in question?

18-3. Initiation of Translation. Figure 18-8 shows diagrammatically how translation begins in prokaryotic cells. Using the text on pages 576–577 as a guide, draw a sketch outlining the steps in *eukaryotic* initiation of translation. What are the main differences between prokaryotic and eukaryotic initiation?

18-4. Prokaryotic and Eukaryotic Protein Synthesis Compared. For each of the following statements, indicate whether it applies to protein synthesis in prokaryotes (P), in eukaryotes (E), in both (B), or in neither (N).

(a) The mRNA has a ribosome-binding site within its leader region.

(b) AUG is a start codon.

(c) The enxyme that catalyzes peptide bond formation is an RNA molecule.

(d) The mRNA is translated in the 3′ → 5′ direction.

(e) The C-terminus of the polypeptide is synthesized last.

(f) Translation is terminated by special tRNA molecules that recognize stop codons.

(g) GTP hydrolysis functions to induce conformational changes in various proteins involved in polypeptide elongation.

(h) ATP hydrolysis is required to attach an amino acid to a tRNA molecule.

(i) The specificity required to link the right amino acids to the right tRNA molecules is a property of the enzymes called aminoacyl-tRNA synthetases.

18-5. An Antibiotic Inhibitor of Translation. Puromycin is a powerful inhibitor of protein synthesis. It is an analogue of the 3′ end of aminoacyl tRNA, as Figure 18-22 reveals. (R represents the functional group of the amino acid; R′ represents the remainder of the tRNA molecule.) When puromycin is added to an in vitro system containing all the necessary machinery for protein synthesis, incomplete polypeptide chains are released from the ribosomes. Each such chain has puromycin covalently attached to one end.

(a) Explain these results.

(b) To which end of the polypeptide chains would you expect the puromycin to be bound? Explain.

(c) Would you expect puromycin to bind to the A or P site on the ribosome, or to both? Explain.

(d) Assuming that it can penetrate into the cell equally well in both cases, would you expect puromycin to be a better inhibitor of protein synthesis in a eukaryotic cell or in a prokaryotic cell? Explain.

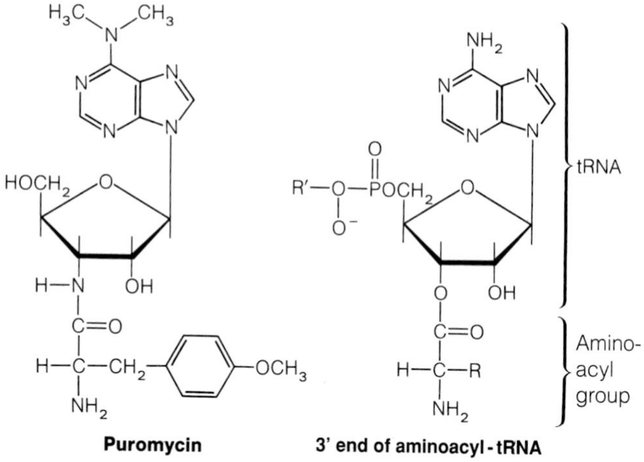

Figure 18-22   The Structure of Puromycin.

18-6. A Fictional Antibiotic. In an in vitro protein synthesis system from *E. coli*, the polyribonucleotide AUGUUUUUUUUUUUU directs the synthesis of the oligopeptide f-Met-Phe-Phe-Phe-Phe. In the presence of Rambomycin, a new antibiotic just developed by Macho Pharmaceuticals, only the dipeptide f-Met-Phe is made.

(a) What step in polypeptide synthesis does Rambomycin inhibit?

(b) Will the oligopeptide product be found attached to tRNA at the end of the uninhibited reaction? Will the dipeptide product be found attached to tRNA at the end of the Rambomycin-inhibited reaction? Explain.

18-7. Frameshift Suppression. As discussed in the chapter, a nonsense mutation can be suppressed by a mutant tRNA in which one of the three anticodon nucleotides has been changed. Describe a mutant tRNA that could suppress a *frameshift* mutation. (Hint: Such a mutant tRNA was used to demonstrate that it is the translocation of tRNA from the A site to the P site that moves the mRNA through the ribosome.)

18-8. The Many Roles of RNA. Consider all the processes involved in the expression of a eukaryotic gene encoding a secretory protein. Explain briefly the roles played by each of the following kinds of RNA:

(a) mRNA

(b) snRNA

(c) rRNA (two roles)

(d) tRNA

(e) SRP RNA

18-9. Signal Peptides. Although initially discovered in studies of secretory proteins, signal peptides (sequences) are now recognized as a common property of almost all proteins that are destined to be segregated into a membranous compartment. What specific property of a signal sequence is suggested by each of the following observations?

(a) Signal sequences are necessary for the transport of polypeptides across membranes.

(b) A number of proteins with different functions and properties are often segregated into the same membrane-bounded compartment.

(c) Although a number of different proteins are segregated into the same compartment, a single kind of peptidase enzyme seems adequate to remove the signal peptide from each of them.

18-10. Two Types of Posttranslational Import. The mechanism by which proteins synthesized in the cytosol are imported into the mitochondrial matrix is different from the mechanism by which proteins enter the nucleus, yet the two mechanisms do share some features. For example, in both cases the protein must cross two membranes. Indicate whether each of the following statements applies to nuclear import (Nu), mitochondrial import (M), both (B), or neither (N). You may want to review the discussion of nuclear import in Chapter 14 before answering this question (see especially Figure 14-26).

(a) The polypeptide to be transported into the organelle has a specific short stretch of amino acids that targets the polypeptide to the organelle.

(b) The signal peptide is always at the polypeptide's N-terminus and is cut off by a peptidase within the organelle.

(c) The signal peptide is recognized and bound by a receptor protein in the organelle's outer membrane.

(d) ATP hydrolysis is known to be required for the translocation process.

(e) GTP hydrolysis is known to be required for the translocation process.

(f) There is strong evidence for the involvement of chaperone proteins in translocation of the protein.

(g) The imported protein enters the organelle through some sort of protein pore.

(h) The pore complex consists of many proteins and, with a total mass of over 100 million Da, is large enough to be readily seen with the electron microscope.

SUGGESTED READING

General References

Alberts, B., D. Bray, J. Lewis, M. Raff, K. Roberts, and J. D. Watson. *Molecular Biology of the Cell,* 3d ed. New York: Garland, 1994.

Gesteland, R. F., and J. F. Atkins, eds. *The RNA World.* Cold Spring Harbor, NY: Cold Spring Harbor Laboratory, 1993.

Griffiths, A., J. Miller, D. Suzuki, R. Lewontin, and W. Gelbart. *An Introduction to Genetic Analysis,* 5th ed. New York: W. H. Freeman, 1993.

Lewin, B. *Genes V.* New York: Oxford University Press, 1994.

Stent, G. S. *Molecular Genetics: An Introductory Narrative.* New York: W. H. Freeman, 1971.

Watson, J. D., N. H. Hopkins, J. W. Roberts, J. A. Steitz, and A. M. Weiner. *Molecular Biology of the Gene,* 4th ed. Menlo Park, CA: Benjamin/Cummings, 1987.

Translation

Carter, C. W., Jr. Cognition, mechanism, and evolutionary relationships in aminoacyl-tRNA synthetases. *Annu. Rev. Biochem.* 62 (1993): 715.

Hendrick, J. P., and F.-U. Hartl. Molecular chaperone functions of heat-shock proteins. *Annu. Rev. Biochem.* 62 (1993): 349.

Hill, W. E., R. A. Dahlberg, R. A. Garrett, P. B. Moore, D. Schlessinger, and J. R. Warner, eds. *The Ribosome: Structure, Function & Evolution.* Washington, DC: Am. Soc. for Microbiol., 1990.

Jimenez, A. Inhibitors of translation. *Trends Biochem. Sci.* 1 (1976): 28.

Lake, J. A. Evolving ribosome structure: Domains in archaebacteria, eubacteria, eocytes and eukaryotes. *Annu. Rev. Biochem.* 54 (1985): 507.

Nierhaus, K. H., F. Franceschi, A. R. Subramanian, V. A. Erdmann, and B. Wittman-Liebold, eds. *The Translational Apparatus: Structure, Function, Regulation, Evolution.* New York: Plenum, 1993.

Noller, H. F. Ribosomal RNA and translation. *Annu. Rev. Biochem.* 60 (1991): 191.

Pace, N. R. New horizons for RNA catalysis. *Science* 256 (1992): 1402.

Cotranslational Protein Import

Blobel, G., P. Walter, G. N. Chang, B. M. Goldman, A. H. Erickson, and V. R. Lingappa. Translocation of proteins across membranes: The signal hypothesis and beyond. *Symp. Soc. Exp. Biol.* 33 (1979): 9.

Crowley, K. S., S. Liao, V. E. Worrell, G. D. Reinhart, and A. E. Johnson. Secretory proteins move through the endoplasmic reticulum membrane via an aqueous, gated pore. *Cell* 78 (1994): 461.

High, S., and B. Dobberstein. Mechanisms that determine the transmembrane disposition of proteins. *Curr. Opin. Cell Biol.* 4 (1992): 581.

Miller, J. D., H. Wilhelm, L. Gierasch, R. Gilmore, and P. Walter. GTP binding and hydrolysis by the signal recognition particle during initiation of protein translocation. *Nature* 366 (1993): 351.

Ng, D. T. W., and P. Walter. Protein translocation across the endoplasmic reticulum. *Curr. Opin. Cell Biol.* 6 (1994): 510.

Nunnari, J., and P. Walter. Protein targeting to and translocation across the membrane of the endoplasmic reticulum. *Curr. Opin. Cell Biol.* 4 (1992): 573.

Sanders, S. L., and R. Schekman. Polypeptide translocation across the endoplasmic reticulum membrane. *J. Biol. Chem.* 267 (1992): 13791.

Siegel, V., and P. Walter. Functional dissection of the signal recognition particle. *Trends Biochem. Sci.* 13 (1988): 314.

Walter, P. and Johnson, A. E. Signal sequence recognition and protein targeting to the endoplasmic reticulum membrane. *Annu. Rev. Cell Biol.* 10 (1994): 211.

Posttranslational Protein Import

Craig, E. A., J. S. Weissman, and A. L. Horwich. Heat shock proteins and molecular chaperones: Mediators of protein conformation and turnover in the cell (Meeting Review). *Cell* 78 (1994): 365.

Hartl, F. U., J. Ostermann, B. Guiard, and W. Neupert. 1987. Successive translocation into and out of the mitochondrial matrix: Targeting of proteins to the intermembrane space by a bipartite signal peptide. *Cell* 51 (1987): 1027.

Höhfeld, J., and F. U. Hartl. Post-translational protein import and folding. *Curr. Opin. Cell Biol.* 6 (1994): 499.

Pain, D., Y. S. Kanwar, and G. Blobel. Identification of a receptor for protein import into chloroplasts and its localization to envelope contact zones. *Nature* 331 (1988): 232.

Schleyer, M., and W. Neupert. Transport of proteins into mitochondria: Translocation intermediates spanning contact sites between outer and inner membranes. *Cell* 43 (1986): 339.

Schmidt, G. W., and M. L. Mishkind. The transport of proteins into chloroplasts. *Annu. Rev. Biochem.* 55 (1986): 879.

Stuart, R. A., D. M. Cyr, E. A. Craig, and W. Neupert. Mitochondrial molecular chaperones: Their role in protein translocation. *Trends Biochem. Sci.* 19 (1994): 87.

19

THE REGULATION OF GENE EXPRESSION

In our exploration of biological information flow, we have met DNA as the main repository of information in cells, we have seen how DNA is replicated, and we have looked at the processes involved in the expression of DNA's genetic information. In this concluding chapter of Part Four, we examine the multiplicity of strategies used by cells to regulate the expression of their genes.

Regulation is an important part of almost every process in nature. Rarely, if ever, is it adequate simply to describe the steps of a process. Instead, for a complete understanding of the process, we must also ask what turns it on, what turns it off, and what determines its rate. This is certainly true of the processes of gene expression. Most genes are not expressed all the time. In some cases, selective gene expression enables cells to be metabolically thrifty, making only those gene products that are of immediate use under the prevailing environmental conditions; this is often the situation with bacteria. In other cases, such as in multicellular organisms, selective gene expression allows cells to fulfill specialized roles.

As you might expect, our first knowledge about the regulation of gene expression came from investigations of prokaryotes. Bacteria were far more amenable to the kinds of genetic and biochemical manipulations that were available for the study of gene control mechanisms. But in recent years, advances in DNA technology have led to considerable progress with eukaryotes as well. We will look first at the prokaryotes, where almost all mechanisms of gene regulation operate at the level of transcription. Then we will turn to the eukaryotes, to explore both their versions of transcriptional control and some of the other levels of control found in these organisms.

Gene Regulation in Prokaryotes

Of the several thousand genes present on the typical bacterial chromosome, some are so important to the ongoing life of the cell that they are turned on at all times; their expression is not regulated. Such **constitutive genes** include, for example, the genes encoding the enzymes of glycolysis. (Such enzymes are called *constitutive enzymes.*) For many other genes, however, expression is regulated so that the amount of the final gene product—protein or RNA—is carefully tuned to the cell's need for that product. A number of these regulated genes encode enzymes for metabolic processes that, unlike glycolysis, are not constantly required. The intracellular concentrations of these enzymes are adjusted by the starting and stopping of gene expression in response to cellular needs, allowing the bacterium to adapt to its environment. As mentioned earlier, the control of prokaryotic gene expression is usually effected by starting and stopping transcription of the gene. Nevertheless, the term often used for the regulation of enzyme-coding genes focuses on the final protein product: **adaptive enzyme synthesis.**

Strategies of Adaptive Enzyme Synthesis

Bacterial cells use two main strategies for regulating enzyme synthesis, depending on whether a given enzyme is involved in a catabolic (degradative) or anabolic (synthetic) pathway. Figures 19-1 and 19-2 illustrate two well-understood metabolic pathways, one of each type. Often the enzymes required to carry out such pathways are regulated coordinately; that is, the synthesis of all the enzymes in the set is turned on and off together, as we will discuss.

Figure 19-1 depicts the steps in a catabolic pathway that lead to the degradation of the disaccharide lactose into sugars that can be further broken down in glycolysis. Lactose is hydrolyzed into the monosaccharides glucose and galactose, a reaction catalyzed by the enzyme *β-galactosidase.* However, before the hydrolysis reaction can occur, lactose must be transported into the cell, across the plasma membrane. A protein called *galactoside permease* is responsible for this transport, and its synthesis is regulated coordinately with *β*-galactosidase. Figure 19-2 shows an anabolic pathway, the

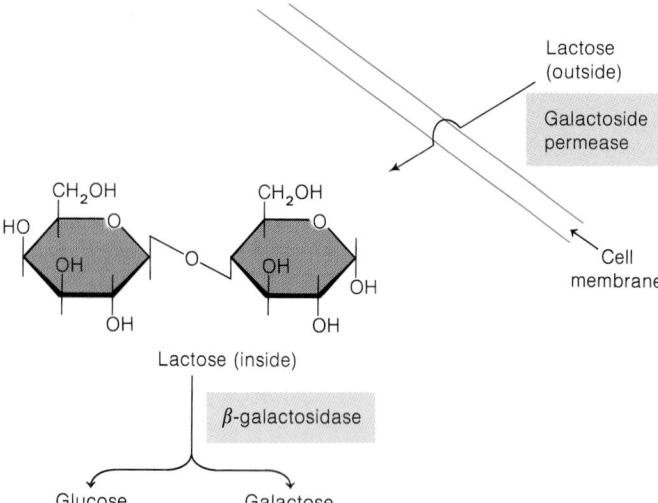

Figure 19-1 A Typical Catabolic Pathway. Breakdown of the disaccharide lactose involves enzymes (pink boxes) whose synthesis is regulated coordinately. See Figures 19-3 and 19-4 for the organization and regulation of the genes that code for these enzymes. (The enzymes responsible for the subsequent catabolism of the monosaccharides glucose and galactose are not a part of the same regulatory unit.)

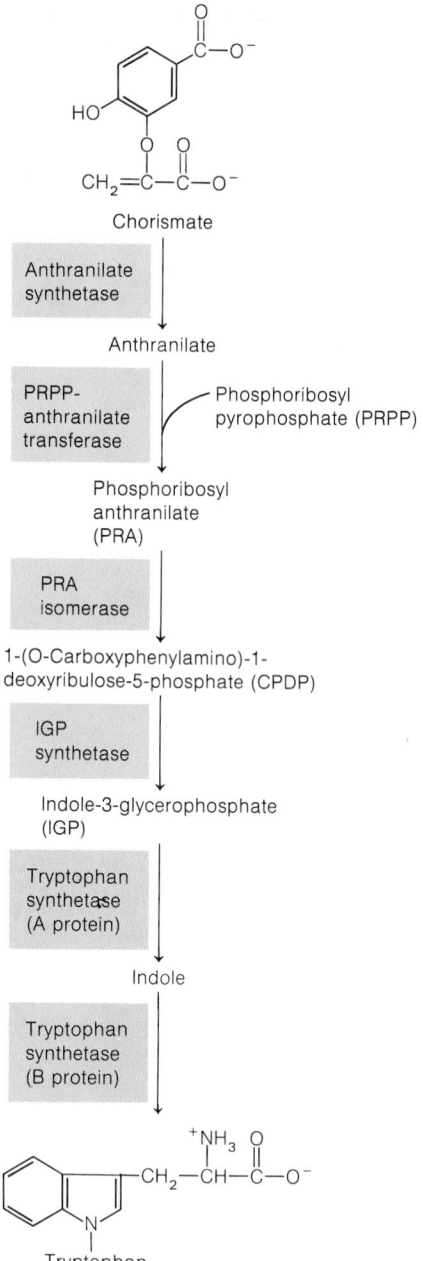

Figure 19-2 A Typical Anabolic Pathway. Synthesis of the amino acid tryptophan from the starting compound chorismate involves a set of enzymes (pink boxes) whose synthesis is regulated in a coordinated way. See Figure 19-6 for the organization and regulation of the genes that code for these enzymes.

synthesis of the amino acid tryptophan from the starting compound chorismate. The enzymes that catalyze the six steps of this pathway, like those involved in the first two steps of lactose catabolism, are regulated coordinately at the genetic level.

Catabolic Pathways. Catabolic enzymes exist for the primary purpose of degrading specific substrates, usually as a means of obtaining energy. Such enzymes are therefore needed only when the cell is confronted by the relevant substrate. The enzyme β-galactosidase, for example, is useful only when the cell is actually using lactose as an energy source; in the absence of lactose, the enzyme is superfluous. Accordingly, it makes sense in terms of cellular economy for the synthesis of β-galactosidase to be turned on, or *induced*, in the presence of lactose, but to be turned off in its absence. This turning on of enzyme synthesis is called **substrate induction**, and enzymes whose synthesis is regulated in this way are referred to as **inducible enzymes**. Most catabolic pathways in bacterial cells are subject to substrate induction of their enzymes.

Anabolic Pathways. The regulation of anabolic pathways is in a sense just the opposite of that for regulated catabolic pathways. For anabolic pathways, enzyme synthesis usually correlates with the cellular level of the end-product of the pathway. Such a relationship makes sense. For example, as the cellular concentration of tryptophan rises, it makes sense for the cell to economize on its metabolic resources by reducing its production of the several enzymes involved in

tryptophan synthesis. But it is equally important that the cell be able to turn the production of these enzymes back on when the level of tryptophan decreases again. This kind of control is made possible by the ability of the end-product of an anabolic pathway—in our example, tryptophan—to somehow *repress* (reduce or stop) the further production of the enzymes involved in its formation. This reduction in the expression of the enzyme-coding genes is called **end-product repression**. Most biosynthetic pathways in bacterial cells are regulated in this way. **Repression** is a general term in molecular genetics, referring to the reduction in expression of any regulated gene.

It is important to note that true genetic repression always has an effect on protein *synthesis*, and not just on protein *activity*. Recall from Chapter 6 that end-products of biosynthetic pathways often have an inhibitory effect on enzyme activity as well. This reduction of enzyme activity is called **feedback inhibition,** and it differs from repression in both mechanism and result. In feedback inhibition by the end-product of a pathway, molecules of enzyme are present but their catalytic activity is blocked; in end-product repression, the enzyme molecules are not even made.

Effector Molecules. One feature common to both repression and induction of enzyme synthesis is that the ultimate control is at the genetic level. Another feature is that the control mechanism is triggered by small organic molecules present in the cell's surroundings. Geneticists call small organic molecules that function in this way **effectors.** As we shall see shortly, effectors induce shape changes in certain allosteric proteins that control gene expression. For catabolic pathways, effectors are almost always substrates (lactose in our example), and they function as inducers of gene expression and, thus, of enzyme synthesis. For anabolic pathways, effectors are usually end-products (tryptophan in our example), and they usually lead to the repression of gene expression and thus repression of enzyme synthesis.

The Lactose System of E. coli and the Operon Concept

The classic example of an inducible enzyme system is the set of enzymes involved in lactose catabolism in the bacterium *Escherichia coli*—the enzymes that catalyze the steps shown in Figure 19-1. Much of what we currently know about the regulation of gene expression in bacteria, and most of the vocabulary used to express that knowledge, have come from studies of this system by French molecular geneticists François Jacob and Jacques Monod. Their classic paper, published in 1961, has probably had more influence on our understanding of gene regulation than any other work.

At the DNA level, the lactose system has separate genes encoding three enzymes (Figure 19-3). The *lacZ* gene codes for the enzyme β-galactosidase, which hydrolyzes lactose and other β-galactosides. A galactoside permease that transports lactose across the bacterial cell membrane is the product of the *lacY* gene. The *lacA* gene encodes a transacetylase, an enzyme that adds an acetyl group to lactose as it is taken up by the cell; the metabolic role of this enzyme is still obscure. Mapping studies showed that the genes *lacZ*, *lacY*, and *lacA* lie adjacent to one another on the genetic map of *E. coli*, as Figure 19-3 illustrates. This discovery led Jacob and Monod to suggest that the three genes might all belong to a single regulatory unit, or, as they called it, an **operon**—a cluster of genes with related functions, regulated in such a way that all the genes in the cluster are turned on and off together. The operon containing the lactose catabolism genes is called the *lac* operon. Operon genes such as *lacZ*, *lacY*, and *lacA* that actually code for enzymes (or other proteins) with related metabolic functions are called **structural genes,** to distinguish them from **regulatory genes,** which control the expression of the structural genes, as we shall see.

The organization of functionally related genes into operons is a feature commonly observed in prokaryotes but not in eukaryotic cells, in which individual genes are typically regulated as independent transcription units. Nevertheless, the operon model proposed by Jacob and Monod established several basic principles that have shaped our understanding of transcription regulation in both prokaryotic and eukaryotic systems.

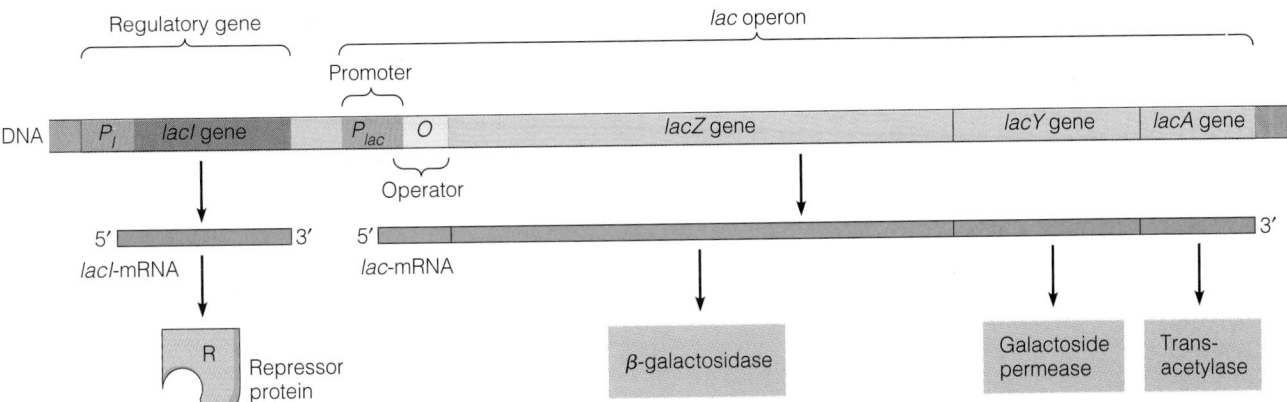

Figure 19-3   The Lactose (*lac*) Operon of *E. coli*. The *lac* operon consists of a segment of DNA that includes three contiguous structural genes (*lacZ*, *lacY*, and *lacA*), which are transcribed and regulated coordinately. The adjacent regulatory gene *lacI* codes for the *lac* repressor protein R. Both the regulatory gene and the *lac* operon itself contain promoters (P_I and P_{lac}) at which RNA polymerase binds and terminators at which transcription halts. P_{lac} overlaps with the operator site (O) to which the active form of the repressor protein binds. The operon is transcribed into a single long molecule of mRNA that codes for all three polypeptides. For details of regulation, see Figure 19-4.

The Operon Model in Detail

Focusing on the *lac* system, Jacob and Monod proposed a mechanism by which coordinated regulation might be accomplished. Their operon model, illustrated in an updated version in Figures 19-3 and 19-4, represents one of the truly important conceptual advances in biology. According to the model, clusters of genes coding for proteins with metabolically related functions are transcribed together into a single messenger RNA molecule. Messenger RNA encoding more than one polypeptide is found only in prokaryotic cells. It is called **polycistronic mRNA,** a term based on "cistron," a segment of DNA encoding a polypeptide—that is, a gene. (*Polygenic mRNA* is a synonym.) Transcription of an operon begins at a **promoter** (here P_{lac}), the site of RNA polymerase attachment, proceeds through all the structural genes, and ends at a terminator sequence.

The essence of the operon is not just that it is a unit of transcription, but that it is a unit of regulation. To understand how the operon is regulated, we need to consider three additional elements of the system: an operator site on the DNA, a repressor protein that binds to DNA, and the regulatory gene that codes for the repressor protein. Figures 19-3 and 19-4 depict each of these elements for the specific case of the *lac* operon.

The **operator** (*O*) is a nucleotide sequence located between the promoter and the first structural gene (actually overlapping the promoter sequence) that is recognized specifically by the **repressor,** a DNA-binding protein designated R in the figures. The repressor is encoded by a **regulatory gene** located outside the operon. In the lactose system, the regulatory gene is called the *lacI* gene (for inducibility) and happens to be located adjacent to the *lac* operon that it regulates. The *lacI* gene has its own promoter and termina-

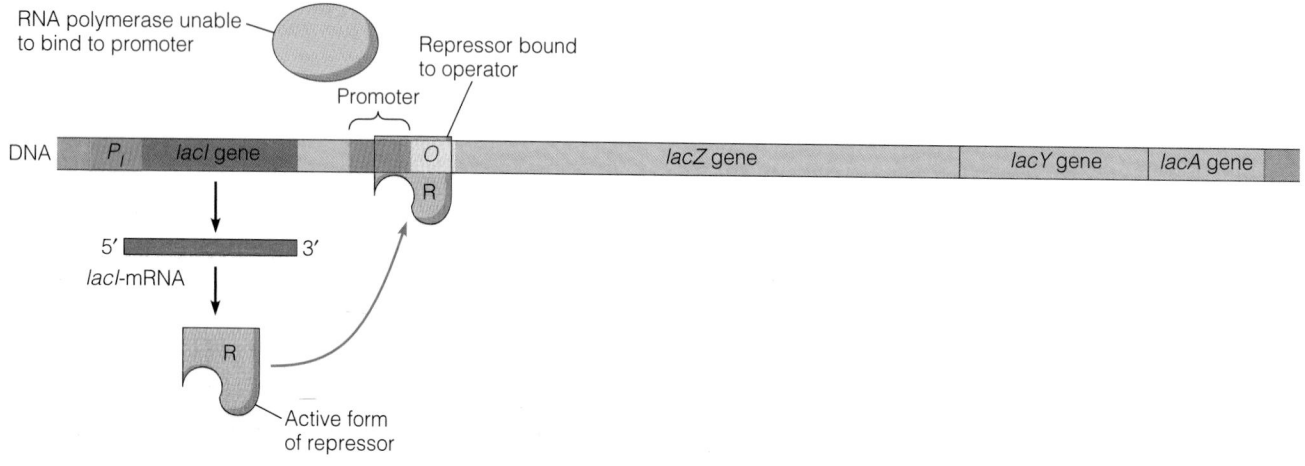

(a) Lactose absent, repressor bound to operator, operon repressed

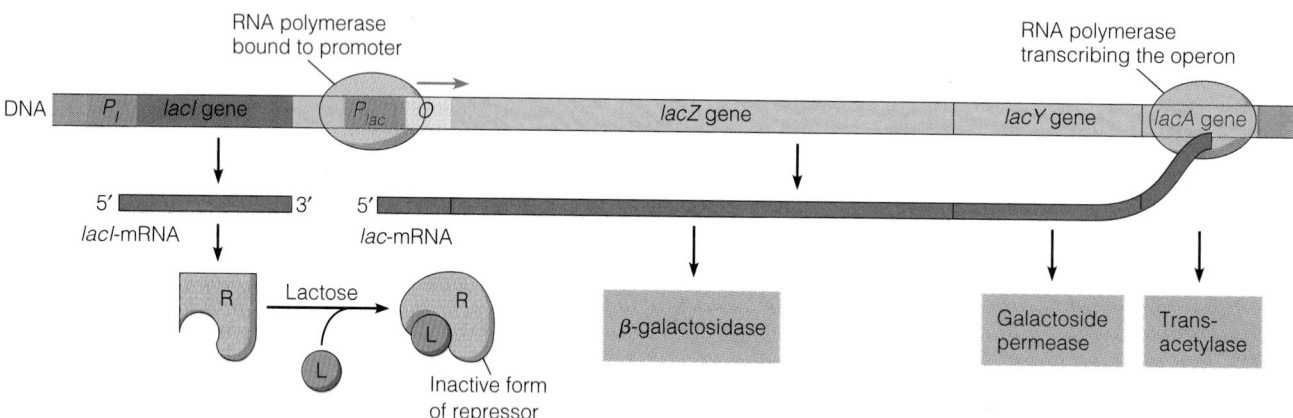

(b) Lactose present, repressor not bound to operator, operon derepressed

Figure 19-4   Regulation of the *lac* Operon.   Transcription of the *lac* operon is regulated by binding of the repressor protein to the operator. **(a)** In the absence of lactose, the repressor remains bound to the operator, and RNA polymerase cannot gain access to the promoter. Transcription is therefore blocked, and the operon remains repressed. **(b)** In the presence of lactose, the repressor is converted to the inactive form, which does not bind to the operator. RNA polymerase is therefore able to bind to the promoter and transcribe the structural genes *lacZ, lacY,* and *lacA* into a single polycistronic mRNA. The form of lactose that binds to the repressor is an isomer called allolactose (L).

tor. These elements—structural genes, promoter, operator, regulatory gene, and repressor protein—are fundamental to the operon model as we now understand it. (The promoter site was not part of the original model of Jacob and Monod because its existence was not recognized until later.)

Binding of the repressor protein to the operator site on the DNA regulates expression of the structural genes in the operon, as Figure 19-4 illustrates. When the repressor is bound to the operator (Figure 19-4a), RNA polymerase cannot bind to the promoter, and transcription of the structural genes is not possible. Binding of the repressor to the operator thus inactivates the operon and keeps its structural genes turned off. Without repressor bound to it, however, the operator site is open (Figure 19-4b). In this case, RNA polymerase can bind to the promoter and proceed down the operon, transcribing the *lacZ*, *lacY*, and *lacA* genes into a single polycistronic mRNA molecule.

The Repressor: An Allosteric Protein

Figure 19-4 reveals a crucial feature of a repressor protein: its ability to exist in two forms, only one of which binds to the operator. In other words, the repressor is an *allosteric protein*. As we learned in Chapter 6, an allosteric protein can exist in either of two conformational states, depending on whether or not the appropriate effector molecule is present. In one state the protein is active, and in the other state the protein is inactive. When the effector molecule binds to the protein, it induces a change in the conformational state of the protein and therefore in its activity. The binding is readily reversible, however, and departure of the effector results in the protein's rapid return to the alternative form.

Figure 19-5 shows the reversible interaction of the *lac* repressor with its effector, which is actually not lactose itself but rather *allolactose*, an isomer of lactose produced after lactose enters the cell. The conformational form assumed by the repressor protein in the absence of allolactose recognizes and binds to the operator, whereas the form with allolactose attached to it does not. The result is that the repressor protein is active (binds to the operator) in the absence of allolactose, when there is no need to produce the catabolic enzymes encoded by the *lac* operon. However, in the presence of allolactose, the repressor converts to its inactive form, which does not recognize the operator and hence does not prevent the transcription of the structural genes by RNA polymerase. In this indirect way, lactose triggers the *induction* of the enzymes encoded by the *lac* operon. Because the *lac* operon is turned off unless induced, it is said to be an **inducible operon**. To borrow a term from the world of computers, the "default state" of an inducible operon is *off*.

Genetic Analysis of the Lactose Operon

Most of the initial evidence in support of the operon model was based on genetic analyses of mutant bacteria that either produced abnormal amounts of the enzymes of the *lac* operon or showed abnormal responses to the addition or removal of lactose. These mutations were found to map either

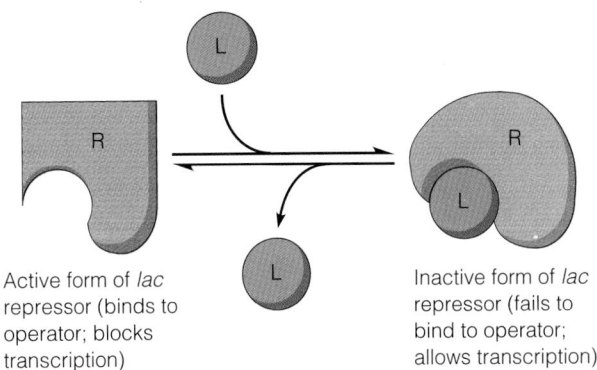

Active form of *lac* repressor (binds to operator; blocks transcription)

Inactive form of *lac* repressor (fails to bind to operator; allows transcription)

Figure 19-5 Allosteric Regulation of the *lac* Repressor.
The *lac* repressor (R) is an allosteric protein, capable of reversible conversion between two alternative forms. In the absence of the effector allolactose (L), the protein assumes the form that is active in binding to the *lac* operator. In the presence of the effector, the protein exists preferentially in the alternative conformational state, which does not recognize the operator and is therefore inactive as a repressor of transcription.

in the structural genes (*lacZ*, *lacY*, or *lacA*) or in the regulatory elements of the system (*O*, P_{lac}, or *lacI*). These two classes of mutations can be readily distinguished, because mutations in a structural gene affect only a single protein, whereas mutations in regulatory regions almost always affect expression of all the structural genes coordinately.

Table 19-1 summarizes these mutations and their phenotypes, starting with the inducible phenotype of the wild-type (nonmutant) cell (line 1). The plus signs in the phenotype columns of the table indicate high enzyme levels. The minus signs indicate very low enzyme levels, although not the complete cessation of enzyme production. Even in the absence of lactose, wild-type cells make small amounts of the *lac* enzymes because the binding of repressor to operator is reversible—the active repressor occasionally "falls off" the operator.

Experiments such as those represented in Table 19-1 are usually carried out using the synthetic β-galactoside *isopropylthiogalactoside* (IPTG) rather than lactose or allolactose as the inducing molecule. IPTG is a good inducer of the system but cannot itself be metabolized by the cell, so its use avoids possible complications due to changes in the level of effector caused by its catabolism. Following common usage, we refer to effectors of inducible operons as **inducers**. But keep in mind that an inducer is not exactly the "opposite" of a repressor, as the words may imply. An inducer is a small molecule that binds to a protein; a repressor is a protein that binds to DNA.

Examination of Table 19-1 should help clarify the operon model and should also illustrate the power of genetic analysis. Simply by observing the phenotypes of various mutants and combining these observations with the results of genetic mapping experiments, Jacob and Monod were able to formulate the operon model of gene regulation. We will now consider in turn each of the six kinds of mutations shown in lines 2–7 of the table.

Table 19-1 Genetic Analysis of Mutations Affecting the *lac* Operon

Line Number	Genotype of Bacterium*	Phenotype with Inducer Absent		Phenotype with Inducer Present	
		β-galactosidase	Permease	β-galactosidase	Permease
1	$I^+ P^+ O^+ Z^+ Y^+$	−	−	+	+
2	$I^+ P^+ O^+ Z^+ Y^-$	−	−	+	−
3	$I^+ P^+ O^+ Z^- Y^+$	−	−	−	+
4	$I^+ P^+ O^c Z^+ Y^+$	+	+	+	+
5	$I^+ P^- O^+ Z^+ Y^+$	−	−	−	−
6	$I^s P^+ O^+ Z^+ Y^+$	−	−	−	−
7	$I^- P^+ O^+ Z^+ Y^+$	+	+	+	+

*$P = P_{lac}$.

Mutations in the Structural Genes.

Mutations in the *lacY* or *lacZ* genes of the *lac* operon can result in the production of altered proteins with little or no biological activity, even in the presence of inducer (Table 19-1, lines 2 and 3). Such mutants are therefore unable to utilize lactose as a carbon source, either because they cannot transport lactose into the cell (Y^- mutants) or because they cannot cleave the glycosidic bond between galactose and glucose in the lactose compound (Z^- mutants). Note that mutations in a bacterial gene or regulatory sequence that render it defective are usually indicated with a superscript minus sign; thus, for example, the genotype of a bacterium carrying a defective *Y* gene is written Y^-. The wild-type allele is indicated with a superscript plus sign.

Mutations in the Operator.

Mutations in the operator can lead to a constitutive phenotype; that is, the mutant cells produce the *lac* enzymes constantly, whether inducer is present or not (Table 19-1, line 4). These mutations change the nucleotide sequence of the operator DNA so that it is no longer recognized by the repressor. The genotype of such *operator-constitutive mutants* is represented as O^c. As is expected for mutations in a regulatory site, O^c mutations simultaneously affect the synthesis of all three structural genes in the same way.

Mutations in the Promoter.

Although the promoter was not a part of the original operon model, it is now known to be the binding site for RNA polymerase and therefore an essential feature of any operon. Promoter mutations decrease the affinity of the polymerase enzyme for the promoter. This change in binding strength can be measured experimentally as a change in the binding constant for the interaction between the promoter and RNA polymerase. As a result, fewer RNA polymerase molecules bind per unit time to the promoter, and the rate of mRNA production decreases. (However, once an RNA polymerase molecule attaches to the DNA and begins transcription, it elongates the mRNA molecule at a normal rate.) Usually, promoter mutations (P^-) in the *lac* operon decrease both the elevated level of enzyme produced in the presence of inducer *and* the already low, basal level of *lac* enzyme production that the cell manages to achieve in the absence of inducer (Table 19-1, line 5).

Mutations in the Regulatory Gene.

Mutations in the *lacI* gene are of two types. Some mutants fail to produce any of the *lac* enzymes, regardless of whether inducer is present, and are therefore called *superrepressor mutants* (I^s in Table 19-1, line 6). The repressor molecule in such mutants either has lost its ability to recognize and bind the inducer but can still recognize the operator, or else it has a high affinity for the operator regardless of whether inducer is bound to it. In either case, the repressor binds tightly to the operator and represses transcription, and hence enzyme synthesis, under all conditions.

The other class of *lacI* mutations causes synthesis of a mutant repressor protein that is not able to recognize the operator (or, in some cases, is not synthesized at all). The *lac* operon in such I^- mutants cannot be turned off, and the enzymes are therefore synthesized constitutively (Table 19-1, line 7).

I^- mutants, along with the O^c and P^- mutants, illustrate the importance of specific recognition of DNA sequences by regulatory proteins in the regulation of gene transcription. A small change—either in the DNA sequence of promoter or operator, or in the regulatory protein that binds to the operator—can dramatically affect expression of all the genes in the operon.

The *Cis-Trans* Test Using Partially Diploid Bacteria.

The existence of two different kinds of constitutive mutants, O^c and I^-, raises the question of how one type might be distinguished from the other. A *cis-trans* test can be used to differentiate between *trans*-acting mutations, which affect proteins (e.g., I^-), and *cis*-acting mutations, which affect DNA binding sites (e.g., O^c). In Latin, *cis* means "on this side" and *trans* means "on the other side"; the meanings of these terms in the present context will be clarified below.

The basis of the *cis-trans* test is a cell (or organism) that has two different copies of the DNA segment of interest. As you know, *E. coli* is usually haploid, but Jacob and Monod constructed partially diploid bacteria by inserting a second copy of the *lac* portion of the bacterial genome into the F-factor plasmid of F⁺ cells. This second copy could be transferred by conjugation into a host bacterium of any desired *lac* genotype to create partial diploids such as the types listed in Table 19-2. If only one copy of the operon contains an I^- or O^c regulatory mutation, one can determine whether the mutation has an effect on both copies of the operon or only on the copy of the operon where it is located. The mutation is said to act in *trans* if the structural genes in both copies of the operon are affected; the effect of the mutation somehow reaches "the other side"—that is, the other copy of the operon. In contrast, the mutation is said to act only in *cis* if the only structural genes affected are those physically linked to—on "the same side" as—the mutant locus. To determine which of the two copies of the *lac* operon is being expressed in each combination, one copy contains a defective *Z* gene while the other contains a defective *Y* gene.

In the examples in Table 19-2, you can see that a diploid with both an I^+ allele and an I^- allele in the same cell is inducible for both β-galactosidase and permease, even though the functional gene for one enzyme is physically linked to a defective *I* gene (Table 19-2, line 3). We now know that this occurs because the one functional *I* gene present in the cell produces active repressor molecules that can diffuse through the cytosol and bind to both operator sites in the absence of lactose. The repressor is said to be a **trans-acting factor.**

Quite different results are obtained with partial diploids containing both the O^+ and O^c alleles (Table 19-2, line 4). In this case, structural genes linked to the O^c allele are constitutively transcribed, whereas those linked to the wild-type allele are inducible. The *O* locus, in other words, acts only in *cis*; it affects the behavior of structural genes only in the operon of which it is physically a part. Such *cis* specificity is characteristic of mutations that affect DNA binding sites rather than protein products of genes. Like other noncoding DNA sequences that are involved in the regulation of gene expression, the *O* site is said to be a **cis-acting element.**

The trp Operon: A Repressible Operon

Although much of the initial work that led to the operon concept was done with the *lac* system of *E. coli*, a number of other regulatory systems in this and other species of bacteria have since been shown to follow the same general pattern: The genes coding for enzymes of a particular metabolic pathway are clustered in a region of the bacterial chromosome that serves as a unit of both transcription and regulation. One or more operators, promoters, and regulatory genes are usually involved, although there is sufficient variation from one operon to another to preclude many generalizations.

You have seen that the *lac* operon regulates the appearance of enzymes involved in a specific catabolic (degradative) pathway. The *lac* operon and many others like it are *inducible;* they are turned *on* by a specific allosteric effector, usually the substrate for that pathway. In contrast, operons that regulate enzymes involved in anabolic (biosynthetic) pathways are **repressible operons;** they are turned *off* allosterically, usually by an effector that is the end-product of the pathway. The tryptophan (*trp*) operon is a good example of a repressible operon (Figure 19-6). The *trp* operon contains the structural genes for the enzymes that catalyze the reactions in the pathway for tryptophan synthesis, as well as the control elements (DNA sequences) necessary to regulate the synthesis of these enzymes. The effector molecule in this case is the amino acid tryptophan.

Expression of the enzymes of the *trp* operon is repressed in the presence of tryptophan (Figure 19-6a) and derepressed in its absence (Figure 19-6b). Thus, unlike the *lac* system, the regulatory gene for this operon, called *trpR*, codes for an allosteric repressor protein that is active (binds to operator DNA) when the effector is attached to it and inactive in its free form. The effector in such systems (in this case, tryptophan) is sometimes called a **corepressor** because it is required, along with the repressor protein, to shut off transcription of the operon.

Table 19-2 Diploid Analysis of Mutations Affecting the *lac* Operon

Line Number	Genotype of Diploid Bacterium*	Phenotype with Inducer Absent		Phenotype with Inducer Present	
		β-galactosidase	Permease	β-galactosidase	Permease
1	$I^+ P^+ O^+ Z^+ Y^+ / I^+ P^+ O^+ Z^+ Y^+$	−	−	+	+
2	$I^+ P^+ O^+ Z^- Y^+ / I^+ P^+ O^+ Z^+ Y^-$	−	−	+	+
3	$I^+ P^+ O^+ Z^- Y^+ / I^- P^+ O^+ Z^+ Y^-$	−	−	+	+
4	$I^+ P^+ O^+ Z^- Y^+ / I^+ P^+ O^c Z^+ Y^-$	+	−	+	+
5	$I^+ P^+ O^+ Z^- Y^+ / I^{s} P^+ O^+ Z^+ Y^-$	−	−	−	−

*$P = P_{lac}.$

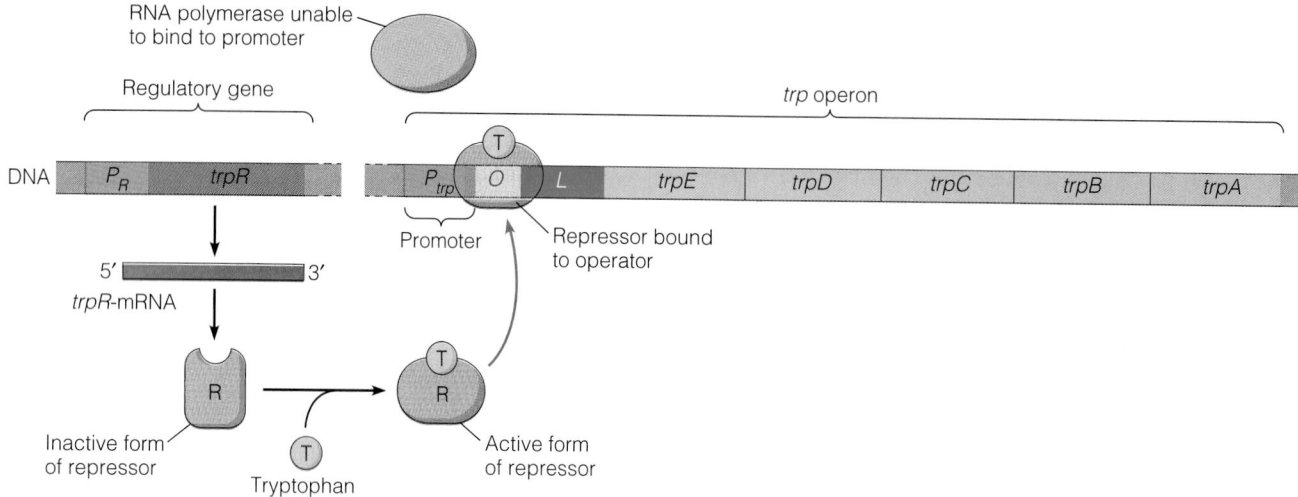

(a) Tryptophan present, repressor bound to operator, operon repressed

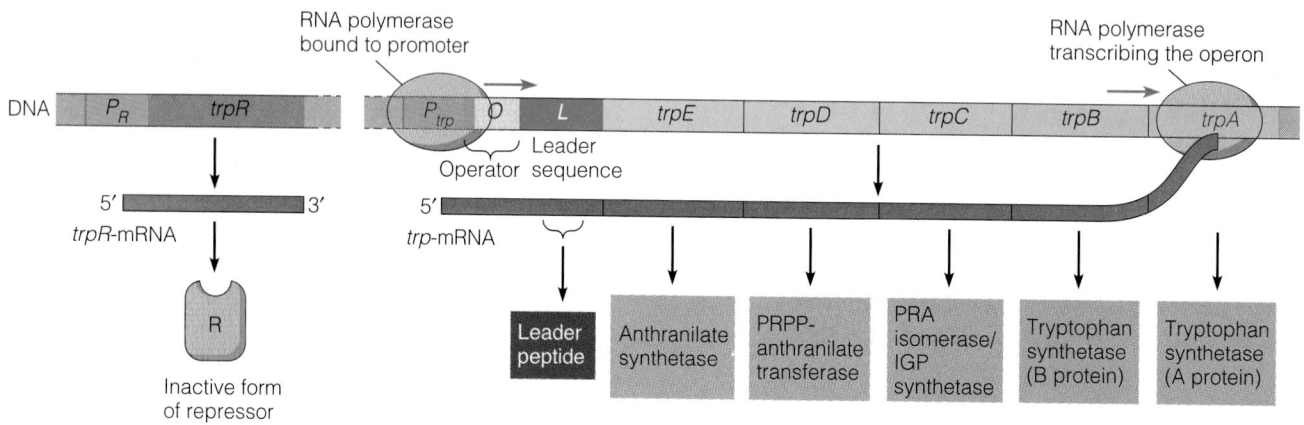

(b) Tryptophan absent, repressor not bound to operator, operon derepressed

Figure 19-6   The Tryptophan (*trp*) Operon of *E. coli*.   (a) The *trp* operon consists of a segment of DNA that includes five contiguous structural genes (*trpE, trpD, trpC, trpB,* and *trpA*) as well as promoter (P_{trp}), operator (*O*), and leader (*L*) sequences. The structural genes are transcribed and regulated as a unit. The resulting polycistronic message codes for the enzymes of the tryptophan biosynthetic pathway (see Figure 19-2). The repressor protein, encoded by the *trpR* gene, is inactive (cannot recognize the operator site) in the free form but is active when complexed with tryptophan. In the presence of tryptophan, the repressor is converted to the active form and binds tightly to the operator, thereby blocking access of RNA polymerase to the promoter and keeping the operon repressed. **(b)** In the absence of tryptophan, the repressor does not bind to the operator site. RNA polymerase is therefore able to bind to the promoter and transcribe the structural genes, giving the cell the capability to synthesize tryptophan. An additional role in regulating expression of this operon is played by the leader segment of the mRNA, as will be explained in Figures 19-8 and 19-9.

The Concept of Negative Control of Transcription

As we have seen, repressor proteins can control the transcription of genes for enzymes of both catabolic and anabolic pathways, but the active (DNA-binding) form of the repressor is the effector-free molecule for a catabolic pathway and the effector-bound molecule for an anabolic pathway. Regardless of which is the active form of the repressor, the effect of repressor action is always the same: The repressor prevents transcription of the operon under appropriate conditions by blocking the proper attachment of RNA polymerase to the DNA. Repressors, in other words, never specifically turn anything on; their effect is always to turn off gene expression (or to keep it turned off). The operon is therefore a system of **negative control** in that the active form of repressor protein, its key regulatory protein, always acts by turning off expression of the operon.

Catabolite Repression: An Example of Positive Control of Transcription

The transcription of some bacterial operons is under **positive control,** meaning that the active form of a key regulatory protein *turns on* expression of the operon. An important example of positive control of transcription is actually a mechanism for regulating operons that are also under negative control. This is **catabolite repression,** a glucose-sensitive mechanism that acts on inducible operons encoding catabolic enzymes. In catabolite repression, the key regulatory protein is a positive regulator whose activity is controlled by the level of glucose in the cell. To understand the value of this sensitivity to glucose, we need to recognize that glucose is the preferred energy source for almost all prokaryotic cells (and for most eukaryotic cells, too). This is because the enzymes of the glycolytic and tricarboxylic acid (TCA) pathways are present constitutively in most cells, so glucose can be catabolized at any time without the synthesis of additional enzymes. Although molecules other than glucose can also be used for energy, their catabolism always requires the synthesis of one or more additional enzymes. Consequently, when presented with a choice of carbon sources, cells usually show a preference for glucose. For example, *E. coli* cells grown in the presence of both glucose and lactose use the glucose preferentially and have very low levels of the enzymes encoded by the *lac* operon, despite the presence of the inducer for that operon.

Catabolite repression guarantees that carbon sources other than glucose will be used only when glucose is not available. Like the other regulatory mechanisms we have encountered, catabolite repression depends on an allosteric regulatory protein and a small organic effector. The actual effector is not glucose, but rather a secondary signal that indirectly reflects the level of glucose in the cell. This secondary signal is a form of AMP called **3′,5′-cyclic AMP,** or **cAMP.** (See Figure 23-7 for more details about cAMP.) The level of cAMP is high in the absence of glucose and low in its presence. Glucose seems to act by indirectly inhibiting adenylate cyclase, the enzyme that catalyzes the synthesis of cAMP from ATP. So the more glucose present, the less cAMP is made, and vice versa.

How does cAMP bring about positive control of an appropriate operon? As already mentioned, cAMP is the effector molecule for catabolite repression. Like other effectors, it acts by binding to an allosteric regulatory protein (Figure 19-7a). In this case, the regulatory protein, called **cAMP receptor protein (CRP),** is an activator of transcription. (An older name for this protein is catabolite activator protein, or CAP.) By itself, CRP is inactive, but when complexed with cAMP, CRP changes to a shape that enables it to bind to a particular nucleotide sequence within the operon. This sequence, the *CRP recognition site,* is located upstream of the promoter. Figure 19-7b shows the location of the CRP recognition site (labeled *C*) in the *lac* operon; similar sites

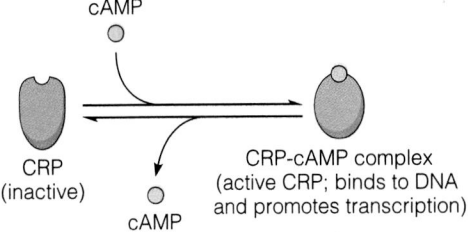

(a) Allosteric activation of CRP

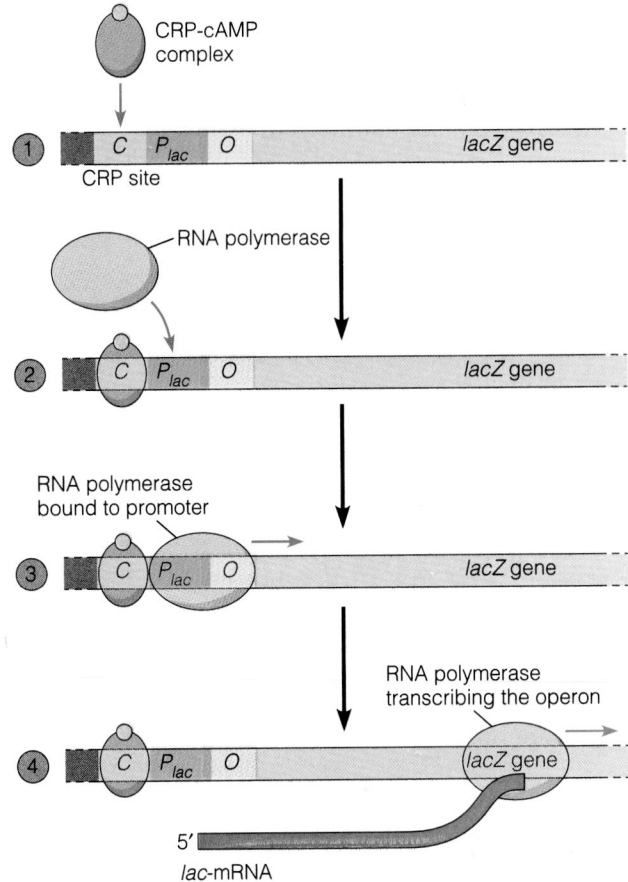

(b) Action of CRP-cAMP complex (active CRP)

Figure 19-7   The cAMP Receptor Protein (CRP) and Its Function.   (a) Allosteric activation of CRP. CRP is an allosteric protein that is inactive in the free form but is converted to the active form by binding to cyclic AMP (cAMP). The CRP-cAMP complex binds at or near the promoter of a variety of inducible operons, increasing the affinity of the promoter for RNA polymerase and thereby stimulating transcription. (Compare with Figure 19-5.) **(b)** The effect of active CRP on the *lac* operon. ① The CRP-cAMP complex binds to the CRP site (*C*) near the promoter region, thereby ② making the promoter more readily bound by RNA polymerase. ③ RNA polymerase binds to the promoter and ④ transcribes the operon.

are found in a variety of inducible operons. To repeat what we said previously, CRP has a *positive* effect on gene regulation. When CRP, in its active form, attaches to its recognition site, the binding of RNA polymerase to the promoter is greatly enhanced, thereby stimulating the initiation of transcription. In the *lac* operon, for example, transcription is increased fiftyfold in this way.

As the level of glucose in the cell rises and the concentration of cAMP falls, CRP is converted to its inactive form, and its stimulatory effect on the transcription of inducible operons is abolished. The cell then resorts to its preferential use of glucose as an energy source. Conversely, when the glucose level falls, the cAMP level rises, activating CRP by binding to it. The CRP-cAMP complex greatly enhances transcription of an inducible operon such as the *lac* operon—*provided,* of course, that the repressor for that operon has been inactivated by the presence of its effector (i.e., allolactose).

Dual Control of Inducible Operons

As we have seen, inducible operons are often under two kinds of control, rendering such operons sensitive to two kinds of signals. The repressor-operator interaction provides a sensitivity to the presence of an alternative energy source (such as lactose) that turns on a particular operon. The effect of CRP, on the other hand, is to render the level of transcription of that operon sensitive to the glucose concentration of the cell, as mediated by the cAMP level. An *E. coli* cell might therefore have very low levels of β-galactosidase either because it is growing in the absence of lactose and its *lac* operon is fully repressed, or because it has access to a large supply of glucose. In the latter case, the glucose will suppress the cAMP level, and CRP will therefore be inactive and incapable of stimulating transcription of the *lac* operon, even if the operon is otherwise derepressed by the presence of lactose.

It is an unfortunate feature of genetic nomenclature that the repressor-operator system of negative control has so positive-sounding a name as "induction," whereas the CRP-cAMP system of positive control has so negative-sounding a name as catabolite "repression." In each case, the name reflects the observed effect of a particular environmental signal: Lactose in the cellular environment induces (turns on) the synthesis of specific enzymes, whereas glucose represses (turns off) the synthesis of the same enzymes. At the molecular level, however, lactose (via allolactose) interferes with the negative effect of a repressor protein, and glucose interferes with the positive effect of an activator protein.

To keep negative and positive types of transcriptional controls clear in your mind, ask about the primary effect of the regulatory protein that binds to the operon DNA. If, in binding to the DNA, the regulatory protein prevents or turns off transcription, then it is part of a negative control mechanism. If, on the other hand, its binding to DNA results in the activation or enhancement of transcription, then the regulatory protein is part of a positive control mechanism.

Regulation of Transcription Initiation by Alternative Sigma Factors

In addition to regulatory proteins that act as activators or repressors of transcription, the bacterial cell makes use of a variety of different sigma (σ) factors in controlling the initiation of transcription. Recall from Chapter 17 that the proper initiation of transcription in bacteria requires the RNA polymerase core enzyme to be combined with a protein called a sigma factor, which is necessary for the recognition of the promoter. Each bacterial cell contains several different types of sigma factors, which are utilized in specific situations. The most common form of prokaryotic RNA polymerase holoenzyme contains a sigma factor called the σ^{70} subunit (molecular weight 70 kDa) and initiates transcription at most prokaryotic promoters. However, changes in a cell's environment, such as an increase in temperature (heat shock), can favor binding of an alternative sigma factor—in this case, σ^{32}—to the RNA polymerase core enzyme. With this sigma factor bound, promoter recognition is slightly altered to favor a shift in initiation of transcription to a special set of genes encoding proteins that help the cell adapt to the altered environment. Yet another sigma factor, σ^{54}, enables RNA polymerase to transcribe preferentially genes involved in nitrogen utilization. For both σ^{32} and σ^{54}, the promoter DNA sequence is somewhat different from the usual (consensus) sequence, and its recognition by an alternative sigma factor enables the cell to respond to the changes in its environment. In addition, most bacteriophages take over transcription in infected cells by using specific sigma factors that recognize only the viral promoters.

Attenuation: Regulation After the Initiation of Transcription

All the regulatory mechanisms discussed so far control the *initiation* of transcription. Prokaryotes also employ some regulatory mechanisms that operate following transcription initiation. One of the most important, attenuation, was discovered when Charles Yanofsky and his colleagues found that the *trp* operon had a novel type of negative regulatory site located between the promoter/operator and the first structural gene of the operon, *trpE*. This stretch of DNA, called the **leader sequence** (or *L*)**,** is transcribed to produce the leader mRNA at the 5′ end of the mRNA (see Figure 19-6b). The *trp* leader mRNA, they found, is 162 nucleotides long.

Analysis of the *trp* operon transcripts produced under various conditions showed that, as expected, the full-length, polycistronic *trp* mRNA is transcribed under conditions of tryptophan scarcity. In this way, more tryptophan can be synthesized as the cell needs it. When tryptophan is plentiful, the genes encoding the enzymes of the tryptophan biosynthetic pathway are not transcribed, also as expected; however, the DNA corresponding to most of the leader sequence *is* transcribed. From these findings, Yanofsky concluded that the *trp* operon must contain within its leader sequence an additional control sequence that is sensitive to tryptophan levels. This control sequence somehow determines not whether *trp* operon RNA synthesis can begin, but whether it can continue to completion. The effect of this control element was called **attenuation** because of its role in attenuating, or reducing, the synthesis of mRNA.

The mechanism of attenuation depends on the tight coupling of transcription and translation in prokaryotes—the fact that protein synthesis begins before the mRNA is completed. To understand how attenuation works, we must start with a closer look at the *trp* operon leader mRNA segment (Figure 19-8). This leader mRNA has two unusual features that enable it to play a regulatory role. First, a portion of the leader sequence is translated into a **leader peptide,** which is 14 amino acids long. Within this coding sequence are two adjacent codons for the amino acid tryptophan; these will prove important. Second, the *trp* leader mRNA also contains four segments (labeled regions 1, 2, 3, and 4) whose nucleotides can base-pair with each other to form several distinctive hairpin (stem-and-loop) structures. As indicated in Figures 19-8 and 19-9a, the part of the RNA comprising regions 3 and 4 and a string of eight U nucleotides is called the **attenuator.** When base pairing between regions 3 and 4 creates a hairpin, the attenuator acts as a transcription termination signal. Recall that the typical eukaryotic termination signal shown in Figure 17-11 was also a hairpin followed by a string of Us. The formation of such a structure causes RNA polymerase and the growing RNA chain to detach from the DNA.

Yanofsky's experiments suggested that translation somehow plays a crucial role in the attenuation mechanism. We are now ready to consider that role. A ribosome attaches to its first binding site on the *trp* mRNA as soon as the site appears, and from there it follows close behind the RNA polymerase. When tryptophan levels are low (Figure 19-9b), the concentration of tryptophanyl tRNA (tRNA molecules carrying tryptophan) is also low. Thus, when the ribosome arrives at the tryptophan codons, it stalls briefly, "awaiting"

the arrival of tryptophanyl tRNA. The stalled ribosome blocks region 1, allowing an alternative hairpin structure to form by pairing of regions 2 and 3. With region 3 tied up in this way, it cannot pair with region 4 to create a termination structure, and the polymerase continues, eventually producing a complete transcript of the *trp* operon. Ribosomes use the mRNA to synthesize the tryptophan pathway enzymes, and production of the amino acid increases.

However, if tryptophan is plentiful and tryptophanyl-tRNA levels are high, the ribosome does not stall at the tryptophan codons (Figure 19-9c). Instead, the ribosome continues to the stop codon at the end of the coding sequence for the leader peptide and pauses there, blocking region 2, before its release from the RNA. This pause permits the formation of the 3–4 hairpin, which is the transcription termination signal. Transcription stops after 141 nucleotides, near the end of the leader sequence. The result is that no additional enzyme synthesis or tryptophan synthesis occurs.

Since the elucidation of attenuation in the *trp* operon, attenuators have been found in numerous other operons, mainly those that code for enzymes involved in amino acid biosynthesis. In some cases, attenuation appears to be the only means of regulation for the operon; in other cases, the attenuator site complements the operator in the regulation of gene expression, as with the *trp* operon. When the cellular level of the relevant amino acid is high, the repressor-effector complex binds to the operator, effectively blocking initiation of transcription. As the level of effector decreases, the operon becomes derepressed, and transcription begins. Fine-tuning of the system then occurs at the attenuator site, allowing greater numbers of RNA polymerase molecules to proceed past the attenuator as the effector becomes scarcer.

While the tight coupling of transcription and translation required for attenuation is the norm in prokaryotic cells, it cannot occur in eukaryotes because transcription in eukaryotic cells occurs in the nucleus, whereas translation is a cytoplasmic event. It appears, then, that attenuation is an exclusively prokaryotic process.

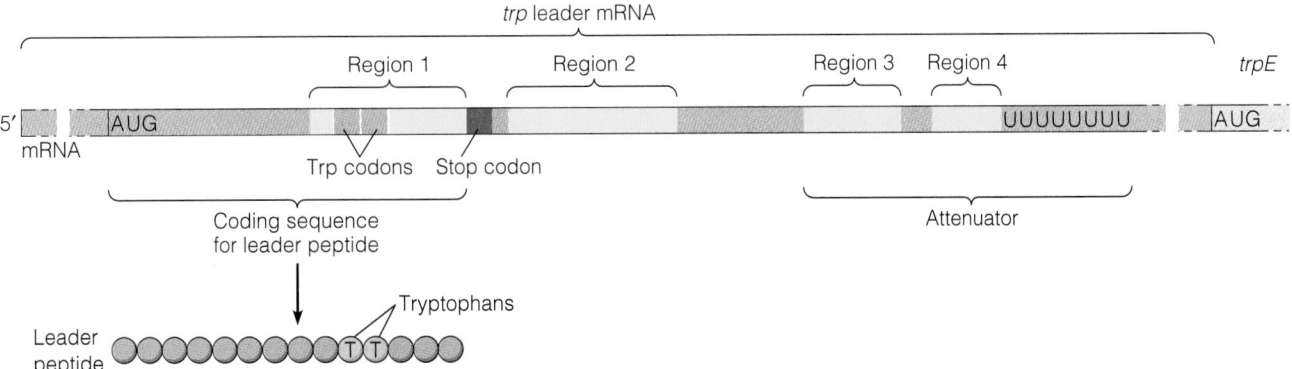

Figure 19-8   The Leader of *trp* mRNA.   The transcript of the *trp* operon includes 162 nucleotides upstream of the initiation codon for *trpE,* the first structural gene. This leader mRNA includes a section encoding a leader peptide of 14 amino acids; two adjacent tryptophan (Trp) codons play an important role in the operon's regulation by attenuation. The leader mRNA also contains four regions capable of base pairing in various combinations to form hairpin (stem-and-loop) structures, as the next figure shows. A part of the leader mRNA containing regions 3 and 4 and a string of eight Us is called the attenuator.

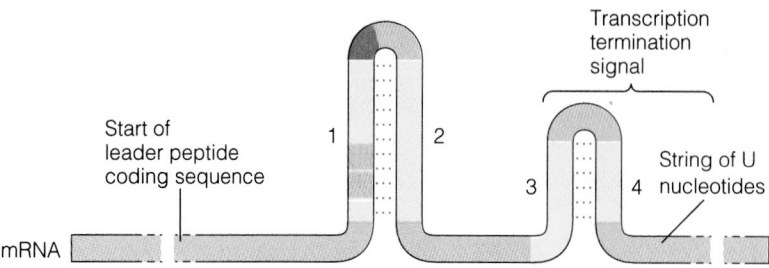

(a) The most stable secondary structures for *trp* leader mRNA

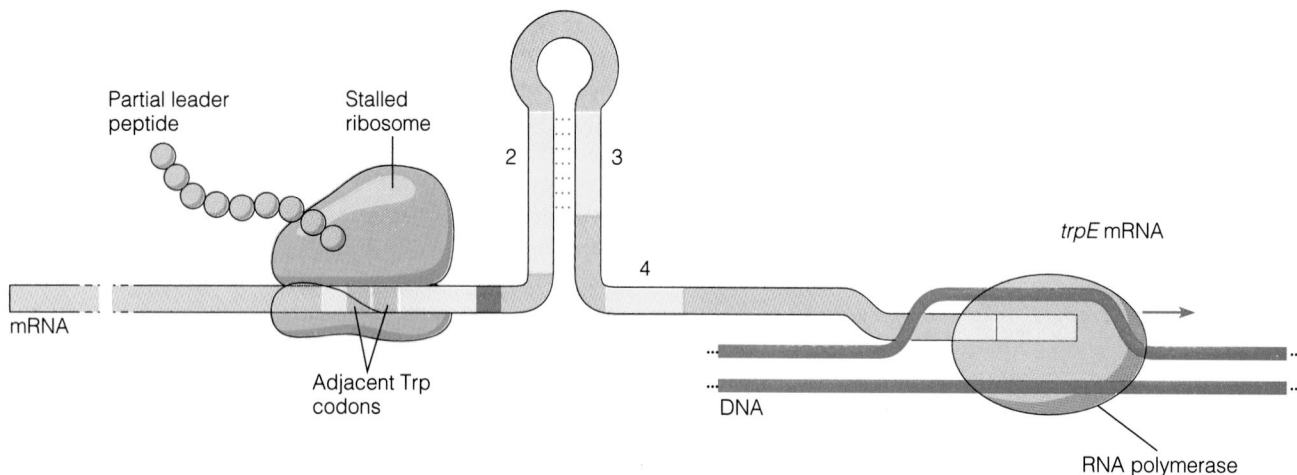

(b) When tryptophan is scarce, ribosome stalls, allowing a 2-3 hairpin to form; RNA polymerase continues transcription

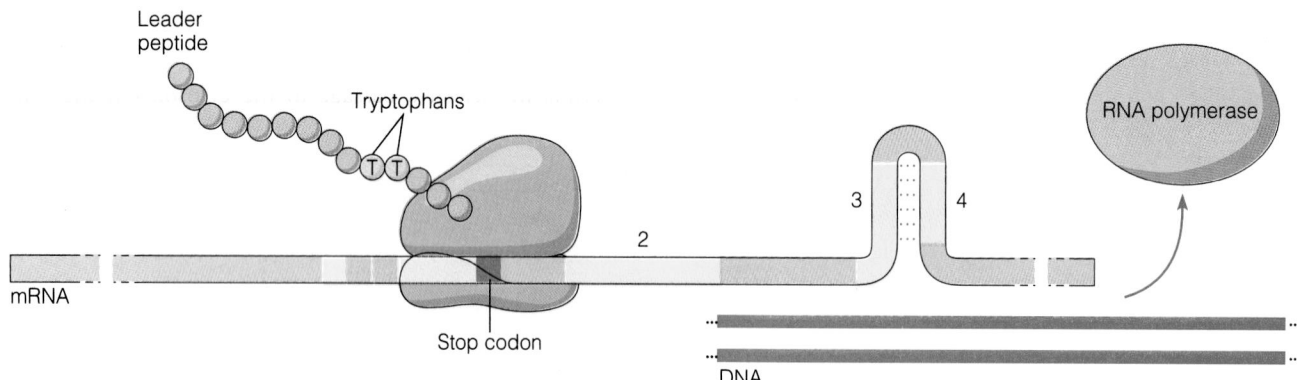

(c) When tryptophan is plentiful, ribosome continues, and RNA polymerase stops transcribing

Figure 19-9   Attenuation in the *trp* Operon.   (a) Attenuation depends upon the ability of regions 1 and 2 and regions 3 and 4 of the *trp* leader sequence to base-pair, forming hairpin secondary structures. The 3–4 hairpin structure acts as a transcription termination signal; as soon as it forms, the RNA and the RNA polymerase are released from the DNA. **(b)** During periods of tryptophan scarcity, a ribosome translating the coding sequence for the leader peptide may stall when it encounters the two tryptophan (Trp) codons because of the shortage of tryptophan-carrying tRNA molecules. Because a stalled ribosome at this site blocks region 1, a 1–2 hairpin cannot form, and an alternative, 2–3 hairpin is created. The 2–3 base pairing prevents formation of the 3–4 transcription termination hairpin, and therefore RNA polymerase can move on to transcribe the entire operon. **(c)** When tryptophan is readily available, a ribosome can complete translation of the leader peptide without stalling. As it pauses at the stop codon, it blocks region 2, preventing it from base pairing. As a result, the 3–4 structure forms and terminates transcription near the end of the leader sequence.

Gene Regulation in Eukaryotes

In the early days of molecular biology, a popular adage claimed that "what is true of *E. coli* is also true of elephants." This maxim aptly expressed the initial conviction of many microbial geneticists that almost everything learned about bacterial function at the molecular level would also be applicable to eukaryotes. Not surprisingly, however, the adage is only partly true. In terms of basic metabolic pathways, mechanisms for the transport of solutes across membranes, and such fundamental features as DNA structure, protein synthesis, and enzyme function, there are many similarities between the prokaryotic and eukaryotic worlds, and findings from studies with prokaryotes can be extrapolated to eukaryotes with considerable confidence. But when the discussion turns to the regulation of gene expression, the comparison needs to be scrutinized.

There are certainly basic similarities between prokaryotic and eukaryotic regulation of gene expression. Chief among these is the importance of transcription initiation as a key control point and the general manner in which control at that point is effected. In eukaryotes as well as prokaryotes, many genes are controlled by regulatory proteins (*trans*-acting factors) that interact with specific DNA sequences (*cis*-acting elements) to turn transcription on or off.

However, there are also dramatic differences between prokaryotic gene regulation and eukaryotic regulation. As we will see shortly, transcriptional regulation is much more elaborate in the eukaryotic cell, involving many more regulatory proteins and DNA control elements. Moreover, while some eukaryotic regulatory proteins interact directly with DNA (as do most prokaryotic regulatory proteins), many others act on DNA only indirectly, via interactions with other proteins.

Furthermore, gene expression in eukaryotes is controlled at a number of levels other than transcription. Eukaryotic cells can regulate gene expression at a variety of points in the flow of genetic information from a gene in the nucleus to an active enzyme (or other gene product) in the cytosol or elsewhere in the cell. The greater diversity of regulatory mechanisms in eukaryotic cells reflects basic differences between prokaryotes and eukaryotes in the structure, organization, and compartmentalization of the genome, as well as some fundamental features of the multicellular way of life. We will now review some of these differences, focusing on ones that are especially relevant to the regulation of gene expression.

Differences Between Prokaryotic and Eukaryotic Cells

Genome Size and Complexity. As discussed in Chapter 14, eukaryotic genomes are almost always much larger than those of prokaryotes—often by two or three orders of magnitude, or more. For most eukaryotes, genes constitute only a small portion of the genome, leaving the role of the bulk of the DNA, including most of the repeated sequences, unexplained. It is thought that at least some of this "extra" DNA might have regulatory functions, but what these functions might be remains obscure.

In addition to the noncoding DNA existing between genes, the noncoding DNA interspersed *within* most eukaryotic genes—introns—provides another genomic feature that may be exploited by the cell for regulatory purposes. Especially intriguing is the finding that the coding sequences of genes—exons—are, in some cases, spliced together in different combinations to generate alternative mRNAs, which, in turn, are translated into related proteins with different properties. RNA splicing was described in Chapter 17. We will consider an example of alternative splicing later in this chapter.

Genomic Compartmentalization. Another fundamental difference between prokaryotes and eukaryotes is that most of the genetic information of eukaryotes is located in the nucleus, segregated from the sites of protein synthesis in the cytoplasm. Transcription and translation in eukaryotic cells are therefore separated in both time and space by the nuclear envelope.

This separation has implications for gene regulation, by allowing for the possibility of selectivity in the processing and transport of nuclear transcripts. It is clear, for example, that primary transcripts are extensively modified, cleaved, and spliced in eukaryotic nuclei, and that most nuclear RNA is degraded and never actually appears in the cytosol. The nuclear envelope may serve as a means of screening transcripts for selective passage to the cytosol. On the other hand, the separation of transcription and translation makes control by attenuation impossible in eukaryotes, since the mechanism of attenuation requires close physical proximity of RNA polymerase molecules and ribosomes.

Structural Organization of the Genome. The chromosomes of a eukaryotic cell differ in both chemistry and structure from the bacterial chromosome. As we saw in Chapter 14, the DNA of eukaryotic chromosomes is intimately associated with histone and nonhistone proteins and is highly folded, with several successive levels of structural organization. Bacterial DNA is complexed with small basic proteins that may interact with DNA in much the same way as histones do in eukaryotic chromatin. However, prokaryotic DNA folding is not based on nucleosomes, the basic packaging units of eukaryotic chromosomes, and the intermediate levels of DNA compaction are not present. Because eukaryotic DNA is packaged in a more elaborate way, it is reasonable to hypothesize that eukaryotic transcription is regulated at two stages: the uncoiling of the appropriate region of the chromosome, and the binding of transcription factors and RNA polymerase to DNA in the uncoiled region (Figure 19-10).

Genetic studies and DNA sequencing have revealed that eukaryotic genes are almost never physically clustered into operonlike units of transcription or regulation. Instead, eukaryotes generally employ a different strategy for coordinating gene transcription. In addition to its own promoter, each eukaryotic gene in a coordinately transcribed group has its own regulatory DNA elements nearby. These sequences are similar for all the genes in the group and are recognized by the same regulatory proteins, which can turn the genes on or off together. (Bacteria also have this sort of regulation, in addition to operons.)

The Stability of Messenger RNA. In Chapter 17, we mentioned that mRNA molecules in prokaryotes and eukaryotes differ in turnover rate—that is, in their average lifetime in the cell before degradation. Although the values vary with the specific mRNA molecule, with the species of organism, and with the tissue type (in the case of multicellular eukaryotes), the average life span of prokaryotic mRNA molecules is about three minutes, while the mRNA of eukaryotes—especially multicellular eukaryotes—may survive for much

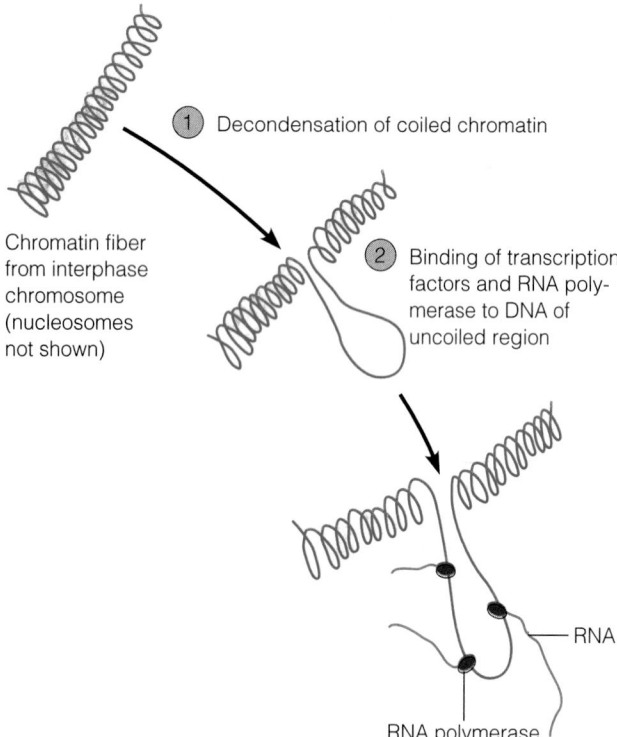

Figure 19-10 Two Stages of Regulation of Eukaryotic Transcription. Transcription in eukaryotes is probably regulated in two stages. ① Localized changes in chromatin structure cause selected regions of the chromatin to decondense. The uncoiled chromatin loops out. ② The DNA is then accessible for the binding of transcription factors and RNA polymerase. Transcription follows.

longer, often many hours. This striking difference can probably best be understood in light of what we might call the "lifestyles" of these organisms.

Prokaryotes are unicellular organisms with little or no assurance of environmental constancy and must therefore be able to adapt readily to new conditions. Consistent with this need is a short life span for most mRNA molecules and rapid transcriptional regulation via effector molecules and allosteric regulatory proteins. The effector molecules are almost always small molecules such as sugars and amino acids—the very environmental factors to which the cells must respond. The prokaryotic cell can rapidly change its population of mRNAs, and therefore the spectrum of proteins that are made, in response to external conditions.

On the other hand, many eukaryotic cells are constituents of multicellular organisms and can depend on a much more predictable environment within the organism. Moreover, cells of multicellular eukaryotes are usually quite highly differentiated (specialized) in function, reflecting the division of labor characteristic of the multicellular way of life. Most cells of multicellular eukaryotes are therefore committed to the synthesis of a relatively small collection of gene products. This commitment to the synthesis of a predictable set of proteins allows such cells to use longer-lived mRNA molecules than are typical for bacterial cells.

While eukaryotic mRNAs are generally longer-lived than their prokaryotic counterparts, there are also numerous examples (some of which we will discuss later in the chapter) of significant variations in the stability of eukaryotic mRNAs. The life spans of different mRNA molecules within the same cell may vary. In addition, the life spans of the same kind of mRNA in cells of a given type may vary with the organism's environmental or developmental state. For single-celled eukaryotes such as yeasts, as for bacteria, environmental conditions are of paramount importance; in general, the mRNAs of these organisms are relatively short-lived.

Posttranslational Modification of Proteins. Proteins may be extensively modified after polypeptide synthesis, as we discussed in Chapter 18, and such modifications are particularly important in eukaryotic cells. For example, some eukaryotic proteins must have blocks of amino acids removed before the protein is biologically active; instances include mammalian insulin and many digestive enzymes. Also, signal sequences that target the protein to specific cellular compartments must often be removed. Other important types of protein modification in eukaryotes include the addition of prosthetic groups and the glycosylation of proteins destined for secretion. All these types of modification reflect the more elaborate structure and more specialized functions of eukaryotic cells, as compared with prokaryotes, and all offer potential opportunities for regulation.

Protein Turnover. Another difference that arises from the contrasting lifestyles of most eukaryotic cells and prokaryotic cells is the relative importance of enzymatic hydrolysis as a means of eliminating proteins that are no longer needed by the cell. Prokaryotes have proteolytic enzymes and can degrade defective or unneeded proteins, but this is not their only option. Because most prokaryotic cells continually grow and divide, they can get rid of proteins simply by stopping their synthesis and allowing them to be diluted out by successive cell divisions. As a result, prokaryotic cells are in general not dependent on proteolytic enzymes as a major means of eliminating proteins.

Eukaryotic cells, on the other hand, are much more likely to cease dividing and to persist as discrete metabolic entities for relatively long periods of time thereafter. The nerve cells in your body are an especially good example: They are as old as you are and will never divide again. Clearly, such cells cannot depend on dilution by division to get rid of unnecessary or defective proteins. Instead, they have specific proteolytic enzymes that degrade proteins selectively. The regulated degradation and replacement of proteins, called **protein turnover,** is a much more prominent feature of eukaryotic cells than of prokaryotes. Thus, protein degradation is yet another potential level of regulation for eukaryotes.

Multiple Levels of Control

The differences described in the preceding section underscore the impossibility of explaining the eukaryotic regulation of gene expression entirely in terms of known prokaryotic mechanisms—elephants are not just large *E. coli* after all! If we are ever to have a thorough understanding of regulation in eukaryotic cells, we must approach the topic from a eukaryotic perspective.

Gene expression in eukaryotic cells, ultimately measured in terms of activity of gene products, is the culmination of controls that act at several different levels. Figure 19-11 traces the flow of genetic information from the DNA of chromatin in the eukaryotic nucleus to functional proteins in the cytoplasm, indicating possibilities for regulation at a number of different levels. At least in theory, there are five main levels at which control might be exerted: (1) the genome, (2) transcription, (3) RNA processing and translocation from nucleus to cytosol, (4) translation, and (5) posttranslational events. Control mechanisms operating at any of the latter three levels are often grouped together under the category *posttranscriptional control,* a term that therefore encompasses a wide variety of quite different phenomena. Experimental means of distinguishing among these levels of control are described in Box 19A (page 635). Next we look at each level of control in turn.

Genomic Control: Exceptions to Genomic Constancy

The basis of gene regulation is almost always the selective utilization of genes present in every cell, rather than any

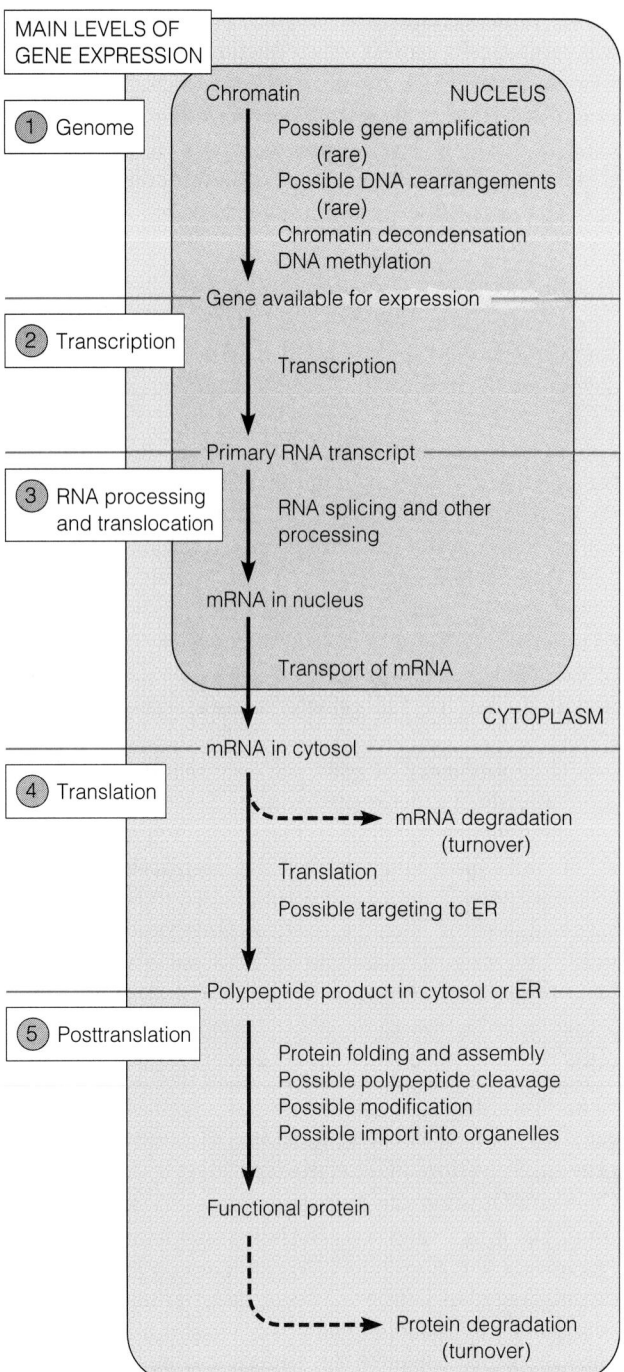

Figure 19-11   The Multiple Levels of Eukaryotic Gene Expression and Regulation.   Gene expression can be regulated by influencing any of the events that occur within any of these levels. ① The genome level, including rarely occurring amplification or rearrangement of DNA segments, chromatin decondensation (and condensation), and DNA methylation. ② Transcription, where critical control mechanisms determine which genes are active at a given time. ③ Processing of RNA and its transport out of the nucleus. ④ Translation, the synthesis of polypeptides. This level includes targeting of some nascent polypeptides to the ER. ⑤ Posttranslational events, including polypeptide folding and assembly, polypeptide cleavage, modifications of polypeptides by the addition of chemical groups, and the import of proteins into organelles (and secretion of some from the cell). Degradation of mRNA and proteins are also subject to regulation.

change in the genome itself. Even in plants and animals, in which most cells express only a fraction of the genes in the genome, virtually all somatic cells contain exactly the same genes. Nevertheless, there are some important types of gene regulation that involve exceptions to this rule. Before discussing a few of these examples of **genomic control,** we will look at the classic experiments that provided evidence for genomic constancy.

The Integrity of the Genome. For animals, strong evidence that even highly specialized somatic cells carry a full complement of genes was provided by John Gurdon and his colleagues. In studies with *Xenopus laevis,* the South African clawed toad, they transplanted nuclei from differentiated cells of tadpoles into oocytes (unfertilized eggs) that had been deprived of their own nuclei (Figure 19-12). Although the frequency of success was low, perhaps reflecting the difficulty of reprogramming a differentiated nucleus, some eggs with transplanted nuclei did give rise to viable, swimming tadpoles. This result indicated that nuclei derived from differentiated tissues are capable of directing the development of the whole organism. Such a nucleus is said to be **totipotent:** It has the full genetic potential expected of the species.

The totipotency of plant cells was demonstrated even more directly in experiments pioneered by F. C. Steward. As we saw in Figure 16-1d, when pieces of differentiated tissue are obtained from a mature plant and dissociated into single cells, the individual cells can often be used to regenerate an entire adult plant. Such regenerated plants, genetically identical to the original plant, are called *clones.* This means of plant propagation avoids the genetic variability inherent in sexual reproduction and is of great commercial interest because it ensures the maintenance of desirable genetic traits.

Gene Amplification. Although an organism's genome generally remains constant in all its cells, one exception is **gene amplification,** the selective replication of certain genes. Gene amplification can be regarded as a mode of genomic control of gene expression. The best-studied example is the amplification of ribosomal RNA genes in *Xenopus laevis,* the organism used by Gurdon in his nuclear transplantation experiments. The haploid genome of *Xenopus* normally contains about 500 copies of the genes that code for 5.8S, 18S, and 28S rRNA. During oogenesis (development of the egg prior to fertilization), the DNA of this entire set of genes is selectively replicated about 4000-fold, such that the mature oocyte contains about 2 million copies of the genes for rRNA. Apparently, this level of amplification is necessary to accommodate the enormous amount of ribosome synthesis that occurs during oogenesis, which in turn is required to sustain the high rate of protein synthesis characteristic of early embryonic development. Note that this example of gene amplification involves genes whose products are *RNA* rather than protein molecules. The expression of genes that encode proteins—even proteins needed in large amounts, such as ribosomal proteins—can usually be increased sufficiently by increasing translation of the mRNA, because each mRNA molecule can be translated numerous times.

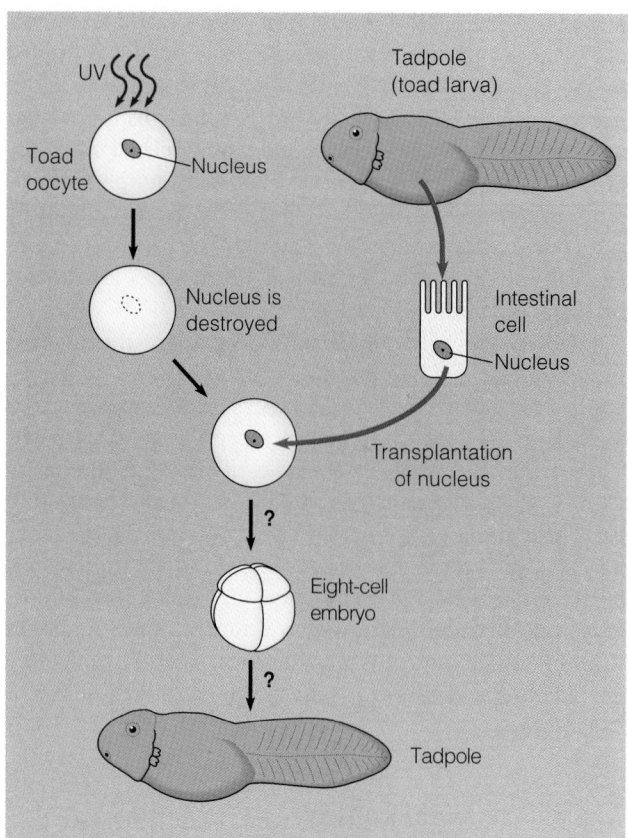

Figure 19-12   Nuclear Transplantation as a Means of Demonstrating Totipotency in Animal Cells. The experiments shown here were done by John Gurdon and his colleagues using *Xenopus laevis,* the South African clawed toad. Unfertilized *Xenopus* eggs (oocytes) were treated with ultraviolet light to inactivate their nuclei. Diploid nuclei were removed from differentiated cells (epithelial cells from tadpole intestine, in the example shown) with a micropipette and were injected into the enucleated eggs. The eggs were then allowed to develop. Some of the eggs never divided at all; other eggs began to develop but gave rise to abnormal embryos. However, some eggs with transplanted nuclei developed into normal embryos or even tadpoles. These results showed that the nucleus of a differentiated cell possesses all the genetic information necessary to direct the development of an entire organism.

The many extra copies of the rRNA genes that result from amplification are present in oocytes as extrachromosomal circles of DNA. These are located in hundreds of nucleoli that appear in the nucleoplasm of the oocyte as amplification progresses. The prominence of these nucleoli in the oocyte nucleus is shown in the micrograph in Figure 19-13.

DNA Rearrangement. A few cases are known in which genes are regulated by means of DNA rearrangement. A well-studied example involves yeast mating types. Haploid cells of *Saccharomyces cerevisiae* exist in two mating types, called *a* and *α,* and mating can occur between cells of opposite types, producing a diploid cell. All haploid cells carry both of the two different alleles for mating type; a cell's mating phenotype depends on which of the two alleles, *a* or *α,* is present at a special site in the genome, called the **MAT locus.**

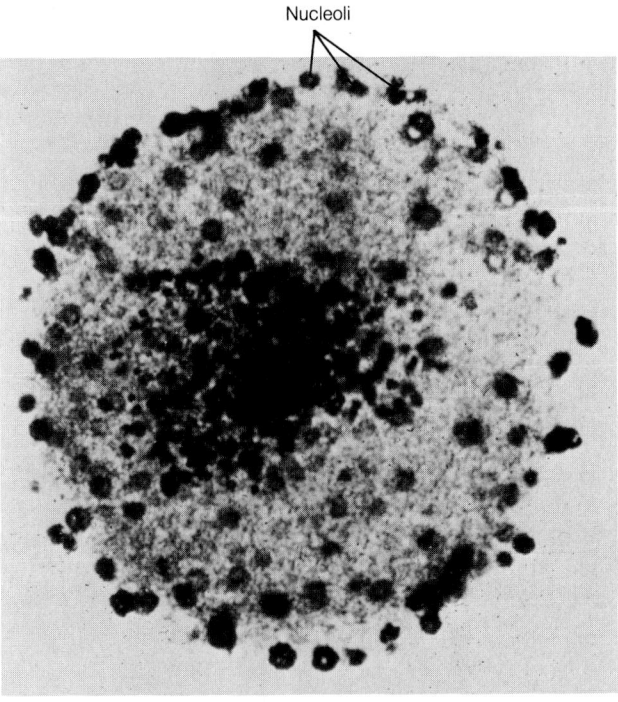

Figure 19-13 Amplification of Genes for rRNA in an Amphibian Oocyte. This micrograph shows an isolated nucleus from a *Xenopus* oocyte, stained to reveal the many nucleoli that are formed during oogenesis by the amplification of genes for ribosomal RNA. Each nucleolus contains multiple copies of the rRNA genes, present as extrachromosomal circles of DNA (LM).

Cells frequently switch mating type, presumably as a means of maximizing opportunities for mating. They do so by moving the alternative allele into the *MAT* locus. This process of DNA rearrangement is called the **cassette mechanism,** because the mating-type locus is like a tape deck into which either the *a* or the *α* "cassette" (allele) can be inserted and "played" (transcribed).

Figure 19-14 describes the yeast cassette mechanism in more detail. The *MAT* locus, existing as either *MATa* or *MATα*, is located on yeast chromosome 3, approximately midway between extra copies of the two alleles. The locus that stores the extra copy of the *α* allele is called *HMLα*; the locus with the extra copy of the *a* allele is called *HMRa*. In switching mating type, a yeast cell makes a DNA copy of the other mating-type allele, either *HMLα* or *HMLa*, and inserts this new cassette into the *MAT* locus. Before the new cassette can be inserted, however, the old DNA cassette at *MAT* must be excised (by a site-specific endonuclease) and discarded. Unlike the DNA sequence at the *MAT* locus, the DNA sequences at *HMLα* and *HMRa* never change.

The DNA of each mating-type allele actually encodes several proteins, including secretory proteins and cell-surface receptors. It is these proteins, encoded by the allele inserted at the *MAT* locus, that give the cell either an "*α*" or an "*a*" mating phenotype. But the extra copies of the alleles at *HMLα* and *HMRa* present a problem: If the cell contains complete copies of *both* the *α* and the *a* alleles, why aren't both sets of proteins made? The answer is that a set of regulatory genes, known as the *silent information regulator (SIR) genes*, act together to prevent expression of the genetic information at *HMLα* and *HMRa*. The proteins encoded by the *SIR* genes block transcription of *HMLα* and *HMRa* by binding to specific DNA sequences surrounding the *α* and *a* DNA cassettes at *HMLα* and *HMRa*. The function of the SIR proteins is thus analogous to that of the *lac* repressor.

Another important example of DNA rearrangement occurs in certain cells of the vertebrate immune system. During differentiation of these cells, functional antibody genes are created by rearrangement of DNA segments within the cell's genome. This topic will be discussed in Chapter 24.

Genomic Control: Chromosome Decondensation and DNA Accessibility

We encounter another aspect of control at the genome level in considering what is involved in making the eukaryotic genome—that is, chromosomal DNA—accessible to the cell's transcription machinery. As mentioned earlier, the

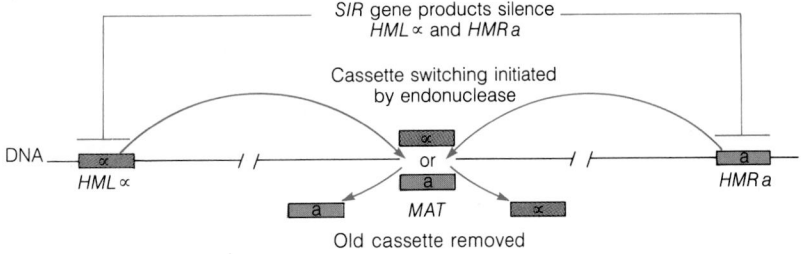

Figure 19-14 The Cassette Mechanism of the Yeast Mating-Type Switch. Chromosome 3 of *Saccharomyces* contains three copies of the mating-type information. The central *MAT* locus contains either the *α*- or the *a*-specific DNA and specifies the mating type of the cell. The *HMRa* locus, 150 kilobase pairs (kbp) to the right (downstream) of the *MAT* locus, contains a complete copy of the *a* mating-type DNA. The *HMLα* locus, 180 kbp to the left of *MAT*, contains a complete copy of the *α* mating-type DNA. *HMRa* and *HMLα* are transcriptionally silenced by the products of the *SIR* genes (located elsewhere). When an *α* or an *a* cell switches mating types, the *α* or *a* DNA at the *MAT* locus is excised, and a DNA "cassette" copy of the alternate mating-type DNA is inserted in its place.

elaborate packaging of DNA with proteins to form the eukaryotic chromosome adds a level of complexity not encountered in our discussion of prokaryotic gene regulation. Recall from Chapter 17 that, to initiate transcription, a eukaryotic RNA polymerase must interact with both DNA and a number of specific proteins (basal regulatory factors) in the promoter region of a gene. Except when a gene is being transcribed, its promoter region is embedded within a highly folded and ordered chromosomal superstructure. Thus, some degree of chromatin decondensation (unfolding) is undoubtedly a critical early event in the expression of eukaryotic genes.

Evidence that transcribed regions of chromatin are decondensed *selectively* comes from a variety of studies. The earliest of such studies involved direct microscopic visualization of certain special kinds of chromosomes caught in the act of transcription.

Visualization of Chromosome Decondensation. Visual evidence for a correlation between chromosome decondensation and transcription comes from several sources. One major source is research on the polytene chromosomes of insects such as the fruit fly *Drosophila melanogaster*. The precise relationship between changes observed in these chromosomes and changes detected biochemically in other experimental systems is unclear, but studies of these unusual chromosomes support the idea that chromosomes are packaged in structural domains that are physically reorganized during periods of active transcription.

A **polytene chromosome** is actually a tightly attached pair of homologous chromosomes with a very large number of precisely aligned, parallel chromatids. Such a multi-stranded chromosome is formed by repeated replication of the chromatin without separation of the daughter chromatids. It is a giant structure, much thicker than an ordinary chromosome and also much more extended in length. The four giant polytene chromosomes found in the salivary glands of *Drosophila* larvae, for example, are generated by ten rounds of chromosome replication, and therefore each has 1024 (or 2^{10}) chromatids in lateral register for each homologue—a total of 2048 chromatids in all. The micrograph in Figure 19-15a shows several polytene chromosomes from the nucleus of a *Drosophila* salivary gland cell. Visible in each of these polytene chromosomes is a characteristic pattern of dark bands, several of which are identified by number. Each such band represents a chromatin domain that is highly condensed compared to the chromatin in the "interband" regions between the bands. The chromatin in a band is thought to be uncoiled as a unit during transcription of the one or more genes it contains. However, genes are not confined to the bands; interband regions also contain genes.

When a polytene chromosome is exposed to fluorescent antibodies that selectively bind RNA polymerase II, transcriptionally active regions light up under the fluorescence microscope (Figure 19-15b). Activation of the genes of a given chromosome band causes the coiled chromatin strands to unwind and expand outward, resulting in a **chromosome puff**. Figure 19-16 shows a clearer view of a

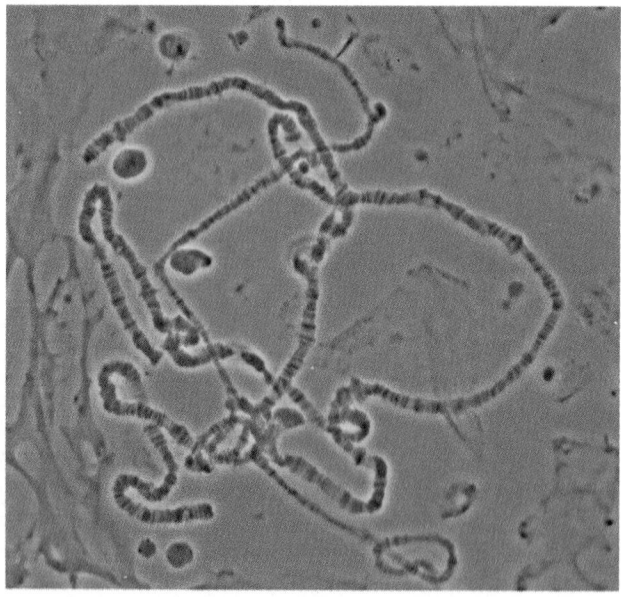

(a)

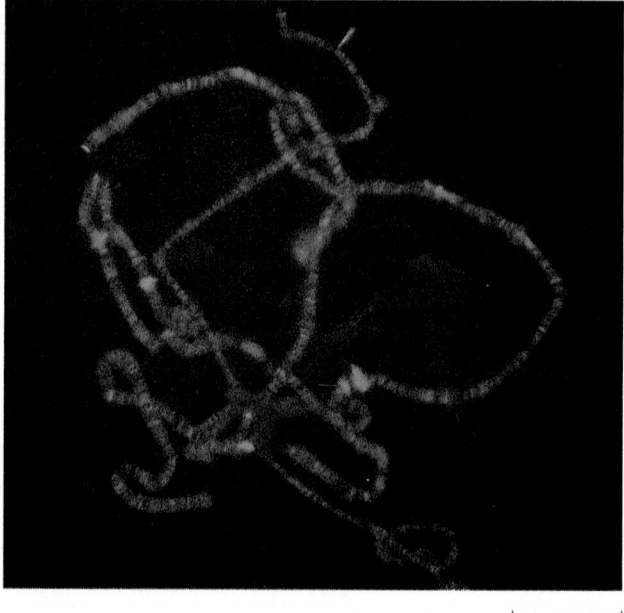

(b)

50 μm

Figure 19-15   Transcriptional Activity of Polytene Chromosomes. **(a)** A phase-contrast micrograph showing several polytene chromosomes from the salivary gland of a *Drosophila* larva. The banding pattern is a characteristic property of each chromosome, such that individual bands can be identified. **(b)** The same chromosomes seen with the fluorescence microscope, after incubation with fluorescent antibodies

that specifically bind to RNA polymerase II. The chromosomes light up brightly wherever RNA polymerase II molecules are located—that is, where transcription is occurring. The larva from which these chromosomes were obtained had been subjected to an elevated temperature for a short time to activate genes that code for heat shock proteins.

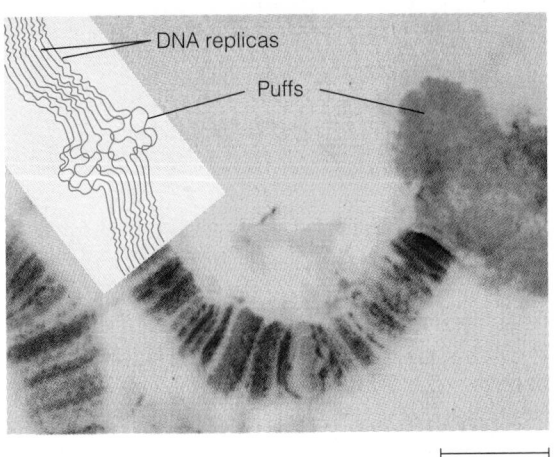

DNA replicas

Puffs

25 μm

Figure 19-16 **Puffs in Polytene Chromosomes.** Puffs are regions in which transcriptionally active chromatin has become less condensed, as indicated diagrammatically. The light micrograph shows part of a polytene chromosome.

puff-containing region of a polytene chromosome. Such puffs consist of loops of DNA that are much less condensed than the DNA of bands elsewhere in the chromosome. While puffs are by no means the only sites of active transcription along the polytene chromosome, the physical change in chromosome condensation at puffs correlates well with an enhancement of transcriptional activity.

As the *Drosophila* larva proceeds through development, each of the polytene chromosomes in salivary gland nuclei undergoes reproducible changes in puffing patterns, under the control of an insect steroid hormone called ecdysone. This hormone functions by binding to and thus activating a regulatory protein that, in turn, stimulates the transcription of certain genes. (This is similar to the action of vertebrate steroid hormones, which will be discussed in Chapter 23.) It appears, in other words, that the characteristic puffing patterns seen during the development of *Drosophila* larvae are direct visual manifestations of the selective decondensation and transcription of specific segments of DNA according to a genetically controlled developmental program.

DNase I Sensitivity Studies. Although decondensation is not readily visualized for most chromosomes, other kinds of studies also demonstrate correlations between chromatin uncoiling and transcription. One particularly useful tool is DNase I, an endonuclease (DNA-cutting enzyme) from the pancreas. At low concentrations, DNase I preferentially degrades transcriptionally active DNA in chromatin. The increased sensitivity of these regions of DNA to DNase I degradation is evidence that the DNA is uncoiled.

Figure 19-17 illustrates a classic DNase I sensitivity experiment focusing on the chicken gene for a globin protein. The gene is expressed at a high level in chicken erythrocyte nuclei. (Avian red blood cells retain nuclei, in contrast to the red blood cells of many other vertebrate species.) In these nuclei, the globin gene is completely digested at DNase I

concentrations that do not affect the same gene in tissues, such as brain, where globin genes are not transcribed. As you might predict, a gene that is *not* active in erythrocytes (e.g., the gene for ovalbumin, an egg white protein) is not digested by DNase I. Such data demonstrate that the transcriptional activity of eukaryotic DNA is at least correlated with DNase accessibility.

The data can be explained by two alternative interpretations: Either chromosome decondensation is necessary to give transcription factors and RNA polymerase access to the DNA, or the selective binding of these proteins has caused the uncoiling. Further experiments have addressed the question of which interpretation is more probable. Studies of the DNase sensitivity of the globin gene during time periods prior to and following actual transcription of the gene suggest that active transcription is not absolutely required for the structural change reflected by DNase sensitivity, and that chromatin decondensation is therefore likely to be a cause of—or at least a prerequisite for—transcriptional activation.

DNase I is also used in other kinds of experiments. When nuclei are treated with very low concentrations of DNase I, it is possible to detect **DNase I hypersensitive sites,** specific locations in the chromatin that are exceedingly vulnerable to digestion. Nicks in the DNA at these regions appear even before the enzyme attacks other regions that are actively being transcribed. DNase I hypersensitive sites can be reproducibly located in the vicinity of particular genes. The positions of hypersensitive sites often correlate with specific DNA-binding sites for regulatory transcription factors (which we discuss later in the chapter). Some hypersensitive sites are constitutive; they are present before, during, and after transcription of their associated genes. Other hypersensitive sites are observed in the chromatin only when the associated genes are transcriptionally active. It is possible that hypersensitive sites of the latter type mark areas where the binding of one or more transcription factors alters normal nucleosome structure in such a way that the DNA helix is less well protected by histones and therefore more vulnerable to nuclease attack.

The Mechanism of Decondensation. We do not yet know what renders specific regions of the chromosome susceptible to decondensation at specific times; nor do we know the details of the decondensation process. Recent studies in yeast, however, have provided a few clues. Yeast cells have a regulatory protein complex, known as *SWI/SNF*, that plays an important role in activating many inducible genes. The SWI/SNF complex seems to function by bringing about the reorganization of nucleosome structure to allow transcription factors to bind DNA in the promoter region of a gene. The human homologue of the SWI/SNF complex has recently been shown to induce an ATP-dependent structural change in nucleosomes. It is not yet clear whether nucleosomes are simply reconfigured during such structural changes, or whether histones are actually removed from the DNA. In any case, once transcription is successfully initiated, the "normal" nucleosome structure reappears. Thus,

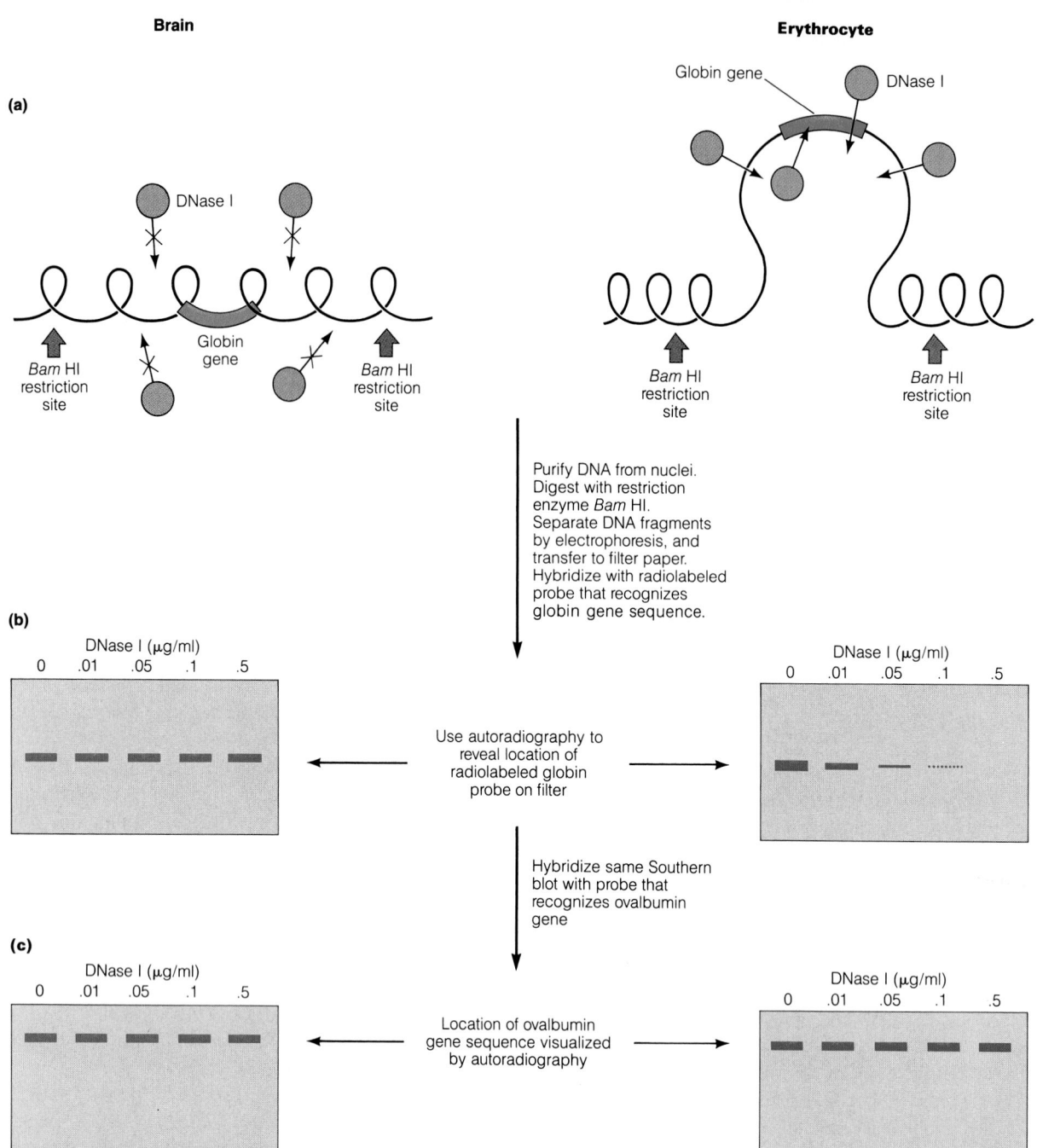

Figure 19-17 Detection of DNase I Sensitivity of Active Genes in Chromatin.   The chromatin configuration of active genes can be assayed by treating cell nuclei with DNase I. As an endonuclease, DNase I digests DNA by repeatedly cutting internal phosphodiester bonds. However, DNA in condensed chromatin is protected from DNase I attack, presumably because it is highly coiled and complexed with proteins. The experiment shown here uses two different kinds of cells, brain cells (neurons) and erythrocytes, and focuses on a gene for globin protein. **(a)** If the gene is inactive in cells of a given type (e.g., a globin gene in brain cells), it will still be present intact in DNA purified from nuclease-treated nuclei. However, if a gene is active (e.g., a globin gene in erythrocytes), the DNA in the region of the gene will be vulnerable to nuclease attack. **(b)** The condition of the gene is assayed by digestion of the puri-

fied nuclear DNA with a restriction enzyme, which should release the globin gene (if it has not been nicked by DNase I) from the rest of the DNA as a restriction fragment of a characteristic size. The presence of that restriction fragment is detected by hybridization of a radioactively labeled DNA probe for the globin gene to the nuclear DNA that is immobilized on a filter paper (Southern blot). DNA isolated from erythrocyte nuclei treated with increasing amounts of DNase I contains increasingly smaller amounts of the intact restriction fragment with the globin gene. However, even high concentrations of DNase I have no effect on the globin gene in brain nuclei. **(c)** An important control experiment shows that the gene for ovalbumin, which is inactive in both erythrocyte and brain nuclei, is resistant to DNase I attack in both cases.

nucleosomes do not seem to interfere with the movement of the RNA polymerase along the DNA during the elongation stage of transcription.

Association of DNA Methylation with Gene Inactivity. An additional means of regulating DNA sequence availability is **DNA methylation,** the addition of methyl groups to selected cytosine groups in the DNA. Many vertebrates have a number of their cytosines methylated, and these nucleotides tend to be clustered in noncoding regions at the 5′ ends of genes. Moreover, the methylation pattern of DNA is heritable, because the enzyme that methylates the DNA is specific for cytosines that are adjacent to guanines (CG) and is active only if the CG sequence is base-paired to a CG sequence in the other strand that is already methylated. Methylation seems to have a negative regulatory effect, somehow interfering with transcription. The mechanism by which methylation affects transcription is not known, but it clearly plays an important role: A recent study found that the gene for the responsible enzyme, DNA methyltransferase, is essential for mouse development.

Many examples of correlations between DNA methylation and gene inactivation have been observed. One important example is the phenomenon of *X*-chromosome inactivation in female mammals. The somatic cells of female mammals each contain two copies of the *X* chromosome, one inherited from each parent. Early in development, one *X* chromosome in each existing cell is inactivated at random. The inactivated chromosome is very tightly condensed as heterochromatin. When an interphase cell is examined under a microscope, the inactivated *X* chromosome is visible as a dark spot called a *Barr body*. In addition, the inactivated chromosome is highly methylated, and it is highly resistant to transcription initiation. Once an embryonic cell has undergone *X* chromosome inactivation, all its cellular descendants will similarly inactivate the same *X* chromosome. The benefit to the organism of X inactivation is unclear, but it may function to prevent the buildup of harmfully high levels of the *X* chromosome's gene products.

Despite the many correlations between methylation and the reduction of gene expression, it often appears that methylation only reinforces the genetic decisions made by other mechanisms and does not itself directly regulate expression. For example, the removal of methyl groups from a gene appears to increase the likelihood of transcription, but demethylation alone does not trigger transcription initiation. Furthermore, the effects of in vitro methylation vary. In some cases, in vitro methylation of a DNA sequence that serves as a transcription factor binding site reduces binding of the factor and subsequent transcription of the gene; but in other cases, in vitro methylation seems to have little effect. Moreover, some invertebrates, such as *Drosophila*, have almost no methylated DNA, although they employ other gene regulatory mechanisms similar to those used by all other eukaryotes. Thus it appears that methylation is a method of gene regulation that evolved relatively recently.

Transcriptional Control

With the exceptions discussed earlier, the coding DNA sequences in a eukaryotic genome generally have the same number and arrangement in all cells. Other types of control at the genomic level, such as those affecting chromatin configuration, are probably highly important but are not yet understood. When we move to **transcriptional control,** however, we come to a level of gene regulation where our knowledge has blossomed in recent years. As already mentioned, much regulation of gene expression—in eukaryotes as well as prokaryotes—is now known to occur at the transcriptional level. We now turn to the regulation of eukaryotic transcription, concentrating on genes that encode proteins.

Evidence for Differential Transcription. Direct evidence for the importance of transcriptional regulation in eukaryotes comes from comparisons of newly synthesized RNA in nuclei of different mammalian tissues. Liver and brain cells, for example, produce different sets of proteins, although there is considerable overlap between the two sets. Is this difference due to *differential transcription* of cellular genes—that is, transcription of different genes to produce different sets of RNAs? An alternative possibility is that the same genes are transcribed, yielding identical populations of primary transcripts, but that some posttranscriptional mechanism prevents much of the RNA from being translated. If the different proteins in different tissues do reflect differential gene transcription in the nucleus, we should see corresponding differences between the populations of nuclear RNA derived from brain and liver tissue. On the other hand, if all genes are equally transcribed in liver and brain cells, we would find few, if any, differences between the populations of nuclear RNA from the two tissues, and we would conclude that the tissue-specific differences in protein synthesis were due to posttranscriptional mechanisms.

The experiment diagrammed in Figure 19-18 is designed to determine which of those alternatives is correct. It demonstrates that there are differences in composition between the mRNA population present in the cytoplasm of liver cells and that in brain cells, and that the difference can be explained by differential transcription of genes in the nuclei of cells of these two organs. As shown in Figure 19-l8a, liver-specific mRNA molecules are isolated from the total mRNA in the cytoplasm of liver cells, by selectively removing all mRNA species common to both liver and brain cells. The remaining liver-specific mRNAs from the cytoplasm are used as templates for the synthesis of complementary DNA (cDNA) by reverse transcription. These cDNAs are then used as probes to test for the presence of the liver-specific sequences in *nuclear* RNA preparations from liver cells (part b) or from brain cells (part c). The liver-specific probes are able to hybridize (base-pair) with complementary RNA in the liver nuclei, but not with complementary mRNA sequences in brain cell nuclei. These results suggest

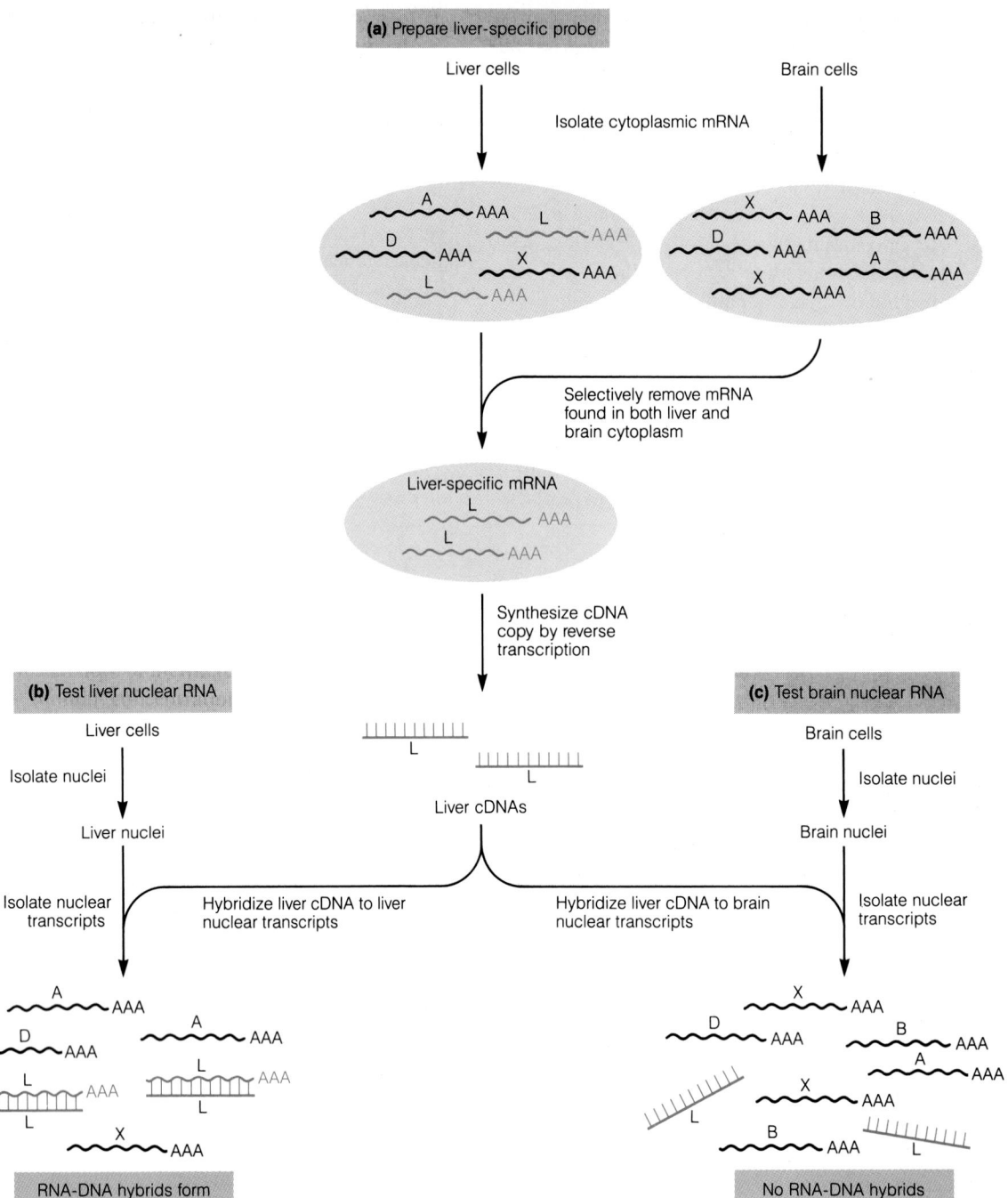

Figure 19-18   Evidence for Transcriptional Control of Eukaryotic Gene Expression.   Hybridization studies can be used to show that specific differences between the populations of mRNA sequences in the cytoplasm of different kinds of cells—liver cells and brain cells, for instance—reflect corresponding differences in nuclear RNA populations. These differences presumably result from differential transcription of genes. **(a)** Complementary DNA (cDNA) probes for liver-specific transcripts are prepared by eliminating from the liver mRNA population the molecules also found in brain tissue, and then reverse-transcribing the remaining (liver-specific) mRNA. (The letters appearing above the RNA molecules represent hypothetical genes.) **(b)** Liver nuclear RNA and **(c)** brain nuclear RNA are then tested for hybridization (binding) to the liver-specific probes. As expected, the liver nuclear RNA contains the liver-specific sequences (positive control experiment). However, the brain nuclear RNA cannot hybridize with liver-specific probes, suggesting that different genes are transcribed in different tissues.

that differential transcription is occurring, and that it is likely to be an important mechanism for generating specific differences between mRNA populations in different tissues.

The same conclusion can also be reached by using a different technique, called *nuclear run-on transcription*. This procedure provides a "snapshot" of the transcriptional activity occurring in a nucleus (Figure 19-19). Transcriptionally active nuclei are gently isolated from cells and allowed to complete synthesis of nascent transcripts in vitro in the presence of radioactively labeled nucleoside triphosphates, under conditions in which new transcripts are not initiated. The presence of labeled nucleotides in an mRNA molecule will thus indicate that it was being transcribed in the nucleus at the time of isolation. When such an experiment is performed using liver and brain cells, the resulting set of newly transcribed (radioactively labeled) mRNAs in the liver nuclei includes transcripts from known liver-specific genes, but these sequences are not detected among the labeled mRNAs synthesized by isolated brain nuclei. The conclusion is that different sets of genes are active in each tissue.

The Main Control Point: Initiation of Transcription. By failing to detect any liver-specific RNA sequences in brain cell nuclei, and vice versa, the experiments described in the preceding section suggest that gene transcription is generally an all-or-none process. In other words, in the cells studied, if transcription starts, it proceeds to completion—just as most often occurs in prokaryotic cells. We now know that control of transcriptional initiation is the usual way that transcription is regulated in eukaryotes.

The Eukaryotic Promoter: A Review. As described in Chapter 17, the specificity of eukaryotic transcription—that is, where on the DNA it initiates—is determined not by RNA polymerase itself, but by a variety of other proteins

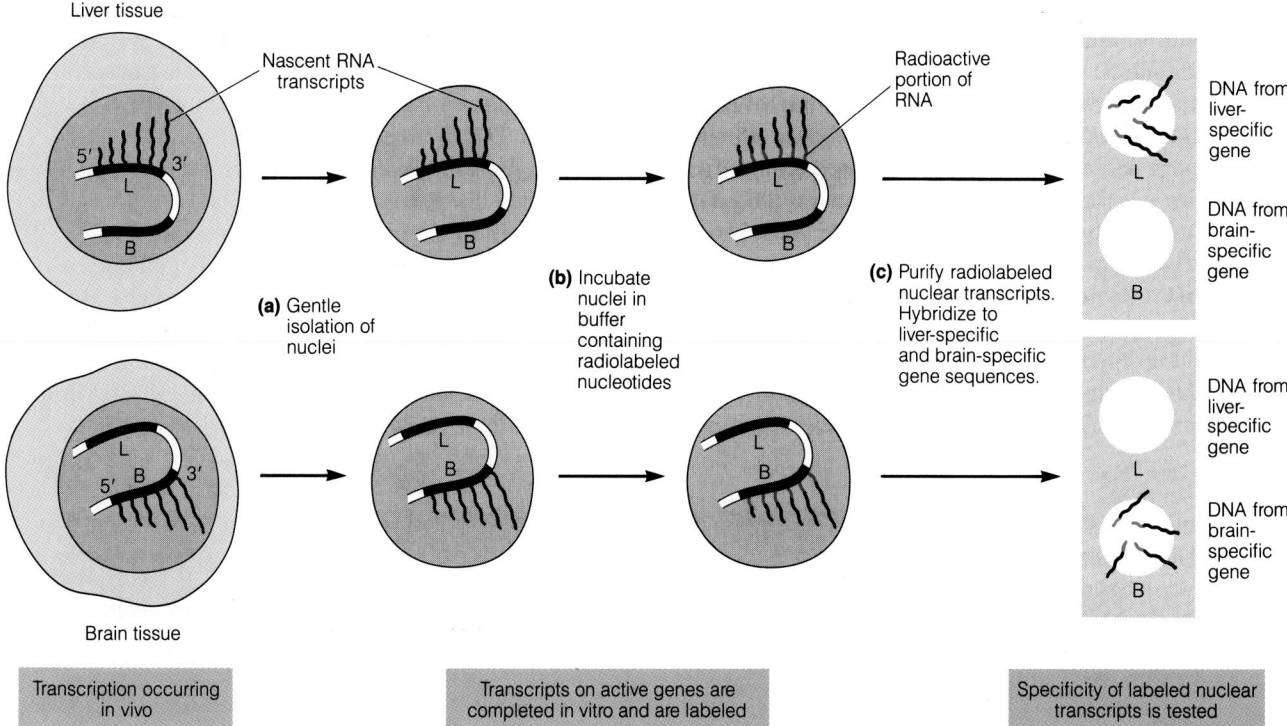

Figure 19-19 Demonstration of Differential Transcription by Nuclear Run-on Transcription Assays. **(a)** Nuclei are gently isolated from brain and liver tissues. On the DNA in the diagram, B is a hypothetical gene expressed only in brain tissue; L is a gene expressed only in liver tissue. **(b)** The isolated nuclei are incubated in a transcription buffer containing radioactively labeled ribonucleotides. The labeled nucleotides can enter the isolated nuclei through nuclear pores and become incorporated in mRNA being synthesized on active genes. If different genes are active in liver and brain tissue, some labeled sequences in the liver nuclear transcripts will not be present in brain transcripts, and vice versa. **(c)** The composition of the labeled RNA population is assayed by allowing the labeled RNA to hybridize with DNA sequences representing different genes that have been attached to a filter paper support. Labeled liver transcripts hybridize with a different set of genes than do labeled brain transcripts, indicating that the identity of active genes in the two tissues differs.

called *transcription factors.* (Unlike the prokaryotic sigma factor, which also determines initiation specificity, none of the eukaryotic transcription factors are considered part of an RNA polymerase molecule.) The transcription factors discussed in Chapter 17 were **basal transcription factors,** proteins that are essential for the transcription of *all* genes transcribed by a given RNA polymerase. Recall that eukaryotic RNA polymerase I transcribes genes for the three largest rRNAs, polymerase II transcribes protein-coding genes, and polymerase III transcribes genes for tRNAs, 5S rRNA, and other small RNAs. Here, as in Chapter 17, we concentrate on the genes transcribed by RNA polymerase II.

For RNA polymerase II, the *core promoter* on the DNA, where the basal transcription factors assemble with the polymerase, typically consists of a short initiator (Inr) sequence surrounding the transcriptional startpoint and, at about 25 nucleotides upstream, a short sequence called a TATA box (Figure 19-20). The transcription factor TFIID, which contains a TATA-binding protein (TBP) subunit, begins the formation of the initiation complex by recognizing and binding to the TATA box. The other basal transcription factors recognize not DNA but other proteins, including each other and RNA polymerase (see Figure 17-14). Thus, protein-protein interactions are crucial to the initiation of eukaryotic transcription.

The interaction of basal transcription factors and RNA polymerase with the core promoter initiates transcription only at a low, "basal" rate, so that few transcripts are produced. However, in addition to a core promoter, most protein-coding genes have short DNA sequences farther upstream to which other transcription factors bind, thereby improving the efficiency of the core promoter. When these DNA elements are deleted or mutated, the frequency and accuracy of transcription initiation are reduced. In discussing

protein-coding genes, we use the term **proximal control elements** to refer to DNA sequences upstream of the core promoter but within about 100 nucleotides of it. The number, identities, and exact locations of these DNA elements vary with the gene, but you can see two of the most common elements in Figure 19-20. The transcription factors that bind to these and other DNA control elements outside the core promoter we call **regulatory transcription factors.** They increase (or, sometimes, decrease) transcription initiation, apparently by interacting with the basal transcription apparatus. Transcription factors that bind specifically to each of these sequences and others have been identified (Table 19-3). Some scientists regard the RNA polymerase II promoter as a region that includes proximal control elements because many of the same elements are associated with a variety of genes. However, the particular combinations and locations of the elements are specific to each gene.

Distal Control Elements: Enhancers and Silencers. Some eukaryotic DNA control elements resemble most prokaryotic control elements in lying very close to the core promoter on its upstream side. However, eukaryotic control elements can be either upstream or downstream from the genes they regulate and are often a great distance along the DNA from that gene. Some can function at distances of 10,000–20,000 bp from the promoter they regulate. Occasionally, regulatory elements are even found *within* genes; one example is a positive regulatory element found within an intron of certain immunoglobulin genes.

DNA elements that increase the initiation of transcription of the associated gene are called **enhancers.** This term was originally applied only to **distal control elements,** sequences located far from the gene. However, the nucleotide sequences of enhancers and their probable mode of influ-

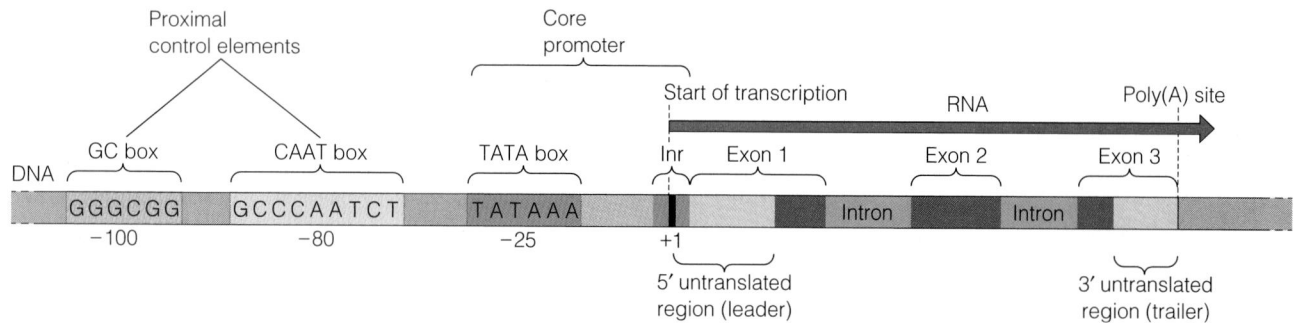

Figure 19-20 Anatomy of a Typical Eukaryotic Gene, with Its Core Promoter and Proximal Control Region. This diagram (not to scale) features a typical protein-coding eukaryotic gene, which is transcribed by RNA polymerase II. The promoter—called the core promoter to distinguish it clearly from the proximal control region—is characterized by an initiator (Inr) sequence surrounding the transcriptional startpoint and a sequence called a TATA box located about 25 bp upstream (to the 5′ side) of the startpoint. The core promoter is where the basal transcription factors and RNA polymerase assemble for the initiation of transcription. Within about 100 nucleotides upstream from the

core promoter lie several proximal control elements, which stimulate transcription of the gene by interacting with regulatory transcription factors. The number, identity, and exact location of the proximal elements vary from gene to gene; here we show a very simple case, with one copy of each of two common elements, the GC box and the CAAT box. The transcription unit includes a 5′ untranslated region (leader) and a 3′ untranslated region (trailer), which are transcribed and included in the mRNA but do not contribute sequence information for the protein product. At the end of the last exon is a site where, in the primary transcript, the RNA will be cleaved and given a poly(A) tail.

Table 19-3 Some Eukaryotic DNA Control Elements and Transcription Factors That Bind Them

DNA Control Element	Genes Where Found	DNA Element Consensus Sequence*	Transcription Factor**
TATA box	Many	TATAAAA (in promoter)	TFIID (basal factor)
CAAT box	Many	GCCCCAATCT	CTF
GC box	Many	GGGCGG	Sp1
Heat shock element	Heat shock genes	CnnGAAnnTTCnnG	HSTF
Octamer	SV40 virus	TGCTTTGCAT	Octamer binding factor
Estrogen response element	Ovalbumin, others	AGGGTCAnnnTGACCT	Estrogen receptor
κB	Immunoglobulin light chain κ	GGGGACTTTCC	NF-κB

*Some researchers give slightly different consensus sequences. Lowercase n signifies that any nucleotide can be located at that position.

**In some cases, other names have also been assigned to the factors listed here.

encing transcription are similar to those of proximal elements. The *GC box* and the *octamer* element (see Table 19-3) are examples of elements that have been found in both proximal and distal positions. For an enhancer to function, the transcription factor or factors that specifically bind to the enhancer sequence must be present; these regulatory transcription factors are called **activators.** The octamer element was one of the first enhancers studied. It was identified in the genome of the virus SV40 (simian virus 40). This enhancer plays an important role in ensuring high levels of synthesis of viral products following infection of a host cell.

Enhancer elements, varying in specific sequence but sharing common properties, have now been identified in association with many eukaryotic genes. Several properties of enhancers have been demonstrated by experiments in which recombinant DNA techniques are used to alter enhancer location and orientation with respect to the regulated gene. As shown in Figure 19-21, enhancers can function properly when relocated at variable distances from the transcription startpoint (nucleotide "+1"), as long as the promoter is present (Figure 19-21a–e). Activity is retained even when the orientation of the enhancer element relative to the beginning of the gene is reversed (Figure 19-21f). Moreover, enhancers usually function properly even when relocated downstream of the 3′ end of the gene (Figure 19-21g). These properties are often used to distinguish enhancers from proximal control elements, whose specific locations in the DNA tend to be more critical for their function.

Subsequent to the discovery and study of enhancers, analogous *negative* regulatory elements, called **silencers,** were also found in eukaryotic DNA. The binding of transcription factors to these sites *reduces* the level of transcription of the associated gene; therefore, the transcription factors involved are called **repressors.** (Like prokaryotic repressors, these proteins turn off transcription, although the eukaryotic mechanisms are probably somewhat different

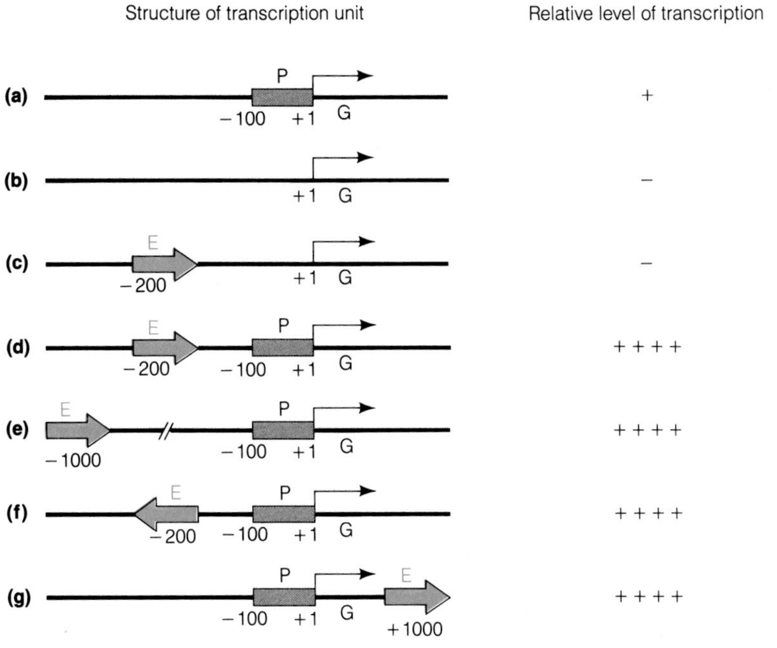

Figure 19-21 Properties of Eukaryotic Enhancer Elements. Recombinant DNA techniques can be used to alter the orientation and location of DNA control elements and study the effect of the change on the level of transcription of the gene. The black arrows indicate the direction of transcription of gene G, with the startpoint (first transcribed nucleotide) labeled +1. The other numbers give the positions of nucleotides relative to the startpoint. **(a)** The core promoter (P) alone, in its typical location just upstream of gene G, allows a basal level of transcription to occur. **(b)** When the core promoter is removed from the gene, no transcription occurs. **(c)** An enhancer element (E) alone cannot substitute for the promoter region, but **(d)** combining an enhancer element with a core promoter results in a significantly higher level of transcription than occurs with the promoter alone. **(e)** This increase in transcription is observed when the enhancer is moved farther upstream, **(f)** when it is inverted in orientation, and **(g)** even when it is moved to the 3′ side of the structural gene.

and also more varied.) Although not yet as well characterized as enhancers, silencers are believed to share the same general features, except for their negative effect. Both enhancers and silencers include a variety of nucleotide sequences, each of which binds a different type of regulatory transcription factor. Shortly, we will discuss the mechanism by which distal control elements are believed to influence the promoters they regulate. But first, let's step back and consider how the variety of regulatory elements and factors could make differential transcription possible.

A Combinatorial Model for Gene Regulation.   The observation that numerous regulatory transcription factors are required for high levels of transcription initiation in eukaryotes has led to a **combinatorial model for gene regulation.** This model proposes that highly specific and precisely controlled patterns of gene expression can be established by

using a relatively small number of different DNA control elements and transcription factors, in different combinations. According to the model, a gene is expressed at a maximum level when the particular set of regulatory transcription factors produced by a cell includes the factors corresponding to all the positive DNA control elements associated with the gene.

As shown schematically in Figure 19-22, this model allows liver cells to produce large amounts of proteins such as albumin but prevents significant production of these proteins in other tissues, such as brain. Some transcription factors are present in many cell types. These include not only the basal transcription factors, which should be present in all cells, but any regulatory factors needed for the transcription of constitutive genes and others that are frequently expressed. On the other hand, transcription of genes that encode tissue-specific proteins depends on the presence of unique transcription factors or unique *combinations* of tran-

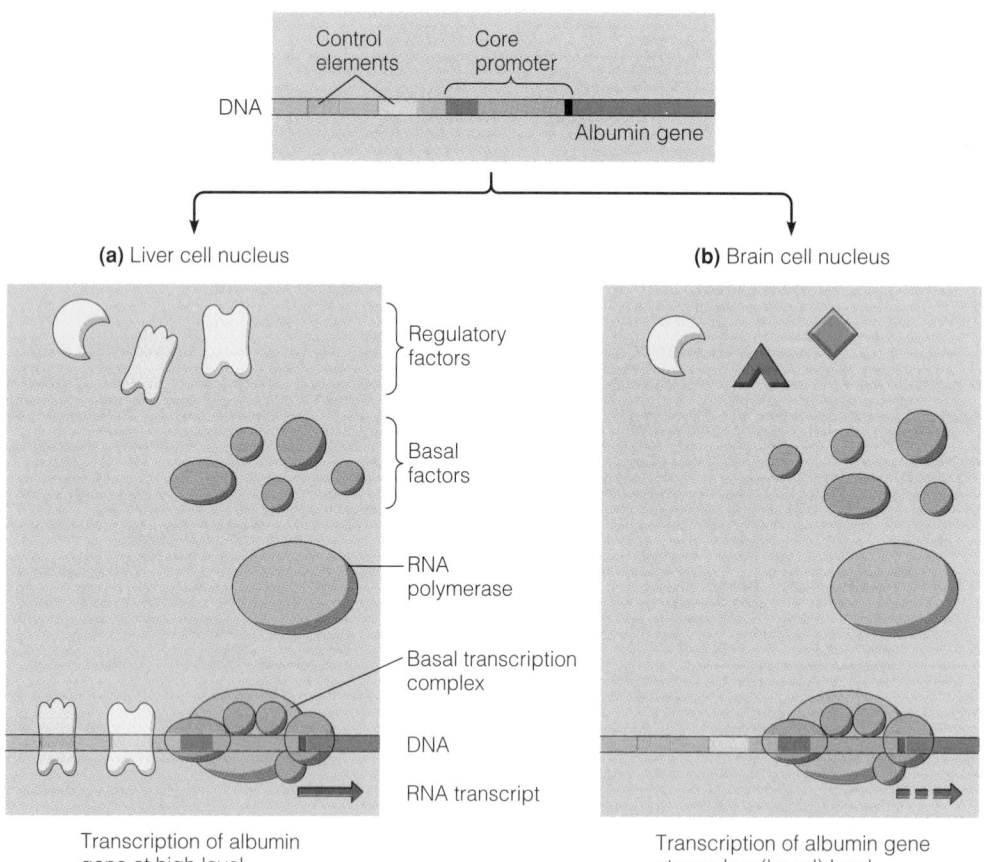

Figure 19-22   A Combinatorial Model for Gene Expression. The gene for the protein albumin, like other genes, is associated with an array of regulatory DNA elements; here we show only two control elements, as well as the core promoter. Cells of all tissues contain RNA polymerase and the basal transcription factors (purple), but the set of regulatory transcription factors (other colors) available varies with the cell type. As shown here, **(a)** liver cells contain a set of regulatory transcription factors that includes the factors for recognizing all the albumin gene

control elements. When these factors bind to the DNA, they facilitate transcription of the albumin gene at a high level. **(b)** Brain cells, however, have a different set of regulatory transcription factors, which does not include all the ones for the albumin gene. Consequently, in brain cells, the basal transcription complex can assemble at the promoter, but not very efficiently. The result is that brain cells transcribe the albumin gene only at a low level.

scription factors in the specialized tissue cells. The original version of the combinatorial model was "all or none," proposing that transcription of a gene would not initiate at all unless the entire set of regulatory factors for the gene was present. It is now considered more likely that there is a continuum of initiation efficiency, ranging from a basal level, occurring when *no* regulatory factors are available, to a maximum level, occurring when the full set of regulatory factors is present.

Structural Motifs of DNA-Binding Proteins. Although not all eukaryotic transcription factors bind directly to DNA, DNA-binding proteins clearly play critical roles in controlling eukaryotic transcription. Proteins of this kind include the basal transcription factor TFIID and the activators and repressors that bind proximal and distal control elements. What features of these proteins enable them to carry out their functions?

Among the multitude of regulatory transcription factors that have now been identified, some common structural patterns have been found. Like many other important proteins, transcription factors possess several distinct functional domains. The portion of a DNA-binding transcription factor that recognizes and interacts with a DNA control element is called the protein's **DNA-binding domain.** This domain is generally distinct from the portion of the protein required for the regulation of transcription, the **transcription regulation domain.** (In activator proteins, which include most of the regulatory proteins studied, the latter domain is called the *transcription activation domain.*) Most DNA-binding transcription factors fall into one of a small number of categories based on the type of protein structural motif present in the DNA-binding domain (Figure 19-23).

A common structural motif, identified in both eukaryotic and prokaryotic regulatory transcription factors, is the **helix-turn-helix motif** (Figure 19-23a). This motif has two α helices separated by a turn in the polypeptide chain, although the specific amino acid sequence differs from one protein to the next. (To review protein secondary structure, see Chapter 3.) The α helices of the helix-turn-helix motif are positioned in such a way that one of them, called the *recognition helix*, fits within the major groove of the DNA. The *lac* and *trp* repressors, the CRP protein, and many phage repressor proteins are prokaryotic examples of proteins with this motif, and transcription factors that regulate certain developmental events in *Drosophila* (the class of factors encoded by homeotic genes, described later) are eukaryotic examples. Figure 19-23b is a model of a repressor protein of phage λ bound to DNA. Like many DNA-binding regulatory proteins, this λ repressor consists of two identical polypeptides. In this case, each DNA-binding domain (white) has a helix-turn-helix motif.

A second structural motif, initially identified in a transcription factor for the 5S rRNA genes (TFIIIA), is the **zinc finger motif** (Figure 19-23c). This motif consists of an α

helix and a two-segment β sheet, held in place by the interaction of precisely positioned cysteine or histidine residues with a zinc atom. The figure shows four zinc fingers in a row. (TFIIIA has nine.) They protrude from the protein and serve as the points of contact with the major groove of the DNA.

A third structural motif, the **leucine zipper motif** (Figure 19-23d), is formed by two polypeptides. Each polypeptide has an α helix with regularly spaced leucine residues, which bond hydrophobically with the leucines of a similar helix in the other polypeptide. The two helices wrap around each other, forming a coiled coil. Some transcription factors whose active form is a dimer of identical polypeptides use leucine zippers to "zip" the polypeptides together. In other cases, leucine zippers bind together two different polypeptides. The parts of the dimeric structure that interact directly with DNA are two other α helices, which fit into DNA's major groove.

A fourth DNA-binding motif is the **helix-loop-helix motif** (Figure 19-23e). It is composed of a short α helix connected by a loop to another, longer α helix. Like leucine zippers, helix-loop-helix structures usually connect two polypeptides, which may be either similar or different. The formation of the four-helix bundle results in the juxtaposition of basic regions of each polypeptide (shown at the right in the diagram) to create a two-part DNA-binding domain. The list of structural motifs of transcription factors will undoubtedly continue to grow as more factors are studied.

Some eukaryotic transcription factors are allosteric proteins that change their DNA-binding affinity in response to certain small molecules, in much the same way as do the prokaryotic regulatory proteins that we discussed earlier in the chapter. Among the eukaryotic transcription factors that function in this way is a family of receptor proteins for steroid hormones such as progesterone, estrogen, and glucocorticoids. Steroid hormones are lipids that are synthesized by cells in endocrine tissues and released into the circulatory system, from which they are taken up by target tissues that contain the appropriate receptor protein. Steroid receptors are intracellular proteins and belong to the zinc finger category of transcription factors. As will be discussed further in Chapter 23 (see Figure 23-3), the steroid hormone enters the cytosol, where it binds to the inactive receptor protein, converting the protein to its active form. The activated protein then enters the nucleus, where it functions as a regulatory transcription factor—an activator. It recognizes and binds to specific enhancers on the DNA, called **hormone-response elements (HRE),** thereby stimulating transcription of the associated genes. In this way, a steroid hormone, made available via the blood to virtually all the cells of the body, can turn on specific genes in specific target tissues.

The sequence of events that starts with the binding of a hormone or other messenger molecule to its receptor and ends with specific changes in cell function is termed *signal transduction* (and will be discussed in detail in Chapter 23). Most such changes in cell function affect protein activity.

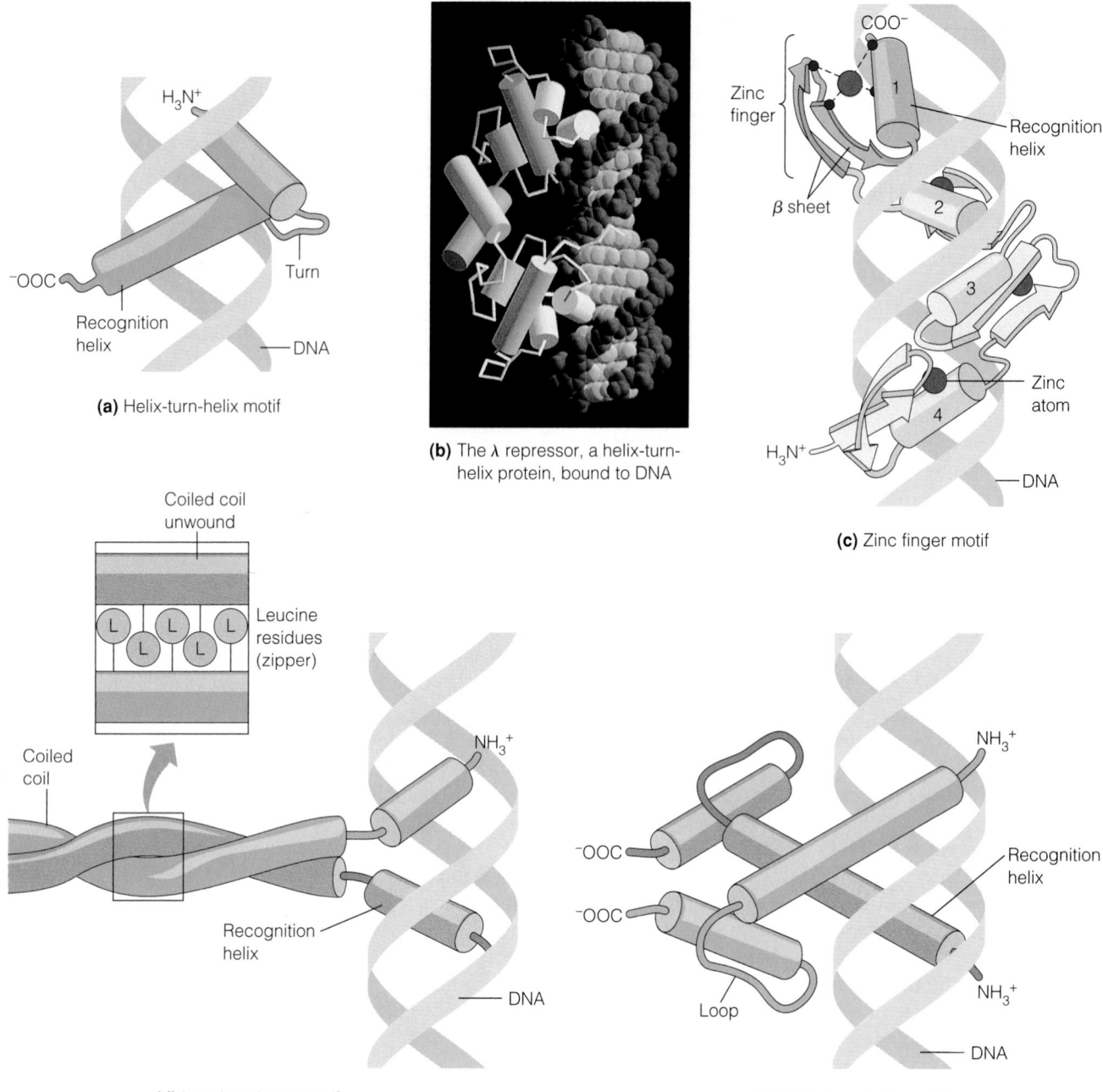

(a) Helix-turn-helix motif

(b) The λ repressor, a helix-turn-helix protein, bound to DNA

(c) Zinc finger motif

(d) Leucine zipper motif

(e) Helix-loop-helix motif

Figure 19-23   Common Structural Motifs in DNA-Binding Transcription Factors.   Several structural motifs are commonly found in the DNA-binding domains of regulatory transcription factors. The parts of these domains that directly interact with specific DNA sequences are usually α helices, called recognition helices, which fit into DNA's major groove. In this figure, all α helices are shown as cylinders. **(a)** The helix-turn-helix motif, in which two α helices are joined by a short flexible turn. **(b)** A computer graphic model showing the λ repressor, a helix-turn-helix protein, bound to DNA. It is a dimer of two identical subunits. The helices of the two DNA-binding domains are white. **(c)** The zinc finger motif. Each zinc finger consists of an α helix and a two-segment, an-

tiparallel β sheet (shown as ribbons), all held together by the interaction of four cysteine residues, or two cysteine and two histidine residues, with a zinc atom. At the top of the diagram, these key residues are shown as small purple balls; the zinc atoms are shown as larger red balls. Zinc-finger proteins typically have several zinc fingers in a row; here we see four. **(d)** The leucine zipper motif, in which an α helix with regularly arranged leucine residues in one polypeptide (green) interacts with a similar region in a second polypeptide (purple). The two helices coil around each other. **(e)** The helix-loop-helix motif, in which a short α helix connected to a longer α helix by a polypeptide loop interacts with a similar region on another polypeptide to create a dimer.

Steroids are unusual messengers in that they affect protein synthesis and do so by directly influencing gene expression at the transcription level.

How Enhancers Act.   We come now to the puzzling question of how DNA control elements at great and variable distances from a promoter can influence the initiation of transcription. We will focus on enhancers, with the under-standing that at least some silencers may act in analogous ways. In recent years, several lines of experimental evidence have led to the general acceptance of the idea that a looping of the DNA brings two linearly distant transcription factor binding sites into close proximity. The looping mechanism is not completely understood yet, but Figure 19-24 illustrates one model that is consistent with much experimental data. The basal transcription factor TFIID plays a central

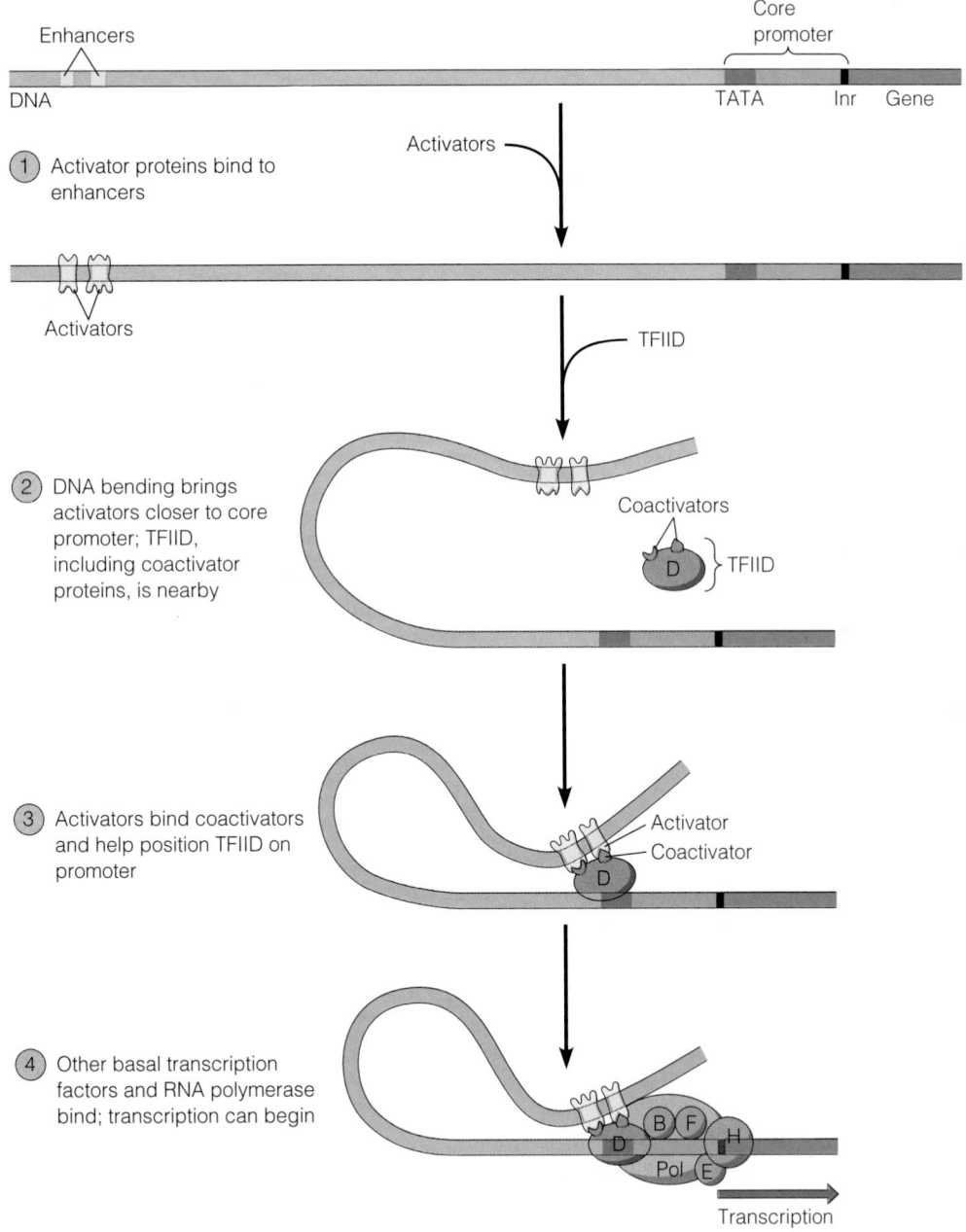

Figure 19-24   A Model for Enhancer Action.   In this model, enhancer elements at great distance along the DNA from the protein-coding gene they regulate get close to the core promoter by a looping of the DNA. The influence of an enhancer on the promoter is mediated by regulatory transcription factors called activators. ① The activator proteins bind to the enhancer elements. ② Bending of the DNA brings the DNA-bound activators closer to the core promoter. The basal transcription fac-tor TFIID is in the promoter's vicinity. For the purpose of this figure, two of the protein subunits of TFIID, which will function as coactivators in step 3, are distinguished from the rest of the factor. ③ The DNA-bound activators interact with specific coactivators that are part of TFIID. This interaction somehow facilitates the correct positioning of TFIID on the promoter. ④ The other basal transcription factors and RNA polymerase join the complex, and transcription is initiated.

role. Recall that this factor consists of TBP (TATA-binding protein) plus a number of other proteins. In the first step of the model, activator proteins bind to the enhancer elements. At the same time, TFIID is in the vicinity of the core promoter, but perhaps not successfully binding to the TATA box there. The DNA bends, bringing the bound activator molecules close to the core promoter. (It is conceivable that the activators induce the bending.) Next, the activators attach to specific protein components of TFIID, called **coactivators,** creating a loop of DNA between the promoter and the enhancers. Somehow this interaction helps position TFIID properly at the TATA box of the promoter. Finally, the other basal transcription factors and RNA polymerase assemble with TFIID at the promoter (see Figure 17-14), and transcription can begin.

The details of this model of enhancer action are less important than the main idea: When activator proteins for a particular gene are present in the cell, their binding to the gene's enhancers stimulates formation of the basal transcription complex at the promoter, resulting in much more frequent initiation of transcription than would otherwise occur. Thus, the differential transcription in a cell probably depends on the activators—and repressors—that the cell makes. What is ultimately responsible for the selective expression of the genes that encode activators and repressors is not yet known.

Coordinate Gene Regulation in Eukaryotes

Like prokaryotic cells, eukaryotic cells often need to activate several different genes in a coordinated fashion. For unicellular eukaryotes, coordinate gene regulation may be required to respond to some signal from the external environment, as for bacteria. For multicellular eukaryotes, the requirement for coordinate gene regulation is especially critical for the differentiation and maintenance of specialized tissues. For example, cells of muscle tissue must synthesize a specific set of proteins required for muscle function. In prokaryotes, genes with related functions are often physically contiguous in an operon, which is regulated and transcribed as a unit, but this is almost never the case in eukaryotes. Instead, eukaryotic genes that must be turned on (or off) at the same time are usually scattered in the genome. Their coordinate regulation must therefore depend on a different strategy.

Heat Shock Genes. Analysis of the heat shock genes of eukaryotes has provided important clues about how eukaryotic cells coordinate the regulation of genes located at many different chromosomal sites. **Heat shock genes** were first defined as genes that are expressed in response to an increase in temperature. They are now known to respond to certain other stressful environmental influences as well, and they are therefore also called *stress-response genes.* There are a number of different heat shock genes in the typical prokaryotic or eukaryotic genome, and the appropriate environmental trigger simultaneously activates all of them. The functions of these genes are not completely understood, but it appears that the protein products of at least some of the heat shock genes help minimize the cellular damage resulting from thermal denaturation of important proteins at various sites within the cell.

Analysis of the upstream regulatory regions of different members of the heat shock gene family has revealed that each gene shares with other members of the set a specific regulatory sequence, the **heat shock element.** Heat-induced activation of a specific regulatory protein, the *heat shock transcription factor,* leads to binding of the factor to the heat shock element for each gene, thereby activating transcription simultaneously from the entire set of different heat shock genes. This system illustrates a basic principle of eukaryotic coordinate regulation: Genes located at different sites can be activated by the same signal if all the genes have copies of the same regulatory element nearby. Although most systems are not as simple as the heat shock case, the same basic principle seems to apply to coordinate transcription regulation in many other situations.

Homeotic Genes. One of the most important examples of coordinate gene expression identified to date can be traced to the 1940s, when Edward B. Lewis discovered in *Drosophila* a cluster of genes, the *bithorax gene complex,* in which certain single mutations caused drastic developmental abnormalities. The mutant fruit flies exhibited such gross changes as two extra wings (Figure 19-25b). Later, Thomas C. Kaufman and his colleagues discovered a second group of genes that, when mutated, led to different but equally bizarre developmental changes—including the growth of legs from the fly's head in place of antennae (Figure 19-25c). This group of genes is called the *antennapedia gene complex.* It was clear that both bithorax and antennapedia genes somehow play key roles in directing embryonic development. Molecular analysis of such master control genes, today known as **homeotic genes,** revealed that each contains a 180-nucleotide sequence termed the **homeobox** (Figure 19-26). (In Greek, *homeo* means "alike," and homeotic genes were so named because of their ability to change one body segment of *Drosophila* to resemble another.)

Homeotic genes encode a family of regulatory transcription factors that bind to particular DNA control elements. By binding to all the copies of the appropriate DNA element in the genome, each of these factors may affect the expression of dozens or even hundreds of genes in the growing embryo. The result of this coordinated gene expression is the development of fundamental body characteristics such as appendage shape and location. The DNA-binding domain of each homeotic protein is encoded by the homeobox sequence and is referred to as the protein's **homeodomain.** Sixty amino acids long, this domain has a helix-turn-helix motif, along with a third α helix (see Figure 19-26).

Homeoboxes have now been found in more than 60 *Drosophila* genes and in a range of other organisms, including fungi, plants, and animals. Most homeotic genes control major developmental pathways, although in vertebrates they

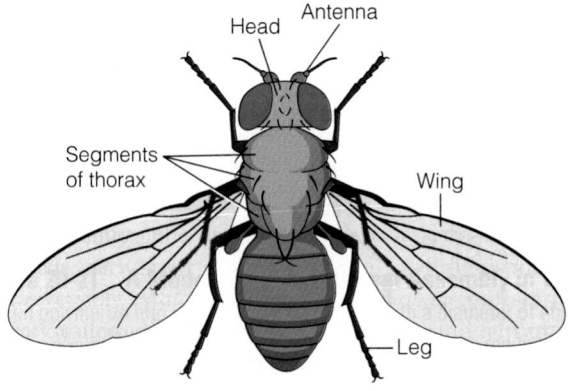

(a) Wild-type *Drosophila*

(b) Bithorax mutant

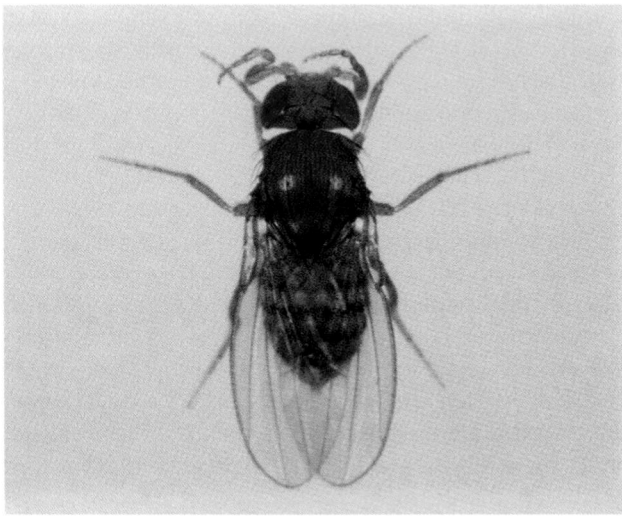

(c) Antennapedia mutant

Figure 19-25   Homeotic Mutants of *Drosophila*.   (a) Wild-type *Drosophila* has two wings and six legs extending from its three thoracic segments. **(b)** Mutations in the bithorax gene complex convert the third thoracic segment to a second thoracic segment (the wing-producing segment), and an additional set of wings is formed. **(c)** A mutation in the antennapedia gene complex causes legs to develop where the insect's antennae should be.

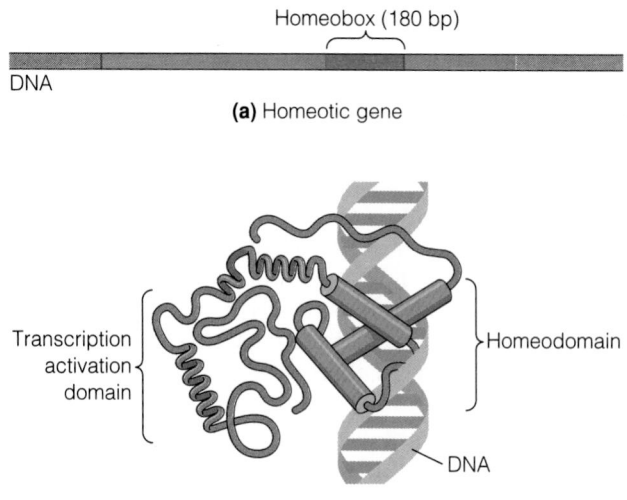

(a) Homeotic gene

(b) Homeotic protein bound to DNA

Figure 19-26   Homeotic Genes and Proteins.   (a) Homeotic mutants like the ones in Figure 19-25 have mutations in homeotic genes. These genes all contain a 180-bp segment, called a homeobox, that encodes a homeodomain in the homeotic protein. **(b)** The homeodomain (red) is a helix-turn-helix DNA-binding domain, which, in combination with a transcription activation domain, enables the protein to function as a regulatory transcription factor. Notice that the homeodomain actually has *three* α helices (shown as cylinders).

may also help regulate other important processes, such as histone production and antibody synthesis. Homeobox sequences have been highly conserved in evolution. Even when the DNA sequence has diverged (homeoboxes of fruit flies and mammals are somewhat different, for example), the structure and function of the homeodomain in the protein have remained remarkably unchanged. It appears that strong evolutionary pressure has been at work to preserve the amino acid sequence of the homeodomain. So widespread and well preserved is this sequence that the C-terminal portion of the typical eukaryotic homeodomain even shows some homology with prokaryotic repressors.

Regulation of Transcription by RNA Polymerases I and III

Regulation of genes that are transcribed by RNA polymerases I and III also depends on specific DNA-protein interactions between DNA control elements and transcription factors. However, the identities of the regulatory proteins and DNA sequences and the arrangements of the sequences relative to the transcription startpoint differ significantly in these two cases from the polymerase II examples we have considered up to now.

As discussed in Chapter 17, RNA polymerase I is responsible for transcribing the genes for three of the four eukaryotic rRNAs: 18S rRNA, 28S rRNA, and 5.8S rRNA. These three products are transcribed as a single precursor, which is then processed to yield equimolar amounts of the individual components (see Figure 17-16). In most eukaryotes, the multiple copies of these rRNA genes are arranged in tandem arrays, separated by nontranscribed DNA spacer regions. A promoter is located in each nontranscribed spacer, spanning nucleotides -45 to $+6$ relative to the transcriptional startpoint. (The promoter slightly overlaps the startpoint.) Each nontranscribed spacer also has a variable number of repetitions of a unit of 60 or 81 nucleotides. In vitro studies have shown that larger numbers of repeats of these units correlate with more transcription of the rRNA gene downstream. Thus, these repeated units seem to act as enhancers, ensuring high levels of transcription of these genes in many different cell types.

RNA polymerase III is responsible for transcription of tRNAs, 5S rRNA, and other small RNAs. This enzyme utilizes yet another type of promoter structure and a different set of transcription factors. Recall that the RNA polymerase III promoter is located internally, *within* the transcribed portion of the gene (see Figure 17-13c). Included in the set of transcription factors that bind to this internal promoter is one of the first transcription factors purified, the basal transcription factor TFIIIA. In general, the transcription factors used by RNA polymerases I, II, and III differ from one another, although recent experiments indicate that all three polymerases associate with basal factors that include TBP, the TATA-binding protein.

Posttranscriptional Controls

The flow of genetic information in eukaryotic cells involves a complex series of posttranscriptional events, any or all of which can also turn out to be regulatory points. Many interesting examples of **posttranscriptional control** have been identified in recent years, ranging from RNA processing, to regulation of translation, to modulation of protein function. Posttranscriptional regulation may be especially useful in providing ways to fine-tune the gene expression pattern rapidly, allowing a cell to respond to the intracellular or extracellular environment without changing overall cellular transcription patterns.

Regulation of RNA Processing. The first possible level of posttranscriptional control is **regulation of RNA processing.** It is clear from previous chapters that virtually all primary transcripts in eukaryotic nuclei undergo substantial processing, most of it occurring before the transcripts leave the nucleus. This processing includes capping at the 5′ end of the RNA, the addition of a poly(A) tail at the 3′ end, splicing together of exons with the removal of introns, and chemical modifications such as methylation.

Regulation of RNA splicing patterns to create alternative mRNA molecules from the same primary transcript— **alternative RNA splicing**—is a mechanism that allows a given gene to give rise to more than one kind of protein. One well-documented example involves the mRNA for a type of antibody molecule called immunoglobulin M (IgM).

This antibody exists in two versions, a secreted version and a version that binds to the plasma membrane of the cell that makes it. Like all antibodies, the basic molecular unit of IgM is a protein with four polypeptide subunits, two heavy chains and two light chains, that combine to form a Y-shaped molecule (see Figure 24-9). It is the heavy chain amino acid sequence that determines whether IgM is secreted or membrane-bound. As shown in Figure 19-27, the gene for the IgM heavy chain has two possible transcription termination, or poly(A), sites, producing primary transcripts that differ at their 3′ ends. It is at this point that alternative RNA splicing occurs. The exons within the primary transcripts are spliced together in two different ways, to produce either the secreted version of the heavy chain or the plasma membrane-bound form. Only the splicing pattern for the membrane-bound version includes the exons that encode the membrane anchor, a hydrophobic amino acid domain that anchors the heavy chain at the cell surface. Alternative splicing also generates variant forms of important cell adhesion molecules, such as the neural cell adhesion molecule (N-CAM) and fibronectin.

Translational Control. Various regulatory possibilities can also be found at the translational level. The differential availability of translational initiation factors or prosthetic groups for the protein product, the involvement of translational regulatory proteins, and variations in the rates of mRNA degradation are some of the mechanisms of **translational control.** As with transcriptional control, translational control appears to occur most often at the initiation stage of the process.

One well-studied example of such translational control involves the dependence of globin synthesis in mammalian reticulocytes (developing red blood cells) on the availability of heme, the iron-containing prosthetic group required for hemoglobin formation. Although reticulocytes have lost their nuclei, they retain large amounts of stable mRNA, almost all of which encodes globin polypeptides. Normally the cells synthesize globin at a high rate. However, globin synthesis would be wasteful if enough heme were not available to complete the formation of hemoglobin molecules. Fortunately, these cells have a mechanism for adjusting polypeptide synthesis to match heme availability.

These red blood cells have a protein kinase, called **heme-controlled inhibitor (HCI),** that responds to heme. Figure 19-28 shows how HCI works. In the presence of heme, it is inactive. In the absence of heme, however, it is active, and it specifically phosphorylates eIF2 (eukaryotic initiation factor 2), one of several proteins required for translation initiation in eukaryotes. Phosphorylated eIF2 cannot form the complex with GTP and methionyl tRNA that is required for initiation (see Chapter 18). Thus, this regulatory mechanism operates at the level of translation initiation. It is a general mechanism for controlling all translation in a cell; but because globins account for more than 90% of the polypeptides made in these cells, it mostly affects globin synthesis. Other translational inhibitors with eIF2 kinase activity have

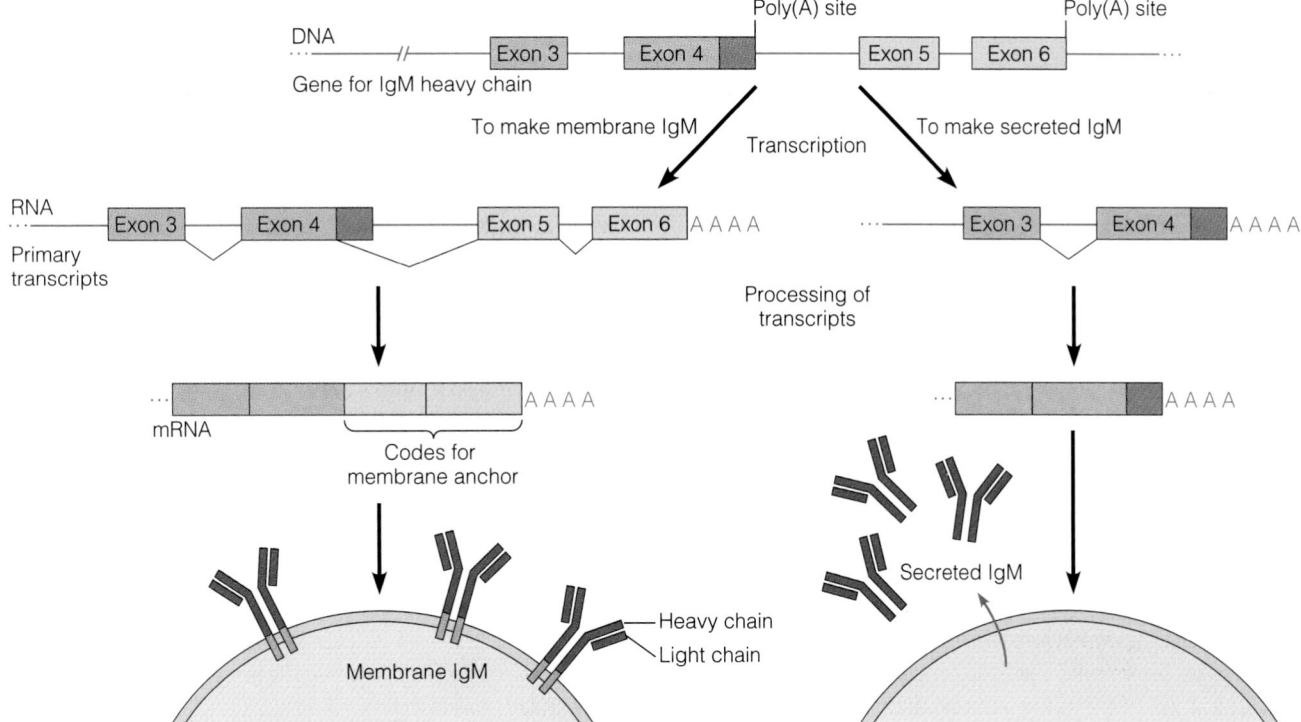

Figure 19-27 Alternative RNA Splicing to Produce Variant Gene Products. The antibody protein immunoglobulin M (IgM) exists in two forms, secreted IgM and membrane-bound IgM, which differ in the carboxyl ends of their heavy chains (each antibody unit consists of two heavy chains and two light chains). A single gene carries the genetic information for both types of IgM heavy chain; the end of the gene corresponding to the polypeptide's C-terminus is shown at the top of the figure. This DNA has two possible poly(A) sites, where RNA transcripts can terminate, and four exons that can be used in two alternative configurations. The splices made in the primary transcripts are indicated by V-shaped symbols. A splicing pattern that uses a splice junction within exon 4 and retains exons 5 and 6 results in the synthesis of IgM heavy chains that are held in the plasma membrane by a membrane anchor encoded by exons 5 and 6. The alternative product is secreted because a splice within exon 4 is not made and the transcript is terminated after exon 4.

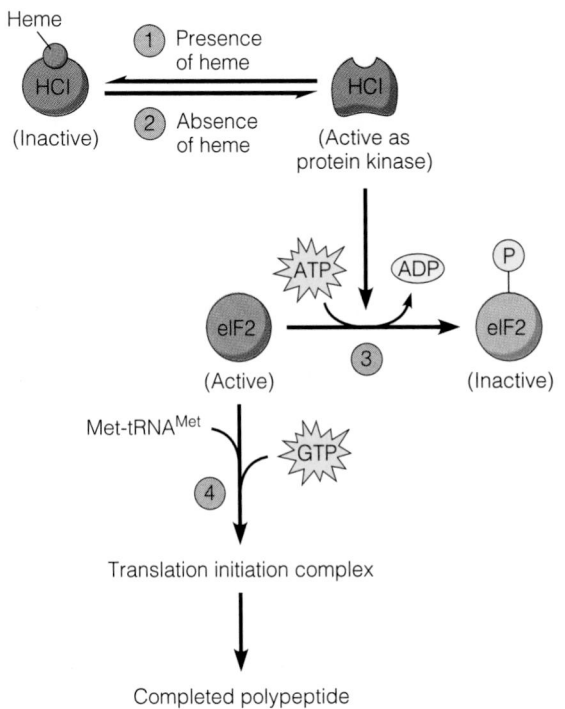

Figure 19-28 Regulation of Translation by Heme in Developing Red Blood Cells. The main function of reticulocytes (developing red blood cells) is to synthesize hemoglobin, which consists of four globin polypeptides and a heme prosthetic group. These cells contain the protein HCI (heme-controlled inhibitor), which regulates this synthesis in response to the availability of heme. ① When heme is present, it binds to HCI, inactivating it. ② When heme is absent, HCI is active. ③ Active HCI functions as a kinase that catalyzes the phosphorylation of eIF2, a key translation initiation factor. Phosphorylated eIF2 is inactive; it cannot combine with methionyl tRNA and GTP to form the translation initiation complex. Thus, in the absence of heme, translation of all mRNA in the cell is inhibited. The main effect is on globin synthesis, because globin mRNA constitutes most of the cell's mRNA. ④ When heme is present, translation of the mRNA proceeds. Newly made globins combine with heme to form hemoglobin molecules (not shown).

been reported in a variety of eukaryotic cells, suggesting that phosphorylation of a translation initiation factor is a widespread means of regulation.

An example of a more specific type of translational control coincidentally also involves iron—in this case, not iron-containing heme in red blood cells, but iron atoms present in other kinds of mammalian cells. When excess iron is present in the cell, synthesis of *ferritin,* the intracellular iron-storage protein, increases dramatically. However, the amount of ferritin mRNA in the cell remains constant, suggesting that translational regulation is occurring. Figure 19-29 illustrates the mechanism of this regulation. When the iron concentration is very low, a regulatory protein binds to a hairpin structure formed in the 5′ untranslated region of the ferritin mRNA. Binding of the protein to this **iron response element (IRE)** in the mRNA blocks the formation of an active polyribosome complex. The protein is allosteric and responds to iron. When more iron is available, the protein binds iron atoms and undergoes a conformational change that prevents binding to the IRE, allowing translation of the ferritin mRNA to occur. This type of response to a change in the cellular environment can be accomplished much faster by translational control than by transcriptional control.

The availability of an mRNA for translation in the cytoplasm can also be regulated by altering mRNA stability. The *half-life,* or time required for 50% of the initial amount of specific transcript to be degraded, varies widely among mRNA species in eukaryotic cells, ranging from 30 minutes or less for some growth factor transcripts to over 10 hours for the mRNA encoding β-globin. The length of the poly(A) tail on a cytoplasmic message is one factor that plays a role in controlling mRNA stability. Messenger RNAs with very short poly(A) tails tend to be less stable. In some cases, the characteristic stability of an mRNA is associated with specific features of the sequence in the 3′ untranslated region. Short-lived mRNAs for several growth factors have a partic-

ular AU-rich sequence in this region. The AU-rich sequence triggers the removal of the poly(A) tail by degradative enzymes. When the AU-rich sequence is transferred to the 3′ end of a normally stable globin message using recombinant DNA techniques, the hybrid mRNA acquires the short half-life of the growth factor mRNA.

Another example of regulation of mRNA stability involves an interesting turn of events in which low levels of iron stabilize an mRNA to protect it from degradation and allow enhanced translation (Figure 19-30). In this case, an IRE similar to the one in ferritin mRNA is located in the 3′ untranslated region of the mRNA for *transferrin receptor,* a plasma-membrane protein important in the uptake of iron from the extracellular fluid. When intracellular iron levels are low and increased uptake of iron is necessary, a regulatory protein (the same allosteric iron-responsive protein mentioned previously) binds to the IRE and thus protects the mRNA from degradation. When iron levels in the cell are high and additional uptake is not necessary, the regulatory protein binds iron and dissociates from the mRNA, allowing the mRNA to be degraded. As a result, the synthesis of transferrin receptor decreases, leading to a decrease in the transport of iron into the cell.

Posttranslational Control. Even after a polypeptide is synthesized, there are many possible means of regulating its function. Mechanisms of **posttranslational control** include processes that modulate protein function in permanent ways, such as proteolysis or glycosylation, or in reversible ways, such as phosphorylation. Other posttranslational processes subject to regulation include the guiding of protein folding by chaperone proteins (Chapter 2), the targeting of proteins to intracellular or extracellular locations (Chapter 18), the interaction of proteins with regulatory molecules or ions, such as cAMP or Ca^{2+} (see Chapter 23), and the turnover of proteins within the cell. These processes are discussed elsewhere in other contexts, but they are mentioned

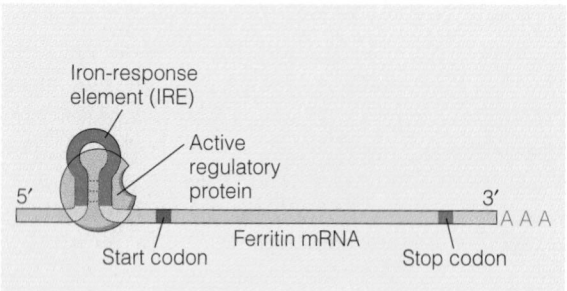

Low iron

Iron-response element (IRE)

Active regulatory protein

5′   3′ A A A

Start codon   Ferritin mRNA   Stop codon

Regulatory protein binds IRE; ferritin synthesis low

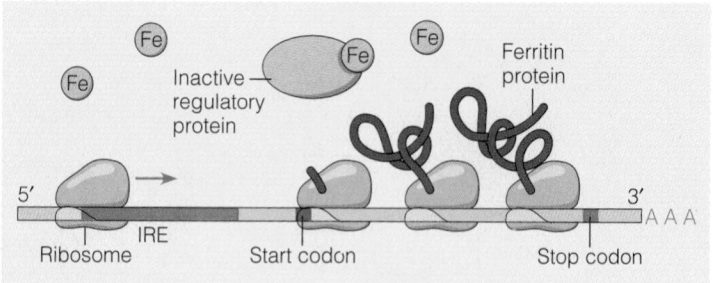

High iron

Fe

Fe   Inactive regulatory protein   Fe   Fe   Ferritin protein

5′   3′ A A A

Ribosome   IRE   Start codon   Stop codon

Regulatory protein cannot bind IRE; ferritin synthesis high

Figure 19-29   Translational Control in Response to Iron.
The initiation of translation of ferritin mRNA is inhibited by the binding of a regulatory protein to the hairpin structure that can be formed by an iron response element (IRE) in the 5′ untranslated leader sequence of the mRNA. The regulatory protein is allosteric. When iron binds to it, it

changes to a conformation that does not recognize the IRE. Therefore, in the presence of iron, ribosomes can assemble on the mRNA and proceed to translate it. Hairpin formation by the IRE does not significantly interfere with ribosome assembly or movement along the mRNA.

here to emphasize that the flow of information in the cell is not complete until a properly functioning gene product is available.

We conclude our discussion of posttranscriptional control with protein turnover. Regulating protein turnover within the cell is an important means of responding to a changing cellular environment or to a particular molecular signal. Many proteins involved in conveying signals within or between cells must be relatively short-lived if the cell is to function appropriately; examples of such proteins include transcription factors and proteins involved in regulating cell division or critical metabolic processes. Furthermore, abnormal proteins of all sorts need to be destroyed.

In the most common mechanism for designating particular proteins for degradation, a small protein called **ubiquitin** is attached to one or more lysine residues of a targeted protein (Figure 19-31). Additional molecules of ubiquitin are then attached to the first, forming a short chain. The

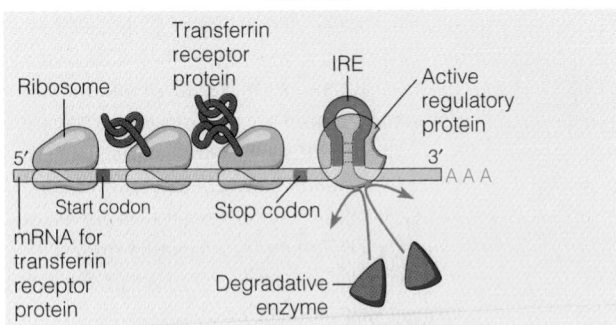

Low iron

Regulatory protein binds IRE; mRNA protected; transferrin receptor synthesis high

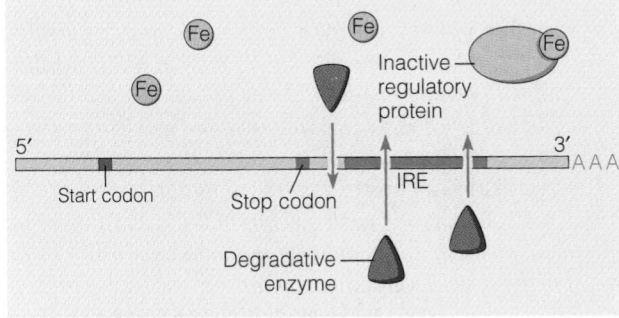

High iron

Regulatory protein cannot bind IRE; mRNA degraded; transferrin receptor synthesis low

Figure 19-30 Control of mRNA Degradation in Response to Iron. Degradation of the mRNA for transferrin receptor, a protein required for iron uptake by the cell, is regulated by the same iron-responsive allosteric protein shown in Figure 19-29. The mRNA for the transferrin receptor has an IRE in its 3′ untranslated region. When the intracellular iron concentration is low, the allosteric regulatory protein remains bound to the IRE, protecting the mRNA from degradation and allowing more transferrin receptor protein to be synthesized. When the iron level is high, iron atoms bind to the regulatory protein, causing it to leave the IRE; the mRNA is left vulnerable to attack by degradative enzymes.

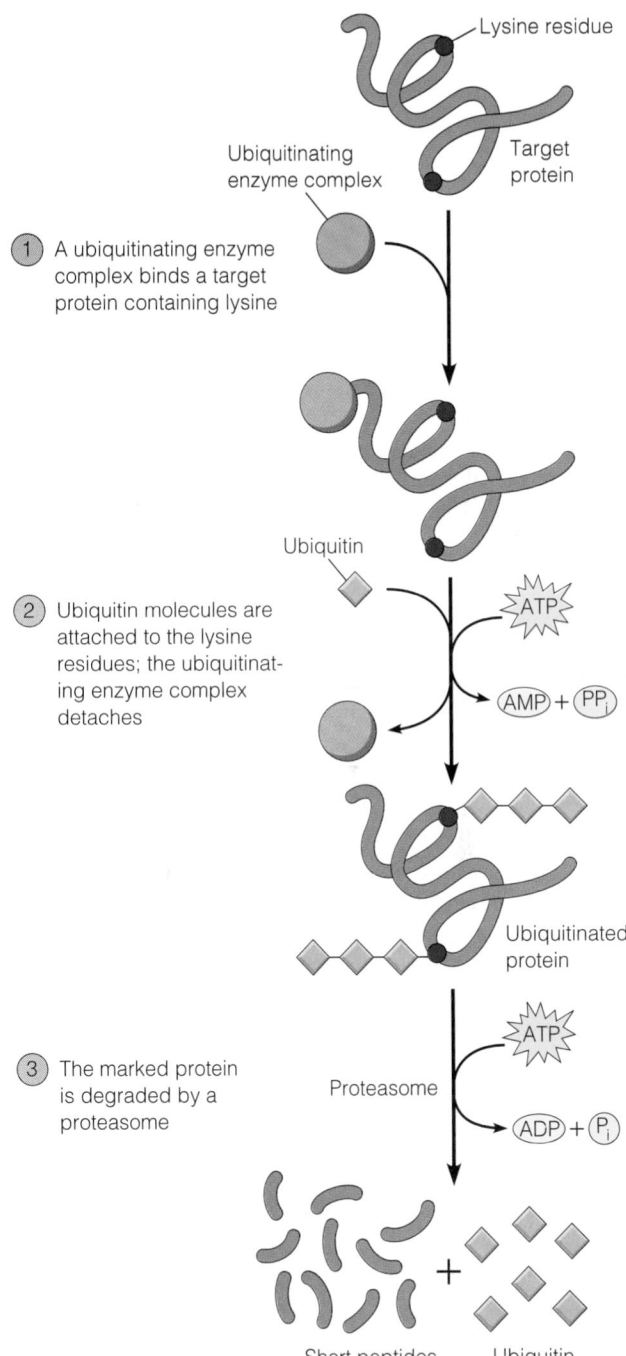

Figure 19-31 Ubiquitin-Dependent Protein Degradation. ① A protein targeted for degradation is bound at its N-terminus by a ubiquitinating enzyme complex. ② In an ATP-dependent series of reactions, ubiquitin molecules are sequentially attached to the protein's lysine residues. The ubiquitinating enzyme complex detaches. ③ A proteasome degrades the ubiquitinated protein into short peptides. The ubiquitin is released and can be recycled.

ubiquitin chains serve as signals that are recognized by large protein complexes called **proteasomes.** Each proteasome has a mass of about 2000 kDa and consists of several proteases (proteolytic enzymes), as well as several ATPases and a binding site for ubiquitin chains. The proteases hydrolyze peptide bonds of ubiquitinated proteins in an ATP-dependent process, degrading the proteins to small peptides (which are later broken down by other enzymes). Proteasomes exist in large numbers throughout the cell; interestingly, given their function, they seem to be shaped somewhat like trash cans! However, the mechanism of the critical step in the ubiquitin pathway—the recognition of a protein to be marked by ubiquitin addition—is not yet well understood.

Most genes are not expressed all the time. Therefore, in addition to knowing how genetic information is expressed in cells, we need to understand how that expression is regulated. In prokaryotes, coordinately regulated genes are often clustered into operons, and most regulation is effected at the level of transcription. Operons are turned on and off in response to cellular needs. In general, those encoding enzymes of catabolic pathways are inducible. Their transcription is specifically activated by the presence of substrate. Operons encoding anabolic enzymes, on the other hand, are subject to repression. Transcription is specifically turned off in the presence of the end-product. Both types of regulation are effected by allosteric repressor proteins that bind to the operator locus and prevent transcription of the associated structural genes.

Some operons coding for catabolic enzymes also have nearby DNA control sites that respond to positive control by the cAMP receptor protein (CRP). This response allows the operon to be shut down in the presence of glucose, ensuring preferential utilization of that sugar. The effect of glucose is mediated by cAMP, the allosteric effector of CRP. Some operons that encode anabolic enzymes, such as those for amino acid biosynthesis, contain an attenuator sequence that renders expression of the operon sensitive to the cellular level of the corresponding aminoacyl tRNA.

Regulation is more complicated in eukaryotes because of the size and organizational complexity of the eukaryotic genome, as well as the progressive, long-term nature of development and tissue differentiation in multicellular organisms. Activation of genes most likely involves a selective decondensation of chromatin. This structural change can be seen microscopically in the giant polytene chromosomes of insect salivary glands, but it is probably a general phenomenon. Once uncoiled, the DNA is more accessible to RNA polymerases and the transcription factors required for initiation. The initiation of gene transcription is subject to regulation by the binding of specific transcription factors, some of which are allosteric proteins, to DNA control elements such as enhancers. In eukaryotes as well as prokaryotes, transcription initiation is perhaps the most important control point in gene expression.

In addition to transcription regulation, eukaryotic cells have a variety of posttranscriptional controls, including the generation of variant forms of a gene product by alternative RNA splicing; general, as well as transcript-specific, mechanisms to regulate translational activity; differences in mRNA stability; and posttranslational modulation of protein activity. Eukaryotic gene regulation is currently an area of intense research and rapid progress, and we are likely to see more exciting developments in the next several years.

(Key Terms for Self-Testing and Problem Set begin on page 638)

Contemporary Techniques

DISCRIMINATING AMONG MULTIPLE LEVELS OF CONTROL

Gene expression in eukaryotic cells is a topic of great interest and excitement, mainly because of relatively new techniques that make it possible to dissect and analyze the complex series of events that lead from a specific gene to the functional protein for which that gene codes. In bacteria, gene expression is regulated primarily at the level of transcription, by varying the kinds and amounts of mRNA molecules that are synthesized. For eukaryotic cells, however, gene expression can be regulated at almost every point along the pathway from DNA to proteins, and many scientists are now working to distinguish carefully among the various levels at which control can be exerted. Especially important in these studies are techniques that make it possible to identify and quantify specific proteins and specific mRNA molecules.

Levels of Gene Expression

The expression of a specific gene usually results in an increase in the level of a particular gene product, commonly an enzyme that can be detected because of its activity. Thus, the first indication of the expression of a particular gene is often an increase in the activity of a specific enzyme. Three examples of induction of enzyme activity are illustrated in

Figure 19A-1. Part a shows the induction of β-galactosidase activity in *E. coli* cells, part b the appearance of tyrosine aminotransferase activity in hormone-treated rat hepatoma cells, and part c the increase in β-amylase activity that occurs in light-grown but not in dark-grown mustard seedlings.

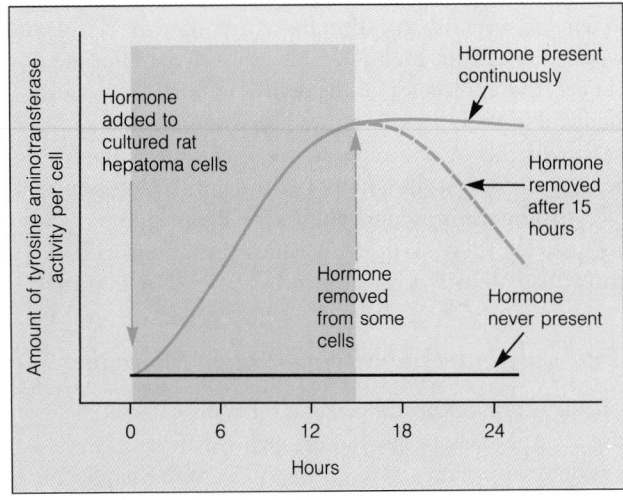

(b) Induction of tyrosine aminotransferase activity in rat hepatoma cells

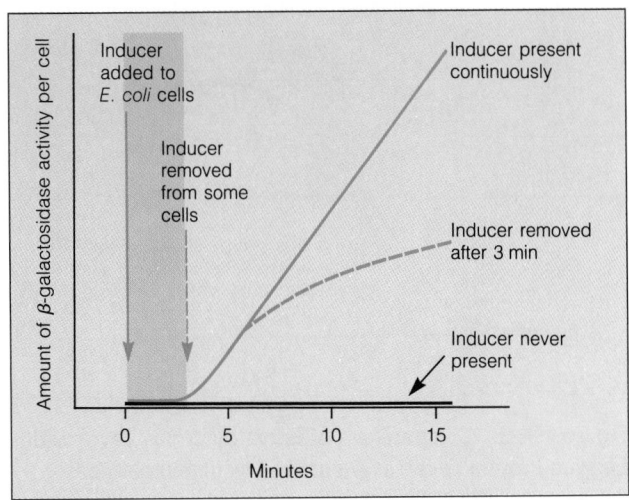

(a) Induction of β-galactosidase activity in *E. coli*

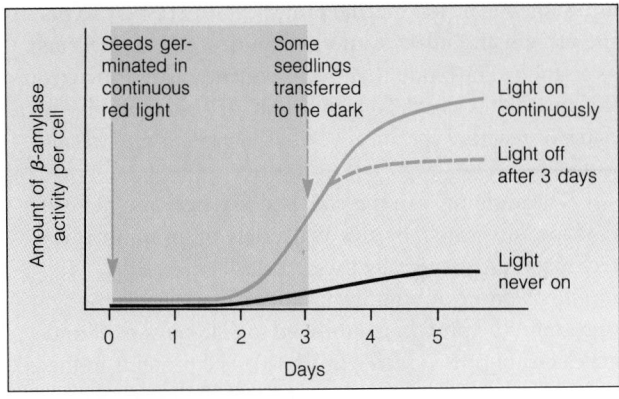

(c) Induction of β-amylase activity in mustard seedlings

Figure 19A-1   Three Examples of Induction of Enzyme Activity.

The time scales vary greatly in the three examples, but the effect is very similar in each case: An enzyme activity that would otherwise remain low (black line) increases many times in response to the appropriate stimulus (blue line). Notice also that if the stimulus is removed, the amount of enzyme stops increasing, or even decreases (dashed line).

An increase in the activity of a specific enzyme indicates the expression of a specific gene, but it tells us little about the level at which that expression is regulated, particularly in eukaryotic cells. From what we know about the control of gene expression in bacterial cells, we might be tempted to regard any increase in enzyme activity as evidence that transcription of a specific gene has been turned on. But other explanations would be equally consistent with the data and need to be considered. For example, the enzyme we are interested in might already be present in the cell in an inactive form, such that the observed increase in activity in response to a stimulus reflects only the activation of existing enzyme molecules. Alternatively, perhaps messenger RNA molecules for the desired enzyme are already on hand in the cell but have not been translated for some reason. In this case, the appearance of enzyme activity would reflect the translation of existing mRNA molecules.

Distinguishing among these several possibilities requires the ability to detect specific proteins and mRNA molecules. We will consider protein and mRNA, in turn.

Detecting and Quantifying Protein Molecules

The detection of specific proteins often depends on the use of antibodies, because of the high specificity of the antibody-antigen reaction. As we will learn in Chapter 24, *antibodies,* also called *immunoglobulins,* are blood proteins that are highly specific with respect to the particular *antigens* (proteins, usually) to which they will bind. To prepare antibodies against a specific protein, it is necessary to purify the protein and inject it into an experimental animal such as a rabbit. The blood that is subsequently withdrawn from the animal has a high concentration, or *titer,* of antibody capable of reacting specifically with the protein that was injected.

The antibody can then be used to detect and quantify that specific protein by any of a variety of immunological techniques. An especially useful technique is called *immunoblotting.* A homogenate or extract of cells thought to contain the protein is subjected to SDS polyacrylamide gel electrophoresis (see Figure 7-18). The proteins in the gel are then transferred to a sheet of nitrocellulose paper, and the paper is exposed to the antibody preparation. Because of the specificity of the antibody-antigen reaction, the anti-

body will bind only to places on the paper where the protein of interest is located. In this way, the protein can be localized to a specific band on the polyacrylamide gel, and the amount of protein in that band can be quantified.

Figure 19A-2 illustrates how we can use the results of an immunological assay such as immunoblotting to analyze gene expression. In addition to assaying for enzyme *activity* at regular time intervals (blue line), we can use the immunoblotting technique to quantify the amount of enzyme *protein* present at the same time points (black lines). If the protein level increases concomitantly with enzyme activity (part a), we can conclude that the appearance of enzyme activity is almost certainly due to the accumulation of newly synthesized enzyme molecules.

However, if the protein is already present at early time points and the amount remains constant as enzyme activity increases (part b), we can conclude instead that the appearance of enzyme activity is due to the activation of preexisting enzyme molecules. This is an example of posttranslational control, because the regulatory events that lead to activation of the enzyme occur at some point after synthesis of the enzyme molecule. Common mechanisms of enzyme activation include the removal of a short peptide from one end of an inactive precursor, phosphorylation of specific amino acid side chains, and transport of the protein into the appropriate organelle.

Detecting and Quantifying mRNA Molecules

Many examples of posttranslational control of enzyme activity are known, but for most enzymes, activity increases

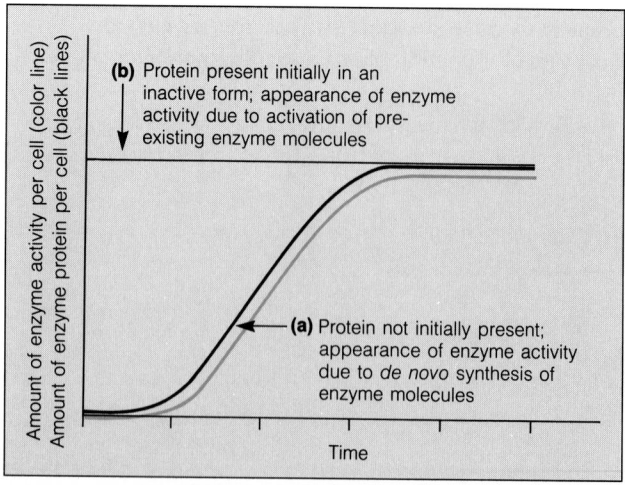

Figure 19A-2 Comparisons of Enzyme Activity Level with Enzyme Protein Level, as Measured by Immunological Assay.

along with enzyme protein. In such cases, the appearance of enzyme activity could be regulated at the level of either transcription or translation, depending on whether the mRNA preexists or must be synthesized as needed. To distinguish between these possibilities, we must be able to detect and quantify specific mRNA molecules. The most commonly used technique takes advantage of the sequence complementarity between the mRNA molecule and a DNA probe.

A complementary DNA (cDNA) probe is synthesized by reverse transcription of mRNA. For detecting a specific mRNA, an RNA sample is size-fractionated by gel electrophoresis and transferred to special blotting paper—a step resembling the size-fractionation step used for proteins in the immunoblotting technique. The paper is exposed to a solution of radioactively labeled cDNA, which binds only to places on the paper that contain RNA molecules with a sequence complementary to that of the probe. The RNA is then quantified by determining the amount of radioactive cDNA bound to a given RNA band. The entire procedure is called a *Northern blot.*

Figure 19A-3 illustrates the results that might be obtained when the same samples used to demonstrate an increase in enzyme activity (blue line) are also assayed for mRNA (black lines). If the mRNA level is low initially and increases just in advance of the enzyme activity (part a), we can conclude that either (1) the appearance of enzyme activity is dependent on newly synthesized mRNA molecules that are beginning to be translated as they accumulate in the cell, an example of *transcriptional control;* or (2) the stabilization of mRNA molecules allows increased accumulation within the cell even though transcription rates have not changed, an example of *posttranscriptional control.* Application of the nuclear run-on transcription technique (see Figure 19-19) would allow us to distinguish between these two possibilities.

Alternatively, the mRNA may already be on hand at early time points, such that the amount remains constant as

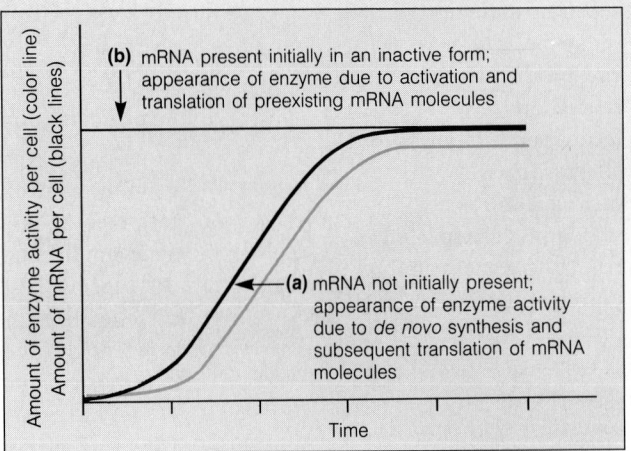

Figure 19A-3   Comparisons of Enzyme Activity Level with the Corresponding mRNA Level.

enzyme level increases (part b). In this case, the appearance of enzyme activity could be due to (1) activation and translation of preexisting mRNA molecules *(translational control)* or (2) stabilization of a protein that initially demonstrated a short half-life *(posttranslational control).*

Thus, the expression of a particular gene may be regulated transcriptionally (neither mRNA nor protein present beforehand), translationally (activation or stabilization of preexisting mRNA), or posttranslationally (activation or stabilization of preexisting protein). Techniques that can detect and quantify specific mRNAs and proteins at very low concentrations are currently being used to study many eukaryotic genes to determine the point at which expression is controlled in each case. If present progress is a fair indication, we can look forward to a fruitful expansion in our understanding of the regulation of gene expression in eukaryotes. ∎

KEY TERMS FOR SELF-TESTING

Gene Regulation in Prokaryotes

constitutive gene (p. 599)

adaptive enzyme synthesis (p. 599)

substrate induction (p. 600)

inducible enzyme (p. 600)

end-product repression (p. 600)

repression (p. 601)

feedback inhibition (p. 601)

effector (p. 601)

operon (p. 601)

structural gene (p. 601)

regulatory gene (p. 601)

polycistronic mRNA (p. 602)

promoter (p. 602)

operator (p. 602)

repressor (p. 602)

regulatory gene (p. 602)

inducible operon (p. 603)

inducer (p. 603)

cis-trans test (p. 604)

trans-acting factor (p. 605)

cis-acting element (p. 605)

repressible operon (p. 605)

corepressor (p. 605)

negative control (p. 606)

positive control (p. 607)

catabolite repression (p. 607)

3′,5′-cyclic AMP (cAMP) (p. 607)

cAMP receptor protein (CRP) (p. 607)

leader sequence (p. 608)

attenuation (p. 609)

leader peptide (p. 609)

attenuator (p. 609)

Gene Regulation in Eukaryotes

protein turnover (p. 613)

genomic control (p. 614)

totipotent (p. 614)

gene amplification (p. 614)

MAT locus (p. 614)

cassette mechanism (p. 615)

polytene chromosome (p. 616)

chromosome puff (p. 616)

DNase I hypersensitive site (p. 617)

DNA methylation (p. 619)

X-chromosome inactivation (p. 619)

transcriptional control (p. 619)

basal transcription factor (p. 622)

proximal control element (p. 622)

regulatory transcription factor (p. 622)

enhancer (p. 622)

distal control element (p. 622)

activator (p. 622)

silencer (p. 623)

repressor (eukaryotic) (p. 623)

combinatorial model for gene regulation (p. 624)

DNA-binding domain (p. 625)

transcription regulation domain (p. 625)

helix-turn-helix motif (p. 625)

zinc finger motif (p. 625)

leucine zipper motif (p. 625)

helix-loop-helix motif (p. 625)

hormone-response element (HRE) (p. 625)

coactivator (p. 628)

heat shock gene (p. 628)

heat shock element (p. 628)

homeotic gene (p. 628)

homeobox (p. 628)

homeodomain (p. 628)

posttranscriptional control (p. 630)

regulation of RNA processing (p. 630)

alternative RNA splicing (p. 630)

translational control (p. 630)

heme-controlled inhibitor (HCI) (p. 630)

iron-response element (IRE) (p. 632)

posttranslational control (p. 632)

ubiquitin (p. 633)

proteasome (p. 634)

PROBLEM SET

19-1. Laboring with *lac*. Most of what we know about the *lac* operon of *E. coli* has come from genetic analysis of various mutants. In the following list are the genotypes of seven strains of *E. coli*. For each strain, indicate whether the *Z* gene product will be expressed in the presence of lactose and whether it will be expressed in the absence of lactose. Explain your reasoning in each case.

(a) $I^+ P^+ O^+ Z^+$

(b) $I^s P^+ O^+ Z^+$

(c) $I^+ P^+ O^c Z^+$

(d) $I^- P^+ O^+ Z^+$

(e) $I^s P^+ O^c Z^+$

(f) $I^+ P^- O^+ Z^+$

(g) Same as (a), but with glucose present

19-2. The Pickled Prokaryote. *Pickelensia hypothetica* is an imaginary prokaryote that converts a wide variety of carbon sources to ethanol when cultured anaerobically in the absence of ethanol. When ethanol is added to the culture medium, however, the organism obligingly shuts off its own production of ethanol and makes lactate instead. Several mutant strains of *Pickelensia* have been isolated that differ in their ability to synthesize ethanol. Class I mutants cannot synthesize ethanol at all. Mutations of this type map at two loci, *A* and *B*. Class II mutants, on the other hand, are constitutive for ethanol synthesis: They continue to produce

ethanol whether it is present in the medium or not. Mutations of this type map at loci *C* and *D*. Strains of *Pickelensia* constructed to be diploid for the ethanol operon have the following genotypes and phenotypes:

$$A^+ B^- C^+ D^+ / A^- B^+ C^+ D^+ \qquad \text{inducible}$$

$$A^+ B^+ C^+ D^- / A^+ B^+ C^+ D^+ \qquad \text{inducible}$$

$$A^+ B^+ C^+ D^+ / A^+ B^+ C^- D^+ \qquad \text{constitutive}$$

(a) Identify each of the four genetic loci of the ethanol operon.

(b) Indicate the expected phenotype for each of the following partially diploid strains:

(i) $A^- B^+ C^+ D^+ / A^- B^+ C^+ D^-$

(ii) $A^+ B^- C^+ D^+ / A^- B^+ C^+ D^-$

(iii) $A^+ B^+ C^+ D^- / A^- B^- C^- D^+$

(iv) $A^- B^+ C^- D^+ / A^+ B^- C^+ D^+$

19-3. Regulation of Bellicose Catabolism. The enzymes bellicose kinase and bellicose phosphate dehydrogenase are coordinately regulated in the bacterium *Hokus focus*. The genes encoding these proteins, *belA* and *belB*, are contiguous segments on the genetic map of the organism. In his pioneering work on this system, Professor Gene X. Pression established that the bacterium can grow with the monosaccharide bellicose as its only carbon and energy source, that the two enzymes involved in bellicose catabolism are synthesized by the bacterium only when bellicose is present in the medium, and that enzyme production is turned off in the presence of glucose. He identified a number of mutations that reduce or eliminate enzyme production and showed that these could be grouped into two classes. Class I mutants are in *cis*-acting elements, whereas those in class II all map at a distance from the structural genes and are *trans*-acting. Thus far, the only constitutive mutations that Pression has found are deletions that connect *belA* and *belB* to new DNA at the upstream side of *belA*. The following is a list of conclusions that he would like to draw from his observations. Indicate in each case whether the conclusion is consistent with the data (C), inconsistent with the data (I), or irrelevant to the data (X).

(a) Bellicose can be metabolized by *H. focus* cells to yield ATP.

(b) Enzyme production by genes *belA* and *belB* is under positive control.

(c) Phosphorylation of bellicose makes the sugar less permeable to transport across the plasma membrane.

(d) The operator for the bellicose operon is located upstream from the promoter.

(e) Some of the mutations in class I may be in the promoter.

(f) Some of the mutations in class II may be in the operator.

(g) Class II mutations may include mutations in the gene that encodes CRP.

(h) The constitutive deletion mutations connect genes *belA* and *belB* to a new promoter.

(i) Expression of the bellicose operon is subject to regulation by attenuation.

19-4. Attenuation in 25 Words or Less. Complete each of the following statements about attenuation in 25 words or less.

(a) Attenuators can also be called *conditional terminators* because . . .

(b) When the gene product of the operon is not needed, the 5′ end of the mRNA forms a secondary structure that . . .

(c) Implicit in our understanding of the mechanism of attenuation is the assumption that ribosomes follow RNA polymerase very closely. This "tailgating" is essential to the model because . . .

(d) Measurements of rates of synthesis indicate that translation normally proceeds faster than transcription. This supports the proposed mechanism for attenuation because . . .

(e) If the operon encoding the enzymes in the biosynthetic pathway for amino acid X is subject to attenuation, the leader sequence probably encodes a polypeptide that . . .

(f) Attenuation makes sense as a mechanism for regulating operons that code for enzymes involved in amino acid biosynthesis because . . .

(g) Attenuation is not a likely mechanism for regulating gene expression in eukaryotes because . . .

19-5. Positive and Negative Control. Assume that you have a culture of *E. coli* cells growing on medium B, which contains both lactose and glucose. At time *t*, 33% of the cells are transferred to medium L, which contains lactose but not glucose; 33% are transferred to medium G, which contains glucose but not lactose; and the remaining cells are left in medium B. For each of the following statements, indicate with an L if it is true of the cells transferred to medium L, with a G if it is true of the cells transferred to medium G, with a B if it is true of the cells left in medium B, and with an N if it is true of none of the cells. In some cases, more than one letter may be appropriate.

(a) The rate of glucose consumption per cell is approximately the same after time *t* as before.

(b) The rate of lactose consumption is higher after time *t* than before.

(c) The intracellular cAMP level is lower after time *t* than before.

(d) Most of the *lac* operator sites have *lac* repressor proteins bound to them.

(e) Most of the cAMP receptor proteins exist as CRP-cAMP complexes.

(f) Most of the *lac* repressor proteins exist as repressor-glucose complexes.

(g) The rate of transcription of the *lac* operon is greater after time *t* than before.

(h) The *lac* operon has both CRP and the repressor protein bound to it.

19-6. Polytene Chromosomes. The fruit fly *Drosophila melanogaster* has about 2×10^8 base pairs of DNA per haploid genome, of which about 75% is unique-sequence DNA. The DNA is distributed over four pairs of homologous chromosomes, which have a total of about 5000 visible bands when in polytene form in the salivary gland. The number of genes estimated from mutational studies is also about 5000, although recent molecular studies indicate that many bands contain more than one gene.

(a) Why was it tempting to speculate that each band corresponds to a single gene? Why might the mutational and molecular studies reach different conclusions about gene number? What does this suggest about the number of different proteins that *Drosophila* can make? Does that seem like a reasonable number to you?

(b) Assuming that all the unique-sequence DNA is uniformly distributed in the chromosomes, how much unique-sequence DNA (in base pairs) is there in the average band?

(c) How much DNA would it take to code for a single polypeptide with a molecular weight of 50,000? (Assume that amino acids have an average molecular weight of 110.) What proportion does that represent of the total unique-sequence DNA in the average band?

(d) How do you account for the discrepancy in part c?

19-7. Enhancers. An enhancer may increase the frequency of transcription initiation for its associated gene when . . . (Indicate true or false for each statement, and explain your answer.)

(a) it is located 1000 nucleotides upstream of the gene's core promoter.

(b) it is in the gene's coding region.

(c) no promoter is present.

(d) it causes looping out of the intervening DNA.

(e) it causes alternative splicing of the DNA.

19-8. Gene Amplification. The best-studied example of gene amplification is that of the ribosomal genes during oogenesis in *Xenopus laevis*. The unamplified number of genes is about 500 per haploid genome. After amplification, the tetraploid (premeiotic) oocyte contains about 2,000,000 genes (i.e., 500,000 genes per haploid genome). This level of amplification is apparently necessary to allow the egg cell to synthesize the 10^{12} ribosomes that accumulate during the two months of oogenesis in this species. Each ribosomal gene consists of about 13,000 nucleotide pairs, and the genome size of *Xenopus* is about 2.7×10^9 nucleotide pairs per haploid genome.

(a) What fraction of the total haploid genome do the 500 copies of the ribosomal genes represent?

(b) What is the total size of the amplified genome (in nucleotide pairs)? What proportion of this do the amplified ribosomal genes represent?

(c) Assume that all the ribosomal genes in the amplified oocyte are transcribed continuously to generate the needed number of ribosomes during the two months of oogenesis. How long would oogenesis have to extend if the genes had not been amplified?

(d) Why do you think genes have to be amplified when the gene product needed by the cell is an RNA, but not usually when the desired gene product is a protein?

19-9. Homeotic Genes. Homeotic genes are considered crucial to early development in *Drosophila* because . . . (Indicate true or false for each statement, and explain your answer.)

(a) they encode proteins containing zinc-finger domains.

(b) mutations in homeotic genes are always lethal.

(c) they control expression of many other genes required for development.

(d) they are identical in all eukaryotes.

(e) homeodomain proteins act by influencing mRNA degradation.

19-10. Levels of Control. Assume that liver and kidney tissues from the same mouse each contain about 10,000 species of cytosolic mRNA, but that only about 25% of these are common to the two tissues.

(a) Suggest an experimental approach that might be used to establish that some of the mRNA molecules are common to both tissues, but others are not.

(b) One possible explanation for the data is that differential transcription occurs in liver and kidney nuclei. What is another

possible explanation? Describe an experiment that would enable you to distinguish between these possibilities.

(c) If all mRNA molecules had been shown to be common to both liver and kidney and yet the two tissues were known to be synthesizing different proteins, what level of control would you have to assume was operating?

19-11. Messenger RNA Complexity in Sea Urchins. The mRNA molecules found in the cytosol in oocytes, embryos, and adult tissues of the sea urchin represent a considerable diversity of sequences. Figure 19-32 depicts the complexity of mRNA molecules found at different stages of sea urchin development and in different adult tissues. Blastula, gastrula, and pluteus are successive stages in embryogenesis; tubefoot, intestine, and coelomocyte are adult tissues. The solid portion of each bar indicates the fraction of the gastrula stage mRNA also found in mRNA from other sources. The open portion of each bar indicates the amount of mRNA not shared in common with the gastrula. Label each of the following conclusions as C if it is consistent with the data in the figure, I if it is inconsistent, and X if you cannot tell from the data.

(a) A small fraction of RNA species is probably common to all tissues examined.

(b) The mRNA molecules common to all tissues probably code for "housekeeping" proteins responsible for metabolic processes common to all cells.

(c) The mRNA molecules present at the gastrula stage of embryogenesis represent less than 3% of the unique-sequence DNA in the genome.

(d) The mRNA molecules present in the coelomocyte of the adult can code for about 30,000 polypeptides with an average molecular weight of 50,000.

(e) Differentiation of adult tissues appears to be accompanied by a decrease in the number of genetic functions being expressed in the cells.

(f) All the genes that are being actively transcribed in the intestine are also turned on at the pluteus stage of embryogenesis.

(g) The oocyte is expressing genetic information that is not being expressed in the intestine.

(h) The intestine is expressing genetic information that is not being expressed in the oocyte.

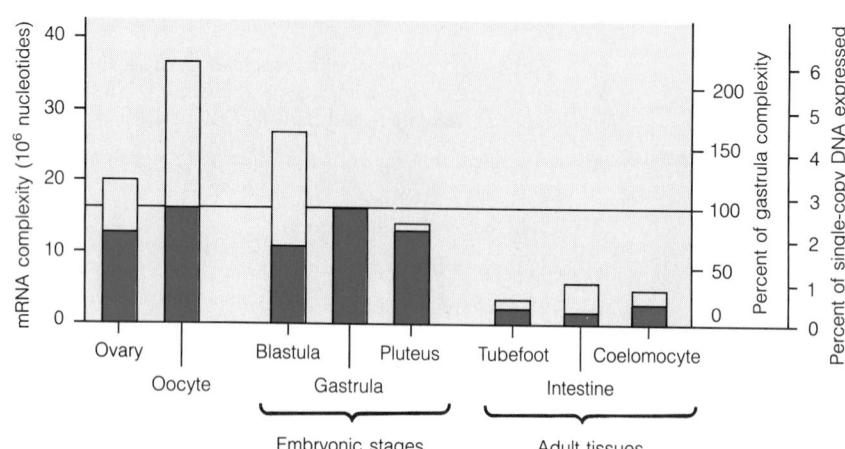

Figure 19-32   Sequence Complexity of mRNA Molecules.

SUGGESTED READING

General References

Lewin, B. *Genes*, 5th ed. New York: Wiley, 1994.

McKnight, S. L., and K. R. Yamamoto, eds. *Transcriptional Regulation.* Cold Spring Harbor, NY: Cold Spring Harbor Laboratory Press, 1992.

Singer, M., and P. Berg. *Genes and Genomes.* Mill Valley, CA: University Science Books, 1991.

Negative Control in Prokaryotes: The Operon Model

Dickson, R., J. Abelson, W. Barnes, and W. Reznikoff. Genetic regulation: The *lac* control region. *Science* 187 (1975): 27.

Gralla, J. D. *lac* Repressor. In *Transcriptional Regulation* (S. L. McKnight and K. R. Yamamoto, eds.). Cold Spring Harbor, NY: Cold Spring Harbor Laboratory Press, 1992.

Jacob, F., and J. Monod. Genetic regulatory mechanisms in the synthesis of proteins. *J. Mol. Biol.* 3 (1961): 318.

Maniatis, T., and M. Ptashne. A DNA operator-repressor system. *Sci. Amer.* 234 (January 1976): 64.

Sigler, P. B. The molecular mechanism of *trp* repression. In *Transcriptional Regulation* (S. L. McKnight and K. R. Yamamoto, eds.). Cold Spring Harbor, NY: Cold Spring Harbor Laboratory Press, 1992.

Positive Control in Prokaryotes: Catabolite Repression

Crothers, K. M., and T. A. Steitz. Transcriptional activation by *Escherichia coli* CAP protein. In *Transcriptional Regulation* (S. L. McKnight, K. R. Yamamoto, eds.). Cold Spring Harbor, NY: Cold Spring Harbor Laboratory Press, 1992.

Kolb, A., S. Busby, H. Buc, S. Garges, and S. Adhya. Transcriptional regulation by cAMP and its receptor protein. *Annu. Rev. Biochem.* 62 (1993): 749.

Bacterial Sigma Factors

Gross, C. A., and M. Lonetto. Bacterial sigma factors. In *Transcriptional Regulation* (S. L. McKnight and K. R. Yamamoto, eds.). Cold Spring Harbor, NY: Cold Spring Harbor Laboratory Press, 1992.

Helmann, J. D., and M. J. Chamberlin. Structure and function of bacterial sigma factors. *Ann. Rev. Biochem.* 57 (1988): 839.

Attenuation

Landick, R., and C. S. Turnbough, Jr. Transcriptional attenuation. In *Transcriptional Regulation* (S. L. McKnight and K. R. Yamamoto, eds.). Cold Spring Harbor, NY: Cold Spring Harbor Laboratory Press, 1992.

Yanofsky, C. Operon specific control by transcription attenuation. *Trends Genet.* 3 (1987): 356.

Gene Amplification in Eukaryotes

Chisholm, R. Gene amplification during development. *Trends Biochem. Sci.* 7 (1982): 161.

Schimke, R. T., ed. *Gene Amplification.* Cold Spring Harbor, NY: Cold Spring Harbor Laboratory, 1982.

Eukaryotic Gene Expression and Chromatin Structure

Ashburner, M. Temporal control of puffing activity in polytene chromosomes. *Cold Spring Harbor Symp. Quant. Biol.* 38 (1973): 655.

Elgin, S. C. R. DNAse l-hypersensitive sites of chromatin. *Cell* 27 (1981): 413.

Kwon, H., A. N. Imbalzano, P. A. Khavarl, R. E. Kingston, and M. R. Green. Nucleosome disruption and enhancement of activator binding by a human SWI/SNF complex. *Nature* 370 (1994): 477.

Li, E., T. H. Bestor, and R. Jaenisch. Targeted mutation of the DNA methyltransferase gene results in embryonic lethality. *Cell* 69 (1992): 915.

Patient, R. K., and J. Allan. Active chromatin. *Curr. Opin. Cell Biol.* 1 (1989): 454.

Wolffe, A. P. Transcription: In tune with the histones. *Cell* 77 (1994):13.

Transcriptional Regulation in Eukaryotes

Derman, E., K. Krauter, L. Walling, C. Weinberger, M. Ray, and J. E. Darnell, Jr. Transcriptional control in the production of liver-specific mRNAs. *Cell* 23 (1981): 731.

Drapkin, R., A. Merino, and D. Reinberg. Regulation of RNA polymerase II transcription. *Curr. Opin. Cell Biol.* 5 (1993): 469.

Herschbach, B. M., and A. D. Johnson. Transcriptional repression in eukaryotes. *Annu. Rev. Cell Biol.* 9 (1993): 479.

Spieth, J., G. Brooke, S. Kuersten, K. Lea, and T. Blumenthal. Operons in *C. elegans:* Polycistronic mRNA precursors are processed by *trans*-splicing of SL2 to downstream coding regions. *Cell* 73 (1993): 521.

Transcription Factors and Enhancers

Atchison, M. L. Enhancers: Mechanisms of action and cell specificity. *Annu. Rev. Cell. Biol.* 4 (1988): 127.

Dynan, W. S., and R. Tijan. Control of eukaryotic messenger RNA synthesis by sequence-specific DNA-binding proteins. *Nature* 316 (1985): 774.

Harrison, S. C. A structural taxonomy of DNA-binding domains. *Nature* 353 (1991): 715.

McKnight, S. L. Molecular zippers in gene regulation. *Sci. Amer.* (April 1991): 54.

Muller, H.-P., and W. Schaffner. Transcriptional enhancers can act in *trans. Trends Genet.* 6 (1990): 300.

Murre, C., and D. Baltimore. The helix-loop-helix motif: Structure and function. In *Transcriptional Regulation* (S. L. McKnight and K. R. Yamamoto, eds.). Cold Spring Harbor, NY: Cold Spring Harbor Laboratory Press, 1992.

Rhodes, D., and A. Klug. Zinc fingers. *Sci. Amer.* (Feb. 1993): 56.

Schleif, R. DNA looping. *Annu. Rev. Biochem.* 61 (1992): 199.

Tjian, R. Molecular machines that control genes. *Sci. Amer.* (Feb. 1995): 54.

Heat Shock Genes and Homeotic Genes

Gehring, W. J., Y. Q. Quan, M. Billeter, K. Furukubo-Tokunaga, A. F. Schier, D. Resendez-Perez, M. Affolter, F. Otting, and K. Wuthrich. Homeodomain-DNA recognition. *Cell* 78 (1994): 211.

Lawrence, P. A., and G. Morata. Homeobox genes: Their function in *Drosophila* segmentation and pattern formation. *Cell* 78 (1994): 181.

McGinnis, W., and M. Kuziora. The molecular architects of body design. *Sci. Amer.* (Feb. 1994): 58.

Welch, W. J. How cells respond to stress. *Sci. Amer.* (May 1993): 56.

Posttranscriptional Control in Eukaryotes

Bernstein, P., and J. Ross. Poly(A), poly(A) binding protein and the regulation of mRNA stability. *Trends Biochem. Sci.* 14 (1989): 373.

Breitbart, R., A. Andreadis, and B. Nadal-Ginard. Alternative splicing: A ubiquitous mechanism for the generation of multiple protein isoforms from single genes. *Annu. Rev. Biochem.* 56 (1987): 467.

Casey, J., M. W. Hentze, D. M. Koeller, S. W. Caughman, T. A. Rouault, R. D. Klausner, and J. B. Harford. Iron responsive elements: Regulatory RNA sequences that control mRNA levels and translation. *Science* 240 (1988): 924.

Early, P., J. Rogers, M. Davis, K. Calame, M. Bond, R. Wall, and L. Hood. Two mRNAs can be produced from a single immunoglobulin mu gene by alternative RNA processing. *Cell* 20 (1980): 313.

Finley, D., and V. Chau. Ubiquitization. *Annu. Rev. Cell Biol.* 7 (1991): 25.

Goldberg, A. L. Functions of the proteasome: The lysis at the end of the tunnel. *Science* 268 (1995): 522.

Hershko, A., and A. Ciechanover. The ubiquitin system for protein degradation. *Annu. Rev. Biochem.* 61 (1992): 761.

Shaw, G., and R. Kamen. A conserved AU sequence from the 3′ untranslated region of GM-CSF mRNA mediates selective mRNA degradation. *Cell* 46 (1986): 659.

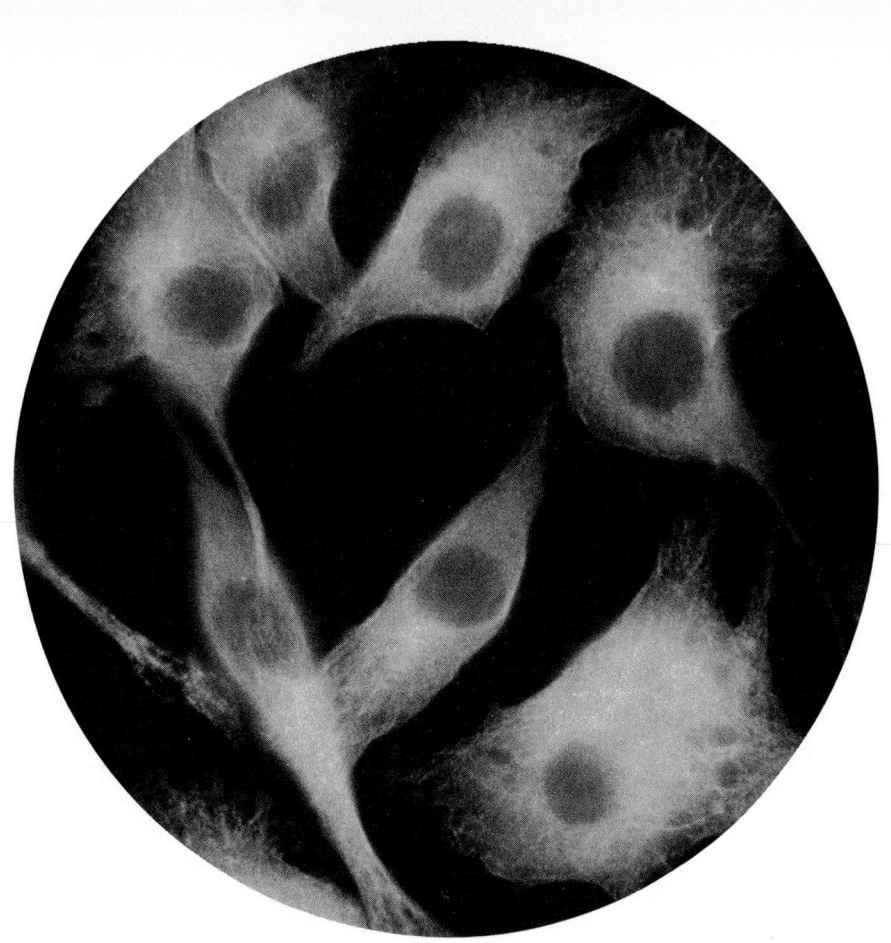

SPECIFIC CELL FUNCTIONS

20

CYTOSKELETAL STRUCTURE AND FUNCTION

I n the preceding chapters, we examined a variety of cellular processes and pathways, many of which occur in the organelles of eukaryotic cells. We now come to the **cytosol**, which is the region of the cytoplasm between and surrounding the organelles. Until a few decades ago, the cytosol of the eukaryotic cell was regarded as the generally uninteresting, gel-like substance in which the nucleus and other organelles were suspended. Cell biologists knew that the cytosol was quite rich in proteins, usually about 20–30%, but these proteins were thought to be soluble and freely diffusible. Except for those of known enzymatic activity, little was understood about the structural or functional significance of the cytosolic proteins. The advent of electron microscopy made us aware of structures such as microtubules and microfilaments in the cytosol, but initially this knowledge did little to change the prevailing view of the cytosol as an amorphous matrix without much fascination for cell biologists. In recent years, however, that view has changed dramatically.

Several new microscopic techniques have revealed that the interior of a eukaryotic cell is highly structured. A complex network of interconnected filaments and tubules called the **cytoskeleton** extends throughout the cytosol, from the nucleus to the inner surface of the plasma membrane. This elaborate array of filaments and tubules forms a highly structured yet very dynamic matrix that helps establish the shape of the cell and plays important roles in cell movement and cell division.

In addition, the cytoskeleton serves as a framework for positioning and actively moving membrane-bounded organelles within the cytosol and may even play a similar role for messenger RNA and ribosomes. For example, clusters of ribosomes are frequently seen in association with filaments, and they remain behind with the cytoskeleton after the extraction of cells with nonionic detergents. The same may be true of enzymes and other soluble proteins. Some researchers think that many, perhaps even most, enzymes in the cytosol are not really soluble at all, but are physically clustered and attached to the cytoskeleton in close proximity to other enzymes involved in the same pathway, thereby facilitating the channeling of intermediates within each pathway.

The term *cytoskeleton* expresses well the role of this proteinaceous matrix in providing an architectural framework for eukaryotic cells. This framework, or internal scaffolding, confers a high level of internal organization on such cells and enables them to assume and maintain complex shapes that would not otherwise be possible. However, the term is less adequate at conveying the functional significance of the cytoskeleton, which is probably more important than its structural role.

The cytoskeleton is a dynamic, changeable matrix. The various components of the cytoskeleton are involved in a great variety of cellular shape changes and movements. Examples of such movements include muscle contraction, the beating of cilia and flagella, amoeboid movement, cell division, chromosome movements, endocytosis and exocytosis, changes in cell shape, and the active movement of vesicles and other organelles within the cell.

Some of these examples obviously involve the movement of whole cells, whereas others entail the motion of intracellular components such as membranes, chromosomes, or organelles. All, however, are mediated by the structural elements of the cytoskeleton, which are the subjects of this and the next chapter.

Structural Elements of the Cytoskeleton

The three major structural elements of the cytoskeleton are *microtubules, microfilaments,* and *intermediate filaments,* each of which is unique to eukaryotic cells, as are their respective monomers. The existence of three distinct systems of filaments and tubules was first revealed by electron microscopy. Biochemical and immunological studies then identified the distinctive proteins of each system, which are

Table 20-1 Properties of Microtubules, Microfilaments, and Intermediate Filaments

	Microtubules	Microfilaments	Intermediate Filaments
Structure	Hollow tube with a wall consisting of 13 protofilaments	Two intertwined chains of F-actin	Eight protofilaments joined end-to-end with staggered overlaps
Diameter	Outer: 25 nm Inner: 15 nm	7 nm	8–12 nm
Monomers	α-tubulin β-tubulin	G-actin	Several proteins; see Table 20–4

Functions	Axonemal: Cell motility
	Cytoplasmic:
	  Organization and maintenance of
	    animal cell shape
	  Chromosome movements
	  Disposition and movement of organelles
		Muscle contraction
		Amoeboid movement
		Cell locomotion
		Cytoplasmic streaming
		Cell division
		Maintenance of animal cell shape
			Structural support
			Maintenance of animal cell shape
			Formation of nuclear lamina and scaffolding
			Strengthening of nerve cell axons (NF protein)
			Keeping muscle fibers in register (desmin)

also unique to eukaryotic cells. The technique of *immuno-fluorescence microscopy,* to be described in the next section, was especially important in localizing specific proteins within the cytoskeletal network.

Each of the structural elements of the cytoskeleton has a characteristic size, structure, and intracellular distribution, and each is formed by the polymerization of a different kind of protein monomer (Table 20-1). Microtubules are composed of the protein *tubulin* and are about 25 nm in diameter. Microfilaments are polymers of the protein *actin* and have a diameter of about 7 nm. Intermediate filaments have diameters in the range of 8–12 nm, and are assembled from different protein monomers, depending on the cell type. We now know that intermediate filament proteins constitute a family of related proteins, and different types of cells utilize different members of this family to assemble intermediate filaments. In addition to its major protein component, each structural element of the cytoskeleton also has a number of other proteins associated with it. These *accessory proteins* account for the remarkable structural and functional diversity of the cytoskeletal elements.

Microtubules and microfilaments are best known for their roles in contraction and motility. Microfilaments are essential components of *muscle fibrils,* and microtubules are the structural elements of *cilia* and *flagella,* appendages that enable certain cells to either propel themselves through a fluid environment or move fluids past the cell. These structures are large enough to be seen with the light microscope and were therefore known and studied long before it became clear that the same structural elements are also integral parts of the cytoskeleton.

Motility and contractility are very important topics in cell biology and will therefore be discussed in detail in Chapter 21. Here, we will focus specifically on the involvement of filaments and tubules in cytoskeletal structure and on the molecular basis for the general functions in which they are involved. We will look first at several of the tech-niques currently used to study cytoskeletal organization and then consider each of these structural elements in detail.

In doing so, we will be discussing microtubules, micro-filaments, and intermediate filaments as though they were separate entities with their own structure and function. But to appreciate their roles in giving cells shape and in mediating various types of cell movement, we need to understand that the components of the cytoskeleton are linked together structurally and function not as independent entities, but as an integral meshwork of filaments, tubules, and their associated proteins. Moreover, the cytoskeleton of one cell can influence that of neighboring cells through intercellular junctions and by means of the extracellular matrix that surrounds many animal cells.

Techniques for Studying the Cytoskeleton

The cytoskeleton is a topic of much current research interest to eukaryotic cell biologists. Most of the recent progress in our understanding of cytoskeleton structure is due to three powerful microscopic techniques: immunofluorescence microscopy, quick-freeze deep-etch microscopy, and digital video microscopy. We will look at each technique briefly, to appreciate its potential for revealing cell structure and to understand the complementary insights into the cytoskeleton that the three techniques provide. In addition, we will consider the use of specific drugs and antibodies in the analysis of cytoskeletal function.

Immunofluorescence Microscopy

Immunofluorescence microscopy combines the power of antibodies to recognize and bind to specific molecules with fluorescence microscopy, one of the most sensitive techniques for seeing where molecules are located inside the cell.

The cytoskeleton is usually not obvious when cells are observed with most types of light microscopes. However, with immunofluorescence microscopy, a cytoskeletal protein such as actin or tubulin can be made to glow vividly against a black background.

To achieve this type of staining, cells are fixed to preserve the cytoskeletal structure. Next, the plasma membrane is partially dissolved with organic solvents or detergents so that the large antibody molecules can gain access to the inside of the cell. The cells are then incubated with a series of two antibody solutions, designated respectively as the *primary antibody* and the *secondary antibody*. The primary antibodies are specific for a particular protein (e.g., actin) and bind to this protein wherever they are found in the cell. If, for example, actin is assembled into microfilaments, primary antibody molecules will bind to each of the actin monomers along the filament. After unbound primary antibody is washed away, the cells are exposed to secondary antibodies that recognize the primary antibodies and are covalently linked to a fluorescent dye such as *fluorescein.*

Thus, the primary antibodies bind to the target protein wherever it occurs in the cell, and the secondary antibodies bind to the primary antibodies, causing the appropriate structures to glow brightly, or "light up," in the fluorescence microscope. Figure 20-1 illustrates this technique for each of the three structural elements of the cytoskeleton, using antibodies against tubulin, actin, and the protein *keratin* to visualize microtubules, microfilaments, and intermediate filaments, respectively.

Quick-Freeze Deep-Etch Electron Microscopy

Quick-freeze deep-etch microscopy, or **deep-etching** for short, is a modification of the freeze-etching technique described in the Appendix. In freeze-etching, cells are frozen very quickly at the temperature of liquid nitrogen ($-196°C$), and the frozen tissue is cracked with a knife blade. Ice is then removed from the fractured surface by sublimation in a vacuum, thus making it possible to see more deeply into the cell. Finally, a platinum replica of the exposed surface is prepared and examined in the electron microscope (see Figure A-30).

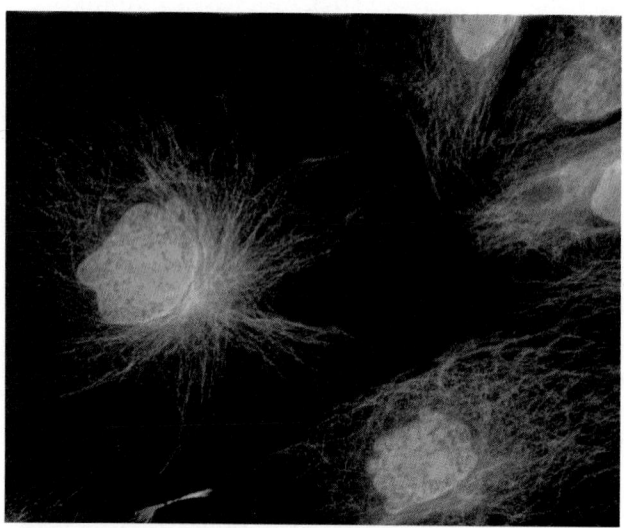

(a) Microtubules

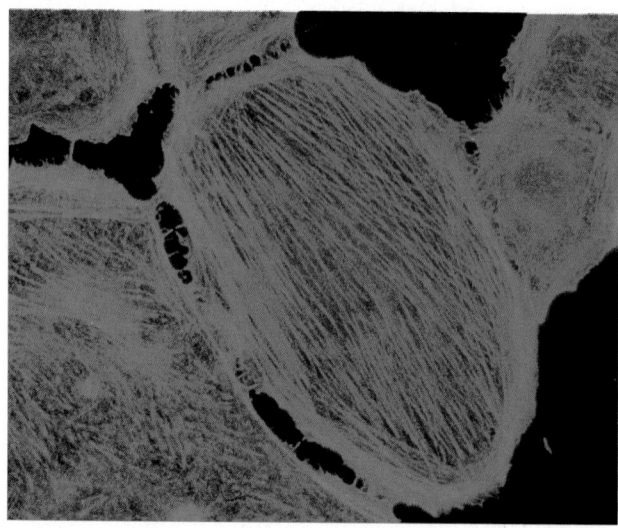

(b) Microfilaments

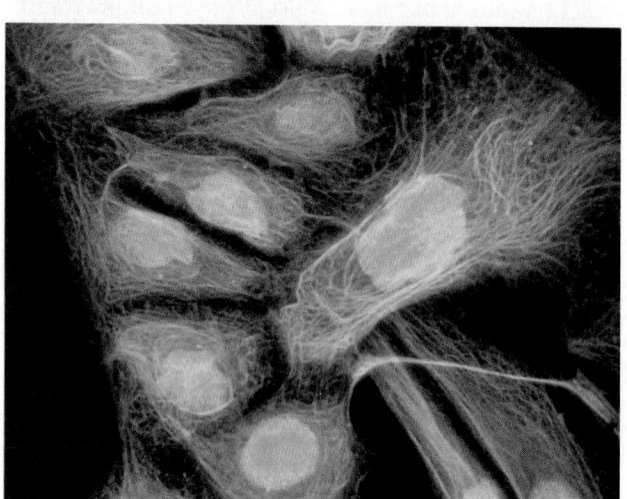

(c) Intermediate filaments

Figure 20-1   The Intracellular Distribution of Microtubules, Microfilaments, and Intermediate Filaments.
(a) The distribution of microtubules in cells of the kangaroo rat kidney cell line (PtK-1) visualized by immunofluorescence staining for tubulin. For reference, the nucleus has been stained with the red fluorescent DNA stain propidium iodide. **(b)** The distribution of microfilaments in a rat kidney cell line visualized by staining actin with a fluorescent derivative of phalloidin. **(c)** The distribution of intermediate filaments in PtK-1 cells, visualized by immunofluorescence staining for keratin. Here the nucleus has also been stained with propidium iodide. $\vdash\!\!\!\!\!\!\!\!\dashv$
                                                                    5 μm

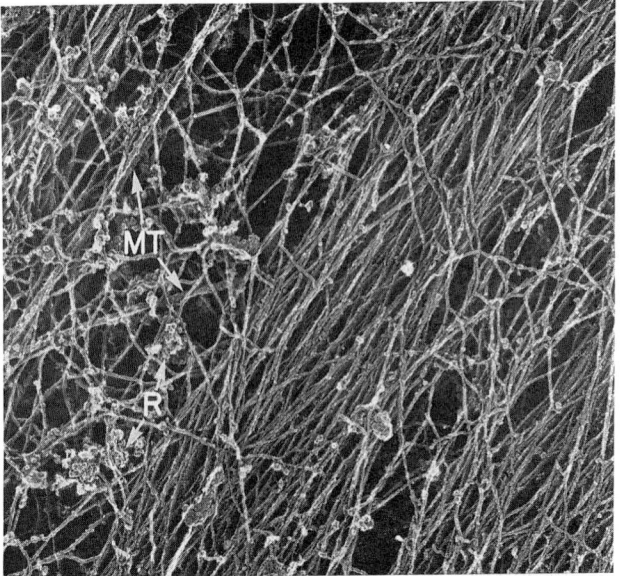

0.5 μm

Figure 20-2 The Cytoskeleton as Revealed by Quick-Freeze Deep-Etch Microscopy. A fibroblast cell was extracted with a nonionic detergent to remove soluble cytoplasmic proteins and integral membrane proteins. A platinum replica of the cytoskeleton was then prepared by the quick-freeze deep-etch technique described in the text. Bundles of actin microfilaments are visible in the lower right, microtubules (MT) are visible in the upper left, and several ribosomes (R) line some of the filaments.

In the past, the major limitation of freeze-etching was the formation of ice crystals during freezing, which distorts the cytoskeletal architecture. In the deep-etching technique, a copper block cooled with liquid helium (−269°C) is used to freeze the sample so quickly (within milliseconds) that ice crystals do not form, thereby avoiding structural distortions. The frozen sample is then subjected to fracturing and sublimation, and coated with a thin molecular layer of platinum to visualize filamentous material. Deep-etching has been very useful in exploring the cytoskeleton and especially in examining connections with other structures of the cell (Figure 20-2).

Digital Video Microscopy

The use of digitizers and computers to acquire, store, and process video images from a microscope is called **digital video microscopy.** This technique was pioneered independently by Robert Allen and Shinya Inoué. (Video microscopy can be either digital or analogue, although most modern video microscopy is digital—see the Appendix for more information.) In digital video microscopy, high-resolution images from a video camera attached to a microscope are digitized, a process that converts the signal from the camera to a two-dimensional array of numbers whose values correspond to the lightness or darkness at a particular location in the image. Once an image is digitized, it can be stored in a computer and processed mathematically to remove background features that obscure the image of interest, reduce noise, and increase contrast. Such enhancement has made it possible to visualize, for example, individual microtubules with the light microscope (Figure 20-3).

Intensified video cameras designed to detect very dim images also greatly extend our ability to study the cytoskeleton. It may take a minute to record a fluorescent image on film, while an intensified video camera in some cases can record the same image at 30 images per second. This increased sensitivity has made it possible to study particular components of the cytoskeleton in living cells. One application of digital video microscopy is a technique called *fluorescence analogue cytochemistry,* where cytoskeletal proteins such as actin or tubulin are purified from cells and covalently labeled or *tagged* with a fluorescent molecule. The fluorescently labeled cytoskeletal proteins can then be reintroduced into the cell using a tiny hollow glass needle, a procedure known as *microinjection.* These fluorescent cytoskeletal proteins become incorporated into the cytoskeleton,

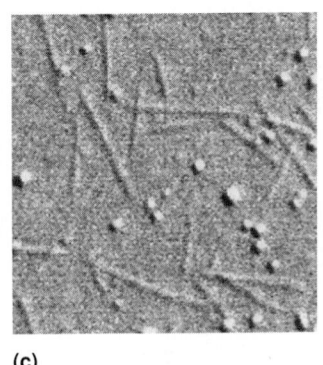

(a)   (b)

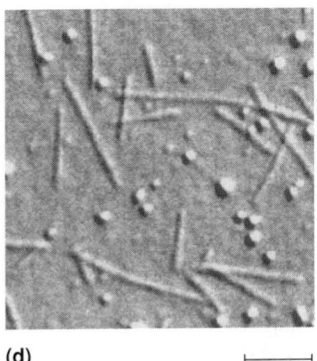

(c)   (d)

2.5 μm

Figure 20-3 Computer-Enhanced Digital Video Microscopy. This series of micrographs shows how computers can be used to enhance images obtained with light microscopy. In this example, an image of several microtubules, which are too small to be seen with unenhanced light microscopy, are processed to make them visible in detail. **(a)** The image resulting from electronic contrast enhancement of the original image (which appeared to be empty). **(b)** The background of the enhanced image in (a), which is then **(c)** subtracted from that image, leaving only the microtubules. **(d)** The final, detailed image resulting from electronic averaging of four separate images processed as shown in (a)–(c).

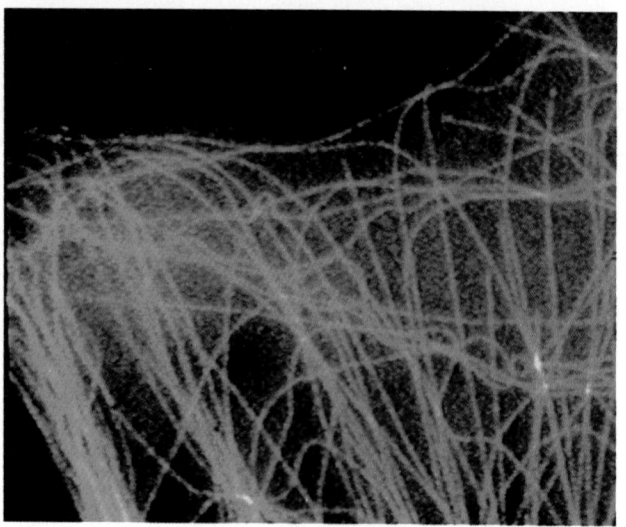

10 μm

Figure 20-4   Fluorescence Analogue Cytochemistry.   The fluorescent molecule rhodamine was covalently attached to purified tubulin, and the fluorescent tubulin molecules were microinjected into living fibroblast cells. Inside the cell, the fluorescent tubulin dimers become incorporated into microtubules, which can be easily seen using a fluorescence microscope.

which then becomes visible when imaged through a fluorescence microscope equipped with an intensified video camera (Figure 20-4).

Drugs and Antibodies as Tools for Studying Cytoskeletal Function

Although microscopic techniques can reveal much about the structure of the cytoskeleton, they do not usually enable us to deduce much about its function because most of the techniques provide only a static view of the cytoskeleton—a glimpse of the cell at the moment it was fixed for observation. Alternatives to microscopic techniques are needed for functional analysis. One such approach is the use of drugs that are known to bind specifically to a particular cytoskeletal protein and to alter its function in a known way. By studying the effects of such drugs on specific cellular processes, it is often possible to identify, at least tentatively, functions in which the particular protein might be involved.

For example, the drugs colchicine and taxol disrupt microtubule function in distinctively different ways. *Colchicine* (an alkaloid from the meadow saffron plant) binds to tubulin monomers, strongly inhibiting their assembly into microtubules and fostering the disassembly of existing ones. In contrast, *taxol* (from the Pacific yew tree) binds tightly to microtubules and stabilizes them, causing much of the free tubulin in the cell to assemble into microtubules. Sensitivity of a cellular process to colchicine or taxol is therefore a good first indication that microtubules may mediate that process

within the cell. In a similar manner, the drug *cytochalasin B* inhibits the polymerization of actin microfilaments, whereas *phalloidin* blocks the depolymerization of actin, thereby stabilizing microfilaments. Obviously, processes that are disrupted in cells treated with either of these drugs are likely to depend in some way on microfilaments. The use and effects of these drugs are discussed in more detail later in the chapter.

Antibodies can also be used to assess the involvement of cytoskeletal proteins in cellular functions because an antibody against a specific protein will often inactivate that protein when bound to it. As with exposure to a drug, microinjection of a specific antibody into a living cell can provide evidence to link a particular protein to a certain cellular function.

With these techniques in mind, we are now ready to look at each of the three major components of the cytoskeleton. In each case, we will consider the chemistry of the monomer, the structure of the polymer, the polymerization process, the role of accessory proteins, and some of the structural and functional roles attributable to that component. We will look first at microtubules, because of their key role in organizing both microfilaments and intermediate filaments in the cytoskeletal array.

Microtubules

With a diameter of about 25 nm, **microtubules (MTs)** are the largest of the cytoskeletal elements (see Table 20-1). Microtubules commonly originate from a structure called a **microtubule-organizing center (MTOC).** An MTOC serves as a site at which microtubule assembly is initiated and as an anchor for one end of these microtubules. Many cells during interphase have an MTOC called the **centrosome** that is positioned near the nucleus (see Figure 20-7b). Some types of cells have many MTOCs. For example, ciliated cells have an MTOC, called the *basal body,* at the base of each cilium.

The microtubules in eukaryotic cells can be classified into two general groups, **axonemal microtubules** and **cytoplasmic microtubules,** which differ in both degree of organization and structural stability. The first group includes the highly organized, stable microtubules found in specific subcellular structures associated with cellular movement, including cilia, flagella, and the basal bodies to which these appendages are attached. The central shaft, or *axoneme,* of a cilium or flagellum consists of a highly ordered bundle of axonemal microtubules and associated proteins. The second group consists of the more loosely organized, dynamic network of cytoplasmic microtubules.

Given their order and stability, it is not surprising that the axonemal microtubules were the first of the two groups to be recognized and studied. We have already encountered an example of such a structure; the axoneme of the sperm tail shown in Figure 4-12 consists of microtubules. As already noted, we will consider axoneme structure and microtubule-mediated motility further in Chapter 21.

The occurrence of cytoplasmic microtubules in eukaryotic cells was not recognized until the early 1960s, when the introduction of better fixation techniques permitted the direct visualization of the network of microtubules now known to pervade the cytosol of most eukaryotic cells. Since then, immunofluorescence microscopy has revealed a diversity and complexity of microtubule networks for different cell types.

Cytoplasmic microtubules are responsible for a variety of functions (see Table 20-1). While all components of the cytoskeleton contribute to the overall shape of the cell, microtubules appear to play a unique role in organizing it. MTs define and maintain the overall shape and architecture of animal cells. For example, they establish both types of nerve cell extensions or processes, called *axons* and *dendrites*. Circular rings of microtubules are also largely responsible for the round shape of red blood cells. In plant cells, MTs are thought to govern the orientation with which cellulose microfibrils are deposited during the growth of plant cell walls.

In addition, cytoplasmic microtubules are involved in many of the changes in cell shape and cell movements that occur during embryonic development in animals. As discussed in Chapter 15, microtubules are also essential for the movement of chromosomes during mitosis and meiosis, and they determine the plane of cell division.

Cytoplasmic microtubules also contribute to the spatial disposition and directional movement of vesicles and other organelles by providing an organized system of fibers to guide their movement. Specifically, MTs may govern the location of organelles, such as the Golgi complex and the endoplasmic reticulum, and the active movement of vesicles.

Structure and Polarity of Microtubules

As mentioned in Chapter 4, MTs are straight, hollow cylinders with an outer diameter of about 25 nm and an inner diameter of about 15 nm (Figure 20-5). Microtubules vary greatly in length. Some are less than 200 nm long, while others, particularly those in nerve cells, can be as long as hundreds of micrometers.

The MT wall consists of longitudinal arrays of linear polymers called **protofilaments.** There are usually 13 protofilaments arranged side by side around the hollow center, or **lumen.** The basic component of the protofilament is a dimer of the protein **tubulin.**

Tubulin is actually a family of similar but distinct polypeptides that can be subdivided into **α-tubulins, β-tubulins,** and **γ-tubulin.** Molecules of α-tubulin and β-tubulin have diameters of about 4–5 nm, molecular weights of about 50,000, and binding sites for the high-energy nucleotide GTP. As soon as individual α-tubulin and β-tubulin molecules are synthesized, they bind tightly to each other to form a dimer that does not dissociate under normal conditions. This **αβ-tubulin dimer** is the basic subunit of the MT protofilament (see Figure 20-5). The γ-tubulin molecule has only recently been discovered. It is found in the centrosome and other MTOCs, where it may initiate the assembly of microtubules.

Within a protofilament, all of the tubulin dimers are oriented in the same direction, such that all of the α-tubulin subunits face the same end. This uniform orientation of tubulin dimers means that one end of the protofilament dif-

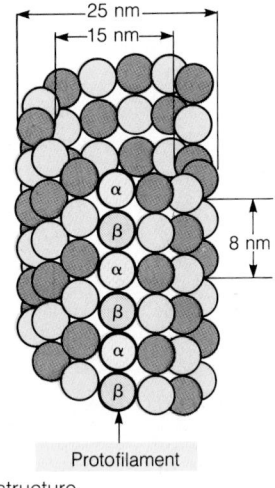

(a) Microtubule structure

(b) Microtubules in an axoneme   0.1 μm

Figure 20-5 Microtubule Structure. **(a)** A schematic diagram, showing a microtubule as a hollow cylinder enclosing a lumen. The outside diameter is about 25 nm, and the inside diameter is about 15 nm. The wall of the cylinder consists of 13 protofilaments, one of which is indicated by an arrow. A protofilament is a linear polymer of tubulin dimers, each of which consists of two polypeptides, α-tubulin and β-tubulin. All dimers in the protofilaments have the same orientation, thereby accounting for the polarity of the microtubule. **(b)** Microtubules in a longitudinal section of an axoneme (TEM).

fers chemically and structurally from the other, giving the protofilament an inherent **polarity.** Because the orientation of the tubulin dimers is the same for all of the protofilaments in a microtubule, the microtubule itself is also a polar structure. Depending on the organization of MTs within a cell, their polarity can confer polarity on the cell as a whole.

The Genetics of Tubulin

Most organisms have several closely related but not necessarily identical genes that encode each of the tubulin subunits, thereby allowing for possible slight variations in the expression of the gene and the structure and function of its protein product. For example, both the fruit fly *Drosophila* and the chicken have six genes that encode different kinds of α-tubulin and another six for different kinds of β-tubulin, whereas the green alga *Chlamydomonas reinhardtii* has two genes for each polypeptide. In the chicken, each of the six genes for β-tubulin is known to be functional, but that does not mean that all six are actually necessary. It is not clear whether all of the various tubulin subunit genes are expressed, nor is it known whether specific genes are expressed only in specific cell types. However, we do know that the expression of a particular gene may change during the development of an organism, or it may be unique to particular tissues in an organism.

Different MT-containing structures appear to contain different kinds of α- and β-tubulin molecules, at least in some species. In *Drosophila,* for example, the tubulin subunits present in the axoneme of the sperm flagellum are different from those in axonemes elsewhere in the organism. It is not yet known, however, whether the observed heterogeneity in tubulin polypeptides is due to the presence of different gene products for each subunit or is instead the result of chemical modifications (such as acetylation) of a single gene product. Nor is it yet clear how significant this heterogeneity is for microtubule function, because some species appear not to require tubulin heterogeneity. Yeast cells, for example, have a single β-tubulin gene, and the two β-tubulin genes of *Chlamydomonas* are known to encode identical proteins. These organisms must therefore be able to assemble all of the necessary MT-containing structures using a single kind of β-tubulin.

When the tubulins of diverse species are compared, they are found to be remarkably similar in sequence and structure—though not as strikingly as in the case of actin, as we will see later. In fact, all known α- and β-tubulins will form MTs if mixed together in vitro, whether they come from different organisms or from different tissues within the same organism.

Microtubule Assembly in Vitro

Microtubules form by the reversible polymerization of tubulin dimers. In order for assembly to occur, these dimers must be bound to GTP. A schematic representation of MT assembly in vitro is shown in Figure 20-6. A critical step in the formation of MTs is the aggregation of tubulin dimers into clusters called **oligomers.** Some of these oligomers are stable enough to form linear chains of protofilaments. The protofilaments then bind laterally to neighboring protofilaments to form sheets. A microtubule is formed as the sheets

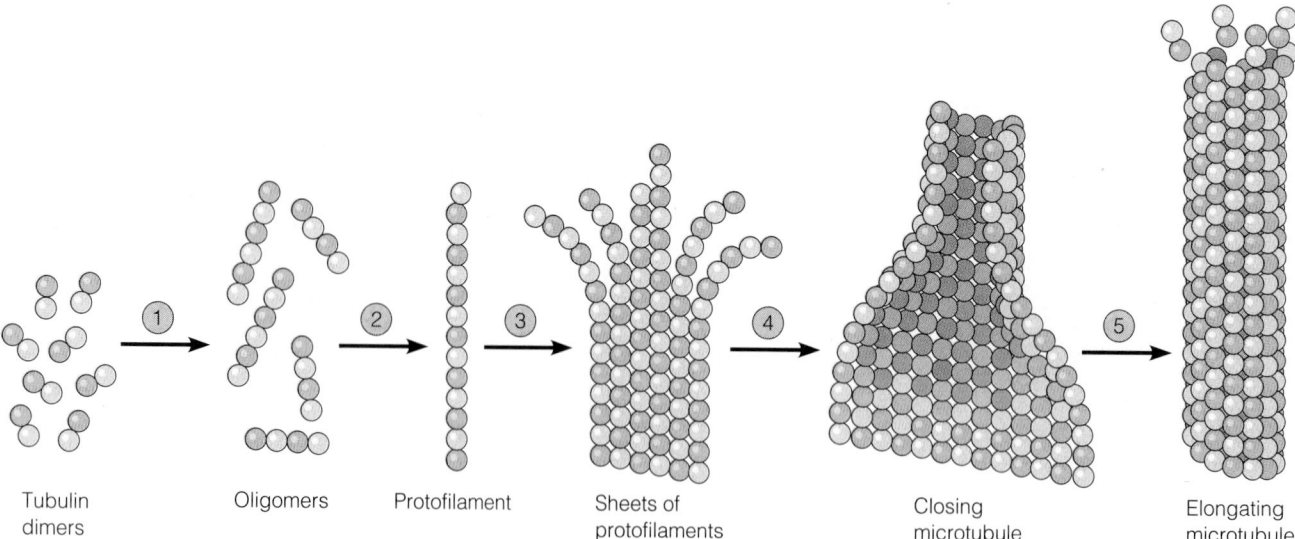

Tubulin dimers Oligomers Protofilament Sheets of protofilaments Closing microtubule Elongating microtubule

Figure 20-6 A Model for Microtubule Assembly in Vitro. Microtubules are assembled from subunits composed of one molecule of α-tubulin and one molecule of β-tubulin bound together tightly as a dimer, called an αβ-tubulin dimer, or simply a tubulin dimer. ① In the start of the nucleation process, several tubulin dimers can aggregate into clusters called oligomers, ② some of which go on to form linear chains of tubulin dimers called protofilaments. ③ The protofilaments can then associate with each other side-by-side to form sheets. ④ Sheets containing thirteen or more protofilaments can close into a tube, forming a microtubule. ⑤ Elongation of the microtubule continues by the addition of tubulin subunits at one or both ends.

curl up and close into a tube. The MT then grows or elongates by the addition of tubulin dimers to one or both ends.

One consequence of microtubule polarity is that the two ends of the microtubule have different properties of assembly. One end can inherently grow or shrink much faster than the other. If, for example, a fragment of a basal body is incubated in vitro with tubulin molecules, MTs growing out from one end of the basal body will elongate more rapidly than those at the other end. This faster-growing end is the **plus end** and the other is the **minus end** (Figure 20-7a). When isolated centrosomes are incubated with tubulin, MT growth occurs in an outward direction such that the plus ends are furthest away from the centrosome (Figure 20-7b). This is similar to the microtubule arrangement seen in fixed cells (Figure 20-1a).

The polymerization process has been studied extensively in vitro. When a solution containing a sufficient concentration of tubulin dimers, GTP, and Mg^{2+} is warmed from 0°C to 37°C, the polymerization reaction begins, forming microtubules. (MT formation in the solution can be readily measured as an increase in light scattering by using a spectrophotometer.)

Microtubule formation is initially slow, a period referred to as the **lag phase** of MT assembly (Figure 20-8). This period reflects the relatively slow process of microtubule **nucleation**—the initiation of MT assembly. Once a microtubule has begun to form, it grows by the addition of tubulin dimers to its ends. This **elongation phase** of MT assembly is relatively fast compared to nucleation. Finally, as the concentration of free tubulin dimers declines and becomes limiting, a plateau phase is reached where assembly is balanced with disassembly.

Microtubule growth in vitro depends on the concentration of tubulin dimers; microtubules grow when tubulin

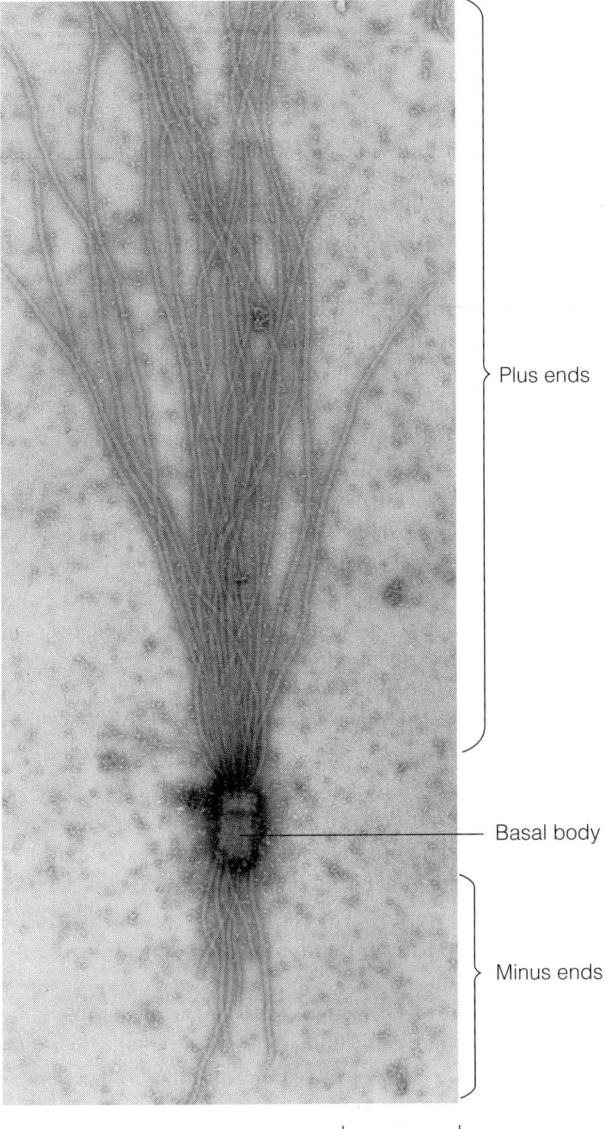

Plus ends

Basal body

Minus ends

(a) Assembly at a basal body |———| 0.5 μm

Centrosome

(b) Assembly at a centrosome |———| 1.4 μm

Figure 20-7 Polar Assembly of Microtubules in Vitro. (a) The polarity of MT assembly can be demonstrated by adding basal bodies to a solution of tubulin dimers. The tubulin dimers add to the plus and minus ends of the microtubules in the basal body. However, MTs that grow from the plus end are much longer than those growing from the minus end. **(b)** MT assembly from isolated centrosomes (TEMs).

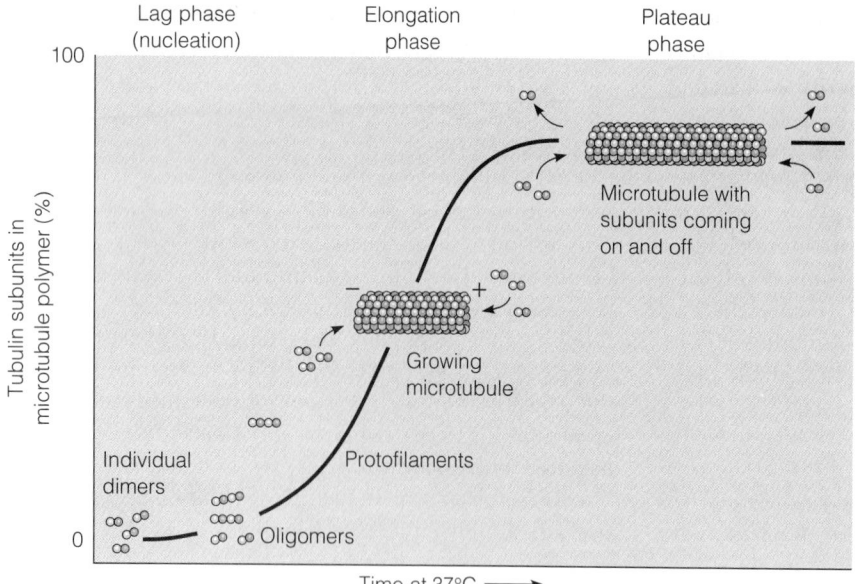

Lag phase (nucleation)   Elongation phase   Plateau phase

Figure 20-8   The Kinetics of Micro-tubule Assembly in Vitro.   The kinetics of MT assembly can be monitored by observing the amount of light scattered by a solution containing GTP-tubulin *after* it is warmed from 0°C to 37°C. (Microtubule assembly is inhibited by cold and activated upon warming.) Such light-scattering measurements reflect changes in the MT population as a whole, not the assembly of individual microtubules. When measured in this way, microtubule assembly exhibits three phases: lag, elongation, and plateau. The lag phase is the period of nucleation. During the elongation phase, MTs grow rapidly, causing the concentration of tubulin subunits in the solution to decline. When this concentration is low enough to limit further assembly, the plateau phase is reached, in which subunits are added and removed from MTs at equal rates.

concentrations are high and depolymerize when tubulin concentrations are low. Somewhere between these two conditions is a tubulin concentration at which assembly is exactly balanced with disassembly. The concentration of

Figure 20-9   The in Vitro Dependence of MT Assembly on Tubulin Concentration.   For a population of microtubules, the minimum tubulin concentration required for growth at the plus end is known as Cc^+. For MTs to grow at the minus end, tubulin must be at a higher concentration, Cc^-. At any concentration of tubulin between Cc^+ and Cc^-, MTs will grow at the plus end and shrink at the minus end, resulting in treadmilling. Cc^o is the tubulin concentration at which the growth rate at the plus end is exactly balanced by the shrinkage rate at the minus end, yielding no net MT growth.

dimers in this condition is called the **overall critical concentration (Cc^o)** (Figure 20-9).

The overall critical concentration reflects the combined assembly properties of the two ends of the MT. However, the critical concentration of tubulin that results in dimer assembly at the plus end (Cc^+) is lower than the critical concentration of tubulin that results in assembly at the minus end (Cc^-). Having a different critical concentration at the two ends of the microtubule requires that there be a difference in the chemical reactions that take place at the two ends. (MT assembly is appropriately regarded as a chemical reaction in which the reactants are the tubulin dimers, and the product is a microtubule.) As with any chemical reaction, there is a direct relationship between the free energy given off during a reaction and the ratio of reactants to product. A microtubule with different critical concentrations at the two ends requires that different amounts of free energy be released when a tubulin dimer adds to either the minus or the plus end. The reason MTs can have different critical concentrations at the two ends is that the chemical reactions at the two ends are not identical. The reason the reactions are different is directly related to GTP and its hydrolysis. GTP-bound tubulin dimers can add quickly to the plus end whose subunits still contain GTP. At the plus end, GTP-bound tubulin dimers add to GTP-bound tubulin already assembled at the tip of the MT. At the minus end, GTP-bound tubulin dimers add to polymerized GDP-bound tubulin subunits. Thus, the chemical reactions are not quite the same at the two ends, and this is necessary for the two ends to have two different critical concentrations.

When tubulin is present at concentrations between Cc^+ and Cc^-, assembly will occur at the plus end while disassembly occurs at the minus end. This simultaneous assembly and disassembly produces the phenomenon known as treadmilling (see Figure 20-9). **Treadmilling** arises when a given tubulin molecule incorporated at the plus end is displaced progressively along the MT and eventually lost by depolymerization at the opposite end. Treadmilling can be demonstrated in vitro but may not occur in vivo. So far, the only situation in which treadmilling might occur in vivo is in the microtubules that connect chromosomes to the spindle pole during mitosis. While controversial, some experiments have detected the movement of tubulin subunits from the chromosome (plus end) to the poles (minus end).

GTP and GDP appear to play a central role in regulating microtubule assembly and disassembly. For example, whether assembly or disassembly occurs at a given plus end may depend on how rapidly the GTP on a newly added tubulin dimer is hydrolyzed to GDP. GTP is needed for MT assembly because the association of GDP-bound tubulin dimers with each other is too weak for polymerization to occur. However, the *hydrolysis* of GTP does not seem to be necessary for assembly because microtubules will polymerize from tubulin-bound to GTPγS, a nonhydrolyzable analogue of GTP. In fact, microtubules assembled using GTPγS-tubulin are more stable than normal. On the one hand, GDP-bound tubulin dimers bind to each other so weakly that they cannot assemble into MTs, and on the other hand, MTs polymerized using tubulin dimers bound to GTPγS are remarkably stable. These observations help us understand the role of GTP hydrolysis in MT assembly (and disassembly). As tubulin dimers within a microtubule hydrolyze GTP to GDP, the MT weakens. It will not dissociate as long as the tubulin dimers at the tip of the plus end still contain GTP. However, if these tubulin dimers at the tip hydrolyze their GTP to GDP, the whole MT dissociates catastrophically. Therefore, the main role of GTP hydrolysis is to make the MT unstable so that it will dissociate quickly under the right conditions. The conditions that determine when a microtubule is stable and when it disassembles

form the basis of the dynamic instability model of microtubule assembly, the topic of the next section.

The Dynamic Instability Model of Microtubule Assembly. In vivo, microtubules constantly grow out from an MTOC and then disassemble. The MTOC nucleates assembly of the MT and anchors the minus end in such a way that only the plus end is free to add or lose tubulin subunits. Therefore, most of the dynamics of microtubule assembly and disassembly are dictated by events at the plus end.

Studies of MT in vitro assembly with isolated centrosomes as nucleation sites show that one set of microtubules can grow by polymerization while another set shrinks by depolymerization. As a result, one set (or perhaps one type of MT-containing structure) effectively enlarges at the expense of another, at least as studied in vitro. That such concomitant growth and shrinkage also occur in vivo can be demonstrated using video-enhanced differential interference contrast microscopy to follow the life cycles of individual microtubules (Figure 20-10).

To explain how both polymerization and depolymerization might occur simultaneously, Tim Mitchison and Marc Kirschner proposed the **dynamic instability model.** This model presumes two populations of MTs, one growing in length by continued polymerization at their plus ends, and the other shrinking in length by depolymerization. The distinction between the two populations is that growing MTs have GTP bound to the tubulin at their plus ends, while shrinking microtubules have GDP instead.

Because bound GTP-tubulin molecules are thought to have a greater affinity for each other than for GDP-tubulin, the hydrolysis of GTP-tubulin to GDP-tubulin weakens the bonds between the subunits, causing the protofilament to change to a less stable conformation (Figure 20-11). Therefore, the presence of a group of such GTP-bound tubulin

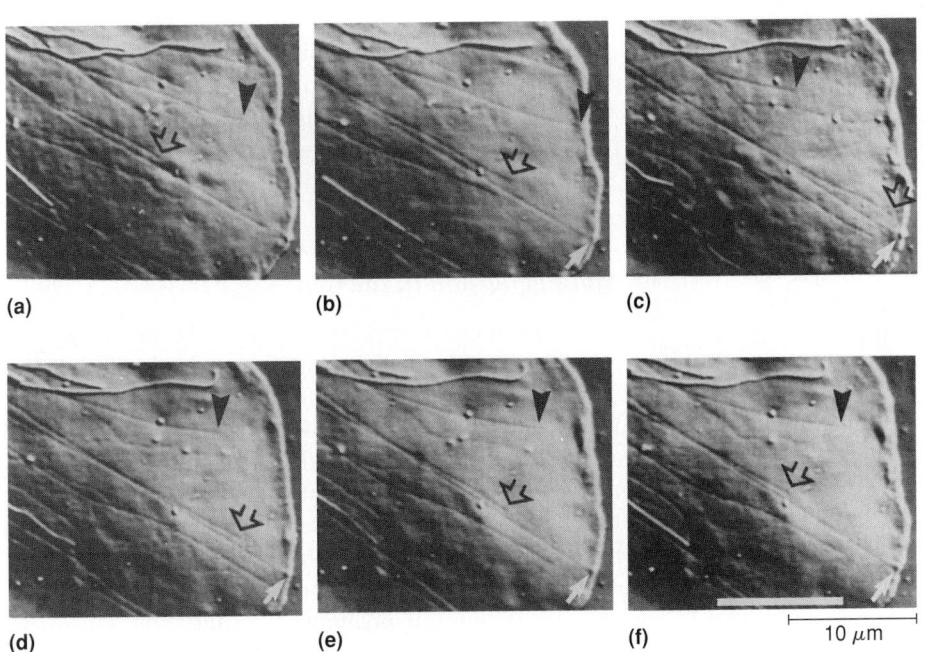

(a) (b) (c)

(d) (e) (f) 10 μm

Figure 20-10 The Dynamic Instability of Microtubules in Vivo. Microtubules visualized in a living cell by video-enhanced differential interference contrast (DIC) microscopy exhibit dynamic instability in vivo. Here, individual microtubules, designated by several types of arrows, are monitored over time, reading from (a) to (f). The MTs grow out to the edge of the cell and then rapidly shorten. For an explanation of DIC, see the Appendix.

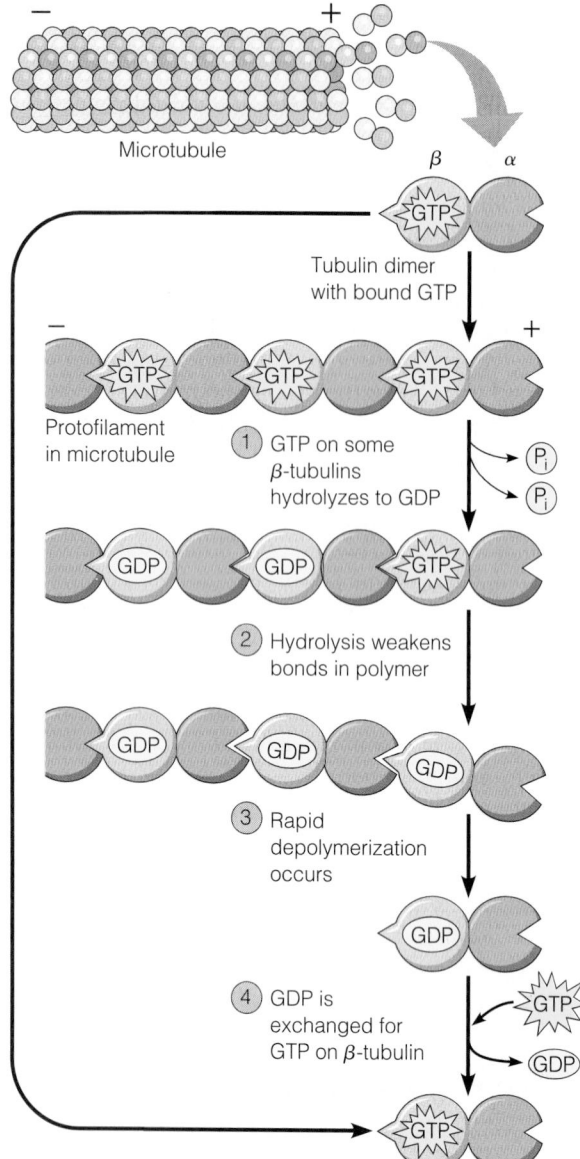

Figure 20-11 The Role of GTP Hydrolysis in Tubulin Assembly and Disassembly. Each of the α- and β-tubulin subunits binds a guanine nucleotide (GTP or GDP). However, only the GTP bound to the β-tubulin is shown here, because only it can be hydrolyzed to GDP and then exchanged for a new GTP. Tubulin dimers bound to GTP can polymerize into linear chains of protofilaments, an early stage of MT formation. Once tubulin dimers have been incorporated into the protofilaments of the microtubule, ① the GTP of the β-tubulin may be hydrolyzed. When the terminal tubulin subunits hydrolyze their GTPs, ② the protofilaments become unstable and ③ depolymerize. ④ The tubulin dimers released from a depolymerizing MT can exchange GDP for a new GTP so that they can again assemble into microtubules.

The concentration of tubulin bound to GTP is crucial to this model. The hydrolysis of GTP-tubulin appears to occur at a fixed rate. When GTP-tubulin is readily available, it is added to the MT quickly, creating a large GTP-tubulin cap. If the concentration of GTP-tubulin falls, however, the rate of tubulin addition decreases, along with the size of the GTP cap. At a low concentration of GTP-tubulin, the MT grows too slowly to maintain the cap, and the entire structure breaks down catastrophically.

Drug Sensitivities of Microtubule Assembly. A number of drugs affect microtubule assembly in different ways. The best-known is colchicine, the plant alkaloid that acts by binding to tubulin dimers, introduced earlier. The resulting tubulin-colchicine complex can still add to the growing end of a microtubule, but it then prevents any further addition of tubulin molecules and destabilizes the structure, thereby promoting microtubule disassembly. *Vinblastine* and *vincristine* are related compounds from the periwinkle plant that cause tubulin to aggregate inside the cell. *Nocodazole* (a synthetic benzimidazole) is another compound that inhibits microtubule assembly and is frequently used instead of colchicine because its effects are more readily reversible when the drug is removed.

These compounds are called *antimitotic drugs* because they disrupt the mitotic spindle of dividing cells, blocking the further progress of mitosis. The sensitivity of the mitotic spindle to these drugs is understandable because the spindle fibers are microtubular structures. Vinblastine and vincristine also find application in medical practice as anticancer drugs. They are useful for this purpose because cancer cells divide rapidly and are therefore preferentially susceptible to drugs that interfere with the mitotic spindle.

Taxol has the opposite effect on microtubules because when it binds to MTs, it stabilizes them. Within cells, it causes free tubulin to assemble into microtubules and arrests dividing cells in mitosis. Thus, both taxol and colchicine block cells in mitosis, but they do so by opposing effects on MTs and hence on the fibers of the mitotic spindle. Taxol has recently been used in the treatment of some cancers and has relatively few undesirable effects.

Microtubule Organization, Function, and Regulation in the Cell

In the previous section, we discussed the properties that tubulin and microtubules exhibit in vitro, providing a foundation for understanding how MTs function in the cell. However, microtubule formation in vivo is a more ordered and regulated process, one that produces sets of microtubules in locations coordinated to specific cell functions. In this section, we will first see how MTs are organized in the cell and then discuss the roles the MTOC (microtubule-organizing center) and the process of dynamic instability play in creating this preliminary organization. Finally, we

molecules at the plus end of an MT forms a *GTP cap* that provides a stable MT tip to which further dimers can be added (Figure 20-12a). A GDP cap, however, results in an unstable tip, at which depolymerization may occur rapidly (Figure 20-12b).

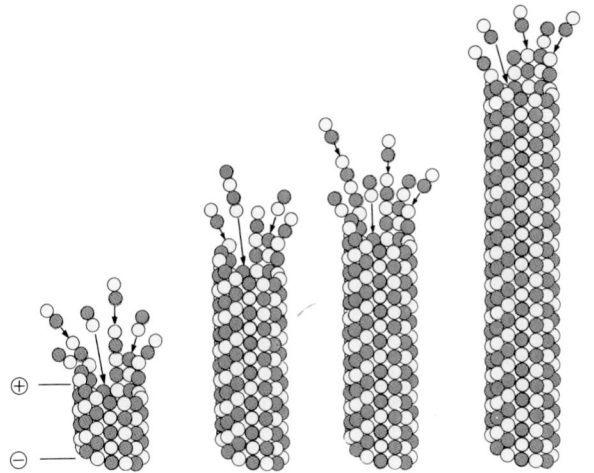

(a) Growing microtubule population. GTP "cap" present at plus end; continued polymerization favored

(b) Shrinking microtubule population. GDP "cap" present at plus end; rapid depolymerization favored

Figure 20-12   The Dynamic Instability Model of Microtubule Assembly and Disassembly.   This model assumes two populations of microtubules, one growing in length and the other shrinking in length. **(a)** The growing MTs have GTP bound to the newly added tubulin dimers at the plus end. Because GTP-tubulin dimers have a high affinity for each other, the GTP cap creates a stable microtubule tip to which further tubulin dimers can add. **(b)** The shrinking MTs have GDP bound to the newly added tubulin dimers instead, because the GTP originally present on each dimer has already hydrolyzed. The presence of a GDP cap makes the tip unstable, and rapid depolymerization is favored.

will look at how microtubule-associated proteins produce a more permanent, functional MT arrangement (sometimes by facilitating the association of MTs with each other to form bundles).

Organization and Maintenance of Cell Shape.   Microtubules make an important contribution to the structure and shape of animal cells but play a less important role in plant cells, where cell shape is determined largely by the rigid cell wall. In animal cells, microtubules appear to determine shape through the initial patterns of MT growth and through the regulation of MT stability.

As already mentioned, the growth of microtubules in the cell generally depends on MTOCs in the cytosol. Most cells have at least one MTOC, the centrosome, which serves as the origination site for microtubules while the cell is in interphase (see Figure 20-7b).

The centrosome in an animal cell normally consists of two *centrioles* surrounded by *pericentriolar material*. In electron micrographs of the centrosome, microtubules originate from the pericentriolar material (Figure 20-13), leaving the role of the centrioles in question. In fact, fertilized mouse eggs and some plant cells have functional centrosomes that lack centrioles. The pericentriolar material has not been completely defined but contains at least two proteins, *pericentrin* and *γ-tubulin*, that are needed for the nucleation of microtubules in vivo. The abundance of pericentrin fluctuates during mitosis, being highest at prophase and

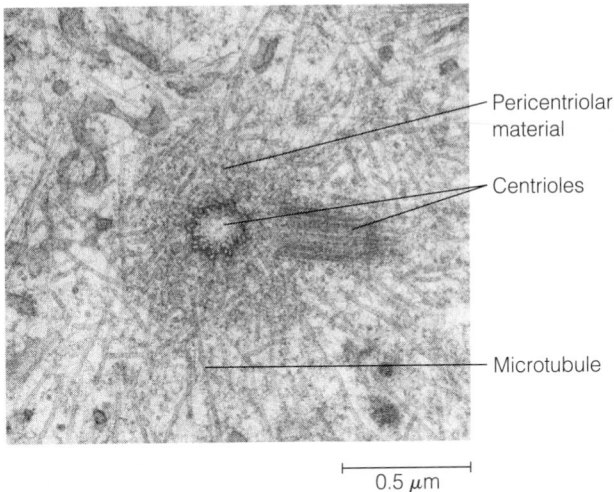

Figure 20-13   A Centrosome.   An electron micrograph of a centrosome showing the centrioles and the pericentriolar material. Notice that microtubules originate from the pericentriolar material (TEM).

metaphase when spindle poles show the greatest MT-nucleating activity. Gamma-tubulin resembles the other tubulins in many respects. However, it will bind to β-tubulin but does not polymerize into microtubules.

The centrosome or MTOC plays an important role in controlling the organization of microtubules in cells. The

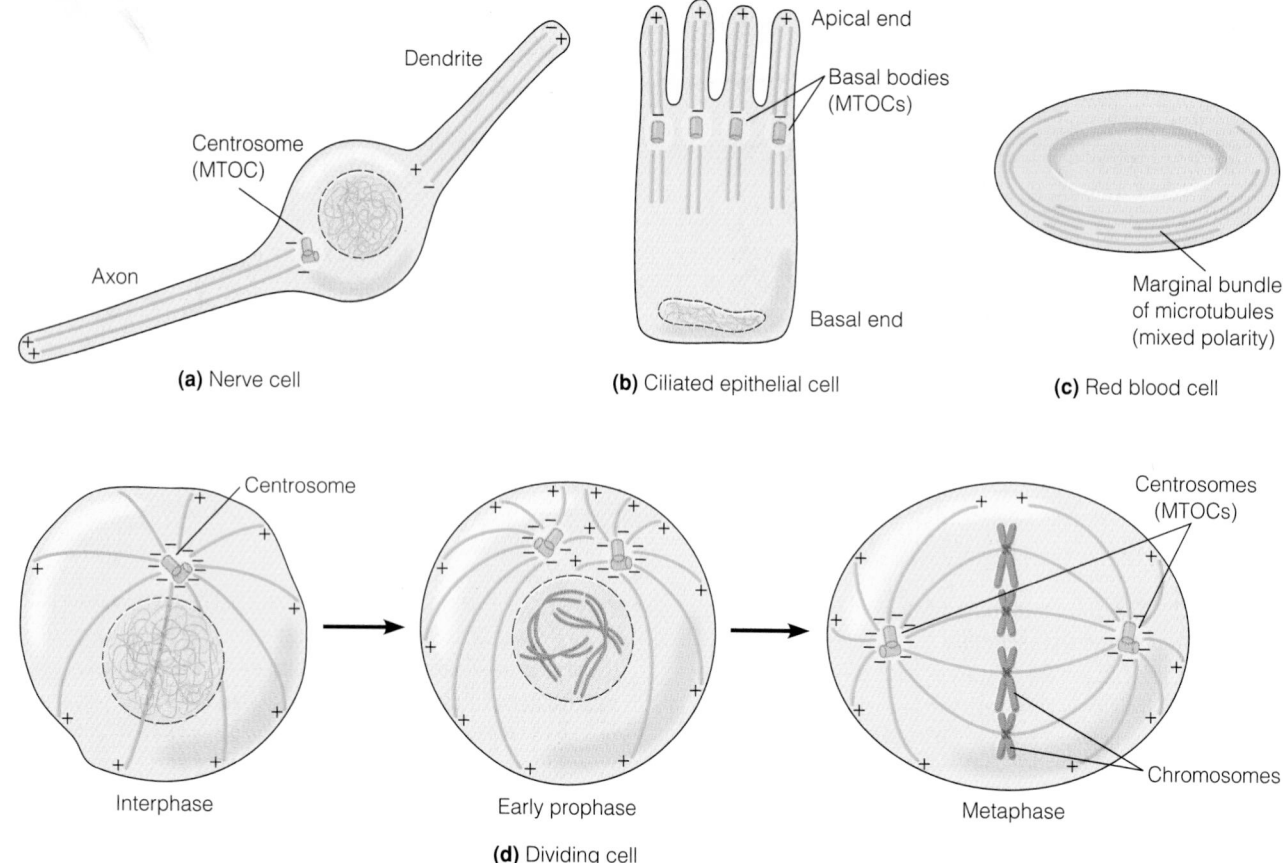

(a) Nerve cell

(b) Ciliated epithelial cell

(c) Red blood cell

Interphase

Early prophase

Metaphase

(d) Dividing cell

Figure 20-14   The Effects of Microtubule Polarity on MT Orientation in Animal Cells.   In the cell, the distribution of most microtubules is determined by the microtubule-organizing center (MTOC), which is sometimes a centrosome. These illustrations show several examples of how microtubule orientation in a cell may vary with that cell's function. **(a)** Nerve cells contain two distinct sets of MTs, those of the axon and those of the dendrite. The axonal microtubules are attached at their minus ends to the centrosome, with their plus ends at the tip of the axon. However, dendritic microtubules are not associated with the centrosome and are of mixed polarities. **(b)** Ciliated epithelial cells have many MTOCs called basal bodies, one at the base of each cilium. Cilia MTs originate with their minus ends in the basal bodies and elongate with their plus ends toward the tip of the cilia. **(c)** Mature human red blood cells have no nucleus or MTOC. However, microtubules of mixed polarities persist as a circular band at the periphery of the cell. This band maintains the cell's round, disklike shape. **(d)** Throughout the process of mitosis, MTs in a dividing cell are oriented with their minus ends anchored in the centrosome and their plus ends pointing away from the centrosome. Cell division is preceded by the division of the centrosome. The two centrosomes then separate, each forming one of the poles of the mitotic spindle. At metaphase, the centrosomes are at opposite sides of the cell. Each centrosome, or spindle pole, forms half of the spindle microtubules, some of which extend from pole to chromosomes, while others extend from one pole to the other pole.

most important aspect of this role is probably the MTOC's ability to nucleate and anchor microtubules. As a result of this ability, microtubules extend out from an MTOC toward the periphery of the cell. Furthermore, they grow out from an MTOC with a fixed polarity—their minus ends are always anchored in the MTOC, and their plus ends always extend out toward the cell membrane. The ability of γ-tubulin to bind only the β-tubulin subunit (and thus only to the minus end of the microtubule) may explain why microtubules grow out from the centrosome with a fixed polarity. The relationship between the MTOC and the distribution and polarity of microtubules is shown in Figure 20-14.

The MTOC also influences the number of microtubules in a cell. Each MTOC has a limited number of nucleation and anchorage sites that seem to control how many microtubules can form. However, the microtubule-nucleating capacity of the MTOC can be modified during certain processes such as mitosis, in which the number of microtubules increases.

The Role of Dynamic Instability and Capping Proteins in Microtubule Organization.   We have seen that cellular microtubules exhibit dynamic instability; they grow out from the centrosome and then disassemble. This process could account for randomly distributed and short-lived MTs but not for the organized and stable arrays of MTs that establish and maintain cell shape. Such microtubules are too unstable to remain intact for long periods of time and will break down unless they are stabilized in some way. One way to do this is to "capture" and protect the growing plus end of the microtubule. *Capping proteins* perform this function: Microtubules with capped plus ends are more stable than those that are uncapped. Thus, the distribution of stable microtubules in a cell is determined by both the position of the MTOC and the distribution of capping proteins that capture and stabilize newly formed MTs.

An example of how dyamic instability and capping proteins can produce a precisely organized set of microtubules is seen during mitosis (see Figure 20-14d), as we discussed in Chapter 15. Prior to prophase, the centrosome replicates and during early prophase the two daughter centrosomes separate and move to opposite sides of the cell. These centrosomes are the poles of the mitotic spindle. When the nuclear envelope breaks down, the chromosomes are connected to the poles by microtubules. To establish these connections, the kinetochore on the chromosomes is thought to serve as a plus-end cap. As MTs grow out from the spindle pole, those that encounter a kinetochore are captured and stabilized. Those microtubules that miss the kinetochore will eventually dissociate through dynamic instability and be replaced by new ones that will go through the same process. Through repeated cycles of MT growth and depolymerization, the kinetochore of each chromosome will eventually capture a MT and become connected to the spindle poles.

Regulation of Microtubules by Microtubule-Associated Proteins. Although tubulin is the main component of microtubules, a variety of other proteins are known to be involved in microtubule structure, assembly, and function. These **microtubule-associated proteins (MAPs)** account for 10–15% of the microtubule mass.

MAPs add another level of regulation to the organization and functions of microtubules in cells. In order to affect MT function, many MAPs bind at regular intervals along the wall of a microtubule and form projections from the wall, allowing interaction with other filaments and cellular structures. MAPs are also important in regulating MT assembly, most likely by binding to the growing plus end of a microtubule, thereby stabilizing it against disassembly. Most MAPs have been shown to increase MT stability, and some also can stimulate MT assembly (Figure 20-15). The many different kinds of MAPs differ mainly in how they link microtubules together or to other structures and how their effects on microtubules are regulated.

MAP function has been studied extensively in brain cells, as they are the most abundant source of these proteins. Two major classes of MAPs are present in microtubules from brain cells: microtubule-associated motor proteins (motor MAPs) and nonmotor MAPs. **Motor MAPs,** which

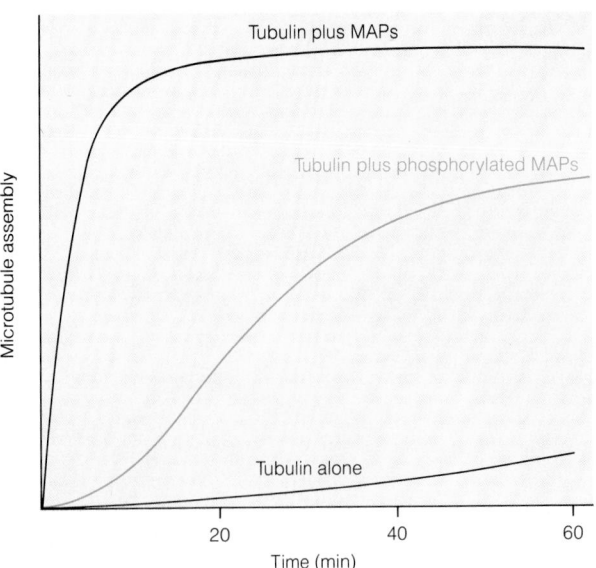

Figure 20-15 Stimulation of Microtubule Assembly by MAPs. MT assembly occurs very slowly in vitro in the presence of purified tubulin only. The process is stimulated significantly by the presence of microtubule-associated proteins (MAPs). When present in phosphorylated form, however, MAPs are less effective at promoting assembly.

include **kinesin** and **dynein,** are so called because they use ATP to drive the transport of vesicles and organelles or to generate sliding forces between microtubules. Kinesin and dynein, along with another type of motor molecule called myosin, which mediates microfilament-based motility, will be discussed in detail in Chapter 21.

Nonmotor MAPs appear to control MT organization in the cytoplasm. A dramatic example of such control is seen in neurons (nerve cells). The proper functioning of the nervous system depends on connections among neurons and with other types of cells. To establish these connections, neurons send out projections called *neurites,* which are reinforced by bundles of microtubules. Neurites eventually become axons and dendrites, the nerve cell processes previously mentioned. *Axons* carry electrical signals away from the nerve cell body, while *dendrites* receive such signals and carry them to the cell body. The MT bundles are characteristically denser in axons than they are in dendrites.

These differences arise because dendrites and axons contain different types of MAPs (Table 20-2). For example, an axon-specific MAP called Tau causes microtubules to form tight bundles. A family of MAPs called MAP2, including the specific MAP called MAP2C, is present in dendrites and causes the formation of looser bundles of microtubules (Figure 20-16).

The importance of MAPs for neurite formation can be demonstrated by introducing the *MAP2C* or *tau* genes into a

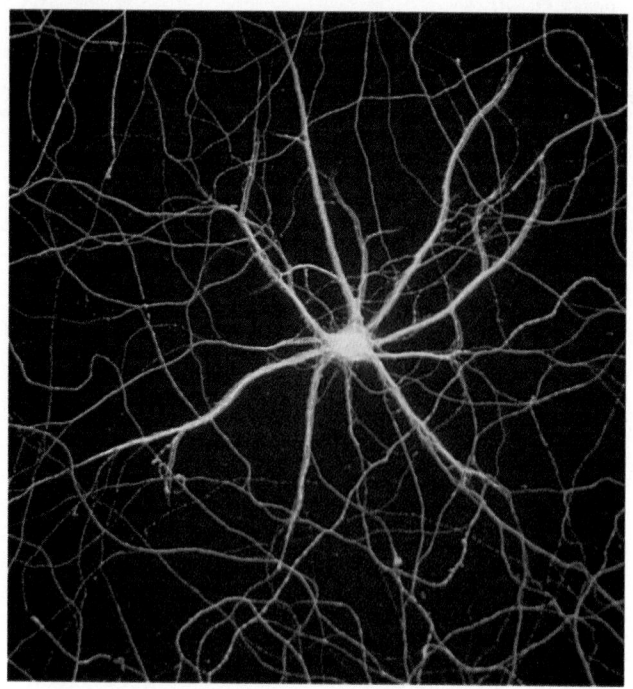

Figure 20-16 Localization of MAP2 and Tau in a Neuron by Immunofluorescence Staining. The unique subcellular distribution of MAP2 and Tau can be demonstrated by staining a cultured hippocampal neuron for both proteins using two different antibodies, one for MAP2 and one for Tau. When stained in this manner, axons exhibit green fluorescence, indicating the presence of Tau, whereas dendrites show an orange fluorescence, indicating the presence of MAP2.

10 μm

Table 20-2 Nonmotor MAPs (Microtubule-Associated Proteins)

MAP	Isoforms	Molecular Mass (kDa)	Tissue	Function
MAP1A (MAP1)	No	350	Nerve dendrites and axons	Induces tubulin polymerization
MAP1B (MAP5)	No	320	Nerve dendrites and axons	Induces neurite elongation
MAP2A	Yes	199	Nerve dendrites	Promotes assembly of tubulin in vitro
MAP2B	Yes	—	Nerve dendrites	Unknown
MAP2C	Yes	70	Nerve dendrites	Bundles MTs, increases MT stability and stiffness
MAP3	Yes	180	Various	Induces tubulin polymerization
MAP4	Yes	200	Various	Induces tubulin polymerization in many cell types during mitosis
Tau	Yes	37–46	Nerve axons	Induces tubulin polymerization, bundles MTs
Radial spoke proteins	—	34–124	Cilia and flagella	Unknown

nonneuronal cell line that cannot normally make either protein. These cells are normally rounded (Figure 20-17a), but when the *tau* gene is introduced and expressed, these cells extend single long processes that look remarkably similar to nerve axons (Figure 20-17b). In contrast, when the *MAP2C* gene is introduced and expressed, several shorter processes form that resemble nerve dendrites (Figure 20-17c).

The diversity of MAPs may help explain how cells can differ in their microtubule organization. In addition, the function of some of these MAPs can be altered by phosphorylation, which provides a means for the cell to rapidly alter MT organization. An example of this kind of regulation is seen with MAP4, a MAP that promotes MT assembly when it is not phosphorylated. As the cell goes through mitosis,

MAP4 becomes phosphorylated and loses its ability to promote assembly.

Phosphorylation of MAPs can also be an important aspect of some diseases. An example is Alzheimer's disease, which causes the progressive deterioration of mental functions due to neuron death in the cerebral cortex and the hippocampus, both of which function in memory. Microscopic examination of these cells reveals abnormal structures called *neurofibrillary tangles* (Figure 20-18). These neurofibrillary tangles are composed largely of abnormally phosphorylated Tau. Under these conditions, Tau loses its ability to bind to microtubules and instead forms paired helical filaments, which clump together with other proteins to form the neurofibrillary tangles.

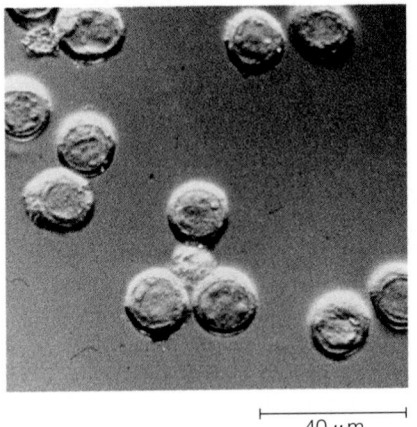

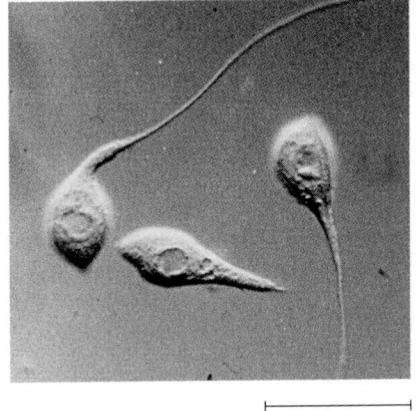

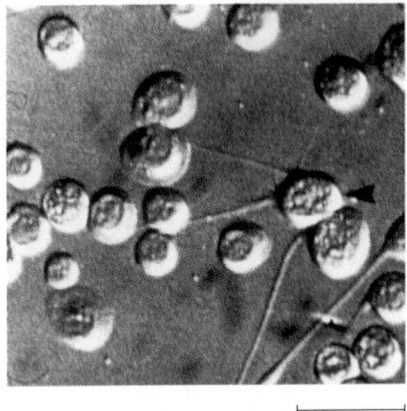

⊢———⊣ 40 μm	⊢———⊣ 40 μm	⊢———⊣ 40 μm
(a) Normal S*f*9 cells	**(b)** S*f*9 cells expressing *tau*	**(c)** S*f*9 cells expressing *MAP2C*

Figure 20-17 Expression of the Genes for MAP2C and Tau in a Nonneuronal Cell Line. Both MAP2C and Tau are microtubule-associated proteins whose genes are expressed in neurons and are important for establishing the dendrites and axons of neurons. One way to demonstrate this is to express the *MAP2C* gene or the *tau* gene in a nonneuronal cell line and observe the changes in cell morphol-ogy that occur. **(a)** Cells of the Sf9 tissue culture line normally exhibit a round morphology and do not express either *MAP2C* or *tau*. **(b)** When *tau* is expressed in Sf9 cells, a single long process resembling an axon grows out from the cell. **(c)** When *MAP2C* is expressed in Sf9 cells, multiple processes grow out from the cell, resembling the formation of dendrites.

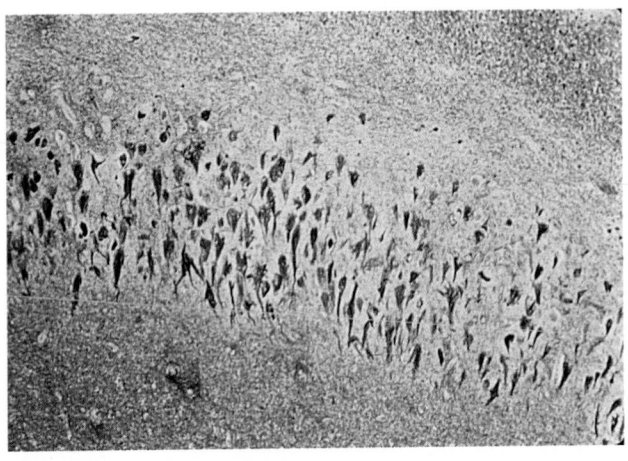

(a) Section of brain tissue affected by Alzheimer's disease

⊢————⊣ 0.1 mm

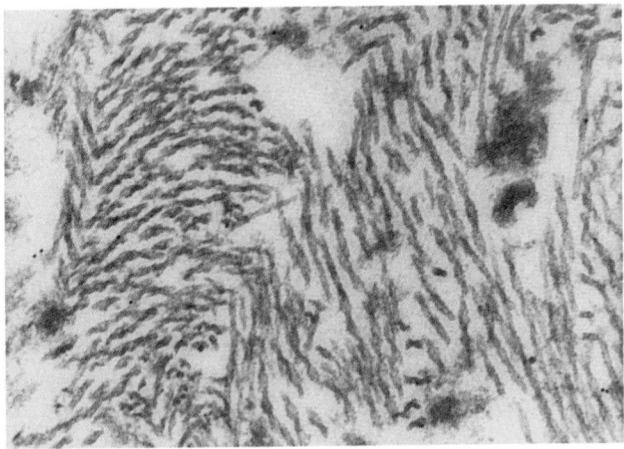

(b) Neurofibrillary tangles

⊢———⊣ 0.1 μm

Figure 20-18 Neurofibrillary Tangles in Neurons Affected by Alzheimer's Disease. Neurons in the hippocampus and cortical regions of the brain of Alzheimer's patients exhibit a characteristic abnormality known as neurofibrillary tangles. These structures contain large amounts of abnormally phosphorylated Tau protein. **(a)** A light microscopic section of the hippocampus from a patient with Alzheimer's disease (LM). **(b)** An electron micrograph of neurofibrillary tangles, which appear as filaments (TEM).

Microfilaments

With a diameter of about 7 nm, **microfilaments (MFs)** are the smallest of the cytoskeletal filaments (see Table 20-1). Microfilaments are best known for their role in the contractile fibrils of muscle cells, where they interact with thicker filaments of myosin to cause the contractions characteristic of muscle. This role will be explored in detail in Chapter 21. However, MFs are not confined to muscle cells; they occur in almost all eukaryotic cells and are involved in numerous other phenomena, including a variety of locomotory and structural functions.

Examples of cell movements in which microfilaments play a role include *amoeboid movement*, locomotion of cultured cells (usually fibroblasts) over a substratum (the surface to which the cell is attached), and *cytoplasmic streaming*, a regular pattern of cytoplasmic flow that includes both the *cyclosis* of algal cells and higher plant cells and the *shuttle streaming* of animal cells. Microfilaments also produce the cleavage furrows that divide the cytoplasm of animal cells after chromosomes have been separated by the spindle. All of these phenomena will also be discussed in Chapter 21.

In addition to mediating a variety of cell movements, MFs are important structurally in the development and maintenance of cell shape. Most animal cells, for example,

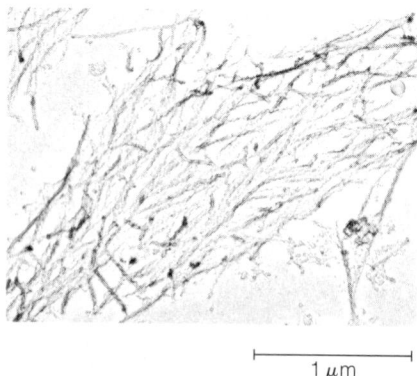

1 μm

Figure 20-19 Stress Fibers. This deep-etch micrograph shows stress fibers in a cultured fibroblast cell. The stress fibers consist of linear bundles of microfilaments with associated proteins and are prominent features of the cytoskeleton in many cultured cells.

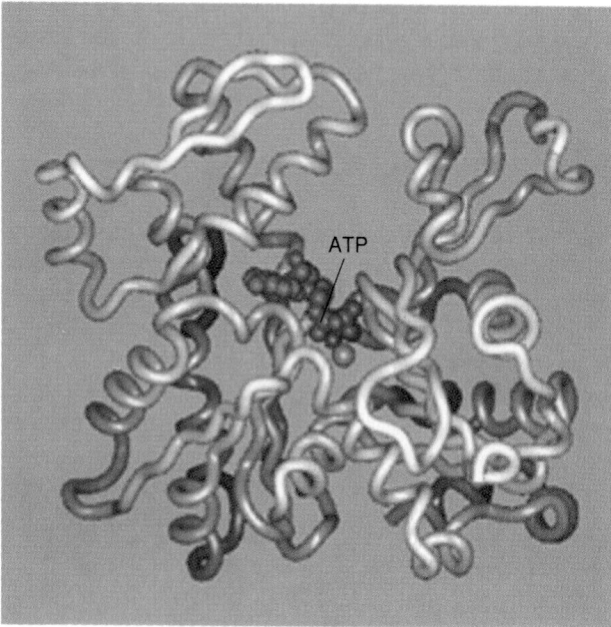

ATP

Figure 20-20 The Molecular Structure of G-Actin Monomer. X-ray crystallography shows that the G-actin monomer is shaped somewhat like a U or a channel. A nucleotide (ATP or ADP) binds reversibly in the groove made by the channel. When G-actin monomers polymerize into F-actin, the mouth of the groove is covered by another G-actin monomer, trapping the bound nucleotide inside. In addition, binding of one G-actin to another forces the mouth of the groove to close more tightly on the bound nucleotide, promoting hydrolysis of the ATP.

have a dense network of microfilaments called the **cell cortex** just beneath the plasma membrane. The cortex confers structural rigidity on the cell surface and facilitates shape changes and cell movement. In fibroblasts growing in culture, microfilaments are ordered into long bundles called **stress fibers** that often span the length of the cell (Figure 20-19). Stress fibers terminate at specialized sites of contact between the plasma membrane and the substrate called **focal contacts** or **focal adhesion plaques.** Parallel bundles of

microfilaments also make up the structural core of microvilli, the fingerlike extensions found on the surface of many animal cells (see Figure 4-2). Several of these structural roles will be discussed later in the chapter.

Microfilament Structure and Assembly

Microfilaments are polymers of the protein **actin.** Although actin is best known for its role in muscle contraction, it is present in virtually all eukaryotic cells, including those of plants, algae, and fungi. In fact, actin is the single most abundant protein in most cells, usually comprising more than 5% of the total cellular protein.

Actin is synthesized as a monomer called **G-actin** (G for globular). The G-actin molecule is a single polypeptide consisting of 375 amino acids, with a molecular weight of about 42,000. G-actin molecules are roughly U-shaped, with a central cavity containing ATP or ADP (Figure 20-20). Several forms of actin can be distinguished, based on slight differences in amino acid sequence. Muscle cells contain α-actin, whereas nonmuscle cells have two similar but distinct forms, β-actin and γ-actin. These different forms of actin are very similar. They can, for example, polymerize into a single microfilament in vitro. Similarly, some antibodies directed against α-actin bind to microfilaments of nonmuscle cells as well.

As with tubulin dimers, G-actin monomers polymerize reversibly into filaments with a *lag phase* corresponding to filament nucleation, followed by a more rapid polymer *elongation phase.* The filaments that form (**F-actin**) appear to be composed of two linear strands of actin wound around each other in a helix, with roughly 13.5 actin monomers per turn (Figure 20-21).

Within a microfilament, all of the actin monomers are oriented in the same direction, such that a microfilament, like a microtubule, has an inherent polarity, with one end differing chemically and structurally from the other end. This polarity can be readily demonstrated by incubating microfilaments with **myosin subfragment 1 (S1),** a proteolytic fragment of myosin containing the myosin head that binds tightly to actin (Figure 20-22a and b). The S1 fragments bind to, or "decorate," the microfilaments to give a distinctive arrowhead pattern, with all the S1 molecules pointing in the same direction (Figure 20-22c). Based on this arrowhead pattern, the terms *pointed end* and *barbed end* are commonly used to identify the two ends of an actin filament. This polarity is important because it provides the basis for the directional protrusion of portions of the plasma membrane and for directional movements of proteins or organelles that associate with actin filaments.

As with microtubules, the polarity of the filament gives rise to preferred assembly of G-actin at one end, the plus end, while disassembly occurs mainly at the other end, the minus end (see Figure 20-21a). If G-actin is polymerized onto short fragments of S1-decorated F-actin, polymerization proceeds much faster at the barbed end. This shows that the barbed end of the filament is the plus assembly end.

As G-actin monomers assemble onto the filament, the bound ATP they carry is hydrolyzed much the same as is the

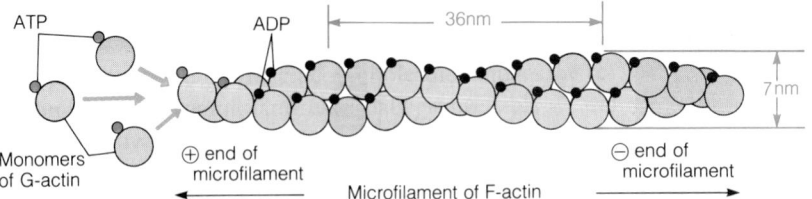

(a) MF assembly

Figure 20-21   A Model for Microfilament Assembly in Vitro.   **(a)** Monomers of G-actin polymerize into long filaments of F-actin with a diameter of about 7 nm. A full turn of the helix occurs every 36–37 nm, with about 13.5 monomers required for a full turn. The addition of each G-actin monomer is usually accompanied or followed by hydrolysis of the ATP molecule so that it is tightly bound to the monomer, although the energy of ATP hydrolysis is not required to drive the polymerization reaction. **(b)** An electron micrograph of purified F-actin (TEM).

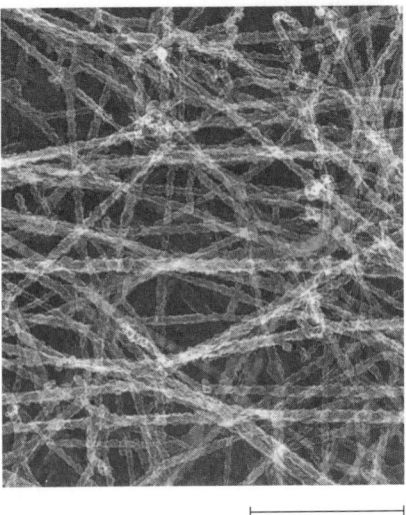

(b) Purified F-actin   0.5 μm

GTP bound to tubulin. The hydrolysis of ATP results in preferential assembly of the filament at the plus end and increases the potential for disassembly at the minus end. When the filament is growing rapidly, there will still be a short segment of ATP-bound monomers at the tip of the plus end. G-actin monomers bound to ATP add preferentially to F-actin

when the already-bound monomers also still contain ATP. If an actin filament is growing rapidly, the tip of the plus end can add ATP-bound monomers faster than hydrolysis can take place. Meanwhile, the minus end will typically contain monomers bound to ADP because they are added more slowly.

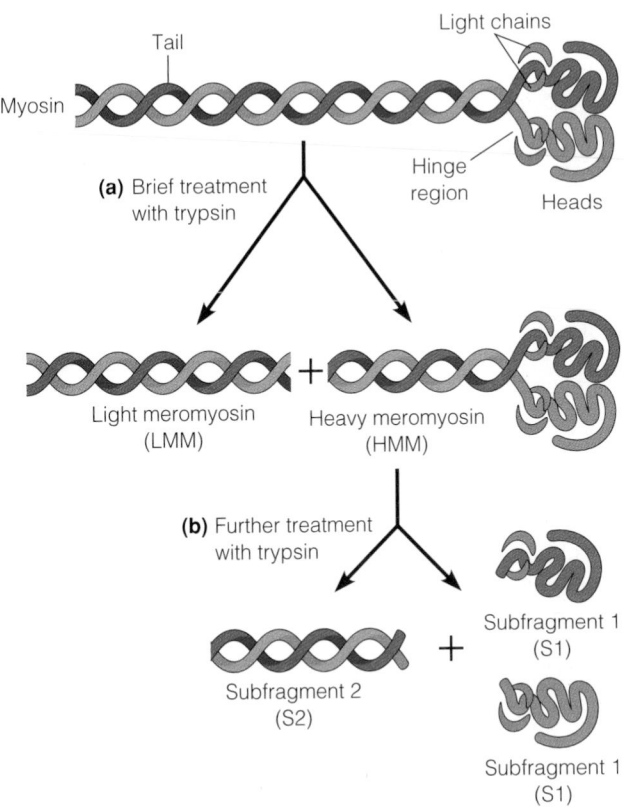

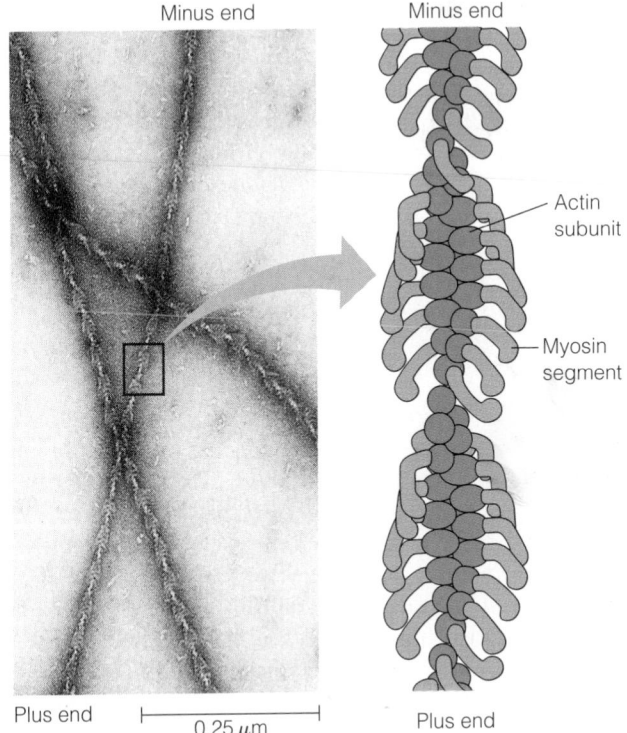

(c) EM and diagram of S1 fragments "decorating" actin microfilaments

Figure 20-22   The Actin-Binding Protein Myosin II.   This protein is part of the contractile machinery found in muscle cells. The globular head of the myosin molecule binds to actin, while the myosin tails can associate with filaments of myosin (the thick myofilaments of muscle cells). **(a)** Myosin II can be cleaved by proteases such as trypsin into two pieces, heavy meromyosin (HMM) and light meromyosin (LMM).

(b) HMM can be further digested, leaving only the globular head. This fragment, called myosin subfragment 1 (S1), retains its actin-binding properties. **(c)** When actin microfilaments are incubated with myosin S1 and then examined in the electron microscope, the S1 fragments appear to "decorate" the microfilaments like arrowheads. All the S1 arrowheads point toward the minus end, indicating the polarity of the MF.

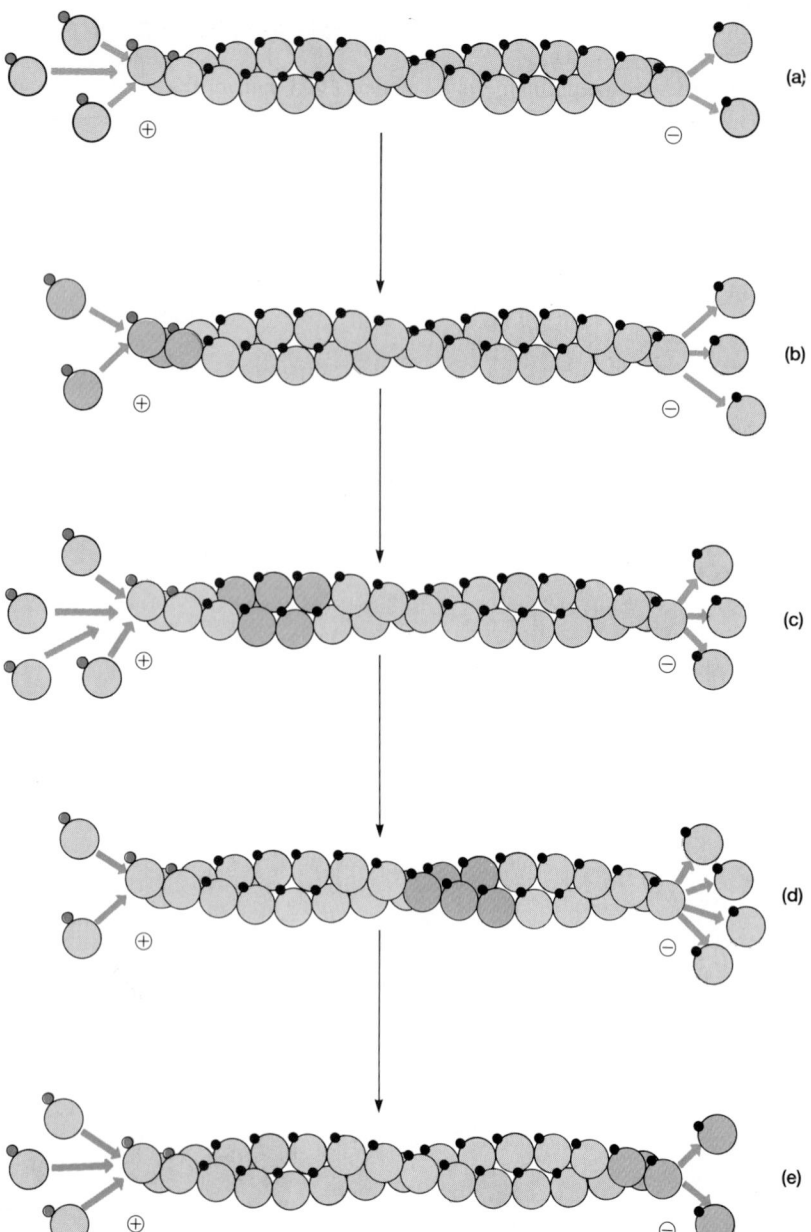

Figure 20-23   The Treadmilling of Actin Microfilaments in Vitro. Even when undergoing no net change in length, MFs can continue to incorporate G-actin monomers at the plus end and lose monomers at the minus end. The result is a treadmilling of actin monomers along the microfilament. Actin monomers added at one point in time **(a)** start out at the plus end, but are **(b)–(d)** displaced progressively along the microfilament as assembly and disassembly continue at the respective end, and **(e)** eventually lost by depolymerization at the minus end.

In this situation, the plus end has a lower critical concentration than the minus end. If G-actin concentrations lie between the critical concentrations for the plus and minus ends, treadmilling can take place (Figure 20-23). If monomer addition slows down or ATP hydrolysis catches up with the growing plus-end tip, the filament suddenly becomes unstable and may start to depolymerize. This behavior is similar to the dynamic instability seen in microtubules.

While the characteristics of microfilament and microtubule assembly appear to be identical, they differ in that actin filaments can assemble from ADP–G-actin, whereas microtubules can only assemble from GTP-tubulin dimers. As a consequence, ADP-actin filaments are relatively more stable than microtubules containing only GDP-tubulin. While actin filaments exhibit dynamic instability, they tend to undergo cycles of shortening and elongation rather than the more catastrophic disassembly seen with microtubules.

This increased stability tends to make treadmilling a more prominent feature of MFs than of MTs. In contrast to microtubules, treadmilling of actin monomers may be a significant part of the mechanism by which microfilaments generate movements in cells.

The Genetics of Actin

Actin varies little in its amino acid sequence from species to species. In fact, actin genes are the most highly conserved of the genes encoding cytoskeletal proteins. For example, the actins from yeast and chicken are identical at more than 90% of their amino acids, whereas tubulins from the two species show only about 70% identity. The minor differences in actin sequences between species do not appear to have much functional significance, because actin molecules from widely divergent species are indistinguishable when used for in vitro assays.

Regulation of Microfilaments by Actin-Binding Proteins

Many proteins are known to bind to actin, thereby conferring specific properties on actin that regulate its assembly and function. These **actin-binding proteins** perform a variety of functions, but they can be grouped into four major classes, as shown in Table 20-3. *Severing and capping proteins* regulate MF length. Under certain conditions, they sever actin filaments and cap one end, preventing further assembly. *Fragmin* is one severing and capping protein.

Monomer-binding proteins such as *profilin* and *thymosin β4* are thought to regulate the assembly of actin monomers into filaments. Although the mechanism is not understood, most of the G-actin in the cell seems to be bound to thymosin β4, which retards its assembly. Profilin can bind to complexes of actin and thymosin β4, releasing the actin and perhaps helping actin release ADP and bind to ATP. In this way, the addition of profilin to a mixture of actin and thymosin β4 stimulates the polymerization of actin.

Cross-linking proteins connect adjacent microfilaments to one another. Depending on the protein, this cross-linkage may result in the formation of loose networks or gels, tight parallel MF bundles (in which all of the filaments have the same polarity), or contractile bundles (containing actin filaments with a mixture of polarities). *Filamin* is an example of a gel-forming protein found in the cortex of cells. *Fimbrin* forms dense parallel bundles of actin that participate in the formation of thin cell protrusions called **filopodia** (singular: **filopodium**) at the front, or **leading edge,** of a moving cell. *Alpha-actinin* links actin into contractile bundles that are found in stress fibers. In moving cells, stress fibers are normally aligned with the direction of movement and may help pull the tail of the cell up toward the front. In addition to the *motor proteins* that bind actin (see Chapter 21), actin-binding proteins are also known that link microfilaments to other cellular structures, including the plasma membrane. We will encounter several different kinds of actin-binding proteins as we consider the involvement of microfilaments in cell structure and shape.

Microfilaments and Cell Shape

In addition to their involvement in the various kinds of cellular movement to be discussed in the next chapter, microfilaments play important roles in the development and maintenance of the asymmetrical shape of most animal cells. This involvement can be demonstrated readily by exposing elongated cells (such as cultured fibroblasts) to the drug cytochalasin B, an alkaloid secreted by some fungi. Like other filament poisons, this drug disrupts the MF bundles found in such cells. The cells lose their characteristic shape and become rounded. Microfilaments stabilize not only the overall shape of the cell but also local projections of the cell surface such as the microvilli mentioned earlier. We will examine both roles briefly, looking first at the cell cortex that underlies and supports the plasma membrane and then at the microvillus as a specific example of a cell-surface projection.

Microfilaments and the Cell Cortex. The cell cortex is a three-dimensional meshwork of actin microfilaments and associated proteins located just under the plasma membrane of most animal cells. The cortex supports the plasma membrane, confers strength on the cell surface, and facilitates shape changes and cellular movement. An important function of certain actin-binding proteins in the cortex is to link MFs into a stable network with gel-like properties. One of

---

Table 20-3 Actin-Binding Proteins

Class	Name	Specific Function(s)
Severing and capping proteins	Fragmin	Severs actin filaments; caps actin filaments
	Gelsolin	Severs actin filaments; caps actin filaments; cross-links actin filaments
Monomer-binding proteins	Profilin	Promotes ATP-ADP exchange
	Thymosin β4	Sequesters G-actin
Cross-linking proteins	α-actinin	Forms contractile bundles of actin filaments
	Filamin	Cross-links actin filaments into a gel
	Fimbrin	Forms parallel bundles of actin filaments
	Villin	Forms parallel bundles of actin filaments
Motor proteins	Myosin I	Moves vesicles along actin filaments
	Myosin II	Generates sliding forces between actin filaments
Other	Spectrin	Attaches actin filaments to plasma membrane
	Talin	Connects actin filaments to transmembrane proteins
	Tropomyosin	Regulates actin-myosin binding in muscle cells; stiffens actin filaments
	Vinculin	Attaches actin filaments to plasma membrane

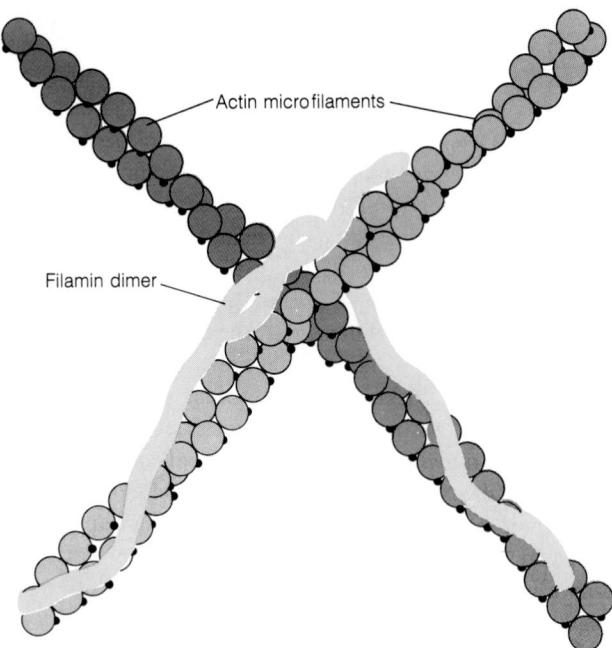

Figure 20-24 Cross-Linking of Actin Filaments by Filamin.
Filamin is one of several cross-linking proteins that link actin filaments into an extensive three-dimensional network. Filamin is a dimer of two identical polypeptides joined head-to-head into a long, flexible molecule. The tail of each polypeptide contains a binding site for actin filaments.

these cross-linking proteins is filamin (introduced earlier), a long molecule consisting of two identical polypeptides joined head-to-head, with an actin-binding site at each tail. Molecules of filamin act as "splices," joining two MFs together where they intersect, as shown in Figure 20-24. In this way, actin microfilaments are linked to form large three-dimensional networks.

Other cortex proteins play the opposite role of breaking up the microfilament network, thereby causing the actin gel

of the cortex to soften and liquefy. These severing and capping proteins are important in mediating the gel-to-sol transition associated with the phenomenon of cytoplasmic streaming, which we will encounter again in Chapter 21. One such protein is *gelsolin,* which functions by breaking actin microfilaments and capping the newly exposed plus ends of the microfilaments, thereby preventing further polymerization. Gelsolin and other severing and capping proteins are activated by Ca^{2+}, which renders them sensitive to extracellular signals that result in transient changes in cytosolic Ca^{2+} levels.

In addition to proteins such as filamin and gelsolin that determine the degree of microfilament cross-linking in the cell cortex, other proteins mediate the association of cortex MFs with the plasma membrane. The erythrocyte proteins *spectrin* and *ankyrin* mentioned in Chapter 7 are examples of this category. As illustrated in Figure 20-25, the plasma membrane of the erythrocyte is supported by a network of spectrin filaments that are cross-linked by very short actin chains. This network is connected to the plasma membrane by molecules of ankyrin and band 4.1 that link the spectrin filaments to specific transmembrane proteins. Although spectrin and ankyrin are found in erythrocytes, the presence of similar or identical proteins in the cortex of other animal cells suggests that the plasma membrane of other, perhaps even all, animal cells is also supported by a cortical network of microfilaments. Actin filaments are also found at specialized sites where the cell attaches to the substrate, to the extracellular matrix, or in some cases to other cells. Central to these connections is a family of proteins called *integrins* that span the plasma membrane, connecting to extracellular proteins on the outside of the cell to actin filaments on the inside of the cell. A protein called *talin* connects the integrin to actin filaments.

Microvilli and Microfilaments. Microvilli (singular: **microvillus**) are cell-surface projections that both increase the

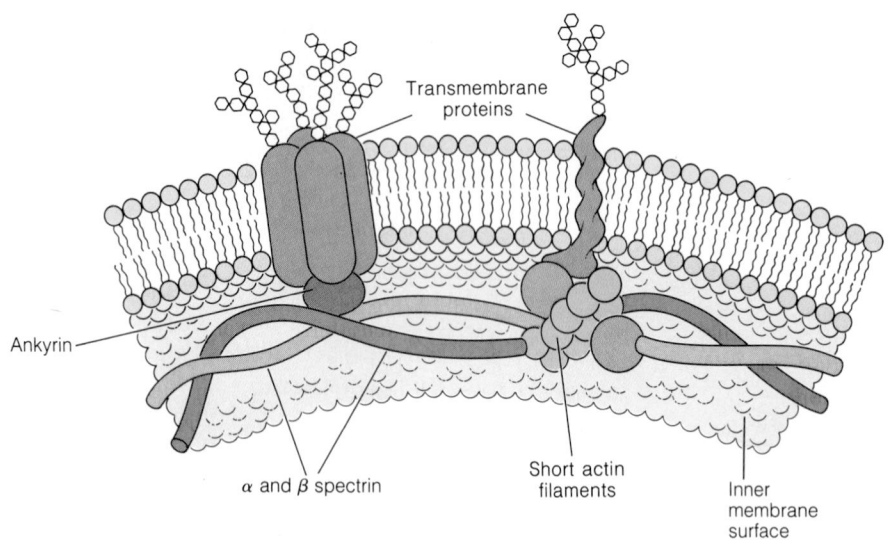

Figure 20-25 Support of the Erythrocyte Plasma Membrane by a Spectrin-Ankyrin-Actin Network. The plasma membrane of a red blood cell is supported on its inner surface by a filamentous network that gives the cell both strength and flexibility. Long filaments of spectrin are cross-linked by short actin filaments. The network is anchored to the band III transmembrane protein by molecules of the protein ankyrin.

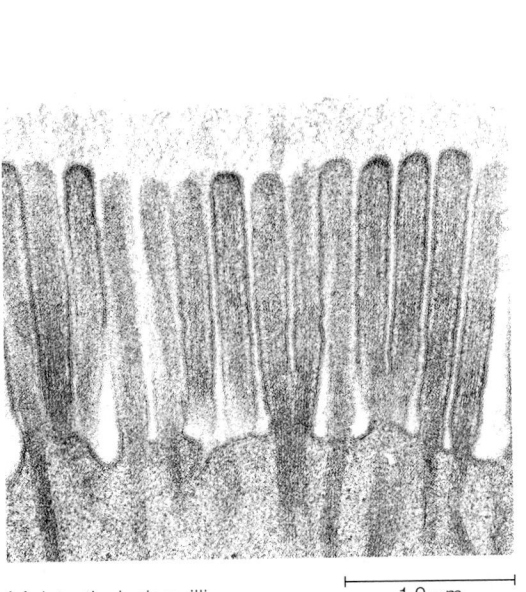

(a) Intestinal microvilli   1.0 μm

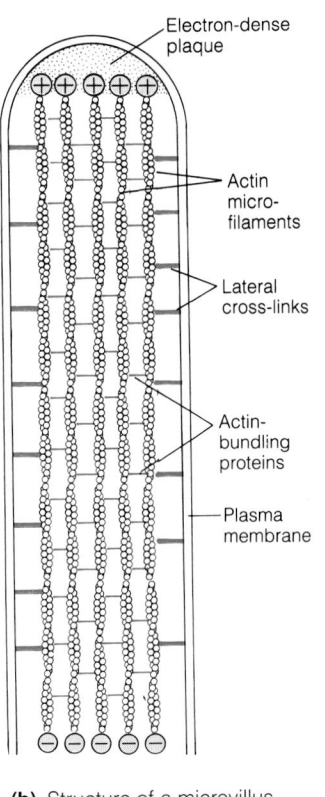

Electron-dense plaque

Actin microfilaments

Lateral cross-links

Actin-bundling proteins

Plasma membrane

(b) Structure of a microvillus

Figure 20-26   Microvillus Structure. **(a)** An electron micrograph of microvilli from intestinal mucosa cells (TEM). **(b)** A schematic diagram of a single microvillus, showing the core of microfilaments that gives the microvillus its characteristic stiffness. The core consists of several dozen microfilaments oriented with their plus ends facing outward toward the tip and their minus ends facing toward the cell. The plus ends are embedded in an amorphous, electron-dense plaque. The MFs are tightly linked together by actin-bundling (cross-linking) proteins and are connected to the inner surface of the plasma membrane by lateral cross-links composed of calmodulin and myosin I, a protein thought to have ATPase activity.

Actin microfilaments

Myosin and spectrin filaments

Intermediate filaments (containing keratin)

0.1 μm

Figure 20-27   The Terminal Web of an Intestinal Epithelial Cell.   The terminal web beneath the plasma membrane is seen in this freeze-etch electron micrograph of an intestinal epithelial cell. Bundles of microfilaments that form the cores of microvilli extend into the terminal web. The intermediate filaments contain keratin, as this is an epithelial cell. (From Nobutaka Hirokawa, *J. Cell Biol.* 94 (1982): 425. Reproduced by copyright permission of The Rockefeller University Press.)

surface area of the cell and mediate changes in cell shape and movement. Microvilli are especially prominent features of intestinal mucosal cells (Figure 20-26a). A single mucosal cell in your small intestine, for example, has several thousand microvilli, each about 1–2 μm long and about 0.1 μm in diameter, which increase the surface area of the cell about twenty-fold. This increase in surface area is essential to intestinal function, because the uptake of digested food depends on an extensive absorptive surface.

As illustrated in Figure 20-26b, the core of the microvillus consists of a tight bundle of microfilaments, oriented with their plus ends pointing toward the tip, where they are attached to the membrane through an amorphous electron-dense plaque. The MFs in the bundle are also connected to the plasma membrane by lateral cross-links composed of the proteins myosin I and calmodulin. These cross-links extend outward about 20–30 nm from the bundle to contact electron-dense patches on the inner membrane surface. The microfilaments in the bundle are held together at regular intervals by the cross-linking proteins fimbrin and villin. These proteins bind adjacent MFs together tightly. For example, each molecule of fimbrin binds two actin monomers, linking the microfilaments into a tightly packed bundle.

At the base of the microvillus, the MF bundle extends into a network of filaments called the terminal web (Figure 20-27). The filaments of the terminal web are composed mainly of myosin and spectrin, which connect the microfilaments to each other, to the plasma membrane, and perhaps

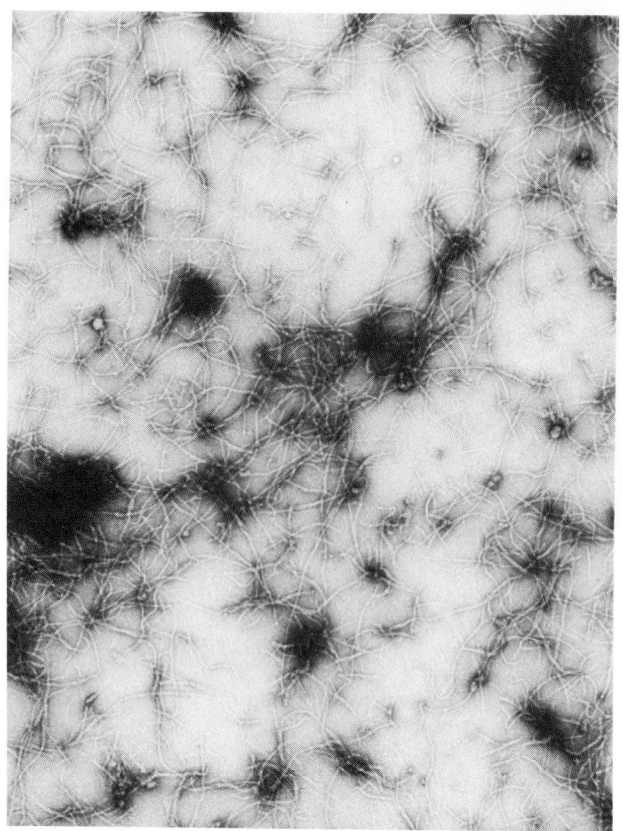

|————————| 0.5 μm

Figure 20-28   Intermediate Filaments.   An electron micrograph of negatively stained intermediate filaments from a cultured human fibroblast cell (TEM).

also to the network of intermediate filaments beneath the terminal web. The purpose of the terminal web is apparently to give rigidity to the microvilli by anchoring their MF bundles securely so that they project straight out from the cell surface.

Intermediate Filaments

Intermediate filaments (IFs) have a diameter of about 8–12 nm, which makes them intermediate in size between microtubules and microfilaments (Table 20-4), or between the thin (actin) and thick (myosin) filaments in muscle cells, where IFs were first discovered. To date, most studies have focused on animal cells, where IFs occur singly or in bundles and appear to play a structural or tension-bearing role. Figure 20-28 is an electron micrograph of IFs from a cultured human fibroblast cell.

Intermediate filaments are the most stable and the least soluble constituents of the cytoskeleton. Treatment of cells with detergents or with solutions of high or low ionic strength removes most of the microtubules, microfilaments, and other proteins of the cytosol, but leaves networks of intermediate filaments that retain their original shape. (In fact, the original structure to which the term *cytoskeleton* was applied was really a residual network of IFs from which MTs and MFs, now considered an integral part of the cytoskeleton, had already been removed.) Because of the stability of the intermediate filaments, some scientists suggest that they serve as a scaffold to support the entire cytoskeletal framework.

Table 20-4   Classes of Intermediate Filaments

Class	IF Protein	Molecular Mass (kDa)	Tissue	Function
I	Acidic cytokeratins	40–56.5	Epithelial cells	Mechanical strength
II	Basic cytokeratins	53–67	Epithelial cells	Mechanical strength
III	Vimentin	54	Fibroblasts; cells of mesenchymal origin; lens of eye	Maintenance of cell shape
III	Desmin	53–54	Muscle cells, especially smooth muscle	Structural support for contractile machinery
III	GFA protein	50	Glial cells and astrocytes	Maintenance of cell shape
IV	Neurofilament proteins		Central and peripheral nerves	Axon strength; determines axon size
	NF-L (major)	62		
	NF-M (minor)	102		
	NF-H (minor)	110		
V	Nuclear lamins		All cell types	Form a nuclear scaffold to give shape to nucleus
	Lamin A	70		
	Lamin B	67		
	Lamin C	60		
VI	Nestin	240	Neuronal stem cells	Unknown

Tissue Specificity and Genetics of Intermediate Filament Proteins

In contrast to microtubules and microfilaments, intermediate filaments differ in their composition from tissue to tissue. Based on the cell type in which they are found, IFs and their proteins can be grouped into six classes (see Table 20-4). Classes I and II comprise the *keratins,* proteins that make up the *tonofilaments* found in the epithelial cells that cover the body surfaces and line its cavities. (The IFs visible beneath the terminal web in the intestinal mucosa cell of Figure 20-27 consist of keratin, because the intestinal lining is an epithelial tissue.) Class I keratins are *acidic keratins,* while class II are *basic* or *neutral keratins.* Each of these classes contains at least 15 different keratins.

Class III IFs include vimentin, desmin, and glial fibrillary acidic (GFA) protein. *Vimentin* is present in connective tissue and other cells of mesenchymal origin. Vimentin-containing filaments are often prominent features in cultured fibroblast cells, in which they form a network that radiates from the center out to the periphery of the cell. *Desmin* is found in muscle cells, and *glial fibrillary acidic (GFA) protein* is characteristic of the glial cells that surround and insulate nerve cells. Class IV IFs are the *neurofilament (NF) proteins* found in the *neurofilaments* of nerve cells. Class V IFs are the *nuclear lamins* (A, B, and C) that form a filamentous scaffold inside the nuclear membrane of virtually all eukaryotic cells. Neurofilaments found in embryonic neural ectoderm cells are made of nestin, which constitutes class VI.

Because of the tissue specificity of intermediate filaments, animal cells from different tissues can be distinguished on the basis of the IF protein present, as determined by immunofluorescence microscopy. This **intermediate filament typing,** as it is called, serves as a diagnostic tool in medicine. IF typing is especially useful in the diagnosis of cancer, because tumor cells are known to retain the IF proteins characteristic of the tissue of origin, regardless of where the tumor occurs in the body. Because the appropriate treatment of cancer often depends on knowing the tissue of origin, IF typing is a valuable diagnostic aid, especially in cases where diagnosis using conventional microscopic techniques is difficult. This medical application of basic research on intermediate filaments is explored in more detail in Box 20A. Classification of IFs by cell type is biologically useful because the genes for IF proteins are expressed in accordance with the pattern of tissue differentiation during embryogenesis.

As IF proteins and their genes have been sequenced, it has become clear that these proteins are encoded by a single (though large) family of related genes and can therefore be classified according to amino acid sequence relatedness as well. The six classes of IF proteins have been distinguished on this basis (see Table 20-4).

Intermediate Filament Structure

As products of a family of related genes, all IF proteins share some common features, although they differ significantly in size and chemical properties. In contrast to actin and tubulin, all IF proteins are fibrous, rather than globular, proteins.

All IF proteins have a homologous central rodlike domain of about 310–318 amino acids that has been remarkably conserved in size, in secondary structure, and to some extent in sequence. As shown in Figure 20-29, this central domain consists of four segments of coiled helices interspersed with three short linker regions. Flanking the central

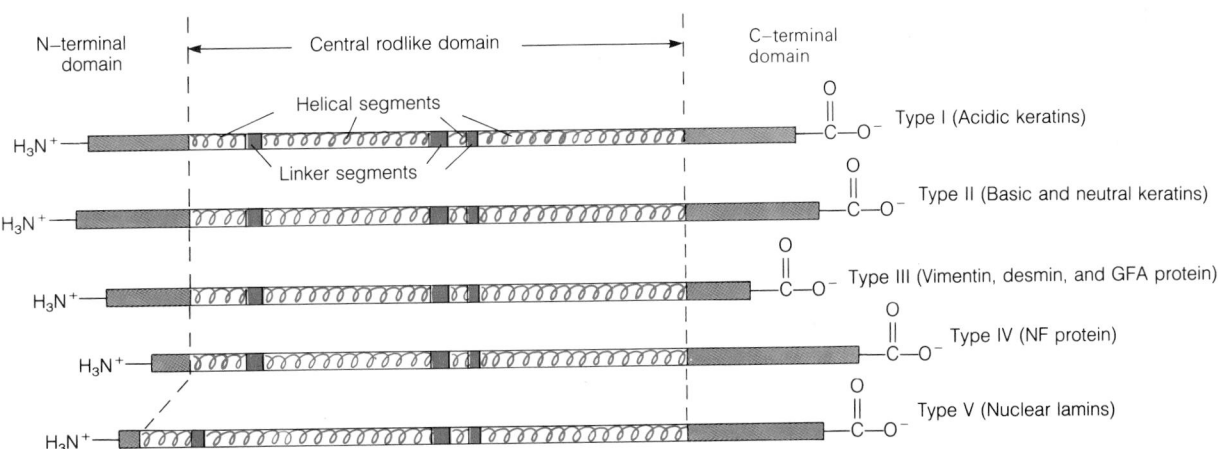

Figure 20-29 Structural Similarities of Intermediate Filament Proteins.   All six types of IF proteins have a central rodlike domain consisting of four helical segments interrupted by three linker segments. The central domain is thought to be important for IF assembly. This domain is highly conserved in size, secondary structure, and sequence, though sequence homologies are confined to the helical regions. In types I–IV, the helical segments contain a total of 276 amino acids and the linker segments are nonhelical. In type V, the helical segments contain 318 amino acids and their linker segments are also helical. The N-terminal and C-terminal domains that flank the central section are nonhelical and are much more variable in size and sequence. (The structure of the sixth type, nestin, is not shown.)

Clinical Applications

INTERMEDIATE FILAMENTS AND
THE DIAGNOSIS OF TUMORS

"How useful is your research? What good will it do anyone?" These kinds of questions are often directed to scientists, usually by citizens whose tax dollars support the research and who therefore deserve honest answers to their questions. The problem with such questions, however, is their inherent implication that research is worthwhile only if it is directed toward the solution of some particular problem or the achievement of some specific benefit, usually defined in human terms. When we invoke such criteria, we fail to distinguish between *applied research,* which is in fact intended to solve a specific problem or to yield a particular benefit, and *basic research,* which is not. Even more important, we may fail to realize that applied research almost always depends critically on basic research, without which we would have no findings to apply.

In reality, both kinds of research are important. We need applied researchers who are motivated by the satisfaction of seeing a problem solved or a benefit realized. But we also need basic researchers whose main reason for studying a specific phenomenon is the same reason people give for climbing Mount Everest—"just because it's there!" Out of basic research intriguing applications often emerge, usually in ways that could not have been anticipated or even guessed in advance.

Research on the cytoskeleton illustrates this point well. Most of what you have been reading about in this chapter is a result of investigations conducted by basic researchers, who study the cytoskeleton and its proteins "just because they're there." But out of their work have come some important applications that could not have been easily predicted beforehand, much less deliberately planned. One such example is the application of basic findings on intermediate filaments to the practical problem of distinguishing tissue types and identifying tumors, thereby providing pathologists with a valuable new tool for cancer diagnosis.

Underlying this application is the recognition, from the results of basic research, that cells from different kinds of vertebrate tissues have remarkably similar microtubules and microfilaments, but distinctively different kinds of intermediate filaments (IFs). In fact, the kind of IF—and hence the kind of IF protein—present in the cells of a particular tissue is a distinguishing characteristic of that tissue. Thus, epithelial cells contain mainly *keratin,* muscle cells have *desmin,* nerve cells have *NF (neurofilament) protein,* glial cells have *GFA (glial fibrillary acidic) protein,* and cells of mesenchymal origin are characterized by the presence of *vimentin* (see Table 20–4). Because of these characteristic differences in IF proteins, various tissues can be differentiated by identifying, or *typing,* the intermediate filaments using *immunofluorescence microscopy.*

IF typing is, in turn, a useful tool for further basic research. Developmental biologists, for example, find the technique useful for tracing cell lineages during embryonic development. But the technique also has important practical implications in the diagnosis of cancer because of its capability for distinguishing between different kinds of tumors. Moreover, IF typing is also a useful diagnostic tool for certain other human diseases as well, including the prenatal detection of congenital defects.

The diagnosis of cancer relies crucially on the ability of pathologists to classify tumors according to the tissue of origin. Such distinctions are important because the appropriate treatment of the cancer depends, sometimes critically, on knowing the cellular origin of the tumor. Yet tumors often appear in the body far from the tissue in which they originated because of the tendency of cancer cells to *metastasize,* or become dislodged from the original tumor and to be carried to distant parts of the body. (As we will learn in Chapter 24, a mass of cancer cells in their original location in the body is called a *primary tumor,* whereas a mass of cancer cells that has spread to a distant part of the body and is proliferating there is called a *metastasis.*)

In most cases, tumors can be successfully diagnosed by light microscopy, using standard staining techniques. However, about 5–10% of all tumors are difficult to diagnose by conventional means. These include small cell tumors in children, tumor samples that contain only a few tumor cells, or poorly differentiated tumors. In such cases, IF typing enables pathologists to characterize the tumor rapidly and unambiguously. Moreover, the technique is sensitive enough to detect tiny tumors containing only a few cells.

IF typing depends not only on the presence of only one kind (or at most two kinds) of IF protein in a given cell type but also on the maintenance of that specificity in *neoplastic* (cancerous) cells. The evidence that tumors maintain the IF specificity of the original tissue is strong. Consider, for example, the carcinomas, which are tumors of epithelial origin and the most common malignant tumors. Virtually all carcinomas react positively with an antibody specific for keratin, the IF protein characteristic of epithelial tissues. Few, if any, carcinomas respond positively when tested with antibodies specific for vimentin, desmin, GFA protein, or NF protein. Thus, carcinomas not only retain the keratin characteristic of epithelial tissues but also do not acquire additional IF types.

Similarly, *sarcomas* of muscle-cell origin stain positively and specifically for desmin, *lymphomas* of mesenchymal origin stain for vimentin, *gliomas* of glial origin stain for GFA protein, and *neuroblastomas* of nervous system origin stain for NF protein. Moreover, these same specificities are generally true for primary tumors and also for metastases. In the case of carcinomas, tumors can be further subtyped using either two-dimensional gel electrophoresis or monoclonal antibodies to identify subsets of keratins, thereby allowing finer diagnostic distinctions. For example, *adenocarcinomas* contain the same subset of keratins as the epithelial cells of the small intestine and colon from which these carcinomas are derived, while *hepatocellular carcinomas* contain only those keratins found in normal liver cells from which these tumors originate.

IF typing has other medical applications as well. Combined with *amniocentesis* (sampling of the *amniotic fluid* surrounding the fetus in the womb), the technique is potentially useful in the prenatal diagnosis of birth defects. Normally, most of the cells in the amniotic fluid react positively with keratin antibodies, presumably because they are of epithelial origin. Most also test positive for vimentin, but not usually for desmin, GFA protein, or NF protein. However, in the case of a fetus with anencephaly, a severe congenital defect in which the brain fails to develop properly, the amniotic fluid contains many cells positive for GFA protein or NF protein, both of which are derived from cell types related to the nervous system.

IF typing can also be used to detect and study abnormalities of IF organization in other diseases. Already it has been used to demonstrate IF anomalies in muscle cells of patients with certain muscle disorders and possibly also in the brains of people with Alzheimer's disease.

Cancer, prenatal diagnosis, muscle disorders, Alzheimer's disease—who could have guessed in advance that basic research on intermediate filaments would lead to medical applications in areas such as these? And who can predict what unexpected future benefits may come from the basic research that is currently under way in other areas of cell biology?

"But how useful is your research? What good will it do anyone?" Good questions to ask, and deserving of answers. For the basic researcher, the appropriate response is to continue climbing Mount Everest, because you never know what you might find when you get to the top—or even along the way. ■

helical domain are N-terminal and C-terminal domains that differ greatly in size, sequence, and function among IF proteins, presumably accounting for the functional diversity of these proteins.

A possible model for IF assembly is shown in Figure 20-30. The basic structural unit of intermediate filaments consists of two IF polypeptides intertwined into a *coiled coil*. The two polypeptides have their central helical domains aligned in parallel, with the N-terminal and C-terminal regions protruding as globular domains at each end. Two such dimers align laterally to form a tetrameric *protofilament*, the basic structural unit of IFs. Protofilaments interact with each other, associating in an overlapping manner to build up a filamentous structure both laterally and longitudinally. When fully assembled, an intermediate filament is thought to be eight protofilaments thick at any point, with protofilaments probably joined end-to-end in staggered overlaps.

Functions of Intermediate Filaments

Intermediate filaments are considered to be the principal structural determinants in many cells and tissues. Because they often occur in areas of the cell that are subject to mechanical stress, IFs are thought to have a tension-bearing role. For example, in epithelial cells, tonofilaments made of keratin loop through plaques called *desmosomes* that provide strong connecting junctions between two neighboring cells or between the cell and the extracellular matrix (see Figure 10-1). This system of desmosomes and tonofilaments bears most of the mechanical stress when the epithelium is stretched. When keratin filaments are genetically modified in the keratinocytes of transgenic mice, the epidermal cells are fragile and rupture easily. In humans, mutations of keratins give rise to a blistering skin disease called *epidermolysis bullosa simplex (EBS)*. IF defects are also suspected in other pathological conditions, including *amyotrophic lateral sclerosis (ALS)* and certain types of inherited *cardiomyopathies*.

A specific function of IFs may be maintenance of the position of the nucleus in the cell. Intermediate filaments form a ring around the nucleus with branches extending outward through the cytosol. These processes extend into the pores of the nuclear envelope, possibly connecting with the filaments of the nuclear lamina that may play a similar structural role within the nucleus. Intermediate filaments may interact with other elements of the cytoskeleton as well. For example, when microtubules are disassembled, the IFs that normally extend throughout the cytosol collapse into a ring around the nucleus.

Figure 20-30 A Model for Intermediate Filament Assembly in Vitro.

(a) The starting point for assembly is a pair of IF polypeptides. The two polypeptides are identical for all IFs except keratin filaments, which are obligate heterodimers with one each of the type I and type II polypeptides. (b) The two polypeptides twist around each other to form a two-chain coiled coil, with their conserved center domain aligned in parallel. (c) Two dimers align laterally to form a tetrameric protofilament. (d) Protofilaments assemble into larger filaments by end-to-end and side-to-side alignment. (e) The fully assembled intermediate filament is thought to be 8 protofilaments thick at any point.

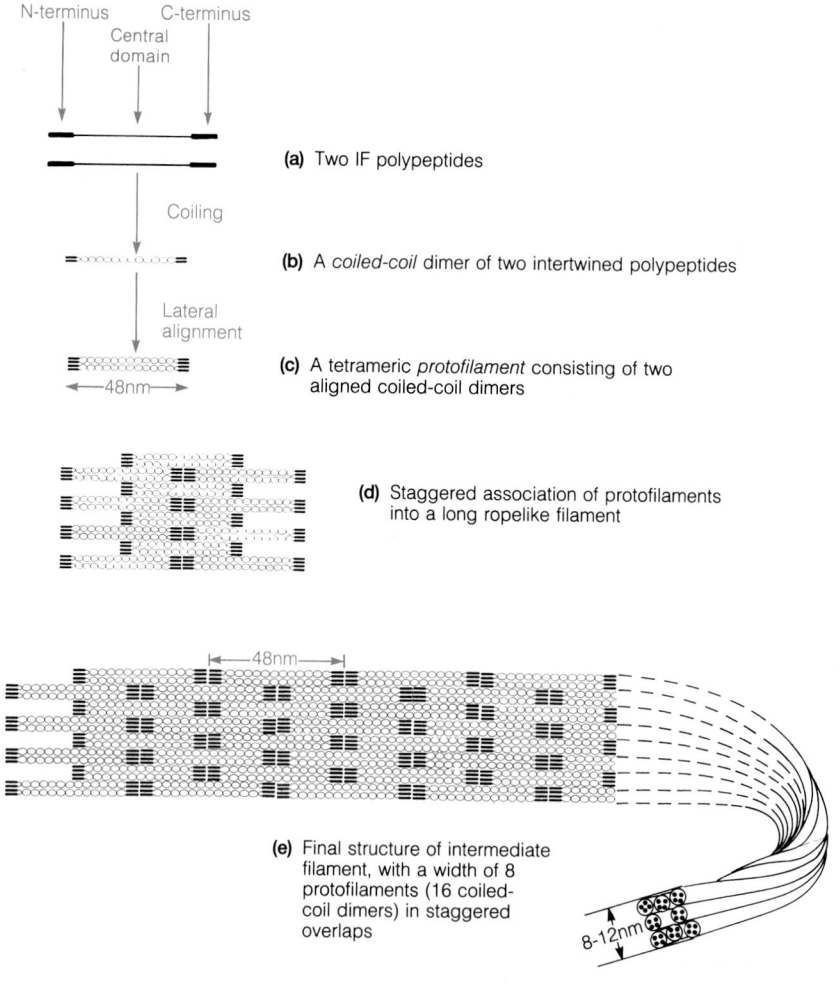

(a) Two IF polypeptides

(b) A *coiled-coil* dimer of two intertwined polypeptides

(c) A tetrameric *protofilament* consisting of two aligned coiled-coil dimers

(d) Staggered association of protofilaments into a long ropelike filament

(e) Final structure of intermediate filament, with a width of 8 protofilaments (16 coiled-coil dimers) in staggered overlaps

The cytoskeleton is a structural feature of eukaryotic cells revealed especially well by digital video microscopy, deep-etching, and immunocytochemistry. It consists of an extensive three-dimensional network of microtubules, microfilaments, and intermediate filaments that determines cell shape and facilitates a variety of cell movements.

Microtubules are hollow tubes with walls consisting of tubulin dimers polymerized linearly into protofilaments. MTs are polar structures and elongate preferentially from one end. First identified as components of the axonemal structures of cilia and flagella and the mitotic spindle of dividing cells, they are now recognized as a general cytoplasmic constituent of most eukaryotic cells. Cytoplasmic microtubules govern the asymmetrical shape of animal cells, the plane of cell division in plant cells, and the changes in cell position and shape that occur during embryonic development.

Microfilaments are double-stranded polymers of actin that were initially discovered because of their role in the contractile fibrils of muscle cells but are now recognized as a component of virtually all eukaryotic cells. Like microtubules, microfilaments are polar structures, with actin monomers added preferentially to one end and removed from the other. Microfilaments are involved in a variety of locomotory functions. They are also important structurally in the development and maintenance of animal cell shape, as evidenced by their presence in the cell cortex, stress fibers, and the cores of microvilli.

Intermediate filaments are the most stable and least soluble constituents of the cytoskeleton. They appear to play a structural or tension-bearing role. Unlike microtubules and microfilaments, IFs are tissue-specific and can be used to identify cell type. Such typing is useful in the diagnosis of cancer, as tumor cells are known to retain the IF proteins of their tissue of origin. Intermediate filaments can also be classified by amino acid sequence, because their constituent proteins are encoded by a family of related genes. All IF proteins have a highly conserved central domain flanked by terminal regions that differ in size and sequence, presumably accounting for the functional diversity of IF proteins.

With this background, we are now ready to proceed to the next chapter, where we will explore in more detail the role of microtubules and microfilaments in cellular motility and contractility.

KEY TERMS FOR SELF-TESTING

cytosol (p. 644)
cytoskeleton (p. 644)

Techniques for Studying the Cytoskeleton

immunofluorescence microscopy (p. 645)
quick-freeze deep-etch microscopy (deep-etching) (p. 646)
digital video microscopy (p. 647)

Microtubules

microtubule (MT) (p. 648)
microtubule-organizing center (MTOC) (p. 648)
centrosome (p. 648)
axonemal microtubule (p. 648)
cytoplasmic microtubule (p. 648)
protofilament (p. 649)
lumen (p. 649)
tubulin (p. 649)

α-tubulin (p. 649)
β-tubulin (p. 649)
γ-tubulin (p. 649)
$\alpha\beta$-tubulin dimer (p. 649)
polarity (p. 650)
oligomer (p. 650)
plus end (p. 651)
minus end (p. 651)
lag phase (p. 651)
nucleation (p. 651)
elongation phase (p.651)
overall critical concentration (Cc^o) (p. 652)
treadmilling (p. 652)
dynamic instability model (p. 653)
microtubule-associated protein (MAP) (p. 657)
motor MAP (p. 657)
kinesin (p. 657)
dynein (p. 657)
nonmotor MAP (p. 657)

Microfilaments

microfilament (MF) (p. 659)
cell cortex (p. 660)
stress fiber (p. 660)
focal contact (focal adhesion plaque) (p. 660)
actin (p. 660)
G-actin (p. 660)
F-actin (p. 660)
myosin subfragment 1 (S1) (p. 660)
actin-binding protein (p. 663)
filopodium (p. 663)
leading edge (p. 663)
microvillus (p. 664)

Intermediate Filaments

intermediate filament (IF) (p. 666)
intermediate filament typing (p. 667)

PROBLEM SET

20-1. Filaments and Tubules. Indicate whether each of the following statements is true of microtubules (MT), microfilaments (MF), intermediate filaments (IF), or none of these (N). More than one response may be appropriate for some statements.

(a) Involved in muscle contraction.

(b) Involved in the movement of cilia and flagella.

(c) More important for chromosome movements than for cell division.

(d) More important for cytokinesis than for chromosome movements in animal cells.

(e) Most likely to remain when cells are treated with solutions of nonionic detergents or high ionic strength.

(f) Found in bacterial cells.

(g) Differ in composition in muscle cells versus nerve cells.

(h) Can be detected by immunofluorescence microscopy.

(i) Play well-documented roles in cell movement.

(j) Assembled from protofilaments.

20-2. True, False, or Maybe. Identify each of the following statements as true (T), false (F), or maybe (M), where M indicates a statement that may be either true or false, depending on the circumstances. If you answer F or M, explain why.

(a) The most abundant cytoskeletal protein in most eukaryotic cells is actin.

(b) The energy required for tubulin and actin polymerization is provided by hydrolysis of a nucleoside triphosphate.

(c) Microtubules, microfilaments, and intermediate filaments all exist in a typical eukaryotic cell in dynamic equilibrium with a pool of subunit proteins.

(d) Cytochalasin B inhibits chromosome movement but not cytokinesis (cell division).

(e) An algal cell contains neither tubulin nor actin.

(f) Despite their structural variability, all of the proteins of intermediate filaments are encoded by genes in the same gene family.

(g) Most eukaryotic cells contain only one kind of intermediate filament.

(h) As long as actin monomers continue to be added to the plus end of a microfilament, the microfilament will continue to elongate.

20-3. Weighty Matters. A microfilament exists as two helical strands of actin molecules twisted around each other. One full turn includes 13.5 monomers in each strand and extends the MF by 72 nm. A microtubule exists as a hollow tube with a wall of 13 protofilaments, each a linear polymer of tubulin dimers. Each dimer extends a protofilament by 8 nm. Assume that intermediate filaments are constructed of tetrameric protofilaments with a length of 48 nm, and that each IF has 8 protofilaments in cross section.

(a) Calculate the mass (in daltons) of a 100-nm microfilament, assuming the presence of actin only.

(b) Calculate the mass (in daltons) of a 100-nm microtubule, assuming the presence of tubulin only.

(c) How much heavier, on the basis of length, is a microtubule compared to a microfilament?

(d) Calculate the mass (in daltons) of a 100-nm intermediate filament from a fibroblast cell, assuming the presence of vimentin only.

(e) How heavy, on the basis of length, is a vimentin filament compared to a microfilament or a microtubule?

20-4. Cytoskeletal Sources. Each of the tissues below is an especially good source of a particular kind of cytoskeletal protein. For each of the tissues, decide for which one of the three possible proteins it is likely to be a good source.

(a) Brain: actin/tubulin/desmin

(b) Intestinal mucosa cells: tubulin/NF protein/keratin

(c) Stomach muscle: tubulin/vimentin/desmin

(d) Skeletal muscle: actin/tubulin/GFA protein

(e) Spinal cord: actin/tubulin/NF protein

20-5. Cytoskeletal Studies. Described below are the results of several recent studies on the proteins of the cytoskeleton. In each case, state the conclusion(s) that can be drawn from the findings.

(a) Filaments of the same polarity were generated by polymerization of purified actin and were attached to a carbon film. Small polystyrene beads with myosin molecules linked to them were placed on the actin. When ATP was added to the system, the beads were observed to move along the filaments.

(b) When an animal cell is treated with colchicine, its microtubules depolymerize and virtually disappear. If the colchicine is then washed away, the microtubules appear again, beginning at the centrosome and elongating outward at about the rate (1 μm/min) at which tubulin polymerizes in vitro.

(c) When a gene for heart muscle actin is introduced into and expressed in a cultured fibroblast cell that normally synthesizes only skeletal muscle actin, the actin produced by the "foreign" gene combines readily with the indigenous actin molecules without any adverse effects on the shape or function of the cell.

(d) Homogenates from dividing clam eggs, but not from nondividing eggs, were found to contain sedimentable structures that could induce the polymerization of tubulin into microtubules in vitro, regardless of whether the tubulin was prepared from dividing or nondividing eggs. When examined with the electron microscope, this structure was shown to consist of the amorphous granular material that surrounds the centrioles in the centrosome.

20-6. Actin Polymerization. Shown in Figure 20-31 is a time course for actin polymerization in vitro. Polymerization was initiated by increasing the ionic strength of a solution of G-actin at time 0, and the concentration of F-actin was determined at time intervals, as shown. Notice that three points on the x-axis are labeled as times A, B, and C. Each of the statements below may apply to any, all, or none of these three time points. Indicate for each statement whether it is true of the actin solution at time A, B, and/or C. Use N if the statement is not true at any point.

(a) The main process occurring in the solution is nucleation.

(b) G-actin is present at its critical concentration.

(c) Actin monomers are polymerizing.

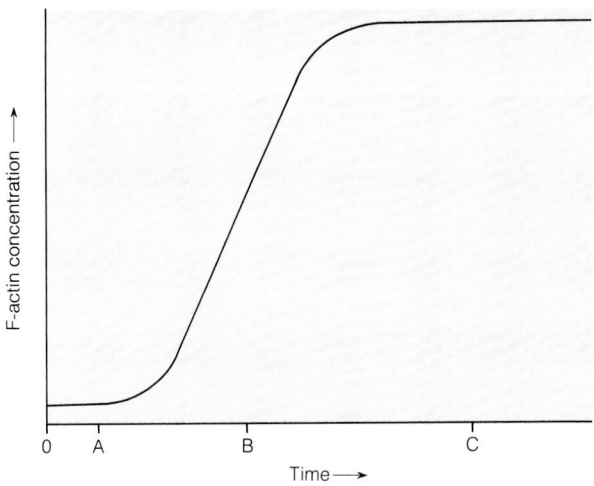

Figure 20-31 Time Course for Actin Polymerization in Vitro.

(d) Actin monomers are adding to the microfilament faster than they are dissociating.

(e) Actin monomers are dissociating from the microfilament faster than they are polymerizing.

(f) If the actin solution is suddenly diluted, the microfilament will begin to depolymerize and will continue to do so until the critical concentration is restored.

20-7. Profilin and the Acrosomal Reaction. Not much is known about how the polymerization of G-actin monomers is initiated except in the case of the *acrosomal reaction* in the sperm of some invertebrates. When a sperm cell makes contact with the outer coat of an egg cell, a long, thin, membrane-covered protrusion called the *acrosomal process* shoots out explosively from the sperm cell, puncturing the egg coat and allowing the sperm and egg membranes to fuse. Prior to such contact, the sperm is crammed with unpolymerized actin molecules, but with little or no F-actin. Also present in the sperm cell is the protein *profilin,* which is equal in abundance to actin on a molecular basis.

(a) Assuming that the formation of F-actin in the core of the acrosomal process drives its rapid elongation, postulate a mechanism to explain the sudden onset of actin polymerization necessary to accomplish the rapid extension of the acrosomal process upon contact with an egg cell. Suggest a way to test your hypothesis.

(b) The first measurable response of sperm upon contact with an egg cell is a very rapid rise in the intracellular pH of the sperm cell, which occurs even before the polymerization of actin begins. Postulate a possible link between the pH change and the initiation of actin polymerization. Suggest a way to test your hypothesis.

20-8. Microtubule Structure. The major proteins making up the outer doublets have been isolated from the cilia of the protozoan *Tetrahymena*. These proteins were found to have the following properties:

Protein	Molecular Mass (kDa)	ATPase Activity
A	50	no
B	450	yes
C	Small multiple of 450	yes

When a solution of protein C is mixed with a suspension of isolated *Tetrahymena* outer doublets from which the sidearms have previously been removed, protein C binds to the doublets, and electron microscopy shows that the sidearms of the outer doublets are thereby restored.

(a) Explain the likely nature and purpose of proteins A, B, and C in outer doublet structure.

(b) Suggest additional experiments to test your explanation.

20-9. Stabilization and the Cc^o. Suppose you have determined the Cc^o for a sample of purified tubulin. Then you add a preparation of centrosomes (microtubule-organizing centers), which nucleate microtubules so that the minus end is bound to the centrosome and stabilized against disassembly. When you again determine the Cc^o, you find it is different. Explain why the Cc^o would change.

SUGGESTED READING

General References

Bray, D. *Cell Movements.* New York: Garland, 1993.
Hyams, J. S., and C. W. Lloyd. *Microtubules.* Wiley-Liss, 1994.
Kreis, T. and R. Vale. *Guidebook to the Cytoskeletal and Motor Proteins.* New York: Oxford University Press, 1993.
Schliwa, M. *The Cytoskeleton: An Introductory Survey.* Cell Biology Monographs, Vol. 13. New York: Springer-Verlag, 1986.

Structural Elements of the Cytoskeleton

Bassell, G. J., R. H. Singer, and K. S. Kosik. Association of poly(A) mRNA with microtubules in cultured neurons. *Neuron* 12 (1994): 571–582.
Fulton, A. B. *The Cytoskeleton: Cellular Architecture and Choreography.* London: Chapman and Hall, 1984.
Luby-Phelps, K., D. L. Taylor, and F. Lanni. Probing the structure of the cytoplasm. *J. Cell Biol.* 102 (1986): 2015.

Techniques for Studying the Cytoskeleton

Bridgman, P. C., and T. S. Reese. The structure of cytoplasm in directly frozen cultured cells. 1. Filamentous meshworks and the cytoplasmic ground substance. *J. Cell Biol.* 99 (1980): 1655.
Heuser, J., and M. W. Kirschner. Filament organization revealed in platinum replicas of freeze-dried cytoskeletons. *J. Cell Biol.* 86 (1980): 212.
Hirokawa, N., and J. E. Heuser. Quick-freeze, deep-etch visualization of the cytoskeleton beneath surface differentiations of intestinal epithelial cells. *J. Cell Biol.* 91 (1981): 399s.
Shotton, D., ed. Electronic light microscopy: Techniques in modern biomedical microscopy. New York: Wiley-Liss, 1993.
Slayter, E. M. *Light and Electron Microscopy.* New York: Cambridge University Press, 1993.

Wang, Y. L. Dynamics of the cytoskeleton in live cells. *Curr. Opin. Cell Biol.* 3 (1991): 27.

Wang, Y. L., and D. L. Taylor. Fluorescence microscopy of living cells in culture. Part A: Fluorescent analogs, labeling cells, and basic microscopy. In *Methods in Cell Biology* (Y. L. Wang and D. L. Taylor, eds.). New York: Academic Press, 1989.

Microtubules

Archer, J., and F. Solomon. Deconstructing the microtubule-organizing center. *Cell* 76 (1994): 589.

Avila, J. Microtubule functions. *Life Sci.* 50 (1992): 327.

Baron, A. T., T. M. Greenwood, C. W. Bazinet, and J. L. Salisbury. Centrin is a component of the pericentriolar lattice. *Biol. Cell* 76 (1992): 383.

Brinkley, B. R. Chromosomes, kinetochores and the microtubule connection. *Bioessays* 13 (1991): 675.

Brinkley, B. R. Microtubule organizing centers. *Annu. Rev. Cell Biol.* 1 (1985): 145.

Caplow, M. Microtubule dynamics. *Curr Opin. Cell Biol.* 4 (1992): 58.

Cassimeris, L. Regulation of microtubule dynamic instability. *Cell Motil. Cytoskel.* 26 (1993): 275.

Lee, G. Non-motor microtubule-associated proteins. *Curr. Opin. Cell Biol.* 5 (1993): 88.

Luduena, R. F., A. Banerjee, and I. A. Khan. Tubulin structure and biochemistry. *Curr. Opin. Cell Biol.* 4 (1992): 53.

Mitchison, T. J. Polewards microtubule flux in the mitotic spindle: Evidence from photoactivation of fluorescence. *J. Cell Biol.* 109 (1989): 637.

Mitchison, T., and M. Kirschner. Dynamic instability of microtubule growth. *Nature* 312 (1984): 237.

Olmstead, J. B. Microtubule-associated proteins. *Annu. Rev. Cell Biol.* 2 (1986): 421.

Olmsted, J. B. Non-motor microtubule-associated proteins. *Curr. Opin. Cell Biol.* 3 (1991): 52.

Microfilaments

Carlier, M. F. Actin polymerization and ATP hydrolysis. *Adv. Biophys.* 26 (1990): 51.

Carlier, M. F. Actin: Protein structure and filament dynamics. *J. Biol. Chem.* 266 (1991): 1.

Dhermy, D. The spectrin super-family. *Biol. Cell* 71 (1991): 249.

Hartwig, J. H., and D. J. Kwiatkowski. Actin-binding proteins. *Curr. Opin. Cell Biol.* 3 (1991): 87.

Mooseker, M. S. Organization, chemistry, and assembly of the cytoskeletal apparatus of the intestinal brush border. *Annu. Rev. Cell Biol.* 1 (1985): 209.

Pollard, T. D. Actin. *Curr. Opin. Cell Biol.* 2 (1990): 33.

Rubenstein, P. A. The functional importance of multiple actin isoforms. *Bioessays* 12 (1990): 309.

Schroder, R. R., D. J. Manstein, W. Jahn, H. Holden, I. Rayment, K. C. Holmes, and J. A. Spudich. Three-dimensional atomic model of F-actin decorated with Dictyostelium myosin S1. *Nature* 364 (1993): 171.

Vandekerckhove, J. Structural principles of actin-binding proteins. *Curr. Opin. Cell Biol.* 1 (1989): 15.

Vandekerckhove, J., and K. Vancompernolle. Structural relationships of actin-binding proteins. *Curr. Opin. Cell Biol.* 4 (1992): 36.

Intermediate Filaments

Albers, K., and E. Fuchs. The molecular biology of intermediate filament proteins. *Int. Rev. Cytol.* 134 (1992): 243.

Klymkowsky, M. W., J. B. Bachant, and A. Domingo. Functions of intermediate filaments. *Cell Motil. Cytoskel.* 14 (1989): 309.

Lazarides, E. Intermediate filaments as mechanical integrators of cellular space. *Nature* 283 (1980): 249.

Osborn, M., and K. Weber. Tumor diagnosis by intermediate filament typing: A novel tool for surgical pathology. *Lab. Invest.* 48 (1983): 372.

Steinert, P. M., and D. R. Roop. Molecular and cellular biology of intermediate filaments. *Annu. Rev. Biochem.* 57 (1988): 593.

Stewart, M. Intermediate filaments: Structure and assembly. *Curr. Opin. Cell Biol.* 5 (1990): 3.

Alzheimer's Disease

Brion, J. P. The pathology of the neuronal cytoskeleton in Alzheimer's disease. *Biochim. Biophys. Acta* 1160 (1992): 134.

Wischik, C. M. Cell biology of the Alzheimer tangle. *Curr. Opin. Cell Biol.* 1 (1989): 115.

21

CELLULAR MOVEMENT: MOTILITY AND CONTRACTILITY

In the previous chapter, we considered cytoskeletal components and their basic functions in eukaryotic cells. We saw that an important function of the cytoskeleton is to provide an intracellular scaffolding that organizes structures within the cell and shapes the cell itself. We also noted that this scaffolding plays a dynamic role in cell motility. In this chapter, we will explore the role of these cytoskeletal elements in cellular movement. This may involve the movement of a cell (or a whole organism) through its environment, the movement of the environment past or through the cell, the movement of components within the cell, or the shortening of the cell itself. In each case, we are dealing with some aspect of **motility. Contractility,** a related term used to describe the shortening in muscle cells, is a specialized form of motility.

Systems of Motility

Motility is an especially intriguing use of energy by cells because it involves the conversion of chemical energy directly to mechanical energy. In contrast, most mechanical devices that produce movement from chemicals (such as a steam engine, which depends on the combustion of coal) require an intermediate form of energy, usually heat or electricity. Motility occurs at the tissue, cellular, and subcellular levels, with the most conspicuous examples at the tissue level, particularly in the animal world. The muscle tissues common to most animals consist of cells specifically adapted for contraction, and the movements produced are often obvious, whether manifested as the bending of an arm or leg, the beating of a heart, or a uterine contraction during childbirth. In humans, about 40% of body weight consists of skeletal muscles, which consume a significant proportion of our total energy budget.

At the cellular level, the emphasis is on the movement of the cell through its environment or, in some cases, the movement of the environment past the cell. Cellular motility is observed most often in organisms that consist of one or only a few cells. It occurs among cell types as diverse as flagellated bacteria, ciliated protozoa, and motile sperm. In each case, movement depends on some sort of cellular appendage adapted for propulsion. In eukaryotic cells, these appendages may be cilia or flagella, which have similar structures and mechanisms of movement. Other examples of motility at the cellular level include amoeboid movement, cell migration during animal embryogenesis, and the invasive action of cancer cells in malignant tumors.

Equally important is the movement of intracellular components, which might be regarded as motility at the subcellular level. For example, highly ordered microtubules of the mitotic spindle play a key role in the separation of chromosomes during cell division, as we saw in Chapter 14. In addition, some cells display a phenomenon called *cytoplasmic streaming,* in which the cytoplasm undergoes rhythmic patterns of flow. Other examples of mechanical work at the subcellular level include the characteristic movements of molecular structures that occur during cell growth and differentiation. An example of such a process is the transport of cellulose myofibrils to the growing wall of a dividing or differentiating plant cell.

The Molecular Basis of Motility

The microfilaments and microtubules of the cytoskeleton provide a basic scaffolding for specialized **motor molecules,** which interact with the cytoskeleton to produce motion at the molecular level (Table 21-1). The molecular motions are summed to produce motion at the cellular level. In cases such as muscle contraction, the cellular motions are then summed to produce motion at the tissue level.

The major motility systems can be classified into three groups. The first type of motility is based on interactions between microfilaments and members of the myosin type of

Table 21–1 Motor Molecules of Eukaryotic Cells

Molecules	Function
Microfilament-associated (actin-binding) proteins	
Myosin I, monomer	Motion along actin filaments.
Myosin II, filament	Thick filament of muscle cell.
Microtubule-associated proteins	
Cytoplasmic dynein	Motion toward minus end of microtubule.
Axonemal dynein	Activation of sliding in flagellar microtubule.
Kinesin	Motion toward plus end of microtubule.

Nonmuscle Microfilament-Based Movement

Actin and myosin are best known as the major components of the thin and thick filaments of muscle cells, which will be discussed later in the chapter. However, actins and myosins have now been discovered in almost all eukaryotic cells and are known to play important roles in various types of **nonmuscle cell movement.** In fact, muscle cells represent only one specialized case of cell movements driven by the interactions of actin and myosin.

Actin and Myosin in Nonmuscle Cells

Actin occurs ubiquitously in eukaryotic cells and usually represents 5–10% of the protein in a nonmuscle cell. Typically, about half of the actin in such cells is present as filaments of polymerized F-actin (see Figure 20-11), whereas the other half is monomeric G-actin that is usually complexed with other proteins. Actins from phylogenetically diverse organisms are remarkably similar in amino acid sequence and structural properties. Such evolutionary conservatism argues for all parts of the molecule being important for its function.

Different forms of **myosin** also occur widely in nonmuscle cells, but usually at a lower concentration than actin. In general, myosin accounts for less than 0.5% of the protein of nonmuscle cells. The greater prominence of actin in such cells probably reflects a dual role: It is involved with myosin in microfilament-based movement, but it also functions in a variety of structural roles. In the cell cortex, for example, actin is crosslinked with filamin to form a gel.

Myosin is most often identified with the myosin found in the thick filaments of muscle cells. In fact, myosin is a family of related proteins, all of which are actin-activated ATPases. All myosins have at least one polypeptide chain, called the *heavy chain,* with a globular head group at one end and a tail of varying length at the other end (Figure 21-1). The globular head binds to actin and uses the energy of ATP hydrolysis to move along an actin filament. The structure of the tail region varies between the different kinds of myosin, giving myosin molecules the ability to bind to a variety of different molecules or cell structures. The tail region in effect specifies what kind of "cargo" the myosin head will carry. Myosins contain one or two small polypeptides bound to the globular head group. These polypeptides, referred to as the *light chains,* often play a role in regulating the activity of the myosin ATPase.

Despite their variability, myosins fall naturally into two groups: **myosin I** (type I myosin) and **myosin II** (type II myosin). Type I myosins (Figure 21-1a) (sometimes called *minimyosins*) contain a single heavy chain and cannot assemble into filaments. While all myosin molecules have head groups that can bind to actin, some type I myosins are unusual in that they also have a binding site for actin on their

motor molecules. A familiar example of **microfilament-based movement** is *muscle contraction.* Contraction occurs when filaments of *actin* slide along thicker filaments composed of *myosin*—a mechanism that is commonly referred to as the *sliding-filament model.* Actin and myosin are also used to generate nonmuscle cell movements, including cell crawling or amoeboid movement, cytoplasmic streaming, and cell division. At the heart of these movements is the ability of a myosin-type motor molecule to propel itself along an actin filament.

The second type of motility involves interactions between specialized motor molecules and microtubules. Microtubules are abundant in interphase cells and are used for a variety of intracellular movements. A specific example of **microtubule-based movement** is *fast axonal transport,* one of the processes by which a nerve cell transports materials between the central part of the cell and the outlying regions. The motor molecule responsible for carrying axoplasmic vesicles from the cell body toward the tips of the axon is *kinesin.* Kinesin binds to axoplasmic vesicles and "walks" along microtubules in the minus-to-plus-end direction. A second example of microtubule-based movement is the motility produced by the *cilia* and *flagella* of certain eukaryotic cells. In this type of movement, a motor MAP called *dynein* causes microtubules in the flagella or cilia to slide past each other. This basic sliding force underlies the beating movements of cilia and flagella.

The third type of motility is unique to prokaryotes and involves **bacterial flagellar rotation.** Despite the similarity in terminology, bacterial flagella are chemically and structurally different from eukaryotic flagella. Bacterial flagellar rotation is driven by the flow of protons through a motor at the base of the flagella. We will now take a closer look at these three motility systems.

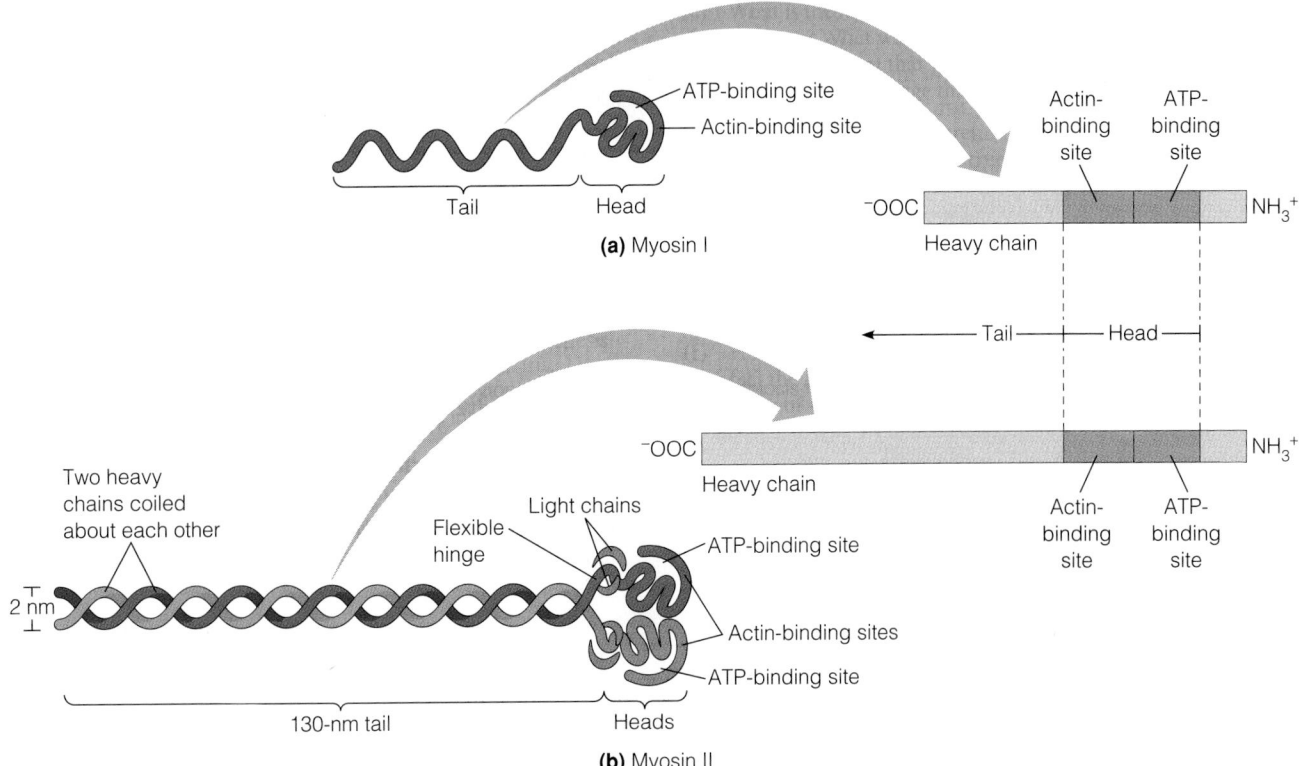

(a) Myosin I

(b) Myosin II

Figure 21-1   The Molecular Structure of Myosins.
(a) Myosin I molecules consist of one heavy chain of approximately 110–140 kDa, which includes a globular head group with an actin-binding site and ATPase and a tail region. The tail region does not permit myosin I to assemble into filaments; instead, it has a second actin-binding site or can bind to phospholipid membranes. **(b)** Myosin II is composed of two heavy chains of about 200 kDa. The heavy chains wrap around each other to form a rodlike tail at one end and two globular heads at the other end. Each head contains an actin-binding site and the ATPase necessary for movement along an actin filament. Typically, the base of each globular head group is also encircled by two light chains, an essential light chain and a regulatory light chain. Myosin II proteins have the distinctive ability of assembling into bipolar filaments, as in the thick filaments of muscle cells.

tail region. In addition, some type I myosins can bind strongly to membranes, suggesting that this form of myosin plays a role in movement of the plasma membrane or in transporting membrane-enclosed organelles inside the cell.

Type II myosins (Figure 21-1b) are composed of two heavy chains, each featuring a globular myosin head, a hinge region, and a long rodlike tail. These myosins are found in skeletal, cardiac (heart), and smooth muscle cells as well as nonmuscle cells. Type II myosins are distinctive in that they can assemble into filaments such as the thick filaments of muscle cells. The basic function of myosin II in all cell types is to convert the energy of ATP to mechanical force that can cause actin filaments to slide past the myosin molecule. While all myosin II molecules exhibit many similarities and carry out the same basic function in cells, their interactions with actin filaments are regulated in different ways. In general, we can distinguish two different types of regulation, both involving calcium but mediated by two different mechanisms. One type is used in skeletal and cardiac muscle, while another type is used for smooth muscle and nonmuscle cells. These distinctive means for regulating muscle contraction will be discussed in the context of each particular type of muscle cell.

Evidence for the Involvement of Actin and Myosin in Nonmuscle Movement

The complexity of cell movement and the many proteins involved make it difficult to identify clearly the function of specific proteins. Often, several approaches are needed to piece together the puzzle. One way to get at the role of a particular protein is to inhibit its function, either by genetic mutation or by microinjecting the cell with antibodies that inhibit the protein. These approaches have been fruitful using unicellular organisms such as the slime mold *Dictyostelium discoideum* and the amoeba *Acanthamoeba castellani* because they exhibit many of the same types of movements seen in animal cells, such as crawling, phagocytosis, and cytokinesis. For example, when the myosin II gene was altered to produce nonfunctional myosin II in *Dictyostelium*, cytokinesis was blocked but the cells were still capable of crawling. This indicates that myosin II is essential for cytokinesis but not absolutely necessary for crawling. In *Acanthamoeba castellani*, antibody-staining experiments showed that one of the myosin I proteins (called myosin IC) was exclusively localized around the contractile vacuole that expels water from these cells. When antibodies that inhibit the function of myosin IC were introduced into the amoeba, the vacuole could no longer contract.

Mutations of particular proteins in a cell does not always provide clear answers to the question of how a protein functions in the cell, however. For example, *Dictyostelium* has an estimated 6–12 genes for myosin. Several of the myosin I genes have been disrupted without blocking the ability of these cells to crawl. In fact, certain mutations seem to have no effect on the cell at all. This has led researchers to believe that some redundancy exists in the myosins or that different myosins overlap in their functions. Thus, various approaches are needed to shed light on how actin, myosin, and related proteins are used by the cell.

Additional information showing how actin is involved in cell motility comes from studies with a family of related drugs called the *cytochalasins,* which were described in Chapter 20. These drugs are produced by various species of molds and have the common property of paralyzing many different kinds of cell movements in vertebrates. At the molecular level, cytochalasins act by binding specifically to the plus end of F-actin filaments, thereby preventing any further addition of G-actin monomers to the filament. The net result is the prevention of actin polymerization.

The disruptive effect of cytochalasins on nonmuscle movement can be explained in terms of the transient nature of the actin filaments in these systems. Unlike the myofibrils of muscle cells, most nonmuscle contractile structures are temporary: They are formed only when needed and are disassembled again thereafter. This explains why at least half of the actin in nonmuscle tissue exists in nonpolymerized form and implies a dynamic relationship between the F-actin filaments and unpolymerized subunits. Clearly, any process that is dependent on the renewed assembly of filaments from subunits will be sensitive to inhibition by cytochalasins.

Cell movements that are sensitive to cytochalasins are in general also inhibited by the drug *phalloidin,* even though the effect of phalloidin is to stabilize actin filaments and prevent their depolymerization, as described in Chapter 20. This dual sensitivity of actin-linked motility to both cytochalasins and phalloidin is further evidence of the importance of the dynamic assembly and disassembly of actin filaments in a variety of nonmuscle movements.

Intracellular Microtubule-Based Movement: Dynein and Kinesin

Microtubules provide a rigid set of tracks for the transport of a variety of membrane-enclosed organelles and vesicles. The polarity of microtubules provides a sense of direction in the same way that the center divider of our roadways separates traffic into two directions. The centrosome provides an organization and orientation to microtubules, because all the minus ends of the microtubules are embedded in the centrosome. The centrosome is generally located near the center of the cell, so traffic toward the minus ends of micro-

tubules might be considered "inbound" traffic. Traffic directed toward the plus ends might likewise be considered "outbound," meaning that it is directed toward the periphery of the cell.

While microtubules provide an organized set of tracks along which organelles can move, they do not directly generate the force necessary for movement. The mechanical work needed for movement depends on **microtubule-associated motor proteins (motor MAPs).** As mentioned earlier, these motor MAPs attach to vesicles or organelles and then "walk" along the microtubule, using ATP to provide the needed energy. Furthermore, these motor MAPs recognize the polarity of the microtubule, with each motor MAP having a preferred direction of movement. At present, we are aware of two families of motor MAPs: **kinesins** and **dyneins.** At least a dozen different proteins similar to kinesin have been identified in various organisms. The dynein family of motor MAPs currently has only two members: *axonemal dynein* and *cytoplasmic dynein.*

With the availability of microtubule-based "roadways" and motor MAPs that function as "trucks" for hauling cargo, the cell clearly has an elaborate transportation system in place. Why is such a transport system necessary, and how it is used? The answers to these questions are beginning to emerge; there are at least a few cellular functions for which we know these transport systems are critical. Examples of such cell functions include fast axonal transport, the processing of material through the Golgi complex, movements and distribution of the endoplasmic reticulum, and mitosis.

Cytoplasmic Microtubules, Motor MAPs, and Axonal Transport

The transport of vesicles and organelles via motor MAPs is particularly important for neurons (nerve cells). A receptor protein or neurotransmitter might be synthesized in the cell body of the neuron and then might have to be transported over distances up to a meter between the cell body and the nerve ending. Some form of energy-dependent transport is clearly required, and microtubule-based movement provides the required mechanism. The need for such transport arises because ribosomes are present only in the cell body, so no protein synthesis occurs in the axons or synaptic knobs. Instead, proteins and membranous vesicles are synthesized in the cell body and transported along the axons to the synaptic knobs. The process is called **fast axonal transport** and appears to involve the movement of protein-containing vesicles and other organelles along "tracks" of microtubules.

The role of microtubules in axonal transport was initially suggested by evidence that this process is inhibited by colchicine and other drugs that impair microtubule function but is insensitive to drugs such as cytochalasin B that affect microfilaments. Since then, microtubules have been visualized along the axon by the deep-etching technique (see Figure 20-2) and have been shown to be prominent features of the axonal cytoskeleton. Moreover, axonal microtubules

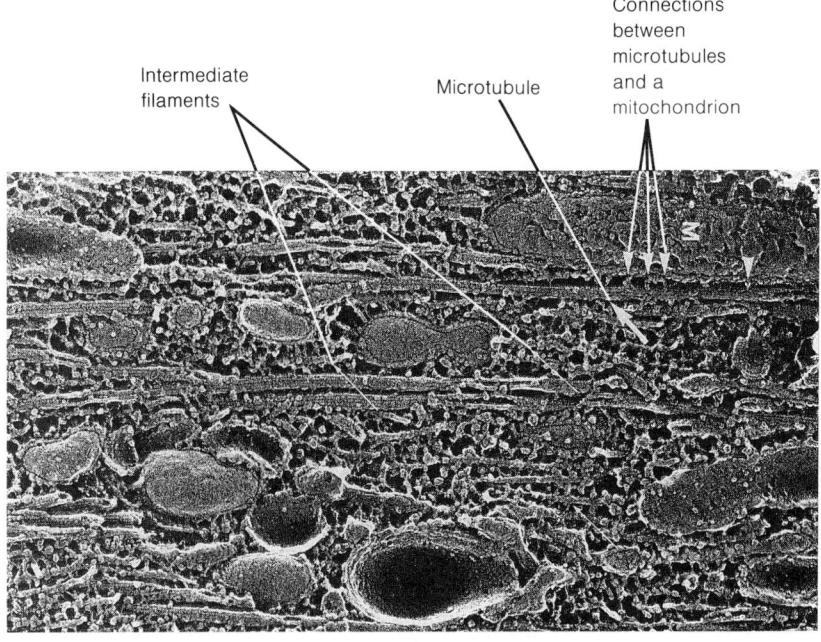

Figure 21-2 The Prominence of Microtubules in Axonal Cytoplasm. The microtubules in a frog axon have been visualized here by quick-freeze deep-etch electron microscopy. Microtubules (large arrow) run longitudinally along the axon, as do intermediate filaments (small arrows). Mitochondria (M) and other membranous vesicles are embedded in the cytoskeletal matrix and appear to be connected to the microtubules (medium arrows). (From Nobutaka Hirokawa, *J. Cell Biol.* 94 (1982): 129. Reproduced by copyright permission of The Rockefeller University Press.)

Intermediate filaments

Microtubule

Connections between microtubules and a mitochondrion

0.2 μm

have small membranous vesicles and mitochondria associated with them (Figure 21-2).

Evidence that a motor MAP drives the movements of organelles was obtained when a group of investigators found that fine filamentous structures present in exuded *axoplasm* (the cytoplasm of axons) could direct the movement of organelles in the presence of ATP. The organelle movement was visualized by video-enhanced differential interference contrast microscopy (a technique described in the Appendix). The rate of organelle movement was shown to be about 2 μm/sec, comparable to the axonal transport rate in intact neurons. The researchers then used a combination of immunofluorescence and electron microscopy to demonstrate that the "tracks" along which the organelles move are single microtubules. They concluded that axonal transport depends on interaction between microtubules and organelles.

Since that time, two motor MAPs responsible for fast axonal transport, kinesin and cytoplasmic dynein, have been purified and characterized in vitro. To determine the direction of transport by these motor MAPs, purified proteins were used experimentally to drive the transport of polystyrene beads along microtubules of known polarity. To obtain microtubules of known polarity, tubulin was polymerized into microtubules using purified centrosomes as MTOCs. This ensured that the microtubules were assembled with their minus ends attached to the centrosome. When polystyrene beads, purified kinesin, and ATP were added to microtubules polymerized by centrosomes, the beads moved toward the plus ends (i.e., away from the centrosome). This finding means that in a nerve cell, kinesin mediates transport from the cell body down the axon to the nerve ending (called *anterograde axonal transport*). When similar experiments were carried out with purified dynein, particles were moved in the opposite direction, toward the minus ends of

the microtubules (called *retrograde axonal transport* when it occurs in neurons).

These results are summarized in Figure 21-3, together with a proposed mechanism for the movement generated by

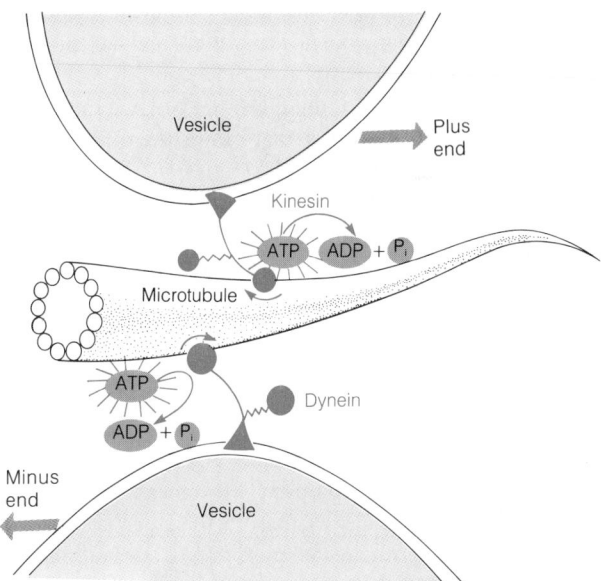

Figure 21-3 Microtubule-Based Motility. Kinesin and dynein are families of molecules that use the energy of ATP hydrolysis to "walk" along microtubules. In the process, they move intracellular structures along microtubules. In general, members of the kinesin family move vesicles or organelles toward the plus ends of microtubules—that is, from the center of the cell to the periphery. Dynein moves in the opposite direction, toward the minus ends of microtubules, and therefore toward the center of the cell where the MTOC is located.

kinesin and dynein. The motion is produced by a cycle of attachment, conformational change, and release that is driven by ATP hydrolysis. As will be described shortly, myosin also undergoes a similar ATP-driven cycle when it interacts with actin filaments to produce muscle contraction.

Motor MAPs and the Transport of Intracellular Vesicles

Movements of vesicles driven by kinesin and cytoplasmic dynein or similar motor MAPs are not restricted to nerve cells. These motor MAPs also play a fundamental role in establishing the distribution and structure of the endoplasmic reticulum, the formation of the Golgi complex, the movements of lysosomes and secretory granules, the internalization of vesicles from the plasma membrane, and a variety of vesicle movements within animal cells. This can be illustrated by looking more closely at how the Golgi complex forms.

As we learned in Chapter 9, the Golgi complex is a series of flattened membrane stacks located in the region of the cell centrosome or microtubule-organizing center (MTOC). The function of the Golgi is to receive proteins made in the endoplasmic reticulum (ER) and to process and package those proteins for distribution to the right cellular destinations. At each step of this process, proteins are transported in vesicles. Thus, there is a continuous flow of vesicles to and from the Golgi. These vesicles are carried by motor MAPs on microtubule tracks (Figure 21-4). If microtubules are depolymerized using nocodazole, a drug that reversibly depolymerizes microtubules, the Golgi complex disperses. When the nocodazole is washed out, the Golgi complex reforms.

Vesicles traveling from the ER to the Golgi complex are carried by a dyneinlike motor that moves them toward the minus ends of microtubules. The minus ends are anchored in the centrosome, so ER vesicles originating from various parts of the cell will all move toward the centrosome. These vesicles fuse together and become part of the flattened membrane stacks that make up one side of the Golgi complex. If the transport of vesicles from the ER to the Golgi complex is blocked, the Golgi again disappears.

Once in the Golgi complex, proteins are processed as they move through the stacks of Golgi membranes. When finished, proteins emerge—still packaged in vesicles—from the other side of the complex. Movement of finished protein-containing vesicles away from the Golgi complex is mediated by a plus end directed motor MAP such as kinesin. This protein carries the vesicles toward the cell periphery and the plasma membrane. Thus motor MAPs and microtubules create what resembles a system of conveyer belts on which vesicles move both toward and away from the Golgi complex. In this respect, the Golgi complex might be considered similar to a waterfall in that it exists only because of the constant flow of vesicles.

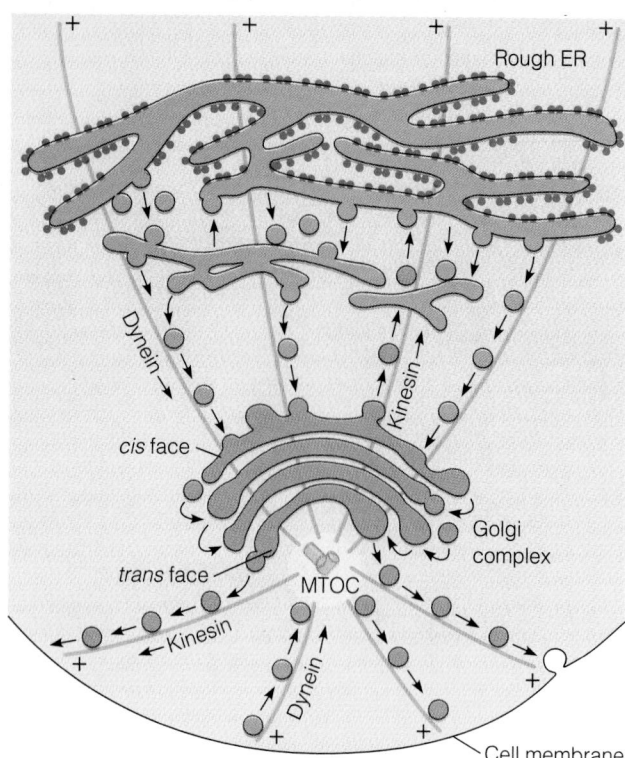

Figure 21-4   The Relationship Between Microtubules, Motor MAPs, and the Golgi Complex. Vesicles going to and from the Golgi complex are attached to microtubules and are thought to be carried by motor MAPs similar or identical to dynein and kinesin. Dynein is a minus end directed motor MAP, while kinesin is plus end directed. Thus, vesicles derived from either the ER or the cell membrane are carried toward the Golgi complex and MTOC by dynein, while vesicles derived from the Golgi complex are carried toward either the ER or the cell periphery by kinesin.

Filament-Based Movement in Muscle

Muscle contraction is the most familiar example of mechanical work mediated by intracellular filaments. Mammals have several different kinds of muscles, including skeletal muscle, cardiac muscle, and smooth muscle. We will first consider skeletal muscle, because much of our knowledge of the contractile process grew out of early investigations of its molecular structure and function.

Skeletal Muscle Cells

Skeletal muscles are responsible for voluntary movement. The structural organization of skeletal muscle is shown in Figure 21-5. A muscle consists of bundles of parallel **muscle fibers** joined by tendons to the bones that the muscle must move. Each fiber is actually a long, thin multinucleate cell, highly specialized for its contractile function. The multi-

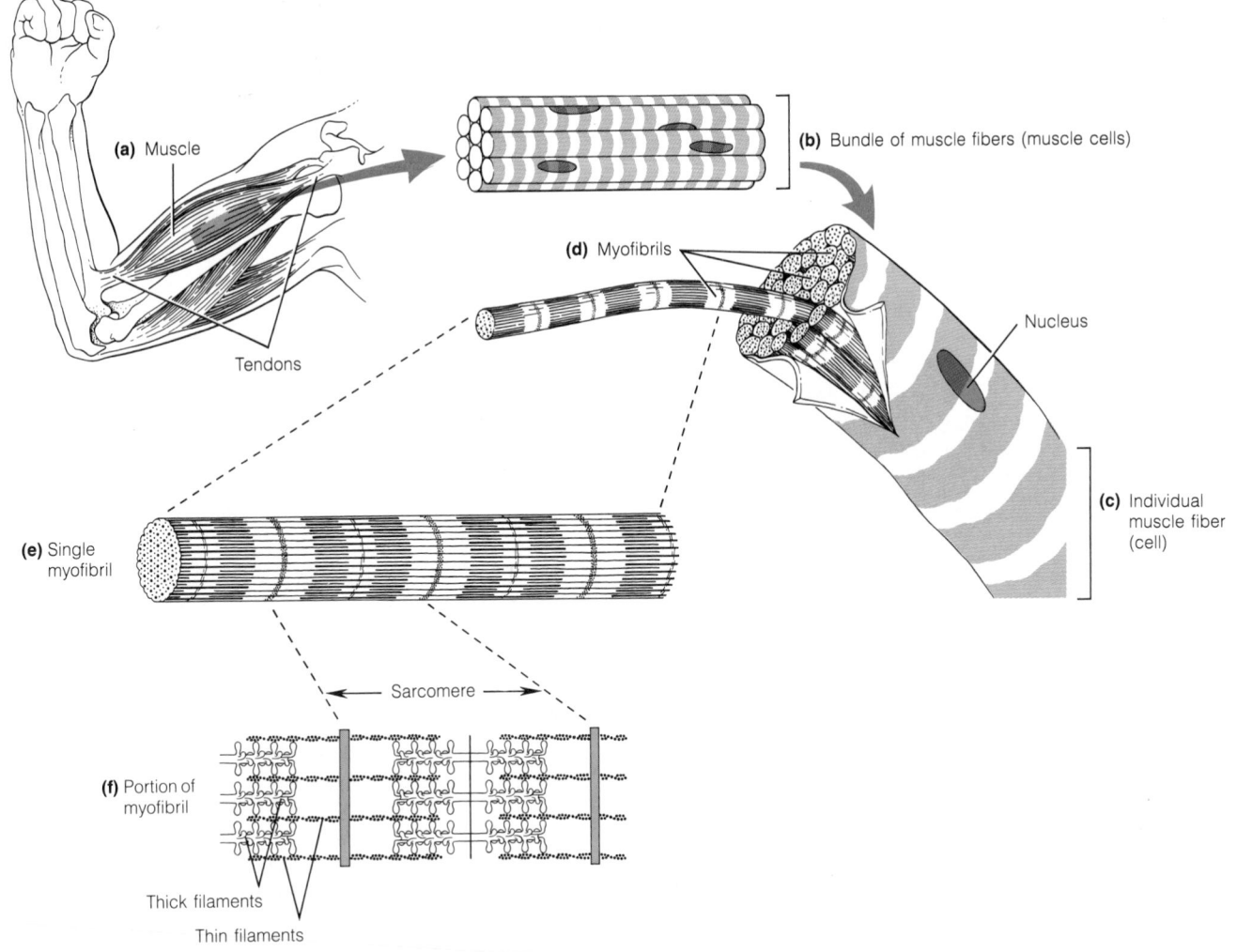

Figure 21-5   Levels of Organization of Skeletal Muscle Tissue. **(a)** Muscle tissue is attached by means of tendons to the specific bones it must move. **(b)** The tissue consists of bundles of muscle fibers, **(c)** each of which is a long, thin multinucleate cell. **(d)** Within each cell are many myofibrils. **(e)** Each myofibril consists of bundles of fila- ments aligned laterally to give skeletal muscle its striated appearance. **(f)** The unit of contraction along each myofibril is the sarcomere, in which thick filaments interdigitate with thin filaments. The thick and thin filaments consist primarily of myosin and actin, respectively.

nucleate state arises from the end-to-end fusion of embryonic cells called *myoblasts* during muscle differentiation. This cell fusion also accounts at least in part for the striking length of muscle cells, which may be measured in centimeters.

At the subcellular level, each muscle fiber (or cell) contains numerous **myofibrils.** These are the functional units of contraction. Myofibrils are 1–2 μm in diameter and may extend the entire length of the cell. Each myofibril is subdivided along its length into repeating units called **sarcomeres.** The sarcomere is the fundamental contractile unit of the muscle cell. Each sarcomere of the myofibril contains bundles of **thick filaments** and **thin filaments.** Thick filaments consist of myosin, whereas thin filaments consist mainly of actin, although several other important proteins are also present. The thin filaments are arranged around the

thick filaments in a hexagonal pattern, as can be seen when the myofibril is viewed in cross section (Figure 21-6).

The filaments in skeletal muscle are aligned in lateral register, giving the myofibrils a pattern of alternating dark and light bands (Figure 21-7a). This pattern of bands, or *striations,* is characteristic of skeletal and cardiac muscle, which are therefore referred to as *striated muscle.*

The dark bands are called **A bands,** and the light bands are called **I bands.** (The terminology for the structure and appearance of muscle myofibrils was developed from observations originally made with the polarizing light microscope. I stands for *isotropic* and A for *anisotropic,* terms related to the appearance of these bands when illuminated with plane-polarized light.)

As illustrated in Figure 21-7b, the lighter region in the middle of each A band is called the **H zone** (from the Ger-

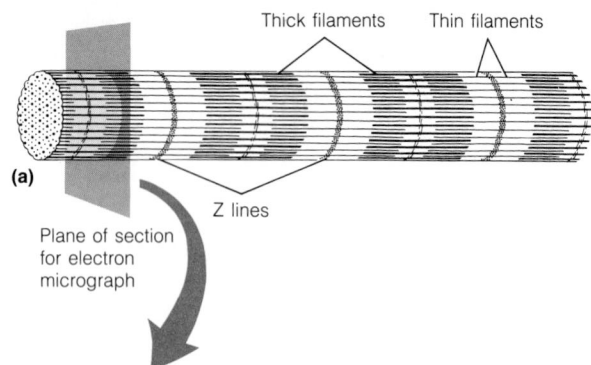

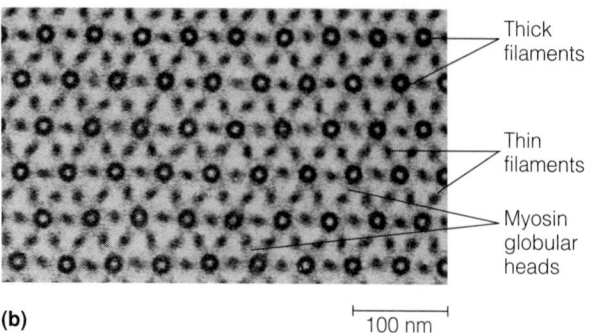

Figure 21-6 Arrangement of Thick and Thin Filaments in a Myofibril. (a) A myofibril consists of interdigitated thick and thin filaments. (b) The thin filaments are arranged around the thick filaments in a hexagonal pattern, as seen in this cross section of a flight muscle from the fruit fly *Drosophila melanogaster* viewed by high-voltage electron microscopy (HVEM).

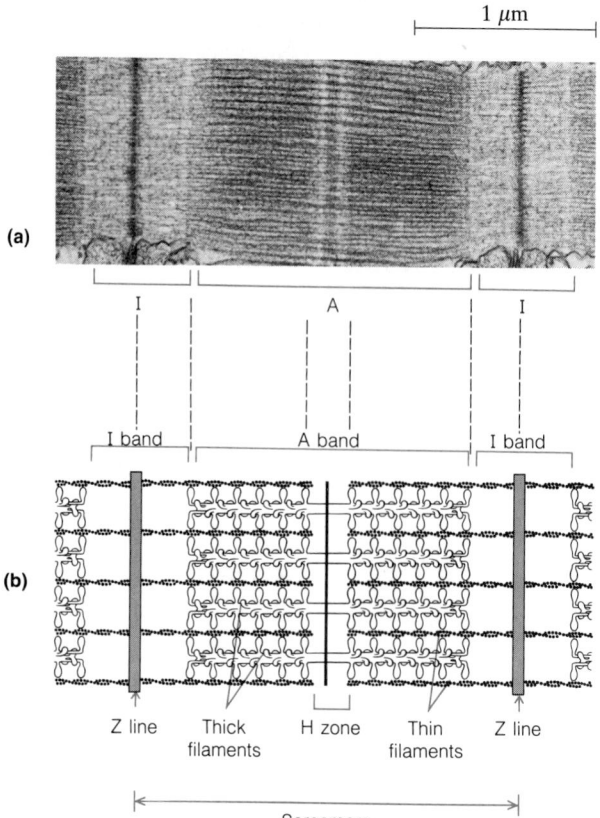

Figure 21-7 Appearance of and Nomenclature for Skeletal Muscle. (a) An electron micrograph of a single sarcomere (TEM). (b) A schematic diagram that can be used to interpret the repeating pattern of bands in striated muscle in terms of the interdigitation of thick and thin filaments. An A band corresponds to the length of the thick filaments, and an I band represents that portion of the thin filaments that does not overlap with thick filaments. The lighter area in the center of the A band is called the H zone. The dense zone in the center of each I band is called the Z line. A sarcomere, the basic repeating unit along the myofibril, is the distance between two successive Z lines.

man word *hell,* meaning "light"). Running down the center of the H zone is the M line, which contains myomesin, a protein that links myosin filaments together. In the middle of each I band appears a dense **Z line** (from the German word *zwischen,* meaning "between"). The distance from one Z line to the next defines a sarcomere. A sarcomere is about 2.5–3.0 μm long in the relaxed state but shortens progressively as the muscle contracts.

Skeletal Muscle Filaments and Their Proteins

The striated pattern of skeletal muscle and the observed shortening of the sarcomeres during contraction can be explained in terms of the thick and thin filaments that make up the myofibrils. We will therefore look in some detail at both types of filaments and then come back to the contraction process in which they play so vital a role.

Thick Filaments. The thick filaments of myofibrils are about 15 nm in diameter and about 1.6 μm long. They lie parallel to one another in the middle of the sarcomere, thus forming the dark A bands of the myofibril (see Figure 21-7).

Each thick filament consists of many molecules of myosin, which are oriented in opposite directions in the two halves of the filament. Each myosin molecule is long and thin, with a molecular weight of about 520,000.

Every thick filament consists of hundreds of myosin molecules organized in a staggered array such that the heads of successive myosin molecules project out of the thick filament in a repeating pattern, as shown in Figure 21-8. The heads occur in pairs, which protrude from the thick filament facing away from the center. Projecting pairs of heads are spaced 14.3 nm apart along the thick filament, with each pair displaced one-third of the way around the filament from the previous pair. These protruding heads can make contact with adjacent thin filaments, forming the crossbridges between thick and thin filaments that are essential to the mechanism of muscle contraction.

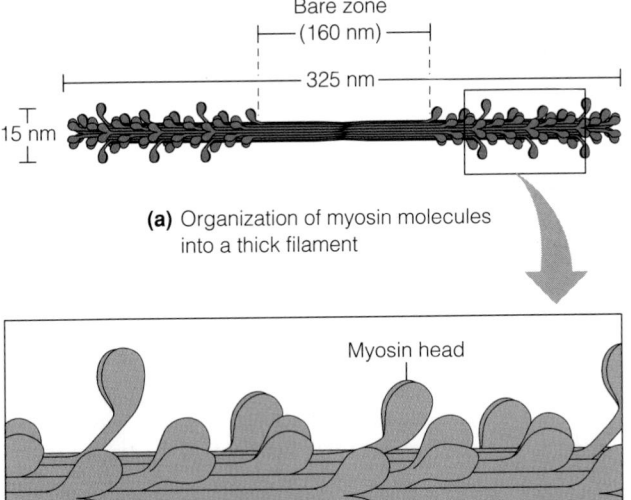

(a) Organization of myosin molecules into a thick filament

Myosin head

14.3 nm

(b) Portion of a thick filament

Figure 21-8 **The Thick Filament of Skeletal Muscle.**
(a) The thick filament of the myofibril consists of hundreds of myosin molecules organized in a repeating, staggered array. A typical thick filament is about 1.6 μm long and about 15 nm in diameter. Individual myosin molecules are integrated into the filament longitudinally, with their ATPase-containing heads oriented away from the center of the filament. The central region of the filament is therefore a bare zone containing no heads.
(b) This enlargement of a portion of the thick filament shows that pairs of myosin heads are spaced 14.3 nm apart.

Thin Filaments. The thin filaments of myofibrils interdigitate with the thick filaments. The thin filaments are about 7 nm in diameter, about 1 μm long, and are the only filaments in the I bands of the myofibril. In fact, each I band consists of *two* sets of thin filaments, one set on either side of the Z line, with each filament attached to the Z line and extending toward and into the A band in the center of the sarcomere. This accounts for the length of almost 2 μm for I bands in extended muscle.

The structure of thin filaments is shown in Figure 21-9. A thin filament consists of at least three proteins, the most prominent of which is actin. Recall from Chapter 20 that actin is synthesized as *G-actin* but polymerizes into long, linear strands of *F-actin*. Thin filaments are strands of F-actin intertwined with the proteins **tropomyosin** and **troponin.** Tropomyosin is a long, rodlike molecule, similar to the myosin tail, that fits in the groove of the actin helix (see Figure 21-9). Each tropomyosin molecule stretches for about 38.5 nm along the filament and associates along its length with 7 actin monomers.

Troponin is actually a complex of three polypeptide chains, called TnT, TnC, and TnI. **TnT** binds to tropomyosin and is thought to be responsible for positioning the complex on the tropomyosin molecule. **TnC** binds calcium ions, and **TnI** binds to actin. (Tn stands for troponin, T for tropomyosin, C for calcium, and I for inhibitory, because TnI inhibits muscle contraction, as we will see later.) One troponin complex is associated with each tropomyosin molecule, so the spacing between successive troponin complexes along the thin filament is 38.5 nm. Troponin and tropomyosin constitute a calcium-sensitive switch that activates contraction in both skeletal and cardiac muscle.

Organization of Muscle Filament Proteins. How can the filamentous proteins of muscle fibers maintain such a precise organization, when in other cells microfilaments are relatively disorganized? The answer has to do with structural proteins that play a central role in maintaining the architectural relationships of muscle proteins (Figure 21-10). For instance, α-**actinin** keeps actin filaments bundled into parallel arrays and attaches them to the Z line. A myosin-binding protein called **myomesin** is present at the H zone of the thick filament arrays and performs the same bundling func-

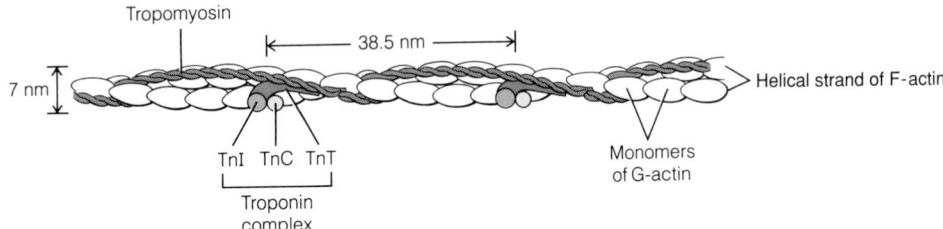

Figure 21-9 **The Thin Filament of Striated Muscle.** Each thin filament is a single strand of F-actin in which the G-actin monomers are staggered and give the appearance of a double-stranded helix. One result of this arrangement is that two grooves run along both sides of the filament. Long, ribbonlike molecules of tropomyosin lie in these grooves.

Each tropomyosin molecule consists of two α helices wound about each other to form a ribbon about 2 nm in diameter and 38.5 nm long. Associated with each tropomyosin molecule is a troponin complex consisting of the three polypeptides TnT, TnC, and TnI.

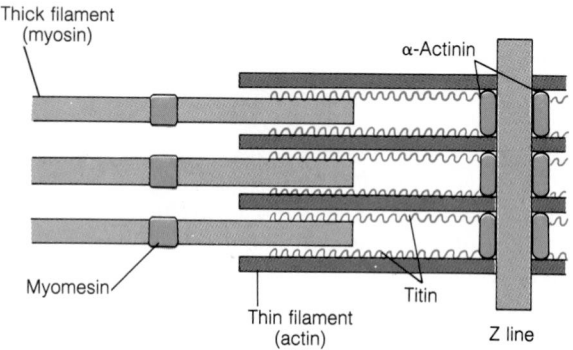

Thick filament (myosin)

α-Actinin

Myomesin

Thin filament (actin)

Titin

Z line

Figure 21-10   Structural Proteins of the Sarcomere. The thick and thin filaments require structural support to maintain their precise organization in the sarcomere. The support is provided by two proteins, α-actinin and myomesin, which bundle actin and myosin filaments, respectively. Titin attaches thick filaments to the Z line, thereby maintaining their position within the thin filament array.

tion for the myosin molecules composing the arrays. A third structural protein, **titin,** attaches the thick filaments to the Z lines. Titin is highly flexible, so during contraction-relaxation cycles, it can keep thick filaments in the correct position relative to thin filaments. The protein components involved in muscle contraction are summarized in Table 21-2. The contractile process involves the complex interaction of all of these proteins.

The Sliding-Filament Model of Muscle Contraction

With our understanding of muscle structure, we can now consider what happens during the contraction process. From electron microscopic studies, it is clear that the A

bands of the myofibrils remain fixed in length during contraction, whereas the I bands shorten progressively and virtually disappear in the fully contracted state. To explain these observations, the **sliding-filament model** illustrated in Figure 21-11 was proposed in 1954. According to this model, muscle contraction is due to thin filaments sliding past thick filaments, with no change in the length of either type of filament. The sliding-filament model not only proved to be correct, but was instrumental in focusing attention on the molecular interactions between thick and thin filaments that underlie the sliding process.

As Figure 21-11a indicates, contraction involves a sliding of thin filaments such that they are drawn progressively into the spaces between adjacent thick filaments, overlapping more and more with the thick filaments and narrowing the I band in the process. The result is a shortening of the individual sarcomeres and myofibrils, and hence a contracting of the muscle cell and the whole tissue. This, in turn, causes the movement of the body parts attached to the muscle.

The interdigitation and sliding of the thin and thick filaments past each other as a means of generating force suggests that there should be a relationship between the force generated during contraction and the degree of shortening of the sarcomere. When the force produced by the muscle at different amounts of sarcomere stretching or shortening is measured, the relationship is exactly what one would predict from the sliding-filament model (Figure 21-11b): The amount of force the muscle can generate during a contraction depends on the number of myosin heads from the thick filament that can make contact with the thin filament. As a result, when the sarcomere is stretched, there is relatively little overlap between thin and thick filaments, so the force generated is small. As the sarcomere shortens, the region of overlap increases and the force of contraction gets correspondingly larger. Finally, a point is reached at which con-

Table 21–2   Major Protein Components of Vertebrate Skeletal Muscle

Protein	Molecular Mass (kDa)	Function
Actin	42	Major component of thin filaments.
Myosin	510	Major component of thick filaments.
Tropomyosin	64	Binds along the length of thin filaments.
Troponin	78	Positioned at regular intervals along thin filaments; mediates calcium regulation of contraction.
Titin	2500	Links thick filaments to Z line.
Myomesin	185	Myosin-binding protein present at the M line of thick filaments.
α-actinin	190	Bundles actin filaments in Z line.
C protein	140	Myosin-binding protein present on either side of the M line of thick filaments.
Ca²⁺ ATPase	115	Major protein of SR; transports Ca²⁺ into SR to relax muscle.
Ryanodine receptor channel	—	A component of the muscle calcium release mechanism.
Dihydropyridine receptor	—	A component of the muscle calcium release mechanism.

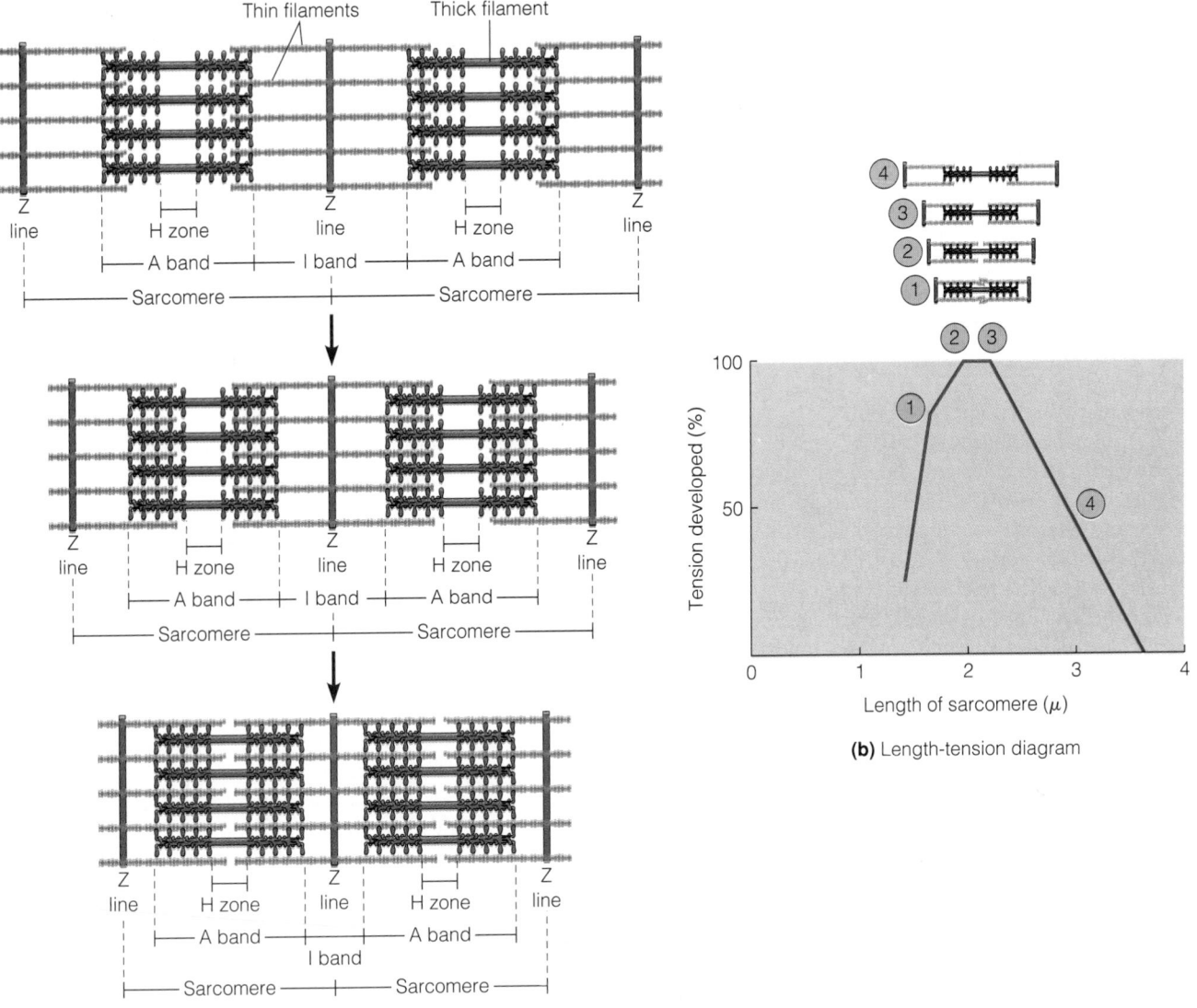

(a) Sliding filament model

(b) Length-tension diagram

Figure 21-11 The Sliding-Filament Model of Muscle Contraction. **(a)** Two sarcomeres of a myofibril during the contraction process. The extended configuration is shown at the top, while the center and bottom views represent progressively more contracted myofibrils. A myofibril shortens by the progressive sliding of thick and thin filaments past each other. The result is a greater interdigitation of filaments without any change in length of individual filaments. The increasing overlap of thick and thin filaments leads to a progressive decrease in the length of the I band as interpenetration continues during contraction. **(b)** This graph shows that the amount of tension developed by the sarcomere is proportional to the amount of overlap between the thin filament and the region of the thick filament containing myosin heads. When the sarcomere begins to shorten, as during a muscle contraction, the Z lines move closer together, increasing the amount of overlap between thin and thick filaments. This overlap allows more of the thick filament to interact with the thin filament. Therefore, the muscle can develop more tension (see 4 to 3). This proportional relationship continues until the ends of the thin filaments move into the H zone. Here they encounter no further myosin heads, so tension remains constant (3 to 2). Any further shortening of the sarcomere results in a dramatic decline in tension (2 to 1) as the filaments crowd into one another.

tinued shortening no longer increases the amount of overlap between thin and thick filaments and the force of contraction stays the same. If the sarcomere shortens further, the strength of contraction will fall off quickly.

The sliding of thin filaments past thick filaments requires energy, which gives rise to three questions: By what mechanism are the thin filaments drawn or pulled progressively into the spaces between thick filaments to cause the actual contraction? How is the energy of ATP used to drive this process? What keeps the partially interdigitated thick and thin filaments associated with each other, so that they do not simply fall apart?

Cross-Bridge Formation. We can begin by taking the third question first and asking what holds interdigitated thick and thin filaments together. This is the role of the heads of the myosin molecules that project outward at regular intervals along each thick filament. These heads represent binding sites for the F-actin of the thin filaments. Regions of overlap between thick and thin filaments, whether extensive (in the contracted muscle) or minimal (in the relaxed muscle), are always characterized by the presence of transient **cross-bridges** between the myosin heads of the thick filaments and the actin of the thin filaments. Numerous cross-bridges can be seen in the electron micrograph shown in Figure 21-12.

Cycles of Cross-Bridge Formation. For actual contraction, it is not enough that cross-bridges be present. They must form and dissociate repeatedly and in such a manner that each cycle of cross-bridge formation causes the thin filaments to interdigitate with the thick filaments more and more, thereby shortening the individual sarcomeres and causing the muscle fiber to contract. A given myosin head on the thick filament repeatedly undergoes a cycle of events in which it binds to a specific actin site on the thin filament, undergoes an energy-requiring change in shape that pulls the thin filament, then breaks that particular association with the thin filament and associates with another actin site farther along the thin filament toward the Z line.

As a result of this **contraction cycle**, the protruding heads of the myosin molecules in the thick filaments draw each thin filament along unidirectionally toward the center of the A band. Overall contraction, then, is the net result of the repeated making and breaking of many such cross-

bridges, with each cycle of cross-bridge formation causing the translocation of a small length of thin filament of a single fibril in a single cell.

What about the direction of contraction? Recall from Chapter 20 that actin filaments have plus and minus ends. We know from the above description that contraction of a muscle cell involves a motor protein—myosin—"walking" along the actin filament in an ATP-dependent process, but does contraction occur toward the plus or minus ends of actin thin filaments? The observation is that myosin always walks toward the *plus* end of the thin filament, thereby establishing the direction of contraction.

ATP and the Contraction Cycle

The driving force for this cyclic formation of cross-bridges in skeletal muscle is the hydrolysis of ATP, catalyzed by the *actin-activated ATPase* located strategically in the myosin head. This requirement for ATP can be demonstrated in vitro because isolated muscle fibers can be shown to contract in response to added ATP.

The mechanism whereby this is accomplished is depicted in Figure 21-13 as a four-step cycle. In step 1, a specific myosin head, in a high-energy configuration containing an ADP and a P_i molecule (a hydrolyzed ATP, in effect), binds loosely to the actin filament. The myosin head then proceeds to a more tightly bound configuration that requires the loss of P_i. Step 2 is then the "power stroke." The transition of myosin to the more tightly bound state triggers a conformational change in myosin. This conformational change is associated with a movement of the head, causing the thick filament to pull against the thin filament, which then moves with respect to the thick filament.

Cross-bridge dissociation follows in step 3, as ATP binds to the myosin head in preparation for the next step. The binding of ATP causes the myosin head to change its conformation in a way that weakens its binding to actin. In the absence of adequate ATP, cross-bridge dissociation does not occur, and the muscle becomes locked in a stiff, rigid state called **rigor.** The rigor mortis associated with death results from the depletion of ATP and the progressive accumulation of cross-bridges in the configuration shown at the end of step 2 in Figure 21-13. Note that once detached, the thick and thin filaments would be free to slip back to their previous positions except that they are held together at all times by the many other cross-bridges along their length at any given moment, just as at least some legs of a millipede are always in contact with the surface on which it is walking. In fact, each thick filament has about 350 myosin heads, and each of these attaches and detaches about five times per second during rapid contraction, so there are always many cross-bridges intact at any time.

Finally, in step 4, the energy of ATP hydrolysis is used to reposition the myosin head for the next cycle by returning it to the high-energy configuration necessary for the next round of cross-bridge formation and filament sliding. This brings us back to where we started, for the myosin head is

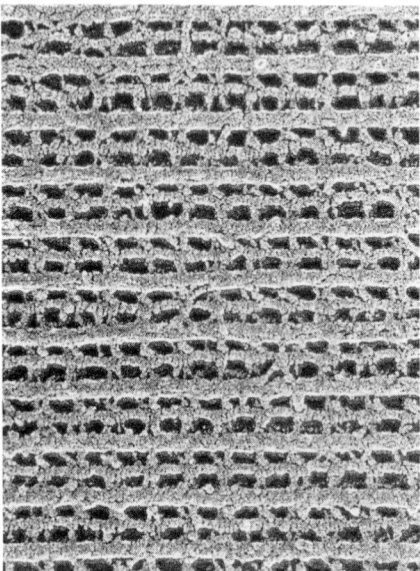

Figure 21-12 Cross-Bridges. The cross-bridges between thick and thin filaments formed by the projecting heads of myosin molecules can be readily seen in this high-resolution electron micrograph (TEM).

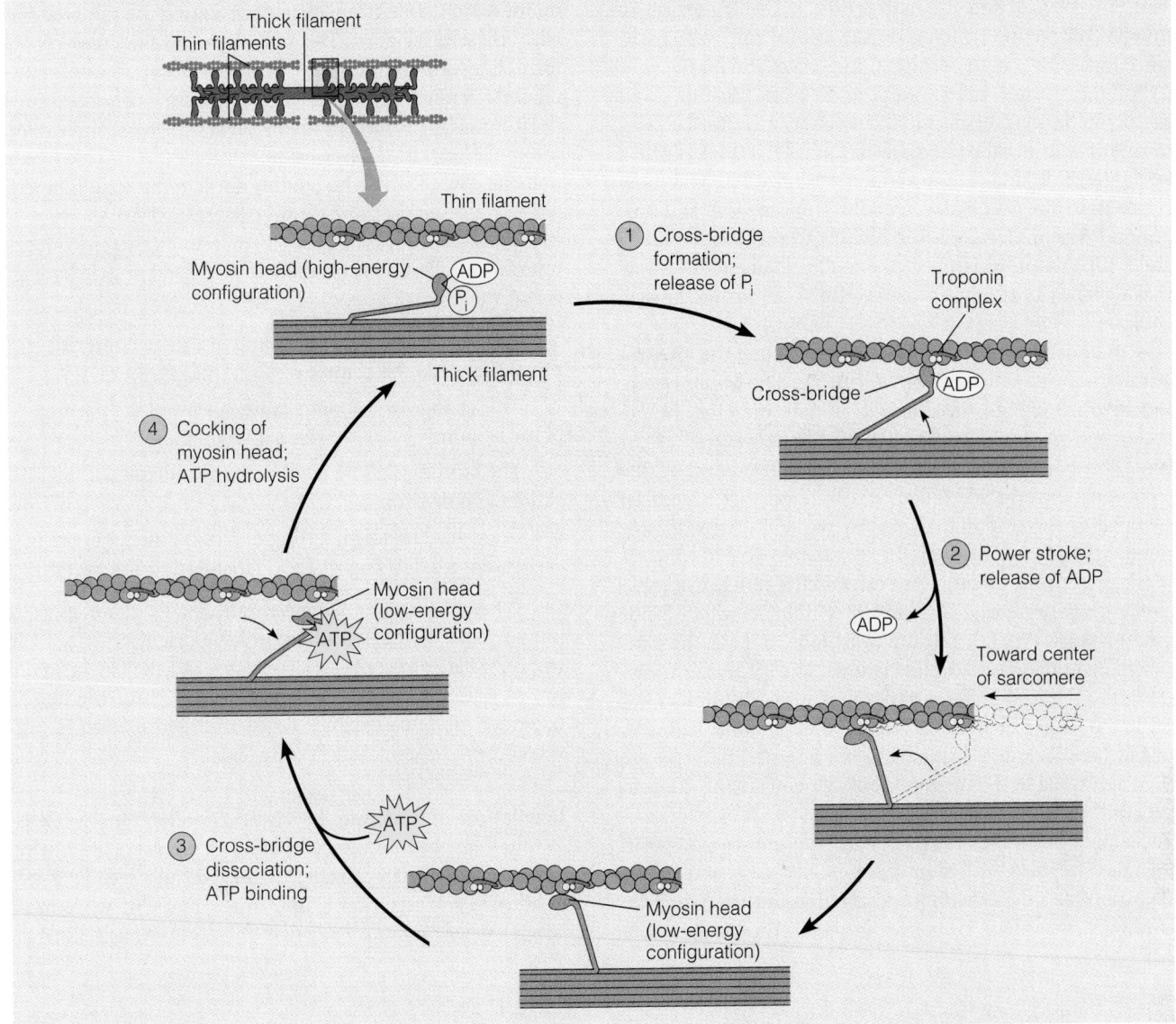

Figure 21-13 The Cyclic Process of Muscle Contraction.
A small segment of adjacent thick and thin filaments (see the orienting diagram) is used to illustrate the series of events in which the cross-bridge formed by a myosin head is used to draw the thin filament toward the center of the sarcomere, thereby causing the myofibril to contract. Step 1 shows the cross-bridge configuration of relaxed muscle, while the end of step 2 shows the configuration of a muscle in rigor. A detailed description of all the steps is given in the text.

now activated and ready to form a bridge to the actin again. But the new bridge will be formed with an actin site further along the thin filament, because the first cycle resulted in a net displacement of the thin filament with respect to the thick filament. In succeeding cycles, the particular myosin head shown in Figure 21-13 will draw the thin filament in the direction of further contraction.

Of particular interest in this contraction cycle are the separation of the actual hydrolysis of ATP (step 4) from the contraction event that it drives (step 2) and the use of an energized configuration of the myosin head as the interim "carrier" of the energy.

The Regulation of Muscle Contraction

The sequence of events depicted in Figure 21-13 shows how muscle contraction is accomplished but provides no insight into how the process is regulated. Indeed, the description provided so far implies that skeletal muscle ought to contract continuously, as long as there is sufficient ATP. Yet experience tells us that most skeletal muscles spend more time in the relaxed state than they spend in contraction. Contraction and relaxation must therefore be regulated to result in the coordinated movements associated with muscle activity.

The Role of Calcium in Contraction. The regulation of muscle contraction depends on the critical role of free calcium ions (Ca²⁺) in the contraction process and on the ability of the muscle cell to raise and lower calcium levels rapidly in the cytosol around the myofibrils. In muscle cells, the cytoplasm is called **myoplasm** and the cytosol is called **sarcoplasm.** In addition to actin, filaments of skeletal muscle contain the regulatory proteins tropomyosin and troponin. These molecules act in concert to regulate the availability of myosin-binding sites on actin filaments in a way that depends critically on the level of calcium in the sarcoplasm.

To understand this, we must recognize that the myosin-binding sites on actin are normally blocked by tropomyosin. For myosin to bind to actin and initiate the cross-bridge cycle, the tropomyosin molecule must be moved out of the way. The positioning of tropomyosin is regulated by the troponins, introduced earlier. Muscle fibers that have been stripped of the troponins and then reconstituted with troponin T (TnT) and troponin I (TnI) are not able to contract in the presence of calcium. The calcium dependence of muscle contraction is due to troponin C (TnC), which binds calcium ions. When a calcium ion binds to TnC, the TnC molecule undergoes a conformational change that is transmitted to the tropomyosin molecule, causing it to move toward the center of the helical groove of the thin filament, out of the blocking position. The binding sites on actin are then accessible to the myosin heads, allowing contraction to proceed.

Figure 21-14 illustrates how the troponin-tropomyosin complex regulates the interaction between actin and myosin. When the calcium concentration in the sarcoplasm is low ($<10^{-4}$ mM), tropomyosin blocks the binding sites on the actin filament, effectively preventing their interaction with myosin (Figure 21-14a). As a result, cross-bridge formation is inhibited, and the muscle becomes or remains relaxed. However, at higher calcium concentrations ($>10^{-3}$ mM), calcium binds to TnC, causing tropomyosin molecules to shift their position, which allows myosin heads to make contact with the binding sites on the actin filament and thereby initiate contraction (Figure 21-14b).

Thus, an increase in the sarcoplasmic calcium concentration stimulates the contraction of skeletal muscle by triggering the following series of events:

1. Calcium binds to troponin and induces a conformational change in the complex.

2. This change in troponin causes a shift in the position of the tropomyosin with which it is complexed.

3. The binding sites on actin become available for interaction with myosin.

4. Cross-bridges form, setting in motion the sequence of events (depicted in Figure 21-13) that lead to contraction.

When the calcium concentration falls again as it is pumped out of the cytosol (as will be discussed later), the troponin-calcium complex dissociates and the tropomyosin moves back to the blocking position. Myosin binding is therefore inhibited, further cross-bridge formation is prevented, and the contraction cycle ends.

Regulation of Calcium Levels in Skeletal Muscle Cells. We now have some understanding of what occurs during contraction and can consider in more detail a second aspect of muscle cell function. Think for a moment what must happen when you move any part of your body, such as when

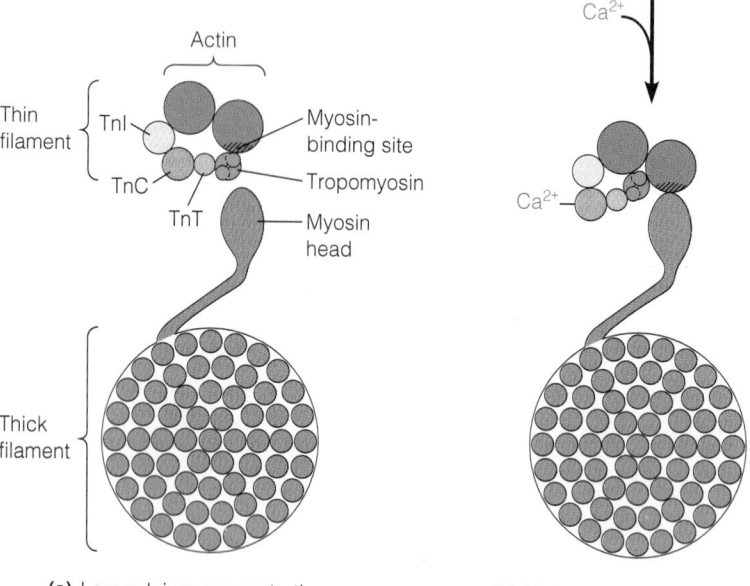

(a) Low calcium concentration   **(b)** High calcium concentration

Figure 21-14 Regulation of Contraction in Striated Muscle. **(a)** At low concentrations ($<10^{-4}$ mM Ca²⁺), calcium is not bound to the TnC subunit of troponin, and tropomyosin blocks the binding sites on actin, preventing access by myosin and thereby maintaining the muscle in the relaxed state. **(b)** At high concentrations ($>10^{-3}$ mM Ca²⁺), calcium binds to the TnC subunit of troponin, inducing a conformational change that is transmitted to tropomyosin. The tropomyosin molecule moves toward the center of the groove in the thin filament, allowing myosin to gain access to the binding sites on actin, thereby triggering contraction.

you flex an index finger. A nerve impulse is generated in the brain and transmitted down the spinal column to the nerve cells, or *motor neurons*, that control a small muscle in your forearm. The motor neurons activate the appropriate muscle cells, which contract and relax, all within about 100 msec. Two main questions arise from this description: How is the muscle cell activated, and how does it relax again after the contraction cycle?

From the previous discussion, you know that calcium ions are required to activate the contraction process. Muscle contraction is regulated by the concentration of calcium ions in the sarcoplasm such that the muscle relaxes when the calcium concentration is low and contracts when calcium levels are elevated. Every time we move, or with every heartbeat, nerve impulses are sent to the appropriate muscles, causing elevation of calcium levels in the sarcoplasm and resulting in contraction. When nerve impulses to the muscle cell cease, calcium levels decline quickly and the muscle relaxes. Therefore, to understand how muscle contraction is regulated, we need to know how nerve impulses cause calcium levels in the sarcoplasm to change and how changes in calcium concentration affect the contractile machinery.

Calcium ions regulate many functions of eukaryotic cells, and cells carefully regulate the concentration of calcium ions in the sarcoplasm. Skeletal muscle cells have many specialized features that facilitate a rapid change in the sarcoplasmic concentration of calcium ions and a rapid response of the contractile machinery. These specializations include the neuromuscular junction, the transverse tubule system, the sarcoplasmic reticulum, and the troponin-tropomyosin proteins. Each of these will be discussed in turn.

Events at the Neuromuscular Junction. The signal for a muscle cell to contract is conveyed by a nerve cell in the form of an electrical impulse called an *action potential* (see Chapter 22). An action potential is carried from the neuron to the muscle cell by nerve axons. The site at which the nerve innervates, or makes contact with, the muscle cell is called the **neuromuscular junction.** At the neuromuscular junction, the axon branches out and forms *axon terminals* that make contact with the muscle cell. These terminals contain a transmitter chemical called *acetylcholine* that is stored in membrane-enclosed vesicles. Acetylcholine is secreted by axon terminals in response to an action potential. The area of the muscle cell plasma membrane under these terminals is called the *motor end plate*. Here, in the plasma membrane, called the **sarcolemma** in muscle cells, are clusters of acetylcholine receptors associated with each axon terminal. Acetylcholine receptors are members of a class of proteins known as *ligand-gated channels* (see Chapter 23). The term *ligand* simply means something that binds, and in the case of acetylcholine receptors, the ligand is acetylcholine. When a ligand-gated channel binds to its ligand, it opens a pore in the plasma membrane through which ions can flow. The acetylcholine receptor is a ligand-gated channel that lets sodium ions enter the muscle cell when acetylcholine is present. As we will see in the next chapter, this causes an electrical impulse to form in the membrane of the muscle cell.

Transmission of an Impulse to the Interior of the Muscle. Once an electrical impulse forms at the the motor end plate, it spreads out over the sarcolemma. This brings us to the next specialization, the **transverse (T) tubule system** (Figure 21-15). The sarcolemma differs from the plasma membrane of most cells in that it has a regular pattern of invaginations, which give rise to a series of tubes called **T tubules** that penetrate the interior of a muscle cell (like fingertips pressed into a balloon). The function of the T tubule is to carry an action potential into the muscle cell. The T tubule system is part of the reason that muscle cells can respond so

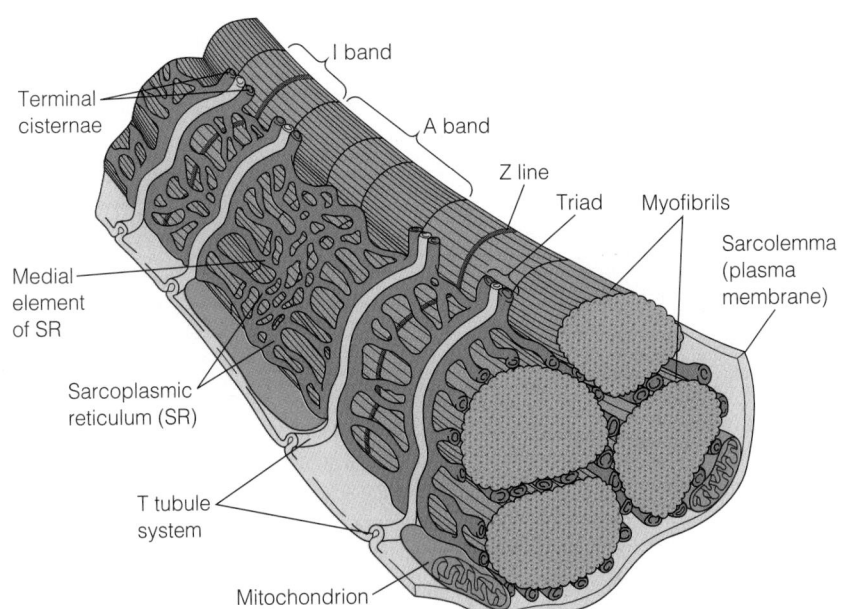

Figure 21-15 The Sarcoplasmic Reticulum and the Transverse Tubule System of Skeletal Muscle Cells. The sarcoplasmic reticulum (SR) is an extensive network of specialized ER that accumulates calcium ions and releases them in response to nerve signals. T tubules are invaginations of the sarcolemma (plasma membrane) that relay the contraction signal to the interior of the cell. Where the T tubule passes near the terminal cisternae of the SR, a triad structure is formed that appears in electron micrographs as the cross section of three adjacent tubes. The T tubule is in the middle, and on each side is one of the SR terminal cisternae. The triad contains the important junctional complex that regulates the release of calcium ions from the SR.

I band
Terminal cisternae
A band
Z line
Triad   Myofibrils
Sarcolemma (plasma membrane)
Medial element of SR
Sarcoplasmic reticulum (SR)
T tubule system
Mitochondrion

quickly to a nerve impulse. For many nonmuscle cells, the rate at which calcium concentrations change in the cytosol is limited by the rate that either calcium itself or other small molecules can diffuse from the plasma membrane into the interior of the cell. However, for large cells like those of skeletal and cardiac muscle, the time it takes for substances to diffuse from the plasma membrane into the interior would greatly limit the cell's rate of response. This limitation is overcome by the T tubule system, by which T tubules carry electrical signals rapidly into the interior of the cell.

Inside the muscle cell, the T tubule system comes into contact with a system of intracellular membranes in the form of flattened sacs or tubes called the **sarcoplasmic reticulum (SR).** As the name suggests, the SR is similar to the endoplasmic reticulum found in nonmuscle cells except that it is highly specialized for accumulating, storing, and releasing calcium ions. The SR runs along the myofibrils where it is poised to release calcium ions directly into the myofibril, causing contraction, and then remove calcium from the myofibril, causing relaxation. This close proximity of the SR to the myofibrils is another specialization that facilitates the rapid response of muscle cells to nerve signals.

SR Function in Calcium Release and Uptake. The SR can be functionally divided into two components, referred to as the **medial element** and the **terminal cisternae** (see Figure 21-15). The medial element of the SR contains a high concentration of ATP-dependent calcium pumps that continually pump calcium into the lumen of the SR. Evidence for the existence and function of the SR was first obtained in the early 1960s, when it was discovered that a particulate fraction obtained from homogenized muscle tissue was able to cause relaxation of specially treated muscle fibers. This *relaxing factor*, as it was then called, later proved to be membrane vesicles containing calcium pump molecules that were derived from the SR of muscle cells.

The ability of the SR to pump calcium ions is crucial for muscle relaxation, but it is also needed for muscle contraction. Calcium pumping produces a high calcium concentration in the lumen of the SR (up to several millimolar). This calcium can then be released from the terminal cisternae of the SR when needed. Figure 21-15 shows how the terminal cisternae of the SR are positioned adjacent to the contractile apparatus of each myofibril. Terminal cisternae are typically found right next to a T tubule, giving rise to a structure called a **triad.** In electron micrographs, a triad appears as three circles in a row. The central circle is the membrane of the T tubule, while the circles on each side are the membranes of the terminal cisternae. Close inspection of the triad reveals that the terminal cisternae appear to be connected to the T tubule by a dense-staining material between the two membranes. This material is referred to as the **junctional complex** (to which we will return shortly).

The proximity of the T tubule, the terminal cisternae of the SR, and the contractile machinery of the myofibril provides the basis for understanding why muscle cells can respond so rapidly to a nerve impulse. The action potential travels from the motor end plate, spreads out over the sarcolemma, and enters the T tubule (Figure 21-16). As the action potential travels down the T tubule, it passes near the terminal cisternae of the SR, causing a special type of calcium channel in the terminal cisternae to open. When this channel opens, calcium rushes into the sarcoplasm, causing contraction.

One of the mysteries of calcium regulation in muscle cells concerns the events that link an action potential traveling down the T tubule to the release of calcium from the terminal cisternae. Researchers have long suspected that the key to this mystery may lie in the junctional complex between the membrane of the T tubule and the terminal cisternae of the SR. While the events at the junctional complex are not fully understood, evidence suggests that the release of calcium from the SR is initiated by a process called

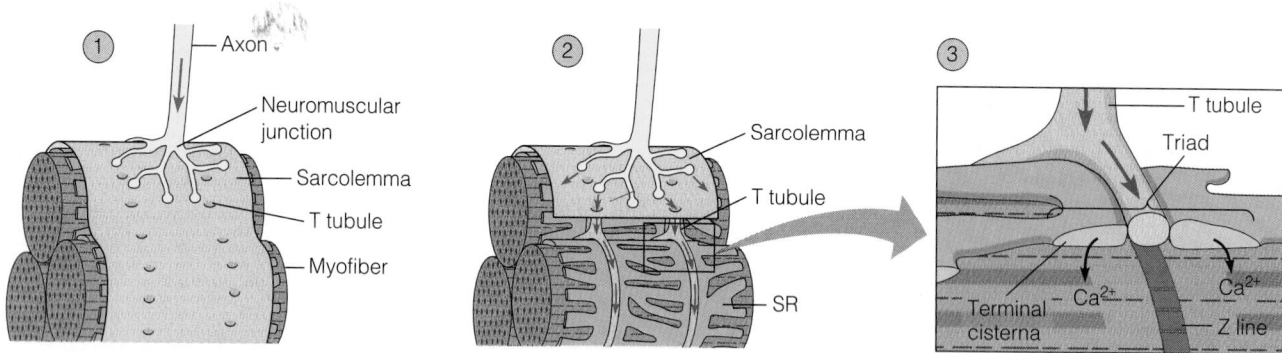

Figure 21-16 Stimulation of a Muscle Cell by a Nerve Impulse. ① An action potential moves down the axon of the neuron until it reaches the end. The ends of the axon branch out over the surface of the muscle cell at the neuromuscular junction to form synapses (contact points) between the neuron and the muscle cell. ② Depolarization of the terminals of the axon causes the release of neurotransmitter molecules, which bind to the acetylcholine receptors on the sarcolemma. Binding of neurotransmitter to the acetylcholine receptors starts an action potential in the muscle cell. As the action potential spreads over the surface of the muscle cell, it travels down into the T tubules. ③ T tubules carry the action potential into the interior of the muscle cell, where it stimulates calcium release from the terminal cisternae of the SR.

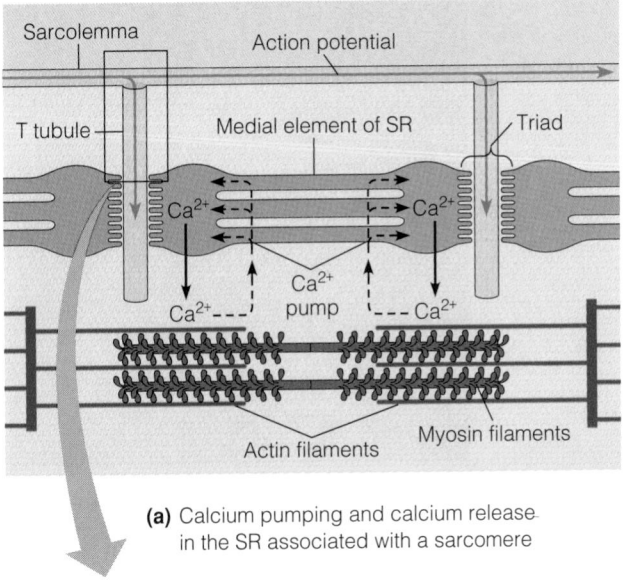

(a) Calcium pumping and calcium release in the SR associated with a sarcomere

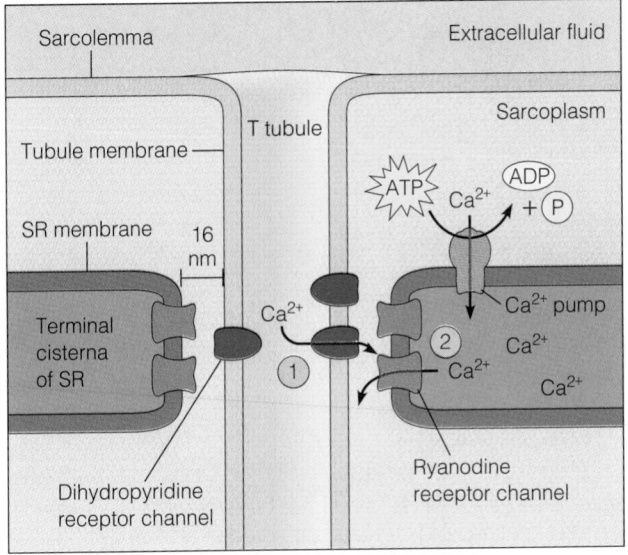

(b) A calcium-dependent calcium channel in a junctional complex

Figure 21-17 The Calcium-Induced Calcium Release Model. **(a)** The SR performs both calcium pumping and calcium release. Calcium pumping lowers the calcium concentration in the sarcomere and causes the muscle cell to relax. Calcium release increases the calcium concentration in the sarcomere, causing the muscle cell to contract. **(b)** The SR releases calcium through a protein called the ryanodine receptor channel. This channel is anchored in the membrane of the terminal cisternae of the SR and protrudes from the membrane toward the T tubule. Whether the channel is open or closed depends on a protein in the T tubule called the dihydropyridine receptor channel, which opens when a depolarization wave travels down into the T tubule. In this model, ① a small amount of calcium moves from the T tubule lumen through the dihydropyridine receptor channel when the T tubule is depolarized. ② This calcium immediately acts on the ryanodine receptor channel, causing it to release calcium from the SR into the sarcoplasm.

calcium-induced calcium release (Figure 21-17). There are two main proteins in the junctional complex, the **dihydropyridine receptor** channel in the T tubule membrane and the **ryanodine receptor** channel in the SR membrane. The dihydropyridine receptor may be similar to a voltage-gated calcium channel, which opens when a membrane is depolarized. In this case, a depolarization traveling down a T tubule would cause this channel to open, letting a small amount of calcium pass from the T tubule into the muscle cell. Calcium passing through the channel would immediately encounter the ryanodine receptor, which acts as a calcium-sensitive channel in the SR. When calcium contacts the ryanodine receptor, the channel opens to let much larger amounts of calcium out of the SR. This causes the calcium concentration in the sarcoplasm of the muscle cell to rise at least an order of magnitude, from 10^{-4} mM to 10^{-3} mM. An alternative model suggests that conformational changes in the dihydropyridine receptor mechanically open the ryanodine receptor channel.

Letting calcium out of the SR causes a muscle cell to contract, but the cell must also relax in order to be ready for another contraction. For the muscle cell to relax, calcium levels must be brought back down to the resting level. This is accomplished by pumping calcium back into the SR. The membrane of the SR contains an active transport protein called a **calcium pump** or **ATPase,** which can pump calcium ions from the sarcoplasm into the cisternae of the SR. These pumps are concentrated in the medial element of the SR. The calcium pump can be seen clearly on freeze-fractured SR membranes (Figure 21-18). In fact, this was the first freeze-fractured particle to be linked to a specific membrane function. The calcium pump from mammalian muscle tissue is a single polypeptide chain with a molecular weight of about 115,000 and a known amino acid sequence. From the sequence and biochemical results, David MacLennan and Michael Green proposed a three-dimensional structure for the protein that suggests a possible pump mechanism, representing the first ATP-driven pump to be analyzed in such detail.

The mechanism is similar to that discussed in Chapter 8 for the sodium-potassium pump. ATP-Mg^{2+} is normally bound to the nucleotide-binding portion of the enzyme, but the ATP is not hydrolyzed in the absence of calcium. The calcium pump has two conformations: E1, which has a high affinity for calcium, and E2, with a low affinity. When a sufficient amount of calcium is present—following a muscle cell contraction, for instance—two calcium ions bind to the high-affinity sites on the ATPase and activate the pump cycle shown in Figure 21-19. A phosphate group is transferred from the bound ATP to the enzyme, shown as E1-P in the figure. Transfer of the phosphate causes a conformational change such that the calcium-binding site simultaneously becomes a low-affinity site (E2-P) and exposes bound calcium to the interior of the SR, releasing the calcium ions. The bound phosphate is then hydrolyzed, causing the ATPase to cycle back to the high-affinity E1 form.

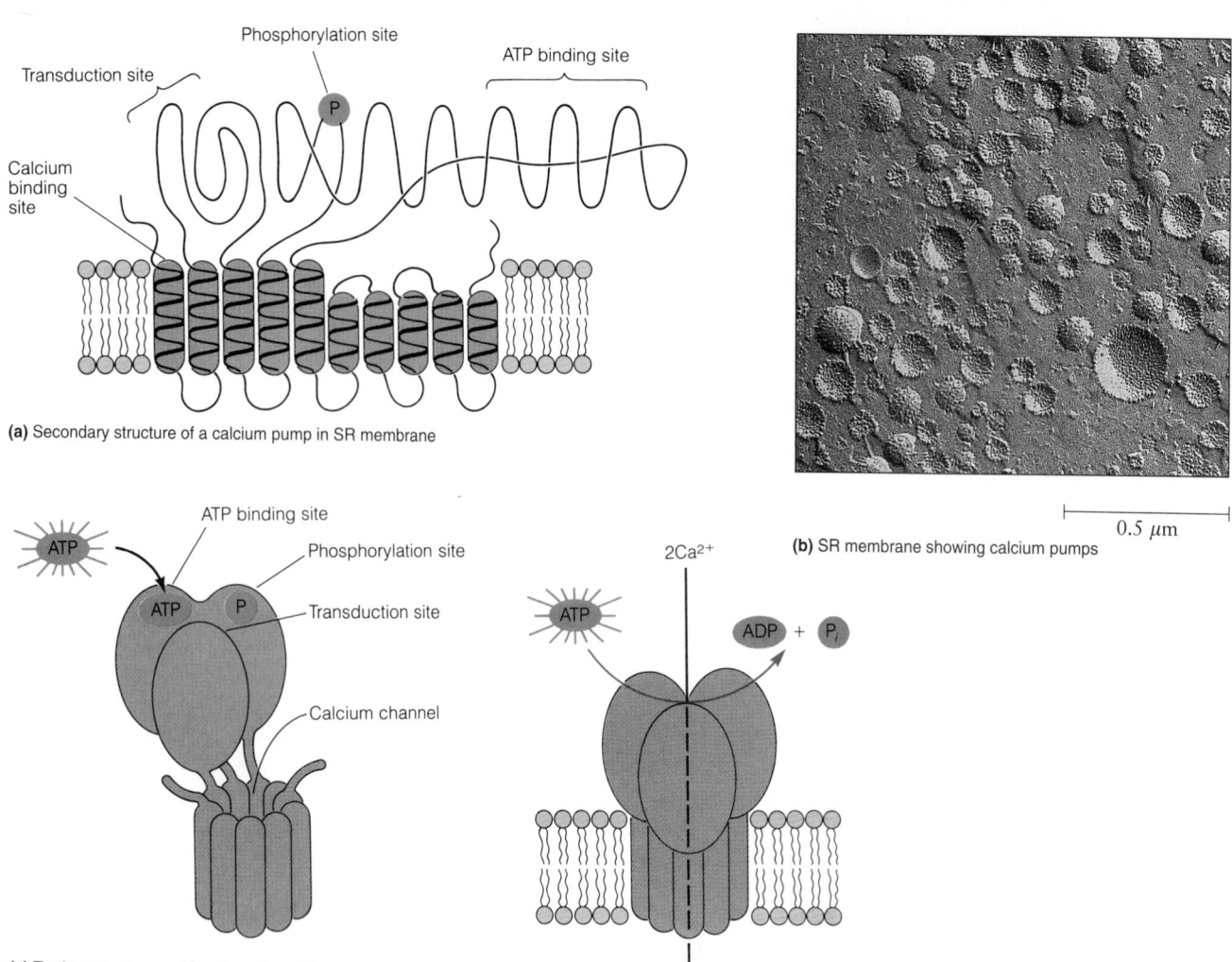

(a) Secondary structure of a calcium pump in SR membrane

(b) SR membrane showing calcium pumps

0.5 μm

(c) Tertiary structure and function of a calcium pump

Figure 21-18   Structure of the Calcium Pump Protein.
The calcium pump protein is a single polypeptide chain containing 1001 amino acids. **(a)** A partially unfolded chain indicates the main features. Ten α helices are embedded in the lipid bilayer, and helices cluster to form a transmembrane calcium channel. **(b)** The pump enzyme, a Ca²⁺ ATPase, can be seen as small particles in the membranes of SR visualized by the freeze-fracture technique (TEM). **(c)** The head group of the calcium pump has three significant features, including an ATP-binding site, a phosphorylation site, and a transduction site, where the energy of ATP hydrolysis is made available for the pumping of calcium ions. When calcium is present, the ATPase activity is activated, and two calcium ions are transported inward for every ATP hydrolyzed.

The continued pumping of calcium from the sarcoplasm back into the SR cisternae quickly lowers the sarcoplasmic calcium level to the point at which troponin releases calcium, tropomyosin again moves back to the blocking position on actin sites, and further cross-bridge formation is prevented. Cross-bridges therefore disappear rapidly as actin dissociates from myosin and becomes blocked by tropomyosin. This leaves the muscle relaxed and free to be reextended, because the absence of cross-bridge contacts allows the thin filaments to slide out from between the thick filaments. In some cases, extension is caused by the contraction of an opposing muscle, and in other cases, by the tension of elastic connective tissue.

Meeting the Energy Needs of Muscle Contraction

Because muscle contraction involves the hydrolysis of an ATP molecule for every cycle of attachment and detachment, muscle cells need ways to regenerate ATP continuously. Muscle cells actually have a variety of mechanisms to ensure maximum ATP availability, even during prolonged periods of intense activity.

ATP Generation: The Aerobic Mode.   The ATP needs of muscle cells are normally met either by glycolysis or by mitochondrial respiration. The extent to which one or the other of these pathways is favored depends on the kind of

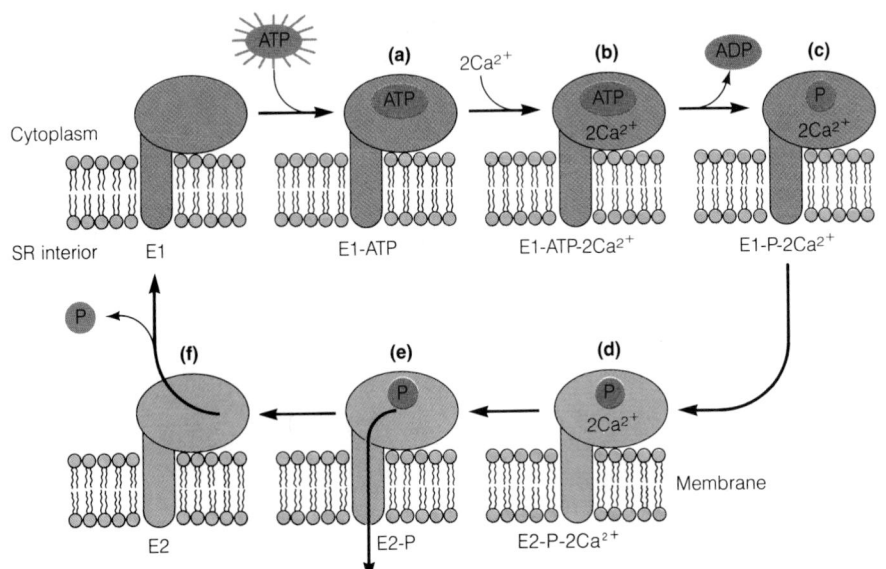

Figure 21-19 A Proposed Mechanism for Active Calcium Ion Transport. The calcium pump has two conformations: E1, high affinity for calcium, and E2, low affinity for calcium (E for enzyme). **(a), (b)** Two calcium ions and one ATP bind to the high-affinity form of the protein on the sarcoplasmic side of the SR membrane, followed by **(c)** phosphorylation as ATP donates its phosphate to an aspartic acid in the protein. **(d)** The protein shifts to its low-affinity form, exposing the bound calcium on the SR interior. **(e)** The calcium ions dissociate, **(f)** the phosphate group is hydrolyzed, and the protein returns to its high-affinity state for the next round of calcium transport.

muscle involved and whether it is functioning under *aerobic* or *hypoxic* (low-oxygen) conditions. Skeletal muscles that are characterized by frequent use and high activity usually rely on complete respiratory metabolism. The flight muscles of birds are a good example. Such muscles draw both glucose and oxygen from the circulatory system, oxidizing the glucose completely to carbon dioxide and water and generating ATP by oxidative phosphorylation in the mitochondrion. These muscles are characterized by an abundance of mitochondria and by a red color. Mitochondria occur in close association with the myofibrils in almost all aerobic muscle cells (see Figure 4-13). The red color of such tissue is due to the high degree of vascularization, to the cytochromes and iron-sulfur proteins present in the mitochondria, and to the presence of **myoglobin**, a protein related in structure and function to hemoglobin but localized in muscle cells and used to bind and store oxygen.

ATP Generation: The Hypoxic Mode. During periods of intense exercise, the demand for ATP regeneration may exceed the rate at which oxygen can be supplied to the tissue by the circulatory system. After depletion of the reserve oxygen available from myoglobin, the tissue begins to function hypoxically, converting glucose to lactate. Because the ATP yield of glucose is greatly reduced under these conditions (from 36 to 2 molecules of ATP per molecule of glucose; see Chapter 11), much more glucose is required per unit time. The extra glucose is supplied by catabolism of *glycogen,* the storage carbohydrate of muscle cells. The lactate formed under hypoxic conditions is usually released into the blood and eventually reaches the liver, where it is either oxidized fully to carbon dioxide and water or converted back into glucose (gluconeogenesis; see Chapter 11). A cycle is therefore established, with glucose moving from the liver via the blood to the hypoxic muscle and lactate returning from the muscle to the liver. This cyclic process, called the *Cori cycle,* is described in Box 11A.

Intense muscular activity cannot be sustained long under hypoxic conditions, because of the rapid depletion of glycogen stores and the accumulation of lactate. However, this option is useful for short bursts of activity when oxygen cannot be supplied fast enough.

The Role of Creatine Phosphate. ATP clearly serves as the immediate source of energy to drive muscle contraction and is also the form in which energy is conserved during glycolysis and respiratory metabolism. However, muscle cells do not use ATP as their major form of stored energy. A surprising observation made early in muscle research was that the ATP content of working muscle remains remarkably constant until the muscle is near exhaustion. What decreases instead during prolonged exertion is the cellular level of **creatine phosphate,** a high-energy compound whose structure is shown in Figure 21-20.

Creatine phosphate represents a reservoir of high-energy phosphate that can be used to recharge ADP, as shown in the reaction of Figure 21-20. The enzyme that catalyzes this reaction is called *creatine kinase.* The hydrolysis of creatine phosphate is even more exergonic than that of ATP ($\Delta G^{\circ\prime} = -10.3$ versus -7.3 kcal/mol), which ensures that the equilibrium for the creatine kinase reaction lies far to the right. The recharging of ADP is therefore driven effectively, maintaining a high ATP/ADP ratio in the muscle cell, even when the creatine phosphate level drops considerably.

The Role of Myokinase. As a final backup system, muscle cells also contain an enzyme called *myokinase,* which is capable of phosphorylating one ADP molecule at the expense of another, as follows:

$$2ADP \longrightarrow ATP + AMP \qquad (21\text{-}1)$$

The myokinase reaction provides a means of extracting energy from the remaining acid anhydride bond of ADP.

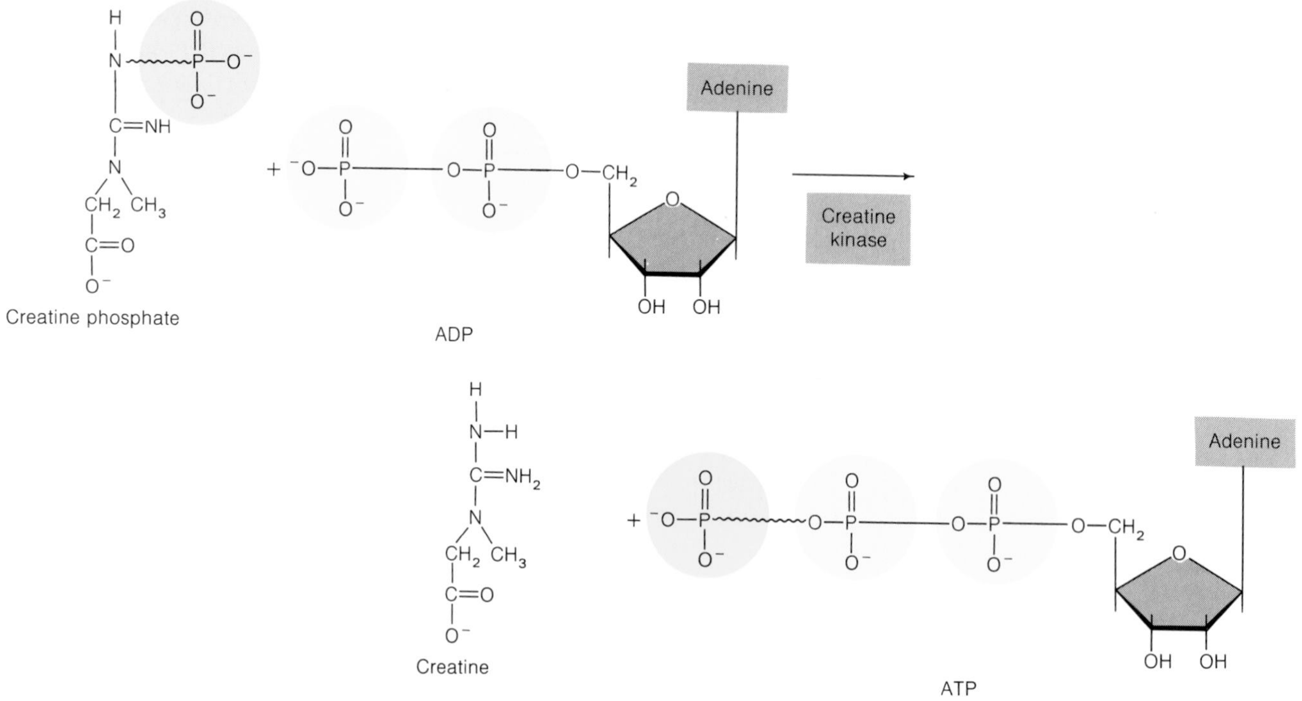

Figure 21-20 The Structure of Creatine Phosphate and the Creatine Kinase Reaction. The transfer of the high-energy phosphate group from creatine phosphate to ADP is thermodynamically favorable, because the $\Delta G^{\circ\prime}$ value for the hydrolysis of creatine phosphate (-10.3 kcal/mol) is substantially more negative than that for the hydrolysis of ATP (-7.3 kcal/mol).

Meeting Energy Needs: A Summary. The variety of mechanisms available to skeletal muscle cells to meet their energy needs for continued contraction under a wide variety of conditions can be summarized as follows:

1. The primary energy storage form in muscle cells is creatine phosphate, and the energetics of phosphate transfer from creatine phosphate to ADP are sufficiently favorable to maintain a high ATP/ADP ratio despite the considerable depletion of creatine phosphate that may occur during muscle activity.

2. With adequate oxygen and glucose available from the circulatory system, complete respiratory metabolism is possible, and the muscle functions aerobically.

3. As a safeguard, oxygen can be stored within the cell as a complex with myoglobin and can be released to prolong respiratory metabolism for a short time, even when the circulatory system is unable to supply oxygen at an adequate rate.

4. When oxygen supplies become limited, the muscle cell can still function hypoxically to meet its needs by lactate formation.

5. If the supply of blood glucose becomes limited, stored glycogen can be used as a source of glucose to continue glycolysis.

6. If all else fails and ADP begins to accumulate, the myokinase reaction provides a backup means of regenerating ATP.

The overall picture that emerges of the skeletal muscle cell is one of almost incredible specialization for its role in contraction—specialization in design of the contractile elements as well as in the mechanism available to ensure that the ATP needed for contraction can be supplied under virtually any condition.

Cardiac Muscle

Cardiac (heart) muscle is responsible for the beating of the heart and the pumping of blood through the body's circulatory system. Cardiac muscle functions continuously; in one year, your heart beats about 40 million times! Cardiac muscle is very similar to skeletal muscle in organization of actin and myosin filaments and has the same striated appearance (Figure 21-21). However, the two kinds of muscle differ significantly in their metabolism. Cardiac muscle is highly dependent on aerobic respiration; only in emergencies is glycogen used as an energy source. Accordingly, cardiac muscle cells have much less glycogen and many more mitochondria than are usually found in skeletal muscle cells. In fact, mitochondria represent about 30–40% of the total volume of heart muscle.

Most of the energy required for the beating of the heart under resting conditions is provided not by blood glucose, but by free fatty acids that are transported from adipose (fat storage) tissue to the heart by serum albumin, a blood protein. These fatty acids are degraded by β oxidation, and the resulting acetyl coenzyme A is oxidized via the TCA cycle (see Figure 12-7). When a heavy workload is suddenly imposed on the heart, the consumption of blood glucose and muscle glycogen increases significantly. Shortly thereafter,

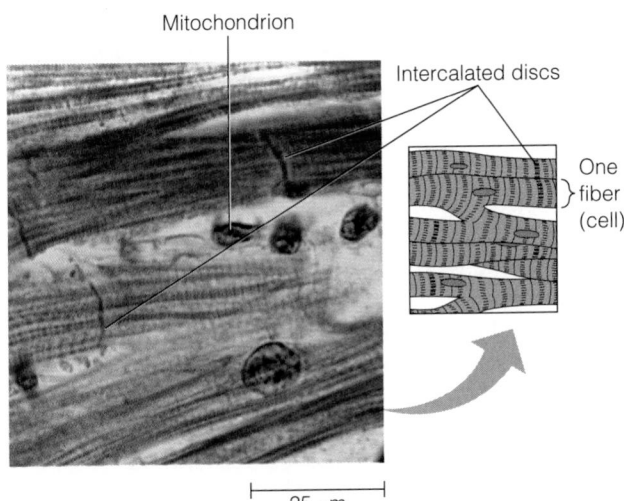

Figure 21-21 **Cardiac Muscle Cells.** Cardiac muscle cells have a contractile mechanism and sarcomeric structure similar to those of skeletal muscle cells. However, unlike skeletal muscle cells, cardiac muscle cells are joined together end-to-end at the intercalated discs, which allow ions and electrical signals to pass from one cell to the next. This ionic permeability enables a contraction stimulus to spread evenly to all the cells of the heart. In addition, cardiac muscle cells exhibit branches that are not seen in skeletal or smooth muscle cells (LM).

however, fatty acid oxidation is greatly accelerated, and the dependence on glucose utilization again becomes minimal.

A second difference between cardiac and skeletal muscle is that heart muscle cells are not multinucleate. Instead, cells are joined end-to-end through structures called **intercalated discs** (see Figure 21-21). The discs have gap junctions that electrically couple neighboring cells, providing a way for depolarization waves to spread throughout the heart during its contraction cycle. The reason for this arrangement is that the heart is not activated by nerve impulses, as skeletal muscle is, but contracts spontaneously once every second or so. The heart rate is controlled by a pacemaker region in an upper portion of the heart (right atrium), which contracts 70–80 times per minute, slightly faster than other heart tissues. The depolarization wave initiated by the pacemaker then spreads to the rest of the heart to produce the heartbeat.

Smooth Muscle

Smooth muscle is responsible for involuntary contractions such as those of the stomach, intestines, uterus, and blood vessels. In general, such contractions are slow, taking up to five seconds to reach maximum tension. Smooth muscle contractions are also of greater duration than those of skeletal or cardiac muscle. While smooth muscle is not able to contract rapidly, it is well adapted to maintain tension for long periods of time, as is required in these organs and tissues.

Smooth muscle cells are long and thin, with pointed ends. Unlike skeletal or heart muscle, smooth muscle has no striations (Figure 21-22a). Thick and thin filaments are both present, but they are not organized into myofibrils and are not regularly aligned. Nor is the number of thick filaments always constant. The thin filaments of smooth muscle cells contain actin and tropomyosin, but no troponin.

Smooth muscle cells do not contain Z lines, which are responsible for the periodic organization of the sarcomeres found in skeletal and cardiac muscle cells. Instead, smooth muscle cells contain **dense bodies,** plaquelike structures in the myoplasm and on the cell membrane (Figure 21-22b). Bundles of actin filaments are anchored at their ends to

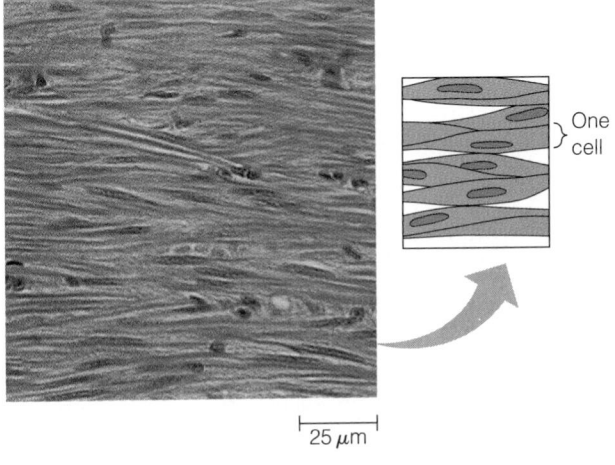

(a) Smooth muscle cells

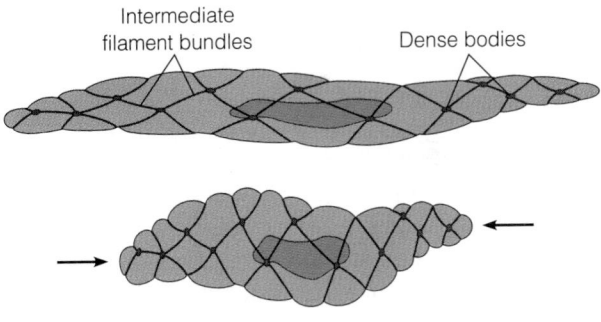

(b) Contraction of smooth muscle cell

Figure 21-22 **Smooth Muscle and Its Contraction.**
(a) Individual smooth muscle cells are long and spindle-shaped, with no Z lines or sarcomeric structure (LM). **(b)** In the smooth muscle cell, contractile bundles of actin and myosin appear to be anchored at one end of the bundle to a plaquelike structure called a dense body on the plasma membrane and at the other end to a dense body in the sarcoplasm. The dense bodies are connected to each other by intermediate filaments, thereby orienting the actin and myosin bundles obliquely to the long axis of the cell. When the actin and myosin bundles contract, they pull on the dense bodies and intermediate filaments, producing the cellular contraction shown here.

these dense bodies. As a result, actin filaments appear in a crisscross pattern, aligned obliquely to the long axis of the cell.

The organization of myosin molecules in the thick filaments of smooth muscle also differs from that of skeletal muscle. Instead of the bare central zone characteristic of thick filaments from skeletal muscle, smooth muscle thick filaments have myosin heads distributed along the entire length of the filament. Cross-bridges connect thick and thin filaments in smooth muscle, but not in the regular, repeating pattern seen in skeletal muscle.

As we will see, the same stimulus that causes a smooth muscle cell to contract can also stimulate the assembly of myosin into thick filaments. In this and other respects, smooth muscle cells resemble nonmuscle cells more closely than they do skeletal muscle cells.

Regulation of Contraction in Smooth Muscle and Nonmuscle Cells. Smooth muscle cell contraction and nonmuscle cell contraction are regulated in a manner distinct from that of skeletal muscle cells. Although skeletal and smooth muscle cells are both stimulated to contract by an increase in the sarcoplasmic concentration of calcium ions, the mechanisms are quite different. When sarcoplasmic calcium concentrations increase in smooth muscle and nonmuscle cells, a cascade of events takes place that includes the activation of **myosin light-chain kinase (MLCK).** Active MLCK then phosphorylates one type of myosin light chain known as a *regulatory light chain.*

Myosin light-chain phosphorylation affects myosin in two ways. First, some myosin molecules are curled up so that they cannot assemble into filaments. When the myosin light chain is phosphorylated, the myosin tail uncurls and becomes capable of assembly. Second, the phosphorylation of the light chains activates myosin, enabling it to interact with actin filaments to undergo the cross-bridge cycle.

The cascade of events involved in the activation of smooth muscle and nonmuscle myosin is shown in Figure 21-23. In response to a nerve impulse or hormonal signal reaching the smooth muscle cell, an influx of extracellular calcium ions occurs, increasing the intracellular calcium concentration and causing contraction. The effect of the increased calcium concentration on contraction is mediated by the binding of calcium to a small cytosolic protein called **calmodulin.** This forms the **calcium-calmodulin complex,** which, unlike free calmodulin, is able to bind to myosin light-chain kinase, thereby activating the enzyme. As a result, myosin light chains become phosphorylated and are therefore able to interact with actin to cause contractions. (Other intracellular effects of the calcium-calmodulin complex are described in Chapter 23.)

The regulation of MLCK by calcium and calmodulin illustrates a common theme in the regulation of protein kinases. MLCK contains a peptide sequence at one end called a pseudosubstrate (Figure 21-23a). A *pseudosubstrate* is a sequence of amino acids similar to the enzyme's normal substrate. In this case, the real substrate is a sequence of amino acids on the myosin light chain that is recognized by MLCK. Within this substrate's sequence of amino acids is the serine that actually accepts the phosphate from ATP. The sequence of the pseudosubstrate differs from that of this real substrate by the replacement of this serine with a different amino acid that cannot accept a phosphate group.

The pseudosubstrate region of MLCK regulates the activity of this enzyme. The pseudosubstrate region can fold around and bind to the active site of the enzyme and thus inhibit its activity. Inhibition of enzyme activity in this manner is referred to as *autoinhibition,* because in effect the enzyme inhibits itself. However, there is also a binding site for calcium-calmodulin near the pseudosubstrate site. The pseudosubstrate region cannot bind to the active site of the enzyme at the same time that calcium-calmodulin is also bound. Therefore, when calcium-calmodulin is available to bind, the pseudosubstrate region is prevented from inhibiting the enzyme, and the enzyme becomes active.

The active MLCK can then phosphorylate myosin. Before phosphorylation, smooth muscle or nonmuscle myosin may be simply inactive, or may be both inactive and in the curled form (Figure 21-23b), preventing assembly, as discussed previously. When the myosin light chain is phosphorylated, myosin in both forms becomes active and able to interact with actin, and the tail of myosin in the curled form straightens out and can assemble with other myosin molecules into filaments.

In addition to phosphorylation by MLCK, myosin-actin interaction in both smooth muscle and nonmuscle cells can be regulated by a protein called *caldesmon.* In the absence of the calcium-calmodulin complex, caldesmon binds to actin filaments and prevents interaction between myosin and actin. When calcium-calmodulin binds to caldesmon, or when caldesmon is phosphorylated, caldesmon detaches from the actin filament, thereby allowing myosin to attach. Caldesmon therefore appears to add a second level of calcium regulation to the interaction between myosin and actin.

Thus, both skeletal muscle and smooth muscle are activated to contract by calcium ions, but from different sources and by different mechanisms. In skeletal muscle, the calcium comes from the sarcoplasmic reticulum. Its effect on actin-myosin interaction is mediated by troponin and is very rapid, because it depends on conformational changes only. In smooth muscle, the calcium comes from outside the cell, and its effect is mediated by calmodulin. The effect is much slower in this case, because it involves a covalent modification of the myosin molecule. The two kinds of muscle also differ in the signals that initiate calcium release from the SR. In skeletal muscle, the signal comes from voluntary nerves, whereas smooth muscle is activated by nerve signals of the autonomic (involuntary) component of the nervous system or by hormonal stimuli.

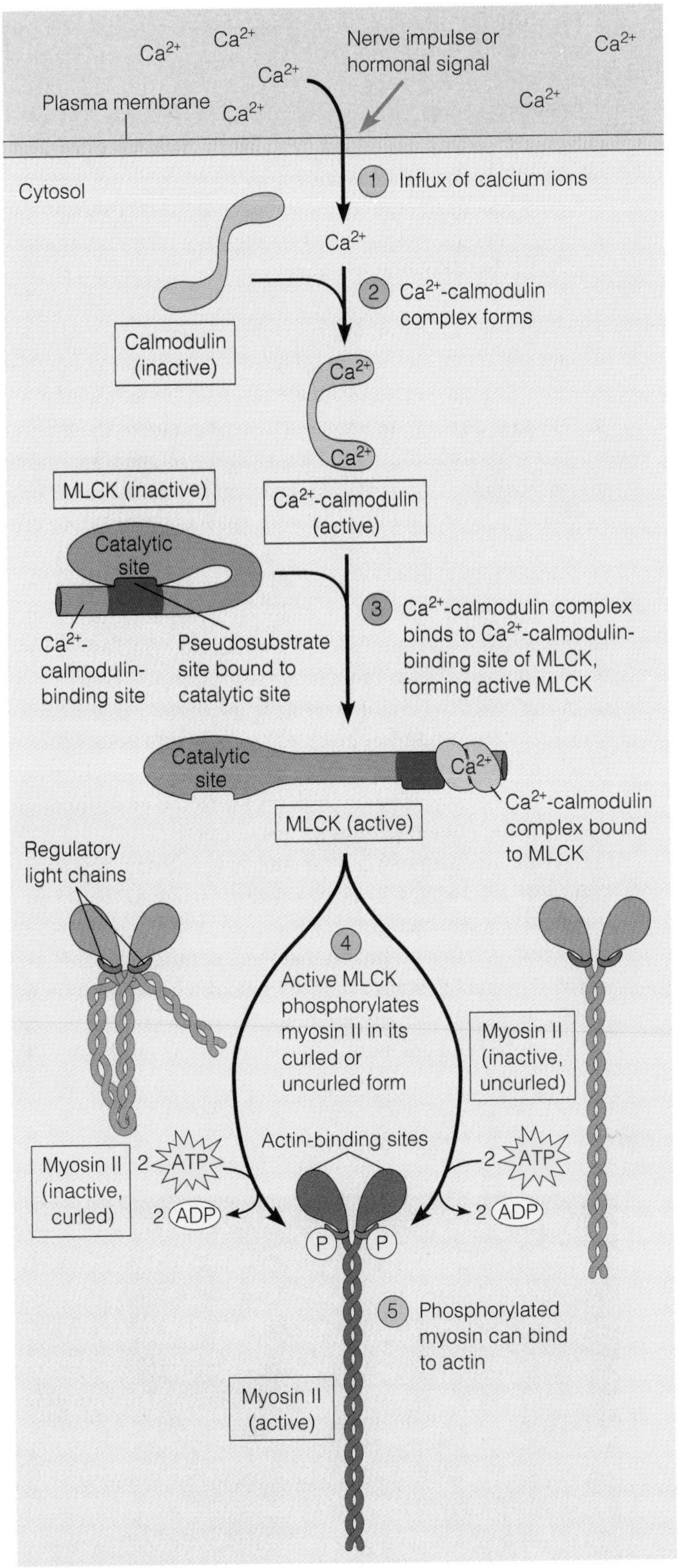

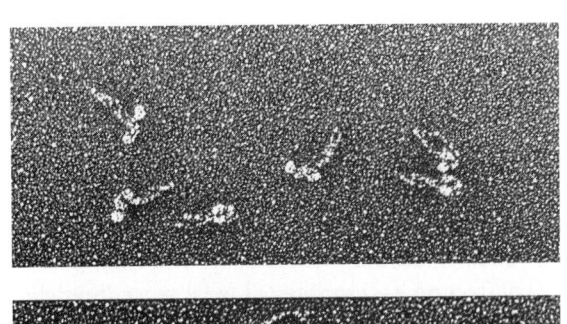

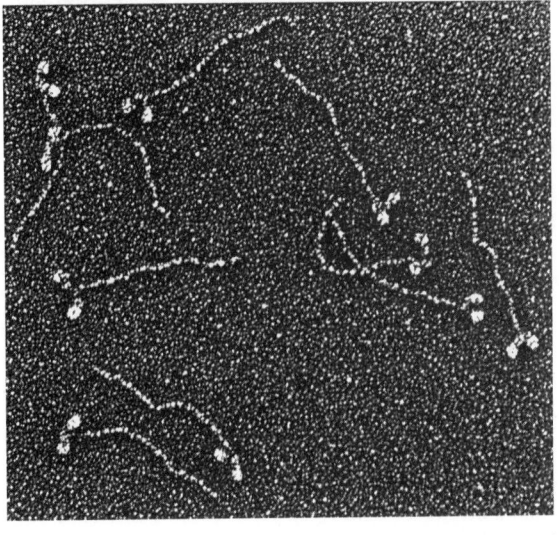

(b) Curled and uncurled myosin II molecules

The figure on the left shows the following labeled pathway:

Nerve impulse or hormonal signal

Ca^{2+} Ca^{2+} Ca^{2+} Ca^{2+}
Plasma membrane Ca^{2+} Ca^{2+}

Cytosol

① Influx of calcium ions

Ca^{2+}

Calmodulin (inactive)

② Ca^{2+}-calmodulin complex forms

Ca^{2+}

Ca^{2+}

Ca^{2+}-calmodulin (active)

MLCK (inactive)

Catalytic site

Ca^{2+}-calmodulin-binding site

Pseudosubstrate site bound to catalytic site

③ Ca^{2+}-calmodulin complex binds to Ca^{2+}-calmodulin-binding site of MLCK, forming active MLCK

Catalytic site

Ca^{2+}

MLCK (active)

Ca^{2+}-calmodulin complex bound to MLCK

Regulatory light chains

④ Active MLCK phosphorylates myosin II in its curled or uncurled form

Myosin II (inactive, uncurled)

Myosin II (inactive, curled)

Actin-binding sites

2 ATP → 2 ADP

2 ATP → 2 ADP

P   P

⑤ Phosphorylated myosin can bind to actin

Myosin II (active)

(a) Phosphorylation of myosin II by myosin light-chain kinase (MLCK)

Figure 21-23   Phosphorylation of Smooth Muscle and Nonmuscle Myosin. **(a)** The functions of both smooth muscle and nonmuscle myosin II are regulated by phosphorylation of the regulatory light chains. ① An influx of calcium ions into the cell is triggered by a nerve impulse or a hormonal signal. ② When present at a sufficiently high concentration, calcium ions bind to calmodulin, forming an active calcium-calmodulin complex. ③ The calcium-calmodulin complex in turn binds to a region of myosin light-chain kinase (MLCK) that overlaps the pseudosubstrate site. When this happens, the pseudosubstrate stretch of amino acids is pulled away and prevented from binding to the active site of MLCK, thus activating MLCK. ④ Activated MLCK phosphorylates the myosin light chains, whether the myosin is curled or uncurled. ⑤ The activated (and uncurled) myosin can then bind to actin and undergo the cross-bridge cycle. **(b)** Electron micrographs of curled and uncurled myosin II molecules. (TEMs)

Specialized Filament-Based Motility Systems

Besides their role in muscle contraction, interactions between myosin and actin microfilaments are also important in other forms of cell motility. The mechanisms are not as well understood, but the lessons learned from skeletal muscle contraction have turned out to be generally applicable in approaching topics as diverse as amoeboid movement, cytoplasmic streaming, and cell division.

Amoeboid Movement

Amoebas, slime molds, leukocytes, and fibroblasts involved in wound healing are all capable of a crawling type of movement generally referred to as **amoeboid movement** (Figure 21-24). In many cases, this type of movement is accompanied by protrusions of the cytoplasm called **pseudopodia** (singular: **pseudopodium**, from the Greek for "false foot"). Pseudopodia is a general term for a variety of cell protrusions. Some types of protrusions, such as those in amoebas, are large and thick and are known as *lobopodia*. Another form of protrusion is the extension of thin sheets of cytoplasm called *lamellipodia*. And some protrusions are thin and pointed structures known as *filopodia*.

Actin appears to be involved in all types of cell protrusions. As mentioned in Chapter 20, actin can be organized in a variety of different states, depending on the presence of different actin-binding proteins. In cells that move, actin is commonly found in three different structures. Stretching from the tail, or trailing edge, of the cell to the front are *stress fibers,* which are *contractile bundles* of actin (Figure 21-25a). In the cell cortex, F-actin is crosslinked into a *gel* (Figure 21-

25b). At the front, or leading edge, and especially in filopodia, F-actin is found in *parallel bundles*. Each actin filament structure appears to be regulated and coordinated during cell crawling. Such changes in the form of actin may contribute to the formation of a protrusion. Movement of a cell over a substrate also depends on its attachment to that surface and on contractile forces acting on its cell protrusions. These events are summarized in Figure 21-26 and are discussed in the following sections.

The Formation of Cell Protrusions. Cell protrusions are prominent features of cell crawling. The mechanism by which cells form protrusions is not yet fully understood. In fact, several different processes may contribute to the formation of a protrusion, including *gelation-solation* and *actin polymerization*. Different cells types may rely more on one mechanism than the other, or cells may use more than one mechanism at the same time.

Cells that undergo amoeboid movement have an outer, or *cortical,* layer of thick, gelatinous cytoplasm called the **ectoplasm** and an inner layer of more fluid cytoplasm called the **endoplasm.** In an amoeba, as a pseudopod is extended from the cell, fluid endoplasm streams forward in the direction of extension and appears to congeal into ectoplasm at the tip of the pseudopod. Meanwhile, at the rear of the moving cell, ectoplasm appears to change into the more fluid endoplasm and stream toward the pseudopod.

The cytoplasm of a cell can alternate between gelatinous and more fluid states. These transitions relate to the state of the actin meshworks in the cytoplasm. When criss-crossed filaments of actin are linked together by other proteins, the result is a **gel.** Proteins such as **gelsolin** that are present within these gels are activated by calcium to convert the gel to the more fluid **sol** state. Alternation between these states

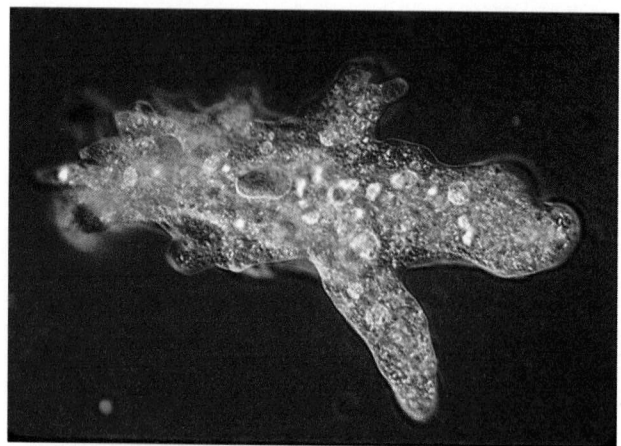

(a) *Amoeba*   50 μm

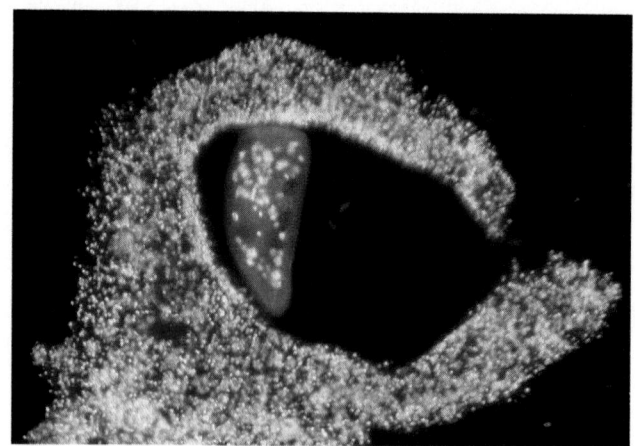

(b) *Amoeba* engulfing prey with pseudopodia   50 μm

Figure 21-24   Amoeboid Movement.   (a) A micrograph of *Amoeba proteus,* a protozoan that moves by extension of pseudopodia. **(b)** This micrograph shows an *Amoeba* cell using its pseudopodia to

engulf a smaller ciliated cell on which *Amoeba* feeds by phagocytosis (LMs).

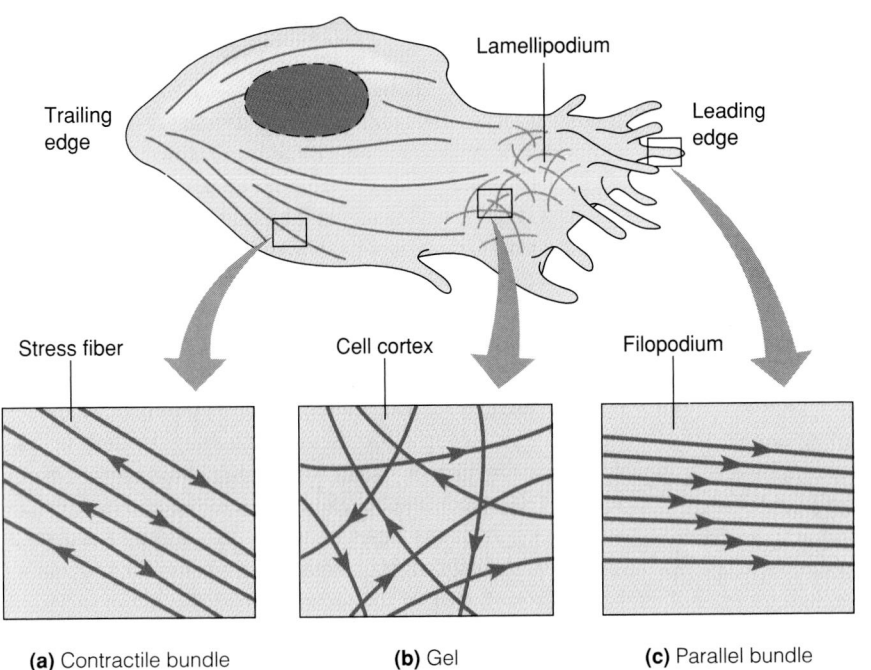

Figure 21-25 The Architecture of Actin in Crawling Cells. Actin is found in a variety of different structures in crawling cells such as this macrophage. **(a)** Running from the trailing edge of the cell to the leading edge are contractile bundles of actin, the stress fibers. **(b)** At the periphery of the cell is the cortex, which contains a three-dimensional meshwork of actin filaments crosslinked into a gel. **(c)** The broad leading edge of the lamellipodia can produce thin, fingerlike projections called filopodia. While the bulk of lamellipodia contain an actin meshwork, filopodia contain parallel bundles of actin filaments.

is called **gelation-solation** (Figure 21-27). The local conversion of cortical actin from a gel to a sol state can result in cell protrusions. Such transitions can weaken a region of the cortex, resulting in the cell bulging out at the site of weakening. Pressure exerted on the endoplasm, possibly due to contraction of actin filaments in the trailing edge of the cell, squeezes the endoplasm forward, forming a protrusion at the site of the bulge.

As important as gelation-solation and internal pressures are, they do not fully account for protrusion forma-

tion. Other factors such as the polymerization of actin are also important. The drugs cytochalasin B and phalloidin both inhibit cell crawling. This inhibition suggests that cell crawling depends not only on gelation but also on the ability of actin to polymerize and depolymerize. Actin is known to polymerize in conjunction with the protrusion of a pseudopod, and polymerization of actin may even cause the protrusion. For example, when actin is encapsulated inside lipid vesicles and then induced to polymerize, the vesicles become distorted and develop protrusions. Furthermore,

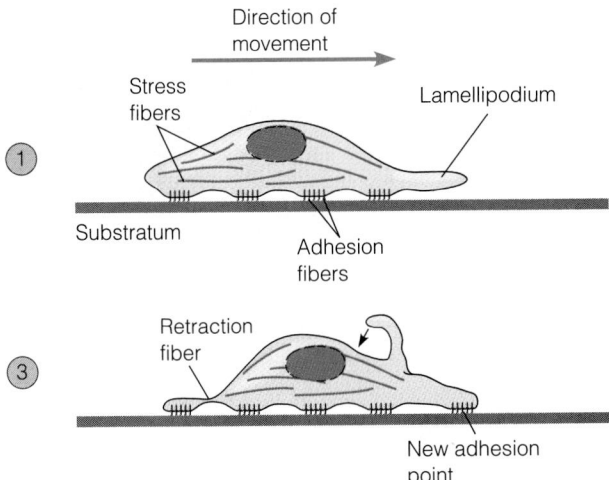

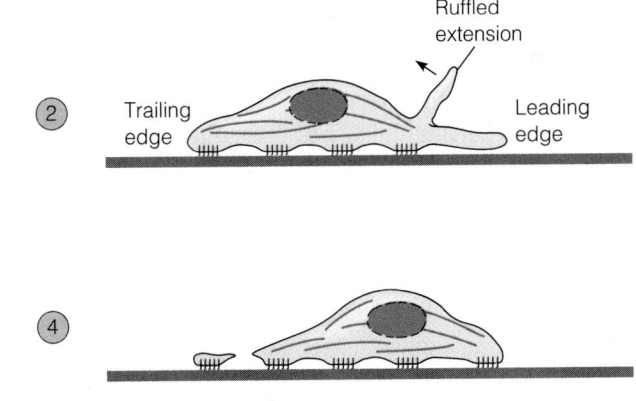

Figure 21-26 The Steps of Cell Crawling. Several different processes are involved in cell crawling, including cell protrusion, attachment, and contractile activities. Illustrated schematically here in a macrophage, ① the leading edge of the cell forms protrusions that extend in the direction of travel. ② These protrusions can adhere to the sub- strate, or they can be pulled upward and back toward the cell body. ③ When the protrusions adhere, they provide anchorage points for actin filaments. ④ Tension on the actin filaments can then cause the rest of the cell to pull forward.

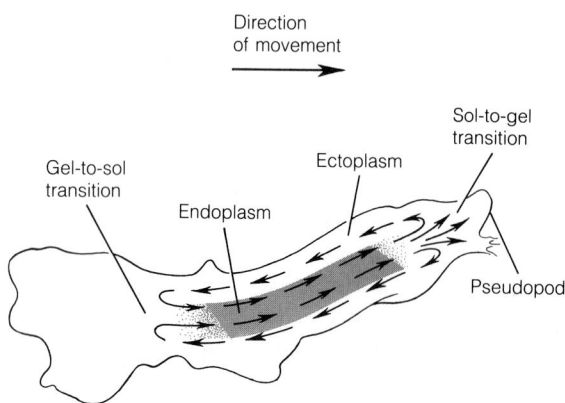

Figure 21-27 Membrane Protrusions and Cell Movement Due to Gelation-Solation. The outer boundary (ectoplasm) of amoeboid cells consists of an actin-rich cortex containing a three-dimensional array of actin filaments crosslinked into a gel. In the interior (endoplasm) of *Amoeba* and perhaps other cells, the cytoplasm is more fluid (sol). Pressure on the endoplasm, perhaps caused by the contraction of actin filaments at the tail, squeezes the endoplasm forward. A local weakening of the cortex at the front of the cell, perhaps caused by the activation of filament-severing proteins like gelsolin, allows the cell to bulge out at the front, forming a pseudopod. As the amoeba contracts at the rear, endoplasm flows into the new pseudopod while the tail pulls forward. The more fluid endoplasm will then convert to a gel.

when G-actin is added to preparations of microvilli, the membrane at the tips of the microvilli is pushed out further by the growth of actin filaments underneath.

Other possible mechanisms for membrane protrusion are closely coupled to the assembly of actin filaments. For example, a type I myosin motor molecule may push actin filaments into the membrane, causing it to protrude. Immunofluorescence staining of migrating *Dicytostelium* cells for myosin I shows that myosin I is localized at the leading edge. However, it is difficult to see how this would work, because myosin can only push actin filaments if they are oriented in parallel, yet actin in the leading edge is organized in a criss-cross three-dimensional array.

Finally, some scientists propose that changes in osmolarity cause local swelling and protrusion of the cell membrane, which is closely followed by the assembly of actin filaments. This observation is supported by studies of the acrosome reaction of sperm from the sea cucumber, *Thyone briareus*. The *acrosome reaction* prepares the sperm to fertilize the egg. In many types of sperm, this reaction is associated with the extension of a protrusion called the *acrosomal process* from the tip of the sperm. The formation of the acrosomal process in *Thyone* sperm depends on the rapid assembly of actin filaments (Figure 21-28). Formation can be slowed or blocked by raising the osmotic strength outside the cell, suggesting that osmotic swelling is required for this process to occur.

Cell Attachment. Membrane protrusions are normally associated with cell movements. However, some cells move without obvious changes in shape or protrusion of pseudopodia. In addition, cells can produce protrusions and still not move anywhere in particular. **Attachment,** or adhesion of the cell to its substrate, is also necessary for cell

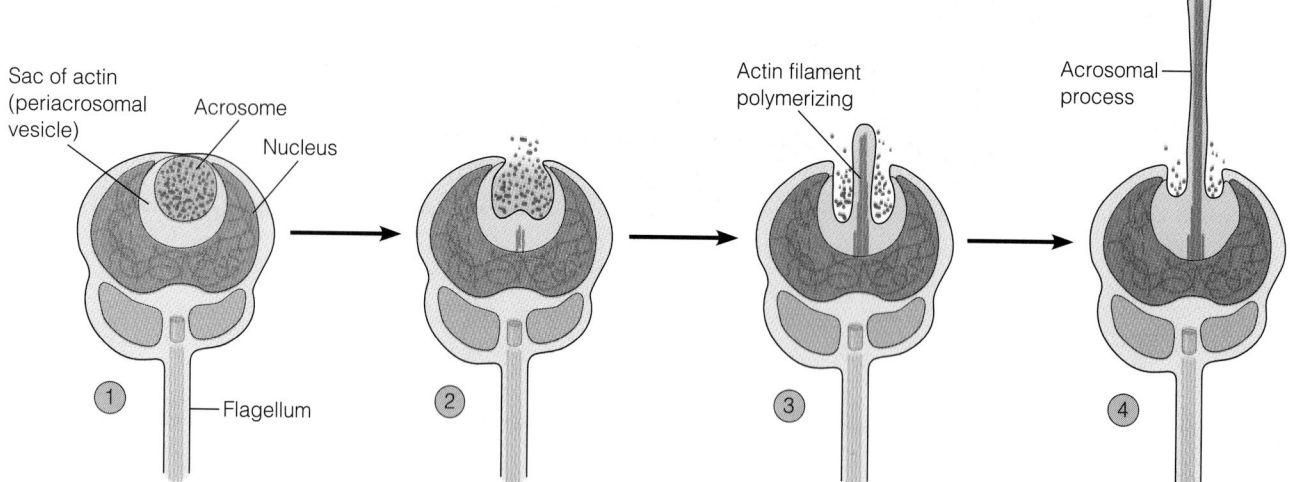

Figure 21-28 Membrane Protrusion Due to Actin Polymerization. The formation of an acrosomal process in *Thyone* (sea cucumber) sperm illustrates another way to generate cell protrusion. In this case, actin polymerization is directly related to the protrusion of the

membrane. However, actin polymerization is thought to occur in conjunction with osmotic swelling of the periacrosomal vesicle, the sac that holds the actin. Actin polymerization follows swelling, suggesting that membrane protrusion may be more directly due to osmotic swelling.

movement. New sites of attachment must be formed at the front of a cell, and contacts at the rear must be broken. Interestingly, sometimes contacts at the rear of the cell are too tight to be broken. In this case, while trying to pull the rear forward, the tail of the cell actually breaks off and is left behind (see Figure 21-26).

Attachment sites between the cell and the substrate are complex structures involving the attachment of proteins in the plasma membrane to other proteins on both the outside and the inside of the cell. One family of such attachment proteins are called *integrins.* On the outside of the cell, integrins attach to extracellular matrix proteins. On the inside of the cell, integrins are connected to actin filaments through a linkage composed of the proteins *talin, vinculin,* and *α-actinin.* The protein talin binds to the membrane and to the integrins, while *α*-actinin binds to actin filaments. Vinculin, through its ability to bind both talin and *α*-actinin, attaches the actin filament to the membrane.

Contractile Forces and Cell Movement. A final component of cell crawling is the ability of a cell to coordinate contraction and exert force at sites where it makes contact with the substrate. As the cell sends out protrusions at its leading edge, these protrusions may or may not adhere to the substrate. If a protrusion does not adhere, it is pulled backward. This retraction often causes the protrusion to lift up and fall backward onto the cell. If the protrusion instead adheres to the substrate, the contractile force pulling on the protrusion will pull the cell forward.

Although the specific molecules responsible for generating contractile forces on the leading edge have not been identified, the best evidence suggests that contraction is due to interactions between actin and myosin I. We know, for example, that myosin I is present at the leading edge of migrating cells, whereas myosin II is localized more toward the trailing edge.

As the cell protrudes, its tail will tend to remain attached to the substrate. In at least some types of moving cells, the mechanism used to detach the tail relies on stress fibers oriented parallel to the direction of movement. These stress fibers are contractile and contain bundles of actin and myosin II. In some ways, their structure resembles the organization of actin and myosin found in sarcomeres. Contraction of these stress fibers will tend to pull the tail of the cell toward the front.

At present, perhaps the best analogy we have for cell crawling is the movement of an inchworm. The worm extends forward, makes contact with the surface, and then pulls up its rear. Unfortunately, even this model falls short of explaining all types of crawling cell movements.

Cytoplasmic Streaming

Cytoplasmic streaming is seen in a variety of organisms that do not display amoeboid movement. In slime molds such as *Physarum polycephalum,* cytoplasm streams back and forth in the branched network of protoplasm that constitutes the *Physarum* cell mass. The flow of cytoplasm reverses direction with predictable periodicity and is therefore called **shuttle streaming.** The purpose of this periodic streaming seems to be both nourishment and locomotion, because the streaming process is correlated with the further extension of fingerlike projections by which the slime mold reaches out into its environment in search of nutrients.

Many plant cells display a circular flow of cell contents around a central vacuole. This streaming process, called **cyclosis,** has been studied most extensively in the giant algal cell *Nitella.* In this case, the movement seems to circulate and mix cell contents.

The unifying feature in these and other cases of cytoplasmic streaming seems to be the presence of filaments of actin in the cytoplasm. These filaments are thought to be integrally involved in the streaming mechanism, and the supporting evidence, as in the case of amoeboid movement, is increasingly convincing.

Cytokinesis

Cell division involves not just the parceling out of chromosomes to the daughter nuclei, but also the physical separation of the cytoplasm into two compartments, each of which gives rise to a daughter cell. This process, called **cytokinesis,** has been studied most extensively in the cleavage, or division, of animal eggs into multicellular embryos, with the sea urchin egg often the system of choice. As Figure 21-29 shows, cytokinesis begins as a slight indentation of the cell surface, which deepens into a **cleavage furrow** that encircles the egg. The plane of the cleavage furrow is perpendicular to the long axis of the mitotic spindle, thereby ensuring that the two sets of chromosomes will be separated into the two daughter cells. The furrow continues to deepen until opposite surfaces make contact and the cell is split in two.

The underlying mechanism depends on a beltlike bundle of actin filaments and myosin called the **contractile ring,** which forms during cell division in the cortical cytoplasm just beneath the plasma membrane. Examination of the cortical cytoplasm of the cleavage furrow with the electron microscope reveals large numbers of microfilaments, all oriented with their long axes parallel to the plane of the furrow. Myosin filaments cannot be seen directly, but their presence has been deduced by the use of antibodies. The filaments in the contractile ring progressively constrict the cleavage furrow, using force generated by the interaction of actin and myosin.

The contractile ring of dividing cells provides one of the most dramatic examples of both the transitory nature of actin-myosin structures in most nonmuscle cells and the rapidity with which such structures can be assembled and disassembled. Polymerization of actin apparently occurs just before the initial indentation becomes visible, and the entire structure is dismantled again shortly after cell division is complete.

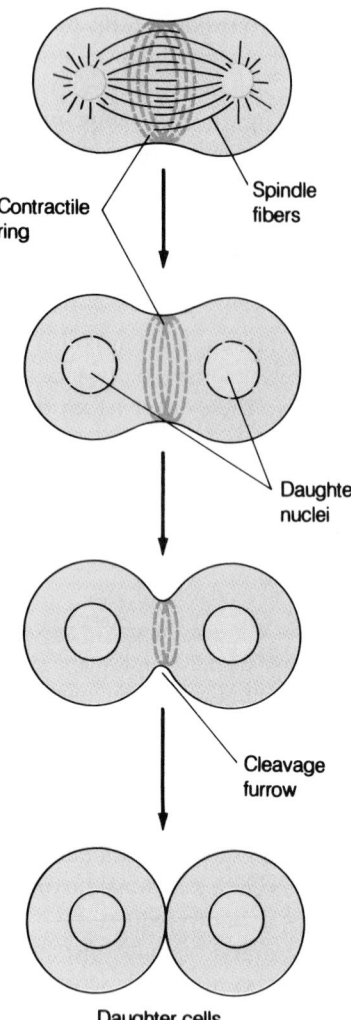

**Fertilized sea urchin egg
toward end of mitotic division**

Contractile ring

Spindle fibers

Daughter nuclei

Cleavage furrow

Daughter cells

Figure 21-29 Cytokinesis. Cytokinesis is illustrated here for the first division of a fertilized sea urchin egg. As the cell approaches the end of mitosis, cytokinesis begins with an indentation on the cell surface created by the contractile ring that encircles the cell in the midregion. The indentation gradually deepens, forming a cleavage furrow, and eventually the cell divides. The contractile ring responsible for the progressive constriction of the cleavage furrow is thought to consist of actin microfilaments and myosin in the cortical cytoplasm just beneath the plasma membrane.

Microtubule-Based Motility: The Motile Appendages of Eukaryotic Cells

We now come to a set of specialized motility mechanisms based on microtubules and the dynein motor molecules we described earlier. We will first discuss flagella and cilia, the motile appendages of eukaryotic cells.

Cilia and Flagella

Cilia and flagella share a common structural basis and differ only in relative length, number per cell, and mode of beating. **Cilia** (singular: **cilium**) have a diameter of about 0.25 μm, are about 2–10 μm long, and tend to occur in large numbers on the cell surface. Each cilium is bounded by an extension of the plasma membrane and is therefore an intracellular structure. Cilia occur in both unicellular and multicellular eukaryotes. Unicellular organisms, such as protozoa, use cilia for both locomotion and the collection of food particles.

In multicellular organisms, cilia serve primarily to move the environment past the cell rather than to propel the cell through the environment. The cells that line the air passages of your respiratory tract, for example, have several hundred cilia each (Figure 21-30a). (This means, incidentally, that every square centimeter of epithelial tissue lining your respiratory tract has about a billion cilia!) It is the coordinated, wavelike beating of these cilia that carries mucus, dust, dead cells, and other foreign matter out of the lungs. One of the health hazards of cigarette smoking lies in the inhibitory effect that smoke has on normal ciliary beating. Moreover, certain respiratory ailments can be traced to defective cilia.

Cilia display an oarlike pattern of beating, with a power stroke perpendicular to the cilium, thereby generating a force parallel to the cell surface. The numerous cilia on the cell surface usually beat in a coordinated manner, ensuring a steady movement of fluid past the cell surface. The cycle of beating for epithelial cilia is shown in Figure 21-30b. Each cycle requires about 0.1–0.2 sec and involves an active power stroke followed by a recovery stroke.

Flagella (singular: **flagellum**), although of the same diameter as cilia, are often much longer—from 1 μm to several millimeters, though usually in the range of 10–200 μm—and may be limited to one or a few per cell (Figure 21-30c). Like cilia, flagella are bounded by an extension of the plasma membrane.

Flagella differ distinctively from cilia in the nature of their beat. Flagella move with a propagated bending motion that is usually symmetrical and undulatory and may even have a helical pattern. This type of beat generates a force parallel to the flagellum, such that the organism moves in approximately the same direction as the axis of the flagellum. The locomotory pattern of most flagellated cells involves propulsion of the cell by the trailing flagellum, but examples are also known in which the flagellum actually precedes the cell. Figure 21-30d illustrates the swimming movement of the unicellular alga *Euglena*.

The Structure of Motile Appendages

Cilia and flagella have a common structure consisting of an **axoneme,** or main cylinder of tubules, about 0.25 μm in diameter, connected to a **basal body** and surrounded by an extension of the cell membrane (Figure 21-31a). Between the axoneme and the basal body is a *transition zone* in which the

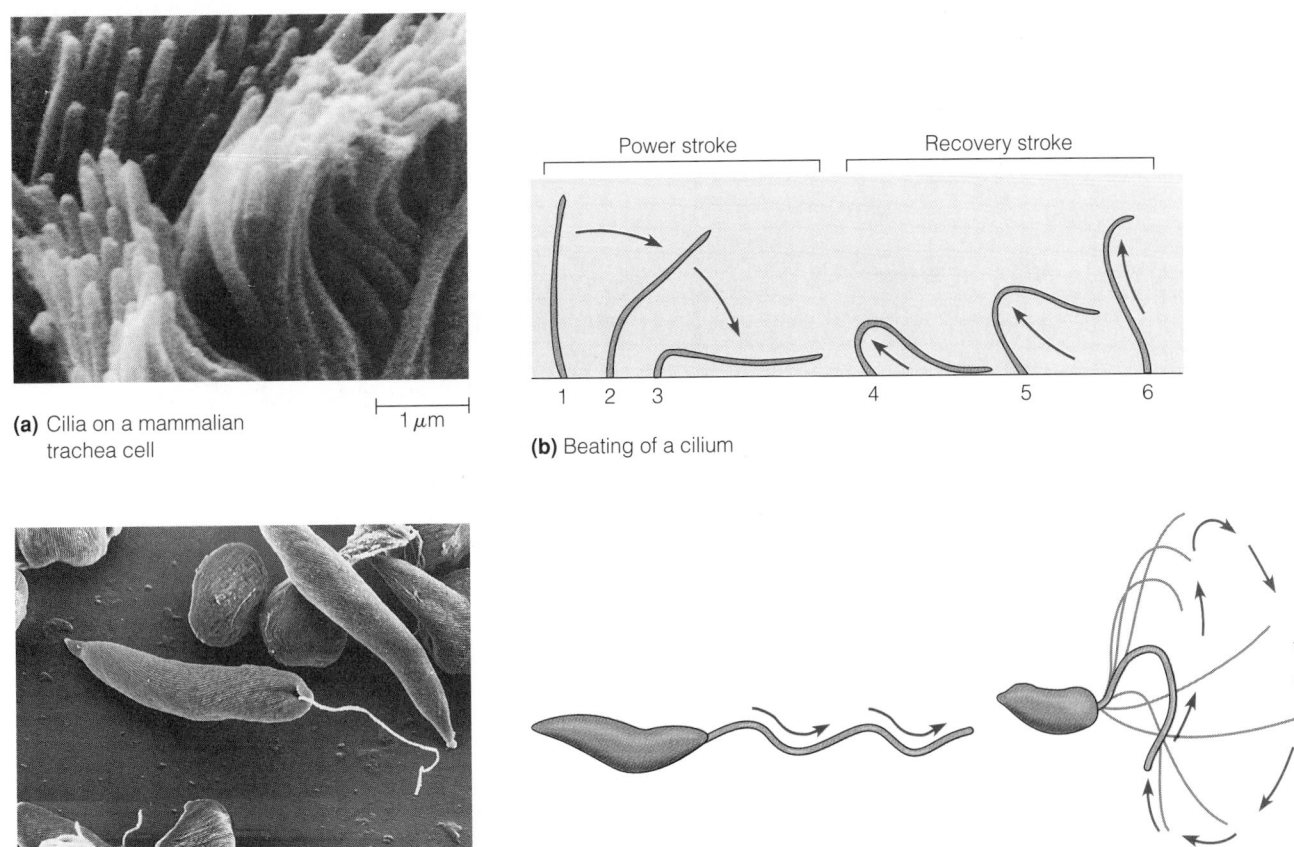

(a) Cilia on a mammalian
trachea cell

1 μm

(b) Beating of a cilium

Power stroke   Recovery stroke

1   2   3   4   5   6

(c) Flagellum on unicellular
alga *Euglena*

1 μm

(d) Movement of flagellated cell

Figure 21-30 Cilia and Flagella. **(a)** A micrograph of cilia on a mammalian tracheal cell (SEM). **(b)** The beating of a cilium on the surface of an epithelial cell from the human respiratory tract. A beat begins with a power stroke that sweeps fluid over the cell surface. A recovery stroke follows, leaving the cilium poised for the next beat. Each cycle requires about 0.1–0.2 sec. **(c)** A micrograph of the flagellated unicellular alga *Euglena* (SEM). **(d)** Movement of a flagellated cell through an aqueous environment.

arrangement of microtubules in the basal body takes on the pattern characteristic of the axoneme. Cross-sectional views of the axoneme, transition zone, and basal body are shown in Figure 21-31 as parts b–d. The basal body is identical in appearance to the centriole, a structure we encountered in our discussion of the mitotic spindle (see Chapter 14). Centrioles and basal bodies are usually considered to be two different functional manifestations of the same structure. A basal body consists of nine sets of tubular structures arranged around its circumference. Each set is called a *triplet* because it consists of three tubules that share common walls. Each triplet has one complete microtubule and two incomplete tubules.

The axoneme, however, has a characteristic "9 + 2" pattern, with nine **outer doublets** of tubules and two additional microtubules in the center, often called the **central pair.** Figure 21-32 illustrates these structural features in greater detail. The nine outer doublets of the axoneme are thought to be extensions of two of the three subfibers from each of the nine triplets of the basal body. Each outer dou-

blet of the axoneme therefore consists of one complete microtubule, called the **A tubule,** and one incomplete MT, the **B tubule.** The A tubule has 13 protofilaments, whereas the B tubule has only 10 or 11. The A tubule shares 5 protofilaments with the otherwise unclosed B tubule. The tubules of the central pair, however, are both complete, with 13 protofilaments each. All of these structures contain the protein *tubulin,* together with a second protein called **tektin.** Tektin is related to intermediate filament proteins (see Chapter 20) and is a necessary component of the axoneme, which cannot be produced from tubulin alone. For instance, the A and B tubules share a wall that appears to contain tektin as a major component.

In addition to microtubules, axonemes contain several other key structures (Figure 21-32b). The most important of these are the sets of **sidearms** that project out from each of the A tubules of the nine outer doublets. Each sidearm reaches out clockwise toward the B tubules of the adjacent doublet. These arms consist of **axonemal dynein,** which has ATPase activity that is critical in motility, just as in

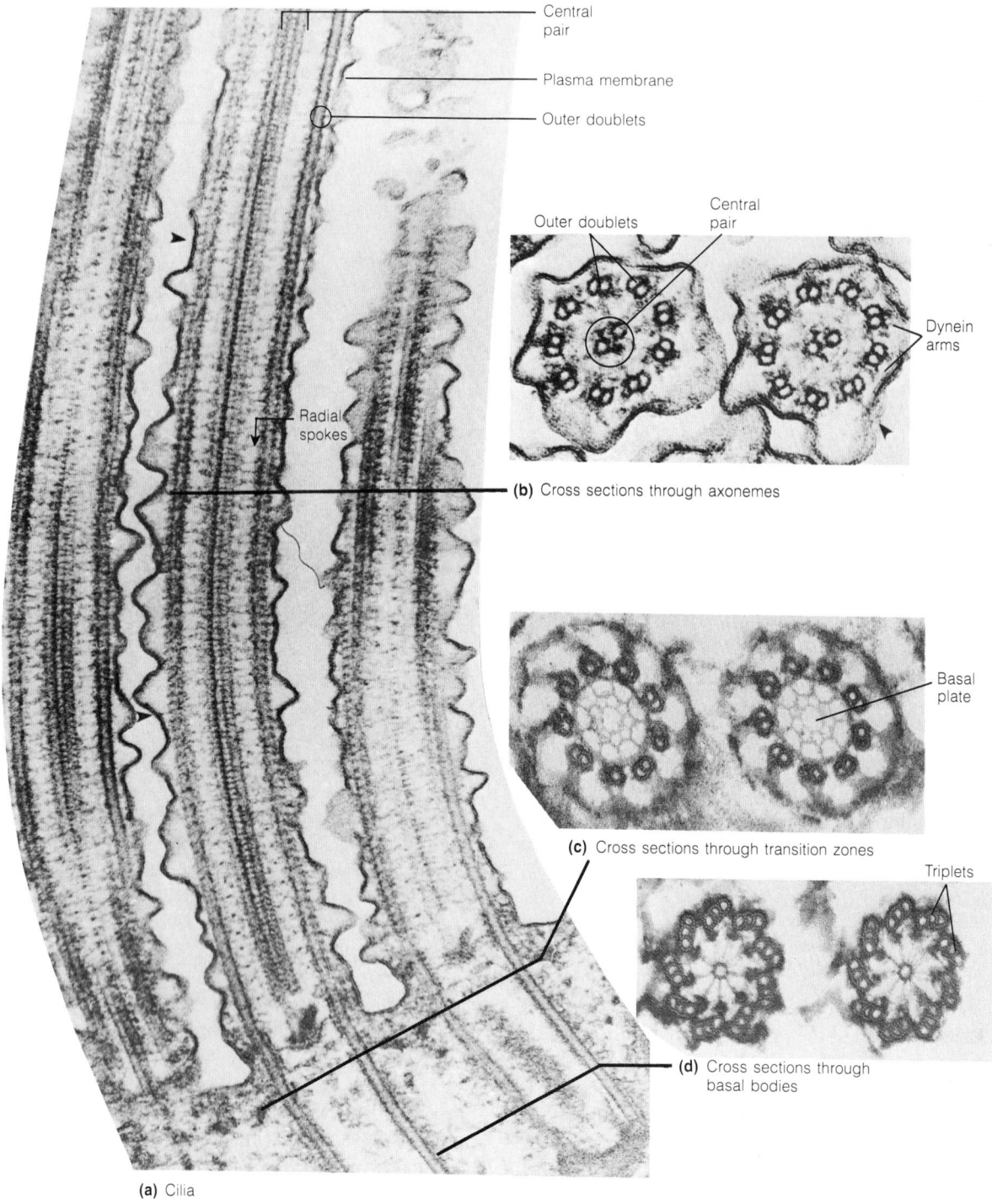

Central pair

Plasma membrane

Outer doublets

Outer doublets

Central pair

Dynein arms

(b) Cross sections through axonemes

Radial spokes

Basal plate

(c) Cross sections through transition zones

Triplets

(d) Cross sections through basal bodies

(a) Cilia

Figure 21-31 The Structure of a Cilium. **(a)** These longitudinal sections of cilia from the protozoan *Tetrahymena thermophila* illustrate several structural features of cilia, including the central pair and outer doublet microtubules, the radial spokes, the dynein arms, and the basal body structures. Cross-sectional views are shown for **(b)** axonemes, **(c)** the transition zone between cilia and basal bodies, and **(d)** basal bodies. Notice the triplet pattern of the basal body and the 9 + 2 pattern of tubule arrangement in the axoneme of the cilium. (All TEMs.)

cytoplasmic dynein function. The dynein arms occur in pairs, one **inner arm** and one **outer arm,** spaced along the MT at regular intervals. At less frequent intervals, adjacent doublets are joined by **interdoublet links.** These links are thought to limit the extent to which doublets can move with respect to each other as the axoneme bends.

At regular intervals, **radial spokes** project inward from each of the nine MT doublets, terminating near a set of projections that extend outward from the central pair of microtubules. Researchers believe that these spokes are important in translating the sliding of adjacent doublets into the bending motion that characterizes the beating of these ap-

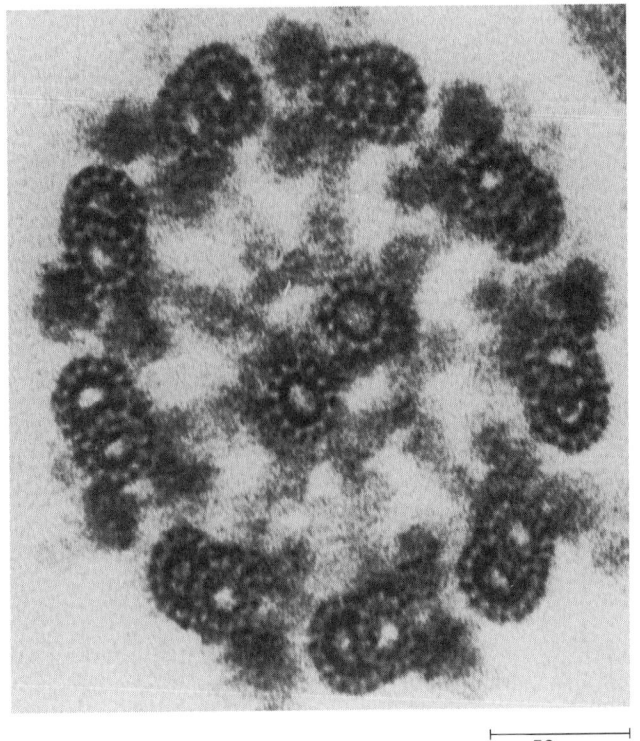

(a)

50 nm

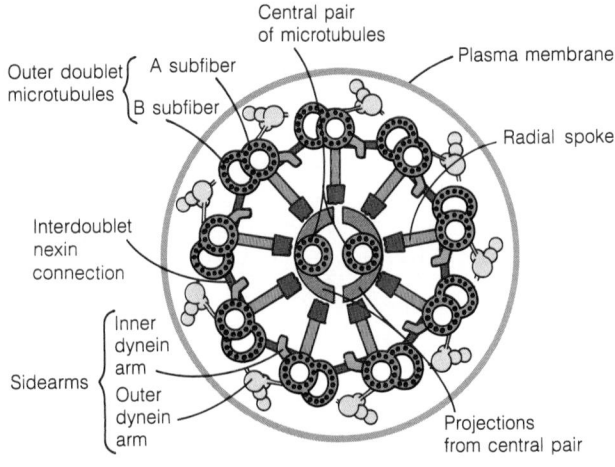

(b)

Figure 21-32 Enlarged Cross Section of an Axoneme.
(a) This micrograph shows an axoneme from a flagellum of *Chlamydomonas*. The microtubules of the central pair have 13 protofilaments each, as do the A tubules of the outer doublets. Each B tubule has 11 protofilaments of its own and shares five protofilaments with the A tubule (TEM). **(b)** This diagram of an axoneme in cross section shows the dynein arms, which have ATPase activity and are thought to be responsible for the sliding of adjacent doublets. The interdoublet links (nexin connections) join adjacent doublets and the radial spokes project inward, terminating near projections that extend outward from the central pair of MTs.

pendages. In addition to the radial spoke attachment to the central pair, a protein called **nexin** links adjacent doublets to one another and probably also plays a role in translating sliding into bending motion.

The Sliding-Microtubule Model for Motile Appendages

The mechanism of motility in microtubule-based systems is not as well understood as muscle contraction, but it is clear that ATP hydrolysis is the driving force in both cases. Axonemes of isolated cilia and flagella that lack their basal bodies and membranes can be induced to beat if ATP is added to the medium, so it appears likely that the appendages are actively involved in their own motility.

The mechanism proposed to explain microtubule-based motility is similar to that of muscle contraction, but modified in a way that accomplishes bending instead of contraction. According to the **sliding-microtubule model** for cilia and flagella, overall MT length remains unchanged, but adjacent outer doublets slide past each other in an ATP-dependent process. Instead of causing contraction, however, this sliding movement is converted to a localized bending, because the doublets of the axoneme are connected radially to the central pair and circumferentially to one another and therefore cannot slide past each other freely. The resultant bending takes the form of a wave that begins at the base of the organelle and proceeds toward the tip.

Dynein Arms Are Responsible for Sliding. The driving force for MT sliding is provided by ATP hydrolysis, catalyzed by the ATPase activity of the dynein arms. The importance of the dynein arms is indicated by two kinds of evidence. When dynein is selectively extracted from isolated axonemes, the arms disappear from the outer doublets, and the axonemes lose both their ability to hydrolyze ATP and their capacity to beat. Furthermore, the effect is reversible: If purified dynein is added to isolated outer doublets, the sidearms reappear and ATP-dependent sliding is restored.

A second kind of evidence comes from studies of nonmotile mutants in such normally motile species as *Chlamydomonas*. In some of these mutants, flagella are still present, but they are nonfunctional. Such nonmotile flagella lack either dynein arms, radial spokes, or the central pair of microtubules. These structures are therefore almost certainly essential to the movement mechanism.

Dynein is a very large protein, with a molecular weight exceeding 1 million. It has multiple subunits, the three largest having ATPase activity and a molecular weight of about 450,000 each. During the sliding process, the stalk of the dynein arm apparently attaches to and detaches from the B tubule in a cyclic manner similar to the contraction cycle of muscle (Figure 21-33). Each cycle requires the hydrolysis of ATP and leaves the dynein arm of one doublet further along the next doublet. In this way, the dynein arms of one doublet move the neighboring doublet, resulting in a relative displacement of the two.

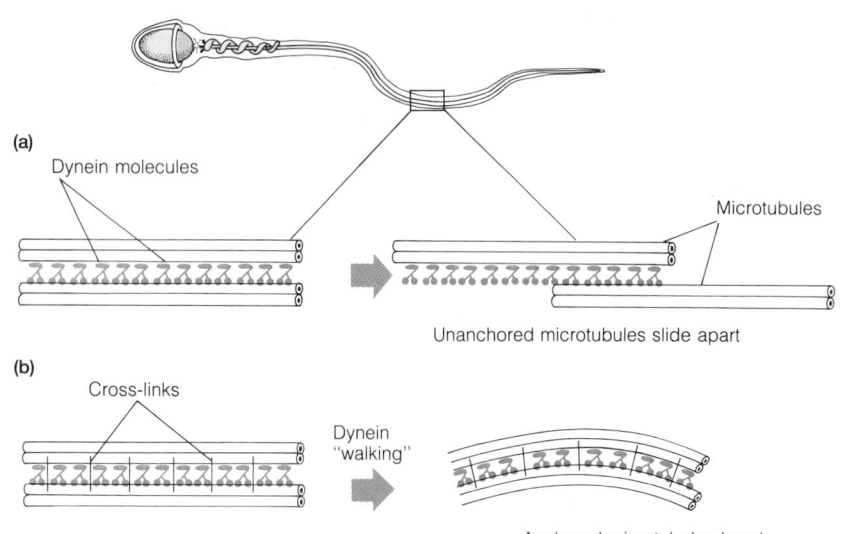

(a)

Dynein molecules

Microtubules

Unanchored microtubules slide apart

(b)

Cross-links

Dynein "walking"

Anchored microtubules bend

Figure 21-33 Dynein and Flagellar Motion. **(a)** If the microtubules on a flagellar axoneme are treated to remove crosslinking proteins, the addition of ATP causes the MT doublets to slide apart. **(b)** This result suggests that crosslinking proteins like nexin translate the sliding motion into a bending motion. The precise mechanism underlying dynein interaction with microtubules at the molecular level is not yet understood, however.

Crosslinks and Spokes Are Responsible for Bending. To convert the dynein-mediated displacement of doublets to a bending motion, the doublets must be restrained in a way that resists sliding but allows deformation. This resistance is provided by the radial spokes that connect the doublets to the central pair of microtubules and possibly also by the nexin crosslinks between doublets. If these crosslinks and spokes are removed (by partial digestion with proteolytic enzymes, for example), the resistance that translates doublet sliding into a bending action is absent, and sliding is uncoupled from bending. Under these conditions, the free doublets move with respect to each other just as actin and myosin filaments do, and the axonemes simply become longer and thinner as the microtubules slide apart (see Figure 21-33a).

The importance of the radial spokes is shown by electron microscopic observations on bent and straight regions of a cilium. The radial spokes are oriented perpendicular to the doublets in straight regions but are tilted at an angle in bent regions. In other words, the spokes are deformed by the sliding process and provide the resistance that translates the sliding of doublets into the bending of the cilium or flagellum. Eventually, though, the sliding movement of the doublets overcomes the resistance and causes displacement of the radial spokes with respect to the central pair of tubules. The progressive displacement of the spokes on the inside of the bend indicates that the two doublets on opposite sides of the axoneme are in fact sliding with respect to each other.

The Bacterial Flagellum

One of the most remarkable motile appendages in all of nature is the **bacterial flagellum,** an appendage strikingly dissimilar to the flagella of eukaryotic cells, both structurally and functionally. A flagellated bacterium is shown in Figure 21-34. Unlike eukaryotic appendages, bacterial fla-

gella are not membrane-bounded and are therefore extracellular structures. As Figure 21-35a and b illustrates, the bacterial flagellum is a spiral filament, usually about 15 nm in diameter and about 10–20 μm long. The filament is attached to a **hook,** which is, in turn, connected by a **rod** to four ringlike structures in the base. The rod penetrates the outer membrane, the peptidoglycan wall, and the inner (plasma) membrane. Two of the rings are anchored in the outer membrane, and two in the plasma membrane. The two rings in the plasma membrane are called the **S (stator) ring** and the **M (motor) ring** (Figure 21-35c).

Chemically, bacterial flagella consist of parallel strands of protein coiled around each other. The common subunit is **flagellin,** with a molecular weight of about 40,000 in the monomeric form. Flagellin is not an ATPase; in fact, it has no known enzymatic activity.

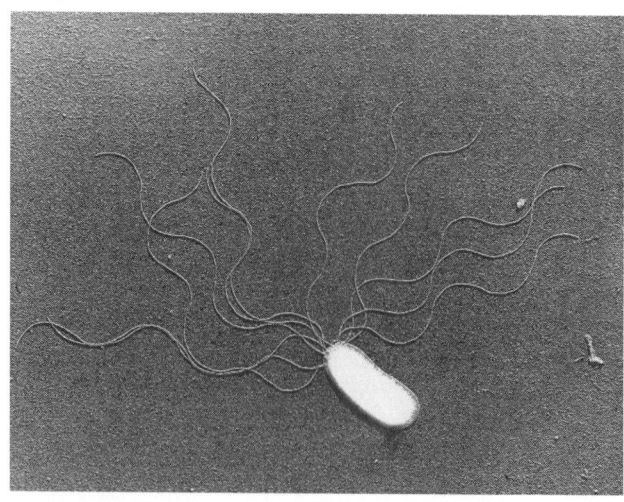

Figure 21-34 A Flagellated Bacterium. An electron micrograph of the flagellated bacterium *Pseudomonas marginalis* (SEM).

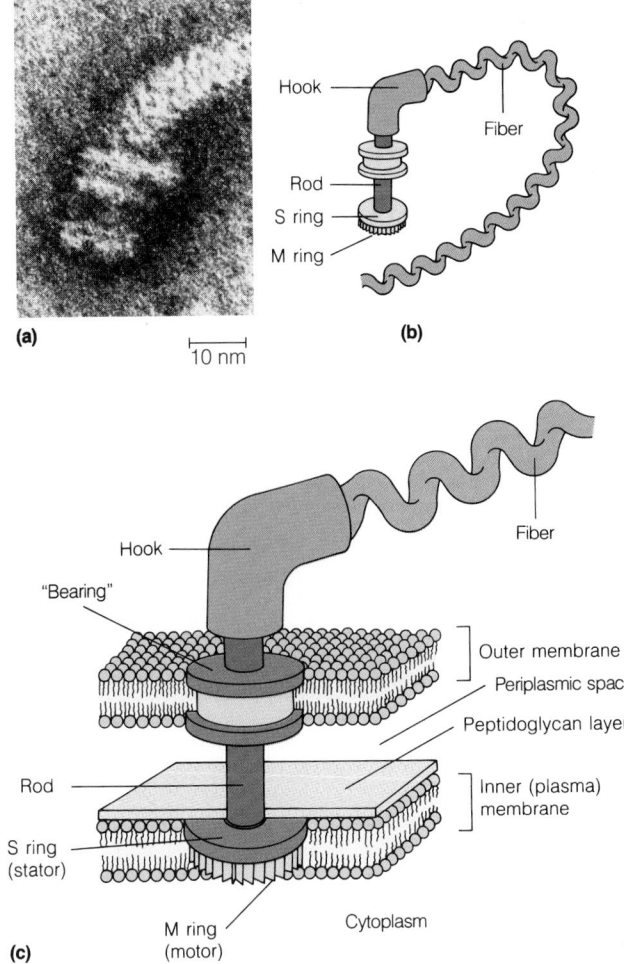

Hook

Fiber

Rod

S ring

M ring

(a)

(b)

10 nm

Hook

"Bearing"

Rod

S ring
(stator)

Fiber

Outer membrane

Periplasmic space

Peptidoglycan layer

Inner (plasma)
membrane

Cytoplasm

M ring
(motor)

(c)

Figure 21-35 Structure of a Bacterial Flagellum and the "Motor" Responsible for Its Rotation. **(a)** A micrograph of a bacterial flagellar hook assembly (TEM). **(b)** The helical filament or fiber of each flagellum is attached to a hook structure, and this, in turn, is joined by a rod to four ringlike structures in the base. **(c)** Two of the rings are anchored in the outer membrane of the bacterial cell and two in the inner (plasma) membrane. The S ring in the plasma membrane is anchored to the peptidoglycan layer and is regarded as the nonrotating stator. The M (motor) ring in the plasma membrane rotates against the S ring, with the torque provided by the proton motive force generated by the electrochemical proton gradient across the plasma membrane. Structures that rotate are the rod, the hook, and the fiber.

Nature's Wheel: Locomotion by Rotation

The bacterial flagellum has been recognized as a locomotory structure for more than a hundred years, but its mode of propulsion has been elucidated only recently. Initially, the characteristic helical motion of the flagellum was thought to result from waves of bending, originating from the base and propagated along the length of the flagellum. It is now quite clear, however, that the flagellum actually *rotates* as a rigid, helical structure, driven by a rotary "motor" at the base of each filament. The S ring in the plasma membrane is anchored to the peptidoglycan layer and is regarded as the

nonrotating stator, as the name suggests. The M ring in the plasma membrane rotates against the S ring and is in effect the motor that drives the flagellum (see Figure 21-35c).

Requiring as it does the structural equivalents of a rotor, a stator, and rotary bearings, such a mechanism was originally considered highly unlikely, and is certainly without precedent in the biological world. But a series of ingenious experiments provided conclusive evidence for just such a propellerlike rotary drive. Much of this evidence is due to the availability of antibodies that react specifically with flagellar filaments. In a particularly clever experiment, Michael Silverman and Melvin Simon used antibodies to link, or "tether," flagellar filaments to a glass surface, thereby preventing normal rotation of the filament. In so doing, they made the remarkable observation that if the filament cannot rotate, the whole body of the cell rotates instead!

The same researchers were also able to visualize flagellar rotation directly. Using antibodies as "glue," they succeeded in linking microscopic beads of polystyrene-latex to individual filaments as visible markers along the filament surface. Viewed under the microscope, the beads remained fixed in location with respect to one another but revolved together around the filament. Apparently, we must credit nature with the invention of the wheel, for one exists in the rotary motor at the base of every flagellar filament in every motile bacterial cell!

Once it became clear that the driving force for flagellar rotation lies in the basal motor rather than in the flagellum itself, it was no longer surprising that the bacterial flagellum, unlike the motile appendages or contractile filaments of eukaryotic cells, possesses no ATPase activity along its length. In fact, ATP is not required at all as an energy source for flagellar rotation. Instead, the driving force appears to be the energy of the electrochemical proton gradient across the plasma membrane, though it is not yet clear how the proton gradient is coupled to flagellar rotation.

Flagellar Rotation and Chemotaxis

Flagellar rotation in the bacterium *E. coli* can be either clockwise or counterclockwise (Figure 21-36). The spiral shape of the flagellum is such that clockwise rotation results in each flagellum acting independently, each pulling the cell in its own direction (Figure 21-36a). Coherent motion of the cell is therefore impossible, and the bacterium displays a chaotic *tumbling* behavior. Counterclockwise rotation, however, results in each flagellum *pushing* on the cell. The flagella associate into a single coherent bundle and function together in concert, causing the bacterium to swim in a straight line (Figure 21-36b).

An *E. coli* cell reverses the rotation of its flagella quickly and frequently, such that the cell alternates between tumbling and smooth swimming. Characteristically, the tumbling lasts for only a few tenths of a second; then the flagella are reversed, and the cell swims off in a randomly chosen direction. In the absence of any changes in environmental stimuli, this results in a characteristic motility pattern in

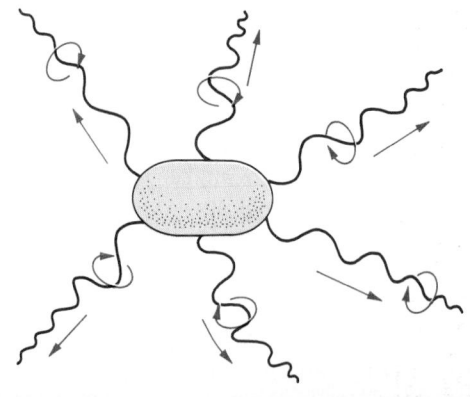

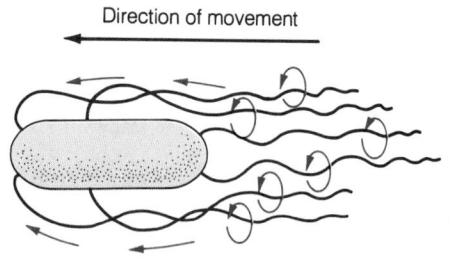

Direction of movement

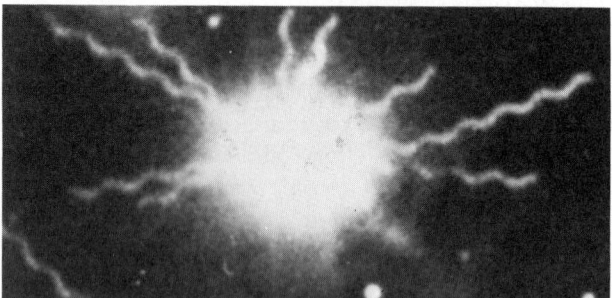

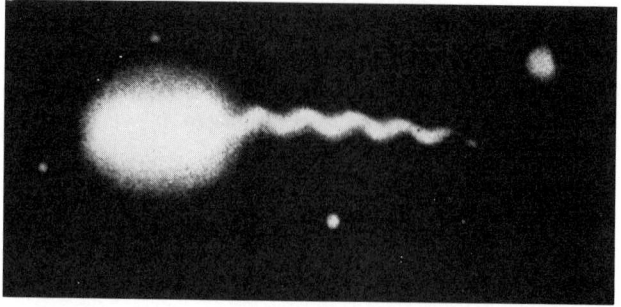

(a) Clockwise rotation of flagella (tumbling movement)

(b) Counterclockwise rotation of flagella (swimming movement)

Figure 21-36 The Effect of Direction of Flagellar Rotation on Cell Movement. **(a)** When the flagella of an *E. coli* cell rotate clockwise (circular arrows), each flagellum pulls outward (straight arrows) because of the helical shape of the flagellum. The bacterium is therefore drawn in several directions at once and displays an undirected tumbling

movement. **(b)** When the flagella of the cell rotate counterclockwise, each flagellum pushes instead of pulls, and the flagella are drawn together into a bundle. Because each flagellum is now pushing in the same direction, the cell is propelled along with a smooth swimming movement.

which short periods of undirected tumbling are interspersed with periods (or "runs") of swimming, but in randomly chosen directions (Figure 21-37).

When exposed to a specific chemical attractant or repellent, however, the bacterial cell modifies this pattern to allow net movement toward the attractant or away from the repellent. *Attractants* include oxygen and nutrients such as sugars or amino acids. *Repellents* include waste substances excreted from the cell and various noxious chemicals such as phenol. The ability to move toward a chemical attractant or away from a repellent is called **chemotaxis.**

A bacterial cell responds not only to the presence or absence of an attractant or repellent, but also to a concentration gradient of the chemical. The chemotactic response involves a relative increase in the length of a run when the cell is moving in the right direction, either toward an attractant (*positive chemotaxis;* Figure 21-38a) or away from a repellent (*negative chemotaxis;* Figure 21-38b). In other words, when a bacterial cell is moving in the right direction with respect to a concentration gradient, it will tend to tumble little and swim much, thereby accomplishing net movement in the desired direction. Chemotaxis therefore depends on the

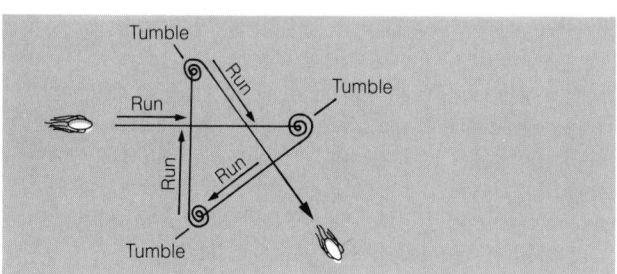

Figure 21-37 Characteristic Motility Pattern of *E. coli.*
In the absence of any changes in environmental stimuli, *E. coli* flagella rotate counterclockwise for a few seconds, for a short period (or run) of smooth swimming, then reverse to clockwise rotation for a few tenths of a second, for a short period of tumbling. This is followed by another run in a randomly chosen direction. The overall pattern is short runs in random directions interspersed with brief periods of tumbling.

Figure 21-38 Chemotaxis in Bacteria. (a) In the presence of a gradient of an attractant, the flagella rotate counterclockwise much longer when the bacterium senses it is moving up the attractant gradient than when it is swimming away from the attractant. Net movement is therefore toward the attractant, even though the direction of swimming chosen after each tumble is still random. **(b)** In the presence of a gradient of a repellent, the runs carrying the bacterium down the concentration gradient are much longer than those that move it up, resulting in net movement away from the repellent.

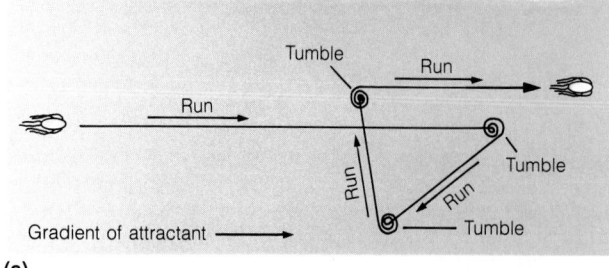

(a)

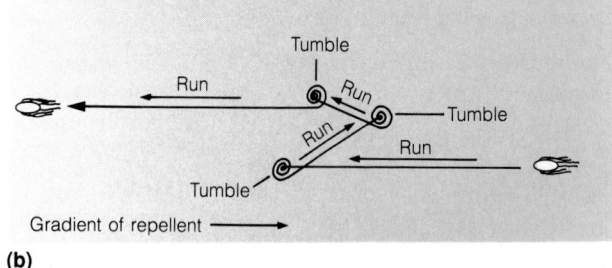

(b)

ability of the bacterial cell to change the direction of flagellar rotation and thereby to either swim smoothly in a desired direction or tumble randomly before striking out in a new direction. And the bacterial cell accomplishes all of this with tiny helical propellers driven by rotary "motors" so small that they can barely be seen with an electron microscope!

PERSPECTIVE

Motility is a major theme in cell biology. Our knowledge of the mechanisms underlying eukaryotic motility has increased considerably over the past decade, and we now understand that it is driven at the molecular level by a set of ATP-dependent motor molecules. These molecules use portions of the cytoskeleton as a kind of track to pull subcellular components into position.

Two major eukaryotic motility systems are known, one based on the interaction of myosin with actin microfilaments and the other on the interaction of dynein or kinesin with microtubules. Actin and myosin are found widely distributed in nonmuscle cells, where they are involved in a variety of motility mechanisms, including amoeboid movement, cytoplasmic streaming, and cytokinesis. Skeletal muscle contraction is a specialization of this more general motility process, and is the best-understood example. Muscle contraction involves a progressive sliding of thin actin filaments past thick myosin filaments, driven by the interaction between the ATPase head of the myosin molecules and successive myosin-binding sites on the actin filaments. Contraction is triggered by the release of calcium from the sarcoplasmic reticulum and ceases again as the calcium is actively pumped back into the SR. In skeletal muscle, calcium binds to troponin and causes a conformational change in tropomyosin, which opens myosin-binding sites on the thin filament. In smooth muscle, the effect of calcium is mediated by calmodulin, which activates myosin light-chain kinase, leading to the phosphorylation of myosin.

The second major motility system in eukaryotic cells is that based on microtubules. Cytoplasmic kinesin and dynein are motor molecules that move intracellular structures in opposite directions along MT tracks. The axoneme present in both cilia and flagella is a highly specialized example of dynein-tubulin interaction. The nine outer doublets of the axoneme are connected laterally to one another and radially to the central pair of single microtubules. Dynein arms project out from one MT doublet to the next. Dynein has ATPase activity and is thought to be involved in the sliding of one set of microtubules past the next. This sliding is opposed by the radial spokes between the doublets and the central pair of tubules and by the connections between adjacent doublets. As a result, the sliding is converted to a bending motion.

Bacterial cells use neither actin and myosin nor tubulin for their motility system. They depend instead on flagella that bear no structural or functional resemblance to the eukaryotic appendages of the same name. Bacterial flagella propel the cell by actually rotating like small propellers, using the energy of an electrochemical proton gradient to drive the rotor. Depending on the direction of rotation, the flagella can cause the cell to move smoothly in a fixed direction or to tumble randomly. This process underlies the phenomenon of bacterial chemotaxis, which involves net movement of a bacterial cell toward a chemical attractant or away from a chemical repellent.

KEY TERMS FOR SELF-TESTING

motility (p. 675)
contractility (p. 675)

Systems of Motility

motor molecule (p. 675)
microfilament-based movement (p. 676)
microtubule-based movement (p. 676)
bacterial flagellar rotation (p. 676)

Nonmuscle Microfilament-Based Movement

nonmuscle cell movement (p. 676)
actin (p. 676)
myosin (p. 676)
myosin I (p. 676)
myosin II (p. 676)

Intracellular Microtubule-Based Movement: Dynein and Kinesin

microtubule-associated motor protein (motor MAP) (p. 678)
kinesin (p. 678)
dynein (p. 678)
fast axonal transport (p. 678)

Filament-Based Movement in Muscle

muscle contraction (p. 680)
skeletal muscle (p. 680)
muscle fiber (p. 680)
myofibril (p. 681)
sarcomere (p. 681)
thick filament (p. 681)
thin filament (p. 681)
A band (p. 681)
I band (p. 681)
H zone (p. 681)
Z line (p. 682)
tropomyosin (p. 683)
troponin (p. 683)
TnT (p. 683)
TnC (p. 683)

TnI (p. 683)
α-actinin (p. 683)
myomesin (p. 683)
titin (p. 684)
sliding-filament model (p. 684)
cross-bridge (p. 686)
contraction cycle (p. 686)
rigor (p. 686)
myoplasm (p. 688)
sarcoplasm (p. 688)
neuromuscular junction (p. 689)
sarcolemma (p. 689)
transverse (T) tubule system (p. 689)
T tubule (p. 689)
sarcoplasmic reticulum (SR) (p. 690)
medial element (p. 690)
terminal cisternae (p. 690)
triad (p. 690)
functional complex (p. 690)
calcium-induced calcium release (p. 691)
dihydropyridine receptor (p. 691)
ryanodine receptor (p. 691)
calcium pump (ATPase) (p. 691)
myoglobin (p. 693)
creatine phosphate (p. 693)
cardiac (heart) muscle (p. 694)
intercalated disc (p. 695)
smooth muscle (p. 695)
dense body (p. 695)
myosin light-chain kinase (MLCK) (p. 696)
calmodulin (p. 696)
calcuim-calmodulin complex (p. 696)

Specialized Filament-Based Motility Systems

amoeboid movement (p. 698)
pseudopodium (p. 698)
ectoplasm (p. 698)
endoplasm (p. 698)

gel (p. 698)
gelsolin (p. 698)
sol (p. 698)
gelation-solation (p. 699)
attachment (p. 700)
cytoplasmic streaming (p. 701)
shuttle streaming (p. 701)
cyclosis (p. 701)
cytokinesis (p. 701)
cleavage furrow (p. 701)
contractile ring (p. 701)

Microtubule-Based Motility: The Motile Appendages of Eukaryotic Cells

cilium (p. 702)
flagellum (p. 702)
axoneme (p. 702)
basal body (p. 702)
outer doublet (p. 703)
central pair (p. 703)
A tubule (p. 703)
B tubule (p. 703)
tektin (p. 703)
sidearm (p. 703)
axonemal dynein (p. 703)
inner arm (p. 704)
outer arm (p. 704)
interdoublet link (p. 704)
radial spoke (p. 704)
nexin (p. 705)
sliding-microtubule model (p. 705)

The Bacterial Flagellum

bacterial flagellum (p. 706)
hook (p. 706)
rod (p. 706)
S (stator) ring (p. 706)
M (motor) ring (p. 706)
flagellin (p. 706)
chemotaxis (p. 708)

PROBLEM SET

21-1. Muscle Structure. Frog skeletal muscle consists of thick filaments with a length of about 1.6 μm and thin filaments with a length of about 1 μm.

(a) What is the length of the A band and the I band in a muscle with a sarcomere length of 3.2 μm? Describe what happens to the length of both bands as the sarcomere length decreases during contraction from 3.2 μm to 2.0 μm.

(b) The H zone is a specific portion of the A band. If the H zone of each A band decreases in length from 1.2 to 0 μm as the sarcomere length contracts from 3.2 to 2.0 μm, what can you deduce about the physical meaning of the H zone?

(c) What can you say about the distance from the Z line to the edge of the H zone during contraction?

21-2. Energy Requirement of Muscle Contraction. During contraction, mammalian skeletal muscle hydrolyzes ATP at the rate of 1 mmol/min per gram of muscle tissue. Muscle concentrations of ATP and creatine phosphate are about 5 μmol/g and 25 μmol/g, respectively.

(a) How long could muscle contraction continue if it depended only on existing ATP supplies in the tissue? How long could it continue if it depended solely on existing supplies of ATP and creatine phosphate? Can you think of circumstances in which these reserves of immediate energy may be essential despite the short times over which they can sustain contraction?

(b) Assuming that the need for ATP is in fact being met by aerobic respiration, at what rate must the tissue be supplied with oxygen (in milliliters of oxygen per minute per gram of tissue) to sustain contraction? (Recall that 1 mole of a gas occupies 22.4 L at standard temperature and pressure, and assume an energy yield of 36 ATP molecules per molecule of catabolized glucose.)

(c) Now assume that the need for ATP is met by glycolysis, with stored glycogen as the only energy source. If muscle glycogen reserves are equal to 1% of the tissue by weight and all of this glycogen is available for catabolism, how long can ATP generation be sustained at the expense of glycogen?

(d) Finally, assume that oxygen is not available, glycogen has been consumed, creatine phosphate stores are depleted, and all ATP has been converted to ADP, leaving the myokinase reaction as the cell's last resort. How long can contraction go on? Are the assumptions of this question realistic?

21-3. Creatine Phosphate. The $\Delta G^{\circ\prime}$ of hydrolysis is -10.3 kcal/mol for creatine phosphate versus -7.3 kcal/mol for ATP. Long after its involvement in muscle energetics was first realized, creatine phosphate was thought to be the immediate source of energy for muscle contraction. It was not until inhibitors of creatine kinase were discovered that the mechanical work of muscle contraction could be correlated directly with ATP consumption and the true nature of creatine phosphate as a reservoir of high-energy phosphate for recharging ADP was realized.

(a) What is $\Delta G^{\circ\prime}$ for the creatine kinase reaction in the direction of ATP generation? What is the value of the equilibrium constant K'_{eq} at 25°C?

(b) Assuming conditions under which $\Delta G' \approx \Delta G^{\circ\prime}$, calculate the ratio ATP/(ATP + ADP) when the following percentages of creatine are phosphorylated: 90%, 50%, 10%, and 1%.

(c) If you were assaying for intracellular creatine phosphate and ATP concentrations during the process of muscle contraction, why might you conclude, as early researchers did, that the creatine phosphate rather than the ATP was the immediate source of the energy needed for contraction?

(d) If, on the other hand, you were to assay for creatine phosphate and ATP levels during contraction in the presence of an inhibitor (such as 2,4-dinitrofluorobenzene) that inactivates creatine kinase in intact muscle, how would your results differ from those of part c?

(e) How does this result support the "reservoir" nature of creatine phosphate?

21-4. Rigor Mortis and the Contraction Cycle. At death, the muscles of the body become very stiff and inextensible, and the corpse is said to go into *rigor*.

(a) Explain the basis of rigor. Where in the contraction cycle is the muscle arrested? Why?

(b) Would you be likely to go into rigor faster if you were to die while racing to class or while sitting in lecture? Explain.

(c) What effect do you think the addition of ATP might have on muscles in rigor?

21-5. AMPPCP and the Contraction Cycle. AMPPCP is the abbreviation for a structural analogue of ATP in which the third phosphate group is linked to the second by a CH_2 group instead of an oxygen atom. AMPPCP binds to the ATP-binding site of virtually all ATPases, including myosin. It differs from ATP, however, in that its terminal phosphate cannot be removed by hydrolysis.

When isolated myofibrils are placed in a flask containing a solution of calcium ions and AMPPCP, contraction is quickly arrested.

(a) Where in the contraction cycle will contraction be arrested by AMPPCP? Draw the arrangement of the thin filament, the thick filament, and a cross-bridge in the arrested configuration.

(b) Do you think contraction would resume if ATP were added to the flask containing the AMPPCP-arrested myofibrils? Explain.

(c) What other processes in a muscle cell do you think are likely to be inhibited by AMPPCP?

21-6. Tetanus. Single nerve impulses to skeletal muscle normally cause separate twitches, interspersed by periods of relaxation. When a string of impulses arrives in rapid succession, however, it results in a summation of twitches and steady contraction. This leads to *tetanus*, a condition of rigidity, with the muscles temporarily "locked" in the fully contracted state. So defined, tetanus is a normal physiological phenomenon, because this kind of extreme contraction occurs naturally under conditions of intense muscular effort and is followed by relaxation as soon as the nerve stimulation stops. However, the same term is used in a pathological sense to describe a disease caused by the bacterium *Clostridium tetani*, a spore-producing anaerobe. *C. tetani* is normally a nonpathogenic inhabitant of the intestinal tract of animals. But if it gains entry into tissue (usually through a wound), it can produce a powerful *exotoxin* (a poison liberated from the bacterial cell) capable of inducing an abnormal state of extreme muscle contraction.

(a) Explain the physiological phenomenon of tetanus in terms of stimulation of muscle contraction by nerve impulses.

(b) What is a likely explanation for the ability of *C. tetani* to produce a pathological tetanus condition?

(c) Why is vaccination with an attenuated strain of *C. tetani* (a "tetanus shot") especially recommended for wounds involving skin puncture?

21-7. A Moving Experience. For each of the following statements, indicate with the appropriate letter(s) whether it is true of the motility system that you use to lift your arm (A), to cause your heart to beat (H), to move ingested food through your intestine (I), or to sweep mucus and debris out of your respiratory tract (R).

(a) It depends on muscles that have a striated appearance when examined with an electron microscope.

(b) It would probably be affected by the same drugs that inhibit motility of a flagellated protozoan.

(c) It requires ATP.

(d) It involves calmodulin-mediated calcium stimulation.

(e) It involves interaction between actin and myosin filaments.

(f) It depends heavily on fatty acid oxidation for energy.

(g) It is under the control of the voluntary nervous system.

21-8. Microtubules and the Endoplasmic Reticulum. Kinesin is an ATPase thought to be involved in the transport of vesicles along axonal microtubules. It may also play a part in the location and shape of the endoplasmic reticulum (ER). Recent studies have demonstrated that kinesin can attach to the ER membrane in vitro and stretch it by pulling it along microtubules. In intact cells, the ER is known to be stretched outward from the centrosome.

(a) Suggest at least two different ways in which you could test whether microtubules are responsible in part for the location and shape of the ER in vivo.

(b) How might you establish more directly the involvement of kinesin in vivo?

21-9. Sterility, Bronchitis, and Sinusitis. Some forms of sterility in human males are due to nonmotile sperm. Upon cytological examination, the sperm of such individuals are found to have tails (i.e., flagella) that lack one or more of the normal structural components. Such individuals are also likely to have histories of respiratory tract disease, especially recurrent bronchitis and sinusitis, caused by an inability to clear mucus from the lungs and sinuses.

(a) What is the likely explanation for nonmotility of sperm in such cases of sterility?

(b) Why is respiratory tract disease linked with sterility in affected individuals?

(c) Would you expect male offspring of an affected individual to have the same defects?

21-10. Pulled in Two Directions. In skeletal muscle sarcomeres, the H zone is in the middle and bounded on each side by a Z line. During contraction, the Z line on each side moves in opposite directions toward the H zone. Myosin, however, can only crawl along an actin filament in one direction. How can you reconcile movements of Z lines in opposite directions with the unidirectional movement of myosin along an actin filament?

21-11. Chlortetracycline Fluorescence. Muscle preparations can be made in which the sarcolemma is removed or rendered permeable to small ions and molecules. Such preparations were used to study the phenomenon of calcium-induced calcium release. The experiment utilized an antibiotic called *chlortetracycline*, which accumulates in membrane-bound organelles or compartments inside the cell if they contain relatively high levels of calcium. If the membrane-bound organelles lose their calcium, chlortetracycline also leaves the organelle. Chlortetracycline is naturally fluorescent, so organelles that accumulate the antibiotic glow brightly under the fluorescence microscope.

(a) Explain how you might use the measurement of chlortetracycline fluorescence to test for calcium-induced calcium release.

(b) Which organelle of the muscle would you expect to accumulate chlortetracycline?

(c) Would ATP have to be present in the medium for chlortetracycline to accumulate?

(d) A tiny pipette containing calcium can be used to deliver small amounts of calcium to localized regions of the muscle cell.

Into which region of the muscle cell should the calcium be delivered?

(e) What should happen to the chlortetracycline fluorescence if calcium-induced calcium release takes place?

21-12. Motility Potpourri. Answer each of the following questions as concisely as possible.

(a) *Melanophores* are cells of teleost fish that contain pigment granules used to control coloration: When the granules are aggregated, the cell appears light; when the pigment granules are dispersed, the cell appears dark. The movement of the granules is under hormonal control. How could you tell if the distribution and movement of granules is an example of microfilament-based or microtubule-based motility?

(b) Stress fibers of nonmuscle cells contain contractile bundles of actin and myosin II. For stress fibers to contract or develop tension, how would actin have to be oriented within the stress fibers?

(c) Protrusion of the acrosomal filament is thought to require osmotic swelling of the membrane, along with actin polymerization. Supporting evidence comes from studies on the acrosome reaction of *Thyone briareus* (sea cucumber) sperm. When sucrose was added to the medium, the extension of the acrosomal process was blocked. Explain this result, and indicate how it supports the hypothesis.

(d) The beating movements of sperm flagella depend on precise control of where and when dynein ATPase molecules exert force on microtubules. Explain why, and indicate what would happen if all the dynein molecules were to exert force at the same time.

21-13. Research with Digital Video Microscopy. Most digital imaging systems used in microscopy digitize the image obtained from the video camera by converting it into an array of numbers. To do this, the image is subdivided into tiny squares called picture elements or pixels, with each pixel corresponding to one element of the array. If an 80-μm square region of the microscope field is digitized into an array of 512 × 512 pixels, what is the width and length of each pixel in μm, or microns? In the chapter, we showed how it was possible to obtain images of individual microtubules in the microscope. It is also known that when plastic beads are coated with kinesin and added to microtubules, the beads will exhibit ATP-dependent movement along microtubules. How could you use digital image processing to measure the rate at which the beads move along microtubules?

Suggested Reading

General References

Bement, W. M., and M. S. Mooseker. Keeping out the rain. *Nature* 365 (1993): 785–786.

Bray, D. *Cell Movements.* New York: Garland, 1992.

Bretscher, M. S. How animal cells move. *Sci. Amer.* 257 (December 1987): 72–90.

Huxley, A. F. *Reflections on Muscle.* Princeton, NJ: Princeton University Press, 1980.

Kreis, T. and R. Vale. *Guidebook to the Cytoskeletal and Motor Proteins.* New York: Oxford University Press, 1993.

Lackie, J. M. *Cell Movement and Cell Behaviour.* London: Allen & Unwin, 1985.

Stossel, T. P. The machinery of cell crawling. *Sci. Amer.* 271 (September 1994): 54–63.

Nonmuscle Microfilament-Based Movement

Condeelis, J. Life at the leading edge. *Annu. Rev. Cell Biol.* 9 (1993): 411–444.

Corthesy-Theulaz, I., A. Pauloin, and S. R. Rfeffer. Cytoplasmic dynein participates in the centrosomal localization of the Golgi complex. *J. Cell Biol.* 118 (1992): 1333–1345.

Egelhoff, T. T., and J. A. Spudich. Molecular genetics of cell migration: Dictyostelium as a model system. *Trends Gen.* 7 (1991): 161–166.

Fath, K. R., and D. R. Burgess. Membrane motility mediated by unconventional myosin. *Curr. Opin. Cell Biol.* 6 (1994): 131–135.

Fukui, Y. Toward a new concept of cell motility: Cytoskeletal dynamics in amoeboid movement and cell division. *Internat. Rev. Cytol.* 144 (1993): 85–127.

Grebecki, A. Membrane and cytoskeleton flow in motile cells with emphasis on the contribution of free living amoeba. *Internat. Rev. Cytol.* 148 (1994): 37–73.

Ho, W. C., V. J. Allan, G. van Meer, E. G. Berger, and T. E. Kreis. Reclustering of scattered Golgi elements occurs along microtubules. *Eur. J. Cell Biol.* 48 (1989): 250–263.

Karecla, P. I., and T. E. Kreis. Interaction of membranes of the Golgi complex with microtubules in vitro. *Eur. J. Cell Biol.* 57 (1992): 139–146.

Kemp, B. E., R. B. Pearson, C. House, P. J. Robinson, and A. R. Means. Regulation of protein kinases by pseudosubstrate prototopes. *Cell Signal* 1 (1989): 303–311.

Kreis, T. E. Role of microtubules in the organisation of the Golgi apparatus. *Cell Motil. and Cytoskel.* 15 (1990): 67–70.

Means, A. R., M. F. VanBerkum, I. Bagchi, K. P. Lu, and C. D. Rasmussen. Regulatory functions of calmodulin. *Pharmacol. Ther.* 50.2 (1991): 255–270.

Olson, N. J., *et al.* Regulatory and structural motifs of chicken gizzard myosin light chain kinase. *Proc. Natl. Acal. Sci. USA* 87 (1990): 2284–2288.

Taylor, D. L., and J. S. Condeelis. Cytoplasmic structure and contractility in amoeboid cells. *Internat. Rev. Cytol.* 56 (1979): 57.

Thyberg, J., and S. Moskalewski. Microtubules and organization of the Golgi apparatus. *Exp. Cell Res.* 159 (1985): 1–16.

Filament-Based Movement in Muscle

Ashcroft, F. M. Ca^{2+} channels and excitation-contraction coupling. *Curr. Opin. Cell Biol.* 3 (1991): 671–675.

Caswell, A. H., and N. R. Brandt. Triadic proteins of skeletal muscle. *J. Bioenerg. Biomembr.* 21 (1989): 149–162.

Eisenberg, E., and L. E. Greene. The relation of muscle biochemistry to muscle physiology. *Annu. Rev. Physiol.* 42 (1980): 293.

Fill, M., J. J. Ma, C. M. Knudson, T. Imagawa, K. P. Campbell, and R. Coronado. Role of the ryanodine receptor of skeletal muscle in excitation-contraction coupling. *Ann. N. Y. Acad. Sci.* 560 (1989): 155–162.

Franzini-Armstrong, C., and L. D. Peachey. Striated muscles: Contractile and control mechanisms. *J. Cell Biol.* 91 (1981): 166.

Huxley, H. E. The mechanism of muscular contraction. *Science* 164 (1969): 1356.

Lamb, G. D. DHP receptors and excitation-contraction coupling. *J. Muscle Res. Cell. Motil.* 13 (1992): 394–405.

MacLennan, D. H., C. J. Brandl, B. Korczak, and N. M. Green. Amino acid sequence of a Ca^{2+}, Mg^{2+}-dependent ATPase from rabbit muscle sarcoplasmic reticulum, deduced from its complementary cDNA sequence. *Nature* 316 (1985): 696.

Numa, S., T. Tanabe, H. Takeshima, A. Mikami, T. Niidome, S. Nishimura, B. A. Adams, and K. G. Beam. Molecular insights into excitation-contraction coupling. *Cold Spring Harb. Symp. Quant. Biol.* 55 (1990): 1–7.

Pollard, T. D. The myosin crossbridge problem. *Cell* 48 (1987): 909.

Squire, J. *The Structural Basis of Muscle Contraction.* New York: Plenum, 1981.

Taylor, K. A., and L. A. Amos. A new model for the geometry of the binding of myosin crossbridges to muscle thin filaments. *J. Mol. Biol.* 147 (1981): 297.

Warrick, H. M., and J. A. Spudich. Myosin structure and function in cell motility. *Annu. Rev. Cell Biol.* 3 (1987): 379–421.

Microtubule-Based Motility

Brokaw, C. J., D. J. L. Luck, and B. Huang. Analysis of the movement of *Chlamydomonas* flagella: The function of the radial-spoke system is revealed by comparison of wild-type and mutant flagella. *J. Cell Biol.* 92 (1982): 722.

Dustin, P. *Microtubules,* 2d ed. New York: Springer-Verlag, 1984.

Gibbons, I. R. Cilia and flagella of eukaryotes. *J. Cell Biol.* 91 (1981): 107.

Goldstein, L. S. B. With apologies to Scheherazade: Tails of 1001 kinesin motors. *Annu. Rev. Genet.* 27 (1993): 319–351.

Goodenough, U. W. Cilia, flagella and the basal apparatus. *Curr. Opin. Cell Biol.* 1 (1989): 58–62.

Goodenough, U. W., and J. E. Heuser. Substructure of the outer dynein arm. *J. Cell Biol.* 95 (1982): 795.

Grigg, G. Discovery of the 9 + 2 subfibrillar structure of flagella/cilia. *Bioessays* 13 (1991): 363–369.

Holwill, M. E., and P. Satir. A physical model of microtubule sliding in ciliary axonemes. *Biophys. J.* 58 (1990): 905–917.

Omoto, C. K. Mechanochemical coupling in cilia. *Internat. Rev. Cytol.* 131 (1991): 255–292.

Porter, M. E., and K. A. Johnson. Dynein structure and function. *Annu. Rev. Cell Biol.* 5 (1989): 119–151.

Satir, P. How cilia move. *Sci. Amer.* 231 (October 1974): 44.

Vallee, R. B., and H. S. Sheptner. Motor proteins of cytoplasmic microtubules. *Annu. Rev. Biochem.* 59 (1990): 909–932.

Warner, F. D., and D. R. Mitchell. Dynein, the mechano-chemical coupling adenosine triphosphatase of microtubule-based sliding filament mechanisms. *Internat. Rev. Cytol.* 66 (1980): 1.

Witman, G. B. Axonemal dyneins. *Curr. Opin. Cell Biol.* 4 (1992): 74–79.

Wilson, L. Action of drugs on microtubules. *Life Sci.* 17 (1975): 303.

Bacterial Flagella

Adler, J. The sensing of chemicals by bacteria. *Sci. Amer.* 234 (April 1976): 40.

Hazelbauer, G. L. Bacterial chemotaxis: Molecular biology of a sensory system. *Endeavour* 4 (1980): 67.

Macnab, R. M. The bacterial flagellar motor. *Trends Biochem. Sci.* 9 (1984): 185.

22

SIGNAL TRANSDUCTION MECHANISMS: I. ELECTRICAL SIGNALS IN NERVE CELLS

In the previous chapter we saw how studies of cytoskeletal structures in muscle cells led to a more general understanding of the molecular mechanisms of motility. This underscores an important principle in biological research: Cellular functions are often best studied in cells that are highly specialized for the function of interest. In the first part of this chapter, we will again be looking at specialized cell functions, but this time in nerve cells. Virtually all cells maintain electrical potentials across their plasma membranes, but nerve cells have special mechanisms for using this potential to transmit information over long distances.

In the second part of the chapter, we will discuss the process by which information is passed between cells in the nervous system. The communication of information always involves at least two components: the sender and the receiver. In the nervous system, the *sender* is a nerve cell or sensory cell, and the *receiver* is a second nerve cell, a gland, or a muscle cell. We will see that nerve cells use specialized processes to deliver information across their junctions with other cells. Such information may be transmitted by direct electrical connection, but more often the process of transmission involves the exocytotic release of chemical *neurotransmitters* at the junction, followed by binding of the neurotransmitters to *receptors* on the plasma membrane of the second cell. Our knowledge of neurotransmitters enables us to understand how pharmaceutical agents interact with the brain and how certain highly toxic compounds affect the nervous system.

The Nervous System

Every animal has a **nervous system,** in which electrical impulses are transmitted along the specialized plasma membranes of nerve cells. The nervous system performs three functions: It *collects* information from the environment ("the light just turned green"), it *processes* that information ("green means go"), and it *elicits responses* to that information by triggering specific effectors, usually muscle tissue or glands ("push on the accelerator").

To accomplish these functions, the nervous system has special components for sensing and processing information and for triggering the appropriate response (Figure 22-1). In vertebrates, the nervous system is divided into two components, the central nervous system and the peripheral nervous system. The **central nervous system (CNS)** consists of the brain and the spinal cord, including both sensory and motor cells; the **peripheral nervous system (PNS)** consists of all other sensory and motor components, including the somatic nervous system and the autonomic nervous system. The **somatic nervous system** controls voluntary movements of skeletal muscles, whereas the **autonomic nervous system** controls the involuntary activities of cardiac muscle, the smooth muscles of the gastrointestinal tract and blood vessels, and a variety of secretory glands.

Cells that make up the nervous system can be broadly divided into two groups: *neurons* or *nerve cells,* and *glial cells.* **Neurons** can be subdivided into three basic types based on function: sensory neurons, motor neurons, and interneurons. **Sensory neurons** are a diverse group of cells specialized for the detection of various types of stimuli. Examples of sensory neurons include the photoreceptors of the retina, olfactory neurons, and the various touch, pressure, pain, and temperature-sensitive neurons located in the skin or joints. Sensory neurons provide a continuous stream of information to the brain about the state of the body and its environment. **Motor neurons** transmit signals from the CNS to muscles or glands, and **interneurons** process signals received from other neurons and relay the information to other parts of the nervous system.

The term **glial cell** (from *glia,* the Greek word for "glue") encompasses a variety of different cell types, including microglia, oligodendrocytes, Schwann cells, and astrocytes. *Microglia* are phagocytic cells that fight infections and remove debris. *Oligodendrocytes* and *Schwann cells* form the

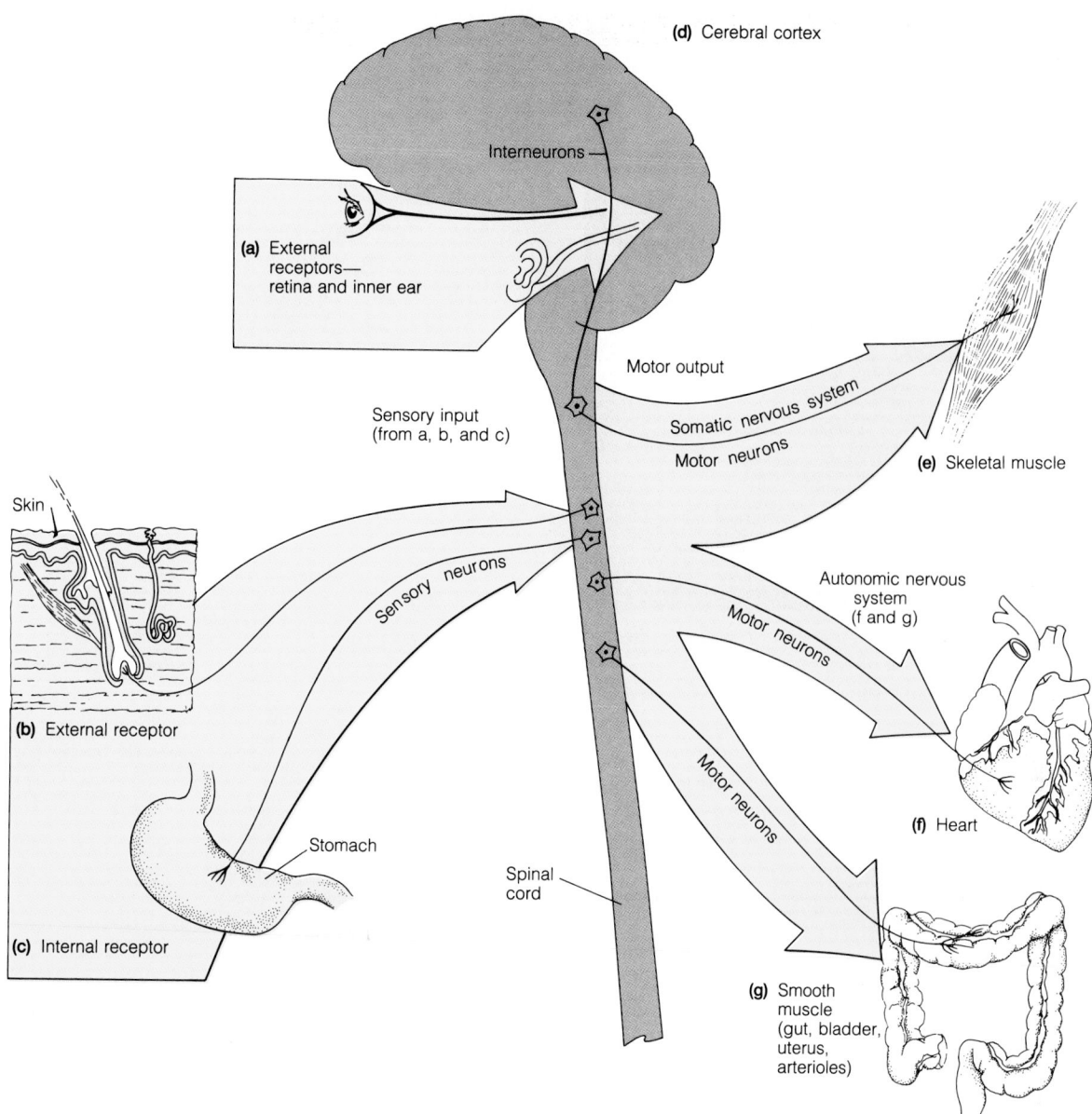

(d) Cerebral cortex

Interneurons

(a) External receptors— retina and inner ear

Motor output

Sensory input (from a, b, and c)

Somatic nervous system

Motor neurons

(e) Skeletal muscle

Skin

Sensory neurons

Autonomic nervous system (f and g)

Motor neurons

(b) External receptor

Stomach

Motor neurons

(f) Heart

Spinal cord

(c) Internal receptor

(g) Smooth muscle (gut, bladder, uterus, arterioles)

Figure 22-1 The Vertebrate Nervous System. The nervous system consists of the central nervous system (gray) and the peripheral nervous system (yellow). The sensory neurons of the peripheral nervous system receive information from **(a, b, c)** external and internal receptors and transmit it to the CNS. Interneurons in the CNS integrate and coordinate response to the sensory input, a primary example being **(d)** cerebral cortical neurons. Motor responses originate in the CNS and are transmitted to **(e)** skeletal muscles by neurons of the somatic nervous system and to **(f, g)** involuntary muscles and glands by means of the autonomic nervous system.

myelin sheath around neurons of the CNS and those of the peripheral nerves, respectively. The function of the *astrocytes* is still obscure but may involve the maintenance of the potassium concentration in the extracellular fluid surrounding nerves and blood vessels.

Intricate networks of neurons make up the complex tissues of the brain that are responsible for coordinating nervous function. About 10 billion nerve cell bodies are involved in the networks, which sometimes form layers of tissue. Each neuron receives input from thousands of other neurons, so the brain's connections easily number in the trillions. By comparison, computers do not even come close to this number of connections, despite their seeming complexity (Figure 22-2).

Like data in a computer, however, neural information does consist of individual "bits," or *signals*. Our purpose here is not to discuss the overall functioning of the nervous system, but to focus on cellular mechanisms by which the

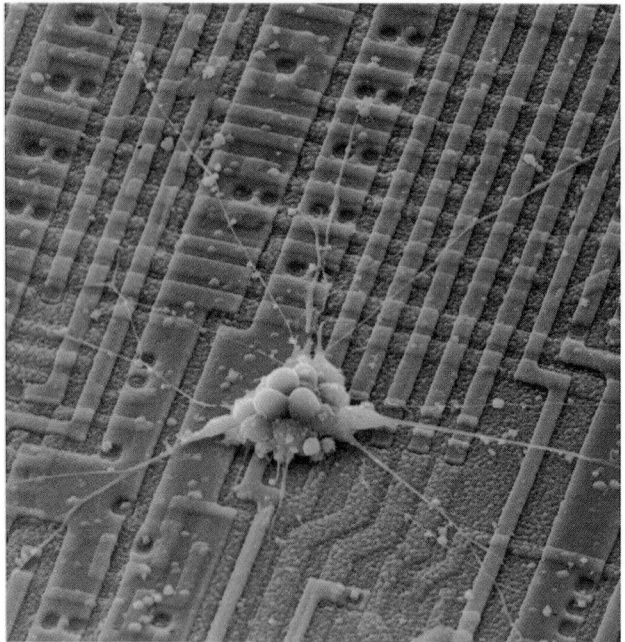

Figure 22-2   **A Neuron on a Microprocessor.**   This micrograph shows a neuron growing on top of a Motorola 68000 microprocessor chip. The neuron is the fundamental information-processing unit of the brain, which might be compared to the transistor as the fundamental processing unit in the computer. However, the brain has 15 billion neurons, whereas microprocessors may have up to a few million transistors (SEM).

series of electrical signals called *nerve impulses* are spread. In addition to understanding the functions of nerve cells specifically, we will also acquire a better appreciation for several general aspects of membrane function, of which nerve cells are a specialized example.

The Structure of a Neuron

The structure of a typical motor neuron is shown in Figure 22-3. The **cell body** of most neurons is similar to that of other cells, consisting of a nucleus and most of the same organelles. In addition, however, neurons contain extensions, or branches, called **processes.** These processes make neurons easy to distinguish from almost any other cell type. There are two types of processes: Those that receive signals and transmit them inward to the cell body are called **dendrites,** and those that conduct signals away from the cell body are called **axons.** The cytoplasm within an axon is commonly referred to as **axoplasm.** A **nerve** is simply a tissue composed of bundles of axons.

Neurons display more structural variability than Figure 22-3 suggests. Some sensory neurons have only one process, which conducts signals both toward and away from the cell body. Moreover, the structure of the dendritic processes is not random; many different classes of neurons in the central

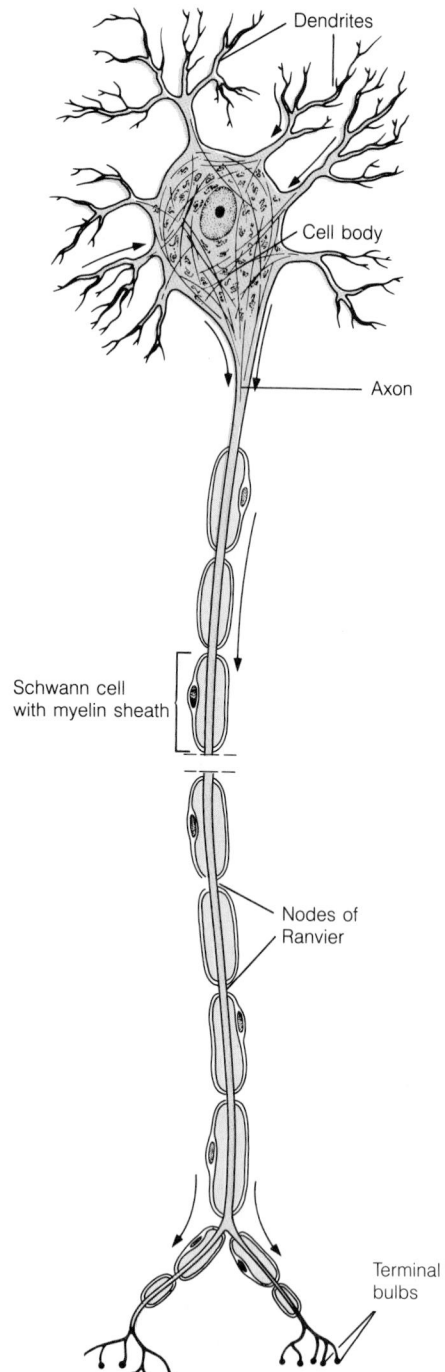

Figure 22-3   **The Structure of a Typical Motor Neuron.**
The cell body contains the nucleus and most of the usual organelles. Dendrites conduct signals passively inward to the cell body, whereas the axon transmits signals actively outward (the direction of transmission is shown by black arrows). At the end of the axon are numerous terminal bulbs. Some, although not all, neurons have a discontinuous myelin sheath around their axons to insulate them electrically. Each segment of the sheath consists of a concentric layer of membranes wrapped around the axon by a Schwann cell (or an oligodendrocyte, in the CNS). The breaks in the myelin sheath, called nodes of Ranvier, are concentrated regions of electrical activity.

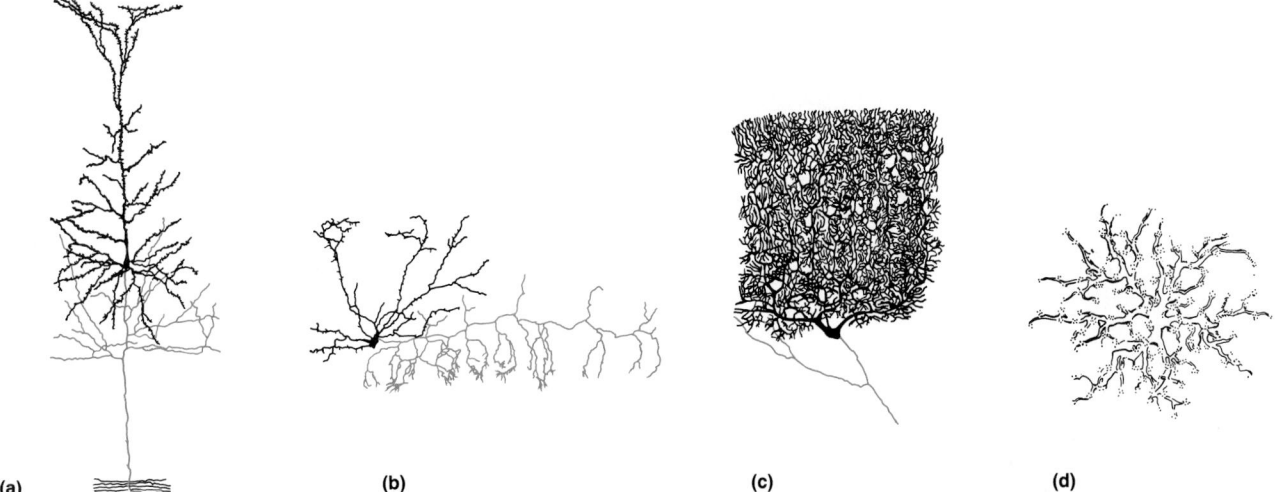

Figure 22-4 Neuron Shapes. Neurons of the central nervous system display a wide variety of characteristic shapes. Axons are shown in pink and dendrites in black. **(a)** A pyramidal neuron from the cerebral cortex. **(b)** Several short-axon cells in the cerebral cortex. **(c)** A Purkinje cell in the cerebellum. **(d)** An axonless horizontal cell in the retina of the eye.

nervous system can be identified by structure alone (Figure 22-4).

As Figure 22-3 illustrates, a motor neuron has multiple, branched dendrites and a single axon leading away from the cell body. The axon of a typical neuron is much longer than the dendrites and forms multiple branches. Each branch terminates in structures called **terminal bulbs,** also known as *synaptic knobs* or *axon terminals.* The terminal bulbs are responsible for transmitting the signal to the next cell, which may be another neuron or a muscle or gland cell. In each case, the junction is called a **synapse.** For neuron-to-neuron junctions, synapses may occur between an axon and a dendrite, between a dendrite and a dendrite, between an axon and a cell body, or even between an axon and an axon. Typically, neurons have synapses with many other neurons.

Axons can be very long—up to several thousand times longer than the diameter of the cell body. For example, a motor neuron that innervates your foot has its cell body in your spinal cord, and its axon extends approximately a meter down a nerve tract in your leg!

The Myelin Sheath

Most axons are surrounded by a discontinuous **myelin sheath** consisting of many concentric layers of membrane. The myelin sheath is a very effective electrical insulation for the segments of the axon that it envelops. As noted, the myelin sheath of neurons in the CNS is formed by **oligodendrocytes,** whereas in the PNS it is formed by **Schwann cells.** Figure 22-5a shows a cross section of a myelinated nerve axon in the peripheral nervous system, and Figure 22-5b illustrates the process of myelination. A Schwann cell envelops an axon and wraps layer after layer of its own plasma membrane around the axon in a tight spiral.

Each Schwann cell is responsible for the myelin sheath around a short segment (about 1 mm) of a single axon. Numerous Schwann cells are therefore required to encase a PNS axon with discontinuous sheaths of myelin. In the CNS, however, a single oligodendrocyte myelinates many axons. The small regions of bare axon between successive segments of the myelin sheath are called **nodes of Ranvier** (see Figure 22-3). These nodes are only about 2 μm long, but they are concentrated regions of electrical activity and therefore play an important role in the transmission of signals, as we shall see shortly.

Electrical Properties of Neurons

A **membrane potential** is a fundamental property of essentially all cells. It results from an excess of negative charge on one side of the plasma membrane and an excess of positive charge on the other side. Cells at rest normally have an excess of negative charge inside and an excess of positive charge outside the cell; this is called the **resting membrane potential (V_m).** The resting membrane potential can be measured by placing one tiny electrode inside the cell and another outside the cell. The electrodes compare the ratio of negative to positive charge inside the cell and outside the cell. Because the inside of a cell typically has an excess of negative charge, we say that the cell has a *negative resting membrane potential.* Cell membrane potentials are measured in millivolts (mV).

Nerve, muscle, and certain other cell types such as the islet cells of the pancreas and cells of a water plant called *Nitella* exhibit a special property called **electrical excitability.** In electrically excitable cells, certain types of stimuli

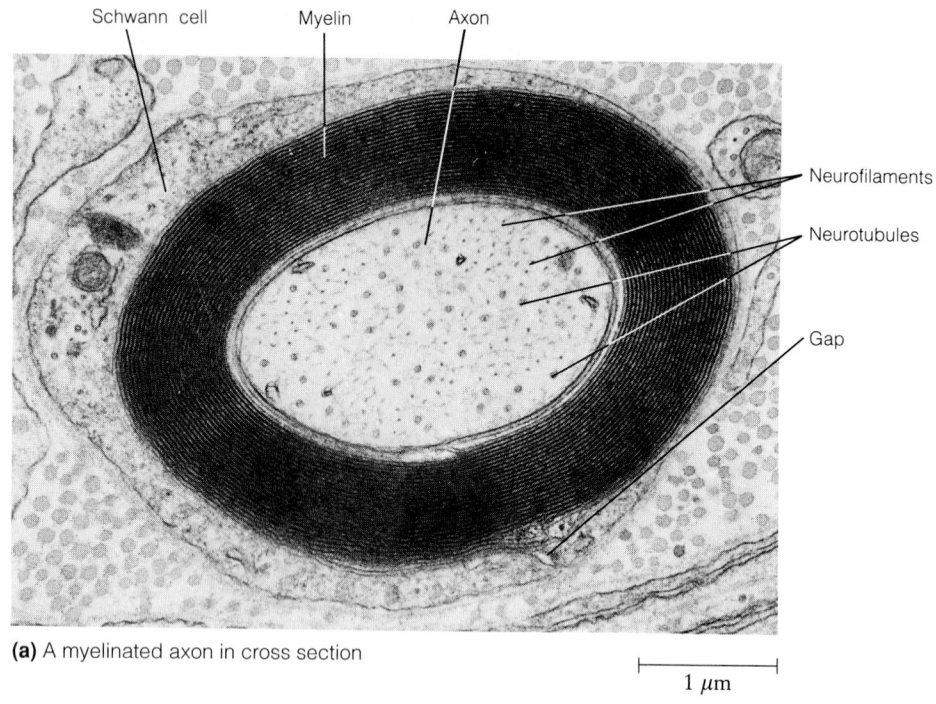

Schwann cell   Myelin   Axon

Neurofilaments

Neurotubules

Gap

(a) A myelinated axon in cross section

1 μm

Figure 22-5   Myelination of Axons. **(a)** This cross-sectional view of a myelinated axon from the nervous system of a cat shows the concentric layers of unit membrane that have been wrapped around the axon by the Schwann cell that envelops it. The gap in the plasma membrane of the Schwann cell is the point at which the membrane initially invaginated to begin enveloping the axon. Notice also the neurotubules and neurofilaments in the axoplasm of the myelinated axon (TEM). **(b)** An axon of the peripheral nervous system being myelinated by a Schwann cell. Each Schwann cell gives rise to one segment of myelin sheath by wrapping its own plasma membrane concentrically around the axon. The myelin layer gets progressively thicker as more layers of unit membrane derived from the plasma membrane of the Schwann cell are added to it, accompanied by the gradual loss of the Schwann cell cytoplasm.

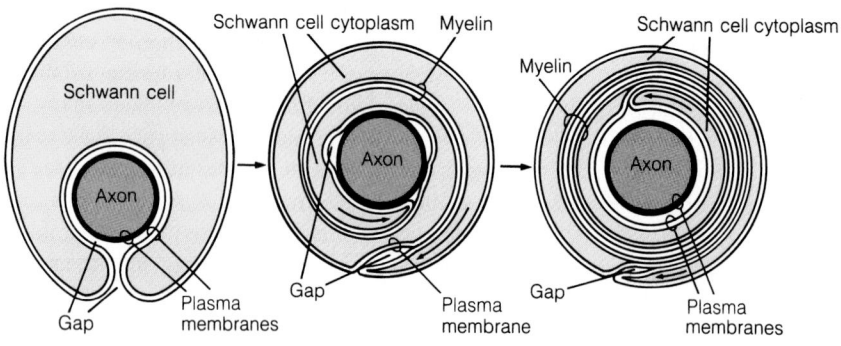

Schwann cell cytoplasm   Myelin   Schwann cell cytoplasm

Myelin

Schwann cell

Axon   Axon   Axon

Gap   Gap   Gap
Plasma   Plasma   Plasma
Gap   membranes   membrane   membranes

(b) The process of axon myelination

trigger a rapid sequence of changes in the membrane potential known as an *action potential*. During an action potential, the membrane potential changes from negative values to positive values and then back to negative values again, all in little over a millisecond. In nerve cells, the action potential has the specific function of transmitting an electrical signal along the axon.

Understanding Membrane Potentials

To understand how nerve cells use action potentials to transmit signals, we must first comprehend how cells generate a resting membrane potential and how the membrane potential changes during an action potential. Then we will examine how an action potential conveys a signal from one neuron to another.

The Resting Membrane Potential

The resting membrane potential develops because the cytosol of the cell and the extracellular fluid contain different compositions of cations and anions. Extracellular fluid is a watery solution of salts, including sodium chloride and lesser amounts of potassium chloride. The cytosol contains potassium rather than sodium as its main cation because of the action of the sodium-potassium pump (described in Chapter 8). The anions in the cytosol consist largely of macromolecules such as proteins, RNA, and a variety of other molecules that are not present outside the cell. These negatively charged macromolecules cannot pass through the cell membrane and therefore remain inside the cell. The presence of impermeable anions in the cytosol is the main reason that cells develop a resting membrane potential.

To understand how a membrane potential forms, we need to recall a few basic physical principles. First, all substances tend to diffuse from a higher concentration to a lower concentration. Cells normally have a high concentration of potassium ions inside and a low concentration of potassium ions outside. We refer to this uneven distribution of potassium ions as a *potassium ion gradient*. By convention, the potassium ion gradient is expressed as the molar concentration of potassium ions outside the cell ($[K^+]_{outside}$) divided by the molar concentration of potassium in the cytoplasm ($[K^+]_{inside}$), or $[K^+]_{outside}/[K^+]_{inside}$. Given the large potassium concentration gradient, potassium ions will tend to diffuse out of the cell.

The second basic principle is that of *electroneutrality*. When ions are in solution, they are always present in pairs, one positive ion for each negative ion, so that there is no charge imbalance. For any given ion, which we will call A, there must be an oppositely charged ion B in the solution; therefore, we can refer to B as the *counterion* for A. In the cytosol, potassium ions (K^+) serve as the counterions for the trapped anions. Outside the cell, sodium (Na^+) is the main cation and chloride (Cl^-) is the counterion.

Although a solution must have an equal number of positive and negative charges overall, these charges can be locally separated so that one region has more positive charges while another region has more negative charges. Because it takes work to separate charges, once they have been separated, they tend to move back toward each other. The tendency of oppositely charged ions to flow back toward each other is called a **potential** or **voltage.** When negative or positive ions are actually moving, one toward the other, we say that current is flowing, and this current is measured in amperes (A). Given these principles, we can understand how a resting membrane potential will form as a result of the ionic compositions of the cytosol and the extracellular fluid, and the characteristics of the plasma membrane.

The plasma membrane is normally permeable to potassium because it contains what are commonly known as *potassium leak channels,* which permit potassium ions to diffuse out of the cell. However, there are no channels for negatively charged macromolecules. As potassium leaves the cytosol, increasing numbers of trapped anions are left behind without counterions. Excess negative charge therefore accumulates in the cytosol and excess positive charge accumulates on the outside of the cell, resulting in a membrane potential.

The formation of a membrane potential is illustrated in Figure 22-6, using a container subdivided into two compartments by a semipermeable membrane. The left compartment represents the cytosol of the cell and the right side the extracellular fluid. The semipermeable membrane is permeable to potassium ions but not to negatively charged macromolecules (represented as M^-). The cytosolic compartment contains a mixture of potassium ions and negatively charged macromolecules. For simplicity, we will assume that the extracellular compartment starts with nothing but water.

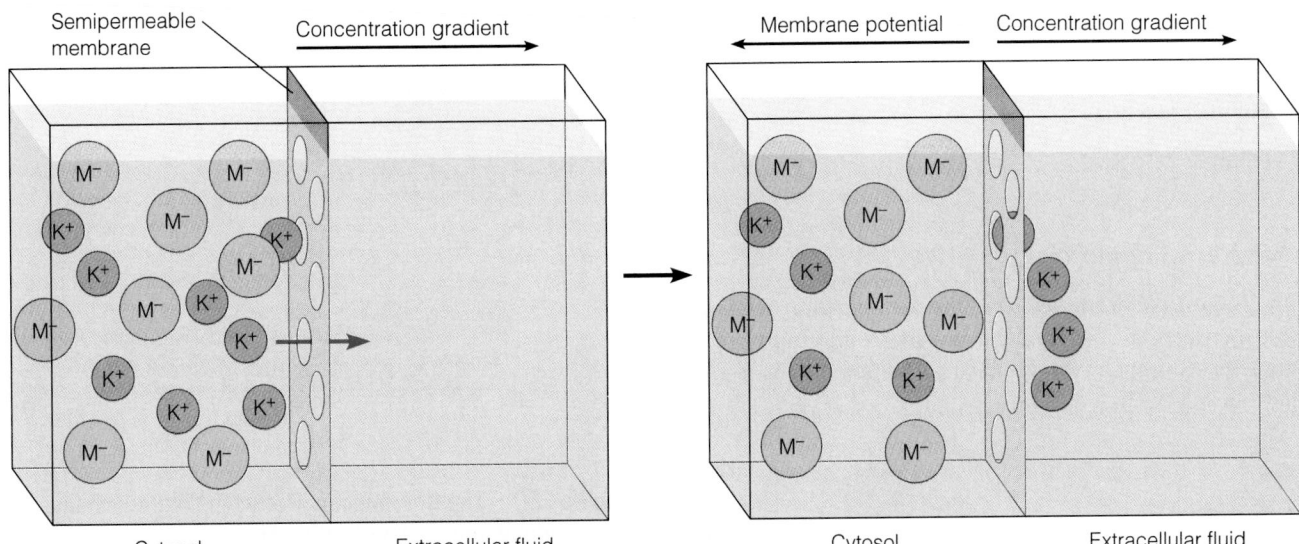

Figure 22-6   Development of the Equilibrium Membrane Potential.   A two-compartment container is used to represent a cell, with a semipermeable membrane separating the compartments. In each container, the left-hand compartment represents the cytosol, and the right-hand compartment represents the extracellular fluid. The cytosol contains a high concentration of potassium ions (K^+) and impermeable anions (M^-) relative to their concentrations in the extracellular fluid. As potassium ions diffuse out of the cell (from left to right), the impermeable anions are left behind, creating a membrane potential. The magnitude of the membrane potential increases until an equilibrium is reached in which the electrical attraction of the anions for potassium ions prevents any further net diffusion of potassium ions out of the cell. At this point the membrane potential is in equilibrium with the potassium ion gradient.

Under the pressure of a concentration gradient, potassium ions diffuse across the membrane from the left side to the right side. However, the negatively charged macromolecules are not free to follow. The result is an accumulation of anions on the left side and of cations on the right side. A membrane potential is created by a separation of negative charge from positive charge. The source of work (defined here as force × distance) needed to produce this charge separation is the potassium ion gradient.

Eventually, the membrane potential becomes great enough to prevent further net diffusion of potassium ions across the membrane. As potassium ions diffuse from left to right in Figure 22-6, the left compartment becomes increasingly more negative. This potential ultimately builds to a point at which the positively charged potassium ions are pulled back into the left compartment as fast as they leave. In this way, an equilibrium is reached in which the force of attraction due to the membrane potential balances the tendency of potassium to diffuse down its concentration gradient. This type of equilibrium, in which a chemical gradient is balanced with an electrical potential, is referred to as an **electrochemical equilibrium**. The membrane potential at the point of equilibrium is known as an **equilibrium membrane potential**.

We have seen that the formation of a membrane potential is an important consequence of the potassium ion gradient when potassium ions are free to diffuse out of the cell while anions remain trapped inside. The potassium ion gradient did the work needed to separate negative and positive charges, resulting in an equilibrium membrane potential. Therefore, the magnitude of the potassium ion gradient is related to the magnitude of the equilibrium membrane potential. The Nernst equation provides a mathematical description of this relationship and enables us to estimate the membrane potential.

The Nernst Equation

The **Nernst equation** describes the mathematical relationship between an ion gradient and the equilibrium membrane potential that will form when the membrane is permeable only to that ion:

$$E_X = RT/zF \ln [X]_{outside}/[X]_{inside} \qquad \textbf{(22–1)}$$

where E_X is the equilibrium membrane potential for ion X (in volts), R is the gas constant (1.987 cal/mol-degree), T is the absolute temperature in kelvin units, z is the valence of the ion, F is the Faraday constant (23,062 cal/V-mol), $[X]_{outside}$ is the concentration of X outside the cell (in M), and X_{inside} is the concentration of X inside the cell (in M). This equation can be simplified if we assume a temperature of 18°C (a value appropriate for marine organisms) and that X is a monovalent cation and therefore has a valence of +1. By substituting into the equation the values for R, T, F, and z

and converting from natural logs to $\log_{10}$ ($\log_{10} = 2.3 \ln$) for illustration purposes, the equation reduces to:

$$E_X = 0.058 \log_{10} [X]_{outside}/[X]_{inside} \qquad \textbf{(22–2)}$$

In this simplified form we can see that for every tenfold increase in the cation gradient, the membrane potential changes by −0.058 V, or −58 mV. This relationship is shown in Figure 22-7 for potassium ions (X = K$^+$).

Diffusible Anions and the Donnan Equilibrium. Figure 22-6 explains the formation of the membrane potential but is incomplete, mainly because it does not take into account diffusible anions that are present on both sides of the membrane. The main anion in the extracellular fluid is chloride, to which the cell membrane is relatively permeable. Now we need to examine how the presence of impermeable anions in the cytosol forces potassium ions and chloride ions to distribute unequally between the cytosol and the extracellular fluid. This unequal distribution is referred to as **Donnan equilibrium,** based on its original description by Frederick Donnan in 1911.

To illustrate how chloride ions distribute between the inside and outside of the cell, we will return to our two-

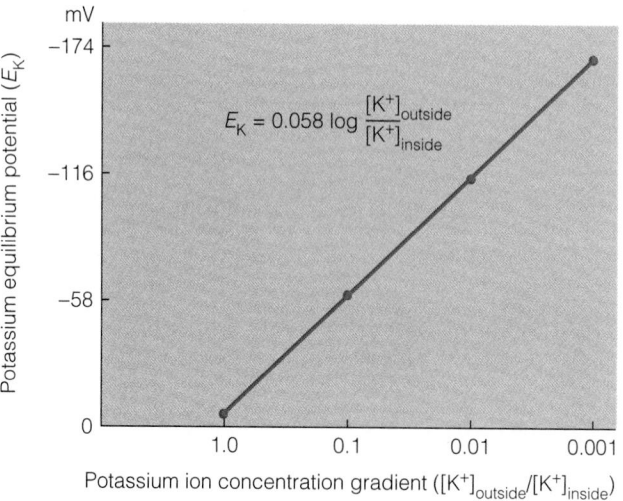

Figure 22-7   The Relationship Between the Potassium Ion Gradient and the Equilibrium Membrane Potential.
The Nernst equation was used to calculate equilibrium membrane potential for tenfold changes in the potassium ion concentration gradient across the membrane. By convention, the potassium ion gradient is expressed as the ratio of [K$^+$] outside the cell to [K$^+$] inside the cell, or [K$^+$]$_{outside}$ /[K$^+$]$_{inside}$. When the ratio is expressed this way, increasing potassium inside the cell relative to outside produces progressively smaller fractions. The result, based on calculations at a temperature of 18°C, shows that for every tenfold change in the potassium ion gradient, there is a 58-mV change in membrane potential.

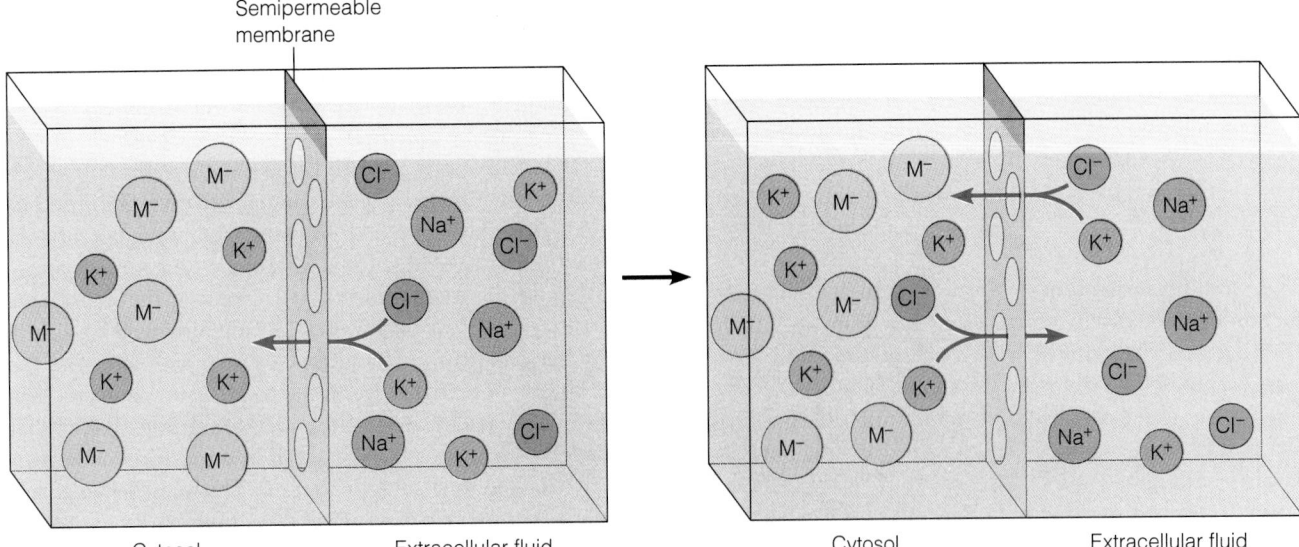

Figure 22-8 Donnan Equilibrium. Donnan equilibrium explains why chloride ions are more concentrated outside the cell than inside, even though the membrane is permeable to both potassium and chloride ions. Chloride ions start out much more concentrated on the outside (right compartment) and will tend to diffuse inward (to the left). Potassium ions are also permeable, and a positively charged potassium ion will diffuse into the cell along with each chloride ion, thereby preserving electrical neutrality. At equilibrium, the chloride ions diffuse across the membrane at the same rate in both directions, but the concentrations of chloride inside and outside the cell are quite different. This is because the high concentration of potassium ions inside the cell leads to the rapid leakage of potassium and chloride ions, even though the concentration of chloride in the cytosol is relatively low.

compartment model (Figure 22-8). As in Figure 22-6, the left compartment represents the cytosol, the right compartment represents the extracellular fluid, and the barrier represents a plasma membrane that is permeable to potassium and chloride ions. Now, however, a solution containing 0.1 M potassium ions and negatively charged macromolecules has been added to the left compartment, and a solution containing 0.1 M potassium chloride has been added to the right compartment. Even though the concentration of potassium ions is the same on both sides, the system is not at equilibrium because of the large chloride ion concentration gradient. Therefore, chloride ions will diffuse from right to left. A potassium ion will accompany each chloride ion that diffuses across the barrier, thereby preserving electrical neutrality. As chloride diffuses down its concentration gradient from right to left, potassium becomes more concentrated on the left side. This imbalance of potassium is one of the features of Donnan equilibrium.

Eventually, the concentration of potassium chloride will build up in the left compartment until an equilibrium is reached, with potassium and chloride ions diffusing in both directions at the same rate. If we were to measure the concentration of potassium and chloride in both compartments at equilibrium, we would find that potassium is more concentrated on the left side and chloride is more concentrated on the right side. In fact, we would find that

$$[K^+]_{left}/[K^+]_{right} = [Cl^-]_{right}/[Cl^-]_{left} \qquad \textbf{(22–3)}$$

In other words, the chloride gradient is the reciprocal or op-posite of the potassium gradient. Reciprocal concentration gradients of chloride and potassium ions are the hallmark of Donnan equilibrium.

Ion Distribution and the Problem of Cell Swelling. Thus far, we have seen that the presence of negatively charged macromolecules in the cytosol lies at the heart of both the negative equilibrium membrane potential and the reciprocal distribution of potassium ions and chloride ions. Their presence also causes the cytosol to be hypertonic with respect to the extracellular fluid. The cytosol is *hypertonic* because it contains more dissolved substances than the solution outside the cell. This can be demonstrated by summing the ions present inside and outside. The cell contains both impermeable and diffusible anions, and to maintain electrical neutrality, it must also contain a matching number of cations. Outside the cell, there will be only diffusible anions (A^-) with a matching number of cations (C^+). We can express this idea mathematically as:

$$[C^+]_{inside} = [M^-]_{inside} + [A^-]_{inside} \qquad \textbf{(22–4)}$$

Outside the cell, electrical neutrality is achieved when:

$$[C^+]_{outside} = [A^-]_{outside} \qquad \textbf{(22–5)}$$

If we were to sum the concentrations of solutes inside the cell, we would find that there are more dissolved substances inside the cell. Therefore, water would tend to enter the cytosol, causing the cell to swell. If this process were to continue unchecked, the cell would eventually burst.

To prevent swelling and rupture, an impermeable solute is needed outside the cell to balance the osmolarity of the cytosol. The high concentration of sodium ions present in the extracellular fluid fulfills this requirement because of the poor permeability of the membrane to sodium ions. In addition, when sodium ions do leak through the membrane, they are removed from the cytosol by the sodium-potassium pump.

The Sodium-Potassium Pump. Although the plasma membrane is relatively impermeable to sodium ions, there is always a small amount of leakage. If sodium continued to leak, eventually the cell would swell and burst. To compensate for this leakage, the **sodium-potassium pump** continually pumps sodium out of the cell while carrying potassium inward (see Figure 8-6). On average, the sodium-potassium pump transports three sodium ions out of the cell and two potassium ions into the cell for every molecule of ATP that is hydrolyzed. This net transport of ions out of the cell dilutes the cytosol, thereby preventing swelling. In addition, the sodium-potassium pump maintains the large potassium ion gradient across the membrane that provides the basis for the resting membrane potential.

Steady-State Ion Concentrations and the Resting Membrane Potential

Sodium, potassium, and chloride ions are the major ionic components present in both the cytosol and the extracellular fluid. Due to their unequal distributions across the cell membrane, each ion has a different impact on the membrane potential. The magnitude of the concentration gradient for each ion is illustrated by the bar graphs in Figure

22-9, in which the height of each bar is proportional to the concentration of a specific ion in the cytosol or the extracellular fluid. Each ion will tend to diffuse down its concentration gradient, and if allowed to do so will produce a change in the membrane potential.

Potassium ions tend to diffuse out of the cell, which makes the membrane potential more negative. Sodium ions tend to flow into the cell, driving the membrane potential in the more positive direction and thereby causing a **depolarization** of the membrane (that is, causing the membrane potential to be less negative). Chloride ions tend to diffuse into the cell slowly which should, in principle, make the membrane potential more negative. However, chloride ions are also repelled by the negative membrane potential, so that chloride ions usually enter the cell in association with positively charged ions such as sodium. This paired movement nullifies the depolarizing effect of sodium entry. Increasing the permeability of cells to chloride can have two effects, both of which decrease neuronal excitability. First, the net entry of chloride ions (chloride entry without a matching cation) causes hyperpolarization of the membrane (that is, the membrane potential becomes more highly negative than usual). Second, when the membrane becomes permeable to sodium ions, some chloride will tend to enter the cell along with sodium. Understanding this effect of chloride entry will be important later when we discuss inhibitory neurotransmitters.

Steady-State Ion Movements and the Goldman Equation

If we want to take into account the relative contributions of each of several ions to the resting membrane potential of a

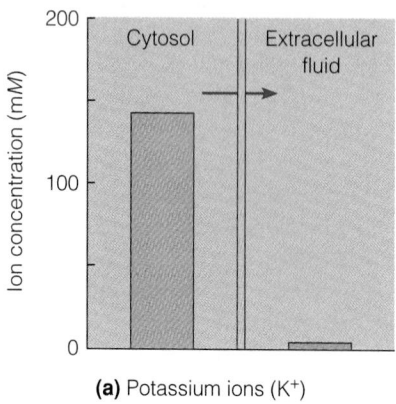 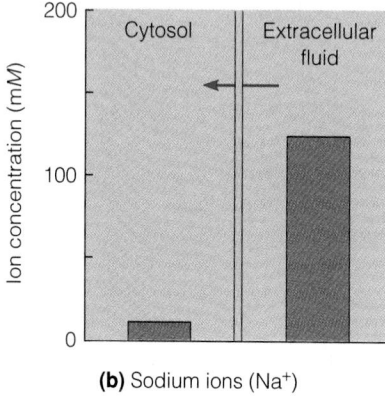 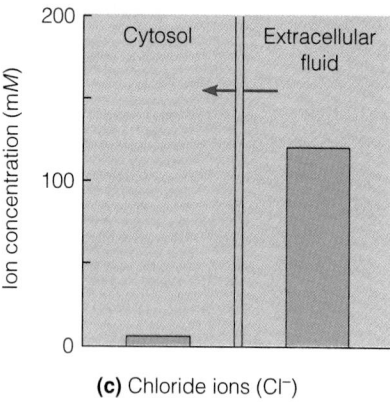

(a) Potassium ions (K⁺)   **(b)** Sodium ions (Na⁺)   **(c)** Chloride ions (Cl⁻)

Figure 22-9   Relative Concentrations of Potassium, Sodium, and Chloride Ions Across the Cell Membrane.
Each of the major ions in the cytosol exists as a concentration gradient across the plasma membrane, and each is capable of having an effect on the membrane potential. **(a)** Potassium ions are more concentrated in the cytosol than in the extracellular fluid. As a result, potassium ions have a tendency to move out of the cell, leaving behind trapped anions. Therefore, the loss of potassium causes the membrane potential to become more negative. **(b)** Sodium ions are much more concentrated outside the cell than inside; therefore, sodium ions tend to enter the cell. As sodium

ions enter, they neutralize some of the excess negative charge in the cytosol. As a result, the membrane potential becomes more positive.
(c) Chloride ions distribute across the membrane in Donnan equilibrium. Chloride ions usually cross the membrane together with a permeable cation (normally a potassium ion). Although chloride ions are concentrated outside the cell, the extracellular potassium ion concentration is low, which limits the rate of chloride entry. In contrast, the potassium concentration in the cytosol is high, allowing chloride to leave the cell even at low cytosolic chloride concentrations.

cell, we cannot use the Nernst equation because it deals with only one type of ion at a time, and it assumes that this ion is in electrochemical equilibrium. These relative contributions are important to an understanding of the actual conditions in the cell because even in its resting state, the cell has some permeability to sodium and chloride ions in addition to potassium ions. To take into account the leakage of sodium and chloride ions into the cell, we must move from the more static concept of an equilibrium membrane potential to a consideration of **steady-state ion movements** across the membrane.

We can illustrate the concept of steady-state ion movements by returning to our model of a cell in electrochemical equilibrium (Figure 22-10). As mentioned in our discussion of the Nernst equation, a cell that is permeable only to potassium will have a membrane potential equal to the equilibrium potential for potassium ions. Under these conditions, there will be no net movement of potassium out of the cell. If we now assume that the membrane is slightly permeable to sodium ions, what will happen? We know that the cell will have both a large sodium gradient across the membrane and a negative membrane potential corresponding to the potassium equilibrium potential. These forces tend to drive sodium ions into the cell. As sodium ions leak inward, the membrane is partially depolarized. At the same time, by neutralizing the membrane potential, there is now less restraining force preventing potassium from leaving the cell, so potassium ions diffuse outward to balance the inward movement of sodium. The inward movement of sodium ions shifts the membrane potential in the positive direction,

while the outward movement of potassium ions shifts the membrane potential in the negative direction.

The membrane potential now becomes a function not only of ion gradients but also of the rate at which ions can flow through the membrane—a very important new concept. Movements of sodium and potassium ions across the membrane have essentially opposite effects on the membrane potential. For mammalian cells, the sodium ion gradient tends toward a cell membrane potential of about +55 mV, while the potassium ion gradient tends toward a membrane potential of about −90 mV. At what value will the membrane potential come to rest? In principle, it could be anywhere between these points. At any given time, the membrane potential will depend on the outward flux of potassium ions relative to the inward flux of sodium ions. Changes in the permeability of the cell to either ion cause corresponding changes in the membrane potential. In a living cell, sodium ions continually leak into the cell and potassium ions leak out, but steady-state concentrations of the two ions are maintained because the sodium-potassium pump acts to move sodium ions outward and potassium ions inward.

The pioneering neurobiologists David E. Goldman, Alan Lloyd Hodgkin, and Bernard Katz were the first to describe how gradients of several different ions each contribute to the membrane potential as a function of relative ionic permeabilities. The Goldman-Hodgkin-Katz equation, more commonly known as the **Goldman equation,** is as follows:

$$V_m = \frac{RT}{F} \ln \frac{(P_K)[K^+]_{outside} + (P_{Na})[Na^+]_{outside} + (P_{Cl})[Cl^-]_{inside}}{(P_K)[K^+]_{inside} + (P_{Na})[Na^+]_{inside} + (P_{Cl})[Cl^-]_{outside}} \qquad \textbf{(22–6)}$$

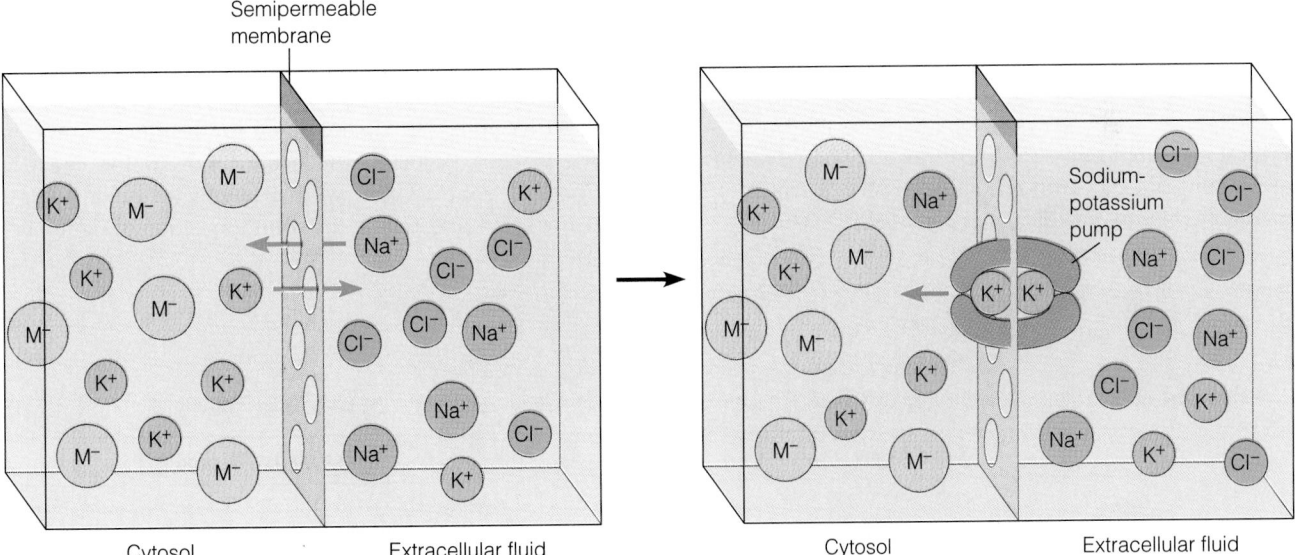

Figure 22-10 Steady-State Ion Movements. The actual membrane potential of a cell depends on the permeability of the membrane to various ions and the steady-state movements of ions across the membrane. As illustrated here with our two-compartment model, a small number of sodium ions continually leak into the cell. This makes the membrane potential more positive, weakening the electrical restraint on the movement of potassium ions. A small number of potassium ions can now leak out of the cell. As sodium ions accumulate in the cytosol, they are pumped outward in exchange for potassium ions by the sodium-potassium pump. The end result is a small, permanent concentration of sodium ions inside the cell. The presence of a small amount of sodium ions in the cytosol causes the membrane potential to be more positive than the equilibrium membrane potential for potassium ions.

One of the key differences between the Nernst equation and the Goldman equation is the incorporation of terms for permeability. Here P_K, P_{Na}, and P_{Cl} are the *relative permeabilities* of the membrane for the respective ions. The use of relative permeabilities circumvents the complicated task of determining the absolute permeability of each ion. Because chloride ions have a negative valence, $[Cl^-]_{inside}$ appears in the numerator and $[Cl^-]_{outside}$ in the denominator. While the equation shown here only takes into account the contributions of potassium, sodium, and chloride ions, other ions could be added as well. Except under special circumstances, however, the permeability of the plasma membrane to other ions is usually so low that their contributions are negligible.

We can use a mammalian neuron to illustrate how one can accurately estimate the resting membrane potential from the known steady-state concentrations of sodium, potassium, and chloride ions as well as their relative permeabilities. To do so, K^+ is assigned a permeability value of 1.0, and the permeability values of all other ions are determined relative to that of K^+. The permeability of sodium ions is only about 1% of that for potassium ions, and for chloride ions, the estimated value is 45%. Relative values of P_K, P_{Na}, and P_{Cl} are therefore 1.0, 0.01, and 0.45, respectively. Using these values, a temperature of 37°C, and the intracellular and extracellular concentrations of Na^+, K^+, and Cl^- from Table 22-1, we can estimate the resting membrane potential of a mammalian neuron as follows:

$$V_m = \frac{RT}{F} \ln \left[\frac{(1.0)(5) + (0.01)(145) + (0.45)(10)}{(1.0)(140) + (0.01)(10) + (0.45)(110)} \right]$$
$$= 0.0267 \ln (0.057) = -0.077V = -77mV \qquad \textbf{(22–7)}$$

Typical measured values for the resting membrane potential of a mammalian neuron are about -80 mV, which is remarkably close to our calculated potential.

Electrical Excitability

The establishment of a resting membrane potential and its dependence on ion gradients and ion permeability are properties of almost all cells and are not what makes electrically excitable cells unique. The unique feature of electrically excitable cells is their response to membrane depolarization. A nonexcitable cell that has been temporarily and slightly depolarized will simply return to its original resting membrane potential. When an electrically excitable cell is depolarized to the same degree, however, it responds with an *action potential.*

Electrically excitable cells produce an action potential because of the presence of particular types of ion channels in the plasma membrane. As we shall see, ion channels are at the heart of electrical signaling in neurons. Thus, to understand how nerve cells communicate signals electrically, we need to know the characteristics of the ion channels that are present in the membrane.

Ion Channels

Ion channels are integral membrane proteins that form ion-conducting pores through the lipid bilayer. Channel types differ in several ways, including their selectivity for a particular ion and the conditions that determine when the channel is opened or closed to the passage of ions. Channels are generally classified according to the kind of ion they conduct, the most common examples being sodium, potassium, calcium, and chloride channels. Channels also differ in what stimulus causes them to open and how long they stay open in response to a particular stimulus. Controlling the opening and closing of a channel is referred to as *gating* the channel.

As the name suggests, **voltage-gated ion channels** respond to changes in the voltage across a membrane. Voltage-gated sodium and potassium channels are responsible for the action potential. **Ligand-gated ion channels,** however, open when a particular molecule binds to the channel. (*Ligand* is from a Latin word meaning "to bind.") These channels are important in the communication of signals be-

---

Table 22-1 Ionic Concentrations Inside and Outside Axons, Concentration Ratios, and Resulting Equilibrium Potentials

Ion	Squid Axon				Mammalian Neuron			
	Outside (mM)	Inside (mM)	Ratio*	Potential (mV)	Outside (mM)	Inside (mM)	Ratio*	Potential (mV)
Na^+	440	50	8.8	+55	145	10	14.4	+68
K^+	20	400	0.05	−75	5	140	0.035	−85
Cl^-	560	50	0.09	−61	110	10	0.09	−60

*Concentration ratios are outside/inside for cations, inside/outside for anions.

tween neurons. Finally, some channels appear to be *ungated* and therefore always open. An example of an ungated channel is the potassium leak channel that makes resting cells somewhat permeable to potassium ions.

The Structure and Function of Voltage-Gated Ion Channels. Understanding the structure and function of voltage-gated sodium and potassium channels provides a basis for understanding the events of the action potential. Therefore, before we discuss the events of an action potential, we will first explore the properties of ion channels.

The structure of voltage-gated ion channels follows two different, though similar, models. *Voltage-gated potassium channels* are multimeric proteins that form from four separate protein subunits present in the membrane (Figure 22-11). When these four subunits come together in the membrane, a central pore is formed through which ions can pass. *Voltage-gated sodium channels,* however, are large, monomeric proteins with four separate domains, each of which is similar to one of the subunits of the voltage-gated potassium channel. In both kinds of channels, each subunit or domain contains six transmembrane α helices. One of these transmembrane α helices has positively charged amino acid groups in the middle of its transmembrane segment. These positively charged amino acids probably serve as the *voltage sensor* that makes these channels voltage-sensitive. In some way, changes in voltage across the membrane cause the positions of these amino acids to shift, thereby opening or closing the channel.

The size of the central pore and, more importantly, the way it interacts with an ion, give a channel its ion specificity (Figure 22-12a). In part, channels select for ions of the right charge through electrostatic attraction to, or repulsion from, charged amino acids at the opening of the pore. Ultimately, however, ions must bind to the channel before they can pass through. When an ion binds to a channel, most of the water molecules bound to the ion are released. A channel thus selects for ions that bind strongly enough to displace the water molecules surrounding the ion. Once this happens, the ion can pass through the pore. It is through binding to an ion that a channel exerts its greatest selectivity.

Gated channels have the ability to open in response to some stimulus and then to close again. This open or closed state is an all-or-none phenomenon: When a channel opens, it conducts ions at a maximum, and when it closes, it does not conduct ions at all. However, a channel can go into either of two different closed states. In the case of *channel gating,* the channel closes but remains capable of opening again in response to the appropriate signal (Figure 22-12b). This is due to the voltage gate swinging open or shut.

The other closed state is referred to as **channel inactivation** which is an important feature of voltage-gated sodium channels (Figure 22-12c). When a channel is inactivated, it is closed in such a way that it cannot reopen immediately, even if stimulated to do so. Inactivation is caused by a portion of

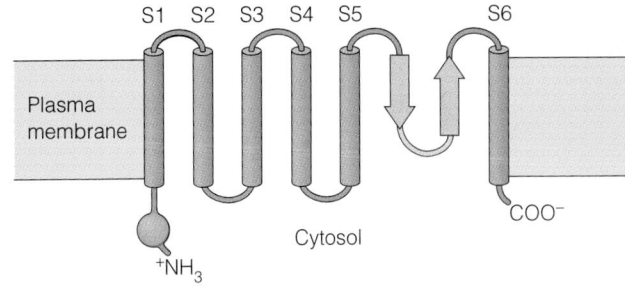

(a) An individual subunit

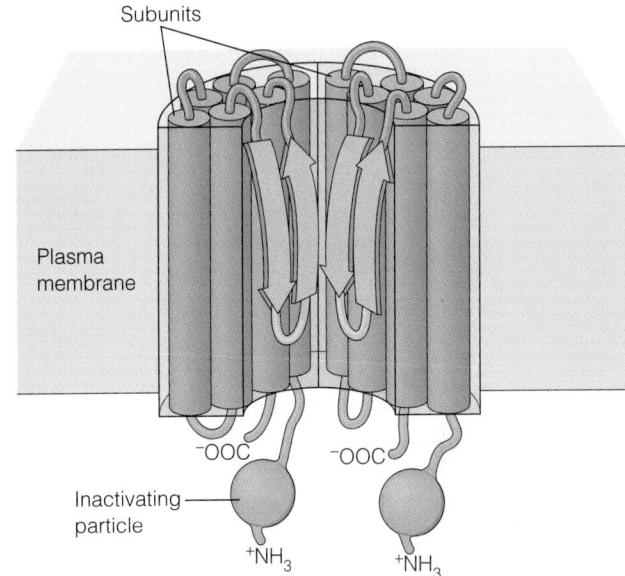

(b) Arrangement of two of four subunits in a channel

Figure 22-11 The General Structure of Voltage-Gated Ion Channels. The voltage-gated channels for sodium, potassium, and calcium ions all share the same basic structural themes. The channel is essentially a rectangular tube whose four walls are formed from either four subunits, or four domains. **(a)** Each subunit or domain contains six transmembrane α helices labeled S1–S6. The fourth transmembrane α helix, S4, contains many positively charged residues, which make it a good candidate for a voltage sensor and part of the gating mechanism. For voltage-gated sodium channels and some types of potassium channels, a region near the N terminus protrudes into the cytosol and forms an inactivating particle. **(b)** Two of the subunits of a voltage-gated potassium channel are brought together to show how the pore forms in the middle. The inactivating particle (when present) causes channel inactivation by extending over the mouth of the channel to block the passage of ions.

the channel protein called the *inactivating particle* that protrudes into the cytosol. During inactivation, this particle covers the opening of the channel pore. For such a channel to reactivate and open in response to a stimulus, the inactivating particle must move away from the pore.

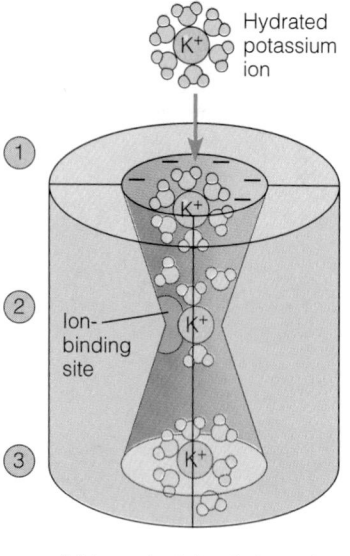

Hydrated potassium ion

①

Ion-binding site

② K⁺

③ K⁺

(a) Ion selectivity of channels

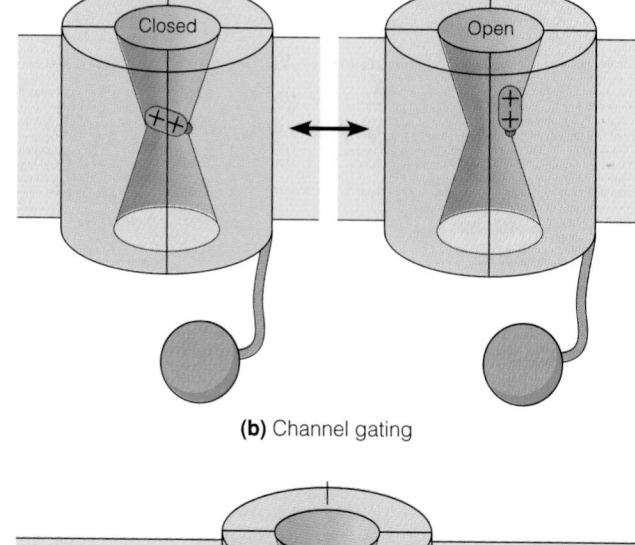

Closed

Open

(b) Channel gating

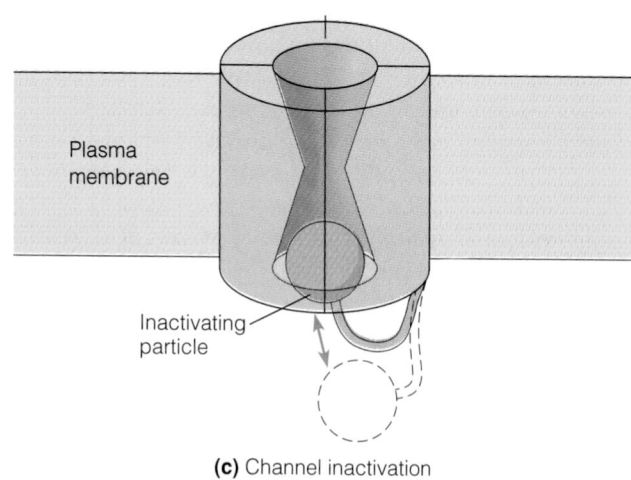

Plasma membrane

Inactivating particle

(c) Channel inactivation

Figure 22-12   The Function of a Voltage-Gated Ion Channel. **(a)** Several ways that a channel can select for different ions are shown here. ① Negative charges at the opening of the channel repel anions and attract cations. ② The pore diameter restricts the size of ions that can pass. ③ Ion-selective binding strips off water molecules so that ions can pass through the pore. **(b)** Channel gating occurs because a portion of the channel changes conformation when the membrane potential changes. Here, a charged gate is depicted as swinging open and shut. **(c)** Inactivation of sodium channels occurs when an inactivating particle blocks the opening of the pore.

Studying a Single Channel: Patch Clamping

Our clearest picture of how channels operate is the result of the development of a technique that permits the recording of ion currents passing through individual channels. This technique, known as *single-channel recording,* or more commonly as **patch clamping** (Figure 22-13), was developed by Erwin Neher and Bert Sackman.

To record single-channel currents, a fire-polished glass micropipette with a tip diameter of approximately 1 μm is carefully pressed up against the surface of a cell such as a neuron (Figure 22-13a). Gentle suction is then applied, so that a tight seal forms between the pipette and the plasma membrane. There is now a "patch" of membrane under the mouth of the micropipette that is sealed off from the surrounding medium (Figure 22-13b).

This patch of membrane is small enough that it will usually contain only one or perhaps a few ion channels. Current can enter and leave the pipette only through these channels, thereby enabling an experimenter to study various

properties of the individual channels. The channels can be studied in the intact cell, or the patch can be pulled away from the cell so that the researcher has access to the cytosolic side of the membrane.

During the experimental process, an amplifier maintains voltage across the membrane with the addition of a sophisticated electronic feedback circuit called a *voltage clamp* (hence the term *patch clamp*). The voltage clamp keeps the cell at a fixed membrane potential, regardless of changes in the electrical properties of the plasma membrane, by injecting current as needed to hold the voltage constant. It then measures tiny changes in current flow—actual ionic currents through individual channels—from the patch pipette.

The patch-clamp method has been used to show that when a particular kind of channel opens, it always conducts the same amount of *current*—that is, the same number of ions per unit of time. There are no partially open or closed states in which the channel conducts more or less current. Therefore, we can characterize a particular channel in terms of its conductance. *Conductance* is the amount of current

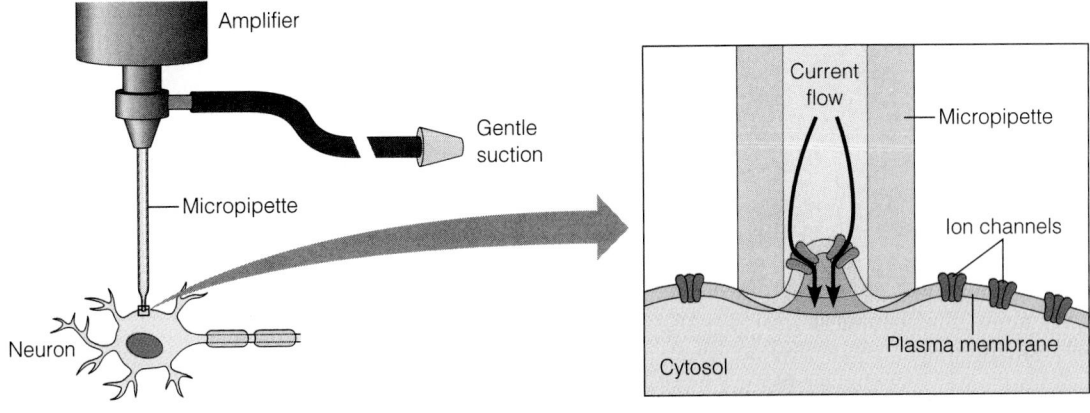

(a) Patch-clamping setup

(b) Membrane patch isolating ion channel(s)

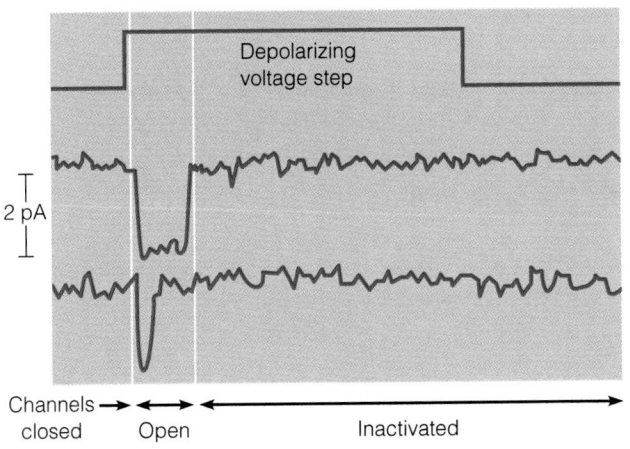

(c) Traces of individual Na⁺ currents during channel openings

Figure 22-13 Patch Clamping. **(a)** A fire-polished micropipette with a diameter of about 1 μm is carefully placed against a cell, such as the neuron shown here, and **(b)** gentle suction is applied to form a tight seal between the pipette and the plasma membrane. The density of channels is usually low enough that only one or a few channels will be in the membrane under the mouth of the pipette. **(c)** When current flow (flow of ions) is recorded while the membrane is subjected to a depolarizing step in voltage, a burst of channel opening occurs. When each channel opens, the amount of current that flows through the channel is always the same (2 pA for the sodium channels shown here). Following the burst of channel opening, a quiescent period occurs because of channel inactivation.

flowing through the channel per unit of time when a specified voltage is applied across the membrane. A voltage-gated sodium channel, for example, conducts an electrical current of approximately 2 picoamps (2×10^{-12} A), which corresponds to about 12 million sodium ions flowing through the channel per second. This can be seen in the traces shown in Figure 22-13c.

In single-channel recording, membrane depolarization (triggered by changing the applied voltage to a more positive potential) increases the probability that a channel will open. Even before the membrane is depolarized, a sodium channel will occasionally flicker open and closed. When the membrane is depolarized, a burst of activity occurs, involving much more frequent opening and closing of sodium channels. This channel activity then dies down and cannot resume unless the membrane potential is restored to a more negative level. The cessation of channel activity that occurs while the membrane is still depolarized is due to channel inactivation.

Mode of Action of Sodium Channels. Much of the current research using patch clamping focuses on the mechanism that enables ion channels of excitable membranes to

sense and respond to changes in membrane potential. Our understanding in this area has been greatly enhanced by studies of **gating currents.** These are small currents that last only about 0.1 msec and precede the opening of the sodium channels. Gating currents are thought to reflect changes in the positions of charged amino acids within the channel. Any movement of charge over a distance is a current. Here, the movement of charge that makes up the gating current is due to the gate of the sodium channel protein swinging open in response to membrane depolarization.

Our understanding of how these channels function has been greatly aided by the isolation and cloning of the gene that encodes the sodium channel protein. This has made it possible to synthesize large amounts of the channel protein and to study its functions in lipid bilayers. In addition, specific molecular modifications or mutations to the channel can be used to determine how various regions of the channel protein are involved in channel function. This approach has been used to study the parts of the channel protein responsible for voltage-gating. Each of the four domains of the sodium channels has six transmembrane helices, called S1–S6. One of these helices, S4, has a conspicuous series of positively charged amino acids. When these positively charged amino acids are replaced with neutral amino acids, gating currents are abolished, and the channel does not open. This suggests that the transmembrane helix S4 functions as the voltage sensor and gate.

Molecular studies have also identified the mechanism of channel inactivation. The current model of inactivation suggests something like a ball tethered on a chain (the inactivating particle) that swings over the cytosolic opening of the channel to block conductance (see Figure 22-12c). Several lines of evidence support this conclusion. When the cytosolic side of the channel is treated with a protease, channels can no longer be inactivated. Treatment of channels with antibodies prepared against the fragment of the channel thought to be responsible for inactivation also prevents inactivation.

Kinds of Potassium Channels. There are many different kinds of potassium channels. The delayed potassium channels play an important role in action potentials. These channels respond to depolarization as the sodium channels do, but they open more slowly. Other potassium channels are activated by internal calcium and may play a role in the regulation of membrane potential.

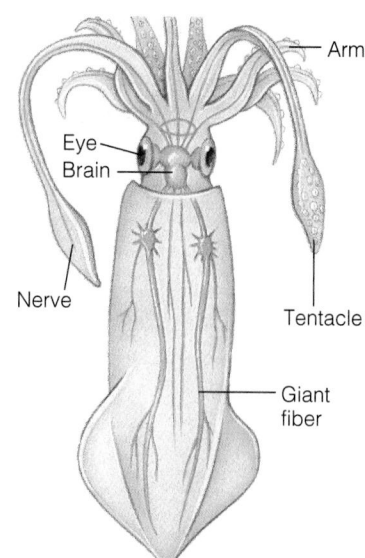

Figure 22-14 Squid Giant Axons. The squid nervous system includes motor nerves that control swimming movements. The nerves contain giant axons (fibers) with diameters ranging up to 1 mm, providing a convenient system for studying resting and action potentials in a biological membrane.

The Action Potential

We have seen how an ion gradient across a selectively permeable membrane can generate a membrane potential, and how, according to the Goldman equation, membrane potential will change in response to changes in ion permeability. We have also examined the nature of ion channels in the membrane, which are responsible for regulating the permeability of the membrane to the different ions. Now we are ready to explore how the coordinated opening and closing of ion channels can lead to an electrical impulse called an action potential. To do this, we will first examine how the membrane potential is measured and how it changes during an action potential, using the squid giant axon as a model system. Then we will discuss how the opening and closing of sodium and potassium channels account for these changes in membrane potential.

The Squid Giant Axon as an Experimental Model

Progress in science is often associated with technological breakthroughs. One of the great advances in understanding the electrical events in neurons came with the discovery of the *giant neurons* of the squid in the 1930s. Since that time, the squid giant axon has become one of the most extensively studied and well characterized of all neurons. In the squid, these giant nerve fibers are used to expel water explosively from the mantle cavity of the animal, enabling it to propel itself quickly backward to escape predators (Figure 22-14). The axons involved in triggering this "jet propulsion" system are very large, with diameters of about 0.5–1.0 mm. This size allows microelectrodes to be inserted readily into the axon,

thereby making it possible to measure and control electrical potentials and ionic currents across the axonal membrane. Because the squid giant axon has been studied so intensively, we will use it to illustrate the properties of an action potential.

Measuring the Resting Membrane Potential

We have seen how the ion currents flowing through a single channel can be measured using patch clamping. To measure the resting membrane potential of the cell, we use a slightly different apparatus, such as that shown in Figure 22-15a. Here, a *recording electrode* is implanted in the axon and a *reference electrode* is placed in the electrolyte solution surrounding the membrane. Wires run from the electrodes to an electronic amplifier to amplify the signals, and then to an oscilloscope or computer screen that displays the strength of the signal over time. As we will discuss later, additional electrodes can be inserted into the neuron for special purposes.

The Sequence of Events During an Action Potential

A resting neuron is a system poised for electrical action. As already mentioned, the membrane potential of the cell is set by a delicate balance of ion gradients and ion permeability. Depolarization of the membrane upsets this balance. If the level of depolarization is small—less than about 20 mV—the membrane potential will normally recover without event. Further depolarization brings the membrane to the

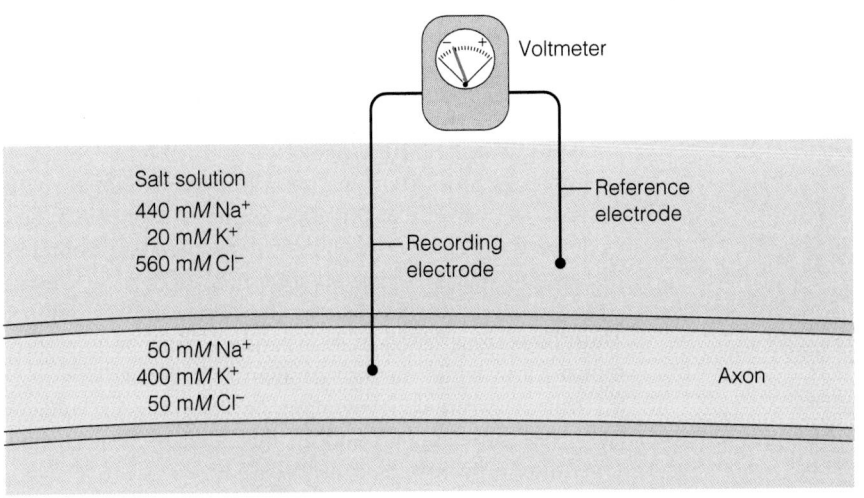

(a) Measuring the resting membrane potential in a squid axon

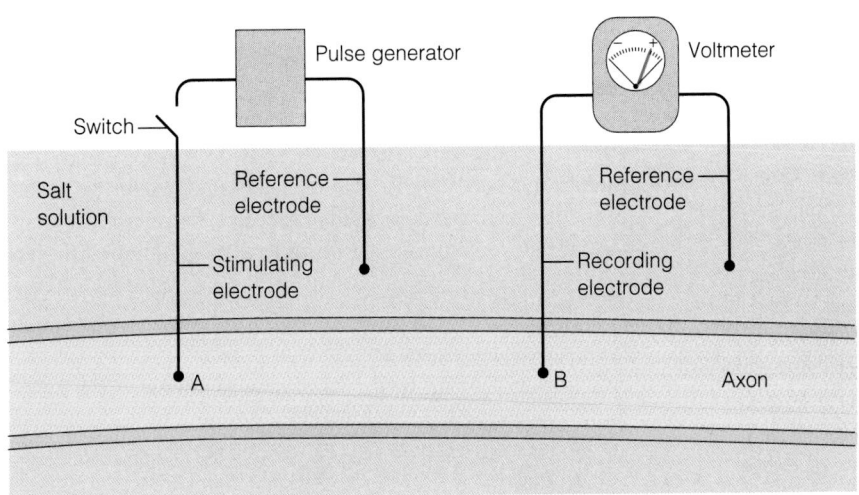

(b) Measuring an action potential in a squid axon

Figure 22-15   An Apparatus for Measuring Membrane Potentials. **(a)** Measurement of the resting membrane potential requires two electrodes, one inserted inside the axon (the recording electrode) and one placed in the fluid surrounding the cell (the reference electrode). Differences in potential between the recording and reference electrodes are amplified by a voltage amplifier and displayed on a voltmeter, an oscilloscope, or a computer monitor. **(b)** Measurement of an action potential requires four electrodes, one in the axon for stimulation, another in the axon for recording, and two in the fluid surrounding the cell for reference. The stimulating electrode is connected to a pulse generator, which delivers a pulse of current to the axon when the switch is momentarily closed. The nerve impulse this generates is propagated down the axon and can be detected a few milliseconds later by the recording electrode. The impulse is detected as a transient change in transmembrane potential, measured with respect to the reference electrodes.

threshold potential, where the events of the action potential take control.

An **action potential** is a brief but large electrical depolarization and repolarization of the neuronal plasma membrane caused by the inward movement of sodium ions followed by the outward movement of potassium ions. These ion movements are, in turn, controlled by the opening and closing of voltage-gated sodium and potassium channels. In fact, we can explain the development of an action potential solely in terms of the behavior of these channels. Once an action potential is initiated in one region of the membrane, it will travel along the membrane away from the site of origin by a process called **propagation.**

Measuring an Action Potential

The development and propagation of an action potential can be readily studied in large axons such as those of the squid. To do so using the apparatus shown in Figure 22-15b,

an additional electrode called the *stimulating electrode* is connected to a power source and inserted into the axon some distance from the recording electrode. A brief impulse from this stimulating electrode can be used to locally depolarize the neuron beyond the threshold potential. This requires an electrical impulse from the stimulating electrode sufficient to depolarize the membrane by about 20 mV (i.e., from -60 to about -40 mV). This triggers an action potential that spreads out, or propagates away from, the stimulating electrode. As the action potential passes the recording electrode, the voltmeter or oscilloscope will display the characteristic pattern of potential changes shown in Figure 22-16.

In less than a millisecond, such an apparatus will record the membrane potential rising dramatically from the resting membrane potential to about $+40$ mV—the interior of the membrane actually becomes positive for a brief period. The potential then falls somewhat more slowly, dropping to about -75 mV (hyperpolarization) before stabilizing again

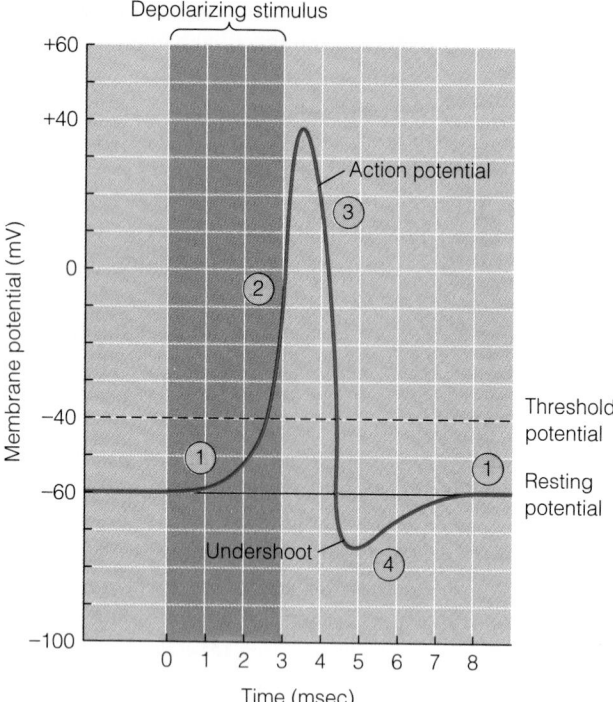

Figure 22-16   The Action Potential of the Squid Axon.
① The resting potential prior to the start of the action potential is approximately −60 mV. ② An action potential begins when the neuron is depolarized by about 20 mV to a point known as the threshold potential. Once the action potential is initiated, the potential swings rapidly in the positive direction. At the peak of the positive swing, the membrane potential reaches a value of about +40 mV. ③ Once the cell reaches the peak positive potential, it begins to repolarize, returning to a negative membrane potential. ④ Repolarization often leads to a membrane potential that is actually hyperpolarized or more negative than the resting potential. This is referred to as the undershoot. The membrane potential then returns to its resting state.

at the resting potential of about −60 mV. As Figure 22-16 indicates, the complete sequence of events during an action potential takes place within a few milliseconds.

The apparatus shown in Figure 22-15 can also be used to measure the ion currents that flow through the membrane at different phases of an action potential. To do so, an additional electrode known as the *holding electrode* is inserted into the cell and connected to a voltage clamp, thereby enabling the investigator to set and hold the membrane at a particular potential, regardless of changes in the membrane's electrical properties. Using the voltage-clamp apparatus, a researcher can measure the current flowing through the membrane at any given membrane potential. Such experiments have contributed fundamentally to our current understanding of the mechanism that causes an action potential.

Understanding the Action Potential in Terms of Ion Channels and Currents

In a resting neuron, the voltage-dependent sodium and potassium channels are usually closed. Therefore, the cell is roughly 100 times more permeable to potassium than to sodium ions because of the potassium leak channels. When a region of the nerve cell is slightly depolarized, a fraction of the sodium channels respond and open. As they do, the increased sodium current acts to depolarize the membrane. Thus, increasing depolarization causes a larger sodium current to flow, which further depolarizes the cell. This relationship between depolarization, the opening of voltage-gated sodium channels, and an increased sodium current constitutes a positive feedback loop known as the *Hodgkin cycle* (Figure 22-17).

Subthreshold and Threshold Depolarization. If the Hodgkin cycle were not opposed by other forces, even a small amount of sodium entry would always lead to complete depolarization of the cell membrane. However, the events of the Hodgkin cycle meet with resistance due to the efflux of potassium ions, which tends to restore the resting membrane potential. As mentioned earlier, when the membrane is depolarized by a small amount, the membrane potential recovers and no action potential is generated. Levels

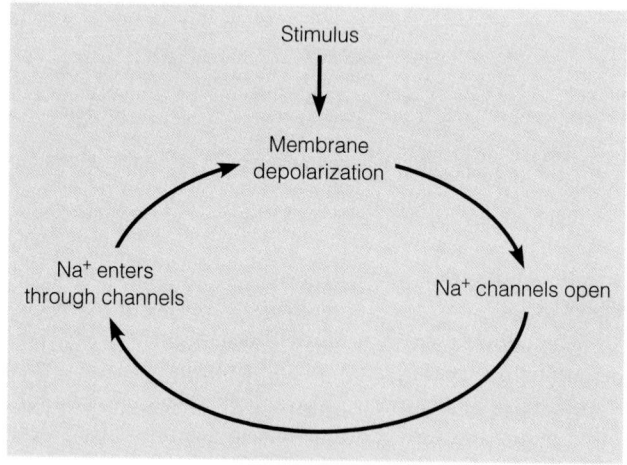

Figure 22-17   The Hodgkin Cycle.   This diagram illustrates the positive feedback relationship between depolarization, the opening of sodium channels, and the corresponding increase in sodium current. A small depolarization causes sodium channels to open. This in turn lets sodium ions flow into the cell, which depolarizes the membrane even more. As long as the membrane's permeability to potassium ions is greater than its permeability to sodium ions, diffusion of potassium ions out of the cell will counteract the Hodgkin cycle. However, when the membrane's permeability to sodium ions approaches its permeability to potassium ions, the Hodgkin cycle will operate unopposed, leading to maximum sodium permeability.

of depolarization that are too small to produce an action potential are referred to as **subthreshold depolarizations.**

To understand why subthreshold depolarizations lead to recovery while larger depolarizations lead to an action potential, we need to consider how potassium ions respond to depolarization. When sodium ions enter a cell and depolarize the membrane, the electrical restraint that keeps potassium ions from diffusing out of the cell is weakened. In response to a depolarization of the membrane, potassium ions diffuse outward, thereby making the membrane potential more negative again.

This efflux of potassium ions can effectively oppose the Hodgkin cycle as long as its rate is greater than or equal to the rate of sodium influx. The rate of sodium influx varies with the degree of depolarization: The greater the depolarization, the faster sodium enters. Up to a point, increasing rates of sodium entry are matched by increasing rates of potassium efflux. For subthreshold levels of depolarization, the rate of potassium efflux can compensate for the rate of sodium entry.

The Depolarizing Phase. If all the voltage-dependent sodium channels in the membrane were to open at once, the cell would suddenly become ten times more permeable to sodium than to potassium. Because sodium would then be the more permeable ion, the membrane potential would be largely a function of the sodium ion gradient. This is effectively what happens when the membrane is depolarized past the threshold potential (Figure 22-18, steps 1 and 2). Once the threshold potential is reached, potassium efflux can no longer compensate for the rate of sodium entry. At this point, the membrane potential shoots rapidly upward, peaking at approximately +40 mV. When the rate of sodium entry slightly exceeds the maximum rate of potassium efflux, an action potential is triggered. Note that at this peak, the action potential approaches, although it does not actually reach, the equilibrium potential for sodium ions (about +55 mV).

The Repolarizing Phase Once the membrane potential has risen to its peak, it quickly repolarizes (Figure 22-18, step 3). This is due to a combination of the inactivation of sodium channels and the opening of voltage-gated potassium channels. When sodium channels are inactivated, they close and remain closed until the membrane potential becomes negative again. Channel inactivation thus stops the inward flow of sodium ions and temporarily blocks the Hodgkin cycle. The cell will now automatically repolarize as potassium ions leak out.

Voltage-gated potassium channels in neurons differ kinetically from voltage-gated sodium channels in that, when the cell is depolarized, the potassium channels open more slowly. As a result, an action potential begins with an increase in the membrane's permeability to sodium, followed by an increased permeability to potassium. The increased permeability to sodium depolarizes the membrane, and the increased permeability to potassium ions that follows repolarizes the membrane.

The Hyperpolarizing Phase (Undershoot). At the end of an action potential, most neurons show a transient **hyperpolarization,** or **undershoot,** in which the membrane potential briefly becomes even more negative than it normally is at rest (Figure 22-18, step 4). The undershoot occurs because of the increased potassium permeability that exists while voltage-gated potassium channels remain open. Note that the potential of the undershoot closely approximates the equilibrium potential for potassium ions (about −75 mV for the squid axon). As the voltage-gated potassium channels close, the membrane potential returns to its original resting state.

The Refractory Periods. For a few milliseconds after an action potential, it is impossible to trigger a new action potential. This interval, the **absolute refractory period,** is due to sodium channel inactivation. As long as the channels are inactivated, they cannot open, even if the membrane is depolarized. In addition, the undershoot makes it more difficult to stimulate an action potential, even after sodium channels have reactivated and are ready to open again. During this interval, the **relative refractory period,** it is possible but difficult to trigger an action potential.

Changes in Ion Concentrations Due to an Action Potential. Our discussion of ion movements might give the impression that an action potential involves large changes in the cytosolic concentrations of sodium and potassium ions. In fact, during a single action potential, the cellular concentrations of sodium and potassium ions hardly change at all. Remember that the membrane potential is due to a slight excess of negative charge on one side and of positive charge on the other side of the membrane. The number of excess charges is a tiny fraction of the total ions in the cell, and the number of ions that must cross the membrane to neutralize or alter the balance of charge is likewise tiny.

However, intense neuronal activity can lead to significant changes in the overall ion concentrations. For example, as a neuron continues to generate large numbers of action potentials, the concentration of potassium outside the cell will begin to rise perceptibly. This can affect the membrane potential of both the neuron itself and surrounding cells. Astrocytes are thought to control this problem by taking up excess potassium ions.

The Propagation of an Action Potential

In order for neurons to transmit signals to one another, the transient depolarization and repolarization that occur during an action potential must travel along the neuronal membrane. The depolarization at one point on the membrane

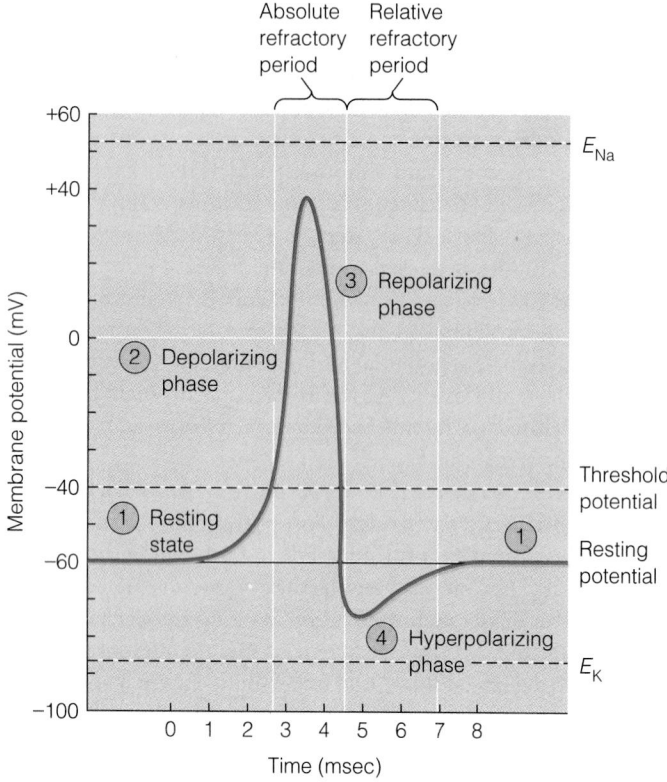

(a) Changes in ion channels and membrane potential

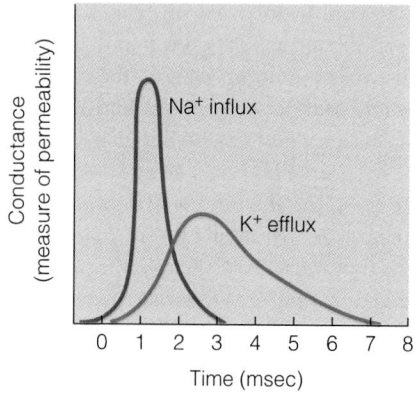

(b) Change in membrane conductance

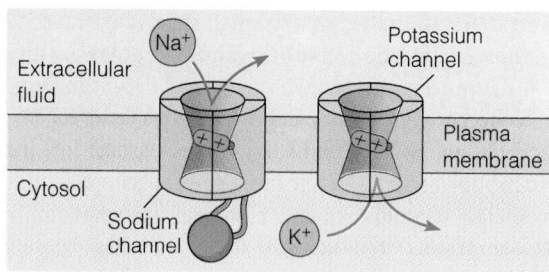

① Resting state: All gated Na⁺ and K⁺ channels closed

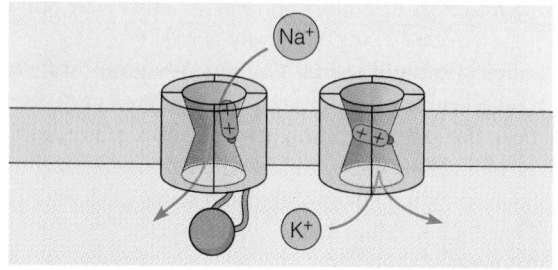

② Depolarizing phase: Na⁺ channels open

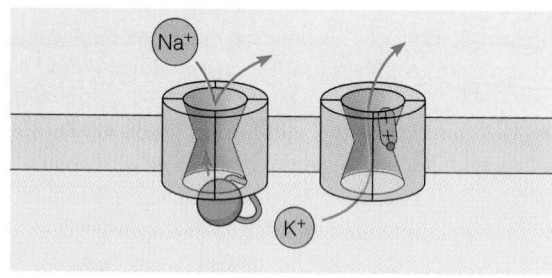

③ Repolarizing phase: Na⁺ channels inactivated and K⁺ channels open

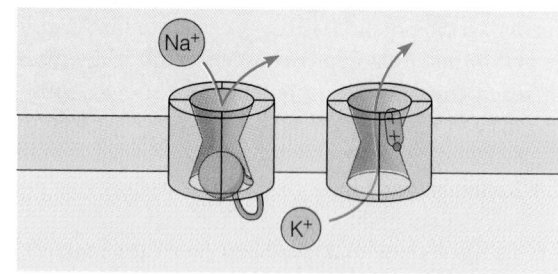

④ Hyperpolarizing phase (undershoot): K⁺ channels remain open and Na⁺ channels inactivated

Figure 22-18   Changes in Ion Channels and Currents in the Membrane of a Squid Axon During an Action Potential.
(a) The change in membrane potential caused by movement of Na⁺ and K⁺ through their voltage-gated channels, which are shown at each step of the action potential at right. The absolute refractory period is caused by sodium channel inactivation. Notice that, at the peak of the action potential, the membrane potential approaches the E_{Na} (sodium equilibrium potential) value of about +55 mV; similarly, the potential undershoots

nearly to the E_K (potassium equilibrium potential) value of about −75 mV. (b) The change in membrane conductance (permeability of the membrane to specific ions). The depolarized membrane initially becomes very permeable to sodium ions, facilitating a large inward rush of sodium. Thereafter, as permeability to sodium declines, the permeability of the membrane to potassium increases transiently, causing the membrane potential to hyperpolarize.

spreads to adjacent regions through a process called the **passive spread of depolarization.** This passive spread is due to the movement of cations (mostly potassium ions) away from the site of depolarization to regions under the membrane where the potential is more negative. As a wave of depolarization spreads passively away from the site of origin, it also decreases in magnitude, however. The fact that the depolarization fades with distance from the source makes it difficult for signals to travel very far by passive means only. For signals to travel longer distances, an action potential must be propagated, or actively regenerated, from point to point along the membrane.

To illustrate the difference between the passive spread of depolarization and the propagation of an action potential, consider how a signal travels along the nerve from the site of origin at the dendrites to the end of the axon (Figure 22-19). Incoming signals are transmitted to a nerve at synapses that

form points of contact between the terminal bulbs of the transmitting neuron and the dendrites of the receiving neuron. When these incoming signals depolarize the dendrites of the receiving neuron, the depolarization spreads passively over the membrane from the dendrites to the base of the axon, the **axon hillock.** The axon hillock is the region where action potentials are initiated most easily. This is because sodium channels are distributed sparsely over the dendrites and cell body but are concentrated at the axon hillock and nodes of Ranvier. A given amount of depolarization will produce the greatest amount of sodium entry at sites where sodium channels are abundant. The depolarization that spreads passively from the dendrites initiates action potentials at the axon hillock, which are then propagated along the axon.

The mechanism for propagating an action potential in nonmyelinated nerve cells is illustrated in Figure 22-20. Stimulation of a resting membrane at point *P* results in a depolarization of the membrane and a sudden rush of sodium ions into the axon at that location. Membrane polarity is temporarily reversed at that point, and this depolarization then spreads passively over a short distance to an adjacent point *Q*. Passive depolarization at point *Q* is sufficient to bring it above the threshold potential, triggering the inward rush of sodium ions. By this time, the membrane at point *P* has become highly permeable to potassium ions. As potassium ions rush out of the cell, negative polarity is restored, and that portion of the membrane returns to its resting state.

Meanwhile, the events at *Q* have stimulated the membrane in the neighboring region at *R*, initiating the same sequence of events there, which then moves on to point *S*. In this way, the signal moves along the membrane as a ripple of depolarization-repolarization events, with the membrane polarity reversed in the immediate vicinity of the signal, but returned to normal again as the signal travels down the axon. The propagation of this cycle of events along the nerve fiber is called a *propagated action potential* or **nerve impulse.** The nerve impulse can only move away from the initial site of depolarization because the sodium channels that have just been depolarized are in the inactivated state and cannot respond immediately to further stimulation.

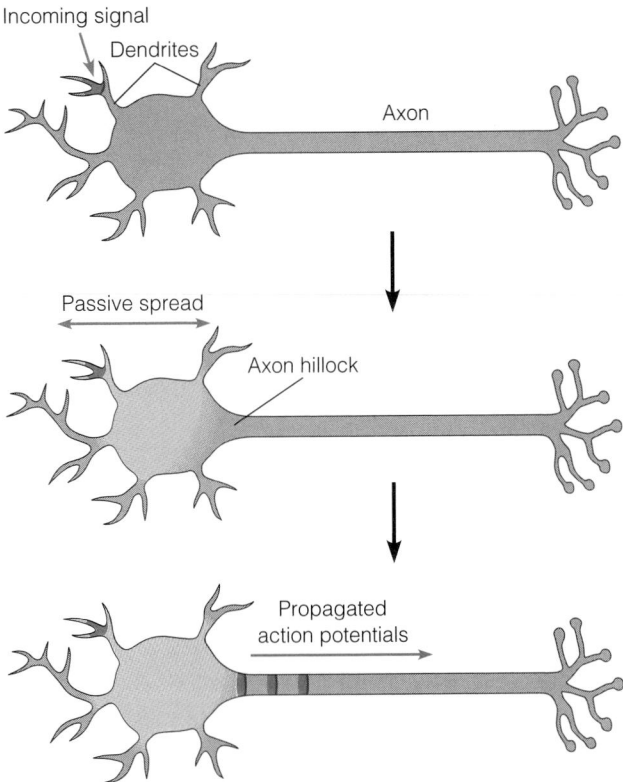

Figure 22-19 The Passive Spread of Depolarization and Propagated Action Potentials in a Neuron. The transmission of a nerve impulse along a neuron depends on both the passive spread of depolarization and the propagation of action potentials. A neuron is stimulated when its dendrites receive a depolarizing stimulus from other neurons. A depolarization starting at a dendrite will spread passively over the cell body to the axon hillock, where an action potential will form. This action potential will then be propagated down the axon.

The Energetics and Rate of Impulse Transmission

We can establish that an action potential is propagated by showing that it does not fade as it travels. This can be demonstrated by measuring changes in membrane potential from two recording electrodes, each inserted at a different distance from a stimulating electrode. The stimulating electrode will trigger an action potential, which will then travel along the axon, first passing by recording electrode 1 and

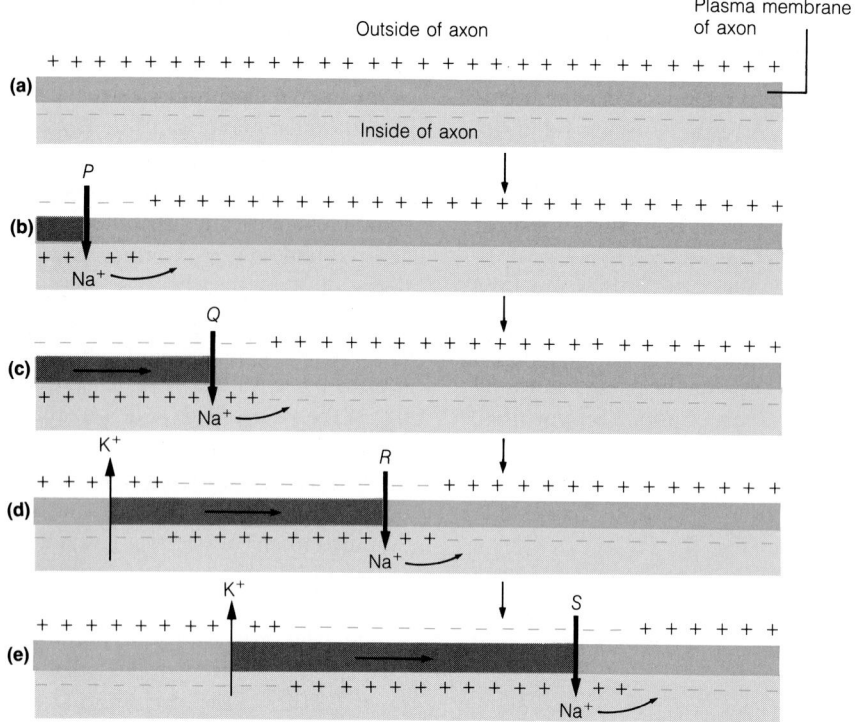

Figure 22-20 The Transmission of an Action Potential Along a Nonmyelinated Axon. A nonmyelinated axon might be viewed as an infinite string of points, each capable of undergoing an action potential. For simplicity, we will consider only the four points *P, Q, R,* and *S,* which represent adjacent regions along the plasma membrane of the axon. **(a)** At the start, the membrane is completely polarized. **(b)** When an action potential is initiated at point *P,* this region of the membrane depolarizes and briefly has a positive potential. The positively charged sodium ions will be drawn along the membrane to adjacent regions where the potential is negative. As this happens, the adjacent regions become depolarized. **(c)** When the adjacent point *Q* is depolarized to its threshold, an action potential starts here. **(d)** Meanwhile, point *P* is recovering from its action potential and has repolarized because of the outward flow of potassium ions. The action potential at point *Q* continues to propagate in the direction of point *R.* (It cannot propagate backward toward *P* because sodium channels are in a refractory, or inactivated, state, and the membrane in this region has become hyperpolarized.) The depolarization spreading from point *Q* will trigger an action potential at *R.* **(e)** Likewise, the depolarization at point *R* will eventually trigger an action potential at point *S.*

then by recording electrode 2. The magnitude of response detected by the two electrodes will be the same, even though the signal has had to travel further along the membrane to reach the second electrode.

A nerve impulse can be propagated along the membrane with no reduction in amplitude because it is constantly being renewed along the way. It is generated anew as an all-or-none event at each successive point along the membrane, using energy provided by the electrochemical ion gradients. Thus, a nerve impulse can be transmitted over essentially any distance with no decrease in strength.

The rate of action potential propagation determines how fast a signal can be transmitted through a nerve. This can be a critical issue in some situations—when an animal needs to respond quickly to danger, for example. The rate-limiting step for the speed of propagation is the passive spread of depolarization. The rate at which a depolarization event spreads passively depends, in turn, on several properties of the plasma membrane and the cytosol, including re-sistance of the cytosol and the capacitance of the plasma membrane (Figure 22-21).

The **resistance** of the cytosol determines how easily positively charged ions can move laterally along the inside of the membrane away from the site of depolarization. Larger axons have less resistance and thus conduct signals more rapidly. This principle is exploited in the squid giant axon, which controls the muscles involved in propelling the squid away from danger. The faster the squid can respond to danger, the better chance it has of escaping.

Membrane capacitance is another factor controlling the rate at which a depolarization event spreads passively along the membrane. **Capacitance** is the ability of the membrane to store or accumulate ions when a potential exists. When a negative membrane potential occurs, the excess negatively charged ions accumulate on the inside of the membrane, and an equal amount of positively charged ions line up along the outside of the membrane. In effect, these oppositely charged ions are attracted to each other even though

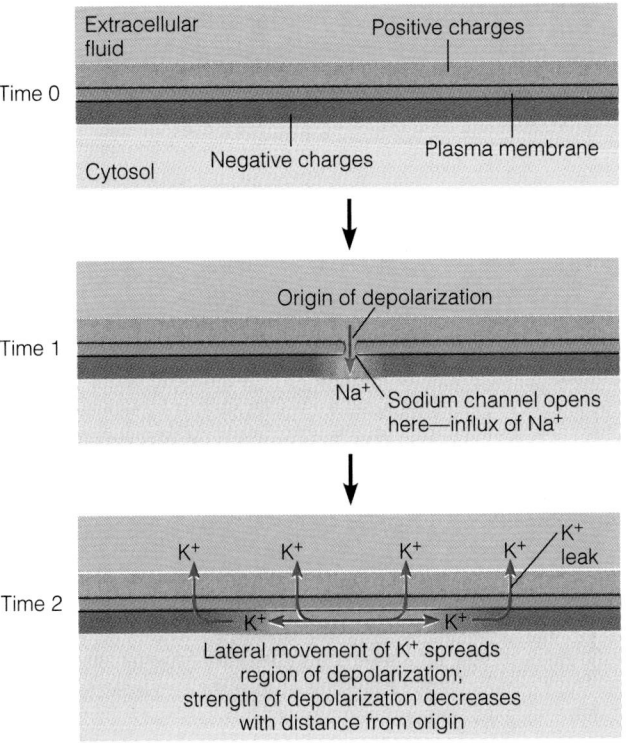

Time 0

Extracellular fluid

Positive charges

Cytosol

Negative charges

Plasma membrane

Time 1

Origin of depolarization

Na⁺

Sodium channel opens here—influx of Na⁺

Time 2

K^+ K^+ K^+ K^+ K^+ leak

K^+ K^+

Lateral movement of K⁺ spreads region of depolarization; strength of depolarization decreases with distance from origin

(a) Passive conduction along the membrane

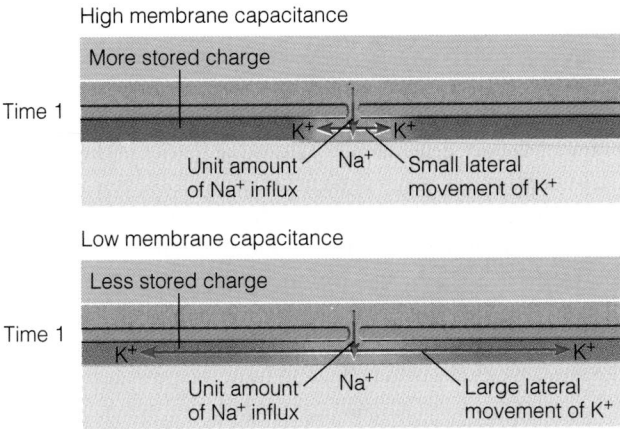

High membrane capacitance

More stored charge

Time 1

K^+ K^+

Unit amount of Na⁺ influx

Na⁺

Small lateral movement of K⁺

Low membrane capacitance

Less stored charge

Time 1

K^+ K^+

Unit amount of Na⁺ influx

Na⁺

Large lateral movement of K⁺

(b) Passive conductance along the membrane with large or small amounts of capacitance

Figure 22-21   Factors Affecting the Rate of the Passive Spread of Depolarization.   The rate at which depolarization spreads passively is a function of the resistance of the cytosol and the capacitance of the membrane. Reducing either resistance or capacitance will allow depolarization to spread passively both further and faster along the membrane. **(a)** A depolarization at one point on the membrane will spread passively to adjacent points on the membrane against the resistance of the cytosol. For example, where sodium enters (blue arrow), the membrane potential is depolarized. In this depolarized region, any positive ions (primarily potassium) will be drawn laterally (pink arrows) to adjacent regions where the membrane potential is more negative. As this occurs, the regions next to the point of sodium entry are also depolarized. However, due to the leakage of potassium ions from the cell and resistance, the strength of the depolarization fades with increasing distance from the site of sodium entry. **(b)** Low amounts of capacitance (charge stored in the membrane) allow a faster passive spread of depolarization.

they are separated by the lipid bilayer. The larger the capacitance, the more ions accumulate per unit area of the membrane at any given membrane potential.

To understand how capacitance affects the passive conduction of depolarization, we can ask what happens during the transition of the membrane from a negative potential to a positive potential. When, for example, sodium channels open and sodium ions rush in, each incoming sodium ion neutralizes one of the excess negative charges. A membrane potential of zero is reached when all of the excess negative charges have been matched with sodium ions. Any excess sodium ions beyond this point will drive the potential to positive values. When the capacitance of a membrane is large, there are more negative charges to neutralize, which

requires a greater amount of sodium influx. Therefore, a given amount of sodium influx will cause less change in membrane potential when the membrane capacitance is large than it will when the capacitance is small. Capacitance, in other words, has the effect of dampening and slowing the passive spread of depolarization.

Accelerating Signal Transmission by Myelination.   Myelination decreases the capacitance of the neuronal membrane. This reduction in membrane capacitance permits a depolarization event to spread further and faster than it would without myelination. However, myelination does not eliminate the need for propagation. For depolarization to spread from

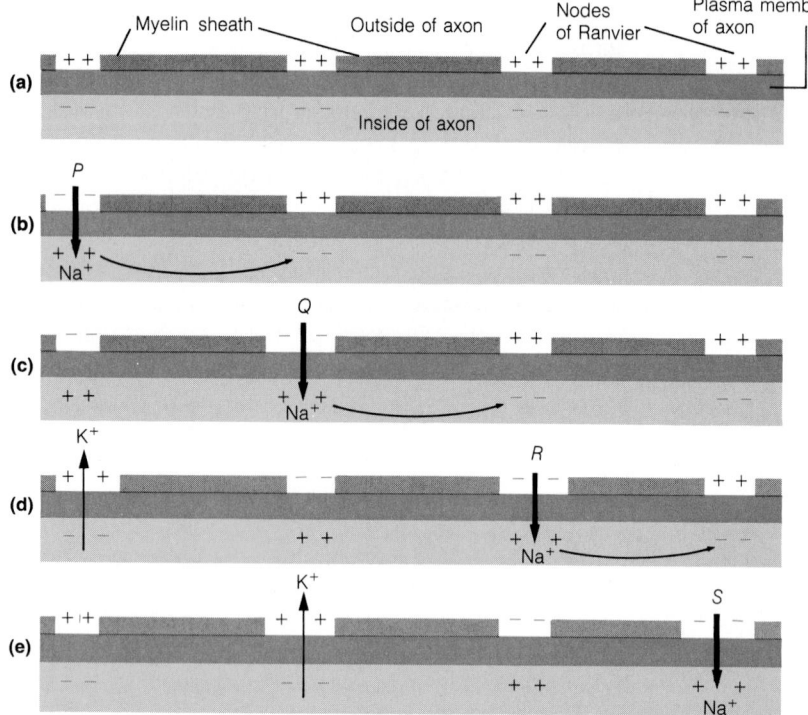

Figure 22-22   The Transmission of an Action Potential Along a Myelinated Axon. Myelination reduces membrane capacitance, thereby allowing a given amount of sodium current, entering at one point of the membrane, to spread much farther along the membrane than it would in the absence of myelin. Thus, there is much less need to regenerate the signal by the relatively slow process of generating an action potential. **(a)** In myelinated neurons, an action potential is usually triggered at the axon hillock, just before the start of the myelin sheath. The depolarization then spreads along the axon. Action potentials can only be generated at nodes of Ranvier, here represented by points *P*, *Q*, *R*, and *S*. **(b)** With myelination, the depolarization at point *P* spreads passively all the way to point *Q* and **(c)** brings *Q* to its threshold. Here a new action potential is generated that **(d)** triggers point *R* to undergo an action potential and **(e)** so on to point *S*. A nerve impulse consists of a wave of depolarization-repolarization events that is propagated along the axon from node to node.

one site to the rest of the neuron, the action potential must still be renewed periodically down the axon. The only places along a myelinated axon that can support an action potential are the nodes of Ranvier.

Nodes of Ranvier are interruptions in the myelin layer that are spaced just close enough together (1–2 mm) to ensure that the depolarization spreading out from an action potential at one node brings an adjacent node above its threshold potential. Voltage-gated sodium channels are concentrated at the nodes and therefore can generate a large response. Thus, action potentials "jump" from node to node along myelinated axons, rather than moving as a steady ripple along the membrane. Nerve impulses are thereby transmitted in a *saltatory* manner along myelinated axons, as illustrated in Figure 22-22. (*Saltatory* is derived from the Latin word for "dancing.") Saltatory propagation is much more rapid than the continuous propagation that occurs in nonmyelinated axons.

Synaptic Transmission

Nerve cells communicate with one another and with glands and muscles at synapses. There are two structurally different types of synapses, electrical and chemical. In an **electrical synapse,** the axon of one neuron, called the **presynaptic neuron,** is connected to another cell, the **postsynaptic neuron,** by gap junctions. These junctions allow ions to move back and forth between the two cells (Figure 22-23). As a result, the depolarization in one cell spreads passively to the connected cell. Electrical synapses provide for transmission

with no delay and tend to occur in places in the nervous system where speed of transmission is of the essence.

In a **chemical synapse,** the presynaptic and postsynaptic neurons are close to each other but not directly connected (Figure 22-24). Typically, the synaptic terminals at the end of an axon are separated from the synaptic membrane of the dendrites by a gap of about 20–50 nm known as the **synaptic cleft.** A nerve signal arriving at the terminals of the presynaptic neuron cannot bridge the synaptic cleft as an electrical impulse. For synaptic transmission to take place, the electrical signal must be converted at the presynaptic neuron to the chemical signal of a **neurotransmitter.** Neurotransmitter molecules are stored in secretory vesicles in the synaptic knobs or terminal bulbs of the presynaptic neuron and can diffuse across the synaptic cleft when released into it. An action potential arriving at the terminal causes the neurotransmitter to be secreted into the synaptic cleft. The chemical signal must then be converted back to an electrical signal at the postsynaptic neuron. This takes place when neurotransmitter molecules bind to their receptors on the postsynaptic neuron.

Understanding how a nerve signal is transmitted across a synapse requires that we know the nature of neurotransmitters and their respective receptors. Specifically, we need to comprehend how the binding of neurotransmitter molecules to their receptors can alter the electrical activity of the postsynaptic cell, either to excite it to the threshold point or to inhibit its electrical activity. Finally, we need to understand how an action potential regulates the secretion of neurotransmitter into the synaptic cleft and the processes that terminate the signal by removing the neurotransmitter from the synaptic cleft.

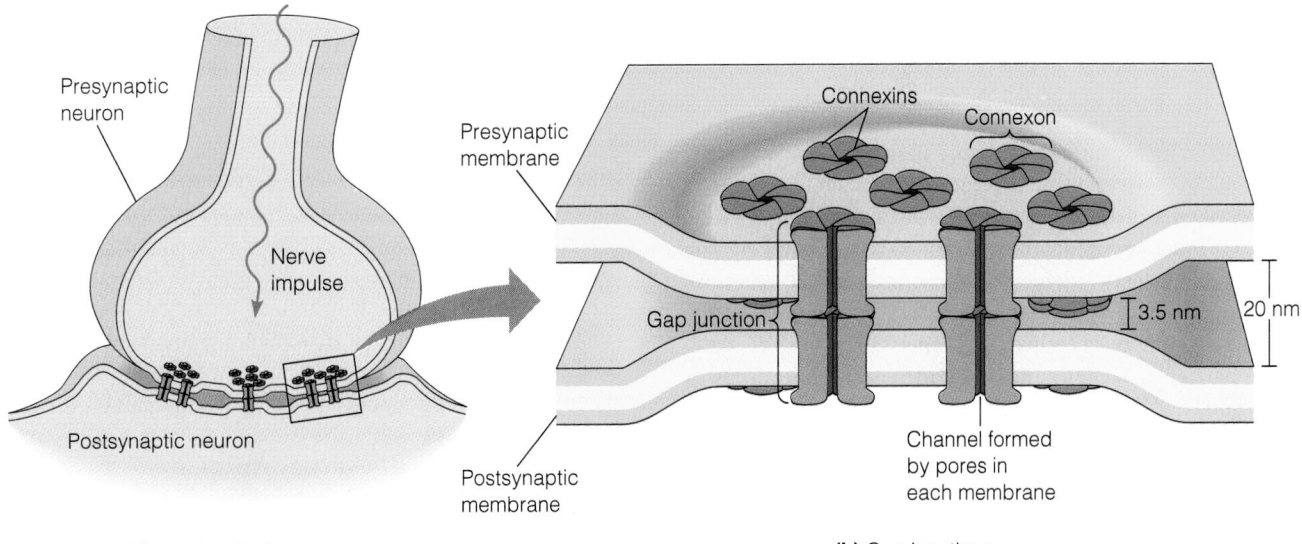

(a) An electrical synapse

(b) Gap junctions

Figure 22-23   An Electrical Synapse. **(a)** In electrical synapses, the presynaptic and postsynaptic neurons are coupled by gap junctions. Gap junctions allow small molecules and ions to pass freely from the cytosol of one cell to the next. Therefore, when an action potential arrives at the presynaptic side of an electrical synapse, the depolarization spreads passively, due to the flow of positively charged ions, across the gap junction. **(b)** The gap junction is composed of sets of channels. Each channel is made up of six protein subunits called connexins. The entire set of six subunits together is called a connexon. Two connexons, one in the presynaptic membrane and one in the postsynaptic membrane, make up a gap junction.

Figure 22-24   A Chemical Synapse. **(a)** When a nerve impulse from the presynaptic axon arrives at the synapse (pink arrow), it causes synaptic vesicles containing neurotransmitter in the synaptic knob to fuse with the presynaptic membrane, releasing their contents into the synaptic cleft. **(b)** Neurotransmitter molecules diffuse across the cleft from the presynaptic (axonal) membrane to the postsynaptic (dendritic) membrane, where they bind to specific membrane receptor sites and change the polarization of the membrane, either exciting or inhibiting the next cell.

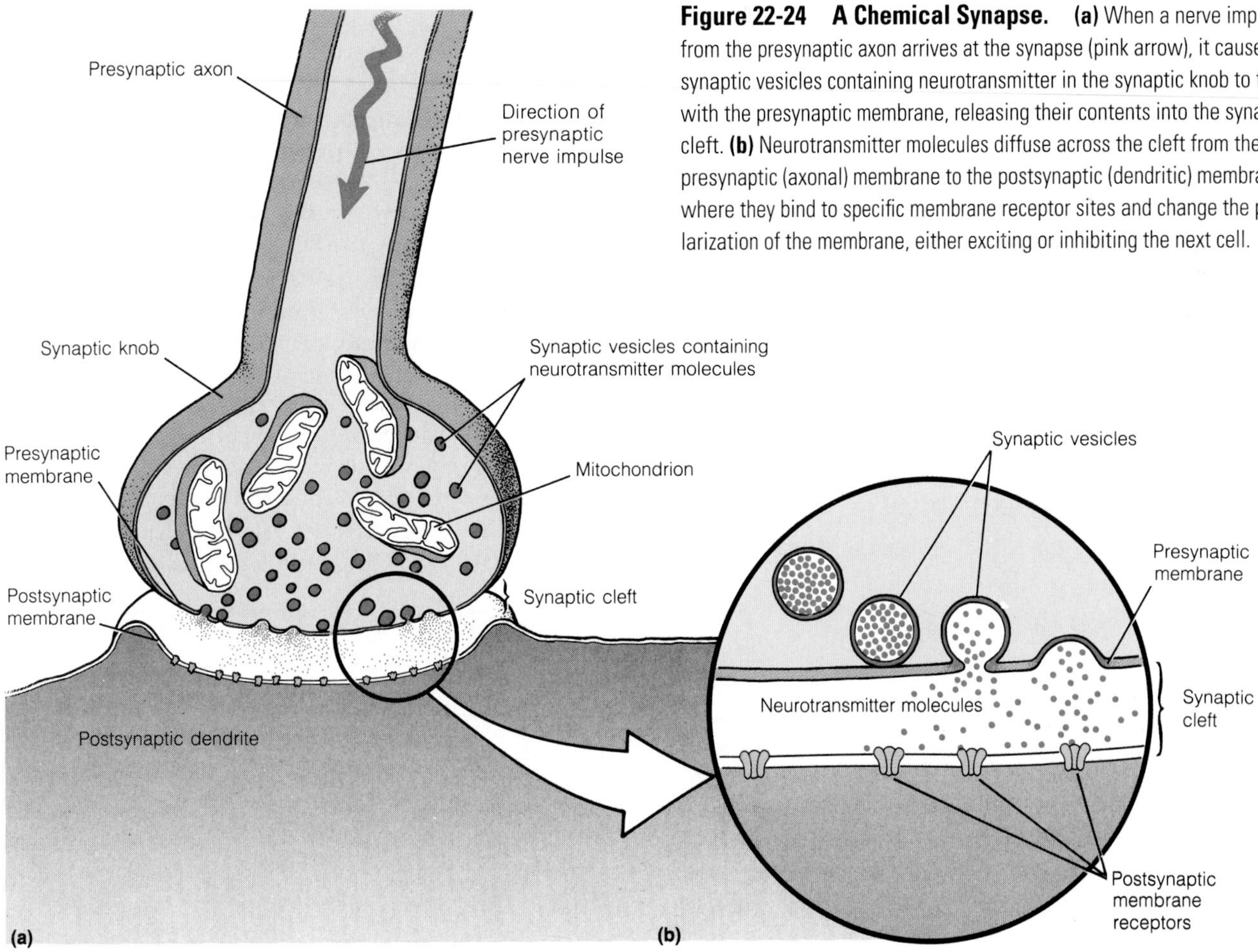

Presynaptic axon

Direction of presynaptic nerve impulse

Synaptic knob

Synaptic vesicles containing neurotransmitter molecules

Presynaptic membrane

Mitochondrion

Postsynaptic membrane

Synaptic cleft

Postsynaptic dendrite

(a)

Synaptic vesicles

Presynaptic membrane

Neurotransmitter molecules

Synaptic cleft

Postsynaptic membrane receptors

(b)

Neurotransmitters

A neurotransmitter is a small molecule whose function is to bind to a receptor. Many kinds of molecules act as neurotransmitters, each with at least one specific type of receptor; some neurotransmitters have more than one type of receptor. When a neurotransmitter molecule binds to its receptor, the properties of the receptor are altered. One might compare receptors and neurotransmitters to a lock and key. The neurotransmitter, by virtue of its unique chemical structure, is the key that fits into the receptor lock. Its function is typically to turn the receptor from an "off" state to an "on" state. For a particular receptor, however, the specific meanings of "off" and "on" lie in the properties of that receptor.

Neurotransmitters can be classified as excitatory or inhibitory, depending on what happens when they bind to their receptors. An *excitatory neurotransmitter* causes depolarization of the postsynaptic neuron. The binding of an *inhibitory neurotransmitter* to its receptor, however, causes the postsynaptic cell to hyperpolarize. This can be accomplished by opening either potassium or chloride ion channels.

To qualify as a neurotransmitter, a compound must satisfy the following three criteria: (1) It must elicit the appropriate response when microinjected into the synaptic cleft, (2) it must occur naturally in the presynaptic axon, and (3) it must be released at the right time when the presynaptic membrane is stimulated. At present, molecules that meet these criteria include acetylcholine, a group of biogenic amines called the catecholamines, certain other amino acids and their derivatives, and perhaps some of the neuropeptides (see below). Figure 22-25 shows several common neurotransmitters.

Acetylcholine. In vertebrates, **acetylcholine** (Figure 22-25a) is the most common neurotransmitter for synapses between neurons outside the central nervous system, as well as for neuromuscular junctions (see Chapter 20). Acetylcholine is an excitatory neurotransmitter. Bernard Katz and his collaborators were the first to make the important observation that acetylcholine increases the permeability of the postsynaptic membrane to sodium within 0.1 msec of binding to its receptor. Synapses that use acetylcholine as their neurotransmitter are called **cholinergic synapses**.

Catecholamines. Another family of neurotransmitters is referred to as the **catecholamines** (Figure 22-25b). Catecholamines include *dopamine* and the hormones *norepinephrine* and *epinephrine,* all derivatives of the amino acid tyrosine. Because the hormones are also synthesized in the adrenal gland, synapses that use them as neurotransmitters are termed **adrenergic synapses.** Adrenergic synapses are found at the junctions between nerves and smooth muscles in internal organs such as the intestines, as well as at nerve-nerve junctions in the brain. The mode of action of the adrenergic hormones will be considered in Chapter 23.

Figure 22-25 The Structure and Synthesis of Neurotransmitters. (a) Acetylcholine is synthesized from acetyl CoA and choline by choline acetyltransferase. (b) The catecholamines dopamine, norepinephrine, and epinephrine are synthesized from the amino acid tyrosine and are inactivated by the enzyme monoamine oxidase. Dopamine can be converted to norepinephrine, and norepinephrine to epinephrine, as indicated by the arrows. (c) Other amino acid derivatives are histamine, serotonin, and γ-aminobutyric acid (GABA). The amino acids glycine and glutamate (not shown) are also neurotransmitters.

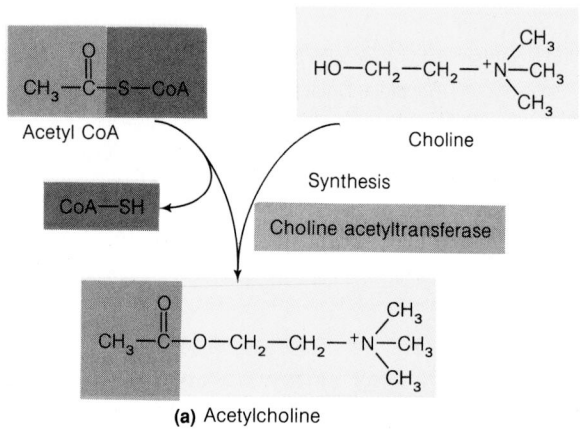

(a) Acetylcholine

(b) Catecholamines

Dopamine   Norepinephrine   Epinephrine

Histamine   Serotonin   γ-aminobutyric acid (GABA)

(c) Amino acid derivatives

Other Amino Acids and Derivatives. Other neurotransmitters that consist of amino acids and derivatives include *histamine, serotonin, γ-aminobutyric acid (GABA), glycine,* and *glutamate* (Figure 22-25c). Serotonin functions in the central nervous system. It is considered an excitatory neurotransmitter because it indirectly causes potassium channels to close, which has an effect similar to opening sodium channels in that the postsynaptic cell is depolarized. However, its effect is exerted much more slowly than that of sodium channels. GABA and glycine are inhibitory neurotransmitters, while glutamate has an excitatory effect.

Neuropeptides. In addition to the neurotransmitters described above, neurons can secrete short chains of amino acids called **neuropeptides.** Neuropeptides are formed by proteolytic cleavage of precursor proteins and can be stored in the secretory vesicles similar to those that hold neurotransmitters. At present, over 50 different neuropeptides have been identified. Some neuropeptides exhibit characteristics similar to neurotransmitters in that they excite, inhibit, or modify the activity of other neurons in the brain. However, they differ from typical neurotransmitters in that they act on groups of neurons and have long-lasting effects. Examples of this group of neuropeptides include *substance P* and the *enkaphalins,* which target regions of the brain involved with the perception of pain. Other neuropeptides act on tissues outside of the brain and are often classified as *endocrine hormones* (see Chapter 23). Examples of this group of neuropeptides are *prolactin, growth hormone,* and *leutinizing hormone.*

Neurotransmitter Receptors

Neurotransmitter receptors fall into two broad groups: ligand-gated ion channels, in which activation has a direct effect on the cell, and receptors that exert their effects indirectly through a system of intracellular messengers. We will focus here on the ligand-gated channels and leave the other category for Chapter 23.

Ligand-gated channels are ion channels in the plasma membrane that open in response to the binding of a neurotransmitter. Functionally, the ligand-gated channel class of neurotransmitter receptors can mediate either excitatory or inhibitory responses in the postsynaptic cell.

The Acetylcholine Receptor. Acetylcholine binds to the ligand-gated sodium channel known as the *nicotinic acetylcholine receptor.* When two molecules of acetylcholine bind, the channel opens and lets sodium ions rush into the postsynaptic neuron, causing a depolarization.

Our understanding of synaptic transmission has been greatly aided by the ease with which membranes rich in acetylcholine receptors can be isolated from the electric organs of the electric ray (*Torpedo californica*), an organism that is also useful as a source of sodium channel protein. The

electric organ consists of *electroplaxes*—stacks of cells that are innervated on one side but not on the other. The innervated side of the stack can undergo a potential change from about −90 mV to about +60 mV upon excitation, whereas the noninnervated side stays at −90 mV. A potential difference of about 150 mV can therefore be built up across a single electroplax at the peak of an action potential. Because the electric organ contains thousands of electroplaxes arranged in series, their voltages are additive, allowing the organism to deliver a jolt of several hundred volts.

When electroplax membranes are examined under the electron microscope, they are found to be rich in rosettelike particles about 8 nm in diameter (Figure 22-26a). Each such particle consists of five subunits arranged around a central axis, which is assumed to be the ion channel. Their size and reaction with antibodies indicate that these structures are the acetylcholine receptors.

The acetylcholine receptor from the electric ray can be purified by solubilizing electroplax membranes with nonionic detergents, followed by several chromatographic procedures. Purification of the acetylcholine receptor was greatly aided by the availability of several substances from snake venom, including *α-bungarotoxin* and *cobratoxin* (see Box 22A, page 744). These toxins serve as a highly specific means of locating and quantifying acetylcholine receptors, because they can be made highly radioactive and they bind to the receptor protein very tightly and specifically. The radioactive toxin can therefore be used as an assay for the acetylcholine receptor after each step in the purification procedure.

The purified acetylcholine receptor has a molecular weight of about 300,000 and consists of four kinds of subunits—α, β, γ, and δ—each containing about 500 amino acids. The transmembrane segment of each subunit includes sequences of relatively hydrophobic amino acids, which probably form α helices grouped together in the plane of the bilayer. The intact receptor contains the subunits in the ratio 2:1:1:1 (Figure 22-26b and c), so the simplest empirical formula for the receptor protein is $\alpha_2\beta\gamma\delta$.

The GABA Receptor. The GABA receptor is also a ligand-gated channel, but when open, it conducts chloride ions rather than sodium ions (Figure 22-27). By opening chloride channels, the activated GABA receptor inhibits formation of an action potential.

To understand this mechanism, we need to see how increasing permeability to chloride opposes the depolarization of the membrane caused by sodium influx. Sodium ions depolarize the membrane as long as they enter the cell unaccompanied by a negatively charged ion. As we discussed previously, if a chloride ion entered at the same time as a sodium ion, there would be no net effect on the membrane potential. Sodium ions normally diffuse across the membrane much faster than chloride ions, but increasing the permeability of the membrane to chloride ions tends to negate this difference. When the membrane is depolarized and sodium ions enter, more chloride ions enter also.

Figure 22-26   The Acetylcholine Receptor. The acetylcholine receptor is the primary excitatory receptor of the central nervous system. **(a)** This micrograph of an electroplax postsynaptic membrane shows the rosettelike particles thought to be the acetylcholine receptors of the membrane (TEM). **(b)** This receptor contains five subunits, including two α subunits with binding sites for acetylcholine and one each of β, γ and δ. The subunits aggregate in the lipid bilayer in such a way that the transmembrane portions form a channel. **(c)** The channel (shown here with the β subunit removed) is normally closed, but when acetylcholine binds to the two sites on the α subunits, the subunits are altered in such a way that the channel opens to allow sodium ions across.

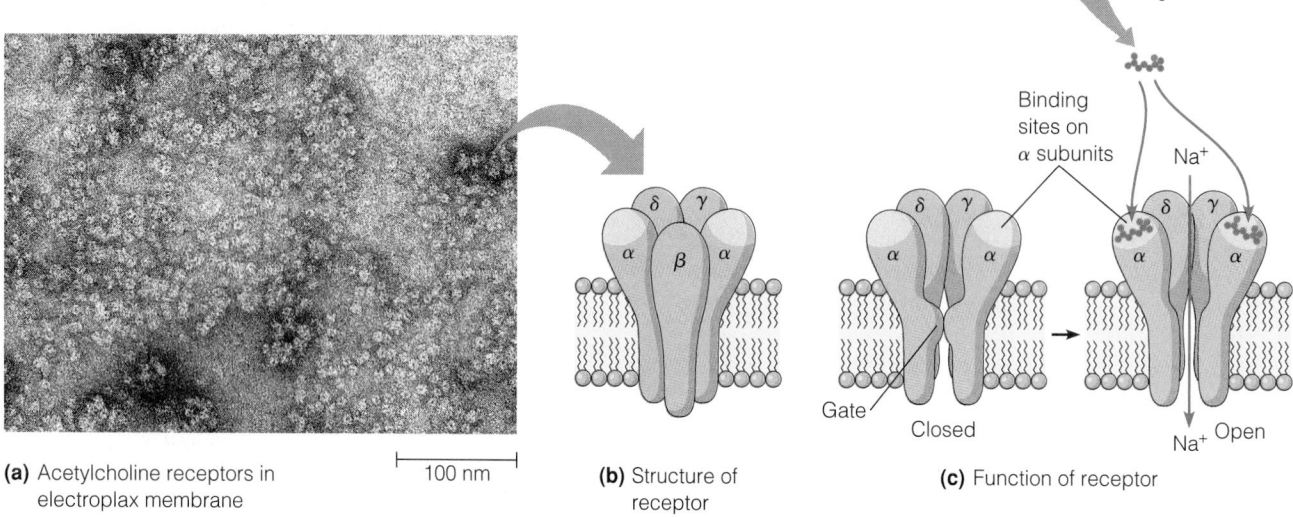

(a) Acetylcholine receptors in electroplax membrane

100 nm

(b) Structure of receptor

(c) Function of receptor

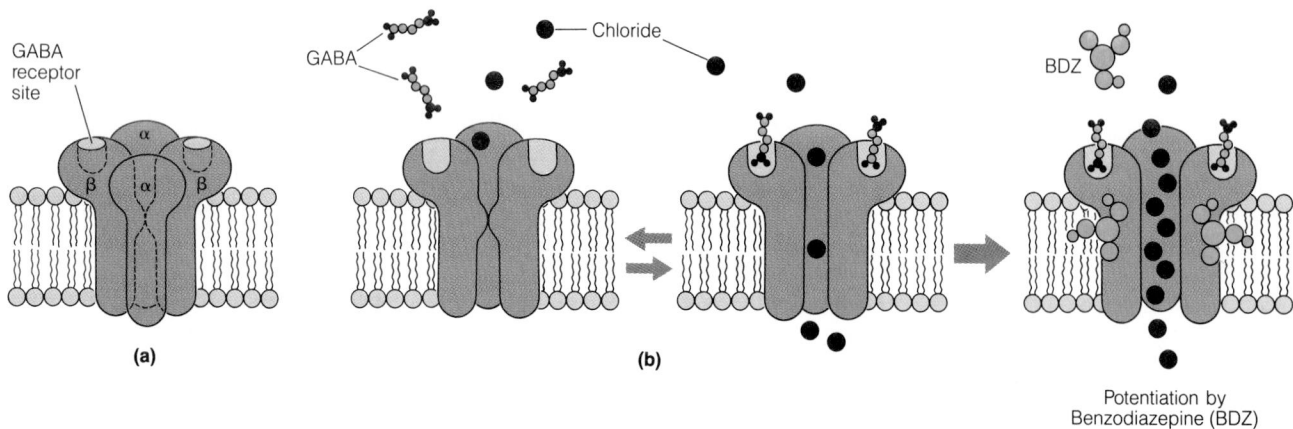

(a)

(b)

Potentiation by Benzodiazepine (BDZ)

(c)

Figure 22-27   The GABA Receptor. The GABA receptor is the primary inhibitory receptor of the central nervous system. **(a)** It is composed of two α subunits and two β subunits, and each β subunit has binding sites for GABA. **(b)** When GABA is present, the channel assumes a configuration that permits chloride ions to enter the cell down their concentration gradient. This increased chloride permeability can lead to both a small hyperpolarization and chloride entry along with sodium ions. Both effects raise the threshold for stimulating an action potential and thus decrease neuronal excitability. Several pharmacologically active agents act at the GABA receptor. For instance, **(c)** when benzodiazepines (BDZ) bind to the receptor, the effect of GABA is enhanced and the overall level of excitability is reduced. Presumably, this produces the tranquilizing effect of benzodiazepine drugs such as Valium and Librium.

Neurotransmitter Secretion

Neurotransmitter molecules are stored in small membrane-bounded **neurosecretory vesicles** in the terminal bulbs. For the neurotransmitter to act on the postsynaptic cell, it must be secreted by the process of exocytosis (see Chapter 9). During this process, the membrane of the vesicles moves into close contact with the plasma membrane of the axon terminal and then fuses with it to release the contents of the vesicle.

As we learned in Chapter 9, secretory processes are either constitutive or regulated. *Constitutive secretion* is an unregulated, ongoing process, whereas *regulated secretion* occurs only in response to a specific signal, which is needed to induce fusion. In a wide variety of cell types, the immediate signal for regulated secretion is an increase in intracellular calcium. This can be demonstrated experimentally using a compound called a *calcium ionophore*. Ionophores are molecules that carry a normally impermeable ion across a lipid bilayer, such as the plasma membrane (see Box 8B, p. 209). Several different ionophores have been isolated from bacteria and fungi, each exhibiting selectivity for a particular kind of ion. For example, a calcium ionophore carries calcium ions into the cytosol from the extracellular medium, causing the cytosolic calcium concentration to increase. In cells capable of regulated secretion, experimental application of a calcium ionophore stimulates such secretion.

The secretion of neurotransmitter by the presynaptic cell is directly controlled by the concentration of calcium ions in the terminal bulb (Figure 22-28). Each time an ac-

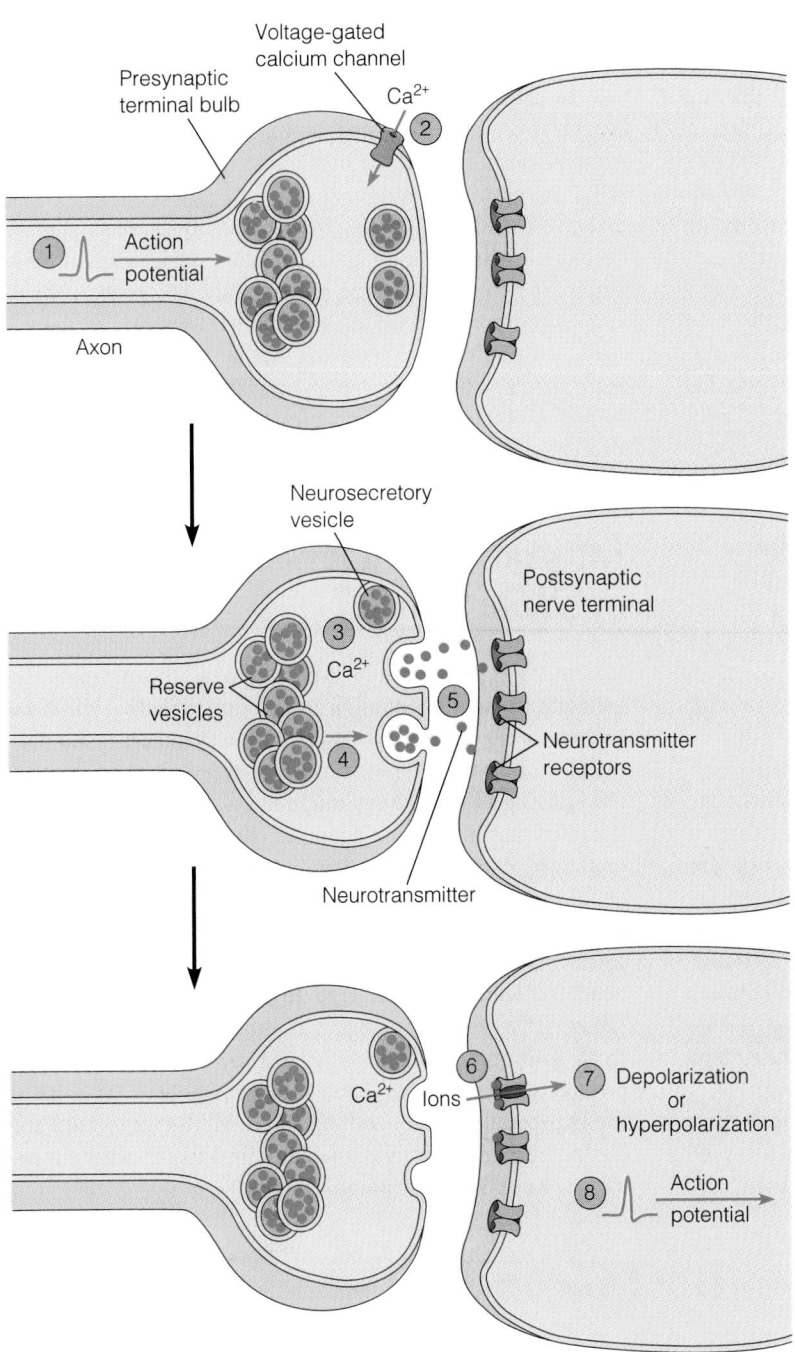

Figure 22-28   The Transmission of a Signal Across a Synapse. ① An action potential arrives at the terminal bulb, resulting in a transient depolarization. ② Depolarization opens voltage-gated calcium channels, allowing calcium ions to rush into the terminal. ③ This increase in the calcium concentration in the terminal bulb induces the secretion of a fraction of the neurosecretory vesicles. ④ Calcium also causes reserve vesicles to be released from the actin cytoskeleton so that they are ready for secretion. ⑤ Secreted neurotransmitter molecules diffuse across the synaptic cleft to receptors on the postsynaptic cell. ⑥ Binding of neurotransmitter to the receptor alters the receptor properties. ⑦ For receptors that are ligand-gated channels, the channel opens, letting ions flow into the postsynaptic cell. Depending on the ion, channel opening leads to either depolarization or hyperpolarization of the postsynaptic cell membrane. ⑧ If depolarization results, a sufficient amount of excitatory neurotransmitter will result in an action potential in the postsynaptic cell.

tion potential arrives, the depolarization causes the calcium concentration in the terminal bulb to increase temporarily due to the opening of voltage-gated calcium channels in the terminal bulbs. Normally, the cell is relatively impermeable to calcium ions, so that the cytosolic calcium concentration remains low (about 10^{-4} mM). However, there is a very large concentration gradient of calcium across the membrane because the calcium concentration outside the cell is about 10,000 times higher than that of the cytosol. As a result, calcium ions will rush into the cell when the calcium channels open.

The details of how calcium stimulates secretion are only beginning to be unraveled. For any given action potential, only a tiny fraction of the total number of vesicles stored in the terminal is secreted. Furthermore, only a relatively small fraction of the total number of vesicles are even available for secretion, the rest being held in reserve. Thus, in the terminal bulb there are at least two pools of vesicles, those that are available for secretion and those held in reserve. Calcium appears to act on these two kinds of neurosecretory vesicles in different ways. Calcium ions appear to trigger the conversion of vesicles from reserve status to those that are ready for secretion. In addition, for those vesicles that are ready, calcium also triggers the actual secretory event.

Neurons hold vesicles in reserve by linking them to actin microfilaments so that they cannot move close to the synaptic membrane for secretion. For the vesicles to become available for secretion, they must become disengaged from the actin meshwork. The key event controlling this transition is the phosphorylation of a protein called *synapsin* by a calcium-calmodulin-regulated kinase called *CAM kinase II*. Synapsin is an integral membrane protein found in the membrane of neurosecretory vesicles. In its unphosphorylated form, synapsin binds to actin filaments and also stimulates the polymerization of G-actin. In doing so, it anchors vesicles to actin filaments and prevents their secretion. When phosphorylated, synapsin no longer binds to actin, freeing the vesicles from the actin cytoskeleton.

Calcium ions also act directly at the secretory step to induce secretion. The key event in secretion is the fusion of the neurosecretory vesicle with the plasma membrane. When this occurs, the contents of the neurosecretory vesicle are discharged into the synaptic cleft. The fusion of neurosecretory vesicles with the plasma membrane appears to be mediated by a complex of proteins that functions as a "secretion machine." The process of vesicle-membrane fusion requires ATP and proceeds through several steps (see Figure 22-28). In the case of neurotransmitter secretion, one of these steps is calcium-dependent.

The calcium dependence of neurotransmitter secretion may be due to a protein called *synaptotagmin*, which is present on the membranes of neurosecretory vesicles. In the ab-sence of calcium, synaptotagmin blocks one of the steps of secretion. When an action potential arrives at an axon terminal and triggers the opening of voltage-gated calcium channels, calcium enters the terminal bulb and binds to synaptotagmin. Once calcium is bound to it, synaptotagmin is no longer inhibitory and secretion proceeds.

Terminating Synaptic Transmission

For neurons to transmit signals effectively, it is just as important to turn off the stimulus as it is to turn it on. Whether excitatory or inhibitory, once the neurotransmitter has been secreted, it must be rapidly removed from the synaptic cleft or neurotransmission becomes effectively paralyzed. In fact, the persistence of an excitatory neurotransmitter such as acetylcholine renders muscles unable to relax, ultimately leading to death.

Neurotransmitters are removed from the synaptic cleft by two specific mechanisms. One is to degrade them into inactive molecules, and the other is to resorb them back into the presynaptic terminals. In the case of acetylcholine, an enzyme called *acetylcholinesterase* hydrolyzes acetylcholine into acetic acid (or acetate ion) and choline. Neither of these products can stimulate the acetylcholine receptor. A much more common method of terminating synaptic transmission is through resorption. This mechanism involves specific transporter proteins in the membrane of the presynaptic terminals that pump the neurotransmitter back into the presynaptic axon terminals.

Integration and Processing of Nerve Signals

Sending a signal across a synapse does not automatically generate an action potential in the postsynaptic cell. If we consider a cholinergic synapse, for example, there is not necessarily a one-to-one relationship between an action potential arriving at the presynaptic neuron and one initiated in the postsynaptic neuron. A single action potential causes the secretion of enough acetylcholine to produce a detectable depolarization in the postsynaptic neuron, but usually not enough to cause the firing of an action potential. These small incremental changes in potential due to the binding of neurotransmitter are referred to as *postsynaptic potentials (PSPs)*. If a neurotransmitter is excitatory, it will cause a small amount of depolarization known as an **excitatory postsynaptic potential (EPSP)** (Figure 22-29a). Likewise, if the neurotransmitter is inhibitory, it will hyperpolarize the postsynaptic neuron by a small amount; this is called an **inhibitory postsynaptic potential (IPSP)**.

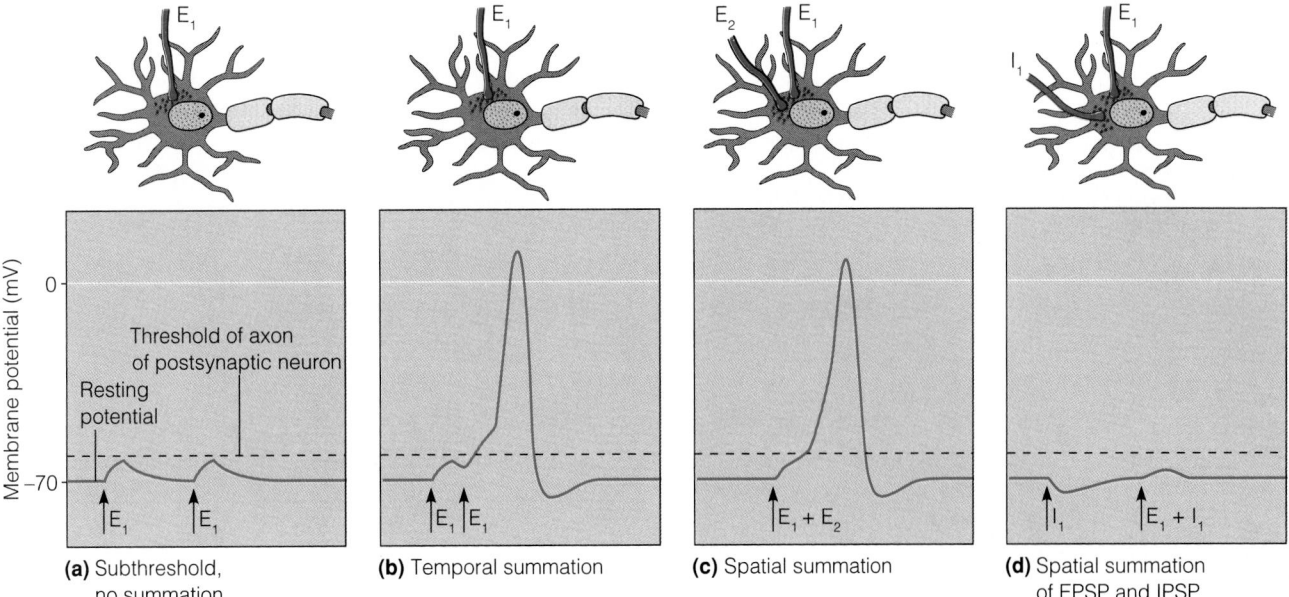

(a) Subthreshold, no summation

(b) Temporal summation

(c) Spatial summation

(d) Spatial summation of EPSP and IPSP

Figure 22-29 Summation of EPSPs and IPSPs. Neurons detect the strength of incoming signals and the strength of excitatory versus inhibitory inputs. **(a)** A single presynaptic action potential does not cause the release of enough neurotransmitter to produce an action potential in the postsynaptic neuron. **(b)** In temporal summation, the response of the postsynaptic neuron is determined by the rate of action potentials arriving at the presynaptic terminal bulb. The strength of a nerve signal is often encoded by varying the frequency of action potentials such that a weak stimulus produces infrequent action potentials and a strong stimulus produces frequent action potentials. When two action potentials arrive at close intervals, the effects of the two overlap and summate to produce an action potential in the postsynaptic cell. **(c)** In spatial summation, even infrequent action potentials can cause an action potential in the postsynaptic cell. This is because many neurons can form synapses with a single postsynaptic cell. If two or more neurons send an action potential at the same time, these action potentials summate. **(d)** One neuron can receive many synaptic inputs, either excitatory or inhibitory. The stimulation of inhibitory inputs makes it more difficult for excitatory transmissions to cause an action potential in the postsynaptic cell.

For a cholinergic presynaptic neuron to stimulate the formation of an action potential in the postsynaptic neuron, the EPSP must build up to a point at which the postsynaptic membrane reaches its threshold for firing an action potential. EPSPs can do so in two different ways, known as temporal and spatial summation. We will examine both means here.

Temporal Summation

As mentioned earlier, an action potential is an all-or-none event, so that neurons do not produce larger or smaller action potentials. Yet sensory neurons, for example, can detect whether the stimulus is weak or strong, and this information must be encoded in the form of action potentials. To encode the strength of a stimulus, neurons fire action potentials at different rates. If a neuron is maintained in a strongly depolarized state, it will fire a train of action potentials in rapid succession.

An individual action potential will only produce a temporary EPSP. However, if two action potentials fire in rapid succession at the presynaptic neuron, the postsynaptic neuron will not have time to recover from the first EPSP before experiencing a second EPSP. The result is that the postsynaptic neuron will be more depolarized. A rapid sequence of action potentials effectively sums EPSPs over time and brings the postsynaptic neuron to its threshold. This is called **temporal summation** (Figure 22-29b).

Spatial Summation

It was noted earlier that neurons can receive literally thousands of synaptic inputs from other neurons, and that the amount of neurotransmitter released at a single synapse during an action potential is usually not sufficient to produce an action potential in the postsynaptic cell. However, when many action potentials occur, causing the release of neurotransmitter, their effects combine, resulting in a large depolarization of the postsynaptic cell. This is known as **spatial summation** because the postsynaptic neuron integrates the numerous small depolarizations that occur over its surface into one large depolarization (Figure 22-29c).

Clinical Applications

POISONED ARROWS, SNAKEBITES, AND NERVE GASES

Because the coherent functioning of the human body depends so critically on its nervous system, anything that disrupts the transmission of nerve impulses is likely to be very harmful. And because of the importance of acetylcholine as a neurotransmitter, any substance that interferes with its function is almost certain to be lethal. Various toxins are known that disrupt nerve and muscle function by specific effects on cholinergic synapses. We will consider several of these substances, not only to underscore the serious threat they pose to human health, but also to illustrate how clearly their modes of action can be explained once the physiology of synaptic transmission is understood. We will also see how useful such compounds can be as research tools in studying the very phenomenon they disrupt so effectively.

Once acetylcholine has been released into the synaptic cleft and depolarization of the postsynaptic membrane has occurred, excess acetylcholine must be rapidly hydrolyzed. If it is not, the membrane cannot be restored to its polarized state, and further transmission will not be possible. The enzyme acetylcholinesterase is therefore essential, and substances that inhibit its activity are usually very toxic.

One such family of acetylcholinesterase inhibitors consists of *carbamoyl esters.* These compounds inhibit acetylcholinesterase by covalently blocking the active site of the enzyme, effectively preventing the breakdown of acetylcholine. An example of such an inhibitor is *physostigmine* (sometimes also called *eserine*), a naturally occurring alkaloid produced by the Calabar bean. Once used as a poison, physostigmine now finds use as an acetylcholinesterase inhibitor in studies of cholinergic transmission. Figure 22A-1 shows the structure of physostigmine and illustrates how it inhibits the enzyme by forming a stable carbamoyl-enzyme complex at the active site.

Many synthetic organic phosphates form even more stable covalent complexes with the active site of acetylcholinesterase and are therefore still more potent inhibitors. Included in this class of compounds are the widely used insecticides *parathion* and *malathion,* as well as nerve gases such as *tabun* and *sarin.* The structures of several such poi-

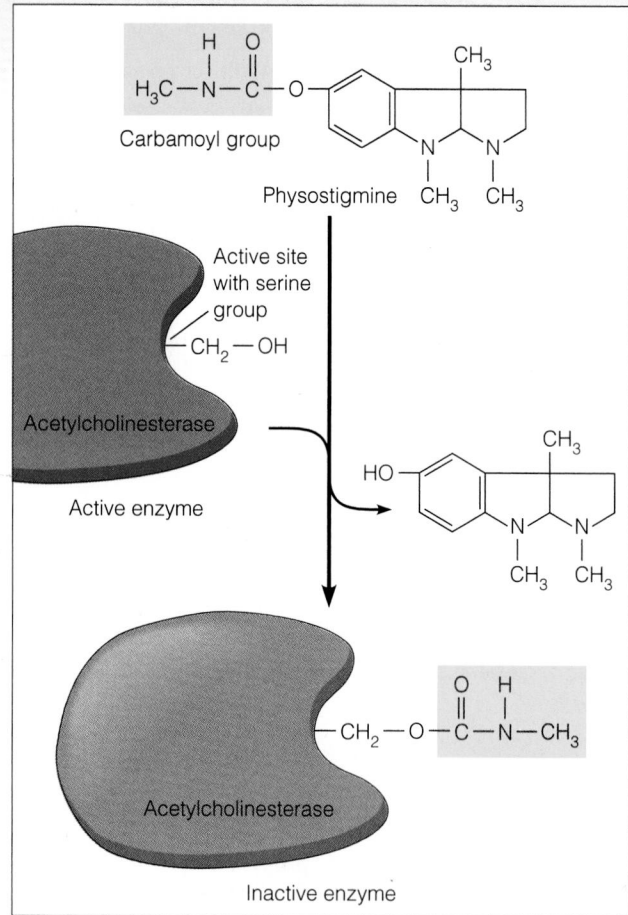

Figure 22A-1   Inactivation of Acetylcholinesterase by Physostigmine.   Physostigmine is a carbamoyl ester that inactivates acetylcholinesterase by carbamoylating the serine group at the active site of the enzyme.

sons are shown in Figure 22A-2. The primary effect of these compounds is muscle paralysis, caused by an inability of the postsynaptic membrane to regain its polarized state.

Nerve transmission at cholinergic synapses can be blocked not only by inhibitors of acetylcholinesterase, but

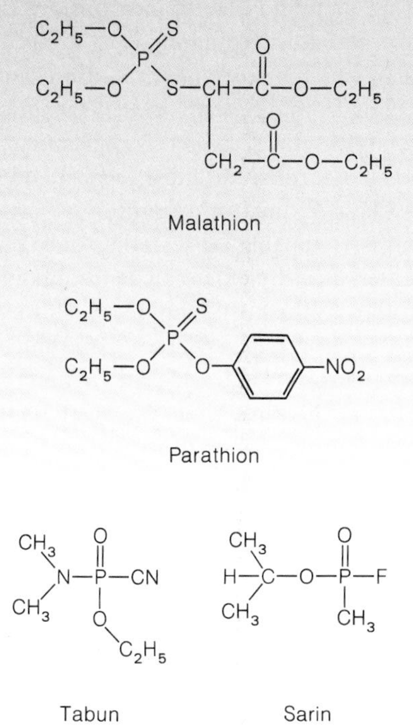

C_2H_5-O
C_2H_5-O —P— S—CH—C—O—C_2H_5
CH_2—C—O—C_2H_5

Malathion

C_2H_5-O
C_2H_5-O —P—O— —NO_2

Parathion

Tabun

Sarin

Figure 22A-2 Structures of Several Organophosphate Inhibitors of Acetylcholinesterase.

also by substances that compete with acetylcholine for binding to its receptor on the postsynaptic membrane. A particularly notorious example of such a poison is *curare,* a plant extract once used by Native South Americans to poison arrows. One of the active factors in curare is *d-tubocurarine* (Figure 22A-3). Snake venoms act in the same way. Both α-bungarotoxin (from snakes of the genus *Bungarus*) and *cobratoxin* (from cobra snakes) are small, basic proteins that bind noncovalently to the acetylcholine receptor, thereby blocking depolarization of the postsynaptic membrane.

Substances that function in this way are referred to as *antagonists* of cholinergic systems. Other compounds, called

agonists, have just the opposite effect. Agonists also bind to the acetylcholine receptor, but in so doing they mimic acetylcholine, causing depolarization of the postsynaptic membrane. Unlike acetylcholine, however, they cannot be rapidly inactivated, so the membrane does not regain its polarized state.

As these effects imply, agonists effectively lock the acetylcholine receptor in its "open" state, whereas antagonists essentially lock it in its "closed" state. These substances have therefore proved inordinately valuable in studying the receptor and especially the effects of its open and closed states on membrane permeability. In addition, several of these toxins have been useful in purification of the receptor protein because of their great specificity for this single protein.

An analogue of acetylcholine called *succinylcholine* is an agonist that is medically useful as a muscle relaxant (Figure 22A-4). Compared with acetylcholine, succinylcholine is hydrolyzed more slowly in vivo, resulting in a persistent depolarization of the postsynaptic membrane. The muscle relaxation that this depolarization produces at neuromuscular junctions is especially useful in surgical procedures.

Though of disparate origins and uses, poisoned arrows, snake venom, nerve gases, and surgical muscle relaxants all turn out to have some features in common. Each interferes in some way with the normal functioning of acetylcholine, and each is therefore a neurotoxin because it disrupts the transmission of nerve impulses, usually with lethal consequences. Each one has turned out to be useful as an investigative tool, illustrating again the strange but powerful arsenal of exotic tools upon which biologists and biochemists are able to draw in their continued probings into the intricacies of cellular function.

Why do you suppose that inhibition is so important for nervous function? A useful analogy is to imagine driving a car with a gas pedal only: In the absence of brakes (inhibition), it would be difficult to control the car's response. You could speed up or slow down but would have trouble making fine adjustments (not to mention stopping entirely). In the same way, most physiological functions, including nervous function, have regulatory processes that include both excitatory and inhibitory controls to "fine-tune" the response.

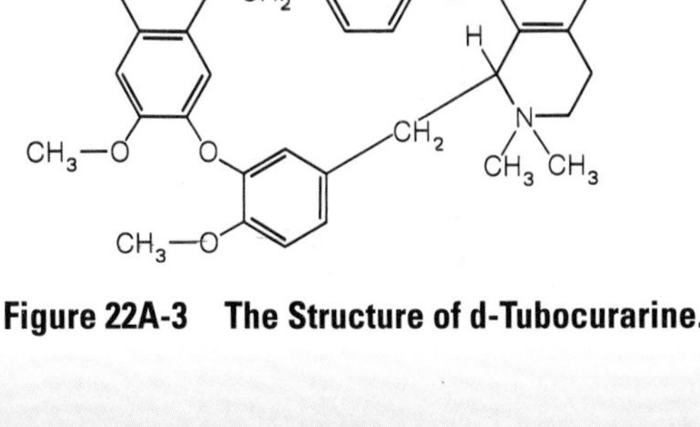

Figure 22A-3 The Structure of d-Tubocurarine.

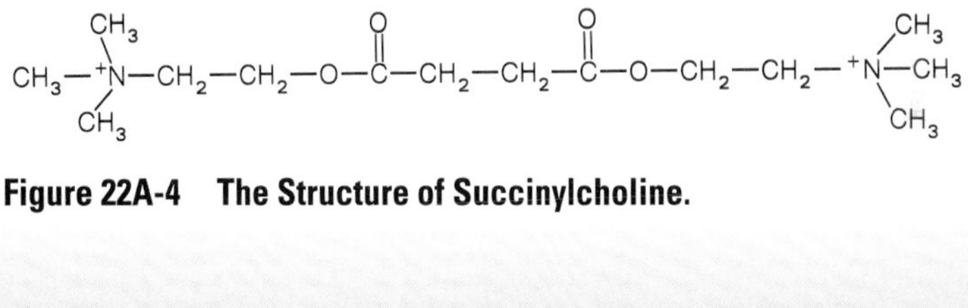

Figure 22A-4 The Structure of Succinylcholine.

A clear example of the importance of inhibition is the effect of *strychnine,* a plant toxin, on motor control. Strychnine blocks the glycine receptors of the spinal cord, so that their normal inhibitory function is lost. As a result, excitatory motor neurons take over, producing uncontrolled convulsions that often lead to death. Another interesting example of an inhibitory response is the effect of *benzodiazepines* on the brain. The benzodiazepines are a family of pharmacologically active compounds that include the drugs Valium and Librium. GABA receptors have highly specific binding sites for benzodiazepines, which produce an enhancement of the hyperpolarizing chloride ion flux that inhibits the excitability of cells with GABA receptors (see Figure 22-27). ∎

Summation of Excitatory and Inhibitory Signals

We have seen the two ways that neurons can assess the strength and significance of an incoming signal. Signal strength can be evaluated in terms of how rapidly a single neuron is firing or by how many neurons are firing all at once. Yet there is a third component to the processing of nerve signals, in which a postsynaptic neuron must respond to signals arriving from different kinds of synaptic inputs.

A postsynaptic neuron can receive synaptic inputs from different kinds of neurons, and these synapses do not necessarily all use the same neurotransmitter. A particular presynaptic neuron is capable of secreting only one type of neurotransmitter, and therefore can send only one kind of chemical stimulus to the postsynaptic cell. However, a postsynaptic neuron can receive synapses from both excitatory and inhibitory neurons (Figure 22-29d). When these different neurons fire at the same time, the postsynaptic neuron has to integrate the two kinds of signals in terms of their combined effect on the membrane potential. Thus, an individual neuron can become a "decision-making center" by virtue of its ability to determine which incoming signal (excitatory or inhibitory) is more significant.

Facilitation: Making Synaptic Transmission Stronger

We can see that synapses give neurons flexibility in how incoming signals are processed. The strength of an incoming signal can be weighed in terms of the number of action potentials received in a unit of time, whether from a few neurons firing in rapid succession or many neurons firing at the same time. When we consider that some incoming signals may be inhibitory and others excitatory, we find another level of complexity. Yet this is only part of the story.

Neurons have yet another level of complexity, referred to as **facilitation**, in which the repeated use of a synapse leads to changes that make synaptic transmission stronger. These changes can affect both the presynaptic and the postsynaptic cells. In principle, facilitation can be achieved by increasing the number of synapses formed between the presynaptic cell and the postsynaptic cell, increasing the pool of neurotransmitter vesicles that are available for secretion, or increasing the responsiveness of the postsynaptic cell. In fact, all of these methods may be involved.

Many neuroscientists believe that facilitation is closely related to learning. If so, this would also suggest that learning relates to connections between neurons. Thus, when we think of a specific memory, we might imagine that a neuron can store that memory. However, studies have shown that memory in a rat that has learned to traverse a maze cannot be localized even to a particular section of the brain. Indeed, large portions of the cerebral cortex can be removed without drastically affecting the memory. This suggests a model of learning and memory involving large numbers of neurons, a model in which memory is somehow due to active neuronal connections distributed over a vast number of neurons.

All cells maintain an electrical potential across their membranes, but neurons have specialized to use membrane potentials as a means of transmitting signals from one part of an organism to another. For this function they possess slender processes (dendrites and axons) that either receive transmitted impulses or conduct them to the next cell. The membrane of an axon may or may not be encased in a myelin sheath.

Cells develop a membrane potential due to the separation of positive and negative charges across the plasma membrane. This potential develops as each ion to which the membrane is permeable moves down its concentration gradient. The maximum membrane potential that an ion gradient can produce is the equilibrium potential for that ion—a theoretical condition that is not met in cells because it requires that the membrane be permeable only to that ion. However, it can be useful to calculate equilibrium potentials for specific ions using the Nernst equation. To calculate the resting membrane potential of a cell, the Goldman equation is used. This equation gives the algebraic sum of the equilibrium potentials for sodium, potassium, and chloride ions, each weighted for the relative permeability of the unstimulated membrane for that ion. The resting potential for the plasma membrane of most animal cells is usually in the range -60 to -75 mV. These values are quite near the equilibrium potential for potassium ion (usually about -75 mV), but very far from that for sodium ion (about $+55$ mV), reflecting the greater permeability of the resting membrane for potassium.

The action potential of a neuron represents a transient depolarization and repolarization of its membrane, due to the sequential opening and closing of sodium and potassium ion channels. These channels have been characterized structurally by molecular techniques and functionally by patch clamping. They are voltage-gated ion channels whose probability of opening, and consequently their conductance, depends on the membrane potential and the state of activation.

An action potential is initiated when the membrane is depolarized to its threshold, a point at which the rate of sodium influx exceeds the maximum rate of potassium efflux under resting conditions (usually about $+20$ mV). The entry of sodium ions drives the membrane potential to approximately $+40$ mV before voltage-gated sodium channels inactivate. Depolarization also stimulates the opening of slower voltage-gated potassium channels, which leads to re-polarization of the membrane, including a short period of hyperpolarization. This sequence of channel opening and closing takes place within a few milliseconds.

The depolarization of the membrane due to an action potential spreads to adjacent regions of the membrane by passive conductance. When an adjacent region is depolarized to its threshold potential, it also undergoes an action potential. This action potential is then propagated along the membrane, eventually reaching a synapse, or junction, between a nerve cell and another cell with which it communicates. Such synapses may be either electrical or chemical. In a chemical synapse, the electrical impulse increases the permeability of the membrane to calcium. As calcium ions cross the presynaptic membrane, they cause synaptic vesicles to fuse with the membrane. The synaptic vesicles contain neurotransmitter molecules, which are released into the synaptic cleft by the fusion event. Neurotransmitter molecules migrate across the cleft to the postsynaptic membrane, where they bind to specific receptors, often ligand-gated ion channels.

The best-understood receptor is the nicotinic acetylcholine receptor of the neuromuscular junction. Binding of acetylcholine stimulates this receptor channel to open, permitting sodium to enter. The resulting sodium influx produces a local depolarization of the postsynaptic membrane, which in turn can initiate an action potential in the postsynaptic cell. Following depolarization, the enzyme acetylcholinesterase hydrolyzes the acetylcholine, thereby returning the synapse to its resting state. An example of an inhibitory receptor is the GABA (γ-aminobutyric acid) receptor, which is a voltage-gated chloride channel. Upon binding GABA, this receptor allow increased chloride influx, leading to hyperpolarization of the postsynaptic membrane and reduced neuronal excitability.

Thirty or more specific neurotransmitters have been identified in the central nervous system that can produce either excitatory or inhibitory postsynaptic potentials. Therefore, transmission of an action potential from one neuron to another requires that the cell body of the postsynaptic neuron integrate the excitatory and inhibitory activity of thousands of synaptic inputs. Because the nerve cell body has a relatively low density of sodium channels, it is less excitable than the axon hillock. Through temporal or spatial summation, however, incoming signals can depolarize the nerve cell body sufficiently to initiate a new action potential at the axon hillock.

KEY TERMS FOR SELF-TESTING

The Nervous System

nervous system (p. 714)
central nervous system (CNS) (p. 714)
peripheral nervous system (PNS) (p. 714)
somatic nervous system (p. 714)
autonomic nervous system (p. 714)
neuron (p. 714)
sensory neuron (p. 714)
motor neuron (p. 714)
interneuron (p. 714)
glial cell (p. 714)
cell body (p. 716)
process (p. 716)
dendrite (p. 716)
axon (p. 716)
axoplasm (p. 716)
nerve (p. 716)
terminal bulb (p. 717)
synapse (p. 717)
myelin sheath (p. 717)
oligodendrocyte (p. 717)
Schwann cell (p. 717)
node of Ranvier (p. 717)

Electrical Properties of Neurons

membrane potential (p. 717)
resting membrane potential (V_m) (p. 717)
electrical excitability (p. 717)

Understanding Membrane Potentials

potential (voltage) (p. 719)
electrochemical equilibrium (p. 720)
equilibrium membrane potential (p. 720)
Nernst equation (p. 720)
Donnan equilibrium (p. 720)
sodium-potassium pump (p. 722)
depolarization (p. 722)
steady-state ion movement (p. 723)
Goldman equation (p. 723)

Electrical Excitability

ion channel (p. 724)
voltage-gated ion channel (p. 724)
ligand-gated ion channel (p. 724)
channel inactivation (p. 725)
patch clamping (p. 726)
gating current (p. 727)

The Action Potential

threshold potential (p. 729)
action potential (p. 729)
propagation (p. 729)
subthreshold depolarization (p. 731)
hyperpolarization (undershoot) (p. 731)
absolute refractory period (p. 731)
relative refractory period (p. 731)

The Propagation of an Action Potential

passive spread of depolarization (p. 733)
axon hillock (p. 733)
nerve impulse (p. 733)
resistance (p. 734)
capacitance (p. 734)

Synaptic Transmission

electrical synapse (p. 736)
presynaptic neuron (p. 736)
postsynaptic neuron (p. 736)
chemical synapse (p. 736)
synaptic cleft (p. 736)
neurotransmitter (p. 736)
acetylcholine (p. 738)
cholinergic synapse (p. 738)
catecholamine (p. 738)
adrenergic synapse (p. 738)
neuropeptide (p. 739)
neurosecretory vesicle (p. 740)

Integration and Processing of Nerve Signals

excitatory postsynaptic potential (EPSP) (p. 742)
inhibitory postsynaptic potential (IPSP) (p. 742)
temporal summation (p. 743)
spatial summation (p. 743)
facilitation (p. 746)

PROBLEM SET

22-1. The Truth About Nerve Cells. For each of the following statements, indicate whether it is true of all nerve cells (A), of some nerve cells (S), or of no nerve cells (N).

(a) The axonal endings make contact with muscle or gland cells.

(b) An electrical potential is maintained across the axonal membrane.

(c) The axon is surrounded by a discontinuous sheath of myelin.

(d) The resting potential of the membrane is much closer to the equilibrium potential for potassium ions than to that for sodium ions because the sodium-potassium pump maintains a much larger transmembrane gradient for potassium than for sodium.

(e) Excitation of the membrane results in a permanent increase in its permeability to sodium ions.

(f) The electrical potential across the membrane of the axon can be easily measured using electrodes.

(g) Both the sodium and potassium concentration gradients completely "collapse" every time a nerve impulse is transmitted along the axon.

(h) Upon arrival at a synapse, a nerve impulse causes the secretion of acetylcholine into the synaptic cleft.

22-2. The Resting Membrane Potential. The Goldman equation is used to calculate V_m, the resting potential of a biological membrane. As presented in the chapter, it contains terms for sodium, potassium, and chloride ions only.

(a) Why do only these three ions appear in the Goldman equation as it applies to nerve impulse transmission?

(b) Suggest a more general formulation for the Goldman equation that would be applicable to membranes that might be selectively permeable to other monovalent ions as well.

(c) Would the version of the Goldman equation you suggested in part b be adequate for calculating the resting potential of the membrane of the sarcoplasmic reticulum in muscle cells? Explain.

(d) How much would the resting potential of the membrane change if the relative permeability for sodium ions were 1.0 instead of 0.01?

(e) Would you expect a plot of V_m versus the relative permeability of the membrane to sodium to be linear? Why or why not?

22-3. Equilibrium Potentials Versus Resting Potentials. Based on the Nernst equation, the equlibrium membrane potential for potassium (E_R) is about −15 mV more negative than the measured resting potential of the cell. For mammalian neurons, the measured resting potential is −75 mV and E_K is −90 mV. However, the equilibrium membrane potential for sodium (E_{Na}) is about +55 mV, far different from the resting membrane potential. Explain why the mammalian resting membrane potential is closer to E_K than to E_{Na}.

22-4. Patch Clamping. Patch-clamp instruments enable researchers to measure the opening and closing of a single channel in a membrane. A typical acetylcholine receptor channel passes about 5 pA of ionic current (1 picoampere = 10^{-12} ampere) at −60 mV over a period of about 5 msec.

(a) Given that an electrical current of 1 A is about 6.2×10^{18} electrical charges per second, how many ions (potassium or sodium) pass through the channel during the time it is open?

(b) Do you think the opening of a single receptor channel would be sufficient to depolarize a postsynaptic membrane? Why or why not?

22-5. The Equilibrium Membrane Potential. Answer each of the following questions with respect to E_{Cl}, the equilibrium membrane potential for chloride ions. The chloride ion concentration inside the squid giant axon can vary from 50 to 150 mM.

(a) Before doing any calculations, predict whether E_{Cl} will be positive or negative. Explain.

(b) Now calculate E_{Cl}, assuming an internal chloride concentration of 50 mM.

(c) How much difference would it make in the value of E_{Cl} if the internal chloride concentration were 150 mM instead?

(d) Why do you suppose the chloride concentration inside the axon is so variable?

22-6. Heart Throbs. An understanding of muscle cell stimulation involves some of the same principles as nerve cell stimulation, except that calcium ions play an important role in the former. The following ion concentrations are typical of those in human heart muscle and in the serum that bathes the muscles:

> [K$^+$]: 150 mM in cell, 4.6 mM in serum
> [Na$^+$]: 10 mM in cell, 145 mM in serum
> [Ca^{2+}]: 0.001 mM in cell, 6 mM in serum

Figure 22-30 depicts the change in membrane potential with time upon stimulation of a cardiac muscle cell.

(a) Calculate the equilibrium membrane potential for each of the three ions, given the concentrations listed.

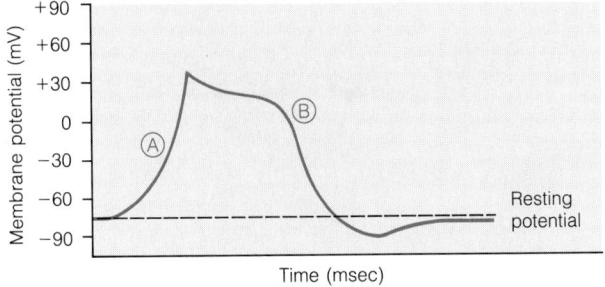

Figure 22-30   The Action Potential of a Muscle Cell of the Human Heart.

(b) Why is the resting membrane potential significantly more negative than that of the squid axon (−75 mV versus −60 mV)?

(c) The increase in membrane potential in the region of the graph marked Ⓐ could in theory be due to the movement across the membrane of one or both of two cations. Which cations are they, and in what direction would you expect each to move across the membrane?

(d) How might you distinguish between the possibilities suggested in part c?

(e) The rapid decrease in membrane potential that is occurring in the region marked Ⓑ is caused by the outward movement of potassium ions. What are the driving forces that cause potassium to leave the cell at this point? Why aren't the same forces operative in the region of the curve marked Ⓐ?

(f) People with heart disease often take drugs that can double or triple their serum potassium levels to about 10 mM without altering intracellular potassium levels. What effect should this increase have on the rate of potassium ion movement across the heart cell membrane during muscle stimulation? What effect should it have on the resting potential of the muscle?

22-7. The All-or-None Response of Membrane Excitation. A nerve cell membrane exhibits an all-or-none response to excitation; that is, the magnitude of the response is independent of the magnitude of the stimulus, once a threshold value is exceeded.

(a) Explain in your own words why this is so.

(b) Why is it necessary that the stimulus exceed a threshold value?

(c) If every neuron exhibits an all-or-none response, how do you suppose the nervous system of an animal can distinguish different intensities of stimulation? How do you think your own nervous system can tell the difference between a warm iron and a hot iron, or between a chamber orchestra and a rock band?

22-8. Excitability of Dendrites and Axons. The dendrites and cell body of a typical motor neuron have a much lower density of sodium channels than does the axon hillock or the nodes of Ranvier. Would the threshold potential be the same in the dendrites, the axon hillock, and the nodes of Ranvier? Why or why not?

22-9. One-Way Propagation. Why does the action potential move in only one direction down the axon?

22-10. Inhibitory Neurotransmitters. Some inhibitory neurotransmitters cause chloride channels to open, while others stimulate the opening of potassium channels. Explain why increasing the permeability of the neuronal membrane to either chloride or potassium would make it more difficult to stimulate an action potential. What generalization can you make about inhibitory neurotransmitters?

22-11. Trouble at the Synapse. Transmission of a nerve impulse across a cholinergic synapse is subject to inhibition by a variety of neurotoxins. Indicate, as specifically as possible, what effect each of the following poisons or drugs has on synaptic transmission and what effect each has on the polarization of the postsynaptic membrane.

(a) The snake poison α-bungarotoxin

(b) The insecticide malathion

(c) Succinylcholine

(d) The carbamoyl ester neostigmine

Suggested Reading

General References

DeCamilli, P. and R. Jahn. Pathways to regulated exocytosis in neurons. *Annu. Rev. Physiol.* 52 (1990): 625.

Hall, Z. W. *An Introduction to Molecular Neurobiology*, Vol. 2. Sunderland, MA: Sinauer, 1992.

Hille, B. *Ionic Channels of Excitable Membranes*. Sunderland, MA: Sinauer, 1992.

Kandel, E. R., J. H. Schwartz, and T. M. Jessell. *Principles of Neural Science*, 3d ed. Norwalk, CT: Appleton & Lange, 1991.

Levitan, I. B., and L. K. Kaczmarek. *The Neuron: Cell and Molecular Biology*. New York: Oxford University Press, 1991.

The Sodium-Potassium Pump

Mercer, R. W. Structure of the Na, K-ATPase. *Internat. Rev. Cytol.* 137C (1993): 139.

Patch Clamping

Neher, E. and B. Sakmann. The patch-clamp technique. *Sci. Amer.* 266 (1992): 28.

Ion Channels and Membrane Excitation

Armstrong, C. M. Voltage-dependent ion channels and their gating. *Physiol. Rev.* (Suppl. on Forty Years of Membrane Current in Nerve) 72 (1992): S5.

Catterall, W. A. Structure and function of voltage-gated ion channels. *Trends Neurosci.* 16 (1993): 500.

Hess, P. Calcium channels in vertebrate cells. *Annu. Rev. Neurosci.* 13 (1990): 337.

Hodgkin, A. L., and A. F. Huxley. A quantitative description of membrane current and its application to conduction and excitation in nerve. *J. Physiol.* 117 (1952): 500.

Karlin, A. The structure of nicotinic acetylcholine receptors. *Curr. Opin. Neurobiol.* 3 (1993): 299.

Unwin, N. Neurotransmitter action: Opening of ligand-gated ion channels. *Cell* 72 (1993): 31.

Unwin, N. The structure of ion channels in membranes of excitable cells. *Neuron* 3 (1989): 665.

Neurotransmitters and Neuropeptides

Bloom, F. E. Neurotransmitters: Past, present, and future directions. *FASEB* 2 (1988): 32.

Hokfelt, T. Neuropeptides in perspective: The last ten years. *Neuron* 7 (1991): 867.

Synaptic Transmission

Brose, N., A. G. Petrenko, T. C. Sudhof, and R. Jahn. Synaptotagmin: A calcium sensor on the synaptic vesicle surface. *Science* 256 (1992): 1021.

Jessell, T. M., and E. R. Kandel. Synaptic transmission: A bidirectional and self-modifiable form of cell-cell communication. *Cell* 72 (1993): 1.

Langosch, D., C. M. Becker, and H. Betz. The inhibitory glycine receptor: A ligand-gated chloride channel of the central nervous system. *Eur. J. Biochem.* 194 (1990): 1.

O'Connor, V., G. Augustine, and G. Betz. Synaptic vesicle exocytosis: Molecules and models. *Cell* 76 (1994): 785.

23

SIGNAL TRANSDUCTION MECHANISMS:
II. MESSENGERS AND RECEPTORS

I n the previous chapter, we learned how nerve cells are able to communicate with one another and with other types of cells by means of electrochemical changes in their membranes. We saw how, in most cases, the arrival of an action potential at a synapse causes the release of *neurotransmitters,* which bind in turn to *receptors* on the postsynaptic cell membrane, thereby passing on the signal. Now we are ready to explore a second major means of intercellular communication that also involves interactions between chemicals and receptors. In this case, however, the signal is transmitted by regulatory chemicals called **messengers,** and the **receptors** are located on cells that may be quite distant from the messenger-secreting cells. Thus, the animal body has two different but complementary systems of communication and control, and receptors play a crucial role in both systems. In this chapter, we will encounter several general mechanisms by which receptors detect and pass on chemical signals that impinge upon the cell.

Chemical Signals and Cellular Receptors

All cells have some ability to sense and respond to specific aspects of their environment. These can be **physical factors** in the environment, such as light, heat, or gravity, or **chemical factors,** such as the presence and concentration of particular molecules. Prokaryotic cells, for instance, have membrane-bound receptor molecules on the cell surface that enable these cells to respond by chemotaxis to chemicals in their environment, as described in Chapter 21. The human body has receptors for light (the rods and cone cells of the retina), sound (the hair cells of the inner ear), and various other physical stimuli. In addition, we have receptors on the tongue and in the nose that detect chemicals in our food and in the environment.

One way that cells communicate with one another is by displaying molecules on their surfaces that are recognized by receptors on the surfaces of other cells. This kind of cell-to-

cell communication requires that cells come into contact with each other. Alternatively, one cell can release chemical signals that are recognized by another cell, either nearby or at a distant location. For example, simple eukaryotic organisms such as amoeboid cells of the slime mold *Dictyostelium* secrete a compound called *cyclic AMP* (cAMP) at one stage in their life cycle (Figure 23-1). The compound binds to receptors on the surfaces of neighboring cells, triggering a process whereby thousands of individual amoeboid cells aggregate and differentiate into a multicellular organism.

In more complex multicellular organisms, the problem of regulating and coordinating the various activities of cells or tissues is particularly important. Here, the whole organism is organized into different tissues containing specialized cells. Furthermore, the specific function of these cells may only be critical for certain occasions, or one tissue may need to perform opposite functions, depending on the circumstances. One way that animal cells regulate the function of tissues is through the nervous system, which is very fast but is confined to tissues that are innervated. Another way to regulate the function of various tissues is through the release of *chemical messengers,* which is the topic of this chapter.

To regulate the function of various cells and tissues, both plants and animals use chemical signals called hormones. **Hormones** are chemical messengers secreted by one tissue in order to regulate the function of cells or tissues in the same organism. In animals, there are hundreds of different hormones that regulate a wide variety of functions. For example, cells of the liver regulate the amount of glucose in the blood. This is critically important because the concentration of glucose in the blood must be kept within narrow limits. Too little glucose in the blood results in unconsciousness because nerve cells rely on glucose for energy. However, too much glucose in the blood causes a number of problems, including damage to red blood cells. To regulate blood glucose concentration, the liver either removes glucose from the blood and stores it as glycogen or breaks down glycogen and releases glucose into the blood. The particular activities

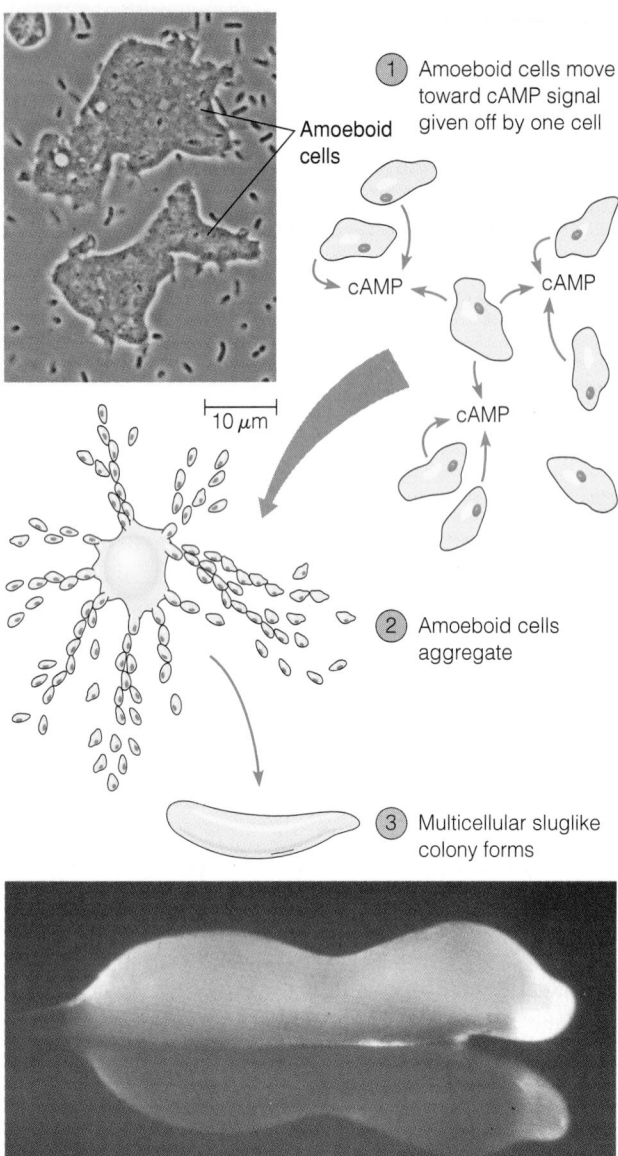

① Amoeboid cells move toward cAMP signal given off by one cell

Amoeboid cells

10 μm

② Amoeboid cells aggregate

③ Multicellular sluglike colony forms

Sluglike colony

0.25 μm

Figure 23-1   The Aggregation of Slime Mold Cells in Response to a Chemical Messenger.   The slime mold *Dictyostelium* uses cyclic AMP (cAMP) as a messenger for communication between cells. Under conditions of nutrient deprivation, some of the cells secrete cAMP. It then diffuses away and binds to receptors on neighboring cells, stimulating them to move toward the source of the cAMP. As a result, a population of *Dictyostelium* cells aggregates and forms a multicellular sluglike colony. Through the interaction between other messenger molecules and receptors, cells of the "slug" are induced to form a fruiting body containing spores. The spores allow *Dictyostelium* to survive until conditions are again favorable for its growth.

of the liver in storing or releasing glucose must be coordinated to the needs of the rest of the body and, in particular, to the levels of glucose in the blood. Hormonal control mechanisms ensure that the liver acts to lower or raise blood glucose levels properly.

Types of Chemical Signals Received by Cells

The kinds of chemical compounds that function as messengers or hormones are very diverse. For example, various messenger molecules can be chemically characterized as amino acids or their derivatives, peptides, proteins, fatty acids, lipids, nucleosides, or nucleotides. However, with one important exception, the chemical composition of the messenger does not have any direct bearing on the kind of signal or message delivered to the target cell. Most messengers are water-soluble compounds; these **water-soluble messengers** bind to receptors on cell surfaces. Furthermore, their function lies entirely in their ability to bind to a receptor. The chemical composition of a messenger molecule is only important in that it provides the correct molecular shape and bonding characteristics to fit in the receptor. In a sense, then, we can compare the interaction between a messenger and its receptor to a key and lock. The key is similar to a messenger molecule in that it has the right shape and grooves needed to fit into the lock.

The one case in which the chemical nature of the messenger is important to how the messenger acts involves certain types of lipid or lipidlike messengers. These **lipid messengers** act on receptors in the nucleus or cytosol whose function is to regulate the transcription of particular genes. For such a messenger molecule to bind to intracellular receptors, it obviously must pass through the plasma membrane. This, in turn, means that the molecule must be sufficiently hydrophobic or lipidlike to cross the lipid bilayer. Among the lipid messengers that bind to intracellular receptors are the *steroid hormones,* which are derived from the compound cholesterol. Cholesterol and the steroid hormones are structurally similar in that they all share a common core chemical structure called a *steroid nucleus.* Other hydrophobic or lipidlike messengers include the thyroid hormones, vitamin D_3, and retinoic acid (Figure 23-2).

Interactions Between Messengers and Receptors

Chemical communication between cells depends on the ability of one cell to secrete or display on its surface a messenger molecule that can be sensed and responded to by another cell. The ability to sense a chemical messenger depends, in turn, on the presence of specific receptor proteins in either the cytosol or the nucleus of a target cell or exposed on its surface. Communication occurs when the messenger binds to the receptor and causes the receptor to undergo a change in its properties that results in its **activation.** Thus, communication depends on a specific chemical messenger sent by one cell and a specific receptor for that messenger on

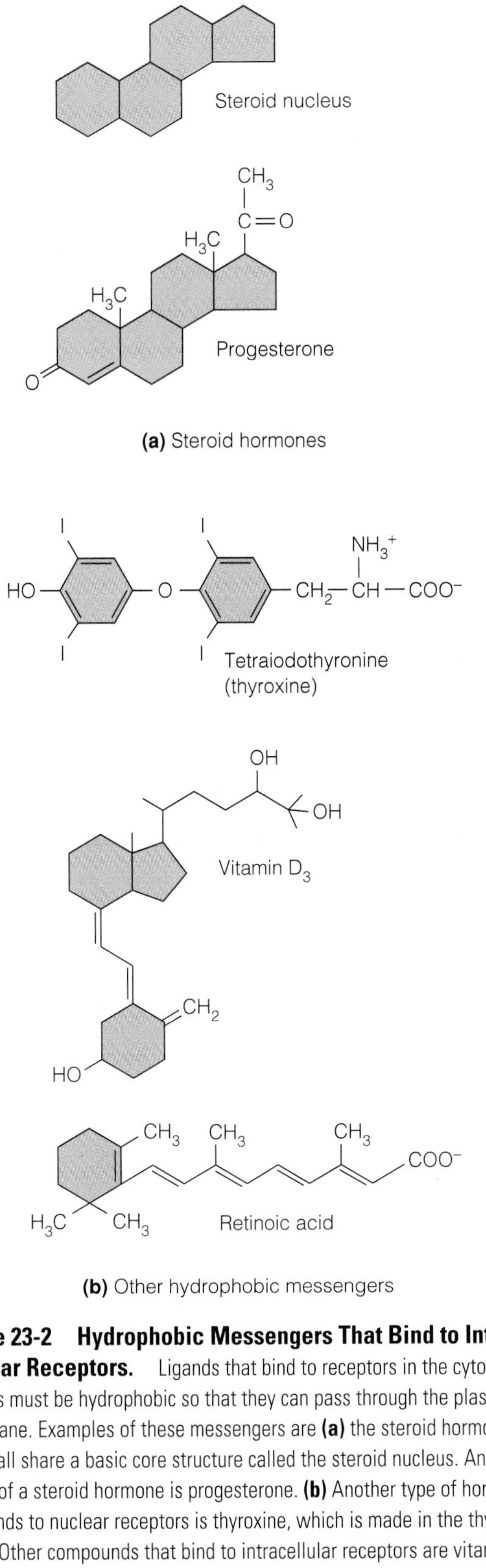

(a) Steroid hormones

(b) Other hydrophobic messengers

Figure 23-2 Hydrophobic Messengers That Bind to Intracellular Receptors. Ligands that bind to receptors in the cytosol or nucleus must be hydrophobic so that they can pass through the plasma membrane. Examples of these messengers are **(a)** the steroid hormones, which all share a basic core structure called the steroid nucleus. An example of a steroid hormone is progesterone. **(b)** Another type of hormone that binds to nuclear receptors is thyroxine, which is made in the thyroid gland. Other compounds that bind to intracellular receptors are vitamin D_3, which induces the synthesis of a protein involved in calcium uptake from the intestines, and retinoic acid, which regulates the transcription of genes during early development.

another cell. For a given messenger-receptor pair, the chemical messenger is often called a **ligand,** or binding molecule. The specific receptor for a particular messenger is called the **cognate receptor** for that particular ligand.

We can now ask how cells distinguish messengers from the myriad of other chemicals in the environment or from messengers intended for other cells. The answer lies in the specific binding of a receptor to the messenger molecule. A messenger binds by forming noncovalent chemical bonds with the receptor protein. Individual noncovalent bonds are generally weak; therefore, several bonds must form to achieve strong bonding. For a receptor to make numerous bonds with its ligand, the receptor must have a *binding site* (or binding pocket) that fits the messenger molecule closely, like a hand in a glove. Furthermore, within the ligand-binding site on the receptor, appropriate amino acid side chains must be positioned so that they can make chemical bonds with the messenger molecule. This combination of binding site shape and the positioning of amino acid side chains within the binding site is what enables the receptor to distinguish its specific ligand from thousands of other chemicals.

In most cases, the binding reaction between a ligand and its cognate receptor is similar to the binding of an enzyme to its substrate. Ligands dissolved in the fluid surrounding the cell can collide with receptors. Some of these collisions lead to binding, in which the ligand adheres to the binding site on the receptor. When a receptor binds its ligand, the receptor is said to be occupied. Similarly, the ligand can be either bound (to a receptor) or free in the solution. The amount of receptor that is occupied by ligand is proportional to the concentration of free ligand in solution. As the ligand concentration increases, an increasingly greater proportion of its cognate receptors will become occupied, until *receptor saturation* is reached, at which point all the receptors are occupied. Further increases in ligand concentration will, in principle, have no further effect on the target cell.

The relationship between the concentration of ligand in solution and the number of receptors occupied can be described qualitatively in terms of **receptor affinity.** When receptors are occupied at low concentrations of free ligand, we say that the receptor has a high affinity for its ligand. Receptor affinity can be described quantitatively in terms of the *dissociation constant,* K_d. The K_d value for a specific ligand-receptor interaction is the concentration of free ligand needed to produce a state in which half the receptors are occupied. Thus, receptors with a high affinity for their ligands have a very small dissociation constant. In general, all hormones bind fairly tightly to their receptors, but the values for K_d range from roughly 10^{-4} to 10^{-9} mM. The importance of the K_d value is that it tells us at what concentration a particular ligand will be effective in producing a cellular response. Specifically, the ligand concentration must be in the range of the K_d value of the receptor in order for the ligand to have an effect on the target tissue.

Although receptors have a characteristic affinity for their ligands, cells are geared to sense *changes* in ligand concentration rather than fixed ligand concentrations. When a ligand is present and receptors are occupied for prolonged periods of time, the cell adapts, or becomes *desensitized,* so that it no longer responds to the ligand. To further stimulate the cell, the ligand concentration must be increased.

Desensitization is due mainly to changes in the properties or cellular location of the receptor, a phenomenon known as **receptor down-regulation.** The down-regulation of receptors occurs in several ways, including (1) removal of the receptor from the cell surface, (2) alterations to the receptor that lower its affinity for ligand, or (3) alterations to the receptor that render it unable to initiate changes in cellular function. The removal of receptors from the cell surface takes place through the process of *receptor-mediated endocytosis,* in which small portions of the plasma membrane containing receptors invaginate and are internalized (see Figure 9-20). Receptor-mediated endocytosis removes the receptor from the cell surface so that it is no longer exposed to ligand. Once the receptor is internalized, it can be either stored or transferred to lysosomes, where it is degraded. In either case, the reduced number of receptors on the cell surface results in a diminished cellular response to ligand. In other cases, the receptors are chemically altered, often by phosphorylation. These modified receptors can remain on the surface yet not respond to the presence of ligand because of either their reduced affinity for ligand or their inability to initiate changes in cellular function.

Desensitization provides a way for cells to adapt to permanent differences in levels of messenger concentration. However, it also leads to the phenomenon of *tolerance.* An example of this is seen in nasal sprays, which contain neosynephrine or other compounds that stimulate particular kinds of receptors called α-adrenergic receptors (discussed later in the chapter). The drug causes blood vessels in the nose to constrict, thereby inhibiting mucous secretions. However, prolonged use leads to a loss of effectiveness due to receptor down-regulation.

Understanding the nature of receptor-ligand binding has provided great opportunities for researchers and pharmaceutical companies. Although receptors have binding sites that fit the messenger molecule quite closely, it is possible to make similar synthetic compounds that bind even more tightly or selectively. This is especially important when more than one type of receptor exists for the same ligand. For example, there are several different receptor proteins that bind acetylcholine as their natural ligand. These acetylcholine receptors are broadly divided into two groups, *nicotinic* and *muscarinic,* so named because the compounds nicotine and muscarine selectively activate different subsets of acetylcholine receptors. Researchers and pharmaceutical companies have developed many synthetic compounds that selectively affect only one type of receptor. In addition, whereas normal messengers cause a change in the receptor when they bind, both synthetic and natural compounds have been discovered that can bind to receptors without triggering such a change. These compounds inhibit the receptor by preventing the naturally occurring messenger from binding and activating the receptor.

The use of drugs that selectively activate or inhibit particular kinds of receptors has become central to the treatment of many medical problems. For example, *isoproterenol* and *propranolol* activate or inhibit the β-adrenergic receptors that will be discussed later in the chapter. Isoproterenol is used to treat asthma or to stimulate the heart, whereas propranolol is used to reduce blood pressure and the strength of cardiac contractions. Another example is *cimetidine,* a compound that selectively binds and inhibits a particular type of histamine receptor found on parietal cells in the stomach. Through its action on histamine receptors, cimetidine blocks the secretion of stomach acid by parietal cells and thus is useful for treating ulcers.

Initiation of Signal Transduction

When a ligand binds to its cognate receptor, the receptor is altered in a way that causes changes in cellular activities. In general, the binding of a ligand either induces a change in receptor conformation or causes receptors to cluster together. Once one of these changes takes place, the receptor initiates a preprogrammed sequence of events inside the cell. By "preprogrammed," we mean that cells have a greater repertoire of functions than are active at any particular time. Some of these cellular processes remain inactive until particular signals are received that trigger them. We might compare the activation effect of a messenger to the stimulation of a reflex. In the knee-jerk reflex, for example, the nerve connections are already in place that can cause the leg to extend in response to a tap just below the knee, but the tap is needed to activate the reflex. In a similar manner, a messenger activates a series of cellular functions that are already in place but are not presently active.

The sequence of biochemical events that are initiated or altered within a cell when a ligand binds to its cognate receptor on the cell surface or within the cell is called **signal transduction.** Different receptors initiate various transduction pathways, but most such pathways nonetheless fall into several general categories. In the following section, we will discuss several of these general pathways, including the control of gene transcription by steroid hormones, the cAMP signaling pathway, the inositol-phospholipid-calcium pathway, and the pathways initiated by tyrosine kinase receptors.

Characteristics and Functions of Receptors

Receptors are the key to understanding how a messenger triggers a change in cellular activities. To understand how a particular hormone or other messenger causes a change in cellular activities, we really need to know what the receptor does when it binds to its ligand. Our purpose here is to classify receptors into several basic categories and to consider the main features of each. The basic types of receptors are *ligand-gated channels* (discussed in Chapter 22), *intracellular receptors*, and *plasma membrane receptors*. Plasma membrane receptors can be classified into two families: those linked to G proteins and those linked to kinase enzymes. The latter family includes receptors that have intrinsic kinase activity and those that stimulate the activity of a kinase.

As we saw previously, intracellular receptors bind lipid messengers and play a specific role in regulating the transcription of specific genes. For the most part, the plasma membrane receptors bind water-soluble messengers that cannot penetrate the plasma membrane. Their effects on cell function are much more diverse than are those of the intracellular receptors. Although the stimulation of plasma membrane receptors can in some cases activate or inhibit the transcription of particular genes, they do so indirectly and usually have a broader range of effects. We will now turn to intracellular receptors and their effects on the cell.

Intracellular Receptors

Intracellular receptors are located primarily within the nucleus, although in some cases the receptor may initially reside in the cytosol and translocate to the nucleus upon ligand binding. As noted earlier, the receptors for the various steroid hormones, thyroxine, vitamin D_3, and retinoic acid are included in this category. Intracellular receptors have certain similarities in their structure and mode of action. Each receptor is a protein with the potential for binding both to a lipid ligand, such as one of the steroid hormones, and to DNA. Once the ligand binds, the receptor-ligand complex becomes a *transcription factor* that can stimulate the transcription of particular genes. Here, we will confine our discussion to the steroid receptors because these are among the best understood.

The Structure and Activation of Steroid Receptors. Analysis of a number of different **steroid receptor** proteins has revealed a remarkably similar basic structure. Each such receptor has a steroid-binding region near the C-terminus and another region in the middle of the protein that binds to specific sequences of DNA. Each of the different steroid receptors is specific for a particular steroid hormone because of unique features of its steroid-binding region. The DNA-binding portions of the various steroid receptors exhibit a similar basic structure called a *zinc finger motif* (see Figure 19-23). However, each different steroid receptor binds to a unique short segment of DNA known as a *response element*. One or more response elements are typically clustered together in regions of DNA known as *enhancers*. As we learned in Chapter 19, an enhancer is a nontranscribed region of DNA that regulates the rate of expression of a specific gene, which may be at a considerable distance from the enhancer. Thus, when a steroid receptor binds to its response element within the enhancer, the rate of gene transcription is increased.

The Effects of Steroid Receptor Activation. The activation of gene expression by a steroid receptor is under hormonal control; in other words, the steroid receptor is only active as a transcription factor when bound to its ligand. Our present understanding of how a steroid regulates the function of its cognate receptor comes largely from work with the receptor for glucocorticoid hormone (cortisol) and follows the model shown in Figure 23-3. In the absence of hormone, the glucocorticoid receptor is predominantly located in the cytosol, where it is apparently anchored by binding to a cytosolic protein. This cytosolic anchor prevents the glucocorticoid receptor from entering the nucleus and binding to its response element. The glucocorticoid receptor is therefore inhibited in the absence of cortisol. The binding of the hormone to its receptor frees the receptor from its anchor and allows translocation of the receptor to the nucleus, where it binds to its response element and stimulates transcription.

This model for hormonal regulation of the glucocorticoid receptor provides a framework for understanding how steroid hormones regulate transcription. However, most steroid receptors are already located in the nucleus even in the absence of hormone. But the nuclear steroid receptors are still incapable of activating transcription. Nuclear steroid receptors are probably also anchored—to nuclear proteins, in this case—so that the receptors cannot bind to their response elements. Here, too, the binding of hormone frees the receptor from its anchor so that it can bind to its response element. The receptor for thyroid hormone deviates even further from the mechanism described for the glucocorticoid receptor. The thyroid receptor is bound to its DNA response element even in the absence of hormone. The binding of thyroid hormone somehow converts the receptor to an active transcription factor, although the mechanism is not yet known.

Plasma Membrane Receptors: G Protein–Linked Receptors

One class of **plasma membrane receptors** is known as the **G protein–linked receptor** family because ligand binding causes a change in receptor conformation that activates a particular **G protein** (*guanine nucleotide–binding protein*). The activated G protein in turn binds to a target protein such as an enzyme or a channel protein, thereby altering its activity. All G protein–linked receptors initiate signal transduction inside the cell by this means.

The Structure and Activation of G Protein–Linked Receptors. The G protein–linked receptors are remarkable in that they all have a similar structure yet differ significantly in their amino acid sequences. In each case, the receptor protein forms seven transmembrane α helices connected by alternating cytosolic or extracellular loops. The N-terminus of the protein is exposed to the extracellular fluid, while the

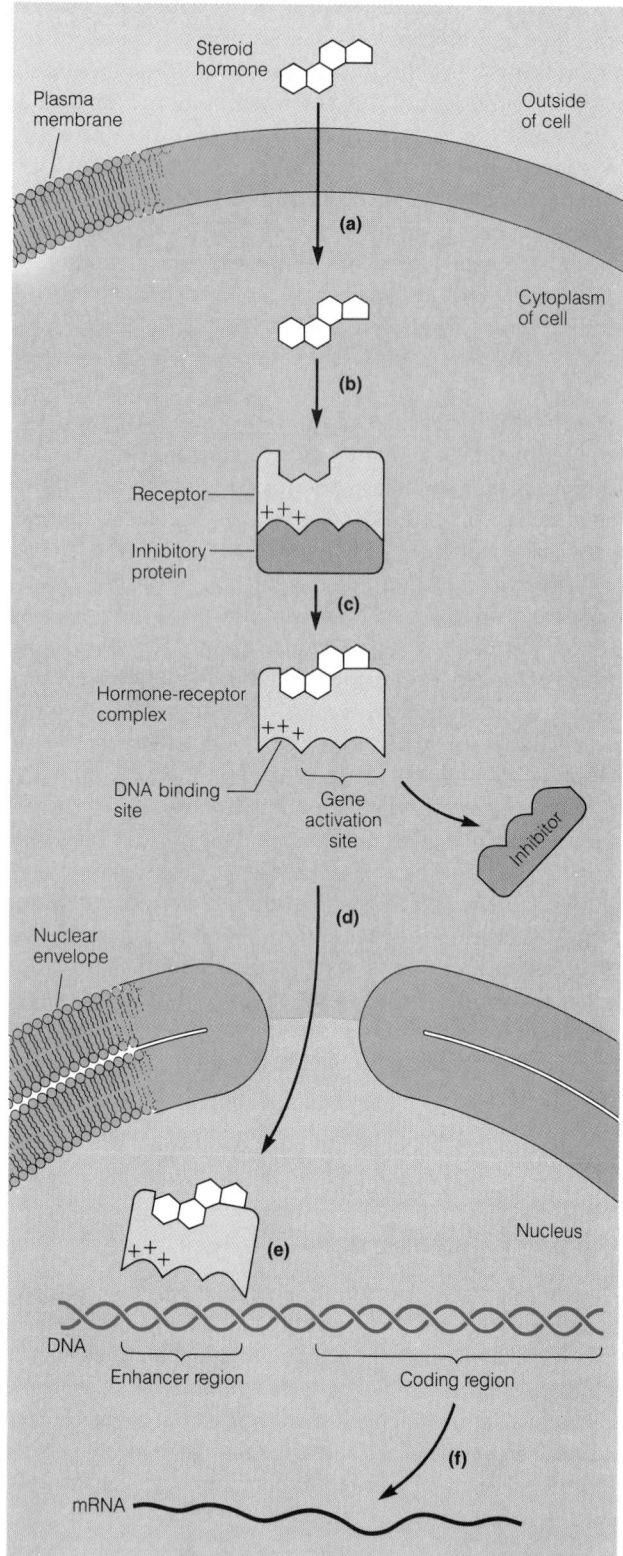

Figure 23-3 The Activation of Glucocorticoid Receptors.
Steroid hormones enter the cell and act at the level of gene regulation.
(a) Because it is lipophilic, glucocorticoid hormone can diffuse through the plasma membrane. **(b)** In the cell, a hormone molecule binds to an intracellular receptor, **(c)** causing the release of an inhibitory protein, which exposes DNA-binding and gene activation sites. **(d)** The activated complex then enters the nucleus, where it **(e)** binds to a DNA enhancer region. **(f)** This binding activates transcription of the RNAs that encode the proteins involved in the cell functions regulated by the hormone.

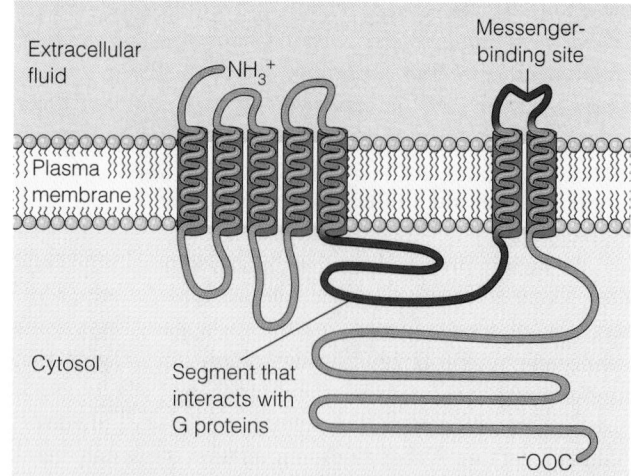

Figure 23-4   The Structure of G Protein–Linked Receptors.
Each G protein–linked receptor has seven transmembrane α helices. The messenger binds to the extracellular portion of the receptor. This binding causes an intracellular portion of the receptor to activate an adjacent G protein.

C-terminus resides in the cytosol (Figure 23-4). The extracellular portion of each G protein–linked receptor has a unique messenger-binding site, whereas a cytosolic loop connecting the fifth and sixth transmembrane α helices is specific for a particular G protein. G protein–linked receptors therefore provide a versatile method for linking different messengers to different signal transduction pathways.

The Structure and Activation of G Proteins.   Because G proteins are central to the function of G protein–linked receptors, we will examine the structure and function of these proteins more closely. We might describe G proteins as a type of molecular switch whose "on" or "off" state depends on whether the G protein is bound to GTP or GDP. There are two distinct classes of G proteins: the *large heterotrimeric G proteins* and the *small monomeric G proteins*. The large heterotrimeric G proteins contain three different subunits called alpha (Gα), beta (Gβ), and gamma (Gγ). These heterotrimeric G proteins mediate signal transduction through G protein–linked receptors as well as the transduction of light into electrical signals by rod cells of the eye (Box 23A). The small monomeric G proteins are represented by Ras, which will be discussed later in the chapter in the context of tyrosine kinase receptors. While the term *G protein* can apply to both trimeric and monomeric types, in our discussion here we will use it only to refer to the trimeric type.

G proteins were first identified in mutant cells that appeared to lack adenylate cyclase, the enzyme that converts ATP to cyclic AMP. In fact, the mutants had the adenylate cyclase enzyme but it could not be activated. Activity was restored, however, when extracts from normal cells were added to extracts from the mutant cells. The component missing in the mutant cells was a 42-kDa protein that turned out to be the α subunit of a G protein. Since then, many different G proteins have been identified in various types of cells.

23A

Further Insights

G Proteins and Vision

At first glance, a relationship between vision and receptors for endocrine hormones might seem highly unlikely. Why would photons of light striking a sensory cell in the retina of the eye have anything to do with hormonal receptors? However, it is now clear that the sensing of light by retinal cells depends directly on a G protein–mediated mechanism that somewhat resembles the hormonal regulatory cascades encountered in this chapter.

The retinal cells that are most sensitive to light are called *rod cells* (Figure 23A-1). The rod cell membrane, like

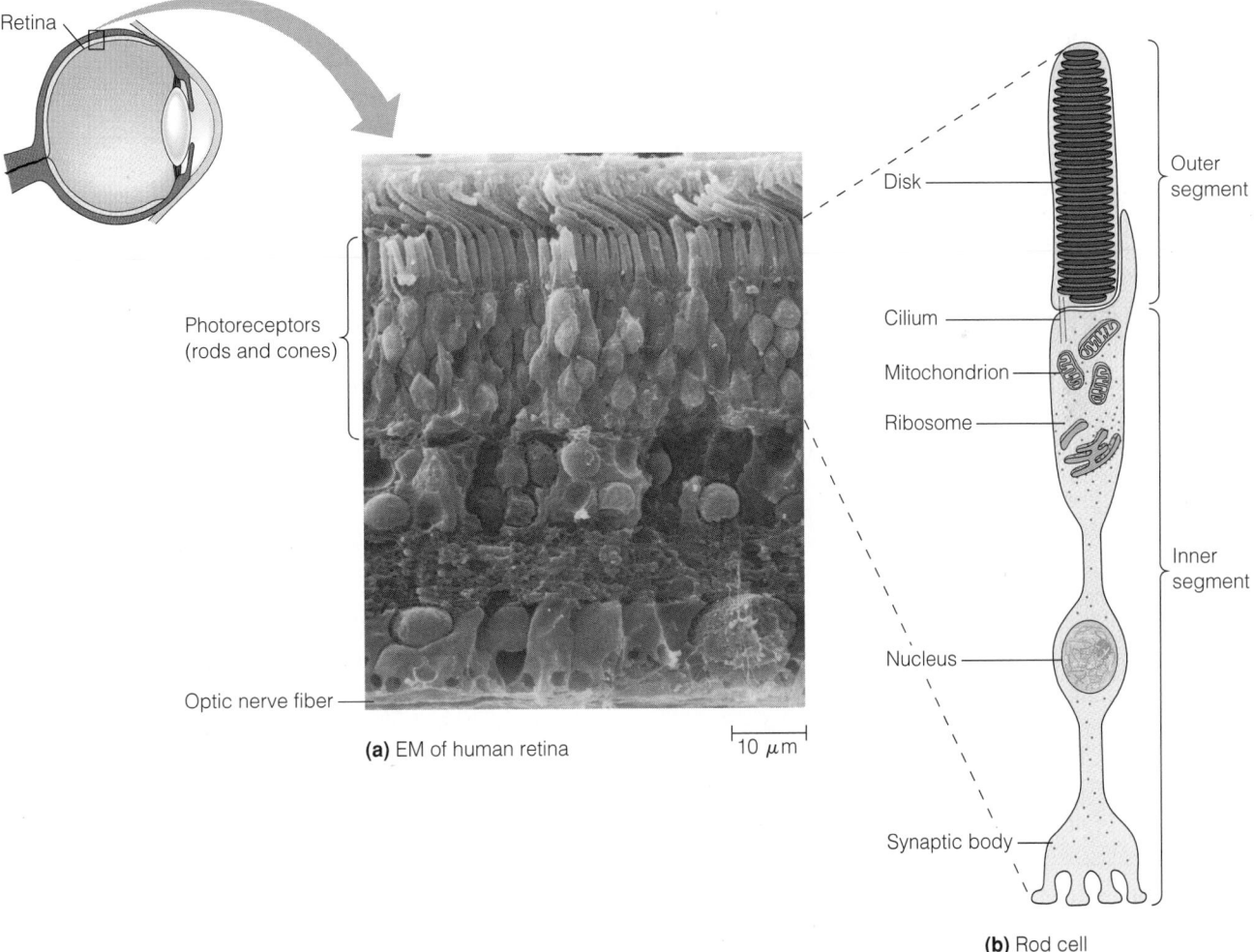

(a) EM of human retina

10 μm

(b) Rod cell

Figure 23A-1 The Structure of Rod Cells. **(a)** An electron micrograph of rod cells, light-sensitive cells in the retina of the human eye. Rods are especially useful for vision in dim light (SEM; R. G. Kessel and R. H. Kardon. *Tissues and Organs: A Text Atlas of Scanning Electron Microscopy.* © 1979 by R. G. Kessel and R. H. Kardon). **(b)** An individual rod cell, showing the outer segment, the inner segment, the nucleus, and the synaptic body (also called the synaptic knob). The outer segment consists of a stack of about one thousand flattened and tightly packed membrane disks. Embedded within these membranes are molecules of rhodopsin, a complex protein consisting of the pigment retinal attached covalently to the transmembrane glycoprotein opsin. The inner segment of the rod cell contains a variety of organelles, predominantly mitochondria. The synaptic body is filled with synaptic vesicles that contain an inhibitory neurotransmitter. Signal transduction via the optic nerve is suppressed in the dark but is activated in the light by a cGMP-mediated decrease in the release of the inhibitory neurotransmitter.

other excitable membranes, contains gated sodium channels. Unlike the channels in the nerve cell membrane, however, the sodium channel of the rod cell membrane has a binding site for *cyclic GMP (cGMP)*. When bound to this site, cGMP keeps the channels open, so that the cell is normally depolarized in the resting state (i.e., in the dark). The rod cell has a synaptic connection to a sensory neuron and continuously releases a neurotransmitter that is inhibitory to the neuron. The neuron is thereby maintained in a state of relatively low activity as long as the rod cell detects no light and the membrane remains depolarized.

What happens when photons of light strike the rod cell? The stacked membrane disks in the outer segment of the cell are packed with molecules of a light-sensitive protein called *rhodopsin*. Rhodopsin has as its prosthetic group a derivative of vitamin A called *retinal* (Figure 23A-2). This molecule is the primary visual pigment of the retina. When

retinal absorbs light energy, it undergoes a conformational change from the 11-*cis* to the all-*trans* configuration, thereby inducing a conformational change in the rhodopsin molecule of which it is a part. The conformationally altered rhodopsin molecule then catalytically activates the heterotrimeric G protein *transducin* (designated Gt) by causing Gtα to exchange GDP for GTP and then to dissociate from Gtβγ. The free GTP-Gtα in turn activates GMP diesterase, a membrane-bound enzyme that hydrolyzes cGMP to GMP. The resulting decrease in the cyclic GMP concentration in the cytosol causes cGMP to be released from its binding sites on the sodium channels. The channels then close, the cell becomes polarized, and less inhibitory neurotransmitter is released, thereby activating the postsynaptic sensory neurons. As a result, increasing numbers of action potentials reach the brain, where they are interpreted as light. ∎

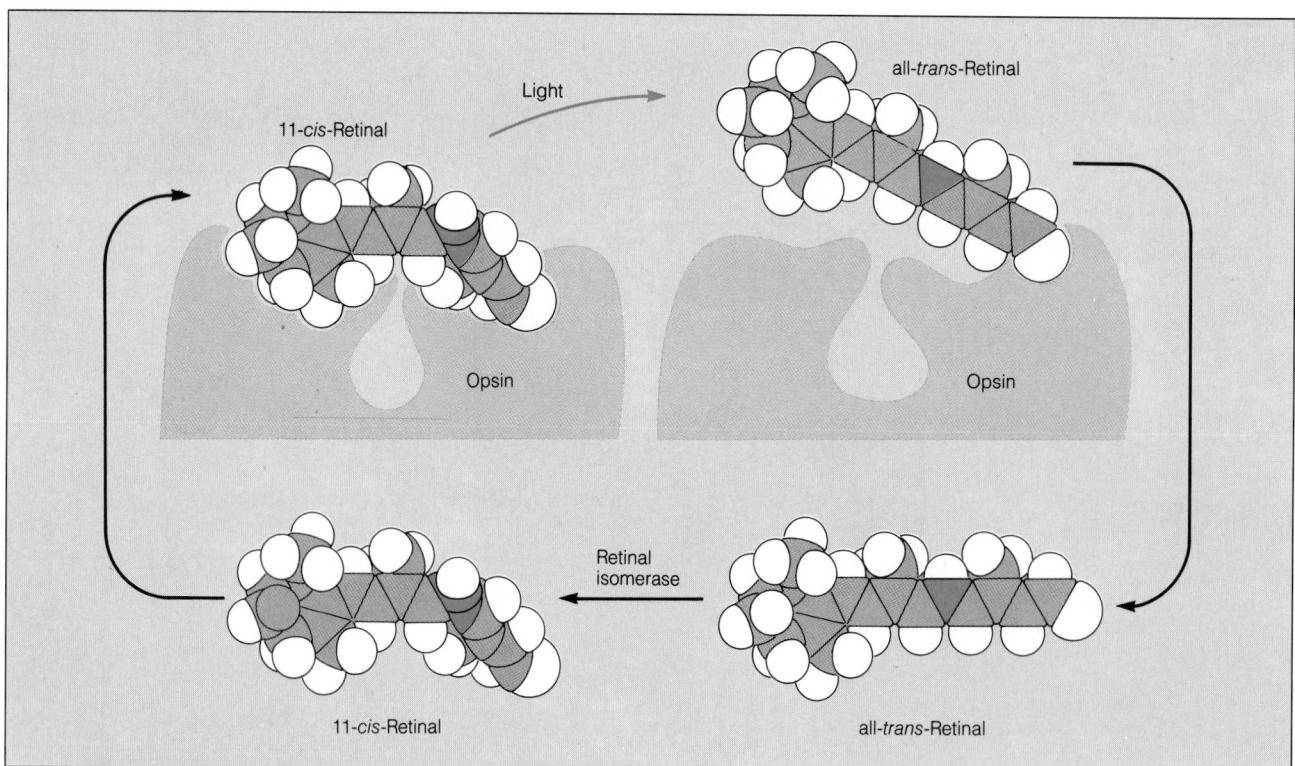

Figure 23A-2   The Effect of Light on Retinal. The primary visual pigment of the rod cell is 11-*cis*-retinal, the light-sensitive prosthetic group of the rhodopsin molecule. When retinal absorbs a photon of light, it isomerizes from the 11-*cis* form to all-*trans*-retinal. As these space-filling models show, 11-*cis*-retinal is attached to the protein opsin, such that the light-induced isomerization to the all-*trans* form leads to a conformational change in the opsin protein. The shape change initiates a signal transduction pathway that includes the G protein transducin, cGMP, and the sodium channels in the plasma membrane of the rod cell. The *trans*-retinal dissociates from the opsin and is released into the cytosol, where it is converted by the action of retinal isomerase to the 11-*cis* form, which is again capable of association with opsin to reinitiate the cycle.

G proteins all have a similar basic structure and mode of activation. Of the three subunits in the $G\alpha\beta\gamma$ heterotrimer, $G\alpha$, the largest, binds to a guanine nucleotide (GDP or GTP). $G\alpha$ also has the ability to hydrolyze GTP, so it only remains in the active GTP-bound form for a short time. When $G\alpha$ binds to GTP, it also detaches from the $G\beta\gamma$ complex. The $G\beta$ and $G\gamma$ subunits, on the other hand, are permanently bound together.

When a messenger binds to a G protein–linked receptor on the exterior surface of the cell, the change in conformation of the receptor causes the $G\alpha$ subunit to release its bound GDP, acquire a GTP, and then detach from the complex (Figure 23-5). Depending on the G protein and the cell type, either the free GTP-$G\alpha$ subunit or the $G\beta\gamma$ complex can then initiate signal transduction events in the cell. Each portion of the G protein exerts its effect by binding to a particular enzyme or other protein in the cell. In some cases, both the GTP-$G\alpha$ and $G\beta\gamma$ subunits simultaneously regulate different processes in the cytosol. However, the activity of the G protein persists only as long as the $G\alpha$ is bound to GTP and the subunits remain separated. Because the $G\alpha$ subunit hydrolyzes GTP, it will only remain active for a short time before reverting to the GDP-bound state and reassociating with $G\beta\gamma$. This feature allows the signal transduction pathway to shut down rapidly when the messenger is removed.

The large number of different G proteins provides for a diversity of G protein–mediated signal transduction events. Some G proteins interact directly with certain types of potassium or calcium ion channels to mediate the actions of specific neurotransmitters. In some organisms, G proteins activate kinases. Perhaps the most important and widespread G protein–mediated signal transduction events are the release or formation of second messengers. *Second messengers* are small molecules or ions that relay the signals from one location in the cell, such as the plasma membrane, where the receptor is located, to the interior of the cell. Two of the most important second messengers are cyclic AMP and calcium ions, both of which will be discussed later.

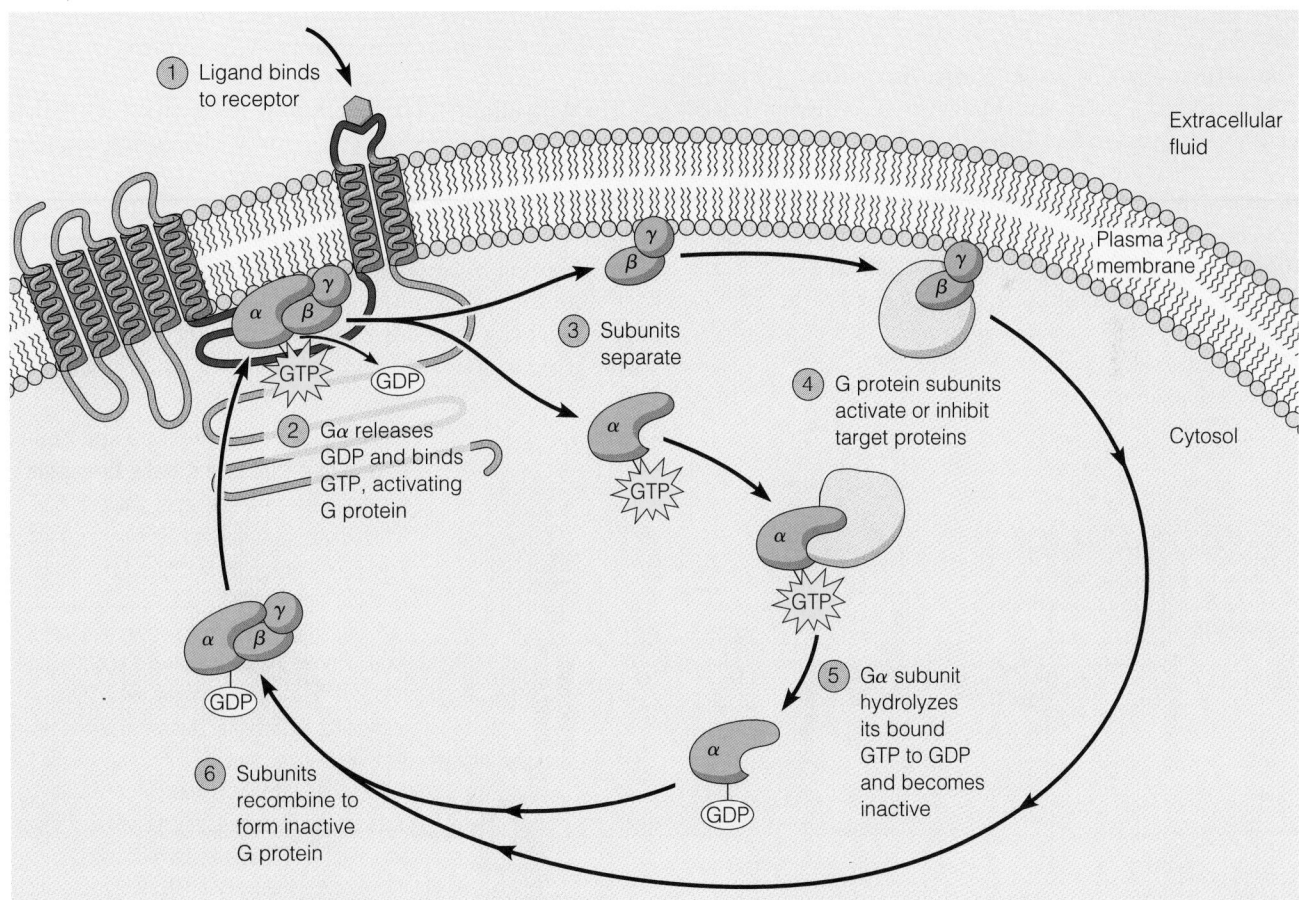

Figure 23-5   The G Protein Activation/Inactivation Cycle.
G protein–linked receptors contain seven transmembrane segments that form a ligand-binding site on the outside of the cell and a G protein–binding site on the inside. ① When the ligand binds, ② the receptor activates a G protein by causing the $G\alpha$ subunit to release GDP and acquire GTP. ③ The $G\alpha$ and $G\beta\gamma$ subunits then separate and ④ initiate different signal transduction events. ⑤ The GTP-$G\alpha$ subunit eventually hydrolyzes its bound GTP, converting the subunit back to its inactive GDP-$G\alpha$ form. ⑥ The inactive GDP-$G\alpha$ subunit then recombines with $G\beta\gamma$ to form the inactive G heterotrimer.

Plasma Membrane Receptors: Tyrosine Kinase Receptors

The next class of receptors is a more diverse group, but it can generally be categorized as receptors that are linked to specific *protein kinases*. Some of the most common and important of these, the **tyrosine kinase receptors,** have their own tyrosine kinase activity. In other words, an intrinsic part of the receptor is an enzyme called a *tyrosine kinase* whose function is to transfer a phosphate from ATP to tyrosine residues in specific substrate proteins. Many cellular enzymes are regulated by kinases, including those that control glycogenolysis (see Figure 11-14). However, most kinases involved in regulating cellular enzymes phosphorylate a serine or threonine residue on the target enzyme and are thus known as *serine/threonine kinases.*

In general, tyrosine kinase receptors are activated by growth factors. Many receptors in this class trigger a chain of signal transduction events inside the cell that ultimately lead to cell growth and proliferation. This process is tightly controlled, so that cells grow when the growth factor is available and stop growing when the stimulus is removed.

The Structure of Tyrosine Kinase Receptors. Tyrosine kinase receptors differ structurally from G protein–linked receptors in many ways. These receptors often consist of a single polypeptide chain with only one transmembrane segment. Within this polypeptide chain are several distinct domains (Figure 23-6a). One end of the polypeptide chain projects out into the extracellular fluid and forms the *ligand-binding domain.* A small segment of the peptide passes through the plasma membrane into the cytosol. On the cytosolic side, a portion of the receptor forms the tyro-sine kinase while the remainder constitutes a cytosolic tail. The cytosolic portion of the receptor contains important tyrosine residues that are in fact substrates or targets for the receptor tyrosine kinase. When the receptor binds to its ligand, these tyrosine residues become phosphorylated.

Nonreceptor Tyrosine Kinases. Nonreceptor tyrosine kinases represent a variant of the theme seen with tyrosine kinase receptors. In many tyrosine kinase receptors, the tyrosine kinase is an integral part of the receptor protein. In other cases, however, the receptor and the tyrosine kinase are two separate proteins. The tyrosine kinase is then referred to as a **nonreceptor tyrosine kinase** because it is a separate protein. However, it can bind to the receptor and be activated when the receptor binds its ligand, so the net effect is quite similar to the activation of a typical tyrosine kinase receptor.

Nonreceptor tyrosine kinases were, in fact, the first example of tyrosine kinases to be discovered. The first nonreceptor tyrosine kinase identified was the Src protein, which is encoded by the *src* gene of the avian sarcoma virus. Since then, many nonreceptor tyrosine kinases have been discovered, most of which are similar to the Src protein.

The Activation of Tyrosine Kinase Receptors. To initiate signal transduction through a tyrosine kinase receptor, a ligand binds, causing the receptors to aggregate (Figure 23-6b). Once the receptors cluster in this way, the tyrosine kinase associated with each receptor phosphorylates the tyrosines of neighboring receptors. Interestingly, one can mimic the action of the real ligand by artificially cross-linking receptors together with antibody molecules that specifically bind to the receptor. While an antibody molecule is quite different

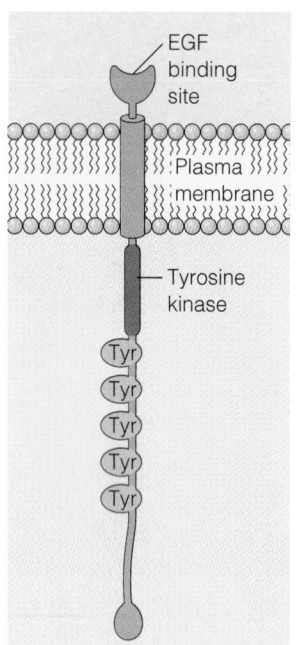
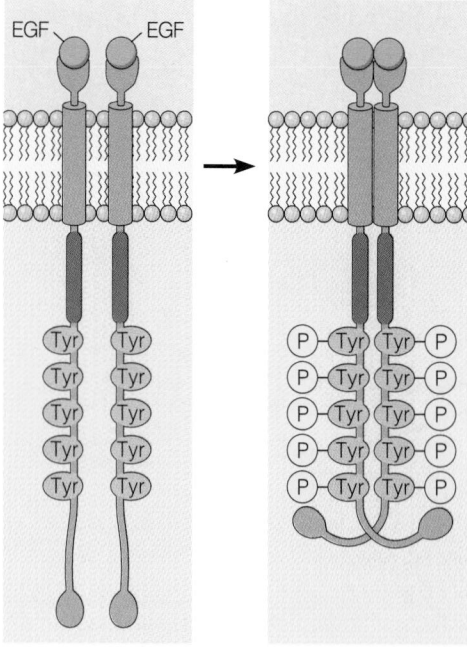

(a) Structure of the epidermal growth factor (EGF) receptor

(b) Activation of the EGF receptor

Figure 23-6 The Structure and Activation of a Tyrosine Kinase Receptor. **(a)** The receptor for epidermal growth factor (EGF), shown here, is typical of many tyrosine kinase receptors. These receptors often have only one transmembrane segment. The extracellular portion of the receptor binds to the growth factor. Inside the cell, a portion of the receptor has tyrosine kinase activity. The remainder of the receptor contains a series of tyrosine residues that are substrates for the tyrosine kinase. **(b)** The activation of tyrosine kinase receptors starts with the binding of a messenger (EGF, in this case), causing receptor aggregation or clustering. Once the receptors aggregate, they cross-phosphorylate each other at a number of tyrosine amino acid residues. The formation of tyrosine phosphate (Tyr-P) residues on the receptor creates binding sites for cytosolic proteins that contain SH2 domains.

from the natural ligand, it can produce the same effect because it causes receptors to aggregate.

Once the cytosolic portion of a receptor becomes phosphorylated, the receptor recruits a number of cytosolic proteins to itself. Each of these recruited proteins binds to the receptor at a phosphorylated tyrosine residue. To bind to the receptor, each cytosolic protein must contain a stretch of amino acids that recognizes the phosphotyrosine and a few neighboring amino acids on the receptor. The portion of a protein that recognizes one of these phosphorylated tyrosines is called an *SH2 domain*. The term SH2, for Src homology (domain) 2, traces its origin to the Src protein, a nonreceptor tyrosine kinase. Proteins with SH2 domains have sequences of amino acids that are strikingly similar to a portion of the Src protein. Several proteins with SH2 domains have been identified and shown to bind to tyrosine phosphate residues, either on a receptor or on other tyrosine-phosphorylated proteins.

The reason the receptor recruits cytosolic proteins is still something of a mystery. In some cases, enzymes seem to be activated merely by binding to the receptor. In other cases, once a protein binds to the receptor, it is also phosphorylated by the receptor tyrosine kinase. In the latter case, binding is required for phosphorylation, but it is the phosphorylation step that activates the enzyme.

Tyrosine kinase receptors can activate several different signal transduction pathways at the same time. These include the Ras pathway, which ultimately activates the expression of genes involved in growth, and the inositol-phospholipid-calcium second messenger pathway, to be discussed shortly.

Signal Transduction Pathways and Second Messenger Systems

As noted earlier, **second messengers** relay signals from the plasma membrane to the interior of the cell, and they mediate many of the effects of receptor activation. The two most widely used second messengers are *cyclic AMP* and *calcium ions*. These second messengers stimulate the activity of target enzymes when their cytosolic concentrations are elevated. G protein–linked receptors can stimulate the elevation of either cyclic AMP or calcium, depending on which G protein is activated. Tyrosine kinase receptors, on the other hand, may elevate cystolic calcium levels but typically do not stimulate cyclic AMP formation. Once the target enzymes are activated in this way, they mediate subsequent steps of the respective signal transduction pathways. In this section, we will consider the cyclic AMP pathway and then the inositol-phospholipid-calcium pathway.

Second Messengers: Cyclic AMP

Cyclic AMP (cAMP) is formed from cytosolic ATP by the enzyme **adenylate cyclase** (Figure 23-7). Adenylate cyclase is anchored in the plasma membrane with its catalytic portion protruding into the cytosol. Normally, the enzyme is inactive until it binds to the α subunit of a specific G protein called Gs (Gs was the first G protein to be discovered). When a G protein–linked receptor is coupled to Gs, the binding of ligand stimulates the Gsα subunit to release GDP and acquire a GTP (Figure 23-8). This in turn causes GTP-Gsα to

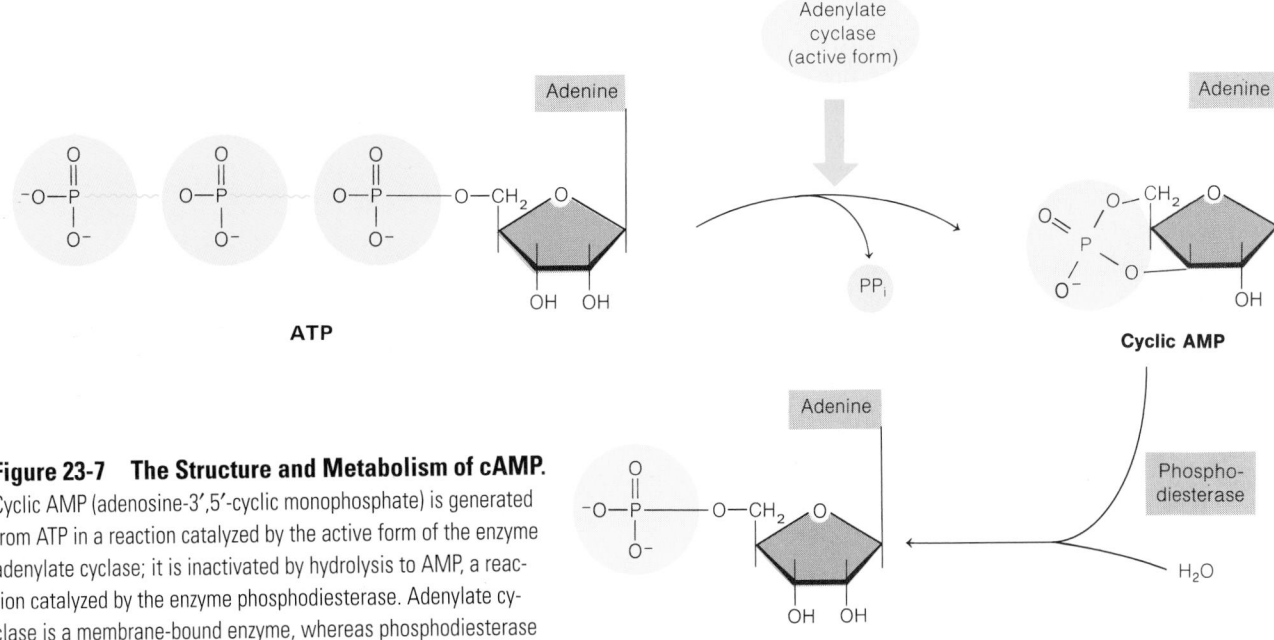

Figure 23-7   The Structure and Metabolism of cAMP. Cyclic AMP (adenosine-3′,5′-cyclic monophosphate) is generated from ATP in a reaction catalyzed by the active form of the enzyme adenylate cyclase; it is inactivated by hydrolysis to AMP, a reaction catalyzed by the enzyme phosphodiesterase. Adenylate cyclase is a membrane-bound enzyme, whereas phosphodiesterase is located in the cytosol.

detach from the Gs$\beta\gamma$ subunits and bind to adenylate cyclase. When GTP-Gsα binds to adenylate cyclase, the enzyme becomes active and converts ATP to cyclic AMP.

We mentioned previously that G proteins respond quickly to changes in ligand concentration because they only remain active for a short period of time before the Gα subunit hydrolyzes its bound GTP and converts to the inactive state. Once the G protein becomes inactive, the adenylate cyclase ceases to make cAMP. However, cAMP levels would still remain elevated in the cell if not for the presence of the enzyme **phosphodiesterase,** which degrades the cAMP. This further ensures that the signal transduction pathway will shut down as soon as the concentration of the ligand outside the cell declines. (For what can happen when this pathway does not shut down, see Box 23B.)

Cyclic AMP appears to have one main intracellular target, an enzyme known as *cAMP-dependent kinase,* or **protein kinase A (PKA).** Protein kinase A phosphorylates a wide variety of cellular proteins, each of which contains a similar short sequence of amino acids that is recognized by the kinase as the phosphorylation site. The kinase transfers a phosphate from ATP to a serine or threonine found within this particular stretch of amino acids.

Cyclic AMP regulates the activity of PKA by removing inhibitory subunits that are normally bound to the kinase. In its inactive state, protein kinase A is a complex of four subunits (Figure 23-9, page 764). Two of the subunits are identical and are known as *catalytic subunits* because they have kinase activity. The other two subunits are called *regulatory subunits* because, in the absence of cAMP, they inhibit the activity of the catalytic subunits. Cyclic AMP binds to the regulatory subunits, causing them to detach from the catalytic subunits. Once they are free, the catalytic subunits can catalyze the phosphorylation of various proteins in the cell. In other words, the effects of elevated cAMP in the cell are due to the activation of PKA.

The next question is how the activated PKA affects the cell. The answer depends on the cell type. In particular, it depends on what enzymes are present in, or what cellular activities are performed by, the cells that serve as targets for PKA. The elevation of cAMP can do many different things. When cAMP is elevated in skeletal muscle and liver cells, the breakdown of glycogen is stimulated. In cardiac muscle, the elevation of cAMP strengthens the heart rate, whereas in smooth muscle, contraction is inhibited. In blood platelets, the elevation of cAMP inhibits the activation of platelets. In intestinal epithelial cells, the elevation of cAMP causes the

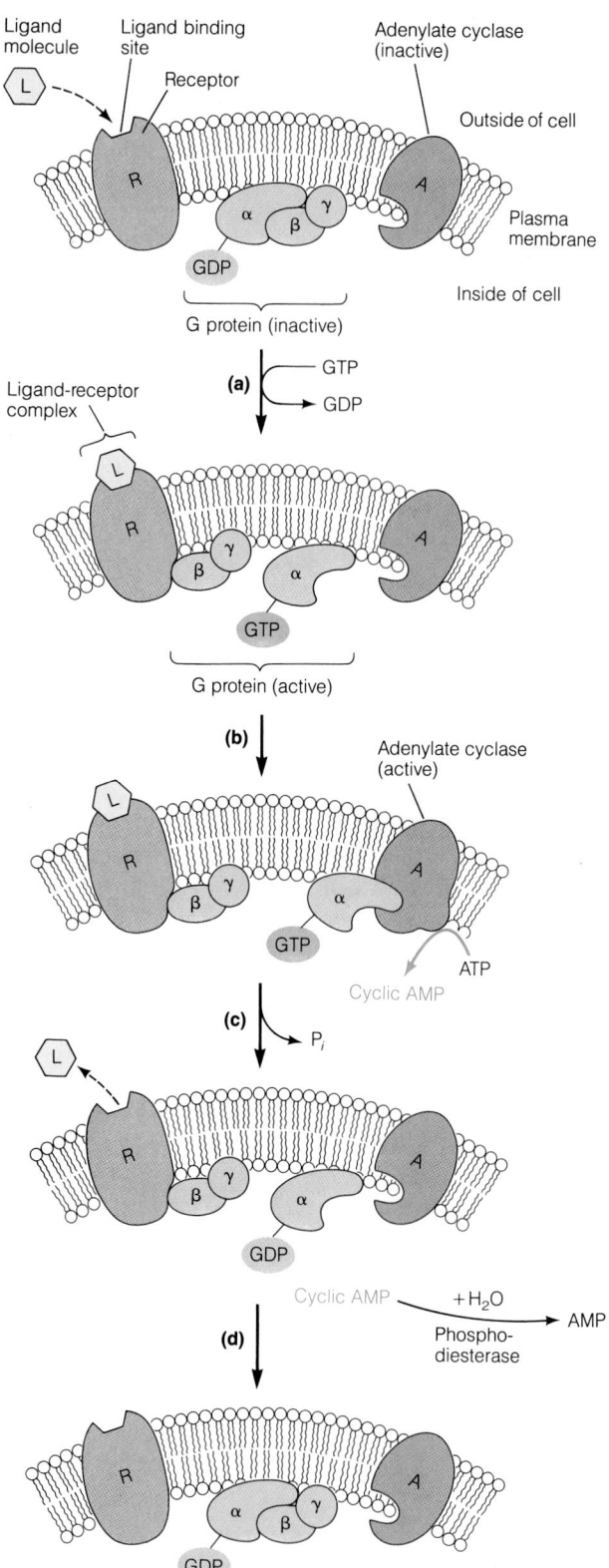

Figure 23-8 The Roles of G Proteins and Cyclic AMP in Signal Transduction. G proteins mediate signal transduction through G protein–linked receptors. When a messenger binds to its receptor, the G protein is activated. In the inactive state, the α, β, and γ subunits are present as a complex, with GDP bound to the α subunit. **(a)** When a receptor is activated by binding of its specific messenger on the outer surface of the plasma membrane, the receptor-messenger complex associates with the G protein, causing the displacement of GDP by GTP and the dissociation of the Gα-GTP complex. **(b)** The GTP-Gα complex then binds tightly to a molecule of membrane-bound adenylate cyclase, activating it for synthesis of cAMP. **(c)** Activation ends when the ligand leaves the receptor, the GTP is hydrolyzed to GDP by the GTPase activity of the Gα subunit, and the Gα dissociates from the adenylate cyclase. **(d)** Adenylate cyclase then reverts to the inactive form, the Gα reassociates with the G$\beta\gamma$ complex, and cAMP molecules in the cytosol are hydrolyzed to AMP by the enzyme phosphodiesterase.

Clinical Applications

HETEROTRIMERIC G PROTEINS AND DISEASE

What would happen if the G protein–adenylate cyclase system could not be shut off? We can answer this question by considering several human diseases that can now be explained at the molecular level because of what we have learned about G proteins. The bacteria *Vibrio cholerae* (which causes cholera) and *Bordetella pertussis* (which causes whooping cough) both elicit pathological responses in epithelial secretory cells. Both bacteria secrete toxins that cause disease by their effect on heterotrimeric G proteins.

Cholera results from the secretion of *cholera toxin* when *V. cholerae* colonize the gut. The toxin alters the secretion of salt (sodium bicarbonate and sodium chloride) and fluid in the intestines. Salt secretion in the intestines is normally regulated by hormones and receptors that act through the G protein Gs to modulate intracellular cAMP concentrations. A portion of the cholera toxin is an enzyme that chemically modifies Gs. This enzyme, known as an *ADP ribosylase*, cleaves NADH and transfers a fragment called ADP ribose to Gs. The ADP ribosylation of Gs inhibits its ability to hydrolyze GTP to GDP. As a result, Gs is unable to shut off and therefore continually stimulates adenylate cyclase to make cAMP. The resulting high concentration of cAMP causes the intestinal cells to secrete large amounts of salt and water into the lumen of the intestines, a condition that, if left untreated, can lead to death by dehydration.

The pertussis toxin secreted by *B. pertussis* ADP ribosylates the inhibitory G protein Gi. The normal function of Gi is to shut off adenylate cyclase. When Gi is ADP ribosylated, it no longer inhibits adenylate cyclase. It is not yet clear why the inability to shut off adenylate cyclase activity leads to whooping cough, however. We do know that the most serious cases of whooping cough are caused by strains of the bacteria that produce the largest amounts of the toxin.

The discovery of these toxins and their modes of action has provided not only advances in medical treatment, but also powerful tools for studying G protein–mediated signal transduction. Each toxin acts on a different set of G proteins, and the ADP-ribosylation reaction can result in activation or inhibition, depending on the particular G protein in question. When ADP ribosylation is carried out using radioactively labeled NADH, the G protein becomes radioactively labeled and therefore readily identifiable. This has enabled researchers to associate a signaling pathway with the activity of a particular G protein. ■

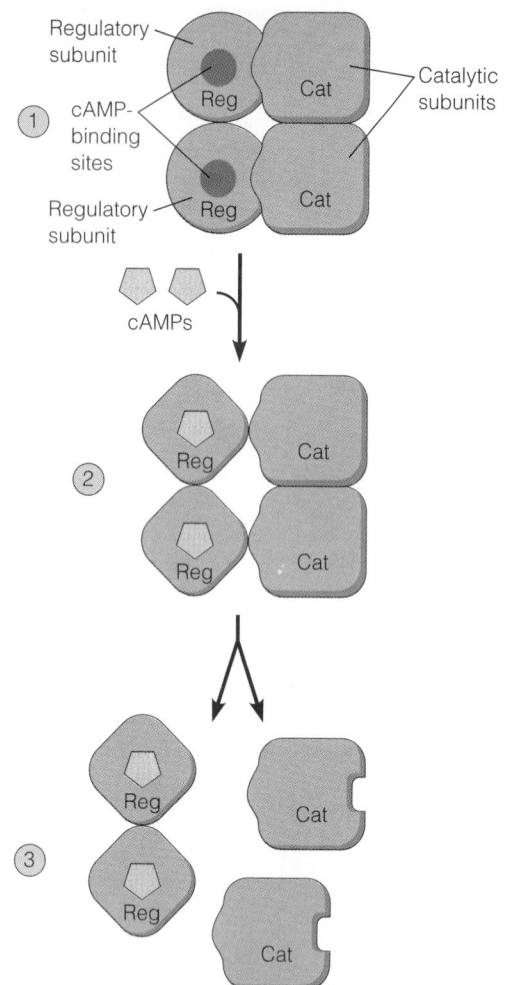

Figure 23-9   The Activation of Protein Kinase A by Cyclic AMP. ① Protein kinase A is composed of four subunits, two catalytic and two regulatory. The regulatory subunits inhibit the catalytic subunits in the absence of cAMP. ② Cyclic AMP activates protein kinase A by binding to the regulatory subunits. When cAMP binds, the regulatory subunits change conformation and ③ detach, freeing the catalytic subunits. Once the catalytic subunits are free, they are activated and become capable of phosphorylating target proteins in the cell.

secretion of salts and water into the lumen of the gut. Each of these reactions is an example of the preprogrammed response discussed earlier. In fact, if we raise the concentration of cAMP artificially in these different types of cells, these same cellular responses can be triggered even in the absence of a receptor ligand. We can do so in two different ways: either by stimulating cAMP production directly or by inhibiting the enzyme phosphodiesterase that degrades cAMP. Inhibition of phosphodiesterase leads to elevated cAMP levels because there is always a low level of cAMP production in cells, but the cAMP is usually degraded before it accumulates. Examples of phosphodiesterase inhibitors are the methylxanthines, compounds such as caffeine and theophylline, found in coffee, tea, and soft drinks. (Theophylline is often used to treat asthma because it relaxes bronchial smooth muscle.)

To summarize, the **cAMP pathway** transduces extracellular signals that affect numerous cell functions. Cyclic AMP is produced when a ligand binds to its membrane receptor, activating adenylate cyclase for cAMP production. Cyclic AMP then activates PKA. The events up through the activation of PKA are present in all cells that use cAMP as a second messenger. The events that follow the activation of PKA are usually specific for each cell type. We will now turn our attention to the other primary pathway of G protein–linked receptors.

Second Messengers: Inositol Trisphosphate and Diacylglycerol

The importance of inositol phospholipids in cell signaling was first brought to light in the pioneering studies of Robert Michell and Michael Berridge. In the early 1980s, Michael Berridge noticed that when the salivary glands of certain insect larvae were stimulated to secrete, changes occurred in membrane inositol phospholipids. We now know that **inositol-1,4,5-trisphosphate (InsP$_3$),** one of the breakdown products of inositol phospholipids, is also a second messenger. InsP$_3$ is generated from *phosphatidylinositol-4,5-bisphosphate (PIP$_2$)*, a relatively uncommon membrane phospholipid, when the enzyme **phospholipase C** is activated. Some forms of phospholipase C (called phospholipase C$_\beta$) are activated by G proteins, whereas others (called phospholipase C$_\gamma$) are activated by tyrosine kinases. Phospholipase C cleaves PIP$_2$ into two molecules: inositol trisphosphate and **diacylglycerol (DAG)** (Figure 23-10).

The roles of InsP$_3$ and DAG as second messengers in the **inositol-phospholipid-calcium pathway** are shown in Figure 23-11 (page 766). The sequence begins with the binding of a ligand to its membrane receptor, leading to the activation of a specific G protein called Gp. Gp then activates phopholipase C$_\beta$, thereby generating both InsP$_3$ and DAG. Inositol trisphosphate is water-soluble and quickly diffuses through the cytosol, binding to a ligand-gated calcium channel known as the *InsP$_3$ receptor channel* in the endoplasmic reticulum. The ER of nonmuscle cells accumulates calcium ions in a manner analogous to that of the sarcoplasmic reticulum of muscle cells. When InsP$_3$ binds, the channel opens, releasing calcium into the cytosol. Calcium then binds to calmodulin, and the calcium-calmodulin complex (discussed later) activates the desired physiological process. The DAG that is also generated by phospholipase C activity remains in the membrane, where it activates the enzyme *protein kinase C*. This enzyme can then phosphorylate specific serine and threonine groups on a variety of target proteins, depending on the cell type.

How do we know that the sequence of events shown in Figure 23-11 actually occurs in cells? The answer begins with the original observation that a specific physiological phenomenon—salivary secretion, in this case—could be activated by the products of phospholipase action: InsP$_3$ and DAG. Evidence for the role of calcium was provided by a different experimental approach involving the injection of a calcium-dependent fluorescent dye into a target cell. Because the fluorescence of the dye changes with the calcium concentration, the dye is a sensitive indicator of the intracel-

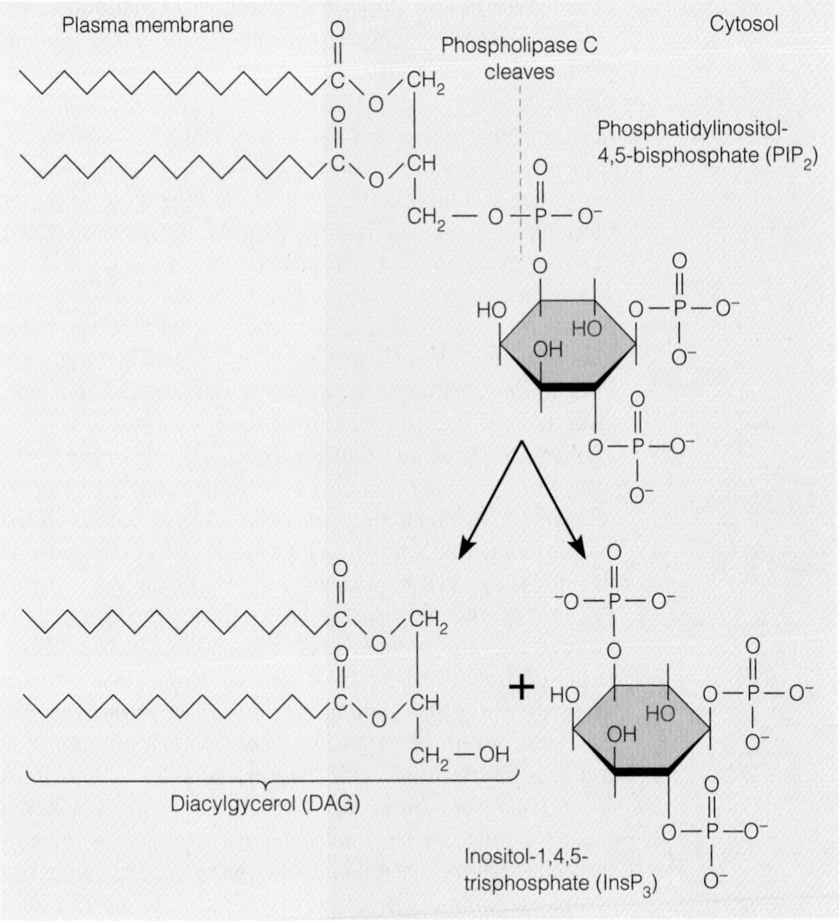

Figure 23-10 The Formation of Inositol Trisphosphate and Diacylglycerol.
Inositol trisphosphate (InsP$_3$) and diacylglycerol (DAG) are formed when phospholipase C cleaves phosphatidylinositol-4,5-bisphosphate (PIP$_2$), one of the phospholipids present in membranes. InsP$_3$ is released into the cytosol, whereas DAG remains within the membrane. Both InsP$_3$ and DAG are second messengers in a variety of signal transduction pathways.

lular calcium concentration. By measuring the increase in fluorescence in response to activation by a ligand or by InsP$_3$ directly, investigators were able to establish a link between ligand binding, InsP$_3$ generation, and increased cytosolic calcium concentration.

To complete the sequence of events, the increase in calcium concentration had to be linked to the actual physiological response of the target cell. This link was established by treating target cells with a *calcium ionophore* (such as *ionomycin*) in the absence of extracellular calcium. The ionophore renders membranes permeable to calcium, thereby releasing intracellular stores of calcium in the absence of a physiological stimulus. Treatment with the calcium ionophore mimicked the effect of InsP$_3$, thus implicating calcium as an intermediary in the InsP$_3$ signal transduction pathway.

The role of DAG was established experimentally by showing that its effects can be mimicked by *phorbol esters,* plant metabolites that bind to protein kinase C, activating it directly. Using ionomycin, phorbol esters, and other pharmacological agents, researchers showed that InsP$_3$-stimulated calcium release and DAG-mediated kinase activity are both required to produce a full response in target cells.

If the sequence of events shown in Figure 23-11 occurred only in insect salivary glands, Berridge's observations would have warranted no more than a modest note in the story of signal transduction. Instead, InsP$_3$ and DAG were quickly shown to be second messengers in a variety of regulated cell functions, some of which are indicated in Table 23-1. To date, more than 25 different cell-surface receptors are known to activate this pathway, with new examples being reported each year.

Table 23-1 Examples of Cell Functions Regulated by Inositol Trisphosphate and Diacylglycerol

Regulated Function	Target Tissue	Messenger
Platelet activation	Blood platelets	Thrombin
Muscle contraction	Smooth muscle	Acetylcholine
Insulin secretion	Pancreas, endocrine	Acetylcholine
Amylase secretion	Pancreas, exocrine	Acetylcholine
Glycogen degradation	Liver	Antidiuretic hormone
Antibody production	B lymphocytes	Foreign antigens

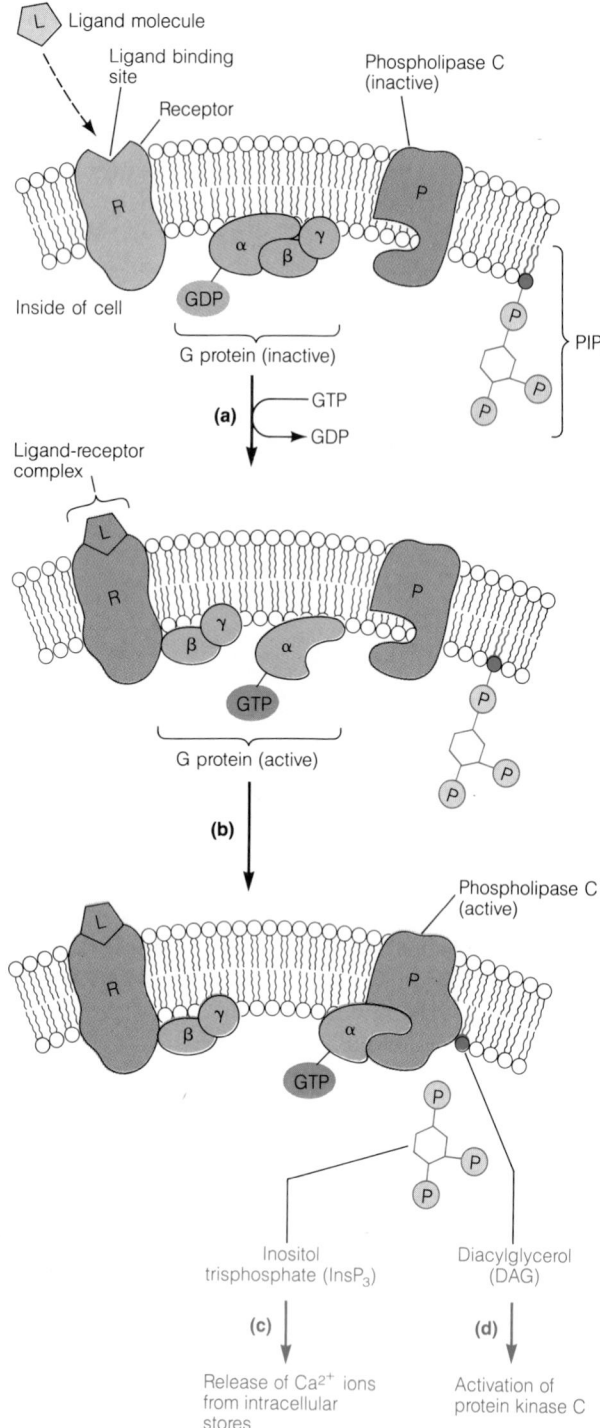

Outside of cell

L Ligand molecule

Ligand binding site

Receptor

Phospholipase C (inactive)

R

α β γ

GDP

Inside of cell

G protein (inactive)

(a) GTP → GDP

PIP₂

P

P

P

Ligand-receptor complex

L

R

β γ

α

GTP

G protein (active)

(b)

P

P

P

L

R

β γ

α

GTP

Phospholipase C (active)

P

P

P

P

Inositol trisphosphate (InsP₃)

Diacylglycerol (DAG)

(c)

(d)

Release of Ca²⁺ ions from intracellular stores

Activation of protein kinase C

Figure 23-11   The Role of InsP₃ and DAG in Signal Transduction.   **(a)** When a receptor (R) is activated by the binding of its ligand (L) on the outer surface of the plasma membrane, the receptor-ligand complex associates with the G protein Gp, causing the displacement of GDP by GTP and the dissociation of the GTP-Gα complex. **(b)** The GTP-Gα complex then binds to phospholipase C (P), activating it and causing cleavage of PIP₂ into one molecule each of InsP₃ and DAG. **(c)** The InsP₃ is released into the cytosol, where it triggers the release of calcium. **(d)** The DAG remains in the membrane, where it activates protein kinase C.

Diacylglycerol and Protein Kinase C.   As mentioned, the second messenger DAG specifically activates the enzyme **protein kinase C (PKC)**. PKC is actually a family of six or more related enzymes. Most members of the PKC family require membrane phospholipids and DAG for activity, and some require calcium as well.

There are many unresolved ambiguities concerning the mechanism of PKC activation. At present, the most widely accepted model depicts PKC as a cytosolic molecule in its inactive state. When phospholipase C generates DAG in the plasma membrane, PKC moves to the membrane and binds to both membrane phospholipids and DAG. The binding of PKC to DAG in the plasma membrane activates the enzyme, leading to the phosphorylation of specific proteins.

A wide variety of cellular effects have been linked to the activation of protein kinase C, including the stimulation of cell growth, the regulation of ion channels, changes in the cytoskeleton, increases in cellular pH, and secretion. The stimulation of cell growth by PKC is probably due to its ability to phosphorylate and activate proteins called *mitogen-activated protein kinases (MAP kinases,* or *MAPKs).* Once phosphorylated, MAP kinases enter the nucleus, where they phosphorylate transcription factors that regulate the transcription of genes involved with cell growth. PKC also phosphorylates specific cytoskeletal proteins known as the *myristolated alanine-rich C-kinase substrate (MARCKS)* and *talin.* Both of these molecules are associated with actin at points where actin filaments make contact with the plasma membrane. Thus, PKC may regulate connections between microfilaments and the plasma membrane.

A recent finding that has generated a great deal of interest is that a variety of compounds known as *tumor promoters* specifically activate PKC. Tumor promoters in general do not cause cancer. Instead, they act on dormant cells that have already undergone the kinds of genetic alterations that give the cells the potential to form tumors. Tumor promoters stimulate the growth of these altered cells, causing them to demonstrate their tumor-forming potential. Structural analysis of tumor promoters suggests that they can act as analogues of diacylglycerol and bind to the DAG-binding site.

Second Messengers: Calcium Ions

Calcium ions (Ca²⁺) also play a very important role in regulating a variety of cellular functions in response to external signals. In this case, regulation is achieved by changes in the calcium concentration of the cytosol.

The concentration of calcium is normally maintained at very low levels in the cytosol because of the presence of calcium pumps in the plasma membrane and the endoplasmic reticulum (Figure 23-12). The calcium pump in the plasma membrane transports calcium out of the cell, whereas the calcium pump in the ER sequesters calcium ions in the lumen of the ER. In addition to the calcium pumps, some cells have sodium-calcium exchangers that further reduce cytosolic calcium concentration. Finally, mitochondria can

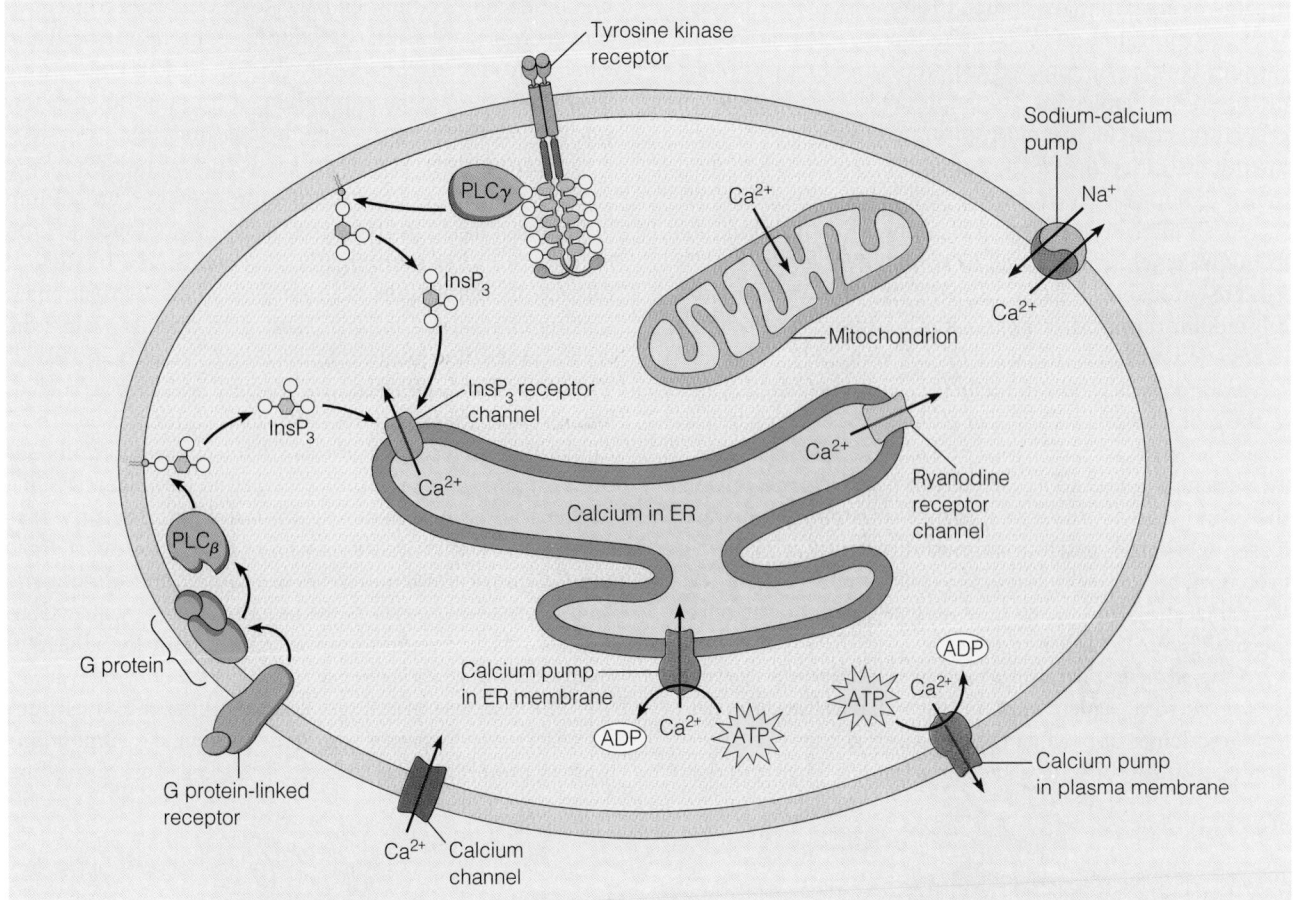

Figure 23-12   An Overview of Calcium Regulation in Cells.
Calcium concentration increases in the cytosol because of the opening of calcium channels in the plasma membrane and the release of calcium through the InsP₃ or ryanodine receptor channels in the ER membrane.

Cytosolic calcium concentration is lowered by the actions of the ER calcium pump, the plasma membrane calcium pump, sodium-calcium exchangers, and the mitochondria.

transport calcium into the mitochondrial matrix. For most cells in their resting state, the action of calcium pumps maintains the calcium concentration in the cytosol at about 10^{-4} mM.

Just as a number of pumps keep calcium concentration low in the cytosol, there are a number of different ways that various stimuli can cause cytosolic calcium concentrations to increase. One way, discussed in Chapter 22 in relation to neurons, is by the opening of calcium channels in the plasma membrane. The calcium concentration in the extracellular fluid and the blood is about 1.2 mM, more than 10,000 times higher than that of the cytosol. As a result, when calcium channels open, calcium ions rush into the cell.

Another way that calcium levels can be elevated depends on the release of calcium from intracellular stores. Calcium ions sequestered in the ER can be released through the InsP₃ receptor channel, introduced earlier, and the ryanodine receptor channel. The *ryanodine receptor channel* is particularly important for calcium release from the sarcoplasmic reticulum of cardiac and skeletal muscle. But

nonmuscle cells such as neurons also have ryanodine calcium-release channels. Surprisingly, the ligand that opens the ryanodine receptor channel appears to be calcium itself. When a neuron is depolarized, calcium channels in the plasma membrane open and allow some calcium to enter the cytosol. Upon exposure to a rapid increase in calcium ions, the ryanodine receptor channel opens, allowing calcium to escape from the ER into the cytosol. This phenomenon has been aptly named *calcium-induced calcium release* (see Figure 21-17).

The Calcium-Calmodulin Complex and Its Intracellular Effects. Calcium can bind directly to some proteins and thereby alter their activity, but more often it works through the protein calmodulin, as introduced in Chapter 21. **Calmodulin**, which is present in eukaryotic, but not in prokaryotic, cells, was discovered in the late 1960s. This protein plays a pivotal role in the regulation of a variety of cellular processes.

The amino acid sequences of calmodulins from many different species of plants, animals, and eukaryotic microorganisms are strikingly similar. In addition, the amino acid sequence of calmodulin is quite similar to that of the calcium-binding subunit of the troponin complex (TnC) discussed in Chapter 21. The importance of calmodulin in cellular function has been demonstrated using yeast cells. By intentionally mutating yeast calmodulin, researchers showed that calmodulin is essential for yeast cell survival. The observed similarity in amino acid sequence of calmodulins from different organisms and its requirement for cell viability are probably related. Apparently, the sequence of calmodulin cannot diverge too much without jeopardizing its function, which is necessary for cell viability.

How does calmodulin mediate calcium-activated events in the cell? The calmodulin molecule has been compared to a flexible "arm" with a "hand" at each end (Figure 23-13a). Two calcium ions bind at each of two hand regions, causing the calmodulin to undergo a change in conformation that forms the active **calcium-calmodulin complex**. When a protein is present that contains a calmodulin-binding site, the hands and arm bind to it by wrapping around the binding site (Figure 23-13b).

One of the important features of calmodulin is its affinity for calcium, which must be such that calmodulin will bind to calcium only when the cytosolic calcium concentration rises in response to a specific stimulus. Thus, calmod-

ulin binds to calcium when the cytosolic calcium concentration increases to about 10^{-3} mM, but it releases calcium when cytosolic calcium levels decline back to the resting level of 10^{-4} mM. For calmodulin to respond to such low calcium concentrations, calcium ions must bind tightly—but not too tightly. Monovalent ions such as potassium and sodium bind to proteins much more weakly than calcium binds to calmodulin. However, heavy metals such as iron typically bind far more tightly to proteins. The binding of calcium to calmodulin is in between these two extremes, making calmodulin uniquely suited to operate within the typical range of cytosolic calcium concentrations.

Most calmodulin-binding proteins are enzymes such as kinases and phosphatases. We have already seen how calmodulin can activate myosin light-chain kinase in our discussion of smooth muscle contraction (see Figure 21-23). As you will recall, the enzyme is autoinhibited by a pseudosubstrate sequence that is normally bound to the catalytic site. Calcium-calmodulin binds to a calmodulin-binding site on the kinase that overlaps the pseudosubstrate sequence. In binding, calcium-calmodulin interferes with the ability of the pseudosubstrate sequence to inhibit the enzyme.

The response of a target cell to an increase in calcium concentration depends on the particular calmodulin-binding proteins that are present in the cell. This means that

(a) Structure of Ca^{2+}-calmodulin complex

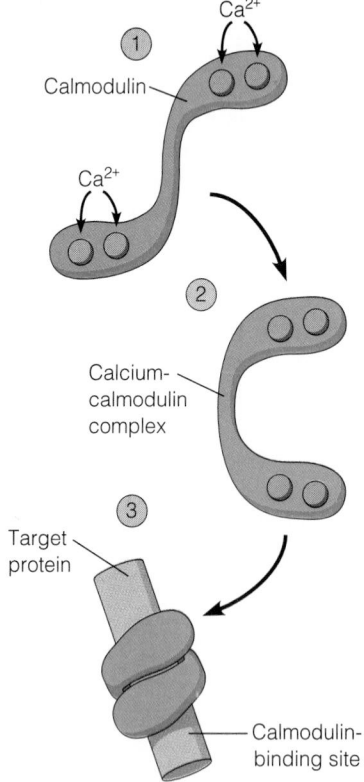

(b) Function of Ca^{2+}-calmodulin complex

Figure 23-13 The Structure and Function of the Calcium-Calmodulin Complex. **(a)** Calmodulin is a cytosolic calcium-binding protein. This model of its molecular structure is based on data from X-ray crystallography. The molecule consists of two globular ends joined by a helical region. Each end has two calcium-binding sites to

which two calcium ions can bind. **(b)** To form a functional calcium-calmodulin complex, ① calmodulin binds four calcium ions. ② Calmodulin changes conformation to active complex. ③ The two globular ends of the complex act like hands that wrap around a calmodulin-binding site on a target protein.

the same change in calcium concentration can produce distinctly different effects in two target cells if each possesses different calmodulin-sensitive enzyme systems.

Among the main targets for calcium-calmodulin in nonmuscle cells are kinases such as the myosin light-chain kinase just mentioned. In some invertebrates, the contraction of striated muscle is also regulated by calmodulin, rather than by troponin. For example, the adductor muscle that closes the valve, or shell, of a clam is under calmodulin control. Another important kinase that is activated by calcium-calmodulin is called *calcium-calmodulin–dependent kinase type II (CAM kinase II)*. This kinase is present in a wide variety of cells, although it is most abundant in neurons. It appears to influence the secretion of neurotransmitter and the transcription of particular genes in neurons. In nonneuronal cells, CAM kinase II seems to regulate one of the steps involved in the breakdown of the nuclear envelope during mitosis.

Thus, calcium plays an important role in a remarkable variety of cellular activities. In each case, the effect is elicited not by the calcium ion itself, but by the calcium-calmodulin complex. In these functions, calcium shares several properties with cAMP: Its concentration is determined by events occurring within a membrane, and its effect depends on allosteric activation of an intracellular protein. Like cAMP, therefore, calcium is a second messenger—a means of transmitting an impinging signal to the interior of the cell, thereby triggering specific intracellular events that are as important as they are diverse.

Signal Transduction Pathways Initiated by Tyrosine Kinase Receptors

Tyrosine kinase receptors use a different method of initiating signal transduction from that seen with G protein–linked receptors. The key initial event is the cross-phosphorylation of tyrosine residues on the cytosolic portion of the receptor that takes place when receptors aggregate in response to ligand binding (Figure 23-14). Once the

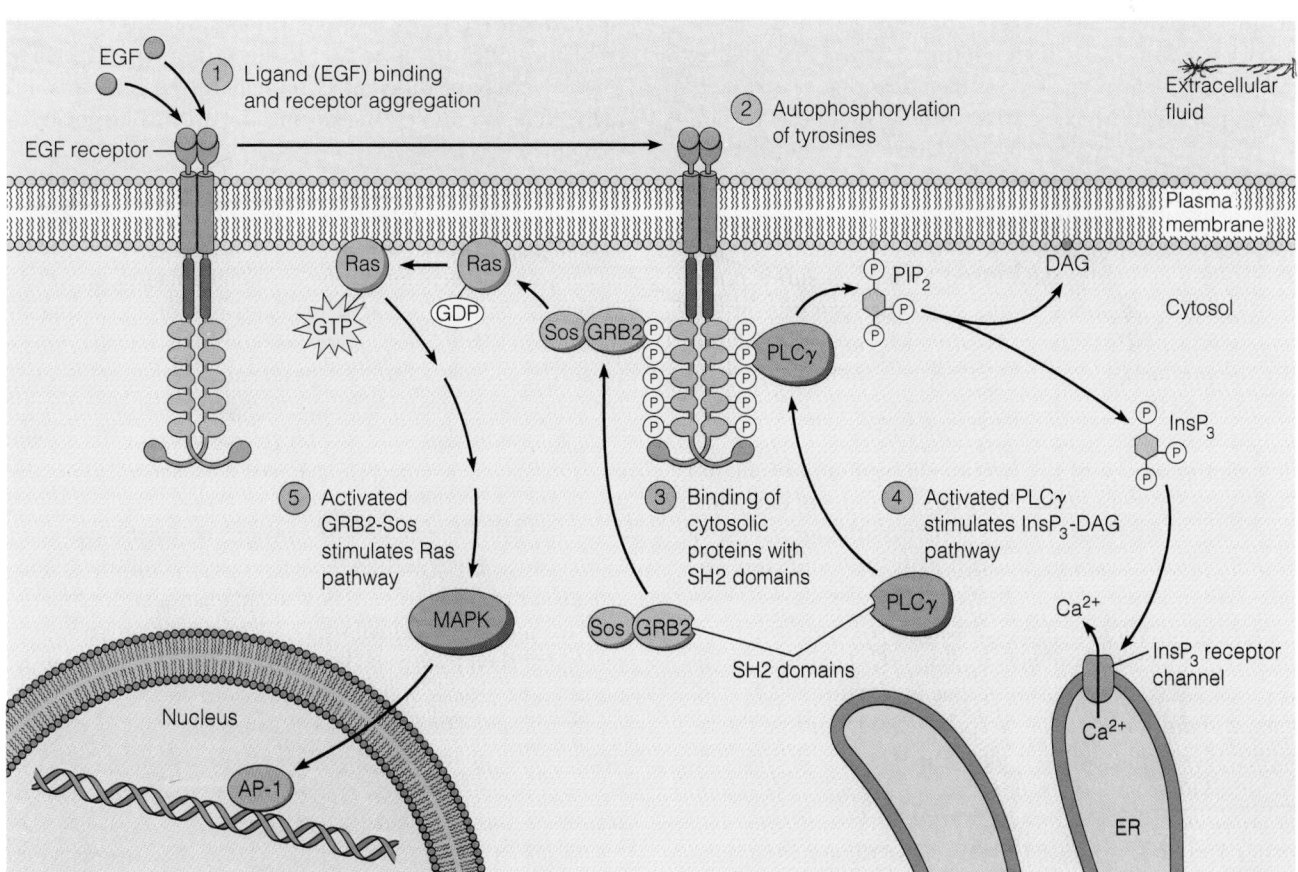

Figure 23-14 Signal Transduction Through Tyrosine Kinase Receptors. ① Upon ligand binding, tyrosine kinase receptors, such as that shown here for the epidermal growth factor (EGF), aggregate and ② cross-phosphorylate each other. Once a receptor is phosphorylated at tyrosine residues on the cytosolic tail, ③ proteins with SH2 domains such as phospholipase C (PLC) and GRB2 bind to the receptor. ④ The binding of phospholipase C results in its activation and the cleavage of PIP$_2$ into InsP$_3$ and DAG. ⑤ The binding of GRB2 causes the activation of Sos, a guanine nucleotide release protein to which it is bound. Sos then causes the activation of the Ras protein by helping it release GDP and acquire GTP. Activated Ras initiates a cascade of events that ultimately results in the formation of AP-1, a transcription factor in the nucleus that stimulates the expression of genes needed for cell growth.

receptors are phosphorylated, these phosphorylated tyrosine residues become binding sites for other proteins in the cytosol that contain SH2 domains. At present, several different important signaling pathways are known to be initiated by tyrosine kinase receptors. These pathways are the activation of the small monomeric G protein called Ras, the activation of phospholipase C_γ, and the activation of other enzymes, including phosphatidylinositol-3-kinase.

The Ras Pathway. Ras is important in regulating the growth of cells. As with other types of G proteins, Ras can be bound to either GDP or GTP, but it is active only when bound to GTP. In the absence of receptor stimulation, Ras is normally in the GDP-bound state. For Ras to become active, it must release GDP and acquire a molecule of GTP. For this to take place, Ras needs the help of another type of protein called a *guanine nucleotide–release protein (GNRP).*

The GNRP that activates Ras was originally identified from a genetic mutation in fruit flies called the Son of Sevenless, so it is also called Sos. For Sos to become active, it must bind indirectly to the tyrosine kinase receptor through another protein called GRB2. GRB2 contains an SH2 domain and binds a tyrosine phosphate on the receptor. Thus, in order to activate Ras, the receptor becomes tyrosine phosphorylated, and GRB2 and Sos form a complex that binds to the receptor, activating Sos. Sos then helps Ras release GDP and acquire GTP, which converts Ras to its active state.

Once Ras is active, it triggers a cascade of cellular events called the **Ras pathway.** One important event in the pathway is the activation of MAPKs. MAPKs are activated when cells receive a stimulus to grow. One of their functions is to phosphorylate a nuclear protein called Jun. When Jun is phosphorylated, it assembles along with other proteins into a transcription factor called AP-1, which appears to stimulate the production of proteins that are needed for cells to grow and divide.

Phospholipase C and the Inositol-Phospholipid-Calcium Pathway. Tyrosine kinase receptors can also activate phospholipase C. We have already seen that the activation of phospholipase C leads to production of $InsP_3$ and DAG and that $InsP_3$ releases calcium from intracellular stores. However, the phospholipase C_γ (activated by tyrosine kinase receptors) is different from phospholipase C_β (activated by the G protein–linked receptors) in that it contains an SH2 domain and must bind to the receptor. Once it binds to the receptor, phospholipase $C\gamma$ is phosphorylated by the receptor tyrosine kinase and becomes active.

Other Enzymes Activated by Tyrosine Kinase Receptors. We have seen that tyrosine kinase receptors can activate Ras, which ultimately stimulates the expression of genes involved in growth, and that they can also activate phospholipase C_γ. They also activate other enzymes such as phosphatidylinositol-3-kinase, which phosphorylates the plasma membrane phospholipid phosphatidylinositol. At present, we know this enzyme is important in regulating cell growth but do not yet understand why.

The Endocrine and Paracrine Hormone Systems

The intracellular receptors, G protein–linked receptors, and the tyrosine kinase receptors we have just discussed represent the three major types of receptors used in chemical communication between cells. Intracellular receptors are primarily involved with events in the nucleus, where they regulate the expression of particular sets of genes. G protein–linked receptors regulate the function of adenylate cyclase, which forms the second messenger cAMP, or phospholipase C_β, which forms $InsP_3$ and DAG. The tyrosine kinase receptors can activate a number of enzymes, including phospholipase C_γ, Ras, and others. We are now ready to examine how the human body uses messengers called hormones and their cognate receptors to regulate the function of various tissues. Despite the diversity of hormones, we will see that they all depend on a few basic types of receptors.

As a group, hormones are quite diverse, including ligands for all the types of receptors just discussed. Some hormones are steroids or lipidlike chemicals that are targeted to intracellular receptors. Other hormones, such as the adrenergic hormones discussed later in this section, are targeted to a wide variety of different G protein–linked receptors. Finally, some hormones, such as insulin, are ligands for tyrosine kinase receptors.

Every hormone can be placed in one of two categories, depending on the distances over which it operates. An **endocrine hormone** is a chemical that is released by one set of cells and travels by means of the circulatory system to other sets of cells, where it regulates one or more specific functions. (The word *endocrine* comes from the Greek meaning "to secrete into.") A **paracrine hormone,** however, is a more local signal that is taken up, destroyed, or immobilized so rapidly that it can act only on cells in the immediate environment. (The word *paracrine* means "to secrete around.")

Endocrine hormones are synthesized by the *endocrine tissues* of the body and are secreted directly into the bloodstream. This mode of secretion distinguishes endocrine tissues from *exocrine tissues*, which secrete their products into ducts that then transport the secretions to other parts of the body. Some organs of the body have both endocrine and exocrine tissues. The pancreas is a good example of such an organ. Cells of the pancreatic endocrine tissue secrete two hormones, glucagon and insulin, that regulate the concentration of glucose circulating in the blood. Cells of the pancreatic exocrine tissue, however, secrete digestive enzymes that are collected and sent to the small intestine through the pancreatic duct. Figure 23-15 depicts the major endocrine tissues of the human body.

Once secreted into the circulatory system, endocrine hormones have a limited life span, ranging from a few seconds for epinephrine (a product of the adrenal gland) to many hours for insulin. As they circulate in the bloodstream, hormone molecules come into contact with cell-surface re-

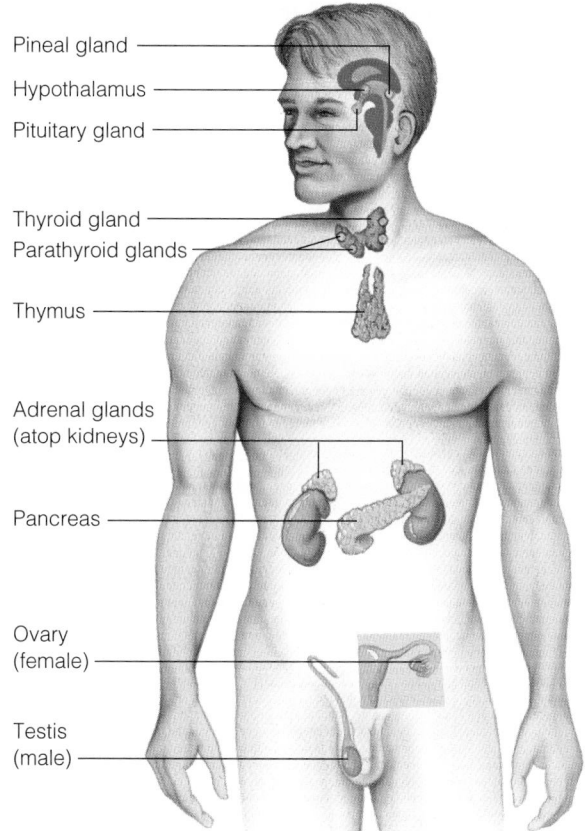

Figure 23-15 The Major Endocrine Tissues of the Human Body. The cells of endocrine tissues secrete hormones directly into the bloodstream, effectively "broadcasting" the hormone throughout the body. The hormones act as chemical messengers, binding to receptors of specific target cells and thereby regulating the function of such cells.

Pineal gland
Hypothalamus
Pituitary gland
Thyroid gland
Parathyroid glands
Thymus
Adrenal glands (atop kidneys)
Pancreas
Ovary (female)
Testis (male)

ceptors in tissues throughout the body. A tissue that is specifically affected by a particular hormone is called a **target tissue** for that hormone (Figure 23-16). For example, the heart and the liver are target tissues for epinephrine, whereas the liver and skeletal muscles are targets for insulin. This contact with cell-surface receptors is the means by which most endocrine hormones regulate cell function in a variety of tissues.

Hormonal Control of Physiological Functions

Hormones regulate a wide range of human physiological functions, including growth and development, rates of body processes, concentrations of substances, and responses to stress and injury (Table 23-2).

For example, *somatotropin* is involved in the regulation of overall growth of the body, whereas *androgens* and *estrogens*, the sex hormones, control the differentiation of tissues into secondary sex characteristics. *Thyroxine* regulates the rate at which the body makes energy available and is therefore an example of a rate-controlling hormone. Hormones that control the concentrations of substances include *insulin*

(control of blood glucose level), aldosterone (control of blood sodium and potassium levels), and *parathyroid hormone* (control of blood calcium level). The body's response to stress is regulated by *epinephrine* and *cortisol,* and its response to local injury is regulated by the release of *histamine* and the production of *prostaglandins.*

The Chemistry of Animal Hormones

Hormones can be classified not only according to function and the distances over which they act, but also according to their chemical properties (Table 23-3). Chemically, the endocrine hormones fall into four categories: amino acid derivatives, peptides, proteins, and the lipid-based hormones such as the steroids (Figure 23-17, page 773). An example of an amino acid derivative is epinephrine, derived from tyrosine. *Antidiuretic hormone* (also called *vasopressin)* is an example of a peptide hormone, whereas insulin is a protein. *Testosterone* is an example of a steroid hormone. The steroid hormones are derivatives of cholesterol that are synthesized either in the gonads (the *sex hormones)* or in the adrenal cortex (the *corticosteroids).*

As already mentioned, paracrine hormones are chemicals that act locally, rather than throughout the body. Histamine is a paracrine hormone, as are the prostaglandins (see Table 23-3). Histamine is produced by decarboxylation of the amino acid histidine and is responsible for local inflammatory responses. Prostaglandins and the closely related thromboxanes and leukotrienes are all locally acting hormones derived from the fatty acid called **arachidonic acid.** These compounds are formed when an enzyme called phospholipase A is activated, resulting in the hydrolysis of membrane phospholipids and the release of arachidonic acid. Depending on which enzymes are present in the cell, the arachidonic acid can then be converted into one or more of the prostaglandins, thromboxanes or leukotrienes.

An Example of Endocrine Regulation: Adrenergic Hormones and Receptors

Hormones are always released for the purpose of modulating the function of particular target tissues. An important aspect of studying a specific hormone is therefore to understand the specific functions of the target tissues on which the hormone acts. To illustrate how endocrine hormones act, we will look more closely at the adrenergic hormones **epinephrine** and **norepinephrine.** (Epinephrine is also called **adrenalin;** the two words are of Greek and Latin derivation, respectively, and mean "above, or near, the kidney," referring to the location in the body of the *adrenal glands,* which synthesize this hormone.) When secreted into the bloodstream, epinephrine and norepinephrine stimulate changes in many different tissues or organs, all aimed at preparing the body for dangerous or stressful situations (the so-called "fight-or-flight syndrome"). Overall, the adrenergic hormones trigger increased cardiac output, shunting blood from the visceral organs to the muscles and the heart, as well as dilation of arterioles to facilitate oxygenation of the blood. In addition,

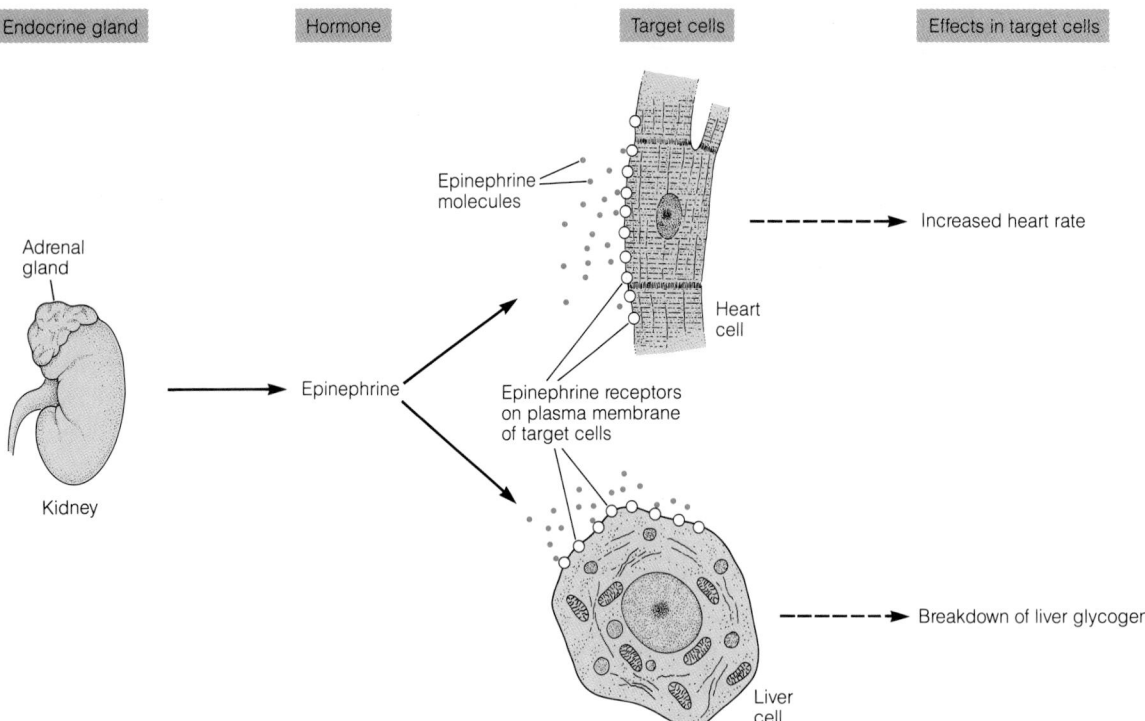

Figure 23-16   Target Tissues for Endocrine Hormones.
Cells in a target tissue have hormone-specific receptors embedded in their plasma membranes (or, in the case of the steroid hormones, present in the nucleus or cytosol). Heart and liver cells can respond to epinephrine synthesized by the adrenal glands because these cells have epinephrine-specific receptors on their outer surfaces. A specific hormone may elicit different responses in different target cells, depending on the enzymes present in the cells. Epinephrine causes an increase in the heart rate but stimulates glycogen breakdown in the liver.

Table 23-2   Physiological Functions of Hormones

Function Under Hormonal Control	Hormone	Source (Endocrine Tissue)
Growth and Development		
Body size	Somatotropin (growth hormone)	Anterior pituitary
Sexual development	Androgens (males)	Testes
	Estrogens (females)	Ovaries
Reproductive cycle	Luteinizing hormone	Anterior pituitary
	Follicle-stimulating hormone	Anterior pituitary
	Chorionic gonadotropin	Follicle
Rates of Body Processes		
Hormone secretion	Tropic hormones	Anterior pituitary
Basal metabolism	Thyroxine	Thyroid
Glucose uptake	Insulin	Pancreas
Kidney filtration	Antidiuretic hormone	Posterior pituitary
Uterine contraction	Oxytocin	Posterior pituitary
Concentrations of Substances		
Blood glucose	Glucagon, insulin	Pancreas
Mineral balance	Corticosteroids (aldosterone)	Adrenal cortex
Blood calcium	Parathyroid hormone	Parathyroid
Responses to Stress and Injury		
Heart rate	Epinephrine	Adrenal medulla
Blood pressure	Epinephrine	Adrenal medulla
Inflammation	Histamine	Mast cells
	Prostaglandins	All tissues
	Corticosteroids (cortisol)	Adrenal cortex

Table 23-3 Chemical Classification and Function of Hormones

Chemical Classification	Examples	Regulated Function
Endocrine Hormones		
Amino acid derivatives	Epinephrine (adrenalin) and norepinephrine (both derived from tyrosine)	Stress responses: regulation of heart rate and blood pressure; release of glucose and fatty acids from storage sites
	Thyroxine (derived from tyrosine)	Regulation of metabolic rate
Peptides	Antidiuretic hormone	Regulation of body water and blood pressure
	Hypothalamic hormones (releasing factors)	Regulation of tropic hormone release from pituitary gland
Proteins	Anterior pituitary hormones	Regulation of other endocrine systems
Steroids	Sex hormones (androgens and estrogens)	Development and control of reproductive capacity
	Corticosteroids	Stress responses; control of blood electrolytes
Paracrine Hormones		
Amino acid derivative	Histamine	Local responses to stress and injury
Arachidonic acid derivatives	Prostaglandins	Local responses to stress and injury

these hormones stimulate the breakdown of glycogen to supply glucose to the muscles.

Adrenergic hormones bind to a family of G protein–linked receptors known as **adrenergic receptors.** The individual members of this family differ mainly in their preference for epinephrine or norepinephrine and in which G protein is linked to the receptor. They can be broadly classified into α- and β-adrenergic receptors. The **α-adrenergic receptors** bind both epinephrine and norepinephrine. These receptors are located on the smooth muscles that regulate blood flow to visceral organs. The **β-adrenergic receptors** bind epinephrine much better than norepinephrine. These

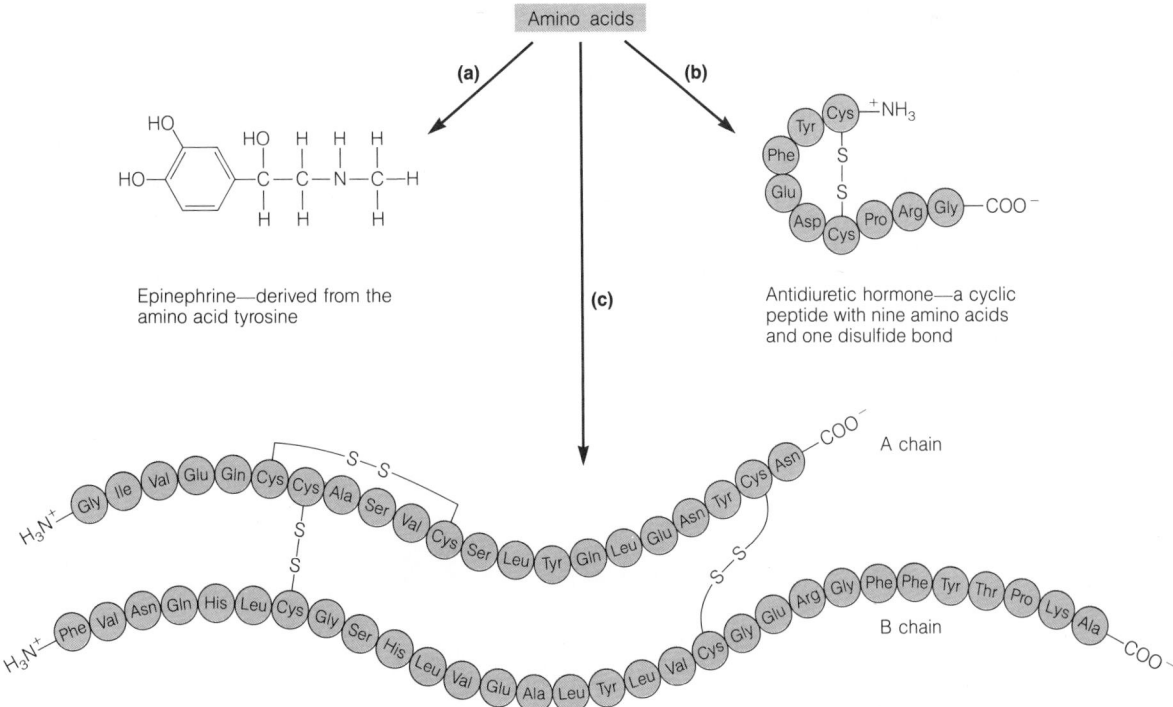

Epinephrine—derived from the amino acid tyrosine

Antidiuretic hormone—a cyclic peptide with nine amino acids and one disulfide bond

Insulin—a protein with an A chain and a B chain linked by two interchain disulfide bonds and one intrachain disulfide bond

Figure 23-17   The Chemistry of Animal Hormones.   Most endocrine hormones are **(a)** derivatives of amino acids, **(b)** peptides, **(c)** proteins, or (see Figure 23-3a) steroids (either sex hormones or corti-costeroids) derived from cholesterol. (Insulin is actually a globular protein and is shown here in extended form only for purposes of illustration.)

receptors are found on smooth muscles associated with arterioles that feed the heart, smooth muscles of the bronchioles in the lungs, and skeletal muscles.

The α- and β-adrenergic receptors stimulate different signal transduction pathways because they are linked to different G proteins (Figure 23-18). The G protein activated by the α-adrenergic receptors is known as Gp (or Gq), while the β-adrenergic receptors bind Gs. Activation of Gp stimulates phospholipase C, leading to the elevation of intracellular calcium levels. Activation of Gs stimulates the cAMP signal transduction pathway. For the moment, we will focus on the cAMP pathway and the β-adrenergic receptors.

β-Adrenergic Receptors and the cAMP Pathway

When epinephrine binds to a β-adrenergic receptor, it activates Gs, and the Gsα subunit converts to the GTP-bound form and then detaches from the Gs$\alpha\beta\gamma$ complex. The free Gsα now binds to adenylate cyclase and stimulates the formation of cAMP from ATP (see Figure 23-7). The Gα subunit stimulates adenylate cyclase only when in the GTP-bound form. As soon as the GTP is hydrolyzed, Gsα

becomes inactive and reassociates with Gs$\beta\gamma$, thus completing a cycle of activation and inactivation. If the receptor continues to stimulate G protein activation, the G proteins will cycle between active and inactive states. Once the messenger is removed, G proteins revert to their inactive form, and the enzyme phosphodiesterase breaks down the remaining cAMP into adenosine monophosphate.

Table 23-4 summarizes some of the major physiological functions that are regulated by cAMP. As a specific example of cAMP-mediated regulation, we will consider the control of glycogen degradation in liver or muscle cells by the hormone epinephrine.

Intracellular Effects of the cAMP Pathway: Control of Glycogen Degradation. One of the actions of the adrenergic hormones is to stimulate the breakdown of glycogen to provide muscle cells with an adequate supply of glucose. The breakdown of glycogen is catalyzed by the enzyme *glycogen phosphorylase*, which cleaves glucose units from glycogen as glucose-1-phosphate by the addition of inorganic phosphate (P_i). Although the sequence of events from hormonal stimulation of the tissue to enhanced glycogen degradation is

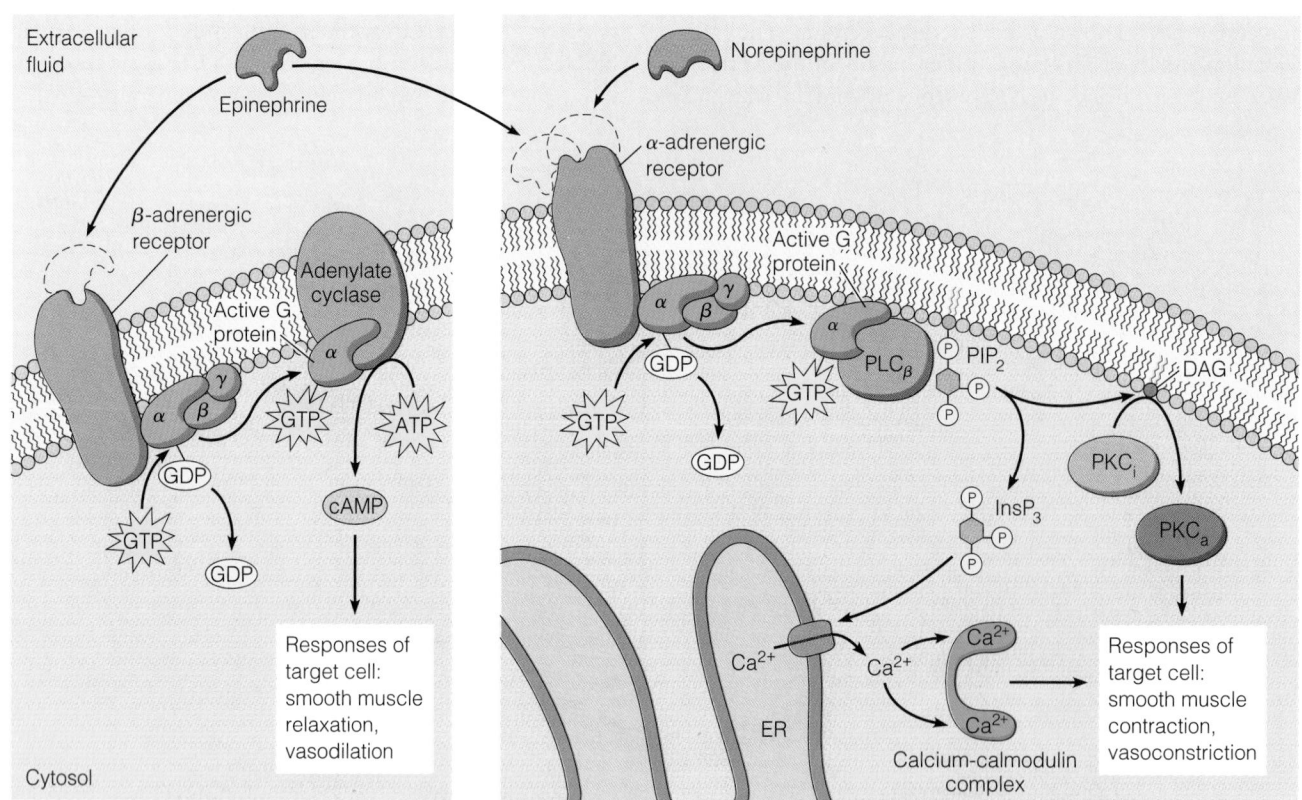

(a) cAMP pathway initiated by activation of β-adrenergic receptor

(b) Inositol-phospholipid-calcium pathway initiated by activation of α-adrenergic receptor

Figure 23-18   The Stimulation of G Protein–Linked Signal Transduction Pathways by α- and β-Adrenergic Receptors. The α- and β-adrenergic receptors are a closely related family of G protein–linked receptors. Each receptor binds to epinephrine or norepinephrine. While both types of receptors can bind the same hormone (epinephrine, for example), they trigger different signal transduction pathways because they activate different G proteins. **(a)** The β receptors activate Gs and stimulate the cAMP signal transduction pathway. **(b)** The α receptors activate Gp and stimulate the inositol-phospholipid-calcium signal transduction pathway.

Table 23-4 Examples of Cell Functions Regulated by cAMP

Regulated Function	Target Tissue	Hormone
Glycogen degradation	Muscle, liver	Epinephrine
Fatty acid production	Adipose	Epinephrine
Heart rate, blood pressure	Cardiovascular	Epinephrine
Water reabsorption	Kidney	Antidiuretic hormone
Bone resorption	Bone	Parathyroid hormone

complex, it is well worth understanding because it serves as a model for a variety of other hormonally regulated processes in mammalian cells. The glycogen phosphorylase system is also of historical interest because it was the first cAMP-mediated regulatory sequence to be elucidated. The original work was published in 1956 by Earl Sutherland, who received a Nobel Prize for this discovery in 1971.

The sequence of events that leads from hormonal stimulation to enhanced glycogen degradation is shown in Figure 23-19. It begins when an epinephrine molecule binds to a β-adrenergic receptor on the plasma membrane of a liver or muscle cell. As described earlier, the receptor activates a neighboring G protein, and the G protein in turn stimulates adenylate cyclase, the membrane-bound enzyme that generates cAMP from ATP (see Figure 23-18a). The resulting

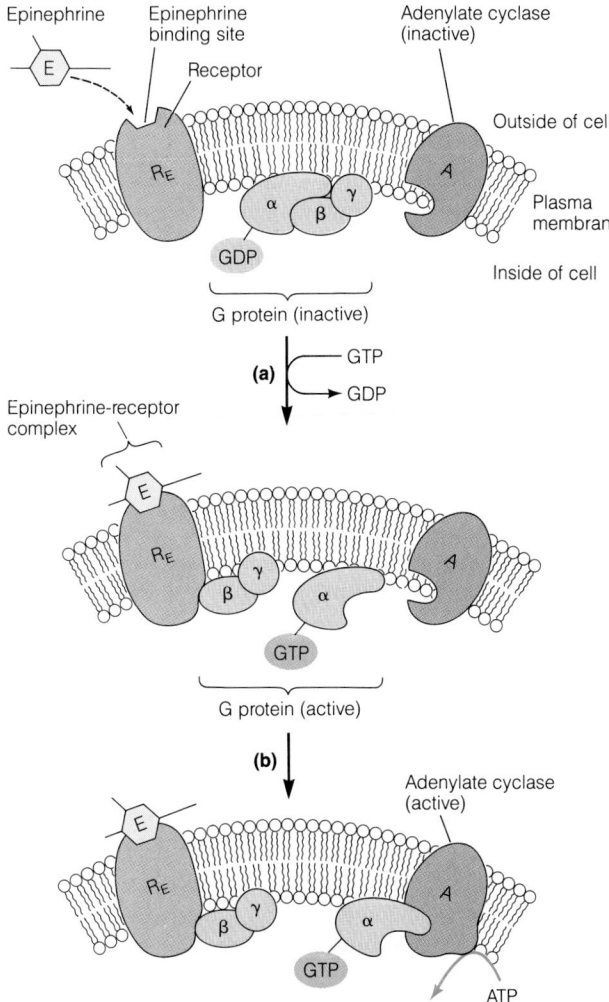

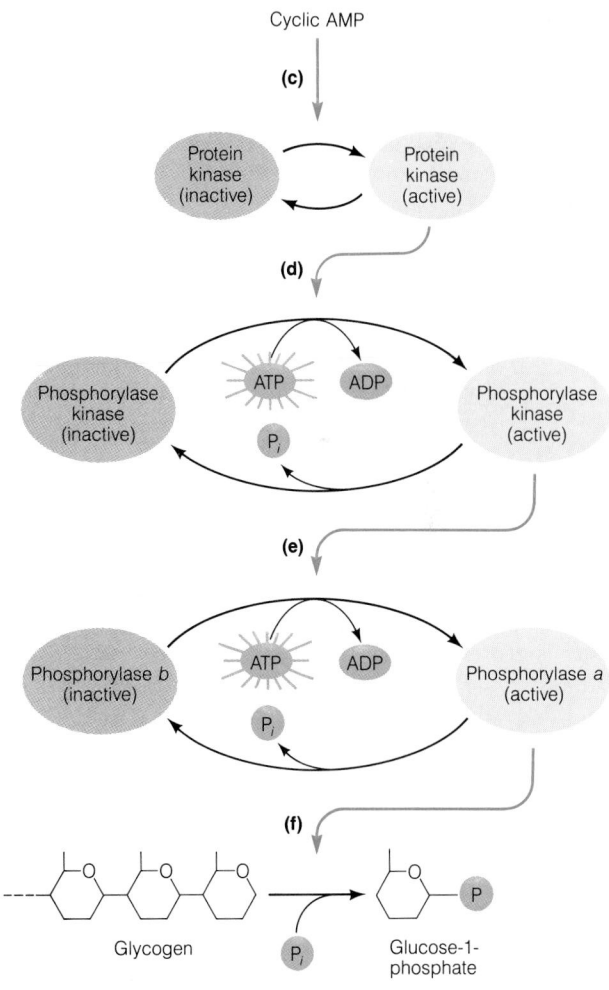

Figure 23-19   Stimulation of Glycogen Breakdown by Epinephrine.   Muscle and liver cells respond to an increased concentration of epinephrine in the blood by increasing their rate of glycogen breakdown. The stimulatory effect of extracellular epinephrine on intracellular glycogen catabolism is mediated by a G protein–cyclic AMP regulatory cascade. In this case, the ligand is epinephrine (E) and the membrane protein to which it binds is the β-adrenergic receptor (R$_E$). **(a)** The binding of epinephrine to its receptor activates the Gs protein, causing dissociation of the GTP-Gα complex. **(b)** The GTP-Gα complex then binds

to and activates adenylate cyclase (A). **(c)** As the intracellular concentration of cyclic AMP rises, cAMP initiates a further cascade of regulatory events that begins with the activation of protein kinase A (see Figure 23-9). **(d)** Protein kinase A then converts inactive phosphorylase kinase to the active form by ATP-dependent phosphorylation. **(e)** Active phosphorylase kinase, in turn, phosphorylates phosphorylase b, converting it to phosphorylase a, the active form of the enzyme. **(f)** Phosphorylase a then catalyzes the phosphorolytic cleavage of glycogen into molecules of glucose-1-phosphate.

transient increase in the concentration of cytosolic cAMP in the cytosol activates protein kinase A. In the case of glycogen degradation, PKA activates another cascade of events that begins with the phosphorylation of the enzyme *phosphorylase kinase*. This leads to the conversion of *glycogen phosphorylase* from phosphorylase *b*, the less active form, to phosphorylase *a*, the more active form, and thus to an increased rate of glycogen breakdown (glycogenolysis).

To shut down the glycogenolytic pathway, glycogen phosphorylase and phosphorylase kinase must be dephosphorylated. These phosphate groups are removed by an enzyme called *phosphoprotein phosphatase*. Protein kinase A inhibits phosphoprotein phosphatase by phosphorylating and activating an inhibitor protein. Once the formation of cAMP ceases and PKA becomes inactive, other phosphatases in the cytosol remove the phosphate from the inhibitor protein, and the phosphoprotein phosphatase becomes active. The active phosphoprotein phosphatase then removes the phosphate groups from glycogen phosphorylase and phosphorylase kinase, causing glycogenolysis to cease.

Thus, the signal transduction pathway for epinephrine-stimulated glycogen breakdown involves the activation, in order, of a membrane-bound G protein, adenylate cyclase, PKA, phosphorylase kinase, and glycogen phosphorylase, which then catalyzes the phosphorolytic cleavage of glycogen into glucose-1-phosphate (see Figure 23-19). Because cAMP also stimulates the inactivation of the enzyme system responsible for glycogen synthesis, the overall effect of cAMP involves both an increase in glycogen breakdown and a decrease in its synthesis.

α-Adrenergic Receptors and the Inositol-Phospholipid-Calcium Pathway

We have seen one important signal transduction pathway in which β-adrenergic stimulation leads to an elevation in the intracellular cAMP concentration. Another important pathway is represented by the α-adrenergic receptors, which stimulate the formation of inositol trisphosphate (InsP$_3$) and diacylglycerol (DAG). The α-adrenergic receptors are found mainly on smooth muscles in the peripheral blood vessels such as those controlling blood flow to the intestines. When α-adrenergic receptors are stimulated, the formation of InsP$_3$ causes an increase in the intracellular calcium concentration. The elevated level of calcium causes smooth muscle contraction, resulting in constriction of the blood vessels and diminished blood flow to the gut. Thus, the activation of α-adrenergic receptors affects smooth muscle cells in a manner opposite to that of β-adrenergic activation, which causes smooth muscles to relax.

Coordination of the Responses to Adrenergic Hormones

We can now consider how the secretion of epinephrine and norepinephrine leads to the coordinated changes in various tissues that are required to prepare the body for stressful situations. The overall strategy of adrenergic hormone actions is to put many of the normal bodily functions on hold and to deliver vital resources to the heart and skeletal muscles instead, as well as to produce a heightened state of alertness. This is effected by the adrenergic hormones and a system of receptors (α- and β-adrenergic receptors) that produce essentially opposite effects on a particular tissue such as smooth muscle.

Using smooth muscle as our example, we can see how the elevation of cAMP levels, which causes smooth muscle to relax, and the elevation of intracellular calcium levels, which causes smooth muscle to contract, coordinate many of the needed changes in blood flow. In the smooth muscles surrounding the blood vessels of the heart and the bronchioles of the lungs, stimulation of β-adrenergic receptors results in muscle relaxation. As a result, blood flow to the heart increases and airway passages widen, facilitating better oxygenation of the blood. At the same time, stimulation of α-adrenergic receptors in the smooth muscles surrounding peripheral veins causes these muscles to contract, constricting the smaller veins. This results in diminished blood flow to the skin, kidneys, and digestive tract, increasing blood pressure and shunting the blood to the tissues where it is needed most: the heart, lungs, and skeletal muscles.

The use of muscles requires a great deal of ATP, which is provided by the catabolism of either glucose or fatty acids. The β-adrenergic receptors on the surfaces of skeletal muscle cells are linked to the formation of cAMP, which stimulates the breakdown of glycogen within the muscle cells, thereby supplying the needed glucose. In addition, lipolysis is stimulated in fat cells, releasing the fatty acids required by cardiac muscle for the generation of ATP. These are just a few effects of adrenergic stimulation, but they enable us to understand that each of these effects is part of a coordinated effort to mobilize the body for stress or strenuous activity.

An Example of Paracrine Regulation: Prostaglandins and Thromboxanes

Paracrine hormones are substances secreted by cells that affect other cells a short distance away. In this case, the messenger clearly has a limited range of action. Examples of messengers commonly classified as paracrine hormones are **histamine** and the **prostaglandins,** as mentioned earlier. Both histamine and the group of prostaglandins known as *leukotrienes* are important in allergic responses and inflammation. In allergies, certain types of foreign substances called *allergens* induce the body to make an antibody protein of the class IgE. Once a person is exposed to the allergen and IgE has been formed, it binds receptors on membranes of cells known as mast cells, which store histamine inside secretory granules. The next time a person is exposed to the allergen, it binds to the antibody and stimulates an increase in calcium in the mast cell, which induces the secretion of histamine. The release of histamine generally acts locally on neighboring tissues and causes many of the symptoms of allergy.

The prostaglandins, which are among the best known examples of paracrine hormones, are so named because they were first identified in human semen as a secretion of the prostate gland. Prostaglandins, including leukotrienes and the related *thromboxanes*, are derived from arachidonic acid, a 20-carbon fatty acid with four double bonds (Figure 23-20). Arachidonic acid is one of the fatty acids normally present in membrane phospholipids and is released from a phospholipid by the enzyme *phospholipase A*. Once arachidonic acid is liberated, the enzyme *cyclooxygenase* converts the arachidonic acid to a prostaglandin precursor, which is then further metabolized to one of the many different prostaglandins.

Prostaglandins have a variety of effects, many of which act on smooth muscle. For example, prostaglandins contained in semen stimulate uterine smooth muscle contraction, which helps transport sperm to the egg. Prostaglandins also initiate smooth muscle contraction during labor and can be used to induce labor clinically if a baby is well past the expected delivery date. Some prostaglandins cause smooth muscle relaxation and can, for example, cause bronchiole dilation or lowering of blood pressure. Typically, the prostaglandins act on G protein–linked receptors to stimulate either the cAMP or the inositol-trisphosphate-calcium second messenger pathway.

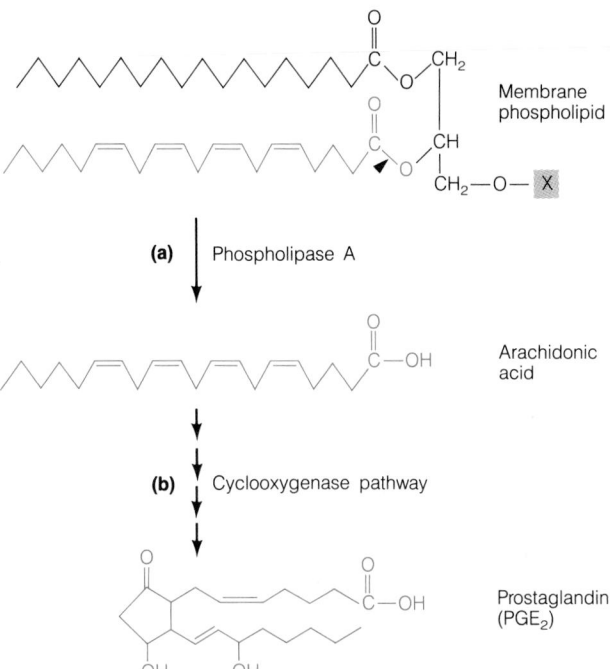

Figure 23-20 Phospholipase A and the Synthesis of Prostaglandins. **(a)** The synthesis of prostaglandins begins with the hydrolysis, by phospholipase A, of phospholipids to release fatty acids, including arachidonic acid. **(b)** Arachidonic acid is then oxidized by cyclooxygenase enzymes to produce the cyclic prostaglandins. Prostaglandins differ in the number and position of their double bonds and the oxidation state of their oxygen-containing functional groups. The prostaglandin shown here is PGE$_2$.

Paracrine Signaling Between Platelets and Blood Clotting. An example of how prostaglandins function in paracrine signaling is seen in the activation of **blood platelets.** Platelets are essential components of the blood-clotting mechanism and act to plug sites where blood vessels are ruptured. Platelets are not true cells, but are produced by the fragmentation of bone marrow cells called *megakaryocytes.* A platelet has no nucleus, but it does contain a cytoskeletal structure, numerous *dense granules*, and a calcium-storing membranous network called the *dense tubular system.* The dense granules contain *platelet factors* such as adenosine diphosphate, which here acts as a paracrine hormone that stimulates other platelets; growth factors that stimulate the formation of scar tissue; and molecules involved in the clotting process.

Clotting involves a complex cascade of enzymatic events that is triggered by injury and leads rapidly to the formation of a *clot*, a tangled mass of red blood cells, platelets, and fibers of a protein called *fibrin.* The clotting mechanism must be carefully regulated, because inappropriate blood clots lead quickly to life-threatening situations. The coronary arteries of the heart are only 1 millimeter or so in diameter, and most heart attacks result from a small blood clot that blocks coronary circulation. Similarly, clots in the venous system can produce *phlebitis* (inflammation of the veins) in the lower limbs and *embolisms* (obstructions) in the lungs. Thus, clot formation is essential, but it must be confined to the area where the damage occurs.

Platelets have many different receptors that sense when a tissue is damaged, including those for thrombin and collagen. *Thrombin* is a protease formed in the blood from *prothrombin* during the clotting reaction. The platelet thrombin receptor is activated when thrombin cleaves a portion of the receptor protein that is exposed on the outside of the cell. Collagen can also activate platelets. Collagen is present within the wall of blood vessels but not in the inner lining, so it is not normally exposed to the blood. Damage to the blood vessel wall that exposes collagen triggers the activation of platelets. Platelets can be activated by several other substances, including the presence of extracellular adenosine diphosphate and members of the prostaglandin family such as thromboxane A2. These substances are released by activated platelets and then function as paracrine hormones to activate other platelets.

The generally accepted platelet activation sequence is shown in Figure 23-21. Most of the known platelet activators act through G protein–linked receptors to activate phospholipase C. This triggers the release of calcium from the dense tubular system. Although the identity of the activator is not known, there is some evidence that a G protein also activates phospholipase A, resulting in the hydrolysis of membrane phospholipids and the release of arachidonic acid. An enzymatic pathway starting with cyclooxygenase then converts the arachidonic acid to thromboxane A2. Thromboxane A2 diffuses out of the platelet and acts on G protein–linked thromboxane receptors of neighboring platelets. The thromboxane receptors in turn activate phos-

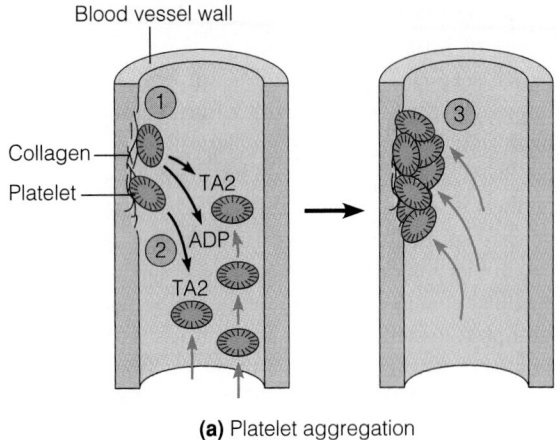

(a) Platelet aggregation

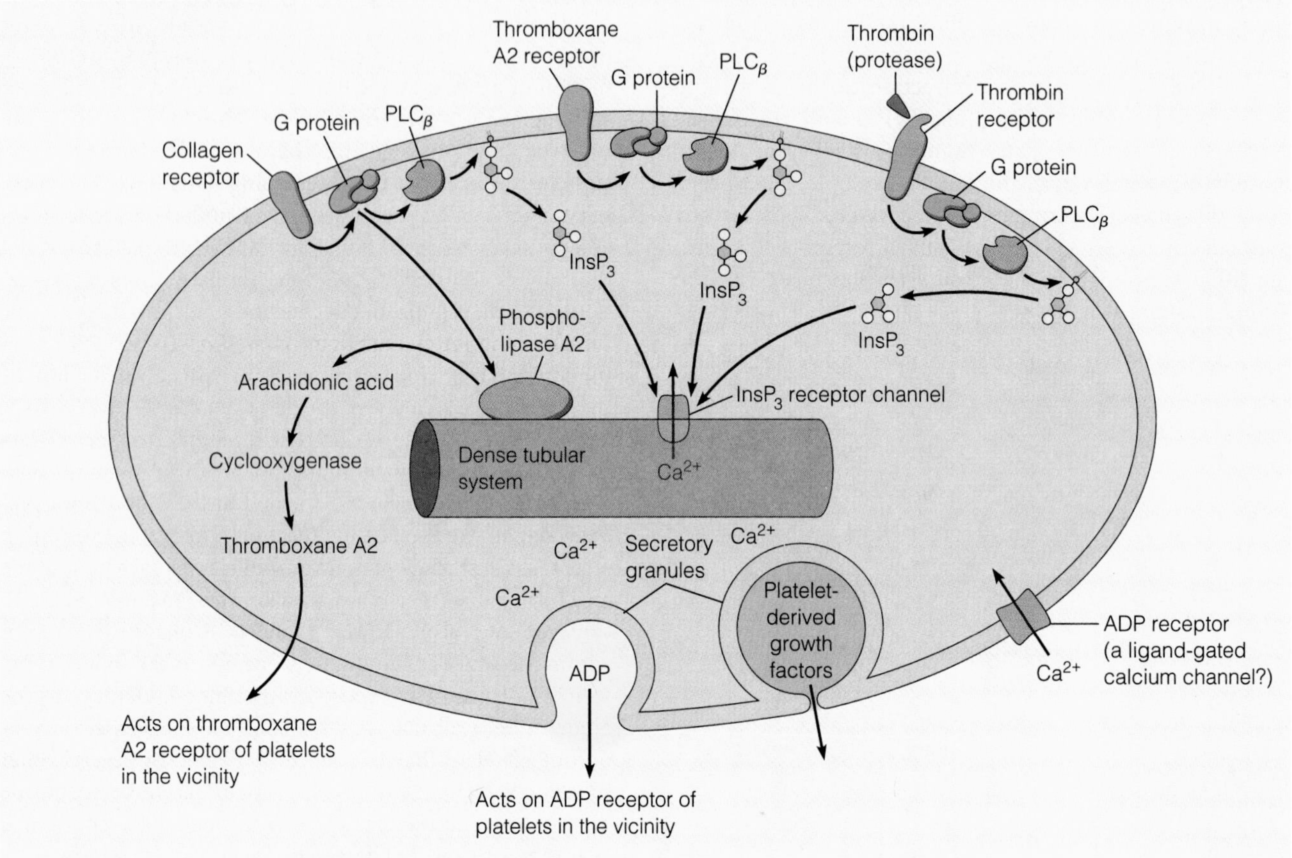

Thromboxane
A2 receptor

G protein

PLC$_\beta$

Thrombin
(protease)

Thrombin
receptor

G protein

Collagen
receptor

G protein

PLC$_\beta$

PLC$_\beta$

InsP$_3$

InsP$_3$

InsP$_3$

Phospho-
lipase A2

Arachidonic acid

InsP$_3$ receptor channel

Cyclooxygenase

Dense tubular
system

Ca^{2+}

Thromboxane A2

Ca^{2+}

Secretory
granules

Ca^{2+}

Ca^{2+}

Platelet-
derived
growth
factors

ADP receptor
(a ligand-gated
calcium channel?)

Ca^{2+}

Acts on thromboxane
A2 receptor of platelets
in the vicinity

ADP

Acts on ADP receptor of
platelets in the vicinity

(b) Platelet activation

Figure 23-21 The Process of Platelet Activation.

(a) Platelets are activated by several substances and aggregate upon activation. ① Collagen exposed in a damaged blood vessel wall stimulates platelet activation and adherence. ② Activated platelets release the platelet factors ADP and thromboxane A2 (TA2), which ③ stimulate the activation and aggregation of more platelets in the vicinity. The release of A2 and ADP is an example of paracrine signaling. **(b)** Paracrine signaling by platelets uses some of the same classes of receptors and signal transduction pathways described for the endocrine hormones. The receptors for thrombin, TA2, and collagen all appear to be G protein–linked receptors that activate phospholipase C. However, the activation of the thrombin receptor is unusual in that it depends on the protease thrombin re-

moving a portion of the receptor protein. When newly exposed collagen in a damaged blood vessel wall binds to a G protein–linked collagen receptor, membrane phospholipase C is activated, causing the production of InsP$_3$ and DAG, and ultimately the release of calcium from the dense tubular system. Increased calcium concentration inside the platelet stimulates the secretion of granules that contain growth factors and ADP. Phospholipase A is thought to be simultaneously activated by a G protein. The activation of phospholipase A2 releases arachidonic acid, which is converted to TA2. TA2 binds to the TA2 receptors on nearby platelets, causing elevated calcium levels, whereas ADP binds to what is probably a ligand-gated calcium channel. Both messengers stimulate platelets to aggregate.

pholipase C, triggering the activation of platelets that are nearby. As a result, platelets in the area are recruited to the site of injury.

Knowledge of the regulatory pathways involved in prostaglandin formation and platelet activation has enabled us to understand the mechanism of action of certain drugs. It has long been known, for example, that aspirin significantly reduces the incidence of heart attacks in older patients, apparently by decreasing the likelihood of small internal blood clots that might block blood flow through the coronary arteries in the heart. We now understand that aspirin specifically blocks cyclooxygenase activity, thereby inhibiting synthesis of thromboxane A2 and other prostaglandins from arachidonic acid. This inhibition slows platelet activation and recruitment, reducing the probability of small blood clot formation. Interestingly, prostaglandins also appear to increase body heat production, causing fever, by their effect on cells of the hypothalamus. By inhibiting the formation of these prostaglandins, aspirin causes the fever to subside.

Growth Factors as Messengers

In order for a cell to grow, it must have all the nutrients needed for synthesizing its component parts. However, the availability of nutrients is usually not enough to ensure that cells will grow. Cells often also need **growth factors,** messengers that act on specific receptors to stimulate cell growth. Biologists encountered the requirements for cell growth when they first tried to culture cells in vitro. Although provided with a growth medium rich in nutrients, including the presence of blood plasma, cells would not grow. A turning point came when blood serum was used instead of plasma: Serum was able to support the growth of cells, whereas plasma would not.

The difference between blood serum and blood plasma held an important clue about growth factors. *Plasma* is whole blood, including unreacted clotting components but without the red and white blood cells. *Serum* is the clear fluid remaining after blood has clotted. When platelets activate, they secrete growth factors into the blood that stimulate the growth of cells called *fibroblasts,* which form the new connective tissue that makes up a scar. Plasma does not contain these growth factors because the clotting reaction has not taken place. After clotting, the resulting serum is full of these **platelet-derived growth factors (PDGFs).**

The PDGF receptor is a tyrosine kinase receptor similar to those described previously. In fact, several growth factors all act by stimulating tyrosine kinase receptors, including insulin, insulinlike growth factor-1, fibroblast growth factor, epidermal growth factor (EGF; see Figure 23-6), and nerve growth factor.

Growth Factors and Oncogenes

Receptors that bind growth factors, such as tyrosine kinase receptors, are especially important because of their relevance to the problem of cancer. As mentioned earlier, receptors that become unregulated can cause cells to grow out of control. Mutant forms of tyrosine kinase receptors and of components of the signal transduction pathways regulated by the receptor are known to occur. Certain types of mutations give rise to a receptor that is always activated and therefore no longer regulated by the growth factor. This results in uncontrolled growth of the affected cells, which is, of course, one of the hallmarks of cancer cells (see Chapter 25).

Some of the proteins that are activated by the receptor can also become mutated so that they are always active. Many of these proteins, when mutated, also stimulate uncontrolled cell growth. Normal cellular proteins that have the potential to cause uncontrolled cell growth when mutated are known as *proto-oncogenes.* The actual mutated form responsible for uncontrolled cell growth is called an **oncogene.** The normal cellular forms of *ras* and *jun* are examples of proto-oncogenes. Mutant forms of *ras* are well known and have, in fact, been detected in approximately 30% of human cervical cancers. (The *ras* oncogene and the cellular chain of events it initiates are discussed further in Box 25A, p. 835.)

PERSPECTIVE

Most cells respond to hormones and other substances present in the extracellular fluid. Such responses are mediated by receptor proteins, either at the cell surface or in the cytosol. Each receptor protein has a binding site that is specific for its particular ligand. In the case of membrane receptors, ligand binding is followed by transmission of the signal to the cytosol, thereby regulating specific intracellular events. Several different mechanisms for signal transduction are known.

Because of their lipophilic nature, steroid hormones can diffuse across the plasma membrane, interacting with receptors in the nucleus or cytosol rather than membrane-bound receptors. The hormone-receptor complex activates the expression of specific genes by binding to their enhancer regions.

Another important signal transduction pathway involves G proteins and G protein–linked receptors. G proteins are activated when ligand binding to a neighboring receptor causes a conformational change in the G protein, resulting in the displacement of GDP by GTP. The G protein then activates an enzyme system that produces intracellular chemical signals called second messengers. One of the most common second messengers is cyclic AMP, synthesized when the enzyme adenylate cyclase is activated by a G protein. In an alternative pathway, the second messengers are inositol trisphosphate and diacylglycerol, which are produced from phosphatidylinositol bisphosphate when a G protein activates the enzyme phospholipase C. Regardless of the pathway, second messengers mediate specific intracellular responses by activating specific enzymes or enzyme cascades.

Calcium ions also play a central role in cellular activation processes and can be considered second messengers as well. Calcium effects are often mediated by calmodulin, a calcium-binding protein that is activated when calcium ions bind to it. Depending on the target cell, the calcium-calmodulin complex can activate any of a variety of enzymes, thereby modulating enzyme activity in response to the calcium concentration.

A third category of receptors are those with kinase activity. Upon binding of the appropriate ligand, such receptors become phosphorylated on specific tyrosines. The phosphorylated receptor becomes a binding site for other proteins that contain SH2 domains. When these proteins bind to the receptor, they are activated either by the binding itself or by subsequent phosphorylation. Each SH2-containing protein initiates a separate signal transduction pathway, including the activation of Ras and phospholipase C.

Hormones are messengers that regulate the activities of tissues in the body. Hormones can be endocrine or paracrine, depending on the mode of delivery. The adrenergic hormones secreted by the adrenal medulla are examples of endocrine hormones. Adrenergic hormone receptors can be divided into two groups: α- and β-adrenergic receptors. The β-adrenergic receptors are linked to a G protein that stimulates the formation of cAMP, whereas the α-adrenergic receptors stimulate phospholipase C, resulting in elevation of the intracellular calcium concentration. The different receptors mediate opposite effects in smooth muscle but work together to coordinate changes in blood flow and other activities, thereby preparing the body for stressful situations or vigorous exercise.

Growth factors are messengers that play a specific role in regulating cell growth. Many growth factors bind to tyrosine kinase receptors. The tyrosine kinase receptors and the signal transduction pathways initiated by these receptors provide an important link to our understanding of cancer. Although growth is normally regulated, mutations can lead to unregulated activity of the receptor or of components of the signal transduction pathway. As a result, cells receive a constant and uncontrolled stimulus for growth, which can lead to cancer.

KEY TERMS FOR SELF-TESTING

messenger (p. 751)
receptor (p. 751)

Chemical Signals and Cellular Receptors

physical factor (p. 751)
chemical factor (p. 751)
hormone (p. 751)
water-soluble messenger (p. 752)
lipid messenger (p. 752)
receptor activation (p. 752)
ligand (p. 753)
cognate receptor (p. 753)
receptor affinity (p. 753)
receptor down-regulation (p. 754)
signal transduction (p. 754)

Characteristics and Functions of Receptors

intracellular receptor (p. 755)
steroid receptor (p. 755)
plasma membrane receptor (p. 755)
G protein–linked receptor (p. 755)

G protein (p. 755)
tyrosine kinase receptor (p. 760)
nonreceptor tyrosine kinase (p. 760)

Signal Transduction Pathways and Second Messenger Systems

second messenger (p. 761)
cyclic AMP (cAMP) (p. 761)
adenylate cyclase (p. 761)
phosphodiesterase (p. 762)
protein kinase A (PKA) (p. 762)
cAMP pathway (p. 764)
inositol-1, 4, 5-trisphosphate ($InsP_3$) (p. 764)
phospholipase C (p. 764)
diacylglycerol (DAG) (p. 764)
inositol-phospholipid-calcium pathway (p. 764)
protein kinase C (PKC) (p. 766)
calcium ion (Ca^{2+}) (p. 766)
calmodulin (p. 767)
calcium-calmodulin complex (p. 768)
Ras pathway (p. 770)

The Endocrine and Paracrine Hormone Systems

endocrine hormone (p. 770)
paracrine hormone (p. 770)
target tissue (p. 771)
arachidonic acid (p. 771)
epinephrine (adrenalin) (p. 771)
norepinephrine (p. 771)
adrenergic receptor (p. 773)
α-adrenergic receptor (p. 773)
β-adrenergic receptor (p. 773)
histamine (p. 776)
prostaglandin (p. 776)
blood platelet (p. 777)

Growth Factors as Messengers

growth factor (p. 779)
platelet-derived growth factor (PGDF) (p. 779)
oncogene (p. 779)

PROBLEM SET

23-1. Chemical Signals and Second Messengers. Fill in the blanks with the appropriate terms:

(a) _____ is an intracellular protein that binds calcium and activates enzymes.

(b) The _____ is a gland that regulates other endocrine tissues of the body.

(c) A substance that fits into a specific binding site on the surface of a protein molecule is called a _____.

(d) Two products of phospholipase C activity that serve as second messengers are _____ and _____.

(e) Cyclic AMP is produced by the enzyme _____ and degraded by the enzyme _____.

23-2. Heterotrimeric and Monomeric G Proteins. G protein–linked receptors activate heterotrimeric G proteins through interactions between the receptor and the $G\beta\gamma$ subunit. Upon binding to the receptor, the $G\beta\gamma$ subunit then catalyzes GDP/GTP exchange by the $G\alpha$ subunit. How is this similar to the activation of Ras by a tyrosine kinase receptor?

23-3. Why Calcium? Compare the use of sodium ions and calcium ions as intracellular signals. What happens when sodium ions rush into a cell? What happens when the cytosolic calcium concentration rises? What is the difference? Why would it be better to use calcium ions as a second messenger rather than sodium ions, if the ion must bind to a protein as part of a signal transduction pathway?

23-4. Calcium Chelators and Ionophores. Two important tools that have aided studies of the role of calcium in triggering different cellular events are calcium chelators and calcium ionophores. Chelators are compounds such as EGTA and EDTA that bind very tightly to calcium ions, thereby effectively reducing the free (or available) calcium ion concentration outside of the cell nearly to zero. Ionophores are compounds that shuttle ions across lipid bilayers and membranes, including the plasma membrane and internal membranes such as the ER. For calcium ions, two of the commonly used ionophores are A23187 and ionomycin. Using these tools, describe how you could demonstrate that a hormone exerts its effect by (1) causing calcium to enter the cell through channels or (2) releasing calcium from intracellular stores such as the ER.

23-5. Anti-Inflammatory vs. Inflammatory Drugs. Anti-inflammatory drugs can generally be classified as nonsteroidal and steroidal. Aspirin is a nonsteroidal anti-inflammatory drug, whereas cortisone is a steroid. Cortisone induces the synthesis of a protein that inhibits phospholipase A. Are the actions of these two kinds of compounds related? Based on how these drugs act, what kind of compound would be a good candidate for stimulating inflammation?

23-6. Neuroendocrine Response. The nervous and endocrine systems often act in concert to regulate a specific physiological function. Imagine that you have just been frightened by something—the sudden thought of a cell biology exam that you forgot to study for, perhaps. Sketch the pathway by which this mental event causes the adrenal gland to release epinephrine. Include the following elements in your pathway: brain, spinal cord, peripheral nervous system, adrenal gland, bloodstream, and receptors on cardiac (heart) cells.

23-7. Membrane Receptors and Medicine. Hypertension, or high blood pressure, is often seen in elderly people. A typical prescription to reduce a patient's blood pressure includes compounds called *beta-blockers,* which block β-adrenergic receptors throughout the body. These receptors bind epinephrine, thereby activating a cellular response. Why do you think that beta-blockers are effective in reducing blood pressure?

23-8. Growth Factors and Wound Healing. A recent survey of the medical literature shows considerable interest in the use of fibroblast growth factors to promote wound healing. These factors are secreted by platelets when they react to form a platelet plug at the site of a wound. One of these platelet growth factors, known as platelet-derived growth factor (PDGF), binds to tyrosine kinase receptors on the surface of fibroblasts and stimulates both the chemotaxis of fibroblasts toward the platelet plug and the growth of fibroblasts. Using this information, explain why growth factors can be effectively used to treat wounds.

23-9. Chemoattractant Receptors on Neutrophils. Neutrophils are normally responsible for killing bacteria at sites of infection. Neutrophils are able to find their way toward sites of infection by the process of chemotaxis. In this process, neutrophils sense the presence of bacterial protein with receptors on their plasma membranes and then follow the trail of these proteins toward the site of infection. Suppose you find that chemotaxis is inhibited by pertussis toxin (see Box 23B). What kind of receptor is more likely to be involved in responding to bacterial proteins?

23-10. Once Is Enough. A person who smokes for the first time tends to experience much more severe effects from a cigarette than someone who smokes frequently. Given that the smoke contains nicotine, provide a reasonable explanation for this difference.

SUGGESTED READING

Chemical Signals and Cellular Receptors

Berridge, M. J. The molecular basis of communication within the cell. *Sci. Amer.* 253 (October 1985): 142.

Hardie, D. G. *Biochemical Messengers: Hormones, Neurotransmitters, and Growth Factors.* London: Chapman & Hall, 1990.

Pawson, T. Signal transduction: A conserved pathway from the membrane to the nucleus. *Dev. Genet.* 14 (1993): 333.

Synder, S. H. The molecular basis of communication between cells. *Sci. Amer.* 253 (October 1985): 132.

Characteristics and Functions of Receptors

Clapham, D. E. Mutations in G protein-linked receptors: Novel insights on disease. *Cell* 75 (1993): 1237.

Gehring, U. Steroid hormone receptors: Biochemistry, genetics, and molecular biology. *Trends Biochem. Sci.* 12 (1987): 399.

Hedin, K. E., K. Duerson, and D. E. Clapham. Specificity of receptor-G protein interactions: Searching for the structure behind the signal. *Cell Signal.* 5 (1993): 505.

Hepler, J., and A. Gilman. G Proteins. *Trends Biochem. Sci.* 17 (1992): 383.

Levitsky, A. *Receptors: A Quantitative Approach.* Menlo Park, CA: Benjamin/Cummings, 1984.

Yamamoto, K. R. Steroid receptor regulated transcription of specific genes and gene networks. *Annu. Rev. Genet.* 19 (1985): 209.

Signal Transduction Pathways and Second Messenger Systems

Asaoka, Y., S. Nakamura, K. Yoshida, and Y. Nishizuka. Protein kinase C, calcium and phospholipid degradation. *Trends Biochem. Sci.* 17 (1992): 414.

Berridge, M. J. The biology and medicine of calcium signalling. *Mol. Cell Endocrinol.* 98 (1994): 119.

Berridge, M. J. Inositol trisphosphate and calcium signalling. *Nature* 361 (1993): 315.

Cole, K., and E. Kohn. Calcium-mediated signal transduction: Biology, biochemistry, and therapy. *Cancer Metastasis Rev.* 13 (1994): 31.

Dissing, S. Cell calcium: Introduction. *Cell Calcium* 14 (1993): 673.

Fantl, W. J., D. E. Johnson, and L. T. Williams. Signalling by receptor tyrosine kinases. *Annu. Rev. Biochem.* 62(1993): 453.

Freissmuth, M., P. J. Casey, and A. G. Gilman. G proteins control diverse pathways of transmembrane signaling. *FASEB* 3 (1989): 2125.

Head, J. F. A better grip on calmodulin. *Curr. Biol.* 2 (1992): 609.

Hinrichsen, R. D. Calcium and calmodulin in the control of cellular behavior and motility. *Biochim. Biophys. Acta* 1155 (1993): 277.

Hokin, L. E., and J. F. Dixon. The phosphoinositide signalling system. I. Historical background. II. Effects of lithium on the accumulation of second messenger inositol 1,4,5-trisphosphate in brain cortex slices. *Prog. Brain Res.* 98 (1993): 309.

Lai, C.-Y. The chemistry and biology of cholera toxin. *CRC Crit. Rev. Biochem.* 9 (1987): 171.

Lambert, D. G. Signal transduction: G proteins and second messengers. *Brit. J. Anaesth.* 71 (1993): 86.

Levitsky, A. From epinephrine to cyclic AMP. *Science* 241 (1988): 800.

Medema, R. H., and J. L. Bos. The role of p21ras in receptor tyrosine kinase signaling. *Crit. Rev. Oncog.* 4 (1993): 615.

Michell, R. Inositol lipids in cellular signalling mechanisms. *Trends Biochem. Sci.* 17 (1992): 274.

Mikoshiba, K. Inositol 1,4,5-trisphosphate receptor. *Trends Pharmacol. Sci.* 14 (1993): 86.

Nishizuka, Y. Intracellular signaling by hydrolysis of phospholipids and activation of protein kinase C. *Science* 258 (1992): 607.

O'Neil, K. T., and W. F. DeGrado. How calmodulin binds its targets: Sequence independent recognition of amphipathic alpha-helices. *Trends Biochem. Sci.* 15 (1990): 59.

Pazin, M., and L. Williams. Triggering signaling cascades by receptor tyrosine kinases. *Trends Biochem. Sci.* 17 (1992): 374.

Putney, J., Jr., and G. S. Bird. The inositol phosphate-calcium signaling system in nonexcitable cells. *Endocr. Rev.* 14 (1993): 610.

Rozakis-Adcock, M., R. Fernley, J. Wade, T. Pawson, and D. Bowtell. The SH2 and SH3 domains of mammalian Grb 2 couple the EGF receptor to the ras activator mSos1. *Nature* 363 (1993): 83.

Schlessinger, J. How receptor tyrosine kinases activate Ras. *Trends Biochem. Sci.* 18 (1993): 273.

Stryer, L. The molecules of visual excitation. *Sci. Amer.* 257 (July 1987): 42.

Sutherland, E. W. Studies on the mechanism of hormone action. *Science* 177 (1972): 401.

Tang, W. J., and A. G. Gilman. Adenylyl cyclases. *Cell* 71 (1992): 1069.

Tang, W. J., J. A. Iniguez-Lluhi, S. Mumby, and A. G. Gilman. Regulation of mammalian adenylyl cyclases by G-protein alpha and beta gamma subunits. *Cold Spring Harb. Symp. Quant. Biol.* 57 (1992): 135.

The Endocrine and Paracrine Hormone Systems

Baulieu, E. E., and P. A. Kelly. *Hormones: From Molecules to Disease.* London: Chapman & Hall, 1990.

Brass, L. F., J. A. Hoxie, and D. R. Manning. Signaling through G proteins and G protein-coupled receptors during platelet activation. *Thromb. Haemost.* 70 (1993): 217.

Norman, A. W., and G. Litwack. *Hormones.* San Diego: Academic Press, 1987.

Growth Factors as Messengers

Bishop, J. M. Molecular themes in oncogenes. *Cell* 64 (1991): 235.

Carpenter, G. Receptors for epidermal growth factor and other polypeptide mitogens. *Annu. Rev. Biochem.* 56 (1987): 881.

Hsuan, J. J. Oncogene regulation by growth factors. *Anticancer Res.* 13 (1993): 2521.

Schlessinger, J., and A. Ullrich. Growth factor signaling by receptor tyrosine kinases. *Neuron* 9 (1992): 383.

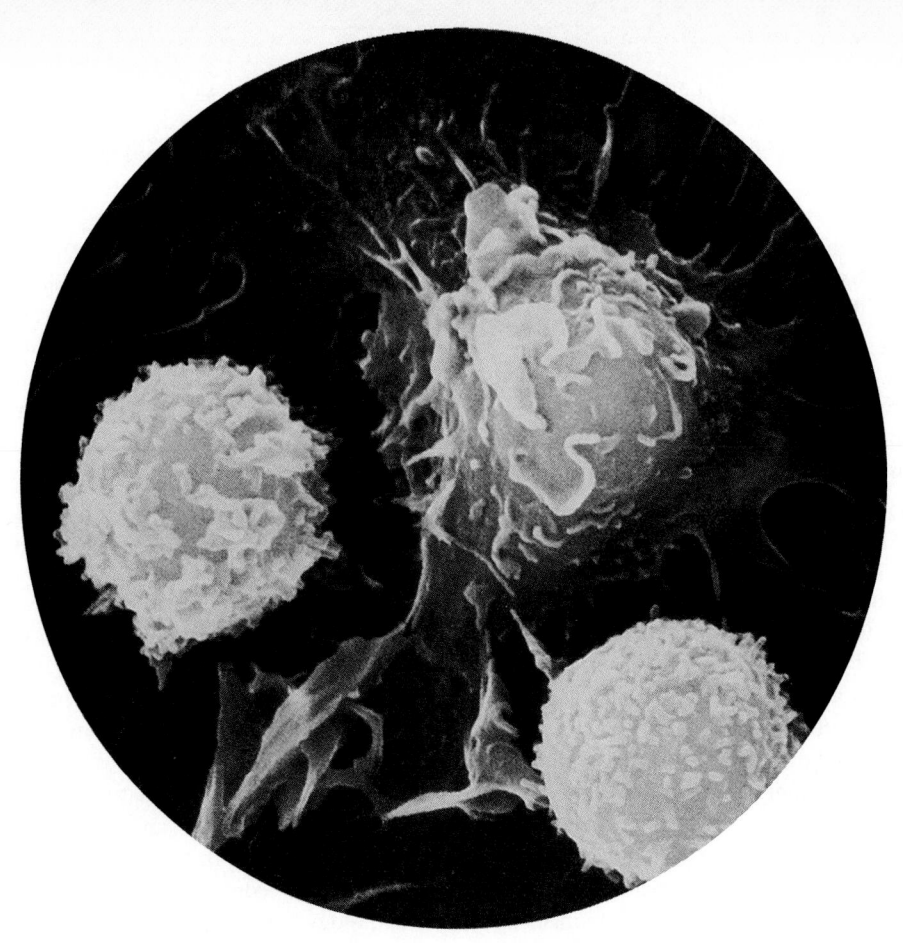

SPECIAL TOPICS IN CELL BIOLOGY

CELLULAR ASPECTS OF THE IMMUNE RESPONSE

Our environment is filled with a host of infectious agents, including bacteria, viruses, parasites, and fungi. As a first line of defense against these pathogenic (disease-causing) invaders, vertebrates depend on **innate immunity,** which is nonspecific and includes physical barriers (e.g., skin and mucous membranes) as well as secretions with antimicrobial activity (e.g., mucus and tears containing the enzyme lysozyme). Pathogens that breach these physical barriers are often destroyed and eliminated by *phagocytes*, white blood cells that engulf foreign agents such as bacteria and fungi by the process of phagocytosis.

But physical barriers and phagocytes alone are not always sufficient to ward off pathogens. Fortunately, vertebrates also have **adaptive immunity,** a specific, inducible system of defense. The adaptive immune system is essentially a means of surveillance intended to protect the organism from infections and cancer by recognizing and destroying the causative agents and abnormal cells. As such, it is indispensable to the health of the organism.

To appreciate how vital the immune system is, we need only consider the fate of individuals with inherited or acquired deficiencies of the immune system. Infants born with a completely defective immune system can survive only if they are maintained in a sterile environment, completely free from microorganisms that might cause infection. Similarly, patients whose immune systems have been medically suppressed to prevent the rejection of transplanted tissues or organs are abnormally susceptible to microbial infections. Such *immunosuppression* has rendered them temporarily defenseless.

The importance of the immune system has been underscored in an especially devastating way by the continuing worldwide epidemic of **acquired immunodeficiency syndrome (AIDS).** People with AIDS have a severely impaired immune system, as a result of infection by **human immunodeficiency virus (HIV).** These individuals are highly susceptible to infections and malignant diseases that can be successfully countered by a functional immune system, but that are often fatal when the immune system is incapacitated.

Immunology, the study of the immune system, arose out of the observation that people who have recovered from certain infections, such as smallpox or chicken pox, rarely, if ever, contract the disease again. This protective immunity is generally long-lasting and is highly specific; that is, an individual who has survived an infection by the virus that causes smallpox is immune to that virus but not to other pathogens, such as the virus that causes mumps. Medical research has taken advantage of this specific immunity to develop vaccines against a variety of pathogens. For example, as of 1980, smallpox, once a major scourge of humankind, has been completely eradicated.

In this chapter, we will first discuss the development of the immune system. Then we will describe what is known about lymphocytes, the major cells of the immune system, and how they and their secreted products account for the specificity and memory of the immune response. Finally, we will describe how lymphocytes interact with one another and with other cells to regulate the immune response and to distinguish between self and foreign substances. As we become acquainted with the field of immunology, we will also see that it has contributed much to our understanding of cell biology, providing model systems of gene regulation, cell-cell communication, and cell-surface interactions that lead to proliferation and differentiation.

The Immune Response

The immune system recognizes and eliminates foreign pathogens or substances by a process known as the **immune response.** A substance capable of inducing an immune response is called an **immunogen** or, more commonly, an **antigen.** The most common antigens are foreign proteins and polysaccharides, although the immune system of an organism may also respond to substances produced by the organism itself (self-antigens), leading to an **autoimmune disease.**

The immune response is directed against small, discrete parts of an antigen called **antigenic determinants** or **epitopes.** An epitope is made up of about five to ten amino acid or sugar residues, so the number of epitopes on an antigen is a function of its molecular size and chemical complexity. Most antigens have multiple epitopes and therefore elicit a complex immune response. Some epitopes of an antigen may be *immunodominant*, which means that they are more effective than others in eliciting an immune response, so they dominate the response.

Characteristics of the Immune Response

When pathogens enter the body, they are carried by the lymphatic and circulatory systems to specialized *lymphoid tissues*. There they encounter lymphocytes that recognize and respond to foreign antigens, culminating in the elimination of the invaders by a number of mechanisms. Later, if the same pathogen is encountered by the immune system, the response is much faster and more intense, tending to prevent full-blown infection. This *adaptive immune system* exhibits four fundamental characteristics: diversity, specificity, memory, and self-nonself recognition.

Diversity is evident in the immune system's capacity to respond to millions of unique antigens. Its *specificity* enables the immune system to distinguish between antigens that are very similar to each other, such as proteins that differ by only a single amino acid. The immune system exhibits *memory* for previously encountered antigens; it can "remember" whether or not it has encountered a particular antigen before. If it has, it produces a secondary response, which is more efficient than the primary response in detecting and eliminating the pathogen. And the immune system is capable of *self-nonself recognition,* the ability to recognize which macromolecules are foreign and which are not. This capability is necessary to prevent the immune response from unleashing its destructive forces on the very organism it serves to protect. If it fails in this task, distinctive autoimmune reactions may occur, possibly leading to autoimmune disease.

These four characteristics set the immune system apart from all other systems of the body. Other systems may share some of these properties, but not all four. For example, the nervous system also displays specific memory, but it cannot distinguish between self and nonself molecules.

Types of Immune Responses

There are two major types of immune responses: humoral and cell-mediated. The **humoral immune response** is mediated by special antigen-binding proteins called **antibodies** that circulate in the bloodstream. In fact, *humoral* comes from the Latin word *humor* (meaning "body fluid"), thus emphasizing that this type of immunity can be transferred from an immune to a nonimmune individual by the admin-istration of serum or plasma (cell-free portions of the blood). When antibodies come in contact with antigens that they recognize, they bind to and trap them, promoting their destruction and removal. The **cell-mediated immune response,** on the other hand, is produced by direct contact between certain lymphocytes and foreign antigens on the surface of self cells, such as viral proteins expressed on the surface of infected cells in the body. Thus, in a viral infection, for example, cell-mediated responses can result in the death of virus-infected cells, thereby stopping further production of viruses, while any escaping viruses may be trapped by antibodies in the bloodstream.

The Cellular Basis of the Immune Response

Both the humoral and the cellular immune responses are mediated by white blood cells called **lymphocytes** that circulate in the blood and the *lymph.* (Lymph is the colorless fluid in the lymphatic system, which consists of vessels that connect the lymphoid organs with one another.) Lymphocytes are also found in specialized **lymphoid tissues** distributed throughout the body (Figure 24-1). The total cell mass of the lymphoid tissue is similar to that of the liver or brain and consists of about 2×10^{12} lymphocytes in humans. This network of lymphoid tissue facilitates the capture of foreign intruders regardless of their route of entry.

The two types of immune response are mediated by two distinct classes of lymphocytes. **B lymphocytes,** or **B cells,** are the antibody-producing cells responsible for the humoral response, while **T lymphocytes,** or **T cells,** are responsible for the cell-mediated response. A cardinal feature of T cells, which will be discussed in detail later, is that they respond only to molecules expressed on the surface of body cells. Therefore, cellular immunity is effective against pathogens that gain access to cells of the body but is ineffective against extracellular microorganisms. On the other hand, antibodies produced in humoral immune responses can attack and eliminate extracellular pathogens. Thus, the two branches of the immune system complement each other. And, as you will see, the B cell and T cell responses themselves are often dependent upon each other.

Lymphocyte Development

Both B cells and T cells develop from **hematopoietic stem cells** located in the hematopoietic ("blood-forming") tissue—mainly the liver in embryos and the bone marrow in adult animals (Figure 24-2, page 787). Hematopoietic stem cells also give rise to all other types of blood cells. These stem cells are *pluripotent*, which means that they can give rise to distinctly different sets of progeny cells, depending on the tissue in which the differentiation occurs. Given a particular *microenvironment*, stem cells become committed to a specific developmental pathway. These committed cells

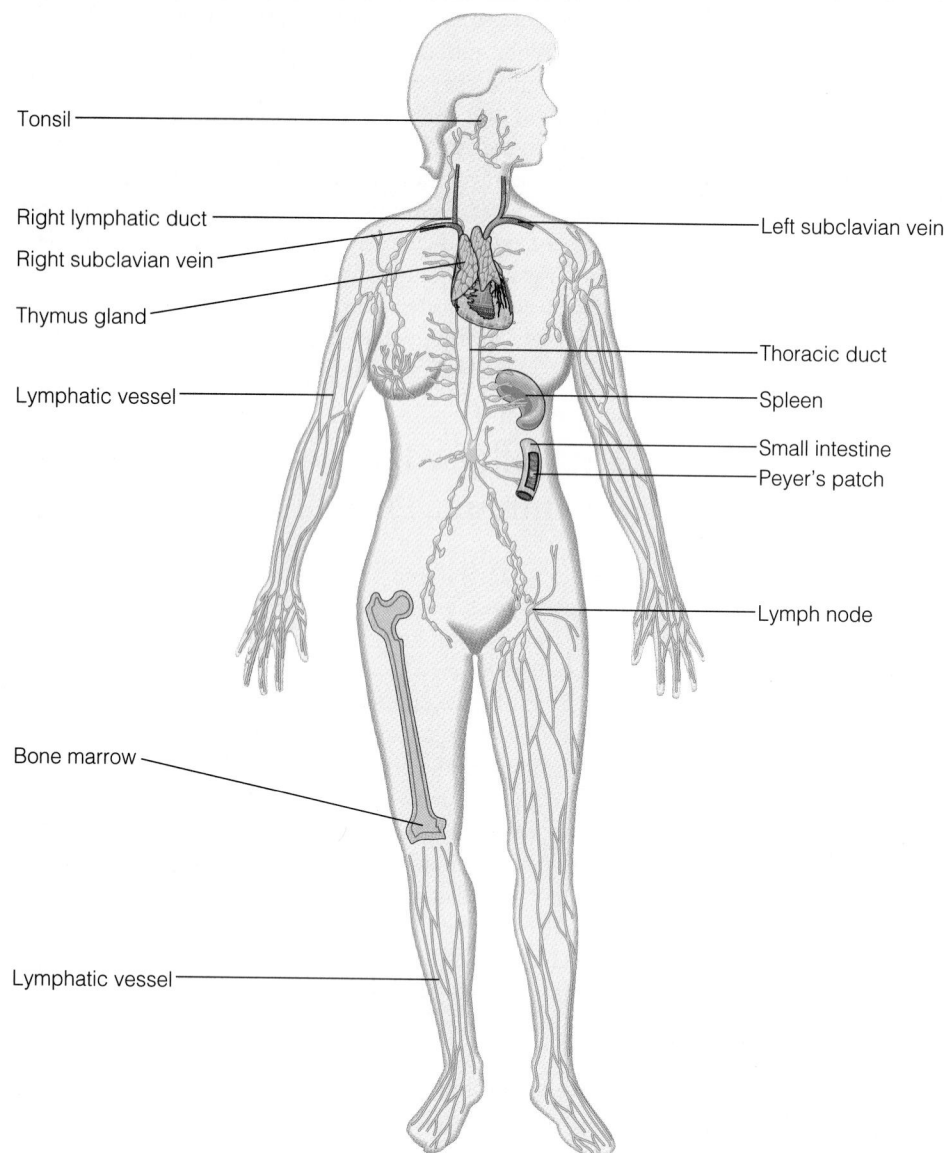

Figure 24-1   The Human Lymphoid System.   The lymphoid system consists of central lymphoid tissues (dark green) and peripheral lymphoid tissues (light green). The central lymphoid tissues are the thymus and the bone marrow (in the long bones, only one of which is shown). The peripheral lymphoid tissues include the lymphatic vessels and lymph nodes, spleen, tonsils, and the Peyer's patches of the small intestine (also the adenoids and appendix, not shown). The lymphatic vessels collect lymphocytes and antibody molecules from the tissues and lymph nodes and return them to the bloodstream at the left subclavian vein.

Tonsil

Right lymphatic duct

Right subclavian vein

Thymus gland

Lymphatic vessel

Left subclavian vein

Thoracic duct

Spleen

Small intestine

Peyer's patch

Lymph node

Bone marrow

Lymphatic vessel

express certain surface receptors, and specific soluble or cell-bound factors then trigger the cells to differentiate further.

For lymphocytes to mature fully, progenitor cells must migrate from the **hematopoietic tissue** to the **central lymphoid tissues,** where they encounter additional cell-bound and soluble factors that induce further differentiation. The central lymphoid tissues are the *thymus,* the *bone marrow,* and, in birds, the *bursa of Fabricius,* an intestine-associated organ. Some lymphoid progenitor cells migrate by way of the blood to the thymus, where they proliferate and differentiate into mature T (for thymus-derived) lymphocytes. In birds, other progenitor cells migrate to the bursa of Fabricius, where they develop into mature B (for bursa-derived) lymphocytes. In mammals, which have no bursa of Fabricius, an as-yet-undefined compartment of the bone marrow is the locus of B cell maturation.

Mature B cells and T cells migrate via the bloodstream to **peripheral lymphoid tissues** such as the *spleen, lymph nodes, tonsils, adenoids, Peyer's patches,* and *appendix.* The B cells seek out specific regions of these tissues called *primary follicles,* while T cells are found principally in regions of the tissues called the *diffuse cortex.* Here, these mature T and B lymphocytes are poised to react with foreign antigens that enter the lymphoid tissues by way of the lymphatic system or bloodstream. Antigen-activated T cells help induce responses in other cells and kill infected cells, while antigen-activated B lymphocytes differentiate into **plasma cells,** which secrete antibodies.

From studies of animals with their central lymphoid tissues removed or absent, it is clear that thymus-derived T cells and bursa-derived B cells have different functions. Removal of the bursa of Fabricius from a newly hatched chick greatly impairs the chick's ability to produce antibodies but has no effect on its cell-mediated responses. On the other hand, removal of the thymus from a newborn animal profoundly impairs the animal's cell-mediated immune re-

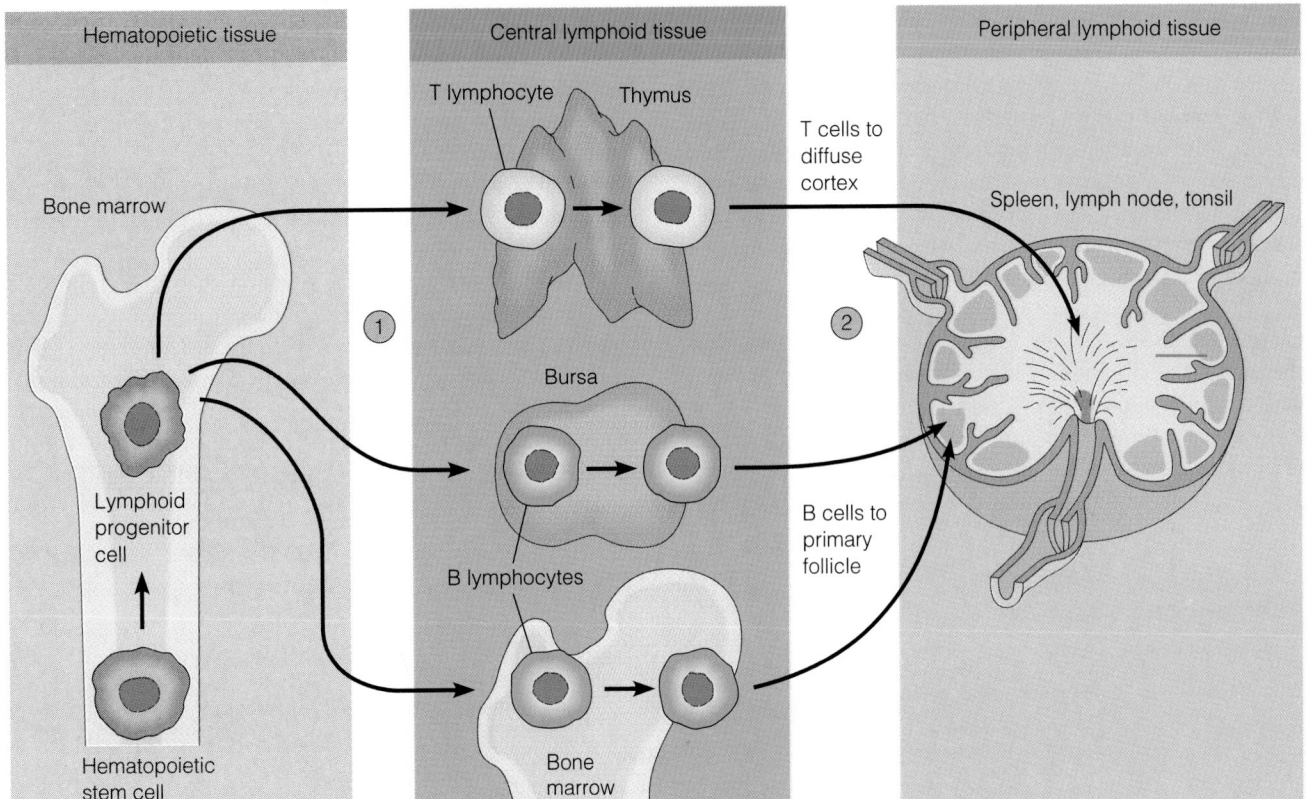

Figure 24-2 Lymphocyte Development. Lymphocytes develop from hematopoietic stem cells into lymphoid progenitor cells in the hematopoietic tissue of the bone marrow. ① Progenitor cells migrate from the bone marrow to the central lymphoid tissue (dark green), where they mature into either T cells (in the thymus) or B cells (in the bone mar-row in mammals or the bursa of Fabricius in birds). ② The T cells and B cells then migrate to the peripheral lymphoid tissue (light green), where they reside mainly in the diffuse cortex (T cells) or the primary follicles (B cells).

---

sponse but has a less dramatic effect on its antibody responses. Moreover, the peripheral lymphoid tissues of an animal that has had its thymus removed at birth have no lymphocytes in the diffuse cortical regions where T cells normally reside. The critical role of the thymus in cell-mediated responses is also clear in children with *DiGeorge syndrome*, a congenital disease characterized by the absence of a thymus. These children lack cell-mediated immunity and are therefore highly susceptible to viral, fungal, and protozoal infections.

Clonal Selection

The most remarkable feature of the immune system is its ability to respond specifically to an enormous variety of antigens. What mechanism could possibly explain such tremendous diversity? A milestone in our understanding of the process was the exposition of the *clonal selection theory* in the 1950s by N. K. Jerne, F. M. Burnet, and others. Clonal selection is amply supported by compelling experimental evidence and is generally accepted by the scientific community. Before considering the theory, however, we must con-

sider the nature of antigen receptors of B lymphocytes and T lymphocytes.

Antigen Receptors. Every lymphocyte carries on its surface about 10^5 membrane-bound molecules that serve as specific **antigen receptors.** The antigen receptors of B cells are a membrane-bound version of antibodies, also called **immunoglobulins (Ig).** Those of T cells, the so-called **T cell receptors (TCR),** are distinct from immunoglobulins in that they are specially designed to bind foreign antigen on self cells. All the receptors on a given lymphocyte are identical in structure and, therefore, in their ability to bind a particular antigen. This commitment to the expression of a particular antigen receptor is acquired by the lymphocyte during its development without any exposure to antigen. The molecular mechanism is described later in the chapter.

Thus, every mature lymphocyte is committed to express a homogeneous set of membrane-bound receptors that have a particular specificity for antigen. The capacity of the immune system to respond to an almost infinite variety of antigens depends on the presence of a diversity of lymphocytes, each having different receptor specificities. In the human body, for example, there are about 1×10^{12} B lymphocytes, representing about 10^8–10^9 different receptor specificities. Because most antigens have many epitopes and a given epitope can be recognized by more than one lymphocyte, the number of lymphocytes that can respond to a given antigen is much larger than the number of cells possessing a certain antigen receptor. In other words, the response to even a single epitope is heterogeneous, involving a number of different lymphocytes.

Clonal Selection. A key feature of the **clonal selection theory** is that the binding of an antigen to a lymphocyte with complementary receptors on its surface selectively activates that lymphocyte to proliferate and differentiate. By binding to the receptor, the antigen *selects* from a very large pool of lymphocytes the set capable of recognizing its epitopes. *Activation* results in clones of lymphocytes, the members of each clone being derived from the same ancestor and therefore all having the same receptor for antigen. Thus, the immune system can be viewed as a vast number of clones of B and T cells, with each clone descended from a single cell and therefore committed to the synthesis of the same surface receptor as the original cell.

According to the clonal selection theory, lymphocytes display two phases of differentiation: antigen-independent differentiation and antigen-dependent differentiation (Figure 24-3). During **antigen-independent differentiation**, hematopoietic stem cells give rise to mature lymphocytes, each bearing surface receptors capable of reacting with a particular epitope or, more precisely, with a small set of

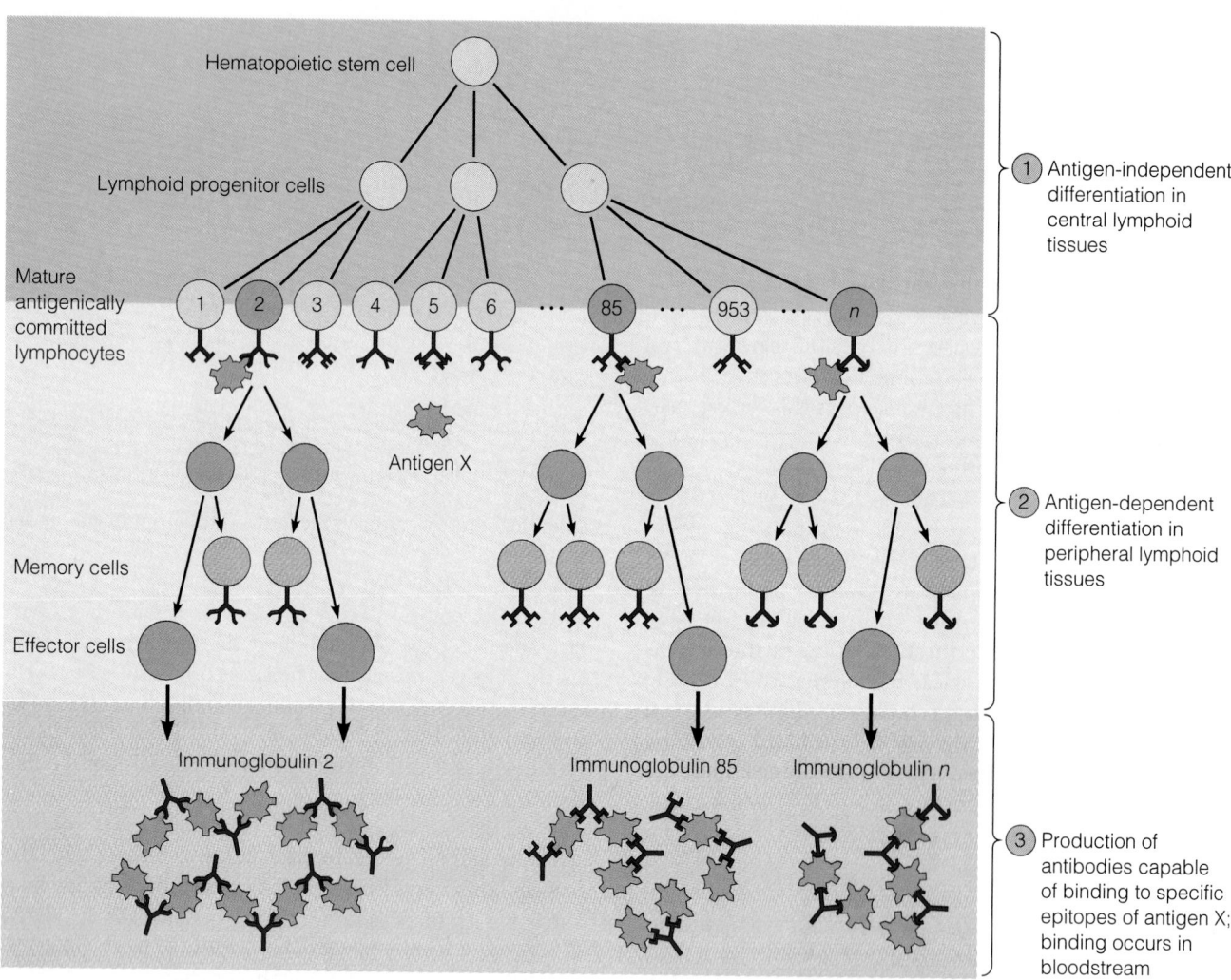

Figure 24-3   Formation of Clones and Their Selection by Antigen.   ① Hematopoietic stem cells undergo antigen-independent differentiation into lymphoid progenitor cells, each of which gives rise to a mature cell committed to the synthesis of one species of surface receptor capable of reacting with a specific antigen. Of the millions of cells with different specificities that can be generated, only a limited number are shown. ② Binding of an antigen (blue) to the surface of specific clones of cells initiates antigen-dependent differentiation, leading to the proliferation of clones of effector cells and clones of memory cells. In the case shown, three B lymphocytes are committed to the production of a surface receptor that recognizes a specific epitope of antigen X, so these cells proliferate and differentiate into effector cells that ③ produce antibodies (immunoglobulins) capable of recognizing and binding to the same epitope of antigen X as they encounter it in the bloodstream.

structurally related epitopes. A few of these mature *antigenically committed* cells are numbered in Figure 24-3; actually, there are millions of them. **Antigen-dependent differentiation** results when a particular antigen comes into contact with lymphocytes that carry surface receptors complementary to one of its epitopes. During this phase, antigen X is shown interacting with cells of three different B cell clones, each specific for a different epitope on the antigen. The cells to which the antigen binds are selectively stimulated to proliferate and to differentiate into functionally active **effector cells.**

As we noted earlier, the effector cell that arises from terminal differentiation of the B lymphocyte is the plasma cell, which secretes antibodies with the same antigen-recognition specificity as its cell-surface receptors. The plasma cell is larger and has a higher ratio of cytoplasm to nucleus than the B lymphocyte from which it originated. Its cytoplasm is loaded with rough ER and the cell is, in essence, an antibody-secreting factory. Indeed, it is estimated that a plasma cell can secrete as many as 2000 antibodies per second.

There are three defined populations of T lymphocytes, each of which is subject to antigen-dependent differentiation. Activated **cytotoxic T (T_c) cells** are responsible for killing cells of the body that are synthesizing foreign antigens. Activated **helper T (T_h) cells** are so named because they help induce other cells of the immune system to proliferate and differentiate. Included among the cells helped by T_h lymphocytes are T_c cells and B cells. **Suppressor T (T_s) cells** are thought to be involved in down-regulating the immune response. However, while it is clear that some T cells can mediate suppression, no distinct T_s subpopulation has yet been identified. It is likely that suppression is mediated by already defined T cell populations, and that such an effect depends upon the state of the T cells involved.

Immunological Memory

An important feature of the immune system is its ability to "remember." If you have ever had a viral disease such as chicken pox, you know that you are now immune to that particular disease and are unlikely to get it again. Somehow, your immune system remembers your initial encounter with that virus and responds accordingly. This phenomenon, called **immunological memory,** can be explained in terms of clonal selection.

Figure 24-4 illustrates the time course for the immune response of an animal that is injected for the first time with antigen X at day 0. The **primary immune response** appears after a lag of about 7–10 days, rises rapidly, reaches a peak, and then falls again gradually. If the animal encounters the same antigen at some later time (day 28, in the example shown in Figure 24-4), the **secondary immune response** that it elicits is more rapid, of greater magnitude, and lasts longer. Clearly, the animal's immune system has remembered antigen X and responds differently in the second encounter. Moreover, the memory is specific for antigen X, be-

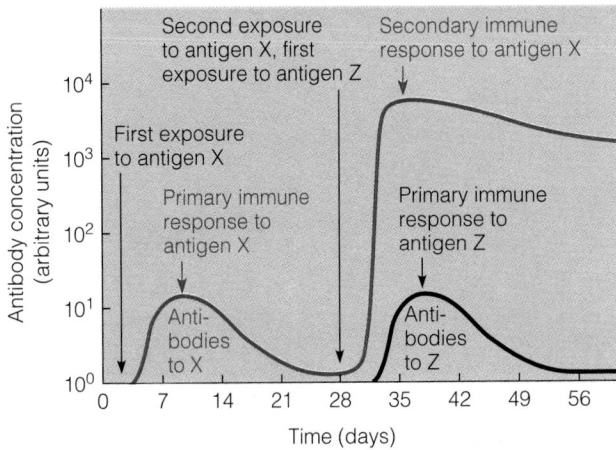

Figure 24-4   Primary and Secondary Immune Responses. Initial exposure to antigen X (at day 0) elicits a primary immune response with a lag phase, a rapid rise in antibody concentration, and then a gradual decrease. A second encounter with the same antigen (at day 28) results in a secondary immune response that is faster, stronger, and longer-lasting than the primary response. The immune response to antigen Z is included at day 28 to demonstrate the specificity of immunological memory for antigen X.

cause a different antigen (e.g., antigen Z) injected at day 28 elicits a primary, not a secondary, immune response. The data in Figure 24-4 represent serum antibody levels and thus refer to a B cell response, but the same scenario applies to T cell responses.

The initial encounter with antigen causes specific B and T cells to proliferate and differentiate. The progeny lymphocytes include not only the effector cells that produce the primary immune response but also large clones of memory B and T cells (Figure 24-5). **Memory cells** have the ability to produce both effector cells and more memory cells when they are stimulated by the same antigen later on. Effector cells have a short life span (a few days, usually), but memory cells survive in the spleen and lymph nodes much longer—decades, in the case of humans. If enough memory cells are produced during initial exposure to an antigen, they may persist for the lifetime of the individual, conferring permanent immunity.

Memory cells thus have three important properties:

1.   They are produced along with effector cells as a result of lymphocyte proliferation in response to encounter with antigen.

2.   They have a much longer life span than effector cells.

3.   Because they are more numerous than the lymphocytes from which they are derived, the secondary and subsequent responses exhibit faster kinetics and greater magnitude than the primary immune response.

Therefore, the difference between primary and secondary immune responses is a function of the number of lymphocytes poised to respond to a given antigen.

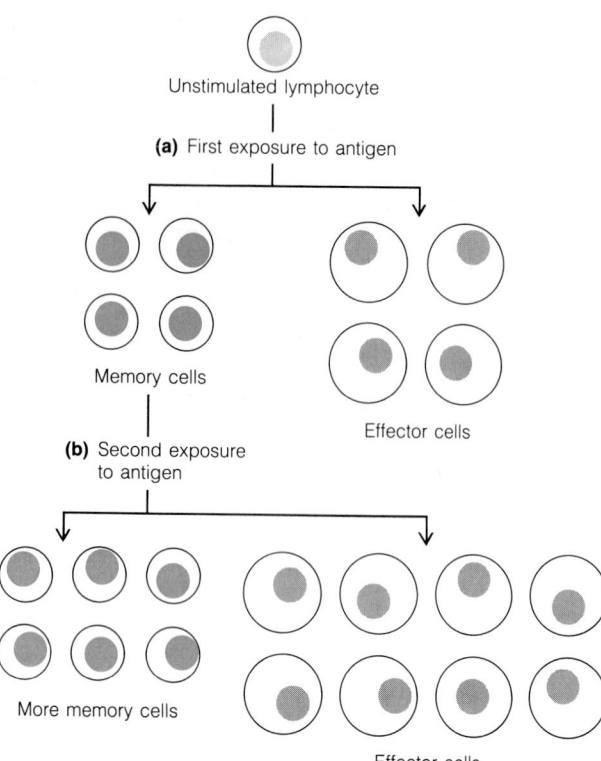

Figure 24-5 The Production of Memory Cells. **(a)** Initial stimulation of a mature resting lymphocyte by its first exposure to antigen results in the proliferation of both effector cells (gray) and memory cells (pink). The effector cells produce the primary immune response and are short-lived. The memory cells last much longer. **(b)** Upon another encounter with the same antigen, the memory cells proliferate, generating still more memory cells and effector cells.

Telling the Players Apart: Differentiation Markers

All lymphocytes, whether B cells or T cells, look alike in the ordinary light microscope. Fortunately for immunologists, however, different lymphocyte types can be identified on the basis of differential expression of certain cell-surface proteins, or **differentiation markers.**

B lymphocytes express surface immunoglobulins (their antigen receptors) and can be distinguished from T lymphocytes on that basis. Similarly, T cells can be identified by the presence of their surface receptor for antigen, the TCR (for "T cell receptor"). Along with the TCR, all T cells express a set of polypeptide chains called the **CD3 complex.** In addition, T_h cells can be distinguished from T_c cells by the expression of the **CD4** or **CD8 glycoprotein.** Typically, T_h cells bear the CD4 marker, while T_c cells have the CD8 marker. Thus T_h cells are phenotypically TCR+, CD3+, CD4+, CD8−, while the T_c population is TCR+, CD3+, CD4−, CD8+.

It is noteworthy that the CD4 glycoprotein has been shown to be a receptor for HIV, the causative agent in AIDS. The virus gains entry into the T cell via the CD4 glycopro-tein, replicates within the cell, and destroys it. This, in part, explains why the CD4+ T cells, which as "helpers" play a central role in the immune response (see Figure 24-8), become depleted in HIV infection, resulting in severe immunodeficiency.

The surface of a lymphocyte actually has many differentiation markers, particular gene products that are expressed only at characteristic stages during lymphocyte development. Some of these glycoproteins function as receptors for molecules on other cells with which they interact, or as receptors for the hormonelike growth factors (cytokines, discussed shortly) that are necessary during a particular stage of differentiation. The importance of these receptors can be appreciated when we consider the complex process whereby progenitor stem cells differentiate into mature functional cells. As lymphoid cells migrate from hematopoietic tissue into central lymphoid tissue and finally into peripheral lymphoid tissue, they encounter several different microenvironments in succession (see Figure 24-2). How do these cells sense these microenvironments, and how do they respond appropriately?

The answers may involve specialized sessile (nonmotile) cells present in each of these different microenvironments. Unlike the motile lymphoid cells, sessile cells are permanent residents of a specific lymphoid tissue. Examples of sessile cells include *stromal cells* of the bone marrow, *epithelial cells* of the thymus, and *dendritic cells* of the lymph nodes. The nonmotile cells in a particular tissue stimulate further differentiation of lymphoid cells either by directly contacting them or by releasing soluble factors. Such cell-cell interactions are as yet poorly defined. In any case, the response of the lymphocyte depends on the presence of cell-surface molecules that serve as receptors for these various signals.

When developing lymphocytes leave the bone marrow or thymus and enter the bloodstream, they must be able to find their way to specific lymphoid tissues, such as the lymph nodes or Peyer's patches. Once again, receptor-mediated cell-cell interactions are involved. The lymphocytes have receptors on their surfaces that target them for preferential entry into particular lymphoid tissues. These receptors bind to complementary structures on specialized endothelial cells that line the tiny veins, or venules, of that tissue. The lymphocytes then migrate across these endothelial walls and enter the lymphoid tissue.

Lymphocyte Activation Pathways

The immune response to an antigen consists of a complex sequence of events involving the interaction of different cells by direct contact or via soluble mediators called **cytokines** (also known as *lymphokines* or *interleukins*) that promote proliferation and/or differentiation. In order to understand how the process works, we must first consider the gene products of the major histocompatibility complex, which play a vital role in antigen recognition by T lymphocytes.

The Major Histocompatibility Complex

As mentioned earlier, T cells respond to antigens only when the antigens are present on the surfaces of other self cells. Indeed, the antigen is *presented* on the surface of cells in association with certain self molecules. These molecules therefore play a critical role in the recognition of foreign antigens by T cells.

The surface molecules involved in T cell recognition were actually discovered from studies of skin grafts. When skin is removed from one individual and grafted onto another individual of either the same species (*allograft*) or a different species (*xenograft*), the grafted skin is usually recognized as foreign and rejected by the immune system of the recipient. This phenomenon is called **graft rejection.** Transplantation of an organ such as a kidney or a heart usually leads to the same result unless the donor and the recipient are genetically identical. In fact, it was the immunological rejection of transplanted tissues, skin grafts in particular, that provided a major driving force for the development of the field of immunology.

Graft rejection is mediated principally by T_c cells that respond to and react against the foreign antigens on the surface of the cells in the grafted tissue or organ. Thus, the cell-surface antigens that elicit this response are called **transplantation** (or **histocompatibility**) **antigens.** (The prefix *histo-* means "tissue.") The most important of these antigens are encoded by a set of genes called the **major histocompatibility complex (MHC).** In humans, the MHC antigens are called **human leukocyte-associated (HLA) antigens** and are encoded by the **HLA complex** on chromosome 6. In mice, they are called **histocompatibility (H-2) antigens** and are encoded by genes in the **H-2 complex** on chromosome 17. Figure 24-6 illustrates the several loci of the HLA and H-2 complexes.

MHC genes are expressed in a *codominant* manner, meaning that all the MHC genes inherited from both parents are expressed in their progeny. In a given species, particularly in outbred populations such as humans, there is great variation in MHC genes. In other words, the MHC genetic loci show extreme **polymorphism** (many forms) such that there are numerous alternative, functional alleles for these loci in the gene pool of the species. In fact, the genes of the MHC are the most polymorphic human genes known. It is therefore quite unlikely that any two unrelated individuals will have the same set of MHC molecules. Since even a mismatch of a single MHC antigen between donor and recipient can result in transplant rejection, the difficulties accompanying organ transplantation are obvious. Graft rejection, then, is a consequence of the gene products of the MHC, which normally serve to display foreign antigen to T cells, as we will describe in detail later.

Classes of MHC Antigens

The MHC genes encode two families of cell surface glycoproteins called **class I MHC antigens** and **class II MHC antigens.** In humans, the loci HLA-A, HLA-B, and HLA-C code for the class I antigens, while a number of loci, including HLA-DP, HLA-DQ, and HLA-DR encode class II antigens (Figure 24-6a). In mice, the H-2K, H-2D, and H-2L loci code for class I antigens, and the IA and IE genes encode class II antigens (Figure 24-6b). Cytotoxic T cells respond primarily to foreign antigens presented by class I MHC molecules, whereas helper T cells respond mainly to those presented by class II MHC molecules. In each case, the necessary cell-cell interactions are greatly enhanced by interaction between class I MHC molecules and CD8 (on the T_c cell surface), and class II MHC molecules and CD4 (on the T_h cell surface).

Class I Molecules.   There are three class I glycoproteins, each encoded by a separate MHC locus. The gene products are all integral membrane proteins ranging in molecular

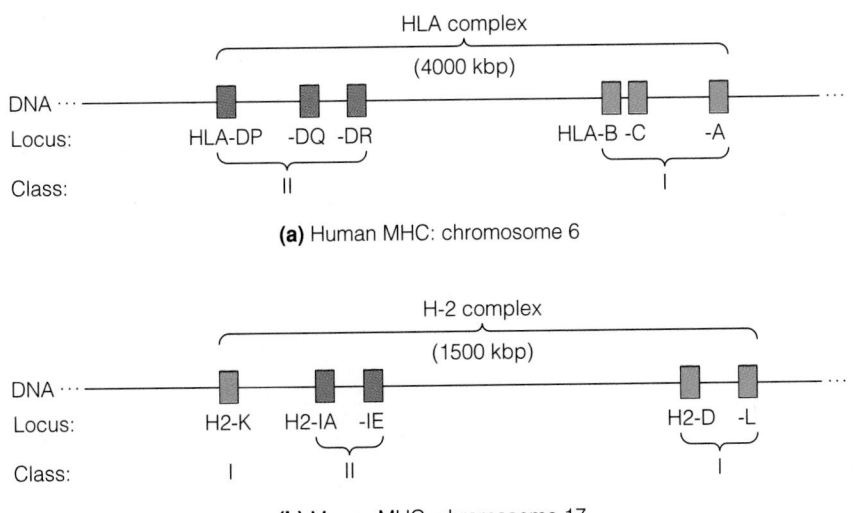

(a) Human MHC: chromosome 6

(b) Mouse MHC: chromosome 17

Figure 24-6   The Major Histocompatibility Complex (MHC) of the Human and Mouse Genomes.   **(a)** The MHC of the human genome is the HLA complex on chromosome 6. **(b)** For the mouse genome, the MHC is the H-2 complex on chromosome 17. Each complex contains multiple loci coding for class I MHC glycoproteins (red) and for class II MHC glycoproteins (blue). Not all of the identified MHC genes are shown.

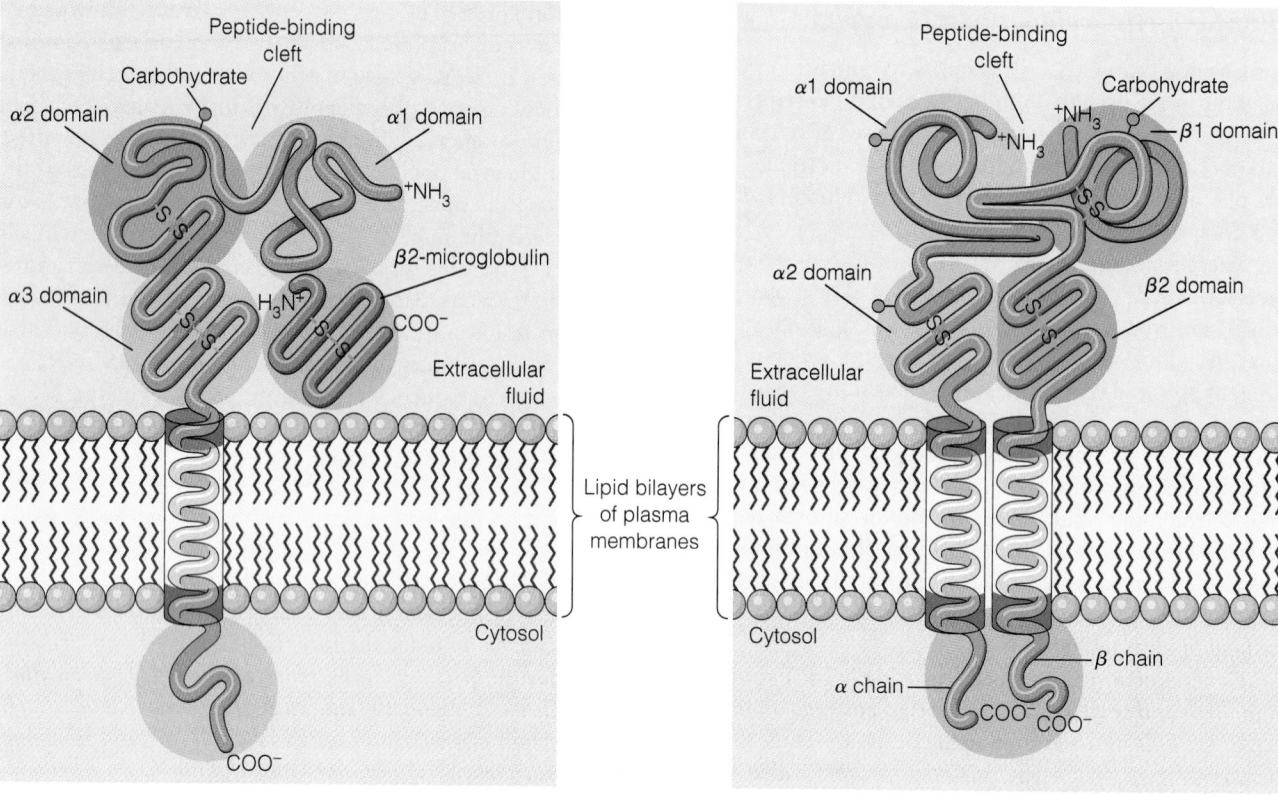

(a) Class I MHC glycoprotein

(b) Class II MHC glycoprotein

Figure 24-7   Structure and Orientation of MHC Glycoproteins.   **(a)** Class I MHC glycoprotein molecules are integral membrane proteins that associate noncovalently with β2-microglobulin on the external side of the membrane. Class I glycoproteins are found on the surface of virtually all nucleated cells, where they account for up to 1% of the total protein of the plasma membrane. **(b)** Class II MHC glycoprotein molecules are integral membrane proteins that consist of an α polypeptide noncovalently associated with a β polypeptide. Class II glycoproteins are found on the surface of B lymphocytes, macrophages, and dendritic cells.

weight from 40,000 to 45,000. As shown in Figure 24-7a, class I MHC molecules have a short hydrophilic cytoplasmic tail and a hydrophobic transmembrane segment. The extracellular portion of class I proteins consists of three domains with two intrachain disulfide bridges.

The exterior domain closest to the membrane associates noncovalently with a small, non-MHC-encoded protein called **β2-microglobulin,** which also has an intrachain disulfide bond. The outer two domains interact to form a groove, or cleft, on the top surface of the molecule. This cleft is likely the binding site for foreign antigen. Indeed, when the structure of the class I molecule was determined by X-ray crystallographic analysis, a small peptide was found in the cleft. Class I MHC antigens are present on virtually all nucleated cells of the body and are involved in presenting peptide antigen to cytotoxic (CD8+) T cells.

Class II Molecules.   Class II MHC molecules are also glycosylated integral membranes proteins (Figure 24-7b). Each class II molecule consists of two MHC-encoded transmembrane chains, an α chain with a molecular weight of about 33,000 and a β chain with a molecular weight of about 28,000. Each chain consists of a short cytoplasmic tail, a hydrophobic transmembrane segment, and two extracellular domains containing intrachain disulfide bonds, one in the α chain (α2 domain) and two in the β chain (β1 and β2 domains). The two noncovalently associated polypeptides fold together to form a cleft at the top surface of the molecule, again a binding site for foreign peptide.

While class I MHC molecules are expressed on all nucleated cells of the body, class II MHC molecules are less widely distributed. Class II MHC molecules are found on macrophages, B cells, dendritic cells, thymic epithelial cells, and, in most species, on activated T cells. Thus, class II molecules, expressed on specialized **antigen-presenting cells (APC),** present antigen to helper (CD4+) T cells.

Both class I and class II molecules contain domains having amino acid sequence homology with certain immunoglobulin domains and thus belong to the so-called *immunoglobulin gene superfamily.* In fact, many cell-surface proteins involved in cell-cell recognition belong to this family, suggesting that they are derived from a common ancestral gene.

Antigen Processing and Presentation

To elicit an immunological response, an antigen must first undergo *processing* and transport to the cell surface. Exogenous and endogenous antigens are processed by separate pathways, each of which is described here.

Processing of Exogenous Antigen. Responses to most antigens require processing of the antigen by APCs. The reason for this is that helper T cells, which regulate the immune response and are essential for its initiation, recognize antigens only in the context of MHC class II proteins on the surface of other cells. Consequently, the first steps in the immune response involve the capture and processing of foreign antigen by an APC and the presentation of processed antigen fragments to T_h cells (Figure 24-8, step 1). Because the antigen originates outside the APC, it is called **exogenous antigen,** and the route it takes on its way to the surface of the APC is called the *exogenous pathway.* This pathway, then, involves internalization of the antigen by phagocytosis (in

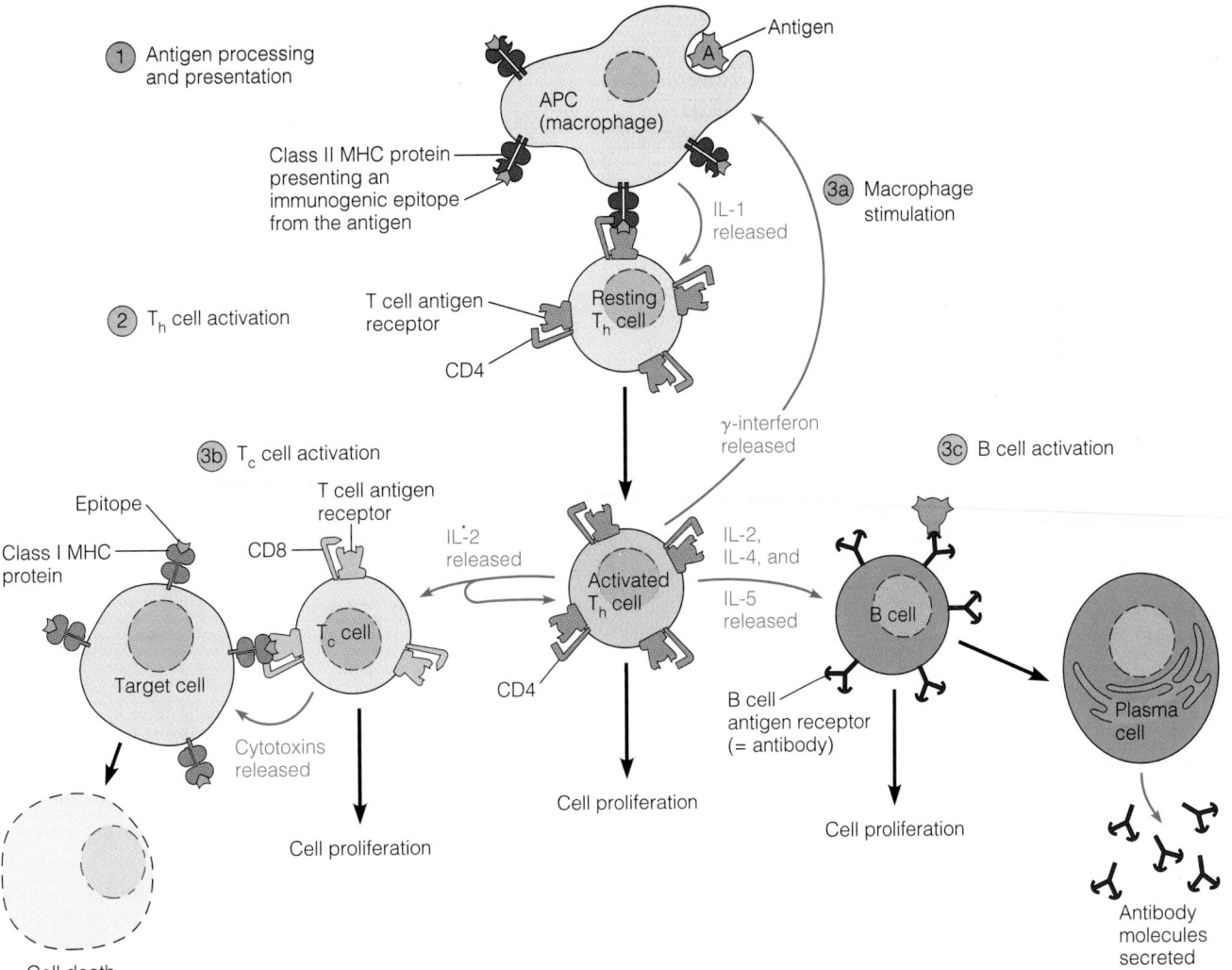

Figure 24-8 Pathways of the Immune Response. The response is initiated by the uptake of antigen by an APC, such as the macrophage shown here. ① *Antigen processing and presentation:* The antigen is partially degraded within the APC, and one or more fragments of it (epitopes) become associated with class II MHC proteins. The epitope-MHC complex is transported to the cell surface, where the epitope is "presented." ② T_h *cell activation:* The epitope-MHC complex can bind to the antigen receptors of T_h cells. This reaction and the cytokine IL-1 released by the APC activate the T_h cell. The activated T_h cell in turn releases IL-2, which stimulates the T_h cell itself to proliferate and to release a number of other cytokines. ③ⓐ *Macrophage stimulation:* Gamma interferon released by T_h cells activates macrophages. ③ⓑ T_c *cell activa-tion:* IL-2 also acts on T_c cells, activating them to kill target cells that present immunogenic epitopes complexed with class I MHC proteins. The cytotoxic effect is caused by cytotoxins released by the activated T_c cells. ③ⓒ *B cell activation:* IL-2, IL-4, and IL-5, released by activated T_h cells, are among the cytokines that activate the B cell to proliferate and to differentiate into antibody-secreting plasma cells. B cells, T_h cells, and T_c cells can be distinguished from one another by certain surface proteins: T_h cells have a surface protein called CD4, T_c cells have the CD8 protein, and B cells have neither. Note that the activation of all three types of lymphocyte requires (1) the action of one or more cytokines and (2) the binding of an antigen (or antigen fragment) to an antigen receptor on the lymphocyte surface.

the case of macrophages) or by receptor-mediated endocytosis (in the case of B cells). Internalization of antigen is followed by its degradation within a cellular compartment called the *endosome*. Some peptide fragments of the original antigen become noncovalently associated with class II molecules that reach the endosome on their way to the cell surface. The class II MHC–peptide complex is then transported to the surface of the APC, where it is accessible to the helper T cell's antigen receptor. Although few cell types express class II MHC proteins, all such types are capable of acting as APCs for helper (CD4+) T cells. The principal antigen-presenting cells are macrophages and B cells.

It may seem curious that B lymphocytes, the precursors of antibody-secreting plasma cells, can present antigens to T helper cells, but their capacity to do so is well documented. Because of its specific receptors (the Ig molecules on its surface), a B cell has a very efficient mechanism for capturing and endocytosing antigen. Moreover, B cells are well designed presenters of exogenous antigen, given their expression of class II MHC molecules. Experimental evidence indicates that B cells are relatively poor activators of naive (previously unactivated) T cells, possibly because such T cells require additional protein mediators (cytokines) that B cells do not adequately supply. However, B cells are more efficient than macrophages in activating memory T cells. Because B cells specific for a particular antigen are rare in an individual who has never encountered that antigen, and naive T cells do not respond well to antigen presented by B cells, it is thought that macrophages play a critical role as APCs in the primary immune response, whereas B cells play a more significant role in secondary responses.

Processing of Endogenous Antigen. Cytotoxic T cells are designed to detect and eliminate "altered self" cells such as virus-infected cells, malignant cells, and grafted cells. T_c cells recognize processed peptide antigens only in association with class I MHC proteins on the surface of such altered cells. Indeed, the wide distribution of class I MHC molecules makes sense; any nucleated cell of the body has the potential of being infected by a virus or becoming malignant. In this case, however, the foreign antigen is synthesized within the cell and is therefore called **endogenous antigen,** and the route by which it is processed and transported is called the *endogenous pathway*. Unlike exogenous antigens, viral or tumor antigens are synthesized along with normal cellular proteins including class I MHC, and peptide fragments of the antigens become associated with class I molecules during assembly in the ER. The class I MHC–peptide complex is then transported to the surface of the altered **target cell** by the exocytic pathway that we encountered in Chapter 9. There it is accessible to the T_c cell's antigen receptor.

An obvious aspect of antigen processing is that T cells do not respond to antigens in their native structural form. Instead, they recognize peptide fragments of the original antigen if, and only if, the peptides are associated with self MHC molecules. T cells that respond to class I–peptide complexes (CD8+ cytotoxic T cells) are said to be **class I-restricted,** while T cells that respond to class II–peptide complexes (CD4+ helper T cells) are **class II-restricted.**

Activation of Helper T Cells

The helper T cell is the conductor of the immunological orchestra, providing signals essential for the activation of macrophages, B cells, and T_c cells, the cells responsible for the elimination of antigens. Activation of T_h cells (Figure 24-8, step 2) occurs early in the immune response and requires at least two signals. One signal is furnished by the binding of the T cell antigen receptors to class II MHC–peptide complexes on an APC. The second signal comes from a cytokine called **interleukin-1 (IL-1),** for which the T_h cell has receptors. Macrophages produce much more IL-1 than do B cells, which may account for the heightened ability of macrophages to activate naive T cells. The two signals induce the T_h cell to express receptors for a cytokine called **interleukin-2 (IL-2)** and to produce IL-2 itself. By binding to IL-2 receptors on the T_h cell, IL-2 stimulates the cell itself to produce other cytokines and to proliferate. IL-2 and other cytokines are, in turn, necessary for full activation of B cells and T_c cells. In addition, as Figure 24-8, step 3a, shows, the helper T cell–derived cytokine γ-**interferon (IFN-γ)** activates macrophages (and exerts potent antiviral activity on other cells). The elaboration of this assortment of cytokines, only some of which are described here, is how the T_h cell orchestrates the effector cells of the immune system. Long-lived memory T cells are also generated in the course of T cell activation. It is not known whether these T cells arise from effector cells themselves or from a separate population of activated helper T cells.

Activation of Cytotoxic T Cells

We have already seen how the activated T_h cell is central to driving the immune response, in part by the production of IL-2 and other cytokines that help activate T_c cells. Thus, T_c cells also require two activation signals (Figure 24-8, step 3b). One signal is provided by specific interaction between its TCR and class I MHC–peptide complexes on a target cell (a virus-infected cell, a tumor cell, or an allograft). The second signal is furnished by helper T cell–derived IL-2. The activated T_c cell kills its target cell by the exocytosis of granules containing lytic enzymes and perforins. *Perforins* act by perforating the target cell membrane, thereby altering the integrity of the membrane and allowing the entry of lytic enzymes. Lysis of the target usually occurs within 15–180 minutes, sometime after the T_c cell has exocytosed its granules and moved on. It is notable that the T_c cell itself is not lysed in the killing process. In fact, a single T_c cell is capable of killing multiple target cells in succession. How the cytotoxic T cell is protected from its own exocytosed cytotoxins is not yet understood.

Table 24-1 Principal Cells of the Immune System

Cell Type	Significant Surface Molecules*	Antigen-Specific	MHC Restriction	Function
B lymphocytes	Ig, class II MHC	Yes	No	Process and present antigen to helper T cells; produce IL-1 and antibody
T lymphocytes				
Helper T cells	TCR, CD3, CD4	Yes	Class II MHC	Produce IL-2, IL-4, IL-5, and other cytokines that help activate other cells
Cytotoxic T cells	TCR, CD3, CD8	Yes	Class I MHC	Produce cytotoxins; kill virus-infected cells, tumor cells, and grafted cells
Macrophages	FcR Class II MHC	No	No	Phagocytose antigen; process and present antigen to helper T cells

*Ig=immunoglobulin. MHC=major histocompatibility complex. FcR=receptor for immunoglobulin. All cells bear Class I MHC.

Activation of B Cells

The production of antibodies requires the activation of B lymphocytes and their differentiation into antibody-secreting plasma cells. The sequence of events in this process may be visualized as follows. While the T_h cell is being activated as described above, relevant B cells are also binding to epitopes on foreign antigens via their antigen receptors, membrane-bound forms of the antibodies they will later secrete (Figure 24-8, step 3c). Antigen binding is followed by endocytosis of the receptor-antigen complex, which appears to serve as a partial activating signal to the B cell. Again, additional signals for full activation are provided by T_h cells in the form of cytokines. These mediators have a short radius of activity, requiring proximity between the relevant antigen-specific T and B cells. Because B cells express class II MHC and can function as APCs, helper T cells directly contact the B cells and can deliver the appropriate cytokines at close range. These include IL-4, which helps drive the cell from the resting stage into the cell cycle; and IL-2, IL-4, IL-5, and IL-6, which together promote proliferation and differentiation of the B cells into plasma cells. B cell activation leads not only to terminally differentiated plasma cells but also to memory B cells (see Figure 24-5). How memory B cells arise is not clear. It has been proposed that the differentiative path taken by the progeny of activated B cells is influenced by the pattern of cytokines they encounter.

Some of the functions and characteristics of the major cell types involved in the immune response are summarized in Table 24-1.

Immunological Tolerance

In a healthy immune system, lymphocytes are activated by foreign but not self antigens; that is, they exhibit **immunological tolerance** to self. There are two mechanisms, both demonstrated experimentally, that are thought to help prevent autoimmune responses. First, many immature T and B lymphocytes with potentially self-reactive receptors may be eliminated during development in the thymus and bone marrow, respectively, by a process termed *clonal deletion*. Thus, in the thymus, for example, a T cell bearing receptors with high affinity for self is signaled to die. Indeed, it is estimated that only about 1% of developing T cells survive to maturity and leave the thymus. Second, mature lymphocytes specific for self components in the periphery may be functionally inactivated, in a process called *clonal anergy*. By this mechanism, a lymphocyte with specificity for a self antigen is not eliminated but is nonetheless unresponsive to that antigen. Together these mechanisms help make the immune system tolerant to self and thus prevent autoimmune reactions.

The Structure and Function of Antibodies

The humoral immune system responds to myriad different antigens by making large numbers (10^8–10^9) of highly specific antibodies. Antibodies account for about 20% of the total plasma (noncellular) protein in the blood and therefore comprise one of the major classes of blood proteins. The capacity to produce a specific antibody is acquired by every B lymphocyte during its development, without prior exposure to antigen.

One of the great mysteries of immunology has been how the immune system makes such large numbers of different antibody molecules. The molecular mechanisms that the immune system uses to meet this challenge are now well understood. But before we can appreciate the elegance of antibody diversity, we must first acquaint ourselves with the structure and function of antibodies.

The Antibody Molecule

Every antibody molecule has two functions: to recognize and bind to an antigen, and to assist in the disposal of that antigen. This duality of function is reflected in antibody structure. Every antibody molecule consists of a set of domains that participate in one of these two functions. The domains of the antibody that bind antigen must display diversity, because antibodies must be able to recognize an almost unlimited number of antigens. These domains differ in amino acid sequence from one antibody to another and are thus called **variable (V) domains.** In contrast, only a few mechanisms are involved in antigen elimination, so only a few different kinds of domains are involved in these functions. These are the **constant (C) domains.**

An antibody is a Y-shaped molecule with a single "leg" and two identical *antigen-binding sites* on its "arms" (Figure 24-9). The antigen-binding sites are formed by the variable domains, whereas the *effector site*, that part of the molecule that facilitates disposal of bound antigen, is within the constant domains. Typically, the arms are called *Fabs* and the leg is called the *Fc*, terms that come from the initial elucidation of antibody structure, which involved studies of fragmented molecules. Antibody fragments found to bind antigen were named Fab (for fragment of antigen binding), and fragments that did not bind to antigen but were easily crystallized were named Fc (for crystallizable fragment). One antibody molecule, then, consists of two Fabs and one Fc connected by flexible *hinge regions.* The hinge regions enable the arms of the molecule to move, thereby accommodating binding to two antigen molecules (or to two antigenic determinants on the same molecule) simultaneously.

The basic antibody molecule consists of four polypeptide chains: two identical **light (L) chains** and two identical **heavy (H) chains,** organized into two H-L pairs (see Figure 24-9). The structure of the molecule can therefore be represented as H_2L_2. The molecular weight of the light chain is about 23,000, while the heavy chain has a molecular weight of about 55,000. Each light chain is linked to one heavy chain by a single disulfide bridge and by noncovalent associations. The heavy chains are glycosylated and are joined to each other by one or more disulfide bonds.

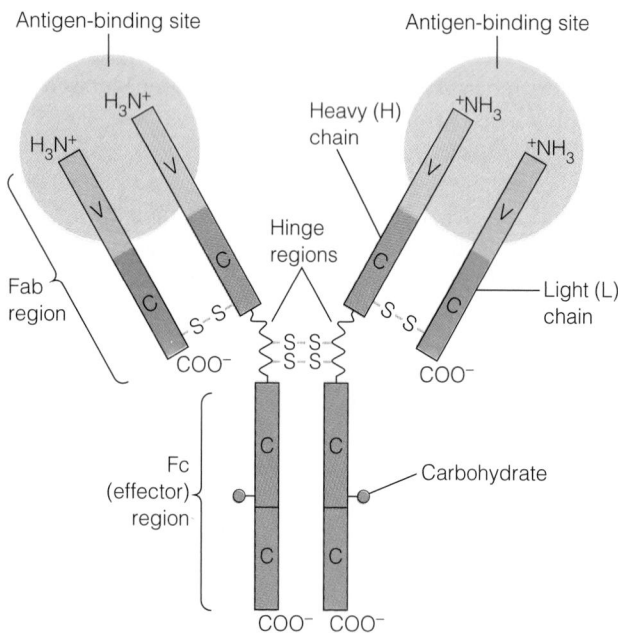

Figure 24-9   Structure of an Immunoglobulin Molecule.
An immunoglobulin (antibody) molecule consists of two identical heavy chains (purple) and two identical light chains (green) associated with one another in a Y-shaped configuration and linked by disulfide bridges (—S—S—). The antigen-binding site at the end of each arm (Fab region) is formed from the N-terminal regions of one light and one heavy chain, whereas the effector site in the leg (Fc region) of the molecule is formed from the C-terminal regions of the heavy chains only. Carbohydrates (oligosaccharides) are attached to the legs of both heavy chains. Also shown are the variable (V) and constant (C) regions (lighter colors) of both light and heavy chains.

Classes of Immunoglobulins

In mammals, there are five major **immunoglobulin classes,** or **isotypes,** each using a particular heavy chain. The five isotypes are IgM, IgD, IgG, IgE, and IgA, and the H chains found in each are μ, δ, γ, ϵ, and α, respectively. There are four distinct types, or subclasses, of IgG and two subclasses of IgA, because of the existence of multiple forms of the γ and α heavy chains. In addition to having their own characteristic H chains, the classes and subclasses of immunoglobulins also differ in the number of disulfide bridges between the H chains.

An antibody molecule of any class can have one of two kinds of light chains, either a κ or a λ chain. The ratio of κ- to λ-containing immunoglobulins varies among species. For example, the κ:λ ratio in humans is about 70:30, whereas in the mouse it is about 95:5. No functional differences between κ and λ have been identified. In general, a given B cell clone produces only one type of light chain and one of the five classes of heavy chains. At some stages in their development, B cells may simultaneously produce more than a single type of heavy chain, however.

Because of the differences in their constant domains, the various isotypes have different biological properties. A good example of a functional difference due to H chain variation is the differential transport of antibodies across membranes. IgA molecules (but not molecules of other classes) are transported across the epithelial lining of the intestine into the mucosa, whereas only IgG molecules can pass across the placenta from the circulatory system of the mother to that of a developing fetus. Specific transport, like most other effector functions of particular antibody classes, is thought to involve binding of the Fc (effector site) of the immunoglobulin to specific **Fc receptors (FcRs)** on the cell surface.

It should be noted that all classes and subclasses of immunoglobulins can be synthesized as either membrane-bound receptors for antigen or as secreted antibodies. The membrane-bound form contains a hydrophobic transmembrane segment to anchor it in the membrane. The secreted form does not contain such a segment. In fact, the secreted form is sufficiently hydrophilic to circulate as a soluble molecule. The difference in solubility is a function of the amino acid sequence near the C-terminal region of the heavy chain. The particular form of heavy chain that is synthesized depends on the stage of the B cell. For example, a mature B cell produces membrane-bound antibody, while a plasma cell makes mostly the secreted form. The molecular mechanism that determines this difference will be discussed later.

The major functional roles of the five classes of immunoglobulins are summarized in Table 24-2. We will discuss each briefly before going on to examine the fine structure of the antibody molecule.

IgM. IgM is the first class of antibody produced during the development of a B cell, although most B cells will later switch to making other classes of antibody. IgM is also the major isotype secreted early in a primary immune response. In fact, low doses of antigen may stimulate IgM production only; higher doses are usually required to stimulate IgG production. Interestingly, IgM is the first class of immunoglobulin to appear in ontogeny (the developing individual) and phylogeny (evolution). Primitive vertebrates, such as sharks,

have only one class of immunoglobulin, and it is IgM. Thus, IgM appears to be the ancestral immunoglobulin from which the other classes descended. IgM is unique among antibody classes in that it is not secreted as a single antibody (monomer) but as a pentamer ring, with five H_2L_2 subunits held together by disulfide bonds. The **IgM pentamer** also contains a polypeptide at its center, called the **J (joining) chain,** which may regulate the extent of polymerization. Because the IgM pentamer has ten antigen-binding sites (although not all are used at one time due to steric hindrance), secreted IgM is very efficient at complexing antigen. Indeed, the IgM pentamer is well designed as the mediator of the primary humoral response, when antibody levels are suboptimal (see Figure 24-4). In addition, when IgM binds to antigen, it can activate a set of serum proteins called the **complement system.** Activated complement has several important properties, including the enhancement of phagocytosis and enzymatic activities that lead to the destruction of bacterial cell membranes or viral envelopes.

IgD. The function of **IgD,** a monomeric antibody, is not yet known. Along with IgM, IgD is a prominent class of antibody on the surface of mature resting B cells, but very few B cells secrete IgD, so it is usually present in the blood only in minute concentrations. Surface IgD molecules may play a role in B cell triggering because B cells do not become re-

---

Table 24-2 Properties of the Five Classes of Human Immunoglobulins

Properties	Isotype				
	IgM	IgD	IgG	IgE	IgA
	J chain				J chain / Secretory component
Formula of monomer	$\mu_2\kappa_2$ or $\mu_2\lambda_2$	$\delta_2\kappa_2$ or $\delta_2\lambda_2$	$\gamma_2\kappa_2$ or $\gamma_2\lambda_2$	$\epsilon_2\kappa_2$ or $\epsilon_2\lambda_2$	$\alpha_2\kappa_2$ or $\alpha_2\lambda_2$
Subclasses of heavy chain	None	None	$\gamma_1, \gamma_2, \gamma_3, \gamma_4$	None	$\alpha 1, \alpha 2$
Oligomeric form	Pentamer	None	None	None	Dimer
Accessory chains	J chain	None	None	None	J chain, secretory component
Serum concentration (mg/mL)	1.0	0.1	12	0.001	3.0
Physiological function	Activates complement	Unknown	Binds to phagocytes; crosses the placenta; activates complement	Binds to mast cells and basophil, leading to degranulation	Crosses epithelial cells

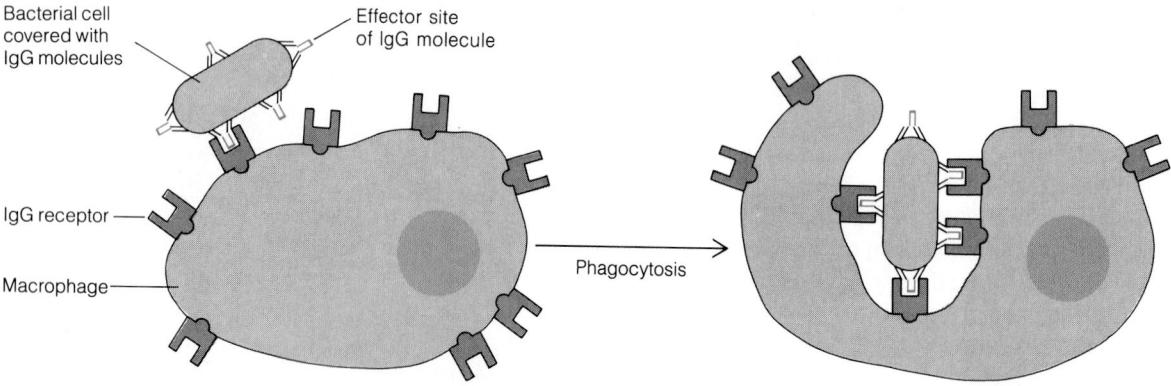

Figure 24-10 The Role of IgG in Phagocytosis of a Bacterial Cell. IgG antibodies bind to epitopes on an invading bacterial cell and deliver the bacterium to a macrophage that has surface receptors (gray) to which the effector site on IgG molecules bind. In this process, the binding of the antibody-coated bacterium to the receptors on the macrophage activates phagocytosis.

sponsive to antigen until they express IgM and IgD together. Immature antigen-unresponsive B cells express only surface IgM.

IgG. IgG is the most abundant antibody produced in late primary and secondary immune responses, and is therefore the major isotype in the blood. It mediates the elimination of antigen to which it is bound in two ways. First, IgG facilitates the destruction of bacterial cells or viruses by phagocytosis. Antigen that is coated with IgG is effectively delivered to macrophages because the cells express IgG receptors specific for Fcγ heavy chain (FcRγ) (Figure 24-10). Second, when IgG is bound to antigen, it activates the complement system, which facilitates antigen disposal.

IgE. Like IgD, IgE is normally present in the blood in low concentrations. IgE is involved in stimulating **mast cells** in connective tissue to release vasoactive amines such as histamine or serotonin that cause dilation and increased permeability of blood vessels, leading to typical symptoms of allergic reactions like hay fever. The IgE effector site has a high affinity for Fc receptors on the surface of mast cells. When adjacent mast cell-bound IgE molecules are cross-linked by antigen, the cells are triggered to exocytose their granule contents. The capacity of IgE to trigger such inflammatory reactions is thought to protect against parasitic infections, although it is not appreciated by hay fever sufferers.

IgA. Although **IgA** is also present in the blood, it is the major class of antibody found in tears, sweat, saliva, milk, and mucus of the intestinal and respiratory tracts. It therefore represents the first line of antibody defense against bacterial and viral antigens. IgA is secreted from plasma cells as a dimer consisting of two H_2L_2 units linked to a single J chain. IgA in the external secretions also contains a polypeptide called **secretory component** that becomes associated with the dimer as a result of its passage across the epithelial cells that line secretory ducts. This process, called *transcyto-*

sis, is initiated by binding of the IgA dimer to receptors on the basal surface of epithelial cells. The IgA-receptor complex is endocytosed, and the IgA is eventually exocytosed into the lumen on the other side of the cell (Figure 24-11). Once transport is complete, the receptor is cleaved, but a

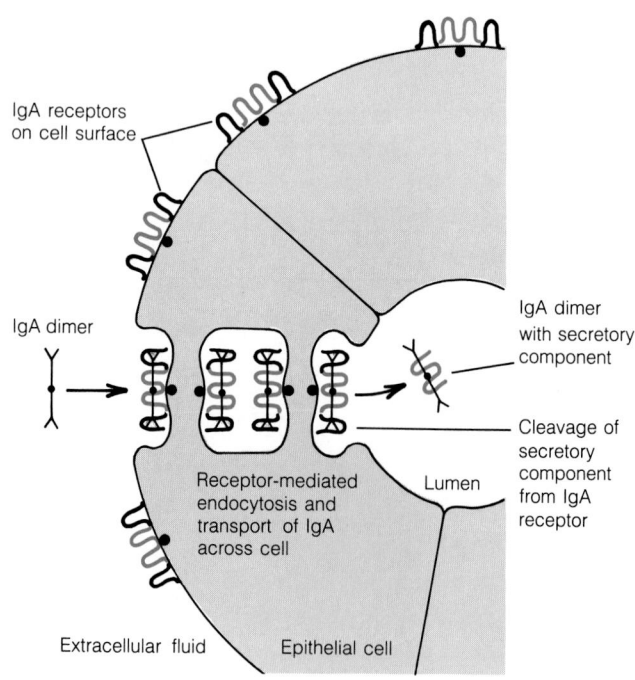

Figure 24-11 The Transport of IgA Molecules Across Epithelial Cells of Secretory Tissue. IgA dimers from the blood are transported across epithelial cells from the extracellular fluid into the lumen of secretory tissue by a process that involves receptor-mediated endocytosis on the basal surface and exocytosis on the luminal surface. IgA dimers bind to specific receptors on the basal surface of the epithelial cells and are transported across the cell as IgA-receptor complexes. On the luminal side, the receptor is cleaved, and a fragment of it, the secretory component, remains associated with the IgA dimer.

fragment of it, the secretory component, remains associated with the IgA dimer. Antigens in the mucosal fluids are complexed by IgA, thereby inhibiting viral infection and bacterial colonization.

Antigen-Antibody Interactions

The binding of antibody to antigen involves a number of noncovalent bonds between the epitope of the antigen and the variable domain of the antibody. While these interactions impart an *affinity* of antibody for antigen, the true strength, or *avidity* of most antibody-antigen reactions arises from the extensive complexing of antigens and antibodies. This is made possible by bivalent (IgG, IgD, IgE) and multivalent (IgM, IgA) antibodies. The extent of antibody-antigen complexing depends on the number of different antibody specificities present, the *valence* of the antigen (the number of epitopes on the antigen that can be recognized by the antibodies present), and the relative concentrations of antibody and antigen. The importance of reactant concentrations can be illustrated by an in vitro *precipitin reaction.* When antigen and specific antibody are mixed in an aqueous solution, a lattice forms, eventually developing into a visible precipitate. However, precipitate formation occurs

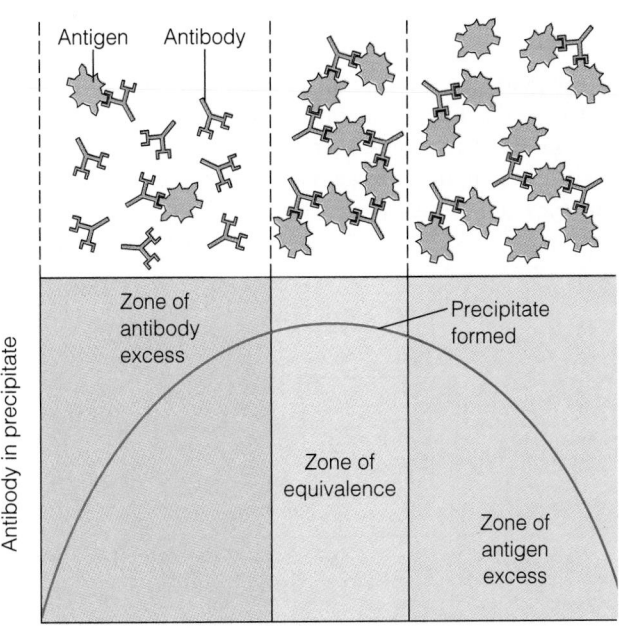

Figure 24-12   The Effects of Antigen and Antibody Concentrations on Complex Formation.   Here, a constant amount of antibody has been mixed with increasing amounts of antigen in a series of reactions. The amount of precipitate in each reaction is quantified, and the data are used to construct a precipitin curve. This plot of the antigen concentration versus precipitate formation reveals that at either antibody excess or antigen excess, precipitate formation is inhibited. However, within a range of relative reactant concentrations, a zone of equivalence is achieved, favoring complex formation (and therefore precipitation).

only at an antibody/antigen ratio of about 2:1 for bivalent antibodies, as shown in Figure 24-12. When either antigen or antibody is present in excess, precipitate formation is inhibited. Most antibody-antigen reactions involve many species of antibodies specific for different epitopes on multivalent antigens and a favorable ratio of reactants, so large complexes are formed. In vivo, the formation of such complexes facilitates phagocytosis and complement activation.

The Fine Structure of Antibodies

What accounts for the specificity of antibodies, and how are so many diverse kinds of antibodies generated? A straightforward approach to answering these questions would be to compare the amino acid sequences of light and heavy chains of antibodies that differ in their specificity. On the face of it, that is an impossible feat, since the immunoglobulin fraction of blood contains millions of different antibodies in small quantities, making them hard to distinguish and separate from one another. How could a protein chemist ever hope to find a single antibody needle in so complex a haystack?

A breakthrough came with the discovery that people with *multiple myeloma* are a good source of homogeneous immunoglobulins. Multiple myeloma is a cancer of antibody-secreting plasma cells. This type of cancer usually results from the uncontrolled proliferation of a single antibody-secreting cell, so that the resulting tumors consist of a clone of cells all secreting a single kind of immunoglobulin, the **myeloma protein.** A large quantity of a single kind of myeloma protein is found in the blood of a myeloma patient, making it possible to isolate homogeneous antibodies from different patients and compare their sequences. Also, myeloma tumors induced in certain inbred strains of mice are very stable and produce large amounts of homogeneous antibody; many of these myelomas can be grown in vitro as cultured cell lines.

Purified homogeneous immunoglobulin derived from a single clone of cells is called a **monoclonal antibody.** Several myeloma cell lines have been selected for special properties and have been used to generate hybrid cells called **hybridomas** that produce specific monoclonal antibodies. Specific monoclonal antibody production by hybridomas is one of the most exciting and important technical advances in immunology this century. In addition to facilitating the study of immunoglobulins of defined specificity, this technology has led to a number of important advances in biological research (Box 24A, page 801).

Variable and Constant Regions of L and H Chains

When the amino acid sequences of light chains from different myeloma proteins are compared, a striking pattern is evident. All such polypeptides differ from one another in amino acid sequence, but virtually all the differences are re-

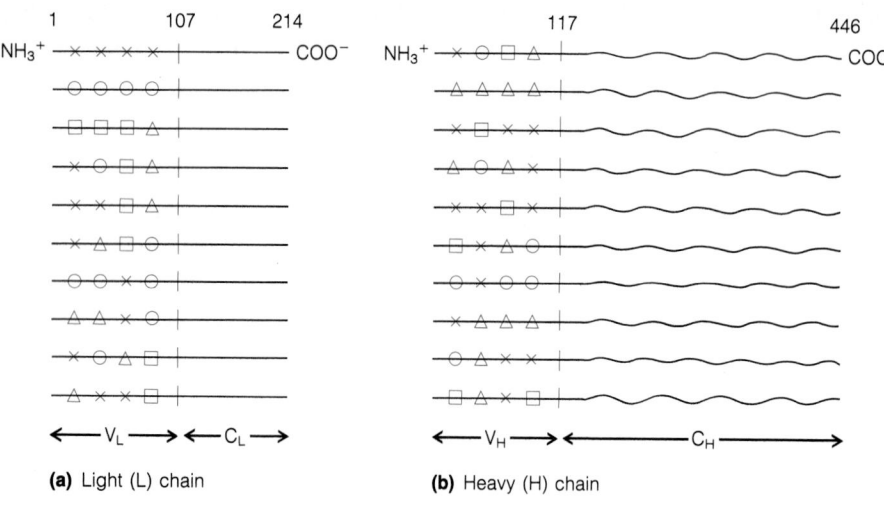

Figure 24-13   Variable and Constant Regions of the Heavy and Light Chains of Human IgA.   Ten human IgA molecules are shown schematically, with differences in their amino acid sequences indicated by the symbols X, △, □, and ○. The lengths of the variable (V) and constant (C) regions are indicated by amino acid number, beginning with amino acid 1 at the N-terminus. **(a)** The light chains. **(b)** The heavy chains.

(a) Light (L) chain

(b) Heavy (H) chain

stricted to the N-terminal half of the proteins (Figure 24-13a). By contrast, the C-terminal halves of all the κ light chains are identical (or nearly so), as are the C-terminal halves of all the λ light chains. The N-terminal half of the L chain is therefore called the *variable (V_L) region* of the chain, and the C-terminal half is called the *constant (C_L) region*. The variable and constant regions of both κ and λ chains are each about 110 amino acids long.

A similar pattern was found when heavy chains were sequenced (Figure 24-13b). Again, the variable region (V_H) is located at the N-terminal end and the constant region (C_H) at the C-terminus. The variable region of the heavy chain is 113 amino acids in length, approximately the same as the V_L region of the light chains. However, the C_H region of the heavy chain is about 330 or 440 amino acids long, depending on the class of immunoglobulin.

The functional significance of these findings is clear and is related to the bifunctional nature of the antibody molecule. The main sequence differences between antibodies are in the domains responsible for antigen binding (see Figure 24-9). Could the sequence heterogeneity in the variable regions of the H and L chains account for the great diversity of antigenic specificities among antibody molecules? Several lines of evidence suggested that this might be the case. But the real proof came when the amino acid sequences of antibody molecules were scrutinized more closely and regions of *hypervariability* were revealed.

Hypervariable Regions and the Antigen-Binding Site

As more and more myeloma proteins were sequenced and compared, it became clear that variable regions of the H and L chains are not uniformly variable. Instead, small segments of the V_H and V_L regions, called **hypervariable regions,** are responsible for most of the variability between the V regions of different antibodies (Figure 24-14). The other portions of the variable regions of both chains, called the **framework regions,** are relatively constant in amino acid sequence. The

V_H and V_L regions both have three hypervariable sites, each consisting of about 5–10 (or 15, in one case) amino acids. Thus, most of the variability in the L and H chains is restricted to about 15–30 amino acids.

Based on these findings, it was suggested in 1970 that the hypervariable regions of the H and L chains come together at one end of the molecule to form a pocket that functions as the antigen-binding site, and that the specificity of the antibody is determined by the particular amino acids in these regions. Two lines of evidence confirmed this suggestion. First, several myeloma protein Fabs have been crystallized and studied by X-ray crystallography. These studies have shown that V_H and V_L regions fold together to form a pocket that can bind antigens and that five of the six hypervariable regions line the walls of this pocket. Moreover, crys-

(Continued on page 804)

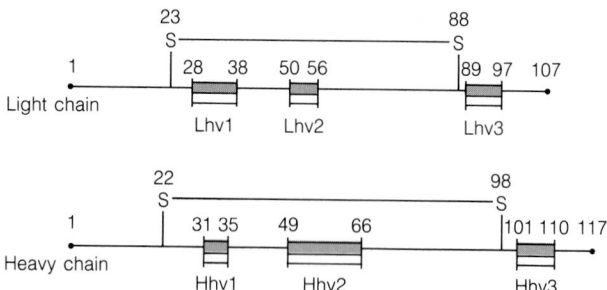

Figure 24-14   Hypervariable Regions of the Light and Heavy Immunoglobulin Chains.   This linear map of the variable regions of the light and heavy chains from human IgG illustrates the localization of sequence variability to three small segments of each chain (blue). These hypervariable (hv) regions are identified by an L (light chain) or an H (heavy chain) and are numbered 1, 2, and 3 from the N-terminal end of each chain. These regions are also called complementarity-determining regions (CDRs), because they make contacts with antigen. Amino acids bracketing each hypervariable region are numbered, as are the cysteines that give rise to the disulfide bond in the variable region of each chain.

24A

Contemporary Techniques

MONOCLONAL ANTIBODIES: A MILESTONE IN IMMUNOLOGY

M ost antibody preparations are highly heterogeneous. Even when an animal is immunized with very pure antigen, it will inevitably produce a great variety of immunoglobulins with a range of affinities against all the immunogenic epitopes of that antigen. Such heterogeneous antibody populations are of obvious advantage to the organism and are often useful to the researcher as well. Frequently, however, it is desirable to have a homogeneous population of identical antibodies, each with the same antigen-binding site. For example, the sequencing of immunoglobulin chains requires such a homogeneous preparation, as do various other research and clinical applications.

Myeloma patients are a valuable source of homogeneous antibodies, as are the myeloma tumors that can be induced in mice and the cell lines derived from them. Myeloma antibodies are of limited use as analytical tools, however, because they are not generated in response to a particular known antigen. Instead, a myeloma tumor develops from the random transformation of a normal lymphocyte, and the antibody produced by the tumor clone is the one that happens to be synthesized by that particular lymphocyte. Myeloma tumors therefore provide large quantities of homogeneous antibodies, but the researcher has no control over the specificity of the immunoglobulin produced.

An exciting breakthrough occurred in 1975, when Cesar Milstein and Georges Köhler discovered a way to obtain homogeneous antibodies of almost any desired specificity. Their approach has already revolutionized the use of antibodies in both research and clinical medicine. For this pioneering work, Milstein and Kohler received the 1984 Nobel Prize. (In fact, they shared the prize with Niels K. Jerne, a leading theoretician in immunology who was responsible in part for the clonal selection theory.)

Milstein and Köhler's technique involves fusing an antibody-producing lymphocyte with a mutant myeloma cell. This approach takes advantage of specific properties of both cell types. The myeloma cell can multiply indefinitely in culture but harbors mutations that prevent it from producing its own antibody. By contrast, a normal lymphocyte

from an immunized animal produces a specific antibody but cannot grow indefinitely in culture. When a lymphocyte is fused with a myeloma cell, the product is a hybrid cell called a **hybridoma** that can both proliferate indefinitely and produce large amounts of a single **monoclonal antibody.**

The fusion technique is illustrated in Figure 24A-1. The antigen of interest is injected into a mouse, causing the proliferation of specific antibody-producing cells. Several weeks later, the spleen of the mouse is removed and a suspension of B lymphocytes is prepared from it. Many of these lymphocytes will be making antibody to the antigen, but they have only a limited life span in culture. However, they can be fused with cultured myeloma cells to generate hybridomas that are, in effect, immortal. Cell fusion is generally induced by the addition of polyethylene glycol, and the resulting hybridomas are then selected by culturing the cells in a medium that does not allow unfused lymphocytes or myeloma cells to grow.

The ability to select hybridomas depends on a special property of the mutant myeloma cell line used in such fusions. The cell line is not only mutant with respect to its own antibody production but is also deficient in the enzyme hypoxanthine-guanine phosphoribosyltransferase (HGPRT). This enzyme catalyzes a reaction in one of two metabolic pathways for the synthesis of guanine monophosphate (GMP), an indispensable substrate for nucleic acid biosynthesis. Cells deficient in HGPRT are nonetheless viable because they can still make GMP by another pathway (which is, in fact, the normal pathway for GMP synthesis in most cells). However, these mutant myeloma cells will not survive if the drug *aminopterin* is added to the medium, because it is a potent inhibitor of the normal pathway for GMP synthesis. By contrast, hybrid cells will be able to grow in the presence of aminopterin, because they contain a functional HGPRT gene supplied by the normal lymphocyte.

This differential sensitivity to aminopterin is used to advantage in the selection procedure. The mixture of hybridoma cells, lymphocytes, and myeloma cells that results

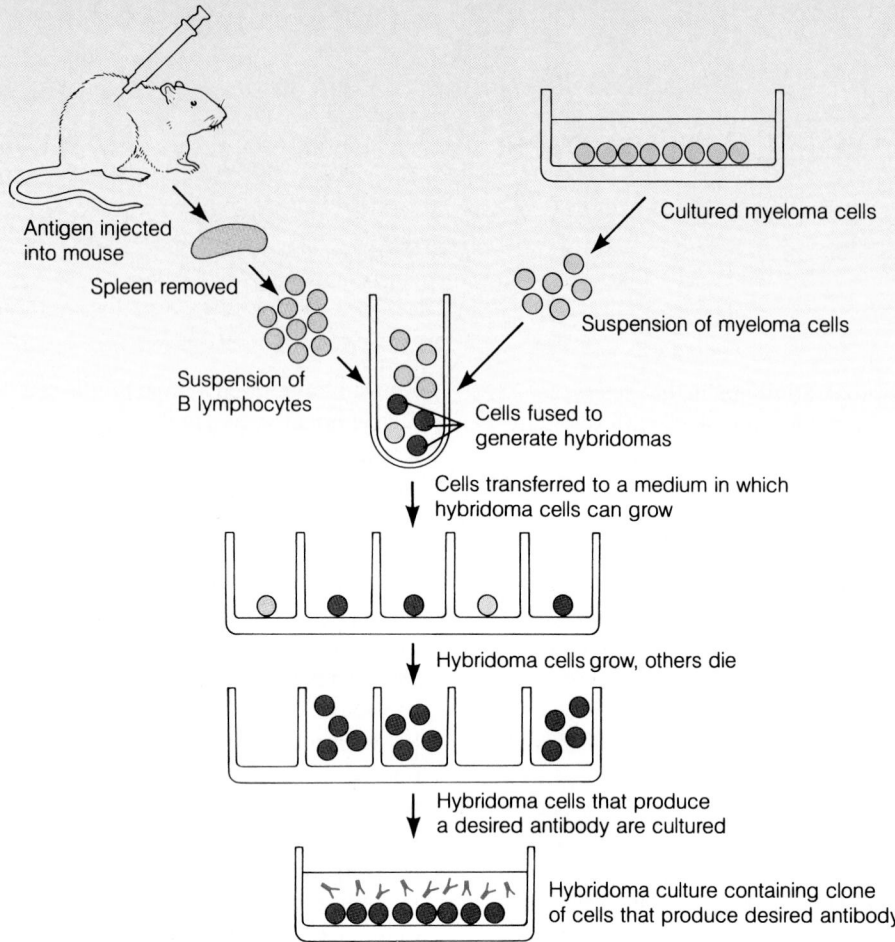

Figure 24A-1 Hybridoma Technique for the Production of Monoclonal Antibodies.   To produce hybridoma cells, the spleen of a mouse injected with the antigen of interest is removed, and B lymphocytes (light gray) are isolated. These are fused with cultured myeloma cells (beige) to generate hybridoma cells (dark gray) that can be cultured indefinitely and produce a single species of antibody. To select for fused hybridoma cells, aminopterin, an inhibitor of the guanine monophosphate (GMP) pathway, is added to the medium. The mutant myeloma cells used in the fusion process cannot survive in the presence of aminopterin because they are deficient in HGPRT, a key enzyme in the alternative pathway to GMP. Hybridoma cells can grow, however, because they have a functional HGPRT gene supplied by the lymphocyte genome.

from the fusion is cultured in the presence of a selective medium containing aminopterin. The myeloma cells cannot grow because of the aminopterin, and the lymphocytes survive but grow poorly. The hybridomas grow rapidly, however, forming large, readily distinguishable colonies. Each such colony is then screened for production of antibody against the antigen of interest, and hybridomas that produce the appropriate antibody can then be cloned in culture.

A tremendous advantage of the hybridoma technique is that monoclonal antibodies can be produced against unpurified molecules, even molecules that make up only a small fraction of a complex mixture of antigens. Regardless of how many different hybridomas are produced in the fusion step and how great a variety of antibodies they produce, individual hybridomas producing the antibody of interest can still be detected and selected. This technique therefore makes it possible to obtain virtually unlimited quantities of

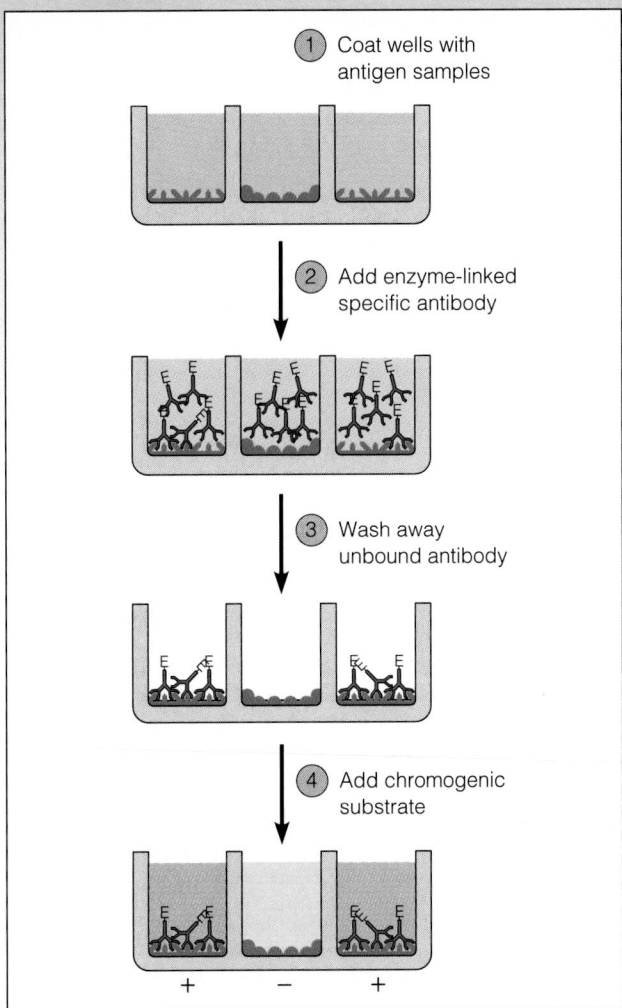

Figure 24A-2 Antigen Detection by the Enzyme-Linked Immunosorbent Assay (ELISA). An ELISA in its simplest form is illustrated here. ① Wells of a microtiter plate are first coated with antigen. ② Antibody specific for the antigen of interest is added to the well. This antibody is linked covalently to an enzyme (E) that reacts with a chromogenic substrate. ③ After an incubation period, the wells are washed to remove any antibody that is unbound, or that is bound at a very low affinity. Antibody that is bound specifically to antigen is not washed out. ④ Uncolored chromogenic substrate is added to all wells. Color develops only in those wells containing enzyme-linked antibody, which is in turn bound to the antigen of interest.

homogeneous antibody with specificity for any desired antigen, without having to purify the antigen. Indeed, that pool of monoclonal antibody is specific for a particular epitope on that antigen, so the specificity of a monoclonal antibody can be precisely defined.

Many kinds of monoclonal antibodies are now being produced for use as highly specific and sensitive analytical reagents. For example, monoclonal antibodies prepared against specific cellular components are very useful in locating such components within the cell by a technique called **immunocytochemistry,** using fluorescence-labeled antibodies to detect the position of molecules of interest. This technique is also useful for the detection of cell surface molecules. In clinical practice, monoclonal antibodies directed against specific drugs, hormones, vitamins, and serum proteins are used to detect and quantify very small amounts of such molecules in body fluids. A typical method is the *enzyme-linked immunosorbent assay,* or *ELISA* (Figure 24A-2). Here, the presence of a particular antigen in a sample is detected because antibody bound specifically to it is linked to an enzyme that reacts with a *chromogenic substrate.* That is, in the presence of enzyme-linked antibody, which is in turn bound to antigen, a colorless substrate becomes a colored reaction product. The amount of reaction product (intensity of color) can be quantified by spectrophotometry and is a measure of antigen concentration. Monoclonal antibodies have also been used in the purification of proteins and have proved particularly valuable in the isolation of rare proteins, so that enough material can be obtained for studies of structure and function.

The hybridoma technique of Milstein and Köhler provides the enormous advantage of an immortal cellular source of homogeneous antibodies with a desired specificity. Such monoclonal antibodies have become essential tools in research and diagnostics and have even been used therapeutically. Clearly, this technology has had and will continue to have a tremendous impact on many areas of biology and medicine. ■

tallographic analysis of several antibodies bound to simple epitopes confirmed that the epitope is bound within the pocket and makes contact with amino acid residues from the hypervariable regions. Figure 24-15 illustrates the binding of an antibody to a simple antigen, a derivative of vitamin K_1. Notice how five of the six hypervariable regions of the antibody molecule interact intimately with the antigen to hold it in the antigen-binding pocket.

A second line of evidence that hypervariable regions form the bulk of the antigen-binding site comes from *affinity labeling* studies. In these studies, small antigens were synthesized with a chemical group that could link covalently to the antibody molecule once the antigen was within the antigen-binding site. Results showed that such synthetic antigens attached only to amino acids in the hypervariable regions of the L and H chains. Because the three hypervariable regions have been shown to make specific contacts with antigen, and thus form the bulk of the antigen-binding site,

they are called **complementarity-determining regions (CDRs)**.

An important feature of antibody structure is that the antigen-binding site involves portions of two separate proteins, the L and H chains. This property contributes significantly to antibody diversity, because different H and L chains can presumably combine to generate different kinds of antigen-binding sites. Theoretically, if any H chain can associate with any L chain to form a functional antibody, then 1000 different H chains and 1000 different L chains could combine to form 1000×1000, or 10^6, different antibody molecules. It is not likely that every possible combination leads to a functional antibody, but even if only a fraction of all possible combinations were active, this ability to form *combinatorial associations* would be a great source of diversity.

Immunoglobulin Fingerprints: Idiotypes

Each clone of B cells produces homogeneous immunoglobulin with unique V_H and V_L sequences that distinguish it from all other monoclonal immunoglobulins. These unique sequences comprise the **idiotype** of that immunoglobulin. An idiotype, then, is a "fingerprint" or "signature" of a single B cell clone. The pool of immunoglobulins in blood is composed of minute quantities of many millions of different clonal products, or idiotypes. When an antigen stimulates an antibody response, the concentration of those antibodies (and their idiotypes) becomes transiently elevated (see Figure 24-4). Although idiotypes are really "self" rather than "foreign" epitopes, this radical change in concentration provides an antigenic stimulus, inducing the production of *anti-idiotype antibodies*. Anti-idiotype antibodies appear during the course of normal immune responses, usually during the declining phase of the original antibody response (see Figure 24-4). They are believed to be involved in the down-regulation of the antibody response.

The Folding of Immunoglobulin Molecules

Analyses of immunoglobulin sequences revealed that both heavy and light chains contain a series of units having similar amino acid sequences. Each of these homologous units, or *domains*, contains about 110 amino acids and includes two conserved cysteine residues. The light chain has one variable domain (V_L) and one constant domain (C_L), whereas the heavy chain has one variable domain (V_H) and three or four constant domains (C_{H1}, C_{H2}, C_{H3}, and C_{H4}), depending on the class of antibody. The constant regions of IgM and IgE heavy chains consist of four C domains, while the other classes contain three.

X-ray crystallographic studies established that immunoglobulin domains assume a characteristic structure consisting of two antiparallel β sheets stabilized by a single intrachain disulfide bond involving the conserved cysteines.

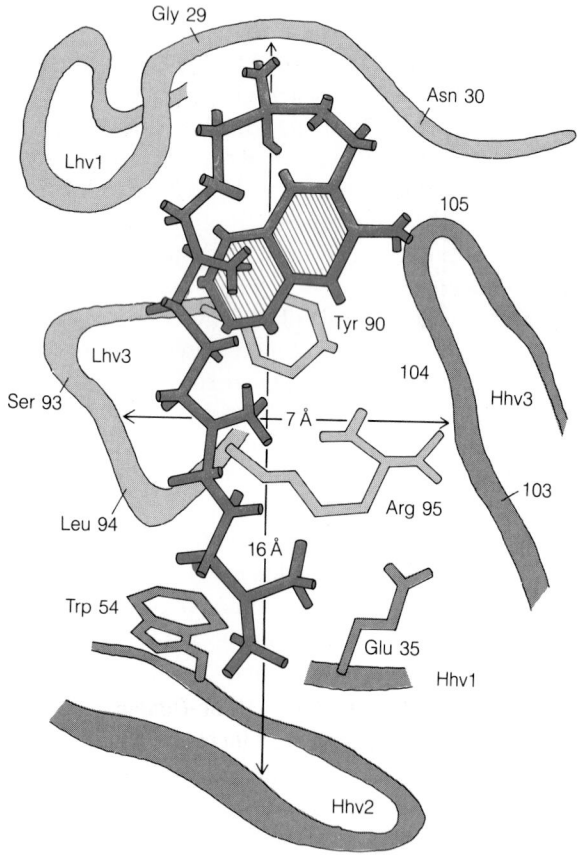

Figure 24-15   The Role of Hypervariable Regions in Antigen Binding.   The binding of an antigen, a derivative of vitamin K_1, to a human IgG molecule. The approximate location of two of the hypervariable regions (CDRs) of the light chain (light gray) and the three CDRs of the heavy chain (dark gray) are indicated with respect to the antigen molecule (blue). Several specific amino acids along each chain are numbered for cross-reference with Figure 24-14. Notice that most of the contacts with antigen are made by the third hypervariable regions of both chains.

This *immunoglobulin fold* structure facilitates noncovalent interactions between adjacent domains. Thus, in an immunoglobulin molecule, the V_L and V_H domains are intimately associated, as are C_L with C_{H1}, C_{H2} with C_{H2}, C_{H3} with C_{H3}, and C_{H4} with C_{H4}. Figure 24-16 illustrates the three-dimensional model of antibody structure that has emerged from X-ray crystallographic studies.

The presence of homologous domains in both heavy and light chains suggests that the genes encoding them arose from a common ancestral gene. Indeed, many cell-surface proteins contain one or more immunoglobulin domains and are therefore members of the *immunoglobulin gene superfamily*. Among the members of the superfamily are the T cell receptor, CD4 and CD8, class I and II MHC molecules, β2-microglobulin, and a number of cellular adhesion molecules.

The Genetic Basis of Antibody Diversity

We already know that the immune system of vertebrates is capable of making an incredibly large number of antibodies—perhaps as many as 10^8–10^9 different molecules. This **antibody repertoire,** as it is called, can recognize virtually any foreign substance the animal will ever encounter. Understanding how this large repertoire is generated has been a major focus of immunology research. From this research has come the remarkable finding that the DNA sequences that encode an immunoglobulin molecule exist at several locations within the genome and are assembled during differentiation of B cells in a way that accounts for much of the diversity of antibodies.

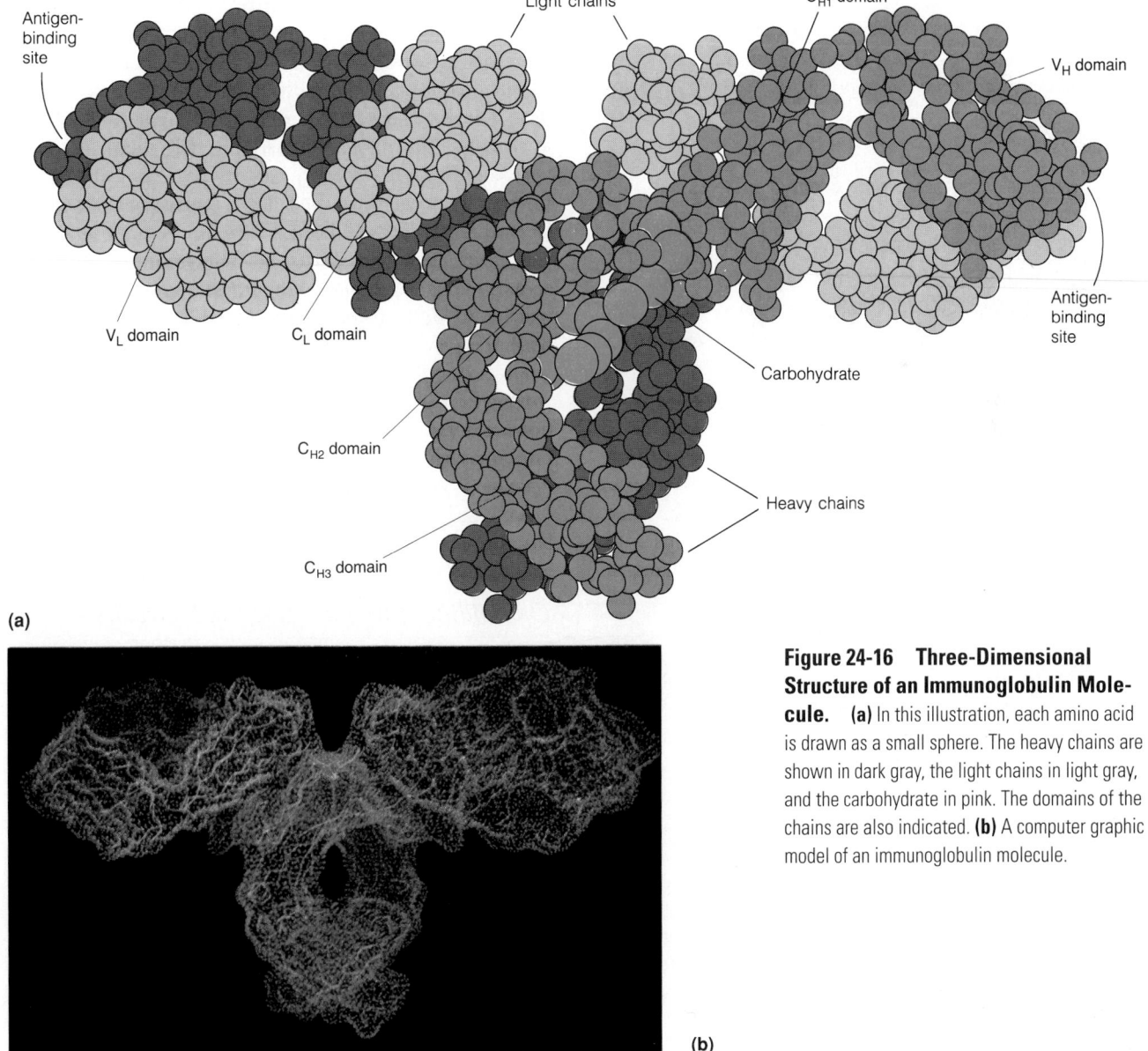

(a)

(b)

Figure 24-16   Three-Dimensional Structure of an Immunoglobulin Molecule.   (a) In this illustration, each amino acid is drawn as a small sphere. The heavy chains are shown in dark gray, the light chains in light gray, and the carbohydrate in pink. The domains of the chains are also indicated. **(b)** A computer graphic model of an immunoglobulin molecule.

Separate V and C Gene Segments Are Brought Together During B Cell Differentiation

We have already encountered an important clue to understanding the genetic basis of antibody diversity. Each heavy and light chain consists of distinct variable (V) and constant (C) regions. In 1965, William Dreyer and J. Claude Bennett proposed that a single immunoglobulin chain is encoded by two separate genes, V and C, that are brought together at the DNA level sometime during early B cell development. Moreover, they proposed that there are hundreds or thousands of different V genes in the genome, but only single copies of genes encoding the constant regions. Thus, one of the many V region genes combines with a particular C region gene, allowing for both variability and constancy in a single polypeptide chain. The suggestion that eukaryotic genes could be separate in the genome and combined or *rearranged* during development was revolutionary at the time, but the hypothesis proved to be essentially correct, albeit more complex than initially thought.

The first direct support for the hypothesis came when it was found that the structure of immunoglobulin genes in mouse myeloma cells (antibody producers) is different from that in embryonic cells. To show this, DNA from the two cell types was first digested with a restriction endonuclease and then hybridized to a radioactive mRNA probe corresponding to the V-C sequence of a specific L chain or to the C region only (Figure 24-17). The V_L and C_L coding sequences were found on different restriction fragments in the embryonic DNA (Figure 24-17a), but on the same fragment when myeloma DNA was used (Figure 24-17b). The conclusion seemed inescapable: Immunoglobulin DNA is rearranged during B cell development. In other words, segments of genes that are distant from one another in the genome of an undifferentiated precursor cell are brought together at the right moment during B cell differentiation. Susumu Tonegawa was awarded a Nobel Prize in 1986 for this landmark research.

The κ light chains, λ light chains, and all classes of heavy chains are each encoded in germ line DNA by separate clusters of gene segments on three different chromosomes. These clusters of gene segments are the κ, λ, and H chain genes, respectively (Figure 24-18). Each gene contains a set of different **V gene segments** (about 300 in κ, two in the λ gene, and an estimated 300–1000 in the H chain gene of the mouse; the size of the gene families varies in different species). In each case, the V gene segments are separated from one or more **C gene segments** by hundreds of kilobase pairs of DNA.

Multiple Gene Segments as a Source of Antibody Diversity

The multiplicity of V gene segments in the germ line can account for a considerable amount of antibody diversity. If 300 κ light chains with different V regions can associate with any of 1000 H chains with different V regions, then 300,000 antibodies with different H-L combinations can be generated. This is already an impressive diversity, but it falls short of accounting for the estimated 10^8–10^9 antibodies in the total repertoire. Clearly, other mechanisms are necessary for generating additional diversity.

Another source of antibody diversity was discovered when the DNA sequences of cloned immunoglobulin genes from embryonic cells and myeloma cells were compared with corresponding amino acid sequences. It quickly became clear that the gene segments encoding the variable domains of immunoglobulin proteins are themselves separated in embryonic cells. For instance, mouse embryonic DNA

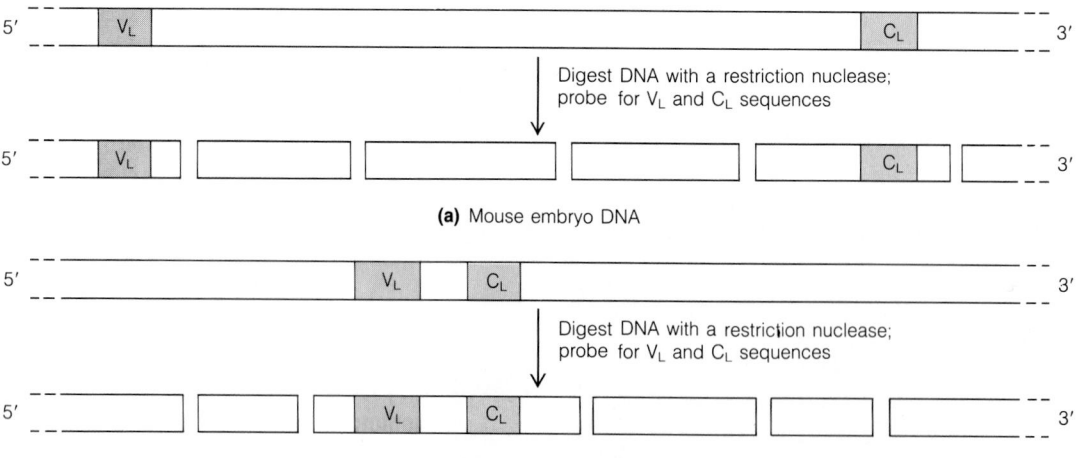

(a) Mouse embryo DNA

(b) Mouse myeloma DNA

Figure 24-17 The Proximity of V and C Genes in Mouse Embryo and Myeloma DNA. To determine the location of the genes coding for the V_L (gray) and C_L (blue) regions of the mouse light chain, mouse DNA was digested with a restriction endonuclease, and the digestion fragments were separated by electrophoresis on agarose. The restriction fragments were then analyzed for the presence of V_L and C_L coding sequences by hybridization with a radioactive mRNA probe corresponding either to the V-C sequences of a specific myeloma chain or to the C sequence only. **(a)** In DNA from a 13-day-old mouse embryo, the V_L and C_L genes are always found on separate fragments. **(b)** In DNA from a mouse myeloma synthesizing a specific light chain, the sequences are always found on the same DNA fragments.

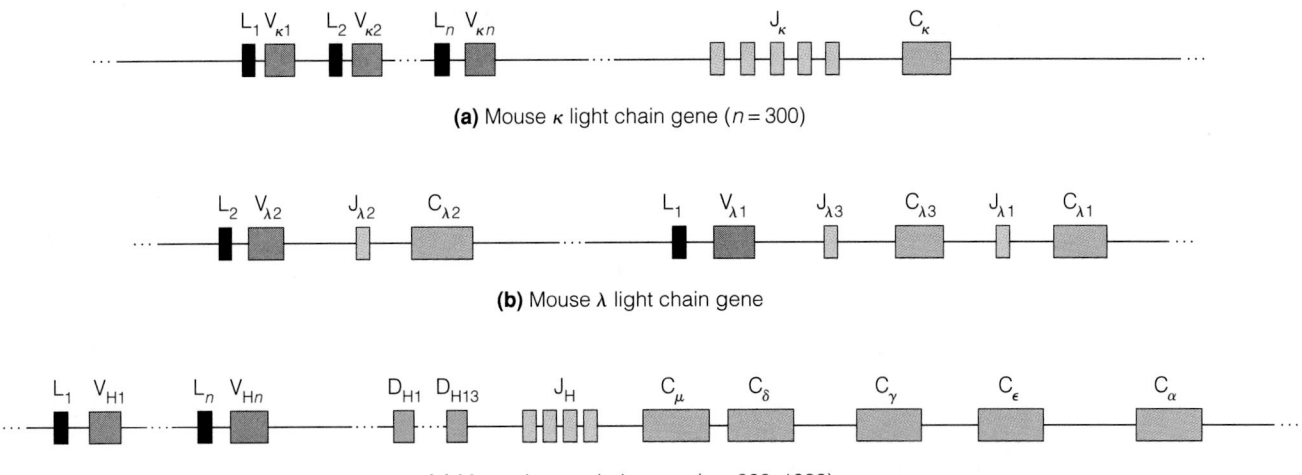

(a) Mouse κ light chain gene (*n* = 300)

(b) Mouse λ light chain gene

(c) Mouse heavy chain gene (*n* = 300–1000)

Figure 24-18   The κ and λ Light Chain and Heavy Chain Genes.   In the mouse, **(a)** the κ light chain gene contains about 300 V gene segments, five J (joining) segments (only four of which can be used), and one C region; **(b)** the λ light chain gene contains two V, three J, and three C gene segments; and **(c)** the heavy chain gene contains as many as 1000 V gene segments, 13 D (diversity) segments, four J segments, and a series of C gene segments encoding the five major heavy chain classes. The four γ subclasses of the mouse are not depicted here. The L (leader) exon located on the 5′ end of each V gene segment encodes the N-terminal signal peptide.

contains a V_λ gene segment that encodes 97 of the 110 amino acids of the λ chain's variable domain. The remaining 13 amino acids are encoded by a DNA sequence located at a great distance downstream of the V gene segment, near the C gene segment. This sequence is called the **J (joining) gene segment** because it joins to V, bringing V closer to C in a differentiated B cell (Figure 24-18b). (Note that the J gene segment is distinct from the J chain polypeptide present in polymeric IgM and IgA molecules; despite the common name, the two are unrelated.) J gene segments encode the last ten or so amino acids of the L and H chain V domains, including part of the third CDR. The heavy chain gene, however, is more complex. The V gene segment encodes amino acids 1–94, while the J gene segment encodes amino acids 98–113. The code for amino acids 95–97 is located at a site between V and J called the **D (diversity) gene segment** (Figure 24-18c).

As shown in Table 24-3, the multiple V domain–encoding segments greatly enhance the diversity of the im-

munoglobulin repertoire. For example, given random VDJ rearrangement in the heavy chain gene of the mouse, 1000 V segments, 13 D segments, and 4 J segments can give rise to about 52,000 possible H chains with different variable domains. At the same time, the potential number of κ light chains with different variable domains is 300 V × 4 J, or 1200. All in all, if any H chain can associate with any L chain, the number of different combinatorial associations is not 300,000, as stated above, but approximately 6×10^7. As we will see shortly, there are still more mechanisms for generating diversity that help account for the estimated repertoire of 10^8–10^9 different antibody specificities.

Gene Segments Are Assembled by Somatic Rearrangement

Once we understand that the L and H chain genes consist of several separate gene segments, we can ask how these

Table 24-3   Antibody Diversity Provided by Germ Line Gene Segments in the Mouse

	Heavy Chain	Light Chain κ	Light Chain λ
Germ line gene segments			
V	300	300–1000	2
D	0	13	0
J	4	4	3
Possible combinations			
VDJ/VJ gene segment combinations	1.2×10^3	5.2×10^4	6
H-κ combinations		6.24×10^7	
H-λ combinations			3.12×10^5

segments are brought together and expressed during B cell development. As Figure 24-19 shows, this is a multistep process, involving both irreversible recombination of gene segments and splicing of RNA transcripts. The initial event is random **gene rearrangement,** which assembles a complete V domain–encoding segment (step 1). For the light chain gene, this rearrangement joins a V segment to a J segment. For the heavy chain gene, two rearrangements are required: A D segment is first linked to a J segment, then a V segment is linked to the DJ segment.

Immunoglobulin gene rearrangements generally involve looping out and deletion of the DNA contained between the randomly selected VJ and VDJ gene segments. These events proceed by way of interactions between special **recombination signal sequences (RSS)** located on the 3′ side of each V gene segment, on the 5′ side of each J segment, and on both sides of each D segment. The recombination signal sequences function as signals for the recombination process and are likely recognized by an undefined enzyme or system of enzymes called a **recombinase.** The RSS and recombinase function together to define the points at which the DNA is cut and rejoined, thus mediating VJ and VDJ recombination.

The cutting and rejoining of gene segments does not always occur at precise locations in the RSS. For example, in the case of the κ light chain gene, the same V_κ and J_κ segments can be joined in slightly different ways, generating several possible nucleotide sequences at the V-J joints, each capable of coding for a different amino acid. The same imprecise joining is seen in the VD and DJ joints of the heavy chain gene. This flexibility of recombination provides additional **junctional diversity.** Moreover, the VDJ joints of heavy chain genes often contain additional nucleotides not found in the corresponding germ line DNA. That is, prior to DJ and VD joining, nucleotides may be added to the existing DNA sequence, further altering the coding at these joints. This so-called **N-region addition,** thought to be catalyzed by the enzyme terminal deoxynucleotidyl transferase (TdT), further contributes to the generation of antibody diversity. It is notable that these mechanisms occur at the coding joints that specify much of the third hypervariable region (CDR3) of the immunoglobulin chains, and that CDR3 regions make up a significant portion of the antigen-binding site of the antibody (see Figure 24-15).

Rearrangement is a prerequisite for high-level transcription of an immunoglobulin gene because it brings a *promoter,* located on the 5′ end of each V gene segment, into functional proximity of an *enhancer,* located within the J-C intron. Therefore, once an immunoglobulin gene is rearranged, it is transcribed (Figure 24-19, step 2). The transcript is then processed to form a functional mRNA (step 3). Processing involves the addition of a 5′ ^7mG cap and a 3′ poly(A) tail, and the removal of introns, as usual. The mature transcript then moves from the nucleus to the cytosol,

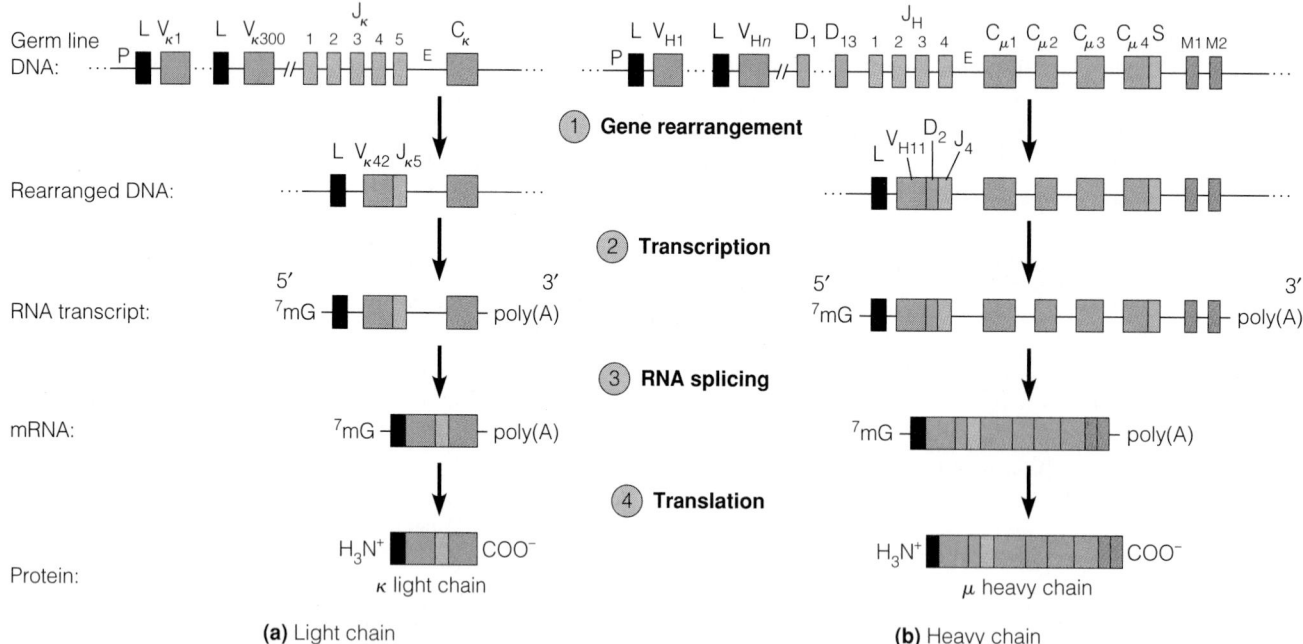

Figure 24-19   Rearrangement and Expression of Immunoglobulin Genes in B Cells. The genes for both **(a)** the light chain and **(b)** the heavy chain of an antibody molecule are assembled from several gene segments by rearrangement of germ line DNA during B cell differentiation. The particular light chain and heavy chain genes shown are those for the κ light chain and the μ heavy chain. ① The light chain gene is assembled from V, J, and C segments; the heavy chain gene requires V, D, J, and C segments. ② The rearranged DNA is then transcribed into an RNA transcript with a ^7mG cap at the 5′ end and a poly(A) sequence at the 3′ end. ③ RNA splicing eliminates introns and generates a functional mRNA molecule. ④ The mRNA is then translated to synthesize the desired polypeptide.

where it is translated into protein (step 4). Each new immunoglobulin chain contains a short N-terminal signal peptide that directs it into the rough ER and then is cleaved as the chain appears in the lumen of the rough ER. Once within the rough ER, two H and two L chains come together to form a functional antibody, H_2L_2.

Recall that the antibodies produced by a single B cell are all identical with respect to antigen binding. This accounts for the monospecificity of a B cell and its progeny cells. Thus, there must be a mechanism limiting B cells to the expression of a single rearranged heavy chain gene and a single rearranged light chain gene; otherwise, a single B cell would make two different H chains and L chains, which could assemble into mixed molecules. The expression of only one H chain gene and one L chain gene is called **allelic exclusion,** which simply means that one of the two alleles fails to be expressed. Thus, although we are diploid organisms, a given B lymphocyte expresses only one of the two genes encoding a particular immunoglobulin chain.

Hypermutation as a Source of Antibody Diversity

So far, we have considered the contributions to antibody diversity made by multiple V, D, and J gene segments and their random assembly, and by imprecise joining of gene segments, N-region addition, and the combinatorial association of H and L chains. Yet another mechanism that contributes significantly to antibody diversity is an unusually high mutation rate within the rearranged VJ and VDJ units. In fact, these mutations tend to occur within the CDR1 and CDR2 hypervariable regions. This process, called **somatic hypermutation,** occurs during the proliferation and differentiation of activated B cells and can increase the antibody repertoire by two to three orders of magnitude. The mechanism that accounts for this V region–directed hypermutation is not yet understood. The mutations accumulate with time and are therefore more numerous in the Ig genes of B cells that have switched from IgM production to the production of one of the other classes of antibodies, such as IgG or IgA. Interestingly, somatic mutation also contributes to an overall increase in the affinity of antibodies for antigen seen in secondary and subsequent immune responses. While the mutation process is random and therefore also generates antibodies of lesser affinity, those B cells bearing high-affinity surface antibody are preferentially selected by antigen (clonally selected) to proliferate and differentiate.

Antibody Class Switching

Having seen how the enormous diversity in antigen-binding sites arises genetically, we turn now to the question of how the switch from one class of immunoglobulin to another is made as a B lymphocyte differentiates. We will find that two distinct mechanisms are involved in the expression of different classes of immunoglobulins that accounts for such **class switching.**

Early in their life span, B lymphocytes synthesize IgM, which is expressed as a membrane-bound antigen receptor. At this stage, the B cell is immature and is unresponsive to antigen. Responsiveness to antigen is acquired with the simultaneous surface expression of IgM and IgD. The simultaneous expression of the μ and δ heavy chains for IgM and IgD, respectively, occurs by **alternative splicing** of large RNA transcripts that contain the assembled V_H sequence and the C_μ and C_δ sequences, which are adjacent to each other on the chromosome (Figure 24-20). The heavy chain transcripts are spliced in two different ways, generating mature mRNAs that have the VDJ sequence linked to either C_μ or C_δ sequences. Translation of these two different mRNAs results in the production of two different H chains, μ and δ, both having the same V region and therefore the same antigen-binding site, but having different C regions.

Following antigen-induced activation, B lymphocytes may switch to the production of one of the other classes of immunoglobulins, usually IgG, but sometimes IgA or IgE. This type of class switching involves a mechanism that is different from alternative splicing but similar to the rearrangement of V gene segments. That is, expression of the classes encoded by C_H gene segments downstream of C_δ requires DNA recombination, in which the VDJ unit is transposed to another C_H gene segment, eliminating the intervening DNA. Class-switch recombination is mediated by **S (switch) regions** that lie just upstream of each C_H gene segment except C_δ. These switch regions are composed of multiple copies of short sequences that apparently serve as recognition elements for recombination. Transcription and translation of the newly structured H chain gene then occurs as usual, generating a new antibody isotype with the original antigen-binding site (Figure 24-21). An important ramification of this process is that class switching in a given B cell can occur sequentially and only in a downstream direction—for example, from μ to γ to α (see Figure 24-18c)—but the B cell cannot return to the synthesis of a previously expressed class because the C_H gene encoding that C_H region has been excised and lost. Class switching provides the immune response with flexibility, inasmuch as antibodies with identical specificity for antigen but with different biological properties can be made against a particular pathogen.

Membrane and Secreted Forms of Antibody

As stated earlier in the chapter, all classes of antibodies can be synthesized as either membrane-bound receptors or as secreted molecules, depending on the stage of the B lymphocyte. Immature, mature, and long-lived memory B cells produce membrane-bound antibodies that serve as receptors for antigen, while plasma cells produce antibodies that are secreted at a high rate and help dispose of antigen. The choice between the production of secreted or membrane forms of the heavy chain depends on alternative processing

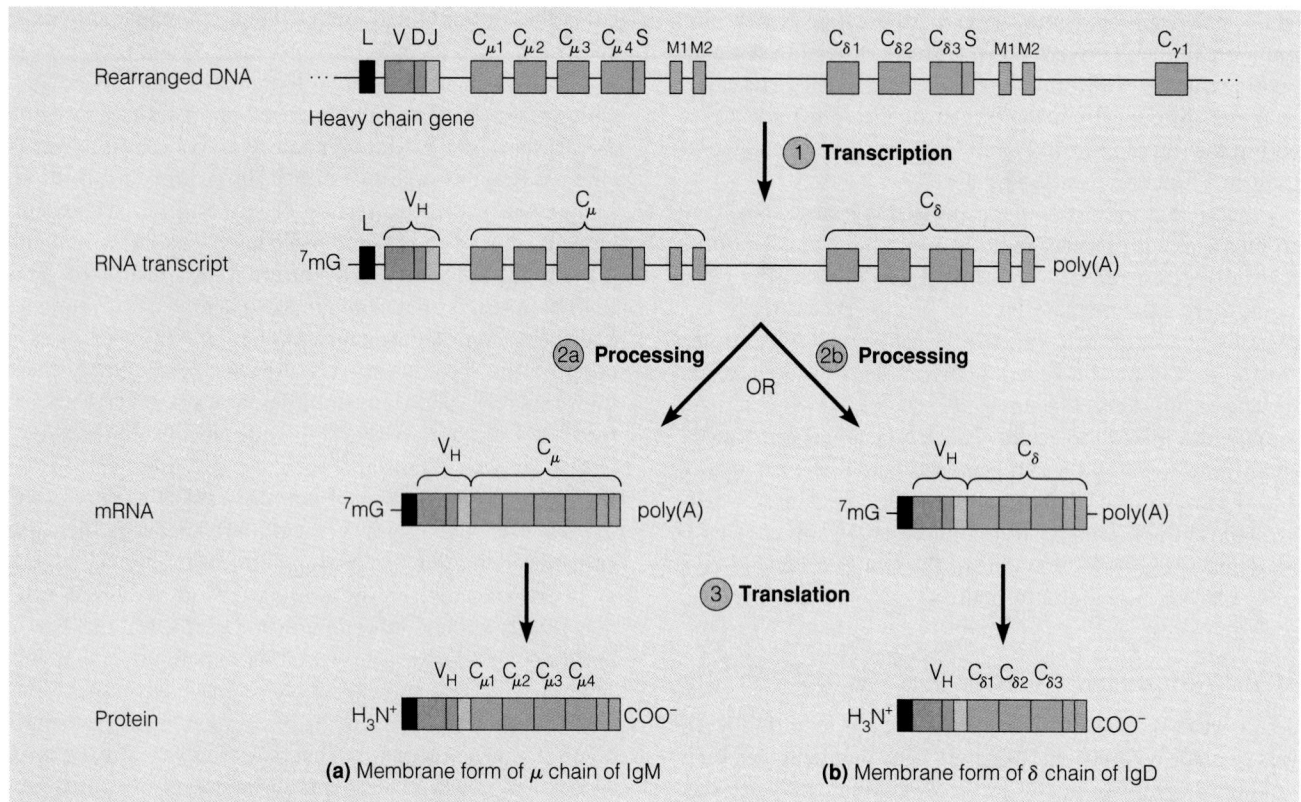

Figure 24-20 Expression of IgM and IgD by Alternative Splicing and Processing of RNA. The coexpression of both IgM and IgD involves the alternative splicing and processing of a single long RNA transcript that encodes both of these C_H regions. As shown here, the choice between IgM and IgD production involves a change in the process- ing pattern of the same primary transcript. **(a)** Initially, processing links the transcript of the assembled V_H sequence (consisting of V, D, and J segments) to the C_μ sequence, generating the mRNA for the μ chain of IgM. **(b)** Later, for some transcripts, processing links the V_H sequence to the C_δ sequence, generating the mRNA for the δ chain of IgD.

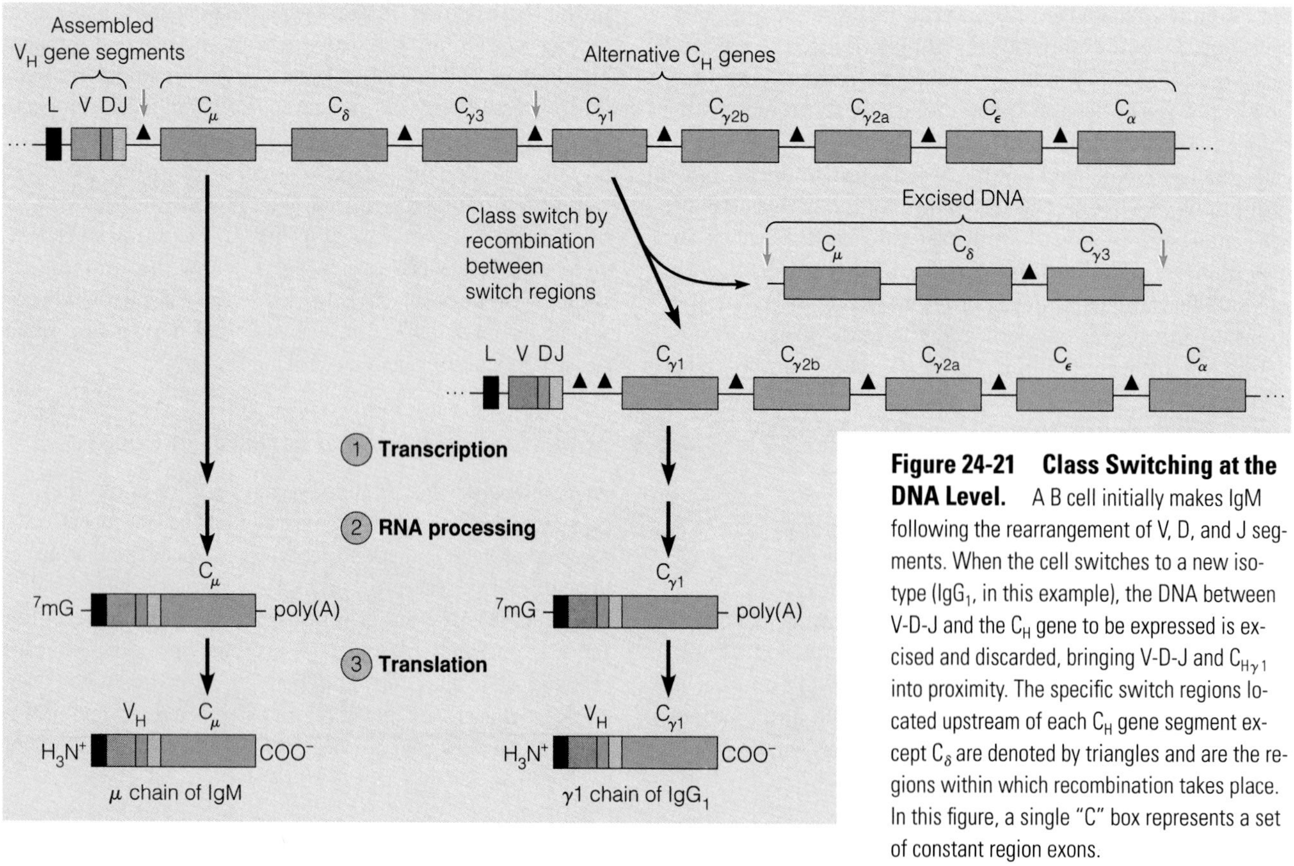

Figure 24-21 Class Switching at the DNA Level. A B cell initially makes IgM following the rearrangement of V, D, and J segments. When the cell switches to a new iso- type (IgG$_1$, in this example), the DNA between V-D-J and the C_H gene to be expressed is ex- cised and discarded, bringing V-D-J and $C_{H\gamma 1}$ into proximity. The specific switch regions lo- cated upstream of each C_H gene segment ex- cept C_δ are denoted by triangles and are the re- gions within which recombination takes place. In this figure, a single "C" box represents a set of constant region exons.

of a common RNA transcript. As shown in Figure 24-20, each heavy chain transcript contains a sequence that encodes the hydrophilic C-terminus of a secreted molecule (S) as well as those that encode the hydrophobic C-terminus of the membrane-bound form (M1 and M2). The particular pattern of RNA processing—poly(A) addition and splicing—presumably regulated in a stage-specific manner, determines which form of heavy chain mRNA, and therefore which immunoglobulin form, is produced in that cell.

The T Cell Antigen Receptor

T lymphocytes, like B lymphocytes, express clonally distributed specific receptors for antigen. The antigen receptors of B and T lymphocytes are similar in many respects, but there are several important differences between them. First, T cell receptors (TCRs) are found only in membrane-bound form, whereas antibodies exist as both membrane-bound and secreted proteins. Second, TCRs have a single antigen-binding site, whereas antibodies have two sites each. Third, TCRs

recognize antigens only when they are presented in association with MHC molecules, while antibodies recognize foreign antigen alone. Consequently, the T cell receptor can recognize a complex epitope composed of foreign peptide antigen and a self MHC molecule (Figure 24-22).

The T cell receptor itself is composed of two disulfide-linked polypeptide chains of about equal size (molecular weights of about 40,000 each). Each of the chains, like immunoglobulin H and L chains, has an N-terminal variable domain and a C-terminal constant region. There are two kinds of TCRs found in the pool of T cells. Most T cells (more than 95%) express a receptor composed of α and β chains. The T cells that bear **$\alpha\beta$ TCRs** are the typical CD4+ helper and CD8+ cytotoxic subpopulations. As shown in Figure 24-22, the $\alpha\beta$ TCR engages peptide antigen and MHC on the surface of an APC or a target cell. This cell-cell interaction is enhanced by the presence of CD4 or CD8, surface molecules that bind to class I and class II MHC molecules, respectively. A small fraction of T cells, found primarily in the thymus, skin, and intestinal epithelia, express a receptor consisting of γ and δ chains. The function of cells

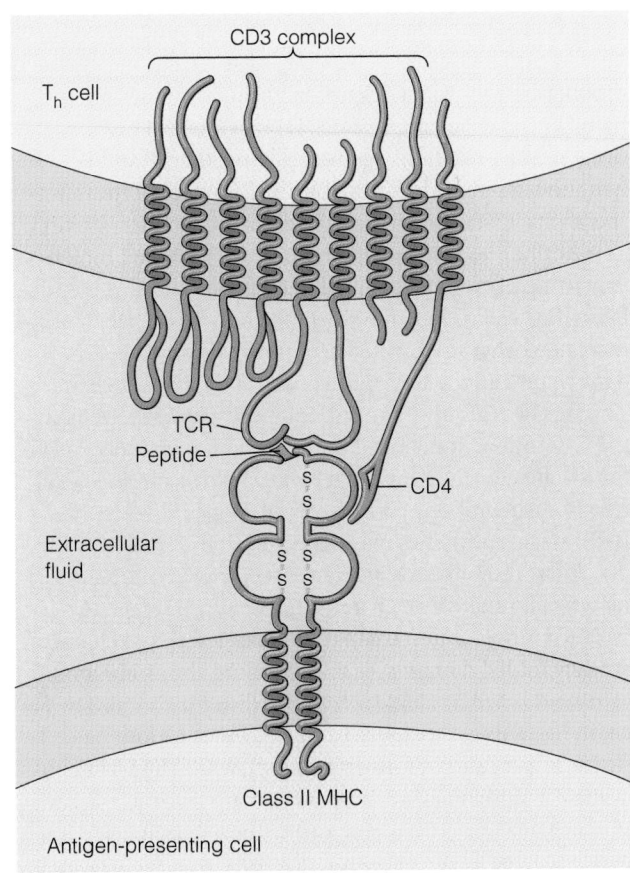

(a) TCR binding to MHC complex on APC

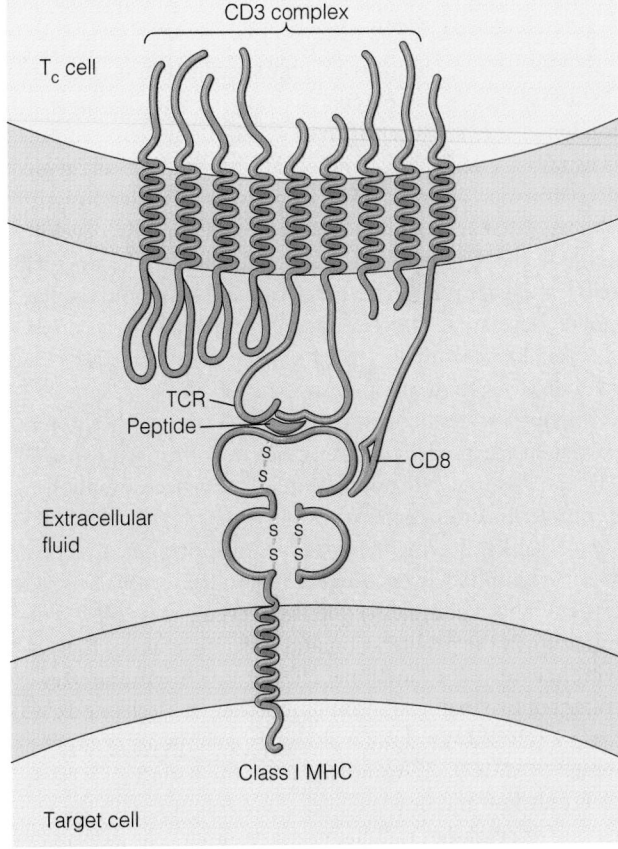

(b) TCR binding to MHC complex on target cell

Figure 24-22 Recognition of a Complex Epitope by the T Cell Antigen Receptor. This schematic diagram shows the TCR binding to the MHC-peptide complex on the surface of either **(a)** an APC or **(b)** a target cell. Also pictured are the CD3 complex, which serves to transduce antigen signals, and the proteins **(a)** CD4 and **(b)** CD8 that help to strengthen these essential cell-cell contacts.

bearing **γδ TCRs** is not fully understood. Because of their distribution and their lack of CD4 and CD8, it is thought that γδ TCRs are involved in non-MHC-restricted cell-mediated immune response against pathogens commonly found at epithelial boundaries.

As in the immunoglobulin genes, functional TCR genes are assembled by the rearrangement of V domain–encoding gene segments. In the α and γ genes, a single V to J recombination is required, while in β and δ genes, recombination involves D to J followed by V to DJ rearrangements. Thus, a diverse repertoire of T cell receptors is achieved in part by the random rearrangement of multiple gene segments. However, TCR genes contain far fewer VJ and VDJ gene segments than do immunoglobulin genes. Additional mechanisms such as imprecise joining of gene segments and N-region nucleotide addition contribute to TCR diversity, but no somatic mutation has been observed in TCR genes. Nonetheless, with the mechanisms available, the repertoire is estimated to be in the range of 10^{15} TCR specificities.

Rearrangements in the TCR genes bring the V domain–encoding region into functional proximity of a constant region (C_α, C_β, C_γ, or C_δ), so the genes are efficiently transcribed. As in B cells, allelic exclusion is thought to operate so that each T cell produces only αβ or γδ TCRs of singular specificity. The αβ and γδ TCRs are always transported to the T cell surface in association with CD3, a complex of five different polypeptide chains (see Table 24-2 and Figure 24-22). This complex is believed to be essential for the transmission of external antigen-induced activation signals to the interior of the cell. Such signals lead to proliferation and cytokine production in helper T cells and to active target cell killing in cytotoxic T cells.

PERSPECTIVE

The immune system is a highly specific defense mechanism that enables vertebrates to recognize and eliminate foreign invaders. The antigen to which the system responds is usually a foreign macromolecule with multiple antigenic determinants (epitopes). The response is mediated by lymphocytes that arise from hematopoietic stem cells, develop either in the thymus (for T cells) or the bone marrow (for B cells), and later migrate to the peripheral lymphoid tissues, where they contact antigen. Each lymphocyte carries on its surface identical antigen receptors that are specific for a structurally related set of epitopes.

Lymphocytes become committed to the synthesis of a particular receptor for antigen prior to antigen exposure. The total lymphocyte population is composed of millions of different clones, each expressing its particular unique receptor, thus endowing the system with enormous diversity. When a lymphocyte encounters an antigen to which its receptors bind, a complex sequence of events takes place, culminating in the proliferation and differentiation of the lymphocyte into a functional effector cell. T lymphocytes consist of two major subpopulations: helper T cells, and cytotoxic T cells. CD4+ T_h cells detect exogenous antigen presented by class II MHC molecules on the surface of B cells, macrophages, and dendritic cells, and respond with the secretion of cytokines that help activate other cells involved in the response. CD8+ T_c cells detect endogenous antigen presented by class I MHC molecules on virus-infected cells, tumor cells, or grafted cells, and respond by becoming active killers. B lymphocytes detect discrete epitopes on whole antigens and become antibody-secreting plasma cells. In addition, through clonal deletion and clonal anergy, potentially self-reactive lymphocytes are taken out of commission, limiting reactions against self. Together, T and B lymphocytes provide for the cardinal features of immunity: specificity, diversity, memory, and self-nonself recognition.

Antibodies (immunoglobulins) are Y-shaped molecules consisting of two identical light (L) chains and two identical heavy (H) chains linked by disulfide bonds. Each arm of the molecule contains an antigen-binding site and consists of a light chain and the N-terminal portion of a heavy chain. The leg (Fc region) of the antibody consists of the remaining C-terminal domains of both H chains and is responsible for the effector functions that determine the physiological role of each antibody class (isotype). Depending on the B cell stage, antibodies may be membrane-bound, serving as cellular receptors for antigen, or secreted, binding to and helping eliminate foreign antigen.

Immunoglobulin chains consist of a series of characteristically folded domains, each containing about 110 amino acids and having a single intrachain disulfide bond. The H chain has a single V domain and a series of C domains (three in δ, γ, and α; four in μ and ε), while the L chain has a single V domain and a single C domain (κ or λ). V_L and V_H domains associate to form an antigen-binding site. The hypervariable CDRs of both V domains in particular are involved in making specific contacts with antigenic epitopes. The large repertoire of antigen-binding specificities arises by a number of mechanisms. The V domain–encoding region is assembled randomly from gene segments (V and J for the light chain gene; V, D, and J for the heavy chain gene). Thus, the V and C regions of both H and L chains are encoded by gene segments that are far apart on the germ line DNA but are brought closer together by somatic recombination during B cell development. These gene segments may be joined imprecisely (junctional diversity), and, in the heavy chain gene, additional nucleotides may be added at the joints (N-region addition). These mechanisms create a great array of immunoglobulin genes encoding chains with different variable domains. And since almost any L chain can associate with almost any H chain, the variety of antibodies that can be made is enormous.

A B lymphocyte expresses the same VDJ and VJ units throughout its life, as do all of its descendants. However, the antigen specificity of the encoded antibodies can be altered by somatic mutation, which occurs with high frequency within regions specifying CDRs 1 and 2. While such mutations can lead to decreased as well as increased affinity for antigen, cells with receptors that gain affinity are preferentially selected to proliferate and differentiate, resulting in an overall increased affinity in secondary and subsequent re-

sponses. Finally, a B cell can switch to making a different class of antibody while maintaining its original antigen specificity. This is achieved by the alternative processing of a common heavy chain transcript (as in IgM and IgD expression) or by class-switch recombination (as in IgG, IgE, or IgA expression).

T cells also express clonally distributed antigen-specific receptors with variable and constant regions. And, as in the case of immunoglobulins, the enormous diversity of T cell receptors (TCRs) is achieved by a number of mechanisms, including the random rearrangement of V, D, and J gene segments. Helper T cells and cytotoxic T cells express the $\alpha\beta$ TCR that recognizes a complex epitope consisting of foreign peptide and self MHC. Helper T cells in particular play a central role in the immune response. These cells detect antigen that has been taken in and degraded by specialized antigen-presenting cells (APCs). The cells then respond by secreting a number of cytokines that drive antigen-activated T_c cells and B cells to become efficient effectors of the immune response.

KEY TERMS FOR SELF-TESTING

innate immunity (p. 784)

adaptive immunity (p. 784)

acquired immunodeficiency syndrome (AIDS) (p. 784)

human immunodeficiency virus (HIV) (p. 784)

immunology (p. 784)

The Immune Response

immune response (p. 784)

immunogen (antigen) (p. 784)

autoimmune disease (p. 785)

antigenic determinant (epitope) (p. 785)

humoral immune response (p. 785)

antibody (p. 785)

cell-mediated immune response (p. 785)

The Cellular Basis of the Immune Response

lymphocyte (p. 785)

lymphoid tissue (p. 785)

B lymphocyte (B cell) (p. 785)

T lymphocyte (T cell) (p. 785)

hematopoietic stem cell (p. 785)

hematopoietic tissue (p. 786)

central lymphoid tissue (p. 786)

peripheral lymphoid tissue (p. 786)

plasma cell (p. 786)

antigen receptor (p. 787)

immunoglobulin (Ig) (p. 787)

T cell receptor (TCR) (p. 787)

clonal selection theory (p. 788)

antigen-independent differentiation (p. 788)

antigen-dependent differentiation (p. 789)

effector cell (p. 789)

cytotoxic T (T_c) cell (p. 789)

helper T (T_h) cell (p. 789)

suppressor T (T_s) cell (p. 789)

immunological memory (p. 789)

primary immune response (p. 789)

secondary immune response (p. 789)

memory cell (p. 789)

differentiation marker (p. 790)

CD3 complex (p. 790)

CD4 glycoprotein (p. 790)

CD8 glycoprotein (p. 790)

Lymphocyte Activation Pathways

cytokine (p. 790)

graft rejection (p. 791)

transplantation (histocompatibility) antigen (p. 791)

major histocompatibility complex (MHC) (p. 791)

human leukocyte-associated (HLA) antigen (p. 791)

HLA complex (p. 791)

histocompatibility (H-2) antigen (p. 791)

H-2 complex (p. 791)

polymorphism (p. 791)

class I MHC antigen (p. 791)

class II MHC antigen (p. 791)

β2-microglobulin (p. 792)

antigen-presenting cell (APC) (p. 792)

exogenous antigen (p. 793)

endogenous antigen (p. 794)

target cell (p. 794)

class I-restricted (p. 794)

class II-restricted (p. 794)

interleukin-1 (IL-1) (p. 794)

interleukin-2 (IL-2) (p. 794)

γ-interferon (IFN-γ) (p. 794)

immunological tolerance (p. 795)

The Structure and Function of Antibodies

variable (V) domain (p. 796)

constant (C) domain (p. 796)

light (L) chain (p. 796)

heavy (H) chain (p. 796)

immunoglobulin class (isotype) (p. 796)

Fc receptor (FcR) (p. 796)

IgM (p. 797)

IgM pentamer (p. 797)

J (joining) chain (p. 797)

complement system (p. 797)

PROBLEM SET

24-1. Immunological True or False. Indicate whether each of the following statements is true (T) or false (F), and explain the error in each false statement.

(a) Most antigens can have many different antibody molecules bound to them.

(b) The removal of the bursa of Fabricius from a newly hatched chick will impair both the humoral and the cell-mediated immune responses of the chick.

(c) Immunoglobulin-producing cells capable of responding to a specific antigen are already present within the lymphoid tissues of an animal prior to exposure to antigen.

(d) A cell expressing class I, but not class II, MHC proteins can present antigen to CD4+ T cells.

(e) The primary humoral response typically results in high-level IgG production.

(f) Immunological memory results from the presence of long-lived lymphocytes generated upon exposure to antigen.

24-2. Lymphocytes and the Immune Response. Figure 24-23 depicts an experiment done in 1954 to establish the central role of lymphocytes in the immune response. ① A mouse that would otherwise show a normal immune response upon immunization with antigen is unable to make a response because it is first heavily irradiated. ② This mouse is given lymphocytes from a normal donor of the same inbred line (they are genetically identical) and the immune response is restored. ③ However, when nonlymphocytes (other white blood cells) are provided, the immune response of the irradiated mouse is not restored.

(a) Why is the irradiated mouse in step 2 unable to respond to antigen?

(b) What conclusions can be drawn from steps 3 and 4 of the experiment?

(c) Which would you expect to find restored in step 3, the humoral immune response, the cell-mediated immune response, or both?

(d) How would the immune response of the mouse in step 2 differ if, instead of being irradiated, the animal had its thymus removed shortly after birth?

(e) How important is it that the donor mouse in steps 3 and 4 be of the same inbred line as the irradiated animal?

(f) What do you think would be the easiest way to detect or measure the immune response in an animal?

24-3. The Major Cellular Players in the Immune Response. Indicate whether each of the following features is descriptive of the macrophage (M), B cell (B), T_c cell (Tc), and/or T_h cell (Th). There may be more than one cell type listed for a single feature.

(a) Presents antigen to T_h cells.

(b) Internalizes antigen by phagocytosis.

(c) Internalizes antigen by receptor-mediated endocytosis.

(d) Produces IL-1.

(e) Produces IL-2.

(f) Responds to IL-1.

(g) Responds to IL-2.

(h) Produces perforins.

(i) Differentiates into a plasma cell.

(j) Functions directly in cell-mediated lysis.

(k) Responds to soluble antigen alone.

(l) Expresses CD3.

(m) Expresses CD8.

(n) Required for B cell activation.

24-4. Antibody Structure. Indicate whether each of the following statements is true of the light chain only (L), of the heavy chain only (H), of both light and heavy chains (LH), or of neither chain (N).

(a) It is found in the Fabs but not in the Fc region of the immunoglobulin molecule.

(b) It contains both a variable and a constant region.

(c) It associates with the secretory component when an IgA dimer crosses an epithelial cell.

(d) It has a length of about 110 amino acids.

(e) It has a length of about 220 amino acids.

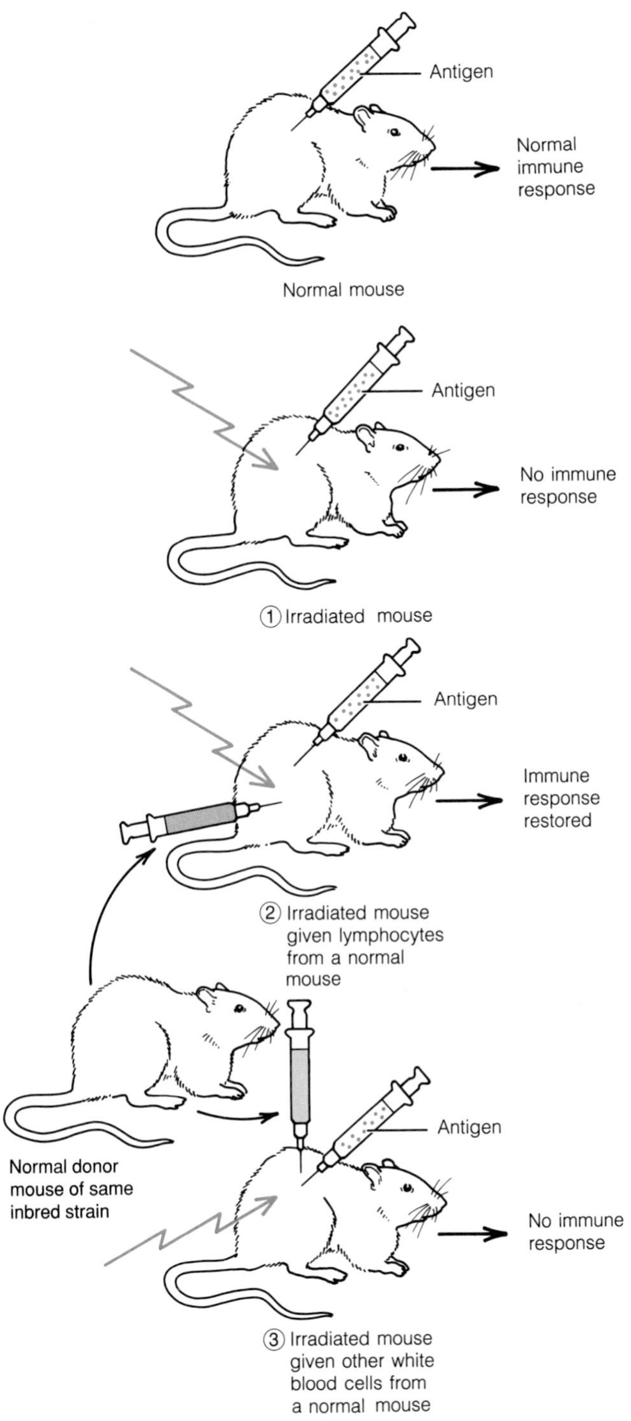

Antigen
Normal immune response
Normal mouse
Antigen
No immune response
① Irradiated mouse
Antigen
Immune response restored
② Irradiated mouse given lymphocytes from a normal mouse
Antigen
Normal donor mouse of same inbred strain
No immune response
③ Irradiated mouse given other white blood cells from a normal mouse

Figure 24-23 Lymphocytes and the Immune Response.

(f) It is involved in the antigen-binding site.

(g) When IgE binds to mast cells, it associates with the Fc receptor.

(h) It associates with β2-microglobin on the surface of B cells.

24-5. The Instruction Hypothesis of Antibody Diversity. Before the clonal selection theory was accepted by immunologists in the 1950s, antibody diversity was explained by the *instruction hypothesis*. According to this theory, antibodies are made as unfolded polypeptide chains, and their final conformation is determined by their association with a specific antigen.

(a) Why do you suppose the instruction hypothesis was an attractive explanation of antibody diversity at one time?

(b) What crucial discovery about the three-dimensional structure of protein molecules do you suppose led to the demise of the instruction hypothesis?

24-6. Antibody Diversity. Consider a genome that contains 500 V_κ and 5 J_κ light chain gene segments and 400 V, 20 D, and 4 J heavy chain gene segments. The genome also contains 2 C_κ sequences and 6 C_H sequences.

(a) Calculate the number of different antibody specificities that can be generated from this genome.

(b) Are the different forms of C_κ and C_H genes considered when calculating the diversity of the antibodies? Why or why not?

(c) Describe two other mechanisms that further enhance antibody diversity.

(d) Compare the mechanisms for the generation of diversity in immunoglobulins and TCR receptors.

24-7. Ordering C_H Gene Segments. Unstimulated lymphocytes from the organism *Vaccinatia hypothetica* make only membrane-bound IgM, though their germ line DNA is known to contain the genes for the constant regions of five different heavy chains, designated m, n, o, p, and q. Myeloma cells secreting IgQ are found to lack the DNA sequences that code for C_m and C_o, while cells secreting IgN lack the sequences that encode all other classes of heavy chains. It has also been shown that some lymphocytes synthesize both IgM and IgO at the same time.

(a) How do you suppose it was determined that the DNA of specific myeloma cells was missing certain sequences?

(b) Suggest an order for the five CH genes and explain why you ordered them as you did.

(c) Suggest a mechanism that could account for the expression of both IgO and IgM in the same cell.

24-8. The Role of MHC Molecules. A singularly important series of experiments performed in the mid-1970s led to our understanding of the role of class I MHC glycoproteins in cellular immune responses. Inbred mice of strain 1 were infected with virus X. One week later, activated T_c cells were recovered from the spleens of these mice and tested for their ability to kill virus-infected mouse fibroblast cells in culture. Explain each of the following findings.

(a) Cultured fibroblasts from strain 1 that were infected with virus X were killed by the T_c cells within a few hours.

(b) Cultured fibroblasts from strain 1 that were infected with an unrelated virus Z were not killed by the T_c cells.

(c) Cultured fibroblasts from strain 2 that were infected with virus X were not killed by the T_c cells.

(d) The experiment in part a was repeated with cultured fibroblasts that differed genetically from the infected mice at all loci except at the class I MHC loci. The virus X–infected fibroblasts were killed.

(e) The experiment in part a was repeated with cultured fibroblasts that were genetically identical to the infected mice at all loci except at the class I MHC loci. The virus X–infected fibroblasts were not killed.

24-9. Acquired Immunodeficiency Syndrome (AIDS). The CD4 glycoprotein is a receptor for human immunodeficiency virus (HIV). Describe how this is related to the severe immunodeficiency seen in AIDS.

SUGGESTED READING

General References

Abbas, A. K., A. H. Lichtman, and J. S. Pober. *Cellular and Molecular Immunology*, 2d ed. Philadelphia: Saunders, 1994.

Kuby, J. *Immunology*, 2d ed. New York: W. H. Freeman, 1994.

Paul, W. E. *Fundamental Immunology*, 3d ed. New York: Raven, 1993.

The Immune System

Ada, G. L., and G. Nossal. The clonal selection theory. *Sci. Amer.* 257 (August 1987): 62.

Burnet, F. M. *The Clonal Selection Theory of Acquired Immunity.* Nashville, TN: Vanderbilt University Press, 1959.

Nossal, G. Life, death and the immune system. *Sci. Amer.* 269 (September 1993): 52.

Lymphocyte Development

Dexter, T. M., and E. Spooncer. Growth and differentiation in the hematopoietic system. *Annu. Rev. Cell Biol.* 3 (1987): 423.

Robey, E. and B. J. Fowlkes. Selective events in T cell development. *Annu. Rev. Immunol.* 12 (1994): 675.

Weissman, I. L. Development switches in the immune response. *Cell* 76 (1994): 207.

The Major Histocompatibility Complex

Bjorkman, P. J., and P. Parham. Structure, function and diversity of class I major histocompatibility complex molecules. *Annu. Rev. Biochem.* 59 (1990): 253.

Bjorkman, P. J., M. A. Saper, B. Samraoui, W. S. Bennett, J. L. Strominger, and D. C. Wiley. Structure of the human class I histocompatibility antigens. *Nature* 329 (1987): 506.

Brown, J. H., T. S. Jardetsky, J. C. Gorga, L. J. Stern, R. G. Urban, J. L. Strominger, and D. C. Wiley. Three-dimensional structure of the human class II histocompatibility antigen HLA-DR1. *Nature* 364 (1993): 33.

Trowsdale, J., J. Ragoussis, and R. D. Campbell. Map of the human MHC. *Immunol. Today* 12 (1991): 443.

Lymphocyte Activation

Arai, K., F. Lee, A. Miyajima, S. Miyatake, N. Arai, and T. Yokota. Cytokines: Coordinators of immune and inflammatory responses. *Annu. Rev. Biochem.* 59 (1990): 783.

Berke, G. The binding and lysis of target cells by cytotoxic lymphocytes: Molecular and cellular aspects. *Annu. Rev. Immunol.* 12 (1994): 735.

Germaine, R. N., and D. H. Marguiles. The biochemistry and cell biology of antigen processing and presentation. *Annu. Rev. Immunol.* 11 (1993): 403.

Janeway, C. A. and K. Bottomly. Signals and signs for lymphocyte responses. *Cell* 76 (1994): 275.

Nossal, G. Molecular and cellular aspects of immunologic tolerance. *Eur. J. Biochem.* 202 (1991): 729.

Parker, D. C. T cell-dependent B cell activation. *Annu. Rev. Immunol.* 11 (1993): 331.

Vitetta, E. S., M. T. Berton, C. Burger, M. Kepron, W. T. Lee, and X. M. Yin. Memory B and T cells. *Annu. Rev. Immunol.* 9 (1991): 193.

Antibody Structure and Function

Alzari, P. M., M. B. Lascombe, and R. J. Polijak. Three dimensional structure of antibodies. *Annu. Rev. Immunol.* 6 (1988): 555.

Greenspan, N. S., and C. A. Bona. Idiotypes: Structure and immunogenicity. *FASEB* 7 (1993): 437.

Stanfield, R. L., T. M. Fieser, R. Lerner, and I. A. Wilson. Crystal structures of an antibody to a peptide and its complex with peptide antigen at 2.8 A. *Science* 248 (1990): 712.

Wilson, I. A., and R. L. Stanfield. Antibody-antigen interactions. *Curr. Opin. Struc. Biol.* 3 (1993): 113.

The Genetic Basis of Diversity and Class Switching

Berek, C. and M. Ziegner. The maturation of the immune response. *Immunol. Today* 14 (1993): 400.

Chen, J., and F. W. Alt. Gene rearrangement and B cell development. *Curr. Opin. Immunol.* 5 (1993): 194.

Dreyer, W. J. and J. C. Bennett. The molecular basis of antibody formation: A paradox. *Proc. Natl. Acad. Sci. USA* 54 (1965): 864.

Harriman, W., H. Volk, N. Defranoux, and M. Wabl. Immunoglobulin class switch recombination. *Annu. Rev. Immunol.* 11 (1993): 361.

Schatz, D. G., M. A. Oettinger, and M. S. Schlissel. V(D)J recombination: Molecular biology and regulation. *Annu. Rev. Immunol.* 10 (1992): 359.

Tonegawa, S. Somatic generation of antibody diversity. *Nature* 302 (1983): 575.

The T Cell Antigen Receptor

Allison, J. P. and L. L. Lanier. Structure, function and serology of the T cell antigen receptor complex. *Annu. Rev. Immunol.* 5 (1987): 503.

Chien, Y. and Davis, M. M. How $\alpha\beta$ T-cell receptors "see" peptide/MHC complexes. *Immunol. Today* 14 (1993): 597.

Davis, M. M. T cell receptor gene diversity and selection. *Annu. Rev. Biochem.* 59 (1990): 475.

Hedrick, S. M., D. I. Cohen, E. A. Nielsen, and M. M. Davis. Isolation of cDNA clones encoding T cell-specific membrane associated proteins. *Nature* 308 (1984): 149.

Raulet, D. H. The structure, function, and molecular genetics of the $\gamma\delta$ T cell receptor. *Annu. Rev. Immunol.* 7 (1989): 175.

Monoclonal Antibodies

Köhler, G., and C. Milstein. Continuous cultures of fused cells secreting antibody of predefined specificity. *Nature* 256 (1975): 495.

Milstein, C. Monoclonal antibodies. *Sci. Amer.* 243 (October 1980): 66.

Reichman, L., M. Clark, H. Waldmann, and G. Winter. Reshaping human antibodies for therapy. *Nature* 323 (1988): 323.

25

CELLULAR ASPECTS OF CANCER

S o far, our discussion of cellular structure and function has focused on normal cells. Knowing how complex cells are, we
can appreciate how remarkable it is that their
appearance and function are normal so
much of the time. This is particularly striking
in the development of higher organisms, which
requires the careful orchestration of the activities
of large numbers of cells that perform different
functions. A good example is the human body. In most
cases, the anatomical organization of the body is stable over
time. The volumes of the different tissues and organs are relatively constant and appropriately proportioned with respect to the size of the body. Furthermore, the organization
of the tissues within organs and the differentiated character
of individual cells of the different tissues are also stable. This
stability of tissue size, differentiation, and organization is essential for the maintenance of function.

However, in all multicellular organisms, from plants to
humans, the regulatory mechanisms that control cell division sometimes go awry. In this situation, uncontrolled
growth may begin, resulting in loss of tissue stability, reduced tissue function, and even death of the organism. In
the last few years we have come to understand some, but by
no means all, of the molecular mechanisms that cause cells
to become transformed in this manner. Interestingly, learning about these abnormal cellular events is also teaching us a
great deal about normal cellular processes. With increased
understanding comes the hope that we may someday be able
to inhibit the transformation process, or, failing that, to control the rampant growth that leads all too often to a fatal
outcome.

Cancer: A Loss of Normal Growth Regulation

If the normal stability of the organization of tissues and organs is disturbed, a variety of disease states can arise. One
example is a tissue in which the control of growth becomes

defective, forming what is called a **tumor** or
neoplasm (literally, "new growth"). Neo-
plasms can be classified as benign or malignant based on their likelihood of spreading.
Benign tumors are typically *encapsulated
nodules* of neoplastic tissue and therefore do
not spread, whereas **malignant tumors** often
invade neighboring tissues and even other parts
of the body, and thus may become lethal. The common term for a malignant tumor is **cancer.** The word is from
the Latin term for "crab" because early physicians noticed
that certain skin cancers had a crablike appearance.

Neoplastic Transformation

In almost every case, malignant tumors are *monoclonal,*
meaning that they develop from a single progenitor cell. The
progenitor cell has undergone a series of permanent, heritable changes in a process called **neoplastic transformation.**
Cells that have been transformed are characterized by two
general features. First, they exhibit uncontrolled growth.
These cancer cells don't necessarily grow faster than normal
cells, but they bypass normal constraints on cell growth and
proliferation. Second, as mentioned, they tend to spread.
The spread of cancer cells to neighboring tissues is called
invasion; the spread to distant organs is termed **metastasis.**
The tumor nodules that implant at sites distant from the
parent tumor are referred to as *metastases.*

In this chapter we will examine the transformed cell
phenotype, investigate the genetic alterations that contribute to transformation, and consider how these genetic
changes arise. Then we will examine the mechanisms by
which these triggers deregulate normal cell growth and tissue organization, leading ultimately to tumor progression
and dissemination.

Classification of Tumors

As we will be discussing numerous cancers throughout the
chapter, here is a brief guide to their nomenclature. Tumors
may arise from mature, differentiated cells or from mitoti-

cally active stem cell populations. **Stem cells** are relatively undifferentiated, actively dividing cells from which some more highly differentiated cells originate, or "stem." An example is the hematopoietic ("blood-forming") stem cell population found in bone marrow that gives rise to red and white blood cells (see Figure 24-2). The stem cells of various tissues are often designated by the suffix *-blast*. For example, *neuroblasts* are the mitotic stem cells for neurons, *myoblasts* are the stem cells for myocytes, and *fibroblasts* are the stem cells for fibrocyte connective tissue cells. Cancers of a stem cell may be designated by the suffix *-oma: neuroblastoma* for a tumor of neuroblastic origin; *fibroma* for a tumor of fibroblastic origin.

Another important element in tumor categorization is the tissue class of origin. *Carcinoma* refers to a tumor derived from an epithelial tissue, and *sarcoma* refers to a tumor derived from connective tissue. As indicated above, the hematopoietic system includes stem cell populations that grow and divide throughout life. All these populations are subject to neoplasia: *Lymphomas* are tumors of the lymphocytic lineage; *leukemias* and *myelomas* are tumors of the leukocytic system (granulocytes, eosinophils, and basophils); and *erythroblastomas* are tumors of the erythroblastic (red blood cell) lineage.

Growth Properties of Cells in Culture

The most useful system for studying the regulation of tissue growth in both normal and cancerous cells has been *monolayer* tissue culture. Cells are placed in culture dishes and provided with nutrient culture medium. Many different types of cells have been studied, including cells freshly isolated from embryonic tissues, cultured cell lines (cells that have been maintained in culture for extended periods and are specifically adapted to cell culture), and cells that have been transformed. Normal cells and transformed cells have been shown to differ in many growth properties when studied this way (Figure 25-1).

Normal Cells

When grown in tissue culture in the laboratory, normal cells exhibit *mortality,* meaning that they grow and divide for a characteristic number of generations and then die spontaneously. While growing, they display the property of **contact inhibition**—they divide and spread until they touch each other, forming a single layer of cells across the surface of the culture dish, and then stop dividing. Normal cells must be regularly provided with nutrients in the form of *serum,* the cell-free portion of blood isolated after the blood has been allowed to clot.

Transformed Cells

In contrast, transformed cells are *immortal*—they continue to grow and divide indefinitely. They are not inhibited by contact with other cells, so that rather than forming a monolayer, they tend to pile on top of one another in the culture dish. Transformed cells do not have a requirement for serum. Cells with this transformed phenotype also display a highly variable karyotype, and their nuclei are often large and abnormally shaped.

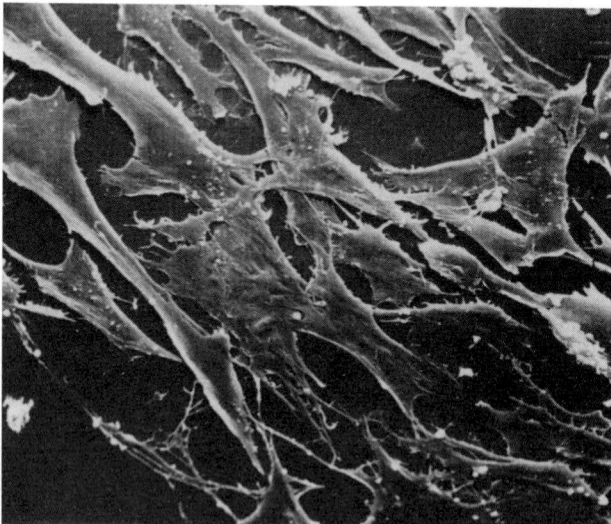

(a) Normal cells

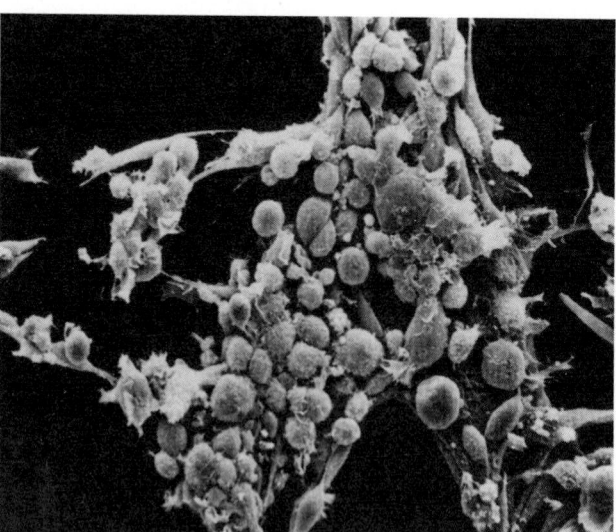

(b) Transformed cells

Figure 25-1 Phenotypic Changes of Transformed Cells. Cells that have undergone neoplastic transformation are characterized by an unusual appearance and growth pattern. **(a)** Chick fibroblasts grown in culture are typically flat and adhere well to their culture dish. **(b)** The same cells, after being experimentally transformed, develop an abnormal rounded shape and tend to cluster together rather than spreading across the culture dish (SEMs). $30 \mu m$

Table 25-1 Examples of Oncogenic Viruses

Virus Family	Specific Virus	Tumor
DNA Tumor Viruses		
Poxviruses	Shope's rabbit fibroma virus	Epidermal carcinomas
Herpesviruses	Marek's disease virus	Marek's disease (a lymphoma of chickens)
	Lucké tumor virus	Lucké renal adenocarcinoma (frog)
	Epstein-Barr virus	Burkitt's lymphoma and nasopharyngeal carcinoma (human)
Adenoviruses	Adenovirus type 12	Multiple tumors result only when injected into newborn rodents
Papovaviruses	Shope's rabbit papilloma virus	Sarcomas (rabbit)
	Human papilloma virus	Cervical carcinoma (human)
	Polyoma viruses	Multiple tumors result only when injected into newborn mice
	Simian virus 40	Multiple tumors result only when injected into newborn hamsters
RNA Tumor Viruses (Retroviruses)		
B-type mammary tumor viruses	Mouse mammary tumor virus	Mammary carcinoma
C-type leukoviruses	Avian (Rous) sarcoma viruses	Leukemias and sarcomas
	Murine sarcoma viruses	
	Feline sarcoma viruses	
	Human T cell leukemia virus-I	

The Genetic Basis of Neoplasia

With only a few exceptions, the neoplastic phenotype is heritable; that is, the daughter cells of a transformed cell also show abnormal growth regulation. This heritability of the neoplastic phenotype suggested to early investigators that genetic alterations play an important role in the establishment and progression of cancers. Currently, one of the most exciting areas of cancer biology is the study of the genetic basis of cancer. The sophisticated tools of virology, genetics, and molecular biology are enabling investigators to clarify the genetics and physiology of cellular growth regulation, resulting in a rapid expansion of our understanding of the details of both normal and neoplastic tissue growth. These studies have defined two categories of genes that are involved in the development of cancer: oncogenes and tumor-suppressor genes. We'll consider each of these in turn in the following sections.

Oncogenes

An **oncogene** is a gene that contributes to neoplastic transformation when introduced into a normal cell. Two means of experimentally introducing oncogenes into cultured cells have been widely used: (1) infection with an **oncogenic** ("tumor-causing") **virus,** a virus that carries an oncogene in its genome; and (2) the introduction of DNA isolated from a naturally occurring tumor into a normal cell, a laboratory technique known as **transfection.** In the first case, the oncogene's identity can be deduced by mutating each viral gene individually and determining which gene's function is required for transformation. In the case of transfection, the oncogene can be identified by isolating and characterizing the particular tumor-derived DNA that transforms the cells.

Oncogenic Viruses. Certain families of viruses are able to trigger a neoplastic transformation in the cells they infect. Members of both the DNA and the RNA classes of viruses (viruses that have, respectively, DNA and RNA as their genetic material) can cause infected cells to become cancerous (Table 25-1). One piece of evidence for a viral etiology is that patients with tumors may have antibodies to antigens associated with the suspected virus (Table 25-2). Further-

---

Table 25-2 Demonstrating a Tumor's Viral Etiology

A. Associate a virus with the tumor by detecting:
  1. Production of viral particles by tumor cells.
    Characterize the particles as viruses by:
    a. Morphology (electron microscopy)
    b. Physical properties:
      Buoyant density
      Molecular weight of nucleic acid
    c. Biochemical properties:
      Viral coat antigens
      Virus-specific enzymes (reverse transcriptase)
    d. Infectivity
  2. Virus-specific antigens within or on the surface of tumor cells
  3. Viral nucleic acid in tumor cells by molecular hybridization
  4. Systemic effects on the diseased host:
    a. Excretion of viral particles by the host
    b. Presence of antiviral antibodies in the serum

B. Demonstrate experimentally by:
  1. Isolating a particular virus from hosts with a specific tumor and demonstrating tumor production by the virus when inoculated into disease-free hosts.
  2. Repeating virus isolation and infection cycle with second set of disease-free hosts, showing that the same tumor is produced.

more, cancerous cells are often found to contain nucleic acid sequences that are diagnostic of the virus. The experimental demonstration of the ability of particular viruses to cause tumors begins with the isolation of viral particles from tumors in mice and other experimental animals. When inoculation of this preparation of viruses into healthy, tumor-free individuals results in the development of tumors, the tumorigenic potential of the virus is confirmed. When the cycle can be repeated by isolating the same type of virus from the experimentally infected individual and using the virus to induce tumors in healthy hosts, the evidence becomes very strong that the particular virus plays a direct role in the initiation of the particular tumor.

The RNA tumor viruses that can cause cancer belong to the *retrovirus* family (Figure 25-2). When one of these viruses infects a cell, it stimulates the production of a complementary DNA copy of the viral genome, called the *provirus,* which becomes incorporated into one of the chromosomes of the host cell (see Box 17A, Figure 17A-1, p. 538). Synthesis of the DNA copy from the viral RNA template is catalyzed by the viral enzyme *reverse transcriptase.* The provirus replicates along with the host cell's DNA, synthesizing RNA copies of the provirus by transcription. These viral RNA copies code for the viral proteins needed to encapsulate new viral particles that contain the RNA as genetic material. The viral particles then bud from the plasma membrane of the infected cell.

The actual oncogene carried by the transforming virus can be identified by mutational analysis. For example, the avian sarcoma virus (Figure 25-3) leads to transformation only when it has a functional *src* gene, indicating that *src* is the oncogene.

Oncogene Transfection Assay. The genes responsible for cancer development in naturally occurring tumors can be isolated and identified using an **oncogene transfection assay** (Figure 25-4). This assay involves three steps: transfection, transformation, and analysis. First, the DNA from the tumor is isolated, cleaved with a restriction endonuclease, and presented to nonneoplastic cultured cells, such as mouse fibroblasts, under conditions that favor the incorporation of foreign DNA fragments into the cells' chromosomes. These transfected cells are allowed to grow in culture and are observed for any type of abnormal growth that might indicate neoplastic transformation had occurred. Most of the transfected cells acquire human DNA sequences that are irrelevant to the neoplastic transformation process, so their growth properties are not affected.

Next, the few cultured cells that display deviant growth characteristics are tested for the ability to cause neoplastic transformation when they are transplanted to experimental animals. If a tumor results from this step, it confirms that the cells indeed carry an oncogene, rather than having some other abnormality responsible for their unusual growth in culture.

Finally, the DNA is isolated from the culture of these neoplastic cells, and the oncogene it carries is identified by comparing it to known oncogenes. To do this, the human DNA is isolated from the cultured mouse cells and fragmented. Meanwhile, known oncogenes from various oncoviruses are isolated and radioactively labeled to create molecular *probes.* When a probe is mixed with the sample

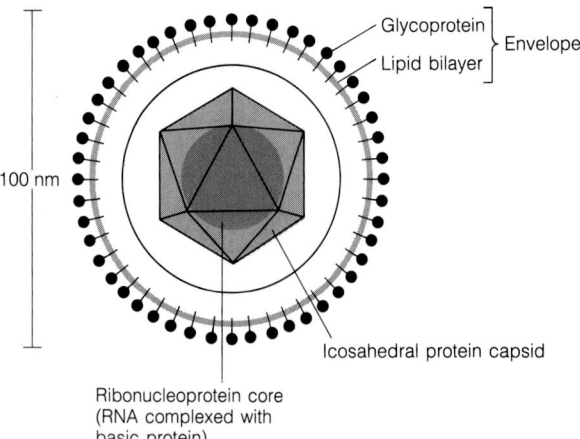

Figure 25-2 The Structure of a Retrovirus. A typical retrovirus consists of a ribonucleoprotein core, an icosahedral protein capsid, and an envelope containing viral glycoprotein molecules in a lipid bilayer that has been derived from the plasma membrane of the host cell. Two copies of the viral RNA genome are packaged in the core, as are molecules of reverse transcriptase.

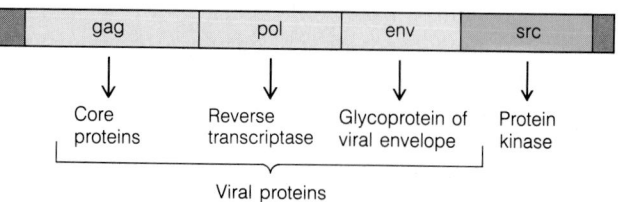

Figure 25-3 Genome Diagram of the Avian Sarcoma Virus, a Transforming Retrovirus. The wild-type progenitor of the avian sarcoma virus presumably contained in its genome only the genes necessary for viral replication. These are *gag,* which codes for core proteins of the virion; *pol,* which encodes reverse transcriptase; and *env,* which codes for glycoproteins of the viral envelope. In addition, the virus contains a fourth gene, *src,* responsible for the transformation of cells infected by the virus. All transformation-defective mutants of the avian sarcoma virus map to the *src* locus; *src* is thus the oncogene carried by avian sarcoma virus.

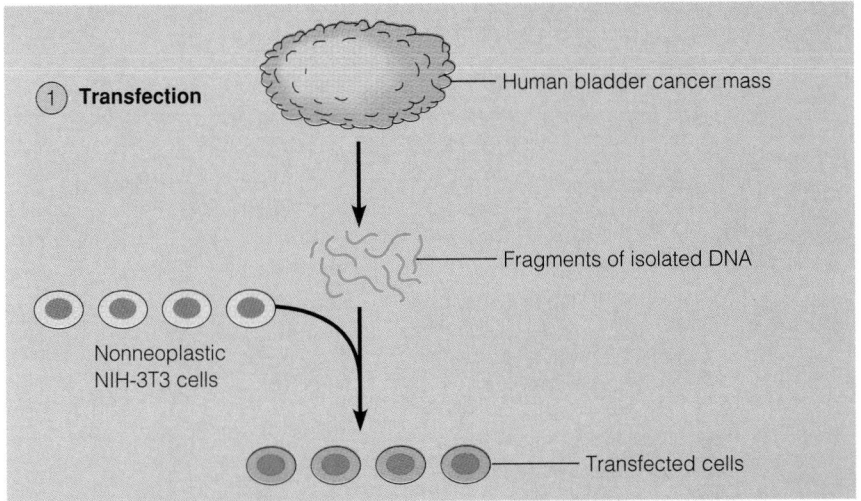

① **Transfection**

Human bladder cancer mass

Fragments of isolated DNA

Nonneoplastic NIH-3T3 cells

Transfected cells

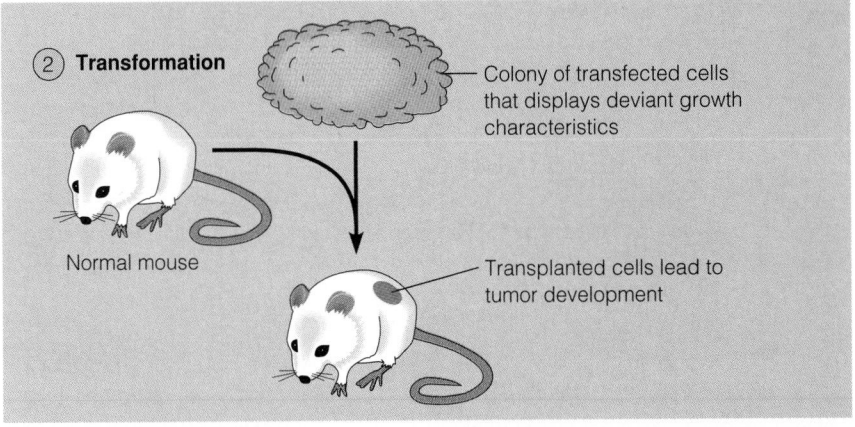

② **Transformation**

Colony of transfected cells that displays deviant growth characteristics

Normal mouse

Transplanted cells lead to tumor development

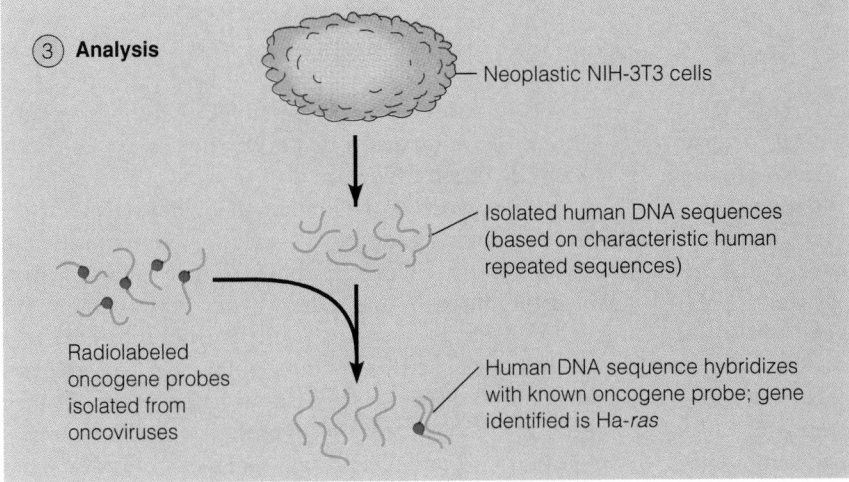

③ **Analysis**

Neoplastic NIH-3T3 cells

Isolated human DNA sequences (based on characteristic human repeated sequences)

Radiolabeled oncogene probes isolated from oncoviruses

Human DNA sequence hybridizes with known oncogene probe; gene identified is Ha-*ras*

Figure 25-4   The Gene Transfection Assay Used for Identifying Oncogenes in a Spontaneous Human Tumor.

① *Transfection.* DNA is isolated from cells obtained from a human bladder carcinoma and is fragmented by *Bam1* endonuclease. Nonneoplastic mouse fibroblasts of the NIH-3T3 cell line are mixed with the DNA under conditions in which the cells incorporate the DNA fragments into their own chromosomes. Most of the transfected cells contain irrelevant bits of the human DNA, but a few contain the oncogene responsible for the bladder tumor. ② *Transformation.* The transfected cells are grown in culture, and the few cell colonies that display deviant growth characteristics are isolated and tested for neoplastic growth behavior by transplantation into healthy mice. If a tumor develops from the transplanted cells, the transfected cells must have carried an oncogene from the original human tumor. ③ *Analysis.* The neoplastic NIH-3T3 cells that triggered the mouse tumor are grown in culture, and the human DNA they harbor is isolated based on the characteristic human repeated sequences it contains. This DNA is analyzed for its ability to hybridize with any of a variety of probes derived from known oncogenes that were isolated from oncoviruses and radioactively labeled. In this case, the oncogenic human DNA bound to a known probe and was thereby identified as Ha-*ras*.

DNA under proper conditions, it will selectively attach to complementary sequences, or *hybridize,* allowing identification of the oncogene in question.

Proto-Oncogenes: Normal Cellular Homologues of Oncogenes.   These two independent means of transforming nonneoplastic cells—infection with an oncovirus and transfection with DNA from a spontaneous tumor—have identi-

fied overlapping sets of genes that can operate as oncogenes (Table 25-3). One of the most unexpected findings of these studies, for which Michael Bishop and Harold Varmus received a Nobel Prize in 1989, is that *most oncogenes isolated from RNA tumor viruses and from spontaneous tumors are variants of genes normally present in the genome of the host animal.* In the usual terminology, the overt oncogene is denoted by an italicized abbreviation, such as *src, myc,* or *ras,*

Table 25-3 Transforming Genes of Transforming Retroviruses

Gene/Gene Product	Prototype Virus	Isolation Source
Oncogenes That Code for Growth Factors		
sis/p28^{sis}	Simian sarcoma virus	Woolly monkey
Oncogenes That Code for Protein Kinases		
src/p60^{src}*	Rous sarcoma virus	Chicken
fps**/p140*	Fujinami sarcoma virus	Chicken
fes**/p85*	Snyder-Theilen feline sarcoma virus	Cat
yes/p90*	Y73 sarcoma virus	Cat
erb-B/p65*	Avian erythroblastosis virus	Chicken
ros/p68*	UR2 sarcoma virus	Cat
fms/p140*	McDonough feline sarcoma virus	Cat
fgr/p70*	Gardner-Rasheed feline sarcoma virus	Cat
mos/p37†	Moloney sarcoma virus	Mouse
raf/?†	3611 murine sarcoma virus	Mouse
abl/p120*	Abelson leukemia virus	Mouse
Oncogenes That Code for Ras Proteins		
Ha-ras/p21-Ha-ras	Harvey sarcoma virus	Rat
Ki-ras/p21-Ki-ras	Kirsten sarcoma virus	Rat
Oncogenes That Code for Nuclear Proteins		
myc/p110	Myelocytomatosis virus	Chicken
myb/p48	Avian myeloblastosis virus	Chicken
fos/p55	FBJ osteosarcoma virus	Mouse

* Tyrosine protein kinase activity.
** Homologous genes.
† Serine, threonine protein kinase activity.

and its normal cellular homologue is denoted by the same abbreviation with the prefix "c-": c-src, c-myc, c-ras. Many of these normal cellular homologues of the oncogenes, also termed **proto-oncogenes,** are widely distributed in eukaryotes. Homologues of the src oncogene have been identified in many vertebrates and insects; homologues of the ras oncogene are present in humans and yeasts. The evolutionary conservation of the proto-oncogenes suggests that their gene products mediate important cellular processes.

The presence of cellular homologues of overt oncogenes raises a key question: What types of alterations transform genes involved in the biology of normal cells into genes that contribute to loss of growth control and cancer development? Two basic categories of alteration have been postulated: expression effects and gene mutations. An *expression effect*, in this case, refers to a genetic change, such as gene amplification or chromosomal rearrangement, that causes overproduction of the gene product. The overproduced protein itself is normal. The second type of alteration, *gene mutation,* results in the synthesis of normal amounts of a protein that has an altered function as a result of a deletion or a point mutation.

The human tumor *Burkitt's lymphoma* provides an example of an expression effect. In Burkitt's lymphoma, a segment of chromosome 8 containing the cellular proto-oncogene c-myc is translocated to chromosome 2, 14, or 22. The sites of translocation are close to the locations of genes coding for subunits of the immunoglobulin proteins. These immunoglobulin genes are transcriptionally very active in the lymphocytic cell lineage: Approximately half of the total protein produced by the mature lymphocyte is immunoglobulin. It has been suggested that spatial proximity to a transcriptionally active immunoglobulin gene causes overexpression of the translocated c-myc gene, which in turn contributes to the establishment of the tumor. This is an example of translocation triggering neoplastic transformation. Other examples of chromosomal abnormalities and their effects will be described later in the chapter.

Genetic mutations can change proto-oncogenes into oncogenes (Figure 25-5), leading to transformation, as demonstrated for the Ha-ras gene (so named for its isolation from the Harvey sarcoma virus). Ha-ras was one of the first oncogenes identified by the cell transfection assay. The parent DNA was from a tissue culture cell line derived from a human bladder carcinoma. Under the experimental conditions used, normal c-Ha-ras was unable to produce neoplastic transformation when transfected into normal host cells. The tumor cell–derived Ha-ras gene responsible for trans-

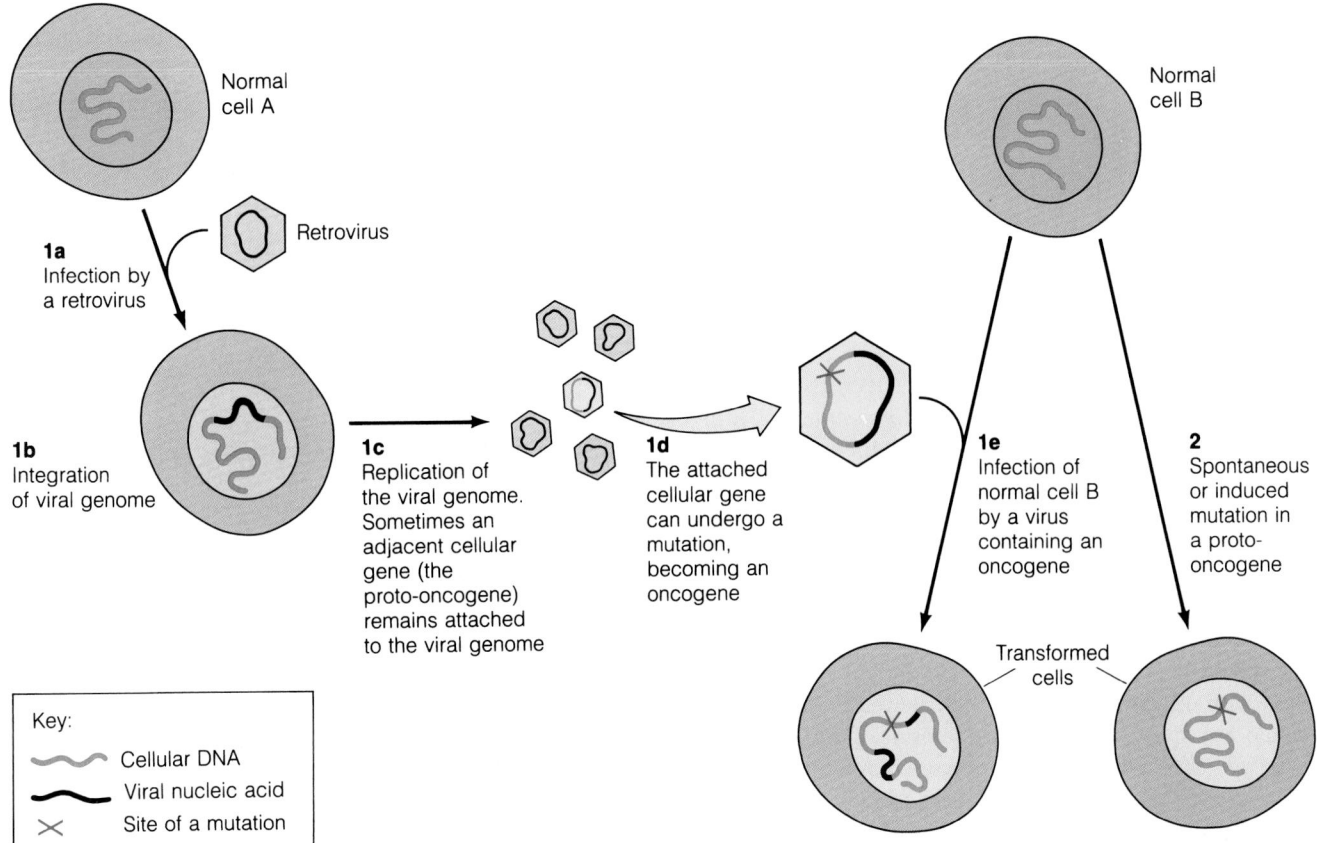

Figure 25-5   Mutation of Proto-Oncogenes to Oncogenes.
In some cases, a mutation can cause a normal cellular gene, known as a proto-oncogene, to become an oncogene. Mutation of a cellular proto-oncogene can occur in a virus that has accidentally picked up the cellular gene, or in a normal cell. ⓐ A retrovirus infects a cell, and ⓑ the viral genome integrates into a host cell chromosome. ⓒ When the virus replicates, it sometimes picks up an adjacent proto-oncogene, which be- comes packaged into new viral particles. ⓓ If the proto-oncogene is mutated and becomes an oncogene while it is carried in the virus, it will ⓔ transform a normal cell when it is introduced by viral infection. ② A proto-oncogene in a normal cell can undergo a mutation, either sponta- neously or as a result of exposure to a carcinogen, that results in an oncogene and therefore a transformed cell.

formation had several point mutations along its length. To determine which of the point mutations of the oncogenic form of Ha-*ras* was responsible for its transforming activity, hybrid DNA molecules were prepared in vitro, in which sec- tions of the normal form of the Ha-*ras* proto-oncogene were replaced by corresponding sections from the tumor-derived homologue. Transformation resulted from transfection with a hybrid molecule containing a short 5' section of the tumor-derived Ha-*ras* that replaced the same 5' segment of the normal homologue. The active 5' segment of the tumor- derived gene differed by only a single base-pair substitution from the normal, nonneoplastic form of the gene. The nor- mal codon GGC, which codes for the amino acid glycine in position 12 of the Ha-*ras* protein, had been replaced by the codon GTC, which codes for valine. These observations offer dramatic proof that point mutations can, in selected cases, transform a normal cellular proto-oncogene into an overt oncogene. For further information on the *ras* onco- gene, see Box 25A (page 835).

Tumor-Suppressor Genes

The neoplastic phenotype of spontaneous tumors is reces- sive. When neoplastic cells are fused in cell culture with nor- mal cells, the progeny, which contain the complete geno- types of both parental cells, are nonneoplastic. They show normal growth behavior in cell culture and, more impor- tantly, fail to form tumors when introduced into appropriate host animals. In these characteristics, they resemble the nor- mal rather than the neoplastic parent cell. However, after ex- tended periods of growth in cell culture, variant clones of cells that have reverted to neoplasia frequently develop. These cells grow abnormally in cell culture and form grow- ing tumors when transplanted into animals. In all cases, the reversion to neoplasia is accompanied by a loss of particular chromosomes initially derived from the normal parent cell. These observations imply that normal cells contain discrete genes on their chromosomes that *suppress* unregulated cell growth. When present, these **tumor-suppressor genes,** also known as *anti-oncogenes*, can slow the uncontrolled growth of even an overtly neoplastic cell. Once these genes are lost, however, the neoplastic cell reverts to uncontrolled growth behavior.

The *Rb* Gene.   One of the best-characterized of the tumor-suppressor genes is the human gene *Rb*. Inactivation of this gene was first associated with the inherited tumor *bilateral retinoblastoma*. The tumor develops in young children from neuroblasts of the developing retina, and both retinas develop one or more metastatic nodules. The tumor is curable by early surgery, and many individuals who have had the disease have survived to maturity and have had children. The children of parents with retinoblastoma show a 50% incidence of the tumor, demonstrating that the tumor is heritable. Development of the tumor is associated with another chromosomal abnormality sometimes involved in neoplastic transformation: a deletion (in this case, at band 14 of the long arm of chromosome 13).

It appears that the *Rb* gene resides in this region of chromosome 13 and is necessary for regulating growth of the rapidly dividing neuroblasts in the retina. Half the children with one parent who has experienced retinoblastoma would be expected to inherit one normal chromosome 13 from the disease-free parent and one chromosome 13 that lacks a functional allele of the *Rb* gene from the parent with retinoblastoma. During the many rounds of cell division in the tissue of the developing retina, occasionally cells are produced in which the remaining normal chromosome 13 is lost or suffers a somatic deletion or mutation involving the *Rb* gene. This cell, which then lacks even one normal *Rb* gene, serves as the founder of a clone of neoplastic tissue that develops into the retinoblastoma tumor. In addition to the heritable, *multifocal* form of the tumor, instances of spontaneous retinoblastoma are also known. This spontaneous condition is *monofocal* and presumably results from the somatic mutation of both copies of the *Rb* gene within a single cell. This double mutation would clearly occur more rarely than the single somatic mutation required for inactivation of the second copy of the *Rb* gene in cells of individuals in whom all cells contained only one normal copy of the gene. This accounts for the rarity and the monofocal character of spontaneous retinoblastoma.

This model received dramatic support with the identification, isolation, and characterization of the *Rb* gene. Cloned fragments of the normal chromosome 13 were tested for their ability to hybridize with fragments of chromosome 13 isolated from retinoblastoma cells. It was expected that the *Rb* gene would be present in DNA segments in the normal chromosome 13 but would be absent in the chromosome from the tumor. A candidate gene was identified that is expressed in normal retinoblasts but not in retinoblastoma cells. When the normal form of this gene was transfected into retinoblastoma cells, the cells lost the neoplastic phenotype and reverted to normal growth behavior. Thus, the *Rb* gene was identified and isolated.

The *p53* Gene.   Another tumor-suppressor gene, *p53*, has been found in recent years to be the most commonly mutated gene in human tumors. About half of the 6.5 million people worldwide who will be diagnosed with some form of cancer this year will have *p53* mutations.

Also known as the "guardian of the genome," the p53 protein normally responds in several ways to DNA damage in the cell. First, p53 acts as a transcription factor, stimulating synthesis of a 21-kDa protein that blocks cyclin-dependent kinases (Cdks) from interacting with cyclin proteins (see Chapter 15). This block serves to halt the cell cycle when DNA damage has occurred, giving the cell time to repair the damage so that genetic errors are not passed on to daughter cells. Second, if the repair fails, p53 can trigger the damaged cells to undergo **apoptosis,** or programmed cell death, before their genetic abnormalities are inherited. Third, recent evidence indicates that p53 also stimulates the DNA repair machinery, acting both directly and indirectly through other proteins. The mechanisms by which p53 performs these functions is an area of intense current investigation. Mutations in the *p53* gene not only cause a lack of these protective effects but also stimulate abnormal cell growth.

Other Causes of Neoplastic Transformation

So far, we have seen that certain genes, the oncogenes, are found in spontaneous tumors and that these genes can cause neoplastic transformation of normal cells when introduced by viral infection or by transfection. Most of these genes have been shown to be variants of normal cellular genes, proto-oncogenes, that exhibit mutations or altered expression. Furthermore, we have seen that normal cells contain tumor-suppressor genes that, when altered, can also lead to neoplastic growth. Before moving on to discuss how these genes are actually involved in regulating tissue growth, let's consider the other events known to trigger neoplasia: chromosomal alterations and exposure to chemical carcinogens.

Chromosomal Abnormalities

Chromosomal alterations such as *deletions, translocations* (exchanges of segments between chromosomes), and *inversions* of chromosomal components are involved in both the initiation and the progression of selected human tumors. An example of this, Burkitt's lymphoma, was discussed earlier. Another very well-studied example is the *Philadelphia chromosomal abnormality,* which plays a role in the initiation of the cancer *chronic myelogenous leukemia* (*CML*). CML is a tumor of a class of white blood cells referred to as *granulocytes* because of the granular appearance of their cytoplasm under the microscope. The mitotically active stem cells of normal granulocytes (known as myeloblasts) are present in bone marrow. The so-called Philadelphia chromosome (named for the city in which it was discovered) results when a portion of the long arm of chromosome 22 is transferred to chromosome 9 (Figure 25-6). The strongest evidence that this chromosomal abnormality initiates CML is that 85% of CML patients carry the Philadelphia chromosome. (The 15% of CML patients lacking the Philadelphia chromosome appear to have a different form of the disease that typically develops in elderly individuals.)

Because patients with CML lack the Philadelphia chromosome in tissues other than the leukemic cells, it can be concluded that CML results from a *somatic chromosomal alteration*. The alteration occurs in the somatic myelogenous stem tissue of the adult and is not transferred to offspring by egg or sperm. Chromosomal abnormalities occur with low frequency in any mitotically active tissue. When they occur in the *germ cell* lineage, they have the potential to be inherited by the next generation; however, when they occur in a somatic tissue cell, they are restricted to the progeny of that cell and do not appear in the germ cells.

Environmental Carcinogens

Environmental carcinogens include physical factors that can provoke neoplastic transformation, such as ionizing radiation and ultraviolet light, which are involved in the development of skin cancer and melanoma. Even more important are chemical carcinogens, which have been implicated in the development of many important human tumors. Evidence for the oncogenic potential of particular chemicals includes experimental studies with animals and epidemiological studies with humans. The extensive body of evidence demonstrating that noxious chemicals in cigarette smoke cause lung cancer is one of the best-publicized examples of an epidemiological study that ties a particular cancer to exposure to particular chemicals. Experimental studies involve determining the incidence of cancers in groups of test animals that have been deliberately exposed to suspected carcinogens. Both experimental and epidemiological studies have indicated an oncogenic potential for a huge array of chemicals (Figure 25-7).

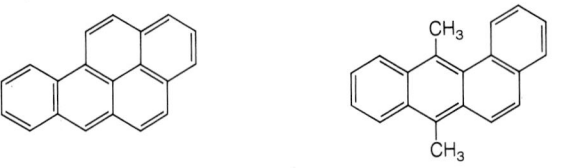

1. Polycyclic aromatic hydrocarbons

Benzo [a] pyrene

7,12-Dimethylbenz [a] anthracene

2. Aromatic amides

2-Naphthylamine

3. Azo dyes

N-dimethylaminoazobenzene (DAB) = butter yellow

4. Nitrosamines

Dimethyl nitrosamine

5. Halogenated hydrocarbons

Carbon tetrachloride

Chloroform

Dichlorodiphenly-trichloroethane (DDT)

6. Alkylating agents

Uracil mustard

7. Metal ions:
Be^{2+}, Ca^{2+}, Co^{2+}, Ni^{2+}, Pb^{2+}

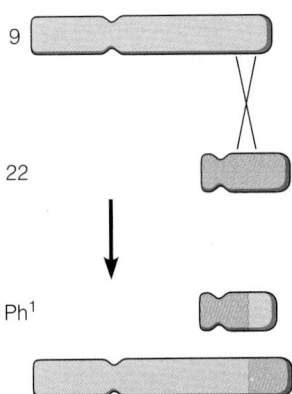

Figure 25-6 The Philadelphia Chromosome. The Philadelphia chromosome is an example of a chromosomal abnormality responsible for a human cancer. It results when a portion of the long arm of chromosome 22 is exchanged with the long arm of chromosome 9 by means of a reciprocal translocation event. The Philadelphia chromosome (Ph[1]) is found in most people with chronic myelogeneous leukemia.

Figure 25-7 Examples of Chemical Carcinogens. Many chemicals have been identified as carcinogens, based on experimental studies with laboratory animals and epidemiological studies with humans. Most important are the aromatic organic compounds (groups 1, 2, and 3), the nitrosamines (group 4), and the halogenated hydrocarbons (group 5).

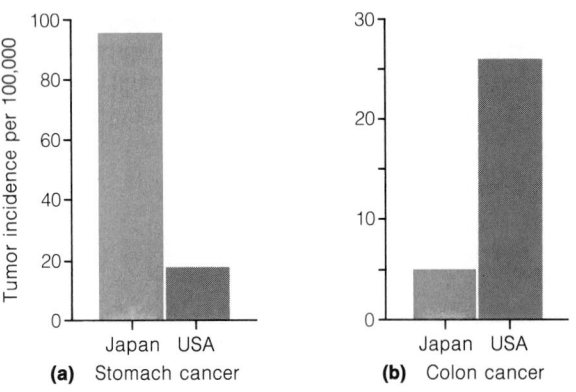

Figure 25-8 National Differences in Cancer Incidence.
National groups show marked differences in incidence rates for various cancers. For example, **(a)** stomach cancer is of high incidence in Japan and of low incidence in the United States. Conversely, **(b)** colon cancer occurs frequently in the United States and relatively rarely in Japan. These national differences in tumor frequency are due mainly to exposure to environmental carcinogens rather than to genetic differences between the populations. (Note difference in scales.)

Exposure to an environmental carcinogenic stimulus is implicated in the initiation of approximately 50–90% of all human cancers. Extensive cross-cultural studies of the differences in tumor incidence among various national groups support these figures. Tumor types that are rare in one national group may be abundant in another. Colon cancer is common in the United States, for example, but rare in Japan, while stomach cancer is common in Japan and comparatively uncommon in the United States (Figure 25-8). Initially, it was unclear whether these profound differences were based on genetic differences between the national groups or whether they resulted from different patterns of exposure to environmental carcinogens.

Indeed, in some cases, genetic differences between human populations are responsible for different rates of certain cancers. For example, dark-skinned people are less susceptible to ultraviolet light–induced cases of skin cancer and melanoma than are fair-skinned individuals. However, differences in exposure to environmental carcinogens are responsible for most of the ethnic differences in tumor incidence. Immigrant populations typically develop the tumor incidence pattern characteristic of the new host country as they adopt its diet and lifestyle, even when these people show only a low frequency of intermarriage with members of the majority population of the host country. As there are profound national differences in the relative frequencies of most important human cancers and because these differences are largely a result of local differences in exposure to environmental carcinogens, environmental carcinogens must be primary factors in the development of most cancers in humans. This implies that the incidence of cancer could be decreased significantly by reducing the exposure to major carcinogens. Certainly, elimination of the smoking of tobacco products would markedly reduce the incidence of lung cancer, a major cancer in many countries.

Most chemical carcinogens require metabolic conversion for the initial chemical, the **procarcinogen,** to be activated to the form that actually produces the neoplastic conversion of the target cell, the **ultimate carcinogen** (Figure 25-9). An important pathway of activation involves the introduction of reactive electrophilic groups, such as epoxide, ester, and carbonium groups, into the procarcinogen by an enzymatic system known as the *mixed-function oxidase* or *aryl hydroxylase system.* This system of enzymes contributes to the ability of the liver to remove toxins from the body by increasing the solubility of toxic organic molecules, thereby facilitating their removal by the kidneys. The liver does this by chemically modifying the organic molecules, adding the electrophilic groups that have the adverse effect of converting the organic procarcinogen molecule to an ultimate carcinogen. The body thus trades safety from acute toxic damage for possible chronic damage due to carcinogenesis.

Chemical Carcinogens Act by Producing Genetic Mutations.   The principal action of most environmental agents in the induction of cancer is the production of *genetic mutations.* Evidence includes the observation that many chemical carcinogens in the form of the ultimate carcinogen bind co-

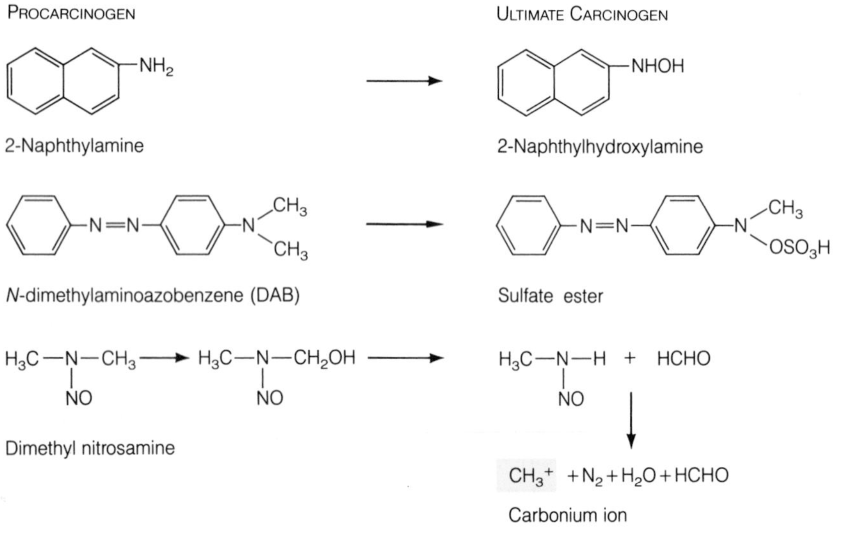

PROCARCINOGEN

2-Naphthylamine

N-dimethylaminoazobenzene (DAB)

$H_3C-\overset{\underset{\textstyle |}{NO}}{N}-CH_3 \longrightarrow H_3C-\overset{\underset{\textstyle |}{NO}}{N}-CH_2OH \longrightarrow$

Dimethyl nitrosamine

ULTIMATE CARCINOGEN

2-Naphthylhydroxylamine

Sulfate ester

$H_3C-\overset{\underset{\textstyle |}{NO}}{N}-H \;+\; HCHO$

$CH_3^+ \;+\; N_2 + H_2O + HCHO$

Carbonium ion

Figure 25-9 Metabolic Activation of Chemical Carcinogens. Most important chemical carcinogens take the form of procarcinogens, chemicals that require metabolic activation into the form of the ultimate carcinogen to acquire carcinogenic potential. The enzymes that catalyze procarcinogen conversion are known as mixed-function oxidases. The normal function of this enzymatic pathway is to add electrophilic groups to organic molecules, in order to render the molecules more soluble in the blood for removal by the kidneys. Acquisition of carcinogenic potential is an unfortunate by-product of this activity.

valently to the nucleotide bases of DNA. This chemical modification of DNA is potentially damaging in a number of ways. Modification may disturb hydrogen bonding between the two DNA strands and may result in incorrect base pairing during replication of the affected DNA strand. Modification may also affect interactions of the DNA with regulatory DNA-binding proteins, which can then affect the level of transcription of the gene.

Evidence for the mutagenic activity of carcinogens is provided by the correlation between the carcinogenic potential of a given agent and its activity as a *mutagen*. A simple and sensitive assay to demonstrate this correlation was developed by Bruce Ames, who compared the mutagenic activity of various chemicals on the bacterium *Salmonella* with

their carcinogenicity, as demonstrated by experimental studies in mice and epidemiological studies with humans. This mutagenesis assay, known as the **Ames test**, involves specially engineered mutant strains of *Salmonella* that require the amino acid histidine for growth (Figure 25-10). When plated on histidine-free culture medium, the only bacteria that grow are those in which a mutant gene in the histidine biosynthetic pathway undergoes a *back-mutation*—a reversion to the normal DNA sequence or a compensating sequence change. In the absence of mutagens in the system, very few bacteria experience the requisite back-mutation to enable them to grow on agar that lacks histidine. When an appropriate mutagen is present, back-mutation occurs with higher frequency, and larger numbers

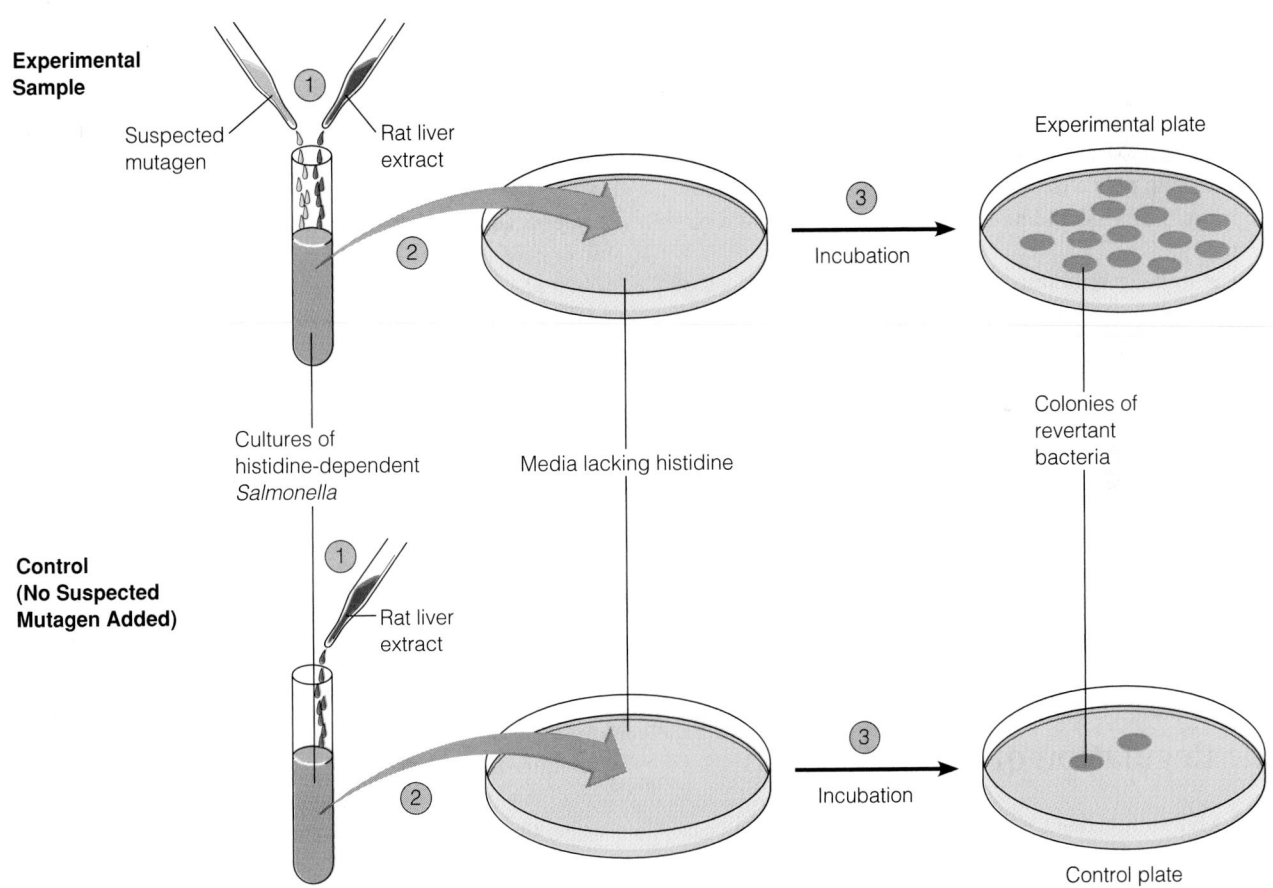

Figure 25-10 The Ames Test for Mutagens. The Ames test depends on *Salmonella* bacteria that have lost the ability to synthesize histidine and therefore depend on the presence of histidine in the medium. However, if a mutation occurs that restores the faulty gene in the histidine biosynthetic pathway to a functional form, the mutant bacteria can synthesize histidine and so grow on a medium lacking histidine. ① Two cultures of histidine-dependent *Salmonella* are prepared, one as the experimental sample to which the suspected mutagen is added (top) and the other as the control without the suspected mutagen (bottom). A membrane-enriched rat liver extract is added to both cultures to ensure the presence of the mixed-function oxidase enzymes that convert pro-

carcinogens to ultimate carcinogens. ② Each sample is then poured onto a plate of medium lacking histidine. ③ The plates are incubated (usually for two days at 37°C), during which time carcinogen-induced mutations may occur in the bacterial cells in the experimental sample. Some of these mutations may restore the capacity to grow in the absence of histidine, leading to the appearance of colonies of revertant bacteria on the medium. The number of colonies on the experimental plate is then determined and corrected for the number of colonies on the control plate, which may have a few spontaneous histidine-synthesizing revertants. The net number of colonies on the experimental plate provides a relative measure of the mutagenicity of the chemical being tested.

of bacterial colonies are present after several days of incubation. The potency of the mutagen can be calculated from the fraction of the bacteria that experience back-mutation and grow into colonies on the histidine-free culture medium.

When using the Ames test to investigate the mutagenic potential of procarcinogens, it is necessary to include both the procarcinogen and a source of the mixed-function oxidase enzyme system to convert the procarcinogen to the ultimate carcinogen. Ames included extracts of rat liver tissue with the bacterial culture for this purpose. The experimental results were striking: In one test, only 13 of 108 "noncarcinogenic" chemicals had mutagenic activity in the bacterial assay (several were subsequently demonstrated to be weak carcinogens), and 158 of 176 of the chemicals known to be carcinogenic were mutagenic. The false-negatives (carcinogens that failed to show mutagenic activity) included a few chemicals, such as the steroid hormone diethylstilbestrol, that do not interact directly with DNA and that stimulate the development of cancer by different pathways. Thus, there is a strong correlation between two properties of these chemicals assessed by two very different assays: The mutagenic activity of a given ultimate carcinogen (assayed by its ability to produce gene mutations in bacteria) correlates with its activity as a carcinogen in mammals. The data indicate that, with only a few exceptions, a chemical's mutagenic activity in an animal underlies its ability to function as a carcinogen.

The mutagenicity tests are important as a first screen of untested chemicals that are suspected to be carcinogens. The test is rapid, relatively inexpensive, readily quantifiable, and results in low levels of false-positives and false-negatives. In contrast, animal tests of potential carcinogens are slow, inaccurate, and prohibitively expensive. Approximately 50,000 synthetic chemicals are produced and used in significant quantities by industry. Testing all these chemicals for carcinogenicity would require enormous expenditures and an army of pathologists to evaluate the histological slides. Development of the Ames test represents a major advance in our ability to prevent cancer by identifying cancer-causing chemicals.

The Regulation of Tissue Growth

As we have seen, a diverse array of events can contribute to neoplastic transformation. Each of these events has at least one of the following genetic effects: It contributes an oncogene to the cell, it activates a proto-oncogene residing in the cell's genome by altering the gene's structural sequence or its expression, or it diminishes the function of a tumor-suppressor gene in the cell. But how do these genetic changes cause cancer? To answer this question, we must consider the recent research that is beginning to shed light on the cellular mechanisms involved in regulating tissue growth in both normal and neoplastic cells.

The stability of anatomical form is, for many tissues, a dynamic situation. This is best illustrated by the regulation of tissue volume. Constancy of tissue volume means that the rates at which cells are added by mitosis are equal to the rates at which cells are lost by cell death or emigration. In many organs, the rates of cell division can increase when injury occurs or when an appropriate physiological demand is made on the system. In rodents, for example, when a portion of the liver is surgically removed, the rate of cell division in the remaining liver tissue increases until the original volume is restored. During embryonic and larval development, and in normal cellular turnover and wound repair in the adult, growth of the tissues is strictly regulated. The volumes of the various tissues are proportional to the size of the organism. Neoplasia results from the establishment of a tumor mass in which this strict regulation of tissue volume is no longer operative. The basic question posed by neoplastic growth is how tissue growth is controlled in the normal state and how this control is lost in the neoplastic state.

Growth Factors: Chemical Regulators of Tissue Growth

One of the most important generalizations to emerge from the study of growth regulation in cell culture is that the growth of normal cells depends on a supply of specific growth-stimulatory proteins known as *growth factors* or *mitogens* (Table 25-4). As described in Chapter 23, growth factors are messenger molecules that act on cell-surface receptors that have kinase activity. Often these receptors specifically phosphorylate the tyrosine residues of proteins. In the usual culture situation, the relevant growth factors are supplied by the serum that is included in the culture medium. One example is the protein **platelet-derived growth factor** (**PDGF**), the most significant growth factor in serum for mesenchymal tissues. Another growth factor is the protein **epidermal growth factor** (**EGF**). This protein, which is widely distributed in embryonic tissues, was initially isolated from the salivary glands of mice by Stanley Cohen, who received a Nobel Prize in 1987 for his pioneering investigations of growth factors.

The usual bioassay for the presence of a growth factor in a sample is to demonstrate the ability of the sample to stimulate cell division in a mitotically quiescent population of cultured cells. Certain growth factors, such as EGF, have broad spectra of activity, stimulating the growth of a wide variety of cell types. Other growth factors are more selective. Growth factors probably function in the stimulation of tissue growth during embryonic and larval development and the reparative growth that contributes to wound repair and regeneration in the adult. For example, release of the growth factor PDGF from blood platelets almost certainly plays a role in the reparative growth of connective tissue cells of injured blood vessels. Platelet aggregation and the exocytotic release of PDGF and other factors take place in an early response to bleeding. The release of PDGF from platelets at sites of vascular wounding makes it available to stimulate reparative growth of the connective tissue of the blood vessel wall.

Table 25-4 Examples of Growth Factors

Growth Factor	Target Cells	Receptor
Epidermal growth factor (EGF)	Wide variety of epithelial and mesenchymal cells	c-*erbB* gene product; 170-kDa tyrosine kinase
Transforming growth factor-α (TGF-α)	Same as EGF	Same as EGF
Platelet-derived growth factor (PDGF)	Mesenchyme, smooth muscle, trophoblast	185-kDa tyrosine kinase
Transforming growth factor-β (TGF-β)	Fibroblastic cells	565–615 kDa complex
Insulin-like growth factor-I (IGF-I)	Epithelium, mesenchyme	450-kDa complex
Insulin-like growth factor-II (IGF-II)	Epithelium, mesenchyme	Single 260-kDa protein
Interleukin-2 (IL-2)	Cytotoxic T lymphocytes	55-kDa glycoprotein
Colony stimulating factor-1 (CSF-1)	Macrophage precursors	c-*fms* gene product; 170-kDa tyrosine kinase

Specificity of Growth Factor Receptors

Peptide growth factors in solution interact with target cells by binding to specific growth factor receptors at the cell surface. The existence of growth factor receptors can be demonstrated by the high affinity and saturable binding of radiolabeled growth factors to living cells (Figure 25-11). Each growth factor receptor binds one or only a few related growth factors. Specificity can be determined by competition assays. The binding of the radiolabeled growth factor is carried out in the presence of a 10-fold to 100-fold excess of unlabeled growth factor. If the labeled and unlabeled growth factors are the same, the unlabeled growth factor molecules will occupy most of the receptor sites, thereby reducing the extent of binding of the radiolabeled molecules and the amount of radioactivity bound to the cells. Thus, for example, excess unlabeled PDGF will reduce the extent of binding to the cell of radiolabeled PDGF. In the presence of high concentrations of other unlabeled growth factors such as EGF, the binding of radiolabeled PDGF remains unaffected. From this, one can conclude that binding to the PDGF receptor is specific: The PDGF receptor binds PDGF but fails to interact with other growth factors. In this way, the growth factor's ability to compete for a given receptor allows the binding specificity of the different growth factor receptors to be characterized experimentally.

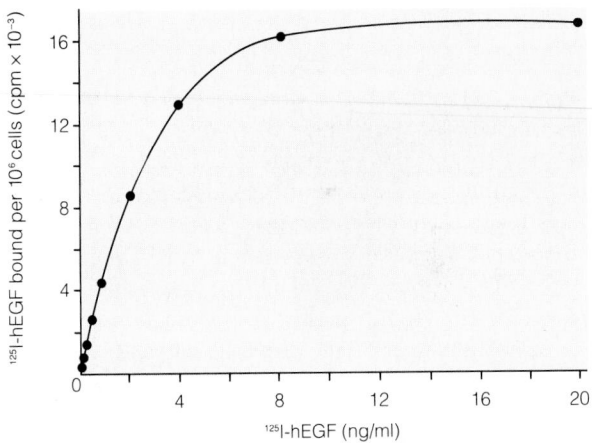

Figure 25-11   Demonstration of the Growth Factor Receptor.   Soluble growth factors interact with target cells by means of specific cell-surface receptors, as shown by binding studies. A radioactive isotope of iodine, ^{125}I, is attached to the purified growth factor molecules before they are mixed with cultures of cells. In this case, the binding of human epidermal growth factor (hEGF) to human fibroblasts was measured. After an incubation period, the unbound molecules are removed by careful washing, and the amount of growth factor associated with the cells is estimated by measuring the remaining ^{125}I. A binding curve is produced by plotting the amount of cell-bound ^{125}I at different doses of ^{125}I-labeled growth factor. When the specific binding curve is known, the number of growth factor receptors per cell and the strength of binding can be calculated. If binding of iodinated molecules of one growth factor is significantly reduced by the presence of unlabeled molecules of another growth factor, it can be concluded that both growth factors use the same receptors. If binding is unaffected, it can be concluded that the two growth factors use different and independent receptor systems.

The Growth Factor Response Pathway

Binding of the growth factor to its receptor initiates a chain of events in the cytoplasm that culminates in the entry of the cell into the S phase of the cell cycle. Growth factor binding frequently initiates activity in one or both of two distinct signal transduction pathways (Figure 25-12). Pathway A: Growth factor binding may activate the **tyrosine-specific protein kinase** (or *tyrosine kinase*) activity of the cytoplasmic domain of the growth factor receptor, leading to the phosphorylation of tyrosine residues of a variety of protein targets (see Figure 23-14). Pathway B: In addition, binding may activate a G protein and in turn the enzyme *phospholipase C,* which catalyzes the hydrolysis of phosphatidylinositol bisphosphate into inositol trisphosphate and diacylglycerol (see Figure 23-11). Binding of the growth factor stimulates the *G protein* associated with the growth factor receptor and with phospholipase C at the inner face of the plasma membrane to bind a molecule of GTP. As a result, the G protein adopts its active configuration, allowing it to activate phospholipase C.

Although some growth factors may activate both of these signal transduction pathways, others activate only one or the other. For example, the Chinese hamster fibroblast cell shows a growth factor response to EGF, thrombin, and bombesin. Binding of EGF activates the tyrosine-specific protein kinase of the cytoplasmic domain of the EGF receptor, whereas thrombin and bombesin activate phospholipase C and the phosphatidylinositol bisphosphate metabolic pathway. Inhibitors of the latter pathway, such as pertussis toxin, have no effect on the growth factor response of hamster fibroblasts to EGF, but block completely the response to thrombin and bombesin. Thus, it appears that activation of the phosphatidylinositol bisphosphate metabolic pathway is essential for the response to bombesin and thrombin but is not essential when EGF is the stimulator of growth.

Although the individual steps further along on the response pathway have not yet been completely defined, an important event in the pathway appears to be induction of the synthesis of a group of nuclear proteins. The mRNAs encoding these proteins appear a short time (0.5 and 3 hours) after stimulation by a growth factor such as EGF or PDGF.

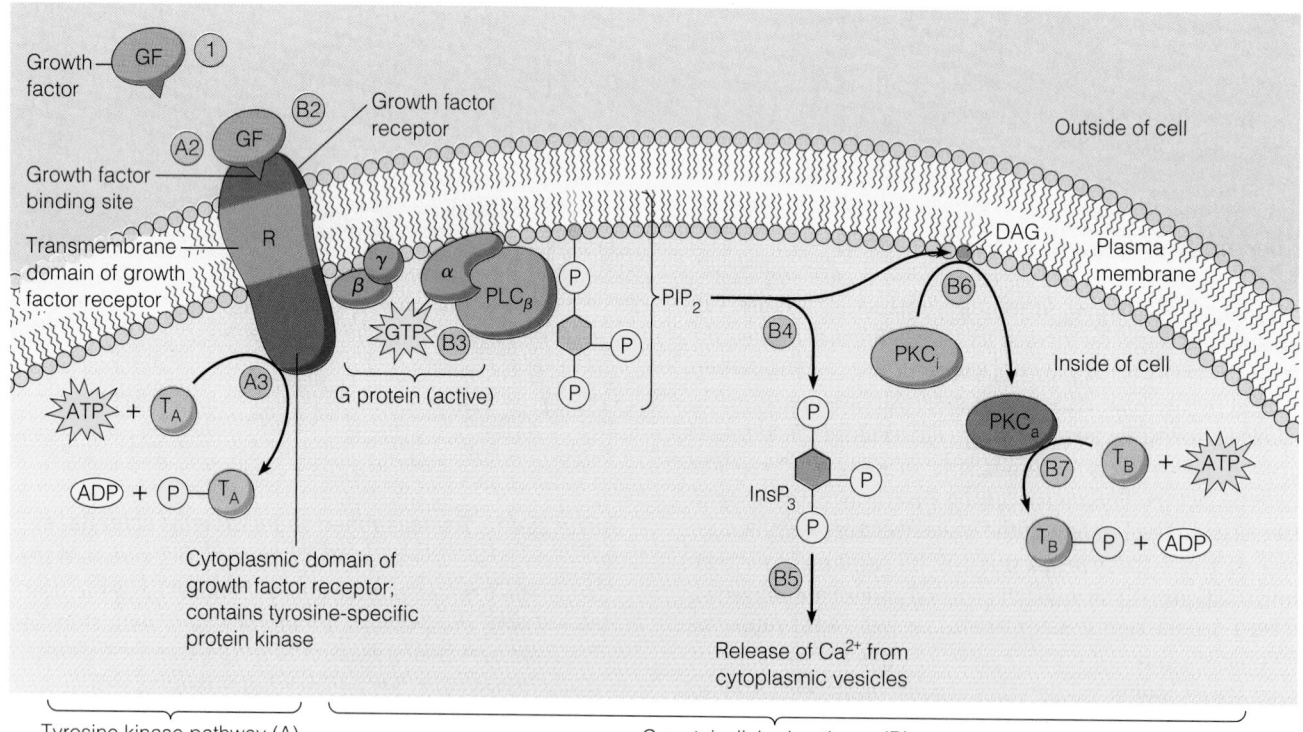

Figure 25-12   Activation of Two Signal Transduction Pathways by the Binding of Growth Factor. ① A peptide growth factor (GF) binds to its receptor (R), a protein embedded in the plasma membrane, and initiates one or both of two signal transduction pathways. In pathway A, Ⓐ2 binding of the growth factor activates the tyrosine-specific protein kinase located in the cytoplasmic domain of the growth factor receptor. Ⓐ3 This kinase can then catalyze the phosphorylation of a target protein (T_A). In pathway B, Ⓑ2 binding of the growth factor to its receptor activates a G protein (see Figure 23-5), which in turn Ⓑ3 converts the inactive form of the membrane enzyme phospholipase C (PLC) to the activated form. Ⓑ4 The PLC enzyme breaks down phosphatidylinositol bisphosphate (PIP_2) into inositol trisphosphate ($InsP_3$) and diacylglycerol (DAG). These two products act as second messengers in the cytoplasm, Ⓑ5 stimulating the release of Ca^{2+} from cytoplasmic storage vesicles and Ⓑ6 activating protein kinase C (PKC_i). Ⓑ7 The active form of PKC (PKC_a), which is dependent on Ca^{2+}, then catalyzes the phosphorylation of its target proteins (T_B).

The nuclear proteins include those encoded by the onco-genes *myc, fos, jun,* and *myb*, suggesting that unregulated expression of one or another of these nuclear proteins produces neoplastic growth. When 3T3 cells are transfected with c-*myc* under the control of a strong promoter (one that directs a high rate of transcription), they become independent of PDGF for growth. Investigation of the function of these nuclear proteins is currently an active area of research.

Dysfunction of the Growth Factor Response Pathway in Neoplasia

One of the most important general findings to emerge from recent investigations of the unregulated growth of neoplastic tissue is the hypothesis that this growth results from a dysfunctional expression of the *growth factor response pathway*. As indicated above, the growth factor response pathway consists of multiple steps. ① The growth factor must be present in the external environment. ② It must bind to specific high-affinity cell-surface receptors. ③ Enzymatic activities are induced, such as that of the tyrosine protein kinase of the cytoplasmic domain of the receptor and the enzyme phospholipase C. ④ Target proteins are phosphorylated. ⑤ The phosphatidylinositol pathway may be activated. ⑥ A group of nuclear proteins is synthesized. ⑦ Ultimately, DNA synthesis is initiated. In theory, if any of these steps is constitutively active (i.e., active in the absence of the growth factor), the cell can be expected to divide continuously and independently of an external supply of the growth factor. In fact, examples have been described in which unregulated cell division is caused by constitutive expression of each of these steps of the growth factor cascade. Furthermore, oncogenes are intimately involved: *The gene products of a number of oncogenes are elements in the growth factor response pathway.* These findings represent a dramatic convergence of molecular genetic and cell biological analysis of neoplasia—one of the outstanding successes of recent biological research.

One way that defective expression of a component of the growth factor response pathway could produce continual cell cycling under conditions in which normal cells are quiescent is for the neoplastic cells to produce their own supply of growth factor. This is the **autocrine secretion mode** of neoplastic growth. While normal cells require an external supply of growth factor and will fail to divide in its absence, cells that can produce and then respond to a growth stimulator will divide even in the absence of an external supply. An example of autocrine growth promotion is the NIH-3T3 fibroblast cell line transformed by infection with the simian sarcoma virus (the SSV-3T3 cell). The 3T3 cell is an example of a normal cultured cell line. In culture, regulated growth of the 3T3 cell is lost following transformation by SSV. Evidence that the SSV-3T3 cell produces autocrine growth factors includes the observations that virus-free culture medium exposed to SSV-3T3 cells contains growth factor activity and that cell division is stimulated when presented to quiescent populations of 3T3 cells.

This activity is embodied in a molecule similar to the growth factor PDGF; SSV-3T3-conditioned medium eliminates binding of radiolabeled authentic PDGF to 3T3 cells, and antibodies to PDGF abolish the growth-stimulatory activity of SSV-3T3-conditioned medium. The precise nature of the relation of the growth-stimulatory activity of SSV-3T3-conditioned medium to PDGF became clear with the analysis of *sis*, the oncogene of the simian sarcoma virus. This gene was isolated and sequenced, and the predicted amino acid sequence was derived. This sequence turned out to be nearly identical to the peptide sequence of the β chain of PDGF. Thus, introducing the *sis* oncogene into a cell by infection with SSV induces the synthesis of an active form of PDGF. The PDGF interacts with the infected cell to stimulate growth, in contrast to the normal situation, in which mesenchymal cells are exposed to PDGF only if blood platelets in their immediate vicinity release PDGF into the environment.

A second way in which the growth factor response pathway could become constitutively active would be for a growth factor receptor to be altered such that the enzymatic activity of the C-terminal tyrosine protein kinase domain would be active even in the absence of growth factor binding. An apparent example of this situation is a cell transformed by the addition of the *erb-B* oncogene. This oncogene is present in the *erythroblastosis virus*, a retrovirus of avian origin that produces tumors of the erythroblastic lineage. When the *erb-B* oncogene was sequenced, it was found that the derived protein sequence was related to that of the EGF receptor. The Erb-B protein included a modified version of the cytoplasmic and transmembrane domains of the EGF receptor but lacked almost all of the extracellular domain. The Erb-B protein may be a constitutively active form of the EGF receptor that conducts the protein phosphorylation reaction independently of EGF. The normal form of the EGF receptor, of course, shows enzymatic activity only when EGF is bound to the appropriate site in the extracellular portion of the receptor protein.

Growth-Inhibitory Proteins

The growth factor response pathway is involved in the stimulation of cell proliferation. In addition to growth stimulators, a number of growth suppressors have been identified. One example is a protein called *transforming growth factor-β* (TGF-β). This is a 25-kDa homodimer with both growth-stimulating and growth-inhibiting properties. Its antiproliferative actions include antagonism of the growth factor effects of peptide growth factors such as EGF and PDGF. Interestingly, cell binding of TGF-β involves a family of three different cell-surface receptors, all three of which are missing in retinoblastoma cells. Clearly, the gene product of the *Rb* gene (discussed on page 824) is a nuclear protein, rather than a cell-surface TGF-β receptor; but it is possible that the loss of the receptors may be important in establishing the neoplastic phenotype of this particular tumor.

Tumor Progression and Dissemination

We have considered the nature of the neoplastic phenotype, the genetic changes that initiate its development, and the ways in which these genetic events alter the regulation of the growth factor response pathway. Although we have generally been describing single genetic alterations, keep in mind that cancer is a multistep process, and several changes are usually required for tumor development. Now let's turn to our final topics: how these changes in a single cell lead to the development of an entire tumor, and how that neoplasia can spread to affect other body tissues.

Tumor Progression

Tumor progression is the incremental development of increasingly malignant states by a tumor. Typically, tumors in the early stages of development are relatively benign. The tumor grows slowly and is either weakly invasive or noninvasive. With time, however, tumors can enter a phase of increasingly rapid growth, becoming highly invasive and metastatic (Figure 25-13). Clinically, tumors are categorized into Classes I–IV according to their progression and likelihood of cure, Class IV (malignant) being the most dangerous.

Tumor progression is exemplified when a hidden metastatic nodule suddenly develops into a life-threatening tumor several years after the primary tumor has been surgically removed. In such cases, the metastatic nodule must have originated before surgery, but it had remained dormant for those several years before progressing to the fully malignant state. Primary tumors presumably also follow this course from a benign state to states of progressively greater malignancy. This course is suggested by the relatively high incidence of benign tumors found upon autopsy of individuals who died of other causes.

Tumor Invasion: Dissemination to Nearby Tissue

As mentioned previously, the anatomical relationships among the tissues of the different organs of the body are highly stable—that is, cells remain within the confines of the parent tissues for the life of the organism. This organizational stability is disrupted by the onset of the invasive and metastatic behavior of cancerous tissue. *Invasion* is the intrusion of malignant tissue into neighboring normal tissues. Typically, the neoplastic tissue invades and replaces adjacent normal tissue, disrupting normal function of the afflicted organ (Figure 25-14).

Several processes contribute to tumor invasion. For example, many tumors release a variety of degradative enzymes into the local environment. These may function to degrade the extracellular matrix of adjacent normal tissues, facilitating penetration by the neoplastic tissue. Regulation of the motile behavior of invasive cells has also been shown

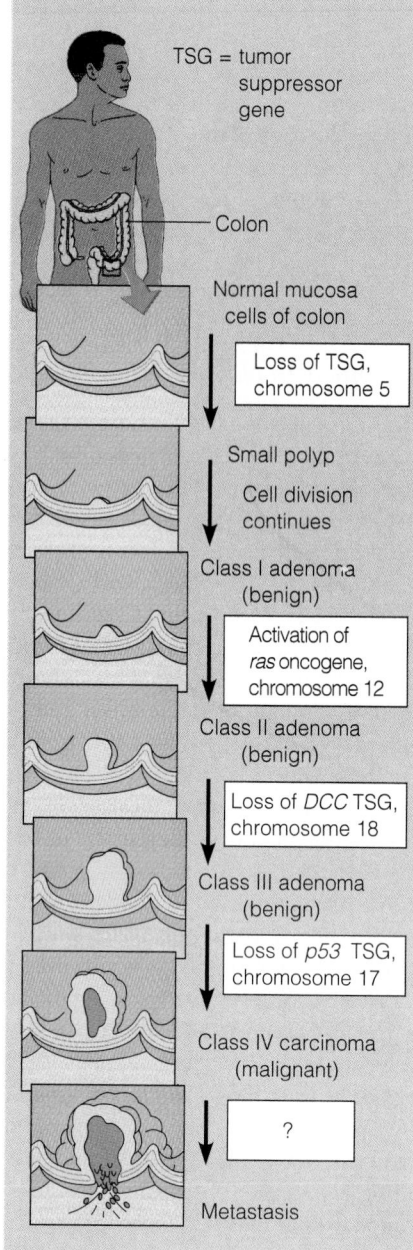

Figure 25-13 Tumor Progression. Colon cancer, like most cancers, develops gradually as specific cellular and genetic changes accumulate. As shown here, a small benign growth, a polyp, develops from apparently normal mucosa cells. As it grows it becomes a Class I benign adenoma, but as various tumor-suppressor genes (TSG), including *DCC*, are inactivated and the *ras* oncogene is activated, the tumor evolves through Classes II and III into a Class IV malignant carcinoma capable of metastasis.

to differ from that of noninvasive cells. Invasive tumor cells have been shown to lack contact inhibition following encounters with normal tissue cells in cell culture, allowing them to migrate freely over the surface of the contacted normal cell. It has been suggested that contact inhibition re-

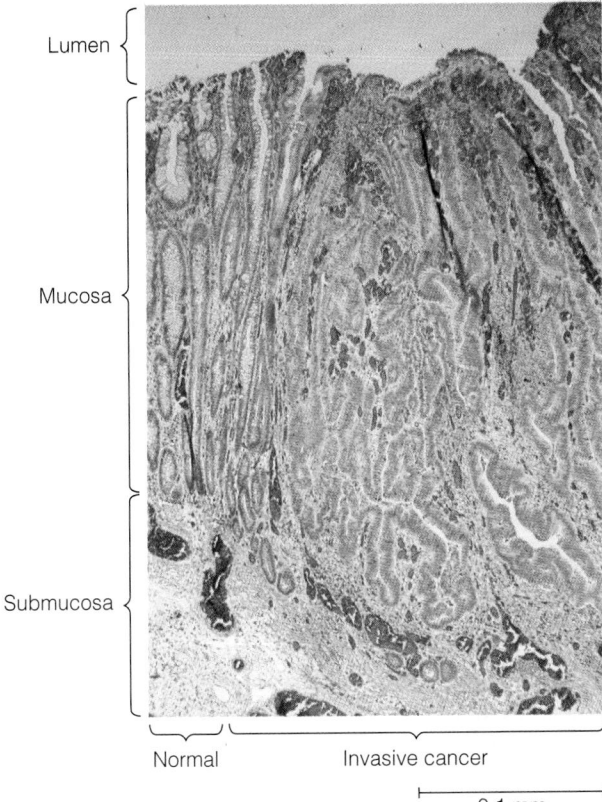

Figure 25-14 Tumor Invasion. Cancerous cells arising in the wall of the human colon can spread from one tissue layer, the mucosa, to the adjacent submucosal tissue. The adenocarcinoma shown here has caused the normally orderly tissue structures (left) to become irregularly shaped, chaotic in distribution, and composed of atypical cells (right) (LM).

strains the active migration of normal cells in coherent tissues and that its absence is a necessary condition for tumor invasion. Contact inhibition is apparently a consequence of a cell's ability to establish specialized adhesive contact junctions following cell-cell collision. Careful electron microscopic studies have determined that the adhesive junctions form within a few minutes after a collision between two noninvasive cells and fail to form when invasive cells collide with normal cells. It follows that the ability or inability to display contact inhibition is a function of the adhesive characteristics of the cells. The inability to display contact inhibition of motility may represent one way in which alterations in cell-cell adhesive interactions can initiate invasive behavior by tumor cells.

Metastasis: Dissemination to Distant Organs

As we have seen, tumor invasion accomplishes local dissemination of the malignant tissue from the parent tumor mass. *Metastasis*, however, disseminates malignant tissue to distant sites in the body by the transport of clumps of tumor cells in the fluids of the body cavities and vasculature (Figure 25-15). Metastatic transport can occur within the peritoneal cavity, the neural canal, the lymphatic system, or the vascular system. The last route is the most important for the dissemination of solid tumors in humans and will therefore be the basis for the discussion that follows.

Vascular metastasis begins with the establishment of a vascular supply for the solid tumor. The endothelial cells that line blood vessels actually invade the tumor from vascular capillaries of adjacent normal tissue. The tumor cells then gain access to the vascular lumen by invasion across the endothelium of these vessels. Small clumps of tumor cells

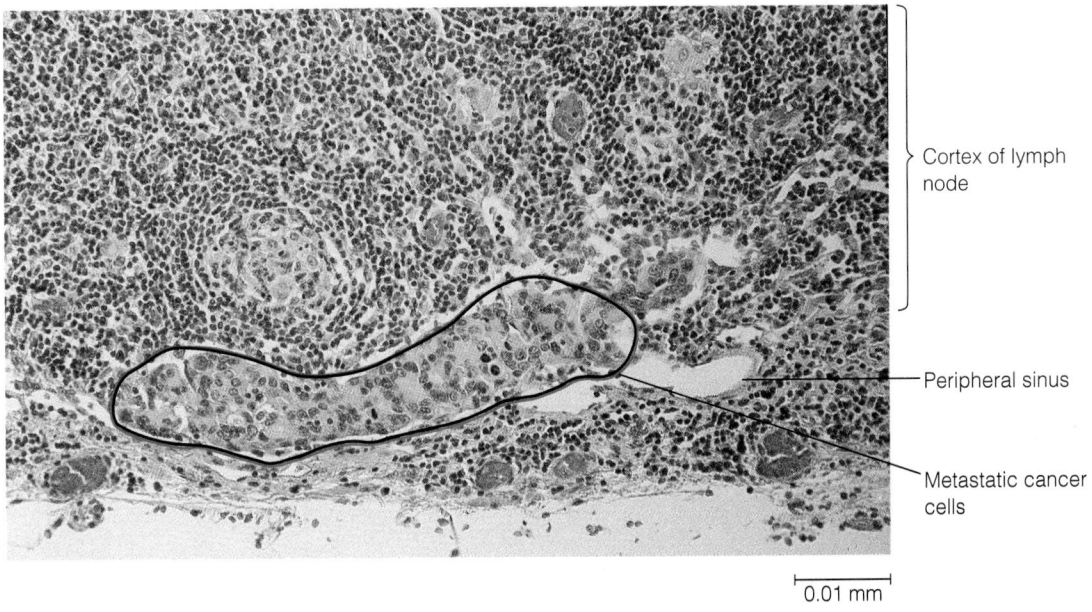

Figure 25-15 Metastasis. This micrograph shows evidence of human stomach cancer having spread to a distant body tissue, a lymph node. Here, the gastric adenocarcinoma cells have metastasized to form a plug (indicated by line) filling the lumen of the peripheral sinus of the lymph node (LM).

that protrude into the vascular lumen dislodge from the main mass and are swept away in the blood. In advanced cancers, many thousands of such cell clumps may be shed every day. Although most of these cells die, the successful clumps must lodge in the capillaries of an appropriate organ and then invade across the capillary endothelium to begin growth in the parenchyma of the new host organ.

As we have just described, metastasis is a complex process of vascularization, invasion, vascular transport, arrest, and invasion. The process of vascularization of solid tumors, termed **tumor angiogenesis,** is essential not only to metastasis but also to malignant progression and growth. In the absence of a vascular supply, solid tumors typically remain small and confined. The typical avascular tumor consists of an actively growing cortex at the surface and dead (necrotic) tissue at the core. The progression from the neoplastic but nonmalignant state of the tumor to the progressively growing state of the malignant tumor requires a vascular bed to supply the interior of the tumor mass with nutrients and oxygen and to remove metabolic wastes. The initiation of tumor vascularization has been shown to depend on the release of diffusible factors from the tumor that stimulate the directed migration of the capillary endothelium from the surrounding normal tissues into the tumor. In this situation, the tumor experiences invasion by capillaries from adjacent tissues.

One of the experimental systems that has proved useful in the analysis of tumor angiogenesis involves implantation of small pieces of tumor tissue in the avascular connective tissue of the cornea of an experimental animal. Following implantation, the tumor establishes itself in the cornea as a tumor in situ—a nonvascularized nodule that remains small with an actively dividing surface layer and a necrotic interior. If placed within a few millimeters of the edge of the cornea, the tumor induces the migration of capillary sprouts from the vascularized tissue peripheral to the cornea. These capillary sprouts progressively invade the connective tissue of the cornea and finally contact and vascularize the tumor. Induction of corneal vascularization can also be produced by cell-free extracts of the tumor, suggesting that chemical factors are released by neoplastic tissue that attract migrating vascular endothelial cells.

Several angiogenic proteins have been identified and characterized, including *angiogenin, basic fibroblast growth factor,* and *tumor necrosis factor-α.* The latter is interesting because it is produced not by the tumor cells themselves but by host macrophages that have colonized the tumor. (Macrophages are scavenging cells derived from a class of circulating blood cells known as *monocytes.*) Most solid tumors contain significant populations of macrophages. Tumor necrosis factor-α is the principal secreted protein that enables the macrophages to kill tumor cells. The factor thus has multiple effects, including the destruction of tumor cells by its direct tumoricidal activity and the promotion of tumor growth by its ability to attract blood vessels into the tumor.

The consequence of the establishment of a vascular supply by a previously avascular tumor in situ is a spectacular increase in the rate of tumor enlargement. In the corneal implantation system, the tumor implant rapidly switches from the dormant state to one of rapid enlargement. This change in the tumor's rate of enlargement indicates that the ability to initiate angiogenesis can play an important role in tumor progression—the avascular solid tumor is generally benign, whereas the vascularized solid tumor is usually overtly malignant. This phenomenon suggests that inhibition of vascularization might produce a reversal of the progressive enlargement of solid tumors. Indeed, combinations of steroidal antiinflammatory agents and the polysaccharide heparin inhibit and reverse the vascularization of tumors in experimental animals and produce a dramatic reversal of tumor growth.

Clinical Applications

ON THE TRAIL OF THE *RAS* ONCOGENE

Although they appeared on the molecular scene only recently, oncogenes already have many investigators on their trail, and intensive molecular analysis is well under way. The research is spurred on by the many significant findings that have already been made and by the obvious implications of these results for our understanding of both cancer and normal growth and differentiation. A particularly intriguing trail is that of the *ras* gene family, so let's follow it to see where it leads.

Our story begins with the RNA tumor viruses, because the *ras* gene was originally identified in two different retroviruses of the rat, the *Harvey sarcoma virus* and the *Kirsten sarcoma virus*. The *ras* genes of these two viruses are referred to as Ha-*ras* and Ki-*ras* (see Table 25-3). The *ras* proto-oncogenes have since been discovered in a variety of organisms, including humans, rodents, fruit flies, and yeast. Like other proto-oncogenes, these *ras* genes are thought to be required in normal cells for the regulation of growth and development.

The *ras* gene codes for a protein that is associated with the inner surface of the plasma membrane. This Ras protein is a G protein that regulates the activities of a number of important enzymes, including phospholipase C and adenylate cyclase, the enzyme responsible for the formation of cyclic AMP from ATP.

An important development in our understanding of oncogenes occurred when the *ras* oncogene trail led to human cancer. DNA isolated from two different human bladder carcinomas was shown to induce neoplastic transformation in the NIH-3T3 line of mouse fibroblast cells. Transforming activity was traced to a single sequence that in both cases turned out to correspond to the *ras* genes of the rat sarcoma viruses. Since then, *ras* oncogenes have also been detected in other human carcinomas, including lung and colon tumors.

Based on prior findings with other viral oncogenes, it was initially presumed that the *ras* gene acted by increasing the cellular level of its product in transformed cells. However, no such increase was found in NIH-3T3 cells following transformation by DNA from the human bladder carcino-

mas. In other words, transformation did not depend critically on the amount of *ras* gene product made by the cells. Instead, when the *ras* genes from the bladder carcinomas were analyzed further in the laboratories of Robert Weinberg and Mariano Barbacid, the trail led to a gene product with an altered amino acid sequence. DNA from both carcinoma and normal tissue of the same patient was digested with restriction enzymes, and the fragments containing the *ras* genes (along with flanking stretches of DNA) were sequenced. The sequences were identical except for a single position, at which the normal DNA fragment (the protooncogene) had a G and the DNA fragments from the carcinomas (the oncogenes) had a T. This was the first time a mutated gene had been found in a primary tumor and not in a normal tissue from the same patient.

This substitution affects the twelfth amino acid from the N-terminus of the protein, which is glycine in the normal cell and valine in the carcinoma. That difference is probably significant, because when the Ha-*ras* and Ki-*ras* oncogenes of the sarcoma viruses were sequenced, the same codon was affected. One of the viruses has an arginine in place of the glycine, and the other has a serine. It appears, therefore, that the glycine in position 12 is critical to the proper functioning of the normal protein, and that any of several amino acid substitutions at this position disrupts cellular function sufficiently to induce malignancy.

More recently, the trail of the *ras* oncogene has broadened as a number of related *ras* genes have been sequenced, including those from lung and colon carcinomas and from a neuroblastoma, a tumor of embryonic nerve cells. In each case, the transforming activity could be attributed to small, specific changes in the nucleotide sequence of the DNA and hence in the amino acid sequence of the protein. It appears that a *ras* gene can be transformed by structural alterations in at least five sites. The sites are clustered in two regions of the gene, affecting the codons for amino acids 12 and 13 and for amino acids 59–63 of the protein.

The implications of this mode of oncogene activation are clear because most human cancers are thought to be caused by chemical carcinogens, all of which are known to

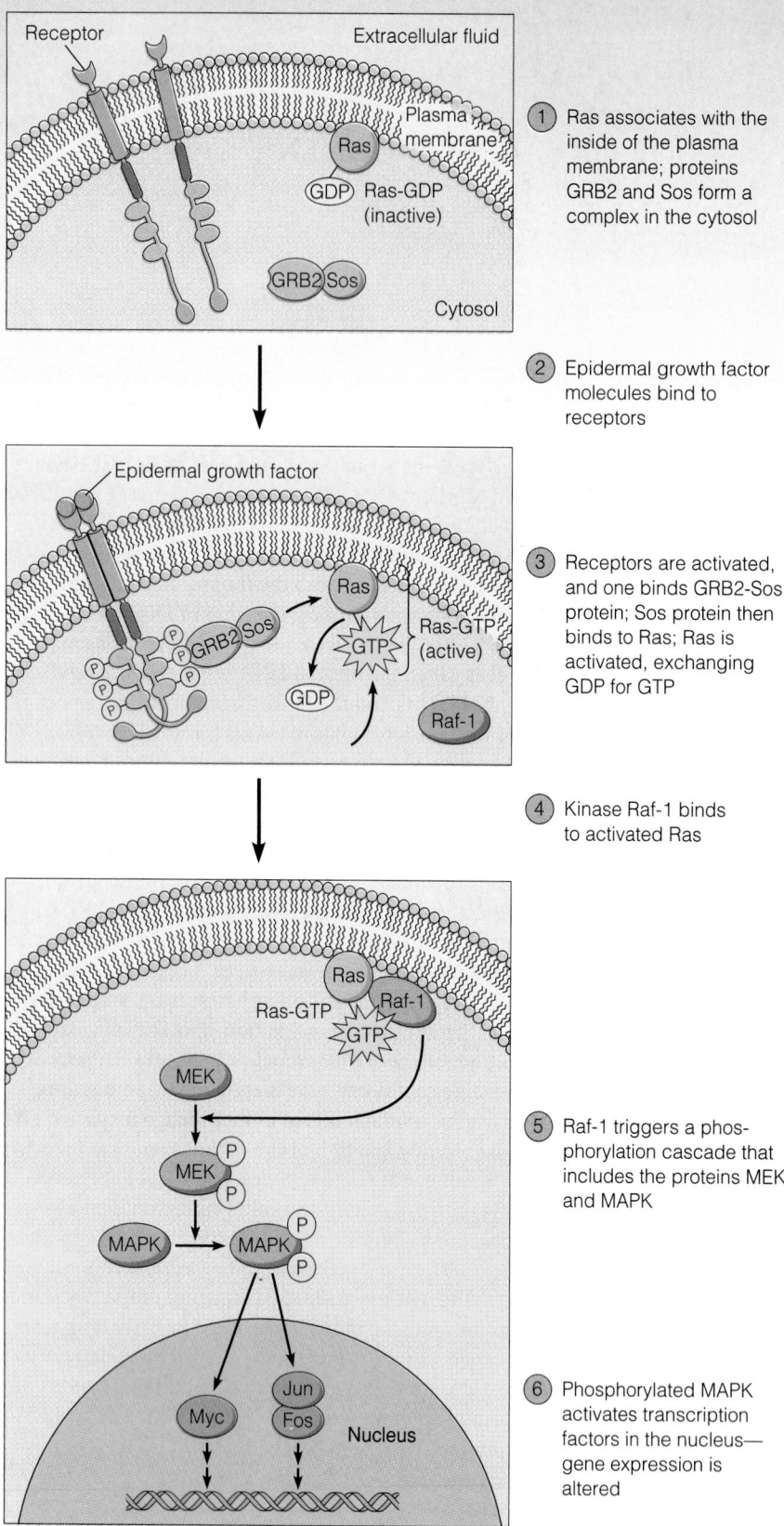

1 Ras associates with the inside of the plasma membrane; proteins GRB2 and Sos form a complex in the cytosol

2 Epidermal growth factor molecules bind to receptors

3 Receptors are activated, and one binds GRB2-Sos protein; Sos protein then binds to Ras; Ras is activated, exchanging GDP for GTP

4 Kinase Raf-1 binds to activated Ras

5 Raf-1 triggers a phosphorylation cascade that includes the proteins MEK and MAPK

6 Phosphorylated MAPK activates transcription factors in the nucleus— gene expression is altered

Figure 25A-1   Transmission of Growth Signals to the Cell Nucleus Via Ras.   Scientists have pieced together most or all of the complex molecular chain of events that starts when a growth factor binds to a receptor on the cell surface, ultimately leading to altered gene expression—such as that directing cell growth and division—in the nucleus. The Ras protein, encoded by the normal proto-oncogene *ras,* is a key intermediate in the chain.

be directly or indirectly mutagenic. In fact, the *ras* trail now seems to be leading to chemical carcinogenesis, if recent results from Barbacid's laboratory are any indication. A single dose of the carcinogen nitrosomethylurea administered to 8-week-old rats was enough to cause mammary tumors within 6–12 months in almost all the animals. In each case, the cancerous tissue contained a transformed *ras* gene, and the active oncogene differed from its proto-oncogene counterpart by a single nucleotide, again in codon 12!

One of the most exciting advances in our understanding of *ras* gene function came as a result of the discovery that *ras* genes are also present in yeast. Identification of oncogenes in yeast cells makes it possible to study these genes using biochemical and genetic methods that cannot be used in more complex organisms. For example, with yeast cells it is possible to replace specific genes with experimentally altered forms. By replacing one or both of the two *ras* genes in the yeast genome with nonfunctional genes, Michael Wigler and his colleagues were able to show that the yeast cell has an absolute requirement for at least one intact *ras* gene for spore germination. This finding strengthens the hypothesis that proto-oncogenes play an indispensable role in regulating cell division or differentiation.

Wigler and his colleagues also showed that gene function can even be restored by substituting a human *ras* gene for an inactivated yeast gene, although the proteins encoded by the *ras* genes are significantly larger in yeast cells than in human cells. They were then able to construct a mutant *ras* yeast gene that mimicked one of the alterations known to activate the transforming potential of mammalian *ras* genes. When the mutant gene was inserted into yeast cells, the cells lost their ability to form spores, suggesting that the transformation of mammalian cells and the inability of yeast cells to form spores may have an explanation in common.

Clues to that explanation came when it was found that the mutant Ras protein in yeast continuously activates adenylate cyclase, an enzyme important for cell regulation because of its role in the formation of cyclic AMP. Discovery of the connection between the Ras protein and adenylate cyclase regulation qualifies as another milestone in cancer research because it links an oncogene to a major regulatory mechanism of eukaryotic cells.

However, an understanding of how extracellular signals are relayed through the Ras protein to the cell nucleus remained elusive until 1993. Then the *ras* oncogene trail, and cancer research in general, took a dramatic turn. Scientists in several laboratories around the world announced results that helped piece together the molecular chain of events in the Ras growth control pathway, extending from growth factors to gene expression effects (Figure 25A-1). A brief overview of the pathway that was uncovered illustrates its complexity. Researchers showed that epidermal growth factor (EGF) binds to its receptor, and the receptor in turn binds to a cytoplasmic protein named GRB2, which is already bound to the protein Sos. Activated Sos then binds to Ras, activating it by causing it to swap its bound GDP for GTP. The activated Ras protein in turn activates another kinase, Raf-1 (perhaps with assistance from other proteins). Raf-1 triggers a phosphorylation cascade, in which one molecule phosphorylates, and thereby activates, another, which phosphorylates another, and so on. In this cascade, the signal is passed from Raf-1 to an enzyme known as MEK, to a group of protein kinases called MAPKs, to transcription factors in the nucleus, including the Myc, Jun, and Fos proteins (encoded by the *myc*, *jun*, and *fos* oncogenes, respectively). Once activated, these proteins stimulate the gene activity necessary for cell growth and division—the ultimate result of the growth factor signal.

More remains to be learned about the *ras* oncogene family. But the trail of discoveries that began with the oncogenes of two rat sarcoma viruses has already led to several human cancers, to some highly specific mutations in the corresponding human genes, and now to signal transduction pathways as well. It seems a promising trail indeed, and it illustrates how the initially separate paths of oncogenes and cellular biology seem to be merging in ways that are almost certain to be as significant as they were unexpected. ∎

The study of the protein products of oncogenes has clarified our understanding of normal growth regulation, and that understanding, in turn, has informed the investigation of neoplastic growth. A basic lesson has been that any of a variety of possible perturbations that make the growth factor response pathway constitutively active can produce the same basic phenotype: neoplastic growth. The constitutive activation of the growth factor cascade can be at the level of the growth factor itself (autocrine stimulation), the growth factor receptor, or postreceptor steps (activation of the relevant G proteins, cytoplasmic protein kinases, or the relevant nuclear proteins).

The abnormal biology of cancerous tissue is largely a consequence of an altered expression of normal cellular processes, rather than the expression of novel processes. Growth factors are important regulators of normal tissue growth during embryonic and larval development and in situations of wound repair. It is only when the growth factor response pathway is activated in tissues in which it is normally silent that unregulated cell proliferation becomes a problem. Certain normal tissues are invasive and even metastatic, especially during embryonic development. However, in the adult, these tissues generally remain in stable anatomical relationships. The reinitiation of invasive and metastatic behavior in the adult is pathological. Angiogenesis is a normal event of embryonic development and of wound healing. Tumor angiogenesis is nothing more than the angiogenic response in an abnormal context. The abnormal behavior of the cancerous tissue has resulted from an exaggerated, unrestrained, or misplaced expression of a family of normal processes. It has been the purpose of this chapter to identify these processes and to show how cancer results from their dysfunctional expression.

A further lesson of this chapter is the utility of studying pathological processes for learning more about normal processes. Understanding the biology of cancerous tissue has provided valuable insights into the biology of normal tissues, since both classes of tissues operate with the same processes. The oncogenes and tumor-suppressor genes were first discovered in the context of neoplasia. Now we know that these genes play essential roles in normal tissues. Most of the early interest in angiogenesis derived from its importance to the growth of solid tumors, but recent interest has shifted to the role of angiogenesis in normal development. Cancer has served as an experiment of nature whose investigation has significantly strengthened our understanding of the important problems of growth control and tissue organizational regulation.

KEY TERMS FOR SELF-TESTING

Cancer: A Loss of Normal Growth Regulation

tumor (neoplasm) (p. 817)
benign tumor (p. 817)
malignant tumor (p. 817)
cancer (p. 817)
neoplastic transformation (p. 817)
invasion (p. 817)
metastasis (p. 817)
stem cell (p. 818)

Growth Properties of Cells in Culture

contact inhibition (p. 818)

The Genetic Basis of Neoplasia

oncogene (p. 819)
oncogenic virus (p. 819)
transfection (p. 819)
oncogene transfection assay (p. 820)
proto-oncogene (p. 822)
tumor-suppressor gene (p. 823)
apoptosis (p. 824)

Other Causes of Neoplastic Transformation

procarcinogen (p. 826)
ultimate carcinogen (p. 826)
Ames test (p. 827)

The Regulation of Tissue Growth

platelet-derived growth factor (PDGF) (p. 828)
epidermal growth factor (EGF) (p. 828)
tyrosine-specific protein kinase (p. 830)
autocrine secretion mode (p. 831)

Tumor Progression and Dissemination

tumor progression (p. 832)
tumor angiogenesis (p. 834)

PROBLEM SET

25-1. Telling the Difference. Suggest an observation or an experimental test to distinguish between the two members of each of the following pairs.

(a) Benign tumor; malignant tumor

(b) Nontransformed 3T3 cells; 3T3 cells transformed with simian sarcoma virus

(c) Viral oncogene; proto-oncogene

(d) Growth factor that binds to the EGF receptor; growth factor that fails to bind to the EGF receptor

(e) Normal granular leukocyte (granulocyte); chronic myelogenous leukocyte

25-2. Oncogenic Viruses. You suspect that a particular tumor found in a laboratory animal is caused by a particular virus. What experiments would you conduct to determine whether that tumor was indeed caused by the virus?

25-3. Activation of Proto-Oncogenes. Describe two general mechanisms by which a cell's proto-oncogene can sustain a mutation that activates it to become an oncogene. Then discuss what types of mutations these might be.

25-4. Oncogene Transfection Assay. In what situation is this assay used? What results can be obtained? What are the three steps of the assay?

25-5. Tumor-Suppressor Genes. Outline the two basic categories of experimental evidence for the involvement of tumor-suppressor genes in the development of the neoplastic phenotype.

25-6. Carcinogens and Mutagens. Present the two basic categories of evidence that can be used to show that a particular substance, such as cigarette smoke, is carcinogenic. How would you show that the same substance is mutagenic?

25-7. Carcinogenic Potential. What evidence leads us to suspect that the carcinogenic potential of most chemical carcinogens depends on their ability to act as mutagens? Cite two different kinds of evidence.

25-8. Tumor Progression. Tumor progression is the development over time of progressively more malignant characteristics in a tumor. What are some of the processes that contribute to tumor progression?

25-9. Medical Possibilities. Recent advances in understanding the basic biology of the cancerous cell have suggested attractive possibilities for the prevention and clinical management of cancer. Discuss the medical possibilities suggested by recent studies of:

(a) Tumor angiogenesis

(b) Techniques to identify chemical carcinogens

SUGGESTED READING

General References

Alberts B., D. Bray, J. Lewis, M. Raff, K. Roberts, and J. D. Watson. Cancer. In *Molecular Biology of the Cell*, 3d ed., Chapter 24, p. 1255. New York: Garland, 1994.

Becker, F. F. *Cancer, a Comprehensive Treatise.* New York: Plenum, 1975–1988.

Lewin, B. *Genes*, 5th ed. New York: Wiley, 1994.

Pitot, H. C. *Fundamentals of Oncology*, 3d ed. New York: Marcel Dekker, 1986.

Oncogenes

Bishop, J. M. The molecular genetics of cancer. *Science* 235 (1987): 305.

Jove, R., and H. Hanafusa. Cell transformation by the viral *src* oncogene. *Annu. Rev. Cell Biol.* 3 (1987): 31.

Lowy, D. R., and B. M. Williamsen. Function and regulation of Ras. *Annu. Rev. Biochem.* 62 (1993): 851.

Reddy, E. P., R. K. Reynolds, E. Santos, and M. Barbacid. A point mutation is responsible for the acquisition of transforming properties by the T24 human bladder carcinoma oncogene. *Nature* 300 (1982): 149.

Tabin, C. J., S. M. Bradley, C. I. Bargmann, R. A. Weinberg, A. G. Papageorge, E. M. Scolnick, R. Dhar, D. R. Lowry, and E. H. Chang. Mechanism of activation of a human oncogene. *Nature* 300 (1982): 143.

Oncogenic Viruses

Gallo, R. C., and F. Wong-Staal. Current thoughts on the viral etiology of certain human cancers. *Cancer Res.* 44 (1984): 2743.

Muhlbock, O., and P. Bentvelzen. The transmission of mammary tumor viruses. *Perspect. Virol.* 6 (1968): 75.

Nazerian, K. Marek's disease: A neoplastic disease of chickens caused by a herpesvirus. *Adv. Cancer Res.* 17 (1973): 279.

Tumor-Suppressor Genes

Culotta, E., and D. E. Koshland, Jr. *p53* sweeps through cancer research. *Science* 262(1993): 1958.

Harris, H. The analysis of malignancy by cell fusion. *Cancer Res.* 48 (1988): 3302.

Huang, H.-J. S., J.-K. Yee, J.-Y. Shew, P.-L. Chen, R. Bookstein, T. Friedmann, E. Y.-H. P. Lee, and W.-H. Lee. Suppression of the neoplastic phenotype by replacement of the RB gene in human cancer cells. *Science* 242 (1988): 1563.

Lee, W.-H., J.-Y. Shew, F. D. Hong, T. W. Sery, L. A. Donoso, L.-J. Young, R. Bookstein, and E. Y.-H. P. Lee. The retinoblastoma susceptibility gene encodes a nuclear phosphoprotein associated with DNA binding activity. *Nature* 329 (1987): 642.

Levine, A. J. The tumor suppressor genes. *Annu. Rev. Biochem.* 62 (1991): 623.

Marx, J. How *p53* suppresses cell growth. *Science* 262 (1993): 1644.

Marx, J. New link found between *p53* and DNA repair. *Science* 266 (1994): 1321.

Riley, D.J., E.Y.-H.P. Lee, and W.-H. Lee. The retinoblastoma protein: More than a tumor suppressor. *Annu. Rev. Cell Biol.* 10 (1994): 1.

Weinberg, R. A. Finding the anti-oncogene. *Sci. Amer.* 259 (March 1988): 44.

Chromosomal Alterations

Croce, C. M., and G. Klein. Chromosome translocations and human cancer. *Sci. Amer.* 252 (March 1985): 54.

Kurzrock, R., J. U. Gutterman, and M. Talpaz. The molecular genetics of Philadelphia chromosome-positive leukemias. *N. Engl. J. Med.* 319 (1988): 990.

Leder, P., J. Battey, G. Lenoir, C. Moulding, W. Murphy, H. Potter, T. Stewart, and R. Taub. Translocations among antibody genes in human cancer. *Science* 222 (1983): 765.

Nowell, P. C. Chromosomal and molecular clues to tumor progression. *Semin. Oncol.* 16 (1989): 116.

Chemical Carcinogens

Ames, B. N. Identifying environmental chemicals causing mutations and cancer. *Science* 204 (1979): 547.

Miller, E. C. Some current perspectives on chemical carcinogenesis in humans and experimental animals. *Cancer Res.* 38 (1978): 1479.

The Regulation of Tissue Growth

Chambard, J. C., S. Paris, G. L'Allemain, and J. Pouyssegur. Two growth factor signalling pathways in fibroblasts distinguished by pertussis toxin. *Nature* 326 (1987): 800.

Cheifetz, S., J. A. Weatherbee, M. L.-S. Tsang, J. K. Anderson, J. E. Mole, R. Lucas, and J. Massague. The transforming growth factor-β system, a complex pattern of cross-reactive ligands and receptors. *Cell* 48 (1987): 409.

Curran, T., and P. K. Vogt. Dangerous liaisons: fos and jun, oncogenic transcription factors. In *Transcriptional Regulation* (S. L. McKnight and K. R. Yamamoto, eds.). Cold Spring Harbor, NY: Cold Spring Harbor Laboratory Press, 1992.

Roberts, A. B., K. C. Flanders, P. Kondaiah, N. L. Thompson, E. Van Obberghen-Schilling, L. Wakefield, P. Rossi, B. De Crombrugghe, U. Heine, and M. B. Sporn. Transforming growth factor β: Biochemistry and roles in embryogenesis, tissue repair and remodeling, and carcinogenesis. *Recent Prog. Hormone Res.* 44 (1988): 157.

Sibley, D. R., J. L. Benovic, M. G. Caron, and R. J. Lefkowitz. Regulation of transmembrane signaling by receptor phosphorylation. *Cell* 48 (1987): 913.

Tumor Progression and Dissemination

Armstrong, P. B. Invasiveness of non-malignant cells. In *Invasion: Experimental and Clinical Implications* (M. Mareel and K. Calman, eds.). Oxford, England: Oxford University Press, 1984.

Bock, G., ed. *Metastasis* (Ciba Foundation Symposium No. 141), Chichester, UK: Wiley, 1988.

Folkman, J., and M. Klagsbrun. Angiogenic factors. *Science* 235 (1987): 235.

Leibovich, J., P. J. Polverini, H. M. Shepard, D. M. Wiseman, V. Shively, and N. Nuseir. Macrophage-induced angiogenesis is mediated by tumor necrosis factor-α. *Nature* 329 (1987): 630.

Stetler-Stevenson, W. G., S. Aznavorian, and L. A. Liotta. Tumor cell interactions with the extracellular matrix during invasion and metastasis. *Annu. Rev. Cell Biol.* 9 (1993): 541.

Box 25A: On the Trail of the *ras* Oncogene

van Aelst, L., M. Barr, S. Marcus, A. Polverino, and M. Wigler. Complex formation between RAS and RAF and other protein kinases. *Proc. Natl. Acad. Sci.* 90 (1993): 6213.

Feig, L. A., and B. Schaffhausen. The hunt for Ras targets. *Nature* 370 (1994): 508.

Kataoka, T., S. Powers, S. Cameron, O. Fasano, M. Goldfarb, J. Broach, and M. Wigler. Functional homology of mammalian and yeast RAS genes. *Cell* 40 (1985): 19.

Kataoka, T., S. Powers, C. McGill, O. Fasano, J. Strathern, J. Broach, and M. Wigler. Genetic analysis of yeast *RAS1* and *RAS2* genes. *Cell* 37 (1984): 437.

Marx, J. Forging a path to the nucleus. *Science* 260 (1993): 1588.

Moodie, S. A., B. M. Willumsen, M. J. Weber, and A. Wolfman. Complexes of Ras•GTP with Raf-1 and mitogen-activated protein kinase kinase. *Science* 260 (1993): 1658.

Reddy, E. P., R. K. Reynold, E. Santos, and M. Barbacid. A point mutation is responsible for the acquisition of transforming properties of the T24 human bladder carcinoma oncogene. *Nature* 300 (1982): 149.

Tabin, C. J., S. M. Bradley, C. I. Bargmann, R. A. Weinberg, A. G. Papageorge, E. M. Scolnick, R. Dhar, D. R. Lowy, and E. H. Chang. Mechanism of activation of a human oncogene. *Nature* 300 (1982): 143.

PRINCIPLES AND TECHNIQUES
OF MICROSCOPY

C ell biologists need to examine the structure of cells and their components. The microscope is an indispensable tool for this purpose because most cellular structures are too small to be seen by the unaided eye. In fact, the beginnings of cell biology can be traced to the invention of the **light microscope,** which made it possible for scientists to see enlarged images of cells for the first time. The first generally useful light microscope was developed in 1590 by Z. Janssen and his nephew H. Janssen. Many important observations in biology were reported during the next century, notably those of Robert Hooke and Antonie van Leeuwenhoek in the last quarter of the seventeenth century. Since then, the light microscope has undergone numerous improvements and modifications, right up to the present time.

By contrast, the **electron microscope** is of much more recent vintage, dating from the early 1930s. Just as the invention of the light microscope heralded a wave of scientific achievement, the development of the electron microscope triggered a revolution in the exploration of cell structure and function and, ultimately, in the way we think about cells.

The past decade has seen a renaissance in light microscopy that has enabled researchers to explore new levels of the subcellular organization and physiological dynamics of cells. These advances have been developed with the merging of technologies from physics, engineering, chemistry, and molecular biology, and they have revolutionized our ability to explore the details of cells using the light microscope. Applications of *electronic imaging* techniques have allowed the mapping of very sparse concentrations of biological molecules. *Confocal microscopy* has greatly improved the resolution and clarity of images obtained with thick samples, and *video microscopy* has opened the door to studying the dynamics of biological processes in living specimens. The development of a large family of new *fluorescent probes* has greatly improved the specificity of the images that can be obtained. We will look at each of these methods in some detail. We will begin by exploring the fundamental principles of both light and electron microscopy. Then we will go on to examine a variety of techniques relevant to cell biology.

Image Formation in Light and Electron Microscopy

Regardless of the kind of microscope being used, three elements are always needed to form an image: a **source of illumination,** the **specimen** to be observed, and a system of **lenses** to focus the illumination on the specimen and to form the image. Figure A-1 illustrates these features for a light microscope and an electron microscope. In a light microscope, the source of illumination is visible light, and the lens system consists of a series of glass lenses. The image can be viewed directly through an eyepiece or focused on a detector, such as photographic film or an electronic camera. In an electron microscope, the illumination source is a beam of electrons emitted by a heated tungsten filament, and the lens system consists of a series of electromagnets. The electron beam is focused either on a fluorescent screen of zinc sulfide for direct visualization of the image or on photographic film.

Despite these differences in illumination source and instrument design, both types of microscope depend on the same principles of optics and form images in a similar manner. When a specimen is placed in the path of the light or electron beam, physical characteristics of the beam are changed in a way that can be interpreted by the human eye or recorded on a photographic plate. The image can often be enhanced by modifying specific properties of the specimen or the illumination source. We will discuss some of these modifications when we examine different types of specimen preparation and microscope design.

The *wavelength* of the illumination source is an important feature of microscopy, because it determines how small an object can be and still be detected. To understand the interaction between the illumination source and the speci-

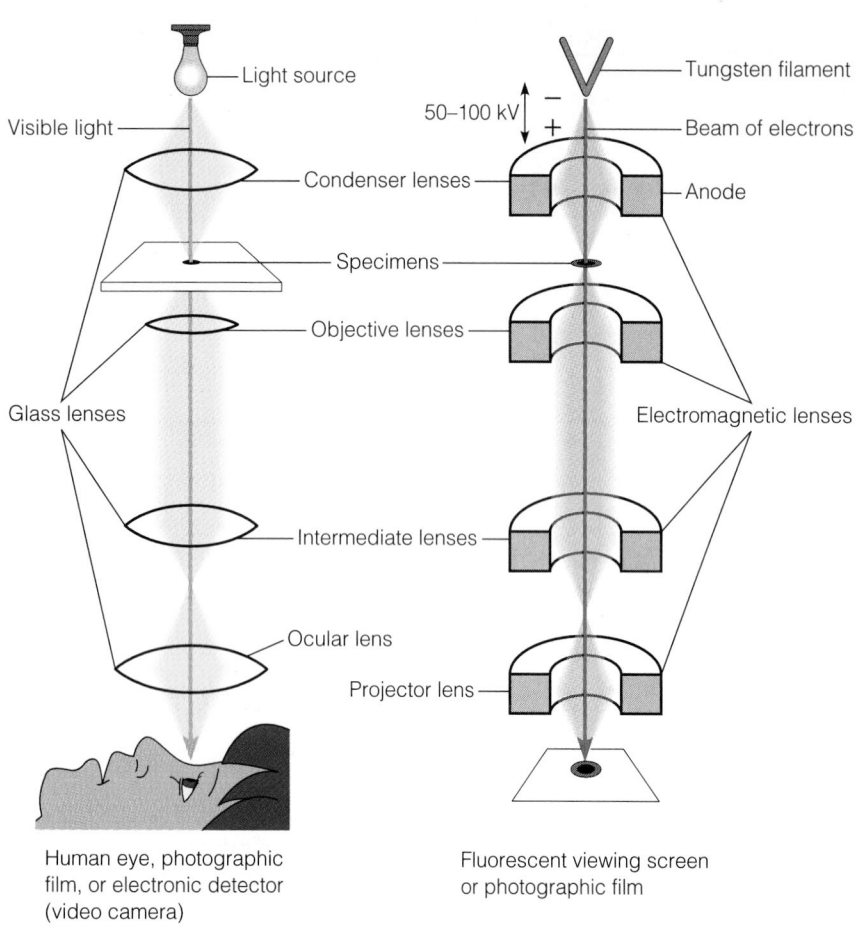

Figure A-1 The Optical Systems of the Light Microscope and the Electron Microscope. **(a)** The light microscope uses visible light and glass lenses to form an image of the specimen that can be seen by the eye, focused on photographic film, or received by an electronic detector such as a video camera. **(b)** The electron microscope uses a beam of electrons emitted by a tungsten filament and focused by electromagnetic lenses to form an image of the specimen on a fluorescent screen or photographic film. (These diagrams have been drawn to emphasize the similarities in overall design between the two types of microscope. In reality, a light microscope is designed with the light source at the bottom and the ocular lens at the top, as shown in Figure A-6b.)

(a) The light microscope

(b) The electron microscope

men, consider the simple analogy depicted in Figure A-2. If two people hold onto opposite ends of a slack rope and wave the rope with a rhythmic up-and-down motion, they will generate a long, regular pattern of movement in the rope called a **wave form** (Figure A-2a). The distance from the crest of one wave to the crest of the next is called the **wavelength.**

If someone standing to one side of the rope tosses a large object such as a beach ball against the rope, the ball may interfere with, or perturb, the wave form of the rope's motion (Figure A-2b). However, if a small object such as a softball is tossed against the rope, the movement of the rope will probably not be affected at all (Figure A-2c). If the rope holders move the rope more rapidly, the motion of the rope will still have a wave form, but the wavelength will be shorter (Figure A-2d). In this case, a softball tossed against the rope is very likely to perturb the rope's movement (Figure A-2e). This simple analogy illustrates an important principle: The ability of an object to perturb a wave motion depends crucially on the size of the object in relation to the wavelength of the motion.

This principle is of great importance in microscopy because it means that the wavelength of the illumination source sets a limit on how small an object can be and still be seen. To understand this relationship, we need to recognize that the moving rope of Figure A-2 is analogous to the beam of photons or electrons that is used as an illumination source in a microscope. When photons or electrons encounter a specimen, the specimen alters the physical characteristics of the illuminating beam, just as the beach ball or softball alters the motion of the rope. And because an object can be detected only by its effect on the wave, the wavelength must be comparable in size to the object that is to be detected.

Once we understand this relationship between wavelength and object size, we can readily appreciate why very small objects can be seen only by electron microscopy: The wavelengths of electrons are very much shorter than those of photons. Thus, objects such as viruses and ribosomes are too small to perturb a wave of photons, but they can readily interact with a wave of electrons. As we discuss different types of microscopes and specimen preparation techniques, you might find it helpful to ask yourself how the source and specimen are interacting and how the characteristics of both are modified to produce an image.

Optical Principles of Microscopy

Although light and electron microscopes differ in many ways, they use similar optical principles to form images. One of the most important underlying principles in image for-

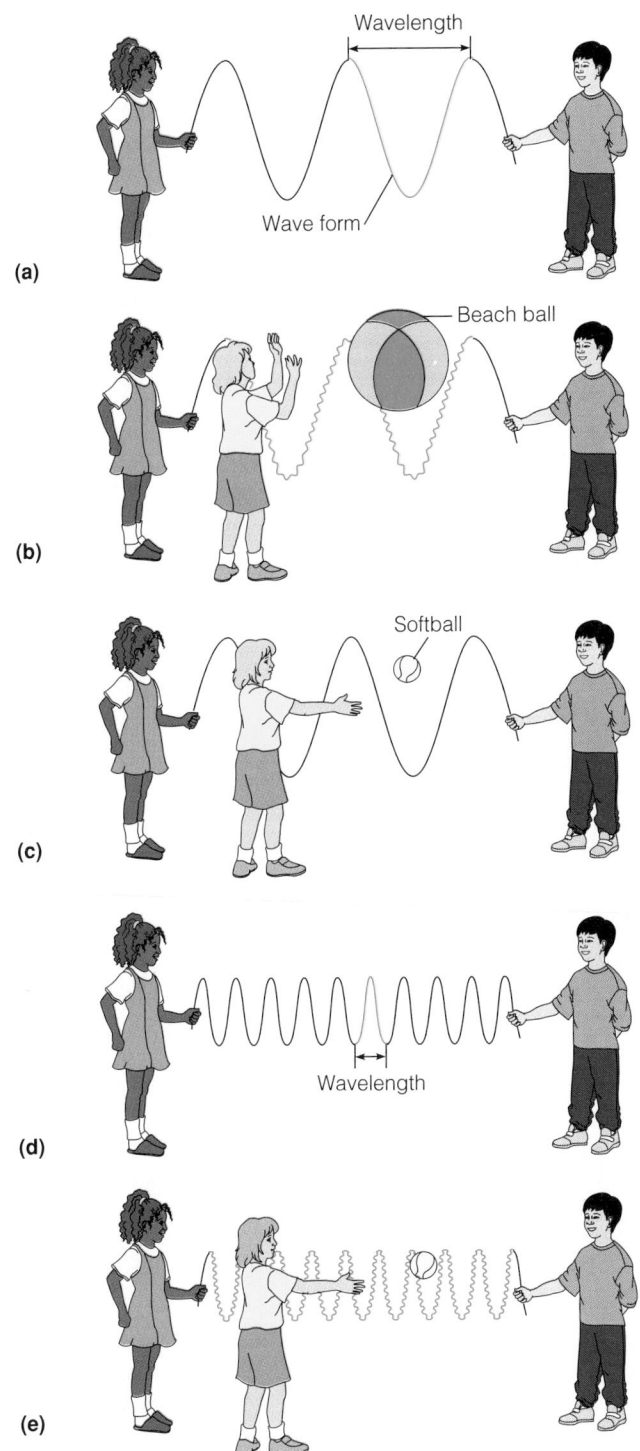

(a)

(b)

(c)

(d)

(e)

Wavelength

Wave form

Beach ball

Softball

Wavelength

Figure A-2 Wave Motion, Wavelength, and Perturbations. The wave motion of a rope held between two people is analogous to the wave form of both photons and electrons, and can be used to illustrate the effect of size of an object on its ability to perturb wave motion. **(a)** Moving a slack rope up and down rhythmically will generate a wave form with a characteristic wavelength. **(b)** When thrown against a rope, a beach ball or other object with a diameter that is comparable to the wavelength of the rope will perturb the motion of the rope. **(c)** A softball or other object with a diameter significantly less than the wavelength of the rope will cause little or no perturbation of the rope. **(d)** If the rope is moved more rapidly, the wavelength will be reduced substantially. **(e)** A softball can now perturb the motion of the rope because its diameter is comparable to the wavelength of the rope.

mation is that both light and electrons behave as *waves.* When light or electrons pass through a lens and come to a focus, the image that is formed results from a property of waves called **interference.** The image that you see when you look at a specimen through a series of lenses is really just a pattern of either additive or cancelling interference of the waves that went through the lenses, a phenomenon known as **diffraction.**

In a light microscope, glass lenses are used to direct the course of photons, whereas an electron microscope uses electromagnets as lenses to direct the course of electrons. Yet both kinds of lenses have two basic properties in common: focal length and angular aperture. The **focal length** is the distance between the midline of the lens and the point at which rays passing through the lens converge to a focus (Figure A-3). The **angular aperture** is the half-angle α of the cone of light entering the objective lens of the microscope from the specimen (Figure A-4). Angular aperture is therefore a measure of how much of the illumination that leaves the specimen actually passes through the lens, which in turn determines the sharpness of the interference pattern, and therefore the ability of the lens to convey information about the specimen. In the best light microscopes, the angular aperture is about 70°.

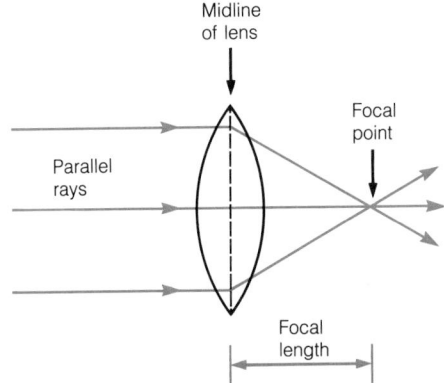

Midline
of lens

Focal
point

Parallel
rays

Focal
length

Figure A-3 The Focal Length of a Lens. Focal length is the distance from the midline of a lens to the point at which rays passing through the lens converge to a focus.

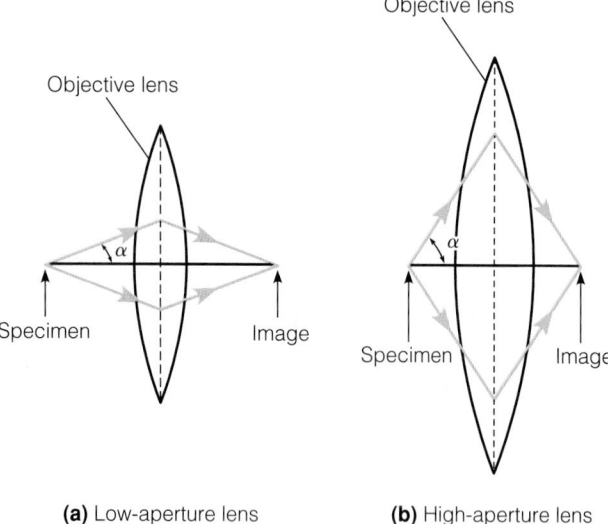

(a) Low-aperture lens   **(b)** High-aperture lens

Figure A-4 The Angular Aperture of a Lens. The angular aperture is the half-angle α of the cone of light entering the objective lens of the microscope from the specimen. **(a)** A low-aperture lens (α is small). **(b)** A high-aperture lens (α is large). The larger the angular aperture, the more information the lens can transmit. The best glass lenses have an angular aperture of about 70°.

Resolution

The most important characteristic of any lens is its **resolution.** To understand the central role that the lens plays in determining the resolution obtained in a microscope, we can ask how an infinitesimally small point of light is imaged by a simple lens. Figure A-5 illustrates the type of pattern that is formed in the image plane under such conditions. Even though the original specimen (the point of light in part a) is infinitesimal, the image formed by a lens is not (parts b and c). The resulting image is a spot whose size and shape are the result of the diffraction of light going through the lens.

The size and shape of the spot that arises from an infinitesimal object will limit how close two small objects can be and still be resolved. Resolution, and hence the size of this *diffraction-limited spot,* is therefore governed by three factors: the wavelength of the light used to illuminate the specimen, the angular aperture, and the refractive index of the medium surrounding the specimen. (**Refractive index** is a measure of the change in the velocity of light as it passes from one medium to another.)

Resolution is described quantitatively by the following equation:

$$r = 0.61\lambda / n \sin \alpha \qquad \textbf{(A-1)}$$

where r is the resolution, λ is the wavelength of the light used for illumination, n is the refractive index of the medium between the specimen and the objective lens of the microscope, and α is the aperture angle as already defined. (The constant 0.61 represents the degree to which image points can overlap and still be recognized as separate points by an observer.)

The quantity $n \sin \alpha$ is called the **numerical aperture** of the objective lens, abbreviated *NA.* An alternative expression for resolution is therefore

$$r = 0.61\lambda / NA \qquad \textbf{(A-2)}$$

Maximizing Resolution

Because r is a measure of how close two points can be and still be distinguished from each other, resolution improves as r becomes smaller. Thus, for the best resolution, the numerator of equation A-2 should be as small as possible and the denominator should be as large as possible.

We will begin by asking how to maximize resolution for a glass lens with visible light as the illumination source. First, we need to make the numerator as small as possible. The wavelength range for visible light is about 400–700 nm, and the minimum value for λ is set by the shortest wavelength in this range that is usable for illumination, which is

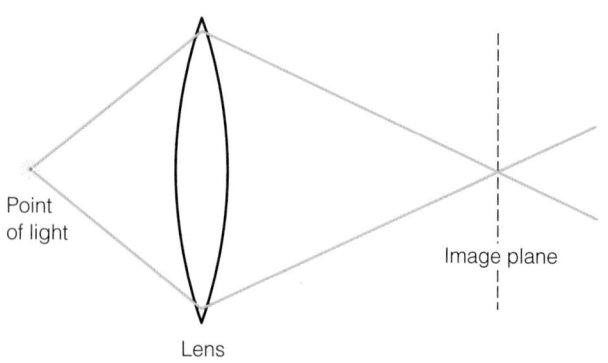
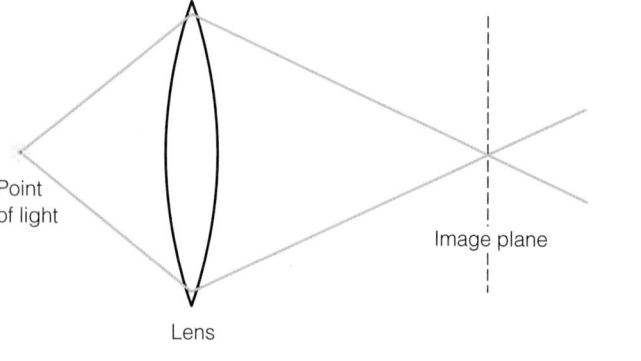

(a) Image formed from infinitesimal point of light

Figure A-5 Image Formation and the Resolution of a Lens.
(a) The image formed by a simple lens of an infinitesimally small point of light. In the plane where the rays come into focus, an interference pattern is formed. **(b)** The pattern formed on a piece of photographic film. **(c)** The

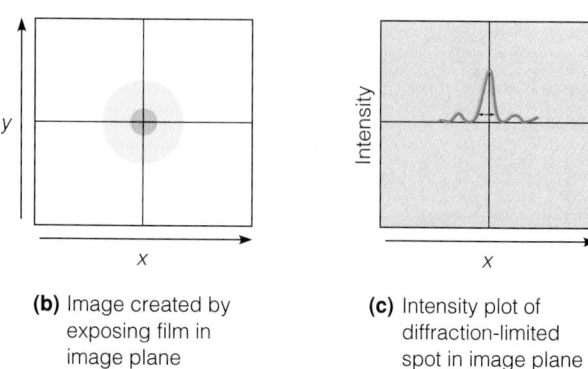

(b) Image created by exposing film in image plane

(c) Intensity plot of diffraction-limited spot in image plane

trace of a scan of the film along the x-axis of part b. Even an infinitesimally small spot will have a finite size when imaged by a lens. The size of the spot determines how close two objects can be and still be resolved.

blue light of about 450 nm. To maximize the denominator of equation A-2, recall that the numerical aperture is the product of the angular aperture and the sine of the refractive index. Both of these values must therefore by maximized for the best possible resolution. Since the angular aperture for the best objective lenses is about 70°, the maximum value for sin α is about 0.94. The refractive index of air is about 1, so for a lens designed for use in air, the maximum numerical aperture is about 0.94.

Thus, for a lens with an aperture angle of 70°, the resolution in air for a sample that is illuminated with blue light of about 450 nm can be calculated as follows:

$$r = 0.61\lambda/NA$$
$$= (0.61)(450)/0.94 = 292 \text{ nm} \approx 0.3 \ \mu\text{m} \qquad \textbf{(A-3)}$$

As a rule of thumb, then, the limit of resolution for a glass lens used in air is about 300 nm, or 0.3 μm.

As a means of increasing the numerical aperture value, some microscope lenses are designed to be used with a layer of *immersion oil* between the lens and the specimen. Immersion oil has a higher refractive index than air and therefore allows the lens to receive more of the light transmitted through the specimen. Since the refractive index of immersion oil is about 1.5, the maximum numerical aperture for an oil immersion lens is about 1.5 × 0.94 = 1.4. The resolution of an oil immersion lens is therefore about 0.2 μm:

$$r = 0.61\lambda/NA$$
$$= (0.61)(450)/1.4 = 196 \text{ nm} \approx 0.2 \ \mu\text{m} \qquad \textbf{(A-4)}$$

Thus, the **limit of resolution** (best possible resolution) for a microscope that uses visible light is about 0.3 μm in air and 0.2 μm with an oil immersion lens. By using ultraviolet light, the resolution can be pushed to about 0.1 μm because of the shorter wavelength (200–300 nm). However, the image must then be recorded by photographic film or another detector because ultraviolet light is invisible to the human eye. Moreover, expensive quartz lenses must be used because ordinary glass is opaque to ultraviolet light. In any case, these calculated values are theoretical limits to the resolution. In actual practice, such limits can rarely be reached because of technical flaws or *aberrations* in the lenses.

The limit of resolution for a lens sets the upper limit on the **useful magnification** that is possible with that lens. In general, the greatest useful magnification that can be achieved with a light microscope is about 1000 times the numerical aperture of the lens used. And because numerical aperture ranges from about 1.0 to about 1.4, the useful magnification of a light microscope is limited to about 1000× in air and about 1400× with immersion oil. Magnification greater than these limits is referred to as "empty magnification" because it provides no additional information about the object being studied.

Because the wavelength of an electron is so much shorter than that of a photon of visible light, the electron microscope has a theoretical limit of resolution much lower than that of the light microscope—about 0.1–0.2 nm instead of 200–300 nm. For biological samples, however, problems with specimen preparation and contrast are such that the practical limit of resolution is almost always much greater than the theoretical limit—generally about 2 nm. Practically speaking, therefore, resolution in an electron microscope is about 100 times greater than in the light microscope. As a result, the useful magnification of an electron microscope is about 100 times that of a light microscope, or about 100,000×.

The Light Microscope

An important name in the history of light microscopy is that of Antonie van Leeuwenhoek, the Dutch shopkeeper who is generally regarded as the father of microscopy. Leeuwenhoek's lenses, which he manufactured himself, were of surprisingly high quality for his time and were capable of 300-fold magnification. His observations, made over a period of more than 25 years, were remarkable, especially in view of the limitations imposed by the single-lens microscopes used at the time.

Today, the instrument of choice for light microscopy uses several lenses in combination and is therefore called a **compound microscope** (Figure A-6). The optical path through a compound microscope is illustrated in Figure A-6b. The path begins with the source of illumination, usually a light source located in the base of the instrument. The light rays from the source first pass through the **condenser lens,** which directs the light toward the specimen that is mounted on a glass slide and positioned on the **stage** of the microscope. The **objective lens** is located immediately above the specimen and is responsible for forming the *primary image.* Most compound microscopes have several objective lenses of differing magnifications mounted on a rotatable turret.

The primary image is further enlarged by the **ocular lens,** or *eyepiece.* In some microscopes, an **intermediate lens** is positioned between the objective and ocular lenses to accomplish still further enlargement. We can calculate the overall magnification of the image by multiplying the enlarging powers of the objective lens, the ocular lens, and the intermediate lens (if present). Thus, a microscope with a 10× objective lens, a 2.5× intermediate lens, and a 10× ocular lens will magnify a specimen 250-fold.

There are many kinds of light microscopy, but all of them are variations on the general theme of image formation that we have just described. Table A-1 illustrates four different techniques by comparing the same human cheek epithelial cell as seen by brightfield microscopy (unstained and stained), phase-contrast microscopy, differential interference contrast microscopy, and confocal microscopy. We will look at these and several other important techniques.

Brightfield Microscopy

The elements of the microscope as they have been described so far represent the basic form of light microscopy, that of **brightfield microscopy.** Compared to other microscopes,

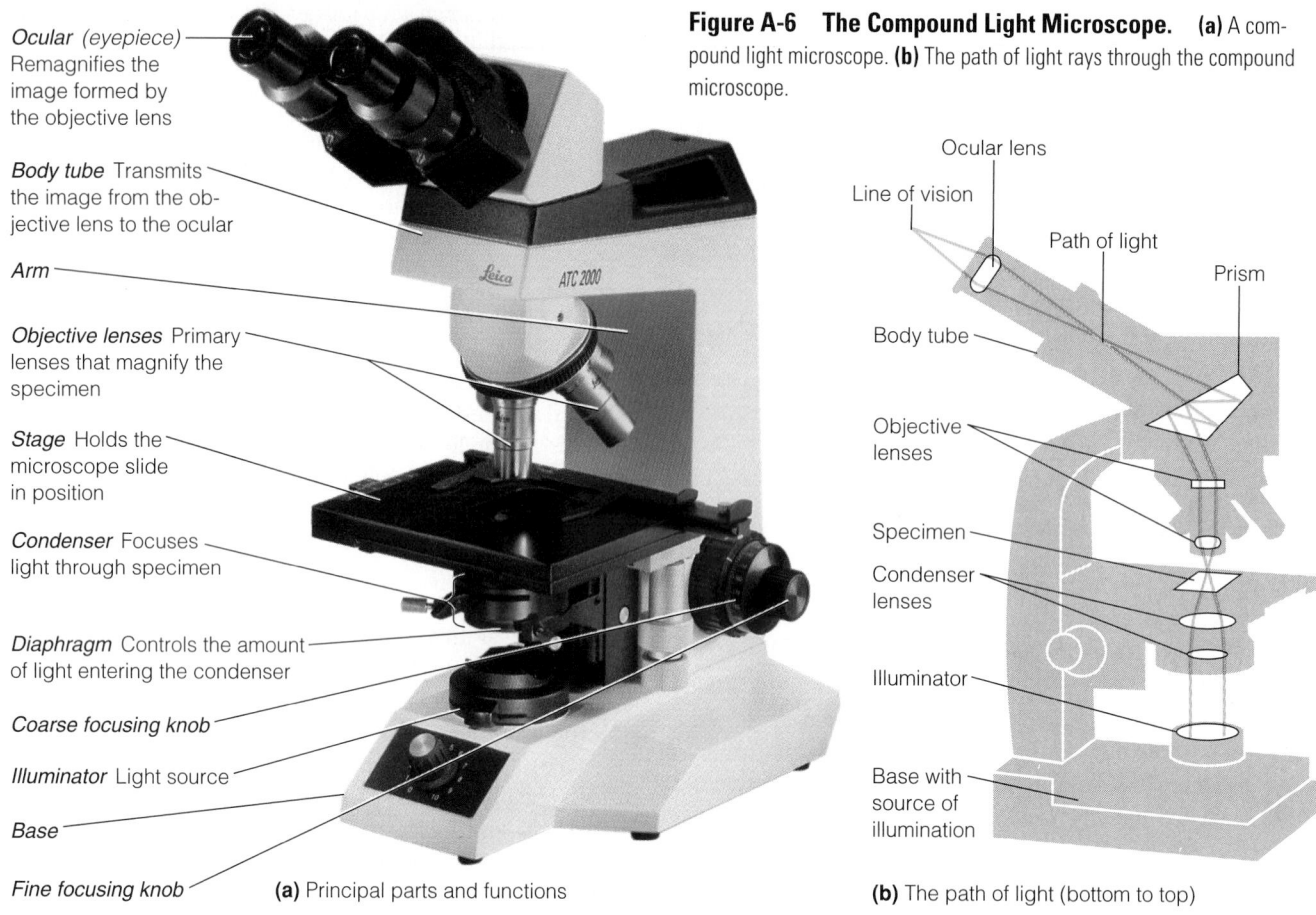

Figure A-6 The Compound Light Microscope. **(a)** A compound light microscope. **(b)** The path of light rays through the compound microscope.

Ocular *(eyepiece)* Remagnifies the image formed by the objective lens

Body tube Transmits the image from the objective lens to the ocular

Arm

Objective lenses Primary lenses that magnify the specimen

Stage Holds the microscope slide in position

Condenser Focuses light through specimen

Diaphragm Controls the amount of light entering the condenser

Coarse focusing knob

Illuminator Light source

Base

Fine focusing knob

(a) Principal parts and functions

Ocular lens
Line of vision
Path of light
Prism
Body tube
Objective lenses
Specimen
Condenser lenses
Illuminator
Base with source of illumination

(b) The path of light (bottom to top)

Table A-1 Different Types of Light Microscopy

Type of Microscopy	Light Micrographs of Human Cheek Epithelial Cells	Type of Microscopy	Light Micrographs of Human Cheek Epithelial Cells
Brightfield (unstained specimen): Passes light directly through specimen; unless cell is naturally pigmented or artificially stained, image has little contrast.		**Differential interference contrast:** Also uses optical modifications to exaggerate differences in density.	
Brightfield (stained specimen): Staining with various dyes enhances contrast, but most staining procedures require that cells be fixed (preserved).		**Confocal:** Uses lasers and special optics for "optical sectioning." Only those regions within a narrow depth of focus are imaged. Regions above and below the selected plane of view appear black rather than blurry.	
Phase-contrast: Enhances contrast in unstained cells by amplifying variations in density within specimen; especially useful for examining living, unpigmented cells.			50 μm

the brightfield microscope is inexpensive and simple to align and use. However, the only specimens that can be seen directly by brightfield microscopy are those that possess color or have some other property that affects the amount of light that passes through. Many biological specimens lack these characteristics and must therefore be stained with dyes before they can be seen by brightfield optics.

Phase-Contrast Microscopy

To understand the basis of **phase-contrast microscopy,** we must recognize that a beam of light is made up of many individual rays of light. As the rays pass from the light source through the specimen, their velocity may be affected by the physical properties of the specimen. Specifically, the rays are *diffracted* and their phase is changed to different extents by different regions of the specimen. As a result, the rays that pass through the specimen get out of phase with respect to rays of light that do not pass through the specimen.

We can exploit this difference by inserting into the light path an optical material capable of bringing the direct or undiffracted rays into phase with those that have been dif-

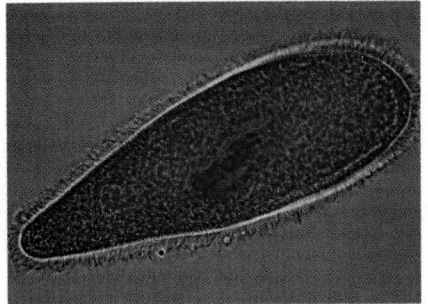

10 μm

Figure A-8   Phase-Contrast Microscopy.   A phase-contrast micrograph of the protozoan *Paramecium.* The cell was observed in an unprocessed and unstained state, which is a major advantage of phase-contrast microscopy.

fracted by the specimen. The resulting pattern of wavelengths intensifies the image. The phase-contrast microscope takes advantage of this effect by inserting a *phase plate* into the light path above the objective lens (Figure A-7). Phase contrast produces an image with highly contrasting bright and dark areas against a neutral gray background (Figure A-8). As a result, internal structures of cells are often better visualized by phase-contrast microscopy than with brightfield optics.

This approach to light microscopy is particularly useful for examining living, unstained specimens because biological materials almost inevitably diffract light. Phase-contrast microscopy is widely used in microbiology and tissue culture research to detect bacteria, cellular organelles, and other small entities in living specimens.

Fluorescence Microscopy

Until now we have dealt with various forms of light microscopy that generate contrast by diffraction, refraction, or interference. Although these techniques are quite powerful, they give little information concerning the distribution of specific molecules within the cell. In contrast, **fluorescence microscopy** involves measurements based on the absorption of light by a specific **fluorescent probe.** A fluorescent probe is either a single synthetic molecule that both fluoresces and indicates, or a combination of a fluorescent molecule and an indicator molecule (most commonly an antibody or toxin). A wide range of fluorescent probes exists, the spectral properties of which can be sensitive indicators of the molecular environment. The direct coupling of fluorescent probes to antibodies provides an especially useful means of mapping the subcellular distribution of cellular proteins.

Occasionally, nature itself provides useful tools. Figure A-9 depicts the fluorescence image of epithelial cells that have been labeled with a fluorescence-tagged mushroom toxin, *phalloidin,* which binds specifically to actin microfilaments. In this way, fluorescence microscopy provides a

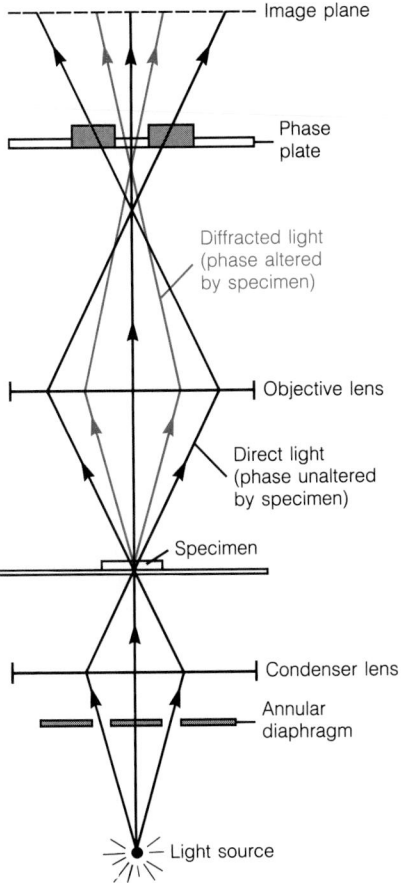

Figure A-7   Optics of the Phase-Contrast Microscope.
The configuration of the optical elements and the path of light rays through the phase-contrast microscope. Pink lines represent light diffracted by the specimen, and black lines represent direct light.

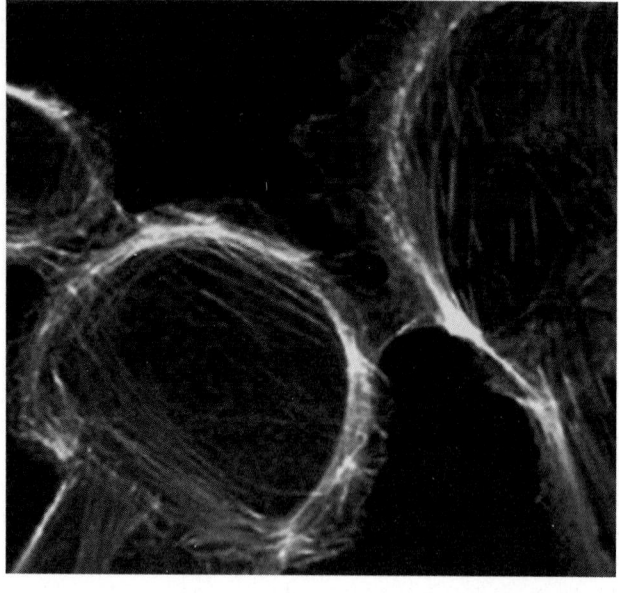

Figure A-9   **Fluorescence Microscopy.**   Cultured dog kidney epithelial cells stained with the fluorescent stain phalloidin, which binds to actin microfilaments.

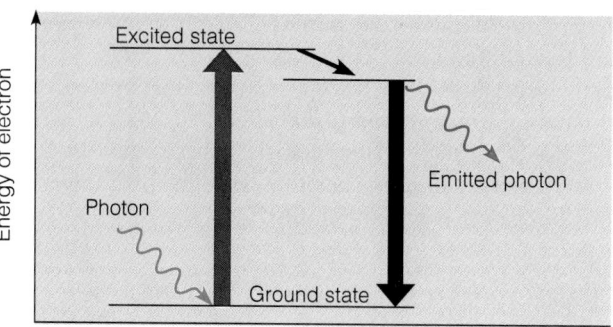

(a) Energy diagram

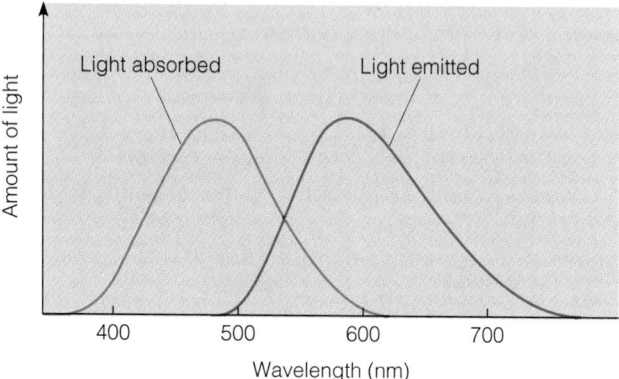

(b) Absorption and emission spectra

Figure A-10   Principles of Fluorescence.   (a) An energy diagram of fluorescence from a simple atom. Light of a certain energy is absorbed (e.g., the blue light shown here). The electron jumps from its ground state to an excited state. It returns to the ground state by emitting a photon of lower energy and hence longer wavelength (e.g., red light). **(b)** The absorption and emission spectra of a typical fluorescent molecule. The blue curve represents the amount of light absorbed as a function of wavelength, while the red curve shows the amount of emitted light as a function of wavelength.

means of making sensitive subcellular chemical and molecular analyses.

Fluorescence and Fluorescent Probes.   To understand fluorescence microscopy and the use of fluorescent probes, it is first necessary to understand fluorescence. **Fluorescence** is a process that begins with the absorption of light and ends with its emission. This phenomenon is best approached by considering the quantum behavior of light, as opposed to its wavelike behavior. Figure A-10a is a diagram of the various energy levels of a simple atom. When an atom absorbs a photon (or *quantum*) of light of the right energy, one of its electrons jumps from its ground state to a higher-energy, or *excited,* state. As the atom jiggles around, this electron often loses some of its energy and drops back down to the original ground state, emitting another photon as it does so. The emitted photon is always of less energy (longer wavelength) than the original photon that was absorbed. Thus, for example, shining blue light on the atom may result in red light being emitted. (Recall that the energy of a photon is inversely proportional to its wavelength; therefore, red light, being longer in wavelength than blue light, is lower in energy.)

Real fluorescent molecules have energy diagrams that are more complicated than that depicted in Figure A-10a. The number of possible energy levels in real molecules is much greater, so the different energies that can be absorbed, and emitted, are correspondingly greater. The absorption and emission spectra of a typical fluorescent molecule are shown in Figure A-10b. Every fluorescent probe has its own characteristic absorption and emission spectra.

In recent years, chemists have synthesized molecules with fluorescent properties that are sensitive to their environment (i.e., to the concentrations of ions such as Ca^{2+}, H^+, Na^+, Zn^{2+}, and Mg^{2+}, as well as to the electrical potential across the plasma membrane). These molecules have wide application in cell biology. When injected into cells or applied in a form that gets trapped in the cytosol or in specific intracellular compartments, such molecules can provide information about the cellular environment.

The Fluorescence Microscope.   A fluorescent microscope has an *exciter filter* between the light source and the condenser lens that transmits only excitation light (Figure A-11). The condenser lens then focuses the rays on the specimen, causing fluorescent compounds in the specimen to emit light of longer wavelength. Both the excitation light from the illuminator and the emitted light generated by fluorescent compounds in the specimen then pass through the

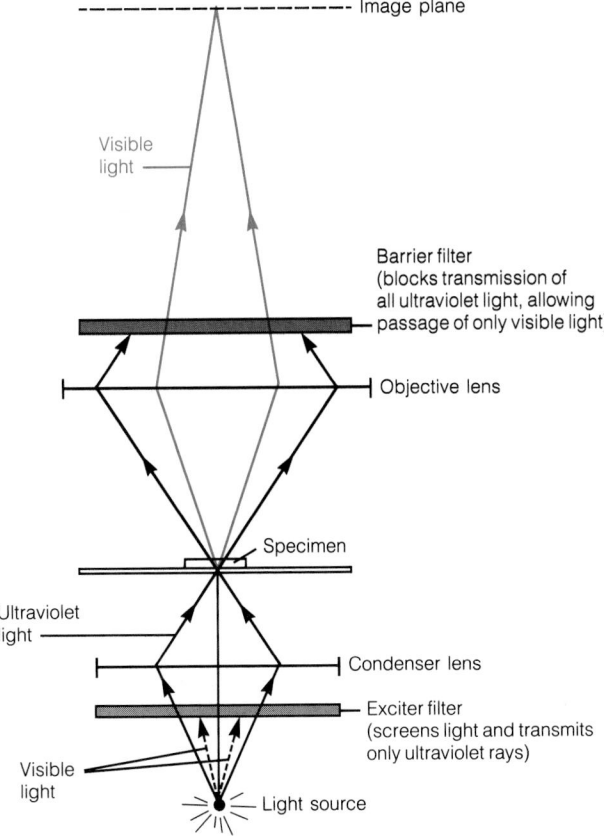

Figure A-11 Optics of the Fluorescence Microscope.
The configuration of the optical elements and the path of light rays through the fluorescence microscope. Light from the source passes through an exciter filter that transmits only excitation light (solid black lines). Illumination of the specimen with this light induces fluorescent molecules in the specimen to emit longer-wavelength light (pink lines). The barrier filter subsequently removes the excitation light, while allowing passage of the emitted light. The image is therefore formed exclusively by light emitted by fluorescent molecules in the specimen.

objective lens. As the light passes through the tube of the microscope above the objective lens, it encounters a *barrier filter* that specifically removes the excitation wavelengths. This leaves only the emission wavelengths to form the final image, which therefore appears bright against a dark background.

Differential Interference Contrast (DIC) Microscopy

Differential interference contrast (DIC) microscopy is an optical method that makes use of the wavelike properties of light and the way in which those properties change when traveling though a specimen. As with phase-contrast microscopy, DIC microscopy depends on changes in the phase of the light wave as it passes through the specimen. In this case, however, gradients in the refractive index in the sample (between the specimen and the surrounding

medium, for example) are detected. In fact, the DIC image is really a map of the differences in phase along one direction, achieved by comparing the phase change of two nearby paths of light (Figure A-12). Because the biggest phase changes usually occur at cell edges (the refractive index is more constant within the cell), the outline of the cell typically gives a strong signal. The image appears three-dimensional as a result of a shadow-casting illusion that arises because differences in phase are positive on one side of the cell but negative on the opposite side of the cell (Figure A-13).

The optical components required for DIC microscopy consist of a *polarizer,* an *analyzer,* and a pair of *Wollaston prisms* (see Figure A-12). The polarizer and the first Wollaston prism split a beam of light, creating two beams that are separated by a small distance along one direction. After traveling through the specimen, the beams are recombined by the second Wollaston prism. If no specimen is present, the beams recombine to form one that is identical to that which initially entered the polarizer and first Wollaston prism. In the presence of a specimen, the two beams do not recombine in the same way (i.e., they interfere with each other), and the resulting beam's polarization becomes rotated slightly compared to the original. This rotation is then de-

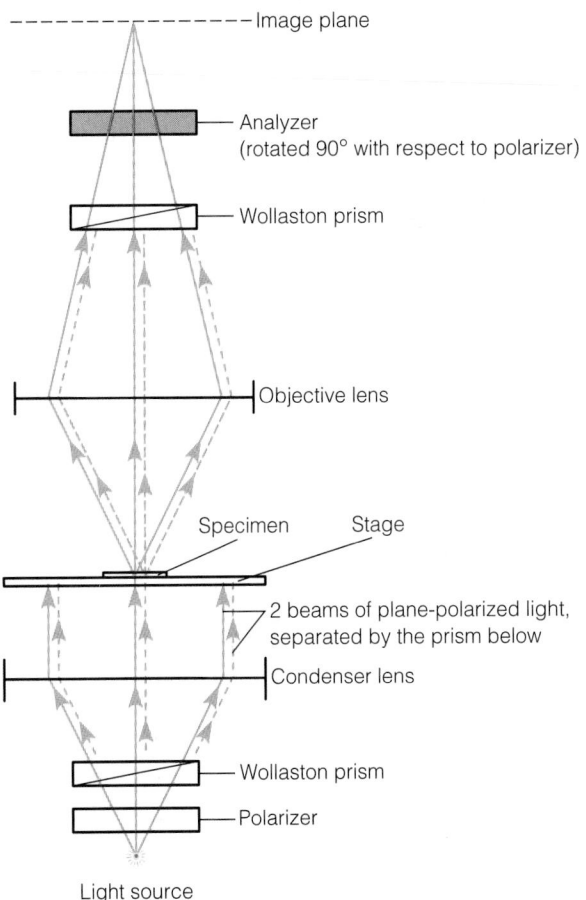

Figure A-12 Optics of the Differential Interference Contrast (DIC) Microscope.   The configuration of the optical elements and the path of light rays through the DIC microscope.

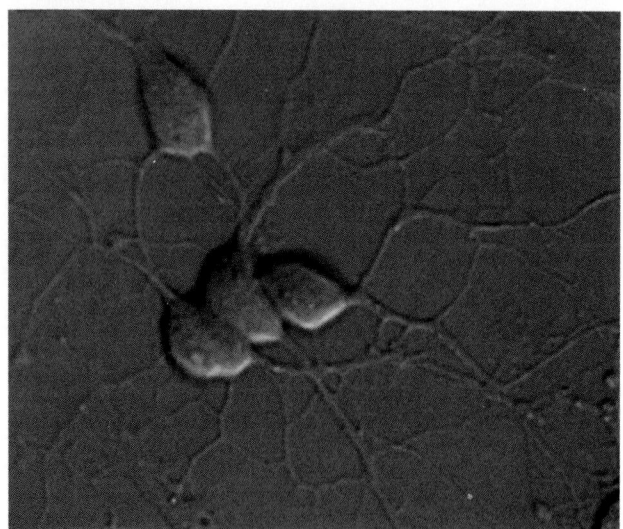

Figure A-13 **DIC Microscopy.** A DIC micrograph of a cluster of rat hippocampal neurons growing in culture. Notice the shadow-casting illusion that makes these cells appear dark at the top and light at the bottom.

$10\ \mu m$

tected with an analyzer. The degree of rotation is related to the difference in phase of the two nearby beams of light.

DIC imaging provides a sensitive means for *optical sectioning* because this interference only occurs in the focal plane of the objective lens. This technique is especially useful for studying living, unstained specimens. By combining this technique with video microscopy, the dynamics of cellular morphogenesis and cell division can be studied, as we are about to see.

Video Microscopy and Electronic Imaging

The advent of solid-state light detectors has in many circumstances made it possible to replace photographic film with an electronic equivalent—that is, with a video camera or a digital-imaging camera. These developments have given rise to the techniques of **video microscopy** and **electronic imaging.** By placing a video camera in the image plane formed by the ocular lens (see Figure A-1), one can record and store the microscope images. Video images, which are in analogue form, can be stored either on videotape or in a computer in digital format. However, images stored on videotape are of low fidelity, so most modern video microscopy is stored digitally, and an increasing number of such images are now produced digitally as well. Digital storage allows images not only to be viewed in higher fidelity, but also to be enhanced with computer-based image processors, increasing the contrast and eliminating distracting defects of the optical system (see Figure 20-3). These techniques have been particularly successful in applications to DIC and brightfield microscopy for the study of cell motility and cell division.

The sensitivity of modern, or intensified, cameras is such that one can use very dim illumination. This, along with the ability to shutter the illumination source, allows time-lapse images of living specimens to be recorded. When coupled with fluorescence microscopy, this technique makes it possible to obtain information on the changes in concentration and subcellular distribution of such cytosolic components as second messengers during physiological processes, and to study cytoskeletal components in living cells, as described in Chapter 20. Thus, video microscopy and electronic imaging have greatly expanded our knowledge of processes that occur in living cells, and have made possible the quantitative analysis of cellular chemistry.

Confocal Microscopy

The methods we have described so far depend on ways of generating contrast to obtain images of subcellular structures. An additional powerful method that also enhances contrast is **confocal microscopy.** This technique is designed to improve the effective resolution along the optical axis of the microscope—that is, to provide depth selection in the specimen. In this way, structures in the middle of a cell may be distinguished from those on the top or bottom. Likewise, a cell in the middle of a piece of tissue can be distinguished from cells above or below it.

To understand this type of microscopy, it is first necessary to consider the paths of light taken through a simple lens. Figure A-14a illustrates how a simple lens forms an image of a point source of light. To understand what your eye would see, imagine placing a piece of photographic film in the plane of focus (see Figure A-5b). Now ask how the images of other points of light placed further away or closer to the lens contribute to the original image (Figure A-14b). As you might guess, there is a precise relationship between o, the distance of the object from the lens, i, the distance from the lens to the image of that object brought into focus, and f, the focal length of the lens. This relationship is given by the equation

$$\frac{1}{f} = \frac{1}{o} + \frac{1}{i} \qquad \textbf{(A-5)}$$

As Figure A-14b shows, light arising from the points that are not in focus covers a greater surface area on the film because the rays are still either converging or diverging. Thus, the image on the film now has the original point source that is in focus, with a superimposed halo of light from the out-of-focus objects.

If we were only interested in seeing the original point source, we could mask out the extraneous light by placing an aperture, or *pinhole*, in the same plane as the film. This principle is used in a confocal microscope to discriminate against out-of-focus rays. In a real specimen, of course, we do not have just a single extraneous source of light on each side of the object we wish to see, but a continuum of points. To understand how this affects our image, imagine that instead of three points of light, our specimen consists of a long thin tube of light, as in Figure A-14c. Now consider obtaining an image of some arbitrary small section, dx. If the tube sends out the same amount of light per unit length, then

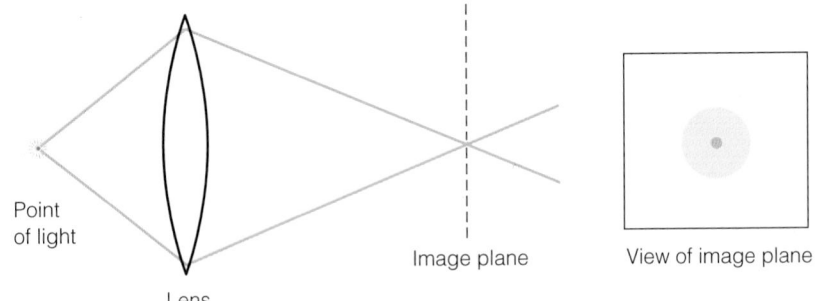

(a) Formation of an image of a single point of light by a lens

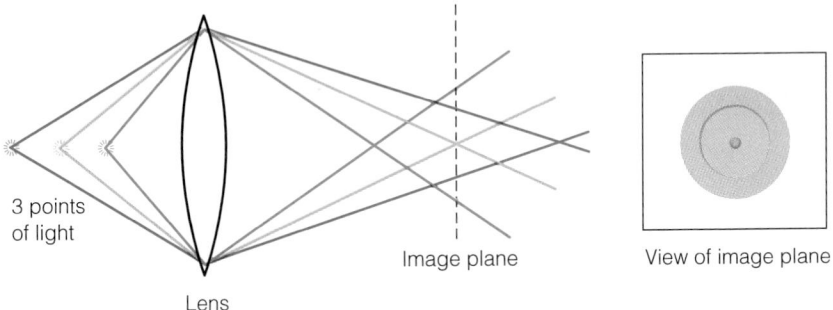

(b) Formation of an image of a point of light in the presence of two other points

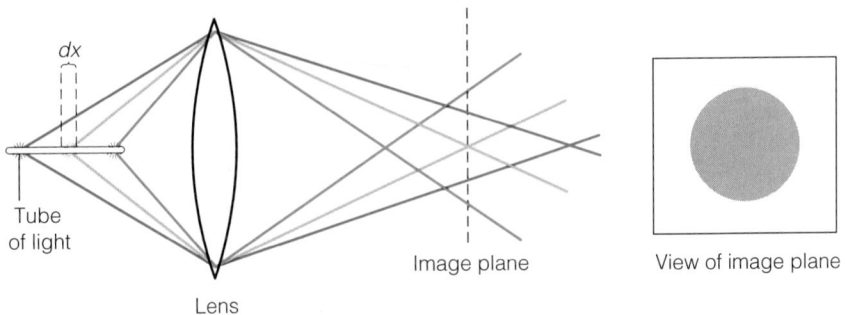

(c) Formation of an image of a section of an equally bright tube of light

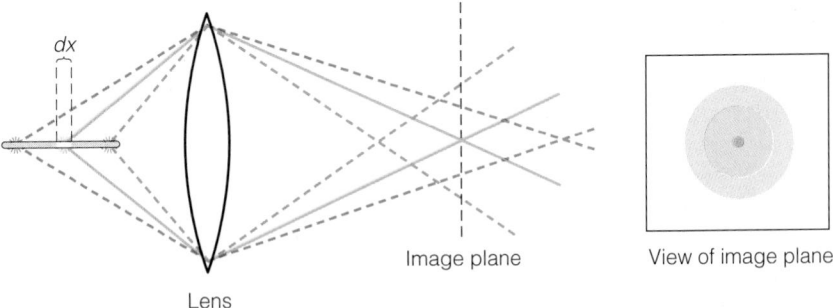

(d) Formation of an image of a brightened section of a tube of light

Figure A-14 Paths of Light Through a Single Lens. **(a)** The image of single point of light formed by a lens. **(b)** The paths of light from three points of light at different distances from the lens. In the image plane, the in-focus image of the central point is superimposed with the out-of-focus rays of the other points. A pinhole or aperture around the central point can be used to discriminate against out-of-focus rays and maximize the contributions from the central point. **(c)** The paths of light originating from a continuum of points, represented as a tube of light. This is similar to a uniformly illuminated sample. In the image plane, the contributions from an arbitrarily small in-focus section, *dx*, are completely obscured by the other out-of-focus rays; here a pinhole does not help. **(d)** By illuminating only a single section of the tube strongly and the rest weakly, we can recover information in the image plane about the section *dx*. Now a pinhole placed around the spot will reject out-of-focus rays. Because the rays in the middle are almost all from *dx*, we have a means of discriminating against the dimmer, out-of-focus points.

even with a pinhole, the image of interest will be obscured by the halos arising from other parts of the tube. This occurs because there is a small contribution from each out-of-focus section, and the sheer number of small sections will create a large background over the section of interest.

This situation is very close to that which we face when dealing with real biological samples that have been stained with a fluorescent probe. In general, the distribution of the probe is three-dimensional, and when we wish to look at the detail of a single object such as a microtubule using conventional fluorescence microscopy, the image is often marred by the halo of background light that arises mostly from microtubules above and below the plane of interest. To circumvent this, we can preferentially illuminate the plane of interest,

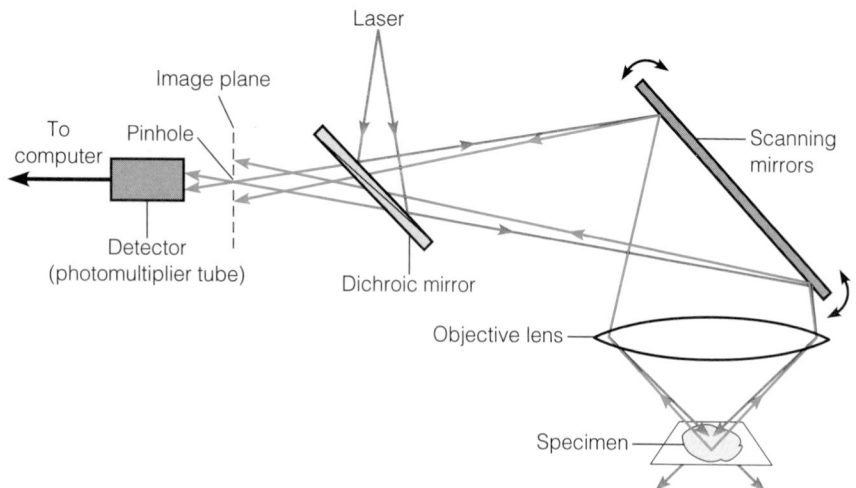

Figure A-15   A Laser Scanning Confocal Microscope.   A laser is used to illuminate one spot at a time in the specimen (blue lines). The scanning mirrors move the spot in a given plane of focus through a precise pattern. The fluorescent light that is emitted from the specimen (red lines) bounces off the same scanning mirrors and returns along the original path of the illumination beam. The emitted light does not return to the laser but instead is transmitted through the dichroic mirror (which in this example reflects blue light but transmits red light). A pinhole in the image plane blocks the extraneous rays that are out of focus. The light is detected by a photomultiplier tube, the signal from which is digitized and stored by a computer.

thereby biasing the contributions in the image plane such that they arise mostly from a single plane (Figure A-14d). Thus, the essence of confocal microscopy is to bring the illumination beam that excites the fluorescence into focus in a single plane, and to use a pinhole to ensure that the light we collect in the image plane arises mainly from that plane of focus (hence the word *confocal*).

One of the most practical applications of these principles is the **laser scanning confocal microscope (LSCM),** which is widely used in fluorescence imaging. As Figure A-15 shows, this type of instrument uses a laser that is focused down to a diffraction-limited spot (see Figure A-5) by the objective lens. The position of the spot is controlled by scanning mirrors, which allow the beam to be swept over the specimen in a precise pattern. The fluorescent light emitted

by the specimen is collected by the objective lens and returns along the same path as the original incoming light. The path of the fluorescent light is separated from the laser light using a dichroic mirror, which reflects one color but transmits another. Because the fluorescent light is lower in energy than the excitation beam, the fluorescence color is shifted toward the red. The light is then detected by a photomultiplier tube. A pinhole is placed at an image plane in front of the detector, and the signal from the photomultiplier tube is digitized for storage and display by a computer. As the beam is scanned over the specimen, an image of the specimen is formed. Shown in Figure A-16 are immunofluorescence images of the same cell (a fertilized sea urchin egg) obtained by conventional fluorescence microscopy and laser scanning confocal microscopy.

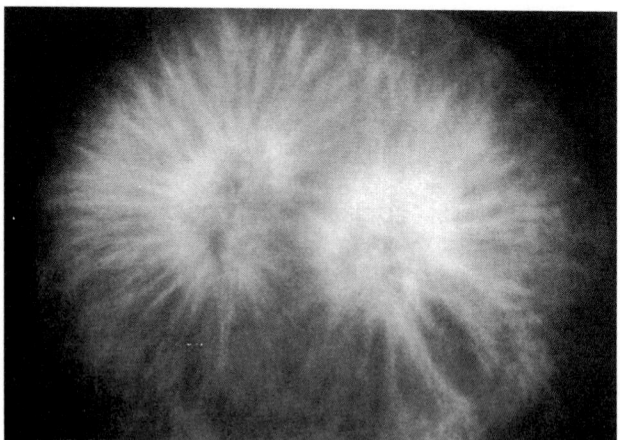

(a) Fluorescence micrograph

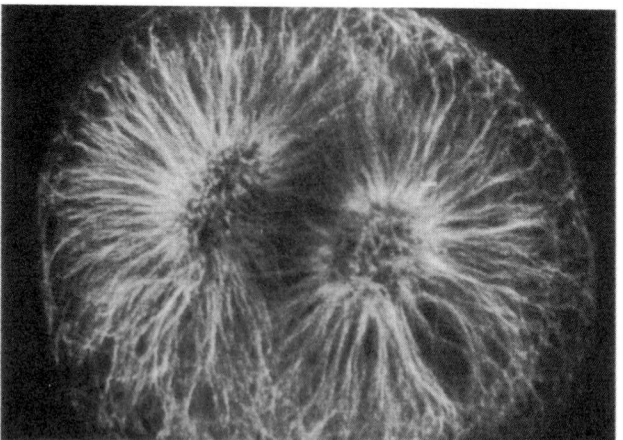

(b) Laser scanning confocal micrograph

Figure A-16   Laser Scanning Confocal and Conventional Fluorescence Microscopy Compared.   Both micrographs show a fertilized sea urchin egg labeled with a fluorescent antibody that binds to microtubules. **(a)** The typical image obtained using a fluorescence microscope. For specimens such as this one that are thicker than a few

microns, the contributions from the out-of-focus portions of the egg blur the image considerably. **(b)** The same specimen viewed with a laser scanning confocal microscope. Notice the considerable improvement in clarity that is achieved when the out-of-focus light from planes above or below the plane of focus is rejected. $\overline{}$
10 μm

Sample Preparation Techniques in Light Microscopy

One of the most attractive features of light microscopy is the ease with which most specimens can be prepared for examination. Preparation often involves nothing more than mounting a small piece of the specimen in a suitable liquid on a glass slide and covering it with a glass coverslip. The slide is then positioned on the specimen stage of the microscope and examined through the ocular lens, or with a camera. Magnification can be changed simply by rotating a turret that holds different objective lenses. Objective lenses capable of high magnification (40× and greater) generally require the application of a small drop of immersion oil to the coverslip for optimal viewing. Specimens that are pathogenic or potentially pathogenic are often killed chemically prior to viewing by the application of a small amount of fixative (such as an aqueous buffered aldehyde solution) to the specimen suspension.

Specimen Processing: Fixation

Often the first step in preparing the specimen is **primary fixation,** generally in a buffered aldehyde fixative. Fixation kills the cells, stabilizes their chemical components, and hardens the specimen in anticipation of further processing and sectioning. One way to fix a specimen is simply to immerse it in the fixative solution. An alternative approach for animal tissues is to pass the fixative through the bloodstream of the animal before removing the organs. This technique, called **perfusion,** may help reduce artifacts, false or inaccurate representations of the specimen that result from chemical treatment or handling of the cells or tissues.

Embedding and Sectioning

In most cases, the next step is to prepare *thin sections* of the fixed specimen. For this purpose, the specimen is embedded in a medium that will hold it rigidly in position while sections are cut. The usual choice of embedding medium is paraffin wax. Since paraffin is insoluble in water, any water in the specimen must first be removed (by dehydration in alcohol, usually) and replaced by an organic solvent such as xylene, in which paraffin is soluble. The processed tissue is then placed in warm, liquefied paraffin and allowed to harden. Dehydration is less critical if the specimen is embedded in a water-soluble medium instead of in paraffin. Specimens may also be embedded in epoxy plastic resin.

Next, the embedded specimen is sliced into thin sections, usually a few (1–10) micrometers thick. Sections are cut with a **microtome,** an instrument that operates somewhat like a meat slicer (Figure A-17). The paraffin or plastic block containing the specimen is mounted on the arm of the microtome, which advances the block by small increments toward a metal or glass blade. As successive sections are cut, they usually adhere to one another, forming a ribbon of thin sections. These sections can be mounted on a glass slide, stained (if desired), and protected with a coverslip.

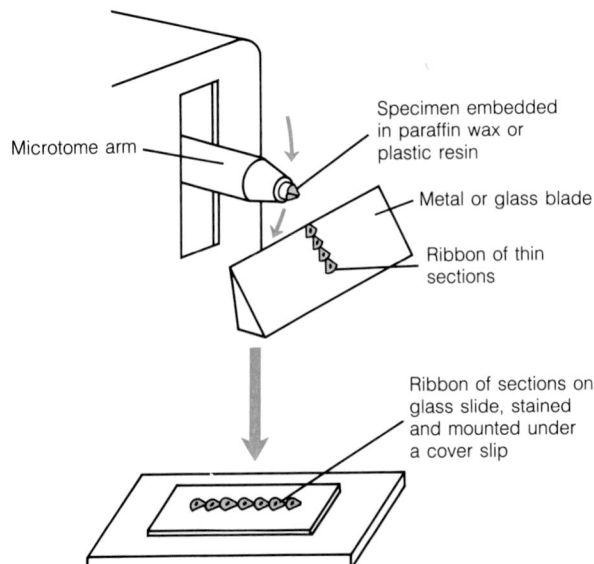

Figure A-17   Sectioning with a Microtome. The fixed specimen is embedded in paraffin wax or plastic resin and mounted on the arm of the microtome. As the arm moves up and down through a circular arc, successive sections are cut. These sections adhere to each other, forming a ribbon of thin sections that can be mounted on a glass slide, stained, and protected with a coverslip.

Staining

The purpose of **staining** is to give distinctive color characteristics to different kinds of cellular components. The general approach is to mount the fixed tissue on a microscope slide and then treat it with any of a variety of dyes and stains that have been adapted for this purpose. Sometimes the tissue is treated with a single stain, but more often a series of stains is used, each with an affinity for a different kind of cellular component.

Autoradiography

Autoradiography is a technique that uses photographic film to determine where within a cell a specific radioactively labeled compound is at the time the cell is fixed and sectioned for microscopy. Autoradiography can therefore be used to localize cellular processes to specific structures and to provide information about cell function. The preparation of a sample for autoradiography is shown in Figure A-18. In essence, the process involves incubating cells, organisms, or tissue with a radioactively labeled compound, fixing the specimen, sectioning it in the conventional way, and mounting the fixed sample on a microscope slide.

The slide is then covered with a thin layer of photographic emulsion and placed in a sealed box for the desired length of time, often for several days or even weeks. During this time, the radioactivity in the sample exposes the emulsion directly above it, providing a photographic record of precisely where in the cell the radioactive compound is located. Thereafter, the slide containing the specimen and the

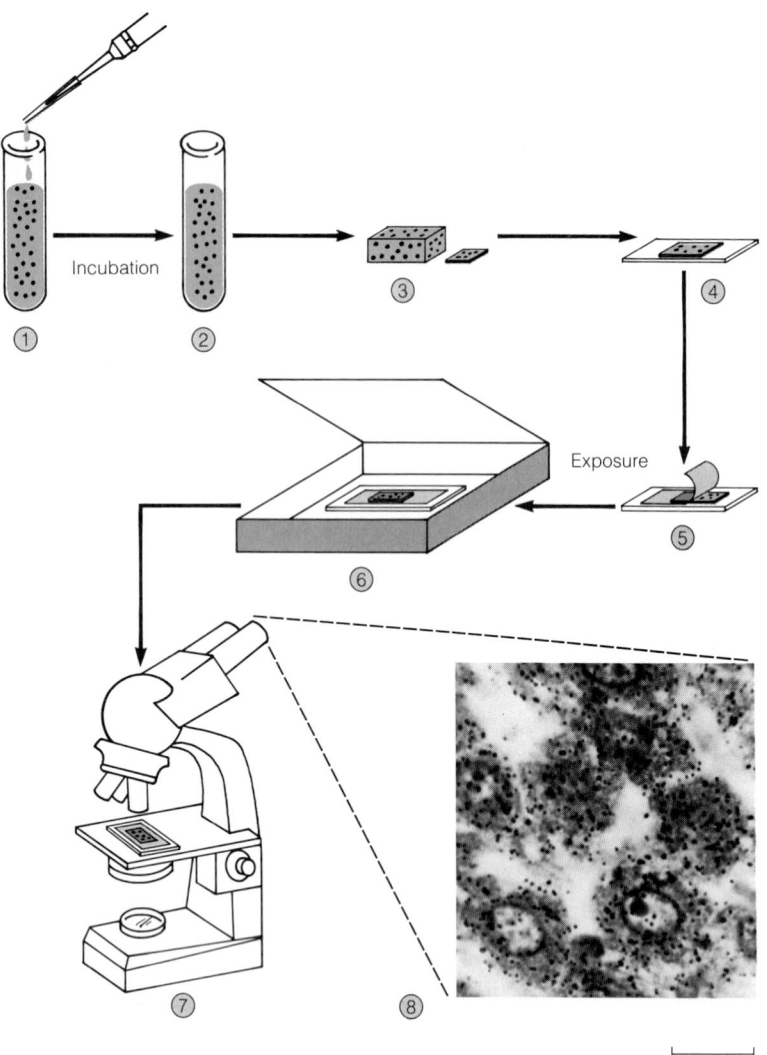

10 μm

emulsion is developed in much the same way as conventional black-and-white film. It is then ready for examination under the microscope.

Autoradiography can be applied to both light microscopy (Figure A-19a) and electron microscopy (Figure A-19b). In either case, it provides valuable information about the localization of specific molecules, structures, or processes within the cell.

The Electron Microscope: Design and Practice

The effect of electron microscopy on our understanding of cells can only be described as revolutionary. Yet, like light microscopy, electron microscopy has both strengths and weaknesses. In light microscopy, most specimens can be prepared easily and examined readily, but resolution of cellular ultrastructure is severely limited by the physical characteristics of light. In electron microscopy, resolution is much better, but specimen preparation and instrument operation are often more difficult.

Electron microscopes are of two basic designs: the *transmission electron microscope* and the *scanning electron microscope*. A third type of instrument, the *scanning transmission electron microscope*, is a hybrid of the two, as the name implies. Scanning and transmission electron microscopes are similar in that each uses a beam of electrons to produce an image. However, the instruments use quite different mechanisms to form the final image.

Transmission Electron Microscopy

A **transmission electron microscope (TEM)** is shown in Figure A-20. Most of the parts of the TEM are similar in name and function to their counterparts in the light microscope, although their physical orientation is reversed. We will look briefly at each of the major features.

The Vacuum System. Because electrons cannot travel very far in air, a strong vacuum must be maintained along the en-

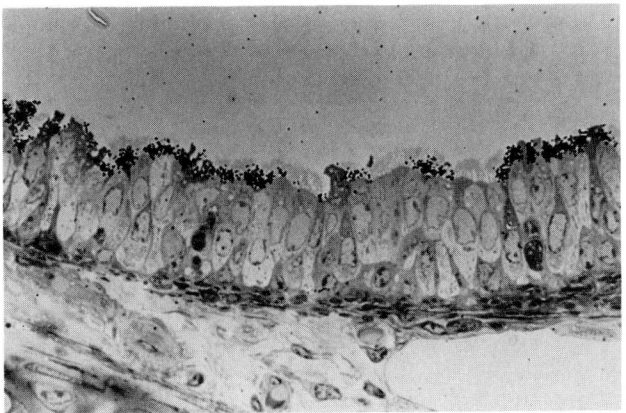

(a) Light autoradiograph  25 μm

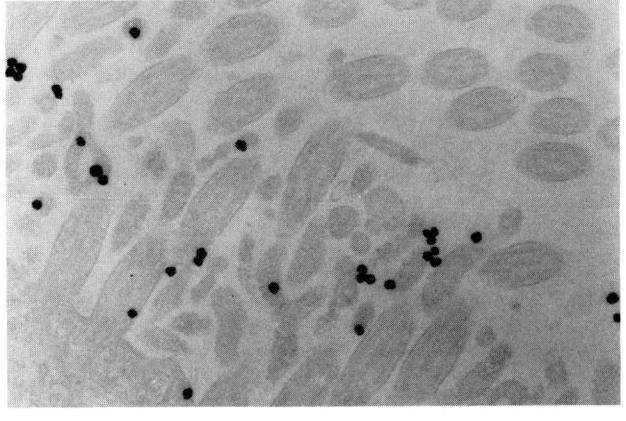

(b) Electron autoradiograph 1 μm

Figure A-19   Autoradiographs Obtained by Light and Electron Microscopy.   Autoradiography can be applied to both light microscopy and electron microscopy. For the autoradiographs shown here, a radioactively labeled microbial pathogen was allowed to infect epithelial cells of the respiratory tract of an animal. The microbes can be identified by the black dots, which represent areas where the radioactive

label exposed silver grains in the overlying photographic emulsion.
(a) The limited resolving power of the light microscope precludes direct observation of the organisms, but their presence is indicated by the numerous black dots corresponding to exposed silver grains. **(b)** The higher magnification and resolution of the electron microscope reveal both the microbial cells and their identifying label (black dots).

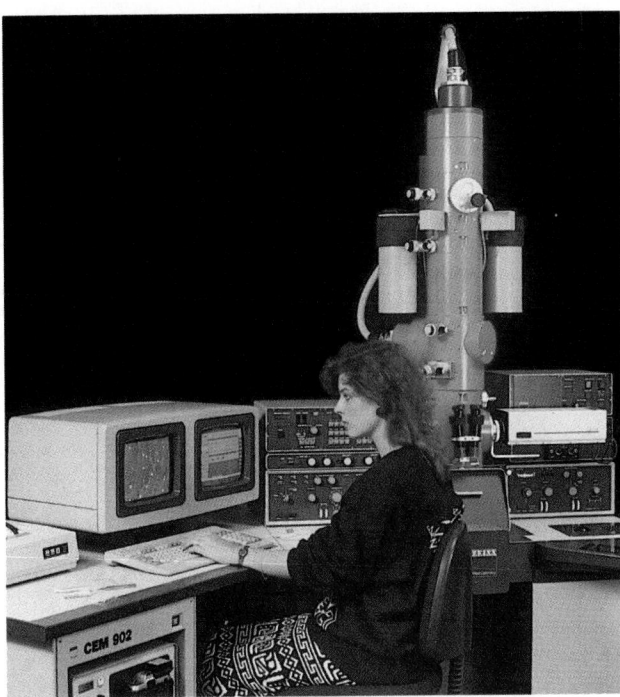

(a)

Figure A-20   A Transmission Electron Microscope.
(a) A photograph and **(b)** a schematic diagram of a TEM.

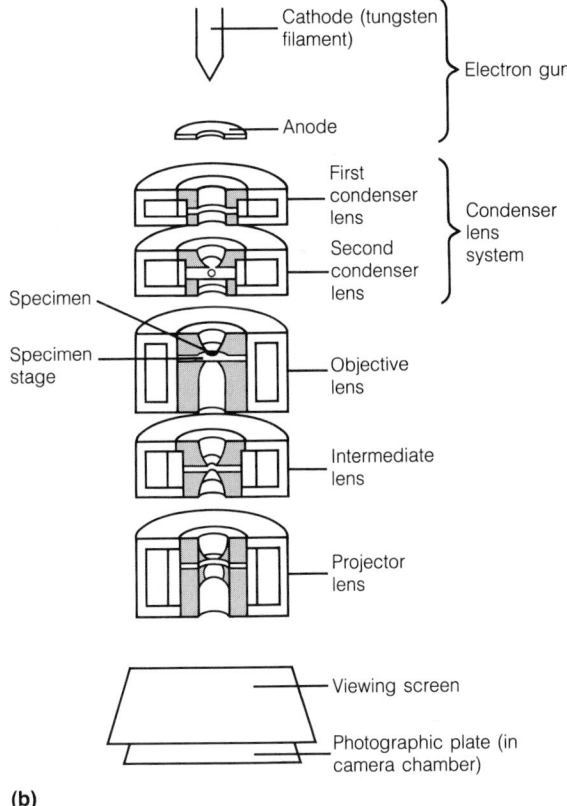

(b)

tire path of the electron beam. Two types of vacuum pumps work together to create the vacuum in the column of an electron microscope. A standard *rotary pump* is used to achieve the initial low vacuum when the instrument is first started up. The high vacuum required for operation is

achieved by an *oil diffusion pump*. The diffusion pump is an oil-filled reservoir in which oil is vaporized by heating. As the oil vapor rises, it traps air molecules and is then condensed by condensing vanes, which are cooled by circulating cold water. The diffusion pump cannot function independently; it requires backup by the rotary pump to remove the trapped air molecules from the system.

On some TEMs, a device called a *coldfinger* is incorporated into the vacuum system to help establish a high vacuum. The coldfinger is a metal insert in the column of the microscope that is cooled by liquid nitrogen. The coldfinger attracts gases and random contaminating molecules, which then solidify on the cold metal surface. When functioning at their best, most modern electron microscopes maintain a vacuum of about 1×10^{-4} mm Hg or less.

The Electron Gun.   The electron beam in a TEM is generated by an **electron gun,** an assembly of several components. The *cathode,* a tungsten filament similar to a light bulb filament, emits electrons from its surface when it is heated. The cathode tip is near a circular opening in a metal housing called the *Wehnelt cylinder.* A negative voltage on the cylinder helps control electron emission and shape the beam. At the other end of the cylinder is the *anode.* The anode is kept at 0 V, while the cathode is maintained at 50–100 kV. This difference in voltage is called the **accelerating voltage** because it causes the electrons to accelerate as they pass through the cylinder.

Electromagnetic Lenses and Image Formation.   The formation of an image using electron microscopy depends on both the wavelike and the particlelike properties of electrons. Just as a glass lens can bend the rays of light that pass through it, the trajectory of an electron can be controlled using electromagnets. Because electrons are charged, they are subject to magnetic forces when they move. This principle can be used to change the direction of an electron beam.

As the electron beam leaves the upper region of the condenser lens system, it enters a series of lenses made of electromagnets (Figure A-20b). The lens itself is simply a space influenced by an electromagnetic field. The focal length of each lens can be increased or decreased by varying the current applied to its energizing coils. Thus, when several lenses are arranged together, they can control illumination, focus, and magnification.

The **condenser lens** is the first lens to affect the electron beam. It functions in the same fashion as its counterpart in the light microscope to focus the beam on the specimen. Most electron microscopes actually use a condenser lens system with two lenses to achieve better focus of the electron beam.

The next component, the **objective lens,** is the most important part of the electron microscope's sophisticated lens system. The specimen is positioned on the specimen stage within the objective lens (see Figure A-20b). The objective lens, in concert with the **intermediate lens** and the **projector lens,** produces a final image on a *viewing screen* of zinc sulfide that fluoresces when struck by the electron beam.

How is an image formed from the action of these lenses on an electron beam? Recall that the electron beam generated by the cathode passes through the condenser lens system and impinges on the specimen. As the beam strikes the specimen, some electrons are scattered by the sample, whereas others continue in their paths relatively unimpeded. This scattering of electrons is the result of properties created in the specimen by the preparation procedure. Specimen preparation, in other words, imparts selective electron density to the specimen; that is, some areas become more opaque to electrons than others. Such electron-dense areas of the specimen will appear dark because few electrons pass through, whereas other areas will appear lighter because they permit the passage of more electrons.

The contrasting light, dark, and intermediate areas of the specimen create the image seen on the screen. The fact that the image is formed by differing extents of electron transmission through the specimen is reflected in the term *transmission electron microscope.*

The Photographic System.   In addition to observing the image on a fluorescent screen, an electron microscopist can also record the image photographically as an **electron micrograph,** which then becomes a permanent photographic record of the specimen.

Most transmission electron microscopes have a *camera chamber* mounted directly beneath the viewing screen. The camera is little more than a box that allows *photographic plates* to be moved manually or automatically to the area immediately beneath the viewing screen. To photograph a specimen, the microscopist simply aligns the image on the screen, focuses the image with the objective lens control, adjusts the illumination to a predetermined intensity with the condenser adjustment, and makes the exposure. The exposure may be made automatically or by lifting the screen with a lever on the microscope console. Once the exposure is made, the plate is advanced out of the exposure position to a container where it is stored until retrieved from the instrument for later development and printing.

Scanning Electron Microscopy

Scanning electron microscopy is a relatively recent development. It is an especially spectacular technique because of the sense of depth it gives to biological structures, thereby allowing surface topography to be studied. As the name implies, a **scanning electron microscope (SEM)** generates an image by scanning the specimen with a beam of electrons.

An SEM and its optical system are shown in Figure A-21. The vacuum system and electron source are similar to those found in the transmission electron microscope, although the accelerating voltage is much lower (about 5–30 kV). The significant difference between the two kinds of instruments lies in the way the image is formed. In the SEM, a magnetic lens system focuses the beam of electrons into an intense spot on the surface of the specimen. The spot is moved back and forth across the specimen by charged plates called *beam deflectors* located between the condenser lens and the specimen. The beam deflectors attract or repel

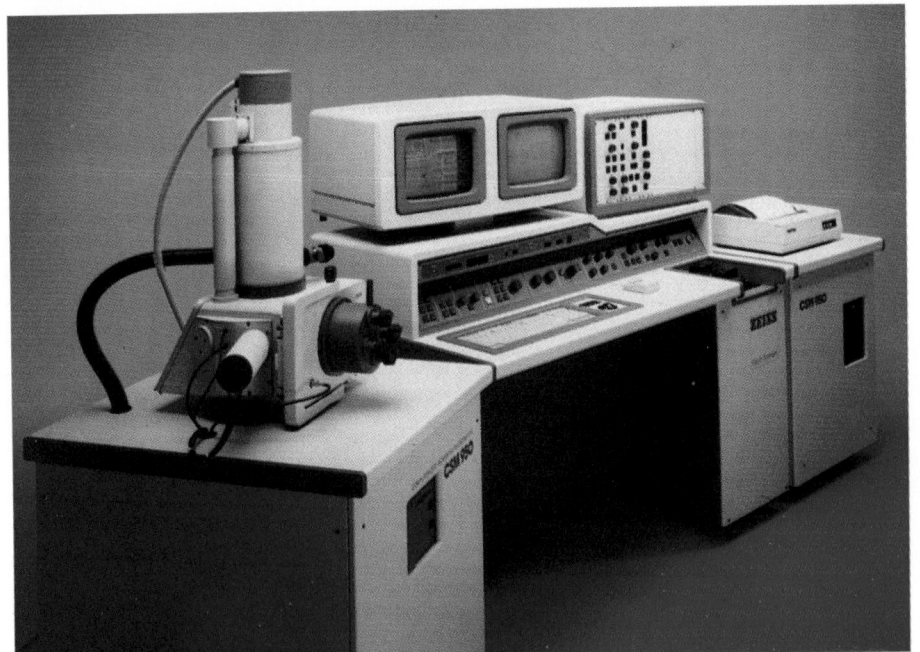

Figure A-21 A Scanning Electron Microscope. **(a)** A photograph and **(b)** schematic diagram of an SEM. The image is generated by secondary electrons (short pink lines) emitted by the specimen as a focused beam of primary electrons (long pink lines) sweeps rapidly over it. The signal to the video screen is synchronized to the movement of the primary electron beam over the specimen by the deflector circuitry of the scan generator.

(a)

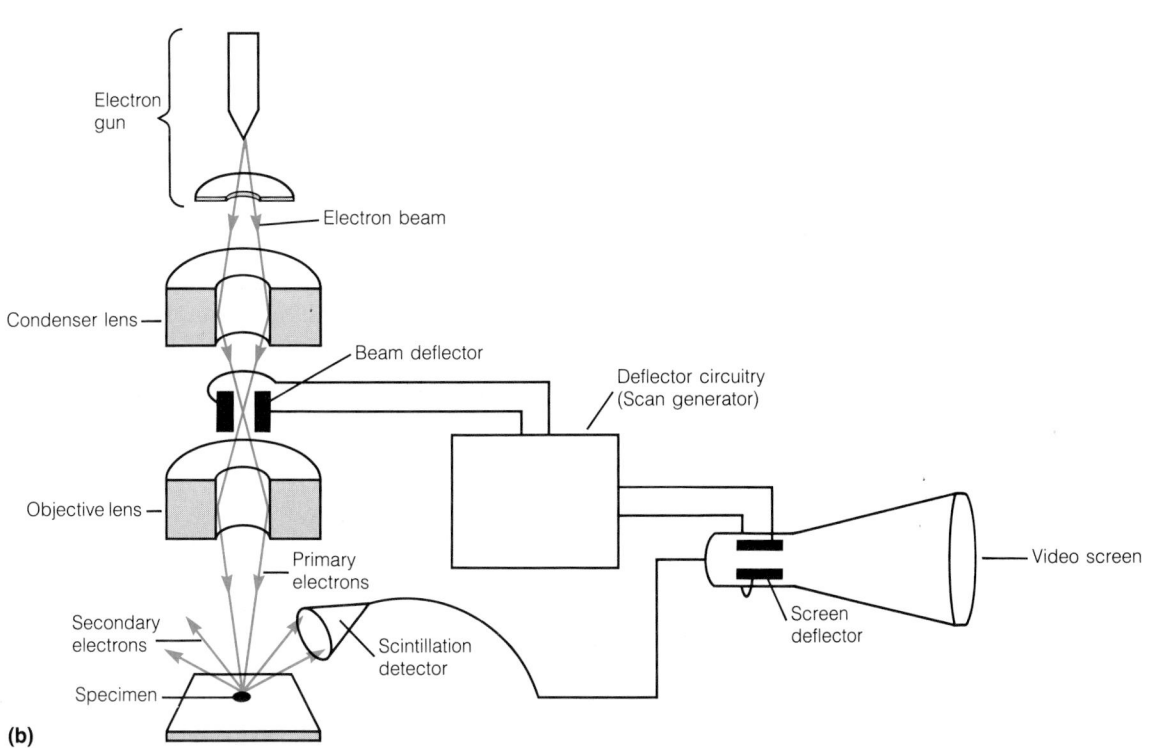

(b)

the beam according to the signals sent to them by the deflector circuitry (Figure A-21b).

As the electron beams sweeps rapidly over the specimen, molecules in the specimen are excited to high energy levels and emit *secondary electrons*. These secondary electrons are used to form an image of the specimen surface. They are then captured by a detector that is located immediately above and to one side of the specimen. The essential component of the detector is a *scintillator*, which emits pho-

tons of light when excited by the electrons incident upon it. The photons are used to generate an electronic signal to a video screen. The image then develops point by point, line by line on the screen as the primary electron beam sweeps over the specimen. Photomicrographs such as the one shown in Figure A-22 are made by directly photographing the video screen, usually with a Polaroid camera. For additional scanning electron micrographs, see Chapter 1, Figure 1-4.

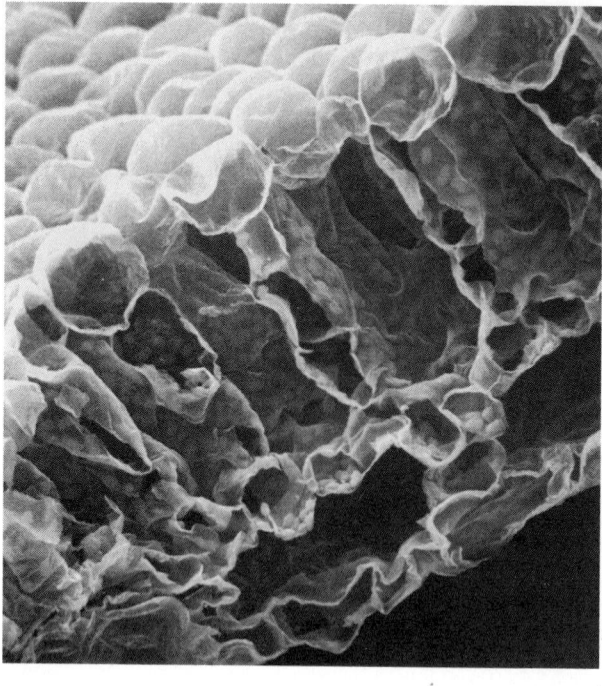

Figure A-22 Scanning Electron Microscopy. A transverse section through the leaf of the water fern *Salvinia,* as seen with a scanning electron microscope. Numerous chloroplasts can be seen inside the palisade cells in the center of the leaf.

Scanning Transmission Electron Microscopy

A **scanning transmission electron microscope (STEM)** contains elements of both transmission and scanning electron microscopes (Figure A-23). Like an SEM, an STEM uses an electron beam that sweeps over the specimen. But the image is then formed by electrons transmitted through the specimen, as with a TEM. An STEM is capable of distinguishing specific characteristics of the electrons that are transmitted by the specimen, thus deriving information about the specimen not obtainable with a conventional TEM. However, an STEM is technically sophisticated, requires a very high vacuum, and is much more electronically complex than a TEM or an SEM.

High-Voltage Electron Microscopy

A **high-voltage electron microscope (HVEM)** is very similar to a transmission electron microscope except that its accelerating voltage is much higher. Whereas a TEM uses accelerating voltages of 50–100 kV, an HVEM uses voltages of about 200–1000 kV. Because of the high voltage and the greatly reduced chromatic aberration that it makes possible, relatively thick specimens can be examined with good resolution. As a result, cellular structure can be studied in sections as thick as 1 μm, about ten times the thickness possible with an ordinary TEM. Figure A-24 shows a million-volt HVEM, and Figure A-25 shows a micrograph of a polytene chromosome from a fruit fly, as visualized by HVEM.

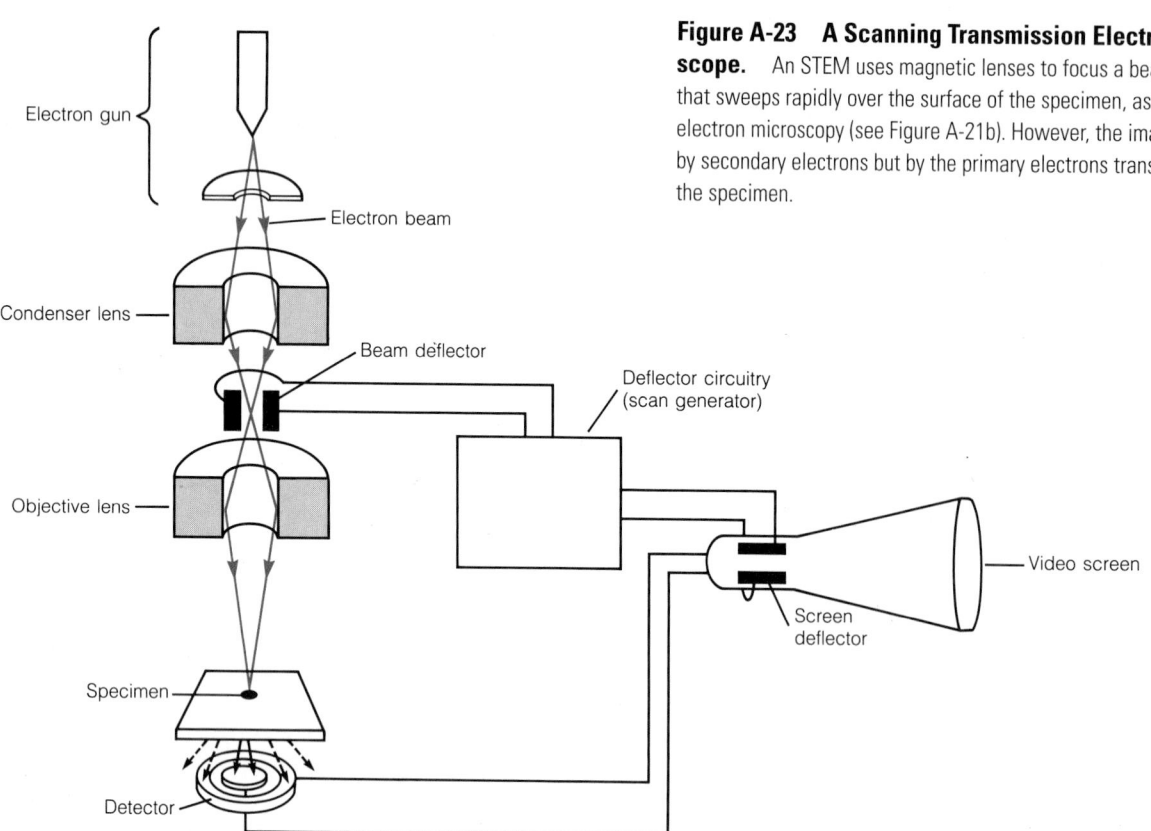

Figure A-23 A Scanning Transmission Electron Microscope. An STEM uses magnetic lenses to focus a beam of electrons that sweeps rapidly over the surface of the specimen, as in scanning electron microscopy (see Figure A-21b). However, the image is formed not by secondary electrons but by the primary electrons transmitted through the specimen.

Sample Preparation Techniques in Transmission Electron Microscopy

Specimens for electron microscopy can be prepared in several different ways, depending on the type of microscope and the kind of information the microscopist wants to obtain. In each case, however, the method is complicated, time-consuming, and costly compared to methods used for light microscopy. Moreover, living specimens cannot be examined because of the vacuum to which specimens are exposed in the electron microscope.

Specimen Processing: Fixation

Specimens to be prepared for electron microscopy must first be chemically fixed and stabilized. This primary fixation kills the cells but keeps the cellular components much as they were in the living cell. Primary fixatives are usually buffered solutions of aldehydes. Glutaraldehyde is the most common fixative. Following primary fixation, the specimen is usually treated with a 1–2% solution of buffered osmium tetroxide (OsO_4). The osmium tetroxide stains specific components of the cell, making them more electron-dense.

Embedding, Sectioning, and Poststaining

The next step in specimen preparation for transmission electron microscopy is to dehydrate the tissue by passing it through a series of alcohol solutions. The specimen is then placed into a fluid such as acetone or propylene oxide to prepare it for embedding in liquefied plastic epoxy resin. After the plastic has infiltrated the specimen, it is put into a mold and heated in an oven to harden the plastic. Thereafter, the

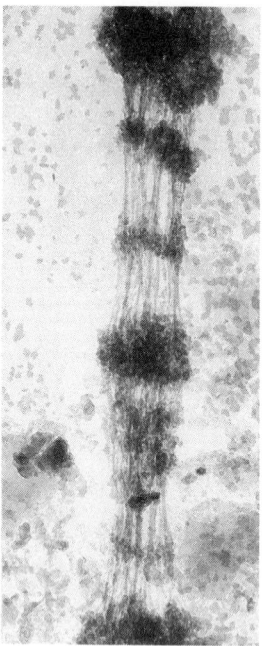

0.5 µm

Figure A-25   High-Voltage Electron Microscopy.   A polytene chromosome from the fruit fly *Drosophila melanogaster,* as seen with an HVEM operated at 1 million volts. For a three-dimensional view of this structure, see the stereo pair in Figure A-33.

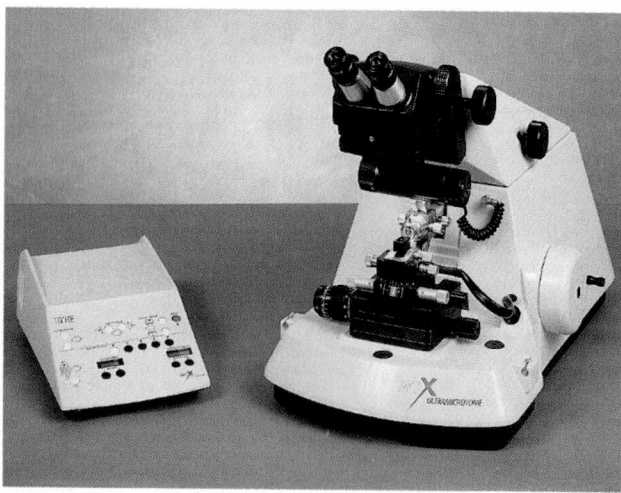

(a) Ultramicrotome

(b) Microtome arm of ultramicrotome

Figure A-26   An Ultramicrotome.   **(a)** A photograph of an ultramicrotome. **(b)** A close-up view of the microtome arm, showing the specimen in a plastic block mounted on the end of the arm. As the microtome arm moves up and down, the block is advanced in small increments, and ultrathin sections are cut from the block face by the diamond knife.

area around the specimen is trimmed to prepare a face that is appropriate for sectioning.

The specimen is then sliced into the ultrathin sections required for examination with the transmission electron microscope. The instrument used for this purpose is an **ultramicrotome** (Figure A-26a). The specimen is mounted firmly on the arm of the ultramicrotome, which then advances the specimen in small increments toward a glass or diamond knife (Figure A-26b). When the block reaches the knife blade, ultrathin sections (about 60–90 nm thick) are cut from the block face. The sections float from the blade onto a water surface, where they can be picked up on a circular copper specimen grid. The grid consists of a meshwork of very thin copper strips, which support the specimen while still allowing "windows" between adjacent strips through which the specimen can be observed.

Once in place on the grid, the sections are usually stained with solutions containing lead and uranium. This procedure, called **poststaining,** enhances the contrast of the specimen because the lead and uranium give still greater electron density to specific parts of the cell. After poststaining, the specimen is ready for viewing or photography with the TEM. Numerous examples of transmission electron micrographs are found throughout this text.

Electron Microscopic Autoradiography

The autoradiographic techniques described in our discussion of light microscopy can be used for transmission elec-

tron microscopy with only minor changes. For the TEM, the specimen containing the radioactively labeled compounds is examined in ultrathin sections on copper specimen grids instead of in thin sections on glass slides.

Immunoelectron Microscopy

In electron microscopy just as in fluorescence microscopy, special techniques have been devised to generate contrast, thereby making it possible to identify and localize specific molecules and subcellular structures. Heavy metal atoms such as iron and gold are commonly used to generate contrast because they are very electron-dense and are therefore readily visible as opaque dots on an electron micrograph. An especially important technique is **immunoelectron microscopy**, which depends on antibody molecules coupled to colloidal gold particles as a means of tagging and visualizing specific antigens. This approach makes it possible to determine the subcellular distribution of specific proteins or protein-containing structures with great precision. Caution is necessary, however, because certain fixation techniques (such as osmium fixation) interfere with the fidelity of antibody-antigen binding. In such cases, milder sample preparation techniques must be used.

Negative Staining

In contrast to the considerable effort necessary to prepare ultrathin sections, **negative staining** is one of the simplest techniques used in transmission electron microscopy. It is the preferred method for examining very small objects, such as viruses or isolated organelles.

For negative staining, the copper specimen grid must first be overlaid with an ultrathin plastic film. The specimen is then suspended in a small drop of liquid, applied to the overlay, and allowed to dry in air. After the specimen has dried on the grid, a drop of stain such as uranyl acetate or phosphotungstic acid is applied to the film surface. The edges of the grid are then blotted in several places with a piece of filter paper to absorb the excess stain. This draws the stain down and around the specimen and its ultrastructural features. When viewed in the TEM, the specimen is seen in **negative contrast** because the background is dark and heavily stained, whereas the specimen itself is lightly stained. Negative staining is illustrated by the electron micrograph of an algal cell wall shown in Figure A-27.

Shadowing

The same cell wall seen in Figure A-27 is shown in Figure A-28, as visualized by the technique of **shadowing.** Shadowing involves the deposition of a thin layer of an electron-dense metal such as gold or platinum on a biological specimen. The metal-emitting electrode is positioned at an angle to the specimen, such that surfaces facing the electrode become coated with the metal, whereas those facing away from the electrode do not.

Figure A-29a illustrates the shadowing technique. Specimens that can be suspended in water-based solutions are especially suitable for shadowing. The specimen is first spread on a clean mica surface and dried (step 1). It is then placed in a **vacuum evaporator,** a bell jar in which a vacuum is created by a system similar to that of an electron microscope (Figure A-29b). Also within the evaporator are two electrodes, one consisting of a carbon rod located directly over the specimen and the other consisting of a metal wire positioned at an angle of about 10°–45° relative to the specimen.

After a vacuum is created in the evaporator, current is applied to the metal electrode, causing the metal to evaporate from the electrode and spray over the surface of the specimen (Figure A-29a, step 2). Because of the angular positioning of the metal electrode, the metal will accumulate as a thin coating on the sides of any surface irregularities that face the electrode, generating a metal **replica** of the surface. These same irregularities prevent deposition of the metal on the side facing away from the electrode, thus producing contrast resulting from the shadow effect.

The carbon electrode is then fired, coating the specimen from directly overhead with evaporated carbon to give stability and support to the metal replicas (step 3): The mica

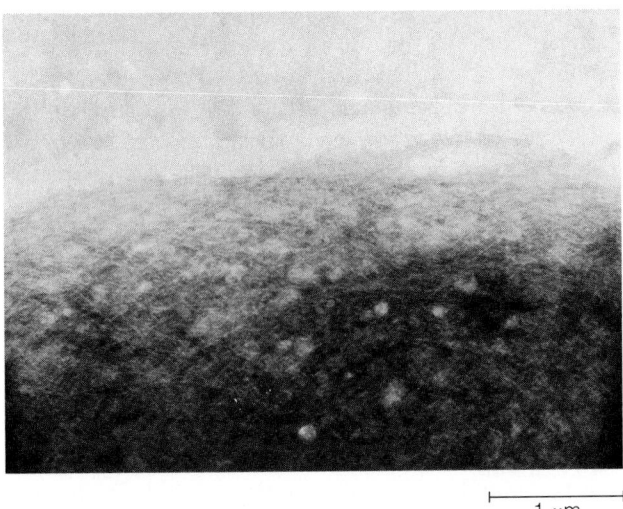

1 µm

Figure A-27   Negative Staining.   An electron micrograph of the cell wall of a flagellated freshwater alga as seen in a negatively stained preparation. For a shadowed preparation of the same specimen, see Figure A-28 (TEM).

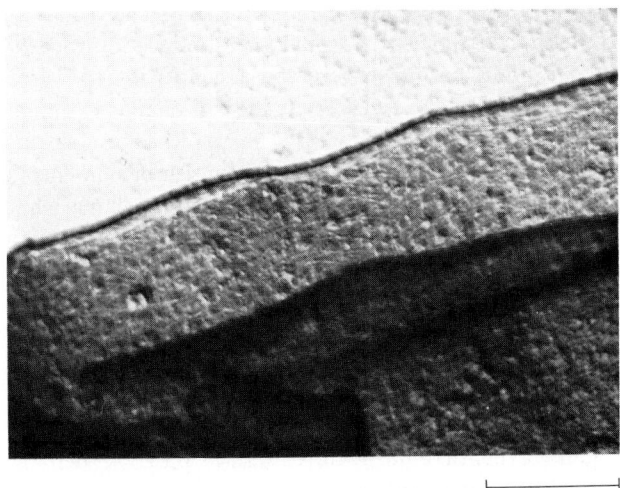

1 µm

Figure A-28   Shadowing.   An electron micrograph of the cell wall of a flagellated freshwater alga as seen in a shadowed preparation. For a negatively stained preparation of the same specimen, see Figure A-27. Notice that the same swirling pattern of cellulose microfibrils is seen in both preparations (TEM).

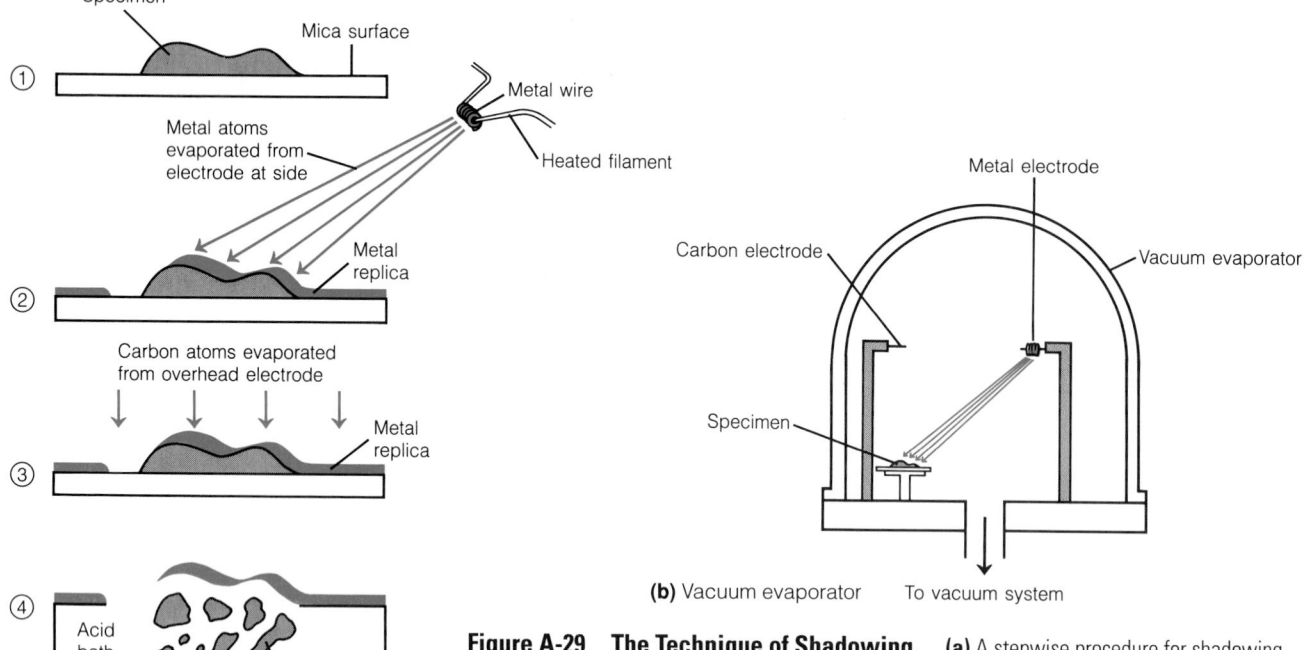

Figure A-29 The Technique of Shadowing. **(a)** A stepwise procedure for shadowing. ① The specimen is spread on a mica surface and dried. ② The specimen is shadowed by coating it with atoms of a heavy metal (platinum or gold, shown in blue) that are evaporated from a heated filament located to the side of the specimen in a vacuum evaporator. This generates a metal replica (blue), the thickness of which reflects the surface contours of the specimen. ③ Next, the specimen is coated with carbon atoms evaporated from an overhead electrode to strengthen and stabilize the metal replica. ④ The replica is then floated onto the surface of an acid bath to dissolve away the specimen, leaving a clean metal replica. ⑤ The replica is washed and picked up on a copper grid for examination in the TEM. **(b)** The vacuum evaporator in which shadowing is done. The carbon electrode is located directly over the specimen, whereas the heavy metal electrode is off to the side.

support containing the specimen is then removed from the vacuum evaporator and lowered gently onto a water surface, causing the replica to float away from the mica surface. The replica is transferred into an acid bath, which dissolves away remaining bits of specimen, leaving a clean metal replica of the specimen (step 4). The replica can then be returned to the water surface and retrieved on a standard copper grid (step 5). Replicas can be viewed in the TEM in the same way as ultrathin sections.

Freeze Fracturing

Freeze fracturing is a relatively recent technique that has proved very useful to cell biologists. It is especially valuable for studying the ultrastructure of biological membranes. Freeze fracturing involves the cleavage of a frozen specimen under a vacuum, followed by platinum/carbon shadowing to create a replica of the fractured surface, which is often the interior of a membrane.

Freeze fracturing is illustrated in Figure A-30. It takes place in a modified vacuum evaporator with an internal microtome knife for fracturing the frozen specimen and with provision for precise control of the temperature of the specimen stage and the microtome arm and knife. Specimens are generally fixed prior to freeze fracturing, although some living tissues can be frozen fast enough to keep them

in almost lifelike condition. Because cells contain a lot of water, fixed specimens are usually treated with an antifreeze such as glycerol to provide **cryoprotection**—that is, to reduce the formation of ice crystals during freezing.

The cryoprotected specimen is mounted on a metal specimen support (Figure A-30, step 1) and immersed rapidly in freon cooled with liquid nitrogen (step 2). This procedure also reduces the formation of ice crystals in the cells. With the frozen specimen positioned on the specimen stage in the vacuum evaporator (step 3), a high vacuum is established, the stage temperature is adjusted to around −100°C, and the frozen specimen is fractured with a blow from the microtome knife (step 4). A replica of the fractured specimen is made by shadowing with platinum and carbon as described in the previous section (step 5), and the replica is then ready to be viewed in the transmission electron microscope (step 6).

Newcomers to the freeze-fracturing technique often misunderstand what a freeze-fracture replica represents. One might think that the fracture plane should pass through the specimen in a straight line, as is clearly the case when a fixed and embedded sample is sectioned conventionally with an ultramicrotome (Figure A-31a). In actuality, however, the fracture line passes through the hydrophobic interior of membranes whenever possible, because this is the line of least resistance through the frozen specimen (Figure A-31b). As a result, a freeze-fracture replica is largely a view of the

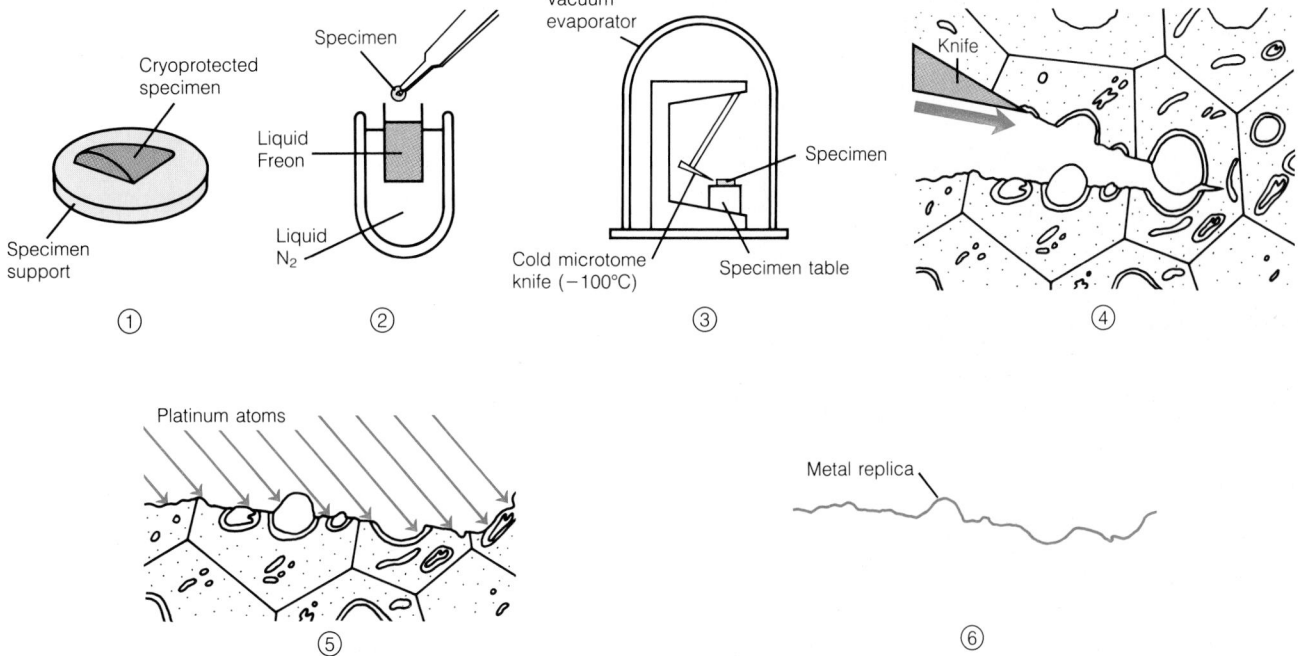

Figure A-30   The Technique of Freeze Fracturing.
① A cryoprotected specimen is mounted on a metal support. ② The mounted specimen is immersed in liquid freon cooled in liquid nitrogen. ③ The frozen specimen is transferred to a vacuum evaporator and adjusted to a temperature of about −100°C. ④ The specimen is fractured with a blow from the microtome knife. The fracture plane passes through the interior of lipid bilayers wherever possible, because this is the line of least resistance through the frozen specimen, as shown in Figure A-31b. ⑤ The fractured specimen is shadowed with platinum and carbon, as in Figure A-29, to make a metal replica of the specimen. ⑥ The metal replica is examined in the TEM.

interiors of membranes, showing the inside of one or the other of the two monolayers of the membrane.

Freeze-fractured membranes appear as smooth surfaces studded with **intramembranous particles (IMPs)** that are either randomly distributed in the membrane or organized into ordered complexes. These are thought to be integral membrane proteins that have remained with one lipid monolayer or the other as the fracture plane passes through the interior of the membrane.

The electron micrograph in Figure A-32 shows the two faces of a plasma membrane as revealed by freeze fracturing. The **P face** is the interior face of the inner monolayer; it is called the P face because this monolayer is on the *proto-plasmic* side of the membrane. The **E face** is the interior face of the outer monolayer; it is called the E face because this monolayer is on the *exterior* side of the membrane. Notice that the P face has far more intramembranous particles than does the E face. In general, most of the particles in the mem-

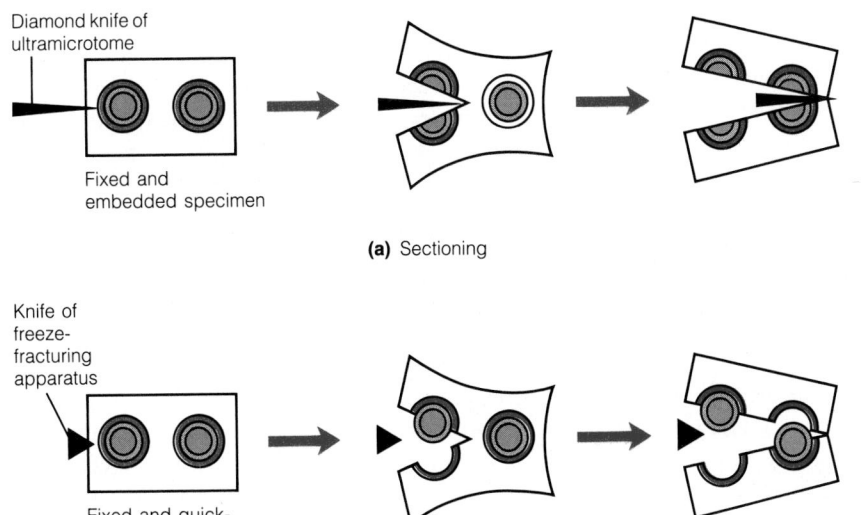

(a) Sectioning

(b) Fracturing

Figure A-31   Sectioning Versus Fracturing of Specimens.   (a) When a fixed and embedded specimen is sectioned for conventional transmission electron microscopy, the edge of the diamond knife makes a clean, linear cut through the tissue. **(b)** When a fixed and quick-frozen specimen is fractured, the blow of the knife generates a fracture plane through the frozen sample that passes through the interiors of membranes whenever possible, because the hydrophobic interior of a phospholipid bilayer is more readily fractured than is the ice that surrounds it. The interiors of membranes are therefore exposed on the fracture surfaces, as Figure A-32 shows.

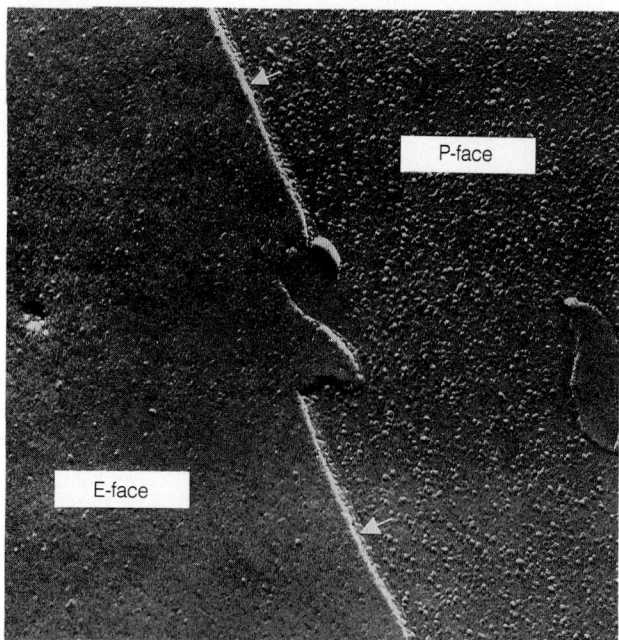

$\overline{0.1 \ \mu m}$

Figure A-32   Freeze Fracturing of the Plasma Membrane.
This electron micrograph shows the exposed faces of the plasma membranes of two adjacent endocrine cells from a rat pancreas as revealed by freeze fracturing. The P face is the inner surface of the lipid monolayer on the protoplasmic side of the plasma membrane. The E face is the inner surface of the lipid monolayer on the exterior side of the plasma membrane. The P face is much more richly studded with intramembranous particles than the E face. The arrows indicate the "step" along which the fracture plane passed from the interior of the plasma membrane of one cell to the interior of the plasma membrane of a neighboring cell. The "step" therefore represents the thickness of the intercellular space (TEM).

brane stay with the inner monolayer when the fracture plane passes down the middle of a membrane.

To have a P face and an E face appear side by side as in Figure A-32, the fracture plane must pass through two neighboring cells, such that one cell has its cytoplasm and the inner monolayer of its plasma membrane removed to reveal the E face, while the other cell has the outer monolayer of its plasma membrane and the associated intercellular space removed to reveal the P face. Accordingly, E faces are always separated from P faces of adjacent cells by a "step" (marked by the arrows in Figure A-32) that represents the thickness of the intercellular space.

Freeze Etching

Freeze etching is related to freeze fracturing, but there is a considerable difference between the two techniques. Freeze etching adds a further step to the conventional freeze-fracture procedure that makes the technique even more informative. Following the fracture of the specimen but prior to shadowing, the microtome arm is placed directly over the specimen for a short time (a few seconds to several min-

utes). This maneuver causes a small amount of water to evaporate (sublime) from the surface of the specimen to the cold knife surface. Where the fracture has passed through a membrane, etching will cause small areas of the true cell surface around the periphery of the fracture face to stand out in relief against the background.

By using ultrarapid freezing techniques and a volatile cryoprotectant such as aqueous methanol, which sublimes very readily to a cold surface, the etching period can be extended and a much deeper layer of ice can be removed, exposing large areas of the specimen surface to view. This modification, called **deep etching**, provides a fascinating new look at cellular structure. For an example, see Figure 20-2, an electron micrograph of the cytoskeleton as revealed by deep etching.

Stereo Electron Microscopy

Electron microscopists frequently want to visualize their specimens in three dimensions. Techniques such as shadowing, freeze fracturing, and scanning electron microscopy are useful for this purpose. However, they can be further enhanced by **stereo electron microscopy**. Specifically, the same specimen is photographed from two different angles to generate a *stereo pair* of photos that are then fused optically. To do this, a special specimen stage is used that can be tilted relative to the electron beam. The specimen is first tilted in one direction and photographed, then tilted an equal amount in the opposite direction and photographed again.

The two micrographs are mounted side by side as a stereo pair. When you view a stereo pair through a stereoscopic viewer, your brain uses the two independent images to construct a three-dimensional view that gives a striking sense of depth to the structure under investigation.

Figure A-33 is a stereo pair of the *Drosophila* polytene chromosome seen earlier in the high-voltage electron micrograph of Figure A-25. Here, however, it appears in three dimensions. A striking view of the chromosome can be achieved either with a stereo viewer or by allowing your eyes to fuse the two images visually.

Sample Preparation Techniques in Scanning Electron Microscopy

When preparing a specimen for scanning electron microscopy, the goal is to preserve the structural features of the cell surface and to treat the tissue in a way that minimizes damage by the electron beam. The procedure is actually quite similar to the preparation of ultrathin sections for transmission electron microscopy. The tissue is fixed in aldehyde, postfixed in osmium tetroxide, and processed for dehydration through a series of alcohol solutions. The tissue is then placed in a fluid such as freon and transferred to a heavy metal canister called a **critical point bomb,** which is used to dry the specimen under conditions of controlled temperature and pressure. This helps keep structures on the surfaces

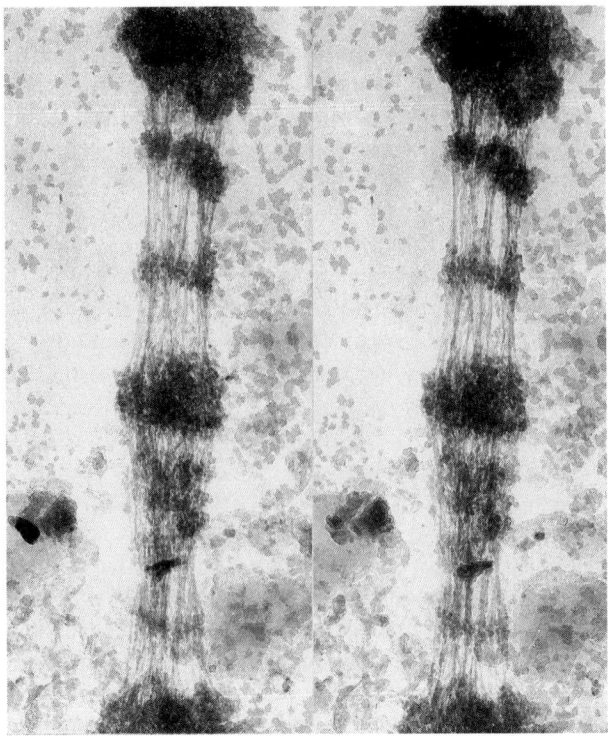

Figure A-33 Stereo Electron Microscopy. The polytene chromosome seen in the high-voltage electron micrograph of Figure A-25 is shown here as a stereo pair of photos that can be fused optically to generate a three-dimensional image. The two photographs were taken by tilting the specimen stage first 5° to the right, then 5° to the left of the electron beam. For a three-dimensional view, a stereo pair can be examined with a stereoscopic viewer. Alternatively, simply let your eyes cross slightly, fusing the two micrographs into a single image (TEM).

of the tissue in almost the same condition they were in before dehydration.

The dried specimen is then attached to a metal specimen mount with a metallic paste. The mounted specimen is coated with a layer of gold or a mixture of gold and palladium, using a modified form of vacuum evaporation called **sputter coating.** These procedures allow electrons to pass through the specimen more readily, thereby minimizing heating of the specimen when struck by the electron beam. Once the specimen has been mounted and coated, it is ready to be examined in the microscope.

As we have seen, preparing specimens for electron microscopy is often expensive and time-consuming. However, the high resolution and unique perspective on the structure and function of cells provide worthwhile insights into the biology of cells and tissues.

Other Imaging Methods

Light and electron microscopy are direct imaging methods in that they use photons or electrons to produce actual images of a specimen. There are other methods of microscopy that are indirect imaging methods. To understand what is meant by indirect imaging, suppose you are given some object to handle with your eyes closed. You might feel six flat surfaces, twelve edges and eight corners, and if you then draw what you have felt, it would turn out to be a box. This is an example of an indirect imaging procedure.

The indirect imaging methods described here are *scanning tunneling microscopy, atomic force microscopy,* and *X-ray diffraction.* Each method has the potential for showing molecular structures at near-atomic resolution, ten times better than the best electron microscope. Each method also has certain characteristics that limit its application to biological material, but when these techniques can be successfully used, the resulting images provide exciting information that cannot be obtained in any other way.

Scanning Tunneling Microscopy

Although "scanning" is involved in both scanning electron microscopy and scanning tunneling microscopy, the two methods are in fact quite different. The **scanning tunneling microscope (STM)** does not use an electron beam, but instead depends on a tip made of a conducting material such as platinum-iridium. The tip is extremely sharp, ideally with its point composed of a single atom. It is under precise control of an electronic circuit that can move it in three dimensions over a surface. The x and y dimensions scan the surface, while the z dimension governs the distance of the tip above the surface (Figure A-34).

The basic principle of the STM is electron tunneling. At the quantum-mechanical level, an electron has both wavelike and particlelike properties. These properties allow the electron to cross barriers that it cannot penetrate as a particle, but that it can penetrate in the form of a wave. This penetration is called *tunneling.* As the tip of the STM is moved across a surface, voltages from a few millivolts to several volts are applied. Under these conditions, if the tip is close enough to the surface and the surface is electrically conductive, electrons will begin to tunnel between tip and surface. The tunneling is highly dependent on the distance, so that even small irregularities in the size range of single atoms will affect the rate of electron conductance. The changes in conductance are used to regulate the distance between tip and surface so that the tunneling current is kept constant. The resulting feedback current is electronically amplified and displayed on a video screen.

An important limitation of the STM is that the specimen must be electrically conductive. Therefore, the technique is better suited to producing images of physical surfaces rather than biological specimens, which are often good insulators. Nonetheless, some progress in biological imaging has been made.

Atomic Force Microscopy

Atomic force microscopy (AFM) is related to STM, in that it uses a tip of atomic dimensions. It has the important advantage that the specimen does not need to be an electrical con-

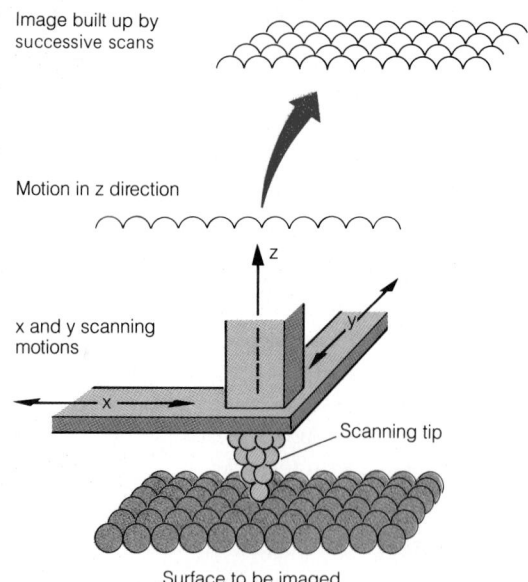

Figure A-34 Scanning Tunneling Electron Microscopy.
The scanning tunneling microscope (STM) uses electronic methods to scan a metallic tip across the surface of a specimen. The tip is not drawn to scale in this illustration, but the point of the tip is ideally composed of one or a few atoms, shown here as balls. An electrical voltage is produced between the tip and the specimen surface. As the tip scans the specimen in the x and y directions, electron tunneling occurs at a rate dependent on the distance between the tip and the first layer of atoms in the surface. The instrument is designed to move the tip in the z direction to maintain a constant current flow. The movement is therefore a function of the tunneling current and is presented on a video screen. Successive scans then build up an image of the surface at atomic resolution.

ductor. This is because the tip is actually moved over the specimen surface, bumping along individual atoms in the specimen. The minute movements of the tip are followed with an optical system that senses deflection.

One of the most important potential applications of the STM and AFM instruments is the measurement of dynamic changes in the conformation of a functioning biomolecule. Consider, for instance, how exciting it would be to "watch" a single enzyme molecule change its shape as it hydrolyzes ATP to provide the energy needed to transport ions across membranes. Such molecular eavesdropping is now entirely within the realm of possibility.

X-Ray Diffraction

X-ray diffraction does not involve microscopy but instead reconstructs images from the diffraction patterns of X rays passing through a crystalline specimen. This method can be used to deduce molecular structure at the atomic level of resolution. X-ray diffraction is, in fact, the only method presently available to analyze the structure of proteins, nucleic acids, and other biological molecules at this resolution.

A good way to understand X-ray diffraction is to draw an analogy with light. As discussed earlier, light has certain properties that are best described as wavelike. Whenever

wave phenomena occur in nature, interaction between waves can occur. If waves from two sources come into phase with one another, their total energy is additive *(constructive interference)*, and if they are out of phase, their energy is reduced *(destructive interference)*. This effect can be seen when light passes through two pinholes in a piece of opaque material and then falls on a white surface. Interference patterns result, with dark regions where light waves are out of phase and bright regions where they are in phase (Figure A-35). If the wavelength of the light is known, one can measure the angle α between the original beam and the first diffraction peak and then calculate the distance d between the two holes with this formula:

$$d = \text{wavelength}/\sin \alpha \qquad \textbf{(A-6)}$$

The same approach can be used to calculate the distance between atoms in crystals of proteins or nucleic acids. Instead of two holes in a sheet of paper, imagine that we have multiple layers of atoms organized in a crystal. And instead of light, which has much too long a wavelength to interact with atoms, we will use a narrow beam of X rays with wavelengths in the range of interatomic distances. As the X rays pass through the crystal, they reflect off planes of atoms, and the reflected beams come into constructive and destructive interference. If the X-ray beams are allowed to fall onto photographic plates behind the crystal, distinctive diffraction patterns are produced. These patterns can be analyzed

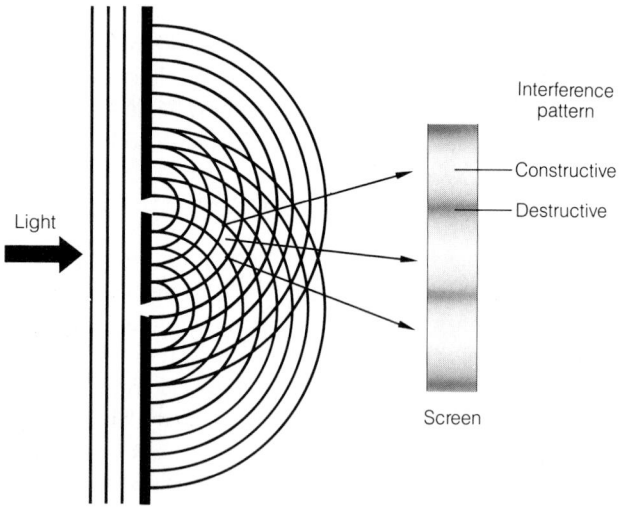

Figure A-35 Understanding Diffraction Patterns. Any energy in the form of waves will produce interference patterns if the waves from two or more sources are superimposed in space. One of the simplest patterns can be seen when monochromatic light passes through two neighboring pinholes and is allowed to fall on a screen. The light is shown on the left as parallel waves with alternating peaks and troughs. When the light passes through the two pinholes, the holes act as light sources, with waves radiating from each and falling on a white surface. Where the waves are in the same phase a bright area appears (constructive interference), but where the waves are out of phase, peaks fall on troughs and cancel each other out, producing dark areas (destructive interference).

Figure A-36 X-Ray Crystallography of Molecular Structure. X-ray crystallography can be used to analyze molecular structure at near-atomic resolution. ① X rays are diffracted by atoms in crystals just as light waves are diffracted by pinholes. In most cases, the crystals of interest to biologists are proteins or nucleic acids. The specific example illustrated in this figure is crystalline DNA. ② The resulting diffraction patterns are recorded photographically and can then be analyzed mathematically to deduce the molecular structure of the crystallized molecules. This photograph depicts the actual X-ray diffraction pattern used by James Watson and Francis Crick to deduce the molecular structure of double-stranded DNA. ③ A computer graphic model of the DNA double helix.

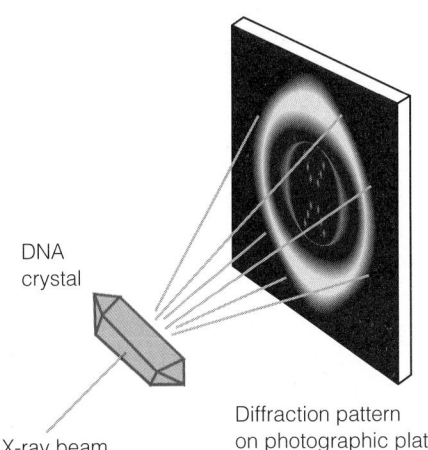
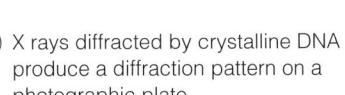
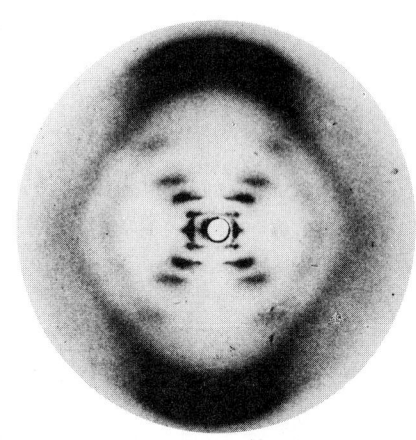
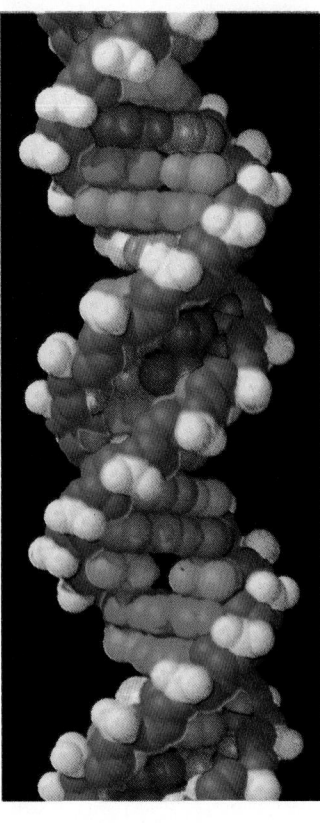

DNA crystal

X-ray beam

Diffraction pattern on photographic plate

① X rays diffracted by crystalline DNA produce a diffraction pattern on a photographic plate

② The resulting diffraction pattern is analyzed mathematically

③ The three-dimensional structure of the molecule is deduced

mathematically to determine the organization of atomic layers in the original crystal and thus to deduce three-dimensional molecular structure. Figure A-36 illustrates the use of this procedure to deduce the double-helical structure of DNA.

The technique of **X-ray crystallography** was developed in 1912 by Sir William Bragg, who went on to establish the structures of relatively simple mineral crystals. Forty years later, Max Perutz and John Kendrew found ways to apply X-ray diffraction to crystals of hemoglobin and myoglobin, providing our first view of the intricacies of protein structure. Since then, many proteins and other biological molecules have been crystallized and subjected to X-ray diffraction. Membrane proteins are much more difficult to crystallize than hemoglobin. In 1985, however, Henri Michel and Johannes Deisenhofer overcame this obstacle by crystallizing the proteins of a bacterial photosynthetic reaction center. They then went on to describe the molecular organization of the reaction center at a resolution of 0.3 nm (see Figure 13-6).

KEY TERMS FOR SELF-TESTING

light microscope (p. 841)
electron microscope (p. 841)

Image Formation in Light and Electron Microscopy

source of illumination (p. 841)
specimen (p. 841)
lens (p. 841)
wave form (p. 842)
wavelength (p. 842)

Optical Principles of Microscopy

interference (p. 843)
diffraction (p. 843)

focal length (p. 843)
angular aperture (p. 843)
resolution (p. 844)
refractive index (p. 844)
numerical aperture (*NA*) (p. 844)
limit of resolution (p. 845)
useful magnification (p. 845)

The Light Microscope

compound microscope (p. 845)
condenser lens (p. 845)
stage (p. 845)
objective lens (p. 845)
ocular lens (p. 845)

intermediate lens (p. 845)
brightfield microscopy (p. 845)
phase-contrast microscopy (p. 847)
fluorescence microscopy (p. 847)
fluorescent probe (p.847)
fluorescence (p. 848)
differential interference contrast (DIC) microscopy (p. 849)
video microscopy (p. 850)
electronic imaging (p. 850)
confocal microscopy (p. 850)
laser scanning confocal microscope (LSCM) (p. 852)

SUGGESTED READING

General References

Bradbury, S. *An Introduction to the Optical Microscope.* New York: Oxford University Press, 1984.

Slayter, E. M., and H. S. Slayter. *Light and Electron Microscopy.* New York: Cambridge University Press, 1992.

Light Microscopy

Haugland, R. P. *Molecular Probes Handbook of Fluorescent Probes and Research Chemicals,* 5th ed. Eugene, OR: Molecular Probes, 1992.

Mason, W. T., ed. *Fluorescent and Luminescent Probes for Biological Activity.* San Diego: Academic Press, 1993.

Pawley, J., ed. *The Handbook of Biological Confocal Microscopy,* 2d ed. University of Wisconsin, Madison: Integrated Microscopy Resources, 1995.

Spencer, M. *Fundamentals of Light Microscopy.* Cambridge, England: Cambridge University Press, 1982.

Stelzer, E. H., I. Wacker, and J. R. De Mey. Confocal fluorescence microscopy in modern cell biology. *Seminars Cell Biol.* 2 (1991): 145.

Tsien, R. Y. Intracellular signal transduction in four dimensions: From molecular design to physiology. *Am. J. Physiol.* 263 (1992): C723.

Video Microscopy

Allen, R. D. New observations on cell architecture and dynamics by video-enhanced contrast optical microscopy. *Annu. Rev. Biophys. Chem.* 14 (1985): 265.

Farkas, D. L., G. Baxter, R. L. DeBiasio, A. Gough, M. A. Nederlof, D. Pane, J. Pane, D. R. Patek, K. W. Ryan, and D. L. Taylor. Multimode light microscopy and the dynamics of molecules, cells, and tissues. *Annu. Rev. Physiol.* 55 (1993): 785.

Inoué, S. *Video Microscopy.* New York: Plenum, 1986.

Transmission and Scanning Electron Microscopy

Flegler, S. L., J. W. Heckman, and K. L. Klomparens. *Scanning and Transmission Microscopy: An Introduction.* New York: Plenum, 1993.

Olson, A. J., and D. S. Goodsell. Visualizing biological molecules. *Sci. Amer.* 267 (November 1992): 76.

Palade, G. E. Albert Claude and the beginning of biological electron microscopy. *J. Cell. Biol.* 50 (1971): 5D.

Pease, D. C., and K. R. Porter. Electron microscopy and ultramicrotomy. *J. Cell. Biol.* 91 (1981): 287s.

Sommerville, J., and U. Scheer, eds. *Electron Microscopy in Molecular Biology: A Practical Approach.* Washington, DC: IRL Press, 1987.

Weakley, B. S. *A Beginner's Handbook in Biological Transmission Microscopy,* 2d ed. New York: Churchill Livingstone, 1981.

Wieschnitzer, S. *Introduction to Electron Microscopy,* 3d ed. Elmsford, NY: Pergamon, 1981.

Sample Preparation Techniques

Griffiths, G. *Fine Structure Immunocytochemistry.* New York: Springer-Verlag, 1993.

Heuser, J. Quick-freeze, deep-etch preparation of samples for 3-D electron microscopy. *Trends Biochem. Sci.* 6 (1981): 64.

Orci, L., and A. Perrelet. *Freeze-Etch Histology: A Comparison Between Thin Sections and Freeze-Etch Replicas.* New York: Springer-Verlag, 1975.

Pinto da Silva, P., and D. Branton. Membrane splitting in freeze-etching. *J. Cell Biol.* 45 (1970): 598.

Other Imaging Methods

Deisenhofer, J., O. Epp, K. Miki, R. Huber, and H. Michel. The structure of the protein subunits in the photosynthetic reaction centre of *Rhodopseudomonas viridis* at 3 Å resolution. *Nature* 318 (1985): 618.

Glusker, J. P., and K. N. Trueblood. *Crystal Structure Analysis: A Primer.* Oxford, England: Oxford University Press, 1985.

Hansma, P. K., V. B. Elings, O. Marti, and C. E. Bracker. Scanning tunneling microscopy and atomic force microscopy: Application to biology and technology. *Science* 242 (1988): 209.

Hoh, J. H., and P. K. Hansma. Atomic force microscopy for high-resolution imaging in cell biology. *Trends Cell Biol.* 2 (1992): 208.

Kendrew, J. C. The three-dimensional structure of a protein molecule. *Sci. Amer.* 205 (December 1961): 96.

Marti, O., and M. Amrein. *STM and SFM in Biology.* San Diego: Academic Press, 1993.

Perutz, M. F. The hemoglobin molecule. *Sci. Amer.* 211 (November 1964): 64.

Photo, Illustration, and Text Credits

Photo Credits

Part and Chapter Openers Part 1, Chapters 1-3: E. H. Newcomb and S. E. Frederick/Biological Photo Service. Part 2, Chapters 4-10: Courtesy of Jerome Gross. Part 3, Chapters 11-13: Courtesy of J. Croxdale. Part 4, Chapters 14-19: © Dr. Gopal Murti/Science Photo Library/Photo Researchers, Inc. Part 5, Chapters 20-23: R. L. Brown, J. P. Horn, L. Wible, R. B. Arlinghaus, and B. R. Brinkley, *Proc. Nat. Acad. Sci.* 78 (1981): 5595. Part 6, Chapters 24-25: © Nina Lampen/Science Source/Photo Researchers.

Chapter 1 1.3a, b: From *Cell Ultrastructure*, by William A. Jensen and Roderic B. Park. Copyright © 1967 by Wadsworth Publishing Co., Inc. Used by permission of the authors and publisher. 1.4a: Courtesy of G. G. Borisy. 1.4b: Courtesy of C. Greaves and J. Croxdale; photo provided by J. Croxdale.

Chapter 2 2.13a: Courtesy of Don Fawcett, © Photo Researchers, Inc. 2.14, upper left: Courtesy of G. F. Bahr. 2.14, upper right: Courtesy of E. F. Newcomb. 2.14, lower right: Courtesy of E. Frei & R. D. Preston. 2.20: Courtesy of IBM U.K. Ltd.

Chapter 3 3.21a: © Jeremy Burgess/Photo Researchers. 3.21b: © Don Fawcett/Photo Researchers. 3.22: Courtesy of E. Frei & R. D. Preston. 3a.1: From J. D. Watson, *The Double Helix*, Atheneum, New York, 1968, p. 215 © 1968 by J. D. Watson.

Chapter 4 4.2: Courtesy of S. Ito. 4.3a: Courtesy of L. D. Simon, Waksman Institute of Microbiology, Rutgers University, New Brunswick, NJ. 4.3b: Courtesy of E. H. Newcomb. 4.4: © David M. Phillips/Visuals Unlimited. 4.5: Courtesy of H. Ris. 4.6: From O. L. Miller, Jr., B. A. Hamkalo, and C. A. Thomas, Jr. *Science*, 169 (1970): 392. Copyright © 1970 by the AAAS. 4.7b: Courtesy of R. Rodewald, University of Virginia/Biological Photo Service. 4.8b: Micrograph by W. P. Wergin; photograph provided by E. H. Newcomb. 4.10b: Courtesy of R. Rodewald, University of Virginia/Biological Photo Service. 4.10c: From J. P. Strafstrom, and L. A. Staehelin, *J. Cell. Biol.* 98 (1984): 699. Reproduced by copyright permission of The Rockefeller University Press. 4.11c: © Keith Porter/Photo Researchers, Inc. 4.12b: From J. B. Rattner and B. R. Brinkley, *J. Ultrastructure Res.* 32 (1970): 316. Copyright © 1970 by Academic Press 4.13: Courtesy of S. M. Wang. 4.14b: Micrograph by W. P. Wergin; photograph provided by E. H. Newcomb. 4.15c: Courtesy of H. S. Pankratz, Michigan State University/Biological Photo Service. 4.15d: Courtesy of J. F. King, University of California School of Medicine/Biological Photo Service. 4.16b: Courtesy of E. H. Newcomb. 4.18b: Courtesy of N. Simionescu. 4.18c: © Don Fawcett/Visuals Unlimited. 4.19: © Barry King, University of California, Davis/Biological Photo Service. 4.20b: From S. E. Frederick and E. H. Newcomb, *J. Cell Biol.* 43 (1969): 343. Reproduced by copyright permission of the Rockefeller University Press; photo provided by E. H. Newcomb. 4.21b: Micrograph by P. J. Gruber; photo provided by E. H. Newcomb. 4.23: From Sigrid Regauer, Werner W. Franke, and Ismo Virtananen, *J. Cell Biol.* 100 (1988): 997-1009. 4.25(2): From P. H. Raven, R. F. Evert, and H. A. Curtis, *Biology of Plants*, 2nd edi-

tion. New York: Worth, 1981. 4.26a: Courtesy of R. C. Williams and H. W. Fisher.

Chapter 5 5.3: From D. A. Cuppels and A. Kelman, *Phytopathology* 70 (1980): 1110. Photo provided by A. Kelman. Copyright © 1980 by the American Phytopathological Soc. 5.6: Courtesy of Dr. Meeuse, University of Washington, Seattle.

Chapter 6 6.2, 6.7: Courtesy of Richard J. Feldmann, National Institutes of Health.

Chapter 7 7.1a: M. Simionescu and N. Simionescu, *J. Cell Biol.* 70 (1976): 622. Reproduced by copyright permission of The Rockefeller University Press. 7.1b: Courtesy of E. H. Newcomb. 7.3: Micrograph courtesy of J. David Robertson. 7.5: Courtesy of John Heuser. 7.9: © Ken Eward/Science Source/Photo Researchers, Inc. 7.16b: Micrographs of E and P faces courtesy of P. Claude. 7.17a, b: Courtesy of David Deamer, University of California, Santa Cruz. 7.18(7): D. Branton et al., *Cell* 24 (1981): 24-32; © Cell Press. 7.21: Courtesy of S. Ito.

Chapter 8 8.13a: © Helen E. Carr/Biological Photo Service.

Chapter 9 9.9: © Don Fawcett/Visuals Unlimited. 9.10: Courtesy of D. S. Friend. 9.11a: © Barry King, U.C. Davis/Biological Photo Service. 9.12b: From R. Wetherbee, *Protoplasma* 95 (1978): 347. Copyright ©1978 by Springer-Verlag. 9.13: From W. J. Brown and M. G. Farquhar, *Cell* 36 (1984): 295. Copyright ©1984 by MIT Press. 9.17 From L. Orci and A. Perrelet, *Freeze-Etch Histology*. Heidelberg: Springer-Verlag, 1975. 9.20b: C. C. F. Blake et al., *J. Mol. Biol.* 88 (1974): 1-12. 9.21: From M. M. Perry and A. B. Gilbert, *J. Cell Sci.* 39 (1979): 257. Copyright © 1979 by The Company of Biologists Ltd. 9.22a: From R. A. Crowther and B. M. F. Pearse, *J. Cell Biol.* 91 (1981): 790. Reproduced by copyright permission of The Rockefeller University Press. 9.23a: Courtesy of E. Ungewickell and D. Branton; photo provided by D. Branton. 9.26a: Courtesy of Z. Hruban. 9.28: Courtesy of H. Shio and P. B. Lazarow. Reproduced by copyright permission of The Rockefeller University Press; photos provided by P. B. Lazarow. 9.29: From S. E. Frederick and E. H. Newcomb, *J. Cell Biol.* 43 (1969): 343. Reproduced by copyright permission of The Rockefeller University Press. Photo provided by E. H. Newcomb. 9.30: 9.28: Courtesy of H. Shio and P. B. Lazarow. Reproduced by copyright permission of The Rockefeller University Press; photos provided by P. B. Lazarow. 9.31:E. Newcomb/Biological Photo Service. 9.33: Keith Wood. 9A.2a: R. G. W. Anderson, M. S. Brown, and J. L. Goldstein, *Cell* 10 (1977): 351-364. 9A.2b: R. G. W. Anderson, M. S. Brown, and J. L. Goldstein, *Cell* 10 (1977): 351-364.

Chapter 10 10.1a From Douglas E. Kelly, *J. Cell. Biol.* 28 (1966): 51. Reproduced by copyright permission of The Rockefeller University Press. 10.2b: From L. Orci and A. Perrelet, *Freeze-Etch Histology*. Heidelberg: Springer-Verlag, 1975. 10.2c, d: Courtesy of P. Claude. 10.4b: From C. Peracchia and A. F. Dulhunty, *J. Cell Biol.* 70 (1976):419. Reproduced by copyright permission of The Rockefeller University Press. 10.4c: Courtesy of P. Claude. 10.5a(1): Courtesy of Jerome Gross. 10.8a: J. A. Buckwalter and L. Rosenberg, *Collagen Relat. Res.* 3 (1983): 489504. 10.12: © Biophoto Associates/Photo Researchers, Inc. 10.13c: Micrograph by B. A. Palevitz; photo provided by E. H. Newcomb. 10.17a(1): Micrograph by W. P. Wergin; photo provided by E. H. Newcomb. 10.17c: Courtesy of E. H. Newcomb. 10A.1: © Leon J. Lebeau/Biological Photo Service.

Chapter 11 11A.1: © Deborah Davis/Photo Edit.

Chapter 12 12.2: Courtesy of Charles R. Hackenbrock. 12.3b: Courtesy of K. R. Porter. 12.4: From L. Packer, *Ann. N. Y. Acad. Sci.* 227 (1974): 166. Copyright © 1974 by the New York Academy of Sciences. Photo provided by H. T. Ngo. 12.5a: Courtesy of A. Tzagoloff from *Mitochondria* (New York: Plenum, 1982). 12.6: Courtesy of E. P. Gogol and R. A. Capaldi. Reprinted with permission from Lucken et al. *Biochemistry* 29 (1990): 5339. Copyright 1990 American Chemical Society. 12A.1: From R. N. Trelease, P. J. Gruber, W. M. Becker, and E. H. Newcomb, *Plant Physiology* 48 (1971): 461.

Chapter 13 13.2a: Micrograph by M. W. Steer; photo provided by E. H. Newcomb. 13.2b: Courtesy of Professor Linda Graham, Department of Botany, University of Wisconsin, Madison. 13.3a: Micrographs by W. P. Wergin; photos provided by E. H. Newcombe.13.4: N. J. Lang, University of California, Davis/Biological Photo Service.13.14: H. Michel and J. Deisenhofer, Max Planck Institute fur Biochemie, Munich. 13.21b: © R. W. Van Norman/Visuals Unlimited. 13A.1: Courtesy of James A. Bassham.

Chapter 14 14.6b: From J. C. Wang, *Sci. Am.* 247 (1982): 94. 14.16: From Donald E. Olins and Ada L. Olins, *Amer. Sci.* 66 (1978): 704. Reprinted by permission of *American Scientist*, the journal of Sigma Xi. Photo Courtesy of Donald and Ada Olins, University of Tennessee—Oak Ridge. 14.18a, top: S. C. Holt, University of Texas Health Science Center, San Antonio/Biological Photo Service. 14.18a, bottom: A. L. Olins, University of Tennessee/Biological Photo Service. 14.18b: Courtesy of Barbara Hamkalo. 14.18c: Courtesy of J. R. Paulsen and U. K. Laemmli, *Cell*. 14.18d, e: G. F. Bahr, Armed Forces Institute of Pathology. 14.19: From D. R. Wolstenholme, K. Koike, and P. Cochran-Fouts, *Cold Spring Harbor Symp. Quant. Biol.* 38 (1973): 267. Copyright © 1973 by Cold Spring Harbor Press. 14.21a: From L. Orci and A. Perrelet, *Freeze-Etch Histology.* Heidelberg: Springer-Verlag, 1975. 14.21b: Micrograph by S. R. Tandon; photo provided by E. H. Newcomb. 14.22a, 14.23: From L. Orci and A. Perrelet, *Freeze-Etch Histology.* Heidelberg: Springer-Verlag, 1975. 14.24a: From A. C. Faberge, *Cell Tiss. Res.* 15 (1974): 403. Heidelberg: Springer-Verlag, 1974. 14.27a: Courtesy of Sheldon Penman from *Science* 259 (1993): 1257; © 1993 by the AAAS. 14.27b: Courtesy of Ueli Aebi. 14.28: © D. Phillips/Photo Researchers, Inc. 14.29: From Sasha Koulish and Ruth G. Kleinfeld, *J. Cell Biol.* 23 (1964): 39. Reproduced by copyright permission of The Rockefeller University Press. 14A.1: © Lee D. Simon/Science Source/Photo Researchers, Inc. 14A.2b: Courtesy of L. D. Simon, Waksman Institute of Microbiology, Rutgers University, New Brunswick, N. J. 14A.3: © Bruce Iverson. 14C.2: Cellmark Diagnostics.

Chapter 15 15.5a: Courtesy of J. Cairns. 15.6b: From D. J. Burks and P. J. Stambrook, *J. Cell Biol.* 77 (1978): 762. Reproduced by permission of The Rockefeller University Press. Photos provided by P. J. Stambrook. 15.17a-e: © Ed Reschke. 15.18a-e: © Carolina Biological Supply/Phototake. 15.19: Courtesy of Richard Macintosh. 15.21: Courtesy of Dr. Matthew Schibler, from *Protoplasma* 137 (1987): 29-44. 15.22c: Courtesy of Jeremy Pickett-Heaps, University of Melbourne. 15.23: From H. W. Beams and R. G. Kessel, *Amer. Sci.* 64 (1976): 279. Reprinted by permission of *American Scientist*, the journal of Sigma Xi. 15.24: Micrograph by B. A. Palevitz. Courtesy of E.H. Newcomb, University of Wisconsin.

Chapter 16 16.7: Courtesy of James L. Walters. 16.9a: From P. B. Moens, *Chromosoma* 23 (1968): 418. Copyright © 1968 by Springer-Verlag. 16.21a: Courtesy of Charles C. Brinton, Jr., and Judith Carnehan. 16.21b: © Omikron/Photo Researchers, Inc. 16.29: Courtesy of Genentech. 16A.1: Courtesy of Dr. R. L. Brinster, School of Veterinary Medicine, Univ. of Pennsylvania.

Chapter 17 17.12: Courtesy of Roger Kornberg, Stanford University. 17.15: Courtesy of D. B. Nikolov and S. K. Burley from Nikolov et al., *Nature* (1992) 360: 40-46. 17.18: From O. L. Miller, Jr., B. A. Hamkalo, and C. A. Thomas, Jr., *Science* 169 (1970): 392. Copyright © 1970 by the AAAS.

Chapter 18 18.1a; Courtesy of J. A. Lake from *Sci. Am.* 245 (1981): 86; © J. A. Lake. 18.3b: Reprinted with permission from S. H. Kim et al. from *Science* 185 (1974): 435; ©1974 by the AAAS. 18.5b: ©M. A. Rould, J. J. Perona, P. Vogt, and T. A. Steitz, *Science* 246 (1 December 1989): cover. Copyright 1989 by the American Association for the Advancement of Science.

Chapter 19 19.13: From D. D. Brown and I. B. Dawid, *Science* 160 (1968): 272. Copyright © 1968 by the AAAS. 19.15a, b: Courtesy of Sarah Elgin/Washington University. 19.16(1): © P. Bryant, University of California, Irvine/Biological Photo Service. 19.23b: Courtesy of IBM U.K. Ltd. 19.25b, c: Courtesy of E. B. Lewis.

Chapter 20 20.1a-c: Courtesy of Mark S. Ladinsky, University of Colorado, Boulder. 20.2: Courtesy of Dr. John Heuser. 20.3: E. D. Salmon, *Trends in Cell Biology* 5: 154-158, figure 3. 20.4: Courtesy of Sammak and Borisy, *Nature* 332: 724-736, fig. 1b. 20.5b: Courtesy of L. E. Roth, Y. Shigenaka, and D. J. Pihlaja/Biological Photo Service. 20.7a: Lester Binder and Joel Rosenbaum, *J. Cell Biol.* 79 (1978): 510. Reproduced by copyright permission of the Rockefeller University Press. 20.7b: Mitchison and Kirschner, *Nature* 312 (1984): 235, fig. 4c. 20.10: L. Casimeris, N. Pryer, and E. Salmon, *J. Cell Biol.* 107 (1988): 2226. Reproduced by copyright permission of the Rockefeller University Press. 20.13: Courtesy of Kent McDonald. 20.16: Courtesy of James Mandell and Gary Banker, University of Virginia. 20.17a, b: J. Knops et al. *J. Cell Biol.* 114 (1991): 725-733. Reproduced by copyright permission of the Rockefeller University Press. 20.17c: Leclerc et al., *Proc. Natl. Acad. Sci.* 90 (13): 6223-6227. 20.18a, b: Courtesy of Alan Snow, University of Washington. 20.19: Courtesy of Dr. John Heuser. 20.20: Adapted from C. E. Schutt et al., 1993, *Nature* 365: 810; courtesy of M. Roozycki. 20.21b: Courtesy of R. Niederman and J. Hartwig. 20.22c: Courtesy of Roger Craig. 20.26a: Courtesy of S. Ito. 20.28: From M. W. Aynardi, P. M. Steinert and R. D. Goldman, *J. Cell Biol.* 98 (1984): 1407. Reproduced by copyright permission of the Rockefeller University Press.

Chapter 21 21.6b: Courtesy of H. Ris. 21.7a: Courtesy of C. Franzini-Armstrong. 21.12: Courtesy of John Heuser. 21.18b: Courtesy of David Deamer. 21.21, 21.22a: © Allen Bell, University of New England/The Benjamin/Cummings Publishing Co. 21.23b: Adapted from Trybus, K. M., and Lowey, S., *J. Biol. Chem.* 259 (1984): 8564-8571. 21.24a: © M. Abbey/Visuals Unlimited. 21.24b: © Peter Parks—OSF/Animals Animals. 21.30a: W. L. Dentler/Biological Photo Service. 21.30c: © Biophoto Associates/Photo Researchers, Inc. 21.31a-d: Courtesy of William L. Dentler. 21.32a: Courtesy of Dr. Lewis Tilney/University of Pennsylvania. 21.34: From D. A. Cuppels and A. Kelman, *Phytopathology* 70 (1980): 1110. Copyright ©1980 by the American Phytopathological Society. Photo provided by. A. Kelman. 21.35a: Courtesy of J. Adler. 21.36a, b: Courtesy of R.M. Macnab and M. K. Ornston, *J. Mol. Biol.* 112 (1977):1-30; © 1977 by Academic Press, Inc.

Chapter 22 22.2: © John Stevens 1992/FPG. 22.5a: Courtesy of G. L. Scott, J. A. Feilbach, and T. A. Duff. 22.26a: From J. Cartaud, E. L. Bendetti, A. Sobel, and J. P. Changeux, *J. Cell Sci.* 29 (1978): 313. Copyright ©1978 by The Company of Biologists, Ltd.

Chapter 23 23.1, top: Courtesy of Matt Springer, Stanford University. 23.1, bottom: Courtesy of Robert Kay, Cambridge University.

Chapter 24 24.16b: © A. J. Olsen, Scripps Clinic and Research Foundation, 1986. 25.1a, b: Courtesy of G. Steven Martin, *Nature* (1970): 227:1021-1023. 25.14, 25.14: John Eichhorn, M.D./Massachusetts General Hospital.

Appendix A-6a: Courtesy of Leica, Inc., Deerfield, IL. A-8: © David M. Phillips/Visuals Unlimited. A-9, A-13: Courtesy of Tim Ryan. A-15a, b: J. G. White, W. B. Amos, and M. Fordham, *J. Cell Biol.* 105 (1987): 41-48. Reproduced by copyright permission of the Rockefeller University Press. A-18: From Sasha Koulish and Ruth G. Kleinfeld, *J. Cell Biol.* 23 (1964): 39. Reproduced by copyright permission of The Rockefeller University Press.

A-19a, b: Courtesy of J. L. Carson, Biological Education Consultants. A-20a, A-21a: Courtesy of Carl Zeiss, Inc. A-22: Courtesy of J. Croxdale. A-24: Courtesy of Lawrence Berkeley Laboratory. A-25: Courtesy of H. Ris. A-26a, b: Courtesy of RMC, Inc. A-27: Courtesy of J. L. Carson, Biological Education Consultants. A-28: Courtesy of J. L. Carson, Biological Education Consultants. A-32: From L. Orci and A. Perrelet, *Freeze-Etch Histology*. Heidelberg: Springer-Verlag, 1975. A-33: Courtesy of H. Ris. A-36(2): Reprinted by permission from *Nature* 171 (1953): 740; © 1953 MacMillan Magazines, Ltd. A-36(3): Richard Wagner, UCSF Graphics.

Illustration Credits

1-2, 14-2, 14A-1(1), 14-2a, 14A-4, 15-5b, 16-26, 17-7, 21-30d, 23-1(2), 24-12, 25-10, A-6b: G. Tortora, B. Funke, and C. Case, *Microbiology, An Introduction,* 5th Ed., Benjamin/Cummings, 1995. ©1995 The Benjamin/Cummings Publishing Company.

3-1a-b, 3-13, 3-29a-f, 6-5a-c, 6-16, 7-3T, 7-14a-b, 7-15a-b, 7-22, 14-18, 14C-1, 15-30, 16-15, 16-16, 16-22, 17-22, 17A-1, 18-3b(1), 18-5a, 19-11, 22-16, 22-29, 23-18: N. Campbell, *Biology,* 3rd Ed., Benjamin/Cummings, 1993. ©1993 The Benjamin/Cummings Publishing Company.

3-8, 3-9a-b, 3-28a-c, 6-7b, 6-15a-b, 6-20, 7-10, 9-1a-b, 13-8, 22-3, 22.5b, 22.15, 23A-1b, 24A-1, A-1: C. K. Mathews and K. E. van Holde, *Biochemistry,* Benjamin/Cummings, 1990. Copyright ©1990 The Benjamin/Cummings Publishing Company.

3-3, 3-10a-c, 7-12, 9-18, 9-19, 9A-1, 10-6, 10-A2, 11A-2, 12-18, 12-21, 12A-3, 13-5, 13-6, 13-13, 13-16, 13-17, 13-20a-b, 17B-1, 18-2a-b, 19-9, 19-23a,c-e, 21-1a-b, 21-13, 21-14, 21-29, 21-32b, 21-35b-c, 21-36a-b(1): C. K. Mathews and K. E. van Holde, *Biochemistry,* 2nd Ed., Benjamin/Cummings, 1996. ©1996 The Benjamin/Cummings Publishing Company.

3-4: R. E. Dickerson and I. Geis, *The Structure and Action of Proteins,* Benjamin/Cummings Publishing, 1969. Illustration ©1969 Irving Geis.

3-5c-d: Copyright ©Irving Geis.

6-2a-b(1): Courtesy of Richard J. Feldmann, National Institutes of Health.

6-3a-b: Reprinted with permission from D. M. Blow and T. A. Steitz, *Ann. Rev. Biochem.* 39:63. ©1970 Annual Reviews, Inc.

6-19, 8-3b, 18-13, 21-8, 21-30b, 24-4, 25-13: E. N. Marieb, *Human Anatomy and Physiology,* 3rd Ed., Benjamin/Cummings, 1995. ©1995 The Benjamin/Cummings Publishing Co.

7-4a-c: Adapted from R. M. Dowben, *Journal of General Physiology,* 1969. Copyright ©1969 Rockefeller University Press.

8-4: Adapted from R. Collander, *Trans. Faraday Soc* 33(1937):986. ©1936 The Royal Society of Chemistry.

9-34: Data from National Cancer Institute Monograph 21, June 1966, *The Development of Zonal Centrifuges and Ancillary Systems for Tissue Fractionation and Analysis,* U.S. Dept HEW, PHS, NCI, Bethesda, MD.

10-5: From data in H. McNeil et al., *Ann. Rev. Biochem.* 53(1984):625.

10-8b: L. Rosenberg, *Dynamics of Connective Tissue Macromolecules,* M. Burleigh and R. Poole, eds., Amsterdam: North Holland Publishing, 1975.

14-3a, 15.17, 15.26, 16-3, 16-6, 18B-1b, 19-12, 19-16(2), 22-14, 23-15: N. Campbell, L. Mitchell, and J. Reece, *Biology: Concepts and Connections,* Benjamin/Cummings,1994. ©1994 The Benjamin/Cummings Publishing Co.

14-20: Reproduced with permission from D. C. Wallace, "Diseases of the Mitochondrial DNA," *Ann. Rev. Biochem.* 61(1992):1175-1212. ©1992 by Annual Reviews, Inc.

18-1b: Adapted from J. A. Lake, *Sci. Am.* 245(1981):86 with permission. ©1981 James A. Lake.

20-8: From Alberts et al., *Molecular Biology of the Cell,* 3rd Ed., Garland Publishing, 1994, fig. 16.23. ©1994 Garland Publishing.

20-11: From Alberts et al., *Molecular Biology of the Cell,* 3rd Ed., Garland Publishing, 1994, fig. 16.33a. ©1994 Garland Publishing.

21-4: Reproduced from Gorthesy-Thevlaz et al., *J. Cell Biol.* 118(1992): 1333-45 by copyright permission of The Rockefeller University Press.

21-11b: Data from Gordon et al., *J. Physiol.* 171(1964):28P.

21-16: D. Bray, *Cell Movements,* Garland Publishing, 1992, fig. 11.10, p. 166. ©1992 Garland Publishing.

21-17; From Guyton et al., *Textbook of Medical Physiology,* 5th Ed., fig. 22.17b, W.B. Saunders, 1996. Copyright ©1996 W.B. Saunders.

21-21(2), 21-22a(2): L. G. Mitchell, J. A. Mutchmor, and W. D. Dolphin, *Zoology,* Benjamin/Cummings, 1988. ©1988 The Benjamin/Cummings Publishing Company.

21-25: From Alberts et al., *Molecular Biology of the Cell,* 3rd Ed., Garland Publishing, 1994, fig. 16.65, p. 835. ©1994 Garland Publishing.

21-28: From D. Bray, *Cell Movements,* Garland Publishing, 1992, fig. 9.8. ©1992 Garland Publishing.

22-4a-d: Adapted from B. B. Boycott, L. Peichl, and H. Wassle, *Proc. Soc. Roy. London B.* 203(1978):229.

23-14: Adapted from M. Berridge, *Nature* 361(1993):315-325.

24-01: G. J. Tortora and S. R. Grabowski, *Principles of Anatomy and Physiology,* 7th Ed. ©1993 Biological Sciences Textbooks, Inc., A&P Textbooks, Inc., Sandra Reynolds Grabowski. Reprinted by permission of Harper-Collins Publishers, Inc.

24-03: Adapted from G. M. Edelman, *Sci. Am.* 223(1970):34. ©1970 Scientific American, Inc. All rights reserved.

24-15: Adapted from L. M. Amzel, R. J. Poljak, F. Saul, J. M. Varga, and F. F. Richards, *Proc. Nat. Acad. Sci.* 71(1974):1427.

24-16a: Adapted from E. W. Silverton, N. A. Navia, and D. R. Davies, *Proc. Nat. Acad. Sci.* 74(1977):5140.

25-06: Reprinted with permission from J. Adams, *Nature* 315(1985):542. ©1985 by Macmillan Magazines Ltd.

Text Credits

p. 62: Box 3-1: From J. D. Watson, *The Double Helix,* pp. 194-197. Copyright © 1969 James D. Watson. Reprinted by permission of the author and Atheneum Publishers.

pp. 160-161: Reprinted from T. R. Cech, "The Chemistry of Self-Splicing RNA and RNA Enzymes," *Science* 236(1987):1533. Copyright © 1987 American Association for the Advancement of Science.

p. 363 Table 12.4:   Data from Futai and Kanazawa, *Microbiol. Rev.* 47(1983):285.

p. 411: From a letter by Oswald Avery in R. D. Hotchkiss, *Phage and the Origins of Molecular Biology,* J. Cairns, G.V.S. Stent, J. D. Watson, eds. Cold Spring Harbor, NY: Cold Spring Harbor Laboratory, 1966. Reprinted by permission.

p. 455: Reprinted with permission from J. D. Watson and F. H. C. Crick, *Nature* 171(1953):964. Copyright © 1953 Macmillan Magazines Ltd.

p. 581 Box 18A: Reproduced with permission from Noller, H., "Ribosomal RNA and Translation." *Ann. Rev. Biochem.* 60(1991):195-196. Copyright © 1991 Annual Reviews Inc.

p. 797: Table 24.2: C. K. Mathews and K.E. van Holde, *Biochemistry,* T7.2, p. 252, Benjamin/Cummings, 1990. Copyright © 1990 The Benjamin/Cummings Publishing Company.

INDEX

microtubules in, 649, 657, 658*t*
myelinated, 717, 718*f*
Axoplasm, 716
MAPs in, 678–680, 679

B

B-DNA, 418
B lymphocytes (B cells), 785, 795*t*
  activation of, 795
  differentiation of, V and C gene segments brought together during, 806
B tubule, 703
Back-mutation, in Ames mutagenesis assay, 827–828
Backcrossing, 509, 510*f*
Bacteria, 82. *See also* Prokaryotic cells
  cell wall of, 104, 279–281
    polysaccharides in, 67, 68*f*
  conjugation in, 518–520
  F₁ complexes/particles in, 334
  recombination in, 516–520
  suppressor strains of, 582
  transformation and transduction in, 517–518
Bacterial chromosome, packaging of, 432–433
Bacterial flagellar rotation, 657, 676, 707
  chemotaxis and, 707–709, 709*f*
Bacterial flagellum, 706–708, 709*f*
  structure of, 706, 707*f*
Bacterial genes, nomenclature for, 486
Bacterial plasmids, 433
Bacteriochlorophyll, 379*f*, 380
Bacteriophages (phages), 105, 106*f*, 411, 412*f*, 413–416
  recombination of, 516–517
  transducing, 517
Bacteriorhodopsin, structure of, 173–175, 175*f*
Bacteriorhodopsin proton pump, 222, 223*f*, 224*f*
Band 4.1, in erythrocyte plasma membrane, 176
Barbed end, of actin filament, 660
Barr body, 619
Barrier filter, in fluorescence microscope, 849
Basal body, 648, 702, 703, 704*f*
  microtubule assembly at, 651*f*
Basal laminae, 285
Basal transcription factors, 554–555, 622
  gene expression regulation and, 622
Base excision repair, 472
Base-pair deletions, mutations caused by, 584*f*, 585
  cancer and, 824–825
Base-pair insertions, mutations caused by, 584*f*, 585
Base-pair substitution, mutations created by, 584–585, 584*f*
Base pairs, 58, 60*f*
  in double helix of DNA, 60*f*, 61, 62–63
  genome size and, 421
  in RNA structure, 61
Basic fibroblast growth factor, 834
Basic keratins, 666*t*, 667
B-DNA, 418
Beam deflectors, for scanning electron microscope, 856–857
Benign tumors, 817
Benzodiazepines, GABA receptors affected by, 746
Beta-adrenergic receptors, 773–774
  cAMP pathway and, 774–776
Beta carotene, 380
  absorption spectrum of, 379*f*
Beta glycosidic bond, 65
Beta2-microglobulin, 792
Beta oxidation, 301, 302–306, 309, 340, 341*f*
  peroxisomes in, 262–263
Beta sheet (beta pleated sheet), 50*f*, 51

Beta (β) tubulin, 102, 649
  genetics of, 650
Bicarbonate, in Hatch-Slack cycle, 401
Bifunctional enzyme, 320
Bilayer, membrane (lipid/phospholipid), 26, 89, 170–171. *See also* Membranes
Binding, substrate, 146–147
Binding site, 753
Binding stage of transcription, 546*f*, 547–549
  for RNA polymerase II, 554, 555*f*
Biochemistry, 18
  as strand in cell biology, 6, 9
Bioenergetics, 119–128
  definition of, 119
  energy systems and, 119–120
  first law of thermodynamics and, 120–122
  heat and, 119–120
  second law of thermodynamics and, 122–128
  work and, 119–120
Biological chemistry. *See* Biochemistry
Biological work, 113–116
  types of, 114*f*
Bioluminescent work, 114*f*, 116
Biopolymers, synthesis of, 31, 32*f*
Biosynthesis, 113
Bithorax gene complex, 628, 629*f*
Bivalents, 500, 501–502, 506*f*
Blood clotting, 777
Blood platelets, 777
  activation of, 777, 778*f*
Blue-green algae. *See* Cyanobacteria
Bond energy, 19
Bone marrow, in lymphocyte development, 786
Bordetella pertussis toxin, G proteins affected by, 763
Brenner, S., 541–542
Brightfield microscope/microscopy, 845–847, 846*f*
Brinster, R., 529
Brown, R., 4
Buchner, E., 9, 11
Buchner, H., 9, 11
Budding, asexual reproduction by, 495, 496*f*
Bulk-phase endocytosis, 247
Bundle sheath cells, 400
α-Bungarotoxin, acetylcholinesterase affected by, 745
Buoyant density (equilibrium density) centrifugation, 232–233, 234*f*
  in DNA analysis, 456, 457*f*
Bursa of Fabricius, in lymphocyte development, 786
Butylene glycol fermentation, 314
Bypass reactions, 316–318

C

C. *See* Carbon atom
C domain (constant domain), antibody, 796, 804–805
C gene segments, 806, 807*f*
c-myc gene, 822
C₃ pathway, 389
C₄ pathway, 400
C₃ plants, 400
C₄ plants, 400
C protein, 684*t*
C regions (constant regions), of light and heavy immunoglobulin chains, 799–800, 806, 807*f*
C-terminus (carboxyl terminus), 46
Ca²⁺. *See* Calcium ions
Ca²⁺ ATPase (calcium pump), 684*t*, 691–692, 767
CAAT box, 554
Cadang-cadang disease, 105

cal. *See* Calorie
Cal/mol. *See* Calories per mole
Calcium
  in contraction, 688
  in exocytosis, 245–246
  inositol trisphosphate signal transduction pathway and, 765
  intracellular, regulation of, 767*f*
  in neurotransmitter secretion, 741–742
  sarcoplasmic reticulum in release and uptake of, 690–692
  as second messenger, 245, 766–769
  in skeletal muscle cells, regulation of, 688–689
Calcium-calmodulin complex, 696, 697*f*, 768–769
Calcium-calmodulin–dependent kinase type II (CAM kinase II), 769
  in neurotransmitter secretion, 742
Calcium-induced calcium release, 691, 767
Calcium ionophores
  inositol trisphosphate signal transduction pathway and, 765
  in neurotransmitter secretion, 741
Calcium ions. *See also* Calcium
  as second messengers, 245, 766–769
Calcium pump (Ca²⁺ ATPase), 684*t*, 691–692, 767
Caldesmon, 696
Calmodulin, 696, 697*f*, 767–768
Calorie, 19, 120
Calorie (nutritional) (kilocalorie), 120
Calories per mole, 19
Calvin cycle, 389–395
  paper chromatography and, 391–392
  3-phosphoglycerate reduction in, 389, 390*f*, 393
  photosynthetic energy transduction and, 395–396
  regulation of, 394–395
  ribulose-1,5-bisphosphate in, 390*f*
    carboxylation of, 389, 389–393
    regeneration of, 389, 393, 394*f*
CAM kinase II. *See* Calcium-calmodulin–dependent kinase type II
CAM plants, photosynthesis in, 401–403
cAMP. *See* Cyclic AMP
cAMP-dependent kinase, 762
cAMP pathway, 764
  β-adrenergic receptors and, 774–776
  intracellular effects of, 774–776
cAMP receptor protein, 607–608
Cancer, 817–840
  chromosomal abnormalities and, 824–825
  classification/nomenclature for, 817–818
  environmental carcinogens and, 825–828
  genetic basis of, 819–824
  growth properties of transformed cells and, 818
  loss of growth regulation and, 817–818
  metastasis and, 817, 833–834
  national differences in incidence of, 826
  neoplastic transformation and, 817
  oncogenes and, 819–823, 835–837
  tissue growth regulation and, 828–831
  tumor invasion and, 817, 832–833
  tumor progression and, 832
  tumor-suppressor genes and, 823–824
Capacitance, impulse transmission and, 734–735
Capping and patching, membrane protein mobility and, 189–190
Capping proteins, in microtubule organization, 656–657
Capping and severing proteins, 663, 664
Caps
  GTP, in microtubule assembly, 653–654
  of mRNA, 559–560, 560*f*, 561*f*

Electron microscope/microscopy (*continued*)
   sample preparation techniques in, 864–865
   scanning transmission, 858
   scanning tunneling, 7–9, 865, 866*f*
   stereo, 864, 865*f*
   transmission, 854–856
      sample preparation techniques in, 859–864
Electron shuttle system, 366
Electron transfer potential, 350
Electron transport, 329, 346–358, 366–369
   coenzyme oxidation and, 346–350
   energetics of, 355–358, 355*f*
   reduction potentials and, 350–352
Electron transport intermediates, 26
Electron transport system/chain, 235, 329, 350, 352–358
   electron carriers in, 353–354
      organization of in membrane, 355–358
      sequence of, 354–355
Electronegative atom, 22
Electroneutrality, 719
Electronic imaging, 850
Electrophilic substitution, 148
Electrophoresis, gel
   for membrane protein analysis, 184, 185*f*
   restriction fragments separated by, 423–424
Electroplaxes, 116
   acetylcholine receptors in, 739, 740*f*
Electrostatic interactions (ionic bonds)
   in protein folding, 35, 48
   in tertiary structure of protein, 51
ELISA. *See* Enzyme-linked immunosorbent assay
Elongation
   in transcription, 546*f*, 549–550
   in translation, 576*f*, 577–579
Elongation phase
   of microfilament assembly, 660
   of microtubule assembly, 651, 652*f*
Embden-Meyerhof pathway, 9, 307. *See also* Glycolysis
Embedding
   for light microscopy, 853
   for transmission electron microscopy, 859–860
Emerson enhancement effect, 381
End-product inhibition (feedback inhibition), 157, 601
End-product repression, 600–601
End-replication problem, 464–467
Endergonic reaction, 124, 296
Endocrine hormones, 770–779. *See also* Hormones
   chemical classification and function of, 771, 773*t*
   neuropeptide activity of, 739
Endocrine tissues, 770, 771*f*
Endocytosis, 84, 169, 246–249. *See also* Phagocytosis
   bulk-phase, 247
   receptor-mediated, 247–249, 250–251, 754
Endocytotic vesicle, 246, 247*f*
Endogenous antigen, processing of, 794
Endogenous pathway, 794
Endomembrane system, 235, 236*f*
   protein sorting in, Golgi complex and, 241–244
Endonucleases, repair, 471
Endopeptidases, 340, 342*f*
Endoplasm, 698
Endoplasmic reticulum, 83, 93, 94*f*, 233–239
   cotranslational import of polypeptides into, 585, 585–589, 586*f*, 590*f*
   membranes of, 235, 236*t*
      insertion of transmembrane proteins into, 589, 590*f*

in protein glycosylation, 241, 242*f*
rough, 93, 94*f*, 235–237, 237*f*, 238–239
smooth, 84*f*, 93, 235–237, 237–238, 237*f*, 239*f*
types of, 235–237
Endosomes, 794
Endosymbiont theory, 100–101
Endothermic reaction, 121
Energetics, of active transport, 212
Energy, 112–137. *See also* Free energy change
   activation, 139
   for active transport, 216
   aerobic production of, 117
   anaerobic production of, 117
   bond, 19
   cellular. *See* Bioenergetics
   conservation of, 190–192
   definition of, 113
   fat as source of, 340, 341*f*
   flow of in biosphere, 117–118
   free, 124
      chemical equilibrium and, 129
   importance of, 112–119
   internal, 121
   need for, 113–116
   protein as source of, 340–343
   sources of, 116*t*
   use of, 116–117
   wavelength of electromagnetic radiation and, 19–20, 19*f*
Energy barrier, activation, 127
Energy metabolism. *See also* Metabolism
   chemotrophic, 116–117, 300–307
      by aerobic respiration, 327–373
      by glycolysis and fermentation, 296–326
Energy systems, 119
Energy transduction, photosynthetic, 374, 375*f*, 377–381, 386–387
   Calvin cycle and, 395–396
Enhancer elements, 554, 622, 623, 627–628
   gene expression regulation and, 622–623, 627–628
   of steroid receptors, 755
Enkephalins, 739
Enolase, in glycolysis, 308*f*, 309
Enthalpy, 121
   changes in, 121, 122*f*
Enthalpy change, 125–126
Entropy, 123, 128
   cellular activities affecting, 118
Entropy change, 123, 125
   as measure of thermodynamic spontaneity, 123–124
Enveloped viruses, 105
Environmental carcinogens, 825–828
Enzymatic method, of DNA sequencing, 425–427
Enzyme activity, for detection and quantification of protein molecules, 636, 637*f*
Enzyme catalysis, 138, 146–148. *See also* Enzymes
   activation energy affected by, 139
Enzyme Commission system, for enzyme nomenclature, 143–144, 145*t*
Enzyme I, 221
Enzyme II, 221
Enzyme III, 221
Enzyme kinetics, 148–160
   double-reciprocal plot and, 152–153, 154*f*
   for hexokinase, 153–155
   Michaelis-Menton, 148–149
   V_{max} and K_m in, 149–152, 153–155
Enzyme-linked immunosorbent assay (ELISA), 803
Enzymes, 26, 138–166
   activation energy and, 139
   activation energy barrier and, 140

active site of, 141, 142*f*
allosteric, 157
allosteric regulation of, 156–158
bifunctional, 320
as catalysts, 140–148
   substrate activation and, 147–148
   substrate binding and, 146–147
concentration of, V_{max} and, 152
core, in transcription, 547
covalent modification of, 158–160
discovery of, 9
diversity of, 143–144
genes coding for, regulation of. *See* Gene expression, regulation of
inducible, 600
inhibition of, 155, 156*f*
marker, 169
membrane, 89
metastable state and, 139
nomenclature for, 143–144, 145*t*
peroxisomal, in fatty acid breakdown, 96–98
pH sensitivity of, 144–146
as proteins, 141–146
regulation of, 155–156
   allosteric, 156–158
   substrate-level, 156
RNA molecules as (ribozymes), 160–162
specificity of, 143
substrate-level regulation of, 156
temperature sensitivity of, 144
Epidermal growth factor, 828, 829*t*
Epinephrine, 738, 738*f*, 771–774
   in cAMP pathway, 774–776
   glycogen degradation affected by, 774–776
   physiologic functions of, 771, 772*t*
Epitopes, 785
EPSP. *See* Excitatory postsynaptic potential
Equilibrium
   Donnan, 720–721
   electrochemical, 720
Equilibrium constant, 125, 128–129
   standard free energy change and, 130–131
Equilibrium density (buoyant density) centrifugation, 232–233, 234*f*
   in DNA analysis, 456, 457*f*
Equilibrium membrane potential, 719*f*, 720
   relationship of to ion gradient (Nernst equation), 720–722
ER. *See* Endoplasmic reticulum
erb-B oncogene, 831
Ergastoplasm, 233
Erythroblastomas, 818
Erythrocyte, 176*f*
   plasma membrane of, 175–177
      spectrin-ankyrin-actin network in, 664
   transport processes of, 202*f*
Escherichia coli
   DNA studied in, 411, 412*f*, 413–416
   lactose system of, 601
Eserine, acetylcholinesterase inhibited by, 744
Ester bonds, 69*f*, 70
Estradiol, 73, 74*f*
Estrogens, 73, 74*f*
   physiologic functions of, 771, 772*t*
Ethanol, fermentation producing, 313
Ethanolamine, 71, 72*f*
Eubacteria, 82
Euchromatin, 434
Eukaryotic cells (eukaryotes), 82, 83–85, 88–104
   DNA in
      expression of, 85
      organization of into chromosomes, 84–85
      packaging of, 433–434, 435*f*
      replication of by meiosis and mitosis, 85
   endocytosis by, 84
   exocytosis by, 84

filaments of, 84, 102–103
genetic information storage by, 83*f*
genomes of, repeated sequences in, 430–431
internal membranes of, 83, 84*f*, 90–104
meiosis, by, 85
mitosis by, 85
nucleus of, 83, 88, 89–90
organelles of, 82, 84*f*, 90–104. *See also specific type*
plasma membrane of, 88–89
prokaryotic cells compared with, gene regulation and, 611–613
properties of, 82*t*
regulation of gene expression in, 611–634. *See also* Gene expression
ribosomes in, 570*t*
rRNA of, 556–558, 557*t*
transcription in, 550–556
promoters in, 552–555
RNA polymerases in, 551–552
termination of, 555–556
translation in, initiation of, 576–577
tubules of, 84, 102–103
Excision repair, of DNA, 471–472
Excitatory neurotransmitter, 738
Excitatory postsynaptic potential, 742, 743*f*
Exciter filter, in fluorescence microscope, 848, 849*f*
Exergonic reaction, 124, 296
Exit site (E site), 571
Exocytosis, 84, 245–246
Exogenous antigen, processing of, 793–794
Exogenous pathway, 793–794
Exons, 561, 562
Exonuclease, 461
Exopeptidases, 340
Exoskeleton, insect, chitin in, 67, 68*f*
Exothermic reaction, 121
Expression effect, 822
Expression vectors, 525
Extensins, 69, 288–289, 290*f*
Extracellular digestion, 257*f*, 258
Extracellular matrix, 84, 88, 103–104, 276–277, 277*t*, 278–287
Extracellular structures, 276–291
Eyepiece (microscope), 845, 846*f*

F

F-actin (filamentous actin), 102*f*, 103, 660, 661*f*, 676, 698
F_0 complex, 333, 362, 364
F_0F_1 complex, 333, 360–361, 362–364
ATP synthesis by, 364
F_1 complexes/particles, 333, 334, 362, 364
F factors (fertility factors), 433, 518
F_1 generation, 508
F_2 generation, 508*f*, 509
f-Met. *See* N-Formylmethionine
Fab regions, antibody, 796
Fabricius, bursa of, in lymphocyte development, 786
Facilitated diffusion, 202–203, 205–206, 205*t*. *See also* Transport proteins
alternating conformation model of, 206, 207*f*
examples of, 207–208
kinetics of, 206*f*
Facilitation, 746
Facultative heterochromatin, 444
Facultative organisms, 306
FAD. *See* Flavin adenine dinucleotide
$FADH_2$, oxidation of, 346–350
Faraday constant, 720
Fast axonal transport, 676, 678
Fats, 70. *See also* Lipids
conversion of to sugars, 347–350

as energy source, 340, 341*f*
Fatty acids, 69*f*, 70, 71*f*. *See also* Lipids
β oxidation of, 340, 341*f*
membrane structure and, 178, 179*t*
oxidation of, peroxisomes in, 262–263
peroxisomal enzymes in breakdown of, 96–98
transition temperature of, 180
Fatty acyl coenzyme A, 340, 341*f*
Fc receptors, 796
Fc region, antibody, 796
FcRs. *See* Fc receptors
Feedback inhibition (end-product inhibition), 157, 601
Fermentation, 116, 312–314
energetics of, 314
vs. respiration, 306
Ferments, 141. *See also* Enzymes
Ferredoxin, 376, 381, 385
reduction of, 384–385, 387
Ferredoxin-NADP$^+$ reductase, in photo-reduction, 385, 387
Ferredoxin-thioredoxin reductase, in Calvin cycle, 395
Ferric ions, in respiration, 327. *See also* Anaerobic respiration
Ferritin, membrane glycoproteins studied with, 186
Fertility factors (F factors), 433, 518
Fertilization, 498
Fibrin, 285, 777
Fibroblasts, 278, 818
Fibroin, 51
Fibroma, 818
Fibronectin receptor, 284, 286, 287*f*
Fibronectins, 284–285
Fibrous proteins, 51–52, 53*f*
Filament-based movement, 676
in muscle, 680–696, 697*f*. *See also* Muscle contraction
nonmuscle, 676–678
specialized, 698–701, 702*f*
Filamentous actin (F-actin), 102*f*, 103, 660, 661*f*, 676, 698
Filaments, intermediate. *See* Intermediate filaments
Filamin, 663, 664
Filopodia, 663, 698
Fimbrin, 663
First law of thermodynamics, 120–122
Fischer projection, 64, 306
Fission, cell, 85
5' cap, 559, 560*f*, 561
Fixation
for light microscopy, 853
primary, 853
for transmission electron microscopy, 859
Flagella (flagellum), 676, 702, 703*f*
bacterial, 706–708, 709*f*
rotation of, 657, 676, 707, 707–709, 709*f*
structure of, 706, 707*f*
sliding-microtubule model for, 705–706
structure of, 702–705
Flagellin, 706
Flavin adenine dinucleotide (FAD), 327, 338, 353
electron shuttle systems for transport of, 366–367
Flavin mononucleotide, 353
Flavoproteins, in electron transport, 353
Flemming, W., 10
Fluid mosaic model, 173, 174*f*, 175–190
membrane lipids and, 177–183
membrane proteins and, 183–190
Fluidity, membrane, 178–183
Fluorescein, cytoskeleton studied with, 646
Fluorescence, principles of, 848

Fluorescence analogue cytochemistry, 647–648
Fluorescence microscope/microscopy, 7, 847–849
Fluorescence recovery after photobleaching, 178, 180*f*
Fluorescent antibodies, membrane protein mobility studied with, 189
Fluorescent probe, 847, 848
FMN. *See* Flavin mononucleotide
Focal contacts (focal adhesion plaques), 660
Focal length, of microscope lens, 843
Food cup, 256
Footprinting, 548–549, 549
Forming face (cis-face), of Golgi stack, 240–241
N-Formylmethionine, in translation, 575, 577*f*
Fragmin, 663
Frameshift mutations, 541–542, 585
Framework regions, 800
Free energy, 124–128
chemical equilibrium and, 129
Free energy change, 124, 126, 128–133
calculation of, 128–130
samples of, 132–133
meaning of, 131–132, 132*t*
as measure of thermodynamic spontaneity, 124
solute transport and, 212–213, 213–214
standard, 130–131
equilibrium constant and, 130–131
meaning of, 131–132, 132*t*
Freeze etching, 864
Freeze fracturing, 862–864, 863*f*, 864*f*
in membrane analysis, 183–184, 863–864, 864*f*
for tight junctions, 274*f*, 275
Fructose-1,6-bisphosphatase, in gluconeogenesis, 316–317
Fructose-2,6-bisphosphatase, 320, 321*f*
Fructose-1,6-bisphosphate, in glycolysis, 307, 308*f*, 309
Fructose-2,6-bisphosphate, glycolysis and gluconeogenesis regulated by, 320, 321*f*
Fructose-6-phosphate
in glycolysis, 307, 308*f*
interconversion of, 122–123, 123*f*
Fumarate, 335*f*, 338
Fumarate hydrolase, 335*f*, 339
Functional groups, 20
van der Waals radii for, 36*t*
Fungi, genomes of, 422*t*
Fusion, cell, membrane protein mobility and, 188–189

G

G. *See* Free energy
G-actin (globular actin), 102*f*, 103, 660–662, 676
G0 phase of cell cycle, 454
variations in, 482
G1 phase of cell cycle, 453, 454*f*
restriction points in, 483
variations in, 482–483
G2 phase of cell cycle, 453, 454*f*
restriction points in, 483
variations in, 482–483
G protein-linked receptors, 755–759
structure and activation of, 755–756, 756*f*
G proteins, 755
α- and β-adrenergic receptors and, 773
cyclic AMP levels affected by, 761–762, 762*f*
diseases and, 763
growth factor binding and, 830
structure and activation of, 756–759, 759*f*
vision and, 757–758
GABA. *See* Gamma-aminobutyric acid
GAGs. *See* Glycosaminoglycans
Galactose, 185, 187*f*, 283

Hydrolytic enzymes, Golgi complex sorting of, 243
Hydrophilic amino acids, 46
Hydrophilic channel, 26
Hydrophilic molecules, 23, 24
Hydrophobic amino acids, 46
Hydrophobic effect, 36, 48
Hydrophobic interactions, in tertiary structure of protein, 51
Hydrophobic messengers, 752, 753f
Hydrophobic molecules, 23, 24
 diffusion across membranes and, 204
Hydroxyl group, 20, 21f
 van der Waals radius for, 36t
Hydroxylation, smooth ER in, 237–238
Hydroxymethyl transferases, in glycolate pathway, 399
Hypercholesterolemia, receptor-mediated endocytosis and, 250–251
Hypermutation, antibody diversity and, 809
Hyperpolarization, 722, 731, 732f
Hypervariable regions, 800–804
Hypothesis, 12
Hypoxic mode, ATP generation for muscle contraction and, 693

I

I. *See* Inosine
I bands, 681, 682f
I-cell disease, 256
Idiotypes, 804
IF. *See* Initiation factors; Intermediate filaments
IF proteins, 103, 666t
 tissue specificity and genetics of, 667
IFN-γ. *See* Interferon, γ
Ig. *See specific type and* Immunoglobulins
IgA, 797t, 798–799, 798f
IgD, 797–798, 797t
IgE, 797t, 798
IGF. *See* Insulin-like growth factors
IgG, 797t, 798, 798f
IgM, 797, 797t
IgM pentamer, 797
IL-1. *See* Interleukin-1
IL-2. *See* Interleukin-2
Illumination source, for microscopy, 841, 846f
 wavelength of, 841–842
Immortal cells, 818
Immune response, 784–816
 adaptive, 784, 785
 antibodies in
  class switching and, 809–811
  fine structure of, 799–805
  genetic basis of diversity and, 805–809
  structure and function of, 795–799
 antigen-antibody interactions in, 799
 antigen presentation and processing in, 793–794
 B cell activation and, 795
 cell-mediated, 785
 cells of, 795t
 cellular basis of, 785–790
 characteristics of, 785
 clonal selection and, 787–789
 cytotoxic T cell activation and, 794
 deficiency in, 784
 differentiation markers and, 790
 helper T cell activation and, 794
 humoral, 785. *See also* Antibodies
 lymphocyte activation pathways and, 790–795
 lymphocyte development and, 785–787
 major histocompatibility complex and, 791–792
 memory and, 785, 789, 790f
 primary, 789

secondary, 789
 T cell antigen receptor in, 811–812
 tolerance and, 795
 types of, 785
Immunity. *See also* Immune response
 adaptive, 784, 785
 innate, 784
Immunoblotting, 636
Immunocytochemistry, 803
Immunodominance, 785
Immunoelectron microscope/microscopy, 860–861
Immunofluorescence microscope/microscopy
 cytoskeleton studied with, 645–646
 for intermediate filament typing, 668
Immunogen, 784. *See also* Antigens
Immunoglobulin chains, 796
 genes coding for, 806, 807f
 hypervariable regions of, 800–804
 variable and constant regions of, 799–800
Immunoglobulin class, 796–799
Immunoglobulin fold structure, 805
Immunoglobulin gene superfamily, 792, 805
Immunoglobulins, 787. *See also specific type under Ig and* Antibodies
 classes of, 796–799
 for detection and quantification of protein molecules, 636
 folding of, 804–805
 idiotypes of, 804
 membrane protein mobility studied with, 189
 structure of, 55f
Immunological assays, 636–637
Immunological memory, 785, 789, 790f
Immunological tolerance, 795
Immunology, 784. *See also* Immune response
Immunosuppression, 784
IMPs. *See* Intramembranous particles
Inactivating particle, for ion channels, 725
Independent assortment, law of, 510, 512–513, 512f
Induced-fit model, for substrate binding, 146
Inducers, 603
Inducible enzymes, 600
Inducible operons, 603
 dual control of, 608
Induction, substrate, 600
Informational macromolecules, 29–30, 30f, 30t
Inheritance
 blending theory of, 509
 chromosome theory of, 511–513
 cytological basis of, discovery of, 510–511
 Mendel's laws of, 509–510
Inhibition
 allosteric, 158, 318
 competitive, 155, 156f
  of transport proteins, 207
 contact, 818
 enzyme, 155, 156f
 feedback (end-product), 157, 601
 irreversible, 155
 noncompetitive, 155, 156f
 reversible, 155, 156f
Inhibitory neurotransmitter, 738
Inhibitory postsynaptic potential, 742, 743f
Initiation
 of transcription, 546f, 549
  alternative sigma factors and, 608
  gene expression regulation and, 608, 621
 of translation, 575, 575–577, 576f
  in eukaryotes, 576–577
  in prokaryotes, 575–576, 577f
Initiation complexes, in translation, 576
Initiation factors, 575, 576
Initiator protein, 459, 461t

Initiator sequence, of promoter for RNA polymerase II, 553f, 554, 622
 gene expression regulation and, 622
Initiator tRNA, 575
Innate immunity, 784
Inner chloroplast membrane, 376
Inner dynein arm, 704, 705f
Inner mitochondrial membrane, 90–91, 331, 332f
 localization of functions in, 332t
 transport systems of, 368–369
Inosine, in tRNA, 571, 573f
Inositol, 71, 72f
Inositol-phospholipid-calcium pathway, 764, 766f
 α-adrenergic receptors in, 776
Inositol-1,4,5-trisphosphate
 cell functions regulated by, 765t
 formation of, 765f
 as second messenger, 764–766, 766f
Inr. *See* Initiator sequence
Insect exoskeleton, chitin in, 67, 68f
InsP₃. *See* Inositol-1,4,5-trisphosphate
InsP₃ receptor channel, 764
Instability, dynamic, microtubule assembly and disassembly and, 653–654, 655f
Insulin, 29, 30f
 amino acid sequencing of, 49–50
 physiologic functions of, 771, 772t
Insulin-like growth factors, 829t
Integral membrane proteins, 173
 Golgi complex sorting of, 243
Integrins, 285–287, 664, 701
Intercalated discs, 695
Interdoublet links, 704, 705f
Interference, 843, 866
Interferon, γ, in helper T cell activation, 794
Interleukin-1, in helper T cell activation, 794
Interleukin-2, 829t
 in helper T cell activation, 794
Interleukins, 790
Intermediate filament proteins, 103, 666t
 structure of, 667–670
 tissue specificity and genetics of, 667
Intermediate filament typing, 667, 668–669
 for tumor diagnosis, 668–669
Intermediate filaments, 84, 102f, 103, 272, 645t, 666–670
 assembly of, 670
 in cell division, 480
 classes of, 666t
 functions of, 670
 structure of, 667–670
Intermediate lens, 845
 of transmission electron microscope, 856
Intermembrane space
 chloroplast, 376, 377f
 mitochondrial, 331
  localization of functions in, 332t
  polypeptide targeting to, 593, 594f
Internal energy, 121
 change in, 121
Interneurons, 714, 715f
Interphase, 453, 474f
 chromosome appearance in, 90
Interphase I, 502f
Interphase II, 504f
Interspersed repeated DNA, 431
Intervening sequences, 160–161, 559. *See also* Introns
Intracellular membranes, 83, 84f, 167, 168f
Intracellular microtubule-based movement, 678–680
Intracellular receptors, 755. *See also* Receptors
Intracellular transport, 198
Intramembranous particles, 863

Polycistronic mRNA (polygenic mRNA), 602
Polymerase chain reaction, 462, 468–469
Polymerases. *See* DNA polymerases; RNA polymerases
Polymerization, 17, 27–31, 32*f*
    actin, in cell protrusion formation, 698, 699–700
    amino acid, 46–48
    in microtubule assembly, 651
    principles of, 31
Polymers
    nucleic acid, 58–61. *See also* DNA; RNA
    polysaccharide, 65–67, 68*f*
    protein and polypeptide, 46–48
    synthesis of, 17, 27–31, 32*f*
Polymorphism, of MHC genetic loci, 791
Polymorphisms, restriction fragment length (RFLPs), 447
    DNA fingerprinting by analysis of, 446–447
Polymorphonuclear leukocytes (neutrophils), in phagocytosis, 256
Polynucleotide phosphorylase, codon assignments and, 544
Polynucleotides, 58
Polypeptide folding, in translation, 580
Polypeptides, 33, 48. *See also* Proteins
    amino acid polymerization producing, 46–48
    cotranslational import of into ER, 585, 585–589, 586*f*, 590*f*
    folding of, 33, 48
    posttranslational import of, 264, 585, 586*f*, 589–593, 594*f*
    posttranslational processing of, 583
    in protein self-assembly, 33, 34*f*
        noncovalent interactions in, 35–37
    synthesis of. *See* Translation
Polyribosomes, 559*f*, 580
Polysaccharides, 30, 61–69
    repeating units of, 30*t*, 64–65
    storage, 30, 31*f*, 65–67
        catabolism of, 314–316
    structural, 30, 31*f*, 67, 68*f*
    structure of, 68–69
    synthesis of, Golgi complex in, 93–95
Polytene chromosomes, 616–617
Pores
    nuclear, 90, 439, 440*f*
    polar, 172
Porins, 104, 281
Porphyrins, 140
Positive control of transcription, 607–608
Poststaining, for transmission electron microscopy, 860
Postsynaptic neuron, 736, 737*f*
Postsynaptic potentials, 742
Posttranscriptional control of gene expression, 613, 630–634, 637
Posttranslational control of gene expression, 632–634
Posttranslational import, 264, 585, 586*f*, 589–593, 594*f*
Posttranslational processing, 583
Potassium, permeability coefficient for, 205*t*
Potassium channels, 728
    voltage-gated, 725
        action potential and, 730–731, 732*f*
Potassium ion gradient, 719
Potassium leak channels, 719
    action potential and, 730
Potential, 719. *See also specific type*
    action, 718, 724, 728–731
        energetics of, 733–736
        events during, 728–729
        ion channels and currents and, 730–731, 732*f*
        ion concentrations and, 731

        measuring, 729–730
        in muscle contraction, 689
        propagated, 733. *See also* Nerve impulses
        propagation of, 729, 731–736
        rate of transmission and, 733–736
    electron transfer, 350
    membrane, 199, 717
        electrochemical proton gradient and, 360
        equilibrium, 719*f*, 720
        formation of, 719–720
        Goldman equation and, 722–724
        measuring, 728
        Nernst equation and, 720–722
        resting, 717, 718–720
        solute transport and, 213–214, 214*f*
        steady-state ion concentrations and, 722
    postsynaptic, 742, 743*f*
    reduction, 350–352
    threshold, 729
Poxviruses, 819*t*
Precipitin reaction, antigen-antibody interactions and, 799
Pre-mRNA, 559
    alternative splicing of, 564
Presynaptic neuron, 736, 737*f*
    neurotransmitter secretion by, 741–742
Pribnow box, 547
Primary cell wall, 287
Primary fixation, for light microscopy, 853
Primary follicles, 786
Primary image, in light microscopy, 845
Primary immune response, 789. *See also* Immune response
Primary lysosome, 256, 257*f*
Primary structure of proteins, 49–50, 49*f*, 49*t*
Primary transcript, modification of (RNA processing), 556–564
Primase, in DNA replication, 461*t*, 463
Primer, RNA, for DNA replication, 462–463
Primosome, 463
Prions, 105
Procarcinogen, 826
    Ames assay of mutagenesis of, 827–828
Processes, neuronal, 716
Profilin, 663
Proflavin, as mutagen, 541
Progesterone, 73, 74*f*
Programmed cell death, 258
Projector lens, of transmission electron microscope, 856
Prokaryotic cells (prokaryotes), 82. *See also* Bacteria
    DNA in
        expression of, 85
        packaging of, 432–433
    eukaryotic cells compared with, gene regulation and, 611–613
    genetic information storage by, 83*f*
    genomes of, 422*t*
    localization of respiratory functions in, 334
    properties of, 82*t*
    regulation of gene expression in, 599–609, 610*f*. *See also* Gene expression
    ribosomes in, 570*t*
    rRNA of, 557*t*, 558
    transcription in, 545–550
        coupling of translation with, 559*f*
        RNA polymerase in, 545–547
    translation in, initiation of, 575–576, 577*f*
Proline, 46*t*, 47*f*
Prometaphase, 473, 474*f*
Promoter
    in eukaryotic transcription, 550, 552–555, 621–622
        gene expression regulation and, 621–622
    mutations in, *lac* operon affected by, 604

    operon transcription beginning at, 602
    in prokaryotic transcription, 547–549
Propagated action potential, 733. *See also* Nerve impulses
Propagation, of action potential, 729, 731–736
Prophage, 415
Prophase, 473, 474*f*, 506*f*
Prophase I, 501, 502*f*, 504*f*, 505*f*, 506*f*
Prophase II, 503, 504*f*
Propionate fermentation, 314
Prostaglandins, 776, 777
    physiologic functions of, 771, 772*t*
Prosthetic groups, 141
Protamine, discovery of, 409
Proteases, 340, 342*f*
Proteasomes, 634
Protein-binding sites, on DNA, footprinting for identification of, 548–549
Protein-coding genes, introns and, 561
Protein factors, in translation, 574–575
Protein kinase, 159
    in cell cycle regulation, 485–486, 487–488
    in exocytosis, 246
    mitogen-activated, 766
    PFK-2 phosphorylation by, 320, 321*f*
Protein kinase A, 762, 764*f*
    subunits of, 762, 764*f*
Protein kinase C, 764
    diacylglycerol and, 766
Protein-protein interactions, in eukaryotic transcription, 551–552
Protein sorting, Golgi complex in, 241–244
Protein synthesis, 48. *See also* Translation
    DNA in, 56*f*
    RNA in, 56*f*
Protein targeting and sorting, 583–593, 594*f*
    cotranslational import and, 585, 585–589, 586*f*, 590*f*
    intracellular, 586*f*
    posttranslational import and, 264, 585, 586*f*, 589–593, 594*f*
Protein turnover, gene expression regulation and, 612–613, 633–634
Proteins, 45–56. *See also specific type and* Polypeptides
    amino acids in, 45–46
    antibodies for detection and quantification of, 636
    conformation of, 49
        native, 51
    as energy source, 340–343
    enzymes as, 141–146
    fibrous, 51–52, 53*f*
    folding of, 33, 35–36, 48
        noncovalent interactions in, 35–37
    growth-inhibitory, 831
    hormone, chemical classification and function of, 771, 773*t*
    as informational macromolecules, 29, 30*f*
    monomeric, 48
    multimeric, 48
        quaternary structure and, 55
    mutations changing, 537–540
    posttranslational import of, 264, 585, 586*f*, 589–593, 594*f*
        regulation of gene expression and, 612
    primary structure of, 49–50, 49*f*, 49*t*
    quaternary structure of, 49*f*, 49*t*, 55–56
    repeating units of, 30*t*, 45–48
    secondary structure of, 49*f*, 49*t*, 50–51, 50*f*
    self-assembly of, 32–33
    space-filling models of, 36
    structure of, 48–56
        level of organization in, 48–49, 49*t*
    tertiary structure of, 49*f*, 49*t*, 51–55
Proteoglycans, 88, 103, 277, 281–283

Restriction mapping, 424–425, 424*f*
Restriction point, 483
Restriction/methylation system, 428
Retinal, 222
    in G protein regulation of vision, 758
Retinoblastoma, *Rb* gene inactivation and, 824
Retrograde axonal transport, 679
Retrotransposons, 539
Retroviruses, 525, 538–539, 819*t*, 820
Reverse transcriptase, 525, 538–539, 820
Reverse transcription, 535, 538–539
Reversible inhibition, 155, 156*f*
RFLPs (restriction fragment length poly-
      morphisms), 447
    DNA fingerprinting by analysis of, 446–447
RGDS sequence, 284, 286
Rho (ρ) factor, 550
Rhodopsin, in G protein regulation of vision,
    758
Riboflavin, 338
Ribonuclease
    amino acid sequencing of, 50
    denaturation and renaturation of, 33, 34*f*
    structure of, 52, 54*f*
    synthesis and self-assembly of, 33, 34*f*
Ribonuclease P, 161
Ribonucleic acid. *See* RNA
Ribonucleoprotein particles, transport of, 442
Ribose, 56
Ribose-5-phosphate, in Calvin cycle, 393, 394*f*
Ribosomal RNA (rRNA), 56, 535, 556–558, 581
    in nucleolus, 444–445
    processing of, 557–558, 557*f*
    promoters for, 554
    sedimentation coefficient for, 556, 557*t*
Ribosome-binding site, mRNA, 576
Ribosomes, 99–101, 569–571
    binding sites on, 571
    eukaryotic, 570*t*
    prokaryotic, 570*f*, 570*t*
    on rough endoplasmic reticulum, 93
    in translation, 569–571
Ribozymes, 141, 160–162
    23S rRNA as, 579
Ribulose-1,5-bisphosphate, in Calvin cycle, 390*f*
    carboxylation of, 389, 389–393
    regeneration of, 389, 393, 394*f*
Ribulose-bisphosphate carboxylase/oxygenase.
    See Rubisco
Ribulose-5-phosphate kinase, in Calvin cycle,
    390*f*, 393, 395
Rigor, 686
RME. *See* Receptor-mediated endocytosis
RNA, 56
    in DNA replication, 462–463
    in flow of genetic information, 535, 536*f*
    guide, 564
    as informational macromolecule, 29, 30*f*
    messenger. *See* Messenger RNA
    in nucleolus, 444–445
    in protein synthesis, 56*f*
    ribosomal. *See* Ribosomal RNA
    structure of, base pairing in, 61
    synthesis of. *See also* Transcription
        in nucleolus, 444–445
        precursors for, 58
        template for, 58
    transfer. *See* Transfer RNA
RNA editing, 564
RNA polymerase I, 551, 552*t*
    in gene expression regulation, 629–630
    promoter for, 552–554, 553*f*
    termination signal for, 555
RNA polymerase II, 551–552, 552*t*
    binding stage of transcription for, 554, 555*f*
    promoter for, 553*f*, 554, 622

gene expression regulation and, 622
    termination signal for, 555–556
    transcription factors in function of,
      554–555
RNA polymerase III, 552, 552*t*
    in gene expression regulation, 629–630
    promoter for, 553*f*
    termination signal for, 556
RNA polymerases
    in eukaryotic transcription, 550, 551–552
    in prokaryotic transcription, 545–547, 548
RNA processing, 551, 556–564
    regulation of, gene expression regulation and,
      630
RNA splicing, 562, 563*f*
    alternative, 564, 630, 631*f*
      in antibody class switching, 809, 810*f*
    self-splicing introns and, 562–564
RNA tumor viruses, 819*t*, 820
RNA viruses, 106*f*
RNP particles, 562
Robertson, J.D., 172
Rod, of bacterial flagellum, 706, 707*f*
Rod cells, 757–758, 757*f*
Rotary pump, for transmission electron micro-
    scope, 855
Rotation
    bacterial flagellar, 657, 676, 707
      chemotaxis and, 707–709, 709*f*
    phospholipid, membrane asymmetry and,
      190
Rotors, centrifuge, 230
Rough endoplasmic reticulum (rough ER), 93,
    94*f*, 235–237, 237*f*, 238–239
Rous sarcoma virus, 106*f*
rRNA. *See* Ribosomal RNA
RSS. *See* Recombination signal sequences
Rubisco, 390, 395
    in Hatch-Slack cycle, 400
    oxygenase activity of, photosynthetic effi-
      ciency affected by, 398–403
Ryanodine receptor channel, 684*t*, 691, 767

S

S. *See* Entropy
S1. *See* Myosin subfragment 1
S phase of cell cycle, 453, 454*f*. *See also* DNA
    replication
    variations in, 482–483
S regions (switch regions), 809
S ring (stator ring), of bacterial flagellum, 706,
    707*f*
S value (sedimentation coefficient), 99, 231
    of ribosome, 99
    for rRNA, 556, 557*t*
Saltatory transmission, 736
Salvage pathway, glycolate pathway as, 398
Sanger method, of DNA sequencing, 425–427
Sarcolemma, 689
Sarcoma, 818
Sarcoma virus, avian, 820
Sarcomeres, 681, 682*f*
    structural proteins of, 683–684
Sarcoplasm, 688
Sarcoplasmic reticulum, 690
    in calcium release and uptake, 690–692
Sarin, 744, 745*f*
Satellite DNA, 430
Saturated fatty acids, 70, 71*f*
Saturation, 149
    membrane fluidity affected by, 180, 181*f*
    receptor, 753
Scanning electron microscope/microscopy, 7, 8*f*,
    856–857, 858*f*
    sample preparation techniques in, 864–865

Scanning transmission electron microscope/
    microscopy, 858
Scanning tunneling microscope/microscopy,
    7–9, 865, 866*f*
Schleiden, M., 4
Schwann, T., 4
Schwann cells, 714–715, 717, 718*f*
Scientific method, 10–13
Scintillator, in scanning electron microscope,
    858
Scrapie, 105
SDS-polyacrylamide gel electrophoresis. *See also*
    Gel electrophoresis
    for membrane protein analysis, 184, 185*f*
Second law of thermodynamics, 122–128
Second messengers, 169, 761–770. *See also* spe-
    cific type
    calcium as, 245, 766–769
    cyclic AMP as, 761–764
    diacylglycerol as, 764–766, 766*f*
    inositol trisphosphate as, 764–766, 766*f*
Secondary cell wall, 287
Secondary electrons, in scanning electron micro-
    scope, 858
Secondary immune response, 789. *See also*
    Immune response
Secondary lysosome, 256–257, 257*f*
Secondary structure of proteins, 49*f*, 49*t*, 50–51,
    50*f*
Secretory component, IgA, 798–799, 798*f*
Secretory granules, 244, 245
Secretory pathways, 244
Secretory proteins
    Golgi complex in, 93–95
    Golgi complex sorting of, 243
    packaging of, smooth ER in, 93
    synthesis of, on rough endoplasmic reticulum,
      93
Secretory vesicles, 95, 244
Sectioning
    for light microscopy, 853
    optical, 850
    for transmission electron microscopy,
      859–860
Sedimentation coefficient (S value), 99, 231
    of ribosome, 99
    for rRNA, 556, 557*t*
Sedoheptulose bisphosphatase, in Calvin cycle,
    395
Segregation
    of alleles, genetic variability and, 508–513
    law of, 510, 511*f*, 512
Selectable markers, 524
Selectively permeable membranes, 24–27
Self, immunologic tolerance to, 795
Self-assembly, 32–39
    assisted, 34, 35
    limits of, 38–39
    strict, 34–35
Self-nonself recognition, immunologic, 785
Self-splicing RNA introns, 562–564
SEM. *See* Scanning electron microscope/
    microscopy
Semiautonomous organelles, 376
Semiconservative DNA replication, 456, 457*f*
Sensory neurons, 714, 715*f*
Serine, 46*t*, 47*f*, 71, 72*f*
Serine/threonine kinases, 760
Serotonin, 738*f*, 739
70S initiation complex, 576
Severing and capping proteins, 663, 664
Sex chromosomes, 497
Sex hormones, 73, 74*f*
    physiologic functions of, 771, 772*t*
Sex pili, 518
Sexual reproduction, 495–498

SH2 domain, 761
Shadowing, for transmission electron microscopy, 861–862, 861f, 862f
Shine-Dalgarno sequence, 576
Shuttle streaming, microfilaments in, 659, 701
Shuttle vesicles, 240
Sialic acid, 185, 187f
Sickle cell anemia
    hemoglobin structure and, 55
    molecular basis of, 540
Sidearms, 703–704, 705f
Sigma (σ) factor, 545–547
    alternative, for regulation of transcription initiation, 608
Signal hypothesis, 585–589, 590f
Signal mechanism, for cotranslational import, 585–589, 590f
Signal peptidase, 587–589, 590f
Signal peptide (signal sequence), in cotranslational import, 585–589, 590f
Signal-recognition particle, 587
    receptor for, 587
Signal transduction, 754
    cellular receptors in, 751–761
    chemical signals in, 751–761
    endocrine and paracrine hormone systems in, 770–779
    G protein-linked receptors in, 755–759, 761–762, 762f
    growth factor binding and, 830–831
    growth factors as messengers in, 779
    initiation of, 754
        by tyrosine kinase receptors, 769–770
    membranes in, 168f, 169
    in nerve cells, 714–750
        action potential and, 728–731, 731–736
        electrical excitability and, 724–728
        integration and processing in, 742–746
        membrane potentials and, 718–724
        synaptic transmission and, 736–742
    pathways of, 761–770
    second messenger systems for, 761–770
Silencers, gene expression regulation and, 623–624
Silent information regulator genes (SIR genes), 615
Silent mutations, 585
Simple diffusion, 169, 202, 203–204
    direction and rate of, 204, 205t, 206f
    membrane permeability and, 203–204
Simple-sequence repeated DNA, 430
Simple transport, 215–216
Singer, S.J., 172–173
Single bonds, 18f, 19
Single-channel recording (patch clamping), ion channels studied with, 726–728
Single-strand binding protein, 461t, 464
SIR genes. See Silent information regulator genes
Sister chromatids, 453, 454f
Size rule, simple diffusion and, 203
Skeletal muscle cells, 680–682
    calcium levels in, regulation of, 688–689
Skeletal muscle contraction, 681–694
    ATP in, 686–687
    energy needs of, 692–694
    regulation of, 687–692
    sliding-filament model of, 684–686
    thick filaments in, 682
    thin filaments in, 683
Skeletal muscle filaments, 682–684. See also Thick filaments; Thin filaments
Skeletal muscles, 680–682
    protein components of, 684t
Sliding-filament model, 676, 684–686
Sliding-microtubule model, 705–706

Small nuclear ribonucleoproteins (snRNPs), 562, 563f
Small nuclear RNA (snRNA), 562
Small subunits, ribosomal, 100
Smooth endoplasmic reticulum (smooth ER), 84f, 93, 235–237, 237–238, 237f, 239f
Smooth muscle, 695–696
    regulation of contraction in, 696, 697f
Smooth muscle contraction, 695–697
Snake venoms, acetylcholinesterase affected by, 745
snRNA, 562
snRNPs, 562, 563f
Sodium, permeability coefficient for, 205t
Sodium channels, 727–728
    voltage-gated, 725
        action potential and, 730–731, 732f
Sodium chloride, solubility of, 24
Sodium cotransport, 215, 219
Sodium dodecyl sulfate (SDS), in membrane protein analysis, 184, 185f
Sodium-potassium pump, 216–219, 722
Sol, 698
Solar radiation (ultraviolet radiation)
    energy of, 19–20, 19f
    hazard to biological molecules from, 20
Solute transport, 197, 199
    charged particles, 213–214
    uncharged particles, 212–213
Solvent, 23
    water as, 23–24
Somatic hypermutation, antibody diversity and, 809
Somatic nervous system, 715, 716f
Somatotropin, physiologic functions of, 771, 772f
Southern blotting, in DNA fingerprinting, 446f, 447
Space-filling models, van der Waals radii in, 36
Spacer DNA (nucleosome linker DNA), 434
Spatial summation, 743
Specific heat, 23
Specificity
    enzyme, 143
    group, 143
    immune response, 785
    substrate, 143
    transport protein, 206–207
Specimen, 841
    preparation of
        for light microscopy, 853–854
        for transmission electron microscopy, 859–864
Spectrin, 176, 185f, 188, 664
Spectrum
    absorption, 378, 379f
        of DNA, 420
    action, 381
Sperm (spermatozoa), 498
    mitochondrion localization within, 91, 92f
Spheres of hydration, 24
Sphingolipids, 71–72, 72, 73f, 177
    source/function of, 177t
Sphingomyelins, 72, 73f
Sphingosine, 72, 73f
Spindle fibers, 102
Spleen, in lymphocyte development, 786
Splice sites, 562
Spliceosome, 562, 563f
Splicing, RNA. See RNA splicing
Spontaneity, 124–128. See also Thermodynamic spontaneity
Spores, haploid, 500
Sporophyte, 500
Sputter coating, 865
Squid giant axon, as experimental model, 728

SR. See Sarcoplasmic reticulum
src oncogene, 820
SRP. See Signal-recognition particle
SRP receptor, 587
SSB. See Single-strand binding protein
Stage (of microscope), 845, 846f
Staining
    for light microscopy, 853
    negative, for transmission electron microscopy, 861, 861f
    for transmission electron microscopy, 859–860
Standard free energy change, 130–131
    equilibrium constant and, 130–131
    meaning of, 131–132, 132t
Standard reduction potentials, 351–352, 351t
Standard state, 130–131
Starch, 30, 31f, 65, 66f, 67
    catabolism of, 315, 316f
    synthesis of, from Calvin cycle products, 396–397, 397
Starch grains, 67
Start codon, 575
Startpoint, of prokaryotic promoter, 547
State, 119
    standard, 130–131
    steady, 127, 133
Stator ring (S ring), of bacterial flagellum, 706, 707f
Steady state, 127, 133
Steady-state ion concentration, resting membrane potential and, 722
Steady-state ion movements, 722–724
Stearate, 70t, 178, 179f
STEM. See Scanning transmission electron microscope/microscopy
Stem cells, 482, 818
    in gene therapy, 527
    hematopoietic, lymphocyte development from, 785–786, 787f
Stereo electron microscope/microscopy, 864, 865f
Stereoisomers, 20–21, 22f
    of amino acids, 45, 46f
Stereospecificity, transport protein, 207
Steroid hormones, 73, 74f, 752
    chemical classification and function of, 771, 773t
Steroid receptors, 755
    activation of, 755, 756f
Steroids, 72–73, 74f
    chemical classification and function of, 771, 773t
Sticky ends, of restriction fragments, 429, 520
STM. See Scanning tunneling microscope/microscopy
Stoma (stomata), 389
Stop codon, 575
Stop signals, 545
Stop-transfer peptide, 589, 590f
Storage macromolecules/polysaccharides, 30, 31f, 65–67
    catabolism of, 314–316
Streaming, cytoplasmic, microfilaments in, 102, 659, 675, 701
Streptococcus pneumoniae ("pneumococcus"), DNA as transforming principle of, 409–411
Stress fibers, 660, 698, 699f
Stress-response genes, gene expression regulation and, 628
Striated muscle, 681. See also Skeletal muscles
Strict aerobes/anaerobes, 306
Stroma, chloroplast, 92, 93f, 376
Stroma lamellae (stroma thylakoids), 91, 93f, 376, 378f